Calculus

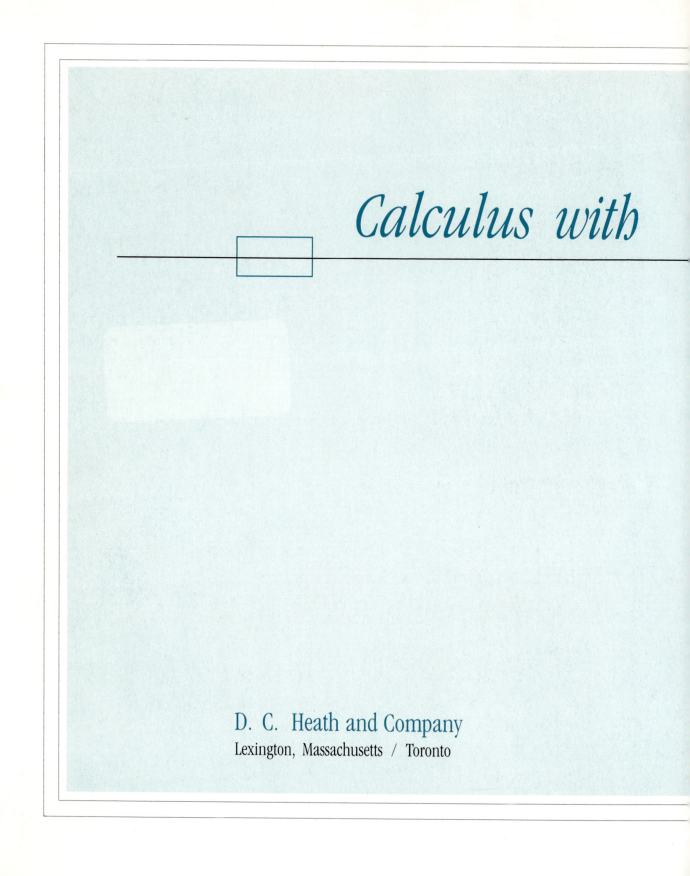

Calculus with

D. C. Heath and Company

Lexington, Massachusetts / Toronto

2427

Arithmetic Operations:

$$ab + ac = a(b + c)$$

$$\frac{a}{b} + \frac{c}{d} = \frac{ad + bc}{bd}$$

$$\frac{a + b}{c} = \frac{a}{c} + \frac{b}{c}$$

$$\frac{\left(\frac{a}{b}\right)}{\left(\frac{c}{d}\right)} = \frac{ad}{bc}$$

$$a\left(\frac{b}{c}\right) = \frac{ab}{c}$$

$$\frac{a - b}{c - d} = \frac{b - a}{d - c}$$

$$\frac{ab + ac}{a} = b + c$$

$$\frac{\left(\frac{a}{b}\right)}{c} = \frac{a}{bc}$$

$$\frac{a}{\left(\frac{b}{c}\right)} = \frac{ac}{b}$$

Exponents and Radicals:

$$a^0 = 1 \ (a \neq 0)$$

$$\frac{a^x}{a^y} = a^{x-y}$$

$$\left(\frac{a}{b}\right)^x = \frac{a^x}{b^x}$$

$$\sqrt[n]{a^m} = a^{m/n} = (\sqrt[n]{a})^m$$

$$a^{-x} = \frac{1}{a^x}$$

$$(a^x)^y = a^{xy}$$

$$\sqrt{a} = a^{1/2}$$

$$\sqrt[n]{ab} = \sqrt[n]{a}\sqrt[n]{b}$$

$$a^x a^y = a^{x+y}$$

$$(ab)^x = a^x b^x$$

$$\sqrt[n]{a} = a^{1/n}$$

$$\sqrt[n]{\left(\frac{a}{b}\right)} = \frac{\sqrt[n]{a}}{\sqrt[n]{b}}$$

Algebraic Errors to Avoid:

$\dfrac{a}{x + b} \neq \dfrac{a}{x} + \dfrac{a}{b}$ To see this error, let $a = b = x = 1$.

$\sqrt{x^2 + a^2} \neq x + a$ To see this error, let $x = 3$ and $a = 4$.

$a - b(x - 1) \neq a - bx - b$ Remember to distribute negative signs. The equation should be $a - b(x - 1) = a - bx + b$.

$\dfrac{\left(\frac{x}{a}\right)}{b} \neq \dfrac{bx}{a}$ To divide fractions, invert and multiply. The equation should be

$$\frac{\left(\frac{x}{a}\right)}{b} = \frac{\left(\frac{x}{a}\right)}{\left(\frac{b}{1}\right)} = \left(\frac{x}{a}\right)\left(\frac{1}{b}\right) = \frac{x}{ab}$$

$\sqrt{-x^2 + a^2} \neq -\sqrt{x^2 - a^2}$ We can't factor a negative sign outside of the square root.

$\dfrac{a + bx}{a} \neq 1 + bx$ This is one of many examples of incorrect cancellation. The equation should be

$$\frac{a + bx}{a} = \frac{a}{a} + \frac{bx}{a} = 1 + \frac{bx}{a}$$

$\dfrac{1}{x^{1/2} - x^{1/3}} \neq x^{-1/2} - x^{-1/3}$ This error is a sophisticated version of the first error.

$(x^2)^3 \neq x^5$ The equation should be $(x^2)^3 = x^2 x^2 x^2 = x^6$.

Conversion Table:

1 centimeter = 0.394 inches	1 joule = 0.738 foot-pounds	1 mile = 1.609 kilometers
1 meter = 39.370 inches	1 gram = 0.035 ounces	1 gallon = 3.785 liters
= 3.281 feet	1 kilogram = 2.205 pounds	1 pound = 4.448 newtons
1 kilometer = 0.621 miles	1 inch = 2.540 centimeters	1 foot-lb = 1.356 joules
1 liter = 0.264 gallons	1 foot = 30.480 centimeters	1 ounce = 28.350 grams
1 newton = 0.225 pounds	= 0.305 meters	1 slug = 14.594 kilograms

analytic geometry

ALTERNATE THIRD EDITION

ROLAND E. LARSON
ROBERT P. HOSTETLER

The Pennsylvania State University
The Behrend College

with the assistance of DAVID E. HEYD

Published simultaneously in Canada.

Printed in the United States of America.

International Standard Book Number: 0-669-09569-9

Library of Congress Catalog Card Number: 85-80735

Preface

Calculus with Analytic Geometry, Alternate Third Edition, is designed for use in a calculus course for students in engineering, economics, life sciences, mathematics, and the physical sciences. In writing this book, we were guided by two primary objectives that have crystallized over many years of teaching calculus. For the student, our objective was to write in a precise, readable manner with the basic concepts and rules of calculus clearly defined and demonstrated. For the instructor, our objective was to design a comprehensive teaching instrument that employs proven pedagogical techniques, thus freeing the instructor to make the most efficient use of classroom time.

Changes in the alternate third edition

We incorporated many suggestions given by instructors and students who used the second edition.

☐ Coverage of the transcendental functions follows the basic order of topics of the second edition, appearing after the presentation of differential and integral calculus of algebraic functions. **However, because several users of the second edition prefer an earlier coverage of trigonometric functions, we have prepared a third edition with that feature.**

☐ The presentation of limits (Chapter 2) has been rewritten. Limits are defined intuitively in Section 2.1, and the section now contains a strategy for finding limits. Section 2.2 discusses techniques for evaluating limits. Continuity is introduced in Section 2.3 and infinite limits in Section 2.4. Finally, the chapter closes with an ε-δ approach to limits. Some of the more technical proofs of theorems in Chapter 2 are given in Appendix A.

☐ In Chapter 3, velocity and acceleration, higher-order derivatives, and the relationship between differentiability and continuity have all been moved ahead of the presentation of differentiation rules. The sections on the Chain Rule and implicit differentiation were completely rewritten, and the chapter now closes with a section on related rate problems.

☐   Applications of the derivative now appear in one chapter (Chapter 4), which begins with a discussion of extrema on an interval. The discussion of curvature has been postponed to Chapter 14.

☐   There are many changes in the chapter introducing integration (Chapter 5). Formal *u*-substitution has been moved from the chapter on techniques of integration. The limit of a sequence is given a brief introduction in Section 5.3. The discussion of area and definite integrals has been divided into two sections.

☐   Chapter 7 begins with an introduction of the natural logarithmic function, followed by a new section on inverse functions.

☐   In Chapter 8 the review of trigonometry has been divided into two sections. Section 8.5 includes a summary of differentiation techniques for elementary functions. This is followed in Section 8.6 with a discussion of completing the square and a summary of integration techniques.

☐   The coverage of techniques of integration (Chapter 9) has been reordered. The primary differences are an early presentation of integration by parts, a greater emphasis on Wallis's formulas, and a later presentation of partial fractions.

☐   The chapter on infinite series has been moved to an earlier position in the text (Chapter 10). The chapter now begins with a discussion of Taylor polynomials and Taylor's Theorem with remainder.

☐   The introduction to parametric equations is now included in the chapter on polar coordinates (Chapter 12). Also, we have added a section on polar equations for conics as well as material on arc length in polar coordinates.

☐   The introduction of vectors in the plane and in space now appears in one chapter (Chapter 13).

☐   Chapter 14 has been completely rewritten to give greater emphasis to vector-valued functions. This presentation now integrates study of vector-valued functions in the plane *and* in space, and ends with a discussion of curvature.

☐   The chapter on functions of several variables has been substantially rewritten and includes four additional sections: a new section on limits and continuity (Section 15.2), two sections on differentials and the Chain Rule, a new section on applications of extrema (Section 15.9), and a new section on Lagrange multipliers (Section 15.10).

☐   In Chapter 16 we have included two new sections on change of variables—one involving polar coordinates and one involving the Jacobian.

☐   Chapter 17 on vector analysis and line integrals has been completely rewritten. The chapter is more applied now and begins with a new section on vector fields. The Fundamental Theorem of Line Integrals, Green's Theorem, surface integrals, the Divergence Theorem, and Stokes's Theorem have all been given greater emphasis.

Features

Order of Topics   The eighteen chapters readily adapt to either semester or quarter systems. In each system both differentiation *and* integration can be introduced in the first course of the sequence. There is some flexibility in the order and depth in which the chapters can be covered. For instance, much of the precalculus material in Chapter 1 can be used as individual review. The ε-δ discussion of limits in Chapter 2 can be given minimal coverage. Sections 6.5, 6.6, and 6.7 can be given later coverage in the course. Chapter 10 can be covered any time after Chapter 9. The coverage of Section 13.7 can be delayed until just before Section 16.8.

Definitions and Theorems   Special care has been taken to state the definitions and theorems simply without sacrificing accuracy.

Proofs   We have chosen to include only those proofs that we have found to be both instructive and within the grasp of a beginning calculus student. Moreover, in presenting proofs, we have found that extensive detail often obscures rather than illuminates. For this reason, many of the proofs are presented in outline form, with an emphasis on the essence of the argument. (See the proof of the Product Rule in Section 3.4.) In some cases, we have included a more complete discussion of proofs in Appendix A. (See the proof of the Chain Rule in Section 3.5.)

Graphics   The alternate third edition contains over 2,250 figures. Of these, over 1,100 are in the examples and exposition, over 625 are in the exercise sets, and over 525 are in the odd-numbered answers. Special care has been taken to produce accurate graphics with the aid of computer-plotting techniques.

Summaries   Many sections have summaries that identify core ideas and procedures—see Sections 3.5, 4.8, 8.6, 10.7, or 14.5. In some instances, an entire section summarizes the preceding topics, for example, Sections 4.6 and 9.1.

Applications   In our choice of applications we have tried for variety, integrity, and a minimal knowledge of other fields.

Exercises   Eight hundred new problems have been added, so the text now contains over 6,800 exercises. The exercises are graded, progressing from skill-development problems to more challenging problems involving applications and proofs. Many exercise sets begin with a set of exercises that provides the graphs of the functions involved. Review exercises are included at the end of each chapter.

Calculators   Special emphasis is given to the use of hand calculators in sections dealing with limits, Newton's Method, numerical integration, and Taylor polynomials. In addition, many of the exercise sets contain problems identified by a colored box as calculator exercises.

Study Aids   For students who need algebraic help, the *Study and Solutions Guide* contains detailed solutions to several representative problems from each exercise set. (These exercises are denoted by exercise numbers set in color.) The solutions to these exercises are given in greater detail than the examples in the text, with special care taken to show the *algebra* involved in the solutions. In addition, the *Study and Solutions Guide* contains a review of algebra. For instructors, the *Complete Solutions Guides* are available.

Examples   Together, the text and the *Study and Solutions Guide* contain over 1,800 examples of solved problems. As a new feature in the alternate third edition, each example in the text has been titled for easy reference, and many examples include color side comments.

Historical Notes   A new feature in the alternate third edition is the addition of short historical notes throughout the text. These are designed to help students gain an appreciation of both the people involved in the development of calculus and the nature of the problems that calculus was designed to solve.

Remarks   Another new feature in the alternate third edition is the use of special instructional notes to students in the form of "Remarks." These appear after definitions, theorems, or examples and are designed to give additional insight, help avoid common errors, or describe generalizations.

Computer Software   Technology Training Associates have produced software for the Apple II® and IBM-PC®*. The software consists of exploratory and directed activities that reinforce the learning and enhance the teaching of calculus. The programs feature a carefully designed user interface and are structured as reusable tools that a student will be able to use in other mathematics courses.

Roland E. Larson
Robert P. Hostetler

---

*Apple is a registered trademark of Apple Computer, Inc. IBM is a registered trademark of International Business Machines Corp.

Acknowledgments

We would like to thank the many people who have helped us at various stages of this project during the past 14 years. Their encouragement, criticisms, and suggestions have been invaluable to us.

The reviewers for the third edition and previous editions are: Dennis Albér, Palm Beach Junior College; Harry L. Baldwin, Jr., San Diego City College; Paul W. Davis, Worcester Polytechnic Institute; Garret J. Etgen, University of Houston; Phillip A. Ferguson, Fresno City College; William R. Fuller, Purdue University; Thomas M. Green, Contra Costa College; Eric R. Immel, Georgia Institute of Technology; Arnold J. Insel, Illinois State University; William J. Keane, Boston College; Timothy J. Kearns, Boston College; Frank T. Kocher, Jr., Pennsylvania State University; Joseph F. Krebs, Boston College; David C. Lantz, Colgate University; Norbert Lerner, State University of New York at Cortland; Robert L. Maynard, Tidewater Community College; Maurice L. Monahan, South Dakota State University; Robert A. Nowlan, Southern Connecticut State University; Barbara L. Osofsky, Rutgers University; Jean E. Rubin, Purdue University; N. James Schoonmaker, University of Vermont; Richard E. Shermoen, Washburn University; Thomas W. Shilgalis, Illinois State University; Lawrence A. Trivieri, Mohawk Valley Community College; J. Philip Smith, Southern Connecticut State University; Bert K. Waits, Ohio State University; and Florence A. Warfel, University of Pittsburgh.

We would like to thank our publisher, D. C. Heath and Company. Several other people worked on this project with us and we appreciate their dedicated and careful help. David E. Heyd assisted us in the text and wrote the *Study and Solutions Guide*. Dianna L. Zook proofread the manuscript and wrote the *Complete Solutions Guide*. Linda L. Matta proofread the galleys and typed much of the manuscript. Timothy R. Larson proofread the galleys and prepared the art. Nancy K. Stout typed the *Complete Solutions Guide*. Linda M. Bollinger typed part of the manuscript. Helen Medley proofread the manuscript. Dee Crenshaw-Crouch rendered the mathematician drawings. A. David Salvia and Gregor M. Olsavsky helped with the exercise solutions.

A special note of thanks goes to the over 250,000 students who have used earlier editions of this text.

On a personal level, we are grateful to our children for their interest and support during the past several years, and to our wives, Deanna Gilbert Larson and Eloise Hostetler, for their love, patience, and understanding.

If you have suggestions for improving this text, please feel free to write to us. Over the past 14 years we have received many useful comments from both instructors and students and we value these very much.

Contents

11 *Conics*   *606*

12 *Plane curves, parametric equations, and polar coordinates*   *640*

13 *Vectors and the geometry of space*   *685*

Vector-valued functions   746

Functions of several variables   786

Multiple integration   872

Vector analysis   930

18

Differential equations   *987*

Computer activities for calculus

Developed by Technology Training Associates, Cambridge, MA

This package, prepared for both the IBM-PC® and Apple® II computers, offers activity-based programs that enhance the learning of calculus. Below is a listing of the six units offered and their contents. Each unit contains both exploratory practice and directed tutorials.

Unit I Functions

- A   Introduction to functions and graphs
- B   Linear functions
- C   Quadratic functions
- D   Trigonometric functions
- E   Exponential and logarithmic functions
- F   Explorer

Unit II Limits

- A   Introduction to limits
- B   Finding limits—Consider the function
- C   Finding limits—Consider the graph
- D   Explorer

Unit III The derivative

- A   Introduction to the derivative
- B   The derivative: Geometric approach
- C   The derivative: Numerical approach
- D   The first and second derivatives
- E   Explorer

What is calculus?

We begin to answer this question by saying that calculus is the reformulation of elementary mathematics through the use of a limit process. If limit processes are unfamiliar to you this answer is, at least for now, somewhat less than illuminating. From an elementary point of view, we may think of calculus as a "limit machine" that generates new formulas from old. Actually, the study of calculus involves three distinct stages of mathematics: *precalculus mathematics* (the length of a line segment, the area of a rectangle, and so forth), the *limit process,* and new *calculus* formulations (derivatives, integrals, and so forth).

Some students try to learn calculus as if it were simply a collection of new formulas. This is unfortunate. When students reduce calculus to the memorization of differentiation and integration formulas, they miss a great deal of understanding, self-confidence, and satisfaction.

On the following two pages we have listed some familiar precalculus concepts coupled with their more powerful calculus versions. Throughout this text, our goal is to show you how precalculus formulas and techniques are used as building blocks to produce the more general calculus formulas and techniques. Don't worry if you are unfamiliar with some of the "old formulas" listed on the following two pages—we will be reviewing all of them.

As you proceed through this text, we suggest that you come back to this discussion repeatedly. Try to keep track of where you are relative to the three stages involved in the study of calculus. For example, the first three chapters break down as follows: precalculus (Chapter 1), the limit process (Chapter 2), and new calculus formulas (Chapter 3). This cycle is repeated many times on a smaller scale throughout the text. We wish you well in your venture into calculus.

WITHOUT CALCULUS	WITH DIFFERENTIAL CALCULUS
value of $f(x)$ when $x = c$	limit of $f(x)$ *as* x approaches c
slope of a line	slope of a curve
secant line to a curve	tangent line to a curve
average rate of change between $t = a$ and $t = b$	instantaneous rate of change at $t = c$
curvature of a circle	curvature of a curve
height of a curve when $x = c$	maximum height of a curve on an interval
tangent plane to a sphere	tangent plane to a surface
direction of motion along a straight line	direction of motion along a curved line

WITHOUT CALCULUS	WITH INTEGRAL CALCULUS
area of a rectangle	area under a curve
work done by a constant force	work done by a variable force
center of a rectangle	centroid of a region
length of a line segment	length of an arc
surface area of a cylinder	surface area of a solid of revolution
mass of a solid of constant density	mass of a solid of variable density
volume of a rectangular solid	volume of a region under a surface
sum of a finite number of terms $\quad a_1 + a_2 + \cdots + a_n = S$	sum of an infinite number of terms $\quad a_1 + a_2 + a_3 \cdots = S$

Calculus

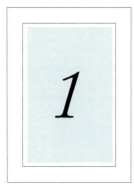

1

The Cartesian plane and functions

1.1 The real line

SECTION TOPICS
- Real numbers
- The real line
- Order and inequalities
- Absolute value
- Distance on the real line
- Intervals on the real line

This first chapter sets the stage for our study of calculus. The props we require include basic algebra and analytic geometry. A good working knowledge of basic algebra is essential for the study of calculus, and we assume that you possess such a knowledge. As for analytic geometry, we assume you have less familiarity with this topic and will discuss its basic concepts as the need arises.

To represent the real numbers we use a coordinate system called the **real line** or x-axis (Figure 1.1). The **positive direction** (to the right) is denoted by an arrowhead and indicates the direction of increasing values of x. The real number corresponding to a particular point on the real line is called the **coordinate** of the point. As shown in Figure 1.1, it is customary to identify those points whose coordinates are integers.

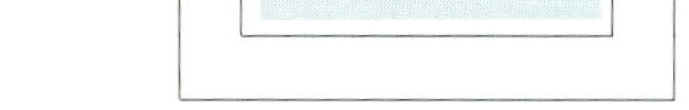

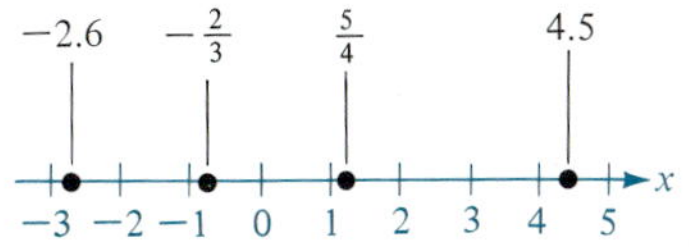

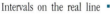

The Real Line

FIGURE 1.1

The point on the real line corresponding to zero is called the **origin** and is denoted by 0. Numbers to the right of the origin are **positive,** and numbers to the left of the origin are **negative.** We use the term **nonnegative** to describe a number that is either positive or zero. Similarly, the term **nonpositive** is used to describe a number that is negative or zero.

The real line provides us with a perfect picture of the real numbers. That is, each point on the real line corresponds to one and only one real number, and each real number corresponds to one and only one point on the real line. This type of relationship is called a **one-to-one correspondence.** (See Figure 1.2.)

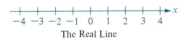

One-to-one correspondence between real numbers and points on the real line.

FIGURE 1.2

Each of the four points in Figure 1.2 corresponds to a real number that can be expressed as the ratio of two integers. (Note that $4.5 = \frac{9}{2}$ and $-2.6 = -\frac{13}{5}$.) We call such numbers **rational.** Rational numbers are represented by either *terminating decimals* (such as $\frac{2}{5} = 0.4$) or *infinite repeating decimals* (such as $\frac{1}{3} = 0.333\ldots = 0.\overline{3}$).

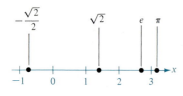

Irrational Numbers

FIGURE 1.3

Real numbers that are not rational are called **irrational,** and they cannot be represented as terminating or infinite repeating decimals. To represent an irrational number, we usually resort to a decimal approximation. For example, $\sqrt{2} \approx 1.4142135623$, $\pi \approx 3.1415926535$, and $e \approx 2.7182818284$.

| **Remark** Note that we use $\approx$ to mean *approximately equal to.* Remember that, even though we cannot represent irrational numbers *exactly* as terminating decimals, we can represent them *exactly* by points on the real line, as shown in Figure 1.3.

Order and inequalities

One important property of real numbers is that they are **ordered.**

| DEFINITION OF ORDER ON THE REAL LINE | If a and b are real numbers, then a is **less than** b if $b - a$ is positive. We denote this order by the **inequality** |

$$a < b$$

The symbol $a \leq b$ means that a is **less than or equal to** b.

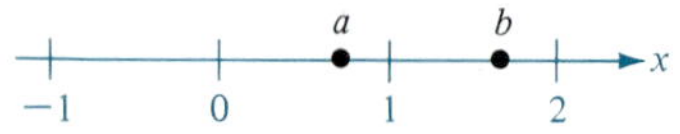

$a < b$ if and only if a lies to the left of b.

FIGURE 1.4

| **Remark** Geometrically, we can show that $a < b$ if and only if a lies to the *left* of b on the real line. (See Figure 1.4.) For example, $1 < 2$ since 1 lies to the left of 2 on the real line.

The following properties are often used in working with inequalities. Similar properties are obtained if $<$ is replaced by $\leq$ and $>$ is replaced by $\geq$.

THEOREM 1.1   **PROPERTIES OF INEQUALITIES**

1. Transitive Property: If $a < b$ and $b < c$, then $a < c$.
2. Additive Property: If $a < b$ and $c < d$, then $a + c < b + d$.
3. If $a < b$ and k is any real number, then $a + k < b + k$.
4. If $a < b$ and k is any real number, then $a - k < b - k$.
5. If $a < b$ and $k > 0$, then $ak < bk$.
6. If $a < b$ and $k < 0$, then $ak > bk$.

| **Remark** Note that we *reverse the inequality* when we multiply by a negative number. For example, if $x < 3$, then $-4x > -12$. This principle also applies to division. Thus, if $-2x > 4$, then $x < -2$.

When three real numbers a, b, and c are ordered such that $a < b$ and $b < c$, we say that b is **between** a and c and we write $a < b < c$.

Occasionally it is convenient to use set notation to describe collections of real numbers. A **set** is a collection of elements. For example, the two major sets we have been discussing are the set of real numbers and the set of points on the real line. Often, we will restrict our interest to a **subset** of one of these two sets, in which case it is convenient to use **set builder notation** of the form

$$\{x: \text{ condition on } x\}$$

The set of all x such that a certain condition is true.

For example, we can describe the set of positive real numbers as $\{x: 0 < x\}$. The **union** of two sets A and B is the set of elements that are members of A or B and is denoted by $A \cup B$. The **intersection** of two sets A and B is the set of elements that are members of A *and* B and is denoted by $A \cap B$.

In calculus, the most common sets we work with are subsets of the real line called **intervals.** For example, the **open** interval $(a, b) = \{x: a < x < b\}$ is the set of all real numbers greater than a and less than b, where a and b are called the **endpoints** of the interval. Note that the endpoints are not included in an open interval. Intervals that include their endpoints are called **closed** and are denoted by $[a, b] = \{x: a \le x \le b\}$. The nine basic types of intervals on the real line are shown in Table 1.1.

TABLE 1.1   INTERVALS ON THE REAL LINE

	Interval notation	*Set notation*	*Graph*
Open interval	(a, b)	$\{x: a < x < b\}$	
Closed interval	$[a, b]$	$\{x: a \le x \le b\}$	
Half-open intervals	$[a, b)$	$\{x: a \le x < b\}$	
	$(a, b]$	$\{x: a < x \le b\}$	
Infinite intervals	$(-\infty, a]$	$\{x: x \le a\}$	
	$(-\infty, a)$	$\{x: x < a\}$	
	(b, ∞)	$\{x: b < x\}$	
	$[b, \infty)$	$\{x: b \le x\}$	
	$(-\infty, \infty)$	$\{x: x \text{ is a real number}\}$	

Remark We use the symbols ∞ and $-\infty$ to refer to positive and negative infinity. These symbols do *not* denote real numbers; they merely enable us to write certain statements more concisely.

EXAMPLE 1   Intervals on the real line

Describe the intervals on the real line that correspond to the temperature ranges (in degrees Fahrenheit) for water in the following two states:
(a) liquid   (b) gas

Solution:

(a) Since water is in a liquid state at temperatures that are greater than 32° and less than 212°, we have the interval

$$(32, 212) = \{x: 32 < x < 212\}$$

as shown in Figure 1.5.

(b) Since water is in a gaseous state (steam) at temperatures that are greater than or equal to 212°, we have the interval

$$[212, \infty) = \{x: 212 \leq x\}$$

as shown in Figure 1.6.

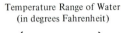

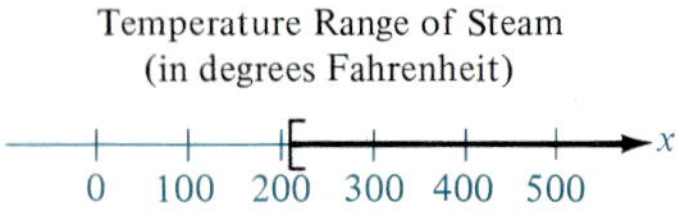

Temperature Range of Water
(in degrees Fahrenheit)

0   50 100 150 200 250

FIGURE 1.5

Temperature Range of Steam
(in degrees Fahrenheit)

0   100 200 300 400 500

FIGURE 1.6

In calculus we are frequently asked to solve inequalities involving variable expressions such as $2x - 5 < 7$. We say $x = a$ is a **solution** to this inequality if the inequality is true when a is substituted for x. The set of all values of x that satisfy the inequality is called the **solution set** of the inequality.

EXAMPLE 2   Solving an inequality

Find the solution set of the inequality $2x - 5 < 7$.

Solution: Using the properties in Theorem 1.1, we have

$$2x - 5 < 7$$
$$2x - 5 + 5 < 7 + 5 \qquad \text{Add 5 to both sides}$$
$$2x < 12$$
$$\frac{1}{2}(2x) < \frac{1}{2}(12) \qquad \text{Multiply both sides by } \frac{1}{2}$$
$$x < 6$$

Thus, the interval representing the solution is $(-\infty, 6)$, as shown in Figure 1.7.

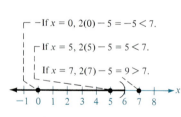

−If $x = 0$, $2(0) - 5 = -5 < 7$.

If $x = 5$, $2(5) - 5 = 5 < 7$.

If $x = 7$, $2(7) - 5 = 9 > 7$.

−1 0 1 2 3 4 5 6 7 8

FIGURE 1.7

Remark In Example 2, all five inequalities listed as steps in the solution have the same solution set and as such are called **equivalent.**

Once you have solved an inequality, it is a good idea to check some x-values in your solution interval to see if they satisfy the original inequality. You also might check some values outside your solution interval to verify that they do not satisfy the inequality. For example, in Figure 1.7 we see that when $x = 0$ or $x = 5$ the inequality is satisfied, but when $x = 7$ the inequality is not satisfied.

EXAMPLE 3   *Finding the intersection of two solution sets*

Find the intersection of the solution sets of the inequalities

$$-3 \leq 2 - 5x \qquad \text{and} \qquad 2 - 5x \leq 12$$

Solution: We could solve both inequalities and then find the intersection of the resulting solution sets. However, since the expression $2 - 5x$ occurs on the left side of one inequality and the right side of the other, it is convenient to work with both inequalities at the same time:

$$-3 \leq \quad 2 - 5x \quad \leq 12$$
$$-3 - 2 \leq 2 - 5x - 2 \leq 12 - 2 \qquad \text{Subtract 2}$$
$$-5 \leq \quad -5x \quad \leq 10$$
$$\frac{-5}{-5} \geq \quad \frac{-5x}{-5} \quad \geq \frac{10}{-5} \qquad \text{Divide by } -5$$
$$1 \geq \quad x \quad \geq -2$$

Thus, the interval representing the solution is $[-2, 1]$.

The inequalities in Examples 2 and 3 involve first-degree polynomials. For inequalities of higher degree we use the principle that a polynomial can change signs *only* at its zeros (the values that make the polynomial zero). Between two consecutive zeros a polynomial must be entirely positive or entirely negative. This means that when the real zeros of a polynomial are put in order, they divide the real line into **test intervals** in which the polynomial has no sign changes. That is, if a polynomial has the factored form

$$(x - r_1)(x - r_2), \ldots, (x - r_n), \qquad r_1 < r_2 < r_3 < \cdots < r_n$$

then the test intervals are

$$(-\infty, r_1), (r_1, r_2), \ldots, (r_{n-1}, r_n), \text{ and } (r_n, \infty)$$

For example, the polynomial

$$x^2 - x - 6 = (x - 3)(x + 2)$$

can change signs only at $x = -2$ and $x = 3$, and we need to test only *one value* from each test interval $(-\infty, -2)$, $(-2, 3)$, and $(3, \infty)$ to solve the inequality.

EXAMPLE 4   *Solving quadratic inequalities*

Find the solution set for the inequality $x^2 < x + 6$.

Solution:

$$x^2 < x + 6 \qquad \text{Given}$$
$$x^2 - x - 6 < 0 \qquad \text{Polynomial form}$$
$$(x - 3)(x + 2) < 0 \qquad \text{Factor}$$

Thus, the quadratic has $x = -2$ and $x = 3$ as its zeros, and we can solve the inequality by testing the sign of the quadratic in each of the following intervals:

$$x < -2, \qquad -2 < x < 3, \qquad x > 3$$

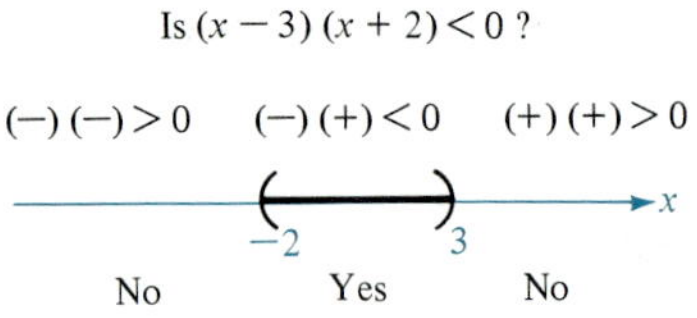

Is $(x - 3)(x + 2) < 0$?

$(-)(-) > 0 \qquad (-)(+) < 0 \qquad (+)(+) > 0$

No   Yes   No

FIGURE 1.8

To test an interval, we choose a representative number in the interval and compute the sign of each factor. For example, for any $x < -2$, both of the factors $(x - 3)$ and $(x + 2)$ are negative. Consequently, the product (of two negatives) is positive and the inequality is *not* satisfied in the interval $x < -2$. We suggest the testing format shown in Figure 1.8.

Since the inequality is satisfied only by the center interval, we conclude that the solution set is

$$-2 < x < 3$$

Absolute value and distance

The absolute value of a real number a is denoted by $|a|$ and defined as follows.

DEFINITION OF ABSOLUTE VALUE	If a is a real number, then the **absolute value** of a is $$|a| = \begin{cases} +a, & \text{if } a \geq 0 \\ -a, & \text{if } a < 0 \end{cases}$$

Remark Note that the absolute value of a number can never be negative. For example, let $a = -4$. Then since $-4 < 0$, we have

$$|a| = |-4| = -(-4) = 4$$

Remember that the symbols $+a$ and $-a$ do not mean that $+a$ is positive or $-a$ is negative.

The following two theorems contain some useful properties of absolute value.

THEOREM 1.2	**OPERATIONS WITH ABSOLUTE VALUE** If a and b are real numbers and n is an integer, then the following properties are true: 1. $\|ab\| = \|a\|\|b\|$ 2. $\left\|\dfrac{a}{b}\right\| = \dfrac{\|a\|}{\|b\|}, \; b \neq 0$ 3. $\|a\| = \sqrt{a^2}$ 4. $\|a^n\| = \|a\|^n$

Proof: We prove Property 3 and leave the proofs of the remaining properties as exercises.

Since $(+a)^2 = a^2$ and $(-a)^2 = a^2$, we know that $+a$ and $-a$ are both square roots of a^2. Moreover, since $\sqrt{a^2}$ denotes the *nonnegative* square root, we have

$$\sqrt{a^2} = \left\{ \begin{array}{ll} +a, & \text{if } a \geq 0 \\ -a, & \text{if } a < 0 \end{array} \right\} = |a|$$

Properties 2 and 4 in the following theorem are also true if $\leq$ is replaced by $<$.

THEOREM 1.3   INEQUALITIES AND ABSOLUTE VALUE
If a and b are real numbers and k is positive, then the following properties are true:

1. $-|a| \leq a \leq |a|$
2. $|a| \leq k$ if and only if $-k \leq a \leq k$.
3. $k \leq |a|$ if and only if $k \leq a$ or $a \leq -k$.
4. Triangle Inequality: $|a + b| \leq |a| + |b|$

Proof: We give a proof of Property 4 and leave the proofs of the first three properties as exercises.

Using Property 1, we have

$$-|a| \leq a \leq |a|$$

and

$$-|b| \leq b \leq |b|$$

Adding these two inequalities produces

$$-(|a| + |b|) \leq a + b \leq |a| + |b|$$

Now, by Property 2 (using $k = |a| + |b|$), we can conclude that

$$|a + b| \leq |a| + |b|$$

| **Remark** Theorem 1.3 uses the phrase "if and only if" as a way of stating two implications at once. For example, in Property 2 one implication is "if $|a| \leq k$, then $-k \leq a \leq k$." The other implication is the *converse*, which says, "if $-k \leq a \leq k$, then $|a| \leq k$."

EXAMPLE 5   Finding an interval defined by an absolute value

Sketch the solution set of $|x - 3| \leq 2$.

Solution: Using Property 2 of Theorem 1.3, we have

$$-2 \leq \quad x - 3 \quad \leq 2$$
$$-2 + 3 \leq x - 3 + 3 \leq 2 + 3$$
$$1 \leq \quad x \quad \leq 5$$

$|x - 3| \leqslant 2$

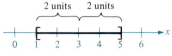

FIGURE 1.9

Thus, the solution set is the closed interval $[1, 5]$, as shown in Figure 1.9.

Example 5 illustrates one of the following general results for intervals defined by absolute value and inequalities.

INTERVALS DEFINED BY ABSOLUTE VALUE

Single interval, $|x - a| \leq d$:

$$a - d \leq x \leq a + d$$

Two intervals, $|x - a| \geq d$:

$$x \leq a - d$$
or
$$x \geq a + d$$

Remark Be sure you see that inequalities of the form $|x - a| \geq d$ have solution sets consisting of *two* intervals.

EXAMPLE 6   A two-interval solution set

Sketch the solution set of $|x + 2| > 3$.

Solution: This inequality can be satisfied in two ways.

    Case 1:   If $x + 2 < 0$, then $|x + 2| = -(x + 2)$ and we have

$$-(x + 2) > 3$$
$$-x - 2 > 3$$
$$-x > 5 \qquad \text{Multiply by } -1 \text{ and reverse the inequality}$$
$$x < -5$$

    Case 2:   If $x + 2 > 0$, then $|x + 2| = x + 2$ and we have

$$x + 2 > 3$$
$$x > 1$$

$|x + 2| > 3$

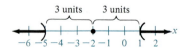

FIGURE 1.10

Therefore, the solution set consists of the union of the disjoint intervals $(-\infty, -5)$ and $(1, \infty)$, as shown in Figure 1.10.

Given two distinct points a and b on the real line, we will in this text make use of each of the distances pictured in Figure 1.11.

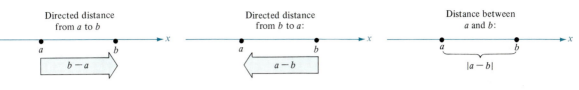

FIGURE 1.11

Using Property 3 of Theorem 1.2, we can write the distance between two points in the following convenient form.

DEFINITION OF DISTANCE BETWEEN TWO POINTS ON THE REAL LINE	The distance d between points x_1 and x_2 on the real line is given by $$d = |x_2 - x_1| = \sqrt{(x_2 - x_1)^2}$$

Remark Note that the order in which x_1 and x_2 are subtracted does not matter since

$$|x_2 - x_1| = |x_1 - x_2| \qquad \text{and} \qquad (x_2 - x_1)^2 = (x_1 - x_2)^2$$

EXAMPLE 7   Distance on the real line

(a) The distance between -3 and 4 is given by

$$|4 - (-3)| = |7| = 7 \qquad \text{or} \qquad |-3 - 4| = |-7| = 7$$

FIGURE 1.12

(See Figure 1.12.)

(b) The directed distance from -3 to 4 is $4 - (-3) = 7$.

(c) The directed distance from 4 to -3 is $-3 - 4 = -7$.

To find the **midpoint** of an interval with endpoints a and b, we simply find the average value of a and b. That is,

$$\text{midpoint of interval } (a, b) = \frac{a + b}{2}$$

To show that this is the midpoint, you need only show that $(a + b)/2$ is equidistant from a and b.

Section Exercises 1.1

In Exercises 1–10, determine whether the real number is rational or irrational.

1. 0.7

2. -3678

3. $\dfrac{3\pi}{2}$

4. $3\sqrt{2} - 1$

5. $4.345\overline{1451}$

6. $\dfrac{22}{7}$

7. *$\sqrt[3]{64}$

8. $0.81\overline{778177}$

9. $4\dfrac{5}{8}$

10. $(\sqrt{2})^3$

*A color number indicates that a detailed solution can be found in the *Student Solution Guide*.

In Exercises 11–14, express the repeating decimal as a ratio of integers using the following procedure. Let $x = 0.6363...$, then

$$100x = 63.6363...$$

When the first equation is subtracted from the second, we have $99x = 63$ or

$$x = \frac{63}{99} = \frac{7}{11}$$

11. $0.36\overline{36}$
12. $0.318\overline{18}$
13. $0.29\overline{7297}$
14. 0.99009900

15. Given $a < b$, determine which of the following are true.
    (a) $a + 2 < b + 2$
    (b) $5b < 5a$
    (c) $5 - a > 5 - b$
    (d) $\frac{1}{a} < \frac{1}{b}$
    (e) $(a - b)(b - a) > 0$
    (f) $a^2 < b^2$

16. For $A = \{x: 0 < x\}$, $B = \{x: -2 \leq x \leq 2\}$, and $C = \{x: x < 1\}$, find the indicated interval.
    (a) $A \cup B$
    (b) $A \cap B$
    (c) $B \cap C$
    (d) $A \cup C$
    (e) $A \cap B \cap C$

In Exercises 17 and 18, complete the given table by filling in the appropriate interval notation, set notation, and graph on the real line.

17.

Interval notation	Set notation	Graph
		(graph: open at -2, solid line to open at 0)
$(-\infty, -4]$		
	$\{x: 3 \leq x \leq \frac{11}{2}\}$	
$(-1, 7)$		

18.

Interval notation	Set notation	Graph
		(graph: bracket at 100, line toward 102)
	$\{x: 10 < x\}$	
$(\sqrt{2}, 8]$		
	$\{x: \frac{1}{3} < x \leq \frac{22}{7}\}$	

In Exercises 19–46, solve the inequality and graph the solution on the real number line.

19. $x - 5 \geq 7$
20. $2x > 3$
21. $4x + 1 < 2x$
22. $2x + 7 < 3$
23. $2x - 1 \geq 0$
24. $3x + 1 \geq 2x + 2$
25. $4 - 2x < 3x - 1$
26. $x - 4 \leq 2x + 1$
27. $-4 < 2x - 3 < 4$
28. $0 \leq x + 3 < 5$
29. $\frac{3}{4}x > x + 1$
30. $-1 < -\frac{x}{3} < 1$
31. $\frac{x}{2} + \frac{x}{3} > 5$
32. $x > \frac{1}{x}$
33. $|x| < 1$
34. $\frac{x}{2} - \frac{x}{3} > 5$
35. $\left|\frac{x - 3}{2}\right| \geq 5$
36. $\left|\frac{x}{2}\right| > 3$
37. $|x - a| < b$
38. $|x + 2| < 5$
39. $|2x + 1| < 5$
40. $|3x + 1| \geq 4$
41. $\left|1 - \frac{2}{3}x\right| < 1$
42. $|9 - 2x| < 1$
43. $x^2 \leq 3 - 2x$
44. $x^4 - x \leq 0$
45. $x^2 + x - 1 \leq 5$
46. $2x^2 + 1 < 9x - 3$

In Exercises 47–50, find the directed distance from a to b, the directed distance from b to a, and the distance between a and b.

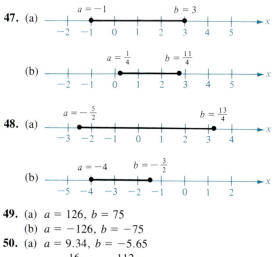

47. (a) $a = -1$, $b = 3$

    (b) $a = \frac{1}{4}$, $b = \frac{11}{4}$

48. (a) $a = -\frac{5}{2}$, $b = \frac{13}{4}$

    (b) $a = -4$, $b = -\frac{3}{2}$

49. (a) $a = 126$, $b = 75$
    (b) $a = -126$, $b = -75$
50. (a) $a = 9.34$, $b = -5.65$
    (b) $a = \frac{16}{5}$, $b = \frac{112}{75}$

In Exercises 51–54, find the midpoint of the given interval.

51. (a) $a = -1$, $b = 3$

    (b) $a = -6$, $b = -1$

52. (a)

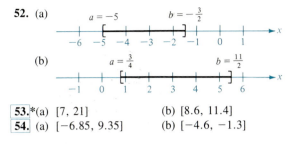

(b)

53. *(a) [7, 21]   (b) [8.6, 11.4]
54. (a) [−6.85, 9.35]   (b) [−4.6, −1.3]

In Exercises 55–60, use absolute values to define each interval (or pair of intervals) on the real line.

55. (a)

56. (a)

(b)

57. (a)

(b)

58. (a)

(b)

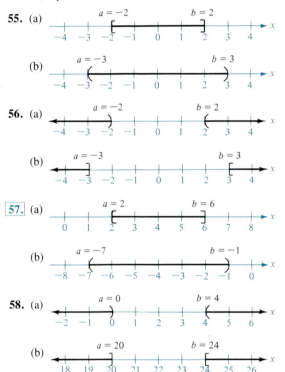

59. (a) All numbers less than 10 units from 12.
   (b) All numbers more than 6 units from 3.
60. (a) y is at most 2 units from a.
   (b) y is less than δ units from c.
61. P dollars invested at r percent simple interest for t years grows to an amount

$$A = P + Prt$$

If an investment of \$1000 is to grow to an amount greater than \$1250 in two years, then the interest rate must be greater than what percentage?

62. In the manufacture and sale of a certain product, the revenue for selling x units is

$$R = 115.95x$$

and the cost of producing x units is

$$C = 95x + 750$$

In order for a profit to be realized, it is necessary that R be greater than C. For what values of x will this product return a profit?

63. A utility company has a fleet of vans for which the annual operating cost per van is estimated to be

$$C = 0.32m + 2300$$

where C is measured in dollars and m is measured in miles. If the company wants the annual operating cost per van to be less than \$10,000, then m must be less than what value?

64. The heights, h, of two-thirds of the members of a certain population satisfy the inequality

$$\left|\frac{h - 68.5}{2.7}\right| \le 1$$

where h is measured in inches. Determine the interval on the real line in which these heights lie.

65. To determine if a coin is fair, an experimenter tosses it 100 times and records the number of heads, x. Through statistical theory, the coin is declared unfair if

$$\left|\frac{x - 50}{5}\right| \ge 1.645$$

For what values of x will the coin be declared unfair?

66. The estimated daily production, p, at a refinery is given by

$$|p - 2,250,000| < 125,000$$

where p is measured in barrels of oil. Determine the high and low production levels.

In Exercises 67 and 68, determine which of the two given real numbers is greater.

67. (a) π or $\dfrac{355}{113}$ **68.** (a) $\dfrac{224}{151}$ or $\dfrac{144}{97}$

   (b) π or $\dfrac{22}{7}$ (b) $\dfrac{73}{81}$ or $\dfrac{6427}{7132}$

In Exercises 69–73, prove each property.

69. $|ab| = |a||b|$
70. $|a - b| = |b - a|$ [Hint: Use Exercise 69 and the fact that $(a - b) = (-1)(b - a)$.]
71. $\left|\dfrac{a}{b}\right| = \dfrac{|a|}{|b|}, \; b \ne 0$ **72.** $\left|\dfrac{1}{a}\right| = \dfrac{1}{|a|}, \; a \ne 0$
73. $|a^n| = |a|^n, \; n = 1, 2, 3, \ldots$

*A color box indicates that a calculator may be helpful.

SECTION TOPICS
- The Cartesian plane
- The Distance Formula
- The Midpoint Formula
- Equations of circles
- Completing the square

1.2 The Cartesian plane, the Distance Formula, and circles

Just as we can represent the real numbers geometrically by points on the real line, we can represent ordered pairs of real numbers by points in a plane. An **ordered pair** (x, y) of real numbers has x as its *first* member and y as its *second* member. The model for representing ordered pairs is called the **rectangular coordinate system,** or the **Cartesian plane.** It is developed by considering two real lines intersecting at right angles (Figure 1.13).

The horizontal real line is traditionally called the **x-axis,** and the vertical real line is called the **y-axis.** Their point of intersection is called the **origin,** and the lines divide the plane into four parts called **quadrants.**

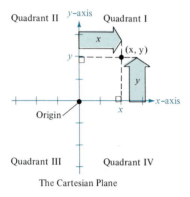

The Cartesian Plane

FIGURE 1.13

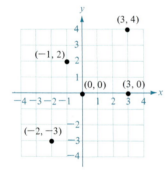

We identify each point in the plane by an ordered pair (x, y) of real numbers x and y, called **coordinates** of the point. The number x represents the directed distance from the y-axis to the point, and y represents the directed distance from the x-axis to the point (Figure 1.13). For the point (x, y), the first coordinate is referred to as the x-coordinate or **abscissa,** and the second or y-coordinate is referred to as the **ordinate.** For example, Figure 1.14 shows the location of the points $(-1, 2)$, $(3, 4)$, $(0, 0)$, $(3, 0)$, and $(-2, -3)$ in the Cartesian plane.

FIGURE 1.14

Remark We use the notation (x, y) to denote both a point in the plane and an open interval on the real line. This should cause no confusion because the nature of a specific problem will show which we are talking about.

The Distance and Midpoint Formulas

In Section 1.1 we defined the distance between two points x_1 and x_2 on the real line. We will now find the distance between two points in the plane. Recall from the Pythagorean Theorem that, for a right triangle with hypotenuse c and sides a and b, we have the relationship $a^2 + b^2 = c^2$. Conversely, if $a^2 + b^2 = c^2$, then the triangle is a right triangle (Figure 1.15).

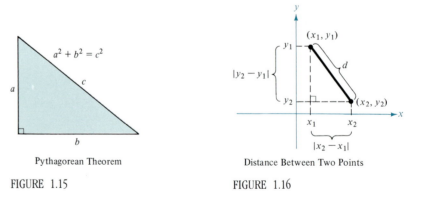

Pythagorean Theorem

FIGURE 1.15

Distance Between Two Points

FIGURE 1.16

Suppose we wish to determine the distance d between two points (x_1, y_1) and (x_2, y_2) in the plane. With these two points, a right triangle can be formed, as shown in Figure 1.16. We see that the length of the vertical side of the triangle is $|y_2 - y_1|$. Similarly, the length of the horizontal side is

$|x_2 - x_1|$. By the Pythagorean Theorem, we then have

$$d^2 = |x_2 - x_1|^2 + |y_2 - y_1|^2$$
$$d = \sqrt{|x_2 - x_1|^2 + |y_2 - y_1|^2}$$

Replacing $|x_2 - x_1|^2$ and $|y_2 - y_1|^2$ by the equivalent expressions $(x_2 - x_1)^2$ and $(y_2 - y_1)^2$, we can write

$$d = \sqrt{(x_2 - x_1)^2 + (y_2 - y_1)^2}$$

We choose the positive square root for d because the distance *between* two points is not a directed distance. We have therefore established the following theorem.

THEOREM 1.4   **DISTANCE FORMULA**
The distance d between the points (x_1, y_1) and (x_2, y_2) is given by

$$d = \sqrt{(x_2 - x_1)^2 + (y_2 - y_1)^2}$$

EXAMPLE 1   The distance between two points

Find the distance between the points $(-2, 1)$ and $(3, 4)$.

Solution: Applying Theorem 1.4, we have

$$d = \sqrt{[3 - (-2)]^2 + (4 - 1)^2}$$
$$= \sqrt{(5)^2 + (3)^2}$$
$$= \sqrt{25 + 9}$$
$$= \sqrt{34} \approx 5.83$$

EXAMPLE 2   Verifying a right triangle

Use the Distance Formula to show that the points $(2, 1)$, $(4, 0)$, and $(5, 7)$ are the vertices of a right triangle.

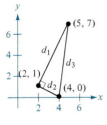

FIGURE 1.17

Solution: Refer to Figure 1.17. The three sides have lengths

$$d_1 = \sqrt{(5 - 2)^2 + (7 - 1)^2} = \sqrt{9 + 36} = \sqrt{45}$$
$$d_2 = \sqrt{(4 - 2)^2 + (0 - 1)^2} = \sqrt{4 + 1} = \sqrt{5}$$
$$d_3 = \sqrt{(5 - 4)^2 + (7 - 0)^2} = \sqrt{1 + 49} = \sqrt{50}$$

Since $d_1^2 + d_2^2 = 45 + 5 = 50 = d_3^2$, we can apply the Pythagorean Theorem to conclude that the triangle must be a right triangle.

| Remark   Although the figure provided in Example 2 was not really essential to the solution, we strongly recommend that you include sketches with your problem solutions.

The formula for the midpoint of a line segment in the plane is similar to that given for intervals on the real line in Section 1.1.

MIDPOINT FORMULA

The midpoint of the line segment joining (x_1, y_1) and (x_2, y_2) is

$$\left(\frac{x_1 + x_2}{2}, \frac{y_1 + y_2}{2} \right)$$

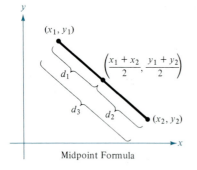

Midpoint Formula

FIGURE 1.18

Proof: Note that the coordinates of the midpoint of a line segment are simply the average values of the respective coordinates of the two endpoints. To prove this we refer to Figure 1.18 and show that $d_1 = d_2$ and $d_1 + d_2 = d_3$. Using the Distance Formula, we obtain

$$d_1 = \sqrt{\left(\frac{x_1 + x_2}{2} - x_1 \right)^2 + \left(\frac{y_1 + y_2}{2} - y_1 \right)^2}$$

$$= \frac{1}{2}\sqrt{(x_2 - x_1)^2 + (y_2 - y_1)^2}$$

$$d_2 = \sqrt{\left(x_2 - \frac{x_1 + x_2}{2} \right)^2 + \left(y_2 - \frac{y_1 + y_2}{2} \right)^2}$$

$$= \frac{1}{2}\sqrt{(x_2 - x_1)^2 + (y_2 - y_1)^2}$$

$$d_3 = \sqrt{(x_2 - x_1)^2 + (y_2 - y_1)^2}$$

Thus, it follows that $d_1 = d_2$ and $d_1 + d_2 = d_3$. You are asked to fill in the algebraic details of this proof in Exercise 63.

EXAMPLE 3   *The midpoint of a line segment*

Find the midpoint of the line segment joining the points $(-5, -3)$ and $(9, 3)$.

Solution: By Theorem 1.5 the midpoint is

$$\left(\frac{-5 + 9}{2}, \frac{-3 + 3}{2} \right) = (2, 0)$$

EXAMPLE 4   *Finding points at a specified distance from a given point*

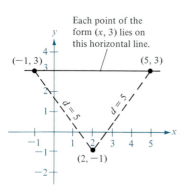

FIGURE 1.19

Find x so that the distance between $(x, 3)$ and $(2, -1)$ is 5.

Solution: Using the Distance Formula, we have

$$d = 5 = \sqrt{(x - 2)^2 + (3 + 1)^2}$$

$$25 = (x^2 - 4x + 4) + 16$$

$$0 = x^2 - 4x - 5$$

$$0 = (x - 5)(x + 1)$$

Therefore, $x = 5$ or $x = -1$, and we conclude that both of the points $(5, 3)$ and $(-1, 3)$ lie 5 units from the point $(2, -1)$, as shown in Figure 1.19.

Circles

One straightforward application of the Distance Formula is in developing an equation for a circle in the plane.

DEFINITION OF A CIRCLE IN THE PLANE

Let (h, k) be a point in the plane and let $r > 0$. The set of all points (x, y) such that r is the distance between (h, k) and (x, y) is called a **circle.** The point (h, k) is the **center** of the circle, and r is the **radius.** (See Figure 1.20.)

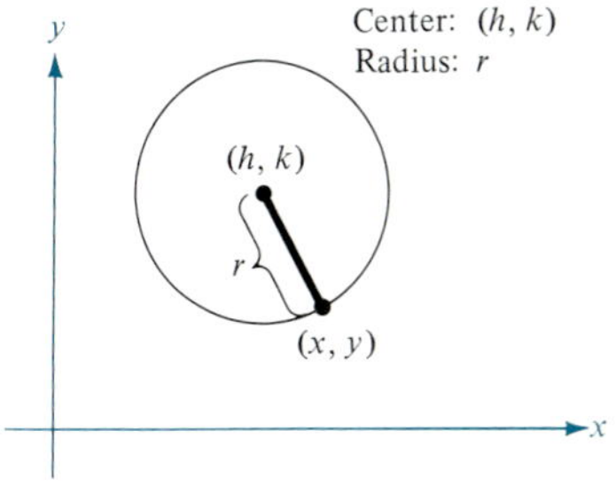

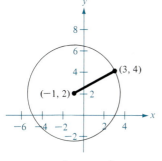

Center: (h, k)
Radius: r

(h, k)

r

(x, y)

FIGURE 1.20

We can use the Distance Formula to write an equation for the circle with center (h, k) and radius r as

$$[\text{distance between } (h, k) \text{ and } (x, y)] = r$$
$$\sqrt{(x - h)^2 + (y - k)^2} = r$$

By squaring both sides of this equation, we obtain the **standard form of the equation of a circle,** which is part of the following theorem.

THEOREM 1.6   STANDARD EQUATION OF A CIRCLE
The point (x, y) lies on the circle of radius r and center (h, k) if and only if

$$(x - h)^2 + (y - k)^2 = r^2$$

It follows from Theorem 1.6 that the equation of a circle with its center at the origin is

$$x^2 + y^2 = r^2$$

y

8
6
4
$(-1, 2)$ 2 $(3, 4)$

-6 -4 -2 2 4 x
-2

$(x + 1)^2 + (y - 2)^2 = 20$

FIGURE 1.21

EXAMPLE 5   Finding the equation of a circle

The point $(3, 4)$ lies on a circle whose center is at $(-1, 2)$, as shown in Figure 1.21. Find an equation for the circle.

Solution: The radius of the circle is the distance between $(-1, 2)$ and $(3, 4)$. Thus,

$$r = \sqrt{[3 - (-1)]^2 + (4 - 2)^2} = \sqrt{16 + 4} = \sqrt{20}$$

Therefore, by Theorem 1.6 the standard equation for this circle is

$$[x - (-1)]^2 + (y - 2)^2 = (\sqrt{20})^2$$
$$(x + 1)^2 + (y - 2)^2 = 20$$

If we remove the parentheses in the standard equation of Example 5, we obtain

$$(x + 1)^2 + (y - 2)^2 = 20$$
$$x^2 + 2x + 1 + y^2 - 4y + 4 = 20$$
$$x^2 + y^2 + 2x - 4y - 15 = 0$$

This equation is written in the **general form of the equation of a circle:**

$$Ax^2 + Ay^2 + Dx + Ey + F = 0, \quad A \neq 0$$

The general form of the equation of a circle is less useful than the equivalent standard form. For instance, it is not immediately apparent from the general equation of the circle in Example 5 that the center is at $(-1, 2)$ and the radius is $\sqrt{20}$. Before graphing the equation of a circle, it is best to write the equation in standard form. This can be accomplished by **completing the square,** as demonstrated in the following example.

EXAMPLE 6   *Completing the square*

Sketch the graph of the circle whose general equation is

$$4x^2 + 4y^2 + 20x - 16y + 37 = 0$$

Solution: To complete the square we will first divide by 4 so that the coefficients of x^2 and y^2 are both 1.

$$4x^2 + 4y^2 + 20x - 16y + 37 = 0 \qquad \text{General form}$$

$$x^2 + y^2 + 5x - 4y + \frac{37}{4} = 0 \qquad \text{Divide by 4}$$

$$(x^2 + 5x + \quad) + (y^2 - 4y + \quad) = -\frac{37}{4} \qquad \text{Group terms}$$

$$\left(x^2 + 5x + \frac{25}{4}\right) + (y^2 - 4y + 4) = -\frac{37}{4} + \frac{25}{4} + 4 \qquad \text{Complete the square}$$

$$\underbrace{\qquad}_{(\text{half})^2} \qquad \underbrace{\qquad}_{(\text{half})^2}$$

$$\left(x + \frac{5}{2}\right)^2 + (y - 2)^2 = 1 \qquad \text{Standard form}$$

Note that we complete the square by adding the square of half the coefficient of x and the square of half the coefficient of y to both sides of the equation. Therefore, the circle is centered at $(-\frac{5}{2}, 2)$, and its radius is 1, as shown in Figure 1.22.

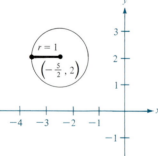

$$\left(x + \tfrac{5}{2}\right)^2 + (y - 2)^2 = 1$$

FIGURE 1.22

EXAMPLE 7   *A single point solution set*

Discuss the graph of $3x^2 + 3y^2 + 24x - 6y + 51 = 0$.

Solution:

First, dividing by 3 we obtain

$$x^2 + y^2 + 8x - 2y + 17 = 0$$

Then we write

$$(x^2 + 8x + \quad) + (y^2 - 2y + \quad) = -17$$
$$(x^2 + 8x + 16) + (y^2 - 2y + 1) = -17 + 16 + 1$$
$$(x + 4)^2 + (y - 1)^2 = 0$$

The sum on the left side of the last equation is zero only if both terms are zero. Since this is true only when $x = -4$ and $y = 1$, the graph consists of the single point $(-4, 1)$.

Example 7 shows that the general equation $Ax^2 + Ay^2 + Dx + Ey + F = 0$ may not always represent a circle. In fact, such an equation may have no solution points if the procedure of completing the square yields the impossible result

$$(x - h)^2 + (y - k)^2 = \text{(negative number)}$$

Circles are one example of a class of curves called **conic sections.** We will study these in detail in Chapter 11.

We have now introduced the basic ideas of *analytic geometry*. Because these ideas are in common use today, it is easy to overlook their revolutionary nature. At the time analytic geometry was being developed by Pierre de Fermat (1601–1655) and René Descartes (1596–1650), the two major branches of mathematics—algebra and geometry—were largely independent of each other. Circles belonged to geometry and equations belonged to algebra. The coordination of the points on a circle and the solutions to an equation belongs to what is now called analytic geometry. It is important that you develop skill in using analytic geometry to move back and forth between algebra and geometry. For instance, in Example 5, we are given a geometric description of a circle and are asked to find an algebraic equation for the circle. Thus, we are moving from geometry to algebra. Similarly, in Example 6 we are given an algebraic equation and asked to sketch a geometric picture. In this case, we are moving from algebra to geometry. These two examples illustrate the two most common problems in analytic geometry.

1. Given a graph, find its equation.

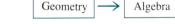

2. Given an equation, find its graph.

In the next two sections, we will look more closely at both of these problems.

Pierre de Fermat

René Descartes

Section Exercises 1.2

In Exercises 1–8, plot the points, find the distance between the points, and find the midpoint of the line segment joining the points.

1. $(2, 1)$, $(4, 5)$

2. $(-3, 2)$, $(3, -2)$

3. $\left(\frac{1}{2}, 1\right)$, $\left(-\frac{3}{2}, -5\right)$

4. $\left(\frac{2}{3}, -\frac{1}{3}\right)$, $\left(\frac{5}{6}, 1\right)$

5. $(2, 2)$, $(4, 14)$

6. $(-3, 7)$, $(1, -1)$

7. $(1, \sqrt{3})$, $(-1, 1)$

8. $(-2, 0)$, $(0, \sqrt{2})$

9. Show that the points $(4, 0)$, $(2, 1)$, and $(-1, -5)$ are vertices of a right triangle.

10. Show that the points $(1, -3)$, $(3, 2)$, and $(-2, 4)$ are vertices of an isosceles triangle.

11. Show that the points $(0, 0)$, $(1, 2)$, $(2, 1)$, and $(3, 3)$ are vertices of a rhombus.

12. Show that the points $(0, 1)$, $(3, 7)$, $(4, 4)$, and $(1, -2)$ are vertices of a parallelogram.

13. Use the Distance Formula to determine if the points $(0, -4)$, $(2, 0)$, and $(3, 2)$ lie on a straight line.

14. Use the Distance Formula to determine if the points $(0, 4)$, $(7, -6)$, and $(-5, 11)$ lie on a straight line.

15. Use the Distance Formula to determine if the points $(-2, 1)$, $(-1, 0)$, and $(2, -2)$ lie on a straight line.

16. Find y so that the distance from the origin to the point $(3, y)$ is 5.

17. Find x so that the distance from the origin to the point $(x, -4)$ is 5.

18. Find x so that the distance from $(2, -1)$ to the point $(x, 2)$ is 5.

19. Find the relationship between x and y so that the point (x, y) is equidistant from $(4, -1)$ and $(-2, 3)$.

20. Find the relationship between x and y so that the point (x, y) is equidistant from $(3, \frac{5}{2})$ and $(-7, -1)$.

21. Use the Midpoint Rule successively to find the three points that divide the line segment joining (x_1, y_1) and (x_2, y_2) into four equal parts.

22. Use the results of Exercise 21 to find the points that divide into four equal parts the line segment joining these points:
    (a) $(1, -2)$ and $(4, -1)$
    (b) $(-2, -3)$ and $(0, 0)$

In Exercises 23 and 24, complete the square on each expression.

23. (a) $x^2 + 5x$
    (b) $x^2 + 8x + 7$

24. (a) $4x^2 - 4x - 39$
    (b) $5x^2 + x$

In Exercises 25–30, match the given equation with its graph. [Graphs are labeled (a), (b), (c), (d), (e), and (f).]

25. $x^2 + y^2 = 1$

26. $(x - 1)^2 + (y - 3)^2 = 4$

27. $(x - 1)^2 + y^2 = 0$

28. $\left(x + \frac{1}{2}\right)^2 + \left(y - \frac{3}{4}\right)^2 = \frac{1}{4}$

29. $(x + 3)^2 + (y - 1)^2 = 16$

30. $x^2 + (y - 1)^2 = 1$

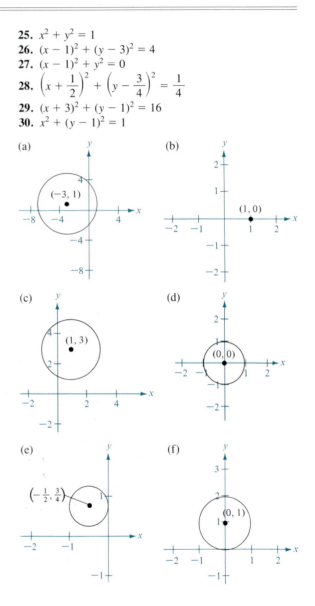

In Exercises 31–40, write the equation of the specified circle in general form.

31. Center: $(0, 0)$; radius: 3

32. Center: $(0, 0)$; radius: 5

33. Center: $(2, -1)$; radius: 4

34. Center: $(-4, 3)$; radius: $\frac{5}{8}$

35. Center: $(-1, 2)$; point on circle: $(0, 0)$

36. Center: $(3, -2)$; point on circle: $(-1, 1)$

37. Endpoints of diameter: (2, 5), (4, −1)
38. Endpoints of diameter: (1, 1), (−1, −1)
39. Points on circle: (0, 0), (0, 8), (6, 0)
40. Points on circle: (1, −1), (2, −2), (0, −2)

In Exercises 41–48, write each equation in standard form and sketch its graph.

41. $x^2 + y^2 - 2x + 6y + 6 = 0$
42. $x^2 + y^2 - 2x + 6y - 15 = 0$
43. $x^2 + y^2 - 2x + 6y + 10 = 0$
44. $3x^2 + 3y^2 - 6y - 1 = 0$
45. $2x^2 + 2y^2 - 2x - 2y - 3 = 0$
46. $4x^2 + 4y^2 - 4x + 2y - 1 = 0$
47. $16x^2 + 16y^2 + 16x + 40y - 7 = 0$
48. $x^2 + y^2 - 4x + 2y + 3 = 0$

49. Find an equation for the path of a communications satellite in a circular orbit 22,000 miles above the surface of the earth. (Assume that the radius of the earth is 4000 miles.)
50. Find the equation of the circle passing through the points (1, 2), (−1, 2), and (2, 1).
51. Find the equation of the circle passing through the points (4, 3), (−2, −5), and (5, 2).
52. Find the equations of the circles passing through the points (4, 1) and (6, 3) and having radius $\sqrt{10}$.

In Exercises 53–56, sketch the set of all points satisfying the given inequality.

53. $x^2 + y^2 - 4x + 2y + 1 \leq 0$
54. $x^2 + y^2 - 4x + 2y + 1 > 0$

55. $(x + 3)^2 + (y - 1)^2 < 9$
56. $(x - 1)^2 + \left(y - \frac{1}{2}\right)^2 > 1$

57. Prove that

$$\left(\frac{2x_1 + x_2}{3}, \frac{2y_1 + y_2}{3}\right)$$

is one of the points of trisection of the line segment joining (x_1, y_1) and (x_2, y_2). Also, find the midpoint of the line segment joining

$$\left(\frac{2x_1 + x_2}{3}, \frac{2y_1 + y_2}{3}\right)$$

and

$$(x_2, y_2)$$

to find the second point of trisection of the line segment joining (x_1, y_1) and (x_2, y_2).
58. Use the results of Exercise 57 to find the points of trisection of the line segment joining these points:
(a) (1, −2) and (4, 1)   (b) (−2, −3) and (0, 0)
59. Prove that the line segments joining the midpoints of the opposite sides of a quadrilateral bisect each other.
60. Prove that the midpoint of the hypotenuse of a right triangle is equidistant from each of the three vertices.
61. Prove that an angle inscribed in a semicircle is a right angle.
62. Prove that the perpendicular bisector of a chord of a circle passes through the center of the circle.
63. Complete the proof of Theorem 1.5 by filling in the missing algebraic details.

SECTION TOPICS ▪
The graph of an equation ▪
Point-plotting method ▪
Intercepts of a graph ▪
Symmetry of a graph ▪
Points of intersection ▪
Mathematical models ▪

1.3
Graphs of equations

The idea of using a graph to show how two quantities are related is familiar. News magazines frequently show graphs that compare the rate of inflation, the gross national product, wholesale prices, or the unemployment rate to the time of year. Industrial firms and businesses use graphs to report their monthly production and sales statistics. The value of such graphs is that they provide a geometric picture of the way one quantity changes with respect to another.

Frequently the relationship between two quantities is expressed in the form of an equation. For instance, degrees on the Fahrenheit scale are related to degrees on the Celsius scale by the equation $F = \frac{9}{5}C + 32$. In this section we introduce a basic procedure for determining the geometric picture associated with such an equation.

TABLE 1.2

x	0	1	2	3	4
y	7	4	1	−2	−5

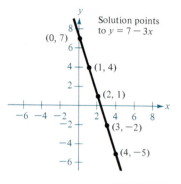

FIGURE 1.23

The graph of an equation

Consider the equation $3x + y = 7$. If $x = 2$ and $y = 1$, the equation is satisfied and we call the point $(2, 1)$ a **solution point** of the equation. Of course, there are other solution points, such as $(1, 4)$ and $(0, 7)$. We can make up a **table of values** for x and y by choosing arbitrary values for x and determining the corresponding values for y. To determine the values for y, it is convenient to write the equation in the form

$$y = 7 - 3x$$

Thus, $(0, 7)$, $(1, 4)$, $(2, 1)$, $(3, -2)$, and $(4, -5)$ are all solution points of the equation $3x + y = 7$, as shown in Table 1.2. There are actually infinitely many solution points for the equation, and the set of all such points is called the **graph** of the equation, as shown in Figure 1.23.

Remark   Even though we refer to the sketch shown in Figure 1.23 as the graph of $y = 7 - 3x$, it actually represents only a *portion* of the graph. The entire graph is a line that would extend off the page.

DEFINITION OF THE GRAPH OF AN EQUATION IN TWO VARIABLES	The **graph of an equation** involving two variables x and y is the set of all points in the plane that are solution points to the equation.

Remark   The graph of an equation is sometimes called its **locus.** Thus, when we refer to the locus of all points satisfying a particular equation, we are referring to its graph.

EXAMPLE 1   The point-plotting method

Sketch the graph of the equation $y = x^2 - 2$.

Solution: First, we make a table of values (Table 1.3) by choosing several convenient values of x and calculating the corresponding values of $y = x^2 - 2$. Next, we locate these points in the plane, as in Figure 1.24. Finally, we connect the points by a *smooth curve,* as shown in Figure 1.25. This particular graph is called a **parabola.** It is one of the conic sections we will study in Chapter 11.

TABLE 1.3

x	−2	−1	0	1	2	3
y	2	−1	−2	−1	2	7

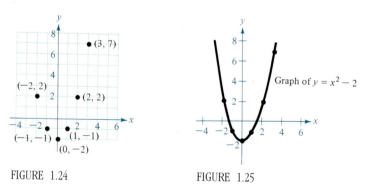

FIGURE 1.24   FIGURE 1.25

We call this method of sketching a graph the **point-plotting method.** Its basic features are given in the following summary.

THE POINT-PLOTTING METHOD OF GRAPHING	1. Make up a table of several solution points of the equation. 2. Plot these points in the plane. 3. Connect the points with a smooth curve.

Steps 1 and 2 of the point-plotting method are usually easy, but step 3 can be the source of some major difficulties. For instance, how would you connect the four points in Figure 1.26? Without additional points or further information about the equation, any one of the three graphs in Figure 1.27 would be reasonable.

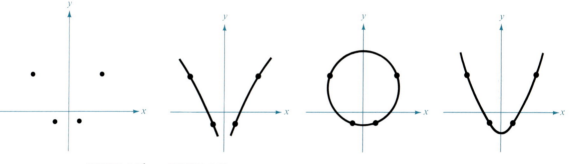

FIGURE 1.26   FIGURE 1.27

Obviously, with too few solution points, we could badly misrepresent the graph of a given equation. How many points should be plotted? For a straight-line graph, two points are sufficient. For more complicated graphs, we need many more points. More sophisticated graphing techniques are discussed in later chapters. In the meantime we suggest that you plot enough points so as to reveal the essential behavior of the graph. The more solution points you plot, the more accurate your graph will be. (A programmable calculator is very useful for determining the solution points.)

Two solution points that are usually easy to determine are those having zero as either their x- or y-coordinate. Such points are called **intercepts,** because they are the points at which the graph intersects the x- or y-axis.

DEFINITION OF INTERCEPTS	The point $(a, 0)$ is called an **x-intercept** of the graph of an equation if it is a solution point of the equation. Such points can be found by letting y be zero and solving the equation for x. The point $(0, b)$ is called a **y-intercept** of the graph of an equation if it is a solution point of the equation. Such points can be found by letting x be zero and solving the equation for y.

| **Remark** Some texts denote the *x*-intercept as the *x*-coordinate of the point $(a, 0)$ rather than the point itself. Unless it is necessary to make a distinction, we will use the term *intercept* to mean either the point or the coordinate.

Of course, it is possible for a graph to have no intercepts, or several. For instance, consider the four graphs in Figure 1.28.

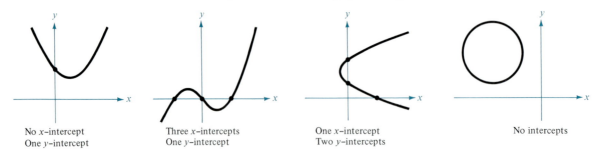

No *x*-intercept   Three *x*-intercepts   One *x*-intercept   No intercepts
One *y*-intercept   One *y*-intercept   Two *y*-intercepts

FIGURE 1.28

EXAMPLE 2 *Finding x- and y-intercepts*

Find the *x*- and *y*-intercepts for the graphs of the following equations:
(a) $y = x^3 - 4x$ (b) $y^2 - 3 = x$

Solution:
(a) Let $y = 0$. Then, $0 = x(x^2 - 4)$ has solutions $x = 0$ and $x = \pm 2$.

$$x\text{-intercepts:}\quad (0, 0), (2, 0), (-2, 0)$$

Let $x = 0$. Then $y = 0$.

$$y\text{-intercept:}\quad (0, 0)$$

(See Figure 1.29.)
(b) Let $y = 0$. Then $-3 = x$.

$$x\text{-intercept:}\quad (-3, 0)$$

Let $x = 0$. Then $y^2 - 3 = 0$ has solutions $y = \pm\sqrt{3}$.

$$y\text{-intercepts:}\quad (0, \sqrt{3}), (0, -\sqrt{3})$$

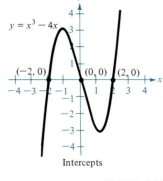

Intercepts

FIGURE 1.29

EXAMPLE 3 *Sketching the graph of an equation*

Sketch the graph of the equation $x^2 + 4y^2 = 16$.

Solution: Sometimes it is beneficial to rewrite an equation before calculating solution points. For example, if we rewrite the equation $x^2 + 4y^2 = 16$ as

$$x = \pm\sqrt{16 - 4y^2} = \pm 2\sqrt{4 - y^2}$$

then we can easily determine several solution points by choosing values for *y* and calculating the corresponding values for *x*. Note that $x = \pm 2\sqrt{4 - y^2}$ is defined only when $|y| \le 2$. By plotting these points and connecting them with a smooth curve, we create the graph shown in Figure 1.30. This particular graph is called an **ellipse**. It is one of the conic sections we will be studying in Chapter 11.

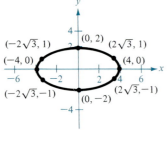

Graph of $x^2 + 4y^2 = 16$

FIGURE 1.30

Symmetry of a graph

The graphs shown in Figures 1.25 and 1.30 are both said to be **symmetric** with respect to the *y*-axis. This means that if the Cartesian plane were folded along the *y*-axis, the portion of the graph to the left of the *y*-axis would coincide with the portion to the right of the *y*-axis. Another way to describe this symmetry is to say that the graph is a reflection of itself with respect to the *y*-axis. Symmetry with respect to the *x*-axis can be described in a similar manner.

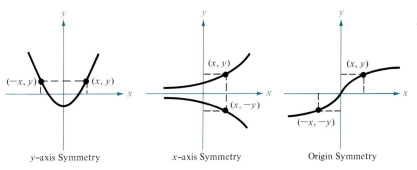

FIGURE 1.31   *y*–axis Symmetry   *x*–axis Symmetry   Origin Symmetry

Knowing that a graph has symmetry *before* attempting to sketch it is helpful because then we need only half as many solution points as we otherwise would. We define three basic types of symmetry, as shown in Figure 1.31.

DEFINITION OF SYMMETRY

A graph is said to be **symmetric with respect to the *y*-axis** if, whenever (x, y) is a point on the graph, $(-x, y)$ is also on the graph.

A graph is said to be **symmetric with respect to the *x*-axis** if, whenever (x, y) is a point on the graph, $(x, -y)$ is also on the graph.

A graph is said to be **symmetric with respect to the origin** if, whenever (x, y) is a point on the graph, $(-x, -y)$ is also on the graph.

Remark Note that a graph is symmetric with respect to the origin if a rotation of 180° leaves the graph unchanged.

Suppose we apply the definition of symmetry to the graph of the equation shown in Figure 1.25.

$$y = x^2 - 2 \qquad \text{Given equation}$$
$$y = (-x)^2 - 2 \qquad \text{Replace } x \text{ by } -x$$
$$y = x^2 - 2 \qquad \text{Equivalent equation}$$

Since substituting $-x$ for x produces an equivalent equation, it follows that if (x, y) is a solution point of the given equation, then $(-x, y)$ must also be a solution point. Therefore, the graph of $y = x^2 - 2$ is symmetric with respect to the *y*-axis.

A similar test can be made for symmetry with respect to the x-axis or to the origin. These three tests are summarized in the following theorem, which we list without proof.

THEOREM 1.7   **TESTS FOR SYMMETRY**
The graph of an equation is symmetric with respect to

1. the y-axis if replacing x by $-x$ yields an equivalent equation.
2. the x-axis if replacing y by $-y$ yields an equivalent equation.
3. the origin if replacing x by $-x$ *and* y by $-y$ yields an equivalent equation.

EXAMPLE 4   Testing for origin symmetry

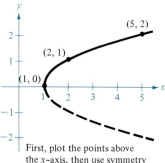

$y = 2x^3 - x$

(1, 1)

(−1, −1)

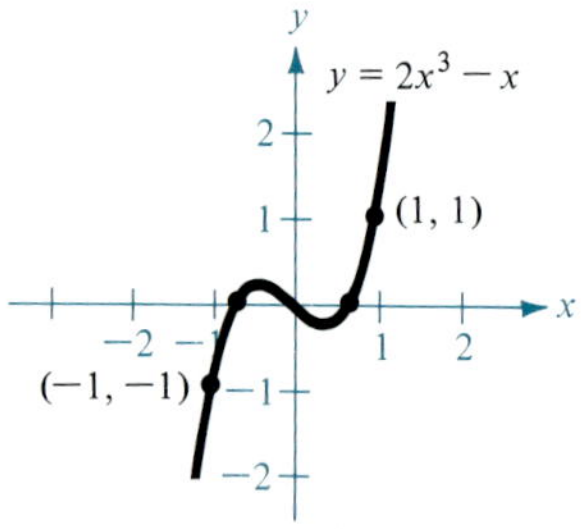

Origin Symmetry

FIGURE 1.32

Show that the graph of $y = 2x^3 - x$ is symmetric with respect to the origin.

Solution: We apply the test for origin symmetry as follows.

$$y = 2x^3 - x \qquad \text{Given equation}$$
$$-y = 2(-x)^3 - (-x) \qquad \text{Replace } x \text{ by } -x \text{ and } y \text{ by } -y$$
$$-y = -2x^3 + x$$
$$y = 2x^3 - x \qquad \text{Equivalent equation}$$

Since the replacement produces an equivalent equation, we conclude that the graph of $y = 2x^3 - x$ is symmetric with respect to the origin, as shown in Figure 1.32.

EXAMPLE 5   Using symmetry to sketch a graph

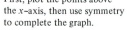

(5, 2)

(2, 1)

(1, 0)

First, plot the points above the x–axis, then use symmetry to complete the graph.

FIGURE 1.33

Sketch the graph of $x - y^2 = 1$.

Solution: The graph is symmetric with respect to the x-axis since replacing y by $-y$ yields

$$x - (-y)^2 = 1$$
$$x - y^2 = 1$$

This means that the graph below the x-axis is a mirror image of the graph above the x-axis. Hence, we can first sketch the graph above the x-axis and then reflect it to obtain the entire graph, as shown in Figure 1.33.

Points of intersection

Since each point of a graph is a solution point of its corresponding equation, a **point of intersection** of two graphs is simply a solution point that satisfies both equations. Moreover, the points of intersection of two graphs can be found by solving the equations simultaneously.

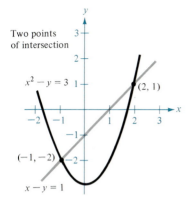

Two points
of intersection

$x^2 - y = 3$

$(2, 1)$

$(-1, -2)$

$x - y = 1$

FIGURE 1.34

EXAMPLE 6   *Finding points of intersection*

Find all points of intersection of the graphs of

$$x^2 - y = 3 \quad \text{and} \quad x - y = 1$$

Solution: It is helpful to begin by making a sketch for each equation on the *same* coordinate plane, as shown in Figure 1.34. Now, from the figure it appears that the two graphs have two points of intersection. To find these two points, we proceed as follows.

$$y = x^2 - 3 \quad \text{and} \quad y = x - 1 \qquad \text{Solve each equation for } y$$

Then

$$x^2 - 3 = x - 1 \qquad \text{Equate } y\text{-values}$$
$$x^2 - x - 2 = 0$$
$$(x - 2)(x + 1) = 0 \qquad \text{Solve for } x$$

The corresponding values of y are obtained by substituting $x = 2$ and $x = -1$ into either of the original equations. For instance, if we choose the equation $y = x - 1$, then the values of y are 1 and -2, respectively. Therefore, the two points of intersection are $(2, 1)$ and $(-1, -2)$.

Mathematical models

In this text, we will frequently be concerned with the use of equations as **mathematical models** of real-world phenomena. In developing a mathematical model to represent actual data, we strive for two (often conflicting) goals: accuracy and simplicity. That is, we would like the model to be simple enough to be workable and at the same time accurate enough to produce meaningful results. Our next example describes a typical mathematical model.

EXAMPLE 7   *A mathematical model*

The median income (between 1955 and 1980) for U.S. families with married couples is given in Table 1.4. A mathematical model for these data is given by

$$y = 0.0333t^2 - 0.1293t + 5.0537$$

where y represents the median income in 1000s of dollars and t represents the year, with $t = 0$ corresponding to 1955. (A technique for finding such a model is discussed in Section 15.9.) Using a graph, compare the data with the model and use the model to predict the median income for 1985.

TABLE 1.4

Year	1955	1960	1965	1970	1975	1980
Income (in 1000s)	4.6	5.9	7.3	10.5	14.9	23.1

Solution: Table 1.5 and Figure 1.35 compare the model's values with the actual values.

TABLE 1.5

t	0	5	10	15	20	25
y	5.1	5.2	7.1	10.6	15.8	22.6
Actual income	4.6	5.9	7.3	10.5	14.9	23.1

To predict the median income for 1985, we let $t = 30$ and calculate y as follows:

$$y = 0.0333(30)^2 - 0.1293(30) + 5.0537 \approx 31.1$$

Thus, we estimate the 1985 median income to be $31,100.

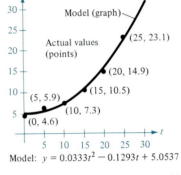

Model: $y = 0.0333t^2 - 0.1293t + 5.0537$

FIGURE 1.35

Section Exercises 1.3

In Exercise 1–6, match the given equation with its graph. [Graphs are labeled (a), (b), (c), (d), (e), and (f).]

1. $y = x - 2$

2. $y = -\dfrac{1}{2}x + 2$

3. $y = x^2 + 2x$

4. $y = \sqrt{9 - x^2}$

5. $y = |x| - 2$

6. $y = x^3 - x$

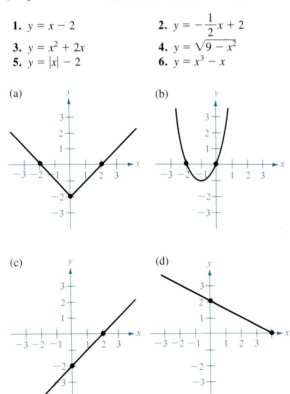

(a)

(b)

(c)

(d)

(e)

(f)

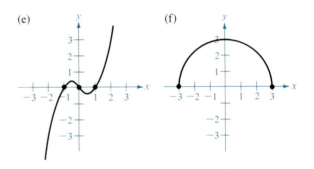

In Exercises 7–16, check for symmetry about both axes and the origin.

7. $y = x^2 - 2$

8. $y = x^4 - x^2 + 3$

9. $x^2y - x^2 + 4y = 0$

10. $x^2y - x^2 - 4y = 0$

11. $y^2 = x^3 - 4x$

12. $xy^2 = -10$

13. $y = x^3 + x$

14. $xy = 1$

15. $y = \dfrac{x}{x^2 + 1}$

16. $y = x^3 + x - 3$

In Exercises 17–26, find the intercepts.

17. $y = 2x - 3$

18. $y = (x - 1)(x - 3)$

19. $y = x^2 + x - 2$

20. $y^2 = x^3 - 4x$

21. $y = x^2\sqrt{9 - x^2}$

22. $xy = 4$

23. $y = \dfrac{x - 1}{x - 2}$

24. $y = \dfrac{x^2 + 3x}{(3x + 1)^2}$

25. $x^2y - x^2 + 4y = 0$

26. $y = 2x - \sqrt{x^2 + 1}$

In Exercises 27–46, use the methods of this section to sketch the graph of each equation. Identify the intercepts and test for symmetry.

27. $y = x$

28. $y = x - 2$

29. $y = x + 3$

30. $y = 2x - 3$

31. $y = -3x + 2$

32. $y = -\frac{1}{2}x + 2$

33. $y = \frac{1}{2}x - 4$

34. $y = x^2 + 3$

35. $y = 1 - x^2$

36. $y = 2x^2 + x$

37. $y = -2x^2 + x + 1$

38. $y = x^3 - 1$

39. $y = x^3 + 2$

40. $y = \sqrt{9 - x^2}$

41. $x^2 + 4y^2 = 4$

42. $x = y^2 - 4$

43. $y = (x + 2)^2$

44. $y = \frac{1}{x^2 + 1}$

45. $y = \frac{1}{x}$

46. $y = 2x^4$

In Exercises 47–56, find the points of intersection of the graphs of the equations; check your results.

47. $x + y = 2$, $2x - y = 1$

48. $2x - 3y = 13$, $5x + 3y = 1$

49. $x + y = 7$, $3x - 2y = 11$

50. $x^2 + y^2 = 25$, $2x + y = 10$

51. $x^2 + y^2 = 5$, $x - y = 1$

52. $x^2 + y = 4$, $2x - y = 1$

53. $y = x^3$, $y = x$

54. $y = x^4 - 2x^2 + 1$, $y = 1 - x^2$

55. $y = x^3 - 2x^2 + x - 1$, $y = -x^2 + 3x - 1$

56. $x = 3 - y^2$, $y = x - 1$

In Exercises 57 and 58, find the sales necessary to break even ($R = C$) for the given cost C of x units and the given revenue R obtained by selling x units.

57. $C = 8650x + 250,000$
$R = 9950x$

58. $C = 5.5\sqrt{x} + 10,000$
$R = 3.29x$

59. Determine whether the points $(1, 2)$, $(1, -1)$, and $(4, 5)$ lie on the graph of $2x - y - 3 = 0$.

60. Determine whether the points $(1, -\sqrt{3})$, $(\frac{1}{2}, -1)$, and $(\frac{3}{2}, \frac{7}{2})$ lie on the graph of $x^2 + y^2 = 4$.

61. Determine whether the points $(1, \frac{1}{5})$, $(2, \frac{1}{2})$, and $(-1, -2)$ lie on the graph of $x^2y - x^2 + 4y = 0$.

62. Determine whether the points $(0, 2)$, $(-2, -\frac{1}{6})$, and $(3, -6)$ lie on the graph of $x^2 - xy + 4y = 3$.

63. For what values of k does the graph of $y = kx^3$ pass through these points?
(a) $(1, 4)$
(b) $(-2, 1)$
(c) $(0, 0)$
(d) $(-1, -1)$

64. For what values of k does the graph of $y^2 = 4kx$ pass through these points?
(a) $(1, 1)$
(b) $(2, 4)$
(c) $(0, 0)$
(d) $(3, 3)$

65. The Consumer Price Index (CPI) for the 1970s is given in the following table.

Year	1970	1971	1972	1973	1974
CPI	116.3	121.3	125.3	133.1	147.7

Year	1975	1976	1977	1978	1979
CPI	161.2	170.5	181.5	195.3	211.1

A mathematical model for the CPI during this 10-year period is

$$y = 0.55t^2 + 5.85t + 114.41$$

where y represents the CPI and t represents the year, with $t = 0$ corresponding to 1970.
(a) Use a graph to compare the CPI with the model.
(b) Use the model to predict the CPI for 1985.

66. From the model in Exercise 65, we obtain the model

$$V = \frac{100}{0.55t^2 + 5.85t + 114.41}$$

where V represents the purchasing power of the dollar (in terms of constant 1967 dollars) and t represents the year, with $t = 0$ corresponding to 1970. Use this model to complete the following table.

t	0	2	4	6	8	10	12
V							

67. The farm population in the United States as a percentage of the total population is given in the following table.

Year	1950	1955	1960	1965	1970	1975	1979
Percentage	15.3	11.6	8.7	6.4	4.8	4.2	3.4

A mathematical model for these data is given by

$$y = \frac{100}{4.90 + 0.79t}$$

where y represents the percentage and t represents the year, with $t = 0$ corresponding to 1950.
(a) Use a graph to compare the actual percentage with that given by the model.
(b) Use the model to predict the farm percentage of the population in 1990.

68. The average number of acres per farm in the United States is given in the following table.

Year	1950	1960	1965	1970	1975	1978
Number of acres	213	297	340	374	391	401

A mathematical model for these data is given by

$$y = -0.13t^2 + 10.43t + 211.3$$

where t represents the year, with $t = 0$ corresponding to 1950.

(a) Use a graph to compare the actual number of acres per farm with that given by the model.

(b) Use the model to predict the average number of acres per farm in the United States in 1985.

69. Prove that if a graph is symmetric with respect to the x-axis and to the y-axis, then it is symmetric with respect to the origin. Give an example to show that the converse is not true.

70. Prove that if a graph is symmetric with respect to one axis and the origin, then it is symmetric with respect to the other axis also.

1.4
Lines in the plane

In this section you can get an inkling of the power and usefulness of calculus in measuring rates of change. Although we will not be introducing calculus techniques for measuring rates of change until Chapter 3, the preliminary material in this section is critical.

Consider an automobile that is traveling at a *constant* rate on a straight highway. At 2:00 P.M. the car has traveled 20 miles from a particular city and at 2:30 the car has traveled 48 miles, as shown in Figure 1.36. How fast is the car traveling?

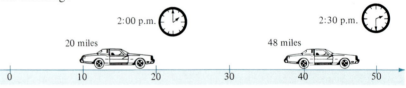

FIGURE 1.36

To measure the rate in *miles per hour,* we divide the distance traveled by the elapsed time.

$$\text{rate (mph)} = \frac{\text{distance (miles)}}{\text{time (hours)}} = \frac{48 - 20}{2.5 - 2} = \frac{28}{0.5} = 56 \text{ mph}$$

If we let s be the distance from the city in miles and t be the time in hours, then the distance is related to the time by the **linear equation**

$$s = 56t - 92, \quad 2 \le t \le 2.5$$

The graph of this equation is a line,* as shown in Figure 1.37. For every unit that t increases, s increases 56 units. Mathematically, we say that this line has a **slope** of 56.

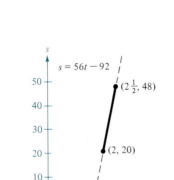

FIGURE 1.37   *In this text, we use the term *line* to mean *straight* line.

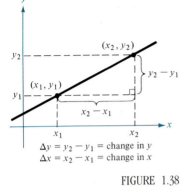

$\Delta y = y_2 - y_1 =$ change in y
$\Delta x = x_2 - x_1 =$ change in x

FIGURE 1.38

The slope of a line

By the **slope** of a (nonvertical) line, we mean the number of units a line rises (or falls) vertically for each unit of horizontal change from left to right. For instance, consider the two points (x_1, y_1) and (x_2, y_2) on the line in Figure 1.38. As we move from left to right along this line, a vertical change of

$$\Delta y = y_2 - y_1 \text{ units}$$

corresponds to a horizontal change of

$$\Delta x = x_2 - x_1 \text{ units}$$

(Δ is the Greek uppercase letter delta, and the symbols Δy and Δx are read ''delta y'' and ''delta x.'') We use the ratio of Δy to Δx to define the slope of a line as follows.

DEFINITION OF THE SLOPE OF A LINE

The **slope** m of a line passing through the points (x_1, y_1) and (x_2, y_2) is

$$m = \frac{\Delta y}{\Delta x} = \frac{y_2 - y_1}{x_2 - x_1}, \quad x_1 \neq x_2$$

Remark Note that

$$\frac{y_2 - y_1}{x_2 - x_1} = \frac{-(y_1 - y_2)}{-(x_1 - x_2)} = \frac{y_1 - y_2}{x_1 - x_2}$$

Hence, it does not matter in which order we subtract *as long as* we are consistent and both ''subtracted coordinates'' come from the same point.

In our first example (Figure 1.39) you will see that, as we move from left to right,

1. a line with positive slope ($m > 0$) *rises*,
2. a line with negative slope ($m < 0$) *falls,* and
3. a line with zero slope ($m = 0$) is *horizontal*.

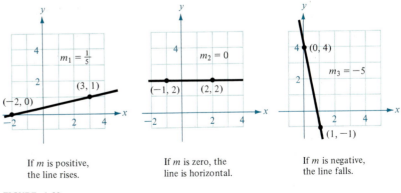

If m is positive, the line rises.

If m is zero, the line is horizontal.

If m is negative, the line falls.

FIGURE 1.39

EXAMPLE 1   The slope of a line passing through two points

(a) The slope of the line containing $(-2, 0)$ and $(3, 1)$ is

$$m = \frac{1 - 0}{3 - (-2)} = \frac{1}{3 + 2} = \frac{1}{5}$$

(b) The slope of the line containing $(-1, 2)$ and $(2, 2)$ is

$$m = \frac{2 - 2}{2 - (-1)} = \frac{0}{3} = 0$$

(c) The slope of the line containing $(0, 4)$ and $(1, -1)$ is

$$m = \frac{-1 - 4}{1 - 0} = \frac{-5}{1} = -5$$

See Figure 1.39.

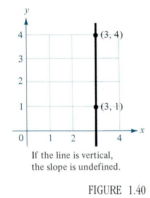

If the line is vertical, the slope is undefined.

FIGURE 1.40

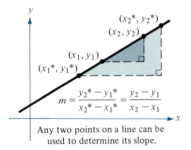

Any two points on a line can be used to determine its slope.

$$m = \frac{y_2{}^* - y_1{}^*}{x_2{}^* - x_1{}^*} = \frac{y_2 - y_1}{x_2 - x_1}$$

FIGURE 1.41

Remark In Example 1 note that we did not consider the slope of a vertical line. This was not an oversight! The definition of slope does not cover vertical lines. For example, consider the points $(3, 4)$ and $(3, 1)$ on the line shown in Figure 1.40. In attempting to find the slope of this line, we obtain

$$m = \frac{4 - 1}{3 - 3} \qquad \text{Division by zero is undefined}$$

The same situation (a zero denominator) arises with any two points on a vertical line. Thus, we say that the slope of a vertical line is undefined.

It is important to realize that *any* two points on a line can be used to calculate the slope. This can be verified from the similar triangles shown in Figure 1.41. (Recall that the ratios of corresponding sides of similar triangles are equal.)

Equations of lines

If we know the slope of a line and one point on the line, how can we determine the equation of the line? Figure 1.41 leads us to the answer to this question. For, if (x_1, y_1) is a point lying on a line of slope m and (x, y) is any *other* point on the line, then

$$\frac{y - y_1}{x - x_1} = m$$

This equation, involving the two variables x and y, can be rewritten in the form

$$y - y_1 = m(x - x_1)$$

which is commonly referred to as the **point-slope equation of a line.**

THEOREM 1.8   **POINT-SLOPE EQUATION OF A LINE**
The equation of the line with slope m passing through the point (x_1, y_1) is given by

$$y - y_1 = m(x - x_1)$$

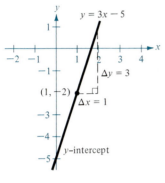

FIGURE 1.42

EXAMPLE 2   The point-slope equation of a line

Find the equation of the line that has a slope of 3 and passes through the point $(1, -2)$.

Solution:

$$y - y_1 = m(x - x_1) \qquad \text{Point-slope form}$$
$$y - (-2) = 3(x - 1)$$
$$y + 2 = 3x - 3$$
$$y = 3x - 5$$

See Figure 1.42.

EXAMPLE 3   An application

The total U.S. sales (including inventories) during the first two quarters of 1978 were 539.9 and 560.2 billion dollars, respectively. Assuming a *linear growth pattern,* predict the total sales during the fourth quarter of 1978.

Solution: Referring to Figure 1.43, we let $(1, 539.9)$ and $(2, 560.2)$ be two points on the line representing the total U.S. sales. We let x represent the quarter and y represent the sales in billions of dollars. The slope of the line passing through these two points is

$$m = \frac{560.2 - 539.9}{2 - 1} = 20.3$$

Thus, the equation of the line is

$$y - y_1 = m(x - x_1)$$
$$y - 539.9 = 20.3(x - 1)$$
$$y = 20.3(x - 1) + 539.9 = 20.3x + 519.6$$

Now, using this linear model, we estimate the fourth quarter sales ($x = 4$) to be

$$y = (20.3)(4) + 519.6 = 600.8 \text{ billion dollars}$$

(In this particular case, the estimate proves to be quite good. The actual fourth quarter sales in 1978 were 600.5 billion dollars.)

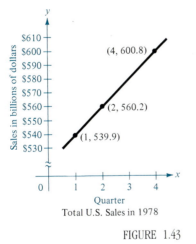

FIGURE 1.43

The prediction method illustrated in Example 3 is called **linear extrapolation.** Note that the extrapolated point does not lie between the given points. (See Figure 1.44.) When the estimated point lies between two given points, we call the procedure **linear interpolation.**

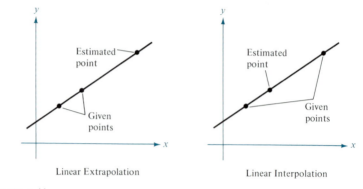

FIGURE 1.44

Graphing lines in the plane

In Section 1.2, we mentioned that many problems in analytic geometry can be classified in two basic categories: (1) Given a graph, what is its equation? and (2) Given an equation, what is its graph? The point-slope equation of a line fits in the first category. However, this form is *not* particularly useful for solving problems in the second category. The form that is best suited to graphing lines is called the **slope-intercept** form for the equation of a line.

THEOREM 1.9   THE SLOPE-INTERCEPT EQUATION OF A LINE
The graph of the equation

$$y = mx + b$$

is a line having a *slope* of m and a *y-intercept* at $(0, b)$.

Remark   For linear equations of the form $y = mx$ ($m \neq 0$), we say y is **proportional** to x and m is the **constant of proportionality.**

EXAMPLE 4   Sketching lines in the plane

Sketch the graphs of the following linear equations:
(a) $y = 2x + 1$
(b) $y = 2$
(c) $3y + x - 6 = 0$

Solution:

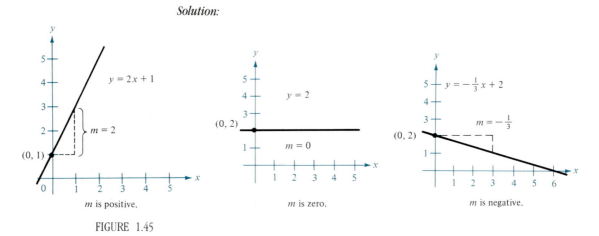

FIGURE 1.45

(a) The y-intercept occurs at $b = 1$, and since the slope is 2 we know that this line rises 2 units for each unit it moves to the right. (See Figure 1.45.)

(b) The y-intercept occurs at $b = 2$, and since the slope is zero we know that the line is horizontal. That is, it doesn't rise or fall. (See Figure 1.45.)

(c) We begin by writing the equation in slope-intercept form:

$$3y + x - 6 = 0$$
$$3y = -x + 6$$
$$y = -\frac{1}{3}x + 2$$

Thus, the y-intercept occurs at $b = 2$ and the slope is $m = -\frac{1}{3}$. This means that the line falls 1 unit for every 3 units it moves to the right. (See Figure 1.45.)

From the slope-intercept equation of a line we can see that a horizontal line ($m = 0$) has an equation of the form

$$y = (0)x + b \quad \text{or} \quad y = b \qquad \text{Horizontal line}$$

This is consistent with the fact that each point on a horizontal line through $(0, b)$ has a y-coordinate of b. (See Figure 1.46.) Similarly, each point on a vertical line through $(a, 0)$ has an x-coordinate of a. (See Figure 1.47.) Hence a vertical line has an equation of the form

$$x = a \qquad \text{Vertical line}$$

This equation cannot be written in the slope-intercept form since the slope of a vertical line is undefined. However, *every* line has an equation that can be written in the **general form**

$$Ax + By + C = 0 \qquad \text{General form}$$

where A and B are not *both* zero. If $A = 0$ (and $B \neq 0$), the equation can be reduced to the form $y = b$, which represents a horizontal line. If $B = 0$ (and $A \neq 0$), the general equation can be reduced to the form $x = a$, which represents a vertical line.

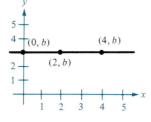

Horizontal Line: $y = b$

FIGURE 1.46

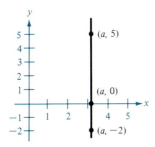

Vertical Line: $x = a$

FIGURE 1.47

Parallel and perpendicular lines

The slope of a line is a convenient tool for determining when two lines are parallel or perpendicular. This is seen in the following two theorems.

THEOREM 1.10   **PARALLEL LINES**
Two distinct nonvertical lines are parallel if and only if their slopes are equal.

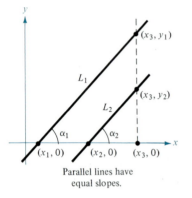

Parallel lines have equal slopes.

FIGURE 1.48

Proof: We will prove one direction of the theorem and leave the proof of its converse as an exercise (see Exercise 76). Assume that we have two parallel lines L_1 and L_2 with slopes m_1 and m_2. If these lines are both horizontal, then $m_1 = m_2 = 0$ and the rule is established. If L_1 and L_2 are not horizontal, then they must intersect the x-axis at points $(x_1, 0)$ and $(x_2, 0)$, as shown in Figure 1.48. Since L_1 and L_2 are parallel, their intersection with the x-axis must produce equal angles α_1 and α_2. (α is the Greek lowercase letter alpha.) Therefore, the two right triangles with vertices

$$(x_1, 0), \ (x_3, 0), \ (x_3, y_1)$$

and

$$(x_2, 0), \ (x_3, 0), \ (x_3, y_2)$$

must be similar. From this we conclude that the ratios of their corresponding sides must be equal, and thus

$$m_1 = \frac{y_1 - 0}{x_3 - x_1} = \frac{y_2 - 0}{x_3 - x_2} = m_2$$

Hence the lines L_1 and L_2 must have equal slopes.

Remark   The angle α_1 shown in the proof of Theorem 1.10 is called the **angle of inclination** of the line L_1. Using trigonometry, we can show that the slope of this line is $m = \tan \alpha_1$.

THEOREM 1.11   **PERPENDICULAR LINES**
Two nonvertical lines are perpendicular if and only if their slopes are related by the equation

$$m_1 = -\frac{1}{m_2}$$

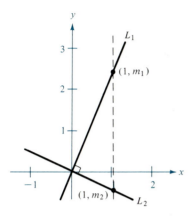

The slopes of perpendicular lines are negative reciprocals of each other.

FIGURE 1.49

Proof: As in the previous theorem, we will prove only one direction of the theorem and leave the other as an exercise (see Exercise 77). Let us assume that we are given two nonvertical perpendicular lines L_1 and L_2 with slopes m_1 and m_2. For simplicity's sake let these two lines intersect at the origin, as shown in Figure 1.49. The vertical line $x = 1$ will intersect L_1 and L_2 at the respective points $(1, m_1)$ and $(1, m_2)$. Since the triangle formed by these two points and the origin is a right triangle, we can apply the Pythagorean Theorem and conclude that

$$\left(\begin{array}{c}\text{distance between}\\ (0,0) \text{ and } (1, m_1)\end{array}\right)^2 + \left(\begin{array}{c}\text{distance between}\\ (0,0) \text{ and } (1, m_2)\end{array}\right)^2 = \left(\begin{array}{c}\text{distance between}\\ (1, m_1) \text{ and } (1, m_2)\end{array}\right)^2$$

Using the Distance Formula, we have

$$(\sqrt{1 + m_1{}^2})^2 + (\sqrt{1 + m_2{}^2})^2 = (\sqrt{0^2 + (m_1 - m_2)^2})^2$$
$$1 + m_1{}^2 + 1 + m_2{}^2 = (m_1 - m_2)^2$$
$$2 + m_1{}^2 + m_2{}^2 = m_1{}^2 - 2m_1m_2 + m_2{}^2$$
$$2 = -2m_1m_2$$
$$-\frac{1}{m_2} = m_1$$

EXAMPLE 5   Finding parallel and perpendicular lines

Find an equation for the line that passes through the point $(2, -1)$ and is
(a) parallel to the line $2x - 3y = 5$
(b) perpendicular to the line $2x - 3y = 5$

Solution: Writing the equation $2x - 3y = 5$ in slope-intercept form, we have

$$y = \frac{2}{3}x - \frac{5}{3}$$

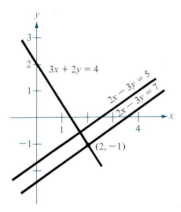

FIGURE 1.50

Therefore, the given line has a slope of $m = \frac{2}{3}$.
(a) The line through $(2, -1)$ that is parallel to the given line has an equation of the form

$$y - (-1) = \frac{2}{3}(x - 2)$$
$$3(y + 1) = 2(x - 2)$$
$$2x - 3y = 7$$

See Figure 1.50. (Note the similarity to the original equation.)
(b) The line through $(2, -1)$ that is perpendicular to the given line has the equation

$$y - (-1) = -\frac{3}{2}(x - 2)$$
$$2(y + 1) = -3(x - 2)$$
$$3x + 2y = 4$$

See Figure 1.50.

Section Exercises 1.4

In Exercises 1–6, estimate the slope of the given line from its graph.

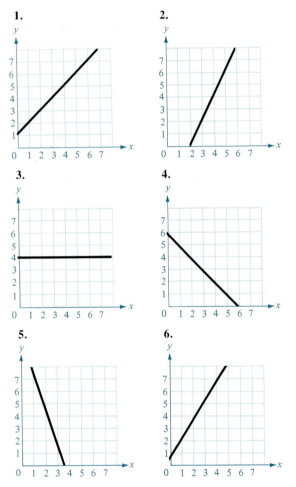

1.

2.

3.

4.

5.

6.

In Exercises 7–14, plot the points and find the slope of the line passing through each pair of points.

7. $(3, -4), (5, 2)$

8. $(-2, 1), (4, -3)$

9. $\left(\frac{1}{2}, 2\right), (6, 2)$

10. $\left(-\frac{3}{2}, -5\right), \left(\frac{5}{6}, 4\right)$

11. $(-6, -1), (-6, -4)$

12. $(2, 1), (2, 5)$

13. $(1, 2), (-2, 2)$

14. $\left(\frac{7}{8}, \frac{3}{4}\right), \left(\frac{5}{4}, -\frac{1}{4}\right)$

In Exercises 15–33, find an equation for the indicated line and sketch its graph.

15. through $(2, 1)$ and $(0, -3)$

16. through $(-3, -4)$ and $(1, 4)$

17. through $(0, 0)$ and $(-1, 3)$

18. through $(-3, 6)$ and $(1, 2)$

19. through $(2, 3)$ and $(2, -2)$

20. through $(6, 1)$ and $(10, 1)$

21. through $(1, -2)$ and $(3, -2)$

22. through $\left(\frac{7}{8}, \frac{3}{4}\right)$ and $\left(\frac{5}{4}, -\frac{1}{4}\right)$

23. through $(0, 3)$ with $m = \frac{3}{4}$

24. through $(-1, 2)$ with m undefined

25. through $(0, 0)$ with $m = \frac{2}{3}$

26. through $(-1, -4)$ with $m = \frac{1}{4}$

27. through $(0, 5)$ with $m = -2$

28. through $(-2, 4)$ with $m = -\frac{3}{5}$

29. y-intercept at 2 with $m = 4$

30. y-intercept at $-\frac{2}{3}$ with $m = \frac{1}{6}$

31. y-intercept at $\frac{2}{3}$ with $m = \frac{3}{4}$

32. y-intercept at 4 with $m = 0$

33. vertical line with x-intercept at 3

34. Show that the line with intercepts $(a, 0)$ and $(0, b)$ has the following equation:

$$\frac{x}{a} + \frac{y}{b} = 1, \quad a \neq 0, b \neq 0$$

In Exercises 35–42, use the result of Exercise 34 to write an equation of the indicated line.

35. x-intercept: $(2, 0)$
y-intercept: $(0, 3)$

36. x-intercept: $(-3, 0)$
y-intercept: $(0, 4)$

37. x-intercept: $\left(-\frac{1}{6}, 0\right)$
y-intercept: $\left(0, -\frac{2}{3}\right)$

38. x-intercept: $\left(-\frac{2}{3}, 0\right)$
y-intercept: $(0, -2)$

39. Point on line: $(1, 2)$
x-intercept: $(a, 0)$
y-intercept: $(0, a)$
$(a \neq 0)$

40. Point on line: $(-3, 4)$
x-intercept: $(a, 0)$
y-intercept: $(0, a)$
$(a \neq 0)$

41. Point on line: $\left(\frac{3}{2}, \frac{1}{2}\right)$
x-intercept: $(2a, 0)$
y-intercept: $(0, a)$
$(a \neq 0)$

42. Point on line: $(-3, 1)$
x-intercept: $(a, 0)$
y-intercept: $(0, -a)$
$(a \neq 0)$

In Exercises 43–48, write an equation of the line through the given point and (a) parallel to the given line and (b) perpendicular to the given line.

43. $(2, 1)$, $4x - 2y = 3$
44. $(-3, 2)$, $x + y = 7$
45. $\left(\dfrac{7}{8}, \dfrac{3}{4}\right)$, $5x + 3y = 0$
46. $(-6, 4)$, $3x + 4y = 7$
47. $(2, 5)$, $x = 4$
48. $(-1, 0)$, $y = -3$

49. Sketch the graph of each equation.
  (a) $y = -3$
  (b) $2x - y - 3 = 0$
  (c) $x + 2y + 6 = 0$

50. Find an equation of the line determined by the points of intersection of

$$y = x^2 - 4x + 3$$
$$y = -x^2 + 2x + 3$$

In Exercises 51–53, use slope to determine if the three given points are collinear (lie on the same straight line). (See Section 1.2, Exercises 13–15.)

51. $(0, -4)$, $(2, 0)$, $(3, 2)$
52. $(0, 4)$, $(7, -6)$, $(-5, 11)$
53. $(-2, 1)$, $(-1, 0)$, $(2, -2)$

In Exercises 54–57, refer to the triangle in Figure 1.51.

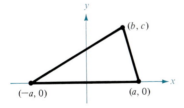

FIGURE 1.51

54. Find the coordinates of the point of intersection of the medians.
55. Find the coordinates of the point of intersection of the perpendicular bisectors of the sides.
56. Find the coordinates of the point of intersection of the altitudes.
57. Show that the points of intersection of Exercises 54, 55, and 56 are collinear.

58. Find the equation of the line giving the relationship between the temperature in degrees Celsius C and degrees Fahrenheit F. Use the fact that water freezes at $0°$ Celsius ($32°$ Fahrenheit) and boils at $100°$ Celsius ($212°$ Fahrenheit).

59. Use the result of Exercise 58 to complete the following table.

C		$-10°$	$10°$			$177°$
F	$0°$			$68°$	$90°$	

60. A manufacturing company pays its assembly line workers $4.50 per hour *plus* an additional piecework rate of $0.75 per unit produced. Write a linear equation for the hourly wages W in terms of the number of units produced per hour x.

61. A small business purchases a piece of equipment for $875. After five years the equipment will be outdated and have no value. Write a linear equation giving the value V of the equipment during the five years it will be used.

62. A company constructs a warehouse for $825,000. It has an estimated useful life of 25 years, after which its value is expected to be $75,000. Use straight-line depreciation to write a linear equation giving the value V of the warehouse during its 25 years of useful life.

63. A real estate office handles an apartment complex with 50 units. When the rent is $280 per month, all 50 units are occupied. However, when the rent is $325, the average number of occupied units drops to 47. Assume that the relationship between the monthly rent p and the demand x is linear.
  (a) Write the equation of the line giving the demand x in terms of the rent p.
  (b) [Linear extrapolation] Use this equation to predict the number of units occupied if the rent is raised to $355.
  (c) [Linear interpolation] Predict the number of units occupied if the rent is lowered to $295.

64. The amount (in billions of dollars) spent by the United States for energy imports between 1975 and 1978 is given in the following table.

Year	1975	1976	1977	1978
t	0	1	2	3
Imports y	$96	$121	$148	$172

  (a) Assuming an approximately linear relation between y and t, write an equation for the line passing through $(0, 96)$ and $(3, 172)$.
  (b) [Linear interpolation] Use this equation to estimate the amount spent in 1976 and 1977. Compare the estimate with the actual amount.

(c) [Linear extrapolation] Predict the amount spent on energy imports in 1980.

(d) What information is given by the slope of the line in part (a)?

65. Consider a plane flying at a constant rate on a direct flight from Los Angeles to Chicago. The distance s (in miles) it has traveled in t hours is given by $s = 560t$.
(a) Sketch the graph of this equation for $t \geq 0$.
(b) What information is given by the slope of this line?

In Exercises 66–71, find the distance between the given point and line (or two lines) using the following formula for the distance between the point (x_1, y_1) and the line $Ax + By + C = 0$,

$$\frac{|Ax_1 + By_1 + C|}{\sqrt{A^2 + B^2}}$$

66. Point: $(0, 0)$; line: $4x + 3y = 10$
67. Point: $(2, 3)$; line: $4x + 3y = 10$

68. Point: $(-2, 1)$; line: $x - y - 2 = 0$
69. Point: $(6, 2)$; line: $x = -1$
70. Lines: $x + y = 1$, $x + y = 5$
71. Lines: $3x - 4y = 1$, $3x - 4y = 10$

72. Find the equation of the line that bisects the acute angle formed by the lines $y = \frac{3}{4}x$ and $y = 2$.

73. Find the distance from the origin to the line $4x + 3y = 10$ by finding the point of intersection of the given line and the line through the origin perpendicular to the given line. Then find the distance between the origin and the point of intersection. Compare the result with that of Exercise 66.

74. Prove analytically that the diagonals of a rhombus intersect at right angles.

75. Prove that the figure formed by connecting consecutive midpoints of the sides of any quadrilateral is a parallelogram.

76. Complete the proof of Theorem 1.10.

77. Complete the proof of Theorem 1.11.

1.5
Functions

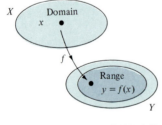

FIGURE 1.52

Many common relationships involve two variables in such a way that the value of one of the variables depends on the value of the other. For example, the sales tax on an item depends on its selling price. The distance an object moves in a given time depends on its speed.

Consider the relationship between the area of a circle and its radius. This relationship can be expressed by the equation

$$A = \pi r^2$$

where the value of A depends on the choice of r. We refer to A as the **dependent variable** and to r as the **independent variable.**

Of particular interest are relationships such that to every value of the independent variable there corresponds *one and only one* value of the dependent variable. We call this type of correspondence a **function.**

DEFINITION OF A FUNCTION	A **function** f from a set X into a set Y is a correspondence that assigns to each element x in X exactly one element y in Y. We call y the **image** of x under f and denote it by $f(x)$. The **domain** of f is the set X, and the **range** consists of all images of elements in X. (See Figure 1.52.)
	If to each value in its range there corresponds exactly one value in its domain, the function is said to be **one-to-one.** Moreover, if the range of f consists of all of Y, then the function is said to be **onto.**

| **Remark**   In this text, we work primarily with functions whose domains and ranges are sets of real numbers. We call such functions **real-valued functions of a real variable.**

Functions can be specified in a variety of ways. However, in this text we concentrate primarily on functions that are given by an equation involving the dependent and independent variables. To evaluate a function described by an equation, we generally isolate the dependent variable to the left side of the equation. For instance, the equation $x + 2y = 1$, written as

$$y = \frac{1 - x}{2}$$

describes y as a function of x, and we can denote this function as

$$f(x) = \frac{1 - x}{2}$$

This functional notation has the advantage of clearly identifying the dependent variable as $f(x)$ while at the same time telling us that x is the independent variable and that the function itself will be called "f." The symbol $f(x)$ is read "f of x." The $f(x)$ notation also allows us to be less wordy. Instead of asking, "What is the value of y that corresponds to $x = 3$?" we can ask, "What is $f(3)$?" In general, to denote the value of the dependent variable when $x = a$, we use the symbol $f(a)$. For example, the value of f when $x = 3$ is

$$f(3) = \frac{1 - (3)}{2} = \frac{-2}{2} = -1$$

In an equation that defines a function, the role of the variable x is simply that of a placeholder. For instance, the function given by

$$f(x) = 2x^2 - 4x + 1$$

can be described by the form

$$f(\ ) = 2(\ )^2 - 4(\ ) + 1$$

where parentheses are used instead of x. Therefore, to evaluate $f(-2)$, we simply place -2 in each set of parentheses:

$$f(-2) = 2(-2)^2 - 4(-2) + 1$$
$$= 2(4) + 8 + 1$$
$$= 17$$

| **Remark**   Although we generally use f as a convenient function name and x as the independent variable, we can use other symbols. For instance, the following equations all define the same function:

$$f(x) = x^2 - 4x + 7$$
$$f(t) = t^2 - 4t + 7$$
$$g(s) = s^2 - 4s + 7$$

EXAMPLE 1   Evaluating a function

For the function f defined by $f(x) = x^2 - 4x + 7$, evaluate

(a) $f(3a)$ (b) $f(b - 1)$ (c) $\dfrac{f(x + \Delta x) - f(x)}{\Delta x}$

Solution: We begin by writing the equation for f in the form

$$f(\ ) = (\ )^2 - 4(\ ) + 7$$

(a) $f(3a) = (3a)^2 - 4(3a) + 7$ Replace x with $3a$

$\qquad\quad = 9a^2 - 12a + 7$ Expand terms

(b) $f(b - 1) = (b - 1)^2 - 4(b - 1) + 7$ Replace x with $(b - 1)$

$\qquad\qquad = b^2 - 2b + 1 - 4b + 4 + 7$ Expand terms

$\qquad\qquad = b^2 - 6b + 12$ Collect like terms

(c) $\dfrac{f(x + \Delta x) - f(x)}{\Delta x} = \dfrac{[(x + \Delta x)^2 - 4(x + \Delta x) + 7] - [x^2 - 4x + 7]}{\Delta x}$

$\qquad\qquad\qquad\quad = \dfrac{x^2 + 2x\Delta x + (\Delta x)^2 - 4x - 4\Delta x + 7 - x^2 + 4x - 7}{\Delta x}$

$\qquad\qquad\qquad\quad = \dfrac{2x\Delta x + (\Delta x)^2 - 4\Delta x}{\Delta x}$

$\qquad\qquad\qquad\quad = \dfrac{\Delta x(2x + \Delta x - 4)}{\Delta x}$

$\qquad\qquad\qquad\quad = 2x + \Delta x - 4$

Remark   The ratio in part (c) of Example 1 is called a difference quotient and has a special significance in calculus.

The domain of a function may be explicitly described along with the function, or it may be *implied* by an equation used to define the function. (The implied domain is the set of all real numbers for which the equation is defined.) For example, the function given by

$$f(x) = \frac{1}{x^2 - 4}, \quad 4 \leq x \leq 5$$

has an explicitly defined domain given by $\{x: 4 \leq x \leq 5\}$. On the other hand, the function given by

$$g(x) = \frac{1}{x^2 - 4}$$

has an implied domain which is the set $\{x: x \neq \pm 2\}$. Another common type of implied domain is that used to avoid even roots of negative numbers. For example, the function given by

$$f(x) = \sqrt{x + 2}$$

has an implied domain $\{x: x \geq -2\}$.

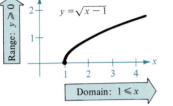

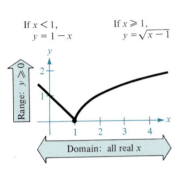

FIGURE 1.53

FIGURE 1.54

EXAMPLE 2   *Finding the domain and range of a function*

Determine the domain and range for the function of x defined by

$$f(x) = \sqrt{x - 1}$$

Solution: Since $\sqrt{x - 1}$ is not defined for $x - 1 < 0$ (that is, for $x < 1$), we must have $x \geq 1$. Therefore, the domain is the interval $\{x\colon x \geq 1\}$.

To find the range, we observe that $f(x) = \sqrt{x - 1}$ is never negative. Moreover, as x takes on the various values in the domain, $f(x)$ takes on all nonnegative values and we find the range to be $f(x) \geq 0$.

The graph of the function lends further support to our conclusions (Figure 1.53).

EXAMPLE 3   *A function defined by more than one equation*

Determine the domain and range for the function of x given by

$$f(x) = \begin{cases} \sqrt{x - 1}, & \text{if } x \geq 1 \\ 1 - x, & \text{if } x < 1 \end{cases}$$

Solution: Since f is defined for $x \geq 1$ and $x < 1$, the domain of the function is the entire set of real numbers.

On the portion of the domain for which $x \geq 1$, the function behaves as in Example 2. For $x < 1$, the value of $1 - x$ is positive, and therefore the range of the function is the interval $[0, \infty)$.

Again, a graph of the function helps to verify our conclusions (Figure 1.54).

Remark Note that the function given in Example 2 is one-to-one whereas the function given in Example 3 is not one-to-one.

The graph of a function

As you study this section, remember that the graph of the function $y = f(x)$ consists of all points $(x, f(x))$ as shown in Figure 1.55, where

$$x = \text{the directed distance from the } y\text{-axis}$$

$$f(x) = \text{the directed distance from the } x\text{-axis}$$

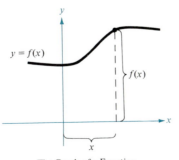

FIGURE 1.55   The Graph of a Function

Since (by the definition of a function) there is exactly one *y*-value for each *x*-value, it follows that a vertical line can intersect the graph of a function at most once. This observation provides us with a convenient visual test for functions. For example, in part (a) of Figure 1.56, we see that the graph does not define *y* as a function of *x* since a vertical line intersects the graph twice.

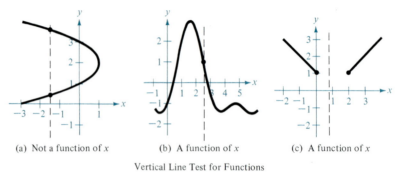

(a) Not a function of *x*   (b) A function of *x*   (c) A function of *x*

Vertical Line Test for Functions

FIGURE 1.56

Functional notation lends itself well to describing transformations of graphs in the plane. Some families of graphs all have the same basic shape. For example, consider the graph of $y = x^2$, as shown in Figure 1.57(a). Now compare this graph to those shown in Figure 1.57(b).

Each of the graphs in Figure 1.57(b) is a **transformation** of the graph of $y = x^2$. The three basic types of transformations involved in these six graphs are (1) horizontal shifts, (2) vertical shifts, and (3) reflections.

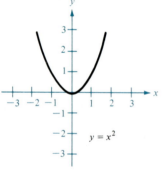

Original Graph

FIGURE 1.57(a)

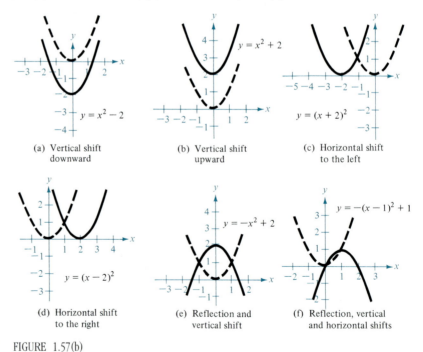

(a) Vertical shift downward   (b) Vertical shift upward   (c) Horizontal shift to the left

(d) Horizontal shift to the right   (e) Reflection and vertical shift   (f) Reflection, vertical and horizontal shifts

FIGURE 1.57(b)

BASIC TYPES OF TRANSFORMATIONS ($c > 0$)		
Original graph:		$y = f(x)$
Horizontal shift c units to the **right:**		$y = f(x - c)$
Horizontal shift c units to the **left:**		$y = f(x + c)$
Vertical shift c units **downward:**		$y = f(x) - c$
Vertical shift c units **upward:**		$y = f(x) + c$
Reflection (about the x-axis):		$y = -f(x)$

Classifications and combinations of functions

Leonhard Euler

The modern notion of a function is due to the efforts of many different mathematicians who lived during the seventeenth and eighteenth centuries. Of particular note was Leonhard Euler (1707–1783), to whom we are indebted for the functional notation $y = f(x)$. By the end of the eighteenth century, mathematicians and scientists had concluded that most real-world phenomena can be represented by mathematical models taken from a basic collection of functions called **elementary functions.** Elementary functions are divided into three categories:

1. algebraic
2. logarithmic and exponential
3. trigonometric

We will study the logarithmic and exponential functions in Chapter 7 and the trigonometric functions in Chapter 8.

The most common type of algebraic function is a **polynomial function.**

DEFINITION OF POLYNOMIAL FUNCTION	
	Let n be a nonnegative integer. The function $$f(x) = a_n x^n + a_{n-1} x^{n-1} + \cdots + a_2 x^2 + a_1 x + a_0$$ is called a **polynomial function** of degree n. The numbers a_i are called **coefficients,** with $a_n \neq 0$ the **leading coefficient** and a_0 the **constant term** of the polynomial function.

Remark It is common practice to use subscript notation for coefficients of general polynomial functions, but for polynomial functions of low degree we often use the following simpler forms:

Zeroth degree:	$f(x) = a$	Constant function
First degree:	$f(x) = ax + b$	Linear function
Second degree:	$f(x) = ax^2 + bx + c$	Quadratic function
Third degree:	$f(x) = ax^3 + bx^2 + cx + d$	Cubic function

Although the graph of a polynomial function can have several turns, eventually the graph will rise or fall without bound as x moves to the right or left. Whether the graph of

$$f(x) = a_n x^n + a_{n-1} x^{n-1} + \cdots + a_2 x^2 + a_1 x + a_0$$

eventually rises or falls can be determined by the function's degree (odd or even) and by the leading coefficient a_n as indicated in Figure 1.58. Note that the dashed portions of the graphs indicate that the **leading coefficient test** determines *only* the right and left behavior of the graph.

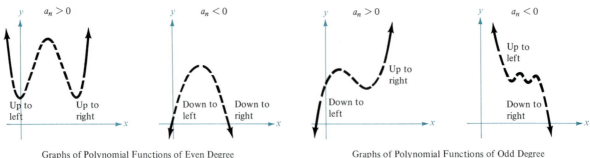

Graphs of Polynomial Functions of Even Degree   Graphs of Polynomial Functions of Odd Degree

FIGURE 1.58   Leading Coefficient Test for Polynomial Functions

Just as a rational number can be written as the quotient of two integers, so can a **rational function** be written as the quotient of two polynomials. Specifically, a function f is **rational** if it has the form

$$f(x) = \frac{P(x)}{Q(x)}, \quad Q(x) \neq 0$$

where $P(x)$ and $Q(x)$ are polynomials.

Polynomial functions and rational functions are two examples of a larger class of functions called **algebraic functions.** An algebraic function is one that can be expressed in terms of finitely many sums, differences, multiples, quotients, and radicals involving x^n. For example, the following functions are algebraic:

$$f(x) = \sqrt{x + 1} \qquad \text{and} \qquad g(x) = x + \frac{1}{\sqrt[3]{x + 1}}$$

Functions that are not algebraic are called **transcendental.**

Two functions can be combined in various ways to create new functions. For example, if

$$f(x) = 2x - 3 \qquad \text{and} \qquad g(x) = x^2 + 1$$

we can form the functions

$$f(x) + g(x) = (2x - 3) + (x^2 + 1) = x^2 + 2x - 2 \qquad \text{Sum}$$
$$f(x) - g(x) = (2x - 3) - (x^2 + 1) = -x^2 + 2x - 4 \qquad \text{Difference}$$
$$f(x)g(x) = (2x - 3)(x^2 + 1) = 2x^3 - 3x^2 + 2x - 3 \qquad \text{Product}$$
$$\frac{f(x)}{g(x)} = \frac{2x - 3}{x^2 + 1} \qquad \text{Quotient}$$

We can combine two functions in yet another way to form what is called a **composite function.**

DEFINITION OF COMPOSITE FUNCTION	Let f and g be functions such that the range of g is in the domain of f. Then the function given by $f \circ g(x) = f(g(x))$ is called the **composite** of f with g.

It is important to realize that the composite of f with g may not be equal to the composite of g with f. This is illustrated in the following example.

EXAMPLE 4   *Composition of functions*

Given $f(x) = 2x - 3$ and $g(x) = x^2 + 1$, find $f(g(x))$ and $g(f(x))$.

Solution: Since $f(x) = 2x - 3$, we have

$$f(g(x)) = 2(g(x)) - 3 = 2(x^2 + 1) - 3 = 2x^2 - 1$$

and since $g(x) = x^2 + 1$, we have

$$g(f(x)) = (f(x))^2 + 1 = (2x - 3)^2 + 1 = 4x^2 - 12x + 10$$

Note that $f(g(x)) \neq g(f(x))$.

In Section 1.3, we defined an x-intercept of a graph to be a point $(x, 0)$ at which the graph crosses the x-axis. When the graph of a function crosses the x-axis, we say that the function has a **zero** at that point. For example, the function $f(x) = x - 4$ has a zero at $(4, f(4))$ since $f(4) = 0$.

In Section 1.3 we also discussed different types of symmetry. In the terminology of functions, we say that a function is **even** if its graph is symmetric with respect to the y-axis, and a function is **odd** if its graph is symmetric with respect to the origin. Thus, the symmetry tests in Section 1.3 yield the following test for even and odd functions.

THEOREM 1.12	**TEST FOR EVEN AND ODD FUNCTIONS** The function $y = f(x)$ is **even** if $f(-x) = f(x)$. The function $y = f(x)$ is **odd** if $f(-x) = -f(x)$.

Remark   Except for trivial cases, such as $f(x) = 0$, the graph of a function cannot have symmetry with respect to the x-axis since it then fails the vertical line test for the graph of a function.

EXAMPLE 5   Even and odd functions

Determine whether the following functions are even, odd, or neither. In each case find the zeros of the function.

(a) $f(x) = x^3 - x$ (b) $g(x) = x^2 + 1$ (c) $h(x) = x^2 - 4$

Solution:

(a) This function is odd since

$$f(-x) = (-x)^3 - (-x) = -x^3 + x = -(x^3 - x) = -f(x)$$

The zeros of f are found as follows:

$$x^3 - x = 0 \qquad \text{Let } f(x) = 0$$
$$x(x^2 - 1) = x(x - 1)(x + 1) = 0 \qquad \text{Factor}$$
$$x = 0, 1, -1 \qquad \text{Zeros of } f$$

(b) This function is even since

$$g(-x) = (-x)^2 + 1 = x^2 + 1 = g(x)$$

It has no zeros since $x^2 + 1$ is positive for all x.

(c) This function is even since

$$h(-x) = (-x)^2 - 4 = x^2 - 4 = h(x)$$

The zeros of f are found as follows:

$$x^2 - 4 = 0$$
$$x^2 = 4$$
$$x = \pm 2$$

Remark Each of the functions in Example 5 is either even or odd. However, some functions such as

$$f(x) = x^2 + x + 1$$

are neither even nor odd.

Section Exercises 1.5

1. Given $f(x) = 2x - 3$, find
   (a) $f(0)$ (b) $f(-3)$
   (c) $f(b)$ (d) $f(x - 1)$

2. Given $f(x) = x^2 - 2x + 2$, find
   (a) $f\left(\dfrac{1}{2}\right)$ (b) $f(-1)$
   (c) $f(c)$ (d) $f(x + \Delta x)$

3. Given $f(x) = \sqrt{x + 3}$, find
   (a) $f(-2)$ (b) $f(6)$
   (c) $f(c)$ (d) $f(x + \Delta x)$

4. Given $f(x) = 1/\sqrt{x}$, find
   (a) $f(2)$ (b) $f\left(\dfrac{1}{4}\right)$
   (c) $f(x + \Delta x)$ (d) $f(x + \Delta x) - f(x)$

5. Given $f(x) = |x|/x$, find
   (a) $f(2)$ (b) $f(-2)$
   (c) $f(x^2)$ (d) $f(x - 1)$

6. Given $f(x) = |x| + 4$, find
   (a) $f(2)$ (b) $f(-2)$
   (c) $f(x^2)$ (d) $f(x + \Delta x) - f(x)$

7. Given $f(x) = x^2 - x + 1$, find

$$\frac{f(2 + \Delta x) - f(2)}{\Delta x}$$

8. Given $f(x) = 1/x$, find

$$\frac{f(1 + \Delta x) - f(1)}{\Delta x}$$

9. Given $f(x) = x^3$, find

$$\frac{f(x + \Delta x) - f(x)}{\Delta x}$$

10. Given $f(x) = 3x - 1$, find

$$\frac{f(x) - f(1)}{x - 1}$$

11. Given $f(x) = 1/\sqrt{x - 1}$, find

$$\frac{f(x) - f(2)}{x - 2}$$

12. Given $f(x) = x^3 - x$, find

$$\frac{f(x) - f(1)}{x - 1}$$

13. Given $f(x) = \sqrt{x}$ and $g(x) = x^2 - 1$, find
  (a) $f(g(1))$ (b) $g(f(1))$
  (c) $g(f(0))$ (d) $f(g(-4))$
  (e) $f(g(x))$ (f) $g(f(x))$
14. Given $f(x) = 1/x$ and $g(x) = x^2 - 1$, find
  (a) $f(g(2))$ (b) $g(f(2))$
  (c) $f\left(g\left(\dfrac{1}{\sqrt{2}}\right)\right)$ (d) $g\left(f\left(\dfrac{1}{\sqrt{2}}\right)\right)$
  (e) $g(f(x))$ (f) $f(g(x))$

In Exercises 15–18, find the (real) zeros of the given function.

15. $f(x) = x^2 - 9$
16. $f(x) = x^3 - x$
17. $f(x) = \dfrac{3}{x - 1} + \dfrac{4}{x - 2}$
18. $f(x) = a + \dfrac{b}{x}$

In Exercises 19–28, find the domain and range of the given function and sketch its graph.

19. $f(x) = \sqrt{x - 1}$ **20.** $f(x) = \sqrt{1 - x}$
21. $f(x) = x^2$ **22.** $f(x) = 4 - x^2$
23. $f(x) = \sqrt{9 - x^2}$ **24.** $f(x) = \sqrt{25 - x^2}$
25. $f(x) = \dfrac{1}{|x|}$ **26.** $f(x) = |x - 2|$

27. $f(x) = \dfrac{|x|}{x}$ **28.** $f(x) = \sqrt{x^2 - 4}$

In Exercises 29–38, use the vertical line test to determine if y is a function of x.

29. $y = x^2$ **30.** $y = x^3 - 1$

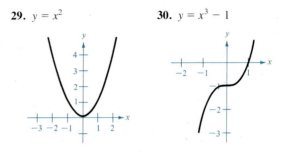

31. $x - y^2 = 0$ **32.** $x^2 + y^2 = 9$

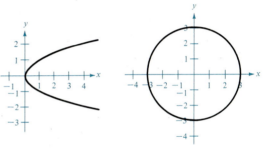

33. $\sqrt{x^2 - 4} - y = 0$ **34.** $x - xy + y + 1 = 0$

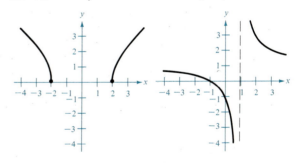

35. $x^2 = xy - 1$ **36.** $x = |y|$

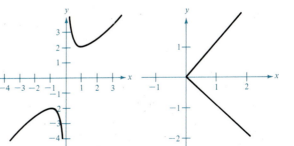

37. $x^2 - 4y^2 + 4 = 0$ **38.** $y = |x - 3| + |x - 1|$

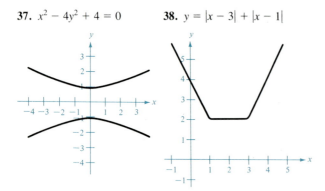

In Exercises 39–46, identify the equations that determine y as a function of x.

39. $x^2 + y^2 = 4$ **40.** $x = y^2$

41. $x^2 + y = 4$ **42.** $x + y^2 = 4$

43. $2x + 3y = 4$

44. $x^2 + y^2 - 2x - 4y + 1 = 0$

45. $y^2 = x^2 - 1$ **46.** $x^2y - x^2 + 4y = 0$

47. Use the accompanying graph of $f(x) = \sqrt{x}$ to sketch the graph of each of the following.
(a) $y = \sqrt{x} + 2$ (b) $y = -\sqrt{x}$
(c) $y = \sqrt{x} - 2$ (d) $y = \sqrt{x + 3}$
(e) $y = \sqrt{x} - 4$ (f) $y = 2\sqrt{x}$

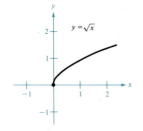

48. Use the accompanying graph of $f(x) = \sqrt[3]{x}$ to sketch the graph of each of the following.
(a) $y = \sqrt[3]{x} - 1$ (b) $y = \sqrt[3]{x + 1}$
(c) $y = \sqrt[3]{x - 1}$ (d) $y = -\sqrt[3]{x} - 2$
(e) $y = \frac{1}{2}\sqrt[3]{x}$ (f) $y = \sqrt[3]{x + 1} - 1$

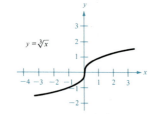

49. Use the accompanying graph of $f(x) = x\sqrt{x + 3}$ to write formulas for the functions whose graphs are shown in parts (a) through (d).

50. Use the accompanying graph of $f(x) = 1/(x^2 + 1)$ to write formulas for the functions whose graphs are shown in parts (a) through (d).

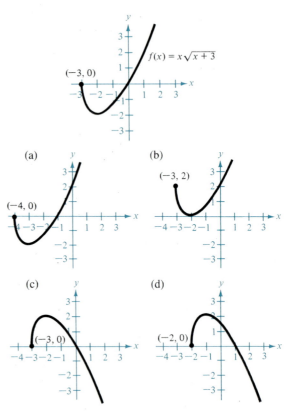

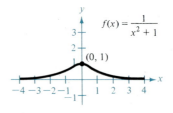

(a) (b)

(c) (d)

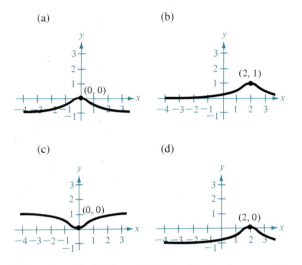

In Exercises 51–56, determine if the function is even, odd, or neither.

51. $f(x) = 4 - x^2$
52. $f(x) = \sqrt[3]{x}$
53. $f(x) = x(4 - x^2)$
54. $f(x) = \sqrt[3]{x} + 1$
55. $f(x) = 4x - x^2$
56. $f(x) = x^{2/3}$

57. What type of symmetry does the graph of an even function possess?

58. What type of symmetry does the graph of an odd function possess?

59. Show that the following function is odd:

$$f(x) = a_{2n+1}x^{2n+1} + \cdots + a_3x^3 + a_1x$$

60. Show that the following function is even:

$$f(x) = a_{2n}x^{2n} + a_{2n-2}x^{2n-2} + \cdots + a_2x^2 + a_0$$

61. Show that the product of two even (or two odd) functions is even.

62. Show that the product of an odd function and an even function is odd.

63. A rectangle as shown in Figure 1.59 has a perimeter of 100 feet. Express the area A of the rectangle as a function of x.

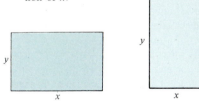

FIGURE 1.59   FIGURE 1.60

64. A rancher has 200 feet of fencing to enclose two adjacent rectangular corrals as shown in Figure 1.60. Express the area A of the enclosures as a function of x.

65. An open box is to be made from a square piece of material 12 inches on a side, by cutting equal squares from each corner and turning up the sides as shown in

Figure 1.61. Express the volume V as a function of x.

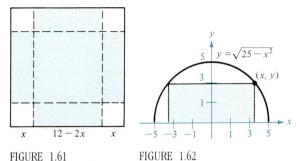

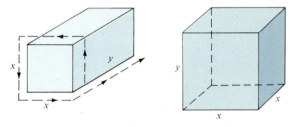

FIGURE 1.61   FIGURE 1.62

66. A rectangle is bounded by the x-axis and the semicircle $y = \sqrt{25 - x^2}$ as shown in Figure 1.62. Write the area A of the rectangle as a function of x.

67. A rectangular package with square cross sections has a combined length and girth (perimeter of a cross section) of 108 inches. Express the volume V as a function of x. (See Figure 1.63.)

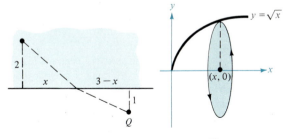

FIGURE 1.63   FIGURE 1.64

68. A closed box with a square base of side x has a surface area of 100 square feet as shown in Figure 1.64. Express the volume V of the box as a function of x.

69. A man is in a boat 2 miles from the nearest point on the coast. He is to go to a point Q, 3 miles down the coast and 1 mile inland as shown in Figure 1.65. He can row at 2 miles per hour and walk at 4 miles per hour. Express the total time T of the trip as a function of x.

FIGURE 1.65   FIGURE 1.66

70. The portion of the vertical line through the point $(x, 0)$ that lies between the x-axis and the graph of $y = \sqrt{x}$ is revolved about the x-axis. Express the area A of the resulting disk as a function of x. (See Figure 1.66.)

Review Exercises for Chapter 1

In Exercises 1–4, sketch the intervals defined by the given inequalities.

1. $|x - 2| \le 3$ **2.** $|3x - 2| \le 0$

3. $4 < (x + 3)^2$ **4.** $\dfrac{1}{|x|} < 1$

In Exercises 5 and 6, find the midpoint of the given interval.

5. $\left[\dfrac{7}{8}, \dfrac{10}{4} \right]$ **6.** $\left[-1, \dfrac{3}{2} \right]$

In Exercises 7 and 8, find the points of trisection for the given interval.

7. $[-2, 6]$ **8.** $[1, 5]$

9. Find the midpoints of the sides of the triangle whose vertices are $(1, 4)$, $(-3, 2)$, and $(5, 0)$.

10. Find the vertices of the triangle whose sides have midpoints $(0, 2)$, $(1, -1)$, and $(2, 1)$.

In Exercises 11–14, determine the radius and center of the given circle and sketch its graph.

11. $x^2 + y^2 + 6x - 2y + 1 = 0$
12. $4x^2 + 4y^2 - 4x + 8y = 11$
13. $x^2 + y^2 + 6x - 2y + 10 = 0$
14. $x^2 - 6x + y^2 + 8y = 0$

15. Determine the value of c so that the given circle has a radius of 2:

$$x^2 - 6x + y^2 + 8y = c$$

16. Find an equation in x and y such that the distance between (x, y) and $(-2, 0)$ is twice the distance between (x, y) and $(3, 1)$.

17. Find an equation for the circle whose center is $(1, 2)$ and whose radius is 3. Then determine if the following points are inside, outside, or on the circle.
(a) $(1, 5)$ (b) $(0, 0)$
(c) $(-2, 1)$ (d) $(0, 4)$

18. Find an equation for the circle whose center is $(2, 1)$ and whose radius is 2. Then determine if the following points are inside, outside, or on the circle.
(a) $(1, 1)$ (b) $(4, 2)$
(c) $(0, 1)$ (d) $(3, 1)$

In Exercises 19–22, sketch the graph of the given equation.

19. $y = \dfrac{-x + 3}{2}$ **20.** $y = 1 + \dfrac{1}{x}$

21. $y = 7 - 6x - x^2$ **22.** $y = 6x - x^2$

In Exercises 23 and 24, determine if the given points lie on the same straight line.

23. $(-1, 3)$, $(2, 9)$, $(3, 1)$
24. $(2, 5)$, $(4, 10)$, $(6, 20)$

In Exercises 25–28, use the slope and y-intercept to sketch the graph of the given line.

25. $4x - 2y = 6$ **26.** $0.02x + 0.15y = 0.25$

27. $-\dfrac{1}{3}x + \dfrac{5}{6}y = 1$ **28.** $51x + 17y = 102$

29. Find equations of the lines passing through $(-2, 4)$ and having the following characteristics:
(a) Slope of $\tfrac{7}{16}$
(b) Parallel to the line $5x - 3y = 3$
(c) Passing through the origin
(d) Parallel to the y-axis

30. Find equations of the lines passing through $(1, 3)$ and having the following characteristics:
(a) Slope of $-\tfrac{2}{3}$
(b) Perpendicular to the line $x + y = 0$
(c) Passing through the point $(2, 4)$
(d) Parallel to the x-axis

31. The midpoint of a line segment is $(-1, 4)$. If one end of the line segment is $(2, 3)$, find the other end.

32. Find the point that is equidistant from $(0, 0)$, $(2, 3)$, and $(3, -2)$.

In Exercises 33 and 34, find the point(s) of intersection of the graphs of the given equations.

33. $3x - 4y = 8$, $x + y = 5$
34. $x - y + 1 = 0$, $y - x^2 = 7$

In Exercises 35–40, find a formula for the given function and find the domain.

35. The value v of a farm at \$850 per acre, with buildings, livestock, and equipment worth \$300,000, is a function of the number of acres a.

36. The value v of wheat at \$3.25 per bushel is a function of the number of bushels b.

37. The surface area s of a cube is a function of the length of an edge x.

38. The surface area s of a sphere is a function of the radius r.

39. The distance d traveled by a car at a speed of 45 miles per hour is a function of the time traveled t.

40. The area a of an equilateral triangle is a function of the length of one of its sides x.

In Exercises 41–44, express the indicated values as functions of x.

41. R and r **42.** R and r

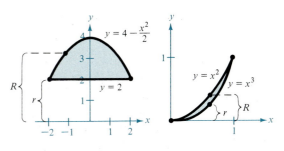

43. h and p **44.** h and p

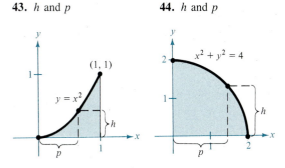

45. The sum of two positive numbers is 500. Let one of the numbers be x, and express the product P of the two numbers as a function of x.

46. The product of two positive numbers is 120. Let one of the numbers be x, and express the sum of the two numbers as a function of x.

In Exercises 47–52, sketch the graph of the given equation and use the vertical line test to determine if the equation expresses y as a function of x.

47. $x^2 - y = 0$ **48.** $x^2 + 4y^2 = 16$
49. $x - y^2 = 0$ **50.** $x^3 - y^2 + 1 = 0$
51. $y = x^2 - 2x$ **52.** $y = 36 - x^2$

53. Given $f(x) = 1 - x^2$ and $g(x) = 2x + 1$, find
    (a) $f(x) + g(x)$ (b) $f(x) - g(x)$

    (c) $f(x)g(x)$ (d) $\dfrac{f(x)}{g(x)}$

    (e) $f(g(x))$ (f) $g(f(x))$

54. Given $f(x) = 2x - 3$ and $g(x) = \sqrt{x + 1}$, find
    (a) $f(x) + g(x)$ (b) $f(x) - g(x)$

    (c) $f(x)g(x)$ (d) $\dfrac{f(x)}{g(x)}$

    (e) $f(g(x))$ (f) $g(f(x))$

55. Sales representatives for a certain company are required to use their own cars for transportation. The cost to the company is $100 per day for lodging and meals plus $0.25 per mile driven. Write a linear equation expressing the daily cost C to the company in terms of x, the number of miles driven.

56. A contractor purchases for $26,500 a piece of equipment that costs an average of $5.25 per hour for fuel and maintenance. The equipment operator is paid $9.50 per hour, and customers are charged $25 per hour.
    (a) Write an equation for the cost C of operating this equipment t hours.
    (b) Write an equation for the revenue R derived from t hours of use.
    (c) Find the break-even point for this equipment by finding the time at which $R = C$.

2 Limits and their properties

2.1
An introduction to limits

The notion of a limit is fundamental to the study of calculus. For this reason it is important for you to acquire a good working knowledge of limits before moving on to other calculus topics. In this chapter, we will discuss limits in two stages. In the first four sections, we build an intuitive understanding of limits by examining various kinds of limits, ways to evaluate limits, and the role of limits in defining continuity. This background is followed, in Section 2.5, by a more rigorous development of the limit concept. This two-stage approach follows closely the historical evolution of limits. In fact, the formal definition of limits in Section 2.5 was developed after much of the early work in calculus was already complete.

Just what do we mean by the term *limit?* In everyday language we refer to the speed limit, a wrestler's weight limit, the limit of one's endurance, the limits of modern technology, or stretching a spring to its limit. These phrases all suggest that a limit is a type of bound, which on some occasions may not be reached but on other occasions may be reached or even exceeded.

We introduce the mathematical notion of a limit by looking at an example. Suppose we are asked to sketch the graph of the function f given by

$$f(x) = \frac{x^3 - 1}{x - 1}, \quad x \neq 1$$

For all points other than those for which $x = 1$, we can use the standard curve-sketching techniques. However, at the undefined point ($x = 1$), we are not sure what to expect. To get an idea of the behavior of the graph of f near $x = 1$, we could use two sets of x-values—one set that approaches 1 from the left and one set that approaches 1 from the right. Table 2.1 shows two such sets with their corresponding functional values. By plotting these points, we find that the graph of f is a parabola that has a hole at the point (1, 3), as shown in Figure 2.1. Though x cannot equal 1, we can move arbitrarily close to 1, and as a result $f(x)$ moves arbitrarily close to 3. Using limit notation, we say that the *limit of f(x) as x approaches 1 is 3,* and we write

$$\lim_{x \to 1} f(x) = 3$$

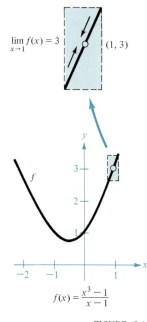

$$\lim_{x \to 1} f(x) = 3 \qquad (1, 3)$$

$$f(x) = \frac{x^3 - 1}{x - 1}$$

FIGURE 2.1

TABLE 2.1

	x approaches 1 from the left						x approaches 1 from the right				
x	0.5	0.75	0.9	0.99	0.999	1	1.001	1.01	1.1	1.25	1.5
f(x)	1.750	2.313	2.710	2.970	2.997	?	3.003	3.030	3.310	3.813	4.750

f(x) approaches 3	f(x) approaches 3

This discussion leads to the following informal definition of a limit.

INFORMAL DEFINITION OF A LIMIT

If $f(x)$ becomes arbitrarily close to a unique number L as x approaches c from either side, then we say that the **limit** of $f(x)$, as x approaches c, is L, and we write

$$\lim_{x \to c} f(x) = L$$

| **Remark** Throughout this text, when we write

$$\lim_{x \to c} f(x) = L$$

we imply two statements—the limit **exists** *and* the limit is L. Some functions do not have a limit as $x \to c$, but those that do cannot have two different limits as $x \to c$. In other words, *if the limit of a function exists, it is unique.*

EXAMPLE 1   Using a calculator to estimate a limit

Evaluate $f(x) = x/(\sqrt{x + 1} - 1)$ at several points near $x = 0$, and use the result to estimate the limit

$$\lim_{x \to 0} \frac{x}{\sqrt{x + 1} - 1}$$

Solution: Table 2.2 lists the values of $f(x)$ for several x-values near 0.

TABLE 2.2

	x approaches 0 from the left					x approaches 0 from the right			
x	−0.1	−0.01	−0.001	−0.0001	0	0.0001	0.001	0.01	0.1
f(x)	1.9487	1.9950	1.9995	1.9999	?	2.0001	2.0005	2.0050	2.0488

f(x) approaches 2	f(x) approaches 2

From the results shown in this table, we estimate the limit to be 2. The graph of f is shown in Figure 2.2.

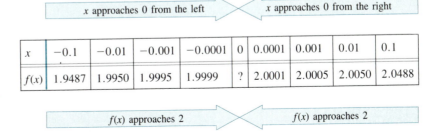

f is undefined at $x = 0$.

$$f(x) = \frac{x}{\sqrt{x + 1} - 1}$$

$$\lim_{x \to 0} \frac{x}{\sqrt{x + 1} - 1} = 2$$

FIGURE 2.2

In Example 1, note that the function is undefined at $x = 0$, and yet $f(x)$ appears to be approaching a limit as $x \to 0$. This often happens, and it is important to realize that the existence or nonexistence of $f(x)$ *at $x = c$* has no bearing on the existence of the limit of $f(x)$ as x approaches c. This is further demonstrated in the next example.

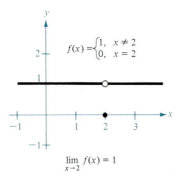

$$f(x) = \begin{cases} 1, & x \neq 2 \\ 0, & x = 2 \end{cases}$$

$$\lim_{x \to 2} f(x) = 1$$

FIGURE 2.3

EXAMPLE 2   *Finding a limit*

Find the limit of $f(x)$ as x approaches 2, where f is defined by

$$f(x) = \begin{cases} 1, & x \neq 2 \\ 0, & x = 2 \end{cases}$$

Solution: Since $f(x) = 1$ for all x other than $x = 2$, and since the value of $f(2)$ is immaterial, we conclude that the limit is 1, as shown in Figure 2.3. ☐

| **Remark**   Note in the first three figures that we use an open "dot" in a graph to mean that the point is not part of the graph, whereas a solid dot means that the point is part of the graph.

Limits that fail to exist

In Examples 1 and 2 we looked at two limits that do exist. We can learn a great deal about what it means for a limit to exist by looking at some situations for which a limit does not exist.

EXAMPLE 3   *Behavior that differs from the right and left*

Show that the following limit does not exist:

$$\lim_{x \to 0} \frac{|x|}{x}$$

Solution: We consider the function $f(x) = |x|/x$. From Figure 2.4, we see that for positive x,

$$\frac{|x|}{x} = 1$$

and for negative x,

$$\frac{|x|}{x} = -1$$

This means that no matter how close we get to zero, we will have x-values that yield $f(x) = 1$ and $f(x) = -1$. Specifically, we have

$$(-\infty, 0) \cup (0, \infty)$$

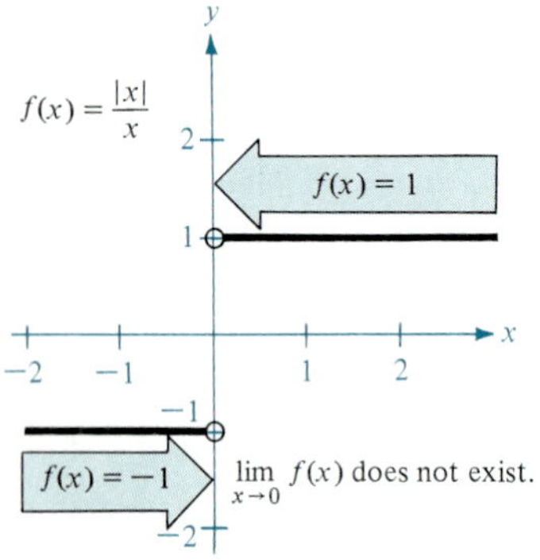

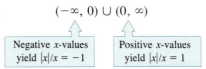

FIGURE 2.4   This implies that the limit does not exist.   ☐

EXAMPLE 4   *Unbounded behavior*

Discuss the existence of the following limit:

$$\lim_{x \to 0} \frac{1}{x^2}$$

Solution: We let $f(x) = 1/x^2$. From Figure 2.5, we see that as x approaches 0 from either the right or the left, $f(x)$ increases without bound. This means that by choosing x close enough to 0, we can force $f(x)$ to be as large as we want. For instance, $f(x)$ will be larger than 100 if we choose x that is within $1/10$ of 0. That is,

$$0 < |x| < \frac{1}{10} \implies f(x) = \frac{1}{x^2} > 100$$

Similarly, we can force $f(x)$ to be larger than 1,000,000, as follows.

$$0 < |x| < \frac{1}{1000} \implies f(x) = \frac{1}{x^2} > 1,000,000$$

Since $f(x)$ is not approaching a real number L as x approaches 0, we say that the limit does not exist.

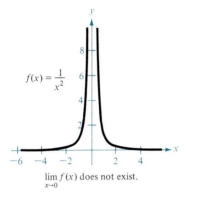

$f(x) = \dfrac{1}{x^2}$

$\lim_{x \to 0} f(x)$ does not exist.

FIGURE 2.5

Examples 3 and 4 show two of the most common types of problems that lead to the nonexistence of a limit of $f(x)$ as x approaches c.

1. $f(x)$ approaches a different number from the right side of c than it approaches from the left side.
2. $f(x)$ increases or decreases without bound as x approaches c.

There are many other interesting functions that have unusual limit behavior. An often-cited one is defined as follows:

$$f(x) = \begin{cases} \dfrac{1}{q}, & x = \dfrac{p}{q}, \text{ rational number in reduced form} \\ 0, & x \text{ is irrational} \end{cases}$$

This function has the property that its limit is 0 as x approaches any real number c. In the early development of calculus, the definition of a function was much more restricted than it is today and "functions" such as this one would not have been considered. The modern definition of a function was given by the German mathematician Peter Gustav Dirichlet (1805–1859). Dirichlet made many contributions to mathematics and, together with Augustin-Louis Cauchy (1789–1857), Georg Friedrich Riemann (1826–1866), and Karl Weierstrauss (1815–1897), developed much of the rigor present in calculus today.

Peter Gustav Dirichlet

A strategy for finding limits

We pointed out that the limit of $f(x)$ as $x \to c$ does not depend upon the value of f at $x = c$. *However,* if it happens that the limit is precisely $f(c)$, then we say that the limit can be evaluated by **direct substitution.** That is,

$$\lim_{x \to c} f(x) = f(c) \qquad \text{Substitute for } x$$

Such *well-behaved* functions are said to be **continuous at** c and we will examine this concept more closely in Section 2.3. An important application of direct substitution is shown in the following theorem.

THEOREM 2.1   **FUNCTIONS THAT AGREE AT ALL BUT ONE POINT**
Let c be a real number and $f(x) = g(x)$ for all $x \neq c$ in an open interval containing c. If the limit of $g(x)$ exists as $x \to c$, then the limit of $f(x)$ also exists, and

$$\lim_{x \to c} f(x) = \lim_{x \to c} g(x)$$

EXAMPLE 5   Two functions that agree at all but one point

Show that the functions

$$f(x) = \frac{x^3 - 1}{x - 1} \qquad \text{and} \qquad g(x) = x^2 + x + 1$$

have the same values for all x other than $x = 1$.

Solution: By factoring the numerator of f, we have

$$f(x) = \frac{x^3 - 1}{x - 1} = \frac{(x - 1)(x^2 + x + 1)}{x - 1}$$

Thus, for $x \neq 1$, we can cancel like factors to obtain

$$f(x) = \frac{\cancel{(x - 1)}(x^2 + x + 1)}{\cancel{x - 1}} = x^2 + x + 1 = g(x), \quad x \neq 1$$

Therefore, for all points other than $x = 1$, the functions f and g are identical. This is shown graphically in Figure 2.6.

Note that in Figure 2.6, there is no gap in the graph of the polynomial function g, and we will see in Theorem 2.4 that its limit as x approaches 1 is simply

$$\lim_{x \to 1} g(x) = \lim_{x \to 1} (x^2 + x + 1) = 1^2 + 1 + 1 = 3 = g(1)$$

Thus, by applying Theorem 2.1, we can conclude that the limit of $f(x)$ as $x \to 1$ is also 3.

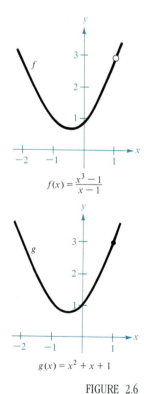

$f(x) = \dfrac{x^3 - 1}{x - 1}$

$g(x) = x^2 + x + 1$

FIGURE 2.6

This discussion identifies the following basic strategy for finding limits.

STRATEGY FOR
FINDING LIMITS

1. Learn to recognize which limits can be evaluated by direct substitution. (Several types will be discussed in the remainder of this section.)
2. If the limit of $f(x)$ as $x \to c$ *cannot* be evaluated by direct substitution, try to find a function g that agrees with f for all x other than $x = c$. (Choose g so that the limit of $g(x)$ *can* be evaluated by direct substitution.)
3. Apply Theorem 2.1 to conclude that

$$\lim_{x \to c} f(x) = \lim_{x \to c} g(x) = g(c)$$

We devote the remainder of this section to the study of limits that can be evaluated by direct substitution. Then in the next section, we will look at techniques for evaluating limits for which direct substitution fails.

Limits of algebraic functions

We begin our discussion of limits that can be evaluated by direct substitution with a look at the limits of some basic functions.

THEOREM 2.2

SOME BASIC LIMITS
If b and c are real numbers and n is an integer,* then the following are true.

1. $\lim_{x \to c} b = b$ 2. $\lim_{x \to c} x = c$ 3. $\lim_{x \to c} x^n = c^n$

EXAMPLE 6   Evaluating a limit

Applying Theorem 2.2, we have

(a) $\lim_{x \to 2} 3 = 3$

(b) $\lim_{x \to 2} x^2 = 2^2 = 4$

By combining Theorem 2.2 with the following theorem, we can find limits for a wide variety of algebraic functions. The proofs of Properties 4 and 5 are given in Appendix A.

*If $c = 0$, then n must be a positive integer.

THEOREM 2.3   **PROPERTIES OF LIMITS**

If b and c are real numbers, n is a positive integer, and the functions f and g have limits as $x \to c$, then the following properties are true.

1. Scalar multiple: $\displaystyle \lim_{x \to c} [b\, f(x)] = b\left[\lim_{x \to c} f(x) \right]$

2. Sum or difference: $\displaystyle \lim_{x \to c} [f(x) \pm g(x)] = \lim_{x \to c} f(x) \pm \lim_{x \to c} g(x)$

3. Product: $\displaystyle \lim_{x \to c} [f(x)g(x)] = \left[\lim_{x \to c} f(x) \right]\left[\lim_{x \to c} g(x) \right]$

4. Quotient: $\displaystyle \lim_{x \to c} \frac{f(x)}{g(x)} = \frac{\lim_{x \to c} f(x)}{\lim_{x \to c} g(x)}, \quad \lim_{x \to c} g(x) \neq 0$

5. Power: $\displaystyle \lim_{x \to c} [f(x)]^n = \left[\lim_{x \to c} f(x) \right]^n$

EXAMPLE 7   The limit of a polynomial

Find the following limit:

$$\lim_{x \to 2} (4x^2 + 3)$$

Solution: Using the results of Example 6 together with Theorem 2.3, we have

$$\lim_{x \to 2} (4x^2 + 3) = \lim_{x \to 2} 4x^2 + \lim_{x \to 2} 3 \qquad \text{Property 2}$$

$$= 4\left[\lim_{x \to 2} x^2 \right] + \lim_{x \to 2} 3 \qquad \text{Property 1}$$

$$= 4(4) + 3 = 19$$

Note that in Example 7, the limit (as $x \to 2$) of the *polynomial function* $p(x) = 4x^2 + 3$ is simply the value of p at $x = 2$.

$$\lim_{x \to 2} p(x) = p(2) = 4(2^2) + 3 = 19$$

This *direct substitution* property is valid for all polynomial functions, as stated in the following theorem.

THEOREM 2.4   **LIMIT OF A POLYNOMIAL FUNCTION**

If p is a polynomial function and c is a real number, then

$$\lim_{x \to c} p(x) = p(c)$$

Proof: Let the polynomial function p be given by

$$p(x) = a_n x^n + \cdots + a_1 x + a_0$$

By repeated applications of the sum and scalar multiple properties, we have

$$\lim_{x \to c} p(x) = a_n \left[\lim_{x \to c} x^n \right] + \cdots + a_1 \left[\lim_{x \to c} x \right] + \lim_{x \to c} a_0$$

Now, using Properties 1, 2, and 3 of Theorem 2.2, we have

$$\lim_{x \to c} p(x) = a_n c^n + \cdots + a_1 c + a_0 = p(c)$$

THEOREM 2.5   LIMIT OF A RATIONAL FUNCTION
If r is a rational function given by $r(x) = p(x)/q(x)$, and c is a real number such that $q(c) \neq 0$, then

$$\lim_{x \to c} r(x) = r(c) = \frac{p(c)}{q(c)}$$

Proof: By Theorem 2.4 we know that for the polynomial functions p and q, we have

$$\lim_{x \to c} p(x) = p(c) \qquad \text{and} \qquad \lim_{x \to c} q(x) = q(c)$$

Now, since $q(c) \neq 0$, we can apply Property 4 of Theorem 2.3 to conclude that

$$\lim_{x \to c} r(x) = \lim_{x \to c} \frac{p(x)}{q(x)} = \frac{\lim_{x \to c} p(x)}{\lim_{x \to c} q(x)} = \frac{p(c)}{q(c)} = r(c)$$

EXAMPLE 8   The limit of a rational function

Find the following limit:

$$\lim_{x \to 1} \frac{x^2 + x + 2}{x + 1}$$

Solution: Since the denominator is not zero when $x = 1$, we can apply Theorem 2.5 to obtain

$$\lim_{x \to 1} \frac{x^2 + x + 2}{x + 1} = \frac{1^2 + 1 + 2}{1 + 1} = \frac{4}{2} = 2$$

Polynomial functions and rational functions constitute two of the three basic types of algebraic functions. The following theorem deals with the limit of a third type of algebraic function—one that involves radicals.

THEOREM 2.6   LIMITS INVOLVING RADICALS

If c is a real number, n and m are positive integers,* and f is a function whose limit exists at c, then the following properties are true.

1. $\lim\limits_{x \to c} \sqrt[n]{x} = \sqrt[n]{c}$ 2. $\lim\limits_{x \to c} \sqrt[n]{f(x)} = \sqrt[n]{\lim\limits_{x \to c} f(x)}$

3. $\lim\limits_{x \to c} x^{m/n} = c^{m/n}$ 4. $\lim\limits_{x \to c} [f(x)]^{m/n} = \left[\lim\limits_{x \to c} f(x)\right]^{m/n}$

Proof: Proofs of the first two properties are given in Appendix A. Property 3 follows from Property 1, together with Theorem 2.2. Property 4 follows from Property 2, together with Theorem 2.3.

EXAMPLE 9   The limit of a radical function

Find the following limit:

$$\lim_{x \to 3} \sqrt{2x^2 - 2}$$

Solution: Since

$$\lim_{x \to 3} (2x^2 - 2) = 2(3^2) - 2 = 16 > 0$$

we have, by Property 2 of Theorem 2.6,

$$\lim_{x \to 3} \sqrt{2x^2 - 2} = \sqrt{\lim_{x \to 3} (2x^2 - 2)} = \sqrt{16} = 4$$

Section Exercises 2.1

In Exercises 1–6, complete the table and use the result to estimate the given limit.

1. $\lim\limits_{x \to 2} \dfrac{x - 2}{x^2 - x - 2}$

x	1.9	1.99	1.999	2.001	2.01	2.1
$f(x)$						

2. $\lim\limits_{x \to 2} \dfrac{x - 2}{x^2 - 4}$

x	1.9	1.99	1.999	2.001	2.01	2.1
$f(x)$						

*If n is *even*, then we assume that $c > 0$ and $\lim\limits_{x \to c} f(x) > 0$.

†A color box indicates that a calculator may be helpful.

✓3. $\lim\limits_{x \to 0} \dfrac{\sqrt{x+3} - \sqrt{3}}{x}$

x	-0.1	-0.01	-0.001	0.001	0.01	0.1
$f(x)$						

4. $\lim\limits_{x \to -3} \dfrac{\sqrt{1-x} - 2}{x+3}$

x	-3.1	-3.01	-3.001	-2.999	-2.99	-2.9
$f(x)$						

✓5. *$\lim\limits_{x \to 3} \dfrac{[1/(x+1)] - (1/4)}{x-3}$

x	2.9	2.99	2.999	3.001	3.01	3.1
$f(x)$						

6. $\lim\limits_{x \to 4} \dfrac{[x/(x+1)] - (4/5)}{x-4}$

x	3.9	3.99	3.999	4.001	4.01	4.1
$f(x)$						

In Exercises 7–12, use the given graph to find the limit (if it exists).

7. $\lim\limits_{x \to 3} (4 - x)$ **8.** $\lim\limits_{x \to 1} (x^2 + 2)$

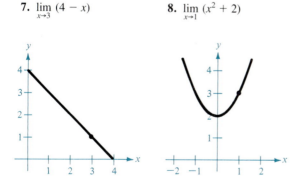

*A color number indicates that a detailed solution can be found in the *Student Solution Guide*.

9. $\lim\limits_{x \to 2} f(x)$ **10.** $\lim\limits_{x \to 1} f(x)$

$$f(x) = \begin{cases} 4 - x, & x \neq 2 \\ 0, & x = 2 \end{cases} \qquad f(x) = \begin{cases} x^2 + 2, & x \neq 1 \\ 1, & x = 1 \end{cases}$$

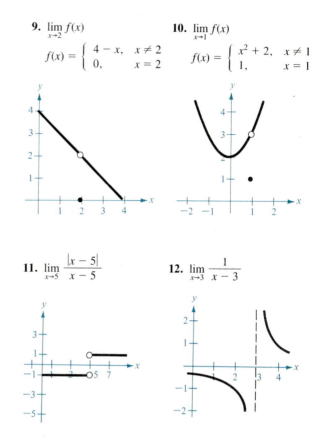

11. $\lim\limits_{x \to 5} \dfrac{|x-5|}{x-5}$ **12.** $\lim\limits_{x \to 3} \dfrac{1}{x-3}$

In Exercises 13–26, find the indicated limit.

13. $\lim\limits_{x \to 2} x^2$ **14.** $\lim\limits_{x \to -3} (3x + 2)$

15. $\lim\limits_{x \to 0} (2x - 1)$ **16.** $\lim\limits_{x \to 1} (-x^2 + 1)$

17. $\lim\limits_{x \to 2} (-x^2 + x - 2)$ **18.** $\lim\limits_{x \to 1} (3x^3 - 2x^2 + 4)$

19. $\lim\limits_{x \to 3} \sqrt{x+1}$ **20.** $\lim\limits_{x \to 4} \sqrt[3]{x+4}$

21. $\lim\limits_{x \to -4} (x+3)^2$ **22.** $\lim\limits_{x \to 0} (2x - 1)^3$

23. $\lim\limits_{x \to 2} \dfrac{1}{x}$ **24.** $\lim\limits_{x \to -3} \dfrac{2}{x+2}$

25. $\lim\limits_{x \to -1} \dfrac{x^2+1}{x}$ **26.** $\lim\limits_{x \to 3} \dfrac{\sqrt{x+1}}{x-4}$

27. If $\lim\limits_{x \to c} f(x) = 2$ and $\lim\limits_{x \to c} g(x) = 3$, find:

(a) $\lim\limits_{x \to c} [f(x) + g(x)]$ (b) $\lim\limits_{x \to c} [f(x)g(x)]$

(c) $\lim\limits_{x \to c} \dfrac{f(x)}{g(x)}$

28. If $\lim\limits_{x \to c} f(x) = 3/2$ and $\lim\limits_{x \to c} g(x) = 1/2$, find:

(a) $\lim\limits_{x \to c} [f(x) - g(x)]$ (b) $\lim\limits_{x \to c} [f(x)g(x)]$

(c) $\lim\limits_{x \to c} \dfrac{f(x)}{g(x)}$

2.2
Techniques for evaluating limits

In Section 2.1, we catalogued several types of limits that can be evaluated by *direct substitution*. We now look at some techniques for reducing other limits to this form. In the first example we use the Factor Theorem from algebra that states that for a polynomial function p, $p(c) = 0$ if and only if $(x - c)$ is a factor of $p(x)$. We can use this result in evaluating the limit of a rational function, as follows. If

$$r(x) = \frac{p(x)}{q(x)}$$

and

$$\lim_{x \to c} p(x) = p(c) = 0 \quad \text{and} \quad \lim_{x \to c} q(x) = q(c) = 0$$

then we may be able to evaluate the limit of $r(x)$ as x approaches c by cancelling the common factor $(x - c)$ out of the numerator and denominator, as shown in the following example.

EXAMPLE 1   Cancellation technique

Find the following limit:

$$\lim_{x \to -3} \frac{x^2 + x - 6}{x + 3}$$

Solution: Even though we are taking the limit of a rational function $p(x)/q(x)$, we *cannot* apply Theorem 2.5, since the limit of the denominator is zero.

$$\lim_{x \to -3} (x^2 + x - 6) = p(-3) = 0$$

$$\lim_{x \to -3} \frac{x^2 + x - 6}{x + 3}$$

Direct substitution fails

$$\lim_{x \to -3} (x + 3) = q(-3) = 0$$

However, since the limit of the numerator is also zero, we know that the numerator and denominator have a *common factor* of $(x + 3)$. Thus, for all $x \neq -3$ we can cancel this factor to obtain

$$\frac{x^2 + x - 6}{x + 3} = \frac{(x + 3)(x - 2)}{x + 3} = x - 2$$

Finally, by Theorem 2.1, it follows that

$$\lim_{x \to -3} \frac{x^2 + x - 6}{x + 3} = \lim_{x \to -3} (x - 2) = -5$$

This result is shown graphically in Figure 2.7.

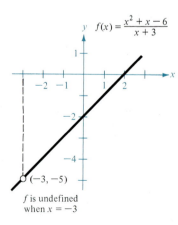

$f(x) = \dfrac{x^2 + x - 6}{x + 3}$

$(-3, -5)$

f is undefined
when $x = -3$

FIGURE 2.7

In Example 1, direct substitution produced the meaningless fractional form 0/0. We call such an expression an **indeterminate form,** since we cannot (from the form alone) determine the limit. When you are trying to evaluate a limit and encounter this form, remember that you must change the fraction so that the new denominator does not have zero as its limit. One way to do this is to *cancel like factors,* as shown in Example 1 and further demonstrated in the next example.

EXAMPLE 2   Cancellation technique

Find the following limit:

$$\lim_{x \to 1} \frac{x - 1}{x^3 - x^2 + x - 1}$$

Solution: By direct substitution, we get the indeterminate form 0/0:

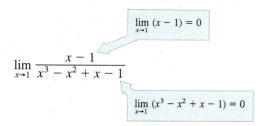

$$\lim_{x \to 1} (x - 1) = 0$$

$$\lim_{x \to 1} \frac{x - 1}{x^3 - x^2 + x - 1}$$

$$\lim_{x \to 1} (x^3 - x^2 + x - 1) = 0$$

Since the limit as x approaches 1 is zero for both the numerator *and* denominator and both are polynomials, we know that $(x - 1)$ is a common factor. Thus, for all $x \neq 1$, we can cancel this factor to obtain

$$\frac{x - 1}{x^3 - x^2 + x - 1} = \frac{x - 1}{(x - 1)(x^2 + 1)} = \frac{1}{x^2 + 1}$$

Now, by Theorem 2.1, it follows that

$$\lim_{x \to 1} \frac{x - 1}{x^3 - x^2 + x - 1} = \lim_{x \to 1} \frac{1}{x^2 + 1} = \frac{1}{2}$$

This result is shown graphically in Figure 2.8.

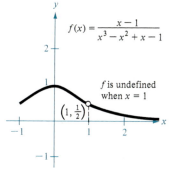

$f(x) = \dfrac{x - 1}{x^3 - x^2 + x - 1}$

f is undefined when $x = 1$

$\left(1, \frac{1}{2}\right)$

FIGURE 2.8

Remark In Example 2 the factorization of the denominator can be obtained by dividing by $(x - 1)$ or by grouping as follows:

$$x^3 - x^2 + x - 1 = x^2(x - 1) + (x - 1)$$
$$= (x - 1)(x^2 + 1)$$

A second way to find the limit of a function for which direct substitution yields the indeterminate form 0/0 is to use a *rationalization technique*. This technique can be used to rationalize either the numerator or the denominator. For instance, in the next example we begin by rationalizing the numerator—then we cancel common factors in the rationalized form.

EXAMPLE 3   Evaluating a limit by rationalizing the numerator

Find the following limit:

$$\lim_{x \to 0} \frac{\sqrt{x + 1} - 1}{x}$$

Solution: By direct substitution, we get the indeterminate form 0/0:

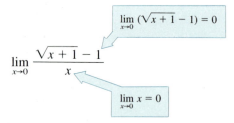

In this case, we change the form of the fraction by rationalizing (eliminating the radical in) the numerator:

$$\frac{\sqrt{x + 1} - 1}{x} = \left(\frac{\sqrt{x + 1} - 1}{x} \right)\left(\frac{\sqrt{x + 1} + 1}{\sqrt{x + 1} + 1} \right)$$

$$= \frac{(x + 1) - 1}{x(\sqrt{x + 1} + 1)}$$

$$= \frac{x}{x(\sqrt{x + 1} + 1)} = \frac{1}{\sqrt{x + 1} + 1}$$

Therefore, by Theorem 2.1, we have

$$\lim_{x \to 0} \frac{\sqrt{x + 1} - 1}{x} = \lim_{x \to 0} \frac{1}{\sqrt{x + 1} + 1} = \frac{1}{1 + 1} = \frac{1}{2}$$

Table 2.3 reinforces our conclusion that this limit is 1/2.

TABLE 2.3

	x approaches 0 from the left					x approaches 0 from the right			
x	−0.25	−0.1	−0.01	−0.001	0	0.001	0.01	0.1	0.25
$f(x)$	0.5359	0.5132	0.5013	0.5001	?	0.4999	0.4988	0.4881	0.4721

$f(x)$ approaches 1/2	$f(x)$ approaches 1/2

Remark The rationalizing technique for evaluating limits is based upon multiplication by a convenient form of 1—in the case of Example 3, the convenient form is

$$1 = \frac{\sqrt{x + 1} + 1}{\sqrt{x + 1} + 1}$$

One-sided limits

In Section 2.1, we saw that one way in which a limit can fail to exist is when a function approaches a different value from the right side of c than it approaches from the left side of c. To further investigate this type of behavior, we first need to look at a different type of "limit" called a **one-sided limit.** For example, when we talk about the limit from the right, we mean that x approaches c from values greater than c. We denote this by

$$\lim_{x \to c^+} f(x) = L \qquad \text{Limit from the right}$$

Similarly, the limit from the left means that x approaches c from values less than c. We denote this by

$$\lim_{x \to c^-} f(x) = L \qquad \text{Limit from the left}$$

One-sided limits possess many of the same properties shared by regular (two-sided) limits. For instance, notice how we use direct substitution to evaluate the limit of a function involving a radical in the next example.

EXAMPLE 4   A one-sided limit

Evaluate the one-sided limit

$$\lim_{x \to 3^-} \sqrt{3 - x}$$

Solution: Using Theorem 2.4 and 2.6 (as applied to one-sided limits), we find that the limit as x approaches 3 *from the left* is given by

$$\lim_{x \to 3^-} \sqrt{3 - x} = \sqrt{3 - 3} = 0$$

This result is shown graphically in Figure 2.9.

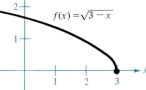

$f(x) = \sqrt{3 - x}$

FIGURE 2.9

Another use of one-sided limits is in investigating the behavior of **step functions.** One common type of step function is shown in the next example.

EXAMPLE 5   The greatest integer function

The **greatest integer function, $[x]$,** is defined by

$$[x] = \text{greatest integer } n \text{ such that } n \le x$$

Find the limit of this function as x approaches 0 from the left and from the right. (See Figure 2.10.)

Solution: The limit as x approaches 0 *from the left* is given by

$$\lim_{x \to 0^-} [x] = \lim_{x \to 0^-} (-1) = -1$$

and the limit as x approaches 0 *from the right* is given by

$$\lim_{x \to 0^+} [x] = \lim_{x \to 0^+} (0) = 0$$

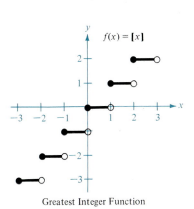

$f(x) = [x]$

Greatest Integer Function

FIGURE 2.10

In Figure 2.10 we see that the greatest integer function approaches a different number from the right of 0 than it approaches from the left of 0. In such cases we say that the (two-sided) limit *does not exist*. The following theorem makes this more explicit. The proof of this theorem follows directly from the definitions of limits and one-sided limits.

THEOREM 2.7   **THE EXISTENCE OF A LIMIT**
If f is a function and c and L are real numbers, then

$$\lim_{x \to c} f(x) = L$$

if and only if

$$\lim_{x \to c^-} f(x) = L \qquad \text{and} \qquad \lim_{x \to c^+} f(x) = L$$

| **Remark**   Theorem 2.7 is particularly useful for determining that a limit does not exist as demonstrated in Example 5.

EXAMPLE 6   Comparing the limits from the left and right

Evaluate the following limit:

$$\lim_{x \to 1} f(x), \quad \text{where } f(x) = \begin{cases} 2x - x^3, & x < 1 \\ 2x^2 - 2, & x \geq 1 \end{cases}$$

Solution: Since f is defined differently for $x < 1$ than for $x \geq 1$, we consider the one-sided limits:

$$\lim_{x \to 1^-} f(x) = \lim_{x \to 1^-} (2x - x^3) = 2 - 1 = 1$$

and

$$\lim_{x \to 1^+} f(x) = \lim_{x \to 1^+} (2x^2 - 2) = 2 - 2 = 0$$

Because these one-sided limits are not equal, we conclude that the limit of $f(x)$ as $x \to 1$ *does not exist*. This is shown graphically in Figure 2.11.

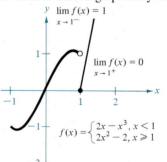

FIGURE 2.11

Section Exercises 2.2   1-20, 35-50

In Exercises 1–6, use the graph to visually determine the limit, if it exists.

1. $f(x) = -2x + 1$
   (a) $\lim_{x \to 0} f(x)$
   (b) $\lim_{x \to -1} f(x)$

2. $f(x) = x^3 - 3x^2$
   (a) $\lim_{x \to 1} f(x)$
   (b) $\lim_{x \to 3} f(x)$

3. $g(x) = \dfrac{-2x^2 + x}{x}$
   (a) $\lim_{x \to 0} g(x)$
   (b) $\lim_{x \to -1} g(x)$

4. $h(x) = \dfrac{x^2 - 3x}{x}$
   (a) $\lim_{x \to -2} h(x)$
   (b) $\lim_{x \to 0} h(x)$

5. $g(x) = \dfrac{x^3 - x}{x - 1}$
   (a) $\lim_{x \to 1} g(x)$
   (b) $\lim_{x \to -1} g(x)$

6. $f(x) = \dfrac{1}{x - 1}$
   (a) $\lim_{x \to 1} f(x)$
   (b) $\lim_{x \to 2} f(x)$

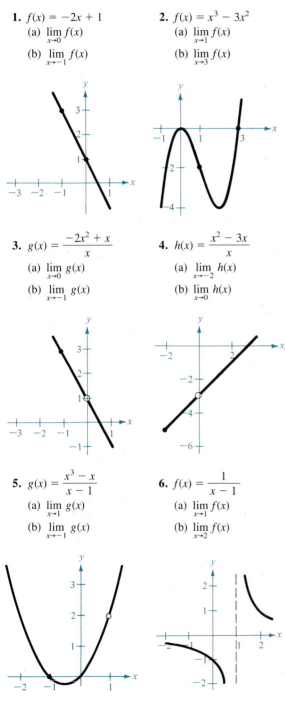

In Exercises 7–24, find the limit (if it exists).

7. $\lim_{x \to -1} \dfrac{x^2 - 1}{x + 1}$

8. $\lim_{x \to -1} \dfrac{2x^2 - x - 3}{x + 1}$

9. $\lim_{x \to 3} \dfrac{x - 3}{x^2 - 9}$

10. $\lim_{x \to -1} \dfrac{x^3 + 1}{x + 1}$

11. $\lim_{x \to -2} \dfrac{x^3 + 8}{x + 2}$

12. $\lim_{\Delta x \to 0} \dfrac{(x + \Delta x)^2 - x^2}{\Delta x}$

13. $\lim_{\Delta x \to 0} \dfrac{2(x + \Delta x) - 2x}{\Delta x}$

14. $\lim_{\Delta x \to 0} \dfrac{(x + \Delta x)^3 - x^3}{\Delta x}$

15. $\lim_{\Delta x \to 0} \dfrac{(x + \Delta x)^2 - 2(x + \Delta x) + 1 - (x^2 - 2x + 1)}{\Delta x}$

16. $\lim_{\Delta x \to 0} \dfrac{(1 + \Delta x)^3 - 1}{\Delta x}$

17. $\lim_{x \to 5} \dfrac{x - 5}{x^2 - 25}$

18. $\lim_{x \to 2} \dfrac{2 - x}{x^2 - 4}$

19. $\lim_{x \to 1} \dfrac{x^2 + x - 2}{x^2 - 1}$

20. $\lim_{x \to 0} \dfrac{\sqrt{2 + x} - \sqrt{2}}{x}$

21. $\lim_{x \to 0} \dfrac{\sqrt{3 + x} - \sqrt{3}}{x}$

22. $\lim_{x \to 0} \dfrac{[1/(x + 4)] - (1/4)}{x}$

23. $\lim_{x \to 0} \dfrac{[1/(2 + x)] - (1/2)}{x}$

24. $\lim_{x \to 3} \dfrac{\sqrt{x + 1} - 2}{x - 3}$

In Exercises 25–28, use a calculator to complete a table of values near $x = c$ to estimate the limit. Then, find the limit by analytic methods and compare the result to your estimated limit.

25. $\lim_{x \to 0} \dfrac{\sqrt{x + 2} - \sqrt{2}}{x}$

26. $\lim_{x \to 1} \dfrac{1 - x}{\sqrt{5 - x^2} - 2}$

27. $\lim_{x \to 0} \dfrac{[1/(2 + x)] - (1/2)}{2x}$

28. $\lim_{x \to 2} \dfrac{x^5 - 32}{x - 2}$

In Exercises 29–34, use the graph to visually determine
(a) $\lim_{x \to c^+} f(x)$ (b) $\lim_{x \to c^-} f(x)$ (c) $\lim_{x \to c} f(x)$

29.

30.

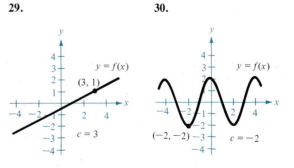

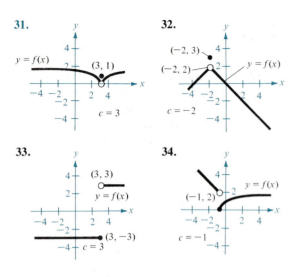

31.

$y = f(x)$

(3, 1)

$c = 3$

32.

(-2, 3)

(-2, 2)

$y = f(x)$

$c = -2$

33.

(3, 3)

$y = f(x)$

(3, -3)

$c = 3$

34.

$y = f(x)$

(-1, 2)

$c = -1$

37. $\lim\limits_{x \to 2^+} \dfrac{x}{\sqrt{x^2 - 4}}$

38. $\lim\limits_{x \to 4^-} \dfrac{\sqrt{x} - 2}{x - 4}$

39. $\lim\limits_{\Delta x \to 0^+} \dfrac{2(x + \Delta x) - 2x}{\Delta x}$

40. $\lim\limits_{x \to 1^-} \dfrac{x^2 - 2x + 1}{x - 1}$

41. $\lim\limits_{x \to 0} \dfrac{|x|}{x}$

42. $\lim\limits_{x \to 2} \dfrac{|x - 2|}{x - 2}$

43. $\lim\limits_{x \to 3} f(x), \quad f(x) = \begin{cases} \dfrac{x + 2}{2}, & x \leq 3 \\ \dfrac{12 - 2x}{3}, & x > 3 \end{cases}$

44. $\lim\limits_{x \to 2} f(x), \quad f(x) = \begin{cases} x^2 - 4x + 6, & x < 2 \\ -x^2 + 4x - 2, & x \geq 2 \end{cases}$

45. $\lim\limits_{x \to 1} f(x), \quad f(x) = \begin{cases} x^3 + 1, & x < 1 \\ x + 1, & x \geq 1 \end{cases}$

46. $\lim\limits_{x \to 1} f(x), \quad f(x) = \begin{cases} x, & x \leq 1 \\ 1 - x, & x > 1 \end{cases}$

47. $\lim\limits_{x \to 3^-} 2[\![x - 3]\!]$

48. $\lim\limits_{x \to 1^+} [\![2x]\!]$

49. $\lim\limits_{x \to 1} \left(\left[\!\!\left[\dfrac{x}{4}\right]\!\!\right] + x \right)$

50. $\lim\limits_{x \to 2} \left[\!\!\left[\dfrac{x - 1}{2}\right]\!\!\right]$

In Exercises 35–50, find the limit (if it exists).

35. $\lim\limits_{x \to 5^+} \dfrac{x - 5}{x^2 - 25}$

36. $\lim\limits_{x \to 2^+} \dfrac{2 - x}{x^2 - 4}$

Continuity at a point ▪
Continuity on an open interval ▪
Continuity on a closed interval ▪
Properties of continuity ▪
Intermediate Value Theorem ▪

$f(c_1)$ is not defined.

$\lim\limits_{x \to c_2} f(x)$ does not exist.

f

$\lim\limits_{x \to c_3} f(x) \neq f(c_3)$

a b

Three Points of Discontinuity

FIGURE 2.12

2.3
Continuity

In mathematics the term *continuous* has much the same meaning as it does in our everyday usage. To say that a function is continuous at $x = c$ means that there is no interruption in the graph of f at c. That is, its graph is unbroken at c and there are no holes, jumps, or gaps. For example, Figure 2.12 identifies three values of x at which the graph of f is not continuous. At all other points of the interval (a, b), the graph of f is uninterrupted, and we say it is **continuous** at such points. Thus, it appears that the continuity of a function at $x = c$ can be destroyed by any one of the following conditions:

1. The function is not defined at $x = c$.
2. The limit of $f(x)$ does not exist at $x = c$.
3. The limit of $f(x)$ exists at $x = c$, but is not equal to $f(c)$.

This brings us to the following definition.

DEFINITION OF CONTINUITY

Continuity at a Point: A function f is said to be **continuous at c** if the following three conditions are met.

1. $f(c)$ is defined.   2. $\lim\limits_{x \to c} f(x)$ exists.   3. $\lim\limits_{x \to c} f(x) = f(c)$

Continuity on an Open Interval: A function is said to be **continuous on an interval (a, b)** if it is continuous at each point in the interval.

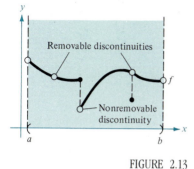

Removable discontinuities

Nonremovable discontinuity

FIGURE 2.13

Remark A function that is continuous on the entire real line $(-\infty, \infty)$ is sometimes called simply a **continuous function.**

A function f is said to be **discontinuous at c** if f is defined on an open interval containing c (except possibly at c) and f is not continuous at c. Discontinuities fall into two categories: **removable** and **nonremovable.** A discontinuity at $x = c$ is called removable if f can be made continuous by redefining f at $x = c$. For example, in Figure 2.13 the function f has two removable discontinuities and one nonremovable discontinuity.

In Section 2.1 we studied several types of functions that meet the three conditions for continuity. In Example 1 we use this knowledge of limits to examine two functions that are continuous and one that is not.

EXAMPLE 1   Testing for continuity

Determine whether the following functions are continuous on the given interval:

(a) $f(x) = \dfrac{1}{x}$, $(0, 1)$

(b) $f(x) = \dfrac{x^2 - 1}{x - 1}$, $(0, 2)$

(c) $f(x) = x^3 - x$, $(-\infty, \infty)$

Solution: The graphs of these three functions are shown in Figure 2.14.

(a) Since f is a rational function whose denominator is not zero in the interval $(0, 1)$, we can apply Theorem 2.5 to conclude that f is continuous on $(0, 1)$.

(b) Since f is undefined at $x = 1$, we conclude that it is discontinuous at $x = 1$. From Theorem 2.5 we know that f is continuous at all other points in the interval $(0, 2)$.

(c) Since polynomial functions are defined over the entire real line, we can apply Theorem 2.4 to conclude that f is continuous on $(-\infty, \infty)$.

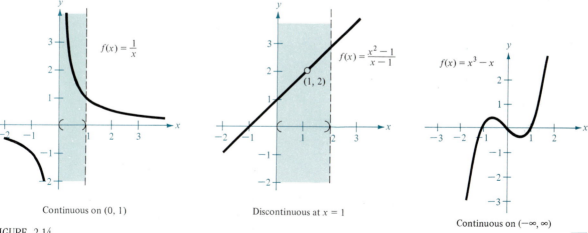

Continuous on $(0, 1)$ Discontinuous at $x = 1$ Continuous on $(-\infty, \infty)$

FIGURE 2.14

| **Remark** In part (b) of Example 1 the discontinuity at $x = 1$ is *removable*. Specifically, by defining $f(1)$ to be 2, we would obtain a function that is continuous on $(0, 1)$. That is, we say that f has a **removable discontinuity** at $x = c$ if f is discontinuous at c and

$$\lim_{x \to c} f(x)$$

exists but does not equal $f(c)$.

Each of the intervals in Example 1 is open. To discuss continuity on a closed interval, we use the concept of one-sided limits, as defined in the previous section.

DEFINITION OF CONTINUITY ON A CLOSED INTERVAL

If f is defined on a closed interval $[a, b]$, continuous on (a, b), and

$$\lim_{x \to a^+} f(x) = f(a) \qquad \text{and} \qquad \lim_{x \to b^-} f(x) = f(b)$$

then f is said to be **continuous on $[a, b]$.**

| **Remark** If a function is continuous on $[a, b]$, we say that **f is continuous from the right at a and f is continuous from the left at b.**

EXAMPLE 2   Continuity at the endpoint of an interval

Discuss the continuity of the function given by $f(x) = \sqrt{x}$.

Solution: The domain of f is $[0, \infty)$. At all points in the interval $(0, \infty)$ the continuity of f follows from Theorems 2.2 and 2.6. Moreover, at the left endpoint of the domain, we have

$$\lim_{x \to 0^+} \sqrt{x} = 0 = f(0) \qquad \text{Limit from the right}$$

Therefore, f is continuous from the right at $x = 0$, and we conclude that f is continuous on its entire domain, as shown in Figure 2.15.

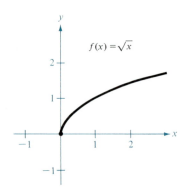

FIGURE 2.15

EXAMPLE 3   Continuity on a closed interval

Discuss the continuity of

$$g(x) = \begin{cases} 5 - x, & -1 \le x \le 2 \\ x^2 - 1, & 2 < x \le 3 \end{cases}$$

Solution: From our work in Section 2.1, we know that the polynomial functions given by $5 - x$ and $x^2 - 1$ are continuous for all real x. Thus, to conclude that g is continuous on the entire interval $[-1, 3]$, we need only check the behavior of g when $x = 2$. By taking the one-sided limits when $x = 2$, we see that

$$\lim_{x \to 2^-} g(x) = \lim_{x \to 2^-} (5 - x) = 3 \qquad \text{Limit from the left}$$

and

$$\lim_{x \to 2^+} g(x) = \lim_{x \to 2^+} (x^2 - 1) = 3 \qquad \text{Limit from the right}$$

Since these two limits are equal, we can apply Theorem 2.7 to conclude that

$$\lim_{x \to 2} g(x) = g(2) = 3$$

Thus, g is continuous at $x = 2$, and consequently it is continuous on the entire interval $[-1, 3]$. The graph of g is shown in Figure 2.16.

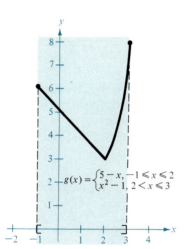

$$g(x) = \begin{cases} 5 - x, -1 \le x \le 2 \\ x^2 - 1, 2 < x \le 3 \end{cases}$$

FIGURE 2.16

Properties of continuous functions

In Section 2.1, we looked at several properties of limits. Each of these properties yields a corresponding property of continuous functions. The proof of Theorem 2.8 follows directly from Theorem 2.3.

THEOREM 2.8   **PROPERTIES OF CONTINUOUS FUNCTIONS**

If b is a real number and f and g are continuous at $x = c$, then the following functions are also continuous at c.

1. Scalar multiple: bf

2. Sum and difference: $f \pm g$

3. Product: fg

4. Quotient: $\dfrac{f}{g}$, if $g(c) \ne 0$

It is important for you to be able to recognize continuous functions, and the following summary lists the basic types of continuous functions we have studied up to this point.

SUMMARY OF   The following functions are continuous at every point in their domain.
CONTINUOUS FUNCTIONS

1. Polynomial functions: $p(x) = a_n x^n + a_{n-1} x^{n-1} + \cdots + a_1 x + a_0$

2. Rational functions: $r(x) = \dfrac{p(x)}{q(x)}, \quad q(x) \neq 0$

3. Radical functions: $f(x) = \sqrt[n]{x}$

By combining Theorem 2.8 with this summary, we can conclude that a wide variety of elementary functions are continuous. For example, the following function is continuous at every point in its domain.

$$f(x) = \frac{x^2 + 1}{\sqrt{x}}$$

The next theorem allows us to determine the continuity of a *composite* function, such as

$$f(x) = \sqrt{x^2 + 1}$$

A proof of this theorem is given in Appendix A.

THEOREM 2.9   CONTINUITY OF A COMPOSITE FUNCTION
If g is continuous at c and f is continuous at $g(c)$, then the composite function given by $f \circ g(x) = f(g(x))$ is continuous at c.

Remark   Note that one consequence of the theorem is that for functions f and g satisfying the given conditions, we can determine the limit as x approaches c to be

$$\lim_{x \to c} f(g(x)) = f\left(\lim_{x \to c} g(x)\right) = f(g(c))$$

EXAMPLE 4   *Testing for continuity*

Find the intervals for which the three functions shown in Figure 2.17 are continuous.

Solution:
(a) The function $f(x) = \sqrt{1 - x^2}$ is continuous on the *closed* interval $[-1, 1]$.
(b) The function $f(x) = 1/\sqrt{1 - x^2}$ is continuous on the *open* interval $(-1, 1)$. (Note that f is *undefined* for all x such that $|x| \geq 1$.)
(c) At $x = \pm 1$, the limits from the right and left are zero. Thus, the function $f(x) = |x^2 - 1|$ is continuous on the entire real line, that is, the interval $(-\infty, \infty)$.

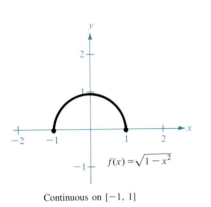

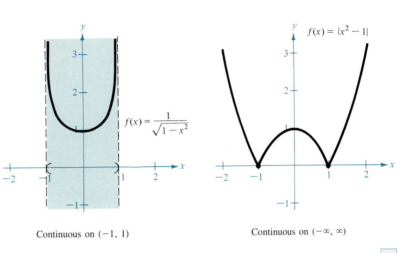

Continuous on $[-1, 1]$

$f(x) = \sqrt{1 - x^2}$

Continuous on $(-1, 1)$

$f(x) = \dfrac{1}{\sqrt{1 - x^2}}$

Continuous on $(-\infty, \infty)$

$f(x) = |x^2 - 1|$

FIGURE 2.17

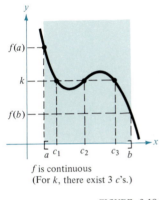

f is continuous
(For k, there exist 3 c's.)

FIGURE 2.18

Intermediate Value Theorem

We conclude this section with an important theorem concerning the behavior of continuous functions. By referring to a text on advanced calculus, you will find that a proof of this theorem is based on a property of real numbers called *completeness*. (The completeness property of real numbers essentially states that there are no holes or gaps on the real number line.) The Intermediate Value Theorem states that for a continuous function f, if x takes on all values between a and b, then $f(x)$ must take on all values between $f(a)$ and $f(b)$. As a practical example of this theorem, consider a person's height. Suppose that a boy is 5 feet tall on his thirteenth birthday and 5 feet 7 inches tall on his fourteenth birthday. Then, for any height h between 5 feet and 5 feet 7 inches, there must have been a time t when his height was exactly h. This seems reasonable, since we believe that normal human growth is continuous and a person's height could not abruptly change from one value to another.

THEOREM 2.10

INTERMEDIATE VALUE THEOREM
If f is continuous on $[a, b]$ and k is any number between $f(a)$ and $f(b)$, then there is at least one number c between a and b such that $f(c) = k$.

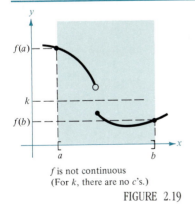

f is not continuous
(For k, there are no c's.)

FIGURE 2.19

The Intermediate Value Theorem states that if the domain of a function is a closed interval and the function is continuous, then there are no holes or gaps in the graph of the function. Discontinuous functions, however, may not possess the intermediate value property, as a comparison of Figures 2.18 and 2.19 indicates.

You may be familiar with one application of the Intermediate Value Theorem in algebra. That is, if we are given a polynomial function f and there are two numbers a and b such that $f(a)$ is negative and $f(b)$ is positive, then the function must have a zero (its graph must cross the x-axis) between a and b.

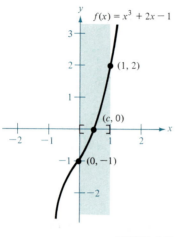

FIGURE 2.20

EXAMPLE 5 An application of the Intermediate Value Theorem

Use the Intermediate Value Theorem to show that the polynomial function $f(x) = x^3 + 2x - 1$ has a zero in the interval $[0, 1]$.

Solution: Since

$$f(0) = 0^3 + 2(0) - 1 = -1 \qquad f(0) < 0$$

and

$$f(1) = 1^3 + 2(1) - 1 = 2 \qquad f(1) > 0$$

we can apply the Intermediate Value Theorem to conclude that there must be some c in $(0, 1)$ such that $f(c) = 0$, as shown in Figure 2.20. ▭

Remark Note that the Intermediate Value Theorem tells us that (at least) one c exists in the interval (a, b). However, the theorem does not give us a method for finding c. Such theorems are called **existence theorems.**

Section Exercises 2.3

In Exercises 1–6, find the points of discontinuity (if any).

1. $f(x) = -\dfrac{x^3}{2}$

2. $f(x) = \dfrac{x^2 - 1}{x}$

3. $f(x) = \dfrac{x^2 - 1}{x + 1}$

4. $f(x) = \dfrac{1}{x^2 - 4}$

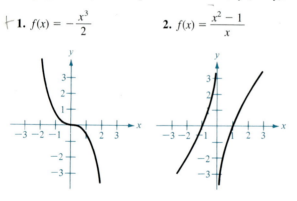

5. $f(x) = \begin{cases} x, & x < 1 \\ 2, & x = 1 \\ 2x - 1, & x > 1 \end{cases}$

6. $f(x) = \dfrac{[x]}{2} + x$

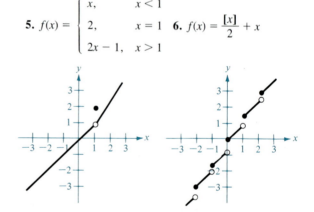

In Exercises 7–24, find the discontinuities (if any) for the given function. Which of the discontinuities are removable?

7. $f(x) = x^2 - 2x + 1$

8. $f(x) = \dfrac{1}{x^2 + 1}$

9. $f(x) = \dfrac{1}{x - 1}$

10. $f(x) = \dfrac{x}{x^2 - 1}$

11. $f(x) = \dfrac{x}{x^2 + 1}$

12. $f(x) = \dfrac{x - 3}{x^2 - 9}$

13. $f(x) = \dfrac{x + 2}{x^2 - 3x - 10}$

14. $f(x) = \dfrac{x - 1}{x^2 + x - 2}$

15. $f(x) = \begin{cases} x, & x \le 1 \\ x^2, & x > 1 \end{cases}$

16. $f(x) = \begin{cases} -2x + 3, & x < 1 \\ x^2, & x \ge 1 \end{cases}$

17. $f(x) = \begin{cases} \dfrac{x}{2} + 1, & x \le 2 \\ 3 - x, & x > 2 \end{cases}$

18. $f(x) = \begin{cases} -2x, & x \le 2 \\ x^2 - 4x + 1, & x > 2 \end{cases}$

19. $f(x) = \dfrac{|x + 2|}{x + 2}$ **20.** $f(x) = \dfrac{|x - 3|}{x - 3}$

21. $f(x) = \begin{cases} |x - 2| + 3, & x < 0 \\ x + 5, & x \ge 0 \end{cases}$

22. $f(x) = \begin{cases} 3 + x, & x \le 2 \\ x^2 + 1, & x > 2 \end{cases}$

23. $f(x) = [x - 1]$ **24.** $f(x) = x - [x]$

In Exercises 25–30, discuss the continuity of the composite function $h(x) = f(g(x))$.

25. $f(x) = x^2, \ g(x) = x - 1$
26. $f(x) = 1/\sqrt{x}, \ g(x) = x - 1$
27. $f(x) = \dfrac{1}{x - 1}, \ g(x) = x^2 + 5$
28. $f(x) = \sqrt{x}, \ g(x) = x^2$
29. $f(x) = \dfrac{1}{x}, \ g(x) = \dfrac{1}{x - 1}$
30. $f(x) = \dfrac{1}{\sqrt{x}}, \ g(x) = \dfrac{1}{x}$

In Exercises 31–34, sketch the graph of the given function to determine any points of discontinuity.

31. $f(x) = \dfrac{x^2 - 16}{x - 4}$ **32.** $f(x) = \dfrac{x^3 - 8}{x - 2}$

33. $f(x) = [x] - x$ **34.** $f(x) = \dfrac{|x^2 - 1|}{x}$

In Exercises 35–38, find the interval(s) for which the function is continuous.

35. $f(x) = \dfrac{x^2}{x^2 - 36}$ **36.** $f(x) = x\sqrt{x + 3}$

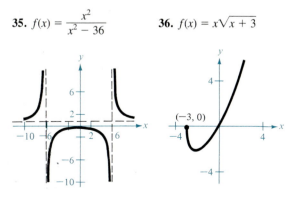

(-3, 0)

37. $f(x) = \dfrac{x}{x^2 + 1}$ **38.** $f(x) = \dfrac{x + 1}{\sqrt{x}}$

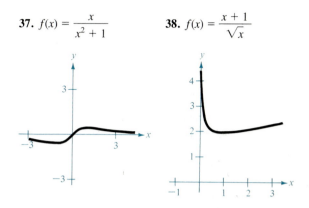

In Exercises 39 and 40, prove that the given function has a zero in the indicated interval.

Function	Interval
39. $f(x) = x^2 - 4x + 3$	$[2, 4]$
40. $f(x) = x^3 + 3x - 2$	$[0, 1]$

In Exercises 41 and 42, use the Intermediate Value Theorem to approximate the zero of the given function in the interval $[0, 1]$. (a) Begin by locating the zero in a subinterval of length 0.1. (b) Refine your approximation by locating the zero in a subinterval of length 0.01.

41. $f(x) = x^3 + x - 1$
42. $f(x) = x^3 + 3x - 2$

In Exercises 43–46, verify the applicability of the Intermediate Value Theorem in the indicated interval and find the value of c guaranteed by the theorem.

43. $f(x) = x^2 + x - 1, \ [0, 5], \ f(c) = 11$
44. $f(x) = x^2 - 6x + 8, \ [0, 3], \ f(c) = 0$
45. $f(x) = x^3 - x^2 + x - 2, \ [0, 3], \ f(c) = 4$
46. $f(x) = \dfrac{x^2 + x}{x - 1}, \ \left[\dfrac{5}{2}, 4\right], \ f(c) = 6$

47. Determine the constant a so that the following function is continuous:

$$f(x) = \begin{cases} x^3, & x \le 2 \\ ax^2, & x > 2 \end{cases}$$

48. Determine the constants a and b so that the following function is continuous:

$$f(x) = \begin{cases} 2, & x \le -1 \\ ax + b, & -1 < x < 3 \\ -2, & x \ge 3 \end{cases}$$

49. Is the function $f(x) = \sqrt{1 - x^2}$ continuous at $x = 1$? Give the reason for your answer.

50. A union contract guarantees a 9 percent annual salary increase for five years. For an initial salary of $28,500, the salary S is given by

$$S = 28,500(1.09)^{[t]}$$

where $t = 0$ corresponds to 1985. Sketch a graph of this function and discuss its continuity.

51. A dial-direct long distance call between two cities costs $1.04 for the first two minutes and $0.36 for each additional minute or fraction thereof. Use the greatest integer function to write the cost C of a call in terms of the time t (in minutes). Sketch a graph of this function and discuss its continuity.

52. The number of units in inventory in a small company is given by

$$N(t) = 25\left(2\left[\frac{t+2}{2}\right] - t\right)$$

where t is the time in months. Sketch the graph of this function and discuss its continuity. How often does this company replenish its inventory?

53. Prove Theorem 2.8.

54. Prove that if f is continuous and has no zeros on $[a, b]$, then either

$$f(x) > 0 \text{ for all } x \text{ in } [a, b]$$

or

$$f(x) < 0 \text{ for all } x \text{ in } [a, b]$$

SECTION TOPICS ▪
Infinite limits ▪
Vertical asymptotes ▪

2.4
Infinite limits

In this section we take a further look at an important way in which a limit can fail to exist—if the function has an infinite discontinuity.

We begin with an example. Let f be the function given by

$$f(x) = \frac{3}{x - 2}$$

From Figure 2.21 and Table 2.4, we can see that $f(x)$ *decreases without bound* as x approaches 2 from the left and $f(x)$ *increases without bound* as x approaches 2 from the right. Symbolically, we write

$$\lim_{x \to 2^-} \frac{3}{x - 2} = -\infty \qquad \text{and} \qquad \lim_{x \to 2^+} \frac{3}{x - 2} = \infty$$

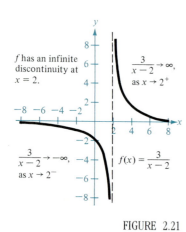

f has an infinite discontinuity at $x = 2$.

$\dfrac{3}{x-2} \to \infty,$ as $x \to 2^+$

$\dfrac{3}{x-2} \to -\infty,$ as $x \to 2^-$

$f(x) = \dfrac{3}{x-2}$

FIGURE 2.21

TABLE 2.4

	x approaches 2 from the left						x approaches 2 from the right				
x	1	1.5	1.9	1.99	1.999	2	2.001	2.01	2.1	2.5	3.0
$f(x)$	-3	-6	-30	-300	-3000	?	3000	300	30	6	3
	f(x) decreases without bound						f(x) increases without bound				

In general, the types of limits where $f(x)$ increases or decreases without bound as x approaches c are called **infinite limits.**

| **Remark** The equal sign in the statement $\lim f(x) = \infty$ does not mean that the limit exists! On the contrary, it tells us how the limit *fails to exist* by denoting the unbounded behavior of $f(x)$ as x approaches c. Thus, when we say "the limit of $f(x)$ is

infinite as x approaches c'' we really mean that ''the limit does not exist *and f* has an infinite discontinuity at $x = c$.''

DEFINITION OF INFINITE LIMITS

The statement

$$\lim_{x \to c} f(x) = \infty$$

means that $f(x)$ *increases* without bound as x approaches c. The statement

$$\lim_{x \to c} f(x) = -\infty$$

means that $f(x)$ *decreases* without bound as x approaches c.

To say that $f(x)$ increases without bound as $x \to c$ means that for each $M > 0$, there exists an open interval I containing c such that $f(x) > M$ for all x in I (other than $x = c$). A similar interpretation is given to define what it means for $f(x)$ to decrease without bound, and both definitions are illustrated in Figure 2.22.

Infinite limits from the right and left are defined similarly. The four possible one-sided infinite limits are

$$\lim_{x \to c^-} f(x) = -\infty, \qquad \lim_{x \to c^-} f(x) = \infty \qquad \text{Infinite limits from the left}$$

$$\lim_{x \to c^+} f(x) = -\infty, \qquad \lim_{x \to c^+} f(x) = \infty \qquad \text{Infinite limits from the right}$$

If $f(x) \to \pm\infty$ from the left or from the right, we say that f has an **infinite discontinuity** at $x = c$.

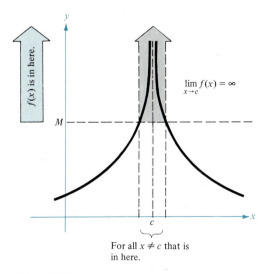

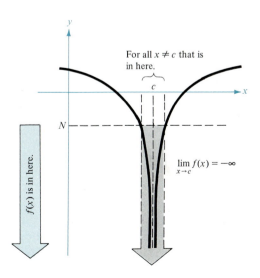

FIGURE 2.22

EXAMPLE 1   Determining infinite limits from a graph

Use Figure 2.23 to determine the limit of each function as $x \to 1$ from the left and from the right.

(a)   (b)   (c)   (d)

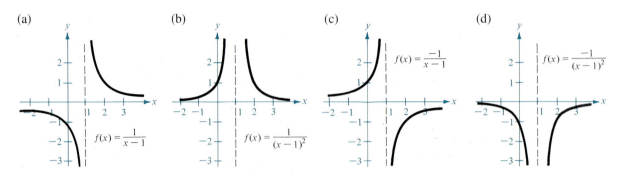

FIGURE 2.23

Solution:

(a) $\displaystyle \lim_{x \to 1^-} \frac{1}{x-1} = -\infty$ and $\displaystyle \lim_{x \to 1^+} \frac{1}{x-1} = \infty$

(b) $\displaystyle \lim_{x \to 1} \frac{1}{(x-1)^2} = \infty$ Limit from both sides is ∞

(c) $\displaystyle \lim_{x \to 1^-} \frac{-1}{x-1} = \infty$ and $\displaystyle \lim_{x \to 1^+} \frac{-1}{x-1} = -\infty$

(d) $\displaystyle \lim_{x \to 1} \frac{-1}{(x-1)^2} = -\infty$ Limit from both sides is $-\infty$

If it were possible to extend the graphs in Figure 2.23 up and down toward infinity, you would see that each graph becomes arbitrarily close to the vertical line $x = 1$. We call this line a **vertical asymptote** of the graph of f.

DEFINITION OF VERTICAL ASYMPTOTE	If $f(x)$ approaches infinity (or negative infinity) as x approaches c from the right or left, then we call the line $x = c$ a **vertical asymptote** of the graph of f.

In Example 1, note that each of the functions is a *quotient* and that the vertical asymptote occurs at the point where the denominator is zero. The following theorem generalizes this observation. (A proof of the theorem is given in Appendix A.)

THEOREM 2.11   VERTICAL ASYMPTOTES

Let f and g be continuous on an open interval containing c. If $f(c) \neq 0$, $g(c) = 0$, and there exists an open interval containing c such that $g(x) \neq 0$ for all $x \neq c$ in the interval, then the graph of the function given by

$$h(x) = \frac{f(x)}{g(x)}$$

has a vertical asymptote at $x = c$.

EXAMPLE 2   Finding vertical asymptotes

Determine all vertical asymptotes for the graphs of the following functions.

(a) $f(x) = \dfrac{1}{2(x + 1)}$ (b) $f(x) = \dfrac{x^2 + 1}{x^2 - 1}$

Solution:

(a) At the point $x = -1$, the denominator is zero and the numerator is not zero. Hence, by Theorem 2.11 we conclude that $x = -1$ is a vertical asymptote, as shown in Figure 2.24.

(b) By factoring the denominator as $(x^2 - 1) = (x - 1)(x + 1)$, we see that the denominator is zero at $x = -1$ and $x = 1$. Moreover, since the numerator is not zero at these two points, we apply Theorem 2.11 to conclude that the graph of f has the two vertical asymptotes shown in Figure 2.24.

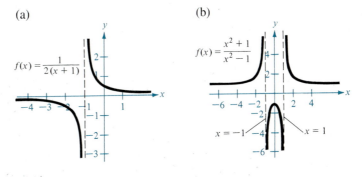

FIGURE 2.24

Theorem 2.11 requires that the value of the numerator at $x = c$ is nonzero. If both the numerator and denominator are zero at $x = c$, then we obtain the *indeterminate form* 0/0 and we cannot determine the limit behavior at $x = c$ without further investigation. One case for which we can determine the limit is when the numerator and denominator are polynomials. We do this by canceling common factors, as shown in the next example.

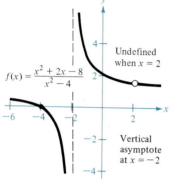

$$f(x) = \frac{x^2 + 2x - 8}{x^2 - 4}$$

Undefined when $x = 2$

Vertical asymptote at $x = -2$

FIGURE 2.25

EXAMPLE 3   *A rational function with common factors*

Determine all vertical asymptotes of the graph of

$$f(x) = \frac{x^2 + 2x - 8}{x^2 - 4}$$

Solution: By factoring both numerator and denominator, we have

$$f(x) = \frac{x^2 + 2x - 8}{x^2 - 4} = \frac{(x + 4)(x - 2)}{(x + 2)(x - 2)} = \frac{x + 4}{x + 2}, \quad x \ne 2$$

Thus, at all points other than $x = 2$, the graph of f coincides with the graph

$$g(x) = \frac{x + 4}{x + 2}$$

and we can apply Theorem 2.11 to g to conclude that there is a vertical asymptote at $x = -2$, as shown in Figure 2.25.

In Example 3, we know that the limit as $x \to -2$ from the right and left is either $+\infty$ or $-\infty$. Specifically, from Figure 2.25, we have the following limits.

$$\lim_{x \to -2^-} \frac{x^2 + 2x - 8}{x^2 - 4} = -\infty \quad \text{and} \quad \lim_{x \to -2^+} \frac{x^2 + 2x - 8}{x^2 - 4} = \infty$$

Once you have determined that the graph of a function has a vertical asymptote at a particular x-value, we suggest that you settle the question of whether $f(x)$ approaches positive or negative infinity, using the graphical approach outlined in the following example.

EXAMPLE 4   *Determining infinite limits*

Find the following limits.

$$\lim_{x \to 1^-} \frac{x^2 - 3x}{x - 1} \quad \text{and} \quad \lim_{x \to 1^+} \frac{x^2 - 3x}{x - 1}$$

Solution:

1. Factor both numerator and denominator. (Cancel any common factors.)

$$f(x) = \frac{x^2 - 3x}{x - 1} = \frac{x(x - 3)}{x - 1}$$

2. The (remaining) factors of the numerator yield the x-intercepts of the function. In this case, the x-intercepts occur at $x = 0$ and $x = 3$.
3. The (remaining) factors of the denominator yield the vertical asymptotes of the function. In this case, a vertical asymptote occurs at $x = 1$.
4. We determine a few additional points on the graph, making sure to choose at least one point between each intercept and asymptote. Then, we plot the information obtained in the first four steps, as shown in Figure 2.26.

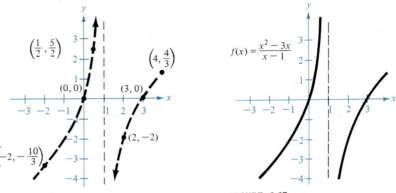

FIGURE 2.26   FIGURE 2.27

x	-2	$\dfrac{1}{2}$	2	4
$f(x)$	$-\dfrac{10}{3}$	$\dfrac{5}{2}$	-2	$\dfrac{4}{3}$

5. Finally, we complete the sketch as shown in Figure 2.27 and conclude that

$$\lim_{x\to 1^-} \frac{x^2 - 3x}{x - 1} = \infty \quad \text{and} \quad \lim_{x\to 1^+} \frac{x^2 - 3x}{x - 1} = -\infty$$

We conclude this section with a theorem involving the limits of sums, products, and quotients of functions.

THEOREM 2.12   **PROPERTIES OF INFINITE LIMITS**
If c and L are real numbers and f and g are functions such that

$$\lim_{x\to c} f(x) = \infty \quad \text{and} \quad \lim_{x\to c} g(x) = L$$

then the following properties are true.

1. Sum or difference: $\displaystyle\lim_{x\to c} [f(x) \pm g(x)] = \infty$

2. Quotient: $\displaystyle\lim_{x\to c} \frac{g(x)}{f(x)} = 0$

3. Product: $\displaystyle\lim_{x\to c} [f(x)g(x)] = \infty, \quad L > 0$

   $\displaystyle\lim_{x\to c} [f(x)g(x)] = -\infty, \quad L < 0$

Similar properties hold if $\displaystyle\lim_{x\to c} f(x) = -\infty$.

Remark Each of the properties in Theorem 2.12 remains true if $x \to c$ is replaced by $x \to c^+$ or $x \to c^-$.

EXAMPLE 5   Determining limits

Find the following limits:

(a) $\lim\limits_{x\to 0} \left(1 + \dfrac{1}{x^2}\right)$ (b) $\lim\limits_{x\to 1^-} \dfrac{x^2 + 1}{1/(x - 1)}$

Solution:

(a) Since

$$\lim\limits_{x\to 0} 1 = 1 \qquad \text{and} \qquad \lim\limits_{x\to 0} \frac{1}{x^2} = \infty$$

we apply Property 1 of Theorem 2.12 to conclude that

$$\lim\limits_{x\to 0} \left(1 + \frac{1}{x^2}\right) = \infty$$

(b) Since

$$\lim\limits_{x\to 1^-} (x^2 + 1) = 2 \qquad \text{and} \qquad \lim\limits_{x\to 1^-} \frac{1}{x - 1} = -\infty$$

we apply Property 2 of Theorem 2.12 to conclude that

$$\lim\limits_{x\to 1^-} \frac{x^2 + 1}{1/(x - 1)} = 0$$

Section Exercises 2.4

In Exercises 1–4, determine the limits as x approaches -3 from the left and from the right.

1. $f(x) = \dfrac{1}{x^2 - 9}$ **2.** $f(x) = \dfrac{x}{x^2 - 9}$

3. $f(x) = \dfrac{x^2}{x^2 - 9}$ **4.** $f(x) = \dfrac{x^3}{x^2 - 9}$

In Exercises 5 and 6, determine the limit as x approaches -2 from the left and from the right.

5. $f(x) = \dfrac{1}{(x + 2)^2}$ **6.** $f(x) = \dfrac{1}{x + 2}$

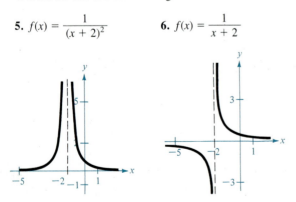

In Exercises 7 and 8, find the vertical asymptotes of the given function.

7. $f(x) = \dfrac{x^2 - 2}{x^2 - x - 2}$ **8.** $f(x) = \dfrac{x^3}{x^2 - 1}$

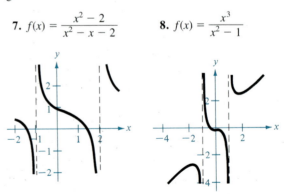

In Exercises 9–18, find the vertical asymptotes (if any) of the given function.

9. $f(x) = \dfrac{1}{x^2}$ **10.** $f(x) = \dfrac{4}{(x - 2)^3}$

11. $f(x) = \dfrac{x^2}{x^2 + x - 2}$ **12.** $f(x) = \dfrac{2 + x}{1 - x}$

13. $f(x) = \dfrac{x^3}{x^2 - 4}$

14. $f(x) = \dfrac{-4x}{x^2 + 4}$

15. $f(x) = 1 - \dfrac{4}{x^2}$

16. $f(x) = \dfrac{-2}{(x - 2)^2}$

17. $f(x) = \dfrac{x}{x^2 + x - 2}$

18. $f(x) = \dfrac{1}{(x + 3)^4}$

In Exercises 19–22, determine whether the given function has a vertical asymptote or a removable discontinuity at $x = -1$.

19. $f(x) = \dfrac{x^2 - 1}{x + 1}$

20. $f(x) = \dfrac{x^2 - 6x - 7}{x + 1}$

21. $f(x) = \dfrac{x^2 + 1}{x + 1}$

22. $f(x) = \dfrac{x - 1}{x + 1}$

In Exercises 23–32, find the indicated limit.

23. $\displaystyle\lim_{x \to 2^+} \dfrac{x - 3}{x - 2}$

24. $\displaystyle\lim_{x \to 1^+} \dfrac{2 + x}{1 - x}$

25. $\displaystyle\lim_{x \to 4} \dfrac{x^2}{x^2 - 16}$

26. $\displaystyle\lim_{x \to 4} \dfrac{x^2}{x^2 + 16}$

27. $\displaystyle\lim_{x \to 0^-} \left(1 + \dfrac{1}{x}\right)$

28. $\displaystyle\lim_{x \to 0^-} \left(x^2 - \dfrac{1}{x}\right)$

29. $\displaystyle\lim_{x \to 1} \dfrac{x^2 - x}{(x^2 + 1)(x - 1)}$

30. $\displaystyle\lim_{x \to 1} \dfrac{x^3 - 1}{x^2 + x + 1}$

31. $\displaystyle\lim_{x \to 1^+} \dfrac{x^2 + x + 1}{x^3 - 1}$

32. $\displaystyle\lim_{x \to 0^-} \dfrac{x^2 - 2x}{x^3}$

33. A 25-foot ladder is leaning against a house, as shown in Figure 2.28. If the base of the ladder is pulled away from the house at a rate of 2 feet per second, the top will move down the wall at a rate of

$$r = \dfrac{2x}{\sqrt{625 - x^2}} \ \text{ft/sec}$$

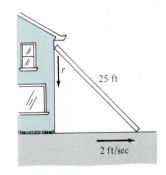

FIGURE 2.28

(a) Find the rate when x is 7 feet.
(b) Find the rate when x is 15 feet.
(c) Find the limit of r as $x \to 25$.

34. The cost in millions of dollars for the federal government to seize x percent of a certain illegal drug as it enters the country is given by

$$C = \dfrac{528x}{100 - x}, \quad 0 \le x < 100$$

(a) Find the cost of seizing 25 percent.
(b) Find the cost of seizing 50 percent.
(c) Find the cost of seizing 75 percent.
(d) Find the limit of C as $x \to 100^-$.

35. The cost in dollars of removing p percent of the air pollutants from the stack emission of a utility company that burns coal to generate electricity is

$$C = \dfrac{80,000p}{100 - p}, \quad 0 \le p < 100$$

(a) Find the cost of removing 15 percent.
(b) Find the cost of removing 50 percent.
(c) Find the cost of removing 90 percent.
(d) Find the limit of C as $x \to 100^-$.

SECTION TOPICS •
Formal definition of limit •
Proofs of limit theorems •
Formal definition of infinite limit •
The Squeeze Theorem •

2.5
ε-δ Definition of limits

We begin this section with another look at our informal description of a limit given at the beginning of this chapter. If $f(x)$ becomes arbitrarily close to a single number L as x approaches c from either side, then we say that the **limit** of $f(x)$, as x approaches c, is L, and we write

$$\lim_{x \to c} f(x) = L$$

At first glance, this description looks fairly technical—so why do we call it informal? The answer lies in the exact meaning of the two phrases

"$f(x)$ becomes arbitrarily close to L"   and   "x approaches c"

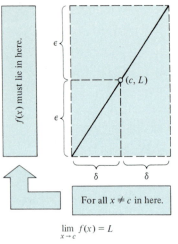

f(x) must lie in here.

For all $x \neq c$ in here.

$$\lim_{x \to c} f(x) = L$$

FIGURE 2.29

The first person to assign a mathematically rigorous meaning to these two phrases was Augustin-Louis Cauchy (1789–1857). His ε-δ definition of a limit is the standard used today. (ε is the lowercase Greek letter *epsilon* and δ is the lowercase Greek *delta*.)

In Figure 2.29, let ε represent a (small) positive number. Then the phrase "$f(x)$ becomes arbitrarily close to L" means that $f(x)$ lies in the interval $(L - \varepsilon, L + \varepsilon)$. In terms of absolute value, we write this as

$$|f(x) - L| < \varepsilon$$

Similarly, the phrase "x approaches c" means that there exists a positive number δ such that x lies in either the interval $(c - \delta, c)$ or the interval $(c, c + \delta)$. In terms of absolute value, we write

$$0 < |x - c| < \delta$$

Putting these two inequalities together, we have the following formal definition of a limit.

DEFINITION OF LIMIT   The statement

$$\lim_{x \to c} f(x) = L$$

means that for each $\varepsilon > 0$ there exists a $\delta > 0$ such that

$$|f(x) - L| < \varepsilon \qquad \text{whenever} \qquad 0 < |x - c| < \delta$$

Augustin-Louis Cauchy

Remark Note that the inequality $0 < |x - c| < \delta$ implies that $x \neq c$. In other words, the value of the limit of $f(x)$ as $x \to c$ does not depend on the value of $f(x)$ at c.

Note that the definition has an *order* to it: "For each $\varepsilon > 0$ there exists a $\delta > 0$." First we are given an ε-value, then we must find an appropriate δ-value. We do not require any specific δ to work for more than one choice of ε. Furthermore, the number δ is not unique, for if a specific δ works, then any smaller positive number will also work. To gain a better understanding of the relationship between ε and δ, consider the graphical description shown in Figure 2.30.

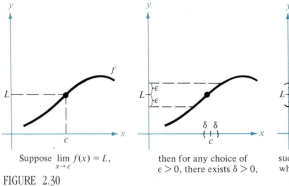
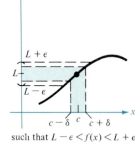

Suppose $\lim_{x \to c} f(x) = L$, then for any choice of $\varepsilon > 0$, there exists $\delta > 0$, such that $L - \varepsilon < f(x) < L + \varepsilon$ whenever $c - \delta < x < c + \delta$.

FIGURE 2.30

The next three examples illustrate some ways to determine a δ-value for a given ε.

EXAMPLE 1   Finding a δ for a given ε

Given the limit

$$\lim_{x \to 3} (2x - 5) = 1$$

find δ such that $\left|(2x - 5) - 1\right| < 0.01$ whenever $0 < |x - 3| < \delta$.

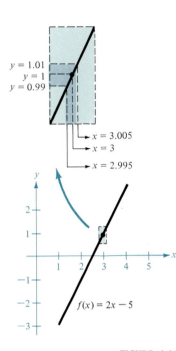

$y = 1.01$
$y = 1$
$y = 0.99$

$x = 3.005$
$x = 3$
$x = 2.995$

$f(x) = 2x - 5$

FIGURE 2.31

Solution: In this problem, we are working with a given value of ε, namely $\varepsilon = 0.01$. To find an appropriate δ, we try to establish a connection between the two absolute values $\left|(2x - 5) - 1\right|$ and $|x - 3|$. By simplifying the first absolute value, we get

$$\left|(2x - 5) - 1\right| = |2x - 6|$$
$$= 2|x - 3|$$

In other words, the inequality $\left|(2x - 5) - 1\right| < 0.01$ is equivalent to $2|x - 3| < 0.01$, and we have

$$|x - 3| < \frac{0.01}{2} = 0.005$$

Thus, we choose $\delta = 0.005$. This choice works because $0 < |x - 3| < 0.005$ implies that

$$\left|(2x - 5) - 1\right| = 2|x - 3| < 2(0.005) = 0.01$$

as indicated in Figure 2.31.

In Example 1, we found a δ-value for a *particular* ε. Figure 2.32 shows how a new ε-value, ε_1, can require a new choice for δ. The original value for δ is now too large, and we must find a smaller value δ_1 that works for ε_1. In other words, finding a δ-value for a particular ε does not prove the existence of the limit. To do that, we must prove that we can find δ for *any* ε, as demonstrated in the next example.

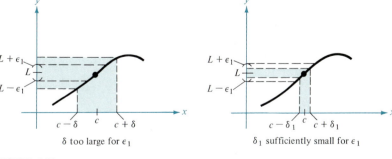

δ too large for ϵ_1 δ_1 sufficiently small for ϵ_1

FIGURE 2.32

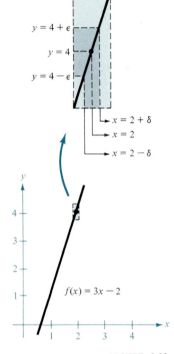

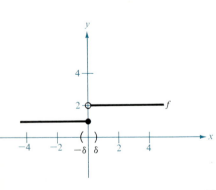

FIGURE 2.33

EXAMPLE 2   Using the ε-δ definition of a limit

Use the ε-δ definition of a limit to prove that

$$\lim_{x \to 2} (3x - 2) = 4$$

Solution: We are required to show that for each $\varepsilon > 0$, there exists a $\delta > 0$ such that

$$|(3x - 2) - 4| < \varepsilon \qquad \text{whenever} \qquad 0 < |x - 2| < \delta$$

Since our choice for δ depends on ε, we try to establish a connection between the absolute values $|(3x - 2) - 4|$ and $|x - 2|$. By simplifying the first absolute value, we get

$$|(3x - 2) - 4| = |3x - 6| = 3|x - 2|$$

Thus, the inequality $|(3x - 2) - 4| < \varepsilon$ requires $3|x - 2| < \varepsilon$, so that we have

$$|x - 2| < \frac{\varepsilon}{3}$$

and we choose $\delta = \varepsilon/3$. This choice works, because

$$0 < |x - 2| < \delta = \frac{\varepsilon}{3}$$

implies that

$$|(3x - 2) - 4| = 3|x - 2| < 3\left(\frac{\varepsilon}{3}\right) = \varepsilon$$

as shown in Figure 2.33.

In the next example we use the ε-δ definition to prove the nonexistence of a limit.

EXAMPLE 3   A limit that does not exist

Show that the limit of $f(x)$ as x approaches 0 *does not* exist for the function given by

$$f(x) = \begin{cases} 1, & x \le 0 \\ 2, & x > 0 \end{cases}$$

Solution: Assume that the limit of $f(x)$ as $x \to 0$ exists and is equal to L. We choose $\varepsilon = 1/2$, which is less than the jump discontinuity at $x = 0$, as shown in Figure 2.34. Then, if there exists a $\delta > 0$ such that

$$|f(x) - L| < \frac{1}{2} \qquad \text{whenever} \qquad 0 < |x - 0| < \delta$$

the limit from the left requires

$$|1 - L| < \frac{1}{2} \qquad \text{whenever} \qquad -\delta < x < 0$$

FIGURE 2.34

and the limit from the right requires

$$|2 - L| < \frac{1}{2} \qquad \text{whenever} \qquad 0 < x < \delta$$

Together these statements imply that

$$-\frac{1}{2} < L - 1 < \frac{1}{2} \qquad \text{and} \qquad -\frac{1}{2} < L - 2 < \frac{1}{2}$$

or

$$\frac{1}{2} < L < \frac{3}{2} \qquad \text{and} \qquad \frac{3}{2} < L < \frac{5}{2}$$

But no single value for L can satisfy both of these inequalities. Therefore, our assumption that the limit exists is false, and we conclude that the limit of $f(x)$ as $x \to 0$ does not exist.

In Example 3 we made use of one-sided limits. For future reference we provide the formal ε-δ definition of one-sided limits.

FORMAL DEFINITION OF ONE-SIDED LIMITS

1. $\lim_{x \to c^-} f(x) = L$ if for each $\varepsilon > 0$, there exists a $\delta > 0$ such that

$$|f(x) - L| < \varepsilon \qquad \text{whenever} \qquad -\delta < x - c < 0$$

2. $\lim_{x \to c^+} f(x) = L$ if for each $\varepsilon > 0$, there exists a $\delta > 0$ such that

$$|f(x) - L| < \varepsilon \qquad \text{whenever} \qquad 0 < x - c < \delta$$

There are several equivalent forms of the limit statement

$$\lim_{x \to c} f(x) = L$$

In the future we will use whichever form is most convenient for the task at hand. Try using the ε-δ definition of a limit to establish the equivalence of the following three forms.

1. $\lim_{x \to c} f(x) = L$

2. $\lim_{x \to c} [f(x) - L] = 0$

3. $\lim_{h \to 0} f(c + h) = L$

Proofs of limit theorems

Your appreciation of the ε-δ definition of a limit should grow as you see how useful it is in proving properties of limits. In the next two examples, we demonstrate the type of argument needed to establish some of the basic theorems identified in Section 2.1.

EXAMPLE 4   *Using the ε-δ definition to prove Theorem 2.1*

Prove Theorem 2.1: Let c be a real number and $f(x) = g(x)$ for all $x \neq c$ in an open interval containing c. If the limit of $g(x)$ exists as $x \to c$, then the limit of $f(x)$ also exists, and

$$\lim_{x \to c} f(x) = \lim_{x \to c} g(x)$$

Proof: Assume that the limit of $g(x)$ as $x \to c$ is L. Then, by definition, for each $\varepsilon > 0$ there exists a $\delta > 0$ such that

$$|g(x) - L| < \varepsilon \qquad \text{whenever} \qquad 0 < |x - c| < \delta$$

However, since $f(x) = g(x)$ for all x in the open interval other than $x = c$, it follows that

$$|f(x) - L| < \varepsilon \qquad \text{whenever} \qquad 0 < |x - c| < \delta$$

Thus, we conclude that the limit of $f(x)$ as $x \to c$ is also L.   ☐

EXAMPLE 5   *Using the ε-δ definition to prove Theorem 2.2*

Prove that if c is a real number, then

$$\lim_{x \to c} x = c$$

Proof: We need to show that for each $\varepsilon > 0$ there exists a $\delta > 0$ such that

$$|x - c| < \varepsilon \qquad \text{whenever} \qquad 0 < |x - c| < \delta$$

The right-hand inequality is similar to the left-hand one, and we simply choose $\delta = \varepsilon$.   ☐

EXAMPLE 6   *Using the ε-δ definition to prove Theorem 2.3*

Prove that if the functions f and g have limits as $x \to c$, then

$$\lim_{x \to c} [f(x) + g(x)] = \lim_{x \to c} f(x) + \lim_{x \to c} g(x)$$

Proof: Assume that

$$\lim_{x \to c} f(x) = L \qquad \text{and} \qquad \lim_{x \to c} g(x) = K$$

Choose $\varepsilon > 0$. Then, since $\varepsilon/2 > 0$, we know that there exists $\delta_1 > 0$ and $\delta_2 > 0$ such that

$$0 < |x - c| < \delta_1 \qquad \text{implies} \qquad |f(x) - L| < \frac{\varepsilon}{2}$$

and

$$0 < |x - c| < \delta_2 \qquad \text{implies} \qquad |g(x) - K| < \frac{\varepsilon}{2}$$

Now, if we let δ be the smaller of δ_1 and δ_2, we have

$$0 < |x - c| < \delta \qquad \text{implies} \qquad |f(x) - L| < \frac{\varepsilon}{2} \qquad \text{and} \qquad |g(x) - K| < \frac{\varepsilon}{2}$$

Finally, we apply the Triangle Inequality to conclude that

$$|[f(x) + g(x)] - (L + K)| < |f(x) - L| + |g(x) - K| < \frac{\varepsilon}{2} + \frac{\varepsilon}{2} = \varepsilon$$

Hence

$$0 < |x - c| < \delta \implies |f(x) + g(x) - (L + K)| < \varepsilon$$

which implies that

$$\lim_{x \to c} [f(x) + g(x)] = L + K = \lim_{x \to c} f(x) + \lim_{x \to c} g(x)$$

In Section 2.4, we discussed vertical asymptotes and their relationship to infinite limits. The following definition formalizes the definition of an infinite limit.

FORMAL DEFINITION OF INFINITE LIMITS

The statement

$$\lim_{x \to c} f(x) = \infty$$

means that for each $M > 0$ there exists a $\delta > 0$ such that $f(x) > M$ whenever $0 < |x - c| < \delta$.

The statement

$$\lim_{x \to c} f(x) = -\infty$$

means that for each $N < 0$ there exists a $\delta > 0$ such that $f(x) < N$ whenever $0 < |x - c| < \delta$.

To define the **infinite limit from the left,** we replace

$$0 < |x - c| < \delta \quad \text{by} \quad c - \delta < x < c$$

To define the **infinite limit from the right,** we replace

$$0 < |x - c| < \delta \quad \text{by} \quad c < x < c + \delta$$

EXAMPLE 7   Using the M-δ definition to prove Theorem 2.12

Prove that if c and L are real numbers and f and g are functions such that

$$\lim_{x \to c} f(x) = \infty \quad \text{and} \quad \lim_{x \to c} g(x) = L$$

then

$$\lim_{x \to c} [f(x) + g(x)] = \infty$$

Proof: To show that the limit of $f(x) + g(x)$ is infinite, we choose $M > 0$, and we need to find $\delta > 0$ such that

$$[f(x) + g(x)] > M \qquad \text{whenever} \qquad 0 < |x - c| < \delta$$

For simplicity's sake, we assume L is positive, and let $M_1 = M + 1$. Since the limit of $f(x)$ is infinite, there exists δ_1 such that

$$f(x) > M_1 \qquad \text{whenever} \qquad 0 < |x - c| < \delta_1$$

Also, since the limit of $g(x)$ is L, there exists δ_2 such that

$$|g(x) - L| < 1 \qquad \text{whenever} \qquad 0 < |x - c| < \delta_2$$

By letting δ be the smaller of δ_1 and δ_2, we conclude that $0 < |x - c| < \delta$ implies

$$f(x) > M + 1 \qquad \text{and} \qquad |g(x) - L| < 1$$

The second of these two inequalities implies that $g(x) > L - 1$, and by adding this to the first inequality, we have

$$f(x) + g(x) > M + L > M$$

Thus,

$$\lim_{x \to c} [f(x) + g(x)] = \infty \qquad \qquad \square$$

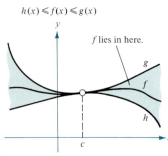

$h(x) \leq f(x) \leq g(x)$

f lies in here.

The Squeeze Theorem

FIGURE 2.35

Later in the text when we encounter the indeterminate form 0/0 in a limit involving trigonometric functions, it will often require a considerable amount of ingenuity to determine the limit. The following theorem is helpful. It concerns the limiting behavior of a function that is squeezed between two other functions, each of which has the same limit at a given point, as shown in Figure 2.35.

THEOREM 2.13   **THE SQUEEZE THEOREM**
If $h(x) \leq f(x) \leq g(x)$ for all x in an open interval containing c, except possibly at c itself, and if

$$\lim_{x \to c} h(x) = L = \lim_{x \to c} g(x)$$

then

$$\lim_{x \to c} f(x) = L$$

Proof: For $\varepsilon > 0$ there exists δ_1 and δ_2 such that

$$|h(x) - L| < \varepsilon \qquad \text{whenever} \qquad 0 < |x - c| < \delta_1$$

and

$$|g(x) - L| < \varepsilon \qquad \text{whenever} \qquad 0 < |x - c| < \delta_2$$

Let δ be the smaller of δ_1 and δ_2. Then, if $0 < |x - c| < \delta$, it follows that $|h(x) - L| < \varepsilon$ and $|g(x) - L| < \varepsilon$, which implies that

$$-\varepsilon < h(x) - L < \varepsilon \qquad \text{and} \qquad -\varepsilon < g(x) - L < \varepsilon$$
$$L - \varepsilon < h(x) \qquad \text{and} \qquad g(x) < L + \varepsilon$$

Now, since $h(x) \le f(x) \le g(x)$, it follows that

$$L - \varepsilon < f(x) < L + \varepsilon$$
$$|f(x) - L| < \varepsilon$$

Therefore,

$$\lim_{x \to c} f(x) = L$$

Section Exercises 2.5

In Exercises 1–6, find the limit L and then find $\delta > 0$ such that $|f(x) - L| < 0.01$ whenever $0 < |x - c| < \delta$.

1. $\lim_{x \to 2} (3x + 2)$

2. $\lim_{x \to 4} \left(4 - \dfrac{x}{2} \right)$

3. $\lim_{x \to 2} (x^2 - 3)$

4. $\lim_{x \to 5} \sqrt{x - 4}$

5. $\lim_{x \to 2} \dfrac{x^2 - 3x + 2}{x - 2}$

6. $\lim_{x \to -2} \dfrac{x^2 + 5x + 6}{x + 2}$

In Exercises 7–22, find the indicated limit L. Then use the definitions given in this section to verify the limit.

7. $\lim_{x \to 2} (x + 3)$

8. $\lim_{x \to -3} (2x + 5)$

9. $\lim_{x \to 6} 3$

10. $\lim_{x \to 2} -1$

11. $\lim_{x \to 0} \sqrt[3]{x}$

12. $\lim_{x \to 3} |x - 3|$

13. $\lim_{x \to 0} x^2$

14. $\lim_{x \to 2} x^3$

15. $\lim_{x \to 2} \dfrac{x^2 + x - 6}{x - 2}$

16. $\lim_{x \to -10} \dfrac{x^2 + 10x}{x + 10}$

17. $\lim_{x \to 2} \dfrac{1}{x}$

18. $\lim_{x \to -1} \dfrac{3}{x + 2}$

19. $\lim_{x \to 2} (x^2 - 2)$

20. $\lim_{x \to -1} x^2$

21. $\lim_{x \to 0^+} \sqrt{x}$

22. $\lim_{x \to 3^-} f(x)$, $f(x) = \begin{cases} 2, & x \le 3 \\ 0, & x > 3 \end{cases}$

In Exercises 23–26, find the infinite limit and use the definition given in this section to verify your result.

23. $\lim_{x \to -1^+} \dfrac{1}{x + 1}$

24. $\lim_{x \to -1^-} \dfrac{1}{x + 1}$

25. $\lim_{x \to 2} \dfrac{1}{(x - 2)^2}$

26. $\lim_{x \to 0} \dfrac{1}{x^2}$

27. Prove that $f(x) = x^2$ is continuous at $x = 3$.

28. Prove that $f(x) = 4 - 3x$ is continuous at $x = 1$.

29. Prove that

$$\lim_{x \to c} f(x) = 0$$

if

$$\lim_{x \to c} |f(x)| = 0$$

30. Prove that

$$\lim_{x \to 0^+} \dfrac{|x|}{x} = 1$$

31. Use the Squeeze Theorem to prove that

$$\lim_{x \to 0} \dfrac{x}{x^2 + 1} = 0$$

32. Prove that

$$\lim_{x \to 1} f(x)$$

does not exist for the function

$$f(x) = \begin{cases} x, & x \le 1 \\ 1 - x, & x > 1 \end{cases}$$

Review Exercises for Chapter 2

In Exercises 1–20, find the given limit (if it exists).

1. $\lim\limits_{x \to 2} (5x - 3)$

2. $\lim\limits_{x \to 2} (3x + 5)$

3. $\lim\limits_{x \to 2} (5x - 3)(3x + 5)$

4. $\lim\limits_{x \to 2} \dfrac{3x + 5}{5x - 3}$

5. $\lim\limits_{t \to 3} \dfrac{t^2 + 1}{t}$

6. $\lim\limits_{t \to 3} \dfrac{t^2 - 9}{t - 3}$

7. $\lim\limits_{t \to -2} \dfrac{t + 2}{t^2 - 4}$

8. $\lim\limits_{x \to 0} \dfrac{\sqrt{4 + x} - 2}{x}$

9. $\lim\limits_{x \to 0} \dfrac{[1/(x + 1)] - 1}{x}$

10. $\lim\limits_{s \to 0} \dfrac{(1/\sqrt{1 + s}) - 1}{s}$

11. $\lim\limits_{x \to -1} \dfrac{x^3 + 1}{x + 1}$

12. $\lim\limits_{x \to -2} \dfrac{x^2 - 4}{x^3 + 8}$

13. $\lim\limits_{x \to 0^+} \left(x - \dfrac{1}{x^3} \right)$

14. $\lim\limits_{x \to 2} \dfrac{1}{\sqrt[3]{x^2 - 4}}$

15. $\lim\limits_{x \to -2^-} \dfrac{2x^2 + x + 1}{x + 2}$

16. $\lim\limits_{x \to 1/2} \dfrac{2x - 1}{6x - 3}$

17. $\lim\limits_{x \to -1} \dfrac{x + 1}{x^3 + 1}$

18. $\lim\limits_{x \to -1} \dfrac{x + 1}{x^4 - 1}$

19. $\lim\limits_{x \to 1} \dfrac{x^2 - 2x + 1}{x + 1}$

20. $\lim\limits_{x \to -1^+} \dfrac{x^2 - 2x + 1}{x + 1}$

21. Estimate the limit

$$\lim\limits_{x \to 1^+} \dfrac{\sqrt{2x + 1} - \sqrt{3}}{x - 1}$$

by completing the following table.

x	1.1	1.01	1.001	1.0001
$f(x)$				

22. Estimate the limit

$$\lim\limits_{x \to 1^+} \dfrac{1 - \sqrt[3]{x}}{x - 1}$$

by completing the following table.

x	1.1	1.01	1.001	1.0001
$f(x)$				

23. Evaluate the limit in Exercise 21 by rationalizing the numerator.

24. Evaluate the limit in Exercise 22 by rationalizing the numerator. [Use $a^3 - b^3 = (a - b)(a^2 + ab + b^2)$.]

In Exercises 25–30, determine if the given limit statement is true or false.

25. $\lim\limits_{x \to 0} \dfrac{|x|}{x} = 1$

26. $\lim\limits_{x \to 0} x^3 = 0$

27. $\lim\limits_{x \to 2} f(x) = 3$, where $f(x) = \begin{cases} 3, & x \le 2 \\ 0, & x > 2 \end{cases}$

28. $\lim\limits_{x \to 3} f(x) = 1$, where $f(x) = \begin{cases} x - 2, & x \le 3 \\ -x^2 + 8x - 14, & x > 3 \end{cases}$

29. $\lim\limits_{x \to 0^+} \sqrt{x} = 0$

30. $\lim\limits_{x \to 0} \sqrt[3]{x} = 0$

In Exercises 31–38, determine the points of discontinuity (if any) of the given function.

31. $f(x) = [x + 3]$

32. $f(x) = \dfrac{3x^2 - x - 2}{x - 1}$

33. $f(x) = \begin{cases} \dfrac{3x^2 - x - 2}{x - 1}, & x \ne 1 \\ 0, & x = 1 \end{cases}$

34. $f(x) = \begin{cases} 5 - x, & x \le 2 \\ 2x - 3, & x > 2 \end{cases}$

35. $f(x) = \dfrac{1}{(x - 2)^2}$

36. $f(x) = \sqrt{\dfrac{x + 2}{x}}$

37. $f(x) = \dfrac{3}{x + 1}$

38. $f(x) = \dfrac{x + 1}{2x + 2}$

39. Determine the value of c so that the following function is continuous.

$$f(x) = \begin{cases} x + 3, & x \le 2 \\ cx + 6, & x > 2 \end{cases}$$

40. Determine the values of b and c so that the following function is continuous.

$$f(x) = \begin{cases} x + 1, & 1 < x < 3 \\ x^2 + bx + c, & |x - 2| \ge 1 \end{cases}$$

41. $5000 is deposited in a savings plan that pays 12 percent compounded semiannually. The amount in the account after t years is given by

$$A = 5000(1.06)^{[2t]}$$

Sketch a graph of this function and discuss its continuity.

42. $1000 is deposited in a savings plan that pays 14 percent compounded annually. The amount in the account after t years is given by

$$A = 1000(1.14)^{[t]}$$

Sketch a graph of this function and discuss its continuity.

3

Differentiation

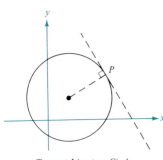

Isaac Newton

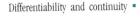

3.1
The derivative and the tangent line problem

The development of calculus grew out of four major problems that European mathematicians were working on during the seventeenth century:

1. The tangent line problem (Section 3.1)
2. The velocity and acceleration problem (Section 3.2)
3. The minimum and maximum problem (Section 4.1)
4. The area problem (Section 5.4)

Each problem involves the notion of a limit, and we could introduce calculus with any one of the four problems. Because of its geometric nature, we choose to begin with the tangent line problem. Partial solutions to this problem were given by Pierre de Fermat, René Descartes, Christian Huygens (1629–1695), and Isaac Barrow (1630–1677). However, credit for the first general solution is usually given to Isaac Newton (1642–1727) and Gottfried Leibniz (1646–1716). Newton's work on the problem stemmed from his interest in optics and light refraction.

The tangent line problem

What do we mean when we say that a line is tangent to a curve at a point? For a circle, we can characterize the tangent line at point P as the line that is perpendicular to the radial line at point P, as shown in Figure 3.1. However, for a general curve the problem is more difficult. For example, how would you define the tangent lines shown in Figure 3.2?

Tangent Line to a Circle

FIGURE 3.1

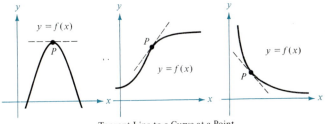

Tangent Line to a Curve at a Point

FIGURE 3.2

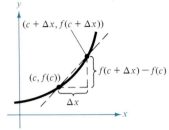

The Secant Line through $(c, f(c))$
and $(c + \Delta x, f(c + \Delta x))$

FIGURE 3.3

Essentially, the problem of finding the tangent line at a point P boils down to the problem of finding the *slope* of the tangent line at P. We can approximate this slope by making use of a line through the point of tangency and a second point on the curve as shown in Figure 3.3. We call such a line a **secant line.** (This use of the word *secant* comes from the Latin *secare* meaning to cut and not from a reference to the trigonometric function of the same name.) If $(c, f(c))$ is the point of tangency and $(c + \Delta x, f(c + \Delta x))$ is a second point on the graph of f, then the slope of the secant line through these two points is given by

$$m_{\text{sec}} = \frac{f(c + \Delta x) - f(c)}{c + \Delta x - c} = \frac{f(c + \Delta x) - f(c)}{\Delta x} \qquad \text{Slope of secant line}$$

This formula is called a **difference quotient.** The denominator Δx is called the **change in x,** and the numerator is called the **change in y,** denoted by

$$\Delta y = f(c + \Delta x) - f(c) \qquad \text{Change in } y$$

The beauty of this procedure is that we can obtain better and better approximations to the slope of the tangent line by choosing a sequence of points closer and closer to the point of tangency as shown in Figure 3.4.

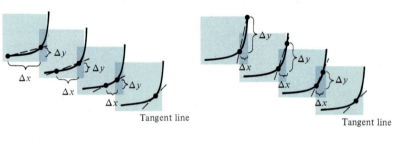

As $\Delta x \to 0$ from the left, the secant lines approach the tangent line.

As $\Delta x \to 0$ from the right, the secant lines approach the tangent line.

FIGURE 3.4

By resorting to the limit process, we can thus define the slope of the tangent line to be the limit of the slopes of the secant lines.

DEFINITION OF TANGENT LINE WITH SLOPE m

If f is defined on an open interval containing c and the following limit exists

$$\lim_{\Delta x \to 0} \frac{\Delta y}{\Delta x} = \lim_{\Delta x \to 0} \frac{f(c + \Delta x) - f(c)}{\Delta x} = m$$

then we call the line passing through $(c, f(c))$ with slope m the **tangent line** to the graph of f at the point $(c, f(c))$.

Remark We often refer to the slope of the tangent line to the graph of f at the point $(c, f(c))$ as simply the **slope of the graph of f at $x = c$.**

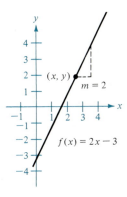

FIGURE 3.5

EXAMPLE 1   The slope of a linear function

Find the slope of $f(x) = 2x - 3$ at the point $(2, 1)$.

Solution:

$$\lim_{\Delta x \to 0} \frac{f(2 + \Delta x) - f(2)}{\Delta x} = \lim_{\Delta x \to 0} \frac{[2(2 + \Delta x) - 3] - [2(2) - 3]}{\Delta x}$$

$$= \lim_{\Delta x \to 0} \frac{4 + 2\Delta x - 3 - 4 + 3}{\Delta x}$$

$$= \lim_{\Delta x \to 0} \frac{2\Delta x}{\Delta x} = \lim_{\Delta x \to 0} 2 = 2$$

Therefore, the slope of f at $(2, 1)$ is $m = 2$, as shown in Figure 3.5.

Remark   Note in Example 1 that the limit definition of the slope of f agrees with our usual definition of the slope of a line as discussed in Section 1.4.

We know that a linear function has the same slope at any point. This is not true of nonlinear functions, as can be seen in the following example.

EXAMPLE 2   Tangent lines to a nonlinear function

Find the slope of the tangent line to $f(x) = x^2 + 3$ at the points $(0, 3)$ and $(-2, 7)$, as shown in Figure 3.6.

Solution: Rather than performing the limit process twice, we let x represent an arbitrary value which we consider to be fixed with respect to Δx. Then the slope of the tangent line at $(x, f(x))$ is given by

$$\lim_{\Delta x \to 0} \frac{f(x + \Delta x) - f(x)}{\Delta x} = \lim_{\Delta x \to 0} \frac{x^2 + 2x(\Delta x) + (\Delta x)^2 + 3 - x^2 - 3}{\Delta x}$$

$$= \lim_{\Delta x \to 0} \frac{2x\Delta x + (\Delta x)^2}{\Delta x}$$

$$= \lim_{\Delta x \to 0} (2x + \Delta x) = 2x$$

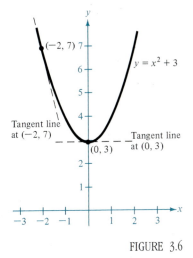

FIGURE 3.6

Therefore, the slope at $(x, f(x))$ is $2x$. At the point $(0, 3)$, the slope is $m = 2(0) = 0$, and at the point $(-2, 7)$, the slope is $2(-2) = -4$.

Remark   In Example 2 note that x is held constant in the limit process (as $\Delta x \to \infty$).

For many functions, the limit procedure used in Example 2 is algebraically complicated. The following **Four-Step Process** is helpful in organizing the procedure.

1. Evaluate f at $x + \Delta x$: $f(x + \Delta x)$
2. Subtract $f(x)$: $f(x + \Delta x) - f(x)$
3. Divide by Δx: $\dfrac{f(x + \Delta x) - f(x)}{\Delta x}$
4. Take the limit as $\Delta x \to 0$: $\lim\limits_{\Delta x \to 0} \dfrac{f(x + \Delta x) - f(x)}{\Delta x} = m$

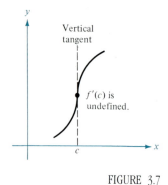

Vertical tangent

$f'(c)$ is undefined.

FIGURE 3.7

The importance of the Four-Step Process is that it gives us a procedure for finding the slope of the tangent line at *any* point for which the limit exists.

Our definition of a tangent line covers only those lines whose slope is defined. In Figure 3.7, the function has a vertical tangent line at $x = c$. To cover this case, we give the following definition. If f is continuous at x and

$$\lim_{\Delta x \to 0} \left| \frac{f(x + \Delta x) - f(x)}{\Delta x} \right| = \infty$$

then the vertical line passing through $(x, f(x))$ is called a **vertical tangent line** to the graph of f. If the domain of f is a closed interval $[a, b]$, then we extend the definition of vertical tangent lines to include the endpoints by considering continuity and limits from the right (for $x = a$) and from the left (for $x = b$).

The derivative of a function

We have now arrived at a crucial point in the study of calculus. The limit used to define the slope of a tangent line is also used to define one of the two fundamental operations of calculus—**differentiation.**

DEFINITION OF THE DERIVATIVE OF A FUNCTION

The **derivative** of f at x is given by

$$f'(x) = \lim_{\Delta x \to 0} \frac{f(x + \Delta x) - f(x)}{\Delta x}$$

provided the limit exists.

A function is said to be **differentiable** at x if its derivative exists at x. A function is called **differentiable on an interval** (a, b) if it is differentiable at every point in the interval. The process of finding the derivative is called **differentiation.** In addition to $f'(x)$, read "f prime of x," various other notations are used to denote the derivative. The most commonly used notations for the derivative of $y = f(x)$ are

$$f'(x), \qquad \frac{dy}{dx}, \qquad y', \qquad \frac{d}{dx}[f(x)], \qquad D_x[y] \qquad \text{Notation for derivatives}$$

The notation dy/dx is read as "the derivative of y *with respect to x*." Using limit notation, we have

$$\frac{dy}{dx} = \lim_{\Delta x \to 0} \frac{\Delta y}{\Delta x}$$

$$= \lim_{\Delta x \to 0} \frac{f(x + \Delta x) - f(x)}{\Delta x} = f'(x)$$

EXAMPLE 3   Finding the derivative by the Four-Step Process

Find the derivative of $f(x) = x^3 + 2x$.

Solution:

1. $f(x + \Delta x) = (x + \Delta x)^3 + 2(x + \Delta x)$
2. $f(x + \Delta x) - f(x) = (x + \Delta x)^3 + 2(x + \Delta x) - (x^3 + 2x)$
$$= 3x^2\Delta x + 3x(\Delta x)^2 + (\Delta x)^3 + 2(\Delta x)$$
3. $\dfrac{f(x + \Delta x) - f(x)}{\Delta x} = \dfrac{\Delta x[3x^2 + 3x\Delta x + (\Delta x)^2 + 2]}{\Delta x}$
4. $\displaystyle\lim_{\Delta x \to 0} \dfrac{f(x + \Delta x) - f(x)}{\Delta x} = \lim_{\Delta x \to 0} [3x^2 + 3x\Delta x + (\Delta x)^2 + 2] = 3x^2 + 2$

Therefore, $f'(x) = 3x^2 + 2$.

Remember that the derivative $f'(x)$ gives us a formula for finding the slope of the tangent line at the point $(x, f(x))$ on the graph of f. This is illustrated in the following example.

EXAMPLE 4   Using the derivative to find the slope at a point

Find $f'(x)$ for $f(x) = \sqrt{x}$, and use the result to find the slope of the graph of f at the points $(1, 1)$, $(4, 2)$, and $(0, 0)$.

Solution: We use the rationalizing procedure discussed in Section 2.2. Since

$$\frac{f(x + \Delta x) - f(x)}{\Delta x} = \frac{\sqrt{x + \Delta x} - \sqrt{x}}{\Delta x}$$

$$= \left(\frac{\sqrt{x + \Delta x} - \sqrt{x}}{\Delta x}\right)\left(\frac{\sqrt{x + \Delta x} + \sqrt{x}}{\sqrt{x + \Delta x} + \sqrt{x}}\right)$$

$$= \frac{(x + \Delta x) - x}{\Delta x(\sqrt{x + \Delta x} + \sqrt{x})}$$

$$= \frac{\Delta x}{\Delta x(\sqrt{x + \Delta x} + \sqrt{x})}$$

$$= \frac{1}{\sqrt{x + \Delta x} + \sqrt{x}}$$

we have

$$f'(x) = \lim_{\Delta x \to 0} \frac{f(x + \Delta x) - f(x)}{\Delta x}$$

$$= \lim_{\Delta x \to 0} \frac{1}{\sqrt{x + \Delta x} + \sqrt{x}} = \frac{1}{2\sqrt{x}}$$

Therefore, at the point $(1, 1)$ the slope is $f'(1) = 1/2$, and at the point $(4, 2)$ the slope is $f'(4) = 1/4$, as shown in Figure 3.8. At the point $(0, 0)$ the slope is undefined, since substituting $x = 0$ in $f'(x)$ produces division by zero. Moreover, since

$$\lim_{\Delta x \to 0^+} \frac{\sqrt{0 + \Delta x} - \sqrt{0}}{\Delta x} = \lim_{\Delta x \to 0^+} \frac{1}{\sqrt{\Delta x}} = \infty$$

the graph of f has a vertical tangent line at $(0, 0)$.

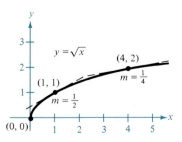

FIGURE 3.8

EXAMPLE 5   Finding the derivative

Find the derivative with respect to t for the function $y = 2/t$.

Solution: Considering $y = f(t)$, we have

$$\Delta y = f(t + \Delta t) - f(t) = \frac{2}{t + \Delta t} - \frac{2}{t}$$

$$= \frac{2t - 2t - 2\Delta t}{t(t + \Delta t)}$$

$$= \frac{-2\Delta t}{t(t + \Delta t)}$$

Now, taking the limit of the difference quotient $\Delta y/\Delta t$, we have

$$\frac{dy}{dt} = \lim_{\Delta t \to 0} \frac{f(t + \Delta t) - f(t)}{\Delta t}$$

$$= \lim_{\Delta t \to 0} \frac{-2}{t(t + \Delta t)} = -\frac{2}{t^2}$$

Differentiability and continuity

There is a close relationship between differentiability and continuity. The following alternative limit form of the derivative is useful in investigating this relationship.

THEOREM 3.1   ALTERNATIVE FORM OF THE DERIVATIVE
The derivative of f at c is given by

$$f'(c) = \lim_{x \to c} \frac{f(x) - f(c)}{x - c}$$

provided this limit exists.

Proof: The derivative of f at c is given by

$$f'(c) = \lim_{\Delta x \to 0} \frac{f(c + \Delta x) - f(c)}{\Delta x}$$

In Figure 3.9, we can see that if $x = c + \Delta x$, then $x \to c$ as $\Delta x \to 0$. Thus, if we replace $c + \Delta x$ by x, we can write

$$f'(c) = \lim_{\Delta x \to 0} \frac{f(c + \Delta x) - f(c)}{\Delta x} = \lim_{x \to c} \frac{f(x) - f(c)}{x - c}$$

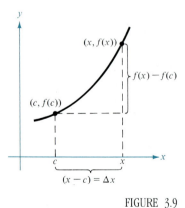

FIGURE 3.9

Note that the existence of the limit in this alternative form requires the equality of the one-sided limits

$$\lim_{x \to c^-} \frac{f(x) - f(c)}{x - c} \qquad \text{and} \qquad \lim_{x \to c^+} \frac{f(x) - f(c)}{x - c}$$

For convenience, we refer to these one-sided limits as the **derivatives from the left and from the right,** respectively. However, keep in mind that if these one-sided limits are not equal at c, then the derivative does not exist at c.

The next three examples will give you a sense of the relationship between differentiability and continuity.

EXAMPLE 6 *A function whose one-sided derivatives are different*

The function $f(x) = |x - 2|$ shown in Figure 3.10 is continuous at $x = 2$. However, the one-sided limits

$$\lim_{x \to 2^-} \frac{f(x) - f(2)}{x - 2} = \lim_{x \to 2^-} \frac{|x - 2| - 0}{x - 2} = -1 \qquad \text{Derivative from the left}$$

and

$$\lim_{x \to 2^+} \frac{f(x) - f(2)}{x - 2} = \lim_{x \to 2^+} \frac{|x - 2| - 0}{x - 2} = 1 \qquad \text{Derivative from the right}$$

are not equal. Therefore, f is not differentiable at $x = 2$ and the graph of f does not have a tangent line at the point $(2, 0)$.

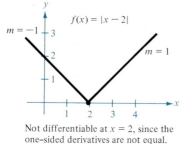

$f(x) = |x - 2|$

$m = -1$ $m = 1$

Not differentiable at $x = 2$, since the one-sided derivatives are not equal.

FIGURE 3.10

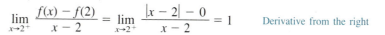

EXAMPLE 7 *A function with a vertical tangent*

The function $f(x) = x^{1/3}$ is continuous at $x = 0$, as shown in Figure 3.11. However, since the following limit is infinite,

$$\lim_{x \to 0} \frac{f(x) - f(0)}{x - 0} = \lim_{x \to 0} \frac{x^{1/3} - 0}{x} = \lim_{x \to 0} \frac{1}{x^{2/3}} = \infty$$

we conclude that the tangent line is vertical at $x = 0$. Therefore, f is not differentiable at $x = 0$.

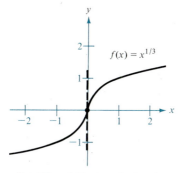

$f(x) = x^{1/3}$

Not differentiable at $x = 0$, since f has a vertical tangent at $x = 0$.

FIGURE 3.11

EXAMPLE 8 *A discontinuous function*

Show that the following function is neither continuous nor differentiable at $x = 0$:

$$f(x) = \begin{cases} x^2 + 1, & x < 0 \\ -x^2 + 4, & x \geq 0 \end{cases}$$

Solution: In Figure 3.12, we see that

$$\lim_{x \to 0^-} f(x) = 1 \neq f(0)$$

which implies that f is not continuous at $x = 0$. Moreover, the derivative from the left does not exist since

$$\lim_{x \to 0^-} \frac{f(x) - f(0)}{x - 0} = \lim_{x \to 0^-} \frac{(x^2 + 1) - 4}{x} = \lim_{x \to 0^-} \frac{x^2 - 3}{x} = \infty$$

Therefore, f is not differentiable at $x = 0$. Note from Figure 3.12 that the function does not have a vertical tangent at $x = 0$ even though it produces an infinite limit. This does not contradict the definition of a vertical tangent line because the function is not continuous at $x = 0$.

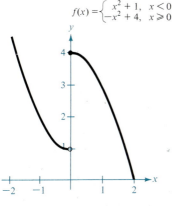

$f(x) = \begin{cases} x^2 + 1, & x < 0 \\ -x^2 + 4, & x \geq 0 \end{cases}$

Not differentiable at $x = 0$, since f is not continuous at $x = 0$.

FIGURE 3.12

From Examples 6, 7, and 8 we list some *characteristics that destroy differentiability at a point:*

1. Sharp turns in the graph (Example 6)
2. Vertical tangents (Example 7)
3. Discontinuities (Example 8)

Thus on one hand, continuity is not sufficient to guarantee differentiability, but on the other hand discontinuity is sufficient to destroy it. This brings us to the following theorem.

THEOREM 3.2   DIFFERENTIABILITY IMPLIES CONTINUITY
If f is differentiable at $x = c$, then f is continuous at $x = c$.

Proof: To prove that f is continuous at $x = c$, we will show that $f(x)$ approaches $f(c)$ as $x \to c$. To do this, we use the differentiability of f at $x = c$ and consider the following limit:

$$\lim_{x \to c} [f(x) - f(c)] = \lim_{x \to c} \left[(x - c) \left(\frac{f(x) - f(c)}{x - c} \right) \right]$$

$$= \left[\lim_{x \to c} (x - c) \right] \left[\lim_{x \to c} \frac{f(x) - f(c)}{x - c} \right] = (0)[f'(c)] = 0$$

Since the difference $[f(x) - f(c)]$ approaches zero as $x \to c$ and $f(c)$ is constant, we conclude that

$$\lim_{x \to c} f(x) = f(c)$$

Therefore, f is continuous at $x = c$.

Section Exercises 3.1

In Exercises 1 and 2, trace the curves on another piece of paper and sketch the tangent line at the point (x, y).

1. (a)   (b)   **2.** (a)   (b)

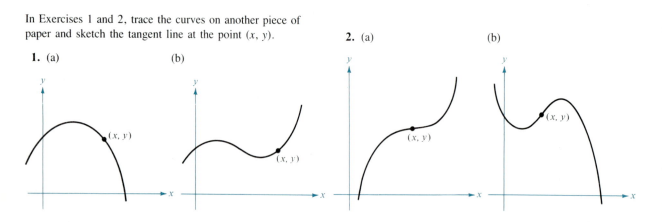

In Exercises 3 and 4, estimate the slope of the curve at the point (x, y).

3. (a)   (b)

4. (a)   (b)

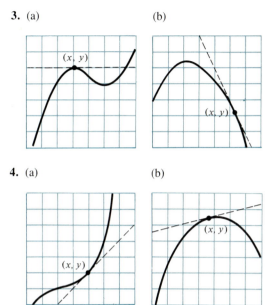

In Exercises 5–14, use the Four-Step Process to find the derivative.

5. $f(x) = 3$ **6.** $f(x) = 3x + 2$

7. $f(x) = -5x$ **8.** $f(x) = 1 - x^2$

9. $f(x) = 2x^2 + x - 1$ **10.** $f(x) = \sqrt{x - 4}$

11. $f(x) = \dfrac{1}{x - 1}$ **12.** $f(x) = \dfrac{1}{x^2}$

13. $f(t) = t^3 - 12t$ **14.** $f(t) = t^3 + t^2$

In Exercises 15–20, find the equation of the tangent line to the graph of f at the indicated point. Then verify your answer by sketching both the graph of f and the tangent line.

Function	*Point of Tangency*
15. $f(x) = x^2 + 1$	$(2, 5)$
16. $f(x) = x^2 + 2x + 1$	$(-3, 4)$
17. $f(x) = x^3$	$(2, 8)$
18. $f(x) = x^3$	$(-2, -8)$
19. $f(x) = \sqrt{x + 1}$	$(3, 2)$
20. $f(x) = \dfrac{1}{(x + 1)}$	$(0, 1)$

In Exercises 21–26, use the alternative limit form to find the derivative at $x = c$ (if it exists).

21. $f(x) = x^2 - 1$, $c = 2$

22. $f(x) = x^3 + 2x$, $c = 1$

23. $f(x) = x^3 + 2x^2 + 1$, $c = -2$

24. $f(x) = \dfrac{1}{x}$, $c = 3$

25. $f(x) = (x - 1)^{2/3}$, $c = 1$

26. $f(x) = |x - 2|$, $c = 2$

In Exercises 27–36, find every point at which the function is not differentiable.

27. $f(x) = |x + 3|$ **28.** $f(x) = |x^2 - 9|$

29. $f(x) = \dfrac{1}{x + 1}$ **30.** $f(x) = \dfrac{2x}{x - 1}$

31. $f(x) = (x - 3)^{2/3}$ **32.** $f(x) = x^{2/5}$

33. $f(x) = \sqrt{x - 1}$ **34.** $f(x) = \dfrac{x^2}{x^2 - 4}$

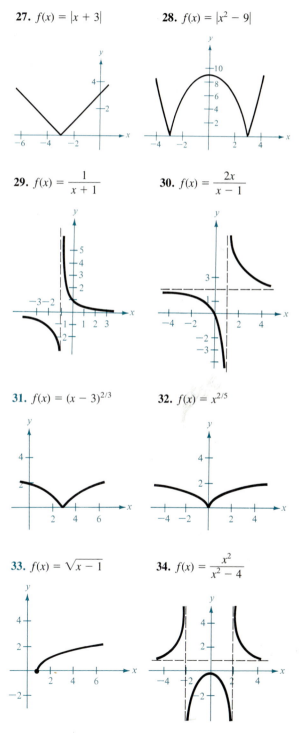

35. $f(x) = \begin{cases} 4 - x^2, & 0 < x \\ x^2 - 4, & x \le 0 \end{cases}$

36. $f(x) = \begin{cases} x^2 - 2x, & x > 1 \\ x^3 - 3x^2 + 3x, & x \le 1 \end{cases}$

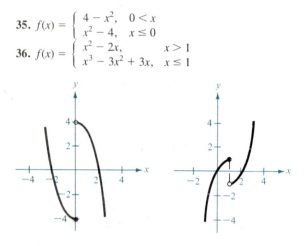

In Exercises 37–40, find the derivatives from the left and from the right at $x = 1$ (if they exist). Is the function differentiable at $x = 1$?

37. $f(x) = \sqrt{1 - x^2}$

38. $f(x) = \begin{cases} x - 1, & x \le 1 \\ (x - 1)^2, & x > 1 \end{cases}$

39. $f(x) = \begin{cases} (x - 1)^3, & x \le 1 \\ (x - 1)^2, & x > 1 \end{cases}$

40. $f(x) = \begin{cases} x, & x \le 1 \\ x^2, & x > 1 \end{cases}$

41. Find the equation of a line that is tangent to the graph of $y = x^3$ and is parallel to the line $3x - y + 1 = 0$.

42. Find the equation of a line that is tangent to the graph of $y = 1/\sqrt{x}$ and is parallel to the line $x + 2y - 6 = 0$.

43. There are two tangent lines to the graph of $y = 4x - x^2$ that pass through the point $(2, 5)$, as shown in Figure 3.13. Find the equations of these two lines.

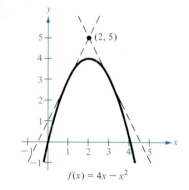

$f(x) = 4x - x^2$

FIGURE 3.13

44. Two lines through the point $(1, -3)$ are tangent to the graph of $y = x^2$. Find the equations of these two lines and make a sketch to verify your results.

3.2
Velocity, acceleration, and other rates of change

We have seen how the derivative is used to determine slope. We now consider another use—to determine the rate of change in one variable with respect to another. Applications involving rates of change can be found in a wide variety of fields. A few examples are population growth rates, unemployment rates, production rates, and the rate of water flow.

Straight line motion

One of the most familiar uses of a rate of change is to describe the motion of an object moving in a straight line. Such motion is called **rectilinear motion.** It is customary to use a horizontal (or vertical) line with a designated origin to represent the line of motion. Movement to the right (or upward) is considered to be in the **positive direction,** and movement to the left (or down) is considered to be in the **negative direction.**

The function s that gives the position (relative to the origin) of an object as a function of time t is called a **position function.** If, over a period of time Δt, the object changes its position by the amount

$$\Delta s = s(t + \Delta t) - s(t) \qquad \text{Change in distance}$$

then, by the familiar formula

$$\text{rate} = \frac{\text{distance}}{\text{time}}$$

the **average rate of change** in distance with respect to time is given by

$$\frac{\text{change in distance}}{\text{change in time}} = \frac{\Delta s}{\Delta t}$$

We call this average rate of change the **average velocity.**

DEFINITION OF AVERAGE VELOCITY	If $s(t)$ gives the position at time t of an object moving in a straight line, then the **average velocity** of the object over the interval $[t, t + \Delta t]$ is given by $$\text{average velocity} = \frac{\Delta s}{\Delta t} = \frac{s(t + \Delta t) - s(t)}{\Delta t}$$

EXAMPLE 1   Finding average velocities for a falling object

If an object is dropped from a height of 100 feet and falls freely (that is, with no air resistance), its height s at time t is given by the position function $s = -16t^2 + 100$, where s is measured in feet and t is measured in seconds. Find the average rate of change of the height over the following intervals:
(a) [1, 2]   (b) [1, 1.5]   (c) [1, 1.1]

Solution: Using the position equation, we determine the heights at $t = 1, 1.1,$ 1.5, and 2, as shown in Table 3.1 and Figure 3.14.

(a) For the interval [1, 2], the object falls from a height of 84 feet to a height of 36 feet. Thus, the average rate of change is

$$\frac{\Delta s}{\Delta t} = \frac{36 - 84}{2 - 1} = \frac{-48}{1} = -48 \text{ ft/sec}$$

(b) For the interval [1, 1.5], the average rate of change is

$$\frac{\Delta s}{\Delta t} = \frac{64 - 84}{1.5 - 1} = \frac{-20}{0.5} = -40 \text{ ft/sec}$$

(c) For the interval [1, 1.1], the average rate of change is

$$\frac{\Delta s}{\Delta t} = \frac{80.64 - 84}{1.1 - 1} = \frac{-3.36}{0.1} = -33.6 \text{ ft/sec}$$

TABLE 3.1

s	84	80.64	64	36
t	1	1.1	1.5	2

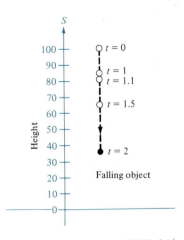

FIGURE 3.14

Remark In Example 1, note that the positive direction is taken to be upward. Thus, the negative average velocities indicate that the object is moving downward. In general, a positive velocity indicates motion in whichever direction is taken as positive and a negative velocity indicates motion in the opposite direction.

Suppose that in Example 1 we wanted to find the velocity at the instant, $t = 1$ second. We call this the **instantaneous velocity,** or simply **velocity** of the object when $t = 1$. Just as we approximate the slope of the tangent line by calculating the slope of the secant line, we can approximate the velocity at $t = 1$ by calculating the average velocity over a small interval $[1, 1 + \Delta t]$, as shown in Table 3.2.

TABLE 3.2

Δt	1	0.5	0.1	0.01	0.001	0.0001
$\dfrac{\Delta s}{\Delta t}$	-48	-40	-33.6	-32.16	-32.016	-32.0016

From this table, it seems reasonable to conclude that the velocity when $t = 1$ is -32 feet per second. We will verify this conclusion after presenting the following definition.

DEFINITION OF (INSTANTANEOUS) VELOCITY	If $s = s(t)$ is the position function for an object moving along a straight line, then the **velocity** of the object at time t is given by $$v(t) = \lim_{\Delta t \to 0} \frac{s(t + \Delta t) - s(t)}{\Delta t} = s'(t)$$

Remark Note that velocity is given by the *derivative* of the position function.

EXAMPLE 2   Using the derivative to find velocity

Find the velocity at $t = 1$ and $t = 2$ of a free-falling object whose position function is

$$s(t) = -16t^2 + 100$$

where s is measured in feet and t is measured in seconds.

Solution: Using the limit definition of the derivative, we have

$$s(t + \Delta t) - s(t) = -16[t^2 + 2t(\Delta t) + (\Delta t)^2] + 100 - [-16t^2 + 100]$$
$$= -32t(\Delta t) - 16(\Delta t)^2$$

$$\lim_{\Delta t \to 0} \frac{s(t + \Delta t) - s(t)}{\Delta t} = \lim_{\Delta t \to 0} (-32t - 16\Delta t) = -32t$$

Therefore, the velocity function is

$$v(t) = s'(t) = -32t$$

and the velocity at $t = 1$ and $t = 2$ is

$$v(1) = -32 \text{ ft/sec} \qquad \text{and} \qquad v(2) = -64 \text{ ft/sec}$$

Sometimes there is confusion between the terms ''speed'' and ''velocity.'' We define **speed** as the absolute value of velocity; as such, it is always nonnegative. This simply means that speed indicates only how fast an object is moving, whereas velocity indicates both the speed and the direction of its motion.

Just as we can obtain the velocity function by differentiating the position function, we can obtain the **acceleration** function by differentiating the velocity function.

DEFINITION OF ACCELERATION

If s is the position function for an object moving along a straight line, then the **acceleration** of the object at time t is given by

$$a(t) = v'(t)$$

where $v(t)$ is the velocity at time t.

| Remark   If the time t is expressed in seconds and the distance s is expressed in feet, then velocity will be expressed in feet per second (ft/sec) and acceleration will be given in feet per second per second (ft/sec^2).

EXAMPLE 3   *Finding acceleration as the derivative of the velocity*

Find the acceleration of a free-falling object whose position function is

$$s(t) = -16t^2 + 100$$

Solution: From Example 2, we know the velocity function for this object is

$$v(t) = -32t$$

Thus, the acceleration is given by

$$a(t) = v'(t) = -32 \, \text{ft/sec}^2$$

| Remark   The acceleration found in Example 3 is called the **acceleration due to gravity.** It is denoted by g, and its exact value depends on one's locaton on the earth. The *standard value* of g is -32.174 feet per second per second (assuming the positive direction is taken as upward).

In general, the position of a free-falling object (neglecting air resistance) under the influence of gravity can be represented by the equation

$$s(t) = \frac{1}{2}gt^2 + v_0 t + s_0$$

where s_0 is the initial height of the object, v_0 is the initial velocity at which the object is thrown, and g is the acceleration due to gravity. Considering the acceleration due to the earth's gravity as $g = -32$ feet per second per second, we have the position function

$$s(t) = -16t^2 + v_0 t + s_0$$

Remember that for free-falling objects, we consider the velocity to be positive for upward motion and negative for downward motion.

EXAMPLE 4   Finding the acceleration of a moving object

Suppose that the velocity of an automobile starting from rest is given by

$$v = \frac{80t}{t + 5} \text{ ft/sec}$$

Find the velocity and acceleration of the automobile when $t = 0, 5, 10, ..., 60$ seconds.

Solution: The position of the car is shown in Figure 3.!5. The acceleration at time t is given by

$$a = \frac{dv}{dt} = \lim_{\Delta t \to 0} \frac{1}{\Delta t}\left[\frac{80(t + \Delta t)}{(t + \Delta t) + 5} - \frac{80t}{t + 5}\right]$$

$$= \lim_{\Delta t \to 0} \frac{80}{\Delta t}\left[\frac{5\Delta t}{(t + \Delta t + 5)(t + 5)}\right]$$

$$= \lim_{\Delta t \to 0} \frac{400}{(t + \Delta t + 5)(t + 5)}$$

$$= \frac{400}{(t + 5)^2} \text{ ft/sec}^2$$

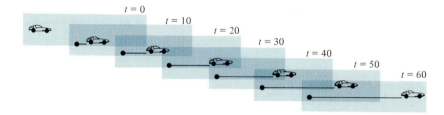

FIGURE 3.15

Table 3.3 compares the velocity and acceleration of the automobile at five-second intervals during its first minute of travel.

TABLE 3.3

t	0	5	10	15	20	25	30	35	40	45	50	55	60
v	0	40	53.3	60.0	64.0	66.7	68.6	70.0	71.1	72.0	72.7	73.3	73.8
$\dfrac{dv}{dt}$	16	4	1.78	1.00	0.64	0.44	0.33	0.25	0.20	0.16	0.13	0.11	0.09

Note from Table 3.3 that the acceleration approaches zero as the velocity levels off. This observation should agree with your experience when riding in an accelerating car—you do not feel the velocity, but you feel the acceleration. In other words, you feel changes in velocity.

Higher-order derivatives

To derive the acceleration function from the position function, we need to differentiate *twice*.

$$s(t) \qquad \text{Position function}$$
$$v(t) = s'(t) \qquad \text{Velocity function}$$
$$a(t) = v'(t) = s''(t) \qquad \text{Acceleration function}$$

We call $a(t)$ the **second derivative** of $s(t)$ and denote it by $s''(t)$.

The second derivative is an example of a **higher-order derivative.** We can define derivatives of any positive integer order. For instance, the **third derivative** is the derivative of the second derivative. We denote higher-order derivatives as follows.

First derivative: $\quad y', \quad f'(x), \quad \dfrac{dy}{dx}, \quad \dfrac{d}{dx}[f(x)], \quad D_x(y)$

Second derivative: $\quad y'', \quad f''(x), \quad \dfrac{d^2y}{dx^2}, \quad \dfrac{d^2}{dx^2}[f(x)], \quad D_x^2(y)$

Third derivative: $\quad y''', \quad f'''(x), \quad \dfrac{d^3y}{dx^3}, \quad \dfrac{d^3}{dx^3}[f(x)], \quad D_x^3(y)$

Fourth derivative: $\quad y^{(4)}, \quad f^{(4)}(x), \quad \dfrac{d^4y}{dx^4}, \quad \dfrac{d^4}{dx^4}[f(x)], \quad D_x^4(y)$

.
.
.

nth derivative: $\quad y^{(n)}, \quad f^{(n)}(x), \quad \dfrac{d^ny}{dx^n}, \quad \dfrac{d^n}{dx^n}[f(x)], \quad D_x^n(y)$

Velocity and acceleration are only two examples of rates of change. In general, we can use the derivative to measure the rate of change of any variable with respect to another [provided the two variables are related by a differentiable function $y = f(x)$]. When determining the rate of change of one variable with respect to another, we must be careful to distinguish between average and instantaneous rates of change. The distinction between these two rates of change is comparable to the distinction between the slope of the secant line through two points on a curve and the slope of the tangent line at one point on a curve.

EXAMPLE 5   Finding the average rate of change over an interval

A drug is administered to a patient. The drug concentration in the patient's bloodstream is monitored over 10-minute intervals for two hours. Find the average rates of change (in milligrams per minute) over the time intervals [0, 10], [0, 20], and [100, 110] for the concentrations in Table 3.4.

TABLE 3.4

t (min)	0	10	20	30	40	50	60	70	80	90	100	110	120
C (mg)	0	2	17	37	55	73	89	103	111	113	113	103	68

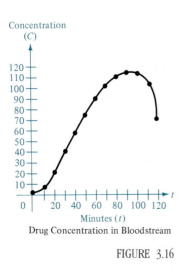

Concentration (C)

Minutes (t)

Drug Concentration in Bloodstream

FIGURE 3.16

Solution: For the interval [0, 10], the average rate of change is

$$\frac{\Delta C}{\Delta t} = \frac{2-0}{10-0} = \frac{2}{10} = 0.2 \text{ mg/min}$$

For the interval [0, 20], the average rate of change is

$$\frac{\Delta C}{\Delta t} = \frac{17-0}{20-0} = \frac{17}{20} = 0.85 \text{ mg/min}$$

For the interval [100, 110], the average rate of change is

$$\frac{\Delta C}{\Delta t} = \frac{103-113}{110-100} = \frac{-10}{10} = -1 \text{ mg/min}$$

Note in Figure 3.16 that the average rate of change is positive when the concentration increases and negative when the concentration decreases.

To conclude this section, we give a summary concerning the derivative and its interpretations.

INTERPRETATIONS OF THE DERIVATIVE

If the function given by $y = f(x)$ is differentiable at x, then its derivative

$$\frac{dy}{dx} = f'(x) = \lim_{\Delta x \to 0} \frac{f(x + \Delta x) - f(x)}{\Delta x}$$

denotes both

1. the *slope* of the graph of f at x and
2. the *instantaneous rate of change* in y with respect to x.

Remark In future work with the derivative, we will use "rate of change" to mean "instantaneous rate of change."

Section Exercises 3.2

In Exercises 1–6, find the average rate of change of the given function over the indicated interval. Compare this to the instantaneous rates of change at the endpoints of the interval.

Function	Interval
1. $f(t) = 2t + 7$	[1, 2]
2. $f(t) = 3t - 1$	$\left[0, \frac{1}{3}\right]$
3. $f(x) = \dfrac{1}{x+1}$	[0, 3]

4. $f(x) = \dfrac{-1}{x}$ [1, 2]

5. $f(t) = t^2 - 3$ [2, 2.1]

6. $f(x) = x^2 - 6x - 1$ [−1, 3]

7. The height s at time t of a silver dollar dropped from the World Trade Center is given by $s(t) = -16t^2 + 1350$, where s is measured in feet and t is measured in seconds [$s'(t) = -32t$].
(a) Find the average velocity on the interval [1, 2].

(b) Find the instantaneous velocity when $t = 1$ and $t = 2$.

(c) How long will it take the dollar to hit the ground?

(d) Find the velocity of the dollar when it hits the ground.

8. The velocity of an automobile starting from rest is given by

$$v = \frac{100t}{2t + 15}$$

where v is measured in feet per second. Find the acceleration at the following times:

(a) 5 seconds   (b) 10 seconds   (c) 20 seconds

In Exercises 9–14, use the following position and velocity functions for free-falling objects:

$$s(t) = -16t^2 + v_0 t + s_0$$
$$s'(t) = -32t + v_0$$

9. A projectile is shot upward from the surface of the earth with an initial velocity of 384 feet per second. What is its velocity after 5 seconds? After 10 seconds?

10. Repeat Exercise 9 for an initial velocity of 256 feet per second.

11. A pebble is dropped from a height of 600 feet. Find the pebble's velocity when it hits the ground.

12. A ball is thrown straight down from the top of a 220-foot building with an initial velocity of -22 feet per second. What is its velocity after 3 seconds? What is its velocity after falling 108 feet?

13. To estimate the height of a building, a stone is dropped from the top of the building into a pool of water at ground level. How high is the building if the splash is seen 6.8 seconds after the stone is dropped? (Ignore the time for light to travel. That is, assume that the splash is seen the instant it occurs.)

14. A ball is dropped from a height of 100 feet. One second later another ball is dropped from a height of 75 feet. Which one hits the ground first?

15. A car is traveling at a rate of 66 feet per second (about 45 miles per hour) when the brakes are applied. The position function for the car is $s(t) = -8.25t^2 + 66t$, where s is measured in feet and t is measured in seconds. Use this function to complete the following table.

t	0	1	2	3	4
$s(t)$					
$v(t)$					
$a(t)$					

16. The position function for an object is given by $s(t) = 10t^2$, $0 \le t \le 10$, where s is measured in feet and t is measured in seconds. Use this function to complete the following table.

t	0	2	4	6	8	10
$s(t)$						
$v(t)$						
$a(t)$						

17. Find the average velocity of the car in Exercise 15 during each given time interval.

18. Find the average velocity of the object in Exercise 16 during each given time interval.

In Exercises 19–24, find the indicated derivative.

19. Given $f'(x) = 2x^2$, find $f''(x)$.

20. Given $f''(x) = 20x^3 - 36x^2$, find $f'''(x)$.

21. Given $f''(x) = (2x - 2)/x$, find $f'''(x)$.

22. Given $f'''(x) = 2\sqrt{x} - 1$, find $f^{(4)}(x)$.

23. Given $f^{(4)}(x) = 2x + 1$, find $f^{(6)}(x)$.

24. Given $f(x) = x^3 - 2x$, find $f''(x)$.

25. The annual inventory cost for a certain manufacturer is given by

$$C = \frac{1,008,000}{Q} + 6.3Q$$

where Q is the order size when the inventory is replenished. Find the change in annual cost when Q is increased from 350 to 351 and compare this with the rate of change

$$\frac{dC}{dQ} = -\frac{1,008,000}{Q^2} + 6.3$$

when $Q = 350$. (Apply the methods of calculus by assuming that the order size Q can be any positive real number.)

26. A car is driven 15,000 miles a year and gets x miles per gallon. Assume that the average fuel cost is $1.50 per gallon. Find the annual cost of fuel C as a function of x and use this function to complete the following table.

x	10	15	20	25	30	35	40
C							
$\dfrac{dC}{dx}$							

27. At 0° Celsius, the wind-chill-corrected temperature is given by

$$T = 33 - (1.43)(10\sqrt{w} - w + 10.45)$$

where T is measured in degrees Celsius and the wind speed w is measured in meters per second. The rate of change of T is given by

$$\frac{dT}{dw} = -1.43\left(\frac{5}{\sqrt{w}} - 1\right)$$

Find the rate at which T is changing when (a) $w = 4$ and (b) $w = 9$.

28. Suppose a guitar string when plucked vibrates with a frequency of

$$F = 200\sqrt{T}$$

where F is measured in vibrations per second and the tension T is measured in pounds. Find the rate of change of the frequency when (a) $T = 4$ and (b) $T = 9$.

SECTION TOPICS •
Constant Rule •
Power Rule •
Constant Multiple Rule •
Sum and Difference Rule •

3.3
Differentiation rules for constant multiples, sums, and powers

Up to this point we have found derivatives by the limit definition of the derivative. This procedure is rather tedious even for simple functions. Fortunately there are rules that greatly simplify the differentiation process. These rules permit us to calculate the derivative without the *direct* use of limits.

THEOREM 3.3   CONSTANT RULE
The derivative of a constant is zero.

$$\frac{d}{dx}[c] = 0, \quad c \text{ is a real number}$$

The slope of a horizontal line is zero.

$f(x) = C$

The derivative of a constant function is zero.

FIGURE 3.17

Proof: Let $f(x) = c$. Then, by the limit definition of the derivative, we have

$$\frac{d}{dx}[f(x)] = f'(x) = \lim_{\Delta x \to 0} \frac{f(x + \Delta x) - f(x)}{\Delta x} = \lim_{\Delta x \to 0} \frac{c - c}{\Delta x} = \lim_{\Delta x \to 0} 0 = 0$$

Therefore,

$$\frac{d}{dx}[c] = 0$$

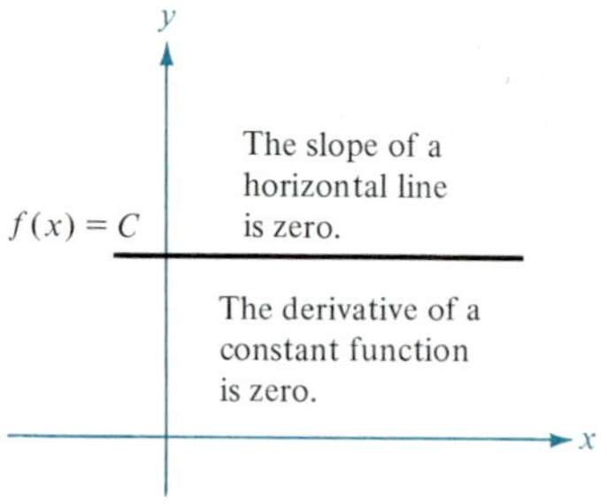

Remark Note in Figure 3.17 that the Constant Rule is equivalent to saying that the slope of a horizontal line is zero.

EXAMPLE 1   The derivative of a constant function

Function	*Derivative*
(a) $y = 7$	$\dfrac{dy}{dx} = 0$
(b) $f(x) = 0$	$f'(x) = 0$
(c) $s'(t) = -3$	$s''(t) = 0$

Before deriving the next rule, we review the procedure for expanding a binomial. Recall that

$$(x + \Delta x)^2 = x^2 + 2x\Delta x + (\Delta x)^2$$
$$(x + \Delta x)^3 = x^3 + 3x^2\Delta x + 3x(\Delta x)^2 + (\Delta x)^3$$

and the general binomial expansion for a positive integer n is

$$(x + \Delta x)^n = x^n + nx^{n-1}\Delta x + \underbrace{\frac{n(n-1)x^{n-2}}{2}(\Delta x)^2 + \cdots + (\Delta x)^n}$$

$(\Delta x)^2$ is a factor of these terms

We use this binomial expansion in proving a special case of the following theorem.

THEOREM 3.4   POWER RULE

If n is a rational number, then

$$\frac{d}{dx}[x^n] = nx^{n-1}$$

Proof: For the time being, we present a proof only for the case in which n is a positive integer. Section 3.4 (Example 7) has a proof for the case in which n is a negative integer, and Section 3.5 (Example 5) has a proof for the case in which n is rational. Using the binomial expansion, we have

$$\frac{d}{dx}[x^n] = \lim_{\Delta x \to 0} \frac{(x + \Delta x)^n - x^n}{\Delta x}$$

$$= \lim_{\Delta x \to 0} \frac{x^n + nx^{n-1}(\Delta x) + [n(n-1)x^{n-2}/2](\Delta x)^2 + \cdots + (\Delta x)^n - x^n}{\Delta x}$$

$$= \lim_{\Delta x \to 0} \left[nx^{n-1} + \left(\frac{n(n-1)x^{n-2}}{2} \right)(\Delta x) + \cdots + (\Delta x)^{n-1} \right]$$

$$= nx^{n-1} + 0 + \cdots + 0 = nx^{n-1}$$

We leave the proof that $D_x[x] = 1$ as an exercise.

Remark The case of the Power Rule for $n = 1$ is worth memorizing as a separate differentiation rule. That is,

$$\frac{d}{dx}[x] = 1$$

EXAMPLE 2   Applying the Power Rule

Function	Derivative
(a) $f(x) = x^3$	$f'(x) = 3x^2$
(b) $y = \dfrac{1}{x^2}$	$\dfrac{dy}{dx} = \dfrac{d}{dx}[x^{-2}] = (-2)x^{-3} = -\dfrac{2}{x^3}$

Remark In part (b) of Example 2, note that *before* differentiating we rewrite $1/x^2$ as x^{-2}. Rewriting is the first step in *many* differentiation problems.

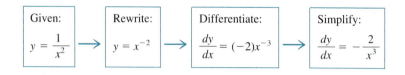

Given:	Rewrite:	Differentiate:	Simplify:
$y = \dfrac{1}{x^2}$	$y = x^{-2}$	$\dfrac{dy}{dx} = (-2)x^{-3}$	$\dfrac{dy}{dx} = -\dfrac{2}{x^3}$

THEOREM 3.5   CONSTANT MULTIPLE RULE

If f is a differentiable function and c is a real number, then

$$\frac{d}{dx}[cf(x)] = cf'(x)$$

Proof:

$$\frac{d}{dx}[cf(x)] = \lim_{\Delta x \to 0} \frac{cf(x + \Delta x) - cf(x)}{\Delta x}$$

$$= \lim_{\Delta x \to 0} c\left[\frac{f(x + \Delta x) - f(x)}{\Delta x}\right]$$

$$= c\left[\lim_{\Delta x \to 0} \frac{f(x + \Delta x) - f(x)}{\Delta x}\right] = cf'(x)$$

Informally, the Constant Multiple Rule states that constants can be factored out of the differentiation process:

$$\frac{d}{dx}[cf(x)] = c\frac{d}{dx}[\,f(x)] = cf'(x)$$

The use of this rule is often overlooked, especially when the constant appears in the denominator, as follows:

$$\frac{d}{dx}\left[\frac{f(x)}{c}\right] = \frac{d}{dx}\left[\left(\frac{1}{c}\right)f(x)\right] = \left(\frac{1}{c}\right)\frac{d}{dx}[\,f(x)] = \left(\frac{1}{c}\right)f'(x)$$

EXAMPLE 3   *Applying the Constant Multiple Rule with the Power Rule*

Function   Derivative

(a) $y = \dfrac{2}{x}$ $\dfrac{dy}{dx} = \dfrac{d}{dx}[2x^{-1}] = 2\dfrac{d}{dx}[x^{-1}] = 2(-1)x^{-2} = -\dfrac{2}{x^2}$

(b) $f(t) = \dfrac{4t^2}{5}$ $f'(t) = \dfrac{d}{dt}\left[\dfrac{4}{5}t^2\right] = \dfrac{4}{5}\dfrac{d}{dt}[t^2] = \dfrac{4}{5}(2t) = \dfrac{8}{5}t$ ☐

Remark It is helpful to see that the Constant Multiple and Power Rules can be combined into one rule. The combination rule is $D_x[cx^n] = cnx^{n-1}$.

The four functions in the next example are very simple, yet errors are frequently made in differentiating functions involving a constant multiple of the first power of x. Keep in mind that $D_x[cx] = c$.

EXAMPLE 4   *Applying the Constant Multiple Rule*

(a) $\dfrac{d}{dx}\left[-\dfrac{3x}{2}\right] = -\dfrac{3}{2}$ (b) If $y = -\dfrac{x}{2}$, then $y' = -\dfrac{1}{2}$.

(c) $\dfrac{d}{dx}[3\pi x] = 3\pi$ (d) If $f(x) = -\dfrac{5x}{6}$, then $f'(x) = -\dfrac{5}{6}$.

The next differentiation rule tells us that we can differentiate the sum of two functions separately and add the resulting derivatives.

THEOREM 3.6   **SUM RULE**
The derivative of the sum of two differentiable functions is the sum of their derivatives:

$$\frac{d}{dx}[f(x) + g(x)] = f'(x) + g'(x)$$

Proof: A proof follows directly from Theorem 2.3 (the limit of a sum).

$$\frac{d}{dx}[f(x) + g(x)] = \lim_{\Delta x \to 0} \frac{[f(x + \Delta x) + g(x + \Delta x)] - [f(x) + g(x)]}{\Delta x}$$

$$= \lim_{\Delta x \to 0} \frac{f(x + \Delta x) + g(x + \Delta x) - f(x) - g(x)}{\Delta x}$$

$$= \lim_{\Delta x \to 0} \left[\frac{f(x + \Delta x) - f(x)}{\Delta x} + \frac{g(x + \Delta x) - g(x)}{\Delta x} \right]$$

$$= \lim_{\Delta x \to 0} \frac{f(x + \Delta x) - f(x)}{\Delta x} + \lim_{\Delta x \to 0} \frac{g(x + \Delta x) - g(x)}{\Delta x}$$

$$= f'(x) + g'(x)$$

By a similar procedure it can be shown that

$$\frac{d}{dx}[f(x) - g(x)] = f'(x) - g'(x) \qquad \text{Difference Rule}$$

Furthermore, the Sum and Difference Rules can be extended to cover the derivative of any finite number of functions. For instance, if

$$F(x) = f(x) + g(x) - h(x) - k(x)$$

then

$$F'(x) = f'(x) + g'(x) - h'(x) - k'(x)$$

EXAMPLE 5   Applying the Sum Rule

Function	Derivative
(a) $f(x) = x^3 - 4x + 5$	$f'(x) = 3x^2 - 4$
(b) $g(x) = -\dfrac{x^4}{2} + 3x^3 - 2x$	$g'(x) = -2x^3 + 9x^2 - 2$

Parentheses can play an important role in the Power Rule and the Constant Multiple Rule, as shown in Example 6.

EXAMPLE 6   Using parentheses when differentiating

Given Function	Rewrite	Differentiate	Simplify
(a) $y = \dfrac{5}{2x^3}$	$y = \dfrac{5}{2}(x^{-3})$	$y' = \dfrac{5}{2}(-3x^{-4})$	$y' = -\dfrac{15}{2x^4}$
(b) $y = \dfrac{5}{(2x)^3}$	$y = \dfrac{5}{8}(x^{-3})$	$y' = \dfrac{5}{8}(-3x^{-4})$	$y' = -\dfrac{15}{8x^4}$
(c) $y = \dfrac{7}{3x^{-2}}$	$y = \dfrac{7}{3}(x^2)$	$y' = \dfrac{7}{3}(2x)$	$y' = \dfrac{14x}{3}$
(d) $y = \dfrac{7}{(3x)^{-2}}$	$y = 63(x^2)$	$y' = 63(2x)$	$y' = 126x$

When differentiating functions involving radicals, we rewrite the function in terms of rational exponents, as shown in the next example.

EXAMPLE 7   Differentiating a function involving a radical

Function

(a) $y = 2\sqrt{x}$

Derivative

$$\frac{dy}{dx} = \frac{d}{dx}[2x^{1/2}] = 2\frac{d}{dx}[x^{1/2}] = 2\left(\frac{1}{2}x^{-1/2}\right)$$

$$= x^{-1/2} = \frac{1}{\sqrt{x}}$$

(b) $y = \dfrac{1}{2\sqrt[3]{x^2}}$

$$\frac{dy}{dx} = \frac{d}{dx}\left[\frac{1}{2}x^{-2/3}\right] = \frac{1}{2}\left(-\frac{2}{3}\right)x^{-5/3}$$

$$= -\frac{1}{3x^{5/3}}$$

(c) $y = \sqrt{2x}$

$$\frac{dy}{dx} = \frac{d}{dx}[\sqrt{2}x^{1/2}] = \sqrt{2}\frac{d}{dx}[x^{1/2}] = \frac{\sqrt{2}}{2}x^{-1/2}$$

$$= \frac{1}{\sqrt{2x}}$$

Applications of the derivative

In the first two sections of this chapter, we looked at two important applications of the derivative—the slope of a curve and rate of change. We conclude this section with two examples of these applications.

EXAMPLE 8   *Using the derivative to find the slope of a curve*

Find the slope of the graph of $f(x) = x^3 - 3x$ at the following points:
(a) $(-2, -2)$ (b) $(0, 0)$ (c) $(1, -2)$

Solution: The derivative of f is $f'(x) = 3x^2 - 3$. Therefore, the slopes at the indicated points are as follows:
(a) At $x = -2$, the slope is

$$f'(-2) = 3(-2)^2 - 3 = 12 - 3 = 9$$

(b) At $x = 0$, the slope is

$$f'(0) = 3(0)^2 - 3 = -3$$

(c) At $x = 1$, the slope is

$$f'(1) = 3(1)^2 - 3 = 0$$

See Figure 3.18.

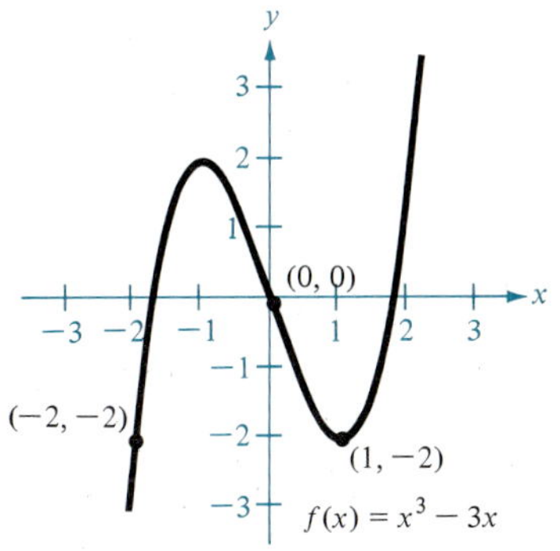

FIGURE 3.18

EXAMPLE 9   *Using the derivative to find velocity*

At time $t = 0$, a diver jumps from a diving board that is 32 feet high. The position of the diver is given by

$$s(t) = -16t^2 + 16t + 32$$

where s is measured in feet and t is measured in seconds. (See Figure 3.19.)
(a) When does the diver hit the water?
(b) What is the diver's velocity at impact?

Solution:
(a) To find the time at which the diver hits the water, we let $s = 0$ and solve for t.

$$-16t^2 + 16t + 32 = -16(t^2 - t - 2) = -16(t + 1)(t - 2) = 0$$
$$t = -1 \text{ or } 2$$

The solution $t = -1$ doesn't make sense since $t \geq 0$, so we conclude that the diver hits the water at $t = 2$ seconds.
(b) The velocity at time t is given by the derivative

$$s'(t) = -32t + 16$$

Therefore, the velocity at time $t = 2$ is

$$s'(2) = -32(2) + 16 = -48 \text{ ft/sec}$$

| Remark   In Figure 3.19, note that the diver moves upward for the first half-second. This corresponds to the fact that the velocity is positive for $0 < t < \frac{1}{2}$.

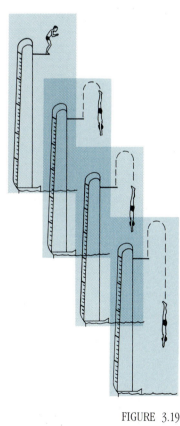

FIGURE 3.19

Section Exercises 3.3

In Exercises 1 and 2, find the slope of the tangent line to $y = x^n$ at the point $(1, 1)$.

1. (a)   (b)

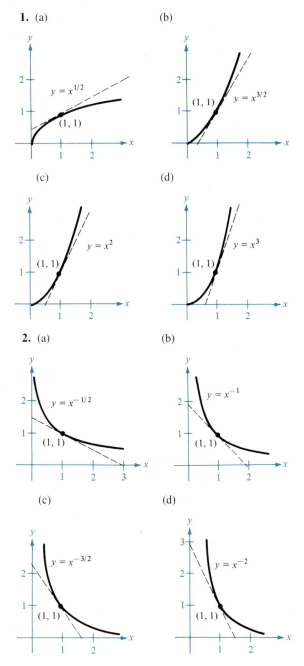

(c)   (d)

2. (a)   (b)

(c)   (d)

In Exercises 3–12, find the derivative of the given function.

3. $y = 3$ **4.** $f(x) = -2$
5. $f(x) = x + 1$ **6.** $g(x) = 3x - 1$

7. $g(x) = x^2 + 4$ **8.** $y = t^2 + 2t - 3$
9. $f(t) = -2t^2 + 3t - 6$ **10.** $y = x^3 - 9$
11. $s(t) = t^3 - 2t + 4$
12. $f(x) = 2x^3 - x^2 + 3x - 1$

In Exercises 13–18, find the value of the derivative of the given function at the indicated point.

Function	Point
13. $f(x) = \dfrac{1}{x}$	$(1, 1)$
14. $f(x) = -\dfrac{1}{2} + \dfrac{7}{5}x^3$	$\left(0, -\dfrac{1}{2}\right)$
15. $f(t) = 3 - \dfrac{3}{5t}$	$\left(\dfrac{3}{5}, 2\right)$
16. $y = 3x\left(x^2 - \dfrac{2}{x}\right)$	$(2, 18)$
17. $y = (2x + 1)^2$	$(0, 1)$
18. $f(x) = 3(5 - x)^2$	$(5, 0)$

In Exercises 19–30, find $f'(x)$.

19. $f(x) = x^2 - \dfrac{4}{x}$

20. $f(x) = x^2 - 3x - 3x^{-2} + 5x^{-3}$

21. $f(x) = x^3 - 3x - \dfrac{2}{x^4}$ **22.** $f(x) = \dfrac{2x^2 - 3x + 1}{x}$

23. $f(x) = \dfrac{x^3 - 3x^2 + 4}{x^2}$ **24.** $f(x) = (x^2 + 2x)(x + 1)$

25. $f(x) = x(x^2 + 1)$ **26.** $f(x) = x + \dfrac{1}{x^2}$

27. $f(x) = x^{4/5}$ **28.** $f(x) = x^{1/3} - 1$

29. $f(x) = \sqrt[3]{x} + \sqrt[5]{x}$ **30.** $f(x) = \dfrac{1}{\sqrt[3]{x^2}}$

In Exercises 31–36, complete the table, using Example 6 as a model.

Function	Rewrite	Derivative	Simplify
31. $y = \dfrac{1}{3x^3}$			
32. $y = \dfrac{2}{3x^2}$			
33. $y = \dfrac{1}{(3x)^3}$			
34. $y = \dfrac{\pi}{(3x)^2}$			
35. $y = \dfrac{\sqrt{x}}{x}$			
36. $y = \dfrac{4}{x^{-3}}$			

In Exercises 37 and 38, find an equation of the tangent line to the given function at the indicated point.

37. $y = x^4 - 3x^2 + 2$, $(1, 0)$

38. $y = x^3 + x$, $(-1, -2)$

In Exercises 39–42, determine the point(s) (if any) at which the given function has a horizontal tangent line.

39. $y = x^4 - 3x^2 + 2$

40. $y = x^3 + x$

41. $y = \dfrac{1}{x^2}$

42. $y = x^2 + 1$

43. Sketch the graphs of the two equations $y = x^2$ and $y = -x^2 + 6x - 5$, and sketch the two lines that are tangent to both graphs. Find the equations of these lines.

44. Show that the graphs of the two equations $y = x$ and $y = 1/x$ have tangent lines that are perpendicular to each other at their points of intersection.

45. The area of a square with sides of length s is given by $A = s^2$. Find the rate of change of the area with respect to s when $s = 4$. →derivative

46. The volume of a cube with sides of length s is given by $V = s^3$. Find the rate of change of the volume with respect to s when $s = 4$.

47. A company finds that if it charges p dollars per unit, its monthly revenue R will be

$$R = 12{,}000p - 1{,}000p^2, \quad 0 \le p \le 12$$

(Note that the revenue is zero when $p = 12$ since no one is willing to pay that much.) Find the rate of change of R with respect to p when

(a) $p = 1$ (b) $p = 4$

(c) $p = 6$ (d) $p = 10$

(Apply the methods of calculus by assuming that the price p can be any positive real number.)

48. Suppose that the profit P obtained in selling x units of a certain item each week is given by

$$P = 50\sqrt{x} - 0.5x - 500, \quad 0 \le x \le 8000$$

Find the rate of change of P with respect to x when

(a) $x = 900$ (b) $x = 1600$

(c) $x = 2500$ (d) $x = 3600$

(Note that the eventual decline in profit occurs because the only way to sell larger quantities is to decrease the price per unit.)

49. Suppose that the effectiveness E of a painkilling drug t hours after entering the bloodstream is given by

$$E = \frac{1}{27}(9t + 3t^2 - t^3), \quad 0 \le t \le 4.5$$

Find the rate of change of E with respect to t when

(a) $t = 1$ (b) $t = 2$

(c) $t = 3$ (d) $t = 4$

50. In a certain chemical reaction, the amount in grams Q of a substance produced in t hours is given by the equation

$$Q = 16t - 4t^2, \quad 0 < t \le 2$$

Find the rate in grams per hour at which the substance is being produced when

(a) $t = \dfrac{1}{2}$ (b) $t = 1$

(c) $t = 2$

51. Prove that $D_x[x] = 1$.

SECTION TOPICS ▪
Product Rule ▪
Quotient Rule ▪

3.4
Differentiation rules for products and quotients

In Section 3.3 we noted that the derivative of a sum or difference of two functions is simply the sum or difference of their derivatives. The rules for the derivative of a product or quotient of two functions are not so simple, and you may find the results surprising.

THEOREM 3.7   **PRODUCT RULE**
The derivative of the product of two differentiable functions $f(x)$ and $g(x)$ is given by

$$\frac{d}{dx}[f(x)g(x)] = f(x)g'(x) + g(x)f'(x)$$

Proof: Some mathematical proofs, such as the proof of the Sum Rule in the previous section, are straightforward. Others involve clever steps that may appear unmotivated to a reader. This proof involves such a step—adding and subtracting the same quantity. Once you see this step (shown in color), you should be able to fill in the details of the proof.

$$\frac{d}{dx}[f(x)g(x)] = \lim_{\Delta x \to 0} \frac{f(x + \Delta x)g(x + \Delta x) - f(x)g(x)}{\Delta x}$$

$$= \lim_{\Delta x \to 0} \frac{f(x + \Delta x)g(x + \Delta x) - f(x + \Delta x)g(x) + f(x + \Delta x)g(x) - f(x)g(x)}{\Delta x}$$

$$= \lim_{\Delta x \to 0} \left[f(x + \Delta x)\frac{g(x + \Delta x) - g(x)}{\Delta x} + g(x)\frac{f(x + \Delta x) - f(x)}{\Delta x} \right]$$

$$= f(x)g'(x) + g(x)f'(x)$$

As part of filling in the details, you should justify the following four limits:

$$\lim_{\Delta x \to 0} f(x + \Delta x) = f(x) \qquad \lim_{\Delta x \to 0} \frac{g(x + \Delta x) - g(x)}{\Delta x} = g'(x)$$

$$\lim_{\Delta x \to 0} g(x) = g(x) \qquad \lim_{\Delta x \to 0} \frac{f(x + \Delta x) - f(x)}{\Delta x} = f'(x)$$

Remark   It is helpful to memorize a *verbal* statement of the Product Rule:

$$\frac{d}{dx}[\text{product}] = (\text{first})\binom{\text{derivative}}{\text{of second}} + (\text{second})\binom{\text{derivative}}{\text{of first}}$$

That is, "The derivative of a product is equal to the first factor times the derivative of the second factor plus the second factor times the derivative of the first factor."

EXAMPLE 1   Differentiation with the Product Rule

Find the derivative of $f(x) = (3x - 2x^2)(5 + 4x)$.

Solution: Using the Product Rule, we have

$$
\overset{\displaystyle(\text{first})}{} \quad \overset{\displaystyle\binom{\text{derivative}}{\text{of second}}}{} \quad + \quad \overset{\displaystyle(\text{second})}{} \quad \overset{\displaystyle\binom{\text{derivative}}{\text{of first}}}{}
$$

$$f'(x) = (3x - 2x^2)\frac{d}{dx}[5 + 4x] + (5 + 4x)\frac{d}{dx}[3x - 2x^2]$$

$$= (3x - 2x^2)(4) + (5 + 4x)(3 - 4x)$$

$$= (12x - 8x^2) + (15 - 8x - 16x^2)$$

$$= -24x^2 + 4x + 15$$

In Example 1, note that the derivative of the product is

$$\frac{d}{dx}[(3x - 2x^2)(5 + 4x)] = -24x^2 + 4x + 15$$

whereas the product of the derivatives is

$$\left(\frac{d}{dx}[3x - 2x^2]\right)\left(\frac{d}{dx}[5 + 4x]\right) = (3 - 4x)(4) = -16x + 12$$

Thus, it is clear that

$$\left(\begin{array}{c}\text{the derivative of the}\\\text{product of two functions}\end{array}\right) \neq \left(\begin{array}{c}\text{the product of the derivatives}\\\text{of the two functions}\end{array}\right)$$

In Example 1, we had the option of finding the derivative with or without the Product Rule. To find the derivative without the Product Rule, we can write

$$D_x[(3x - 2x^2)(5 + 4x)] = D_x[-8x^3 + 2x^2 + 15x]$$
$$= -24x^2 + 4x + 15$$

After we introduce the Chain Rule in the next section, the benefit of the Product Rule will become more apparent.

EXAMPLE 2   Using the Product Rule

Find the derivative of $y = (1 + x^{-1})(x - 1)$.

Solution:

$$D_x[(1\text{st})(2\text{nd})] = (1\text{st})(D_x[2\text{nd}]) + (2\text{nd})(D_x[1\text{st}])$$

$$f'(x) = (1 + x^{-1})\frac{d}{dx}[x - 1] + (x - 1)\frac{d}{dx}[1 + x^{-1}]$$

$$= (1 + x^{-1})(1) + (x - 1)(-x^{-2})$$

$$= 1 + \frac{1}{x} - \frac{x - 1}{x^2}$$

$$= \frac{x^2 + x - x + 1}{x^2} = \frac{x^2 + 1}{x^2}$$

EXAMPLE 3   Comparing the Product Rule and the Constant Multiple Rule

Find the derivative of the following:
(a) $y = \sqrt{x}\,g(x)$ (b) $y = \sqrt{2}\,g(x)$

Solution:
(a) Using the Product Rule, we have

$$\frac{dy}{dx} = \sqrt{x}\left(\frac{d}{dx}[g(x)]\right) + g(x)\left(\frac{d}{dx}[\sqrt{x}]\right)$$

$$= \sqrt{x}\,g'(x) + g(x)\left(\frac{1}{2}x^{-1/2}\right) = \sqrt{x}\,g'(x) + g(x)\frac{1}{2\sqrt{x}}$$

(b) Using the Constant Multiple Rule, we have

$$\frac{dy}{dx} = \sqrt{2}\,g'(x)$$

| **Remark** In Example 3, notice that we use the Product Rule when both factors of the product are variable, and we use the Constant Multiple Rule when one of the factors is a constant.

The Product Rule can be extended to cover products involving more than two factors. For example, if f, g, and h are differentiable functions of x, then

$$\frac{d}{dx}[f(x)g(x)h(x)] = f'(x)g(x)h(x) + f(x)g'(x)h(x) + f(x)g(x)h'(x)$$

THEOREM 3.8   QUOTIENT RULE

The derivative of the quotient of two differentiable functions $f(x)$ and $g(x)$ is given by

$$\frac{d}{dx}\left[\frac{f(x)}{g(x)}\right] = \frac{g(x)f'(x) - f(x)g'(x)}{[g(x)]^2}, \quad g(x) \neq 0$$

Proof: As in the proof of the Product Rule, a key step in this proof involves adding and subtracting the same quantity.

$$\frac{d}{dx}\left[\frac{f(x)}{g(x)}\right] = \lim_{\Delta x \to 0} \frac{\dfrac{f(x + \Delta x)}{g(x + \Delta x)} - \dfrac{f(x)}{g(x)}}{\Delta x}$$

$$= \lim_{\Delta x \to 0} \frac{g(x)f(x + \Delta x) - f(x)g(x + \Delta x)}{\Delta x g(x)g(x + \Delta x)}$$

$$= \lim_{\Delta x \to 0} \frac{g(x)f(x + \Delta x) - f(x)g(x) + f(x)g(x) - f(x)g(x + \Delta x)}{\Delta x g(x)g(x + \Delta x)}$$

$$= \frac{\displaystyle\lim_{\Delta x \to 0} \frac{g(x)[f(x + \Delta x) - f(x)]}{\Delta x} - \lim_{\Delta x \to 0} \frac{f(x)[g(x + \Delta x) - g(x)]}{\Delta x}}{\displaystyle\lim_{\Delta x \to 0} [g(x)g(x + \Delta x)]}$$

$$= \frac{g(x)\left[\displaystyle\lim_{\Delta x \to 0} \frac{f(x + \Delta x) - f(x)}{\Delta x}\right] - f(x)\left[\displaystyle\lim_{\Delta x \to 0} \frac{g(x + \Delta x) - g(x)}{\Delta x}\right]}{\displaystyle\lim_{\Delta x \to 0} [g(x)g(x + \Delta x)]}$$

$$= \frac{g(x)f'(x) - f(x)g'(x)}{[g(x)]^2}$$

| **Remark** As with the Product Rule, it is helpful to memorize a verbal statement of the Quotient Rule:

$$\frac{d}{dx}[\text{quotient}] = \frac{(\text{denominator})\dfrac{d}{dx}[\text{numerator}] - (\text{numerator})\dfrac{d}{dx}[\text{denominator}]}{(\text{denominator})^2}$$

That is, "The derivative of a quotient is equal to the denominator times the derivative of the numerator minus the numerator times the derivative of the denominator all divided by the square of the denominator."

Be sure you see that

$$\left(\begin{array}{c}\text{the derivative of the}\\\text{quotient of two functions}\end{array}\right) \neq \left(\begin{array}{c}\text{the quotient of the derivatives}\\\text{of the two functions}\end{array}\right)$$

EXAMPLE 4   *Using the Quotient Rule*

Differentiate

$$y = \frac{2x^2 - 4x + 3}{2 - 3x}$$

Solution:

$$y' = \frac{(2 - 3x)\dfrac{d}{dx}[2x^2 - 4x + 3] - (2x^2 - 4x + 3)\dfrac{d}{dx}[2 - 3x]}{(2 - 3x)^2}$$

$$= \frac{(2 - 3x)(4x - 4) - (2x^2 - 4x + 3)(-3)}{(2 - 3x)^2}$$

$$= \frac{(-12x^2 + 20x - 8) - (-6x^2 + 12x - 9)}{(2 - 3x)^2}$$

$$= \frac{-6x^2 + 8x + 1}{(2 - 3x)^2}$$

Remark Note the use of parentheses in Example 4. A liberal use of parentheses is recommended for *all* types of differentiation problems. For instance, with the Quotient Rule, it is a good idea to enclose all factors and derivatives in parentheses and to pay special attention to the subtraction required in the numerator.

When we introduced differentiation rules in the previous section, we emphasized the need for rewriting *before* differentiating. The next example illustrates this point with the Quotient Rule.

EXAMPLE 5   *Rewriting before differentiating*

$$y = \frac{3 - (1/x)}{x + 5} \qquad \text{Given function}$$

$$= \frac{(3x - 1)/x}{x + 5} = \frac{3x - 1}{x(x + 5)} = \frac{3x - 1}{x^2 + 5x} \qquad \text{Rewrite}$$

$$\frac{dy}{dx} = \frac{(x^2 + 5x)(3) - (3x - 1)(2x + 5)}{(x^2 + 5x)^2} \qquad \text{Quotient Rule}$$

$$= \frac{(3x^2 + 15x) - (6x^2 + 13x - 5)}{(x^2 + 5x)^2}$$

$$= \frac{-3x^2 + 2x + 5}{(x^2 + 5x)^2} \qquad \text{Simplify}$$

Not every quotient needs to be differentiated by the Quotient Rule. For example, the quotients in the next example can be considered as products of a constant times a function of x. In such cases the Constant Multiple Rule is more convenient than the Quotient Rule.

EXAMPLE 6   Differentiating quotients with the Constant Multiple Rule

	Given Function	Rewrite	Differentiate	Simplify
(a)	$y = \dfrac{x^2 + 3x}{6}$	$y = \dfrac{1}{6}(x^2 + 3x)$	$y' = \dfrac{1}{6}(2x + 3)$	$y' = \dfrac{2x + 3}{6}$
(b)	$y = \dfrac{5x^4}{8}$	$y = \dfrac{5}{8}x^4$	$y' = \dfrac{5}{8}(4x^3)$	$y' = \dfrac{5}{2}x^3$
(c)	$y = \dfrac{-3(3x - 2x^2)}{7x}$	$y = -\dfrac{3}{7}(3 - 2x)$	$y' = -\dfrac{3}{7}(-2)$	$y' = \dfrac{6}{7}$
(d)	$y = \dfrac{9}{5x^2}$	$y = \dfrac{9}{5}(x^{-2})$	$y' = \dfrac{9}{5}(-2x^{-3})$	$y' = -\dfrac{18}{5x^3}$

In Section 3.3, we claimed that the Power Rule, $D_x[x^n] = nx^{n-1}$, is valid for any rational number n. However, we proved only the case where n is a positive integer. In the next example, we prove the rule for the case where n is a negative integer.

EXAMPLE 7   Proof of the Power Rule for negative integers

Use the Quotient Rule to prove the Power Rule for the case when n is a negative integer.

Solution: If n is a negative integer, then there exists a positive integer k such that $n = -k$. Thus, by the Quotient Rule, we have

$$\frac{d}{dx}[x^n] = \frac{d}{dx}\left[\frac{1}{x^k}\right]$$

$$= \frac{x^k(0) - (1)(kx^{k-1})}{(x^k)^2}$$

$$= \frac{0 - kx^{k-1}}{x^{2k}}$$

$$= -kx^{-k-1}$$

$$= nx^{n-1}$$

The summary in Table 3.5 shows that much of the work in obtaining the simplified form of a derivative occurs *after* differentiating. Note that two characteristics of a simplified form are the absence of negative exponents and the combining of like terms.

TABLE 3.5

	$f'(x)$ after differentiating	$f'(x)$ after simplifying
Example 1	$(3x - 2x^2)(4) + (5 + 4x)(3 - 4x)$	$-24x^2 + 4x + 15$
Example 2	$(1 + x^{-1})(1) + (x - 1)(-x^{-2})$	$\dfrac{x^2 + 1}{x^2}$
Example 4	$\dfrac{(2 - 3x)(4x - 4) - (2x^2 - 4x + 3)(-3)}{(2 - 3x)^2}$	$\dfrac{-6x^2 + 8x + 1}{(2 - 3x)^2}$
Example 5	$\dfrac{(x^2 + 5x)(3) - (3x - 1)(2x + 5)}{(x^2 + 5x)^2}$	$\dfrac{-3x^2 + 2x + 5}{(x^2 + 5x)^2}$

We conclude this section with a summary of the differentiation rules we have studied so far. To become skilled at differentiation, it is essential that you memorize each rule.

GENERAL DIFFERENTIATION RULES

For u and v, differentiable functions of x:

Constant Multiple Rule: $\dfrac{d}{dx}[cu] = cu'$

Sum or Difference Rule: $\dfrac{d}{dx}[u \pm v] = u' \pm v'$

Product Rule: $\dfrac{d}{dx}[uv] = uv' + vu'$

Quotient Rule: $\dfrac{d}{dx}\left[\dfrac{u}{v}\right] = \dfrac{vu' - uv'}{v^2}$

DERIVATIVES OF ALGEBRAIC FUNCTIONS

Constant Rule: $\dfrac{d}{dx}[c] = 0$

Power Rule: $\dfrac{d}{dx}[x^n] = nx^{n-1} \qquad \dfrac{d}{dx}[x] = 1$

Section Exercises 3.4

In Exercises 1–8, find $f'(x)$ and $f'(c)$.

1. $f(x) = \dfrac{1}{3}(2x^3 - 4)$, $c = 0$

2. $f(x) = \dfrac{5 - 6x^2}{7}$, $c = 1$

3. $f(x) = 5x^{-2}(x + 3)$, $c = 1$

4. $f(x) = (x^2 - 2x + 1)(x^3 - 1)$, $c = 1$

5. $f(x) = (x^3 - 3x)(2x^2 + 3x + 5)$, $c = 0$

6. $f(x) = (x - 1)(x^2 - 3x + 2)$, $c = 0$

7. $f(x) = (x^5 - 3x)\left(\dfrac{1}{x^2}\right)$, $c = -1$

8. $f(x) = \dfrac{x + 1}{x - 1}$, $c = 2$

In Exercises 9–24, differentiate the given function.

9. $f(x) = \dfrac{3x - 2}{2x - 3}$

10. $f(x) = \dfrac{x^3 + 3x + 2}{x^2 - 1}$

11. $f(x) = \dfrac{3 - 2x - x^2}{x^2 - 1}$

12. $f(x) = x^4\left(1 - \dfrac{2}{x + 1}\right)$

13. $f(x) = \dfrac{x + 1}{\sqrt{x}}$

14. $f(x) = \sqrt[3]{x}(\sqrt{x} + 3)$

15. $h(t) = \dfrac{t + 1}{t^2 + 2t + 2}$

16. $h(x) = (x^2 - 1)^2$

17. $h(s) = (s^3 - 2)^2$

18. $f(x) = \left(\dfrac{x^2 - x - 3}{x^2 + 1}\right)(x^2 + x + 1)$

19. $g(x) = \left(\dfrac{x + 1}{x + 2}\right)(2x - 5)$

20. $f(x) = (x^2 - x)(x^2 + 1)(x^2 + x + 1)$

21. $f(x) = (3x^3 + 4x)(x - 5)(x + 1)$

22. $f(x) = \dfrac{x^2 + c^2}{x^2 - c^2}$, c is a constant

23. $f(x) = \dfrac{c^2 - x^2}{c^2 + x^2}$

24. $f(x) = \dfrac{x(x^2 - 1)}{x + 3}$

In Exercises 25–28, find the second derivative of the given function.

25. $f(x) = 4x^{3/2}$

26. $f(x) = \dfrac{x^2 + 2x - 1}{x}$

27. $f(x) = \dfrac{x}{x - 1}$

28. $f(x) = x + \dfrac{32}{x^2}$

In Exercises 29–34, complete the table without using the Quotient Rule.

Function	Rewrite	Derivative	Simplify
29. $y = \dfrac{x^2 + 2x}{x}$			
30. $y = \dfrac{4x^{3/2}}{x}$			
31. $y = \dfrac{7}{3x^3}$			
32. $y = \dfrac{4}{5x^2}$			
33. $y = \dfrac{3x^2 - 5}{7}$			
34. $y = \dfrac{x^2 - 4}{x + 2}$			

In Exercises 35–38, find an equation of the tangent line to the graph of the given function at the indicated point.

Function	Point
35. $f(x) = \dfrac{x}{x - 1}$	$(2, 2)$
36. $f(x) = (x - 1)(x^2 - 2)$	$(0, 2)$
37. $f(x) = (x^3 - 3x + 1)(x + 2)$	$(1, -3)$
38. $f(x) = \dfrac{(x - 1)}{(x + 1)}$	$\left(2, \dfrac{1}{3}\right)$

39. At what point(s) does the graph of

$$f(x) = \dfrac{x^2}{x - 1}$$

slope = 0

have a horizontal tangent?

40. At what point(s) does the graph of

$$f(x) = \dfrac{x^2}{x^2 + 1}$$

slope = 0

have a horizontal tangent?

41. The function

$$f(t) = \dfrac{t^2 - t + 1}{t^2 + 1}$$

measures the percentage of the normal level of oxygen in a pond, where t is the time in weeks after organic waste is dumped into the pond. Find the rate of change of f with respect to t when (a) $t = 0.5$, (b) $t = 2$, and (c) $t = 8$.

42. Suppose that the temperature T of food placed in a refrigerator drops according to the equation

$$T = 10\left(\dfrac{4t^2 + 16t + 75}{t^2 + 4t + 10}\right)$$

where t is the time in hours. What is the initial temperature of the food? Find the rate of change of T with respect to t when (a) $t = 1$, (b) $t = 3$, (c) $t = 5$, and (d) $t = 10$.

43. A population of 500 bacteria is introduced into a culture and grows in number according to the equation

$$P(t) = 500\left(1 + \dfrac{4t}{50 + t^2}\right)$$

where t is measured in hours. Find the rate at which the population is growing when $t = 2$.

44. Prove that if f, g, and h are differentiable functions of x, then

$$\dfrac{d}{dx}[f(x)g(x)h(x)]$$
$$= f'(x)g(x)h(x) + f(x)g'(x)h(x) + f(x)g(x)h'(x)$$

3.5
The Chain Rule

We have yet to discuss one of the most powerful rules in differential calculus—the **Chain Rule.** This differentiation rule deals with composite functions and adds a surprising versatility to the rules we discussed in the two previous sections. For example, compare the following functions. Those on the left can be differentiated without the Chain Rule and those on the right are best done with the Chain Rule.

Without the Chain Rule	*With the Chain Rule*
$y = x^2 + 1$	$y = \sqrt{x^2 + 1}$
$y = x + 1$	$y = (x + 1)^{-1/2}$
$y = 3x + 2$	$y = (3x + 2)^5$

We begin with a physical example that should give you an intuitive idea about the nature of the Chain Rule. Basically, the rule says that if y changes dy/du times as fast as u and u changes du/dx times as fast as x, then y changes $(dy/du)(du/dx)$ times as fast as x.

EXAMPLE 1   *The derivative of a composite function*

Let y, u, and x represent the number of revolutions per minute for the first, second, and third axles shown in Figure 3.20. Find dy/du, du/dx, and dy/dx, and show that

$$\frac{dy}{dx} = \left(\frac{dy}{du}\right)\left(\frac{du}{dx}\right)$$

Solution: To find the rate of change of one axle with respect to an adjacent axle, we compare the circumference of their connecting gears.

$$\left. \begin{array}{l} \text{circumference of gear } \#1 = 2\pi(1) = 2\pi \\ \text{circumference of gear } \#2 = 2\pi(3) = 6\pi \end{array} \right\} \quad y = 3u$$

$$\left. \begin{array}{l} \text{circumference of gear } \#3 = 2\pi(1) = 2\pi \\ \text{circumference of gear } \#4 = 2\pi(2) = 4\pi \end{array} \right\} \quad u = 2x$$

Now, to find y as a function of x, we substitute $2x$ for u and obtain

$$y = 3u = 3(2x) = 6x$$

Therefore, we have

$$\frac{dy}{du} = \frac{d}{du}[3u] = 3, \qquad \frac{du}{dx} = \frac{d}{dx}[2x] = 2, \qquad \frac{dy}{dx} = \frac{d}{dx}[6x] = 6$$

and it follows that

$$\frac{dy}{dx} = \left(\frac{dy}{du}\right)\left(\frac{du}{dx}\right) = (3)(2) = 6$$

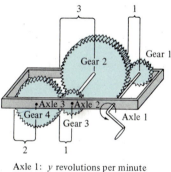

Axle 1: y revolutions per minute
Axle 2: u revolutions per minute
Axle 3: x revolutions per minute

FIGURE 3.20

Example 1 illustrates a simple case of the Chain Rule. The general case is stated as follows.

THEOREM 3.9   **CHAIN RULE**

If $y = f(u)$ is a differentiable function of u and $u = g(x)$ is a differentiable function of x, then $y = f(g(x))$ is a differentiable function of x and

$$\frac{dy}{dx} = \frac{dy}{du} \cdot \frac{du}{dx}$$

or, equivalently,

$$\frac{d}{dx}[f(g(x))] = f'(g(x))g'(x)$$

Proof: Let $F(x) = f(g(x))$. Then, using the alternative form of the derivative (Theorem 3.1), we need to show that for $x = c$

$$F'(c) = f'(g(c))g'(c)$$

An important consideration in this proof is the behavior of g as x approaches c. A problem occurs if there are values of x, other than c, such that $g(x) = g(c)$. In Appendix A we show how to use the differentiability of f and g to overcome this problem. For now, we assume that $g(x) \neq g(c)$ for values of x other than c.

In the proof of the Product and Quotient Rules, we added and subtracted the same quantity to obtain the desired form. In this proof we use a similar technique—multiplying and dividing by the same (nonzero) quantity. Note that since g is differentiable, it is also continuous, and it follows that $g(x) \to g(c)$ as $x \to c$.

$$F'(c) = \lim_{x \to c} \frac{f(g(x)) - f(g(c))}{x - c}$$

$$= \lim_{x \to c} \left[\frac{f(g(x)) - f(g(c))}{g(x) - g(c)} \cdot \frac{g(x) - g(c)}{x - c} \right], \quad g(x) \neq g(c)$$

$$= \left[\lim_{x \to c} \frac{f(g(x)) - f(g(c))}{g(x) - g(c)} \right] \left[\lim_{x \to c} \frac{g(x) - g(c)}{x - c} \right]$$

$$= f'(g(c))g'(c)$$

In applying the Chain Rule, it is helpful to think of the composite function $f \circ g$ as having two parts—an *inside* and an *outside*, as follows:

$$y = f(\underbrace{g(x)}) = \overbrace{f(u)}^{\text{outside}}$$

$$\underbrace{u = g(x)}_{\text{inside}}$$

The next example illustrates two cases.

EXAMPLE 2   Decomposition of a composite function

$y = f(g(x))$	$u = g(x)$	$y = f(u)$
(a) $y = \dfrac{1}{x + 1}$	$u = x + 1$	$y = \dfrac{1}{u}$
(b) $y = \sqrt{3x^2 - x + 1}$	$u = 3x^2 - x + 1$	$y = \sqrt{u}$

EXAMPLE 3   Using the Chain Rule

Find dy/dx for $y = (x^2 + 1)^3$.

Solution: To apply the Chain Rule, we identify the inside function u as follows:

$$y = \underbrace{(x^2 + 1)}_{u}{}^3 = u^3$$

Now, by the Chain Rule, we have

$$\frac{dy}{dx} = \underbrace{3(x^2 + 1)^2}_{\frac{dy}{du}}\underbrace{(2x)}_{\frac{du}{dx}} = 6x(x^2 + 1)^2$$

The function in Example 3 is an instance of one of the most common types of composite functions, $y = [u(x)]^n$. The rule for differentiating such *power functions* is called the **General Power Rule,** and it is a special case of the Chain Rule.

THEOREM 3.10   **GENERAL POWER RULE**

If $y = [u(x)]^n$, where u is a differentiable function of x and n is a rational number, then

$$\frac{dy}{dx} = n[u(x)]^{n-1}\frac{du}{dx}$$

or, equivalently,

$$\frac{d}{dx}[u^n] = nu^{n-1}u'$$

Proof: Since $y = u^n$, we apply the Chain Rule to obtain

$$\frac{dy}{dx} = \left(\frac{dy}{du}\right)\left(\frac{du}{dx}\right) = \frac{d}{du}[u^n]\frac{du}{dx}$$

Now, by the (simple) Power Rule in Section 3.3, we have $D_u[u^n] = nu^{n-1}$, and it follows that

$$\frac{dy}{dx} = nu^{n-1}\frac{du}{dx}$$

EXAMPLE 4   Applying the General Power Rule

Find the derivative of $f(x) = (3x - 2x^2)^3$.

Solution: If we let $u = 3x - 2x^2$, then $f(x) = (3x - 2x^2)^3 = u^3$, and by the General Power Rule the derivative is

$$\overset{n}{\overbrace{}} \quad \overset{u^{n-1}}{\overbrace{}} \qquad \overset{u'}{\overbrace{}}$$

$$f'(x) = 3(3x - 2x^2)^2\frac{d}{dx}[3x - 2x^2]$$

$$= 3(3x - 2x^2)^2(3 - 4x) = (9 - 12x)(3x - 2x^2)^2$$

In Section 3.3 we claimed that the Power Rule is valid for *any* rational number n. However, at this point we have given proofs for only two cases—positive integer powers (Section 3.3) and negative integer powers (Section 3.4). Using the Chain Rule, we can now prove the case for rational powers.

EXAMPLE 5   A proof of the Power Rule for rational exponents

Use the Chain Rule to prove the Power Rule for rational exponents:

$$\frac{d}{dx}[x^n] = nx^{n-1}$$

Solution: Let n be a rational number with denominator m. Then nm is an integer and we consider the equation

$$u^m = (x^n)^m = x^{nm}, \quad m \text{ and } nm \text{ are integers}$$

Using the Chain Rule on the left and the Power Rule (for integers) on the right, we differentiate both sides of this equation to obtain

$$\overset{m}{\overbrace{}} \quad \overset{u^{m-1}}{\overbrace{}} \qquad \overset{u'}{\overbrace{}}$$

$$m(x^n)^{m-1}\frac{d}{dx}[x^n] = nmx^{nm-1}$$

$$\frac{d}{dx}[x^n] = nx^{nm-1}x^{-nm+n}$$

$$= nx^{n-1}$$

EXAMPLE 6   *Differentiating radical functions*

Find the derivative of $y = \sqrt[3]{(x^2 + 2)^2}$.

Solution: We rewrite the function as $y = (x^2 + 2)^{2/3}$. Then, by the Power Rule (with $u = x^2 + 2$), we have

$$y' = \overset{n}{\overbrace{\frac{2}{3}}}\overset{u^{n-1}}{\overbrace{(x^2 + 2)^{-1/3}}}\overset{u'}{\overbrace{(2x)}} = \frac{4x}{3\sqrt[3]{x^2 + 2}}$$

EXAMPLE 7   *Differentiating quotients with constant numerators*

Differentiate

$$g(t) = \frac{-7}{(2t - 3)^2}$$

Solution: If we rewrite the function as $g(t) = -7(2t - 3)^{-2}$, then by the General Power Rule we obtain

$$g'(t) = \underbrace{(-7)}_{\substack{\text{Constant} \\ \text{Multiple Rule}}}\overset{n}{\overbrace{(-2)}}\overset{u^{n-1}}{\overbrace{(2t - 3)^{-3}}}\overset{u'}{\overbrace{(2)}} = 28(2t - 3)^{-3} = \frac{28}{(2t - 3)^3}$$

Simplifying derivatives

The next three examples illustrate some useful simplification techniques for the "raw derivatives" of functions involving products, quotients, and composites.

EXAMPLE 8   *Simplification by factoring least powers*

Differentiate $f(x) = x^2\sqrt{1 - x^2}$.

Solution:

$$f(x) = x^2(1 - x^2)^{1/2} \qquad \text{Rewrite}$$

$$f'(x) = x^2\frac{d}{dx}[(1 - x^2)^{1/2}] + (1 - x^2)^{1/2}\frac{d}{dx}[x^2] \qquad \text{Product Rule}$$

$$= x^2\left[\frac{1}{2}(1 - x^2)^{-1/2}(-2x)\right] + (1 - x^2)^{1/2}(2x) \qquad \text{Power Rule}$$

$$= -x^3(1 - x^2)^{-1/2} + 2x(1 - x^2)^{1/2}$$

$$= x(1 - x^2)^{-1/2}[-x^2(1) + 2(1 - x^2)] \qquad \text{Factor}$$

$$= \frac{x(2 - 3x^2)}{\sqrt{1 - x^2}} \qquad \text{Simplify}$$

EXAMPLE 9   Simplifying the derivative of a quotient

Differentiate

$$f(x) = \frac{x}{\sqrt[3]{x^2 + 4}}$$

Solution:

$$f(x) = \frac{x}{(x^2 + 4)^{1/3}} \qquad \text{Rewrite}$$

$$f'(x) = \frac{(x^2 + 4)^{1/3}(1) - x(1/3)(x^2 + 4)^{-2/3}(2x)}{(x^2 + 4)^{2/3}} \qquad \text{Quotient Rule}$$

$$= \frac{1}{3}(x^2 + 4)^{-2/3}\left[\frac{3(x^2 + 4) - (2x^2)(1)}{(x^2 + 4)^{2/3}}\right] \qquad \text{Factor}$$

$$= \frac{x^2 + 12}{3(x^2 + 4)^{4/3}} \qquad \text{Simplify}$$

EXAMPLE 10   Differentiating a quotient raised to a power

Find the derivative of

$$y = \left(\frac{3x - 1}{x^2 + 3}\right)^2$$

Solution:

$$\frac{dy}{dx} = \overset{n}{2}\overset{u^{n-1}}{\left(\frac{3x - 1}{x^2 + 3}\right)}\overset{u'}{\frac{d}{dx}\left[\frac{3x - 1}{x^2 + 3}\right]}$$

$$= \left[\frac{2(3x - 1)}{x^2 + 3}\right]\left[\frac{(x^2 + 3)(3) - (3x - 1)(2x)}{(x^2 + 3)^2}\right]$$

$$= \frac{2(3x - 1)(3x^2 + 9 - 6x^2 + 2x)}{(x^2 + 3)^3}$$

$$= \frac{2(3x - 1)(-3x^2 + 2x + 9)}{(x^2 + 3)^3}$$

Try finding y' using the Quotient Rule on $y = (3x - 1)^2/(x^2 + 3)^2$ and compare the results.

Section Exercises 3.5

In Exercises 1–6, complete the table using Example 2 as a model.

$y = f(g(x))$	$u = g(x)$	$y = f(u)$
1. $y = \sqrt{x^2 - 1}$		
2. $y = \left(\frac{3x}{2}\right)^2$		
3. $y = (x^2 - 3x + 4)^6$		
4. $y = (6x - 5)^4$		

$y = f(g(x))$	$u = g(x)$	$y = f(u)$
5. $y = \dfrac{1}{\sqrt{x^2 + 1}}$		
6. $y = \dfrac{1}{\sqrt{x + 1}}$		

In Exercises 7–44, find the derivative.

7. $y = (2x - 7)^3$ **8.** $y = (3x^2 + 1)^4$

9. $f(x) = 2(x^2 - 1)^3$ **10.** $g(x) = 3(9x - 4)^4$

11. $y = \dfrac{1}{x - 2}$

12. $s(t) = \dfrac{1}{t^2 + 3t - 1}$

13. $f(t) = \left(\dfrac{1}{t - 3}\right)^2$

14. $y = -\dfrac{4}{(t + 2)^2}$

15. $f(x) = \dfrac{3}{x^3 - 4}$

16. $f(x) = \dfrac{1}{(x^2 - 3x)^2}$

17. $f(x) = x^2(x - 2)^4$

18. $f(x) = x(3x - 9)^3$

19. $f(t) = \sqrt{1 - t}$

20. $g(x) = \sqrt{3 - 2x}$

21. $s(t) = \sqrt{t^2 + 2t - 1}$

22. $y = \sqrt[3]{3x^3 + 4x}$

23. $y = \sqrt[3]{9x^2 + 4}$

24. $g(x) = \sqrt{x^2 - 2x + 1}$

25. $y = 2\sqrt{4 - x^2}$

26. $f(x) = -3\sqrt[4]{2 - 9x}$

27. $f(x) = (9 - x^2)^{2/3}$

28. $f(t) = (9t + 2)^{2/3}$

29. $y = \dfrac{1}{\sqrt{x + 2}}$

30. $g(t) = \sqrt{\dfrac{1}{t^2 - 2}}$

31. $g(x) = \dfrac{3}{\sqrt[3]{x^3 - 1}}$

32. $s(x) = \dfrac{1}{\sqrt{x^2 - 3x + 4}}$

33. $y = \dfrac{-1}{\sqrt{x + 1}}$

34. $y = \dfrac{1}{2\sqrt{t - 3}}$

35. $y = \dfrac{\sqrt{x + 1}}{x^2 + 1}$

36. $f(x) = \dfrac{x + 1}{2x - 3}$

37. $f(t) = \dfrac{3t + 2}{t - 1}$

38. $y = \sqrt{\dfrac{2x}{x + 1}}$

39. $g(t) = \dfrac{3t^2}{\sqrt{t^2 + 2t - 1}}$

40. $f(x) = \sqrt{x}(2 - x)^2$

41. $y = \sqrt{\dfrac{x + 1}{x}}$

42. $y = (t^2 - 9)\sqrt{t + 2}$

43. $s(t) = \dfrac{-2(2 - t)\sqrt{1 + t}}{3}$

44. $g(x) = \sqrt{x - 1} + \sqrt{x + 1}$

In Exercises 45–46, find an equation of the tangent line to the graph of f at the given point.

45. $f(x) = \sqrt{3x^2 - 2}$, $(3, 5)$

46. $f(x) = x\sqrt{x^2 + 5}$, $(2, 6)$

In Exercises 47–50, find the second derivative of the function.

47. $f(x) = 2(x^2 - 1)^3$

48. $f(x) = \dfrac{1}{x - 2}$

49. $f(x) = \sqrt{x^2 + x + 1}$

50. $f(t) = \dfrac{\sqrt{t^2 + 1}}{t}$

51. (Doppler effect) The frequency F of a fire truck siren heard by a stationary observer is given by

$$F = \dfrac{132{,}400}{331 \pm v}$$

where $\pm v$ represents the velocity of the fire truck, as shown in Figure 3.21. Find the rate of change of F with respect to v when

(a) the fire truck is approaching at a velocity of 30 meters per second [use $-v$]

(b) the fire truck is moving away at a velocity of 30 meters per second [use $+v$]

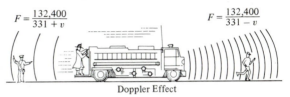

$F = \dfrac{132{,}400}{331 + v}$ $F = \dfrac{132{,}400}{331 - v}$

Doppler Effect

FIGURE 3.21

52. A certain automobile depreciates according to the formula

$$V = \dfrac{7500}{1 + 0.4t + 0.1t^2}$$

where $t = 0$ represents the time of purchase (in years). At what rate is the car depreciating

(a) 1 year after purchase?

(b) 2 years after purchase?

53. The speed S of blood that is r centimeters from the center of an artery is given by

$$S = C(R^2 - r^2)$$

where C is a constant, R is the radius of the artery, and S is measured in centimeters per second. Suppose a drug is administered and the artery begins dilating at the rate of dR/dt. At a constant distance r, find the rate at which S changes with respect to t for $C = 1.76 \times 10^5$, $R = 1.2 \times 10^{-2}$, and $dR/dt = 10^{-5}$.

3.6
Implicit differentiation

SECTION TOPICS ▪
Implicit and explicit functions ▪
Implicit differentiation ▪

Up to this point, our functions involving two variables were generally expressed in the **explicit form** $y = f(x)$. That is, one of the two variables was explicitly given in terms of the other. For example

$$y = 3x - 5, \qquad s = -16t^2 + 20t, \qquad u = 3w - w^2$$

are all written in explicit form, and we say that y, s, and u are functions of x, t, and w, respectively.

However, many functions are not given explicitly and are only implied by a given equation. For instance, the function $y = 1/x$ is defined **implicitly** by the equation $xy = 1$. Suppose that you were asked to find dy/dx in this equation. As it turns out, this is a relatively simple task, and you would probably begin by solving the equation for y.

Implicit form	*Explicit form*	*Derivative*
$xy = 1$	$y = \dfrac{1}{x} = x^{-1}$	$\dfrac{dy}{dx} = -x^{-2} = -\dfrac{1}{x^2}$

This procedure works well whenever we can easily solve for the function explicitly. However, we cannot use this procedure in cases in which we are unable to solve for y as a function of x. For instance, how would we find dy/dx in the equation

$$x^2 - 2y^3 + 4y = 2$$

where it is very difficult to express y as a function of x explicitly? To do this, we use a procedure called **implicit differentiation** in which we assume y is a differentiable function of x.

To understand how to find dy/dx implicitly you must realize that the differentiation is taking place *with respect to x*. This means that when we differentiate terms involving x alone, we can differentiate as usual. *But* when we differentiate terms involving y, we must apply the Chain Rule because we are assuming that y is defined implicitly as a function of x. Study the next example carefully. Note in particular how the Chain Rule is used to introduce the dy/dx factors.

EXAMPLE 1   Applying the Chain Rule

Differentiate the following with respect to x.
(a) $3x^2$ (b) $2y^3$ (c) $x + 3y$ (d) xy^2

Solution:

(a) $\dfrac{d}{dx}[3x^2] = 6x$

(b) $\dfrac{d}{dx}[2y^3] = \overset{u^n}{\overbrace{}}\,\overset{nu^{n-1}\quad u'}{\overbrace{2(3)y^2 \dfrac{dy}{dx}}} = 6y^2 \dfrac{dy}{dx}$ Chain Rule

(c) $\dfrac{d}{dx}[x + 3y] = 1 + 3\dfrac{dy}{dx}$

(d) $\dfrac{d}{dx}[xy^2] = x\dfrac{d}{dx}[y^2] + y^2\dfrac{d}{dx}[x]$ Product Rule

$\qquad\qquad = x\left(2y\dfrac{dy}{dx}\right) + y^2(1)$ Chain Rule

$\qquad\qquad = 2xy\dfrac{dy}{dx} + y^2$

For equations involving x and y, we suggest the following procedure for finding dy/dx implicitly.

IMPLICIT DIFFERENTIATION   Given an equation involving x and y and assuming y is a differentiable function of x, we can find dy/dx as follows.

1. Differentiate both sides of the equation *with respect to x*.
2. Collect all terms involving dy/dx on the left side of the equation and move all other terms to the right side of the equation.
3. Factor dy/dx out of the left side of the equation.
4. Solve for dy/dx by dividing both sides of the equation by the left-hand factor that does not contain dy/dx.

EXAMPLE 2   Implicit differentiation

Find dy/dx given that $y^3 + y^2 - 5y - x^2 = -4$.

Solution:

1. Differentiating both sides of the equation with respect to x, we have

$$\frac{d}{dx}[y^3 + y^2 - 5y - x^2] = \frac{d}{dx}[-4]$$

$$\frac{d}{dx}[y^3] + \frac{d}{dx}[y^2] - \frac{d}{dx}[5y] - \frac{d}{dx}[x^2] = \frac{d}{dx}[-4]$$

$$3y^2\frac{dy}{dx} + 2y\frac{dy}{dx} - 5\frac{dy}{dx} - 2x = 0$$

2. Collecting the dy/dx terms on the left, we have

$$3y^2\frac{dy}{dx} + 2y\frac{dy}{dx} - 5\frac{dy}{dx} = 2x$$

3. Factoring dy/dx out of the left side, we have

$$\frac{dy}{dx}(3y^2 + 2y - 5) = 2x$$

4. Finally, we divide by $(3y^2 + 2y - 5)$ to obtain

$$\frac{dy}{dx} = \frac{2x}{3y^2 + 2y - 5}$$

| **Remark** In Example 2, note that implicit differentiation can produce an expression for dy/dx that contains both x and y.

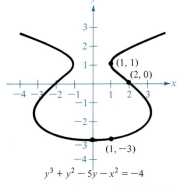

$y^3 + y^2 - 5y - x^2 = -4$

FIGURE 3.22

To see how we can use an *implicit derivative,* consider the graph of the equation $y^3 + y^2 - 5y - x^2 = -4$ as shown in Figure 3.22. The derivative found in Example 2 gives us a formula for the slope of the tangent line at a point on this graph. The slopes at several points on the graph are as follows:

Point on Graph	(2, 0)	(1, −3)	$x = 0$	(1, 1)
Slope of Graph	$-\dfrac{4}{5}$	$\dfrac{1}{8}$	0	undefined

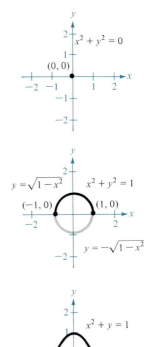

When using implicit differentiation, keep in mind that it is meaningless to solve for dy/dx in an equation that has no solution points. (For example, $x^2 + y^2 = -4$ has no solution points.) If a segment of a graph can be represented by a differentiable function, then dy/dx will have meaning as the slope at each point on the segment. Recall that a function is not differentiable at (1) points with vertical tangents and (2) points of discontinuity. The following example illustrates this idea.

EXAMPLE 3   Representing a graph by differentiable functions

If possible, represent the graphs of the following equations by differentiable functions. (See Figure 3.23.)
(a) $x^2 + y^2 = 0$ (b) $x^2 + y^2 = 1$ (c) $x^2 + y = 1$

Solution:
(a) The graph of this equation is a single point. Therefore, it does not define y as a differentiable function of x.
(b) The graph of this equation is the unit circle, centered at $(0, 0)$. The upper semi-circle is given by the differentiable function

$$y = \sqrt{1 - x^2}, \quad -1 < x < 1$$

and the lower semi-circle is given by

$$y = -\sqrt{1 - x^2}, \quad -1 < x < 1$$

FIGURE 3.23

At the points $(-1, 0)$ and $(1, 0)$, the slope of the graph is undefined.
(c) This parabola is given by the differentiable function

$$y = -x^2 + 1, \quad -\infty < x < \infty$$

EXAMPLE 4   Finding the slope of a curve implicitly

Determine the slope of the tangent line to the graph of $x^2 + 4y^2 = 4$ at the point $(\sqrt{2}, -1/\sqrt{2})$. (See Figure 3.24.)

Solution: Implicit differentiation of the equation $x^2 + 4y^2 = 4$ with respect to x yields

$$2x + 8y\frac{dy}{dx} = 0$$

$$\frac{dy}{dx} = \frac{-2x}{8y} = \frac{-x}{4y}$$

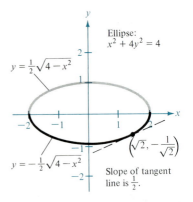

Therefore, at $(\sqrt{2}, -1/\sqrt{2})$ the slope is

$$\frac{dy}{dx} = \frac{-\sqrt{2}}{-4/\sqrt{2}} = \frac{1}{2}$$

Slope of tangent line is $\frac{1}{2}$.

FIGURE 3.24

Remark To see the benefit of implicit differentiation, we suggest you do Example 4 using the explicit function

$$y = -\frac{1}{2}\sqrt{4 - x^2}$$

The graph of this function is the lower half of the ellipse.

EXAMPLE 5   Finding the slope of a curve implicitly

Determine the slope of the graph of

$$3(x^2 + y^2)^2 = 100xy$$

at the point (3, 1).

Solution:

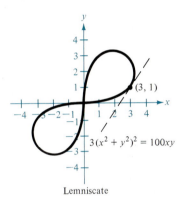

$3(x^2 + y^2)^2 = 100xy$

Lemniscate

FIGURE 3.25

1.
$$\frac{d}{dx}[3(x^2 + y^2)^2] = \frac{d}{dx}[100xy]$$

$$3(2)(x^2 + y^2)\left(2x + 2y\frac{dy}{dx}\right) = 100\left[x\frac{dy}{dx} + y(1)\right]$$

2.
$$12y(x^2 + y^2)\frac{dy}{dx} - 100x\frac{dy}{dx} = 100y - 12x(x^2 + y^2)$$

3.
$$[12y(x^2 + y^2) - 100x]\frac{dy}{dx} = 100y - 12x(x^2 + y^2)$$

4.
$$\frac{dy}{dx} = \frac{100y - 12x(x^2 + y^2)}{-100x + 12y(x^2 + y^2)} = \frac{25y - 3x(x^2 + y^2)}{-25x + 3y(x^2 + y^2)}$$

Now, at the point (3, 1) the slope of the graph is

$$\frac{dy}{dx} = \frac{25(1) - 3(3)(3^2 + 1^2)}{-25(3) + 3(1)(3^2 + 1^2)} = \frac{25 - 90}{-75 + 30} = \frac{-65}{-45} = \frac{13}{9}$$

as shown in Figure 3.25. This graph is called a **lemniscate.**

EXAMPLE 6   Finding a differentiable function

Find dy/dx implicitly for the equation $4x - y^3 + 12y = 0$, and use Figure 3.26 to find the largest interval of the form $-a < y < a$ such that y is a differentiable function of x.

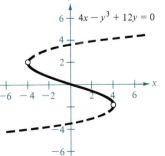

$4x - y^3 + 12y = 0$

FIGURE 3.26

Solution:

$$4x - y^3 + 12y = 0$$

$$4 - 3y^2y' + 12y' = 0$$

$$y'(-3y^2 + 12) = -4$$

$$y' = \frac{4}{3(y^2 - 4)}$$

From Figure 3.26 we see that the largest interval about the origin for which y is a differentiable function of x is $-2 < y < 2$.

In implicit differentiation, the form of the derivative can often be simplified or changed to a single variable by an appropriate use of the *original* equation. A similar technique can be used to find and simplify higher-order derivatives obtained implicitly. This is demonstrated in the next example.

EXAMPLE 7   Finding the second derivative implicitly

Given $x^2 + y^2 = 25$, find y''.

Solution: Differentiating each term with respect to x, we obtain

$$2x + 2yy' = 0$$
$$2yy' = -2x$$
$$y' = \frac{-2x}{2y} = -\frac{x}{y}$$

Differentiating a second time with respect to x yields

$$y'' = -\frac{(y)(1) - (x)(y')}{y^2}$$

$$= -\frac{y - (x)(-x/y)}{y^2} \qquad \text{Substitute } y' = -x/y$$

$$= -\frac{y^2 + x^2}{y^3}$$

$$= -\frac{25}{y^3} \qquad \text{Substitute } x^2 + y^2 = 25 \qquad \square$$

Isaac Barrow

The graph in Example 8 is called the *kappa curve* because of its resemblance to the Greek letter kappa. The general solution for the tangent line to this particular curve was first discovered by the English mathematician Isaac Barrow (1630–1677). Barrow and Newton were contemporaries and corresponded frequently regarding their work in the early development of calculus.

EXAMPLE 8   Finding a tangent line to a graph

Find the tangent line to the **kappa curve** given by

$$x^2(x^2 + y^2) = y^2$$

at the point $(\sqrt{2}/2, \sqrt{2}/2)$ as shown in Figure 3.27.

Solution: By rewriting and differentiating implicitly, we have

$$x^4 + x^2y^2 - y^2 = 0$$
$$4x^3 + x^2(2yy') + 2xy^2 - 2yy' = 0$$
$$2y(x^2 - 1)y' = -2x(2x^2 + y^2)$$
$$y' = \frac{x(2x^2 + y^2)}{y(1 - x^2)}$$

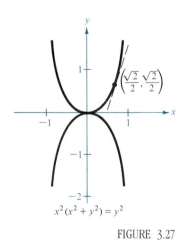

$$x^2(x^2 + y^2) = y^2$$

FIGURE 3.27

Now, at the point $(\sqrt{2}/2, \sqrt{2}/2)$, the slope is

$$m = \frac{(\sqrt{2}/2)[2(1/2) + (1/2)]}{(\sqrt{2}/2)[1 - (1/2)]} = \frac{3/2}{1/2} = 3$$

and the equation of the tangent line at this point is

$$y - \frac{\sqrt{2}}{2} = 3\left(x - \frac{\sqrt{2}}{2}\right)$$

$$y = 3x - \sqrt{2}$$

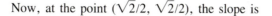

Section Exercises 3.6

In Exercises 1–16, find dy/dx by implicit differentiation and evaluate the derivative at the indicated point.

Equation	Point
1. $x^2 + y^2 = 16$	$(3, \sqrt{7})$
2. $x^2 - y^2 = 16$	$(4, 0)$
3. $xy = 4$	$(-4, -1)$
4. $x^2 - y^3 = 0$	$(1, 1)$
5. $x^{1/2} + y^{1/2} = 9$	$(16, 25)$
6. $x^3 + y^3 = 8$	$(0, 2)$
7. $x^3 - xy + y^2 = 4$	$(0, -2)$
8. $x^2y + y^2x = -2$	$(2, -1)$
9. $y^2 = \dfrac{x^2 - 9}{x^2 + 9}$	$(3, 0)$
10. $(x + y)^3 = x^3 + y^3$	$(-1, 1)$
11. $x^3y^3 - y = x$	$(0, 0)$
12. $\sqrt{xy} = x - 2y$	$(4, 1)$
13. $x^{2/3} + y^{2/3} = 5$	$(8, 1)$
14. $x^3 + y^3 = 2xy$	$(1, 1)$
15. $x^3 - 2x^2y + 3xy^2 = 38$	$(2, 3)$
16. $x^3 - y^3 = x - y$	$(1, 1)$

In Exercises 17–20, sketch the graph of the given equation. Then find dy/dx implicitly and explicitly and show that the two results are equivalent.

17. $x^2 + y^2 = 16$
18. $x^2 + y^2 - 4x + 6y + 9 = 0$
19. $9x^2 + 16y^2 = 144$
20. $4y^2 - x^2 = 4$

$f''(x) = y''$

In Exercises 21–26, find d^2y/dx^2 in terms of x and y.

21. $x^2 + xy = 5$ **22.** $x^2y^2 - 2x = 3$
23. $x^2 - y^2 = 16$ **24.** $1 - xy = x - y$
25. $y^2 = x^3$ **26.** $y^2 = 4x$

In Exercises 27 and 28, find equations for the tangent line and normal line to the given circle at the indicated points. (The **normal line** at a point is perpendicular to the tangent line at the point.)

27. $x^2 + y^2 = 25$, $(4, 3)$ and $(-3, 4)$
28. $x^2 + y^2 = 9$, $(0, 3)$ and $(2, \sqrt{5})$

In Exercises 29 and 30, find the points at which the graph of the given equation has a vertical or horizontal tangent line.

29. $25x^2 + 16y^2 + 200x - 160y + 400 = 0$
30. $4x^2 + y^2 - 8x + 4y + 4 = 0$

In Exercises 31–34, sketch the intersecting graphs of the given equations and show that they are **orthogonal.** (Two graphs are orthogonal if at their point(s) of intersection, their tangent lines are perpendicular to each other.)

31. $2x^2 + y^2 = 6$ and $y^2 = 4x$
32. $y^2 = x^3$ and $2x^2 + 3y^2 = 5$
33. $x + y = 0$ and $x^2 + y^2 = 4$
34. $x^3 = 3(y - 1)$ and $x(3y - 29) = 3$

35. Two circles of radius 4 are tangent to the graph of $y^2 = 4x$ at the point $(1, 2)$. Find equations for these two circles.

36. Show that the normal line (the line perpendicular to the tangent line) at any point on the circle $x^2 + y^2 = r^2$ passes through the origin.

37. Show that

$$\frac{dy}{dx} = -\frac{x - h}{y - k}$$

for the circle

$$(x - h)^2 + (y - k)^2 = r^2$$

Volume is related to
radius and height.

FIGURE 3.28

3.7
Related rates

We have seen how the Chain Rule can be used to find dy/dx implicitly. Another important use of the Chain Rule is to find the rates of change of two or more related variables that are changing with respect to time.

For example, when water is drained out of a conical tank (see Figure 3.28), the volume V, the radius r, and the height h of the water level are all functions of time t. Knowing these variables are related by the equation

$$V = \frac{\pi}{3} r^2 h$$

we can differentiate implicitly with respect to t to obtain the **related rate** equation

$$\frac{dV}{dt} = \frac{\pi}{3}\left[r^2 \frac{dh}{dt} + h\left(2r \frac{dr}{dt}\right)\right] = \frac{\pi}{3}\left[r^2 \frac{dh}{dt} + 2rh \frac{dr}{dt}\right]$$

From this equation we see that the rate of change of V is related to the rates of change of both h and r. This concept is further demonstrated in Example 1.

EXAMPLE 1   Two rates that are related

Suppose that x and y are both differentiable functions of t and are related by the equation $y = x^2 + 3$. Find dy/dt when $x = 1$, given that $dx/dt = 2$ when $x = 1$.

Solution: Using the Chain Rule, we can differentiate both sides of the equation *with respect to* t.

$$y = x^2 + 3 \qquad \text{Given equation}$$

$$\frac{d}{dt}[y] = \frac{d}{dt}[x^2 + 3] \qquad \text{Differentiate with respect to } t$$

$$\frac{dy}{dt} = 2x \frac{dx}{dt} \qquad \text{Chain Rule}$$

Now, when $x = 1$ and $dx/dt = 2$, we have

$$\frac{dy}{dt} = 2(1)(2) = 4$$

In Example 1, we are *given* the following mathematical model.

Given equation: $y = x^2 + 3$

Given rate: $\dfrac{dx}{dt} = 2$ when $x = 1$

Find: $\dfrac{dy}{dt}$ when $x = 1$

Now we look at an example in which we must *create* the model from a verbal description.

EXAMPLE 2   An application involving related rates

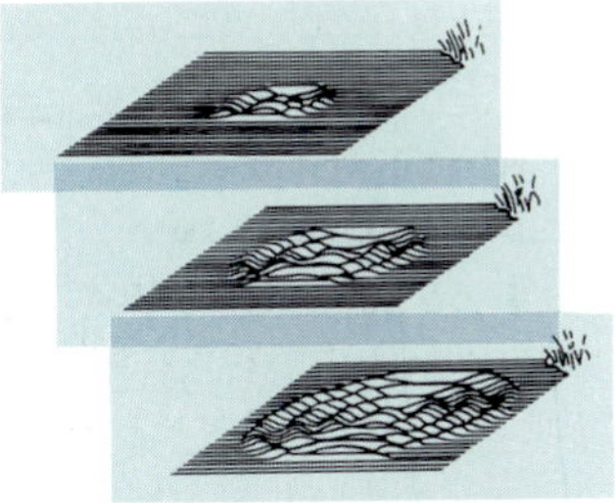

A pebble is dropped into a calm pond, causing ripples in the form of concentric circles, as shown in Figure 3.29. The radius r of the outer ripple is increasing at a constant rate of 1 foot per second. When this radius is 4 feet, at what rate is the total area A of the disturbed water increasing?

Solution: The variables r and A are related by the equation for the area of a circle, $A = \pi r^2$. To solve this problem, we must remember that the rate of change of the radius r is given by its derivative, dr/dt. Thus, the problem can be summarized by the following model.

Expanding ripples in pond

FIGURE 3.29

Given equation: $A = \pi r^2$

Given rate: $\dfrac{dr}{dt} = 1$ when $r = 4$

Find: $\dfrac{dA}{dt}$ when $r = 4$

With this information, we proceed as in Example 1.

$$\frac{d}{dt}[A] = \frac{d}{dt}[\pi r^2]$$

$$\frac{dA}{dt} = 2\pi r \frac{dr}{dt}$$

Thus, when $r = 4$, we have

$$\frac{dA}{dt} = 2\pi(4)(1) = 8\pi \ \text{ft}^2/\text{sec}$$

Following the procedure shown in Example 2, we suggest the following four solution steps for related rate problems.

PROCEDURE FOR SOLVING RELATED RATE PROBLEMS	1. Assign symbols to all *given* quantities and *quantities to be determined.* Make a sketch and label the quantities if feasible. 2. Write an equation involving the variables whose rates of change either are given or are to be determined. 3. Using the Chain Rule, implicitly differentiate both sides of the equation *with respect to time t.* 4. Substitute into the resulting equation all known values for the variables and their rates of change. Then solve for the required rate of change.

Table 3.6 shows the mathematical models for some common rates of change, which can be used in the first step of the solution to related rate problems.

TABLE 3.6

Verbal statement	Mathematical model
The velocity of a car after traveling one hour is 50 miles per hour.	x = distance traveled $\dfrac{dx}{dt} = 50$ when $t = 1$
Water is being pumped into a swimming pool at the rate of 10 cubic feet per minute.	V = volume of water in pool $\dfrac{dV}{dt} = 10$ ft^3/min
A gear is revolving at the rate of 25 revolutions per minute (1 rev = 2π rad).	θ = angle of revolution $\dfrac{d\theta}{dt} = 25(2\pi)$ rad/min

EXAMPLE 3 An inflating balloon

Air is being pumped into a spherical balloon at the rate of 4.5 cubic inches per minute, as shown in Figure 3.30. Find the rate of change of the radius when the radius is 2 inches.

Solution:

1. We let V be the volume of the balloon and let r be its radius. Since the volume is increasing at the rate of 4.5 in^3/min, we know that at time t the rate of change of the volume is $dV/dt = 9/2$. Thus, the problem can be stated as follows.

$$\text{Given:} \quad \frac{dV}{dt} = \frac{9}{2} \quad \text{(constant rate)}$$

$$\text{Find:} \quad \frac{dr}{dt} \quad \text{when} \quad r = 2$$

2. To find the rate of change of the radius, we must find an equation that relates the radius r to the volume V. This is given by the formula for the volume of a sphere.

$$V = \frac{4}{3}\pi r^3$$

3. Now we implicitly differentiate *with respect to t* to obtain

$$\frac{dV}{dt} = 4\pi r^2 \frac{dr}{dt}$$

$$\frac{dr}{dt} = \frac{1}{4\pi r^2}\left(\frac{dV}{dt}\right)$$

4. Finally, when $r = 2$, the rate of change of the radius is

$$\frac{dr}{dt} = \frac{1}{16\pi}\left(\frac{9}{2}\right) \approx 0.09 \text{ in/min}$$

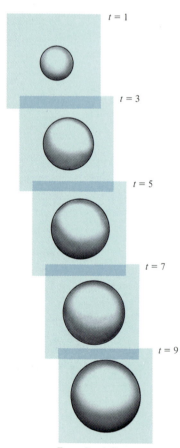

$t = 1$

$t = 3$

$t = 5$

$t = 7$

$t = 9$

Expanding balloon

FIGURE 3.30

In Example 3, note that the volume is increasing at a *constant* rate but the radius is increasing at a *variable* rate. Thus, when we say that two rates are related, we do not mean that they are proportional. In this particular case, the radius is growing more and more slowly as t increases. This is illustrated in Figure 3.30 and Table 3.7.

TABLE 3.7

t	1	3	5	7	9	11
V	4.5	13.5	22.5	31.5	40.5	49.5
r	1.02	1.48	1.75	1.96	2.13	2.28
dr/dt	0.34	0.16	0.12	0.09	0.08	0.07

EXAMPLE 4   The velocity of an airplane tracked by radar

An airplane is flying at an elevation of 6 miles on a flight path that is directly over a radar tracking station. Let s represent the distance (measured in miles) between the radar station and the plane. If s is decreasing at a rate of 400 miles per hour when s is 10 miles, what is the velocity of the plane?

Solution:

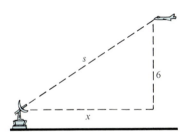

FIGURE 3.31

1. We label the distances x and s in Figure 3.31.

$$\text{Given:} \quad \frac{ds}{dt} = -400 \quad \text{when} \quad s = 10$$

$$\text{Find:} \quad \frac{dx}{dt} \quad \text{when} \quad s = 10$$

2. We can use the Pythagorean Theorem to form an equation relating s and x.

$$x^2 + 6^2 = s^2$$

3. Differentiating implicitly with respect to t, we have

$$2x\frac{dx}{dt} = 2s\frac{ds}{dt}$$

$$\frac{dx}{dt} = \frac{s}{x}\left(\frac{ds}{dt}\right)$$

4. To find dx/dt, we first must find x when $s = 10$:

$$x = \sqrt{s^2 - 36} = \sqrt{100 - 36} = \sqrt{64} = 8$$

Finally, when $s = 10$ we have

$$\frac{dx}{dt} = \frac{10}{8}(-400) = -500 \text{ mph}$$

Since the velocity is -500 miles per hour, the *speed* is 500 miles per hour.

EXAMPLE 5   *Gravel falling in a conical pile*

Gravel is falling in a conical pile at the rate of 100 cubic feet per minute. Find the rate of change of the height of the pile when the height is 10 feet. (Assume that the coarseness of the gravel is such that the radius of the cone is equal to its height.)

Solution:

1. We label the height and radius on the cone in Figure 3.32. Note that the given rate is a change in *volume*.

$$\text{Given:} \quad \frac{dV}{dt} = 100 \quad \text{(constant rate)}$$

$$\text{Find:} \quad \frac{dh}{dt} \quad \text{when} \quad h = 10$$

FIGURE 3.32

2. The volume of a cone is given by the formula

$$V = \frac{\pi r^2 h}{3}$$

and since we are given that $r = h$, this equation simplifies to

$$V = \frac{\pi}{3} h^3$$

3. Implicit differentiation with respect to t yields

$$\frac{dV}{dt} = \frac{\pi}{3} 3h^2 \frac{dh}{dt}$$

$$\frac{dh}{dt} = \frac{1}{\pi h^2}\left(\frac{dV}{dt}\right)$$

4. Finally, when $h = 10$ we have

$$\frac{dh}{dt} = \frac{1}{100\pi}(100) = \frac{1}{\pi} \approx 0.318 \text{ ft/min}$$

EXAMPLE 6   *Tracking an accelerating object*

A television camera at ground level is filming the lift-off of a space shuttle that is rising vertically according to the position equation $s = 50\,t^2$, where s is measured in feet and t is measured in seconds. The camera is 2000 feet from the launch pad. Find the rate of change in the distance between the camera and the base of the shuttle 10 seconds after lift-off. (Assume that the camera and the base of the shuttle are level with each other when $t = 0$.)

Solution:

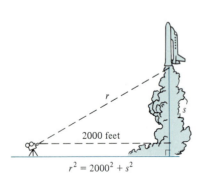

2000 feet

$r^2 = 2000^2 + s^2$

FIGURE 3.33

1. We let r be the distance between the camera and the base of the shuttle, as shown in Figure 3.33. Then we can find the velocity of the rocket by differentiating s with respect to t to obtain $ds/dt = 100t$. Thus, we have

Given: $\dfrac{ds}{dt} = 100t = $ velocity

Find: $\dfrac{dr}{dt}$ when $t = 10$

2. Using Figure 3.33, we relate s and r by the equation

$$r^2 = 2000^2 + s^2$$

3. Implicit differentiation with respect to t yields

$$2r\,\frac{dr}{dt} = 2s\,\frac{ds}{dt}$$

$$\frac{dr}{dt} = \frac{s}{r} \cdot \frac{ds}{dt} = \frac{s}{r}(10t)$$

4. Now, when $t = 10$ we know that $s = 50(10^2) = 5000$, and we have

$$r = \sqrt{2000^2 + 5000^2} = 1000\sqrt{29}$$

Finally, the rate of change of r when $t = 10$ is

$$\frac{dr}{dt} = \frac{5000}{1000\sqrt{29}}(100)(10) = 928.48 \text{ ft/sec}$$

Section Exercises 3.7

In Exercises 1–4, assume that x and y are both differentiable functions of t and find the indicated values of dy/dt and dx/dt.

Equation	Find	Given
1. $y = \sqrt{x}$	(a) $\dfrac{dy}{dt}$ when $x = 4$	$\dfrac{dx}{dt} = 3$
	(b) $\dfrac{dx}{dt}$ when $x = 25$	$\dfrac{dy}{dt} = 2$
2. $y = x^2 - 3x$	(a) $\dfrac{dy}{dt}$ when $x = 3$	$\dfrac{dx}{dt} = 2$
	(b) $\dfrac{dx}{dt}$ when $x = 1$	$\dfrac{dy}{dt} = 5$
3. $xy = 4$	(a) $\dfrac{dy}{dt}$ when $x = 8$	$\dfrac{dx}{dt} = 10$
	(b) $\dfrac{dx}{dt}$ when $x = 1$	$\dfrac{dy}{dt} = -6$
4. $x^2 + y^2 = 25$	(a) $\dfrac{dy}{dt}$ when $x = 3$, $y = 4$	$\dfrac{dx}{dt} = 8$
	(b) $\dfrac{dx}{dt}$ when $x = 4$, $y = 3$	$\dfrac{dy}{dt} = -2$

5. The radius r of a circle is increasing at a rate of 2 inches per minute. Find the rate of change of the area when (a) $r = 6$ inches and (b) $r = 24$ inches.

6. The radius r of a sphere is increasing at a rate of 2 inches per minute. Find the rate of change of the volume when (a) $r = 6$ inches and (b) $r = 24$ inches.

7. Let A be the area of a circle of radius r that is changing with respect to time.. If dr/dt is constant, is dA/dt constant? Explain why or why not.

8. Let V be the volume of a sphere of radius r that is changing with respect to time. If dr/dt is constant, is dV/dt constant? Explain why or why not.

9. A spherical balloon is inflated with gas at the rate of 20 cubic feet per minute. How fast is the radius of the balloon increasing at the instant the radius is (a) 1 foot and (b) 2 feet?

10. The formula for the volume of a cone is

$$V = \frac{1}{3}\pi r^2 h$$

Find the rate of change of the volume if dr/dt is 2 inches per minute and $h = 3r$ when (a) $r = 6$ inches and (b) $r = 24$ inches.

11. At a sand and gravel plant, sand is falling off a conveyor and onto a conical pile at the rate of 10 cubic feet per minute. The diameter of the base of the cone is approximately three times the altitude. At what rate is the height of the pile changing when it is 15 feet high?

12. A conical tank (with vertex down) is 10 feet across the top and 12 feet deep. If water is flowing into the tank at the rate of 10 cubic feet per minute, find the rate of change of the depth of the water the instant it is 8 feet deep.

13. All edges of a cube are expanding at the rate of 3 centimeters per second. How fast is the volume changing when each edge is (a) 1 centimeter and (b) 10 centimeters?

14. The conditions are the same as in Exercise 13. Now measure how fast the *surface* area is changing when each edge is (a) 1 centimeter and (b) 10 centimeters.

15. A point is moving along the graph of $y = x^2$ so that dx/dt is 2 centimeters per minute. Find dy/dt when (a) $x = 0$ and (b) $x = 3$.

16. The conditions are the same as in Exercise 15 but now measure the rate of change of the distance between the point and the origin.

17. A point is moving along the graph of $y = 1/(1 + x^2)$ so that $dx/dt = 2$ centimeters per minute. Find dy/dt when
   (a) $x = -2$ (b) $x = 0$
   (c) $x = 2$ (d) $x = 10$

18. A point is moving along the graph of $y = x^2$ so that $dx/dt = 2$ centimeters per minute. Find dy/dt when
   (a) $x = -2$ (b) $x = 2$
   (c) $x = 0$ (d) $x = 10$

19. A swimming pool is 40 feet long, 20 feet wide, 4 feet deep at the shallow end, and 9 feet deep at the deep end, as shown in Figure 3.34. Water is being pumped into the pool at 10 cubic feet per minute, and there is 4 feet of water at the deep end.
   (a) What percentage of the pool is filled?
   (b) At what rate is the water level rising?

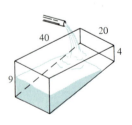

FIGURE 3.34

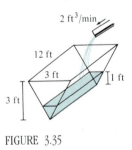

FIGURE 3.35

20. A trough is 12 feet long and 3 feet across the top, as shown in Figure 3.35. Its ends are isosceles triangles with an altitude of 3 feet. If water is being pumped into the trough at 2 cubic feet per minute, how fast is the water level rising when it is 1 foot deep?

21. A ladder 25 feet long is leaning against a house, as shown in Figure 3.36. The base of the ladder is pulled away from the house wall at a rate of 2 feet per second. How fast is the top moving down the wall when the base of the ladder is (a) 7 feet, (b) 15 feet, and (c) 24 feet from the wall?

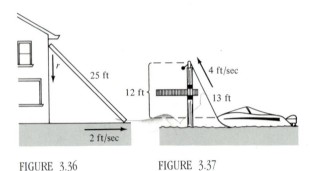

FIGURE 3.36   FIGURE 3.37

22. A boat is pulled in by means of a winch on the dock 12 feet above the deck of the boat as shown in Figure 3.37. The winch pulls in rope at the rate of 4 feet per second. Determine the speed of the boat when there is 13 feet of rope out. What happens to the speed of the boat as it gets closer to the dock?

23. An air traffic controller spots two planes at the same altitude converging on a point as they fly at right angles to one another, as shown in Figure 3.38. One plane is 150 miles from the point and is moving at 450 miles per hour. The other plane is 200 miles from the point and has a speed of 600 miles per hour.
   (a) At what rate is the distance between the planes decreasing?
   (b) How much time does the traffic controller have to get one of the planes on a different flight path?

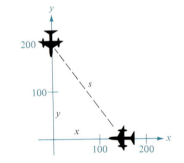

FIGURE 3.38

24. The point $(0, y)$ moves along the y-axis at a constant rate of R feet per second, while the point $(x, 0)$ moves along the x-axis at a constant rate of r feet per second. Find an expression for the rate of change of the distance between the two points.

25. A baseball diamond has the shape of a square with sides 90 feet long, as shown in Figure 3.39. A player 30 feet from third base is running at a speed of 28 feet per second. At what rate is the player's distance from home plate changing?

28. An airplane is flying at an altitude of 6 miles and passes directly over a radar antenna, as shown in Figure 3.41. When the plane is 10 miles away, the radar detects that the distance s is changing at a rate of 240 miles per hour. What is the speed of the plane?

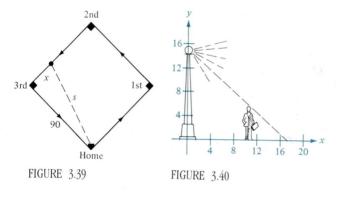

FIGURE 3.39   FIGURE 3.40

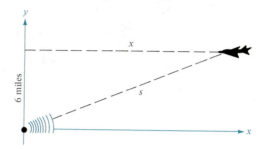

FIGURE 3.41

26. For the baseball diamond in Exercise 25, suppose the player is running from first to second at a speed of 28 feet per second. Find the rate at which the distance from home plate is changing when the player is 30 feet from second.

27. A man 6 feet tall walks at a rate of 5 feet per second away from a light that is 15 feet above the ground, as shown in Figure 3.40. When he is 10 feet from the base of the light,
(a) at what rate is the tip of his shadow moving?
(b) at what rate is the length of his shadow changing?

29. As a spherical raindrop falls, it reaches a layer of dry air and begins to evaporate at a rate that is proportional to its surface area ($S = 4\pi r^2$). Show that the radius decreases at a constant rate.

30. The combined electrical resistance R of R_1 and R_2, connected in parallel, is given by

$$\frac{1}{R} = \frac{1}{R_1} + \frac{1}{R_2}$$

where R, R_1, and R_2 are measured in ohms. R_1 and R_2 are increasing at rates of 1 and 1.5 ohms per second, respectively. At what rate is R changing when $R_1 = 50$ ohms and $R_2 = 75$ ohms?

Review Exercises for Chapter 3

In Exercises 1–24, find the derivative of the given function.

1. $f(x) = x^3 - 3x^2$

2. $f(x) = \dfrac{2x^3 - 1}{x^2}$

3. $f(x) = x^{1/2} - x^{-1/2}$

4. $f(x) = \dfrac{x + 1}{x - 1}$

5. $g(t) = \dfrac{2}{3t^2}$

6. $h(x) = \dfrac{2}{(3x)^2}$

7. $f(x) = \sqrt{x^3 + 1}$

8. $f(x) = \sqrt[3]{x^2 - 1}$

9. $f(x) = (3x^2 + 7)(x^2 - 2x + 3)$

10. $f(x) = \left(x^2 + \dfrac{1}{x}\right)^5$ $5(x^2+\frac{1}{x})^4 \ (2x-\frac{1}{x^2})$

11. $f(s) = (s^2 - 1)^{5/2}(s^3 + 5)^{5/3}$

12. $h(\theta) = \dfrac{\theta}{(1 - \theta)^3}$ $\dfrac{2\theta+1}{(1-\theta)^4}$

13. $g(x) = \sqrt{x}\sqrt{x^2 + 1}$

14. $f(x) = \dfrac{6x - 5}{x^2 + 1}$ $\dfrac{2(3+5x-3x^2)}{(x^2+1)^2}$

15. $f(x) = \dfrac{x^2 + x - 1}{x^2 - 1}$

16. $f(t) = t^2(t - 1)^5(t + 2)^3$ $t(t-1)^4 (t+2)^2 (10t^2+9t-4)$

17. $f(x) = -2(1 - 4x^2)^2$

18. $f(x) = [(x - 2)(x + 4)]^2$

19. $f(x) = \dfrac{1}{4 - 3x^2}$

20. $f(x) = \dfrac{9}{3x^2 - 2x}$

21. $g(x) = \dfrac{2x}{\sqrt{x + 1}}$

22. $f(x) = \dfrac{-2x^2}{x - 1}$

23. $f(t) = \sqrt{t + 1}\sqrt[3]{t + 1}$

24. $y = \sqrt{3x}(x + 2)^3$

In Exercises 25–30, find the second derivative of the given function.

25. $f(x) = \sqrt{x^2 + 9}$

26. $h(x) = x\sqrt{x^2 - 1}$

27. $f(t) = \dfrac{t}{(1 - t)^2}$

28. $h(x) = x^2 + \dfrac{3}{x}$

29. $g(x) = \dfrac{6x - 5}{x^2 + 1}$

30. $f(x) = (3x^2 + 7)(x^2 - 2x + 3)$

In Exercises 31–36, use implicit differentiation to find dy/dx.

31. $x^2 + 3xy + y^3 = 10$

32. $x^2 + 9y^2 - 4x + 3y - 7 = 0$

33. $y\sqrt{x} - x\sqrt{y} = 16$

34. $y^2 + x^2 - 6y - 2x - 5 = 0$

35. $y^2 - x^2 = 25$ **36.** $y^2 = (x - y)(x^2 + y)$

In Exercises 37–42, find the equation of the tangent line and the normal line to the graph of the given equation at the indicated point.

Equation	*Point*
37. $y = (x + 3)^3$	$(-2, 1)$
38. $y = (x - 2)^2$	$(2, 0)$
39. $x^2 + y^2 = 20$	$(2, 4)$
40. $x^2 - y^2 = 16$	$(5, 3)$
41. $y = \sqrt[3]{(x - 2)^2}$	$(3, 1)$
42. $y = \dfrac{2x}{1 - x^2}$	$(0, 0)$

43. Find the points on the graph of

$$f(x) = \frac{1}{3}x^3 + x^2 - x - 1$$

at which the slope is
   (a) -1 (b) 2 (c) 0

44. Find the points on the graph of

$$f(x) = x^2 + 1$$

at which the slope is
   (a) -1 (b) 0 (c) 1

In Exercises 45–48, find the derivative of the given function by the Four-Step Process.

45. $f(x) = \dfrac{1}{x^2}$ **46.** $f(x) = \dfrac{x + 1}{x - 1}$

47. $f(x) = \sqrt{x + 2}$ **48.** $f(x) = \dfrac{1}{\sqrt{x}}$

49. Derive the equations for the velocity and acceleration of a particle whose position function is

$$s(t) = t + \frac{1}{t + 1}$$

50. Derive the equations for the velocity and acceleration of a particle whose position equation is

$$s(t) = \frac{1}{t^2 + 2t + 1}$$

51. Suppose that the temperature T of food placed in a freezer drops according to the equation

$$T = \frac{700}{t^2 + 4t + 10}$$

where t is the time in hours. Find the rate of change of T with respect to t when
   (a) $t = 1$ (b) $t = 3$ (c) $t = 5$ (d) $t = 10$

52. The emergent velocity v of a liquid flowing from a hole in the bottom of a tank is given by $v = \sqrt{2gh}$, where g is the acceleration due to gravity (32 ft/s^2) and h is the depth of the liquid in the tank. Find the rate of change of v with respect to h when
   (a) $h = 9$ (b) $h = 4$

53. What is the smallest initial velocity that is required to throw a stone to the top of a 49-foot silo?

54. A bomb is dropped from a plane at an altitude of 14,400 feet. How long will it take to reach the ground? (Even though it will not be a vertical fall because of the motion of the plane, the time will be the same as that for a vertical fall.) The plane is moving at 600 miles per hour. How far will the bomb move horizontally after it is released from the plane?

55. A ball is thrown and follows a path described by $y = x - 0.02x^2$.
   (a) Sketch a graph of the path.
   (b) Find the total horizontal distance the ball was thrown.
   (c) For what x-value does the ball reach its maximum height? (Use the symmetry of the path.)
   (d) Find the equation that gives the instantaneous rate of change in the height of the ball with respect to the horizontal change and evaluate this equation at $x = 0, 10, 25, 30, 50$.
   (e) What is the instantaneous rate of change of the height when the ball reaches its maximum height?

56. The path of a projectile thrown at an angle of 45° with level ground is given by

$$y = x - \frac{32}{v_0^2}(x^2)$$

where the initial velocity is v_0 feet per second.
   (a) Sketch the path followed by the projectile.
   (b) Find the x-coordinate of the point where the projectile strikes the ground. Use the symmetry of the path of the projectile to locate the abscissa of the point where the projectile reaches its maximum height.
   (c) What is the instantaneous rate of change of the height when the projectile is at its maximum height?

57. The path of a projectile, thrown at an angle of $45°$ with the ground, is given by

$$y = x - \frac{32}{v_0^2}(x^2)$$

where the initial velocity is v_0 feet per second. Show that doubling the initial velocity of the projectile multiplies both the maximum height and the range by a factor of 4.

58. Use the equation given in Exercise 57 to find the maximum height and range of a projectile thrown with an initial velocity of 70 feet per second.

59. A point moves along the curve $y = \sqrt{x}$ in such a way that the y-value is increasing at the rate of 2 units per second. At what rate is x changing for the following values?

(a) $x = \frac{1}{2}$ (b) $x = 1$ (c) $x = 4$

60. The same conditions exist as in Exercise 59. Find the rate the distance between (x, y) and the origin is changing for the following:

(a) $x = \frac{1}{2}$ $\frac{dL}{dt} = 8/\sqrt{6}$ (b) $x = 1$ $3\sqrt{2}$ (c) $x = 4$ $18/\sqrt{5}$

61. The cross section of a 5-foot trough is an isosceles trapezoid with a 2-foot lower base, a 3-foot upper base, and an altitude of 2 feet. Water is running into the trough at the rate of 1 cubic foot per minute. How fast is the water level rising when the water is 1 foot deep?

62. The pressure p and the volume V of an ideal gas are related by the adiabatic compression formula $pV^k = C$. The constant C is determined by an observation that $p = p_1$ when $V = V_1$, and k is a constant determined by the heat capacity of the gas. If V is changing at the rate

$$\frac{dp}{dt} = -\frac{Kp_1}{V_1}$$

of 1 cubic centimeter per second, at what rate is p changing?

63. A population of 500 bacteria is introduced into a culture and grows in number according to the equation

$$P(t) = 500\left(1 + \frac{4t}{50 + t^2}\right)$$

where t is measured in hours. Find the rate at which the population is growing when $t = 2$.

64. The geometric mean of x and $x + n$ is

$$g = \sqrt{x(x + n)}$$

and the arithmetic mean is

$$a = \frac{x + (x + n)}{2}$$

Show that $dg/dx = a/g$.

65. For $y = 1/x$, show that

$$y^{(n)} = \frac{(-1)^n n!}{x^{n+1}}$$

[Note: $n! = n(n - 1)(n - 2)\cdots3 \cdot 2 \cdot 1.$]

66. Sketch the graph of $f(x) = 4 - |x - 2|$.
   (a) Is f continuous at $x = 2$?
   (b) Is f differentiable at $x = 2$? Why or why not?

67. Sketch the graph of

$$f(x) = \begin{cases} x^2 + 4x + 2, & x < -2 \\ 1 - 4x - x^2, & x \geq -2 \end{cases}$$

   (a) Is f continuous at $x = -2$?
   (b) Is f differentiable at $x = -2$? Why or why not?

4

Applications of differentiation

4.1
Extrema on an interval

In calculus a great deal of effort is devoted to determining the behavior of a function on an interval. For instance, we are interested in the following questions: Does it have a maximum value, a minimum value? Where is the function increasing? Where is it decreasing? In this chapter we will show how the derivative can be used to answer these questions.

We begin by looking at the minimum and maximum values of a function on an interval.

DEFINITION OF EXTREMA

Let f be defined on an interval I containing c.

1. $f(c)$ is the **minimum of f on I** if $f(c) \leq f(x)$ for all x in I.
2. $f(c)$ is the **maximum of f on I** if $f(c) \geq f(x)$ for all x in I.

The minimum and maximum of a function on an interval are called the **extreme values,** or **extrema,** of the function on the interval.

| **Remark** The minimum and maximum of a function on an interval are sometimes called the **absolute minimum** and **absolute maximum,** respectively.

It may be surprising to learn that a function need not have a minimum and a maximum on an interval. In fact, it may lack both. Figure 4.1 shows several possibilities regarding the existence of extreme values on an interval. By comparing the second graph to the first, we see that a maximum is lost by changing from the closed interval $[-1, 2]$ to the open interval $(-1, 2)$. Moreover, in the third and fourth graphs we see that a discontinuity (at $x = 0$) can affect the existence of an extremum on a closed or open interval. This suggests the following theorem, which identifies conditions that guarantee the existence of both a minimum and a maximum of a function on an interval.

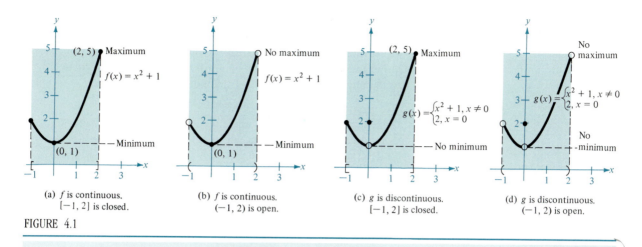

FIGURE 4.1

THEOREM 4.1   THE EXTREME VALUE THEOREM
If f is continuous on a closed interval, then f has both a minimum and a maximum in the interval.

Remark Note that the Extreme Value Theorem (like the Intermediate Value Theorem) is an *existence theorem*. That is, it tells of the existence of minimum and maximum values but does not give a way to find these values.

Relative extrema

From Figure 4.1 we can see that extrema can occur at interior points or endpoints of an interval. Extrema that occur at the endpoints are called **endpoint extrema,** and those that occur at interior points are called *relative extrema.*

EXAMPLE 1   *Extrema on an open interval*

Use the graph of $f(x) = (x^3 - 3x^2)/2$ in Figure 4.2 to locate the extrema on the open intervals
(a) $(-2, 4)$ (b) $(-2, 3)$ (c) $(-1, 3)$ (d) $(-1, 4)$

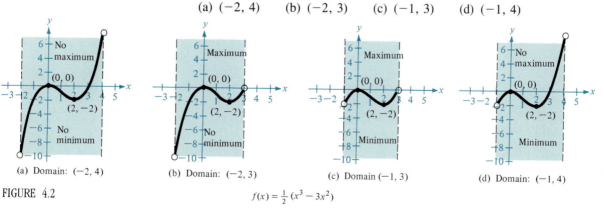

FIGURE 4.2 $f(x) = \frac{1}{2}(x^3 - 3x^2)$

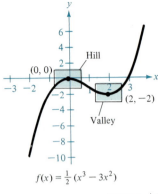

$$f(x) = \tfrac{1}{2}(x^3 - 3x^2)$$

FIGURE 4.3

Solution:

(a) On the interval $(-2, 4)$, f has neither a minimum nor a maximum.

(b) On the interval $(-2, 3)$, f has a maximum but no minimum.

(c) On the interval $(-1, 3)$, f has both a minimum and a maximum.

(d) On the interval $(-1, 4)$, f has a minimum but no maximum.

In Example 1 we see that the function f has a maximum at $(0, 0)$ relative to some open intervals but not to others. Similarly, f has a minimum at $(2, -2)$ relative to some open intervals but not to others. Moreover, it appears in Figure 4.3 that the way to force these two points to yield extreme values is to choose intervals that are small enough to produce a local hill or valley on the graph. This is the essence of the following definition.

DEFINITION OF
RELATIVE EXTREMA

1. If there is an open interval in which $f(c)$ is a maximum, then $f(c)$ is called a **relative maximum** of f.
2. If there is an open interval in which $f(c)$ is a minimum, then $f(c)$ is called a **relative minimum** of f.

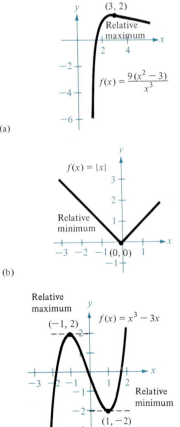

FIGURE 4.4

| Remark In Figure 4.3, note that a relative maximum occurs at the point $(0, 0)$ and a relative minimum occurs at the point $(2, -2)$.

EXAMPLE 2   A property of relative extrema

Find the value of the derivative at each of the relative extrema shown in Figure 4.4.

Solution:

(a) The derivative of this function is

$$f'(x) = \frac{x^3(18x) - (9)(x^2 - 3)(3x^2)}{(x^3)^2} = \frac{9(9 - x^2)}{x^4}$$

Thus, at the point $(3, 2)$, the value of the derivative is

$$f'(3) = \frac{9(9 - 3^2)}{3^4} = 0$$

(b) At $x = 0$, the derivative of $f(x) = |x|$ *does not exist* since the following one-sided limits differ:

$$\lim_{x \to 0^-} \frac{f(x) - f(0)}{x - 0} = \lim_{x \to 0^-} \frac{|x|}{x} = -1 \qquad \text{Limit from the left}$$

$$\lim_{x \to 0^+} \frac{f(x) - f(0)}{x - 0} = \lim_{x \to 0^+} \frac{|x|}{x} = +1 \qquad \text{Limit from the right}$$

(c) Since $f'(x) = 3x^2 - 3$, we have

$$f'(-1) = 3(-1)^2 - 3 = 0 \qquad \text{and} \qquad f'(1) = 3(1)^2 - 3 = 0$$

Note in Example 2 that the relative extrema occur when the derivative is either zero or undefined. We call the *x*-values at these special points **critical numbers.**

DEFINITION OF CRITICAL NUMBER	If *f* is defined at *c*, then *c* is called a **critical number** of *f* if $f'(c) = 0$ or if f' is undefined at *c*.

Remark Figure 4.5 illustrates the two types of critical numbers given in this definition.

THEOREM 4.2	**RELATIVE EXTREMA OCCUR ONLY AT CRITICAL NUMBERS** If *f* has a relative minimum or relative maximum at $x = c$, then *c* is a critical number of *f*.

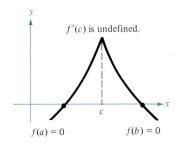

f'(*c*) is undefined.

f(*a*) = 0   *f*(*b*) = 0

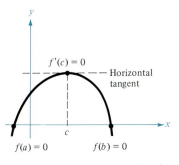

f'(*c*) = 0 — Horizontal tangent

f(*a*) = 0   *f*(*b*) = 0

FIGURE 4.5

Proof:

  Case 1: If *f* is *not* differentiable at $x = c$, then by definition *c* is a critical number of *f* and we have nothing to prove.

  Case 2: If *f* is* differentiable at $x = c$, then $f'(c)$ must be either positive, negative, or zero. Suppose $f'(c)$ is positive, Then we have

$$f'(c) = \lim_{x \to c} \frac{f(x) - f(c)}{x - c} > 0$$

But this implies that there exists an interval (a, b) containing *c* such that

$$\frac{f(x) - f(c)}{x - c} > 0, \quad \text{for all } x \neq c \text{ in } (a, b)$$

Since this quotient is positive, the signs of the denominator and numerator must agree; this produces the following inequalities for *x*-values in the interval (a, b).

Left of *c*: $x < c$ and $f(x) < f(c)$ ⟹ $f(c)$ is not a relative minimum
Right of *c*: $x > c$ and $f(x) > f(c)$ ⟹ $f(c)$ is not a relative maximum

Thus, our assumption that $f'(c) > 0$ contradicts the fact that $f(c)$ is a relative extremum. Assuming that $f'(c) < 0$ produces a similar contradiction, we are left with only one possibility—$f'(c) = 0$.

Theorem 4.2 tells us that the relative extrema of a function can occur *only* at the critical numbers of the function. By combining this observation with our knowledge of endpoint extrema, we obtain the following guidelines for finding extrema on a closed interval.

GUIDELINES FOR FINDING EXTREMA ON A CLOSED INTERVAL	To find the extrema of a continuous function f on a closed interval $[a, b]$, we suggest the following steps.

1. Evaluate f at each of its critical numbers in (a, b).
2. Evaluate f at each endpoint of $[a, b]$.
3. The least of these values is the minimum, and the greatest is the maximum.

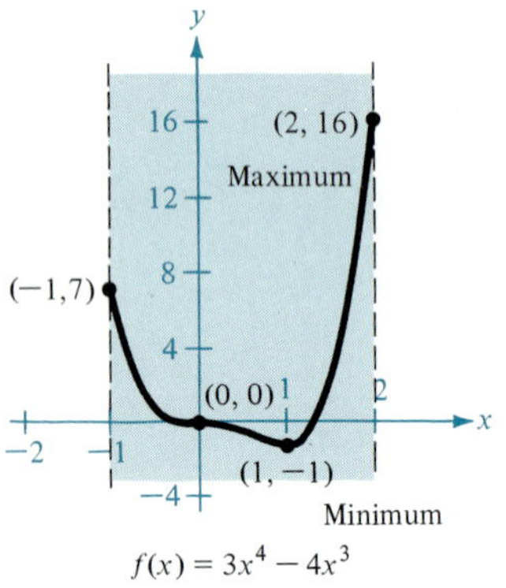

$f(x) = 3x^4 - 4x^3$

FIGURE 4.6

EXAMPLE 3   *Finding extrema on a closed interval*

Find the extrema of $f(x) = 3x^4 - 4x^3$ on the interval $[-1, 2]$.

Solution: To find the critical numbers, we differentiate to obtain

$$f'(x) = 12x^3 - 12x^2 = 0 \qquad \text{Set } f'(x) = 0$$
$$12x^2(x - 1) = 0 \qquad \text{Factor}$$
$$x = 0, 1 \qquad \text{Critical numbers}$$

Since f' is defined for all x, we conclude that these are the only critical numbers of f. Finally, by evaluating f at these two critical numbers and at the endpoints of $[-1, 2]$, we determine that the maximum is $f(2) = 16$ and the minimum is $f(1) = -1$, as indicated in Table 4.1. The graph of f is shown in Figure 4.6.

TABLE 4.1

Endpoint	Critical number	Critical number	Endpoint
$f(-1) = 7$	$f(0) = 0$	$f(1) = -1$ Minimum	$f(2) = 16$ Maximum

Remark   It is important that you realize in Figure 4.6 that the critical number $x = 0$ does not yield a relative minimum or maximum. This tells us that the converse of Theorem 4.2 is not true. In other words, the critical numbers of a function need not produce relative extrema.

EXAMPLE 4   *Finding extrema on a closed interval*

Find the extrema of $f(x) = 2x - 3x^{2/3}$ on the interval $[-1, 3]$.

Solution: Differentiating, we have

$$f'(x) = 2 - \frac{2}{x^{1/3}} = 2\left(\frac{x^{1/3} - 1}{x^{1/3}}\right)$$

This gives us the following critical numbers:

$$x = 1 \qquad f'(1) = 0$$
$$x = 0 \qquad f' \text{ is undefined}$$

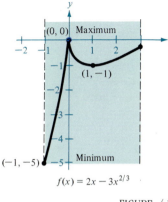

$f(x) = 2x - 3x^{2/3}$

FIGURE 4.7

Finally, by evaluating f at these two points and at the endpoints of the interval, we conclude that the minimum is $f(-1) = -5$ and the maximum is $f(0) = 0$, as indicated in Table 4.2. The graph of f is shown in Figure 4.7.

TABLE 4.2

Endpoint	Critical number	Critical number	Endpoint
$f(-1) = -5$ Minimum	$f(0) = 0$ Maximum	$f(1) = -1$	$f(3) = 6 - 3\sqrt[3]{9} \approx -0.24$

Section Exercises 4.1

In Exercises 1–4, determine from the graph whether or not f possesses a minimum in the interval (a, b).

1. (a)   (b)

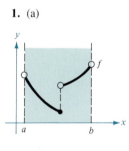

2. (a)   (b)

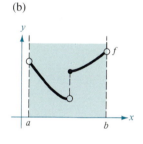

3. (a)   (b)

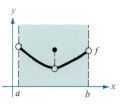

4. (a)   (b)

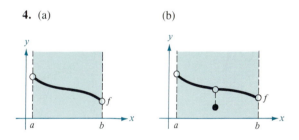

In Exercises 5–10, locate the extrema of the function (if any exist) over the indicated interval.

5. $f(x) = 2x - 3$
    (a) $[0, 2]$
    (b) $[0, 2)$
    (c) $(0, 2]$
    (d) $(0, 2)$

6. $f(x) = 5 - x$
    (a) $[1, 4]$
    (b) $[1, 4)$
    (c) $(1, 4]$
    (d) $(1, 4)$

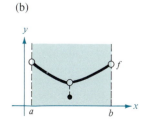

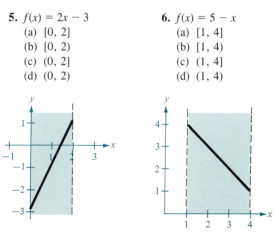

7. $f(x) = x^2 - 2x$
  (a) $[-1, 2]$
  (b) $(1, 3]$
  (c) $(0, 2)$
  (d) $[1, 3]$

8. $f(x) = \sqrt{4 - x^2}$
  (a) $[-2, 2]$
  (b) $[-2, 0)$
  (c) $(-2, 2)$
  (d) $[1, 2)$

9. $f(x) = \dfrac{1}{2}(x - 1)^2(4 - x)$
  (a) $(-1, 5)$
  (b) $(-1, 4)$
  (c) $(0, 4)$
  (d) $(0, 5)$

10. $f(x) = \dfrac{4}{x^2 + 1}$
  (a) $[-1, 2]$
  (b) $(-1, 2)$
  (c) $[0, 2]$
  (d) $(0, 2)$

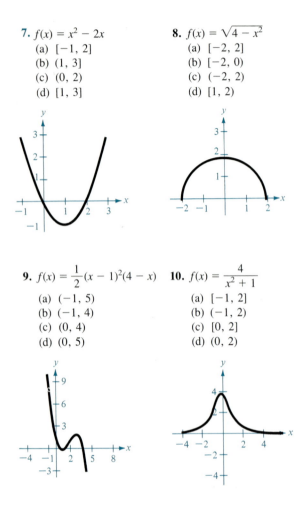

In Exercises 11–16, find the value of the derivative (if it exists) at each of the indicated critical points.

11. $f(x) = \dfrac{x^2}{x^2 + 4}$

12. $f(x) = -x^2 + 4x$

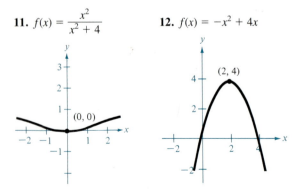

13. $f(x) = x + \dfrac{32}{x^2}$

14. $f(x) = -3x\sqrt{x + 1}$

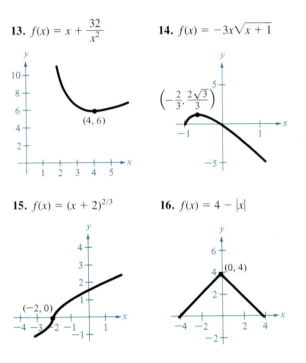

15. $f(x) = (x + 2)^{2/3}$

16. $f(x) = 4 - |x|$

In Exercises 17–26, locate the extrema of the given function on the indicated interval.

Function	Interval
17. $f(x) = 2(3 - x)$	$[-1, 2]$
18. $f(x) = \dfrac{2x + 5}{3}$	$[0, 5]$
19. $f(x) = -x^2 + 3x$	$[0, 3]$
20. $f(x) = x^2 + 2x - 4$	$[-1, 1]$
21. $f(x) = x^3 - 3x^2$	$[-1, 3]$
22. $f(x) = x^3 - 12x$	$[0, 4]$
23. $f(x) = 3x^{2/3} - 2x$	$[-1, 1]$
24. $g(t) = \dfrac{t^2}{t^2 + 3}$	$[-1, 1]$
25. $h(s) = \dfrac{1}{s - 2}$	$[0, 1]$
26. $h(t) = \dfrac{t}{t - 2}$	$[3, 5]$

27. Explain why the function given by $f(x) = 1/x^2$ has a maximum on $[1, 2]$, but not on $[0, 2]$.

28. Explain why the function given by $y = 1/(x + 1)$ has a minimum on $[0, 2]$, but not on $[-2, 0]$.

The error estimate for the Trapezoidal Rule (given in Section 5.7) involves the maximum of the absolute value of the second derivative in an interval. In Exercises 29–32, find the maximum value of $|f''(x)|$ in the indicated interval.

Function	Interval
29. $f(x) = \dfrac{1}{x^2 + 1}$	$[0, 3]$
30. $f(x) = \dfrac{1}{x^2 + 1}$	$\left[\dfrac{1}{2}, 3\right]$
31. $f(x) = \sqrt{1 + x^3}$	$[0, 2]$
32. $f(x) = x^3(3x^2 - 10)$	$[0, 1]$

The error estimate for Simpson's Rule (given in Section 5.7) involves the maximum of the absolute value of the fourth derivative in an interval. In Exercises 33–36, find the maximum value of $|f^{(4)}(x)|$ in the indicated interval.

Function	Interval
33. $f(x) = 15x^4 - \left(\dfrac{2x - 1}{2}\right)^6$	$[0, 1]$
34. $f(x) = x^5 - 5x^4 + 20x^3 + 600$	$\left[0, \dfrac{3}{2}\right]$

35. $f(x) = (x + 1)^{2/3}$ $[0, 2]$

36. $f(x) = \dfrac{1}{x^2 + 1}$ $[-1, 1]$

37. The formula for the output P of a battery is given by

$$P = VI - RI^2$$

where V is the voltage, R is the resistance, and I is the current. Find the current (measured in amperes) that corresponds to a maximum value of P in a battery for which $V = 12$ volts and $R = 0.5$ ohms. (Assume that a 15-ampere fuse bounds the output in the interval $0 \le I \le 15$.)

38. A retailer has determined that the cost C for ordering and storing x units of a certain product is

$$C = 2x + \frac{300{,}000}{x}, \quad 0 < x \le 300$$

Find the order size that will minimize the cost if the delivery truck can bring a maximum of 300 units per order.

SECTION TOPICS ▪
Rolle's Theorem ▪
Mean Value Theorem ▪
Applications ▪

4.2
The Mean Value Theorem

The Extreme Value Theorem tells us that a continuous function on a closed interval must have both a minimum and maximum on the interval. However, both of these values can occur at the endpoints. We now present a theorem called Rolle's Theorem, named after the French mathematician Michel Rolle (1652–1719), that implies the existence of extrema in the interior of a closed interval.

THEOREM 4.3   **ROLLE'S THEOREM**
Let f be continuous on $[a, b]$ and differentiable on (a, b). If

$$f(a) = f(b)$$

then there is at least one number c in (a, b) such that $f'(c) = 0$.

Proof: Let $f(a) = d = f(b)$.

Case 1: If $f(x) = d$ for all x in $[a, b]$, then f is constant on the interval, and by Theorem 3.3 $f'(x) = 0$ for all x in (a, b).

Case 2: If $f(x) > d$ for some x in (a, b), then by the Extreme Value Theorem we know that f has a maximum at some c in the interval. Moreover,

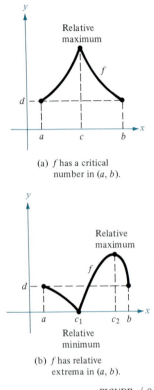

(a) f has a critical
number in (a, b).

(b) f has relative
extrema in (a, b).

FIGURE 4.8

since $f(c) > d$, this maximum does not occur at either endpoint. Therefore, f has a maximum in the *open* interval (a, b). This implies that $f(c)$ is a *relative* maximum, and by Theorem 4.2 we know c is a critical number of f. Finally, since f is differentiable at c, we can conclude that $f'(c) = 0$.

Case 3: If $f(x) < d$ for some x in (a, b), then we can use an argument similar to that in Case 2.

If we drop the differentiability requirement from Rolle's Theorem, then f will still have a critical number in (a, b) but it need not yield a horizontal tangent. This is shown in Figure 4.8 and stated in the following two corollaries to Rolle's Theorem.

Corollary 1 to Rolle's Theorem: Let f be continuous on $[a, b]$. If $f(a) = f(b)$, then f has a critical number in (a, b).

Corollary 2 to Rolle's Theorem: Let f be continuous on $[a, b]$ such that $f(a) = d = f(b)$.

1. If $f(x) > d$ for some x in (a, b), then f has a relative maximum in (a, b).
2. If $f(x) < d$ for some x in (a, b), then f has a relative minimum in (a, b).

EXAMPLE 1   *An example of Rolle's Theorem*

Find the two x-intercepts of $f(x) = x^2 - 3x + 2$ and show that $f'(x) = 0$ at some point between the two intercepts.

Solution: Note that f is differentiable on the entire real line. Setting $f(x)$ equal to zero, we have

$$x^2 - 3x + 2 = 0$$
$$(x - 1)(x - 2) = 0$$

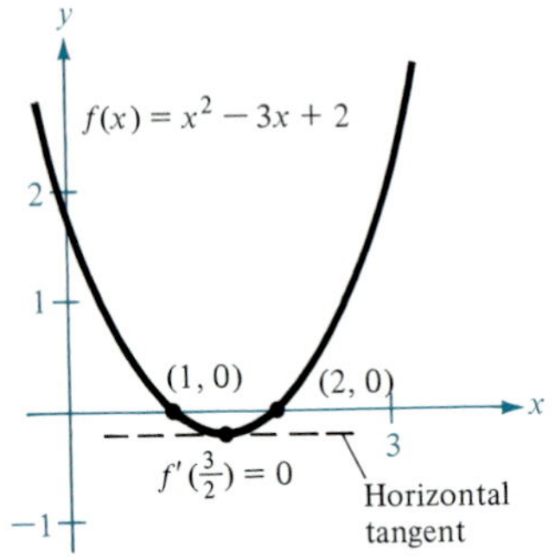

Thus $f(1) = f(2) = 0$, and from Rolle's Theorem we know that there exists c in the interval $(1, 2)$ such that $f'(c) = 0$. To find c we solve the equation $f'(x) = 0$ as follows:

$$f'(x) = 2x - 3 = 0$$
$$2x = 3$$
$$x = \frac{3}{2}$$

Since $\frac{3}{2}$ is in the interval $(1, 2)$ and $f'(\frac{3}{2}) = 0$, we let $c = \frac{3}{2}$, as shown in Figure 4.9.

FIGURE 4.9

Rolle's Theorem states that if f satisfies the conditions of the theorem, then there must be *at least* one point between a and b at which the derivative is zero. There may of course be more than one such point, as illustrated in the next example.

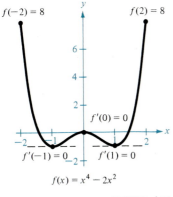

$f(-2) = 8$ $f(2) = 8$

$f'(0) = 0$

$f'(-1) = 0$ $f'(1) = 0$

$f(x) = x^4 - 2x^2$

FIGURE 4.10

EXAMPLE 2   An example of Rolle's Theorem

Let $f(x) = x^4 - 2x^2$. Find all c in the interval $(-2, 2)$ such that $f'(c) = 0$.

Solution: Since $f(-2) = 8 = f(2)$ and f is differentiable, Rolle's Theorem guarantees the existence of at least one c in $(-2, 2)$ such that $f'(c) = 0$. Setting the derivative equal to zero, we have

$$f'(x) = 4x^3 - 4x = 0$$
$$4x(x^2 - 1) = 0$$
$$x = 0, 1, -1$$

Thus, in the interval $(-2, 2)$ the derivative is zero at each of these three points, as shown in Figure 4.10.

The Mean Value Theorem

Rolle's Theorem can be used to prove one of the most important theorems in calculus—the **Mean Value Theorem.** It can be viewed as a generalization of Rolle's Theorem [in which $f(a) \neq f(b)$].

THEOREM 4.4   **THE MEAN VALUE THEOREM**
If f is continuous on $[a, b]$ and differentiable on (a, b), then there exists a number c in (a, b) such that

$$f'(c) = \frac{f(b) - f(a)}{b - a}$$

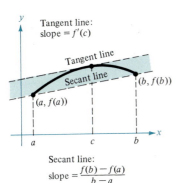

Tangent line:
slope $= f'(c)$

Tangent line

Secant line

$(b, f(b))$

$(a, f(a))$

a c b

Secant line:
slope $= \dfrac{f(b) - f(a)}{b - a}$

FIGURE 4.11

Proof: Referring to Figure 4.11, the equation of the secant line containing the points $(a, f(a))$ and $(b, f(b))$ is given by

$$y = \left[\frac{f(b) - f(a)}{b - a} \right](x - a) + f(a)$$

Let $g(x)$ be the difference between $f(x)$ and y. Then

$$g(x) = f(x) - y$$
$$= f(x) - \left[\frac{f(b) - f(a)}{b - a} \right](x - a) - f(a)$$

Now, by evaluating g at a and b, we see that

$$g(a) = 0 = g(b)$$

Furthermore, since f is differentiable, g is also differentiable, and we can apply Rolle's Theorem to the function g. Thus, there exists a point c in (a, b) such that $g'(c) = 0$. This means that

$$0 = g'(c) = f'(c) - \frac{f(b) - f(a)}{b - a}$$

Therefore, there exists a point c in (a, b) such that

$$f'(c) = \frac{f(b) - f(a)}{b - a}$$

Joseph-Louis Lagrange

Remark The "mean" in the Mean Value Theorem refers to the mean (or average) rate of change of f in the interval $[a, b]$.

Although the Mean Value Theorem can be used directly in problem solving, it is more often used in proving other theorems. It was proved by a famous French mathematician Joseph-Louis Lagrange (1736–1813), and it is closely related to the Fundamental Theorem of Calculus discussed in Chapter 5. For now, you can get an idea of the versatility of this theorem by looking at the results stated in Exercises 27–31 in this section.

The Mean Value Theorem has implications for both of our basic interpretations of the derivative. Geometrically, the theorem guarantees the existence of a tangent line that is parallel to the secant line through the points $(a, f(a))$ and $(b, f(b))$, as shown in Figure 4.11. The next example illustrates this geometrical interpretation of the Mean Value Theorem.

EXAMPLE 3   An example of the Mean Value Theorem

Given $f(x) = 5 - (4/x)$, find all c in the interval $(1, 4)$ such that

$$f'(c) = \frac{f(4) - f(1)}{4 - 1}$$

Solution: The slope of the secant line through $(1, f(1))$ and $(4, f(4))$ is

$$\frac{f(4) - f(1)}{4 - 1} = \frac{4 - 1}{4 - 1} = 1$$

Since f satisfies the conditions of the Mean Value Theorem, there exists at least one c in $(1, 4)$ such that $f'(c) = 1$. Solving the equation $f'(x) = 1$, we obtain

$$f'(x) = \frac{4}{x^2} = 1$$

$$4 = x^2$$

$$x = \pm 2$$

FIGURE 4.12   Finally, in the interval $(1, 4)$ we choose $c = 2$, as shown in Figure 4.12.

In terms of rates of change, the Mean Value Theorem tells us that there must be a point in (a, b) at which the instantaneous rate of change is equal to the average rate of change over the interval $[a, b]$. This is illustrated in the next example.

EXAMPLE 4 *An application involving the Mean Value Theorem*

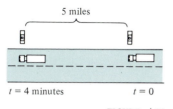

5 miles

t = 4 minutes   *t* = 0

FIGURE 4.13

Two stationary patrol cars equipped with radar are located 5 miles apart on a highway, as shown in Figure 4.13. As a truck passes the first patrol car, its speed is clocked at 55 miles per hour. Four minutes later, when the truck passes the second patrol car, its speed is clocked at 50 miles per hour. Prove that the truck must have exceeded the speed limit at some time during the four minutes.

Solution: We let $t = 0$ be the time (in hours) when the truck passes the first patrol car. Then, the time when the truck passes the second patrol car is

$$t = \frac{4}{60} = \frac{1}{15} \text{ hr}$$

Now, if we let $s(t)$ represent the distance (in miles) traveled by the truck, we have $s(0) = 0$ and $s(\frac{1}{15}) = 5$. Therefore, the average velocity of the truck over the five-mile stretch of highway is given by

$$\text{average velocity} = \frac{s(1/15) - s(0)}{(1/15) - 0} = \frac{5}{1/15} = 75 \text{ mph}$$

Assuming that the position function is differentiable, we can apply the Mean Value Theorem to conclude that the truck must have been traveling at a rate of 75 miles per hour sometime during the four minutes.

Remark   A useful alternative form of the Mean Value Theorem is as follows: If f is continuous on $[a, b]$ and differentiable on (a, b), then there exists a number c in (a, b) such that

$$f(b) = f(a) + (b - a)f'(c)$$

When working the exercises for this section, keep in mind that polynomial and rational functions are differentiable everywhere they are defined.

Section Exercises 4.2

In Exercises 1–12, find all the intervals $[a, b]$ on which $f(a) = f(b) = 0$ and Rolle's Theorem can be applied. For each interval find all values of c such that $f'(c) = 0$.

1. $f(x) = x^2 - 2x$ **2.** $f(x) = x^2 - 3x + 2$

3. $f(x) = x^3 - 6x^2 + 11x - 6$

4. $f(x) = x(x^2 - x - 2)$

5. $f(x) = |x| - 1$ **6.** $f(x) = 3 - |x - 3|$

7. $f(x) = x^{2/3} - 1$ **8.** $f(x) = x - x^{1/3}$

9. $f(x) = \dfrac{x^2 - 2x - 3}{x + 2}$ **10.** $f(x) = (x - 3)(x + 1)^2$

11. $f(x) = x^2 - 6x + 10$ **12.** $f(x) = \dfrac{x + 1}{x}$

In Exercises 13–20, apply the Mean Value Theorem to f on the indicated interval. In each case, find all values of c in the interval (a, b) such that

$$f'(c) = \frac{f(b) - f(a)}{b - a}$$

Function	Interval
13. $f(x) = x^2$	$[-2, 1]$
14. $f(x) = x(x^2 - x - 2)$	$[-1, 1]$
15. $f(x) = x^{2/3}$	$[0, 1]$
16. $f(x) = \dfrac{x + 1}{x}$	$\left[\dfrac{1}{2}, 2\right]$

17. $f(x) = \dfrac{x}{x + 1}$ $\qquad$ $\left[-\dfrac{1}{2}, 2\right]$

18. $f(x) = \sqrt{x - 2}$ $\qquad$ $[2, 6]$

19. $f(x) = x^3$ $\qquad$ $[0, 1]$

20. $f(x) = x^3 - 2x$ $\qquad$ $[0, 2]$

21. The height of a ball t seconds after it is thrown is given by

$$f(t) = -16t^2 + 48t + 32$$

(a) Verify that $f(1) = f(2)$.

(b) According to Rolle's Theorem, what must the velocity be at some time in the interval $[1, 2]$?

22. The ordering and transportation cost C of components used in a manufacturing process is approximated by

$$C(x) = 10\left(\dfrac{1}{x} + \dfrac{x}{x + 3}\right)$$

where C is measured in thousands of dollars and x is the order size in hundreds.

(a) Verify that $C(3) = C(6)$.

(b) According to Rolle's Theorem, the rate of change of cost must be zero for some order size in the interval $[3, 6]$. Find that order size.

23. The height of an object t seconds after it was dropped from a height of 500 feet is given by

$$s(t) = -16t^2 + 500$$

(a) Find the average velocity of the object during the first 3 seconds.

(b) Use the Mean Value Theorem to verify that at some time during the first three seconds of fall the instantaneous velocity equals the average velocity. Find that time.

24. A company introduces a new product for which the number of sales S is given by

$$S(t) = 200\left(5 - \dfrac{9}{2 + t}\right)$$

where t is the time in months.

(a) Find the average value of $S'(t)$ during the first year.

(b) During what month does $S'(t)$ equal its average value during the first year?

25. Prove Corollary 1 to Rolle's Theorem.

26. Prove Corollary 2 to Rolle's Theorem.

27. If $a > 0$ and n is any integer, prove that the polynomial function

$$p(x) = x^{2n+1} + ax + b$$

cannot have two real roots.

28. Let p be a *nonconstant* polynomial function. Prove that between any two consecutive zeros of p', there is at most one zero of p.

29. Prove that if $f'(x) = 0$ for all x in an interval $[a, b]$, then f is constant on the interval.

30. Let $p(x) = Ax^2 + Bx + C$. Prove that for any interval $[a, b]$, the value c guaranteed by the Mean Value Theorem is the midpoint of the interval.

31. Prove that if two functions f and g have the same derivatives on an interval, then they must differ only by a constant on the interval. [Hint: Let $h(x) = f(x) - g(x)$ and use the result of Exercise 29.]

SECTION TOPICS ▪
Increasing and decreasing functions ▪
The First Derivative Test ▪
Monotonic function ▪

4.3
Increasing and decreasing functions and the First Derivative Test

You have already seen that the derivative is useful in *locating* the relative extrema of a function. In this section we will show that the derivative can also be used to *classify* relative extrema as either relative minima or relative maxima. We begin by defining what is meant when we say a function increases (or decreases).

DEFINITION OF INCREASING
AND DECREASING FUNCTIONS

A function f is said to be **increasing** on an interval if for any two numbers x_1 and x_2 in the interval,

$$x_1 < x_2 \quad \text{implies} \quad f(x_1) < f(x_2)$$

A function f is said to be **decreasing** on an interval if for any two numbers x_1 and x_2 in the interval,

$$x_1 < x_2 \quad \text{implies} \quad f(x_1) > f(x_2)$$

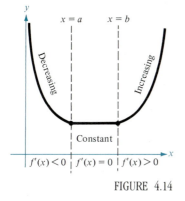

FIGURE 4.14

From this definition we can see that a function is increasing if its graph moves up as x moves to the right and a function is decreasing if its graph moves down as x moves to the right. For example, the function in Figure 4.14 is decreasing on the interval $(-\infty, a)$, is constant on the interval (a, b), and is increasing on the interval (b, ∞).

The derivative is useful in determining whether a function is increasing or decreasing on a given interval. Specifically, as Figure 4.14 shows, a positive derivative implies that the graph slopes upward and the function is increasing. Similarly, a negative derivative implies that the function is decreasing. Finally, a zero derivative on an entire interval implies that the function is constant on the interval.

THEOREM 4.5

TEST FOR INCREASING OR DECREASING FUNCTIONS
Let f be a function that is differentiable on the interval (a, b).

1. If $f'(x) > 0$ for all x in (a, b), then f is increasing on (a, b).
2. If $f'(x) < 0$ for all x in (a, b), then f is decreasing on (a, b).
3. If $f'(x) = 0$ for all x in (a, b), then f is constant on (a, b).

Proof: To prove the first case, we assume that $f'(x) > 0$ for all x in the interval (a, b) and let $x_1 < x_2$ be any two points in the interval. By the Mean Value Theorem, we know there exists a number c such that $x_1 < c < x_2$, and

$$f'(c) = \frac{f(x_2) - f(x_1)}{x_2 - x_1}$$

Since $f'(c) > 0$ and $x_2 - x_1 > 0$, we know that $f(x_2) - f(x_1) > 0$, which implies that

$$f(x_1) < f(x_2)$$

Thus, f is increasing on the interval. The second case has a similar proof, and the third case was given as Exercise 29 in Section 4.2.

| **Remark** Theorem 4.5 is also valid if the interval (a, b) is replaced by an interval of the form $(-\infty, b)$, (a, ∞), or $(-\infty, \infty)$.

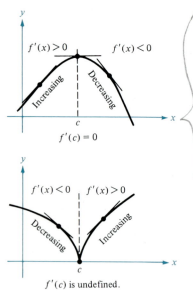

$f'(x) > 0$ $f'(x) < 0$

Increasing   Decreasing

c

$f'(c) = 0$

$f'(x) < 0$ $f'(x) > 0$

Decreasing   Increasing

c

$f'(c)$ is undefined.

FIGURE 4.15

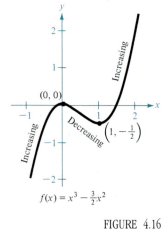

$f(x) = x^3 - \frac{3}{2}x^2$

FIGURE 4.16

To see how to apply Theorem 4.5, we note that for a continuous function, $f'(x)$ can change sign only at a critical number, as shown in Figure 4.15. Thus, to determine the intervals on which a continuous function is increasing or decreasing, we suggest using the following guidelines.

1. Locate the critical numbers of f and use these numbers to determine test intervals.
2. Test the sign of $f'(x)$ at an arbitrary point in each of the intervals determined in the first step.
3. Use Theorem 4.5 to decide whether f is increasing or decreasing on each interval.

EXAMPLE 1   Determining intervals on which f is increasing or decreasing

Find the intervals on which $f(x) = x^3 - \frac{3}{2}x^2$ is increasing or decreasing.

Solution: We begin by setting $f'(x)$ equal to zero.

$$f'(x) = 3x^2 - 3x = 0 \qquad \text{Let } f'(x) = 0$$
$$3(x)(x - 1) = 0 \qquad \text{Factor}$$
$$x = 0, 1 \qquad \text{Critical numbers}$$

Since there are no points for which f' is undefined, we conclude that $x = 0$ and $x = 1$ are the only critical numbers. Table 4.3 summarizes the testing of the three intervals determined by these critical numbers.

TABLE 4.3

Interval	$-\infty < x < 0$	$0 < x < 1$	$1 < x < \infty$
Test value	$x = -1$	$x = \dfrac{1}{2}$	$x = 2$
Sign of $f'(x)$	$f'(-1) = 6 > 0$	$f'\left(\dfrac{1}{2}\right) = -\dfrac{3}{4} < 0$	$f'(2) = 6 > 0$
Conclusion	increasing	decreasing	increasing

The graph of f is shown in Figure 4.16. Note that the test values in Table 4.3 were chosen for convenience; other points could have been used.

Not only is the function in Example 1 continuous, it is differentiable on the entire real line. For such functions, the only critical numbers are those for which $f'(x) = 0$. In the next example we look at a continuous function that has two types of critical numbers—those for which $f'(x) = 0$ and those for which f' is undefined.

EXAMPLE 2   Determining intervals on which f is increasing or decreasing

Find the intervals on which $f(x) = (x^2 - 4)^{2/3}$ is increasing or decreasing.

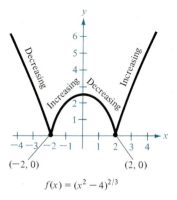

$f(x) = (x^2 - 4)^{2/3}$

FIGURE 4.17

Solution:

$$f(x) = (x^2 - 4)^{2/3}$$

$$f'(x) = \frac{2}{3}(x^2 - 4)^{-1/3}(2x) = \frac{4x}{3(x^2 - 4)^{1/3}}$$

Since $f'(x)$ is zero at $x = 0$ and f' is undefined at $x = \pm 2$, the critical numbers are

$$x = -2, \qquad x = 0, \qquad \text{and} \qquad x = 2$$

Table 4.4 summarizes the testing of the resulting four intervals. The graph of the function is shown in Figure 4.17.

TABLE 4.4

Interval	$-\infty < x < -2$	$-2 < x < 0$	$0 < x < 2$	$2 < x < \infty$
Test value	$x = -3$	$x = -1$	$x = 1$	$x = 3$
Sign of $f'(x)$	$f'(-3) < 0$	$f'(-1) > 0$	$f'(1) < 0$	$f'(3) > 0$
Conclusion	decreasing	increasing	decreasing	increasing

Remark In Table 4.4, note that we do not need to *evaluate* $f'(x)$ at the test values—we only need to determine its sign. For example, we can determine that $f'(-3)$ is negative as follows:

$$f'(-3) = \frac{4(-3)}{3[(-3)^2 - 4]^{1/3}} = \frac{4(-3)}{3(9 - 4)^{1/3}} = \frac{\text{negative}}{\text{positive}} = \text{negative}$$

The functions in Examples 1 and 2 are continuous over the entire real line. If the domain of a function includes points of discontinuity, then these points should be used along with the critical numbers to determine the test intervals. This is demonstrated in the next example.

EXAMPLE 3   Test intervals for a discontinuous function

Find the intervals on which $f(x) = (x^4 + 1)/x^2$ is increasing or decreasing.

Solution:

$$f(x) = \frac{x^4 + 1}{x^2} = x^2 + \frac{1}{x^2}$$

$$f'(x) = 2x - \frac{2}{x^3}$$

$$= \frac{2(x^4 - 1)}{x^3} = \frac{2(x^2 + 1)(x - 1)(x + 1)}{x^3}$$

Since $f'(x)$ is zero at $x = \pm 1$ and f is discontinuous at $x = 0$, we use these three x-values to determine the test intervals:

$$x = -1 \qquad \text{and} \qquad x = 1 \qquad \text{Critical numbers}$$

$$x = 0 \qquad\qquad\qquad\qquad \text{Point of discontinuity}$$

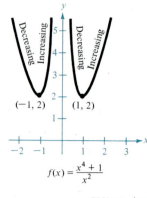

$$f(x) = \frac{x^4 + 1}{x^2}$$

FIGURE 4.18

Table 4.5 summarizes the testing of the four intervals determined by these three points, and the graph of f is shown in Figure 4.18.

TABLE 4.5

Interval	$-\infty < x < -1$	$-1 < x < 0$	$0 < x < 1$	$1 < x < \infty$
Test value	$x = -2$	$x = -\dfrac{1}{2}$	$x = \dfrac{1}{2}$	$x = 2$
Sign of $f'(x)$	$f'(-2) < 0$	$f'\left(-\dfrac{1}{2}\right) > 0$	$f'\left(\dfrac{1}{2}\right) < 0$	$f'(2) > 0$
Conclusion	decreasing	increasing	decreasing	increasing

From the functions in the first three examples, you might conclude that on adjacent test intervals a function must alternate between increasing and decreasing. This is not necessarily true, as is shown in the next example.

EXAMPLE 4   A function that increases on both sides of a critical number

Find the intervals on which $f(x) = x^3$ is increasing or decreasing.

Solution:

$$f'(x) = 3x^2 = 0$$

$$x = 0 \qquad \text{Critical number}$$

Even though the derivative is zero at $x = 0$, it is positive for all other x-values:

$$f'(x) = 3x^2 > 0, \quad x \neq 0$$

Therefore, f is increasing on $(-\infty, 0)$ and on $(0, \infty)$.

| Remark   In Example 4, we used the test for increasing functions (Theorem 4.5) to conclude that f is increasing on the intervals $(-\infty, 0)$ and $(0, \infty)$. In Figure 4.19, we can see that this function is actually increasing on the entire real line. Look back at the definition of an increasing function to convince yourself that this is true.

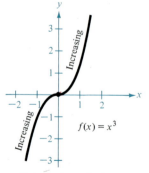

FIGURE 4.19

f is increasing on both sides of a critical number.

The First Derivative Test

Once we have determined the intervals on which a function is increasing or decreasing, we can easily locate the relative extrema of the function. This procedure is described in the First Derivative Test.

THEOREM 4.6   **THE FIRST DERIVATIVE TEST**

Let c be a critical number of a function f that is continuous on an open interval I containing c. If f is differentiable on the interval, except possibly at c, then $f(c)$ can be classified as follows.

1. If f' changes from negative to positive at c, then $f(c)$ is a **relative minimum** of f.
2. If f' changes from positive to negative at c, then $f(c)$ is a **relative maximum** of f.
3. If f' does not change signs at c, then $f(c)$ is neither a relative minimum nor a relative maximum.

Proof: We prove the first case and leave the other two cases as exercises. Assume that f' changes from negative to positive at c. Then there exist a and b in I such that

$$f'(x) < 0 \text{ for all } x \text{ in } (a, c)$$

and

$$f'(x) > 0 \text{ for all } x \text{ in } (c, b)$$

By Theorem 4.5, f is decreasing on (a, c) and increasing on (c, b). Therefore, $f(c)$ is a minimum of f on the open interval (a, b) and consequently a relative minimum of f.

The First Derivative Test is illustrated graphically in Figure 4.20.

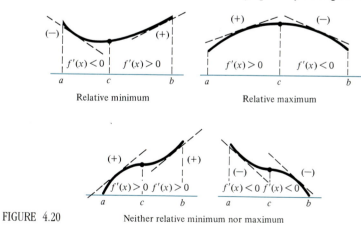

FIGURE 4.20

To apply the First Derivative Test, we use the tabular format shown in Examples 1, 2, and 3. This is demonstrated in the next example.

EXAMPLE 5   Applying the First Derivative Test

Use the First Derivative Test to find all relative maxima and minima for the function given by

$$f(x) = 2x^3 - 3x^2 - 36x + 14$$

Solution:

$$f'(x) = 6x^2 - 6x - 36 = 0 \qquad \text{Let } f'(x) = 0$$
$$6(x^2 - x - 6) = 0$$
$$6(x - 3)(x + 2) = 0$$
$$x = -2, 3 \qquad \text{Critical numbers}$$

Table 4.6 shows a practical format for applying the First Derivative Test.

$f(x) = 2x^3 - 3x^2 - 36x + 14$

TABLE 4.6

Interval	$-\infty < x < -2$	$-2 < x < 3$	$3 < x < \infty$
Test value	$x = -3$	$x = 0$	$x = 4$
Sign of $f'(x)$	$f'(-3) > 0$	$f'(0) < 0$	$f'(4) > 0$
Conclusion	Increasing	Decreasing	Increasing

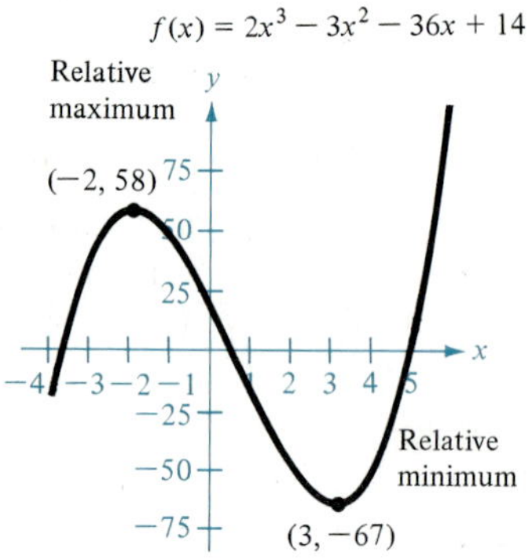

FIGURE 4.21

From Table 4.6, we conclude that a relative maximum occurs at $x = -2$ and a relative minimum occurs at $x = 3$. A graph of the function is shown in Figure 4.21.

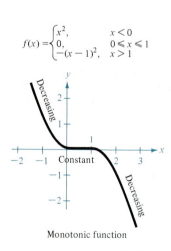

Monotonic function

FIGURE 4.22

A function is called **monotonic** on an interval if it is either nondecreasing or nonincreasing on the interval. For example, the function given by

$$f(x) = \begin{cases} x^2, & x < 0 \\ 0, & 0 \leq x \leq 1 \\ -(x-1)^2, & x > 1 \end{cases}$$

is monotonic since it is never increasing, as seen in Figure 4.22.

A function is called **strictly monotonic** on an interval if it is either increasing on the entire interval or decreasing on the entire interval. For instance, in Example 4 we saw that the function given by $f(x) = x^3$ is strictly monotonic on the entire real line. Functions that are strictly monotonic play a special role in calculus. You will see why this is true when we study inverse functions in Chapter 7.

Section Exercises 4.3

In Exercises 1–6, identify the intervals on which the function is increasing or decreasing.

1. $f(x) = x^2 - 6x + 8$

2. $y = -(x + 1)^2$

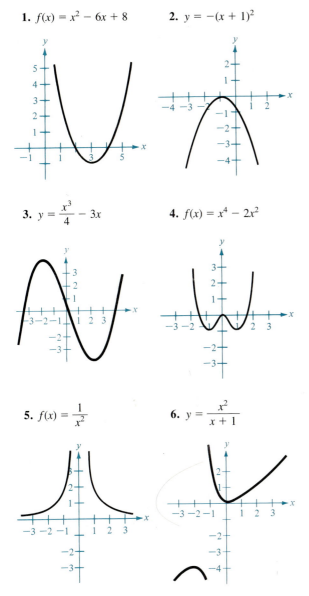

3. $y = \dfrac{x^3}{4} - 3x$

4. $f(x) = x^4 - 2x^2$

5. $f(x) = \dfrac{1}{x^2}$

6. $y = \dfrac{x^2}{x + 1}$

In Exercises 7–26, find the critical numbers of f (if any), find the intervals on which f is increasing or decreasing, and use the First Derivative Test to locate all relative extrema.

7. $f(x) = -2x^2 + 4x + 3$

8. $f(x) = x^2 + 8x + 10$

9. $f(x) = x^2 - 6x$

10. $f(x) = (x - 1)^2(x + 2)$

11. $f(x) = 2x^3 + 3x^2 - 12x$

12. $f(x) = (x - 3)^3$

13. $f(x) = x^3 - 6x^2 + 15$

14. $f(x) = (x - 1)^{2/3}$

15. $f(x) = x^4 - 2x^3$

16. $f(x) = (x - 1)^{1/3}$

17. $f(x) = x^{1/3} + 1$

18. $f(x) = x^{2/3}(x - 5)$

19. $f(x) = x + \dfrac{1}{x}$

20. $f(x) = \dfrac{x}{x + 1}$

21. $f(x) = \dfrac{x^2}{x^2 - 9}$

22. $f(x) = \dfrac{x + 3}{x^2}$

23. $f(x) = \dfrac{x^5 - 5x}{5}$

24. $f(x) = x^4 - 32x + 4$

25. $f(x) = \dfrac{x^2 - 2x + 1}{x + 1}$

26. $f(x) = \dfrac{x^2 - 3x - 4}{x - 2}$

27. The height (in feet) of a ball at time t (in seconds) is given by the position function

$$s(t) = 96t - 16t^2$$

Find the time interval on which the ball is moving up and the interval on which it is moving down. What is the maximum height of the ball?

28. Repeat Exercise 27 using the position function

$$s(t) = -16t^2 + 64t$$

29. Coughing forces the trachea (windpipe) to contract, which in turn affects the velocity v of the air through the trachea. Suppose the velocity of the air during coughing is

$$v = k(R - r)r^2$$

where k is a constant, R is the normal radius of the trachea, and r is the radius during coughing. What radius will produce the maximum air velocity?

30. The concentration C of a certain chemical in the bloodstream t hours after injection into muscle tissue is given by

$$C = \dfrac{3t}{27 + t^3}$$

When is the concentration greatest?

31. After a drug is administered to a patient, the drug concentration in the patient's bloodstream over a two-hour period is given by

$$C = 0.29483t + 0.04253t^2 - 0.00035t^3$$

where C is measured in milligrams and t is the time in minutes. Find the interval on which C is increasing or decreasing.

32. A fast-food restaurant sells x hamburgers to make a profit P given by

$$P = 2.44x - \dfrac{x^2}{20,000} - 5000, \quad 0 \le x \le 35,000$$

Find the intervals on which P is increasing or decreasing.

33. After birth, an infant will normally lose weight for a few days and then start gaining. A model for the average weight W of infants over the first two weeks following birth is

$$W = 0.033t^2 - 0.3974t + 7.3032$$

Find the intervals on which W is increasing or decreasing.

34. The ordering and transportation cost C of components used in a manufacturing firm is given by

$$C = 10\left(\frac{1}{x} + \frac{x}{x+3}\right), \quad 1 \le x$$

where C is measured in thousands of dollars and x is the order size in hundreds. Find the intervals on which C is increasing or decreasing.

35. The resistance R of a certain type of resistor is given by

$$R = \sqrt{0.001T^4 - 4T + 100}$$

where R is measured in ohms and the temperature T is measured in degrees Celsius. What temperature produces a minimum resistance for this type of resistor?

36. The electric power P in watts in a direct-current circuit with two resistors R_1 and R_2 connected in series is

$$P = \frac{vR_1R_2}{(R_1 + R_2)^2}$$

where v is the voltage. If v and R_1 are held constant, what resistance R_2 produces maximum power?

37. Find a, b, c, and d so that the function given by

$$f(x) = ax^3 + bx^2 + cx + d$$

has a relative minimum at $(0, 0)$ and a relative maximum at $(2, 2)$.

38. Find a, b, and c so that the function given by

$$f(x) = ax^2 + bx + c$$

has a relative maximum at $(5, 20)$ and passes through the point $(2, 10)$.

In Exercises 39–44, assume that f is differentiable for all x. The sign of f' is as follows:

$$f'(x) > 0 \text{ on } (-\infty, -4)$$
$$f'(x) < 0 \text{ on } (-4, 6)$$
$$f'(x) > 0 \text{ on } (6, \infty)$$

In each exercise, supply the appropriate inequality for the indicated value of c.

Function	*Sign of $g'(c)$*
39. $g(x) = f(x) + 5$	$g'(0)$ ___ 0
40. $g(x) = 3f(x) - 3$	$g'(-5)$ ___ 0
41. $g(x) = -f(x)$	$g'(-6)$ ___ 0
42. $g(x) = -f(x)$	$g'(0)$ ___ 0
43. $g(x) = f(x - 10)$	$g'(0)$ ___ 0
44. $g(x) = f(x - 10)$	$g'(8)$ ___ 0

45. Prove the second case of Theorem 4.5.

46. Prove the second and third cases of Theorem 4.6.

SECTION TOPICS ▪
Concavity ▪
Points of inflection ▪
The Second Derivative Test ▪

4.4
Concavity and the Second Derivative Test

We have already seen that locating the intervals in which a function f increases or decreases is helpful in determining its graph. In this section we show that, by locating the intervals in which f' increases or decreases, we can determine where the graph of f is *curving upward* or *curving downward*. We define this notion of curving upward or downward as **concavity.**

DEFINITION OF CONCAVITY   Let f be differentiable on an open interval. We say that the graph of f is **concave upward** if f' is increasing on the interval and **concave downward** if f' is decreasing on the interval.

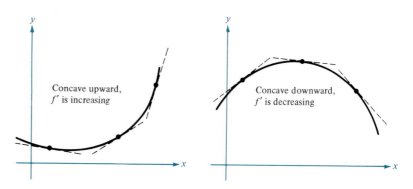

FIGURE 4.23

From Figure 4.23, we can get the following graphical interpretation of concavity.

1. If a curve lies *above* its tangent lines, then it is concave upward.
2. If a curve lies *below* its tangent lines, then it is concave downward.

To determine concavity without seeing a graph of f, we can use the second derivative to determine the intervals in which f' is increasing or decreasing (just as we used the first derivative to determine the intervals in which f is increasing or decreasing).

THEOREM 4.7   **TEST FOR CONCAVITY**
Let f be a function whose second derivative exists on an open interval I.

1. If $f''(x) > 0$ for all x in I, then the graph of f is concave upward.
2. If $f''(x) < 0$ for all x in I, then the graph of f is concave downward.

Proof: This theorem follows directly from Theorem 4.5 and the definition of concavity. Note that we do not define concavity for straight lines. In other words, a straight line is neither concave upward nor concave downward. In this context, we could add a third conclusion to the theorem. If $f''(x) = 0$ for all x in I, then f is linear.

EXAMPLE 1   Comparing the graphs of f and f'

For each of the following functions, sketch the graph of f and f' to show that f' is increasing on the intervals for which the graph of f is concave upward and f' is decreasing on the intervals for which the graph of f is concave downward.

(a) $f(x) = \dfrac{x^2}{2}$ (b) $f(x) = \dfrac{x^3 - 3x}{3}$

Solution: The solution is shown in Figure 4.24.

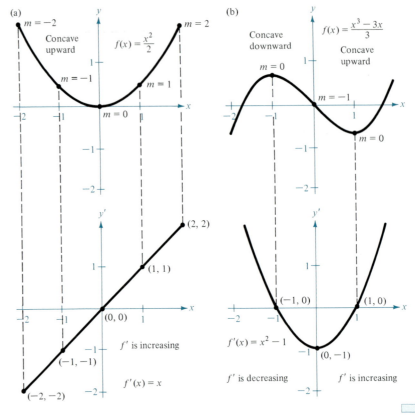

FIGURE 4.24

Remark In Example 1(a), f' is increasing on the interval $(-\infty, 0)$ even though f is decreasing there. Be sure you see that the increasing or decreasing of f' does not necessarily correspond to the increasing or decreasing of f.

For a *continuous* function f, we can find the intervals on which the graph of f is concave upward and concave downward as follows.

1. Locate the x-values at which $f''(x) = 0$ or f'' is undefined.
2. Use these x-values to determine the test intervals.
3. Test the sign of $f''(x)$ in each of the test intervals.

We illustrate this procedure in Example 2. For *discontinuous* functions, the test intervals should be formed using the points of discontinuity (along with the points at which f'' is zero or undefined). This is demonstrated in Example 3.

EXAMPLE 2   Determining concavity

Determine the intervals in which the graph of

$$f(x) = \frac{6}{x^2 + 3}$$

is concave upward or downward.

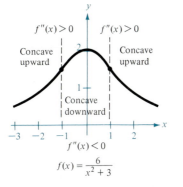

$$f(x) = \frac{6}{x^2 + 3}$$

FIGURE 4.25

Solution: We begin by finding the second derivative.

$$f(x) = 6(x^2 + 3)^{-1}$$

$$f'(x) = (-6)(2x)(x^2 + 3)^{-2} = \frac{-12x}{(x^2 + 3)^2}$$

$$f''(x) = \frac{(x^2 + 3)^2(-12) - (-12x)(2)(2x)(x^2 + 3)}{(x^2 + 3)^4}$$

$$= \frac{36(x^2 - 1)}{(x^2 + 3)^3}$$

Since $f''(x) = 0$ when $x = \pm 1$ and f'' is defined on the entire real line, we test f'' in the intervals $(-\infty, -1)$, $(-1, 1)$, and $(1, \infty)$. The results are shown in Table 4.7 and Figure 4.25.

TABLE 4.7

Interval	$-\infty < x < -1$	$-1 < x < 1$	$1 < x < \infty$
Test value	$x = -2$	$x = 0$	$x = 2$
Sign of $f''(x)$	$f''(-2) > 0$	$f''(0) < 0$	$f''(2) > 0$
Conclusion	Concave upward	Concave downward	Concave upward

EXAMPLE 3   *Determining concavity for a discontinuous function*

Determine the intervals in which the graph of

$$f(x) = \frac{x^2 + 1}{x^2 - 4}$$

is concave upward or downward.

Solution:

$$f'(x) = \frac{(x^2 - 4)(2x) - (x^2 + 1)(2x)}{(x^2 - 4)^2} = \frac{-10x}{(x^2 - 4)^2}$$

$$f''(x) = -10\frac{(x^2 - 4)^2(1) - (x)(2)(2x)(x^2 - 4)}{(x^2 - 4)^4} = \frac{10(3x^2 + 4)}{(x^2 - 4)^3}$$

There are no points at which f'' is zero, but at $x = \pm 2$ the function f is discontinuous, so we test for concavity in the intervals $(-\infty, -2)$, $(-2, 2)$, and $(2, \infty)$, as shown in Table 4.8. The graph of f is shown in Figure 4.26.

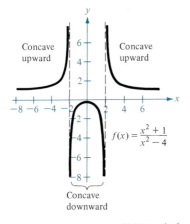

$$f(x) = \frac{x^2 + 1}{x^2 - 4}$$

FIGURE 4.26

TABLE 4.8

Interval	$-\infty < x < -2$	$-2 < x < 2$	$2 < x < \infty$
Test value	$x = -3$	$x = 0$	$x = 3$
Sign of $f''(x)$	$f''(-3) > 0$	$f''(0) < 0$	$f''(3) > 0$
Conclusion	Concave upward	Concave downward	Concave upward

Points of inflection

The graph in Figure 4.25 has two points at which the concavity changes. If at such a point the tangent line to the graph exists, we call the point a **point of inflection.** Three types of inflection points are shown in Figure 4.27.

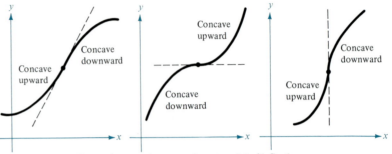

The graph *crosses* its tangent line at a point of inflection.

FIGURE 4.27

| DEFINITION OF POINT OF INFLECTION | If the graph of a continuous function possesses a tangent line at a point where its concavity changes from upward to downward (or vice versa), we call the point a **point of inflection.** |

Remark Note in Figure 4.27 that at a point of inflection, the graph crosses its tangent line.

Since a point of inflection occurs where the concavity of a graph changes, it must be true that the sign of f'' changes at such points. Thus, to locate possible points of inflection, we need only determine the values of x for which $f''(x) = 0$ or for which f'' is undefined. This parallels the procedure for locating relative extrema of f.

| THEOREM 4.8 | POINTS OF INFLECTION
If $(c, f(c))$ is a point of inflection of the graph of f, then either $f''(c) = 0$ or f'' is undefined at $x = c$. |

critical pts for the derivative

EXAMPLE 4   Finding points of inflection

Determine the points of inflection and discuss the concavity of the graph of

$$f(x) = x^4 + x^3 - 3x^2 + 1$$

Solution: Differentiating twice, we have

$$f'(x) = 4x^3 + 3x^2 - 6x$$
$$f''(x) = 12x^2 + 6x - 6 = 6(2x^2 + x - 1) = 6(2x - 1)(x + 1)$$

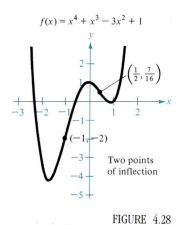

$f(x) = x^4 + x^3 - 3x^2 + 1$

$\left(\frac{1}{2}, \frac{7}{16}\right)$

$(-1, -2)$

Two points
of inflection

FIGURE 4.28

Possible points of inflection occur at $x = -1$ and $x = \frac{1}{2}$. By testing the intervals determined by these x-values, we conclude that they are both points of inflection. A summary of this testing is shown in Table 4.9, and the graph of f is shown in Figure 4.28.

TABLE 4.9

Interval	$-\infty < x < -1$	$-1 < x < \dfrac{1}{2}$	$\dfrac{1}{2} < x < \infty$
Test value	$x = -2$	$x = 0$	$x = 1$
Sign of $f''(x)$	$f''(-2) > 0$	$f''(0) < 0$	$f''(1) > 0)$
Conclusion	Concave upward	Concave downward	Concave upward

EXAMPLE 5   Finding points of inflection

Determine the points of inflection of the graph of

$$f(x) = \frac{x - 2}{\sqrt{x^2 + 2}}$$

Solution:

$$f(x) = (x - 2)(x^2 + 2)^{-1/2}$$

$$f'(x) = (x - 2)(2x)\left(-\frac{1}{2}\right)(x^2 + 2)^{-3/2} + (x^2 + 2)^{-1/2}(1)$$

$$= 2(x + 1)(x^2 + 2)^{-3/2}$$

$$f''(x) = 2(x + 1)(2x)\left(-\frac{3}{2}\right)(x^2 + 2)^{-5/2} + 2(x^2 + 2)^{-3/2}(1)$$

$$= 2(x^2 + 2)^{-5/2}[(-3x)(x + 1) + (x^2 + 2)]$$

$$= -2(2x - 1)(x + 2)(x^2 + 2)^{-5/2}$$

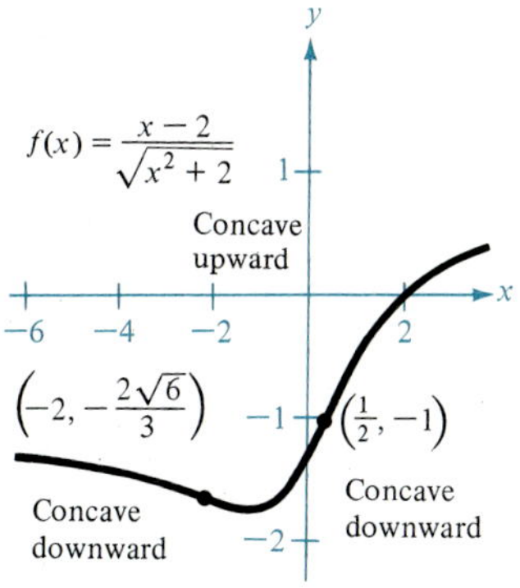

Possible points of inflection occur at $x = -2$ and $x = \frac{1}{2}$. By testing the intervals determined by these x-values, we conclude that they are both points of inflection. A summary of this testing is shown in Table 4.10, and the graph of f is shown in Figure 4.29.

$f(x) = \dfrac{x - 2}{\sqrt{x^2 + 2}}$

Concave
upward

$\left(-2, -\dfrac{2\sqrt{6}}{3}\right)$

$\left(\dfrac{1}{2}, -1\right)$

Concave
downward

Concave
downward

FIGURE 4.29

TABLE 4.10

Interval	$-\infty < x < -2$	$-2 < x < \dfrac{1}{2}$	$\dfrac{1}{2} < x < \infty$
Test value	$x = -3$	$x = 0$	$x = 1$
Sign of $f''(x)$	$f''(-3) < 0$	$f''(0) > 0$	$f''(1) < 0$
Conclusion	Concave downward	Concave upward	Concave downward

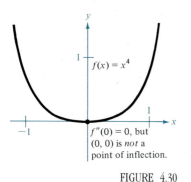

$f''(0) = 0$, but
$(0, 0)$ is *not* a
point of inflection.

FIGURE 4.30

Remark Note that it is possible for the second derivative to be zero at a point that is *not* a point of inflection. For example, the graph of $f(x) = x^4$ is shown in Figure 4.30. The second derivative is zero when $x = 0$, but the point $(0, 0)$ is not a point of inflection. Before concluding that a point of inflection exists at a value of x for which $f''(x) = 0$, you should test to be certain that the concavity actually changes there.

The Second Derivative Test

If the second derivative exists, we can often use it as a simple test for relative minima and maxima. The test is based on the fact that if $f(c)$ is a relative maximum of a *differentiable* function, then its graph is concave downward in some interval containing c. Similarly, if $f(c)$ is a relative minimum, its graph is concave upward in some interval containing c, as shown in Figure 4.31.

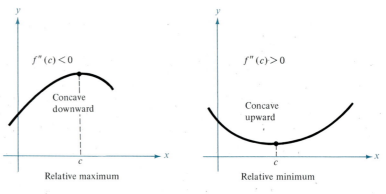

FIGURE 4.31

THEOREM 4.9   SECOND DERIVATIVE TEST

Let f be a function such that $f'(c) = 0$ and the second derivative of f exists on an open interval containing c.

1. If $f''(c) > 0$, then $f(c)$ is a relative minimum.
2. If $f''(c) < 0$, then $f(c)$ is a relative maximum.
3. If $f''(c) = 0$, then the test fails.

Proof: This theorem follows from Theorem 4.7. We outline a proof for the first case. If $f''(c) > 0$, then f is concave upward in some interval I containing c. This implies that the graph of f lies above its tangent lines in I. Since $f'(c) = 0$, the tangent line must be horizontal at $(c, f(c))$. Thus, we can conclude that $f(c)$ is a minimum of f in the interval I, and consequently $f(c)$ must be a relative minimum of f.

Remark Be sure that you see that if $f''(c) = 0$, the Second Derivative Test does not apply. In such cases we can use the First Derivative Test.

EXAMPLE 6   Using the Second Derivative Test

Find the relative extrema for $f(x) = -3x^5 + 5x^3$.

Solution: We begin by finding the critical numbers of f.

$$f'(x) = -15x^4 + 15x^2 = 15x^2(1 - x^2) = 0$$
$$x = -1, 0, 1 \quad \text{potential extrema} \quad \text{Critical numbers}$$

These are the only critical numbers, since f' exists on the entire real line. Applying the Second Derivative Test, we have

$$f''(x) = 15(-4x^3 + 2x)$$

Point	Sign of f''		Conclusion
$(-1, -2)$	$f''(-1) = 30 > 0$	⇨	Relative minimum
$(1, 2)$	$f''(1) = -30 < 0$	⇨	Relative maximum
$(0, 0)$	$f''(0) = 0$	⇨	Test fails

Since the Second Derivative Test fails at $(0, 0)$, we use the First Derivative Test and observe that f increases to the right and left of $x = 0$. Thus, $(0, 0)$ is neither a relative minimum nor a relative maximum. The graph of f is shown in Figure 4.32.

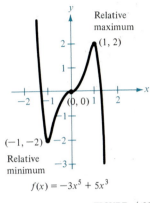

Relative maximum (1, 2)

(0, 0)

(−1, −2) Relative minimum

$f(x) = -3x^5 + 5x^3$

FIGURE 4.32

In this and the three previous sections we have studied several related concepts: extrema, relative extrema, critical numbers, increasing functions, decreasing functions, concavity, and points of inflection. Although we have seen that there are relationships among these concepts, the relationships are not simple. Moreover, the theorems in these four sections are fussy in that their conclusions can depend on continuity, differentiability, and closed or open intervals. Remember that when you are applying a theorem it is not enough to know the conclusion of the theorem. You must also know its hypotheses. For example, the Second Derivative Test can only be applied if all of the following hypotheses are satisfied:

1. f is twice differentiable on an open interval containing c.
2. $f'(c) = 0$
3. $f''(c) \neq 0$

Section Exercises 4.4

In Exercises 1–4, use the given graph to sketch the graph of f'. Find the intervals (if any) on which

    (a) $f'(x)$ is positive
    (b) $f'(x)$ is negative
    (c) f' is increasing
    (d) f' is decreasing

For each of these intervals, describe the corresponding behavior of f.

1. $f(x) = 4 - x^2$

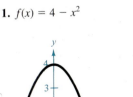

2. $f(x) = \dfrac{|x|}{x}$

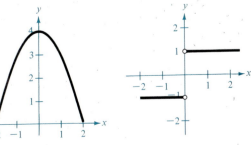

3. $f(x) = -\dfrac{1}{3}x^3 + x^2$ **4.** $f(x) = \dfrac{3}{2}x^{2/3}$

11. $y = -x^3 + 3x^2 - 2$ **12.** $f(x) = \dfrac{x^2 - 1}{2x + 1}$

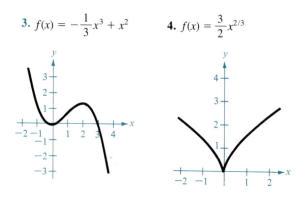

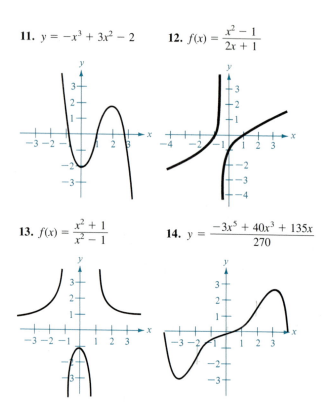

In Exercises 5–8, trace the given graph of f. On the same set of axes sketch the graph of f' and f''.

13. $f(x) = \dfrac{x^2 + 1}{x^2 - 1}$ **14.** $y = \dfrac{-3x^5 + 40x^3 + 135x}{270}$

5.   **6.**

7.   **8.**

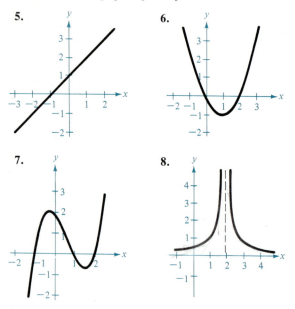

In Exercises 15–26, identify all relative extrema. Use the Second Derivative Test when applicable.

15. $f(x) = 6x - x^2$ **16.** $f(x) = x^2 + 3x - 8$
17. $f(x) = (x - 5)^2$ **18.** $f(x) = -(x - 5)^2$
19. $f(x) = x^3 - 3x^2 + 3$ **20.** $f(x) = 5 + 3x^2 - x^3$
21. $f(x) = x^4 - 4x^3 + 2$
22. $f(x) = x^3 - 9x^2 + 27x - 26$
23. $f(x) = x^{2/3} - 3$ **24.** $f(x) = \sqrt{x^2 + 1}$
25. $f(x) = x + \dfrac{4}{x}$ **26.** $f(x) = \dfrac{x}{x - 1}$

In Exercises 9–14, find the intervals on which the given function is concave upward and those on which it is concave downward.

In Exercises 27–40, sketch the graph of the given function and identify all relative extrema and points of inflection.

27. $f(x) = x^3 - 12x$ **28.** $f(x) = x^3 + 1$
29. $f(x) = x^3 - 6x^2 + 12x - 8$
30. $f(x) = 2x^3 - 3x^2 - 12x + 8$
31. $f(x) = \dfrac{1}{4}x^4 - 2x^2$ **32.** $f(x) = 2x^4 - 8x + 3$
33. $f(x) = x^2 + \dfrac{1}{x^2}$ **34.** $f(x) = \dfrac{x^2}{x^2 - 1}$
35. $f(x) = x\sqrt{x + 3}$ **36.** $f(x) = x\sqrt{x + 1}$
37. $f(x) = \dfrac{x}{x^2 - 4}$ **38.** $f(x) = \dfrac{1}{x^2 - x - 2}$
39. $f(x) = \dfrac{x - 2}{x^2 - 4x + 3}$ **40.** $f(x) = \dfrac{x + 1}{x^2 + x + 1}$

9. $y = x^2 - x - 2$ **10.** $f(x) = \dfrac{24}{x^2 + 12}$

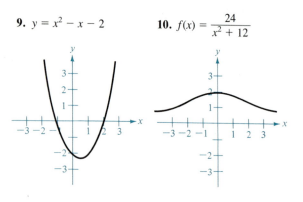

41. Sketch the graph of a function f having the following characteristics:
    (a) $f(2) = f(4) = 0$
    (b) $f'(x) < 0$ if $x < 3$
        $f'(3)$ is undefined.
        $f'(x) > 0$ if $x > 3$
    (c) $f''(x) < 0$

42. Sketch the graph of a function f having the following characteristics:
    (a) $f(2) = f(4) = 0$
    (b) $f'(x) > 0$ if $x < 3$
        $f'(3)$ is undefined.
        $f'(x) < 0$ if $x > 3$
    (c) $f''(x) > 0$

43. Find a cubic polynomial function that has a relative maximum at $(3, 3)$, a relative minimum at $(5, 1)$, and a point of inflection at $(4, 2)$.

44. Find a cubic polynomial function that has a relative maximum at $(2, 4)$, a relative minimum at $(4, 2)$, and a point of inflection at $(3, 3)$.

45. Show that the point of inflection of

$$f(x) = x(x - 6)^2$$

    lies midway between the relative extrema of f.

46. Prove that a cubic function with three real zeros has a point of inflection whose x-coordinate is the average of the three zeros.

47. The deflection D of a particular beam of length L is given by

$$D = 2x^4 - 5Lx^3 + 3L^2x^2$$

    where x is the distance from one end of the beam. Find the value of x that yields the maximum deflection.

48. The equation

$$E = \frac{T}{(x^2 + a^2)^{3/2}}$$

    gives the electric field intensity on the axis of a uniformly charged ring, where T is the total charge on the ring and a is the radius of the ring. At what value of x is E maximum?

49. A manufacturer has determined that the total cost C of operating a certain facility is given by

$$C = 0.5x^2 + 15x + 5000$$

    where x is the number of units produced. At what level of production will the average cost per unit be minimum? (The average cost per unit is given by C/x.)

50. The total cost C for ordering and storing x units is

$$C = 2x + \frac{300,000}{x}$$

    What order size will produce a minimum cost?

SECTION TOPICS ▪
Limits at infinity ▪
Horizontal asymptotes ▪

4.5
Limits at infinity

In this section we take a further look at the behavior of a function on *infinite* intervals. Consider the graph of

$$f(x) = \frac{3x^2}{x^2 + 1}$$

as shown in Figure 4.33. Table 4.11 on page 178 suggests that the value of $f(x)$ approaches 3 as x increases without bound ($x \to \infty$). Similarly, $f(x)$ approaches 3 as $x \to -\infty$. We denote these **limits at infinity** by

$$\lim_{x \to -\infty} f(x) = 3$$

and

$$\lim_{x \to \infty} f(x) = 3$$

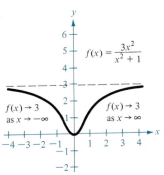

FIGURE 4.33

TABLE 4.11

	x decreases without bound					x increases without bound			
x	$-\infty$ $\leftarrow$	-100	-10	-1	0	1	10	100	$\rightarrow$ ∞
$f(x)$	3 $\leftarrow$	2.9997	2.97	1.5	0	1.5	2.97	2.9997	$\rightarrow$ 3

$f(x)$ approaches 3	$f(x)$ approaches 3

To say that a statement is true as x increases *without bound* means that for any (large) real number M, the statement is true for *all* x in the interval $\{x: x > M\}$. The following definition makes use of this concept.

DEFINITION OF LIMITS AT INFINITY

The statement

$$\lim_{x\to\infty} f(x) = L$$

means that for each $\varepsilon > 0$ there exists an $M > 0$ such that $|f(x) - L| < \varepsilon$ whenever $x > M$. The statement

$$\lim_{x\to-\infty} f(x) = L$$

means that for each $\varepsilon > 0$ there exists an $N < 0$ such that $|f(x) - L| < \varepsilon$ whenever $x < N$.

This definition is illustrated graphically in Figure 4.34. Remember that when we write

$$\lim_{x\to-\infty} f(x) = L \quad \text{or} \quad \lim_{x\to\infty} f(x) = L$$

we mean that the limit exists *and* the limit is equal to L.

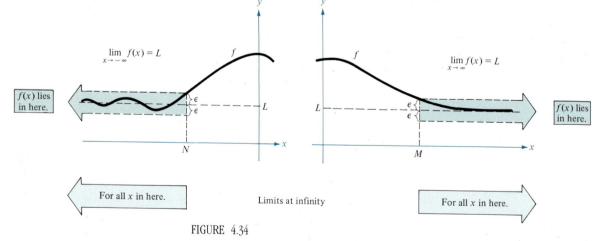

FIGURE 4.34

In Figure 4.34, the graph of f approaches the line $y = L$. We call this line a **horizontal asymptote** of the graph of f.

DEFINITION OF HORIZONTAL ASYMPTOTE

If

$$\lim_{x \to -\infty} f(x) = L \qquad \text{or} \qquad \lim_{x \to \infty} f(x) = L$$

then the line $y = L$ is called a **horizontal asymptote** of the graph of f.

Remark From this definition it follows that the graph of a *function* of x can have at most two horizontal asymptotes—one to the right and one to the left.

Limits at infinity share many of the properties of limits discussed in Section 2.1. We list these, without proof, in the following theorem. Although the properties are stated for limits as $x \to \infty$, they are also valid for limits as $x \to -\infty$.

THEOREM 4.10

PROPERTIES OF LIMITS AT INFINITY

If b is a real number, n is a positive integer, and the functions f and g have limits as $x \to \infty$, then the following properties are true.

1. Constant:

$$\lim_{x \to \infty} b = b$$

2. Scalar multiple:

$$\lim_{x \to \infty} [b\, f(x)] = b \left[\lim_{x \to \infty} f(x) \right]$$

3. Sum or difference:

$$\lim_{x \to \infty} [f(x) \pm g(x)] = \lim_{x \to \infty} f(x) \pm \lim_{x \to \infty} g(x)$$

4. Product:

$$\lim_{x \to \infty} [f(x)g(x)] = \left[\lim_{x \to \infty} f(x) \right] \left[\lim_{x \to \infty} g(x) \right]$$

5. Quotient:

$$\lim_{x \to \infty} \frac{f(x)}{g(x)} = \frac{\lim\limits_{x \to \infty} f(x)}{\lim\limits_{x \to \infty} g(x)} \quad \text{if } \lim_{x \to \infty} g(x) \neq 0$$

6. Power:

$$\lim_{x \to \infty} [f(x)]^n = \left[\lim_{x \to \infty} f(x) \right]^n$$

7. Root:

$$\lim_{x \to \infty} \sqrt[n]{f(x)} = \sqrt[n]{\lim_{x \to \infty} f(x)}$$

In evaluating limits at infinity, the following theorem is helpful.

THEOREM 4.11

LIMITS AT INFINITY

If r is positive and rational and c is any real number, then $\lim\limits_{x \to \infty} \dfrac{c}{x^r} = 0$.

Furthermore, if x^r is defined when $x < 0$, then $\lim\limits_{x \to -\infty} \dfrac{c}{x^r} = 0$.

Proof: We prove the first case and leave the second as an exercise. We begin by proving that

$$\lim_{x \to \infty} \frac{1}{x} = 0$$

For $\varepsilon > 0$, let $M = 1/\varepsilon$. Then for $x > M$, we have

$$x > M = \frac{1}{\varepsilon} \quad \Rightarrow \quad \frac{1}{x} < \varepsilon \quad \Rightarrow \quad \left| \frac{1}{x} - 0 \right| < \varepsilon$$

Therefore, by the definition, we conclude that the limit of $1/x$ as $x \to \infty$ is 0. Now, applying Theorem 4.10, we let $r = m/n$ and write

$$
\begin{aligned}
\lim_{x \to \infty} \frac{c}{x^r} &= \lim_{x \to \infty} \frac{c}{x^{m/n}} \\
&= c \left[\lim_{x \to \infty} \left(\frac{1}{\sqrt[n]{x}} \right)^m \right] \qquad \text{Property 2} \\
&= c \left[\lim_{x \to \infty} \sqrt[n]{\frac{1}{x}} \right]^m \qquad \text{Property 6} \\
&= c \left[\sqrt[n]{\lim_{x \to \infty} \frac{1}{x}} \right]^m \qquad \text{Property 7} \\
&= c \left[\sqrt[n]{0} \right]^m \\
&= 0
\end{aligned}
$$

EXAMPLE 1   Evaluating a limit at infinity

Find

$$\lim_{x \to \infty} \left(5 - \frac{2}{x^2} \right)$$

Solution: Using Theorem 4.11 and Properties 1 and 3 of Theorem 4.10, we have

$$\lim_{x \to \infty} \left(5 - \frac{2}{x^2} \right) = \lim_{x \to \infty} 5 - \lim_{x \to \infty} \frac{2}{x^2} = 5 - 0 = 5$$

You cannot apply Theorem 4.10 unless the limits of both f and g exist. However, if you are given functions that do not meet this requirement, you can sometimes form algebraic combinations of f and g to produce functions that do meet the hypotheses of the theorem. This procedure is demonstrated in the next example.

EXAMPLE 2   Evaluating a limit at infinity

Find

$$\lim_{x \to \infty} \frac{2x - 1}{x + 1}$$

Solution: We cannot apply Theorem 4.10 directly, since neither the numerator nor the denominator possesses a limit as $x \to \infty$.

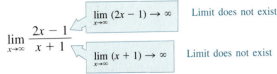

$$\lim_{x \to \infty} \frac{2x - 1}{x + 1}$$

$$\lim_{x \to \infty} (2x - 1) \to \infty \qquad \text{Limit does not exist}$$

$$\lim_{x \to \infty} (x + 1) \to \infty \qquad \text{Limit does not exist}$$

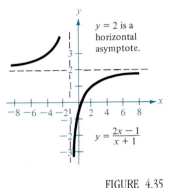

$y = 2$ is a horizontal asymptote.

$y = \frac{2x - 1}{x + 1}$

FIGURE 4.35

However, we can apply Theorem 4.10 if we first divide both the numerator and denominator by x. This division yields

$$\lim_{x \to \infty} \frac{2x - 1}{x + 1} = \lim_{x \to \infty} \frac{2 - (1/x)}{1 + (1/x)} = \frac{\displaystyle\lim_{x \to \infty} 2 - \lim_{x \to \infty} \frac{1}{x}}{\displaystyle\lim_{x \to \infty} 1 + \lim_{x \to \infty} \frac{1}{x}} = \frac{2 - 0}{1 + 0} = 2$$

Thus, the line $y = 2$ is a horizontal asymptote to the right. By taking the limit as $x \to -\infty$, we can see that $y = 2$ is also a horizontal asymptote to the left. The graph of this function is shown in Figure 4.35.

| **Remark** You may wish to test the reasonability of the limit found in Example 2 by evaluating $f(x)$ for large positive values of x.

In Example 2, our first attempt to apply Theorem 4.10 resulted in the **indeterminate form** ∞/∞. We were able to resolve the difficulty by rewriting the given expression in an equivalent form. Specifically, we divided both the numerator and the denominator by x. In general, we suggest dividing by the highest power of x in the *denominator*. This is illustrated in the following example.

EXAMPLE 3   A comparison of three rational functions
_____

Find the following limits:

(a) $\displaystyle\lim_{x \to \infty} \frac{2x + 5}{3x^2 + 1}$ (b) $\displaystyle\lim_{x \to \infty} \frac{2x^2 + 5}{3x^2 + 1}$ (c) $\displaystyle\lim_{x \to \infty} \frac{2x^3 + 5}{3x^2 + 1}$

Solution:
(a) To apply Theorem 4.10, we divide both the numerator and denominator by x^2 to obtain

$$\lim_{x \to \infty} \frac{2x + 5}{3x^2 + 1} = \lim_{x \to \infty} \frac{(2/x) + (5/x^2)}{3 + (1/x^2)} = \frac{0 + 0}{3 + 0} = \frac{0}{3} = 0$$

(b) In this case, we also divide by x^2 to obtain

$$\lim_{x \to \infty} \frac{2x^2 + 5}{3x^2 + 1} = \lim_{x \to \infty} \frac{2 + (5/x^2)}{3 + (1/x^2)} = \frac{2 + 0}{3 + 0} = \frac{2}{3}$$

(c) Again, we divide by x^2 to obtain

$$\lim_{x \to \infty} \frac{2x^3 + 5}{3x^2 + 1} = \lim_{x \to \infty} \frac{2x + (5/x^2)}{3 + (1/x^2)}$$

$$\lim_{x \to \infty} \left[2x + \frac{5}{x^2} \right] \to \infty$$

$$\lim_{x \to \infty} \left[3 + \frac{1}{x^2} \right] \to 3$$

In this case, we conclude that the limit *does not exist* because the numerator increases without bound while the modified denominator approaches 3.

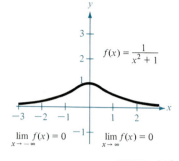

$$f(x) = \frac{1}{x^2 + 1}$$

$$\lim_{x \to -\infty} f(x) = 0 \qquad \lim_{x \to \infty} f(x) = 0$$

FIGURE 4.36

Maria Agnesi

It is instructive to compare the three rational functions in Example 3. In part (a) the degree of the numerator is *less* than the degree of the denominator and the limit of the rational function is zero. In part (b) the degrees of the numerator and denominator are *equal* and the limit is simply the ratio of the two leading coefficients 2 and 3. Finally, in part (c) the degree of the numerator is *greater* than that of the denominator and the limit does not exist. This seems reasonable when we realize that for large values of x the highest-powered term of the rational function is the most "influential" in determining the limit. For instance, the limit as x approaches infinity of the function given by

$$f(x) = \frac{1}{x^2 + 1}$$

is zero since the denominator overpowers the numerator as x increases or decreases without bound, as shown in Figure 4.36.

The function shown in Figure 4.36 is a special case of a type of curve studied by the Italian mathematician Maria Gaetana Agnesi (1717–1783). The general form of this function is $f(x) = a^3/(x^2 + a^2)$ and through a mistranslation of the Italian word *vertéré*, the curve has come to be known as the witch of Agnesi. Agnesi's work with this curve first appeared in a comprehensive text on calculus that was published in 1748.

In Figure 4.36, we can see that the function $f(x) = 1/(x^2 + 1)$ approaches the same horizontal asymptote to the right and to the left. That is,

$$\lim_{x \to -\infty} f(x) = 0 = \lim_{x \to \infty} f(x)$$

This is always the case with rational functions. However, functions that are not rational may approach different horizontal asymptotes to the right and to the left, as shown in Example 4.

EXAMPLE 4   *A function with two horizontal asymptotes*

Determine the following:

(a) $\displaystyle\lim_{x \to \infty} \frac{3x - 2}{\sqrt{2x^2 + 1}}$ (b) $\displaystyle\lim_{x \to -\infty} \frac{3x - 2}{\sqrt{2x^2 + 1}}$

Solution:

(a) To apply Theorem 4.10, we divide numerator and denominator by x. Since $\sqrt{x^2} = x$ for $x > 0$, we have

$$\frac{3x - 2}{\sqrt{2x^2 + 1}} = \frac{(3x - 2)/x}{\sqrt{2x^2 + 1}/\sqrt{x^2}} = \frac{3 - (2/x)}{\sqrt{(2x^2 + 1)/x^2}} = \frac{3 - (2/x)}{\sqrt{2 + (1/x^2)}}$$

Therefore,

$$\lim_{x \to \infty} \frac{3x - 2}{\sqrt{2x^2 + 1}} = \lim_{x \to \infty} \frac{3 - (2/x)}{\sqrt{2 + (1/x^2)}}$$

$$= \frac{3 - 0}{\sqrt{2 + 0}} = \frac{3}{\sqrt{2}} \approx 2.12$$

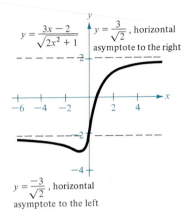

$y = \dfrac{3x - 2}{\sqrt{2x^2 + 1}}$

$y = \dfrac{3}{\sqrt{2}}$, horizontal asymptote to the right

$y = \dfrac{-3}{\sqrt{2}}$, horizontal asymptote to the left

FIGURE 4.37

(b) For $x \to -\infty$, we have $x < 0$ and $x = -\sqrt{x^2}$. Thus,

$$\frac{3x - 2}{\sqrt{2x^2 + 1}} = \frac{(3x - 2)/x}{\sqrt{2x^2 + 1}/(-\sqrt{x^2})} = \frac{3 - (2/x)}{-\sqrt{2 + (1/x^2)}}$$

Therefore,

$$\lim_{x \to -\infty} \frac{3x - 2}{\sqrt{2x^2 + 1}} = \lim_{x \to -\infty} \frac{3 - (2/x)}{-\sqrt{2 + (1/x^2)}}$$

$$= \frac{3 - 0}{-\sqrt{2 + 0}}$$

$$= -\frac{3}{\sqrt{2}} \approx -2.12$$

See Figure 4.37.

There are many examples of asymptotic behavior in the physical sciences. For instance, the following example describes the asymptotic recovery of oxygen in a pond.

EXAMPLE 5   An application

Suppose that $f(t)$ measures the level of oxygen in a pond, where $f(t) = 1$ is the normal (unpolluted) level and the time t is measured in weeks. When $t = 0$, organic waste is dumped into the pond, and as the waste material oxidizes, the amount of oxygen in the pond is given by

$$f(t) = \frac{t^2 - t + 1}{t^2 + 1}$$

What percentage of the normal level of oxygen exists in the pond after 1 week? After 2 weeks? After 10 weeks? What is the limit as t approaches infinity?

Solution: When $t = 1$, 2, and 10, the levels of oxygen are

$$f(1) = \frac{1^2 - 1 + 1}{1^2 + 1} = \frac{1}{2} = 50\% \qquad \text{1 week}$$

$$f(2) = \frac{2^2 - 2 + 1}{2^2 + 1} = \frac{3}{5} = 60\% \qquad \text{2 weeks}$$

$$f(10) = \frac{10^2 - 10 + 1}{10^2 + 1} = \frac{91}{101} \approx 90.1\% \qquad \text{10 weeks}$$

To take the limit as t approaches infinity, we divide the numerator and denominator by t^2 to obtain

$$\lim_{t \to \infty} \frac{t^2 - t + 1}{t^2 + 1} = \lim_{t \to \infty} \frac{1 - (1/t) + (1/t^2)}{1 + (1/t^2)}$$

$$= \frac{1 - 0 + 0}{1 + 0} = 1 = 100\%$$

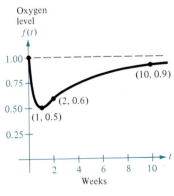

Oxygen level
$f(t)$

(10, 0.9)

(2, 0.6)

(1, 0.5)

Weeks

FIGURE 4.38

See Figure 4.38.

Section Exercises 4.5

In Exercises 1–8, match the given function to one of the graphs (a)–(h), using horizontal asymptotes as an aid.

1. $f(x) = \dfrac{3x^2}{x^2 + 2}$

2. $f(x) = \dfrac{2x}{\sqrt{x^2 + 2}}$

3. $f(x) = \dfrac{x}{x^2 + 2}$

4. $f(x) = 2 + \dfrac{x^2}{x^4 + 1}$

5. $f(x) = \dfrac{-6x}{\sqrt{4x^2 + 5}}$

6. $f(x) = 5 - \dfrac{1}{x^2 + 1}$

7. $f(x) = \dfrac{4}{x^2 + 1}$

8. $f(x) = \dfrac{2x^2 - 3x + 5}{x^2 + 1}$

(g)   (h)

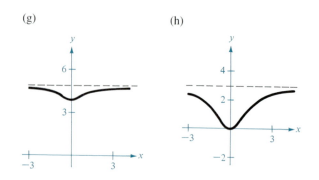

(a)   (b)

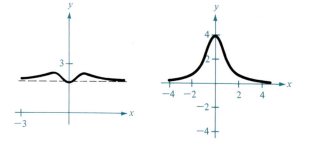

(c)   (d)

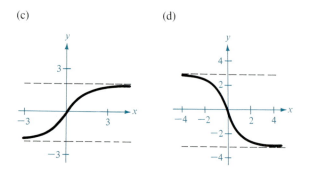

(e)   (f)

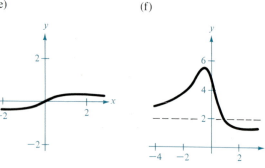

In Exercises 9–28, find the indicated limit.

9. $\displaystyle\lim_{x \to \infty} \dfrac{2x - 1}{3x + 2}$

10. $\displaystyle\lim_{x \to \infty} \dfrac{5x^3 + 1}{10x^3 - 3x^2 + 7}$

11. $\displaystyle\lim_{x \to \infty} \dfrac{x}{x^2 - 1}$

12. $\displaystyle\lim_{x \to \infty} \dfrac{2x^{10} - 1}{10x^{11} - 3}$

13. $\displaystyle\lim_{x \to -\infty} \dfrac{5x^2}{x + 3}$

14. $\displaystyle\lim_{x \to \infty} \dfrac{x^3 - 2x^2 + 3x + 1}{x^2 - 3x + 2}$

15. $\displaystyle\lim_{x \to \infty} \left(2x - \dfrac{1}{x^2}\right)$

16. $\displaystyle\lim_{x \to \infty} (x + 3)^{-2}$

17. $\displaystyle\lim_{x \to -\infty} \left(\dfrac{2x}{x - 1} + \dfrac{3x}{x + 1}\right)$

18. $\displaystyle\lim_{x \to \infty} \left(\dfrac{2x^2}{x - 1} + \dfrac{3x}{x + 1}\right)$

19. $\displaystyle\lim_{x \to -\infty} (x + \sqrt{x^2 + 3})$

20. $\displaystyle\lim_{x \to \infty} (2x - \sqrt{4x^2 + 1})$

21. $\displaystyle\lim_{x \to \infty} (x - \sqrt{x^2 + x})$

22. $\displaystyle\lim_{x \to -\infty} (3x + \sqrt{9x^2 - x})$

23. $\displaystyle\lim_{x \to -\infty} \dfrac{x}{\sqrt{x^2 - x}}$

24. $\displaystyle\lim_{x \to \infty} \dfrac{x}{\sqrt{x^2 + 1}}$

25. $\displaystyle\lim_{x \to \infty} \dfrac{2x + 1}{\sqrt{x^2 - x}}$

26. $\displaystyle\lim_{x \to -\infty} \dfrac{-3x + 1}{\sqrt{x^2 + x}}$

27. $\displaystyle\lim_{x \to \infty} \dfrac{x^2 - x}{\sqrt{x^4 + x}}$

28. $\displaystyle\lim_{x \to \infty} \dfrac{2x}{\sqrt{4x^2 + 1}}$

In Exercises 29–40, sketch the graph of each equation. As a sketching aid examine each equation for intercepts, symmetry, and asymptotes.

29. $y = \dfrac{2 + x}{1 - x}$

30. $y = \dfrac{x - 3}{x - 2}$

31. $y = \dfrac{x^2}{x^2 + 9}$

32. $y = \dfrac{x^2}{x^2 - 9}$

33. $xy^2 = 4$

34. $x^2 y = 4$

35. $y = \dfrac{2x}{1 - x}$

36. $y = \dfrac{2x}{1 - x^2}$

37. $y = 2 - \dfrac{3}{x^2}$

38. $y = 1 + \dfrac{1}{x}$

39. $y = \dfrac{x^3}{\sqrt{x^2 - 4}}$ **40.** $y = \dfrac{x}{\sqrt{x^2 - 4}}$

In Exercises 41–44, complete the given table and estimate the limit of $f(x)$ as x approaches infinity. Then find the limit analytically and compare your results.

41. $f(x) = \dfrac{x + 1}{x\sqrt{x}}$

x	10^0	10^1	10^2	10^3	10^4	10^5	10^6
$f(x)$							

42. $f(x) = x - \sqrt{x(x - 1)}$

x	10^0	10^1	10^2	10^3	10^4	10^5	10^6
$f(x)$							

43. $f(x) = 2x - \sqrt{4x^2 + 1}$

x	10^0	10^1	10^2	10^3	10^4	10^5	10^6
$f(x)$							

44. $f(x) = x^2 - x\sqrt{x(x - 1)}$

x	10^0	10^1	10^2	10^3	10^4	10^5	10^6
$f(x)$							

45. A business has a cost of $C = 0.5x + 500$ for producing x units. The average cost per unit is given by $\overline{C} = C/x$. Find the limit of $\overline{C}$ as x approaches infinity.

46. According to the theory of relativity, the mass m of a particle depends on its velocity v. That is,

$$m = \frac{m_0}{\sqrt{1 - (v^2/c^2)}}$$

where m_0 is the mass when the particle is at rest and c is the speed of light. Find the limit of the mass as v approaches c.

47. The efficiency of an internal combustion engine is defined to be

$$\text{efficiency } (\%) = 100\left[1 - \frac{1}{(v_1/v_2)^c}\right]$$

where v_1/v_2 is the ratio of the uncompressed gas to the compressed gas and c is a constant dependent upon the engine design. Find the limit of the efficiency as the compression ratio approaches infinity.

48. The game commission in a certain state introduces 50 deer into some newly acquired state game land. It is believed that the size of the herd will increase according to the model

$$N = \frac{10(5 + 3t)}{1 + 0.04t}$$

where t is the time in years. Find the number in the herd when $t = 5$, 10, and 25 years. What is the limit of N as t approaches infinity?

49. Prove that if

$$p(x) = a_n x^n + \cdots + a_1 x + a_0$$
$$q(x) = b_m x^m + \cdots + b_1 x + b_0$$

then

$$\lim_{x \to \infty} \frac{p(x)}{q(x)} = \begin{cases} 0, & n < m \\ \dfrac{a_n}{b_m}, & n = m \\ \pm\infty, & n > m \end{cases}$$

50. Complete the proof of Theorem 4.11.

SECTION TOPIC ▪
Summary of curve sketching techniques ▪

4.6
A summary of curve sketching

It would be difficult to overstate the importance of curve sketching in mathematics. Descartes's introduction of this concept not only preceded, but to a great extent contributed to, the rapid advances in mathematics since the mid-seventeenth century. In the words of Lagrange, "As long as algebra and geometry traveled separate paths their advance was slow and their applications limited. But when these two sciences joined company, they drew from each other fresh vitality and thenceforth marched on at a rapid pace towards perfection."

Today, government, science, industry, business, education, and the social and health sciences all make widespread use of graphs to describe and predict relationships between variables. However, we have seen that sometimes sketching a graph can require considerable ingenuity.

Up to this point in the text, we have discussed several concepts that are useful in sketching the graph of a function.

- Domain and range
- *x*-intercepts and *y*-intercepts
- Symmetry
- Points of discontinuity
- Horizontal and vertical asymptotes
- Points of nondifferentiability
- Relative extrema
- Concavity
- Points of inflection

In this section we show some examples that incorporate these concepts into an effective procedure for sketching curves.

SUGGESTIONS FOR SKETCHING THE GRAPH OF A FUNCTION

1. Make a rough preliminary sketch that includes any easily determined intercepts and asymptotes.
2. Locate the *x*-values where $f'(x)$ and $f''(x)$ are either zero or undefined.
3. Test the behavior of f at and between each of these *x*-values.
4. Sharpen the accuracy of the final sketch by plotting the relative extrema, the points of inflection, and a few points between.

| **Remark** Note in these guidelines the importance of *algebra* (as well as calculus) for solving the equations $f(x) = 0$, $f'(x) = 0$, and $f''(x) = 0$.

EXAMPLE 1   Sketching the graph of a rational function

Sketch the graph of

$$f(x) = \frac{2(x^2 - 9)}{x^2 - 4}$$

Solution:

$$f'(x) = \frac{20x}{(x^2 - 4)^2} \qquad \text{and} \qquad f''(x) = -\frac{20(3x^2 + 4)}{(x^2 - 4)^3}$$

x-intercepts: $(-3, 0)$, $(3, 0)$
y-intercept: $(0, \frac{9}{2})$
Vertical asymptotes: $x = -2$ and $x = 2$
Horizontal asymptote: $y = 2$
Critical number: $x = 0$
Possible points of inflection: None
Test intervals: $(-\infty, -2)$, $(-2, 0)$, $(0, 2)$, $(2, \infty)$ [See Table 4.12.]

The intercepts and asymptotes can be used to give preliminary clues regarding the nature of the graph of f as shown in Figure 4.39. Then, we can complete the graph as shown in Figure 4.40 using a few additional points together with the conclusions summarized in Table 4.12.

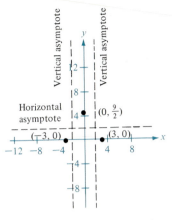

FIGURE 4.39

TABLE 4.12

	$f(x)$	$f'(x)$	$f''(x)$	*Shape of graph*
$-\infty < x < -2$		$-$	$-$	decreasing, concave down
$x = -2$	undefined	undefined	undefined	vertical asymptote
$-2 < x < 0$		$-$	$+$	decreasing, concave up
$x = 0$	$\dfrac{9}{2}$	0	$+$	relative minimum
$0 < x < 2$		$+$	$+$	increasing, concave up
$x = 2$	undefined	undefined	undefined	vertical asymptote
$2 < x < \infty$		$+$	$-$	increasing, concave down

EXAMPLE 2   *Sketching the graph of a rational function*

Sketch the graph of

$$f(x) = \frac{x^2 - 2x + 4}{x - 2}$$

Solution:

$$f'(x) = \frac{x(x - 4)}{(x - 2)^2} \quad \text{and} \quad f''(x) = \frac{8}{(x - 2)^3}$$

x-intercepts: None, since $x^2 - 2x + 4 = (x - 1)^2 + 3 > 0$ for all x.
y-intercept: $(0, -2)$
Vertical asymptote: $x = 2$
Critical numbers: $x = 0$ and $x = 4$
Possible points of inflection: None
Test intervals: $(-\infty, 0)$, $(0, 2)$, $(2, 4)$, $(4, \infty)$ [See Table 4.13.]
The graph of f is shown in Figure 4.41.

TABLE 4.13

	$f(x)$	$f'(x)$	$f''(x)$	*Shape of graph*
$-\infty < x < 0$		$+$	$-$	increasing, concave down
$x = 0$	-2	0	$-$	relative maximum
$0 < x < 2$		$-$	$-$	decreasing, concave down
$x = 2$	undefined	undefined	undefined	vertical asymptote
$2 < x < 4$		$-$	$+$	decreasing, concave up
$x = 4$	6	0	$+$	relative minimum
$4 < x < \infty$		$+$	$+$	increasing, concave up

$$f(x) = \frac{2(x^2 - 9)}{x^2 - 4}$$

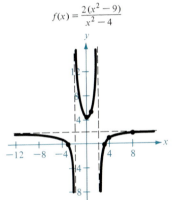

FIGURE 4.40

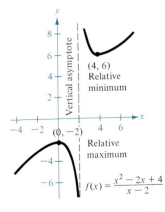

$$f(x) = \frac{x^2 - 2x + 4}{x - 2}$$

FIGURE 4.41

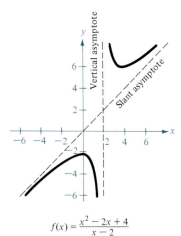

$$f(x) = \frac{x^2 - 2x + 4}{x - 2}$$

FIGURE 4.42

A rational function is called **improper** if the degree of the numerator exceeds the degree of the denominator. If the degree of the numerator exceeds the degree of the denominator by one, then the function has a **slant asymptote.** To find the slant asymptote of such a function, use division to rewrite the (improper) rational expression as the sum of a first-degree polynomial and a (proper) rational expression. For instance, in Example 2 the slant asymptote is determined as follows:

$$f(x) = \frac{x^2 - 2x + 4}{x - 2} = x + \frac{4}{x - 2}$$

y = x is a slant asymptote

In Figure 4.42, we can see that the graph of f approaches the slant asymptote $y = x$ as x approaches $\pm\infty$.

EXAMPLE 3   Sketching a rational function with a slant asymptote

Sketch the graph of

$$f(x) = \frac{-x^3 + x^2 + 4}{x^2}$$

Solution:

$$f'(x) = -\frac{x^3 + 8}{x^3}$$

$$f''(x) = \frac{24}{x^4}$$

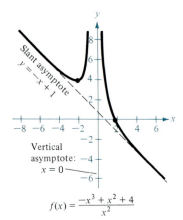

$$f(x) = \frac{-x^3 + x^2 + 4}{x^2}$$

FIGURE 4.43

x-intercept: $(2, 0)$ since $-x^3 + x^2 + 4 = -(x - 2)(x^2 + x + 2)$
Vertical asymptote: $x = 0$

Slant asymptote: $y = -x + 1$ since $\dfrac{-x^3 + x^2 + 4}{x^2} = -x + 1 + \dfrac{4}{x^2}$

Critical number: $x = -2$
Test intervals: $(-\infty, -2), (-2, 0), (0, \infty)$ [See Table 4.14.]
The graph of f is shown in Figure 4.43.

TABLE 4.14

	$f(x)$	$f'(x)$	$f''(x)$	*Shape of graph*
$-\infty < x < -2$		−	+	decreasing, concave up
$x = -2$	4	0	+	relative minimum
$-2 < x < 0$		+	+	increasing, concave up
$x = 0$	undefined	undefined	undefined	vertical asymptote
$0 < x < \infty$		−	+	decreasing, concave up

EXAMPLE 4   *Sketching the graph of a function involving a radical*

Sketch the graph of

$$f(x) = \frac{x}{\sqrt{x^2 + 2}}$$

Solution:

$$f'(x) = \frac{2}{(x^2 + 2)^{3/2}}$$

$$f''(x) = \frac{-6x}{(x^2 + 2)^{5/2}}$$

x-intercepts: $(0, 0)$
y-intercept: $(0, 0)$
Vertical asymptotes: None
Horizontal asymptotes: $y = 1$ (to the right)
$\quad\quad\quad\quad\quad\quad\quad\quad\quad\quad\quad y = -1$ (to the left)
Critical numbers: None
Possible points of inflection: $x = 0$
Points of discontinuity: None
Test intervals: $(-\infty, 0), (0, \infty)$ [See Table 4.15.]
The graph of f is shown in Figure 4.44.

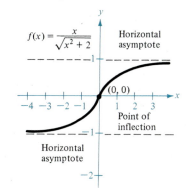

FIGURE 4.44

TABLE 4.15

	$f(x)$	$f'(x)$	$f''(x)$	*Shape of graph*
$-\infty < x < 0$		$+$	$+$	increasing, concave up
$x = 0$	0	$\dfrac{1}{\sqrt{2}}$	0	point of inflection
$0 < x < \infty$		$+$	$-$	increasing, concave down

| Remark   Although the function in Example 4 is defined over the entire real line, a square root often indicates a restricted domain. For example,

$$f(x) = \sqrt{x^2 - 4}$$

has $(-\infty, -2] \cup [2, \infty)$ as its domain.

EXAMPLE 5   Sketching the graph of a function involving cube roots

Sketch the graph of $f(x) = 2x^{5/3} - 5x^{4/3}$.

Solution:

$$f'(x) = \frac{10}{3}x^{1/3}(x^{1/3} - 2)$$

$$f''(x) = \frac{20(x^{1/3} - 1)}{9x^{2/3}}$$

x-intercepts: $(0, 0)$ and $\left(\frac{125}{8}, 0\right)$

y-intercept: $(0, 0)$

Critical number: $x = 0$ and $x = 8$

Possible points of inflection: $x = 0$ and $x = 1$

Points of discontinuity: None

Test intervals: $(-\infty, 0)$, $(0, 1)$, $(1, 8)$, $(8, \infty)$ [See Table 4.16.]

The graph of f is shown in Figure 4.45.

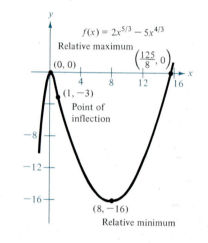

FIGURE 4.45

TABLE 4.16

	$f(x)$	$f'(x)$	$f''(x)$	*Shape of graph*
$-\infty < x < 0$		+	−	increasing, concave down
$x = 0$	0	0	undefined	relative maximum
$0 < x < 1$		−	−	decreasing, concave down
$x = 1$	−3	−	0	point of inflection
$1 < x < 8$		−	+	decreasing, concave up
$x = 8$	−16	0	+	relative minimum
$8 < x < \infty$		+	+	increasing, concave up

Remark   Although the function in Example 5 is differentiable over the entire real line, cube roots often indicate that there are points where the derivative is undefined. For example, $f(x) = x^{1/3}$ has a vertical tangent at $(0, 0)$.

EXAMPLE 6   Sketching the graph of a polynomial function

Sketch the graph of $f(x) = x^4 - 12x^3 + 48x^2 - 64x = x(x - 4)^3$.

Solution:

$$f'(x) = 4x^3 - 36x^2 + 96x - 64 = 4(x - 1)(x - 4)^2$$
$$f''(x) = 12x^2 - 72x + 96 = 12(x - 4)(x - 2)$$

x-intercepts: $(0, 0)$ and $(4, 0)$
y-intercept: $(0, 0)$
Critical numbers: $x = 1$ and $x = 4$
Possible points of inflection: $x = 2$ and $x = 4$
Points of discontinuity: None
Test intervals: $(-\infty, 1)$, $(1, 2)$, $(2, 4)$, $(4, \infty)$ [See Table 4.17.]
The graph of f is shown in Figure 4.46.

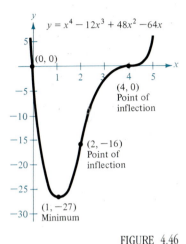

$y = x^4 - 12x^3 + 48x^2 - 64x$

(0, 0)

(4, 0)
Point of inflection

(2, −16)
Point of inflection

(1, −27)
Minimum

FIGURE 4.46

TABLE 4.17

	$f(x)$	$f'(x)$	$f''(x)$	*Shape of graph*
$-\infty < x < 1$		−	+	decreasing, concave up
$x = 1$	−27	0	+	relative minimum
$1 < x < 2$		+	+	increasing, concave up
$x = 2$	−16	+	0	point of inflection
$2 < x < 4$		+	−	increasing, concave down
$x = 4$	0	0	0	point of inflection
$4 < x < \infty$		+	+	increasing, concave up

Remark   The fourth degree polynomial function in Example 6 has one relative minimum and no relative maxima. In general, a polynomial function of degree n can have *at most* $n - 1$ relative extrema. Moreover, polynomial functions of even degree must have *at least* one relative extremum.

Section Exercises 4.6

In Exercises 1–24, sketch the graph of the given function. Choose a scale that allows all relative extrema and points of inflection to be identified on the sketch.

1. $y = x^3 - 3x^2 + 3$

2. $y = -\dfrac{1}{3}(x^3 - 3x + 2)$

3. $y = 2 - x - x^3$

4. $y = x^3 + 3x^2 + 3x + 2$

5. $f(x) = 3x^3 - 9x + 1$

6. $f(x) = (x + 1)(x - 2)(x - 5)$

7. $f(x) = -x^3 + 3x^2 + 9x - 2$

8. $f(x) = \dfrac{1}{3}(x - 1)^3 + 2$

9. $y = 3x^4 + 4x^3$
10. $y = 3x^4 - 6x^2$
11. $f(x) = x^4 - 4x^3 + 16x$
12. $f(x) = x^4 - 8x^3 + 18x^2 - 16x + 5$
13. $f(x) = x^4 - 4x^3 + 16x - 16$
14. $f(x) = x^5 + 1$
15. $y = x^5 - 5x$ 16. $y = (x - 1)^5$
17. $y = |2x - 3|$ 18. $y = |x^2 - 6x + 5|$
19. $y = \dfrac{x^2}{x^2 + 3}$ 20. $y = \dfrac{x}{x^2 + 1}$
21. $y = 3x^{2/3} - 2x$ 22. $y = 3x^{2/3} - x^2$
23. $f(x) = \dfrac{x}{\sqrt{x^2 + 7}}$ 24. $f(x) = \dfrac{4x}{\sqrt{x^2 + 15}}$

In Exercises 25–36, sketch the graph of the given function. In each case label the intercepts, relative extrema, points of inflection, and asymptotes.

25. $y = \dfrac{1}{x - 2} - 3$ 26. $y = \dfrac{x^2 + 1}{x^2 - 2}$
27. $y = \dfrac{2x}{x^2 - 1}$ 28. $y = \dfrac{x^2 - 6x + 12}{x - 4}$
29. $y = x\sqrt{4 - x}$ 30. $y = x\sqrt{4 - x^2}$
31. $f(x) = \dfrac{x + 2}{x}$ 32. $f(x) = x + \dfrac{32}{x^2}$
33. $f(x) = \dfrac{x^2 + 1}{x}$ 34. $f(x) = \dfrac{x^3}{x^2 - 1}$
35. $y = \dfrac{x^3}{2x^2 - 8}$ 36. $y = \dfrac{2x^2 - 5x + 5}{x - 2}$

In Exercises 37–42, determine conditions on the coefficients of

$$f(x) = ax^3 + bx^2 + cx + d$$

such that the graph of f will resemble the given graph.

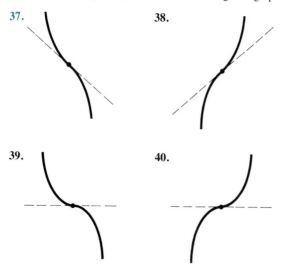

37.

38.

39.

40.

41.

42.

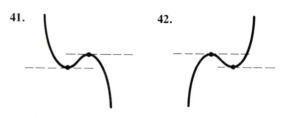

In Exercises 43–48, use the given graph (of f' or f'') to sketch a graph of the function f. [Hint: The solutions are not unique.]

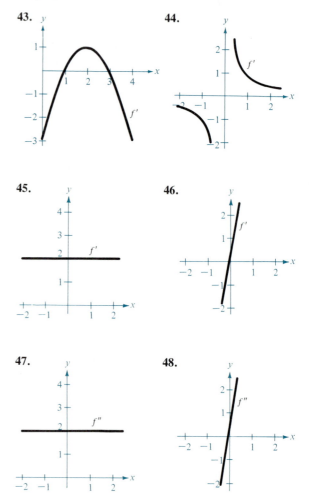

43.

44.

45.

46.

47.

48.

49. Find the relative extrema and points of inflection of the graph of

$$f(x) = \dfrac{3x - 2}{\sqrt{2x^2 + 1}}$$

(The graph of f is shown in Example 4 of Section 4.5.)

SECTION TOPIC ▪
Applied minimum and
maximum problems ▪

4.7
Optimization problems

One of the most common applications of calculus involves the determination of minimum or maximum values. Consider how frequently we hear or read terms like greatest profit, least cost, least time, greatest voltage, optimum size, least area, greatest strength, or greatest distance. Before outlining a general method of solution for such problems, we present an example.

EXAMPLE 1   Finding the maximum volume

An open box having a square base is to be constructed from 108 square inches of material. What dimensions will produce a box with maximum volume?

Solution: Since the box (in Figure 4.47) has a square base, its volume is

$$V = x^2h \qquad \text{Primary equation}$$

Furthermore, since the box is open at the top, its surface area is

$$S = (\text{area of base}) + (\text{area of four sides})$$
$$S = x^2 + 4xh = 108 \qquad \text{Secondary equation}$$

Since V is to be maximized, we express it as a function of just one variable. To do this, we solve the equation $108 = x^2 + 4xh$ for h in terms of x to obtain

$$h = \frac{108 - x^2}{4x}$$

Substituting in the equation for volume, we have

$$V = x^2h = x^2\left(\frac{108 - x^2}{4x}\right) = 27x - \frac{x^3}{4} \qquad \text{Function of one variable}$$

Before we try to find which x-value will yield a maximum value of V, we should determine the *feasible domain*. That is, what values of x make sense in this problem? We know that x should be nonnegative and that the area of the base $(A = x^2)$ is at most 108. Thus, we have

$$0 \le x \le \sqrt{108} \qquad \text{Feasible domain}$$

Now, to maximize V we find the critical numbers as follows.

$$\frac{dV}{dx} = 27 - \frac{3x^2}{4} = 0$$
$$3x^2 = 108 \;\Longrightarrow\; x = \pm 6 \qquad \text{Critical numbers}$$

Evaluating V at the critical numbers in the domain and at the endpoints of the domain, we have

$$V(0) = 0, \qquad V(6) = 108, \qquad \text{and} \qquad V(\sqrt{108}) = 0$$

We conclude that V is maximum when $x = 6$ and the dimensions of the box are $6 \times 6 \times 3$.

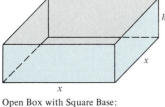

Open Box with Square Base:
$S = x^2 + 4xh = 108$

FIGURE 4.47

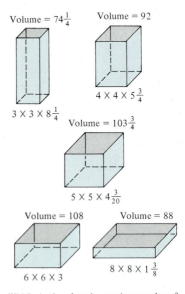

Volume = $74\frac{1}{4}$ Volume = 92

$3 \times 3 \times 8\frac{1}{4}$ $4 \times 4 \times 5\frac{3}{4}$

Volume = $103\frac{3}{4}$

$5 \times 5 \times 4\frac{3}{20}$

Volume = 108   Volume = 88

$6 \times 6 \times 3$ $8 \times 8 \times 1\frac{3}{8}$

Which size box has the maximum volume?

FIGURE 4.48

| **Remark** One might argue that the feasible domain in Example 1 is $0 < x < \sqrt{108}$ since the values $x = 0$ and $x = \sqrt{108}$ cannot be used to construct an actual "box". With such an interpretation, we can still conclude that a maximum occurs when $x = 6$ by using the First Derivative Test.

Many students have trouble with word problems because they are too eager to use a pat formula. For instance, in Example 1 you should realize that there are infinitely many open boxes having 108 square inches of surface area. You might begin by asking yourself which basic shape would seem to yield a maximum volume. Should the box be tall, squat, or more cubical? You might even try calculating a few volumes, as shown in Figure 4.48, to see if you can get a better feeling for what the optimum dimensions should be. Remember that you are not ready to begin solving a problem until you have clearly identified what the problem is.

Example 1 illustrates the following five-step procedure for solving applied minimum and maximum problems.

PROCEDURES FOR SOLVING APPLIED MINIMUM AND MAXIMUM PROBLEMS

1. Assign symbols to all given quantities and quantities to be determined. If applicable, make a sketch.
2. Write a **primary equation** for the quantity to be maximized or minimized.
3. Reduce the primary equation to one having a single independent variable. This may involve the use of **secondary equations** relating the independent variables of the primary equation.
4. Determine the domain of the primary equation. That is, determine the values for which the stated problem makes sense.
5. Determine the desired maximum or minimum value by the techniques discussed in Sections 4.1 through 4.4.

| **Remark** When performing step 5, recall that to determine the maximum or minimum value of a continuous function f on a closed interval, we compare the values of f at its relative extrema to the values of f at the endpoints of the interval.

EXAMPLE 2 *Finding the minimum sum*

Find two positive numbers $(0, \infty)$ that minimize the sum of twice the first number plus the second if the product of the two numbers is 288.

Solution:

1. Let x be the first number, y the second, and S the sum to be minimized.
2. Since we wish to minimize S, we write

$$S = 2x + y \qquad \text{Primary equation}$$

3. Since the product of the two numbers is 288, we have

$$xy = 288 \implies y = \frac{288}{x} \qquad \text{Secondary equation}$$

Thus, we can rewrite the primary equation in terms of x alone as

$$S = 2x + \frac{288}{x} \qquad \text{Function of one variable}$$

4. Since x must be positive, the feasible domain is $(0, \infty)$.
5. To find the critical numbers, we set the derivative to zero.

$$\frac{dS}{dx} = 2 - \frac{288}{x^2} = 0$$

$$x^2 = 144 \implies x = \pm 12$$

Choosing the positive x-value, we use the First Derivative Test to conclude that S is decreasing in the interval $(0, 12)$ and increasing in the interval $(12, \infty)$. Therefore, $x = 12$ yields a minimum and the two numbers are

$$x = 12 \qquad \text{and} \qquad y = \frac{288}{12} = 24$$

EXAMPLE 3   Finding the minimum distance

Find the points on the graph of $y = 4 - x^2$ that are closest to the point $(0, 2)$.

Solution:

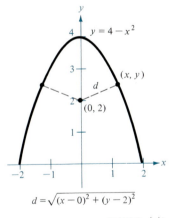

$$d = \sqrt{(x - 0)^2 + (y - 2)^2}$$

FIGURE 4.49

1. The symmetry of the graph in Figure 4.49 indicates that there are two points at a minimum distance from $(0, 2)$.
2. To obtain an equation for d, we use the *Distance Formula:*

$$d = \sqrt{(x - 0)^2 + (y - 2)^2} \qquad \text{Primary equation}$$

3. Using the secondary equation $y = 4 - x^2$, we can rewrite the primary equation as

$$d = \sqrt{x^2 + (4 - x^2 - 2)^2} = \sqrt{x^2 + (2 - x^2)^2} = \sqrt{x^4 - 3x^2 + 4}$$

Since d is smallest when the expression inside the radical is smallest, we need only find the critical numbers of

$$f(x) = x^4 - 3x^2 + 4$$

4. The domain of f is the entire real line.
5. Differentiation yields

$$f'(x) = 4x^3 - 6x = 2x(2x^2 - 3) = 0$$

$$x = 0, \ \sqrt{\frac{3}{2}}, \ -\sqrt{\frac{3}{2}}$$

The First Derivative Test verifies that $x = 0$ yields a relative maximum, while both $x = \sqrt{3/2}$ and $x = -\sqrt{3/2}$ yield a minimum distance. Hence, the closest points are $(\sqrt{3/2}, 5/2)$ and $(-\sqrt{3/2}, 5/2)$.

EXAMPLE 4   *Finding the minimum area*

A rectangular page is to contain 24 square inches of print. The margins at the top and bottom of the page are each $1\frac{1}{2}$ inches wide. The margins on each side are 1 inch. What should the dimensions of the page be so that the least amount of paper is used?

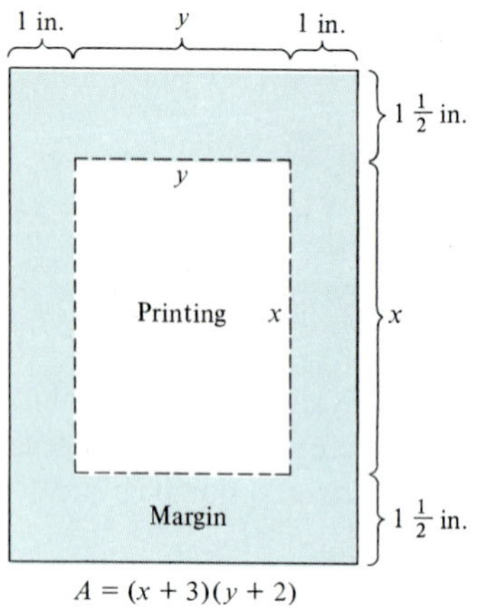

Solution: Letting A be the area to be minimized, we have, from Figure 4.50,

$$A = (x + 3)(y + 2) \qquad \text{Primary equation}$$

The printed area inside the margins is given by

$$24 = xy \qquad \text{Secondary equation}$$

By solving this equation for y, we have

$$y = \frac{24}{x}$$

and the primary equation becomes

$$A = (x + 3)\left(\frac{24}{x} + 2\right) = 30 + 2x + \frac{72}{x} \qquad \text{Function of one variable}$$

Since x must be positive, we are only interested in values of A when $x > 0$. Now, to find the minimum area, we differentiate with respect to x and obtain

$$\frac{dA}{dx} = 2 - \frac{72}{x^2} = 0 \implies x^2 = 36$$

Thus, the critical numbers are $x = \pm 6$. Since -6 is outside the domain, we choose $x = 6$. The First Derivative Test confirms that A is a minimum at this point. Therefore, $y = \frac{24}{6} = 4$ and the dimensions of the page should be

$$x + 3 = 9 \text{ inches} \qquad \text{by} \qquad y + 2 = 6 \text{ inches}$$

1 in. y 1 in.

$1\frac{1}{2}$ in.

y

Printing x x

Margin $1\frac{1}{2}$ in.

$A = (x + 3)(y + 2)$

FIGURE 4.50

EXAMPLE 5   *Finding the minimum length*

Two posts, one 12 feet high and the other 28 feet high, stand 30 feet apart. They are to be stayed by wires, attached to a single stake, running from ground level to the tops of the posts. Where should the stake be placed to use the least wire?

Solution: Let W be the wire length to be minimized. Using Figure 4.51, we have

$$W = y + z \qquad \text{Primary equation}$$

In this problem, rather than solving for y in terms of z (or vice versa), we solve for both y and z in terms of a third variable x, as shown in Figure 4.51. From the Pythagorean Theorem, we have

$$x^2 + 12^2 = y^2 \implies y = \sqrt{x^2 + 144}$$
$$(30 - x)^2 + 28^2 = z^2 \implies z = \sqrt{x^2 - 60x + 1684}$$

Thus, we have

$$W = y + z = \sqrt{x^2 + 144} + \sqrt{x^2 - 60x + 1684}, \quad 0 \le x \le 30$$

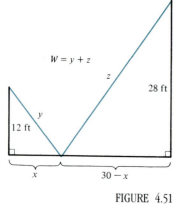

$W = y + z$

z

28 ft

y

12 ft

x $30 - x$

FIGURE 4.51

Differentiating W with respect to x, we have

$$\frac{dW}{dx} = \frac{x}{\sqrt{x^2 + 144}} + \frac{x - 30}{\sqrt{x^2 - 60x + 1684}}$$

By letting $dW/dx = 0$, we obtain

$$x^2(x^2 - 60x + 1684) = (30 - x)^2(x^2 + 144)$$

which simplifies to

$$320(x - 9)(2x + 45) = 0$$

Since $x = -22.5$ is not in the interval $[0, 30]$, we compare the values of W at the critical number $x = 9$ with the values of W at the endpoints $x = 0$ and $x = 30$ as follows:

$$W(0) \approx 53.04, \qquad W(9) = 50, \qquad \text{and} \qquad W(30) \approx 60.31$$

This implies that W is a minimum when $x = 9$, and we conclude that the wire should be stayed at a point 9 feet from the 12-foot pole.

In each of the first five examples, the extreme value occurred at a critical number. Although this is often the case, you must keep in mind that an extreme value can occur at an endpoint of an interval. The next example illustrates this.

EXAMPLE 6   An endpoint maximum

Four feet of wire is to be used to form a square and a circle. How much of the wire should be used for the square and how much should be used for the circle to enclose the maximum total area?

Solution: From Figure 4.52, we see that the total area is given by

$$A = (\text{area of square}) + (\text{area of circle})$$
$$A = x^2 + \pi r^2 \qquad\qquad \text{Primary equation}$$

Since the total amount of wire is 4 feet, we have

$$4 = (\text{perimeter of square}) + (\text{circumference of circle})$$
$$4 = 4x + 2\pi r$$

Thus, $r = 2(1 - x)/\pi$, and by substituting in the primary equation we have

$$A = x^2 + \pi\left[\frac{2(1 - x)}{\pi}\right]^2 = x^2 + \frac{4(1 - x)^2}{\pi} = \frac{1}{\pi}[(\pi + 4)x^2 - 8x + 4]$$

The feasible domain is $0 \le x \le 1$. Since $dA/dx = [2(\pi + 4)x - 8]/\pi$, the only critical number is $x = 4/(\pi + 4) \approx 0.56$. Therefore, since

$$A(0) \approx 1.273, \qquad A(0.56) \approx 0.56, \qquad \text{and} \qquad A(1) = 1$$

we conclude that the maximum area occurs when $x = 0$. That is, all of the wire is used for the circle.

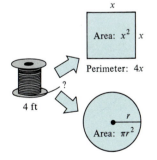

x

Area: x^2 | x

Perimeter: $4x$

r

Area: πr^2

Circumference: $2\pi r$

$4x + 2\pi r = 4$

$A = x^2 + \pi r^2$

4 ft

?

FIGURE 4.52

Let's review the primary equations developed in the first six examples. As applications go, these six examples are fairly simple, and yet the resulting primary equations are quite complicated.

$$V = 27x - \frac{x^3}{4} \qquad S = 2x + \frac{288}{x}$$

$$d = \sqrt{x^4 - 3x^2 + 4} \qquad W = \sqrt{x^2 + 144} + \sqrt{x^2 - 60x + 1684}$$

$$A = 30 + 2x + \frac{72}{x} \qquad A = \frac{1}{\pi}[(\pi + 4)x^2 - 8x + 4]$$

Real applications often involve equations that are at least as messy as these six, and you must expect that. Remember that one of the main goals of this course is to learn to use the power of calculus to analyze equations that at first glance seem formidable.

Section Exercises 4.7

1. A rectangle has a perimeter of 100 feet. What length and width should it have so that its area is maximum?
2. What positive number x minimizes the sum of x and its reciprocal?
3. The sum of one number and two times a second number is 24. What numbers should be selected so that their product is as large as possible?
4. The difference of two numbers is 50. Find the two numbers so that their product is as small as possible.
5. Find two positive numbers whose sum is 110 and whose product is a maximum.
6. The sum of one positive number and twice a second positive number is 100. Find the two numbers so that their product is a maximum.
7. Find two positive numbers whose product is 192 and whose sum is a minimum.
8. The product of two positive numbers is 192. What numbers should be chosen so that the sum of the first plus three times the second is a minimum?
9. A rancher has 200 feet of fencing to enclose two adjacent rectangular corrals, as shown in Figure 4.53. What dimensions should be used so that the enclosed area will be maximum?

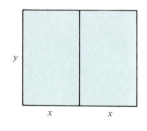

FIGURE 4.53

10. A dairy farmer plans to fence in a rectangular pasture adjacent to a river. The pasture must contain 180,000 square meters in order to provide enough grass for the herd. What dimensions would require the least amount of fencing if no fencing is needed along the river?
11. An open box is to be made from a square piece of material, 12 inches on a side, by cutting equal squares from each corner and turning up the sides, as shown in Figure 4.54. Find the volume of the largest box that can be made in this manner.

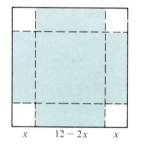

FIGURE 4.54

12. (a) Solve Exercise 11 if the square of material is s inches on a side.
    (b) If the dimensions of the square piece of material are doubled, how does the volume change?
13. An open box is to be made from a rectangular piece of material by cutting equal squares from each corner and turning up the sides. Find the dimensions of the box of maximum volume if the material has dimensions of 2 feet by 3 feet.

14. A net enclosure for golf practice is open at one end, as shown in Figure 4.55. Find the dimensions that require the least amount of netting if the volume of the enclosure is to be $83\frac{1}{3}$ cubic meters.

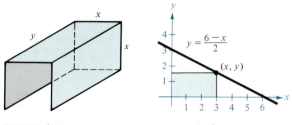

FIGURE 4.55   FIGURE 4.56

15. A rectangle is bounded by the *x*- and *y*-axes and the graph of $y = (6 - x)/2$, as shown in Figure 4.56. What length and width should the rectangle have so that its area is maximum?

16. Find the dimensions of the largest isosceles triangle that can be inscribed in a circle of radius *r*.

17. An indoor physical fitness room consists of a rectangular region with a semicircle on each end. If the perimeter of the room is to be a 200-meter running track, find the dimensions that will make the area of the rectangular region as large as possible.

18. A page is to contain 30 square inches of print. The margins at the top and the bottom of the page are each 2 inches wide. The margins on each side are only 1 inch wide. Find the dimensions of the page so that the least paper is used.

19. A right triangle is formed in the first quadrant by the *x*- and *y*-axes and a line through the point (2, 3). Find the vertices of the triangle so that its area is minimum.

20. A right triangle is formed in the first quadrant by the *x*- and *y*-axes and a line through the point (1, 2), as shown in Figure 4.57. Find the vertices of the triangle so that the length of the hypotenuse is minimum.

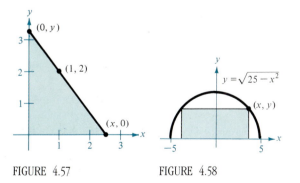

FIGURE 4.57   FIGURE 4.58

21. A rectangle is bounded by the *x*-axis and the semicircle $y = \sqrt{25 - x^2}$, as shown in Figure 4.58. What length

and width should the rectangle have so that its area is maximum?

22. Find the dimensions of the largest rectangle that can be inscribed in a semicircle of radius *r*. (See Exercise 21.)

23. Find the point on the graph of $y = \sqrt{x}$ that is closest to the point (4, 0).

24. Find the point on the graph of $y = x^2$ that is closest to the point $(2, \frac{1}{2})$.

25. A right circular cylinder is to be designed to hold 12 fluid ounces of a soft drink and to use a minimal amount of material in its construction. Find the required dimensions for the container (1 fl oz ≈ 1.80469 in³.)

26. Find the dimensions of the trapezoid of greatest area that can be inscribed in a semicircle of radius *r*.

27. A rectangular package to be sent by a postal service can have a maximum combined length and girth (perimeter of a cross section) of 108 inches, as shown in Figure 4.59. Find the dimensions of the package of maximum volume that can be sent. (Assume the cross section is square.)

28. Rework Exercise 27 for a cylindrical package. (The cross sections are circular.)

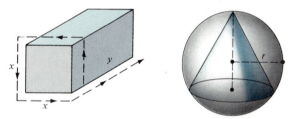

FIGURE 4.59   FIGURE 4.60

29. Find the volume of the largest right circular cone that can be inscribed in a sphere of radius *r*. (See Figure 4.60.)

30. Find the volume of the largest right circular cylinder that can be inscribed in a sphere of radius *r*.

31. A figure is formed by adjoining two hemispheres to each end of a right circular cylinder. The total volume of the figure is 12 cubic inches. Find the radius of the cylinder that produces the minimum surface area.

32. Show that among all positive numbers *x* and *y* with $x^2 + y^2 = r^2$, the sum $x + y$ is largest when $x = y$.

33. The combined perimeter of an equilateral triangle and a square is 10. Find the dimensions of the triangle and square that produce a minimum total area.

34. The combined perimeter of a circle and a square is 16. Find the dimensions of the circle and square that produce a minimum total area.

35. Ten feet of wire is to be used to form an isosceles right triangle and a circle. How much of the wire should be used on the circle if the total area enclosed is to be (a) minimum and (b) maximum?

36. Twenty feet of wire is to be used to form two figures. In each of the following cases, how much should be used for each figure so that the total enclosed area is a maximum?
 (a) equilateral triangle and square
 (b) square and regular pentagon
 (c) regular pentagon and hexagon

37. A wooden beam has a rectangular cross section of height h and width w, as shown in Figure 4.61. The strength S of the beam is directly proportional to the width and the square of the height. What are the dimensions of the strongest beam that can be cut from a round log of diameter 24 inches? [Hint: $S = kh^2w$, where k is the proportionality constant.]

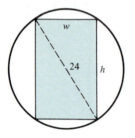

FIGURE 4.61

38. The illumination from a light source is directly proportional to the strength of the source and inversely proportional to the square of the distance from the source. Two light sources of intensities I_1 and I_2 are d units apart. At what point on the line segment joining the two sources is the illumination least?

39. A man is in a boat 2 miles from the nearest point on the coast. He is to go to a point Q, 3 miles down the coast and 1 mile inland, as shown in Figure 4.62. If he can row at 2 miles per hour and walk at 4 miles per hour, toward what point on the coast should he row in order to reach point Q in the least time?

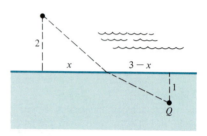

FIGURE 4.62

40. The conditions are the same as in Exercise 39 except that the man can row at 4 miles per hour. How does this change the solution?

4.8
Business and economics applications

In Section 3.8 we mentioned that one of the most common ways to measure change is with respect to time. In this section we study some important rates of change in the field of economics that are not measured with respect to time. For example, economists refer to **marginal profit, marginal revenue,** and **marginal cost** as the rates of change of the profit, revenue, and cost with respect to the number of units sold, which we represent by the variable x. In many problems in business and economics, the number of units produced or sold is restricted to positive integer values, as shown in Figure 4.63. (Of course, it could happen that a sale involves half or quarter units, but it is hard to conceive of a sale involving $\sqrt{2}$ units.) We call such a variable **discrete.** To analyze a function of a discrete variable, we temporarily assume that x is able to take on any real value in a given interval, as shown in Figure 4.64. Then we use the standard methods of calculus to find the x-value that corresponds to the marginal revenue, maximum profit, least cost, or whatever is called for. Finally, we round the solution off to the nearest sensible x-value—cents, dollars, units, or days, depending on the context of the problem.

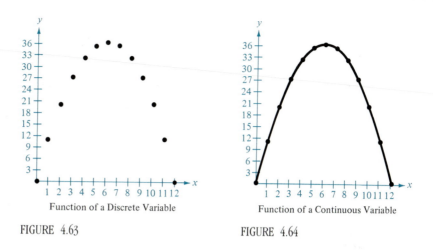

Function of a Discrete Variable

FIGURE 4.63

Function of a Continuous Variable

FIGURE 4.64

We begin with a summary of some basic terms and formulas.

SUMMARY OF BUSINESS TERMS AND FORMULAS

Basic Terms   **Basic Formulas**

x is the number of units produced (or sold)
p is the price per unit
R is the total revenue from selling x units $R = xp$
C is the total cost of producing x units
$\overline{C}$ is the average cost per unit $\overline{C} = C/x$
P is the total profit from selling x units $P = R - C$

The **break-even point** is the number of units for which $R = C$.

Marginals

$\dfrac{dR}{dx}$ = marginal revenue ≈ the *extra* revenue for selling one additional unit

$\dfrac{dC}{dx}$ = marginal cost ≈ the *extra* cost of producing one additional unit

$\dfrac{dP}{dx}$ = marginal profit ≈ the *extra* profit for selling one additional unit

---

Remark   In this summary note that marginals can be used to approximate the *extra* revenue, cost, or profit associated with selling or producing one additional unit. This is illustrated graphically for marginal revenue in Figure 4.65.

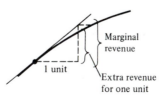

FIGURE 4.65   Revenue Function

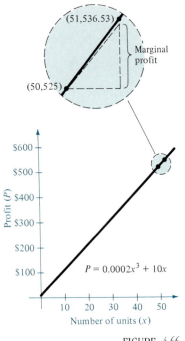

(51,536.53)

Marginal
profit

(50,525)

$600
$500
$400
$300
$200
$100

Profit (P)

$P = 0.0002x^3 + 10x$

10   20   30   40   50
Number of units (x)

FIGURE 4.66

EXAMPLE 1   Using marginals as approximations

A manufacturer determines that the profit derived from selling x units of a certain item is given by $P = 0.0002x^3 + 10x$.

(a) Find the marginal profit for a production level of 50 units.

(b) Compare this to the actual gain in profit obtained by increasing the production from 50 to 51 units. (See Figure 4.66.)

Solution:

(a) The marginal profit is given by

$$\frac{dP}{dx} = 0.0006x^2 + 10$$

When $x = 50$, the marginal profit is

$$\frac{dP}{dx} = (0.0006)(50)^2 + 10 = \$11.50 \qquad \text{Marginal profit}$$

(b) For $x = 50$ and 51, the actual profit is

$$P = (0.0002)(50)^3 + 10(50) = 25 + 500 = \$525.00$$
$$P = (0.0002)(51)^3 + 10(51) = 26.53 + 510 = \$536.53$$

Thus, the additional profit obtained by increasing the production level from 50 to 51 units is

$$536.53 - 525.00 = \$11.53 \qquad \text{Extra profit for one unit} \qquad \square$$

The profit function in Example 1 is quite unusual in that the profit continues to increase as long as the number of units sold increases. In practice, it is more common to encounter situations in which sales can be increased only by lowering the price per item, and such reductions in price will ultimately cause the profit to decline. We define the number of units x that consumers are willing to purchase at a given price per unit p as the **demand function** $p = f(x)$.

EXAMPLE 2   Finding the demand function

A business sells 2000 items per month at a price of $10 each. It is predicted that monthly sales will increase by 250 items for each $0.25 reduction in price. Find the demand function corresponding to this prediction.

Solution: From the given prediction, x increases 250 units each time p drops $0.25 from the original cost of $10. This is described by the equation

$$x = 2000 + 250\left(\frac{10 - p}{0.25}\right) = 12{,}000 - 1000p$$

or

$$p = 12 - \frac{x}{1000} \qquad \text{Demand function} \qquad \square$$

EXAMPLE 3   Finding the marginal revenue

A fast-food restaurant has determined that the monthly demand for their hamburgers is given by

$$p = \frac{60,000 - x}{20,000}$$

Find the increase in revenue per hamburger for monthly sales of 20,000 hamburgers. In other words, find the marginal revenue when $x = 20,000$.

Solution: Since the total revenue is given by $R = xp$, we have

$$R = xp = x\left(\frac{60,000 - x}{20,000}\right) = \frac{1}{20,000}(60,000x - x^2)$$

and the marginal revenue is

$$\frac{dR}{dx} = \frac{1}{20,000}(60,000 - 2x)$$

Finally, when $x = 20,000$, the marginal revenue is

$$\frac{dR}{dx} = \frac{1}{20,000}[60,000 - 2(20,000)] = \frac{20,000}{20,000} = \$1/\text{unit}$$

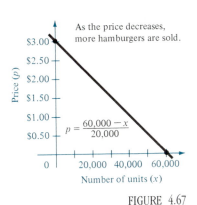

As the price decreases, more hamburgers are sold.

$$p = \frac{60,000 - x}{20,000}$$

Number of units (x)

FIGURE 4.67

Remark   Note that the demand function in Example 3 is typical in that a high demand corresponds to a low price, as seen in Figure 4.67 and Table 4.18.

TABLE 4.18

p	$0.00	$0.50	$1.00	$1.50	$2.00	$2.50	$3.00
x	60,000	50,000	40,000	30,000	20,000	10,000	0

EXAMPLE 4   Finding the marginal profit

Suppose that in Example 3 the cost of producing x hamburgers is

$$C = 5000 + 0.56x$$

Find the total profit and the marginal profit for (a) 20,000, (b) 24,400, and (c) 30,000 units.

Solution: Since $P = R - C$, we use the revenue function in Example 3 to obtain

$$P = \frac{1}{20,000}(60,000x - x^2) - 5,000 - 0.56x$$

$$= 2.44x - \frac{x^2}{20,000} - 5,000$$

Thus, the marginal profit is

$$\frac{dP}{dx} = 2.44 - \frac{x}{10,000}$$

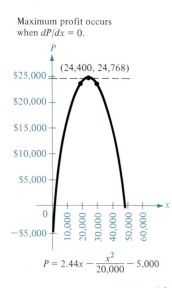

Maximum profit occurs
when $dP/dx = 0$.

(24,400, 24,768)

$P = 2.44x - \dfrac{x^2}{20,000} - 5,000$

FIGURE 4.68

Table 4.19 shows the total profit and the marginal profit for the three indicated demands.

TABLE 4.19

Demand	20,000	24,400	30,000
Profit	$23,800	$24,768	$23,200
Marginal Profit	$0.44	$0.00	−$0.56

Remark   In Example 4 we can see that when more than 24,400 hamburgers are sold, the marginal profit is negative. This means that increasing production beyond this point will *reduce* rather than increase profit. Figure 4.68 shows that the maximum profit corresponds to the point where the marginal profit is zero.

It is worthwhile to note that many problems in economics are minimum and maximum problems. For such problems, the procedure used in Section 4.7 is an appropriate model to follow.

EXAMPLE 5   *Finding the maximum profit*

In marketing a certain item, a business has discovered that the demand for the item is represented by

$$p = \frac{50}{\sqrt{x}} \qquad \text{Demand function}$$

The cost of producing x items is given by $C = 0.5x + 500$. Find the price per unit that yields a maximum profit.

Solution: From the given cost function, we have

$$P = R - C = xp - (0.5x + 500) \qquad \text{Primary equation}$$

Substituting for p (from the demand function), we have

$$P = x\left(\frac{50}{\sqrt{x}}\right) - (0.5x + 500) = 50\sqrt{x} - 0.5x - 500$$

Setting the marginal profit equal to zero, we obtain

$$\frac{dP}{dx} = \frac{25}{\sqrt{x}} - 0.5 = 0$$

$$\sqrt{x} = \frac{25}{0.5} = 50 \implies x = 2500$$

Finally, we conclude that the maximum profit occurs when the price is

$$p = \frac{50}{\sqrt{2500}} = \frac{50}{50} = \$1.00$$

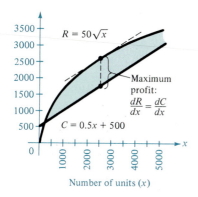

$R = 50\sqrt{x}$

Maximum profit:
$\dfrac{dR}{dx} = \dfrac{dC}{dx}$

$C = 0.5x + 500$

Number of units (x)

FIGURE 4.69

Remark   To find the maximum profit in Example 5, we differentiated the profit function, $P = R - C$, and set dP/dx equal to zero. From the equation

$$\frac{dP}{dx} = \frac{dR}{dx} - \frac{dC}{dx} = 0$$

it follows that the maximum profit occurs when the marginal revenue is equal to the marginal cost, as shown in Figure 4.69.

To study the effect of production levels on cost, economists use the average cost function $\overline{C}$ defined as $\overline{C} = C/x$. A use of this function is illustrated in the next example.

EXAMPLE 6   *Minimizing the average cost*

A company estimates that the cost (in dollars) of producing x units of a certain product is given by $C = 800 + 0.04x + 0.0002x^2$. Find the production level that minimizes the average cost per unit.

Solution: Substituting from the given equation for C, we have

$$\overline{C} = \frac{C}{x} = \frac{800 + 0.04x + 0.0002x^2}{x} = \frac{800}{x} + 0.04 + 0.0002x$$

Setting the derivative $d\overline{C}/dx$ equal to zero yields

$$\frac{d\overline{C}}{dx} = -\frac{800}{x^2} + 0.0002 = 0$$

$$x^2 = \frac{800}{0.0002} = 4{,}000{,}000 \implies x = 2{,}000 \text{ units}$$

See Figure 4.70.

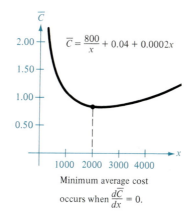

$$\overline{C} = \frac{800}{x} + 0.04 + 0.0002x$$

1000 2000 3000 4000

Minimum average cost

FIGURE 4.70

occurs when $\dfrac{d\overline{C}}{dx} = 0$.

Economists often describe the behavior of a demand function by a term called the **price elasticity of demand.** It describes the relative responsiveness of consumers to a change in the price of an item. If $p = f(x)$ is a differentiable demand function, then the price elasticity of demand is given by

$$\eta = \frac{p/x}{dp/dx}$$

(where η is the lowercase Greek letter eta). For a given price, if $|\eta| < 1$, the demand is **inelastic.** If $|\eta| > 1$, the demand is **elastic.**

EXAMPLE 7   *Test for elasticity*

Show that the demand function in Example 5, $p = 50x^{-1/2}$, is elastic.

Solution:

$$\eta = \frac{p/x}{dp/dx} = \frac{(50x^{-1/2})/x}{-25x^{-3/2}} = \frac{50x^{-3/2}}{-25x^{-3/2}} = -2$$

Since $|\eta| = 2$, we conclude that the demand is elastic for any price.   ☐

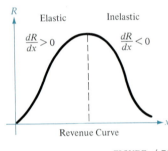

R

Elastic   Inelastic

$\frac{dR}{dx} > 0$ $\frac{dR}{dx} < 0$

x

Revenue Curve

FIGURE 4.71

Elasticity has an interesting relationship to the total revenue, as shown in Figure 4.71.

1. If the demand is *elastic,* then a decrease in the price per unit is accompanied by a sufficient increase in unit sales to cause the total revenue to increase.
2. If the demand is *inelastic,* then the increase in unit sales accompanying a decrease in price per unit is *not* sufficient to increase revenue, and the total revenue decreases.

EXAMPLE 8   *Elasticity and marginal revenue*

Show that for a differentiable demand function, the marginal revenue is positive when the demand is elastic and negative when the demand is inelastic. (Assume that the quantity demanded increases as the price decreases, and thus dp/dx is negative.)

Solution: Since the revenue is given by $R = xp$, we calculate the marginal revenue to be

$$\frac{dR}{dx} = x\left(\frac{dp}{dx}\right) + p = p\left[\left(\frac{x}{p}\right)\left(\frac{dp}{dx}\right) + 1\right] = p\left[\frac{dp/dx}{p/x} + 1\right] = p\left(\frac{1}{\eta} + 1\right)$$

If the demand is inelastic, then $|\eta| < 1$ implies that $1/\eta < -1$, since dp/dx is negative. (We assume that x and p are positive.) Therefore,

$$\frac{dR}{dx} = p\left(\frac{1}{\eta} + 1\right)$$

is negative. Similarly, if the demand is elastic, then $|\eta| > 1$ and $-1 < 1/\eta$, which implies that dR/dx is positive.   ☐

We conclude this section with a summary of the properties of a *typical demand function* and its corresponding revenue, cost, and profit functions.

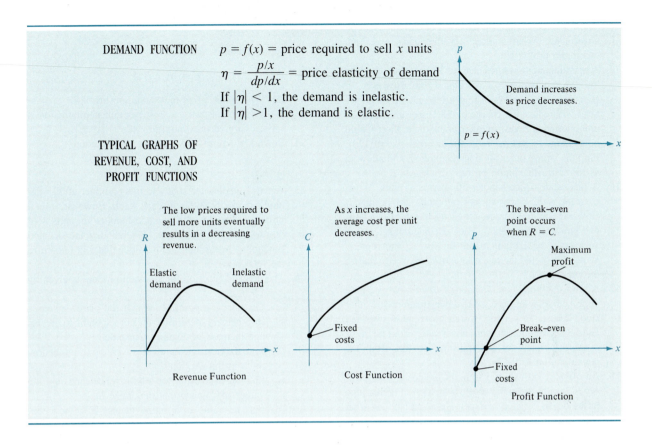

DEMAND FUNCTION $p = f(x)$ = price required to sell x units

$$\eta = \frac{p/x}{dp/dx} = \text{price elasticity of demand}$$

If $|\eta| < 1$, the demand is inelastic.
If $|\eta| > 1$, the demand is elastic.

Demand increases
as price decreases.

$p = f(x)$

TYPICAL GRAPHS OF
REVENUE, COST, AND
PROFIT FUNCTIONS

The low prices required to
sell more units eventually
results in a decreasing
revenue.

Elastic
demand

Inelastic
demand

Revenue Function

As x increases, the
average cost per unit
decreases.

Fixed
costs

Cost Function

The break–even
point occurs
when $R = C$.

Maximum
profit

Break–even
point

Fixed
costs

Profit Function

Section Exercises 4.8

In Exercises 1–4, find the number of units x that produce a maximum revenue R.

1. $R = 900x - 0.1x^2$
2. $R = 600x^2 - 0.02x^3$
3. $R = \dfrac{1,000,000x}{0.02x^2 + 1800}$
4. $R = 30x^{2/3} - 2x$

In Exercises 5–8, find the number of units x that produce the minimum average cost per unit $\overline{C}$.

5. $C = 0.125x^2 + 20x + 5000$
6. $C = 0.001x^3 - 5x + 250$
7. $C = 3000x - x^2\sqrt{300 - x}$
8. $C = \dfrac{2x^3 - x^2 + 5000x}{x^2 + 2500}$

In Exercises 9–12, find the price per unit p that produces the maximum profit P.

	Cost function	*Demand function*
9.	$C = 100 + 30x$	$p = 90 - x$
10.	$C = 2400x + 5200$	$p = 6000 - 0.4x^2$
11.	$C = 4000 - 40x + 0.02x^2$	$p = 50 - \dfrac{x}{100}$
12.	$C = 35x + 2\sqrt{x - 1}$	$p = 40 - \sqrt{x - 1}$

13. A manufacturer of lighting fixtures has daily production costs of

$$C = 800 - 10x + \frac{1}{4}x^2$$

How many fixtures x should be produced each day to minimize costs?

14. The profit for a certain company is given by

$$P = 230 + 20s - \frac{1}{2}s^2$$

where s is the amount (in hundreds of dollars) spent on advertising. What amount of advertising gives the maximum profit?

15. A manufacturer of radios charges $90 per unit when the average production cost per unit is $60. However, to encourage large orders from distributors, the manufacturer will reduce the charge by $0.10 per unit for each unit ordered in excess of 100 (for example, there would be a charge of $88 per radio for an order size of 120). Find the largest order the manufacturer should allow so as to realize maximum profit.

16. A real estate office handles 50 apartment units. When the rent is $270 per month, all units are occupied. However, on the average, for each $15 increase in rent, one unit becomes vacant. Each occupied unit requires an average of $18 per month for service and repairs. What rent should the real estate office charge to realize the most profit?

17. A power station is on one side of a river that is $\frac{1}{2}$ mile wide, and a factory is 6 miles downstream on the other side. It costs $6 per foot to run power lines overland and $8 per foot to run them underwater. Find the most economical path for the transmission line from the power station to the factory.

18. An offshore oil well is 1 mile off the coast. The refinery is 2 miles down the coast. If the cost of laying pipe in the ocean is twice as expensive as on land, in what path should the pipe be constructed in order to minimize the cost?

19. Assume that the amount of money deposited in a bank is proportional to the square of the interest rate the bank pays on this money. Furthermore, the bank can reinvest this money at 12 percent. Find the interest rate the bank should pay to maximize profit. (Use the simple interest formula.)

20. Prove that the average cost is minimum at the value of x where the average cost equals the marginal cost.

21. Given the cost function

$$C = 2x^2 + 5x + 18$$

   (a) Find the value of x where the average cost is minimum.
   (b) For the value of x found in part (a), show that the marginal cost and average cost are equal (see Exercise 20).

22. Given the cost function

$$C = x^3 - 6x^2 + 13x$$

   (a) Find the value of x where the average cost function is minimum.
   (b) For the value of x found in part (a), show that the marginal cost and average cost are equal.

23. A given commodity has a demand function given by

$$p = 100 - \tfrac{1}{2}x^2$$

and a total cost function of

$$C = 4x + 375$$

   (a) What price gives the maximum profit?
   (b) What is the average cost per unit if production is set to give maximum profit?

24. Rework Exercise 23 using the cost function

$$C = 26.5x + 375$$

25. When a wholesaler sold a certain product at $25 per unit, sales were 800 units each week. However, after a price raise of $5, the average number of units sold dropped to 775 per week. Assume that the demand function is linear and find the price that will maximize the total revenue.

26. The demand function for a certain product is given by $x = 20 - 2p^2$.
   (a) Consider the point (2, 12). If the price decreases by 5 percent, determine the corresponding percentage increase in quantity demanded.
   (b) Find the exact elasticity at (2, 12) by using the formula in this section.
   (c) Find an expression for total revenue, and find the values of x and p that maximize R.
   (d) For the value of x found in part (c), show that $|\eta| = 1$.

27. The demand equation is $p^3 + x^3 = 9$.
   (a) Find η when $x = 2$.
   (b) Find the values of x and p that maximize the total revenue.
   (c) Show that $|\eta| = 1$ for the value of x found in part (b).

28. The demand function for a particular commodity is given by

$$p = (16 - x)^{1/2}, \quad 0 \leq x \leq 16$$

Determine the price and quantity for which revenue is maximum.

29. The demand function is given by the equation

$$x = a/p^m, \text{ where } m > 1$$

Show that $\eta = -m$. (In other words, in terms of approximate price changes, a 1 percent increase in price results in an m percent decrease in the quantity demanded.)

4.9
Newton's Method

In the first eight sections of this chapter, we frequently needed to find the zeros of a function. Until now our functions have been carefully chosen so that elementary algebraic techniques suffice for finding the zeros. For example, the zeros of

$$f(x) = x^2 - 6x + 8$$
$$g(x) = 2x^2 - 3x - 7$$
and
$$h(x) = x^3 - 2x^2 - x + 2$$

can all be found by factoring or by the quadratic formula. However, in practice we frequently encounter functions whose zeros are more difficult to find. For example, the zeros of a function as simple as

$$f(x) = x^3 - x + 1$$

cannot be found by elementary algebraic methods.* In such cases the tangent line may serve as a convenient tool for approximating the zeros.

To illustrate this, suppose we wish to find a zero for a function f, which is continuous on the interval $[a, b]$ and differentiable on the interval (a, b). If $f(a)$ and $f(b)$ differ in sign, then by the Intermediate Value Theorem f must possess at least one zero ($x = c$) in (a, b). Suppose we estimate this zero to occur at $x = x_1$, as shown in Figure 4.72.

Newton's Method for approximating zeros is based on the assumption that the graph of f and the tangent line at $(x_1, f(x_1))$ both cross the x-axis at *about* the same point. Since we can easily calculate the x-intercept for this tangent line, we use it as our second (and, we hope, better) estimate for the zero of f. The tangent passes through the point $(x_1, f(x_1))$ with a slope of $f'(x_1)$. In point-slope form, the equation of the tangent line is therefore

$$y - f(x_1) = f'(x_1)(x - x_1)$$
$$y = f'(x_1)(x - x_1) + f(x_1)$$

Letting $y = 0$ and solving for x, we have

$$x = x_1 - \frac{f(x_1)}{f'(x_1)}$$

Thus, from our initial guess we arrive at a new estimate

$$x_2 = x_1 - \frac{f(x_1)}{f'(x_1)}$$

We may improve upon x_2 and calculate yet a third estimate:

$$x_3 = x_2 - \frac{f(x_2)}{f'(x_2)}$$

Repeated application of this process is called **Newton's Method.**

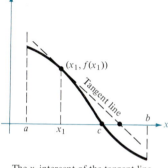

$(x_1, f(x_1))$

Tangent line

The x-intercept of the tangent line approximates the zero of f.

FIGURE 4.72

*See the note regarding algebraic solutions of polynomial equations at the end of this section.

| NEWTON'S METHOD FOR APPROXIMATING THE ZEROS OF A FUNCTION | Let $f(c) = 0$, where f is differentiable on an open interval containing c. Then, to approximate c we use the following steps.

1. Make an initial estimate x_1 that is "close" to c. (A graph is helpful.)
2. Determine a new approximation

$$x_{n+1} = x_n - \frac{f(x_n)}{f'(x_n)}$$

3. If $|x_n - x_{n+1}|$ is less than the desired accuracy, let x_{n+1} serve as the final approximation. Otherwise, return to step 2 and calculate a new approximation.

Each successive approximation is called an **iteration.** |

For many functions, just a few iterations of Newton's Method will produce results that are surprisingly accurate. To demonstrate this, we begin with an example in which we know the exact zero.

EXAMPLE 1   Using Newton's Method

Calculate three iterations of Newton's Method to approximate a zero of $f(x) = x^2 - 2$. Use $x_1 = 1$ as the initial guess.

Solution: Since $f(x) = x^2 - 2$, we have $f'(x) = 2x$ and the iterative process is given by the formula

$$x_{n+1} = x_n - \frac{f(x_n)}{f'(x_n)}$$

$$= x_n - \frac{x_n^2 - 2}{2x_n}$$

The calculations for three iterations are shown in Table 4.20.

TABLE 4.20

n	x_n	$f(x_n)$	$f'(x_n)$	$\dfrac{f(x_n)}{f'(x_n)}$	$x_n - \dfrac{f(x_n)}{f'(x_n)}$
1	1.000000	−1.000000	2.000000	−0.500000	1.500000
2	1.500000	0.250000	3.000000	0.083333	1.416667
3	1.416667	0.006945	2.833334	0.002451	1.414216
4	1.414216				

Of course, in this case we know that the two zeros of the function are $\pm\sqrt{2}$. To six decimal places, $\sqrt{2} = 1.414214$. Thus, after only three iterations of Newton's Method, we have obtained an approximation that is within 0.000002 of an actual root. The iterations shown in Table 4.20 are graphically illustrated in Figure 4.73.

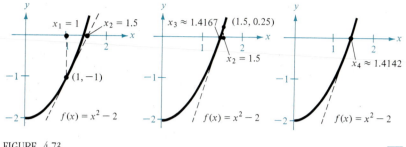

FIGURE 4.73

| **Remark** From the calculations in Example 1, you can see that Newton's Method has become more practical with the easy access to calculators and computers.

EXAMPLE 2   *Using Newton's Method*

Use Newton's Method to approximate the zeros of $f(x) = 2x^3 + x^2 - x + 1$. Continue the iterations until two successive approximations differ by less than 0.0001.

Solution: We begin by sketching a graph of f and observing that it has only one zero, which occurs near $x = -1.2$, as shown in Figure 4.74. Now, we differentiate f to form the iterative formula

$$x_{n+1} = x_n - \frac{f(x_n)}{f'(x_n)}$$

$$= x_n - \frac{2x_n^3 + x_n^2 - x_n + 1}{6x_n^2 + 2x_n - 1}$$

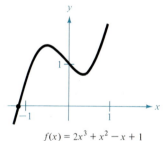

$f(x) = 2x^3 + x^2 - x + 1$

FIGURE 4.74

The calculations are shown in Table 4.21 on the following page.

TABLE 4.21

n	x_n	$f(x_n)$	$f'(x_n)$	$\dfrac{f(x_n)}{f'(x_n)}$	$x_n - \dfrac{f(x_n)}{f'(x_n)}$
1	-1.20000	0.18400	5.24000	0.03511	-1.23511
2	-1.23511	-0.00771	5.68276	-0.00136	-1.23375
3	-1.23375	0.00001	5.66533	0.00000	-1.23375
4	-1.23375				

Thus, we estimate the zero of f to be -1.23375, since two successive approximations differ by less than the required 0.0001.

    When, as in Examples 1 and 2, the approximations approach a limit, we say that the sequence

$$x_1, \; x_2, \; x_3, \; \ldots, \; x_n, \; \ldots$$

converges. Moreover, if the limit is c, then it can be shown that c must be a zero of f.

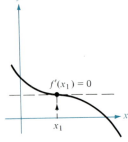

$f'(x_1) = 0$

x_1

FIGURE 4.75

It is important to realize that Newton's Method does not always yield a convergent sequence. There are two potential problems. The first type is shown in Figure 4.75. Since Newton's Method involves division by $f'(x_n)$, it is clear that the method will fail if the derivative is zero for any x_n in the sequence. If you encounter this problem, you can usually overcome it with a better choice for x_1.

The problem illustrated in the next example is usually more serious and may be independent of the choice of x_1. Figure 4.76 shows that Newton's Method does not converge for any choice of x_1 (other than the actual zero) for the function $f(x) = x^{1/3}$.

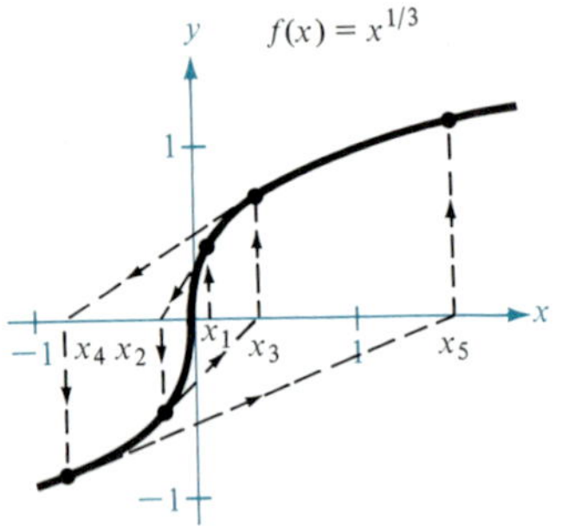

$f(x) = x^{1/3}$

-1 x_4 x_2 x_1 x_3 x_5

FIGURE 4.76

EXAMPLE 3   An example in which Newton's Method fails

Using $x_1 = 0.1$, show that Newton's Method fails to converge for the function $f(x) = x^{1/3}$.

Solution: Since $f'(x) = \frac{1}{3}x^{-2/3}$, the iterative formula is

$$x_{n+1} = x_n - \frac{f(x_n)}{f'(x_n)} = x_n - \frac{x_n^{1/3}}{\frac{1}{3}x_n^{-2/3}} = x_n - 3x_n = -2x_n$$

The calculations are given in Table 4.22. This table and Figure 4.76 indicate that x_n increases in magnitude as $n \to \infty$, and thus the limit of the sequence does not exist.

TABLE 4.22

n	x_n	$f(x_n)$	$f'(x_n)$	$\dfrac{f(x_n)}{f'(x_n)}$	$x_n - \dfrac{f(x_n)}{f'(x_n)}$
1	0.10000	0.46416	1.54720	0.30000	−0.20000
2	−0.20000	−0.58480	0.97467	−0.60000	0.40000
3	0.40000	0.73681	0.61401	1.20000	−0.80000
4	−0.80000	−0.92832	0.38680	−2.40000	1.60000

It can be shown that a sufficient condition to produce convergence of Newton's Method to a zero of f is that

$$\left| \frac{f(x)f''(x)}{[f'(x)]^2} \right| < 1$$

on an open interval containing the zero. For instance, in Example 1 this test would yield: $f(x) = x^2 - 2$, $f'(x) = 2x$, $f''(x) = 2$, and

$$\left| \frac{f(x)f''(x)}{[f'(x)]^2} \right| = \left| \frac{(x^2 - 2)(2)}{4x^2} \right| = \left| \frac{1}{2} - \frac{1}{x^2} \right| \qquad \text{Example 1}$$

On the interval $(1, 3)$ this quantity is less than 1 and therefore the convergence of Newton's Method is guaranteed. On the other hand, in Example 3, we have $f(x) = x^{1/3}$, $f'(x) = \frac{1}{3}x^{-2/3}$, $f''(x) = -\frac{2}{9}x^{-5/3}$, and

$$\left| \frac{f(x)f''(x)}{[f'(x)]^2} \right| = \left| \frac{x^{1/3}(-2/9)(x^{-5/3})}{(1/9)(x^{-4/3})} \right| = 2 \qquad \text{Example 3}$$

which is not less than 1 for any value of x.

EXAMPLE 4   Using Newton's Method to find a point of intersection

Estimate the point of intersection of the graphs of

$$y = \frac{1}{x^2 + 1} \qquad \text{and} \qquad y = 2x$$

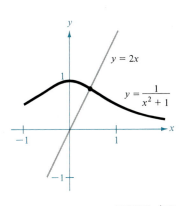

FIGURE 4.77

as shown in Figure 4.77. Use Newton's Method and continue the iterations until two successive approximations differ by less than 0.0001.

Solution: We let $2x = 1/(x^2 + 1)$. Since this implies that $2x(x^2 + 1) - 1 = 0$, we need to find the zero of the function given by

$$f(x) = 2x(x^2 + 1) - 1 = 2x^3 + 2x - 1$$

Thus, the iterative formula for Newton's Method takes the form

$$x_{n+1} = x_n - \frac{f(x_n)}{f'(x_n)} = x_n - \frac{2x_n^3 + 2x_n - 1}{6x_n^2 + 2}$$

The calculations are shown in Table 4.23, beginning with an initial guess of $x_1 = 0.5$.

TABLE 4.23

n	x_n	$f(x_n)$	$f'(x_n)$	$\dfrac{f(x_n)}{f'(x_n)}$	$x_n - \dfrac{f(x_n)}{f'(x_n)}$
1	0.50000	0.25957	3.50000	0.07143	0.42857
2	0.42857	0.01458	3.10204	0.00470	0.42387
3	0.42387	0.00006	3.07801	0.00002	0.42385

Thus, we approximate the point of intersection to occur when $x = 0.42385$.

Algebraic solutions of polynomial equations

At the beginning of this section we mentioned that the zeros of the cubic function $f(x) = x^3 - x + 1$ cannot be found by *elementary* algebraic methods. This particular function happens to have only one real zero, and using more advanced algebraic techniques we can determine this value to be

$$x = -\sqrt[3]{\frac{3 - \sqrt{23/3}}{6}} - \sqrt[3]{\frac{3 + \sqrt{23/3}}{6}}$$

Since the *exact* solution is written in terms of square roots and cube roots, we call it a **solution by radicals.**

Neils Henrik Abel

Evariste Galois

The determination of radical solutions to a polynomial equation is one of the fundamental problems of algebra. The earliest result in this category is the well-known Quadratic Formula, and it dates back at least to Babylonian times. The general formula for the zeros of a cubic function came much later in history. In the sixteenth century an Italian mathematician, Jerome Cardan, published a method for finding radical solutions to cubic and quartic equations. Then, for three hundred years the problem of finding a general quintic formula remained open. Finally, in the nineteenth century the problem was answered independently by two young mathematicians. Niels Henrik Abel (1802–1829), a Norwegian mathematician, and Evariste Galois (1811–1832), a French mathematician, proved that it is not possible to solve a *general* fifth (or higher) degree polynomial equation by radicals. Of course, we can solve particular fifth-degree equations such as $x^5 - 1 = 0$, but Abel and Galois were able to show that no general *radical* solution exists.

Section Exercises 4.9

In Exercises 1–4, complete one iteration of Newton's Method for the given function using the indicated initial guess.

Function	*Initial Guess*
1. $f(x) = x^2 - 3$	$x_1 = 1.7$
2. $f(x) = 3x^2 - 2$	$x_1 = 1$
3. $f(x) = 3x^3 - 2$	$x_1 = 1$
4. $f(x) = x^3 - x^2 - 2x - 2$	$x_1 = 2.5$

In Exercises 5–10, approximate the indicated zero(s) of the function. Use Newton's Method and continue the process until two successive approximations differ by less than 0.001.

5. $f(x) = x^3 + x - 1$

6. $f(x) = x^5 + x - 1$

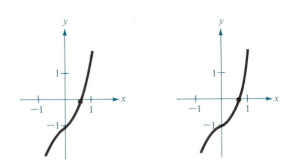

7. $f(x) = 3\sqrt{x-1} - x$

8. $f(x) = x^3 - 3.9x^2 + 4.79x - 1.881$

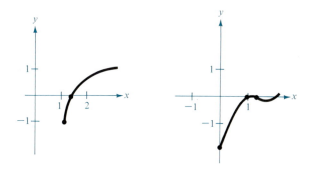

13. $f(x) = \dfrac{4}{x}$;

$g(x) = x^2 + 1$

14. $f(x) = x^3$;

$g(x) = x^2 + 2$

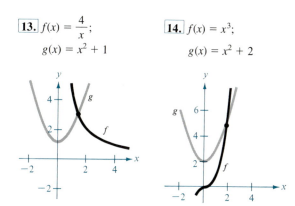

9. $f(x) = x^4 - 10x^2 - 11$

10. $f(x) = x^3 + 3$

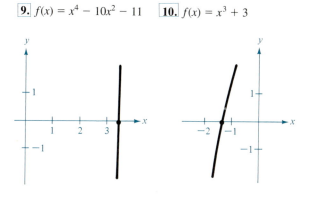

In Exercises 15–18, apply Newton's Method, using the indicated initial guess, and explain why the method fails.

15. $y = 2x^3 - 6x^2 + 6x - 1$; $x_1 = 1$

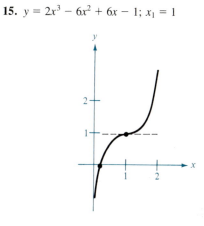

In Exercises 11–14, apply Newton's Method to approximate the x-value of the point of intersection of the two graphs. Continue the process until two successive approximations differ by less than 0.001.

11. $f(x) = 2x + 1$;

$g(x) = \sqrt{x+4}$

12. $f(x) = 3 - x$;

$g(x) = \dfrac{1}{x^2 + 1}$

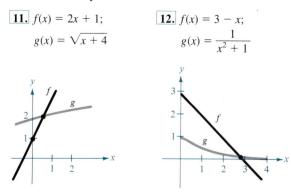

16. $y = 4x^3 - 12x^2 + 12x - 3$; $x_1 = \dfrac{3}{2}$

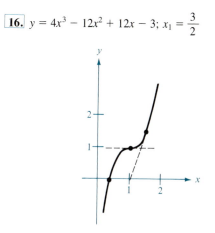

17. $y = -x^3 + 3x^2 - x + 1$; $x_1 = 1$

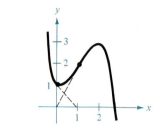

18. $f(x) = \sqrt[3]{x - 1}$; $x_1 = 2$

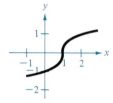

19. Use Newton's Method to obtain a general formula for approximating $\sqrt[n]{a}$. [Hint: Consider $f(x) = x^n - a$.]

In Exercises 20–22, use the result of Exercise 19 to approximate the indicated radical to three decimal places.

20. $\sqrt[3]{7}$ **21.** $\sqrt[4]{6}$
22. $\sqrt{5}$

In Exercises 23–28, we review some typical problems from the previous sections of this chapter. In each case, use Newton's Method to approximate the solution.

23. Find the point on the graph of $f(x) = 4 - x^2$ that is closest to the point $(1, 0)$.
24. Find the point on the graph of $f(x) = x^2$ that is closest to the point $(4, -3)$.

25. A woman is in a boat 2 miles from the nearest point on the coast, as shown in Figure 4.78. She is to go to a point Q, which is 3 miles down the coast and 1 mile inland. She can row at 3 miles per hour and walk at 4 miles per hour. Toward what point on the coast should she row in order to reach Q in the least time?

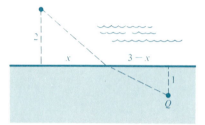

FIGURE 4.78

26. A company estimates that the cost (in dollars) of producing x units of a certain product is given by the model

$$C = 800 + 0.4x + 0.02x^2 + 0.0001x^3$$

Find the production level that minimizes the average cost per unit.

27. The concentration C of a certain chemical in the bloodstream t hours after injection into muscle tissue is given by

$$C = \frac{3t^2 + t}{50 + t^3}$$

When is the concentration greatest?

28. The ordering and transportation cost C of the components used in manufacturing a certain product is given by

$$C = 100\left(\frac{200}{x^2} + \frac{x}{x + 30}\right), \quad 1 \leq x$$

where C is measured in thousands of dollars and x is the order size in hundreds. Find the order size that minimizes the costs.

SECTION TOPICS ▪
Differentials ▪
Error propagation ▪
Differential formulas ▪

4.10
Differentials

We previously defined the derivative as the limit ($\Delta x \to 0$) of the ratio $\Delta y / \Delta x$, and it seemed natural to retain the quotient symbolism for the limit itself. Thus, we denoted the derivative by

$$\frac{dy}{dx} = \lim_{\Delta x \to 0} \frac{\Delta y}{\Delta x}$$

even though we did not think of dy/dx as the quotient of the two separate quantities dy and dx. In this section we give separate meanings to dy and dx in such a way that their quotient, when $dx \neq 0$, is equal to the derivative of y with respect to x. We do this in the following manner.

DEFINITION OF DIFFERENTIALS

Let $y = f(x)$ represent a differentiable function. The **differential of x** (denoted by dx) is any nonzero real number. The **differential of y** (denoted by dy) is given by

$$dy = f'(x)\, dx$$

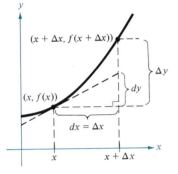

FIGURE 4.79

Remark Note that in this definition dx can have any nonzero value. However, in most applications of differentials we choose dx to be small and we denote this choice by $dx = \Delta x$.

One use of differentials is in approximating the change in $f(x)$ that corresponds to a change in x, as shown in Figure 4.79. We denote this change by

$$\Delta y = f(x + \Delta x) - f(x)$$

The following example compares the values of dy and Δy.

EXAMPLE 1   *Comparing Δy and dy*

Let $y = x^2$. Find dy when $x = 1$ and $dx = 0.01$. Compare this value to Δy when $x = 1$ and $\Delta x = 0.01$.

Solution: Since $y = f(x) = x^2$, we have $f'(x) = 2x$ and the differential dy is given by

$$
\begin{aligned}
dy &= f'(x)\, dx \\
   &= f'(1)(0.01) \qquad \text{Differential of } y\\
   &= 2(0.01)\\
   &= 0.02
\end{aligned}
$$

Now, since $\Delta x = 0.01$, the change in y is

$$
\begin{aligned}
\Delta y &= f(x + \Delta x) - f(x) \qquad \text{Change in } y\\
         &= f(1.01) - f(1)\\
         &= (1.01)^2 - 1^2\\
         &= 0.0201
\end{aligned}
$$

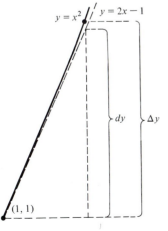

FIGURE 4.80

Figure 4.80 shows the geometric comparison of dy and Δy.

For the function in Example 1, we can see in Table 4.24 that dy approximates Δy more and more closely as dx approaches zero.

TABLE 4.24

$dx = \Delta x$	dy	Δy	$\Delta y - dy$
0.1000	0.2000	0.21000000	0.01000000
0.0100	0.0200	0.02010000	0.00010000
0.0010	0.0020	0.00200100	0.00000100
0.0001	0.0002	0.00020001	0.00000001

Remark   Note that for the sequence shown in Table 4.24, as Δx decreases by tenths, $\Delta y - dy$ decreases by hundredths.

Actually, this is not the first time we have encountered an approximation of this type. In Section 4.8 we used the *marginal profit* to approximate the extra profit gained by selling one more unit. In Section 4.9 on Newton's Method, we used the tangent line to approximate the x-intercept of the graph of a function. Each of these cases is an example of a **linear approximation.** More specifically, these particular examples are called **tangent line approximations,** since we are using the tangent line at a point to approximate the graph of the function near the point.

The validity of using a tangent line to approximate a curve stems from the limit definition of the tangent line. That is, the existence of the limit

$$f'(x) = \lim_{\Delta x \to 0} \frac{f(x + \Delta x) - f(x)}{\Delta x}$$

implies that when Δx is close to zero (within δ units of zero), then $f'(x)$ is close to the difference quotient (within ε units) and we have

$$\frac{f(x + \Delta x) - f(x)}{\Delta x} \approx f'(x), \quad \Delta x \neq 0$$

$$f(x + \Delta x) - f(x) \approx f'(x)\,\Delta x$$

$$\Delta y \approx f'(x)\,\Delta x$$

Substituting dx for Δx, we have

$$\Delta y \approx f'(x)\,dx = dy, \quad dx \neq 0$$

We formalize this result in the following theorem. The proof follows directly from the definition of the derivative.

THEOREM 4.12   THE RELATIVE SIZE OF dy AND Δy
Let $y = f(x)$ be differentiable at x such that $f'(x) \neq 0$. If $dx = \Delta x$, then

$$\lim_{\Delta x \to 0} \frac{\Delta y}{dy} = 1$$

Error propagation

Physicists and engineers tend to make liberal use of the approximation of Δy by dy. One way this occurs in practice is in the estimation of errors propagated by physical measuring devices. For example, if we let x represent the measured value of a variable and let $x + \Delta x$ represent the exact value, then Δx is the *error in measurement*. Finally, if the measured value x is used to compute another value $f(x)$, then the difference between $f(x + \Delta x)$ and $f(x)$ is called the **propagated error.**

$$\underbrace{f(x + \Delta x)}_{\substack{\text{exact} \\ \text{value}}} - \underbrace{f(x)}_{\substack{\text{measured} \\ \text{value}}} = \underbrace{\Delta y}_{\substack{\text{propagated} \\ \text{error}}}$$

with the brace labeled:
$$\overbrace{}^{\substack{\text{measurement} \\ \text{error}}}$$

EXAMPLE 2   Estimation of error

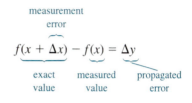

The diameter of a ball bearing is measured to be 1.4 inch, as shown in Figure 4.81. If the measurement is correct to within 0.01 inch, estimate the propagated error in the volume V of the ball bearing.

Solution: The formula for the volume of a sphere is $V = \frac{4}{3}\pi r^3$, where r is the radius of the sphere. Thus, we have

$$r = 0.7 \qquad \text{Measured radius}$$

and

$$-0.01 \le \Delta r \le 0.01 \qquad \text{Possible error}$$

To approximate the propagated error in the volume, we differentiate V to obtain $dV/dr = 4\pi r^2$ and write

$$\Delta V \approx dV = 4\pi r^2 dr = 4\pi(0.7)^2(\pm 0.01) \approx \pm 0.06158 \text{ in}^3 \qquad \square$$

FIGURE 4.81

Would you say that the propagated error in Example 2 is large or small? The answer is best given in *relative* terms by comparing dV to V. We obtain the ratio

$$\frac{dV}{V} = \frac{4\pi r^2 dr}{\frac{4}{3}\pi r^3} = \frac{3 dr}{r}$$

$$\approx \frac{3}{0.7}(\pm 0.01) \approx \pm 0.0429$$

which is called the **relative error.** The corresponding **percentage error** is

$$\frac{dV}{V}(100) \approx 4.29\%$$

We can use the definition of a differential to rewrite each of the derivative rules in **differential form.** For example, suppose u and v are differentiable functions of x. By the definition of differentials, we have $du = u'dx$ and

$dv = v'dx$. Therefore, we can write the differential form of the Product Rule as follows:

$$d[uv] = \frac{d}{dx}[uv]\,dx \qquad \text{Differential of } uv$$

$$= [uv' + vu']\,dx \qquad \text{Product Rule}$$

$$= uv'\,dx + vu'\,dx$$

$$= u\,dv + v\,du$$

The following summary lists the differential forms corresponding to the derivative rules we have studied so far in the text.

GENERAL DIFFERENTIAL FORMULAS

For u and v, differentiable functions of x:

Constant multiple: $d[cu] = c\,du$
Sum or difference: $d[u \pm v] = du \pm dv$
Product: $d[uv] = u\,dv + v\,du$

Quotient: $d\left[\dfrac{u}{v}\right] = \dfrac{v\,du - u\,dv}{v^2}$

DIFFERENTIALS OF ALGEBRAIC FUNCTIONS

Constant Rule: $d[c] = 0$
Power Rule: $d[x^n] = nx^{n-1}\,dx$
   $d[x] = dx$

In the next example, we compare the derivatives and differentials of several functions.

EXAMPLE 3   *Finding differentials*

Function	*Derivative*	*Differential*
(a) $y = x^2$	$\dfrac{dy}{dx} = 2x$	$dy = 2x\,dx$
(b) $y = 2x^3 + x$	$\dfrac{dy}{dx} = 6x^2 + 1$	$dy = (6x^2 + 1)\,dx$
(c) $y = \sqrt{2x + 1}$	$\dfrac{dy}{dx} = \dfrac{1}{\sqrt{2x+1}}$	$dy = \left(\dfrac{1}{\sqrt{2x+1}}\right)dx$
(d) $y = \dfrac{1}{x}$	$\dfrac{dy}{dx} = -\dfrac{1}{x^2}$	$dy = -\dfrac{dx}{x^2}$

Gottfried Wilhelm Leibniz

The notation in Example 3 is called the **Leibniz notation** for derivatives and differentials, named after the German mathematician Gottfried Wilhelm Leibniz (1646–1716). The beauty of this notation is that it provides us with an easy way to remember several important calculus formulas by making it seem as though the formulas were derived from algebraic manipulations of differentials. We will encounter several instances of this later in the text (in substitutions, inverse functions, parametric equations, and polar coordinates). For now, we simply compare the the *Chain Rule* in Leibniz notation to other possible notations.

$$\frac{dy}{dx} = \frac{dy}{du} \cdot \frac{du}{dx}$$ Leibniz notation

$$f'(g(x)) = f'(u)g'(x), \quad \text{where } u = g(x)$$ Function notation

$$D_x[y] = D_u[y]D_x[u]$$ Operator notation

The Leibniz form of the rule would probably appear to be true to a student in elementary algebra. Of course, it appears to be true for the *wrong reason*— that we have simply canceled the *du*'s. Even so, the notation's many advantages overshadow the potential problem of drawing valid conclusions from invalid arguments.

EXAMPLE 4   *Differential of a composite function*

Use the Chain Rule to find *dy* for $y = (3x^2 + x)^4$.

Solution:

$$y = f(x) = (3x^2 + x)^4$$
$$f'(x) = 4(6x + 1)(3x^2 + x)^3$$
$$dy = f'(x)\ dx = 4(6x + 1)(3x^2 + x)^3\ dx$$

EXAMPLE 5   *Differential of a composite function*

Use the Chain Rule to find *dy* for $y = \sqrt{x^2 + 1}$.

Solution:

$$y = f(x) = (x^2 + 1)^{1/2}$$
$$f'(x) = \frac{1}{2}(2x)(x^2 + 1)^{-1/2}$$
$$= \frac{x}{\sqrt{x^2 + 1}}$$
$$dy = f'(x)\ dx = \frac{x}{\sqrt{x^2 + 1}}\ dx$$

Differentials can be used to approximate function values. To do this for the function given by $y = f(x)$, we make use of the formula

$$f(x + \Delta x) \approx f(x) + dy = f(x) + f'(x)\ dx$$

which arises from the approximation $\Delta y = f(x + \Delta x) - f(x) \approx dy$. The key to the use of this formula lies in choosing a value for *x* that makes the calculations easier. For instance, to approximate $\sqrt{101}$ we would let $x = 100$ and $\Delta x = 1$. This is illustrated in the next example.

EXAMPLE 6   Approximating function values

Use differentials to approximate $\sqrt{16.5}$.

Solution: We use the function $f(x) = \sqrt{x}$ and choose $x = 16$. Then $dx = 0.5$, and we obtain

$$f(x + \Delta x) \approx f(x) + f'(x)\, dx = \sqrt{x} + \frac{1}{2\sqrt{x}}\, dx$$

$$\sqrt{16.5} \approx \sqrt{16} + \frac{1}{2\sqrt{16}}(0.5) = 4 + \left(\frac{1}{8}\right)\left(\frac{1}{2}\right)$$

$$= 4 + \frac{1}{16} = 4.0625$$

Remark The use of differentials to approximate function values has diminished because of the availability of calculators and computers. Using a calculator, we obtain the value (accurate to four decimal places)

$$\sqrt{16.5} \approx 4.0620$$

which indicates the accuracy of the differential method in Example 6.

Section Exercises 4.10

In Exercises 1–10, find the differential dy of each function.

1. $y = 3x^2 - 4$

2. $y = 2x^{3/2}$

3. $y = 4x^3$

4. $y = 3$

5. $y = \dfrac{x + 1}{2x - 1}$

6. $y = \dfrac{x}{x + 5}$

7. $y = \sqrt{x}$

8. $y = \sqrt{x^2 - 4}$

9. $y = x\sqrt{1 - x^2}$

10. $y = \sqrt{x} + \dfrac{1}{\sqrt{x}}$

In Exercises 11–16, let $x = 2$ and use the given function and value of $\Delta x = dx$ to complete the table:

$dx = \Delta x$	dy	Δy	$\Delta y - dy$	$\dfrac{dy}{\Delta y}$
1.000				
0.500				
0.100				
0.010				
0.001				

11. $y = x - 1$

12. $y = 2x$

13. $y = x^2$

14. $y = \dfrac{1}{x^2}$

15. $y = x^5$

16. $y = \sqrt{x}$

17. The area of a square of side x is given by $A(x) = x^2$.
   (a) Compute dA and ΔA in terms of x and Δx.
   (b) In Figure 4.82, identify the region whose area is dA.
   (c) In Figure 4.82 identify the region whose area is $\Delta A - dA$.

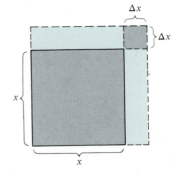

FIGURE 4.82

18. The measurement of the side of a square is found to be 12 inches, with a possible error of $\frac{1}{64}$ inch. Use differentials to approximate the possible error in computing the area of the square.

19. The measurement of the diameter of the end of a log is found to be 28 inches, with a possible error of $\frac{1}{4}$ inch. Use differentials to approximate the possible error in computing the area of the end of the log.

20. The measurement of the edge of a cube is found to be 12 inches, with a possible error of 0.03 inch. Use differentials to approximate the maximum possible error in computing the following:
    (a) the volume of the cube
    (b) the surface area of the cube

21. The radius of a sphere is claimed to be 6 inches, with a possible error of 0.02 inch. Use differentials to approximate the maximum possible error in calculating (a) the volume of the sphere and (b) the surface area of the sphere. (c) What is the relative error in parts (a) and (b)?

22. The profit P for a company is given by

$$P = (500x - x^2) - \left(\frac{1}{2}x^2 - 77x + 3000\right)$$

Approximate the change in profit as production changes from $x = 115$ to $x = 120$ units. Approximate the percentage change in profit when x changes from $x = 115$ to $x = 120$ units.

23. The period of a pendulum is given by

$$T = 2\pi\sqrt{\frac{L}{g}}$$

where L is the length of the pendulum in feet, g is the acceleration due to gravity, and T is time in seconds. Suppose that the pendulum has been subjected to an increase in temperature so that the length increases by $\frac{1}{2}$ percent.
(a) Find the approximate percentage change in the period.
(b) Using the result of part (a), find the approximate error in this pendulum clock in one day.

24. The acceleration of gravity varies somewhat from point to point on the earth's surface. Suppose a pendulum is to be used to find the acceleration of gravity at a given point on the earth's surface. Use the formula

$$T = 2\pi\sqrt{\frac{L}{g}}$$

to approximate the percentage error in the value of g if you can identify the period T to within 0.1 percent of its true value.

25. Examine the completed tables for Exercises 11–16 and explain why $\Delta y - dy$ decreases as $\Delta x \to 0$.

Review Exercises for Chapter 4

In Exercises 1–24, make use of domain, range, symmetry, asymptotes, intercepts, relative extrema, or points of inflection to obtain an accurate graph of each function.

1. $f(x) = 4x - x^2$
2. $f(x) = 4x^3 - x^4$
3. $f(x) = x\sqrt{16 - x^2}$
4. $f(x) = x + \dfrac{4}{x^2}$
5. $f(x) = \dfrac{x + 1}{x - 1}$
6. $f(x) = x^2 + \dfrac{1}{x}$
7. $f(x) = x^3 + x + \dfrac{4}{x}$
8. $f(x) = x^3(x + 1)$
9. $f(x) = (x - 1)^3(x - 3)^2$

10. $f(x) = (x - 3)(x + 2)^3$
11. $f(x) = (5 - x)^3$
12. $f(x) = (x^2 - 4)^2$
13. $f(x) = x^{1/3}(x + 3)^{2/3}$
14. $f(x) = (x - 2)^{1/3}(x + 1)^{2/3}$
15. $f(x) = x^3 + \dfrac{243}{x}$
16. $f(x) = \dfrac{2x}{1 + x^2}$

17. $f(x) = \dfrac{4}{1 + x^2}$
18. $f(x) = \dfrac{x^2}{1 + x^4}$
19. $f(x) = |x^2 - 9|$
20. $f(x) = |9 - x^2|$
21. $f(x) = |x^3 - 3x^2 + 2x|$
22. $f(x) = |x - 1| + |x - 3|$
23. $f(x) = \dfrac{1}{|x - 1|}$
24. $f(x) = \dfrac{x - 1}{1 + 3x^2}$

25. Find the maximum and minimum points on the graph of

$$x^2 + 4y^2 - 2x - 16y + 13 = 0$$

(a) without using calculus and (b) using calculus.

26. Consider the function $f(x) = x^n$ for positive integer values of n.
    (a) For what values of n does the function have a relative minimum at the origin?
    (b) For what values of n does the function have a point of inflection at the origin?

In Exercises 27–32, find the point(s) guaranteed by the Mean Value Theorem for the indicated interval.

Function	Interval
27. $f(x) = \dfrac{2x + 3}{3x + 2}$	$1 \leq x \leq 5$
28. $f(x) = \dfrac{1}{x}$	$1 \leq x \leq 4$
29. $f(x) = x^{2/3}$	$1 \leq x \leq 8$
30. $f(x) = |x^2 - 9|$	$0 \leq x \leq 2$
31. $f(x) = x - \dfrac{1}{x}$	$1 \leq x \leq 4$
32. $f(x) = \sqrt{x} - 2x$	$0 \leq x \leq 4$

33. Can the Mean Value Theorem be applied to the function $f(x) = 1/x^2$ on the interval $[-2, 1]$?

34. If $f(x) = 3 - |x - 4|$, verify that $f(1) = f(7)$ and yet $f'(x)$ is not equal to zero for any x in $[1, 7]$. Does this contradict Rolle's Theorem?

35. For the function

$$f(x) = Ax^2 + Bx + C$$

determine the value of c guaranteed by the Mean Value Theorem on the interval $[x_1, x_2]$.

36. Demonstrate the result of Exercise 35 for

$$f(x) = 2x^2 - 3x + 1$$

on the interval $[0, 4]$.

37. At noon ship A was 100 miles due east of ship B. Ship A is sailing west at 12 miles per hour, and ship B is sailing south at 10 miles per hour. At what time will the ships be nearest to each other, and what will this distance be?

38. The cost C of producing x units per day is

$$C = \frac{1}{4}x^2 + 62x + 125$$

and the price p per unit is

$$p = 75 - \frac{1}{3}x$$

(a) What should be the daily output to obtain maximum profit?
(b) What should be the daily output to obtain minimum average cost?
(c) Find the elasticity of demand.

39. Find the maximum profit if the demand equation is

$$p = 36 - 4x$$

and the total cost is

$$C = 2x^2 + 6$$

40. Find the dimensions of the rectangle of maximum area, with sides parallel to the coordinate axes, that can be inscribed in the ellipse given by

$$\frac{x^2}{144} + \frac{y^2}{16} = 1$$

41. A right triangle in the first quadrant has the coordinate axes as sides, and the hypotenuse passes through the point $(1, 8)$. Find the vertices of the triangle so that the length of the hypotenuse is minimum.

42. The wall of a building is to be braced by a beam that must pass over a parallel fence 5 feet high and 4 feet from the building. Find the length of the shortest beam that can be used.

43. For groups of 80 or more, a charter bus company determines the rate per person according to the following formula:

$$\text{Rate} = \$8.00 - \$0.05(n - 80), \quad n \geq 80$$

What number of passengers will give the bus company maximum revenue?

44. Show that the greatest area of any rectangle inscribed in a triangle is one-half that of the triangle.

45. Three sides of a trapezoid have the same length s. Of all such possible trapezoids, show that the one of maximum area has a fourth side of length $2s$.

46. The cost of fuel for running a locomotive is proportional to the $\frac{3}{2}$ power of the speed $(s^{3/2})$ and is $50 per hour for a speed of 25 miles per hour. Other fixed costs amount to an average of $100 per hour. Find the speed that will minimize the cost per mile.

47. Find the length of the longest pipe that can be carried level around a right-angle corner if the two intersecting corridors are of width 4 feet and 6 feet.

48. Rework Exercise 47, given corridors of width a feet and b feet.

49. The cost of inventory depends on ordering cost and storage cost, according to the following inventory model:

$$C = \left(\frac{Q}{x}\right)s + \left(\frac{x}{2}\right)r$$

Determine the order size that will minimize the cost, assuming that sales occur at a constant rate, Q is the number of units sold per year, r is the cost of storing one unit for one year, s is the cost of placing an order, and x is the number of units per order.

50. The demand and cost equations for a certain product are

$$p = 600 - 3x$$

and

$$C = 0.3x^2 + 6x + 600$$

where p is the price per unit, x is the number of units, and C is the cost of producing x units. If t is the excise tax per unit, the profit for producing x units is $P = xp - C - xt$. Find the maximum profit for (a) $t = 5$, (b) $t = 10$, and (c) $t = 20$.

51. Approximate, to three decimal places, the zero of

$$f(x) = x^3 - 3x - 1$$

in the interval $[-1, 0]$.

52. Approximate to three decimal places, the x-value of the points of intersection of the equations $y = x^4$ and $y = x + 3$.

53. Approximate, to three decimal places, the real root of

$$x^3 + 2x + 1 = 0$$

54. The diameter of a sphere is measured to be 18 inches with a maximum possible error of 0.05 inch. Use the differential to approximate the possible error in the surface area and the volume of the sphere.

55. If a 1 percent error is made in measuring the edge of a cube, approximately what percentage error will be made in calculating the surface area and the volume of the cube?

56. A company finds that the demand for its commodity is given by

$$p = 75 - \frac{1}{4}x$$

If x changes from 7 to 8, find the corresponding change in p. Compare the values of Δp and dp.

5 Integration

5.1

Antiderivatives and indefinite integration

Up to this point in our study of calculus, we have been concerned primarily with this problem: *given a function, find its derivative*. However, many important applications of calculus involve the inverse problem: *given the derivative of a function, find the original function*. For example, suppose you were asked to find a function F that has the following derivative:

$$F'(x) = 3x^2$$

From your knowledge of derivatives, you would probably say that

$$F(x) = x^3 \quad \text{because} \quad \frac{d}{dx}[x^3] = 3x^2$$

We call the function F an **antiderivative** of F'. For convenience, we will use the phrase *$F(x)$ is an antiderivative of $f(x)$* synonymously with *F is an antiderivative of f*. For instance, it is convenient to say that x^3 is an antiderivative of $3x^2$.

DEFINITION OF ANTIDERIVATIVE

A function F is called an **antiderivative** of the function f if for every x in the domain of f,

$$F'(x) = f(x)$$

In this definition, we call F *an* antiderivative of f, rather than *the* antiderivative of f. To see why, consider the fact that $F_1(x) = x^3$, $F_2(x) = x^3 - 5$, and $F_3(x) = x^3 + 97$ are all antiderivatives of $f(x) = 3x^2$. This suggests that for any constant C, the function given by $F(x) = x^3 + C$ is an antiderivative of f. This result is part of the following theorem.

THEOREM 5.1

REPRESENTATION OF ANTIDERIVATIVES

If F is an antiderivative of f on an interval I, then G is an antiderivative of f on the interval I if and only if it is of the form

$$G(x) = F(x) + C, \quad \text{for all } x \text{ in } I$$

where C is a constant.

Proof: There are two directions in this theorem. The proof of one direction is straightforward. That is, if $F'(x) = f(x)$ and C is a constant, then

$$G'(x) = \frac{d}{dx}[F(x) + C] = F'(x) + 0 = f(x)$$

The proof of the other direction takes a little more work, but it can be accomplished with the Mean Value Theorem, by defining a function H such that

$$H(x) = G(x) - F(x)$$

Now, if H is not constant on the interval I, then there must exist a and b ($a < b$) in the interval such that $H(a) \neq H(b)$. Moreover, since H is differentiable on $[a, b]$, we can apply the Mean Value Theorem to conclude that there exists some c in (a, b) such that

$$H'(c) = \frac{H(b) - H(a)}{b - a}$$

Since $H(b) \neq H(a)$, it follows that $H'(c) \neq 0$. However, since $G'(c) = F'(c)$, we know that

$$H'(c) = G'(c) - F'(c) = 0$$

and we have arrived at a contradiction. Consequently, our assumption that $H(x)$ is not constant must be false, and we conclude that $H(x) = C$. Therefore, $G(x) - F(x) = C$ and we have

$$G(x) = F(x) + C$$

Remark The crucial point of Theorem 5.1 is that we can represent the entire family of antiderivatives of a function by adding a constant to a *known* antiderivative. For example, knowing that $D_x[x^2] = 2x$, we can represent the family of *all* antiderivatives of $f(x) = 2x$ by $G(x) = x^2 + C$, where C is a constant. We call G the **general antiderivative** of f and we call $G(x) = x^2 + C$ the **general solution** of the equation $G'(x) = 2x$.

Notation for antiderivatives

If $y = F(x)$ is an antiderivative of f, then we say $F(x)$ is a solution of the equation

$$\frac{dy}{dx} = f(x)$$

When solving such an equation, it is convenient to write it in the equivalent differential form

$$dy = f(x)\ dx$$

The operation of finding all solutions (the general antiderivative of f) of this equation is called **antidifferentiation** and is denoted by the symbol $\int$. The general solution of the equation $dy = f(x)\ dx$ is denoted by

$$y = \int f(x)\ dx = F(x) + C$$

where x is called the **variable of integration,** $f(x)$ is called the **integrand,** and C is called the **constant of integration.** We read the symbol $\int f(x)\ dx$ as the *antiderivative of f with respect to x.* Thus the differential dx serves to identify x as the variable of integration. The term **indefinite integral** is a synonym for antiderivative, and the term **integration** is a synonym for antidifferentiation. We put these ideas together in the following definition.

DEFINITION OF INTEGRAL NOTATION FOR ANTIDERIVATIVES

The notation

$$\int f(x)\ dx = F(x) + C$$

where C is an arbitrary constant, means that F is an antiderivative of f. That is, $F'(x) = f(x)$ for all x in the domain of f.

The inverse nature of integration and differentiation can be seen from the fact that by substituting $F'(x)$ for $f(x)$ in this definition, we get

$$\int F'(x)\ dx = F(x) + C \qquad \text{Integration is the inverse of differentiation}$$

Moreover, if $\int f(x)\ dx = F(x) + C$, then

$$\frac{d}{dx}\left[\int f(x)\ dx\right] = \frac{d}{dx}[F(x) + C] \qquad \text{Differentiation is the inverse of integration}$$
$$= F'(x) = f(x)$$

This inverse characteristic allows us to obtain integration formulas directly from differentiation formulas. In the following theorem, we summarize the most common integration and differentiation formulas.

THEOREM 5.2   BASIC INTEGRATION RULES

| **Differentiation formula** | **Integration formula** |

Constant: $\dfrac{d}{dx}[kx] = k$ $\displaystyle\int k\,dx = kx + C$

Constant Multiple: $\dfrac{d}{dx}[kf(x)] = kf'(x)$ $\displaystyle\int kf(x)\,dx = k\int f(x)\,dx$

Sum or Difference: $\dfrac{d}{dx}[f(x) \pm g(x)] = f'(x) \pm g'(x)$ $\displaystyle\int [f(x) \pm g(x)]\,dx = \int f(x)\,dx \pm \int g(x)\,dx$

Power: $\dfrac{d}{dx}[x^n] = nx^{n-1}$ $\displaystyle\int x^n\,dx = \dfrac{x^{n+1}}{n+1} + C,\ n \neq -1$

| Remark Be sure you see that the Power Rule for integration has the restriction that $n \neq -1$. Hence the evaluation of

$$\int \frac{1}{x}\,dx$$

must wait until the natural logarithm function is introduced in Section 5.5.

EXAMPLE 1   Applying the basic integration rules

Evaluate $\int 3x\,dx$.

Solution:

$$\int 3x\,dx = 3\int x\,dx \qquad\qquad \text{Constant Multiple Rule}$$

$$= 3\int x^1\,dx \qquad\qquad \text{Rewrite } (x = x^1)$$

$$= 3\left(\frac{x^2}{2}\right) + C \qquad\qquad \text{Power Rule } (n = 1)$$

$$= \frac{3}{2}x^2 + C \qquad\qquad \text{Simplify}$$

| Remark When indefinite integrals are being evaluated, a strict application of the basic integration rules tends to produce messy constants of integration. For instance, in Example 1, we could have written

$$\int 3x\,dx = 3\int x\,dx$$

$$= 3\left[\frac{x^2}{2} + C\right]$$

$$= \frac{3}{2}x^2 + 3C$$

However, since C represents *any* constant, it is both cumbersome and unnecessary to write $3C$ as the constant of integration, and we opt for the simpler $\frac{3}{2}x^2 + C$.

EXAMPLE 2   *Applying the basic integration rules*

Evaluate $\int (3x^2 + 2x)\, dx$.

Solution:

$$\int (3x^2 + 2x)\, dx = \int 3x^2\, dx + \int 2x\, dx \qquad \text{Sum Rule}$$

$$= 3 \int x^2\, dx + 2 \int x\, dx \qquad \text{Constant Multiple Rule}$$

$$= 3\left(\frac{x^3}{3}\right) + 2\left(\frac{x^2}{2}\right) + C \qquad \text{Power Rule}$$

$$= x^3 + x^2 + C$$

In Examples 1 and 2 note that the general pattern of integration is similar to that of differentiation.

Given Integral $\longrightarrow$ Rewrite $\longrightarrow$ Integrate $\longrightarrow$ Simplify

This pattern is further demonstrated in the next example.

EXAMPLE 3   *Applying the basic integration rules*

Given integral	*Rewrite*	*Integrate*	*Simplify*
(a) $\int \dfrac{1}{x^3}\, dx$	$\int x^{-3}\, dx$	$\dfrac{x^{-2}}{-2} + C$	$-\dfrac{1}{2x^2} + C$
(b) $\int \sqrt{x}\, dx$	$\int x^{1/2}\, dx$	$\dfrac{x^{3/2}}{3/2} + C$	$\dfrac{2}{3}x^{3/2} + C$

The basic integration rules listed in Theorem 5.2 allow us to integrate *any* polynomial function. This is demonstrated in the next example.

EXAMPLE 4   *Integrating polynomial functions*

(a) $\int 1\, dx = x + C$

(b) $\int (x + 2)\, dx = \int x\, dx + \int 2\, dx$

$$= \frac{x^2}{2} + 2x + C$$

(c) $\int (3x^4 - 5x^2 + x)\, dx = 3\left(\frac{x^5}{5}\right) - 5\left(\frac{x^3}{3}\right) + \frac{x^2}{2} + C$

$$= \frac{3}{5}x^5 - \frac{5}{3}x^3 + \frac{1}{2}x^2 + C$$

Remark   The integral in Example 4(a) is usually simplified to the form

$$\int 1\, dx = \int dx$$

EXAMPLE 5   *Rewriting before integrating*

Evaluate

$$\int \frac{x + 1}{\sqrt{x}}\, dx$$

Solution:

$$\int \frac{x + 1}{\sqrt{x}}\, dx = \int \left(\frac{x}{\sqrt{x}} + \frac{1}{\sqrt{x}} \right) dx$$

$$= \int (x^{1/2} + x^{-1/2})\, dx$$

$$= \frac{x^{3/2}}{3/2} + \frac{x^{1/2}}{1/2} + C = \frac{2}{3}x^{3/2} + 2x^{1/2} + C$$

> Remark   When integrating quotients, don't make the mistake of integrating the numerator and denominator separately. This is no more valid in integration than it is in differentiation. For instance, in Example 5, be sure you see that

$$\int \frac{x + 1}{\sqrt{x}}\, dx \neq \frac{\int (x + 1)\, dx}{\int \sqrt{x}\, dx}$$

EXAMPLE 6   *Rewriting before integrating*

Evaluate

$$\int \frac{3x^2 - 4}{x^2}\, dx$$

Solution:

$$\int \frac{3x^2 - 4}{x^2}\, dx = \int (3 - 4x^{-2})\, dx$$

$$= 3x - 4\left(\frac{x^{-1}}{-1} \right) + C = 3x + \frac{4}{x} + C$$

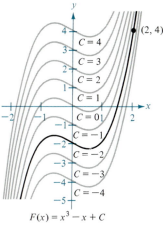

$F(x) = x^3 - x + C$

FIGURE 5.1

Particular solutions

We have already mentioned that the equation $y = \int f(x)\, dx$ can have many solutions (each differing by a constant). This means that the graphs of any two antiderivatives of f are vertical translations of each other. For example, Figure 5.1 shows the graphs of several antiderivatives of the form

$$y = \int (3x^2 - 1)\, dx = x^3 - x + C$$

for various integer values of C. Each of these antiderivatives is a solution of the equation

$$\frac{dy}{dx} = 3x^2 - 1$$

In many applications of integration, we are given enough information to determine a **particular solution.** To do so, we need only know the value of $F(x)$ for one value of x. (This information is called an **initial condition.**) For example, in Figure 5.1, there is only one curve that passes through the point $(2, 4)$. To find this particular curve, we use the information

$$F(x) = x^3 - x + C \qquad \text{General solution}$$
$$F(2) = 4 \qquad \text{Initial condition}$$

By using the initial condition in the general solution, we determine that $F(2) = 8 - 2 + C = 4$, which implies that $C = -2$. Thus, we obtain

$$F(x) = x^3 - x - 2 \qquad \text{Particular solution}$$

EXAMPLE 7   *Finding a particular solution*

Find the general solution of the equation

$$F'(x) = \frac{1}{x^2}$$

and find the particular solution that satisfies the initial condition $F(1) = 2$.

Solution: To find the general solution, we have

$$F(x) = \int \frac{1}{x^2}\, dx = \int x^{-2}\, dx = \frac{x^{-1}}{-1} + C = -\frac{1}{x} + C$$

Now, since $F(1) = 2$, we write

$$F(1) = -\frac{1}{1} + C = 2 \;\Longrightarrow\; C = 3$$

Thus, the particular solution, as shown in Figure 5.2, is

$$F(x) = -\frac{1}{x} + 3$$

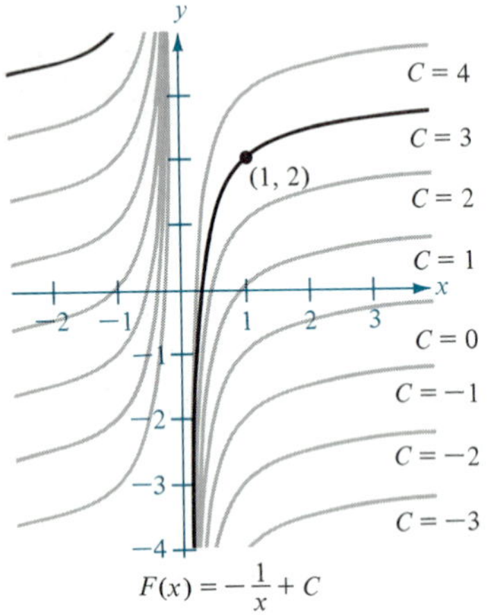

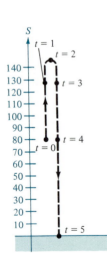

$C = 4$
$C = 3$
$(1, 2)$
$C = 2$
$C = 1$
$C = 0$
$C = -1$
$C = -2$
$C = -3$

$F(x) = -\dfrac{1}{x} + C$

FIGURE 5.2

EXAMPLE 8   *An application involving gravity*

A ball is thrown upward with an initial velocity of 64 feet per second from a height of 80 feet, as shown in Figure 5.3. Find the position function for this motion.

Solution: We let $t = 0$ represent the initial time. Then the two stated initial conditions in the problem can be written as follows:

$$s(0) = 80 \qquad \text{Initial height is 80 ft}$$
$$s'(0) = 64 \qquad \text{Initial velocity is 64 ft/sec}$$

Furthermore, we know that the acceleration due to gravity is -32 feet per second per second, so

$$s''(t) = -32$$

Now integrating $s''(t)$, we have

FIGURE 5.3

$$s'(t) = \int s''(t) \, dt = \int -32 \, dt = -32t + C_1$$

where $C_1 = s'(0) = 64$. Similarly, by integrating $s'(t)$, we have

$$s(t) = \int s'(t) \, dt = \int (-32t + 64) \, dt = -16t^2 + 64t + C_2$$

where $C_2 = s(0) = 80$. Therefore, we have

$$s(t) = -16t^2 + 64t + 80$$

Before beginning the exercise set for this section, be sure you realize that one of the most important steps in integration is *rewriting the integrand* in a form that fits the basic integration rules. To further illustrate this point, we list several additional examples in Table 5.1.

TABLE 5.1

Given	Rewrite	Integrate	Simplify
$\int \dfrac{2}{\sqrt{x}} \, dx$	$2 \int x^{-1/2} \, dx$	$2\left(\dfrac{x^{1/2}}{1/2}\right) + C$	$4x^{1/2} + C$
$\int (x^2 + 1)^2 \, dx$	$\int (x^4 + 2x^2 + 1) \, dx$	$\dfrac{x^5}{5} + 2\left(\dfrac{x^3}{3}\right) + x + C$	$\dfrac{1}{5}x^5 + \dfrac{2}{3}x^3 + x + C$
$\int \dfrac{x^3 + 3}{x^2} \, dx$	$\int (x + 3x^{-2}) \, dx$	$\dfrac{x^2}{2} + 3\left(\dfrac{x^{-1}}{-1}\right) + C$	$\dfrac{1}{2}x^2 - \dfrac{3}{x} + C$
$\int \sqrt[3]{x}(x - 4) \, dx$	$\int (x^{4/3} - 4x^{1/3}) \, dx$	$\dfrac{x^{7/3}}{7/3} - 4\left(\dfrac{x^{4/3}}{4/3}\right) + C$	$\dfrac{3}{7}x^{4/3}(x - 7) + C$

Section Exercises 5.1

In Exercises 1–6, complete the table using Table 5.1 as a model.

Given	Rewrite	Integrate	Simplify
1. $\int \sqrt[3]{x} \, dx$			
2. $\int \dfrac{1}{x^2} \, dx$			
3. $\int \dfrac{1}{x\sqrt{x}} \, dx$			
4. $\int x(x^2 + 3) \, dx$			
5. $\int \dfrac{1}{2x^3} \, dx$			
6. $\int \dfrac{1}{(2x)^3} \, dx$			

In Exercises 7–26, evaluate the indefinite integral and check your result by differentiation.

7. $\int (x^3 + 2) \, dx$ **8.** $\int (x^2 - 2x + 3) \, dx$

9. $\int (x^{3/2} + 2x + 1) \, dx$

10. $\int \left(\sqrt{x} + \dfrac{1}{2\sqrt{x}}\right) dx$

11. $\int \sqrt[3]{x^2} \, dx$ **12.** $\int (\sqrt[4]{x^3} + 1) \, dx$

13. $\int \dfrac{1}{x^3} \, dx$ **14.** $\int \dfrac{1}{x^4} \, dx$

15. $\int \dfrac{1}{4x^2} \, dx$ **16.** $\int (2x + x^{-1/2}) \, dx$

17. $\int \dfrac{x^2 + x + 1}{\sqrt{x}} \, dx$ **18.** $\int \dfrac{x^2 + 1}{x^2} \, dx$

19. $\int (x + 1)(3x - 2)\, dx$

20. $\int (2t^2 - 1)^2\, dt$

21. $\int \dfrac{t^2 + 2}{t^2}\, dt$

22. $\int (1 - 2y + 3y^2)\, dy$

23. $\int y^2 \sqrt{y}\, dy$

24. $\int (1 + 3t)t^2\, dt$

25. $\int dx$

26. $\int 3\, dt$

In Exercises 27–30, find the equation of the curve, given the derivative and the indicated point on the curve.

27. $\dfrac{dy}{dx} = 2x - 1$

28. $\dfrac{dy}{dx} = 2(x - 1)$

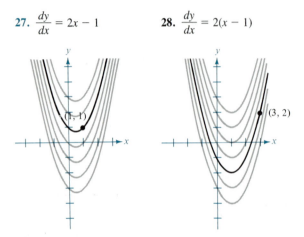

(1, 1)

(3, 2)

29. $\dfrac{dy}{dx} = 3x^2 - 1$

30. $\dfrac{dy}{dx} = -\dfrac{1}{x^2}$

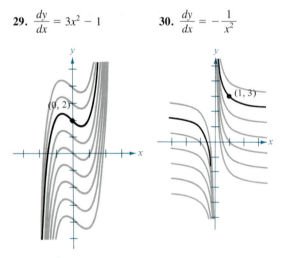

(0, 2)

(1, 3)

In Exercises 31–34, find $y = f(x)$ satisfying the given conditions.

31. $f''(x) = 2,\ f'(2) = 5,\ f(2) = 10$
32. $f''(x) = x^2,\ f'(0) = 6,\ f(0) = 3$
33. $f''(x) = x^{-3/2},\ f'(4) = 2,\ f(0) = 0$
34. $f''(x) = x^{-3/2},\ f'(1) = 2,\ f(9) = -4$

In Exercises 35–39, use

$$a(t) = -32 \text{ ft/s}^2$$

as the acceleration due to gravity. (Neglect air resistance.)

35. An object is dropped from a balloon which is stationary at 1600 feet. Express its height above the ground as a function of t. How long does it take the object to reach the ground?

36. A ball is thrown vertically upward from the ground with an initial velocity of 60 feet per second. How high will the ball go?

37. With what initial velocity must an object be thrown upward from the ground to reach a maximum height of 550 feet (approximate height of the Washington Monument)?

38. Show that the height above the ground of an object thrown upward from a point s_0 feet above the ground with an initial velocity of v_0 feet per second is given by the function

$$f(t) = -16t^2 + v_0 t + s_0$$

39. A balloon, rising vertically with a velocity of 16 feet per second, releases a sandbag at an instant when the balloon is 64 feet above the ground.
  (a) How many seconds after its release will the bag strike the ground?
  (b) With what velocity will it reach the ground?

40. Assume that a fully loaded plane starting from rest has a constant acceleration while moving down the runway. Find this acceleration if the plane requires, on the average, 0.7 mile of runway and a speed of 160 miles per hour before lifting off.

41. The makers of a certain automobile advertise that it will accelerate from 15 to 50 miles per hour in high gear in 13 seconds. Assuming constant acceleration, compute
  (a) the acceleration in feet per second per second
  (b) the distance the car travels in the given time

42. A car traveling at 45 miles per hour was brought to a stop, at constant deceleration, 132 feet from where the brakes were applied. How far had the car moved by the time its speed was reduced to (a) 30 miles per hour and (b) 15 miles per hour? (c) Draw the real number line from 0 to 132, and plot the points found in parts (a) and (b). What conclusions can you draw?

43. At the instant the traffic light turns green, an automobile that has been waiting at an intersection starts ahead with a constant acceleration of 6 feet per second per second. At the same instant a truck traveling with a constant velocity of 30 feet per second overtakes and passes the car.
  (a) How far beyond its starting point will the automobile overtake the truck?

(b) How fast will the automobile be traveling?

44. A ball is released from rest and rolls down an inclined plane, requiring 4 seconds to cover a distance of 100 centimeters. What is its acceleration in centimeters per second per second?

45. Galileo Galilei (1564–1642) stated the following proposition: The time in which any space is traversed by a body starting from rest and uniformly accelerated is equal to the time in which that same space would be traversed by the same body moving at a uniform speed whose value is the mean of the highest speed and the speed just before acceleration began. Use the techniques of this section to verify the statement.

46. If marginal cost is constant, show that the cost function is a straight line.

47. The marginal cost for production is

$$\frac{dC}{dx} = 2x - 12$$

and the fixed costs are $50. Find the total cost function and the average cost function.

48. If marginal revenue is

$$\frac{dR}{dx} = 100 - 5x$$

find the revenue and demand functions.

49. If marginal revenue is

$$\frac{dR}{dx} = 10 - 6x - 2x^2$$

find the revenue and demand functions.

5.2
Integration by substitution

SECTION TOPICS ▪
u-substitution ▪
Pattern recognition ▪
Change of variables ▪

In this section we demonstrate techniques for integrating composite functions. We will split the discussion into two parts—*pattern recognition* and *change of variables*. Both techniques involve a **u-substitution.** With pattern recognition we perform the substitution mentally, and with change of variables we actually write the substitution steps.

The role of substitution in integration is comparable to the role of the Chain Rule in differentiation. Recall that for differentiable functions given by $y = F(u)$ and $u = g(x)$, the Chain Rule states that

$$\frac{d}{dx}[F(g(x))] = F'(g(x))g'(x)$$

From our definition of an antiderivative, it follows that

$$\int F'(g(x))g'(x)\, dx = F(g(x)) + C = F(u) + C$$

We state these results in the following theorem.

THEOREM 5.3   **ANTIDIFFERENTIATION OF A COMPOSITE FUNCTION**
Let f and g be functions that satisfy the conditions of the Chain Rule for the composite function $y = f(g(x))$. If F is an antiderivative of f, then

$$\int f(g(x))g'(x)\, dx = F(g(x)) + C$$

If $u = g(x)$, then $du = g'(x)\, dx$ and $\int f(u)\, du = F(u) + C$

Pattern recognition

There are several techniques for applying substitution, each differing slightly from the other. However, you should remember that the goal is the same with every technique—*we are trying to find an antiderivative of the integrand.*

To apply Theorem 5.3 you must develop the ability to recognize what part of the integrand is $g(x)$. When you find the right function, you will see that the integrand is composed of the composite function $f(g(x))$ multiplied by $g'(x)$. Of course, part of the key is a familiarity with differentiation rules.

The first two examples show how to apply the theorem *directly*, by recognizing the presence of $f(g(x))$ and $g'(x)$. Note that the composite function in the integrand has an *outside function f* and an *inside function g*. Moreover, the derivative $g'(x)$ is present as a factor of the integrand.

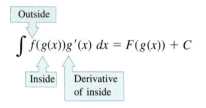

$$\int f(g(x))g'(x)\ dx = F(g(x)) + C$$

___

EXAMPLE 1   Recognizing the $f(g(x))g'(x)$ pattern

Evaluate $\int (x^2 + 1)^2(2x)\ dx$.

Solution: Letting $g(x) = x^2 + 1$, we have $g'(x) = 2x$, $f(g(x)) = [g(x)]^2$, and we recognize that the integrand follows the $f(g(x))g'(x)$ pattern. Moreover, by the Power Rule, we know that $F(g(x)) = \frac{1}{3}[g(x)]^3$ is an antiderivative of f. Thus, by Theorem 5.3, we have

$$\int (x^2 + 1)^2(2x)\ dx = F(x^2 + 1) + C = \frac{(x^2 + 1)^3}{3} + C$$

$\qquad\quad [g(x)]^2 \quad g'(x) \qquad\qquad F(g(x)) \qquad\qquad \frac{1}{3}[g(x)]^3$

___

EXAMPLE 2   Recognizing the $f(g(x))g'(x)$ pattern

Evaluate $\int 5\sqrt{5x + 1}\ dx$.

Solution: By letting $g(x) = 5x + 1$, we have $g'(x) = 5$, and we recognize that the integrand follows the $f(g(x))g'(x)$ pattern. Moreover, by the Power Rule, we know that

$$F(g(x)) = \frac{[g(x)]^{3/2}}{3/2} = \frac{2}{3}[g(x)]^{3/2}$$

is an antiderivative of f. Thus, by Theorem 5.3, we have

$$\int (5x + 1)^{1/2}(5)\ dx = F(5x + 1) + C = \frac{2}{3}(5x + 1)^{3/2} + C$$

$\qquad\quad [g(x)]^{1/2} \quad g'(x) \qquad\qquad F(g(x)) \qquad\qquad \frac{2}{3}[g(x)]^{3/2}$

Both of the integrands in Examples 1 and 2 fit the $f(g(x))g'(x)$ pattern exactly—all we had to do was recognize the pattern. We can extend this technique considerably with the Constant Multiple Rule

$$\int kf(x)\,dx = k \int f(x)\,dx$$

Many integrands contain the essential part (the variable part) of $g'(x)$, but are missing a constant multiple. In such cases we can multiply and divide by the necessary constant multiple, as shown in the following example.

EXAMPLE 3   *Multiplying and dividing by a constant*

Evaluate $\int x(x^2 + 1)^2\,dx$.

Solution: This is similar to the integral given in Example 1, except there is a missing factor of 2. Recognizing that $2x$ is the derivative of $x^2 + 1$, we let $u = x^2 + 1$ and supply the $2x$ as follows.

$$\int x(x^2 + 1)^2\,dx = \int (x^2 + 1)^2 \left(\frac{1}{2}\right)(2x)\,dx \qquad \text{Multiply and divide by 2}$$

$$= \frac{1}{2} \int (x^2 + 1)^2(2x)\,dx \qquad \text{Constant Multiple Rule}$$

$$= \frac{1}{2} \int [g(x)]^2 g'(x)\,dx$$

$$= \frac{1}{2} \frac{[g(x)]^3}{3} + C \qquad \text{Integrate}$$

$$= \frac{(x^2 + 1)^3}{6} + C \qquad \Box$$

EXAMPLE 4   *Multiplying and dividing by a constant*

Evaluate $\int \sqrt{5x + 1}\,dx$.

Solution: By recognizing that 5 is the derivative of $5x + 1$, we let $g(x) = 5x + 1$ and supply the 5 as follows.

$$\int (5x + 1)^{1/2}\,dx = \int (5x + 1)^{1/2} \left(\frac{1}{5}\right)(5)\,dx \qquad \text{Multiply and divide by 5}$$

$$= \frac{1}{5} \int (5x + 1)^{1/2}(5)\,dx \qquad \text{Constant Multiple Rule}$$

$$= \frac{1}{5}\left[\frac{(5x + 1)^{3/2}}{3/2}\right] + C \qquad \text{Integrate}$$

$$= \frac{2}{15}(5x + 1)^{3/2} + C \qquad \Box$$

Remark Be sure you see that the *Constant* Multiple Rule applies only to constants. You cannot multiply and divide by a variable and then move the variable outside the integral sign. For instance,

$$\int (x^2 + 1)^2\,dx \neq \frac{1}{2x} \int (x^2 + 1)^2(2x)\,dx$$

After all, if it were legitimate to move variable quantities outside the integral sign, you could move the entire integrand out and simplify the whole process!

Change of variables

The integration technique used in Examples 1 through 4 depends on the ability to recognize (or create) integrands of the form $f(g(x))g'(x)$. With a formal *change of variables,* we completely rewrite the integral in terms of u and du. Although this procedure involves more written steps, it is useful for complicated integrands. The change of variable technique uses the Leibniz notation for the differential. That is, if $u = g(x)$, we write $du = g'(x)\,dx$, and the integral in Theorem 5.3 takes the form

$$\int f(g(x))g'(x)\,dx = \int f(u)\,du = F(u) + C$$

We summarize this procedure in the following way.

INTEGRATION BY SUBSTITUTION

1. Choose a substitution $u = g(x)$. Usually, it is best to choose the *inner* part of a composite function, such as a quantity raised to a power.
2. Evaluate the differential $du = g'(x)\,dx$. Note any constant factor k of $g'(x)$ that is not a factor of the given integrand.
3. Rewrite the integrand in the form $f(u)\dfrac{du}{k}$.
4. Evaluate the resulting integral in terms of u.
5. Backsubstitute to obtain an antiderivative in terms of x.

There are other ways to accomplish the third step. We can multiply and divide by a constant as we did with pattern recognition in Examples 3 and 4, or we can solve for dx as demonstrated in the next example.

EXAMPLE 5   Change of variable

Evaluate $\int \sqrt{2x - 1}\,dx$.

Solution: First, we let u be the inner function, $u = 2x - 1$. Then, we obtain

$$du = 2\,dx \qquad \text{Solve for } du$$

Now, since $\sqrt{2x - 1} = \sqrt{u}$ and $dx = du/2$, we substitute to obtain

$$\int \sqrt{2x - 1}\,dx = \int \sqrt{u}\left(\frac{du}{2}\right) \qquad \text{Integral in terms of } u$$

$$= \frac{1}{2}\int u^{1/2}\,du$$

$$= \frac{1}{2}\left(\frac{u^{3/2}}{3/2}\right) + C \qquad \text{Antiderivative in terms of } u$$

$$= \frac{1}{3}u^{3/2} + C$$

Backsubstitution of $u = 2x - 1$ yields

$$\int \sqrt{2x - 1}\,dx = \frac{1}{3}(2x - 1)^{3/2} + C \qquad \text{Antiderivative in terms of } x$$

EXAMPLE 6   *Change of variable*

Evaluate $\int x\sqrt{2x - 1}\ dx$.

Solution: As in the previous example, we let $u = 2x - 1$ and obtain $dx = du/2$. Since the integrand contains a factor of x, we must also solve for x in terms of u as follows:

$$u = 2x - 1 \implies x = \frac{u + 1}{2} \qquad \text{Solve for } x \text{ in terms of } u$$

Thus, the integral becomes

$$\int x\sqrt{2x - 1}\ dx = \int \left(\frac{u + 1}{2}\right) u^{1/2} \left(\frac{du}{2}\right)$$

$$= \frac{1}{4} \int (u^{3/2} + u^{1/2})\ du$$

$$= \frac{1}{4} \left[\frac{u^{5/2}}{5/2} + \frac{u^{3/2}}{3/2}\right] + C$$

Backsubstitution yields

$$\int x\sqrt{2x - 1}\ dx = \frac{1}{10}(2x - 1)^{5/2} + \frac{1}{6}(2x - 1)^{3/2} + C$$

To complete the change of variable in Example 6, we solved for x in terms of u. Sometimes this is very difficult. Fortunately it is not always necessary, as shown in the next example.

EXAMPLE 7   *Change of variable*

Evaluate $\int x\sqrt{x^2 - 1}\ dx$.

Solution: Since $\sqrt{x^2 - 1} = (x^2 - 1)^{1/2}$, we let $u = x^2 - 1$. Then

$$du = (2x)\ dx$$

Now, since $x\ dx$ is part of the given integral, we write

$$\frac{du}{2} = x\ dx$$

Substituting u and $du/2$ in the given integral yields

$$\int x\sqrt{x^2 - 1}\ dx = \int u^{1/2} \frac{du}{2}$$

$$= \frac{1}{2} \int u^{1/2}\ du$$

$$= \frac{1}{2} \left(\frac{u^{3/2}}{3/2}\right) + C = \frac{1}{3} u^{3/2} + C$$

Backsubstitution yields

$$\int x\sqrt{x^2 - 1}\ dx = \frac{1}{3}(x^2 - 1)^{3/2} + C$$

One of the most common u-substitutions involves quantities in the integrand that are raised to a power. Because of the importance of this type of substitution, we give it a special name—the **General Power Rule.**

THEOREM 5.4   THE GENERAL POWER RULE

If g is a differentiable function of x, then

$$\int [g(x)]^n g'(x)\, dx = \frac{[g(x)]^{n+1}}{n+1} + C, \quad n \neq -1$$

If $u = g(x)$, then

$$\int u^n\, du = \frac{u^{n+1}}{n+1} + C, \quad n \neq -1$$

Proof: This theorem follows directly from the (simple) Power Rule for integration together with Theorem 5.3.

Study the following example carefully to see the variety of integrals that can be evaluated with the General Power Rule.

EXAMPLE 8   Substitution and the General Power Rule

(a) $\displaystyle \int 3(3x-1)^4\, dx = \int \overbrace{(3x-1)^4}^{u^4}\overbrace{(3)\, dx}^{du} = \overbrace{\frac{(3x-1)^5}{5}}^{u^5/5} + C$

(b) $\displaystyle \int (2x+1)(x^2+x)\, dx = \int \overbrace{(x^2+x)^1}^{u^1}\overbrace{(2x+1)\, dx}^{du} = \overbrace{\frac{(x^2+x)^2}{2}}^{u^2/2} + C$

(c) $\displaystyle \int 3x^2\sqrt{x^3-2}\, dx = \int \overbrace{(x^3-2)^{1/2}}^{u^{1/2}}\overbrace{(3x^2)\, dx}^{du} = \overbrace{\frac{(x^3-2)^{3/2}}{3/2}}^{u^{3/2}/(3/2)} + C$

(d) $\displaystyle \int \frac{-4x}{(1-2x^2)^2}\, dx = \int \overbrace{(1-2x^2)^{-2}}^{u^{-2}}\overbrace{(-4x)\, dx}^{du} = \overbrace{\frac{(1-2x^2)^{-1}}{-1}}^{u^{-1}/(-1)} + C$ ▢

Remark   Example 8(b) shows a use of the General Power Rule that is often overlooked—the case when $n=1$. Remember that

$$\int u\, du = \int u^1\, du = \frac{u^2}{2} + C$$

Just because an integrand involves a quantity raised to a power *does not* mean that the integral can be evaluated by the General Power Rule. Consider the two integrals

$$\int x(x^2 + 1)^2 \, dx \qquad \text{and} \qquad \int (x^2 + 1)^2 \, dx$$

Can both be evaluated by the General Power Rule? We evaluated the first integral in Example 3 by letting $u = x^2 + 1$, from which we obtained $du = 2x \, dx$. However, in the second integral the substitution $u = x^2 + 1$ fails, since the integrand lacks the critical factor x needed for du. Fortunately, *for this particular integral,* we can expand into the polynomial form

$$(x^2 + 1)^2 = x^4 + 2x^2 + 1$$

and use the Simple Power Rule on each term.

Section Exercises 5.2

In Exercises 1–4, complete the table by identifying u and du for the given integral.

$$\int f(g(x))g'(x) \, dx \qquad u = g(x) \qquad du = g'(x) \, dx$$

1. $\displaystyle\int (5x^2 + 1)^2(10x) \, dx$

2. $\displaystyle\int x^2 \sqrt{x^3 + 1} \, dx$

3. $\displaystyle\int \frac{x}{\sqrt{x^2 + 1}} \, dx$

4. $\displaystyle\int (x^3 + 3)3x^2 \, dx$

In Exercises 5–34, find the indefinite integral and check the result by differentiation.

5. $\displaystyle\int (1 + 2x)^4 2 \, dx$

6. $\displaystyle\int (x^2 - 1)^3 2x \, dx$

7. $\displaystyle\int x^2(x^3 - 1)^4 \, dx$

8. $\displaystyle\int x(1 - 2x^2)^3 \, dx$

9. $\displaystyle\int x(x^2 - 1)^7 \, dx$

10. $\displaystyle\int \frac{x^2}{(x^3 - 1)^2} \, dx$

11. $\displaystyle\int \frac{4x}{\sqrt{1 + x^2}} \, dx$

12. $\displaystyle\int \frac{6x}{(1 + x^2)^3} \, dx$

13. $\displaystyle\int 5x\sqrt[3]{1 + x^2} \, dx$

14. $\displaystyle\int 3(x - 3)^{5/2} \, dx$

15. $\displaystyle\int \frac{-3}{\sqrt{2x + 3}} \, dx$

16. $\displaystyle\int \frac{4x + 6}{(x^2 + 3x + 7)^3} \, dx$

17. $\displaystyle\int \frac{x + 1}{(x^2 + 2x - 3)^2} \, dx$

18. $\displaystyle\int u^3 \sqrt{u^4 + 2} \, du$

19. $\displaystyle\int \frac{1}{\sqrt{x}(1 + \sqrt{x})^2} \, dx$

20. $\displaystyle\int \left(1 + \frac{1}{t}\right)^3 \left(\frac{1}{t^2}\right) dt$

21. $\displaystyle\int \frac{x^2}{(1 + x^3)^2} \, dx$

22. $\displaystyle\int \frac{x^2}{\sqrt{1 + x^3}} \, dx$

23. $\displaystyle\int \frac{x^3}{\sqrt{1 + x^4}} \, dx$

24. $\displaystyle\int \frac{t + 2t^2}{\sqrt{t}} \, dt$

25. $\displaystyle\int \frac{1}{2\sqrt{x}} \, dx$

26. $\displaystyle\int \frac{1}{(3x)^2} \, dx$

27. $\displaystyle\int \frac{1}{\sqrt{2x}} \, dx$

28. $\displaystyle\int \frac{1}{3x^2} \, dx$

29. $\displaystyle\int \frac{x^2 + 3x + 7}{\sqrt{x}} \, dx$

30. $\displaystyle\int \frac{x^{5/2} + 5x^{1/2}}{x^{5/2}} \, dx$

31. $\displaystyle\int t^2\left(t - \frac{2}{t}\right) dt$

32. $\displaystyle\int \left(\frac{t^3}{3} + \frac{1}{4t^2}\right) dt$

33. $\displaystyle\int (9 - y)\sqrt{y} \, dy$

34. $\displaystyle\int 2\pi y(8 - y^{3/2}) \, dy$

In Exercises 35 and 36, perform the integration in two ways and explain the difference in appearance of the answers.

35. $\displaystyle\int (2x - 1)^2 \, dx$

36. $\displaystyle\int x(x^2 - 1)^2 \, dx$

In Exercises 37–46, find the indefinite integral by the method of Example 6.

37. $\displaystyle\int x\sqrt{x - 3} \, dx$

38. $\displaystyle\int x\sqrt{2x + 1} \, dx$

39. $\displaystyle\int x^2\sqrt{1 - x} \, dx$

40. $\displaystyle\int \frac{2x - 1}{\sqrt{x + 3}} \, dx$

41. $\displaystyle\int \frac{x^2 - 1}{\sqrt{2x - 1}} \, dx$

42. $\displaystyle\int x^3\sqrt{x + 2} \, dx$

43. $\displaystyle\int \frac{-x}{(x + 1) - \sqrt{x + 1}} \, dx$

44. $\displaystyle\int t\sqrt[3]{t + 1} \, dt$

45. $\displaystyle\int \frac{x}{\sqrt{2x + 1}} \, dx$

46. $\displaystyle\int (x + 1)\sqrt{2 - x} \, dx$

47. Which of the following integrals can be evaluated by the General Power Rule?

(a) $\int \sqrt{x^2 + 1}\, dx$ (b) $\int x\sqrt{x^2 + 1}\, dx$

(c) $\int \dfrac{x}{\sqrt{x^2 + 1}}\, dx$

48. Which of the following integrals can be evaluated by the General Power Rule?

(a) $\int \sqrt{x}\sqrt{1 + x^{3/2}}\, dx$ (b) $\int \dfrac{1}{\sqrt{x}(1 + \sqrt{x})^2}\, dx$

(c) $\int \dfrac{\sqrt{x}}{(1 + \sqrt{x})^2}\, dx$

49. Find the equation of the curve passing through the point $(0, \frac{4}{3})$, given the derivative $f'(x) = x\sqrt{1 - x^2}$.

50. Find the equation of the function f whose graph passes through the point $(0, \frac{7}{3})$ and whose derivative is $f'(x) = x\sqrt{1 - x^2}$.

51. A lumber company is seeking a model that yields the average weight loss W per log as a function of the number of days of drying time t. The model is to be reliable up to 100 days after the log is cut. Based on the weight loss during the first 30 days, it was determined that

$$\frac{dW}{dt} = \frac{12}{\sqrt{16t + 9}}$$

(a) Find W as a function of t. Note that no weight loss occurs until the tree is cut.

(b) Find the total weight loss after 100 days.

52. A particular company has determined their marginal cost to be

$$\frac{dC}{dx} = \frac{4}{\sqrt{x + 1}}$$

(a) Find the cost function if $C = 50$ when $x = 15$.

(b) Graph the marginal cost function and the cost function on the same set of axes.

53. The marginal cost for a certain commodity has been determined to be

$$\frac{dC}{dx} = \frac{12}{\sqrt[3]{12x + 1}}$$

(a) Find the cost function if $C = 100$ when $x = 13$.

(b) Graph the marginal cost function and the cost function on the same set of axes.

5.3
Sigma notation and the limit of a sequence

SECTION TOPICS ▪
Sigma notation ▪
The limit of a sequence ▪

In Section 3.1 we mentioned that one of the major problems in calculus is determining the area of a bounded region in the plane. In the next section we will see that one approach to the area problem involves the sum of many terms. In preparation for Section 5.4, we introduce a concise notation for such summations. This notation is called **sigma notation** because it uses the upper case Greek letter sigma, written as Σ.

DEFINITION OF SIGMA NOTATION

The sum of n terms $\{a_1, a_2, a_3, \ldots, a_n\}$ is written as

$$\sum_{i=1}^{n} a_i = a_1 + a_2 + a_3 + \cdots + a_n$$

where i is called the **index of summation**, a_i is called the **ith term** of the sum, and the **upper and lower bounds of summation** are n and 1, respectively.

Remark The upper and lower bounds of summation must be constant *with respect to the index of summation*. However, the lower bound does not have to be 1—any integer value less than (or equal to) the upper bound is legitimate.

EXAMPLE 1   Examples of sigma notation

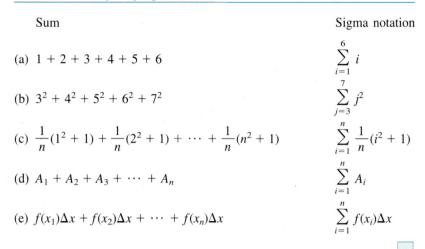

Sum	Sigma notation
(a) $1 + 2 + 3 + 4 + 5 + 6$	$\displaystyle\sum_{i=1}^{6} i$
(b) $3^2 + 4^2 + 5^2 + 6^2 + 7^2$	$\displaystyle\sum_{j=3}^{7} j^2$
(c) $\dfrac{1}{n}(1^2 + 1) + \dfrac{1}{n}(2^2 + 1) + \cdots + \dfrac{1}{n}(n^2 + 1)$	$\displaystyle\sum_{i=1}^{n} \dfrac{1}{n}(i^2 + 1)$
(d) $A_1 + A_2 + A_3 + \cdots + A_n$	$\displaystyle\sum_{i=1}^{n} A_i$
(e) $f(x_1)\Delta x + f(x_2)\Delta x + \cdots + f(x_n)\Delta x$	$\displaystyle\sum_{i=1}^{n} f(x_i)\Delta x$

Example 1 illustrates several characteristics of sigma notation. Though any variable can be used as the index of summation, we prefer i, j, and k because they are usually associated with integers. Note that the index of summation does not appear in the terms of the expanded sum.

The following properties of sigma notation are useful. To verify these two properties, we suggest that you write them in expanded form and apply the associative, commutative, and distributive properties of arithmetic.

THEOREM 5.5   SUMMATION PROPERTIES

1. $\displaystyle\sum_{i=1}^{n} ka_i = k \sum_{i=1}^{n} a_i,$ k is a constant   2. $\displaystyle\sum_{i=1}^{n} [a_i \pm b_i] = \sum_{i=1}^{n} a_i \pm \sum_{i=1}^{n} b_i$

The next theorem gives us some useful formulas for sums of powers.

THEOREM 5.6   SUMMATION FORMULAS

1. $\displaystyle\sum_{i=1}^{n} c = cn$

2. $\displaystyle\sum_{i=1}^{n} i = \dfrac{n(n + 1)}{2}$

3. $\displaystyle\sum_{i=1}^{n} i^2 = \dfrac{n(n + 1)(2n + 1)}{6}$

4. $\displaystyle\sum_{i=1}^{n} i^3 = \dfrac{n^2(n + 1)^2}{4}$

Proof: We give a proof for Property 2 and leave the proofs of the other properties as exercises. We write the sum in increasing *and* decreasing order and add corresponding terms.

$$\sum_{i=1}^{n} i = 1 + 2 + 3 + \cdots + (n-1) + n$$

$$\downarrow \quad \downarrow \quad \downarrow \quad \downarrow \quad \downarrow$$

$$\sum_{i=1}^{n} i = n + (n-1) + (n-2) + \cdots + 2 + 1$$

$$\downarrow \quad \downarrow \quad \downarrow \quad \downarrow \quad \downarrow$$

$$2\sum_{i=1}^{n} i = (n+1) + (n+1) + (n+1) + \cdots + (n+1) + (n+1)$$

$$n \text{ terms}$$

Therefore,

$$\sum_{i=1}^{n} i = \frac{n(n+1)}{2}$$

The importance of Theorem 5.6 is that it provides us with a simple method of evaluating certain sums. For example,

$$\sum_{i=1}^{100} i = 1 + 2 + \cdots + 100 = \frac{100(101)}{2} = 5050$$

This is further illustrated in the next example.

EXAMPLE 2   Evaluating a sum

Evaluate

$$\sum_{i=1}^{n} \frac{i+1}{n^2}$$

when $n = 10, 100, 1000,$ and $10,000$.

Solution: Applying Theorems 5.5 and 5.6, we have

$$\sum_{i=1}^{n} \frac{i+1}{n^2} = \frac{1}{n^2}\sum_{i=1}^{n}(i+1) = \frac{1}{n^2}\left[\sum_{i=1}^{n} i + \sum_{i=1}^{n} 1\right] = \frac{1}{n^2}\left[\frac{n(n+1)}{2} + n\right]$$

$$= \frac{1}{n^2}\left[\frac{n^2 + 3n}{2}\right] = \frac{1}{2} + \frac{3}{2n} = \frac{n+3}{2n}$$

Now, with the formula

$$\sum_{i=1}^{n} \frac{i+1}{n^2} = \frac{n+3}{2n}$$

we can find the sum of n terms by simply substituting different values of n as shown in Table 5.2.

TABLE 5.2

n	10	100	1,000	10,000
$\dfrac{(n+3)}{2n}$	0.65000	0.51500	0.50150	0.50015

Remark In Table 5.2, note that the sum appears to be approaching a limit as n increases. We will say more about this after introducing the concept of the limit of a sequence.

The limit of a sequence

Up to this point in the text, we have dealt only with functions whose domains are intervals on the real line. In the next section we will have need for a different type of function called a sequence. A **sequence** is a function whose domain is the set of positive integers. For example, the sequence given by $s(n) = 1/2^n$ has the following values:

1	2	3	4	5	$\cdots$	n	$\cdots$
↓	↓	↓	↓	↓		↓	
$\dfrac{1}{2^1}$	$\dfrac{1}{2^2}$	$\dfrac{1}{2^3}$	$\dfrac{1}{2^4}$	$\dfrac{1}{2^5}$	$\cdots$	$\dfrac{1}{2^n}$	$\cdots$

Sequences have played an important role in the history of calculus and we will study them extensively in Chapter 10. In the meantime, we preview an important theorem dealing with the limit of a sequence as $n \to \infty$. The definition of this limit parallels that for a function of a real variable. That is, if for $\varepsilon > 0$ there exists $M > 0$ such that $|s(n) - L| < \varepsilon$ whenever $n > M$, then we say that the limit of $s(n)$ is L and write

$$\lim_{n \to \infty} s(n) = L$$

The next theorem compares the limit of a sequence to the limit of a function of a real variable.

THEOREM 5.7   **LIMIT OF A SEQUENCE**
Let f be a function of a real variable such that

$$\lim_{x \to \infty} f(x) = L$$

If s is a sequence such that $s(n) = f(n)$ for every positive integer n, then

$$\lim_{n \to \infty} s(n) = L$$

The importance of this theorem is that it allows us to apply many of the techniques used for functions of a real variable to find the limit of a sequence. This is demonstrated in Example 3.

EXAMPLE 3   Finding the limit of a sequence

Find the following limits.

(a) $\lim\limits_{x \to \infty} \dfrac{1}{x}$ (b) $\lim\limits_{n \to \infty} \dfrac{1}{n}$

Solution:

(a) By Theorem 4.11, we know that

$$\lim_{x \to \infty} \frac{1}{x} = 0$$

(b) If x is any integer n, then

$$\frac{1}{x} = \frac{1}{n}$$

and we can conclude from Theorem 5.7 that

$$\lim_{n \to \infty} \frac{1}{n} = 0$$

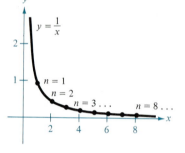

FIGURE 5.4   Figure 5.4 graphically depicts these two limits.

> Remark   When we study sequences in Chapter 10, we will see that the converse of Theorem 5.7 is not true.

EXAMPLE 4   The limit of a sequence

Let the sequence s be defined by

$$s(n) = \sum_{i=1}^{n} \frac{i + 1}{n^2}$$

and find the limit of $s(n)$ as $n \to \infty$.

Solution: From Example 2, we know that $s(n) = (n + 3)/2n$, and by applying Theorem 5.7, we have

$$\lim_{n \to \infty} \left[\sum_{i=1}^{n} \frac{i + 1}{n^2} \right] = \lim_{n \to \infty} \frac{n + 3}{2n}$$

$$= \lim_{n \to \infty} \left[\frac{1}{2} + \frac{3}{2n} \right]$$

$$= \frac{1}{2}$$

Section Exercises 5.3

In Exercises 1–8, find the given sum.

1. $\displaystyle\sum_{i=1}^{5} (2i + 1)$

2. $\displaystyle\sum_{i=1}^{6} 2i$

3. $\displaystyle\sum_{k=0}^{4} \frac{1}{k^2 + 1}$

4. $\displaystyle\sum_{j=3}^{5} \frac{1}{j}$

5. $\displaystyle\sum_{k=1}^{4} c$

6. $\displaystyle\sum_{n=1}^{10} \frac{3}{n + 1}$

7. $\displaystyle\sum_{i=1}^{4} [(i - 1)^2 + (i + 1)^3]$

8. $\displaystyle\sum_{k=2}^{5} (k + 1)(k - 3)$

In Exercises 9–18, use sigma notation to write the given sum.

9. $\dfrac{1}{3(1)} + \dfrac{1}{3(2)} + \dfrac{1}{3(3)} + \cdots + \dfrac{1}{3(9)}$

10. $\dfrac{5}{1 + 1} + \dfrac{5}{1 + 2} + \dfrac{5}{1 + 3} + \cdots + \dfrac{5}{1 + 15}$

11. $\left[2\left(\frac{1}{8}\right) + 3\right] + \left[2\left(\frac{2}{8}\right) + 3\right] + \cdots + \left[2\left(\frac{8}{8}\right) + 3\right]$

12. $\left[1 - \left(\frac{1}{4}\right)^2\right] + \left[1 - \left(\frac{2}{4}\right)^2\right] + \cdots + \left[1 - \left(\frac{4}{4}\right)^2\right]$

13. $\left[\left(\frac{1}{6}\right)^2 + 2\right]\left(\frac{1}{6}\right) + \cdots + \left[\left(\frac{6}{6}\right)^2 + 2\right]\left(\frac{1}{6}\right)$

14. $\left[\left(\frac{1}{n}\right)^2 + 2\right]\left(\frac{1}{n}\right) + \cdots + \left[\left(\frac{n}{n}\right)^2 + 2\right]\left(\frac{1}{n}\right)$

15. $\left[\left(\frac{2}{n}\right)^3 - \frac{2}{n}\right]\left(\frac{2}{n}\right) + \cdots + \left[\left(\frac{2n}{n}\right)^3 - \frac{2n}{n}\right]\left(\frac{2}{n}\right)$

16. $\left[1 - \left(\frac{2}{n} - 1\right)^2\right]\left(\frac{2}{n}\right) + \cdots + \left[1 - \left(\frac{2n}{n} - 1\right)^2\right]\left(\frac{2}{n}\right)$

17. $\left[2\left(1 + \frac{3}{n}\right)^2\right]\left(\frac{3}{n}\right) + \cdots + \left[2\left(1 + \frac{3n}{n}\right)^2\right]\left(\frac{3}{n}\right)$

18. $\left(\frac{1}{n}\right)\sqrt{1 - \left(\frac{0}{n}\right)^2} + \cdots + \left(\frac{1}{n}\right)\sqrt{1 - \left(\frac{n-1}{n}\right)^2}$

In Exercises 19–24, use the properties of sigma notation and summation formulas to evaluate the given sums.

19. $\displaystyle\sum_{i=1}^{20} 2i$

20. $\displaystyle\sum_{i=1}^{10} i(i^2 + 1)$

21. $\displaystyle\sum_{i=1}^{20} (i - 1)^2$

22. $\displaystyle\sum_{i=1}^{15} (2i - 3)$

23. $\displaystyle\sum_{i=1}^{15} \frac{1}{n^3}(i - 1)^2$

24. $\displaystyle\sum_{i=1}^{10} (i^2 - 1)$

In Exercises 25–30, find the limit of the given sequence as $n \to \infty$.

25. $s(n) = \left(\dfrac{4}{3n^3}\right)(2n^3 + 3n^2 + n)$

26. $s(n) = \left(\dfrac{8}{3} + \dfrac{4}{n} + \dfrac{4}{3n^2}\right)$

27. $s(n) = \dfrac{81}{n^4}\left[\dfrac{n^2(n + 1)^2}{4}\right]$

28. $s(n) = \dfrac{64}{n^3}\left[\dfrac{n(n + 1)(2n + 1)}{6}\right]$

29. $s(n) = \dfrac{18}{n^2}\left[\dfrac{n(n + 1)}{2}\right]$

30. $s(n) = \dfrac{1}{n^2}\left[\dfrac{n(n + 1)}{2}\right]$

In Exercises 31–36, use the properties of sigma notation to find a formula for the given sum of n terms. Then use the formula to find the limit as $n \to \infty$.

31. $\displaystyle\lim_{n\to\infty} \sum_{i=1}^{n} \frac{1}{n^3}(i - 1)^2$

32. $\displaystyle\lim_{n\to\infty} \sum_{i=1}^{n} \left(1 + \frac{2i}{n}\right)^2\left(\frac{2}{n}\right)$

33. $\displaystyle\lim_{n\to\infty} \sum_{i=1}^{n} \frac{16i}{n^2}$

34. $\displaystyle\lim_{n\to\infty} \sum_{i=1}^{n} \left(\frac{2i}{n}\right)\left(\frac{2}{n}\right)$

35. $\displaystyle\lim_{n\to\infty} \sum_{i=1}^{n} \left(1 + \frac{2i}{n}\right)^3\left(\frac{2}{n}\right)$

36. $\displaystyle\lim_{n\to\infty} \sum_{i=1}^{n} \left(1 + \frac{i}{n}\right)\left(\frac{2}{n}\right)$

37. (Learning Curve) Psychologists have developed models to predict performance as a function of the number of trials for a certain task. One such model is

$$P = \frac{b + \theta a(n - 1)}{1 + \theta(n - 1)}$$

where P is the percentage of correct responses after n trials and a, b, and θ are constants depending upon the actual learning situation. Find the limit of P as n approaches infinity.

38. Consider the learning curve given by

$$P = \frac{0.5 + 0.9(n - 1)}{1 + 0.9(n - 1)}$$

Complete the following table and find the limit as n approaches infinity.

n	1	2	3	4	5	6	7	8	9	10
P										

5.4
Area

In Euclidean geometry, the simplest type of plane region is a rectangle. Although we often say that the *formula* for the area of a rectangle is $A = bh$ as shown in Figure 5.5, it is actually more proper to say that this is the *definition* of the area of a rectangle.

DEFINITION OF THE AREA OF A RECTANGLE	The **area of a rectangle** with height h and a base of length b is $$A = bh$$

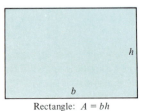

Rectangle: $A = bh$

FIGURE 5.5

From this definition we can develop formulas for the areas of many other plane regions. For example, to determine the area of a triangle, we can form a rectangle whose area is twice that of the triangle as shown in Figure 5.6. Once we know how to find the area of a triangle, we can determine the area of any polygon by subdividing the polygon into triangular regions as shown in Figure 5.7.

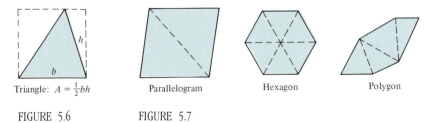

Triangle: $A = \frac{1}{2}bh$ Parallelogram   Hexagon   Polygon

FIGURE 5.6   FIGURE 5.7

When we move from polygons to more general plane regions, finding the area becomes more difficult. The ancient Greeks were able to determine formulas for the area of some general regions (principally those bounded by conics) by the *exhaustion* method. The clearest description of this method was given by Archimedes (287–212 B.C.). Essentially, the method is a limiting process in which the area is squeezed between two polygons—one inscribed in the region and one circumscribed about the region. The process we use to determine the area of a plane region is similar to that used by Archimedes.

The area of a plane region

We introduce the general problem of finding the area of a region in the plane with an example.

EXAMPLE 1   Approximating the area of a plane region

Archimedes

Use the five rectangles in parts (a) and (b) of Figure 5.8 to find two approximations for the area of the region lying between the graph of $f(x) = -x^2 + 5$ and the x-axis between $x = 0$ and $x = 2$.

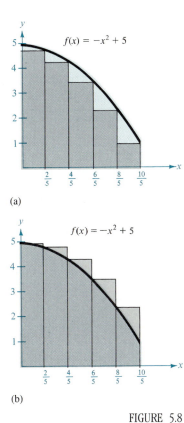

(a)

(b)

FIGURE 5.8

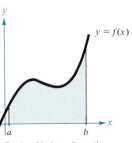

"Region Under a Curve"

FIGURE 5.9

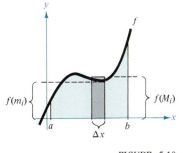

FIGURE 5.10

Solution: We can find the height of the five rectangles shown in Figure 5.8(a) by evaluating the function f at the right endpoint of each of the following intervals:

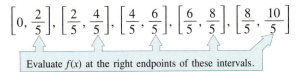

$$\left[0, \frac{2}{5}\right], \left[\frac{2}{5}, \frac{4}{5}\right], \left[\frac{4}{5}, \frac{6}{5}\right], \left[\frac{6}{5}, \frac{8}{5}\right], \left[\frac{8}{5}, \frac{10}{5}\right]$$

Evaluate $f(x)$ at the right endpoints of these intervals.

Since the width of each rectangle is $\frac{2}{5}$, the sum of the areas of the five rectangles is

$$\sum_{i=1}^{5} f\left(\frac{2i}{5}\right)\left(\frac{2}{5}\right) = \sum_{i=1}^{5} \left[-\left(\frac{2i}{5}\right)^2 + 5\right]\left(\frac{2}{5}\right) = \frac{2}{5} \sum_{i=1}^{5} \frac{-4i^2 + 125}{25}$$

$$= \frac{2}{125}[121 + 109 + 89 + 61 + 25] = \frac{162}{25} = 6.48$$

Since each of these five rectangles lies inside the given region, we conclude that the area of the region is *greater* than 6.48. To find the sum of the areas of the five rectangles in Figure 5.8(b), we use the same basic procedure, except we evaluate f at the left endpoints of the intervals and obtain

$$\sum_{i=1}^{5} f\left(\frac{2i-2}{5}\right)\left(\frac{2}{5}\right) = \sum_{i=1}^{5} \left[-\left(\frac{2i-2}{5}\right)^2 + 5\right]\left(\frac{2}{5}\right) = \frac{202}{25} = 8.08$$

From Figure 5.8(b), we see that the area of the given region is *less* than 8.08. By combining these two results, we conclude that the area of the given region lies between the two bounds

$$6.48 < \text{area of region} < 8.08$$

Remark By increasing the number of rectangles used in this approximation procedure, we can obtain closer and closer approximations for the area of the region. For instance, using 25 rectangles of width $\frac{2}{25}$ each, we could conclude that the area lies between the bounds

$$7.17 < \text{area of region} < 7.49$$

We now generalize the procedure described in Example 1. To begin, we look at a plane region bounded above by the graph of a nonnegative, continuous function $y = f(x)$ as shown in Figure 5.9. The region is bounded below by the x-axis and the left and right boundaries of the region are the vertical lines $x = a$ and $x = b$. We subdivide the interval $[a, b]$ into n subintervals, each of width $\Delta x = (b - a)/n$ as shown in Figure 5.10. The endpoints of the intervals are:

$$\underbrace{a + 0(\Delta x)}_{a = x_0} < \underbrace{a + 1(\Delta x)}_{x_1} < \underbrace{a + 2(\Delta x)}_{x_2} < \cdots < \underbrace{a + n(\Delta x)}_{x_n = b}$$

We call this subdivision a **regular partition** of the interval $[a, b]$. (We will look at more general partitions in the next section.)

Since f is continuous, the Extreme Value Theorem guarantees the existence of a minimum and a maximum value of $f(x)$ in *each* subinterval:

$$f(m_i) = \text{minimum value of } f(x) \text{ in } i\text{th subinterval}$$
$$f(M_i) = \text{maximum value of } f(x) \text{ in } i\text{th subinterval}$$

Thus, we can define an **inscribed rectangle** lying *inside* the ith subregion and a **circumscribed rectangle** extending *outside* the ith subregion. We use the minimum value as the height of the inscribed rectangle and the maximum value as the height of the circumscribed rectangle as shown in Figure 5.10. The areas of these two rectangles are related as follows:

$$\left(\begin{array}{c}\text{area of inscribed}\\\text{rectangle}\end{array}\right) = f(m_i)\Delta x \le f(M_i)\Delta x = \left(\begin{array}{c}\text{area of circumscribed}\\\text{rectangle}\end{array}\right)$$

Summing these areas, we have

$$\textbf{lower sum} = s(n) = \sum_{i=1}^{n} f(m_i)\Delta x \qquad \text{Area of inscribed rectangles}$$

$$\textbf{upper sum} = S(n) = \sum_{i=1}^{n} f(M_i)\Delta x \qquad \text{Area of circumscribed rectangles}$$

From Figure 5.11 we can see that the lower sum $s(n)$ is less than or equal to the upper sum $S(n)$ and we have $s(n) \le S(n)$.

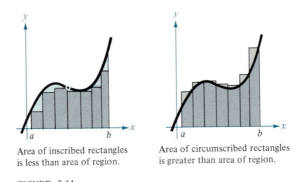

Area of inscribed rectangles is less than area of region.

Area of circumscribed rectangles is greater than area of region.

FIGURE 5.11

EXAMPLE 2 Finding upper and lower sums for a region

Find the upper and lower sums for the region bounded by the graph of

$$f(x) = x^2$$

and the x-axis between $x = 0$ and $x = 2$.

Solution: To begin we partition the interval $[0, 2]$ into n subintervals, each of length

$$\Delta x = \frac{b - a}{n} = \frac{2 - 0}{n} = \frac{2}{n}$$

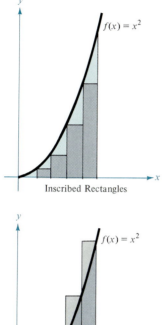

Inscribed Rectangles

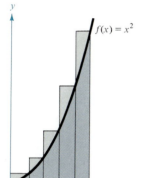

Circumscribed Rectangles

FIGURE 5.12

Figure 5.12 shows the endpoints of the subintervals and several inscribed and circumscribed rectangles. Since f is increasing on the interval $[0, 2]$, the minimum value on each subinterval occurs at the left endpoint and the maximum value occurs at the right endpoint. Thus, we have

Left Endpoints

$$m_i = 0 + (i - 1)\left(\frac{2}{n}\right) = \frac{2(i - 1)}{n}$$

Right Endpoints

$$M_i = 0 + i\left(\frac{2}{n}\right) = \frac{2i}{n}$$

Using the left endpoints, the lower sum is

$$s(n) = \sum_{i=1}^{n} f(m_i)\Delta x = \sum_{i=1}^{n} f\left[\frac{2(i - 1)}{n}\right]\left(\frac{2}{n}\right)$$

$$= \sum_{i=1}^{n} \left[\frac{2(i - 1)}{n}\right]^2\left(\frac{2}{n}\right) = \sum_{i=1}^{n} \left(\frac{8}{n^3}\right)(i^2 - 2i + 1)$$

$$= \frac{8}{n^3}\left[\sum_{i=1}^{n} i^2 - 2\sum_{i=1}^{n} i + \sum_{i=1}^{n} 1\right]$$

$$= \frac{8}{n^3}\left(\frac{n(n + 1)(2n + 1)}{6} - 2\left[\frac{n(n + 1)}{2}\right] + n\right)$$

$$= \frac{4}{3n^3}(2n^3 - 3n^2 + n)$$

$$= \frac{8}{3} - \frac{4}{n} + \frac{4}{3n^2}$$

Using the right endpoints, the upper sum is

$$S(n) = \sum_{i=1}^{n} f(M_i)\Delta x = \sum_{i=1}^{n} f\left(\frac{2i}{n}\right)\left(\frac{2}{n}\right)$$

$$= \sum_{i=1}^{n} \left(\frac{2^2 i^2}{n^2}\right)\left(\frac{2}{n}\right)$$

$$= \sum_{i=1}^{n} \left(\frac{8}{n^3}\right)i^2$$

$$= \frac{8}{n^3}\left[\frac{n(n + 1)(2n + 1)}{6}\right]$$

$$= \frac{4}{3n^3}(2n^3 + 3n^2 + n)$$

$$= \frac{8}{3} + \frac{4}{n} + \frac{4}{3n^2}$$

Remark You can see from Example 2 that the factor $(i - 1)$ makes the calculations for the left endpoints more complicated than those for the right endpoints. This is often true, since for any interval $[a, b]$ the n equal subintervals have the following endpoints:

$$\text{left endpoints} = a + (i - 1)\Delta x$$

$$\text{right endpoints} = a + i(\Delta x)$$

Example 2 contains a great deal of algebra, but we will now see that it is worth the effort! Although it is true that the lower sum in Example 2 is always less than the upper sum,

$$s(n) = \frac{8}{3} - \frac{4}{n} + \frac{4}{3n^2} < \frac{8}{3} + \frac{4}{n} + \frac{4}{3n^2} = S(n)$$

we can see that the difference between these two sums lessens as n increases. In fact, if we take the limit as $n \to \infty$, both the upper and lower sums approach 8/3.

$$\lim_{n \to \infty} s(n) = \lim_{n \to \infty} \left[\frac{8}{3} - \frac{4}{n} + \frac{4}{3n^2} \right] = \frac{8}{3} \qquad \text{Lower sum limit}$$

$$\lim_{n \to \infty} S(n) = \lim_{n \to \infty} \left[\frac{8}{3} + \frac{4}{n} + \frac{4}{3n^2} \right] = \frac{8}{3} \qquad \text{Upper sum limit}$$

Since both sums approach the same limit, it is natural to *define* the area of the region to be this limit.

The next theorem shows that the equivalence of the limit (as $n \to \infty$) of the upper and lower sums is not mere coincidence. It is true for all functions that satisfy the conditions stated in the theorem.

THEOREM 5.8   **LIMIT OF THE LOWER AND UPPER SUMS**

Let f be continuous and nonnegative on the interval $[a, b]$. The limits as $n \to \infty$ of both the lower and upper sums exist and are equal to each other. That is,

$$\lim_{n \to \infty} s(n) = \lim_{n \to \infty} \sum_{i=1}^{n} f(m_i)\Delta x = \lim_{n \to \infty} S(n) = \lim_{n \to \infty} \sum_{i=1}^{n} f(M_i)\Delta x$$

where $\Delta x = (b - a)/n$ and $f(m_i)$ and $f(M_i)$ are the minimum and maximum values of f on the ith subinterval.

The proof of this theorem is best left to a course in advanced calculus. However, from this theorem we can deduce an important result. Since the same limit is attained for both the minimum value $f(m_i)$ and the maximum value $f(M_i)$, it follows from the Squeeze Theorem (Theorem 2.13) that the choice of x in the ith subinterval does not affect the limit. This means that we are free to choose an *arbitrary* x-value in the ith subinterval, and we do this in the following definition of the *area under a curve*.

DEFINITION OF AREA OF A REGION IN THE PLANE

Let f be continuous and nonnegative on the interval $[a, b]$. The **area** of the region bounded by the graph of f, the x-axis, and the vertical lines $x = a$ and $x = b$ is

$$\text{area} = \lim_{n \to \infty} \sum_{i=1}^{n} f(c_i)\Delta x, \quad x_{i-1} \le c_i \le x_i$$

where $\Delta x = (b - a)/n$.

We can use this definition of area in two different ways: (1) to *compute exact areas,* and (2) to *approximate areas.* Actually, with the techniques we have available now, there are only a few types of functions for which we can find exact areas. We are basically limited to the lower-degree polynomial functions covered by Theorems 5.5 and 5.6. In Section 5.6, we will considerably extend this list by a procedure described in the Fundamental Theorem of Calculus.

EXAMPLE 3   *Finding area by the limit definition*

Find the area of the region bounded by the graph of $f(x) = x^3$, the x-axis, and the vertical lines $x = 0$ and $x = 1$, as shown in Figure 5.13.

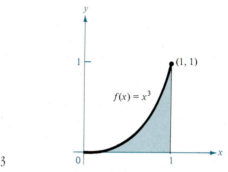

FIGURE 5.13

Solution: We note that f is continuous and nonnegative on the interval $[0, 1]$. Thus, we begin by partitioning the interval $[0, 1]$ into n equal subintervals, each of width $\Delta x = 1/n$. According to the definition of area, we can choose any x-value in the ith subinterval, and for the sake of convenience we choose the right endpoint $c_i = i/n$. Thus, we have

$$\text{area} = \lim_{n \to \infty} \sum_{i=1}^{n} f(c_i)\Delta x = \lim_{n \to \infty} \sum_{i=1}^{n} \left(\frac{i}{n}\right)^3 \left(\frac{1}{n}\right)$$

$$= \lim_{n \to \infty} \frac{1}{n^4} \sum_{i=1}^{n} i^3$$

$$= \lim_{n \to \infty} \frac{n^2(n + 1)^2}{4n^4}$$

$$= \lim_{n \to \infty} \left(\frac{1}{4} + \frac{1}{2n} + \frac{1}{4n^2}\right) = \frac{1}{4}$$

In the next example we look at a region that is bounded by the y-axis (rather than the x-axis).

EXAMPLE 4   *Finding area by the limit definition*

Find the area of the region bounded by the graph of $f(x) = x^2$, the x-axis, and the vertical lines $x = 1$ and $x = 3$, as shown in Figure 5.14.

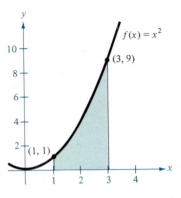

FIGURE 5.14

Solution: Since f is continuous and nonnegative on the interval $[1, 3]$, we can apply the definition of area to the region. We begin by partitioning the interval $[1, 3]$ into n equal subintervals, each of width $\Delta x = (3 - 1)/n = 2/n$. For convenience, we choose to evaluate f at the right endpoint

$$c_i = a + (i)(\Delta x) = 1 + (i)\left(\frac{2}{n}\right) = 1 + \frac{2i}{n}$$

of each interval. Thus, the area of the region is

$$\text{area} = \lim_{n \to \infty} \sum_{i=1}^{n} f(c_i) \, \Delta x$$

$$= \lim_{n \to \infty} \sum_{i=1}^{n} \left(1 + \frac{2i}{n}\right)^2 \left(\frac{2}{n}\right)$$

$$= \lim_{n \to \infty} \sum_{i=1}^{n} \left(1 + \frac{4i}{n} + \frac{4i^2}{n^2}\right)\left(\frac{2}{n}\right)$$

$$= \lim_{n \to \infty} \frac{2}{n} \sum_{i=1}^{n} 1 + \frac{8}{n^2} \sum_{i=1}^{n} i + \frac{8}{n^3} \sum_{i=1}^{n} i^2$$

$$= \lim_{n \to \infty} \frac{2}{n}(n) + \frac{8}{n^2}\left(\frac{n(n+1)}{2}\right) + \frac{8}{n^3}\left(\frac{n(n+1)(2n+1)}{6}\right)$$

$$= \lim_{n \to \infty} 2 + 4 + \frac{4}{n} + \frac{8}{3} + \frac{4}{n} + \frac{4}{3n^2}$$

$$= \frac{26}{3}$$

EXAMPLE 5   A region bounded by the y-axis

Find the area of the region bounded by the graph of $f(y) = y^2$ and the y-axis for $0 \le y \le 1$, as shown in Figure 5.15.

Solution: When f is a continuous, nonnegative function of y, we can still use the same basic procedure illustrated in Example 3. We partition $[0, 1]$ into n equal subintervals, each of width $\Delta y = 1/n$. Using the left (upper) endpoints $c_i = i/n$, we obtain the following limit.

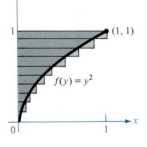

FIGURE 5.15

$$\text{area} = \lim_{n \to \infty} \sum_{i=1}^{n} f(c_i)\Delta y$$

$$= \lim_{n \to \infty} \sum_{i=1}^{n} \left(\frac{i}{n}\right)^2 \left(\frac{1}{n}\right)$$

$$= \lim_{n \to \infty} \frac{1}{n^3} \sum_{i=1}^{n} i^2$$

$$= \lim_{n \to \infty} \frac{n(n+1)(2n+1)}{6n^3}$$

$$= \lim_{n \to \infty} \left(\frac{1}{3} + \frac{1}{2n} + \frac{1}{6n^2}\right) = \frac{1}{3}$$

Note in Examples 3, 4, and 5 that we chose c_i in the ith interval to be a value that is *convenient* for calculating the limit. Since the limit yields the exact area for *any* c_i in the ith interval, we didn't have to try to find values that gave good approximations when n is small. However, if we are using the area definition for *approximation* purposes, then we try to find values of c_i that give good approximations of the area of the ith subregion. In general, a good value to choose is the midpoint, $c_i = (x_i + x_{i-1})/2$, of the interval. This yields the following **Midpoint Rule:**

$$\text{area} \approx \sum_{i=1}^{n} f\left(\frac{x_i + x_{i-1}}{2}\right)\Delta x \qquad \text{Midpoint Rule}$$

(We will look at other approximation strategies in Section 5.7. One called the *Trapezoidal Rule* is similar to the Midpoint Rule.)

EXAMPLE 6   Approximating area with the Midpoint Rule

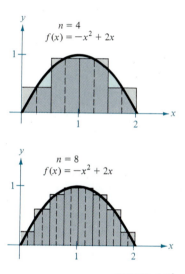

$n = 4$
$f(x) = -x^2 + 2x$

$n = 8$
$f(x) = -x^2 + 2x$

FIGURE 5.16

Approximate the area of the region bounded by the graph of $f(x) = -x^2 + 2x$ and the x-axis for $0 \le x \le 2$. Use the Midpoint Rule with $n = 4$ and $n = 8$, as shown in Figure 5.16.

Solution: For $n = 4$, we have $\Delta x = 1/2$, and the midpoints are given by

$$c_1 = \frac{0 + (1/2)}{2} = \frac{1}{4}$$

$$c_2 = \frac{(1/2) + 1}{2} = \frac{3}{4}$$

$$c_3 = \frac{1 + (3/2)}{2} = \frac{5}{4}$$

$$c_4 = \frac{(3/2) + 2}{2} = \frac{7}{4}$$

Thus, for $n = 4$, the area is approximated by

$$\text{area} \approx \sum_{i=1}^{4} f(c_i)\Delta x$$

$$= \sum_{i=1}^{n} [-(c_i)^2 + 2c_i]\left(\frac{1}{2}\right)$$

$$= \frac{1}{2}\left[-\left(\frac{1}{4}\right)^2 + 2\left(\frac{1}{4}\right) - \left(\frac{3}{4}\right)^2 + 2\left(\frac{3}{4}\right) - \left(\frac{5}{4}\right)^2 + 2\left(\frac{5}{4}\right) - \left(\frac{7}{4}\right)^2 + 2\left(\frac{7}{4}\right)\right]$$

$$\approx 1.375$$

Using the same type of procedure with $n = 8$, we have

$$\text{area} \approx 1.344$$

Remark   In Exercise 11 you are asked to show that the exact area of this region is $4/3 \approx 1.333$.

Section Exercises 5.4

In Exercises 1–6, use the upper and lower sums to approximate the area of the given region using the indicated number of subdivisions.

1. $y = \sqrt{x}$

2. $y = \sqrt{x} + 1$

3. $y = \dfrac{1}{x}$

4. $y = \dfrac{1}{x - 2}$

5. $y = \sqrt{1 - x^2}$

6. $y = \sqrt{x + 1}$

7. Consider the triangle of area 2 bounded by the graphs of $y = x$, $y = 0$, and $x = 2$.

(a) Sketch the graph of the region.

(b) Divide the interval $[0, 2]$ into n equal subintervals and show that the endpoints are

$$0 < 1\left(\frac{2}{n}\right) < \cdots < (n - 1)\left(\frac{2}{n}\right) < n\left(\frac{2}{n}\right)$$

(c) Show that $s(n) = \displaystyle\sum_{i=1}^{n} \left[(i - 1)\left(\frac{2}{n}\right)\right]\left(\frac{2}{n}\right)$.

(d) Show that $S(n) = \displaystyle\sum_{i=1}^{n} \left[i\left(\frac{2}{n}\right)\right]\left(\frac{2}{n}\right)$.

(e) Complete the following table:

n	5	10	50	100
$s(n)$				
$S(n)$				

(f) Show that $\lim\limits_{n \to \infty} s(n) = \lim\limits_{n \to \infty} S(n) = 2$.

8. Consider the trapezoid of area 4 bounded by the graphs of $y = x$, $y = 0$, $x = 1$, and $x = 3$.

(a) Sketch the graph of the region.

(b) Divide the interval $[1, 3]$ into n equal subintervals and show that the endpoints are

$$1 < 1 + 1\left(\frac{2}{n}\right) < \cdots < 1 + (n - 1)\left(\frac{2}{n}\right) < 1 + n\left(\frac{2}{n}\right)$$

(c) Show that $s(n) = \displaystyle\sum_{i=1}^{n} \left[1 + (i - 1)\left(\frac{2}{n}\right)\right]\left(\frac{2}{n}\right)$.

(d) Show that $S(n) = \displaystyle\sum_{i=1}^{n} \left[1 + i\left(\frac{2}{n}\right)\right]\left(\frac{2}{n}\right)$.

(e) Complete the following table:

n	5	10	50	100
$s(n)$				
$S(n)$				

(f) Show that $\lim\limits_{n \to \infty} s(n) = \lim\limits_{n \to \infty} S(n) = 4$.

In Exercises 9–18, use the limit process to find the area of the region between the graph of each function and the x-axis over the given interval. Sketch each region.

Function	Interval
9. $y = -2x + 3$	$[0, 1]$
10. $y = 1 - x^2$	$[-1, 1]$
11. $y = -x^2 + 2x$	$[0, 2]$
12. $y = 3x - 4$	$[2, 5]$
13. $y = 2x^2$	$[1, 3]$
14. $y = 2x - x^3$	$[0, 1]$
15. $y = 1 - x^3$	$[0, 1]$
16. $y = x^2 - x^3$	$[0, 1]$
17. $y = x^2 - x^3$	$[-1, 0]$
18. $y = 2x^2 - x + 1$	$[0, 2]$

In Exercises 19 and 20, use the limit process to find the area of the region between the graph of each function and the y-axis over the given interval. Sketch each region.

Function	Interval
19. $f(y) = 3y$	$[0, 2]$
20. $f(y) = y^2$	$[0, 3]$

In Exercises 21–24, use the Midpoint Rule to approximate the integral of the given function over the indicated interval. (Use $n = 4$.)

Function	Interval
21. $f(x) = x^2 + 3$	$0 \le x \le 2$
22. $f(x) = x^2 + 4x$	$0 \le x \le 4$
23. $f(x) = \sqrt{x - 1}$	$1 \le x \le 2$
24. $f(x) = \dfrac{1}{x^2 + 1}$	$0 \le x \le 2$

5.5
Riemann sums and the definite integral

SECTION TOPICS ▪
Riemann sums ▪
The definite integral ▪
Properties of definite integrals ▪

In the definition of area in Section 5.4, we required that the partitions have subintervals of *equal width*. This was done for computational convenience. We begin this section with an example to show that it is not necessary to have subintervals of equal width. (In fact, a partition with equal subintervals would make the following example more difficult!)

EXAMPLE 1   A partition with subintervals of unequal widths

Consider the region bounded by the graph of $f(x) = \sqrt{x}$ and the x-axis for $0 \le x \le 1$, as shown in Figure 5.17. Evaluate the limit

$$\lim_{n\to\infty} \sum_{i=1}^{n} f(c_i)\Delta x_i$$

where c_i is the right endpoint of the partition given by $x_i = i^2/n^2$ and Δx_i is the width of the ith interval.

Solution: The width of the ith interval is given by

$$\Delta x_i = \frac{i^2}{n^2} - \frac{(i-1)^2}{n^2} = \frac{i^2 - i^2 + 2i - 1}{n^2} = \frac{2i - 1}{n^2}$$

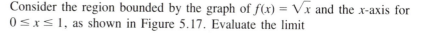

FIGURE 5.17   Thus, we have

$$\lim_{n\to\infty} \sum_{i=1}^{n} f(c_i)\Delta x_i = \lim_{n\to\infty} \sum_{i=1}^{n} \sqrt{\frac{i^2}{n^2}}\left(\frac{2i-1}{n^2}\right) = \lim_{n\to\infty} \frac{1}{n^3} \sum_{i=1}^{n} (2i^2 - i)$$

$$= \lim_{n\to\infty} \frac{1}{n^3}\left[2\frac{n(n+1)(2n+1)}{6} - \frac{n(n+1)}{2}\right]$$

$$= \lim_{n\to\infty} \frac{4n^3 - 3n^2 - n}{6n^3} = \frac{2}{3}$$

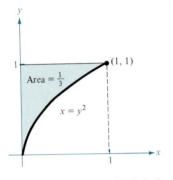

y

Area = $\frac{1}{3}$

$x = y^2$

(1, 1)

1

x

FIGURE 5.18

It is important to make an observation at this point. The two regions shown in Figures 5.17 and 5.18 can be united to form a square whose area is 1. Furthermore, we already know from Example 5 in Section 5.4 that the region in Figure 5.18 has an area of $\frac{1}{3}$. Thus, it is reasonable to conclude that the region in Figure 5.17 has an area of $\frac{2}{3}$. This agrees with the limit found in Example 1 above, even though we used a partition having subintervals of unequal widths. The reason we were able to use this particular partition is that as n increases, the *width of the largest subinterval approaches zero*. This is a key feature of the following development of the definite integral.

Riemann sums

In the previous section we used the limit of a sum to define the area of a special type of region in the plane. Finding area by this means is only one of the *many* applications involving the limit of a sum. We will soon see that a similar approach can be used to determine quantities as diverse as arc length, average value, centroids, volumes, work, and surface areas. The following development is named after Georg Friedrich Bernhard Riemann (1826–1866). Although the definite integral had been defined and used long before the time of Riemann, he generalized the concept to cover a broader category of functions.

To begin, we assume that f is defined but not necessarily continuous on an interval $[a, b]$. We let Δ be an arbitrary partition of the interval

$$a = x_0 < x_1 < x_2 < \cdots < x_{n-1} < x_n = b$$

and let Δx_i be the width of the ith subinterval. Now, if we let c_i be *any* point in the ith subinterval, the sum

$$\sum_{i=1}^{n} f(c_i)\Delta x_i, \quad x_{i-1} \leq c_i \leq x_i$$

is called a **Riemann sum** of f for the partition Δ. The sums in Section 5.4 are examples of Riemann sums, but there are more general Riemann sums than those covered in Section 5.4.

For a given partition Δ, we call the width of the largest subinterval the **norm** of the partition, and we denote it by $\|\Delta\|$. If every subinterval is of equal width, then we call the partition **regular** and denote the norm by

$$\|\Delta\| = \Delta x = \frac{b - a}{n} \qquad \text{Regular partition}$$

For a general partition, the norm is related to the number of subintervals of $[a, b]$ in the following way:

$$\frac{b - a}{\|\Delta\|} \leq n \qquad \text{General partition}$$

Thus, we can see that the number of subintervals in a partition approaches infinity as the norm of the partition approaches zero.

$$\|\Delta\| \to 0 \quad \text{implies that} \quad n \to \infty$$

Georg Friedrich Riemann

| Remark   The converse of this statement is not true. For example, let Δ_n be the partition given by

$$0 < \frac{1}{2^n} < \frac{1}{2^{n-1}} < \cdots < \frac{1}{8} < \frac{1}{4} < \frac{1}{2} < 1$$

For any positive value of n, the norm of the partition Δ_n is $\frac{1}{2}$. Thus, letting n approach infinity does not force $\|\Delta\|$ to approach zero.

To define the definite integral, we use the following limit:

$$\lim_{\|\Delta\| \to 0} \sum_{i=1}^{n} f(c_i)\Delta x_i = L$$

To say that this limit exists means that for $\varepsilon > 0$ there exists a $\delta > 0$ such that for every partiton with $\|\Delta\| < \delta$ it follows that

$$\left| L - \sum_{i=1}^{n} f(c_i)\Delta x_i \right| < \varepsilon$$

(This must be true for any choice of c_i in the ith subinterval of Δ.)

DEFINITION OF THE DEFINITE INTEGRAL

If f is defined on the interval $[a, b]$ and the limit of the Riemann sum of f exists, then we say f is **integrable** on $[a, b]$ and we denote the limit by

$$\lim_{\|\Delta\| \to 0} \sum_{i=1}^{n} f(c_i)\Delta x_i = \int_a^b f(x)\,dx$$

The limit is called the **definite integral** of f from a to b. The number a is the **lower limit** of integration, and the number b is the **upper limit** of integration.

| Remark   By this definition, we see that a definite integral is a number. This means that once we have evaluated a definite integral, the variable of integration is no longer present. For this and other reasons, we call x a **dummy variable.**

It is not a coincidence that the notation for definite integrals is similar to that used for indefinite integrals. We will see why in the next section when we discuss the Fundamental Theorem of Calculus. For now it is important to see that definite integrals and indefinite integrals are very different entities. A definite integral is a *number,* whereas an indefinite integral is a *family of functions*.

A sufficient condition for f to be integrable on $[a, b]$ is given in the following theorem, which we state without proof.

THEOREM 5.9   CONTINUITY IMPLIES INTEGRABILITY
If a function f is continuous on the closed interval $[a, b]$, then f is integrable on $[a, b]$.

EXAMPLE 2   *Evaluating a definite integral as a limit*

Evaluate the definite integral

$$\int_{-2}^{1} 2x \, dx$$

Solution: Since $f(x) = 2x$ is continuous on the interval $[-2, 1]$, we know it is integrable. Moreover, the definition of integrability implies that any partition whose norm approaches zero can be used to determine the limit. Thus, for computational convenience, we define Δ by subdividing $[-2, 1]$ into n subintervals of equal width

$$\Delta x_i = \Delta x = \frac{b - a}{n} = \frac{3}{n}$$

Choosing c_i as the right endpoint of each subinterval, we have

$$c_i = a + i(\Delta x) = -2 + \frac{3i}{n}$$

Thus, the definite integral is given by

$$\int_{-2}^{1} 2x \, dx = \lim_{\|\Delta\| \to 0} \sum_{i=1}^{n} f(c_i)\Delta x_i$$

$$= \lim_{n \to \infty} \sum_{i=1}^{n} f(c_i)\Delta x$$

$$= \lim_{n \to \infty} \sum_{i=1}^{n} 2\left(-2 + \frac{3i}{n}\right)\left(\frac{3}{n}\right)$$

$$= \lim_{n \to \infty} \frac{6}{n} \sum_{i=1}^{n} \left(-2 + \frac{3i}{n}\right)$$

$$= \lim_{n \to \infty} \frac{6}{n}\left[-2n + \frac{3}{n}\left(\frac{n(n + 1)}{2}\right)\right]$$

$$= \lim_{n \to \infty} \left(-12 + 9 + \frac{9}{n}\right)$$

$$= -3$$

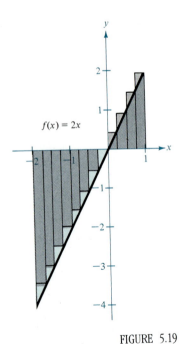

$f(x) = 2x$

FIGURE 5.19

Note that since the particular definite integral in Example 2 is negative, it *does not* represent the area of the region shown in Figure 5.19. Definite integrals can be positive, negative, or zero. In order for a definite integral to be interpreted as an area (as defined in Section 5.4), the function f must satisfy the conditions given in the following theorem.

Start here →

THEOREM 5.10	**THE DEFINITE INTEGRAL AS THE AREA OF A REGION**

If f is continuous and nonnegative on the closed interval $[a, b]$, then the area of the region bounded by the graph of f, the x-axis, and the vertical lines $x = a$ and $x = b$ is given by

$$\text{area} = \int_a^b f(x)\,dx$$

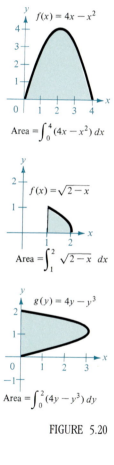

Area $= \int_0^4 (4x - x^2)\,dx$

Area $= \int_1^2 \sqrt{2 - x}\,dx$

Area $= \int_0^2 (4y - y^3)\,dy$

FIGURE 5.20

Proof: We can obtain a proof by combining Theorems 5.8 and 5.9. From Theorem 5.9, we know that f is integrable on $[a, b]$. This means that the definite integral of f from a to b can be written as the limit of a Riemann sum. When such a limit exists, we can use any convenient partition so long as its norm approaches zero. Thus, we choose a regular partition and conclude from Theorem 5.8 that the limit is equal to the area of the given region.

As examples of this theorem, the regions shown in Figure 5.20 are given by the indicated definite integrals.

From geometry we know several formulas for areas of plane regions. Since at this stage in our development evaluation of definite integrals is not particularly easy, it is a good idea to learn to recognize the definite integrals that correspond to areas of common geometric figures. We give some of these in the next example.

EXAMPLE 3   Areas of common geometric figures

Sketch the region corresponding to each of the following definite integrals. Then evaluate each integral using a geometric formula.

(a) $\int_1^3 4\,dx$ (b) $\int_0^3 (x + 2)\,dx$ (c) $\int_{-2}^2 \sqrt{4 - x^2}\,dx$

Solution: A sketch of each region is shown in Figure 5.21.

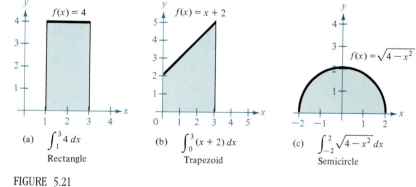

(a) $\int_1^3 4\,dx$
Rectangle

(b) $\int_0^3 (x + 2)\,dx$
Trapezoid

(c) $\int_{-2}^2 \sqrt{4 - x^2}\,dx$
Semicircle

FIGURE 5.21

(a) This region is a rectangle of height 4 and width 2. Thus,

$$\int_1^3 4 \, dx = 4(2) = 8$$

(b) This region is a trapezoid with an altitude of 3 and parallel bases of lengths 2 and 5. The formula for the area of a trapezoid is $\frac{1}{2}h(b_1 + b_2)$, and so we have

$$\int_0^3 (x + 2) \, dx = \frac{1}{2}(3)(2 + 5) = \frac{21}{2}$$

(c) This region is a semicircle of radius 2. Thus, the area is $\frac{1}{2}\pi r^2$, and we have

$$\int_{-2}^2 \sqrt{4 - x^2} \, dx = \frac{1}{2}\pi(2^2) = 2\pi$$

Properties of definite integrals

In the definition of the definite integral of f on the interval $[a, b]$, we assumed that $a < b$. For future work it will be convenient to extend this definition somewhat. Geometrically, the following two special definitions seem reasonable. For instance, it makes sense to define the area of a region of zero width and finite height to be zero.

SOME SPECIAL
DEFINITE INTEGRALS

1. If f is defined at $x = a$, then

$$\int_a^a f(x) \, dx = 0$$

2. If f is integrable on $[a, b]$, then

$$\int_b^a f(x) \, dx = -\int_a^b f(x) \, dx$$

EXAMPLE 4   Evaluating definite integrals

Evaluate the following definite integrals:

(a) $\int_2^2 \sqrt{x^2 + 1} \, dx$

(b) $\int_3^0 (x + 2) \, dx$

Solution:
(a) Since the integrand is defined at $x = 2$, and since the upper and lower limits of integration are equal, we have

$$\int_2^2 \sqrt{x^2 + 1} \, dx = 0$$

(b) This integral is the same as that given in Example 3(b) except that the upper and lower limits are interchanged. Since the integral in Example 3(b) has a value of $\frac{21}{2}$, we have

$$\int_3^0 (x + 2)\, dx = -\int_0^3 (x + 2)\, dx$$

$$= -\frac{21}{2}$$

It is sometimes convenient to break a definite integral into two (or more) parts, as shown in the following theorem.

THEOREM 5.11   **ADDITIVE INTERVAL PROPERTY**

If f is integrable on the three intervals determined by a, b, and c, then

$$\int_a^b f(x)\, dx = \int_a^c f(x)\, dx + \int_c^b f(x)\, dx$$

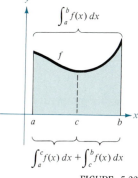

FIGURE 5.22

This theorem is valid for any integrable function and for any three numbers a, b, and c. However, rather than give a formal proof for the general case, it seems more instructive to give a geometric argument for the case in which $a < c < b$ and f is continuous and nonnegative. From Figure 5.22, we can see that the larger region can be divided at $x = c$ into two subregions whose intersection is a line segment. Since the line segment has zero area, it follows that the area of the larger region is equal to the sum of the areas of the two smaller regions.

Since the definite integral is defined as the limit of a sum, it inherits the two properties of summation given in Theorem 5.5 in Section 5.3. We state this result, without proof, in the following theorem.

THEOREM 5.12   **PROPERTIES OF DEFINITE INTEGRALS**

If f and g are integrable on $[a, b]$ and k is a constant, then the following properties are true.

1. $\displaystyle\int_a^b kf(x)\, dx = k \int_a^b f(x)\, dx$

2. $\displaystyle\int_a^b [f(x) \pm g(x)]\, dx = \int_a^b f(x)\, dx \pm \int_a^b g(x)\, dx$

Remark   Property 2 in Theorem 5.12 can be extended to cover any finite number of functions. For example,

$$\int_a^b [f(x) + g(x) + h(x)]\, dx = \int_a^b f(x)\, dx + \int_a^b g(x)\, dx + \int_a^b h(x)\, dx$$

EXAMPLE 5   Evaluation of a definite integral

Evaluate

$$\int_1^3 (-x^2 + 4x - 3)\, dx$$

using the following values:

$$\int_1^3 x^2\, dx = \frac{26}{3}$$

$$\int_1^3 x\, dx = 4$$

and

$$\int_1^3 dx = 2$$

Solution: Using Theorem 5.12, we have

$$\int_1^3 (-x^2 + 4x - 3)\, dx = \int_1^3 (-x^2)\, dx + \int_1^3 4x\, dx + \int_1^3 (-3)\, dx$$

$$= -\int_1^3 x^2\, dx + 4\int_1^3 x\, dx - 3\int_1^3 dx$$

$$= -\left(\frac{26}{3}\right) + 4(4) - 3(2) = \frac{4}{3}$$

Another property that definite integrals inherit from summation is preservation of inequality. For instance, if

$$f(x_1) < g(x_1) \qquad \text{and} \qquad f(x_2) < g(x_2)$$

and Δx_1 and Δx_2 are positive, then we know from the properties of inequalities that

$$f(x_1)\Delta x_1 + f(x_2)\Delta x_2 < g(x_1)\Delta x_1 + g(x_2)\Delta x_2$$

This is the basis of the following theorem.

THEOREM 5.13   PRESERVATION OF INEQUALITY

If f and g are integrable on $[a, b]$ and $f(x) \le g(x)$ for every x in the interval, then

$$\int_a^b f(x)\, dx \le \int_a^b g(x)\, dx$$

Remark A useful corollary of Theorem 5.13 is that if f is integrable and nonnegative on $[a, b]$, then

$$\int_a^b f(x)\, dx \ge 0$$

Section Exercises 5.5

In Exercises 1–10, set up a definite integral that yields the area of the given region. (Do not evaluate the integral.)

1. $f(x) = 3$

2. $f(x) = 4 - 2x$

3. $f(x) = 4 - |x|$

4. $f(x) = x^2$

5. $f(x) = 4 - x^2$

6. $f(x) = (y - 2)^2$

7. $g(y) = y^3$

8. $f(x) = (x^2 + 1)^3$

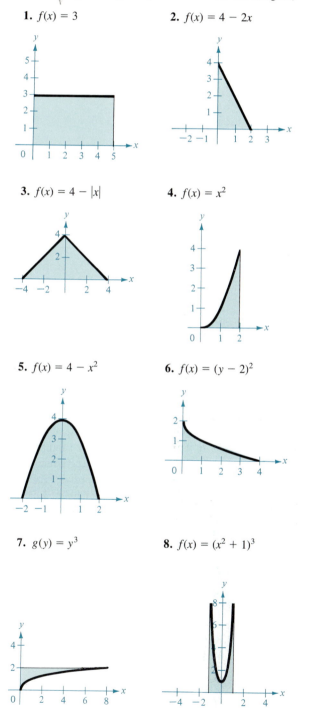

9. $f(x) = \sqrt{x + 1}$

10. $f(x) = \dfrac{1}{x^2 + 1}$

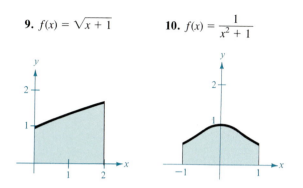

In Exercises 11–20, sketch the region whose area is indicated by the given definite integral. Then use a geometric formula to evaluate the integral.

11. $\displaystyle\int_0^3 4 \, dx$

12. $\displaystyle\int_{-a}^a 4 \, dx$

13. $\displaystyle\int_0^4 x \, dx$

14. $\displaystyle\int_0^4 \frac{x}{2} \, dx$

15. $\displaystyle\int_0^2 (2x + 5) \, dx$

16. $\displaystyle\int_0^5 (5 - x) \, dx$

17. $\displaystyle\int_{-1}^1 (1 - |x|) \, dx$

18. $\displaystyle\int_{-a}^a (a - |x|) \, dx$

19. $\displaystyle\int_{-3}^3 \sqrt{9 - x^2} \, dx$

20. $\displaystyle\int_{-r}^r \sqrt{r^2 - x^2} \, dx$

21. If $\displaystyle\int_0^5 f(x) \, dx = 10$ and $\displaystyle\int_5^7 f(x) \, dx = 3$, find

(a) $\displaystyle\int_0^7 f(x) \, dx$

(b) $\displaystyle\int_5^0 f(x) \, dx$

(c) $\displaystyle\int_5^5 f(x) \, dx$

(d) $\displaystyle\int_0^5 3f(x) \, dx$

22. If $\displaystyle\int_0^3 f(x) \, dx = 4$ and $\displaystyle\int_3^6 f(x) \, dx = -1$, find

(a) $\displaystyle\int_0^6 f(x) \, dx$

(b) $\displaystyle\int_6^3 f(x) \, dx$

(c) $\displaystyle\int_4^4 f(x) \, dx$

(d) $\displaystyle\int_3^6 -5f(x) \, dx$

23. If $\displaystyle\int_2^6 f(x) \, dx = 10$ and $\displaystyle\int_2^6 g(x) \, dx = -2$, find

(a) $\displaystyle\int_2^6 [f(x) + g(x)] \, dx$

(b) $\displaystyle\int_2^6 [g(x) - f(x)] \, dx$

(c) $\displaystyle\int_2^6 [2f(x) - 3g(x)] \, dx$

(d) $\displaystyle\int_2^6 3f(x) \, dx$

24. If $\int_{-1}^{1} f(x)\,dx = 0$ and $\int_{0}^{1} f(x)\,dx = 5$, find

(a) $\int_{-1}^{0} f(x)\,dx$

(b) $\int_{0}^{1} f(x)\,dx - \int_{-1}^{0} f(x)\,dx$

(c) $\int_{-1}^{1} 3f(x)\,dx$

(d) $\int_{0}^{1} 3f(x)\,dx$

In Exercises 25–30, evaluate the definite integral by the limit definition.

25. $\int_{4}^{10} 6\,dx$

26. $\int_{-2}^{3} x\,dx$

27. $\int_{-1}^{1} x^3\,dx$

28. $\int_{0}^{1} x^3\,dx$

29. $\int_{1}^{2} (x^2 + 1)\,dx$

30. $\int_{1}^{2} 4x^2\,dx$

In Exercises 31 and 32, use Example 1 as a model to evaluate the limit

$$\lim_{n \to \infty} \sum_{i=1}^{n} f(c_i) \Delta x_i$$

over the given region bounded by the graphs of the given equations.

31. $f(x) = \sqrt{x}$, $y = 0$, $x = 0$, $x = 2$

[Hint: Let $c_i = 2i^2/n^2$.]

32. $f(x) = \sqrt[3]{x}$, $y = 0$, $x = 0$, $x = 1$

[Hint: Let $c_i = i^3/n^3$.]

33. Prove Theorem 5.12.

34. Prove Theorem 5.13.

35. Prove that for a continuous function f on a closed interval $[a, b]$,

$$\left| \int_{a}^{b} f(x)\,dx \right| \le \int_{a}^{b} |f(x)|\,dx$$

SECTION TOPICS ▪
The Fundamental Theorem of Calculus ▪
The Mean Value Theorem for Integrals ▪
Average value of a function
over an interval ▪

5.6
The Fundamental Theorem of Calculus

We have now looked at the two major branches of calculus: differential calculus, which we introduced with the tangent line problem, and integral calculus, which we introduced with the area problem. Initially, there seems no reason to assume that these two problems are related. But as it turns out, there is a close connection. This connection was discovered independently by Isaac Newton and Gottfried Leibniz, and for this reason these two men are usually credited with the discovery of calculus. The connection is stated in a theorem that is appropriately called the **Fundamental Theorem of Calculus.**

Roughly, the theorem tells us that differentiation and (definite) integration are inverse operations, in somewhat the same sense as division and multiplication are inverse operations. To see how Newton and Leibniz might have anticipated this relationship, let's look at the approximations used in the development of the two operations. In Figure 5.23 we use the *quotient* $\Delta y/\Delta x$ (the slope of the secant line) to approximate the slope of the tangent line at (x, y). Similarly, in Figure 5.24 we use the *product* $\Delta y \Delta x$ (the area of a rectangle) to approximate the area under the curve. Thus, at least in the

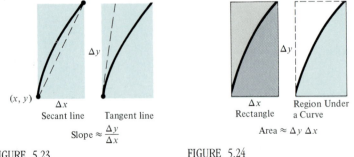

(x, y) Δx Δy

Secant line   Tangent line

Slope $\approx \dfrac{\Delta y}{\Delta x}$

Δx Δy

Rectangle   Region Under a Curve

Area $\approx \Delta y\,\Delta x$

FIGURE 5.23   FIGURE 5.24

primitive approximation stage, the two operations appear to have an inverse relationship. The Fundamental Theorem of Calculus tells us that the limit processes (used to define the derivative and the definite integral) preserve this inverse relationship.

THEOREM 5.14

THE FUNDAMENTAL THEOREM OF CALCULUS
If a function f is continuous on the interval $[a, b]$, then

$$\int_a^b f(x)\, dx = F(b) - F(a)$$

where F is any function such that $F'(x) = f(x)$ for all x in $[a, b]$.

Proof: The key to the proof is in writing the difference $F(b) - F(a)$ in a convenient form. Let Δ be the following partition of $[a, b]$:

$$a = x_0 < x_1 < x_2 < \cdots < x_{n-1} < x_n = b$$

By pairwise subtraction and addition of like terms, we can write

$$F(b) - F(a) = F(x_n) - F(x_{n-1}) + F(x_{n-1}) - \cdots - F(x_1) + F(x_1) - F(x_0)$$

$$= \sum_{i=1}^n [F(x_i) - F(x_{i-1})]$$

Now, by the Mean Value Theorem, we know that there exists a number c_i in the ith subinterval such that

$$F'(c_i) = \frac{F(x_i) - F(x_{i-1})}{x_i - x_{i-1}}$$

Since $F'(c_i) = f(c_i)$, we let $\Delta x_i = x_i - x_{i-1}$ and write

$$F(b) - F(a) = \sum_{i=1}^n f(c_i)\Delta x_i$$

This surprising equation tells us that by applying the Mean Value Theorem we can always find a collection of c_i's such that the *constant* $F(b) - F(a)$ is a Riemann sum of f on $[a, b]$. Taking the limit (as $\|\Delta\| \to 0$), we have

$$F(b) - F(a) = \int_a^b f(x)\, dx$$

Several comments are in order regarding the Fundamental Theorem of Calculus. First, *provided we can find* an antiderivative of f, we now have a way to evaluate a definite integral without having to use the limit of a sum. Second, in applying this theorem it is helpful to use the notation

$$\int_a^b f(x)\, dx = \left[F(x) \right]_a^b = F(b) - F(a)$$

For instance, we write

$$\int_1^3 x^3 \, dx = \left[\frac{x^4}{4}\right]_1^3 = \frac{3^4}{4} - \frac{1^4}{4} = \frac{81}{4} - \frac{1}{4} = 20$$

Finally, we observe that the constant of integration C can be dropped from the antiderivative, because

$$\int_a^b f(x) \, dx = \left[F(x) + C\right]_a^b = [F(b) + C] - [F(a) + C] = F(b) - F(a)$$

EXAMPLE 1   Using the Fundamental Theorem

(a) $\displaystyle\int_1^2 (x^2 - 3) \, dx = \left[\frac{x^3}{3} - 3x\right]_1^2$

$$= \left(\frac{8}{3} - 6\right) - \left(\frac{1}{3} - 3\right) = -\frac{2}{3}$$

(b) $\displaystyle\int_0^1 (4t + 1)^2 \, dt = \frac{1}{4} \int_0^1 (4t + 1)^2(4) \, dt \qquad u = 4t + 1$

$$= \frac{1}{4}\left[\frac{(4t + 1)^3}{3}\right]_0^1$$

$$= \frac{1}{4}\left[\frac{125}{3} - \frac{1}{3}\right] = \frac{31}{3}$$

EXAMPLE 2   Using the Fundamental Theorem

Evaluate

$$\int_1^4 3\sqrt{x} \, dx$$

Solution:

$$\int_1^4 3\sqrt{x} \, dx = 3 \int_1^4 x^{1/2} \, dx = 3\left[\frac{x^{3/2}}{3/2}\right]_1^4 = 2(4)^{3/2} - 2(1)^{3/2} = 14$$

EXAMPLE 3   Using the Fundamental Theorem

Evaluate

$$\int_{-8}^{-1} \frac{x + 2x^2}{\sqrt[3]{x}} \, dx$$

Solution:

$$\int_{-8}^{-1} \frac{x + 2x^2}{\sqrt[3]{x}} \, dx = \int_{-8}^{-1} \left(\frac{x}{x^{1/3}} + \frac{2x^2}{x^{1/3}}\right) dx$$

$$= \int_{-8}^{-1} (x^{2/3} + 2x^{5/3}) \, dx = \left[\frac{x^{5/3}}{5/3} + \frac{2x^{8/3}}{8/3}\right]_{-8}^{-1}$$

$$= \left(-\frac{3}{5} + \frac{3}{4}\right) - \left(-\frac{96}{5} + 192\right) = -172.65$$

EXAMPLE 4   Integrating absolute value

Evaluate

$$\int_0^2 |2x - 1|\, dx$$

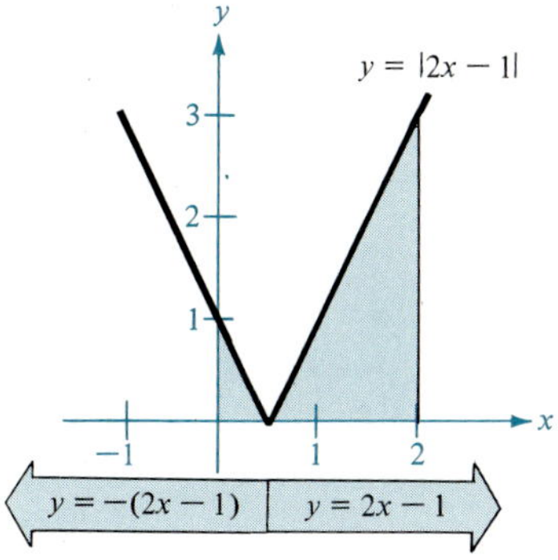

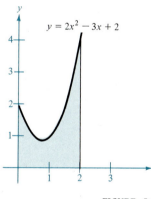

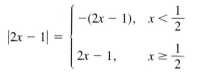

Solution: From Figure 5.25 and the definition of absolute value, we note that

$$|2x - 1| = \begin{cases} -(2x - 1), & x < \dfrac{1}{2} \\[2mm] 2x - 1, & x \geq \dfrac{1}{2} \end{cases}$$

Hence, we rewrite the integral in two parts as

$$\int_0^2 |2x - 1|\, dx = \int_0^{1/2} -(2x - 1)\, dx + \int_{1/2}^2 (2x - 1)\, dx$$

$$= \left[-x^2 + x \right]_0^{1/2} + \left[x^2 - x \right]_{1/2}^2$$

$$= \left(-\frac{1}{4} + \frac{1}{2} \right) - (0 + 0) + (4 - 2) - \left(\frac{1}{4} - \frac{1}{2} \right)$$

$$= \frac{5}{2}$$

y = |2x − 1|

y = −(2x − 1)   *y = 2x − 1*

FIGURE 5.25

EXAMPLE 5   Using the Fundamental Theorem to find area

Find the area of the region bounded by the graph of $y = 2x^2 - 3x + 2$, the x-axis, and the vertical lines $x = 0$ and $x = 2$, as shown in Figure 5.26.

Solution:

$$\text{area} = \int_0^2 (2x^2 - 3x + 2)\, dx$$

$$= \left[\frac{2x^3}{3} - \frac{3x^2}{2} + 2x \right]_0^2$$

$$= \frac{16}{3} - 6 + 4$$

$$= \frac{10}{3}$$

y = 2x² − 3x + 2

FIGURE 5.26

In Examples 1 and 3, we used a simple u-substitution to find the antiderivative in terms of x. For more complicated substitutions, it is often more convenient to determine the limits of integration for the variable u than to convert back to the variable x and evaluate the antiderivative at the original limits. This change of variables is stated explicitly in the next theorem. The proof follows from the substitution theorem for indefinite integrals (Theorem 5.3) combined with the Fundamental Theorem of Calculus.

THEOREM 5.15   **CHANGE OF VARIABLES FOR DEFINITE INTEGRALS**
If the function $u = g(x)$ has a continuous derivative on $[a, b]$ and f has an antiderivative over the range of g, then

$$\int_a^b f(g(x))g'(x)\, dx = \int_{g(a)}^{g(b)} f(u)\, du$$

EXAMPLE 6   *Change of variables*

Evaluate

$$\int_0^1 x(x^2 + 1)^3\, dx$$

Solution: To evaluate this integral, we let $u = x^2 + 1$. Then we have

$$u = x^2 + 1 \implies du = 2x\, dx$$

Before substituting, we determine the new upper and lower limits of integration.

Lower limit   *Upper limit*

When $x = 0$, $u = 0^2 + 1 = 1$ When $x = 1$, $u = 1^2 + 1 = 2$

Now, we substitute to obtain

$$\int_0^1 x(x^2 + 1)^3\, dx = \frac{1}{2} \int_0^1 (x^2 + 1)^3(2x)\, dx = \frac{1}{2} \int_1^2 u^3\, du$$

$$= \frac{1}{2}\left[\frac{u^4}{4}\right]_1^2 = \frac{1}{2}\left(4 - \frac{1}{4}\right) = \frac{15}{8}$$

EXAMPLE 7   *Change of variables*

Evaluate

$$A = \int_1^5 \frac{x}{\sqrt{2x - 1}}\, dx$$

Solution: To evaluate this integral, we let $u = \sqrt{2x - 1}$. Then

$$u^2 = 2x - 1 \implies x = \frac{u^2 + 1}{2} \implies dx = u\, du$$

Before substituting, we determine the new upper and lower limits of integration.

Lower limit   *Upper limit*

When $x = 1$, $u = \sqrt{2 - 1} = 1$.   When $x = 5$, $u = \sqrt{10 - 1} = 3$.

Now, we substitute to obtain

$$\int_1^5 \frac{x}{\sqrt{2x-1}}\, dx = \int_1^3 \frac{1}{2}\left(\frac{u^2+1}{u}\right)u\, du = \frac{1}{2}\int_1^3 (u^2+1)\, du$$

$$= \frac{1}{2}\left[\frac{u^3}{3}+u\right]_1^3 = \frac{1}{2}\left(9+3-\frac{1}{3}-1\right) = \frac{16}{3}$$

Remark   Geometrically, we can interpret the equation

$$\int_1^5 \frac{x}{\sqrt{2x-1}}\, dx = \int_1^3 \frac{u^2+1}{2}\, du$$

to mean that the two *different* regions shown in Figures 5.27 and 5.28 have the *same* area.

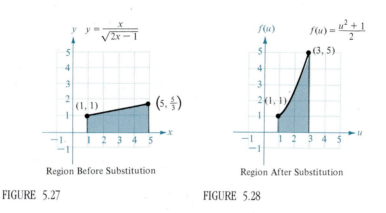

FIGURE 5.27	FIGURE 5.28

When evaluating definite integrals by substitution, don't be surprised if the upper limit of integration of the *u*-variable form is smaller than the lower limit. If this happens, don't rearrange the limits. Simply evaluate as usual. For example, after substituting $u = \sqrt{1-x}$ in the integral

$$\int_0^1 x^2(1-x)^{1/2}\, dx$$

we have $u = \sqrt{1-1} = 0$ when $x = 1$ and $u = \sqrt{1-0} = 1$ when $x = 0$. Thus, the correct *u*-variable form of this integral is

$$-2 \int_1^0 (1-u^2)^2 u\, du$$

The Mean Value Theorem for Integrals

Recall from our discussion in Section 5.5 that we used rectangles to approximate the area of a region under a curve. In that discussion we observed that the actual area of the region was greater than the area of an inscribed rectangle and less than the area of a circumscribed rectangle. The Mean Value Theorem for Integrals states that somewhere "between" the inscribed and circumscribed rectangles there is a rectangle whose area is precisely equal to the area of the region under the curve, as shown in Figure 5.29.

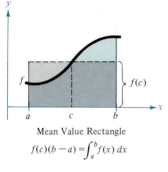

Mean Value Rectangle

$$f(c)(b-a) = \int_a^b f(x)\, dx$$

FIGURE 5.29

THEOREM 5.16

MEAN VALUE THEOREM FOR INTEGRALS

If f is continuous on $[a, b]$, then there exists a number c in (a, b) such that

$$\int_a^b f(x)\ dx = f(c)(b - a)$$

Proof:

Case 1: If f is constant over the interval $[a, b]$, the result is trivial, since c can be any point in (a, b).

Case 2: If f is not constant on $[a, b]$, then by the Extreme Value Theorem we choose $f(m)$ and $f(M)$ to be the minimum and maximum values of f on $[a, b]$. Since $f(m) \leq f(x) \leq f(M)$ for all x in $[a, b]$, we conclude from Theorem 5.13 that

$$\int_a^b f(m)\ dx \leq \int_a^b f(x)\ dx \leq \int_a^b f(M)\ dx$$

This inequality is graphically depicted in Figure 5.30.

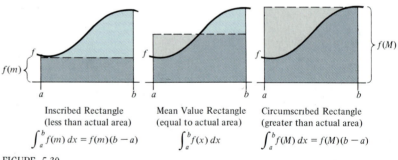

Inscribed Rectangle (less than actual area)	Mean Value Rectangle (equal to actual area)	Circumscribed Rectangle (greater than actual area)
$\int_a^b f(m)\ dx = f(m)(b-a)$	$\int_a^b f(x)\ dx$	$\int_a^b f(M)\ dx = f(M)(b-a)$

FIGURE 5.30

Thus, we have

$$f(m)(b - a) \leq \int_a^b f(x)\ dx \leq f(M)(b - a)$$

$$f(m) \leq \frac{1}{b - a} \int_a^b f(x)\ dx \leq f(M)$$

Finally, by the Intermediate Value Theorem we conclude that there exists some c in (a, b) such that

$$f(c) = \frac{1}{b - a} \int_a^b f(x)\ dx$$

$$f(c)(b - a) = \int_a^b f(x)\ dx$$

| Remark   Note that the Mean Value Theorem for Integrals does not specify how to determine c. It merely guarantees the existence of the number c.

The value of $f(c)$, given in the Mean Value Theorem for Integrals, is called the **average value** of f in the interval $[a, b]$.

DEFINITION OF THE AVERAGE VALUE OF A FUNCTION	If f is continuous on $[a, b]$, then the **average value** of f on this interval is given by

$$\frac{1}{b - a} \int_a^b f(x) \, dx$$

To see why we call this the average value of f, suppose that we partition $[a, b]$ into n subintervals of equal width $\Delta x = (b - a)/n$. If c_i is any point in the ith subinterval, then the arithmetic average (or mean) of the function values at the c_i's is given by

$$a_n = \frac{1}{n}[f(c_1) + f(c_2) + \cdots + f(c_n)] \qquad \text{Average of } f(c_i), \ldots, f(c_n)$$

By multiplying and dividing by $(b - a)$, we can write the average as

$$a_n = \frac{1}{n} \sum_{i=1}^{n} f(c_i)\left(\frac{b - a}{b - a}\right) = \frac{1}{b - a} \sum_{i=1}^{n} f(c_i)\left(\frac{b - a}{n}\right)$$

$$= \frac{1}{b - a} \sum_{i=1}^{n} f(c_i)\Delta x$$

Finally, by taking the limit as $n \to \infty$, we obtain the average value of f on the interval $[a, b]$, as given in the above definition.

| Remark   This development of the average value of a function on an interval is only one of many practical uses of definite integrals to represent summation processes. In Chapter 6 we will study other applications, such as volume, arc length, centers of mass, and work.

EXAMPLE 8   Finding the average value of a function

Find the average value of $f(x) = 3x^2 - 2x$ on the interval $[1, 4]$.

Solution: The average value is given by

$$\frac{1}{b - a} \int_a^b f(x) \, dx = \frac{1}{3} \int_1^4 (3x^2 - 2x) \, dx = \frac{1}{3}\left[x^3 - x^2\right]_1^4$$

$$= \frac{1}{3}[64 - 16 - (1 - 1)] = \frac{48}{3} = 16 \qquad \square$$

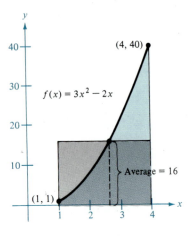

$f(x) = 3x^2 - 2x$

(4, 40)

Average = 16

(1, 1)

FIGURE 5.31

| Remark Note in Figure 5.31 that the area of the region is equal to the area of the rectangle whose height is the average value.

Even with the Fundamental Theorem, integration is not a simple task, because antiderivatives are not always easy to find. However, from the previous section we know that we can sometimes simplify the process by using area formulas from geometry. The next theorem describes another special category of definite integrals whose intervals are symmetric with respect to the origin.

THEOREM 5.17   **INTEGRATION OF EVEN AND ODD FUNCTIONS**
Let f be integrable on the interval $[a, b]$.

1. If f is an *even* function, then

$$\int_{-a}^{a} f(x)\, dx = 2 \int_{0}^{a} f(x)\, dx$$

2. If f is an *odd* function, then

$$\int_{-a}^{a} f(x)\, dx = 0$$

Proof: We prove Property 1 and leave the proof of Property 2 as an exercise. Since f is even, we know that $f(x) = f(-x)$. Now we use Theorem 5.15 with the substitution $u = -x$ to obtain

$$\int_{-a}^{0} f(x)\, dx = -\int_{a}^{0} f(x)\, dx = \int_{0}^{a} f(x)\, dx$$

Finally, using Theorem 5.11, we have

$$\int_{-a}^{a} f(x)\, dx = \int_{-a}^{0} f(x)\, dx + \int_{0}^{a} f(x)\, dx$$

$$= \int_{0}^{a} f(x)\, dx + \int_{0}^{a} f(x)\, dx = 2 \int_{0}^{a} f(x)\, dx$$

EXAMPLE 9   *Integration of an odd function*

Evaluate

$$\int_{-2}^{2} (x^5 - 4x^3 + 6x)\, dx$$

Solution: By letting $f(x) = x^5 - 4x^3 + 6x$, we have

$$f(-x) = (-x)^5 - 4(-x)^3 + 6(-x) = -x^5 + 4x^3 - 6x = -f(x)$$

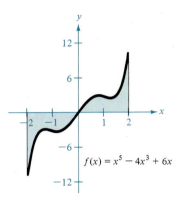

$f(x) = x^5 - 4x^3 + 6x$

FIGURE 5.32

Thus, f is an odd function, and since $[-2, 2]$ is symmetric about the origin, we can apply Theorem 5.17 to conclude that

$$\int_{-2}^{2} (x^5 - 4x^3 + 6x)\, dx = 0$$

Remark From Figure 5.32, we see that the two regions on either side of the y-axis have the same area. However, since one lies below the x-axis and one lies above, integration produces a cancellation effect. (We will say more about finding the area of a region below the x-axis in Section 6.1.)

Section Exercises 5.6

In Exercises 1–38, evaluate the definite integrals.

1. $\int_{0}^{1} 2x\, dx$

2. $\int_{2}^{7} 3\, dv$

3. $\int_{-1}^{0} (x - 2)\, dx$

4. $\int_{2}^{5} (-3v + 4)\, dv$

5. $\int_{-1}^{1} (t^2 - 2)\, dt$

6. $\int_{0}^{3} (3x^2 + x - 2)\, dx$

7. $\int_{0}^{1} (2t - 1)^2\, dt$

8. $\int_{-1}^{1} (t^3 - 9t)\, dt$

9. $\int_{1}^{2} \left(\frac{3}{x^2} - 1\right) dx$

10. $\int_{0}^{1} (3x^3 - 9x + 7)\, dx$

11. $\int_{1}^{2} (5x^4 + 5)\, dx$

12. $\int_{-3}^{3} v^{1/3}\, dv$

13. $\int_{-1}^{1} (\sqrt[3]{t} - 2)\, dt$

14. $\int_{-2}^{-1} \sqrt{\frac{-2}{x}}\, dx$

15. $\int_{1}^{4} \frac{u - 2}{\sqrt{u}}\, du$

16. $\int_{-2}^{-1} \left(u - \frac{1}{u^2}\right) du$

17. $\int_{0}^{1} \frac{x - \sqrt{x}}{3}\, dx$

18. $\int_{0}^{2} (2 - t)\sqrt{t}\, dt$

19. $\int_{-1}^{0} (t^{1/3} - t^{2/3})\, dt$

20. $\int_{-8}^{-1} \frac{x - x^2}{2\sqrt[3]{x}}\, dx$

21. $\int_{0}^{4} \frac{1}{\sqrt{2x + 1}}\, dx$

22. $\int_{0}^{1} x\sqrt{1 - x^2}\, dx$

23. $\int_{-1}^{1} x(x^2 + 1)^3\, dx$

24. $\int_{0}^{2} \frac{x}{\sqrt{1 + 2x^2}}\, dx$

25. $\int_{0}^{2} x\sqrt[3]{4 + x^2}\, dx$

26. $\int_{1}^{9} \frac{1}{\sqrt{x}(1 + \sqrt{x})^2}\, dx$

27. $\int_{-1}^{1} |x|\, dx$

28. $\int_{0}^{3} |2x - 3|\, dx$

29. $\int_{0}^{4} |x^2 - 4x + 3|\, dx$

30. $\int_{-1}^{1} |x^3|\, dx$

31. $\int_{1}^{2} (x - 1)\sqrt{2 - x}\, dx$

32. $\int_{0}^{4} \frac{x}{\sqrt{2x + 1}}\, dx$

33. $\int_{3}^{7} x\sqrt{x - 3}\, dx$

34. $\int_{0}^{1} \frac{1}{\sqrt{x} + \sqrt{x + 1}}\, dx$

35. $\int_{0}^{7} x\sqrt[3]{x + 1}\, dx$

36. $\int_{-2}^{6} x^2\sqrt[3]{x + 2}\, dx$

37. $\int_{1}^{5} x^2\sqrt{x - 1}\, dx$

38. $\int_{0}^{2} x^2\sqrt{2x + 1}\, dx$

In Exercises 39–44, determine the area of the indicated region.

39. $y = x - x^2$

40. $y = -x^2 + 2x + 3$

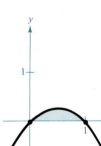

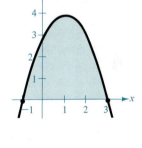

41. $y = 1 - x^4$

42. $y = \frac{1}{x^2}$

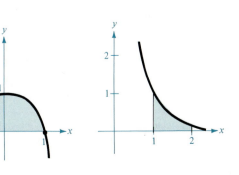

43. $y = \sqrt[3]{2x}$ **44.** $y = (3 - x)\sqrt{x}$

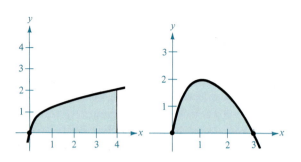

In Exercises 45–48, determine the area of each region having the given boundaries.

45. $y = 3x^2 + 1$, $x = 0$, $x = 2$, $y = 0$
46. $y = 1 + \sqrt{x}$, $x = 0$, $x = 4$, $y = 0$
47. $y = x^3 + x$, $x = 2$, $y = 0$
48. $y = -x^2 + 3x$, $y = 0$

In Exercises 49–54, sketch the graph of each function over the given interval. Find the average value of each function over the given interval and all values of x where the function equals its average.

Function	Interval
49. $f(x) = 4 - x^2$	$[-2, 2]$
50. $f(x) = x^2 - 2x + 1$	$[0, 1]$
51. $f(x) = x\sqrt{4 - x^2}$	$[0, 2]$
52. $f(x) = \dfrac{x^2 + 1}{x^2}$	$\left[\dfrac{1}{2}, 2\right]$
53. $f(x) = x - 2\sqrt{x}$	$[0, 4]$
54. $f(x) = 1/(x - 3)^2$	$[0, 2]$

55. Use the fact that

$$\int_0^2 x^2 \, dx = \frac{8}{3}$$

to evaluate the following definite integrals without using the Fundamental Theorem of Calculus.

(a) $\int_{-2}^0 x^2 \, dx$ (b) $\int_{-2}^2 x^2 \, dx$

(c) $\int_0^2 -x^2 \, dx$ (d) $\int_0^2 (x^2 + 1) \, dx$

(e) $\int_{-2}^0 3x^2 \, dx$

56. Use the fact that

$$\int_0^2 x^3 \, dx = 4$$

to evaluate the following definite integrals without using the Fundamental Theorem of Calculus.

(a) $\int_{-2}^0 x^3 \, dx$ (b) $\int_{-2}^2 x^3 \, dx$

(c) $\int_0^2 -x^3 \, dx$ (d) $\int_0^2 (x^3 + 1) \, dx$

(e) $\int_{-2}^0 3x^3 \, dx$

57. The volume V in liters of air in the lungs during a 5-second respiratory cycle is approximated by the model

$$V = 0.1729t + 0.1522t^2 - 0.0374t^3$$

where t is the time in seconds. Approximate the average volume of air in the lungs during one cycle.

58. The velocity v of the flow of blood at a distance r from the central axis of an artery of radius R is given by

$$v = k(R^2 - r^2)$$

where k is the constant of proportionality. Find the average rate of flow of blood along a radius of the artery. (Use zero and R as the limits of integration.)

59. The air temperature during a period of 12 hours is given by the model

$$T = 53 + 5t - 0.3t^2, \quad 0 \le t \le 12$$

where t is measured in hours and T in degrees Fahrenheit, as shown in Figure 5.33. Find the average temperature during (a) the first 6 hours of the period and (b) the entire period.

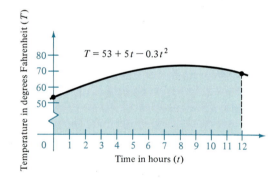

FIGURE 5.33

60. Prove Property 2 of Theorem 5.17.

5.7
Numerical integration

Occasionally we encounter functions for which we cannot find antiderivatives. Of course, that may be due to a lack of cleverness on our part. On the other hand, some elementary functions simply do not possess antiderivatives that are elementary functions. For example, there is no elementary function that has either of the following functions as its derivative.

$$\sqrt[3]{x}\,\sqrt{1-x} \quad \text{or} \quad \sqrt{1-x^3}$$

If we wish to evaluate a definite integral involving a function whose antiderivative we cannot find, then the Fundamental Theorem of Calculus cannot be applied, and we must resort to an approximation technique. We describe two such techniques in this section.

Development of the Trapezoidal Rule

One way we can approximate a definite integral is by the use of n trapezoids, as shown in Figure 5.34. In the development of this method, we assume that f is continuous and positive on the interval $[a, b]$, and that the definite integral $\int_a^b f(x)\,dx$ represents the area of the region bounded by the graph of f and the x-axis, from $x = a$ to $x = b$.

First, we partition the interval $[a, b]$ into n equal subintervals, each of width $\Delta x = (b - a)/n$, such that

$$a = x_0 < x_1 < x_2 < \cdots < x_n = b$$

We then form trapezoids for each subinterval, as shown in Figure 5.35. The areas of these trapezoids are given by

$$\text{area of the 1st trapezoid} = \left[\frac{f(x_0) + f(x_1)}{2}\right]\left(\frac{b - a}{n}\right)$$

$$\text{area of the 2nd trapezoid} = \left[\frac{f(x_1) + f(x_2)}{2}\right]\left(\frac{b - a}{n}\right)$$

$$\vdots$$

$$\text{area of the } n\text{th trapezoid} = \left[\frac{f(x_{n-1}) + f(x_n)}{2}\right]\left(\frac{b - a}{n}\right)$$

Finally, the sum of the areas of the n trapezoids is

$$\left(\frac{b - a}{n}\right)\left[\frac{f(x_0) + f(x_1)}{2} + \frac{f(x_1) + f(x_2)}{2} + \cdots + \frac{f(x_{n-1}) + f(x_n)}{2}\right]$$

$$= \left(\frac{b - a}{2n}\right)[f(x_0) + \underbrace{f(x_1) + f(x_1)}_{2f(x_1)} + \underbrace{f(x_2) + \cdots}_{2f(x_2)} + \underbrace{f(x_{n-1})}_{2f(x_{n-1})} + f(x_n)]$$

$$= \left(\frac{b - a}{2n}\right)[f(x_0) + 2f(x_1) + 2f(x_2) + \cdots + 2f(x_{n-1}) + f(x_n)]$$

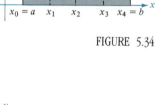

FIGURE 5.34

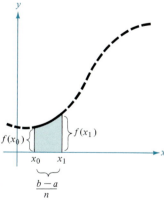

FIGURE 5.35

By letting $\Delta x = (b - a)/n$, we can take the limit as $n \to \infty$, to obtain

$$\lim_{n \to \infty} \left(\frac{b - a}{2n}\right)[f(x_0) + 2f(x_1) + \cdots + 2f(x_{n-1}) + f(x_n)]$$

$$= \lim_{n \to \infty} \left[\frac{[f(a) - f(b)]\, \Delta x}{2} + \sum_{i=1}^{n} f(x_i)\, \Delta x\right]$$

$$= \lim_{n \to \infty} \frac{[f(a) - f(b)](b - a)}{2n} + \lim_{n \to \infty} \sum_{i=1}^{n} f(x_i)\, \Delta x$$

$$= 0 + \int_a^b f(x)\, dx$$

This brings us to the following theorem, which we call the Trapezoidal Rule.

THEOREM 5.18   **TRAPEZOIDAL RULE**

Let f be continuous on $[a, b]$. The Trapezoidal Rule for approximating $\int_a^b f(x)\, dx$ is given by

$$\int_a^b f(x)\, dx \approx \frac{b - a}{2n}[f(x_0) + 2f(x_1) + 2f(x_2) + \cdots + 2f(x_{n-1}) + f(x_n)]$$

Moreover, as $n \to \infty$, the right-hand side approaches $\int_a^b f(x)\, dx$.

| **Remark** Although we assumed f to be positive and continuous in the development of the Trapezoidal Rule, the rule is valid for any continuous function on the closed interval $[a, b]$.

EXAMPLE 1   *Approximation with the Trapezoidal Rule*

Use the Trapezoidal Rule to approximate the definite integral

$$A = \int_0^1 \sqrt{x + 1}\, dx$$

Compare the results for $n = 4$ and $n = 8$, as shown in Figure 5.36.

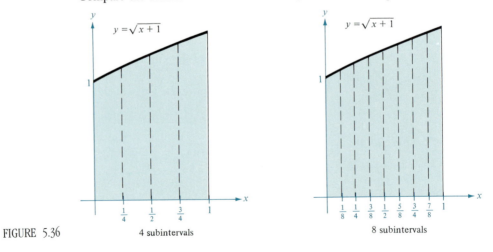

FIGURE 5.36   4 subintervals   8 subintervals

Solution: When $n = 4$, we have $\Delta x = 1/4$, and by the Trapezoidal Rule we have

$$A \approx \frac{1}{8}\left[\sqrt{1} + 2\sqrt{\frac{5}{4}} + 2\sqrt{\frac{6}{4}} + 2\sqrt{\frac{7}{4}} + \sqrt{2}\right]$$

$$= \frac{1}{8}\left[1 + 2\left(\frac{\sqrt{5}}{2}\right) + 2\left(\frac{\sqrt{6}}{2}\right) + 2\left(\frac{\sqrt{7}}{2}\right) + \sqrt{2}\right]$$

$$\approx 1.2182$$

When $n = 8$, we have $\Delta x = 1/8$, and

$$A \approx \frac{1}{16}\left[\sqrt{1} + 2\sqrt{\frac{9}{8}} + 2\sqrt{\frac{10}{8}}\right.$$

$$+ 2\sqrt{\frac{11}{8}} + 2\sqrt{\frac{12}{8}} + 2\sqrt{\frac{13}{8}}$$

$$\left. + 2\sqrt{\frac{14}{8}} + 2\sqrt{\frac{15}{8}} + \sqrt{2}\right]$$

$$\approx 1.2188$$

For this particular integral we could have found an antiderivative and determined that the exact area of the region is $\frac{2}{3}(2^{3/2} - 1) \approx 1.2190$. ☐

Remark It is interesting to compare the Trapezoidal Rule for approximating definite integrals to the Midpoint Rule discussed in Section 5.4. For the Trapezoidal Rule, we average the functional values at the endpoints of the subintervals, but for the Midpoint Rule we take the functional value of the subinterval midpoints.

Midpoint Rule   *Trapezoidal Rule*

$$\int_a^b f(x)\, dx \approx \sum_{i=1}^n f\left(\frac{x_i + x_{i-1}}{2}\right)\Delta x \qquad \int_a^b f(x)\, dx \approx \sum_{i=1}^n \left(\frac{f(x_i) + f(x_{i-1})}{2}\right)\Delta x$$

There are two important points that should be made concerning the Trapezoidal Rule or the Midpoint Rule. First, the approximation tends to become more accurate as n increases. For instance, in Example 1, if $n = 16$, the Trapezoidal Rule yields an approximation of 1.2189. Second, though we could have used the Fundamental Theorem to evaluate the integral in Example 1, this theorem cannot be used to evaluate an integral as simple looking as $\int_0^1 \sqrt{x^3 + 1}\, dx$, because $\sqrt{x^3 + 1}$ has no elementary antiderivative. Yet, the Trapezoidal Rule can be readily applied to this integral.

Development of Simpson's Rule

One way to view the trapezoidal approximation of a definite integral is to say that on each subinterval we approximate f by a *first*-degree polynomial. In Simpson's Rule, named after the English mathematician Thomas Simpson (1710–1761), we take this procedure one step further and we approximate f by *second*-degree polynomials.

Before presenting Simpson's Rule, we give a theorem for evaluating integrals of second-degree polynomials.

THEOREM 5.19

INTEGRAL OF A QUADRATIC FUNCTION

If $p(x) = Ax^2 + Bx + C$, then

$$\int_a^b p(x)\, dx = \left(\frac{b-a}{6}\right)\left[p(a) + 4p\left(\frac{a+b}{2}\right) + p(b)\right]$$

Proof:

$$\int_a^b p(x)\, dx = \int_a^b (Ax^2 + Bx + C)\, dx$$

$$= \left[\frac{Ax^3}{3} + \frac{Bx^2}{2} + Cx\right]_a^b$$

$$= \frac{A(b^3 - a^3)}{3} + \frac{B(b^2 - a^2)}{2} + C(b - a)$$

$$= \left(\frac{b-a}{6}\right)[2A(a^2 + ab + b^2) + 3B(b + a) + 6C]$$

Now, by expanding and collecting terms, the expression inside the brackets becomes

$$\underbrace{(Aa^2 + Ba + C)}_{p(a)} + \underbrace{4\left[A\left(\frac{a+b}{2}\right)^2 + B\left(\frac{b+a}{2}\right) + C\right]}_{4p\left(\frac{a+b}{2}\right)} + \underbrace{(Ab^2 + Bb + C)}_{p(b)}$$

and we have

$$\int_a^b p(x)\, dx = \left(\frac{b-a}{6}\right)\left[p(a) + 4p\left(\frac{a+b}{2}\right) + p(b)\right]$$

To develop Simpson's Rule for approximating a definite integral, we again partition the interval $[a, b]$ into n equal subintervals, each of width $\Delta x = (b - a)/n$. However, this time we require n to be even, and group the subintervals into pairs such that

$$a = \underbrace{x_0 < x_1 < x_2}_{[x_0,\, x_2]} < \underbrace{x_3 < x_4}_{[x_2,\, x_4]} < \cdots < \underbrace{x_{n-2} < x_{n-1} < x_n}_{[x_{n-2},\, x_n]} = b$$

Then on each (double) subinterval $[x_{i-2}, x_i]$ we approximate f by a polynomial p of degree less than or equal to 2. For example, on the subinterval $[x_0, x_2]$ we choose the polynomial of least degree passing through the points (x_0, y_0), (x_1, y_1), and (x_2, y_2), as shown in Figure 5.37. Now, using p as an approximation for f on this subinterval, we have

$$\int_{x_0}^{x_2} f(x)\, dx \approx \int_{x_0}^{x_2} p(x)\, dx = \frac{x_2 - x_0}{6}\left[p(x_0) + 4p\left(\frac{x_2 + x_0}{2}\right) + p(x_2)\right]$$

$$= \frac{2[(b - a)/n]}{6}[p(x_0) + 4p(x_1) + p(x_2)]$$

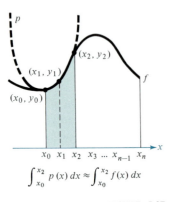

$$\int_{x_0}^{x_2} p(x)\, dx \approx \int_{x_0}^{x_2} f(x)\, dx$$

FIGURE 5.37

$$= \frac{b-a}{3n}[f(x_0) + 4f(x_1) + f(x_2)]$$

Repeating this procedure on the entire interval $[a, b]$ results in the following theorem.

THEOREM 5.20   **SIMPSON'S RULE** (n is even)
Let f be continuous on $[a, b]$. Simpson's Rule for approximating $\int_a^b f(x)\, dx$ is given by

$$\int_a^b f(x)\, dx \approx \frac{b-a}{3n}[f(x_0) + 4f(x_1) + 2f(x_2) + 4f(x_3) + \cdots + 4f(x_{n-1}) + f(x_n)]$$

Moreover, as $n \to \infty$, the right-hand side approaches $\int_a^b f(x)\, dx$.

Remark   Note that the coefficients in Simpson's Rule follow the pattern

$$1 \quad 4 \quad 2 \quad 4 \quad 2 \quad 4 \quad \cdots \quad 4 \quad 2 \quad 4 \quad 1$$

In Example 1 we used the Trapezoidal Rule to estimate $\int_0^1 \sqrt{x+1}\, dx$. In the next example we see how well Simpson's Rule does for the same integral.

EXAMPLE 2   Approximation with Simpson's Rule

Use Simpson's Rule to approximate the definite integral

$$A = \int_0^1 \sqrt{x+1}\, dx$$

Compare the results for $n = 4$ and $n = 8$.

Solution: When $n = 4$, we have $\Delta x = \frac{1}{4}$, and by Simpson's Rule we have

$$A \approx \frac{1}{12}\left[\sqrt{1} + 4\sqrt{\frac{5}{4}} + 2\sqrt{\frac{6}{4}} + 4\sqrt{\frac{7}{4}} + \sqrt{2}\right]$$

$$= \frac{1}{12}\left[1 + 4\left(\frac{\sqrt{5}}{2}\right) + 2\left(\frac{\sqrt{6}}{2}\right) + 4\left(\frac{\sqrt{7}}{2}\right) + \sqrt{2}\right] \approx 1.2189$$

When $n = 8$, we have $\Delta x = \frac{1}{8}$, and

$$A \approx \frac{1}{24}\left[\sqrt{1} + 4\sqrt{\frac{9}{8}} + 2\sqrt{\frac{10}{8}} + 4\sqrt{\frac{11}{8}} + 2\sqrt{\frac{12}{8}} + 4\sqrt{\frac{13}{8}} + 2\sqrt{\frac{14}{8}} + 4\sqrt{\frac{15}{8}} + \sqrt{2}\right]$$

$$\approx 1.2190$$

In Examples 1 and 2 we were able to calculate the exact value of the integral and compare that to our approximations to see how good they were. Of course, in practice we would not bother with an approximation if it were possible to evaluate the integral exactly. However, if we must use an approximation technique, then it is important to know how good we can expect the approximation to be. The following theorem, which we list without proof, gives the formulas for estimating the error involved in the use of Simpson's Rule and the Trapezoidal Rule.

THEOREM 5.21   **ERROR IN TRAPEZOIDAL AND SIMPSON'S RULES**

If f has a continuous second derivative on $[a, b]$, then the error E in approximating $\int_a^b f(x)\,dx$ by the Trapezoidal Rule is

$$0 \leq E \leq \frac{(b-a)^3}{12n^2}[\max |f''(x)|], \quad a \leq x \leq b \qquad \text{Trapezoidal Rule}$$

Moreover, if f has a continuous fourth derivative on $[a, b]$, then the error E in approximating $\int_a^b f(x)\,dx$ by Simpson's Rule is

$$0 \leq E \leq \frac{(b-a)^5}{180n^4}[\max |f^{(4)}(x)|], \quad a \leq x \leq b \qquad \text{Simpson's Rule}$$

Theorem 5.21 states that the errors generated by the Trapezoidal Rule and Simpson's Rule have upper bounds dependent upon the extreme values of $f''(x)$ and $f^{(4)}(x)$, respectively, in the interval $[a, b]$. Furthermore, it is evident from this theorem that these errors can be made arbitrarily small by *increasing* n, provided that f'' and $f^{(4)}$ are continuous and therefore bounded in $[a, b]$. The next example shows how to determine a value of n that will bound the error within a predetermined tolerance.

EXAMPLE 3   The error in the Trapezoidal Rule

Use the Trapezoidal Rule to estimate the value of

$$\int_0^1 \sqrt{1 + x^2}\,dx$$

Determine n so that the approximation error is less than 0.01.

Solution: If $f(x) = \sqrt{1 + x^2}$, then

$$f'(x) = x(1 + x^2)^{-1/2}$$

$$f''(x) = x\left(-\frac{1}{2}\right)(2x)(1 + x^2)^{-3/2} + (1 + x^2)^{-1/2}$$

$$= (1 + x^2)^{-3/2}$$

Now, we can determine that the maximum value of $|f''(x)|$ on this interval is $|f''(0)| = 1$. By Theorem 5.21 we have

$$E \leq \frac{(b-a)^3}{12n^2}|f''(0)| \leq \frac{1}{12n^2}(1) = \frac{1}{12n^2}$$

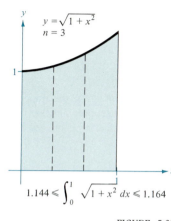

$1.144 \leqslant \int_0^1 \sqrt{1 + x^2}\, dx \leqslant 1.164$

FIGURE 5.38

To ensure that our approximation has an error less than 0.01, we must choose n so that $(1/12n^2) \leq 1/100$. Thus,

$$100 \leq 12n^2 \implies 2.89 \approx \sqrt{\frac{100}{12}} \leq n$$

Therefore, we choose $n = 3$ (since n must be greater than or equal to 2.89) and apply the Trapezoidal Rule, as shown in Figure 5.38, to obtain

$$\int_0^1 \sqrt{1 + x^2}\, dx$$
$$\approx \frac{1}{6}\left[\sqrt{1 + 0^2} + 2\sqrt{1 + (1/3)^2} + 2\sqrt{1 + (2/3)^2} + \sqrt{1 + 1^2} \right]$$
$$\approx 1.154$$

Finally, with an error no larger than 0.01, we know that

$$1.144 \leq \int_0^1 \sqrt{1 + x^2}\, dx \leq 1.164$$

EXAMPLE 4   The error in Simpson's Rule

Use Simpson's Rule to estimate the value of

$$\int_1^3 \frac{1}{x}\, dx$$

Determine n so that the approximation error is less than 0.01.

Solution: According to Theorem 5.21, the error in Simpson's Rule involves the fourth derivative. Hence, by successive differentiation, we have

$$f(x) = x^{-1}$$
$$f'(x) = -x^{-2}$$
$$f''(x) = 2x^{-3}$$
$$f'''(x) = -6x^{-4}$$
$$f^{(4)}(x) = 24x^{-5}$$

Since, in the interval [1, 3], the maximum of $|f^{(4)}(x)|$ is $|f^{(4)}(1)| = 24$, we have

$$E \leq \frac{(b - a)^5}{180n^4}|f^{(4)}(1)| = \frac{32}{180n^4}|f^{(4)}(1)| = \frac{32}{180n^4}(24) = \frac{64}{15n^4}$$

Now by choosing n so that $(64/15n^4) < 1/100$, we have

$$\frac{6400}{15} < n^4 \implies 4.54 < n$$

Therefore, we choose $n = 6$, as shown in Figure 5.39, and obtain

$$\int_1^3 \frac{1}{x}\, dx$$
$$\approx \frac{2}{18}\left[\frac{1}{1} + 4\left(\frac{1}{4/3}\right) + 2\left(\frac{1}{5/3}\right) + 4\left(\frac{1}{6/3}\right) + 2\left(\frac{1}{7/3}\right) + 4\left(\frac{1}{8/3}\right) + \frac{1}{3} \right]$$
$$\approx 1.0989$$

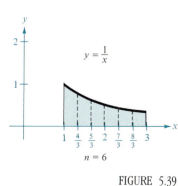

$n = 6$

FIGURE 5.39

and we conclude that

$$1.0889 \le \int_1^3 \frac{1}{x} \le 1.1089$$

| Remark From Examples 1 and 2, you may wonder why we introduced the Trapezoidal Rule, since for a fixed n, Simpson's Rule usually gives a better approximation. The main reason is that the error in the Trapezoidal Rule can be more easily estimated than the error involved in Simpson's Rule. For instance, if

$$f(x) = \sqrt{x}\,\sqrt[3]{x+1}$$

then, to estimate the error in Simpson's Rule, we would need to determine the fourth derivative of f—a monumental task! Therefore, we sometimes prefer to use the Trapezoidal Rule, even though we may have to use a larger n to obtain the desired accuracy.

Section Exercises 5.7

In Exercises 1–10, use the Trapezoidal Rule and Simpson's Rule to approximate the value of the definite integral for the indicated value of n. Round your answer to four decimal places and compare the results with the exact value of the definite integral.

1. $\int_0^2 x^2\, dx$

2. $\int_0^1 \left(\frac{x^2}{2} + 1\right) dx$

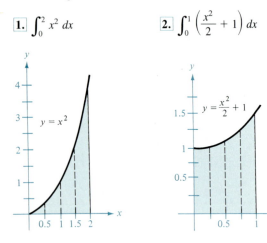

3. $\int_0^2 x^3\, dx$

4. $\int_1^2 \frac{1}{x^2}\, dx$

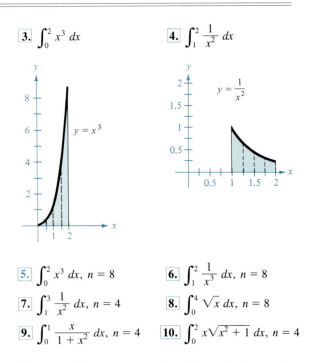

5. $\int_0^2 x^3\, dx,\ n = 8$

6. $\int_1^2 \frac{1}{x^3}\, dx,\ n = 8$

7. $\int_1^3 \frac{1}{x^2}\, dx,\ n = 4$

8. $\int_0^4 \sqrt{x}\, dx,\ n = 8$

9. $\int_0^1 \frac{x}{1 + x^2}\, dx,\ n = 4$

10. $\int_0^2 x\sqrt{x^2 + 1}\, dx,\ n = 4$

In Exercises 11–20, approximate each integral using (a) the Trapezoidal Rule, and (b) Simpson's Rule with $n = 4$.

11. $\int_0^4 \frac{1}{x+1}\, dx$

12. $\int_0^4 \sqrt{1 + x^2}\, dx$

13. $\int_0^2 \sqrt{1 + x^3}\, dx$

14. $\int_0^2 \frac{1}{\sqrt{1 + x^3}}\, dx$

15. $\int_0^1 \sqrt{x}\,\sqrt{1 - x}\, dx$

16. $\int_0^1 \frac{1}{x^2 + 1}\, dx$

17. $\int_{-1}^{1} \dfrac{1}{x^2 + 1}\, dx$ **18.** $\int_{-1}^{1} x\sqrt{x + 1}\, dx$

19. $\int_{1}^{5} \dfrac{\sqrt{x - 1}}{x}\, dx$ **20.** $\int_{2}^{5} \dfrac{1}{1 + \sqrt{x - 1}}\, dx$

In Exercises 21–26, find the maximum possible error in approximating the given integral by (a) the Trapezoidal Rule, and (b) Simpson's Rule.

21. $\int_{0}^{2} x^3\, dx,\; n = 4$ **22.** $\int_{0}^{2} x^4\, dx,\; n = 4$

23. $\int_{0}^{1} \dfrac{1}{x + 1}\, dx,\; n = 4$ **24.** $\int_{0}^{1} \dfrac{1}{x^2 + 1}\, dx,\; n = 4$

25. $\int_{0}^{1} \sqrt{1 + x^2}\, dx,\; n = 2$ **26.** $\int_{0}^{1} \sqrt{1 + x^3}\, dx,\; n = 2$

27. Find n so that the Trapezoidal Rule will have an error less than 0.00001 in the approximation of

$$\int_{1}^{5} \frac{1}{x}\, dx$$

28. Find n so that Simpson's Rule will have an error less than 0.00001 in the approximation of

$$\int_{1}^{5} \frac{1}{x}\, dx$$

29. Approximate the area, accurate to one decimal place, of the region bounded by the graph of $y = x\sqrt{4 - x}$ and the x-axis for $0 \leq x \leq 4$.

30. Approximate the area, accurate to one decimal place, of the region bounded by the graph of $y = \sqrt{1 - x^3}$ and the x-axis for $0 \leq x \leq 1$.

31. Use Simpson's Rule on the integral

$$\pi = \int_{0}^{1} \frac{4}{1 + x^2}\, dx$$

to obtain an approximation of π that is correct to five decimal places. (Use $n = 6$.)

32. Prove that Simpson's Rule is exact when approximating the integral of a cubic polynomial function, and demonstrate the result for

$$\int_{0}^{1} x^3\, dx, \quad n = 2$$

In Exercises 33 and 34, use Simpson's Rule to estimate the number of square feet of land in the given lot where x and y are measured in feet, as shown in Figures 5.40 and 5.41. In each case the land is bounded by a stream and two straight roads that meet at right angles.

33. See Figure 5.40.

x	0	100	200	300	400	500
y	125	125	120	112	90	90

x	600	700	800	900	1000
y	95	88	75	35	0

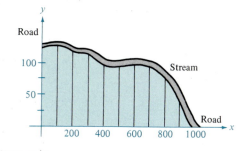

FIGURE 5.40

34. See Figure 5.41.

x	0	10	20	30	40	50
y	75	81	84	76	67	68

x	60	70	80	90	100	110	120
y	69	72	68	56	42	23	0

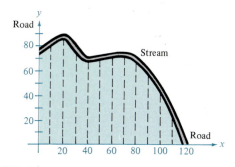

FIGURE 5.41

Review Exercises for Chapter 5

In Exercises 1–20, find the indefinite integral.

1. $\int \frac{2}{3\sqrt[3]{x}}\,dx$

2. $\int \frac{2}{\sqrt[3]{3x}}\,dx$

3. $\int (2x^2 + x - 1)\,dx$

4. $\int \frac{x^3 - 2x^2 + 1}{x^2}\,dx$

5. $\int \frac{(1 + x)^2}{\sqrt{x}}\,dx$

6. $\int x^2\sqrt{x^3 + 3}\,dx$

7. $\int \frac{x^2}{\sqrt{x^3 + 3}}\,dx$

8. $\int \frac{x^2 + 2x}{(x + 1)^2}\,dx$

9. $\int (x^2 + 1)^3\,dx$

10. $\int \sqrt{2 - 5x}\,dx$

11. $\int x(x^2 + 1)^3\,dx$

12. $\int \frac{x}{(x^2 + 1)^3}\,dx$

13. $\int \frac{2x}{(1 + 4x^2)^4}\,dx$

14. $\int \frac{(x^2 + 1)^3 - 1}{x}\,dx$

15. $\int \frac{x}{\sqrt{25 - 9x^2}}\,dx$

16. $\int \frac{x + 2}{\sqrt{x^2 + 4x + 1}}\,dx$

17. $\int x^2\sqrt{x + 5}\,dx$

18. $\int x\sqrt{x + 5}\,dx$

19. $\int \frac{1}{\sqrt{x}(5 + \sqrt{x})^2}\,dx$

20. $\int \frac{(6 + x^{2/3})^3}{\sqrt[3]{x}}\,dx$

In Exercises 21–24, write the given sum in sigma notation.

21. The sum of the first ten positive odd integers
22. The sum of the cubes of the first n positive integers
23. $6 + 10 + 14 + 18 + \cdots + 42$
24. $1 + 7 + 13 + 19 + \cdots + 43$

In Exercises 25–28, evaluate the given sums if $x_1 = 2$, $x_2 = -1$, $x_3 = 5$, $x_4 = 3$, and $x_5 = 7$.

25. $\frac{1}{5}\sum_{i=1}^{5} x_i$

26. $\sum_{i=1}^{5} \frac{1}{x_i}$

27. $\sum_{i=1}^{5} (2x_i - x_i^2)$

28. $\sum_{i=2}^{5} (x_i - x_{i-1})$

In Exercises 29–40, use the Fundamental Theorem of Calculus to evaluate the definite integral.

29. $\int_0^4 (2 + x)\,dx$

30. $\int_{-1}^1 (t^2 + 2)\,dt$

31. $\int_{-1}^1 (4t^3 - 2t)\,dt$

32. $\int_3^6 \frac{x}{3\sqrt[3]{x^2 - 8}}\,dx$

33. $\int_0^3 \frac{1}{\sqrt{1 + x}}\,dx$

34. $\int_0^1 x^2(x^3 + 1)^3\,dx$

35. $\int_4^9 x\sqrt{x}\,dx$

36. $\int_1^2 \left(\frac{1}{x^2} - \frac{1}{x^3}\right)\,dx$

37. $\int_0^4 x\sqrt{4 - x}\,dx$

38. $\int_0^4 x\sqrt[4]{4 - x}\,dx$

39. $\int_0^1 \frac{x}{\sqrt{x + 3}}\,dx$

40. $\int_0^1 \frac{x}{\sqrt{x^2 + 3}}\,dx$

41. Find the function f whose derivative is $f'(x) = -2x$ and whose graph passes through the point $(-1, 1)$.
42. A function f has a second derivative $f''(x) = 6(x - 1)$. Find the function if its graph passes through the point $(2, 1)$ and at that point is tangent to the line $3x - y - 5 = 0$.
43. An airplane taking off from a runway travels 3600 feet before lifting off. If it starts from rest, moves with constant acceleration, and makes the run in 30 seconds, with what velocity does it lift off?
44. The speed of a car traveling in a straight line is reduced from 45 to 30 miles per hour in a distance of 264 feet. Find the distance in which the car can be brought to rest from 30 miles per hour, assuming the same constant acceleration.
45. A ball is thrown vertically upward from the ground with velocity of 96 feet per second.
   (a) How long will it take the ball to rise to its maximum height?
   (b) What is the maximum height?
   (c) When is the velocity of the ball one-half the initial velocity?
   (d) What is the height of the ball when its velocity is one-half the initial velocity?
46. Repeat Exercise 45, assuming that the ball has an initial velocity of 128 feet per second.
47. Given the region bounded by $y = mx$, $y = 0$, $x = 0$ and $x = b$, find the following:
   (a) The upper and lower sum to approximate the area of the region when $\Delta x = b/4$.
   (b) The upper and lower sum to approximate the area of the region when $\Delta x = b/n$.
   (c) The area of the region by letting n approach infinity in both sums of part (b). Show that in each case you obtain the formula for the area of a triangle.
   (d) The area of the region by using the Fundamental Theorem of Calculus.
48. (a) Find the area of the region bounded by the graphs of $y = x^3$, $y = 0$, $x = 1$, and $x = 3$, by using the limit definition.

(b) Find the area of the given region by using the Fundamental Theorem of Calculus.

In Exercises 49–56, sketch the graph of the region whose area is given by the integral, and find the area.

49. $\int_1^3 (2x - 1)\, dx$

50. $\int_0^2 (x + 4)\, dx$

51. $\int_3^4 (x^2 - 9)\, dx$

52. $\int_{-1}^2 (-x^2 + x + 2)\, dx$

53. $\int_0^1 (x - x^3)\, dx$

54. $\int_0^1 \sqrt{x}(1 - x)\, dx$

55. $\int_0^1 (y - y^2)\, dy$

56. $\int_1^4 \sqrt{y}\, dy$

In Exercises 57–60, find the average value of the function over the given interval. Find the values of x where the function assumes its mean value, and sketch the graph of the function.

57. $f(x) = \dfrac{1}{\sqrt{x - 1}}$, [5, 10]

58. $f(x) = x^3$, [0, 2]

59. $f(x) = x$, [0, 4]

60. $f(x) = x^2 - \dfrac{1}{x^2}$, [1, 2]

In Exercises 61–64, approximate each integral using (a) the Trapezoidal Rule and (b) Simpson's Rule.

61. $\int_1^5 \dfrac{\sqrt{1 + x^2}}{x}\, dx$, $n = 4$

62. $\int_0^3 x\sqrt{1 + x^3}\, dx$, $n = 6$

63. $\int_0^2 \sqrt[3]{1 + x^2}\, dx$, $n = 8$

64. $\int_0^2 \dfrac{5}{t^3 + 1}\, dt$, $n = 4$

65. Suppose that gasoline is increasing in price according to the equation

$$p = 1 + 0.1t + 0.02t^2$$

where p is the dollar price per gallon and $t = 0$ represents the year 1983. If an automobile is driven 15,000 miles a year and gets M miles per gallon, then the annual fuel cost is

$$C = \frac{15,000}{M} \int_t^{t+1} p\, dt$$

Find the annual fuel cost for the years (a) 1985 and (b) 1990.

In Exercises 66 and 67, consider the function

$$f(x) = kx^n(1 - x)^m, \quad 0 \le x \le 1$$

where n, $m > 0$ and k is a constant. This function can be used to represent various percentage distributions. If k is chosen so that

$$\int_0^1 f(x)\, dx = 1$$

then the probability that x will fall between a and b is given by

$$p_{a,b} = \int_a^b f(x)\, dx$$

66. The probability of recall in a certain experiment is found to be

$$p_{a,b} = \int_a^b \frac{15}{4} x\sqrt{1 - x}\, dx$$

where x represents the percentage of recall. (See Figure 5.42.)

(a) For a randomly chosen individual, what is the probability that he or she will recall between 50% and 75% of the material?

(b) What is the median percentage recall? That is, for what value of b is it true that the probability from 0 to b is 0.5?

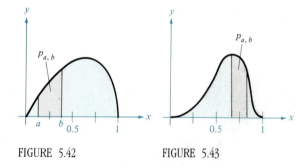

FIGURE 5.42   FIGURE 5.43

67. The probability of finding between a and b percentage of iron in ore samples taken from a certain region is given by

$$p_{a,b} = \int_a^b \frac{1155}{32} x^3(1 - x)^{3/2}\, dx$$

(See Figure 5.43.) What is the probability that a sample will contain between:

(a) 0% and 25%?   (b) 50% and 100%?

6

Applications of integration

6.1
Area of a region between two curves

With a few modifications we can extend the application of definite integrals from finding the area of a region *under* a curve to finding the area of a region *between* two curves. If, as in Figure 6.1, the graphs of both f and g lie above the x-axis, we can geometrically interpret the area of the region between the graphs as the area of the region under the graph of g subtracted from the area of the region under the graph of f as shown in Figure 6.2.

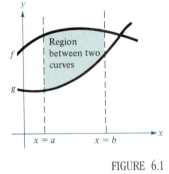

FIGURE 6.1

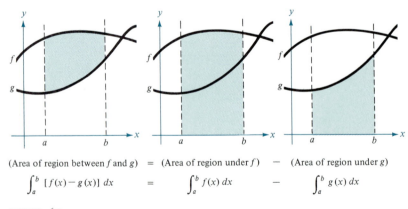

(Area of region between f and g) $=$ (Area of region under f) $-$ (Area of region under g)

$$\int_a^b [f(x) - g(x)]\, dx \quad = \quad \int_a^b f(x)\, dx \quad - \quad \int_a^b g(x)\, dx$$

FIGURE 6.2

Although Figure 6.1 depicts the graphs of f and g above the x-axis, this is not necessary and the same integrand $[f(x) - g(x)]$ can be used as long as f and g are continuous and $g(x) \le f(x)$ on the interval $[a, b]$. This result is summarized in the following theorem.

THEOREM 6.1

AREA OF A REGION BETWEEN TWO CURVES

If f and g are continuous on $[a, b]$ and $g(x) \leq f(x)$ for all x in $[a, b]$, then the area of the region bounded by the graphs of f and g and the vertical lines $x = a$ and $x = b$ is

$$A = \int_a^b [f(x) - g(x)] \, dx$$

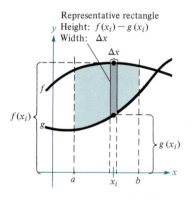

Representative rectangle
Height: $f(x_i) - g(x_i)$
Width: Δx

FIGURE 6.3

Proof: We partition the interval $[a, b]$ into n subintervals, each of width Δx and sketch a **representative rectangle** of width Δx and height $f(x_i) - g(x_i)$, where x_i is in the ith interval as shown in Figure 6.3. The area of this representative rectangle is

$$\Delta A_i = (\text{height})(\text{width}) = [f(x_i) - g(x_i)] \, \Delta x$$

By adding the areas of the n rectangles and taking the limit as $\|\Delta\| \to 0$ $(n \to \infty)$, we have

$$\lim_{n \to \infty} \sum_{i=1}^n [f(x_i) - g(x_i)] \, \Delta x$$

Since f and g are continuous on $[a, b]$, $f - g$ is also continuous on this interval and the limit exists. Therefore, the area A of the given region is

$$A = \lim_{n \to \infty} \sum_{i=1}^n [f(x_i) - g(x_i)] \, \Delta x = \int_a^b [f(x) - g(x)] \, dx$$

| **Remark** Representative rectangles are used throughout this chapter in various applications of integration. A vertical rectangle (of width Δx) implies integration with respect to x, while a horizontal rectangle (of width Δy) implies integration with respect to y.

It is important to realize that the development of the area formula in Theorem 6.1 depends *only* on the continuity of f and g and the assumption that $g(x) \leq f(x)$. It does not depend on the positions of the graphs of f and g with respect to the x-axis. Try to convince yourself of this by referring to Figure 6.4.

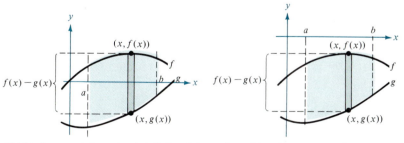

FIGURE 6.4   Height of representative rectangle is $f(x) - g(x)$ regardless of the relative position of the x-axis.

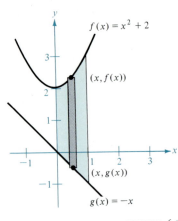

$f(x) = x^2 + 2$

$(x, f(x))$

$(x, g(x))$

$g(x) = -x$

FIGURE 6.5

EXAMPLE 1   *Finding the area of a region between two curves*

Find the area of the region bounded by the graphs of $y = x^2 + 2$, $y = -x$, $x = 0$, and $x = 1$.

Solution: If we let $g(x) = -x$ and $f(x) = x^2 + 2$, then $g(x) \le f(x)$ for all x in $[0, 1]$ as shown in Figure 6.5. Thus, the area of the representative rectangle is

$$\Delta A = [f(x) - g(x)]\, \Delta x = [(x^2 + 2) - (-x)]\, \Delta x$$

Therefore, by Theorem 6.1 we find the area of the region to be

$$A = \int_a^b [f(x) - g(x)]\, dx = \int_0^1 [(x^2 + 2) - (-x)]\, dx$$

$$= \left[\frac{x^3}{3} + \frac{x^2}{2} + 2x \right]_0^1$$

$$= \frac{1}{3} + \frac{1}{2} + 2 = \frac{17}{6}$$

In Example 1, the graphs of $f(x) = x^2 + 2$ and $g(x) = -x$ do not intersect, and the values of a and b are given explicitly. A more common type of problem involves the area of a region bounded by two *intersecting* graphs, where the values of a and b must be calculated.

EXAMPLE 2   *A region lying between two intersecting graphs*

Find the area of the region bounded by the graphs of $f(x) = 2 - x^2$ and $g(x) = x$.

Solution: From Figure 6.6, we see that the graphs of f and g have two points of intersection. To find the x-coordinates of these points, we set $f(x)$ and $g(x)$ equal to each other and solve for x.

$$2 - x^2 = x \qquad \text{Set } f(x) \text{ equal to } g(x)$$

$$-x^2 - x + 2 = 0$$

$$-(x + 2)(x - 1) = 0 \qquad x = -2 \text{ and } x = 1$$

Thus, we have $a = -2$ and $b = 1$. Since $g(x) \le f(x)$ on the interval $[-2, 1]$, the representative rectangle has an area of

$$\Delta A = [f(x) - g(x)]\, \Delta x = [(2 - x^2) - x]\, \Delta x$$

and the area of the region is

$$A = \int_{-2}^1 [(2 - x^2) - x]\, dx = \left[-\frac{x^3}{3} - \frac{x^2}{2} + 2x \right]_{-2}^1$$

$$= \left(-\frac{1}{3} - \frac{1}{2} + 2 \right) - \left(\frac{8}{3} - 2 - 4 \right)$$

$$= \frac{9}{2}$$

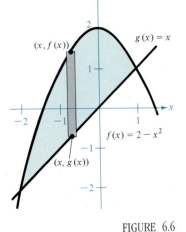

$(x, f(x))$

$g(x) = x$

$f(x) = 2 - x^2$

$(x, g(x))$

FIGURE 6.6

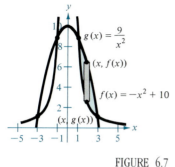

FIGURE 6.7

EXAMPLE 3   A region lying between two intersecting graphs

The graphs of

$$f(x) = -x^2 + 10 \qquad \text{and} \qquad g(x) = \frac{9}{x^2}$$

intersect four times, bounding two regions of equal areas, as shown in Figure 6.7. Find the area of one of these regions.

Solution: To find the points of intersection of the graphs of $f(x) = -x^2 + 10$ and $g(x) = 9/x^2$, we set the two functions equal to each other and solve for x.

$$-x^2 + 10 = \frac{9}{x^2} \qquad \text{Set } f(x) \text{ equal to } g(x)$$

$$-x^4 + 10x^2 - 9 = 0$$

$$-(x^2 - 9)(x^2 - 1) = 0 \qquad x = -3, -1, 1, 3$$

We choose to find the area of the region in the first quadrant. Thus, $a = 1$ and $b = 3$. Since $-x^2 + 10 \geq (9/x^2)$ on the interval $[1, 3]$, the area of the region is

$$A = \int_1^3 \left[(-x^2 + 10) - \frac{9}{x^2} \right] dx$$

$$= \left[-\frac{x^3}{3} + 10x + \frac{9}{x} \right]_1^3$$

$$= (-9 + 30 + 3) - \left(-\frac{1}{3} + 10 + 9 \right)$$

$$= \frac{16}{3}$$

If two curves intersect in *more* than two points, then to find the area of the region between the curves, we must find all points of intersection and check to see which curve is above the other in each interval determined by these points.

EXAMPLE 4   *Curves that intersect at more than two points*

Find the area of the region between the graphs of $f(x) = x^3 - 2x^2 + x - 1$ and $g(x) = -x^2 + 3x - 1$.

Solution: Solving for x in the equation $f(x) = g(x)$, we have

$$f(x) - g(x) = x^3 - x^2 - 2x = 0$$

$$x(x^2 - x - 2) = 0$$

$$x(x - 2)(x + 1) = 0$$

Thus, the zeros are $x = -1$, $x = 0$, and $x = 2$. In Figure 6.8, we see that $g(x) \leq f(x)$ on $[-1, 0]$. However, the two curves switch at the point $(0, -1)$, and $f(x) \leq g(x)$ on $[0, 2]$. Hence, we need two integrals—one for the interval $[-1, 0]$ and one for $[0, 2]$.

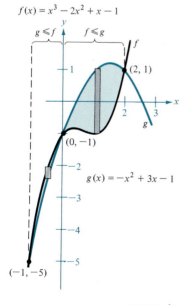

FIGURE 6.8

$$A = \int_{-1}^{0} [f(x) - g(x)]\, dx + \int_{0}^{2} [g(x) - f(x)]\, dx$$

$$= \int_{-1}^{0} (x^3 - x^2 - 2x)\, dx + \int_{0}^{2} (-x^3 + x^2 + 2x)\, dx$$

$$= \left[\frac{x^4}{4} - \frac{x^3}{3} - x^2 \right]_{-1}^{0} + \left[\frac{-x^4}{4} + \frac{x^3}{3} + x^2 \right]_{0}^{2}$$

$$= -\left(\frac{1}{4} + \frac{1}{3} - 1 \right) + \left(-4 + \frac{8}{3} + 4 \right) = \frac{37}{12}$$

If the graph of a function of y is a boundary of a region, it is often convenient to use representative rectangles that are *horizontal* and find the area by integrating with respect to y. In general, to determine the area between two curves, we have

$$A = \int_{x_1}^{x_2} \underbrace{[(\text{top curve}) - (\text{bottom curve})]}_{\text{in variable } x}\, dx \qquad \text{Vertical rectangles}$$

$$A = \int_{y_1}^{y_2} \underbrace{[(\text{right curve}) - (\text{left curve})]}_{\text{in variable } y}\, dy \qquad \text{Horizontal rectangles}$$

where (x_1, y_1) and (x_2, y_2) are either adjacent points of intersection of the two curves involved or points on the specified boundary lines.

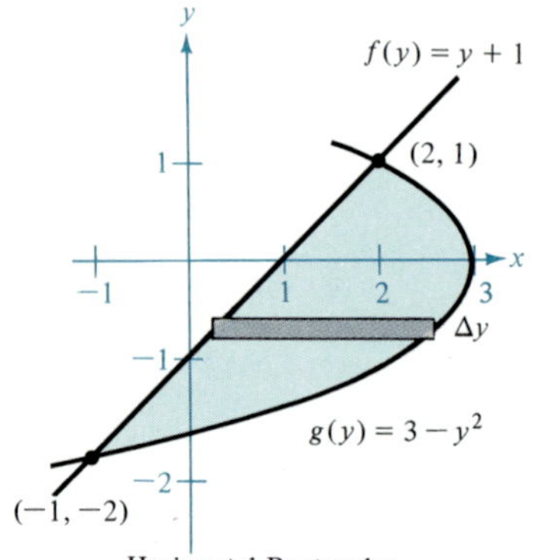

EXAMPLE 5   *Horizontal representative rectangles*

Find the area of the region bounded by the graphs of $x = 3 - y^2$ and $y = x - 1$.

Solution: We consider $g(y) = 3 - y^2$ and $f(y) = y + 1$. These two curves intersect when $y = -2$ and $y = 1$ as shown in Figure 6.9. Since $f(y) \le g(y)$ on this interval, we have

$$\Delta A = [g(y) - f(y)]\, \Delta y = [(3 - y^2) - (y + 1)]\, \Delta y$$

Hence the area is

$$A = \int_{-2}^{1} [(3 - y^2) - (y + 1)]\, dy$$

$$= \int_{-2}^{1} (-y^2 - y + 2)\, dy$$

$$= \left[\frac{-y^3}{3} - \frac{y^2}{2} + 2y \right]_{-2}^{1}$$

$$= \left(\frac{-1}{3} - \frac{1}{2} + 2 \right) - \left(\frac{8}{3} - 2 - 4 \right) = \frac{9}{2}$$

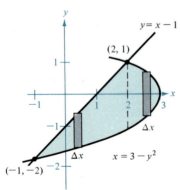

Horizontal Rectangles
(Integration with respect to y)

FIGURE 6.9

Vertical Rectangles
(Integration with respect to x)

FIGURE 6.10

Remark Note in Example 5 that by integrating with respect to y we need only one integral. If we had integrated with respect to x, we would have needed two integrals

$$A = \int_{-2}^{2} [(x - 1) - (-\sqrt{3 - x})]\, dx + \int_{2}^{3} [\sqrt{3 - x} - (-\sqrt{3 - x})]\, dx$$

$$= \int_{-2}^{2} (x - 1 + \sqrt{3 - x})\, dx + \int_{2}^{3} 2\sqrt{3 - x}\, dx$$

as shown in Figure 6.10.

In this section, we developed the integration formula for area between two curves by using a rectangle as the *representative element*. For each new application in the remaining sections of this chapter, we will construct an appropriate representative element using precalculus formulas you already know. Each integration formula will then be obtained by summing these representative elements.

For example, in this section we developed the area formula as

$$A = \text{(height)(width)} \longrightarrow \Delta A = [f(x) - g(x)] \Delta x \longrightarrow A = \int_a^b [f(x) - g(x)] \, dx$$

If you set up the form of each representative element in this chapter, you will see how integration works as a summation process. Moreover, by setting up representative elements, you will minimize the need to memorize integration formulas.

EXAMPLE 6   An application

The total consumption of petroleum fuel for transportation in the United States from 1960 to 1979 followed a growth pattern described by the equation

$$f(t) = 0.000433t^2 + 0.0962t + 2.76, \quad -10 \le t \le 9$$

where $f(t)$ is measured in billions of barrels and t in years, with $t = 0$ corresponding to January 1, 1970. With the onset of dramatic increases in crude oil prices in the late 1970s, the growth pattern for consumption changed and began following the pattern described by the model

$$g(t) = -0.00831t^2 + 0.152t + 2.81, \quad 9 \le t \le 16$$

as shown in Figure 6.11. Find the total amount of fuel saved from 1979 through 1985 as a result of consuming fuel at the post-1979 rate rather than the pre-1979 rate.

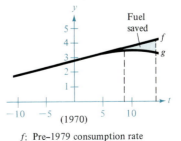

f: Pre–1979 consumption rate
g: Post–1979 consumption rate

FIGURE 6.11

Solution: Since the graph of the pre-1979 model lies above the post-1979 graph on the interval [9, 16], the amount of gasoline saved is given by the following integral:

$$\int_9^{16} [\overbrace{(0.000433t^2 + 0.0962t + 2.76)}^{f(t)} - \overbrace{(-0.00831t^2 + 0.152t + 2.81)}^{g(t)}]\ dt$$

$$= \int_9^{16} (0.008743t^2 - 0.0558t - 0.05)\ dt$$

$$= \left[\frac{0.008743t^3}{3} - \frac{0.0558t^2}{2} - 0.05t \right]_9^{16}$$

$$\approx 4.58 \text{ billion barrels}$$

Therefore, approximately 4.58 billion barrels of fuel were saved. (At 42 gallons per barrel, that is a savings of about 200 billion gallons!)

Section Exercises 6.1

In Exercises 1–6, find the area of the given region.

1. $f(x) = x^2 - 6x$
$g(x) = 0$

2. $f(x) = x^2 + 2x + 1$
$g(x) = 2x + 5$

3. $f(x) = x^2 - 4x + 3$
$g(x) = -x^2 + 2x + 3$

4. $f(x) = x^2$
$g(x) = x^3$

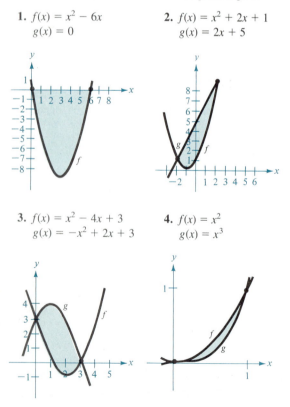

5. $f(x) = 3(x^3 - x)$
$g(x) = 0$

6. $f(x) = (x - 1)^3$
$g(x) = x - 1$

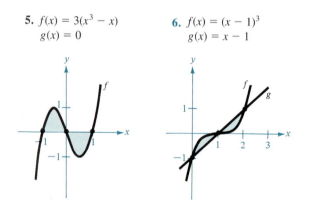

In Exercises 7–24, sketch the region bounded by the graphs of the given functions and find the area of each region.

7. $f(x) = x^2 - 4x$, $g(x) = 0$
8. $f(x) = 3 - 2x - x^2$, $g(x) = 0$
9. $f(x) = x^2 + 2x + 1$, $g(x) = 3x + 3$
10. $f(x) = -x^2 + 4x + 2$, $g(x) = x + 2$
11. $y = x$, $y = 2 - x$, $y = 0$
12. $y = \dfrac{1}{x^2}$, $y = 0$, $x = 1$, $x = 5$
13. $f(x) = 3x^2 + 2x$, $g(x) = 8$
14. $f(x) = x(x^2 - 3x + 3)$, $g(x) = x^2$
15. $f(x) = x^3 - 2x + 1$, $g(x) = -2x$, $x = 1$
16. $f(x) = \sqrt[3]{x}$, $g(x) = x$

17. $f(x) = \sqrt{3x} + 1$, $g(x) = x + 1$
18. $f(x) = x^2 + 5x - 6$, $g(x) = 6x - 6$
19. $y = x^2 - 4x + 3$, $y = 3 + 4x - x^2$
20. $y = x^4 - 2x^2$, $y = 2x^2$
21. $f(y) = y^2$, $g(y) = y + 2$
22. $f(y) = y(2 - y)$, $g(y) = -y$
23. $f(y) = y^2 + 1$, $g(y) = 0$, $y = -1$, $y = 2$
24. $f(y) = \dfrac{y}{\sqrt{16 - y^2}}$, $g(y) = 0$, $y = 3$

In Exercises 25–28, use integration to find the areas of the triangles having the given vertices.

25. $(0, 0)$, $(4, 0)$, $(4, 4)$
26. $(0, 0)$, $(4, 0)$, $(6, 4)$
27. $(0, 0)$, $(a, 0)$, (b, c)
28. $(2, -3)$, $(4, 6)$, $(6, 1)$

In Exercises 29 and 30, find b so that $y = b$ divides the region bounded by the graphs of the two equations into two regions of equal area.

29. $y = 9 - x^2$, $y = 0$
30. $y = 9 - |x|$, $y = 0$

31. The graphs of $y = x^4 - 2x^2 + 1$ and $y = 1 - x^2$ intersect at three points. However, the area between the curves *can* be found by a single integral. Explain why this is so and write an integral for this area.

32. The graphs of $y = x^3$ and $y = x$ intersect at three points, but the area between the curves *cannot* be found by the single integral

$$\int_{-1}^{1} (x^3 - x)\, dx$$

Explain why this is so. Use symmetry to write a single integral that does represent the area.

In Exercises 33 and 34, two models, R_1 and R_2, are given for revenue (in billions of dollars) for a large corporation. The model R_1 gives projected annual revenues from 1985 to

1990, with $t = 0$ corresponding to 1985, and R_2 gives projected revenues if there is a decrease in growth of corporate sales over the period. Approximate the total reduction in revenue if corporate sales are actually closer to the model R_2.

33. $R_1 = 7.21 + 0.58t$
    $R_2 = 7.21 + 0.45t$
34. $R_1 = 7.21 + 0.26t + 0.02t^2$
    $R_2 = 7.21 + 0.1t + 0.01t^2$

In Exercises 35–38, find the consumer surplus and producer surplus for each of the given supply and demand curves. The consumer surplus and producer surplus are represented by the areas shown in Figure 6.12.

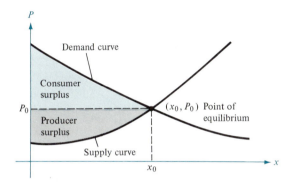

FIGURE 6.12

35. Demand function: $p_1(x) = 50 - 0.5x$
    Supply function: $p_2(x) = 0.125x$
36. Demand function: $p_1(x) = 1000 - 0.4x^2$
    Supply function: $p_2(x) = 42x$
37. Demand function: $p_1(x) = \dfrac{10{,}000}{\sqrt{x + 100}}$
    Supply function: $p_2(x) = 100\sqrt{0.05x + 10}$
38. Demand function: $p_1(x) = \sqrt{25 - 0.1x}$
    Supply function: $p_2(x) = \sqrt{9 + 0.1x} - 2$

SECTION TOPICS •
Solid of revolution •
The disc method •
The washer method •
Solids with known cross sections •

6.2
Volume: The disc method

When we introduced integration in Chapter 5, we mentioned that area is only *one* of the many applications of the definite integral. Another important application is its use in finding the volume of a three-dimensional solid. In this section we consider a particular type of three-dimensional solid—one whose

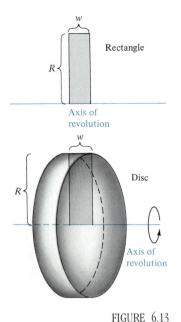

FIGURE 6.13

cross sections are similar. We begin with solid figures with *circular* (or annular) cross sections. Such solids are called **solids of revolution** and are commonly used in engineering and manufacturing. Some examples are axles, funnels, pills, bottles, and pistons.

If a region in the plane is revolved about a line that does not cross the interior of the region, the resulting solid is called a **solid of revolution** and the line is called the **axis of revolution.** The simplest such solid is a right circular cylinder which is formed by revolving a rectangle about an axis adjacent to one of the sides of the rectangle as shown in Figure 6.13. We call this cylinder a **disc** since its width is small compared to its radius. The volume of such a disc is given by

$$\text{volume of disc} = \pi R^2 w$$

where R is the radius of the disc and w is the width.

To see how to use the volume of a disc to find the volume of a general solid of revolution, consider a solid of revolution formed by revolving the plane region in Figure 6.14 about the indicated axis. To determine the volume of this solid, we consider a representative rectangle in the plane region. When this rectangle is revolved about the axis of revolution, it generates a representative disc whose volume is

$$\Delta V = \pi R^2 \, \Delta x$$

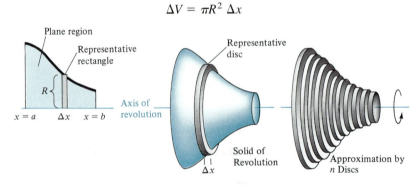

FIGURE 6.14

If we approximate the volume of the solid by n such discs of width Δx and radius $R(x_i)$, we have

$$\text{volume of solid} \approx \sum_{i=1}^{n} \pi [R(x_i)]^2 \, \Delta x = \pi \sum_{i=1}^{n} [R(x_i)]^2 \, \Delta x$$

By taking the limit as $\|\Delta\| \to 0$ $(n \to \infty)$, we have

$$\text{volume of solid} = \lim_{n \to \infty} \pi \sum_{i=1}^{n} [R(x_i)]^2 \, \Delta x = \pi \int_{a}^{b} [R(x)]^2 \, dx$$

Schematically, the disc method looks like this:

Known precalculus   *Representative*   *New integration*
formula   *element*   *formula*

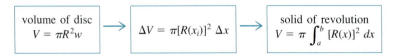

A similar formula can be derived if the axis of revolution is vertical.

THE DISC METHOD

To find the volume of a solid of revolution with the **disc method,** use one of the following as indicated in Figure 6.15.

Horizontal axis of revolution

$$\text{volume} = V = \pi \int_a^b [R(x)]^2 \, dx$$

Vertical axis of revolution

$$\text{volume} = V = \pi \int_c^d [R(y)]^2 \, dy$$

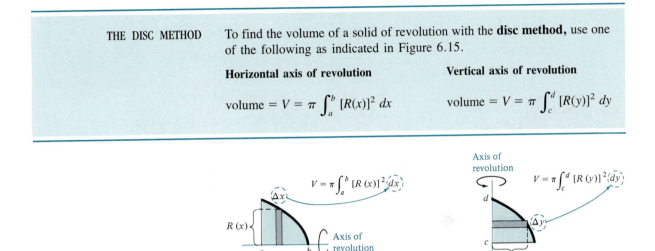

Horizontal Axis of Revolution   Vertical Axis of Revolution

FIGURE 6.15

| Remark In Figure 6.15 note that we can determine the variable of integration by placing a representative rectangle in the *plane* region "perpendicular" to the axis of revolution. If the width of the rectangle is Δx, we integrate with respect to x, and if the width of the rectangle is Δy, we integrate with respect to y.

The simplest application of the disc method involves a plane region bounded by the graph of f and the x-axis. If the axis of revolution is the x-axis, then the radius $R(x)$ is simply $f(x)$ as shown in Example 1.

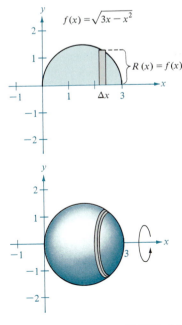

EXAMPLE 1   Finding the volume of a solid of revolution: the disc method

Find the volume of the solid formed by revolving the region bounded by the graph of $f(x) = \sqrt{3x - x^2}$ and the x-axis $(0 \le x \le 3)$ about the x-axis.

Solution: From the representative rectangle in Figure 6.16, we see that the radius of this solid is given by

$$R(x) = f(x) = \sqrt{3x - x^2}$$

and it follows that its volume is

$$V = \pi \int_a^b [R(x)]^2 \, dx = \pi \int_0^3 (\sqrt{3x - x^2})^2 \, dx$$

$$= \pi \int_0^3 (3x - x^2) \, dx$$

$$= \pi \left(\frac{3x^2}{2} - \frac{x^3}{3} \right) \Big]_0^3 = \frac{9\pi}{2}$$

FIGURE 6.16

| Remark Note in Example 1 that the problem was solved *without* any reference to the three-dimensional portion of the sketch in Figure 6.16. In general, to set up an integral to find the volume of a solid of revolution, a sketch of the plane region is more useful than a sketch of the solid, since the radius is more easily visualized in the plane region.

EXAMPLE 2   *Finding the volume of a solid of revolution: the disc method*

Find the volume of the solid formed by revolving the region bounded by $f(x) = 2 - x^2$ and $g(x) = 1$ about the line $y = 1$.

Solution: We begin by finding the points of intersection of f and g:

$$2 - x^2 = 1 \qquad \text{Set } f(x) \text{ equal to } g(x)$$
$$x^2 = 1$$
$$x = \pm 1$$

Now, to find the radius, we subtract $g(x)$ from $f(x)$ as shown in Figure 6.17 to obtain

$$R(x) = f(x) - g(x) = (2 - x^2) - 1 = 1 - x^2$$

Finally, we integrate between -1 and 1 to find the volume.

$$V = \pi \int_a^b [R(x)]^2 \, dx = \pi \int_{-1}^1 (1 - x^2)^2 \, dx$$

$$= \pi \int_{-1}^1 (1 - 2x^2 + x^4) \, dx$$

$$= \left[\pi \left(x - 2\frac{x^3}{3} + \frac{x^5}{5} \right) \right]_{-1}^1 = \frac{16\pi}{15} \approx 3.35 \quad \square$$

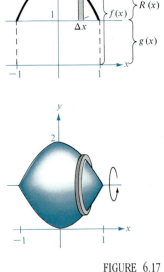

$f(x) = 2 - x^2$

$R(x)$

$f(x)$

Δx

$g(x)$

FIGURE 6.17

The washer method

The disc method can easily be extended to cover solids of revolution with a hole by replacing the representative disc with a representative **washer.** The washer is formed by revolving a rectangle about an axis as shown in Figure 6.18. If r and R are the inner and outer radii of the washer and w is the width of the washer, then the volume is given by

$$\text{volume of washer} = \pi(R^2 - r^2)w$$

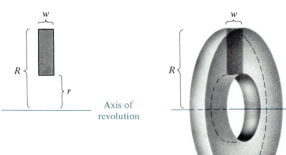

w

R

r

Axis of revolution

w

R

r

Axis of revolution

FIGURE 6.18

Now, suppose that a region is bounded by an **outer radius** $R(x)$ and an **inner radius** $r(x)$ as shown in Figure 6.19. If this region is revolved about its axis of revolution, then the volume of the resulting solid is given by

$$V = \pi \int_a^b [R(x)]^2 \, dx - \pi \int_a^b [r(x)]^2 \, dx$$

$$= \pi \int_a^b ([R(x)]^2 - [r(x)]^2) \, dx$$

Note that the integral involving the inner radius represents the volume of the hole and is *subtracted* from the integral involving the outer radius.

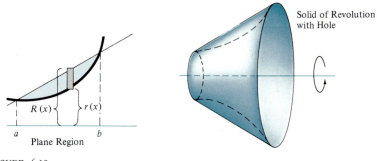

Solid of Revolution with Hole

$R(x)$ $r(x)$

a b

Plane Region

FIGURE 6.19

EXAMPLE 3   A solid of revolution with a hole

Find the volume of the solid formed by revolving the region bounded by the graphs of $y = \sqrt{x}$ and $y = x^2$ about the *x*-axis as shown in Figure 6.20.

Solution: From Figure 6.20, we have

$$R(x) = \sqrt{x} \qquad \text{Outer radius}$$
$$r(x) = x^2 \qquad \text{Inner radius}$$

Now, integrating between 0 and 1, we have

$$V = \pi \int_a^b ([R(x)]^2 - [r(x)]^2) \, dx$$

$$= \pi \int_0^1 (x - x^4) \, dx$$

$$= \pi \left[\frac{x^2}{2} - \frac{x^5}{5} \right]_0^1$$

$$= \frac{3\pi}{10}$$

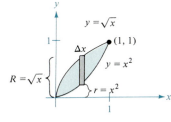

$y = \sqrt{x}$

Δx (1, 1)

$y = x^2$

$R = \sqrt{x}$

$r = x^2$

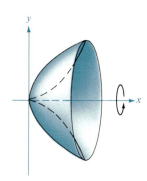

FIGURE 6.20

In each example so far, the axis of revolution has been *horizontal* and we integrated with respect to *x*. In the next example the axis of revolution is *vertical* and (as with the disc method) we must integrate with respect to *y*. In this particular example, we need two separate integrals to compute the volume.

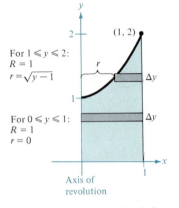

For $1 \leq y \leq 2$:
$R = 1$
$r = \sqrt{y-1}$

For $0 \leq y \leq 1$:
$R = 1$
$r = 0$

Axis of
revolution

FIGURE 6.21

EXAMPLE 4   Two-integral case, integrating with respect to y

Find the volume of the solid formed by revolving the region bounded by the graphs of $y = x^2 + 1$, $y = 0$, $x = 0$, and $x = 1$ about the y-axis, as shown in Figure 6.21.

Solution: We use two integrals because we do not have a single formula to represent the inner radius r. When $0 \leq y \leq 1$, then $r = 0$, but when $1 \leq y \leq 2$, r is determined by the equation $y = x^2 + 1$, which implies that $r = \sqrt{y-1}$. Since $R = 1$,

$$V = \pi \int_0^1 (1^2 - 0^2)\, dy + \pi \int_1^2 [1^2 - (\sqrt{y-1})^2]\, dy$$

$$= \pi \int_0^1 1\, dy + \pi \int_1^2 (2 - y)\, dy$$

$$= \pi y \Big]_0^1 + \pi \left[2y - \frac{y^2}{2} \right]_1^2$$

$$= \pi \left(1 + 4 - 2 - 2 + \frac{1}{2} \right) = \frac{3}{2}\pi$$

Note that the first integral $\pi \int_0^1 1\, dy$ represents the volume of a right circular cylinder of radius 1 and height 1. This portion of the volume could have been determined without using calculus.

EXAMPLE 5   An application

A manufacturer drills a hole through the center of a metal sphere of radius 5 inches, as shown in Figure 6.22. The hole has a radius of 3 inches. What is the volume of the resulting metal ring?

Solution: We imagine the ring to be generated by a segment of the circle whose equation is $x^2 + y^2 = 25$, as shown in Figure 6.22. Since the radius of the hole is 3 inches, its intersection with the circle occurs at $x = \pm 4$. Thus, we have

$$r(x) = 3 \qquad \text{and} \qquad R(x) = \sqrt{25 - x^2}$$

and the volume is given by

$$V = \pi \int_a^b ([R(x)]^2 - [r(x)]^2)\, dx$$

$$= \pi \int_{-4}^4 [(\sqrt{25 - x^2})^2 - (3)^2]\, dx$$

$$= \pi \int_{-4}^4 (16 - x^2)\, dx$$

$$= \pi \left[16x - \frac{x^3}{3} \right]_{-4}^4$$

$$= \frac{256\pi}{3} \text{ in}^3$$

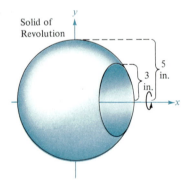

Solid of
Revolution

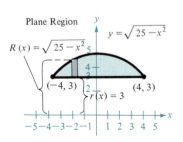

Plane Region

$y = \sqrt{25 - x^2}$

$R(x) = \sqrt{25 - x^2}$

$(-4, 3)$ $(4, 3)$

$r(x) = 3$

FIGURE 6.22

Solids with known cross sections

With the disc method, we can find the volume of a solid having a circular cross section whose area is $\Delta A = \pi R^2$. We can generalize this method to solids of any shape, so long as we know a formula for the area of an arbitrary cross section. Some common cross sections are squares, rectangles, triangles, semicircles, and trapezoids.

VOLUMES OF SOLIDS WITH KNOWN CROSS SECTIONS

1. For cross sections of area $A(x)$, taken perpendicular to the x-axis:

$$\text{volume} = \int_a^b A(x)\, dx$$

2. For cross sections of area $A(y)$, taken perpendicular to the y-axis:

$$\text{volume} = \int_c^d A(y)\, dy$$

See Figure 6.23.

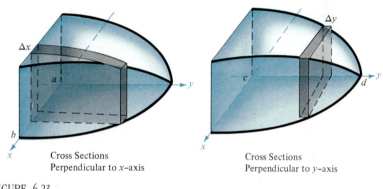

Cross Sections Perpendicular to x-axis

Cross Sections Perpendicular to y-axis

FIGURE 6.23

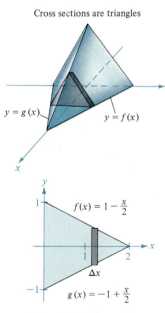

Cross sections are triangles

$y = g(x)$ $y = f(x)$

$f(x) = 1 - \dfrac{x}{2}$

Δx

$g(x) = -1 + \dfrac{x}{2}$

Triangular Base in xy-plane

FIGURE 6.24

EXAMPLE 6   Triangular cross sections

Find the volume of the solid whose base is the area bounded by the lines

$$f(x) = 1 - \frac{x}{2}, \qquad g(x) = -1 + \frac{x}{2}, \qquad \text{and} \qquad x = 0$$

and whose cross sections perpendicular to the x-axis are equilateral triangles, as shown in Figure 6.24.

Solution: Each triangular cross section has a base of

$$b = \left(1 - \frac{x}{2}\right) - \left(-1 + \frac{x}{2}\right) = 2 - x \qquad \text{Length of base}$$

and since the area of an equilateral triangle with base b is

$$\text{area} = \frac{1}{2}(\text{height})(\text{base}) = \frac{1}{2}\left(\frac{\sqrt{3}}{2}b\right)(b) = \frac{\sqrt{3}}{4}b^2$$

the area of each cross section is

$$A(x) = \frac{\sqrt{3}}{4}(2 - x)^2 \qquad \text{Area of cross section}$$

Since x ranges from 0 to 2, the volume of the solid is

$$V = \int_a^b A(x)\, dx = \int_0^2 \frac{\sqrt{3}}{4}(2 - x)^2\, dx$$

$$= \frac{-\sqrt{3}}{4}\left[\frac{(2 - x)^3}{3}\right]_0^2$$

$$= \frac{2\sqrt{3}}{3}$$

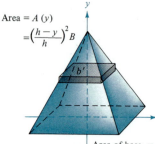

Area = $A(y)$

$= \left(\dfrac{h - y}{h}\right)^2 B$

Area of base $= B$

EXAMPLE 7   An application to geometry

Prove that the volume of a pyramid with a square base is $V = \frac{1}{3}hB$, where h is the height of the pyramid and B is the area of the base.

Solution: In Figure 6.25, we intersect the pyramid with a plane parallel to the base at height y to form a square cross section whose sides are of length b'. Using similar triangles, we can show that

$$\frac{b'}{b} = \frac{h - y}{h} \qquad \text{or} \qquad b' = \frac{b}{h}(h - y)$$

where b is the length of the sides of the base of the pyramid. Thus,

$$A(y) = (b')^2 = \frac{b^2}{h^2}(h - y)^2$$

Integrating between 0 and h, we have

$$\text{volume} = \int_0^h \frac{b^2}{h^2}(h - y)^2\, dy = \frac{b^2}{h^2}\int_0^h (h - y)^2\, dy$$

$$= -\left(\frac{b^2}{h^2}\right)\frac{(h - y)^3}{3}\bigg]_0^h = \frac{b^2}{h^2}\left(\frac{h^3}{3}\right) = \frac{b^2 h}{3} = \frac{1}{3}hB$$

FIGURE 6.25

An interesting generalization of the formula in Example 7 is that the volume of *any cone* can be determined by the formula $V = \frac{1}{3}hB$, where B is the area of the base and h is the height of the cone, as shown in Figure 6.26.

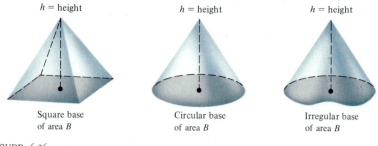

h = height   h = height   h = height

Square base   Circular base   Irregular base
of area B of area B of area B

FIGURE 6.26

Section Exercises 6.2

In Exercises 1–8, find the volume of the solid formed by revolving the given region about the x-axis.

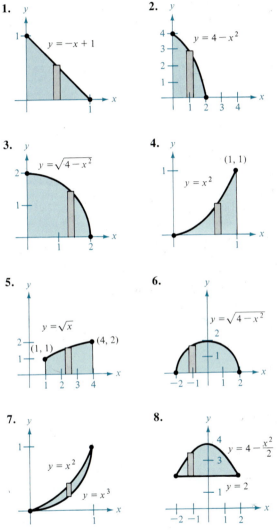

1. $y = -x + 1$

2. $y = 4 - x^2$

3. $y = \sqrt{4 - x^2}$

4. $y = x^2$ (1, 1)

5. $y = \sqrt{x}$ (1, 1), (4, 2)

6. $y = \sqrt{4 - x^2}$

7. $y = x^2$, $y = x^3$

8. $y = 4 - \dfrac{x^2}{2}$, $y = 2$

In Exercises 9–12, find the volume of the solid formed by revolving the given region about the y-axis.

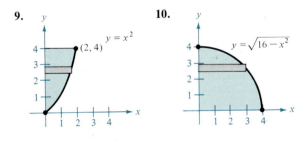

9. $y = x^2$ (2, 4)

10. $y = \sqrt{16 - x^2}$

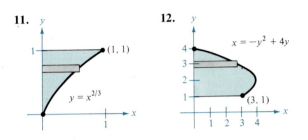

11. $y = x^{2/3}$ (1, 1)

12. $x = -y^2 + 4y$ (3, 1)

In Exercises 13–16, find the volume of the solid generated by revolving the region bounded by the graphs of the given equations about the indicated axis.

13. $y = \sqrt{x}$, $y = 0$, $x = 4$
    (a) the x-axis   (b) the y-axis
    (c) the line $x = 4$ (d) the line $x = 6$

14. $y = 2x^2$, $y = 0$, $x = 2$
    (a) the y-axis   (b) the x-axis
    (c) the line $y = 8$ (d) the line $x = 2$

15. $y = x^2$, $y = 4x - x^2$
    (a) the x-axis   (b) the line $y = 6$

16. $y = 6 - 2x - x^2$, $y = x + 6$
    (a) the x-axis   (b) the line $y = 3$

In Exercises 17–20, find the volume of the solid generated by revolving the region bounded by the graphs of the given equations about the x-axis.

17. $y = x\sqrt{4 - x^2}$, $y = 0$

18. $y = 2x^{2/3} - x$, $y = 0$

19. $y = \dfrac{1}{x}$, $y = 0$, $x = 1$, $x = 4$

20. $y = \dfrac{3}{x + 1}$, $y = 0$, $x = 0$, $x = 8$

In Exercises 21–24, find the volume of the solid generated by revolving the region bounded by the graphs of the given equations about the line $y = 4$.

21. $y = x$, $y = 3$, $x = 0$

22. $y = x^2$, $y = 4$

23. $y = \dfrac{1}{x^2}$, $y = 0$, $x = 1$, $x = 4$

24. $y = \sqrt{x}$, $y = 0$, $0 \le x \le 4$

In Exercises 25–28, find the volume of the solid generated by revolving the region bounded by the graphs of the given equations about the line $x = 6$.

25. $y = x$, $y = 0$, $y = 4$, $x = 6$

26. $y = 6 - x$, $y = 0$, $y = 4$, $x = 0$

27. $x = y^2$, $x = 4$

28. $xy^2 = 12$, $y = 2$, $y = 6$, $x = 0$

29. The region bounded by the parabola $y = 4x - x^2$ and the x-axis is revolved about the x-axis. Find the volume of the resulting solid.

30. If the equation of the parabola in Exercise 29 were changed to $y = 4 - x^2$, would the volume of the solid generated be different? Why or why not?

31. The upper half of the ellipse $9x^2 + 25y^2 = 225$ is revolved about the x-axis to form a prolate spheroid (shaped like a football). Find the volume of the spheroid.

32. The right half of the ellipse $9x^2 + 25y^2 = 225$ is revolved about the y-axis to form an oblate spheroid (shaped like an M&M candy). Find the volume of the spheroid.

33. If the portion of the line $y = \frac{1}{2}x$ lying in the first quadrant is revolved about the x-axis, a cone is generated. Find the volume of the cone extending from $x = 0$ to $x = 6$.

34. Use the disc method to verify that the volume of a right circular cone is $\frac{1}{3}\pi r^2 h$, where r is the radius of the base and h is the height.

35. Use the disc method to verify that the volume of a sphere of radius r is $\frac{4}{3}\pi r^3$.

36. A sphere of radius r is cut by a plane h ($h < r$) units above the equator. Find the volume of the solid (spherical segment) above the plane.

37. A cone with a base of radius r and height H is cut by a plane parallel to and h units above the base. Find the volume of the solid (frustrum of a cone) below the plane.

38. The region bounded by $y = \sqrt{x}$, $y = 0$, $x = 0$, and $x = 4$ is revolved about the x-axis. Find the value of x in the interval $[0, 4]$ that divides the solid into two parts of equal volume.

39. Find the values of x in the interval $[0, 4]$ that divide the solid in Exercise 38 into three parts of equal volume.

40. A tank on a water tower is a sphere of radius 50 feet. Determine the depth of the water when the tank is filled to 21.6 percent of its total capacity.

41. A tank on the wing of a jet is formed by revolving the region bounded by the graph of $y = \frac{1}{8}x^2\sqrt{2 - x}$ and the x-axis about the x-axis, as shown in Figure 6.27, where x and y are measured in meters. Find the volume of the tank.

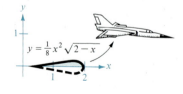

$y = \frac{1}{8}x^2\sqrt{2 - x}$

FIGURE 6.27

42. Match each of the following integrals with the solid whose volume it represents, and give the dimensions of each solid.
    (a) right circular cylinder   (b) ellipsoid
    (c) sphere   (d) right circular cone
    (e) torus

____ $\pi \int_0^h \left(\dfrac{rx}{h}\right)^2 dx$

____ $\pi \int_{-r}^{r} (\sqrt{r^2 - x^2})^2 \, dx$

____ $\pi \int_0^h r^2 \, dx$

____ $\pi \int_{-r}^{r} [(R + \sqrt{r^2 - x^2})^2 - (R - \sqrt{r^2 - x^2})^2] \, dx$

____ $\pi \int_{-b}^{b} \left(a\sqrt{1 - \dfrac{x^2}{b^2}}\right)^2 dx$

43. Find the volume of the solid whose base is bounded by the circle $x^2 + y^2 = 4$, with the indicated cross sections taken perpendicular to the x-axis.
    (a) squares   (b) equilateral triangles

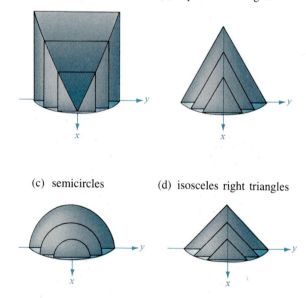

    (c) semicircles   (d) isosceles right triangles

44. Find the volume of the solid whose base is bounded by the graphs of $y = x + 1$ and $y = x^2 - 1$, with the indicated cross sections taken perpendicular to the x-axis.
    (a) squares   (b) rectangles of height 1
    (c) semiellipses of   (d) isosceles right
        height 2 ($A = \pi ab$)   triangles

45. The base of a solid is bounded by $y = x^3$, $y = 0$, and $x = 1$. Find the volume of the solid if the cross sections perpendicular to the y-axis are
    (a) squares   (b) semicircles

(c) equilateral triangles

(d) trapezoids for which $h = b_1 = \frac{1}{2}b_2$, where b_1 and b_2 are the lengths of the upper and lower bases

(e) semiellipses whose heights are twice the lengths of their bases

46. Find the volume of the solid of intersection (the solid common to both) of the two right circular cylinders of radius r whose axes meet at right angles, as shown in Figure 6.28.

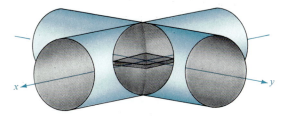

FIGURE 6.28

47. A wedge is cut from a right circular cylinder of radius r inches by a plane through the diameter of the base, making an angle of 45° with the plane of the base, as shown in Figure 6.29. Find the volume of the wedge.

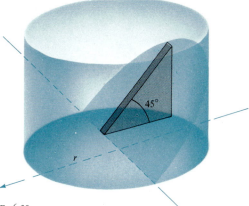

FIGURE 6.29

6.3
Volume: The shell method

In this section, we look at an alternative method for finding the volume of a solid of revolution, a method that uses cylindrical shells. Both the disc method and the shell method are important and we will compare the advantages of one method over the other later in this section.

To introduce the **shell method,** we consider a representative rectangle as shown in Figure 6.30, where

w = width of the rectangle

h = height of the rectangle

p = distance between axis of revolution and *center* of the rectangle

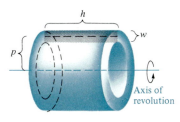

FIGURE 6.30

When this rectangle is revolved about its axis of revolution, it forms a cylindrical shell (or tube) of thickness w. To find the volume of this shell we consider two cylinders. The radius of the larger cylinder corresponds to the outer radius of the shell and the radius of the smaller cylinder corresponds to the inner radius of the shell. Since p is the average radius of the shell, we know the outer radius is $p + (w/2)$ and the inner radius is $p - (w/2)$. Thus, the volume of the shell is given by the difference:

volume of shell = (volume of cylinder) − (volume of hole)

$$= \pi\left(p + \frac{w}{2}\right)^2 h - \pi\left(p - \frac{w}{2}\right)^2 h$$

$$= \pi\left(p^2 + pw + \frac{w^2}{4} - p^2 + pw - \frac{w^2}{4}\right)h$$

$$= \pi 2phw = 2\pi(\text{average radius})(\text{height})(\text{thickness})$$

We use this formula to find the volume of a solid of revolution as follows. Assume that the plane region in Figure 6.31 is revolved about a line to form the indicated solid. If we consider a horizontal rectangle of width Δy, then as the plane region is revolved about its axis of revolution, the rectangle generates a representative shell whose volume is

$$\Delta V = 2\pi[p(y)h(y)] \, \Delta y$$

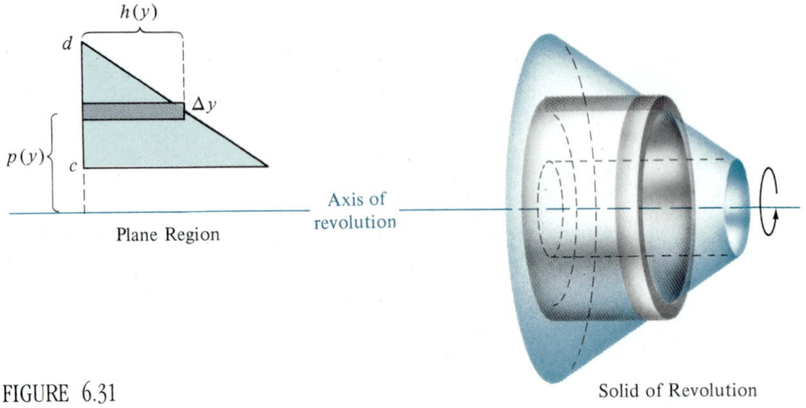

FIGURE 6.31

Plane Region

Axis of revolution

Solid of Revolution

Now, if we approximate the volume of the solid by n such shells of thickness Δy, height $h(y_i)$, and average radius $p(y_i)$, we have

$$\text{volume of solid} \approx \sum_{i=1}^{n} 2\pi[p(y_i)h(y_i)] \, \Delta y$$

$$= 2\pi \sum_{i=1}^{n} [p(y_i)h(y_i)] \, \Delta y$$

By taking the limit as $\|\Delta\| \to 0$ $(n \to \infty)$, we have

$$\text{volume of solid} = \lim_{n \to \infty} 2\pi \sum_{i=1}^{n} [p(y_i)h(y_i)] \, \Delta y$$

$$= 2\pi \int_{c}^{d} [p(y)h(y)] \, dy$$

THE SHELL METHOD

To find the volume of a solid of revolution with the **shell method,** use one of the following, as shown in Figure 6.32.

Horizontal axis of revolution

$$\text{volume} = V = 2\pi \int_{c}^{d} p(y)h(y) \, dy$$

Vertical axis of revolution

$$\text{volume} = V = 2\pi \int_{a}^{b} p(x)h(x) \, dx$$

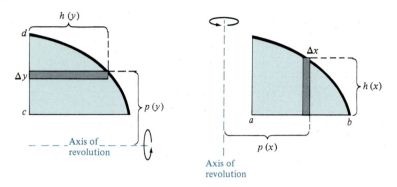

FIGURE 6.32

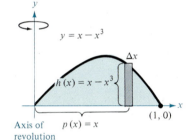

FIGURE 6.33

EXAMPLE 1   *Using the shell method to find volume*

Find the volume of the solid of revolution formed by revolving the region bounded by $y = x - x^3$ and the x-axis ($0 \le x \le 1$) about the y-axis.

Solution: Since the axis of revolution is vertical, we use a vertical representative rectangle as shown in Figure 6.33. The width Δx indicates that x is the variable of integration. The distance from the center of the rectangle to the axis of revolution is $p(x) = x$ and the height of the rectangle is $h(x) = x - x^3$. Since x ranges from 0 to 1, the volume of the solid is

$$V = 2\pi \int_a^b p(x)h(x)\, dx = 2\pi \int_0^1 x(x - x^3)\, dx = 2\pi \int_0^1 (-x^4 + x^2)\, dx$$

$$= 2\pi \left[-\frac{x^5}{5} + \frac{x^3}{3} \right]_0^1 = \frac{4\pi}{15}$$

EXAMPLE 2   *Using the shell method to find volume*

Find the volume of the solid of revolution formed by revolving the region bounded by

$$y = \frac{1}{(x^2 + 1)^2}$$

and the x-axis ($0 \le x \le 1$) about the y-axis.

Solution: Since the axis of revolution is vertical, we use a vertical representative rectangle, as shown in Figure 6.34. The width Δx indicates that x is the variable of integration. The distance from the center of the rectangle to the axis of revolution is $p(x) = x$, and the height of the rectangle is $h(x) = 1/(x^2 + 1)^2$. Since x ranges from 0 to 1, the volume of the solid is

$$V = 2\pi \int_a^b p(x)h(x)\, dx = 2\pi \int_0^1 \frac{x}{(x^2 + 1)^2}\, dx$$

$$= \left[-\frac{\pi}{x^2 + 1} \right]_0^1 = \frac{\pi}{2}$$

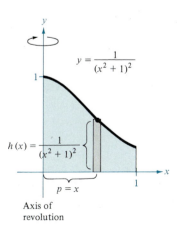

FIGURE 6.34

Remark   The disc and shell methods can be distinguished as follows. For the disc method the representative rectangle is always *perpendicular* to the axis of revolution, whereas for the shell method the representative rectangle is always *parallel* to the axis of revolution, as shown in Figure 6.35.

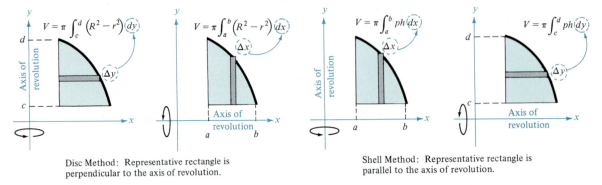

Disc Method: Representative rectangle is perpendicular to the axis of revolution.

Shell Method: Representative rectangle is parallel to the axis of revolution.

FIGURE 6.35

Often, one method is more convenient to use than the other. The following example illustrates a case in which the shell method is preferable.

EXAMPLE 3   *Shell method preferable*

Find the volume of the solid formed by revolving the region bounded by the graphs of $y = x^2 + 1$, $y = 0$, $x = 0$, and $x = 1$ about the y-axis.

Solution: To begin, look back at Example 4 in Section 6.2. There, in Figure 6.21, you can see that with the disc method two integrals are needed to find the volume.

$$V = \pi \int_0^1 (1^2 - 0^2)\, dy + \pi \int_1^2 [1^2 - (\sqrt{y - 1})^2]\, dy \qquad \text{Disc method}$$

$$= \pi \int_0^1 1\, dy + \pi \int_1^2 (2 - y)\, dy$$

$$= \pi y \Big]_0^1 + \pi \Big[2y - \frac{y^2}{2} \Big]_1^2$$

$$= \pi \Big(1 + 4 - 2 - 2 + \frac{1}{2} \Big) = \frac{3}{2}\pi$$

From Figure 6.36 we can see that the shell method requires only one integral to find the volume.

$$V = 2\pi \int_a^b p(x)h(x)\, dx \qquad \text{Shell method}$$

$$= 2\pi \int_0^1 x(x^2 + 1)\, dx$$

$$= 2\pi \Big[\frac{x^4}{4} + \frac{x^2}{2} \Big]_0^1$$

$$= 2\pi \Big(\frac{3}{4} \Big) = \frac{3\pi}{2}$$

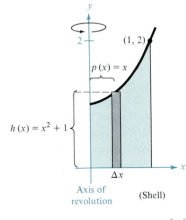

FIGURE 6.36

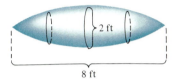

FIGURE 6.37

EXAMPLE 4   Disc method preferable

A pontoon is to be made in the shape shown in Figure 6.37. The pontoon is designed by rotating the graph of

$$y = 1 - \frac{x^2}{16}, \quad -4 \le x \le 4$$

about the x-axis, where x and y are measured in feet. Find the volume of the pontoon.

Solution: In Figure 6.38 we compare the disc and shell methods.

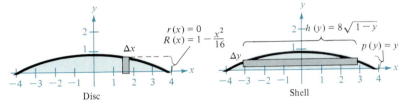

FIGURE 6.38

Using the shell method, we see that Δy is the width of the rectangle, $p(y) = y$, and $h(y) = 2x$. Solving for x in the equation $y = 1 - (x^2/16)$, we obtain

$$x = 4\sqrt{1-y} \qquad \text{or} \qquad h(y) = 2x = 8\sqrt{1-y}$$

Hence, the shell method results in the integral

$$V = 2\pi \int_0^1 8y\sqrt{1-y}\; dy \qquad \text{Shell method}$$

Although this is not a particularly difficult integral, its evaluation does require a *u*-substitution. By contrast, the disc method yields the relatively simple integral

$$V = \pi \int_{-4}^4 \left(1 - \frac{x^2}{16}\right)^2 dx = \pi \int_{-4}^4 \left(1 - \frac{x^2}{8} + \frac{x^4}{256}\right) dx \qquad \text{Disc method}$$

$$= \pi \left[x - \frac{x^3}{24} + \frac{x^5}{1280}\right]_{-4}^4$$

$$= \frac{64\pi}{15} \approx 13.4 \text{ ft}^3$$

Note that to use the shell method in Example 4 we had to solve for x in terms of y in the equation $y = 1 - (x^2/16)$. Sometimes, as in Example 1 in this section, solving for x is very difficult (or even impossible). In such cases we must use a vertical rectangle (of width Δx), thus making x the variable of integration. The position (horizontal or vertical) of the axis of revolution then determines the method to be used. This is illustrated in Example 5.

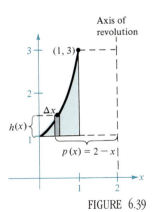

FIGURE 6.39

EXAMPLE 5   Shell method necessary

Find the volume of the solid formed by revolving the region bounded by the graphs of $y = x^3 + x + 1$, $y = 1$, and $x = 1$ about the line $x = 2$, as shown in Figure 6.39.

Solution: In the equation $y = x^3 + x + 1$, we cannot easily solve for x in terms of y. (See the discussion at the end of Section 4.9.) Therefore, the variable of integration must be x, and we choose a vertical representative rectangle. Now, since the rectangle is parallel to the axis of revolution, we use the shell method, and the volume is

$$V = 2\pi \int_a^b p(x)h(x)\, dx = 2\pi \int_0^1 (2-x)(x^3+x+1-1)\, dx$$

$$= 2\pi \int_0^1 (-x^4 + 2x^3 - x^2 + 2x)\, dx$$

$$= 2\pi \left[-\frac{x^5}{5} + \frac{x^4}{2} - \frac{x^3}{3} + x^2 \right]_0^1$$

$$= 2\pi \left(-\frac{1}{5} + \frac{1}{2} - \frac{1}{3} + 1 \right) = \frac{29\pi}{15}$$

Section Exercises 6.3

In Exercises 1–20, use the shell method to find the volume of the solid generated by revolving the given plane region about the indicated line.

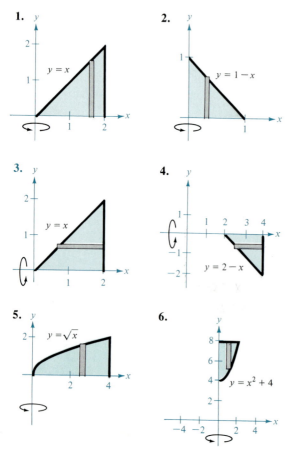

1. $y = x$

2. $y = 1 - x$

3. $y = x$

4. $y = 2 - x$

5. $y = \sqrt{x}$

6. $y = x^2 + 4$

7. $y = x^2$, $y = 0$, $x = 2$, about the y-axis
8. $y = x^2$, $y = 0$, $x = 4$, about the y-axis
9. $y = x^2$, $y = 4x - x^2$, about the y-axis
10. $y = x^2$, $y = 4x - x^2$, about the line $x = 2$
11. $y = x^2$, $y = 4x - x^2$, about the line $x = 4$
12. $y = \dfrac{1}{x}$, $x = 1$, $x = 2$, $y = 0$, about the x-axis
13. $y = 4x - x^2$, $y = 0$, about the line $x = 5$
14. $x + y^2 = 9$, $x = 0$, about the x-axis
15. $y = 4x - x^2$, $x = 0$, $y = 4$, about the y-axis
16. $y = 4 - x^2$, $y = 0$, about the y-axis
17. $y = \sqrt{x}$, $y = 0$, $x = 4$, about the line $x = 6$
18. $y = 2x$, $y = 4$, $x = 0$, about the y-axis
19. $x = y - y^2$, $x = 0$, about the line $y = 2$
20. $x = y - y^2$, $x = 0$, about the line $y = -4$

21. Use the disc or shell method to find the volume of the solid generated by revolving the region bounded by the graphs of $y = x^3$, $y = 0$, and $x = 2$ about
(a) the x-axis   (b) the y-axis
(c) the line $x = 4$ (d) the line $y = 8$

22. Find the volume of the solid generated by revolving the region bounded by the graphs of $y = 1/x^2$, $y = 0$, $x = 1$, and $x = 4$ about
(a) the x-axis   (b) the y-axis
(c) the line $x = 4$

23. Find the volume of the solid generated by revolving the plane region bounded by $x^{1/2} + y^{1/2} = a^{1/2}$, $x = 0$, and $y = 0$ about
(a) the x-axis   (b) the y-axis
(c) the line $x = a$

24. Find the volume of the solid generated by revolving the region bounded by $x^{2/3} + y^{2/3} = a^{2/3}$ (hypocycloid of four cusps) about the x-axis.

25. A solid is generated by revolving the region bounded by $y = \frac{1}{2}x^2$ and $y = 2$ about the y-axis. A hole, centered along the axis of revolution, is drilled through this solid so that one-quarter of the volume is removed. Find the diameter of the hole.

26. A solid is generated by revolving the region bounded by $y = \sqrt{9 - x^2}$ and $y = 0$ about the y-axis. A hole, centered along the axis of revolution, is drilled through this solid so that one-third of the volume is removed. Find the diameter of the hole.

27. Let a sphere of radius r be cut by a plane, thus forming a segment of height h. Show that the volume of this segment is $\frac{1}{3}\pi h^2(3r - h)$.

28. Match each of the following integrals with the solid whose volume it represents and give the dimensions of each solid.

_____ $2\pi \int_0^r hx \, dx$ (a) right circular cone

_____ $2\pi \int_0^r hx\left(1 - \dfrac{x}{r}\right) dx$ (b) torus

_____ $2\pi \int_0^r 2x\sqrt{r^2 - x^2} \, dx$ (c) sphere

_____ $2\pi \int_0^b 2ax\sqrt{1 - \dfrac{x^2}{b^2}} \, dx$ (d) right circular cylinder

_____ $2\pi \int_{-r}^r (R - x)(2\sqrt{r^2 - x^2}) \, dx$ (e) ellipsoid

In Exercises 29 and 30, use Simpson's Rule to approximate the volume of the solid accurate to one decimal place. The solid is generated by revolving the region bounded by the graphs of the given equations about the y-axis.

29. $y = \sqrt{1 - x^3}$, $y = 0$, $x = 0$ **30.** $x^{4/3} + y^{4/3} = 1$

6.4
Work

SECTION TOPICS ▪
Work done by a constant force ▪
Work done by a variable force ▪

The concept of work is important to scientists and engineers in determining the energy needed to perform various physical tasks. For instance, it is useful to know the amount of work done when a crane lifts a steel girder, when a spring is compressed, when a rocket is propelled into the air, or when a truck pulls a load along a highway.

In general, we say that work is done when a force moves an object. If the force applied to the object is *constant*, we have the following definition of work.

DEFINITION OF WORK DONE BY A CONSTANT FORCE   If an object is moved a distance D in the direction of an applied constant force F, then the **work** W done by the force is defined as $W = FD$.

There are many types of forces—centrifugal, electromotive, and gravitational, to name a few. Generally, we measure forces according to their ability to accelerate a unit of mass over time. For gravitational forces on the earth, we commonly use units of measure corresponding to the weight of an object, as illustrated in the following example.

EXAMPLE 1   *Work done by a constant force*

Determine the work done in lifting a 150-pound object a height of 4 feet.

Solution: The magnitude of the required force F is the weight of the object, as shown in Figure 6.40. Thus, the work done in lifting the object 4 feet is

$$\text{work} = W = FD = 150(4) = 600 \text{ ft} \cdot \text{lb}$$

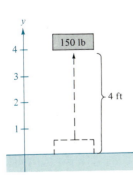

FIGURE 6.40

Remark   In the U.S. system of measurement, work is typically expressed in foot-pounds (ft · lb), inch-pounds, or foot-tons. In the Standard International system, the basic unit of force is the **dyne**—the force required to produce an acceleration of one centimeter per second per second in a gram mass. In this system work is typically expressed in dyne-centimeters (erg) or newton-meters (joule), where one joule $= 10^7$ ergs.

In Example 1, the force involved is *constant*. If a *variable* force is applied to an object, then calculus is needed to determine the work done, because the amount of force changes as the object changes positions. For instance, the force required to compress a spring increases as the spring is compressed.

Suppose that an object is moved along a straight line from $x = a$ to $x = b$ by a continuously varying force $F(x)$. Let Δ be a partition that divides the interval $[a, b]$ into n subintervals determined by

$$a = x_0 < x_1 < x_2 < \cdots < x_n = b$$

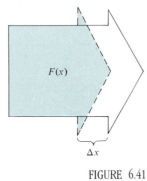

$F(x)$

Δx

FIGURE 6.41

and let $\Delta x_i = x_i - x_{i-1}$. For each i, choose c_i such that $x_{i-1} \leq c_i \leq x_i$. Then at c_i the force is given by $F(c_i)$. Since F is continuous, we conclude, assuming that Δx_i is small, that the work done in moving the object through the *i*th subinterval is approximated by the increment $\Delta W_i = F(c_i)\,\Delta x_i$, as shown in Figure 6.41. By adding the work done in each subinterval, we can approximate the total work done as the object moves from a to b by

$$W \approx \sum_{i=1}^{n} \Delta W_i = \sum_{i=1}^{n} F(c_i)\,\Delta x_i$$

Taking the limit of this sum as $\|\Delta\| \to 0$, $(n \to \infty)$, we have

$$W = \lim_{n \to \infty} \sum_{i=1}^{n} F(c_i)\,\Delta x_i = \int_a^b F(x)\,dx$$

We therefore define work done by a variable force as follows.

DEFINITION OF WORK DONE BY A VARIABLE FORCE

If an object is moved along a straight line by a continuously varying force $F(x)$, then the **work** W done by the force as the object is moved from $x = a$ to $x = b$ is given by

$$W = \lim_{n \to \infty} \sum_{i=1}^{n} \Delta W_i = \int_a^b F(x)\,dx$$

In the remaining examples in this section, we will make use of some well-known physical laws. The discovery of many of these laws occurred during the same period of history in which calculus was being developed. In fact, during the seventeenth and eighteenth centuries, there was little difference between physicists and mathematicians. One such physicist-mathematician was Emilie de Breteuil (1706–1749). Breteuil was instrumental in syn-

Emilie de Breteuil

thesizing the work of many other scientists including Newton, Leibniz, Huygens, Kepler, and Descartes. Her physics text *Institutions* was widely used for many years.

The following three laws were developed by Robert Hooke (1635–1703), Isaac Newton (1642–1727), and Charles Coulomb (1736–1806). Although Hooke and Coulomb are known primarily for their work in physics, they were also gifted mathematicians.

1. **Hooke's Law:** The force F required to compress or stretch a spring (within its elastic limits) is proportional to the distance d that the spring is compressed or stretched from its original length. That is,

$$F = kd$$

where the constant of proportionality k depends on the specific nature of the spring.

2. **Law of Universal Gravitation:** The force F of attraction between two particles of mass m_1 and m_2 is proportional to the product of the masses and inversely proportional to the square of the distance d between the two particles. That is,

$$F = k\frac{m_1 m_2}{d^2}$$

(If m_1 and m_2 are given in grams and d in centimeters, then F will be in dynes for a value of $k = 6.670 \times 10^{-8}$.)

3. **Coulomb's Law:** The force between two charges q_1 and q_2 in a vacuum is proportional to the product of the charges and inversely proportional to the square of the distance d between the two charges. That is,

$$F = k\frac{q_1 q_2}{d^2}$$

(If q_1 and q_2 are given in electrostatic units and d in centimeters, then F will be in dynes for a value of $k = 1$.)

EXAMPLE 2   An application of Hooke's Law

A force of 750 pounds compresses a spring 3 inches from its natural length of 15 inches. Find the work done in compressing the spring an additional 3 inches.

Solution: Using Hooke's Law, the force $F(x)$ required to compress the spring x units (from its natural length) is given by $F(x) = kx$. Using the given data, we have

$$F(3) = 750 = (k)(3)$$

and thus $k = 250$ and $F(x) = 250x$, as shown in Figure 6.42. To find the increment of work, we assume that the force required to compress the spring over a small increment Δx is nearly constant. Thus, the increment of work is

$$\Delta W = (\text{force})(\text{distance increment}) = (250x)\,\Delta x$$

Natural length ($F = 0$)

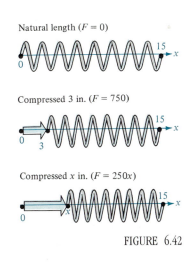

Compressed 3 in. ($F = 750$)

Compressed x in. ($F = 250x$)

FIGURE 6.42

Now, since the spring is compressed from $x = 3$ to $x = 6$ inches less than its natural length, the work required is

$$W = \int_3^6 250x \, dx = 125x^2 \Big]_3^6 = 4500 - 1125 = 3375 \text{ in} \cdot \text{lb}$$

Note that we do not integrate from $x = 0$ to $x = 6$ because we were asked to determine the work done in compressing the spring an *additional* 3 inches, not including the first 3 inches.

EXAMPLE 3   An application involving gravitational force

If a space module weighs 15 tons on the surface of the earth, how much work is done in propelling the module to a height of 800 miles above the earth, as shown in Figure 6.43? (Do not consider the effect of air resistance or the weight of the propellant.)

Solution: Since the weight of a body varies inversely as the square of its distance from the center of the earth, the force $F(x)$ exerted by gravity is

$$F(x) = \frac{k}{x^2}$$

Since the module weighs 15 tons on the surface of the earth and the radius of the earth is approximately 4000 miles, we obtain

$$15 = \frac{k}{(4000)^2} \quad \Longrightarrow \quad k = 240,000,000$$

Thus, the increment of work is

$$\Delta x = (\text{force})(\text{distance increment})$$
$$= \frac{240,000,000}{x^2} \, \Delta x$$

Finally, since the module is propelled from $x = 4000$ to $x = 4800$ miles, the total work done is

$$W = \int_{4000}^{4800} \frac{240,000,000}{x^2} \, dx = \frac{-240,000,000}{x} \Big]_{4000}^{4800}$$
$$= -50,000 + 60,000 = 10,000 \text{ mile-tons}$$
$$\approx 1.056 \times 10^{11} \text{ ft} \cdot \text{lb}$$

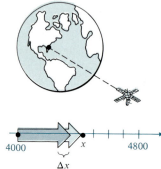

Space module is moved 800 miles above the earth.

FIGURE 6.43

The solutions to Examples 2 and 3 conform to our development of work as the summation of increments in the form

$$\Delta W = (\text{force})(\text{distance increment}) = F(\Delta x)$$

An equally useful way to formulate the increment of work is

$$\Delta W = (\text{force increment})(\text{distance}) = (\Delta F)x$$

This second interpretation of ΔW is useful in work problems involving the movement of nonrigid substances such as fluids or chains, as seen in the next two examples.

EXAMPLE 4   *Work required to move a liquid*

A spherical tank of radius 8 feet is half full of oil that weighs 50 pounds per cubic foot. Find the work required to pump the oil out through a hole in the top of the tank.

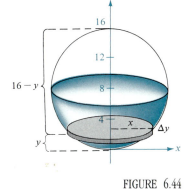

FIGURE 6.44

Solution: We consider the oil to be subdivided into discs of thickness Δy and radius x, as shown in Figure 6.44. Since the increment of force for each disc is given by its weight, we have

$$\Delta F = \text{weight} = \left(\frac{50 \text{ lb}}{\text{ft}^3}\right)(\text{volume}) = 50(\pi x^2 \, \Delta y)$$

Now, for a circle of radius 8 and center at $(0, 8)$, we have

$$x^2 + (y - 8)^2 = 8^2$$
$$x^2 = 16y - y^2$$

and we can write the force increment as

$$\Delta F = 50(\pi x^2 \, \Delta y) = 50\pi(16y - y^2) \, \Delta y$$

In Figure 6.44, we see that a disc y feet from the bottom of the tank must be moved a distance of $(16 - y)$ feet. Therefore, the increment of work is

$$\Delta W = \Delta F (16 - y) = 50\pi(16y - y^2) \, \Delta y \, (16 - y)$$
$$= 50\pi(256y - 32y^2 + y^3) \, \Delta y$$

Finally, since the tank is half full, y ranges from 0 to 8, and the work required to empty the tank is

$$W = \int_0^8 50\pi(256y - 32y^2 + y^3) \, dy = 50\pi\left[128y^2 - \frac{32}{3}y^3 + \frac{y^4}{4}\right]_0^8$$
$$= 50\pi\left(\frac{11264}{3}\right) \approx 589{,}782 \text{ ft} \cdot \text{lb}$$

EXAMPLE 5   *Work done in lifting a chain*

A 20-foot chain, weighing 5 pounds per foot, is lying coiled on the ground. How much work is required to raise one end of the chain to a height of 20 feet so that it is fully extended, as shown in Figure 6.45?

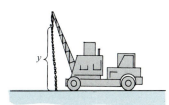

FIGURE 6.45

Solution: Imagine that the chain is divided into small sections, each of length Δy. Then the weight of each section is the increment of force

$$\Delta F = (\text{weight}) = \left(\frac{5 \text{ lb}}{\text{ft}}\right)(\text{length}) = 5 \, \Delta y$$

Since a typical section (initially on the ground) is raised to a height of y, we conclude that the increment of work is

$$\Delta W = (\text{force increment})(\text{distance}) = (5 \, \Delta y)y = 5y \, \Delta y$$

Finally, since y ranges from 0 to 20, the total work is

$$W = \int_0^{20} 5y \, dy = \frac{5y^2}{2}\bigg]_0^{20} = \frac{5(400)}{2} = 1000 \text{ ft} \cdot \text{lb}$$

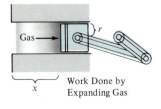

Work Done by
Expanding Gas

FIGURE 6.46

In the next example we consider a piston of radius r in a cylindrical casing, as shown in Figure 6.46. As the gas in the cylinder expands, the piston moves and work is done. If p represents the pressure of the gas (in pounds per square foot) against the piston head and V represents the volume of the gas (in cubic feet), then the work increment involved in moving the piston Δx feet is

$$\Delta W = (\text{force})(\text{distance increment}) = F(\Delta x) = p(\pi r^2)\,\Delta x = p\,\Delta V$$

Thus, as the volume of the gas expands from V_0 to V_1, the work done in moving the piston is

$$W = \int_{V_0}^{V_1} p\,dV$$

Assuming the pressure of the gas to be inversely proportional to its volume, we have $p = k/V$ and the integral for work becomes

$$W = \int_{V_0}^{V_1} \frac{k}{V}\,dV$$

EXAMPLE 6   *Work done by a gas*

A quantity of gas with an initial volume of 1 cubic foot and pressure of 500 pounds per square foot expands to a volume of 2 cubic feet. Find the work done by the gas. (Assume the pressure is inversely proportional to the volume.)

Solution: Since $p = k/V$ and $p = 500$ when $V = 1$, we have $k = 500$. Thus, the work is

$$W = \int_{V_0}^{V_1} \frac{k}{V}\,dV = \int_1^2 \frac{500}{V}\,dV$$

Using Simpson's Rule (see Example 4, Section 5.7), we can approximate the work to be

$$W \approx 346.6 \text{ ft} \cdot \text{lb}$$

Section Exercises 6.4

1. Determine the work done in lifting a 100-pound bag of sugar 10 feet.
2. Determine the work done by a hoist in lifting a 2400-pound car 6 feet.
3. A force of 25 pounds is required to slide a cement block on a plank in a construction project. The plank is 12 feet long. Determine the work done in sliding the block along the length of the plank.
4. The locomotive of a freight train pulls its cars with a constant force of 9 tons while traveling at a constant rate of 55 miles per hour on a level track. How many foot-pounds of work does the locomotive do in a distance of one-half mile?
5. A force of 5 pounds compresses a 15-inch spring a total of 4 inches. How much work is done in compressing the spring a total of 7 inches?
6. How much work is done in compressing the spring in Exercise 5 from a length of 10 inches to a length of 6 inches?
7. A force of 60 pounds stretches a spring 1 foot. How much work is done in stretching the spring from 9 inches to 15 inches?

8. A force of 200 pounds stretches a spring 2 feet on a mechanical device for driving fence posts. Find the work done by the tractor in stretching the spring the required 2 feet.

9. A force of 15 pounds stretches a spring 6 inches in an exercise machine. Find the work done in stretching the spring 1 foot from its natural position.

10. An overhead garage door has two springs, one on each side of the door. A force of 15 pounds is required to stretch each spring 1 foot. Because of the pulley system, the springs only stretch one-half the distance the door travels. Find the work done by the pair of springs if the door moves a total of 8 feet and the springs are at their natural length when the door is open.

11. A rectangular tank with a base 4 feet by 5 feet and a height of 4 feet is filled with water, as shown in Figure 6.47. (The water weighs 62.4 pounds per cubic foot.) How much work is done in pumping water out over the top edge if
(a) half of the tank is emptied?
(b) the tank is completely emptied?

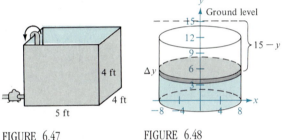

FIGURE 6.47   FIGURE 6.48

12. Repeat Exercise 11 for a tank filled with gasoline that weighs 42 pounds per cubic foot.

13. A cylindrical water tank 12 feet high with a radius of 8 feet is buried so that the top of the tank is 3 feet below ground level, as shown in Figure 6.48. How much work is done in pumping a full tank of water up to ground level?

14. Suppose the tank in Exercise 13 is located on a tower so that the bottom of the tank is 20 feet above the level of a stream as shown in Figure 6.49. How much work is

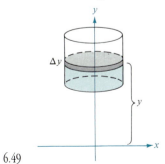

FIGURE 6.49

done in filling the tank half full of water through a hole in the bottom, using water from the stream?

15. A hemispherical tank of radius 6 feet is positioned so that its base is circular. How much work is required to fill the tank with water through a hole in the base if the water source is at the base?

16. Suppose the tank in Exercise 15 is inverted and the top 2 feet of water is pumped out through a hole in the top. How much work is done?

17. An open tank has the shape of a right circular cone as shown in Figure 6.50. The tank is 8 feet across the top and has a height of 6 feet. How much work is done in emptying the tank by pumping the water over the top edge?

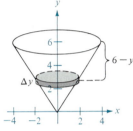

FIGURE 6.50

18. If water is pumped in through the bottom of the tank in Exercise 17, how much work is done to fill the tank
(a) to a depth of 2 feet?
(b) from a depth of 4 feet to a depth of 6 feet?

19. A cylindrical gasoline tank with a diameter of 3 feet and length of 4 feet is carried on the back of a truck and is used to fuel tractors in the field. The axis of the tank is horizontal. Find the work done to pump the entire contents of the full tank into a tractor if the opening on the tractor tank is 5 feet above the top of the tank in the truck. Assume gasoline weighs 42 pounds per cubic foot. (Hint: Evaluate one integral by a geometric formula and the other by observing that the integrand is an odd function.)

20. The top of a cylindrical storage tank for gasoline at a service station is 4 feet below ground level. The axis of the tank is horizontal and its diameter and length are 5 feet and 12 feet, respectively. Find the work done in pumping the entire contents of the full tank to a height of 3 feet above ground level. (Hint: Evaluate one integral by a geometric formula and the other by observing that the integrand is an odd function.)

21. Neglecting air resistance and the weight of the propellant, determine the work done in propelling a 4-ton satellite to a height of (a) 200 miles and (b) 400 miles above the earth.

22. Neglecting air resistance, determine the work done in propelling a 10-ton satellite to a height of (a) 11,000 miles and (b) 22,000 miles above the earth.

23. If a lunar module weighs 12 tons on the surface of the earth, how much work is done in propelling the module from the surface of the moon to a height of 50 miles? Consider the radius of the moon to be 1100 miles and its force of gravity to be one-sixth that of the earth's.

24. Two electrons repel each other with a force that varies inversely as the square of the distance between them. If one electron is fixed at the point (2, 4), find the work done in moving a second electron from (−2, 4) to (1, 4).

In Exercises 25–28, consider a 15-foot chain hanging from a winch 15 feet above ground level. Find the work done by the winch in winding up the required amount of chain, if the chain weighs 3 pounds per foot.

25. Wind up the entire chain.
26. Wind up one-third of the chain.
27. Run the winch until the bottom of the chain is at the 10-foot level.
28. Wind up the entire chain with a 100-pound load attached.

In Exercises 29 and 30, consider a 15-foot hanging chain that weighs 3 pounds per foot. Find the work done in lifting the chain vertically to the specified position.

29. Take the bottom of the chain and raise it to the 15-foot level, leaving the chain doubled and still hanging vertically as shown in Figure 6.51.
30. Repeat Exercise 29 raising the bottom of the chain to the 12-foot level.

In Exercises 31 and 32, consider a demolition crane with a 500-pound ball suspended from a 40-foot cable that weighs 1 pound per foot.

31. Find the work required to wind up 15 feet of the apparatus.

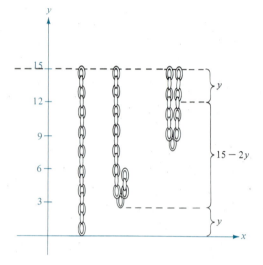

FIGURE 6.51

32. Find the work required to wind up all 40 feet of the apparatus.

In Exercise 33 and 34, find the work done by the gas for the given volumes and pressures. Assume the pressure is inversely proportional to the volume. (Use Example 6 as a model.)

33. A quantity of gas with an initial volume of 2 cubic feet and pressure of 1000 pounds per square foot expands to a volume of 3 cubic feet.

34. A quantity of gas with an initial volume of 1 cubic foot and pressure of 2000 pounds per square foot expands to a volume of 4 cubic feet.

SECTION TOPICS ▪
Fluid pressure ▪
Fluid force ▪

6.5
Fluid pressure and fluid force

Swimmers know that the deeper an object is submerged into a fluid, the greater the pressure on the object. We define **pressure** to be the force per unit of area over the surface of a body. For example, since a ten-foot column of water (one-inch square) weighs 4.3 pounds, we say that the *fluid* pressure at a depth of 10 feet of water is 4.3 pounds per square inch.* At 20 feet, this

*The total pressure on an object in ten feet of water would also include the pressure due to the earth's atmosphere. At sea level, this is approximately 14.7 pounds per square inch.

would increase to 8.6 pounds per square inch, and in general the pressure is proportional to the depth of the object in the fluid, as indicated in the following definition.

DEFINITION OF FLUID
PRESSURE
The **pressure** on an object at depth h in a liquid is given by

$$\text{pressure} = P = wh$$

where w is the weight of the liquid per unit of volume.

Blaise Pascal

Listed below are some commonly used weights of fluids.

Fluid	Pounds per cubic foot
Water	62.4
Ethyl alcohol	49.4
Gasoline	41.0–43.0
Kerosene	51.2
Mercury	849.0
Glycerin	78.6
Sea water	64.0

When calculating fluid pressure, we use an important (and rather surprising) physical law called **Pascal's Principle,** named after the French mathematician Blaise Pascal (1623–1662). Pascal is well known for his work in many areas of mathematics and physics and also for his influence on Leibniz. Although much of Pascal's work in calculus was intuitive and lacked the rigor of modern mathematics, he nevertheless anticipated many important results. Pascal's Principle states that the pressure exerted by a fluid at a depth h is *equal in all directions*. For example, in Figure 6.52, the pressure at the indicated depth is the same for all three objects.

Since fluid pressure is given in terms of force per unit area ($P = F/A$), the fluid force on a *submerged horizontal* surface of area A is given by

$$\text{fluid force} = F = PA = (\text{pressure})(\text{area})$$

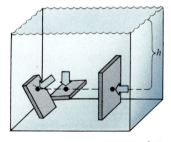

FIGURE 6.52

EXAMPLE 1   *Fluid force on a horizontal surface*

Find the fluid force on a rectangular metal sheet 3 feet by 4 feet submerged in 6 feet of water, as shown in Figure 6.53.

Solution: Since the weight of water is 62.4 pounds per cubic foot and the sheet is submerged in 6 feet of water, the fluid pressure is

$$P = (62.4)(6) = 374.4 \text{ lb/ft}^2$$

Now, since the total area of the sheet is $A = (3)(4) = 12$ square feet, the fluid force is given by

$$F = PA = (374.4)(12) = 4492.8 \text{ lb}$$

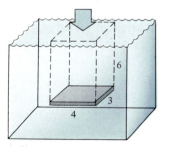

FIGURE 6.53

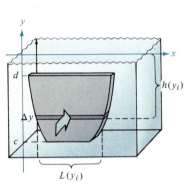

$h(y_i)$

$L(y_i)$

FIGURE 6.54

The answer to Example 1 is independent of the size of the body of water. The fluid force would be the same in a swimming pool or a lake.

In Example 1, the fact that the sheet is rectangular and horizontal means that we do not need the methods of calculus to solve the problem. We now look at a surface that is submerged vertically in a fluid. This problem is more difficult because the pressure is not constant over the surface.

Suppose a vertical plane region is submerged in a fluid of weight w (per unit of volume), as shown in Figure 6.54. We wish to determine the total force against one side of this region from depth c to depth d. First, we subdivide the interval $[c, d]$ into n subintervals, each of width Δy. Now consider the representative rectangle of width Δy and length $L(y_i)$, where y_i is in the ith subinterval. The force against this representative rectangle is

$$\Delta F_i = w(\text{depth})(\text{area}) = wh(y_i)L(y_i)\ \Delta y$$

The force against n such rectangles is

$$\sum_{i=1}^{n} \Delta F_i = w \sum_{i=1}^{n} h(y_i)L(y_i)\ \Delta y$$

Note that w is considered to be constant and is factored out of the summation. Therefore, taking the limit as $\|\Delta\| \to 0$ ($n \to \infty$), we find the total force against the region to be

$$F = w \lim_{n \to \infty} \sum_{i=1}^{n} h(y_i)L(y_i)\ \Delta y = w \int_{c}^{d} h(y)L(y)\ dy$$

DEFINITION OF FORCE
EXERTED BY A FLUID

The **force F exerted by a fluid** of constant weight w (per unit of volume) against a submerged vertical plane region from $y = c$ to $y = d$ is given by

$$F = w \int_{c}^{d} h(y)L(y)\ dy$$

where $h(y)$ is the depth of the fluid at y and $L(y)$ is the horizontal length of the region at y.

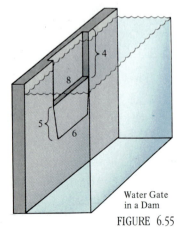

Water Gate
in a Dam

FIGURE 6.55

EXAMPLE 2   Fluid force on a vertical surface

A vertical gate in a dam has the shape of an isosceles trapezoid 8 feet across the top and 6 feet across the bottom, with a height of 5 feet, as shown in Figure 6.55. What is the fluid force against the gate if the top of the gate is 4 feet below the surface of the water?

Solution: In setting up a mathematical model for this problem, we are at liberty to locate the x- and y-axes in a number of different ways. A convenient approach is to let the y-axis bisect the gate and place the x-axis at the surface of the water, as shown in Figure 6.56. Thus, the depth of the water at y is

$$\text{depth} = h(y) = -y$$

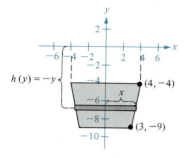

FIGURE 6.56

To find the length $L(y)$ of the region at y, we find the equation of the line forming the right side of gate. Since this line passes through the two points $(3, -9)$ and $(4, -4)$, its equation is

$$y - (-9) = \frac{-4 - (-9)}{4 - 3}(x - 3) = 5(x - 3)$$

$$y = 5x - 24 \implies x = \frac{y + 24}{5}$$

From Figure 6.56 we can see that the length of the region at y is given by

$$\text{length} = 2x = \frac{2}{5}(y + 24) = L(y)$$

Finally, by integrating from $y = -9$ to $y = -4$, we find the fluid force to be

$$F = w \int_c^d h(y)L(y) \, dy = 62.4 \int_{-9}^{-4} (-y)\left(\frac{2}{5}\right)(y + 24) \, dy$$

$$= -62.4\left(\frac{2}{5}\right) \int_{-9}^{-4} (y^2 + 24y) \, dy$$

$$= -62.4\left(\frac{2}{5}\right)\left[\frac{y^3}{3} + 12y^2\right]_{-9}^{-4}$$

$$= -62.4\left(\frac{2}{5}\right)\left(\frac{-1675}{3}\right) = 13{,}936 \text{ lb}$$

| **Remark** In Example 2, we let the x-axis coincide with the surface of the water. This was convenient, but arbitrary. In choosing a coordinate system to represent a physical situation, you ought to consider various possibilities. Often, you can simplify the calculations in a problem by locating the coordinate system so as to take advantage of special characteristics of the problem, such as symmetry.

EXAMPLE 3   Fluid force on a vertical surface

A circular observation window on a marine science ship has a radius of 1 foot, and the center of the window is 8 feet below water level, as shown in Figure 6.57. What is the fluid force on the window?

Solution: To take advantage of symmetry, we locate a coordinate system so that the origin coincides with the center of the window, as shown in Figure 6.58 on page 322. The depth at y is then given by

$$\text{depth} = h(y) = 8 - y$$

The horizontal length of the window is $2x$, and we can use the equation for the circle, $x^2 + y^2 = 1$, to solve for x as follows:

$$\text{length} = 2x = 2\sqrt{1 - y^2} = L(y)$$

Finally, since y ranges from -1 to 1, we have, using the weight of sea water as 64 pounds per cubic foot,

$$F = w \int_c^d h(y)L(y) \, dy = 64 \int_{-1}^1 (8 - y)(2)\sqrt{1 - y^2} \, dy$$

FIGURE 6.57

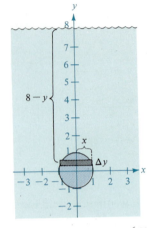

FIGURE 6.58

Initially it looks as if this integral would be difficult to solve. However, if we break the integral into two parts and apply symmetry, the solution is actually quite simple.

$$F = 64(16) \int_{-1}^{1} \sqrt{1 - y^2} \, dy - 64(2) \int_{-1}^{1} y\sqrt{1 - y^2} \, dy$$

The second integral is zero (since the integrand is odd and the limits of integration are symmetric to the origin). Moreover, recognizing that the first integral represents the area of a semicircle of radius 1, we have

$$F = 64(16)\left(\frac{\pi}{2}\right) - 64(2)(0) = 512\pi \approx 1608.5 \text{ lb}$$

EXAMPLE 4   Fluid force on a vertical surface

A swimming pool is 2 feet deep at one end and 10 feet deep at the other, as shown in Figure 6.59. The pool is 40 feet long and 30 feet wide with vertical sides. Find the fluid force against one of the 40-foot sides.

FIGURE 6.59

Solution: By placing the x- and y-axes as shown in Figure 6.60, we see that the depth at y is

$$\text{depth} = h(y) = 10 - y$$

The equation representing the base of the side is given by

$$y = mx + b = \frac{1}{5}x \implies x = 5y$$

However, we must note that the equation $x = 5y$ is only valid for $0 \le y \le 8$. When y is between 8 and 10, x is a constant 40. Thus, the length of the side of the pool is

$$\text{length} = L(y) = \begin{cases} 5y, & 0 \le y \le 8 \\ 40, & 8 \le y \le 10 \end{cases}$$

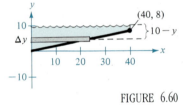

FIGURE 6.60

Therefore, the fluid force on the side of the pool is given by the two integrals

$$F = \int_{0}^{8} 62.4(10 - y)(5y) \, dy + \int_{8}^{10} 62.4(10 - y)(40) \, dy$$

$$= 312 \int_{0}^{8} (10y - y^2) \, dy + 2496 \int_{8}^{10} (10 - y) \, dy$$

$$= 312\left[5y^2 - \frac{y^3}{3}\right]_{0}^{8} + 2496\left[10y - \frac{y^2}{2}\right]_{8}^{10}$$

$$= 312\left(\frac{448}{3}\right) + 2496(2)$$

$$= 51{,}584 \text{ lb}$$

Remark In Example 4, note that the 30-foot width of the swimming pool is a "red herring" in the sense that the fluid force against the 40-foot sides is not dependent on the width of the pool.

Section Exercises 6.5

In Exercises 1–6, find the fluid force on the indicated vertical side of a tank that is full of water.

1. Rectangle

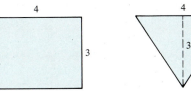

2. Triangle

3. Trapezoid

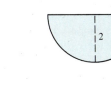

4. Semicircle

5. Parabola

$y = x^2$

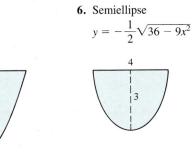

6. Semiellipse

$y = -\frac{1}{2}\sqrt{36 - 9x^2}$

In Exercises 7–10, find the fluid force on the given vertical plates submerged in water.

7. Square

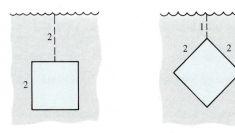

8. Diamond

9. Triangle

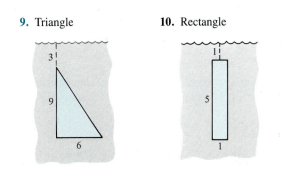

10. Rectangle

In Exercises 11–14, the given vertical plate is the side of a form for poured concrete that weighs 140.7 pounds per cubic foot. Determine the fluid force on the plate.

11. Rectangle

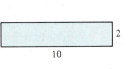

12. Rectangle

13. Semiellipse

14. Triangle

15. A cylindrical gasoline tank is placed so that the axis of the cylinder is horizontal. Find the fluid force on a circular end of the tank if the tank is half full, assuming that the diameter is 3 feet and the gasoline weighs 42 pounds per cubic foot.

16. Repeat Exercise 15 for a tank that is full. (Evaluate one integral by a geometric formula and the other by observing that the integrand is an odd function.)

17. A circular plate of radius r feet is submerged vertically in a tank of fluid that weighs w pounds per cubic foot. The center of the circle is k feet below the surface of the fluid. Show that the fluid force on the surface of the circle is given by

$$F = wk(\pi r^2)$$

(Evaluate one integral by a geometric formula and the other by observing that the integrand is an odd function.)

18. A rectangular plate of height h feet and base b feet is submerged vertically in a tank of fluid that weighs w pounds per cubic foot. The center is k feet below the surface of the fluid, where $h \leq k/2$. Show that the fluid force on the surface of the rectangle is given by

$$F = wkhb$$

19. A porthole on a vertical side of a submarine (in sea water) is 1 foot square. Find the fluid force on the port-

hole, assuming that the center of the square is 15 feet below the surface.

20. Repeat Exercise 19 for a circular porthole that has a diameter of 1 foot. The center is 15 feet below the surface.

21. A swimming pool is 20 feet wide, 40 feet long, 4 feet deep at one end, and 8 feet deep at the other. The bottom is an inclined plane. Find the fluid force on each of the vertical walls.

6.6
Moments, centers of mass, and centroids

In this section we look at several important applications of integration that are related to the concept of **mass.** Mass is considered an *absolute* measure of the amount of matter in a body, since it is independent of the particular gravitational system in which the body is located. However, because so many applications involving mass occur on the earth's surface, we tend to equate an object's mass with its *weight*. This is not technically correct. Weight is a type of force, and as such is dependent on gravity. Force and mass are related by the equation

$$\text{force} = (\text{mass})(\text{acceleration})$$

In Table 6.1 we list some commonly used (and misused) measures of mass and force, together with their conversion factors.

TABLE 6.1

System of Measurement	Measure of Mass	Measure of Force
U.S.	slug	pound = (slug) (ft/sec^2)
International	kilogram	newton = (kilogram) (m/sec^2)
C-G-S	gram	dyne = (gram) (cm/sec^2)
Conversions: 1 pound = 4.448 newtons 1 newton = 0.2248 pound 1 dyne = 0.02248 pound		1 slug = 14.59 kilograms 1 kilogram = 0.06854 slug 1 gram = 0.00006854 slug

EXAMPLE 1   An example of mass on the surface of the earth

Find the mass (in slugs) of an object whose weight at sea level is one pound.

Solution: Using 32 feet per second per second as the acceleration due to gravity, we have

$$\text{mass} = \frac{\text{force}}{\text{acceleration}} = \frac{1 \text{ lb}}{32 \text{ ft/sec}^2} \approx 0.03125 \frac{\text{lb}}{\text{ft/sec}^2} = 0.03125 \text{ slug}$$

Because many applications occur on the earth's surface, this amount of mass is called a **pound mass.**

The moment of a mass

We shall consider two types of moments of a mass—the **moment about a point** and the **moment about a line.** To define these two moments, we consider an idealized situation in which a mass m is concentrated at a point. If x is the distance between this point-mass and another point P, then the **moment of m about the point P** is given by

$$\text{moment} = mx$$

where x is the **length of the moment arm.**

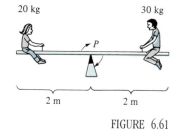

20 kg   30 kg

P

2 m   2 m

FIGURE 6.61

The concept of moment can be simply demonstrated by a seesaw, as shown in Figure 6.61. Suppose a child of mass 20 kilograms sits 2 meters to the left of fulcrum P, and an older child of mass 30 kilograms sits 2 meters to the right of P. From experience, we know that the seesaw would begin to rotate clockwise, moving the larger child down. This rotation occurs because the moment produced by the child on the left is less than the moment produced by the child on the right:

$$\text{left moment} = (20)(2) = 40 \text{ kg} \cdot \text{m}$$
$$\text{right moment} = (30)(2) = 60 \text{ kg} \cdot \text{m}$$

For the seesaw to balance, the two moments must be equal. For example, if the larger child moved to a position $\tfrac{4}{3}$ meters from the fulcrum, then the seesaw would balance, since both children would produce a moment of 40 kilogram-meters.

If $x_1 m_1 + x_2 m_2 + \ldots + x_n m_n = 0$, then the system is in equilibrium.

FIGURE 6.62

To generalize this situation we introduce a coordinate line on which the origin corresponds to the fulcrum, as shown in Figure 6.62. Suppose several point masses are located on the x-axis. The measure of the tendency of this system to rotate about the origin is called the **moment about the origin,** and it is defined to be the sum of the n products $m_i x_i$:

$$M_0 = m_1 x_1 + m_2 x_2 + \cdots + m_n x_n$$

If M_0 is zero, then the system is said to be in **equilibrium.**

For a system that is not in equilibrium, we define the **center of mass** to be the point $\bar{x}$ at which the fulcrum could be relocated to attain equilibrium. If the system were translated $\bar{x}$ units, each coordinate x_i would become $(x_i - \bar{x})$, and since the moment of the translated system is zero, we have

$$\sum_{i=1}^{n} m_i(x_i - \bar{x}) = \sum_{i=1}^{n} m_i x_i - \sum_{i=1}^{n} m_i \bar{x} = 0$$

Solving for $\bar{x}$, we have

$$\bar{x} = \frac{\displaystyle\sum_{i=1}^{n} m_i x_i}{\displaystyle\sum_{i=1}^{n} m_i} = \frac{\text{moment of system about origin}}{\text{total mass of system}}$$

DEFINITION OF THE MOMENT
AND CENTER OF MASS
OF A LINEAR SYSTEM

Given the point masses $m_1, m_2, \ldots, m_n$ located at $x_1, x_2, \ldots, x_n$,

1. the **moment about the origin** is $M_0 = m_1 x_1 + m_2 x_2 + \cdots + m_n x_n$

2. the **center of mass** is $\bar{x} = \dfrac{M_0}{m}$

where $m = m_1 + m_2 + \cdots + m_n$ is the **total mass** of the system.

EXAMPLE 2   *The center of mass of a linear system*

Find the center of mass of the following linear system:

$$m_1 = 10, \quad m_2 = 15, \quad m_3 = 5, \quad m_4 = 10$$
$$x_1 = -5, \quad x_2 = 0, \quad x_3 = 4, \quad x_4 = 7$$

Solution: The moment about the origin is given by

$$M_0 = m_1 x_1 + m_2 x_2 + m_3 x_3 + m_4 x_4$$
$$= 10(-5) + 15(0) + 5(4) + 10(7) = -50 + 0 + 20 + 70 = 40$$

Since the total mass of the system is $m = 10 + 15 + 5 + 10 = 40$, the center of mass is

$$\bar{x} = \frac{M_0}{m} = \frac{40}{40} = 1$$

Rather than defining the moment of a *mass,* we could define the moment of a *force*. In this context, the center of mass is called the **center of gravity.** Suppose that a system of point masses $m_1, m_2, \ldots, m_n$ is located at $x_1, x_2, \ldots, x_n$. Then, since force = (mass)(acceleration), the total force of the system is

$$F = m_1 a + m_2 a + \cdots + m_n a = ma$$

The **torque** (moment) about the origin is given by

$$T_0 = (m_1 a)x_1 + (m_2 a)x_2 + \cdots + (m_n a)x_n = M_0 a$$

and the **center of gravity** is

$$\frac{T_0}{F} = \frac{M_0 a}{ma} = \frac{M_0}{m} = \bar{x}$$

Therefore, the center of gravity and the center of mass have the same location.

We can extend the concept of moment to two dimensions by considering a system of masses located in the *xy*-plane at the points (x_1, y_1), (x_2, y_2), $\ldots, (x_n, y_n)$ as shown in Figure 6.63. Rather than defining a single moment (with respect to the origin), we define two moments—one with respect to the *x*-axis and one with respect to the *y*-axis.

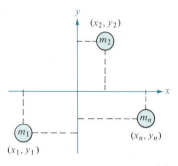

FIGURE 6.63

DEFINITION OF THE MOMENTS AND CENTER OF MASS OF A TWO-DIMENSIONAL SYSTEM

Given the point masses $m_1, m_2, \ldots, m_n$ located at (x_1, y_1), (x_2, y_2), $\ldots$, and (x_n, y_n),

1. the **moment about the y-axis** is $M_y = m_1x_1 + m_2x_2 + \cdots + m_nx_n$
2. the **moment about the x-axis** is $M_x = m_1y_1 + m_2y_2 + \cdots + m_ny_n$
3. the **center of mass** $(\bar{x}, \bar{y})$ (or **center of gravity**) is given by

$$\bar{x} = \frac{M_y}{m} \quad \text{and} \quad \bar{y} = \frac{M_x}{m}$$

where $m = m_1 + m_2 + \cdots + m_n$ is the **total mass** of the system.

Remark The moment of a system of masses in the plane can be calculated about any horizontal line or vertical line. In such cases, the moment about the line is the sum of the product of the masses and the *directed distances* from the points to the line. For the horizontal line $y = b$,

$$\text{moment} = m_1(y_1 - b) + m_2(y_2 - b) + \cdots + m_n(y_n - b)$$

For the vertical line $x = a$,

$$\text{moment} = m_1(x_1 - a) + m_2(x_2 - a) + \cdots + m_n(x_n - a)$$

EXAMPLE 3   The center of mass of a two-dimensional system

Find the center of mass of a system of point masses $m_1 = 6$, $m_2 = 3$, $m_3 = 2$, and $m_4 = 9$, located at $(3, -2)$, $(0, 0)$, $(-5, 3)$, and $(4, 2)$, as shown in Figure 6.64.

Solution:

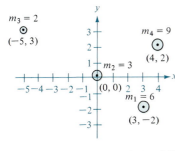

$$
\begin{array}{llllll}
m = 6 & +\ 3 & +\ 2 & +\ 9 & = 20 & \text{Mass} \\
M_y = 6(3) & +\ 3(0) & +\ 2(-5) & +\ 9(4) & = 44 & \text{Moment about } y\text{-axis} \\
M_x = 6(-2) & +\ 3(0) & +\ 2(3) & +\ 9(2) & = 12 & \text{Moment about } x\text{-axis}
\end{array}
$$

FIGURE 6.64

Therefore,

$$\bar{x} = \frac{M_y}{m} = \frac{44}{20} = \frac{11}{5} \quad \text{and} \quad \bar{y} = \frac{M_x}{m} = \frac{12}{20} = \frac{3}{5}$$

and we conclude that the center of mass is $(\frac{11}{5}, \frac{3}{5})$.

The center of mass of a planar lamina

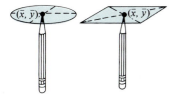

FIGURE 6.65

In the preceding discussion we assumed the total mass of a system to be distributed at discrete points in the plane (or on a line). We now consider a thin flat plate of material of uniform density called a **planar lamina**. Intuitively, we think of the center of mass $(\bar{x}, \bar{y})$ of a lamina as its balancing point. For example, the center of mass of a circular lamina is located at the center of the circle, and the center of mass of a rectangular lamina is located at the center of the rectangle, as shown in Figure 6.65.

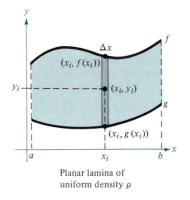

Planar lamina of
uniform density ρ

FIGURE 6.66

| **Remark**   **Density** is a measure of mass per unit of volume, such as grams per cubic centimeter. We denote density by ρ, the lowercase Greek letter rho.

Consider the irregularly shaped planar lamina of density ρ, bounded by the graphs of $y = f(x)$, $y = g(x)$, and $a \le x \le b$, as shown in Figure 6.66. The mass of this region is given by

$$m = (\text{density})(\text{area}) = \rho \int_a^b [f(x) - g(x)] \, dx = \rho A$$

where A is the area of the region. To find the center of mass of this lamina, we partition the interval $[a, b]$ into n equal subintervals of width Δx. If x_i is the center of the ith subinterval, then we can approximate the portion of the lamina lying in the ith subinterval by a rectangle whose height is $h = f(x_i) - g(x_i)$. Since the density of the rectangle is ρ, we know that its mass is

$$m_i = (\text{density})(\text{area}) = \underbrace{\rho}_{\text{density}}\,\underbrace{[f(x_i) - g(x_i)]}_{\text{height}}\,\underbrace{\Delta x}_{\text{width}}$$

Now, considering this mass to be located at the center (x_i, y_i) of the rectangle, we know that the directed distance from the x-axis to (x_i, y_i) is $y_i = [f(x_i) + g(x_i)]/2$. Thus, the moment of m_i about the x-axis is

$$\text{moment} = (\text{mass})(\text{distance}) = m_i y_i = \rho[f(x_i) - g(x_i)]\,\Delta x \left[\frac{f(x_i) + g(x_i)}{2}\right]$$

Summing these moments and taking the limit as $n \to \infty$, we have the moment about the x-axis defined as

$$M_x = \rho \int_a^b \left[\frac{f(x) + g(x)}{2}\right][f(x) - g(x)] \, dx$$

For the moment about the y-axis, the directed distance from the y-axis to (x_i, y_i) is x_i and we have

$$M_y = \rho \int_a^b x[f(x) - g(x)] \, dx$$

DEFINITION OF MOMENTS AND CENTER OF MASS OF A PLANAR LAMINA

Let f and g be continuous functions such that $f(x) \ge g(x)$ on $[a, b]$. Given the planar lamina of uniform density ρ bounded by the graphs of $y = f(x)$, $y = g(x)$, $a \le x \le b$,

1. the **moments about the x- and y-axes** are

$$M_x = \rho \int_a^b \left[\frac{f(x) + g(x)}{2}\right][f(x) - g(x)] \, dx$$
$$M_y = \rho \int_a^b x[f(x) - g(x)] \, dx$$

2. the **center of mass** $(\bar{x}, \bar{y})$ is given by $\bar{x} = \dfrac{M_y}{m}$ and $\bar{y} = \dfrac{M_x}{m}$

where $m = \rho \int_a^b [f(x) - g(x)] \, dx$ is the mass of the lamina.

| Remark   Note that the integrals for both moments can be formed by inserting into the integral for mass the directed distance from the axis (or line) about which the moment is taken to the center of the representative rectangle.

EXAMPLE 4   The center of mass of a planar lamina

Find the center of mass of the lamina of uniform density ρ bounded by the graph of $f(x) = 4 - x^2$ and the x-axis.

Solution: First, since the center of mass lies on the axis of symmetry, we know that $\bar{x} = 0$. Moreover, the mass of the lamina is given by

$$m = \rho \int_{-2}^{2} (4 - x^2)\, dx$$

$$= \rho \left[4x - \frac{x^3}{3} \right]_{-2}^{2}$$

$$= \frac{32\rho}{3}$$

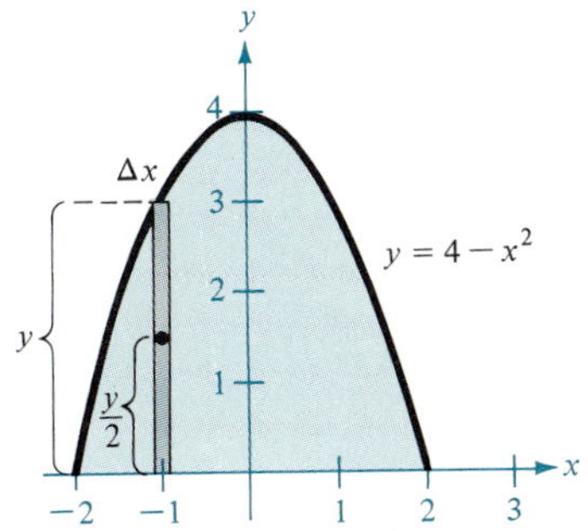

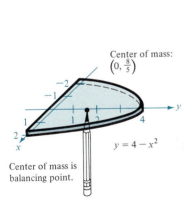

FIGURE 6.67

To find the moment about the x-axis, we place a representative rectangle in the region, as shown in Figure 6.67. The distance from the x-axis to the center of this rectangle is

$$y_i = \frac{f(x)}{2} = \frac{4 - x^2}{2}$$

Since the mass of the representative rectangle is

$$\rho f(x)\, \Delta x = \rho (4 - x^2)\, \Delta x$$

we have

$$M_x = \rho \int_{-2}^{2} \frac{4 - x^2}{2} (4 - x^2)\, dx$$

$$= \frac{\rho}{2} \int_{-2}^{2} (16 - 8x^2 + x^4)\, dx$$

$$= \frac{\rho}{2} \left[16x - \frac{8x^3}{3} + \frac{x^5}{5} \right]_{-2}^{2} = \frac{256\rho}{15}$$

and $\bar{y}$ is given by

$$\bar{y} = \frac{M_x}{m} = \frac{256\rho/15}{32\rho/3} = \frac{8}{5}$$

Thus, the center of mass (the balancing point) of the lamina is $(0, \frac{8}{5})$, as shown in Figure 6.68.

Center of mass:
$\left(0, \frac{8}{5} \right)$

$y = 4 - x^2$

Center of mass is balancing point.

FIGURE 6.68

The density ρ in Example 4 is a common factor of both the moments and the mass, and as such cancels out of the quotients representing the coordinates of the center of mass. In other words, the center of mass of a lamina of *uniform* density depends only on the shape of the lamina and not on its density. From this observation, we use the center of mass of a lamina of uniform density to define the **centroid** of the plane region.

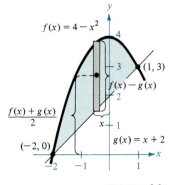

FIGURE 6.69

EXAMPLE 5 The centroid of a plane region

Find the centroid of the region bounded by the graphs of $f(x) = 4 - x^2$ and $g(x) = x + 2$.

Solution: The two graphs intersect at the points $(-2, 0)$ and $(1, 3)$, as shown in Figure 6.69. Thus, the area of the region is given by

$$A = \int_{-2}^{1} [f(x) - g(x)]\, dx = \int_{-2}^{1} (2 - x - x^2)\, dx = \frac{9}{2}$$

The centroid $(\bar{x}, \bar{y})$ of the region has coordinates

$$\begin{aligned}
\bar{x} &= \frac{1}{A}\int_{-2}^{1} x[(4 - x^2) - (x + 2)]\, dx = \frac{2}{9}\int_{-2}^{1} (-x^3 - x^2 + 2x)\, dx \\
&= \frac{2}{9}\left[-\frac{x^4}{4} - \frac{x^3}{3} + x^2\right]_{-2}^{1} = -\frac{1}{2} \\
\bar{y} &= \frac{1}{A}\int_{-2}^{1} \left[\frac{(4 - x^2) + (x + 2)}{2}\right][(4 - x^2) - (x + 2)]\, dx \\
&= \frac{2}{9}\left(\frac{1}{2}\right)\int_{-2}^{1} (-x^2 + x + 6)(-x^2 - x + 2)\, dx \\
&= \frac{1}{9}\int_{-2}^{1} (x^4 - 9x^2 - 4x + 12)\, dx \\
&= \frac{1}{9}\left[\frac{x^5}{5} - 3x^3 - 2x^2 + 12x\right]_{-2}^{1} = \frac{12}{5}
\end{aligned}$$

Thus, the centroid of the region is $(\bar{x}, \bar{y}) = (-\frac{1}{2}, \frac{12}{5})$.

For simple plane regions you may be able to find the centroid without resorting to integration. Example 6 presents such a case.

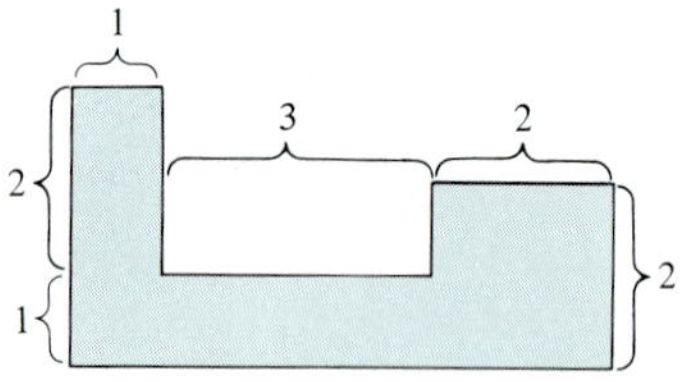

FIGURE 6.70

EXAMPLE 6 The centroid of a simple plane region

Find the centroid of the region shown in Figure 6.70.

Solution: By superimposing a coordinate system on the region, as indicated in Figure 6.71, we locate the centroids of the three rectangles as

$$\left(\frac{1}{2}, \frac{3}{2}\right), \quad \left(\frac{5}{2}, \frac{1}{2}\right), \quad \text{and} \quad (5, 1)$$

FIGURE 6.71

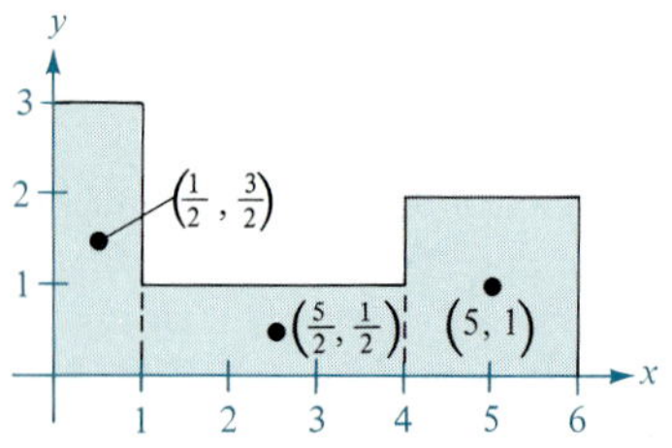

Now, we can calculate the centroid as follows:

$$\begin{aligned}
A &= \text{area of region} = 3 + 3 + 4 = 10 \\
\bar{x} &= \frac{(1/2)(3) + (5/2)(3) + (5)(4)}{10} = \frac{29}{10} = 2.9 \\
\bar{y} &= \frac{(3/2)(3) + (1/2)(3) + (1)(4)}{10} = \frac{10}{10} = 1
\end{aligned}$$

Thus, the centroid of the region is (2.9, 1).

As a final topic in this section, we list a useful theorem that is credited to Pappus of Alexandria (c. 300 A.D.), a Greek mathematician whose *Mathematical Collection* contains a record of much of classical Greek mathematics. We state the theorem without proof.

THEOREM 6.2   **THE THEOREM OF PAPPUS**
Let R be a plane region, and let V be the volume of the solid of revolution formed by revolving R about an axis. If the axis of revolution does not intersect the interior of R, then the volume is given by

$$V = 2\pi r A$$

where A is the area of R and r is the distance between the centroid of R and the axis of revolution.

EXAMPLE 7   Finding volume by the Theorem of Pappus

Find the volume of the torus formed by revolving the circular region bounded by $(x - 2)^2 + y^2 = 1$ about the y-axis, as shown in Figure 6.72.

Solution: From Figure 6.73 we see that the centroid of the circular region is $(2, 0)$. Thus, the distance between the centroid and the axis of revolution is $r = 2$. Since the area of the circular region is $A = \pi$, the volume of the torus is given by

$$V = 2\pi r A = 2\pi(2)(\pi) = 4\pi^2 \approx 39.5$$

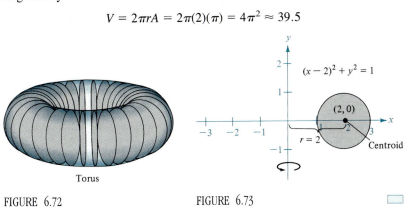

FIGURE 6.72 — Torus

FIGURE 6.73

Section Exercises 6.6

In Exercises 1–4, find the center of the given point masses lying on the x-axis.

1. $m_1 = 6$, $m_2 = 3$, $m_3 = 5$
  $x_1 = -5$, $x_2 = 1$, $x_3 = 3$

2. $m_1 = 7$, $m_2 = 4$, $m_3 = 3$, $m_4 = 8$
  $x_1 = -3$, $x_2 = -2$, $x_3 = 5$, $x_4 = 6$

3. $m_1 = 1$, $m_2 = 1$, $m_3 = 1$, $m_4 = 1$, $m_5 = 1$
  $x_1 = 7$, $x_2 = 8$, $x_3 = 12$, $x_4 = 15$, $x_5 = 18$

4. $m_1 = 12$, $m_2 = 1$, $m_3 = 6$, $m_4 = 3$, $m_5 = 11$
  $x_1 = -3$, $x_2 = -2$, $x_3 = -1$, $x_4 = 0$, $x_5 = 4$

5. Notice in Exercise 3 that $\bar{x}$ is the arithmetic mean of the x-coordinates. Translate each point mass to the right 5 units and compare the resulting center of mass with that obtained in Exercise 3.

6. Translate each point mass in Exercise 4 to the left 3 units and compare the resulting center of mass with that obtained in Exercise 4.

In Exercises 7–10, find the center of mass of the given system of point masses.

7.

m_i	5	1	3
(x_i, y_i)	(2, 2)	(−3, 1)	(1, −4)

8.

m_i	10	2	5
(x_i, y_i)	(1, −1)	(5, 5)	(−4, 0)

9.

m_i	3	4	2	1	6
(x_i, y_i)	(−2, −3)	(−1, 0)	(7, 1)	(0, 0)	(−3, 0)

10.

m_i	4	2	5/2	5
(x_i, y_i)	(2, 3)	(−1, 5)	(6, 8)	(2, −2)

In Exercises 11–14, introduce an appropriate coordinate system and find the coordinates of the center of mass of the given planar lamina.

11.   **12.**

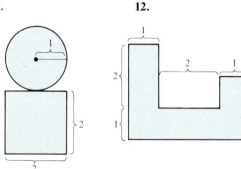

13.   **14.**

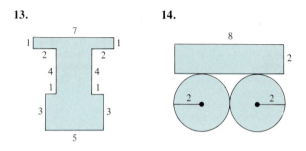

15. Suppose that the circular lamina in Exercise 11 has twice the density of the square lamina, and find the resulting center of mass.

16. Suppose that the square lamina in Exercise 11 has twice the density of the circular lamina, and find the resulting center of mass.

In Exercises 17–28, find M_x, M_y, and $(\bar{x}, \bar{y})$ for the laminas of uniform density ρ bounded by the graphs of the given equations.

17. $y = \sqrt{x}$, $y = 0$, $x = 4$
18. $y = x^2$, $y = 0$, $x = 4$
19. $y = -x^2 + 4x + 2$, $y = x + 2$
20. $y = x^2$, $y = x^3$
21. $y = \sqrt{x}$, $y = x$, $x \geq 0$
22. $y = \sqrt{3x + 1}$, $y = x + 1$
23. $x = 4 - y^2$, $x = 0$
24. $x = 2y - y^2$, $x = 0$
25. $x = -y$, $x = 2y - y^2$
26. $x = y + 2$, $x = y^2$

In Exercises 27–30, find the centroids of the regions bounded by the graphs of the given equations.

27. $y = \sqrt{1^2 - x^2}$, $y = 0$
28. $y = x^3$, $y = x$, $0 \leq x \leq 1$
29. $y = x^2$, $y = x$
30. $y = 2x + 4$, $y = 0$, $0 \leq x \leq 3$
31. Find the centroid of the triangular region with vertices $(-a, 0)$, $(a, 0)$, and (b, c), as shown in Figure 6.74. Show that it is the point of intersection of the medians of the triangle.

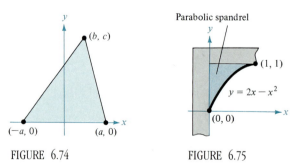

FIGURE 6.74   FIGURE 6.75

32. Find the centroid of the **parabolic spandrel** shown in Figure 6.75.

In Exercises 33–36, use the Theorem of Pappus to find the volume of the solid of revolution.

33. Torus formed by revolving the circle $(x - 5)^2 + y^2 = 16$ about the y-axis.

34. Torus formed by revolving the circle $x^2 + (y - 3)^2 = 4$ about the x-axis.

35. Solid formed by revolving the region bounded by the graphs of $y = x$, $y = 4$, and $x = 0$ about the x-axis.

36. Solid formed by revolving the region bounded by the graphs of $y = \sqrt{x - 1}$, $y = 0$, and $x = 5$ about the y-axis.

SECTION TOPICS •
Arc length •
Surface of revolution •
Area of surface of revolution •

Christian Huygens

6.7
Arc length and surfaces of revolution

In this section we use the summation character of the definite integral to find the arc length of a plane curve and the area of a surface of revolution. In both cases, we approximate an arc (a segment of a curve) by straight line segments whose lengths are given by the familiar Distance Formula

$$d = \sqrt{(x_2 - x_1)^2 + (y_2 - y_1)^2}$$

If a segment of a curve has a finite arc length, we say that it is **rectifiable.** The problem of calculating the length of a curve has motivated a wide range of mathematical work. Some early contributions to the problem were by the Dutch mathematician Christian Huygens (1629–1695), who invented the pendulum clock. Another pioneer in the work of rectifiable curves was James Gregory (1638–1675), a Scottish mathematician. Both of these men were important in the early development of calculus.

We will see in the development of the formula for arc length that a sufficient condition for the graph of a function f to be rectifiable between $(a, f(a))$ and $(b, f(b))$ is that f' is continuous on $[a, b]$. Such a function is said to be **smooth** or **continuously differentiable** on $[a, b]$.

Assume that a function given by $y = f(x)$ is continuously differentiable on the interval $[a, b]$, and let s denote the length of its graph on this interval. We approximate the graph of f by n line segments whose endpoints are determined by the partition

$$a = x_0 < x_1 < x_2 < \cdots < x_n = b$$

as shown in Figure 6.76.

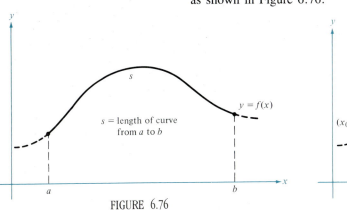

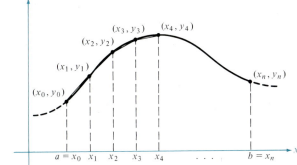

FIGURE 6.76

By letting $\Delta x_i = x_i - x_{i-1}$ and $\Delta y_i = y_i - y_{i-1}$, we approximate the total arc length by

$$s \approx \sum_{i=1}^{n} \sqrt{(\Delta x_i)^2 + (\Delta y_i)^2}$$

Taking the limit as $\|\Delta\| \to 0$, $(n \to \infty)$, we obtain

$$s = \lim_{n \to \infty} \sum_{i=1}^{n} \sqrt{(\Delta x_i)^2 + (\Delta y_i)^2}$$

$$= \lim_{n \to \infty} \sum_{i=1}^{n} \sqrt{1 + \left(\frac{\Delta y_i}{\Delta x_i}\right)^2} (\Delta x_i)$$

Since $f'(x)$ exists for each x in $(x_i\ x_{i-1})$, the Mean Value Theorem guarantees the existence of c_i in (x_i, x_{i-1}) such that

$$f(x_i) - f(x_{i-1}) = f'(c_i)(x_i - x_{i-1}) \implies \frac{\Delta y_i}{\Delta x_i} = f'(c_i)$$

Moreover, since f' is continuous on $[a, b]$, we know that $\sqrt{1 + [f'(x)]^2}$ is also continuous (and hence integrable) on $[a, b]$ and we have

$$s = \lim_{n \to \infty} \sum_{i=1}^{n} \sqrt{1 + [f'(c_i)]^2} (\Delta x_i)$$

$$= \int_{a}^{b} \sqrt{1 + [f'(x)]^2} \, dx$$

We call s the **arc length** of f between a and b.

DEFINITION OF ARC LENGTH

If the function given by $y = f(x)$ has a continuous derivative f' on the interval $[a, b]$, then the **arc length** of f between a and b is given by

$$s = \int_{a}^{b} \sqrt{1 + [f'(x)]^2} \, dx$$

Similarly, for a smooth curve given by $x = g(y)$, the **arc length** of g between c and d is given by

$$s = \int_{c}^{d} \sqrt{1 + [g'(y)]^2} \, dy$$

Remark Definite integrals representing arc length are often very difficult to evaluate. In this section we present a few examples. In the next chapter, with more advanced integration techniques, we will be able to tackle more difficult arc length problems.

Since the definition for arc length can be applied to a linear function, we should check to see that this new definition agrees with the standard distance formula for the length of a line segment. To do this, we let (x_1, y_1) and (x_2, y_2) be any two points on the graph of $f(x) = mx + b$, as shown in Figure 6.77.

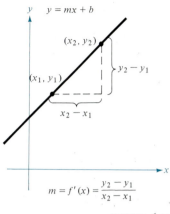

$y = mx + b$

(x_2, y_2)

(x_1, y_1)

$y_2 - y_1$

$x_2 - x_1$

$m = f'(x) = \dfrac{y_2 - y_1}{x_2 - x_1}$

FIGURE 6.77

Since

$$m = f'(x) = \frac{y_2 - y_1}{x_2 - x_1}$$

it follows that

$$s = \int_{x_1}^{x_2} \sqrt{1 + [f'(x)]^2}\, dx = \int_{x_1}^{x_2} \sqrt{1 + \left(\frac{y_2 - y_1}{x_2 - x_1}\right)^2}\, dx$$

$$= \left[\sqrt{\frac{(x_2 - x_1)^2 + (y_2 - y_1)^2}{(x_2 - x_1)^2}}\,(x) \right]_{x_1}^{x_2}$$

$$= \sqrt{\frac{(x_2 - x_1)^2 + (y_2 - y_1)^2}{(x_2 - x_1)^2}}\,(x_2 - x_1)$$

$$= \sqrt{(x_2 - x_1)^2 + (y_2 - y_1)^2}$$

which is the formula for the distance between two points in the plane.

EXAMPLE 1   *Finding arc length*

Find the arc length of the graph of

$$f(x) = \frac{x^3}{6} + \frac{1}{2x}$$

on the interval $[\tfrac{1}{2}, 2]$, as shown in Figure 6.78.

Solution: Since

$$f'(x) = \frac{3x^2}{6} - \frac{1}{2x^2} = \frac{1}{2}\left(x^2 - \frac{1}{x^2}\right)$$

the arc length is

$$s = \int_a^b \sqrt{1 + \left(\frac{dy}{dx}\right)^2}\, dx = \int_{1/2}^{2} \sqrt{1 + \left[\frac{1}{2}\left(x^2 - \frac{1}{x^2}\right)\right]^2}\, dx$$

$$= \int_{1/2}^{2} \sqrt{\frac{1}{4}\left(x^4 + 2 + \frac{1}{x^4}\right)}\, dx$$

$$= \int_{1/2}^{2} \frac{1}{2}\left(x^2 + \frac{1}{x^2}\right) dx$$

$$= \left[\frac{1}{2}\left(\frac{x^3}{3} - \frac{1}{x}\right)\right]_{1/2}^{2}$$

$$= \frac{1}{2}\left(\frac{13}{6} + \frac{47}{24}\right) = \frac{99}{48} = \frac{33}{16}$$

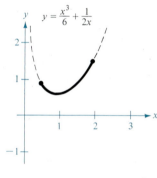

$y = \dfrac{x^3}{6} + \dfrac{1}{2x}$

FIGURE 6.78

EXAMPLE 2   *Finding arc length*

Find the arc length of the graph of $(y - 1)^3 = x^2$ on the interval $[0, 8]$, as shown in Figure 6.79.

Solution: We can solve for either x or y in the equation $(y - 1)^3 = x^2$. For this example we choose to solve for x, and we obtain

$$x = \pm(y - 1)^{3/2}$$

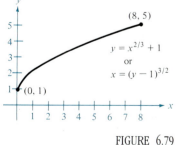

$(8, 5)$

$y = x^{2/3} + 1$

or

$x = (y - 1)^{3/2}$

$(0, 1)$

FIGURE 6.79

Choosing the positive value for x, we have

$$\frac{dx}{dy} = \frac{3}{2}(y-1)^{1/2}$$

Therefore, the arc length is

$$s = \int_c^d \sqrt{1 + \left(\frac{dx}{dy}\right)^2}\, dy = \int_1^5 \sqrt{1 + \left[\frac{3}{2}(y-1)^{1/2}\right]^2}\, dy$$

$$= \int_1^5 \sqrt{\frac{9}{4}y - \frac{5}{4}}\, dy$$

$$= \frac{1}{2}\int_1^5 \sqrt{9y - 5}\, dy$$

$$= \frac{1}{18}\left[\frac{(9y-5)^{3/2}}{3/2}\right]_1^5 = \frac{1}{27}(40^{3/2} - 4^{3/2})$$

$$= \frac{8}{27}(10^{3/2} - 1) \approx 9.0734$$

EXAMPLE 3   An application of arc length

Approximate the distance traveled by a projectile that travels along the path given by

$$y = x - 0.005x^2$$

where x and y are measured in feet, as shown in Figure 6.80.

Solution: Since

$$\frac{dy}{dx} = 1 - 0.01x = 1 - \frac{x}{100}$$

we can use the following arc length to find the distance traveled.

$$s = \int_a^b \sqrt{1 + \left(\frac{dy}{dx}\right)^2}\, dx = \int_0^{200} \sqrt{1 + \left(1 - \frac{x}{100}\right)^2}\, dx$$

$$= \int_0^{200} \sqrt{2 - \frac{x}{50} + \frac{x^2}{10,000}}\, dx$$

Using Simpson's Rule (with $n = 10$), we can approximate this distance to be 229.56 feet.

In Sections 6.2 and 6.3 we used integration to calculate the volume of a solid of revolution. We now look at a procedure for finding the area of a surface of revolution.

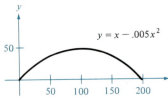

$y = x - .005x^2$

FIGURE 6.80

DEFINITION OF SURFACE OF REVOLUTION	If the graph of a continuous function is revolved about a line, the resulting surface is called a **surface of revolution**.

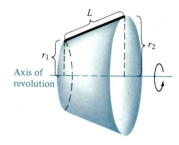

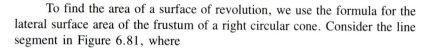

FIGURE 6.81

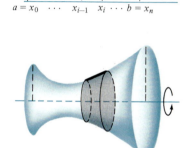

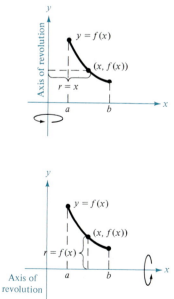

FIGURE 6.82

FIGURE 6.83

To find the area of a surface of revolution, we use the formula for the lateral surface area of the frustum of a right circular cone. Consider the line segment in Figure 6.81, where

$$L = \text{length of line segment}$$
$$r_1 = \text{radius at left end of line segment}$$
$$r_2 = \text{radius at right end of line segment}$$

When the line segment is revolved about its axis of revolution, it forms a frustum of a right circular cone, with

$$S = 2\pi r L \qquad \text{Lateral surface area of frustum}$$
$$r = \frac{1}{2}(r_1 + r_2) \qquad \text{Average radius of frustum}$$

Now, suppose the graph of a function f, having a continuous derivative on the interval $[a, b]$, is revolved about the x-axis to form a surface of revolution, as shown in Figure 6.82. Let Δ be a partition of $[a, b]$, with subintervals of width Δx_i. Then the line segment of length

$$\Delta L_i = \sqrt{\Delta x_i^2 + \Delta y_i^2}$$

generates a frustum of a cone.

By the Intermediate Value Theorem, there exists a point d_i such that $r_i = f(d_i)$ is the average radius of this frustum. Finally, the lateral surface area, ΔS_i, of the frustum is given by

$$\Delta S_i = 2\pi r_i \Delta L_i = 2\pi f(d_i)\sqrt{\Delta x_i^2 + \Delta y_i^2} = 2\pi f(d_i)\sqrt{1 + \left(\frac{\Delta y_i}{\Delta x_i}\right)^2}\, \Delta x_i$$

By the Mean Value Theorem, there exists a point c_i in (x_{i-1}, x_i) such that

$$f'(c_i) = \frac{f(x_i) - f(x_{i-1})}{x_i - x_{i-1}} = \frac{\Delta y_i}{\Delta x_i}$$

Therefore, $\Delta S_i = 2\pi f(d_i)\sqrt{1 + [f'(c_i)]^2}\, \Delta x_i$, and the total surface area can be approximated by

$$S \approx 2\pi \sum_{i=1}^{n} f(d_i)\sqrt{1 + [f'(c_i)]^2}\, \Delta x_i$$

Taking the limit as $\|\Delta\| \to 0$ $(n \to \infty)$, we have

$$S = 2\pi \int_a^b f(x)\sqrt{1 + [f'(x)]^2}\, dx$$

In a similar manner, it follows that if the graph of f is revolved about the y-axis, then S is given by

$$S = 2\pi \int_a^b x\sqrt{1 + [f'(x)]^2}\, dx$$

In both formulas for S, we can regard the products $2\pi f(x)$ and $2\pi x$ as the circumference of the circle traced by a point (x, y) on the graph of f as it is revolved about the x- or y-axis (Figure 6.83). In one case the radius is $r = f(x)$, and in the other case the radius is $r = x$. Moreover, by appropriately adjusting r, we can generalize this formula for surface area to cover *any* horizontal or vertical axis of revolution, as indicated in the following definition.

DEFINITION OF THE AREA OF A SURFACE OF REVOLUTION	If $y = f(x)$ has a continuous derivative on the interval $[a, b]$, then the area S of the surface of revolution formed by revolving the graph of f about a horizontal or vertical axis is $$S = 2\pi \int_a^b r(x)\sqrt{1 + [f'(x)]^2}\, dx$$ where $r(x)$ is the distance between the graph of f and the axis of revolution.

Remark If $x = g(y)$ on the interval $[c, d]$, then the surface area is

$$S = 2\pi \int_c^d r(y)\sqrt{1 + [g'(y)]^2}\, dy$$

where $r(y)$ is the distance between the graph of g and the axis of revolution.

EXAMPLE 4   The area of a surface of revolution

Find the area of the surface formed by revolving the graph of $f(x) = x^3$ on the interval $[0, 1]$ about the x-axis, as shown in Figure 6.84.

Solution: The distance between the x-axis and the graph of f is $r(x) = f(x)$, and since $f'(x) = 3x^2$, the surface area is given by

$$S = 2\pi \int_a^b r(x)\sqrt{1 + [f'(x)]^2}\, dx = 2\pi \int_0^1 x^3 \sqrt{1 + (3x^2)^2}\, dx$$

$$= \frac{2\pi}{36} \int_0^1 (36x^3)(1 + 9x^4)^{1/2}\, dx$$

$$= \frac{\pi}{18} \left[\frac{(1 + 9x^4)^{3/2}}{3/2} \right]_0^1$$

$$= \frac{\pi}{27}(10^{3/2} - 1) \approx 3.563$$

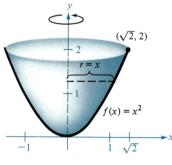

FIGURE 6.84

EXAMPLE 5   The area of a surface of revolution

Find the area of the surface formed by revolving the graph of $f(x) = x^2$ on the interval $[0, \sqrt{2}]$ about the y-axis, as shown in Figure 6.85.

Solution: In this case, the distance between the graph of f and the y-axis is $r(x) = x$, and since $f'(x) = 2x$, the surface area is

$$S = 2\pi \int_a^b r(x)\sqrt{1 + [f'(x)]^2}\, dx = 2\pi \int_0^{\sqrt{2}} x\sqrt{1 + (2x)^2}\, dx$$

$$= \frac{2\pi}{8} \int_0^{\sqrt{2}} (1 + 4x^2)^{1/2}(8x)\, dx$$

$$= \frac{2\pi}{8} \left[\frac{(1 + 4x^2)^{3/2}}{3/2} \right]_0^{\sqrt{2}}$$

$$= \frac{\pi}{6}[(1 + 8)^{3/2} - (1)^{3/2}] = \frac{13\pi}{3}$$

FIGURE 6.85

Section Exercises 6.7

In Exercises 1 and 2, find the distance between the given points by (a) using the Distance Formula and (b) determining the equation of the line through the points and using the formula for arc length.

1. (0, 0), (5, 12)   **2.** (1, 2), (7, 10)

In Exercises 3–8, find the arc length of the graph of the given function over the indicated interval.

Function	Interval
3. $y = \dfrac{2}{3}x^{3/2} + 1$	[0, 1]
4. $y = x^{3/2} - 1$	[0, 4]
5. $y = \dfrac{x^4}{8} + \dfrac{1}{4x^2}$	[1, 2]
6. $y = \dfrac{x^5}{10} + \dfrac{1}{6x^3}$	[1, 2]
7. $y = \dfrac{3}{2}x^{2/3}$	[1, 8]
8. $y = 2 - \dfrac{2}{3}x$	[0, 3]

In Exercises 9–18, find a definite integral that represents the arc length of the curves over the indicated interval. (Do not evaluate the integral.)

Function	Interval
9. $y = \dfrac{1}{x}$	[1, 3]
10. $y = x^2$	[0, 1]
11. $y = x^2 + x - 2$	[-2, 1]
12. $y = \dfrac{1}{x+1}$	[0, 1]
13. $x = 4 - y^2$	[0, 2]
14. $x = \sqrt{a^2 - y^2}$	$\left[0, \dfrac{a}{2}\right]$

15. A fleeing object leaves the origin and moves up the y-axis, as shown in Figure 6.86. At the same time a

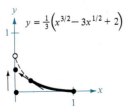

$y = \frac{1}{3}\left(x^{3/2} - 3x^{1/2} + 2\right)$

FIGURE 6.86

pursuer leaves the point (1, 0) and moves always toward the fleeing object. If the pursuer's speed is twice that of the fleeing object, the equation of the path is

$$y = \frac{1}{3}(x^{3/2} - 3x^{1/2} + 2)$$

How far has the fleeing object traveled when it is caught?

16. In Exercise 15, show that the pursuer travels twice as far as the fleeing object.

17. Find the arc length from (0, 3) clockwise to (2, $\sqrt{5}$) along the circle $x^2 + y^2 = 9$.

18. Find the arc length from (-3, 4) clockwise to (4, 3) along the circle $x^2 + y^2 = 25$. Show that the result is one-fourth of the circumference of the circle.

In Exercises 19–22, find the area of the surface of revolution generated by revolving the given plane curve about the x-axis.

Function	Interval
19. $y = \dfrac{x^3}{3}$	[0, 3]
20. $y = \sqrt{x}$	[1, 4]
21. $y = \dfrac{x^3}{6} + \dfrac{1}{2x}$	[1, 2]
22. $y = \dfrac{x}{2}$	[0, 6]

In Exercises 23 and 24, find the area of the surface of revolution generated by revolving the given plane curve over the indicated interval about the y-axis.

Function	Interval
23. $y = \sqrt[3]{x} + 2$	[1, 8]
24. $y = 4 - x^2$	[0, 2]

25. A right circular cone is generated by revolving the region bounded by $y = hx/r$, $y = h$, and $x = 0$ about the y-axis. Verify that the lateral surface area of the cone is

$$S = \pi r \sqrt{r^2 + h^2}$$

26. A sphere of radius r is generated by revolving the graph of $y = \sqrt{r^2 - x^2}$ about the x-axis. Verify that the surface area of the sphere is $4\pi r^2$.

27. Find the area of the zone of a sphere formed by revolving the graph of $y = \sqrt{9 - x^2}$, $0 \le x \le 2$, about the y-axis.

28. Find the area of the zone of a sphere formed by revolving the graph of $y = \sqrt{r^2 - x^2}$, $0 \le x \le a$, about the y-axis. Assume that $a < r$.

29. An ornamental light bulb is designed by revolving the graph of

$$y = \frac{1}{3}x^{1/2} - x^{3/2}, \quad 0 \le x \le \frac{1}{3}$$

about the x-axis, where x and y are measured in feet, as shown in Figure 6.87. Find the surface area of the bulb and use the result to approximate the amount of glass

needed to make the bulb. (Assume the glass is 0.015 inch thick.)

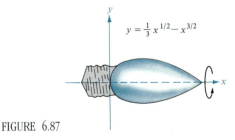

FIGURE 6.87

Review Exercises for Chapter 6

In Exercises 1–18, sketch the region bounded by the graphs of the given equations and determine the area of the region.

1. $y = \dfrac{1}{x^2}$, $y = 0$, $x = 1$, $x = 5$

2. $y = \dfrac{1}{x^2}$, $y = 4$, $x = 5$

3. $y = \dfrac{x}{(x^2 + 1)^2}$, $y = 0$, $x = 0$, $x = 1$

4. $y = 1 - \dfrac{x}{2}$, $y = x - 2$, $y = 1$

5. $x = y^2 - 2y$, $x = 0$

6. $x = y^2 - 2y$, $x = -1$, $y = 0$

7. $y = x$, $y = x^3$

8. $x = y^2 + 1$, $x = y + 3$

9. $y = x^2 - 8x + 3$, $y = 3 + 8x - x^2$

10. $y = x^2 - 4x + 3$, $y = x^3$, $x = 0$

11. $y = \sqrt{x - 1}$, $y = 2$, $y = 0$, $x = 0$

12. $y = \sqrt{x - 1}$, $y = \dfrac{x - 1}{2}$

13. $\sqrt{x} + \sqrt{y} = 1$, $y = 0$, $x = 0$

14. $y = x^4 - 2x^2$, $y = 2x^2$

In Exercises 15–22, find the volume of the solid generated by revolving the plane region bounded by the given equations about the indicated line.

15. $y = x$, $y = 0$, $x = 4$
   (a) the x-axis   (b) the y-axis
   (c) the line $x = 4$ (d) the line $x = 6$

16. $y = \sqrt{x}$, $y = 2$, $x = 0$
   (a) the x-axis   (b) the line $y = 2$
   (c) the y-axis   (d) the line $x = -1$

17. $\dfrac{x^2}{16} + \dfrac{y^2}{9} = 1$
   (a) the y-axis (oblate spheroid)
   (b) the x-axis (prolate spheroid)

18. $\dfrac{x^2}{a^2} + \dfrac{y^2}{b^2} = 1$
   (a) the y-axis (oblate spheroid)
   (b) the x-axis (prolate spheroid)

19. $y = \dfrac{1}{(x^2 + 1)^2}$, $y = 0$, $x = 0$, $x = 1$, revolved about the y-axis

20. $y = \dfrac{1}{(x + 1)^{3/2}}$, $y = 0$, $x = -1$, $x = 1$, revolved about the x-axis

21. $y = -x^2 + 6x - 5$, $y = 0$
   (a) the x-axis   (b) the y-axis

22. $y = x^3 + 1$, $y = 2$, $x = 0$, revolved about the x-axis

23. Find the work done in stretching a spring from its natural length of 10 inches to a length of 15 inches, if a force of 4 pounds is needed to stretch it 1 inch from its natural position.

24. Find the work done in stretching a spring from its natural length of 9 inches to double that length, if a force of 50 pounds is required to hold the spring at double its natural length.

25. A water well has an 8-inch casing (diameter) and is 175 feet deep. If the water is 25 feet from the top of the well, determine the amount of work done in pumping it dry, assuming that no water enters the well while it is being pumped.

26. Repeat Exercise 25, assuming that water enters the well at the rate of 4 gallons per minute and the pump works at the rate of 12 gallons per minute. How many gallons are pumped in this case?

27. A chain 10 feet long weighs 5 pounds per foot and is suspended from a platform 20 feet above the ground. How much work is required to raise the entire chain to the 20-foot level?

28. A windlass, situated 200 feet above ground level on the top of a building, uses a cable weighing 4 pounds per foot. Find the work done in winding up the cable if
(a) one end is at ground level
(b) there is a 300-pound load attached to the end of the cable

29. A swimming pool is 5 feet deep at one end and 10 feet deep at the other, and the bottom is an inclined plane. The length and width of the pool are 40 feet and 20 feet, respectively. If the pool is full of water, what is the fluid force on each of the vertical walls?

30. Show that the fluid force against any vertical region in a liquid is the product of the weight per cubic volume of the liquid, the area of the region, and the depth of the centroid of the region.

31. Using the result of Exercise 30, find the fluid force on one side of a vertical circular plate of radius 4 feet that is submerged in water so that its center is 5 feet below the surface.

32. How much must the water level be raised to double the fluid force on one side of the plate in Exercise 31?

In Exercises 33–36, find the centroid of the region bounded by the graphs of the given equations.

33. $\sqrt{x} + \sqrt{y} = \sqrt{a}$, $x = 0$, $y = 0$
34. $y = x^2$, $y = 2x + 3$
35. $y = a^2 - x^2$, $y = 0$
36. $y = x^{2/3}$, $y = \dfrac{1}{2}x$

37. Use integration to show that the arc length of the circle $x^2 + y^2 = 4$ from $(-\sqrt{3}, 1)$ clockwise to $(\sqrt{3}, 1)$ is one-third the circumference of the circle.

38. Find the length of the graph of

$$y = \frac{1}{6}x^3 + \frac{1}{2x}$$

from $x = 1$ to $x = 3$.

39. Using integration, find the lateral surface area of a right circular cone of height 4 and radius 3.

40. A gasoline tank is an oblate spheroid generated by revolving the region bounded by the graph of

$$\frac{x^2}{16} + \frac{y^2}{9} = 1$$

about the y-axis, where x and y are measured in feet. Find the depth of the gasoline in the tank when it is filled to one-fourth its capacity.

41. Find the area of the region bounded by $y = x\sqrt{x + 1}$ and $y = 0$.

42. The region defined in Exercise 41 is revolved around the x-axis. Find the volume of the solid generated.

43. Find the volume of the solid generated by revolving the region defined in Exercise 41 about the y-axis.

44. The region bounded by $y = 2\sqrt{x}$, $y = 0$, and $x = 3$ is revolved around the x-axis. Find the surface area of the solid generated.

45. Find the arc length of the graph of $f(x) = \frac{4}{5}x^{5/4}$ from $x = 0$ to $x = 4$.

7

Logarithmic and exponential functions

7.1
The natural logarithmic function

When we defined the definite integral of f on the interval $[a, b]$, we used the constant b as the upper limit of integration and x as the variable of integration. In this section we look at a slightly different situation in which the variable x is used as the upper limit of integration. To avoid the confusion of using x in two different ways, we temporarily switch to using t as the variable of integration. (Remember that the definite integral is *not* a function of its variable of integration—any dummy variable can be used.)

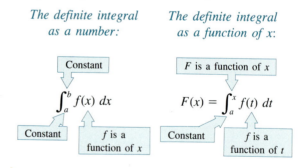

The definite integral as a number:

Constant

$$\int_a^b f(x)\, dx$$

Constant f is a function of x

The definite integral as a function of x:

F is a function of x

$$F(x) = \int_a^x f(t)\, dt$$

Constant f is a function of t

EXAMPLE 1   *The definite integral as a function*

Evaluate the function

$$F(x) = \int_0^x (3 - 3t^2)\, dt$$

at $x = 0$, 1/4, 1/2, 3/4, and 1.

Solution: We could evaluate five different definite integrals, one for each of the given upper limits. However, it is much simpler to temporarily fix x (as a constant) and apply the Fundamental Theorem once, to obtain

$$\int_0^x (3 - 3t^2) \, dt = \left[3t - t^3 \right]_0^x = [3x - x^3] - [3(0) - 0^3] = 3x - x^3$$

Now, using $F(x) = 3x - x^3$ we have the result shown in Figure 7.1.

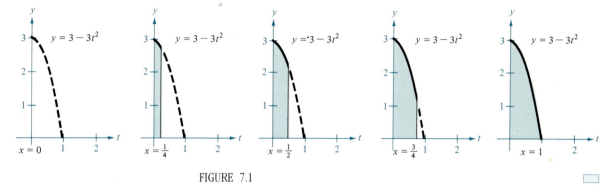

FIGURE 7.1

In Example 1, the function F defined by a definite integral with a variable upper limit x happened to be differentiable *and* its derivative turned out to be the original integrand. That is,

$$\frac{d}{dx} [3x - x^3] = \frac{d}{dx} \left[\int_0^x (3 - 3t^2) \, dt \right] = 3 - 3x^2$$

We generalize this result in the following theorem.

THEOREM 7.1   **THE DERIVATIVE OF A FUNCTION DEFINED BY A DEFINITE INTEGRAL**
If f is continuous on an open interval I containing a, then for every x in the interval,

$$\frac{d}{dx} \left[\int_a^x f(t) \, dt \right] = f(x)$$

Proof: We define F as

$$F(x) = \int_a^x f(t) \, dt$$

Then by the definition of the derivative, we have

$$F'(x) = \lim_{\Delta x \to 0} \frac{F(x + \Delta x) - F(x)}{\Delta x}$$

$$= \lim_{\Delta x \to 0} \frac{1}{\Delta x} \left[\int_a^{x+\Delta x} f(t) \, dt - \int_a^x f(t) \, dt \right]$$

$$= \lim_{\Delta x \to 0} \frac{1}{\Delta x} \left[\int_a^{x+\Delta x} f(t) \, dt + \int_x^a f(t) \, dt \right]$$

$$= \lim_{\Delta x \to 0} \frac{1}{\Delta x} \left[\int_x^{x+\Delta x} f(t) \, dt \right]$$

Now, from the Mean Value Theorem for integrals, we know there exists a number c in the interval $[x, x + \Delta x]$ such that the integral in the above expression is equal to $f(c)\Delta x$. Moreover, since $x \le c \le x + \Delta x$, it follows that $c \to x$ as $\Delta x \to 0$. Thus, we have

$$F'(x) = \lim_{\Delta x \to 0} \left[\frac{1}{\Delta x} f(c)\Delta x \right] = \lim_{\Delta x \to 0} f(c) = f(x)$$

f(t)

Δx

$f(x)$

t

$f(x) \, \Delta x \approx \int_x^{x + \Delta x} f(t) \, dt$

FIGURE 7.2

Remark Using the area model for definite integrals, we can view the approximation

$$f(x)\Delta x \approx \int_x^{x+\Delta x} f(t) \, dt$$

as saying that the area of the rectangle of height $f(x)$ and width Δx is approximately equal to the area of the region lying between the graph of f and the x-axis on the interval $[x, x + \Delta x]$ as shown in Figure 7.2.

This theorem has important consequences for the Fundamental Theorem. In fact, it is sometimes called the **Second Fundamental Theorem of Calculus.** It tells us that if a function is continuous, then we can be sure that it has an antiderivative. Since the Fundamental Theorem requires the existence of an antiderivative, we now know that it can be applied to *any* continuous function.

The natural logarithmic function

John Napier

One of the reasons that Theorem 7.1 is important is that it is a **constructive theorem** as opposed to an existence theorem. That is, it actually gives us a recipe for defining an antiderivative of any continuous function. We have a special use for just such a procedure. Recall that the Power Rule

$$\int x^n \, dx = \frac{x^{n+1}}{n + 1} + C, \quad n \ne -1$$

has a rather glaring disclaimer—it doesn't apply when $n = -1$. The result is that, up to this point, we have not found an antiderivative for the function $f(x) = 1/x$. Now, using Theorem 7.1, we can *define* such a function. This function turns out to be one we have not encountered previously in this text. It is not algebraic, but falls into a new class of functions called *logarithmic functions*. We call this particular function the **natural logarithmic function.** Logarithms were invented by the Scottish mathematician John Napier (1550–1617). Although Napier did not actually introduce the *natural* logarithmic function, it is often called the *Naperian* logarithm.

DEFINITION OF NATURAL LOGARITHMIC FUNCTION	The **natural logarithmic function** is $$\ln x = \int_1^x \frac{1}{t} \, dt, \quad 0 < x$$

| Remark   Note that the function $f(x) = 1/x$ is continuous over the entire real line, except at $x = 0$. Thus, we can apply Theorem 7.1 to define an antiderivative on the interval $(-\infty, 0)$ or $(0, \infty)$. By convention, we choose the domain of the natural logarithmic function to be the set of positive reals.

We can study the properties of the natural logarithmic function from two points of view. As the *definite integral* of $f(t) = 1/t$ from 1 to x, we can tell that ln x is positive for $x > 1$ and negative for $0 < x < 1$ as shown in Figure 7.3. Moreover, we can see that ln $(1) = 0$ since the upper and lower limits of integration are equal when $x = 1$.

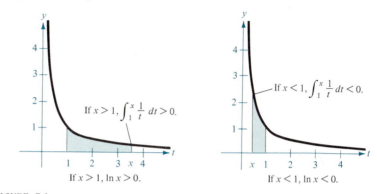

If $x > 1$, $\int_1^x \frac{1}{t} dt > 0$.

If $x < 1$, $\int_1^x \frac{1}{t} dt < 0$.

If $x > 1$, ln $x > 0$.   If $x < 1$, ln $x < 0$.

FIGURE 7.3

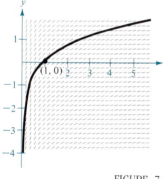

FIGURE 7.4

However, from Figure 7.3, it is difficult to actually picture the graph of $y = \ln x$. To do that, we can think of the natural logarithmic function as an *antiderivative* given by the differential equation

$$\frac{dy}{dx} = \frac{1}{x}$$

Figure 7.4 is a computer-generated graph showing small tangent segments. The particular solution we want is given by the initial condition $y = 0$ when $x = 1$ as shown in the colored portion of the graph.

EXAMPLE 2   Evaluating the slope of the graph of $y = \ln x$

Table 7.1 shows the results of applying Simpson's Rule to the integral

$$\int_0^x \frac{1}{t} dt$$

where $x = 1/4, 1/3, 1/2, 1, 2, 3$, and 4. Use this table to sketch the graph of $y = \ln x$ and sketch a small portion of the tangent line at each of the given points.

TABLE 7.1

x	1/4	1/3	1/2	1	2	3	4
$y = \ln x$	−1.386	−1.099	−0.693	0	0.693	1.099	1.386

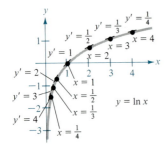

FIGURE 7.5

Solution: Since the slope of the graph of $y = \ln x$ is given by $y' = 1/x$, we can see that the slope of the tangent line at each point is the reciprocal of the x-coordinate of the point as shown in Table 7.2 and Figure 7.5.

TABLE 7.2

x	1/4	1/3	1/2	1	2	3	4
$y' = 1/x$	4	3	2	1	1/2	1/3	1/4

We list some basic properties of the natural logarithmic function in the following theorem.

THEOREM 7.2   **PROPERTIES OF THE NATURAL LOGARITHMIC FUNCTION**
The natural logarithmic function has the following properties.

1. The domain is $(0, \infty)$ and the range is $(-\infty, \infty)$.
2. The function is continuous, increasing, and one-to-one.
3. The graph is concave downward.

Proof: The domain of $f(x) = \ln x$ is $(0, \infty)$ by definition. Moreover, by Theorem 3.2 the function is continuous since it is differentiable. It is increasing since its derivative $f'(x) = 1/x$ is positive for $x > 0$, and concave downward since $f''(x) = -1/x^2$ is negative. (We leave the proof that f is one-to-one as an exercise.) Since f is continuous, we can see that its range is the entire real line by verifying the following limits:

$$\lim_{x \to 0^+} \ln x = -\infty \quad \text{and} \quad \lim_{x \to \infty} \ln x = \infty$$

We leave the verification of these two limits as an exercise.

Using the definition of the natural logarithmic function, we are able to prove some important properties involving operations with natural logarithms. You will notice that these properties are the same as those of common logarithms. In fact, these properties are characteristic of all logarithms.

THEOREM 7.3   **LOGARITHMIC PROPERTIES**
If a and b are positive numbers and n is rational, then the following properties are true.

1. $\ln (1) = 0$ 2. $\ln (ab) = \ln a + \ln b$

3. $\ln (a^n) = n \ln a$ 4. $\ln \left(\dfrac{a}{b}\right) = \ln a - \ln b$

Proof: We have already discussed the first property. The second property has a nice proof that follows from the fact that two antiderivatives of the same function differ at most by a constant. From Theorem 7.1 together with the definition of the natural logarithmic function, we know that

$$\frac{d}{dx} [\ln x] = \frac{1}{x}$$

Thus, we consider the two derivatives

$$\frac{d}{dx} [\ln (ax)] = \frac{a}{ax} = \frac{1}{x}$$

and

$$\frac{d}{dx} [\ln a + \ln x] = 0 + \frac{1}{x} = \frac{1}{x}$$

Since $\ln (ax)$ and $(\ln a + \ln x)$ are both antiderivatives of $1/x$, they must differ at most by a constant

$$\ln (ax) = \ln a + \ln x + C$$

Letting $x = 1$, we see that $C = 0$. The third property can be proved in a similar way by comparing the derivatives of $\ln (x^n)$ and $n \ln x$. Try filling in the details suggested by the following two derivatives.

$$\frac{d}{dx}[\ln (x^n)] = \frac{nx^{n-1}}{x^n} = n\left(\frac{1}{x}\right)$$

$$\frac{d}{dx}[n \ln x] = n\left(\frac{d}{dx}[\ln x]\right) = n\left(\frac{1}{x}\right)$$

Finally, using the second and third properties, we have

$$\ln \left(\frac{a}{b}\right) = \ln [a(b^{-1})]$$

$$= \ln a + \ln (b^{-1})$$

$$= \ln a - \ln b$$

EXAMPLE 3   *Expanding logarithmic expressions*

(a) $\ln \dfrac{10}{9} = \ln 10 - \ln 9$ Property 4

(b) $\ln \sqrt{3x + 2} = \ln (3x + 2)^{1/2} = \dfrac{1}{2} \ln (3x + 2)$ Property 3

(c) $\ln \dfrac{xy}{5} = \ln (xy) - \ln 5 = \ln x + \ln y - \ln 5$ Properties 2 and 4

(d) $\ln \dfrac{(x - 3)^2}{x \sqrt[3]{x - 1}} = \ln (x - 3)^2 - \ln (x \sqrt[3]{x - 1})$

$$= 2 \ln (x - 3) - [\ln x + \ln (x - 1)^{1/3}]$$
$$= 2 \ln (x - 3) - \ln x - \ln (x - 1)^{1/3}$$
$$= 2 \ln (x - 3) - \ln x - \frac{1}{3} \ln (x - 1)$$

EXAMPLE 4   Condensing logarithmic expressions

(a) $\ln x + 2 \ln y = \ln x + \ln y^2 = \ln xy^2$

(b) $\ln (x + 1) - \dfrac{1}{2} \ln x - \ln (x^2 - 1) = \ln (x + 1) - [\ln \sqrt{x} + \ln (x^2 - 1)]$

$$= \ln (x + 1) - \ln [\sqrt{x}(x^2 - 1)]$$

$$= \ln \left[\dfrac{x + 1}{\sqrt{x}(x^2 - 1)} \right]$$

$$= \ln \left[\dfrac{1}{\sqrt{x}(x - 1)} \right]$$

The base of the natural logarithmic function

It is likely that you have previously studied logarithms in an algebra course. There, without the benefit of calculus, logarithms would have been defined in terms of a **base** number. (For example, common logarithms have a base of 10 since $\log_{10} (10) = 1$. We will say more about this in Section 7.5.) Leonhard Euler was the first to recognize that the base for natural logarithms is a special number that he denoted by e. We define e as follows.

DEFINITION OF e

$$e = \lim_{n \to \infty} \left(1 + \dfrac{1}{n} \right)^n$$

$$= \lim_{n \to \infty} \left(\dfrac{n + 1}{n} \right)^n$$

In Chapter 10 we will give a formal proof that this limit exists. For now, we can compute the value of $[(n + 1)/n]^n$ for several values of n to get a decimal approximation for e as shown in Table 7.3.

TABLE 7.3

n	10	100	1,000	10,000	100,000	1,000,000
$\left(\dfrac{n + 1}{n} \right)^n$	2.59374	2.70481	2.71692	2.71815	2.71827	2.71828

Remark   To eleven decimal places the value of e is $e \approx 2.71828182846$.

Since the natural logarithmic function is one-to-one and continuous and has $(-\infty, \infty)$ as its range, we know that there exists a unique number such that $\ln x = 1$ as shown in Figure 7.6. We call this number the natural logarithmic **base.** The next theorem tells us that this number is e. For a proof of this theorem, see Appendix A.

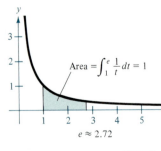

$e \approx 2.72$

Area $= \displaystyle\int_1^e \dfrac{1}{t} dt = 1$

FIGURE 7.6

THEOREM 7.4   NATURAL LOGARITHMIC BASE IS e

The number e is the base for the natural logarithmic function. That is,

$$\ln e = \int_1^e \frac{1}{t}\, dt = 1$$

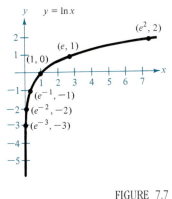

FIGURE 7.7

Once we know that $\ln e = 1$, we can use logarithmic properties to evaluate the natural logarithm of several other numbers. For example, using the property

$$\ln (e^n) = n \ln e = n(1) = n$$

we can evaluate $\ln (e^n)$ for various powers of n as shown in Table 7.4 and Figure 7.7.

TABLE 7.4

x	$\dfrac{1}{e^3} \approx 0.050$	$\dfrac{1}{e^2} \approx 0.135$	$\dfrac{1}{e} \approx 0.368$	$e^0 = 1$	$e \approx 2.718$	$e^2 \approx 7.389$
$\ln x$	-3	-2	-1	0	1	2

Remark The logarithms given in Table 7.4 are convenient because the x-values are integer powers of e. Of course, for most logarithms you would use a calculator. For example,

$$\ln (2) \approx 0.693 \qquad \text{and} \qquad \ln (0.1) \approx -2.303$$

Section Exercises 7.1

In Exercises 1–6, (a) integrate to find F as a function of x and (b) demonstrate Theorem 7.1 by differentiating the result of part (a).

1. $F(x) = \displaystyle\int_0^x (t + 2)\, dt$ **2.** $F(x) = \displaystyle\int_0^x (3t - 1)\, dt$

3. $F(x) = \displaystyle\int_4^x \sqrt{t}\, dt$ **4.** $F(x) = \displaystyle\int_7^x \sqrt[3]{t + 1}\, dt$

5. $F(x) = \displaystyle\int_0^x t(t^2 + 1)^3\, dt$ **6.** $F(x) = \displaystyle\int_1^x (t^2 + 2)\, dt$

In Exercises 7–10, use Theorem 7.1 to find $F'(x)$.

7. $F(x) = \displaystyle\int_{-2}^x (t^2 - 2t + 5)\, dt$

8. $F(x) = \displaystyle\int_{-1}^x \sqrt{t^4 + 1}\, dt$

9. $F(x) = \displaystyle\int_1^x \sqrt[4]{t}\, dt$ **10.** $F(x) = \displaystyle\int_1^x \frac{t^2}{t^2 + 1}\, dt$

In Exercises 11–16, sketch the graph of the function.

11. $f(x) = \ln 2x$ **12.** $f(x) = \ln |x|$
13. $f(x) = \ln x^2$ **14.** $f(x) = 2 \ln x$
15. $f(x) = \ln (x - 1)$ **16.** $f(x) = -2 \ln x$

In Exercises 17–24, use the properties of logarithms and the fact that $\ln 2 \approx 0.6931$ and $\ln 3 \approx 1.0986$, to approximate the given logarithm.

17. $\ln 6$ **18.** $\ln (2/3)$
19. $\ln 81$ **20.** $\ln \sqrt{3}$
21. $\ln 0.25$ **22.** $\ln 24$
23. $\ln \sqrt[3]{12}$ **24.** $\ln (1/72)$

In Exercises 25–34, use the properties of logarithms to write each as a sum, difference, or multiple of logarithms.

25. $\ln \dfrac{2}{3}$ **26.** $\ln (xyz)$

27. $\ln \dfrac{xy}{z}$ **28.** $\ln \sqrt{a-1}$

29. $\ln \sqrt{2^3}$ **30.** $\ln \dfrac{1}{5}$

31. $\ln \left(\dfrac{x^2-1}{x^3}\right)^3$ **32.** $\ln 3e^2$

33. $\ln z(z-1)^2$ **34.** $\ln \dfrac{1}{e}$

In Exercises 35–40, write each expression as a logarithm of a single expression in x.

35. $\ln (x-2) - \ln (x+2)$

36. $3 \ln x + 2 \ln y - 4 \ln z$

37. $\dfrac{1}{3}[2 \ln (x+3) + \ln x - \ln (x^2-1)]$

38. $2[\ln x - \ln (x+1) - \ln (x-1)]$

39. $2 \ln 3 - \dfrac{1}{2} \ln (x^2+1)$

40. $\dfrac{3}{2}[\ln (x^2+1) - \ln (x+1) - \ln (x-1)]$

SECTION TOPICS ▪
The derivative of the natural logarithmic function ▪
Logarithmic differentiation ▪

7.2
The natural logarithmic function and differentiation

Since the natural logarithmic function is defined as an antiderivative, the first part of the following theorem is an immediate consequence of that definition together with Theorem 7.1. The second part of the theorem is simply the Chain Rule version of the first part.

THEOREM 7.5   **DERIVATIVE OF THE NATURAL LOGARITHMIC FUNCTION**
If u is a differentiable function of x, then

1. $\dfrac{d}{dx}[\ln x] = \dfrac{1}{x}, \quad x > 0$ 2. $\dfrac{d}{dx}[\ln u] = \dfrac{1}{u}\dfrac{du}{dx}, \quad u > 0$

EXAMPLE 1   Differentiation of logarithmic functions

(a) $\dfrac{d}{dx}[\ln (2x)] = \dfrac{u'}{u} = \dfrac{2}{2x} = \dfrac{1}{x}$ $u = 2x$

(b) $\dfrac{d}{dx}[\ln (x^2+1)] = \dfrac{u'}{u} = \dfrac{2x}{x^2+1}$ $u = x^2+1$

(c) $\dfrac{d}{dx}[x \ln x] = x\left(\dfrac{d}{dx}[\ln x]\right) + (\ln x)\left(\dfrac{d}{dx}[x]\right)$ Product Rule

$\qquad = x\left(\dfrac{1}{x}\right) + (\ln x)(1) = 1 + \ln x$

When Napier introduced logarithms, he used logarithmic properties to simplify *calculations* involving products, quotients, and powers. Of course, with the availability of calculators, we have little need for this particular application of logarithms today. However, we do find a great value in using logarithmic properties to simplify *differentiation* involving products, quotients, and powers. The following examples illustrate this point.

EXAMPLE 2   *Logarithmic properties as an aid to differentiation*

Differentiate $f(x) = \ln \sqrt{x + 1}$.

Solution: Since

$$f(x) = \ln \sqrt{x + 1} = \ln (x + 1)^{1/2} = \frac{1}{2} \ln (x + 1)$$

we have

$$f'(x) = \frac{1}{2}\left(\frac{1}{x + 1}\right) = \frac{1}{2(x + 1)}$$

EXAMPLE 3   *Logarithmic properties as an aid to differentiation*

Differentiate $f(x) = \ln [x\sqrt{1 - x^2}]$.

Solution: Since

$$f(x) = \ln [x\sqrt{1 - x^2}] = \ln x + \ln (1 - x^2)^{1/2}$$

$$= \ln x + \frac{1}{2} \ln (1 - x^2)$$

we have

$$f'(x) = \frac{1}{x} + \frac{1}{2}\left(\frac{-2x}{1 - x^2}\right) = \frac{1}{x} - \frac{x}{1 - x^2}$$

$$= \frac{1 - x^2 - x^2}{x(1 - x^2)}$$

$$= \frac{1 - 2x^2}{x(1 - x^2)}$$

EXAMPLE 4   *Logarithmic properties as an aid to differentiation*

Differentiate

$$f(x) = \ln \frac{x(x^2 + 1)^2}{\sqrt{2x^3 - 1}}$$

Solution: Since

$$f(x) = \ln \frac{x(x^2 + 1)^2}{\sqrt{2x^3 - 1}} = \ln x + 2 \ln (x^2 + 1) - \frac{1}{2} \ln (2x^3 - 1)$$

we have

$$f'(x) = \frac{1}{x} + 2\left(\frac{2x}{x^2 + 1}\right) - \frac{1}{2}\left(\frac{6x^2}{2x^3 - 1}\right)$$

$$= \frac{1}{x} + \frac{4x}{x^2 + 1} - \frac{3x^2}{2x^3 - 1}$$

Remark In Examples 2, 3, and 4, be sure you see the great benefit in applying logarithmic properties *before* differentiating. For instance, consider the difficulty of direct differentiation of the function given in Example 4.

On occasion, it is convenient to use logarithms as an aid in differentiating *non*logarithmic functions. We call this procedure **logarithmic differentiation,** and we illustrate its use in Examples 5 and 6.

LOGARITHMIC DIFFERENTIATION

To differentiate the function $y = u$, use the following steps.

1. Take the natural logarithm of both sides: $\ln y = \ln u$
2. Use logarithmic properties to rid $\ln u$ of as many products, quotients, and exponents as possible.

3. Differentiate *implicitly:* $\dfrac{y'}{y} = \dfrac{d}{dx}[\ln u]$

4. Solve for y': $y' = y\dfrac{d}{dx}[\ln u]$

5. Substitute for y and simplify: $y' = u\dfrac{d}{dx}[\ln u]$

EXAMPLE 5   Logarithmic differentiation

Find the derivative of $y = x\sqrt{x^2 + 1}$.

Solution: We begin by taking the natural logarithms of both sides of the equation. Then, we apply logarithmic properties and differentiate. Finally, we solve for y'.

$$y = x\sqrt{x^2 + 1}$$

$$\ln y = \ln [x\sqrt{x^2 + 1}] \qquad \text{Take log of both sides}$$

$$= \ln x + \frac{1}{2}\ln(x^2 + 1) \qquad \text{Logarithmic properties}$$

$$\frac{y'}{y} = \frac{1}{x} + \frac{1}{2}\left(\frac{2x}{x^2 + 1}\right) \qquad \text{Differentiate}$$

$$= \frac{x^2 + 1 + x^2}{x(x^2 + 1)}$$

$$y' = y\left[\frac{2x^2 + 1}{x(x^2 + 1)}\right] \qquad \text{Solve for } y'$$

$$y' = x\sqrt{x^2 + 1}\left[\frac{2x^2 + 1}{x(x^2 + 1)}\right] \qquad \text{Substitute for } y$$

$$= \frac{2x^2 + 1}{\sqrt{x^2 + 1}} \qquad \text{Simplify}$$

EXAMPLE 6   Logarithmic differentiation

Find the derivative of

$$y = \frac{(x - 2)^2}{\sqrt{x^2 + 1}}$$

Solution: Following the procedure of Example 5, we have

$$\ln y = \ln \frac{(x-2)^2}{\sqrt{x^2+1}} \qquad\qquad \text{Take ln of both sides}$$

$$\ln y = 2 \ln (x-2) - \frac{1}{2} \ln (x^2+1) \qquad\qquad \text{Logarithmic properties}$$

$$\frac{y'}{y} = 2\left(\frac{1}{x-2}\right) - \frac{1}{2}\left(\frac{2x}{x^2+1}\right) \qquad\qquad \text{Differentiate}$$

$$= \frac{2}{x-2} - \frac{x}{x^2+1}$$

$$y' = y\left(\frac{2}{x-2} - \frac{x}{x^2+1}\right) \qquad\qquad \text{Solve for } y'$$

$$= \frac{(x-2)^2}{\sqrt{x^2+1}} \left[\frac{x^2+2x+2}{(x-2)(x^2+1)}\right] \qquad\qquad \text{Substitute for } y$$

$$= \frac{(x-2)(x^2+2x+2)}{(x^2+1)^{3/2}} \qquad\qquad \text{Simplify}$$

Since the natural logarithm is undefined for negative numbers, we often encounter expressions of the form $\ln |u|$. The following theorem tells us that we can differentiate functions of the form $y = \ln |u|$ as if the absolute value sign were not present.

THEOREM 7.6 **DERIVATIVE INVOLVING ABSOLUTE VALUE**
If u is a differentiable function of x such that $u \neq 0$, then

$$\frac{d}{dx} [\ln |u|] = \frac{u'}{u}$$

Proof: If $u > 0$, then $|u| = u$, and the result follows from Theorem 7.5. If $u < 0$, then $|u| = -u$, and we have

$$\frac{d}{dx} [\ln |u|] = \frac{d}{dx} [\ln (-u)] = \frac{-u'}{-u} = \frac{u'}{u}$$

EXAMPLE 7   Derivative involving absolute value

Find the derivative of $f(x) = \ln |2x - 1|$.

Solution: Using Theorem 7.6, we let $u = 2x - 1$ and write

$$\frac{d}{dx} [\ln |2x - 1|] = \frac{u'}{u} = \frac{2}{2x - 1}$$

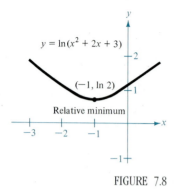

FIGURE 7.8

EXAMPLE 8   *Finding relative extrema*

Locate the relative extrema of $y = \ln(x^2 + 2x + 3)$.

Solution: Differentiating y, we obtain

$$\frac{dy}{dx} = \frac{2x + 2}{x^2 + 2x + 3}$$

Now, since $dy/dx = 0$ when $x = -1$, we apply the First Derivative Test to conclude that the point $(-1, \ln 2)$ is a relative minimum. Since there are no other critical points, we conclude that this is the only relative extrema as shown in Figure 7.8.

Section Exercises 7.2

In Exercises 1–4, find the slope of the tangent line to the given logarithmic function at the point (1, 0).

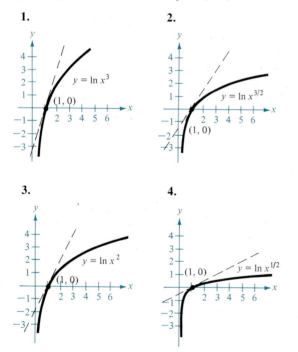

1.

2.

3.

4.

In Exercises 5–24, find dy/dx.

5. $y = \ln(x^2)$

6. $y = \ln(x^2 + 3)$

7. $y = \ln\sqrt{x^4 - 4x}$

8. $y = \ln(1 - x)^{3/2}$

9. $y = (\ln x)^4$

10. $y = x(\ln x)$

11. $y = \ln(x\sqrt{x^2 - 1})$

12. $y = \ln\left(\frac{x}{x + 1}\right)$

13. $y = \ln\left(\frac{x}{x^2 + 1}\right)$

14. $y = \frac{\ln x}{x}$

15. $y = \frac{\ln x}{x^2}$

16. $y = \ln(\ln x)$

17. $y = \ln(\ln x^2)$

18. $y = \ln\sqrt{\frac{x - 1}{x + 1}}$

19. $y = \ln\sqrt{\frac{x + 1}{x - 1}}$

20. $y = \ln\sqrt{x^2 - 4}$

21. $y = \ln\left(\frac{\sqrt{4 + x^2}}{x}\right)$

22. $y = \ln(x + \sqrt{4 + x^2})$

23. $y = \frac{-\sqrt{x^2 + 1}}{x} + \ln(x + \sqrt{x^2 + 1})$

24. $y = \frac{-\sqrt{x^2 + 4}}{2x^2} - \left(\frac{1}{4}\right)\ln\left(\frac{2 + \sqrt{x^2 + 4}}{x}\right)$

In Exercises 25–30, find dy/dx using logarithmic differentiation.

25. $y = x\sqrt{x^2 - 1}$

26. $y = \sqrt{(x - 1)(x - 2)(x - 3)}$

27. $y = \frac{x^2\sqrt{3x - 2}}{(x - 1)^2}$

28. $y = \sqrt[3]{\frac{x^2 + 1}{x^2 - 1}}$

29. $y = \frac{x(x - 1)^{3/2}}{\sqrt{x + 1}}$

30. $y = \frac{(x + 1)(x + 2)}{(x - 1)(x - 2)}$

In Exercises 31 and 32, show that the given equation is a solution to the differential equation.

31. $y = 2(\ln x) + 3$, $x(y'') + y' = 0$

32. $y = x(\ln x) - 4x$, $(x + y) - x(y') = 0$

In Exercises 33–38, find any relative extrema and inflection points, and sketch the graph of the function.

33. $y = \dfrac{x^2}{2} - \ln x$ 　　　**34.** $y = x - \ln x$

35. $y = x\,(\ln x)$ 　　　**36.** $y = \dfrac{\ln x}{x}$

37. $y = \dfrac{x}{\ln x}$ 　　　**38.** $y = x^2\,(\ln x)$

39. Use Newton's Method to approximate, to three decimal places, the value of x satisfying the equation $\ln x = -x$.

40. Use Newton's Method to approximate, to three decimal places, the x-coordinate of the point of intersection of the graphs of $y = 3 - x$ and $y = \ln x$.

41. Show that $f(x) = (\ln x^n)/x$ is a decreasing function for $x > e$ and $n > 0$.

42. A person walking along a dock drags a boat by a 10-foot rope. The boat travels along a path known as a tractrix as shown in Figure 7.9. The equation of this path is

$$y = 10 \ln \left(\frac{10 + \sqrt{100 - x^2}}{x} \right) - \sqrt{100 - x^2}$$

What is the slope of this path when
(a) $x = 10$? 　　　(b) $x = 5$?

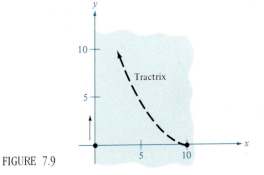

FIGURE 7.9

43. There are 25 prime numbers less than 100. The **Prime Number Theorem** states that the number of primes less than x approaches

$$p(x) \approx \frac{x}{\ln x}$$

Use this approximation to estimate the rate (in primes per 100 integers) at which the prime numbers occur when

(a) $x = 1,000$ 　　　(b) $x = 1,000,000$
(c) $x = 1,000,000,000$

SECTION TOPIC ▪
Log Rule for integration ▪

7.3
The natural logarithmic function and integration

In the previous section we defined the natural logarithmic function as an antiderivative of the function $y = 1/x$. With this definition, we are able to integrate several important functions that are not covered by other integration rules. Specifically, the differentiation rules

$$\frac{d}{dx}\,[\ln |x|] = \frac{1}{x} \quad \text{and} \quad \frac{d}{dx}\,[\ln |u|] = \frac{u'}{u}$$

produce the following integration formulas.

THEOREM 7.7　　**LOG RULE FOR INTEGRATION**
Let u be a differentiable function of x.

1. $\displaystyle \int \frac{1}{x}\,dx = \ln |x| + C$ 　　　　　2. $\displaystyle \int \frac{1}{u}\,du = \ln |u| + C$

EXAMPLE 1   *Using the Log Rule for integration*

Evaluate

$$\int \frac{2}{x} \, dx$$

Solution:

$$\int \frac{2}{x} \, dx = 2 \int \frac{1}{x} \, dx = 2 \ln |x| + C = \ln (x^2) + C$$

Remark In Example 1, the absolute value is unnecessary in the final form of the solution, since x^2 cannot be negative.

EXAMPLE 2   *Using the Log Rule with u-substitution*

Evaluate

$$\int \frac{1}{2x - 1} \, dx$$

Solution: Letting $u = 2x - 1$, we have $du = 2 \, dx$. Multiplying and dividing by 2, we have

$$\int \frac{1}{2x - 1} \, dx = \frac{1}{2} \int \left(\frac{1}{2x - 1} \right) 2 \, dx$$

$$= \frac{1}{2} \int \frac{1}{u} \, du$$

$$= \frac{1}{2} \ln |u| + C$$

$$= \frac{1}{2} \ln |2x - 1| + C$$

EXAMPLE 3   *Finding area with the Log Rule*

Find the area of the region bounded by the graph of $y = x/(x^2 + 1)$, the x-axis, and the line $x = 3$.

Solution: From Figure 7.10 we see that the area of the specified region is given by the definite integral

$$\int_0^3 \frac{x}{x^2 + 1} \, dx$$

Letting $u = x^2 + 1$, we have $u' = 2x$. Thus, to apply the Log Rule, we multiply and divide by 2 and write

$$\int_0^3 \frac{x}{x^2 + 1} \, dx = \frac{1}{2} \int_0^3 \frac{2x}{x^2 + 1} \, dx$$

$$= \frac{1}{2} \left[\ln (x^2 + 1) \right]_0^3$$

$$= \frac{1}{2} (\ln 10 - \ln 1) = \frac{1}{2} \ln 10 \approx 1.151$$

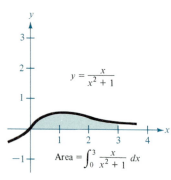

$y = \dfrac{x}{x^2 + 1}$

Area $= \displaystyle\int_0^3 \frac{x}{x^2 + 1} \, dx$

FIGURE 7.10

| **Remark** In Example 3 note that we used an alternative form of the Log Rule:

$$\int \frac{u'}{u}\, dx = \ln |u| + C$$

This form of the Log Rule is convenient, especially for simpler integrals. To apply this rule, be on the lookout for quotients in which the numerator is the derivative of the denominator.

EXAMPLE 4   *Recognizing quotient forms of the Log Rule*
_____

(a) $\displaystyle \int \frac{3x^2 + 1}{x^3 + x}\, dx = \ln |x^3 + x| + C$ $u = x^3 + x$

(b) $\displaystyle \int \frac{x + 1}{x^2 + 2x}\, dx = \frac{1}{2} \int \frac{2x + 2}{x^2 + 2x}\, dx$ $u = x^2 + 2x$

$$= \frac{1}{2} \ln |x^2 + 2x| + C$$

Integrals to which the Log Rule can be applied often occur in disguised form. For instance, if a rational function has a *numerator of degree greater than or equal to that of the denominator*, division may reveal a form to which we can apply the Log Rule. The next example illustrates this.

EXAMPLE 5   *Using long division before integrating*
_____

Evaluate

$$\int \frac{x^2 + x + 1}{x^2 + 1}\, dx$$

Solution: First, by dividing we have

$$\frac{x^2 + x + 1}{x^2 + 1} = 1 + \frac{x}{x^2 + 1}$$

Now, we integrate to obtain

$$\int \frac{x^2 + x + 1}{x^2 + 1}\, dx = \int \left(1 + \frac{x}{x^2 + 1}\right) dx$$

$$= \int dx + \frac{1}{2} \int \frac{2x}{x^2 + 1}\, dx$$

$$= x + \frac{1}{2} \ln (x^2 + 1) + C$$

The next example gives another instance in which the use of the Log Rule is disguised. In this case, a change of variables helps us recognize the Log Rule.

EXAMPLE 6   Change of variables with the Log Rule

Evaluate

$$\int \frac{2x}{(x+1)^2}\, dx$$

Solution: If we let $u = x + 1$, then $du = dx$ and $x = u - 1$. Changing variables, we have

$$\int \frac{2x}{(x+1)^2}\, dx = \int \frac{2(u-1)}{u^2}\, du$$

$$= 2 \int \left[\frac{u}{u^2} - \frac{1}{u^2} \right] du$$

$$= 2 \int \frac{du}{u} - 2 \int u^{-2}\, du$$

$$= 2 \ln |u| - 2\left(\frac{u^{-1}}{-1} \right) + C$$

$$= 2 \ln |u| + \frac{2}{u} + C$$

$$= 2 \ln |x+1| + \frac{2}{x+1} + C \qquad \text{Substitute for } u$$

As you study the two different methods shown in Examples 5 and 6, be
aware that both methods involve rewriting a disguised integrand into a form
that fits one or more of the basic integration formulas. In Chapters 8 and 9 we
will be devoting a great deal of time to integration techniques. To master these
techniques you must recognize the "form fitting" nature of integration. In
this sense integration is not nearly as straightforward as differentiation. Differentiation takes the form "Here is the question; what is the answer?" Integration is more like "Here is the answer; what is the question?" We suggest
the following general approach to integration.

**GUIDELINES FOR
INTEGRATION**

1. Memorize a basic list of integration formulas. (At this point our list
   consists only of the Power Rule and the Log Rule. By the end of
   Chapter 8, we will have expanded this list to nineteen basic formulas.)
2. Find an integration formula that resembles all or part of the integrand,
   and by trial and error find a choice of u that will make the integrand
   conform to the formula.
3. If you cannot find a u-substitution that works, try altering the integrand.
   You might try a trigonometric identity, division, addition and subtraction of the same quantity—be creative.

EXAMPLE 7   u-Substitution and the Log Rule

Evaluate

$$\int \frac{1}{x \ln x}\, dx$$

Solution: Since neither the Power Rule nor the Log Rule apply directly, we consider a *u*-substitution. There are three basic choices for *u*. The choices $u = x$ and $u = x \ln x$ both fail to fit the u'/u form of the Log Rule. However, the third choice does fit. Letting $u = \ln x$, we have $u' = 1/x$ and write

$$\int \frac{1}{x \ln x}\, dx = \int \frac{(1/x)}{\ln x}\, dx = \int \frac{u'}{u}\, dx = \ln |u| + C$$

$$= \ln |\ln x| + C$$

EXAMPLE 8   *u-Substitution and the Log Rule*

Evaluate

$$\int \frac{1}{\sqrt{x} + 1}\, dx$$

Solution: Since neither the Power Rule nor the Log Rule apply to the integral as given, we consider the substitution $u = \sqrt{x}$. Then,

$$u^2 = x \qquad \text{and} \qquad 2u\, du = dx$$

Substitution in the original integral yields

$$\int \frac{1}{\sqrt{x} + 1}\, dx = \int \frac{1}{u + 1}(2u\, du) = 2 \int \frac{u}{u + 1}\, du$$

Since the degree of the numerator is equal to the degree of the denominator, we divide *u* by $(u + 1)$ to obtain

$$\int \frac{1}{\sqrt{x} + 1}\, dx = 2 \int \left(1 - \frac{1}{u + 1} \right) du$$

$$= 2(u - \ln |u + 1|) + C$$

$$= 2\sqrt{x} - 2 \ln (\sqrt{x} + 1) + C$$

In Section 6.4, Example 6, we used Simpson's Rule to find the work done by an expanding gas. Now, with the Log Rule, we can evaluate this work using the Fundamental Theorem of Calculus.

EXAMPLE 9   *An application*

A quantity of gas with an initial volume of 1 cubic foot and pressure of 500 pounds per square foot expands to a volume of 2 cubic feet. Find the work done by the gas. (Assume the pressure is inversely proportional to the volume.)

Solution: Since $p = k/V$ and $p = 500$ when $V = 1$, we have $k = 500$. Thus, the work is

$$W = \int_{V_0}^{V_1} \frac{k}{V}\, dV = \int_1^2 \frac{500}{V}\, dV = 500 \Big[\ln V \Big]_1^2$$

$$= 500(\ln 2 - \ln 1) \approx 346.6 \text{ ft} \cdot \text{lb}$$

Section Exercises 7.3

In Exercises 1–28, evaluate each integral.

1. $\int \dfrac{1}{x+1}\,dx$

2. $\int \dfrac{1}{x-5}\,dx$

3. $\int \dfrac{1}{3-2x}\,dx$

4. $\int \dfrac{1}{6x+1}\,dx$

5. $\int \dfrac{x}{x^2+1}\,dx$

6. $\int \dfrac{x^2}{3-x^3}\,dx$

7. $\int \dfrac{x^2-4}{x}\,dx$

8. $\int \dfrac{x+5}{x}\,dx$

9. $\int_1^e \dfrac{\ln x}{2x}\,dx$

10. $\int_e^{e^2} \dfrac{1}{x(\ln x)}\,dx$

11. $\int_1^e \dfrac{(1+\ln x)^2}{x}\,dx$

12. $\int_0^1 \dfrac{x-1}{x+1}\,dx$

13. $\int_0^2 \dfrac{x^2-2}{x+1}\,dx$

14. $\int \dfrac{1}{(x+1)^2}\,dx$

15. $\int \dfrac{1}{\sqrt{x}+1}\,dx$

16. $\int \dfrac{x+3}{x^2+6x+7}\,dx$

17. $\int \dfrac{x^2+2x+3}{x^3+3x^2+9x+1}\,dx$

18. $\int \dfrac{(\ln x)^2}{x}\,dx$

19. $\int \dfrac{1}{x^{2/3}(1+x^{1/3})}\,dx$

20. $\int \dfrac{1}{x\ln(x^2)}\,dx$

21. $\int \dfrac{1}{1+\sqrt{x}}\,dx$

22. $\int \dfrac{1-\sqrt{x}}{1+\sqrt{x}}\,dx$

23. $\int \dfrac{\sqrt{x}}{\sqrt{x}-3}\,dx$

24. $\int_0^2 \dfrac{1}{1+\sqrt{2x}}\,dx$

25. $\int \dfrac{\sqrt{x}}{1-x\sqrt{x}}\,dx$

26. $\int \dfrac{2x}{(x-1)^2}\,dx$

27. $\int \dfrac{x(x-2)}{(x-1)^3}\,dx$

28. $\int \dfrac{x\sqrt{x}}{1+x^2\sqrt{x}}\,dx$

In Exercises 29 and 30, find the area of the region bounded by the graphs of the given equations.

29.

30.

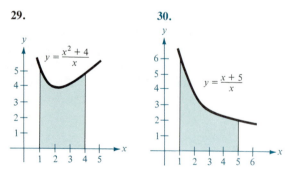

In Exercises 31–35, find the volume of the solid generated by revolving the regions bounded by the graphs of the given equations about the indicated axis.

31. $y = \dfrac{1}{\sqrt{x}+1}$, $y=0$, $x=0$, $x=3$, about the x-axis

32. $y = \dfrac{1}{x(x+1)}$, $y=0$, $x=1$, $x=3$, about the y-axis

33. $xy=1$, $y=0$, $x=1$, $x=4$, about the line $y=4$

34. $xy=6$, $y=2$, $y=6$, $x=6$, about the line $x=6$

35. $y = \dfrac{1}{1+\sqrt{x+2}}$, $y=0$, $x=2$, $x=6$, about the y-axis

36. Find the centroid of the region bounded by the graphs of $y=1/x$, $y=0$, $x=1$, and $x=4$.

In Exercises 37 and 38 find the work done by the gas for the given volumes and pressures. Assume the pressure is inversely proportional to the volume. (Use Example 9 as a model.)

37. A quantity of gas with an initial volume of 2 cubic feet and pressure of 1000 pounds per square foot expands to a volume of 3 cubic feet.

38. A quantity of gas with an initial volume of 1 cubic foot and pressure of 2000 pounds per square foot expands to a volume of 4 cubic feet.

39. A population of bacteria is growing at the rate of

$$\frac{dP}{dt} = \frac{3000}{1+0.25t}$$

where t is the time in days. Assuming that the initial population (when $t=0$) is 1000, write an equation that gives the population at any time t, and then find the population when $t=3$ days.

40. The demand equation for a product is given by

$$p = \frac{90,000}{400+3x}$$

Find the *average* price p on the interval $40 \le x \le 50$.

7.4
Inverse functions

When we introduced the notion of a composite function in Section 1.5, we noted that composition is not commutative. That is, it is not necessarily true that $f(g(x))$ and $g(f(x))$ are equal. We now look at a special case for which composition is commutative—when f and g are inverses of each other.

DEFINITION OF INVERSE FUNCTION

Two functions f and g are **inverses** of each other if

$$f(g(x)) = x \quad \text{for each } x \text{ in the domain of } g$$

and

$$g(f(x)) = x \quad \text{for each } x \text{ in the domain of } f$$

We denote g by f^{-1} (read "f inverse").

| **Remark**   There are some important observations to be made about this definition.

1. If g is the inverse of f, then f is also the inverse of g. That is, $(f^{-1})^{-1} = f$.
2. The domain of f must be equal to the range of f^{-1} (and vice versa), as indicated in Figure 7.11.
3. Even though the notation used to denote an inverse function resembles *exponential notation*, it is a different use of the -1 superscript. That is, in general $f^{-1}(x) \neq 1/f(x)$.

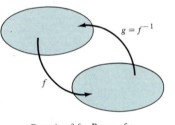

FIGURE 7.11

Domain of f = Range of g
Domain of g = Range of f

To understand the concept of an inverse function, it is helpful to think of f^{-1} as *undoing* what has been done by f. For example, subtraction can be used to undo addition, and division can be used to undo multiplication. To see this, try using the definition of an inverse function to check the following inverses.

1. $f(x) = x + c \qquad$ and $\qquad f^{-1}(x) = x - c$

2. $f(x) = cx \qquad$ and $\qquad f^{-1}(x) = \dfrac{x}{c}, \quad c \neq 0$

EXAMPLE 1   Example of inverse functions

Show that the following functions are inverses of each other:

$$f(x) = 2x^3 - 1 \qquad \text{and} \qquad g(x) = \sqrt[3]{\frac{x + 1}{2}}$$

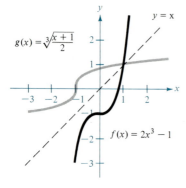

$g(x) = \sqrt[3]{\dfrac{x+1}{2}}$

$y = x$

$f(x) = 2x^3 - 1$

FIGURE 7.12

Solution: First, note that both composite functions exist, since the domain and range of both f and g consist of the set of all real numbers. The composite of f with g is given by

$$f(g(x)) = 2\left(\sqrt[3]{\frac{x+1}{2}}\right)^3 - 1 = 2\left(\frac{x+1}{2}\right) - 1 = x + 1 - 1 = x$$

The composite of g with f is given by

$$g(f(x)) = \sqrt[3]{\frac{(2x^3 - 1) + 1}{2}} = \sqrt[3]{\frac{2x^3}{2}} = \sqrt[3]{x^3} = x$$

Since $f(g(x)) = g(f(x)) = x$, we conclude that f and g are inverses of each other. (See Figure 7.12.)

In Figure 7.12, the graphs of f and f^{-1} appear to be mirror images of each other with respect to the line $y = x$. We say that the graph of f^{-1} is a **reflection** of the graph of f in the line $y = x$. This idea is generalized in the following theorem.

THEOREM 7.8   REFLECTIVE PROPERTY OF INVERSES
The graph of f contains the point (a, b) if and only if the graph of f^{-1} contains the point (b, a).

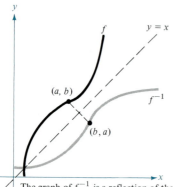

f

$y = x$

(a, b)

f^{-1}

(b, a)

The graph of f^{-1} is a reflection of the graph of f in the line $y = x$.

FIGURE 7.13

Proof: If (a, b) is on the graph of f, then $f(a) = b$ and we have

$$f^{-1}(b) = f^{-1}(f(a)) = a$$

Thus, (b, a) is on the graph of f^{-1}, as shown in Figure 7.13. A similar argument will prove the theorem in the other direction.

Not every function has an inverse, and Theorem 7.8 suggests a graphical test for those that do. It is called the **horizontal line test** for an inverse function, and it follows directly from the vertical line test for functions together with the reflective property of the graphs of f and f^{-1}. The test states that a function f has an inverse if and only if every horizontal line intersects the graph of f at most once. Theorem 7.9 formally states why the horizontal line test is valid. (We leave the proof of this theorem as an exercise.)

THEOREM 7.9   THE EXISTENCE OF AN INVERSE
A function possesses an inverse if and only if the function is one-to-one.

In Section 4.3 we called a function *strictly monotonic* if it is either increasing on its entire domain or decreasing on its entire domain. Functions that are strictly monotonic must be one-to-one.

THEOREM 7.10   **STRICTLY MONOTONIC FUNCTIONS ARE ONE-TO-ONE**
If f is strictly monotonic on an interval, then it is one-to-one on the interval.

Proof: Recall from Section 1.5 that f is one-to-one if for x_1, x_2 in its domain

$$f(x_1) = f(x_2) \implies x_1 = x_2$$

The *contrapositive* of this implication is logically equivalent and it states that

$$x_1 \neq x_2 \implies f(x_1) \neq f(x_2)$$

Now, choose x_1 and x_2 in the given interval. If $x_1 \neq x_2$, then since f is strictly monotonic it follows that either $f(x_1) < f(x_2)$ or $f(x_1) > f(x_2)$. In either case, $f(x_1) \neq f(x_2)$. Thus, f is one-to-one on the interval.

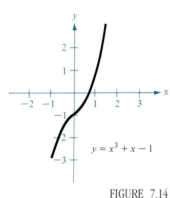

$y = x^3 + x - 1$

FIGURE 7.14

| Remark   An immediate consequence of Theorems 7.9 and 7.10 is that a strictly monotonic function possesses an inverse.

EXAMPLE 2   The existence of an inverse

Determine which of the following functions has an inverse.
(a) $f(x) = x^3 + x - 1$ (b) $f(x) = x^3 - x + 1$

Solution:
(a) From the graph of f given in Figure 7.14, it appears that f is increasing over its entire domain. To verify this, we note that the derivative, $f'(x) = 3x^2 + 1$, is positive for all real values of x. Therefore, f is strictly monotonic and it must have an inverse.
(b) From the graph given in Figure 7.15, we can see that the function does not pass the horizontal line test. In other words, it is not one-to-one. For instance, f has the same value when $x = -1$, 0, and 1.

$$f(-1) = f(1) = f(0) = 1 \qquad \text{Not one-to-one}$$

Therefore, by Theorem 7.9, f does not have an inverse.

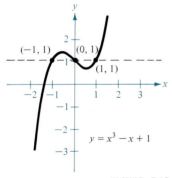

$y = x^3 - x + 1$

FIGURE 7.15

| Remark   It is often easier to prove that a function has an inverse than to actually find the inverse. For instance, it would be difficult to algebraically determine the inverse of the function in Example 2(a).

FINDING AN INVERSE FUNCTION

Use Theorem 7.9 to determine whether the function given by $y = f(x)$ has an inverse. If it does, use the following steps:

1. Solve for x as a function of y: $x = g(y) = f^{-1}(y)$.

2. Define the domain of f^{-1} to be the range of f.

To avoid the confusion that could arise from using y as the independent variable for f^{-1}, it is customary to write f^{-1} as a *function of x* by simply interchanging the variables x and y after solving for x. This is illustrated in the next example.

EXAMPLE 3   *Finding the inverse of a function*

Find the inverse of the function given by $f(x) = \sqrt{2x - 3}$.

Solution: We begin by observing that f is increasing on its entire domain, as shown in Figure 7.16. To find an equation for this inverse, we let $y = f(x)$ and solve for x in terms of y.

$$\sqrt{2x - 3} = y \qquad \text{Let } y = f(x)$$

$$2x - 3 = y^2$$

$$x = \frac{y^2 + 3}{2} \qquad \text{Solve for } x$$

$$f^{-1}(y) = \frac{y^2 + 3}{2}$$

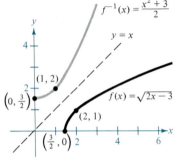

FIGURE 7.16

Since the range of f is $[0, \infty)$, we define this interval to be the domain of f^{-1}. Finally, using x as the independent variable, we have

$$f^{-1}(x) = \frac{x^2 + 3}{2}, \quad 0 \le x \qquad \text{Determine domain} \qquad \square$$

| **Remark** Remember that any letter can be used to represent the independent variable. Thus,

$$f^{-1}(y) = \frac{y^2 + 3}{2}, \qquad f^{-1}(x) = \frac{x^2 + 3}{2}, \qquad \text{and} \qquad f^{-1}(s) = \frac{s^2 + 3}{2}$$

all represent the same function.

Theorem 7.10 is useful in the following type of problem. Suppose you are given a function that is *not* one-to-one on its domain. If you restrict the domain to an interval on which the function is strictly monotonic, you can conclude that the new function *is* one-to-one on the restricted domain. The next example illustrates this procedure.

EXAMPLE 4   *Finding an interval on which a function is one-to-one*

Show that the function, $f(x) = x^3 - 3x$, is not one-to-one on the entire real line. Then show that $[-1, 1]$ is the largest interval, centered at the origin, for which f is strictly monotonic.

Solution: It is clear that f is not one-to-one since different x-values yield the same y-value. For instance, $f(-\sqrt{3}) = f(0) = f(\sqrt{3}) = 0$. Moreover, f is decreasing on the open interval $(-1, 1)$ since its derivative

$$f'(x) = 3x^2 - 3$$

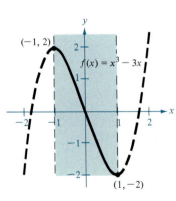

FIGURE 7.17

is negative there. Finally, since the left and right endpoints correspond to relative extrema of f, we can conclude that f is decreasing on the closed interval $[-1, 1]$ *and* that in any larger interval containing the origin the function would not be strictly monotonic. (See Figure 7.17.)

The next two theorems discuss the derivative of an inverse function. Although we state the first theorem without proof, its reasonableness follows from the reflective property of inverse functions as shown in Figure 7.13.

THEOREM 7.11   **CONTINUITY AND DIFFERENTIABILITY OF INVERSE FUNCTIONS**
Let f be a function that possesses an inverse.

1. If f is continuous, then f^{-1} is continuous.
2. If f is increasing, then f^{-1} is increasing.
   If f is decreasing, then f^{-1} is decreasing.
3. If f is differentiable at c and $f'(c) \neq 0$, then f^{-1} is differentiable at $f(c)$.

THEOREM 7.12   **THE DERIVATIVE OF AN INVERSE FUNCTION**
If f is a differentiable function that possesses an inverse function g, then

$$g'(x) = \frac{1}{f'(g(x))}, \quad f'(g(x)) \neq 0$$

Proof: From Theorem 7.11 we know that g is differentiable. Using the Chain Rule we differentiate both sides of the equation

$$x = f(g(x))$$

to obtain

$$1 = f'(g(x))\frac{d}{dx}[g(x)]$$

Since $f'(g(x)) \neq 0$, we can divide by this quantity to obtain

$$\frac{d}{dx}[g(x)] = \frac{1}{f'(g(x))}$$

Geometrically, Theorem 7.12 tells us that the graphs of inverse functions have reciprocal slopes at the points (a, b) and (b, a) as illustrated in the next example. If we let $y = g(x)$, then the equation

$$\frac{dy}{dx} = \frac{1}{dx/dy}$$

gives us an easy way to remember this reciprocal relationship.

EXAMPLE 5   *Graphs of inverse functions have reciprocal slopes*

Let $f(x) = x^2$ (for $x \geq 0$) and let $f^{-1}(x) = \sqrt{x}$. Show that the slopes of the graphs of f and f^{-1} are reciprocals at the following points:
(a) (2, 4) and (4, 2)
(b) (3, 9) and (9, 3)

Solution: The derivatives of f and f^{-1} are given by

$$f'(x) = 2x$$

and

$$(f^{-1})'(x) = \frac{1}{2\sqrt{x}}$$

(a) At (2, 4) the slope of the graph of f is $f'(2) = 2(2) = 4$. At (4, 2), the slope of the graph of f^{-1} is

$$(f^{-1})'(4) = \frac{1}{2\sqrt{4}} = \frac{1}{2(2)} = \frac{1}{4}$$

(b) At (3, 9) the slope of the graph of f is $f'(3) = 2(3) = 6$. At (9, 3), the slope of the graph of f^{-1} is

$$(f^{-1})'(9) = \frac{1}{2\sqrt{9}} = \frac{1}{2(3)} = \frac{1}{6}$$

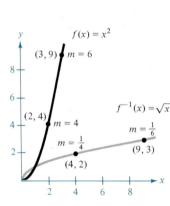

FIGURE 7.18   Thus, in both cases, the slopes are reciprocals as shown in Figure 7.18.

| **Remark**   In Example 5, note that f^{-1} is not differentiable at (0, 0). This is consistent with Theorem 7.11 since the derivative of f at (0, 0) is zero.

Section Exercises 7.4

In Exercises 1–8, (a) show that f and g are inverse functions by showing that $f(g(x)) = x$ and $g(f(x)) = x$, and (b) graph f and g on the same set of coordinate axes.

1. $f(x) = x^3$; $g(x) = \sqrt[3]{x}$

2. $f(x) = \dfrac{1}{x}$; $g(x) = \dfrac{1}{x}$

3. $f(x) = 5x + 1$; $g(x) = \dfrac{x - 1}{5}$

4. $f(x) = 3 - 4x$; $g(x) = \dfrac{3 - x}{4}$

5. $f(x) = \sqrt{x - 4}$; $g(x) = x^2 + 4,\ x \geq 0$

6. $f(x) = 9 - x^2,\ x \geq 0$; $g(x) = \sqrt{9 - x},\ x \leq 9$

7. $f(x) = 1 - x^3$; $g(x) = \sqrt[3]{1 - x}$

8. $f(x) = \dfrac{1}{1 + x^2},\ x \geq 0$

     $g(x) = \sqrt{\dfrac{1 - x}{x}},\ 0 < x \leq 1$

In Exercises 9–22, find the inverse of f. Then graph both f and f^{-1}.

9. $f(x) = 2x - 3$ **10.** $f(x) = 3x$

11. $f(x) = x^5$ **12.** $f(x) = x^3 + 1$

13. $f(x) = \sqrt{x}$ **14.** $f(x) = x^2,\ x \geq 0$

15. $f(x) = \sqrt{4 - x^2},\ 0 \leq x \leq 2$

16. $f(x) = \sqrt{x^2 - 4},\ x \geq 2$

17. $f(x) = \sqrt[3]{x - 1}$

18. $f(x) = 3\sqrt[3]{2x - 1}$

19. $f(x) = x^{2/3},\ x \geq 0$ **20.** $f(x) = x^{3/5}$

21. $f(x) = \dfrac{x}{\sqrt{x^2 + 7}}$ **22.** $f(x) = \dfrac{x + 2}{x}$

In Exercises 23–30, use the horizontal line test to determine if the function is one-to-one (strictly monotonic) on its entire domain and therefore has an inverse.

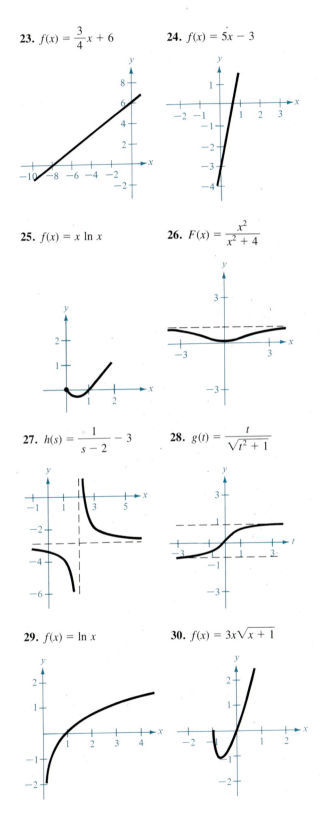

23. $f(x) = \dfrac{3}{4}x + 6$

24. $f(x) = 5x - 3$

25. $f(x) = x \ln x$

26. $F(x) = \dfrac{x^2}{x^2 + 4}$

27. $h(s) = \dfrac{1}{s - 2} - 3$

28. $g(t) = \dfrac{t}{\sqrt{t^2 + 1}}$

29. $f(x) = \ln x$

30. $f(x) = 3x\sqrt{x + 1}$

In Exercises 31–36, use the derivative to determine if the continuous function is strictly monotonic on its entire domain and therefore has an inverse.

31. $f(x) = (x + a)^3 + b$ **32.** $f(x) = x^3 + x + 1$

33. $f(x) = \dfrac{x^4}{4} - 2x^2$

34. $f(x) = x^3 - 6x^2 + 12x - 8$
35. $f(x) = 2 - x - x^3$ **36.** $f(x) = \ln(x - 3)$

In Exercises 37–40, show that f is strictly monotonic on the given interval and therefore has an inverse on that interval.

Function	Interval
37. $f(x) = (x - 4)^2$	$[4, \infty)$
38. $f(x) = \|x + 2\|$	$[-2, \infty)$
39. $f(x) = \dfrac{4}{x^2}$	$(0, \infty)$
40. $f(x) = 2x^3 - 3x^2$	$[0, 1]$

In Exercises 41–44, show that the slopes of the graphs of f and f^{-1} are reciprocals at the given points.

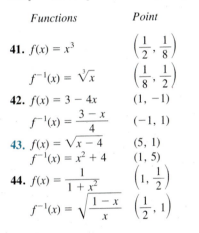

Functions	Point
41. $f(x) = x^3$	$\left(\dfrac{1}{2}, \dfrac{1}{8}\right)$
$f^{-1}(x) = \sqrt[3]{x}$	$\left(\dfrac{1}{8}, \dfrac{1}{2}\right)$
42. $f(x) = 3 - 4x$	$(1, -1)$
$f^{-1}(x) = \dfrac{3 - x}{4}$	$(-1, 1)$
43. $f(x) = \sqrt{x - 4}$	$(5, 1)$
$f^{-1}(x) = x^2 + 4$	$(1, 5)$
44. $f(x) = \dfrac{1}{1 + x^2}$	$\left(1, \dfrac{1}{2}\right)$
$f^{-1}(x) = \sqrt{\dfrac{1 - x}{x}}$	$\left(\dfrac{1}{2}, 1\right)$

In Exercises 45 and 46, the derivative of the function has the same sign for all x in its domain, but the function is not monotonic. Explain why.

45. $f(x) = \dfrac{1}{x}$ **46.** $f(x) = \dfrac{x}{x^2 - 4}$

47. Prove that if a function has an inverse, then the inverse is unique.
48. Prove that if f has an inverse, then $(f^{-1})^{-1} = f$.
49. Prove Theorem 7.9.

7.5
Exponential functions and differentiation

We know what it means to raise a real number x to a *rational* power n. However, we have yet to give a meaning to the exponential expression x^n for *irrational n*. In this section we will accomplish this by means of the inverse of the natural logarithmic function.

In Section 7.1 we saw that the function $f(x) = \ln x$ is increasing on its entire domain, and hence it has an inverse f^{-1}. Moreover, from the domain and range of f, we can conclude that the domain of f^{-1} is the set of all reals, and the range is the set of positive reals, as shown in Figure 7.19. Thus, for any real number x,

$$f(f^{-1}(x)) = \ln (f^{-1}(x)) = x \qquad x \text{ is any real number}$$

We also know that if x happens to be rational, then

$$\ln (e^x) = x \ln e = x(1) = x \qquad x \text{ is a rational number}$$

Since the natural logarithmic function is one-to-one, we conclude that $f^{-1}(x)$ and e^x agree for *rational* values of x. Because of this agreement, we extend the definition of e^x to cover *all* real numbers.

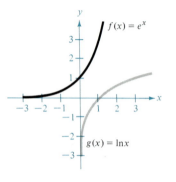

FIGURE 7.19

DEFINITION OF THE NATURAL EXPONENTIAL FUNCTION

The inverse of the natural logarithmic function $f(x) = \ln x$ is called the **natural exponential function** and is denoted by

$$f^{-1}(x) = e^x$$

That is,

$$y = e^x \quad \text{if and only if} \quad x = \ln y$$

We summarize the inverse relationship between the natural logarithmic function and the natural exponential function as follows.

INVERSE RELATIONSHIP OF NATURAL LOGARITHMIC AND EXPONENTIAL FUNCTIONS

$$\ln (e^x) = x \qquad \text{and} \qquad e^{\ln x} = x$$

These two properties are useful in solving equations involving exponential and logarithmic functions, as we demonstrate in the following example.

EXAMPLE 1   Solving exponential and logarithmic equations

Solve for x in the following equations:
(a) $7 = e^{x+1}$ (b) $\ln (2x - 3) = 5$

Solution:

(a) We can convert from exponential form to logarithmic form by *taking the natural log of both sides* of the exponential equation.

$$7 = e^{x+1} \Rightarrow \ln 7 = \ln (e^{x+1}) \Rightarrow \ln 7 = x + 1$$

Thus,

$$x = -1 + \ln 7 \approx 0.946$$

(b) To convert from logarithmic form to exponential form, we *apply the exponential function to both sides* of the logarithmic equation.

$$\ln (2x - 3) = 5 \Rightarrow e^{\ln (2x-3)} = e^5 \Rightarrow 2x - 3 = e^5$$

Thus,

$$x = \frac{e^5 + 3}{2} \approx 75.707$$

The familiar rules for operating with rational exponents can be extended to the natural exponential function, as indicated in the following theorem.

THEOREM 7.13   **OPERATIONS WITH EXPONENTIAL FUNCTIONS**

Let a and b be any real numbers. Then the following properties are true.

1. $e^a e^b = e^{a+b}$ 2. $\dfrac{e^a}{e^b} = e^{a-b}$ 3. $(e^a)^b = e^{ab}$

Proof: We prove Property 1 and leave the proofs of the other two properties as exercises.

$$\ln (e^a e^b) = \ln (e^a) + \ln (e^b) = a + b = \ln (e^{a+b})$$

Thus, since the natural log function is one-to-one, we conclude that

$$e^a e^b = e^{a+b}$$

In Section 7.4, we saw that an inverse function f^{-1} inherits many properties from f. Thus, the natural exponential function inherits the following properties from the natural logarithmic function. (See Figure 7.20.)

1. The domain of $f(x) = e^x$ is $(-\infty, \infty)$, and the range is $(0, \infty)$.
2. The function $f(x) = e^x$ is continuous, increasing, and one-to-one on its entire domain.
3. The graph of $f(x) = e^x$ is concave upward on its entire domain.
4. $\lim\limits_{x \to -\infty} e^x = 0$ and $\lim\limits_{x \to \infty} e^x = \infty$

One of the most intriguing (and useful) characteristics of the natural exponential function is that *it is its own derivative*. In other words, it is a solution to the differential equation $y' = y$. This result is stated in the next theorem.

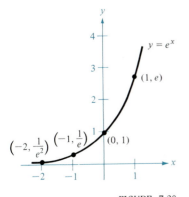

FIGURE 7.20

THEOREM 7.14   THE DERIVATIVE OF THE NATURAL EXPONENTIAL FUNCTION
If u is a differentiable function of x, then

$$1. \ \frac{d}{dx}[e^x] = e^x \qquad 2. \ \frac{d}{dx}[e^u] = e^u \frac{du}{dx}$$

Proof: Let $f(x) = \ln x$ and $g(x) = e^x$. Then $f'(x) = 1/x$, and by Theorem 7.12 we have

$$\frac{d}{dx}[e^x] = \frac{d}{dx}[g(x)] = \frac{1}{f'(g(x))} = \frac{1}{f'(e^x)} = \frac{1}{(1/e^x)} = e^x$$

The derivative of e^u follows from the Chain Rule.

| **Remark** We can interpret this theorem geometrically by saying that the slope of the graph of $f(x) = e^x$ at any point (x, e^x) is numerically equal to the y-coordinate of the point.

EXAMPLE 2   Differentiating exponential functions

(a) $\dfrac{d}{dx}[e^{2x-1}] = \dfrac{du}{dx}e^u = 2e^{2x-1}$ $u = 2x - 1$

(b) $\dfrac{d}{dx}[e^{-3/x}] = \dfrac{du}{dx}e^u = \left(\dfrac{3}{x^2}\right)e^{-3/x} = \dfrac{3e^{-3/x}}{x^2}$ $u = -\dfrac{3}{x}$

EXAMPLE 3   Locating relative extrema

Find the relative extrema of $f(x) = xe^x$.

Solution: The derivative of f is given by

$$f'(x) = x(e^x) + e^x(1) \qquad \text{Product Rule}$$
$$= e^x(x + 1)$$

Now, since e^x is never zero, the derivative is zero only when $x = -1$. Moreover, by the First Derivative Test, we can determine that this corresponds to a relative minimum, as shown in Figure 7.21. Since the derivative $f'(x) = e^x(x + 1)$ is defined for all x, there are no other critical points.

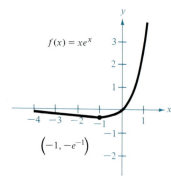

$f(x) = xe^x$

$(-1, -e^{-1})$

FIGURE 7.21

EXAMPLE 4   The normal probability density function

Show that the *normal probability density function*

$$f(x) = \frac{1}{\sqrt{2\pi}}e^{-x^2/2}$$

has points of inflection when $x = \pm 1$.

Solution: To locate possible points of inflection, we find the *x*-values for which the second derivative is zero.

$$f'(x) = \frac{1}{\sqrt{2\pi}}(-x)e^{-x^2/2}$$

$$f''(x) = \frac{1}{\sqrt{2\pi}}[(-x)(-x)e^{-x^2/2} + (-1)e^{-x^2/2}] \quad \text{Product Rule}$$

$$= \frac{1}{\sqrt{2\pi}}(e^{-x^2/2})(x^2 - 1)$$

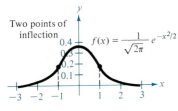

Two points of inflection

$$f(x) = \frac{1}{\sqrt{2\pi}}e^{-x^2/2}$$

Bell–Shaped Curve Given By Normal Probability Density Function

FIGURE 7.22

Therefore, $f''(x) = 0$ when $x = \pm 1$, and we can apply the techniques of Chapter 4 to conclude that these values yield the two points of inflection shown in Figure 7.22.

Remark The general form of a normal probability density function is given by

$$f(x) = \frac{1}{\sigma\sqrt{2\pi}}e^{-x^2/2\sigma^2}$$

where σ is the standard deviation (σ is the lowercase Greek letter sigma). By following the procedure of Example 4, we can show that the bell-shaped curve of this function has points of inflection when $x = \pm\sigma$.

Bases other than *e*

The **base** of the natural exponential function is *e*. Using the third property in Theorem 7.13, we now assign a meaning to the exponential function $y = a^x$.

DEFINITION OF EXPONENTIAL FUNCTION TO BASE *a*	If *a* is a positive real number ($a \neq 1$) and *x* is any real number, then $f(x) = a^x$ is given by $$f(x) = a^x = (e^{\ln a})^x = e^{(\ln a)x}$$ We call *f* the **exponential function to the base *a*.**

We can give a similar definition for logarithmic functions to other bases.

DEFINITION OF LOGARITHMIC FUNCTION TO BASE *a*	If *a* is a positive real number ($a \neq 1$) and *x* is any positive real number, then $f(x) = \log_a x$ is given by $$f(x) = \log_a x = \frac{1}{\ln a}\ln x$$ We call *f* the **logarithmic function to the base *a*.**

When you studied exponential and logarithmic functions in algebra, you encountered the following relationship:

$$y = a^x \quad \text{if and only if} \quad x = \log_a y$$

We can now prove this relationship for the functions a^x and $\log_a x$ *as we have defined them.* (We leave the proof to you.)

> **Remark**   The logarithmic function to the base 10 is called the **common logarithmic function.** Thus, for common logarithms, we have
>
> $$y = 10^x \quad \text{if and only if} \quad x = \log_{10} y$$

To differentiate exponential and logarithmic functions to other bases, you have two options: (1) use the definition of a^x and $\log_a x$ and differentiate using the rules for the natural exponential and logarithmic functions, or (2) use the following differentiation rules for bases other than e.

THEOREM 7.15   **DERIVATIVES FOR BASES OTHER THAN e**
Let a be a positive real number ($a \neq 1$) and let u be a differentiable function of x.

1. $\dfrac{d}{dx}[a^x] = (\ln a)a^x$ 2. $\dfrac{d}{dx}[a^u] = (\ln a)a^u \dfrac{du}{dx}$

3. $\dfrac{d}{dx}[\log_a x] = \dfrac{1}{(\ln a)x}$ 4. $\dfrac{d}{dx}[\log_a u] = \dfrac{1}{(\ln a)u}\dfrac{du}{dx}$

Proof: By definition, $a^x = e^{(\ln a)x}$. Therefore, we can prove the first property by letting $u = (\ln a)x$ and differentiating with base e to obtain

$$\frac{d}{dx}[a^x] = \frac{d}{dx}[e^{(\ln a)x}] = e^u \frac{du}{dx} = e^{(\ln a)x}(\ln a) = (\ln a)a^x$$

The other three parts of this theorem can be proved in a similar way.

> **Remark**   Note that these differentiation rules are similar to those for the natural exponential function and natural logarithmic function. In fact, they differ only by the constant factors $\ln a$ and $1/\ln a$, respectively. This points out one reason why, for calculus, e is the most convenient base.

EXAMPLE 5   *Differentiating functions to other bases*

Find the derivatives of
(a) $y = 2^x$
(b) $y = 2^{\sqrt{x}}$

Solution:

(a) $y' = \dfrac{d}{dx}[2^x] = (\ln 2)2^x$

(b) Let $u = \sqrt{x}$, then $u' = 1/(2\sqrt{x})$ and we have

$$y' = \dfrac{d}{dx}[2^{\sqrt{x}}] = (\ln 2)2^{\sqrt{x}}\left(\dfrac{1}{2\sqrt{x}}\right) = \dfrac{\ln 2}{2\sqrt{x}}2^{\sqrt{x}}$$

Now that we have defined real exponents, we can extend the Power Rule to cover this case. Try proving this theorem using logarithmic differentiation.

THEOREM 7.16   **THE POWER RULE FOR REAL EXPONENTS**
Let n be any real number and let u be a differentiable function of x.

1. $\dfrac{d}{dx}[x^n] = nx^{n-1}$

2. $\dfrac{d}{dx}[u^n] = nu^{n-1}\dfrac{du}{dx}$

In the next example, we differentiate four different types of functions. Each function uses a different differentiation formula, depending on whether the base and exponent are constants or variables.

EXAMPLE 6   Comparing variables and constants

(a) $\dfrac{d}{dx}[e^e] = 0$ Constant Rule

(b) $\dfrac{d}{dx}[e^x] = e^x$ Exponential Rule

(c) $\dfrac{d}{dx}[x^e] = ex^{e-1}$ Power Rule

(d) $y = x^x$ Logarithmic differentiation
       $\ln y = x \ln x$
       $\dfrac{y'}{y} = x\left(\dfrac{1}{x}\right) + (\ln x)(1) = 1 + \ln x$
       $y' = y(1 + \ln x) = x^x(1 + \ln x)$

Applications of exponential functions

We now look at a business problem in which the limit definition of e plays an important role. Suppose P dollars are deposited in a savings account at an annual interest rate r. If accumulated interest is deposited into the account, what is the balance in the account at the end of one year? Of course, the answer depends on the number of times the interest is compounded according to the formula.

$$A = P\left(1 + \frac{r}{n}\right)^n$$

where n is the number of compoundings per year. For instance, the results for a deposit of $1000 at 8 percent interest compounded n times a year are as shown in Table 7.5.

TABLE 7.5

n	1	2	4	12	365
A	$1080.00	$1081.60	$1082.43	$1083.00	$1083.28

It may surprise you to discover that as n increases, the balance A approaches a limit, as indicated in the following development. In this development, we let $x = n/r$. Then $x \to \infty$ as $n \to \infty$, and we have

$$A = \lim_{n\to\infty} P\left(1 + \frac{r}{n}\right)^n = P \lim_{n\to\infty} \left[\left(1 + \frac{1}{n/r}\right)^{n/r}\right]^r$$

$$= P\left[\lim_{x\to\infty}\left(1 + \frac{1}{x}\right)^x\right]^r$$

$$= Pe^r$$

We call this limit the balance after one year of **continuous compounding.** Thus, for a deposit of $1000 at 8 percent interest, compounded continuously, the balance at the end of one year would be

$$A = 1000e^{0.08} \approx \$1083.29$$

We summarize these results as follows.

SUMMARY OF COMPOUND INTEREST FORMULAS

Let P = amount of deposit, t = number of years, A = balance after t years, and r = annual interest rate (decimal form).

1. Compounded n times per year: $A = P\left(1 + \frac{r}{n}\right)^{nt}$

2. Compounded continuously: $A = Pe^{rt}$

EXAMPLE 7   Comparing continuous compounding to quarterly compounding

A deposit of $2500 is made in an account that pays an annual interest rate of 10 percent. Find the balance in the account at the end of 5 years if the interest is compounded (a) quarterly and (b) continuously.

Solution:

(a) The balance after quarterly compounding is

$$A = P\left(1 + \frac{r}{n}\right)^{nt} = 2500\left(1 + \frac{0.1}{4}\right)^{4(5)}$$
$$= 2500(1.025)^{20} = \$4096.54$$

(b) The balance after continuous compounding is

$$A = Pe^{rt} = 2500[e^{0.1(5)}]$$
$$= 2500e^{0.5} = \$4121.80$$

EXAMPLE 8   An application to biology

A bacterial culture is growing according to the *logistics growth function*

$$y = \frac{1.25}{1 + 0.25e^{-0.4t}}, \quad 0 \le t$$

where y is the weight of the culture in grams and t is the time in hours. Find the weight of the culture after (a) 0 hours, (b) 1 hour, and (c) 10 hours. (d) What is the limit as t approaches infinity?

Solution:

(a) When $t = 0$,

$$y = \frac{1.25}{1 + 0.25e^0} = \frac{1.25}{1.25} = 1 \text{ g}$$

(b) When $t = 1$,

$$y = \frac{1.25}{1 + 0.25e^{-0.4}} \approx 1.070 \text{ g}$$

(c) When $t = 10$,

$$y = \frac{1.25}{1 + 0.25e^{-4}} \approx 1.244 \text{ g}$$

(d) Finally, taking the limit as t approaches infinity, we have

$$\lim_{t \to \infty} \frac{1.25}{1 + 0.25e^{-0.4t}} = \frac{1.25}{1 + 0} = 1.25 \text{ g}$$

(See Figure 7.23.)

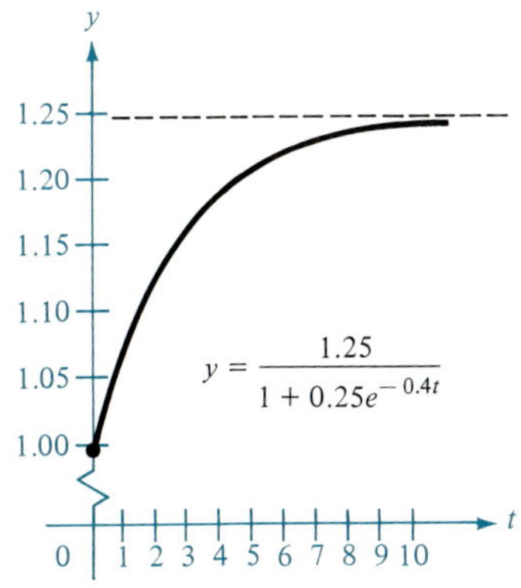

FIGURE 7.23

$$y = \frac{1.25}{1 + 0.25e^{-0.4t}}$$

Section Exercises 7.5

In Exercises 1–6, write the logarithmic equation as an exponential equation and vice versa.

1. (a) $2^3 = 8$ (b) $3^{-1} = \dfrac{1}{3}$

2. (a) $27^{2/3} = 9$ (b) $16^{3/4} = 8$

3. (a) $\log_{10} 0.01 = -2$ (b) $\log_{0.5} 8 = -3$
4. (a) $e^0 = 1$ (b) $e^2 = 7.389...$
5. (a) $\ln 2 = 0.6931...$ (b) $\ln 8.4 = 2.128...$
6. (a) $\ln 0.5 = -0.6931...$ (b) $49^{1/2} = 7$

In Exercises 7–14, solve for x (or b).

7. (a) $\log_{10} 1000 = x$ (b) $\log_{10} 0.1 = x$

8. (a) $\log_4 \dfrac{1}{64} = x$ (b) $\log_5 25 = x$

9. (a) $\log_3 x = -1$ (b) $\log_2 x = -4$

10. (a) $\log_b 27 = 3$ (b) $\log_b 125 = 3$

11. (a) $\log_{27} x = -\dfrac{2}{3}$ (b) $\ln e^x = 3$

12. (a) $e^{\ln x} = 4$ (b) $\ln x = 2$

13. (a) $x^2 - x = \log_5 25$

     (b) $3x + 5 = \log_2 64$

14. (a) $\log_3 x + \log_3 (x - 2) = 1$

     (b) $\log_{10} (x + 3) - \log_{10} x = 1$

In Exercises 15–20, sketch the graph of the given function.

15. $y = 3^x$ **16.** $y = 3^{x-1}$

17. $y = \left(\dfrac{1}{3}\right)^x$ **18.** $y = 2^{x^2}$

19. $y = e^{-x^2}$ **20.** $y = e^{-x/2}$

In Exercises 21–26, show that the given functions are inverses of each other by sketching their graphs on the same coordinate system.

21. $f(x) = 4^x$, $g(x) = \log_4 x$

22. $f(x) = 3^x$, $g(x) = \log_3 x$

23. $f(x) = e^{2x}$, $g(x) = \ln \sqrt{x}$

24. $f(x) = e^{x/3}$, $g(x) = \ln x^3$

25. $f(x) = e^x - 1$, $g(x) = \ln (x + 1)$

26. $f(x) = e^{x-1}$, $g(x) = 1 + \ln x$

27. Complete the accompanying table to demonstrate that e can also be defined as

$$\lim_{x \to 0} (1 + x)^{1/x}$$

x	1	10^{-1}	10^{-2}	10^{-4}	10^{-6}
$(1 + x)^{1/x}$					

28. Show that

$$\log_{10} 2 = \frac{\ln 2}{\ln 10}$$

In Exercises 29 and 30, compare the given number to e.

29. (a) $\dfrac{271{,}801}{99{,}990}$ (b) $\dfrac{299}{110}$

30. (a) $1 + 1 + \dfrac{1}{2} + \dfrac{1}{6} + \dfrac{1}{24}$

     (b) $1 + 1 + \dfrac{1}{2} + \dfrac{1}{6} + \dfrac{1}{24} + \dfrac{1}{120} + \dfrac{1}{720} + \dfrac{1}{5040}$

31. Find the amount of time necessary for P dollars to double if it is compounded continuously at $7\frac{1}{2}$-percent interest. Find the time necessary for it to triple.

32. Complete the accompanying table to find the time t necessary for P dollars to double if interest is compounded continuously at the rate r.

r	2%	4%	6%	8%	10%	12%
t						

33. If \$1,000 is invested at $7\frac{1}{2}$-percent interest, find the amount after 10 years if interest is compounded

  (a) annually   (b) semiannually

  (c) quarterly   (d) monthly

  (e) daily   (f) continuously

34. If \$2,500 is invested at 12-percent interest, find the amount after 20 years if interest is compounded

  (a) annually   (b) semiannually

  (c) quarterly   (d) monthly

  (e) daily   (f) continuously

In Exercises 35 and 36 find the slope of the tangent line to the given exponential function at the point (0, 1).

35. (a)   (b)

36. (a)   (b)

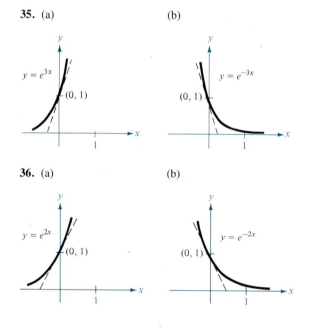

In Exercises 37–68, find dy/dx.

37. $y = e^{2x}$ **38.** $y = e^{1-x}$

39. $y = e^{-2x+x^2}$ **40.** $y = e^{-x^2}$

41. $y = e^{\sqrt{x}}$ **42.** $y = x^2 e^{-x}$

43. $y = (e^{-x} + e^x)^3$

44. $y = e^{-1/x^2}$

45. $y = \ln (e^{x^2})$

46. $y = \ln \left(\dfrac{1 + e^x}{1 - e^x} \right)$

47. $y = \ln (1 + e^{2x})$

48. $y = \dfrac{2}{e^x + e^{-x}}$

49. $y = \ln \dfrac{e^x + e^{-x}}{2}$

50. $y = xe^x - e^x$

51. $y = x^2 e^x - 2xe^x + 2e^x$

52. $y = \dfrac{e^x - e^{-x}}{2}$

53. $y = 5^{x-2}$

54. $y = x(7^{-3x})$

55. $y = e^{-x} \ln x$

56. $y = 2^{x^2} 3^{-x}$

57. $y = \ln e^x$

58. $y = \ln (e^{-x})$

59. $y = \log_3 x$

60. $y = \log_3 \dfrac{x\sqrt{x - 1}}{2}$

61. $y = \log_5 \sqrt{x^2 - 1}$

62. $y = \log_{10} \dfrac{x^2 - 1}{x}$

63. $y = x^{2/x}$

64. $y = x^{x-1}$

65. $y = (x - 2)^{x+1}$

66. $y = (1 + x)^{1/x}$

In Exercises 67–72, find the extrema and the points of inflection (if any exist), and sketch the graph of the function.

67. $f(x) = \dfrac{1}{\sqrt{2\pi}} e^{-(x^2/2)}$

68. $f(x) = \dfrac{e^x - e^{-x}}{2}$

69. $f(x) = \dfrac{e^x + e^{-x}}{2}$

70. $f(x) = xe^{-x}$

71. $f(x) = x^2 e^{-x}$

72. $f(x) = -2 + e^{3x}(4 - 2x)$

73. Find an equation of the line normal to the graph of $y = e^{-x}$ at $(0, 1)$.

74. Find the point on the graph of $y = e^{-x}$ where the normal line to the curve will pass through the origin.

75. Find the area of the largest rectangle that can be inscribed under the curve $y = e^{-x^2}$ in the first and second quadrants.

76. Find, to three decimal places, the value of x so that $e^{-x} = x$. [Use Newton's Method.]

77. The yield V (in millions of cubic feet per acre) for a forest stand at age t is given by

$$V = 6.7 e^{(-48.1)/t}$$

where t is measured in years.
(a) Find the limiting volume of wood per acre as t approaches infinity.
(b) Find the rate at which the yield is increasing when $t = 20$ and $t = 60$ years.

78. The average typing speed (in the number of words per minute) after t weeks of lessons is given by

$$N = \dfrac{157}{1 + 5.4 e^{-0.12t}}$$

(a) Find the limiting number of words per minute as t approaches infinity.

(b) Find the rate at which typing speed is increasing when $t = 5$ and $t = 25$ weeks.

79. In a group project in learning theory, a mathematical model for the proportion P of correct responses after n trials was found to be

$$P = \dfrac{0.83}{1 + e^{-0.2n}}$$

(a) Find the limiting proportion of correct responses as n approaches infinity.
(b) Find the rate at which P is increasing after $n = 3$ and $n = 10$ trials.

80. A lake is stocked with 500 fish, and their population increases according to the **logistics curve**

$$p(t) = \dfrac{10,000}{1 + 19 e^{-t/5}}$$

where t is measured in months. At what rate is the fish population increasing at the end of 1 month and at the end of 10 months? After how many months is the population increasing most rapidly? (See Figure 7.24.)

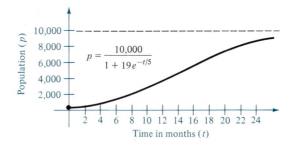

FIGURE 7.24

81. The balance in a certain savings account is given by

$$A = (5000)e^{0.08t}$$

where A is measured in dollars and t is measured in years. Find the rate at which the balance is increasing when
(a) $t = 1$ year   (b) $t = 10$ years
(c) $t = 50$ years

82. Prove Properties 2 and 3 of Theorem 7.13.

7.6
Integration of exponential functions: Growth and decay

Each of the differentiation formulas for exponential functions has its corresponding integration formula, as shown next.

THEOREM 7.17   **INTEGRATION RULES FOR EXPONENTIAL FUNCTIONS**
Let u be a differentiable function of x.

$$1. \int e^x \, dx = e^x + C \qquad 2. \int e^u \, du = e^u + C$$

EXAMPLE 1   *Integrating exponential functions*

Evaluate $\int e^{3x+1} \, dx$.

Solution: Considering $u = 3x + 1$, we have $du = 3 \, dx$. Then

$$\int e^{3x+1} \, dx = \frac{1}{3} \int e^{3x+1}(3) \, dx = \frac{1}{3} \int e^u \, du$$

$$= \frac{1}{3} e^u + C$$

$$= \frac{e^{3x+1}}{3} + C$$

Remark   In Example 1, we introduced the missing *constant* factor 3 to create $du = 3 \, dx$. However, remember that you cannot introduce a missing *variable* factor in the integrand. For instance,

$$\int e^{-x^2} \, dx \neq \frac{1}{x} \int e^{-x^2}(x \, dx)$$

EXAMPLE 2   *Integrating exponential functions*

Evaluate $\int 5xe^{-x^2} \, dx$.

Solution: If we let $u = -x^2$, then $du = -2x \, dx$ or $x \, dx = -du/2$. Thus, we have

$$\int 5xe^{-x^2} \, dx = \int 5e^{-x^2}(x \, dx) = \int 5e^u\left(-\frac{du}{2}\right)$$

$$= -\frac{5}{2} \int e^u \, du$$

$$= -\frac{5}{2} e^u + C$$

$$= -\frac{5}{2} e^{-x^2} + C$$

EXAMPLE 3   Integrating exponential functions

(a) $\displaystyle\int \frac{e^{1/x}}{x^2}\,dx = -\int \underbrace{e^{1/x}}_{e^u}\underbrace{\left(-\frac{1}{x^2}\right)dx}_{du} = -e^{1/x} + C$ $u = \dfrac{1}{x}$

(b) $\displaystyle\int (1 + e^x)^2\,dx = \int (1 + 2e^x + e^{2x})\,dx = x + 2e^x + \frac{1}{2}e^{2x} + C$

EXAMPLE 4   Finding area bounded by exponential functions

(a) $\displaystyle\int_0^1 e^{-x}\,dx = \left[-e^{-x}\right]_0^1 = -e^{-1} - (-1) = 1 - \frac{1}{e} \approx 0.632$

(b) $\displaystyle\int_0^1 \frac{e^x}{1 + e^x}\,dx = \left[\ln (1 + e^x)\right]_0^1 = \ln (1 + e) - \ln 2 \approx 0.620$

(See Figure 7.25.)

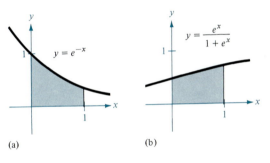

(a)   (b)

FIGURE 7.25

Occasionally, an integrand involves an exponential function to a base other than e. When this occurs, you have two options: (1) convert to base e using the formula $a^x = e^{(\ln a)x}$ and then integrate, or (2) integrate directly, using the following integration formula:

$$\int a^x\,dx = \left(\frac{1}{\ln a}\right)a^x + C$$

EXAMPLE 5   Integrating an exponential function to another base

Evaluate $\int 2^x\,dx$.

Solution:

$$\int 2^x\,dx = \frac{1}{\ln 2}2^x + C$$

| Remark   The integral in Example 5 can also be evaluated by converting to base e as follows

$$\int 2^x\,dx = \int e^{(\ln 2)x}\,dx = \frac{1}{\ln 2}e^{(\ln 2)x} + C = \frac{1}{\ln 2}2^x + C$$

Exponential growth and decay

One very practical application of exponential functions involves mathematical models in which the rate of change of a variable y is proportional to the value of y. If y is a function of time t, then we can write

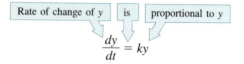

$$\frac{dy}{dt} = ky$$

The general solution to this **differential equation** is given in the following theorem.

THEOREM 7.18   LAW OF EXPONENTIAL GROWTH AND DECAY
If y is a quantity whose rate of change (with respect to time) is proportional to the quantity present at time t, then y is of the form

$$y = Ce^{kt}$$

where C is called the **initial value** and k is called the **constant of proportionality. Exponential growth** is indicated by $k > 0$ and **exponential decay** by $k < 0$.

Proof: Since the rate of change of y is proportional to y, we have $y' = ky$. An obvious solution to this differential equation is $y = 0$. To find any other solutions, we assume $y \neq 0$ and use a solution technique called **separation of variables.** (We study this technique in detail in Chapter 18.)

$$y' = ky \qquad \text{\small y' is proportional to y}$$

$$\frac{y'}{y} = k \qquad \text{\small Divide by y since $y \neq 0$}$$

$$\int \frac{y'}{y}\, dt = \int k\, dt \qquad \text{\small Integrate with respect to t}$$

$$\int \frac{1}{y}\, dy = \int k\, dt \qquad \text{\small $dy = y'\, dt$}$$

$$\ln |y| = kt + C_1$$

Now, solving for y, we have

$$|y| = e^{kt+C_1} = e^{C_1}e^{kt} \implies y = \pm e^{C_1}e^{kt}$$

Since this value for y represents all nonzero solutions (and $y = 0$ is already known to be a solution), we can write the general solution as

$$y = Ce^{kt}, \quad C \text{ is a real number}$$

The next example involves radioactive decay, which is measured in terms of **half-life**—the number of years required for half of the atoms in a sample of radioactive material to decay. For example, the half-lives of some common radioactive isotopes are as follows:

Uranium (U^{238})	4,510,000,000 years
Plutonium (Pu^{230})	24,360 years
Carbon (C^{14})	5730 years
Radium (Ra^{226})	1620 years
Einsteinium (Es^{254})	276 days
Nobelium (No^{257})	23 seconds

EXAMPLE 6   *Radioactive decay*

Let y represent the mass (in grams) of a particular radioactive element whose half-life is 25 years as shown in Figure 7.26. How much of 1 gram would remain after 15 years?

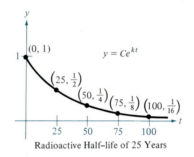

Radioactive Half–life of 25 Years

FIGURE 7.26

Solution: Since the rate of decay is proportional to y, we apply the Law of Exponential Decay to conclude that y is of the form $y = Ce^{kt}$, where t is measured in years. We are given the following initial (or boundary) conditions: $y = 1$ when $t = 0$, and $y = \frac{1}{2}$ when $t = 25$. Applying the first of these conditions, we have

$$y = 1 = Ce^{k(0)} = C \qquad t = 0$$

Thus, $C = 1$. Now, applying the second condition, we have

$$y = \frac{1}{2} = e^{k(25)} \qquad t = 25$$

To solve for k, we take the log of both sides of this equation to obtain $\ln \frac{1}{2} = 25k$. Therefore, $k = (\ln \frac{1}{2})/25 \approx -0.0277$, and the equation for y as a function of time is

$$y = Ce^{kt} \approx e^{-0.0277t}$$

Finally, when $t = 15$ years, the amount remaining is

$$y \approx e^{-0.0277(15)} \approx 0.660 \text{ g} \qquad t = 15$$

Remark Example 6 demonstrates that finding exponential growth functions basically centers around solving for the constants C and k. Furthermore, we can see in Example 6 that it is easy to solve for C when we are given the value at $t = 0$.

In the next example, we demonstrate a procedure for solving for C and k in the growth equation $y = Ce^{kt}$ when we do not know the value of y at $t = 0$.

EXAMPLE 7   Population growth

Suppose that an experimental population of fruit flies increases according to the law of exponential growth. If there are 100 flies after the second day of the experiment and 300 flies after the fourth day, how many flies were in the original population?

Solution: Let $y = Ce^{kt}$ be the number of flies at time t, where t is measured in days. Since $y = 100$ when $t = 2$ and $y = 300$ when $t = 4$, we have

$$100 = Ce^{2k} \qquad \text{and} \qquad 300 = Ce^{4k}$$

From the first equation, we have $C = 100e^{-2k}$. Substituting this value into the second equation, we have $300 = 100e^{-2k}e^{4k} = 100e^{2k}$. Now we can solve for k as follows:

$$\frac{300}{100} = e^{2k} \implies \ln 3 = 2k \implies k = \frac{\ln 3}{2} \approx 0.5493$$

Therefore, the equation for exponential growth is $y = Ce^{0.5493t}$. To solve for C, we can reapply the condition $y = 100$ when $t = 2$ and obtain

$$100 = Ce^{0.5493(2)} \implies C = 100e^{-1.0986} \approx 33$$

Thus, the original population (when $t = 0$) consisted of approximately $y = C = 33$ flies, as indicated in Figure 7.27.

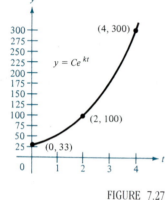

FIGURE 7.27

EXAMPLE 8   Continuously compounded interest

Money is deposited in an account for which the interest is compounded continuously. If the balance doubles in 6 years, what is the annual percentage rate?

Solution: The formula for the balance A using continuously compounded interest is given by the exponential growth function

$$A = Pe^{rt}$$

where P is the original deposit, r is the annual percentage rate, and t is the time in years. When $t = 6$, we know that $A = 2P$, as shown in Figure 7.28. Thus, we have

$$2P = Pe^{6r} \implies 2 = e^{6r} \implies \ln 2 = 6r$$

and we conclude that the annual percentage rate is

$$r = \frac{1}{6} \ln 2 \approx 0.1155 = 11.55\%$$

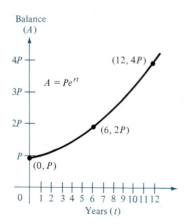

FIGURE 7.28

Number of
units sold

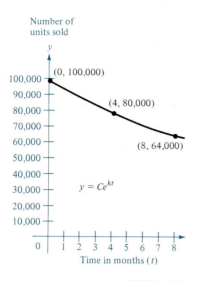

$y = Ce^{kt}$

FIGURE 7.29

EXAMPLE 9   Declining sales

Four months after it discontinued its advertising campaign, a manufacturing company noticed that its sales had dropped from 100,000 units per month to 80,000 units per month. If the sales follow an exponential pattern of decline, what will they be after another two months?

Solution: We use the exponential model $y = Ce^{kt}$, where t is measured in months, as shown in Figure 7.29. From the initial condition ($t = 0$), we know that $C = 100{,}000$. Moreover, since $y = 80{,}000$ when $t = 4$, we have

$$80{,}000 = 100{,}000e^{4k} \implies 0.8 = e^{4k} \implies \ln{(0.8)} = 4k$$

Thus, $k = [\ln{(0.8)}]/4 \approx -0.0558$. Finally, after two more months ($t = 6$), we can expect the monthly sales rate to be

$$y = 100{,}000e^{-0.0558(6)} \approx 71{,}500 \text{ units}$$

In each of the four examples describing exponential growth or decay, we did not actually have to solve the differential equation $y' = ky$. (We did that once in the proof of Theorem 7.18.) In the last example in this section, we demonstrate a problem whose solution involves the separation of variables technique used in proving the Law of Exponential Growth. The example involves **Newton's Law of Cooling,** which states that the rate of change in the temperature of an object is proportional to the difference between the object's temperature and the temperature of the surrounding air.

EXAMPLE 10   Newton's Law of Cooling

Let y represent the temperature (in degrees) of an object in a room whose temperature is kept at a constant 60°. If the object cools from 100° to 90° in 10 minutes, how much longer will it take for its temperature to decrease to 80°?

Solution: From Newton's Law of Cooling, we know that the rate of change in y is proportional to the difference between y and 60. This gives us

$$\frac{dy}{dt} = k(y - 60), \quad 80 \le y \le 100$$

$$\left(\frac{1}{y - 60}\right)\frac{dy}{dt} = k$$

$$\int \frac{1}{y - 60}\, dy = \int k\, dt \qquad \text{Separation of variables}$$

$$\ln{|y - 60|} = kt + C$$

Since $y > 60$, we can omit the absolute value signs. Now, using $y = 100$ when $t = 0$, we have $C = \ln 40$, which implies that

$$kt = \ln{(y - 60)} - \ln 40 = \ln\left(\frac{y - 60}{40}\right)$$

Furthermore, since $y = 90$ when $t = 10$, we have $k = \frac{1}{10} \ln \frac{3}{4}$, and thus

$$t = \left(\frac{10}{\ln (3/4)}\right) \ln \left(\frac{y - 60}{40}\right)$$

Finally, when $y = 80$,

$$t = \frac{10 \ln (1/2)}{\ln (3/4)} \approx 24.09 \text{ min}$$

Therefore, it will take approximately 14.09 *more* minutes for the object to cool to a temperature of $80°$.

Section Exercises 7.6

In Exercises 1–20, evaluate the integral.

1. $\int_0^1 e^{-2x} dx$

2. $\int_1^2 e^{1-x} dx$

3. $\int_0^2 (x^2 - 1)e^{x^3 - 3x + 1} dx$

4. $\int x^2 e^{x^3} dx$

5. $\int \frac{e^{-x}}{1 + e^{-x}} dx$

6. $\int \frac{e^{2x}}{1 + e^{2x}} dx$

7. $\int xe^{ax^2} dx$

8. $\int_0^{\sqrt{2}} xe^{-(x^2/2)} dx$

9. $\int_1^3 \frac{e^{3/x}}{x^2} dx$

10. $\int (e^x - e^{-x})^2 dx$

11. $\int_{-1}^2 2^x dx$

12. $\int x5^{-x^2} dx$

13. $\int e^x \sqrt{1 - e^x} dx$

14. $\int \frac{e^x - e^{-x}}{e^x + e^{-x}} dx$

15. $\int \frac{e^x + e^{-x}}{e^x - e^{-x}} dx$

16. $\int \frac{2e^x - 2e^{-x}}{(e^x + e^{-x})^2} dx$

17. $\int \frac{5 - e^x}{e^{2x}} dx$

18. $\int (3 - x)7^{(3-x)^2} dx$

19. $\int_{-2}^0 (3^3 - 5^2) dx$

20. $\int \frac{e^{2x} + 2e^x + 1}{e^x} dx$

In Exercises 21–24, find the area of the region bounded by the graphs of the equations.

21. $y = e^x$, $y = 0$, $x = 0$, $x = 5$
22. $y = e^{-x}$, $y = 0$, $x = a$, $x = b$
23. $y = xe^{-(x^2/2)}$, $y = 0$, $x = 0$, $x = \sqrt{2}$
24. $y = 3^x$, $y = 0$, $x = 0$, $x = 3$

In Exercises 25 and 26 find the maximum possible error if the integral is approximated by (a) the Trapezoidal Rule and (b) Simpson's Rule.

25. $\int_0^1 e^{x^3} dx$, $n = 4$

26. $\int_0^1 e^{-x^2} dx$, $n = 4$

In Exercises 27 and 28, use the standard normal probability density function

$$f(z) = \frac{1}{\sqrt{2\pi}} e^{-z^2/2}$$

The probability that z is in the interval $[a, b]$ is the area of the region bounded by $y = f(z)$, $y = 0$ on the interval $a \leq z \leq b$ and is denoted by $\Pr(a \leq z \leq b)$. Approximate the given probability. (Use Simpson's Rule with $n = 6$.)

27. $\Pr(0 \leq z \leq 1)$
28. $\Pr(0 \leq z \leq 2)$

29. Given $e^x \geq 1$ for $x \geq 0$, it follows that

$$\int_0^x e^t dt \geq \int_0^x 1 dt$$

Perform this integration to derive the inequality $e^x \geq 1 + x$ for $x \geq 0$.

30. Integrate each term of the following inequalities in a manner similar to that of Exercise 29 to obtain each succeeding inequality for $x \geq 0$. Then evaluate both sides of each inequality when $x = 1$.
(a) $e^x \geq 1 + x$
(b) $e^x \geq 1 + x + \frac{x^2}{2}$
(c) $e^x \geq 1 + x + \frac{x^2}{2} + \frac{x^3}{6}$
(d) $e^x \geq 1 + x + \frac{x^2}{2} + \frac{x^3}{6} + \frac{x^4}{24}$

In Exercises 31 and 32, find the exponential function $y = Ce^{kt}$ that passes through the two given points.

31. (a)   (b)

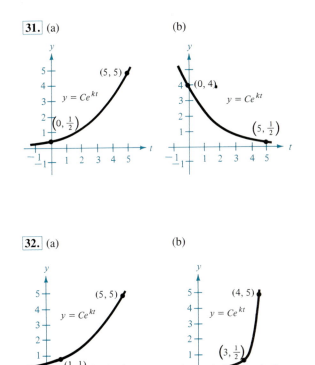

32. (a)   (b)

33. A deposit of \$750 is made in a savings account for which the interest is compounded continuously. The balance will double in $7\frac{3}{4}$ years.
(a) What is the annual percentage rate for this account?
(b) Find the balance in the account after 10 years.

34. A deposit of \$10,000 is made in a savings account for which the interest is compounded continuously. The balance will double in 5 years.
(a) What is the annual percentage rate for this account?
(b) Find the balance after 1 year.

35. Because of a slump in the economy, a company finds that its annual revenues have dropped from \$742,000 in 1980 to \$632,000 in 1982. If the revenue is following an exponential pattern of decline, what is the expected revenue for 1983? (Let $t = 0$ represent 1980.)

36. Atmospheric pressure P (measured in millimeters of mercury) decreases exponentially with increasing altitude x (measured in meters). If the pressure is 760 millimeters of mercury at sea level ($x = 0$) and 672.71 millimeters of mercury at an altitude of 1000 meters, find the pressure at an altitude of 3000 meters.

37. The management at a certain factory has found that the maximum number of units a worker can produce in a day is 30. The learning curve for the number of units N produced per day after a new employee has worked t days is given by

$$N = 30(1 - e^{kt})$$

After 20 days on the job, a particular worker produced 19 units.
(a) Find the learning curve for this worker.
(b) How many days should pass before this worker is producing 25 units per day?

38. If in Exercise 37 the management requires that a new employee be producing at least 20 units per day after 30 days on the job, find
(a) the learning curve describing this minimum requirement
(b) the number of days before a minimal achiever is producing 25 units per day

39. The sales S (in thousands of units) of a new product after it is on the market t years is given by

$$S = Ce^{k/t}$$

(a) Find S as a function of t if 5000 units have been sold after 1 year and the saturation point for the market is 30,000 (that is, $\lim_{t \to \infty} S = 30$).
(b) How many units have been sold after 5 years?
(c) Sketch a graph of this sales function.

40. The number of sales S (in thousands of units) of a new product after it has been on the market t years is given by

$$S = 30(1 - e^{kt})$$

(a) Find S as a function of t if 5000 units have been sold after 1 year.
(b) How many units will saturate this market?
(c) How many units will have been sold after 5 years?
(d) Sketch a graph of this sales function.

41. A certain type of bacteria increases continuously at a rate proportional to the number present. If there are 100 present at a given time and 300 present 5 hours later, how many will there be 10 hours after the initial given time?

42. Given the conditions of Exercise 41, how long would it take for the number of bacteria to double?

43. In 1960 the population of a town was 2500, and in 1970 it was 3350. Assuming the population increases continuously at a constant rate proportional to the existing population, estimate the population in 1990.

44. Given the conditions of Exercise 43, how many years would be necessary for the population to double?

45. Radioactive radium has a half-life of approximately 1600 years. What percentage of a present amount remains after 100 years?

46. If radioactive material decays continuously at a rate proportional to the amount present, find the half-life of the material if after 1 year 99.57 percent of an initial amount still remains.

47. Using Newton's Law of Cooling (see Example 10), determine the reading on a thermometer 5 minutes after it is taken from a room at 72° Fahrenheit to the outdoors

where the temperature is 20°, if the reading dropped to 48° after 1 minute.

48. A body in a room at 70° cools from 350° to 150° in 45 minutes. Using Newton's Law of Cooling, find the time necessary for the body to cool to 80°.

49. Using Newton's Law of Cooling, determine the outdoor temperature if a thermometer is taken from a room where the temperature is 68° to the outdoors, where after $\frac{1}{2}$ minute and 1 minute the thermometer reads 53° and 42°, respectively.

7.7
Indeterminate forms and L'Hôpital's Rule

In Chapter 2, we described the forms 0/0, ∞/∞, and ∞ − ∞ as *indeterminate*, because they do not guarantee that a limit exists, nor do they indicate what the limit is, if one exists. When we encountered one of these indeterminate forms, we attempted to rewrite the expression by using various algebraic techniques, as illustrated by the examples in Table 7.6.

TABLE 7.6   INDETERMINATE FORMS

0/0	$\lim\limits_{x \to -1} \dfrac{2x^2 - 2}{x + 1} = \lim\limits_{x \to -1} 2(x - 1) = -4$	Divide numerator and denominator by $(x + 1)$.
∞/∞	$\lim\limits_{x \to \infty} \dfrac{3x^2 - 1}{2x^2 + 1} = \lim\limits_{x \to \infty} \dfrac{3 - (1/x^2)}{2 + (1/x^2)} = \dfrac{3}{2}$	Divide numerator and denominator by x^2.
∞ − ∞	$\lim\limits_{x \to \infty} (x - \sqrt{x^2 + x}) = \lim\limits_{x \to \infty} \dfrac{-x}{x + \sqrt{x^2 + x}}$ $= \lim\limits_{x \to \infty} \dfrac{-1}{1 + \sqrt{1 + (1/x)}} = -\dfrac{1}{2}$	Rationalize: Then divide numerator and denominator by x.

Occasionally, we can extend these algebraic techniques to find limits of transcendental functions. For instance, the limit

$$\lim_{x \to 0} \frac{e^{2x} - 1}{e^x - 1}$$

produces the indeterminate form 0/0. Factoring and then dividing, we have

$$\lim_{x \to 0} \frac{e^{2x} - 1}{e^x - 1} = \lim_{x \to 0} \frac{(e^x + 1)(e^x - 1)}{e^x - 1}$$
$$= \lim_{x \to 0} (e^x + 1) = 2$$

However, not all indeterminate forms can be evaluated by algebraic manipulation. This is particularly true when *both* algebraic and transcendental functions are involved. For instance, the limit

$$\lim_{x \to 0} \frac{e^{2x} - 1}{x}$$

produces the indeterminate form 0/0. Dividing the numerator and denominator by x to obtain

$$\lim_{x \to 0} \left[\frac{e^{2x}}{x} - \frac{1}{x} \right]$$

merely produces another indeterminate form, $\infty - \infty$. Of course, we could use a calculator to estimate this limit, as shown in Table 7.7. From the table, the limit appears to be 2. (This limit will be verified in Example 1.)

TABLE 7.7

x	-1.0	-0.1	-0.01	-0.001	0	0.001	0.01	0.1	1.0
$\dfrac{e^{2x} - 1}{x}$	0.865	1.813	1.980	1.998	?	2.002	2.020	2.214	6.389

Guillaume L'Hôpital

To find the limit illustrated in Table 7.7, we introduce a theorem called **L'Hôpital's Rule.** This theorem states that under certain conditions the limit of the quotient $f(x)/g(x)$ is determined by the limit of $f'(x)/g'(x)$. This theorem is named after the French mathematician Guillaume Francois Antoine De L'Hôpital (1661–1704), who published the first calculus text in 1696. To prove this theorem, we use a more general result called the **Extended Mean-Value Theorem,** which states the following. If f and g are differentiable on an open interval (a, b) and continuous on $[a, b]$ such that $g(x) \neq 0$ for any x in (a, b), then there exists a point c in (a, b) such that

$$\frac{f'(c)}{g'(c)} = \frac{f(b) - f(a)}{g(b) - g(a)}$$

We prove the Extended Mean-Value Theorem and L'Hôpital's Rule in Appendix A.

| **Remark** In the following theorem, we use the symbol "lim" to represent any of the following types of limits:

$$\lim_{x \to a} \qquad \lim_{x \to a^+} \qquad \lim_{x \to a^-} \qquad \lim_{x \to \infty} \qquad \lim_{x \to -\infty}$$

THEOREM 7.19

L'HÔPITAL'S RULE

If $\lim f(x)/g(x)$ results in the indeterminate form 0/0 or ∞/∞,* then

$$\lim \frac{f(x)}{g(x)} = \lim \frac{f'(x)}{g'(x)}$$

provided the latter limit exists (or is infinite).

*The indeterminate form ∞/∞ actually comes in four forms: ∞/∞, $(-\infty)/\infty$, $\infty/(-\infty)$, and $(-\infty)/(-\infty)$. L'Hôpital's Rule can be applied to each of these forms.

EXAMPLE 1   *Indeterminate form 0/0*

Evaluate

$$\lim_{x \to 0} \frac{e^{2x} - 1}{x}$$

Solution: Since direct substitution results in the indeterminate form 0/0, we apply L'Hôpital's Rule to obtain

$$\lim_{x \to 0} \frac{e^{2x} - 1}{x} = \lim_{x \to 0} \frac{\dfrac{d}{dx}[e^{2x} - 1]}{\dfrac{d}{dx}[x]} = \lim_{x \to 0} \frac{2e^{2x}}{1} = 2$$

| **Remark**   In writing the string of equations in Example 1, we actually do not know that the first limit is equal to the second until we have shown that the second limit exists. In other words, if the second limit had not existed, it would not have been permissible to apply L'Hôpital's Rule.

Occasionally it is necessary to apply L'Hôpital's Rule more than once to remove an indeterminate form. This is illustrated in the next example.

EXAMPLE 2   *Indeterminate form ∞/∞*

Evaluate

$$\lim_{x \to -\infty} \frac{x^2}{e^{-x}}$$

Solution: Since direct substitution results in the indeterminate form ∞/∞, we apply L'Hôpital's Rule to obtain

$$\lim_{x \to -\infty} \frac{x^2}{e^{-x}} = \lim_{x \to -\infty} \frac{\dfrac{d}{dx}[x^2]}{\dfrac{d}{dx}[e^{-x}]} = \lim_{x \to -\infty} \frac{2x}{-e^{-x}}$$

Since this limit yields the indeterminate form $(-\infty)/(-\infty)$, we apply L'Hôpital's Rule again to obtain

$$\lim_{x \to -\infty} \frac{2x}{-e^{-x}} = \lim_{x \to -\infty} \frac{\dfrac{d}{dx}[2x]}{\dfrac{d}{dx}[-e^{-x}]} = \lim_{x \to -\infty} \frac{2}{e^{-x}} = 0$$

In addition to the forms 0/0, ∞/∞, and ∞ − ∞, there are other indeterminate forms, such as $0 \cdot \infty$, 1^{∞}, ∞^0, and 0^0. For example, consider the following four limits that lead to the indeterminate form $0 \cdot \infty$:

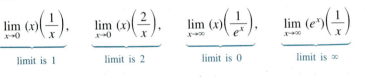

$$\lim_{x \to 0} (x)\left(\frac{1}{x}\right), \qquad \lim_{x \to 0} (x)\left(\frac{2}{x}\right), \qquad \lim_{x \to \infty} (x)\left(\frac{1}{e^x}\right), \qquad \lim_{x \to \infty} (e^x)\left(\frac{1}{x}\right)$$

     limit is 1   limit is 2   limit is 0   limit is ∞

Since each limit is different, it is clear that the form $0 \cdot \infty$ is indeterminate in the sense that it does not determine the value (or even the existence) of the limit. The following examples indicate methods to evaluate these forms. Basically, we attempt to convert each of these forms to those for which L'Hôpital's Rule is applicable.

EXAMPLE 3   Indeterminate form $0 \cdot \infty$

Evaluate

$$\lim_{x \to \infty} e^{-x} \sqrt{x}$$

Solution: Since direct substitution produces the indeterminate form $0 \cdot \infty$, we rewrite the limit to fit the form $0/0$ or ∞/∞. In this case, we choose the second form and write

$$\lim_{x \to \infty} e^{-x} \sqrt{x} = \lim_{x \to \infty} \frac{\sqrt{x}}{e^x}$$

Now, by L'Hôpital's Rule, we have

$$\lim_{x \to \infty} \frac{\sqrt{x}}{e^x} = \lim_{x \to \infty} \frac{1/(2\sqrt{x})}{e^x} = \lim_{x \to \infty} \frac{1}{2\sqrt{x}\, e^x} = 0$$

Remark   If rewriting a limit in one of the forms $0/0$ or ∞/∞ does not seem to work, try the other form. For instance, in Example 3 we could have written the limit as

$$\lim_{x \to \infty} e^{-x} \sqrt{x} = \lim_{x \to \infty} \frac{e^{-x}}{x^{-1/2}}$$

which yields the indeterminate form $0/0$. However, in applying L'Hôpital's Rule to this limit, we obtain

$$\lim_{x \to \infty} \frac{e^{-x}}{x^{-1/2}} = \lim_{x \to \infty} \frac{-e^{-x}}{-1/(2x^{3/2})}$$

which also yields the indeterminate form $0/0$. Moreover, since the quotient seems to be getting more complicated, we abandon this approach and try the ∞/∞ form, as shown in Example 3.

The indeterminate forms 1^∞, ∞^0, and 0^0 arise from limits of functions that have a variable base and a variable exponent. When we encountered this type of function in Section 7.5, we used logarithmic differentiation to find the derivative. We use a similar procedure when taking limits, as indicated in the next example.

EXAMPLE 4   Indeterminate form 1^∞

Evaluate

$$\lim_{x \to \infty} \left(1 + \frac{2}{x}\right)^x$$

Solution: Since direct substitution yields the indeterminate form 1^∞, we proceed in the following manner. We assume the limit exists, and we represent it by

$$y = \lim_{x \to \infty} \left(1 + \frac{2}{x}\right)^x$$

Now, taking the natural logarithm of both sides, we have

$$\ln y = \ln \left[\lim_{x \to \infty} \left(1 + \frac{2}{x}\right)^x\right]$$

and using the fact that the natural logarithmic function is continuous, we write

$$\ln y = \lim_{x \to \infty} \left[x \ln \left(1 + \frac{2}{x}\right)\right] \qquad \text{Indeterminate form: } \infty \cdot 0$$

$$= \lim_{x \to \infty} \left[\frac{\ln \left[1 + (2/x)\right]}{1/x}\right] \qquad \text{Indeterminate form: } 0/0$$

$$= \lim_{x \to \infty} \left[\frac{(-2/x^2)(1/[1 + (2/x)])}{-1/x^2}\right] \qquad \text{L'Hôpital's Rule}$$

$$= \lim_{x \to \infty} \frac{2}{1 + (2/x)} = 2$$

Finally, since $\ln y = 2$, we know that $y = e^2$ and we conclude that

$$\lim_{x \to \infty} \left(1 + \frac{2}{x}\right)^x = e^2 \qquad \Box$$

EXAMPLE 5   *Indeterminate form 0^0*

Evaluate

$$\lim_{x \to 0^+} x^x$$

Solution: Since direct substitution produces the indeterminate form 0^0, we proceed as follows.

$$y = \lim_{x \to 0^+} x^x \qquad \text{Indeterminate form: } 0^0$$

$$\ln y = \ln \left[\lim_{x \to 0^+} x^x\right] \qquad \text{Take log of both sides}$$

$$= \lim_{x \to 0^+} [\ln (x^x)] \qquad \text{Continuity}$$

$$= \lim_{x \to 0^+} [x \ln x] \qquad \text{Indeterminate form: } 0 \cdot (-\infty)$$

$$= \lim_{x \to 0^+} \frac{\ln x}{1/x} \qquad \text{Indeterminate form: } -\infty/\infty$$

$$= \lim_{x \to 0^+} \frac{1/x}{-1/x^2} \qquad \text{L'Hôpital's Rule}$$

$$= \lim_{x \to 0^+} -x = 0$$

Now, since $\ln y = 0$, we conclude that $y = e^0 = 1$, and it follows that

$$\lim_{x \to 0^+} x^x = 1 \qquad \Box$$

| Remark   When evaluating complicated limits like the one in Example 5, it is helpful to check the reasonableness of the solution with a calculator. For instance, the calculations shown in Table 7.8 are consistent with our conclusion that x^x approaches 1 as x approaches 0 from the right.

TABLE 7.8

x	1.0	0.1	0.01	0.001	0.0001	0.00001
x^x	1.0000	0.7943	0.9550	0.9931	0.9991	0.9999

EXAMPLE 6   Indeterminate form $\infty - \infty$

Evaluate

$$\lim_{x \to 1^+} \left(\frac{1}{\ln x} - \frac{1}{x - 1} \right)$$

Solution: Since direct substitution yields the indeterminate form $\infty - \infty$, we try to rewrite the expression to produce a form to which we can apply L'Hôpital's Rule. In this case, we combine the two fractions to obtain

$$\lim_{x \to 1^+} \left(\frac{1}{\ln x} - \frac{1}{x - 1} \right) = \lim_{x \to 1^+} \left[\frac{x - 1 - \ln x}{(x - 1) \ln x} \right]$$

Now, since direct substitution produces the indeterminate form 0/0, we can apply L'Hôpital's Rule to obtain

$$\lim_{x \to 1^+} \left(\frac{1}{\ln x} - \frac{1}{x - 1} \right) = \lim_{x \to 1^+} \left[\frac{1 - (1/x)}{(x - 1)(1/x) + \ln x} \right]$$

$$= \lim_{x \to 1^+} \left[\frac{x - 1}{x - 1 + x \ln x} \right]$$

This limit also yields the indeterminate form 0/0, so we apply L'Hôpital's Rule again to obtain

$$\lim_{x \to 1^+} \left(\frac{1}{\ln x} - \frac{1}{x - 1} \right) = \lim_{x \to 1^+} \left[\frac{1}{1 + x(1/x) + \ln x} \right]$$

$$= \frac{1}{2}$$

| Remark   We have identified the forms 0/0, ∞/∞, $\infty - \infty$, $0 \cdot \infty$, 0^0, 1^∞, and ∞^0 as *indeterminate*. There are similar forms that you should recognize as *determinate*, such as

$$\infty + \infty \to \infty \qquad -\infty - \infty \to -\infty \qquad 0^\infty \to 0 \qquad 0^{-\infty} \to \infty$$

In each of the examples so far in this section, we have used L'Hôpital's Rule to find a limit that exists. L'Hôpital's Rule can also be used to conclude that a limit is infinite and this is demonstrated in the last example.

EXAMPLE 7   An infinite limit

Evaluate

$$\lim_{x\to\infty} \frac{e^x}{x}$$

Solution: Direct substitution produces the indeterminate form ∞/∞, so we apply L'Hôpital's Rule to obtain

$$\lim_{x\to\infty} \frac{e^x}{x} = \lim_{x\to\infty} \frac{e^x}{1} = \lim_{x\to\infty} e^x = \infty$$

Now, since $e^x \to \infty$ as $x \to \infty$, we conclude that the limit of e^x/x as $x \to \infty$ is also infinite.

As a final comment, we remind you that L'Hôpital's Rule can only be applied to quotients leading to the indeterminate forms $0/0$ or ∞/∞. For instance, the following application of L'Hôpital's Rule is *incorrect*.

$$\lim_{x\to 0} \frac{e^x}{x} = \lim_{x\to 0} \frac{e^x}{1} = 1 \qquad \text{Incorrect use of L'Hôpital's Rule}$$

The reason this application is incorrect is that, even though the limit of the denominator is 0, the limit of the numerator is 1—which means that the hypotheses of L'Hôpital's Rule have not been satisfied.

Section Exercises 7.7

In Exercises 1–34, evaluate each limit, using L'Hôpital's Rule where necessary.

1. $\lim\limits_{x\to 2} \dfrac{x^2 - x - 2}{x - 2}$

2. $\lim\limits_{x\to -1} \dfrac{x^2 - x - 2}{x + 1}$

3. $\lim\limits_{x\to 0} \dfrac{\sqrt{4 - x^2} - 2}{x}$

4. $\lim\limits_{x\to 2^-} \dfrac{\sqrt{4 - x^2}}{x - 2}$

5. $\lim\limits_{x\to 0} \dfrac{e^x - (1 - x)}{x}$

6. $\lim\limits_{x\to 0^+} \dfrac{e^x - (1 + x)}{x^3}$

7. $\lim\limits_{x\to 0^+} \dfrac{e^x - (1 + x)}{x^n}, \quad n = 1, 2, 3, \ldots$

8. $\lim\limits_{x\to 1} \dfrac{\ln x}{x^2 - 1}$

9. $\lim\limits_{x\to\infty} \dfrac{\ln x}{x}$

10. $\lim\limits_{x\to\infty} \dfrac{e^x}{x}$

11. $\lim\limits_{x\to\infty} \dfrac{3x^2 - 2x + 1}{2x^2 + 3}$

12. $\lim\limits_{x\to\infty} \dfrac{x - 1}{x^2 + 2x + 3}$

13. $\lim\limits_{x\to\infty} \dfrac{x^2 + 2x + 3}{x - 1}$

14. $\lim\limits_{x\to\infty} \dfrac{x^2}{e^x}$

15. $\lim\limits_{x\to 0^+} x^2 \ln x$

16. $\lim\limits_{x\to 0} \left(\dfrac{1}{x} - \dfrac{1}{x^2} \right)$

17. $\lim\limits_{x\to 2} \left(\dfrac{8}{x^2 - 4} - \dfrac{x}{x - 2} \right)$

18. $\lim\limits_{x\to 2} \left(\dfrac{1}{x^2 - 4} - \dfrac{\sqrt{x - 1}}{x^2 - 4} \right)$

19. $\lim\limits_{x\to\infty} \dfrac{x}{\sqrt{x^2 + 1}}$

20. $\lim\limits_{x\to 1^+} \left(\dfrac{3}{\ln x} - \dfrac{2}{x - 1} \right)$

21. $\lim\limits_{x\to 0^+} x^{1/x}$

22. $\lim\limits_{x\to 0^+} (e^x + x)^{1/x}$

23. $\lim\limits_{x\to\infty} x^{1/x}$

24. $\lim\limits_{x\to\infty} \left(1 + \dfrac{1}{x} \right)^x$

25. $\lim\limits_{x\to\infty} (1 + x)^{1/x}$

26. $\lim\limits_{x\to\infty} \dfrac{x}{x - 1}$

In Exercises 27–32, use L'Hôpital's Rule to determine the comparative rates of increase of the functions

$$f(x) = x^m$$
$$g(x) = e^{nx}$$
$$h(x) = (\ln x)^n$$

where $n > 0$, $m > 0$, and $x \to \infty$. The limits obtained in these exercises suggest that $(\ln x)^n$ tends toward infinity more slowly than x^m, which in turn tends toward infinity more slowly than e^{nx}.

27. $\lim\limits_{x\to\infty} \dfrac{x^2}{e^{5x}}$

28. $\lim\limits_{x\to\infty} \dfrac{x^3}{e^{2x}}$

29. $\lim\limits_{x\to\infty} \dfrac{(\ln x)^3}{x}$

30. $\lim\limits_{x\to\infty} \dfrac{(\ln x)^2}{x^3}$

31. $\lim\limits_{x\to\infty} \dfrac{(\ln x)^n}{x^m}$, where $0 < n$, m

32. $\lim\limits_{x\to\infty} \dfrac{x^m}{e^{nx}}$, where $0 < n$, m

In Exercises 33 and 34, L'Hôpital's Rule is used *incorrectly*. Describe the error.

33. $\lim\limits_{x\to 0} \dfrac{e^{2x} - 1}{e^x} = \lim\limits_{x\to 0} \dfrac{2e^{2x}}{e^x} = \lim\limits_{x\to 0} 2e^x = 2$

34. $\lim\limits_{x\to\infty} \dfrac{e^{-x}}{1 + e^{-x}} = \lim\limits_{x\to\infty} \dfrac{-e^{-x}}{-e^{-x}} = \lim\limits_{x\to\infty} 1 = 1$

35. Complete the following table to show that x eventually "overpowers" $(\ln x)^4$.

x	10	10^2	10^4	10^6	10^8	10^{10}
$\dfrac{(\ln x)^4}{x}$						

36. Complete the following table to show that e^x eventually "overpowers" x^5.

x	1	5	10	20	30	40	50	100
$\dfrac{e^x}{x^5}$								

37. Prove that if $f(x) \geq 0$, $\lim\limits_{x\to a} f(x) = 0$, and $\lim\limits_{x\to a} g(x) = \infty$, then

$$\lim\limits_{x\to a} f(x)^{g(x)} = 0$$

38. Prove that if $f(x) \geq 0$, $\lim\limits_{x\to a} f(x) = 0$, and $\lim\limits_{x\to a} g(x) = -\infty$, then

$$\lim\limits_{x\to a} f(x)^{g(x)} = \infty$$

39. The Prime Number Theorem states that as n increases, the number of primes less than n approaches $n/\ln n$. Use this approximation together with L'Hôpital's Rule to show that there are infinitely many prime numbers.

40. The Gamma Function $\Gamma(n)$ is defined in terms of the integral of the function

$$f(x) = x^{n-1}e^{-x}, \quad n > 0$$

Show that for any fixed value of n, the limit of $f(x)$ as x approaches infinity is zero.

41. The velocity of an object falling in a resisting medium such as air or water is given by

$$v = \frac{32}{k}\left(1 - e^{-kt} + \frac{v_0 k e^{-kt}}{32}\right)$$

where v_0 is the initial velocity, t is the time, and k is the resistance constant of the medium. Use L'Hôpital's Rule to find the formula for the velocity of a falling body in a vacuum by fixing v_0 and t and letting k approach zero. (Assume the downward direction is positive.)

42. The formula for the amount A in a savings account compounded n times a year for t years at an interest rate of r and an initial deposit of P is

$$A = P\left(1 + \frac{r}{n}\right)^{nt}$$

Use L'Hôpital's Rule to show that the limiting formula as the number of compoundings per year becomes infinite is

$$A = Pe^{rt}$$

(Note that this limiting formula is used for continuous compounding of interest.)

Review Exercises for Chapter 7

In Exercises 1–6, (a) find the inverse, f^{-1}, of the given function, (b) sketch the graphs of f and f^{-1} on the same axes, and (c) verify that $f^{-1}[f(x)] = f[f^{-1}(x)] = x$.

1. $f(x) = \dfrac{1}{2}x - 3$

2. $f(x) = 5x - 7$

3. $f(x) = \sqrt{x + 1}$

4. $f(x) = x^3 + 2$

5. $f(x) = x^2 - 5$, $x \geq 0$

6. $f(x) = \sqrt[3]{x + 1}$

In Exercises 7 and 8, the function does not have an inverse. Give a restriction on the domain so that the restricted function has an inverse, and then find the inverse.

7. $f(x) = 2(x - 4)^2$

8. $f(x) = |x - 2|$

In Exercises 9–12, solve the given equations for x.

9. $e^{\ln x} = 3$

10. $\ln x + \ln (x - 3) = 0$

11. $\log_3 x + \log_3 (x - 1) - \log_3 (x - 2) = 2$

12. $\log_x 125 = 3$

In Exercises 13–34, find dy/dx.

13. $y = \ln \sqrt{x}$

14. $y = \ln \dfrac{x(x-1)}{x-2}$

15. $y = x\sqrt{\ln x}$

16. $y = \ln [x(x^2 - 2)^{2/3}]$

17. $y \ln x + y^2 = 0$

18. $\ln (x + y) = x$

19. $\ln y = x \ln x$

20. $\ln y = x \ln \sqrt{x^2 + 1}$

21. $y = \dfrac{1}{b^2}\left[\ln (a + bx) + \dfrac{a}{a + bx}\right]$

22. $y = \dfrac{1}{b^2}[a + bx - a \ln (a + bx)]$

23. $y = -\dfrac{1}{a} \ln \left(\dfrac{a + bx}{x}\right)$

24. $y = -\dfrac{1}{ax} + \dfrac{b}{a^2} \ln \left(\dfrac{a + bx}{x}\right)$

25. $y = \ln (e^{-x^2})$

26. $y = \ln \left(\dfrac{e^x}{1 + e^x}\right)$

27. $y = x^2 e^x$

28. $y = e^{-x^2/2}$

29. $y = \sqrt{e^{2x} + e^{-2x}}$

30. $y = x^{2x+1}$

31. $y = 3^{x-1}$

32. $y = (4e)^x$

33. $ye^x + xe^y = xy$

34. $y = \dfrac{x^2}{e^x}$

In Exercises 35–38, find the derivative of the given function. (Assume that a is constant.)

35. $y = x^a$

36. $y = a^x$

37. $y = x^x$

38. $y = a^a$

In Exercises 39–56, find the indefinite integral.

39. $\displaystyle\int \dfrac{1}{7x - 2}\, dx$

40. $\displaystyle\int \dfrac{x}{x^2 - 1}\, dx$

41. $\displaystyle\int \dfrac{1}{x \ln (3x)}\, dx$

42. $\displaystyle\int \dfrac{\ln \sqrt{x}}{x}\, dx$

43. $\displaystyle\int \dfrac{x^2 + 3}{x}\, dx$

44. $\displaystyle\int \dfrac{x^3 + 1}{x^2}\, dx$

45. $\displaystyle\int \dfrac{x^2}{x^3 - 1}\, dx$

46. $\displaystyle\int \left(x + \dfrac{1}{x}\right)^2 dx$

47. $\displaystyle\int \dfrac{1}{x\sqrt{\ln x}}\, dx$

48. $\displaystyle\int \dfrac{x + 2}{2x + 3}\, dx$

49. $\displaystyle\int xe^{-3x^2}\, dx$

50. $\displaystyle\int \dfrac{e^{1/x}}{x^2}\, dx$

51. $\displaystyle\int \dfrac{e^{4x} - e^{2x} + 1}{e^x}\, dx$

52. $\displaystyle\int \dfrac{e^{2x} - e^{-2x}}{e^{2x} + e^{-2x}}\, dx$

53. $\displaystyle\int \dfrac{e^x}{e^x - 1}\, dx$

54. $\displaystyle\int x^2 e^{x^3+1}\, dx$

55. $\displaystyle\int xe^{-x^2/2}\, dx$

56. $\displaystyle\int \dfrac{x - 1}{3x^2 - 6x - 1}\, dx$

In Exercises 57–60, evaluate the definite integral.

57. $\displaystyle\int_1^4 \dfrac{x + 1}{x}\, dx$

58. $\displaystyle\int_1^4 \dfrac{\ln x}{x}\, dx$

59. $\displaystyle\int_3^4 \dfrac{1}{x - 2}\, dx - \int_3^4 \dfrac{1}{x + 2}\, dx$

60. $\displaystyle\int_0^2 xe^{-x^2}\, dx$

In Exercises 61 and 62, sketch the graph of the region whose area is given by the integral, and find the area.

61. $\displaystyle\int_0^2 4e^{-2x}\, dx$

62. $\displaystyle\int_0^1 \dfrac{1}{x + 1}\, dx$

In Exercises 63 and 64, find the average value of the function over the given interval. Find the values of x where the function assumes its mean value and sketch the graph of the function.

	Function	Interval
63.	$f(x) = \dfrac{1}{x - 1}$	[5, 10]
64.	$f(x) = e^{-x}$	[0, 2]

In Exercises 65 and 66, find the area of the region bounded by the graphs of the equations.

65. $y = xe^{-x^2}$, $y = 0$, $x = 0$, $x = 4$

66. $y = 3e^{-x/2}$, $y = 0$, $x = 0$, $x = 4$

67. A deposit of $500 earns interest at the rate of 5% compounded continuously. Find its value after:
(a) 1 year
(b) 10 years
(c) 100 years

68. A deposit earns interest at the rate r compounded continuously and doubles in 10 years. Find r.

69. How large a deposit, at 7% interest compounded continuously, must be made to obtain a balance of $10,000 in 15 years?

70. A deposit of $2500 is made in a savings account at an annual percentage rate of 12% compounded continuously. Find the average balance in this account during the first five years.

71. A population is growing continuously at the rate of $2\tfrac{1}{2}\%$ per year. Find the time necessary for the population to (a) double and (b) triple in size.

72. Under ideal conditions the rate of change of the air pressure with respect to the height above sea level, is proportional to the pressure at the given height. The air pressure is 30 inches at sea level and 15 inches at 18,000 feet. Find the air pressure at 35,000 inches.

In Exercises 73 and 74, use the following model for human memory

$$p(t) = 80e^{-0.5t} + 20$$

where $p(t)$ is the percentage retained after t weeks as shown in Figure 7.30.

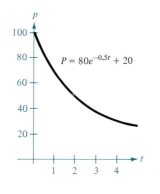

FIGURE 7.30

73. At what rate is information being retained after (a) 1 week? (b) 2 weeks?

74. Find the average percentage retained during (a) the first 2 weeks and (b) the second 2 weeks.

In Exercises 75 and 76, the **exponential density function** is given by $P = (1/\mu)e^{-t/\mu}$, where t is the time in minutes and μ is the average time between successive events. The definite integral

$$\int_{t_1}^{t_2} \frac{1}{\mu} e^{-t/\mu}\, dt$$

gives the probability that the elapsed time before the next occurrence lies between t_1 and t_2 units of time.

75. Trucks arrive at a terminal at an average rate of 3 per hour ($\mu = 20$ minutes is the average time between arrivals). If a truck has just arrived, find the probability that the next arrival will be:
(a) within 10 minutes
(b) within 30 minutes
(c) between 15 and 30 minutes
(d) within 60 minutes

76. The average time between incoming calls at a switchboard is 3 minutes. Complete the accompanying table to show the probabilities of time elapsed between incoming calls.

Time elapsed between calls in minutes	0–2	2–4	4–6	6–8	8–10
Probability					

77. Two numbers between 0 and 10 are chosen at random. The probability that their product is less than n $(0 < n < 100)$ is given by

$$P = \frac{1}{100}\left(n + \int_{10/n}^{10} \frac{n}{x}\, dx\right)$$

(a) What is the probability that the product is less than 25?
(b) What is the probability that the product is less than 50?

78. A solution of a certain drug contains 500 units per milliliter when it is prepared. After 40 days it contains 300 units per milliliter. Assuming that the rate of decomposition is proportional to the amount present, find an equation giving the amount A after t days.

79. An object is projected horizontally with an initial velocity of v_0 feet per second. Assume that the air resistance is proportional to the velocity of the object. Show that the distance traveled in t seconds is given by

$$s = \frac{v_0}{k}(1 - e^{-kt})$$

where k is a constant.

80. Assume that an object falling from rest encounters air resistance that is proportional to its velocity. Considering the acceleration due to gravity to be -32 feet per second per second, the net change in velocity is given by

$$\frac{dv}{dt} = kv - 32$$

Find the velocity of the object as a function of time.

81. Solve Exercise 80 for an object that is projected downward with an initial velocity of 20 feet per second.

82. For the falling object in Exercise 80, find the limit of the velocity as t approaches infinity.

83. Integrate the velocity function found in Exercise 80 to find the position function s.

84. A certain automobile gets 28 miles per gallon of gasoline for speeds up to 50 miles per hour. Over 50 miles per hour the miles per gallon drop at the rate of 12% for each 10 miles per hour.
(a) If s is the speed and y is the miles per gallon, find y as a function of s by solving the differential equation

$$\frac{dy}{ds} = -0.012y \ (s > 50)$$

(b) Use the function in part (a) to complete the following table.

Speed	50	55	60	65	70
Miles per gallon					

In Exercises 85–92, use L'Hôpital's Rule to evaluate the given limit.

85. $\displaystyle \lim_{x \to 1} \frac{(\ln x)^2}{x - 1}$

86. $\displaystyle \lim_{x \to k} \frac{\sqrt[3]{x} - \sqrt[3]{k}}{x - k}$

87. $\displaystyle \lim_{x \to \infty} \frac{e^{2x}}{x^2}$

88. $\displaystyle \lim_{x \to 1^+} \left(\frac{2}{\ln x} - \frac{2}{x - 1} \right)$

89. $\displaystyle \lim_{x \to \infty} (\ln x)^{2/x}$

90. $\displaystyle \lim_{x \to 1} (x - 1)^{\ln x}$

91. $\displaystyle \lim_{n \to \infty} 1000\left(1 + \frac{0.09}{n} \right)^n$

92. $\displaystyle \lim_{x \to \infty} xe^{-x^2}$

8

Trigonometric functions and inverse trigonometric functions

8.1
Review of trigonometric functions

Standard Position of
an Angle

FIGURE 8.1

The concept of an angle is central to the study of trigonometry. As shown in Figure 8.1, an **angle** has three parts: an **initial ray,** a **terminal ray,** and a **vertex** (the point of intersection of the two rays). We say that an angle is in **standard position** if its initial ray coincides with the positive x-axis and its vertex is at the origin. We assume that you are familiar* with the degree measure of an angle. It is common practice to use θ (the Greek lowercase letter theta) to represent both an angle and its measure. We classify angles between 0° and 90° as **acute** and angles between 90° and 180° as **obtuse.** Positive angles are measured *counterclockwise* beginning with the initial ray. Negative angles are measured *clockwise*. For instance, Figure 8.2 shows an angle whose measure is $-45°$. We cannot assign a measure to an angle merely by knowing where its initial and terminal rays are located. To measure an angle, we must also know how the terminal ray was revolved. For example, Figure 8.2 shows that the angle measuring $-45°$ has the same terminal ray as the angle measuring 315°. We call such angles **coterminal.**

An angle that is larger than 360° is one whose terminal ray has revolved more than one full revolution counterclockwise. Figure 8.3 shows an angle measuring more than 360°. Similarly, we can generate angles whose measure is less than $-360°$ by revolving a terminal ray more than one full revolution clockwise.

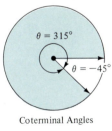

Coterminal Angles

FIGURE 8.2

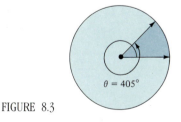

FIGURE 8.3

*For a more complete review of trigonometry, see *Algebra and Trigonometry*, Larson and Hostetler (Lexington, Massachusetts: D.C. Heath and Company, 1985).

397

The arc
length of the
sector is the
radian measure
of θ.

Unit Circle

FIGURE 8.4

Circle of radius r

FIGURE 8.5

Radian measure

A second way to measure angles is in terms of radians. We will see in Section 8.2 that in calculus this measure is preferable to degree measure. To assign a radian measure to an angle θ, we consider θ to be the central angle of a circular sector of radius 1, as shown in Figure 8.4. The **radian measure** of θ is then defined to be the length of the arc of this sector. Recall that the total circumference of a circle is $2\pi r$. Thus, the circumference of a **unit circle** (that is, a circle of radius 1) is simply 2π, and we may conclude that the radian measure of an angle measuring $360°$ is 2π. In other words, $360° = 2\pi$ radians.

Using radian measure, we have a simple formula for the length s of a circular arc of radius r as shown in Figure 8.5.

$$\boxed{\text{arclength} = s = r\theta}\qquad (\theta \text{ measured in radians})$$

It is helpful to memorize the common conversions (such as those pictured in Figure 8.6). For other angles, you can use one of the following conversion rules.

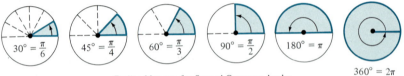

Radian Measure for Several Common Angles

FIGURE 8.6

CONVERSION RULES	$180° = \pi$ radians	
	Degrees $\Rightarrow$ **Radians**	**Radians** $\Rightarrow$ **Degrees**
	$1° = \dfrac{\pi}{180}$ radians	$1 \text{ radian} = \dfrac{180°}{\pi}$

EXAMPLE 1   Conversions between degrees and radians

(a) $40° = (40)\left(\dfrac{\pi}{180}\right) = \dfrac{2\pi}{9}$ radians

(b) $-270° = (-270)\left(\dfrac{\pi}{180}\right) = -\dfrac{3\pi}{2}$ radians

(c) $-\dfrac{\pi}{2}$ radians $= \left(-\dfrac{\pi}{2}\right)\left(\dfrac{180}{\pi}\right) = -90°$

(d) $\dfrac{9\pi}{2}$ radians $= \left(\dfrac{9\pi}{2}\right)\left(\dfrac{180}{\pi}\right) = 810°$

The trigonometric functions

There are two common approaches to the study of trigonometry. In one case the trigonometric functions are defined as ratios of two sides of a right triangle. In the other case these functions are defined in terms of a point on the terminal side of an angle in standard position. The first approach is the one generally used in surveying, navigation, and astronomy, where a typical problem involves a fixed triangle having three of its six parts (sides and angles) known and three to be determined. The second approach is the one normally used in physics, electronics, and biology, where the periodic nature of the trigonometric functions is emphasized. In the following definition we define the six trigonometric functions, **sine, cosine, tangent, cotangent, secant,** and **cosecant** (abbreviated as sin, cos, etc.) from both viewpoints.

DEFINITION OF THE SIX TRIGONOMETRIC FUNCTIONS

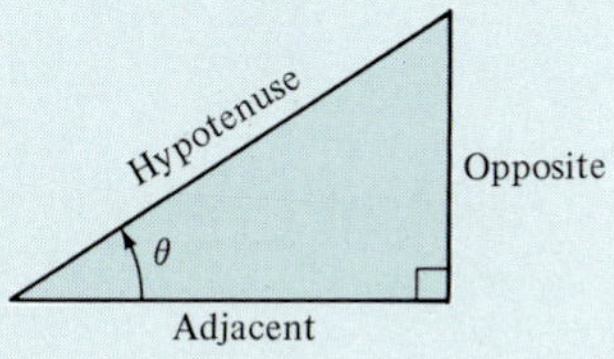

Right Triangle Definitions: $0 < \theta < \dfrac{\pi}{2}$. Refer to Figure 8.7.

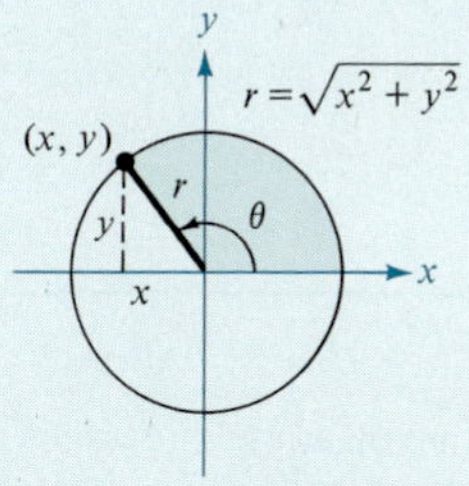

$$\sin \theta = \frac{\text{opp.}}{\text{hyp.}} \qquad \csc \theta = \frac{\text{hyp.}}{\text{opp.}}$$

$$\cos \theta = \frac{\text{adj.}}{\text{hyp.}} \qquad \sec \theta = \frac{\text{hyp.}}{\text{adj.}}$$

$$\tan \theta = \frac{\text{opp.}}{\text{adj.}} \qquad \cot \theta = \frac{\text{adj.}}{\text{opp.}}$$

FIGURE 8.7

Circular Function Definitions: θ is any angle. Refer to Figure 8.8.

$$\sin \theta = \frac{y}{r} \qquad \csc \theta = \frac{r}{y}$$

$$\cos \theta = \frac{x}{r} \qquad \sec \theta = \frac{r}{x}$$

$$\tan \theta = \frac{y}{x} \qquad \cot \theta = \frac{x}{y}$$

FIGURE 8.8

The following formulas are direct consequences of the definitions.

$$\csc \theta = \frac{1}{\sin \theta} \qquad \sec \theta = \frac{1}{\cos \theta} \qquad \cot \theta = \frac{1}{\tan \theta}$$

$$\tan \theta = \frac{\sin \theta}{\cos \theta} \qquad \cot \theta = \frac{\cos \theta}{\sin \theta}$$

Furthermore, since

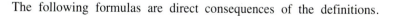

$$\sin^2 \theta + \cos^2 \theta = \left(\frac{y}{r}\right)^2 + \left(\frac{x}{r}\right)^2 = \frac{x^2 + y^2}{r^2} = \frac{r^2}{r^2} = 1$$

we can readily obtain the Pythagorean Identity

$$\sin^2 \theta + \cos^2 \theta = 1$$

Note that we use $\sin^2 \theta$ to mean $(\sin \theta)^2$. Additional trigonometric identities are listed next.

TRIGONOMETRIC IDENTITIES

Pythagorean Identities:

$$\sin^2 \theta + \cos^2 \theta = 1$$
$$\tan^2 \theta + 1 = \sec^2 \theta$$
$$\cot^2 \theta + 1 = \csc^2 \theta$$

Sum or Difference of Two Angles:*

$$\sin (\theta \pm \phi) = \sin \theta \cos \phi \pm \cos \theta \sin \phi$$
$$\cos (\theta \pm \phi) = \cos \theta \cos \phi \mp \sin \theta \sin \phi$$
$$\tan (\theta \pm \phi) = \frac{\tan \theta \pm \tan \phi}{1 \mp \tan \theta \tan \phi}$$

Double Angle:

$$\sin 2\theta = 2 \sin \theta \cos \theta$$
$$\cos 2\theta = 2 \cos^2 \theta - 1$$
$$= 1 - 2 \sin^2 \theta = \cos^2 \theta - \sin^2 \theta$$

Reduction Formula:

$$\sin (-\theta) = -\sin \theta$$
$$\cos (-\theta) = \cos \theta$$
$$\tan (-\theta) = -\tan \theta$$
$$\sin \theta = -\sin (\theta - \pi)$$
$$\cos \theta = -\cos (\theta - \pi)$$
$$\tan \theta = \tan (\theta - \pi)$$

Half Angle:

$$\sin^2 \theta = \frac{1}{2}(1 - \cos 2\theta)$$
$$\cos^2 \theta = \frac{1}{2}(1 + \cos 2\theta)$$

Law of Cosines:

$$a^2 = b^2 + c^2 - 2bc \cos A$$

Law of Cosines

| **Remark** All angles in the remainder of this text are measured in radians unless stated otherwise. In other words, when we write sin 3, we mean the sine of three radians, and when we write sin 3°, we mean the sine of three degrees.

Evaluation of trigonometric functions

There are two common methods of evaluating trigonometric functions: decimal approximations with a calculator (or a table of trigonometric values) and exact evaluations using trigonometric identities and formulas from geometry. We demonstrate the latter method first.

EXAMPLE 2   Evaluating trigonometric functions

Evaluate the sine, cosine, and tangent of $\pi/3$.

*ϕ is the lowercase Greek letter phi.

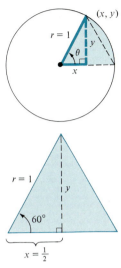

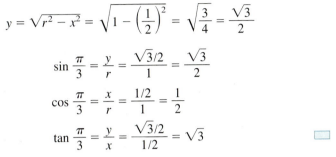

Solution: We begin by drawing the angle $\theta = \pi/3$ in the standard position, as shown in Figure 8.9. Then since $60° = \pi/3$ radians, we obtain an equilateral triangle whose sides have a length of 1 and with θ as one of its angles. Since the altitude of this triangle bisects its base, we know that $x = 1/2$. Now, using the Pythagorean Theorem, we have

$$y = \sqrt{r^2 - x^2} = \sqrt{1 - \left(\frac{1}{2}\right)^2} = \sqrt{\frac{3}{4}} = \frac{\sqrt{3}}{2}$$

Thus,

$$\sin\frac{\pi}{3} = \frac{y}{r} = \frac{\sqrt{3}/2}{1} = \frac{\sqrt{3}}{2}$$

$$\cos\frac{\pi}{3} = \frac{x}{r} = \frac{1/2}{1} = \frac{1}{2}$$

$$\tan\frac{\pi}{3} = \frac{y}{x} = \frac{\sqrt{3}/2}{1/2} = \sqrt{3}$$

FIGURE 8.9

The degree and radian measure of several common angles are given in Table 8.1 with the corresponding values of the sine, cosine, and tangent.

TABLE 8.1   COMMON FIRST QUADRANT ANGLES

Degrees	0	30°	45°	60°	90°
Radians	0	$\dfrac{\pi}{6}$	$\dfrac{\pi}{4}$	$\dfrac{\pi}{3}$	$\dfrac{\pi}{2}$
$\sin\theta$	0	$\dfrac{1}{2}$	$\dfrac{\sqrt{2}}{2}$	$\dfrac{\sqrt{3}}{2}$	1
$\cos\theta$	1	$\dfrac{\sqrt{3}}{2}$	$\dfrac{\sqrt{2}}{2}$	$\dfrac{1}{2}$	0
$\tan\theta$	0	$\dfrac{1}{\sqrt{3}}$	1	$\sqrt{3}$	undef.

In Figure 8.8, note that r is always positive. Thus, the quadrant signs of x and y determine the quadrant signs of the various trigonometric functions as indicated in Figure 8.10. To extend the use of Table 8.1 to angles in quadrants other than the first quadrant, we can use the concept of a **reference angle** (as shown in Figure 8.11), together with the appropriate quadrant sign.

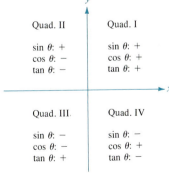

FIGURE 8.10

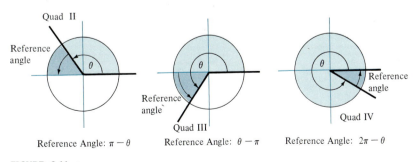

FIGURE 8.11

EXAMPLE 3   *Evaluating trigonometric functions with reference angles*

Function	Quadrant	Sign	Reference Angle	Value
(a) $\sin \dfrac{3\pi}{4}$	II	$+$	$\pi - \dfrac{3\pi}{4} = \dfrac{\pi}{4}$	$\sin \dfrac{3\pi}{4} = \sin \dfrac{\pi}{4} = \dfrac{\sqrt{2}}{2}$
(b) $\tan 330°$	IV	$-$	$360° - 330° = 30°$	$\tan 330° = -\tan 30°$ $= -\dfrac{\sqrt{3}}{3}$
(c) $\cos \dfrac{7\pi}{6}$	III	$-$	$\dfrac{7\pi}{6} - \pi = \dfrac{\pi}{6}$	$\cos \dfrac{7\pi}{6} = -\cos \dfrac{\pi}{6}$ $= -\dfrac{\sqrt{3}}{2}$

EXAMPLE 4   *Trigonometric identities and calculators*

(a) Using the reduction formula $\sin(-\theta) = -\sin\theta$, we have

$$\sin\left(-\frac{\pi}{3}\right) = -\sin\frac{\pi}{3}$$
$$= -\frac{\sqrt{3}}{2}$$

(b) Using the reciprocal formula $\sec\theta = 1/\cos\theta$, we have

$$\sec 60° = \frac{1}{\cos 60°} = \frac{1}{1/2} = 2$$

(c) Using a calculator, we have

$$\cos(1.2) \approx 0.3624$$

Remember that 1.2 is given in *radian* measure, so your calculator must be set in radian mode.

Solving trigonometric equations

In Examples 2, 3, and 4 we looked at techniques for evaluating trigonometric functions for given values of θ. In the next two examples, we look at the reverse problem. That is, if we are given the value of a trigonometric function, how can we solve for θ? For example, consider the equation

$$\sin\theta = 0$$

We know $\theta = 0$ is one solution. But this is not the only solution. Any one of the following values will work:

$$\ldots, \ -3\pi, \ -2\pi, \ -\pi, \ 0, \ \pi, \ 2\pi, \ 3\pi, \ \ldots$$

We can write this infinite solution set as $\{n\pi: n \text{ is an integer}\}$.

EXAMPLE 5   *Solving trigonometric equations*

Solve for θ in the following equation:

$$\sin \theta = -\frac{\sqrt{3}}{2}$$

Solution: To solve the equation, we make two observations: (1) the sine is negative in Quadrants III and IV, and (2) $\sin (\pi/3) = \sqrt{3}/2$. By combining these two observations, we conclude that we are seeking values of θ in the third and fourth quadrants that have a reference angle of $\pi/3$. In the interval $[0, 2\pi]$, the two angles fitting these criteria are

$$\theta = \pi + \frac{\pi}{3} = \frac{4\pi}{3}$$

and

$$\theta = 2\pi - \frac{\pi}{3} = \frac{5\pi}{3}$$

Finally, we can add $2n\pi$ to either of these angles to obtain the solution set

$$\theta = \frac{4\pi}{3} + 2n\pi, \ n \text{ is an integer}$$

or

$$\theta = \frac{5\pi}{3} + 2n\pi, \ n \text{ is an integer}$$

EXAMPLE 6   *Solving trigonometric equations*

Solve the following equation for θ:

$$\cos 2\theta = 2 - 3 \sin \theta, \quad 0 \le \theta \le 2\pi$$

Solution: Using the double angle identity $\cos 2\theta = 1 - 2 \sin^2 \theta$, we obtain the following polynomial (in $\sin \theta$):

$$1 - 2 \sin^2 \theta = 2 - 3 \sin \theta$$
$$0 = 2 \sin^2 \theta - 3 \sin \theta + 1$$
$$0 = (2 \sin \theta - 1)(\sin \theta - 1)$$

If $2 \sin \theta - 1 = 0$, we have $\sin \theta = 1/2$ and $\theta = \pi/6$ or $\theta = 5\pi/6$. If $\sin \theta - 1 = 0$, we have $\sin \theta = 1$ and $\theta = \pi/2$. Thus, for $0 \le \theta \le 2\pi$, there are three solutions to the given equation:

$$\theta = \frac{\pi}{6}, \ \frac{5\pi}{6}, \ \text{ or } \ \frac{\pi}{2}$$

Section Exercises 8.1

In Exercises 1 and 2, determine two coterminal angles (one positive and one negative) for the given angle. Give your answers in degrees.

1. (a)   (b)

    $\theta = 36°$ $\theta = -45°$

(c)   (d)

    $\theta = -120°$ $\theta = 390°$

2. (a) $\theta = 300°$ (b) $\theta = 740°$

(c) $\theta = -420°$ (d) $\theta = 230°$

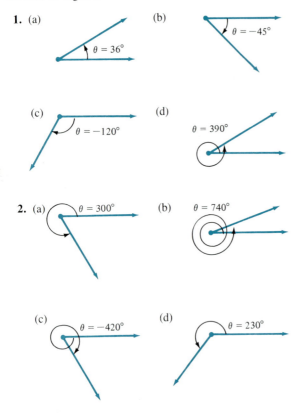

In Exercises 3 and 4, determine two coterminal angles (one positive and one negative) for the given angle. (Give your answers in radians.)

3. (a) $\theta = \frac{\pi}{9}$ (b) $\theta = \frac{4\pi}{3}$

(c) $\theta = \frac{11\pi}{6}$ (d) $\theta = -\frac{7\pi}{6}$

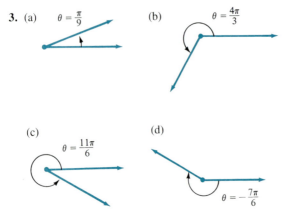

4. (a) $\theta = -\frac{9\pi}{4}$ (b) $\theta = -\frac{2\pi}{15}$

(c) $\theta = \frac{8\pi}{9}$ (d) $\theta = \frac{8\pi}{45}$

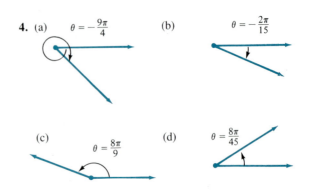

In Exercises 5 and 6, express the given angle in radian measure as a multiple of π.

5. (a) 30°   (b) 150°
   (c) 315°   (d) 120°

6. (a) $-20°$ (b) $-240°$
   (c) $-270°$ (d) 144°

In Exercises 7 and 8, express the given angle in degree measure.

7. (a) $\frac{3\pi}{2}$ (b) $\frac{7\pi}{6}$

   (c) $-\frac{7\pi}{12}$ (d) $\frac{\pi}{9}$

8. (a) $\frac{7\pi}{3}$ (b) $-\frac{11\pi}{30}$

   (c) $\frac{11\pi}{6}$ (d) $\frac{34\pi}{15}$

9. Let r represent the radius of a circle, θ the central angle (measured in radians), and s the length of the arc subtended by the angle. Use the relationship $\theta = s/r$ to complete the following table.

r	8 ft	15 in	85 cm		
s	12 ft			96 in	8642 mi
θ		1.6	$\frac{3\pi}{4}$	4	$\frac{2\pi}{3}$

10. The minute hand on a clock is $3\frac{1}{2}$ inches long as shown in Figure 8.12. Through what distance does the tip of the minute hand move in 25 minutes?

11. A man bends his elbow through 75°. The distance from his elbow to the top of his index finger is $18\frac{3}{4}$ inches as shown in Figure 8.13. (a) Find the radian measure of

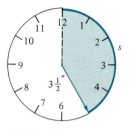

FIGURE 8.12

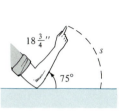

FIGURE 8.13

this angle. (b) Find the distance the tip of the index finger moves.

12. A tractor tire, 5 feet in diameter, is partially filled with a liquid ballast for additional traction. To check the air pressure, the tractor operator rotates the tire until the valve stem is at the top so that the liquid will not enter the gauge. On a given occasion, the operator notes that the tire must be rotated 80° to have the stem in the proper position. (a) Find the radian measure of this rotation. (b) How far must the tractor be moved to get the valve stem in the proper position?

In Exercises 13 and 14, determine all six trigonometric functions for the given angle θ.

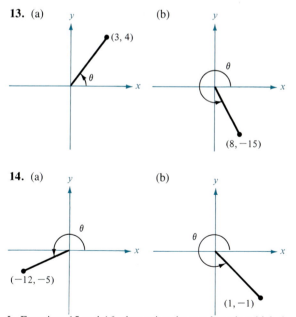

13. (a)   (b)

14. (a)   (b)

In Exercises 15 and 16, determine the quadrant in which θ lies.

15. $\sin \theta < 0$ and $\cos \theta < 0$
16. $\sin \theta > 0$ and $\cos \theta < 0$

In Exercises 17–22, find the indicated trigonometric function from the given one.

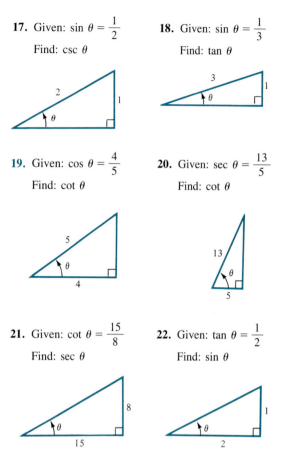

17. Given: $\sin \theta = \dfrac{1}{2}$

    Find: $\csc \theta$

18. Given: $\sin \theta = \dfrac{1}{3}$

    Find: $\tan \theta$

19. Given: $\cos \theta = \dfrac{4}{5}$

    Find: $\cot \theta$

20. Given: $\sec \theta = \dfrac{13}{5}$

    Find: $\cot \theta$

21. Given: $\cot \theta = \dfrac{15}{8}$

    Find: $\sec \theta$

22. Given: $\tan \theta = \dfrac{1}{2}$

    Find: $\sin \theta$

In Exercises 23–26, evaluate the sine, cosine, and tangent of the given angles *without* using a calculator.

23. (a) $60°$ (b) $\dfrac{2\pi}{3}$

     (c) $\dfrac{\pi}{4}$ (d) $\dfrac{5\pi}{4}$

24. (a) $-\dfrac{\pi}{6}$ (b) $150°$

     (c) $-\dfrac{\pi}{2}$ (d) $\dfrac{\pi}{2}$

25. (a) $225°$ (b) $-225°$
     (c) $300°$ (d) $330°$

26. (a) $750°$ (b) $510°$
     (c) $\dfrac{10\pi}{3}$ (d) $\dfrac{17\pi}{3}$

In Exercises 27–30, use a calculator to evaluate the given trigonometric functions to four significant figures.

27. (a) $\sin 10°$ (b) $\csc 10°$
28. (a) $\sec 225°$ (b) $\sec 135°$

29. (a) $\tan \dfrac{\pi}{9}$ (b) $\tan \dfrac{10\pi}{9}$

30. (a) $\cot (1.35)$ (b) $\tan (1.35)$

In Exercises 31–34, find two values of θ corresponding to the given functions. List the measure of θ in radians ($0 \leq \theta < 2\pi$). Do not use a calculator.

31. (a) $\cos \theta = \dfrac{\sqrt{2}}{2}$ (b) $\cos \theta = -\dfrac{\sqrt{2}}{2}$

32. (a) $\sec \theta = 2$ (b) $\sec \theta = -2$

33. (a) $\tan \theta = 1$ (b) $\cot \theta = -\sqrt{3}$

34. (a) $\sin \theta = \dfrac{\sqrt{3}}{2}$ (b) $\sin \theta = -\dfrac{\sqrt{3}}{2}$

In Exercises 35–42, solve the given equation for θ ($0 \leq \theta < 2\pi$). For some of the equations, you should use the trigonometric identities listed in this section.

35. $2 \sin^2 \theta = 1$
36. $\tan^2 \theta = 3$
37. $\tan^2 \theta - \tan \theta = 0$
38. $2 \cos^2 \theta - \cos \theta = 1$
39. $\sec \theta \csc \theta = 2 \csc \theta$
40. $\sin \theta = \cos \theta$
41. $\cos^2 \theta + \sin \theta = 1$
42. $\cos (\theta/2) - \cos \theta = 1$

In Exercises 43–46, solve for x, y, or r as indicated.

43. Solve for y.

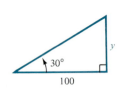

44. Solve for x.

45. Solve for x.

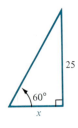

46. Solve for r.

47. A 20-foot ladder leaning against the side of a house makes a $75°$ angle with the ground as shown in Figure 8.14. How far up the side of the house does the ladder reach?

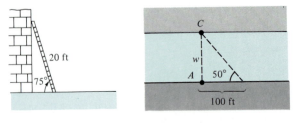

FIGURE 8.14   FIGURE 8.15

48. A biologist wants to know the width w of a river in order to properly set instruments to study the pollutants in the water. From point A, the biologist walks downstream 100 feet and sights to point C. For this sighting, it is determined that $\theta = 50°$ as shown in Figure 8.15. How wide is the river?

49. From a 150-foot observation tower on the coast, a Coast Guard officer sights a boat in difficulty. The angle of depression of the boat is $4°$ as shown in Figure 8.16. How far is the boat from the shoreline?

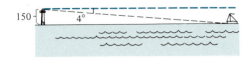

FIGURE 8.16

50. A ramp $17\frac{1}{2}$ feet in length rises to a loading platform that is $3\frac{1}{2}$ feet off the ground as shown in Figure 8.17. Find the angle that the ramp makes with the ground. (Hint: Find the sine of θ and then use the table of trigonometric values in Appendix D to estimate θ.)

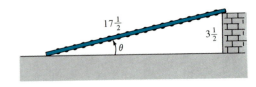

FIGURE 8.17

8.2
Graphs and limits of trigonometric functions

One of the first things we notice about the graphs of all six trigonometric functions is that they are periodic. We call a function f **periodic** if there exists a nonzero number p such that $f(x + p) = f(x)$ for all x in the domain of f. The smallest such positive value of p is called the **period** of f. Both the sine and cosine functions have a period of 2π and by plotting several values in the interval $0 \le x \le 2\pi$, we obtain the graphs shown in Table 8.2.

TABLE 8.2   GRAPHS OF THE SINE AND COSINE FUNCTIONS (PERIOD: 2π)

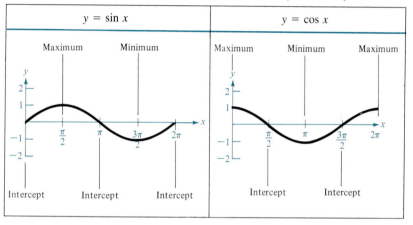

Note in Table 8.2 that the maximum value of $\sin x$ is 1 and the minimum value is -1. The **amplitude** of the sine function (or of the cosine function) is defined to be half the difference between the maximum and minimum values. Thus the amplitude of $f(x) = \sin x$ is 1. Figure 8.18 (on page 408) shows the graphs of all six trigonometric functions. Familiarity with these six basic graphs will serve as a valuable aid in graphing more complicated trigonometric functions.

The graph of the function $y = a \sin bx$ oscillates between $-a$ and a and hence has an amplitude of $|a|$. Furthermore, since

$$bx = 0 \text{ when } x = 0 \quad \text{and} \quad bx = 2\pi \text{ when } x = 2\pi/b$$

we may conclude that the function $y = a \sin bx$ has a period of $2\pi/|b|$. Table 8.3 summarizes the amplitudes and periods for some general types of trigonometric functions.

TABLE 8.3   PERIODS AND AMPLITUDES OF TRIGONOMETRIC FUNCTIONS

Function	Period	Amplitude
$y = a \sin bx$ or $y = a \cos bx$	$2\pi/|b|$	$|a|$
$y = a \tan bx$ or $y = a \cot bx$	$\pi/|b|$	—
$y = a \sec bx$ or $y = a \csc bx$	$2\pi/|b|$	—

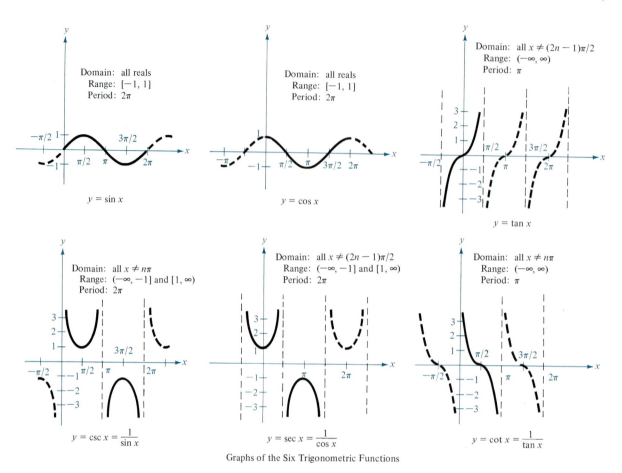

Graphs of the Six Trigonometric Functions

FIGURE 8.18

EXAMPLE 1   *Sketching the graph of a trigonometric function*

Sketch the graph of $f(x) = 3 \cos 2x$.

Solution: The graph of $f(x) = 3 \cos 2x$ has the following characteristics.

$$\text{amplitude: } 3 \qquad \text{period: } \frac{2\pi}{2} = \pi$$

Using the basic shape of the graph of the cosine function, we sketch one period of the function on the interval $[0, \pi]$, following the pattern

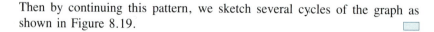

maximum: $(0, 3)$ maximum: $(\pi, 3)$

minimum: $\left(\dfrac{\pi}{2}, -3\right)$

Then by continuing this pattern, we sketch several cycles of the graph as shown in Figure 8.19.

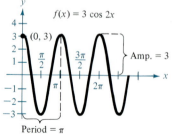

FIGURE 8.19

The discussion of horizontal shifts, vertical shifts, and reflections given in Section 1.5 can be applied nicely to the graphs of trigonometric functions.

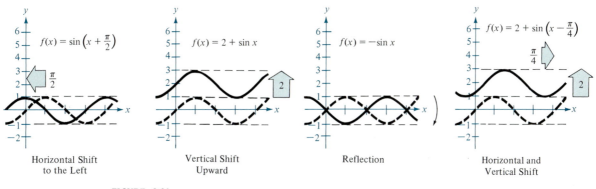

FIGURE 8.20

For instance, Figure 8.20 shows four different shifted or reflected graphs of sine functions.

EXAMPLE 2   Sketching the graph of a trigonometric function

Sketch the graph of

$$f(x) = 2 \sin \left(3x - \frac{\pi}{2} \right)$$

Solution: We write

$$f(x) = 2 \sin \left[3 \left(x - \frac{\pi}{6} \right) \right]$$

and observe that the graph of f has the following characteristics.

$$\text{amplitude: } 2$$

$$\text{period: } \frac{2\pi}{3}$$

$$\text{right shift: } \frac{\pi}{6}$$

Since the shift is $\pi/6$, we start one cycle at $x = \pi/6$. Then, since the period is $2\pi/3$, this cycle ends at $x = (\pi/6) + (2\pi/3) = 5\pi/6$. Two complete cycles are shown in Figure 8.21.

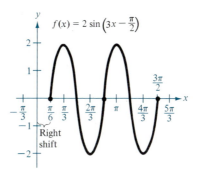

FIGURE 8.21

To sketch the graph of a function that combines an algebraic function with a trigonometric function, it is helpful to use a technique called **addition of ordinates** as demonstrated in the next example.

EXAMPLE 3   Addition of ordinates

Sketch the graph of $f(x) = x + \cos x$ on the interval $0 \le x \le 2\pi$.

Solution: We first make an accurate sketch of the graphs of $y = x$ and $y = \cos x$ on the same coordinate plane and then geometrically add the ordinates (y-values) for each x. This addition is aided by the use of a compass to measure the displacement of one graph from the x-axis and then mark off an equal displacement from the other graph as shown in Figure 8.22.

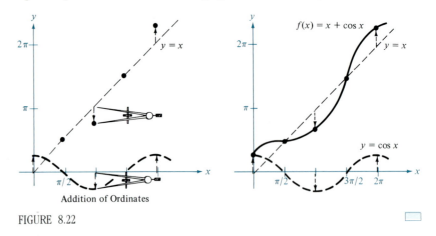

Addition of Ordinates

FIGURE 8.22

Limits of trigonometric functions

In the next section we will show how the derivatives of the trigonometric functions can assist us in sketching more complicated graphs. However, before we discuss these derivatives, we need to obtain some results regarding the limits of trigonometric functions. In Chapter 2, we saw that the limits of many algebraic functions can be evaluated by direct substitution. The following theorem (which we present without proof) tells us that each of the six basic trigonometric functions also possesses this desirable quality.

THEOREM 8.1   LIMITS OF TRIGONOMETRIC FUNCTIONS
If c is a real number then the following properties are true.

1. $\lim\limits_{x \to c} \sin x = \sin c$ 2. $\lim\limits_{x \to c} \cos x = \cos c$

*3. $\lim\limits_{x \to c} \tan x = \tan c$ *4. $\lim\limits_{x \to c} \cot x = \cot c$

*5. $\lim\limits_{x \to c} \sec x = \sec c$ *6. $\lim\limits_{x \to c} \csc x = \csc c$

*If c is not in the domain of the given function, then the limit does not exist.

| Remark   From Theorem 8.1, it follows that each of the six trigonometric functions is *continuous* at every point in its domain.

EXAMPLE 4   *Limits involving trigonometric functions*

(a) By Theorem 8.1, we have

$$\lim_{x \to 0} \sin x = \sin (0) = 0$$

(b) By Theorem 8.1, and Property 3 of Theorem 2.3, we have

$$\lim_{x \to \pi} (\theta \cos \theta) = \left[\lim_{x \to \pi} \theta\right]\left[\lim_{x \to \pi} \cos \theta\right] = \pi \cos (\pi) = -\pi$$

EXAMPLE 5   *A limit involving trigonometric functions*

Find the following limit:

$$\lim_{x \to 0} \frac{\tan x}{\sin x}$$

Solution: Direct substitution produces the indeterminate form 0/0. However, by using the fact that tan x = (sin x)/(cos x), we can rewrite the function as

$$\frac{\tan x}{\sin x} = \frac{(\sin x)/(\cos x)}{\sin x} = \frac{\cancel{\sin x}}{(\cos x)(\cancel{\sin x})} = \frac{1}{\cos x}$$

Thus, by Theorem 2.1, we have

$$\lim_{x \to 0} \frac{\tan x}{\sin x} = \lim_{x \to 0} \frac{1}{\cos x} = \frac{1}{1} = 1$$

| Remark   After we learn how to differentiate trigonometric functions, we can handle problems like the one in Example 5 with L'Hôpital's Rule.

EXAMPLE 6   *A limit that does not exist*

Discuss the existence of the limit:

$$\lim_{x \to 0} \sin \left(\frac{1}{x}\right)$$

Solution: We let $f(x) = \sin (1/x)$. In Figure 8.23, we see that as x approaches 0, $f(x)$ oscillates between -1 and 1. Therefore, the limit does not exist since no matter how small we choose δ, it is possible to choose x_1 and x_2 within δ units of 0 such that $\sin (1/x_1) = 1$ and $\sin (1/x_2) = -1$ as indicated in Table 8.4.

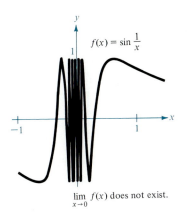

$f(x) = \sin \dfrac{1}{x}$

$\lim_{x \to 0} f(x)$ does not exist.

FIGURE 8.23

TABLE 8.4

x	$\dfrac{2}{\pi}$	$\dfrac{2}{3\pi}$	$\dfrac{2}{5\pi}$	$\dfrac{2}{7\pi}$	$\dfrac{2}{9\pi}$	$\dfrac{2}{11\pi}$	$x \longrightarrow 0$
$\sin \dfrac{1}{x}$	1	-1	1	-1	1	-1	Limit does not exist.

EXAMPLE 7   *Testing for continuity*

Find the intervals for which the two functions shown in Figure 8.24 are continuous.

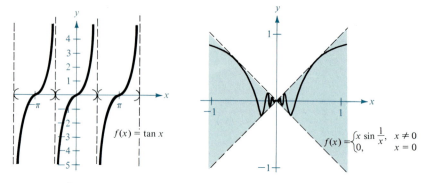

FIGURE 8.24

Solution:

(a) The tangent function is undefined at $x = (\pi/2) + n\pi$. At all other points it is continuous. Thus, $f(x) = \tan x$ is continuous on the open intervals

$$\ldots, \left(-\frac{3\pi}{2}, -\frac{\pi}{2}\right), \left(-\frac{\pi}{2}, \frac{\pi}{2}\right), \left(\frac{\pi}{2}, \frac{3\pi}{2}\right), \ldots$$

(b) This function is similar to that in Example 6 except that the oscillations are damped by the factor x. Using the Squeeze Theorem, we have

$$-|x| \le \sin \frac{1}{x} \le |x|$$

and we conclude that the limit as $x \to 0$ is zero. Thus, f is continuous on the entire real line.

In the next section, we will see that the following two important limits are useful in determining the derivatives of trigonometric functions.

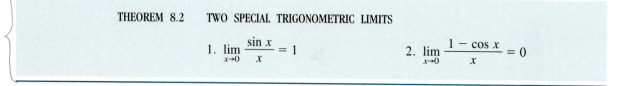

THEOREM 8.2   TWO SPECIAL TRIGONOMETRIC LIMITS

1. $\lim\limits_{x \to 0} \dfrac{\sin x}{x} = 1$ 2. $\lim\limits_{x \to 0} \dfrac{1 - \cos x}{x} = 0$

Proof: We prove the first limit and leave the proof of the second as an exercise. (Note that direct substitution produces the indeterminate form 0/0 in both cases.) To avoid the confusion of two different uses of x, we present the proof using the variable θ, where θ is an acute positive angle (measured in radians). Figure 8.25 shows a circular section that is squeezed between two triangles. The areas of these three figures satisfy the following inequality.

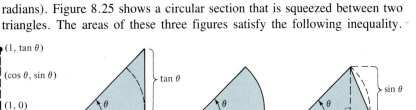

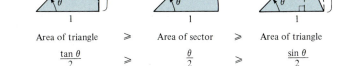

Area of triangle	$\geqslant$	Area of sector	$\geqslant$	Area of triangle
$\dfrac{\tan \theta}{2}$	$\geqslant$	$\dfrac{\theta}{2}$	$\geqslant$	$\dfrac{\sin \theta}{2}$

FIGURE 8.25

If we multiply each inequality by $2/\sin \theta$, then we have

$$\frac{1}{\cos \theta} \geq \frac{\theta}{\sin \theta} \geq 1$$

Now, by taking reciprocals and reversing the inequalities, we have

$$\cos \theta \leq \frac{\sin \theta}{\theta} \leq 1$$

Since $\cos \theta = \cos(-\theta)$ and $(\sin \theta)/\theta = [\sin(-\theta)]/(-\theta)$, we can conclude that this inequality is valid for all nonzero θ in the open interval $(-\pi/2, \pi/2)$. Finally, since

$$\lim_{\theta \to 0} \cos \theta = 1 \qquad \text{and} \qquad \lim_{\theta \to 0} 1 = 1$$

we can apply the Squeeze Theorem to conclude that

$$\lim_{\theta \to 0} \frac{\sin \theta}{\theta} = 1$$

EXAMPLE 8   *A limit involving trigonometric functions*

Find the following limit:

$$\lim_{x \to 0} \frac{\tan x}{x}$$

Solution: Direct substitution gives the indeterminate form 0/0. To solve this problem, we write $\tan x = (\sin x)/(\cos x)$ and obtain

$$\lim_{x \to 0} \frac{\tan x}{x} = \lim_{x \to 0} \left(\frac{\sin x}{x}\right)\left(\frac{1}{\cos x}\right)$$

Now, since

$$\lim_{x \to 0} \frac{\sin x}{x} = 1 \qquad \text{and} \qquad \lim_{x \to 0} \frac{1}{\cos x} = 1$$

we have

$$\lim_{x \to 0} \frac{\tan x}{x} = \left[\lim_{x \to 0} \frac{\sin x}{x}\right]\left[\lim_{x \to 0} \sec x\right] = (1)(1) = 1$$

EXAMPLE 9   *A limit involving trigonometric functions*

Find the following limit:

$$\lim_{x \to 0} \frac{\sin 2x}{x}$$

Solution: Direct substitution gives the indeterminate form 0/0. To solve this problem, we use the identity $\sin 2x = 2 \sin x \cos x$ and obtain

$$\lim_{x \to 0} \frac{\sin 2x}{x} = \lim_{x \to 0} \frac{2 \sin x \cos x}{x}$$

$$= 2 \left[\lim_{x \to 0} \frac{\sin x}{x} \right] \left[\lim_{x \to 0} \cos x \right] = 2(1)(1) = 2$$

There is a simpler way to evaluate the limit in Example 9. Suppose that we rewrite the limit as

$$\lim_{x \to 0} \frac{\sin 2x}{x} = 2 \left[\lim_{x \to 0} \frac{\sin 2x}{2x} \right]$$

Now by letting $y = 2x$ and observing that $x \to 0$ if and only if $y \to 0$, we can write

$$\lim_{x \to 0} \frac{\sin 2x}{x} = 2 \left[\lim_{x \to 0} \frac{\sin 2x}{2x} \right] = 2 \left[\lim_{y \to 0} \frac{\sin y}{y} \right] = 2[1] = 2$$

Section Exercises 8.2

In Exercises 1–12, determine the period and amplitude of the given function.

1. $y = 2 \sin 2x$

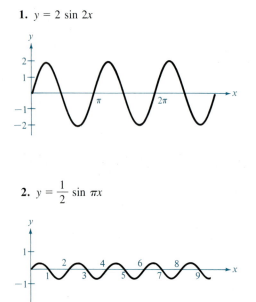

2. $y = \dfrac{1}{2} \sin \pi x$

3. $y = 3 \cos 3x$

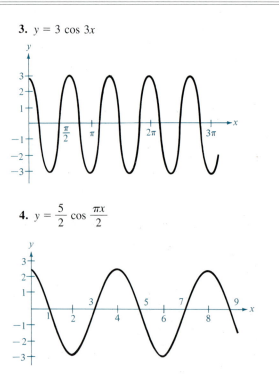

4. $y = \dfrac{5}{2} \cos \dfrac{\pi x}{2}$

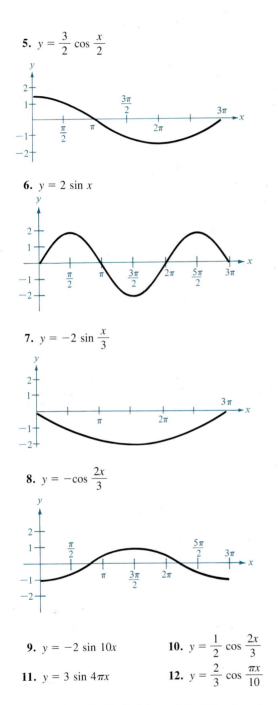

5. $y = \dfrac{3}{2} \cos \dfrac{x}{2}$

6. $y = 2 \sin x$

7. $y = -2 \sin \dfrac{x}{3}$

8. $y = -\cos \dfrac{2x}{3}$

9. $y = -2 \sin 10x$ **10.** $y = \dfrac{1}{2} \cos \dfrac{2x}{3}$

11. $y = 3 \sin 4\pi x$ **12.** $y = \dfrac{2}{3} \cos \dfrac{\pi x}{10}$

In Exercises 13–16, find the period of the given function.

13. $y = 5 \tan 2x$ **14.** $y = 7 \tan 2\pi x$
15. $y = \sec 5x$ **16.** $y = \csc 4x$

In Exercises 17–30, sketch the graph of the given function.

17. $y = \sin \dfrac{x}{2}$ **18.** $y = 2 \cos 2x$

19. $y = -2 \sin 6x$ **20.** $y = \cos 2\pi x$

21. $y = -\sin \dfrac{2\pi x}{3}$ **22.** $y = 2 \tan x$

23. $y = \csc \dfrac{x}{2}$ **24.** $y = \tan 2x$

25. $y = 2 \sec 2x$ **26.** $y = \csc 2\pi x$

27. $y = \sin(x + \pi)$ **28.** $y = \cos\left(2x - \dfrac{\pi}{3}\right)$

29. $y = 1 + \cos\left(x - \dfrac{\pi}{2}\right)$ **30.** $y = -1 + \sin\left(x + \dfrac{\pi}{2}\right)$

In Exercises 31–34, use addition of ordinates to sketch the graph of the given function.

31. $y = x + \cos x$ **32.** $y = x + 2 \sin x$
33. $f(x) = \sin x + \sin 2x$ **34.** $f(x) = \sin x + \cos 2x$

In Exercises 35–50, determine the limit (if it exists).

35. $\lim\limits_{x \to 0} \dfrac{\sin x}{5x}$ **36.** $\lim\limits_{x \to 0} \dfrac{3(1 - \cos x)}{x}$

37. $\lim\limits_{\theta \to 0} \dfrac{\sec \theta - 1}{\theta \sec \theta}$ **38.** $\lim\limits_{\theta \to 0} \dfrac{\cos \theta \tan \theta}{\theta}$

39. $\lim\limits_{x \to 0} \dfrac{\sin^2 x}{x}$ **40.** $\lim\limits_{\phi \to \pi} \phi \sec \phi$

41. $\lim\limits_{x \to \pi/2} \dfrac{\cos x}{\cot x}$ **42.** $\lim\limits_{x \to \pi/4} \dfrac{1 - \tan x}{\sin x - \cos x}$

43. $\lim\limits_{t \to 0} \dfrac{\sin^2 t}{t^2}$ $\left[\text{Hint: Find } \lim\limits_{t \to 0} \left(\dfrac{\sin t}{t}\right)^2.\right]$

44. $\lim\limits_{t \to 0} \dfrac{\sin 3t}{t}$ $\left[\text{Hint: Find } \lim\limits_{t \to 0} 3\left(\dfrac{\sin 3t}{3t}\right).\right]$

45. $\lim\limits_{x \to 0} \dfrac{\sin 2x}{\sin 3x}$

$\left[\text{Hint: Find } \lim\limits_{x \to 0} \left(\dfrac{2 \sin 2x}{2x}\right)\left(\dfrac{3x}{3 \sin 3x}\right).\right]$

46. $\lim\limits_{x \to 0} \dfrac{\tan^2 x}{x}$

47. $\lim\limits_{h \to 0} \dfrac{(1 - \cos h)^2}{h}$ **48.** $\lim\limits_{h \to 0} (1 + \cos 2h)$

49. $\lim\limits_{x \to \pi} \cot x$ **50.** $\lim\limits_{x \to \pi/2} \sec x$

51. $\lim\limits_{x \to 0^+} \dfrac{2}{\sin x}$ **52.** $\lim\limits_{x \to \pi/2^+} \dfrac{-2}{\cos x}$

In Exercises 53–58, find the discontinuities (if any) for the given function. Which of the discontinuities are removable?

53. $f(x) = x + \sin x$ **54.** $f(x) = \cos \dfrac{\pi x}{2}$

55. $f(x) = \csc 2x$ **56.** $f(x) = \tan \dfrac{\pi x}{2}$

57. $f(x) = \begin{cases} \csc \dfrac{\pi x}{6}, & |x - 3| < 2 \\ 2, & |x - 3| \geq 2 \end{cases}$

58. $f(x) = \begin{cases} \tan \dfrac{\pi x}{4}, & |x| < 1 \\ x, & |x| \geq 1 \end{cases}$

59. For a person at rest, the velocity v (in liters per second) of air flow during a respiratory cycle is

$$v = 0.85 \sin \frac{\pi t}{3}$$

where t is the time in seconds. Inhalation occurs when $v > 0$ and exhalation occurs when $v < 0$. (a) Find the time for one full respiratory cycle. (b) Find the number of cycles per minute. (c) Sketch the graph of the velocity function.

60. After exercising for a few minutes, a person has a respiratory cycle for which the velocity of air flow is approximated by

$$v = 1.75 \sin \frac{\pi t}{2}$$

Use this model to repeat Exercise 59.

61. When tuning a piano, a technician strikes a tuning fork for the A above middle C and sets up wave motion that can be approximated by

$$y = 0.001 \sin 880 \pi t$$

where t is the time in seconds. (a) What is the period p of this function? (b) What is the frequency f of this note? ($f = 1/p$) (c) Sketch the graph of this function.

62. Prove the second part of Theorem 8.2.

8.3

Derivatives of trigonometric functions

In the previous section, we obtained the following limits:

$$\lim_{\Delta x \to 0} \frac{\sin \Delta x}{\Delta x} = 1 \quad \text{and} \quad \lim_{\Delta x \to 0} \frac{1 - \cos \Delta x}{\Delta x} = 0$$

These two limits turn out to be crucial in the proofs of the derivatives of the sine and cosine functions. The derivatives of the other four trigonometric functions follow easily from these two.

THEOREM 8.3   DERIVATIVES OF SINE AND COSINE

$$\frac{d}{dx}[\sin x] = \cos x \qquad\qquad \frac{d}{dx}[\cos x] = -\sin x$$

Proof: We prove the first of these two rules and leave the proof of the second as an exercise.

$$\frac{d}{dx}[\sin x] = \lim_{\Delta x \to 0} \frac{\sin(x + \Delta x) - \sin x}{\Delta x}$$

$$= \lim_{\Delta x \to 0} \frac{\sin x \cos \Delta x + \cos x \sin \Delta x - \sin x}{\Delta x}$$

$$= \lim_{\Delta x \to 0} \frac{\cos x \sin \Delta x - \sin x(1 - \cos \Delta x)}{\Delta x}$$

$$= \lim_{\Delta x \to 0} \left[\cos x \left(\frac{\sin \Delta x}{\Delta x} \right) - \sin x \left(\frac{1 - \cos \Delta x}{\Delta x} \right) \right]$$

$$= \cos x \left[\lim_{\Delta x \to 0} \frac{\sin \Delta x}{\Delta x} \right] - \sin x \left[\lim_{\Delta x \to 0} \frac{1 - \cos \Delta x}{\Delta x} \right]$$

$$= (\cos x)(1) - (\sin x)(0)$$

$$= \cos x$$

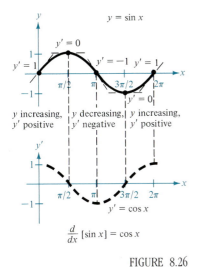

$$\frac{d}{dx}[\sin x] = \cos x$$

FIGURE 8.26

This differentiation formula is shown graphically in Figure 8.26. Note that for each x the *slope* of the sine curve determines the ordinate *value* of the cosine curve.

When taking the derivatives of trigonometric functions, the standard differentiation rules still apply—the Sum Rule, Constant Multiple Rule, the Product Rule, and so on.

EXAMPLE 1   Derivatives involving sines and cosines

Function	*Derivative*
(a) $y = 3 \sin x$	$y' = 3 \cos x$
(b) $y = x + \cos x$	$y' = 1 - \sin x$
(c) $y = \dfrac{\sin x}{2} = \dfrac{1}{2} \sin x$	$y' = \dfrac{1}{2} \cos x = \dfrac{\cos x}{2}$

EXAMPLE 2   A derivative involving the Product Rule

Find the derivative of $y = 2x \cos x - 2 \sin x$.

Solution:

$$\frac{dy}{dx} = \overbrace{(2x)\left(\frac{d}{dx}[\cos x]\right) + (\cos x)\left(\frac{d}{dx}[2x]\right)}^{\text{Product Rule}} - \overbrace{2\,\frac{d}{dx}[\sin x]}^{\text{Constant Multiple Rule}}$$

$$= (2x)(-\sin x) + (\cos x)(2) - 2(\cos x)$$

$$= -2x \sin x$$

EXAMPLE 3   Using the derivative to find the slope of a curve

Find the slope of the graph of $f(x) = 2 \cos x$ at the following points:

(a) $\left(-\dfrac{\pi}{2}, 0\right)$ (b) $\left(\dfrac{\pi}{3}, 1\right)$ (c) $(\pi, -2)$

Solution: The derivative of f is $f'(x) = -2 \sin x$. Therefore, the slopes at the indicated points are as follows.

(a) At $x = -\dfrac{\pi}{2}$, the slope is

$$f'\left(-\frac{\pi}{2}\right) = -2 \sin\left(-\frac{\pi}{2}\right) = -2(-1) = 2$$

(b) At $x = \dfrac{\pi}{3}$, the slope is

$$f'\left(\frac{\pi}{3}\right) = -2 \sin \frac{\pi}{3} = -2\left(\frac{\sqrt{3}}{2}\right) = -\sqrt{3}$$

(c) At $x = \pi$, the slope is

$$f'(\pi) = -2 \sin \pi = -2(0) = 0$$

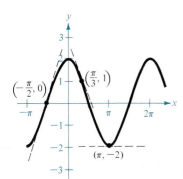

FIGURE 8.27   See Figure 8.27.

Derivatives of the tangent, cotangent, secant, and cosecant functions

Knowing the derivatives of the sine and cosine functions, we can use the Quotient Rule to find the derivatives of the four remaining trigonometric functions.

THEOREM 8.4   DERIVATIVES OF TANGENT, COTANGENT, SECANT, AND COSECANT

$$\frac{d}{dx}[\tan x] = \sec^2 x \qquad\qquad \frac{d}{dx}[\cot x] = -\csc^2 x$$

$$\frac{d}{dx}[\sec x] = \sec x \tan x \qquad\qquad \frac{d}{dx}[\csc x] = -\csc x \cot x$$

Proof: We give a proof for the derivative of the tangent function and leave the proofs of the remaining three formulas as exercises. Considering $\tan x = (\sin x)/(\cos x)$ and applying the Quotient Rule, we obtain

$$\frac{d}{dx}[\tan x] = \frac{(\cos x)(\cos x) - (\sin x)(-\sin x)}{\cos^2 x}$$

$$= \frac{\cos^2 x + \sin^2 x}{\cos^2 x}$$

$$= \frac{1}{\cos^2 x}$$

$$= \sec^2 x$$

EXAMPLE 4   *Differentiating trigonometric functions*

Function	*Derivative*
(a) $y = x - \tan x$	$\dfrac{dy}{dx} = 1 - \sec^2 x$
(b) $y = x \sec x$	$y' = x(\sec x \tan x) + (\sec x)(1)$
	$= \sec x(1 + x \tan x)$

Because of the abundance of trigonometric identities, the derivative of a trigonometric function can take many forms. This presents a challenge when you are trying to match your answer to one given in the back of the text. To help you meet this challenge, we will occasionally list more than one form of the answer to trigonometric problems. The next example illustrates the possible diversity of trigonometric forms.

EXAMPLE 5   *Different forms of a derivative*

Differentiate the function

$$y = \frac{1 - \cos x}{\sin x} = \csc x - \cot x$$

in both forms and show that the derivatives are equal.

Solution: For the first form,

$$y = \frac{1 - \cos x}{\sin x}$$

$$y' = \frac{(\sin x)(\sin x) - (1 - \cos x)(\cos x)}{\sin^2 x}$$

$$= \frac{\sin^2 x + \cos^2 x - \cos x}{\sin^2 x}$$

$$= \frac{1 - \cos x}{\sin^2 x}$$

For the second form,

$$y = \csc x - \cot x$$

$$y' = -\csc x \cot x + \csc^2 x$$

To show that these two derivatives are equal, we write

$$\frac{1 - \cos x}{\sin^2 x} = \frac{1}{\sin^2 x} - \frac{\cos x}{\sin^2 x}$$

$$= \frac{1}{\sin^2 x} - \left(\frac{1}{\sin x}\right)\left(\frac{\cos x}{\sin x}\right)$$

$$= \csc^2 x - \csc x \cot x$$

With the Chain Rule, we can extend the six trigonometric differentiation rules to cover composite functions and we summarize the "Chain Rule Versions" of the six basic formulas as follows.

DERIVATIVES OF TRIGONOMETRIC FUNCTIONS

$$\frac{d}{dx}[\sin u] = \cos u \, \frac{du}{dx} \qquad \frac{d}{dx}[\cos u] = -\sin u \, \frac{du}{dx}$$

$$\frac{d}{dx}[\tan u] = \sec^2 u \, \frac{du}{dx} \qquad \frac{d}{dx}[\cot u] = -\csc^2 u \, \frac{du}{dx}$$

$$\frac{d}{dx}[\sec u] = \sec u \tan u \, \frac{du}{dx} \qquad \frac{d}{dx}[\csc u] = -\csc u \cot u \, \frac{du}{dx}$$

EXAMPLE 6   *Applying the Chain Rule to trigonometric functions*

Function

Derivative

$$u$$

$$\cos u \qquad u'$$

(a) $y = \sin 2x$ $\qquad y' = \cos 2x \dfrac{d}{dx}[2x] = (\cos 2x)(2) = 2\cos 2x$

(b) $y = \cos(x-1)$ $\qquad y' = -\sin(x-1)$

(c) $y = \tan e^x$ $\qquad y' = e^x \sec^2 e^x$

Be sure that you understand the mathematical conventions regarding parentheses and trigonometric functions. For instance, in part (a) of Example 6 we write $\sin 2x$ to mean $\sin(2x)$. The next example shows the effect of different placements of parentheses.

EXAMPLE 7   *Parentheses and trigonometric functions*

Function

Derivative

(a) $y = \cos 3x^2 = \cos(3x^2)$ $\qquad y' = (-\sin 3x^2)(6x) = -6x\sin 3x^2$

(b) $y = (\cos 3)x^2$ $\qquad y' = (\cos 3)(2x) = 2x\cos 3$

(c) $y = \cos(3x)^2 = \cos(9x^2)$ $\qquad y' = (-\sin 9x^2)(18x) = -18x\sin 9x^2$

(d) $y = \cos^2 3x = (\cos 3x)^2$ $\qquad y' = 2(\cos 3x)(-\sin 3x)(3)$
$\qquad\qquad\qquad\qquad = -6\cos 3x\sin 3x$

EXAMPLE 8   *Differentiating a composite function*

Differentiate $f(t) = \sqrt{\sin 4t}$.

Solution: First we write

$$f(t) = (\sin 4t)^{1/2}$$

Then, by the Power Rule we have

$$f'(t) = \frac{1}{2}(\sin 4t)^{-1/2}\frac{d}{dx}[\sin 4t]$$

$$= \frac{1}{2}(\sin 4t)^{-1/2}(4\cos 4t)$$

$$= \frac{2\cos 4t}{\sqrt{\sin 4t}}$$

Applications

In the remainder of this section, we review some applications of the derivative in the context of trigonometric functions. We begin with an application to minimum and maximum values of a function.

EXAMPLE 9   Finding extrema on a closed interval

Find the extrema of $f(x) = 2 \sin x - \cos 2x$ on the interval $[0, 2\pi]$.

Solution: This function is differentiable for all real x, so we can find all critical numbers by setting $f'(x)$ equal to zero, as follows:

$$f'(x) = 2 \cos x + 2 \sin 2x = 0$$

$$2 \cos x + 4 \cos x \sin x = 0 \qquad \text{\color{blue}{$\sin 2x = 2 \cos x \sin x$}}$$

$$2(\cos x)(1 + 2 \sin x) = 0 \qquad \text{\color{blue}{Factor}}$$

By setting the two factors equal to zero and solving for x in the interval $[0, 2\pi]$, we have

$$\cos x = 0 \;\Longrightarrow\; x = \frac{\pi}{2}, \frac{3\pi}{2} \qquad \text{\color{blue}{Critical numbers}}$$

$$\sin x = -\frac{1}{2} \;\Longrightarrow\; x = \frac{7\pi}{6}, \frac{11\pi}{6} \qquad \text{\color{blue}{Critical numbers}}$$

Finally, by evaluating f at these four critical numbers and at the endpoints of the interval, we conclude that the maximum is $f(\pi/2) = 3$ and the minimum occurs at *two* points, $f(7\pi/6) = -3/2$ and $f(11\pi/6) = -3/2$ as indicated in Table 8.5. The graph is shown in Figure 8.28.

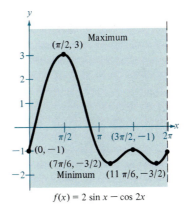

$$f(x) = 2 \sin x - \cos 2x$$

FIGURE 8.28

TABLE 8.5

Endpoint	Critical number	Critical number	Critical number	Critical number	Endpoint
$f(0) = -1$	$f\left(\dfrac{\pi}{2}\right) = 3$ Maximum	$f\left(\dfrac{7\pi}{6}\right) = -\dfrac{3}{2}$ Minimum	$f\left(\dfrac{3\pi}{2}\right) = -1$	$f\left(\dfrac{11\pi}{6}\right) = -\dfrac{3}{2}$ Minimum	$f(2\pi) = -1$

EXAMPLE 10   Applying the First Derivative Test

Use the First Derivative Test to find all relative maxima and minima for the function given by

$$f(x) = \frac{x}{2} - \sin x$$

on the interval $(0, 2\pi)$.

Solution:

$$f'(x) = \frac{1}{2} - \cos x = 0 \implies \cos x = \frac{1}{2}$$

Therefore, the critical numbers, in the interval $(0, 2\pi)$, are $x = \pi/3$ and $x = 5\pi/3$. Table 8.6 summarizes the application of the First Derivative Test to these critical numbers.

TABLE 8.6

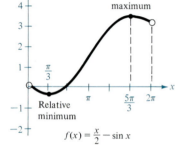

Interval	$0 < x < \dfrac{\pi}{3}$	$\dfrac{\pi}{3} < x < \dfrac{5\pi}{3}$	$\dfrac{5\pi}{3} < x < 2\pi$
Test value	$x = \dfrac{\pi}{4}$	$x = \pi$	$x = \dfrac{7\pi}{4}$
Sign of $f'(x)$	$f'\left(\dfrac{\pi}{4}\right) < 0$	$f'(\pi) > 0$	$f'\left(\dfrac{7\pi}{4}\right) < 0$
Conclusion	decreasing	increasing	decreasing

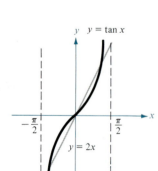

$$f(x) = \frac{x}{2} - \sin x$$

FIGURE 8.29

From Table 8.6, we conclude that a relative minimum occurs at $x = \pi/3$ and a relative maximum occurs at $x = 5\pi/3$ as shown in Figure 8.29.

EXAMPLE 11   Using Newton's Method to find a point of intersection

Estimate the point of intersection of the graphs of $y = \tan x$ and $y = 2x$ as shown in Figure 8.30. Use Newton's Method and continue the iterations until two successive approximations differ by less than 0.0001.

Solution: We let $\tan x = 2x$. Since this implies that $2x - \tan x = 0$, we need to find the zeros of the function given by

$$f(x) = 2x - \tan x$$

Thus, the iterative formula for Newton's Method takes the form

$$x_{n+1} = x_n - \frac{f(x_n)}{f'(x_n)} = x_n - \frac{2x_n - \tan x_n}{2 - \sec^2 x_n}$$

The calculations are shown in Table 8.7, beginning with an initial guess of $x_1 = 1.25$.

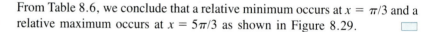

FIGURE 8.30

TABLE 8.7

n	x_n	$f(x_n)$	$f'(x_n)$	$\dfrac{f(x_n)}{f'(x_n)}$	$x_n - \dfrac{f(x_n)}{f'(x_n)}$
1	1.25000	−0.50957	−8.05751	0.06324	1.18676
2	1.18676	−0.10110	−5.12374	0.01973	1.16703
3	1.16703	−0.00653	−4.47832	0.00146	1.16557
4	1.16557	−0.00003	−4.43435	0.00001	1.16556

Thus, we approximate the point of intersection to occur when $x = 1.16556$.

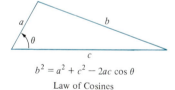

EXAMPLE 12   The velocity of a piston

For the engine shown in Figure 8.31, a connecting rod of a length of 7 inches is fastened to a crank of radius 3 inches. The crank shaft rotates counterclockwise at a constant rate of 200 revolutions per minute. Find the velocity of the piston when $\theta = 60°$.

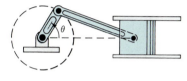

Solution:

1. We label the distances as shown in Figure 8.31. Furthermore, since a complete revolution corresponds to 2π radians, we can determine that $d\theta/dt = 200(2\pi) = 400\pi$ rad/min.

FIGURE 8.31

$$\text{Given: } \frac{d\theta}{dt} = 400\pi \qquad \text{Constant rate}$$

$$\text{Find: } \frac{dx}{dt} \text{ when } \theta = 60°$$

2. From the Law of Cosines (see Figure 8.32), we have

$$7^2 = 3^2 + x^2 - 2(3)(x)\cos\theta$$

3. Implicit differentiation with respect to t yields

$$0 = 2x\frac{dx}{dt} - 6\left[-x\sin\theta\left(\frac{d\theta}{dt}\right) + \cos\theta\left(\frac{dx}{dt}\right)\right]$$

$$(6\cos\theta - 2x)\frac{dx}{dt} = 6x\sin\theta\left(\frac{d\theta}{dt}\right)$$

$$\frac{dx}{dt} = \frac{6x\sin\theta}{6\cos\theta - 2x}\left(\frac{d\theta}{dt}\right)$$

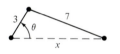

$b^2 = a^2 + c^2 - 2ac\cos\theta$

Law of Cosines

FIGURE 8.32

4. When $\theta = 60°$, we can solve for x as follows:

$$7^2 = 3^2 + x^2 - 2(3)(x)\cos(60°)$$

$$49 = 9 + x^2 - 6x\left(\frac{1}{2}\right)$$

$$0 = x^2 - 3x - 40$$

$$0 = (x - 8)(x + 5)$$

Choosing the positive solution, we have $x = 8$. Finally, when $\theta = 60°$ and $x = 8$, the velocity of the piston is

$$\frac{dx}{dt} = \frac{6(8)(\sqrt{3}/2)}{6(1/2) - 16}(400\pi) = \frac{9600\pi\sqrt{3}}{-13} \approx -4018 \text{ in/min} \qquad \square$$

| **Remark**   Note that the velocity in Example 12 is negative since the piston is moving to the left.

Section Exercises 8.3

In Exercises 1–44, find the derivative of the given function.

1. $y = x^2 - \dfrac{1}{2}\cos x$ **2.** $y = 5 + \sin x$

3. $y = \dfrac{1}{x} - 3\sin x$ **4.** $g(t) = \pi \cos t$

5. $f(x) = 4\sqrt{x} + 3\cos x$
6. $f(x) = 2\sin x + 3\cos x$

7. $f(t) = t^2 \sin t$ **8.** $f(x) = \dfrac{\sin x}{x}$

9. $g(t) = \dfrac{\cos t}{t}$ **10.** $f(\theta) = (\theta + 1)\cos \theta$

11. $y = \tan x - x$ **12.** $y = x + \cot x$

13. $y = 5x \csc x$ **14.** $y = \dfrac{\sec x}{x}$

15. $f(\theta) = -\csc \theta - \sin \theta$
16. $h(x) = x \sin x + \cos x$
17. $g(t) = t^2 \sin t + 2t \cos t - 2 \sin t$
18. $h(\theta) = 5\sec 3\theta + \tan 3\theta$
19. $f(x) = \sin \pi x \cos \pi x$ **20.** $f(x) = \tan 2x \cot 2x$

21. $y = \dfrac{1 + \csc x}{1 - \csc x}$ **22.** $y = \dfrac{\sin \theta}{1 - \cos \theta}$

23. $y = \cos 3x$ **24.** $y = \sin 2x$

25. $y = 3\tan 4x$ **26.** $y = 2\cos \dfrac{x}{2}$

27. $y = \sin \pi x$ **28.** $y = \sec x^2$

29. $y = \dfrac{1}{4}\sin^2 x$ **30.** $y = 5\cos^2 \pi x$

31. $y = \dfrac{1}{4}\sin^2 2x$ **32.** $y = 5\cos(\pi x)^2$

33. $y = \sqrt{\sin x}$ **34.** $y = \csc^2 4x$

35. $y = \sec^3 2x$ **36.** $y = x^2 \sin \dfrac{1}{x}$

37. $y = \ln |\csc x - \cot x|$ **38.** $y = \ln |\sec x + \tan x|$
39. $y = e^x(\sin x + \cos x)$ **40.** $y = \tan^2 e^x$
41. $y = e^{\tan x}$ **42.** $y = \ln |\cot x|$
43. $y = \ln |\tan x|$ **44.** $y = \ln |\sin x|$

In Exercises 45–50, use implicit differentiation to find dy/dx and evaluate the derivative at the indicated point.

	Equation	*Point*
45.	$\sin x + \cos 2y = 2$	$\left(\dfrac{\pi}{2}, 0\right)$
46.	$2\sin x \cos y = 1$	$\left(\dfrac{\pi}{4}, \dfrac{\pi}{4}\right)$
47.	$\tan(x + y) = x$	$(0, 0)$
48.	$\cot y = x - y$	$\left(\dfrac{\pi}{2}, \dfrac{\pi}{2}\right)$
49.	$x \cos y = 1$	$\left(2, \dfrac{\pi}{3}\right)$
50.	$x = \sec \dfrac{1}{y}$	$\left(\sqrt{2}, \dfrac{\pi}{4}\right)$

In Exercises 51 and 52, show that the function satisfies the differential equation.

51. $y = 2\sin x + 3\cos x$
$\quad y'' + y = 0$
52. $y = e^x(\cos \sqrt{2}x + \sin \sqrt{2}x)$
$\quad y'' - 2y' + 3y = 0$

In Exercises 53 and 54, find the slope of the tangent line to the given sine function at the origin. Compare this value to the number of complete cycles in the interval $[0, 2\pi]$.

53. (a) $y = \sin x$ (b) $y = \sin 2x$

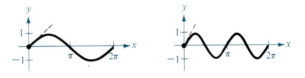

54. (a) $y = \sin 3x$ (b) $y = \sin \dfrac{x}{2}$

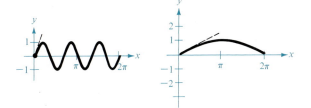

In Exercises 55 and 56, find an equation of the tangent line to the graph of the function at the indicated point.

Function	*Point*
55. $f(x) = \tan x$	$\left(-\dfrac{\pi}{4}, -1\right)$
56. $f(x) = \sec x$	$\left(\dfrac{\pi}{3}, 2\right)$

In Exercises 57–64, evaluate each limit, using L'Hôpital's Rule when necessary.

57. $\lim\limits_{x \to 0} \dfrac{\sin 2x}{\sin 3x}$ **58.** $\lim\limits_{x \to \pi} \dfrac{\sin x}{x - \pi}$

59. $\lim\limits_{x \to 0} \dfrac{x - \tan x}{x - \sin x}$ **60.** $\lim\limits_{\theta \to 0} \dfrac{1 - \cos \theta}{\theta}$

61. $\lim\limits_{\theta \to 0} \dfrac{1 - \cos 2\theta}{4\theta^2}$ **62.** $\lim\limits_{x \to \pi/4} (\tan 2x - \sec 2x)$

63. $\lim\limits_{x \to \infty} x \sin \dfrac{1}{x}$ **64.** $\lim\limits_{x \to 0} \dfrac{1 - e^x}{\sin x}$

In Exercises 65–68, sketch the graph of each function on the indicated interval, making use of relative extrema and points of inflection.

Function	*Interval*
65. $f(x) = 2 \sin x + \sin 2x$	$[0, 2\pi]$
66. $f(x) = 2 \sin x + \cos 2x$	$[0, 2\pi]$
67. $f(x) = x - \sin x$	$[0, 4\pi]$
68. $f(x) = \cos x - x$	$[0, 4\pi]$

In Exercises 69 and 70, sketch the graph of the function on the interval $[-\pi, \pi]$. In each case use Newton's Method to approximate the critical number to two decimal places.

69. $f(x) = x \sin x$ **70.** $f(x) = x \cos x$

71. The height of a weight oscillating on a spring is given by the equation $y = \frac{1}{3} \cos 12t - \frac{1}{4} \sin 12t$, where y is measured in inches and t is measured in seconds.

(a) Calculate the height and velocity of the weight when $t = \pi/8$ seconds.

(b) Show that the maximum displacement of the weight is $\frac{5}{12}$ inches.

(c) Find the period P of y. Find the frequency f (number of oscillations per second) if $f = 1/P$.

72. The general equation giving the height of an oscillating weight attached to a spring is

$$y = A \sin \left(\sqrt{\dfrac{k}{m}}\, t\right) + B \cos \left(\sqrt{\dfrac{k}{m}}\, t\right)$$

where k is the spring constant and m is the mass of the weight.

(a) Show that the maximum displacement of the weight is $\sqrt{A^2 + B^2}$.

(b) Show that the frequency (number of oscillations per second) is $(1/2\pi)\sqrt{k/m}$. How is the frequency changed if the stiffness k of the spring is increased? How is the frequency changed if the mass m of the weight is increased?

73. A component is designed to slide a block of steel of weight W across a table and into a chute as shown in Figure 8.33. The motion of the block is resisted by a frictional force proportional to its net weight. (Let k be the constant of proportionality.) Find the minimum force F needed to slide the block and find the corresponding value of θ. [Hint: $F \cos \theta$ is the force in the direction of motion and $F \sin \theta$ is the amount of force tending to lift the block. Therefore, the net weight of the block is $W - F \sin \theta$.]

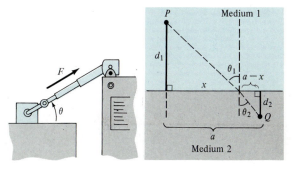

FIGURE 8.33   FIGURE 8.34

74. When light waves, traveling in a transparent medium, strike the surface of a second transparent medium they tend to "bend." This tendency is called refraction, and is given by **Snell's Law of Refraction,**

$$\dfrac{\sin \theta_1}{v_1} = \dfrac{\sin \theta_2}{v_2}$$

where θ_1 and θ_2 are the magnitudes of the angles shown in Figure 8.34, and v_1 and v_2 are the velocities of light in the two media. Show that light waves traveling from *P* to *Q* follow the path of minimum time.

75. An airplane flys at an altitude of 5 miles toward a point directly over an observer as shown in Figure 8.35. The speed of the plane is 600 miles per hour. Find the rate at which the angle of elevation θ is changing when the angle is
(a) $\theta = 30°$ (b) $\theta = 60°$ (c) $\theta = 75°$

of horizontal movement of the dot for the following angles:
(a) $\theta = 0°$ (b) $\theta = 30°$ (c) $\theta = 60°$

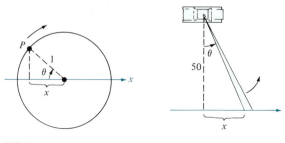

FIGURE 8.37   FIGURE 8.38

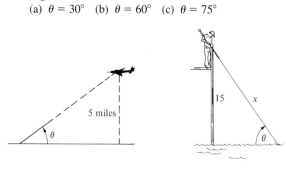

FIGURE 8.35   FIGURE 8.36

76. A fish is reeled in at a rate of 1 foot per second from a bridge 15 feet above the water as shown in Figure 8.36. At what rate is the angle between the line and the water changing when there is 25 feet of line out? (Assume the fish stays near the surface of the water.)

77. A wheel of radius 1 foot revolves at a rate of 10 revolutions per second. A dot is painted at a point *P* on the rim of the wheel as shown in Figure 8.37. Find the rate

78. A patrol car is parked 50 feet from a long warehouse as shown in Figure 8.38. The light on the car turns at a rate of 30 revolutions per minute. How fast is the light beam moving along the wall when the beam makes the following angles with the line perpendicular from the light to the wall:
(a) $\theta = 30°$ (b) $\theta = 60°$ (c) $\theta = 70°$

79. For $f(x) = \sec^2 x$ and $g(x) = \tan^2 x$, show that $f'(x) = g'(x)$.

80. Derive the following differentiation rules.
(a) $D_x[\sec x] = \sec x \tan x$
(b) $D_x[\csc x] = -\csc x \cot x$
(c) $D_x[\cot x] = -\csc^2 x$

8.4
Integrals of trigonometric functions

Corresponding to each trigonometric differentiation formula is an integration formula. For instance, the differentiation formula

$$\frac{d}{dx}[\cos u] = -\sin u \frac{du}{dx}$$

corresponds to the integration formula

$$\int \sin u \, du = -\cos u + C$$

The following list summarizes all six integration formulas corresponding to the derivatives of the basic trigonometric functions.

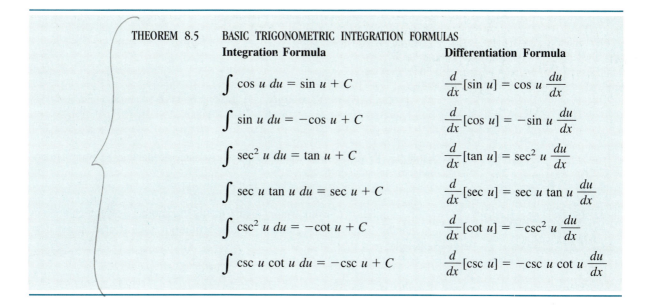

THEOREM 8.5   BASIC TRIGONOMETRIC INTEGRATION FORMULAS

Integration Formula

$$\int \cos u \, du = \sin u + C$$

$$\int \sin u \, du = -\cos u + C$$

$$\int \sec^2 u \, du = \tan u + C$$

$$\int \sec u \tan u \, du = \sec u + C$$

$$\int \csc^2 u \, du = -\cot u + C$$

$$\int \csc u \cot u \, du = -\csc u + C$$

Differentiation Formula

$$\frac{d}{dx}[\sin u] = \cos u \, \frac{du}{dx}$$

$$\frac{d}{dx}[\cos u] = -\sin u \, \frac{du}{dx}$$

$$\frac{d}{dx}[\tan u] = \sec^2 u \, \frac{du}{dx}$$

$$\frac{d}{dx}[\sec u] = \sec u \tan u \, \frac{du}{dx}$$

$$\frac{d}{dx}[\cot u] = -\csc^2 u \, \frac{du}{dx}$$

$$\frac{d}{dx}[\csc u] = -\csc u \cot u \, \frac{du}{dx}$$

EXAMPLE 1   Integration of trigonometric functions

(a) $\displaystyle\int 2 \cos x \, dx = 2 \int \cos x \, dx = 2 \sin x + C$ $u = x$

(b) $\displaystyle\int 3x^2 \sin x^3 \, dx = \int \underbrace{\sin x^3}_{\sin u}\underbrace{(3x^2) \, dx}_{du} = -\cos x^3 + C$ $u = x^3$

(c) $\displaystyle\int \sec^2 3x \, dx = \frac{1}{3} \int \underbrace{(\sec^2 3x)}_{\sec^2 u}\underbrace{(3) \, dx}_{du} = \frac{1}{3} \tan 3x + C$ $u = 3x$

The integrals in Example 1 are easily recognized as fitting one of the basic integration formulas in Theorem 8.5. However, because of the variety of trigonometric identities, it often happens that an integrand that fits one of the basic formulas will come in a disguised form. This is illustrated in the next two examples.

EXAMPLE 2   Using a trigonometric identity

Evaluate $\int \tan^2 x \, dx$.

Solution:

$$\int \tan^2 x \, dx = \int (-1 + \sec^2 x) \, dx = -x + \tan x + C$$

<u>EXAMPLE 3   *Using a trigonometric identity*</u>

Evaluate $\int (\csc x + \sin x)(\csc x)\, dx$.

Solution:

$$\int (\csc x + \sin x)(\csc x)\, dx = \int (\csc^2 x + 1)\, dx = -\cot x + x + C$$

| Remark   We will say more about the use of trigonometric identities to evaluate trigonometric integrals in Section 9.3.

In addition to using trigonometric identities, another useful technique in evaluating trigonometric integrals is u-substitution as illustrated in the next example.

<u>EXAMPLE 4   *Integration by u-substitution*</u>

Evaluate

$$\int \frac{\sec^2 \sqrt{x}}{\sqrt{x}}\, dx$$

Solution: Let $u = \sqrt{x}$. Then we have

$$u = \sqrt{x} \implies du = \frac{1}{2\sqrt{x}}\, dx \implies 2\,du = \frac{1}{\sqrt{x}}\, dx$$

$$\int \frac{\sec^2 \sqrt{x}}{\sqrt{x}}\, dx = \int \sec^2 \sqrt{x}\left(\frac{1}{\sqrt{x}}\, dx\right) = \int \sec^2 u(2\,du)$$

$$= 2 \int \sec^2 u\, du = 2 \tan u + C = 2 \tan \sqrt{x} + C$$

One of the most common u-substitutions involves quantities in the integrand that are raised to a power as illustrated in the next two examples.

<u>EXAMPLE 5   *Change of variable*</u>

Evaluate $\int \sin^2 3x \cos 3x\, dx$.

Solution: Since $\sin^2 3x = (\sin 3x)^2$, we let $u = \sin 3x$. Then

$$du = (\cos 3x)(3)\, dx \implies \frac{du}{3} = \cos 3x\, dx$$

Substituting u and $du/3$ in the given integral yields

$$\int \sin^2 3x \cos 3x\, dx = \int u^2 \frac{du}{3} = \frac{1}{3} \int u^2\, du$$

$$= \frac{1}{3}\left(\frac{u^3}{3}\right) + C = \frac{1}{9} \sin^3 3x + C$$

EXAMPLE 6   Substitution and the Power Rule

(a) $\displaystyle\int \frac{\sin x}{\cos^2 x}\, dx = -\int \overbrace{(\cos x)^{-2}}^{u^{-2}}\overbrace{(-\sin x)\, dx}^{du} = -\overbrace{\frac{(\cos x)^{-1}}{-1}}^{u^{-1}/(-1)} + C$

$\qquad\qquad\qquad = \sec x + C$

(b) $\displaystyle\int 4\cos^2 4x \sin 4x \, dx = -\int \overbrace{(\cos 4x)^2}^{u^2}\overbrace{(-4\sin 4x)\, dx}^{du}$

$\qquad\qquad\qquad = -\overbrace{\frac{(\cos 4x)^3}{3}}^{u^3/3} + C$

(c) $\displaystyle\int \frac{\sec^2 x}{\sqrt{\tan x}}\, dx = \int \overbrace{(\tan x)^{-1/2}}^{u^{-1/2}}\overbrace{(\sec^2 x)\, dx}^{du} = \overbrace{\frac{(\tan x)^{1/2}}{1/2}}^{u^{1/2}/(1/2)} + C$

In Theorem 8.5 we listed six trigonometric integration formulas—the six that correspond directly to differentiation rules. We now complete our set of basic trigonometric integration formulas by making use of the Log Rule.

EXAMPLE 7   The antiderivative of the tangent

Evaluate $\int \tan x \, dx$.

Solution: This integral doesn't seem to fit any formulas on our basic list. However, by a trigonometric identity, we obtain the following quotient form.

$$\int \tan x \, dx = \int \frac{\sin x}{\cos x}\, dx$$

Now, knowing that $D_x[\cos x] = -\sin x$, we consider $u = \cos x$ and write

$$\int \tan x \, dx = -\int \frac{(-\sin x)}{\cos x}\, dx = -\int \frac{u'}{u}\, dx$$

$$= -\ln |u| + C = -\ln |\cos x| + C$$

In Example 7, we used a trigonometric identity together with the Log Rule to derive an integration formula for the tangent function. In the next example we use a rather unusual step (multiplying and dividing by the same quantity) to derive an integration formula for the secant function.

EXAMPLE 8   Derivation of the secant formula

Evaluate $\int \sec x \, dx$.

Solution: Consider the following procedure:

$$\int \sec x \, dx = \int \sec x\left(\frac{\sec x + \tan x}{\sec x + \tan x}\right) dx = \int \frac{\sec^2 x + \sec x \tan x}{\sec x + \tan x} \, dx$$

Now, letting u be the denominator of this quotient, we have

$$u = \sec x + \tan x \implies u' = \sec x \tan x + \sec^2 x$$

Therefore, we conclude that

$$\int \sec x \, dx = \int \frac{\sec^2 x + \sec x \tan x}{\sec x + \tan x} \, dx = \int \frac{u'}{u} \, dx = \ln |u| + C$$

$$= \ln |\sec x + \tan x| + C \quad \square$$

With Examples 7 and 8 we now have integration formulas for $\sin x$, $\cos x$, $\tan x$, and $\sec x$. We leave the derivations for $\csc x$ and $\cot x$ as exercises and summarize all six trigonometric formulas in the following theorem.

THEOREM 8.6   INTEGRALS OF THE SIX BASIC TRIGONOMETRIC FUNCTIONS

$$\int \sin u \, du = -\cos u + C \qquad \int \cos u \, du = \sin u + C$$

$$\int \tan u \, du = -\ln |\cos u| + C \qquad \int \cot u \, du = \ln |\sin u| + C$$

$$\int \sec u \, du = \ln |\sec u + \tan u| + C \qquad \int \csc u \, du = -\ln |\csc u + \cot u| + C$$

Remark As you memorize these formulas, note that the three formulas on the right follow the pattern set by the three on the left.

EXAMPLE 9   Integrating trigonometric functions

Evaluate

$$\int_0^{\pi/4} \sqrt{1 + \tan^2 x} \, dx$$

Solution: Since $1 + \tan^2 x = \sec^2 x$, we have

$$\int_0^{\pi/4} \sqrt{1 + \tan^2 x} \, dx = \int_0^{\pi/4} \sqrt{\sec^2 x} \, dx = \int_0^{\pi/4} \sec x \, dx$$

$$= \left[\ln |\sec x + \tan x| \right]_0^{\pi/4}$$

$$= \ln (\sqrt{2} + 1) - \ln (1) \approx 0.8814 \quad \square$$

Remark Note in Example 9 that $\sec x > 0$ for $0 \le x \le \pi/4$ and thus it is valid to replace $\sqrt{\sec^2 x}$ by $\sec x$.

Applications

$f(x) = \sin^3 x \cos x + \sin x \cos x$

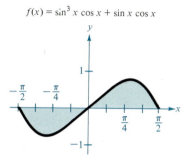

FIGURE 8.39

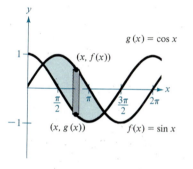

FIGURE 8.40

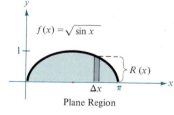

Plane Region

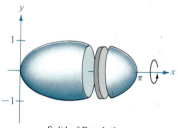

Solid of Revolution

FIGURE 8.41

EXAMPLE 10   *Finding the area of a plane region*

Find the area of the region bounded by the graph of

$$f(x) = \sin^3 x \cos x + \sin x \cos x$$

and the *x*-axis over the interval $0 \le x \le \pi/2$ as shown in Figure 8.39.

Solution: Using the Power Rule, we find the area of this region to be

$$A = \int_0^{\pi/2} (\sin^3 x \cos x + \sin x \cos x)\, dx = \left[\frac{\sin^4 x}{4} + \frac{\sin^2 x}{2} \right]_0^{\pi/2}$$

$$= \frac{1}{4} + \frac{1}{2} = \frac{3}{4}$$

EXAMPLE 11   *A region lying between two intersecting graphs*

The sine and cosine curves intersect infinitely many times bounding regions of equal areas as shown in Figure 8.40. Find the area of one of these regions.

Solution: To find the points of intersection of the graphs of $f(x) = \sin x$ and $g(x) = \cos x$, we set the two functions equal to each other and solve for *x*.

$$\sin x = \cos x \qquad \text{Set } f(x) \text{ equal to } g(x)$$

$$\frac{\sin x}{\cos x} = 1$$

$$\tan x = 1 \qquad x = \frac{\pi}{4}, \frac{5\pi}{4}; \ 0 \le x \le 2\pi$$

Thus, $a = \pi/4$ and $b = 5\pi/4$. Since $\sin x \ge \cos x$ on the interval $[\pi/4, 5\pi/4]$, the area of the region is

$$A = \int_{\pi/4}^{5\pi/4} [\sin x - \cos x]\, dx = \left[-\cos x - \sin x \right]_{\pi/4}^{5\pi/4}$$

$$= \left(\frac{\sqrt{2}}{2} + \frac{\sqrt{2}}{2} \right) - \left(-\frac{\sqrt{2}}{2} - \frac{\sqrt{2}}{2} \right)$$

$$= 2\sqrt{2}$$

EXAMPLE 12   *Finding the volume of a solid of revolution*

Find the volume of the solid formed by revolving the region bounded by the graph of $f(x) = \sqrt{\sin x}$ and the *x*-axis $(0 \le x \le \pi)$ about the *x*-axis.

Solution: From the representative rectangle in Figure 8.41, we see that the radius of this solid is given by

$$R(x) = f(x) = \sqrt{\sin x}$$

and it follows that its volume is

$$V = \pi \int_0^\pi [R(x)]^2 \, dx = \pi \int_0^\pi (\sqrt{\sin x})^2 \, dx$$

$$= \pi \int_0^\pi \sin x \, dx = -\pi \cos x \Big]_0^\pi$$

$$= \pi(1 + 1) = 2\pi$$

EXAMPLE 13   *Finding the average value of a function on an interval*

The electromotive force E of a particular electrical circuit is given by

$$E = 3 \sin 2t$$

where E is measured in volts and t is measured in seconds. Find the average value of E as t ranges from 0 to 0.5 second.

Solution: The average value of E is given by

$$\frac{1}{b - a} \int_a^b f(t) \, dt = \frac{1}{0.5 - 0} \int_0^{0.5} 3 \sin 2t \, dt$$

$$= 6 \int_0^{0.5} \sin 2t \, dt$$

Letting $u = 2t$ implies that $du = 2 \, dt$. Thus, we have

$$\text{average value} = 6\left(\frac{1}{2}\right) \int_0^{0.5} (\sin 2t)(2) \, dt$$

$$= \left[3(-\cos 2t) \right]_0^{0.5}$$

$$= 3[-\cos (1) + 1] \approx 1.379 \text{ volts}$$

EXAMPLE 14   *Finding arc length*

Find the arc length of the graph of $y = \ln (\cos x)$ from $x = 0$ to $x = \pi/4$ as shown in Figure 8.42.

Solution: Since

$$y' = -\frac{\sin x}{\cos x} = -\tan x$$

the arc length is given by

$$s = \int_a^b \sqrt{1 + (y')^2} \, dx = \int_0^{\pi/4} \sqrt{1 + \tan^2 x} \, dx$$

$$= \int_0^{\pi/4} \sqrt{\sec^2 x} \, dx = \int_0^{\pi/4} \sec x \, dx$$

$$= \left[\ln |\sec x + \tan x| \right]_0^{\pi/4}$$

$$= \ln (\sqrt{2} + 1) - \ln (1) \approx 0.8814$$

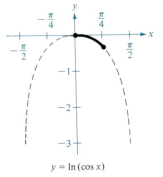

$y = \ln (\cos x)$

FIGURE 8.42

Section Exercises 8.4

In Exercises 1–36, evaluate the given integral.

1. $\int (2 \sin x + 3 \cos x)\, dx$

2. $\int (t^2 - \sin t)\, dt$

3. $\int (1 - \csc t \cot t)\, dt$

4. $\int (\theta^2 + \sec^2 \theta)\, d\theta$

5. $\int (\sec^2 \theta - \sin \theta)\, d\theta$

6. $\int \sec y(\tan y - \sec y)\, dy$

7. $\int \sin 2x\, dx$

8. $\int \cos 6x\, dx$

9. $\int x \cos x^2\, dx$

10. $\int x \sin x^2\, dx$

11. $\int \sec^2 \dfrac{x}{2}\, dx$

12. $\int \csc^2 \dfrac{x}{2}\, dx$

13. $\int \dfrac{\csc^2 x}{\cot^3 x}\, dx$

14. $\int \dfrac{\sin x}{\cos^2 x}\, dx$

15. $\int \cot^2 x\, dx$

16. $\int \csc 2x \cot 2x\, dx$

17. $\int \tan^4 x \sec^2 x\, dx$

18. $\int \sqrt{\cot x}\, \csc^2 x\, dx$

19. $\int \cot \pi x\, dx$

20. $\int \tan 5x\, dx$

21. $\int \csc 2x\, dx$

22. $\int \sec \dfrac{x}{2}\, dx$

23. $\int \dfrac{\sec^2 x}{\tan x}\, dx$

24. $\int \dfrac{\tan^2 2x}{\sec 2x}\, dx$

25. $\int \dfrac{\sec x \tan x}{\sec x - 1}\, dx$

26. $\int \dfrac{\sin x}{1 + \cos x}\, dx$

27. $\int \dfrac{\cos t}{1 + \sin t}\, dt$

28. $\int \dfrac{\sin^2 x - \cos^2 x}{\cos x}\, dx$

29. $\int \dfrac{1 - \cos \theta}{\theta - \sin \theta}\, d\theta$

30. $\int \dfrac{1 - \sin^2 \theta}{\cos^2 \theta}\, d\theta$

31. $\int e^x \cos e^x\, dx$

32. $\int e^{\sin x} \cos x\, dx$

33. $\int e^{-x} \tan (e^{-x})\, dx$

34. $\int e^{\sec x} \sec x \tan x\, dx$

35. $\int (\sin 2x + \cos 2x)^2\, dx$

36. $\int (\csc 2\theta - \cot 2\theta)^2\, d\theta$

In Exercises 37–44, evaluate the definite integral.

37. $\int_0^{\pi/2} \cos \dfrac{2x}{3}\, dx$

38. $\int_0^{\pi/2} \sin 2x\, dx$

39. $\int_{\pi/2}^{2\pi/3} \sec^2 \dfrac{x}{2}\, dx$

40. $\int_{\pi/2}^{\pi/2} (x + \cos x)\, dx$

$\pi/3$

41. $\int_{\pi/12}^{\pi/4} \csc 2x \cot 2x\, dx$

42. $\int_0^{\pi/8} \sin 2x \cos 2x\, dx$

43. $\int_0^1 \sec (1 - x) \tan (1 - x)\, dx$

44. $\int_0^{\pi/4} (\sec x)^3 (\sec x \tan x)\, dx$

In Exercises 45–50, determine the area of the given region.

45. $y = \cos \dfrac{x}{2}$

46. $y = x + \sin x$

47. $y = 2 \sin x + \sin 2x$

48. $y = \sin x + \cos 2x$

49. $f(x) = 2$
$g(x) = \sec x$

50. $f(x) = 2 \sin x$
$g(x) = \tan x$

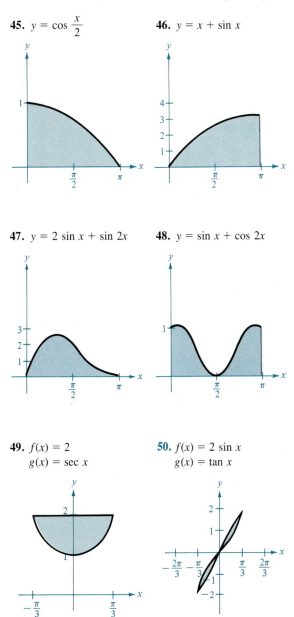

In Exercises 51–54, find the volume of the solid generated by revolving the region bounded by the graphs of the given equations about the *x*-axis.

51. $y = \sqrt{\sin x}$, $y = 0$, $0 \le x \le \pi$
52. $y = \sqrt{\cos x}$, $y = 0$, $0 \le x \le \pi/2$
53. $y = \sec x$, $y = 0$, $x = 0$, $x = \pi/4$
54. $y = \csc x$, $y = 0$, $x = \pi/6$, $x = 5\pi/6$

In Exercises 55 and 56, find the arc length of the graphs of the given function over the indicated interval. Use Simpson's Rule with $n = 4$ to approximate the definite integral.

Function	Interval
55. $y = \sin x$	$[0, \pi]$
56. $y = 4 \cos \dfrac{x}{2}$	$[0, \pi]$

57. The sales of a seasonal product are given by the model

$$S = 74.50 + 43.75 \sin \frac{\pi t}{6}$$

where S is measured in thousands of units and t is the time in months, with $t = 1$ corresponding to January. Find the average sales during:
(a) the first quarter $(0 \le t \le 3)$
(b) the second quarter $(3 \le t \le 6)$
(c) the entire year $(0 \le t \le 12)$

58. The minimum stockpile level of gasoline in the United States can be approximated by the model

$$Q = 217 + 13 \cos \frac{\pi(t - 3)}{6}$$

where Q is measured in millions of barrels of gasoline and t is the time in months, with $t = 1$ corresponding to January. Find the average minimum level given by this model during:
(a) the first quarter $(0 \le t \le 3)$
(b) the second quarter $(3 \le t \le 6)$
(c) the entire year $(0 \le t \le 12)$

59. The oscillating current in an electrical circuit is given by

$$I = 2 \sin (60\pi t) + \cos (120\pi t)$$

where I is measured in amperes and t is measured in seconds. Find the average current for the following time intervals:
(a) $0 \le t \le 1/60$
(b) $0 \le t \le 1/240$
(c) $0 \le t \le 1/30$

60. A horizontal plane is ruled with parallel lines 2 inches apart. If a 2-inch needle is randomly tossed onto the plane, it can be shown that the probability of the needle touching a line is given by

$$P = \frac{2}{\pi} \int_0^{\pi/2} \sin \theta \; d\theta$$

where θ is the acute angle between the needle and any one of the parallel lines. Find this probability.

61. Evaluate $\int \sin x \cos x \, dx$ two ways. First, make the substitution $u = \sin x$, and second, integrate by letting $u = \cos x$. Explain the difference in the results.

62. The graphs of the sine and cosine are symmetric to the origin and the *y*-axis respectively. Use this fact to aid in the evaluation of the following definite integrals:

(a) $\displaystyle\int_{-\pi/4}^{\pi/4} \sin x \, dx$ (b) $\displaystyle\int_{-\pi/4}^{\pi/4} \cos x \, dx$

(c) $\displaystyle\int_{-\pi/2}^{\pi/2} \cos x \, dx$ (d) $\displaystyle\int_{-1.32}^{1.32} \sin 2x \, dx$

(e) $\displaystyle\int_{-\pi/2}^{\pi/2} \sin x \cos x \, dx$

In Exercises 63–66, show the equivalence of each pair of formulas.

63. $\displaystyle\int \tan x \, dx = -\ln |\cos x| + C$

$\displaystyle\int \tan x \, dx = \ln |\sec x| + C$

64. $\displaystyle\int \cot x \, dx = \ln |\sin x| + C$

$\displaystyle\int \cot x \, dx = -\ln |\csc x| + C$

65. $\displaystyle\int \sec x \, dx = \ln |\sec x + \tan x| + C$

$\displaystyle\int \sec x \, dx = -\ln |\sec x - \tan x| + C$

66. $\displaystyle\int \csc x \, dx = \ln |\csc x - \cot x| + C$

$\displaystyle\int \csc x \, dx = -\ln |\csc x + \cot x| + C$

In Exercises 67–72, approximate the given integral using (a) the Trapezoidal Rule and (b) Simpson's Rule.

67. $\displaystyle\int_0^{\sqrt{\pi/2}} \cos x^2 \, dx$, $n = 4$

68. $\displaystyle\int_0^{\sqrt{\pi/4}} \tan x^2 \, dx$, $n = 4$

69. $\displaystyle\int_0^1 \sin x^2 \, dx$, $n = 2$

70. $\displaystyle\int_0^{\pi} \sqrt{x} \sin x \, dx$, $n = 4$

71. $\displaystyle\int_0^{\pi/4} x \tan x \, dx$, $n = 4$

72. $\displaystyle\int_0^{\pi/2} \sqrt{1 + \cos^2 x} \, dx$, $n = 2$

8.5
Inverse trigonometric functions and differentiation

We begin this section with a rather startling statement: *None of the six basic trigonometric functions has an inverse*. This statement follows from the fact that all six functions are periodic, and hence not one-to-one. It seems, then, that we have little to do in this section, but of course that is not the case. Rather, we need to examine the six basic trigonometric functions to see if we can redefine their domains in such a way that they will have inverses on the *restricted domains*.

We begin by looking at the sine function.

EXAMPLE 1   *Finding an interval on which the sine function is one-to-one*

Show that the sine function $f(x) = \sin x$ is not one-to-one on the entire real line. Then show that $[-\pi/2, \pi/2]$ is the largest interval, centered at the origin, for which f is strictly monotonic.

Solution: It is clear that f is not one-to-one since many different x-values yield the same y-value. For instance,

$$\sin (0) = 0 = \sin (\pi)$$

Moreover, f is increasing on the open interval $(-\pi/2, \pi/2)$ since its derivative

$$f'(x) = \cos x$$

is positive there. Finally, since the left and right endpoints correspond to relative extrema of the sine, we can conclude that f is increasing on the closed interval $[-\pi/2, \pi/2]$ *and* that in any larger interval the function would not be strictly monotonic. (See Figure 8.43.)

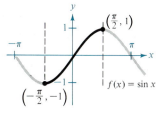

FIGURE 8.43

From Example 1, we can apply Theorem 7.10 to conclude that the *restricted* sine function, whose domain is $[-\pi/2, \pi/2]$, has an inverse function. On this interval, we define the inverse sine function to be

$$y = \arcsin x \qquad \text{if and only if} \qquad \sin y = x$$

where $-1 \le x \le 1$ and $-\pi/2 \le \arcsin x \le \pi/2$.

Under suitable restrictions, each of the six trigonometric functions is one-to-one and so possesses an inverse, as indicated in the following definition. (The term iff is used to represent the phrase "if and only if.")

DEFINITION OF INVERSE TRIGONOMETRIC FUNCTIONS	**Function**	**Domain**	**Range**
	$y = \textbf{arcsin } x$ iff $\sin y = x$	$-1 \le x \le 1$	$-\dfrac{\pi}{2} \le y \le \dfrac{\pi}{2}$
	$y = \textbf{arccos } x$ iff $\cos y = x$	$-1 \le x \le 1$	$0 \le y \le \pi$
	$y = \textbf{arctan } x$ iff $\tan y = x$	$-\infty < x < \infty$	$-\dfrac{\pi}{2} < y < \dfrac{\pi}{2}$
	$y = \textbf{arccot } x$ iff $\cot y = x$	$-\infty < x < \infty$	$0 < y < \pi$
	$y = \textbf{arcsec } x$ iff $\sec y = x$	$x \le -1,\ 1 \le x$	$0 \le y \le \pi,\ y \ne \dfrac{\pi}{2}$
	$y = \textbf{arccsc } x$ iff $\csc y = x$	$x \le -1,\ 1 \le x$	$-\dfrac{\pi}{2} \le y \le \dfrac{\pi}{2},\ y \ne 0$

Remark   The term arcsin x is read as the "inverse sine of x" or sometimes the "angle whose sine is x." An alternative notation for the inverse sine function is $\sin^{-1} x$.

The graphs of these six inverse functions are shown in Figure 8.44.

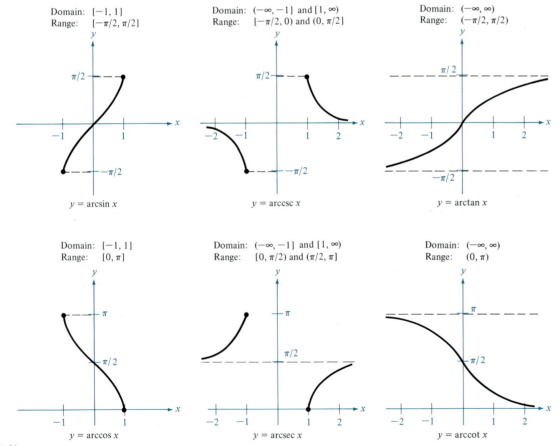

FIGURE 8.44

When evaluating inverse trigonometric functions, remember that they denote *angles in radian measure.*

EXAMPLE 2   *Evaluating inverse trigonometric functions*

Evaluate the following:

(a) $\arcsin\left(-\dfrac{1}{2}\right)$ (b) $\arccos 0$ (c) $\arctan \sqrt{3}$ (d) $\arcsin (0.3)$

Solution:

(a) By definition, $y = \arcsin(-\frac{1}{2})$ implies that $\sin y = -\frac{1}{2}$. In the interval $[-\pi/2, \pi/2]$, we choose $y = -\pi/6$. Therefore,

$$\arcsin\left(-\frac{1}{2}\right) = -\frac{\pi}{6}$$

(b) By definition, $y = \arccos 0$ implies that $\cos y = 0$. In the interval $[0, \pi]$, we choose $y = \pi/2$. Therefore,

$$\arccos 0 = \frac{\pi}{2}$$

(c) By definition, $y = \arctan \sqrt{3}$ implies that $\tan y = \sqrt{3}$. In the interval $(-\pi/2, \pi/2)$, we choose $y = \pi/3$. Therefore,

$$\arctan \sqrt{3} = \frac{\pi}{3}$$

(d) Using the table at the back of the book or a calculator set in *radian mode,* we obtain

$$\arcsin (0.3) \approx 0.3047$$

Inverse functions possess the properties

$$f(f^{-1}(x)) = x \qquad \text{and} \qquad f^{-1}(f(x)) = x$$

When you apply these properties to inverse trigonometric functions, remember that the trigonometric functions possess inverses only in restricted domains. For *x*-values outside these domains, these two properties do not hold. For example,

$$\arcsin (\sin \pi) = \arcsin 0 = 0 \neq \pi$$

INVERSE PROPERTIES   If $-1 \leq x \leq 1$ and $-\pi/2 \leq y \leq \pi/2$, then

$$\sin (\arcsin x) = x \qquad \text{and} \qquad \arcsin (\sin y) = y$$

If $-\pi/2 < y < \pi/2$, then

$$\tan (\arctan x) = x \qquad \text{and} \qquad \arctan (\tan y) = y$$

If $|x| \geq 1$ and $0 \leq y < \pi/2$ or $\pi/2 < y \leq \pi$, then

$$\sec (\operatorname{arcsec} x) = x \qquad \text{and} \qquad \operatorname{arcsec} (\sec y) = y$$

| Remark Similar properties hold for the other three inverse trigonometric functions.

Notice how we use one of these inverse properties to solve the equation in the next example.

EXAMPLE 3   *Solving an equation*

Solve for x in the equation

$$\arctan (2x - 3) = \frac{\pi}{4}$$

Solution:

$$\arctan (2x - 3) = \frac{\pi}{4}$$

$$\tan [\arctan (2x - 3)] = \tan \frac{\pi}{4}$$

$$2x - 3 = 1$$

$$x = 2$$

There are some important types of problems in calculus in which we evaluate expressions like sec (arctan x). In solving this type of problem, it is helpful to use right triangles, as demonstrated in the next example.

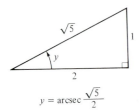

$$y = \operatorname{arcsec} \frac{\sqrt{5}}{2}$$

FIGURE 8.45

EXAMPLE 4   *Using right triangles*

(a) Given $y = \operatorname{arcsec} (\sqrt{5}/2)$, find tan y.
(b) Given $y = \arcsin x$, find cos y.

Solution:
(a) Since y is the angle whose secant is $\sqrt{5}/2$, we can sketch this angle as part of a triangle, as shown in Figure 8.45. Therefore,

$$\tan y = \tan \left[\operatorname{arcsec} \left(\frac{\sqrt{5}}{2} \right) \right] \qquad \tan y = \frac{\text{opp.}}{\text{adj.}}$$

$$= \frac{1}{2}$$

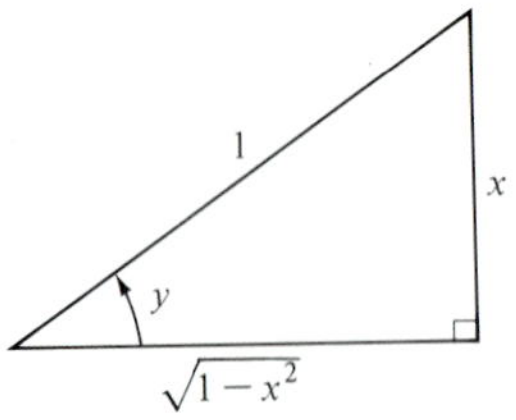

$$y = \arcsin x$$

FIGURE 8.46

(b) Since y is the angle whose sine is x, we form a right triangle having an acute angle $y = \arcsin x$, as shown in Figure 8.46. Therefore,

$$\cos y = \cos (\arcsin x) \qquad \cos y = \frac{\text{adj.}}{\text{hyp.}}$$

$$= \sqrt{1 - x^2}$$

Derivatives of inverse trigonometric functions

In Section 7.2 we saw that the derivative of the *transcendental* function $f(x) = \ln x$ is the *algebraic* function $f'(x) = 1/x$. We will now see that the derivatives of the inverse trigonometric functions are also algebraic, even though the inverse trigonometric functions are themselves transcendental.

The following theorem lists the derivative of each of the six inverse trigonometric functions. Note that the derivatives of arccos u, arccot u, and arccsc u are the *negatives* of the derivatives of arcsin u, arctan u, and arcsec u, respectively.

THEOREM 8.7   **DERIVATIVES OF THE INVERSE TRIGONOMETRIC FUNCTIONS**
Let u be a differentiable function of x.

$$\frac{d}{dx}[\arcsin u] = \frac{u'}{\sqrt{1 - u^2}} \qquad \frac{d}{dx}[\arccos u] = \frac{-u'}{\sqrt{1 - u^2}}$$

$$\frac{d}{dx}[\arctan u] = \frac{u'}{1 + u^2} \qquad \frac{d}{dx}[\text{arccot } u] = \frac{-u'}{1 + u^2}$$

$$\frac{d}{dx}[\text{arcsec } u] = \frac{u'}{|u|\sqrt{u^2 - 1}} \qquad \frac{d}{dx}[\text{arccsc } u] = \frac{-u'}{|u|\sqrt{u^2 - 1}}$$

Proof: We prove the first of these formulas and leave the proofs of the others as exercises. Let $f(x) = \sin x$ and $f^{-1}(x) = g(x) = \arcsin x$. Then, using Theorem 7.12, we have

$$\frac{d}{dx}[g(x)] = \frac{1}{f'[g(x)]}$$

$$= \frac{1}{\cos(\arcsin x)}$$

Now, using the result of Example 3(b), we have

$$\frac{d}{dx}[\arcsin x] = \frac{1}{\cos(\arcsin x)}$$

$$= \frac{1}{\sqrt{1 - x^2}}$$

If u is a differentiable function of x, then the Chain Rule gives us

$$\frac{d}{dx}[\arcsin u] = \frac{u'}{\sqrt{1 - u^2}}, \quad \text{where } u' = \frac{du}{dx}$$

There is no common agreement on the definition of arcsec x (or arccsc x) for negative values of x. When we defined the range of the arcsecant, we chose to preserve the reciprocal identity

$$\text{arcsec } x = \arccos \frac{1}{x}$$

One of the consequences of this choice is that the slope of the graph of the inverse secant function is always positive (see Figure 8.44), which accounts for the absolute value sign in the formula for the derivative of arcsec x.

EXAMPLE 5 *Differentiating inverse trigonometric functions*

(a) $\dfrac{d}{dx}[\arctan (3x)] = \dfrac{3}{1 + (3x)^2}$ $u = 3x$

$= \dfrac{3}{1 + 9x^2}$

(b) $\dfrac{d}{dx}[\arcsin \sqrt{x}] = \dfrac{(1/2)x^{-1/2}}{\sqrt{1 - x}}$ $u = \sqrt{x}$

$= \dfrac{1}{2\sqrt{x}\sqrt{1 - x}}$

$= \dfrac{1}{2\sqrt{x - x^2}}$

(c) $\dfrac{d}{dx}[\operatorname{arcsec} e^{2x}] = \dfrac{2e^{2x}}{e^{2x}\sqrt{(e^{2x})^2 - 1}}$ $u = e^{2x}$

$= \dfrac{2e^{2x}}{e^{2x}\sqrt{e^{4x} - 1}}$

$= \dfrac{2}{\sqrt{e^{4x} - 1}}$

Note that we omit the absolute value sign, since $e^{2x} > 0$.   ▢

Derivatives of inverse trigonometric functions enable us to expand our list of integration formulas to cover some important algebraic functions. Example 6 shows a typical case.

EXAMPLE 6 *A derivative that can be simplified*

Differentiate $y = \arcsin x + x\sqrt{1 - x^2}$.

Solution:

$$y' = \frac{1}{\sqrt{1 - x^2}} + x\left(\frac{1}{2}\right)(-2x)(1 - x^2)^{-1/2} + \sqrt{1 - x^2}$$

$$= \frac{1}{\sqrt{1 - x^2}} - \frac{x^2}{\sqrt{1 - x^2}} + \sqrt{1 - x^2}$$

$$= \sqrt{1 - x^2} + \sqrt{1 - x^2} = 2\sqrt{1 - x^2} \quad ▢$$

| **Remark** From Example 6 we can conclude that

$$\int \sqrt{1 - x^2}\, dx = \frac{1}{2}[\arcsin x + x\sqrt{1 - x^2}] + C$$

We are not suggesting that you memorize this "integration rule." In Section 9.4 we will introduce a general integration technique (called trigonometric substitution) that covers this particular integral.

EXAMPLE 7 *Graphing an inverse trigonometric function*

Sketch the graph of $y = (\arctan x)^2$.

Solution: From the derivative

$$y' = 2(\arctan x)\left(\frac{1}{1 + x^2}\right) = \frac{2 \arctan x}{1 + x^2}$$

we can see that the only critical number is $x = 0$. By the First Derivative Test, this value corresponds to a relative minimum. From the second derivative

$$y'' = \frac{(1 + x^2)\left(\dfrac{2}{1 + x^2}\right) - (2 \arctan x)(2x)}{(1 + x^2)^2} = \frac{2(1 - 2x \arctan x)}{(1 + x^2)^2}$$

we see that points of inflection occur when $2x \arctan x = 1$. Using Newton's Method, we find these points to occur when $x \approx \pm 0.765$. Finally, the graph has a horizontal asymptote at $y = \pi^2/4$ since

$$\lim_{x \to \pm \infty} (\arctan x)^2 = \frac{\pi^2}{4}$$

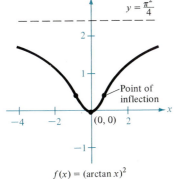

The graph is shown in Figure 8.47.

FIGURE 8.47

We conclude this section with an example describing an application involving the derivative of an inverse trigonometric function.

EXAMPLE 8   An application

A photographer is taking a picture of a four-foot painting hung in an art gallery. The camera lens is one foot below the lower edge of the painting, as shown in Figure 8.48. How far back should the camera be to maximize the angle subtended by the camera lens?

Solution: In Figure 8.48, we let β be the angle we wish to maximize. Then we have

$$\beta = \theta - \alpha = \operatorname{arccot} \frac{x}{5} - \operatorname{arccot} x$$

Differentiating, we have

$$\frac{d\beta}{dx} = \frac{-1/5}{1 + (x^2/25)} - \frac{-1}{1 + x^2}$$

$$= \frac{-5}{25 + x^2} + \frac{1}{1 + x^2}$$

$$= \frac{4(5 - x^2)}{(25 + x^2)(1 + x^2)}$$

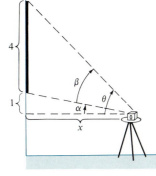

FIGURE 8.48

Since $d\beta/dx = 0$ when $x = \sqrt{5}$, we conclude that this distance yields a maximum value for β.

Review of basic differentiation formulas

In the 1600s, Europe was ushered into the scientific age by such great thinkers as Descartes, Galileo, Huygens, Newton, and Kepler. These men believed that nature is governed by basic laws—laws that can, for the most part, be

Galileo Galilei

written in terms of mathematical equations. One of the most influential publications during this period was written by Galileo Galilei (1564–1642). It is called *Dialogue on the Great World Systems,* and it became a classic description of modern scientific thought.

As mathematics has developed during the past few hundred years, a basic list of elementary functions has proven to be sufficient to model most phenomena that occur in physics, chemistry, biology, engineering, economics, and a variety of other fields. An **elementary function** is one that can be formed as the sum, product, quotient, or composition of functions from the following list.

Algebraic functions	*Transcendental functions*
Polynomial functions	Logarithmic functions
Rational functions	Exponential functions
Functions involving radicals	Trigonometric functions
	Inverse trigonometric functions

There are isolated cases of the use of other types of functions, but this list is fairly comprehensive. Moreover, with the differentiation formulas introduced in this section, we can differentiate *any* elementary function. For convenience, we summarize these differentiation rules here.

BASIC DIFFERENTIATION RULES FOR ELEMENTARY FUNCTIONS

1. $\dfrac{d}{dx}[cu] = cu'$

2. $\dfrac{d}{dx}[u \pm v] = u' \pm v'$

3. $\dfrac{d}{dx}[uv] = uv' + vu'$

4. $\dfrac{d}{dx}\left[\dfrac{u}{v}\right] = \dfrac{vu' - uv'}{v^2}$

5. $\dfrac{d}{dx}[c] = 0$

6. $\dfrac{d}{dx}[u^n] = nu^{n-1}u'$

7. $\dfrac{d}{dx}[x] = 1$

8. $\dfrac{d}{dx}[|u|] = \dfrac{u}{|u|}(u'), \ u \neq 0$

9. $\dfrac{d}{dx}[\ln u] = \dfrac{u'}{u}$

10. $\dfrac{d}{dx}[e^u] = e^u u'$

11. $\dfrac{d}{dx}[\sin u] = (\cos u)u'$

12. $\dfrac{d}{dx}[\cos u] = -(\sin u)u'$

13. $\dfrac{d}{dx}[\tan u] = (\sec^2 u)u'$

14. $\dfrac{d}{dx}[\cot u] = -(\csc^2 u)u'$

15. $\dfrac{d}{dx}[\sec u] = (\sec u \tan u)u'$

16. $\dfrac{d}{dx}[\csc u] = -(\csc u \cot u)u'$

17. $\dfrac{d}{dx}[\arcsin u] = \dfrac{u'}{\sqrt{1 - u^2}}$

18. $\dfrac{d}{dx}[\arccos u] = \dfrac{-u'}{\sqrt{1 - u^2}}$

19. $\dfrac{d}{dx}[\arctan u] = \dfrac{u'}{1 + u^2}$

20. $\dfrac{d}{dx}[\text{arccot } u] = \dfrac{-u'}{1 + u^2}$

21. $\dfrac{d}{dx}[\text{arcsec } u] = \dfrac{u'}{|u|\sqrt{u^2 - 1}}$

22. $\dfrac{d}{dx}[\text{arccsc } u] = \dfrac{-u'}{|u|\sqrt{u^2 - 1}}$

Section Exercises 8.5

In Exercises 1–10, evaluate the given expression.

1. $\arcsin \dfrac{1}{2}$

2. $\arcsin 0$

3. $\arccos \dfrac{1}{2}$

4. $\arccos 0$

5. $\arctan \dfrac{\sqrt{3}}{3}$

6. $\operatorname{arccot}(-1)$

7. $\operatorname{arccsc}\sqrt{2}$

8. $\arcsin(-0.39)$

9. $\operatorname{arcsec} 1.269$

10. $\arctan(-3)$

In Exercises 11–16, evaluate the given expression without the use of a calculator. [Hint: Make a sketch of a right triangle, as illustrated in Example 4.]

11. (a) $\sin\left(\arcsin \dfrac{1}{2}\right)$ (b) $\cos\left(2 \arcsin \dfrac{1}{2}\right)$

12. (a) $\tan\left(\arccos \dfrac{\sqrt{2}}{2}\right)$ (b) $\cot\left[\arcsin\left(-\dfrac{1}{2}\right)\right]$

13. (a) $\sin\left(\arctan \dfrac{3}{4}\right)$ (b) $\sec\left(\arcsin \dfrac{4}{5}\right)$

14. (a) $\tan(\operatorname{arccot} 2)$ (b) $\cos(\operatorname{arcsec}\sqrt{5})$

15. (a) $\cos\left(\arcsin \dfrac{5}{13}\right)$ (b) $\csc\left[\arctan\left(-\dfrac{5}{12}\right)\right]$

16. (a) $\sec\left[\arctan\left(-\dfrac{3}{5}\right)\right]$ (b) $\tan\left[\arcsin\left(-\dfrac{5}{6}\right)\right]$

In Exercises 17–26, use a right triangle to write the given expression in algebraic form.

17. $\tan(\arctan x)$

18. $\sin(\arccos x)$

19. $\cos(\arcsin 2x)$

20. $\sec(\arctan 3x)$

21. $\sin(\operatorname{arcsec} x)$

22. $\cos(\operatorname{arccot} x)$

23. $\tan\left(\operatorname{arcsec} \dfrac{x}{3}\right)$

24. $\sec[\arcsin(x-1)]$

25. $\csc\left(\arctan \dfrac{x}{\sqrt{2}}\right)$

26. $\cos\left(\arcsin \dfrac{x-h}{r}\right)$

In Exercises 27 and 28, fill in the blanks.

27. $\arctan \dfrac{9}{x} = \arcsin \underline{\qquad}$

28. $\arcsin \dfrac{\sqrt{36-x^2}}{6} = \arccos \underline{\qquad}$

In Exercises 29 and 30, verify each identity.

29. (a) $\operatorname{arccsc} x = \arcsin \dfrac{1}{x}$, $|x| \geq 1$

   (b) $\operatorname{arccot} x = \arctan \dfrac{1}{x}$, $x > 0$

30. (a) $\arcsin(-x) = -\arcsin x$, $|x| \leq 1$

   (b) $\arccos(-x) = \pi - \arccos x$, $|x| \leq 1$

In Exercises 31–34, sketch the graph of the function.

31. $f(x) = \arcsin(x-1)$

32. $f(x) = \dfrac{\pi}{2} + \arctan x$

33. $f(x) = \operatorname{arcsec} 2x$

34. $f(x) = \arccos \dfrac{x}{4}$

In Exercises 35–38, solve the given equation for x.

35. $\arcsin(3x-\pi) = \dfrac{1}{2}$

36. $\arctan 2x = -1$

37. $\arcsin\sqrt{2x} = \arccos\sqrt{x}$

38. $\arccos x = \operatorname{arcsec} x$

In Exercises 39–58, find the derivative of the given function.

39. $f(x) = \arcsin 2x$

40. $f(x) = \arcsin x^2$

41. $f(x) = 2\arcsin(x-1)$

42. $f(x) = \arccos\sqrt{x}$

43. $f(x) = 3\arccos \dfrac{x}{2}$

44. $f(x) = \arctan\sqrt{x}$

45. $f(x) = \arctan 5x$

46. $f(x) = x\arctan x$

47. $f(x) = \arccos \dfrac{1}{x}$

48. $f(x) = \operatorname{arcsec} 2x$

49. $f(x) = \arcsin x + \arccos x$

50. $f(x) = \operatorname{arcsec} x + \operatorname{arccsc} x$

51. $h(t) = \sin(\arccos t)$

52. $g(t) = \tan(\arcsin t)$

53. $f(t) = \dfrac{1}{\sqrt{6}}\arctan \dfrac{\sqrt{6}t}{2}$

54. $f(x) = \dfrac{1}{2}\left(\dfrac{1}{2}\ln \dfrac{x+1}{x-1} - \arctan x\right)$

55. $f(x) = \dfrac{1}{2}\left(\dfrac{1}{2}\ln \dfrac{x+1}{x-1} + \arctan x\right)$

56. $f(x) = \dfrac{1}{2}(x\sqrt{1-x^2} + \arcsin x)$

57. $f(x) = x\arcsin x + \sqrt{1-x^2}$

58. $f(x) = x\arctan 2x - \dfrac{1}{4}\ln(1+4x^2)$

In Exercises 59 and 60, find the point of inflection of the graph of the given function.

59. $f(x) = \arcsin x$

60. $f(x) = \operatorname{arccot} 2x$

In Exercises 61 and 62, find any relative extrema of the function.

61. $f(x) = \operatorname{arcsec} x - x$

62. $f(x) = \arcsin x - 2x$

In Exercises 63 and 64, find the point of intersection of the graphs of the given functions.

63. $y = \arccos x$, $y = \arctan x$

64. $y = \arcsin x, y = \arccos x$

65. A boat is being pulled toward a dock that is 10 feet above the level of the water. The rope (attached to the boat at water level) is being pulled in at a rate of 1.5 feet per second. Find the rate at which the angle the rope makes with the horizontal is changing when there is 20 feet of rope out.

66. An observer is standing 300 feet from the point at which a balloon is released. The balloon rises at a rate of 5 feet per second. How fast is the angle of elevation of the observer's line of sight increasing when the balloon is 100 feet above the observer's eye?

SECTION TOPICS ▪
Integration formulas ▪
Completing the square ▪
Review of basic integration formulas ▪

8.6
Inverse trigonometric functions:
Integration and completing the square

The derivatives of the six inverse trigonometric functions occur in three pairs. In each pair the derivative of one function is the negative of the other. For example,

$$\frac{d}{dx}[\arcsin x] = \frac{1}{\sqrt{1 - x^2}} \quad \text{and} \quad \frac{d}{dx}[\arccos x] = -\frac{1}{\sqrt{1 - x^2}}$$

When listing the *antiderivative* that corresponds to each of the inverse trigonometric functions, we need use only one member from each pair. For example, we choose to use the arcsin x as the antiderivative of $1/\sqrt{1 - x^2}$, rather than $-\arccos x$. The next theorem gives one antiderivative formula for each of the three pairs.

THEOREM 8.8   **INTEGRALS INVOLVING INVERSE TRIGONOMETRIC FUNCTIONS**
Let u be a differentiable function of x.

$$\int \frac{du}{\sqrt{a^2 - u^2}} = \arcsin \frac{u}{a} + C$$

$$\int \frac{du}{a^2 + u^2} = \frac{1}{a} \arctan \frac{u}{a} + C$$

$$\int \frac{du}{u\sqrt{u^2 - a^2}} = \frac{1}{a} \operatorname{arcsec} \frac{|u|}{a} + C$$

Proof: We prove the first formula for $a > 0$ and leave the remaining proofs as exercises. Let $y = \arcsin (u/a)$. Then

$$y' = \frac{1}{\sqrt{1 - (u/a)^2}} \left(\frac{u'}{a} \right) = \frac{u'}{a\sqrt{(a^2 - u^2)/a^2}}$$

$$= \frac{u'}{\sqrt{a^2 - u^2}}$$

Consequently, by the definition of an antiderivative, we have

$$\int \frac{du}{\sqrt{a^2 - u^2}} = \int \frac{u'}{\sqrt{a^2 - u^2}} \, dx$$

$$= \arcsin \frac{u}{a} + C$$

EXAMPLE 1   Integration with inverse trigonometric functions

(a) $\displaystyle\int \frac{dx}{\sqrt{4 - x^2}} = \arcsin \frac{x}{2} + C$

(b) $\displaystyle\int \frac{dx}{2 + 9x^2} = \frac{1}{3} \int \frac{3 \, dx}{(\sqrt{2})^2 + (3x)^2}$ $u = 3x, \; a = \sqrt{2}$

$$= \frac{1}{3\sqrt{2}} \arctan \frac{3x}{\sqrt{2}} + C$$

The integrals in Example 1 are fairly straightforward applications of integration formulas. Unfortunately, this is not typical. The inverse trigonometric integration formulas can be disguised in many ways. In this section we give some typical ways in which these integrals can be disguised, and we illustrate techniques for removing the disguise.

EXAMPLE 2   Integration by substitution

Evaluate

$$\int \frac{dx}{\sqrt{e^{2x} - 1}}$$

Solution: As it stands, this integral doesn't fit any of the three inverse trigonometric formulas, so we try the substitution $u = e^x$.

$$u = e^x \; \Longrightarrow \; du = e^x \, dx \; \Longrightarrow \; dx = \frac{du}{e^x} = \frac{du}{u}$$

Now, we have

$$\int \frac{dx}{\sqrt{(e^x)^2 - 1}} = \int \frac{du/u}{\sqrt{u^2 - 1}} = \int \frac{du}{u\sqrt{u^2 - 1}}$$

This integral now fits the arcsecant formula, and we have

$$\int \frac{dx}{\sqrt{e^{2x} - 1}} = \operatorname{arcsec} \frac{|u|}{1} + C = \operatorname{arcsec} (e^x) + C$$

Another common technique is to rewrite the integrand as the sum of two quotients.

EXAMPLE 3   Rewriting the integrand as the sum of two quotients

Evaluate

$$\int \frac{x+2}{\sqrt{4-x^2}}\,dx$$

Solution: This integral does not appear to fit any of our integration formulas, but by splitting the integrand into two parts we obtain

$$\int \frac{x+2}{\sqrt{4-x^2}}\,dx = \int \frac{x}{\sqrt{4-x^2}}\,dx + \int \frac{2}{\sqrt{4-x^2}}\,dx$$

Now, the first integral on the right can be evaluated with the Power Rule, and the second integral will yield an inverse sine function. Therefore, we write

$$\int \frac{x+2}{\sqrt{4-x^2}}\,dx = -\frac{1}{2}\int (4-x^2)^{-1/2}(-2x)\,dx + 2\int \frac{1}{\sqrt{4-x^2}}\,dx$$

$$= -\frac{1}{2}\left[\frac{(4-x^2)^{1/2}}{1/2}\right] + 2\,\arcsin\frac{x}{2} + C$$

$$= -\sqrt{4-x^2} + 2\,\arcsin\frac{x}{2} + C$$

Another way integration formulas can be disguised is when the integrand is an improper rational function. (That is, the integrand is a rational function in which the degree of the numerator is greater than or equal to the degree of the denominator.) For such functions, we use long division to write the integrand as the sum of a polynomial and a proper rational function.

EXAMPLE 4   Integrating an improper rational function

Evaluate

$$\int \frac{3x^3-2}{x^2+4}\,dx$$

Solution: Since the degree of the numerator is greater than the degree of the denominator, we divide to obtain

$$\int \frac{3x^3-2}{x^2+4}\,dx = \int \left(3x - \frac{12x+2}{x^2+4}\right)\,dx$$

$$= \int 3x\,dx - 6\int \frac{2x}{x^2+4}\,dx - 2\int \frac{dx}{x^2+4}$$

$$= \frac{3x^2}{2} - 6\ln(x^2+4) - \arctan\frac{x}{2} + C$$

Completing the square

Completing the square is a versatile technique that is helpful when quadratic functions are involved in the integrand. For example, the quadratic $x^2 + bx + c$ can be written as

$$x^2 + bx + c = x^2 + bx + \left(\frac{b}{2}\right)^2 - \left(\frac{b}{2}\right)^2 + c = \left(x + \frac{b}{2}\right)^2 + \left(\frac{4c-b^2}{4}\right)$$

Thus, we have written $x^2 + bx + c$ in the form $u^2 \pm a^2$, where $u = x + (b/2)$ and $a = \sqrt{|4c - b^2|}/2$. Notice that the key step in completing the square is to *add and subtract* $(b/2)^2$.

EXAMPLE 5   Completing the square

Evaluate

$$\int \frac{dx}{x^2 - 4x + 7}$$

Solution: By completing the square, we obtain

$$x^2 - 4x + 7 = (x^2 - 4x + 4) - 4 + 7 = (x - 2)^2 + 3 = u^2 + a^2$$

Therefore,

$$\int \frac{dx}{x^2 - 4x + 7} = \int \frac{dx}{(x - 2)^2 + 3} \qquad u = x - 2, \ a = \sqrt{3}$$

$$= \frac{1}{\sqrt{3}} \arctan \frac{x - 2}{\sqrt{3}} + C$$

In Example 5, we completed the square for a polynomial whose leading coefficient is 1. If the leading coefficient is not 1, we suggest factoring before completing the square, as demonstrated in the next example.

EXAMPLE 6   Completing the square when the leading coefficient is not 1

Evaluate

$$\int \frac{dx}{2x^2 - 8x + 9}$$

Solution: To complete the square, we write

$$2x^2 - 8x + 9 = 2(x^2 - 4x) + 9$$
$$= 2(x^2 - 4x + 4 - 4) + 9$$
$$= 2(x^2 - 4x + 4) - 2(4) + 9$$
$$= 2(x - 2)^2 + 1 = u^2 + a^2 \qquad u = \sqrt{2}(x - 2), \ a = 1$$

Therefore,

$$\int \frac{dx}{2x^2 - 8x + 9} = \int \frac{dx}{2(x - 2)^2 + 1}$$

$$= \frac{1}{\sqrt{2}} \int \frac{\sqrt{2} \, dx}{[\sqrt{2}(x - 2)]^2 + 1}$$

$$= \frac{1}{\sqrt{2}} \int \frac{du}{u^2 + 1} = \frac{1}{\sqrt{2}} \arctan u + C$$

$$= \frac{1}{\sqrt{2}} \arctan [\sqrt{2}(x - 2)] + C$$

In the next example, notice how we complete the square when the coefficient of x^2 is negative.

EXAMPLE 7   Completing the square for a negative leading coefficient

Find the area bounded by the graph of $f(x) = 1/\sqrt{3x - x^2}$ and the x-axis for $\frac{3}{2} \le x \le \frac{9}{4}$.

Solution: From Figure 8.49, we see that the area is given by

$$\text{area} = \int_{3/2}^{9/4} \frac{1}{\sqrt{3x - x^2}} \, dx$$

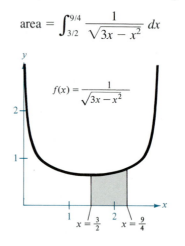

$$f(x) = \frac{1}{\sqrt{3x - x^2}}$$

FIGURE 8.49

To complete the square, we write

$$3x - x^2 = -(x^2 - 3x) = -\left[x^2 - 3x + \left(\frac{3}{2}\right)^2 - \left(\frac{3}{2}\right)^2 \right]$$

$$= \left(\frac{3}{2}\right)^2 - \left(x - \frac{3}{2}\right)^2$$

$$= a^2 - u^2$$

Therefore,

$$\int_{3/2}^{9/4} \frac{dx}{\sqrt{3x - x^2}} = \int_{3/2}^{9/4} \frac{dx}{\sqrt{(3/2)^2 - [x - (3/2)]^2}}$$

$$= \left[\arcsin \frac{x - (3/2)}{3/2} \right]_{3/2}^{9/4}$$

$$= \arcsin \frac{1}{2} - \arcsin 0$$

$$= \frac{\pi}{6} - 0 = \frac{\pi}{6}$$

Review of basic integration formulas

We have now completed our list of the **basic integration formulas.** It is important that you memorize each formula in the list.

BASIC INTEGRATION FORMULAS

1. $\int kf(u)\ du = k \int f(u)\ du$

2. $\int [f(u) \pm g(u)]\ du = \int f(u)\ du \pm \int g(u)\ du$

3. $\int du = u + C$

4. $\int u^n\ du = \dfrac{u^{n+1}}{n+1} + C,\ n \neq -1$

5. $\int \dfrac{du}{u} = \ln |u| + C$

6. $\int e^u\ du = e^u + C$

7. $\int \sin u\ du = -\cos u + C$

8. $\int \cos u\ du = \sin u + C$

9. $\int \tan u\ du = -\ln |\cos u| + C$

10. $\int \cot u\ du = \ln |\sin u| + C$

11. $\int \sec u\ du = \ln |\sec u + \tan u| + C$

12. $\int \csc u\ du = -\ln |\csc u + \cot u| + C$

13. $\int \sec^2 u\ du = \tan u + C$

14. $\int \csc^2 u\ du = -\cot u + C$

15. $\int \sec u \tan u\ du = \sec u + C$

16. $\int \csc u \cot u\ du = -\csc u + C$

17. $\int \dfrac{du}{\sqrt{a^2 - u^2}} = \arcsin \dfrac{u}{a} + C$

18. $\int \dfrac{du}{a^2 + u^2} = \dfrac{1}{a} \arctan \dfrac{u}{a} + C$

19. $\int \dfrac{du}{u\sqrt{u^2 - a^2}} = \dfrac{1}{a} \operatorname{arcsec} \dfrac{|u|}{a} + C$

You can learn quite a bit about the nature of integration by comparing this list to the summary of differentiation formulas given in the previous section. For differentiation, we now have formulas that allow us to differentiate *any* elementary function. For integration, this is certainly not true. The above integration formulas are those that we happened upon when we were developing differentiation formulas. We do not have a formula for the antiderivative of a general product or quotient, the natural logarithmic function, or the inverse trigonometric functions. More importantly, we cannot apply any of the formulas in this list unless we can create the proper *du* corresponding to the *u* in the formula. For example, try integrating $\int e^{x^2}\ dx$ with the substitution $u = x^2$. You will see that it cannot be done. The point is that we need to work some more on integration techniques, which we will do in Chapter 9. To give you a better feeling for the integration problems that we can *and* cannot do with the techniques and formulas we now have, we conclude this section with two examples.

EXAMPLE 8   Comparing integration problems

Evaluate as many of the following integrals as you can using the formulas and techniques we have studied up to this point in the text.

(a) $\int \dfrac{dx}{x\sqrt{x^2 - 1}}$ (b) $\int \dfrac{x\ dx}{\sqrt{x^2 - 1}}$ (c) $\int \dfrac{dx}{\sqrt{x^2 - 1}}$

Solution:
(a) We *can* evaluate this integral (it fits the arcsecant formula):

$$\int \frac{dx}{x\sqrt{x^2 - 1}} = \operatorname{arcsec} |x| + C$$

(b) We *can* also evaluate this integral (it fits the Power Rule):

$$\int \frac{x\,dx}{\sqrt{x^2 - 1}} = \frac{1}{2} \int (x^2 - 1)^{-1/2}(2x)\,dx$$

$$= \frac{1}{2}\left(\frac{(x^2 - 1)^{1/2}}{1/2}\right) + C$$

$$= \sqrt{x^2 - 1} + C$$

(c) We *cannot* evaluate this integral using our present techniques. (You should scan the list of Basic Integration Formulas to verify this conclusion.)

EXAMPLE 9   *Comparing integration problems*

Evaluate as many of the following integrals as you can using the formulas and techniques we have studied up to this point in the text.

(a) $\displaystyle\int \frac{dx}{x \ln x}$ (b) $\displaystyle\int \frac{\ln x\,dx}{x}$ (c) $\displaystyle\int \ln x\,dx$

Solution:
(a) We *can* evaluate this integral (it fits the log formula):

$$\int \frac{dx}{x \ln x} = \int \frac{1/x}{\ln x}\,dx = \ln |\ln x| + C$$

(b) We *can* also evaluate this integral (it fits the Power Rule):

$$\int \frac{\ln x}{x}\,dx = \int \left(\frac{1}{x}\right)(\ln x)^1\,dx = \frac{(\ln x)^2}{2} + C$$

(c) We *cannot* evaluate this integral using our present techniques.

| **Remark**   Note in Examples 8 and 9 that it is the *simplest* function that we cannot yet integrate.

It may seem unusual that we have summarized differentiation and integration of the elementary functions before completing the chapter. The reason for this is that the *hyperbolic functions* introduced in the next section are actually just special combinations of exponential functions. Thus, we already have the tools for differentiating and integrating these functions. The hyperbolic and inverse hyperbolic functions occur in some important applications, and they have a mathematical appeal because of the way in which they parallel the trigonometric functions. Yet, because time is a critical factor in most calculus courses, we consider Section 8.7 to be *optional,* and we will not assume its coverage in later chapters in the text.

Section Exercises 8.6

In Exercises 1–44, evaluate the given integral.

1. $\displaystyle\int_0^{1/6} \frac{1}{\sqrt{1-9x^2}}\, dx$

2. $\displaystyle\int_0^1 \frac{dx}{\sqrt{4-x^2}}$

3. $\displaystyle\int_0^{\sqrt{3}/2} \frac{1}{1+4x^2}\, dx$

4. $\displaystyle\int_{\sqrt{3}}^3 \frac{1}{9+x^2}\, dx$

5. $\displaystyle\int \frac{1}{x\sqrt{4x^2-1}}\, dx$

6. $\displaystyle\int \frac{1}{4+(x-1)^2}\, dx$

7. $\displaystyle\int \frac{x^3}{x^2+1}\, dx$

8. $\displaystyle\int \frac{x^4-1}{x^2+1}\, dx$

9. $\displaystyle\int \frac{1}{\sqrt{1-(x+1)^2}}\, dx$

10. $\displaystyle\int \frac{t}{t^4+16}\, dt$

11. $\displaystyle\int \frac{t}{\sqrt{1-t^4}}\, dt$

12. $\displaystyle\int \frac{1}{x\sqrt{x^4-4}}\, dx$

13. $\displaystyle\int \frac{\arctan x}{1+x^2}\, dx$

14. $\displaystyle\int \frac{1}{(x-1)\sqrt{(x-1)^2-4}}\, dx$

15. $\displaystyle\int_0^{1/\sqrt{2}} \frac{\arcsin x}{\sqrt{1-x^2}}\, dx$

16. $\displaystyle\int_0^{1/\sqrt{2}} \frac{\arccos x}{\sqrt{1-x^2}}\, dx$

17. $\displaystyle\int_{-1/2}^0 \frac{x}{\sqrt{1-x^2}}\, dx$

18. $\displaystyle\int_{-\sqrt{3}}^0 \frac{x}{1+x^2}\, dx$

19. $\displaystyle\int \frac{e^x}{\sqrt{1-e^{2x}}}\, dx$

20. $\displaystyle\int \frac{\cos x}{\sqrt{4-\sin^2 x}}\, dx$

21. $\displaystyle\int \frac{1}{9+(x-3)^2}\, dx$

22. $\displaystyle\int \frac{x+1}{x^2+1}\, dx$

23. $\displaystyle\int \frac{1}{\sqrt{x}(1+x)}\, dx$

24. $\displaystyle\int_1^2 \frac{1}{3+(x-2)^2}\, dx$

25. $\displaystyle\int_{\pi/2}^{\pi} \frac{\sin x}{1+\cos^2 x}\, dx$

26. $\displaystyle\int \frac{e^{2x}}{4+e^{4x}}\, dx$

27. $\displaystyle\int_0^2 \frac{dx}{x^2-2x+2}$

28. $\displaystyle\int_{-3}^{-1} \frac{dx}{x^2+6x+13}$

29. $\displaystyle\int \frac{2x}{x^2+6x+13}\, dx$

30. $\displaystyle\int \frac{2x-5}{x^2+2x+2}\, dx$

31. $\displaystyle\int \frac{1}{\sqrt{-x^2-4x}}\, dx$

32. $\displaystyle\int \frac{x+2}{\sqrt{-x^2-4x}}\, dx$

33. $\displaystyle\int \frac{1}{\sqrt{-x^2+2x}}\, dx$

34. $\displaystyle\int \frac{x-1}{\sqrt{x^2-2x}}\, dx$

35. $\displaystyle\int_2^3 \frac{2x-3}{\sqrt{4x-x^2}}\, dx$

36. $\displaystyle\int \frac{1}{(x-1)\sqrt{x^2-2x}}\, dx$

37. $\displaystyle\int \frac{x}{x^4+2x^2+2}\, dx$

38. $\displaystyle\int \frac{x}{\sqrt{9+8x^2-x^4}}\, dx$

39. $\displaystyle\int \frac{1}{\sqrt{-16x^2+16x-3}}\, dx$

40. $\displaystyle\int \frac{1}{(x-1)\sqrt{9x^2-18x+5}}\, dx$

41. $\displaystyle\int \frac{\sqrt{x}-1}{x}\, dx$

42. $\displaystyle\int \frac{\sqrt{x}-2}{x+1}\, dx$

43. $\displaystyle\int \sqrt{e^t-3}\, dt$

44. $\displaystyle\int \frac{1}{t\sqrt{t}+\sqrt{t}}\, dt$

In Exercises 45–48, find the area of the region bounded by the graphs of the given equations.

45. $y = \dfrac{1}{1+x^2}$, $y = 0$, $x = 0$, $x = 1$

46. $y = \dfrac{1}{\sqrt{4-x^2}}$, $y = 0$, $x = 0$, $x = 1$

47. $y = \dfrac{1}{x^2-2x+5}$, $y = 0$, $x = 1$, $x = 3$

48. $y = \dfrac{1}{\sqrt{3+2x-x^2}}$, $y = 0$, $x = 0$, $x = 2$

49. An object is projected upward with an initial velocity of 500 feet per second. If the air resistance is proportional to the square of the velocity, we obtain the equation

$$\frac{dv}{dt} = -(32 + kv^2)$$

where 32 feet per second is the acceleration due to gravity and k is a constant. (We assume the positive direction to be up.) Find the velocity as a function of time by solving the equation

$$\int \frac{dv}{32 + kv^2} = -\int dt$$

50. A weight of mass m is attached to a spring and oscillates with simple harmonic motion, as shown in Figure 8.50. By Hooke's Law, we can determine that

$$\int \frac{dy}{\sqrt{A^2 - y^2}} = \int \sqrt{\frac{k}{m}}\, dt$$

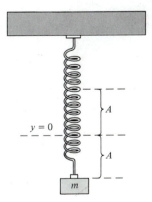

FIGURE 8.50

where A is the maximum displacement, t is the time, and k a constant. Find y as a function of t, given that $y = 0$ when $t = 0$.

In Exercises 51–54, determine which of the given integrals can be evaluated using the Basic Integration Formulas we have studied so far.

51. (a) $\displaystyle\int \frac{1}{\sqrt{1 - x^2}}\, dx$ (b) $\displaystyle\int \frac{x}{\sqrt{1 - x^2}}\, dx$

(c) $\displaystyle\int \frac{1}{x\sqrt{1 - x^2}}\, dx$

52. (a) $\displaystyle\int e^{x^2}\, dx$ (b) $\displaystyle\int x\, e^{x^2}\, dx$

(c) $\displaystyle\int \frac{1}{x^2} e^{1/x}\, dx$

53. (a) $\displaystyle\int \sqrt{x - 1}\, dx$ (b) $\displaystyle\int x\sqrt{x - 1}\, dx$

(c) $\displaystyle\int \frac{x}{\sqrt{x - 1}}\, dx$

54. (a) $\displaystyle\int \frac{1}{1 + x^4}\, dx$ (b) $\displaystyle\int \frac{x}{1 + x^4}\, dx$

(c) $\displaystyle\int \frac{x^3}{1 + x^4}\, dx$

55. Verify the following rules by differentiating.

(a) $\displaystyle\int \frac{du}{a^2 + u^2} = \frac{1}{a}\arctan\frac{u}{a} + C$

(b) $\displaystyle\int \frac{du}{u\sqrt{u^2 - a^2}} = \frac{1}{a}\operatorname{arcsec}\frac{|u|}{a} + C$

SECTION TOPICS •
Hyperbolic functions •
Differentiation and integration of
hyperbolic functions •
Inverse hyperbolic functions •
Differentiation and integration involving
inverse hyperbolic functions •

Johann Heinrich Lambert

8.7
Hyperbolic functions

In this section we take a brief look at a special class of exponential functions called the **hyperbolic functions.** The first person to publish a comprehensive study on hyperbolic functions was Johann Lambert (1728–1777), a Swiss-German mathematician and colleague of Euler.

The name, *hyperbolic function,* arose from comparing the area of a semicircular region, as shown in Figure 8.51, to the area of a region bounded by a hyperbola, as shown in Figure 8.52. The integral for the semicircular region involves an *inverse trigonometric (circular) function:*

$$\int_{-1}^{1} \sqrt{1 - x^2}\, dx = \frac{1}{2}\left[x\sqrt{1 - x^2} + \arcsin x\right]_{-1}^{1} \approx 1.571$$

The integral for the hyperbolic region involves an inverse hyperbolic function:

$$\int_{-1}^{1} \sqrt{1 + x^2}\, dx = \frac{1}{2}\left[x\sqrt{1 + x^2} + \sinh^{-1} x\right]_{-1}^{1} \approx 2.296$$

This is only one of many ways in which the hyperbolic functions parallel the trigonometric functions.

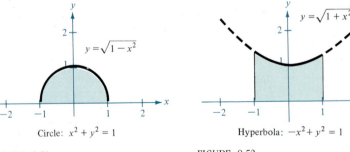

Circle: $x^2 + y^2 = 1$

FIGURE 8.51

Hyperbola: $-x^2 + y^2 = 1$

FIGURE 8.52

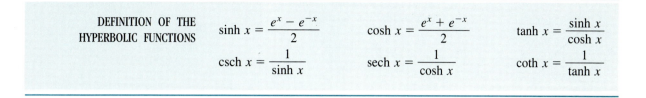

DEFINITION OF THE HYPERBOLIC FUNCTIONS	$\sinh x = \dfrac{e^x - e^{-x}}{2}$	$\cosh x = \dfrac{e^x + e^{-x}}{2}$	$\tanh x = \dfrac{\sinh x}{\cosh x}$
	$\operatorname{csch} x = \dfrac{1}{\sinh x}$	$\operatorname{sech} x = \dfrac{1}{\cosh x}$	$\coth x = \dfrac{1}{\tanh x}$

| **Remark**   We read $\sinh x$ as "the hyperbolic sine of x," $\cosh x$ as "the hyperbolic cosine of x," and so on.

The graphs of the six hyperbolic functions are shown in Figure 8.53, along with the domain and range of each function. Note that you can obtain the graphs of $\sinh x$ and $\cosh x$ by *addition of ordinates* using the exponential functions $f(x) = \frac{1}{2}e^x$ and $g(x) = \frac{1}{2}e^{-x}$.

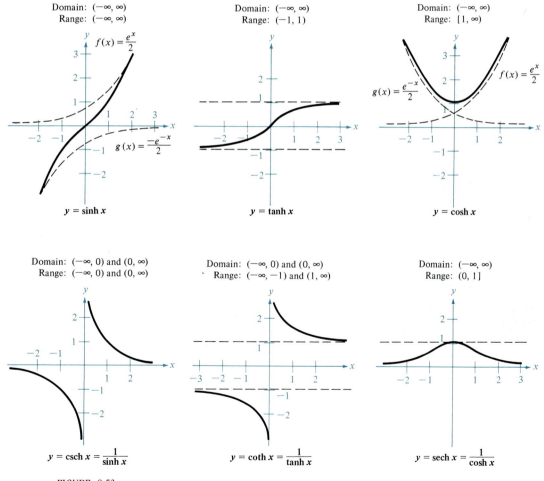

FIGURE 8.53

Many of the trigonometric identities have corresponding *hyperbolic identities*. For instance,

$$\cosh^2 x - \sinh^2 x = \left(\frac{e^x + e^{-x}}{2}\right)^2 - \left(\frac{e^x - e^{-x}}{2}\right)^2$$

$$= \frac{e^{2x} + 2 + e^{-2x}}{4} - \frac{e^{2x} - 2 + e^{-2x}}{4} = \frac{4}{4} = 1$$

or

$$2 \sinh x \cosh x = 2\left(\frac{e^x + e^{-x}}{2}\right)\left(\frac{e^x - e^{-x}}{2}\right)$$

$$= \frac{e^{2x} - e^{-2x}}{2} = \sinh 2x$$

In the following list, notice the similarity to trigonometric identities.

HYPERBOLIC IDENTITIES

$\cosh^2 x - \sinh^2 x = 1$	$\sinh (x + y) = \sinh x \cosh y + \cosh x \sinh y$
$\tanh^2 x + \text{sech}^2 x = 1$	$\sinh (x - y) = \sinh x \cosh y - \cosh x \sinh y$
$\coth^2 x - \text{csch}^2 x = 1$	$\cosh (x + y) = \cosh x \cosh y + \sinh x \sinh y$
	$\cosh (x - y) = \cosh x \cosh y - \sinh x \sinh y$

$$\sinh^2 x = \frac{-1 + \cosh 2x}{2} \qquad \sinh 2x = 2 \sinh x \cosh x$$

$$\cosh^2 x = \frac{1 + \cosh 2x}{2} \qquad \cosh 2x = \cosh^2 x + \sinh^2 x$$

Since the hyperbolic functions are written in terms of e^x and e^{-x}, we can easily derive formulas for their derivatives. In the following theorem, we list these derivatives along with the corresponding integration formulas.

THEOREM 8.9   DERIVATIVES AND INTEGRALS OF HYPERBOLIC FUNCTIONS
Let u be a differentiable function of x.

$$\frac{d}{dx}[\sinh u] = (\cosh u)u' \qquad\qquad \int \cosh u \, du = \sinh u + C$$

$$\frac{d}{dx}[\cosh u] = (\sinh u)u' \qquad\qquad \int \sinh u \, du = \cosh u + C$$

$$\frac{d}{dx}[\tanh u] = (\text{sech}^2 u)u' \qquad\qquad \int \text{sech}^2 u \, du = \tanh u + C$$

$$\frac{d}{dx}[\coth u] = -(\text{csch}^2 u)u' \qquad\qquad \int \text{csch}^2 u \, du = -\coth u + C$$

$$\frac{d}{dx}[\text{sech } u] = -(\text{sech } u \tanh u)u' \qquad \int \text{sech } u \tanh u \, du = -\text{sech } u + C$$

$$\frac{d}{dx}[\text{csch } u] = -(\text{csch } u \coth u)u' \qquad \int \text{csch } u \coth u \, du = -\text{csch } u + C$$

Proof: We give proofs for the derivatives of the hyperbolic sine and tangent, and leave the proofs of the remaining formulas as exercises.

$$\frac{d}{dx}[\sinh x] = \frac{d}{dx}\left[\frac{e^x - e^{-x}}{2}\right] = \frac{e^x + e^{-x}}{2} = \cosh x$$

$$\frac{d}{dx}[\tanh x] = \frac{d}{dx}\left[\frac{\sinh x}{\cosh x}\right] = \frac{\cosh x(\cosh x) - \sinh x(\sinh x)}{\cosh^2 x}$$

$$= \frac{1}{\cosh^2 x} = \text{sech}^2\, x$$

EXAMPLE 1   Differentiation of hyperbolic functions

Differentiate the following:
(a) $\sinh (x^2 - 3)$ (b) $\ln (\cosh x)$ (c) $x \sinh x - \cosh x$

Solution:

(a) $\dfrac{d}{dx}[\sinh (x^2 - 3)] = 2x \cosh (x^2 - 3)$

(b) $\dfrac{d}{dx}[\ln (\cosh x)] = \dfrac{\sinh x}{\cosh x} = \tanh x$

(c) $\dfrac{d}{dx}[x \sinh x - \cosh x] = x \cosh x + \sinh x - \sinh x = x \cosh x$ ☐

EXAMPLE 2   Finding relative extrema

Find the relative extrema of $f(x) = (x - 1) \cosh x - \sinh x$.

Solution: Letting $f'(x) = 0$,

$$f'(x) = (x - 1) \sinh x + \cosh x - \cosh x = (x - 1) \sinh x = 0$$

we see that the critical numbers are $x = 1$ and $x = 0$. By the Second Derivative Test, we can verify that the point $(0, -1)$ is a relative maximum and the point $(1, -\sinh 1)$ is a relative minimum, as shown in Figure 8.54.   ☐

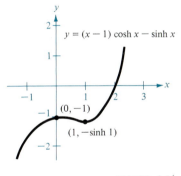

FIGURE 8.54

When a uniform flexible cable, such as a telephone wire, is suspended from two points, it takes the shape of a **catenary**, as shown in Figure 8.55(a). The equation for such a curve is

$$y = a \cosh \left(\frac{x}{a}\right)$$

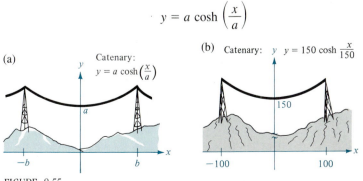

FIGURE 8.55

EXAMPLE 3   An application of arc length

An electric cable is hung between two towers that are 200 feet apart as shown in Figure 8.55(b) on page 455. The cable takes the shape of a catenary whose equation is

$$y = 75(e^{x/150} + e^{-x/150})$$

$$= 150 \cosh \left(\frac{x}{150} \right)$$

Find the arc length of the cable between the two towers.

Solution: Since

$$y' = \sinh \left(\frac{x}{150} \right)$$

the arc length is given by

$$s = \int_{-100}^{100} \sqrt{1 + (y')^2} \, dx$$

$$= \int_{-100}^{100} \sqrt{1 + \sinh^2 (x/150)} \, dx$$

$$= \int_{-100}^{100} \cosh \left(\frac{x}{150} \right) \, dx$$

$$= \left[150 \sinh \left(\frac{x}{150} \right) \right]_{-100}^{100}$$

$$= 300 \sinh \left(\frac{2}{3} \right)$$

$$\approx 215 \text{ ft}$$

EXAMPLE 4   Integrating a hyperbolic function

Evaluate $\int \cosh 2x \sinh^2 2x \, dx$.

Solution: If we let $u = \sinh 2x$, then by the Power Rule it follows that

$$\int \cosh 2x \sinh^2 2x \, dx = \frac{1}{2} \int (\sinh 2x)^2 (2 \cosh 2x) \, dx$$

$$= \frac{1}{2} \left[\frac{(\sinh 2x)^3}{3} \right] + C$$

$$= \frac{\sinh^3 2x}{6} + C$$

Unlike the trigonometric functions, the hyperbolic functions are *not* periodic. In fact, by looking back at Figure 8.53, we can see that four of the six hyperbolic functions are actually one-to-one (the hyperbolic sine, tangent, cosecant, and cotangent). Thus, we can apply Theorem 7.9 to conclude that these four functions have inverse functions. The other two (the hyperbolic cosine and secant) are one-to-one if their domains are restricted to the positive real numbers, and for this restricted domain they also have inverse functions. Since the hyperbolic functions are defined in terms of exponential functions, it is not surprising to find that the inverse hyperbolic functions can be written in terms of logarithmic functions, as shown in Theorem 8.10.

THEOREM 8.10   **INVERSE HYPERBOLIC FUNCTIONS**

Domain

$$\sinh^{-1} x = \ln (x + \sqrt{x^2 + 1}) \qquad\qquad (-\infty, \infty)$$

$$\cosh^{-1} x = \ln (x + \sqrt{x^2 - 1}) \qquad\qquad [1, \infty)$$

$$\tanh^{-1} x = \frac{1}{2} \ln \frac{1 + x}{1 - x} \qquad\qquad (-1, 1)$$

$$\coth^{-1} x = \frac{1}{2} \ln \frac{x + 1}{x - 1} \qquad\qquad (-\infty, -1) \cup (1, \infty)$$

$$\operatorname{sech}^{-1} x = \ln \frac{1 + \sqrt{1 - x^2}}{x} \qquad\qquad (0, 1]$$

$$\operatorname{csch}^{-1} x = \ln \frac{1 + \sqrt{1 + x^2}}{|x|} \qquad\qquad (-\infty, 0) \cup (0, \infty)$$

Proof: The proof of this theorem is a straightforward application of the properties of the exponential and logarithmic functions. For example, if

$$f(x) = \sinh x = \frac{e^x - e^{-x}}{2}$$

and

$$g(x) = \ln (x + \sqrt{x^2 + 1})$$

then

$$f(g(x)) = \frac{(x + \sqrt{x^2 + 1}) - [1/(x + \sqrt{x^2 + 1})]}{2}$$

$$= \frac{2x(x + \sqrt{x^2 + 1})}{2(x + \sqrt{x^2 + 1})}$$

$$= x$$

A similar argument can show that $g(f(x)) = x$, and we can conclude that g is the inverse function of f.

The graphs of the inverse hyperbolic functions are shown in Figure 8.56.

The inverse hyperbolic secant can be used to define a curve called a **tractrix,** or *pursuit curve*. This is described in the following example.

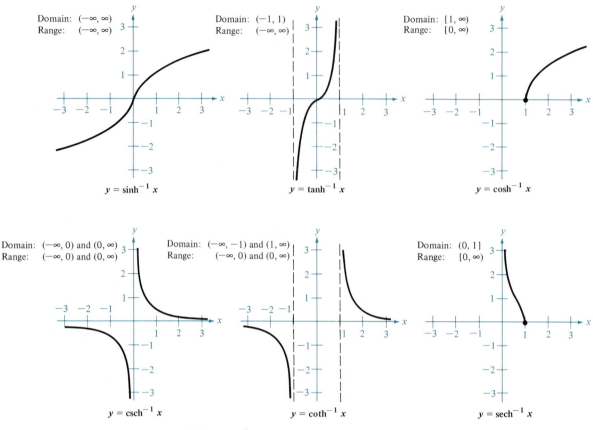

$y = \sinh^{-1} x$

Domain: $(-\infty, \infty)$
Range: $(-\infty, \infty)$

$y = \tanh^{-1} x$

Domain: $(-1, 1)$
Range: $(-\infty, \infty)$

$y = \cosh^{-1} x$

Domain: $[1, \infty)$
Range: $[0, \infty)$

$y = \operatorname{csch}^{-1} x$

Domain: $(-\infty, 0)$ and $(0, \infty)$
Range: $(-\infty, 0)$ and $(0, \infty)$

$y = \coth^{-1} x$

Domain: $(-\infty, -1)$ and $(1, \infty)$
Range: $(-\infty, 0)$ and $(0, \infty)$

$y = \operatorname{sech}^{-1} x$

Domain: $(0, 1]$
Range: $[0, \infty)$

FIGURE 8.56

EXAMPLE 5   An application: the tractrix

A person is holding a rope that is tied to a boat, as shown in Figure 8.57. As the person walks along the dock, the boat travels along a tractrix given by the equation

$$y = a \operatorname{sech}^{-1} \frac{x}{a} - \sqrt{a^2 - x^2}$$

where a is the length of the rope. If $a = 20$ feet, find the distance the person must walk to bring the boat 5 feet from the dock. (We assume that the water is calm, with no current to affect the boat's position.)

Solution: In Figure 8.57, we see that the distance the person has walked is given by

$$y_1 = y + \sqrt{20^2 - x^2} = \left(20 \operatorname{sech}^{-1} \frac{x}{20} - \sqrt{20^2 - x^2} \right) + \sqrt{20^2 - x^2}$$

$$= 20 \operatorname{sech}^{-1} \frac{x}{20}$$

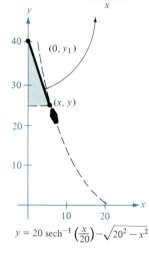

$y = 20 \operatorname{sech}^{-1}\left(\frac{x}{20}\right) - \sqrt{20^2 - x^2}$

FIGURE 8.57

When $x = 5$, this distance is

$$y_1 = 20 \text{ sech}^{-1} \frac{5}{20} = 20 \ln \frac{1 + \sqrt{1 - (1/4)^2}}{1/4}$$

$$= 20 \ln (4 + \sqrt{15}) \approx 41.27 \text{ ft}$$

The derivatives of the inverse hyperbolic functions resemble the derivatives of the inverse trigonometric functions. We list these in Theorem 8.11 with the corresponding integration formulas. Note that for convenience we also list the integration formulas in logarithmic form.

THEOREM 8.11   **DIFFERENTIATION AND INTEGRATION WITH INVERSE HYPERBOLIC FUNCTIONS**
Let u be a differentiable function of x.

$$\frac{d}{dx}[\sinh^{-1} u] = \frac{u'}{\sqrt{u^2 + 1}} \qquad \int \frac{du}{\sqrt{u^2 + 1}} = \sinh^{-1} u + C$$

$$\frac{d}{dx}[\cosh^{-1} u] = \frac{u'}{\sqrt{u^2 - 1}} \qquad \int \frac{du}{\sqrt{u^2 - 1}} = \cosh^{-1} u + C$$

$$\frac{d}{dx}[\tanh^{-1} u] = \frac{u'}{1 - u^2} \qquad \int \frac{du}{1 - u^2} = \tanh^{-1} u + C, \ |u| < 1$$

$$\frac{d}{dx}[\coth^{-1} u] = \frac{u'}{1 - u^2} \qquad \int \frac{du}{1 - u^2} = \coth^{-1} u + C, \ |u| > 1$$

$$\frac{d}{dx}[\text{sech}^{-1} u] = \frac{-u'}{u\sqrt{1 - u^2}} \qquad \int \frac{du}{u\sqrt{1 - u^2}} = -\text{sech}^{-1} u + C$$

$$\frac{d}{dx}[\text{csch}^{-1} u] = \frac{-u'}{|u|\sqrt{1 + u^2}} \qquad \int \frac{du}{|u|\sqrt{1 + u^2}} = -\text{csch}^{-1} u + C$$

$$\int \frac{du}{\sqrt{u^2 \pm a^2}} = \ln (u + \sqrt{u^2 \pm a^2}) + C$$

$$\int \frac{du}{a^2 - u^2} = \frac{1}{2a} \ln \left| \frac{a + u}{a - u} \right| + C$$

$$\int \frac{du}{u\sqrt{a^2 \pm u^2}} = -\frac{1}{a} \ln \frac{a + \sqrt{a^2 \pm u^2}}{|u|} + C$$

Proof: As with the previous theorem, we can prove these differentiation and integration formulas with a straightforward application of the properties of exponential and logarithmic functions. For example,

$$\frac{d}{dx}[\tanh^{-1} x] = \frac{d}{dx}\left[\frac{1}{2} \ln \frac{1 + x}{1 - x}\right] = \frac{1}{2}\left[\frac{d}{dx}[\ln (1 + x) - \ln (1 - x)]\right]$$

$$= \frac{1}{2}\left[\frac{1}{1 + x} - \frac{-1}{1 - x}\right]$$

$$= \frac{1}{2}\left[\frac{2}{1 - x^2}\right] = \frac{1}{1 - x^2}$$

EXAMPLE 6   *Differentiation of inverse hyperbolic functions*

For the tractrix given in Example 5, show that the boat is always pointing toward the person.

Solution: For a point (x, y) on the tractrix, the slope of the graph gives us the direction of the boat, as shown in Figure 8.57. This slope is given by

$$y' = \frac{d}{dx}\left[20\,\text{sech}^{-1}\frac{x}{20} - \sqrt{20^2 - x^2}\right]$$

$$= -20\left(\frac{1}{20}\right)\left(\frac{1}{(x/20)\sqrt{1 - (x/20)^2}}\right) - \left(\frac{1}{2}\right)\left(\frac{-2x}{\sqrt{20^2 - x^2}}\right)$$

$$= \frac{-20^2}{x\sqrt{20^2 - x^2}} + \frac{x}{\sqrt{20^2 - x^2}} = -\frac{\sqrt{20^2 - x^2}}{x}$$

However, from Figure 8.57 we can see that the slope of the line segment connecting the point $(0, y_1)$ with the point (x, y) is also

$$m = -\frac{\sqrt{20^2 - x^2}}{x}$$

Thus, we can conclude that the boat is pointing toward the person. (It is because of this property that a tractrix is called a *pursuit curve*.)

EXAMPLE 7   *Integration using inverse hyperbolic functions*

Evaluate

$$\int \frac{dx}{x\sqrt{4 - 9x^2}}$$

Solution: Since the denominator contains the form $\sqrt{a^2 - u^2}$, we let $u = 3x$ and $a = 2$. Thus, $du = 3\,dx$ and we have

$$\int \frac{dx}{x\sqrt{4 - 9x^2}} = \int \frac{3\,dx}{(3x)\sqrt{4 - 9x^2}} = \int \frac{du}{u\sqrt{a^2 - u^2}}$$

$$= -\frac{1}{a}\ln\frac{a + \sqrt{a^2 - u^2}}{|u|} + C$$

$$= -\frac{1}{2}\ln\frac{2 + \sqrt{4 - 9x^2}}{|3x|} + C$$

EXAMPLE 8   *Integration using inverse hyperbolic functions*

Evaluate

$$\int_0^1 \frac{dx}{5 - 4x^2}$$

Solution: Since the denominator is of the form $a^2 - u^2$, we let $u = 2x$ and $a = \sqrt{5}$. Then $du = 2\,dx$, and we have

$$\int_0^1 \frac{dx}{5 - 4x^2} = \frac{1}{2}\int_0^1 \frac{2\,dx}{(\sqrt{5})^2 - (2x)^2} = \frac{1}{2}\,\frac{1}{2\sqrt{5}}\left[\ln\frac{\sqrt{5} + 2x}{\sqrt{5} - 2x}\right]_0^1$$

$$= \frac{1}{4\sqrt{5}}\ln\frac{\sqrt{5} + 2}{\sqrt{5} - 2} \approx 0.3228$$

Section Exercises 8.7

In Exercises 1–6, evaluate the given function. If the function value is not a rational number, give the answer to three-decimal-place accuracy.

1. (a) $\sinh 3$ (b) $\tanh(-2)$
2. (a) $\cosh 0$ (b) $\text{sech } 1$
3. (a) $\text{csch}(\ln 2)$ (b) $\coth(\ln 5)$
4. (a) $\sinh^{-1} 0$ (b) $\tanh^{-1} 0$
5. (a) $\cosh^{-1} 2$ (b) $\text{sech}^{-1} \dfrac{2}{3}$
6. (a) $\text{csch}^{-1} 2$ (b) $\coth^{-1} 3$

In Exercises 7–12, verify the given identity.

7. $\tanh^2 x + \text{sech}^2 x = 1$ **8.** $\cosh^2 x = \dfrac{1 + \cosh 2x}{2}$

9. $\sinh(x + y) = \sinh x \cosh y + \cosh x \sinh y$
10. $\sinh 2x = 2 \sinh x \cosh x$
11. $\sinh 3x = 3 \sinh x + 4 \sinh^3 x$
12. $\cosh x + \cosh y = 2 \cosh \dfrac{x + y}{2} \cosh \dfrac{x - y}{2}$

In Exercises 13–34, find y' and simplify.

13. $y = \sinh(1 - x^2)$ **14.** $y = \coth 3x$
15. $y = \ln(\sinh x)$ **16.** $y = \ln(\cosh x)$
17. $y = \ln\left(\tanh \dfrac{x}{2}\right)$ **18.** $y = x \sinh x - \cosh x$
19. $y = \dfrac{1}{4} \sinh 2x - \dfrac{x}{2}$ **20.** $y = x - \coth x$
21. $y = \arctan(\sinh x)$ **22.** $y = e^{\sinh x}$
23. $y = x^{\cosh x}$ **24.** $y = \text{sech}^2 3x$
25. $y = (\cosh x - \sinh x)^2$ **26.** $y = \text{sech}(x + 1)$
27. $y = \cosh^{-1}(3x)$ **28.** $y = \tanh^{-1} \dfrac{x}{2}$
29. $y = \sinh^{-1}(\tan x)$
30. $y = \text{sech}^{-1}(\cos 2x),\ 0 < x < \dfrac{\pi}{4}$
31. $y = \coth^{-1}(\sin 2x)$ **32.** $y = (\text{csch}^{-1} x)^2$
33. $y = 2x \sinh^{-1}(2x) - \sqrt{1 + 4x^2}$
34. $y = x \tanh^{-1} x + \ln \sqrt{1 - x^2}$

In Exercises 35–62, evaluate the integral.

35. $\displaystyle\int \sinh(1 - 2x)\, dx$ **36.** $\displaystyle\int \dfrac{\cosh \sqrt{x}}{\sqrt{x}}\, dx$

37. $\displaystyle\int \cosh^2(x - 1) \sinh(x - 1)\, dx$

38. $\displaystyle\int \dfrac{\sinh x}{1 + \sinh^2 x}\, dx$

39. $\displaystyle\int \dfrac{\cosh x}{\sinh x}\, dx$ **40.** $\displaystyle\int \text{sech}^2(2x - 1)\, dx$

41. $\displaystyle\int x \, \text{csch}^2 \dfrac{x^2}{2}\, dx$ **42.** $\displaystyle\int \text{sech}^3 x \tanh x\, dx$

43. $\displaystyle\int \dfrac{\text{csch}(1/x) \coth(1/x)}{x^2}\, dx$

44. $\displaystyle\int \sinh^2 x\, dx$

45. $\displaystyle\int_0^4 \dfrac{1}{25 - x^2}\, dx$ **46.** $\displaystyle\int_0^4 \dfrac{1}{\sqrt{25 - x^2}}\, dx$

47. $\displaystyle\int_0^{\sqrt{2}/4} \dfrac{2}{\sqrt{1 - 4x^2}}\, dx$ **48.** $\displaystyle\int \dfrac{2}{x\sqrt{1 + 4x^2}}\, dx$

49. $\displaystyle\int \dfrac{x}{x^4 + 1}\, dx$ **50.** $\displaystyle\int \dfrac{\cosh x}{\sqrt{9 - \sinh^2 x}}\, dx$

51. $\displaystyle\int \dfrac{1}{\sqrt{1 + e^{2x}}}\, dx$ **52.** $\displaystyle\int \dfrac{e^x}{1 - e^{2x}}\, dx$

53. $\displaystyle\int \dfrac{1}{\sqrt{x}\sqrt{1 + x}}\, dx$ **54.** $\displaystyle\int \dfrac{\sqrt{x}}{\sqrt{1 + x^3}}\, dx$

55. $\displaystyle\int \dfrac{1}{(x - 1)\sqrt{x^2 - 2x + 2}}\, dx$

56. $\displaystyle\int \dfrac{-1}{4x - x^2}\, dx$

57. $\displaystyle\int \dfrac{1}{1 - 4x - 2x^2}\, dx$

58. $\displaystyle\int \dfrac{1}{(x + 1)\sqrt{2x^2 + 4x + 8}}\, dx$

59. $\displaystyle\int \dfrac{1}{\sqrt{80 + 8x - 16x^2}}\, dx$

60. $\displaystyle\int \dfrac{1}{(x - 1)\sqrt{-4x^2 + 8x - 1}}\, dx$

61. $\displaystyle\int \dfrac{x^3 - 21x}{5 + 4x - x^2}\, dx$

62. $\displaystyle\int \dfrac{1 - 2x}{4x - x^2}\, dx$

In Exercises 63 and 64, find any relative extrema and points of inflection of the function. Sketch the graph.

63. $f(x) = x \cosh x - \sinh x$
64. $f(x) = x - \tanh x$

In Exercises 65 and 66, show that the given equation satisfies the differential equation.

65. $y = a \sinh x,\ y''' - y' = 0$
66. $y = a \cosh x,\ y'' - y = 0$

In Exercises 67 and 68, use the equation of the tractrix

$$y = a \, \text{sech}^{-1} \dfrac{x}{a} - \sqrt{a^2 - x^2}$$

67. Find dy/dx.

68. Let L be the tangent line at the point P to the tractrix. If L intersects the y-axis at the point Q, show that the distance between P and Q is a.

69. Suppose that two chemicals A and B combine in a 3-to-1 ratio to form a compound. The amount of compound x being produced at any time t is proportional to the unchanged amounts of A and B remaining in the solution. Thus, if 3 kilograms of A is mixed with 2 kilograms of B, we have

$$\frac{dx}{dt} = k\left(3 - \frac{3x}{4}\right)\left(2 - \frac{x}{4}\right) = \frac{3k}{16}(x^2 - 12x + 32)$$

If 1 kilogram of the compound is formed after 10 minutes, find the amount formed after 20 minutes by solving the integral

$$\int \frac{3k}{16}\, dt = \int \frac{dx}{x^2 - 12x + 32}$$

70. A barn has length 100 feet and width 40 feet, as shown in Figure 8.58. A cross section of the roof is the inverted catenary

$$y = 31 - 20 \cosh \frac{x}{20}$$

Find the number of cubic feet of storage space in the barn.

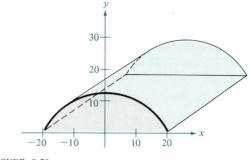

FIGURE 8.58

In Exercises 71–76, verify each derivative formula.

71. $\dfrac{d}{dx}[\tanh x] = \operatorname{sech}^2 x$

72. $\dfrac{d}{dx}[\operatorname{sech} x] = -\operatorname{sech} x \tanh x$

73. $\dfrac{d}{dx}[\cosh^{-1} x] = \dfrac{1}{\sqrt{x^2 - 1}}$

74. $\dfrac{d}{dx}[\tanh^{-1} x] = \dfrac{1}{1 - x^2}$

75. $\dfrac{d}{dx}[\operatorname{sech}^{-1} x] = \dfrac{-1}{x\sqrt{1 - x^2}}$

76. $\dfrac{d}{dx}[\sinh^{-1} x] = \dfrac{1}{\sqrt{x^2 + 1}}$

Review Exercises for Chapter 8

In Exercises 1–24, find dy/dx.

1. $y = \dfrac{\sin x}{x^2}$

2. $y = \csc 3x + \cot 3x$

3. $y = -x \tan x$

4. $y = \dfrac{\cos (x - 1)}{x - 1}$

5. $y = \dfrac{1}{4} \sin 4x + x$

6. $y = x \cos x - \sin x$

7. $y = \dfrac{1}{2}x - \dfrac{1}{4} \sin 2x$

8. $y = \dfrac{1}{7} \sec^7 x - \dfrac{1}{5} \sec^5 x$

9. $y = 2 \csc^3 (\sqrt{x})$

10. $y = \dfrac{1}{2} \sec 2x$

11. $y = \tan \sqrt{1 - x}$

12. $y = \dfrac{1}{2} e^{\sin 2x}$

13. $y = \tan (\arcsin x)$

14. $y = \arctan (x^2 - 1)$

15. $y = x \operatorname{arcsec} x$

16. $y = \dfrac{1}{2} \arctan e^{2x}$

17. $y = x(\arcsin x)^2 - 2x + 2\sqrt{1 - x^2} \arcsin x$

18. $y = \sqrt{x^2 - 4} - 2 \operatorname{arcsec} \dfrac{x}{2}$

19. $y = \left(\dfrac{x^2 + 1}{2}\right) \arctan x$

20. $x \sin y = y \cos x$

21. $x = 2 + \sin y$

22. $\sin (x + y) = x$

23. $\cos x^2 = xe^y$

24. $\cos (x + y) = x$

In Exercises 25–30, find the second derivative of the function.

25. $f(x) = \cot x$

26. $g(t) = \sin^2 t$

27. $h(x) = \dfrac{\cos x}{x}$

28. $f(x) = x \tan x$

29. $f(x) = \arcsin 2x$

30. $f(x) = \arctan \dfrac{x}{2}$

In Exercises 31–50, evaluate the indefinite integral.

31. $\displaystyle\int \dfrac{\cos x}{1 + \sin^2 x}\, dx$

32. $\displaystyle\int \sin^3 x \cos x\, dx$

33. $\displaystyle\int \dfrac{\arctan 2x}{1 + 4x^2}\, dx$

34. $\displaystyle\int \dfrac{\cos x}{\sqrt{\sin x}}\, dx$

35. $\displaystyle\int \tan^n x \sec^2 x\, dx,\ n \neq -1$

36. $\displaystyle\int \frac{\sin\theta}{\sqrt{1-\cos\theta}}\,d\theta$

37. $\displaystyle\int \frac{e^{-2x}}{1+e^{-2x}}\,dx$

38. $\displaystyle\int \frac{x-1}{3x^2-6x-1}\,dx$

39. $\displaystyle\int \frac{1}{e^{2x}+e^{-2x}}\,dx$

40. $\displaystyle\int \frac{\tan(1/x)}{x^2}\,dx$

41. $\displaystyle\int x\tan x^2\,dx$

42. $\displaystyle\int \sec 2x\tan 2x\,dx$

43. $\displaystyle\int \frac{x}{\sqrt{1-x^4}}\,dx$

44. $\displaystyle\int \frac{1}{3+25x^2}\,dx$

45. $\displaystyle\int \frac{x}{16+x^2}\,dx$

46. $\displaystyle\int \frac{1}{16+x^2}\,dx$

47. $\displaystyle\int \frac{\arctan(x/2)}{4+x^2}\,dx$

48. $\displaystyle\int \frac{\arcsin x}{\sqrt{1-x^2}}\,dx$

49. $\displaystyle\int \frac{4-x}{\sqrt{4-x^2}}\,dx$

50. $\displaystyle\int x\sin 3x^2\,dx$

In Exercises 51 and 52, show that the given equation satisfies the differential equation.

51. $y = [a + \ln|\cos x|]\cos x + (b+x)\sin x$
$y'' + y = \sec x$

52. $y = a\cos 2x + b\sin 2x + 2x^2 - 1$
$y'' + 4y = 8x^2$

In Exercises 53 and 54, find an equation of the tangent line to the graph of the equation at the specified point.

Function	Point

53. $y = x\cos x$ $\qquad\qquad \left(\dfrac{\pi}{2}, 0\right)$

54. $y = \arctan x$ $\qquad\qquad \left(1, \dfrac{\pi}{4}\right)$

55. Find the area of the largest rectangle that can be inscribed between the graph of $y = \cos x$ and the x-axis.

56. A certain lawn sprinkler is constructed in such a way that $d\theta/dt$ is constant, where θ ranges between $45°$ and $135°$ as shown in Figure 8.59. The distance the water travels horizontally is given by

$$x = \frac{v^2\sin 2\theta}{32}$$

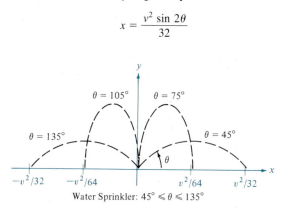

Water Sprinkler: $45° \leqslant \theta \leqslant 135°$

FIGURE 8.59

where v is the speed of the water. Find dx/dt and explain why this model of lawn sprinkler does not water evenly. Which portion of the lawn receives the most water?

57. A hallway of width 6 feet meets a hallway of width 9 feet at right angles. Find the length of the longest pipe that can be carried horizontally around this corner. (Hint: If L is the length of the pipe, show that

$$L = 6\csc\theta + 9\csc\left(\frac{\pi}{2} - \theta\right)$$

where θ is the angle between the pipe and the wall of the narrower hallway.)

58. Rework Exercise 57 if one of the hallways is of width a and the other is of width b.

59. Let θ be the angle of displacement (from the vertical) of a pendulum that is L feet long. Find the maximum rate of change of θ if

$$\theta(t) = A\sin\frac{32t}{L} + B\cos\frac{32t}{L}$$

where t is measured in seconds.

60. Electricity sales in the United States have had both an increasing annual sales pattern and a cyclical monthly sales pattern. For the years 1978 and 1979, the sales pattern can be approximated by the model

$$S(t) = 164.68 + 0.56t + 13.60\cos\frac{\pi t}{3}$$

where S is the sales (per month) in billions of kilowatt hours and t is the time in months as shown in Figure 8.60.

(a) Find the relative extrema of this function for 1978 and 1979.

(b) Use this model to predict the sales in August, 1980. (Use $t = 31.5$.)

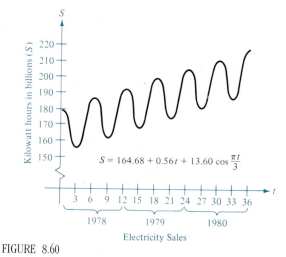

FIGURE 8.60

61. Domestic energy consumption in the United States is seasonal. Suppose the consumption is approximated by the model

$$Q(t) = 6.9 + \cos\left[\frac{\pi(2t - 1)}{12}\right]$$

where Q is the total consumption in quads (quadrillion BTUs) and t is the time in months, with $0 \le t \le 1$ corresponding to January as shown in Figure 8.61.

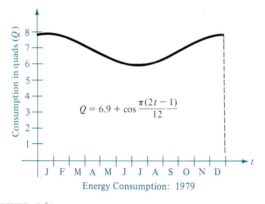

FIGURE 8.61

(a) On which day does this model predict the greatest consumption, and what is the monthly rate for that day?
(b) On which day does this model predict the least consumption, and what is the monthly rate for that day?

62. For a person at rest, the velocity v, in liters per second, of air flow during a respiratory cycle is

$$v = 0.85 \sin\frac{\pi t}{3}$$

where t is the time in seconds. Find the volume, in liters, of air inhaled during one cycle by integrating this function over the interval [0, 3].

63. The temperature in degrees Fahrenheit is given by

$$T = 72 + 12 \sin\left[\frac{\pi(t - 8)}{12}\right]$$

where t is the time in hours, with $t = 0$ representing midnight. Suppose the hourly cost of cooling a house is $0.10 per degree.

(a) Find the cost C of cooling this house if its thermostat is set at 72° by evaluating the integral

$$C = 0.1 \int_8^{20}\left[72 + 12 \sin\frac{\pi(t - 8)}{12} - 72\right] dt$$

(See Figure 8.62.)

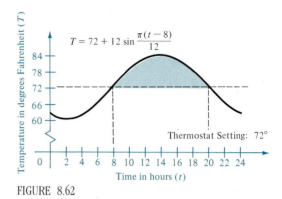

FIGURE 8.62

(b) Find the savings in resetting the thermostat to 78° by evaluating the integral

$$C = 0.1 \int_{10}^{18}\left[72 + 12 \sin\frac{\pi(t - 8)}{12} - 78\right] dt$$

(See Figure 8.63.)

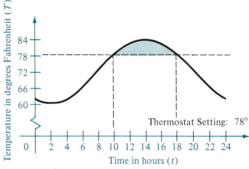

FIGURE 8.63

64. A fertilizer manufacturer finds that national sales of fertilizer roughly follows the seasonal pattern

$$F(t) = 100,000\left[1 + \sin\frac{2\pi(t - 60)}{365}\right]$$

where F is measured in pounds and t is the time in days, with $t = 1$ representing January 1. The manufacturer wants to set up a schedule to produce a uniform amount each day. What should this amount be?

In Exercises 65 and 66, find the volume of the solid generated by revolving the plane region bounded by the graphs of the given equations about the indicated line.

65. $y = \dfrac{1}{x^4 + 1}$, $y = 0$, $x = 0$, $x = 1$,

revolved about the y-axis

66. $y = \dfrac{1}{\sqrt{1 + x^2}}$, $y = 0$, $x = -1$, $x = 1$,

revolved about the x-axis

9
Integration techniques and improper integrals

9.1
Review of basic integration formulas

By now it should be clear to you that integration is not nearly as straightforward as differentiation. To apply the basic integration formulas developed so far, we have often had to rewrite the integrand in an equivalent form. In this chapter we study several new rewriting techniques, which we call *techniques of integration*. These techniques greatly expand the set of integrals to which the basic integration formulas can be applied.

BASIC INTEGRATION FORMULAS

1. $\displaystyle\int kf(u)\,du = k\int f(u)\,du$

2. $\displaystyle\int [f(u) \pm g(u)]\,du = \int f(u)\,du \pm \int g(u)\,du$

3. $\displaystyle\int du = u + C$

4. $\displaystyle\int u^n\,du = \frac{u^{n+1}}{n+1} + C,\ n \neq -1$

5. $\displaystyle\int \frac{du}{u} = \ln |u| + C$

6. $\displaystyle\int e^u\,du = e^u + C$

7. $\displaystyle\int \sin u\,du = -\cos u + C$

8. $\displaystyle\int \cos u\,du = \sin u + C$

9. $\displaystyle\int \tan u\,du = -\ln |\cos u| + C$

10. $\displaystyle\int \cot u\,du = \ln |\sin u| + C$

11. $\displaystyle\int \sec u\,du = \ln |\sec u + \tan u| + C$

12. $\displaystyle\int \csc u\,du = -\ln |\csc u + \cot u| + C$

13. $\displaystyle\int \sec^2 u\,du = \tan u + C$

14. $\displaystyle\int \csc^2 u\,du = -\cot u + C$

15. $\displaystyle\int \sec u \tan u\,du = \sec u + C$

16. $\displaystyle\int \csc u \cot u\,du = -\csc u + C$

17. $\displaystyle\int \frac{du}{\sqrt{a^2 - u^2}} = \arcsin \frac{u}{a} + C$

18. $\displaystyle\int \frac{du}{a^2 + u^2} = \frac{1}{a} \arctan \frac{u}{a} + C$

19. $\displaystyle\int \frac{du}{u\sqrt{u^2 - a^2}} = \frac{1}{a} \operatorname{arcsec} \frac{|u|}{a} + C$

A major step in solving any integration problem is recognizing which basic integration formula to use to solve the problem. Skill in recognizing which formula to use requires memorization of the basic formulas and lots of practice in using them. In this first section we review the use of the basic integration formulas.

One of the challenging things about integration is the fact that slight differences in the integrand can lead to quite different solution techniques.

EXAMPLE 1   A comparison of three similar integrals

Evaluate the following integrals:

(a) $\int \dfrac{4}{x^2 + 9}\, dx$ (b) $\int \dfrac{4x}{x^2 + 9}\, dx$ (c) $\int \dfrac{4x^2}{x^2 + 9}\, dx$

Solution:

(a) Considering the Arctangent Rule, we let $u = x$ and $a = 3$. Then

$$\int \frac{4}{x^2 + 9}\, dx = 4 \int \frac{1}{x^2 + 3^2}\, dx = 4\left[\frac{1}{3} \arctan \frac{x}{3}\right] + C$$

$$= \frac{4}{3} \arctan \frac{x}{3} + C$$

(b) Here the Arctangent Rule does not apply because the numerator contains a factor of x. Considering the Log Rule, we let $u = x^2 + 9$. Then, $du = 2x\, dx$, and we have

$$\int \frac{4x}{x^2 + 9}\, dx = 2 \int \frac{2x\, dx}{x^2 + 9} = 2 \int \frac{du}{u}$$

$$= 2 \ln |u| + C$$

$$= 2 \ln (x^2 + 9) + C$$

(c) Since the degree of the numerator is equal to the degree of the denominator, we first divide to obtain

$$\frac{4x^2}{x^2 + 9} = 4 - \frac{36}{x^2 + 9}$$

Thus,

$$\int \frac{4x^2}{x^2 + 9}\, dx = \int \left(4 - \frac{36}{x^2 + 9}\right) dx = \int 4\, dx - 36 \int \frac{1}{x^2 + 9}\, dx$$

$$= 4x - 36\left[\frac{1}{3} \arctan \frac{x}{3}\right] + C$$

$$= 4x - 12 \arctan \frac{x}{3} + C$$

Notice in part (c) of Example 1 that some preliminary algebra was required before application of the rules for integration, and that subsequently more than one formula was needed to evaluate the resulting integral. This is also the case in the next example.

EXAMPLE 2   Using two basic formulas to solve a single integral

Evaluate

$$\int \frac{x + 3}{\sqrt{4 - x^2}} \, dx$$

Solution: We rewrite the integral as the sum of two integrals and then apply the Power Rule and the Arcsine Rule as follows.

$$\int \frac{x + 3}{\sqrt{4 - x^2}} \, dx = \int \frac{x}{\sqrt{4 - x^2}} \, dx + \int \frac{3}{\sqrt{4 - x^2}} \, dx$$

$$= -\frac{1}{2} \int \underbrace{(4 - x^2)^{-1/2}}_{u^n} \underbrace{(-2x) \, dx}_{du} + 3 \int \frac{1}{\sqrt{\underbrace{2^2 - x^2}_{a = 2, \ u = x}}} \, dx$$

$$= -\frac{1}{2} \int u^{-1/2} \, du + 3 \int \frac{1}{\sqrt{a^2 - u^2}} \, du$$

$$= -\frac{1}{2} \left(\frac{u^{1/2}}{1/2} \right) + 3 \arcsin \frac{u}{a} + C$$

$$= -(4 - x^2)^{1/2} + 3 \arcsin \frac{x}{2} + C$$

EXAMPLE 3   The exponential rule

Evaluate

$$\int_0^{\pi/4} (\sin 2x) e^{-\cos 2x} \, dx$$

Solution: Using the Exponential Rule with $u = -\cos 2x$, we have $du = 2 \sin 2x \, dx$. Moreover, when $x = \pi/4$, $u = -\cos (\pi/2) = 0$, and when $x = 0$, $u = -\cos (0) = -1$. Thus, we have

$$\int_0^{\pi/4} (\sin 2x) e^{-\cos 2x} \, dx = \frac{1}{2} \int_0^{\pi/4} e^{-\cos 2x} (2 \sin 2x) \, dx$$

$$= \frac{1}{2} \int_{-1}^0 e^u \, du = \frac{1}{2} \left[e^u \right]_{-1}^0$$

$$= \frac{1}{2} (1 - e^{-1}) \approx 0.3161$$

When the integrand is a quotient whose numerator is a sum (or difference), you should consider breaking the quotient into two or more parts, as in Example 2. The next example further demonstrates this useful procedure.

EXAMPLE 4   Breaking a quotient into two parts

Evaluate

$$\int \frac{1 + \cos (e^{-2x})}{e^{2x}} \, dx$$

Solution: As it stands, the integrand does not appear to fit any of the basic formulas. However, since the numerator consists of a sum, we break the integrand into two fractions and proceed as follows.

$$\int \frac{1 + \cos e^{-2x}}{e^{2x}} \, dx = \int \frac{dx}{e^{2x}} + \int \frac{\cos e^{-2x}}{e^{2x}} \, dx$$

$$= \underbrace{\int e^{-2x} \, dx}_{u = -2x} + \underbrace{\int (\cos e^{-2x})(e^{-2x}) \, dx}_{u = e^{-2x}}$$

$$= -\frac{1}{2} \int e^{-2x}(-2) \, dx - \frac{1}{2} \int (\cos e^{-2x})(-2e^{-2x}) \, dx$$

$$= -\frac{1}{2} \int e^u \, du - \frac{1}{2} \int \cos u \, du$$

$$= -\frac{1}{2} e^u - \frac{1}{2} \sin u + C$$

$$= -\frac{1}{2} e^{-2x} - \frac{1}{2} \sin e^{-2x} + C$$

Formulas 17, 18, and 19 all have expressions involving the sum or difference of two squares:

$$a^2 - [f(x)]^2$$
$$a^2 + [f(x)]^2$$
$$[f(x)]^2 - a^2$$

When you encounter such an expression, you should consider the substitution $u = f(x)$, as shown in the next example.

EXAMPLE 5   A substitution involving $a^2 - u^2$

Evaluate

$$\int \frac{x^2}{\sqrt{16 - x^6}} \, dx$$

Solution: The radical in the denominator can be written in the form $\sqrt{a^2 - u^2} = \sqrt{4^2 - (x^3)^2}$ and we try the substitution $u = x^3$. Then $du = 3x^2 \, dx$, and we have

$$\int \frac{x^2}{\sqrt{16 - x^6}} \, dx = \frac{1}{3} \int \frac{3x^2 \, dx}{\sqrt{16 - (x^3)^2}}$$

$$= \frac{1}{3} \int \frac{du}{\sqrt{4^2 - u^2}} = \frac{1}{3} \arcsin \frac{u}{4} + C$$

$$= \frac{1}{3} \arcsin \frac{x^3}{4} + C$$

Surprisingly, two of the most commonly overlooked integration formulas are the Log Rule and the Power Rule. Notice in the next two examples how these two integration formulas can take disguised forms.

EXAMPLE 6   *A disguised form of the Log Rule*

Evaluate

$$\int \frac{1}{1 + e^x}\, dx$$

Solution: The integral does not appear to fit any of the basic formulas. However, the quotient form suggests the Log Rule. If we let $u = 1 + e^x$, then $du = e^x\, dx$. We can obtain the required du by adding and subtracting e^x in the numerator as follows.

$$\int \frac{1}{1 + e^x}\, dx = \int \frac{1 + e^x - e^x}{1 + e^x}\, dx \qquad \text{Add and subtract } e^x \text{ in numerator}$$

$$= \int \left(\frac{1 + e^x}{1 + e^x} - \frac{e^x}{1 + e^x} \right) dx$$

$$= \int dx - \int \frac{e^x\, dx}{1 + e^x}$$

$$= \int dx - \int \frac{du}{u}$$

$$= x - \ln |u| + C$$

$$= x - \ln (1 + e^x) + C$$

> **Remark** An alternative way to solve Example 6 would be to multiply the numerator and denominator by e^{-x}. See if you can get the same answer by this procedure. (Be careful; the answer will appear in a different form.)

EXAMPLE 7   *A disguised form of the Power Rule*

Evaluate $\int (\cot x)[\ln (\sin x)]\, dx$.

Solution: Again, this integral does not appear to fit any of the basic formulas. However, considering the two primary choices for u ($u = \cot x$ or $u = \ln \sin x$), we see that the second choice is the appropriate one, since

$$u = \ln \sin x \ \Longrightarrow\ du = \frac{\cos x}{\sin x}\, dx = \cot x\, dx$$

Thus, we have

$$\int (\cot x)[\ln (\sin x)]\, dx = \int u\, du = \frac{u^2}{2} + C$$

$$= \frac{1}{2}[\ln (\sin x)]^2 + C$$

Trigonometric identities can often be used to fit integrals to one of the basic integration formulas. The last example in this review section demonstrates this procedure.

EXAMPLE 8   *Using trigonometric identities*

Evaluate $\int \tan^2 2x \, dx$.

Solution: By looking over the list of basic integration formulas, we see that $\tan^2 u$ is not on the list, but $\sec^2 u$ is on the list. This observation suggests the trigonometric identity $\tan^2 u = \sec^2 u - 1$. If we let $u = 2x$, then $du = 2 \, dx$ and

$$\int \tan^2 2x \, dx = \frac{1}{2} \int \tan^2 u \, du = \frac{1}{2} \int (\sec^2 u - 1) \, du$$

$$= \frac{1}{2} \int \sec^2 u \, du - \frac{1}{2} \int du = \frac{1}{2} \tan u - \frac{u}{2} + C$$

$$= \frac{1}{2} \tan 2x - x + C$$

We conclude with a summary of some common procedures for fitting integrands to the basic integration formulas.

PROCEDURES FOR FITTING INTEGRANDS TO BASIC FORMULAS	Technique	Example
	Expand	$(1 + e^x)^2 = 1 + 2e^x + e^{2x}$
	Separate numerator	$\dfrac{1 + x}{x^2 + 1} = \dfrac{1}{x^2 + 1} + \dfrac{x}{x^2 + 1}$
	Complete the square	$\dfrac{1}{\sqrt{2x - x^2}} = \dfrac{1}{\sqrt{1 - (x - 1)^2}}$
	Divide if rational function is improper	$\dfrac{x^2}{x^2 + 1} = 1 - \dfrac{1}{x^2 + 1}$
	Add and subtract terms in numerator	$\dfrac{2x}{x^2 + 2x + 1} = \dfrac{2x + 2 - 2}{x^2 + 2x + 1}$
		$= \dfrac{2x + 2}{x^2 + 2x + 1} - \dfrac{2}{(x + 1)^2}$
	Use trigonometric identities	$\cot^2 x = \csc^2 x - 1$
	Multiply and divide by Pythagorean conjugate	$\dfrac{1}{1 + \sin x} = \left(\dfrac{1}{1 + \sin x} \right)\left(\dfrac{1 - \sin x}{1 - \sin x} \right)$
		$= \dfrac{1 - \sin x}{1 - \sin^2 x} = \dfrac{1 - \sin x}{\cos^2 x}$
		$= \sec^2 x - \dfrac{\sin x}{\cos^2 x}$

Remark   Watch out for this common error when fitting integrands to basic formulas.

$$\frac{1}{x^2 + 1} \ne \frac{1}{x^2} + \frac{1}{1} \qquad \text{Do } not \text{ separate denominators}$$

Section Exercises 9.1

In Exercises 1–50, evaluate the indefinite integral.

1. $\int (3x - 2)^4 \, dx$

2. $\int \dfrac{2}{(t - 9)^2} \, dt$

3. $\int (-2x + 5)^{3/2} \, dx$

4. $\int x\sqrt{4 - 2x^2} \, dx$

5. $\int \left[v + \dfrac{1}{(3v - 1)^3} \right] dv$

6. $\int \dfrac{2t - 1}{t^2 - t + 2} \, dt$

7. $\int \dfrac{t^2 - 3}{-t^3 + 9t + 1} \, dt$

8. $\int \dfrac{2x}{x - 4} \, dx$

9. $\int \dfrac{x^2}{x - 1} \, dx$

10. $\int \dfrac{x + 1}{\sqrt{x^2 + 2x - 4}} \, dx$

11. $\int \left(\dfrac{1}{3x - 1} - \dfrac{1}{3x + 1} \right) dx$

12. $\int e^{5x} \, dx$

13. $\int t \sin t^2 \, dt$

14. $\int \sec 4u \, du$

15. $\int \cos x \, e^{\sin x} \, dx$

16. $\int \dfrac{e^x}{1 + e^x} \, dx$

17. $\int \dfrac{(1 + e^t)^2}{e^t} \, dt$

18. $\int \dfrac{1 + \sin x}{\cos x} \, dx$

19. $\int \sec 3x \tan 3x \, dx$

20. $\int \dfrac{1}{\sqrt{x}(1 - 2\sqrt{x})} \, dx$

21. $\int \dfrac{2}{e^{-x} + 1} \, dx$

22. $\int \dfrac{1}{2e^x - 3} \, dx$

23. $\int \dfrac{1}{1 - \cos x} \, dx$

24. $\int \dfrac{1}{\sec x - 1} \, dx$

25. $\int \dfrac{2t - 1}{t^2 + 4} \, dt$

26. $\int \dfrac{2}{(2t - 1)^2 + 4} \, dt$

27. $\int \dfrac{3}{t^2 + 1} \, dt$

28. $\int \dfrac{3}{\sqrt{1 - t^2}} \, dt$

29. $\int \dfrac{1}{x\sqrt{x^2 - 4}} \, dx$

30. $\int \dfrac{-2x}{\sqrt{x^2 - 4}} \, dx$

31. $\int \dfrac{-1}{\sqrt{1 - (2t - 1)^2}} \, dt$

32. $\int \dfrac{1}{4 + 3x^2} \, dx$

33. $\int \csc \pi x \cot \pi x \, dx$

34. $\int \dfrac{1}{x\sqrt{4x^2 - 1}} \, dx$

35. $\int \dfrac{t}{\sqrt{1 - t^4}} \, dt$

36. $\int \dfrac{\sin x}{\sqrt{\cos x}} \, dx$

37. $\int \dfrac{\sec^2 x}{4 + \tan^2 x} \, dx$

38. $\int \tan^2 2x \, dx$

39. $\int \dfrac{\tan (2/t)}{t^2} \, dt$

40. $\int \dfrac{e^{1/t}}{t^2} \, dt$

41. $\int (1 + 2x^2)^2 \, dx$

42. $\int (2 - x^2)^2 \, dx$

43. $\int x \left(1 + \dfrac{1}{x} \right)^3 dx$

44. $\int (x + x^2)^3 \, dx$

45. $\int (1 + e^x)^2 \, dx$

46. $\int \left(\dfrac{e^x + e^{-x}}{2} \right)^2 dx$

47. $\int \dfrac{3}{\sqrt{6x - x^2}} \, dx$

48. $\int \dfrac{1}{(x - 1)\sqrt{4x^2 - 8x + 3}} \, dx$

49. $\int \dfrac{4}{4x^2 + 4x + 65} \, dx$

50. $\int \dfrac{1}{\sqrt{2 - 2x - x^2}} \, dx$

In Exercises 51–60, evaluate the definite integrals.

51. $\int_0^1 xe^{-x^2} \, dx$

52. $\int_0^\pi \sin^2 t \cos t \, dt$

53. $\int_1^e \dfrac{1 - \ln x}{x} \, dx$

54. $\int_1^2 \dfrac{x - 2}{x} \, dx$

55. $\int_0^4 \dfrac{2x}{\sqrt{x^2 + 9}} \, dx$

56. $\int_0^{\pi/4} \cos 2x \, dx$

57. $\int_{\pi/4}^{\pi/2} \cot x \, dx$

58. $\int_0^4 \dfrac{1}{\sqrt{25 - x^2}} \, dx$

59. $\int_1^2 \dfrac{1}{2x\sqrt{4x^2 - 1}} \, dx$

60. $\int_0^{2/\sqrt{3}} \dfrac{1}{4 + 9x^2} \, dx$

61. Find the area of the region bounded by the graph of $y^2 = x^2(1 - x^2)$.

62. Find the area of the region bounded by $y = \sin 2x$, $y = 0$, $0 \le x \le \pi/2$.

63. The graphs of $f(x) = x$ and $g(x) = ax^2$ intersect at the points $(0, 0)$ and $(1/a, 1/a)$. Find a so that the area of the region bounded by the graphs of these two functions is $\frac{2}{3}$.

64. Find the volume of the solid generated by revolving the region bounded by $y = e^{-x^2}$, $y = 0$, $x = 0$, and $x = 1$ about the y-axis.

65. The region bounded by $y = e^{-x^2}$, $y = 0$, $x = 0$, and $x = b$ is revolved around the y-axis. Find b so that the volume of the generated solid is $\frac{4}{3}$ cubic units.

66. Compute the average value of each of the functions over the indicated interval.

  (a) $f(x) = \sin nx$, $0 \le x \le \pi/n$, n is a positive integer

  (b) $f(x) = \dfrac{1}{1 + x^2}$, $-3 \le x \le 3$

67. Find the x-coordinate of the centroid of the region bounded by the graphs of

$$y = \dfrac{5}{\sqrt{25 - x^2}}, \quad y = 0, \quad x = 0, \quad x = 4$$

68. Find the area of the surface formed by revolving the graph of $y = 2\sqrt{x}$ on the interval $[0, 9]$ about the x-axis.

9.2
Integration by parts

The first new integration technique we present in this chapter is called **integration by parts.** This technique applies to a wide variety of functions and is particularly useful for integrands involving a *product* of algebraic and transcendental functions. For instance, integration by parts works well with integrals like $\int x \ln x \, dx$, $\int x^2 e^x \, dx$, and $\int e^x \sin x \, dx$.

Integration by parts is based on the formula for the derivative of a product

$$\frac{d}{dx}[uv] = u \frac{dv}{dx} + v \frac{du}{dx} = uv' + vu'$$

where both u and v are differentiable functions of x. If u' and v' are continuous, then we can integrate both sides of this equation to obtain

$$uv = \int uv' \, dx + \int vu' \, dx = \int u \, dv + \int v \, du$$

By rewriting this equation, we obtain the following theorem.

THEOREM 9.1   **INTEGRATION BY PARTS**
If u and v are functions of x and have continuous derivatives, then

$$\int u \, dv = uv - \int v \, du$$

This formula expresses the original integrand in terms of another integral. Depending on the choices for u and dv, it may be easier to evaluate the second integral than the original one. Since the choices of u and dv are critical in the integration by parts process, we suggest the following guidelines.

**GUIDELINES FOR
INTEGRATION BY PARTS**

1. Try letting dv be the most complicated portion of the integrand that fits a basic integration formula. Then u will be the remaining factor(s) of the integrand.
2. Try letting u be the portion of the integrand whose derivative is a simpler function than u. Then dv will be the remaining factor(s) of the integrand.

| **Remark**   These are only suggested guidelines, and they should not be followed blindly or without some thought. Furthermore, it is usually best to consider assigning dv first.

EXAMPLE 1   Integration by parts

Evaluate $\int xe^x \, dx$.

Solution: To apply integration by parts, we want to write the integral in the form $\int u \, dv$. There are several ways to do this.

$$\underbrace{\int (x)(e^x \, dx)}_{u \quad dv}, \quad \underbrace{\int (e^x)(x \, dx)}_{u \quad dv}, \quad \underbrace{\int (1)(xe^x \, dx)}_{u \quad dv}, \quad \underbrace{\int (xe^x)(dx)}_{u \quad dv}$$

Following our guidelines, we choose the first option, since e^x is the most complicated portion of the integrand that fits a basic integration formula. Thus, we have

$$dv = e^x \, dx \implies v = \int dv = \int e^x \, dx = e^x$$

$$u = x \qquad \implies du = dx$$

Now, by the integration by parts formula, we have

$$\int u \, dv = uv - \int v \, du$$

$$\int x \, e^x \, dx = xe^x - \int e^x \, dx = xe^x - e^x + C$$

> **Remark** Note in Example 1 that it is not necessary to include a constant of integration when solving $v = \int e^x \, dx = e^x + C_1$. To illustrate this, we replace v by $v + C_1$ in the general formula to obtain
>
> $$\int u \, dv = u(v + C_1) - \int (v + C_1) \, du = uv + C_1 u - \int C_1 \, du - \int v \, du$$
>
> $$= uv + C_1 u - C_1 u - \int v \, du = uv - \int v \, du$$

EXAMPLE 2   Integration by parts

Evaluate $\int x^2 \ln x \, dx$.

Solution: In this case x^2 is more easily integrated than $\ln x$. Furthermore, the derivative of $\ln x$ is simpler than $\ln x$. Therefore, we let $dv = x^2 \, dx$.

$$dv = x^2 \, dx \implies v = \int x^2 \, dx = \frac{x^3}{3}$$

$$u = \ln x \quad \implies du = \frac{1}{x} \, dx$$

Therefore, we have

$$\int x^2 \ln x \, dx = \frac{x^3}{3} \ln x - \int \left(\frac{x^3}{3}\right)\left(\frac{1}{x}\right) dx$$

$$= \frac{x^3}{3} \ln . x - \frac{1}{3} \int x^2 \, dx$$

$$= \frac{x^3}{3} \ln x - \frac{x^3}{9} + C$$

One unusual application of integration by parts involves integrands consisting of a single factor, such as $\int \ln x \, dx$ or $\int \arcsin x \, dx$. In such cases, we let $dv = dx$, as illustrated in the next example.

EXAMPLE 3   *An integrand with a single term*

Evaluate

$$\int_0^1 \arcsin x \, dx$$

Solution: Letting $dv = dx$, we have

$$dv = dx \quad \Longrightarrow \quad v = \int dx = x$$

$$u = \arcsin x \quad \Longrightarrow \quad du = \frac{1}{\sqrt{1 - x^2}} \, dx$$

Therefore, we have

$$\int \arcsin x \, dx = x \arcsin x - \int \frac{x}{\sqrt{1 - x^2}} \, dx$$

$$= x \arcsin x + \frac{1}{2} \int (1 - x^2)^{-1/2}(-2x) \, dx$$

$$= x \arcsin x + \sqrt{1 - x^2} + C$$

Now, using this antiderivative, we evaluate the definite integral as follows:

$$\int_0^1 \arcsin x \, dx = \left[x \arcsin x + \sqrt{1 - x^2} \right]_0^1$$

$$= \frac{\pi}{2} - 1$$

The area represented by this definite integral is shown in Figure 9.1.

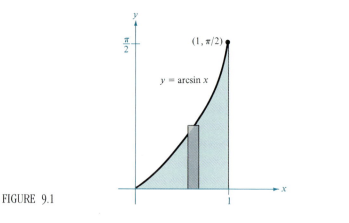

FIGURE 9.1

It may happen that an integral requires repeated application of the integration by parts formula. This is demonstrated in the next example.

EXAMPLE 4   Repeated application of integration by parts

Evaluate $\int x^2 \sin x \, dx$.

Solution: We may consider x^2 and $\sin x$ to be equally easy to integrate. However, the derivative of x^2 becomes simpler, whereas the derivative of $\sin x$ does not. Therefore, we let $u = x^2$ and write

$$dv = \sin x \, dx \implies v = \int \sin x \, dx = -\cos x$$

$$u = x^2 \implies du = 2x \, dx$$

and it follows that

$$\int x^2 \sin x \, dx = -x^2 \cos x + \int 2x \cos x \, dx$$

Now, we apply integration by parts to the new integral. We let $u = 2x$ and write

$$dv = \cos x \, dx \implies v = \int \cos x \, dx = \sin x$$

$$u = 2x \implies du = 2 \, dx$$

and it follows that

$$\int 2x \cos x \, dx = 2x \sin x - \int 2 \sin x \, dx = 2x \sin x + 2 \cos x + C$$

Combining these two results, we have

$$\int x^2 \sin x \, dx = -x^2 \cos x + 2x \sin x + 2 \cos x + C$$

When making repeated application of integration by parts, you need to be careful not to interchange the substitutions in successive applications. For instance, in Example 4 our first substitution was $u = x^2$ and $dv = \sin x \, dx$. If, in the second application, we had switched the substitution to

$$dv = 2x \, dx \implies v = \int 2x \, dx = x^2$$

$$u = \cos x \implies du = -\sin x \, dx$$

we would have obtained

$$\int x^2 \sin x \, dx = -x^2 \cos x + \int 2x \cos x \, dx$$

$$= -x^2 \cos x + x^2 \cos x + \int x^2 \sin x \, dx$$

$$= \int x^2 \sin x \, dx$$

thus undoing the previous integration and returning to the *original* integral.

When making repeated applications of integration by parts, you should also watch for the appearance of a *constant multiple* of the original integral. This is illustrated in the next example.

EXAMPLE 5   Repeated application of integration by parts

Evaluate $\int e^x \cos 2x \, dx$.

Solution: Our guidelines fail to help with a choice of u and dv, so we arbitrarily choose $dv = e^x \, dx$ and $u = \cos 2x$. (You might try verifying that the choice of $dv = \cos 2x \, dx$ and $u = e^x$ works equally well.)

$$dv = e^x \, dx \quad \Longrightarrow \quad v = \int e^x \, dx = e^x$$

$$u = \cos 2x \quad \Longrightarrow \quad du = -2 \sin 2x \, dx$$

and it follows that

$$\int e^x \cos 2x \, dx = e^x \cos 2x + 2 \int e^x \sin 2x \, dx$$

Making the same type of substitutions for the next application of integration by parts, we have

$$dv = e^x \, dx \quad \Longrightarrow \quad v = \int e^x \, dx = e^x$$

$$u = \sin 2x \quad \Longrightarrow \quad du = 2 \cos 2x \, dx$$

and it follows that

$$\int e^x \sin 2x \, dx = e^x \sin 2x - 2 \int e^x \cos 2x \, dx$$

Therefore, we have

$$\int e^x \cos 2x \, dx = e^x \cos 2x + 2 e^x \sin 2x - 4 \int e^x \cos 2x \, dx$$

Now, since the right-hand integral is a constant multiple of the original integral, we add it to the left side of the equation to obtain

$$5 \int e^x \cos 2x \, dx = e^x \cos 2x + 2 e^x \sin 2x$$

$$\int e^x \cos 2x \, dx = \frac{1}{5} e^x \cos 2x + \frac{2}{5} e^x \sin 2x + C \qquad \text{Divide by 5}$$

The integral in the next example is an important one. In Section 9.4 we will see that it is used in finding the arc length of a parabolic segment. (See Example 8 in Section 9.4.)

EXAMPLE 6   Integration by parts

Evaluate $\int \sec^3 x \, dx$.

Solution: The most complicated portion of the integrand that can be easily integrated is $\sec^2 x$. Letting $dv = \sec^2 x \, dx$ and $u = \sec x$, we have

$$dv = \sec^2 x \, dx \;\Longrightarrow\; v = \int \sec^2 x \, dx = \tan x$$

$$u = \sec x \;\Longrightarrow\; du = \sec x \tan x \, dx$$

Therefore, we have

$$\int \sec^3 x \, dx = \sec x \tan x - \int \sec x \tan^2 x \, dx$$

$$= \sec x \tan x - \int \sec x (\sec^2 x - 1) \, dx$$

$$= \sec x \tan x - \int \sec^3 x \, dx + \int \sec x \, dx$$

$$2 \int \sec^3 x \, dx = \sec x \tan x + \int \sec x \, dx \qquad \text{Collect like integrals}$$

$$\int \sec^3 x \, dx = \frac{1}{2} \sec x \tan x + \frac{1}{2} \ln |\sec x + \tan x| + C \qquad \square$$

Since we developed the integration by parts formula from the Product Rule for derivatives, we would expect many of our examples of this technique to involve a product. However, integration by parts is also useful in cases where the integrand is a quotient, as demonstrated in the next example.

EXAMPLE 7   An integrand involving a quotient

Evaluate

$$\int \frac{xe^x}{(x + 1)^2} \, dx$$

Solution: Since $1/(x + 1)^2$ is easily integrated, we make the following choices:

$$dv = \frac{dx}{(x + 1)^2} \;\Longrightarrow\; v = \int \frac{dx}{(x + 1)^2} = -\frac{1}{x + 1}$$

$$u = xe^x \;\Longrightarrow\; du = (xe^x + e^x) \, dx = e^x(x + 1) \, dx$$

Thus, we have

$$\int \frac{xe^x}{(x + 1)^2} \, dx = xe^x \left(\frac{-1}{x + 1} \right) - \int (x + 1)e^x \left(\frac{-1}{x + 1} \right) dx$$

$$= -\frac{xe^x}{x + 1} + \int e^x \, dx$$

$$= -\frac{xe^x}{x + 1} + e^x + C = \frac{e^x}{x + 1} + C \qquad \square$$

EXAMPLE 8   *An application of integration by parts*

Find the centroid of the region bounded by the graph of $y = \sin x$ and the x-axis, $0 \leq x \leq \pi/2$.

Solution: Using the formulas presented in Section 6.6, together with Figure 9.2, we have

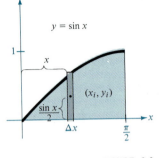

$y = \sin x$

(x_i, y_i)

$\dfrac{\sin x}{2}$

$\Delta x \qquad \dfrac{\pi}{2}$

FIGURE 9.2

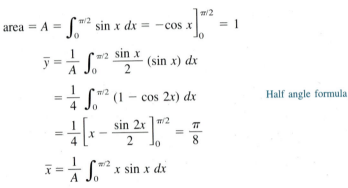

$$\text{area} = A = \int_0^{\pi/2} \sin x \, dx = -\cos x \Big]_0^{\pi/2} = 1$$

$$\bar{y} = \frac{1}{A} \int_0^{\pi/2} \frac{\sin x}{2} (\sin x) \, dx$$

$$= \frac{1}{4} \int_0^{\pi/2} (1 - \cos 2x) \, dx \qquad \text{Half angle formula}$$

$$= \frac{1}{4} \left[x - \frac{\sin 2x}{2} \right]_0^{\pi/2} = \frac{\pi}{8}$$

$$\bar{x} = \frac{1}{A} \int_0^{\pi/2} x \sin x \, dx$$

Now, using *integration by parts* on this integral, we let $dv = \sin x \, dx$, $u = x$, and obtain $v = -\cos x$ and $du = dx$. Thus, we have

$$\int x \sin x \, dx = -x \cos x + \int \cos x \, dx = -x \cos x + \sin x + C$$

Now, we determine $\bar{x}$ to be

$$\bar{x} = \left[-x \cos x + \sin x \right]_0^{\pi/2} = 1$$

Therefore, we conclude that the centroid of the region is $(1, \pi/8)$.

As you gain experience in using integration by parts, your skill in determining u and dv will increase. In the following summary, we list several common integrals with suggestions for the choice of u and dv.

SUMMARY OF COMMON INTEGRALS USING INTEGRATION BY PARTS	
SUMMARY OF COMMON INTEGRALS USING INTEGRATION BY PARTS	1. $\displaystyle\int x^n e^{ax} \, dx, \qquad \int x^n \sin ax \, dx, \qquad \int x^n \cos ax \, dx$ Let $u = x^n$ and $dv = e^{ax} \, dx$, $\sin ax \, dx$, or $\cos ax \, dx$. (Examples 1, 4) 2. $\displaystyle\int x^n \ln x \, dx, \qquad \int x^n \arcsin ax \, dx, \qquad \int x^n \arctan ax \, dx$ Let $u = \ln x$, $\arcsin ax$, or $\arctan ax$ and $dv = x^n \, dx$. (Examples 2, 3) 3. $\displaystyle\int e^{ax} \sin bx \, dx, \qquad \int e^{ax} \cos bx \, dx$ Let $u = \sin bx$ or $\cos bx$ and $dv = e^{ax} \, dx$. (Example 5)

Section Exercises 9.2

In Exercises 1–34, evaluate the given integral. (Note: Solve by the simplest method—not all require integration by parts.)

1. $\int xe^{2x}\,dx$

2. $\int x^2 e^{2x}\,dx$

3. $\int xe^{x^2}\,dx$

4. $\int x^2 e^{x^3}\,dx$

5. $\int xe^{-2x}\,dx$

6. $\int \dfrac{x}{e^x}\,dx$

7. $\int x^3 e^x\,dx$

8. $\int \dfrac{e^{1/t}}{t^2}\,dt$

9. $\int x^3 \ln x\,dx$

10. $\int x^2 \ln x\,dx$

11. $\int t \ln (t+1)\,dt$

12. $\int \dfrac{1}{x(\ln x)^3}\,dx$

13. $\int (\ln x)^2\,dx$

14. $\int \ln 3x\,dx$

15. $\int \dfrac{(\ln x)^2}{x}\,dx$

16. $\int \dfrac{\ln x}{x^2}\,dx$

17. $\int \dfrac{xe^{2x}}{(2x+1)^2}\,dx$

18. $\int \dfrac{x^3 e^{x^2}}{(x^2+1)^2}\,dx$

19. $\int x\sqrt{x-1}\,dx$

20. $\int x^2\sqrt{x-1}\,dx$

21. $\int (x^2-1)e^x\,dx$

22. $\int \dfrac{\ln 2x}{x^2}\,dx$

23. $\int \ln x\,dx$

24. $\int \dfrac{x}{\sqrt{2+3x}}\,dx$

25. $\int x \cos x\,dx$

26. $\int x^2 \cos x\,dx$

27. $\int x \sec^2 x\,dx$

28. $\int \theta \sec \theta \tan \theta\,d\theta$

29. $\int \arcsin 2x\,dx$

30. $\int \arccos x\,dx$

31. $\int \arctan x\,dx$

32. $\int \arctan \dfrac{x}{2}\,dx$

33. $\int e^{2x} \sin x\,dx$

34. $\int e^x \cos 2x\,dx$

In Exercises 35–40, evaluate the definite integral.

35. $\int_0^{\pi} x \sin 2x\,dx$

36. $\int_0^1 x \arcsin x^2\,dx$

37. $\int_0^1 e^x \sin x\,dx$

38. $\int_0^1 x^2 e^x\,dx$

39. $\int_0^{\pi/2} x \cos x\,dx$

40. $\int_0^1 \ln (1+x^2)\,dx$

41. Integrate $\int 2x\sqrt{2x-3}\,dx$
  (a) by parts, letting $dv = \sqrt{2x-3}\,dx$
  (b) by substitution, letting $u = \sqrt{2x-3}$

42. Integrate $\int x\sqrt{4+x}\,dx$
  (a) by parts, letting $dv = \sqrt{4+x}\,dx$
  (b) by substitution, letting $u = \sqrt{4+x}$

43. Integrate

$$\int \dfrac{x^3}{\sqrt{4+x^2}}\,dx$$

  (a) by parts, letting $dv = \dfrac{x}{\sqrt{4+x^2}}\,dx$
  (b) by substitution, letting $u = \sqrt{4+x^2}$

44. Integrate $\int x\sqrt{4-x}\,dx$
  (a) by parts, letting $dv = \sqrt{4-x}\,dx$
  (b) by substitution, letting $u = \sqrt{4-x}$

In Exercises 45–50, use integration by parts to verify the given formula.

45. $\int x^n \sin x\,dx = -x^n \cos x + n \int x^{n-1} \cos x\,dx$

46. $\int x^n \cos x\,dx = x^n \sin x - n \int x^{n-1} \sin x\,dx$

47. $\int x^n \ln x\,dx = \dfrac{x^{n+1}}{(n+1)^2}[-1 + (n+1)\ln x] + C$,

  $n \neq -1$

48. $\int x^n e^{ax}\,dx = \dfrac{x^n e^{ax}}{a} - \dfrac{n}{a}\int x^{n-1} e^{ax}\,dx$

49. $\int e^{ax} \sin bx\,dx = \dfrac{e^{ax}(a \sin bx - b \cos bx)}{a^2+b^2} + C$

50. $\int e^{ax} \cos bx\,dx = \dfrac{e^{ax}(a \cos bx + b \sin bx)}{a^2+b^2} + C$

In Exercises 51–54, find the area of the region bounded by the graphs of the given equations.

51. $y = xe^{-x}$, $y = 0$, $x = 4$

52. $y = \dfrac{1}{9} xe^{-x/3}$, $y = 0$, $x = 0$, $x = 3$

53. $y = e^{-x} \sin \pi x$, $y = 0$, $x = 0$, $x = 1$

54. $y = x \sin x$, $y = 0$, $x = 0$, $x = \pi$

55. Given the region bounded by the graphs of $y = \ln x$, $y = 0$, and $x = e$, find
  (a) the area of the region
  (b) the volume of the solid generated by revolving the region about the x-axis
  (c) the volume of the solid generated by revolving the region about the y-axis
  (d) the centroid of the region

56. Find the volume of the solid generated by revolving the region bounded by $y = e^x$, $y = 0$, $x = 0$, and $x = 1$ about the y-axis.

57. A model for the ability M of a child to memorize, measured on a scale from 0 to 10, is given by

$$M = 1 + 1.6t \ln t, \quad 0 < t \leq 4$$

where t is the child's age in years. Find the average value of this function

(a) between the child's first and second birthdays

(b) between the child's third and fourth birthdays

58. A company sells a seasonal product, and the model for the daily revenue from the product is

$$R = 410.5t^2 e^{-t/30} + 25{,}000, \quad 0 \leq t \leq 365$$

where t is the time in days.

(a) Find the average daily receipts during the first quarter, $0 \leq t \leq 91$.

(b) Find the average daily receipts during the fourth quarter, $274 \leq t \leq 365$.

59. A string stretched between two points $(0, 0)$ and $(0, 2)$ is plucked by displacing the string h units at its midpoint. The motion of the string is modeled by a **Fourier Sine Series** whose coefficients are given by

$$b_n = h \int_0^1 x \sin \frac{n\pi x}{2} \, dx + h \int_1^2 (-x + 2) \sin \frac{n\pi x}{2} \, dx$$

Evaluate b_n.

60. A damping force affects the vibration of a spring so that the displacement of the spring is given by

$$y = e^{-4t}(\cos 2t + 5 \sin 2t)$$

Find the average value of y on the interval from $t = 0$ to $t = \pi$.

61. Find the fallacy in the following argument that $0 = 1$.

$$dv = dx \implies v = x$$

$$u = \frac{1}{x} \implies du = -\frac{1}{x^2} \, dx$$

$$0 + \int \frac{dx}{x} = \left(\frac{1}{x}\right)(x) - \int \left(-\frac{1}{x^2}\right)(x) \, dx$$

$$= 1 + \int \frac{dx}{x}$$

Hence $0 = 1$.

62. Is there a fallacy in the following evaluation of $\int \ln(x + 5) \, dx$?

$$dv = dx \implies v = x + 5$$

$$u = \ln(x + 5) \implies du = \frac{1}{x + 5} \, dx$$

$$\int \ln(x + 5) \, dx = (x + 5) \ln(x + 5) - \int dx$$

$$= (x + 5) \ln(x + 5) - x + C$$

9.3
Trigonometric integrals

In this section we introduce techniques for evaluating integrals of the form

$$\int \sin^m x \cos^n x \, dx \quad \text{and} \quad \int \sec^m x \tan^n x \, dx$$

where either m or n is a positive integer. To find antiderivatives for these forms, we try to break them into combinations of trigonometric integrals to which we can apply the Power Rule. For instance, we can evaluate $\int \sin^5 x \cos x \, dx$ with the Power Rule by letting $u = \sin x$. Then,

$$du = \cos x \, dx$$

and we have

$$\int \sin^5 x \cos x \, dx = \int u^5 \, du = \frac{u^6}{6} + C = \frac{\sin^6 x}{6} + C$$

Similarly, to evaluate $\int \sec^4 x \tan x \, dx$ by the Power Rule, we let $u = \sec x$. Then,

$$du = \sec x \tan x \, dx$$

and we have

$$\int \sec^4 x \tan x \, dx = \int \sec^3 x \, (\sec x \tan x) \, dx = \frac{\sec^4 x}{4} + C$$

To break up $\int \sin^m x \cos^n x \, dx$ into forms to which we can apply the Power Rule, we use the identities

$$\sin^2 x + \cos^2 x = 1 \qquad \text{Pythagorean identity}$$

$$\sin^2 x = \frac{1 - \cos 2x}{2} \qquad \text{Half-angle identity for } \sin^2 x$$

$$\cos^2 x = \frac{1 + \cos 2x}{2} \qquad \text{Half-angle identity for } \cos^2 x$$

as indicated in the following guidelines.

INTEGRALS INVOLVING SINE AND COSINE

1. If the power of the sine is odd and positive, save one sine factor and convert the remaining factors to cosine. Then, expand and integrate.

$$\int \sin^{\overbrace{2k+1}^{\text{odd}}} x \cos^n x \, dx = \int \overbrace{(\sin^2 x)^k}^{\text{convert to cosine}} \cos^n x \overbrace{\sin x \, dx}^{\text{save for } du}$$

$$= \int (1 - \cos^2 x)^k \cos^n x \sin x \, dx$$

2. If the power of the cosine is odd and positive, save one cosine factor and convert the remaining factors to sine. Then, expand and integrate.

$$\int \sin^m x \cos^{\overbrace{2k+1}^{\text{odd}}} x \, dx = \int \sin^m x \overbrace{(\cos^2 x)^k}^{\text{convert to sine}} \overbrace{\cos x \, dx}^{\text{save for } du}$$

$$= \int \sin^m x \, (1 - \sin^2 x)^k \cos x \, dx$$

3. If the powers of *both* the sine and cosine are even and nonnegative, make repeated use of the identities

$$\sin^2 x = \frac{1 - \cos 2x}{2} \qquad \text{and} \qquad \cos^2 x = \frac{1 + \cos 2x}{2}$$

to convert the integrand to odd powers of the cosine. Then, proceed as in case 2.

EXAMPLE 1   *Power of sine is odd and positive*

Evaluate $\int \sin^3 x \cos^4 x \, dx$.

Solution: With the expectation of using the Power Rule with $u = \cos x$, we *save one sine factor* to form du and convert the remaining sine factors to cosines, as follows.

$$\int \sin^3 x \cos^4 x \, dx = \int \sin^2 x \cos^4 x \, (\sin x) \, dx \qquad \text{Save } \sin x$$

$$= \int (1 - \cos^2 x) \cos^4 x \sin x \, dx \qquad \text{Identity for } \sin^2 x$$

$$= \int (\cos^4 x - \cos^6 x) \sin x \, dx$$

$$= \int \cos^4 x \sin x \, dx - \int \cos^6 x \sin x \, dx$$

$$= -\int \cos^4 x \, (-\sin x) \, dx + \int \cos^6 x \, (-\sin x) \, dx$$

$$= -\frac{\cos^5 x}{5} + \frac{\cos^7 x}{7} + C \qquad \text{Power Rule}$$

EXAMPLE 2   *Power of cosine is odd and positive*

Evaluate $\int \sin^2 x \cos^5 x \, dx$.

Solution: Since the power of the cosine is odd and positive, we have

$$\int \sin^2 x \cos^5 x \, dx = \int \sin^2 x \cos^4 x \cos x \, dx \qquad \text{Save } \cos x$$

$$= \int \sin^2 x \, (\cos^2 x)^2 \cos x \, dx$$

$$= \int \sin^2 x \, (1 - \sin^2 x)^2 \cos x \, dx \qquad \text{Identity for } \cos^2 x$$

$$= \int \sin^2 x \, (1 - 2 \sin^2 x + \sin^4 x) \cos x \, dx$$

$$= \int [\sin^2 x - 2 \sin^4 x + \sin^6 x] \cos x \, dx$$

$$= \frac{\sin^3 x}{3} - \frac{2 \sin^5 x}{5} + \frac{\sin^7 x}{7} + C \qquad \text{Power Rule}$$

In Examples 1 and 2, *both* of the powers m and n happened to be positive integers. However, this is not necessary. The method will work as long as either m or n is odd and positive. For instance, in the next example the power of the sine is 3, but the power of the cosine is $-\frac{1}{2}$.

EXAMPLE 3   Power of sine is odd and positive

Evaluate
$$\int_0^{\pi/3} \frac{\sin^3 x}{\sqrt{\cos x}} \, dx$$

Solution: Since the power of the sine is odd and positive, we have

$$\int_0^{\pi/3} \frac{\sin^3 x}{\sqrt{\cos x}} \, dx = \int_0^{\pi/3} \frac{(1 - \cos^2 x)(\sin x)}{\sqrt{\cos x}} \, dx$$

$$= \int_0^{\pi/3} [(\cos x)^{-1/2} \sin x - (\cos x)^{3/2} \sin x] \, dx$$

$$= \int_0^{\pi/3} [-(\cos x)^{-1/2} (-\sin x) + (\cos x)^{3/2} (-\sin x)] \, dx$$

$$= \left[-\frac{(\cos x)^{1/2}}{1/2} + \frac{(\cos x)^{5/2}}{5/2} \right]_0^{\pi/3}$$

$$= \left(-\frac{2}{\sqrt{2}} + \frac{2}{5(2^{5/2})} + 2 - \frac{2}{5} \right) = \frac{32 - 19\sqrt{2}}{20} \approx 0.256$$

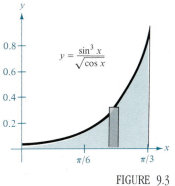

$$y = \frac{\sin^3 x}{\sqrt{\cos x}}$$

FIGURE 9.3

The region whose area is represented by this integral is shown in Figure 9.3.

EXAMPLE 4   Powers of sine and cosine are even and nonnegative

Evaluate $\int \cos^4 x \, dx$.

Solution: Since m and n are both even and nonnegative ($m = 0$), we replace $\cos^4 x$ by $[(1 + \cos 2x)/2]^2$. Then, we have

$$\int \cos^4 x \, dx = \int \left(\frac{1 + \cos 2x}{2} \right)^2 dx \qquad \text{Identity for } \cos^2 x$$

$$= \int \left[\frac{1}{4} + \frac{\cos 2x}{2} + \frac{\cos^2 2x}{4} \right] dx$$

$$= \int \left[\frac{1}{4} + \frac{\cos 2x}{2} + \frac{1}{4} \left(\frac{1 + \cos 4x}{2} \right) \right] dx \qquad \text{Identity for } \cos^2 2x$$

$$= \frac{3}{8} \int dx + \frac{1}{4} \int 2 \cos 2x \, dx + \frac{1}{32} \int 4 \cos 4x \, dx$$

$$= \frac{3x}{8} + \frac{\sin 2x}{4} + \frac{\sin 4x}{32} + C$$

In Example 4, suppose that we were to evaluate the *definite* integral from 0 to $\pi/2$. We would obtain

$$\int_0^{\pi/2} \cos^4 x \, dx = \left[\frac{3x}{8} + \frac{\sin 2x}{4} + \frac{\sin 4x}{32} \right]_0^{\pi/2} = \frac{3\pi}{16}$$

John Wallis

Note that the only term that contributes to the solution is $3x/8$. This observation is generalized in the following formulas developed by John Wallis (1616–1703). Wallis did much of his work prior to Newton and Leibniz, and he influenced the thinking of both of these men.

Wallis's Formulas: If n is odd $(n \geq 1)$, then

$$\int_0^{\pi/2} \cos^n x \, dx = \left(\frac{2}{3}\right)\left(\frac{4}{5}\right)\left(\frac{6}{7}\right) \cdots \left(\frac{n-1}{n}\right)$$

If n is even $(n \geq 2)$, then

$$\int_0^{\pi/2} \cos^n x \, dx = \left(\frac{1}{2}\right)\left(\frac{3}{4}\right)\left(\frac{5}{6}\right) \cdots \left(\frac{n-1}{n}\right)\left(\frac{\pi}{2}\right)$$

These formulas are also valid if $\cos^n x$ is replaced by $\sin^n x$.

To evaluate integrals of the form

$$\int \sec^m x \tan^n x \, dx$$

we suggest the following guidelines.

INTEGRALS INVOLVING SECANTS AND TANGENTS

1. If the power of the secant is even and positive, save a secant squared factor and convert the remaining factors to tangents. Then, expand and integrate.

$$\int \overbrace{\sec^{2k}}^{\text{even}} x \tan^n x \, dx = \int \overbrace{(\sec^2 x)^{k-1}}^{\text{convert to tangents}} \tan^n x \overbrace{\sec^2 x \, dx}^{\text{save for } du}$$

$$= \int (1 + \tan^2 x)^{k-1} \tan^n x \sec^2 x \, dx$$

2. If the power of the tangent is odd and positive, save a secant-tangent factor and convert the remaining factors to secants. Then, expand and integrate.

$$\int \sec^m x \overbrace{\tan^{2k+1}}^{\text{odd}} x \, dx = \int \sec^{m-1} x \overbrace{(\tan^2 x)^k}^{\text{convert to secants}} \overbrace{\sec x \tan x \, dx}^{\text{save for } du}$$

$$= \int \sec^{m-1} x (\sec^2 x - 1)^k \sec x \tan x \, dx$$

3. If there are no secant factors and the power of the tangent is even and positive, convert a tangent squared factor to secants; then expand and repeat if necessary.

$$\int \tan^n x \, dx = \int \tan^{n-2} x \overbrace{(\tan^2 x)}^{\text{convert to secants}} dx = \int \tan^{n-2} x (\sec^2 x - 1) \, dx$$

$$= \int \tan^{n-2} x (\sec^2 x) \, dx - \int \tan^{n-2} x \, dx$$

4. If none of the first three cases apply, try converting to sines and cosines.

| Remark   For integrals of the form $\int \sec^m x \, dx$ with m odd and positive, use integration by parts as illustrated in Example 6 in the previous section.

EXAMPLE 5   *Power of tangent is odd and positive*

Evaluate

$$\int \frac{\tan^3 x}{\sqrt{\sec x}} \, dx$$

Solution: Since the power of the tangent is odd and positive, we write

$$\int \frac{\tan^3 x}{\sqrt{\sec x}} \, dx = \int (\sec x)^{-1/2} \tan^3 x \, dx$$

$$= \int (\sec x)^{-3/2} (\tan^2 x)(\sec x \tan x) \, dx \qquad \text{Save } \sec x \tan x$$

$$= \int (\sec x)^{-3/2} (\sec^2 x - 1)(\sec x \tan x) \, dx$$

$$= \int [(\sec x)^{1/2} - (\sec x)^{-3/2}](\sec x \tan x) \, dx$$

$$= \frac{2}{3}(\sec x)^{3/2} + 2(\sec x)^{-1/2} + C \qquad \text{Power Rule}$$

EXAMPLE 6   *Power of secant is even and positive*

Evaluate $\int \sec^4 3x \tan^3 3x \, dx$.

Solution: Since the power of the secant is even, we save a secant squared factor to create du. If $u = \tan 3x$, then

$$du = 3 \sec^2 3x \, dx$$

and we write

$$\int \sec^4 3x \tan^3 3x \, dx = \int \sec^2 3x \tan^3 3x \, (\sec^2 3x) \, dx \qquad \text{Save } \sec^2 3x$$

$$= \int (1 + \tan^2 3x) \tan^3 3x \, (\sec^2 3x) \, dx$$

$$= \frac{1}{3} \int [\tan^3 3x + \tan^5 3x](3 \sec^2 3x) \, dx$$

$$= \frac{1}{3} \left[\frac{\tan^4 3x}{4} + \frac{\tan^6 3x}{6} \right] + C \qquad \text{Power Rule}$$

| Remark   In Example 6, the power of the tangent is odd and positive. Thus, we could have evaluated the integral with the procedure described in case 2. In Exercise 61, you are asked to show that the results obtained by these two procedures differ only by a constant.

EXAMPLE 7   Power of tangent is even

Evaluate

$$\int_0^{\pi/4} \tan^4 x \, dx$$

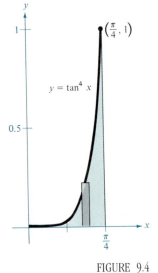

FIGURE 9.4

Solution: Since there are no secant factors, we convert a tangent squared factor to secants.

$$\int \tan^4 x \, dx = \int \tan^2 x \, (\tan^2 x) \, dx = \int \tan^2 x \, (\sec^2 x - 1) \, dx$$

$$= \int \tan^2 x \, \sec^2 x \, dx - \int \tan^2 x \, dx$$

$$= \int \tan^2 x \, \sec^2 x \, dx - \int (\sec^2 x - 1) \, dx$$

$$= \frac{\tan^3 x}{3} - \tan x + x + C$$

Thus, the definite integral (representing the area shown in Figure 9.4) has the following value.

$$\int_0^{\pi/4} \tan^4 x \, dx = \left[\frac{\tan^3 x}{3} - \tan x + x \right]_0^{\pi/4} = \frac{\pi}{4} - \frac{2}{3} \approx 0.119 \quad \square$$

For integrals involving powers of cotangents and cosecants, we follow a strategy similar to that used for powers of tangents and secants, as illustrated in the next example.

EXAMPLE 8   Powers of cotangent and cosecant

Evaluate $\int \csc^4 x \cot^4 x \, dx$.

Solution: Since the power of the cosecant is even, we write

$$\int \csc^4 x \cot^4 x \, dx = \int \csc^2 x \cot^4 x \, (\csc^2 x) \, dx$$

$$= \int (1 + \cot^2 x) \cot^4 x \, (\csc^2 x) \, dx$$

$$= -\int (\cot^4 x + \cot^6 x)(-\csc^2 x) \, dx$$

$$= -\left(\frac{\cot^5 x}{5} + \frac{\cot^7 x}{7} \right) + C = -\frac{\cot^5 x}{5} - \frac{\cot^7 x}{7} + C$$

$$\square$$

EXAMPLE 9   Converting to sines and cosines

Evaluate

$$\int \frac{\sec x}{\tan^2 x} \, dx$$

Solution: Since the first two cases do not apply, we try converting the integrand to sines and cosines. In this case, we are able to integrate the resulting powers of sine and cosine as follows.

$$\int \frac{\sec x}{\tan^2 x}\, dx = \int \left(\frac{1}{\cos x}\right)\left(\frac{\cos x}{\sin x}\right)^2 dx = \int (\sin x)^{-2} (\cos x)\, dx$$

$$= -(\sin x)^{-1} + C = -\csc x + C \qquad \square$$

Occasionally, we encounter integrals involving the product of sines and cosines of two *different* angles. In such instances we use the following product-to-sum identities:

$$\sin mx \sin nx = \frac{1}{2}(\cos [(m-n)x] - \cos [(m+n)x])$$

$$\sin mx \cos nx = \frac{1}{2}(\sin [(m-n)x] + \sin [(m+n)x])$$

$$\cos mx \cos nx = \frac{1}{2}(\cos [(m-n)x] + \cos [(m+n)x])$$

EXAMPLE 10   Using product-to-sum identities

Evaluate $\int \sin 5x \cos 4x\, dx$.

Solution: Considering the second product-to-sum identity, we write

$$\int \sin 5x \cos 4x\, dx = \frac{1}{2} \int (\sin 9x + \sin x)\, dx$$

$$= \frac{1}{2}\left[-\frac{\cos 9x}{9} - \cos x\right] + C$$

$$= -\frac{\cos 9x}{18} - \frac{\cos x}{2} + C \qquad \square$$

As you review this section, concentrate on the *general pattern* followed in evaluating trigonometric integrals. Study the examples and guidelines and then try to work the exercises by reasoning each one out rather than looking back to the boxed-in rules.

Section Exercises 9.3

In Exercises 1–50, evaluate the given integral.

1. $\int \cos^3 x \sin x\, dx$

2. $\int \cos^3 x \sin^2 x\, dx$

3. $\int \sin^5 2x \cos 2x\, dx$

4. $\int \sin^3 x\, dx$

5. $\int \sin^5 x \cos^2 x\, dx$

6. $\int \cos^3 \frac{x}{3}\, dx$

7. $\int \cos^2 3x\, dx$

8. $\int \sin^4 2x\, dx$

9. $\int \sin^2 \pi x \cos^4 \pi x\, dx$

10. $\int \frac{\sin^3 3\theta}{\sqrt{\cos 3\theta}}\, d\theta$

11. $\int \sin^4 x \cos^2 x\, dx$

12. $\int \sin^2 \frac{x}{2} \cos^2 \frac{x}{2}\, dx$

13. $\int x \sin^2 x\, dx$ (integration by parts)

14. $\int x^2 \sin^2 x\, dx$ (integration by parts)

15. $\int \sec 3x\, dx$

16. $\int \sec^2 (2x - 1)\, dx$

17. $\displaystyle\int \sec^4 5x \; dx$ **18.** $\displaystyle\int \sec^6 \frac{x}{2} \; dx$

19. $\displaystyle\int \sec^3 \pi x \; dx$ (integration by parts)

20. $\displaystyle\int \sec^5 \pi x \; dx$ (integration by parts)

21. $\displaystyle\int \tan^3 (1 - x) \; dx$ **22.** $\displaystyle\int \tan^2 x \; dx$

23. $\displaystyle\int \tan^5 \frac{x}{4} \; dx$

24. $\displaystyle\int \tan^3 \frac{\pi x}{2} \sec^2 \frac{\pi x}{2} \; dx$

25. $\displaystyle\int \sec^2 x \tan x \; dx$ **26.** $\displaystyle\int \csc^2 3x \cot 3x \; dx$

27. $\displaystyle\int \tan^2 x \sec^2 x \; dx$ **28.** $\displaystyle\int \tan^5 2x \sec^2 2x \; dx$

29. $\displaystyle\int \sec^5 \pi x \tan \pi x \; dx$

30. $\displaystyle\int \sec^4 (1 - x) \tan (1 - x) \; dx$

31. $\displaystyle\int \sec^6 4x \tan 4x \; dx$ **32.** $\displaystyle\int \sec^2 \frac{x}{2} \tan \frac{x}{2} \; dx$

33. $\displaystyle\int \sec^3 x \tan x \; dx$ **34.** $\displaystyle\int \tan^3 3x \; dx$

35. $\displaystyle\int \tan^3 3x \sec 3x \; dx$ **36.** $\displaystyle\int \sqrt{\tan x} \sec^4 x \; dx$

37. $\displaystyle\int \cot^3 2x \; dx$ **38.** $\displaystyle\int \tan^4 \frac{x}{2} \sec^4 \frac{x}{2} \; dx$

39. $\displaystyle\int \csc^4 \theta \; d\theta$ **40.** $\displaystyle\int \tan^3 t \sec^3 t \; dt$

41. $\displaystyle\int \frac{\cot^2 t}{\csc t} \; dt$ **42.** $\displaystyle\int \frac{\cot^3 t}{\csc t} \; dt$

43. $\displaystyle\int \sin 3x \cos 2x \; dx$ **44.** $\displaystyle\int \cos 3\theta \cos (-2\theta) \; d\theta$

45. $\displaystyle\int \sin \theta \sin 3\theta \; d\theta$ **46.** $\displaystyle\int x \tan^2 x^2 \; dx$

47. $\displaystyle\int \frac{1}{\sec x \tan x} \; dx$ **48.** $\displaystyle\int \frac{\sin^2 x - \cos^2 x}{\cos x} \; dx$

49. $\displaystyle\int (\tan^4 t - \sec^4 t) \; dt$

50. $\displaystyle\int \frac{1 - \sec t}{\cos t - 1} \; dt$

In Exercises 51–60, evaluate the given definite integral.

51. $\displaystyle\int_{-\pi}^{\pi} \sin^2 x \; dx$ **52.** $\displaystyle\int_0^{\pi/4} \tan^2 x \; dx$

53. $\displaystyle\int_0^{\pi/4} \tan^3 x \; dx$ **54.** $\displaystyle\int_0^{\pi/4} \sec^2 t \sqrt{\tan t} \; dt$

55. $\displaystyle\int_0^{\pi/2} \frac{\cos t}{1 + \sin t} \; dt$ **56.** $\displaystyle\int_{-\pi}^{\pi} \sin 3\theta \cos \theta \; d\theta$

57. $\displaystyle\int_0^{\pi/4} \sin 2\theta \sin 3\theta \; d\theta$ **58.** $\displaystyle\int_0^{\pi/2} (1 - \cos \theta)^2 \; d\theta$

59. $\displaystyle\int_{-\pi/2}^{\pi/2} \cos^3 x \; dx$ **60.** $\displaystyle\int_{-\pi/2}^{\pi/2} (\sin^2 x + 1) \; dx$

In Exercises 61 and 62, find the indefinite integral in two different ways and show that the results differ only by a constant.

61. $\displaystyle\int \sec^4 3x \tan^3 3x \; dx$ **62.** $\displaystyle\int \sec^2 x \tan x \; dx$

63. The **inner product** of two functions f and g on $[a, b]$ is the number $\langle f, g \rangle$ defined by

$$\langle f, g \rangle = \int_a^b f(x)g(x) \; dx$$

The distinct functions f and g are said to be **orthogonal** if $\langle f, g \rangle = 0$.

(a) Show that the functions given by

$$f_n(x) = \cos nx, \quad n = 0, 1, 2, \ldots$$

form an orthogonal family on $[0, \pi]$.

(b) Show that the functions given by

$$f_n(x) = \sin nx, \quad n = 1, 2, 3, \ldots$$
$$g_n(x) = \cos nx, \quad n = 0, 1, 2, \ldots$$

form an orthogonal family on $[-\pi, \pi]$.

64. Find the area of the region bounded by the graphs of $y = \tan^2 x$, $y = 4x/\pi$, $x = -\pi/4$, and $x = \pi/4$.

In Exercises 65 and 66, find the volume of the solid generated by revolving the region bounded by the graphs of the given equations about the x-axis.

65. $y = \tan x$, $y = 0$, $x = -\dfrac{\pi}{4}$, $x = \dfrac{\pi}{4}$

66. $y = \sin x \cos^2 x$, $y = 0$, $x = 0$, $x = \dfrac{\pi}{2}$

In Exercises 67 and 68, for the region bounded by the graphs of the given equations, find (a) the volume of the solid formed by revolving the region about the x-axis and (b) the centroid of the region.

67. $y = \sin x$, $y = 0$, $x = 0$, $x = \pi$

68. $y = \cos x$, $y = 0$, $x = 0$, $x = \dfrac{\pi}{2}$

In Exercises 69–74, verify Wallis's Formulas for the given integral.

69. $\displaystyle\int_0^{\pi/2} \cos^3 x \; dx = \frac{2}{3}$ **70.** $\displaystyle\int_0^{\pi/2} \cos^5 x \; dx = \frac{8}{15}$

71. $\displaystyle\int_0^{\pi/2} \cos^7 x \; dx = \frac{16}{35}$ **72.** $\displaystyle\int_0^{\pi/2} \sin^2 x \; dx = \frac{\pi}{4}$

73. $\displaystyle\int_0^{\pi/2} \sin^4 x \; dx = \frac{3\pi}{16}$ **74.** $\displaystyle\int_0^{\pi/2} \sin^6 x \; dx = \frac{5\pi}{32}$

In Exercises 75–77, use integration by parts to verify the given reduction formula.

75. $\displaystyle \int \sin^n x \, dx = -\frac{\sin^{n-1} x \cos x}{n}$

$$+ \frac{n-1}{n} \int \sin^{n-2} x \, dx$$

76. $\displaystyle \int \cos^n x \, dx = \frac{\cos^{n-1} x \sin x}{n}$

$$+ \frac{n-1}{n} \int \cos^{n-2} x \, dx$$

77. $\displaystyle \int \cos^m x \sin^n x \, dx = -\frac{\cos^{m+1} x \sin^{n-1} x}{m+n}$

$$+ \frac{n-1}{m+n} \int \cos^m x \sin^{n-2} x \, dx$$

SECTION TOPIC ▪
Trigonometric substitution ▪

9.4
Trigonometric substitution

Now that we can evaluate integrals involving powers of trigonometric functions, we can use the method of **trigonometric substitution** to evaluate integrals involving the radicals

$$\sqrt{a^2 - u^2}, \qquad \sqrt{a^2 + u^2}, \qquad \text{and} \qquad \sqrt{u^2 - a^2}$$

Our objective with trigonometric substitution is to eliminate the radical in the integrand. We do this with the Pythagorean identities

$$\cos^2 \theta = 1 - \sin^2 \theta$$
$$\sec^2 \theta = 1 + \tan^2 \theta$$
$$\tan^2 \theta = \sec^2 \theta - 1$$

For example, if $a > 0$, we let $u = a \sin \theta$, where $-\pi/2 \le \theta \le \pi/2$. Then

$$\sqrt{a^2 - u^2} = \sqrt{a^2 - a^2 \sin^2 \theta} = \sqrt{a^2(1 - \sin^2 \theta)}$$
$$= \sqrt{a^2 \cos^2 \theta} = a \cos \theta$$

Note that $\cos \theta \ge 0$, since $-\pi/2 \le \theta \le \pi/2$.

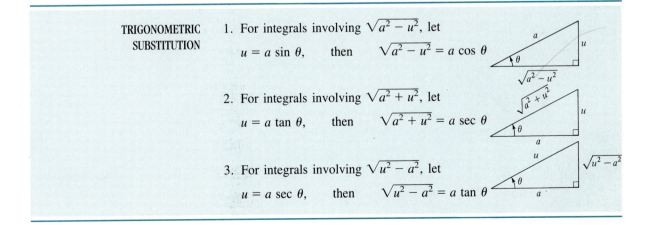

TRIGONOMETRIC SUBSTITUTION

1. For integrals involving $\sqrt{a^2 - u^2}$, let

   $u = a \sin \theta$,   then $\sqrt{a^2 - u^2} = a \cos \theta$

2. For integrals involving $\sqrt{a^2 + u^2}$, let

   $u = a \tan \theta$,   then $\sqrt{a^2 + u^2} = a \sec \theta$

3. For integrals involving $\sqrt{u^2 - a^2}$, let

   $u = a \sec \theta$,   then $\sqrt{u^2 - a^2} = a \tan \theta$

EXAMPLE 1   Trigonometric substitution: u = a sin θ

Evaluate

$$\int \frac{dx}{x^2\sqrt{9-x^2}}$$

Solution: First, we observe that none of the basic integration formulas in Section 9.1 apply. To use trigonometric substitution, we observe that $\sqrt{9-x^2}$ is of the form $\sqrt{a^2-u^2}$, with the corresponding triangle shown in Figure 9.5. Hence, we use the substitution

$$x = a \sin\theta = 3 \sin\theta$$

which implies that

$$dx = 3 \cos\theta\,d\theta, \qquad \sqrt{9-x^2} = 3 \cos\theta, \qquad \text{and} \qquad x^2 = 9 \sin^2\theta$$

Therefore,

$$\int \frac{dx}{x^2\sqrt{9-x^2}} = \int \frac{3 \cos\theta\,d\theta}{(9 \sin^2\theta)(3 \cos\theta)}$$

$$= \frac{1}{9} \int \frac{d\theta}{\sin^2\theta}$$

$$= \frac{1}{9} \int \csc^2\theta\,d\theta = -\frac{1}{9} \cot\theta + C$$

$$= -\frac{1}{9}\left(\frac{\sqrt{9-x^2}}{x}\right) + C \qquad \text{Substitute for cot } \theta$$

$$= -\frac{\sqrt{9-x^2}}{9x} + C$$

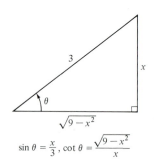

$$\sin\theta = \tfrac{x}{3}, \ \cot\theta = \frac{\sqrt{9-x^2}}{x}$$

FIGURE 9.5

In Section 8.7, we saw how the inverse hyperbolic functions can be used to evaluate the integrals

$$\int \frac{du}{\sqrt{u^2 \pm a^2}}, \qquad \int \frac{du}{a^2 - u^2}, \qquad \text{and} \qquad \int \frac{du}{u\sqrt{a^2 \pm u^2}}$$

If you did not cover that section (or if you prefer not to memorize the formulas for these integrals), you can solve them quite easily using trigonometric substitution. This is illustrated in the next example.

EXAMPLE 2   Trigonometric substitution: u = a tan θ

Evaluate

$$\int \frac{dx}{\sqrt{4x^2 + 1}}$$

Solution: We let $u = 2x$, $a = 1$, and $2x = \tan\theta$, as shown in Figure 9.6. Then,

$$dx = \frac{1}{2} \sec^2\theta\,d\theta \qquad \text{and} \qquad \sqrt{4x^2 + 1} = \sec\theta$$

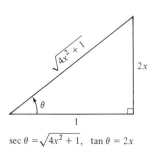

$$\sec\theta = \sqrt{4x^2 + 1}, \ \tan\theta = 2x$$

FIGURE 9.6

Therefore,

$$\int \frac{1}{\sqrt{4x^2 + 1}}\, dx = \frac{1}{2} \int \frac{\sec^2 \theta\, d\theta}{\sec \theta}$$

$$= \frac{1}{2} \int \sec \theta\, d\theta$$

$$= \frac{1}{2} \ln |\sec \theta + \tan \theta| + C$$

$$= \frac{1}{2} \ln |\sqrt{4x^2 + 1} + 2x| + C$$

We can extend the use of trigonometric substitution to cover integrals involving expressions such as $(a^2 - u^2)^{n/2}$ by writing the expression as

$$(a^2 - u^2)^{n/2} = (\sqrt{a^2 - u^2})^n$$

This procedure is demonstrated in the next example.

EXAMPLE 3   Trigonometric substitution: rational powers

Evaluate

$$\int \frac{dx}{(x^2 + 1)^{3/2}}$$

Solution: We write $(x^2 + 1)^{3/2}$ in the form $(\sqrt{x^2 + 1})^3$, and let $a = 1$ and $u = x = \tan \theta$, as shown in Figure 9.7. Then,

$$dx = \sec^2 \theta\, d\theta$$

and

$$\sqrt{x^2 + 1} = \sec \theta$$

Therefore,

$$\int \frac{dx}{(x^2 + 1)^{3/2}} = \int \frac{dx}{(\sqrt{x^2 + 1})^3}$$

$$= \int \frac{\sec^2 \theta\, d\theta}{\sec^3 \theta}$$

$$= \int \frac{d\theta}{\sec \theta}$$

$$= \int \cos \theta\, d\theta = \sin \theta + C$$

$$= \frac{x}{\sqrt{x^2 + 1}} + C \qquad \text{Substitute for } \sin \theta$$

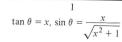
$$\tan \theta = x,\ \sin \theta = \frac{x}{\sqrt{x^2 + 1}}$$

FIGURE 9.7

For definite integrals, it is often convenient to determine the integration limits for θ and thus avoid converting back to x. You might want to review this procedure in Section 5.6, Examples 6 and 7.

EXAMPLE 4   Converting the limits of integration

Evaluate

$$\int_{\sqrt{3}}^{2} \frac{\sqrt{x^2 - 3}}{x} \, dx$$

Solution: Since $\sqrt{x^2 - 3}$ has the form $\sqrt{u^2 - a^2}$, we consider $u = x$, $a = \sqrt{3}$, and let $x = \sqrt{3} \sec \theta$. Then,

$$dx = \sqrt{3} \sec \theta \tan \theta \, d\theta \qquad \text{and} \qquad \sqrt{x^2 - 3} = \sqrt{3} \tan \theta$$

To determine the upper and lower limits of integration, we use the substitution $x = \sqrt{3} \sec \theta$ as follows.

Lower limit   *Upper limit*

When $x = \sqrt{3}$, $\sec \theta = 1$ When $x = 2$, $\sec \theta = \dfrac{2}{\sqrt{3}}$

and $\theta = 0$.   and $\theta = \dfrac{\pi}{6}$.

Therefore, we have

Integration limits for x

Integration limits for θ

$$\int_{\sqrt{3}}^{2} \frac{\sqrt{x^2 - 3}}{x} \, dx = \int_{0}^{\pi/6} \frac{(\sqrt{3} \tan \theta)(\sqrt{3} \sec \theta \tan \theta) \, d\theta}{\sqrt{3} \sec \theta}$$

$$= \int_{0}^{\pi/6} \sqrt{3} \tan^2 \theta \, d\theta = \sqrt{3} \int_{0}^{\pi/6} (\sec^2 \theta - 1) \, d\theta$$

$$= \sqrt{3} \Big[\tan \theta - \theta \Big]_{0}^{\pi/6} = \sqrt{3} \left(\frac{1}{\sqrt{3}} - \frac{\pi}{6} \right)$$

$$= 1 - \frac{\sqrt{3} \pi}{6} \approx 0.0931$$

| **Remark**   In Example 4, we leave as an exercise the alternative procedure of converting back to variable x and evaluating the antiderivative at the original limits of integration. (See Exercise 37.)

We can further expand the range of problems to which trigonometric substitution applies by employing the technique of completing the square. (You can review this procedure as covered in Examples 5, 6, and 7 in Section 8.6.) We further demonstrate the technique of completing the square in the next two examples.

EXAMPLE 5   Completing the square

Evaluate

$$\int \frac{dx}{(x^2 - 4x)^{3/2}}$$

Solution: Completing the square, we have

$$x^2 - 4x = x^2 - 4x + 4 - 4 = (x - 2)^2 - 2^2 = u^2 - a^2$$

We let $a = 2$ and $u = x - 2 = 2 \sec \theta$ as shown in Figure 9.8. Then

$$dx = 2 \sec \theta \tan \theta \, d\theta \qquad \text{and} \qquad \sqrt{(x-2)^2 - 2^2} = 2 \tan \theta$$

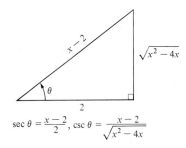

$$\sec \theta = \frac{x-2}{2}, \ \csc \theta = \frac{x-2}{\sqrt{x^2 - 4x}}$$

FIGURE 9.8

Therefore,

$$\int \frac{dx}{(x^2 - 4x)^{3/2}} = \int \frac{2 \sec \theta \tan \theta \, d\theta}{(2 \tan \theta)^3} = \frac{2}{8} \int \frac{\sec \theta}{\tan^2 \theta} \, d\theta$$

$$= \frac{1}{4} \int (\sin \theta)^{-2}(\cos \theta) \, d\theta = -\frac{1}{4} \csc \theta + C$$

$$= -\frac{x - 2}{4\sqrt{x^2 - 4x}} + C \qquad\qquad \text{Substitute for } \csc \theta$$

EXAMPLE 6   *Completing the square*

Evaluate

$$\int \frac{dx}{\sqrt{x^2 - 6x + 10}}$$

Solution: By completing the square, we have

$$x^2 - 6x + 10 = x^2 - 6x + 9 - 9 + 10$$
$$= (x^2 - 6x + 9) + 1 = (x - 3)^2 + 1^2$$

Now, we let $a = 1$ and $u = x - 3 = \tan \theta$. Then

$$dx = \sec^2 \theta \, d\theta \qquad \text{and} \qquad \sqrt{(x-3)^2 + 1^2} = \sec \theta$$

Therefore,

$$\int \frac{dx}{\sqrt{x^2 - 6x + 10}} = \int \frac{\sec^2 \theta \, d\theta}{\sec \theta} = \int \sec \theta \, d\theta$$

$$= \ln |\sec \theta + \tan \theta| + C$$

$$= \ln \left| \sqrt{(x-3)^2 + 1^2} + x - 3 \right| + C$$

$$= \ln \left| \sqrt{x^2 - 6x + 10} + x - 3 \right| + C$$

In Section 9.5, we will encounter integrals involving rational functions. Some of these important integral forms can be solved using trigonometric substitution as illustrated in the next example.

EXAMPLE 7   Trigonometric substitution for rational functions

Evaluate

$$\int \frac{dx}{(x^2 + 1)^2}$$

Solution: Letting $a = 1$ and $u = x = \tan \theta$ as shown in Figure 9.9, we have

$$dx = \sec^2 \theta \, d\theta \qquad \text{and} \qquad x^2 + 1 = \sec^2 \theta$$

Therefore,

$$\int \frac{dx}{(x^2 + 1)^2} = \int \frac{\sec^2 \theta \, d\theta}{\sec^4 \theta}$$

$$= \int \cos^2 \theta \, d\theta$$

$$= \frac{1}{2} \int (1 + \cos 2\theta) \, d\theta$$

$$= \frac{\theta}{2} + \frac{\sin 2\theta}{4} + C$$

$$= \frac{\theta}{2} + \frac{1}{2} \sin \theta \cos \theta + C$$

$$= \frac{\arctan x}{2} + \frac{1}{2}\left(\frac{x}{\sqrt{x^2 + 1}}\right)\left(\frac{1}{\sqrt{x^2 + 1}}\right) + C$$

$$= \frac{1}{2}\left(\arctan x + \frac{x}{x^2 + 1}\right) + C$$

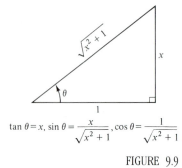

$\tan \theta = x,\ \sin \theta = \dfrac{x}{\sqrt{x^2 + 1}},\ \cos \theta = \dfrac{1}{\sqrt{x^2 + 1}}$

FIGURE 9.9

Trigonometric substitution can be used to evaluate the three integrals listed in the following theorem. We will have occasion to use these integrals several times in the remainder of the text and when we do, we will simply refer to this theorem. If you would rather not memorize these formulas, you should remember that they can be generated by trigonometric substitution.

THEOREM 9.2   SPECIAL INTEGRATION FORMULAS

1. $\displaystyle\int \sqrt{a^2 - u^2} \, du = \frac{1}{2}\left(a^2 \arcsin \frac{u}{a} + u\sqrt{a^2 - u^2}\right) + C$

2. $\displaystyle\int \sqrt{u^2 - a^2} \, du = \frac{1}{2}(u\sqrt{u^2 - a^2} - a^2 \ln |u + \sqrt{u^2 - a^2}|) + C$

3. $\displaystyle\int \sqrt{u^2 + a^2} \, du = \frac{1}{2}(u\sqrt{u^2 + a^2} + a^2 \ln |u + \sqrt{u^2 + a^2}|) + C$

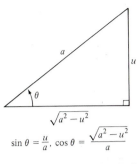

$$\sin \theta = \frac{u}{a}, \quad \cos \theta = \frac{\sqrt{a^2 - u^2}}{a}$$

FIGURE 9.10

Proof: We outline the proof of part 1 and leave the second and third parts as exercises. For the first integral, we let $u = a \sin \theta$. Then $du = a \cos \theta \, d\theta$ and $\sqrt{a^2 - u^2} = a \cos \theta$ and we have

$$\int \sqrt{a^2 - u^2} \, du = \int a^2 \cos^2 \theta \, d\theta = a^2 \int \frac{1 + \cos 2\theta}{2} \, d\theta$$

$$= \frac{a^2}{2} \left(\theta + \frac{1}{2} \sin 2\theta \right) + C = \frac{a^2}{2} (\theta + \sin \theta \cos \theta) + C$$

From Figure 9.10, it follows that

$$\int \sqrt{a^2 - u^2} \, du = \frac{a^2}{2} \left[\arcsin \frac{u}{a} + \left(\frac{u}{a} \right) \left(\frac{\sqrt{a^2 - u^2}}{a} \right) \right] + C$$

$$= \frac{a^2}{2} \left[\frac{a^2 \arcsin (u/a) + u\sqrt{a^2 - u^2}}{a^2} \right] + C$$

$$= \frac{1}{2} \left[a^2 \arcsin \frac{u}{a} + u\sqrt{a^2 - u^2} \right] + C$$

There are many practical applications involving trigonometric substitution, as illustrated in the last two examples.

EXAMPLE 8 *An application involving arc length*

Find the arc length of the graph of $f(x) = \frac{1}{2}x^2$ from $x = 0$ to $x = 1$, as shown in Figure 9.11.

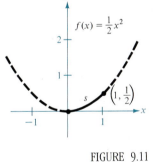

FIGURE 9.11

Solution: From the formula for arc length s, we have

$$s = \int_0^1 \sqrt{1 + [f'(x)]^2} \, dx = \int_0^1 \sqrt{1 + x^2} \, dx$$

Letting $a = 1$ and $x = \tan \theta$, we have

$$dx = \sec^2 \theta \, d\theta \qquad \text{and} \qquad \sqrt{1 + x^2} = \sec \theta$$

Moreover, the upper and lower limits are as follows.

Lower limit   *Upper limit*

When $x = \tan \theta = 0$,   When $x = \tan \theta = 1$,

$$\theta = 0.$$ $$\theta = \frac{\pi}{4}.$$

Therefore, the arc length is given by

$$s = \int_0^1 \sqrt{1 + x^2} \, dx = \int_0^{\pi/4} \sec^3 \theta \, d\theta \qquad \text{Example 6, Section 9.2}$$

$$= \frac{1}{2} \left[\sec \theta \tan \theta + \ln |\sec \theta + \tan \theta| \right]_0^{\pi/4}$$

$$= \frac{1}{2} [\sqrt{2} + \ln (\sqrt{2} + 1)] \approx 1.148$$

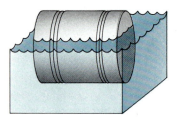

FIGURE 9.12

EXAMPLE 9   A comparison of two fluid forces

A sealed barrel of oil (weighing 48 pounds per cubic foot) is floating in sea water (weighing 64 pounds per cubic foot), as shown in Figures 9.12 and 9.13. Compare the fluid force against one end of the barrel from the inside and from the outside.

Solution: In Figure 9.13 we locate the coordinate system with the origin at the center of the circle given by $x^2 + y^2 = 1$. Then, to find the fluid force against an end of the barrel *from the inside,* we integrate between -1 and 0.8 (using a weight of $w = 48$) to obtain

$$F_{\text{inside}} = 48 \int_{-1}^{0.8} (0.8 - y)(2)\sqrt{1 - y^2}\; dy$$

$$= 76.8 \int_{-1}^{0.8} \sqrt{1 - y^2}\; dy - 96 \int_{-1}^{0.8} y\sqrt{1 - y^2}\; dy$$

To find the fluid force *from the outside,* we integrate between -1 and 0.4 (using a weight of $w = 64$) to obtain

$$F_{\text{outside}} = 64 \int_{-1}^{0.4} (0.4 - y)(2)\sqrt{1 - y^2}\; dy$$

$$= 51.2 \int_{-1}^{0.4} \sqrt{1 - y^2}\; dy - 128 \int_{-1}^{0.4} y\sqrt{1 - y^2}\; dy$$

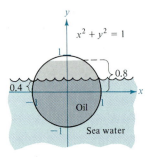

FIGURE 9.13

We leave the details of integration for you to complete in Exercises 51 and 52. Intuitively, would you say that the force from the oil (the inside) or the force from the sea water (the outside) is greater? By solving these two integrals, you can determine that

$$F_{\text{inside}} \approx 121.3 \text{ lb} \qquad \text{and} \qquad F_{\text{outside}} \approx 93.0 \text{ lb}$$

Section Exercises 9.4

In Exercises 1–36, evaluate the indefinite integral.

1. $\displaystyle\int \frac{1}{(25 - x^2)^{3/2}}\; dx$

2. $\displaystyle\int \frac{1}{x^2\sqrt{25 - x^2}}\; dx$

3. $\displaystyle\int \frac{\sqrt{25 - x^2}}{x}\; dx$

4. $\displaystyle\int \frac{1}{\sqrt{25 - x^2}}\; dx$

5. $\displaystyle\int \frac{x}{\sqrt{x^2 + 9}}\; dx$

6. $\displaystyle\int \frac{1}{x\sqrt{4x^2 + 16}}\; dx$

7. $\displaystyle\int_0^2 \sqrt{16 - 4x^2}\; dx$

8. $\displaystyle\int_0^2 x\sqrt{16 - 4x^2}\; dx$

9. $\displaystyle\int \frac{1}{\sqrt{x^2 - 9}}\; dx$

10. $\displaystyle\int \frac{t}{(1 - t^2)^{3/2}}\; dt$

11. $\displaystyle\int_0^{\sqrt{3}/2} \frac{t^2}{(1 - t^2)^{3/2}}\; dt$

12. $\displaystyle\int \frac{1}{(1 - t^2)^{5/2}}\; dt$

13. $\displaystyle\int \frac{\sqrt{1 - x^2}}{x^4}\; dx$

14. $\displaystyle\int \frac{\sqrt{4x^2 + 9}}{x^4}\; dx$

$\displaystyle -\frac{(4x^2+9)^{3/2}}{27x^3}$

15. $\displaystyle\int \frac{1}{x\sqrt{4x^2 + 9}}\; dx$

16. $\displaystyle\int \frac{1}{(x^2 + 3)^{3/2}}\; dx \quad \frac{x}{3\sqrt{x^2+3}}$

17. $\displaystyle\int \frac{x}{(x^2 + 3)^{3/2}}\; dx$

18. $\displaystyle\int \frac{x^3}{\sqrt{x^2 - 4}}\; dx \quad \frac{\sqrt{x^2-4}}{3}(x^2+8)$

19. $\displaystyle\int x^3\sqrt{x^2 - 4}\; dx$

20. $\displaystyle\int \frac{\sqrt{x^2 - 4}}{x}\; dx \quad \sqrt{x^2-4} - 2\arccos$

21. $\displaystyle\int e^{2x}\sqrt{1 + e^{2x}}\; dx$

22. $\displaystyle\int_{-1}^{1} \frac{1}{(1 + x^2)^3}\; dx \quad \frac{3\pi}{16} + \frac{1}{2}$

23. $\displaystyle\int_0^1 \frac{x^2}{(4 + x^2)^2}\; dx$

24. $\displaystyle\int \frac{1}{\sqrt{4x - x^2}}\; dx \quad \arcsin \frac{x-2}{2}$

25. $\displaystyle\int (x + 1)\sqrt{x^2 + 2x + 2}\; dx$

26. $\displaystyle\int \frac{x^2}{\sqrt{2x - x^2}}\; dx$

27. $\displaystyle\int e^x\sqrt{1 - e^{2x}}\; dx$

28. $\displaystyle\int \frac{\sqrt{1 - x}}{\sqrt{x}}\; dx$

29. $\int \dfrac{1}{4 + 4x^2 + x^4}\, dx$

30. $\int \dfrac{x^3 + x + 1}{x^4 + 2x^2 + 1}\, dx$

31. $\int \sqrt{4 + 9x^2}\, dx$

32. $\int \sqrt{1 + x^2}\, dx$

33. $\int \dfrac{x^2}{\sqrt{x^2 - 1}}\, dx$

34. $\int x^2\sqrt{x^2 - 4}\, dx$

35. $\int \operatorname{arcsec} 2x\, dx$

36. $\int x \arcsin x\, dx$

37. Find the value of the definite integral of Example 4 by converting back to the variable x and evaluating the antiderivative at the original limits.

In Exercises 38 and 39, evaluate the given integral using (a) the given integration limits and (b) the limits obtained by trigonometric substitution.

38. $\displaystyle\int_0^3 \dfrac{x^3}{\sqrt{x^2 + 9}}\, dx$

39. $\displaystyle\int_0^{5/3} \sqrt{25 - 9x^2}\, dx$

40. The field strength H of a magnet of length $2L$ on a particle r units from the center of the magnet is given by

$$H = \frac{2mL}{(r^2 + L^2)^{3/2}}$$

where $\pm m$ are the poles of the magnet, as shown in Figure 9.14. Find the average field strength as the particle moves from 0 to R units away from the center by evaluating the integral

$$\frac{1}{R}\int_0^R \frac{2mL}{(r^2 + L^2)^{3/2}}\, dr$$

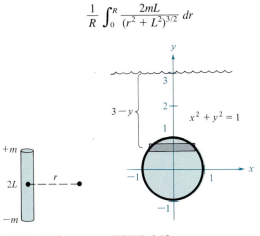

FIGURE 9.14   FIGURE 9.15

In Exercises 41 and 42, find the fluid force on the circular observation window of radius 1 foot in a vertical wall of a large water-filled tank at a fish hatchery.

41. The center of the window is 3 feet below the surface of the water, as shown in Figure 9.15.

42. The center of the window is d feet below the surface of the water $(d > 1)$.

In Exercises 43 and 44, find the volume of the torus generated by revolving the region bounded by the graph of the given circle about the y-axis.

43. $(x - 3)^2 + y^2 = 1$ (Figure 9.16)

44. $(x - h)^2 + y^2 = r^2, \ h > r$

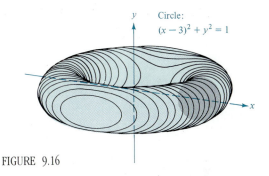

FIGURE 9.16

In Exercises 45 and 46, find the arc length of the given plane curve over the indicated interval.

Function	Interval
45. $y = \ln x$	$[1, 5]$
46. $y = x^2$	$[0, 3]$

In Exercises 47 and 48, find the distance that a projectile travels when it follows the path given by the graph of the equation.

Function	Interval
47. $y = x - 0.005x^2$	$[0, 200]$
48. $y = x - \dfrac{x^2}{72}$	$[0, 72]$

49. Find the surface area of the solid generated by revolving the region bounded by the graphs of $y = x^2$, $y = 0$, $x = 0$, and $x = \sqrt{2}$ about the x-axis.

50. Find the centroid of the region bounded by $y = 3/\sqrt{x^2 + 9}$, $y = 0$, $x = -4$, and $x = 4$.

51. Evaluate the first integral for the fluid force given in Example 9:

$$F = 48 \int_{-1}^{0.8} (0.8 - y)(2)\sqrt{1 - y^2}\, dy$$

52. Evaluate the second integral for the fluid force given in Example 9:

$$F = 64 \int_{-1}^{0.4} (0.4 - y)(2)\sqrt{1 - y^2}\, dy$$

9.5
Partial fractions

In this section we look at a procedure for decomposing a rational function into simpler rational functions to which we can apply the basic integration formulas. We call this procedure the method of **partial fractions.** This technique was introduced in 1702 by John Bernoulli (1667–1748), a Swiss mathematician who was instrumental in the early development of calculus. John Bernoulli was a professor at the University of Basel and taught many outstanding students, the most famous of whom was Leonhard Euler.

To introduce the method of partial fractions, let's consider the integral of $1/(x^2 - 5x + 6)$. To evaluate this integral *without* partial fractions, we complete the square and use trigonometric substitution (see Figure 9.17) to obtain the following.

John Bernoulli

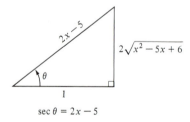

$\sec \theta = 2x - 5$

FIGURE 9.17

$$\int \frac{1}{x^2 - 5x + 6} \, dx = \int \frac{dx}{(x - 5/2)^2 - (1/2)^2} \qquad a = \frac{1}{2}, \ x - \frac{5}{2} = \frac{1}{2} \sec \theta$$

$$= \int \frac{(1/2) \sec \theta \tan \theta \, d\theta}{(1/4) \tan^2 \theta} \qquad dx = \frac{1}{2} \sec \theta \tan \theta \, d\theta$$

$$= 2 \int \csc \theta \, d\theta$$

$$= 2 \ln |\csc \theta - \cot \theta| + C$$

$$= 2 \ln \left| \frac{2x - 5}{2\sqrt{x^2 - 5x + 6}} - \frac{1}{2\sqrt{x^2 - 5x + 6}} \right| + C$$

$$= 2 \ln \left| \frac{x - 3}{\sqrt{x^2 - 5x + 6}} \right| + C$$

$$= 2 \ln \left| \frac{\sqrt{x - 3}}{\sqrt{x - 2}} \right| + C$$

$$= \ln \left| \frac{x - 3}{x - 2} \right| + C$$

$$= \ln |x - 3| - \ln |x - 2| + C$$

Now, to see the benefit of the partial fraction method, suppose we knew that

$$\frac{1}{x^2 - 5x + 6} = \frac{1}{x - 3} - \frac{1}{x - 2}$$

Then we could easily evaluate the integral as follows:

$$\int \frac{1}{x^2 - 5x + 6} \, dx = \int \left(\frac{1}{x - 3} - \frac{1}{x - 2} \right) dx$$

$$= \ln |x - 3| - \ln |x - 2| + C$$

This method is clearly preferable to trigonometric substitution. However, its use depends on the ability to factor the denominator, $x^2 - 5x + 6$, and to find the **partial fractions** $1/(x - 3)$ and $-1/(x - 2)$.

Recall from algebra that every polynomial with real coefficients can be factored into linear and irreducible quadratic factors.* For instance, the polynomial $x^5 + x^4 - x - 1$ can be written as

$$x^5 + x^4 - x - 1 = (x - 1)(x + 1)^2(x^2 + 1)$$

where $(x - 1)$ is a linear factor, $(x + 1)^2$ is a repeated linear factor, and $(x^2 + 1)$ is an irreducible quadratic factor. Using this factorization, we can write the partial fraction decomposition of the rational expression

$$\frac{N(x)}{x^5 + x^4 - x - 1}$$

where $N(x)$ is a polynomial of degree less than 5, as follows.

$$\frac{N(x)}{(x - 1)(x + 1)^2(x^2 + 1)} = \frac{A}{x - 1} + \frac{B}{x + 1} + \frac{C}{(x + 1)^2} + \frac{Dx + E}{x^2 + 1}$$

We summarize the steps for partial fraction decomposition as follows.

DECOMPOSITION OF $N(x)/D(x)$ INTO PARTIAL FRACTIONS

1. **Divide if improper:** If $N(x)/D(x)$ is an improper fraction (that is, if the degree of the numerator is greater than or equal to the degree of the denominator), then divide the denominator into the numerator to obtain

$$\frac{N(x)}{D(x)} = (\text{a polynomial}) + \frac{N_1(x)}{D(x)}$$

where the degree of $N_1(x)$ is less than the degree of $D(x)$. Then apply steps 2, 3, and 4 to the proper rational expression $N_1(x)/D(x)$.

2. **Factor denominator:** Completely factor the denominator into factors of the form

$$(px + q)^m$$

and

$$(ax^2 + bx + c)^n$$

where $ax^2 + bx + c$ is irreducible.

3. **Linear factors:** For *each* factor of the form $(px + q)^m$, the partial fraction decomposition must include the following sum of m fractions:

$$\frac{A_1}{(px + q)} + \frac{A_2}{(px + q)^2} + \cdots + \frac{A_m}{(px + q)^m}$$

4. **Quadratic factors:** For *each* factor of the form $(ax^2 + bx + c)^n$, the partial fraction decomposition must include the following sum of n fractions:

$$\frac{B_1x + C_1}{ax^2 + bx + c} + \frac{B_2x + C_2}{(ax^2 + bx + c)^2} + \cdots + \frac{B_nx + C_n}{(ax^2 + bx + c)^n}$$

---

*For a review of factorization techniques, see *Precalculus,* Larson and Hostetler, D. C. Heath and Company, 1985.

Algebraic techniques for determining the constants in the numerators are demonstrated in the examples that follow.

EXAMPLE 1   Distinct linear factors

Write the partial fraction decomposition for

$$\frac{1}{(x^2 - 5x + 6)}$$

Solution: Since

$$x^2 - 5x + 6 = (x - 3)(x - 2)$$

we include one partial fraction for each factor and write

$$\frac{1}{x^2 - 5x + 6} = \frac{A}{x - 3} + \frac{B}{x - 2}$$

where A and B are to be determined so that the equality holds for all x. Multiplying this equation by the lowest common denominator, $(x - 3)$ $(x - 2)$, leads to the **basic equation**

$$1 = A(x - 2) + B(x - 3) \qquad \text{Basic equation}$$

Since this equation is to be true for all x, we can substitute *convenient* values for x to obtain equations in A and B. These values are the ones that make particular factors zero.

To solve for A, we let $x = 3$ and obtain

$$1 = A(3 - 2) + B(3 - 3) \qquad \text{Let } x = 3 \text{ in basic equation.}$$
$$1 = A(1) + B(0)$$
$$A = 1$$

To solve for B, we let $x = 2$ and obtain

$$1 = A(2 - 2) + B(2 - 3) \qquad \text{Let } x = 2 \text{ in basic equation.}$$
$$1 = A(0) + B(-1)$$
$$B = -1$$

Therefore, the decomposition is

$$\frac{1}{x^2 - 5x + 6} = \frac{1}{x - 3} - \frac{1}{x - 2}$$

as indicated at the beginning of this section.

Remark The substitutions for x in Example 1 were chosen for their convenience in determining values for A and B. We chose $x = 2$ so as to eliminate the term $A(x - 2)$, and $x = 3$ was chosen to eliminate the term $B(x - 3)$. The goal is to make *convenient* substitutions whenever possible.

EXAMPLE 2   Repeated linear factors

Evaluate

$$\int \frac{5x^2 + 20x + 6}{x^3 + 2x^2 + x} \, dx$$

Solution: Since

$$x^3 + 2x^2 + x = x(x^2 + 2x + 1) = x(x + 1)^2$$

we include one fraction for *each power* of x and $(x + 1)$ and write

$$\frac{5x^2 + 20x + 6}{x(x + 1)^2} = \frac{A}{x} + \frac{B}{x + 1} + \frac{C}{(x + 1)^2}$$

Multiplying by the least common denominator, $x(x + 1)^2$, leads to the *basic equation*

$$5x^2 + 20x + 6 = A(x + 1)^2 + Bx(x + 1) + Cx$$

To solve for A, let $x = 0$. This eliminates the B and C terms and yields

$$6 = A(1) + 0 + 0$$
$$A = 6$$

To solve for C, let $x = -1$. This eliminates the A and B terms and yields

$$5 - 20 + 6 = 0 + 0 - C$$
$$C = 9$$

We have exhausted the most convenient choices for x, so to find the value of B we use *any other value* for x along with the calculated values of A and C. Thus, using $x = 1$, $A = 6$, and $C = 9$, we have

$$5 + 20 + 6 = A(4) + B(2) + C$$
$$= 6(4) + 2B + 9$$
$$B = -1$$

Therefore, it follows that

$$\int \frac{5x^2 + 20x + 6}{x(x + 1)^2} \, dx = \int \left(\frac{A}{x} - \frac{1}{x + 1} + \frac{9}{(x + 1)^2} \right) dx$$

$$= \int \left(\frac{6}{x} - \frac{1}{x + 1} + \frac{9}{(x + 1)^2} \right) dx$$

$$= 6 \ln |x| - \ln |x + 1| + 9 \frac{(x + 1)^{-1}}{-1} + C$$

$$= \ln \left| \frac{x^6}{x + 1} \right| - \frac{9}{x + 1} + C$$

Remark   Note that it is necessary to make as many substitutions for x as there are unknowns $(A, B, C, \ldots)$ to be determined. For instance, in Example 2 we made three substitutions $(x = -1, x = 0, \text{ and } x = 1)$ to solve for C, A, and B, respectively.

EXAMPLE 3   Distinct linear and quadratic factors

Evaluate

$$\int \frac{2x^3 - 4x - 8}{(x^2 - x)(x^2 + 4)} \, dx$$

Solution: Since

$$(x^2 - x)(x^2 + 4) = x(x - 1)(x^2 + 4)$$

we include one partial fraction for each factor and write

$$\frac{2x^3 - 4x - 8}{x(x - 1)(x^2 + 4)} = \frac{A}{x} + \frac{B}{x - 1} + \frac{Cx + D}{x^2 + 4}$$

Multiplying by the least common denominator, $x(x - 1)(x^2 + 4)$, yields the *basic equation*

$$2x^3 - 4x - 8 = A(x - 1)(x^2 + 4) + Bx(x^2 + 4) + (Cx + D)(x)(x - 1)$$

To solve for A, let $x = 0$ and obtain

$$-8 = A(-1)(4) + 0 + 0 \implies A = 2$$

To solve for B, let $x = 1$ and obtain

$$-10 = 0 + B(5) + 0 \implies B = -2$$

At this point C and D are yet to be determined. We can find these remaining constants by choosing two other values for x and solving the resulting system of linear equations.

If $x = -1$, then, since $A = 2$ and $B = -2$, we have

$$-6 = (2)(-2)(5) + (-2)(-1)(5) + (-C + D)(-1)(-2)$$
$$2 = -C + D$$

If $x = 2$, then we have

$$0 = (2)(1)(8) + (-2)(2)(8) + (2C + D)(2)(1)$$
$$8 = 2C + D$$

Solving this system of two equations with two unknowns, we have

$$
\begin{array}{rl}
-C + D = 2 & \\
\underline{2C + D = 8} & \\
-3C \quad\quad = -6 & \text{\small Subtract second equation from first}
\end{array}
$$

which yields $C = 2$. Consequently, $D = 4$, and it follows that

$$\int \frac{2x^3 - 4x - 8}{x(x - 1)(x^2 + 4)} \, dx$$

$$= \int \left(\frac{2}{x} - \frac{2}{x - 1} + \frac{2x}{x^2 + 4} + \frac{4}{x^2 + 4} \right) dx$$

$$= 2 \ln |x| - 2 \ln |x - 1| + \ln (x^2 + 4) + 2 \arctan \frac{x}{2} + C$$

When integrating rational expressions, keep in mind that for *improper* rational expressions like

$$\frac{N(x)}{D(x)} = \frac{2x^3 + x^2 - 7x + 7}{x^2 + x - 2}$$

you must first divide

$$\begin{array}{r} 2x - 1 \\ x^2 + x - 2 \overline{)2x^3 +\ \ x^2 - 7x + 7} \\ \underline{2x^3 + 2x^2 - 4x} \\ -x^2 - 3x + 7 \\ \underline{-x^2 -\ \ x + 2} \\ -2x + 5 \end{array}$$

to obtain the form

$$\frac{N(x)}{D(x)} = 2x - 1 + \frac{-2x + 5}{x^2 + x - 2}$$

The proper rational expression is then decomposed into its partial fractions by the usual methods.

Quadratic factors

In Examples 1, 2, and 3, we began the solution of the basic equation by substituting values of x that made the linear factors zero. This method works well when the partial fraction decomposition involves *only* linear factors. However, if the decomposition involves a quadratic factor, then an alternative procedure is often more convenient. Both methods are outlined in the following summary.

GUIDELINES FOR SOLVING
THE BASIC EQUATION

Linear factors

1. Substitute the *roots* of the distinct linear factors into the basic equation.
2. For repeated linear factors use the coefficients determined in part 1 to rewrite the basic equation. Then substitute *other* convenient values of x and solve for the remaining coefficients.

Quadratic factors

1. Expand the basic equation.
2. Collect terms according to powers of x.
3. Equate the coefficients of like powers to obtain a system of linear equations involving A, B, C, and so on.
4. Solve the system of linear equations.

The second procedure for solving the basic equation is demonstrated in the next two examples.

EXAMPLE 4   Repeated quadratic factors

Evaluate

$$\int \frac{8x^3 + 13x}{(x^2 + 2)^2}\, dx$$

Solution: We include one partial fraction for each power of $(x^2 + 2)$ and write

$$\frac{8x^3 + 13x}{(x^2 + 2)^2} = \frac{Ax + B}{x^2 + 2} + \frac{Cx + D}{(x^2 + 2)^2}$$

Multiplying by the least common denominator, $(x^2 + 2)^2$, yields the *basic equation*

$$8x^3 + 13x = (Ax + B)(x^2 + 2) + Cx + D$$

Expanding the basic equation and collecting like terms, we have

$$8x^3 + 13x = Ax^3 + 2Ax + Bx^2 + 2B + Cx + D$$
$$8x^3 + 13x = Ax^3 + Bx^2 + (2A + C)x + (2B + D)$$

Now, we can equate the coefficients of like terms on opposite sides of the equation

$$8 = A$$
$$13 = 2A + C$$
$$8x^3 + 0x^2 + 13x + 0 = Ax^3 + Bx^2 + (2A + C)x + (2B + D)$$
$$0 = 2B + D$$
$$0 = B$$

Using the known values $A = 8$ and $B = 0$, we have

$$13 = 2A + C = 2(8) + C \implies C = -3$$
$$0 = 2B + D = 2(0) + D \implies D = 0$$

Finally, we conclude that

$$\int \frac{8x^3 + 13x}{(x^2 + 2)^2}\, dx = \int \left(\frac{8x}{x^2 + 2} + \frac{-3x}{(x^2 + 2)^2} \right) dx$$

$$= 4 \ln (x^2 + 2) + \frac{3}{2(x^2 + 2)} + C$$

EXAMPLE 5   Repeated quadratic factors

Evaluate

$$\int \frac{x^2}{(x^2 + 1)^2}\, dx$$

Solution: We write

$$\frac{x^2}{(x^2 + 1)^2} = \frac{Ax + B}{x^2 + 1} + \frac{Cx + D}{(x^2 + 1)^2}$$

and multiply by $(x^2 + 1)^2$ to obtain

$$x^2 = (Ax + B)(x^2 + 1) + Cx + D$$
$$= Ax^3 + Bx^2 + (A + C)x + (B + D)$$

Therefore, by equating coefficients we obtain

$$A = 0, \qquad B = 1, \qquad C = 0, \qquad D = -1$$

and we have

$$\int \frac{x^2}{(x^2 + 1)^2}\, dx = \int \left[\frac{1}{x^2 + 1} - \frac{1}{(x^2 + 1)^2} \right] dx$$

The first integral can be evaluated by the Arctangent Rule, and the second integral was solved in Section 9.4 (see Example 7). Therefore, we have

$$\int \frac{x^2\, dx}{(x^2 + 1)^2} = \int \frac{dx}{x^2 + 1} - \int \frac{dx}{(x^2 + 1)^2}$$

$$= \arctan x - \frac{1}{2}\left(\arctan x + \frac{x}{x^2 + 1} \right) + C$$

$$= \frac{1}{2}\left(\arctan x - \frac{x}{x^2 + 1} \right) + C$$

Our last example shows how to combine the two methods effectively to solve a basic equation.

EXAMPLE 6   Linear and quadratic factors

Evaluate

$$\int \frac{3x + 4}{x^3 - 2x - 4}\, dx$$

Solution: Since

$$x^3 - 2x - 4 = (x - 2)(x^2 + 2x + 2)$$

we write

$$\frac{3x + 4}{x^3 - 2x - 4} = \frac{A}{x - 2} + \frac{Bx + C}{x^2 + 2x + 2}$$

$$3x + 4 = A(x^2 + 2x + 2) + (Bx + C)(x - 2) \qquad \text{Basic equation}$$

$$3x + 4 = (A + B)x^2 + (2A - 2B + C)x + (2A - 2C)$$

If $x = 2$, then in the basic equation we have $10 = 10A$ and $A = 1$. Now, from the expanded equation, we have the system

$$A + B \qquad\quad = 0$$
$$2A - 2B + C = 3$$
$$2A \qquad\quad - 2C = 4$$

Using $A = 1$, we have $B = -1$ and $C = -1$. Therefore,

$$\int \frac{3x + 4}{x^3 - 2x - 4} \, dx = \int \left(\frac{A}{x - 2} + \frac{Bx + C}{x^2 + 2x + 2} \right) dx$$

$$= \int \left(\frac{1}{x - 2} + \frac{-x - 1}{x^2 + 2x + 2} \right) dx$$

$$= \ln |x - 2| - \frac{1}{2} \ln (x^2 + 2x + 2) + C$$

Before we conclude this section, a few comments are in order. First, it is not necessary to use the partial fractions technique on all rational functions. For instance, the following integral is more easily evaluated by the Log Rule.

$$\int \frac{x^2 + 1}{x^3 + 3x - 4} \, dx = \frac{1}{3} \int \frac{3x^2 + 3}{x^3 + 3x - 4} \, dx = \frac{1}{3} \ln |x^3 + 3x - 4| + C$$

Second, if the integrand is not in reduced form, reducing it may eliminate the need for partial fractions, as shown in the following integral.

$$\int \frac{x^2 - x - 2}{x^3 - 2x - 4} \, dx = \int \frac{(x + 1)(x - 2)}{(x - 2)(x^2 + 2x + 2)} \, dx = \int \frac{x + 1}{x^2 + 2x + 2} \, dx$$

$$= \frac{1}{2} \ln |x^2 + 2x + 2| + C$$

Finally, partial fractions can be used with some quotients involving transcendental functions. For instance, the substitution $u = \sin x$ allows us to write

$$\int \frac{\cos x}{\sin x (\sin x - 1)} \, dx = \int \frac{du}{u(u - 1)} = \int \left(-\frac{1}{u} + \frac{1}{u - 1} \right) du$$

$$= -\ln |u| + \ln |u - 1| + C = \ln \left| \frac{\sin x - 1}{\sin x} \right| + C$$

Section Exercises 9.5

In Exercises 1–30, find the indefinite integral.

1. $\int \frac{1}{x^2 - 1} \, dx$

2. $\int \frac{1}{4x^2 - 9} \, dx$

3. $\int \frac{3}{x^2 + x - 2} \, dx$

4. $\int \frac{x + 1}{x^2 + 4x + 3} \, dx$

5. $\int \frac{5 - x}{2x^2 + x - 1} \, dx$

6. $\int \frac{3x^2 - 7x - 2}{x^3 - x} \, dx$

7. $\int \frac{x^2 + 12x + 12}{x^3 - 4x} \, dx$

8. $\int \frac{x^3 - x + 3}{x^2 + x - 2} \, dx$

9. $\int \frac{2x^3 - 4x^2 - 15x + 5}{x^2 - 2x - 8} \, dx$

10. $\int \frac{x + 2}{x^2 - 4x} \, dx$

11. $\int \frac{4x^2 + 2x - 1}{x^3 + x^2} \, dx$

12. $\int \frac{2x - 3}{(x - 1)^2} \, dx$

13. $\int \frac{x^4}{(x - 1)^3} \, dx$

14. $\int \frac{4x^2 - 1}{(2x)(x^2 + 2x + 1)} \, dx$

15. $\int \frac{3x}{x^2 - 6x + 9} \, dx$

16. $\int \frac{6x^2 + 1}{x^2(x - 1)^3} \, dx$

17. $\int \frac{x^2 - 1}{x^3 + x} \, dx$

18. $\int \frac{x}{x^3 - 1} \, dx$

19. $\int \frac{x^2}{x^4 - 2x^2 - 8} \, dx$

20. $\int \frac{2x^2 + x + 8}{(x^2 + 4)^2} \, dx$

21. $\int \dfrac{x}{16x^4 - 1}\, dx$

22. $\int \dfrac{x^2 - 4x + 7}{x^3 - x^2 + x + 3}\, dx$

23. $\int \dfrac{x^2 + x + 2}{(x^2 + 2)^2}\, dx$

24. $\int \dfrac{x^3}{(x^2 - 4)^2}\, dx$

25. $\int \dfrac{x^2 + 5}{x^3 - x^2 + x + 3}\, dx$

26. $\int \dfrac{x^2 + x + 3}{x^4 + 6x^2 + 9}\, dx$

27. $\int \dfrac{6x^2 - 3x + 14}{x^3 - 2x^2 + 4x - 8}\, dx$

28. $\int \dfrac{x(2x - 9)}{x^3 - 6x^2 + 12x - 8}\, dx$

29. $\int \dfrac{2x^2 - 2x + 3}{x^3 - x^2 - x - 2}\, dx$

30. $\int \dfrac{-x^4 + 5x^3 - 9x^2 + 3x - 6}{x^5 - x^4 - x + 1}\, dx$

In Exercises 31–36, evaluate the definite integral.

31. $\int_0^1 \dfrac{3}{2x^2 + 5x + 2}\, dx$

32. $\int_3^4 \dfrac{1}{x^2 - 4}\, dx$

33. $\int_1^2 \dfrac{x + 1}{x(x^2 + 1)}\, dx$

34. $\int_1^5 \dfrac{x - 1}{x^2(x + 1)}\, dx$

35. $\int_2^3 \dfrac{x^4 + 4x^3 - x^2 + 5x + 1}{x^5 + x^4 + x^3 - x^2 - 2}\, dx$

36. $\int_0^1 \dfrac{x^2 - x}{x^2 + x + 1}\, dx$

In Exercises 37–41, find the indefinite integral by using the indicated substitution.

37. $\int \dfrac{\sin x}{\cos x\, (\cos x - 1)}\, dx$; let $u = \cos x$.

38. $\int \dfrac{\sin x}{\cos x + \cos^2 x}\, dx$; let $u = \cos x$.

39. $\int \dfrac{e^x}{(e^x - 1)(e^x + 4)}\, dx$; let $u = e^x$.

40. $\int \dfrac{e^x}{(e^{2x} + 1)(e^x - 1)}\, dx$; let $u = e^x$.

41. $\int \dfrac{3 \cos x}{\sin^2 x + \sin x - 2}\, dx$; let $u = \sin x$.

In Exercises 42–44, use the method of partial fractions to verify the given indefinite integral.

42. $\int \dfrac{1}{a^2 - x^2}\, dx = \dfrac{1}{2a} \ln \left| \dfrac{a + x}{a - x} \right| + C$

43. $\int \dfrac{x}{(a + bx)^2}\, dx = \dfrac{1}{b^2}\left(\dfrac{a}{a + bx} + \ln |a + bx| \right) + C$

44. $\int \dfrac{1}{x^2(a + bx)}\, dx = -\dfrac{1}{ax} - \dfrac{b}{a^2} \ln \left| \dfrac{x}{a + bx} \right| + C$

45. Find the area of the region bounded by the graphs of $y = 7/(16 - x^2)$ and $y = 1$.

46. Find the centroid of the region bounded by the graphs of $y = 2x/(x^2 + 1)$, $y = 0$, $x = 0$, and $x = 3$.

47. Find the volume of the solid generated by revolving the region in Exercise 46 about the x-axis.

48. In Section 7.6 the exponential growth equation was derived under the assumption that the rate of growth is proportional to the existing quantity. In practice, there often exists some upper limit L past which growth cannot occur. In such cases, we assume the rate of growth is proportional not only to the existing quantity, but also to the difference between the existing quantity y and the upper limit L. That is,

$$\frac{dy}{dt} = ky(L - y)$$

In integral form we can express this relationship as

$$\int \frac{dy}{y(L - y)} = \int k\, dt$$

(a) Use partial fractions to evaluate the integral on the left, and solve for y as a function of t, where y_0 is the initial quantity.

(b) The graph of the function y is called a **logistics curve** (Figure 9.18). Show that the rate of growth is a maximum at the point of inflection, and that this occurs when $y = L/2$.

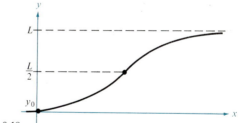

FIGURE 9.18   Logistics Curve

49. A single infected individual enters a community of n susceptible individuals. Let x be the number of newly infected individuals at time t. The common **epidemic model** assumes that the disease spreads at a rate proportional to the product of the total number infected and the number not yet infected. Thus

$$\frac{dx}{dt} = k(x + 1)(n - x)$$

and we obtain

$$\int \frac{1}{(x + 1)(n - x)}\, dx = \int k\, dt$$

Solve for x as a function of t.

9.6
Summary and integration by tables

So far in this chapter we have discussed a number of integration techniques to use with the basic integration formulas. Certainly we have not considered every possible method for finding antiderivatives, but we have considered the most important ones.

But merely knowing *how* to use the various techniques is not enough. You also need to know *when* to use them. Integration is first and foremost a problem of recognition. That is, you must recognize which formula or technique to apply to obtain an antiderivative. Frequently, a slight alteration of an integrand will require the use of a different integration technique, as shown below.

$$\int x \ln x \, dx = \frac{x^2}{2} \ln x - \frac{x^2}{4} + C \qquad \text{Integration by parts}$$

$$\int \frac{\ln x}{x} \, dx = \frac{(\ln x)^2}{2} + C \qquad \text{Power Rule}$$

$$\int \frac{1}{x \ln x} \, dx = \ln |\ln x| + C \qquad \text{Log Rule}$$

In Appendix G of this text is a table of 91 common integrals. As we add new integration formulas to the basic list of 19, two things occur. On the one hand, it becomes increasingly difficult to memorize, or even to become familiar with, the entire list of formulas. On the other hand, we need fewer integration techniques to fit most integrals to one of the formulas on the list. We call the procedure of integrating by means of a long unmemorized list of formulas **integration by tables.**

Integration by tables requires considerable thought and insight and often involves substitution. Many people find a table of integrals to be a valuable supplement to the integration techniques discussed in this chapter. We encourage you to gain competence in the use of a table of integrals. A combination of techniques and tables is the most versatile approach to integration.

Each integration formula in the table in this text can be developed using one or more of the techniques we have studied. We encourage you to verify several of the formulas. For instance, Formula 4

$$\int \frac{u}{(a + bu)^2} \, du = \frac{1}{b^2} \left(\frac{a}{a + bu} + \ln |a + bu| \right) + C \qquad \text{Formula 4}$$

can be verified using the method of partial fractions, and Formula 19

$$\int \frac{\sqrt{a + bu}}{u} \, du = 2\sqrt{a + bu} + a \int \frac{du}{u\sqrt{a + bu}} \qquad \text{Formula 19}$$

can be verified using integration by parts.

The next four examples demonstrate the use of integration tables. Note that the integrals in the tables in this text are classified according to forms involving the following: u^n, $(a + bu)$, $(a + bu + u^2)$, $\sqrt{a + bu}$, $(a^2 \pm u^2)$, $\sqrt{u^2 \pm a^2}$, $\sqrt{a^2 - u^2}$, trigonometric functions, inverse trigonometric functions, exponential functions, and logarithmic functions.

EXAMPLE 1   Integration by tables

Evaluate

$$\int \frac{dx}{x\sqrt{x-1}}$$

Solution: Since the expression inside the radical is linear, we consider forms involving $\sqrt{a+bu}$.

$$\int \frac{du}{u\sqrt{a+bu}} = \frac{2}{\sqrt{-a}} \arctan \sqrt{\frac{a+bu}{-a}} + C \qquad \text{Formula 17}$$

We let $a = -1$, $b = 1$, and $u = x$. Then $du = dx$, and we write

$$\int \frac{dx}{x\sqrt{x-1}} = 2 \arctan \sqrt{x-1} + C$$

EXAMPLE 2   Integration by tables

Evaluate $\int x\sqrt{x^4-9}\, dx$.

Solution: Since the radical has the form $\sqrt{u^2 - a^2}$, we consider the following formula:

$$\int \sqrt{u^2 - a^2}\, du = \frac{1}{2}(u\sqrt{u^2-a^2} - a^2 \ln|u + \sqrt{u^2-a^2}|) + C \quad \text{Formula 26}$$

We let $u = x^2$, and $a = 3$. Then $du = 2x\, dx$, and we write

$$\int x\sqrt{x^4-9}\, dx = \frac{1}{2} \int \sqrt{(x^2)^2 - 3^2}\ (2x)\, dx$$

$$= \frac{1}{4}(x^2\sqrt{x^4-9} - 9 \ln|x^2 + \sqrt{x^4-9}|) + C$$

EXAMPLE 3   Integration by tables

Evaluate

$$\int \frac{x}{1 + e^{-x^2}}\, dx$$

Solution: Of the forms involving e^u, we consider the following:

$$\int \frac{du}{1 + e^u} = u - \ln(1 + e^u) + C \qquad \text{Formula 84}$$

We let $u = -x^2$. Then, $du = -2x\, dx$, and we write

$$\int \frac{x}{1 + e^{-x^2}}\, dx = -\frac{1}{2} \int \frac{-2x\, dx}{1 + e^{-x^2}}$$

$$= -\frac{1}{2}[-x^2 - \ln(1 + e^{-x^2})] + C$$

$$= \frac{1}{2}[x^2 + \ln(1 + e^{-x^2})] + C$$

EXAMPLE 4   Integration by tables

Evaluate

$$\int \frac{\sin 2x}{2 + \cos x} \, dx$$

Solution: Substituting $2 \sin x \cos x$ for $\sin 2x$, we have

$$\int \frac{\sin 2x}{2 + \cos x} \, dx = 2 \int \frac{\sin x \cos x}{2 + \cos x} \, dx$$

A check of the forms involving $\sin u$ or $\cos u$ shows that none of those listed apply. Therefore, we consider forms involving $a + bu$. For example,

$$\int \frac{u \, du}{a + bu} = \frac{1}{b^2}(bu - a \ln |a + bu|) + C \qquad \text{Formula 3}$$

We let $a = 2$, $b = 1$, and $u = \cos x$. Then, $du = -\sin x \, dx$ and we write

$$2 \int \frac{\sin x \cos x}{2 + \cos x} \, dx = -2 \int \frac{\cos x \, (-\sin x \, dx)}{2 + \cos x}$$

$$= -2(\cos x - 2 \ln |2 + \cos x|) + C$$

$$= -2 \cos x + 4 \ln |2 + \cos x| + C \qquad \square$$

You will notice that a number of integrals in the integration tables have the form

$$\int f(x) \, dx = g(x) + \int h(x) \, dx$$

Such integration formulas are called **reduction formulas,** since they reduce a given integral to the sum of a function and a simpler integral. We demonstrate the use of a reduction formula in the next two examples.

EXAMPLE 5   Use of a reduction formula

Evaluate

$$\int x^3 \sin x \, dx$$

Solution: From the integration table, we have the following three formulas.

$$\int u \sin u \, du = \sin u - u \cos u + C \qquad \text{Formula 52}$$

$$\int u^n \sin u \, du = -u^n \cos u + n \int u^{n-1} \cos u \, du \qquad \text{Formula 54}$$

$$\int u^n \cos u \, du = u^n \sin u - n \int u^{n-1} \sin u \, du \qquad \text{Formula 55}$$

Using Formula 54 followed by Formula 55, we have

$$\int x^3 \sin x \, dx = -x^3 \cos x + 3 \int x^2 \cos x \, dx$$

$$= -x^3 \cos x + 3\left(x^2 \sin x - 2 \int x \sin x \, dx\right)$$

Now, by Formula 52, we have

$$\int x^3 \sin x \, dx = -x^3 \cos x + 3x^2 \sin x + 6x \cos x - 6 \sin x + C \quad \square$$

EXAMPLE 6   Use of a reduction formula

Evaluate

$$\int \frac{\sqrt{3 - 5x}}{2x} \, dx$$

Solution: From the integration table, we have the following two formulas.

$$\int \frac{du}{u\sqrt{a + bu}} = \frac{1}{\sqrt{a}} \ln \left| \frac{\sqrt{a + bu} - \sqrt{a}}{\sqrt{a + bu} + \sqrt{a}} \right| + C \qquad \text{Formula 17 } (0 < a)$$

$$\int \frac{\sqrt{a + bu}}{u} \, du = 2\sqrt{a + bu} + a \int \frac{du}{u\sqrt{a + bu}} \qquad \text{Formula 19}$$

Using Formula 19 with $a = 3$, $b = -5$, and $u = x$, it follows that

$$\frac{1}{2} \int \frac{\sqrt{3 - 5x}}{x} \, dx = \frac{1}{2}\left(2\sqrt{3 - 5x} + 3 \int \frac{dx}{x\sqrt{3 - 5x}} \right)$$

$$= \sqrt{3 - 5x} + \frac{3}{2} \int \frac{dx}{x\sqrt{3 - 5x}}$$

Now, by Formula 17, with $a = 3$, $b = -5$, and $u = x$, we conclude that

$$\int \frac{\sqrt{3 - 5x}}{2x} \, dx = \sqrt{3 - 5x} + \frac{3}{2}\left(\frac{1}{\sqrt{3}} \ln \left| \frac{\sqrt{3 - 5x} - \sqrt{3}}{\sqrt{3 - 5x} + \sqrt{3}} \right| \right) + C$$

$$= \sqrt{3 - 5x} + \frac{\sqrt{3}}{2} \ln \left| \frac{\sqrt{3 - 5x} - \sqrt{3}}{\sqrt{3 - 5x} + \sqrt{3}} \right| + C \quad \square$$

Rational functions of sine and cosine

Example 4 involves a rational expression of $\sin x$ and $\cos x$. If you are unable to find an integral of this form in the integration tables, use the following special substitution to convert the trigonometric expression to a standard rational expression.

SUBSTITUTION FOR RATIONAL FUNCTIONS OF SINE AND COSINE

For integrals involving rational functions of the sine and cosine, use the substitution

$$u = \frac{\sin x}{1 + \cos x} = \tan \frac{x}{2}$$

which implies that

$$\cos x = \frac{1 - u^2}{1 + u^2}, \qquad \sin x = \frac{2u}{1 + u^2}, \qquad \text{and} \qquad dx = \frac{2 \, du}{1 + u^2}$$

Proof: From the substitution for u, it follows that

$$u^2 = \frac{\sin^2 x}{(1 + \cos x)^2} = \frac{1 - \cos^2 x}{(1 + \cos x)^2} = \frac{1 - \cos x}{1 + \cos x}$$

Solving for $\cos x$ in this equation, we have

$$\cos x = \frac{1 - u^2}{1 + u^2}$$

To find $\sin x$, we write $u = \sin x/(1 + \cos x)$ as

$$\sin x = u(1 + \cos x) = u\left(1 + \frac{1 - u^2}{1 + u^2}\right) = \frac{2u}{1 + u^2}$$

Finally, to find dx, we consider $u = \tan(x/2)$. Then $\arctan u = x/2$ and

$$dx = \frac{2\ du}{1 + u^2}$$

EXAMPLE 7   *Substitution for rational functions of sine and cosine*

Evaluate

$$\int \frac{dx}{1 + \sin x - \cos x}$$

Solution: Let $u = \sin x/(1 + \cos x)$, then

$$\int \frac{dx}{1 + \sin x - \cos x} = \int \frac{2\ du/(1 + u^2)}{1 + [2u/(1 + u^2)] - [(1 - u^2)/(1 + u^2)]}$$

$$= \int \frac{2\ du}{(1 + u^2) + 2u - (1 - u^2)} = \int \frac{2\ du}{2u + 2u^2}$$

$$= \int \frac{du}{u(1 + u)} = \int \frac{1}{u}\ du - \int \frac{1}{1 + u}\ du \qquad \text{\color{blue}Partial fractions}$$

$$= \ln|u| - \ln|1 + u| + C = \ln\left|\frac{u}{1 + u}\right| + C$$

$$= \ln\left|\frac{\sin x/(1 + \cos x)}{1 + [\sin x/(1 + \cos x)]}\right| + C$$

$$= \ln\left|\frac{\sin x}{1 + \sin x + \cos x}\right| + C$$

Section Exercises 9.6

In Exercises 1–52, use the integration tables at the end of the text to evaluate the given integral.

1. $\displaystyle \int \frac{x^2}{1 + x}\ dx$

2. $\displaystyle \int \frac{x}{\sqrt{1 + x}}\ dx$

3. $\displaystyle \int \frac{1}{x^2\sqrt{1 - x^2}}\ dx$

4. $\displaystyle \int x \sin x\ dx$

5. $\displaystyle \int x^2 \ln x\ dx$

6. $\displaystyle \int \operatorname{arcsec} 2x\ dx$

7. $\displaystyle \int \frac{1}{x^2\sqrt{x^2 - 4}}\ dx$

8. $\displaystyle \int \frac{\sqrt{x^2 - 4}}{x}\ dx$

9. $\displaystyle \int x\ e^{x^2}\ dx$

10. $\displaystyle \int \frac{x}{\sqrt{9 - x^4}}\ dx$

11. $\displaystyle\int \frac{2x}{(1-3x)^2}\,dx$

12. $\displaystyle\int \frac{1}{x^2+2x+2}\,dx$

13. $\displaystyle\int e^x \arccos e^x\,dx$

14. $\displaystyle\int \frac{\theta^2}{1-\sin\theta^3}\,d\theta$

15. $\displaystyle\int x^3 \ln x\,dx$

16. $\displaystyle\int \cot^3\theta\,d\theta$

17. $\displaystyle\int \frac{x^2}{(3x-5)^2}\,dx$

18. $\displaystyle\int \frac{1}{2x^2(2x-1)^2}\,dx$

19. $\displaystyle\int \frac{x}{1-\sec x^2}\,dx$

20. $\displaystyle\int \frac{e^x}{1-\tan e^x}\,dx$

21. $\displaystyle\int \frac{\cos x}{1+\sin^2 x}\,dx$

22. $\displaystyle\int \frac{1}{t[1+(\ln t)^2]}\,dt$

23. $\displaystyle\int \frac{1}{1+e^{2x}}\,dx$

24. $\displaystyle\int \frac{1}{\sqrt{x}(1+2\sqrt{x})}\,dx$

25. $\displaystyle\int \frac{\cos\theta}{3+2\sin\theta+\sin^2\theta}\,d\theta$

26. $\displaystyle\int x^2\sqrt{2+9x^2}\,dx$

27. $\displaystyle\int \frac{1}{x^2\sqrt{2+9x^2}}\,dx$

28. $\displaystyle\int \sqrt{3+x^2}\,dx$

29. $\displaystyle\int e^x\sqrt{1+e^{2x}}\,dx$

30. $\displaystyle\int \frac{1}{\sqrt{x}(x-4)^{3/2}}\,dx$

31. $\displaystyle\int \sin^4 2x\,dx$

32. $\displaystyle\int \frac{\cos^3\sqrt{x}}{\sqrt{x}}\,dx$

33. $\displaystyle\int \frac{1}{\sqrt{x}(1-\cos\sqrt{x})}\,dx$

34. $\displaystyle\int \frac{1}{1-\tan 5x}\,dx$

35. $\displaystyle\int t^4 \cos t\,dt$

36. $\displaystyle\int \sqrt{x}\arctan x^{3/2}\,dx$

37. $\displaystyle\int x\,\mathrm{arcsec}\,(x^2+1)\,dx$

38. $\displaystyle\int (\ln x)^3\,dx$

39. $\displaystyle\int \frac{\ln x}{x(3+2\ln x)}\,dx$

40. $\displaystyle\int \frac{e^x}{(1-e^{2x})^{3/2}}\,dx$

41. $\displaystyle\int \frac{\sqrt{2-2x-x^2}}{x+1}\,dx$

42. $\displaystyle\int \frac{1}{(x^2-6x+10)^2}\,dx$

43. $\displaystyle\int \frac{x}{x^4-6x^2+10}\,dx$

44. $\displaystyle\int (2x-3)^2\sqrt{(2x-3)^2+4}\,dx$

45. $\displaystyle\int \frac{x}{\sqrt{x^4-6x^2+5}}\,dx$

46. $\displaystyle\int \frac{\cos x}{\sqrt{\sin^2 x+1}}\,dx$

47. $\displaystyle\int \frac{x^3}{\sqrt{4-x^2}}\,dx$

48. $\displaystyle\int \sqrt{\frac{3-x}{3+x}}\,dx$

49. $\displaystyle\int \frac{1}{x^{3/2}\sqrt{1-x}}\,dx$

50. $\displaystyle\int x\sqrt{x^2+2x}\,dx$

51. $\displaystyle\int \frac{e^{3x}}{(1+e^x)^3}\,dx$

52. $\displaystyle\int \sec^5\theta\,d\theta$

In Exercises 53–58, verify the given integration formula.

53. $\displaystyle\int \frac{u^2}{(a+bu)^2}\,du =$
$$\frac{1}{b^3}\left(bu-\frac{a^2}{a+bu}-2a\ln|a+bu|\right)+C$$

54. $\displaystyle\int \frac{u^n}{\sqrt{a+bu}}\,du =$
$$\frac{2}{(2n+1)b}\left(u^n\sqrt{a+bu}-na\int\frac{u^{n-1}}{\sqrt{a+bu}}\,du\right)$$

55. $\displaystyle\int \frac{1}{(u^2\pm a^2)^{3/2}}\,du = \frac{\pm u}{a^2\sqrt{u^2\pm a^2}}+C$

56. $\displaystyle\int u^n\cos u\,du = u^n\sin u - n\int u^{n-1}\sin u\,du$

57. $\displaystyle\int \arctan u\,du = u\arctan u - \ln\sqrt{1+u^2}+C$

58. $\displaystyle\int (\ln u)^n\,du = u(\ln u)^n - n\int(\ln u)^{n-1}\,du$

In Exercises 59–68, evaluate the given integral.

59. $\displaystyle\int \frac{1}{2-3\sin\theta}\,d\theta$

60. $\displaystyle\int \frac{\sin\theta}{1+\cos^2\theta}\,d\theta$

61. $\displaystyle\int_0^{\pi/2} \frac{1}{1+\sin\theta+\cos\theta}\,d\theta$

62. $\displaystyle\int_0^{\pi/2} \frac{1}{3-2\cos\theta}\,d\theta$

63. $\displaystyle\int \frac{\sin\theta}{3-2\cos\theta}\,d\theta$

64. $\displaystyle\int \frac{\sin\theta}{1+\sin\theta}\,d\theta$

65. $\displaystyle\int \frac{\cos\sqrt{\theta}}{\sqrt{\theta}}\,d\theta$

66. $\displaystyle\int \frac{\sin\theta}{(\cos\theta)(1+\sin\theta)}\,d\theta$

67. $\displaystyle\int \frac{1}{\sin\theta\tan\theta}\,d\theta$

68. $\displaystyle\int \frac{1}{\sec\theta-\tan\theta}\,d\theta$

69. A hydraulic cylinder on an industrial machine pushes a steel block a distance of x feet $(0\le x\le 5)$, where the variable force required is
$$F(x)=2000xe^{-x}\text{ lb}$$
Find the work done in pushing the block the full 5 feet through the machine.

70. Repeat Exercise 69, using a force of
$$F(x)=\frac{500x}{\sqrt{26-x^2}}\text{ lb}$$

9.7
Improper integrals

The definition of the definite integral $\int_a^b f(x)\, dx$ requires that the interval $[a, b]$ be finite and that f be bounded on $[a, b]$. Furthermore, the Fundamental Theorem of Calculus, by which we have been evaluating definite integrals, requires that f be continuous on $[a, b]$. In this section we discuss a limit procedure for evaluating integrals that do not satisfy these requirements because either (1) one or both of the limits of integration are infinite or (2) f has a finite number of infinite discontinuities on the interval $[a, b]$. Integrals that possess either of these properties are called **improper integrals.**

EXAMPLE 1   Examples of improper integrals

(a) The integrals

$$\int_1^\infty \frac{dx}{x} \qquad \text{and} \qquad \int_{-\infty}^\infty \frac{dx}{x^2 + 1}$$

are improper because one or both of the limits of integration are infinite.

(b) The integrals

$$\int_1^5 \frac{dx}{\sqrt{x - 1}} \qquad \text{and} \qquad \int_{-2}^2 \frac{dx}{(x + 1)^2}$$

are improper because the integrands have infinite discontinuities somewhere in the interval of integration. $1/\sqrt{x - 1}$ has an infinite discontinuity at $x = 1$, and $1/(x + 1)^2$ has an infinite discontinuity at $x = -1$.

To get an idea of how we can evaluate an improper integral, consider the integral

$$\int_1^b \frac{dx}{x^2} = \left[-\frac{1}{x} \right]_1^b$$

$$= -\frac{1}{b} + 1$$

$$= 1 - \frac{1}{b}$$

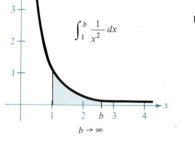

$$\int_1^b \frac{1}{x^2}\, dx$$

$b \to \infty$

FIGURE 9.19

which we can interpret as the area of the shaded region shown in Figure 9.19. Taking the limit as $b \to \infty$, we have

$$\int_1^\infty \frac{dx}{x^2} = \lim_{b \to \infty} \left[\int_1^b \frac{dx}{x^2} \right]$$

$$= \lim_{b \to \infty} \left(1 - \frac{1}{b} \right) = 1$$

and we can interpret this improper integral as the area of the *unbounded* region between the graph of $f(x) = 1/x^2$ and the x-axis (to the right of $x = 1$). We define improper integrals with infinite limits of integration as follows.

DEFINITION OF IMPROPER INTEGRAL WITH INFINITE LIMITS OF INTEGRATION	1. If f is continuous on the interval $[a, \infty)$, then $$\int_a^\infty f(x)\, dx = \lim_{b \to \infty} \int_a^b f(x)\, dx$$ 2. If f is continuous on the interval $(-\infty, b]$, then $$\int_{-\infty}^b f(x)\, dx = \lim_{a \to -\infty} \int_a^b f(x)\, dx$$ 3. If f is continuous on the interval $(-\infty, \infty)$, then $$\int_{-\infty}^\infty f(x)\, dx = \int_{-\infty}^c f(x)\, dx + \int_c^\infty f(x)\, dx$$ where c is any real number.

Remark In each case, if the limit exists, then the improper integral is said to **converge;** otherwise, the improper integral **diverges.** This means that in the third case the integral will diverge if either one of the integrals on the right diverges.

EXAMPLE 2   An improper integral that diverges

Evaluate

$$\int_1^\infty \frac{dx}{x}$$

Solution:

$$\int_1^\infty \frac{dx}{x} = \lim_{b \to \infty} \int_1^b \frac{dx}{x} = \lim_{b \to \infty} \left[\ln x\right]_1^b = \lim_{b \to \infty} (\ln b - 0) = \infty$$

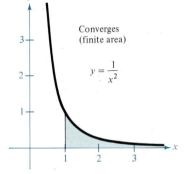

FIGURE 9.20

A comparison of the improper integral $\int_1^\infty (1/x^2)\, dx$, which converges to 1, and $\int_1^\infty (1/x)\, dx$, which diverges, suggests the rather unpredictable nature of improper integrals. The functions $f(x) = 1/x^2$ and $g(x) = 1/x$ have similar graphs, as shown in Figures 9.20 and 9.21. Yet, the shaded region shown in Figure 9.20 has a finite area, whereas the shaded region shown in Figure 9.21 has an infinite area.

EXAMPLE 3   Evaluating improper integrals

(a) $\displaystyle\int_0^\infty e^{-x}\, dx = \lim_{b \to \infty} \int_0^b e^{-x}\, dx = \lim_{b \to \infty} \left[-e^{-x}\right]_0^b = \lim_{b \to \infty} (-e^{-b} + 1) = 1$

(b) $\displaystyle\int_0^\infty \frac{1}{x^2 + 1}\, dx = \lim_{b \to \infty} \int_0^b \frac{1}{x^2 + 1}\, dx = \lim_{b \to \infty} \left[\arctan x\right]_0^b$

$$= \lim_{b \to \infty} \arctan b = \frac{\pi}{2}$$

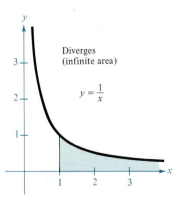

FIGURE 9.21

EXAMPLE 4 *An improper integral that diverges*

Evaluate

$$\int_0^\infty \sin x \, dx$$

Solution:

$$\int_0^\infty \sin x \, dx = \lim_{b \to \infty} \int_0^b \sin x \, dx = \lim_{b \to \infty} \left[-\cos x \right]_0^b = \lim_{b \to \infty} (1 - \cos b)$$

Since $\cos b$ does not approach a limit as b approaches infinity, we conclude that the given improper integral diverges.

In the following example, we illustrate a technique for using L'Hôpital's Rule to evaluate an improper integral.

EXAMPLE 5 *Using L'Hôpital's Rule with an improper integral*

Evaluate

$$\int_1^\infty (1 - x)e^{-x} \, dx$$

Solution: Using integration by parts, with $dv = e^{-x} \, dx$ and $u = (1 - x)$, we have

$$\int u \, dv = uv \quad - \quad \int v \, du$$

$$\int (1 - x)e^{-x} \, dx = -e^{-x}(1 - x) - \int e^{-x} \, dx$$
$$= -e^{-x} + xe^{-x} + e^{-x} + C = xe^{-x} + C$$

Therefore, we have

$$\int_1^\infty (1 - x)e^{-x} \, dx = \lim_{b \to \infty} \left[xe^{-x} \right]_1^b = \left(\lim_{b \to \infty} \frac{b}{e^b} \right) - \frac{1}{e}$$

Now, using L'Hôpital's Rule on the right-hand limit, we have

$$\lim_{b \to \infty} \frac{b}{e^b} = \lim_{b \to \infty} \frac{1}{e^b} = 0$$

Finally, we conclude that

$$\int_1^\infty (1 - x)e^{-x} \, dx = -\frac{1}{e}$$

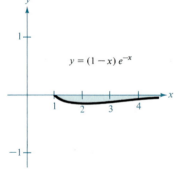

$y = (1 - x)\,e^{-x}$

FIGURE 9.22   See Figure 9.22.

The definition of the improper integral $\int_{-\infty}^\infty f(x) \, dx$ requires the evaluation of two limits—one for the upper limit of integration and one for the lower limit of integration. However, in practice it is often possible to combine these two limits, as indicated in the following example.

Evaluate

$$\int_{-\infty}^{\infty} \frac{e^x}{1 + e^{2x}} \, dx$$

Solution:

$$\int_{-\infty}^{\infty} \frac{e^x}{1 + e^{2x}} \, dx = \lim_{b \to \infty} \left[\arctan e^x \right]_{-b}^{b}$$

$$= \lim_{b \to \infty} [\arctan e^b - \arctan e^{-b}]$$

$$= \frac{\pi}{2} - 0 = \frac{\pi}{2}$$

See Figure 9.23.

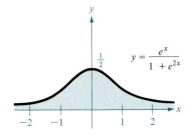

FIGURE 9.23

EXAMPLE 7 An application involving an improper integral

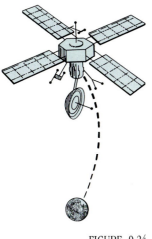

FIGURE 9.24

In Example 3 of Section 6.4, we determined that it would take 10,000 mile-tons of work to propel a 15-ton space module to a height of 800 miles above the earth. How much work is required to propel this same module an unlimited distance away from the earth's surface? (See Figure 9.24.)

Solution: At first you might think that an infinite amount of work would be required. But if this were the case, then it would be impossible to send rockets into outer space. Since this has been done, the work required must be finite, and we can determine the work in the following manner. Using the integral of Example 3, Section 6.4, we replace the upper bound of 4800 miles by ∞ and write

$$W = \int_{4000}^{\infty} \frac{240,000,000}{x^2} \, dx$$

$$= \lim_{b \to \infty} \left[-\frac{240,000,000}{x} \right]_{4000}^{b}$$

$$= 60,000 \text{ mile-tons}$$

The second basic type of improper integral is one that has an infinite discontinuity *at or between* the limits of integration.

DEFINITION OF IMPROPER
INTEGRAL WITH AN
INFINITE DISCONTINUITY

1. If f is continuous on $[a, b)$ and has an infinite discontinuity at b, then

$$\int_a^b f(x)\,dx = \lim_{c \to b^-} \int_a^c f(x)\,dx$$

2. If f is continuous on $(a, b]$ and has an infinite discontinuity at a, then

$$\int_a^b f(x)\,dx = \lim_{c \to a^+} \int_c^b f(x)\,dx$$

3. If f is continuous on $[a, b]$, except for some c in (a, b) at which f has an infinite discontinuity, then

$$\int_a^b f(x)\,dx = \int_a^c f(x)\,dx + \int_c^b f(x)\,dx$$

provided *both* intervals on the right converge. If either integral on the right diverges, we say that the integral on the left diverges.

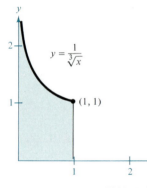

FIGURE 9.25

EXAMPLE 8   An improper integral with an infinite discontinuity

Evaluate

$$\int_0^1 \frac{dx}{\sqrt[3]{x}}$$

Solution: The integrand has an infinite discontinuity at $x = 0$, as shown in Figure 9.25. To evaluate this integral, we write

$$\int_0^1 x^{-1/3}\,dx = \lim_{b \to 0^+} \left[\frac{x^{2/3}}{2/3} \right]_b^1$$

$$= \lim_{b \to 0^+} \frac{3}{2}(1 - b^{2/3})$$

$$= \frac{3}{2}$$

EXAMPLE 9   An improper integral that diverges

Evaluate

$$\int_0^2 \frac{dx}{x^3}$$

Solution: Since the integrand has an infinite discontinuity at $x = 0$, we write

$$\int_0^2 \frac{dx}{x^3} = \lim_{b \to 0^+} \left[-\frac{1}{2x^2} \right]_b^2$$

$$= \lim_{b \to 0^+} \left[-\frac{1}{8} + \frac{1}{2b^2} \right]$$

$$= \infty$$

Thus, we conclude that this improper integral diverges.

EXAMPLE 10   An improper integral with an interior discontinuity

Evaluate

$$\int_{-1}^{2} \frac{dx}{x^3}$$

Solution: This integral is improper because the integrand has an infinite discontinuity at the interior point $x = 0$, as shown in Figure 9.26. Thus, we write

$$\int_{-1}^{2} \frac{dx}{x^3} = \int_{-1}^{0} \frac{dx}{x^3} + \int_{0}^{2} \frac{dx}{x^3}$$

From Example 9 we know that the second integral diverges. Therefore, the original improper integral also diverges.

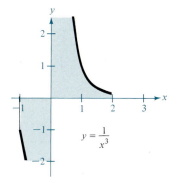

FIGURE 9.26

| Remark Remember that you must check for infinite discontinuities at interior points as well as endpoints when determining whether or not an integral is improper. For instance, if we had not recognized that the integral in Example 10 is improper, we would have obtained the *incorrect* result

$$\int_{-1}^{2} \frac{dx}{x^3} = \left[\frac{-1}{2x^2} \right]_{-1}^{2} = -\frac{1}{8} + \frac{1}{2} = \frac{3}{8} \qquad \text{Incorrect evaluation}$$

The integral in the next example is improper for *two* reasons. One limit of integration is infinite, and the other has an infinite discontinuity, as shown in Figure 9.27.

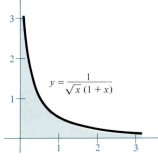

FIGURE 9.27

EXAMPLE 11   A doubly improper integral

Evaluate

$$\int_{0}^{\infty} \frac{dx}{\sqrt{x}(x+1)}$$

Solution: To evaluate this integral, we split it at a convenient point (say $x = 1$) and write

$$\int_{0}^{\infty} \frac{dx}{\sqrt{x}(x+1)} = \int_{0}^{1} \frac{dx}{\sqrt{x}(x+1)} + \int_{1}^{\infty} \frac{dx}{\sqrt{x}(x+1)}$$

$$= \lim_{b \to 0^{+}} \left[2 \arctan \sqrt{x} \right]_{b}^{1} + \lim_{b \to \infty} \left[2 \arctan \sqrt{x} \right]_{1}^{b}$$

$$= 2\left(\frac{\pi}{4} \right) - 0 + 2\left(\frac{\pi}{2} \right) - 2\left(\frac{\pi}{4} \right) = \pi$$

EXAMPLE 12   An improper integral involving L'Hôpital's Rule

Evaluate

$$\int_0^1 \ln x \, dx$$

Solution: Using integration by parts, with $dv = dx$ and $u = \ln x$, we have

$$\int_0^1 \ln x \, dx = \lim_{b \to 0^+} \left[x \ln x - x \right]_b^1 = \lim_{b \to 0^+} [0 - 1 - b \ln b + b]$$

By L'Hôpital's Rule, we have

$$\lim_{b \to 0^+} b \ln b = \lim_{b \to 0^+} \frac{\ln b}{1/b} = \lim_{b \to 0^+} \frac{1/b}{-1/b^2} = \lim_{b \to 0^+} -b = 0$$

Finally, we conclude that

$$\int_0^1 \ln x \, dx = \lim_{b \to 0^+} [0 - 1 - b \ln b + b] = 0 - 1 - 0 + 0 = -1$$

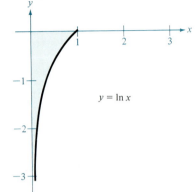

$y = \ln x$

FIGURE 9.28   See Figure 9.28.

EXAMPLE 13   An application involving arc length

Use the formula for arc length to show that the circumference of the circle $x^2 + y^2 = 1$ is 2π.

Solution: To simplify the work, we consider the quarter circle given by $y = \sqrt{1 - x^2}$, where $0 \le x \le 1$. The arc length of this quarter circle is given by

$$s = \int_0^1 \sqrt{1 + (y')^2} \, dx = \int_0^1 \sqrt{1 + \left(\frac{-x}{\sqrt{1 - x^2}} \right)^2} \, dx = \int_0^1 \frac{dx}{\sqrt{1 - x^2}}$$

This integral is improper, since it has an infinite discontinuity at $x = 1$. Thus, we write

$$s = \int_0^1 \frac{dx}{\sqrt{1 - x^2}} = \lim_{b \to 1^-} \left[\arcsin x \right]_0^b = \frac{\pi}{2} - 0 = \frac{\pi}{2}$$

Finally, multiplying by 4, we conclude that the circumference of the circle is $4s = 2\pi$, as shown in Figure 9.29.

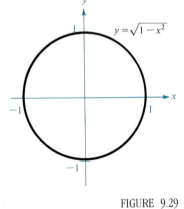

$y = \sqrt{1 - x^2}$

FIGURE 9.29

We conclude this section with a useful theorem describing the convergence (or divergence) of a common type of improper integral. The proof of this theorem is left as an exercise.

THEOREM 9.3   A SPECIAL TYPE OF IMPROPER INTEGRAL

$$\int_1^\infty \frac{dx}{x^p} = \begin{cases} \dfrac{1}{p - 1}, & \text{if } p > 1 \\[2mm] \text{diverges}, & \text{if } p \le 1 \end{cases}$$

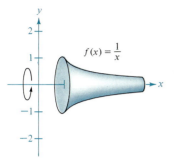

FIGURE 9.30

EXAMPLE 14 *An application involving a solid of revolution*

The solid formed by revolving the *unbounded* region lying between the graph of $f(x) = 1/x$ and the *x*-axis ($x \geq 1$) is called **Gabriel's Horn.** (See Figure 9.30.) Show that this solid has a finite volume and an infinite surface area.

Solution: Using the disc method, with Theorem 9.3, we determine the volume to be

$$V = \pi \int_1^\infty \left(\frac{1}{x}\right)^2 dx$$

$$= \pi\left(\frac{1}{2-1}\right) = \pi$$

The surface area is given by

$$S = 2\pi \int_1^\infty f(x)\sqrt{1 + [f'(x)]^2}\, dx$$

$$= 2\pi \int_1^\infty \frac{1}{x}\sqrt{1 + \frac{1}{x^4}}\, dx$$

$$= 2\pi \int_1^\infty \frac{\sqrt{1+x^4}}{x^3}\, dx$$

Now, using the substitution $u = x^2$ together with Formula 30 in the integration table, we have

$$S = \lim_{b\to\infty} \pi \left[\frac{-\sqrt{1+x^4}}{x^2} + \ln\left|x^2 + \sqrt{1+x^4}\right|\right]_1^b$$

We leave it up to you to show that this limit is infinite.

Section Exercises 9.7

In Exercises 1–32, determine the divergence or convergence of the given improper integral. Evaluate the integrals that converge.

1. $\int_0^4 \frac{1}{\sqrt{x}}\, dx$

2. $\int_3^4 \frac{1}{\sqrt{x-3}}\, dx$

3. $\int_0^2 \frac{1}{(x-1)^{2/3}}\, dx$

4. $\int_0^2 \frac{1}{(x-1)^2}\, dx$

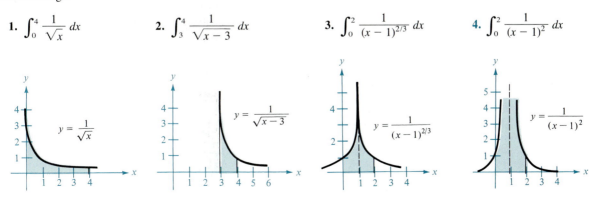

5. $\int_0^\infty e^{-x}\,dx$

6. $\int_{-\infty}^0 e^{2x}\,dx$

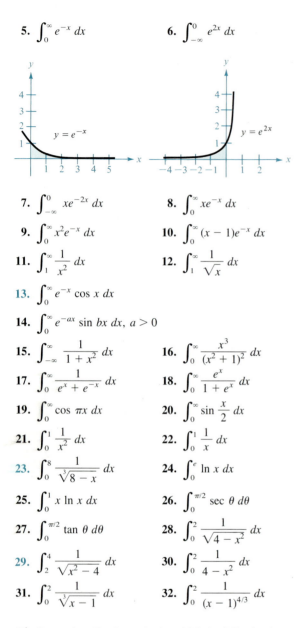

7. $\int_{-\infty}^0 xe^{-2x}\,dx$

8. $\int_0^\infty xe^{-x}\,dx$

9. $\int_0^\infty x^2 e^{-x}\,dx$

10. $\int_0^\infty (x-1)e^{-x}\,dx$

11. $\int_1^\infty \frac{1}{x^2}\,dx$

12. $\int_1^\infty \frac{1}{\sqrt{x}}\,dx$

13. $\int_0^\infty e^{-x}\cos x\,dx$

14. $\int_0^\infty e^{-ax}\sin bx\,dx,\; a>0$

15. $\int_{-\infty}^\infty \frac{1}{1+x^2}\,dx$

16. $\int_0^\infty \frac{x^3}{(x^2+1)^2}\,dx$

17. $\int_0^\infty \frac{1}{e^x+e^{-x}}\,dx$

18. $\int_0^\infty \frac{e^x}{1+e^x}\,dx$

19. $\int_0^\infty \cos \pi x\,dx$

20. $\int_0^\infty \sin \frac{x}{2}\,dx$

21. $\int_0^1 \frac{1}{x^2}\,dx$

22. $\int_0^1 \frac{1}{x}\,dx$

23. $\int_0^8 \frac{1}{\sqrt[3]{8-x}}\,dx$

24. $\int_0^e \ln x\,dx$

25. $\int_0^1 x \ln x\,dx$

26. $\int_0^{\pi/2} \sec \theta\,d\theta$

27. $\int_0^{\pi/2} \tan \theta\,d\theta$

28. $\int_0^2 \frac{1}{\sqrt{4-x^2}}\,dx$

29. $\int_2^4 \frac{1}{\sqrt{x^2-4}}\,dx$

30. $\int_0^2 \frac{1}{4-x^2}\,dx$

31. $\int_0^2 \frac{1}{\sqrt[3]{x-1}}\,dx$

32. $\int_0^2 \frac{1}{(x-1)^{4/3}}\,dx$

33. Determine all values of p for which the following integral converges:
$$\int_1^\infty \frac{1}{x^p}\,dx$$

34. Given continuous functions f and g such that $0 \le f(x) \le g(x)$ on the interval (a, ∞), prove the following.
(a) If $\int_a^\infty g(x)\,dx$ converges, then $\int_a^\infty f(x)\,dx$ converges.
(b) If $\int_a^\infty f(x)\,dx$ diverges, then $\int_a^\infty g(x)\,dx$ diverges.

In Exercises 35–40, use the results of Exercises 33 and 34 to determine whether or not the improper integral converges.

35. $\int_1^\infty \frac{1}{x^2+5}\,dx$

36. $\int_2^\infty \frac{1}{\sqrt{x-1}}\,dx$

37. $\int_2^\infty \frac{1}{\sqrt[3]{x(x-1)}}\,dx$

38. $\int_1^\infty \frac{1}{\sqrt{x}(x+1)}\,dx$

39. $\int_0^\infty e^{-x^2}\,dx$

40. $\int_2^\infty \frac{1}{\sqrt{x}\ln x}\,dx$

41. Given the region bounded by the graphs of $y = 1/x^2$ and $y = 0$ ($x \ge 1$), find the following:
(a) the area of the region
(b) the volume of the solid generated by revolving the region about the x-axis
(c) the volume of the solid generated by revolving the region about the y-axis

42. Given the region bounded by the graphs of $y = e^{-x}$ and $y = 0$ ($x \ge 0$), find the following:
(a) the area of the region
(b) the volume of the solid generated by revolving the region about the y-axis

43. Sketch the graph of the hypocycloid of four cusps
$$x^{2/3} + y^{2/3} = 1$$
and find its perimeter.

44. The region bounded by
$$(x-2)^2 + y^2 = 1$$
is revolved about the y-axis to form a torus. Find the surface area of this torus.

45. The **Gamma Function** $\Gamma(n)$ is defined by
$$\Gamma(n) = \int_0^\infty x^{n-1}e^{-x}\,dx, \quad n > 0$$
Evaluate this function at $n = 1$, 2, and 3. Then use integration by parts to show that $\Gamma(n+1) = n\Gamma(n)$.

46. A 5-ton rocket is fired from the surface of the earth into outer space. How much work is required to overcome the earth's gravitational force? How far has the rocket traveled when half the total work has occurred?

In Exercises 47 and 48, use the following definitions. A nonnegative function f is called a **probability density function** if
$$\int_{-\infty}^\infty f(t)\,dt = 1$$

The **probability** that t lies between a and b is given by
$$P(a \le t \le b) = \int_a^b f(t)\,dt$$

The **expected value** of t is given by
$$E(t) = \int_{-\infty}^\infty tf(t)\,dt$$

47. For
$$f(t) = \begin{cases} \frac{1}{7}e^{-t/7}, & t \geq 0 \\ 0, & t < 0 \end{cases}$$

(a) Show that f is a probability density function.
(b) Find $P(0 \leq t \leq 5)$.
(c) Find $E(t)$.

48. For
$$f(t) = \begin{cases} \frac{2}{5}e^{-2t/5}, & t \geq 0 \\ 0, & t < 0 \end{cases}$$

(a) Show that f is a probability density function.
(b) Find $P(0 \leq t \leq 3)$.
(c) Find $E(t)$.

In Exercises 49 and 50, find the capitalized cost C of an asset (a) for $n = 5$ years, (b) for $n = 10$ years, and (c) forever. The **capitalized cost** is given by

$$C = C_0 + \int_0^n c(t)e^{-rt}\, dt$$

where C_0 is the original investment, t is the time in years, r is the annual rate compounded continuously, and $c(t)$ is annual cost of maintenance.

49. $C_0 = \$650,000$, $c(t) = \$25,000$, $r = 0.12$
50. $C_0 = \$650,000$, $c(t) = \$25,000(1 + 0.08t)$, $r = 0.12$

Review Exercises for Chapter 9

In Exercises 1–40, evaluate the given integral

1. $\displaystyle\int \frac{x^2}{x^2 + 2x - 15}\, dx$

2. $\displaystyle\int \frac{\sqrt{x^2 - 9}}{x}\, dx$

3. $\displaystyle\int \frac{1}{1 - \sin \theta}\, d\theta$

4. $\displaystyle\int x^2 \sin 2x\, dx$

5. $\displaystyle\int e^{2x} \sin 3x\, dx$

6. $\displaystyle\int (x^2 - 1)e^x\, dx$

7. $\displaystyle\int \frac{\ln (2x)}{x^2}\, dx$

8. $\displaystyle\int 2x\sqrt{2x - 3}\, dx$

9. $\displaystyle\int \sqrt{4 - x^2}\, dx$

10. $\displaystyle\int \frac{\sqrt{4 - x^2}}{2x}\, dx$

11. $\displaystyle\int \frac{-12}{x^2\sqrt{4 - x^2}}\, dx$

12. $\displaystyle\int \tan \theta \sec^4 \theta\, d\theta$

13. $\displaystyle\int \sec^4 \frac{x}{2}\, dx$

14. $\displaystyle\int \sec \theta \cos 2\theta\, d\theta$

15. $\displaystyle\int \frac{9}{x^2 - 9}\, dx$

16. $\displaystyle\int \frac{\sec^2 \theta}{\tan \theta (\tan \theta - 1)}\, d\theta$

17. $\displaystyle\int \frac{x^2 + 2x}{x^3 - x^2 + x - 1}\, dx$

18. $\displaystyle\int \frac{4x - 2}{3(x - 1)^2}\, dx$

19. $\displaystyle\int \frac{3x^3 + 4x}{(x^2 + 1)^2}\, dx$

20. $\displaystyle\int \sqrt{\frac{x - 2}{x + 2}}\, dx$

21. $\displaystyle\int \frac{16}{\sqrt{16 - x^2}}\, dx$

22. $\displaystyle\int \frac{\sin \theta}{1 + 2\cos^2 \theta}\, d\theta$

23. $\displaystyle\int \frac{e^x}{4 + e^{2x}}\, dx$

24. $\displaystyle\int \frac{x}{x^2 - 4x + 8}\, dx$

25. $\displaystyle\int \frac{x}{x^2 + 4x + 8}\, dx$

26. $\displaystyle\int \frac{3}{2x\sqrt{9x^2 - 1}}\, dx$

27. $\displaystyle\int \theta \sin \theta \cos \theta\, d\theta$

28. $\displaystyle\int \frac{\csc \sqrt{2x}}{\sqrt{x}}\, dx$

29. $\displaystyle\int (\sin \theta + \cos \theta)^2\, d\theta$

30. $\displaystyle\int \cos 2\theta (\sin \theta + \cos \theta)^2\, d\theta$

31. $\displaystyle\int \frac{x^{1/4}}{1 + x^{1/2}}\, dx$

32. $\displaystyle\int \sqrt{1 - \cos x}\, dx$

33. $\displaystyle\int \sqrt{1 + \cos x}\, dx$

34. $\displaystyle\int \ln \sqrt{x^2 - 1}\, dx$

35. $\displaystyle\int \ln (x^2 + x)\, dx$

36. $\displaystyle\int x \arcsin 2x\, dx$

37. $\displaystyle\int \cos x \ln (\sin x)\, dx$

38. $\displaystyle\int e^x \arctan e^x\, dx$

39. $\displaystyle\int \frac{x^4 + 2x^2 + x + 1}{(x^2 + 1)^2}\, dx$

40. $\displaystyle\int \sqrt{1 + \sqrt{x}}\, dx$

In Exercises 41–44, evaluate the given integral using the indicated methods.

41. $\displaystyle\int \frac{1}{x^2\sqrt{4 + x^2}}\, dx$
(a) trigonometric substitution
(b) substitution: $x = 2/u$

42. $\displaystyle\int \frac{1}{x\sqrt{4 + x^2}}\, dx$
(a) trigonometric substitution
(b) substitution: $u^2 = 4 + x^2$

43. $\displaystyle\int \frac{x^3}{\sqrt{4 + x^2}}\, dx$
(a) trigonometric substitution
(b) substitution: $u^2 = 4 + x^2$
(c) by parts: $dv = (x/\sqrt{4 + x^2})\, dx$

44. $\int x\sqrt{4+x}\,dx$

(a) trigonometric substitution
(b) substitution: $u^2 = 4 + x$
(c) substitution: $u = 4 + x$
(d) by parts: $dv = \sqrt{4+x}\,dx$

In Exercises 45–52, evaluate each definite integral to two-decimal-place accuracy.

45. $\displaystyle\int_0^{\pi/2} \frac{1}{2 - \cos\theta}\,d\theta$ 46. $\displaystyle\int_0^1 \frac{x^{3/2}}{2 - x^2}\,dx$

47. $\displaystyle\int_0^2 \frac{1}{\sqrt{1 + x^3}}\,dx$ 48. $\displaystyle\int_0^{\pi/4} \theta\tan\theta\,d\theta$

49. $\displaystyle\int_1^2 \frac{1}{1 + \ln x}\,dx$ 50. $\displaystyle\int_0^2 \sqrt{1 + x^3}\,dx$

51. (a) $\displaystyle\int_0^1 e^x\,dx$ (b) $\displaystyle\int_0^1 xe^x\,dx$

(c) $\displaystyle\int_0^1 xe^{x^2}\,dx$ (d) $\displaystyle\int_0^1 e^{x^2}\,dx$

52. (a) $\displaystyle\int_0^{\pi/2} \cos x\,dx$ (b) $\displaystyle\int_0^{\pi/2} \cos^2 x\,dx$

(c) $\displaystyle\int_0^{\pi/2} \cos x^2\,dx$ (d) $\displaystyle\int_0^{\pi/2} \cos\sqrt{x}\,dx$

53. Approximate

$$\int_2^\infty \frac{1}{x^5 - 1}\,dx$$

using the inequality

$$\frac{1}{x^5} + \frac{1}{x^{10}} + \frac{1}{x^{15}} < \frac{1}{x^5 - 1} < \frac{1}{x^5} + \frac{1}{x^{10}} + \frac{2}{x^{15}}$$

54. Let

$$I_n = \int \frac{x^{2n-1}}{(x^2 + 1)^{n+3}}\,dx, \quad n \geq 1$$

Prove that

$$I_n = \int \left(\frac{n-1}{n+2}\right) I_{n-1}$$

and then evaluate the following:

(a) $\displaystyle\int_0^\infty \frac{x^3}{(x^2 + 1)^5}\,dx$

(b) $\displaystyle\int_0^\infty \frac{x^5}{(x^2 + 1)^6}\,dx$

(c) $\displaystyle\int_0^\infty \frac{x^7}{(x^2 + 1)^7}\,dx$

10 Infinite series

10.1
Introduction: Taylor polynomials and approximations

At this point it might appear that our discussion of the elementary functions of a single variable is complete. We have learned to sketch graphs, evaluate limits, differentiate, integrate, and use elementary functions in a variety of practical applications. However, a gap remains in our study—we do not yet have a way to *evaluate* many elementary functions. Polynomial functions can be evaluated by simple arithmetic operations, but evaluation of functions such as $f(x) = \sin x$ or $g(x) = \ln x$ is more complicated. We can use tables, a calculator, or a computer. But how does a computer calculate these values and how are the table values determined?

Our goal in this section is to show how special types of polynomials can be used as approximations for other elementary functions. To simplify some of the formulas we will develop, we use the symbol $n!$ (read "n factorial"). If n is a positive integer, then ***n* factorial** is defined as

$$n! = 1 \cdot 2 \cdot 3 \cdot 4 \cdots (n - 1) \cdot n$$

For the sake of writing formulas concisely, we make a special definition for $n = 0$ and define $0! = 1$. Thus, we have

$$0! = 1 \qquad\qquad 3! = 1 \cdot 2 \cdot 3 = 6$$
$$1! = 1 \qquad\qquad 4! = 1 \cdot 2 \cdot 3 \cdot 4 = 24$$
$$2! = 1 \cdot 2 = 2$$

| **Remark** Factorials follow the same conventions for order of operation as exponents. That is, just as $2x^3$ and $(2x)^3$ imply different orders of operations, $2n!$ and $(2n)!$ imply the following orders:

$$2n! = 2(n!) = 2(1 \cdot 2 \cdot 3 \cdot 4 \cdots n)$$

and

$$(2n)! = 1 \cdot 2 \cdot 3 \cdot 4 \cdots n \cdot (n + 1) \cdots 2n$$

525

Polynomial approximations

To find a polynomial function P that approximates another function f, we begin by choosing a number c in the domain of f at which we require that f and P have the same functional value. That is,

$$P(c) = f(c) \qquad \text{Graphs of } f \text{ and } P \text{ pass through } (c, f(c))$$

We say that the approximating polynomial is **expanded about c or centered at c**. Geometrically, the requirement that $P(c) = f(c)$ means that the graph of p passes through the point $(c, f(c))$. Of course, there are many polynomials whose graph passes through the point $(c, f(c))$. Our task is to find a polynomial whose graph resembles the graph of f near this point. One way to do this is to impose the additional requirement that the slope of the polynomial function be the same as the slope of the graph of f at the point $(c, f(c))$. That is, we require that

$$P'(c) = f'(c) \qquad \text{Graphs of } f \text{ and } P \text{ have same slope at } (c, f(c))$$

With these two requirements, we can obtain a simple linear approximation of f—using a first-degree polynomial function, as shown in Figure 10.1. This procedure is demonstrated in Example 1.

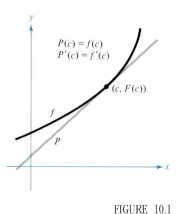

$P(c) = f(c)$
$P'(c) = f'(c)$

$(c, F(c))$

f

P

FIGURE 10.1

EXAMPLE 1   First-degree polynomial approximation

For the function $f(x) = e^x$, find a first-degree polynomial function

$$P(x) = a_1 x + a_0$$

whose value and slope agree with the value and slope of f at $x = 0$.

Solution: Since $f(x) = e^x$ and $f'(x) = e^x$, the value and slope of f, at $x = 0$, are given by

$$f(0) = e^0 = 1 \qquad \text{and} \qquad f'(0) = e^0 = 1$$

Now, since $P(x) = a_1 x + a_0$ we can impose the condition that $P(0) = f(0)$ to conclude that $a_0 = 1$. Moreover, since $P'(x) = a_1$, we can use the condition that $P'(0) = f'(0)$ to conclude that $a_1 = 1$. Therefore, we have

$$P(x) = x + 1$$

Figure 10.2 shows the graphs of $P(x) = x + 1$ and $f(x) = e^x$.

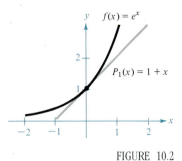

$f(x) = e^x$

$P_1(x) = 1 + x$

FIGURE 10.2

From Figure 10.2, we can see that at points near $(0, 1)$, the graph of $P(x) = x + 1$ is reasonably close to the graph of $f(x) = e^x$. However, as we move away from $(0, 1)$, the graphs move farther apart from each other and the approximation is not good. To improve the approximation, we can impose yet another requirement—that the value of the second derivatives of P and f agree when $x = 0$. The polynomial of least degree that satisfies all three requirements $P_2(0) = f(0)$, $P_2'(0) = f'(0)$, and $P_2''(0) = f''(0)$ is

$$P_2(x) = 1 + x + \frac{1}{2}x^2$$

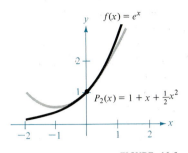

$f(x) = e^x$

$P_2(x) = 1 + x + \frac{1}{2}x^2$

FIGURE 10.3

From Figure 10.3, we can see that P_2 is a better approximation to f than P. If we continue this pattern, requiring

$$P_n(x) = a_0 + a_1 x + a_2 x^2 + a_3 x^3 + \cdots + a_n x^n$$

and that the first n derivatives of $P_n(x)$ match the value of $f(x) = e^x$ and its first n derivatives at $x = 0$, we obtain the following.

$$P_n(x) = 1 + x + \frac{1}{2}x^2 + \frac{1}{3!}x^3 + \cdots + \frac{1}{n!}x^n \approx e^x$$

For instance, in Table 10.1 we compare the values of $P_3(x)$ and e^x near $x = 0$. Note that the closer x is to 0, the better the approximation is.

TABLE 10.1 $P_3(x) = 1 + x + \frac{1}{2}x^2 + \frac{1}{3!}x^3 \approx e^x$

x	-1.0	-0.2	-0.1	0.0	0.1	0.2	1.0
e^x	0.367879	0.818731	0.904837	1.000000	1.105171	1.221403	2.718282
$P_3(x)$	0.333333	0.818667	0.904833	1.000000	1.105167	1.221333	2.666667

The polynomial approximation of $f(x) = e^x$ in Table 10.1 is expanded about $c = 0$. For expansions about an arbitrary value of c, it is convenient to write the polynomial in the form

$$P_n(x) = a_0 + a_1(x - c) + a_2(x - c)^2 + a_3(x - c)^3 + \cdots + a_n(x - c)^n$$

In this form, repeated differentiation produces

$$P_n'(x) = a_1 + 2a_2(x - c) + 3a_3(x - c)^2 + \cdots + na_n(x - c)^{n-1}$$
$$P_n''(x) = 2a_2 + 2 \cdot 3a_3(x - c) + \cdots + n(n - 1)a_n(x - c)^{n-2}$$
$$P_n'''(x) = 2 \cdot 3a_3 + \cdots + n(n - 1)(n - 2)a_n(x - c)^{n-3}$$
$$\vdots$$
$$P_n^{(n)}(x) = n(n - 1)(n - 2) \cdots (2)(1)a_n$$

Letting $x = c$, we then obtain

$$P_n(c) = a_0, \qquad P_n'(c) = a_1, \qquad P_n''(c) = 2a_2, \qquad \ldots, \qquad P_n^{(n)}(c) = n!a_n$$

and using the requirements that the value of f and its first n derivatives agree with the value of P_n and its first n derivatives at $x = c$, it follows that

$$f(c) = a_0, \qquad f'(c) = a_1, \qquad \frac{f''(c)}{2!} = a_2, \qquad \ldots, \qquad \frac{f^{(n)}(c)}{n!} = a_n$$

With these coefficients we obtain the following definition of **Taylor polynomials,** named after the English mathematician Brook Taylor (1685–1731). Although Taylor was not the first to seek polynomial approximations of transcendental functions, his published account in 1715 was the first comprehensive work on the subject.

Brook Taylor

DEFINITION OF nTH DEGREE TAYLOR POLYNOMIAL	If f has n derivatives at c, then the polynomial $$P_n(x) = f(c) + f'(c)(x - c) + \frac{f''(c)}{2!}(x - c)^2 + \cdots + \frac{f^{(n)}(c)}{n!}(x - c)^n$$ is called the **nth Taylor polynomial for f at c.**

Remark If $c = 0$, then
$$P_n(x) = f(0) + f'(0)x + \frac{f''(0)}{2!}x^2 + \frac{f'''(0)}{3!}x^3 + \cdots + \frac{f^{(n)}(0)}{n!}x^n$$
is called the **nth Maclaurin polynomial for f,** named after another English mathematician, Colin Maclaurin (1698–1746).

EXAMPLE 2   Finding Taylor polynomials for ln x

Find the Taylor polynomials P_0, P_1, P_2, P_3, and P_4 for $f(x) = \ln x$ at $c = 1$.

Solution: Expanding about $x = c = 1$, we have

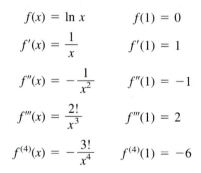

$$f(x) = \ln x \qquad\qquad f(1) = 0$$
$$f'(x) = \frac{1}{x} \qquad\qquad f'(1) = 1$$
$$f''(x) = -\frac{1}{x^2} \qquad\qquad f''(1) = -1$$
$$f'''(x) = \frac{2!}{x^3} \qquad\qquad f'''(1) = 2$$
$$f^{(4)}(x) = -\frac{3!}{x^4} \qquad\qquad f^{(4)}(1) = -6$$

Therefore,

$$P_0(x) = f(1) = 0$$
$$P_1(x) = P_0(x) + f'(1)(x - 1) = (x - 1)$$
$$P_2(x) = P_1(x) + \frac{f''(1)}{2!}(x - 1)^2 = (x - 1) - \frac{1}{2}(x - 1)^2$$
$$P_3(x) = P_2(x) + \frac{f'''(1)}{3!}(x - 1)^3 = (x - 1) - \frac{1}{2}(x - 1)^2 + \frac{1}{3}(x - 1)^3$$
$$P_4(x) = P_3(x) + \frac{f^{(4)}(1)}{4!}(x - 1)^4$$
$$= (x - 1) - \frac{1}{2}(x - 1)^2 + \frac{1}{3}(x - 1)^3 - \frac{1}{4}(x - 1)^4$$

FIGURE 10.4   (See Figure 10.4.)

EXAMPLE 3   Finding Maclaurin polynomials for cos x

Find the Maclaurin polynomials P_0, P_2, P_4, and P_6 for $f(x) = \cos x$. Use $P_6(x)$ to approximate the value of $\cos(0.1)$.

Solution: Expanding about $x = c = 0$, we have

$$
\begin{aligned}
f(x) &= \cos x & f(0) &= 1 \\
f'(x) &= -\sin x & f'(0) &= 0 \\
f''(x) &= -\cos x & f''(0) &= -1 \\
f'''(x) &= \sin x & f'''(0) &= 0
\end{aligned}
$$

Through repeated differentiation, we can see that the pattern 1, 0, -1, 0 continues, and we have

$$P_0(x) = f(0) = 1$$

$$P_2(x) = P_0(x) + f'(0)x + \frac{f''(0)}{2!}x^2 = 1 - \frac{1}{2!}x^2$$

$$P_4(x) = P_2(x) + \frac{f'''(0)}{3!}x^3 + \frac{f^{(4)}(0)}{4!}x^4 = 1 - \frac{1}{2!}x^2 + \frac{1}{4!}x^4$$

$$P_6(x) = P_4(x) + \frac{f^{(5)}(0)}{5!}x^5 + \frac{f^{(6)}(0)}{6!}x^6 = 1 - \frac{1}{2!}x^2 + \frac{1}{4!}x^4 - \frac{1}{6!}x^6$$

Using $P_6(x)$, we obtain the approximation $\cos(0.1) \approx 0.995004165$, which coincides with the calculator value to nine decimal places.   ☐

Remark   Note in Example 3 that the Maclaurin polynomials for $\cos x$ have only even powers of x. Similarly, the Maclaurin polynomials for $g(x) = \sin x$ have only odd powers of x (see Exercise 5). This is not generally true for the Taylor polynomials for $\sin x$ and $\cos x$, expanded about $c \neq 0$, as we will see in the next example.

EXAMPLE 4   Finding Taylor polynomials for sin x

Find the fourth Taylor polynomial for $f(x) = \sin x$, expanded about $c = \pi/6$.

Solution: We have

$$
\begin{aligned}
f(x) &= \sin x & f\left(\frac{\pi}{6}\right) &= \frac{1}{2} \\[2mm]
f'(x) &= \cos x & f'\left(\frac{\pi}{6}\right) &= \frac{\sqrt{3}}{2} \\[2mm]
f''(x) &= -\sin x & f''\left(\frac{\pi}{6}\right) &= -\frac{1}{2} \\[2mm]
f'''(x) &= -\cos x & f'''\left(\frac{\pi}{6}\right) &= -\frac{\sqrt{3}}{2} \\[2mm]
f^{(4)}(x) &= \sin x & f^{(4)}\left(\frac{\pi}{6}\right) &= \frac{1}{2}
\end{aligned}
$$

Therefore, expansion about $x = \pi/6$ yields

$$P_4(x) = f\left(\frac{\pi}{6}\right) + f'\left(\frac{\pi}{6}\right)\left(x - \frac{\pi}{6}\right) + \frac{f''\left(\frac{\pi}{6}\right)}{2!}\left(x - \frac{\pi}{6}\right)^2$$

$$+ \frac{f'''\left(\frac{\pi}{6}\right)}{3!}\left(x - \frac{\pi}{6}\right)^3 + \frac{f^{(4)}\left(\frac{\pi}{6}\right)}{4!}\left(x - \frac{\pi}{6}\right)^4$$

$$= \frac{1}{2} + \frac{\sqrt{3}}{2}\left(x - \frac{\pi}{6}\right) - \frac{1}{2 \cdot 2!}\left(x - \frac{\pi}{6}\right)^2 - \frac{\sqrt{3}}{2 \cdot 3!}\left(x - \frac{\pi}{6}\right)^3$$

$$+ \frac{1}{2 \cdot 4!}\left(x - \frac{\pi}{6}\right)^4$$

To approximate the value of a function at a specific point, we can often use *either* Taylor or Maclaurin polynomials. For instance, to approximate the value of ln (1.1), we can use Taylor polynomials for $f(x) = \ln x$ expanded about $x = 1$, as in Example 2. Or we can use Maclaurin polynomials, as shown in the next example.

EXAMPLE 5   Approximation using Maclaurin polynomials

Use a fourth Maclaurin polynomial to approximate the value of ln (1.1).

Solution: Since $x = 1.1$ is closer to 1 than to 0, we consider Maclaurin polynomials for the function $g(x) = \ln (1 + x)$. We have

$$
\begin{aligned}
g(x) &= \ln (1 + x) & g(0) &= 0 \\
g'(x) &= (1 + x)^{-1} & g'(0) &= 1 \\
g''(x) &= -(1 + x)^{-2} & g''(0) &= -1 \\
g'''(x) &= 2(1 + x)^{-3} & g'''(0) &= 2 \\
g^{(4)}(x) &= -6(1 + x)^{-4} & g^{(4)}(0) &= -6
\end{aligned}
$$

Note that we get the same coefficients as we did in Example 2. Therefore, the fourth Maclaurin polynomial for $g(x) = \ln (1 + x)$ is

$$P_4(x) = g(0) + g'(0)x + \frac{g''(0)}{2!}x^2 + \frac{g'''(0)}{3!}x^3 + \frac{g^{(4)}(0)}{4!}x^4$$

$$= 0 + x - \frac{1}{2}x^2 + \frac{1}{3}x^3 - \frac{1}{4}x^4$$

Consequently,

$$\ln (1.1) = \ln (1 + 0.1) \approx P_4(0.1)$$

$$= (0.1) - \frac{1}{2}(0.1)^2 + \frac{1}{3}(0.1)^3 - \frac{1}{4}(0.1)^4$$

$$\approx 0.0953083$$

Check to see that the fourth Taylor polynomial (from Example 2), evaluated at $x = 1.1$, yields the same result.

Table 10.2 shows the accuracy of the Taylor polynomial approximation to the calculator value of ln (1.1). We can see that as n becomes larger, $P_n(1.1)$ is closer to the calculator value of 0.0953102.

TABLE 10.2   Approximations of ln (1.1), Using Taylor Polynomials

n	1	2	3	4
$P_n(1.1)$	0.1000000	0.0950000	0.0953333	0.0953083

On the other hand, we can see from Table 10.3 that as we move away from the expansion point $c = 1$, the accuracy of the approximation decreases.

TABLE 10.3   Fourth Taylor Polynomial Approximations of ln x

x	1.0	1.1	1.5	1.75	2.0
ln x	0.0000000	0.0953102	0.4054651	0.5596158	0.6931472
$P_4(x)$	0.0000000	0.0953083	0.4010417	0.5302734	0.5833333

Tables 10.2 and 10.3 illustrate two very important points about the accuracy of Taylor (or Maclaurin) polynomials for use in approximations.

1. The approximation is usually better at x-values close to c than at x-values far from c.
2. The approximation is usually better for higher degree Taylor (or Maclaurin) polynomials than for those of lower degree.

The remainder of a Taylor polynomial

An approximation technique is of little value unless we have some idea of its accuracy. To measure the accuracy of approximating a functional value $f(x)$ by the Taylor polynomial $P_n(x)$, we use the concept of a **remainder, $R_n(x)$,** defined as follows:

$$f(x) = P_n(x) + R_n(x)$$

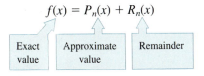

| Exact value | Approximate value | Remainder |

Thus, $R_n(x) = f(x) - P_n(x)$, and we call the absolute value of $R_n(x)$ the **error** associated with the approximation. That is,

$$\text{error} = |R_n(x)| = |f(x) - P_n(x)|$$

The next theorem gives a general procedure for estimating the remainder associated with a Taylor polynomial. This important theorem is called **Taylor's Theorem** and the remainder stated in the theorem is called the **Lagrange form of the remainder.**

THEOREM 10.1   **TAYLOR'S THEOREM**

If a function f is differentiable up through order $n + 1$ in an interval I containing c, then for each x in I, there exists z between x and c such that

$$f(x) = f(c) + f'(c)(x - c) + \frac{f''(c)}{2!}(x - c)^2 + \cdots + \frac{f^{(n)}(c)}{n!}(x - c)^n + R_n(x)$$

where

$$R_n(x) = \frac{f^{(n+1)}(z)}{(n + 1)!}(x - c)^{n+1}$$

Proof: To find $R_n(x)$ we fix x in I ($x \neq c$) and write

$$R_n(x) = f(x) - P_n(x)$$

where $P_n(x)$ is the nth Taylor polynomial for $f(x)$. Then we let g be a function of t defined by

$$g(t) = f(x) - f(t) - f'(t)(x - t) - \cdots - \frac{f^{(n)}(t)}{n!}(x - t)^n - R_n(x)\frac{(x - t)^{n+1}}{(x - c)^{n+1}}$$

The reason for defining g this way is that differentiation with respect to t has a telescoping effect. For example, we have

$$\frac{d}{dt}[-f(t) - f'(t)(x - t)] = -f'(t) + f'(t) - f''(t)(x - t) = -f''(t)(x - t)$$

The result is that the derivative $g'(t)$ simplifies to

$$g'(t) = -\frac{f^{(n+1)}(t)}{n!}(x - t)^n + (n + 1)R_n(x)\frac{(x - t)^n}{(x - c)^{n+1}}$$

for all t between c and x. Moreover, for a fixed x

$$g(c) = f(x) - [P_n(x) + R_n(x)] = f(x) - f(x) = 0$$

and

$$g(x) = f(x) - f(x) - 0 \cdots - 0 = f(x) - f(x) = 0$$

Therefore, g satisfies the conditions of Rolle's Theorem, and it follows that there is a number z between c and x such that $g'(z) = 0$. Substituting z for t in the equation for $g'(t)$ and then solving for $R_n(x)$, we obtain

$$g'(z) = -\frac{f^{(n+1)}(z)}{n!}(x - z)^n + (n + 1)R_n(x)\frac{(x - z)^n}{(x - c)^{n+1}} = 0$$

$$R_n(x) = \frac{f^{(n+1)}(z)}{(n + 1)!}(x - c)^{n+1}$$

Finally, since $g(c) = 0$, we have

$$0 = f(x) - f(c) - f'(c)(x - c) - \cdots - \frac{f^{(n)}(c)}{n!}(x - c)^n - R_n(x)$$

$$f(x) = f(c) + f'(c)(x - c) + \cdots + \frac{f^{(n)}(c)}{n!}(x - c)^n + R_n(x)$$

EXAMPLE 6   Determining the accuracy of an approximation

The third Maclaurin polynomial for $\sin x$ is given by

$$P_3(x) = x - \frac{x^3}{3!}$$

Use Taylor's Theorem to approximate $\sin (0.1)$ by $P_3(0.1)$ and determine the accuracy of the approximation.

Solution: Using Taylor's Theorem, we have

$$\sin x = x - \frac{x^3}{3!} + R_3(x) = x - \frac{x^3}{3!} + \frac{f^{(4)}(z)}{4!} x^4$$

where $0 < z < 0.1$. Therefore,

$$\sin (0.1) \approx 0.1 - \frac{(0.1)^3}{3!} \approx 0.1 - 0.000167 = 0.099833$$

Since $f^{(4)}(z) = \sin z$ and the sine function is increasing on the interval $[0, 0.1]$, it follows that $0 < \sin z < 1$ and we have

$$0 \le R_n(z) = \frac{\sin z}{4!}(0.1)^4 < \frac{0.0001}{4!} \approx 0.000004$$

and we conclude that

$$0.099833 \le \sin (0.1) \le 0.099837$$
☐

EXAMPLE 7   Approximating a functional value to a desired accuracy

Determine the degree of the Taylor polynomial $P_n(x)$ expanded about $c = 1$ that should be used to approximate $\ln (1.2)$ so that the error is less than 0.001.

Solution: Following the pattern of Example 2, we see that the $(n + 1)$st derivative of $f(x) = \ln x$ is given by

$$f^{(n+1)}(x) = (-1)^{n+1}\frac{n!}{x^{n+1}}$$

Using Taylor's Theorem, we know that the error $|R_n(1.2)|$ is given by

$$|R_n(1.2)| = \left| \frac{f^{(n+1)}(z)}{(n + 1)!}(1.2 - 1)^{n+1} \right| = \frac{n!}{z^{n+1}}\left(\frac{1}{(n + 1)!} \right)(0.2)^{n+1}$$

$$= \frac{(0.2)^{n+1}}{z^{n+1}(n + 1)}$$

where $1 < z < 1.2$. In this interval, $|f^{(n+1)}(z)|$ is largest when $z = 1$, thus we are seeking a value of n such that

$$\frac{(0.2)^{n+1}}{(1)^{n+1}(n + 1)} < 0.001 \implies 1000 < (n + 1)5^{n+1}$$

By trial and error, we can determine that the smallest value of n satisfying this inequality is $n = 3$. Thus, we would need the third Taylor polynomial to achieve the desired accuracy in approximating $\ln (1.2)$.
☐

Ultimately, we want to be able to determine the values of x for which the **infinite sequence**

$$P_1(x), P_2(x), P_3(x), \ldots, P_n(x), \ldots$$

has a limiting value as $n \to \infty$. We will study infinite sequences in Section 10.2. From a different (but related) perspective, we will also study ways to determine the values of x for which an nth Taylor polynomial

$$P_n(x) = f(c) + f'(c)(x - c) + \frac{f''(c)}{2!}(x - c)^2 + \cdots + \frac{f^{(n)}(c)}{n!}(x - c)^n$$

expanded to infinitely many terms

$$P(x) = f(c) + f'(c)(x - c) + \frac{f''(c)}{2!}(x - c)^2 + \cdots + \frac{f^{(n)}(c)}{n!}(x - c)^n + \cdots$$

will have a limiting value (a finite sum) as $n \to \infty$. Such infinite summations are called **infinite series** and we will study them in detail in Sections 10.3 through 10.10.

Section Exercises 10.1

In Exercises 1–14, find the Maclaurin polynomial of degree n for the given function.

1. $f(x) = e^{-x}$, $n = 3$
2. $f(x) = e^{-x}$, $n = 5$
3. $f(x) = e^{2x}$, $n = 4$
4. $f(x) = e^{3x}$, $n = 4$
5. $f(x) = \sin x$, $n = 5$
6. $f(x) = \sin \pi x$, $n = 3$
7. $f(x) = xe^x$, $n = 4$
8. $f(x) = x^2 e^{-x}$, $n = 4$
9. $f(x) = \dfrac{1}{x + 1}$, $n = 4$
10. $f(x) = \dfrac{1}{x^2 + 1}$, $n = 4$
11. $f(x) = \sec x$, $n = 2$
12. $f(x) = \tan x$, $n = 3$
13. $f(x) = 2 - 3x^3 + x^4$, $n = 4$
14. $f(x) = 3 - 2x^2 + x^3$, $n = 3$

In Exercises 15–18, find the Taylor polynomial of degree n centered at $x = c$.

15. $f(x) = \dfrac{1}{x}$, $n = 4$, $c = 1$
16. $f(x) = \sqrt{x}$, $n = 4$, $c = 4$
17. $f(x) = x^2 \cos x$, $n = 2$, $c = \pi$
18. $f(x) = \ln x$, $n = 4$, $c = 1$

19. Use the Maclaurin polynomials $P_1(x)$, $P_3(x)$, and $P_5(x)$ for $f(x) = \sin x$ to complete the following table. (See Exercise 5.)

x	0	0.25	0.50	0.75	1.00
$\sin x$	0	0.2474	0.4794	0.6816	0.8415
$P_1(x)$					
$P_3(x)$					
$P_5(x)$					

20. Use the Taylor polynomials $P_1(x)$ and $P_4(x)$ for $f(x) = \ln x$ centered at $x = 1$ to complete the following table. (See Exercise 18.)

x	1.00	1.25	1.50	1.75	2.00
$\ln x$	0	0.2231	0.4055	0.5596	0.6931
$P_1(x)$					
$P_4(x)$					

In Exercises 21–24, approximate the function at the given value of x, using the polynomial found in the indicated exercise.

21. $f(x) = e^{-x}$, $f\left(\frac{1}{2}\right)$, Exercise 1

22. $f(x) = x^2 e^{-x}$, $f\left(\frac{1}{4}\right)$, Exercise 8

23. $f(x) = x^2 \cos x$, $f\left(\frac{7\pi}{8}\right)$, Exercise 17

24. $f(x) = \sqrt{x}$, $f(5)$, Exercise 16

25. Compare the Maclaurin polynomials of degree 4 for the functions

$$f(x) = e^x \qquad \text{and} \qquad g(x) = xe^x$$

What is the relationship between them?

26. Differentiate the Maclaurin polynomial of degree 5 for $f(x) = \sin x$ and compare the result with the Maclaurin polynomial of degree 4 for $g(x) = \cos x$.

In Exercises 27 and 28, determine the values of x for which the given function can be replaced by the Taylor Polynomial if the error cannot exceed 0.001.

27. $f(x) = e^x \approx 1 + x + \frac{x^2}{2!} + \frac{x^3}{3!}$, $x < 0$

28. $f(x) = \sin x \approx x - \frac{x^3}{3!}$

29. Estimate the error in approximating $e^{0.6}$ by the fifth-degree polynomial

$$1 + x + \frac{x^2}{2!} + \frac{x^3}{3!} + \frac{x^4}{4!} + \frac{x^5}{5!}$$

30. What degree Maclaurin polynomial for $\ln (x + 1)$ should be used to guarantee finding $\ln 1.5$ if the error cannot exceed 0.0001?

In Exercises 31 and 32,
(a) Find the Taylor polynomial $P_3(x)$ of degree three for $f(x)$.
(b) Complete the accompanying table for $f(x)$ and $P_3(x)$.
(c) Sketch the graphs of $f(x)$ and $P_3(x)$ on the same axes.

31. $f(x) = \arcsin x$ **32.** $f(x) = \arctan x$

x	-1	-0.75	-0.50	-0.25	0	0.25	0.50	0.75	1
$f(x)$									
$P_3(x)$									

33. Prove that if f is an odd function then the nth Maclaurin polynomial for the function will contain only terms with odd powers of x.

34. Prove that if f is an even function then the nth Maclaurin polynomial for the function will contain only terms with even powers of x.

35. Let $P_n(x)$ be the nth Taylor polynomial for f at c. Prove that $P_n(c) = f(c)$ and $P^{(k)}(c) = f^{(k)}(c)$ for $1 \le k \le n$.

10.2
Sequences

SECTION TOPICS •
Infinite sequence •
Limit of a sequence •
Pattern recognition for sequences •
Monotonic sequences •

In mathematics the word "sequence" is used in much the same way as in ordinary English. When we say that a collection of objects or events is *in sequence* we usually mean that the collection is ordered so that it has an identified first member, second member, third member, and so on.

We define a sequence mathematically as a *function whose domain is the set of positive integers*. Even though a sequence is a function, we usually represent sequences by subscript notation, rather than the standard function notation. For instance, in the sequence

$$1, \quad 2, \quad 3, \quad 4, \quad \ldots, \quad n, \quad \ldots$$
$$\downarrow \quad \downarrow \quad \downarrow \quad \downarrow \qquad\qquad \downarrow$$
$$a_1, \quad a_2, \quad a_3, \quad a_4, \quad \ldots, a_n, \ldots$$

1 is mapped onto a_1, 2 is mapped onto a_2, and so on. We call a_n the **nth term** of the sequence and we denote the sequence by $\{a_n\}$.

DEFINITION OF A SEQUENCE	A **sequence** $\{a_n\}$ is a function whose domain is the set of positive integers. The functional values $a_1, a_2, a_3, \ldots, a_n, \ldots$ are called the **terms** of the sequence.

| Remark   On some occasions it is convenient to begin a sequence with a_0, so that the terms of the sequence become

$$a_0, a_1, a_2, a_3, a_4, \ldots, a_n, \ldots$$

EXAMPLE 1   Listing the terms of a sequence

(a) For the sequence $\{a_n\} = \{3 + (-1)^n\}$, the first four terms are

$$3 + (-1)^1, \; 3 + (-1)^2, \; 3 + (-1)^3, \; 3 + (-1)^4, \ldots$$
$$2, 4, 2, 4, \ldots$$

(b) For the sequence $\{b_n\} = \{2n/(1 + n)\}$, the first four terms are

$$\frac{2 \cdot 1}{1 + 1}, \; \frac{2 \cdot 2}{1 + 2}, \; \frac{2 \cdot 3}{1 + 3}, \; \frac{2 \cdot 4}{1 + 4}, \ldots$$
$$\frac{2}{2}, \frac{4}{3}, \frac{6}{4}, \frac{8}{5}, \ldots$$

(c) For the sequence $\{c_n\} = \{n^2/(2^n - 1)\}$, the first four terms are

$$\frac{1^2}{2^1 - 1}, \; \frac{2^2}{2^2 - 1}, \; \frac{3^2}{2^3 - 1}, \; \frac{4^2}{2^4 - 1}, \ldots$$
$$\frac{1}{1}, \frac{4}{3}, \frac{9}{7}, \frac{16}{15}, \ldots$$

In this chapter our primary interest is in sequences whose terms approach a limiting value. Such sequences are said to **converge.** For instance, the sequence $\{1/2^n\}$

$$\frac{1}{2}, \frac{1}{4}, \frac{1}{8}, \frac{1}{16}, \frac{1}{32}, \ldots$$

converges to 0 as indicated in the following definition.

| DEFINITION OF THE LIMIT OF A SEQUENCE | If for $\varepsilon > 0$ there exists $M > 0$ such that $|a_n - L| < \varepsilon$ whenever $n > M$, then we say that the limit of the sequence $\{a_n\}$ is L and write

$$\lim_{n \to \infty} a_n = L$$

Sequences that have a (finite) limit are said to **converge,** and sequences that do not have a limit are said to **diverge.** |

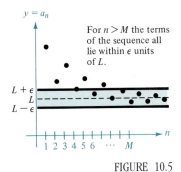

For $n > M$ the terms of the sequence all lie within ϵ units of L.

FIGURE 10.5

Graphically, this definition says that eventually (for $n > M$) the terms of a sequence that converges to L will lie within the band between the lines

$$y = L + \varepsilon \text{ and } y = L - \varepsilon$$

as illustrated in Figure 10.5.

In Section 5.3, we compared the limit of a sequence to the limit of a function of a real variable x as $x \to \infty$. There, we saw that if a sequence happens to agree with a function f of a real variable x at every positive integer, and if $f(x)$ approaches a limit as $x \to \infty$, then the sequence approaches the same limit as $n \to \infty$. For convenience, we give this result as Theorem 10.2.

THEOREM 10.2

LIMIT OF A SEQUENCE

Let f be a function of a real variable such that

$$\lim_{x \to \infty} f(x) = L$$

If $\{a_n\}$ is a sequence such that $f(n) = a_n$ for every positive integer n, then

$$\lim_{n \to \infty} a_n = L$$

Remark Recall from Example 3 in Section 5.3 that the converse to this theorem is not true. That is, it is possible for the sequence $\{a_n\}$ to converge to a limit L even though $f(x)$ does not converge to L.

The following properties of limits of sequences parallel those given for limits of functions of a real variable in Section 4.5.

PROPERTIES OF LIMITS OF SEQUENCES

If

$$\lim_{n \to \infty} a_n = L \qquad \text{and} \qquad \lim_{n \to \infty} b_n = K$$

then the following properties are true.

1. $\lim\limits_{n \to \infty} (a_n \pm b_n) = L \pm K$ 2. $\lim\limits_{n \to \infty} c a_n = cL$, c is any real number

3. $\lim\limits_{n \to \infty} (a_n b_n) = LK$ 4. $\lim\limits_{n \to \infty} \dfrac{a_n}{b_n} = \dfrac{L}{K}$, $b_n \neq 0$ and $K \neq 0$

EXAMPLE 2   Determining the convergence or divergence of a sequence

Determine the convergence or divergence of the sequences given by

(a) $a_n = 3 + (-1)^n$ (b) $b_n = \dfrac{n}{1 - 2n}$

Solution:

(a) Since the sequence $\{a_n\} = \{3 + (-1)^n\}$ has terms

$$2, 4, 2, 4, \ldots$$

that oscillate between 2 and 4, the limit does not exist, and we conclude that the sequence diverges.

(b) For $\{b_n\}$, we can divide the numerator and denominator by n to obtain

$$\lim_{n \to \infty} \frac{n}{1 - 2n} = \lim_{n \to \infty} \left[\frac{1}{(1/n) - 2} \right] = -\frac{1}{2}$$

and we conclude that the sequence converges to $-\frac{1}{2}$.

Theorem 10.2 opens up the possibility of using L'Hôpital's Rule to determine the limit of a sequence, as demonstrated in the next example.

EXAMPLE 3   *Using L'Hôpital's Rule to determine convergence*

Show that the sequence given by $a_n = n^2/(2^n - 1)$ converges.

Solution: We consider the function of a real variable

$$f(x) = \frac{x^2}{2^x - 1}$$

Then, applying L'Hôpital's Rule twice, we have

$$\lim_{x \to \infty} \frac{x^2}{2^x - 1} = \lim_{x \to \infty} \frac{2x}{(\ln 2)2^x} = \lim_{x \to \infty} \frac{2}{(\ln 2)^2 2^x} = 0$$

Since $f(n) = a_n$ for every positive integer, we can apply Theorem 10.2 to conclude that

$$\lim_{n \to \infty} \frac{n^2}{2^n - 1} = 0$$

Another useful limit theorem that can be rewritten for sequences is the Squeeze Theorem of Section 2.5.

THEOREM 10.3   **SQUEEZE THEOREM FOR SEQUENCES**

If

$$\lim_{n \to \infty} a_n = L = \lim_{n \to \infty} b_n$$

and there exists an integer N such that $a_n \leq c_n \leq b_n$ for all $n > N$, then

$$\lim_{n \to \infty} c_n = L$$

The usefulness of the Squeeze Theorem is seen in the next example where we squeeze a factorial sequence between two geometric sequences.

EXAMPLE 4   Using the Squeeze Theorem

Show that the following sequence converges and find its limit:

$$c_n = (-1)^n \frac{1}{n!}$$

Solution: To apply the Squeeze Theorem, we must find two convergent sequences that can be related to the given factorial sequence. Two possibilities are $a_n = -1/2^n$ and $b_n = 1/2^n$, both of which converge to zero. By comparing the term $n!$ with 2^n, we see that

$$n! = 1 \cdot 2 \cdot 3 \cdot 4 \cdot 5 \cdot 6 \cdots n = 24 \cdot 5 \cdot 6 \cdots n$$

and $$2^n = 2 \cdot 2 \cdot 2 \cdot 2 \cdot 2 \cdot 2 \cdots 2 = 16 \cdot \underbrace{2 \cdot 2 \cdots 2}_{n-4 \text{ factors}}$$

This implies that for $n \geq 4$, $2^n < n!$, and we have

$$\frac{-1}{2^n} \leq (-1)^n \frac{1}{n!} \leq \frac{1}{2^n}$$

Therefore, by the Squeeze Theorem it follows that

$$\lim_{n \to \infty} (-1)^n \frac{1}{n!} = 0$$

Table 10.4 lists the first six terms of each of the three sequences a_n, b_n, and c_n, and they are shown visually in Figure 10.6.

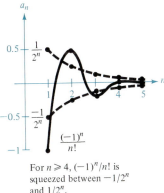

For $n \geq 4$, $(-1)^n/n!$ is squeezed between $-1/2^n$ and $1/2^n$.

FIGURE 10.6

TABLE 10.4

n	1	2	3	4	5	6
$-\dfrac{1}{2^n}$	$-\dfrac{1}{2}$	$-\dfrac{1}{4}$	$-\dfrac{1}{8}$	$-\dfrac{1}{16}$	$-\dfrac{1}{32}$	$-\dfrac{1}{64}$
$(-1)^n \dfrac{1}{n!}$	$-\dfrac{1}{1}$	$\dfrac{1}{2}$	$-\dfrac{1}{6}$	$\dfrac{1}{24}$	$-\dfrac{1}{120}$	$\dfrac{1}{720}$
$\dfrac{1}{2^n}$	$\dfrac{1}{2}$	$\dfrac{1}{4}$	$\dfrac{1}{8}$	$\dfrac{1}{16}$	$\dfrac{1}{32}$	$\dfrac{1}{64}$

In Example 4, the sequence $\{c_n\}$ has both positive and negative terms. For this particular sequence, it happens that the sequence of absolute values, $\{|c_n|\}$, also converges to zero. You can show this by the Squeeze Theorem using the inequality

$$0 \leq \frac{1}{n!} \leq \frac{1}{2^n}, \quad n \geq 4$$

In such cases it is often convenient to consider the sequence of positive terms, then apply Theorem 10.4, which tells us that if the positive term sequence converges to zero, the original signed sequence must also converge to zero.

THEOREM 10.4   ABSOLUTE VALUE THEOREM
For the sequence $\{a_n\}$, if

$$\lim_{n \to \infty} |a_n| = 0 \qquad \text{then} \qquad \lim_{n \to \infty} a_n = 0$$

Proof: Consider the two sequences $\{|a_n|\}$ and $\{-|a_n|\}$. Since both of these sequences converge to 0 and since

$$-|a_n| \le a_n \le |a_n|$$

we can apply the Squeeze Theorem and conclude that $\{a_n\}$ converges to 0.

The result in Example 4 suggests something about the rate at which $n!$ increases as $n \to \infty$. From Figure 10.6 we can see that both $\{1/2^n\}$ and $\{1/n!\}$ approach zero as $n \to \infty$. Yet $\{1/n!\}$ approaches 0 so much faster than $\{1/2^n\}$ does that

$$\lim_{n \to \infty} \frac{1/n!}{1/2^n} = \lim_{n \to \infty} \frac{2^n}{n!} = 0$$

In fact, it can be shown that for any fixed number k,

$$\lim_{n \to \infty} \frac{k^n}{n!} = 0$$

This means that *the factorial function grows faster than any exponential function*. We will find this to be a very useful result as we work with limits of sequences.

Pattern recognition for sequences

Sometimes the first several terms of a sequence are listed without the nth term, or the terms may be generated by some rule that does not explicitly identify the nth term of the sequence. In such cases, we are required to discover a *pattern* in the sequence and to describe the nth term. Once the nth term is specified, we can discuss the convergence or divergence of the sequence. This is demonstrated in the next two examples.

EXAMPLE 5   Finding the nth term of a sequence

Find a sequence $\{a_n\}$ whose first five terms are

$$\frac{2}{1}, \frac{4}{3}, \frac{8}{5}, \frac{16}{7}, \frac{32}{9}, \cdots$$

and then determine whether or not the particular sequence you have chosen converges or diverges.

Solution: First, we note that the numerators are successive powers of 2 and the denominators form the sequence of positive odd integers. Then, by comparing a_n with n, we have

$$1, \quad 2, \quad 3, \quad 4, \quad 5, \quad \cdots$$
$$\downarrow \quad \downarrow \quad \downarrow \quad \downarrow \quad \downarrow$$

$$\frac{2}{1}, \frac{4}{3}, \frac{8}{5}, \frac{16}{7}, \frac{32}{9}, \cdots$$

$$\frac{2^1}{1}, \frac{2^2}{3}, \frac{2^3}{5}, \frac{2^4}{7}, \frac{2^5}{9}, \cdots$$

Using the fact that the odd integer denominators are generated by $2n - 1$, we can conclude that $a_n = 2^n/(2n - 1)$. Now, using L'Hôpital's Rule to evaluate the limit of $f(x) = 2^x/(2x - 1)$, we obtain

$$\lim_{x\to\infty} \frac{2^x}{2x - 1} = \lim_{x\to\infty} \frac{2^x(\ln 2)}{2} = \infty \implies \lim_{n\to\infty} \frac{2^n}{2n - 1} = \infty$$

Hence the sequence *diverges*.

EXAMPLE 6   *Finding the nth term of a sequence*

Find the nth term for the sequence given by

$$a_n = 1 - f^{(n-1)}(0)$$

where $f(x) = e^{x/3}$, and determine whether the sequence converges or diverges.

Solution: The first several terms of the sequence are listed in Table 10.5, along with the first several derivatives of $f(x) = e^{x/3}$. Comparing terms with term numbers, we conclude that

$$a_n = 1 - \frac{1}{3^{n-1}}$$

The sequence *converges* because

$$\lim_{n\to\infty}\left(1 - \frac{1}{3^{n-1}}\right) = 1 - 0 = 1$$

TABLE 10.5

n	1	2	3	4	5	$\cdots$	n
$f^{(n-1)}(x)$	$e^{x/3}$	$\dfrac{e^{x/3}}{3}$	$\dfrac{e^{x/3}}{3^2}$	$\dfrac{e^{x/3}}{3^3}$	$\dfrac{e^{x/3}}{3^4}$	$\cdots$	$\dfrac{e^{x/3}}{3^{n-1}}$
$1 - f^{(n-1)}(0)$	$1 - 1$	$1 - \dfrac{1}{3}$	$1 - \dfrac{1}{3^2}$	$1 - \dfrac{1}{3^3}$	$1 - \dfrac{1}{3^4}$	$\cdots$	$1 - \dfrac{1}{3^{n-1}}$

Without a specific rule for generating the terms of a sequence or some knowledge of the context in which the terms of the sequence are obtained, it is not possible to determine the convergence or divergence of the sequence merely from its first several terms. For instance, although the first three terms of the four sequences given next are identical, the first two sequences converge to 0, the third sequence converges to $\frac{1}{9}$, and the fourth one diverges.

$$\{a_n\}: \frac{1}{2}, \frac{1}{4}, \frac{1}{8}, \frac{1}{16}, \cdots, \frac{1}{2^n}, \cdots$$

$$\{b_n\}: \frac{1}{2}, \frac{1}{4}, \frac{1}{8}, \frac{1}{15}, \cdots, \frac{6}{(n+1)(n^2-n+6)}, \cdots$$

$$\{c_n\}: \frac{1}{2}, \frac{1}{4}, \frac{1}{8}, \frac{7}{62}, \cdots, \frac{n^2-3n+3}{9n^2-25n+18}, \cdots$$

$$\{d_n\}: \frac{1}{2}, \frac{1}{4}, \frac{1}{8}, 0, \cdots, \frac{-n(n+1)(n-4)}{6(n^2+3n-2)}, \cdots$$

The process of determining an nth term from the pattern observed in the first several terms of a sequence is an example of *inductive reasoning*.

EXAMPLE 7   Finding the nth term of a sequence

Determine an nth term for the sequence whose first five terms are

$$-\frac{2}{1}, \frac{8}{2}, -\frac{26}{6}, \frac{80}{24}, -\frac{242}{120}, \cdots$$

and then decide if the sequence converges or diverges.

Solution: We observe that the numerators are one less than 3^n. Hence, we reason that the numerators are given by the rule $3^n - 1$. If we factor the denominators, we have

$$1 = 1$$
$$2 = 1 \cdot 2$$
$$6 = 1 \cdot 2 \cdot 3$$
$$24 = 1 \cdot 2 \cdot 3 \cdot 4$$
$$120 = 1 \cdot 2 \cdot 3 \cdot 4 \cdot 5 \ \dots$$

This suggests that the denominators are represented by $n!$. Finally, since the signs alternate, we can write the nth term as

$$a_n = (-1)^n\left(\frac{3^n - 1}{n!}\right)$$

From Theorem 10.4 and the discussion following Example 4 about the growth of $n!$, it follows that

$$\lim_{n\to\infty}|a_n| = \lim_{n\to\infty}\frac{3^n-1}{n!} = 0 = \lim_{n\to\infty}a_n$$

and we conclude that $\{a_n\}$ converges to 0.

Table 10.6 lists some common sequence patterns.

TABLE 10.6   SEQUENCE PATTERNS

1. Changes in sign:

$$\{(-1)^n\} = \{-1, 1, -1, 1, -1, 1, \ldots\}$$
$$\{(-1)^{n+1}\} = \{1, -1, 1, -1, 1, -1, \ldots\}$$
$$\{(-1)^{n(n+1)/2}\} = \{-1, -1, 1, 1, -1, -1, \ldots\}$$

2. Arithmetic sequences (successive terms differ by a constant value):

$$\{2n\} = \{2, 4, 6, 8, 10, 12, \ldots\}$$
$$\{2n - 1\} = \{1, 3, 5, 7, 9, 11, \ldots\}$$
$$\{an + b\} = \{a + b, 2a + b, 3a + b, 4a + b, 5a + b, \ldots\}$$

3. Binary (powers of 2) and geometric (powers of r) sequences:

$$\{2^{n-1}\} = \{1, 2, 4, 8, 16, 32, \ldots\}$$
$$\{ar^{n-1}\} = \{a, ar, ar^2, ar^3, ar^4, ar^5, \ldots\}$$

4. Power sequences:

$$\{n^2\} = \{1, 4, 9, 16, 25, 36, \ldots\}$$
$$\{n^m\} = \{1, 2^m, 3^m, 4^m, 5^m, 6^m, \ldots\}$$

5. Sequences of products:

$$\{n!\} = \{1, 1 \cdot 2, 1 \cdot 2 \cdot 3, 1 \cdot 2 \cdot 3 \cdot 4, 1 \cdot 2 \cdot 3 \cdot 4 \cdot 5, \ldots\}$$
$$\{2^n n!\} = \{2, 2 \cdot 4, 2 \cdot 4 \cdot 6, 2 \cdot 4 \cdot 6 \cdot 8, 2 \cdot 4 \cdot 6 \cdot 8 \cdot 10, \ldots\}$$
$$\left\{\frac{(2n)!}{2^n n!}\right\} = \{1, 1 \cdot 3, 1 \cdot 3 \cdot 5, 1 \cdot 3 \cdot 5 \cdot 7, 1 \cdot 3 \cdot 5 \cdot 7 \cdot 9, \ldots\}$$

Monotonic sequences

So far we have determined the convergence of a sequence by finding its limit. Even if we cannot determine the limit of a particular sequence, it may still be useful to know whether or not the sequence converges. Theorem 10.5 identifies a test for convergence of sequences without involving the determination of the limit. First, we look at some preliminary definitions.

DEFINITION OF A MONOTONIC SEQUENCE

A sequence $\{a_n\}$ is **monotonic** if its terms are nondecreasing

$$a_1 \leq a_2 \leq a_3 \leq \ldots \leq a_n \leq \ldots$$

or if its terms are nonincreasing

$$a_1 \geq a_2 \geq a_3 \geq \ldots \geq a_n \geq \ldots$$

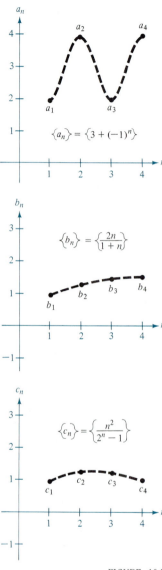

FIGURE 10.7

EXAMPLE 8   Determining if a sequence is monotonic

Determine whether the following sequences are monotonic.

(a) $a_n = 3 + (-1)^n$ (b) $b_n = \dfrac{2n}{1 + n}$ (c) $c_n = \dfrac{n^2}{2^n - 1}$

Solution:

(a) This sequence alternates between 2 and 4. Therefore, it is not monotonic.
(b) This sequence is monotonic because each successive term is larger than its predecessor. To see this, we compare the terms b_n and b_{n+1} as follows. (Note that we use the fact that n is positive to be able to multiply both sides of the inequality by $(1 + n)$ and $(2 + n)$ without reversing the inequality sign.)

$$b_n = \frac{2n}{1 + n} \overset{?}{<} \frac{2(n + 1)}{1 + (n + 1)} = b_{n+1}$$

$$2n(2 + n) \overset{?}{<} (1 + n)(2n + 2)$$

$$4n + 2n^2 \overset{?}{<} 2 + 4n + 2n^2$$

$$0 < 2$$

Since the final inequality is valid, we can reverse the steps to conclude that the original inequality is also valid.

(c) This sequence is not monotonic since the second term is larger than the first term, but smaller than the third. (Note that if we drop the first term, the remaining sequence $c_2, c_3, c_4, \ldots$ is monotonic.) Figure 10.7 graphically illustrates these three sequences.

| Remark   In Example 8(b), another way to see that the sequence is monotonic is to consider the derivative of the corresponding function of a real variable $f(x) = 2x/(1 + x)$. Since the derivative

$$f'(x) = \frac{2}{(1 + x)^2}$$

is positive for all x, we can conclude that f is increasing. This implies that $\{a_n\}$ is increasing.

DEFINITION OF A BOUNDED SEQUENCE   A sequence $\{a_n\}$ is **bounded** if there is a positive real number M such that $|a_n| \le M$ for all n.

| Remark   We call M an **upper bound** for the sequence. In Figure 10.7, all three sequences are bounded, since

$$|3 + (-1)^n| \le 4, \qquad \left|\frac{2n}{1 + n}\right| \le 2, \qquad \text{and} \qquad \left|\frac{n^2}{2^n - 1}\right| \le \frac{4}{3}$$

One important property of real numbers is that they are **complete.** Geometrically, this means that there are no holes or gaps on the real number line.

(The set of rational numbers does not have the completeness property.) The completeness axiom for real numbers can be used to conclude that if a sequence has an upper bound, then there must exist a *least* upper bound. For example, the least upper bound of the sequence $\{a_n\} = \{n/(n + 1)\}$,

$$\frac{1}{2}, \frac{2}{3}, \frac{3}{4}, \frac{4}{5}, \ldots, \frac{n}{n + 1}, \ldots$$

is 1, since the sequence is monotonic and converges to 1 and $|n/(n + 1)| \leq 1$. We use the completeness axiom in the proof of Theorem 10.5.

THEOREM 10.5   **BOUNDED MONOTONIC SEQUENCES**
If a sequence $\{a_n\}$ is bounded and monotonic, then it converges.

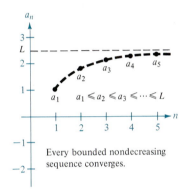

Every bounded nondecreasing sequence converges.

FIGURE 10.8

Proof: We give the proof for a nondecreasing sequence as shown in Figure 10.8, and leave the nonincreasing case as an exercise. For the sake of simplicity, we assume that each term in the sequence is positive. Then, since the sequence is bounded, there must exist an upper bound M such that

$$a_1 \leq a_2 \leq a_3 \ldots \leq a_n \leq \ldots \leq M$$

From the completeness axiom, it follows that there is a least upper bound L such that

$$a_1 \leq a_2 \leq a_3 \ldots \leq a_n \leq \ldots \leq L$$

Now, we will show that $\{a_n\}$ converges to L. For $\varepsilon > 0$, it follows that $L - \varepsilon < L$ and $L - \varepsilon$ cannot be an upper bound for the sequence. Consequently, at least one term of $\{a_n\}$ is greater than $L - \varepsilon$. That is, $L - \varepsilon < a_N$ for some positive integer N. Since the terms of $\{a_n\}$ are nondecreasing we conclude that $a_N \leq a_n$ for $n > N$. We now know that

$$L - \varepsilon < a_N < a_n \leq L < L + \varepsilon, \quad \text{for every } n > N$$

It follows that $|a_n - L| < \varepsilon$ for $n > N$, which by definition means that $\{a_n\}$ converges to L.

Remark   Theorem 10.5 tells us something about divergent sequences—they are either *not bounded* or *not monotonic*. For example,

$$\frac{1}{2}, \frac{4}{3}, \frac{9}{4}, \frac{16}{5}, \ldots, \frac{n^2}{n + 1}, \ldots$$

is monotonic but is not bounded, since

$$\lim_{n \to \infty} a_n = \infty$$

On the other hand, the divergent sequence

$$2, 4, 2, 4, \ldots, [3 + (-1)^n], \ldots$$

is bounded but not monotonic.

Section Exercises 10.2

In Exercises 1–8, write out the first five terms of the sequence with the given nth term.

1. $a_n = 2^n$

2. $a_n = \dfrac{n}{n+1}$

3. $a_n = \left(-\dfrac{1}{2}\right)^n$

4. $a_n = \sin \dfrac{n\pi}{2}$

5. $a_n = \dfrac{3^n}{n!}$

6. $a_n = 5 - \dfrac{1}{n} + \dfrac{1}{n^2}$

7. $a_n = \dfrac{(-1)^{n(n+1)/2}}{n^2}$

8. $a_n = \dfrac{3n!}{(n-1)!}$

In Exercises 9–24, write an expression for the nth term of the sequence.

9. 1, 4, 7, 10, . . .

10. 3, 7, 11, 15, . . .

11. -1, 2, 7, 14, 23, . . .

12. $1, \dfrac{1}{4}, \dfrac{1}{9}, \dfrac{1}{16}, \ldots$

13. $\dfrac{2}{3}, \dfrac{3}{4}, \dfrac{4}{5}, \dfrac{5}{6}, \ldots$

14. $2, \dfrac{3}{3}, \dfrac{4}{5}, \dfrac{5}{7}, \dfrac{6}{9}, \ldots$

15. $2, -1, \dfrac{1}{2}, -\dfrac{1}{4}, \dfrac{1}{8}, \ldots$

16. $\dfrac{1}{2}, \dfrac{1}{3}, \dfrac{2}{9}, \dfrac{4}{27}, \dfrac{8}{81}, \ldots$

17. $2, 1 + \dfrac{1}{2}, 1 + \dfrac{1}{3}, 1 + \dfrac{1}{4}, 1 + \dfrac{1}{5}, \ldots$

18. $1 + \dfrac{1}{2}, 1 + \dfrac{3}{4}, 1 + \dfrac{7}{8}, 1 + \dfrac{15}{16}, 1 + \dfrac{31}{32}, \ldots$

19. $\dfrac{1}{2 \cdot 3}, \dfrac{2}{3 \cdot 4}, \dfrac{3}{4 \cdot 5}, \dfrac{4}{5 \cdot 6}, \ldots$

20. $1, \dfrac{1}{2}, \dfrac{1}{6}, \dfrac{1}{24}, \dfrac{1}{120}, \ldots$

21. $1, \dfrac{1}{1 \cdot 3}, \dfrac{1}{1 \cdot 3 \cdot 5}, \dfrac{1}{1 \cdot 3 \cdot 5 \cdot 7}, \ldots$

22. 2, -4, 6, -8, 10, . . .

23. 1, -1, -1, 1, 1, -1, -1, . . .

24. $1, x, \dfrac{x^2}{2}, \dfrac{x^3}{6}, \dfrac{x^4}{24}, \dfrac{x^5}{120}, \ldots$

In Exercises 25–46, determine the convergence or divergence of each sequence. If the sequence converges, find its limit.

25. $a_n = \dfrac{n+1}{n}$

26. $a_n = \dfrac{1}{n^{3/2}}$

27. $a_n = (-1)^n \left(\dfrac{n}{n+1}\right)$

28. $a_n = \dfrac{n-1}{n} - \dfrac{n}{n-1}$, $n \geq 2$

29. $a_n = \dfrac{3n^2 - n + 4}{2n^2 + 1}$

30. $a_n = \dfrac{\sqrt{n}}{\sqrt{n}+1}$

31. $a_n = \dfrac{n^2 - 1}{n+1}$

32. $a_n = 1 + (-1)^n$

33. $a_n = \dfrac{1 + (-1)^n}{n}$

34. $a_n = \dfrac{\ln (n^2)}{n}$

35. $a_n = \cos \dfrac{n\pi}{2}$

36. $a_n = \dfrac{n}{\sqrt{n^2 + 1}}$

37. $a_n = \dfrac{3^n}{4^n}$

38. $a_n = \dfrac{(n-2)!}{n!}$

39. $a_n = f^{(n-1)}(2)$, $f(x) = \ln x$

40. $a_n = \dfrac{n^2}{2n+1} - \dfrac{n^2}{2n-1}$

41. $a_n = 3 - \dfrac{1}{2^n}$

42. $a_n = \dfrac{n!}{n^n}$

43. $a_n = \left(1 + \dfrac{k}{n}\right)^n$

44. $a_n = 2^{1/n}$

45. $a_n = \dfrac{n^p}{e^n}$ $(p > 0)$

46. $a_n = n \sin \dfrac{1}{n}$

In Exercises 47–56, determine if the sequence is monotonic.

47. $a_n = 4 - \dfrac{1}{n}$

48. $a_n = \dfrac{4n}{n+1}$

49. $a_n = \dfrac{\cos n}{n}$

50. $a_n = ne^{-n/2}$

51. $a_n = (-1)^n \left(\dfrac{1}{n}\right)$

52. $a_n = \left(-\dfrac{2}{3}\right)^n$

53. $a_n = \left(\dfrac{2}{3}\right)^n$

54. $a_n = \left(\dfrac{3}{2}\right)^n$

55. $a_n = \sin \dfrac{n\pi}{6}$

56. $a_n = \dfrac{n}{2^{n+2}}$

In Exercises 57–60, use Theorem 10.5 to show that the sequence converges, and find the limit.

57. $a_n = 5 + \dfrac{1}{n}$

58. $a_n = 3 - \dfrac{4}{n}$

59. $a_n = \dfrac{1}{3}\left(1 - \dfrac{1}{3^n}\right)$

60. $a_n = 4 + \dfrac{1}{2^n}$

61. Consider the sequence $\{A_n\}$, whose nth term is given by

$$A_n = P\left(1 + \dfrac{r}{12}\right)^n$$

where P is the principal, A_n is the amount at compound interest after n months, and r is the annual percentage rate.

(a) Is $\{A_n\}$ a convergent series?

(b) Find the first ten terms of the sequence if $P =$ \$9,000 and $r = 0.115$.

62. A deposit of $100 is made each month in an account that earns 12-percent interest compounded monthly. The balance in the account after n months is given by

$$A_n = 100(101)[(1.01)^n - 1]$$

(a) Compute the first 6 terms of this sequence.
(b) Find the balance after 5 years by computing the 60th term of the sequence.
(c) Find the balance after 20 years by computing the 240th term of the sequence.

63. A government program that currently costs taxpayers $2.5 billion per year is to be cut back by 20 percent per year.
(a) Write an expression for the amount budgeted for this program after n years.
(b) Compute the budgets for the first 4 years.
(c) Determine the convergence or divergence of the sequence of reduced budgets. If the sequence converges, find its limit.

64. If the average price of a new car increases $5\frac{1}{2}$ percent per year and the average price is currently $11,000, then the average price after n years is

$$P_n = \$11,000(1.055)^n$$

Compute the average price for the first 5 years of increases.

65. Consider an idealized population with the characteristic that each population member produces 1 offspring at the end of every time period. If each population member has a life span of 3 time periods and the population begins with 10 newborn members, then the following table gives the population during the first 5 time periods.

Age bracket	Time period				
	1	2	3	4	5
0–1	10	10	20	40	70
1–2		10	10	20	40
2–3			10	10	20
Total	10	20	40	70	130

The sequence for the total population has the property that

$$S_n = S_{n-1} + S_{n-2} + S_{n-3}, \quad n > 3$$

Find the total population during the next 5 time periods.

66. Consider the sequence $\{x_n\}$ defined by

$$x_n = x_{n-1} - \frac{f(x_{n-1})}{f'(x_{n-1})}$$

If $f(x) = x^2 + x - 1$ and $x_1 = 0.5$, find the next three terms of the sequence.

67. Show that $\{ar^n/(1 - r)\}$ diverges if $|r| \geq 1$, and converges to 0 if $|r| < 1$.

68. Prove that if $\{s_n\}$ converges to L and $L > 0$, then there exists a number N such that $s_n > 0$ for $n > N$.

69. If $\{s_n\}$ is a convergent sequence, show that

$$\lim_{n \to \infty} s_{n-1} = \lim_{n \to \infty} s_n$$

10.3
Series and convergence

SECTION TOPICS •
Infinite series •
Convergence of an infinite series •
nth-Term Test for divergence •
Telescoping series •
Geometric series •

One important application of infinite sequences is their use in representing infinite summations.

DEFINITION OF AN INFINITE SERIES

If $\{a_n\}$ is an infinite sequence, then

$$\sum_{n=1}^{\infty} a_n = a_1 + a_2 + a_3 + \cdots + a_n + \cdots$$

is called an **infinite series** (or simply a **series**). The numbers a_1, a_2, a_3, . . . are called the **terms** of the series.

> **Remark** For some series it is convenient to begin the index at $n = 0$. As a typesetting convention, it is common to represent an infinite series as simply Σa_n. In such cases the starting point for the index ($n = 0$ or $n = 1$) must be taken from the context of the statement.

To find the sum of an infinite series, we consider the following **sequence of partial sums:**

$$S_1 = a_1$$
$$S_2 = a_1 + a_2$$
$$S_3 = a_1 + a_2 + a_3$$
$$\vdots$$
$$S_n = a_1 + a_2 + a_3 + \cdots + a_n$$

If this sequence converges, then we say that the series also converges and has the sum indicated in the following definition.

DEFINITION OF CONVERGENT AND DIVERGENT SERIES

For the infinite series Σa_n, the **nth partial sum** is given by

$$S_n = a_1 + a_2 + \cdots + a_n$$

If the sequence of partial sums $\{S_n\}$ converges to S, then we say that the series Σa_n **converges.** We call S the **sum of the series** and write

$$S = a_1 + a_2 + \cdots + a_n + \cdots$$

If $\{S_n\}$ diverges, then we say that the series also **diverges.**

This definition implies that a series can be identified with its sequence of partial sums. Thus, the following properties of infinite series are direct consequences of the corresponding properties of limits of sequences.

THEOREM 10.6

PROPERTIES OF INFINITE SERIES
If $\Sigma a_n = A$, $\Sigma b_n = B$, and c is a real number, then the following series converge to the indicated sums.

1. $\displaystyle\sum_{n=1}^{\infty} ca_n = cA$ 2. $\displaystyle\sum_{n=1}^{\infty} (a_n + b_n) = A + B$

3. $\displaystyle\sum_{n=1}^{\infty} (a_n - b_n) = A - B$

If the first N terms of a series are dropped, this will not destroy the convergence (or divergence) of the series. This is the substance of our next theorem.

THEOREM 10.7   DELETING THE FIRST N TERMS OF A SERIES
For any positive integer N, the series

$$\sum_{n=1}^{\infty} a_n = a_1 + a_2 + \cdots \qquad \text{and} \qquad \sum_{n=N+1}^{\infty} a_n = a_{N+1} + a_{N+2} + \cdots$$

both converge or both diverge.

Remark If both series in Theorem 10.7 converge, their sums differ by the partial sum S_N.

As you study this chapter, you will see that there are two basic questions involving infinite series. Does a series converge or does it diverge? If a series converges, what is its sum? These questions are not always easy to answer, especially the second one. We begin our pursuit for answers with a simple test for *divergence*.

THEOREM 10.8 nTH-TERM TEST FOR DIVERGENCE
If the sequence $\{a_n\}$ does not converge to 0, then the series Σa_n diverges.

Proof: For convenience, we prove the equivalent contrapositive—that is, if the series Σa_n converges, then the sequence $\{a_n\}$ converges to 0. Assume that the given series converges and that

$$\sum_{n=1}^{\infty} a_n = \lim_{n \to \infty} S_n = L$$

Then, because

$$S_n = S_{n-1} + a_n \qquad \text{and} \qquad \lim_{n \to \infty} S_n = \lim_{n \to \infty} S_{n-1} = L$$

it follows that

$$L = \lim_{n \to \infty} S_n = \lim_{n \to \infty} (S_{n-1} + a_n) = \lim_{n \to \infty} S_{n-1} + \lim_{n \to \infty} a_n = L + \lim_{n \to \infty} a_n$$

which requires that $\{a_n\}$ converges to 0.

Remark Be sure you see that this theorem does *not* state that the series Σa_n converges if $\{a_n\}$ converges to 0. Rather, the theorem states that the series diverges if $\{a_n\}$ does not converge to 0. In other words, the condition that the nth term of a series must approach zero as $n \to \infty$ is *necessary* for convergence to occur, but it is *not sufficient* to guarantee convergence.

EXAMPLE 1   Using the nth-Term Test

Using the nth-Term Test, determine which of the following series diverge.

(a) $\displaystyle\sum_{n=0}^{\infty} 2^n$ (b) $\displaystyle\sum_{n=0}^{\infty} \frac{1}{2^n}$ (c) $\displaystyle\sum_{n=1}^{\infty} \frac{n!}{2n! + 1}$ (d) $\displaystyle\sum_{n=1}^{\infty} \frac{10}{n}$

Solution: For the series in parts (b) and (d), we have, respectively,

$$\lim_{n\to\infty} \frac{1}{2^n} = 0 \quad \text{and} \quad \lim_{n\to\infty} \frac{10}{n} = 0$$

Hence, the nth-Term Test does not apply and we can draw *no* conclusions about convergence or divergence. (We will see later that one of these series diverges and the other converges.) For the series in parts (a) and (c), we have

$$\lim_{n\to\infty} 2^n = \infty \quad \text{and} \quad \lim_{n\to\infty} \frac{n!}{2n! + 1} = \lim_{n\to\infty} \frac{1}{2 + (1/n!)} = \frac{1}{2}$$

which imply, by the nth-Term Test, that both of the series diverge.   ☐

One way to find the sum of an infinite series is to find an expression for its nth partial sum S_n. This can require ingenuity, as demonstrated in the next example.

EXAMPLE 2   Using partial sums to test for convergence

Use the sequence $\{S_n\}$ to determine the sum of

$$\sum_{n=1}^{\infty} \frac{2}{4n^2 - 1}$$

Solution: If we had sufficient patience, we could compute several partial sums, such as

$$S_{10} \approx 0.952, \qquad S_{50} \approx 0.990, \qquad S_{100} \approx 0.995, \qquad S_{500} \approx 0.999$$

From these partial sums, we might suspect that this series converges to 1. But how can we be sure? In this case we can develop a formula for S_n by observing that the nth term of the series has the following partial fraction decomposition:

$$a_n = \frac{2}{4n^2 - 1} = \frac{2}{(2n - 1)(2n + 1)} = \frac{1}{2n - 1} - \frac{1}{2n + 1}$$

Now S_n can be written in **telescoping form,** in which each term after the first is canceled by its successor

$$S_n = \left(\frac{1}{1} - \frac{1}{3}\right) + \left(\frac{1}{3} - \frac{1}{5}\right) + \cdots + \left(\frac{1}{2n - 1} - \frac{1}{2n + 1}\right)$$

$$= 1 - \frac{1}{2n + 1}$$

Thus, the series converges to 1, since

$$\sum_{n=1}^{\infty} \frac{2}{4n^2 - 1} = \lim_{n\to\infty} S_n = \lim_{n\to\infty} \left(1 - \frac{1}{2n + 1}\right) = 1 \qquad ☐$$

In Example 2 we were able to use partial fractions to write the given series in **telescoping form.** When this is possible, it is usually a simple matter to derive a general formula for S_n from which we can then determine the sum of the series.

Geometric series

In the remaining portion of this section we will discuss a special type of series that has a simple test for convergence.

DEFINITION OF GEOMETRIC SERIES

The series given by

$$\sum_{n=0}^{\infty} ar^n = a + ar + ar^2 + \cdots + ar^n + \cdots, \quad a \neq 0$$

is called a **geometric series** with ratio r.

The conditions for the convergence or divergence of a geometric series are given in the following theorem.

THEOREM 10.9

CONVERGENCE OF A GEOMETRIC SERIES
A geometric series with ratio r diverges if $|r| \geq 1$. If $0 < |r| < 1$, then the series converges to the sum

$$\sum_{n=0}^{\infty} ar^n = \frac{a}{1-r}, \quad 0 < |r| < 1$$

Proof: To begin, we note that if $|r| \geq 1$, then $\{ar^n\}$ does not converge to 0, and by the *n*th-Term Test the series diverges. If $0 < |r| < 1$, then we let

$$S_n = a + ar + ar^2 + \cdots + ar^{n-1}$$

Multiplication by r yields

$$rS_n = ar + ar^2 + ar^3 + \cdots + ar^n$$

By subtracting these two equations, we obtain the telescoping sum

$$S_n = a + ar + ar^2 + \cdots + ar^{n-1}$$
$$-rS_n = \quad - ar - ar^2 - \cdots - ar^{n-1} - ar^n$$
$$S_n - rS_n = a - ar^n$$

Therefore, $S_n(1-r) = a(1-r^n)$, and we have

$$S_n = \frac{a}{1-r}(1-r^n)$$

Now, since $0 < |r| < 1$, it follows that $r^n \to 0$ as $n \to \infty$, and we obtain

$$\lim_{n \to \infty} S_n = \lim_{n \to \infty} \left[\frac{a}{1-r}(1 - r^n) \right] = \frac{a}{1-r}[\lim_{n \to \infty}(1 - r^n)] = \frac{a}{1-r}$$

which means the series *converges* to $a/(1 - r)$.

Remark Note that it is convenient to begin the index of a geometric series at $n = 0$.

EXAMPLE 3   Geometric series

(a) The geometric series

$$\sum_{n=0}^{\infty} \frac{3}{2^n} = \sum_{n=0}^{\infty} 3\left(\frac{1}{2}\right)^n = 3(1) + 3\left(\frac{1}{2}\right) + 3\left(\frac{1}{2}\right)^2 + \cdots$$

has a ratio of $r = \frac{1}{2}$ with $a = 3$. Since $0 < |r| < 1$, the series converges to

$$\frac{a}{1-r} = \frac{3}{1 - (1/2)} = 6$$

(b) The geometric series

$$\sum_{n=0}^{\infty} \left(\frac{3}{2}\right)^n = 1 + \frac{3}{2} + \frac{9}{4} + \frac{27}{8} + \cdots$$

has a ratio of $r = \frac{3}{2}$. Since $|r| \geq 1$, the series diverges.

EXAMPLE 4   A geometric series for a repeating decimal

Use a geometric series to express $0.08\overline{0808}$ as a rational number.

Solution: For

$$0.080808\ldots = \frac{8}{10^2} + \frac{8}{10^4} + \frac{8}{10^6} + \frac{8}{10^8} + \cdots = \sum_{n=0}^{\infty} \left(\frac{8}{10^2}\right)\left(\frac{1}{10^2}\right)^n$$

we have $a = 8/10^2$ and $r = 1/10^2$. Thus,

$$0.080808\ldots = \frac{a}{1-r} = \frac{8/10^2}{1 - (1/10^2)} = \frac{8}{99}$$

EXAMPLE 5   An application of geometric series

A ball is dropped from a height of 6 feet and begins bouncing as shown in Figure 10.9. The height of each bounce is three-fourths the height of the previous bounce. Find the total vertical distance traveled by the ball.

Solution: When the ball hits the ground for the first time, it has traveled a distance of $D_1 = 6$. For subsequent bounces, we let D_n be the distance traveled up *and* down, and obtain

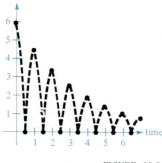

D

FIGURE 10.9

$$D_2 = 6\left(\frac{3}{4}\right) + 6\left(\frac{3}{4}\right) = 12\left(\frac{3}{4}\right)$$

$$\underbrace{\phantom{6\left(\frac{3}{4}\right)}}_{up} \quad \underbrace{\phantom{6\left(\frac{3}{4}\right)}}_{down}$$

$$D_3 = 6\left(\frac{3}{4}\right)\left(\frac{3}{4}\right) + 6\left(\frac{3}{4}\right)\left(\frac{3}{4}\right) = 12\left(\frac{3}{4}\right)^2$$

$$\underbrace{\phantom{6\left(\frac{3}{4}\right)\left(\frac{3}{4}\right)}}_{up} \quad \underbrace{\phantom{6\left(\frac{3}{4}\right)\left(\frac{3}{4}\right)}}_{down}$$

By continuing this process, we have a total vertical distance of

$$D = 6 + 12\left(\frac{3}{4}\right) + 12\left(\frac{3}{4}\right)^2 + 12\left(\frac{3}{4}\right)^3 + \cdots$$

To find the sum of this series, we rewrite the first term in geometric form as $6 = -6 + 12 = -6 + 12(\frac{3}{4})^0$. Then we obtain

$$D = -6 + 12\left(\frac{3}{4}\right)^0 + 12\left(\frac{3}{4}\right)^1 + 12\left(\frac{3}{4}\right)^2 + 12\left(\frac{3}{4}\right)^3 + \cdots$$

$$= -6 + \sum_{n=0}^{\infty} 12\left(\frac{3}{4}\right)^n = -6 + \frac{12}{1 - (3/4)} = -6 + 48 = 42 \text{ ft}$$

Section Exercises 10.3

In Exercises 1–6, for the given series find the first five terms of the sequence of partial sums.

1. $1 + \dfrac{1}{4} + \dfrac{1}{9} + \dfrac{1}{16} + \dfrac{1}{25} + \cdots$

2. $\dfrac{1}{2 \cdot 3} + \dfrac{2}{3 \cdot 4} + \dfrac{3}{4 \cdot 5} + \dfrac{4}{5 \cdot 6} + \dfrac{5}{6 \cdot 7} + \cdots$

3. $3 - \dfrac{9}{2} + \dfrac{27}{4} - \dfrac{81}{8} + \dfrac{243}{16} - \cdots$

4. $\dfrac{1}{1} + \dfrac{1}{3} + \dfrac{1}{5} + \dfrac{1}{7} + \dfrac{1}{9} + \dfrac{1}{11} + \cdots$

5. $\displaystyle\sum_{n=1}^{\infty} \dfrac{3}{2^{n-1}}$

6. $\displaystyle\sum_{n=1}^{\infty} \dfrac{(-1)^{n+1}}{n!}$

In Exercises 7–18, verify that the infinite series diverges.

7. $\dfrac{1}{2} + \dfrac{2}{3} + \dfrac{3}{4} + \dfrac{4}{5} + \cdots$

8. $3 - \dfrac{9}{2} + \dfrac{27}{4} - \dfrac{81}{8} + \dfrac{243}{16} - \cdots$

9. $\displaystyle\sum_{n=1}^{\infty} \dfrac{3n}{n+1} = \dfrac{3}{2} + \dfrac{6}{3} + \dfrac{9}{4} + \dfrac{12}{5} + \cdots$

10. $\displaystyle\sum_{n=1}^{\infty} \dfrac{n}{2n+3} = \dfrac{1}{5} + \dfrac{2}{7} + \dfrac{3}{9} + \dfrac{4}{11} + \cdots$

11. $\displaystyle\sum_{n=1}^{\infty} \dfrac{n^2}{n^2 + 1}$

12. $\displaystyle\sum_{n=1}^{\infty} \dfrac{n}{\sqrt{n^2 + 1}}$

13. $\displaystyle\sum_{n=0}^{\infty} 3\left(\dfrac{3}{2}\right)^n$

14. $\displaystyle\sum_{n=0}^{\infty} \left(\dfrac{4}{3}\right)^n$

15. $\displaystyle\sum_{n=0}^{\infty} 1000(1.055)^n$

16. $\displaystyle\sum_{n=0}^{\infty} 2(-1.03)^n$

17. $\displaystyle\sum_{n=1}^{\infty} \dfrac{2^n + 1}{2^{n+1}}$

18. $\displaystyle\sum_{n=1}^{\infty} \dfrac{n!}{2^n}$

In Exercises 19–24, verify that the infinite series converges.

19. $2 + \dfrac{3}{2} + \dfrac{9}{8} + \dfrac{27}{32} + \dfrac{81}{128} + \cdots$

20. $2 - 1 + \dfrac{1}{2} - \dfrac{1}{4} + \dfrac{1}{8} - \cdots$

21. $\displaystyle\sum_{n=0}^{\infty} (0.9)^n = 1 + 0.9 + 0.81 + 0.729 + \cdots$

22. $\displaystyle\sum_{n=0}^{\infty} (-0.6)^n = 1 - 0.6 + 0.36 - 0.216 + \cdots$

23. $\displaystyle\sum_{n=1}^{\infty} \dfrac{1}{n(n + 1)}$ (Use partial fractions.)

24. $\displaystyle\sum_{n=1}^{\infty} \frac{1}{n(n + 2)}$ (Use partial fractions.)

In Exercises 25–40, find the sum of the convergent series.

25. $\displaystyle\sum_{n=0}^{\infty} \left(\frac{1}{2}\right)^n$

26. $\displaystyle\sum_{n=0}^{\infty} 2\left(\frac{2}{3}\right)^n$

27. $\displaystyle\sum_{n=0}^{\infty} \left(\frac{-1}{2}\right)^n$

28. $\displaystyle\sum_{n=0}^{\infty} 2\left(\frac{-2}{3}\right)^n$

29. $\displaystyle\sum_{n=0}^{\infty} 2\left(\frac{1}{\sqrt{2}}\right)^n = 2 + \sqrt{2} + 1 + \frac{1}{\sqrt{2}} + \cdots$

30. $4 + 1 + \dfrac{1}{4} + \dfrac{1}{16} + \cdots$

31. $1 + 0.1 + 0.01 + 0.001 + \cdots$

32. $8 + 6 + \dfrac{9}{2} + \dfrac{27}{8} + \cdots$

33. $3 - 1 + \dfrac{1}{3} - \dfrac{1}{9} + \cdots$

34. $4 - 2 + 1 - \dfrac{1}{2} + \cdots$

35. $\displaystyle\sum_{n=2}^{\infty} \frac{1}{n^2 - 1}$

36. $\displaystyle\sum_{n=1}^{\infty} \frac{1}{n(n + 1)}$

37. $\displaystyle\sum_{n=1}^{\infty} \frac{4}{n(n + 2)}$

38. $\displaystyle\sum_{n=1}^{\infty} \frac{1}{(2n + 1)(2n + 3)}$

39. $\displaystyle\sum_{n=0}^{\infty} \left(\frac{1}{2^n} - \frac{1}{3^n}\right)$

40. $\displaystyle\sum_{n=1}^{\infty} [(0.7)^n + (0.9)^n]$

In Exercises 41–44, express the repeated decimal as a geometric series, and write its sum as the ratio of two integers.

41. $0.6\overline{666}$

42. $0.2\overline{323}$

43. $0.07\overline{575}$

44. $0.215\overline{15}$

In Exercises 45–56, determine the convergence or divergence of the series.

45. $\displaystyle\sum_{n=1}^{\infty} \frac{n + 10}{10n + 1}$

46. $\displaystyle\sum_{n=0}^{\infty} \frac{4}{2^n}$

47. $\displaystyle\sum_{n=1}^{\infty} \left(\frac{1}{n} - \frac{1}{n + 2}\right)$

48. $\displaystyle\sum_{n=1}^{\infty} \frac{n + 1}{2n - 1}$

49. $\displaystyle\sum_{n=1}^{\infty} \frac{3n - 1}{2n + 1}$

50. $\displaystyle\sum_{n=0}^{\infty} \frac{1}{4^n}$

51. $\displaystyle\sum_{n=0}^{\infty} (1.075)^n$

52. $\displaystyle\sum_{n=1}^{\infty} \frac{2^n}{100}$

53. $\displaystyle\sum_{n=2}^{\infty} \frac{n}{\ln n}$

54. $\displaystyle\sum_{n=1}^{\infty} \frac{2^n}{n^2}$

55. $\displaystyle\sum_{n=1}^{\infty} \left(1 + \frac{k}{n}\right)^n$

56. $\displaystyle\sum_{n=1}^{\infty} \frac{1}{n(n + 3)}$

57. A company producing a new product estimates the annual sales to be 8000 units. Suppose that in any given year 10 percent of the units (regardless of age) will become inoperative. How many units will be in use after n years?

58. Repeat Exercise 57 with the assumption that 25 percent of the units will become inoperative each year.

59. A ball is dropped from a height of 16 feet. Each time it drops h feet, it rebounds vertically $0.81h$ feet. Find the total distance traveled by the ball.

60. Find the total time it takes for the ball in Exercise 59 to come to rest.

61. Find the fraction of the total area of the square that is eventually shaded if the pattern of shading shown in Figure 10.10 is continued. (Note that each side of the shaded corner squares is one-fourth that of the square in which it is placed.)

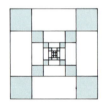

FIGURE 10.10

In Exercises 62–66, use the formula for the nth partial sum of a geometric series:

$$\sum_{i=0}^{n-1} ar^i = \frac{a(1 - r^n)}{1 - r}$$

62. A deposit of $100 is made each month for 5 years in an account that pays 10-percent interest compounded monthly. What is the balance A in the account at the end of the 5 years?

$$A = 100\left(1 + \frac{0.10}{12}\right) + \cdots + 100\left(1 + \frac{0.10}{12}\right)^{60}$$

63. A deposit of $50 is made each month in an account that pays 12-percent interest compounded monthly. What is the balance in the account at the end of 10 years?

64. A deposit of P dollars is made each month for t years in an account that pays interest at an annual rate of r, compounded monthly. Let $N = 12t$ be the total number of deposits, and show that the balance in the account after t years is

$$A = P\left[\left(1 + \frac{r}{12}\right)^N - 1\right]\left(1 + \frac{12}{r}\right)$$

65. Use the formula in Exercise 64 to find the amount in an account earning 9-percent interest compounded monthly after deposits of $50 have been made monthly for 40 years.

66. The number of direct ancestors a person has had is given by

$$2 + 2^2 + 2^3 + 2^4 + \cdots + 2^n + \cdots$$

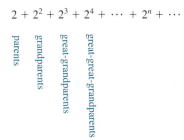

This formula is valid *provided* the person has had no common ancestors. [A common ancestor is one to whom you are related in more than one way. For example, one of your great-grandmothers on your father's side might also be one of your great-grandmothers on your mother's side, as shown in Figure 10.11.] How

Have you had common ancestors?

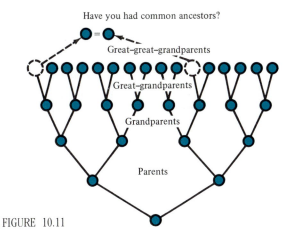

FIGURE 10.11

many direct ancestors have you had who lived since the year 1 A.D.? Assume that the average time between generations was 30 years (resulting in 66 generations) so that the total is given by

$$2 + 2^2 + 2^3 + 2^4 + \cdots + 2^{66}$$

Considering your total, is it reasonable to assume that you have had no common ancestors in the past 2000 years?

67. Prove that $0.75 = 0.749999\ldots$.

68. Prove that every decimal with a repeating pattern of digits is a rational number.

69. Show that the series

$$\sum_{n=1}^{\infty} a_n$$

can be written in the telescoping form

$$\sum_{n=1}^{\infty} [(c - S_{n-1}) - (c - S_n)]$$

where $S_0 = 0$ and S_n is the nth partial sum.

70. Let Σa_n be a convergent series, and let

$$R_N = a_{N+1} + a_{N+2} + \cdots$$

be the remainder of the series after the first N terms. Prove that

$$\lim_{N \to \infty} R_N = 0$$

10.4
The Integral Test and *p*-Series

In this and the following section, we look at several convergence tests that apply to series with *positive* terms.

Since a definite integral is defined as the limit of a sum, it seems logical to expect that we might be able to use integrals to test for convergence or divergence of an infinite series. This is indeed the case, and we show how this is done in the proof of the following theorem.

THEOREM 10.10

INTEGRAL TEST

If f is positive, continuous, and decreasing for $x \geq 1$ and $a_n = f(n)$, then

$$\sum_{n=1}^{\infty} a_n \quad \text{and} \quad \int_1^{\infty} f(x)\, dx$$

either both converge or both diverge. If the series converges to S, then the remainder $R_N = S - S_N$ is bounded by

$$0 \leq R_N \leq \int_N^{\infty} f(x)\, dx$$

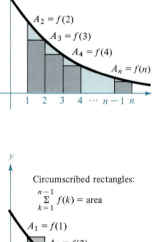

Inscribed rectangles:
$$\sum_{k=2}^{n} f(k) = \text{area}$$

$A_2 = f(2)$
$A_3 = f(3)$
$A_4 = f(4)$
$A_n = f(n)$

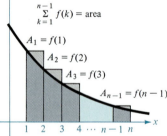

Circumscribed rectangles:
$$\sum_{k=1}^{n-1} f(k) = \text{area}$$

$A_1 = f(1)$
$A_2 = f(2)$
$A_3 = f(3)$
$A_{n-1} = f(n-1)$

FIGURE 10.12

Proof: We begin by partitioning the interval $[1, n]$ into $n - 1$ unit intervals, as shown in Figure 10.12. The total area of the inscribed rectangles is given by

$$\sum_{i=2}^{n} f(i) = f(2) + f(3) + \cdots + f(n) \qquad \text{Inscribed area}$$

Similarly, the total area of the circumscribed rectangles is

$$\sum_{i=1}^{n-1} f(i) = f(1) + f(2) + \cdots + f(n - 1) \qquad \text{Circumscribed area}$$

The exact area under the graph of f from $x = 1$ to $x = n$, $\int_1^n f(x)\, dx$, lies between the inscribed and circumscribed areas, and we have

$$\sum_{i=2}^{n} f(i) \leq \int_1^n f(x)\, dx \leq \sum_{i=1}^{n-1} f(i)$$

Using the nth partial sum, $S_n = f(1) + f(2) + \cdots + f(n)$, we can write this inequality as

$$S_n - f(1) \leq \int_1^n f(x)\, dx \leq S_{n-1}$$

Now assuming that $\int_1^{\infty} f(x)\, dx$ converges to $L > 0$, it follows that for $n \geq 1$

$$S_n - f(1) \leq L \implies S_n \leq L + f(1)$$

Consequently, $\{S_n\}$ is bounded and monotonic, and by Theorem 10.5 it converges. Thus, Σa_n converges.

For the other direction of the proof, we assume the improper integral diverges. Then $\int_1^n f(x)\, dx$ approaches infinity as $n \to \infty$, and the inequality

$$S_{n-1} \geq \int_1^n f(x)\, dx$$

implies that $\{S_n\}$ diverges. Thus, Σa_n diverges.

Finally, for $n > N$, we have

$$S_n - f(1) \leq \int_1^n f(x)\, dx \quad \text{and} \quad S_N - f(1) \leq \int_1^N f(x)\, dx$$

Subtracting the second inequality from the first, we have

$$S_n - S_N \leq \int_N^n f(x) \, dx$$

Assuming that both the integral and the series converge, we can take the limit as $n \to \infty$ to conclude that

$$R_N = S - S_N \leq \int_N^\infty f(x) \, dx$$

| **Remark** We noted earlier that the convergence or divergence of Σa_n is not affected by deleting the first N terms. Similarly, if the conditions for the Integral Test are satisfied for all $x \geq N > 1$, we simply use the integral $\int_N^\infty f(x) \, dx$ to test for convergence or divergence. (This is illustrated in Example 4.)

EXAMPLE 1   *Using the Integral Test*

Apply the Integral Test to the series

$$\sum_{n=1}^\infty \frac{n}{n^2 + 1}$$

Solution: Since $f(x) = x/(x^2 + 1)$ satisfies the conditions for the Integral Test, we integrate to obtain

$$\int_1^\infty \frac{x}{x^2 + 1} \, dx = \frac{1}{2} \int_1^\infty \frac{2x}{x^2 + 1} \, dx$$

$$= \frac{1}{2} \lim_{b \to \infty} \left[\ln (x^2 + 1) \right]_1^b$$

$$= \frac{1}{2} \lim_{b \to \infty} [\ln (b^2 + 1) - \ln 2] = \infty$$

Thus, the series *diverges*.

EXAMPLE 2   *Using the Integral Test*

Apply the Integral Test to the series

$$\sum_{n=1}^\infty \frac{1}{n^2 + 1}$$

Solution: Since $f(x) = 1/(x^2 + 1)$ satisfies the conditions for the Integral Test, we integrate to obtain

$$\int_1^\infty \frac{1}{x^2 + 1} \, dx = \lim_{b \to \infty} \left[\arctan x \right]_1^b$$

$$= \lim_{b \to \infty} [\arctan b - \arctan 1] = \frac{\pi}{4}$$

Thus, the series *converges*.

EXAMPLE 3   Approximating the sum of a series by the Integral Test

Approximate the sum of the following series using its first ten terms:

$$\sum_{n=1}^{\infty} \frac{n}{e^n}$$

Use the Integral Test remainder to determine the maximum error in the approximation.

Solution: Since $f(x) = x/e^x$ satisfies the conditions for the Integral Test, we integrate by parts, to obtain

$$\int_1^{\infty} \frac{x}{e^x}\, dx = \lim_{b \to \infty} \left[-e^{-x}(x+1) \right]_1^b = \lim_{b \to \infty} \left[\frac{b+1}{-e^b} + \frac{2}{e} \right] = \frac{2}{e}$$

Thus, the series converges and has a sum S. Furthermore, since

$$R_{10} \le \int_{10}^{\infty} \frac{x}{e^x}\, dx = \frac{11}{e^{10}} \approx 0.00050$$

and

$$S_{10} = \frac{1}{e} + \frac{2}{e^2} + \frac{3}{e^3} + \cdots + \frac{10}{e^{10}} \approx 0.92037$$

Now, since the terms of the series are positive, we know that

$$S_{10} < S \le S_{10} + 0.00050$$

from which we conclude that

$$0.92037 < S \le 0.92087$$

In the remainder of this section, we investigate a second type of series that has a simple arithmetic test for convergence or divergence.

DEFINITION OF *p*-SERIES

A series of the form

$$\sum_{n=1}^{\infty} \frac{1}{n^p} = \frac{1}{1^p} + \frac{1}{2^p} + \frac{1}{3^p} + \cdots$$

is called a **p-series,** where p is a positive constant. For $p = 1$, the series

$$\sum_{n=1}^{\infty} \frac{1}{n} = 1 + \frac{1}{2} + \frac{1}{3} + \cdots$$

is called the **harmonic series.**

Remark A general harmonic series is of the form $\Sigma[1/(an + b)]$. In music, strings of the same material, diameter, and tension, whose lengths form a harmonic series, produce harmonic tones.

The Integral Test is convenient for establishing the convergence or divergence of p-series. This is shown in the proof of Theorem 10.11.

THEOREM 10.11   CONVERGENCE OF p-SERIES
The p-series

$$\sum_{n=1}^{\infty} \frac{1}{n^p} = \frac{1}{1^p} + \frac{1}{2^p} + \frac{1}{3^p} + \cdots$$

1. Diverges if $0 < p \leq 1$.

2. Converges if $p > 1$, and its sum S has a remainder $R_N = S - S_N$ bounded by

$$0 < R_N < \frac{N^{1-p}}{p-1}$$

Proof: The proof that the series diverges if $p \leq 1$ and converges if $p > 1$ follows from Theorem 9.3, together with the Integral Test. Moreover, if $p > 1$, then

$$\int_N^{\infty} \frac{1}{x^p}\,dx = \lim_{b \to \infty} \left[\frac{x^{1-p}}{1-p} \right]_N^b$$

$$= \left(\frac{1}{1-p} \right) \lim_{b \to \infty} [b^{1-p} - N^{1-p}]$$

$$= \frac{-N^{1-p}}{1-p} = \frac{N^{1-p}}{p-1}$$

Therefore, from the Integral Test we obtain

$$0 < R_N < \int_N^{\infty} \frac{1}{x^p}\,dx = \frac{N^{1-p}}{p-1}$$

It follows from Theorem 10.11 that the *harmonic series*

$$\sum_{n=1}^{\infty} \frac{1}{n} = \frac{1}{1} + \frac{1}{2} + \frac{1}{3} + \cdots \qquad p = 1$$

diverges, whereas the series

$$\sum_{n=1}^{\infty} \frac{1}{n^{1.1}} = \frac{1}{1^{1.1}} + \frac{1}{2^{1.1}} + \frac{1}{3^{1.1}} + \cdots \qquad p = 1.1$$

whose terms are only *slightly* smaller, converges. The next example shows another series whose terms are also slightly smaller than those of the harmonic series.

EXAMPLE 4   A comparison to the harmonic series

Determine if the following series converges or diverges:

$$\sum_{n=2}^{\infty} \frac{1}{n \ln n}$$

Solution: This series is similar to the divergent harmonic series. If its terms were larger than those of the harmonic series, we would expect it to diverge. However, since its terms are smaller, we are not sure what to expect. Using the Integral Test, with $f(x) = 1/(x \ln x)$, we see that the series diverges.

$$\int_{2}^{\infty} \frac{1}{x \ln x}\, dx = \int_{2}^{\infty} \frac{1/x}{\ln x}\, dx$$

$$= \lim_{b \to \infty} \left[\ln (\ln x) \right]_{2}^{b}$$

$$= \lim_{b \to \infty} [\ln (\ln b) - \ln (\ln 2)] = \infty$$

EXAMPLE 5   Approximation of the sum of a p-series

Approximate the sum of the series

$$\frac{2}{1} + \frac{2}{4} + \frac{2}{9} + \frac{2}{16} + \cdots$$

using its first ten terms. Use the *p*-series remainder to determine the maximum error of the approximation.

Solution: The *n*th term of this series is $a_n = 2/n^2$. Therefore, by Theorem 10.11, it is a convergent *p*-series whose sum is

$$S = \sum_{n=1}^{\infty} \frac{2}{n^2}$$

$$= 2 \sum_{n=1}^{\infty} \frac{1}{n^2}$$

The tenth partial sum of this series is

$$S_{10} = \frac{2}{1} + \frac{2}{2^2} + \frac{2}{3^2} + \cdots + \frac{2}{10^2} \approx 3.09954$$

Finally, using the *p*-series remainder (with $p = 2$), we have a maximum error of

$$R_{10} < 2\left(\frac{10^{1-2}}{2 - 1} \right) = 0.2$$

and we conclude that $S_{10} < S < S_{10} + 0.2$. That is,

$$3.09954 < S < 3.29954$$

Section Exercises 10.4

In Exercises 1–10, determine the convergence or divergence of the given *p*-series.

1. $\displaystyle\sum_{n=1}^{\infty} \frac{1}{n^3}$

2. $\displaystyle\sum_{n=1}^{\infty} \frac{1}{n^{1/3}}$

3. $\displaystyle\sum_{n=1}^{\infty} \frac{1}{\sqrt[5]{n}}$

4. $\displaystyle\sum_{n=1}^{\infty} \frac{1}{n^{4/3}}$

5. $1 + \dfrac{1}{\sqrt{2}} + \dfrac{1}{\sqrt{3}} + \dfrac{1}{\sqrt{4}} + \cdots$

6. $1 + \dfrac{1}{4} + \dfrac{1}{9} + \dfrac{1}{16} + \dfrac{1}{25} + \cdots$

7. $1 + \dfrac{1}{2\sqrt{2}} + \dfrac{1}{3\sqrt{3}} + \dfrac{1}{4\sqrt{4}} + \dfrac{1}{5\sqrt{5}} + \cdots$

8. $1 + \dfrac{1}{\sqrt[3]{4}} + \dfrac{1}{\sqrt[3]{9}} + \dfrac{1}{\sqrt[3]{16}} + \dfrac{1}{\sqrt[3]{25}} + \cdots$

9. $\displaystyle\sum_{n=1}^{\infty} \frac{1}{n^{1.04}}$

10. $\displaystyle\sum_{n=1}^{\infty} \frac{1}{n^{\pi}}$

In Exercises 11–20, determine the convergence or divergence of the given series using the Integral Test.

11. $\displaystyle\sum_{n=1}^{\infty} \frac{1}{n+1}$

12. $\displaystyle\sum_{n=1}^{\infty} e^{-n}$

13. $\displaystyle\sum_{n=1}^{\infty} ne^{-n}$

14. $\displaystyle\sum_{n=1}^{\infty} e^{-n} \cos n$

15. $\dfrac{1}{2} + \dfrac{1}{5} + \dfrac{1}{10} + \dfrac{1}{17} + \dfrac{1}{26} + \cdots$

16. $\dfrac{1}{3} + \dfrac{1}{5} + \dfrac{1}{7} + \dfrac{1}{9} + \dfrac{1}{11} + \cdots$

17. $\dfrac{\ln 2}{2} + \dfrac{\ln 3}{3} + \dfrac{\ln 4}{4} + \dfrac{\ln 5}{5} + \dfrac{\ln 6}{6} + \cdots$

18. $\dfrac{1}{4} + \dfrac{2}{7} + \dfrac{3}{12} + \cdots + \dfrac{n}{n^2 + 3} + \cdots$

19. $\displaystyle\sum_{n=1}^{\infty} \frac{n^{k-1}}{n^k + c}$, *k* is a positive integer

20. $\displaystyle\sum_{n=1}^{\infty} n^k e^{-n}$, *k* is a positive integer

In Exercises 21–30, determine the convergence or divergence of the given series.

21. $\displaystyle\sum_{n=1}^{\infty} \frac{1}{2n-1}$

22. $\displaystyle\sum_{n=2}^{\infty} \frac{1}{n\sqrt{n^2 - 1}}$

23. $\displaystyle\sum_{n=1}^{\infty} \frac{1}{n\sqrt[3]{n}}$

24. $3 \displaystyle\sum_{n=1}^{\infty} \frac{1}{n^{0.95}}$

25. $\displaystyle\sum_{n=0}^{\infty} \left(\frac{2}{3}\right)^n$

26. $\displaystyle\sum_{n=1}^{\infty} \frac{n}{\sqrt{n^2 + 1}}$

27. $\displaystyle\sum_{n=0}^{\infty} (1.075)^n$

28. $\displaystyle\sum_{n=1}^{\infty} \left(\frac{1}{n^2} - \frac{1}{n^3}\right)$

29. $\displaystyle\sum_{n=1}^{\infty} \left(1 + \frac{1}{n}\right)^n$

30. $\displaystyle\sum_{n=2}^{\infty} \ln n$

In Exercises 31–36, approximate the sum of the given convergent series using the indicated number of terms. Include an estimate of the maximum error for your approximation.

31. $\displaystyle\sum_{n=1}^{\infty} (0.075)^n$, six terms

32. $\displaystyle\sum_{n=1}^{\infty} \frac{1}{n^5}$, four terms

33. $\displaystyle\sum_{n=1}^{\infty} \frac{1}{n^2 + 1}$, ten terms

34. $\displaystyle\sum_{n=1}^{\infty} \frac{1}{(n+1)[\ln (n+1)]^3}$, ten terms

35. $\displaystyle\sum_{n=1}^{\infty} ne^{-n^2}$, four terms

36. $\displaystyle\sum_{n=1}^{\infty} e^{-n}$, four terms

In Exercises 37–40, find *N* so that $R_N < 0.001$ for the given convergent series.

37. $\displaystyle\sum_{n=1}^{\infty} \frac{1}{n^4}$

38. $\displaystyle\sum_{n=1}^{\infty} \frac{1}{n^{3/2}}$

39. $\displaystyle\sum_{n=1}^{\infty} e^{-5n}$

40. $\displaystyle\sum_{n=1}^{\infty} \frac{1}{n^2 + 1}$

10.5
Comparisons of series

SECTION TOPICS ▪
Direct Comparison Test ▪
Limit Comparison Test ▪

In the previous two sections we developed tests for convergence or divergence of special types of series. In each case the terms of the series were quite simple, and the series *had to possess* special characteristics in order for the convergence tests to be applied. The slightest deviation from these special characteristics could make a test nonapplicable. For example, notice in the following pairs that the second series cannot be tested by the same convergence test as the first series even though it is similar to the first.

1. $\displaystyle\sum_{n=0}^{\infty} \frac{1}{2^n}$ is geometric, but $\displaystyle\sum_{n=0}^{\infty} \frac{n}{2^n}$ is not.

2. $\displaystyle\sum_{n=1}^{\infty} \frac{1}{n^3}$ is a *p*-series, but $\displaystyle\sum_{n=1}^{\infty} \frac{1}{n^3 + 1}$ is not.

3. $a_n = \dfrac{n}{(n^2 + 3)^2}$ is easily integrated, but $b_n = \dfrac{n^2}{(n^2 + 3)^2}$ is not.

In this section we discuss two additional tests for positive term series. These two tests greatly expand the variety of series we are able to test for convergence or divergence by allowing us to *compare* a series, having similar but more complicated terms, to a simpler series whose convergence or divergence is known.

THEOREM 10.12   **DIRECT COMPARISON TEST**
Let $0 \le a_n \le b_n$ for all n.

1. If $\displaystyle\sum_{n=1}^{\infty} b_n$ converges, then $\displaystyle\sum_{n=1}^{\infty} a_n$ converges.

2. If $\displaystyle\sum_{n=1}^{\infty} a_n$ diverges, then $\displaystyle\sum_{n=1}^{\infty} b_n$ diverges.

Proof: To prove the first property, let

$$L = \sum_{n=1}^{\infty} b_n \quad \text{and} \quad S_n = a_1 + a_2 + \cdots + a_n$$

Since $0 \le a_n \le b_n$, we know that the sequence

$$S_1, S_2, S_3, \ldots$$

is nondecreasing and bounded above by L; hence it must converge. Moreover, since

$$\lim_{n \to \infty} S_n = \sum_{n=1}^{\infty} a_n$$

it follows that Σa_n converges. The second property is logically equivalent to the first, and the proof is complete.

| Remark   As stated, the Direct Comparison Test requires that $0 \leq a_n \leq b_n$ for all n. Since the convergence of a series is not dependent on its first several terms, we could modify the test to require only that $0 \leq a_n \leq b_n$ for all n greater than some integer N.

EXAMPLE 1   *Using the Direct Comparison Test*

Use the Direct Comparison Test on

$$\sum_{n=1}^{\infty} \frac{1}{2 + 3^n}$$

Solution: This series resembles

$$\sum_{n=1}^{\infty} \frac{1}{3^n} \qquad \text{Convergent geometric series}$$

Term-by-term comparison yields

$$a_n = \frac{1}{2 + 3^n} < \frac{1}{3^n} = b_n, \quad n \geq 1$$

Thus, it follows by the Direct Comparison Test that the series converges.

EXAMPLE 2   *Using the Direct Comparison Test*

Use the Direct Comparison Test on

$$\sum_{n=1}^{\infty} \frac{1}{2 + \sqrt{n}}$$

Solution: This series resembles

$$\sum_{n=1}^{\infty} \frac{1}{n^{1/2}} \qquad \text{Divergent } p\text{-series}$$

Term-by-term comparison yields

$$\frac{1}{2 + \sqrt{n}} \leq \frac{1}{\sqrt{n}}, \qquad n \geq 1$$

which *does not* meet the requirements for divergence. (Remember that if term-by-term comparison reveals a series that is *smaller* than a divergent series, the Direct Comparison Test tells us nothing.) Still expecting the series to diverge, we compare the given series to

$$\sum_{n=1}^{\infty} \frac{1}{n} \qquad \text{Divergent harmonic series}$$

In this case term-by-term comparison yields

$$a_n = \frac{1}{n} \leq \frac{1}{2 + \sqrt{n}} = b_n, \quad n \geq 4$$

and, by the Direct Comparison Test, the given series diverges.

Remark   To convince yourself of the validity of the last inequality in Example 2, try showing that $2 + \sqrt{n} \le n$ whenever $n \ge 4$.

Be careful when using the Direct Comparison Test. Keep in mind that $0 \le a_n \le b_n$ for both parts of the theorem. Informally the test says that for series with nonnegative terms:

1. If the "larger" series Σb_n converges, then the "smaller" series Σa_n must also converge.
2. If the "smaller" series Σa_n diverges, then the "larger" series Σb_n must also diverge.

The Direct Comparison Test can be used to approximate the sum of a convergent series. For instance, consider the two series

$$\sum_{n=0}^{\infty} \frac{1}{n!} = L \quad \text{and} \quad \sum_{n=0}^{\infty} \frac{1}{2^n} = 2$$

Since

$$\frac{1}{n!} \le \frac{1}{2^n}, \quad n > 3$$

it follows that for $N > 3$ the remainder for the factorial series is less than the remainder for the geometric series. Now, if we want to approximate the sum of the factorial series (with an error of less than 0.001), we find an N such that $R_N < 0.001$ for the geometric series. For this geometric series, we have $a = 1$ and $r = \frac{1}{2}$. Thus,

$$R_N = 2 - S_N = 2 - \frac{a(1 - r^N)}{1 - r} = 2 - \frac{(1)[1 - (1/2)^N]}{1 - (1/2)} = 2\left(\frac{1}{2}\right)^N < 0.001$$

which implies that we can choose $N = 11$. Therefore, we approximate the sum of the factorial series, with an error of less than 0.001, to be

$$\sum_{n=0}^{11} \frac{1}{n!} = \frac{1}{0!} + \frac{1}{1!} + \frac{1}{2!} + \cdots + \frac{1}{11!} \approx 2.718$$

Later in this chapter we will see that the exact sum of this series is e.

Limit Comparison Test

Often a given series closely resembles a p-series or a geometric series, yet we cannot establish the term-by-term comparison necessary to apply the Direct Comparison Test. Under these circumstances we may be able to apply a second comparison test, called the **Limit Comparison Test.**

THEOREM 10.13   LIMIT COMPARISON TEST
Suppose a_n, $b_n > 0$, and

$$\lim_{n \to \infty} \left(\frac{a_n}{b_n}\right) = L$$

where L is *finite and positive*. Then the two series Σa_n and Σb_n either both converge or both diverge.

Proof: Since a_n, $b_n > 0$, and $(a_n/b_n) \to L$ as $n \to \infty$, there exists $N > 0$ such that for $n \geq N$, $0 < (a_n/b_n) < (L + 1)$. This implies that

$$0 < a_n < (L + 1)b_n$$

Hence, by the Direct Comparison Test, the convergence of Σb_n implies the convergence of Σa_n. Similarly, the fact that $(b_n/a_n) \to (1/L)$ as $n \to \infty$ can be used to show that the convergence of Σa_n implies the convergence of Σb_n.

The versatility of the Limit Comparison Test is seen in the next example, where we show that a **general harmonic series** diverges.

| **Remark** As with the Direct Comparison Test, the Limit Comparison Test could be modified to require only that a_n and b_n be positive for all n greater than some integer M.

EXAMPLE 3   Using the Limit Comparison Test

Show that the following general harmonic series diverges:

$$\sum_{n=1}^{\infty} \frac{1}{an + b}, \quad a > 0$$

Solution: By comparison to

$$\sum_{n=1}^{\infty} \frac{1}{n} \qquad \text{Divergent harmonic series}$$

we have

$$\lim_{n \to \infty} \frac{1/(an + b)}{1/n} = \lim_{n \to \infty} \frac{n}{an + b} = \frac{1}{a}$$

Since this limit is greater than 0, we conclude from the Limit Comparison Test that the given series diverges.

The Limit Comparison Test works well for comparing "messy" algebraic series to a *p*-series. In choosing an appropriate *p*-series, we must choose one with an *n*th term of the same magnitude as the *n*th term of the given series. For example:

1. Given $\displaystyle\sum \frac{1}{3n^2 - 4n + 5}$, choose $\displaystyle\sum \frac{1}{n^2}$.

2. Given $\displaystyle\sum \frac{1}{\sqrt{3n - 2}}$, choose $\displaystyle\sum \frac{1}{\sqrt{n}}$.

3. Given $\displaystyle\sum \frac{n^2 - 10}{4n^5 + n^3}$, choose $\displaystyle\sum \frac{n^2}{n^5} = \sum \frac{1}{n^3}$.

4. Given $\displaystyle\sum \frac{\sqrt{n}}{\sqrt{n^3 + 1}}$, choose $\displaystyle\sum \frac{n^{1/2}}{n^{3/2}} = \sum \frac{1}{n}$.

In other words, when choosing a series to which we want to make a comparison, we disregard all but the *highest powers of n* in both the numerator and the denominator.

EXAMPLE 4   Using the Limit Comparison Test

Use the Limit Comparison Test on

$$\sum_{n=1}^{\infty} \frac{4\sqrt{n} - 1}{n^2 + 2\sqrt{n}}$$

Solution: Disregarding all but the highest powers of n in the numerator and the denominator, we compare the series to

$$\sum_{n=1}^{\infty} \frac{\sqrt{n}}{n^2} = \sum_{n=1}^{\infty} \frac{1}{n^{3/2}} \qquad \text{Convergent } p\text{-series}$$

Since

$$\lim_{n \to \infty} \frac{a_n}{b_n} = \lim_{n \to \infty} \left(\frac{4\sqrt{n} - 1}{n^2 + 2\sqrt{n}} \right) \left(\frac{n^{3/2}}{1} \right)$$

$$= \lim_{n \to \infty} \frac{4n^2 - n^{3/2}}{n^2 + 2\sqrt{n}} = 4$$

we conclude by the Limit Comparison Test that the given series converges.

EXAMPLE 5   Using the Limit Comparison Test

Use the Limit Comparison Test on

$$\sum_{n=1}^{\infty} \frac{n2^n + 5}{4n^3 + 3n}$$

Solution: A reasonable comparison would be to the series

$$\sum_{n=1}^{\infty} \frac{2^n}{n^2}$$

Note that this series diverges by the nth-Term Test, since

$$\lim_{n \to \infty} \frac{2^n}{n^2} \neq 0$$

From the limit

$$\lim_{n \to \infty} \frac{a_n}{b_n} = \lim_{n \to \infty} \left(\frac{n2^n + 5}{4n^3 + 3n} \right) \left(\frac{n^2}{2^n} \right)$$

$$= \lim_{n \to \infty} \frac{1 + [5/(2^n n)]}{4 + (3/n^2)} = \frac{1}{4}$$

we conclude that the given series diverges.

Section Exercises 10.5

In Exercises 1–12, use the Direct Comparison Test to determine the convergence or divergence of the series.

1. $\sum_{n=1}^{\infty} \dfrac{1}{n^2+1}$

2. $\sum_{n=1}^{\infty} \dfrac{1}{3n^2+2}$

3. $\sum_{n=2}^{\infty} \dfrac{1}{n-1}$

4. $\sum_{n=2}^{\infty} \dfrac{1}{\sqrt{n}-1}$

5. $\sum_{n=0}^{\infty} \dfrac{1}{3^n+1}$

6. $\sum_{n=0}^{\infty} \dfrac{2^n}{3^n+5}$

7. $\sum_{n=2}^{\infty} \dfrac{\ln n}{n+1}$

8. $\sum_{n=1}^{\infty} \dfrac{1}{\sqrt{n^3+1}}$

9. $\sum_{n=0}^{\infty} \dfrac{1}{n!}$

10. $\sum_{n=1}^{\infty} \dfrac{1}{3\sqrt[4]{n}-1}$

11. $\sum_{n=0}^{\infty} e^{-n^2}$

12. $\sum_{n=0}^{\infty} \dfrac{4^n}{3^n-1}$

In Exercises 13–26, use the Limit Comparison Test to determine the convergence or divergence of the series.

13. $\sum_{n=1}^{\infty} \dfrac{n}{n^2+1}$

14. $\sum_{n=2}^{\infty} \dfrac{1}{\sqrt{n^2-1}}$

15. $\sum_{n=0}^{\infty} \dfrac{1}{\sqrt{n^2+1}}$

16. $\sum_{n=1}^{\infty} \dfrac{1}{2^n-5}$

17. $\sum_{n=1}^{\infty} \dfrac{2n^2-1}{3n^5+2n+1}$

18. $\sum_{n=1}^{\infty} \dfrac{5n-3}{n^2-2n+5}$

19. $\sum_{n=1}^{\infty} \dfrac{n+3}{n(n+2)}$

20. $\sum_{n=1}^{\infty} \dfrac{1}{n(n^2+1)}$

21. $\sum_{n=1}^{\infty} \dfrac{1}{n\sqrt{n^2+1}}$

22. $\sum_{n=1}^{\infty} \dfrac{n}{(n+1)2^{n-1}}$

23. $\sum_{n=1}^{\infty} \dfrac{n^{k-1}}{n^k+1}, k>2$

24. $\sum_{n=1}^{\infty} \dfrac{1}{n+\sqrt{n^2+1}}$

25. $\sum_{n=1}^{\infty} \sin \dfrac{1}{n}$

26. $\sum_{n=1}^{\infty} \tan \dfrac{1}{n}$

In Exercises 27–34, test for convergence or divergence, using each of the seven tests at least once.
(a) nth-Term Test
(b) Geometric Series Test
(c) p-Series Test
(d) telescoping series
(e) Integral Test
(f) Direct Comparison Test
(g) Limit Comparison test

27. $\sum_{n=1}^{\infty} \dfrac{\sqrt{n}}{n}$

28. $\sum_{n=0}^{\infty} 5\left(\dfrac{-1}{5}\right)^n$

29. $\sum_{n=1}^{\infty} \dfrac{1}{3^n+2}$

30. $\sum_{n=4}^{\infty} \dfrac{1}{3n^2-2n-15}$

31. $\sum_{n=1}^{\infty} \dfrac{n}{2n+3}$

32. $\sum_{n=1}^{\infty} \left(\dfrac{1}{n+1}-\dfrac{1}{n+2}\right)$

33. $\sum_{n=1}^{\infty} \dfrac{n}{(n^2+1)^2}$

34. $\sum_{n=1}^{\infty} \dfrac{3}{n(n+3)}$

35. Determine the values of p for which the following series converges:
$$\sum_{n=2}^{\infty} \dfrac{\ln n}{n^p}$$

36. Estimate the error in approximating the sum of the following series by its first ten terms:
$$\sum_{n=0}^{\infty} 2\left(\dfrac{1}{3n}\right)^n$$

37. Estimate the error in approximating the sum of the following series by its first five terms:
$$\sum_{n=1}^{\infty} \dfrac{1}{n^n}$$

38. Prove that if $P(n)$ and $Q(n)$ are polynomials of degree j and k, respectively, then the series
$$\sum_{n=1}^{\infty} \dfrac{P(n)}{Q(n)}$$
converges if $j<k-1$, and diverges if $j\geq k-1$.

In Exercises 39–42, use the polynomial test as given in Exercise 38 to determine if the series converges or diverges.

39. $\dfrac{1}{2}+\dfrac{2}{5}+\dfrac{3}{10}+\dfrac{4}{17}+\dfrac{5}{26}+\cdots$

40. $\dfrac{1}{3}+\dfrac{1}{8}+\dfrac{1}{15}+\dfrac{1}{24}+\dfrac{1}{35}+\cdots$

41. $\sum_{n=1}^{\infty} \dfrac{1}{n^3+1}$

42. $\sum_{n=1}^{\infty} \dfrac{n^2}{n^3+1}$

43. Use the Limit Comparison Test with the harmonic series to show that the series Σa_n $(a_n>0)$ diverges if
$$\lim_{n\to\infty} na_n \neq 0$$

In Exercises 44 and 45, use the divergence test used in Exercise 43 to show that each series diverges.

44. $\sum_{n=1}^{\infty} \dfrac{n^3}{5n^4+3}$

45. $\sum_{n=2}^{\infty} \dfrac{1}{\ln n}$

10.6
Alternating series

Most of the results we have studied so far have been restricted to series with positive terms. In this and the following section we consider series that contain both positive and negative terms. The simplest such series is an **alternating series,** whose terms alternate in sign. For example, the geometric series

$$\sum_{n=1}^{\infty} \frac{1}{(-2)^{n-1}} = \sum_{n=1}^{\infty} (-1)^{n-1}\left(\frac{1}{2^{n-1}}\right) = 1 - \frac{1}{2} + \frac{1}{4} - \frac{1}{8} + \frac{1}{16} - \cdots$$

is an *alternating geometric series* with $r = -\frac{1}{2}$.

Alternating series occur in two ways:

$$\sum_{n=1}^{\infty} (-1)^n a_n = -a_1 + a_2 - a_3 + a_4 - \cdots$$

and

$$\sum_{n=1}^{\infty} (-1)^{n-1} a_n = a_1 - a_2 + a_3 - a_4 + \cdots$$

where $a_n > 0$. In one case the *odd* terms are negative, and in the other case the *even* terms are negative. The conditions for the convergence of an alternating series are given in the following theorem.

THEOREM 10.14   **ALTERNATING SERIES TEST**
If $a_n > 0$, then the alternating series

$$\sum_{n=1}^{\infty} (-1)^n a_n \qquad \text{and} \qquad \sum_{n=1}^{\infty} (-1)^{n-1} a_n$$

converge, provided both of the following conditions are met:

1. $0 < a_{n+1} \le a_n$, for all n
2. $\lim_{n\to\infty} a_n = 0$

Proof: The proof follows the same pattern for either form of the alternating series. In this proof, we use the form

$$\sum_{n=1}^{\infty} (-1)^{n-1} a_n = a_1 - a_2 + a_3 - a_4 + \cdots$$

For this series, the even partial sum

$$S_{2n} = (a_1 - a_2) + (a_3 - a_4) + (a_5 - a_6) + \cdots + (a_{2n-1} - a_{2n})$$

has all nonnegative terms, and therefore $\{S_{2n}\}$ is a nondecreasing sequence.

But we can also write

$$S_{2n} = a_1 - (a_2 - a_3) - (a_4 - a_5) - \cdots - (a_{2n-2} - a_{2n-1}) - a_{2n}$$

which implies that $S_{2n} \le a_1$ for every integer n. Thus $\{S_{2n}\}$ is a bounded, nondecreasing sequence that converges to some value L.

Now, since $S_{2n-1} = S_{2n} - a_{2n}$ and $a_{2n} \to 0$, we have

$$\lim_{n\to\infty} S_{2n-1} = \lim_{n\to\infty} S_{2n} - \lim_{n\to\infty} a_{2n} = L - \lim_{n\to\infty} a_{2n} = L$$

Since both the odd and even sequences of partial sums converge to the same limit L, it follows that $\{S_n\}$ also converges to L. Consequently, the given alternating series converges.

| **Remark** The first condition in the Alternating Series Test can be modified to require only that $0 < a_{n+1} \le a_n$ for all n greater than some integer N.

EXAMPLE 1   Using the Alternating Series Test

Use the Alternating Series Test on

$$\sum_{n=1}^{\infty} \frac{n}{(-2)^{n-1}} = \frac{1}{1} - \frac{2}{2} + \frac{3}{4} - \frac{4}{8} + \cdots$$

Solution: The inequality

$$a_{n+1} = \frac{n+1}{2^n} \le \frac{n}{2^{n-1}} = a_n$$

is satisfied if

$$(n+1)2^{n-1} \le n2^n$$

or if

$$\frac{1}{2} \le \frac{n}{n+1}$$

But this latter inequality is satisfied by $n \ge 1$, hence $a_{n+1} \le a_n$ for all n. Furthermore, by L'Hôpital's Rule,

$$\lim_{x\to\infty} \frac{x}{2^{x-1}} = \lim_{x\to\infty} \frac{1}{2^{x-1}(\ln 2)} = 0 \implies \lim_{n\to\infty} \frac{n}{2^{n-1}} = 0$$

Therefore, by the Alternating Series Test, the given series converges. ☐

EXAMPLE 2   Using the Alternating Series Test

Determine whether the following series converge or diverge.

(a) $\displaystyle\sum_{n=1}^{\infty} \frac{(-1)^n n}{\ln 2n}$ (b) $\displaystyle\sum_{n=1}^{\infty} (-1)^{n+1}\left(\frac{3n+2}{4n^2-3}\right)$

Solution:

(a) By L'Hôpital's Rule, we have

$$\lim_{x \to \infty} \frac{x}{\ln 2x} = \lim_{x \to \infty} \frac{1}{1/x} = \lim_{x \to \infty} x = \infty \neq 0$$

Thus, $\{a_n\}$ does not converge to 0, and, by the nth-Term Test for divergence, the given series diverges.

(b) Sometimes it is convenient to use differentiation to establish that $a_{n+1} \leq a_n$. In this case, we let

$$f(x) = \frac{3x + 2}{4x^2 - 3}$$

Then

$$f'(x) = \frac{-12x^2 - 16x - 9}{(4x^2 - 3)^2}$$

is always negative. Hence, f is a decreasing function, and it follows that $a_{n+1} \leq a_n$ for $n \geq 1$. Furthermore, since

$$\lim_{n \to \infty} \frac{3n + 2}{4n^2 - 3} = 0$$

the series converges by the Alternating Series Test.

For a convergent alternating series, the partial sum S_N can be a useful approximation for the sum S of the series. Just how close S_N is to S is stated in the following theorem.

THEOREM 10.15

ALTERNATING SERIES REMAINDER
If a convergent alternating series satisfies the condition $0 < a_{n+1} \leq a_n$, then the remainder R_N involved in approximating the sum S by S_N is less in magnitude than the first neglected term. That is,

$$|S - S_N| = |R_N| \leq a_{N+1}$$

Proof: Let $\Sigma(-1)^{n-1}a_n = S$. Then the series obtained by deleting the first N terms satisfies the conditions of the Alternating Series Test and has the sum

$$R_N = S - S_N = \sum_{n=1}^{\infty} (-1)^{n-1}a_n - \sum_{n=1}^{N} (-1)^{n-1}a_n$$

$$= (-1)^N a_{N+1} + (-1)^{N+1}a_{N+2} + (-1)^{N+2}a_{N+3} + \cdots$$

$$= (-1)^N (a_{N+1} - a_{N+2} + a_{N+3} - \cdots)$$

$$|R_N| = a_{N+1} - a_{N+2} + a_{N+3} - a_{N+4} + a_{N+5} - \cdots$$

$$= a_{N+1} - (a_{N+2} - a_{N+3}) - (a_{N+4} - a_{N+5}) - \cdots$$

$$|R_N| \leq a_{N+1}$$

Consequently, $|S - S_N| = |R_N| \leq a_{N+1}$, which establishes the theorem.

EXAMPLE 3   *Approximating the sum of an alternating series*

Approximate the sum of the following series by its first six terms:

$$\sum_{n=1}^{\infty} (-1)^{n-1} \left(\frac{1}{n!}\right) = \frac{1}{1!} - \frac{1}{2!} + \frac{1}{3!} - \frac{1}{4!} + \frac{1}{5!} - \frac{1}{6!} + \cdots$$

Solution: The Alternating Series Test establishes that the series converges, because

$$\frac{1}{(n+1)!} \le \frac{1}{n!} \qquad \text{and} \qquad \lim_{n \to \infty} \frac{1}{n!} = 0$$

Now, the sum of the first six terms is

$$S_6 = 1 - \frac{1}{2} + \frac{1}{6} - \frac{1}{24} + \frac{1}{120} - \frac{1}{720} \approx 0.63194$$

and, by the Alternating Series Remainder, we have

$$|S - S_6| = |R_6| \le a_7 = \frac{1}{5040} \approx 0.0002$$

Therefore, the sum S lies between $0.63194 - 0.0002$ and $0.63194 + 0.0002$, and we have $0.63174 \le S \le 0.63214$. (Later, we will see that the actual sum is $(e-1)/e \approx 0.63212$.)

Absolute and conditional convergence

On occasion a series may have both positive and negative terms and not be an alternating series. For instance, the series

$$\sum_{n=1}^{\infty} \frac{\sin n}{n^2} = \frac{\sin 1}{1} + \frac{\sin 2}{4} + \frac{\sin 3}{9} + \cdots$$

has both positive and negative terms, yet it is not an alternating series. One way to obtain some information about the convergence of this series is to investigate the convergence of the series

$$\sum_{n=1}^{\infty} \left| \frac{\sin n}{n^2} \right|$$

By direct comparison, we have $|\sin n| \le 1$ for all n, so

$$\left| \frac{\sin n}{n^2} \right| \le \frac{1}{n^2}, \quad n \ge 1$$

Thus, by the Direct Comparison Test, the series $\Sigma |(\sin n)/n^2|$ converges. But the question still is "Does the original series converge or not?" The next theorem gives us the answer.

THEOREM 10.16   **ABSOLUTE CONVERGENCE**

If the series $\Sigma |a_n|$ converges, then the series Σa_n also converges.

Proof: Since $0 \le a_n + |a_n| \le 2|a_n|$ for all n, the series

$$\sum_{n=1}^{\infty} (a_n + |a_n|)$$

converges by comparison to the convergent series

$$\sum_{n=1}^{\infty} 2|a_n|$$

Furthermore, since $a_n = (a_n + |a_n|) - |a_n|$, we can write

$$\sum_{n=1}^{\infty} a_n = \sum_{n=1}^{\infty} (a_n + |a_n|) - \sum_{n=1}^{\infty} |a_n|$$

where both series on the right converge. Hence it follows that Σa_n converges.

It should be noted that the converse of Theorem 10.16 is not true. For instance, the **alternating harmonic series**

$$\sum_{n=1}^{\infty} \frac{(-1)^{n+1}}{n} = \frac{1}{1} - \frac{1}{2} + \frac{1}{3} - \frac{1}{4} + \cdots$$

converges by the Alternating Series Test. Yet the harmonic series diverges. We call this type of convergence **conditional.**

DEFINITION OF ABSOLUTE AND CONDITIONAL CONVERGENCE

1. Σa_n is **absolutely convergent** if $\Sigma |a_n|$ converges.
2. Σa_n is **conditionally convergent** if Σa_n converges but $\Sigma |a_n|$ diverges.

EXAMPLE 4   Absolute and conditional convergence

Determine whether the following series are convergent or divergent. If convergent, classify the series as absolutely or conditionally convergent.

(a) $\displaystyle\sum_{n=1}^{\infty} \frac{(-1)^{n(n+1)/2}}{3^n} = -\frac{1}{3} - \frac{1}{9} + \frac{1}{27} + \frac{1}{81} - \cdots$

(b) $\displaystyle\sum_{n=1}^{\infty} \frac{(-1)^n}{\ln(n+1)}$

Solution:

(a) This is *not* an alternating series, since the signs change in pairs. However, we note that

$$\sum_{n=1}^{\infty} \left| \frac{(-1)^{n(n+1)/2}}{3^n} \right| = \sum_{n=1}^{\infty} \frac{1}{3^n}$$

is a convergent geometric series with $r = \frac{1}{3} < 1$. Consequently, by Theorem 10.16, we can conclude that the given series is *absolutely* convergent, hence convergent.

(b) In this case, the Alternating Series Test indicates that the given series converges. However, the series

$$\sum_{n=1}^{\infty} \left| \frac{(-1)^n}{\ln (n+1)} \right| = \frac{1}{\ln 2} + \frac{1}{\ln 3} + \frac{1}{\ln 4} + \cdots$$

diverges by direct comparison with the terms of the harmonic series. Therefore, we conclude that the given series is *conditionally* convergent.

Section Exercises 10.6

In Exercises 1–22, use the Alternating Series Test to determine the convergence or divergence of the series.

1. $\sum_{n=1}^{\infty} \frac{(-1)^{n+1}}{n}$

2. $\sum_{n=1}^{\infty} \frac{(-1)^{n+1}n}{2n-1}$

3. $\sum_{n=1}^{\infty} \frac{(-1)^{n+1}}{2n-1}$

4. $\sum_{n=2}^{\infty} \frac{(-1)^n}{\ln n}$

5. $\sum_{n=1}^{\infty} \frac{(-1)^n n^2}{n^2 + 1}$

6. $\sum_{n=1}^{\infty} \frac{(-1)^{n+1}n}{n^2 + 1}$

7. $\sum_{n=1}^{\infty} \frac{(-1)^n}{\sqrt{n}}$

8. $\sum_{n=1}^{\infty} \frac{(-1)^{n+1}n^3}{n^3 + 6}$

9. $\sum_{n=1}^{\infty} \frac{(-1)^{n+1}(n+1)}{\ln (n+1)}$

10. $\sum_{n=1}^{\infty} \frac{(-1)^{n+1} \ln (n+1)}{n+1}$

11. $\sum_{n=0}^{\infty} (-1)^n e^{-n}$

12. $\sum_{n=1}^{\infty} \frac{(-1)^{n+1}\sqrt{n+2}}{\sqrt{n(n+2)}}$

13. $\sum_{n=1}^{\infty} \sin \frac{(2n-1)\pi}{2}$

14. $\sum_{n=1}^{\infty} \cos n\pi$

15. $\sum_{n=1}^{\infty} \frac{1}{n} \sin \frac{(2n-1)\pi}{2}$

16. $\sum_{n=1}^{\infty} \frac{1}{n} \cos n\pi$

17. $\sum_{n=0}^{\infty} \frac{(-1)^n}{n!}$

18. $\sum_{n=0}^{\infty} \frac{(-1)^n}{(2n)!}$

19. $\sum_{n=1}^{\infty} \frac{(-1)^{n+1}\sqrt{n}}{n+2}$

20. $\sum_{n=1}^{\infty} \frac{(-1)^{n+1}\sqrt{n}}{\sqrt[3]{n}}$

21. $\sum_{n=1}^{\infty} \frac{2(-1)^{n+1}}{e^n - e^{-n}} = \sum_{n=1}^{\infty} (-1)^{n+1} \operatorname{csch} n$

22. $\sum_{n=1}^{\infty} \frac{2(-1)^{n+1}}{e^n + e^{-n}} = \sum_{n=1}^{\infty} (-1)^{n+1} \operatorname{sech} n$

In Exercises 23–38, examine the series for conditional convergence or absolute convergence.

23. $\sum_{n=1}^{\infty} \frac{(-1)^{n+1}}{(n+1)^2}$

24. $\sum_{n=1}^{\infty} \frac{(-1)^{n+1}}{n+1}$

25. $\sum_{n=1}^{\infty} \frac{(-1)^{n+1}}{\sqrt{n}}$

26. $\sum_{n=1}^{\infty} \frac{(-1)^{n+1}}{n\sqrt{n}}$

27. $\sum_{n=2}^{\infty} \frac{(-1)^n}{\ln n}$

28. $\sum_{n=0}^{\infty} (-1)^n e^{-n^2}$

29. $\sum_{n=2}^{\infty} \frac{(-1)^n n}{n^3 - 1}$

30. $\sum_{n=1}^{\infty} \frac{(-1)^{n+1}}{n^{1.5}}$

31. $\sum_{n=0}^{\infty} \frac{(-1)^n}{(2n+1)!}$

32. $\sum_{n=0}^{\infty} \frac{(-1)^n}{\sqrt[3]{n+1}}$

33. $\sum_{n=0}^{\infty} \frac{\cos n\pi}{n+1}$

34. $\sum_{n=1}^{\infty} (-1)^{n+1} \arctan n$

35. $\sum_{n=1}^{\infty} \frac{\cos n}{n^2}$

36. $\sum_{n=1}^{\infty} \frac{\cos n}{n\sqrt{n}}$

37. $\sum_{n=1}^{\infty} \frac{\sin [(2n-1)\pi/2]}{\sqrt{n}}$

38. $\sum_{n=1}^{\infty} \frac{\sin [(2n-1)\pi/2]}{n}$

In Exercises 39–46, approximate the sum of the series with an error of less than 0.001.

39. $\sum_{n=1}^{\infty} \frac{(-1)^{n+1}}{2n^3 - 1}$

40. $\sum_{n=1}^{\infty} \frac{(-1)^{n+1}}{n^4}$

41. $\sum_{n=0}^{\infty} \frac{(-1)^n}{n!}$ $\left(\text{the sum is } \frac{1}{e} \right)$

42. $\sum_{n=0}^{\infty} \frac{(-1)^n}{2^n n!}$ $\left(\text{the sum is } \frac{1}{\sqrt{e}} \right)$

43. $\sum_{n=0}^{\infty} \frac{(-1)^n}{(2n+1)!}$ (the sum is sin 1)

44. $\sum_{n=0}^{\infty} \frac{(-1)^n}{(2n)!}$ (the sum is cos 1)

45. $\displaystyle\sum_{n=1}^{\infty} \frac{(-1)^{n+1}}{n2^n}$ $\left(\text{the sum is } \ln \dfrac{3}{2}\right)$

46. $\displaystyle\sum_{n=1}^{\infty} \frac{(-1)^{n+1}}{n4^n}$ $\left(\text{the sum is } \ln \dfrac{5}{4}\right)$

In Exercises 47 and 48, find the number of terms necessary to approximate the sum of the series with an error less than 0.001.

47. $\displaystyle\sum_{n=0}^{\infty} \frac{(-1)^n}{2n+1}$ $\left(\text{the sum is } \dfrac{\pi}{4}\right)$

48. $\displaystyle\sum_{n=1}^{\infty} \frac{(-1)^{n+1}}{n^2}$ $\left(\text{the sum is } \dfrac{\pi^2}{12}\right)$

49. Prove that the alternating p-series

$$\sum_{n=1}^{\infty} (-1)^n \left(\frac{1}{n^p}\right)$$

converges if $p > 0$.

50. Prove that if $\Sigma |a_n|$ converges, then Σa_n^2 converges.

SECTION TOPICS ▪
Ratio Test ▪
Root Test ▪
Summary of tests for convergence and divergence ▪

10.7
The Ratio and Root Tests

We begin this section with a test for absolute convergence—the **Ratio Test.**

THEOREM 10.17

RATIO TEST

Let Σa_n be a series with nonzero terms.

1. Σa_n converges if

$$\lim_{n \to \infty} \left| \frac{a_{n+1}}{a_n} \right| < 1 \qquad \text{Absolute convergence}$$

2. Σa_n diverges if

$$\lim_{n \to \infty} \left| \frac{a_{n+1}}{a_n} \right| > 1$$

3. The Ratio Test is inconclusive if

$$\lim_{n \to \infty} \left| \frac{a_{n+1}}{a_n} \right| = 1$$

Proof: For the proof of the first part, we let

$$\lim_{n \to \infty} \left| \frac{a_{n+1}}{a_n} \right| = r < 1$$

and choose R such that $0 \le r < R < 1$. By the definition of the limit of a sequence, there exists some $N > 0$ such that $|a_{n+1}/a_n| < R$ for all $n > N$. Therefore, we can write

$$|a_{N+1}| < |a_N|R$$
$$|a_{N+2}| < |a_{N+1}|R < |a_N|R^2$$
$$|a_{N+3}| < |a_{N+2}|R < |a_{N+1}|R^2 < |a_N|R^3$$
$$\vdots$$

The geometric series

$$\sum_{n=1}^{\infty} a_N R^n = a_N R + a_N R^2 + \cdots + a_N R^n + \cdots$$

converges, so, by the Direct Comparison Test, the series

$$\sum_{n=1}^{\infty} |a_{N+n}| = |a_{N+1}| + |a_{N+2}| + \cdots + |a_{N+n}| + \cdots$$

also converges. This in turn implies that the series $\Sigma|a_n|$ converges, since discarding a finite number of terms ($n = N - 1$) does not affect convergence. Consequently, by Theorem 10.16, the series Σa_n converges.

The proof of the second part is similar, except that we choose R such that

$$\lim_{n \to \infty} \left| \frac{a_{n+1}}{a_n} \right| = r > R > 1$$

and show that there exists some $M > 0$ such that $|a_{M+n}| > |a_M| R^n$.

The fact that the Ratio Test fails to give us any useful information when $|a_{n+1}/a_n| \to 1$ can be seen by comparing the two series

$$\sum_{n=1}^{\infty} \frac{1}{n} \quad \text{and} \quad \sum_{n=1}^{\infty} \frac{1}{n^2}$$

The first series diverges and the second one converges, but in both cases

$$\lim_{n \to \infty} \left| \frac{a_{n+1}}{a_n} \right| = 1$$

Though the Ratio Test is not a cure for all ills related to tests for convergence, it is particularly useful for series that *converge rapidly*. Series involving factorials or exponentials are frequently of this type.

EXAMPLE 1   *Using the Ratio Test*

Use the Ratio Test on

$$\sum_{n=0}^{\infty} \frac{2^n}{n!}$$

Solution: Since $a_n = 2^n/n!$, we have

$$\lim_{n \to \infty} \left| \frac{a_{n+1}}{a_n} \right| = \lim_{n \to \infty} \left[\frac{2^{n+1}}{(n + 1)!} \div \frac{2^n}{n!} \right] = \lim_{n \to \infty} \left[\frac{2^{n+1}}{(n + 1)!} \cdot \frac{n!}{2^n} \right]$$

$$= \lim_{n \to \infty} \frac{2}{n + 1} = 0$$

Therefore, the series converges.

EXAMPLE 2   A failure of the Ratio Test

Determine the convergence of

$$\sum_{n=1}^{\infty} (-1)^n \frac{\sqrt{n}}{n+1}$$

Solution: Using the Ratio Test, we have

$$\lim_{n\to\infty} \left| \frac{a_{n+1}}{a_n} \right| = \lim_{n\to\infty} \left[\left(\frac{\sqrt{n+1}}{n+2} \right) \left(\frac{n+1}{\sqrt{n}} \right) \right]$$

$$= \lim_{n\to\infty} \left[\sqrt{\frac{n+1}{n}} \left(\frac{n+1}{n+2} \right) \right]$$

$$= \sqrt{1}(1) = 1$$

Thus, the Ratio Test gives us no useful information, so we use the Alternating Series Test. To show that $a_{n+1} \le a_n$, we let $f(x) = \sqrt{x}/(x+1)$. Then the derivative is

$$f'(x) = \frac{-x+1}{2\sqrt{x}(x+1)^2}$$

and since this is negative for $x > 1$, we know that f is a decreasing function. Also, by L'Hôpital's Rule,

$$\lim_{x\to\infty} \frac{\sqrt{x}}{x+1} = \lim_{x\to\infty} \frac{1/(2\sqrt{x})}{1}$$

$$= \lim_{x\to\infty} \frac{1}{2\sqrt{x}} = 0$$

Therefore, the series converges.

Remark   Note that the series in Example 2 is *conditionally convergent*, since $\Sigma|a_n|$ diverges (by the Limit Comparison Test with $\Sigma 1/\sqrt{n}$), but Σa_n converges.

EXAMPLE 3   Using the Ratio Test

Determine whether the following series converge or diverge:

(a) $\displaystyle\sum_{n=0}^{\infty} \frac{n^2 2^{n+1}}{3^n}$ (b) $\displaystyle\sum_{n=0}^{\infty} \frac{n^n}{n!}$

Solution:
(a) Since

$$\lim_{n\to\infty} \left| \frac{a_{n+1}}{a_n} \right| = \lim_{n\to\infty} \left[(n+1)^2 \left(\frac{2^{n+2}}{3^{n+1}} \right) \left(\frac{3^n}{n^2 2^{n+1}} \right) \right]$$

$$= \lim_{n\to\infty} \frac{2(n+1)^2}{3n^2} = \frac{2}{3} < 1$$

we conclude by the Ratio Test that the series converges.

(b) Since

$$\lim_{n \to \infty} \left| \frac{a_{n+1}}{a_n} \right| = \lim_{n \to \infty} \left[\frac{(n+1)^{n+1}}{(n+1)!} \left(\frac{n!}{n^n} \right) \right]$$

$$= \lim_{n \to \infty} \left[\frac{(n+1)^{n+1}}{(n+1)} \left(\frac{1}{n^n} \right) \right]$$

$$= \lim_{n \to \infty} \frac{(n+1)^n}{n^n}$$

$$= \lim_{n \to \infty} \left(1 + \frac{1}{n} \right)^n = e > 1$$

we conclude by the Ratio Test that the series diverges.

The next test for convergence or divergence of series with positive and negative terms works especially well for series involving an nth power. The proof of this theorem is similar to that given for the Ratio Test, and we leave it as an exercise.

THEOREM 10.18   **ROOT TEST**

1. Σa_n converges if

$$\lim_{n \to \infty} \sqrt[n]{|a_n|} < 1$$

2. Σa_n diverges if

$$\lim_{n \to \infty} \sqrt[n]{|a_n|} > 1$$

3. The Root Test is inconclusive if

$$\lim_{n \to \infty} \sqrt[n]{|a_n|} = 1$$

EXAMPLE 4   Using the Root Test

Determine the convergence or divergence of

$$\sum_{n=1}^{\infty} \frac{e^{2n}}{n^n}$$

Solution: Since

$$\lim_{n \to \infty} \sqrt[n]{|a_n|} = \lim_{n \to \infty} \sqrt[n]{\frac{e^{2n}}{n^n}} = \lim_{n \to \infty} \frac{e^{2n/n}}{n^{n/n}}$$

$$= \lim_{n \to \infty} \frac{e^2}{n} = 0 < 1$$

we conclude by the Root Test that the series converges.

EXAMPLE 5   Using the Root Test

Determine the convergence or divergence of

$$\sum_{n=1}^{\infty} \frac{n^3}{3^n}$$

Solution: First, we have

$$\lim_{n\to\infty} \sqrt[n]{|a_n|} = \lim_{n\to\infty} \sqrt[n]{\frac{n^3}{3^n}}$$

$$= \lim_{n\to\infty} \frac{n^{3/n}}{3}$$

Since the limit in the numerator yields the indeterminate form ∞^0, we apply L'Hôpital's Rule as follows:

$$y = \lim_{x\to\infty} x^{3/x}$$

$$\ln y = \ln\left(\lim_{x\to\infty} x^{3/x}\right)$$

$$= \lim_{x\to\infty} (\ln x^{3/x})$$

$$= \lim_{x\to\infty} \left(\frac{3}{x} \ln x\right)$$

$$= \lim_{x\to\infty} \frac{3 \ln x}{x}$$

$$= \lim_{x\to\infty} \frac{3/x}{1} = 0$$

Now, since $\ln y = 0$, we conclude that

$$y = \lim_{x\to\infty} x^{3/x} = 1$$

from which it follows that

$$\lim_{n\to\infty} \frac{n^{3/n}}{3} = \frac{1}{3} < 1$$

Thus by the Root Test, the given series converges.

In the last five sections we have discussed ten different tests for determining the convergence or divergence of an infinite series. Skill in choosing and applying the various tests will come only with practice. We summarize in Table 10.7 the various tests we have studied. Below is a useful checklist for choosing an appropriate test.

1. Does the *n*th term approach zero? If not, the series diverges.
2. Is the series one of the special types—geometric, *p*-series, telescoping, alternating?
3. Can the Integral, Root, or Ratio Test be applied?
4. Can the series be compared favorably to one of the special types?

In some instances more than one test is applicable. However, your objective should be to learn to choose the most efficient way to test a series.

TABLE 10.7   SUMMARY OF TESTS FOR SERIES

Test	Series	Converges	Sum or Remainder	Diverges
nth-Term	$\sum\limits_{n=1}^{\infty} a_n$			$\lim\limits_{n\to\infty} a_n \neq 0$
Geometric Series	$\sum\limits_{n=0}^{\infty} ar^n$	$\lvert r \rvert < 1$	$S = \dfrac{a}{1-r}$	$\lvert r \rvert \geq 1$
Telescoping Series	$\sum\limits_{n=1}^{\infty} (b_n - b_{n+1})$	$\lim\limits_{n\to\infty} b_n = 0$	$S = b_1$	
p-Series	$\sum\limits_{n=1}^{\infty} \dfrac{1}{n^p}$	$p > 1$	$0 < R_N < \dfrac{N^{1-p}}{p-1}$	$p \leq 1$
Alternating Series	$\sum\limits_{n=1}^{\infty} (-1)^{n-1} a_n$	$0 < a_{n+1} \leq a_n$ and $\lim\limits_{n\to\infty} a_n = 0$	$\lvert R_N \rvert \leq a_{N+1}$	
Integral (f is continuous, positive, and decreasing)	$\sum\limits_{n=1}^{\infty} a_n,\ a_n = f(n) \geq 0$	$\int_{1}^{\infty} f(x)\, dx$ converges	$0 < R_N < \int_{N}^{\infty} f(x)\, dx$	$\int_{1}^{\infty} f(x)\, dx$ diverges
Root	$\sum\limits_{n=1}^{\infty} a_n$	$\lim\limits_{n\to\infty} \sqrt[n]{\lvert a_n \rvert} < 1$		$\lim\limits_{n\to\infty} \sqrt[n]{\lvert a_n \rvert} > 1$
Ratio	$\sum\limits_{n=1}^{\infty} a_n$	$\lim\limits_{n\to\infty} \left\lvert \dfrac{a_{n+1}}{a_n} \right\rvert < 1$		$\lim\limits_{n\to\infty} \left\lvert \dfrac{a_{n+1}}{a_n} \right\rvert > 1$
Direct Comparison	$\sum\limits_{n=1}^{\infty} a_n$	$0 \leq a_n \leq b_n$ and $\sum\limits_{n=1}^{\infty} b_n$ converges	$0 \leq R_N \leq \sum\limits_{n=N}^{\infty} b_n$	$0 \leq b_n \leq a_n$ and $\sum\limits_{n=1}^{\infty} b_n$ diverges
Limit Comparison ($a_n,\ b_n > 0$)	$\sum\limits_{n=1}^{\infty} a_n$	$\lim\limits_{n\to\infty} \dfrac{a_n}{b_n} = L > 0$ and $\sum\limits_{n=1}^{\infty} b_n$ converges		$\lim\limits_{n\to\infty} \dfrac{a_n}{b_n} = L > 0$ and $\sum\limits_{n=1}^{\infty} b_n$ diverges

Section Exercises 10.7

In Exercises 1–20, use the Ratio Test to test for convergence or divergence of the series.

1. $\displaystyle\sum_{n=0}^{\infty} \frac{n!}{3^n}$ **2.** $\displaystyle\sum_{n=1}^{\infty} n\left(\frac{2}{3}\right)^n$

3. $\displaystyle\sum_{n=0}^{\infty} \frac{3^n}{n!}$ **4.** $\displaystyle\sum_{n=1}^{\infty} n\left(\frac{3}{2}\right)^n$

5. $\displaystyle\sum_{n=1}^{\infty} \frac{n}{2^n}$ **6.** $\displaystyle\sum_{n=1}^{\infty} \frac{n^2}{2^n}$

7. $\displaystyle\sum_{n=0}^{\infty} \frac{(-1)^{n+1}n!}{1 \cdot 3 \cdot 5 \cdots (2n+1)}$

8. $\displaystyle\sum_{n=1}^{\infty} \frac{(-1)^n 2 \cdot 4 \cdot 6 \cdots (2n)}{2 \cdot 5 \cdot 8 \cdots (3n-1)}$

9. $\displaystyle\sum_{n=1}^{\infty} \frac{2^n}{n^2}$

10. $\displaystyle\sum_{n=1}^{\infty} \frac{(-1)^{n+1}(n+2)}{n(n+1)}$

11. $\displaystyle\sum_{n=0}^{\infty} \frac{(-1)^n 2^n}{n!}$ **12.** $\displaystyle\sum_{n=1}^{\infty} \frac{(-1)^{n-1}(3/2)^n}{n^2}$

13. $\displaystyle\sum_{n=0}^{\infty} \frac{n!}{n3^n}$ **14.** $\displaystyle\sum_{n=1}^{\infty} \frac{(2n)!}{n^5}$

15. $\displaystyle\sum_{n=0}^{\infty} \frac{4^n}{n!}$ **16.** $\displaystyle\sum_{n=1}^{\infty} \frac{n^n}{n!}$

17. $\displaystyle\sum_{n=0}^{\infty} \frac{3^n}{(n+1)^n}$ **18.** $\displaystyle\sum_{n=0}^{\infty} \frac{(n!)^2}{(3n)!}$

19. $\displaystyle\sum_{n=0}^{\infty} \frac{4^n}{3^n+1}$ **20.** $\displaystyle\sum_{n=0}^{\infty} \frac{(-1)^n 2^{4n}}{(2n+1)!}$

In Exercises 21–28, use the Root Test to test for convergence or divergence of the series.

21. $\displaystyle\sum_{n=1}^{\infty} \left(\frac{n}{2n+1}\right)^n$ **22.** $\displaystyle\sum_{n=1}^{\infty} \left(\frac{2n}{n+1}\right)^n$

23. $\displaystyle\sum_{n=1}^{\infty} (2\sqrt[n]{n}+1)^n$ **24.** $\displaystyle\sum_{n=1}^{\infty} (\sqrt[n]{n}-1)^n$

25. $\displaystyle\sum_{n=2}^{\infty} \left(\frac{\ln n}{n}\right)^n$ **26.** $\displaystyle\sum_{n=0}^{\infty} e^{-n}$

27. $\displaystyle\frac{1}{(\ln 3)^3} + \frac{1}{(\ln 4)^4} + \frac{1}{(\ln 5)^5} + \frac{1}{(\ln 6)^6} + \cdots$

28. $\displaystyle 1 + \frac{2}{3} + \frac{3}{3^2} + \frac{4}{3^3} + \frac{5}{3^4} + \frac{6}{3^5} + \cdots$

In Exercises 29–40, test for convergence or divergence using any appropriate test from this chapter. Identify the test used.

29. $\displaystyle\sum_{n=1}^{\infty} \frac{(-1)^n 3^{n-2}}{2^n}$ **30.** $\displaystyle\sum_{n=1}^{\infty} \frac{10}{3\sqrt[3]{n}}$

31. $\displaystyle\sum_{n=1}^{\infty} \frac{10n+3}{n2^n}$ **32.** $\displaystyle\sum_{n=1}^{\infty} \frac{2^n}{4n^2-1}$

33. $\displaystyle\sum_{n=1}^{\infty} (-1)^n \ln\left(\frac{n+2}{n}\right)$ **34.** $\displaystyle\sum_{n=1}^{\infty} \frac{1}{\sqrt{n}+2}$

35. $\displaystyle\sum_{n=1}^{\infty} \frac{\cos n}{2^n}$ **36.** $\displaystyle\sum_{n=2}^{\infty} \frac{(-1)^n}{n \ln n}$

37. $\displaystyle\sum_{n=1}^{\infty} \frac{n7^n}{n!}$ **38.** $\displaystyle\sum_{n=1}^{\infty} \frac{\ln n}{n^2}$

39. $\displaystyle\sum_{n=1}^{\infty} \frac{(-1)^n 3^{n-1}}{n!}$ **40.** $\displaystyle\sum_{n=1}^{\infty} \frac{(-1)^n 3^n}{n2^n}$

41. Prove Theorem 10.18.

42. Show that the Ratio Test is inconclusive for a *p*-series.

10.8
Power series

Now that we have described the machinery for dealing with series, let's return to the original problem introduced in the beginning of the chapter. We begin by considering series that are different for different values of *x* as described in the following definition.

DEFINITION OF
POWER SERIES

If x is a variable, then an infinite series of the form

$$\sum_{n=0}^{\infty} a_n x^n = a_0 + a_1 x + a_2 x^2 + a_3 x^3 + \cdots + a_n x^n + \cdots$$

is called a **power series.** More generally, we call a series of the form

$$\sum_{n=0}^{\infty} a_n(x - c)^n = a_0 + a_1(x - c) + a_2(x - c)^2 + \cdots + a_n(x - c)^n + \cdots$$

a **power series centered at c,** where c is a constant.

| Remark   Note that we begin a power series with $n = 0$, and to simplify the nth term we agree that $(x - c)^0 = 1$, even if $x = c$.

Since a power series has variable terms, it can be viewed as a function of x,

$$f(x) = \sum_{n=0}^{\infty} a_n(x - c)^n$$

where the *domain of f* is the set of all x for which the power series converges. The determination of this domain is one of the primary problems associated with power series. Of course, every power series converges at its center c, since for $x = c$,

$$f(c) = \sum_{n=0}^{\infty} a_n(c - c)^n = a_0(1) + 0 + 0 + \cdots + 0 + \cdots = a_0$$

Thus, c always lies in the domain of f. Sometimes the domain of f consists *only* of the single point $x = c$. In fact, this is one of the three basic types of domains for power series, as shown in Figure 10.13. All three types are described in the following theorem.

Interval: single point $x = c$
Radius : 0

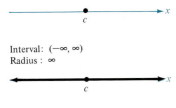

Interval: $(-\infty, \infty)$
Radius : ∞

Interval: $(c - R, c + R)$
Radius : R

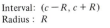

FIGURE 10.13

THEOREM 10.19

CONVERGENCE OF A POWER SERIES
For a power series centered at c, precisely one of the following is true.

1. The series converges only at c.
2. The series converges for all x.
3. There exists an $R > 0$ such that the series converges for $|x - c| < R$, and diverges for $|x - c| > R$.

| Remark   In part 3 of Theorem 10.19 we call R the **radius of convergence** of the power series, and in parts 1 and 2 we consider the radius of convergence to be 0 and ∞, respectively. The corresponding domain is called the **interval of convergence.**

The three types of intervals of convergence are illustrated in Examples 1, 2, and 3. These examples show how the Ratio Test can be used to determine the interval (domain) of convergence.

EXAMPLE 1   *A power series that converges only at a point*

Find the interval of convergence for

$$f(x) = \sum_{n=0}^{\infty} n! x^n$$

Solution: When $x = 0$,

$$f(0) = \sum_{n=0}^{\infty} n! 0^n = 1 + 0 + 0 + \cdots = 1$$

For any fixed value of x such that $|x| > 0$, let $u_n = n! x^n$. Then

$$\lim_{n \to \infty} \left| \frac{u_{n+1}}{u_n} \right| = \lim_{n \to \infty} \left| \frac{(n+1)! x^{n+1}}{n! x^n} \right| = |x| \lim_{n \to \infty} (n+1) = \infty$$

Therefore, by the Ratio Test we conclude that the series diverges for $|x| > 0$, and converges only when $x = 0$.

EXAMPLE 2   *Finding the interval of convergence*

Find the interval of convergence for

$$f(x) = \sum_{n=0}^{\infty} 3(x-2)^n$$

Solution: Since this series is geometric for each fixed value of x, we know that it converges only when $|r| = |x - 2| < 1$. Therefore, the interval of convergence consists of all x such that $-1 < x - 2 < 1$, and we have

$$1 < x < 3 \qquad \text{Interval of convergence}$$

In Figure 10.14, we see that the radius of convergence for this series is 1.

Interval: $(1, 3)$
Radius : $R = 1$

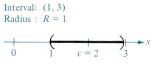

FIGURE 10.14

EXAMPLE 3   *An infinite interval of convergence*

Find the interval of convergence for

$$\sum_{n=0}^{\infty} \frac{(-1)^n x^{2n+1}}{(2n+1)!}$$

Solution: If we let $u_n = (-1)^n x^{2n+1}/(2n+1)!$, then

$$\lim_{n \to \infty} \left| \frac{u_{n+1}}{u_n} \right| = \lim_{n \to \infty} \left| \frac{[(-1)^{n+1} x^{2n+3}]/(2n+3)!}{[(-1)^n x^{2n+1}]/(2n+1)!} \right|$$

$$= \lim_{n \to \infty} \frac{x^2}{(2n+3)(2n+2)}$$

For any *fixed* value for x, this limit is 0, and, by the Ratio Test, the series converges for all x.

Endpoint convergence

Note that for a power series whose radius of convergence is R, Theorem 10.19 says nothing about the convergence at the *endpoints* of the interval of convergence. Each endpoint must be tested separately for convergence or divergence. As a result, an interval of convergence can be any one of the four types shown in Figure 10.15.

Examples 1 and 3 show the usefulness of the Ratio Test in determining the radius of convergence for a power series. From the radius of convergence, we can specify the exact interval of convergence after testing for convergence at the endpoints. We summarize the Ratio Test as applied to a power series as follows:

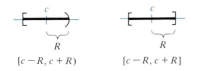

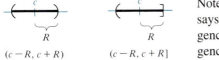

$(c - R, c + R)$ $(c - R, c + R]$

$[c - R, c + R)$ $[c - R, c + R]$

Intervals of Convergence

FIGURE 10.15

Ratio Test Applied to Power Series: For a power series $\Sigma a_n(x - c)^n$, if

1. $\lim\limits_{n \to \infty} \left| \dfrac{a_{n+1}}{a_n} \right| = L \neq 0$, then the radius of convergence is $R = 1/L$.

2. $\lim\limits_{n \to \infty} \left| \dfrac{a_{n+1}}{a_n} \right| = 0$, then the radius of convergence is infinite.

3. $\lim\limits_{n \to \infty} \left| \dfrac{a_{n+1}}{a_n} \right| = \infty$, then the radius of convergence is $R = 0$.

EXAMPLE 4   Finding the interval of convergence and testing endpoints

Find the interval of convergence for

$$\sum_{n=1}^{\infty} \frac{x^n}{n}$$

Solution: Since $a_n = 1/n$ and

$$\lim_{n \to \infty} \left| \frac{a_{n+1}}{a_n} \right| = \lim_{n \to \infty} \left| \frac{1/(n + 1)}{1/n} \right| = \lim_{n \to \infty} \left| \frac{n}{n + 1} \right| = 1$$

the radius of convergence is $R = 1$. Moreover, since the series is centered at $x = 0$, it will converge in the interval $(-1, 1)$, and we proceed to test for convergence at the endpoints. When $x = 1$, we obtain the *divergent* harmonic series

$$\sum_{n=1}^{\infty} \frac{1}{n} = \frac{1}{1} + \frac{1}{2} + \frac{1}{3} + \cdots \qquad \text{Diverges when } x = 1$$

When $x = -1$, we obtain the *convergent* alternating harmonic series

$$\sum_{n=1}^{\infty} \frac{(-1)^n}{n} = -1 + \frac{1}{2} - \frac{1}{3} + \frac{1}{4} - \cdots \qquad \text{Converges when } x = -1$$

Therefore, the interval of convergence for the given series is $[-1, 1)$.

EXAMPLE 5   Finding the interval of convergence and testing endpoints

Find the interval of convergence for the series

$$\sum_{n=0}^{\infty} \frac{(-1)^n (x+1)^n}{2^n}$$

Solution: Since

$$\lim_{n\to\infty} \left| \frac{a_{n+1}}{a_n} \right| = \lim_{n\to\infty} \left| \frac{(-1)^{n+1}/2^{n+1}}{(-1)^n/2^n} \right| = \lim_{n\to\infty} \left| \frac{2^n}{2^{n+1}} \right| = \frac{1}{2}$$

the radius of convergence is $R = 1/L = 2$. Since the series is centered at $x = -1$, it will converge in the interval $(-3, 1)$. Furthermore, at the endpoints we have

$$\sum_{n=0}^{\infty} \frac{(-1)^n(-2)^n}{2^n} = \sum_{n=0}^{\infty} \frac{2^n}{2^n} = \sum_{n=0}^{\infty} 1 \qquad \text{Diverges when } x = -3$$

and

$$\sum_{n=0}^{\infty} \frac{(-1)^n(2)^n}{2^n} = \sum_{n=0}^{\infty} (-1)^n \qquad \text{Diverges when } x = 1$$

both of which diverge. Thus, the interval of convergence is $(-3, 1)$. ▢

EXAMPLE 6   Finding the interval of convergence and testing endpoints

Find the interval of convergence for

$$\sum_{n=1}^{\infty} \frac{x^n}{n^2}$$

Solution: Since $a_n = 1/n^2$, and

$$\lim_{n\to\infty} \left| \frac{a_{n+1}}{a_n} \right| = \lim_{n\to\infty} \left| \frac{1/(n+1)^2}{1/n^2} \right|$$

$$= \lim_{n\to\infty} \left| \frac{n^2}{(n+1)^2} \right| = 1$$

the radius of convergence is $R = 1$. Moreover, since the series is centered at $x = 0$, it will converge in the interval $(-1, 1)$ and we proceed to test for convergence at the endpoints. When $x = 1$, we obtain the *convergent p*-series

$$\sum_{n=1}^{\infty} \frac{1}{n^2} = \frac{1}{1^2} + \frac{1}{2^2} + \frac{1}{3^2} + \frac{1}{4^2} + \cdots \qquad \text{Converges when } x = 1$$

When $x = -1$, we obtain the *convergent* alternating series

$$\sum_{n=1}^{\infty} \frac{(-1)^n}{n^2} = -\frac{1}{1^2} + \frac{1}{2^2} - \frac{1}{3^2} + \frac{1}{4^2} - \cdots \qquad \text{Converges when } x = -1$$

Therefore, the interval of convergence for the given series is $[-1, 1]$. ▢

James Gregory

Differentiation and integration of power series

Power series representation of functions has played an important role in the development of calculus. In fact, much of Newton's work with differentiation and integration was done in the context of power series—especially his work with complicated algebraic functions and transcendental functions. Euler, Lagrange, Leibniz, and the Bernoullis all used power series extensively in calculus. One of the earliest mathematicians to work with power series was a Scotsman, James Gregory (1638–1675). He developed a power series method for interpolating table values—a method that was later used by Brook Taylor in the development of Taylor polynomials and Taylor series.

Once we have defined a function with a power series, it is natural to wonder how we can determine the characteristics of the function. Is it continuous? Differentiable? Integrable? Theorem 10.20, which we state without proof, answers these questions.

THEOREM 10.20   **PROPERTIES OF A FUNCTION DEFINED BY A POWER SERIES**
If the function given by

$$f(x) = \sum_{n=0}^{\infty} a_n(x - c)^n = a_0 + a_1(x - c) + a_2(x - c)^2 + a_3(x - c)^3 + \cdots$$

has a radius of convergence of $R > 0$, then on the interval $(c - R, c + R)$ f is continuous, differentiable, and integrable. Moreover, the derivative and antiderivative of f are given by

1. $f'(x) = a_1 + 2a_2(x - c) + 3a_3(x - c)^2 + \cdots$
2. $\displaystyle \int f(x)\,dx = C + a_0(x - c) + a_1\frac{(x - c)^2}{2} + a_2\frac{(x - c)^3}{3} + \cdots$

Remark Theorem 10.20 tells us that, in many ways, a function defined by a power series behaves like a polynomial. It is continuous in its interval of convergence, and both its derivative and antiderivative can be determined by differentiating and integrating each term of the given power series.

The *radius of convergence* for the series obtained by differentiating or integrating a power series is the same as for the original power series. For example, suppose the power series

$$f(x) = a_0 + a_1(x - c) + a_2(x - c)^2 + a_3(x - c)^3 + \cdots$$

has a finite radius of convergence of $R \neq 0$. Then, by the Ratio Test, we know that

$$\lim_{n \to \infty} \left| \frac{a_{n+1}}{a_n} \right| = L, \quad \text{where } R = \frac{1}{L}$$

Now, for the series

$$f'(x) = a_1 + 2a_2(x - c) + 3a_3(x - c)^2 + \cdots$$

it follows that the radius of convergence is also R since

$$\lim_{n\to\infty} \left| \frac{(n+1)a_{n+1}}{na_n} \right| = \lim_{n\to\infty} \left(\frac{n+1}{n} \right) \left| \frac{a_{n+1}}{a_n} \right| = L, \quad \text{where } R = \frac{1}{L}.$$

However, the *interval of convergence* may differ due to the behavior at the endpoints. This is demonstrated in Example 7.

EXAMPLE 7   *Intervals of convergence for $f(x)$, $f'(x)$, and $\int f(x)\, dx$*

Find the intervals of convergence of $\int f(x)\, dx$, $f(x)$, and $f'(x)$, where

$$f(x) = \sum_{n=1}^{\infty} \frac{x^n}{n} = x + \frac{x^2}{2} + \frac{x^3}{3} + \cdots$$

Solution: By Theorem 10.20 we have

$$f'(x) = \sum_{n=1}^{\infty} x^{n-1} = 1 + x + x^2 + x^3 + \cdots$$

and

$$\int f(x)\, dx = C + \sum_{n=1}^{\infty} \frac{x^{n+1}}{n(n+1)} = C + \frac{x^2}{1\cdot 2} + \frac{x^3}{2\cdot 3} + \frac{x^4}{3\cdot 4} + \cdots$$

By the Ratio Test we know that each series has a radius of convergence of $R = 1$. Now considering the interval $(-1, 1)$, we have the following.

For $\int f(x)\, dx$, the series

$$\sum_{n=1}^{\infty} \frac{x^{n+1}}{n(n+1)} \qquad \text{Interval of convergence: } [-1, 1]$$

converges for $x = \pm 1$, and its interval of convergence is $[-1, 1]$.

For $f(x)$, the series

$$\sum_{n=1}^{\infty} \frac{x^n}{n} \qquad \text{Interval of convergence: } [-1, 1)$$

converges for $x = -1$ and diverges for $x = 1$, and its interval of convergence is $[-1, 1)$.

For $f'(x)$, the series

$$\sum_{n=1}^{\infty} x^{n-1} \qquad \text{Interval of convergence: } (-1, 1)$$

diverges for $x = \pm 1$ and its interval of convergence is $(-1, 1)$.

Remark   From Example 7, it appears that of the three series, the one for the derivative, $f'(x)$, is the least likely to converge at the endpoints. In fact, it can be shown that if the series for $f'(x)$ converges at the endpoints $x = c \pm R$, then the series for $f(x)$ will also converge there (see Exercise 34).

The next example shows how to approximate the value of the definite integral of a function represented by a power series. In Section 10.10, we will see that this technique is useful for a function such as e^{x^2}, which has no elementary antiderivative but can be readily represented by a power series.

EXAMPLE 8   *A power series approximation of a definite integral*

Using $R_N \leq 0.001$, approximate the definite integral $\int_0^1 f(x)\, dx$, where

$$f(x) = \sum_{n=0}^{\infty} \frac{(-1)^n x^{2n}}{(2n)!}$$

Solution: The Ratio Test shows that the series converges for all x ($R = \infty$), so f is integrable on $[0, 1]$. By Theorem 10.20 we have

$$\int_0^1 f(x)\, dx = \int_0^1 \left[\sum_{n=0}^{\infty} \frac{(-1)^n x^{2n}}{(2n)!} \right] dx = \left[\sum_{n=0}^{\infty} \frac{(-1)^n x^{2n+1}}{(2n + 1)(2n)!} \right]_0^1$$

$$= \left[\sum_{n=0}^{\infty} \frac{(-1)^n x^{2n+1}}{(2n + 1)!} \right]_0^1 = \frac{1}{1!} - \frac{1}{3!} + \frac{1}{5!} - \frac{1}{7!} + \frac{1}{9!} - \cdots$$

$$= \sum_{n=0}^{\infty} (-1)^n \frac{1}{(2n + 1)!}$$

In order to use the Alternating Series Remainder, we must choose N such that

$$|R_N| \leq a_{N+1} = \frac{1}{(2N + 3)!} < 0.001$$

If $N = 2$, then $a_{N+1} = 1/7! = 1/5040 < 0.001$. Therefore,

$$\int_0^1 f(x)\, dx \approx S_2 = 1 - \frac{1}{3!} + \frac{1}{5!} \approx 0.8417$$

(The actual value of this integral is $\sin 1 \approx 0.8415$. See Exercise 39 of Section 10.10.)

Section Exercises 10.8

In Exercises 1–24, find the interval of convergence of the given power series. Include a check for convergence at the endpoints of the interval.

1. $\displaystyle\sum_{n=0}^{\infty} \left(\frac{x}{2} \right)^n$

2. $\displaystyle\sum_{n=0}^{\infty} \left(\frac{x}{k} \right)^n$

3. $\displaystyle\sum_{n=1}^{\infty} \frac{(-1)^n x^n}{n}$

4. $\displaystyle\sum_{n=0}^{\infty} (-1)^{n+1} n x^n$

5. $\displaystyle\sum_{n=0}^{\infty} \frac{x^n}{n!}$

6. $\displaystyle\sum_{n=0}^{\infty} \frac{(3x)^n}{(2n)!}$

7. $\displaystyle\sum_{n=0}^{\infty} (2n)! \left(\frac{x}{2} \right)^n$

8. $\displaystyle\sum_{n=0}^{\infty} \frac{(-1)^n x^n}{(n + 1)(n + 2)}$

9. $\displaystyle\sum_{n=1}^{\infty} \frac{(-1)^{n+1} x^n}{4^n}$

10. $\displaystyle\sum_{n=0}^{\infty} \frac{(-1)^n n! (x - 4)^n}{3^n}$

11. $\displaystyle\sum_{n=1}^{\infty} \frac{(-1)^{n+1} (x - 5)^n}{n 5^n}$

12. $\displaystyle\sum_{n=0}^{\infty} \frac{(x - 2)^{n+1}}{(n + 1) 3^{n+1}}$

13. $\displaystyle\sum_{n=0}^{\infty} \frac{(-1)^{n+1} (x - 1)^{n+1}}{n + 1}$

14. $\displaystyle\sum_{n=1}^{\infty} \frac{(-1)^{n+1} (x - c)^n}{n c^n}$

15. $\displaystyle\sum_{n=1}^{\infty} \frac{(x - c)^{n-1}}{c^{n-1}}, \quad 0 < c$

16. $\displaystyle\sum_{n=1}^{\infty} \left(\frac{2 \cdot 4 \cdot 6 \cdots 2n}{3 \cdot 5 \cdot 7 \cdots (2n + 1)} \right) x^{2n+1}$

17. $\displaystyle\sum_{n=1}^{\infty} \frac{(-1)^{n+1} x^{2n-1}}{2n - 1}$

18. $\displaystyle\sum_{n=1}^{\infty} \frac{n! (x - c)^n}{1 \cdot 3 \cdot 5 \cdots (2n - 1)}$

19. $\displaystyle\sum_{n=1}^{\infty} \frac{n}{n + 1} (-2x)^{n-1}$

20. $\displaystyle\sum_{n=0}^{\infty} \frac{(-1)^n x^{2n}}{n!}$

21. $\displaystyle\sum_{n=0}^{\infty} \frac{x^{2n+1}}{(2n + 1)!}$

22. $\displaystyle\sum_{n=1}^{\infty} \frac{n! x^n}{(2n)!}$

23. $\displaystyle\sum_{n=1}^{\infty} \frac{k(k + 1)(k + 2) \cdots (k + n - 1) x^n}{n!}, \quad k \geq 1$

24. $\displaystyle\sum_{n=1}^{\infty} \frac{(-1)^n 2^{2n-1} x^{2n}}{(2n)!}$

In Exercises 25–28, find the interval of convergence of (a) $f(x)$, (b) $f'(x)$, (c) $f''(x)$, and (d) $\int f(x)\,dx$. Include a check for convergence at the endpoints.

25. $\displaystyle f(x) = \sum_{n=0}^{\infty} \left(\frac{x}{2}\right)^n$

26. $\displaystyle f(x) = \sum_{n=1}^{\infty} \frac{(-1)^{n+1}(x-5)^n}{n5^n}$

27. $\displaystyle f(x) = \sum_{n=0}^{\infty} \frac{(-1)^{n+1}(x-1)^{n+1}}{n+1}$

28. $\displaystyle f(x) = \sum_{n=1}^{\infty} \frac{(-1)^{n+1}(x-1)^n}{n}$

In Exercises 29–32, use the given series to approximate $\int_0^1 f(x)\,dx$ using $R_N \le 0.001$.

29. $\displaystyle f(x) = \sum_{n=0}^{\infty} \frac{(-1)^n x^{2n}}{(2n+1)!}$ 30. $\displaystyle f(x) = \sum_{n=0}^{\infty} \frac{(-1)^n x^n}{(2n)!}$

31. $\displaystyle f(x) = \sum_{n=1}^{\infty} \frac{(-1)^{n+1} n x^{n-1}}{3^n n!}$

32. $\displaystyle f(x) = \sum_{n=0}^{\infty} \frac{(-1)^n x^{2n}}{2^n n!}$

33. Let

$$f(x) = \sum_{n=0}^{\infty} \frac{(-1)^n x^{2n+1}}{(2n+1)!}$$

and

$$g(x) = \sum_{n=0}^{\infty} \frac{(-1)^n x^{2n}}{(2n)!}$$

(a) Find the interval of convergence of f and g.
(b) Show that $f'(x) = g(x)$.
(c) Show that $g'(x) = -f(x)$.
(d) From parts (b) and (c), identify the functions f and g.

34. Prove that if the series for f' converges at the endpoints of its interval of convergence, then the series for f does also.

SECTION TOPICS ▪
Geometric power series ▪
Operations with power series ▪
Power series representation of a function ▪

10.9
Representation of functions by power series

In the previous section we established some properties of power series and of functions represented by power series. You could say that we were seeking an answer to the question "Given a power series, what function does it represent?" In this and the next section, we want to turn this question around and ask: "Given a function, can it be represented by a power series, and if it can, what is the power series?" One person who gave some important answers to this question was the French mathematician Joseph Fourier (1768–1830). Fourier's work occupies an important place in the history of calculus, partly because it forced eighteenth century mathematicians to question the then prevailing narrow concept of a function. Both Cauchy and Dirichlet were motivated by Fourier's work with series and in 1837 Dirichlet published the general definition of function that is used today.

In this section we begin to look at the question of representing a given function by a power series by deriving power series for some of the basic elementary functions. Also, we will study techniques for producing power series for other functions that are combinations of the basic functions. Then, in the next section we will look at necessary conditions for representing a function by a power series.

Joseph Fourier

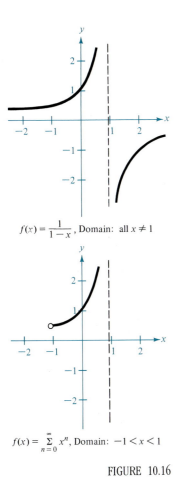

$f(x) = \dfrac{1}{1-x}$, Domain: all $x \neq 1$

$f(x) = \sum\limits_{n=0}^{\infty} x^n$, Domain: $-1 < x < 1$

FIGURE 10.16

Consider the function given by $f(x) = 1/(1 - x)$. The form of f closely resembles the sum of a geometric series

$$\sum_{n=0}^{\infty} ar^n = \frac{a}{1 - r}$$

where $|r| < 1$. In other words, if we let $a = 1$ and $r = x$, then a power series representation for $1/(1 - x)$, centered at 0, is

$$\frac{1}{1 - x} = \sum_{n=0}^{\infty} x^n = 1 + x + x^2 + x^3 + \cdots \qquad (-1, 1)$$

Of course, this series represents $f(x) = 1/(1 - x)$ only on the interval $(-1, 1)$, whereas f is defined for all $x \neq 1$, as shown in Figure 10.16. To represent f on another interval, we must develop a different series. For instance, to obtain the power series centered at -1, we write

$$\frac{1}{1 - x} = \frac{1}{2 - (x + 1)} = \frac{1/2}{1 - [(x + 1)/2]}$$

$$= \frac{a}{1 - r}$$

which implies that $a = \tfrac{1}{2}$ and $r = (x + 1)/2$. Thus, for $|x + 1| < 2$, we have

$$\frac{1}{1 - x} = \sum_{n=0}^{\infty} \left(\frac{1}{2}\right)\left(\frac{x + 1}{2}\right)^n$$

$$= \frac{1}{2} \sum_{n=0}^{\infty} \left(\frac{x + 1}{2}\right)^n$$

$$= \frac{1}{2}\left[1 + \frac{(x + 1)}{2} + \frac{(x + 1)^2}{4} + \frac{(x + 1)^3}{8} + \cdots\right] \qquad (-3, 1)$$

which converges on the interval $(-3, 1)$. This result is generalized in the following theorem.

THEOREM 10.21   **GEOMETRIC POWER SERIES**

The power series for the function $f(x) = a/(x - b)$ is

1. Centered at 0: $\dfrac{a}{x - b} = -\dfrac{a}{b} \sum\limits_{n=0}^{\infty} \left(\dfrac{x}{b}\right)^n$

2. Centered at c: $\dfrac{a}{x - b} = -\dfrac{a}{b - c} \sum\limits_{n=0}^{\infty} \left(\dfrac{x - c}{b - c}\right)^n$

The radius of convergence is $R = |b - c|$.

EXAMPLE 1   Finding a geometric power series

Find a power series for $f(x) = 3/(x + 2)$, centered at 0.

Solution: Writing $f(x)$ in the form $a/(x - b)$, we have

$$\frac{3}{x + 2} = \frac{3}{x - (-2)} = \frac{a}{x - b}$$

which implies that $a = 3$ and $b = -2$. Therefore, by Theorem 10.21,

$$\frac{3}{x + 2} = \frac{-3}{-2} \sum_{n=0}^{\infty} \left(\frac{x}{-2}\right)^n = \frac{3}{2}\left[1 - \frac{x}{2} + \frac{x^2}{4} - \frac{x^3}{8} + \cdots\right]$$

The radius of convergence is $R = |b - c| = 2$ and the interval of convergence is $(-2, 2)$.

EXAMPLE 2   *Finding a geometric power series*

Find a power series for $f(x) = 1/x$ centered at 1.

Solution: First, we note that $f(x)$ has the form

$$\frac{1}{x} = \frac{1}{x - 0} = \frac{a}{x - b}$$

which implies that $a = 1$ and $b = 0$. Therefore, by Theorem 10.21,

$$\frac{1}{x} = \frac{-1}{0 - 1} \sum_{n=0}^{\infty} \left(\frac{x - 1}{0 - 1}\right)^n = \sum_{n=0}^{\infty} (-1)^n (x - 1)^n$$

$$= 1 - (x - 1) + (x - 1)^2 - (x - 1)^3 + \cdots$$

The radius of convergence is $R = |b - c| = 1$, and the interval of convergence is $(0, 2)$.

The versatility of Theorem 10.21 will be seen later in this section, following a discussion of power series operations. These operations, used in conjunction with Theorem 10.21 with differentiation and integration, will give us a means for developing power series for a wide variety of elementary functions. For instance, we will find power series for functions as diverse as

$$y = \frac{5x - 1}{x^2 - x - 2}, \qquad y = \ln x, \qquad \text{and} \qquad y = x \arctan x$$

We summarize the basic operations with power series in the following theorem. For simplicity, we state the results for a series centered at 0.

THEOREM 10.22   **OPERATIONS WITH POWER SERIES**
Given $f(x) = \Sigma a_n x^n$ and $g(x) = \Sigma b_n x^n$, the following properties are true.

1. $f(kx) = \displaystyle\sum_{n=0}^{\infty} a_n k^n x^n$ 2. $f(x^N) = \displaystyle\sum_{n=0}^{\infty} a_n x^{nN}$

3. $f(x) \pm g(x) = \displaystyle\sum_{n=0}^{\infty} (a_n \pm b_n) x^n$ 4. $f(x)g(x) = \left(\displaystyle\sum_{n=0}^{\infty} a_n x^n\right)\left(\displaystyle\sum_{n=0}^{\infty} b_n x^n\right)$

> **Remark** The operations described in Theorem 10.22 can change the interval of convergence for the resulting series. For example, in the following addition, the interval of convergence for the sum is the *intersection* of the intervals of convergence of the two original series:

$$\sum_{n=0}^{\infty} x^n + \sum_{n=0}^{\infty} \left(\frac{x}{2}\right)^n = \sum_{n=0}^{\infty} \left(1 + \frac{1}{2^n}\right)x^n$$

$$(-1, 1) \cap (-2, 2) \quad = \quad (-1, 1)$$

EXAMPLE 3   *Finding a power series by addition*

Find a power series, centered at 0, for

$$f(x) = \frac{5x - 1}{x^2 - x - 2}$$

Solution: Using partial fractions, we can rewrite $f(x)$ as

$$\frac{5x - 1}{x^2 - x - 2} = \frac{2}{x + 1} + \frac{3}{x - 2}$$

Now, applying Theorem 10.21, we obtain

$$\frac{2}{x + 1} = \frac{2}{x - (-1)}$$

$$= \frac{-2}{-1} \sum_{n=0}^{\infty} \left(\frac{x}{-1}\right)^n$$

$$= \sum_{n=0}^{\infty} 2(-1)^n x^n \qquad (-1, 1)$$

and

$$\frac{3}{x - 2} = -\frac{3}{2} \sum_{n=0}^{\infty} \left(\frac{x}{2}\right)^n$$

$$= \sum_{n=0}^{\infty} \frac{-3x^n}{2^{n+1}} \qquad (-2, 2)$$

Finally, adding these two series, we have

$$\frac{5x - 1}{x^2 - x - 2} = \sum_{n=0}^{\infty} \left[2(-1)^n - \frac{3}{2^{n+1}}\right]x^n$$

$$= \sum_{n=0}^{\infty} \left[\frac{(-1)^n 2^{n+2} - 3}{2^{n+1}}\right]x^n$$

$$= \frac{1}{2} - \frac{11}{4}x + \frac{13}{8}x^2 - \frac{35}{16}x^3 + \cdots \qquad (-1, 1)$$

It is possible to determine the power series representation for a rational function such as the one in Example 3 by long division. For instance,

$$
-2 - x + x^2 \overline{\smash{\big)}\, -1 + 5x} \quad \genfrac{}{}{0pt}{}{}{\dfrac{1}{2} - \dfrac{11}{4}x + \dfrac{13}{8}x^2 - \dfrac{35}{16}x^3 + \cdots}
$$

$$
\underline{-1 - \dfrac{x}{2} + \dfrac{x^2}{2}}
$$

$$
\dfrac{11x}{2} - \dfrac{x^2}{2}
$$

$$
\underline{\dfrac{11x}{2} + \dfrac{11x^2}{4} - \dfrac{11x^3}{4}}
$$

$$
-\dfrac{13x^2}{4} + \dfrac{11x^3}{4}
$$

$$
\underline{-\dfrac{13x^2}{4} - \dfrac{13x^3}{8} + \dfrac{13x^4}{8}}
$$

$$
\dfrac{35x^3}{8} - \dfrac{13x^4}{8}
$$

$$
\underline{\dfrac{35x^3}{8} + \dfrac{35x^4}{16} - \dfrac{35x^5}{16}}
$$

One disadvantage of this alternative procedure is that the general term of the series is not given and we may be hard pressed to determine it. And without knowing the general term of the series, we cannot readily determine the interval of convergence.

EXAMPLE 4   Finding a power series by integration

Find a power series for $f(x) = \ln x$, centered at 1.

Solution: From Example 2 we know that

$$
\frac{1}{x} = \sum_{n=0}^{\infty} (-1)^n (x - 1)^n \qquad (0, 2)
$$

By integration, we have

$$
\ln x = \int \frac{1}{x}\, dx + C = C + \sum_{n=0}^{\infty} (-1)^n \frac{(x - 1)^{n+1}}{n + 1}
$$

To determine C, we let $x = 1$ and conclude that $C = 0$. Therefore,

$$
\ln x = \sum_{n=0}^{\infty} (-1)^n \frac{(x - 1)^{n+1}}{n + 1}
$$

$$
= \frac{(x - 1)}{1} - \frac{(x - 1)^2}{2} + \frac{(x - 1)^3}{3} - \frac{(x - 1)^4}{4} + \cdots \qquad (0, 2]
$$

Note that the series converges at $x = 2$. This is consistent with our observation in the previous section that integration of a power series may alter the convergence at the endpoints of the interval of convergence.

EXAMPLE 5   Finding a power series by integration

Find a power series for $f(x) = x \arctan x$, centered at 0.

Solution: Since $D_x[\arctan x] = 1/(1 + x^2)$, we use the series

$$f(x) = \frac{1}{1 + x} = \sum_{n=0}^{\infty} (-1)^n x^n \qquad (-1, 1)$$

Now, substituting x^2 for x and then integrating, we have

$$f(x^2) = \frac{1}{1 + x^2}$$

$$= \sum_{n=0}^{\infty} (-1)^n x^{2n}$$

$$\arctan x = \int \frac{1}{1 + x^2}\, dx + C$$

$$= C + \sum_{n=0}^{\infty} (-1)^n \frac{x^{2n+1}}{2n + 1}$$

$$= \sum_{n=0}^{\infty} (-1)^n \frac{x^{2n+1}}{2n + 1} \qquad \text{Let } x = 0, \text{ then } C = 0.$$

$$x \arctan x = \sum_{n=0}^{\infty} (-1)^n \frac{x^{2n+2}}{2n + 1} \qquad \text{Multiply by } x$$

$$= \frac{x^2}{1} - \frac{x^4}{3} + \frac{x^6}{5} - \frac{x^8}{7} + \cdots \qquad [-1, 1]$$

Again, note that the series converges at the endpoints $x = \pm 1$.

In Section 10.1, we approximated ln (1.1) using the fourth Taylor polynomial for the natural logarithmic function.

$$\ln x \approx (x - 1) - \frac{(x - 1)^2}{2} + \frac{(x - 1)^3}{3} - \frac{(x - 1)^4}{4}$$

$$\ln (1.1) \approx (0.1) - \frac{1}{2}(0.1)^2 + \frac{1}{3}(0.1)^3 - \frac{1}{4}(0.1)^4 \approx 0.0953083$$

We now know from Example 4 that this polynomial represents the first four terms of the power series for ln x. Moreover, using the Alternating Series Remainder, we can determine that the error in this approximation is less than

$$|R_3| \le |a_4| = \frac{1}{5}(0.1)^5 \approx 0.000002$$

During the seventeenth and eighteenth centuries, mathematical tables for logarithms and values of other transcendental functions were computed in this manner. Moreover, the use for such numerical techniques is far from outdated, since it is precisely by such means that modern calculating devices are programmed to evaluate transcendental functions.

Section Exercises 10.9

In Exercises 1–14, find a power series for the given function, centered at c, and determine the interval of convergence.

1. $f(x) = \dfrac{1}{2 - x}$, $c = 0$

2. $f(x) = \dfrac{3}{4 - x}$, $c = 0$

3. $f(x) = \dfrac{1}{2 - x}$, $c = 5$

4. $f(x) = \dfrac{3}{4 - x}$, $c = -2$

5. $f(x) = \dfrac{3}{2x - 1}$, $c = 0$

6. $f(x) = \dfrac{3}{2x - 1}$, $c = 2$

7. $f(x) = \dfrac{1}{2x - 5}$, $c = -3$

8. $f(x) = \dfrac{1}{2x - 5}$, $c = 0$

9. $f(x) = \dfrac{3}{x + 2}$, $c = 0$

10. $f(x) = \dfrac{4}{3x + 2}$, $c = 2$

11. $f(x) = \dfrac{3x}{x^2 + x - 2}$, $c = 0$

12. $f(x) = \dfrac{4x - 7}{2x^2 + 3x - 2}$, $c = 0$

13. $f(x) = \dfrac{2}{1 - x^2}$, $c = 0$

14. $f(x) = \dfrac{4}{4 + x^2}$, $c = 0$

In Exercises 15–22, use the power series

$$\frac{1}{1 + x} = \sum_{n=0}^{\infty} (-1)^n x^n$$

to determine a power series representation, centered at 0, for the given function. List the interval of convergence with your solution.

15. $f(x) = -\dfrac{1}{(x + 1)^2}$

16. $f(x) = \dfrac{2}{(x + 1)^3}$

17. $f(x) = \ln (x + 1)$

18. $f(x) = \ln (x^2 + 1)$

19. $f(x) = \dfrac{1}{4x^2 + 1}$

20. $f(x) = \arctan 2x$

21. $f(x) = \ln \sqrt{\dfrac{1 + x}{1 - x}}$

22. $f(x) = \dfrac{1}{x^2 - x + 1} = \dfrac{x + 1}{x^3 + 1}$

23. Complete the following table to demonstrate the inequalities

$$x - \frac{x^2}{2} \leq \ln (x + 1) \leq x - \frac{x^2}{2} + \frac{x^3}{3}$$

x	0.0	0.2	0.4	0.6	0.8	1.0
$x - \dfrac{x^2}{2}$						
$\ln (x + 1)$						
$x - \dfrac{x^2}{2} + \dfrac{x^3}{3}$						

24. Complete a table for the same values of x as in Exercise 23 to demonstrate the inequalities

$$x - \frac{x^2}{2} + \frac{x^3}{3} - \frac{x^4}{4} \leq \ln (x + 1)$$

$$\leq x - \frac{x^2}{2} + \frac{x^3}{3} - \frac{x^4}{4} + \frac{x^5}{5}$$

In Exercises 25–28, use the series for $f(x) = \arctan x$ to approximate the given value, using $R_N \leq 0.001$.

25. $\arctan \dfrac{1}{4}$

26. $\displaystyle\int_0^{3/4} \arctan x^2 \, dx$

27. $\displaystyle\int_0^{1/2} \dfrac{\arctan x^2}{x} \, dx$

28. $\displaystyle\int_0^{1/2} x^2 \arctan x \, dx$

29. Use long division to develop the power series for $1/(1 + x)$.

10.10
Taylor and Maclaurin series

In Section 10.9 we derived power series for several functions using geometric series together with term by term differentiation or integration. In this section we develop a *general* procedure for deriving the power series for a function that has derivatives of all orders.

The development of power series to represent functions is due to the combined work of many seventeenth and eighteenth century mathematicians. Gregory, Newton, John and James Bernoulli, Leibniz, Euler, Lagrange, Wallis, and Fourier all contributed to this work. However, the two names that are most commonly associated with power series are Brook Taylor (1685–1731) and Colin Maclaurin (1698–1746).

We begin with a theorem that gives us the form that *every* (convergent) power series must take.

THEOREM 10.23   **THE FORM OF A CONVERGENT POWER SERIES**
If f is represented by a power series

$$f(x) = a_0 + a_1(x - c) + a_2(x - c)^2 + a_3(x - c)^3 + \cdots$$

for all x in an open interval I containing c, then $a_n = f^{(n)}(c)/n!$ and

$$f(x) = f(c) + f'(c)(x - c) + \frac{f''(c)}{2!}(x - c)^2 + \cdots + \frac{f^{(n)}(c)}{n!}(x - c)^n + \cdots$$

Colin Maclaurin

Proof: Suppose the power series $\Sigma a_n(x - c)^n$ has a radius of convergence of R. Then by Theorem 10.20 we know that the nth derivative of f exists for $|x - c| < R$ and by successive differentiation we obtain

$$f^{(0)}(x) = a_0 + a_1(x - c) + a_2(x - c)^2 + a_3(x - c)^3 + a_4(x - c)^4 + \cdots$$
$$f^{(1)}(x) = a_1 + 2a_2(x - c) + 3a_3(x - c)^2 + 4a_4(x - c)^3 + \cdots$$
$$f^{(2)}(x) = 2a_2 + 3!a_3(x - c) + 4 \cdot 3a_4(x - c)^2 + \cdots$$
$$f^{(3)}(x) = 3!a_3 + 4!a_4(x - c) + \cdots$$
$$\vdots$$
$$f^{(n)}(x) = n!a_n + (n + 1)!a_{n+1}(x - c) + \cdots$$

Now, evaluating each of these derivatives at $x = c$ yields

$$f^{(0)}(c) = 0!a_0, \qquad f^{(1)}(c) = 1!a_1, \qquad f^{(2)}(c) = 2!a_2, \qquad f^{(3)}(c) = 3!a_3$$

and, in general, $f^{(n)}(c) = n!a_n$. By solving for a_n we find that the coefficients of the power series representation of $f(x)$ are

$$a_n = \frac{f^{(n)}(c)}{n!}$$

Notice that the coefficients of the power series in Theorem 10.23 are precisely the coefficients of the Taylor polynomials for $f(x)$ at c as defined in Section 10.1. For this reason, the series is called the **Taylor series** for $f(x)$ at c.

| DEFINITION OF TAYLOR | If a function f has derivatives of all orders at $x = c$, then the series |

$$\sum_{n=0}^{\infty} \frac{f^{(n)}(c)}{n!}(x-c)^n = f(c) + f'(c)(x-c) + \cdots + \frac{f^{(n)}(c)}{n!}(x-c)^n + \cdots$$

is called the **Taylor series for $f(x)$ at c.** Moreover, if $c = 0$, then this series is called the **Maclaurin series for f.**

If we know the pattern for the coefficients of the Taylor polynomials for a function, then we can easily extend the pattern to form the corresponding Taylor series. For instance, in Example 2 of Section 10.1, we found the fourth Taylor polynomial for $\ln x$, centered at 1, to be

$$P_4(x) = (x-1) - \frac{1}{2}(x-1)^2 + \frac{1}{3}(x-1)^3 - \frac{1}{4}(x-1)^4$$

Following this pattern, we obtain the series

$$(x-1) - \frac{1}{2}(x-1)^2 + \cdots + \frac{(-1)^{n+1}}{n}(x-1)^n + \cdots$$

EXAMPLE 1   Forming a power series

Use the function $f(x) = \sin x$ to form the power series

$$\sum_{n=0}^{\infty} \frac{f^{(n)}(0)}{n!}x^n = f(0) + f'(0)x + \frac{f''(0)}{2!}x^2 + \frac{f^{(3)}(0)}{3!}x^3 + \frac{f^{(4)}(0)}{4!}x^4 + \cdots$$

and determine the interval of convergence.

Solution: Successive differentiation of $f(x)$ yields

$$
\begin{aligned}
f(x) &= \sin x & f(0) &= 0 \\
f'(x) &= \cos x & f'(0) &= 1 \\
f''(x) &= -\sin x & f''(0) &= 0 \\
f^{(3)}(x) &= -\cos x & f^{(3)}(0) &= -1 \\
f^{(4)}(x) &= \sin x & f^{(4)}(0) &= 0 \\
f^{(5)}(x) &= \cos x & f^{(5)}(0) &= 1
\end{aligned}
$$

and so on. The pattern repeats after the third derivative. Hence, the power series is

$$\sum_{n=0}^{\infty} \frac{f^{(n)}(0)}{n!}x^n = f(0) + f'(0)x + \frac{f''(0)}{2!}x^2 + \frac{f^{(3)}(0)}{3!}x^3 + \frac{f^{(4)}(0)}{4!}x^4 + \cdots$$

$$\sum_{n=0}^{\infty} \frac{(-1)^n x^{2n+1}}{(2n+1)!} = x - \frac{x^3}{3!} + \frac{x^5}{5!} - \frac{x^7}{7!} + \cdots$$

By the Ratio Test, we can conclude that this series converges for all x.

Notice that we did not conclude that the power series in Example 1 converges to sin x for all x. We simply concluded that the power series converges to some function, but we are not sure what function it is. This is a subtle, but important, point in dealing with Taylor or Maclaurin series. To convince yourself that the series

$$f(c) + f'(c)(x - c) + \frac{f''(c)}{2!}(x - c)^2 + \cdots + \frac{f^{(n)}(c)}{n!}(x - c)^n + \cdots$$

might converge to a function other than f, remember that the derivatives are being evaluated at a single point. It can easily happen that another function will agree with the values of $f^{(n)}(x)$ when $x = c$ and disagree at other x-values. For instance, if we formed the power series (centered at 0) for the function

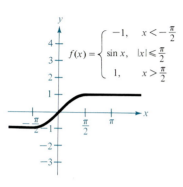

$$f(x) = \begin{cases} -1, & x < -\dfrac{\pi}{2} \\ \sin x, & |x| \le \dfrac{\pi}{2} \\ 1, & x > \dfrac{\pi}{2} \end{cases} \qquad \text{(See Figure 10.17.)}$$

FIGURE 10.17

we would obtain the same series we obtained in Example 1. We know that the series converges for all x, and yet it obviously cannot converge to both $f(x)$ and sin x for all x.

| **Remark** Don't confuse this observation with the result given in Theorem 10.23. That theorem says that *if a power series converges to $f(x)$, then the series must be a Taylor series.* The theorem does not say that every series formed with the Taylor coefficients $a_n = f^{(n)}(c)/n!$ will converge to $f(x)$.

To determine whether or not a power series formed with Taylor coefficients actually converges to $f(x)$, it is helpful to consider the series as the sum of the nth Taylor polynomial for $f(x)$ and a remainder term. That is,

$$\underbrace{f(c) + \cdots + \frac{f^{(n)}(c)}{n!}(x - c)^n}_{P_n(x)} + \underbrace{\frac{f^{(n+1)}(c)}{(n + 1)!}(x - c)^{n+1} + \cdots}_{R_n(x)}$$

Recall that Taylor's Theorem, given in Section 10.1, tells us that the remainder for a Taylor polynomial is given by

$$R_n(x) = \frac{f^{(n+1)}(z)}{(n + 1)!}(x - c)^{n+1}$$

where z lies between x and c. The problem we now want to tackle is to find those values of x for which the Taylor polynomials $P_n(x)$ actually converge to $f(x)$ as $n \to \infty$. In other words, we want to find the values of x for which

$$f(x) = \lim_{n \to \infty} P_n(x) = \lim_{n \to \infty} \sum_{k=0}^{n} \frac{f^{(k)}(c)}{k!}(x - c)^k = \sum_{k=0}^{\infty} \frac{f^{(k)}(c)}{k!}(x - c)^k$$

The conditions that guarantee this convergence are given in Theorem 10.24.

THEOREM 10.24   **CONVERGENCE OF TAYLOR SERIES**

If a function f has derivatives of all orders in an interval I centered at c, then the equality

$$f(x) = \sum_{n=0}^{\infty} \frac{f^{(n)}(c)}{n!}(x - c)^n$$

holds if and only if

$$\lim_{n \to \infty} R_n(x) = 0$$

for every x in I.

Proof: For a Taylor series, the nth partial sum coincides with the nth Taylor polynomial. That is, $S_n(x) = P_n(x)$. Moreover, since

$$P_n(x) = f(x) - R_n(x)$$

it follows that

$$\lim_{n \to \infty} S_n(x) = \lim_{n \to \infty} P_n(x) = \lim_{n \to \infty} [f(x) - R_n(x)] = f(x) - \lim_{n \to \infty} R_n(x)$$

Hence, for a given x, the Taylor series (the sequence of partial sums) converges to $f(x)$ if and only if $R_n(x) \to 0$ as $n \to \infty$.

Remark   Stated another way, this theorem says that a power series formed with the Taylor coefficients $a_n = f^{(n)}(c)/n!$ converges to the function from which it was derived at precisely those values for which the remainder approaches zero as $n \to \infty$.

In Example 2, we take another look at the series formed in Example 1. We derived that series from the sine function and we also concluded that the series converges to some function on the entire real line. We now show that the series actually converges to $\sin x$.

EXAMPLE 2   A convergent Maclaurin series

Show that the Maclaurin series for $f(x) = \sin x$ converges to $\sin x$ for all x.

Solution: Using the result in Example 1, we need to show that

$$\sin x = x - \frac{x^3}{3!} + \frac{x^5}{5!} - \frac{x^7}{7!} + \cdots + \frac{(-1)^n x^{2n+1}}{(2n+1)!} + \cdots$$

is true for all x. Since

$$f^{(n+1)}(x) = \pm \sin x \qquad \text{or} \qquad f^{(n+1)}(x) = \pm \cos x$$

we know that $|f^{(n+1)}(z)| \le 1$ for every real number z. Therefore, for any fixed x, we can apply Taylor's Theorem (Theorem 10.1) to conclude that

$$0 \le |R_n(x)| \le \left| \frac{f^{(n+1)}(z)}{(n+1)!} x^{n+1} \right| \le \frac{|x|^{n+1}}{(n+1)!}$$

Now, from our discussion in Section 10.2 regarding the relative rate of convergence of exponential and factorial sequences, it follows that for a fixed x

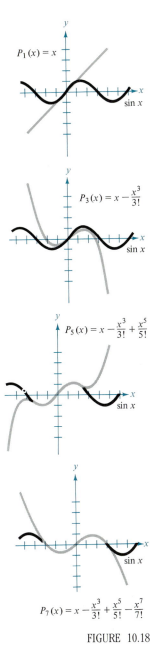

$P_1(x) = x$

$P_3(x) = x - \dfrac{x^3}{3!}$

$P_5(x) = x - \dfrac{x^3}{3!} + \dfrac{x^5}{5!}$

$P_7(x) = x - \dfrac{x^3}{3!} + \dfrac{x^5}{5!} - \dfrac{x^7}{7!}$

FIGURE 10.18

$$\lim_{n \to \infty} \frac{|x|^{n+1}}{(n+1)!} = 0$$

Finally, by the Squeeze Theorem, it follows that for all x, $R_n(x) \to 0$ as $n \to \infty$ for all x. Hence, by Theorem 10.24, the Maclaurin series for $\sin x$ converges to $\sin x$ for all x. ☐

Figure 10.18 visually illustrates the convergence of the Maclaurin series for $\sin x$ by comparing the graphs of the Maclaurin polynomials $P_1(x)$, $P_3(x)$, $P_5(x)$, and $P_7(x)$ with the graph of the sine function. Notice that as the degree of the polynomial increases, its graph more closely resembles that of the sine function.

We summarize the steps for finding a Taylor series for $f(x)$ at c as follows.

1. Differentiate $f(x)$ several times and evaluate each derivative at c.
$$f(c), f'(c), f''(c), f'''(c), \ldots, f^{(n)}(c), \ldots$$
Try to recognize a pattern for these numbers.

2. Use the sequence developed in the first step to form the Taylor coefficients $a_n = f^{(n)}(c)/n!$, and determine the interval of convergence for the resulting power series
$$f(c) + f'(c)(x - c) + \frac{f''(c)}{2!}(x - c)^2 + \cdots + \frac{f^{(n)}(c)}{n!}(x - c)^n + \cdots$$

3. Within this interval of convergence, determine whether or not the series converges to $f(x)$.

The direct determination of Taylor or Maclaurin coefficients using successive differentiation can be difficult, and the next example illustrates a shortcut for finding the coefficients indirectly—using the coefficients of a known Taylor or Maclaurin series.

EXAMPLE 3   *Maclaurin series for a composite function*

Find the Maclaurin series for $f(x) = \sin x^2$.

Solution: To find the coefficients for this Maclaurin series *directly* we must calculate successive derivatives of $f(x) = \sin x^2$. By calculating just the first two,
$$f'(x) = 2x \cos x^2 \qquad \text{and} \qquad f''(x) = -4x^2 \sin x^2 + 2 \cos x^2$$
we recognize this to be a rather cumbersome task. Fortunately there is an alternative. Suppose we first consider the Maclaurin series for $\sin x$ found in Example 1.
$$g(x) = \sin x = x - \frac{x^3}{3!} + \frac{x^5}{5!} - \frac{x^7}{7!} + \cdots$$

Now since $\sin x^2 = g(x^2)$, we can apply Theorem 10.22 and substitute x^2 for x in the series for $\sin x$ to obtain
$$\sin x^2 = g(x^2) = x^2 - \frac{x^6}{3!} + \frac{x^{10}}{5!} - \frac{x^{14}}{7!} + \cdots$$
☐

Be sure that you understand the point being illustrated in Example 3. Since direct computation of Taylor or Maclaurin coefficients can be very tedious, the most practical way to find a Taylor or Maclaurin series is in developing power series for a *basic list* of elementary functions. From this list, we can determine power series for other functions by the operations of addition, subtraction, multiplication, division, differentiation, integration, or composition with known power series. Before presenting the basic list for elementary functions we develop one additional formula—for a function of the form $f(x) = (1 + x)^k$. This produces the **binomial series.**

EXAMPLE 4   Binomial series

Find the Maclaurin series for $f(x) = (1 + x)^k$ and determine its radius of convergence.

Solution: By successive differentiation, we have

$$f(x) = (1 + x)^k \qquad\qquad f(0) = 1$$
$$f'(x) = k(1 + x)^{k-1} \qquad\qquad f'(0) = k$$
$$f''(x) = k(k - 1)(1 + x)^{k-2} \qquad\qquad f''(0) = k(k - 1)$$
$$f'''(x) = k(k - 1)(k - 2)(1 + x)^{k-3} \qquad\qquad f'''(0) = k(k - 1)(k - 2)$$
$$\vdots \qquad\qquad\qquad\qquad \vdots$$
$$f^{(n)}(x) = k \cdots (k - n + 1)(1 + x)^{k-n} \qquad f^{(n)}(0) = k(k - 1) \cdots (k - n + 1)$$

which produces the series

$$1 + kx + \frac{k(k - 1)x^2}{2} + \cdots + \frac{k(k - 1) \cdots (k - n + 1)x^n}{n!} + \cdots$$

For all x in an open interval I containing c, $a_n = f^{(n)}(c)/n!$ and

$$f(x) = f(c) + f'(c)(x - c) + \frac{f''(c)}{2!}(x - c)^2 + \cdots + \frac{f^{(n)}(c)}{n!}(x - c)^n + \cdots$$

By the Ratio Test, we have $a_{n+1}/a_n \to 1$ as $n \to \infty$. This implies that the radius of convergence is $R = 1$, and thus the series converges to *some* function in the interval $(-1, 1)$.

Note that in Example 4 we showed that the Taylor series for $(1 + x)^k$ converges to *some* function in the interval $(-1, 1)$. However, we did not show that the series actually converges to $(1 + x)^k$. To do this, we could show that the remainder $R_n(x)$ converges to 0, as we did in Example 2. Another way to show that the Taylor series for f converges to $f(x)$ is to use the series to form a differential equation. Then, the verification of convergence to $f(x)$ can be obtained by solving the differential equation. For example, if we let y be the binomial series obtained in Example 4,

$$y = 1 + kx + \frac{k(k - 1)x^2}{2} + \cdots + \frac{k(k - 1) \cdots (k - n + 1)x^n}{n!} + \cdots$$

then we can derive the differential equation

$$y' + xy' = ky \qquad \text{Initial condition } y = 1 \text{ if } x = 0$$

(For a detailed discussion of this procedure, see Example 4 in Section 18.7.)

In the following list, we provide the power series for several elementary functions together with the corresponding intervals of convergence.

POWER SERIES FOR ELEMENTARY FUNCTIONS

$$\frac{1}{x} = 1 - (x - 1) + (x - 1)^2 - (x - 1)^3 + (x - 1)^4 - \cdots + (-1)^n(x - 1)^n + \cdots, \qquad 0 < x < 2$$

$$\frac{1}{1 + x} = 1 - x + x^2 - x^3 + x^4 - x^5 + \cdots + (-1)^n x^n + \cdots, \qquad -1 < x < 1$$

$$\ln x = (x - 1) - \frac{(x - 1)^2}{2} + \frac{(x - 1)^3}{3} - \frac{(x - 1)^4}{4} + \cdots + \frac{(-1)^{n-1}(x - 1)^n}{n} + \cdots, \qquad 0 < x \le 2$$

$$e^x = 1 + x + \frac{x^2}{2!} + \frac{x^3}{3!} + \frac{x^4}{4!} + \frac{x^5}{5!} + \cdots + \frac{x^n}{n!} + \cdots, \qquad -\infty < x < \infty$$

$$\sin x = x - \frac{x^3}{3!} + \frac{x^5}{5!} - \frac{x^7}{7!} + \frac{x^9}{9!} - \cdots + \frac{(-1)^n x^{2n+1}}{(2n + 1)!} + \cdots, \qquad -\infty < x < \infty$$

$$\cos x = 1 - \frac{x^2}{2!} + \frac{x^4}{4!} - \frac{x^6}{6!} + \frac{x^8}{8!} - \cdots + \frac{(-1)^n x^{2n}}{(2n)!} + \cdots, \qquad -\infty < x < \infty$$

$$\arctan x = x - \frac{x^3}{3} + \frac{x^5}{5} - \frac{x^7}{7} + \frac{x^9}{9} - \cdots + \frac{(-1)^n x^{2n+1}}{2n + 1} + \cdots, \qquad -1 \le x \le 1$$

$$\arcsin x = x + \frac{x^3}{2 \cdot 3} + \frac{1 \cdot 3x^5}{2 \cdot 4 \cdot 5} + \frac{1 \cdot 3 \cdot 5x^7}{2 \cdot 4 \cdot 6 \cdot 7} + \cdots + \frac{(2n)!x^{2n+1}}{(2^n n!)^2(2n + 1)} + \cdots, \qquad -1 \le x \le 1$$

$$(1 + x)^k = 1 + kx + \frac{k(k - 1)x^2}{2!} + \frac{k(k - 1)(k - 2)x^3}{3!} + \frac{k(k - 1)(k - 2)(k - 3)x^4}{4!} + \cdots, \qquad -1 < x < 1*$$

$$(1 + x)^{-k} = 1 - kx + \frac{k(k + 1)x^2}{2!} - \frac{k(k + 1)(k + 2)x^3}{3!} + \frac{k(k + 1)(k + 2)(k + 3)x^4}{4!} - \cdots, \qquad -1 < x < 1*$$

| **Remark** The binomial series is valid for noninteger values of k. Moreover, if k happens to be a positive integer, then the binomial series reduces to a simple binomial expansion.

---

*The convergence at $x = \pm 1$ depends on the value k.

In the next three examples we show how to use the basic list of power series to obtain series for functions not on the list.

EXAMPLE 5   *Deriving a new power series from a given series*

Find the power series for $f(x) = \cos \sqrt{x}$.

Solution: Using the power series

$$\cos x = 1 - \frac{x^2}{2!} + \frac{x^4}{4!} - \frac{x^6}{6!} + \frac{x^8}{8!} - \cdots$$

we replace x by $\sqrt{x}$ to obtain the series

$$\cos \sqrt{x} = 1 - \frac{x}{2!} + \frac{x^2}{4!} - \frac{x^3}{6!} + \frac{x^4}{8!} - \cdots$$

which converges for all x in the domain of $\cos \sqrt{x}$, that is, for $x \geq 0$.

EXAMPLE 6   *Using the binomial series*

Find the power series for $g(x) = \sqrt[3]{1 + x}$.

Solution: Using the binomial series

$$(1 + x)^k = 1 + kx + \frac{k(k - 1)x^2}{2!} + \frac{k(k - 1)(k - 2)x^3}{3!} + \cdots$$

we let $k = \frac{1}{3}$ and write

$$(1 + x)^{1/3} = 1 + \frac{x}{3} - \frac{2x^2}{3^2 2!} + \frac{2 \cdot 5x^3}{3^3 3!} - \frac{2 \cdot 5 \cdot 8x^4}{3^4 4!} + \cdots$$

which converges for $-1 \leq x \leq 1$.

EXAMPLE 7   *A power series for $\sin^2 x$*

Find the power series for $f(x) = \sin^2 x$.

Solution: Instead of squaring the power series for $\sin x$, we write

$$\sin^2 x = \frac{1 - \cos 2x}{2} = \frac{1}{2} - \frac{\cos 2x}{2}$$

Now, using the series for $\cos x$, we proceed as follows.

$$\cos x = 1 - \frac{x^2}{2!} + \frac{x^4}{4!} - \frac{x^6}{6!} + \frac{x^8}{8!} - \cdots$$

$$\cos 2x = 1 - \frac{2^2}{2!}x^2 + \frac{2^4}{4!}x^4 - \frac{2^6}{6!}x^6 + \frac{2^8}{8!}x^8 - \cdots$$

$$-\frac{1}{2}\cos 2x = -\frac{1}{2} + \frac{2}{2!}x^2 - \frac{2^3}{4!}x^4 + \frac{2^5}{6!}x^6 - \frac{2^7}{8!}x^8 + \cdots$$

$$\sin^2 x = \frac{1}{2} - \frac{1}{2}\cos 2x = \frac{1}{2} - \frac{1}{2} + \frac{2}{2!}x^2 - \frac{2^3}{4!}x^4 + \frac{2^5}{6!}x^6 - \frac{2^7}{8!}x^8 + \cdots$$

$$= \frac{2}{2!}x^2 - \frac{2^3}{4!}x^4 + \frac{2^5}{6!}x^6 - \frac{2^7}{8!}x^8 + \cdots$$

This series converges for $-\infty < x < \infty$.

As mentioned in the previous section, power series can be used to obtain tables of values for transcendental functions. They are also useful for estimating the value of definite integrals for which antiderivatives cannot be found. The last example demonstrates this use.

EXAMPLE 8 *Power series approximation of a definite integral*

Use a power series to approximate

$$\int_0^1 e^{-x^2}\, dx$$

with an error of less than 0.01.

Solution: Replacing x with $-x^2$ in the series for e^x, we have

$$e^{-x^2} = 1 - x^2 + \frac{x^4}{2!} - \frac{x^6}{3!} + \frac{x^8}{4!} - \cdots$$

$$\int_0^1 e^{-x^2}\, dx = \left[x - \frac{x^3}{3} + \frac{x^5}{5\cdot 2!} - \frac{x^7}{7\cdot 3!} + \frac{x^9}{9\cdot 4!} - \cdots \right]_0^1$$

$$= 1 - \frac{1}{3} + \frac{1}{10} - \frac{1}{42} + \frac{1}{216} - \cdots$$

Summing the first *four* terms, we have

$$\int_0^1 e^{-x^2}\, dx \approx 0.74$$

which, by the Alternating Series Test, has an error of less than $\frac{1}{216} \approx 0.005$.

Section Exercises 10.10

In Exercises 1–10, find the Taylor series (centered at c) for the given function.

1. $f(x) = e^{2x}$, $c = 0$
2. $f(x) = e^{-2x}$, $c = 0$
3. $f(x) = \cos x$, $c = \dfrac{\pi}{4}$
4. $f(x) = \sin x$, $c = \dfrac{\pi}{4}$

5. $f(x) = \ln x$, $c = 1$
6. $f(x) = e^x$, $c = 1$
7. $f(x) = \sin 2x$, $c = 0$
8. $f(x) = \tan x$, $c = 0$ (first three terms)
9. $f(x) = \sec x$, $c = 0$ (first three terms)
10. $f(x) = \ln (x^2 + 1)$, $c = 0$

In Exercises 11–16, use the binomial series to find the Maclaurin series for the given function.

11. $f(x) = \dfrac{1}{(1 + x)^2}$ **12.** $f(x) = \dfrac{1}{\sqrt{1 - x}}$

13. $f(x) = \dfrac{1}{\sqrt{4 + x^2}}$ **14.** $f(x) = \sqrt{1 + x}$

15. $f(x) = \sqrt{1 + x^2}$ **16.** $f(x) = \sqrt{1 + x^3}$

In Exercises 17–22, find the Maclaurin series for the given function using the indicated method.

17. $f(x) = \sin x$; differentiate the series for $\cos x$.
18. $f(x) = \cos^2 x$; use the identity

$$\cos^2 x = \frac{1}{2}(1 + \cos 2x)$$

19. $f(x) = e^{x^2/2}$; use the series for e^x.
20. $f(x) = \sec^2 x$; differentiate the series for $\tan x$.

21. $f(x) = \dfrac{e^x - e^{-x}}{2} = \sinh x$; use the series for e^x.

22. $f(x) = \dfrac{e^x + e^{-x}}{2} = \cosh x$; differentiate the series for $\sinh x$.

23. Use the Maclaurin series for $\sin x$, $\cos x$, and e^x, together with the fact that $i^2 = -1$, to prove the following:

(a) $\sin x = \dfrac{e^{ix} - e^{-ix}}{2i}$

(b) $\cos x = \dfrac{e^{ix} + e^{-ix}}{2}$

24. Integrate the Maclaurin series for $f(x) = 1/\sqrt{x^2 + 1}$ to find the Maclaurin series for

$$\sinh^{-1} x = \ln (x + \sqrt{x^2 + 1})$$

25. Find the Maclaurin series for

$$f(x) = \frac{\sin x}{x}$$

26. Use the series in Exercise 25 to show that

$$\lim_{x \to 0} \frac{\sin x}{x} = 1$$

In Exercises 27–34, use power series to approximate the given integral with an error of less than 0.0001.

27. $\int_0^{\pi/2} \dfrac{\sin x}{x}\, dx$ **28.** $\int_0^1 \cos x^2\, dx$

29. $\int_0^{\pi/2} \sqrt{x} \cos x\, dx$ **30.** $\int_0^{1/2} \dfrac{\ln (x + 1)}{x}\, dx$

31. $\int_0^{1/2} \dfrac{\arctan x}{x}\, dx$ **32.** $\int_1^2 e^{-x^2}\, dx$

33. $\int_{0.1}^{0.3} \sqrt{1 + x^3}\, dx$ **34.** $\int_{0.5}^1 \cos \sqrt{x}\, dx$

In Exercises 35 and 36, find a Maclaurin series for $f(x)$ defined by the given integral.

35. $\int_0^x (e^{-t^2} - 1)\, dt$ **36.** $\int_0^x \sqrt{1 + t^3}\, dt$

In Exercises 37 and 38, approximate the given probability with an error of less than 0.0001 for the standard normal probability density function

$$P(a < x < b) = \frac{1}{\sqrt{2\pi}} \int_a^b e^{-x^2/2}\, dx$$

37. $P(0 < x < 1)$ **38.** $P(1 < x < 2)$

39. Show that

$$\sum_{n=0}^{\infty} (-1)^n \left[\frac{1}{(2n + 1)!}\right] = \sin 1$$

40. Show that

$$\sum_{n=1}^{\infty} (-1)^{n-1}\left(\frac{1}{n!}\right) = \frac{e - 1}{e}$$

Review Exercises for Chapter 10

In Exercises 1 and 2, find the general term of the sequence.

1. $1, \dfrac{1}{2}, \dfrac{1}{6}, \dfrac{1}{24}, \dfrac{1}{120}, \ldots$

2. $\dfrac{1}{2}, \dfrac{2}{5}, \dfrac{3}{10}, \dfrac{4}{17}, \ldots$

In Exercises 3–10, determine the convergence or divergence of the sequence with the given general term.

3. $a_n = \dfrac{n + 1}{n^2}$ **4.** $a_n = \dfrac{1}{\sqrt{n}}$

5. $a_n = \dfrac{n^3}{n^2 + 1}$ **6.** $a_n = \dfrac{n}{\ln n}$

7. $a_n = \sqrt{n+1} - \sqrt{n}$

8. $a_n = \left(1 + \dfrac{1}{2n}\right)^n$

9. $a_n = \dfrac{\sin \sqrt{n}}{\sqrt{n}}$

10. $a_n = (b^n + c^n)^{1/n}$
(b and c are positive real numbers)

In Exercises 11–14, find the first five terms of the sequence of partial sums for the given series.

11. $\displaystyle\sum_{n=0}^{\infty} \left(\dfrac{3}{2}\right)^n$

12. $\displaystyle\sum_{n=1}^{\infty} \dfrac{(-1)^{n+1}}{2n}$

13. $\displaystyle\sum_{n=1}^{\infty} \dfrac{(-1)^{n+1}}{(2n)!}$

14. $\displaystyle\sum_{n=1}^{\infty} \dfrac{1}{n(n+1)}$

In Exercises 15–18, find the sum of the given series.

15. $\displaystyle\sum_{n=0}^{\infty} \left(\dfrac{2}{3}\right)^n$

16. $\displaystyle\sum_{n=0}^{\infty} \dfrac{2^{n+2}}{3^n}$

17. $\displaystyle\sum_{n=0}^{\infty} \left(\dfrac{1}{2^n} - \dfrac{1}{3^n}\right)$

18. $\displaystyle\sum_{n=0}^{\infty} \left[\left(\dfrac{2}{3}\right)^n \dfrac{1}{(n+1)(n+2)}\right]$

In Exercises 19 and 20, express the repeating decimal as the ratio of two integers.

19. $0.090909\ldots$

20. $0.923076923076\ldots$

In Exercises 21–32, determine the convergence or divergence of the given series.

21. $\displaystyle\sum_{n=1}^{\infty} \dfrac{2^n}{n^3}$

22. $\displaystyle\sum_{n=1}^{\infty} \dfrac{1}{\sqrt{n^2 + 2n}}$

23. $\displaystyle\sum_{n=1}^{\infty} \dfrac{1}{\sqrt{n^3 + 2n}}$

24. $\displaystyle\sum_{n=1}^{\infty} \dfrac{n+1}{n(n+2)}$

25. $\displaystyle\sum_{n=1}^{\infty} \dfrac{1}{(n^3 + 2n)^{1/3}}$

26. $\displaystyle\sum_{n=1}^{\infty} \dfrac{n!}{e^n}$

27. $\displaystyle\sum_{n=1}^{\infty} \dfrac{(-1)^n n}{\ln n}$

28. $\displaystyle\sum_{n=1}^{\infty} \dfrac{(-1)^n \sqrt{n}}{n+1}$

29. $\displaystyle\sum_{n=1}^{\infty} \dfrac{(-1)^n 1 \cdot 3 \cdot 5 \cdots (2n-1)}{2 \cdot 4 \cdot 6 \cdots (2n)(2n+1)}$

30. $\displaystyle\sum_{n=1}^{\infty} \dfrac{1 \cdot 3 \cdot 5 \cdots (2n-1)}{2 \cdot 5 \cdot 8 \cdots (3n-1)}$

31. $\displaystyle\sum_{n=1}^{\infty} \left(\dfrac{1}{n^2} - \dfrac{1}{n}\right)$

32. $\displaystyle\sum_{n=1}^{\infty} \left(\dfrac{1}{n^2} - \dfrac{1}{2^n}\right)$

In Exercises 33–36, find the interval of convergence of the power series.

33. $\displaystyle\sum_{n=0}^{\infty} \dfrac{(-1)^n (x-2)^n}{(n+1)^2}$

34. $\displaystyle\sum_{n=0}^{\infty} (2x)^n$

35. $\displaystyle\sum_{n=0}^{\infty} n!(x-2)^n$

36. $\displaystyle\sum_{n=0}^{\infty} \dfrac{(x-2)^n}{2^n}$

In Exercises 37–42, find the power series for $f(x)$ centered at c.

37. $f(x) = \sin x, \ c = \dfrac{3\pi}{4}$

38. $f(x) = \sqrt{x}, \ c = 4$

39. $f(x) = 3^x, \ c = 0$

40. $f(x) = \csc x, \ c = \dfrac{\pi}{2}$ (first three terms)

41. $f(x) = \dfrac{1}{x}, \ c = -1$

42. $f(x) = \cos x, \ c = -\dfrac{\pi}{4}$

In Exercises 43–46, find the series representation of the function defined by the given integral.

43. $\displaystyle\int_0^x \dfrac{\sin t}{t}\, dt$

44. $\displaystyle\int_0^x \cos \dfrac{\sqrt{t}}{2}\, dt$

45. $\displaystyle\int_0^x \dfrac{\ln(t+1)}{t}\, dt$

46. $\displaystyle\int_0^x \dfrac{e^t - 1}{t}\, dt$

In Exercises 47–50, use a Taylor polynomial to approximate the given value with an error of less than 0.001.

47. $\sin 95°$

48. $\cos (0.75)$

49. $\ln (1.75)$

50. $e^{-0.25}$

51. Sketch the graphs of $f(x) = e^x$ and its fifth Taylor polynomial on the same axes.

52. A ball is dropped from a height of 8 feet. Each time it drops h feet, it rebounds $0.7h$ feet. Find the total distance traveled by the ball.

11 Conics

Conic sections were discovered sometime during the classical Greek period, which lasted from 600 to 300 B.C. By the beginning of the Alexandrian period, enough was known of conics for Apollonius (262–190 B.C.) to produce an eight-volume work on the subject. Later, toward the end of the Alexandrian period, Hypatia (370–415) wrote a textbook titled *On the Conics of Apollonius*. Her death marked the end of any major mathematical discoveries for several hundred years.

The early Greeks were largely concerned with the geometrical properties of conics. It was not until the early seventeenth century that the broad applicability of conics became apparent, and they then played a prominent role in the early development of calculus.

Each **conic section** (or simply **conic**) can be described as the intersection of a plane and a double-napped cone. Notice from Figure 11.1 that in the formation of the four basic conics, the intersecting plane does not pass through the vertex of the cone. When the plane does pass through the vertex, we call the resulting figure a **degenerate conic,** as shown in Figure 11.2.

There are several ways to approach a study of the conics. We could begin as the Greeks did by defining the conics in terms of the intersections of planes and cones. Or we could define them algebraically in terms of the general second-degree equation

$$Ax^2 + Bxy + Cy^2 + Dx + Ey + F = 0 \qquad \text{General second-degree equation}$$

a procedure to be discussed in Section 11.4. However, a third approach, in which each of the conics is defined as a **locus** (collection) of points satisfying a certain geometric property, suits our needs best. For example, in Section 1.2 we defined a circle as the collection of all points (x, y) that are equidistant

Hypatia

606

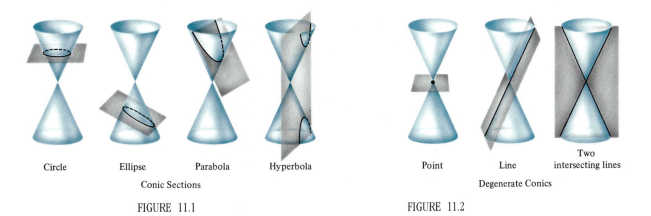

Circle   Ellipse   Parabola   Hyperbola   Point   Line   Two intersecting lines

Conic Sections   Degenerate Conics

FIGURE 11.1   FIGURE 11.2

from a fixed point (h, k). This locus definition easily produced the standard equation of a circle,

$$(x - h)^2 + (y - k)^2 = r^2 \qquad \text{Standard equation of a circle}$$

In this and the following two sections, we give similar definitions to the other three types of conics. We will also identify practical geometric properties used in the construction of objects such as bridges, searchlights, telescopes, and radar detectors. Later, in multivariable calculus, we will make extensive use of conics in illustrative examples.

DEFINITION OF A PARABOLA

A **parabola** is the set of all points (x, y) that are equidistant from a fixed line (**directrix**) and a fixed point (**focus**) not on the line.

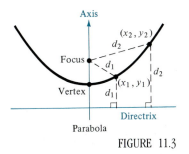

Parabola

FIGURE 11.3

Remark The midpoint between the focus and the directrix is called the **vertex,** and the line passing through the focus and the vertex is called the **axis** of the parabola. Note in Figure 11.3 that a parabola is symmetric with respect to its axis.

Using the definition of a parabola, we derive the following theorem, which gives the **standard form** of the equation of a parabola whose directrix is parallel to the x-axis or to the y-axis.

THEOREM 11.1

STANDARD EQUATION OF A PARABOLA
The **standard form** of the equation of a parabola with vertex (h, k) and directrix $y = k - p$ is

$$(x - h)^2 = 4p(y - k) \qquad \text{Vertical axis}$$

For directrix $x = h - p$, the equation is

$$(y - k)^2 = 4p(x - h) \qquad \text{Horizontal axis}$$

The focus lies on the axis p units (*directed distance*) from the vertex.

Proof: We prove only the case for which the directrix is parallel to the *x*-axis and the focus lies above the vertex, as shown in Figure 11.4(a). If (x, y) is any point on the parabola, then by definition it is equidistant from the focus $(h, k + p)$ and the directrix $y = k - p$, and we have

$$\sqrt{(x - h)^2 + [y - (k + p)]^2} = y - (k - p)$$
$$(x - h)^2 + [y - (k + p)]^2 = [y - (k - p)]^2$$
$$(x - h)^2 + y^2 - 2y(k + p) + (k + p)^2 = y^2 - 2y(k - p) + (k - p)^2$$
$$(x - h)^2 - 2py + 2pk = 2py - 2pk$$
$$(x - h)^2 = 4p(y - k)$$

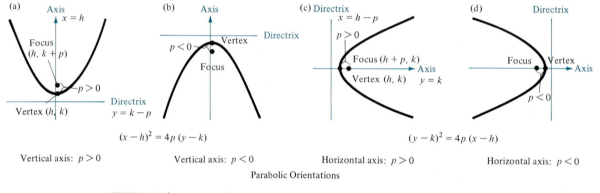

(a) Vertical axis: $p > 0$

(b) $(x - h)^2 = 4p(y - k)$ Vertical axis: $p < 0$

(c) $(y - k)^2 = 4p(x - h)$ Horizontal axis: $p > 0$

(d) Horizontal axis: $p < 0$

Parabolic Orientations

FIGURE 11.4

EXAMPLE 1   *Finding the standard equation for a parabola*

Find the standard form of the equation of the parabola with vertex $(2, 1)$ and focus $(2, 4)$.

Solution: Since the axis of the parabola is vertical, we consider the equation

$$(x - h)^2 = 4p(y - k)$$

where $h = 2$, $k = 1$, and $p = 3$. Thus the standard form is

$$(x - 2)^2 = 12(y - 1)$$

The graph of this parabola is shown in Figure 11.5.

| Remark   By expanding the standard equation in Example 1, we obtain the more common quadratic form

$$y = \frac{1}{12}(x^2 - 4x + 16)$$

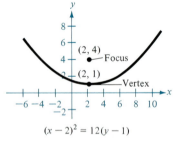

$(x - 2)^2 = 12(y - 1)$

FIGURE 11.5

EXAMPLE 2   *Finding the vertex and focus*

Find the focus of the parabola given by $y = -\frac{1}{2}x^2 - x + \frac{1}{2}$.

Solution: To find the focus, we convert to standard form by completing the square.

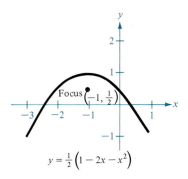

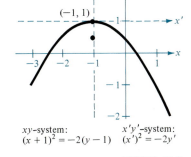

$y = \frac{1}{2}\left(1 - 2x - x^2\right)$

FIGURE 11.6

$$y = \frac{1}{2}(1 - 2x - x^2)$$

$$2y = 1 - (x^2 + 2x + \quad)$$

$$2y = 2 - (x^2 + 2x + 1)$$

$$(x + 1)^2 = -2(y - 1) \qquad \text{Standard form}$$

Comparing this equation to $(x - h)^2 = 4p(y - k)$, we conclude that

$$h = -1, \qquad k = 1, \qquad \text{and} \qquad p = -\frac{1}{2}$$

Since p is negative, the parabola opens downward, as shown in Figure 11.6. Therefore, the focus of the parabola is

$$(h, k + p) = \left(-1, \frac{1}{2}\right) \qquad \text{Focus}$$

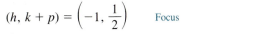

xy-system:
$(x + 1)^2 = -2(y - 1)$

x'y'-system:
$(x')^2 = -2y'$

FIGURE 11.7

If the vertex of a parabola is at the origin, then the standard form

$$(x - h)^2 = 4p(y - k) \qquad \text{Vertex at } (h, k)$$

simplifies to

$$x^2 = 4py \qquad \text{Vertex at origin}$$

This is consistent with the discussion of *horizontal* and *vertical shifts* in Section 1.5. That is, the factor $(x - h)$ indicates a horizontal shift, and $(y - k)$ represents a vertical shift from the origin. We can view this as a **translation of axes** from an xy-system to an $x'y'$-system related by the equations

$$x' = x - h \qquad \text{and} \qquad y' = y - k$$

as shown in Figure 11.7.

EXAMPLE 3 Vertex at the origin

Find the standard equation of the parabola with vertex at the origin and focus at $(2, 0)$.

Solution: The axis of the parabola is horizontal, passing through $(0, 0)$ and $(2, 0)$, as shown in Figure 11.8. Thus, we consider the standard form

$$y^2 = 4px$$

where $h = k = 0$ and $p = 2$. Therefore, the equation is

$$y^2 = 8x$$

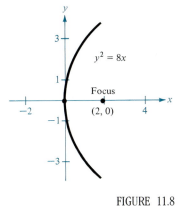

FIGURE 11.8

A line segment that passes through the focus of a parabola and has endpoints on the parabola is called a **focal chord.** The specific focal chord perpendicular to the axis of the parabola is called the **latus rectum.** In our next example we determine the length of the latus rectum and the length of the corresponding intercepted arc.

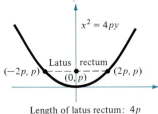

Length of latus rectum: $4p$
Arc length: $4.59p$

Latus Rectum and Intercepted Arc

FIGURE 11.9

EXAMPLE 4   *Focal chord length and arc length*

Find the length of the latus rectum of the parabola given by $x^2 = 4py$. Then find the length of the parabolic arc intercepted by the latus rectum.

Solution: Since the latus rectum passes through the focus $(0, p)$ and is perpendicular to the y-axis, the coordinates of its endpoints are $(-x, p)$ and (x, p). Substituting p for y in the equation of the parabola, we have

$$x^2 = 4p(p) = 4p^2 \implies x = \pm 2p$$

Thus, the endpoints of the latus rectum are $(-2p, p)$ and $(2p, p)$, and we conclude that its length is $4p$, as shown in Figure 11.9. In contrast, the length of the intercepted arc is given by

$$s = \int_{-2p}^{2p} \sqrt{1 + (y')^2}\, dx = 2 \int_0^{2p} \sqrt{1 + \left(\frac{x}{2p}\right)^2}\, dx$$

$$= \frac{1}{p} \int_0^{2p} \sqrt{4p^2 + x^2}\, dx$$

$$= \frac{1}{2p} \left[x\sqrt{4p^2 + x^2} + 4p^2 \ln \left| x + \sqrt{4p^2 + x^2} \right| \right]_0^{2p}$$

$$= \frac{1}{2p} [2p\sqrt{8p^2} + 4p^2 \ln (2p + \sqrt{8p^2}) - 4p^2 \ln (2p)]$$

$$= 2p[\sqrt{2} + \ln (1 + \sqrt{2})] \approx 4.59p$$

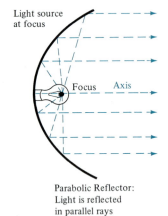

Parabolic Reflector:
Light is reflected
in parallel rays

FIGURE 11.10

Applications

One of the most widely used properties of a parabola is its reflective property. In physics, a surface is called **reflective** if the tangent line at any point on the surface makes equal angles with an incoming ray and the resulting outgoing ray. The angle corresponding to the incoming ray is called the **angle of incidence,** and the angle corresponding to the outgoing ray is called the **angle of reflection.** One example of a reflective surface is a flat mirror. Another type of reflective surface is that formed by revolving a parabola about its axis. A special property of parabolic reflectors is that they allow us to direct all incoming rays parallel to the axis through the focus of the parabola—this is the principle behind the construction of the parabolic mirrors used in reflecting telescopes. Conversely, the light rays emanating from the focus of a parabolic reflector used in a flashlight are all parallel to one another, as shown in Figure 11.10.

THEOREM 11.2   REFLECTIVE PROPERTY OF A PARABOLA
The tangent line to a parabola at the point P makes equal angles with the following two lines:

1. the line passing through P and the focus and
2. the line passing through P parallel to the axis of the parabola.

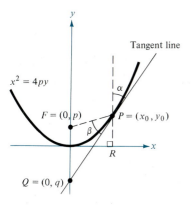

$x^2 = 4py$

Tangent line

$F = (0, p)$

$P = (x_0, y_0)$

R

$Q = (0, q)$

FIGURE 11.11

Proof: We consider the point $P = (x_0, y_0)$ on the parabola given by $x^2 = 4py$, as shown in Figure 11.11. Using differentiation, we can determine the tangent line at P to have a slope of $x_0/2p$, which implies that the equation of the tangent line is

$$y - y_0 = \frac{x_0}{2p}(x - x_0) \qquad \text{Tangent line at } P$$

If P lies at the vertex of the parabola, then the other two lines referred to in the theorem coincide and the theorem is trivially valid. Thus, we assume P is a point other than the vertex. This implies that the tangent line intersects the y-axis at a point $Q = (0, q)$, different from P. Using Figure 11.11, we can show that the angles α and β are equal by showing that the triangle with vertices at F, P, and Q is isosceles. Using the equation of the tangent line at $x = 0$, we see that

$$q = y_0 - \frac{x_0^2}{2p} = \frac{x_0^2}{4p} - \frac{x_0^2}{2p} = -\frac{x_0^2}{4p}$$

which implies that the length of $\overline{FQ}$ is

$$p - q = p + \frac{x_0^2}{4p} \qquad \text{Length of } \overline{FQ}$$

Moreover, using the distance formula, we find the length of $\overline{FP}$ to be

$$\sqrt{x_0^2 + (y_0 - p)^2} = \sqrt{x_0^2 + \left(\frac{x_0^2}{4p} - p\right)^2} = \sqrt{\left(\frac{x_0^2}{4p} + p\right)^2}$$

$$= \frac{x_0^2}{4p} + p \qquad \text{Length of } \overline{FP}$$

Therefore, ΔFQP is isosceles, and we conclude that $\alpha = \beta$.

Section Exercises 11.1

In Exercises 1–6, match the equation with the correct graph.

1. $y^2 = 4x$
2. $x^2 = -2y$
3. $x^2 = 8y$
4. $y^2 = -12x$
5. $(y - 1)^2 = 4(x - 2)$
6. $(x + 3)^2 = -2(y - 2)$

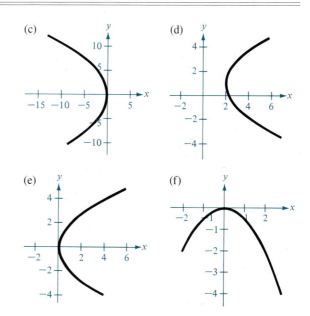

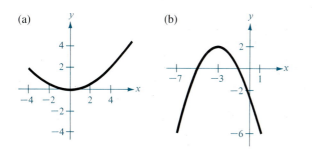

In Exercises 7–26, find the vertex, focus, and directrix of the parabola, and sketch its graph.

7. $y = 4x^2$ **8.** $y = 2x^2$
9. $y^2 = -6x$ **10.** $y^2 = 3x$
11. $x^2 + 8y = 0$ **12.** $x + y^2 = 0$
13. $(x - 1)^2 + 8(y + 2) = 0$
14. $(x + 3) + (y - 2)^2 = 0$
15. $\left(y + \dfrac{1}{2}\right)^2 = 2(x - 5)$
16. $\left(x + \dfrac{1}{2}\right)^2 - 4(y - 3) = 0$
17. $y = \dfrac{1}{4}(x^2 - 2x + 5)$
18. $y = -\dfrac{1}{6}(x^2 + 4x - 2)$
19. $4x - y^2 - 2y - 33 = 0$
20. $y^2 + x + y = 0$
21. $y^2 + 6y + 8x + 25 = 0$
22. $x^2 - 2x + 8y + 9 = 0$
23. $y^2 - 4y - 4x = 0$
24. $y^2 - 4x - 4 = 0$
25. $x^2 + 4x + 4y - 4 = 0$
26. $y^2 + 4y + 8x - 12 = 0$

In Exercises 27–40, find an equation of the specified parabola.

27. Vertex $(0, 0)$, focus $\left(0, -\dfrac{3}{2}\right)$
28. Vertex $(0, 0)$, focus $(2, 0)$
29. Vertex $(3, 2)$, focus $(1, 2)$
30. Vertex $(-1, 2)$, focus $(-1, 0)$
31. Vertex $(0, -4)$, directrix $y = 2$
32. Vertex $(-2, 1)$, directrix $x = 1$
33. Focus $(0, 0)$, directrix $y = 4$
34. Focus $(2, 2)$, directrix $x = -2$
35. Axis parallel to y-axis, graph passes through $(0, 3)$, $(3, 4)$, and $(4, 11)$
36. Axis parallel to x-axis, graph passes through $(4, -2)$, $(0, 0)$, and $(3, -3)$
37.   **38.**

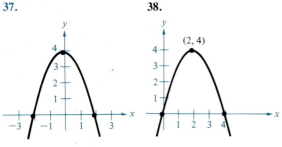

39. Directrix $y = 1$, length of latus rectum is 8, opens downward

40. Directrix $y = -2$, endpoints of latus rectum $(0, 2)$ and $(8, 2)$

41. The filament of a flashlight bulb is $\frac{3}{8}$ inch from the vertex of the parabolic reflector and is located at the focus. Find an equation of a cross section of the reflector. (Assume that it is directed toward the right and the vertex is at the origin.)

42. The receiver in a parabolic television dish antenna is 3 feet from the vertex and is located at the focus. Find an equation of a cross section of the reflector. (Assume that the dish is directed upward and the vertex is at the origin.)

In Exercises 43–46, find the arc length of the parabola over the given interval.

Function	Interval
43. $x^2 + 8y = 0$	$0 \le x \le 4$
44. $x + y^2 = 0$	$0 \le y \le 2$
45. $4x - y^2 = 0$	$0 \le y \le 4$
46. $x^2 - 2y = 0$	$0 \le x \le 1$

47. A cable of a parabolic suspension bridge is suspended between two towers that are 400 feet apart and 50 feet above the roadway, as shown in Figure 11.12. The cable touches the roadway midway between the towers. Find the length of the cable.

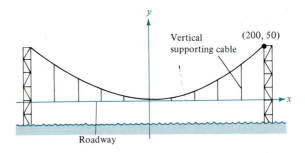

FIGURE 11.12

48. An earth satellite in a circular orbit 100 miles high has a speed of approximately 17,500 miles per hour. If this speed is multiplied by $\sqrt{2}$, then the satellite will have the minimum velocity necessary to escape the earth's gravitational force and will follow a parabolic path with the center of the earth as the focus, as shown in Figure 11.13. Find an equation for the resulting parabolic path. (Assume that the radius of the earth is 4000 miles.)

49. Water is flowing from a horizontal pipe 48 feet above the ground with a horizontal velocity of 10 feet per

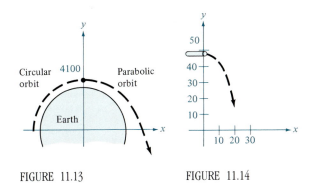

FIGURE 11.13   FIGURE 11.14

second. The falling stream of water has the shape of a parabola whose vertex, (0, 48), is at the end of the pipe, as shown in Figure 11.14. Where does the water hit the ground?

50. Prove that the area enclosed by a parabola and a line parallel to its directrix is two-thirds the area of the circumscribed rectangle.

51. Find an equation of the tangent line to the parabola $y = ax^2$ at $x = x_0$. Prove that the x-intercept of this tangent line is $(x_0/2, 0)$.

52. Prove that any two distinct tangent lines to a parabola intersect.

53. Prove that if any two tangent lines to a parabola intersect at right angles, then their point of intersection must lie on the directrix.

54. Prove that two tangent lines to a parabola intersect at right angles if and only if the focus of the parabola lies on the line segment connecting the two points of tangency.

55. Demonstrate the results of Exercises 53 and 54 for the parabola

$$y = \frac{1}{4}(x^2 - 4x + 8)$$

11.2
Ellipses

SECTION TOPICS ▪
Ellipses ▪
Standard equation of an ellipse ▪
Reflective property ▪
Eccentricity ▪

More than a thousand years after the close of the Alexandrian period of Greek mathematics, Western civilization was finally ushered into the Renaissance of mathematical and scientific discovery. One of the principal figures in this rebirth was the Polish astronomer Nicholas Copernicus (1473–1543). In his work *On the Revolutions of the Heavenly Spheres,* Copernicus claimed that all of the planets, including the earth, revolved about the sun in circular orbits. Although many of Copernicus's claims were invalid, the controversy set off by his heliocentric theory motivated astronomers to search for a mathematical model to explain the observed movements of the sun and planets. The first to find the correct model was the German astronomer Johannes Kepler (1571–1630). Kepler discovered that the planets move about the sun in elliptical orbits, with the sun not as the center but as a focal point of the orbit.

The use of ellipses to explain the movement of the planets is only one of many practical and aesthetic uses. As in the previous section, we begin our study of this second type of conic by defining it as a locus of points. However, in this case, we use *two* focal points rather than one.

Nicholas Copernicus

DEFINITION OF AN ELLIPSE	An **ellipse** is the set of all points (x, y) the sum of whose distances from two distinct fixed points (**foci**) is constant. (See Figure 11.15.)

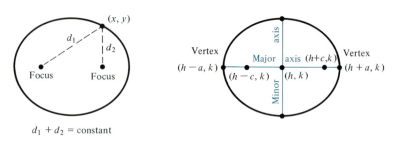

$d_1 + d_2 = \text{constant}$

FIGURE 11.15

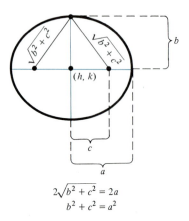

FIGURE 11.16

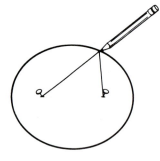

$2\sqrt{b^2 + c^2} = 2a$
$b^2 + c^2 = a^2$

FIGURE 11.17

The line through the foci intersects the ellipse at two points called the **vertices.** The chord joining the vertices is called the **major axis,** and its midpoint is called the **center** of the ellipse. The chord perpendicular to the major axis at the center is called the **minor axis** of the ellipse.

Remark You can visualize the definition of an ellipse by imagining two thumbtacks placed at the foci, as shown in Figure 11.16. If the ends of a fixed length of string are fastened to the thumbtacks and the string is drawn taut with a pencil, the path traced by the pencil will be an ellipse.

To derive the standard form of the equation of an ellipse, consider the ellipse in Figure 11.17 with the following points:

center: (h, k) vertices: $(h \pm a, k)$ foci: $(h \pm c, k)$

The sum of the distances from any point on the ellipse to the two foci is constant. At a vertex, this constant sum is

$$(a + c) + (a - c) = 2a \qquad \text{Length of major axis}$$

or simply the length of the major axis. Now, for *any* point (x, y) on the ellipse, the sum of the distances between (x, y) and the two foci must also be $2a$. That is,

$$\sqrt{[x - (h - c)]^2 + (y - k)^2} + \sqrt{[x - (h + c)]^2 + (y - k)^2} = 2a$$

which, after expanding and regrouping, reduces to

$$(a^2 - c^2)(x - h)^2 + a^2(y - k)^2 = a^2(a^2 - c^2)$$

Finally, in Figure 11.17 we can see that $b^2 = a^2 - c^2$, which implies that the equation of the ellipse is

$$b^2(x - h)^2 + a^2(y - k)^2 = a^2b^2$$

$$\frac{(x - h)^2}{a^2} + \frac{(y - k)^2}{b^2} = 1$$

Had we chosen a vertical major axis, we would have obtained a similar equation. Both results are summarized in the following theorem.

THEOREM 11.3   STANDARD EQUATION OF AN ELLIPSE

The standard form of the equation of an ellipse, with center (h, k) and major and minor axes of lengths $2a$ and $2b$, where $a > b$, is

$$\frac{(x - h)^2}{a^2} + \frac{(y - k)^2}{b^2} = 1 \qquad \text{Major axis is horizontal}$$

$$\frac{(x - h)^2}{b^2} + \frac{(y - k)^2}{a^2} = 1 \qquad \text{Major axis is vertical}$$

The foci lie on the major axis, c units from the center, with $c^2 = a^2 - b^2$.

Figure 11.18 shows both the vertical and horizontal orientations for an ellipse.

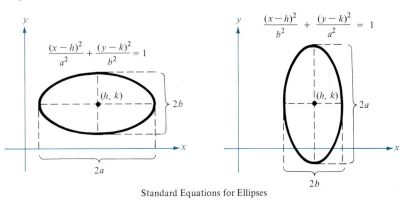

Standard Equations for Ellipses

FIGURE 11.18

EXAMPLE 1   Finding the standard equation for an ellipse

Find the standard form of the equation of the ellipse having foci at $(0, 1)$ and $(4, 1)$, with a major axis of length 6.

Solution: Since the foci occur at $(0, 1)$ and $(4, 1)$, the center of the ellipse is $(2, 1)$. This implies that the distance from the center to one of the foci is $c = 2$, and since $2a = 6$ we know that $a = 3$. Now, using $c^2 = a^2 - b^2$, we have

$$b = \sqrt{a^2 - c^2} = \sqrt{9 - 4} = \sqrt{5}$$

Since the major axis is parallel to the x-axis, the standard equation is

$$\frac{(x - 2)^2}{9} + \frac{(y - 1)^2}{5} = 1$$

Remark By translating the axes in Figure 11.19 according to the equations $x' = x - 2$ and $y' = y - 1$, we obtain the equation

$$\frac{(x')^2}{9} + \frac{(y')^2}{5} = 1$$

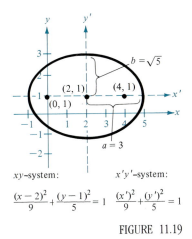

xy-system: $x'y'$-system:

$\dfrac{(x-2)^2}{9} + \dfrac{(y-1)^2}{5} = 1$ $\dfrac{(x')^2}{9} + \dfrac{(y')^2}{5} = 1$

FIGURE 11.19

EXAMPLE 2   *Completing the square to find the standard equation*

Find the center, vertices, and foci of the ellipse given by

$$4x^2 + y^2 - 8x + 4y - 8 = 0$$

Solution: By completing the square, we can write the given equation in standard form.

$$4x^2 + y^2 - 8x + 4y - 8 = 0$$

$$4(x^2 - 2x + 1) + (y^2 + 4y + 4) = 8 + 4 + 4$$

$$4(x - 1)^2 + (y + 2)^2 = 16$$

$$\frac{(x - 1)^2}{4} + \frac{(y + 2)^2}{16} = 1$$

Thus, the major axis is parallel to the y-axis, where $h = 1$, $k = -2$, $a = 4$, $b = 2$, and $c = \sqrt{16 - 4} = 2\sqrt{3}$. Therefore, we have

center: $(1, -2)$ vertices: $(1, -6)$ foci: $(1, -2 - 2\sqrt{3})$

$(1, 2)$ $(1, -2 + 2\sqrt{3})$

The graph of the ellipse is shown in Figure 11.20.

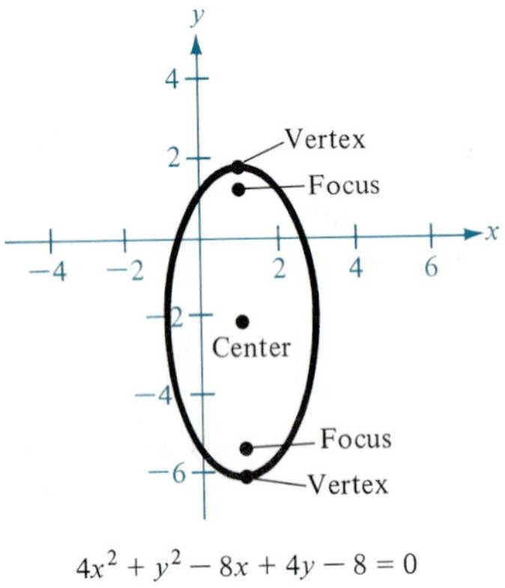

FIGURE 11.20

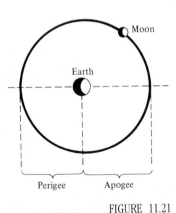

Vertex
Focus

Center

Focus
Vertex

$4x^2 + y^2 - 8x + 4y - 8 = 0$

| **Remark** If the constant term -8 in the equation in Example 2 had been $F \geq 8$, we would have obtained one of the following degenerate cases:

1. Single point: $\dfrac{(x - 1)^2}{4} + \dfrac{(y + 2)^2}{16} = 0$ $F = 8$

2. No solution points: $\dfrac{(x - 1)^2}{4} + \dfrac{(y + 2)^2}{16} < 0$ $F > 8$

EXAMPLE 3   *An application involving an elliptical orbit*

The moon travels about the earth in an elliptical orbit with the earth at one focus, as shown in Figure 11.21. The major and minor axes of the orbit have lengths of 768,806 kilometers and 767,746 kilometers, respectively. Find the greatest and least distances (the apogee and perigee) from the earth's center to the moon's center.

Solution: Since $2a = 768,806$ and $2b = 767,746$, we have $a = 384,403$, $b = 383,873$, and

$$c = \sqrt{a^2 - b^2} \approx 20,179$$

Therefore, the greatest distance between the centers of the earth and the moon is

$$a + c \approx 404,582 \text{ km}$$

and the least distance is

$$a - c \approx 364,224 \text{ km}$$

Moon

Earth

Perigee   Apogee

FIGURE 11.21

EXAMPLE 4   *Finding the area of an ellipse*

Find the area of an ellipse whose major and minor axes have lengths of $2a$ and $2b$, respectively.

Solution: For simplicity, we choose an ellipse centered at the origin

$$\frac{x^2}{a^2} + \frac{y^2}{b^2} = 1$$

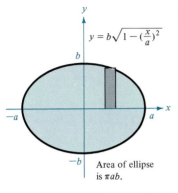

$y = b\sqrt{1 - (\frac{x}{a})^2}$

Area of ellipse
is πab.

FIGURE 11.22

Then, using symmetry, we can find the area of the entire region lying within the ellipse by finding the area of the region in the first quadrant and multiplying by 4 as indicated in Figure 11.22. In the first quadrant, we have

$$y = \frac{b}{a}\sqrt{a^2 - x^2}$$

which implies that the entire area is

$$A = 4 \int_0^a \frac{b}{a}\sqrt{a^2 - x^2}\, dx$$

Using the trigonometric substitution $x = a \sin \theta$, we have

$$A = \frac{4b}{a} \int_0^{\pi/2} a^2 \cos^2 \theta\, d\theta$$

$$= 4ab \int_0^{\pi/2} \frac{1 + \cos 2\theta}{2}\, d\theta$$

$$= 2ab \left[\theta + \frac{\sin 2\theta}{2} \right]_0^{\pi/2}$$

$$= 2ab \left(\frac{\pi}{2}\right) = \pi ab$$

| **Remark**   Note that if $a = b$, then the formula for the area of an ellipse reduces to the area of a circle.

In the previous section we looked at a reflective property of parabolas. Ellipses have a similar reflective property.

THEOREM 11.4   **REFLECTIVE PROPERTY OF AN ELLIPSE**
The tangent line to an ellipse at the point P makes equal angles with the lines through P and the foci.

Proof: Let $P = (x_0, y_0)$ be a point on the ellipse

$$\frac{x^2}{a^2} + \frac{y^2}{b^2} = 1$$

We prove the case in which $0 < x_0 < c$ and $0 < y_0$. Let α and β be the angles made by the tangent line and the lines through P and the foci, as shown in

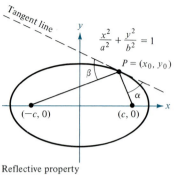

Reflective property
of an ellipse: $\alpha = \beta$

FIGURE 11.23

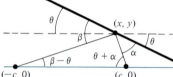

FIGURE 11.24

Figure 11.23. Also let θ be the acute angle between the tangent line and the horizontal axis. Then, since $dy/dx = -b^2x/a^2y$, we have

$$\tan \theta = \left| \frac{-b^2x_0}{a^2y_0} \right| = \frac{b^2x_0}{a^2y_0}$$

and

$$\tan (\theta + \alpha) = \left| \frac{y_0}{x_0 - c} \right| = \frac{y_0}{c - x_0}$$

Now, we apply the trigonometric identity

$$\tan (\theta + \alpha) = \frac{\tan \theta + \tan \alpha}{1 - \tan \theta \tan \alpha}$$

to obtain

$$\frac{y_0}{c - x_0} = \frac{(b^2x_0/a^2y_0) + \tan \alpha}{1 - (b^2x_0/a^2y_0)(\tan \alpha)}$$

Using $b^2x_0^2 + a^2y_0^2 = a^2b^2$ and $a^2 - b^2 = c^2$, we solve this equation for $\tan \alpha$ and conclude that

$$\tan \alpha = \frac{b^2}{cy_0}$$

In a similar manner, using angle $\beta - \theta$ in Figure 11.24, we have

$$\tan (\beta - \theta) = \frac{y_0}{x_0 + c} = \frac{\tan \beta - \tan \theta}{1 + \tan \beta \tan \theta}$$

$$= \frac{\tan \beta - (b^2x_0/a^2y_0)}{1 + (\tan \beta)(b^2x_0/a^2y_0)}$$

Solving this equation for $\tan \beta$, we have

$$\tan \beta = \frac{b^2}{cy_0}$$

and we conclude that $\alpha = \beta$.

Eccentricity

One of the reasons that it was difficult for astronomers to detect that the orbits of the planets are ellipses is that the foci of the planetary orbits are relatively close to their centers, making the orbits nearly circular. To measure the ovalness of an ellipse, we use the concept of **eccentricity.**

DEFINITION OF
ECCENTRICITY

The **eccentricity** e of an ellipse is given by the ratio

$$e = \frac{c}{a}$$

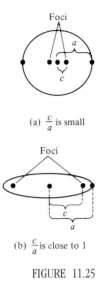

Foci

(a) $\frac{c}{a}$ is small

Foci

(b) $\frac{c}{a}$ is close to 1

FIGURE 11.25

To see how this ratio is used to describe the shape of an ellipse, note that since the foci of an ellipse are located along the major axis between the vertices and the center, it follows that

$$0 < c < a$$

For an ellipse that is nearly circular, the foci are close to the center and the ratio c/a is small, as shown in Figure 11.25. On the other hand, for an elongated ellipse, the foci are close to the vertices and the ratio c/a is close to 1.

| **Remark** Note that $0 < e < 1$ for every ellipse.

The orbit of the moon has an eccentricity of $e = 0.0549$, and the eccentricities of the nine planetary orbits are as follows:

Mercury:	$e = 0.2056$	Saturn:	$e = 0.0543$
Venus:	$e = 0.0068$	Uranus:	$e = 0.0460$
Earth:	$e = 0.0167$	Neptune:	$e = 0.0082$
Mars:	$e = 0.0934$	Pluto:	$e = 0.2481$
Jupiter:	$e = 0.0484$		

In Example 4, we found that the formula for the *area* of an ellipse is $A = \pi ab$. However, it is not so simple to find the *circumference* of an ellipse. The next example shows how to use eccentricity to set up an **elliptic integral** for the circumference of an ellipse.

EXAMPLE 5   *Finding the circumference of an ellipse*

Show that the circumference of the ellipse given by $(x^2/a^2) + (y^2/b^2) = 1$ is

$$4a \int_0^{\pi/2} \sqrt{1 - e^2 \sin^2 \theta}\, d\theta \qquad e = c/a$$

Solution: Since the given ellipse is symmetric with respect to both the x-axis and y-axis, we know that its circumference C is four times the arc length of $y = (b/a)\sqrt{a^2 - x^2}$ in the first quadrant. Thus

$$C = 4 \int_0^a \sqrt{1 + (y')^2}\, dx = 4 \int_0^a \sqrt{1 + \frac{b^2 x^2}{a^2(a^2 - x^2)}}\, dx$$

Using the trigonometric substitution $x = a \sin \theta$, we have

$$C = 4 \int_0^{\pi/2} \sqrt{1 + \frac{b^2 \sin^2 \theta}{a^2 \cos^2 \theta}}\,(a \cos \theta)\, d\theta$$

$$= 4 \int_0^{\pi/2} \sqrt{a^2 \cos^2 \theta + b^2 \sin^2 \theta}\, d\theta$$

$$= 4 \int_0^{\pi/2} \sqrt{a^2(1 - \sin^2 \theta) + b^2 \sin^2 \theta}\, d\theta$$

$$= 4 \int_0^{\pi/2} \sqrt{a^2 - (a^2 - b^2) \sin^2 \theta}\, d\theta$$

Since $e^2 = c^2/a^2 = (a^2 - b^2)/a^2$, we can rewrite this integral as

$$C = 4a \int_0^{\pi/2} \sqrt{1 - e^2 \sin^2 \theta}\, d\theta$$

A great deal of time has been devoted to the study of elliptic integrals. Such integrals do not generally have elementary antiderivatives. To find the circumference of an ellipse, we usually have to resort to an approximation technique, as illustrated in the following example.

EXAMPLE 6   Approximating the value of an elliptic integral

Use the elliptic integral in Example 5 to approximate the circumference of the ellipse

$$\frac{x^2}{25} + \frac{y^2}{16} = 1$$

Solution: Since $e^2 = c^2/a^2 = (a^2 - b^2)/a^2 = 9/25$, we have

$$C = (4)(5) \int_0^{\pi/2} \sqrt{1 - \frac{9 \sin^2 \theta}{25}}\, d\theta$$

Applying Simpson's Rule with $n = 4$, we have

$$C \approx 20\left(\frac{\pi}{6}\right)\left(\frac{1}{4}\right)[1 + 4(0.9733) + 2(0.9055) + 4(0.8323) + 0.8] \approx 28.36$$

Section Exercises 11.2

In Exercises 1–6, match the equation with the correct graph.

1. $\dfrac{x^2}{1} + \dfrac{y^2}{9} = 1$ **2.** $\dfrac{x^2}{9} + \dfrac{y^2}{1} = 1$

3. $\dfrac{x^2}{9} + \dfrac{y^2}{4} = 1$ **4.** $\dfrac{x^2}{9} + \dfrac{y^2}{9} = 1$

5. $\dfrac{(x - 2)^2}{16} + \dfrac{(y + 1)^2}{4} = 1$

6. $\dfrac{(x + 2)^2}{4} + \dfrac{(y + 2)^2}{25} = 1$

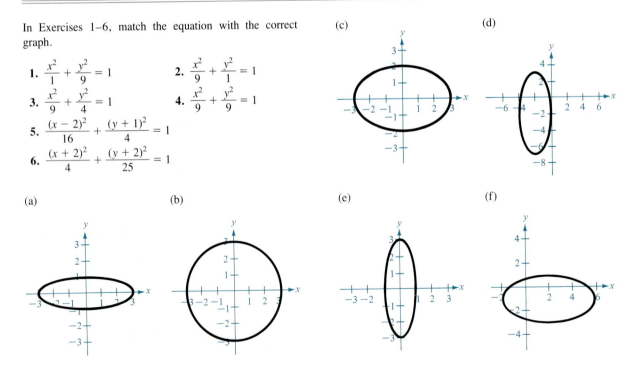

(a)   (b)   (c)   (d)   (e)   (f)

In Exercises 7–26, find the center, foci, vertices, and eccentricity of the ellipse, and sketch its graph.

7. $\dfrac{x^2}{25} + \dfrac{y^2}{16} = 1$

8. $\dfrac{x^2}{144} + \dfrac{y^2}{169} = 1$

9. $\dfrac{x^2}{16} + \dfrac{y^2}{25} = 1$

10. $\dfrac{x^2}{169} + \dfrac{y^2}{144} = 1$

11. $\dfrac{x^2}{9} + \dfrac{y^2}{5} = 1$

12. $\dfrac{x^2}{28} + \dfrac{y^2}{64} = 1$

13. $x^2 + 4y^2 = 4$

14. $5x^2 + 3y^2 = 15$

15. $3x^2 + 2y^2 = 6$

16. $5x^2 + 7y^2 = 70$

17. $4x^2 + y^2 = 1$

18. $16x^2 + 25y^2 = 1$

19. $\dfrac{(x-1)^2}{9} + \dfrac{(y-5)^2}{25} = 1$

20. $(x+2)^2 + 4(y+4)^2 = 1$

21. $9x^2 + 4y^2 + 36x - 24y + 36 = 0$

22. $9x^2 + 4y^2 - 36x + 8y + 31 = 0$

23. $16x^2 + 25y^2 - 32x + 50y + 31 = 0$

24. $9x^2 + 25y^2 - 36x - 50y + 61 = 0$

25. $12x^2 + 20y^2 - 12x + 40y - 37 = 0$

26. $36x^2 + 9y^2 + 48x - 36y + 43 = 0$

In Exercises 27–36, find an equation for the specified ellipse.

27. Center $(0, 0)$
focus $(2, 0)$
vertex $(3, 0)$

28. Center $(0, 0)$
vertex $(2, 0)$
minor axis length 3

29. Vertices $(\pm 5, 0)$, eccentricity $\dfrac{3}{5}$

30. Vertices $(0, \pm 8)$, eccentricity $\dfrac{1}{2}$

31. Vertices $(0, 2)$ and $(4, 2)$, minor axis length 2

32. Foci $(\pm 2, 0)$, major axis length 8

33. Vertices $(3, 1)$ and $(3, 9)$, minor axis length 6

34. Center $(0, 0)$, major axis horizontal, points on ellipse $(3, 1)$ and $(4, 0)$

35. Foci $(0, \pm 5)$, major axis length 14

36. Center $(1, 2)$, major axis vertical, points on ellipse $(1, 6)$ and $(3, 2)$

In Exercises 37 and 38, use the following definition. The **latera recta** (plural of **latus rectum**) of an ellipse are the chords passing through the foci and perpendicular to the major axis.

37. Show that the length of a latus rectum of an ellipse is $2b^2/a$.

38. With the result of Exercise 37, sketch the graph of each of the following, making use of the endpoints of the latera recta.

(a) $\dfrac{x^2}{4} + \dfrac{y^2}{1} = 1$

(b) $5x^2 + 3y^2 = 15$

In Exercises 39 and 40, determine the points at which the derivative is zero or undefined to locate the endpoints of the major and minor axes of the ellipse.

39. $16x^2 + 9y^2 + 96x + 36y + 36 = 0$

40. $9x^2 + 4y^2 + 36x - 24y + 36 = 0$

In Exercises 41 and 42, consider a particle traveling clockwise on the given elliptical path. The particle leaves the orbit at the indicated point and travels in a straight line tangent to the ellipse. At what point will the particle cross the y-axis?

41. $\dfrac{x^2}{100} + \dfrac{y^2}{25} = 1,\ (-8, 3)$

42. $\dfrac{x^2}{16} + \dfrac{y^2}{25} = 1,\ \left(3, \dfrac{5\sqrt{7}}{4}\right)$

In Exercises 43–46, use Simpson's Rule with $n = 8$ to approximate the elliptic integral representing the circumference of the given ellipse.

43. $\dfrac{x^2}{9} + \dfrac{y^2}{16} = 1$

44. $\dfrac{x^2}{9} + \dfrac{y^2}{1} = 1$

45. $\dfrac{x^2}{3} + \dfrac{y^2}{2} = 1$

46. $\dfrac{(x-3)^2}{4} + \dfrac{(y+1)^2}{3} = 1$

In Exercises 47 and 48, find (a) the area of the region bounded by the given ellipse, (b) the volume and surface area of the solid generated by revolving the region about its major axis (prolate spheroid), and (c) the volume and surface area of the solid generated by revolving the region about its minor axis (oblate spheroid).

47. $\dfrac{x^2}{4} + \dfrac{y^2}{1} = 1$

48. $\dfrac{x^2}{16} + \dfrac{y^2}{9} = 1$

49. Find the dimensions of the rectangle of maximum area that can be inscribed in the ellipse

$$\dfrac{x^2}{a^2} + \dfrac{y^2}{b^2} = 1$$

(Assume that the sides of the rectangle are parallel to the coordinate axes.)

50. A solid has an elliptical base given by

$$\dfrac{x^2}{25} + \dfrac{y^2}{16} = 1$$

Cross sections taken perpendicular to the major axis are isosceles triangles of height 6. Find the volume of the solid.

51. Show that the tangent line to the ellipse

$$\frac{x^2}{a^2} + \frac{y^2}{b^2} = 1$$

at the point (x_0, y_0) is given by

$$\frac{x_0}{a^2}x + \frac{y_0}{b^2}y = 1$$

52. The equation of an ellipse with its center at the origin can be written as

$$\frac{x^2}{a^2} + \frac{y^2}{a^2(1 - e^2)} = 1$$

Show that as $e \to 0$, with a remaining fixed, the ellipse approaches a circle.

53. Show that the eccentricity of the ellipse

$$\frac{x^2}{a^2} + \frac{y^2}{b^2} = 1$$

is identical to the eccentricity of

$$\frac{(tx)^2}{a^2} + \frac{(ty)^2}{b^2} = 1$$

for any real t. Give a geometrical explanation of this result.

54. A line segment 9 inches long moves so that one endpoint is always on the y-axis and the other always on the x-axis. Find the equation of the curve traced by a point on the line segment 6 inches from the endpoint that is on the y-axis.

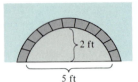

FIGURE 11.26

55. A fireplace arch is to be constructed in the shape of a semiellipse, as shown in Figure 11.26. The opening is to have a height of 2 feet at the center and a width of 5 feet along the base. To sketch the outline of the fireplace, the contractor uses a 5-foot string tied to two thumbtacks. Where should the thumbtacks be placed?

56. The length of half of the major axis of the earth's orbit is 14,957,000 kilometers, and the eccentricity is 0.0167. Find the least and greatest distances of the earth from the sun.

57. If the distances to the apogee and the perigee of an elliptical orbit of an earth satellite are measured from the center of the earth, show that the eccentricity of the orbit is given by

$$e = \frac{A - P}{A + P}$$

where A and P are the apogee and perigee distances, respectively.

58. The first artificial satellite to orbit the earth was Sputnik I (launched by Russia in 1957). Its highest point above the earth's surface was 583 miles, and its lowest point was 132 miles. Find the eccentricity of its orbit.

11.3
Hyperbolas

The definition of a hyperbola parallels that of an ellipse. For an ellipse the *sum* of the distances between the foci and a point on the ellipse is fixed, whereas for a hyperbola the *difference* of these distances is fixed.

DEFINITION OF A HYPERBOLA   A **hyperbola** is the set of all points (x, y) the difference of whose distances from two distinct fixed points (foci) is constant. (See Figure 11.27.)

The line through the two foci intersects a hyperbola at two points called the **vertices.** The line segment connecting the vertices is called the **transverse**

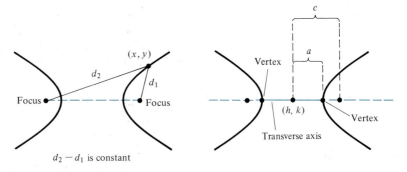

FIGURE 11.27

axis, and the midpoint of the transverse axis is called the **center** of the hyperbola. One distinguishing feature of hyperbolas is that their graphs have two separate *branches*.

The development of the standard form of the equation of a hyperbola is similar to that for an ellipse, so we list the following theorem without proof.

THEOREM 11.5   **STANDARD EQUATION OF A HYPERBOLA**

The standard form of the equation of a hyperbola with center at (h, k) is

$$\frac{(x - h)^2}{a^2} - \frac{(y - k)^2}{b^2} = 1 \qquad \text{Transverse axis is } \textit{horizontal}$$

$$\frac{(y - k)^2}{a^2} - \frac{(x - h)^2}{b^2} = 1 \qquad \text{Transverse axis is } \textit{vertical}$$

The vertices are a units from the center, and the foci are c units from the center. Moreover, $b^2 = c^2 - a^2$.

Figure 11.28 shows both the vertical and horizontal orientations for a hyperbola.

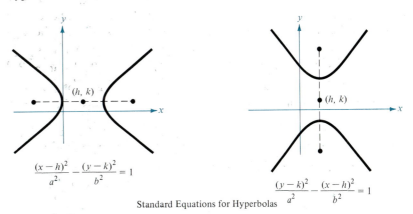

$$\frac{(x - h)^2}{a^2} - \frac{(y - k)^2}{b^2} = 1$$

$$\frac{(y - k)^2}{a^2} - \frac{(x - h)^2}{b^2} = 1$$

Standard Equations for Hyperbolas

FIGURE 11.28

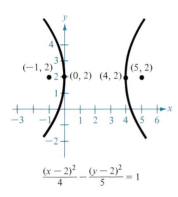

$$\frac{(x-2)^2}{4} - \frac{(y-2)^2}{5} = 1$$

FIGURE 11.29

EXAMPLE 1   Finding the standard equation for a hyperbola

Find the standard form of the equation of the hyperbola with foci at $(-1, 2)$ and $(5, 2)$ and vertices at $(0, 2)$ and $(4, 2)$.

Solution: By the Midpoint Formula, the center of the hyperbola occurs at the point $(2, 2)$. Furthermore, $c = 3$ and $a = 2$, and it follows that

$$b^2 = 3^2 - 2^2 = 9 - 4 = 5$$

Thus, the equation of the hyperbola is

$$\frac{(x-2)^2}{4} - \frac{(y-2)^2}{5} = 1$$

Figure 11.29 shows the graph of the hyperbola.

An important aid in sketching the graph of a hyperbola is the determination of its **asymptotes,** as shown in Figure 11.30. Each hyperbola has two asymptotes that intersect at the center of the hyperbola. The asymptotes pass through the vertices of a rectangle of dimension $2a$ by $2b$, with its center at (h, k). The line segment of length $2b$ joining $(h, k + b)$ and $(h, k - b)$ is referred to as the **conjugate axis** of the hyperbola. The following theorem identifies the equation for the asymptotes.

THEOREM 11.6   **ASYMPTOTES OF A HYPERBOLA**
For a *horizontal* transverse axis, the equations of the asymptotes are

$$y = k + \frac{b}{a}(x - h) \qquad \text{and} \qquad y = k - \frac{b}{a}(x - h)$$

For a *vertical* transverse axis, the equations of the asymptotes are

$$y = k + \frac{a}{b}(x - h) \qquad \text{and} \qquad y = k - \frac{a}{b}(x - h)$$

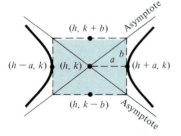

FIGURE 11.30

Proof: Solving for y in the equation

$$\frac{(x-h)^2}{a^2} - \frac{(y-k)^2}{b^2} = 1$$

gives us

$$y = k \pm \frac{b}{a}\sqrt{(x-h)^2 - a^2}$$

If $y = k + (b/a)(x - h)$ is to be an asymptote to the hyperbola, then the difference of the y-values of a point on the hyperbola and a point on the asymptote must approach zero as $x \to \infty$. To prove this, observe that

$$\left[k + \frac{b}{a} \sqrt{(x-h)^2 - a^2} \right] - \left[k + \frac{b}{a}(x-h) \right]$$

$$= \frac{b}{a}[\sqrt{(x-h)^2 - a^2} - (x-h)]$$

$$= \frac{b}{a}\left(\frac{[(x-h)^2 - a^2] - (x-h)^2}{\sqrt{(x-h)^2 - a^2} + (x-h)} \right)$$

$$= \frac{-ab}{\sqrt{(x-h)^2 - a^2} + (x-h)}$$

Therefore,

$$\lim_{x \to \infty} \frac{-ab}{\sqrt{(x-h)^2 - a^2} + (x-h)} = 0$$

The asymptotic behavior of the other three portions of the hyperbola can be established in a similar manner.

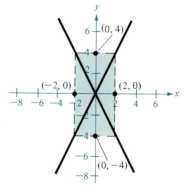

FIGURE 11.31

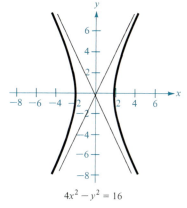

$4x^2 - y^2 = 16$

FIGURE 11.32

Remark In Figure 11.30 we can see that the asymptotes coincide with the diagonals of the rectangle with dimensions $2a$ and $2b$, centered at (h, k). This provides us with a quick means of sketching the asymptotes, which in turn aids in sketching the hyperbola.

EXAMPLE 2   Using asymptotes to sketch a hyperbola

Sketch the graph of the hyperbola whose equation is

$$4x^2 - y^2 = 16$$

Solution: Rewriting this equation in standard form, we have

$$\frac{4x^2}{16} - \frac{y^2}{16} = \frac{16}{16}$$

$$\frac{x^2}{2^2} - \frac{y^2}{4^2} = 1$$

From this, we conclude that the transverse axis is horizontal and the vertices occur at $(-2, 0)$ and $(2, 0)$. Moreover, the ends of the conjugate axis occur at $(0, -4)$ and $(0, 4)$, and we are able to sketch the rectangle shown in Figure 11.31. Finally, by drawing the asymptotes through the corners of this rectangle, we complete the sketch shown in Figure 11.32.

EXAMPLE 3   Finding the asymptotes of a hyperbola

Sketch the hyperbola given by

$$4x^2 - 3y^2 + 8x + 16 = 0$$

and find the equations of its asymptotes.

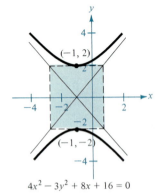

$$4x^2 - 3y^2 + 8x + 16 = 0$$

FIGURE 11.33

Solution: Rewriting this equation in standard form, we have

$$4x^2 - 3y^2 + 8x + 16 = 0$$
$$4(x^2 + 2x) - 3y^2 = -16$$
$$-4(x^2 + 2x + 1) + 3y^2 = 16 - 4$$
$$-4(x + 1)^2 + 3y^2 = 12$$
$$\frac{y^2}{4} - \frac{(x + 1)^2}{3} = 1$$

From this equation we conclude that the hyperbola is centered at $(-1, 0)$ and has vertices at $(-1, 2)$ and $(-1, -2)$, and the ends of the conjugate axis occur at $(-1 - \sqrt{3}, 0)$ and $(-1 + \sqrt{3}, 0)$. To sketch the graph of the hyperbola, we draw a rectangle through these four points. The asymptotes are the lines passing through the corners of the rectangle as shown in Figure 11.33. Finally, using $a = 2$ and $b = \sqrt{3}$, we apply Theorem 11.6 to conclude that the equations of the asymptotes are

$$y = \frac{2}{\sqrt{3}}(x + 1) \qquad \text{and} \qquad y = -\frac{2}{\sqrt{3}}(x + 1) \qquad \square$$

Remark If the constant term in the equation in Example 3 had been $F = -4$ instead of 16, then we would have obtained the following degenerate case:

$$\text{Two Intersecting Lines: } \frac{y^2}{4} - \frac{(x + 1)^2}{3} = 0$$

EXAMPLE 4 *Using asymptotes to find the standard equation*

Find the standard form of the equation of the hyperbola having vertices at $(3, -5)$ and $(3, 1)$ and asymptotes $y = 2x - 8$ and $y = -2x + 4$, as shown in Figure 11.34.

Solution: By the Midpoint Formula the center of the hyperbola is at $(3, -2)$. Furthermore, the hyperbola has a vertical transverse axis with $a = 3$. By Theorem 11.6 the asymptotes have equations whose slopes are

$$m_1 = \frac{a}{b} \qquad \text{and} \qquad m_2 = -\frac{a}{b}$$

From the given equations of the asymptotes, we know that

$$\frac{a}{b} = 2 \qquad \text{and} \qquad -\frac{a}{b} = -2$$

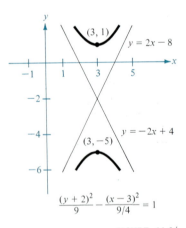

$$\frac{(y + 2)^2}{9} - \frac{(x - 3)^2}{9/4} = 1$$

FIGURE 11.34

and since $a = 3$, we conclude that $b = \frac{3}{2}$. Therefore, the standard equation is

$$\frac{(y + 2)^2}{9} - \frac{(x - 3)^2}{9/4} = 1 \qquad \square$$

As with ellipses, the **eccentricity** of an hyperbola is $e = c/a$. Because $c > a$, it follows that $e > 1$. If the eccentricity is large, then the branches of the hyperbola are nearly flat. If the eccentricity is close to 1, then the branches of the hyperbola are more pointed, as shown in Figure 11.35.

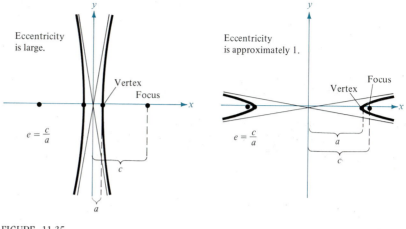

FIGURE 11.35

| DEFINITION OF ECCENTRICITY | The eccentricity e of a hyperbola is given by the ratio $$e = \frac{c}{a}$$ |

Applications

The following application was developed during World War II. It shows how the properties of hyperbolas can be used in radar or other detection systems.

EXAMPLE 5   An application involving hyperbolas

Two microphones, 1 mile apart, record an explosion. Microphone A received the sound 2 seconds before microphone B. Where did the explosion come from?

Solution: Assuming that sound travels at 1100 feet per second, we know that the explosion took place 2200 feet farther from B than from A as shown in Figure 11.36. The locus of all points that are 2200 feet closer to A than to B is one branch of the hyperbola $(x^2/a^2) - (y^2/b^2) = 1$, where

$$c = \frac{5280}{2} = 2640 \quad \text{and} \quad a = \frac{2200}{2} = 1100$$

Thus, $b^2 = c^2 - a^2 = 5{,}759{,}600$, and we conclude that the explosion occurred somewhere on the right branch of the hyperbola given by

$$\frac{x^2}{1{,}210{,}000} - \frac{y^2}{5{,}759{,}600} = 1$$

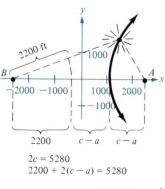

$$2c = 5280$$
$$2200 + 2(c - a) = 5280$$

FIGURE 11.36

In Example 5 we were only able to determine the hyperbola on which the explosion occurred but not the exact location of the explosion. If, how-

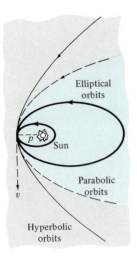

Elliptical
orbits

Sun

Parabolic
orbits

Hyperbolic
orbits

FIGURE 11.37

ever, we had received the sound at a third position C, then two other hyperbolas would be determined. The exact location of the explosion would be the point where these three hyperbolas intersect.

Another interesting application of conic sections involves the orbits of comets in our solar system. Of the 610 comets identified prior to 1970, 245 have elliptical orbits, 295 have parabolic orbits, and 70 have hyperbolic orbits. The center of the sun is a focus point of each of these orbits, and each orbit has a vertex at the point where the comet is closest to the sun, as shown in Figure 11.37. There undoubtedly have been many comets with parabolic or hyperbolic orbits that were not identified. We only get to see such comets *once*. Comets with elliptical orbits such as Halley's Comet are the only ones that remain in our solar system.

If p is the distance between the vertex and the focus in meters, and v is the velocity of the comet at the vertex in meters per second, then the type of orbit is determined as follows.

1. Ellipse: $v < \sqrt{2GM/p}$
2. Parabola: $v = \sqrt{2GM/p}$
3. Hyperbola: $v > \sqrt{2GM/p}$

where $M \approx 1.991 \times 10^{30}$ kilograms (the mass of the sun) and $G \approx 6.67 \times 10^{-11}$ cubic meters per gram-second squared.

In the next theorem we identify ways to classify a conic using the coefficients in the general form of its equation.

THEOREM 11.7   **CLASSIFYING A CONIC FROM ITS GENERAL EQUATION**
The graph of $Ax^2 + Cy^2 + Dx + Ey + F = 0$ is one of the following (except in degenerate cases).

1. Circle: $A = C$
2. Parabola: $AC = 0$ $A = 0$ or $C = 0$, but not both.
3. Ellipse: $AC > 0$ A and C have like signs.
4. Hyperbola: $AC < 0$ A and C have unlike signs.

Proof: To change the general equation to standard form, we complete the square as follows.

$$Ax^2 + Cy^2 + Dx + Ey + F = 0$$

$$A\left(x^2 + \frac{D}{A}x\right) + C\left(y^2 + \frac{E}{C}y\right) = -F$$

$$A\left(x^2 + \frac{D}{A}x + \frac{D^2}{4A^2}\right) + C\left(y^2 + \frac{E}{C}y + \frac{E^2}{4C^2}\right) = \frac{D^2}{4A} + \frac{E^2}{4C} - F = R$$

$$\frac{[x + (D/2A)]^2}{C} + \frac{[y + (E/2C)]^2}{A} = \frac{R}{AC} \qquad \text{Standard form}$$

The nondegenerate cases are

For $A = C$: $\left(x + \dfrac{D}{2A}\right)^2 + \left(y + \dfrac{E}{2C}\right)^2 = \dfrac{R}{A}$ Circle

For $A \neq 0$, $C = 0$: $A\left(x + \dfrac{D}{2A}\right)^2 = -F - Ey + \dfrac{D^2}{4A}$ Parabola

For $A = 0$, $C \neq 0$: $C\left(y + \dfrac{E}{2C}\right)^2 = -F - Dx + \dfrac{E^2}{4C}$ Parabola

For $AC > 0$: $\dfrac{[x + (D/2A)]^2}{R/A} + \dfrac{[y + (E/2C)]^2}{R/C} = 1$ Ellipse

For $AC < 0$: $\dfrac{[x + (D/2A)]^2}{R/A} - \dfrac{[y + (E/2C)]^2}{R/C} = \pm 1$ Hyperbola

The degenerate cases are

For $R < 0$:   No graph

For $R = 0$, $AC > 0$: $\left(-\dfrac{D}{2A}, -\dfrac{E}{2C}\right)$ Single point

For $R = 0$, $AC < 0$: $y = \pm\sqrt{-\dfrac{A}{C}}\left(x + \dfrac{D}{2A}\right) - \dfrac{E}{2C}$ Two lines

---

EXAMPLE 6   *Classifying conics from general equations*

(a) For the general equation

$$4x^2 - 9x + y - 5 = 0$$

we have $AC = 4(0) = 0$. Thus, the graph is a parabola.

(b) For the general equation

$$4x^2 - y^2 + 8x - 6y + 4 = 0$$

we have $AC = 4(-1) < 0$ and

$$R = \frac{D^2}{4A} + \frac{E^2}{4C} - F = \frac{64}{16} + \frac{36}{-4} - 4 = -9 \neq 0$$

Thus, the graph is a hyperbola.

(c) For the general equation

$$2x^2 + 4y^2 - 4x + 12y = 0$$

we have $AC = (2)(4) > 0$ and

$$R = \frac{16}{8} + \frac{144}{16} - 0 > 0$$

Thus, the graph is an ellipse.

Section Exercises 11.3

In Exercises 1–6, match the equation with the correct graph.

1. $\dfrac{x^2}{9} - \dfrac{y^2}{4} = 1$

2. $\dfrac{y^2}{9} - \dfrac{x^2}{4} = 1$

3. $\dfrac{y^2}{1} - \dfrac{x^2}{16} = 1$

4. $\dfrac{y^2}{16} - \dfrac{x^2}{1} = 1$

5. $\dfrac{(x-2)^2}{9} - \dfrac{y^2}{4} = 1$

6. $\dfrac{(x+1)^2}{16} - \dfrac{(y-3)^2}{9} = 1$

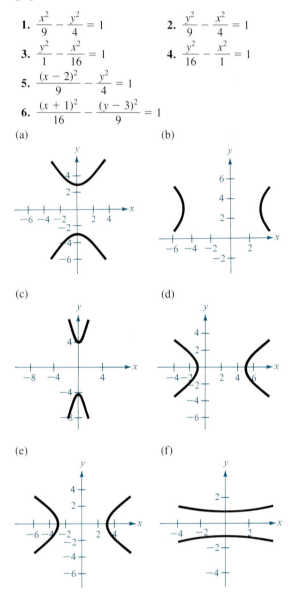

(a)   (b)

(c)   (d)

(e)   (f)

In Exercises 7–26, find the center, vertices, and foci of the hyperbola and sketch its graph, using asymptotes as an aid.

7. $x^2 - y^2 = 1$

8. $\dfrac{x^2}{9} - \dfrac{y^2}{16} = 1$

9. $y^2 - \dfrac{x^2}{4} = 1$

10. $\dfrac{y^2}{9} - x^2 = 1$

11. $\dfrac{y^2}{25} - \dfrac{x^2}{144} = 1$

12. $\dfrac{x^2}{36} - \dfrac{y^2}{4} = 1$

13. $2x^2 - 3y^2 = 6$

14. $3y^2 = 5x^2 + 15$

15. $5y^2 = 4x^2 + 20$

16. $7x^2 - 3y^2 = 21$

17. $\dfrac{(x-1)^2}{4} - (y+2)^2 = 1$

18. $\dfrac{(x+1)^2}{144} - \dfrac{(y-4)^2}{25} = 1$

19. $(y+6)^2 - (x-2)^2 = 1$

20. $4(y-1)^2 - 9(x+3)^2 = 1$

21. $9x^2 - y^2 - 36x - 6y + 18 = 0$

22. $x^2 - 9y^2 + 36y - 72 = 0$

23. $9y^2 - x^2 + 2x + 54y + 62 = 0$

24. $16y^2 - x^2 + 2x + 64y + 63 = 0$

25. $x^2 - 9y^2 + 2x - 54y - 80 = 0$

26. $9x^2 - y^2 + 54x + 10y + 55 = 0$

In Exercises 27–36, find an equation for the specified hyperbola.

27. Center $(0, 0)$, one vertex $(0, 2)$, one focus $(0, 4)$

28. Center $(0, 0)$, one vertex $(3, 0)$, one focus $(5, 0)$

29. Vertices $(\pm 1, 0)$, asymptotes $y = \pm 3x$

30. Vertices $(0, \pm 3)$, asymptotes $y = \pm 3x$

31. Vertices $(0, 2)$ and $(6, 2)$, asymptotes $y = \frac{2}{3}x$ and $y = 4 - \frac{2}{3}x$

32. Vertices $(2, \pm 3)$, foci $(2, \pm 5)$

33. Vertices $(2, \pm 3)$, point on graph $(0, 5)$

34. Focus $(10, 0)$, asymptotes $y = \pm\frac{3}{4}x$

35. For any point on the hyperbola, the difference of its distances from the points $(2, 2)$ and $(10, 2)$ is 6.

36. For any point on the hyperbola, the difference of its distances from the points $(-3, 0)$ and $(-3, 3)$ is 2.

In Exercises 37 and 38, find equations for (a) the tangent lines and (b) the normal lines to the hyperbola for the given value of x.

37. $\dfrac{x^2}{9} - y^2 = 1$, $x = 6$

38. $\dfrac{y^2}{4} - \dfrac{x^2}{2} = 1$, $x = 4$

In Exercises 39 and 40, the region bounded by the graphs of the given equations is revolved about the x-axis. Find the volume and surface area of the resulting solid.

39. $x^2 - y^2 = 1$, $y = 0$, $x = 2$

40. $\dfrac{x^2}{16} - \dfrac{y^2}{9} = 1$, $y = 0$, $x = 5$

41. A rifle, positioned at point $(-c, 0)$, is fired at a target positioned at point $(c, 0)$. A person hears the sound of the rifle and the sound of the bullet hitting the target at the same time. Show that the person is positioned on one branch of the hyperbola given by

$$\frac{x^2}{c^2 v_s^2/v_m^2} - \frac{y^2}{c^2(v_m^2 - v_s^2)/v_m^2} = 1$$

where v_m is the muzzle velocity of the rifle and the speed of sound is $v_s = 1100$ feet per second.

42. Show that the equation of the tangent line to

$$\frac{x^2}{a^2} - \frac{y^2}{b^2} = 1$$

at the point (x_0, y_0) is

$$\frac{x_0}{a^2}x - \frac{y_0}{b^2}y = 1$$

43. Show that the ellipse

$$\frac{x^2}{a^2} + \frac{2y^2}{b^2} = 1$$

and the hyperbola

$$\frac{x^2}{a^2 - b^2} - \frac{2y^2}{b^2} = 1$$

intersect at right angles.

In Exercises 44–56, classify the graph of each equation as a circle, a parabola, an ellipse, or a hyperbola.

44. $x^2 + y^2 - 6x + 4y + 9 = 0$
45. $x^2 + 4y^2 - 6x + 16y + 21 = 0$
46. $4x^2 - y^2 - 4x - 3 = 0$
47. $y^2 - 4y - 4x = 0$
48. $4x^2 + 3y^2 + 8x - 24y + 51 = 0$
49. $4y^2 - 2x^2 - 4y - 8x - 15 = 0$
50. $25x^2 - 10x - 200y - 119 = 0$
51. $4x^2 + 4y^2 - 16y + 15 = 0$
52. $y^2 - 4y = x + 5$
53. $9x^2 + 9y^2 - 36x + 6y + 34 = 0$
54. $2x(x - y) = y(3 - y - 2x)$
55. $3(x - 1)^2 = 6 + 2(y + 1)^2$
56. $9(x + 3)^2 = 36 - 4(y - 2)^2$

11.4
Rotation and the general second-degree equation

In previous sections we have shown that the equation of a conic with axes parallel to one of the coordinate axes has a standard form that can be written in the general form

$$Ax^2 + Cy^2 + Dx + Ey + F = 0 \qquad \text{Horizontal or vertical axes}$$

In this section we investigate the equations of conics whose axes are rotated so that they are not parallel to either the x-axis or the y-axis. The general equation for such conics contains an *xy term*:

$$Ax^2 + Bxy + Cy^2 + Dx + Ey + F = 0 \qquad \text{Equation in } xy\text{-plane}$$

To eliminate this xy term, we use a procedure called **rotation of axes.** Our objective is to rotate the x- and y-axes until they are parallel to the axes of the conic. We denote the rotated axes as the x'-axis and the y'-axis, as shown in Figure 11.38. Once this has been accomplished, the equation of the conic in the new $x'y'$-plane will have the form

$$A'(x')^2 + C'(y')^2 + D'x' + E'y' + F' = 0 \qquad \text{Equation in } x'y'\text{-plane}$$

Since this equation has no $x'y'$ term, we can obtain a standard form by completing the square.

The following theorem identifies how much to rotate the axes to eliminate an xy term, and also the equations for determining the new coefficients A', C', D', E', and F'.

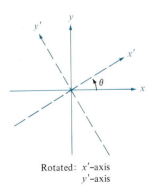

Rotated: x'-axis
y'-axis

FIGURE 11.38

THEOREM 11.8   ROTATION OF AXES TO ELIMINATE AN *XY* TERM

$Ax^2 + Bxy + Cy^2 + Dx + Ey + F = 0$ can be rewritten as

$$A'(x')^2 + C'(y')^2 + D'x' + E'y' + F' = 0$$

by rotating the coordinate axes through an angle θ, where

$$\cot 2\theta = \frac{A - C}{B}$$

The coefficients of the new equation are obtained by making the substitutions

$$x = x' \cos \theta - y' \sin \theta$$
$$y = x' \sin \theta + y' \cos \theta$$

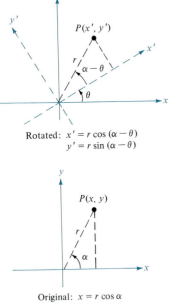

Rotated: $x' = r \cos (\alpha - \theta)$
$y' = r \sin (\alpha - \theta)$

Original: $x = r \cos \alpha$
$y = r \sin \alpha$

FIGURE 11.39

Proof: We need to discover how the coordinates in the *xy*-system are related to the coordinates in the *x'y'*-system. To do this, we choose a point $P = (x, y)$ in the original system and attempt to find its coordinates (x', y') in the rotated system. In either system the distance r between the point P and the origin is the same, thus the equations for x, y, x', and y' are those given in Figure 11.39. Using the formulas for the sine and cosine of the difference of two angles, we have

$$x' = r \cos (\alpha - \theta) = r(\cos \alpha \cos \theta + \sin \alpha \sin \theta)$$
$$= r \cos \alpha \cos \theta + r \sin \alpha \sin \theta = x \cos \theta + y \sin \theta$$
$$y' = r \sin (\alpha - \theta) = r(\sin \alpha \cos \theta - \cos \alpha \sin \theta)$$
$$= r \sin \alpha \cos \theta - r \cos \alpha \sin \theta = y \cos \theta - x \sin \theta$$

Solving this system for x and y yields

$$x = x' \cos \theta - y' \sin \theta$$
$$y = x' \sin \theta + y' \cos \theta$$

Finally, by substituting these values for x and y into the original equation and collecting terms, we have

$$A' = A \cos^2 \theta + B \cos \theta \sin \theta + C \sin^2 \theta$$
$$C' = A \sin^2 \theta - B \cos \theta \sin \theta + C \cos^2 \theta$$
$$D' = D \cos \theta + E \sin \theta$$
$$E' = -D \sin \theta + E \cos \theta$$
$$F' = F$$

Now, in order to eliminate the *x'y'* term, we must select θ so that $B' = 0$ as follows:

$$B' = 2(C - A) \sin \theta \cos \theta + B(\cos^2 \theta - \sin^2 \theta)$$
$$= (C - A) \sin 2\theta + B \cos 2\theta$$
$$= B(\sin 2\theta)\left(\frac{C - A}{B} + \cot 2\theta\right) = 0, \quad \sin 2\theta \neq 0$$

If $B = 0$, no rotation is necessary since the xy term is not present in the original equation. If $B \neq 0$, then the only way to make $B' = 0$ is to let

$$\cot 2\theta = \frac{A - C}{B}, \quad B \neq 0$$

Thus, we have established the desired results.

EXAMPLE 1   Rotation of a hyperbola

Write the equation $xy - 1 = 0$ in standard form.

Solution: Since $A = 0$, $B = 1$, and $C = 0$, we have

$$\cot 2\theta = \frac{A - C}{B} = 0 \implies 2\theta = \frac{\pi}{2} \implies \theta = \frac{\pi}{4}$$

Therefore, the equation in the $x'y'$-system is given by making the following substitution:

$$x = x' \cos \frac{\pi}{4} - y' \sin \frac{\pi}{4} = x'\left(\frac{\sqrt{2}}{2}\right) - y'\left(\frac{\sqrt{2}}{2}\right) = \frac{x' - y'}{\sqrt{2}}$$

$$y = x' \sin \frac{\pi}{4} + y' \cos \frac{\pi}{4} = x'\left(\frac{\sqrt{2}}{2}\right) + y'\left(\frac{\sqrt{2}}{2}\right) = \frac{x' + y'}{\sqrt{2}}$$

Substituting these expressions into the equation $xy - 1 = 0$, we have

$$\left(\frac{x' - y'}{\sqrt{2}}\right)\left(\frac{x' + y'}{\sqrt{2}}\right) - 1 = 0$$

$$\frac{(x')^2 - (y')^2}{2} - 1 = 0$$

$$\frac{(x')^2}{(\sqrt{2})^2} - \frac{(y')^2}{(\sqrt{2})^2} = 1 \qquad \text{Standard form}$$

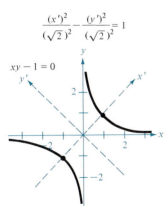

$$\frac{(x')^2}{(\sqrt{2})^2} - \frac{(y')^2}{(\sqrt{2})^2} = 1$$

$xy - 1 = 0$

Vertices:
$(\sqrt{2}, 0), (-\sqrt{2}, 0)$ in $x'y'$-system
$(1, 1), (-1, -1)$ in xy-system

FIGURE 11.40

This is the equation of a hyperbola centered at the origin with vertices at $(\pm\sqrt{2}, 0)$ in the $x'y'$-system as shown in Figure 11.40. To find the coordinates of the vertices in the xy-system, we substituted the $x'y'$-coordinates into the equations

$$x = \frac{x' - y'}{\sqrt{2}} \qquad \text{and} \qquad y = \frac{x' + y'}{\sqrt{2}}$$

From this substitution, we find the vertices to be $(1, 1)$ and $(-1, -1)$ in the xy-system. Note also that the asymptotes of the hyperbola have equations $y' = \pm x'$, which correspond to the original x- and y-axes.

EXAMPLE 2   Rotation of an ellipse

Sketch the graph of the equation

$$7x^2 - 6\sqrt{3}xy + 13y^2 - 16 = 0$$

Solution: Since $A = 7$, $B = -6\sqrt{3}$, and $C = 13$, we have

$$\cot 2\theta = \frac{A - C}{B} = \frac{7 - 13}{-6\sqrt{3}} = \frac{1}{\sqrt{3}} \implies \theta = \frac{\pi}{6}$$

Therefore, the equation in the $x'y'$-system is given by making the following substitution

$$x = x' \cos \frac{\pi}{6} - y' \sin \frac{\pi}{6} = x'\left(\frac{\sqrt{3}}{2}\right) - y'\left(\frac{1}{2}\right) = \frac{\sqrt{3}x' - y'}{2}$$

$$y = x' \sin \frac{\pi}{6} + y' \cos \frac{\pi}{6} = x'\left(\frac{1}{2}\right) + y'\left(\frac{\sqrt{3}}{2}\right) = \frac{x' + \sqrt{3}y'}{2}$$

Substituting these expressions into the original equation, we have

$$7x^2 - 6\sqrt{3}xy + 13y^2 - 16 = 0$$

$$7\left(\frac{\sqrt{3}x' - y'}{2}\right)^2 - 6\sqrt{3}\left(\frac{\sqrt{3}x' - y'}{2}\right)\left(\frac{x' + \sqrt{3}y'}{2}\right)$$

$$+ 13\left(\frac{x' + \sqrt{3}y'}{2}\right)^2 - 16 = 0$$

which simplifies to

$$4(x')^2 + 16(y')^2 - 16 = 0$$

$$\frac{(x')^2}{4} + \frac{(y')^2}{1} = 1 \qquad \text{Standard form}$$

This is the equation of an ellipse centered at the origin with vertices at $(\pm 2, 0)$ in the $x'y'$-system as shown in Figure 11.41.

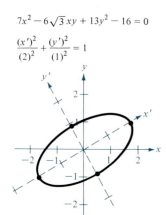

$$7x^2 - 6\sqrt{3}xy + 13y^2 - 16 = 0$$

$$\frac{(x')^2}{(2)^2} + \frac{(y')^2}{(1)^2} = 1$$

Vertices:
$(\pm 2, 0)$, $(0, \pm 1)$ in $x'y'$-system

$(\pm\sqrt{3}, \pm 1)$, $\left(\pm\frac{1}{2}, \mp\frac{\sqrt{3}}{2}\right)$ in xy-system

FIGURE 11.41

> **Remark**   Remember that the substitution
>
> $$x = x' \cos \theta - y' \sin \theta \qquad \text{and} \qquad y = x' \sin \theta + y' \cos \theta$$
>
> was developed to eliminate the $x'y'$ term in the rotated system. You can use this as a check on your work. In other words, if your final equation contains an $x'y'$ term, you know that you made a mistake.

In constructing Examples 1 and 2 we carefully chose the equations so that θ would turn out to be one of the common angles 30°, 45°, and so forth. Of course, many second-degree equations do not yield such common solutions to the equation $\cot 2\theta = (A - C)/B$. Example 3 illustrates such a case.

EXAMPLE 3   *Rotation of a parabola*

Sketch the graph of $x^2 - 4xy + 4y^2 + 5\sqrt{5}y + 1 = 0$.

Solution: Since $A = 1$, $B = -4$, and $C = 4$, we have

$$\cot 2\theta = \frac{A - C}{B} = \frac{1 - 4}{-4} = \frac{3}{4}$$

Using the trigonometric identity $\cot 2\theta = (\cot^2 \theta - 1)/(2 \cot \theta)$, we have

$$\cot 2\theta = \frac{3}{4} = \frac{\cot^2 \theta - 1}{2 \cot \theta}$$

from which we obtain the equation

$$6 \cot \theta = 4 \cot^2 \theta - 4 \quad \Longrightarrow \quad 4 \cot^2 \theta - 6 \cot \theta - 4 = 0$$

$$(2 \cot \theta - 4)(2 \cot \theta + 1) = 0$$

Considering $0 < \theta < \pi/2$, we have $2 \cot \theta = 4$. Thus,

$$\cot \theta = 2 \quad \Longrightarrow \quad \theta \approx 26.6°$$

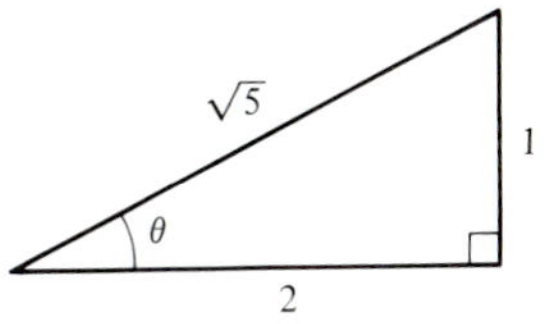

From the triangle in Figure 11.42 we obtain $\sin \theta = 1/\sqrt{5}$ and $\cos \theta = 2/\sqrt{5}$. Consequently, we use the substitution

$$x = x' \cos \theta - y' \sin \theta$$
$$= x'\left(\frac{2}{\sqrt{5}}\right) - y'\left(\frac{1}{\sqrt{5}}\right)$$
$$= \frac{2x' - y'}{\sqrt{5}}$$
$$y = x' \sin \theta + y' \cos \theta$$
$$= x'\left(\frac{1}{\sqrt{5}}\right) + y'\left(\frac{2}{\sqrt{5}}\right)$$
$$= \frac{x' + 2y'}{\sqrt{5}}$$

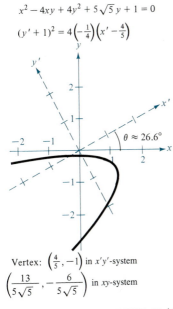

FIGURE 11.42

Substituting these expressions into the original equation, we have

$$x^2 - 4xy + 4y^2 + 5\sqrt{5}y + 1 = 0$$

$$\left(\frac{2x' - y'}{\sqrt{5}}\right)^2 - 4\left(\frac{2x' - y'}{\sqrt{5}}\right)\left(\frac{x' + 2y'}{\sqrt{5}}\right) + 4\left(\frac{x' + 2y'}{\sqrt{5}}\right)^2$$

$$+ 5\sqrt{5}\left(\frac{x' + 2y'}{\sqrt{5}}\right) + 1 = 0$$

which simplifies to

$$5(y')^2 + 5x' + 10y' + 1 = 0$$

By completing the square, we can obtain the standard form

$$5(y' + 1)^2 = -5x' + 4$$

$$(y' + 1)^2 = (-1)\left(x' - \frac{4}{5}\right) \qquad \text{Standard form}$$

The graph of the equation is a parabola with its vertex at $(\frac{4}{5}, -1)$ and its axis parallel to the x'-axis in the $x'y'$-system, as shown in Figure 11.43.

$x^2 - 4xy + 4y^2 + 5\sqrt{5}\,y + 1 = 0$

$(y' + 1)^2 = 4\left(-\frac{1}{4}\right)\left(x' - \frac{4}{5}\right)$

Vertex: $\left(\frac{4}{5}, -1\right)$ in $x'y'$-system

$\left(\frac{13}{5\sqrt{5}}, -\frac{6}{5\sqrt{5}}\right)$ in xy-system

FIGURE 11.43

Invariants under rotation

In Theorem 11.8, note that the constant term $F' = F$ is the same in both equations. We say that it is **invariant under rotation.** Theorem 11.9 lists some other rotation invariants. We leave the proof of this theorem as an exercise.

THEOREM 11.9   ROTATION INVARIANTS
The rotation of coordinate axes through an angle θ that transforms the equation

$$Ax^2 + Bxy + Cy^2 + Dx + Ey + F = 0$$

into the form

$$A'(x')^2 + C'(y')^2 + D'x' + E'y' + F' = 0$$

has the following rotation invariants

1. $F = F'$ 2. $A + C = A' + C'$ 3. $B^2 - 4AC = (B')^2 - 4A'C'$

We can use the results of this theorem to classify the graph of a second-degree equation *with* an xy term in much the same way we did for second-degree equations *without* an xy term. Note that since $B' = 0$, the invariant $B^2 - 4AC$ reduces to

$$B^2 - 4AC = -4A'C' \qquad \text{Discriminant}$$

We call this quantity the **discriminant** of the equation

$$Ax^2 + Bxy + Cy^2 + Dx + Ey + F = 0$$

Now, using Theorem 11.7, we know that the sign of $A'C'$ determines the type of graph for the equation

$$A'(x')^2 + C'(y')^2 + D'x' + E'y' + F' = 0$$

Consequently, the sign of $B^2 - 4AC$ will determine the type of graph for the original equation as given in Theorem 11.10.

THEOREM 11.10   CLASSIFICATION OF CONICS BY THE DISCRIMINANT
The graph of the equation $Ax^2 + Bxy + Cy^2 + Dx + Ey + F = 0$ is, except in degenerate cases, determined by its discriminant as follows.

1. Ellipse or circle: $B^2 - 4AC < 0$
2. Parabola: $B^2 - 4AC = 0$
3. Hyperbola: $B^2 - 4AC > 0$

EXAMPLE 4   *Using the discriminant*

Classify the graph of each of the following equations:

(a) $4xy - 9 = 0$

(b) $2x^2 - 3xy + 2y^2 - 2x = 0$

(c) $x^2 - 6xy + 9y^2 - 2y + 1 = 0$

(d) $3x^2 + 8xy + 4y^2 - 7 = 0$

Solution:

(a) Since

$$B^2 - 4AC = 16 - 0 > 0$$

the graph is a hyperbola.

(b) Since

$$B^2 - 4AC = 9 - 16 < 0$$

the graph is a circle or an ellipse.

(c) Since

$$B^2 - 4AC = 36 - 36 = 0$$

the graph is a parabola.

(d) Since

$$B^2 - 4AC = 64 - 48 > 0$$

the graph is a hyperbola.

Section Exercises 11.4

In Exercises 1–16, rotate the axes to eliminate the xy term. Sketch the graph of the resulting equation, showing both sets of axes.

1. $xy + 1 = 0$
2. $xy - 4 = 0$
3. $9x^2 + 24xy + 16y^2 + 90x - 130y = 0$
4. $9x^2 + 24xy + 16y^2 + 80x - 60y = 0$
5. $x^2 - 10xy + y^2 + 1 = 0$
6. $xy + x - 2y + 3 = 0$
7. $xy - 2y - 4x = 0$
8. $2x^2 - 3xy - 2y^2 + 10 = 0$
9. $5x^2 - 2xy + 5y^2 - 12 = 0$
10. $13x^2 + 6\sqrt{3}xy + 7y^2 - 16 = 0$
11. $3x^2 - 2\sqrt{3}xy + y^2 + 2x + 2\sqrt{3}y = 0$
12. $16x^2 - 24xy + 9y^2 - 60x - 80y + 100 = 0$
13. $17x^2 + 32xy - 7y^2 = 75$
14. $40x^2 + 36xy + 25y^2 = 52$
15. $32x^2 + 50xy + 7y^2 = 52$
16. $4x^2 - 12xy + 9y^2 + (4\sqrt{13} - 12)x - (6\sqrt{13} + 8)y = 91$

In Exercises 17–24, use the discriminant to determine if the graph of the equation is a parabola, ellipse, or hyperbola.

17. $16x^2 - 24xy + 9y^2 - 30x - 40y = 0$
18. $x^2 - 4xy - 2y^2 - 6 = 0$
19. $13x^2 - 8xy + 7y^2 - 45 = 0$
20. $2x^2 + 4xy + 5y^2 + 3x - 4y - 20 = 0$
21. $x^2 - 6xy - 5y^2 + 4x - 22 = 0$
22. $36x^2 - 60xy + 25y^2 + 9y = 0$
23. $x^2 + 4xy + 4y^2 - 5x - y - 3 = 0$
24. $x^2 + xy + 4y^2 + x + y - 4 = 0$

25. Show that the equation $x^2 + y^2 = r^2$ is invariant under rotation of axes.
26. Prove Theorem 11.9.

Review Exercises for Chapter 11

In Exercises 1–10, match the equation with the correct graph.

1. $4x^2 + y^2 = 4$

2. $x^2 = 4y$

3. $4x^2 - y^2 = 4$

4. $y^2 = -4x$

5. $x^2 + 4y^2 = 4$

6. $y^2 - 4x^2 = 4$

7. $x^2 = -6y$

8. $x^2 + 5y^2 = 10$

9. $x^2 - 5y^2 = -5$

10. $y^2 - 8x = 0$

(i)   (j)

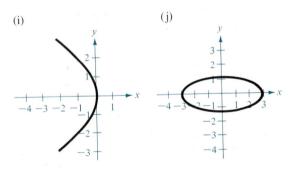

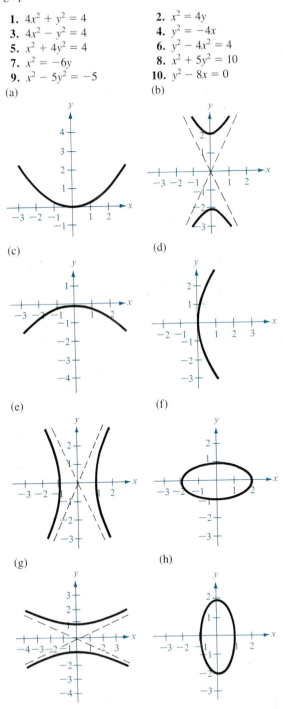

(a)   (b)

(c)   (d)

(e)   (f)

(g)   (h)

In Exercises 11–22, analyze each equation and sketch its graph.

11. $16x^2 + 16y^2 - 16x + 24y - 3 = 0$

12. $y^2 - 12y - 8x + 20 = 0$

13. $3x^2 - 2y^2 + 24x + 12y + 24 = 0$

14. $4x^2 + y^2 - 16x + 15 = 0$

15. $3x^2 + 2y^2 - 12x + 12y + 29 = 0$

16. $4x^2 - 4y^2 - 4x + 8y - 11 = 0$

17. $x^2 - 6x + 2y + 9 = 0$

18. $x^2 + y^2 - 2x - 4y + 5 = 0$

19. $x^2 + y^2 + 2xy + 2\sqrt{2}x - 2\sqrt{2}y + 2 = 0$

20. $9x^2 + 4xy + 6y^2 - 20 = 0$

21. $4x^2 + 9y^2 - 8x + 9y + 4 = 0$

22. $9x^2 + 4y^2 - 36x + 8y + 31 = 0$

In Exercises 23–32, find an equation of the specified conic.

23. Hyperbola; vertices $(0, \pm1)$, foci $(0, \pm3)$

24. Hyperbola; vertices $(\pm2, 2)$, foci $(\pm4, 2)$

25. Ellipse; foci $(0, 0)$ and $(4, 0)$, sum of the distances from a point on the ellipse to the foci is 10

26. Parabola; vertex $(4, 2)$, focus $(4, 0)$

27. Parabola; vertex $(0, 0)$, focus $(1, 1)$

28. Ellipse; vertices $(2, 0)$ and $(2, 4)$, foci $(2, 1)$ and $(2, 3)$

29. Ellipse; points on graph $(1, 2)$ and $(2, 0)$, center at $(0, 0)$

30. Hyperbola; foci $(\pm4, 0)$, asymptotes $y = \pm2x$

31. Hyperbola; foci $(\pm4, 0)$, absolute value of the difference of the distances from a point on the hyperbola to the foci is 4

32. Parabola; vertex $(0, 2)$, point on graph $(-1, 0)$

33. Find the equation of the line tangent to the parabola $y = x^2 - 2x + 2$ and perpendicular to the line $y = x - 2$.

34. Find the equations of the lines tangent to the ellipse

$$x^2 + 4y^2 - 4x - 8y - 24 = 0$$

and parallel to the line

$$x - 2y - 12 = 0$$

35. Find a so that the hyperbola given by

$$\frac{x^2}{a^2} - \frac{y^2}{4} = 1$$

is tangent to the line $2x - y - 4 = 0$.

36. A large parabolic antenna is described as the surface formed by revolving the parabola $y = x^2/200$ on the interval $[0, 100]$ about the y-axis. The receiving and transmitting equipment is positioned at the focus.
(a) Find the focus.
(b) Find the antenna's surface area.

In Exercises 37–39, consider the region bounded by the ellipse

$$\frac{x^2}{a^2} + \frac{y^2}{b^2} = 1$$

with eccentricity $e = c/a$.

37. Show that the area of the region is πab.

38. Show that the solid (oblate spheroid) generated by revolving the region about the minor axis of the ellipse has a volume of $V = 4\pi a^2 b/3$ and a surface area of

$$S = 2\pi a^2 + \pi\left(\frac{b^2}{e}\right)\ln\left(\frac{1 + e}{1 - e}\right)$$

39. Show that the solid (prolate spheroid) generated by revolving the region about the major axis of the ellipse

has a volume of $V = 4\pi ab^2/3$ and a surface area of

$$S = 2\pi b^2 + 2\pi\left(\frac{ab}{e}\right)\arcsin e$$

40. Use the results of Exercises 38 and 39 to find the volume and surface areas of the prolate and oblate spheroids generated by

$$\frac{x^2}{9} + \frac{y^2}{4} = 1$$

41. Approximate, to two decimal places, the perimeter of the ellipse

$$\frac{x^2}{9} + \frac{y^2}{4} = 1$$

In Exercises 42–45, consider a firetruck with a water tank 16 feet long whose vertical cross sections are ellipses described by the equation

$$\frac{x^2}{16} + \frac{y^2}{9} = 1$$

42. Find the volume of the tank.

43. Find the depth of the water in the tank if it is $\frac{3}{4}$ full (by volume) and the truck is on level ground.

44. Find the force on the end of the tank when it is full of water.

45. Approximate, to two decimal places, the surface area of the tank.

12

Plane curves, parametric equations, and polar coordinates

12.1
Plane curves and parametric equations

Up to this point we have been representing a graph by a single equation involving the *two* variables x and y. In this section we look at situations in which it is useful to introduce a *third* variable to represent a curve in the plane.

To see the usefulness of this procedure, consider the path followed by an object that is propelled into the air at an angle of 45°. If the initial velocity of the object is 48 feet per second, it can be shown that it follows the parabolic path given by

$$y = -\frac{x^2}{72} + x \qquad \text{Rectangular equation}$$

as shown in Figure 12.1. However, this equation does not tell the whole story. Although it does tell us *where* the object has been, it doesn't tell us *when* the object was at a given point (x, y) on the path. To determine this time, we introduce a third variable t, which we call a **parameter.** By writing both x and y as functions of t, we obtain the **parametric equations**

$$x = 24\sqrt{2}t \qquad \text{and} \qquad y = -16t^2 + 24\sqrt{2}t \qquad \text{Parametric equations}$$

Now, from this set of equations we can determine that at time $t = 0$, the object is at the point $(0, 0)$. Similarly, at time $t = 1$, the object is at the point $(24\sqrt{2}, 24\sqrt{2} - 16)$, and so on.*

For this particular motion problem x and y are continuous functions of t, and we call the resulting path a **plane curve.**

Rectangular equation:

$$y = -\frac{x^2}{72} + x$$

Parametric equations:

$$x = 24\sqrt{2}\,t, \; y = -16t^2 + 24\sqrt{2}\,t$$

Curvilinear Motion:

Two variables for position
One variable for time

FIGURE 12.1

---

*We will discuss a method for determining this particular set of parametric equations— the equations of motion—later in the text.

DEFINITION OF A PLANE CURVE	If f and g are continuous functions of t on an interval I, then the set of ordered pairs $(f(t), g(t))$ is called a **plane curve** C. The equations $x = f(t)$ and $y = g(t)$ are called **parametric equations** for C, and t is called the **parameter**.

> **Remark**   The set of points $(x, y) = (f(t), g(t))$ in the plane is called the **graph of the curve** C. For simplicity, we will not distinguish between a curve and its graph.

When sketching a curve represented by a pair of parametric equations, we still plot points in the xy-plane. Each set of coordinates (x, y) is determined from a value chosen for the parameter t. By plotting the resulting points in the order of *increasing* values of t, we trace the curve out in a specific direction. This is called the **orientation** of the curve.

EXAMPLE 1   *Sketching a curve*

Sketch the curve described by the parametric equations

$$x = t^2 - 4 \qquad \text{and} \qquad y = \frac{t}{2}, \quad -2 \le t \le 3$$

Solution: For values of t on the given interval, the parametric equations yield the points (x, y) shown in Table 12.1.

TABLE 12.1

t	-2	-1	0	1	2	3
x	0	-3	-4	-3	0	5
y	-1	$-\dfrac{1}{2}$	0	$\dfrac{1}{2}$	1	$\dfrac{3}{2}$

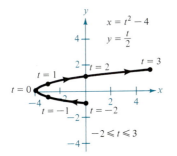

FIGURE 12.2

By plotting these points in the order of increasing t and using the continuity of f and g, we obtain the curve C shown in Figure 12.2. Note that the arrows on the curve indicate its orientation as t increases from -2 to 3.

> **Remark**   By the vertical line test, we can see that the graph shown in Figure 12.2 does not define y as a function of x. This points out one of the benefits of parametric equations—they can be used to represent graphs that are more general than graphs of functions.

It often happens that two different sets of parametric equations have the same graph. For example, the set of parametric equations

$$x = 4t^2 - 4 \qquad \text{and} \qquad y = t, \quad -1 \le t \le \frac{3}{2}$$

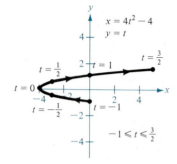

FIGURE 12.3

has the same graph as the set given in Example 1. However, comparing the values of t in Figures 12.2 and 12.3, we see that this second graph is traced out more *rapidly* (considering t as time) than the first graph. Thus, in applications, different parametric representations can be used to represent various *speeds* at which objects travel along a given path.

Eliminating the parameter

In Example 1 we used a simple point-plotting method to sketch the given curve. This tedious process can sometimes be simplified by finding a rectangular equation (in x and y) that has the same graph. We call this process **eliminating the parameter.**

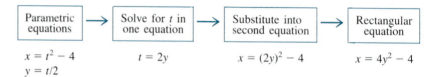

Parametric equations	Solve for t in one equation	Substitute into second equation	Rectangular equation
$x = t^2 - 4$ $y = t/2$	$t = 2y$	$x = (2y)^2 - 4$	$x = 4y^2 - 4$

Now, we recognize that the equation $x = 4y^2 - 4$ represents a parabola with a horizontal axis and vertex at $(-4, 0)$.

One word of caution is in order with regard to converting equations from parametric to rectangular form—the range of x and y implied by the parametric equations may be altered by the change to rectangular form. In such instances it is necessary to adjust the domain of the rectangular equation so that its graph matches the graph of the parametric equations. Such a situation is demonstrated in the next example.

EXAMPLE 2   Adjusting the domain after eliminating the parameter

Sketch the curve represented by the equations

$$x = \frac{1}{\sqrt{t+1}} \quad \text{and} \quad y = \frac{t}{t+1}$$

by eliminating the parameter and adjusting the domain of the resulting rectangular equation.

Solution: Solving for t in the equation for x, we have

$$x = \frac{1}{\sqrt{t+1}} \quad \Longrightarrow \quad x^2 = \frac{1}{t+1}$$

which implies that

$$t = \frac{1 - x^2}{x^2}$$

Now, substituting into the equation for y, we obtain

$$y = \frac{t}{t+1} = \frac{(1 - x^2)/x^2}{[(1 - x^2)/x^2] + 1} = 1 - x^2$$

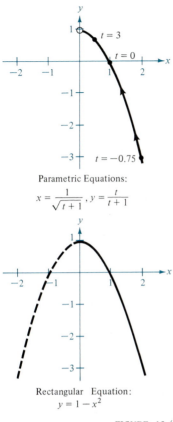

Parametric Equations:

$$x = \frac{1}{\sqrt{t+1}}, \, y = \frac{t}{t+1}$$

Rectangular Equation:

$$y = 1 - x^2$$

FIGURE 12.4

The rectangular equation, $y = 1 - x^2$, is defined for all values of x, but from the parametric equation for x we see that the curve is defined only when $-1 < t$. This implies that we should restrict the domain of x to positive values, as shown in Figure 12.4.

It is not necessary for the parameter in a set of parametric equations to represent time. Our next example uses an *angle* as the parameter. In this example we use *trigonometric identities* to eliminate the parameter.

EXAMPLE 3   *Using a trigonometric identity to eliminate a parameter*

Sketch the curve represented by

$$x = 3 \cos \theta \quad \text{and} \quad y = 4 \sin \theta, \quad 0 \leq \theta \leq 2\pi$$

by eliminating the parameter and finding the corresponding rectangular equation.

Solution: We begin by solving for $\cos \theta$ and $\sin \theta$ in the given equations.

$$\cos \theta = \frac{x}{3} \quad \text{and} \quad \sin \theta = \frac{y}{4} \qquad \text{Solve for } \sin \theta \text{ and } \cos \theta$$

Now we make use of the identity $\sin^2 \theta + \cos^2 \theta = 1$ to form an equation involving only x and y.

$$\cos^2 \theta + \sin^2 \theta = 1 \qquad \text{Trigonometric identity}$$

$$\cos^2 \theta + \sin^2 \theta = \left(\frac{x}{3}\right)^2 + \left(\frac{y}{4}\right)^2 = 1 \qquad \text{Substitute}$$

$$\frac{x^2}{9} + \frac{y^2}{16} = 1 \qquad \text{Rectangular equation}$$

From this rectangular equation we see that the graph is an ellipse centered at $(0, 0)$, with vertices at $(0, 4)$ and $(0, -4)$ and minor axis of length $2b = 6$, as shown in Figure 12.5. Note that the elliptic curve is traced out *counterclockwise* as θ varies from 0 to 2π.

In Examples 2 and 3 it is important to realize that eliminating the parameter is primarily an *aid to curve sketching*. If the parametric equations represent the path of a moving object, the graph alone is not sufficient to describe the object's motion. We still need the parametric equations to tell us the *position, direction,* and *speed* at a given time.

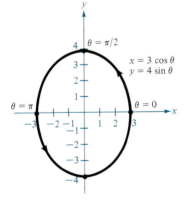

$$x = 3 \cos \theta$$
$$y = 4 \sin \theta$$

FIGURE 12.5

Finding parametric equations for a graph

We have been looking at techniques for sketching the graph represented by a set of parametric equations. We now look at the reverse problem. How can we determine a set of parametric equations for a given graph or a given physical description? From the discussion following Example 1, we know that such a representation is not unique. This is further demonstrated in the following example, in which we find two different parametric representations for a given graph.

EXAMPLE 4   Finding parametric equations for a given graph

Find a set of parametric equations to represent the graph of $y = 1 - x^2$, using the following parameters:

(a) $t = x$ (b) the slope $m = dy/dx$ at the point (x, y)

Solution:

(a) Letting $x = t$, we obtain the parametric equations

$$x = t \qquad \text{and} \qquad y = 1 - x^2 = 1 - t^2$$

(b) To express x and y in terms of the parameter m, we proceed as follows:

$$m = \frac{dy}{dx} = -2x \qquad\qquad \text{Differentiate } y = 1 - x^2$$

$$x = -\frac{m}{2} \qquad\qquad \text{Solve for } x$$

$$y = 1 - x^2 = 1 - \left(-\frac{m}{2}\right)^2 = 1 - \frac{m^2}{4} \qquad\qquad \text{Substitute}$$

Thus, the parametric equations are

$$x = -\frac{m}{2} \qquad \text{and} \qquad y = 1 - \frac{m^2}{4}$$

In Figure 12.6, note that the resulting curve has a right-to-left orientation as determined by the direction of increasing values of slope m. For part (a) the curve would have the opposite orientation.

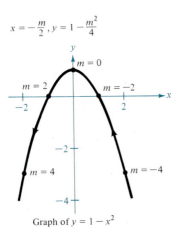

$x = -\frac{m}{2}, y = 1 - \frac{m^2}{4}$

$m = 0$

$m = 2$ $m = -2$

$m = 4$ $m = -4$

Graph of $y = 1 - x^2$

FIGURE 12.6

EXAMPLE 5   Parametric equations for a cycloid

Determine the curve traced out by a point P on the circumference of a circle of radius a as the circle rolls along a straight line in a plane. Such a curve is called a **cycloid.**

Solution: As our parameter we let θ be the measure of the circle's rotation, and we let the point $P = (x, y)$ begin at the origin. When $\theta = 0$, P is at the origin; when $\theta = \pi$, P is at a maximum point $(\pi a, 2a)$; and when $\theta = 2\pi$, P is back on the x-axis at $(2\pi a, 0)$. From Figure 12.7 we see that $\angle APC = 180° - \theta$. Hence,

$$\sin \theta = \sin (180° - \theta) = \sin (\angle APC) = \frac{AC}{a} = \frac{BD}{a}$$

$$\cos \theta = -\cos (180° - \theta) = -\cos (\angle APC) = \frac{AP}{-a}$$

which implies that

$$AP = -a \cos \theta \qquad \text{and} \qquad BD = a \sin \theta$$

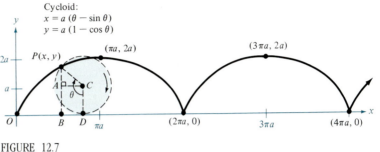

FIGURE 12.7

Now since the circle rolls along the x-axis, we know that $OD = \overset{\frown}{PD} = a\theta$. Furthermore, since $BA = DC = a$, we have

$$x = OD - BD = a\theta - a \sin \theta$$
$$y = BA + AP = a - a \cos \theta$$

Therefore, the parametric equations are

$$x = a(\theta - \sin \theta) \qquad \text{and} \qquad y = a(1 - \cos \theta) \qquad \square$$

The cycloid in Figure 12.7 has sharp corners at the values $x = 2n\pi a$. Between these points, it is called **smooth**.

DEFINITION OF A SMOOTH CURVE

A curve C represented by $x = f(t)$ and $y = g(t)$ on an interval I is called **smooth** if f' and g' are continuous on I and not simultaneously zero, except possibly at the endpoints of I. The curve C is called **piecewise smooth** if it is smooth on each subinterval of some partition of I.

The brachistochrone problem

The type of curve described in Example 5 is related to one of the most famous pairs of problems in the history of calculus. The first problem began with Galileo's discovery that the time required to complete a full swing of a given pendulum is *approximately* the same, whether it makes a large movement at high speeds or a small movement at lower speeds. Late in his life, Galileo (1564–1642) realized that he could use this principle to construct a clock. However, he was not able to conquer the mechanics of actual construction. Christian Huygens (1629–1695) was the first to actually design and construct a working model. In his work with pendulums, Huygens realized that a pendulum does not take exactly the same time to complete swings of varying lengths. (This doesn't affect a pendulum clock, since the length of the circular arc is kept constant by giving the pendulum a slight boost each time it passes its lowest point.) But, in studying the problem, Huygens discovered that a ball rolling back and forth on an inverted cycloid does complete each cycle in exactly the same time.

Inverted Cycloid

FIGURE 12.8

The second problem, posed by John Bernoulli in 1696, is called the **brachistochrone problem**—in Greek *brachys* means short and *chronos* means time. The problem was to determine the path down which a particle will slide from point A to point B in the *shortest time*. Several mathematicians took up the challenge, and the following year the problem was solved by Newton, Leibniz, L'Hôpital, John Bernoulli, and James Bernoulli. As it turns out, the solution is not a straight line from A to B, but an inverted cycloid passing through the points A and B, as shown in Figure 12.8. The amazing part of the solution is that a particle starting at rest at *any* other point C of the cycloid between A and B will take exactly the same time to reach B.

Properties of an inverted cycloid

1. It is the curve of fastest descent from point A to point B—faster than a straight-line path from A to B.
2. The time required to slide to B from any point C between A and B is the same as the time required to slide from A to B.

James Bernoulli (1654–1705), also called Jacques, was the older brother of John. They are only two of several accomplished mathematicians produced by the Swiss Bernoulli family. James's mathematical accomplishments have given him a prominent place in the early development of calculus.

James Bernoulli

Section Exercises 12.1

In Exercises 1–30, sketch the curve represented by the parametric equations and write the corresponding rectangular equation by eliminating the parameter.

1. $x = 3t - 1$, $y = 2t + 1$

2. $x = 3 - 2t$, $y = 2 + 3t$

3. $x = \sqrt{t}$, $y = 1 - t$

4. $x = \sqrt[3]{t}$, $y = 1 - t$

5. $x = t + 1$, $y = t^2$

6. $x = t + 1$, $y = t^3$

7. $x = t^3$, $y = \dfrac{t^2}{2}$

8. $x = 1 + \dfrac{1}{t}$, $y = t - 1$

9. $x = t - 1$, $y = \dfrac{t}{t - 1}$

10. $x = t^2 + t$, $y = t^2 - t$

11. $x = 2t$, $y = |t - 2|$

12. $x = |t - 1|$, $y = t + 2$

13. $x = \sec \theta$, $y = \cos \theta$

14. $x = \tan^2 \theta$, $y = \sec^2 \theta$

15. $x = 3 \cos \theta$, $y = 3 \sin \theta$

16. $x = \cos \theta$, $y = 3 \sin \theta$

17. $x = 4 \sin 2\theta$, $y = 2 \cos 2\theta$

18. $x = \cos \theta$, $y = 2 \sin 2\theta$

19. $x = \cos \theta$, $y = 2 \sin^2 \theta$

20. $x = 4 \cos^2 \theta$, $y = 2 \sin \theta$

21. $x = 4 + 2 \cos \theta$, $y = -1 + \sin \theta$

22. $x = 4 + 2 \cos \theta$, $y = -1 + 2 \sin \theta$

23. $x = 4 + 2 \cos \theta$, $y = -1 + 4 \sin \theta$

24. $x = \sec \theta$, $y = \tan \theta$

25. $x = 4 \sec \theta$, $y = 3 \tan \theta$ **26.** $x = \cos^3 \theta$, $y = \sin^3 \theta$

27. $x = t^3$, $y = 3 \ln t$ **28.** $x = e^{2t}$, $y = e^t$

29. $x = e^{-t}$, $y = e^{3t}$ **30.** $x = \ln 2t$, $y = t^2$

In Exercises 31 and 32, determine how the plane curves differ from each other.

31. (*a*) $x = t$, $y = 2t + 1$
(b) $x = \cos \theta$, $y = 2 \cos \theta + 1$
(c) $x = e^{-t}$, $y = 2e^{-t} + 1$
(d) $x = e^t$, $y = 2e^t + 1$

32. (*a*) $x = 2 \cos \theta$, $y = 2 \sin \theta$
(b) $x = \dfrac{\sqrt{4t^2 - 1}}{|t|}$, $y = \dfrac{1}{t}$

(c) $x = \sqrt{t}$, $y = \sqrt{4-t}$
(d) $x = -\sqrt{4-e^{2t}}$, $y = e^{t}$

In Exercises 33–38, sketch the curve represented by the parametric equations.

33. Cycloid:
$x = 2(\theta - \sin\theta)$, $y = 2(1 - \cos\theta)$
34. Cycloid:
$x = \theta + \sin\theta$, $y = 1 - \cos\theta$
35. Witch of Agnesi:
$x = 2\cot\theta$, $y = 2\sin^2\theta$
36. Curtate cycloid:
$x = 2\theta - \sin\theta$, $y = 2 - \cos\theta$
37. Prolate cycloid:
$x = \theta - \dfrac{3}{2}\sin\theta$, $y = 1 - \dfrac{3}{2}\cos\theta$
38. Folium of Descartes:
$x = \dfrac{3t}{1+t^3}$, $y = \dfrac{3t^2}{1+t^3}$

In Exercises 39–44, eliminate the parameter and obtain the standard form of the rectangular equation of the curve.

39. Line through (x_1, y_1) and (x_2, y_2):
$x = x_1 + t(x_2 - x_1)$, $y = y_1 + t(y_2 - y_1)$
40. Circle:
$x = r\cos\theta$, $y = r\sin\theta$
41. Circle:
$x = h + r\cos\theta$, $y = k + r\sin\theta$
42. Ellipse:
$x = h + a\cos\theta$, $y = k + b\sin\theta$
43. Hyperbola:
$x = h + a\sec\theta$, $y = k + b\tan\theta$
44. Hyperbola:
$x = h + a\sqrt{t+1}$, $y = k + b\sqrt{t}$

In Exercises 45–52, find a set of parametric equations for the given graph.

45. Line: passes through $(0, 0)$ and $(5, -2)$
46. Line: passes through $(1, 4)$ and $(5, -2)$

47. Circle: center at $(2, 1)$ and radius of 4
48. Circle: center at $(-3, 1)$ and radius of 3
49. Ellipse: vertices at $(\pm5, 0)$ and foci at $(\pm4, 0)$
50. Ellipse: vertices at $(4, 7)$ and $(4, -3)$ and foci at $(4, 5)$ and $(4, -1)$
51. Hyperbola: vertices at $(\pm4, 0)$ and foci at $(\pm5, 0)$
52. Hyperbola: vertices at $(0, \pm1)$ and foci at $(0, \pm2)$

In Exercises 53 and 54, find two different sets of parametric equations for the given rectangular equation.

53. $y = x^3$ **54.** $y = x^2$

55. A wheel of radius a rolls along a straight line without slipping. The curve traced by a point P that is b units from the center $(b < a)$ is called a **curtate cycloid,** as shown in Figure 12.9. Use the angle θ shown in the figure to find a set of parametric equations for the curve.

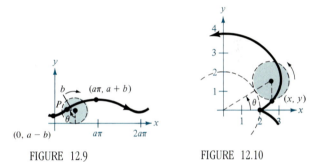

FIGURE 12.9   FIGURE 12.10

56. A circle of radius 1 rolls around the outside of a circle of radius 2 without slipping. The curve traced by a point on the circumference of the smaller circle is called an **epicycloid,** as shown in Figure 12.10. Use the angle θ shown in the figure to find a set of parametric equations for the curve.

12.2
Parametric equations and calculus

Now that we can represent a curve in the plane by a set of parametric equations, it is natural to ask how we can apply the techniques of calculus to study curves. To begin, let's take another look at the projectile represented by the parametric equations

$$x = 24\sqrt{2}t \qquad \text{and} \qquad y = -16t^2 + 24\sqrt{2}t$$

as shown in Figure 12.11 below. From the previous section we know that these parametric equations enable us to locate the position of the object at a given time. We also know that the object is initially projected at an angle of 45°. But, how can we find the angle θ representing the object's direction at some other time t? The following theorem answers this question by giving us a formula for the slope of the tangent line as a function of t.

THEOREM 12.1

PARAMETRIC FORM OF THE DERIVATIVE

If a smooth curve C is given by the equations $x = f(t)$ and $y = g(t)$, then the slope of C at (x, y) is

$$\frac{dy}{dx} = \frac{dy/dt}{dx/dt}, \quad dx/dt \neq 0$$

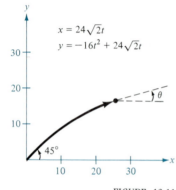

$x = 24\sqrt{2}t$

$y = -16t^2 + 24\sqrt{2}t$

FIGURE 12.11

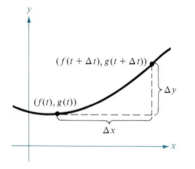

FIGURE 12.12

Proof: In Figure 12.12 consider $\Delta t > 0$ and let

$$\Delta y = g(t + \Delta t) - g(t) \quad \text{and} \quad \Delta x = f(t + \Delta t) - f(t)$$

Since

$$\frac{dy}{dx} = \lim_{\Delta x \to 0} \frac{\Delta y}{\Delta x}$$

and $\Delta x \to 0$ as $\Delta t \to 0$, we can write

$$\frac{dy}{dx} = \lim_{\Delta t \to 0} \frac{g(t + \Delta t) - g(t)}{f(t + \Delta t) - f(t)}$$

Finally, dividing both the numerator and denominator by Δt, we can use the differentiability of f and g to conclude that

$$\frac{dy}{dx} = \lim_{\Delta t \to 0} \frac{[g(t + \Delta t) - g(t)]/\Delta t}{[f(t + \Delta t) - f(t)]/\Delta t}$$

$$= \frac{\displaystyle\lim_{\Delta t \to 0} \frac{g(t + \Delta t) - g(t)}{\Delta t}}{\displaystyle\lim_{\Delta t \to 0} \frac{f(t + \Delta t) - f(t)}{\Delta t}} = \frac{g'(t)}{f'(t)} = \frac{dy/dt}{dx/dt}$$

Since dy/dx is a function of t, we can use Theorem 12.1 repeatedly to find *higher-order* derivatives. For instance,

$$\frac{d^2y}{dx^2} = \frac{d}{dx}\left[\frac{dy}{dx}\right] = \frac{\dfrac{d}{dt}\left[\dfrac{dy}{dx}\right]}{dx/dt} \qquad \text{Second derivative}$$

$$\frac{d^3y}{dx^3} = \frac{d}{dx}\left[\frac{d^2y}{dx^2}\right] = \frac{\dfrac{d}{dt}\left[\dfrac{d^2y}{dx^2}\right]}{dx/dt} \qquad \text{Third derivative}$$

EXAMPLE 1   *Finding slope and concavity*

For the curve given by

$$x = \sqrt{t} \qquad \text{and} \qquad y = \frac{1}{4}(t^2 - 4)$$

find the slope and concavity at the point (2, 3).

Solution: Since

$$\frac{dy}{dx} = \frac{dy/dt}{dx/dt} = \frac{(1/2)t}{(1/2)t^{-1/2}} = t^{3/2}$$

we have

$$\frac{d^2y}{dx^2} = \frac{\dfrac{d}{dt}\left[\dfrac{dy}{dx}\right]}{dx/dt} = \frac{(3/2)t^{1/2}}{(1/2)t^{-1/2}} = 3t$$

At $(x, y) = (2, 3)$, it follows that $t = 4$ and the slope is

$$\frac{dy}{dx} = (4)^{3/2} = 8$$

Moreover, when $t = 4$, the second derivative is

$$\frac{d^2y}{dx^2} = 3(4) = 12 > 0$$

and we conclude that the graph is concave upward at (2, 3), as shown in Figure 12.13.

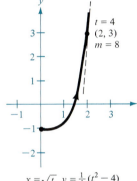

$x = \sqrt{t}, y = \frac{1}{4}(t^2 - 4)$

FIGURE 12.13

Since the parametric equations $x = f(t)$ and $y = g(t)$ need not define y as a function of x, it follows that a curve defined parametrically can loop and cross itself in the plane. At such points the curve may have more than one tangent line, as shown in the next example.

EXAMPLE 2   *A curve with two tangent lines at a point*

The **prolate cycloid** given by

$$x = 2t - \pi \sin t \qquad \text{and} \qquad y = 2 - \pi \cos t$$

crosses itself at the point (0, 2), as shown in Figure 12.14. Find the equations of both tangent lines at this point.

Solution: Since $x = 0$ and $y = 2$ when $t = \pm \pi/2$, and since

$$\frac{dy}{dx} = \frac{dy/dt}{dx/dt} = \frac{\pi \sin t}{2 - \pi \cos t}$$

we have $dy/dx = -\pi/2$ when $t = -\pi/2$ and $dy/dx = \pi/2$ when $t = \pi/2$. Therefore, the two tangent lines at (0, 2) are

$$y - 2 = -\left(\frac{\pi}{2}\right)x \qquad \text{Tangent line when } t = -\frac{\pi}{2}$$

$$y - 2 = \left(\frac{\pi}{2}\right)x \qquad \text{Tangent line when } t = \frac{\pi}{2}$$

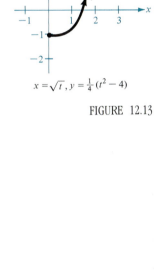

$x = 2t - \pi \sin t$
$y = 2 - \pi \cos t$

FIGURE 12.14

Arc length

We have some idea how parametric equations can be used to describe the path of a particle moving in the plane. We now present a formula for determining the *distance* traveled by the particle along its path.

Recall from Section 6.7 that the rectangular formula for arc length is

$$s = \int_{x_0}^{x_1} \sqrt{1 + \left(\frac{dy}{dx}\right)^2}\, dx, \qquad x_0 \le x \le x_1$$

Converted to variable t, this integral has the form

$$s = \int_{x_0}^{x_1} \sqrt{1 + \left(\frac{dy/dt}{dx/dt}\right)^2}\, dx$$

$$= \int_a^b \frac{\sqrt{(dx/dt)^2 + (dy/dt)^2}}{dx/dt} \left(\frac{dx}{dt}\right) dt, \qquad a \le t \le b$$

$$= \int_a^b \sqrt{\left(\frac{dx}{dt}\right)^2 + \left(\frac{dy}{dt}\right)^2}\, dt$$

This brings us to Theorem 12.2.

THEOREM 12.2   **ARC LENGTH IN PARAMETRIC FORM**

If a smooth curve C is given by $x = f(t)$ and $y = g(t)$ such that C does not intersect itself on the interval $a \le t \le b$, then the arc length of C over this interval is given by

$$s = \int_a^b \sqrt{\left(\frac{dx}{dt}\right)^2 + \left(\frac{dy}{dt}\right)^2}\, dt$$

In the previous section we saw that if a circle rolls along a line, then a point on its circumference will trace out a path that is called a cycloid. If the circle rolls around the circumference of another circle, then the path of the point is called an epicycloid. In the next example, we find the arc length of such a curve.

EXAMPLE 3   Finding arc length

A circle of radius 1 rolls around the circumference of a larger circle of radius 4, as shown in Figure 12.15. The epicycloid traced by a point on the circumference of the smaller circle is given by

$$x = 5 \cos t - \cos 5t \qquad \text{and} \qquad y = 5 \sin t - \sin 5t$$

Find the distance traveled by the point in one complete trip about the larger circle.

Solution: Before applying Theorem 12.2, we note from Figure 12.15 that the curve comes to sharp points when $t = 0$ and $t = \pi/2$. Between these two points, dx/dt and dy/dt are not both simultaneously zero. Thus, the portion of

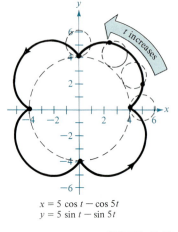

$x = 5 \cos t - \cos 5t$
$y = 5 \sin t - \sin 5t$

FIGURE 12.15

the curve generated from $t = 0$ to $t = \pi/2$ is smooth. To find the total distance traveled by the point, we find the arc length of that portion lying in the first quadrant and multiply by 4.

$$s = 4 \int_0^{\pi/2} \sqrt{\left(\frac{dx}{dt}\right)^2 + \left(\frac{dy}{dt}\right)^2}\, dt$$

$$= 4 \int_0^{\pi/2} \sqrt{(-5\sin t + 5\sin 5t)^2 + (5\cos t - 5\cos 5t)^2}\, dt$$

$$= 20 \int_0^{\pi/2} \sqrt{2 - 2\sin t \sin 5t - 2\cos t \cos 5t}\, dt$$

$$= 20 \int_0^{\pi/2} \sqrt{2 - 2\cos 4t}\, dt = 20 \int_0^{\pi/2} \sqrt{4\sin^2 2t}\, dt$$

$$= 40 \int_0^{\pi/2} \sin 2t\, dt = -20\left[\cos 2t\right]_0^{\pi/2} = 40$$

EXAMPLE 4   *An application involving arc length*

A recording tape 0.001 inch thick is wound around a reel whose inner radius is 0.5 inch and outer radius is 2 inches, as shown in Figure 12.16. How much tape is required to fill the reel?

Solution: To set up a model for this problem, we assume that as the tape is wound around the reel its distance r from the center increases linearly at the rate of 0.001 inch per revolution. In other words,

$$r = (0.001)\frac{\theta}{2\pi} = \frac{\theta}{2000\pi}, \qquad 1000\pi \le \theta \le 4000\pi$$

where θ is measured in radians. Now, we can determine the coordinates of the point (x, y) corresponding to a given radius to be

$$x = r\cos\theta \qquad \text{and} \qquad y = r\sin\theta$$

Substituting for r, we obtain the parametric equations

$$x = \left(\frac{\theta}{2000\pi}\right)\cos\theta \qquad \text{and} \qquad y = \left(\frac{\theta}{2000\pi}\right)\sin\theta$$

Now, we use the arc length formula to determine the total length of the tape to be

$$s = \int_{1000\pi}^{4000\pi} \sqrt{\left(\frac{dx}{d\theta}\right)^2 + \left(\frac{dy}{d\theta}\right)^2}\, d\theta$$

$$= \frac{1}{2000\pi} \int_{1000\pi}^{4000\pi} \sqrt{(-\theta\sin\theta + \cos\theta)^2 + (\theta\cos\theta + \sin\theta)^2}\, d\theta$$

$$= \frac{1}{2000\pi} \int_{1000\pi}^{4000\pi} \sqrt{\theta^2 + 1}\, d\theta \qquad \text{Integration tables, Formula 26}$$

$$= \frac{1}{2000\pi}\left(\frac{1}{2}\right)\left[\theta\sqrt{\theta^2 + 1} + \ln\left|\theta + \sqrt{\theta^2 + 1}\right|\right]_{1000\pi}^{4000\pi}$$

$$\approx 11{,}781 \text{ in} \approx 982 \text{ ft}$$

Remark The graph of $r = a\theta$ is called the **Spiral of Archimedes.**

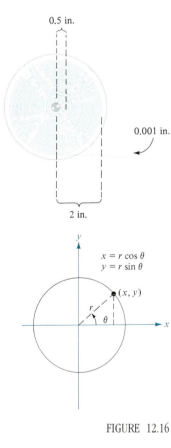

0.5 in.

0.001 in.

2 in.

$x = r\cos\theta$
$y = r\sin\theta$

(x, y)

r

θ

FIGURE 12.16

Area of a surface of revolution

As we did for arc length, we can use the formula for the area of a surface of revolution in rectangular form to develop a formula for surface area in parametric form.

THEOREM 12.3   AREA OF A SURFACE OF REVOLUTION
If a smooth curve C given by $x = f(t)$ and $y = g(t)$ does not cross itself on an interval $a \leq t \leq b$, then the area S of the surface of revolution formed by revolving C about the coordinate axes is given by

1. $S = 2\pi \displaystyle\int_a^b g(t)\sqrt{\left(\dfrac{dx}{dt}\right)^2 + \left(\dfrac{dy}{dt}\right)^2}\, dt$ Revolution about x-axis: $g(t) \geq 0$

2. $S = 2\pi \displaystyle\int_a^b f(t)\sqrt{\left(\dfrac{dx}{dt}\right)^2 + \left(\dfrac{dy}{dt}\right)^2}\, dt$ Revolution about y-axis: $f(t) \geq 0$

EXAMPLE 5   Finding the area of a surface of revolution

Let C be the arc of the circle $x^2 + y^2 = 9$ from $(3, 0)$ to $(3/2, 3\sqrt{3}/2)$, as shown in Figure 12.17. Find the area of the surface formed by revolving C about the x-axis.

Solution: We can represent C parametrically by the equations

$$x = 3 \cos t \quad \text{and} \quad y = 3 \sin t, \quad 0 \leq t \leq \pi/3$$

(Note that we determine the interval for t by observing that $t = 0$ when $x = 3$ and $t = \pi/3$ when $x = 3/2$.) On this interval, C is smooth and y is nonnegative, and we can apply Theorem 12.3 to obtain a surface area of

$$S = 2\pi \int_0^{\pi/3} (3 \sin t)\sqrt{(-3 \sin t)^2 + (3 \cos t)^2}\, dt$$

$$= 6\pi \int_0^{\pi/3} \sin t \sqrt{9(\sin^2 t + \cos^2 t)}\, dt$$

$$= 6\pi \int_0^{\pi/3} 3 \sin t\, dt$$

$$= -18\pi \left[\cos t\right]_0^{\pi/3}$$

$$= -18\pi \left[\frac{1}{2} - 1\right]$$

$$= 9\pi$$

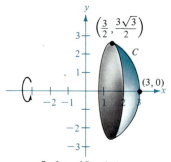

$\left(\dfrac{3}{2}, \dfrac{3\sqrt{3}}{2}\right)$

C

$(3, 0)$

Surface of Revolution

FIGURE 12.17

Area in parametric form

We have seen that the parametric equations $x = f(t)$ and $y = g(t)$ need not define y as a function of x. Nevertheless, for those curves that do define y as a function of x, by the following theorem we can use a definite integral to find the area of a region lying between the curve and the x-axis.

THEOREM 12.4   **DEFINITE INTEGRAL IN PARAMETRIC FORM**
If y is a continuous function of x on the interval $a \leq x \leq b$, where $x = f(t)$ and $y = g(t)$, then

$$\int_a^b y \, dx = \int_{t_1}^{t_2} g(t) f'(t) \, dt, \quad f(t_1) = a, f(t_2) = b$$

provided both g and f' are continuous on $[t_1, t_2]$.

EXAMPLE 6   Area of a region

Find the area of the region bounded by the x-axis and one arch of the cycloid given by

$$x = a(t - \sin t) \quad \text{and} \quad y = a(1 - \cos t)$$

Solution: In Figure 12.18, we see that one arch is defined on the interval $0 \leq x \leq 2\pi a$. The corresponding interval for t is $0 \leq t \leq 2\pi$. Therefore, using Theorem 12.4, we find the area under one arch to be

$$A = \int_0^{2\pi a} y \, dx = \int_0^{2\pi} a(1 - \cos t)(a)(1 - \cos t) \, dt$$

$$= a^2 \int_0^{2\pi} (1 - 2\cos t + \cos^2 t) \, dt$$

$$= a^2 \int_0^{2\pi} \left(1 - 2\cos t + \frac{1 + \cos 2t}{2}\right) dt$$

$$= \frac{a^2}{2} \int_0^{2\pi} (3 - 4\cos t + \cos 2t) \, dt$$

$$= \frac{a^2}{2} \left[3t - 4\sin t + \frac{\sin 2t}{2}\right]_0^{2\pi} = 3a^2\pi$$

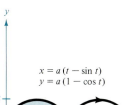

$x = a(t - \sin t)$
$y = a(1 - \cos t)$

$\pi a \quad 2\pi a \quad 3\pi a \quad 4\pi a$

FIGURE 12.18

Section Exercises 12.2

In Exercises 1–10, find dy/dx and d^2y/dx^2, and evaluate each at the specified value of the parameter.

Parametric equations	Point
1. $x = 2t, \ y = 3t - 1$	$t = 3$
2. $x = \sqrt{t}, \ y = 3t - 1$	$t = 1$
3. $x = t + 1, \ y = t^2 + 3t$	$t = -1$
4. $x = t^2 + 3t, \ y = t + 1$	$t = 0$
5. $x = 2\cos\theta, \ y = 2\sin\theta$	$\theta = \dfrac{\pi}{4}$
6. $x = \cos\theta, \ y = 3\sin\theta$	$\theta = 0$
7. $x = 2 + \sec\theta, \ y = 1 + 2\tan\theta$	$\theta = \dfrac{\pi}{6}$
8. $x = \sqrt{t}, \ y = \sqrt{t - 1}$	$t = 2$
9. $x = \cos^3\theta, \ y = \sin^3\theta$	$\theta = \dfrac{\pi}{4}$
10. $x = \theta - \sin\theta, \ y = 1 - \cos\theta$	$\theta = \pi$

In Exercises 11–16, find an equation of the tangent line to the curve at the specified value of the parameter.

Parametric equations	Point
11. $x = 2t, \ y = t^2 - 1$	$t = 2$
12. $x = t - 1, \ y = \dfrac{1}{t} + 1$	$t = 1$
13. $x = t^2 - t + 2, \ y = t^3 - 3t$	$t = -1$
14. $x = 4\cos\theta, \ y = 3\sin\theta$	$\theta = \dfrac{3\pi}{4}$

15. $x = 2 \cot \theta$, $y = 2 \sin^2 \theta$ $\theta = \dfrac{\pi}{4}$

16. $x = 2 - 3 \cos \theta$, $y = 3 + 2 \sin \theta$ $\theta = \dfrac{5\pi}{3}$

In Exercises 17–24, find all points of horizontal tangency on the indicated curve.

17. $x = 1 - t$, $y = t^2$
18. $x = t + 1$, $y = t^2 + 3t$
19. $x = 1 - t$, $y = t^3 - 3t$
20. $x = t^2 - t + 2$, $y = t^3 - 3t$
21. $x = \cot \theta$, $y = 2 \sin \theta \cos \theta$
22. $x = \dfrac{3t}{1 + t^3}$, $y = \dfrac{3t^2}{1 + t^3}$
23. $x = 2\theta$, $y = 2(1 - \cos \theta)$
24. Involute of a circle:
$x = \cos \theta + \theta \sin \theta$, $y = \sin \theta - \theta \cos \theta$

In Exercises 25–30, find the arc length of the given curve.

Parametric equations	*Interval*
25. $x = e^{-t} \cos t$, $y = e^{-t} \sin t$	$0 \le t \le \dfrac{\pi}{2}$
26. $x = t^2$, $y = 4t^3 - 1$	$-1 \le t \le 1$
27. $x = t^2$, $y = 2t$	$0 \le t \le 2$
28. $x = \arcsin t$, $y = \ln \sqrt{1 - t^2}$	$0 \le t \le \dfrac{1}{2}$
29. $x = \sqrt{t}$, $y = 3t - 1$	$0 \le t \le 1$
30. $x = t$, $y = \dfrac{t^5}{10} + \dfrac{1}{6t^3}$	$1 \le t \le 2$

In Exercises 31–34, find the arc length of the given curve.

31. Perimeter of a hypocycloid:
$x = a \cos^3 \theta$, $y = a \sin^3 \theta$, $0 \le \theta \le 2\pi$
32. Circumference of a circle:
$x = a \cos \theta$, $y = a \sin \theta$, $0 \le \theta \le 2\pi$
33. Length of one arch of a cycloid:
$x = a(\theta - \sin \theta)$, $y = a(1 - \cos \theta)$, $0 \le \theta \le 2\pi$
34. Circumference of an ellipse:
$x = 3 \cos \theta$, $y = 4 \sin \theta$, $0 \le \theta < 2\pi$
(Use Simpson's Rule with $n = 6$.)

In Exercises 35–40, find the area of the surface generated by revolving the curve about the given axis.

35. $x = t$, $y = 2t$, $0 \le t \le 4$
(a) x-axis   (b) y-axis
36. $x = t$, $y = 4 - 2t$, $0 \le t \le 2$
(a) x-axis   (b) y-axis
37. $x = 4 \cos \theta$, $y = 4 \sin \theta$, $0 \le \theta \le \dfrac{\pi}{2}$

y-axis

38. $x = t^3$, $y = t + 2$, $1 \le t \le 2$
y-axis
39. $x = a \cos^3 \theta$, $y = a \sin^3 \theta$, $0 \le \theta \le \pi$
x-axis
40. $x = a \cos \theta$, $y = b \sin \theta$, $0 \le \theta \le 2\pi$
(a) x-axis   (b) y-axis

In Exercises 41–44, find the area of the indicated region.

41. $x = \cos^3 \theta$
$y = \sin^3 \theta$

42. $x = a \cos \theta$
$y = b \sin \theta$

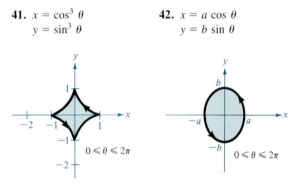

$0 \le \theta \le 2\pi$ $0 \le \theta \le 2\pi$

43. $x = 2 \sin^2 \theta$
$y = 2 \sin^2 \theta \tan \theta$

44. $x = 2 \cot \theta$
$y = 2 \sin^2 \theta$

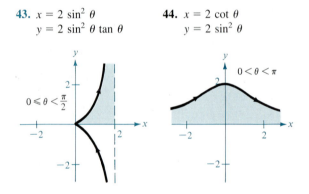

$0 \le \theta < \dfrac{\pi}{2}$ $0 < \theta < \pi$

In Exercises 45 and 46, find the centroid of the region bounded by the graph of the parametric equations and the coordinate axes.

45. $x = \sqrt{t}$, $y = 4 - t$
46. $x = \sqrt{4 - t}$, $y = \sqrt{t}$

In Exercises 47 and 48, find the volume of the solid formed by revolving the region bounded by the graphs of the given equations about the x-axis.

47. $x = 3 \cos \theta$, $y = 3 \sin \theta$
48. $x = \cos \theta$, $y = 3 \sin \theta$
49. A portion of a sphere is removed by a circular cone with its vertex at the center of the sphere. Find the surface area removed from the sphere if the vertex of the cone forms an angle 2θ.

12.3
Polar coordinates and polar graphs

So far, we have been representing plane curves as collections of points (x, y) in the rectangular coordinate system, where x and y represent the directed distances from the coordinate axes to the point (x, y). The corresponding equations for these curves have been in either rectangular or parametric form. In this section we introduce a third system called the **polar coordinate system.**

To form the polar coordinate system in the plane, we fix a point O, called the **pole** (or **origin**), and construct from O an initial ray called the **polar axis,** as shown in Figure 12.19. Then each point P in the plane can be assigned **polar coordinates** (r, θ), as follows:

1. $r = $ *directed distance* from O to P
2. $\theta = $ *directed angle,* counterclockwise from the polar axis to segment $\overline{OP}$

In the polar coordinate system, it is convenient to locate points with respect to a grid of concentric circles intersected by **radial lines** through the pole. This procedure is shown in the following example.

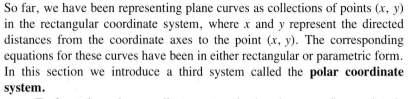

Polar Coordinates

FIGURE 12.19

EXAMPLE 1   Plotting points in the polar coordinate system

(a) The point $(r, \theta) = (2, \pi/3)$ lies at the intersection of a circle of radius $r = 2$ and a ray that is the terminal side of the angle $\theta = \pi/3$, as shown in Figure 12.20(a).
(b) The point $(r, \theta) = (3, -\pi/6)$ lies in the fourth quadrant, 3 units from the pole. Note that we measure negative angles *clockwise,* as shown in Figure 12.20(b).
(c) The point $(r, \theta) = (3, 11\pi/6)$ coincides with the point $(3, -\pi/6)$, as shown in Figure 12.20(c).

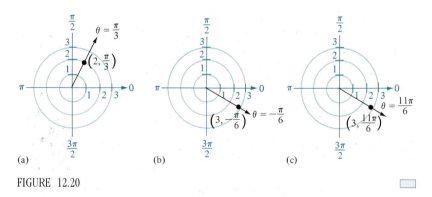

(a)   (b)   (c)

FIGURE 12.20

With rectangular coordinates, each point (x, y) has a unique representation. This is not true with polar coordinates. For instance, the coordinates (r, θ) and $(r, 2\pi + \theta)$ represent the same point, as illustrated in Example 1.

Another way to obtain multiple representations of a point is to use negative values for r. Since r is a *directed distance,* the coordinates (r, θ) and $(-r, \theta + \pi)$ represent the same point. In general, the point (r, θ) has the following different polar coordinate representations

1. $(r, \theta) = (r, \theta + 2n\pi)$ 2. $(r, \theta) = (-r, \theta + (2n + 1)\pi)$

where n is any integer. Moreover, the pole is represented by $(0, \theta)$, where θ is any angle.

EXAMPLE 2   *Multiple representations of points*

Plot the point $(3, -3\pi/4)$ and find three additional polar coordinate representations of this point, using $-2\pi < \theta < 2\pi$.

Solution: The point is shown in Figure 12.21. Three other representations are

$$\left(3, -\frac{3\pi}{4} + 2\pi\right) = \left(3, \frac{5\pi}{4}\right)   \quad\text{Add } 2\pi \text{ to } \theta$$

$$\left(-3, -\frac{3\pi}{4} - \pi\right) = \left(-3, -\frac{7\pi}{4}\right)   \quad\text{Replace } r \text{ by } -r \text{ and subtract } \pi \text{ from } \theta$$

$$\left(-3, -\frac{3\pi}{4} + \pi\right) = \left(-3, \frac{\pi}{4}\right)   \quad\text{Replace } r \text{ by } -r \text{ and add } \pi \text{ to } \theta \qquad \square$$

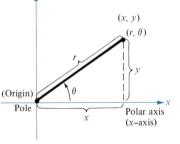

$\left(3, -\frac{3\pi}{4}\right)$

$\theta = -\frac{3\pi}{4}$

$\left(3, -\frac{3\pi}{4}\right) = \left(3, \frac{5\pi}{4}\right) = \left(-3, \frac{\pi}{4}\right) = \ldots$

FIGURE 12.21

To establish the relationship between polar and rectangular coordinates, we let the polar axis coincide with the positive x-axis and the pole with the origin, as shown in Figure 12.22. Since (x, y) lies on a circle of radius r, it follows that $r^2 = x^2 + y^2$. Moreover, for $r > 0$, the definition of the trigonometric functions implies that

$$\tan \theta = \frac{y}{x} \quad \cos \theta = \frac{x}{r} \quad \sin \theta = \frac{y}{r}$$

If $r < 0$, we can show that the same relationships hold. For example, consider the point (r, θ), where $r < 0$. Then, since $(-r, \theta + \pi)$ represents the same point and $-r > 0$, we have $-\sin \theta = \sin(\theta + \pi) = -y/r$, which implies that $\sin \theta = y/r$.

These relationships allow us to convert *coordinates* or *equations* from one system to the other, as indicated in the following theorem.

Relating Polar
and Rectangular Coordinates

FIGURE 12.22

THEOREM 12.5   COORDINATE CONVERSION
The polar coordinates (r, θ) are related to the rectangular coordinates (x, y) as follows:

$$x = r \cos \theta \qquad \text{and} \qquad y = r \sin \theta$$

$$\tan \theta = \frac{y}{x} \qquad \text{and} \qquad r^2 = x^2 + y^2$$

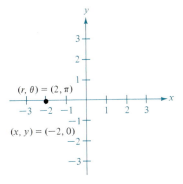

FIGURE 12.23

EXAMPLE 3   Polar-to-rectangular conversion

(a) For the point $(r, \theta) = (2, \pi)$, we have

$$x = r \cos \theta = 2 \cos \pi = -2 \qquad \text{and} \qquad y = r \sin \theta = 2 \sin \pi = 0$$

Thus, the rectangular coordinates are $(x, y) = (-2, 0)$, as shown in Figure 12.23.

(b) For the point $(r, \theta) = (\sqrt{3}, \pi/6)$, we have

$$x = \sqrt{3} \cos \frac{\pi}{6} = \sqrt{3}\left(\frac{\sqrt{3}}{2}\right) = \frac{3}{2}$$

and

$$y = \sqrt{3} \sin \frac{\pi}{6} = \sqrt{3}\left(\frac{1}{2}\right) = \frac{\sqrt{3}}{2}$$

and the rectangular coordinates are $(x, y) = (3/2, \sqrt{3}/2)$.

EXAMPLE 4   Rectangular-to-polar conversion

(a) For the second quadrant point $(x, y) = (-1, 1)$, we have

$$\tan \theta = \frac{y}{x} = -1 \implies \theta = \frac{3\pi}{4}$$

Since our choice of θ lies in the same quadrant as (x, y), we use positive r:

$$r = \sqrt{x^2 + y^2} = \sqrt{(-1)^2 + (1)^2} = \sqrt{2}$$

Thus, *one* set of polar coordinates is $(r, \theta) = (\sqrt{2}, 3\pi/4)$.

(b) Since the point $(x, y) = (0, 2)$ lies on the positive y-axis, we choose $\theta = \pi/2$ and $r = 2$, and one set of polar coordinates is $(r, \theta) = (2, \pi/2)$.

By comparing Examples 3 and 4, we see that point conversion from the polar to the rectangular system is straightforward, whereas point conversion from the rectangular to the polar system is more involved. For equations, the opposite is true. To convert a rectangular equation to polar form, we simply replace x by $r \cos \theta$ and y by $r \sin \theta$. For instance, the rectangular equation $y = x^2$ has the polar form

$$r \sin \theta = (r \cos \theta)^2$$

On the other hand, to convert a polar equation to rectangular form can require considerable ingenuity.

Graphs of polar equations

Our approach to curve sketching in polar coordinates is similar to that for rectangular coordinates. We rely heavily on point-by-point plotting, aided by intercepts and symmetry. Converting to rectangular coordinates and then making the sketch is also a viable option, as demonstrated in the next example.

EXAMPLE 5   *Graphing polar equations*

Describe the graphs of the following polar equations and find the corresponding rectangular equations.

(a) $r = 2$ (b) $\theta = \dfrac{\pi}{3}$ (c) $r = \sec\theta$

Solution:

(a) The graph of the polar equation $r = 2$ consists of all points that are 2 units from the pole. In other words, this graph is a circle centered at the origin and having a radius of 2, as shown in Figure 12.24(a). We can confirm this by converting to rectangular coordinates, using the relationship $r^2 = x^2 + y^2$:

$$r = 2 \implies r^2 = 2^2 \implies x^2 + y^2 = 2^2$$

Polar equation   Rectangular equation

(b) The graph of the polar equation $\theta = \pi/3$ consists of all points on the line that makes an angle of $\pi/3$ with the positive x-axis, as shown in Figure 12.24(b). To convert to rectangular form, we make use of the relationship $\tan\theta = y/x$:

$$\theta = \frac{\pi}{3} \implies \tan\theta = \sqrt{3} \implies y = \sqrt{3}x$$

Polar equation   Rectangular equation

(c) The graph of the polar equation $r = \sec\theta$ is not evident by simple inspection, so we convert to rectangular form by using the relationship $r\cos\theta = x$:

$$r = \sec\theta \implies r\cos\theta = 1 \implies x = 1$$

Polar equation   Rectangular equation

Now we see that the graph is a vertical line, as shown in Figure 12.24(c).

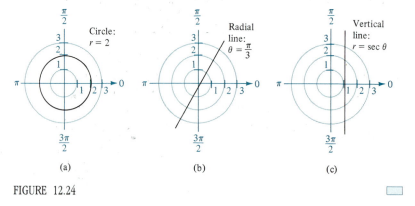

(a)   (b)   (c)

FIGURE 12.24

Curve sketching by converting to rectangular form is not always convenient, and in the next example we demonstrate a straightforward point-plotting technique.

EXAMPLE 6   Graphing a polar equation by point-plotting

Sketch the graph of the polar equation

$$r = 4 \sin \theta$$

Solution: Since the sine function is periodic, we can get a full range of r values by considering values of θ in the interval $0 \le \theta \le 2\pi$, as shown in Table 12.2.

TABLE 12.2

θ	0	$\dfrac{\pi}{6}$	$\dfrac{\pi}{3}$	$\dfrac{\pi}{2}$	$\dfrac{2\pi}{3}$	$\dfrac{5\pi}{6}$	π	$\dfrac{7\pi}{6}$	$\dfrac{3\pi}{2}$	$\dfrac{11\pi}{6}$	2π
r	0	2	$2\sqrt{3}$	4	$2\sqrt{3}$	2	0	-2	-4	-2	0

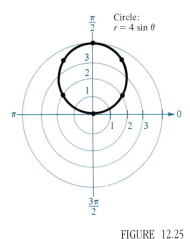

Circle:
$r = 4 \sin \theta$

FIGURE 12.25

By plotting these points as shown in Figure 12.25, it appears that the graph is a circle of radius 2 whose center is at the point $(x, y) = (0, 2)$. (In Exercise 33 you are asked to confirm this by converting to rectangular form.)

Remark   The points on the quadrant rays $\theta = 0$, $\pi/2$, π, $3\pi/2$, and 2π are referred to as the **intercepts.** They are shown in color in Table 12.2.

Symmetry

When making up a table of points, it is often a good idea to use at least one θ-value from each of the four quadrants. However, in Figure 12.25 we see that no points are plotted in Quadrants III and IV because r is negative for these values of θ. Thus, on the interval from $\theta = 0$ to $\theta = 2\pi$ the curve is traced out twice. Moreover, we also see that this curve is *symmetric with respect to the line $\theta = \pi/2$*. Had we known about this symmetry and retracing ahead of time we could have reduced our table of points to half its size.

Symmetry with respect to the line $\theta = \pi/2$ is one of three important types of symmetry to consider in polar curve sketching. (See Figure 12.26.) The following theorem lists a test for each of the three types.

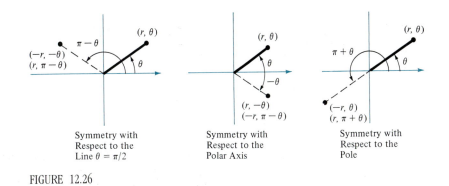

Symmetry with
Respect to the
Line $\theta = \pi/2$

Symmetry with
Respect to the
Polar Axis

Symmetry with
Respect to the
Pole

FIGURE 12.26

THEOREM 12.6   **TESTS FOR SYMMETRY IN POLAR COORDINATES**
The graph of a polar equation is symmetric with respect to the following if the indicated substitution produces an equivalent equation.

1. *The line $\theta = \pi/2$:*   Replace (r, θ) by $(r, \pi - \theta)$ or $(-r, -\theta)$.
2. *The polar axis:*   Replace (r, θ) by $(r, -\theta)$ or $(-r, \pi - \theta)$.
3. *The pole:*   Replace (r, θ) by $(r, \pi + \theta)$ or $(-r, \theta)$.

EXAMPLE 7   Using symmetry to sketch the graph of a polar equation

Use symmetry to sketch the graph of $r = 3 + 2 \cos \theta$.

Solution: Since the cosine is an even function, we replace (r, θ) by $(r, -\theta)$:

$$r = 3 + 2 \cos (-\theta) \implies r = 3 + 2 \cos \theta$$

Since the substitution produces an equivalent equation, the curve is symmetric with respect to the polar axis. This means that we need only use θ-values from the first two quadrants, as shown in Table 12.3.

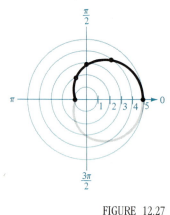

FIGURE 12.27

TABLE 12.3

θ	0	$\dfrac{\pi}{3}$	$\dfrac{\pi}{2}$	$\dfrac{2\pi}{3}$	π
r	5	4	3	2	1

Plotting these points and using polar axis symmetry, we obtain the graph shown in Figure 12.27.

The equations discussed in Examples 6 and 7 are of the form

$$r = 4 \sin \theta = f(\sin \theta) \qquad r \text{ is a function of } \sin \theta$$
$$r = 3 + 2 \cos \theta = g(\cos \theta) \qquad r \text{ is a function of } \cos \theta$$

From Figures 12.25 and 12.27 we see that the graphs of these equations are symmetric with respect to the line $\theta = \pi/2$ and the polar axis, respectively. This observation can be generalized to form the following *quick test for symmetry.*

1. The graph of $r = f(\sin \theta)$ is symmetric with respect to the line $\theta = \pi/2$.
2. The graph of $r = g(\cos \theta)$ is symmetric with respect to the polar axis.

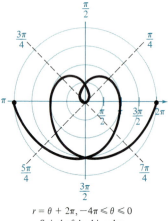

$r = \theta + 2\pi, \; -4\pi \leqslant \theta \leqslant 0$
Spiral of Archimedes

FIGURE 12.28

When applying Theorem 12.6, note that we are using only two of the many possible polar representations of the point (r, θ). Consequently, the conditions of Theorem 12.6 are *sufficient* to guarantee symmetry, but they are not necessary. For instance, Figure 12.28 shows the graph of $r = \theta + 2\pi$ to be symmetric with respect to the line $\theta = \pi/2$. Yet the test fails to indicate symmetry because neither of the following replacements yields an equivalent equation.

Original equation	Replacement	New equation
$r = \theta + 2\pi$	(r, θ) by $(-r, -\theta)$	$-r = -\theta + 2\pi$
$r = \theta + 2\pi$	(r, θ) by $(r, \pi - \theta)$	$r = -\theta + 3\pi$

Section Exercises 12.3

In Exercises 1–8, the polar coordinates of a point are given. Plot the point and find the corresponding rectangular coordinates.

1. $\left(4, \dfrac{3\pi}{6}\right)$

2. $\left(4, \dfrac{3\pi}{2}\right)$

3. $\left(-1, \dfrac{5\pi}{4}\right)$

4. $(0, -\pi)$

5. $\left(4, -\dfrac{\pi}{3}\right)$

6. $\left(-1, -\dfrac{3\pi}{4}\right)$

7. $(\sqrt{2}, 2.36)$

8. $(-3, -1.57)$

In Exercises 9–16, the rectangular coordinates of a point are given. Find two sets of polar coordinates for the point, using $0 \le \theta < 2\pi$.

9. $(1, 1)$

10. $(0, -5)$

11. $(-3, 4)$

12. $(3, -1)$

13. $(-\sqrt{3}, -\sqrt{3})$

14. $(-2, 0)$

15. $(4, 6)$

16. $(5, 12)$

In Exercises 17–32, convert the given rectangular equation to polar form.

17. $x^2 + y^2 = 9$

18. $x^2 + y^2 = a^2$

19. $x^2 + y^2 - 2ax = 0$

20. $x^2 + y^2 - 2ay = 0$

21. $y = 4$

22. $y = b$

23. $x = 10$

24. $x = a$

25. $3x - y + 2 = 0$

26. $4x + 7y - 2 = 0$

27. $xy = 4$

28. $y = x$

29. $y^2 = 9x$

30. $y^2 - 8x - 16 = 0$

31. $x^2 - 4ay - 4a^2 = 0$

32. $(x^2 + y^2)^2 - 9(x^2 - y^2) = 0$

In Exercises 33–44, convert the given polar equation to rectangular form.

33. $r = 4\sin\theta$

34. $r = 4\cos\theta$

35. $\theta = \dfrac{\pi}{6}$

36. $r = 4$

37. $r = 2\csc\theta$

38. $r^2 = \sin 2\theta$

39. $r = 1 - 2\sin\theta$

40. $r = \dfrac{1}{1 - \cos\theta}$

41. $r = \dfrac{6}{2 - 3\sin\theta}$

42. $r = 2\sin 3\theta$

43. $r^2 = 4\sin\theta$

44. $r = \dfrac{6}{2\cos\theta - 3\sin\theta}$

In Exercises 45–54, sketch the graph of the polar equation. Indicate any symmetry possessed by the graph.

45. $r = 5$

46. $r = -2$

47. $r = \theta$

48. $r = \dfrac{\theta}{\pi}$

49. $r = \sin\theta$

50. $r = 3\cos\theta$

51. $r = 2\sec\theta$

52. $r = 3\csc\theta$

53. $r = 4(2 + \sin\theta)$

54. $r = 2 + \cos\theta$

SECTION TOPICS ▪
Relative extrema of r ▪
Tangent lines to polar graphs ▪
Special polar graphs ▪

12.4
Tangent lines and curve sketching in polar coordinates

We have seen that polar curve sketching can be simplified by considering the periodic nature and symmetry of the graph. We begin this section by using calculus to develop two additional curve sketching aids.

If r is a differentiable function of θ, then we can determine the **relative extrema of r** using the procedures discussed in Chapter 4. However, since the

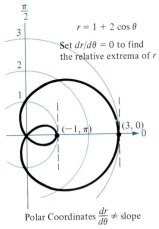

$$r = 1 + 2 \cos \theta$$

Set $dr/d\theta = 0$ to find the relative extrema of r

Polar Coordinates $\dfrac{dr}{d\theta} \neq$ slope

FIGURE 12.29

distance between a point (r, θ) and the pole is $|r|$, it is possible that both the relative maxima and relative minima of r yield points whose distance from the pole is a relative maximum. We illustrate this with an example.

EXAMPLE 1   Relative extrema of r

Find the relative extrema of the function given by $r = 1 + 2 \cos \theta$ and locate the corresponding points on the graph of the function.

Solution: By setting the derivative equal to zero, we have

$$\frac{dr}{d\theta} = -2 \sin \theta = 0$$

and we conclude that on the interval $0 \le \theta < 2\pi$ there are two critical numbers

$$\theta = 0 \qquad \text{and} \qquad \theta = \pi$$

The corresponding points $(3, 0)$ and $(-1, \pi)$ are shown in Figure 12.29.

Remark   In Figure 12.29 we can see that the graph has vertical tangent lines at the points $(3, 0)$ and $(-1, \pi)$. Since $dr/d\theta = 0$ at these points, it is clear that $dr/d\theta$ does *not* represent the slope of a polar graph.

Slope and tangent lines

To find the slope of a tangent line to a polar graph, consider a differentiable function given by $r = f(\theta)$. If α is the angle from the polar axis to the tangent line at the point (r, θ), then $dy/dx = \tan \alpha$, as shown in Figure 12.30. To convert to polar form, we use the parametric equations

$$x = r \cos \theta = f(\theta) \cos \theta \qquad \text{and} \qquad y = r \sin \theta = f(\theta) \sin \theta$$

Using the parametric form of dy/dx given in Theorem 12.1, we have

$$\frac{dy}{dx} = \frac{dy/d\theta}{dx/d\theta} = \frac{f(\theta) \cos \theta + f'(\theta) \sin \theta}{-f(\theta) \sin \theta + f'(\theta) \cos \theta}$$

which establishes the following theorem.

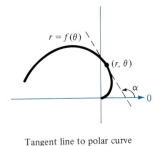

$r = f(\theta)$

(r, θ)

α

Tangent line to polar curve

FIGURE 12.30

THEOREM 12.7   **SLOPE IN POLAR FORM**
If f is a differentiable function of θ, then the *slope* of the tangent line to the graph of $r = f(\theta)$ at the point (r, θ) is

$$\frac{dy}{dx} = \frac{dy/d\theta}{dx/d\theta} = \frac{f(\theta) \cos \theta + f'(\theta) \sin \theta}{-f(\theta) \sin \theta + f'(\theta) \cos \theta}$$

provided $dx/d\theta \neq 0$ at (r, θ).

| Remark   Rather than try to memorize the complete formula for dy/dx, we suggest that you simply remember the parametric form

$$\frac{dy}{dx} = \frac{dy/d\theta}{dx/d\theta}$$

where $y = f(\theta) \sin \theta$ and $x = f(\theta) \cos \theta$. This form is convenient because

1. Solutions to $dy/d\theta = 0$ yield horizontal tangents, provided $dx/d\theta \neq 0$.
2. Solutions to $dx/d\theta = 0$ yield vertical tangents, provided $dy/d\theta \neq 0$.

If $dy/d\theta$ and $dx/d\theta$ are *simultaneously* zero, then no conclusions can be drawn about tangent lines.

EXAMPLE 2   *Finding horizontal and vertical tangents to polar curves*

Find the horizontal and vertical tangents to the graph of $r = 2(1 - \cos \theta)$.

Solution: Using $y = r \sin \theta$, we differentiate and set $dy/d\theta$ equal to zero:

$$y = r \sin \theta = 2(1 - \cos \theta) \sin \theta$$

$$\frac{dy}{d\theta} = 2[(1 - \cos \theta)(\cos \theta) + \sin \theta(\sin \theta)]$$

$$= 2(1 + \cos \theta - 2 \cos^2 \theta) = -2(2 \cos \theta + 1)(\cos \theta - 1) = 0$$

Thus, $\cos \theta = -\frac{1}{2}$ and $\cos \theta = 1$, and we conclude that $dy/d\theta = 0$ when $\theta = 2\pi/3$, $4\pi/3$, and 0. Similarly, using $x = r \cos \theta$, we have

$$x = r \cos \theta = 2 \cos \theta - 2 \cos^2 \theta$$

$$\frac{dx}{d\theta} = -2 \sin \theta + 4 \cos \theta \sin \theta$$

$$= 2 \sin \theta(2 \cos \theta - 1) = 0$$

Thus, $\sin \theta = 0$ and $\cos \theta = \frac{1}{2}$, and we conclude that $dx/d\theta = 0$ when $\theta = 0$, π, $\pi/3$, and $5\pi/3$. Finally, we eliminate $\theta = 0$ because both derivatives are zero there, and conclude that the graph has horizontal and vertical tangents at

$$\left(3, \frac{2\pi}{3}\right), \left(3, \frac{4\pi}{3}\right) \qquad \text{Horizontal tangents}$$

$$\left(1, \frac{\pi}{3}\right), \left(1, \frac{5\pi}{3}\right), (4, \pi) \qquad \text{Vertical tangents}$$

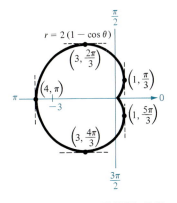

FIGURE 12.31   as shown in Figure 12.31.

Theorem 12.7 has an important consequence. Suppose that the graph of $r = f(\theta)$ passes through the pole when $\theta = \alpha$ and $f'(\alpha) \neq 0$. Then the formula for dy/dx simplifies as follows:

$$\frac{dy}{dx} = \frac{f'(\alpha) \sin \alpha + f(\alpha) \cos \alpha}{f'(\alpha) \cos \alpha - f(\alpha) \sin \alpha} = \frac{f'(\alpha) \sin \alpha + 0}{f'(\alpha) \cos \alpha - 0} = \frac{\sin \alpha}{\cos \alpha} = \tan \alpha$$

This means that the line $\theta = \alpha$ is tangent to the graph at the pole, $(0, \alpha)$. We summarize this result in the following theorem.

THEOREM 12.8	**TANGENT LINES AT THE POLE**
	If $f(\alpha) = 0$ and $f'(\alpha) \neq 0$, then the line $\theta = \alpha$ is tangent at the pole to the graph of $r = f(\theta)$.

Theorem 12.8 is useful because it tells us that the zeros of $r = f(\theta)$ can be used to find the tangent lines at the pole. Note that since a polar curve can cross the pole more than once, it can have more than one tangent line at the pole. Study the next example carefully to see how the various aids to polar curve sketching are combined.

EXAMPLE 3   *Sketching a polar graph*

Sketch the graph of $r = 2 \cos 3\theta$.

Solution: The graph is symmetric with respect to the polar axis since

$$\cos 3\theta = \cos (-3\theta)$$

Setting $dr/d\theta$ equal to zero, we have

$$\frac{dr}{d\theta} = -6 \sin 3\theta = 0$$

Thus, in the interval $0 \leq \theta \leq \pi$, the extrema of r occur when $\theta = 0$, $\pi/3$, and $2\pi/3$. Moreover, by setting r equal to zero, we have

$$2 \cos 3\theta = 0$$

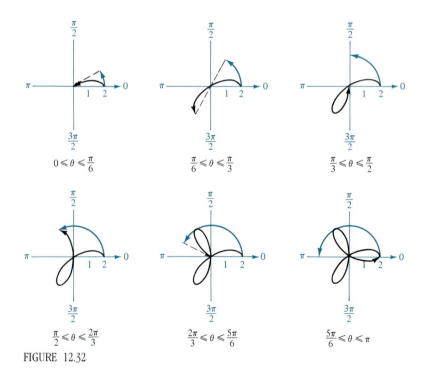

$$0 \leqslant \theta \leqslant \frac{\pi}{6} \qquad \frac{\pi}{6} \leqslant \theta \leqslant \frac{\pi}{3} \qquad \frac{\pi}{3} \leqslant \theta \leqslant \frac{\pi}{2}$$

$$\frac{\pi}{2} \leqslant \theta \leqslant \frac{2\pi}{3} \qquad \frac{2\pi}{3} \leqslant \theta \leqslant \frac{5\pi}{6} \qquad \frac{5\pi}{6} \leqslant \theta \leqslant \pi$$

FIGURE 12.32

and in the interval $0 \le \theta \le \pi$ the tangent lines to the pole occur when $\theta = \pi/6$, $\pi/2$, and $5\pi/6$. Finally, by plotting a few points, we obtain the graph shown in Figure 12.32. It is interesting to note in this figure that both the upper and lower halves of the graph are traced out as θ increases from 0 to π. Were we to continue plotting points as θ increases from π to 2π, the graph would be traced out a second time.

Special polar graphs

Several important types of graphs have equations that are simpler in polar form than in rectangular form. For example, the polar equation of a circle having a radius of a and centered at the origin is simply $r = a$. Later in the text you will come to appreciate this benefit. For now, we summarize some other types of graphs that have simpler equations in polar form. (Conics are considered in the following section.)

Limaçons
$r = a \pm b \cos \theta$
$r = a \pm b \sin \theta$
$(0 < a, \; 0 < b)$

Rose Curves
n petals if n is odd
$2n$ petals if n is even
$(n \ge 2)$

Circles and Lemniscates

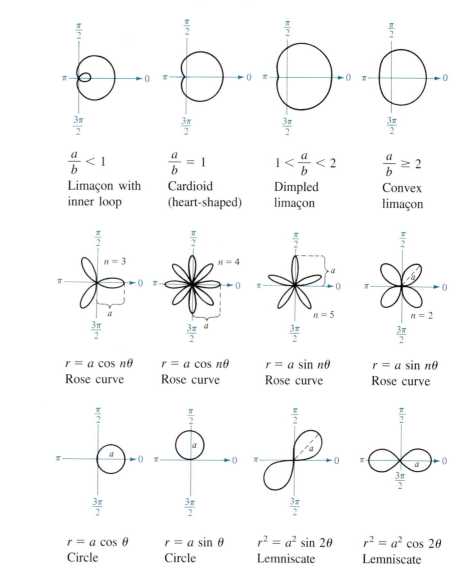

$\dfrac{a}{b} < 1$

Limaçon with inner loop

$\dfrac{a}{b} = 1$

Cardioid (heart-shaped)

$1 < \dfrac{a}{b} < 2$

Dimpled limaçon

$\dfrac{a}{b} \ge 2$

Convex limaçon

$r = a \cos n\theta$
Rose curve

$r = a \cos n\theta$
Rose curve

$r = a \sin n\theta$
Rose curve

$r = a \sin n\theta$
Rose curve

$r = a \cos \theta$
Circle

$r = a \sin \theta$
Circle

$r^2 = a^2 \sin 2\theta$
Lemniscate

$r^2 = a^2 \cos 2\theta$
Lemniscate

Miscellaneous

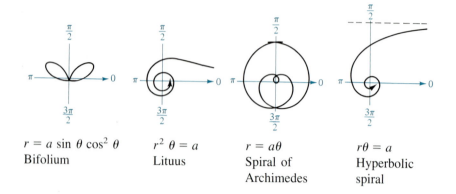

$r = a \sin \theta \cos^2 \theta$
Bifolium

$r^2 \theta = a$
Lituus

$r = a\theta$
Spiral of
Archimedes

$r\theta = a$
Hyperbolic
spiral

EXAMPLE 4   *Sketching a polar graph*

Sketch the graph of $r = 3 \cos 2\theta$.

Solution:

Type of curve: Rose curve with $2n = 4$ petals
Symmetry: With respect to the polar axis and the line $\theta = \pi/2$
Extrema of r: $(3, 0)$, $(-3, \pi/2)$, $(3, \pi)$, and $(-3, 3\pi/2)$
Tangents at pole: $r = 0$ when $\theta = \pi/4$, $3\pi/4$

Using this information together with the additional points shown in Table 12.4, we obtain the graph shown in Figure 12.33.

TABLE 12.4

θ	0	$\dfrac{\pi}{6}$	$\dfrac{\pi}{4}$	$\dfrac{\pi}{3}$
r	3	$\dfrac{3}{2}$	0	$-\dfrac{3}{2}$

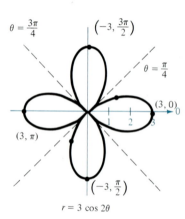

$r = 3 \cos 2\theta$

FIGURE 12.33

EXAMPLE 5   *Sketching a polar graph*

Sketch the graph of $r^2 = 9 \sin 2\theta$.

Solution:

Type of curve: Lemniscate
Symmetry: With respect to the pole
Extrema of r: $r = \pm 3$ when $\theta = \pi/4$
Tangents at pole: $r = 0$ when $\theta = 0$, $\pi/2$

If $\sin 2\theta < 0$, then this equation has no solution points. Thus, we restrict the values of θ to those for which $\sin 2\theta \geq 0$:

$$0 \leq \theta \leq \frac{\pi}{2} \qquad \text{or} \qquad \pi \leq \theta \leq \frac{3\pi}{2}$$

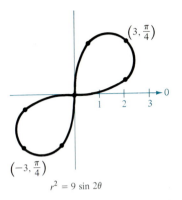

$(3, \frac{\pi}{4})$

$(-3, \frac{\pi}{4})$

$r^2 = 9 \sin 2\theta$

FIGURE 12.34

Moreover, using symmetry, we need only consider the first of these two intervals. In Table 12.5 we find a few additional points, and the graph is shown in Figure 12.34.

TABLE 12.5

θ	0	$\dfrac{\pi}{12}$	$\dfrac{\pi}{4}$	$\dfrac{5\pi}{12}$	$\dfrac{\pi}{2}$
$r = \pm 3\sqrt{\sin 2\theta}$	0	$\pm \dfrac{3}{\sqrt{2}}$	± 3	$\pm \dfrac{3}{\sqrt{2}}$	0

In Examples 4 and 5 we were able to recognize the type of graph from the given polar equation. Of course, there are many other polar graphs, not included in our summary. For instance, the graph in the next example is called a **conchoid**.

EXAMPLE 6   A polar graph with a horizontal asymptote

Show that $y = -1$ is a horizontal asymptote to the graph of

$$r = 2 - \csc \theta, \quad 0 < \theta < \pi$$

Solution: To show that $y = -1$ is a horizontal asymptote, we could show that $y \to -1$ as $x \to \pm\infty$. However, that would necessitate converting to rectangular coordinates, and for this equation the conversion is cumbersome. Another approach is to find the limit of y as $r \to \pm\infty$. To find this limit, we use the fact that $y = r \sin \theta$ to conclude that $\csc \theta = r/y$, and by substituting in the given equation, we have

$$r = 2 - \csc \theta = 2 - \frac{r}{y}$$

$$\frac{r}{y} = 2 - r$$

Solving for y, we have

$$y = \frac{r}{2 - r}$$

$$= -1 + \frac{2}{2 - r}$$

Now, we can determine the limit of y as $r \to \pm\infty$ to be

$$\lim_{r \to \pm\infty} y = \lim_{r \to \pm\infty} \left(-1 + \frac{2}{2 - r} \right) = -1$$

and we conclude that $y = -1$ is a horizontal asymptote to the graph, as shown in Figure 12.35.

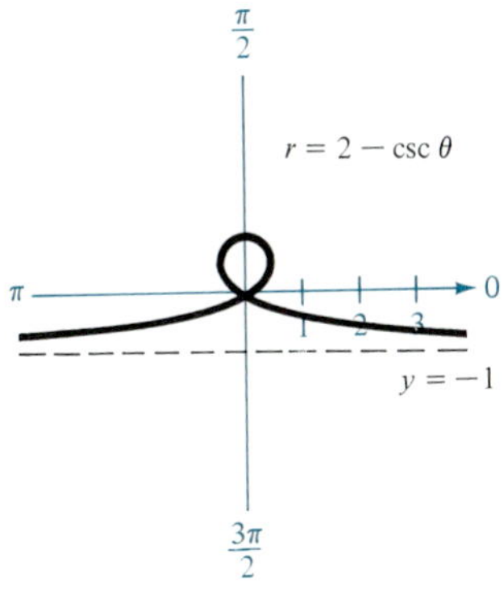

$\dfrac{\pi}{2}$

$r = 2 - \csc \theta$

π

0

$y = -1$

$\dfrac{3\pi}{2}$

FIGURE 12.35

Section Exercises 12.4

In Exercises 1–10, find dy/dx and the slope of the graph of the polar curve at the given value of θ.

Polar equation	Point
1. $r = 3(1 - \cos\theta)$	$\theta = \dfrac{\pi}{2}$
2. $r = 2(1 - \sin\theta)$	$\theta = \dfrac{\pi}{6}$
3. $r = 2 + 3\sin\theta$	$\theta = \dfrac{\pi}{2}$
4. $r = 3 - 2\cos\theta$	$\theta = 0$
5. $r = \theta$	$\theta = \pi$
6. $r = 3\cos\theta$	$\theta = \dfrac{\pi}{4}$
7. $r = 3\sin\theta$	$\theta = \dfrac{\pi}{3}$
8. $r = 4$	$\theta = \dfrac{\pi}{4}$
9. $r = 2\sec\theta$	$\theta = \dfrac{\pi}{4}$
10. $r = \dfrac{6}{2\sin\theta - 3\cos\theta}$	$\theta = \pi$

In Exercises 11 and 12, find the points of horizontal and vertical tangency (if any) to the polar curve.

11. $r = 1 + \sin\theta$ **12.** $r = a\sin\theta$

In Exercises 13 and 14, find the points of horizontal tangency (if any) to the polar curve.

13. $r = 2\csc\theta + 3$ **14.** $r = a\sin\theta\cos^2\theta$

In Exercises 15–48, identify and sketch the graph of the given equation. In each case, find the tangents at the pole.

15. $r = 4$
16. $r = -2$
17. $r = 3\sin\theta$
18. $r = 3\cos\theta$
19. $r = 3(1 - \cos\theta)$
20. $r = 2(1 - \sin\theta)$
21. $r = 4(1 + \sin\theta)$
22. $r = 1 + \cos\theta$
23. $r = 2 + 3\sin\theta$
24. $r = 4 + 5\cos\theta$
25. $r = 3 - 4\cos\theta$
26. $r = 2(1 - 2\sin\theta)$
27. $r = 3 - 2\cos\theta$
28. $r = 5 - 4\sin\theta$
29. $r = 2 + \sin\theta$
30. $r = 4 + 3\cos\theta$
31. $r = 2\cos 3\theta$
32. $r = -\sin 5\theta$
33. $r = 3\sin 2\theta$
34. $r = 3\cos 2\theta$
35. $r = 2\sec\theta$
36. $r = 3\csc\theta$
37. $r = \dfrac{3}{\sin\theta - 2\cos\theta}$
38. $r = \dfrac{6}{2\sin\theta - 3\cos\theta}$
39. $r^2 = 4\cos 2\theta$
40. $r^2 = 4\sin\theta$
41. $r^2 = 4\sin 2\theta$
42. $r^2 = \cos 3\theta$

43. $r = \sin\theta\cos^2\theta$
44. $r^2 = \dfrac{1}{\theta}$
45. $r = 2\theta$
46. $r = \dfrac{1}{\theta}$
47. $r = 2\cos\left(\dfrac{3\theta}{2}\right)$
48. $r = 3\sin\left(\dfrac{5\theta}{2}\right)$

In Exercises 49–52, sketch the graph of the given equation and show that the indicated line is an asymptote to the graph.

Polar equation	Asymptote
49. $r = 2 - \sec\theta$	$x = -1$
50. $r = 2 + \csc\theta$	$y = 1$
51. $r = \dfrac{2}{\theta}$	$y = 2$
52. $r = 2\cos 2\theta\sec\theta$	$x = -2$

53. Verify that if the curve whose polar equation is $r = f(\theta)$ is rotated about the pole through an angle ϕ, then an equation for the rotated curve is $r = f(\theta - \phi)$.

54. If the polar form of an equation for a curve is $r = f(\sin\theta)$, show that the form becomes
 (a) $r = f(-\cos\theta)$ if the curve is rotated counterclockwise $\pi/2$ radians about the pole.
 (b) $r = f(-\sin\theta)$ if the curve is rotated counterclockwise π radians about the pole.
 (c) $r = f(\cos\theta)$ if the curve is rotated counterclockwise $3\pi/2$ radians about the pole.

In Exercises 55–58, use the results of Exercises 53 and 54.

55. Write an equation for the limaçon
$$r = 2 - \sin\theta$$
after it has been rotated by the given amount:
 (a) $\dfrac{\pi}{4}$ (b) $\dfrac{\pi}{2}$ (c) π (d) $\dfrac{3\pi}{2}$

56. Write an equation for the rose curve
$$r = 2\sin 2\theta$$
after it has been rotated by the given amount:
 (a) $\dfrac{\pi}{6}$ (b) $\dfrac{\pi}{2}$ (c) $\dfrac{2\pi}{3}$ (d) π

57. Sketch the graphs of the equations.
 (a) $r = 1 - \sin\theta$ (b) $r = 1 - \sin\left(\theta - \dfrac{\pi}{4}\right)$

58. Sketch the graphs of the equations.
 (a) $r = 3\sec\theta$ (b) $r = 3\sec\left(\theta - \dfrac{\pi}{4}\right)$
 (c) $r = 3\sec\left(\theta + \dfrac{\pi}{3}\right)$

12.5
Polar equations for conics

In Chapter 11 we saw that the rectangular equations of ellipses and hyperbolas take simple forms when the origin lies at their *center*. As it happens, there are many important applications of conics in which it is more convenient to use one of the *foci* as the reference point (the origin) for the coordinate system. For example, the sun lies at the focus of the earth's orbit. Similarly, the light source of a parabolic reflector lies at its focus. In this section we will see that polar equations of conics take simple forms if one of the foci lies at the pole.

We begin with an alternative definition of a conic using the concept of eccentricity.

**ALTERNATIVE
DEFINITION OF CONIC**

The locus of a point in the plane that moves so that its distance from a fixed point (focus) is in constant ratio to its distance from a fixed line (directrix) is called a **conic**. The constant ratio is called the **eccentricity** of the conic and is denoted by e. Moreover, the conic is an **ellipse** if $0 < e < 1$, a **parabola** if $e = 1$, and a **hyperbola** if $e > 1$.

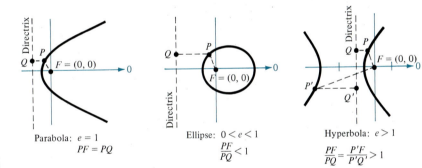

Parabola: $e = 1$
$PF = PQ$

Ellipse: $0 < e < 1$
$\dfrac{PF}{PQ} < 1$

Hyperbola: $e > 1$
$\dfrac{PF}{PQ} = \dfrac{P'F}{P'Q'} > 1$

FIGURE 12.36

In Figure 12.36, note that for each type of conic, the pole corresponds to the fixed point (focus) given in the definition. The benefit of this location is seen in the proof of the following theorem.

THEOREM 12.9

POLAR EQUATIONS FOR CONICS
The graph of a polar equation of the form

1. $r = \dfrac{ep}{1 \pm e \cos \theta}$ 2. $r = \dfrac{ep}{1 \pm e \sin \theta}$

is a conic, where $e > 0$ is the eccentricity and $|p|$ is the distance between the focus (pole) and the directrix.

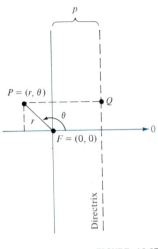

FIGURE 12.37

Proof: We give a proof for $r = ep/(1 + e \cos \theta)$ with $p > 0$. The proofs of the other cases are similar. In Figure 12.37, consider a vertical directrix, p units to the right of the focus $F = (0, 0)$. If $P = (r, \theta)$ is a point on the graph of $r = ep/(1 + e \cos \theta)$, then the distance between P and the directrix is

$$PQ = |p - x| = |p - r \cos \theta| = \left| p - \left(\frac{ep}{1 + e \cos \theta} \right) \cos \theta \right|$$

$$= \left| p \left(1 - \frac{e \cos \theta}{1 + e \cos \theta} \right) \right| = \left| \frac{p}{1 + e \cos \theta} \right| = \left| \frac{r}{e} \right|$$

Moreover, since the distance between P and the pole is simply $PF = |r|$, the ratio of PF to PQ is

$$\frac{PF}{PQ} = \frac{|r|}{|r/e|} = |e| = e$$

and, by definition, the graph of the equation must be a conic.

| **Remark** By completing the proofs of the other three cases, you can see that the equations $r = ep/(1 \pm e \cos \theta)$ correspond to conics with vertical directrices, and the equations $r = ep/(1 \pm e \sin \theta)$ correspond to conics with horizontal directrices. Moreover, the converse is also true—that is, any conic with a focus at the pole and having a horizontal or vertical directrix can be represented by one of the equations in Theorem 12.9.

EXAMPLE 1   Determining a conic from its equation

Sketch the graph of the conic given by

$$r = \frac{15}{3 - 2 \cos \theta}$$

Solution: To determine the type of conic, we rewrite the equation as

$$r = \frac{15}{3 - 2 \cos \theta} = \frac{5}{1 - (2/3) \cos \theta}$$

From this form we conclude that the graph is an ellipse with $e = \frac{2}{3}$. We sketch the upper half of the ellipse by plotting points from $\theta = 0$ to $\theta = \pi$, as shown in Figure 12.38. Then, using symmetry with respect to the polar axis, we sketch the lower half.

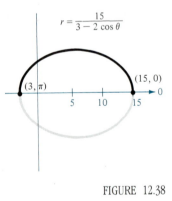

FIGURE 12.38

For the ellipse in Figure 12.38, the major axis is horizontal and the vertices lie at $(15, 0)$ and $(3, \pi)$. Thus, the length of the *major* axis is $2a = 18$. To find the length of the *minor* axis, we use the equations $e = c/a$ and $b^2 = a^2 - c^2$ and conclude that

$$b^2 = a^2 - c^2 = a^2 - (ea)^2 = a^2(1 - e^2) \qquad \text{Ellipse}$$

Since $e = \frac{2}{3}$, we have $b^2 = 9^2[1 - (\frac{2}{3})^2] = 45$, which implies that $b = \sqrt{45} = 3\sqrt{5}$. Thus, the length of the minor axis is $2b = 6\sqrt{5}$. A similar analysis for hyperbolas yields

$$b^2 = c^2 - a^2 = (ea)^2 - a^2 = a^2(e^2 - 1) \qquad \text{Hyperbola}$$

EXAMPLE 2   Sketching a conic from its polar equation

Sketch the graph of the polar equation

$$r = \frac{-32}{3 - 5 \sin \theta}$$

Solution: Dividing each term by 3, we have

$$r = \frac{-32/3}{1 - (5/3) \sin \theta}$$

Since $e = \frac{5}{3} > 1$, the graph is a hyperbola. The transverse axis of the hyperbola lies on the line $\theta = \pi/2$, and the vertices occur at

$$\left(16, \frac{\pi}{2}\right) \qquad \text{and} \qquad \left(-4, \frac{3\pi}{2}\right)$$

Since the length of the transverse axis is 12, we see that $a = 6$. To find b, we write

$$b^2 = a^2(e^2 - 1) = 6^2\left[\left(\frac{5}{3}\right)^2 - 1\right] = 64$$

Therefore, $b = 8$. Finally, we use a and b to determine the asymptotes of the hyperbola and obtain the sketch shown in Figure 12.39.

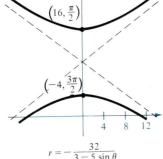

$$r = -\frac{32}{3 - 5 \sin \theta}$$

FIGURE 12.39

The next example uses the converse of Theorem 12.9 to find a polar equation for a specified conic. To do this we let p be the *distance between* the pole and the directrix. With this interpretation of p, we suggest the following guidelines for finding a polar equation for a conic.

1. Horizontal directrix above the pole: $r = ep/(1 + e \sin \theta)$
2. Horizontal directrix below the pole: $r = ep/(1 - e \sin \theta)$
3. Vertical directrix to the right of the pole: $r = ep/(1 + e \cos \theta)$
4. Vertical directrix to the left of the pole: $r = ep/(1 - e \cos \theta)$

EXAMPLE 3   Finding the polar equation for a conic

Find a polar equation for the parabola whose focus is the pole and directrix is the line $y = 3$.

Solution: From Figure 12.40 we see that the directrix is horizontal, and we choose an equation of the form

$$r = \frac{ep}{1 + e \sin \theta}$$

Moreover, since the eccentricity of a parabola is $e = 1$ and the distance between the pole and the directrix is $p = 3$, we have the equation

$$r = \frac{3}{1 + \sin \theta}$$

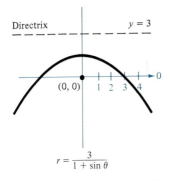

$$r = \frac{3}{1 + \sin \theta}$$

FIGURE 12.40

| Remark   In Example 3 we could have used equation $r = -3/(1 - \sin \theta)$. However, this form is not as convenient, since it yields negative r-values.

Johannes Kepler

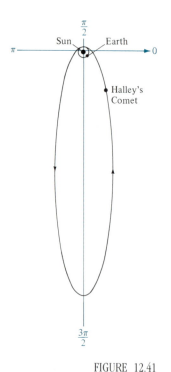

FIGURE 12.41

Applications

Kepler's Laws, named after the German astronomer Johannes Kepler (1571–1630), can be used to describe the orbits of the planets about the sun.

1. Each planet moves in an elliptical orbit with the sun as a focus.
2. The ray from the sun to the planet sweeps out equal areas of the ellipse in equal times.
3. The square of the period is proportional to the cube of the mean distance between the planet and the sun.*

Although Kepler simply stated these laws on the basis of observation, they were later validated by Newton. In fact, Newton was able to show that each law can be deduced from a set of universal laws of motion and gravitation which govern the movement of all heavenly bodies, including comets and satellites. This is illustrated in the next example, involving the comet named after the English mathematician and physicist Edmund Halley (1656–1742).

EXAMPLE 4   An application

Halley's comet has an elliptical orbit with an eccentricity of $e \approx 0.97$. The length of the major axis of the orbit is approximately 36.18 astronomical units. (An astronomical unit is defined to be the mean distance between the earth and the sun, 93 million miles.) Find a polar equation for the orbit. How close does Halley's comet come to the sun?

Solution: Using a vertical axis, we choose an equation of the form $r = ep/(1 + e \sin \theta)$. Since the vertices of the ellipse occur when $\theta = \pi/2$ and $\theta = 3\pi/2$, we can determine the length of the major axis to be the sum of the r-values of the vertices, as shown in Figure 12.41. That is,

$$2a = \frac{0.97p}{1 + 0.97} + \frac{0.97p}{1 - 0.97} \approx 32.83p \approx 36.18$$

Thus, $p \approx 1.102$ and $ep \approx (0.97)(1.102) \approx 1.069$. Using this value in the equation, we have

$$r = \frac{1.069}{1 + 0.97 \sin \theta}$$

where r is measured in astronomical units. To find the closest point to the sun (the focus), we write $c = ea \approx (0.97)(18.09) \approx 17.55$. Since c is the distance between the focus and the center, the closest point is

$$a - c \approx 18.09 - 17.55$$
$$\approx 0.54 \text{ AU}$$
$$\approx 50,000,000 \text{ mi}$$

*If the earth is used as a reference with a period of 1 year and a distance of 1 astronomical unit, the proportionality constant is 1. For example, since Mars has a mean distance to the sun of $d = 1.523$ AU, its period P is given by $d^3 = P^2$. Thus, the period for Mars is $P = 1.88$ years.

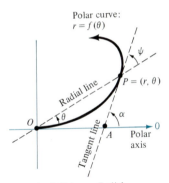

Polar curve:
$r = f(\theta)$

$P = (r, \theta)$

Radial line

Tangent line

Polar
axis

**Angle Between Radial
and Tangent Line**

FIGURE 12.42

There are several applications using polar graphs that involve the angle ψ between the radial line and the tangent line at the point (r, θ), as shown in Figure 12.42. (The tangent and radial lines actually form two angles, but we define ψ so that $0 \le \psi \le \pi/2$.) Using the triangle $\triangle AOP$, we can relate the angles θ, α, and ψ as follows:

$$\theta + \psi + (\pi - \alpha) = \pi \implies \psi = \alpha - \theta$$

Using the identity for the tangent of a difference of two angles together with the slope of the tangent line $dy/dx = \tan \alpha$, we have

$$\tan \psi = \tan (\alpha - \theta) = \frac{\tan \alpha - \tan \theta}{1 + \tan \alpha \tan \theta} = \frac{(dy/dx) - \tan \theta}{1 + (dy/dx) \tan \theta}$$

From the previous section we know that

$$\frac{dy}{dx} = \frac{r \cos \theta + (dr/d\theta) \sin \theta}{-r \sin \theta + (dr/d\theta) \cos \theta}$$

and by substituting this into the equation for $\tan \psi$, we have

$$\tan \psi = \frac{r}{dr/d\theta} \qquad \text{Angle between radial and tangent line}$$

We leave the details of this substitution and simplification as an exercise.

EXAMPLE 5   *Finding the angle between the radial and tangent lines*

Find the angle between the radial and tangent lines at the point $(2/3, \pi/3)$ on the parabola given by $r = 1/(1 + \cos \theta)$.

Solution: Differentiating, we have

$$\frac{dr}{d\theta} = \frac{d}{d\theta}[(1 + \cos \theta)^{-1}]$$

$$= (-1)(-\sin \theta)(1 + \cos \theta)^{-2}$$

$$= \frac{\sin \theta}{(1 + \cos \theta)^2}$$

Thus, at the point $(2/3, \pi/3)$, the value of the derivative is

$$\frac{dr}{d\theta} = \frac{\sqrt{3}/2}{[1 + (1/2)]^2} = \frac{\sqrt{3}/2}{9/4} = \frac{2}{3\sqrt{3}}$$

Now, using the formula for $\tan \psi$, we obtain

$$\tan \psi = \frac{r}{dr/d\theta} = \frac{2/3}{2/(3\sqrt{3})} = \sqrt{3}$$

and we conclude that $\psi = \pi/3$, as shown in Figure 12.43.

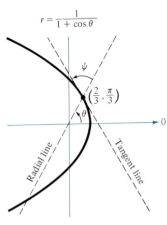

$r = \dfrac{1}{1 + \cos \theta}$

$\left(\dfrac{2}{3}, \dfrac{\pi}{3}\right)$

Radial line

Tangent line

FIGURE 12.43

| **Remark** The fact that the angles θ and ψ in Example 5 are the same illustrates the reflective property of a parabola. Of course, for a general polar curve it is not necessarily true that these two angles agree.

Section Exercises 12.5

In Exercises 1–16, sketch the graph of the equation and identify the curve.

1. $r = \dfrac{2}{1 - \cos \theta}$

2. $r = \dfrac{4}{1 + \sin \theta}$

3. $r = \dfrac{5}{1 + \sin \theta}$

4. $r = \dfrac{6}{1 + \cos \theta}$

5. $r = \dfrac{2}{2 - \cos \theta}$

6. $r = \dfrac{3}{3 + 2 \sin \theta}$

7. $r(2 + \sin \theta) = 4$

8. $r(3 - 2 \cos \theta) = 6$

9. $r = \dfrac{-1}{1 - \sin \theta}$

10. $r = \dfrac{-3}{2 + 4 \sin \theta}$

11. $r = \dfrac{5}{-1 + 2 \cos \theta}$

12. $r = \dfrac{3}{-4 + 2 \cos \theta}$

13. $r = \dfrac{3}{2 - 6 \cos \theta}$

14. $r = \dfrac{4}{1 - 2 \cos \theta}$

15. $r = \dfrac{3}{2 + 6 \sin \theta}$

16. $r = \dfrac{4}{1 + 2 \cos \theta}$

In Exercises 17–28, find a polar equation for the specified conic. In each case consider the focus to be at the pole. (For convenience, the equation for the directrix is given in rectangular form.)

Conic	Eccentricity	Directrix
17. Ellipse	$e = \dfrac{1}{2}$	$y = 1$
18. Ellipse	$e = \dfrac{3}{4}$	$y = -2$
19. Parabola	$e = 1$	$x = -1$
20. Parabola	$e = 1$	$y = 1$
21. Hyperbola	$e = 2$	$x = 1$
22. Hyperbola	$e = \dfrac{3}{2}$	$x = -1$

23. Parabola: vertex at $\left(1, -\dfrac{\pi}{2}\right)$

24. Parabola: directrix is $x = 8$

25. Ellipse: vertices at $\left(6, \dfrac{\pi}{2}\right)$ and $\left(2, \dfrac{3\pi}{2}\right)$

26. Ellipse: vertex at $(2, 0)$, directrix is $x = 6$

27. Hyperbola: vertex at $\left(2, \dfrac{3\pi}{2}\right)$, directrix is $y = -3$

28. Hyperbola: vertices at $(2, 0)$ and $(6, 0)$

29. Show that the polar equation for the ellipse

$$\frac{x^2}{a^2} + \frac{y^2}{b^2} = 1$$

is

$$r^2 = \frac{b^2}{1 - e^2 \cos^2 \theta}$$

30. Show that the polar equation for the hyperbola

$$\frac{x^2}{a^2} - \frac{y^2}{b^2} = 1$$

is

$$r^2 = \frac{-b^2}{1 - e^2 \cos^2 \theta}$$

In Exercises 31–34, use the results of Exercises 29 and 30 to find the polar equation of the given conic.

31. Ellipse: focus at $(4, 0)$, vertices at $(5, 0)$, $(5, \pi)$
32. Hyperbola: focus at $(5, 0)$, vertices at $(4, 0)$, $(4, \pi)$

33. $\dfrac{x^2}{9} - \dfrac{y^2}{16} = 1$

34. $\dfrac{x^2}{4} + y^2 = 1$

In Exercises 35–38, sketch the graph of the rotated conic.

35. $r = \dfrac{2}{1 - \cos (\theta - \pi/4)}$ (See Exercise 1)

36. $r = \dfrac{4}{1 + \sin (\theta - \pi/3)}$ (See Exercise 2)

37. $r = \dfrac{4}{2 + \sin (\theta + \pi/6)}$ (See Exercise 7)

38. $r = \dfrac{4}{1 + 2 \cos (\theta + 2\pi/3)}$ (See Exercise 16)

In Exercises 39–44, find the angle ψ between the radial and tangent lines to the given graph at the indicated point.

Polar equation	Point
39. $r = 2(1 - \cos \theta)$	$\theta = \pi$
40. $r = 3(1 - \cos \theta)$	$\theta = \dfrac{3\pi}{4}$
41. $r = 2 \cos 3\theta$	$\theta = \dfrac{\pi}{6}$
42. $r = 4 \sin 2\theta$	$\theta = \dfrac{\pi}{6}$
43. $r = \dfrac{6}{1 - \cos \theta}$	$\theta = \dfrac{2\pi}{3}$
44. $r = 5$	$\theta = \dfrac{\pi}{6}$

12.6
Area and arc length in polar coordinates

The development of a formula for the area of a polar region parallels that for the area of regions in the rectangular coordinate system, but with *sectors* of a circle used instead of rectangles as the basic element of area. In Figure 12.44, note that the area of a circular sector of radius r is given by

$$\text{area of circular sector} = \frac{1}{2}\theta r^2$$

provided θ is measured in radians.

Consider the function given by $r = f(\theta)$, where f is continuous and nonnegative in the interval $[\alpha, \beta]$. The region bounded by the graph of f and the radial lines $\theta = \alpha$ and $\theta = \beta$ is shown in Figure 12.45. To find the area of this region, we partition the interval $[\alpha, \beta]$ into n equal subintervals,

$$\alpha = \theta_0 < \theta_1 < \theta_2 < \cdots < \theta_{n-1} < \theta_n = \beta$$

Then we approximate the area of the region by the sum of the areas of the n sectors, where

$$\text{radius of } i\text{th sector} = f(\theta_i)$$

$$\text{central angle of } i\text{th sector} = \frac{\beta - \alpha}{n} = \Delta\theta$$

$$A \approx \sum_{i=1}^{n} \left(\frac{1}{2}\right)\Delta\theta f(\theta_i)^2$$

Taking the limit as $n \to \infty$, we have

$$A = \lim_{n\to\infty} \frac{1}{2} \sum_{i=1}^{n} f(\theta_i)^2 \,\Delta\theta = \frac{1}{2} \int_{\alpha}^{\beta} [f(\theta)]^2 \, d\theta$$

which leads to the following theorem.

Area of sector: $\frac{1}{2}\theta r^2$

FIGURE 12.44

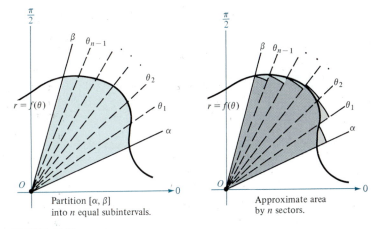

Partition $[\alpha, \beta]$ into n equal subintervals.

Approximate area by n sectors.

FIGURE 12.45

THEOREM 12.10   **AREA IN POLAR COORDINATES**

If f is continuous and nonnegative on the interval $[\alpha, \beta]$, then the area of the region bounded by the graph of $r = f(\theta)$ between the radial lines $\theta = \alpha$ and $\theta = \beta$ is given by

$$A = \frac{1}{2} \int_{\alpha}^{\beta} [f(\theta)]^2 \, d\theta = \frac{1}{2} \int_{\alpha}^{\beta} r^2 \, d\theta$$

Remark We can use the same formula to find the area of a region bounded by the graph of a continuous *nonpositive* function. However, the formula is not necessarily valid if f takes on both positive *and* negative values in the interval $[\alpha, \beta]$.

Sometimes the most difficult part of finding the area of a polar region is determining the limits of integration. A good sketch of the region helps.

EXAMPLE 1   Finding the area of a polar region

Find the area of *one petal* of the rose curve given by

$$r = 3 \cos 3\theta$$

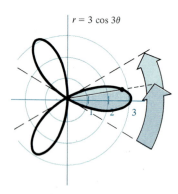

$r = 3 \cos 3\theta$

FIGURE 12.46

Solution: In Figure 12.46 we see that the right petal is traced out as θ increases from $-\pi/6$ to $\pi/6$. Thus, the area is

$$A = \frac{1}{2} \int_{\alpha}^{\beta} r^2 \, d\theta = \frac{1}{2} \int_{-\pi/6}^{\pi/6} (3 \cos 3\theta)^2 \, d\theta$$

$$= \frac{9}{2} \int_{-\pi/6}^{\pi/6} \frac{1 + \cos 6\theta}{2} \, d\theta$$

$$= \frac{9}{4} \left[\theta + \frac{\sin 6\theta}{6} \right]_{-\pi/6}^{\pi/6}$$

$$= \frac{9}{4} \left[\frac{\pi}{6} + \frac{\pi}{6} \right]$$

$$= \frac{3\pi}{4}$$

Suppose you were asked to find the area of the region lying inside all three petals of the rose curve in Example 1. You could not simply integrate between 0 and 2π. In doing this you would obtain $9\pi/2$, which is twice the area of the three petals—the duplication occurs because the rose curve is traced out *twice* as θ increases from 0 to 2π. To avoid this type of error, be sure that the graph of $r = f(\theta)$ does not cross the pole between your chosen limits of integration.

EXAMPLE 2   *Finding the area bounded by a single curve*

Find the area of the region lying between the inner and outer loops of the limaçon $r = 1 - 2 \sin \theta$.

Solution: In Figure 12.47 we see that the inner loop is traced as θ increases from $\pi/6$ to $5\pi/6$. Thus, the area of the region lying inside the *inner loop* is given by

$r = 1 - 2 \sin \theta$

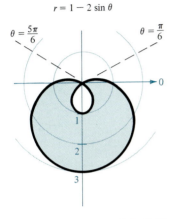

$\theta = \dfrac{5\pi}{6}$ $\theta = \dfrac{\pi}{6}$

FIGURE 12.47

$$A_1 = \frac{1}{2} \int_{\pi/6}^{5\pi/6} (1 - 2 \sin \theta)^2 \, d\theta$$

$$= \frac{1}{2} \int_{\pi/6}^{5\pi/6} (1 - 4 \sin \theta + 4 \sin^2 \theta) \, d\theta$$

$$= \frac{1}{2} \int_{\pi/6}^{5\pi/6} \left[1 - 4 \sin \theta + 4 \left(\frac{1 - \cos 2\theta}{2} \right) \right] d\theta$$

$$= \frac{1}{2} \int_{\pi/6}^{5\pi/6} (3 - 4 \sin \theta - 2 \cos 2\theta) \, d\theta$$

$$= \frac{1}{2} \left[3\theta + 4 \cos \theta - \sin 2\theta \right]_{\pi/6}^{5\pi/6}$$

$$= \frac{1}{2} [2\pi - 3\sqrt{3}] = \pi - \frac{3\sqrt{3}}{2}$$

In a similar way, we can integrate from $5\pi/6$ to $13\pi/6$ to find that the area of the region lying inside the *outer loop* is $A_2 = 2\pi + (3\sqrt{3}/2)$. Finally, to find the area of the region lying between the two loops, we subtract the smaller area from the larger to obtain

$$A = A_2 - A_1 = \left(2\pi + \frac{3\sqrt{3}}{2} \right) - \left(\pi - \frac{3\sqrt{3}}{2} \right) = \pi + 3\sqrt{3} \approx 8.34$$

Points of intersection

Because a point may be represented in different ways in polar coordinates, special care must be taken in determining the points of intersection of two polar graphs. For example, consider the points of intersection of the graphs of $r = 1 - 2 \cos \theta$ and $r = 1$, as shown in Figure 12.48. If, as with rectangular equations, we attempted to find the points of intersection by solving the two equations simultaneously, we would obtain

$$1 = 1 - 2 \cos \theta \implies \cos \theta = 0 \implies \theta = \frac{\pi}{2}, \frac{3\pi}{2}$$

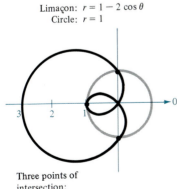
Limaçon: $r = 1 - 2 \cos \theta$
Circle: $r = 1$

Three points of
intersection:
$(1, \pi/2), (-1, 0), (1, 3\pi/2)$

FIGURE 12.48

The corresponding points of intersection are $(1, \pi/2)$ and $(1, 3\pi/2)$. However, from Figure 12.48 we see that there is a *third* point of intersection that did not show up when we set the two polar equations equal to each other. (This is one of the reasons we stress sketching a graph when finding the area of a polar region.) The reason we didn't find the third point is that it does not occur with the same coordinates in the two graphs. On the graph of $r = 1$, the point occurs with the coordinates $(1, \pi)$, but on the graph of $r = 1 - 2 \cos \theta$, the point occurs with coordinates $(-1, 0)$.

FIGURE 12.49

We can compare the problem of finding points of intersection of two polar graphs with that of finding collision points of two satellites in intersecting orbits about the earth, as shown in Figure 12.49. The satellites will not collide as long as they reach the points of intersection at different times (θ values). A collision will occur only at the points of intersection that are "simultaneous points"—those reached at the same time (θ-value).

| **Remark** Because the pole can be represented by $O = (0, \theta)$, where θ is *any* angle, it is a good idea to make a separate check for the pole when hunting for points of intersection.

EXAMPLE 3   *Finding the area of a region between two curves*

Find the area of the region common to the two regions bounded by the circle $r = -6 \cos \theta$ and the cardioid $r = 2 - 2 \cos \theta$.

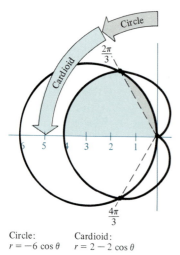

Circle: Cardioid:
$r = -6 \cos \theta$ $r = 2 - 2 \cos \theta$

FIGURE 12.50

Solution: By symmetry we can work with the upper half-plane, so we begin by dividing the region into two parts, as shown in Figure 12.50. The gray shaded region lies between the circle and the radial line $\theta = 2\pi/3$. Since the circle has coordinates $(0, \pi/2)$ at the pole, we integrate between $\pi/2$ and $2\pi/3$ to obtain the area of this region. The colored region lies between the radial lines $\theta = 2\pi/3$ and $\theta = \pi$ and the cardioid. Thus, we can find the area of this second region by integrating between $2\pi/3$ and π. The sum of these two integrals gives the area of the common region lying *above* the polar axis.

$$\underbrace{}_{\substack{\text{Region between circle} \\ \text{and radial line } \theta = 2\pi/3}} \qquad \underbrace{}_{\substack{\text{Region between cardioid and} \\ \text{radial lines } \theta = 2\pi/3 \text{ and } \theta = \pi}}$$

$$\frac{A}{2} = \frac{1}{2} \int_{\pi/2}^{2\pi/3} (-6 \cos \theta)^2 \, d\theta + \frac{1}{2} \int_{2\pi/3}^{\pi} (2 - 2 \cos \theta)^2 \, d\theta$$

$$= 18 \int_{\pi/2}^{2\pi/3} \cos^2 \theta \, d\theta + \frac{1}{2} \int_{2\pi/3}^{\pi} (4 - 8 \cos \theta + 4 \cos^2 \theta) \, d\theta$$

$$= 9 \int_{\pi/2}^{2\pi/3} (1 + \cos 2\theta) \, d\theta + \int_{2\pi/3}^{\pi} (3 - 4 \cos \theta + \cos 2\theta) \, d\theta$$

$$= 9 \left[\theta + \frac{\sin 2\theta}{2} \right]_{\pi/2}^{2\pi/3} + \left[3\theta - 4 \sin \theta + \frac{\sin 2\theta}{2} \right]_{2\pi/3}^{\pi}$$

$$= 9 \left(\frac{2\pi}{3} - \frac{\sqrt{3}}{4} - \frac{\pi}{2} \right) + \left(3\pi - 2\pi + 2\sqrt{3} + \frac{\sqrt{3}}{4} \right)$$

$$= \frac{5\pi}{2}$$

Finally, multiplying by 2, we conclude that the total area is 5π. ☐

Arc length

The formula for the length of a polar arc can be obtained from the arc length formula for a curve described by parametric equations.

THEOREM 12.11

ARC LENGTH OF A POLAR CURVE
Let f be a function whose derivative is continuous in an interval $\alpha \leq \theta \leq \beta$. The length of the graph of $r = f(\theta)$ from $\theta = \alpha$ to $\theta = \beta$ is

$$s = \int_{\alpha}^{\beta} \sqrt{[f(\theta)]^2 + [f'(\theta)]^2}\, d\theta$$

Proof: Considering the parametric form

$$x = r \cos \theta = f(\theta) \cos \theta \qquad \text{and} \qquad y = r \sin \theta = f(\theta) \sin \theta$$

we have

$$\frac{dx}{d\theta} = f'(\theta) \cos \theta - f(\theta) \sin \theta \qquad \text{and} \qquad \frac{dy}{d\theta} = f'(\theta) \sin \theta + f(\theta) \cos \theta$$

and it follows that

$$\left(\frac{dx}{d\theta}\right)^2 + \left(\frac{dy}{d\theta}\right)^2 = [f(\theta)]^2 + [f'(\theta)]^2$$

Consequently, we can apply Theorem 12.2, using θ in place of t, to obtain the desired result.

EXAMPLE 4   *Finding the length of a polar curve*

Find the length of the arc from $\theta = 0$ to $\theta = 2\pi$ for the cardioid

$$r = f(\theta) = 2 - 2 \cos \theta$$

Solution: Since $f'(\theta) = 2 \sin \theta$, we have

$$s = \int_{\alpha}^{\beta} \sqrt{[f(\theta)]^2 + [f'(\theta)]^2}\, d\theta$$

$$= \int_{0}^{2\pi} \sqrt{(2 - 2 \cos \theta)^2 + (2 \sin \theta)^2}\, d\theta$$

$$= 2\sqrt{2} \int_{0}^{2\pi} \sqrt{1 - \cos \theta}\, d\theta$$

$$= 2\sqrt{2} \int_{0}^{2\pi} \frac{\sqrt{1 - \cos^2 \theta}}{\sqrt{1 + \cos \theta}}\, d\theta$$

$$= 4\sqrt{2} \int_{0}^{\pi} \sin \theta\, (1 + \cos \theta)^{-1/2}\, d\theta \qquad \sin \theta = \sqrt{1 - \cos^2 \theta},\ 0 \leq \theta \leq \pi$$

$$= -8\sqrt{2}\left[\sqrt{1 + \cos \theta}\,\right]_{0}^{\pi} = -8\sqrt{2}[0 - \sqrt{2}] = 16$$

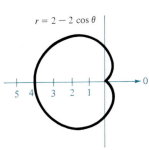

$r = 2 - 2 \cos \theta$

FIGURE 12.51

Using Figure 12.51, we can determine the reasonableness of this answer by comparing it to the circumference of a circle. For example, a circle of radius $\frac{5}{2}$ has a circumference of $5\pi \approx 15.7$.

Area of a surface of revolution

The polar coordinate version of the formulas for the area of a surface of revolution can be obtained from the parametric versions given in Theorem 12.3, using the equations $x = r \cos \theta$ and $y = r \sin \theta$.

THEOREM 12.12   **AREA OF A SURFACE OF REVOLUTION**
Let f be a function whose derivative is continuous in an interval $\alpha \le \theta \le \beta$. The area of the surface formed by revolving the graph of $r = f(\theta)$ from $\theta = \alpha$ to $\theta = \beta$ about the indicated line is

1. $S = 2\pi \displaystyle\int_{\alpha}^{\beta} f(\theta) \sin \theta \sqrt{[f(\theta)]^2 + [f'(\theta)]^2} \, d\theta$ About polar axis

2. $S = 2\pi \displaystyle\int_{\alpha}^{\beta} f(\theta) \cos \theta \sqrt{[f(\theta)]^2 + [f'(\theta)]^2} \, d\theta$ About line $\theta = \dfrac{\pi}{2}$

Remark When using Theorems 12.11 or 12.12, you must check to see that the graph of $r = f(\theta)$ is traced out only once on the interval $\alpha \le \theta \le \beta$. For example, the circle given by $r = \sin \theta$ is traced out once on the interval $0 \le \theta \le \pi$.

EXAMPLE 5   Finding the area of a surface of revolution

Find the area of the surface formed by revolving the circle given by

$$r = f(\theta) = \cos \theta$$

about the line $\theta = \pi/2$ as shown in Figure 12.52.

Solution: We use the second formula given in Theorem 12.12 with $f'(\theta) = -\sin \theta$. Moreover, since the circle is traced out once as θ increases from 0 to π, we have the following integral.

$$S = 2\pi \int_{\alpha}^{\beta} f(\theta) \cos \theta \sqrt{[f(\theta)]^2 + [f'(\theta)]^2} \, d\theta$$

$$= 2\pi \int_{0}^{\pi} \cos \theta \, (\cos \theta) \sqrt{\cos^2 \theta + \sin^2 \theta} \, d\theta$$

$$= 2\pi \int_{0}^{\pi} \cos^2 \theta \, d\theta = \pi \int_{0}^{\pi} (1 + \cos 2\theta) \, d\theta$$

$$= \pi \left[\theta + \frac{\sin 2\theta}{2} \right]_{0}^{\pi} = \pi^2$$

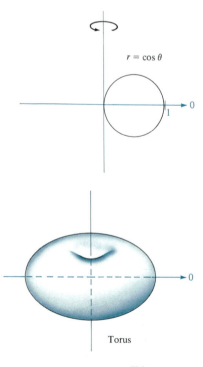

$r = \cos \theta$

Torus

FIGURE 12.52

Applications

Kepler's Second Law states that as a planet moves about the sun, a ray from the sun to the planet sweeps out equal areas in equal times. This law can also be applied to comets or asteroids with elliptical orbits. For example, Figure 12.53 shows the orbit of the asteroid Apollo about the sun. Applying Kepler's Second Law to this asteroid, we know that the closer it is to the sun, the

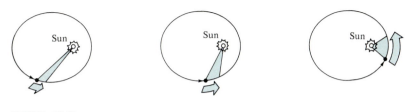

FIGURE 12.53

greater its velocity, since a short ray must be moving quickly to sweep out as much area as a long ray.

EXAMPLE 6   *An application to elliptical orbits*

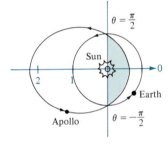

FIGURE 12.54

The asteroid Apollo has a period of 478 earth days, and its orbit is approximated by the ellipse

$$r = \frac{1}{1 + (5/9)\cos\theta} = \frac{9}{9 + 5\cos\theta}$$

where r is measured in astronomical units. How long does it take Apollo to move from the position given by $\theta = -\pi/2$ to $\theta = \pi/2$, as shown in Figure 12.54?

Solution: We begin by finding the area swept out as θ increases from $-\pi/2$ to $\pi/2$.

$$A = \frac{1}{2}\int_\alpha^\beta r^2\,d\theta = \frac{1}{2}\int_{-\pi/2}^{\pi/2}\left(\frac{9}{9 + 5\cos\theta}\right)^2 d\theta$$

Using the substitution $u = \tan(\theta/2)$ as discussed in Section 9.6, we can obtain

$$A = \frac{81}{112}\left[\frac{-5\sin\theta}{9 + 5\cos\theta} + \frac{18}{\sqrt{56}}\arctan\frac{\sqrt{56}\tan(\theta/2)}{14}\right]_{-\pi/2}^{\pi/2} \approx 0.90429$$

Now, since the major axis of the ellipse has length $2a = 81/28$ and the eccentricity is $e = 5/9$, we can determine that $b = a\sqrt{1 - e^2} = 9/\sqrt{56}$. Thus, the area of the ellipse is

$$\text{area of ellipse} = \pi ab = \pi\left(\frac{81}{56}\right)\left(\frac{9}{\sqrt{56}}\right) \approx 5.46507$$

Since the time required to complete the entire orbit is 478 days, we can apply Kepler's Second Law to conclude that the time t required to move from the position $\theta = -\pi/2$ to $\theta = \pi/2$ is given by

$$\frac{t}{478} = \frac{\text{area of elliptical segment}}{\text{area of ellipse}} \approx \frac{0.90429}{5.46507}$$

which implies that

$$t = 478\left(\frac{0.90429}{5.46507}\right) \approx 79 \text{ days}$$

Section Exercises 12.6

In Exercises 1–12, find the points of intersection of the graphs of the given equations.

1. $r = 1 + \cos \theta$
$r = 1 - \cos \theta$

2. $r = 3(1 + \sin \theta)$
$r = 3(1 - \sin \theta)$

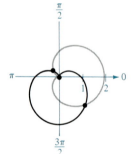

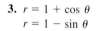

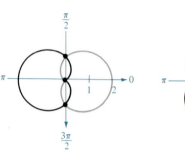

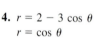

3. $r = 1 + \cos \theta$
$r = 1 - \sin \theta$

4. $r = 2 - 3 \cos \theta$
$r = \cos \theta$

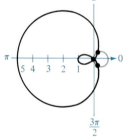

5. $r = 4 - 5 \sin \theta$
$r = 3 \sin \theta$

6. $r = 1 + \cos \theta$
$r = 3 \cos \theta$

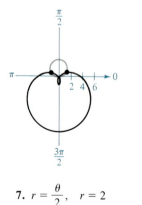

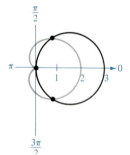

7. $r = \dfrac{\theta}{2}, \quad r = 2$

8. $r = \dfrac{\pi}{4}, \quad r = 2$

9. $r = 4 \sin 2\theta, \quad r = 2$

10. $r = 3 + \sin \theta, \quad r = 2 \csc \theta$

11. $r = 2 + 3 \cos \theta, \quad r = \dfrac{\sec \theta}{2}$

12. $r = 3(1 - \cos \theta), \quad r = \dfrac{6}{1 - \cos \theta}$

In Exercises 13–34, find the area of the given region.

13. One petal of $r = 2 \cos 3\theta$

14. One petal of $r = 4 \sin 2\theta$

15. One petal of $r = \cos \dfrac{3\theta}{2}$

16. One petal of $r = \cos 5\theta$

17. Interior of $r = 1 - \sin \theta$

18. Interior of $r = 1 - \sin \theta$
(above the polar axis)

19. Interior of $r = \dfrac{3}{2 - \cos \theta}$

20. Interior of $r = \dfrac{2}{3 - 2 \sin \theta}$

21. Inner loop of $r = 1 + 2 \cos \theta$

22. Inner loop of $r = 3 + 4 \sin \theta$

23. Between the loops of $r = 1 + 2 \cos \theta$

24. Between the loops of $r = 2(1 + 2 \sin \theta)$

25. Common interior of $r = 4 \sin 2\theta$ and $r = 2$

26. Common interior of $r = 3(1 + \sin \theta)$ and
$r = 3(1 - \sin \theta)$

27. Common interior of $r = 3 - 2 \sin \theta$ and
$r = -3 + 2 \sin \theta$

28. Common interior of $r = 3 - 2 \sin \theta$ and
$r = 3 - 2 \cos \theta$

29. Common interior of $r = 4 \sin \theta$ and $r = 2$

30. Inside $r = 3 \sin \theta$ and outside $r = 2 - \sin \theta$

31. Inside $r = a(1 + \cos \theta)$ and outside $r = a \cos \theta$

32. Inside $r = 2a \cos \theta$ and outside $r = a$

33. Common interior of $r = a(1 + \cos \theta)$ and $r = a \sin \theta$

34. Region bounded by the graphs of

$$r = \dfrac{ab}{a \sin \theta + b \cos \theta}, \quad \theta = 0, \quad \text{and} \quad \theta = \dfrac{\pi}{2}$$

In Exercises 35–42, find the length of the given graph over the indicated interval.

35. $r = a, \quad 0 \le \theta < 2\pi$

36. $r = a \cos \theta, \quad 0 \le \theta < 2\pi$

37. $r = 1 + \sin \theta, \quad 0 \le \theta < 2\pi$

38. $r = 5(1 + \cos \theta), \quad 0 \le \theta < 2\pi$

39. $r = 2\theta$, $0 \le \theta \le \dfrac{\pi}{2}$

40. $r = \sec\theta$, $0 \le \theta \le \dfrac{\pi}{3}$

41. $r = \dfrac{1}{\theta}$, $\pi \le \theta \le 2\pi$

42. $r = e^{\theta}$, $0 \le \theta \le \pi$

In Exercises 43–48, find the area of the surface formed by revolving the indicated curve about the given line.

43. $r = 2\cos\theta$, $0 \le \theta \le \dfrac{\pi}{2}$

revolved about the polar axis

44. $r = a\cos\theta$, $0 \le \theta \le \dfrac{\pi}{2}$

revolved about the line $\theta = \dfrac{\pi}{2}$

45. $r = e^{a\theta}$, $0 \le \theta \le \dfrac{\pi}{2}$

revolved about the line $\theta = \dfrac{\pi}{2}$

46. $r = a(1 + \cos\theta)$, $0 \le \theta \le \pi$

revolved about the polar axis

[Hint: Use a half-angle identity.]

47. $r = 4\cos 2\theta$, $0 \le \theta \le \dfrac{\pi}{4}$

revolved about the polar axis

[Hint: Use Simpson's Rule with $n = 4$.]

48. $r = \theta$, $0 \le \theta \le \pi$

revolved about the polar axis

[Hint: Use Simpson's Rule with $n = 4$.]

49. The elliptical orbit of a satellite is given by

$$r = \frac{ep}{1 + e\cos\theta}$$

where the center of the earth lies at the pole. Show that the area swept out by the ray from the center of the earth to the satellite as it moves from $\theta = \theta_1$ to $\theta = \theta_2$ is

$$A = \left[\frac{e\sin\theta}{(e^2 - 1)(1 + e\cos\theta)} + \frac{2}{(1 - e^2)^{3/2}} \arctan\frac{\sqrt{1 - e^2}\,\tan(\theta/2)}{1 + e} \right]_{\theta_1}^{\theta_2}$$

50. *Explorer 1* had an orbit that ranged from 220 to 1580 miles over the surface of the earth. Sketch its orbit and use the result of Exercise 49 and 4000 miles as the radius of the earth to find the following:

(a) the area the ray from the earth to the satellite sweeps out from $\theta = 0$ to $\theta = \pi/12$

(b) the angle θ_1, such that the area from $\theta = \theta_1$ to $\theta = \pi$ is equal to that of part (a)

(c) the approximate distance traveled and average speed of the satellite in parts (a) and (b)

Review Exercises for Chapter 12

In Exercises 1–10, (a) find dy/dx and all points of horizontal tangency, (b) eliminate the parameter where possible, and (c) sketch the curve represented by the parametric equations.

1. $x = 1 + 4t$, $y = 2 - 3t$

2. $x = e^{t}$, $y = e^{-t}$

3. $x = 3 + 2\cos\theta$, $y = 2 + 5\sin\theta$

4. $x = t^2 - 3t + 2$, $y = t^3 - 3t^2 + 2$

5. $x = \dfrac{1}{t}$, $y = 2t + 3$

6. $x = 2t - 1$, $y = \dfrac{1}{t^2 - 2t}$

7. $x = \dfrac{1}{2t + 1}$, $y = \dfrac{2t(t + 1)}{2t + 1}$

8. $x = \cot\theta$, $y = \sin 2\theta$

9. $x = \cos^3\theta$, $y = 4\sin^3\theta$

10. $x = 2\theta - \sin\theta$, $y = 2 - \cos\theta$

In Exercises 11 and 12, find a parametric representation of the conic.

11. Ellipse: center at $(-3, 4)$, horizontal major axis of length 8, and minor axis of length 6

12. Hyperbola: vertices at $(0, \pm 4)$ and foci at $(0, \pm 5)$

13. Eliminate the parameter from

$$x = a(\theta - \sin\theta)$$

and

$$y = a(1 - \cos\theta)$$

to show that the rectangular equation of a cycloid is

$$x = a \arccos\left(\frac{a - y}{a}\right) \pm \sqrt{2ay - y^2}$$

14. The **involute of a circle** is described by the endpoint P of a string that is held taut as it is unwound from a spool that does not turn, as shown in Figure 12.55. Show that a parametric representation of the involute is given by

$$x = r(\cos \theta + \theta \sin \theta)$$

and

$$y = r(\sin \theta - \theta \cos \theta)$$

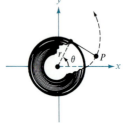

FIGURE 12.55

In Exercises 15 and 16, find the length of the curve represented by the parametric equations over the given interval.

15. $x = r(\cos \theta + \theta \sin \theta)$, $y = r(\sin \theta - \theta \cos \theta)$, $0 \le \theta \le \pi$
16. $x = 6 \cos \theta$, $y = 6 \sin \theta$, $0 \le \theta \le \pi$

In Exercises 17–32, sketch the graph of the equation.

17. $r = -2(1 + \cos \theta)$ **18.** $r = 3 - 4 \cos \theta$
19. $r = 4 - 3 \cos \theta$ **20.** $r = \cos 5\theta$
21. $r = -3 \cos 2\theta$ **22.** $r^2 = \cos 2\theta$
23. $r = 4$ **24.** $r = 2\theta$
25. $r = -\sec \theta$ **26.** $r = 3 \csc \theta$
27. $r^2 = 4 \sin^2 2\theta$ **28.** $r = 2 \sin \theta \cos^2 \theta$
29. $r = \dfrac{2}{1 - \sin \theta}$ **30.** $r = \dfrac{4}{5 - 3 \cos \theta}$
31. $r = 4 \cos 2\theta \sec \theta$ **32.** $r = 4(\sec \theta - \cos \theta)$

In Exercises 33–38, convert the given polar equation to rectangular form.

33. $r = 3 \cos \theta$ **34.** $r = 4 \sec \left(\theta - \dfrac{\pi}{3} \right)$
35. $r = -2(1 + \cos \theta)$ **36.** $r = 1 + \tan \theta$
37. $r = 4 \cos 2\theta \sec \theta$ **38.** $\theta = \dfrac{3\pi}{4}$

In Exercises 39–42, convert the given rectangular equation to polar form.

39. $(x^2 + y^2)^2 = ax^2 y$
40. $x^2 + y^2 - 4x = 0$
41. $x^2 + y^2 = a^2 \left(\arctan \dfrac{y}{x} \right)^2$
42. $(x^2 + y^2) \left(\arctan \dfrac{y}{x} \right)^2 = a^2$

In Exercises 43–48, find a polar equation for the given graph.

43. Parabola: focus at the pole, vertex at $(2, \pi)$
44. Ellipse: a focus at the pole, vertices at $(5, 0)$ and $(1, \pi)$
45. Hyperbola: a focus at the pole, vertices at $(1, 0)$ and $(7, 0)$
46. Circle: center $(0, 5)$, passing through the origin
47. Line: intercepts at $(3, 0)$ and $(0, 4)$
48. Line: through the origin, slope $\sqrt{3}$

In Exercises 49 and 50, (a) find the tangents at the pole, (b) find all points of horizontal and vertical tangency, and (c) sketch the graph of the equation.

49. $r = 1 - 2 \cos \theta$ **50.** $r^2 = 4 \sin 2\theta$

In Exercises 51 and 52, show that the graphs of the given polar equations are orthogonal at the points of intersection.

51. $r = 1 + \cos \theta$ $r = 1 - \cos \theta$
52. $r = a \sin \theta$ $r = a \cos \theta$

In Exercises 53–60, find the area of the given region.

53. Interior of $r = 2 + \cos \theta$
54. Interior of $r = 5(1 - \sin \theta)$
55. Interior of $r = \sin \theta \cos^2 \theta$
56. Interior of $r = 4 \sin 3\theta$
57. Interior of $r^2 = a^2 \sin 2\theta$
58. Common interior of $r = a$ and $r^2 = 2a^2 \sin 2\theta$
59. Common interior of $r = 4 \cos \theta$ and $r = 2$
60. Region bounded by the polar axis and $r = e^\theta$ for $0 \le \theta \le \pi$

In Exercises 61 and 62, find the perimeter of the curve represented by the polar equation.

61. $r = a(1 - \cos \theta)$ **62.** $r = a \cos 2\theta$

63. Find the angle between the circle $r = 3 \sin \theta$ and the limaçon $r = 4 - 5 \sin \theta$ at the point of intersection $(3/2, \pi/6)$.

13

Vectors and the geometry of space

13.1
Vectors in the plane

Many quantities in geometry and physics, such as area, volume, temperature, mass, and time, can be characterized by a single real number scaled to an appropriate unit of measure. We call these **scalar quantities,** and the real number associated with each is called a **scalar.**

Other quantities, such as force and velocity, involve both magnitude and direction and cannot be completely characterized by a single real number. To represent such a quantity, we use a **directed line segment,** as shown in Figure 13.1. The directed line segment $\overrightarrow{PQ}$ has **initial point** P and **terminal point** Q, and we denote its **length** by $\|\overrightarrow{PQ}\|$. Two directed line segments that have the same length and direction are called **equivalent.** For example, the directed line segments in Figure 13.2 are all equivalent. We call the set of all directed line segments that are equivalent to a given directed line segment $\overrightarrow{PQ}$ a **vector in the plane** and write $\mathbf{v} = \overrightarrow{PQ}$.

Q
●
Terminal
point

P
●
Initial
point

$\vec{PQ}$

FIGURE 13.1

| Remark We denote vectors by lowercase, boldface letters such as **u, v,** and **w.**

Be sure you see that a vector in the plane can be represented by many different directed line segments. This is illustrated in the following example.

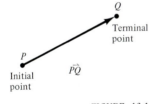

Equivalent Directed
Line Segments

FIGURE 13.2

EXAMPLE 1   Vector representation by directed line segments

Let **u** be represented by the directed line segment from $(0, 0)$ to $(3, 2)$, and let **v** be represented by the directed line segment from $(1, 2)$ to $(4, 4)$. Show that **u = v.**

685

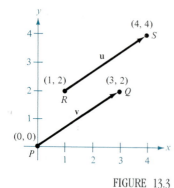

FIGURE 13.3

Solution: Let $P = (0, 0)$, $Q = (3, 2)$, $R = (1, 2)$, and $S = (4, 4)$, as shown in Figure 13.3. We can show that $\overrightarrow{PQ}$ and $\overrightarrow{RS}$ have the *same length* by using the Distance Formula in the plane:

$$\|\overrightarrow{PQ}\| = \sqrt{(3 - 0)^2 + (2 - 0)^2} = \sqrt{13}$$
$$\|\overrightarrow{RS}\| = \sqrt{(4 - 1)^2 + (4 - 2)^2} = \sqrt{13}$$

Moreover, both line segments have the *same direction*, since they are both directed toward the upper right on lines having the same slope.

$$\text{slope of } \overrightarrow{PQ} = \frac{2 - 0}{3 - 0} = \frac{2}{3} \quad \text{and} \quad \text{slope of } \overrightarrow{RS} = \frac{4 - 2}{4 - 1} = \frac{2}{3}$$

Thus, $\overrightarrow{PQ}$ and $\overrightarrow{RS}$ have the same length and direction, and we conclude that $\mathbf{v} = \mathbf{u}$.

Remark The directed line segment whose initial point is the origin is often the most convenient representative of a set of equivalent directed line segments such as those shown in Figure 13.3. We say that this representation of $\mathbf{v}$ is in **standard position.**

A directed line segment whose initial point is at the origin can be uniquely represented by the coordinates of its terminal point $Q = (v_1, v_2)$. We call this the **component form of a vector** and write $\mathbf{v} = \langle v_1, v_2 \rangle$.

DEFINITION OF COMPONENT FORM OF A VECTOR IN THE PLANE	If $\mathbf{v}$ is a vector in the plane whose initial point is the origin and whose terminal point is (v_1, v_2), then the **component form of v** is given by $$\mathbf{v} = \langle v_1, v_2 \rangle$$ The coordinates v_1 and v_2 are called the **components of v.** If both the initial point and terminal point lie at the origin, then $\mathbf{v}$ is called the **zero vector** and is denoted by $\mathbf{0} = \langle 0, 0 \rangle$.

Remark This definition implies that two vectors $\mathbf{u} = \langle u_1, u_2 \rangle$ and $\mathbf{v} = \langle v_1, v_2 \rangle$ are **equal** if and only if $u_1 = v_1$ and $u_2 = v_2$.

To convert directed line segments to component form or vice versa, we use the following procedures.

1. If $P = (p_1, p_2)$ and $Q = (q_1, q_2)$, then the component form of the vector $\mathbf{v}$ represented by $\overrightarrow{PQ}$ is $\langle v_1, v_2 \rangle = \langle q_1 - p_1, q_2 - p_2 \rangle$. Moreover, the length of $\mathbf{v}$ is given by

$$\|\mathbf{v}\| = \sqrt{(q_1 - p_1)^2 + (q_2 - p_2)^2} = \sqrt{v_1^2 + v_2^2} \qquad \text{Length of a vector}$$

2. If $\mathbf{v} = \langle v_1, v_2 \rangle$, then $\mathbf{v}$ can be represented by the directed line segment, in standard position, from $P = (0, 0)$ to $Q = (v_1, v_2)$.

EXAMPLE 2   Finding the component form and length of a vector

Find the component form and length of the vector $\mathbf{v}$ that has initial point $(3, -7)$ and terminal point $(-2, 5)$.

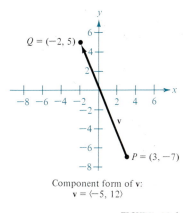

$Q = (-2, 5)$

$P = (3, -7)$

$\mathbf{v}$

Component form of $\mathbf{v}$:
$\mathbf{v} = \langle -5, 12 \rangle$

FIGURE 13.4

Solution: We let $P = (3, -7) = (p_1, p_2)$ and $Q = (-2, 5) = (q_1, q_2)$. Then the components of $\mathbf{v} = \langle v_1, v_2 \rangle$ are given by

$$v_1 = q_1 - p_1 = -2 - 3 = -5$$
$$v_2 = q_2 - p_2 = 5 - (-7) = 12$$

Thus, $\mathbf{v} = \langle -5, 12 \rangle$, and the length of $\mathbf{v}$ is

$$\|\mathbf{v}\| = \sqrt{(-5)^2 + 12^2} = \sqrt{169} = 13$$

as shown in Figure 13.4.

| Remark   The length of $\mathbf{v}$ is also called the **norm of v.** If $\|\mathbf{v}\| = 1$, then $\mathbf{v}$ is called a **unit vector.** Moreover, $\|\mathbf{v}\| = 0$ if and only if $\mathbf{v}$ is the zero vector **0.**

Vector operations

Vectors are relative newcomers to mathematics. Some of the earliest work with vectors was done by the Irish mathematician William Rowan Hamilton (1805–1865). Hamilton spent many years developing a system of vector-like quantities that he called quaternions. This work paved the way for the development of the modern notion of a vector. Although Hamilton was convinced of the benefits of quaternions, the operations he defined did not produce good models for physical phenomena. It wasn't until the latter half of the nineteenth century that the Scottish physicist James Maxwell (1831–1879) restructured Hamilton's quaternions in a form useful for representing physical quantities such as force, velocity, and acceleration.

William Hamilton

The two basic vector operations are **vector addition** and **scalar multiplication.** Geometrically, the product of a vector $\mathbf{v}$ and a scalar k is the vector that is k times as long as $\mathbf{v}$. If k is positive, then $k\mathbf{v}$ has the same direction as $\mathbf{v}$, and if k is negative, then $k\mathbf{v}$ has the direction opposite that of $\mathbf{v}$, as shown in Figure 13.5.

To add two vectors geometrically, we position them (without changing their magnitude or direction) so that the initial point of one coincides with the terminal point of the other. The sum $\mathbf{u} + \mathbf{v}$ (or $\mathbf{v} + \mathbf{u}$), called the **resultant vector,** is formed by joining the initial point of the first vector with the terminal point of the second, as shown in Figure 13.6. The vector $\mathbf{u} + \mathbf{v}$ is the diagonal of a parallelogram having $\mathbf{u}$ and $\mathbf{v}$ as its adjacent sides.

Vector addition and scalar multiplication can also be defined using components of vectors.

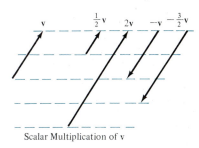

$\mathbf{v}$ $\frac{1}{2}\mathbf{v}$ $2\mathbf{v}$ $-\mathbf{v}$ $-\frac{3}{2}\mathbf{v}$

Scalar Multiplication of $\mathbf{v}$

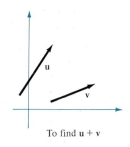

To find $\mathbf{u} + \mathbf{v}$

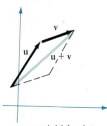

$\mathbf{v}$

$\mathbf{u}$

$\mathbf{u} + \mathbf{v}$

move initial point of $\mathbf{v}$ to the terminal point of $\mathbf{u}$

or

$\mathbf{u} + \mathbf{v}$

$\mathbf{u}$

$\mathbf{v}$

move initial point of $\mathbf{u}$ to the terminal point of $\mathbf{v}$.

FIGURE 13.5   FIGURE 13.6

DEFINITION OF VECTOR ADDITION AND SCALAR MULTIPLICATION	For vectors $\mathbf{u} = \langle u_1, u_2 \rangle$ and $\mathbf{v} = \langle v_1, v_2 \rangle$ and scalar k, we define the following operations. 1. The **vector sum** of $\mathbf{u}$ and $\mathbf{v}$ is the vector $\mathbf{u} + \mathbf{v} = \langle u_1 + v_1, u_2 + v_2 \rangle$. 2. The **scalar multiple** of k and $\mathbf{u}$ is the vector $k\mathbf{u} = \langle ku_1, ku_2 \rangle$. 3. The **negative** of $\mathbf{v}$ is the vector $-\mathbf{v} = (-1)\mathbf{v} = \langle -v_1, -v_2 \rangle$. 4. The **difference** of $\mathbf{u}$ and $\mathbf{v}$ is $\mathbf{u} - \mathbf{v} = \mathbf{u} + (-\mathbf{v}) = \langle u_1 - v_1, u_2 - v_2 \rangle$.

In Figure 13.7, we show that the geometrical and algebraic definitions of vector addition and scalar multiplication are the same. Note that to represent $\mathbf{u} - \mathbf{v}$ graphically, we use directed line segments with the *same* initial points. The difference $\mathbf{u} - \mathbf{v}$ is the vector from the terminal point of $\mathbf{v}$ to the terminal point of $\mathbf{u}$.

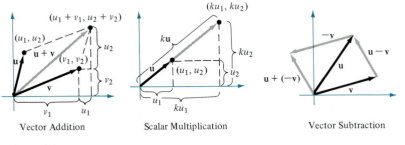

Vector Addition   Scalar Multiplication   Vector Subtraction

FIGURE 13.7

EXAMPLE 3   *Vector operations*

Given the vectors $\mathbf{v} = \langle -2, 5 \rangle$ and $\mathbf{w} = \langle 3, 4 \rangle$, find the following vectors:

(a) $\dfrac{1}{2}\mathbf{v}$ (b) $\mathbf{w} - \mathbf{v}$ (c) $\mathbf{v} + 2\mathbf{w}$

Solution:
(a) Since $\mathbf{v} = \langle -2, 5 \rangle$, we have

$$\frac{1}{2}\mathbf{v} = \left\langle \frac{1}{2}(-2), \frac{1}{2}(5) \right\rangle = \left\langle -1, \frac{5}{2} \right\rangle$$

(b) $\mathbf{w} - \mathbf{v} = \langle w_1 - v_1, w_2 - v_2 \rangle = \langle 3 - (-2), 4 - 5 \rangle = \langle 5, -1 \rangle$
(c) Since $2\mathbf{w} = \langle 6, 8 \rangle$, it follows that

$$\begin{aligned} \mathbf{v} + 2\mathbf{w} &= \langle -2, 5 \rangle + \langle 6, 8 \rangle \\ &= \langle -2 + 6, 5 + 8 \rangle \\ &= \langle 4, 13 \rangle \end{aligned}$$

Vector addition and scalar multiplication share many of the properties of ordinary arithmetic, as shown in the following theorem.

THEOREM 13.1

PROPERTIES OF VECTOR ADDITION AND SCALAR MULTIPLICATION
Let **u, v,** and **w** be vectors in the plane, and let c and d be scalars.

1. $\mathbf{u} + \mathbf{v} = \mathbf{v} + \mathbf{u}$
2. $(\mathbf{u} + \mathbf{v}) + \mathbf{w} = \mathbf{u} + (\mathbf{v} + \mathbf{w})$
3. $\mathbf{u} + \mathbf{0} = \mathbf{u}$
4. $\mathbf{u} + (-\mathbf{u}) = \mathbf{0}$
5. $c(d\mathbf{u}) = (cd)\mathbf{u}$
6. $(c + d)\mathbf{u} = c\mathbf{u} + d\mathbf{u}$
7. $c(\mathbf{u} + \mathbf{v}) = c\mathbf{u} + c\mathbf{v}$
8. $1(\mathbf{u}) = \mathbf{u},\ 0(\mathbf{u}) = \mathbf{0}$

Proof: We prove only the second and sixth properties and leave the others as exercises. Note that the proofs rely on the corresponding properties of real numbers. For example, the proof of the *associative property* of vector addition uses the associative property of addition of real numbers, as follows.

$$
\begin{aligned}
(\mathbf{u} + \mathbf{v}) + \mathbf{w} &= [\langle u_1, u_2\rangle + \langle v_1, v_2\rangle] + \langle w_1, w_2\rangle \\
&= \langle u_1 + v_1, u_2 + v_2\rangle + \langle w_1, w_2\rangle \\
&= \langle (u_1 + v_1) + w_1, (u_2 + v_2) + w_2\rangle \\
&= \langle u_1 + (v_1 + w_1), u_2 + (v_2 + w_2)\rangle \\
&= \langle u_1, u_2\rangle + \langle v_1 + w_1, v_2 + w_2\rangle = \mathbf{u} + (\mathbf{v} + \mathbf{w})
\end{aligned}
$$

Similarly, the proof of the following distributive property depends on the distributive property of real numbers.

$$
\begin{aligned}
(c + d)\mathbf{u} &= (c + d)\langle u_1, u_2\rangle \\
&= \langle (c + d)u_1, (c + d)u_2\rangle \\
&= \langle cu_1 + du_1, cu_2 + du_2\rangle \\
&= \langle cu_1, cu_2\rangle + \langle du_1, du_2\rangle = c\mathbf{u} + d\mathbf{u}
\end{aligned}
$$

THEOREM 13.2

LENGTH OF A SCALAR MULTIPLE
Let **v** be a vector and c be a scalar, then

$$\|c\mathbf{v}\| = |c|\,\|\mathbf{v}\|$$

where $|c|$ is the absolute value of c.

Proof: Since $c\mathbf{v} = \langle cv_1, cv_2\rangle$, it follows that

$$
\begin{aligned}
\|c\mathbf{v}\| = \|\langle cv_1, cv_2\rangle\| &= \sqrt{(cv_1)^2 + (cv_2)^2} = \sqrt{c^2 v_1^2 + c^2 v_2^2} \\
&= \sqrt{c^2(v_1^2 + v_2^2)} = |c|\sqrt{v_1^2 + v_2^2} = |c|\,\|\mathbf{v}\|
\end{aligned}
$$

In many applications of vectors it is useful to find a unit vector that has the same direction as a given vector. The following theorem gives us a procedure for doing this.

THEOREM 13.3   UNIT VECTOR IN THE DIRECTION OF v

If **v** is a nonzero vector in the plane, then the vector

$$\mathbf{u} = \frac{\mathbf{v}}{\|\mathbf{v}\|}$$

has length 1 and the same direction as **v**.

Proof: Since $1/\|\mathbf{v}\|$ is positive and

$$\mathbf{u} = \left(\frac{1}{\|\mathbf{v}\|}\right)\mathbf{v}$$

we conclude that **u** has the same direction as **v**. We leave the proof that **u** has a length of 1 as an exercise.

| **Remark** In Theorem 13.3, we call **u** a **unit vector in the direction of v**. The process of multiplying **v** by $1/\|\mathbf{v}\|$ to get a unit vector is called **normalization of v**.

EXAMPLE 4   Finding a unit vector

Find a unit vector in the direction of $\mathbf{v} = \langle -2, 5 \rangle$ and verify that the result has length 1.

Solution: From Theorem 13.3, the unit vector in the direction of **v** is

$$\frac{\mathbf{v}}{\|\mathbf{v}\|} = \frac{\langle -2, 5 \rangle}{\sqrt{(-2)^2 + (5)^2}} = \frac{1}{\sqrt{29}}\langle -2, 5 \rangle = \left\langle \frac{-2}{\sqrt{29}}, \frac{5}{\sqrt{29}} \right\rangle$$

This vector has length 1, since

$$\sqrt{\left(\frac{-2}{\sqrt{29}}\right)^2 + \left(\frac{5}{\sqrt{29}}\right)^2} = \sqrt{\frac{4}{29} + \frac{25}{29}} = \sqrt{\frac{29}{29}} = 1$$

It is *not* generally true that the length of the sum of two vectors is equal to the sum of their lengths. To see this, consider the vectors **u** and **v** as shown in Figure 13.8. By considering **u** and **v** as two sides of a triangle, we see that the length of the third side is $\|\mathbf{u} + \mathbf{v}\|$, and we have

$$\|\mathbf{u} + \mathbf{v}\| \le \|\mathbf{u}\| + \|\mathbf{v}\|$$

Equality occurs only if the vectors **u** and **v** have the *same* direction. We call this result the **Triangle Inequality** for vectors.

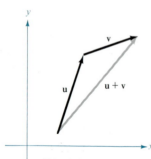

Triangle Inequality

FIGURE 13.8

> **THEOREM 13.4   TRIANGLE INEQUALITY**
> If **u** and **v** are vectors in the plane, then
> $$\|\mathbf{u} + \mathbf{v}\| \leq \|\mathbf{u}\| + \|\mathbf{v}\|$$

Standard unit vectors

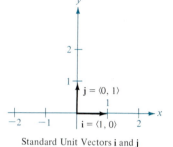

Standard Unit Vectors **i** and **j**

FIGURE 13.9

The unit vectors $\langle 1, 0 \rangle$ and $\langle 0, 1 \rangle$ are called the **standard unit vectors** in the plane and are denoted by

$$\mathbf{i} = \langle 1, 0 \rangle \qquad \text{and} \qquad \mathbf{j} = \langle 0, 1 \rangle$$

as shown in Figure 13.9. These vectors can be used to represent any vector

$$\mathbf{v} = \langle v_1, v_2 \rangle = v_1 \langle 1, 0 \rangle + v_2 \langle 0, 1 \rangle = v_1 \mathbf{i} + v_2 \mathbf{j}$$

We call

$$\mathbf{v} = v_1 \mathbf{i} + v_2 \mathbf{j}$$

a **linear combination** of **i** and **j**. The scalars v_1 and v_2 are called the **horizontal** and **vertical components of v**, respectively.

EXAMPLE 5   Representing a vector as a linear combination of unit vectors

Let **u** be the vector with initial point $(2, -5)$ and terminal point $(-1, 3)$, and let $\mathbf{v} = 2\mathbf{i} - \mathbf{j}$. Write the following vectors as a linear combination of the standard unit vectors **i** and **j**.
(a) **u**   (b) $\mathbf{w} = 2\mathbf{u} - 3\mathbf{v}$

Solution:
(a) For the vector **u**, we have

$$\mathbf{u} = \langle q_1 - p_1, q_2 - p_2 \rangle = \langle -1 - 2, 3 + 5 \rangle = \langle -3, 8 \rangle = -3\mathbf{i} + 8\mathbf{j}$$

(b) For the vector **w**, we have

$$\mathbf{w} = 2\mathbf{u} - 3\mathbf{v} = 2(-3\mathbf{i} + 8\mathbf{j}) - 3(2\mathbf{i} - \mathbf{j})$$
$$= -6\mathbf{i} + 16\mathbf{j} - 6\mathbf{i} + 3\mathbf{j} = -12\mathbf{i} + 19\mathbf{j}$$

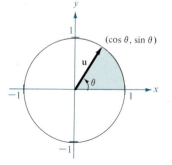

FIGURE 13.10

If **u** is a unit vector such that θ is the angle (measured counterclockwise) from the positive x-axis to **u**, then the terminal point of **u** lies on the unit circle, and we have

$$\mathbf{u} = \langle \cos \theta, \sin \theta \rangle = \cos \theta \, \mathbf{i} + \sin \theta \, \mathbf{j}$$

as shown in Figure 13.10. Moreover, it follows that any other nonzero vector **v** making an angle θ with the positive x-axis has the same direction as **u**, and we can write

$$\mathbf{v} = \|\mathbf{v}\| \langle \cos \theta, \sin \theta \rangle = \|\mathbf{v}\| \cos \theta \, \mathbf{i} + \|\mathbf{v}\| \sin \theta \, \mathbf{j}$$

For instance, the vector **v** of length 3 making an angle of 30° with the positive x-axis is given by

$$\mathbf{v} = 3 \cos \frac{\pi}{6} \mathbf{i} + 3 \sin \frac{\pi}{6} \mathbf{j} = \frac{3\sqrt{3}}{2} \mathbf{i} + \frac{3}{2} \mathbf{j}$$

Applications of vectors

A host of applications dealing with vectors are found in physics and engineering. We conclude this section by considering two examples.

EXAMPLE 6   An application involving force

Two tugboats are pushing an ocean liner, as shown in Figure 13.11. Each boat is exerting a force of 400 pounds. What is the resultant force on the ocean liner?

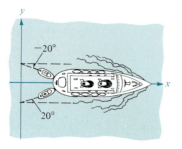

FIGURE 13.11

Solution: From Figure 13.11, we represent the forces exerted by the first and second tugboats as

$$\mathbf{F}_1 = 400\langle \cos 20°, \sin 20° \rangle = 400 \cos (20°) \mathbf{i} + 400 \sin (20°) \mathbf{j}$$
$$\mathbf{F}_2 = 400\langle \cos (-20°), \sin (-20°) \rangle = 400 \cos (20°) \mathbf{i} - 400 \sin (20°) \mathbf{j}$$

To obtain the resultant force on the ocean liner, we add these two forces.

$$\mathbf{F} = \mathbf{F}_1 + \mathbf{F}_2$$
$$= 400 \cos (20°) \mathbf{i} + 400 \sin (20°) \mathbf{j} + 400 \cos (20°) \mathbf{i} - 400 \sin (20°) \mathbf{j}$$
$$= 800 \cos (20°) \mathbf{i} \approx 752\mathbf{i}$$

Thus, the resultant force on the ocean liner is approximately 752 pounds in the direction of the positive x-axis.

EXAMPLE 7   An application involving velocity

An airplane is traveling at a fixed altitude with a negligible wind factor. The plane is headed N 30° W at a speed of 500 miles per hour, as shown in Figure 13.12. As the plane reaches a certain point, it encounters wind with a velocity of 70 miles per hour in the direction E 45° N. What is the resultant speed and direction of the plane?

Solution: Using Figure 13.12, we can represent the velocity of the plane by the vector

$$\mathbf{v}_1 = 500\langle \cos (120°), \sin (120°) \rangle = 500 \cos (120°) \mathbf{i} + 500 \sin (120°) \mathbf{j}$$

The velocity of the wind is represented by the vector

$$\mathbf{v}_2 = 70\langle \cos (45°), \sin (45°) \rangle = 70 \cos (45°) \mathbf{i} + 70 \sin (45°) \mathbf{j}$$

The resultant velocity of the plane is

$$\mathbf{v} = 500 \cos (120°) \mathbf{i} + 500 \sin (120°) \mathbf{j} + 70 \cos (45°) \mathbf{i} + 70 \sin (45°) \mathbf{j}$$
$$= [500 \cos (120°) + 70 \cos (45°)] \mathbf{i} + [500 \sin (120°) + 70 \sin (45°)] \mathbf{j}$$
$$\approx -200.5\mathbf{i} + 482.5\mathbf{j}$$

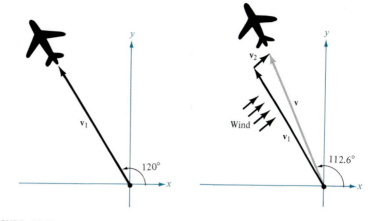

FIGURE 13.12

Now, to find the speed and direction, we write $\mathbf{v} = \|\mathbf{v}\|(\cos\theta\,\mathbf{i} + \sin\theta\,\mathbf{j})$. Since

$$\|\mathbf{v}\| = \sqrt{(-200.5)^2 + (482.5)^2} \approx 522.5$$

we have

$$\mathbf{v} \approx 522.5\left[\frac{-200.5}{522.5}\mathbf{i} + \frac{482.5}{522.5}\mathbf{j}\right] \approx 522.5[-0.3837\mathbf{i} + 0.9234\mathbf{j}]$$

$$\approx 522.5[\cos(112.6°)\,\mathbf{i} + \sin(112.6°)\,\mathbf{j}]$$

Therefore, the speed of the plane, as altered by the wind, is approximately 522.5 miles per hour in a flight path that makes an angle of 112.6° with the positive x-axis.

Section Exercises 13.1

In Exercises 1–4, (a) find the component form of the vector **v** and (b) sketch the vector with its initial point at the origin.

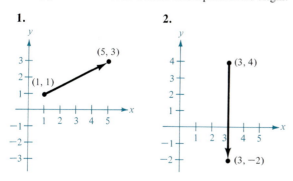

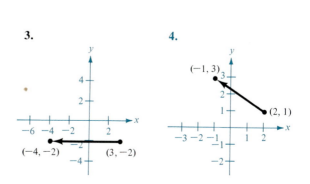

In Exercises 5–12, the initial and terminal points of a vector **v** are given. (a) Sketch the given directed line segment, (b) write the vector in component form, and (c) sketch the vector with its initial point at the origin.

5. Initial point $(1, 2)$, terminal point $(5, 5)$
6. Initial point $(3, -5)$, terminal point $(4, 7)$
7. Initial point $(10, 2)$, terminal point $(6, -1)$
8. Initial point $(0, -4)$ terminal point $(-5, -1)$
9. Initial point $(6, 2)$, terminal point $(6, 6)$
10. Initial point $(7, -1)$, terminal point $(-3, -1)$
11. Initial point $\left(\frac{3}{2}, \frac{4}{3}\right)$, terminal point $\left(\frac{1}{2}, 3\right)$
12. Initial point $(0.12, 0.60)$, terminal point $(0.84, 1.25)$

In Exercises 13 and 14, sketch the scalar multiple of **v**.

13. $\mathbf{v} = \langle 2, 3 \rangle$
  (a) $2\mathbf{v}$ (b) $-3\mathbf{v}$
  (c) $\frac{7}{2}\mathbf{v}$ (d) $\frac{2}{3}\mathbf{v}$
14. $\mathbf{v} = \langle -1, 5 \rangle$
  (a) $4\mathbf{v}$ (b) $-\frac{1}{2}\mathbf{v}$
  (c) $0\mathbf{v}$ (d) $-6\mathbf{v}$

In Exercises 15–20, find the vector **v** and illustrate the indicated vector operations geometrically, where $\mathbf{u} = \langle 2, -1 \rangle$ and $\mathbf{w} = \langle 1, 2 \rangle$.

15. $\mathbf{v} = \frac{3}{2}\mathbf{u}$ **16.** $\mathbf{v} = \mathbf{u} + \mathbf{w}$
17. $\mathbf{v} = \mathbf{u} + 2\mathbf{w}$ **18.** $\mathbf{v} = -\mathbf{u} + \mathbf{w}$
19. $\mathbf{v} = \frac{1}{2}(3\mathbf{u} + \mathbf{w})$ **20.** $\mathbf{v} = \mathbf{u} - 2\mathbf{w}$

In Exercises 21–26, find a and b such that $\mathbf{v} = a\mathbf{u} + b\mathbf{w}$, where $\mathbf{u} = \langle 1, 2 \rangle$ and $\mathbf{w} = \langle 1, -1 \rangle$.

21. $\mathbf{v} = \langle 2, 1 \rangle$ **22.** $\mathbf{v} = \langle 0, 3 \rangle$
23. $\mathbf{v} = \langle 3, 0 \rangle$ **24.** $\mathbf{v} = \langle 3, 3 \rangle$
25. $\mathbf{v} = \langle 1, 1 \rangle$ **26.** $\mathbf{v} = \langle -1, 7 \rangle$

In Exercises 27 and 28, the vector **v** and its initial point are given. Find the terminal point.

27. $\mathbf{v} = \langle -1, 3 \rangle$, initial point $(4, 2)$
28. $\mathbf{v} = \langle 4, -9 \rangle$, initial point $(3, 2)$

In Exercises 29–34, find the magnitude of **v**.

29. $\mathbf{v} = \langle 4, 3 \rangle$ **30.** $\mathbf{v} = \langle 12, -5 \rangle$
31. $\mathbf{v} = 6\mathbf{i} - 5\mathbf{j}$ **32.** $\mathbf{v} = -10\mathbf{i} + 3\mathbf{j}$
33. $\mathbf{v} = 4\mathbf{j}$ **34.** $\mathbf{v} = \mathbf{i} - \mathbf{j}$

In Exercises 35 and 36, find

(a) $\|\mathbf{u}\|$ (b) $\|\mathbf{v}\|$ (c) $\|\mathbf{u} + \mathbf{v}\|$
(d) $\left\|\dfrac{\mathbf{u}}{\|\mathbf{u}\|}\right\|$ (e) $\left\|\dfrac{\mathbf{v}}{\|\mathbf{v}\|}\right\|$ (f) $\left\|\dfrac{\mathbf{u} + \mathbf{v}}{\|\mathbf{u} + \mathbf{v}\|}\right\|$

35. $\mathbf{u} = \left\langle 1, \frac{1}{2} \right\rangle$, $\mathbf{v} = \langle 2, 3 \rangle$
36. $\mathbf{u} = \langle 2, -4 \rangle$, $\mathbf{v} = \langle 5, 5 \rangle$

In Exercises 37 and 38, demonstrate the triangle inequality using the vectors **u** and **v**.

37. $\mathbf{u} = \langle 2, 1 \rangle$, $\mathbf{v} = \langle 5, 4 \rangle$
38. $\mathbf{u} = \langle -3, 2 \rangle$, $\mathbf{v} = \langle 1, -2 \rangle$

In Exercises 39–42, find the vector **v** with the given magnitude and the same direction as **u**.

	Magnitude	Direction
39.	$\|\mathbf{v}\| = 4$	$\mathbf{u} = \langle 1, 1 \rangle$
40.	$\|\mathbf{v}\| = 4$	$\mathbf{u} = \langle -1, 1 \rangle$
41.	$\|\mathbf{v}\| = 2$	$\mathbf{u} = \langle \sqrt{3}, 3 \rangle$
42.	$\|\mathbf{v}\| = 3$	$\mathbf{u} = \langle 0, 3 \rangle$

In Exercises 43–46, find a unit vector (a) parallel to and (b) normal to the graph of $f(x)$ at the indicated point.

	Graph	Point
43.	$f(x) = x^3$	$(1, 1)$
44.	$f(x) = x^3$	$(-2, -8)$
45.	$f(x) = \sqrt{25 - x^2}$	$(3, 4)$
46.	$f(x) = \tan x$	$\left(\frac{\pi}{4}, 1\right)$

In Exercises 47–50, find the component form of **v** given its magnitude and the angle it makes with the positive x-axis.

	Magnitude	Angle
47.	$\|\mathbf{v}\| = 3$	$\theta = 0°$
48.	$\|\mathbf{v}\| = 1$	$\theta = 45°$
49.	$\|\mathbf{v}\| = 2$	$\theta = 150°$
50.	$\|\mathbf{v}\| = 1$	$\theta = 3.5°$

In Exercises 51–54, find the component form of $\mathbf{u} + \mathbf{v}$ given the magnitudes of **u** and **v** and the angles **u** and **v** make with the positive x-axis.

51. $\|\mathbf{u}\| = 1$, $\theta = 0°$; $\|\mathbf{v}\| = 3$, $\theta = 45°$
52. $\|\mathbf{u}\| = 4$, $\theta = 0°$; $\|\mathbf{v}\| = 2$, $\theta = 60°$
53. $\|\mathbf{u}\| = 2$, $\theta = 4$; $\|\mathbf{v}\| = 1$, $\theta = 2$
54. $\|\mathbf{u}\| = 5$, $\theta = -0.5$; $\|\mathbf{v}\| = 5$, $\theta = 0.5$

In Exercises 55 and 56, find the component form of **v** given the magnitudes of **u** and **u** + **v** and the angles **u** and **u** + **v** make with the positive *x*-axis.

55. $\|\mathbf{u}\| = 1$, $\theta = 45°$; $\|\mathbf{u} + \mathbf{v}\| = \sqrt{2}$, $\theta = 90°$
56. $\|\mathbf{u}\| = 4$, $\theta = 30°$; $\|\mathbf{u} + \mathbf{v}\| = 6$, $\theta = 120°$

57. A force of 150 pounds in a direction 30° above the horizontal is applied to a bolt. Find the horizontal and vertical components of the force (see Figure 13.13).

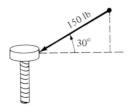

FIGURE 13.13

58. To carry a 100-pound cylindrical weight, two men lift on the ends of short ropes that are tied to an eyelet on the top center of the cylinder. If one rope makes a 20° angle away from the vertical and the other a 30° angle, find the following:
(a) the tension in each rope if the resultant force is vertical
(b) the vertical component of each man's force (see Figure 13.14)

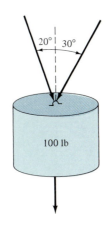

FIGURE 13.14

59. An airplane's velocity with respect to the air is 580 miles per hour, and it is headed 32° north of west. The wind at the altitude of the plane is from the southwest and has a velocity of 60 miles per hour. What is the true direction of the plane, and what is its speed with respect to the ground? (See Figure 13.15.)

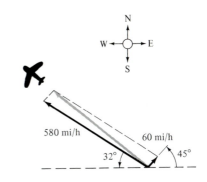

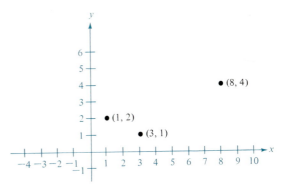

FIGURE 13.15

60. A ball is thrown into the air with an initial velocity of 80 feet per second and at an angle of 50° with the horizontal. Find the vertical and horizontal components of the velocity.

61. Three vertices of a parallelogram are (1, 2), (3, 1), and (8, 4). Find the three possible fourth vertices. (See Figure 13.16.)

FIGURE 13.16

62. Using vectors, prove that the diagonals of a parallelogram bisect each other.

63. Use vectors to find the points of trisection of the line segment with endpoints (1, 2) and (7, 5).

64. Prove that the vectors

$$\mathbf{u} = (\cos \theta)\,\mathbf{i} - (\sin \theta)\,\mathbf{j}$$
and
$$\mathbf{v} = (\sin \theta)\,\mathbf{i} + (\cos \theta)\,\mathbf{j}$$

are unit vectors for any angle θ.

65. Using vectors, prove that the line segment joining the midpoints of two sides of a triangle is parallel to and one-half the length of the third side.

13.2
Space coordinates and vectors in space

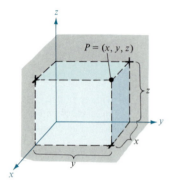

FIGURE 13.17

Physical forces and velocities are not confined to the plane, and hence it is natural that we extend our discussion from vectors in the plane to vectors in space. We do this with the aid of a **three-dimensional coordinate system.** We construct this system by passing a z-axis perpendicular to both the x- and y-axes at the origin. Figure 13.17 shows the positive portion of each coordinate axis. Taken as pairs, the axes determine three **coordinate planes** called the **xy-plane,** the **xz-plane,** and the **yz-plane.** These three coordinate planes separate three-space into eight **octants.** The first octant is the one for which all three coordinates are positive. In this three-dimensional system, a point P in space is determined by an ordered triple (x, y, z), where

$$x = \text{directed distance from } yz\text{-plane to } P$$
$$y = \text{directed distance from } xz\text{-plane to } P$$
$$z = \text{directed distance from } xy\text{-plane to } P$$

Several points are shown in Figure 13.18.

A three-dimensional coordinate system can have either a **left-handed** or a **right-handed** orientation. To determine the orientation of a system, imagine that you are standing at the origin, with your arms pointing in the direction of the positive x- and y-axes, as shown in Figure 13.19. The system is right-handed or left-handed depending on which hand points along the x-axis. In this text we work exclusively with the right-handed system.

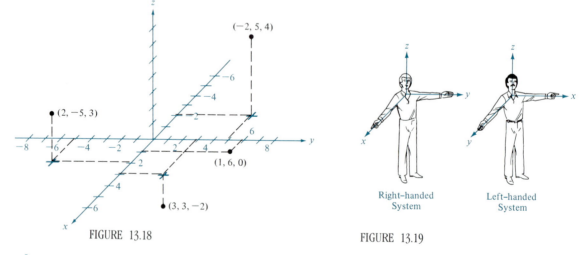

FIGURE 13.18

FIGURE 13.19

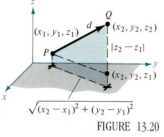

FIGURE 13.20

Many of the formulas established for the two-dimensional coordinate system can be extended to three dimensions. For example, to find the distance between two points in space, we use the Pythagorean Theorem twice, as shown in Figure 13.20, to obtain the formula for the distance between the points (x_1, y_1, z_1) and (x_2, y_2, z_2).

$$d = \sqrt{(x_2 - x_1)^2 + (y_2 - y_1)^2 + (z_2 - z_1)^2} \qquad \text{Distance Formula}$$

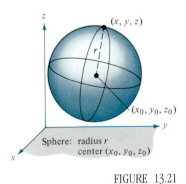

Sphere: radius r
center (x_0, y_0, z_0)

FIGURE 13.21

EXAMPLE 1   *Finding the distance between two points in space*

Find the distance between $(2, -1, 3)$ and $(1, 0, -2)$.

Solution: By the Distance Formula, we have

$$d = \sqrt{(1 - 2)^2 + (0 + 1)^2 + (-2 - 3)^2} = \sqrt{1 + 1 + 25} = \sqrt{27} = 3\sqrt{3}$$

We can use the Distance Formula to find the **standard equation of a sphere** of radius r, centered at (x_0, y_0, z_0). If (x, y, z) is an arbitrary point on the sphere, then we have

$$(x - x_0)^2 + (y - y_0)^2 + (z - z_0)^2 = r^2 \qquad \text{Equation of sphere}$$

as shown in Figure 13.21. Moreover, the midpoint of the line segment joining the points (x_1, y_1, z_1) and (x_2, y_2, z_2) has coordinates

$$\left(\frac{x_1 + x_2}{2}, \frac{y_1 + y_2}{2}, \frac{z_1 + z_2}{2} \right) \qquad \text{Midpoint Rule}$$

The following example makes use of both of these formulas.

EXAMPLE 2   *Finding the equation of a sphere*

Find the standard equation for the sphere that has the points $(5, -2, 3)$ and $(0, 4, -3)$ as endpoints of a diameter.

Solution: By the Midpoint Rule, the center of the sphere is

$$\left(\frac{5 + 0}{2}, \frac{-2 + 4}{2}, \frac{3 - 3}{2} \right) = \left(\frac{5}{2}, 1, 0 \right)$$

By the Distance Formula, the radius is

$$r = \sqrt{\left(0 - \frac{5}{2} \right)^2 + (4 - 1)^2 + (-3 - 0)^2} = \sqrt{\frac{97}{4}} = \frac{\sqrt{97}}{2}$$

Therefore, the standard equation of the sphere is

$$\left(x - \frac{5}{2} \right)^2 + (y - 1)^2 + (z - 0)^2 = \frac{97}{4}$$

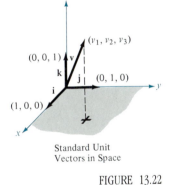

Standard Unit
Vectors in Space

FIGURE 13.22

Vectors in space

In space, we denote vectors by ordered triples

$$\mathbf{v} = \langle v_1, v_2, v_3 \rangle$$

with the **zero vector** denoted by $\mathbf{0} = \langle 0, 0, 0 \rangle$. Using the unit vectors $\mathbf{i}$, $\mathbf{j}$, and a unit vector $\mathbf{k}$ in the direction of the positive z-axis, the **standard unit vector notation** for $\mathbf{v}$ is

$$\mathbf{v} = v_1 \mathbf{i} + v_2 \mathbf{j} + v_3 \mathbf{k}$$

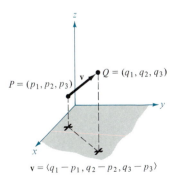

$\mathbf{v} = \langle q_1 - p_1, q_2 - p_2, q_3 - p_3 \rangle$

FIGURE 13.23

as shown in Figure 13.22. If $\mathbf{v}$ is represented by the directed line segment from $P = (p_1, p_2, p_3)$ to $Q = (q_1, q_2, q_3)$ as shown in Figure 13.23, then the component form of **v** is given by subtracting the coordinates of the initial point from the coordinates of the terminal point as follows

$$\mathbf{v} = \langle v_1, v_2, v_3 \rangle = \langle q_1 - p_1, q_2 - p_2, q_3 - p_3 \rangle$$

The following box summarizes the basic definitions and operations of vectors in space.

VECTORS IN SPACE

Let $\mathbf{u} = \langle u_1, u_2, u_3 \rangle$ and $\mathbf{v} = \langle v_1, v_2, v_3 \rangle$ be vectors in space and let c be a scalar.

1. *Equality of Vectors:* $\mathbf{u} = \mathbf{v}$ if and only if $u_1 = v_1$, $u_2 = v_2$, and $u_3 = v_3$
2. *Component Form:* If **v** is represented by the directed line segment from $P = (p_1, p_2, p_3)$ to $Q = (q_1, q_2, q_3)$, then

$$\mathbf{v} = \langle v_1, v_2, v_3 \rangle = \langle q_1 - p_1, q_2 - p_2, q_3 - p_3 \rangle$$

3. *Length:* $\|\mathbf{v}\| = \sqrt{v_1^2 + v_2^2 + v_3^2}$
4. *Unit Vector in the Direction of* **v**: $\dfrac{\mathbf{v}}{\|\mathbf{v}\|} = \left(\dfrac{1}{\|\mathbf{v}\|}\right)\langle v_1, v_2, v_3 \rangle$
5. *Vector Addition:* $\mathbf{v} + \mathbf{u} = \langle v_1 + u_1, v_2 + u_2, v_3 + u_3 \rangle$
6. *Scalar Multiplication:* $c\mathbf{v} = \langle cv_1, cv_2, cv_3 \rangle$

Remark The properties of vector addition and scalar multiplication given in Theorem 13.1 are also valid for vectors in space. We leave the proof of this as an exercise.

EXAMPLE 3   Finding the component form of a vector in space

Find the component form and length of the vector **v** having initial point $(-2, 3, 1)$ and terminal point $(0, -4, 4)$. Then find a unit vector in the direction of **v**.

Solution: The component form of **v** is

$$\begin{aligned}\mathbf{v} &= \langle q_1 - p_1, q_2 - p_2, q_3 - p_3 \rangle \\ &= \langle 0 - (-2), -4 - 3, 4 - 1 \rangle \\ &= \langle 2, -7, 3 \rangle\end{aligned}$$

which implies that its length is

$$\|\mathbf{v}\| = \sqrt{(2)^2 + (-7)^2 + (3)^2} = \sqrt{62}$$

Now, using this length, we find that the unit vector in the direction of **v** is

$$\mathbf{u} = \frac{\mathbf{v}}{\|\mathbf{v}\|} = \frac{1}{\sqrt{62}}\langle 2, -7, 3 \rangle$$

Recall from the definition of scalar multiplication that positive scalar multiples of a vector **v** all have the same direction as **v** whereas negative multiples have the direction opposite that of **v**. In general, we say that vector **u** is **parallel** to vector **v** if there is some scalar $c \neq 0$ such that $\mathbf{u} = c\mathbf{v}$.

DEFINITION OF PARALLEL VECTORS

The vector **u** is **parallel** to vector **v** if there is some scalar $c \neq 0$ such that

$$\mathbf{u} = c\mathbf{v}$$

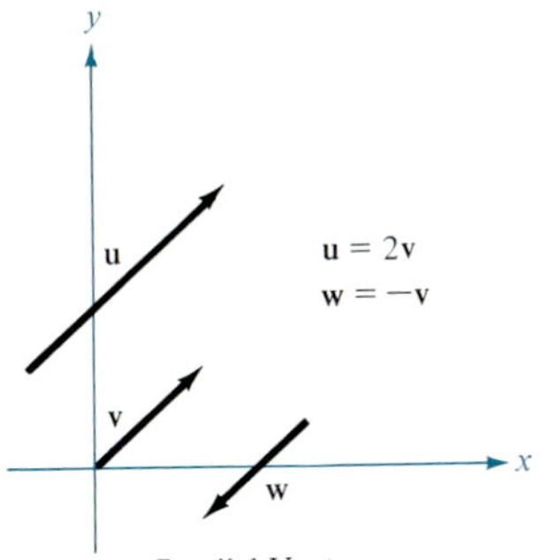

Parallel Vectors

FIGURE 13.24

For example, in Figure 13.24 the vectors **u**, **v**, and **w** are parallel, since **u** = 2**v** and **w** = −**v**.

EXAMPLE 4   *Parallel vectors*

Vector **w** has initial point $(2, -1, 3)$ and terminal point $(-4, 7, 5)$. Which of the following vectors is parallel to **w**?
(a) $\mathbf{u} = \langle 3, -4, -1 \rangle$ (b) $\mathbf{v} = \langle 12, -16, 4 \rangle$

Solution: First, we determine **w** to have the component form

$$\mathbf{w} = \langle -4 - 2, 7 - (-1), 5 - 3 \rangle = \langle -6, 8, 2 \rangle$$

(a) Since $\mathbf{u} = \langle 3, -4, -1 \rangle = -\frac{1}{2}\langle -6, 8, 2 \rangle = -\frac{1}{2}\mathbf{w}$, we conclude that **u** *is* parallel to **w.**
(b) In this case, we want to find a scalar c such that

$$\langle 12, -16, 4 \rangle = c\langle -6, 8, 2 \rangle$$

However, when we equate corresponding components, we get $c = -2$ for the first two components and $c = 2$ for the third. Hence, the equation has no solution, and we conclude that the vectors are *not* parallel.   ☐

EXAMPLE 5   *Using vectors to determine collinear points*

Determine if the points $P = (1, -2, 3)$, $Q = (2, 1, 0)$, and $R = (4, 7, -6)$ lie on the same line.

Solution: We can solve this problem by finding the component forms of the vectors $\overrightarrow{PQ}$ and $\overrightarrow{PR}$. Since these two vectors have a common initial point, it follows that P, Q, and R lie on the same line if and only if $\overrightarrow{PQ}$ and $\overrightarrow{PR}$ are parallel.

$$\overrightarrow{PQ} = \langle 2 - 1, 1 - (-2), 0 - 3 \rangle = \langle 1, 3, -3 \rangle$$
$$\overrightarrow{PR} = \langle 4 - 1, 7 - (-2), -6 - 3 \rangle = \langle 3, 9, -9 \rangle$$

Since $\overrightarrow{PR} = 3\overrightarrow{PQ}$, we conclude that these two vectors are parallel, and therefore the three given points are collinear.   ☐

EXAMPLE 6   *Standard unit vector notation*

(a) Write the vector $\mathbf{v} = 4\mathbf{i} - 5\mathbf{k}$ in component form.
(b) Find the terminal point of the vector $\mathbf{v} = 7\mathbf{i} - \mathbf{j} + 3\mathbf{k}$, given that the initial point is $P = (-2, 3, 5)$.

Solution:

(a) Since **j** is missing, its component is zero and $\mathbf{v} = 4\mathbf{i} - 5\mathbf{k} = \langle 4, 0, -5 \rangle$.

(b) We seek $Q = (x, y, z)$ such that $\mathbf{v} = \overrightarrow{PQ} = 7\mathbf{i} - \mathbf{j} + 3\mathbf{k}$. This implies that

$$x - (-2) = 7, \qquad y - 3 = -1, \qquad z - 5 = 3$$

with solutions $x = 5$, $y = 2$, and $z = 8$. Therefore, $Q = (5, 2, 8)$. ▭

EXAMPLE 7 *An application involving force*

An object weighing 120 pounds is supported by a tripod, as shown in Figure 13.25. Represent the force exerted on each leg of the tripod as a vector. (Assume that the weight is distributed equally on each of the three legs.)

Solution: We let the vectors $\mathbf{F}_1$, $\mathbf{F}_2$, and $\mathbf{F}_3$ represent the force exerted on the three legs. From Figure 13.25, we can determine the direction of $\mathbf{F}_1$, $\mathbf{F}_2$, and $\mathbf{F}_3$ to be given by

$$\overrightarrow{PQ_1} = \langle 0 - 0, -1 - 0, 0 - 4 \rangle = \langle 0, -1, -4 \rangle$$

$$\overrightarrow{PQ_2} = \left\langle \frac{\sqrt{3}}{2} - 0, \frac{1}{2} - 0, 0 - 4 \right\rangle = \left\langle \frac{\sqrt{3}}{2}, \frac{1}{2}, -4 \right\rangle$$

$$\overrightarrow{PQ_3} = \left\langle -\frac{\sqrt{3}}{2} - 0, \frac{1}{2} - 0, 0 - 4 \right\rangle = \left\langle -\frac{\sqrt{3}}{2}, \frac{1}{2}, -4 \right\rangle$$

Since the total force is equally distributed on the three legs, we know that

$$\mathbf{F}_1 = c\langle 0, -1, -4 \rangle, \mathbf{F}_2 = c\left\langle \frac{\sqrt{3}}{2}, \frac{1}{2}, -4 \right\rangle, \text{ and } \mathbf{F}_3 = c\left\langle -\frac{\sqrt{3}}{2}, \frac{1}{2}, -4 \right\rangle$$

Now, we let the total force exerted by the object be given by $\mathbf{F} = -120\mathbf{k}$, and using the fact that

$$\mathbf{F} = \mathbf{F}_1 + \mathbf{F}_2 + \mathbf{F}_3$$

we can conclude that $\mathbf{F}_1$, $\mathbf{F}_2$, and $\mathbf{F}_3$ each have a vertical component of -40. This implies that $c(-4) = -40$ and $c = 10$. Therefore, we have

$$\mathbf{F}_1 = \langle 0, -10, -40 \rangle, \qquad \mathbf{F}_2 = \langle 5\sqrt{3}, 5, -40 \rangle, \qquad \mathbf{F}_3 = \langle -5\sqrt{3}, 5, -40 \rangle$$
▭

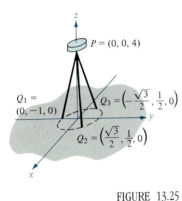

FIGURE 13.25

Section Exercises 13.2

In Exercises 1–4, plot the points on the same three-dimensional coordinate system.

1. (a) $(2, 1, 3)$ (b) $(-1, 2, 1)$

2. (a) $(3, -2, 5)$ (b) $\left(\frac{3}{2}, 4, -2 \right)$

3. (a) $(5, -2, 2)$ (b) $(5, -2, -2)$

4. (a) $(0, 4, -5)$ (b) $(4, 0, 5)$

In Exercises 5–8, find the lengths of the sides of the triangle with the indicated vertices, and determine whether the triangle is a right triangle, an isosceles triangle, or neither of these.

5. $(0, 0, 0)$, $(2, 2, 1)$, $(2, -4, 4)$
6. $(5, 3, 4)$, $(7, 1, 3)$, $(3, 5, 3)$
7. $(1, -3, -2)$, $(5, -1, 2)$, $(-1, 1, 2)$

8. $(5, 0, 0)$, $(0, 2, 0)$, $(0, 0, -3)$

In Exercises 9 and 10, find the coordinates of the midpoint of the line segment joining the given points.

9. $(5, -9, 7)$ and $(-2, 3, 3)$
10. $(4, 0, -6)$ and $(8, 8, 20)$

In Exercises 11–14, find the general form of the equation of the sphere.

11. Center $(0, 2, 5)$, radius 2
12. Center $(4, -1, 1)$, radius 5
13. Endpoints of a diameter are $(2, 0, 0)$ and $(0, 6, 0)$
14. Center $(-2, 1, 1)$, tangent to the xy-coordinate plane

In Exercises 15–18, find the center and radius of the sphere.

15. $x^2 + y^2 + z^2 - 2x + 6y + 8z + 1 = 0$
16. $x^2 + y^2 + z^2 + 9x - 2y + 10z + 19 = 0$
17. $9x^2 + 9y^2 + 9z^2 - 6x + 18y + 1 = 0$
18. $4x^2 + 4y^2 + 4z^2 - 4x - 32y + 8z + 33 = 0$

In Exercises 19–22, (a) find the component form of the vector **v** and (b) sketch the vector with its initial point at the origin.

19.   **20.**

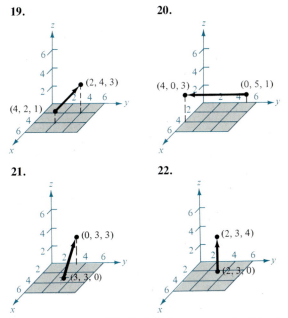

21.   **22.**

In Exercises 23 and 24, the initial and terminal points of a vector **v** are given. (a) Sketch the directed line segment, (b) find the component form of the vector, and (c) sketch the vector with its initial point at the origin.

23. Initial point $(-1, 2, 3)$, terminal point $(3, 3, 4)$
24. Initial point $(2, -1, -2)$, terminal point $(-4, 3, 7)$

In Exercises 25 and 26, sketch each scalar multiple of **v**.

25. $\mathbf{v} = \langle 1, 2, 2 \rangle$
   (a) $2\mathbf{v}$ (b) $-\mathbf{v}$
   (c) $\dfrac{3}{2}\mathbf{v}$ (d) $0\mathbf{v}$

26. $\mathbf{v} = \langle 2, -2, 1 \rangle$
   (a) $-\mathbf{v}$ (b) $2\mathbf{v}$
   (c) $\dfrac{1}{2}\mathbf{v}$ (d) $\dfrac{5}{2}\mathbf{v}$

In Exercises 27–32, find the indicated vector, given $\mathbf{u} = \langle 1, 2, 3 \rangle$, $\mathbf{v} = \langle 2, 2, -1 \rangle$, and $\mathbf{w} = \langle 4, 0, -4 \rangle$.

27. $\mathbf{u} - \mathbf{v}$ **28.** $\mathbf{u} - \mathbf{v} + 2\mathbf{w}$
29. $2\mathbf{u} + 4\mathbf{v} - \mathbf{w}$ **30.** $5\mathbf{u} - 3\mathbf{v} - \dfrac{1}{2}\mathbf{w}$
31. $\mathbf{z}$, where $2\mathbf{z} - 3\mathbf{u} = \mathbf{w}$
32. $\mathbf{z}$, where $2\mathbf{u} + \mathbf{v} - \mathbf{w} + 3\mathbf{z} = \mathbf{0}$

In Exercises 33–36, determine which of the vectors are parallel to **z**.

33. $\mathbf{z} = \langle 3, 2, -5 \rangle$:
   (a) $\langle -6, -4, 10 \rangle$ (b) $\left\langle 2, \dfrac{4}{3}, -\dfrac{10}{3} \right\rangle$
   (c) $\langle 6, 4, 10 \rangle$ (d) $\langle 1, -4, 2 \rangle$

34. $\mathbf{z} = \dfrac{1}{2}\mathbf{i} - \dfrac{2}{3}\mathbf{j} + \dfrac{3}{4}\mathbf{k}$:
   (a) $6\mathbf{i} - 4\mathbf{j} + 9\mathbf{k}$ (b) $-\mathbf{i} + \dfrac{4}{3}\mathbf{j} - \dfrac{3}{2}\mathbf{k}$
   (c) $12\mathbf{i} + 9\mathbf{k}$ (d) $\dfrac{3}{4}\mathbf{i} - \mathbf{j} + \dfrac{9}{8}\mathbf{k}$

35. $\mathbf{z}$ has initial point $(1, -1, 3)$ and terminal point $(-2, 3, 5)$:
   (a) $-6\mathbf{i} + 8\mathbf{j} + 4\mathbf{k}$ (b) $4\mathbf{j} + 2\mathbf{k}$
36. $\mathbf{z}$ has initial point $(3, 2, -1)$ and terminal point $(-1, -3, 5)$:
   (a) $\langle 0, 5, -6 \rangle$ (b) $\langle 8, 10, -12 \rangle$

In Exercises 37–40, use vectors to determine whether or not the given points lie on a straight line.

37. $(0, -2, -5)$, $(3, 4, 4)$, $(2, 2, 1)$
38. $(1, -1, 5)$, $(0, -1, 6)$, $(3, -1, 3)$
39. $(1, 2, 4)$, $(2, 5, 0)$, $(0, 1, 5)$
40. $(0, 0, 0)$, $(1, 3, -2)$, $(2, -6, 4)$

In Exercises 41 and 42, use vectors to show that the given points form the vertices of a parallelogram.

41. $(2, 9, 1)$, $(3, 11, 4)$, $(0, 10, 2)$, $(1, 12, 5)$
42. $(1, 1, -3)$, $(9, -1, 2)$, $(11, 2, 1)$, $(3, 4, -4)$

In Exercises 43 and 44 on page 702, the vector **v** and its initial point are given. Find the terminal point.

43. $\mathbf{v} = \langle 3, -5, 6 \rangle$, initial point $(0, 6, 2)$

44. $\mathbf{v} = \left\langle 0, \dfrac{1}{2}, -\dfrac{1}{3} \right\rangle$, initial point $\left(3, 0, -\dfrac{2}{3}\right)$

In Exercises 45–50, find the magnitude of **v**.

45. $\mathbf{v} = \langle 0, 0, 0 \rangle$ **46.** $\mathbf{v} = \langle 1, 0, 3 \rangle$

47. **v** has $(1, -3, 4)$ and $(1, 0, -1)$ as its initial and terminal points, respectively.

48. **v** has $(0, -1, 0)$ and $(1, 2, -2)$ as its initial and terminal points, respectively.

49. $\mathbf{v} = \mathbf{i} - 2\mathbf{j} - 3\mathbf{k}$ **50.** $\mathbf{v} = -4\mathbf{i} + 3\mathbf{j} + 7\mathbf{k}$

In Exercises 51–54, find a unit vector (a) in the direction of **u** and (b) in the direction opposite that of **u**.

51. $\mathbf{u} = \langle 2, -1, 2 \rangle$ **52.** $\mathbf{u} = \langle 6, 0, 8 \rangle$

53. $\mathbf{u} = \langle 3, 2, -5 \rangle$ **54.** $\mathbf{u} = \langle 8, 0, 0 \rangle$

In Exercises 55–58, determine the values of c that satisfy the given equation. Let $\mathbf{u} = \mathbf{i} + 2\mathbf{j} + 3\mathbf{k}$ and $\mathbf{v} = 2\mathbf{i} + 2\mathbf{j} - \mathbf{k}$.

55. $\|c\mathbf{u}\| = 1$ **56.** $\|c\mathbf{v}\| = 1$

57. $\|c\mathbf{v}\| = 5$ **58.** $\|c\mathbf{u}\| = 3$

In Exercises 59–62, use vectors to find the point that lies two-thirds of the way from P to Q.

59. $P = (4, 3, 0)$, $Q = (1, -3, 3)$

60. $P = (-2, 1, 6)$, $Q = (6, 1, 4)$

61. $P = (1, 2, 5)$, $Q = (6, 8, 2)$

62. $P = (-9, -8, 5)$, $Q = (12, 3, -1)$

In Exercises 63 and 64, write the component form of **v** and sketch it.

63. **v** lies in the yz-plane, has magnitude 2, and makes an angle of $30°$ with the positive y-axis.

64. **v** lies in the xz-plane, has magnitude 5, and makes an angle of $45°$ with the positive z-axis.

In Exercises 65 and 66, write the component form of **v**.

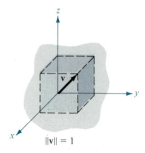

FIGURE 13.26

65. **v** is a unit vector representing the diagonal of a cube, as shown in Figure 13.26.

66. **v** is a vector of magnitude 5 forming the diagonal of a cube, as shown in Figure 13.26.

67. Let $\mathbf{u} = \mathbf{i} + \mathbf{j}$, $\mathbf{v} = \mathbf{j} + \mathbf{k}$, and $\mathbf{w} = a\mathbf{u} + b\mathbf{v}$.
  (a) Sketch **u** and **v**.
  (b) If $\mathbf{w} = \mathbf{0}$, show that a and b must both be zero.
  (c) Find a and b such that $\mathbf{w} = \mathbf{i} + 2\mathbf{j} + \mathbf{k}$.
  (d) Show that no choice of a and b yields $\mathbf{w} = \mathbf{i} + 2\mathbf{j} + 3\mathbf{k}$.

68. The initial and terminal points of the vector **v** are (x_1, y_1, z_1) and (x, y, z), respectively. Describe the set of all points (x, y, z) such that $\|\mathbf{v}\| = 4$.

69. The guy wire to a 100-foot tower has a tension of 550 pounds. Using the distances shown in Figure 13.27, write the component form of the vector **F** representing the tension in the wire.

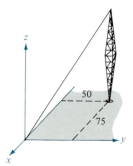

FIGURE 13.27

70. The lights in an auditorium are in the form of 25-pound disks of radius 18 inches. Each disk is supported by three equally spaced wires to the ceiling. (See Figure 13.28.) Find the tension in each wire if they are 48 inches in length.

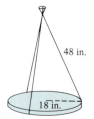

FIGURE 13.28

13.3
The dot product of two vectors

So far we have studied two operations with vectors—vector addition and multiplication by a scalar—each of which yields another vector. In this section we introduce a third vector operation, called the **dot product.** This product yields a scalar, rather than a vector.

DEFINITION OF DOT PRODUCT

Vectors in the Plane: The **dot product** of $\mathbf{u} = \langle u_1, u_2 \rangle$ and $\mathbf{v} = \langle v_1, v_2 \rangle$ is

$$\mathbf{u} \cdot \mathbf{v} = u_1 v_1 + u_2 v_2$$

Vectors in Space: The **dot product** of $\mathbf{u} = \langle u_1, u_2, u_3 \rangle$ and $\mathbf{v} = \langle v_1, v_2, v_3 \rangle$ is

$$\mathbf{u} \cdot \mathbf{v} = u_1 v_1 + u_2 v_2 + u_3 v_3$$

| Remark   Since the dot product of two vectors is a scalar, it is sometimes called the **scalar product.**

The properties listed in the next theorem follow readily from the definition of the dot product.

THEOREM 13.5

PROPERTIES OF THE DOT PRODUCT
If $\mathbf{u}$, $\mathbf{v}$, and $\mathbf{w}$ are vectors in the plane or in space and c is a scalar, then the following properties are true.

1. $\mathbf{u} \cdot \mathbf{v} = \mathbf{v} \cdot \mathbf{u}$
2. $\mathbf{u} \cdot (\mathbf{v} + \mathbf{w}) = \mathbf{u} \cdot \mathbf{v} + \mathbf{u} \cdot \mathbf{w}$
3. $c(\mathbf{u} \cdot \mathbf{v}) = (c\mathbf{u}) \cdot \mathbf{v} = \mathbf{u} \cdot (c\mathbf{v})$
4. $\mathbf{0} \cdot \mathbf{v} = 0$
5. $\mathbf{v} \cdot \mathbf{v} = \|\mathbf{v}\|^2$

Proof: We prove the first and fifth properties for vectors in space and leave the remaining proofs as exercises. To prove the first property, let $\mathbf{u} = \langle u_1, u_2, u_3 \rangle$ and $\mathbf{v} = \langle v_1, v_2, v_3 \rangle$. Then

$$\mathbf{u} \cdot \mathbf{v} = u_1 v_1 + u_2 v_2 + u_3 v_3 = v_1 u_1 + v_2 u_2 + v_3 u_3 = \mathbf{v} \cdot \mathbf{u}$$

For the fifth property, we let $\mathbf{v} = \langle v_1, v_2, v_3 \rangle$. Then

$$\mathbf{v} \cdot \mathbf{v} = v_1^2 + v_2^2 + v_3^2$$
$$= \left(\sqrt{v_1^2 + v_2^2 + v_3^2} \right)^2 = \|\mathbf{v}\|^2$$

EXAMPLE 1   Finding dot products

Given $\mathbf{u} = \langle 2, -2 \rangle$, $\mathbf{v} = \langle 5, 8 \rangle$, and $\mathbf{w} = \langle -4, 3 \rangle$, find each of the following:
(a) $\mathbf{u} \cdot \mathbf{v}$ (b) $(\mathbf{u} \cdot \mathbf{v})\mathbf{w}$ (c) $\mathbf{u} \cdot (2\mathbf{v})$ (d) $\|\mathbf{w}\|^2$

Solution:
(a) $\mathbf{u} \cdot \mathbf{v} = \langle 2, -2 \rangle \cdot \langle 5, 8 \rangle = 2(5) + (-2)(8) = -6$
(b) $(\mathbf{u} \cdot \mathbf{v})\mathbf{w} = -6\langle -4, 3 \rangle = \langle 24, -18 \rangle$
(c) $\mathbf{u} \cdot (2\mathbf{v}) = 2(\mathbf{u} \cdot \mathbf{v}) = 2(-6) = -12$
(d) $\|\mathbf{w}\|^2 = \mathbf{w} \cdot \mathbf{w} = \langle -4, 3 \rangle \cdot \langle -4, 3 \rangle = (-4)(-4) + (3)(3) = 25$

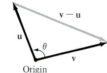

Origin

Angle between Two Vectors

FIGURE 13.29

The **angle between two nonzero vectors** is the angle θ, $0 \le \theta \le \pi$, between their respective standard position vectors, as shown in Figure 13.29. The next theorem shows how to find this angle using the dot product. (Note that we do not define the angle between the zero vector and another vector.)

THEOREM 13.6   ANGLE BETWEEN TWO VECTORS
If θ is the angle between two nonzero vectors $\mathbf{u}$ and $\mathbf{v}$, then

$$\cos \theta = \frac{\mathbf{u} \cdot \mathbf{v}}{\|\mathbf{u}\| \, \|\mathbf{v}\|}$$

Proof: Consider the triangle determined by vectors $\mathbf{u}$, $\mathbf{v}$, and $\mathbf{v} - \mathbf{u}$ as shown in Figure 13.29. By the Law of Cosines, we have

$$\|\mathbf{v} - \mathbf{u}\|^2 = \|\mathbf{u}\|^2 + \|\mathbf{v}\|^2 - 2\|\mathbf{u}\| \, \|\mathbf{v}\| \cos \theta$$

Using the properties of the dot product, we can rewrite the left side as

$$\|\mathbf{v} - \mathbf{u}\|^2 = (\mathbf{v} - \mathbf{u}) \cdot (\mathbf{v} - \mathbf{u}) = (\mathbf{v} - \mathbf{u}) \cdot \mathbf{v} - (\mathbf{v} - \mathbf{u}) \cdot \mathbf{u}$$
$$= \mathbf{v} \cdot \mathbf{v} - \mathbf{u} \cdot \mathbf{v} - \mathbf{v} \cdot \mathbf{u} + \mathbf{u} \cdot \mathbf{u} = \|\mathbf{v}\|^2 - 2\mathbf{u} \cdot \mathbf{v} + \|\mathbf{u}\|^2$$

and substitution back into the Law of Cosines yields

$$\|\mathbf{v}\|^2 - 2\mathbf{u} \cdot \mathbf{v} + \|\mathbf{u}\|^2 = \|\mathbf{u}\|^2 + \|\mathbf{v}\|^2 - 2\|\mathbf{u}\| \, \|\mathbf{v}\| \cos \theta$$
$$-2\mathbf{u} \cdot \mathbf{v} = -2\|\mathbf{u}\| \, \|\mathbf{v}\| \cos \theta$$
$$\cos \theta = \frac{\mathbf{u} \cdot \mathbf{v}}{\|\mathbf{u}\| \, \|\mathbf{v}\|}$$

Remark If the angle between two vectors is known, then rewriting Theorem 13.6 in the form

$$\mathbf{u} \cdot \mathbf{v} = \|\mathbf{u}\| \, \|\mathbf{v}\| \cos \theta$$

gives us an alternative way to calculate the dot product. This form also shows us that because $\|\mathbf{u}\|$ and $\|\mathbf{v}\|$ are always positive, $\mathbf{u} \cdot \mathbf{v}$ and $\cos \theta$ will always have the same sign.

Figure 13.30 shows the five possible orientations of two vectors.

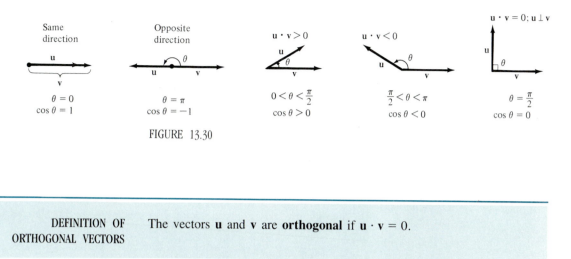

FIGURE 13.30

DEFINITION OF ORTHOGONAL VECTORS

The vectors **u** and **v** are **orthogonal** if $\mathbf{u} \cdot \mathbf{v} = 0$.

From this definition it follows that the zero vector is orthogonal to every vector **u,** since $\mathbf{0} \cdot \mathbf{u} = 0$. Moreover, for $0 \leq \theta \leq \pi$, we know that $\cos \theta = 0$ if and only if $\theta = \pi/2$. Thus, we can use Theorem 13.6 to conclude that two *nonzero* vectors are orthogonal if and only if the angle between them is $\pi/2$.

| **Remark** The terms perpendicular, orthogonal, and normal all mean essentially the same thing—meeting at right angles. However, preference is given to saying that two vectors are *orthogonal,* two lines or planes are *perpendicular,* and a vector is *normal* to a given line or plane.

EXAMPLE 2   *Finding the angle between two vectors*

For $\mathbf{u} = \langle 3, -1, 2 \rangle$, $\mathbf{v} = \langle -4, 0, 2 \rangle$, $\mathbf{w} = \langle 1, -1, -2 \rangle$, and $\mathbf{z} = \langle 2, 0, -1 \rangle$, find the angle between
(a) **u** and **v**   (b) **u** and **w**   (c) **v** and **z**

Solution:

(a) $\cos \theta = \dfrac{\mathbf{u} \cdot \mathbf{v}}{\|\mathbf{u}\|\,\|\mathbf{v}\|} = \dfrac{-12 + 4}{\sqrt{14}\sqrt{20}} = \dfrac{-8}{2\sqrt{14}\sqrt{5}} = \dfrac{-4}{\sqrt{70}}$

Since $\mathbf{u} \cdot \mathbf{v} < 0$, $\theta = \arccos \dfrac{-4}{\sqrt{70}} \approx 2.069$ radians.

(b) $\cos \theta = \dfrac{\mathbf{u} \cdot \mathbf{w}}{\|\mathbf{u}\|\,\|\mathbf{w}\|} = \dfrac{3 + 1 - 4}{\sqrt{14}\sqrt{6}} = \dfrac{0}{\sqrt{84}} = 0$

Since $\mathbf{u} \cdot \mathbf{w} = 0$, **u** and **w** are *orthogonal* vectors, and furthermore, $\theta = \pi/2$.

(c) $\cos \theta = \dfrac{\mathbf{v} \cdot \mathbf{z}}{\|\mathbf{v}\|\,\|\mathbf{z}\|} = \dfrac{-8 + 0 - 2}{\sqrt{20}\sqrt{5}} = \dfrac{-10}{\sqrt{100}} = -1$

Consequently, $\theta = \pi$.

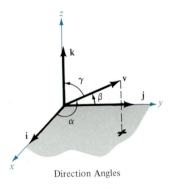

Direction Angles

FIGURE 13.31

Direction cosines

For a vector in the plane, we found it convenient to measure direction in terms of the angle, measured counterclockwise, *from* the positive x-axis to the vector. In space it is more convenient to measure direction in terms of the angles *between* the vector $\mathbf{v}$ and the three unit vectors $\mathbf{i}$, $\mathbf{j}$, and $\mathbf{k}$, as shown in Figure 13.31. We call the angles α, β, and γ the **direction angles of v**, and we call $\cos \alpha$, $\cos \beta$, and $\cos \gamma$ the **direction cosines of v**. Since

$$\mathbf{v} \cdot \mathbf{i} = \langle v_1, v_2, v_3 \rangle \cdot \langle 1, 0, 0 \rangle = v_1$$

it follows that $\cos \alpha = v_1/\|\mathbf{v}\|$. By similar reasoning with the unit vectors $\mathbf{j}$ and $\mathbf{k}$, we have

$$\cos \alpha = \frac{v_1}{\|\mathbf{v}\|}, \qquad \cos \beta = \frac{v_2}{\|\mathbf{v}\|}, \qquad \text{and} \qquad \cos \gamma = \frac{v_3}{\|\mathbf{v}\|}$$

Consequently, any nonzero vector $\mathbf{v}$ in space has the normalized form

$$\frac{\mathbf{v}}{\|\mathbf{v}\|} = \frac{v_1}{\|\mathbf{v}\|}\mathbf{i} + \frac{v_2}{\|\mathbf{v}\|}\mathbf{j} + \frac{v_3}{\|\mathbf{v}\|}\mathbf{k} = \cos \alpha \, \mathbf{i} + \cos \beta \, \mathbf{j} + \cos \gamma \, \mathbf{k}$$

and since $\mathbf{v}/\|\mathbf{v}\|$ is a unit vector, it follows that

$$\cos^2 \alpha + \cos^2 \beta + \cos^2 \gamma = 1$$

EXAMPLE 3   *Finding direction angles*

Find the direction cosines and angles for the vector $\mathbf{v} = 2\mathbf{i} + 3\mathbf{j} + 4\mathbf{k}$, and show that $\cos^2 \alpha + \cos^2 \beta + \cos^2 \gamma = 1$.

Solution: Since $\|\mathbf{v}\| = \sqrt{2^2 + 3^2 + 4^2} = \sqrt{29}$, we have

$$\cos \alpha = \frac{v_1}{\|\mathbf{v}\|} = \frac{2}{\sqrt{29}} \quad \Longrightarrow \quad \alpha \approx 68.2°$$

$$\cos \beta = \frac{v_2}{\|\mathbf{v}\|} = \frac{3}{\sqrt{29}} \quad \Longrightarrow \quad \beta \approx 56.1°$$

$$\cos \gamma = \frac{v_3}{\|\mathbf{v}\|} = \frac{4}{\sqrt{29}} \quad \Longrightarrow \quad \gamma \approx 42.0°$$

and

$$\cos^2 \alpha + \cos^2 \beta + \cos^2 \gamma = \frac{4}{29} + \frac{9}{29} + \frac{16}{29} = \frac{29}{29} = 1$$

(See Figure 13.32.)

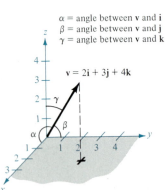

α = angle between $\mathbf{v}$ and $\mathbf{i}$
β = angle between $\mathbf{v}$ and $\mathbf{j}$
γ = angle between $\mathbf{v}$ and $\mathbf{k}$

$\mathbf{v} = 2\mathbf{i} + 3\mathbf{j} + 4\mathbf{k}$

Direction Angles of $\mathbf{v}$

FIGURE 13.32

Vector components

We have already seen applications in which two vectors are added to produce a resultant vector. There are many applications in physics and engineering in which we want to look at the reverse problem—that is, to decompose a given vector into the sum of two **vector components.** To see the usefulness of this procedure, we look at a physical example.

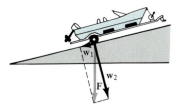

FIGURE 13.33

Consider a boat on an inclined ramp, as shown in Figure 13.33. The force **F** due to gravity pulls the boat down the ramp and against the ramp. These two forces, $\mathbf{w}_1$ and $\mathbf{w}_2$, are orthogonal, and we call them the vector components of **F** and write

$$\mathbf{F} = \mathbf{w}_1 + \mathbf{w}_2$$

The forces $\mathbf{w}_1$ and $\mathbf{w}_2$ help us analyze the effect of gravity on the boat. For example, $\mathbf{w}_1$ indicates the force necessary to keep the boat from rolling down the ramp. On the other hand, $\mathbf{w}_2$ indicates the force that the tires must withstand.

DEFINITION OF VECTOR COMPONENTS

Let **u** and **v** be nonzero vectors. Moreover, let $\mathbf{u} = \mathbf{w}_1 + \mathbf{w}_2$, where $\mathbf{w}_1$ is parallel to **v** and $\mathbf{w}_2$ is orthogonal to **v,** as shown in Figure 13.34.

1. $\mathbf{w}_1$ is called the **vector component of u along v.** $\mathbf{w}_1$ is also called the projection of **u** onto **v** and is denoted by $\mathbf{w}_1 = \text{proj}_\mathbf{v}\,\mathbf{u}.$
2. $\mathbf{w}_2 = \mathbf{u} - \mathbf{w}_1$ is called the **vector component of u orthogonal to v.**

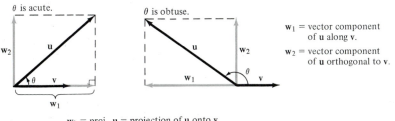

FIGURE 13.34

$\mathbf{w}_1 = \text{proj}_\mathbf{v}\,\mathbf{u} = \text{projection of } \mathbf{u} \text{ onto } \mathbf{v}.$

From this definition we see that it is easy to find the vector component $\mathbf{w}_2$ once we have found the projection of **u** onto **v.** To find the projection, we use the dot product, as indicated in the next theorem.

THEOREM 13.7

ORTHOGONAL PROJECTION USING THE DOT PRODUCT
If **u** and **v** are nonzero vectors, then the projection of **u** onto **v** is given by

$$\text{proj}_\mathbf{v}\mathbf{u} = \left(\frac{\mathbf{u} \cdot \mathbf{v}}{\|\mathbf{v}\|^2} \right)\mathbf{v}$$

Proof: From Figure 13.34, we let $\mathbf{w}_1 = \text{proj}_\mathbf{v}\mathbf{u}.$ Since $\mathbf{w}_1$ is a scalar multiple of **v,** we can write

$$\mathbf{u} = \mathbf{w}_1 + \mathbf{w}_2 = c\mathbf{v} + \mathbf{w}_2$$

Now, taking the dot product of both sides with **v,** we have

$$\mathbf{u} \cdot \mathbf{v} = (c\mathbf{v} + \mathbf{w}_2) \cdot \mathbf{v} = c\mathbf{v} \cdot \mathbf{v} + \mathbf{w}_2 \cdot \mathbf{v} = c\|\mathbf{v}\|^2 + \mathbf{w}_2 \cdot \mathbf{v}$$

Since $\mathbf{w}_2$ and $\mathbf{v}$ are orthogonal, $\mathbf{w}_2 \cdot \mathbf{v} = 0$ and we have

$$\mathbf{u} \cdot \mathbf{v} = c\|\mathbf{v}\|^2 \implies c = \frac{\mathbf{u} \cdot \mathbf{v}}{\|\mathbf{v}\|^2}$$

Therefore, we can write $\mathbf{w}_1$ as

$$\mathbf{w}_1 = \left(\frac{\mathbf{u} \cdot \mathbf{v}}{\|\mathbf{v}\|^2} \right) \mathbf{v}$$

The projection of $\mathbf{u}$ onto $\mathbf{v}$ can be written as a scalar multiple of a unit vector in the direction of $\mathbf{v}$. That is,

$$\left(\frac{\mathbf{u} \cdot \mathbf{v}}{\|\mathbf{v}\|^2} \right) \mathbf{v} = \left(\frac{\mathbf{u} \cdot \mathbf{v}}{\|\mathbf{v}\|} \right) \frac{\mathbf{v}}{\|\mathbf{v}\|} = (k) \frac{\mathbf{v}}{\|\mathbf{v}\|}$$

We call k the **component of u in the direction of v.** Thus,

$$k = \frac{\mathbf{u} \cdot \mathbf{v}}{\|\mathbf{v}\|} = \|\mathbf{u}\| \cos \theta \qquad \text{Component of u in the direction of v}$$

| **Remark**   Be sure you see the distinction between the terms component and vector component. For example, using the standard unit vectors with $\mathbf{u} = u_1 \mathbf{i} + u_2 \mathbf{j}$, we have

$$u_1 \text{ is the } component \text{ of } \mathbf{u} \text{ in the direction of } \mathbf{i}$$

and

$$u_1 \mathbf{i} \text{ is the } vector\ component \text{ in the direction of } \mathbf{i}$$

EXAMPLE 4   Decomposing a vector into vector components

Find the vector component of $\mathbf{u}$ along $\mathbf{v}$ and the vector component of $\mathbf{u}$ orthogonal to $\mathbf{v}$ for the vectors

$$\mathbf{u} = 3\mathbf{i} - 5\mathbf{j} + 2\mathbf{k} \qquad \text{and} \qquad \mathbf{v} = 7\mathbf{i} + \mathbf{j} - 2\mathbf{k}$$

Solution: The vector component of $\mathbf{u}$ along $\mathbf{v}$ is

$$\mathbf{w}_1 = \left(\frac{\mathbf{u} \cdot \mathbf{v}}{\|\mathbf{v}\|^2} \right) \mathbf{v} = \left(\frac{12}{54} \right) (7\mathbf{i} + \mathbf{j} - 2\mathbf{k}) = \frac{14}{9}\mathbf{i} + \frac{2}{9}\mathbf{j} - \frac{4}{9}\mathbf{k}$$

The vector component of $\mathbf{u}$ orthogonal to $\mathbf{v}$ is the vector

$$\mathbf{w}_2 = \mathbf{u} - \mathbf{w}_1 = (3\mathbf{i} - 5\mathbf{j} + 2\mathbf{k}) - \left(\frac{14}{9}\mathbf{i} + \frac{2}{9}\mathbf{j} - \frac{4}{9}\mathbf{k} \right)$$

$$= \frac{13}{9}\mathbf{i} - \frac{47}{9}\mathbf{j} + \frac{22}{9}\mathbf{k}$$

(See Figure 13.35.)

Now, let's return to the problem involving the boat and the ramp to see how we can use vector components in a physical problem.

EXAMPLE 5   An application using a vector component

A 600-pound boat sits on a ramp inclined at 30°, as shown in Figure 13.36. What force is required to keep the boat from rolling down the ramp?

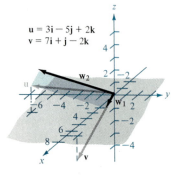

$\mathbf{u} = 3\mathbf{i} - 5\mathbf{j} + 2\mathbf{k}$
$\mathbf{v} = 7\mathbf{i} + \mathbf{j} - 2\mathbf{k}$

FIGURE 13.35

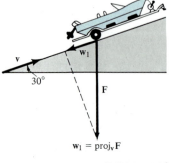

$\mathbf{w}_1 = \text{proj}_\mathbf{v} \mathbf{F}$

FIGURE 13.36

Solution: Since the force due to gravity is vertical and downward, we represent the gravitational force by the vector

$$\mathbf{F} = -600\mathbf{j} \qquad \text{Force due to gravity}$$

To find the force required to keep the boat from rolling down the ramp, we project $\mathbf{F}$ onto a unit vector $\mathbf{v}$ in the direction of the ramp, as follows.

$$\mathbf{v} = \cos 30° \, \mathbf{i} + \sin 30° \, \mathbf{j} = \frac{\sqrt{3}}{2}\mathbf{i} + \frac{1}{2}\mathbf{j} \qquad \text{Unit vector along ramp}$$

Therefore, the projection of $\mathbf{F}$ onto $\mathbf{v}$ is given by

$$\mathbf{w}_1 = \text{proj}_\mathbf{v}\mathbf{F} = \left(\frac{\mathbf{F} \cdot \mathbf{v}}{\|\mathbf{v}\|^2}\right)\mathbf{v} = (\mathbf{F} \cdot \mathbf{v})\mathbf{v} = (-600)\left(\frac{1}{2}\right)\mathbf{v}$$

$$= -300\left(\frac{\sqrt{3}}{2}\mathbf{i} + \frac{1}{2}\mathbf{j}\right)$$

The magnitude of this force is 300, and therefore a force of 300 pounds is required to keep the boat from rolling down the ramp.

Work

The work W done by a constant force $\mathbf{F}$ acting along the lines of motion of an object is given by

$$W = (\text{magnitude of force})(\text{distance}) = \|\mathbf{F}\| \, \|\overrightarrow{PQ}\|$$

as shown in Figure 13.37(a). If the constant force $\mathbf{F}$ is not directed along the line of motion, then we can see from Figure 13.37(b) that the work W done by the force is

$$W = \|\text{proj}_{\overrightarrow{PQ}}\mathbf{F}\| \, \|\overrightarrow{PQ}\| = (\cos \theta)\|\mathbf{F}\| \, \|\overrightarrow{PQ}\| = \mathbf{F} \cdot \overrightarrow{PQ}$$

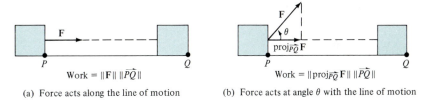

(a) Force acts along the line of motion   (b) Force acts at angle θ with the line of motion

FIGURE 13.37

We summarize this notion of work in the following definition.

DEFINITION OF WORK	The work W done by a constant force $\mathbf{F}$ as its point of application moves along the vector $\overrightarrow{PQ}$ is given by either of the following:
	1. $W = \|\text{proj}_{\overrightarrow{PQ}}\mathbf{F}\| \, \|\overrightarrow{PQ}\|$ Projection form
	2. $W = \mathbf{F} \cdot \overrightarrow{PQ}$ Dot product form

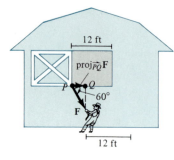

12 ft

proj$_{\vec{PQ}}$F

P Q

60°

F

12 ft

FIGURE 13.38

EXAMPLE 6   An application involving work

To close a sliding door, a person pulls on a rope with a constant force of 50 pounds at a constant angle of 60°, as shown in Figure 13.38. Find the work done in moving the door 12 feet to its closed position.

Solution: Using a projection, we calculate the work as follows:

$$W = \|\text{proj}_{\vec{PQ}}\mathbf{F}\| \, \|\vec{PQ}\|$$
$$= \cos (60°)\|\mathbf{F}\| \, \|\vec{PQ}\|$$
$$= \frac{1}{2}(50)(12) = 300 \text{ ft-lb}$$

Section Exercises 13.3

In Exercises 1–6, find (a) $\mathbf{u} \cdot \mathbf{v}$, (b) $\mathbf{u} \cdot \mathbf{u}$, (c) $\|\mathbf{u}\|^2$, (d) $(\mathbf{u} \cdot \mathbf{v})\mathbf{v}$, and (e) $\mathbf{u} \cdot (2\mathbf{v})$.

1. $\mathbf{u} = \langle 3, 4 \rangle$, $\mathbf{v} = \langle 2, -3 \rangle$
2. $\mathbf{u} = \langle 5, 12 \rangle$, $\mathbf{v} = \langle -3, 2 \rangle$
3. $\mathbf{u} = \langle 2, -3, 4 \rangle$, $\mathbf{v} = \langle 0, 6, 5 \rangle$
4. $\mathbf{u} = \mathbf{i}$, $\mathbf{v} = \mathbf{i}$
5. $\mathbf{u} = 2\mathbf{i} - \mathbf{j} + \mathbf{k}$, $\mathbf{v} = \mathbf{i} - \mathbf{k}$
6. $\mathbf{u} = 2\mathbf{i} + \mathbf{j} - 2\mathbf{k}$, $\mathbf{v} = \mathbf{i} - 3\mathbf{j} + 2\mathbf{k}$

In Exercises 7–14, find the angle θ between the given vectors.

7. $\mathbf{u} = \langle 1, 1 \rangle$, $\mathbf{v} = \langle 2, -2 \rangle$
8. $\mathbf{u} = \langle 3, 1 \rangle$, $\mathbf{v} = \langle 2, -1 \rangle$
9. $\mathbf{u} = 3\mathbf{i} + \mathbf{j}$, $\mathbf{v} = -2\mathbf{i} + 4\mathbf{j}$
10. $\mathbf{u} = \cos \left(\dfrac{\pi}{6} \right) \mathbf{i} + \sin \left(\dfrac{\pi}{6} \right) \mathbf{j}$

   $\mathbf{v} = \cos \left(\dfrac{3\pi}{4} \right) \mathbf{i} + \sin \left(\dfrac{3\pi}{4} \right) \mathbf{j}$
11. $\mathbf{u} = \langle 1, 1, 1 \rangle$, $\mathbf{v} = \langle 2, 1, -1 \rangle$
12. $\mathbf{u} = 2\mathbf{i} + 3\mathbf{j} + \mathbf{k}$, $\mathbf{v} = -3\mathbf{i} + 2\mathbf{j}$
13. $\mathbf{u} = 3\mathbf{i} + 4\mathbf{j}$, $\mathbf{v} = -2\mathbf{j} + 3\mathbf{k}$
14. $\mathbf{u} = 2\mathbf{i} - 3\mathbf{j} + \mathbf{k}$, $\mathbf{v} = \mathbf{i} - 2\mathbf{j} + \mathbf{k}$

In Exercises 15–22, determine whether $\mathbf{u}$ and $\mathbf{v}$ are orthogonal, parallel, or neither.

15. $\mathbf{u} = \langle 4, 0 \rangle$, $\mathbf{v} = \langle 1, 1 \rangle$
16. $\mathbf{u} = \langle 2, 18 \rangle$, $\mathbf{v} = \left\langle \dfrac{3}{2}, -\dfrac{1}{6} \right\rangle$
17. $\mathbf{u} = \langle 4, 3 \rangle$, $\mathbf{v} = \left\langle \dfrac{1}{2}, -\dfrac{2}{3} \right\rangle$
18. $\mathbf{u} = -\dfrac{1}{3}(\mathbf{i} - 2\mathbf{j})$, $\mathbf{v} = 2\mathbf{i} - 4\mathbf{j}$

19. $\mathbf{u} = \mathbf{j} + 6\mathbf{k}$, $\mathbf{v} = \mathbf{i} - 2\mathbf{j} - \mathbf{k}$
20. $\mathbf{u} = -2\mathbf{i} + 3\mathbf{j} - \mathbf{k}$, $\mathbf{v} = 2\mathbf{i} + \mathbf{j} - \mathbf{k}$
21. $\mathbf{u} = \langle 2, -3, 1 \rangle$, $\mathbf{v} = \langle -1, -1, -1 \rangle$
22. $\mathbf{u} = \langle \cos \theta, \sin \theta, -1 \rangle$
    $\mathbf{v} = \langle \sin \theta, -\cos \theta, 0 \rangle$

In Exercises 23–26, find the direction cosines of $\mathbf{u}$ and demonstrate that the sum of the squares of the direction cosines is 1.

23. $\mathbf{u} = \mathbf{i} + 2\mathbf{j} + 2\mathbf{k}$ 24. $\mathbf{u} = 3\mathbf{i} - \mathbf{j} + 5\mathbf{k}$
25. $\mathbf{u} = \langle 0, 6, -4 \rangle$ 26. $\mathbf{u} = \langle a, b, c \rangle$

In Exercises 27–34, (a) find the vector component of $\mathbf{u}$ along $\mathbf{v}$, and (b) find the vector component of $\mathbf{u}$ orthogonal to $\mathbf{v}$.

27. $\mathbf{u} = \langle 2, 3 \rangle$, $\mathbf{v} = \langle 5, 1 \rangle$
28. $\mathbf{u} = \langle 2, -3 \rangle$, $\mathbf{v} = \langle 3, 2 \rangle$
29. $\mathbf{u} = \langle 2, 1, 2 \rangle$, $\mathbf{v} = \langle 0, 3, 4 \rangle$
30. $\mathbf{u} = \langle 0, 4, 1 \rangle$, $\mathbf{v} = \langle 0, 2, 3 \rangle$
31. $\mathbf{u} = \langle 1, 1, 1 \rangle$, $\mathbf{v} = \langle -2, -1, 1 \rangle$
32. $\mathbf{u} = \langle -2, -1, 1 \rangle$, $\mathbf{v} = \langle 1, 1, 1 \rangle$
33. $\mathbf{u} = \langle 5, -4, 3 \rangle$, $\mathbf{v} = \langle 1, 0, 0 \rangle$
34. $\mathbf{u} = \langle 5, -4, 3 \rangle$, $\mathbf{v} = \langle 0, 1, 0 \rangle$

35. A truck with a gross weight of 32,000 pounds is parked on a 15° slope. (See Figure 13.39.) Assuming the only force to overcome is that due to gravity, find
    (a) the force required to keep the truck from rolling down the hill
    (b) the force perpendicular to the hill
36. The guy wire to a 100-foot tower has a tension of 550 pounds. Using the distances shown in Figure 13.40, find the direction cosines and direction angles for the vector $\mathbf{F}$ representing the tension in the wire.

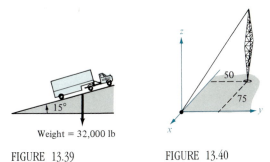

Weight = 32,000 lb

FIGURE 13.39   FIGURE 13.40

37. An implement is dragged 10 feet across a floor, using a force of 85 pounds. Find the work done if the direction of the force is 60° above the horizontal. (See Figure 13.41.)

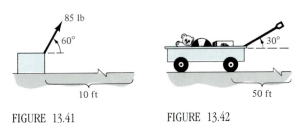

85 lb

60°

10 ft   50 ft

FIGURE 13.41   FIGURE 13.42

38. A toy wagon is pulled by exerting a force of 15 pounds on a handle that makes a 30° angle with the horizontal. Find the work done in pulling the wagon 50 feet. (See Figure 13.42.)

In Exercises 39 and 40, find the work done in moving a particle from P to Q if the magnitude and direction of the force is given by $\mathbf{v}$.

39. $P = (0, 0, 0)$, $Q = (4, 7, 5)$,
    $\mathbf{v} = \langle 1, 4, 8 \rangle$

40. $P = (1, 3, 0)$, $Q = (-3, 5, 10)$,
    $\mathbf{v} = -2\mathbf{i} + 3\mathbf{j} + 6\mathbf{k}$

41. The vector $\mathbf{u} = \langle 3240, 1450, 2235 \rangle$ gives the number of bushels of corn, oats, and wheat raised by a farmer in a certain year. The vector $\mathbf{v} = \langle 2.22, 1.85, 3.25 \rangle$ gives the price (in dollars) per bushel of each of the crops. Find the dot product, $\mathbf{u} \cdot \mathbf{v}$, and explain what information it gives.

42. Find the angle between the diagonal of a cube and one of its edges.

43. What is known about θ, the angle between two vectors $\mathbf{u}$ and $\mathbf{v}$, if
    (a) $\mathbf{u} \cdot \mathbf{v} = 0$?   (b) $\mathbf{u} \cdot \mathbf{v} > 0$?
    (c) $\mathbf{u} \cdot \mathbf{v} < 0$?

44. What can be said about the vectors $\mathbf{u}$ and $\mathbf{v}$ if
    (a) the projection of $\mathbf{u}$ onto $\mathbf{v}$ equals $\mathbf{u}$?
    (b) the projecton of $\mathbf{u}$ onto $\mathbf{v}$ equals $\mathbf{0}$?

45. Use vectors to prove that the diagonals of a rhombus are perpendicular.

46. Prove that

$$\|\mathbf{u} - \mathbf{v}\|^2 = \|\mathbf{u}\|^2 + \|\mathbf{v}\|^2 - 2\mathbf{u} \cdot \mathbf{v}$$

47. Prove Properties 2, 3, and 4 of Theorem 13.5.

48. Prove that if $\mathbf{u}$ is orthogonal to $\mathbf{v}$ and $\mathbf{w}$, then $\mathbf{u}$ is orthogonal to $c\mathbf{v} + d\mathbf{w}$ for any scalars c and d.

13.4

The cross product of two vectors in space

SECTION TOPICS ▪
The cross product ▪
Triple scalar product ▪
Applications ▪

Many applications in physics, engineering, and geometry involve finding a vector in space that is orthogonal to two given vectors. In this section we look at a product that will yield such a vector. It is called the **cross product,** and it is most conveniently defined and calculated using the standard unit vector form.

DEFINITION OF CROSS
PRODUCT OF TWO VECTORS
IN SPACE

Let
$$\mathbf{u} = u_1\mathbf{i} + u_2\mathbf{j} + u_3\mathbf{k} \qquad \text{and} \qquad \mathbf{v} = v_1\mathbf{i} + v_2\mathbf{j} + v_3\mathbf{k}$$
be vectors in space. The **cross product** of $\mathbf{u}$ and $\mathbf{v}$ is the vector
$$\mathbf{u} \times \mathbf{v} = (u_2v_3 - u_3v_2)\,\mathbf{i} - (u_1v_3 - u_3v_1)\,\mathbf{j} + (u_1v_2 - u_2v_1)\,\mathbf{k}$$

| Remark   Be sure you see that this definition applies only to three-dimensional vectors. We do not define the cross product of two-dimensional vectors.

A convenient way to calculate $\mathbf{u} \times \mathbf{v}$ is to use the following *determinant form* with cofactor expansion:

$$\mathbf{u} \times \mathbf{v} = \begin{vmatrix} \mathbf{i} & \mathbf{j} & \mathbf{k} \\ u_1 & u_2 & u_3 \\ v_1 & v_2 & v_3 \end{vmatrix}$$

$$= \begin{vmatrix} \mathbf{i} & \mathbf{j} & \mathbf{k} \\ u_1 & u_2 & u_3 \\ v_1 & v_2 & v_3 \end{vmatrix} \mathbf{i} - \begin{vmatrix} \mathbf{i} & \mathbf{j} & \mathbf{k} \\ u_1 & u_2 & u_3 \\ v_1 & v_2 & v_3 \end{vmatrix} \mathbf{j} + \begin{vmatrix} \mathbf{i} & \mathbf{j} & \mathbf{k} \\ u_1 & u_2 & u_3 \\ v_1 & v_2 & v_3 \end{vmatrix} \mathbf{k}$$

$$= \begin{vmatrix} u_2 & u_3 \\ v_2 & v_3 \end{vmatrix} \mathbf{i} - \begin{vmatrix} u_1 & u_3 \\ v_1 & v_3 \end{vmatrix} \mathbf{j} + \begin{vmatrix} u_1 & u_2 \\ v_1 & v_2 \end{vmatrix} \mathbf{k}$$

where

$$\begin{vmatrix} u_2 & u_3 \\ v_2 & v_3 \end{vmatrix} = u_2 v_3 - u_3 v_2,$$

$$\begin{vmatrix} u_1 & u_3 \\ v_1 & v_3 \end{vmatrix} = u_1 v_3 - u_3 v_1,$$

$$\begin{vmatrix} u_1 & u_2 \\ v_1 & v_2 \end{vmatrix} = u_1 v_2 - u_2 v_1$$

(Note the minus sign in front of the $\mathbf{j}$-component.) This 3×3 determinant form is used simply to help remember the formula for the cross product—it is technically not a determinant since the elements of the corresponding matrix are not all real numbers.

EXAMPLE 1   Finding the cross product

Given $\mathbf{u} = \mathbf{i} - 2\mathbf{j} + \mathbf{k}$ and $\mathbf{v} = 3\mathbf{i} + \mathbf{j} - 2\mathbf{k}$, find the following:
(a) $\mathbf{u} \times \mathbf{v}$ (b) $\mathbf{v} \times \mathbf{u}$ (c) $\mathbf{v} \times \mathbf{v}$

Solution:

(a) $\mathbf{u} \times \mathbf{v} = \begin{vmatrix} \mathbf{i} & \mathbf{j} & \mathbf{k} \\ 1 & -2 & 1 \\ 3 & 1 & -2 \end{vmatrix} = \begin{vmatrix} -2 & 1 \\ 1 & -2 \end{vmatrix} \mathbf{i} - \begin{vmatrix} 1 & 1 \\ 3 & -2 \end{vmatrix} \mathbf{j} + \begin{vmatrix} 1 & -2 \\ 3 & 1 \end{vmatrix} \mathbf{k}$

$$= (4 - 1)\,\mathbf{i} - (-2 - 3)\,\mathbf{j} + (1 + 6)\,\mathbf{k}$$

$$= 3\mathbf{i} + 5\mathbf{j} + 7\mathbf{k}$$

(b) $\mathbf{v} \times \mathbf{u} = \begin{vmatrix} \mathbf{i} & \mathbf{j} & \mathbf{k} \\ 3 & 1 & -2 \\ 1 & -2 & 1 \end{vmatrix} = \begin{vmatrix} 1 & -2 \\ -2 & 1 \end{vmatrix} \mathbf{i} - \begin{vmatrix} 3 & -2 \\ 1 & 1 \end{vmatrix} \mathbf{j} + \begin{vmatrix} 3 & 1 \\ 1 & -2 \end{vmatrix} \mathbf{k}$

$$= (1 - 4)\,\mathbf{i} - (3 + 2)\,\mathbf{j} + (-6 - 1)\,\mathbf{k}$$

$$= -3\mathbf{i} - 5\mathbf{j} - 7\mathbf{k}$$

Note that this result is the negative of that in part (a).

$$(c) \quad \mathbf{v} \times \mathbf{v} = \begin{vmatrix} \mathbf{i} & \mathbf{j} & \mathbf{k} \\ 3 & 1 & -2 \\ 3 & 1 & -2 \end{vmatrix} = \mathbf{0}$$

The results obtained in Example 1 suggest some interesting *algebraic* properties of the cross product. For instance,

$$\mathbf{u} \times \mathbf{v} = -(\mathbf{v} \times \mathbf{u}) \qquad \text{and} \qquad \mathbf{v} \times \mathbf{v} = \mathbf{0}$$

These properties, along with several others, are given in the following theorem.

THEOREM 13.8   **ALGEBRAIC PROPERTIES OF THE CROSS PRODUCT**
If **u**, **v**, and **w** are vectors in space and c is any scalar, then the following properties are true.

1. $\mathbf{u} \times \mathbf{v} = -(\mathbf{v} \times \mathbf{u})$
2. $\mathbf{u} \times (\mathbf{v} + \mathbf{w}) = (\mathbf{u} \times \mathbf{v}) + (\mathbf{u} \times \mathbf{w})$
3. $c(\mathbf{u} \times \mathbf{v}) = (c\mathbf{u}) \times \mathbf{v} = \mathbf{u} \times (c\mathbf{v})$
4. $\mathbf{u} \times \mathbf{0} = \mathbf{0} \times \mathbf{u} = \mathbf{0}$
5. $\mathbf{u} \times \mathbf{u} = \mathbf{0}$
6. $\mathbf{u} \cdot (\mathbf{v} \times \mathbf{w}) = (\mathbf{u} \times \mathbf{v}) \cdot \mathbf{w}$

Proof: We give details only for the proof of the second property. The remaining proofs are left as exercises. Let

$$\mathbf{u} = u_1\mathbf{i} + u_2\mathbf{j} + u_3\mathbf{k}$$
$$\mathbf{v} = v_1\mathbf{i} + v_2\mathbf{j} + v_3\mathbf{k}$$
and
$$\mathbf{w} = w_1\mathbf{i} + w_2\mathbf{j} + w_3\mathbf{k}$$

Then $\mathbf{u} \times (\mathbf{v} + \mathbf{w})$ is given by

$$\begin{vmatrix} \mathbf{i} & \mathbf{j} & \mathbf{k} \\ u_1 & u_2 & u_3 \\ v_1 + w_1 & v_2 + w_2 & v_3 + w_3 \end{vmatrix}$$

$$= [u_2(v_3 + w_3) - u_3(v_2 + w_2)]\, \mathbf{i} - [u_1(v_3 + w_3) - u_3(v_1 + w_1)]\, \mathbf{j}$$
$$+ [u_1(v_2 + w_2) - u_2(v_1 + w_1)]\, \mathbf{k}$$
$$= [(u_2v_3 - u_3v_2)\, \mathbf{i} - (u_1v_3 - u_3v_1)\, \mathbf{j} + (u_1v_2 - u_2v_1)\, \mathbf{k}]$$
$$+ [(u_2w_3 - u_3w_2)\, \mathbf{i} - (u_1w_3 - u_3w_1)\, \mathbf{j} + (u_1w_2 - u_2w_1)\, \mathbf{k}]$$
$$= (\mathbf{u} \times \mathbf{v}) + (\mathbf{u} \times \mathbf{w})$$

Note that the first property listed in Theorem 13.8 indicates that the cross product is *not commutative*. In particular, this property indicates that the vectors $\mathbf{u} \times \mathbf{v}$ and $\mathbf{v} \times \mathbf{u}$ have equal lengths but opposite directions. The geometric implication of this result will be discussed after we establish some other interesting *geometric* properties of the cross product of two vectors.

THEOREM 13.9	**GEOMETRIC PROPERTIES OF THE CROSS PRODUCT**
If $\mathbf{u}$ and $\mathbf{v}$ are nonzero vectors in space and θ is the angle between $\mathbf{u}$ and $\mathbf{v}$, then the following properties are true.

1. $\mathbf{u} \times \mathbf{v}$ is orthogonal to both $\mathbf{u}$ and $\mathbf{v}$.
2. $\|\mathbf{u} \times \mathbf{v}\| = \|\mathbf{u}\| \, \|\mathbf{v}\| \sin \theta$.
3. $\mathbf{u} \times \mathbf{v} = \mathbf{0}$ if and only if $\mathbf{u}$ and $\mathbf{v}$ are scalar multiples of each other.
4. $\|\mathbf{u} \times \mathbf{v}\|$ = area of parallelogram having $\mathbf{u}$ and $\mathbf{v}$ as adjacent sides.

FIGURE 13.43

Proof: We prove the second and fourth properties and leave the proofs of the first and third properties as exercises. For the second property, we note that since $\cos \theta = (\mathbf{u} \cdot \mathbf{v})/(\|\mathbf{u}\| \, \|\mathbf{v}\|)$, it follows that

$$\|\mathbf{u}\| \, \|\mathbf{v}\| \sin \theta = \|\mathbf{u}\| \, \|\mathbf{v}\| \sqrt{1 - \cos^2 \theta}$$

$$= \|\mathbf{u}\| \, \|\mathbf{v}\| \sqrt{1 - \frac{(\mathbf{u} \cdot \mathbf{v})^2}{\|\mathbf{u}\|^2 \|\mathbf{v}\|^2}}$$

$$= \sqrt{\|\mathbf{u}\|^2 \|\mathbf{v}\|^2 - (\mathbf{u} \cdot \mathbf{v})^2}$$

$$= \sqrt{(u_1^2 + u_2^2 + u_3^2)(v_1^2 + v_2^2 + v_3^2) - (u_1 v_1 + u_2 v_2 + u_3 v_3)^2}$$

$$= \sqrt{(u_2 v_3 - u_3 v_2)^2 + (u_1 v_3 - u_3 v_1)^2 + (u_1 v_2 - u_2 v_1)^2}$$

$$= \|\mathbf{u} \times \mathbf{v}\|$$

To prove the fourth property, we use Figure 13.43, which is a parallelogram having $\mathbf{v}$ and $\mathbf{u}$ as adjacent sides. Since the height of the parallelogram is $\|\mathbf{v}\| \sin \theta$, the area is

$$\text{area} = (\text{base})(\text{height}) = \|\mathbf{u}\| \, \|\mathbf{v}\| \sin \theta = \|\mathbf{u} \times \mathbf{v}\|$$

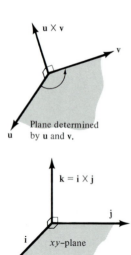

Plane determined by u and v.

$\mathbf{k} = \mathbf{i} \times \mathbf{j}$

Right-handed Systems

FIGURE 13.44

Remark It follows from the second property listed in Theorem 13.9 that if $\mathbf{n}$ is a unit vector orthogonal to both $\mathbf{u}$ and $\mathbf{v}$, then

$$\mathbf{u} \times \mathbf{v} = \pm(\|\mathbf{u}\| \, \|\mathbf{v}\| \sin \theta)\mathbf{n}$$

Both $\mathbf{u} \times \mathbf{v}$ and $\mathbf{v} \times \mathbf{u}$ are perpendicular to the plane determined by $\mathbf{u}$ and $\mathbf{v}$. One way to remember the orientation of the vectors $\mathbf{u}$, $\mathbf{v}$, and $\mathbf{u} \times \mathbf{v}$ is to compare them with the unit vectors $\mathbf{i}$, $\mathbf{j}$, and $\mathbf{k} = \mathbf{i} \times \mathbf{j}$, as shown in Figure 13.44. The three vectors $\mathbf{u}$, $\mathbf{v}$, and $\mathbf{u} \times \mathbf{v}$ form a *right-handed system*, whereas the three vectors $\mathbf{u}$, $\mathbf{v}$, and $\mathbf{v} \times \mathbf{u}$ form a *left-handed system*. Using a right-handed system (or the determinant form of the cross product), we can verify each of the following:

$$\mathbf{i} \times \mathbf{j} = \mathbf{k} \qquad \mathbf{j} \times \mathbf{k} = \mathbf{i} \qquad \mathbf{k} \times \mathbf{i} = \mathbf{j}$$
$$\mathbf{j} \times \mathbf{i} = -\mathbf{k} \qquad \mathbf{k} \times \mathbf{j} = -\mathbf{i} \qquad \mathbf{i} \times \mathbf{k} = -\mathbf{j}$$
$$\mathbf{i} \times \mathbf{i} = \mathbf{0} \qquad \mathbf{j} \times \mathbf{j} = \mathbf{0} \qquad \mathbf{k} \times \mathbf{k} = \mathbf{0}$$

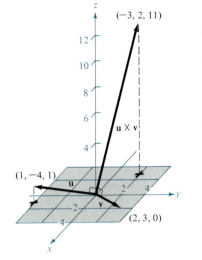

FIGURE 13.45

EXAMPLE 2   Applications of the cross product

Find a unit vector that is orthogonal to both

$$\mathbf{u} = \mathbf{i} - 4\mathbf{j} + \mathbf{k} \qquad \text{and} \qquad \mathbf{v} = 2\mathbf{i} + 3\mathbf{j}$$

Solution: The cross product $\mathbf{u} \times \mathbf{v}$, as shown in Figure 13.45, is orthogonal to both $\mathbf{u}$ and $\mathbf{v}$.

$$\mathbf{u} \times \mathbf{v} = \begin{vmatrix} \mathbf{i} & \mathbf{j} & \mathbf{k} \\ 1 & -4 & 1 \\ 2 & 3 & 0 \end{vmatrix} = -3\mathbf{i} + 2\mathbf{j} + 11\mathbf{k}$$

Since $\|\mathbf{u} \times \mathbf{v}\| = \sqrt{(-3)^2 + 2^2 + 11^2} = \sqrt{134}$, a unit vector orthogonal to both $\mathbf{u}$ and $\mathbf{v}$ is

$$\frac{\mathbf{u} \times \mathbf{v}}{\|\mathbf{u} \times \mathbf{v}\|} = -\frac{3}{\sqrt{134}}\mathbf{i} + \frac{2}{\sqrt{134}}\mathbf{j} + \frac{11}{\sqrt{134}}\mathbf{k}$$

EXAMPLE 3   An application of the cross product

Show that the quadrilateral with vertices at the following points is a parallelogram, and find its area:

$$A = (5, 2, 0) \qquad B = (2, 6, 1)$$
$$C = (2, 4, 7) \qquad D = (5, 0, 6)$$

Solution: From Figure 13.46 we see that the sides of the quadrilateral correspond to the four vectors

$$\overrightarrow{AB} = -3\mathbf{i} + 4\mathbf{j} + \mathbf{k} \qquad \overrightarrow{CD} = 3\mathbf{i} - 4\mathbf{j} - \mathbf{k} = -\overrightarrow{AB}$$
$$\overrightarrow{AD} = 0\mathbf{i} - 2\mathbf{j} + 6\mathbf{k} \qquad \overrightarrow{CB} = 0\mathbf{i} + 2\mathbf{j} - 6\mathbf{k} = -\overrightarrow{AD}$$

Thus, $\overrightarrow{AB}$ is parallel to $\overrightarrow{CD}$ and $\overrightarrow{AD}$ is parallel to $\overrightarrow{CB}$, and we conclude that the quadrilateral is a parallelogram with $\overrightarrow{AB}$ and $\overrightarrow{AD}$ as adjacent sides. Moreover, since

$$\overrightarrow{AB} \times \overrightarrow{AD} = \begin{vmatrix} \mathbf{i} & \mathbf{j} & \mathbf{k} \\ -3 & 4 & 1 \\ 0 & -2 & 6 \end{vmatrix} = 26\mathbf{i} + 18\mathbf{j} + 6\mathbf{k}$$

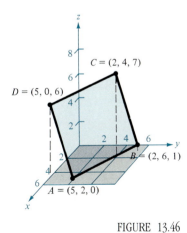

FIGURE 13.46

the area of the parallelogram is

$$\|\overrightarrow{AB} \times \overrightarrow{AD}\| = \sqrt{1036} \approx 32.19$$

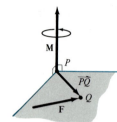

FIGURE 13.47

In physics, the cross product can be used to measure **torque**—the **moment M of a force F about a point *P*,** as shown in Figure 13.47. If the point of application of the force is Q, then the moment of **F** about *P* is given by

$$\mathbf{M} = \overrightarrow{PQ} \times \mathbf{F} \qquad \text{Moment of } \mathbf{F} \text{ about } P$$

The magnitude of the moment **M** measures the tendency of the vector $\overrightarrow{PQ}$ to rotate counterclockwise (using the right-hand rule) about an axis directed along the vector **M**.

EXAMPLE 4   *An application of the cross product*

A vertical force of 50 pounds is applied to the end of a 1-foot lever that is attached to an axle at point *P*, as shown in Figure 13.48. Find the moment of this force about the point *P* when $\theta = 60°$.

Solution: If we represent the 50-pound force as $\mathbf{F} = -50\mathbf{k}$ and the lever as

$$\overrightarrow{PQ} = \cos(60°)\,\mathbf{j} + \sin(60°)\,\mathbf{k} = \frac{1}{2}\mathbf{j} + \frac{\sqrt{3}}{2}\mathbf{k}$$

the moment of **F** about *P* is given by

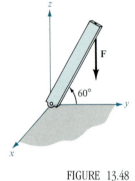

FIGURE 13.48

$$\mathbf{M} = \overrightarrow{PQ} \times \mathbf{F} = \begin{vmatrix} \mathbf{i} & \mathbf{j} & \mathbf{k} \\ 0 & \dfrac{1}{2} & \dfrac{\sqrt{3}}{2} \\ 0 & 0 & -50 \end{vmatrix} = -25\mathbf{i}$$

and the magnitude of this moment is 25 foot-pounds.

Remark   In Example 4, note that the moment (the tendency of the lever to rotate about its axle) is dependent upon the angle θ. When $\theta = \pi/2$, the moment is 0, and the moment is greatest when $\theta = 0$.

The triple scalar product

For vectors **u, v,** and **w** in space, the dot product of **u** and $\mathbf{v} \times \mathbf{w}$

$$\mathbf{u} \cdot (\mathbf{v} \times \mathbf{w})$$

is called the **triple scalar product.** This product has a practical geometric interpretation and it can be evaluated by the determinant given in the following theorem. We leave the proof of this theorem as an exercise.

THEOREM 13.10   **THE TRIPLE SCALAR PRODUCT**
For $\mathbf{u} = u_1\mathbf{i} + u_2\mathbf{j} + u_3\mathbf{k}$, $\mathbf{v} = v_1\mathbf{i} + v_2\mathbf{j} + v_3\mathbf{k}$, and $\mathbf{w} = w_1\mathbf{i} + w_2\mathbf{j} + w_3\mathbf{k}$, the triple scalar product is given by

$$\mathbf{u} \cdot (\mathbf{v} \times \mathbf{w}) = \begin{vmatrix} u_1 & u_2 & u_3 \\ v_1 & v_2 & v_3 \\ w_1 & w_2 & w_3 \end{vmatrix}$$

| Remark From the properties of determinants we know that the value of a determinant is multiplied by -1 if two rows are interchanged. After two such interchanges, the value of the determinant will be unchanged. Thus, the following triple scalar products are equivalent:

$$\mathbf{u} \cdot (\mathbf{v} \times \mathbf{w}) = \mathbf{v} \cdot (\mathbf{w} \times \mathbf{u}) = \mathbf{w} \cdot (\mathbf{u} \times \mathbf{v})$$

If the vectors **u, v,** and **w** do not lie in the same plane, then the triple scalar product $\mathbf{u} \cdot (\mathbf{v} \times \mathbf{w})$ can be used to determine the volume of the parallelepiped with **u, v,** and **w** as adjacent sides, as shown in Figure 13.49. This is established in the following theorem.

THEOREM 13.11

GEOMETRIC PROPERTY OF TRIPLE SCALAR PRODUCT
The volume of V of a parallelepiped with vectors **u, v,** and **ẇ** as adjacent sides is given by

$$V = |\mathbf{u} \cdot (\mathbf{v} \times \mathbf{w})|$$

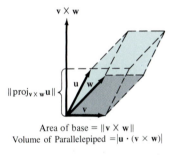

Area of base $= \|\mathbf{v} \times \mathbf{w}\|$
Volume of Parallelepiped $= |\mathbf{u} \cdot (\mathbf{v} \times \mathbf{w})|$

FIGURE 13.49

Proof: In Figure 13.49, we note that

$$\|\mathbf{v} \times \mathbf{w}\| = \text{area of base}$$
$$\|\text{proj}_{\mathbf{v} \times \mathbf{w}}\mathbf{u}\| = \text{height of parallelepiped}$$

Therefore, the volume is

$$V = (\text{height})(\text{area of base}) = \|\text{proj}_{\mathbf{v} \times \mathbf{w}}\mathbf{u}\| \, \|\mathbf{v} \times \mathbf{w}\|$$
$$= \left| \frac{\mathbf{u} \cdot (\mathbf{v} \times \mathbf{w})}{\|\mathbf{v} \times \mathbf{w}\|} \right| \|\mathbf{v} \times \mathbf{w}\|$$
$$= |\mathbf{u} \cdot (\mathbf{v} \times \mathbf{w})|$$

EXAMPLE 5   Volume by the triple scalar product

Find the volume of the parallelepiped having $\mathbf{u} = 3\mathbf{i} - 5\mathbf{j} + \mathbf{k}$, $\mathbf{v} = 2\mathbf{j} - 2\mathbf{k}$, and $\mathbf{w} = 3\mathbf{i} + \mathbf{j} + \mathbf{k}$ as adjacent edges.

Solution: By Theorem 13.11, we have

$$V = |\mathbf{u} \cdot (\mathbf{v} \times \mathbf{w})| = \begin{vmatrix} 3 & -5 & 1 \\ 0 & 2 & -2 \\ 3 & 1 & 1 \end{vmatrix} = 36$$

A natural consequence of Theorem 13.11 is that the volume of the parallelepiped is zero if and only if the three vectors are coplanar. This gives us the following test.

Test for Coplanar Vectors: If the vectors $\mathbf{u} = \langle u_1, u_2, u_3 \rangle$, $\mathbf{v} = \langle v_1, v_2, v_3 \rangle$, $\mathbf{w} = \langle w_1, w_2, w_3 \rangle$ have the same initial point, then they lie in the same plane if and only if

$$\mathbf{u} \cdot (\mathbf{v} \times \mathbf{w}) = \begin{vmatrix} u_1 & u_2 & u_3 \\ v_1 & v_2 & v_3 \\ w_1 & w_2 & w_3 \end{vmatrix} = 0$$

Section Exercises 13.4

In Exercises 1–6, find the cross product of the given unit vectors and sketch your result.

1. $\mathbf{j} \times \mathbf{i}$ **2.** $\mathbf{i} \times \mathbf{j}$
3. $\mathbf{j} \times \mathbf{k}$ **4.** $\mathbf{k} \times \mathbf{j}$
5. $\mathbf{i} \times \mathbf{k}$ **6.** $\mathbf{k} \times \mathbf{i}$

In Exercises 7–14, find $\mathbf{u} \times \mathbf{v}$ and show that it is orthogonal to both $\mathbf{u}$ and $\mathbf{v}$.

7. $\mathbf{u} = \langle 2, -3, 1 \rangle$, $\mathbf{v} = \langle 1, -2, 1 \rangle$
8. $\mathbf{u} = \langle -1, 1, 2 \rangle$, $\mathbf{v} = \langle 0, 1, 0 \rangle$
9. $\mathbf{u} = \langle 12, -3, 0 \rangle$, $\mathbf{v} = \langle -2, 5, 0 \rangle$
10. $\mathbf{u} = \langle -10, 0, 6 \rangle$, $\mathbf{v} = \langle 7, 0, 0 \rangle$
11. $\mathbf{u} = \mathbf{i} + \mathbf{j} + \mathbf{k}$, $\mathbf{v} = 2\mathbf{i} + \mathbf{j} - \mathbf{k}$
12. $\mathbf{u} = \mathbf{j} + 6\mathbf{k}$, $\mathbf{v} = \mathbf{i} - 2\mathbf{j} + \mathbf{k}$
13. $\mathbf{u} = -3\mathbf{i} + 2\mathbf{j} - 5\mathbf{k}$, $\mathbf{v} = \dfrac{1}{2}\mathbf{i} - \dfrac{3}{4}\mathbf{j} + \dfrac{1}{10}\mathbf{k}$
14. $\mathbf{u} = \dfrac{2}{3}\mathbf{k}$, $\mathbf{v} = \dfrac{1}{2}\mathbf{i} + 6\mathbf{k}$

In Exercises 15–18, find the area of the parallelogram that has the given vectors as adjacent sides.

15. $\mathbf{u} = \mathbf{j}$, $\mathbf{v} = \mathbf{j} + \mathbf{k}$
16. $\mathbf{u} = \mathbf{i} + \mathbf{j} + \mathbf{k}$, $\mathbf{v} = \mathbf{j} + \mathbf{k}$
17. $\mathbf{u} = \langle 3, 2, -1 \rangle$, $\mathbf{v} = \langle 1, 2, 3 \rangle$
18. $\mathbf{u} = \langle 2, -1, 0 \rangle$, $\mathbf{v} = \langle -1, 2, 0 \rangle$

In Exercises 19 and 20, find the area of the parallelogram with the given vertices.

19. $(1, 1, 1)$, $(2, 3, 4)$, $(6, 5, 2)$, $(7, 7, 5)$
20. $(2, -1, 1)$, $(5, 1, 4)$, $(0, 1, 1)$, $(3, 3, 4)$

In Exercises 21–24, find the area of the triangle with the given vertices. ($\frac{1}{2}\|\mathbf{u} \times \mathbf{v}\|$ is the area of the triangle having $\mathbf{u}$ and $\mathbf{v}$ as adjacent sides.)

21. $(0, 0, 0)$, $(1, 2, 3)$, $(-3, 0, 0)$
22. $(2, -3, 4)$, $(0, 1, 2)$, $(-1, 2, 0)$

23. $(1, 3, 5)$, $(3, 3, 0)$, $(-2, 0, 5)$
24. $(1, 2, 0)$, $(-2, 1, 0)$, $(0, 0, 0)$

In Exercises 25–28, find $\mathbf{u} \cdot (\mathbf{v} \times \mathbf{w})$.

25. $\mathbf{u} = \mathbf{i}$, $\mathbf{v} = \mathbf{j}$, $\mathbf{w} = \mathbf{k}$
26. $\mathbf{u} = \langle 1, 1, 1 \rangle$, $\mathbf{v} = \langle 2, 1, 0 \rangle$, $\mathbf{w} = \langle 0, 0, 1 \rangle$
27. $\mathbf{u} = \langle 2, 0, 1 \rangle$, $\mathbf{v} = \langle 0, 3, 0 \rangle$, $\mathbf{w} = \langle 0, 0, 1 \rangle$
28. $\mathbf{u} = \langle 2, 0, 0 \rangle$, $\mathbf{v} = \langle 1, 1, 1 \rangle$, $\mathbf{w} = \langle 0, 2, 2 \rangle$

In Exercises 29 and 30, use the triple scalar product to find the volume of the parallelepiped having adjacent edges $\mathbf{u}$, $\mathbf{v}$, and $\mathbf{w}$.

29. $\mathbf{u} = \mathbf{i} + \mathbf{j}$, $\mathbf{v} = \mathbf{j} + \mathbf{k}$, $\mathbf{w} = \mathbf{i} + \mathbf{k}$
  (See Figure 13.50.)
30. $\mathbf{u} = \langle 1, 3, 1 \rangle$, $\mathbf{v} = \langle 0, 5, 5 \rangle$, $\mathbf{w} = \langle 4, 0, 4 \rangle$
  (See Figure 13.51.)

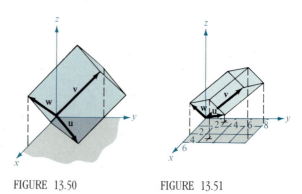

FIGURE 13.50   FIGURE 13.51

In Exercises 31 and 32, find the volume of the parallelepiped with the given vertices.

31. $(0, 0, 0)$, $(3, 0, 0)$, $(0, 5, 1)$, $(3, 5, 1)$, $(2, 0, 5)$, $(5, 0, 5)$, $(2, 5, 6)$, $(5, 5, 6)$
  (See Figure 13.52.)

32. $(0, 0, 0)$, $(1, 1, 0)$, $(1, 0, 2)$, $(0, 1, 1)$, $(2, 1, 2)$,
$(1, 1, 3)$, $(1, 2, 1)$, $(2, 2, 3)$
(See Figure 13.53.)

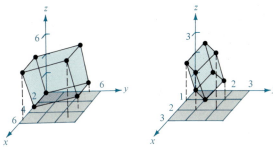

FIGURE 13.52   FIGURE 13.53

In Exercises 33–39, prove the property of the cross product.

33. $\mathbf{u} \times \mathbf{v} = -(\mathbf{v} \times \mathbf{u})$
34. $(c\mathbf{u}) \times \mathbf{v} = c(\mathbf{u} \times \mathbf{v})$
35. $\mathbf{u} \times (\mathbf{v} \times \mathbf{w}) = (\mathbf{u} \cdot \mathbf{w})\mathbf{v} - (\mathbf{u} \cdot \mathbf{v})\mathbf{w}$
36. $\mathbf{u} \cdot (\mathbf{v} \times \mathbf{w}) = (\mathbf{u} \times \mathbf{v}) \cdot \mathbf{w}$
37. $\mathbf{u} \times \mathbf{v}$ is orthogonal to both $\mathbf{u}$ and $\mathbf{v}$.
38. $\mathbf{u} \times \mathbf{v} = \mathbf{0}$ if and only if $\mathbf{u}$ and $\mathbf{v}$ are scalar multiples of each other.
39. $\|\mathbf{u} \times \mathbf{v}\| = \|\mathbf{u}\|\,\|\mathbf{v}\|$ if $\mathbf{u}$ and $\mathbf{v}$ are orthogonal.

SECTION TOPICS ▪
Lines in space ▪
Planes in space ▪
Distance between points,
lines, and planes ▪

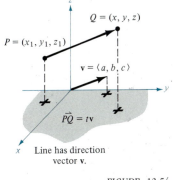

FIGURE 13.54

13.5
Lines and planes in space

In the plane we use *slope* to determine an equation of the line. In space it is more convenient to use *vectors* to determine the equation of a line.

In Figure 13.54, consider the line L through the point $P = (x_1, y_1, z_1)$ and parallel to the vector $\mathbf{v} = \langle a, b, c \rangle$. We call $\mathbf{v}$ the **direction vector** for line L, and a, b, and c the **direction numbers.** One way of describing line L is to say that it consists of all points $Q = (x, y, z)$ for which the vector $\overrightarrow{PQ}$ is parallel to $\mathbf{v}$. This means that $\overrightarrow{PQ}$ is a scalar multiple of $\mathbf{v}$, and we write $\overrightarrow{PQ} = t\mathbf{v}$, where t is a scalar. In component form, we have

$$\overrightarrow{PQ} = \langle x - x_1, y - y_1, z - z_1 \rangle$$
$$= \langle at, bt, ct \rangle = t\mathbf{v}$$

By equating corresponding components, we obtain the equations

$$x = x_1 + at, \qquad y = y_1 + bt, \qquad \text{and} \qquad z = z_1 + ct$$

These are called **parametric equations** of a line in space.

THEOREM 13.12   **PARAMETRIC EQUATIONS OF A LINE IN SPACE**
A line L parallel to the vector $\mathbf{v} = \langle a, b, c \rangle$ and passing through the point $P = (x_1, y_1, z_1)$ is represented by the **parametric equations**

$$x = x_1 + at, \qquad y = y_1 + bt, \qquad \text{and} \qquad z = z_1 + ct$$

| **Remark** If the direction numbers a, b, and c are all nonzero, then we can eliminate the parameter t and obtain the following **symmetric equations** for a line:

$$\frac{x - x_1}{a} = \frac{y - y_1}{b} = \frac{z - z_1}{c}$$

EXAMPLE 1   *Finding parametric and symmetric equations for a line*

Find a set of parametric equations and a set of symmetric equations for the line L that passes through the point $(1, -2, 4)$ and is parallel to $\mathbf{v} = \langle 2, 4, -4 \rangle$.

Solution: For the coordinates $x_1 = 1$, $y_1 = -2$, and $z_1 = 4$ and direction numbers $a = 2$, $b = 4$, and $c = -4$, a set of parametric equations for the line L is

$$x = 1 + 2t, \qquad y = -2 + 4t, \qquad \text{and} \qquad z = 4 - 4t$$

Since a, b, and c are all nonzero, a set of symmetric equations is

$$\frac{x - 1}{2} = \frac{y + 2}{4} = \frac{z - 4}{-4}$$

Neither the parametric equations nor the symmetric equations for a given line are unique. For instance, in Example 1, by letting $t = 1$ in the parametric equations we obtain the point $(3, 2, 0)$. Using this point with the direction numbers $a = 2$, $b = 4$, and $c = -4$, we have the parametric equations

$$x = 3 + 2t, \qquad y = 2 + 4t, \qquad \text{and} \qquad z = -4t$$

EXAMPLE 2   *Finding an equation for a line through two points*

Find a set of parametric equations for the line that passes through the points $(-2, 1, 0)$ and $(1, 3, 5)$.

Solution: We begin by letting $P = (-2, 1, 0)$ and $Q = (1, 3, 5)$. Then a direction vector for the line passing through P and Q is given by

$$\mathbf{v} = \overrightarrow{PQ} = \langle 1 - (-2), 3 - 1, 5 - 0 \rangle = \langle 3, 2, 5 \rangle = \langle a, b, c \rangle$$

Now, using the direction numbers $a = 3$, $b = 2$, and $c = 5$, with the point $P = (-2, 1, 0)$, we obtain the parametric equations

$$x = -2 + 3t, \qquad y = 1 + 2t, \qquad \text{and} \qquad z = 5t$$

Planes in space

We have seen how an equation for a line in space can be obtained from a point on the line and a vector *parallel* to it. We now look at how to obtain an equation for a plane in space from a point in the plane and a vector *normal* (perpendicular) to it.

Consider the plane containing the point $P = (x_1, y_1, z_1)$ having a nonzero normal vector $\mathbf{n} = \langle a, b, c \rangle$, as shown in Figure 13.55. This plane con-

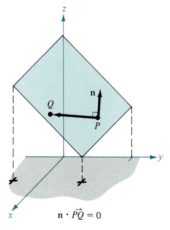

$\mathbf{n} \cdot \overrightarrow{PQ} = 0$

FIGURE 13.55

sists of all points $Q = (x, y, z)$ for which vector $\overrightarrow{PQ}$ is orthogonal to **n.** Using the dot product, we have

$$\mathbf{n} \cdot \overrightarrow{PQ} = 0$$
$$\langle a, b, c \rangle \cdot \langle x - x_1, y - y_1, z - z_1 \rangle = 0$$
$$a(x - x_1) + b(y - y_1) + c(z - z_1) = 0$$

The third equation of the plane is said to be in **standard form.**

THEOREM 13.13   STANDARD EQUATION OF A PLANE IN SPACE
The plane containing the point (x_1, y_1, z_1) and having a normal vector $\mathbf{n} = \langle a, b, c \rangle$ can be represented, in **standard form,** by the equation

$$a(x - x_1) + b(y - y_1) + c(z - z_1) = 0$$

Remark   By regrouping terms, we obtain the **general form** of the equation of a plane in space,

$$ax + by + cz + d = 0$$

If you are given this general form, it is easy to find a normal vector to the plane. Just use the coefficients of x, y, and z and write $\mathbf{n} = \langle a, b, c \rangle$.

EXAMPLE 3   Finding an equation of a plane in three-space

Find the general equation of the plane containing the points $(2, 1, 1)$, $(0, 4, 1)$, and $(-2, 1, 4)$.

Solution: To apply Theorem 13.13 we need a point in the plane and a vector that is normal to the plane. There are three choices for the point, but no normal vector is given. To obtain a normal vector, we use the cross product of vectors **u** and **v** extending from the point $(2, 1, 1)$ to the points $(0, 4, 1)$ and $(-2, 1, 4)$, as shown in Figure 13.56. We have

$$\mathbf{u} = \langle 0 - 2, 4 - 1, 1 - 1 \rangle = \langle -2, 3, 0 \rangle$$
$$\mathbf{v} = \langle -2 - 2, 1 - 1, 4 - 1 \rangle = \langle -4, 0, 3 \rangle$$

and it follows that

$$\mathbf{n} = \mathbf{u} \times \mathbf{v} = \begin{vmatrix} \mathbf{i} & \mathbf{j} & \mathbf{k} \\ -2 & 3 & 0 \\ -4 & 0 & 3 \end{vmatrix} = 9\mathbf{i} + 6\mathbf{j} + 12\mathbf{k} = \langle a, b, c \rangle$$

is normal to the given plane. Using the direction numbers for **n** and the point $(x_1, y_1, z_1) = (2, 1, 1)$, we determine an equation of the plane to be

$$a(x - x_1) + b(y - y_1) + c(z - z_1) = 0$$
$$9(x - 2) + 6(y - 1) + 12(z - 1) = 0$$
$$9x + 6y + 12z - 36 = 0$$
$$3x + 2y + 4z - 12 = 0$$

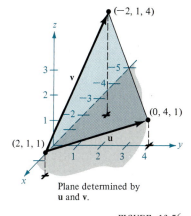

FIGURE 13.56

Plane determined by **u** and **v**.

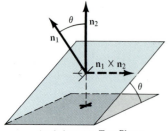

θ n_2

n_1

$n_1 \times n_2$

θ

Angle between Two Planes

FIGURE 13.57

Two planes in three-space either are parallel or they intersect in a line. If they intersect we can determine the angle between them from the angle between their normal vectors, as shown in Figure 13.57. Specifically, if vectors $\mathbf{n}_1$ and $\mathbf{n}_2$ are normal to two intersecting planes, then the angle θ between the normal vectors is equal to the angle between the two planes and is given by

$$\cos \theta = \frac{|\mathbf{n}_1 \cdot \mathbf{n}_2|}{\|\mathbf{n}_1\| \, \|\mathbf{n}_2\|}$$

Consequently, two planes with normal vectors $\mathbf{n}_1$ and $\mathbf{n}_2$ are

1. *perpendicular* if $\mathbf{n}_1 \cdot \mathbf{n}_2 = 0$
2. *parallel* if $\mathbf{n}_1$ is a scalar multiple of $\mathbf{n}_2$

EXAMPLE 4   *Finding the line of intersection of two planes*

Find the angle between the two planes given by

$$x - 2y + z = 0 \qquad \text{and} \qquad 2x + 3y - 2z = 0$$

and find parametric equations for their line of intersection.

Solution: The normal vectors for the planes are $\mathbf{n}_1 = \langle 1, -2, 1 \rangle$ and $\mathbf{n}_2 = \langle 2, 3, -2 \rangle$. Consequently, the angle between the two planes is

$$\cos \theta = \frac{|\mathbf{n}_1 \cdot \mathbf{n}_2|}{\|\mathbf{n}_1\| \, \|\mathbf{n}_2\|} = \frac{|-6|}{\sqrt{6} \, \sqrt{17}} = \frac{6}{\sqrt{102}} \approx 0.59409$$

$$\theta \approx 53.55°$$

We can find the line of intersection of the two planes by simultaneously solving the two linear equations representing the planes.

$$\begin{array}{l} x - 2y + z = 0 \\ 2x + 3y - 2z = 0 \end{array} \implies \begin{array}{l} -2x + 4y - 2z = 0 \\ \underline{2x + 3y - 2z = 0} \\ 7y - 4z = 0 \implies y = 4z/7 \end{array}$$

Now, substituting $y = 4z/7$ back into one of the original equations, we find that $x = z/7$. Finally, letting $t = z/7$, we have the parametric equations

$$x = t, \qquad y = 4t, \qquad \text{and} \qquad z = 7t$$

which indicate that the direction numbers are 1, 4, and 7.

| **Remark** Note that the direction numbers in Example 3 can be obtained from the cross product of the two normal vectors:

$$\mathbf{n}_1 \times \mathbf{n}_2 = \begin{vmatrix} \mathbf{i} & \mathbf{j} & \mathbf{k} \\ 1 & -2 & 1 \\ 2 & 3 & -2 \end{vmatrix} = \mathbf{i} + 4\mathbf{j} + 7\mathbf{k}$$

This means that the line of intersection of the two planes is parallel to the cross product of their normal vectors.

Sketching planes in space

If a plane in space intersects one of the coordinate planes, we call the line of intersection the **trace** of the given plane in the coordinate plane. To sketch a plane in space, it is helpful to find its points of intersection with the coordinate axes and its traces in the coordinate planes. For example, consider the plane given by

$$3x + 2y + 4z = 12 \qquad \text{Equation of plane}$$

We find the xy-trace by letting $z = 0$ and sketching the line

$$3x + 2y = 12 \qquad \text{xy-trace}$$

in the xy-plane. This line intersects the x-axis at $(4, 0, 0)$ and the y-axis at $(0, 6, 0)$. In Figure 13.58, we continue this process by finding the yz-trace and the xz-trace, and then shading in the triangular region lying in the first octant.

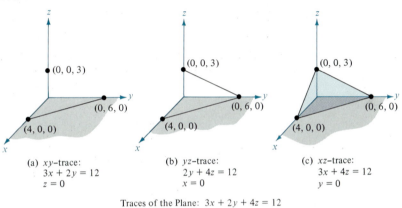

(a) xy-trace:
  $3x + 2y = 12$
  $z = 0$

(b) yz-trace:
  $2y + 4z = 12$
  $x = 0$

(c) xz-trace:
  $3x + 4z = 12$
  $y = 0$

Traces of the Plane: $3x + 2y + 4z = 12$

FIGURE 13.58

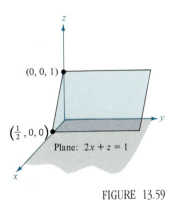

Plane: $2x + z = 1$

FIGURE 13.59

If the equation of a plane has a missing variable such as $2x + z = 1$, then the plane must be *parallel to the axis* represented by the missing variable, as shown in Figure 13.59. If two variables are missing from the equation of a plane, then it is *parallel to the coordinate plane* represented by the missing variables, as shown in Figure 13.60.

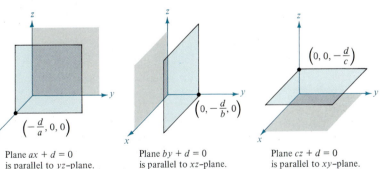

Plane $ax + d = 0$
is parallel to yz-plane.

Plane $by + d = 0$
is parallel to xz-plane.

Plane $cz + d = 0$
is parallel to xy-plane.

FIGURE 13.60

Distances between points, planes, and lines

We conclude this section with two basic distance problems in space.

1. Finding the distance between a point and a plane
2. Finding the distance between a point and a line

The solutions to these problems illustrate the versatility and usefulness of vectors in coordinate geometry. In the first problem we use the *dot product* of two vectors, and in the second problem we use the *cross product*.

By the distance between a point Q and a plane, we mean the length of the shortest line segment connecting Q to the plane. If we knew which point in the plane was closest to Q, we could simply apply the formula for the distance between two points in space. However, this closest point is not easy to find, and so we use a different approach. In Figure 13.61, let D be the distance between Q and the given plane. Then if P is *any* point on the plane, we can find the distance by projecting the vector $\overrightarrow{PQ}$ onto the normal vector $\mathbf{n}$. The length of this projection gives us the desired distance. This result is stated in the following theorem.

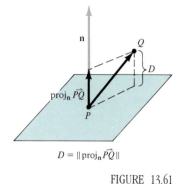

$D = \|\operatorname{proj}_{\mathbf{n}} \overrightarrow{PQ}\|$

FIGURE 13.61

THEOREM 13.14   **DISTANCE BETWEEN A POINT AND A PLANE**
The distance between a plane and a point Q (not on the plane) is

$$D = \|\operatorname{proj}_{\mathbf{n}} \overrightarrow{PQ}\| = \frac{|\overrightarrow{PQ} \cdot \mathbf{n}|}{\|\mathbf{n}\|}$$

where P is a point on the plane and $\mathbf{n}$ is normal to the plane.

Remark To find a point in the plane given by $ax + by + cz + d = 0$, where $a \neq 0$, let $y = 0$ and $z = 0$. Then from the equation $ax + d = 0$, you can conclude that the point $(-d/a, 0, 0)$ lies in the plane.

From this theorem, we see that the distance between the point $Q = (x_0, y_0, z_0)$ and the plane given by $ax + by + cz + d = 0$ is

$$D = \frac{|a(x_0 - x_1) + b(y_0 - y_1) + c(z_0 - z_1)|}{\sqrt{a^2 + b^2 + c^2}}$$

$$= \frac{|ax_0 + by_0 + cz_0 + d|}{\sqrt{a^2 + b^2 + c^2}}$$

where $d = -(ax_1 + by_1 + cz_1)$ and $P = (x_1, y_1, z_1)$ is a point on the plane. (Note the similarity between this formula and the corresponding formula for the distance between a point and a line in the plane, as given in Exercises 66–71 in Section 1.4.)

EXAMPLE 5   Finding the distance between a point and a plane

Find the distance between the point $Q = (1, 5, -4)$ and the plane given by $3x - y + 2z = 6$.

Solution: We know that $\mathbf{n} = \langle 3, -1, 2 \rangle$ is normal to the given plane. To find a point in the plane, we let $y = 0$ and $z = 0$ and obtain the point $P = (2, 0, 0)$. Now, the vector from P to Q is given by

$$\overrightarrow{PQ} = \langle 1 - 2, 5 - 0, -4 - 0 \rangle = \langle -1, 5, -4 \rangle$$

Finally, using the distance formula given in Theorem 13.14, we have

$$D = \frac{|\overrightarrow{PQ} \cdot \mathbf{n}|}{\|\mathbf{n}\|} = \frac{|\langle -1, 5, -4 \rangle \cdot \langle 3, -1, 2 \rangle|}{\sqrt{9 + 1 + 4}} = \frac{|-3 - 5 - 8|}{\sqrt{14}} = \frac{16}{\sqrt{14}}$$

Remark The choice of the point P in Example 5 is arbitrary. Try choosing a different point to verify that you obtain the same distance.

EXAMPLE 6 *Finding the distance between two parallel planes*

Find the distance between the two parallel planes given by

$$3x - y + 2z = 6 \qquad \text{and} \qquad -6x + 2y - 4z = 4$$

Solution: Note that the normal vectors

$$\mathbf{n}_1 = \langle 3, -1, 2 \rangle \qquad \text{and} \qquad \mathbf{n}_2 = \langle -6, 2, -4 \rangle$$

are parallel, since $\mathbf{n}_2 = -2\mathbf{n}_1$. Hence the two planes are parallel. To find the distance between the planes we choose a point in each one. For instance, choosing

$$P = (2, 0, 0) \qquad \text{Point in plane: } 3x - y + 2z = 6$$
$$Q = (0, 0, -1) \qquad \text{Point in plane: } -6x + 2y - 4z = 4$$

we have $\overrightarrow{PQ} = \langle -2, 0, -1 \rangle$. Thus, the distance between the planes is

$$D = \frac{|\overrightarrow{PQ} \cdot \mathbf{n}_1|}{\|\mathbf{n}_1\|} = \frac{|\langle -2, 0, -1 \rangle \cdot \langle 3, -1, 2 \rangle|}{\sqrt{9 + 1 + 4}} = \frac{|-6 - 2|}{\sqrt{14}} = \frac{8}{\sqrt{14}}$$

The formula for the distance between a point and a line in space resembles that for the distance between a point and a plane. For this distance, we replace the dot product by the cross product and replace the normal vector $\mathbf{n}$ by a direction vector for the given line. The validity of this procedure is shown in the proof of the following theorem.

THEOREM 13.15   DISTANCE BETWEEN A POINT AND A LINE IN SPACE
The distance between a point Q and a line in space is given by

$$D = \frac{\|\overrightarrow{PQ} \times \mathbf{u}\|}{\|\mathbf{u}\|}$$

where $\mathbf{u}$ is the direction vector for the line and P is any point on the line.

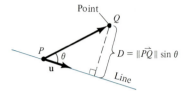

Point

$D = \|\overrightarrow{PQ}\| \sin \theta$

Line

Distance between a point and a line

FIGURE 13.62

Proof: In Figure 13.62, we let D be the distance between the point Q and the given line. Then

$$D = \|\overrightarrow{PQ}\| \sin \theta$$

where θ is the angle between $\mathbf{u}$ and $\overrightarrow{PQ}$. By Theorem 13.9, we have

$$\|\mathbf{u}\| \, \|\overrightarrow{PQ}\| \sin \theta = \|\mathbf{u} \times \overrightarrow{PQ}\| = \|\overrightarrow{PQ} \times \mathbf{u}\|$$

Consequently,

$$D = \|\overrightarrow{PQ}\| \sin \theta = \frac{\|\overrightarrow{PQ} \times \mathbf{u}\|}{\|\mathbf{u}\|}$$

EXAMPLE 7   Finding the distance between a point and a line

Find the distance between the point $Q = (3, -1, 4)$ and the line given by

$$x = -2 + 3t, \qquad y = -2t, \qquad \text{and} \qquad z = 1 + 4t$$

Solution: Using the direction numbers 3, -2, and 4, we find the direction vector for the line to be

$$\mathbf{u} = \langle 3, -2, 4 \rangle \qquad \text{Direction vector for line}$$

To find a point on the line, we let $t = 0$ and obtain

$$P = (-2, 0, 1) \qquad \text{Point on the line}$$

Thus,

$$\overrightarrow{PQ} = \langle 3 - (-2), -1 - 0, 4 - 1 \rangle = \langle 5, -1, 3 \rangle$$

and we form the cross product

$$\overrightarrow{PQ} \times \mathbf{u} = \begin{vmatrix} \mathbf{i} & \mathbf{j} & \mathbf{k} \\ 5 & -1 & 3 \\ 3 & -2 & 4 \end{vmatrix} = 2\mathbf{i} - 11\mathbf{j} - 7\mathbf{k} = \langle 2, -11, -7 \rangle$$

Finally, using Theorem 13.15, we find the distance to be

$$D = \frac{\|\overrightarrow{PQ} \times \mathbf{u}\|}{\|\mathbf{u}\|} = \frac{\sqrt{174}}{\sqrt{29}} = \sqrt{6}$$

Section Exercises 13.5

In Exercises 1–10, find a set of (a) parametric equations and (b) symmetric equations of the specified line. (For each line, express the direction numbers as integers.)

1. The line passes through the origin and is parallel to $\mathbf{v} = \langle 1, 2, 3 \rangle$.

2. The line passes through the origin and is parallel to $\mathbf{v} = \langle -2, \frac{5}{2}, 1 \rangle$.

3. The line passes through the point $(-2, 0, 3)$ and is parallel to $\mathbf{v} = 2\mathbf{i} + 4\mathbf{j} - 2\mathbf{k}$.

4. The line passes through the point $(-2, 0, 3)$ and is parallel to $\mathbf{v} = 6\mathbf{i} + 3\mathbf{j}$.

5. The line passes through the points $(5, -3, -2)$ and $(-\frac{2}{3}, \frac{2}{3}, 1)$.

6. The line passes through the points $(1, 0, 1)$ and $(1, 3, -2)$.

7. The line passes through the point $(1, 0, 1)$ and is parallel to the line given by

$$x = 3 + 3t$$
$$y = 5 - 2t$$
$$z = -7 + t$$

8. The line passes through the point $(-3, 5, 4)$ and is parallel to the line given by

$$\frac{x - 1}{3} = \frac{y + 1}{-2} = z - 3$$

9. The line passes through the point $(2, 3, 4)$ and is parallel to the xz-plane and the yz-plane.

10. The line passes through the point $(2, 3, 4)$ and is perpendicular to the plane given by $3x + 2y - z = 6$.

In Exercises 11 and 12, determine which of the points lie on the line L.

11. The line L passes through the point $(-2, 3, 1)$ and is parallel to the vector $\mathbf{v} = 4\mathbf{i} - \mathbf{k}$.
   (a) $(2, 3, 0)$ (b) $(-6, 3, 2)$
   (c) $(2, 1, 0)$ (d) $(10, 3, -2)$
   (e) $(6, 3, -2)$

12. The line L passes through the points $(2, 0, -3)$ and $(4, 2, -2)$.
   (a) $(4, 1, -2)$ (b) $\left(3, 1, -\dfrac{5}{2}\right)$
   (c) $\left(\dfrac{5}{2}, \dfrac{1}{2}, -\dfrac{11}{4}\right)$ (d) $(-1, -3, -4)$
   (e) $(0, -2, -4)$

In Exercises 13–16, determine if the lines intersect, and if so, find the point of intersection and the cosine of the angle of intersection.

13. $x = 4t + 2$ $x = 2s + 2$
   $y = 3$ $y = 2s + 3$
   $z = -t + 1$ $z = s + 1$

14. $x = -3t + 1$ $x = 3s + 1$
   $y = 4t + 1$ $y = 2s + 4$
   $z = 2t + 4$ $z = -s + 1$

15. $\dfrac{x}{3} = \dfrac{y - 2}{-1} = z + 1$

   $\dfrac{x - 1}{4} = y + 2 = \dfrac{z + 3}{-3}$

16. $\dfrac{x - 2}{-3} = \dfrac{y - 2}{6} = z - 3$

   $\dfrac{x - 3}{2} = y + 5 = \dfrac{z + 2}{4}$

In Exercises 17–32, find the equation of the specified plane.

17. The plane passes through the point $(2, 1, 2)$ and has normal vector $\mathbf{n} = \mathbf{i}$.

18. The plane passes through the point $(1, 0, -3)$ and has normal vector $\mathbf{n} = \mathbf{k}$.

19. The plane passes through the point $(3, 2, 2)$ and has normal vector $\mathbf{n} = 2\mathbf{i} + 3\mathbf{j} - \mathbf{k}$.

20. The plane passes through the point $(3, 2, 2)$ and is perpendicular to the line given by

$$\frac{x - 1}{4} = y + 2 = \frac{z + 3}{-3}$$

21. The plane passes through the points $(0, 0, 0)$, $(1, 2, 3)$, and $(-2, 3, 3)$.

22. The plane passes through the points $(1, 2, -3)$, $(2, 3, 1)$, and $(0, -2, -1)$.

23. The plane passes through the points $(1, 2, 3)$, $(3, 2, 1)$, and $(-1, -2, 2)$.

24. The plane passes through the point $(1, 2, 3)$ and is parallel to the yz-plane.

25. The plane passes through the point $(1, 2, 3)$ and is parallel to the xy-plane.

26. The plane contains the y-axis and makes an angle of $\pi/6$ with the positive x-axis.

27. The plane contains lines given by

$$\frac{x - 1}{-2} = y - 4 = z$$

$$\frac{x - 2}{-3} = \frac{y - 1}{4} = \frac{z - 2}{-1}$$

28. The plane passes through the point $(2, 2, 1)$ and contains the line given by

$$\frac{x}{2} = \frac{y - 4}{-1} = z$$

29. The plane passes through the points $(2, 2, 1)$ and $(-1, 1, -1)$ and is perpendicular to the plane $2x - 3y + z = 3$.

30. The plane passes through the points $(3, 2, 1)$ and $(3, 1, -5)$ and is perpendicular to the plane $6x + 7y + 2z = 10$.

31. The plane passes through the points $(1, -2, -1)$ and $(2, 5, 6)$ and is parallel to the x-axis.

32. The plane passes through the points $(4, 2, 1)$ and $(-3, 5, 7)$ and is parallel to the z-axis.

In Exercises 33–40, determine if the planes are parallel, orthogonal, or neither. If they are neither parallel nor orthogonal, find the angle of intersection.

33. $5x - 3y + z = 4,\ x + 4y + 7z = 1$
34. $3x + y - 4z = 3,\ -9x - 3y + 12z = 4$
35. $x - 3y + 6z = 4,\ 5x + y - z = 4$
36. $3x + 2y - z = 7,\ x - 4y + 2z = 0$

37. $x - 5y - z = 1$, $5x - 25y - 5z = -3$
38. $2x - z = 1$, $4x + y + 8z = 10$
39. $x + 3y + z = 7$, $x - 5z = 0$
40. $2x + y = 3$, $x - 5z = 0$

In Exercises 41–48, mark the intercepts and sketch the graph of the plane.

41. $4x + 2y + 6z = 12$ **42.** $3x + 6y + 2z = 6$
43. $2x - y + 3z = 4$ **44.** $2x - y + z = 4$
45. $y + z = 5$ **46.** $x + 2y = 4$
47. $2x + y - z = 6$ **48.** $x - 3z = 3$

In Exercises 49 and 50, find a set of parametric equations for the line of intersection of the planes.

49. $3x + 2y - z = 7$, $x - 4y + 2z = 0$
50. $x - 3y + 6z = 4$, $5x + y - z = 4$

In Exercises 51–54, find the point of intersection (if any) of the plane and the line. Also determine if the line lies in the plane.

51. $2x - 2y + z = 12$
$$x - \frac{1}{2} = \frac{y + (3/2)}{-1} = \frac{z + 1}{2}$$
52. $2x + 3y = -5$
$$\frac{x - 1}{4} = \frac{y}{2} = \frac{z - 3}{6}$$
53. $2x + 3y = 10$
$$\frac{x - 1}{3} = \frac{y + 1}{-2} = z - 3$$

54. $5x + 3y = 17$
$$\frac{x - 4}{2} = \frac{y + 1}{-3} = \frac{z + 2}{5}$$

In Exercises 55 and 56, find the distance between the point and the line.

55. $(10, 3, -2)$; $x = 4t - 2$, $y = 3$, $z = -t + 1$
56. $(4, 1, -2)$; $x = 2t + 2$, $y = 2t$, $z = t - 3$

In Exercises 57 and 58, find the distance between the point and the plane.

57. $(0, 0, 0)$, $2x + 3y + z = 12$
58. $(1, 2, 3)$, $2x - y + z = 4$

In Exercises 59 and 60, find the distance between the planes.

59. $x - 3y + 4z = 10$, $x - 3y + 4z = 6$
60. $2x - 4z = 4$, $2x - 4z = 10$

In Exercises 61 and 62, find the distance between the lines.

61. $x = \frac{y}{2} = \frac{z}{3}$, $\frac{x - 1}{-1} = y - 4 = z + 1$

62. $x = 3t$ $x = 4s + 1$
      $y = -t + 2$ $y = s - 2$
      $z = t - 1$ $z = -3s - 3$

63. If a_1, b_1, c_1 and a_2, b_2, c_2 are two sets of direction numbers for the same line, show that there exists a scalar d such that $a_1 = a_2 d$, $b_1 = b_2 d$, and $c_1 = c_2 d$.

Surfaces in space

SECTION TOPICS ▪
Cylindrical surfaces ▪
Quadric surfaces ▪
Surfaces of revolution ▪

In the first five sections of this chapter, we introduced the vector portion of the preliminary work necessary to study vector calculus and the calculus of space. In this and the next section we complete this preliminary development.

We begin with the classification of several important surfaces in space. So far, we have studied two special types of surfaces.

1. Spheres: $(x - x_0)^2 + (y - y_0)^2 + (z - z_0)^2 = r^2$ Section 13.2
2. Planes: $ax + by + cz + d = 0$ Section 13.5

A third type of surface in space is called a **cylindrical surface,** or, more simply, a **cylinder.** To see how we define a cylinder, consider the familiar right circular cylinder shown in Figure 13.63. We can imagine that this cylinder is generated by a vertical line moving around the circle $x^2 + y^2 = a^2$ in the xy-plane. We call this circle a **generating curve** for the cylinder, as indicated in the following definition.

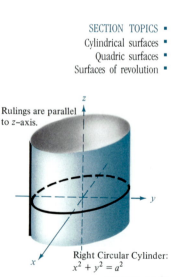

Rulings are parallel to z-axis.

Right Circular Cylinder:
$x^2 + y^2 = a^2$

FIGURE 13.63

DEFINITION OF A CYLINDER   Let C be a curve in a plane and L be a line not in that plane. The set of all lines parallel to L and intersecting C is called a **cylinder**. C is called the **generating curve** (or **directrix**) of the cylinder, and the parallel lines are called **rulings.**

Remark   Without loss of generality, we can assume that C lies in one of the three coordinate planes. Moreover, in this text we restrict our discussion to *right* cylinders—cylinders whose rulings are perpendicular to the coordinate plane containing C, as shown in Figure 13.64.

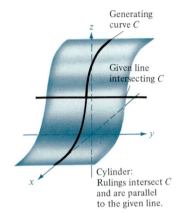

FIGURE 13.64

For the right circular cylinder shown in Figure 13.63, the equation of the generating curve is

$$x^2 + y^2 = a^2 \qquad \text{Equation of generating curve in } xy\text{-plane}$$

To find an equation for the cylinder, note that we can generate any one of the rulings by fixing the values of x and y and then allowing z to take on all real values. In this sense the value of z is arbitrary and is therefore not included in the equation. In other words, the equation of this cylinder is simply the equation of its generating curve:

$$x^2 + y^2 = a^2 \qquad \text{Equation of cylinder in space}$$

This result is generalized in the following theorem.

THEOREM 13.16   EQUATIONS OF CYLINDERS
In space, the graph of an equation in two of the three variables x, y, and z is a cylinder whose rulings are parallel to the axis of the missing variable.

EXAMPLE 1   Sketching a cylinder

Sketch the surfaces represented by the following equations:
(a) $z = y^2$ (b) $z = \sin x, \quad 0 \le x \le 2\pi$

Solution:

(a) The graph is a cylinder whose generating curve, $z = y^2$, is a parabola in the yz-plane. The rulings of the cylinder are parallel to the x-axis, as shown in Figure 13.65.

(b) The graph is a cylinder generated by the sine curve in the xz-plane. The rulings are parallel to the y-axis, as shown in Figure 13.66.

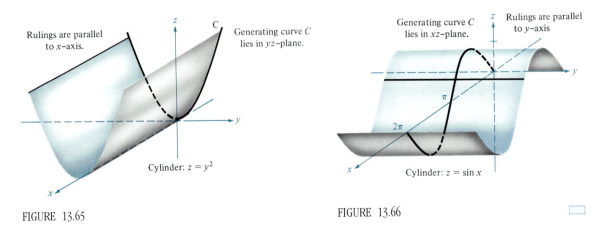

Rulings are parallel to x-axis.

Generating curve C lies in yz-plane.

Cylinder: $z = y^2$

FIGURE 13.65

Generating curve C lies in xz-plane.

Rulings are parallel to y-axis

Cylinder: $z = \sin x$

FIGURE 13.66

The intersection of a surface with a plane is called the **trace of the surface** in the plane. To visualize a surface in space, it is helpful to determine its traces in some well-chosen planes. This will become evident as we study the fourth common type of surface—a **quadric surface.** Quadric surfaces are the three-dimensional analogues of conic sections, as represented by the general second-degree equation

$$Ax^2 + Bxy + Cy^2 + Dx + Ey + F = 0 \qquad \text{Conic section in } xy\text{-plane}$$

The general second-degree equation in three variables follows a similar pattern.

DEFINITION OF A QUADRIC SURFACE

In space, the graph of a second-degree equation of the form

$$Ax^2 + By^2 + Cz^2 + Dxy + Exz + Fyz + Gx + Hy + Iz + J = 0$$

is called a **quadric surface.**

There are six basic types of quadric surfaces: **ellipsoid, hyperboloid of one sheet, hyperboloid of two sheets, elliptic cone, elliptic paraboloid,** and **hyperbolic paraboloid.** From the definition of a quadric surface, we can see that the traces taken in each of the coordinate planes are conic sections. These traces, together with the **standard form** of the equation of each quadric surface, are shown in Table 13.1.

TABLE 13.1   QUADRIC SURFACES

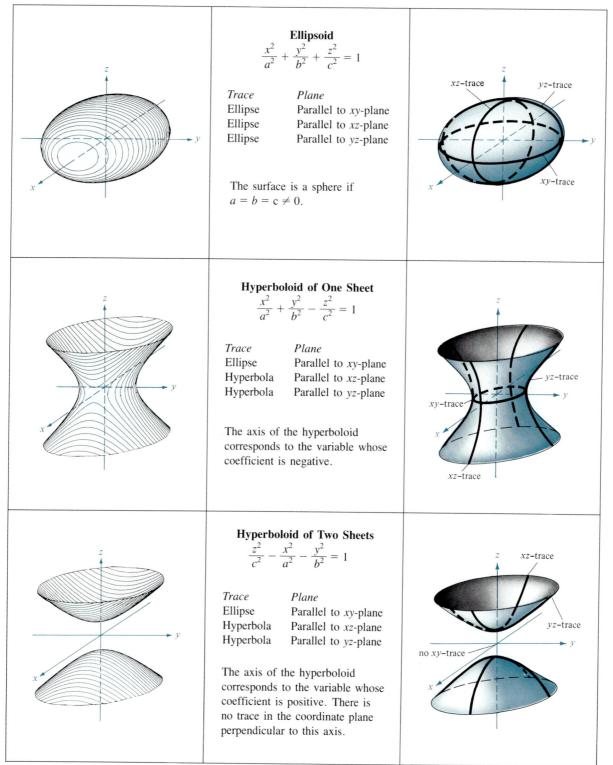

Ellipsoid

$$\frac{x^2}{a^2} + \frac{y^2}{b^2} + \frac{z^2}{c^2} = 1$$

Trace	*Plane*
Ellipse	Parallel to *xy*-plane
Ellipse	Parallel to *xz*-plane
Ellipse	Parallel to *yz*-plane

The surface is a sphere if
$a = b = c \neq 0$.

Hyperboloid of One Sheet

$$\frac{x^2}{a^2} + \frac{y^2}{b^2} - \frac{z^2}{c^2} = 1$$

Trace	*Plane*
Ellipse	Parallel to *xy*-plane
Hyperbola	Parallel to *xz*-plane
Hyperbola	Parallel to *yz*-plane

The axis of the hyperboloid
corresponds to the variable whose
coefficient is negative.

Hyperboloid of Two Sheets

$$\frac{z^2}{c^2} - \frac{x^2}{a^2} - \frac{y^2}{b^2} = 1$$

Trace	*Plane*
Ellipse	Parallel to *xy*-plane
Hyperbola	Parallel to *xz*-plane
Hyperbola	Parallel to *yz*-plane

The axis of the hyperboloid
corresponds to the variable whose
coefficient is positive. There is
no trace in the coordinate plane
perpendicular to this axis.

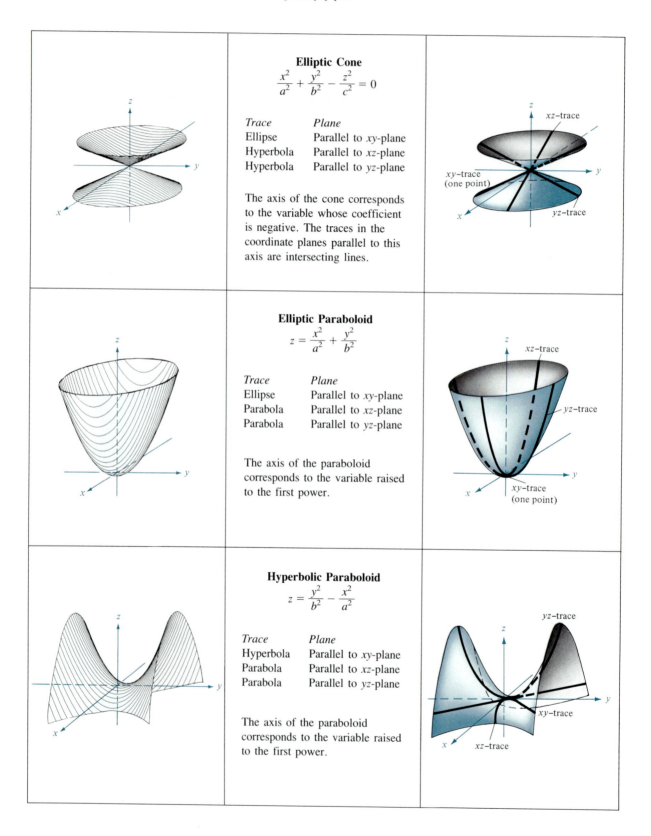

Elliptic Cone

$$\frac{x^2}{a^2} + \frac{y^2}{b^2} - \frac{z^2}{c^2} = 0$$

Trace	Plane
Ellipse	Parallel to xy-plane
Hyperbola	Parallel to xz-plane
Hyperbola	Parallel to yz-plane

The axis of the cone corresponds to the variable whose coefficient is negative. The traces in the coordinate planes parallel to this axis are intersecting lines.

Elliptic Paraboloid

$$z = \frac{x^2}{a^2} + \frac{y^2}{b^2}$$

Trace	Plane
Ellipse	Parallel to xy-plane
Parabola	Parallel to xz-plane
Parabola	Parallel to yz-plane

The axis of the paraboloid corresponds to the variable raised to the first power.

Hyperbolic Paraboloid

$$z = \frac{y^2}{b^2} - \frac{x^2}{a^2}$$

Trace	Plane
Hyperbola	Parallel to xy-plane
Parabola	Parallel to xz-plane
Parabola	Parallel to yz-plane

The axis of the paraboloid corresponds to the variable raised to the first power.

In Table 13.1 only one of several orientations of each quadric surface is shown. If the surface is oriented along a different axis, then its standard equation will change accordingly, as illustrated in the next two examples. The fact that the two types of paraboloids have one variable raised to the first power can be helpful in classifying quadric surfaces. The other four types of basic quadric surfaces have equations that are of *second degree* in all three variables.

EXAMPLE 2   *Sketching a quadric surface*

Describe and sketch the surface given by $4x^2 - 3y^2 + 12z^2 + 12 = 0$.

Solution: We express the given equation in standard form as follows.

$$\frac{x^2}{-3} + \frac{y^2}{4} - z^2 - 1 = 0 \qquad \text{Divide by } -12$$

$$\frac{y^2}{4} - \frac{x^2}{3} - \frac{z^2}{1} = 1 \qquad \text{Standard form}$$

From Table 13.1 we conclude that the surface is a hyperboloid of two sheets with the y-axis as its axis. To help sketch the graph of this surface, we find the following traces.

$$\begin{array}{lll} xy\text{-trace:} & \dfrac{y^2}{4} - \dfrac{x^2}{3} = 1 & \text{Hyperbola} \\ (z = 0) & & \\[2mm] xz\text{-trace:} & \dfrac{x^2}{3} + \dfrac{z^2}{1} = -1 & \text{No trace} \\ (y = 0) & & \\[2mm] yz\text{-trace:} & \dfrac{y^2}{4} - \dfrac{z^2}{1} = 1 & \text{Hyperbola} \\ (x = 0) & & \end{array}$$

The graph is shown in Figure 13.67.

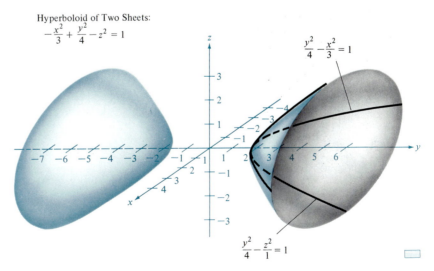

Hyperboloid of Two Sheets:
$$-\frac{x^2}{3} + \frac{y^2}{4} - z^2 = 1$$

$$\frac{y^2}{4} - \frac{x^2}{3} = 1$$

$$\frac{y^2}{4} - \frac{z^2}{1} = 1$$

FIGURE 13.67

EXAMPLE 3   Sketching a quadric surface

Classify and sketch the surface given by $x - y^2 - 4z^2 = 0$.

Solution: Since x is raised only to the first power, the surface will be a paraboloid, and in this case its axis is the x-axis. In the standard form, we have

$$x = y^2 + 4z^2$$

Some convenient traces are

$$xy\text{-trace:}\quad x = y^2 \qquad\qquad \text{Parabola}$$
$$(z = 0)$$

$$xz\text{-trace:}\quad x = 4z^2 \qquad\qquad \text{Parabola}$$
$$(y = 0)$$

$$\text{parallel to } yz\text{-plane:}\quad \frac{y^2}{4} + \frac{z^2}{1} = 1 \qquad \text{Ellipse}$$
$$(x = 4)$$

Thus, the surface is an *elliptic* paraboloid, as shown in Figure 13.68.

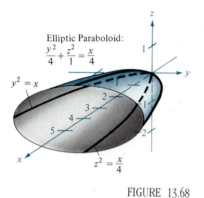

Elliptic Paraboloid:
$$\frac{y^2}{4} + \frac{z^2}{1} = \frac{x}{4}$$

$y^2 = x$

$z^2 = \dfrac{x}{4}$

FIGURE 13.68

For a quadric surface not centered at the origin, we can form the standard equation by completing the square, as we did for conics. A translation of axes then makes sketching easier. This is demonstrated in the next example.

EXAMPLE 4   A quadric surface not centered at the origin

Classify and sketch the surface given by

$$x^2 + 2y^2 + z^2 - 4x + 4y - 2z + 3 = 0$$

Solution: Completing the square for each variable, we have

$$(x^2 - 4x + \quad) + 2(y^2 + 2y + \quad) + (z^2 - 2z + \quad) = -3$$
$$(x^2 - 4x + 4) + 2(y^2 + 2y + 1) + (z^2 - 2z + 1) = -3 + 4 + 2 + 1$$
$$(x - 2)^2 + 2(y + 1)^2 + (z - 1)^2 = 4$$
$$\frac{(x - 2)^2}{4} + \frac{(y + 1)^2}{2} + \frac{(z - 1)^2}{4} = 1$$

For this equation, we see that the quadric surface is centered at $(2, -1, 1)$, and if we let

$$x' = x - 2, \qquad y' = y + 1, \qquad \text{and} \qquad z' = z - 1$$

we obtain the standard form

$$\frac{(x')^2}{4} + \frac{(y')^2}{2} + \frac{(z')^2}{4} = 1$$

of an ellipsoid. Its graph is shown in Figure 13.69.

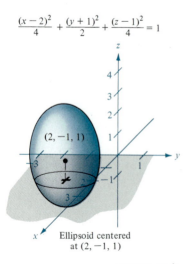

$$\frac{(x - 2)^2}{4} + \frac{(y + 1)^2}{2} + \frac{(z - 1)^2}{4} = 1$$

$(2, -1, 1)$

Ellipsoid centered at $(2, -1, 1)$

FIGURE 13.69

Surfaces of revolution

The fifth special type of surface we will study is called a **surface of revolution.** In Section 6.7 we looked at a method for finding the *area* of such a surface. We now look at a procedure for finding its *equation*.

Consider the graph of the **radius function**

$$y = r(z) \qquad \text{Generating curve}$$

in the *yz*-plane. If this graph is revolved about the *z*-axis, it forms a surface of revolution as shown in Figure 13.70. The trace of the surface in the plane $z = z_0$ is a circle whose radius is $r(z_0)$ and whose equation is

$$x^2 + y^2 = [r(z_0)]^2 \qquad \text{Circular trace in plane: } z = z_0$$

Replacing z_0 by z, we obtain an equation that is valid for all values of z. In a similar manner we can obtain equations for surfaces of revolution for the other two axes, and we summarize the results in the following theorem.

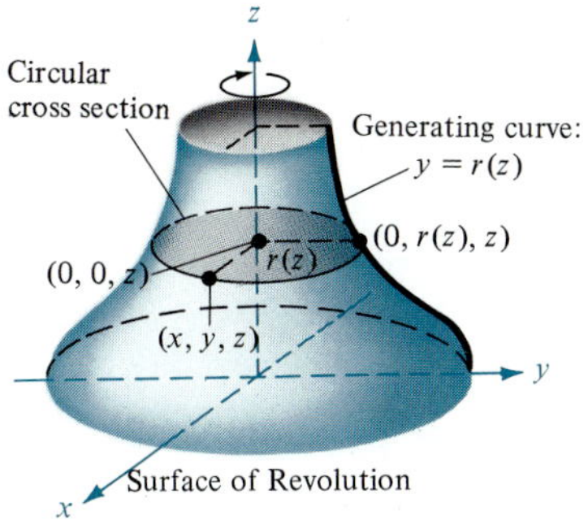

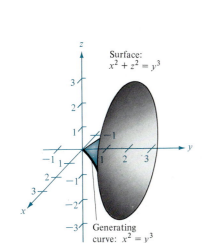

Circular cross section

Generating curve: $y = r(z)$

$(0, r(z), z)$

$(0, 0, z)$ $r(z)$

(x, y, z)

Surface of Revolution

FIGURE 13.70

THEOREM 13.17   **SURFACE OF REVOLUTION**

If the graph of a radius function r is revolved about one of the coordinate axes, then the equation of the resulting surface of revolution has one of the following forms.

1. Revolved about the *x*-axis: $y^2 + z^2 = [r(x)]^2$
2. Revolved about the *y*-axis: $x^2 + z^2 = [r(y)]^2$
3. Revolved about the *z*-axis: $x^2 + y^2 = [r(z)]^2$

EXAMPLE 5   Finding an equation for a surface of revolution

(a) An equation for the surface of revolution formed by revolving the graph of $y = 1/z$ about the *z*-axis is

$$x^2 + y^2 = [r(z)]^2 = \left(\frac{1}{z}\right)^2$$

(b) To find an equation for the surface formed by revolving the graph of $x^2 = y^3$ about the *y*-axis, we solve for *x* in terms of *y* to obtain

$$x = y^{3/2} = r(y) \qquad \text{Radius function}$$

Thus, the equation for this surface is

$$x^2 + z^2 = [r(y)]^2 = [y^{3/2}]^2 = y^3$$

The graph is shown in Figure 13.71.

Surface: $x^2 + z^2 = y^3$

Generating curve: $x^2 = y^3$

FIGURE 13.71

The generating curve for a surface of revolution is not unique. For instance, the surface $x^2 + z^2 = e^{-2y}$ can be formed by revolving either the graph of $x = e^{-y}$ about the y-axis or the graph of $z = e^{-y}$ about the y-axis, as shown in Figure 13.72.

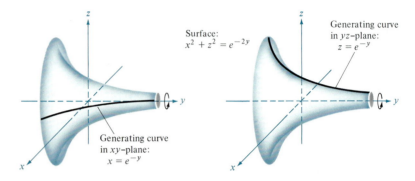

Surface:
$x^2 + z^2 = e^{-2y}$

Generating curve in xy-plane:
$x = e^{-y}$

Generating curve in yz-plane:
$z = e^{-y}$

FIGURE 13.72

EXAMPLE 6   Finding a generating curve for a surface of revolution

Find a generating curve and the axis of revolution for the surface given by $x^2 + 3y^2 + z^2 = 9$.

Solution: From Theorem 13.17 we know that the equation has one of the following forms:

$$x^2 + y^2 = [r(z)]^2 \qquad y^2 + z^2 = [r(x)]^2 \qquad x^2 + z^2 = [r(y)]^2$$

Since the coefficients of x^2 and z^2 are equal, we choose the third form and write

$$x^2 + z^2 = 9 - 3y^2$$

We conclude that the y-axis is the axis of revolution. We can choose a generating curve from either of the following traces:

$$x^2 = 9 - 3y^2 \qquad \text{Trace in xy-plane}$$
$$z^2 = 9 - 3y^2 \qquad \text{Trace in yz-plane}$$

For example, using the first trace, we choose the semiellipse given by

$$x = \sqrt{9 - 3y^2} \qquad \text{Generating curve}$$

The graph of this surface is shown in Figure 13.73.

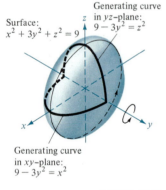

Surface:
$x^2 + 3y^2 + z^2 = 9$

Generating curve in yz-plane:
$9 - 3y^2 = z^2$

Generating curve in xy-plane:
$9 - 3y^2 = x^2$

FIGURE 13.73

Section Exercises 13.6

In Exercises 1–8, match the equation with its graph. (Note that the graphs show only the portion of each surface that lies above the xy-plane.)

1. $\dfrac{x^2}{9} + \dfrac{y^2}{16} + \dfrac{z^2}{9} = 1$

2. $15x^2 - 4y^2 + 15z^2 = -4$
3. $4x^2 - y^2 + 4z^2 = 4$
4. $y^2 = 4x^2 + 9z^2$
5. $4x^2 - 4y + z^2 = 0$
6. $12z = -3y^2 + 4x^2$
7. $4x^2 - y^2 + 4z = 0$
8. $x^2 + y^2 + z^2 = 9$

(a)

(b)

(c)

(d)

(e)

(f)

(g)

(h)

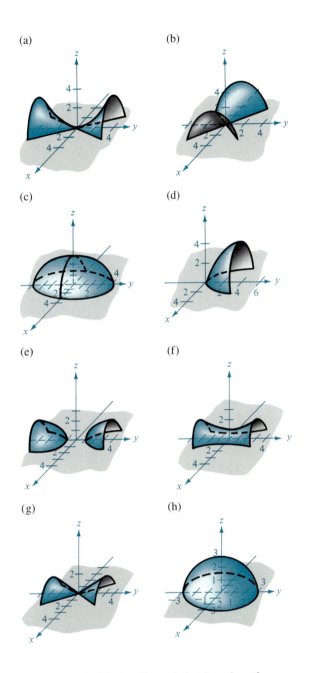

In Exercises 9–18, describe and sketch each surface.

9. $z = 3$

10. $x = 4$

11. $y^2 + z^2 = 9$

12. $x^2 + z^2 = 16$

13. $x^2 - y = 0$

14. $y^2 + z = 4$

15. $4x^2 + y^2 = 4$

16. $z - \sin y = 0$

17. $y^2 - z^2 = 4$

18. $z - e^y = 0$

In Exercises 19–34, identify and sketch the given quadric surface.

19. $x^2 + \dfrac{y^2}{4} + z^2 = 1$

20. $\dfrac{x^2}{9} + \dfrac{y^2}{16} + \dfrac{z^2}{16} = 1$

21. $16x^2 - y^2 + 16z^2 = 4$

22. $9x^2 + 4y^2 - 8z^2 = 72$

23. $x^2 - y + z^2 = 0$

24. $z = 4x^2 + y^2$

25. $x^2 - y^2 + z = 0$

26. $z^2 - x^2 - \dfrac{y^2}{4} = 1$

27. $4x^2 - y^2 + 4z^2 = -16$

28. $z^2 = x^2 + \dfrac{y^2}{4}$

29. $z^2 = x^2 + 4y^2$

30. $4y = x^2 + z^2$

31. $3z = -y^2 + x^2$

32. $z^2 = 2x^2 + 2y^2$

33. $16x^2 + 9y^2 + 16z^2 - 32x - 36y + 36 = 0$

34. $4x^2 + y^2 - 4z^2 - 16x - 6y - 16z + 9 = 0$

In Exercises 35–40, sketch the region bounded by the graphs of the equations.

35. $z = 2\sqrt{x^2 + y^2}, \ z = 2$

36. $z = \sqrt{4 - x^2}, \ y = \sqrt{4 - x^2}, \ x = 0, \ y = 0, \ z = 0$

37. $x^2 + y^2 = 1, \ x + z = 2, \ z = 0$

38. $x^2 + y^2 + z^2 = 4, \ z = \sqrt{x^2 + y^2}, \ z = 0$

39. $z = \sqrt{4 - x^2 - y^2}, \ y = 2z, \ z = 0$

40. $z = \sqrt{x^2 + y^2}, \ z = 4 - x^2 - y^2$

In Exercises 41–46, find an equation for the surface of revolution generated by revolving the given curve about the specified axis.

41. $z^2 = 4y$ in the yz-plane about the y-axis

42. $z = 2y$ in the yz-plane about the y-axis

43. $z = 2y$ in the yz-plane about the z-axis

44. $2z = \sqrt{4 - x^2}$ in the xz-plane about the x-axis

45. $xy = 2$ in the xy-plane about the x-axis

46. $z = \ln y$ in the yz-plane about the z-axis

In Exercises 47 and 48, find an equation of a generating curve given the equation of its surface of revolution.

47. $x^2 + y^2 - 2z = 0$

48. $x^2 + z^2 = \sin^2 y$

In Exercises 49 and 50, analyze the trace when the surface given by

$$z = \dfrac{x^2}{2} + \dfrac{y^2}{4}$$

is intersected by the given planes.

49. Find the length of the major and minor axes and the coordinates of the foci of the ellipse generated when the surface is intersected by the planes given by (a) $z = 2$ and (b) $z = 8$.

50. Find the coordinates of the focus of the parabola formed when the surface is intersected by the planes given by (a) $y = 4$ and (b) $x = 2$.

13.7
Cylindrical and spherical coordinates

We have seen that in the plane some graphs are easier to represent in polar coordinates than in rectangular coordinates. In this section we will see that a similar situation exists for surfaces in space. We will look at two new space coordinate systems. The first is an extension of polar coordinates to space and is called the **cylindrical coordinate system.**

THE CYLINDRICAL
COORDINATE SYSTEM

In a **cylindrical coordinate system,** a point P in space is represented by an ordered triple (r, θ, z).

1. (r, θ) is a polar representation of the projection of P in the xy-plane.
2. z is the directed distance from (r, θ) to P.

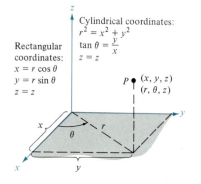

Cylindrical coordinates:
$$r^2 = x^2 + y^2$$
$$\tan \theta = \frac{y}{x}$$
$$z = z$$

Rectangular
coordinates:
$$x = r \cos \theta$$
$$y = r \sin \theta$$
$$z = z$$

$P \bullet \;\; (x, y, z)$
(r, θ, z)

FIGURE 13.74

To convert from rectangular to cylindrical coordinates (or vice versa), we follow the conversion guidelines for polar coordinates as illustrated in Figure 13.74.

Cylindrical to rectangular:

$$x = r \cos \theta, \qquad y = r \sin \theta, \qquad z = z$$

Rectangular to cylindrical:

$$r^2 = x^2 + y^2, \qquad \tan \theta = \frac{y}{x}, \qquad z = z$$

We call the point $(0, 0, 0)$ the **pole.** Moreover, since the representation of a point in the polar coordinate system is not unique, it follows that the representation in the cylindrical coordinate system is also not unique.

EXAMPLE 1   Changing from cylindrical to rectangular coordinates

Express the point $(r, \theta, z) = (4, 5\pi/6, 3)$ in rectangular coordinates.

Solution: Using the *cylindrical-to-rectangular* conversion equations, we have

$$x = 4 \cos \frac{5\pi}{6} = 4\left(-\frac{\sqrt{3}}{2}\right) = -2\sqrt{3}$$

$$y = 4 \sin \frac{5\pi}{6} = 4\left(\frac{1}{2}\right) = 2$$

$$z = 3$$

Thus, the point is $(-2\sqrt{3}, 2, 3)$ in rectangular coordinates, as shown in Figure 13.75.

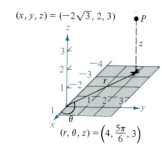

$(x, y, z) = (-2\sqrt{3}, 2, 3)$

$(r, \theta, z) = \left(4, \frac{5\pi}{6}, 3\right)$

FIGURE 13.75

EXAMPLE 2   Changing from rectangular to cylindrical coordinates

Express the point $(x, y, z) = (1, \sqrt{3}, 2)$ in cylindrical coordinates.

Solution: Using the rectangular-to-cylindrical conversion equations, we have

$$r = \pm\sqrt{1 + 3} = \pm 2$$

$$\tan \theta = \sqrt{3} \implies \theta = \arctan(\sqrt{3}) + n\pi = \frac{\pi}{3} + n\pi$$

$$z = 2$$

We have two choices for r and infinitely many choices for θ. As shown in Figure 13.76, two convenient representations of the point are

$$\left(2, \frac{\pi}{3}, 2\right) \qquad r > 0 \text{ and } \theta \text{ in Quadrant I}$$

$$\left(-2, \frac{4\pi}{3}, 2\right) \qquad r < 0 \text{ and } \theta \text{ in Quadrant III}$$

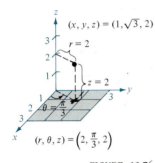

FIGURE 13.76

Cylindrical coordinates are especially convenient for representation of cylindrical surfaces and surfaces of revolution with the z-axis as the axis of symmetry, as shown in Figure 13.77.

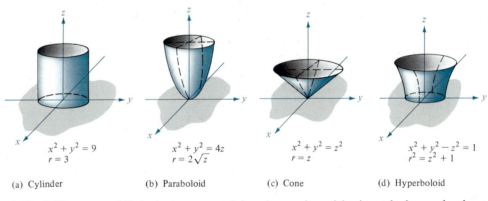

$$x^2 + y^2 = 9$$
$$r = 3$$
(a) Cylinder

$$x^2 + y^2 = 4z$$
$$r = 2\sqrt{z}$$
(b) Paraboloid

$$x^2 + y^2 = z^2$$
$$r = z$$
(c) Cone

$$x^2 + y^2 - z^2 = 1$$
$$r^2 = z^2 + 1$$
(d) Hyperboloid

FIGURE 13.77   Vertical planes containing the z-axis and horizontal planes also have simple cylindrical coordinate equations, as shown in Figure 13.78.

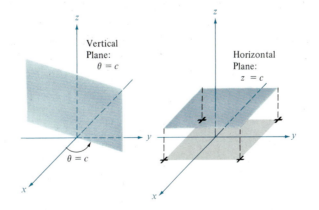

Vertical Plane: $\theta = c$

$\theta = c$

Horizontal Plane: $z = c$

FIGURE 13.78

EXAMPLE 3 Rectangular-to-cylindrical conversion

Find equations in cylindrical coordinates for the surfaces whose rectangular equations are

(a) $x^2 + y^2 = 4z^2$ (b) $y^2 = x$

Solution:

(a) From the previous section we know that the graph of $x^2 + y^2 = 4z^2$ is a cone with its axis along the z-axis. If we replace $x^2 + y^2$ by r^2, the equation in cylindrical coordinates is

$$r^2 = 4z^2 \implies r = 2z$$

(b) The graph of the surface $y^2 = x$ is a parabolic cylinder with rulings parallel to the z-axis, as shown in Figure 13.79. If we replace y^2 by $r^2 \sin^2 \theta$ and x by $r \cos \theta$, the equation in cylindrical coordinates is

$$r^2 \sin^2 \theta = r \cos \theta$$
$$r \sin^2 \theta = \cos \theta, \qquad r \neq 0$$
$$r = \csc^2 \theta \cos \theta$$
$$r = \csc \theta \cot \theta$$

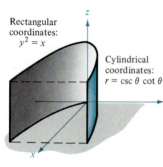

Rectangular
coordinates:
$y^2 = x$

Cylindrical
coordinates:
$r = \csc \theta \cot \theta$

FIGURE 13.79

EXAMPLE 4 Cylindrical-to-rectangular conversion

Find a rectangular equation for the graph represented by the cylindrical equation $r^2 \cos 2\theta + z^2 + 1 = 0$.

Solution: If we replace $\cos 2\theta$ by $\cos^2 \theta - \sin^2 \theta$, we obtain

$$r^2 (\cos^2 \theta - \sin^2 \theta) + z^2 + 1 = 0$$
$$r^2 \cos^2 \theta - r^2 \sin^2 \theta + z^2 = -1$$

Now, replacing $r \cos \theta$ by x and $r \sin \theta$ by y, we have $x^2 - y^2 + z^2 = -1$, or

$$y^2 - x^2 - z^2 = 1$$

which is a hyperboloid of two sheets whose axis lies along the y-axis, as shown in Figure 13.80.

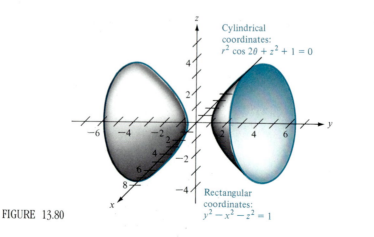

Cylindrical
coordinates:
$r^2 \cos 2\theta + z^2 + 1 = 0$

Rectangular
coordinates:
$y^2 - x^2 - z^2 = 1$

FIGURE 13.80

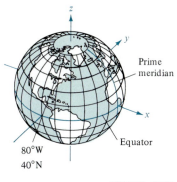

80°W
40°N

FIGURE 13.81

Spherical coordinates

In the **spherical coordinate system,** each point is represented by an ordered triple—the first coordinate is a directed distance, and the second and third coordinates are angles. This system is similar to the latitude-longitude system used to identify points on the surface of the earth. For example, the point on the surface of the earth whose latitude is 40° North (of the equator) and whose longitude is 80° West (of the prime meridian) is shown in Figure 13.81. Assuming that the earth is spherical with a radius of 4000 miles, we would label this point as

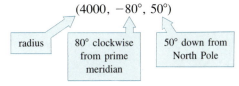

$$(4000, -80°, 50°)$$

| radius | 80° clockwise from prime meridian | 50° down from North Pole |

THE SPHERICAL
COORDINATE SYSTEM

In a **spherical coordinate system,** a point P in space is represented by an ordered triple (ρ, θ, ϕ).

1. ρ is the directed distance from P to the origin.
2. θ is the same angle used in cylindrical coordinates.
3. ϕ is the angle *between* the positive z-axis and the line segment $\overrightarrow{OP}$.

Remark ρ is the lowercase Greek letter rho, and ϕ is the lowercase Greek letter phi. By definition, we have $0 \le \phi \le \pi$.

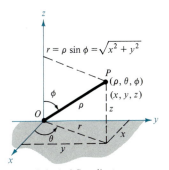

Spherical Coordinates

FIGURE 13.82

The relationship between the rectangular and the spherical coordinates is illustrated in Figure 13.82. To convert from one system to the other, we use the following conversion equations.

Spherical to rectangular:

$$x = \rho \sin \phi \cos \theta, \qquad y = \rho \sin \phi \sin \theta, \qquad z = \rho \cos \phi$$

Rectangular to spherical:

$$\rho^2 = x^2 + y^2 + z^2, \qquad \tan \theta = \frac{y}{x}, \qquad \phi = \arccos \left(\frac{z}{\sqrt{x^2 + y^2 + z^2}} \right)$$

The spherical coordinate system is useful primarily for surfaces in space that have a *point* or *center* of symmetry. For example, Figure 13.83 shows three surfaces with simple spherical equations.

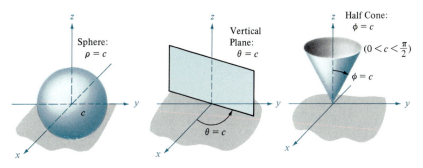

FIGURE 13.83

EXAMPLE 5   Rectangular-to-spherical conversion

Find an equation in spherical coordinates for the surfaces represented by the following rectangular coordinate equations:

(a) Cone: $x^2 + y^2 = z^2$ (b) Sphere: $x^2 + y^2 + z^2 - 4z = 0$

Solution:

(a) Making the appropriate replacements for x, y, and z in the given equation yields

$$x^2 + y^2 = z^2$$

$$\rho^2 \sin^2 \phi \cos^2 \theta + \rho^2 \sin^2 \phi \sin^2 \theta = \rho^2 \cos^2 \phi$$

$$\rho^2 \sin^2 \phi (\cos^2 \theta + \sin^2 \theta) = \rho^2 \cos^2 \phi$$

$$\rho^2 \sin^2 \phi = \rho^2 \cos^2 \phi$$

$$\frac{\sin^2 \phi}{\cos^2 \phi} = 1 \quad \rho > 0$$

$$\tan^2 \phi = 1 \quad \phi = \frac{\pi}{4} \quad \text{or} \quad \phi = \frac{3\pi}{4}$$

The equation $\phi = \pi/4$ represents the *upper* half-cone, and $\phi = 3\pi/4$ represents the *lower* half-cone.

(b) Since $\rho^2 = x^2 + y^2 + z^2$ and $z = \rho \cos \phi$, the given equation has the spherical form

$$\rho^2 - 4\rho \cos \phi = 0$$

$$\rho(\rho - 4 \cos \phi) = 0$$

Temporarily discarding the possibility that $\rho = 0$, we have the spherical equation

$$\rho - 4 \cos \phi = 0 \quad \text{or} \quad \rho = 4 \cos \phi$$

Note that the solution set for this equation includes a point for which $\rho = 0$, so nothing is lost by discarding the factor ρ.   ▢

Section Exercises 13.7

In Exercises 1–6, convert the given point from rectangular to cylindrical coordinates.

1. $(0, 5, 1)$

3. $(1, \sqrt{3}, 4)$

5. $(2, -2, -4)$

2. $(2\sqrt{2}, -2\sqrt{2}, 4)$

4. $(\sqrt{3}, -1, 2)$

6. $(-3, 2, -1)$

In Exercises 7–12, convert the given point from cylindrical to rectangular coordinates.

7. $(5, 0, 2)$

8. $\left(4, \frac{\pi}{2}, -2\right)$

9. $\left(2, \frac{\pi}{3}, 2\right)$

10. $\left(3, -\frac{\pi}{4}, 1\right)$

11. $\left(4, \frac{7\pi}{6}, 3\right)$

12. $\left(1, \frac{3\pi}{2}, 1\right)$

In Exercises 13–18, convert the given point from rectangular to spherical coordinates.

13. $(4, 0, 0)$

14. $(1, 1, 1)$

15. $(-2, 2\sqrt{3}, 4)$

16. $(2, 2, 4\sqrt{2})$

17. $(\sqrt{3}, 1, 2\sqrt{3})$

18. $(-4, 0, 0)$

In Exercises 19–24, convert the given point from spherical to rectangular coordinates.

19. $\left(4, \dfrac{\pi}{6}, \dfrac{\pi}{4}\right)$

20. $\left(12, \dfrac{3\pi}{4}, \dfrac{\pi}{9}\right)$

21. $\left(12, -\dfrac{\pi}{4}, 0\right)$

22. $\left(9, \dfrac{\pi}{4}, \pi\right)$

23. $\left(5, \dfrac{\pi}{4}, \dfrac{3\pi}{4}\right)$

24. $\left(6, \pi, \dfrac{\pi}{2}\right)$

In Exercises 25–30, convert the given point from cylindrical to spherical coordinates.

25. $\left(4, \dfrac{\pi}{4}, 0\right)$

26. $\left(2, \dfrac{2\pi}{3}, -2\right)$

27. $\left(4, -\dfrac{\pi}{6}, 6\right)$

28. $\left(-4, \dfrac{\pi}{3}, 4\right)$

29. $(12, \pi, 5)$

30. $\left(4, \dfrac{\pi}{2}, 3\right)$

In Exercises 31–36, convert the given point from spherical to cylindrical coordinates.

31. $\left(10, \dfrac{\pi}{6}, \dfrac{\pi}{2}\right)$

32. $\left(4, \dfrac{\pi}{18}, \dfrac{\pi}{2}\right)$

33. $\left(6, -\dfrac{\pi}{6}, \dfrac{\pi}{3}\right)$

34. $\left(5, -\dfrac{5\pi}{6}, \pi\right)$

35. $\left(8, \dfrac{7\pi}{6}, \dfrac{\pi}{6}\right)$

36. $\left(7, \dfrac{\pi}{4}, \dfrac{3\pi}{4}\right)$

In Exercises 37–42, match the equation (expressed in terms of cylindrical or spherical coordinates) to the correct graph.

37. $r = 5$

38. $\theta = \dfrac{\pi}{4}$

39. $\rho = 5$

40. $\phi = \dfrac{\pi}{4}$

41. $r^2 = z$

42. $\rho = 4 \sec \phi$

(a)

(b)

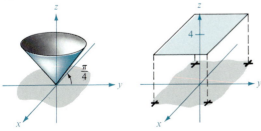

(c)

(d)

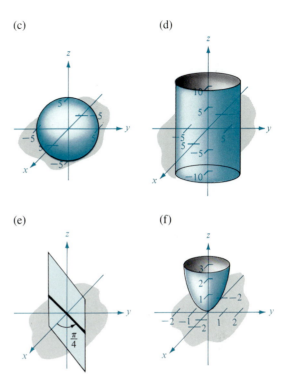

(e)

(f)

In Exercises 43–50, find an equation in rectangular coordinates for the equation in cylindrical coordinates, and sketch its graph.

43. $r = 2$

44. $z = 2$

45. $\theta = \dfrac{\pi}{6}$

46. $r = \dfrac{z}{2}$

47. $r = 2 \sin \theta$

48. $r = 2 \cos \theta$

49. $r^2 + z^2 = 4$

50. $z = r^2 \sin^2 \theta$

In Exercises 51–58, find an equation in rectangular coordinates for the equation in spherical coordinates and sketch its graph.

51. $\rho = 2$

52. $\theta = \dfrac{3\pi}{4}$

53. $\phi = \dfrac{\pi}{6}$

54. $\phi = \dfrac{\pi}{2}$

55. $\rho = 4 \cos \phi$

56. $\rho = 2 \sec \phi$

57. $\rho = \csc \phi$

58. $\rho = 4 \csc \phi \sec \theta$

In Exercises 59–66, find an equation of the given surface in (a) cylindrical coordinates and (b) spherical coordinates.

59. $x^2 + y^2 + z^2 = 16$

60. $4(x^2 + y^2) = z^2$

61. $x^2 + y^2 + z^2 - 2z = 0$

62. $x^2 + y^2 = z$

63. $x^2 + y^2 = 4y$

64. $x^2 + y^2 = 16$

65. $x^2 - y^2 = 9$

66. $y = 4$

In Exercises 67 and 68, sketch the solid that has the given description in cylindrical coordinates.

67. $0 \leq \theta \leq 2\pi,\ 0 \leq r \leq a,\ r \leq z \leq a$
68. $0 \leq \theta \leq 2\pi,\ 2 \leq r \leq 4,\ z^2 \leq -r^2 + 6r - 8$

In Exercises 69 and 70, sketch the solid that has the given description in spherical coordinates.

69. $0 \leq \theta \leq 2\pi,\ 0 \leq \phi \leq \pi/6,\ 0 \leq \rho \leq a \sec \phi$
70. $0 \leq \theta \leq 2\pi,\ \pi/4 \leq \phi \leq \pi/2,\ 0 \leq \rho \leq 1$

Review Exercises for Chapter 13

In Exercises 1 and 2, let $\mathbf{u} = \overrightarrow{PQ}$ and $\mathbf{v} = \overrightarrow{PR}$, and find (a) the component forms of $\mathbf{u}$ and $\mathbf{v}$, (b) the magnitude of $\mathbf{v}$, (c) $\mathbf{u} \cdot \mathbf{v}$, (d) $2\mathbf{u} + \mathbf{v}$, (e) the vector component of $\mathbf{u}$ in the direction of $\mathbf{v}$, and (f) the vector component of $\mathbf{u}$ orthogonal to $\mathbf{v}$.

1. $P = (1, 2),\ Q = (4, 1),\ R = (5, 4)$
2. $P = (-2, -1),\ Q = (5, -1),\ R = (2, 4)$

In Exercises 3 and 4, let $\mathbf{u} = \overrightarrow{PQ}$ and $\mathbf{v} = \overrightarrow{PR}$, and find (a) the component forms of $\mathbf{u}$ and $\mathbf{v}$, (b) $\mathbf{u} \cdot \mathbf{v}$, (c) $\mathbf{u} \times \mathbf{v}$, (d) an equation of the plane containing P, Q, and R, and (e) a set of parametric equations of the line through P and Q.

3. $P = (5, 0, 0),\ Q = (4, 4, 0),\ R = (2, 0, 6)$
4. $P = (2, -1, 3),\ Q = (0, 5, 1),\ R = (5, 5, 0)$

In Exercises 5–8, find the angle θ between the vectors $\mathbf{u}$ and $\mathbf{v}$.

5. $\mathbf{u} = 5\left[\cos\left(\dfrac{3\pi}{4}\right)\mathbf{i} + \sin\left(\dfrac{3\pi}{4}\right)\mathbf{j}\right]$

$\mathbf{v} = 2\left[\cos\left(\dfrac{2\pi}{3}\right)\mathbf{i} + \sin\left(\dfrac{2\pi}{3}\right)\mathbf{j}\right]$

6. $\mathbf{u} = \langle 4, -1, 5 \rangle,\ \mathbf{v} = \langle 3, 2, -2 \rangle$
7. $\mathbf{u} = \langle 10, -5, 15 \rangle,\ \mathbf{v} = \langle -2, 1, -3 \rangle$
8. $\mathbf{u} = \langle 1, 0, -3 \rangle,\ \mathbf{v} = \langle 2, -2, 1 \rangle$

In Exercises 9–12, find $\mathbf{u}$.

9. The angle, measured counterclockwise, from $\mathbf{u}$ to the positive x-axis is $135°$, and $\|\mathbf{u}\| = 4$.
10. The angle between $\mathbf{u}$ and the positive x-axis is $180°$, and $\|\mathbf{u}\| = 8$.
11. $\mathbf{u}$ is perpendicular to the plane $x - 3y + 4z = 0$, and $\|\mathbf{u}\| = 3$.
12. $\mathbf{u}$ is a unit vector perpendicular to the lines

$$
\begin{array}{ll}
x = 4 - t & x = -3 + 7t \\
y = 3 + 2t & y = -2 + t \\
z = 1 + 5t & z = 1 + 2t
\end{array}
$$

In Exercises 13–20, let $\mathbf{u} = \langle 3, -2, 1 \rangle$, $\mathbf{v} = \langle 2, -4, -3 \rangle$, and $\mathbf{w} = \langle -1, 2, 2 \rangle$.

13. Find $\|\mathbf{u}\|$.
14. Find the angle between $\mathbf{u}$ and $\mathbf{v}$.
15. Show that $\mathbf{u} \cdot \mathbf{u} = \|\mathbf{u}\|^2$.
16. Determine a unit vector perpendicular to the plane containing $\mathbf{v}$ and $\mathbf{w}$.
17. Show that $\mathbf{u} \cdot (\mathbf{v} + \mathbf{w}) = \mathbf{u} \cdot \mathbf{v} + \mathbf{u} \cdot \mathbf{w}$.
18. Show that

$$\mathbf{u} \times (\mathbf{v} + \mathbf{w}) = (\mathbf{u} \times \mathbf{v}) + (\mathbf{u} \times \mathbf{w})$$

19. Find the volume of the solid whose edges are $\mathbf{u}$, $\mathbf{v}$, and $\mathbf{w}$.
20. Find the work done in moving an object along the vector $\mathbf{u}$, if the applied force is $\mathbf{w}$.

In Exercises 21–24, find (a) a set of parametric equations, and (b) a set of symmetric equations for the given line.

21. The line passes through $(1, 2, 3)$ and is perpendicular to the xz-coordinate plane.
22. The line passes through $(1, 2, 3)$ and is parallel to the line given by $x = y = z$.
23. The line is the intersection of the planes given by

$$
\begin{aligned}
3x - 3y - 7z &= -4 \\
x - y + 2z &= 3
\end{aligned}
$$

24. The line passes through the point $(0, 1, 4)$ and is perpendicular to $\mathbf{u} = \langle 2, -5, 1 \rangle$ and $\mathbf{v} = \langle -3, 1, 4 \rangle$.

In Exercises 25–28, find an equation of the plane.

25. The plane passes through the point $(1, 2, 3)$ and is orthogonal to the line given by $x = y = z$.
26. The plane passes through the point $(4, 2, 1)$ and is parallel to the yz-coordinate plane.
27. The plane contains the lines

$$
\frac{x - 1}{-2} = y = z + 1
$$

$$
\frac{x + 1}{-2} = y - 1 = z - 2
$$

28. The plane passes through the points $(-3, -4, 2)$, $(-3, 4, 1)$, and $(1, 1, -2)$.

In Exercises 29 and 30, find the distance from the point to the plane.

29. $(1, 0, 2)$, $2x - 3y + 6z = 6$
30. $(0, 0, 0)$, $4x - 7y + z = 2$

In Exercises 31 and 32, find the distance between the parallel planes.

31. $5x - 3y + z = 2$, $5x - 3y + z = -3$
32. $3x + 2y - z = -1$, $6x + 4y - 2z = 1$

In Exercises 33–36, find the distance between the lines.

33.
$$\begin{array}{ll} x = 4 & x = -1 - s \\ y = -2t & y = 2 + s \\ z = 1 + t & z = 3 \end{array}$$

34.
$$\begin{array}{ll} x = 4 + t & x = -3 - 5s \\ y = 3 - t & y = 7 + 2s \\ z = 7 + 3t & z = -5 - 6s \end{array}$$

35. $\dfrac{x}{1} = \dfrac{y}{2} = \dfrac{z}{3}$

$\dfrac{x+1}{-1} = \dfrac{y}{3} = \dfrac{z+2}{2}$

36. $\dfrac{x-2}{1} = \dfrac{y+3}{-2} = \dfrac{z-1}{4}$

$\dfrac{x}{3} = \dfrac{y-2}{1} = \dfrac{z-3}{1}$

In Exercises 37–48, sketch the graph of the specified surface.

37. $x + 2y + 3z = 6$
38. $y = z^2$
39. $x^2 + z^2 = 4$
40. $x^2 + y^2 + z^2 - 2x + 4y - 6z + 5 = 0$

41. $16x^2 + 16y^2 - 9z^2 = 0$
42. $\dfrac{x^2}{16} + \dfrac{y^2}{9} + z^2 = 1$
43. $y = \dfrac{1}{2}z$
44. $\dfrac{x^2}{16} + \dfrac{y^2}{9} - z^2 = 1$
45. $\dfrac{x^2}{16} - \dfrac{y^2}{9} + z^2 = -1$
46. $\dfrac{x^2}{25} + \dfrac{y^2}{4} - \dfrac{z^2}{100} = 1$
47. $y^2 - 4x^2 = z$
48. $y = \cos z$

In Exercises 49 and 50, convert the given point from rectangular to (a) cylindrical coordinates and (b) spherical coordinates.

49. $(-2\sqrt{2}, 2\sqrt{2}, 2)$
50. $\left(\dfrac{\sqrt{3}}{4}, \dfrac{3}{4}, \dfrac{3\sqrt{3}}{2} \right)$

In Exercises 51 and 52, find an equation of the given surface in (a) cylindrical coordinates and (b) spherical coordinates.

51. $x^2 - y^2 = 2z$
52. $x^2 + y^2 + z^2 = 16$

In Exercises 53 and 54, find an equation in rectangular coordinates for the equation in cylindrical coordinates.

53. $r^2(\cos^2 \theta - \sin^2 \theta) + z^2 = 1$
54. $r^2 = 16z$

In Exercises 55 and 56, find an equation in rectangular coordinates for the equation in spherical coordinates.

55. $\rho = \csc \phi$
56. $\rho = 5$

14 *Vector-valued functions*

14.1
Vector-valued functions

In Section 12.1, we defined a *plane curve* as the set of ordered pairs $(f(t), g(t))$ satisfying the parametric equations

$$x = f(t) \quad \text{and} \quad y = g(t)$$

where f and g are continuous functions of t on an interval I. This definition can be extended to three-dimensional space as follows.

DEFINITION OF A SPACE CURVE

A **space curve** C is the set of all ordered triples $(f(t), g(t), h(t))$ satisfying the parametric equations

$$x = f(t), \quad y = g(t), \quad \text{and} \quad z = h(t)$$

where f, g, and h are continuous functions of t on an interval I.

Before looking at examples of space curves, we introduce a new type of function, called a **vector-valued function,** that maps real numbers onto vectors.

DEFINITION OF A VECTOR-VALUED FUNCTION

A function of the form

$$\mathbf{r}(t) = f(t)\,\mathbf{i} + g(t)\,\mathbf{j} \qquad \text{Plane}$$

or

$$\mathbf{r}(t) = f(t)\,\mathbf{i} + g(t)\,\mathbf{j} + h(t)\,\mathbf{k} \qquad \text{Space}$$

is called a **vector-valued function,** where the **component functions** f, g, and h are real-valued functions of the parameter t.

| Remark   Note the distinction between the vector-valued function **r** and the real-valued functions f, g, and h. All are functions of the real variable t, but $\mathbf{r}(t)$ is a vector, whereas $f(t)$, $g(t)$ and $h(t)$ are real numbers. Unless stated otherwise, we consider the **domain** of **r** to be the intersection of the domains of f, g, and h.

Vector-valued functions serve a dual role in the representation of curves. By letting the parameter t represent time, we can use a vector-valued function to represent *motion* along a curve. Or, in the more general case, we can use a vector function to *trace the graph* of a curve. In either case, the terminal point of the position vector $\mathbf{r}(t)$ coincides with the point (x, y) or (x, y, z) on the curve given by the parametric equations, as shown in Figure 14.1.

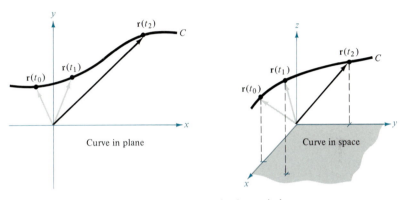

Curve in plane   Curve in space

FIGURE 14.1

Curve C is traced out by the terminal point of position vector $\mathbf{r}(t)$.

EXAMPLE 1   A plane curve represented by a vector-valued function

Sketch the plane curve represented by the vector-valued function

$$\mathbf{r}(t) = 2 \cos t \, \mathbf{i} - 3 \sin t \, \mathbf{j}, \quad 0 \le t \le 2\pi$$

Solution: From the position vector $\mathbf{r}(t)$, we obtain the parametric equations

$$x = 2 \cos t \qquad \text{and} \qquad y = -3 \sin t$$

Solving for $\cos t$ and $\sin t$, we can use the identity $\cos^2 t + \sin^2 t = 1$ to obtain the rectangular equation

$$\frac{x^2}{2^2} + \frac{y^2}{3^2} = 1$$

The graph of this rectangular equation is the ellipse shown in Figure 14.2. Note that the negative coefficient of the sine produces a *clockwise* orientation. That is, as t increases from 0 to 2π, the position vector $\mathbf{r}(t)$ moves clockwise, and its terminal point traces the ellipse.

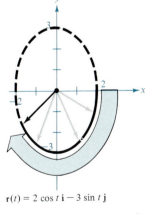

$\mathbf{r}(t) = 2 \cos t \, \mathbf{i} - 3 \sin t \, \mathbf{j}$

FIGURE 14.2

EXAMPLE 2   A space curve represented by a vector-valued function

Sketch the space curve represented by the vector-valued function

$$\mathbf{r}(t) = 4 \sin t \, \mathbf{i} + 4 \cos t \, \mathbf{j} + t\mathbf{k}, \quad 0 \le t \le 4\pi$$

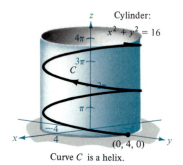

Curve *C* is a helix.

FIGURE 14.3

Solution: From the first two parametric equations $x = 4 \sin t$ and $y = 4 \cos t$ we obtain

$$x^2 + y^2 = 16$$

This means that the curve lies on a right circular cylinder of radius 4, centered about the *z*-axis. To locate the curve on this cylinder, we use the third parametric equation $z = t$. Then, as *t* increases from 0 to 4π, the point (x, y, z) spirals counterclockwise up the cylinder to produce the **helix** shown in Figure 14.3.

In Examples 1 and 2 we were given a vector-valued function and asked to sketch the corresponding curve. In the next two examples we look at the reverse problem—finding a vector-valued function to represent a given curve. Of course, if the curve is described parametrically, then representation by a vector-valued function is straightforward. For instance, to represent the line in space given by

$$x = 2 + t, \qquad y = 3t, \qquad z = 4 - t$$

we simply use the vector-valued function given by

$$\mathbf{r}(t) = (2 + t)\,\mathbf{i} + 3t\mathbf{j} + (4 - t)\,\mathbf{k}$$

If a set of parametric equations for the curve *C* is not given, then the problem of representing *C* by a vector-valued function boils down to finding such a set of equations.

EXAMPLE 3   Representing a plane curve by a vector-valued function

Represent the curve given by $y = x^2 + 1$ by a vector-valued function.

Solution: Although there are many ways to choose the parameter *t*, a natural choice is to let $x = t$. Then $y = t^2 + 1$ and we have

$$\mathbf{r}(t) = t\mathbf{i} + (t^2 + 1)\,\mathbf{j}$$

In the next example, we look at a space curve that is defined as the intersection of two surfaces in space.

EXAMPLE 4   Representing a space curve by a vector-valued function

Sketch the curve *C* represented by the intersection of the semiellipsoid

$$\frac{x^2}{12} + \frac{y^2}{24} + \frac{z^2}{4} = 1, \quad z \ge 0$$

and the parabolic cylinder $y = x^2$. Then, find a vector-valued function to represent *C*.

Solution: The intersection of the two surfaces is shown in Figure 14.4. As in Example 3, a natural choice of parameter is $x = t$. For this choice, we use the given equation $y = x^2$ to obtain $y = t^2$. Then, it follows that

$$\frac{z^2}{4} = 1 - \frac{x^2}{12} - \frac{y^2}{24} = 1 - \frac{t^2}{12} - \frac{t^4}{24} = \frac{24 - 2t^2 - t^4}{24}$$

Since the curve lies above the *xy*-plane, we choose the positive square root for z and obtain the following parametric equations

$$x = t, \qquad y = t^2, \qquad \text{and} \qquad z = 2\sqrt{\frac{24 - 2t^2 - t^4}{6}}$$

The resulting vector-valued function is

$$\mathbf{r}(t) = t\mathbf{i} + t^2\mathbf{j} + 2\sqrt{\frac{24 - 2t^2 - t^4}{6}}\ \mathbf{k}$$

From the points $(-2, 4, 0)$ and $(2, 4, 0)$ shown in Figure 14.4, we see that the curve is traced out as t increases from -2 to 2.

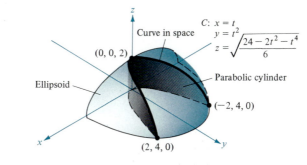

FIGURE 14.4

| Remark   Notice that the orientation of the curve in Example 4 is dependent upon the choice of parameter. For example, by letting $x = -t$, we would orient the curve in the opposite direction.

Limits and continuity

Before we can see the real power of representing curves by vector-valued functions, we must do some additional preliminary work. As it turns out, many of the techniques and definitions used in the calculus of real-valued functions can be applied to vector-valued functions. For instance, we can add and subtract vector-valued functions, multiply a vector-valued function by a scalar, take the limit of a vector-valued function, differentiate a vector-valued function, and so on. The basic approach is to capitalize on the linearity of vector operations by extending the definitions on a component by component basis. For example, to add two vector-valued functions (in the plane), we write

$$\mathbf{r}_1(t) + \mathbf{r}_2(t) = [f_1(t)\ \mathbf{i} + g_1(t)\ \mathbf{j}] + [f_2(t)\ \mathbf{i} + g_2(t)\ \mathbf{j}]$$
$$= [f_1(t) + f_2(t)]\ \mathbf{i} + [g_1(t) + g_2(t)]\ \mathbf{j}$$

Similarly, to multiply a vector-valued function by a scalar, we write

$$c\mathbf{r}(t) = c[f_1(t)\ \mathbf{i} + g_1(t)\ \mathbf{j}] = cf_1(t)\ \mathbf{i} + cg_1(t)\ \mathbf{j}$$

This component-by-component extension of operations with real-valued functions to vector-valued functions is further illustrated in the following definition of the limit of a vector-valued function.

DEFINITION OF THE LIMIT OF A VECTOR-VALUED FUNCTION

1. If **r** is a vector-valued function such that $\mathbf{r}(t) = f(t)\ \mathbf{i} + g(t)\ \mathbf{j}$, then

$$\lim_{t \to a} \mathbf{r}(t) = \left[\lim_{t \to a} f(t)\right] \mathbf{i} + \left[\lim_{t \to a} g(t)\right] \mathbf{j} \qquad \text{Plane}$$

provided f and g have limits as $t \to a$.

2. If **r** is a vector-valued function such that $\mathbf{r}(t) = f(t)\ \mathbf{i} + g(t)\ \mathbf{j} + h(t)\ \mathbf{k}$, then

$$\lim_{t \to a} \mathbf{r}(t) = \left[\lim_{t \to a} f(t)\right] \mathbf{i} + \left[\lim_{t \to a} g(t)\right] \mathbf{j} + \left[\lim_{t \to a} h(t)\right] \mathbf{k} \qquad \text{Space}$$

provided f, g, and h have limits as $t \to a$.

If $\mathbf{r}(t)$ approaches the vector **L**, then the length of the vector $\mathbf{r}(t) - \mathbf{L}$ approaches zero. That is,

$$\|\mathbf{r}(t) - \mathbf{L}\| \to 0 \qquad \text{as} \qquad t \to a$$

This is graphically illustrated in Figure 14.5. With this definition of the limit of a vector-valued function we can develop vector versions of most of the limit theorems given in Chapter 2. For example, the limit of the sum of two vector-valued functions is the sum of their individual limits. Also, we can use the orientation of t to define one-sided limits of vector-valued functions. In the following definition, we extend the notion of continuity to vector-valued functions.

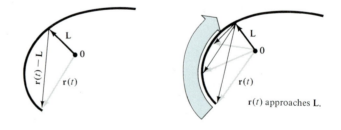

FIGURE 14.5

DEFINITION OF CONTINUITY

A vector-valued function **r** is **continuous** at $t = a$ if the limit of $\mathbf{r}(t)$ exists as $t \to a$ and

$$\lim_{t \to a} \mathbf{r}(t) = \mathbf{r}(a)$$

| Remark A vector-valued function is called **continuous** on an interval I if it is continuous at every point in the interval.

EXAMPLE 5 Continuity of vector-valued functions

Discuss the continuity of the vector-valued function given by

$$\mathbf{r}(t) = (\tan t)\,\mathbf{i} + \sqrt{t+1}\,\mathbf{j} + \ln(t^2 + 1)\,\mathbf{k}$$

at $t = 0$ and $t = -2$.

Solution: As t approaches 0, the limit is

$$\lim_{t \to 0} \mathbf{r}(t) = \left[\lim_{t \to 0} \tan t\right]\mathbf{i} + \left[\lim_{t \to 0}\sqrt{t+1}\right]\mathbf{j} + \left[\lim_{t \to 0}\ln(t^2 + 1)\right]\mathbf{k}$$

$$= 0\mathbf{i} + \mathbf{j} + 0\mathbf{k} = \mathbf{j}$$

Since $\mathbf{r}(0) = (\tan 0)\,\mathbf{i} + \sqrt{1}\,\mathbf{j} + \ln(1)\,\mathbf{k} = 0\mathbf{i} + \mathbf{j} + 0\mathbf{k} = \mathbf{j}$, we conclude that $\mathbf{r}$ is continuous at $t = 0$. At $t = -2$, the component $g(t) = \sqrt{t+1}$ is not defined; hence $\mathbf{r}(-2)$ is not defined and $\mathbf{r}$ is not continuous at $t = -2$.

Section Exercises 14.1

In Exercises 1–10, sketch the curve represented by the given vector-valued function.

1. $\mathbf{r}(t) = 3t\mathbf{i} + (t-1)\mathbf{j}$
2. $\mathbf{r}(t) = 2\cos t\,\mathbf{i} + 2\sin t\,\mathbf{j}$
3. $\mathbf{r}(t) = (-t+1)\mathbf{i} + (4t+2)\mathbf{j} + (2t+3)\mathbf{k}$
4. $\mathbf{r}(t) = t\mathbf{i} + (2t-5)\mathbf{j} + 3t\mathbf{k}$
5. $\mathbf{r}(t) = 2\cos t\,\mathbf{i} + 2\sin t\,\mathbf{j} + t\mathbf{k}$
6. $\mathbf{r}(t) = 3\cos t\,\mathbf{i} + 4\sin t\,\mathbf{j} + \dfrac{t}{2}\mathbf{k}$
7. $\mathbf{r}(t) = 2\cos t\,\mathbf{i} + 2\sin t\,\mathbf{j} + e^{-t}\mathbf{k}$
8. $\mathbf{r}(t) = t\mathbf{i} + t^2\mathbf{j} + \dfrac{3t}{2}\mathbf{k}$
9. $\mathbf{r}(t) = \left\langle t,\, t^2,\, \dfrac{2}{3}t^3\right\rangle$
10. $\mathbf{r}(t) = \langle \cos t + t\sin t,\, \sin t - t\cos t,\, t\rangle$

In Exercises 11–18, sketch the space curve represented by the intersection of the given surfaces, and represent the curve by a vector-valued function using the given parameter.

Surfaces	Parameter
11. $z = x^2 + y^2,\ x + y = 0$	$x = t$
12. $z = x^2 + y^2,\ z = 4$	$x = 2\cos t$
13. $x^2 + y^2 = 4,\ z = x^2$	$x = 2\sin t$
14. $4x^2 + y^2 + 4z^2 = 16,\ x = y^2$	$y = t$

15. $x^2 + y^2 + z^2 = 4,\ x + z = 2$	$x = 1 + \sin t$
16. $x^2 + y^2 + z^2 = 10,\ x + y = 4$	$x = 2 + \sin t$
17. $x^2 + z^2 = 4,\ y^2 + z^2 = 4$	$x = t$
(first octant)	
18. $x^2 + y^2 + z^2 = 16,\ xy = 4$	$x = t$
(first octant)	

In Exercises 19–24, evaluate the limit.

19. $\displaystyle\lim_{t \to 2}\left(t\mathbf{i} + \frac{t^2 - 4}{t^2 - 2t}\mathbf{j} + \frac{1}{t}\mathbf{k}\right)$

20. $\displaystyle\lim_{t \to 0}\left(e^t\mathbf{i} + \frac{\sin t}{t}\mathbf{j} + e^{-t}\mathbf{k}\right)$

21. $\displaystyle\lim_{t \to 0}\left(t^2\mathbf{i} + 3t\mathbf{j} + \frac{1 - \cos t}{t}\mathbf{k}\right)$

22. $\displaystyle\lim_{t \to 1}\left(\sqrt{t}\,\mathbf{i} + \frac{\ln t}{t^2 - 1}\mathbf{j} + 2t^2\mathbf{k}\right)$

23. $\displaystyle\lim_{t \to 0}\left(\frac{1}{t}\mathbf{i} + \cos t\,\mathbf{j} + \sin t\,\mathbf{k}\right)$

24. $\displaystyle\lim_{t \to \infty}\left(e^{-t}\mathbf{i} + \frac{1}{t}\mathbf{j} + \frac{t}{t^2 + 1}\mathbf{k}\right)$

In Exercises 25–30, discuss the continuity of the given vector-valued function.

25. $\mathbf{r}(t) = t\mathbf{i} + \dfrac{1}{t}\mathbf{j}$

26. $\mathbf{r}(t) = \sqrt{t}\,\mathbf{i} + \sqrt{t-1}\,\mathbf{j}$

27. $\mathbf{r}(t) = t\mathbf{i} + \arcsin t\,\mathbf{j} + (t - 1)\,\mathbf{k}$

28. $\mathbf{r}(t) = \sin t\,\mathbf{i} + \cos t\,\mathbf{j} + \ln t\,\mathbf{k}$

29. $\mathbf{r}(t) = \langle e^{-t},\ t^2,\ \tan t \rangle$

30. $\mathbf{r}(t) = \langle 8,\ \sqrt{t},\ \sqrt[3]{t} \rangle$

31. Let $\mathbf{r}(t)$ and $\mathbf{u}(t)$ be vector functions whose limits exist as $t \to c$. Prove that

$$\lim_{t \to c} [\mathbf{r}(t) \times \mathbf{u}(t)] = \lim_{t \to c} \mathbf{r}(t) \times \lim_{t \to c} \mathbf{u}(t)$$

SECTION TOPICS ▪
Differentiation of vector-valued functions ▪
Integration of vector-valued functions ▪

14.2
Differentiation and integration of vector-valued functions

In Sections 14.3–14.5, we will look at several important applications involving the calculus of vector-valued functions. As a prelude to that discussion, we devote this section to the mechanics of differentiation and integration of vector-valued functions.

The definition of the derivative of a vector-valued function parallels that given for real-valued functions.

DEFINITION OF THE DERIVATIVE OF A VECTOR-VALUED FUNCTION

The **derivative of a vector-valued function r** is defined by

$$\mathbf{r}'(t) = \lim_{\Delta t \to 0} \frac{\mathbf{r}(t + \Delta t) - \mathbf{r}(t)}{\Delta t}$$

for all t for which the limit exists.

Differentiation of vector-valued functions can be done on a *component-by-component basis*. To see why this is true, consider the function given by $\mathbf{r}(t) = f(t)\,\mathbf{i} + g(t)\,\mathbf{j} + h(t)\,\mathbf{k}.$ Applying the definition of the derivative, we have

$$\mathbf{r}'(t) = \lim_{\Delta t \to 0} \frac{\mathbf{r}(t + \Delta t) - \mathbf{r}(t)}{\Delta t}$$

$$= \lim_{\Delta t \to 0} \frac{f(t + \Delta t)\,\mathbf{i} + g(t + \Delta t)\,\mathbf{j} + h(t + \Delta t)\,\mathbf{k} - f(t)\,\mathbf{i} - g(t)\,\mathbf{j} - h(t)\,\mathbf{k}}{\Delta t}$$

$$= \lim_{\Delta t \to 0} \left[\left(\frac{f(t + \Delta t) - f(t)}{\Delta t} \right) \mathbf{i} \right.$$

$$\left. + \left(\frac{g(t + \Delta t) - g(t)}{\Delta t} \right) \mathbf{j} + \left(\frac{h(t + \Delta t) - h(t)}{\Delta t} \right) \mathbf{k} \right]$$

$$= f'(t)\,\mathbf{i} + g'(t)\,\mathbf{j} + h'(t)\,\mathbf{k}$$

We list this important result in the following theorem. Note that the derivative of the vector-valued function $\mathbf{r}$ is itself a vector-valued function. (A geometrical interpretation of the derivative of a vector-valued function will be given in the next section.)

THEOREM 14.1   **DIFFERENTIATION OF VECTOR-VALUED FUNCTIONS**

1. If $\mathbf{r}(t) = f(t)\,\mathbf{i} + g(t)\,\mathbf{j}$ where f and g are differentiable functions of t, then

$$\mathbf{r}'(t) = f'(t)\,\mathbf{i} + g'(t)\,\mathbf{j} \qquad \text{Plane}$$

2. If $\mathbf{r}(t) = f(t)\,\mathbf{i} + g(t)\,\mathbf{j} + h(t)\,\mathbf{k}$, where f, g, and h are differentiable functions of t, then

$$\mathbf{r}'(t) = f'(t)\,\mathbf{i} + g'(t)\,\mathbf{j} + h'(t)\,\mathbf{k} \qquad \text{Space}$$

| **Remark**   Higher-order derivatives are obtained by successive differentiation of each component function. Moreover, we call the curve represented by $\mathbf{r}$ **smooth** if f', g', and h' are continuous and $\mathbf{r}'(t) \neq \mathbf{0}$ for any value of t.

EXAMPLE 1   *Differentiation of vector-valued functions*

Find the derivative of the following vector-valued functions:

(a) $\mathbf{r}(t) = t^2\mathbf{i} - 4\mathbf{j}$ (b) $\mathbf{r}(t) = \dfrac{1}{t}\,\mathbf{i} + \ln t\,\mathbf{j}$

Solution: Differentiating on a component-by-component basis, we obtain the following.

(a) $\mathbf{r}'(t) = 2t\mathbf{i} - 0\mathbf{j} = 2t\mathbf{i}$

(b) $\mathbf{r}'(t) = -\dfrac{1}{t^2}\,\mathbf{i} + \dfrac{1}{t}\,\mathbf{j}$

EXAMPLE 2   *Differentiation of vector-valued functions*

For the vector-valued function given by

$$\mathbf{r}(t) = e^{2t}\mathbf{i} - \sin t\,\mathbf{j} + 2t\mathbf{k}$$

find

(a) $\mathbf{r}'(t)$ (b) $\mathbf{r}''(t)$ (c) $\mathbf{r}'(t) \cdot \mathbf{r}''(t)$

Solution:

(a) $\mathbf{r}'(t) = 2e^{2t}\mathbf{i} - \cos t\,\mathbf{j} + 2\mathbf{k}$

(b) $\mathbf{r}''(t) = 4e^{2t}\mathbf{i} + \sin t\,\mathbf{j} + 0\mathbf{k} = 4e^{2t}\mathbf{i} + \sin t\,\mathbf{j}$

(c) $\mathbf{r}'(t) \cdot \mathbf{r}''(t) = 8e^{4t} - \cos t \sin t$

| **Remark**   As with real-valued functions, with vector-valued functions it is occasionally convenient to use Leibniz or operator notation for the derivative. Thus, for the derivative of $\mathbf{r}(t) = f(t)\,\mathbf{i} + g(t)\,\mathbf{j} + h(t)\,\mathbf{k}$, we can write

$$\mathbf{r}'(t), \quad \frac{d\mathbf{r}}{dt}, \quad \frac{d}{dt}[\mathbf{r}(t)], \quad \text{or} \quad D_t[\mathbf{r}(t)]$$

Most of the differentiation rules in Chapter 3 have counterparts for vector-valued functions. We list several in the following theorem.

THEOREM 14.2   **PROPERTIES OF THE DERIVATIVE**

If $\mathbf{r}$ and $\mathbf{u}$ are differentiable vector-valued functions of t, f is a differentiable real-valued function of t, and c is a scalar, then the following properties are true.

1. $D_t[c\mathbf{r}(t)] = c\mathbf{r}'(t)$
2. $D_t[\mathbf{r}(t) \pm \mathbf{u}(t)] = \mathbf{r}'(t) \pm \mathbf{u}'(t)$
3. $D_t[f(t)\mathbf{r}(t)] = f(t)\mathbf{r}'(t) + f'(t)\mathbf{r}(t)$
4. $D_t[\mathbf{r}(t) \cdot \mathbf{u}(t)] = \mathbf{r}(t) \cdot \mathbf{u}'(t) + \mathbf{r}'(t) \cdot \mathbf{u}(t)$
5. $D_t[\mathbf{r}(t) \times \mathbf{u}(t)] = \mathbf{r}(t) \times \mathbf{u}'(t) + \mathbf{r}'(t) \times \mathbf{u}(t)$

Proof: We prove the fourth property for vectors in the plane and leave the proofs of the other properties as exercises. Let

$$\mathbf{r}(t) = f_1(t)\,\mathbf{i} + g_1(t)\,\mathbf{j} \qquad \text{and} \qquad \mathbf{u}(t) = f_2(t)\,\mathbf{i} + g_2(t)\,\mathbf{j}$$

where f_i and g_i are differentiable functions of t. Then,

$$\mathbf{r}(t) \cdot \mathbf{u}(t) = f_1(t)f_2(t) + g_1(t)g_2(t)$$

and it follows that

$$\begin{aligned} D_t[\mathbf{r}(t) \cdot \mathbf{u}(t)] &= f_1(t)f_2'(t) + f_1'(t)f_2(t) + g_1(t)g_2'(t) + g_1'(t)g_2(t) \\ &= [f_1(t)f_2'(t) + g_1(t)g_2'(t)] + [f_1'(t)f_2(t) + g_1'(t)g_2(t)] \\ &= \mathbf{r}(t) \cdot \mathbf{u}'(t) + \mathbf{r}'(t) \cdot \mathbf{u}(t) \end{aligned}$$

Remark Note that Theorem 14.2 contains three different versions of "product rules." Property 3 gives the derivative of the product of a real-valued function f and a vector-valued function $\mathbf{r}$. Property 4 gives the derivative of the dot product of two vector-valued functions, and Property 5 gives the derivative of the cross product of two vector-valued functions. Of course, Property 5 applies only in three dimensions since we do not define the cross product of two-dimensional vectors.

EXAMPLE 3   *Using properties of the derivative*

For the vector-valued functions given by

$$\mathbf{r}(t) = \frac{1}{t}\,\mathbf{i} - \mathbf{j} + (\ln t)\,\mathbf{k} \qquad \text{and} \qquad \mathbf{u}(t) = t^2\mathbf{i} - 2t\mathbf{j} + \mathbf{k}$$

find
(a) $D_t[\mathbf{r}(t) \cdot \mathbf{u}(t)]$ (b) $D_t[\mathbf{u}(t) \times \mathbf{u}'(t)]$

Solution:
(a) Since

$$\mathbf{r}'(t) = -\frac{1}{t^2}\,\mathbf{i} + \frac{1}{t}\,\mathbf{k} \qquad \text{and} \qquad \mathbf{u}'(t) = 2t\mathbf{i} - 2\mathbf{j}$$

we have

$$D_t[\mathbf{r}(t) \cdot \mathbf{u}(t)] = \mathbf{r}(t) \cdot \mathbf{u}'(t) + \mathbf{r}'(t) \cdot \mathbf{u}(t)$$

$$= \left[\frac{1}{t} \mathbf{i} - \mathbf{j} + \ln t \, \mathbf{k} \right] \cdot [2t\mathbf{i} - 2\mathbf{j}]$$

$$+ \left[-\frac{1}{t^2} \mathbf{i} + \frac{1}{t} \mathbf{k} \right] \cdot [t^2\mathbf{i} - 2t\mathbf{j} + \mathbf{k}]$$

$$= 2 + 2 + (-1) + \frac{1}{t} = 3 + \frac{1}{t}$$

(b) Since $\mathbf{u}'(t) = 2t\mathbf{i} - 2\mathbf{j}$ and $\mathbf{u}''(t) = 2\mathbf{i}$, we have

$$D_t[\mathbf{u}(t) \times \mathbf{u}'(t)] = \mathbf{u}(t) \times \mathbf{u}''(t) + \mathbf{u}'(t) \times \mathbf{u}'(t)$$

$$= \begin{vmatrix} \mathbf{i} & \mathbf{j} & \mathbf{k} \\ t^2 & -2t & 1 \\ 2 & 0 & 0 \end{vmatrix} + \mathbf{0}$$

$$= \begin{vmatrix} -2t & 1 \\ 0 & 0 \end{vmatrix} \mathbf{i} - \begin{vmatrix} t^2 & 1 \\ 2 & 0 \end{vmatrix} \mathbf{j} + \begin{vmatrix} t^2 & -2t \\ 2 & 0 \end{vmatrix} \mathbf{k}$$

$$= 0\mathbf{i} - (-2)\,\mathbf{j} + 4t\mathbf{k} = 2\mathbf{j} + 4t\mathbf{k}$$

Integration of vector-valued functions

The following definition is a natural consequence of the definition of the derivative of a vector-valued function.

DEFINITION OF INTEGRATION OF VECTOR-VALUED FUNCTIONS

1. If $\mathbf{r}(t) = f(t)\,\mathbf{i} + g(t)\,\mathbf{j}$ where f and g are continuous on $[a, b]$, then the **indefinite integral (antiderivative)** of $\mathbf{r}$ is

$$\int \mathbf{r}(t)\,dt = \left[\int f(t)\,dt \right]\mathbf{i} + \left[\int g(t)\,dt \right]\mathbf{j} \qquad \text{Plane}$$

and its **definite integral** over the interval $a \le t \le b$ is

$$\int_a^b \mathbf{r}(t)\,dt = \left[\int_a^b f(t)\,dt \right]\mathbf{i} + \left[\int_a^b g(t)\,dt \right]\mathbf{j}$$

2. If $\mathbf{r}(t) = f(t)\,\mathbf{i} + g(t)\,\mathbf{j} + h(t)\,\mathbf{k}$, where f, g, and h are continuous on $[a, b]$, then the **indefinite integral (antiderivative)** of $\mathbf{r}$ is

$$\int \mathbf{r}(t)\,dt = \left[\int f(t)\,dt \right]\mathbf{i} + \left[\int g(t)\,dt \right]\mathbf{j} + \left[\int h(t)\,dt \right]\mathbf{k} \qquad \text{Space}$$

and its **definite integral** over the interval $a \le t \le b$ is

$$\int_a^b \mathbf{r}(t)\,dt = \left[\int_a^b f(t)\,dt \right]\mathbf{i} + \left[\int_a^b g(t)\,dt \right]\mathbf{j} + \left[\int_a^b h(t)\,dt \right]\mathbf{k}$$

The antiderivative of a vector-valued function is a family of vector-valued functions all differing by a constant vector **C**. For instance, if $\mathbf{r}(t)$ is a three-dimensional vector-valued function, then for the indefinite integral $\int \mathbf{r}(t)\, dt$, we obtain three constants of integration:

$$\int f(t)\, dt = F(t) + C_1, \qquad \int g(t)\, dt = G(t) + C_2, \qquad \int h(t)\, dt = H(t) + C_3$$

where $F'(t) = f(t)$, $G'(t) = g(t)$, and $H'(t) = h(t)$. These three *scalar* constants produce one *vector* constant of integration,

$$\begin{aligned}
\int \mathbf{r}(t)\, dt &= [F(t) + C_1]\,\mathbf{i} + [G(t) + C_2]\,\mathbf{j} + [H(t) + C_3]\,\mathbf{k} \\
&= [F(t)\,\mathbf{i} + G(t)\,\mathbf{j} + H(t)\,\mathbf{k}] + [C_1\mathbf{i} + C_2\mathbf{j} + C_3\mathbf{k}] \\
&= \mathbf{R}(t) + \mathbf{C}
\end{aligned}$$

where $\mathbf{R}'(t) = \mathbf{r}(t)$.

EXAMPLE 4   Integrating a vector-valued function

Evaluate the indefinite integral

$$\int (t\mathbf{i} + 3\mathbf{j})\, dt$$

Solution: Integrating on a component-by-component basis, we have

$$\int (t\mathbf{i} + 3\mathbf{j})\, dt = \frac{t^2}{2}\,\mathbf{i} + 3t\mathbf{j} + \mathbf{C}$$

EXAMPLE 5   Definite integral of a vector-valued function

Evaluate the integral

$$\int_0^1 \mathbf{r}(t)\, dt = \int_0^1 \left(\sqrt[3]{t}\,\mathbf{i} + \frac{1}{t+1}\,\mathbf{j} + e^{-t}\mathbf{k} \right) dt$$

Solution:

$$\begin{aligned}
\int_0^1 \mathbf{r}(t)\, dt &= \left[\int_0^1 t^{1/3}\, dt \right]\mathbf{i} + \left[\int_0^1 \frac{1}{t+1}\, dt \right]\mathbf{j} + \left[\int_0^1 e^{-t}\, dt \right]\mathbf{k} \\
&= \left[\left(\frac{3}{4}\right)t^{4/3} \right]_0^1 \mathbf{i} + \left[\ln|t+1| \right]_0^1 \mathbf{j} + \left[-e^{-t} \right]_0^1 \mathbf{k} \\
&= \frac{3}{4}\,\mathbf{i} + (\ln 2)\,\mathbf{j} + \left(1 - \frac{1}{e} \right)\mathbf{k}
\end{aligned}$$

As with real-valued functions, we can narrow the family of antiderivatives of a vector-valued function $\mathbf{r}'$ down to a single antiderivative by imposing an initial (or boundary) condition on the vector-valued function $\mathbf{r}$. This is demonstrated in the next example.

EXAMPLE 6   *Finding the antiderivative of a vector-valued function*

Find the antiderivative of

$$\mathbf{r}'(t) = \cos 2t \,\mathbf{i} - 2 \sin t \,\mathbf{j} + \frac{1}{1+t^2} \,\mathbf{k}$$

that satisfies the initial condition $\mathbf{r}(0) = 3\mathbf{i} - 2\mathbf{j} + \mathbf{k}$.

Solution:

$$\mathbf{r}(t) = \int \mathbf{r}'(t) \, dt$$

$$= \left[\int \cos 2t \, dt \right] \mathbf{i} + \left[\int -2 \sin t \, dt \right] \mathbf{j} + \left[\int \frac{1}{1+t^2} \, dt \right] \mathbf{k}$$

$$= \left[\frac{1}{2} \sin 2t + C_1 \right] \mathbf{i} + [2 \cos t + C_2] \mathbf{j} + [\arctan t + C_3] \mathbf{k}$$

Letting $t = 0$ and using the fact that $\mathbf{r}(0) = 3\mathbf{i} - 2\mathbf{j} + \mathbf{k}$, we have

$$\mathbf{r}(0) = [0 + C_1] \mathbf{i} + [2 + C_2] \mathbf{j} + [0 + C_3] \mathbf{k}$$
$$= 3\mathbf{i} + (-2)\mathbf{j} + \mathbf{k}$$

Equating corresponding components, we have $C_1 = 3$, $2 + C_2 = -2$, and $C_3 = 1$. Thus, the antiderivative that satisfies the given initial condition is

$$\mathbf{r}(t) = \left[\frac{1}{2} \sin 2t + 3 \right] \mathbf{i} + [2 \cos t - 4] \mathbf{j} + [\arctan t + 1] \mathbf{k}$$

Section Exercises 14.2

In Exercises 1–6, find $\mathbf{r}'(t)$.

1. $\mathbf{r}(t) = a \cos^3 t \,\mathbf{i} + a \sin^3 t \,\mathbf{j} + \mathbf{k}$
2. $\mathbf{r}(t) = \sqrt{t}\,\mathbf{i} + t\sqrt{t}\,\mathbf{j} + \ln t \,\mathbf{k}$
3. $\mathbf{r}(t) = e^{-t}\mathbf{i} + 4\mathbf{j}$
4. $\mathbf{r}(t) = \langle \sin t - t \cos t, \cos t + t \sin t, t^2 \rangle$
5. $\mathbf{r}(t) = \langle t \sin t, t \cos t, t \rangle$
6. $\mathbf{r}(t) = \langle \arcsin t, \arccos t, 0 \rangle$

In Exercises 7 and 8, find
(a) $\mathbf{r}'(t)$
(b) $\mathbf{r}''(t)$
(c) $D_t[\mathbf{r}(t) \cdot \mathbf{u}(t)]$
(d) $D_t[3\mathbf{r}(t) - \mathbf{u}(t)]$
(e) $D_t[\mathbf{r}(t) \times \mathbf{u}(t)]$
(f) $D_t[\|\mathbf{r}(t)\|]$

7. $\mathbf{r}(t) = t\mathbf{i} + 3t\mathbf{j} + t^2\mathbf{k}$
   $\mathbf{u}(t) = 4t\mathbf{i} + t^2\mathbf{j} + t^3\mathbf{k}$
8. $\mathbf{r}(t) = t^2\mathbf{i} + \sin t \,\mathbf{j} + \cos t \,\mathbf{k}$

   $\mathbf{u}(t) = \frac{1}{t^2}\mathbf{i} + \sin t \,\mathbf{j} + \cos t \,\mathbf{k}$

In Exercises 9–14, evaluate the indefinite integral.

9. $\displaystyle\int (2t\mathbf{i} + \mathbf{j} + \mathbf{k}) \, dt$

10. $\displaystyle\int [(2t - 1)\,\mathbf{i} + 4t^3\mathbf{j} + 3\sqrt{t}\,\mathbf{k}] \, dt$

11. $\displaystyle\int [e^t\mathbf{i} + \sin t \,\mathbf{j} + \cos t \,\mathbf{k}] \, dt$

12. $\displaystyle\int \left[\ln t \,\mathbf{i} + \frac{1}{t}\mathbf{j} + \mathbf{k} \right] dt$

13. $\displaystyle\int \left[\sec^2 t \,\mathbf{i} + \frac{1}{1+t^2}\mathbf{j} \right] dt$

14. $\displaystyle\int [e^{-t} \sin t \,\mathbf{i} + e^{-t} \cos t \,\mathbf{j}] \, dt$

In Exercises 15–20, find $\mathbf{r}(t)$ for the given conditions.

15. $\mathbf{r}'(t) = 4e^{2t}\mathbf{i} + 3e^t\mathbf{j}$
   $\mathbf{r}(0) = 2\mathbf{i}$
16. $\mathbf{r}'(t) = 2t\mathbf{j} + \sqrt{t}\,\mathbf{k}$
   $\mathbf{r}(0) = \mathbf{i} + \mathbf{j}$

17. $\mathbf{r}''(t) = -32\mathbf{j}$
$\mathbf{r}'(0) = 600\sqrt{3}\mathbf{i} + 600\mathbf{j}$
$\mathbf{r}(0) = \mathbf{0}$

18. $\mathbf{r}''(t) = -4\cos t\,\mathbf{j} - 3\sin 5\,\mathbf{k}$
$\mathbf{r}'(0) = 3\mathbf{k}$
$\mathbf{r}(0) = 4\mathbf{j}$

19. $\mathbf{r}'(t) = te^{-t^2}\mathbf{i} - e^{-t}\mathbf{j} + \mathbf{k}$
$\mathbf{r}(0) = \dfrac{1}{2}\,\mathbf{i} - \mathbf{j} + \mathbf{k}$

20. $\mathbf{r}'(t) = \dfrac{1}{1+t^2}\,\mathbf{i} + \dfrac{1}{t^2}\,\mathbf{j} + \dfrac{1}{t}\,\mathbf{k}$
$\mathbf{r}(1) = 2\mathbf{i}$

In Exercises 21–24, evaluate the definite integral.

21. $\displaystyle\int_0^1 (8t\mathbf{i} + t\mathbf{j} - \mathbf{k})\,dt$

22. $\displaystyle\int_{-1}^1 (t\mathbf{i} + t^3\mathbf{j} + \sqrt[3]{t}\mathbf{k})\,dt$

23. $\displaystyle\int_0^{\pi/2} [(a\cos t)\,\mathbf{i} + (a\sin t)\,\mathbf{j} + \mathbf{k}]\,dt$

24. $\displaystyle\int_0^3 (e^t\mathbf{i} + te^t\mathbf{k})\,dt$

25. Complete the proof of Theorem 14.2.

SECTION TOPICS ▪
Velocity ▪
Acceleration ▪
Projectile motion ▪

14.3
Velocity and acceleration

The stage is now set, and we are ready to combine our study of parametric equations, curves, vectors, and vector-valued functions to form a model for **curvilinear motion.** We begin by looking at the motion of an object in the plane. The motion of an object in space could be developed in a similar manner.

As an object moves along a curve in the plane, the coordinates x and y of its center of mass are each functions of time t. Rather than using f and g to represent these two functions, it is convenient to write $x = x(t)$ and $y = y(t)$. Thus, the position vector $\mathbf{r}(t)$ takes the form

$$\mathbf{r}(t) = x(t)\,\mathbf{i} + y(t)\,\mathbf{j} \qquad \text{Position vector}$$

The beauty of this vector model for representing motion is that we can use the first and second derivatives of the vector-valued function $\mathbf{r}$ to find the object's velocity and acceleration. (Recall from the previous chapter that velocity and acceleration are both vector quantities having magnitude and direction.)

To find the velocity and acceleration vectors at a given time t, we consider a point $Q = (x(t + \Delta t),\ y(t + \Delta t))$ that is approaching the point $P = (x(t),\ y(t))$ along the curve C is given by

$$x = x(t) \qquad \text{and} \qquad y = y(t)$$

as shown in Figure 14.6. As $\Delta t \to 0$, the direction of the vector $\overrightarrow{PQ}$ (denoted by $\Delta\mathbf{r}$) approaches the *direction of motion* at the time t and we write

$$\Delta\mathbf{r} = \mathbf{r}(t + \Delta t) - \mathbf{r}(t)$$

$$\frac{\Delta\mathbf{r}}{\Delta t} = \frac{\mathbf{r}(t + \Delta t) - \mathbf{r}(t)}{\Delta t}$$

$$\lim_{\Delta t \to 0} \frac{\Delta\mathbf{r}}{\Delta t} = \lim_{\Delta t \to 0} \frac{\mathbf{r}(t + \Delta t) - \mathbf{r}(t)}{\Delta t}$$

We define this limit, if it exists, as the **tangent vector** to the curve at the point P. Note that this is the same limit used to define $\mathbf{r}'(t)$. Thus, the direction of

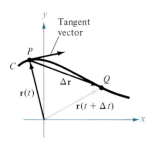

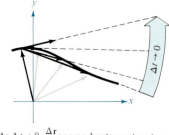

As $\Delta t \to 0$, $\dfrac{\Delta\mathbf{r}}{\Delta t}$ approaches tangent vector.

FIGURE 14.6

$\mathbf{r}'(t)$ gives us the direction of motion at the time t. Moreover, the magnitude of the vector $\mathbf{r}'(t)$ is

$$\|\mathbf{r}'(t)\| = \|x'(t)\,\mathbf{i} + y'(t)\,\mathbf{j}\|$$
$$= \sqrt{[x'(t)]^2 + [y'(t)]^2} = \left|\frac{ds}{dt}\right|$$

where ds/dt is the rate of change in the arc length s with respect to time. We define $|ds/dt|$ to be the **speed** of the object at time t. Thus, the vector $\mathbf{r}'(t)$ gives us both the direction and the speed of the object at any time t. In a similar way, we can use $\mathbf{r}''(t)$ to represent acceleration, as indicated in the following definition.

DEFINITION OF VELOCITY AND ACCELERATION

If x and y are twice differentiable functions of t and $\mathbf{r}$ is a vector-valued function given by $\mathbf{r}(t) = x(t)\,\mathbf{i} + y(t)\,\mathbf{j}$, then the velocity vector and acceleration vector at time t are

$$\textbf{velocity} = \mathbf{v}(t) = \mathbf{r}'(t) = x'(t)\,\mathbf{i} + y'(t)\,\mathbf{j}$$
$$\textbf{acceleration} = \mathbf{a}(t) = \mathbf{r}''(t) = x''(t)\,\mathbf{i} + y''(t)\,\mathbf{j}$$

| Remark   The **speed** of the object is

$$\text{speed} = \|\mathbf{v}(t)\| = \|\mathbf{r}'(t)\| = \sqrt{[x'(t)]^2 + [y'(t)]^2}$$

For motion along a space curve, the definitions are similar. For instance, the velocity vector of the function given by $\mathbf{r}(t) = x(t)\,\mathbf{i} + y(t)\,\mathbf{j} + z(t)\,\mathbf{k}$ is

$$\textbf{velocity} = \mathbf{v}(t) = \mathbf{r}'(t) = x'(t)\,\mathbf{i} + y'(t)\,\mathbf{j} + z'(t)\,\mathbf{k}$$

EXAMPLE 1   Finding velocity and acceleration along a plane curve

Find the velocity vector, speed, and acceleration vector of a particle that moves along the plane curve C described by

$$\mathbf{r}(t) = 2\sin\frac{t}{2}\,\mathbf{i} + 2\cos\frac{t}{2}\,\mathbf{j}$$

Solution: The velocity vector is given by

$$\mathbf{v}(t) = \mathbf{r}'(t) = \cos\frac{t}{2}\,\mathbf{i} - \sin\frac{t}{2}\,\mathbf{j}$$

and the speed (at any time) is

$$\|\mathbf{r}'(t)\| = \sqrt{\cos^2\frac{t}{2} + \sin^2\frac{t}{2}} = 1$$

Finally, the acceleration vector is given by

$$\mathbf{a}(t) = \mathbf{r}''(t) = -\frac{1}{2}\sin\frac{t}{2}\,\mathbf{i} - \frac{1}{2}\cos\frac{t}{2}\,\mathbf{j}$$

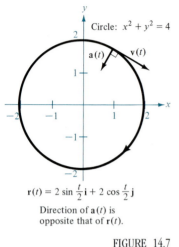

Circle: $x^2 + y^2 = 4$

$$\mathbf{r}(t) = 2 \sin \frac{t}{2}\mathbf{i} + 2 \cos \frac{t}{2}\mathbf{j}$$

Direction of $\mathbf{a}(t)$ is
opposite that of $\mathbf{r}(t)$.

FIGURE 14.7

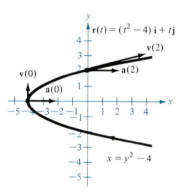

$$\mathbf{r}(t) = (t^2 - 4)\mathbf{i} + t\mathbf{j}$$

$x = y^2 - 4$

FIGURE 14.8

The parametric equations for the curve in Example 1 are $x = 2 \sin(t/2)$ and $y = 2 \cos(t/2)$. By eliminating the parameter t, we can obtain the rectangular equation

$$x^2 + y^2 = 4 \qquad \text{Rectangular equation}$$

Thus, the curve is a circle of radius 2 centered at the origin, as shown in Figure 14.7. Since the velocity vector $\mathbf{v}(t) = \cos(t/2)\mathbf{i} - \sin(t/2)\mathbf{j}$ has a constant magnitude but a changing direction as t increases, the particle moves around this circle at a constant speed.

| Remark   It is interesting to note that the velocity and acceleration vectors in Example 1 are orthogonal at any point in time. (Remember that two vectors are orthogonal if their dot product is zero.) This is characteristic of motion at a constant speed.

EXAMPLE 2   Finding velocity and acceleration along a plane curve

Sketch the path of an object moving along the plane curve given by

$$\mathbf{r}(t) = (t^2 - 4)\mathbf{i} + t\mathbf{j}$$

and find the velocity and acceleration vectors when $t = 0$ and $t = 2$.

Solution: Using the parametric equations $x = t^2 - 4$ and $y = t$, we see that the curve is a parabola given by $x = y^2 - 4$ as shown in Figure 14.8. The velocity vector (at any time) is given by

$$\mathbf{v}(t) = \mathbf{r}'(t) = 2t\mathbf{i} + \mathbf{j}$$

and the acceleration vector (at any time) is given by

$$\mathbf{a}(t) = \mathbf{r}''(t) = 2\mathbf{i}$$

Therefore, when $t = 0$, the velocity and acceleration vectors are given by

$$\mathbf{v}(0) = 2(0)\mathbf{i} + \mathbf{j} = \mathbf{j} \qquad \text{and} \qquad \mathbf{a}(0) = 2\mathbf{i}$$

and when $t = 2$, the velocity and acceleration vectors are given by

$$\mathbf{v}(2) = 2(2)\mathbf{i} + \mathbf{j} = 4\mathbf{i} + \mathbf{j} \qquad \text{and} \qquad \mathbf{a}(2) = 2\mathbf{i}$$

EXAMPLE 3   Finding velocity and acceleration along a space curve

Find the velocity vector, speed, and acceleration vector of a particle that moves along the space curve given by

$$\mathbf{r}(t) = e^t\mathbf{i} + e^{2t}\mathbf{j} + e^{2t}\mathbf{k}$$

Solution: The velocity vector is given by

$$\mathbf{v}(t) = \mathbf{r}'(t) = e^t\mathbf{i} + 2e^{2t}\mathbf{j} + 2e^{2t}\mathbf{k}$$

and the magnitude of this vector gives the speed

$$\|\mathbf{v}(t)\| = \sqrt{(e^t)^2 + (2e^{2t})^2 + (2e^{2t})^2} = e^t\sqrt{1 + 8e^{2t}}$$

By differentiating the velocity vector, we obtain the acceleration vector

$$\mathbf{a}(t) = \mathbf{v}'(t) = e^t\mathbf{i} + 4e^{2t}\mathbf{j} + 4e^{2t}\mathbf{k}$$

EXAMPLE 4   *Sketching velocity and acceleration vectors in space*

Sketch the path of an object moving along the space curve C given by

$$\mathbf{r}(t) = t\mathbf{i} + t^3\mathbf{j} + 3t\mathbf{k}, \quad t \geq 0$$

and find the velocity and acceleration vectors when $t = 1$.

Solution: Using the parametric equations $x = t$ and $y = t^3$, we see that the path of the object lies on the cubic cylinder given by $y = x^3$. Moreover, since $z = 3t$, we see that the object starts at $(0, 0, 0)$ and moves upward as t increases, as shown in Figure 14.9. Since $\mathbf{r}(t) = t\mathbf{i} + t^3\mathbf{j} + 3t\mathbf{k}$, we have

$$\mathbf{v}(t) = \mathbf{r}'(t) = \mathbf{i} + 3t^2\mathbf{j} + 3\mathbf{k} \qquad \text{and} \qquad \mathbf{a}(t) = \mathbf{r}''(t) = 6t\mathbf{j}$$

Therefore, the velocity and acceleration vectors when $t = 1$ are

$$\mathbf{v}(1) = \mathbf{r}'(1) = \mathbf{i} + 3\mathbf{j} + 3\mathbf{k} \qquad \text{and} \qquad \mathbf{a}(1) = \mathbf{r}''(1) = 6\mathbf{j}$$

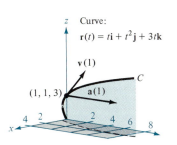

z   Curve:
$\mathbf{r}(t) = t\mathbf{i} + t^2\mathbf{j} + 3t\mathbf{k}$
$\mathbf{v}(1)$
C
$(1, 1, 3)$ $\mathbf{a}(1)$

FIGURE 14.9

So far in this section we have concentrated on finding the velocity and acceleration by differentiating the position function. Many practical applications involve the reverse problem—finding the position function for a given velocity or acceleration. This is demonstrated in the next example.

EXAMPLE 5   *Finding the position function by integration*

An object starts from rest at the point $P = (1, 2, 0)$ and moves with an acceleration of

$$\mathbf{a}(t) = \mathbf{j} + 2\mathbf{k} \qquad \text{\small Acceleration vector}$$

where the $\|\mathbf{a}(t)\|$ is measured in feet per second per second. Find the location of the object after $t = 2$ seconds.

Solution: From the description of the object's motion, we deduce the following *initial conditions*. Since the object starts from rest, we have

$$\mathbf{v}(0) = \mathbf{0}$$

Moreover, since the object starts at the point $(x, y, z) = (1, 2, 0)$, we have

$$\mathbf{r}(0) = x(0)\,\mathbf{i} + y(0)\,\mathbf{j} + z(0)\,\mathbf{k} = 1\mathbf{i} + 2\mathbf{j} + 0\mathbf{k} = \mathbf{i} + 2\mathbf{j}$$

Now, to find the position function, we integrate twice, each time using one of the initial conditions to solve for the constant of integration.

$$\mathbf{v}(t) = \int \mathbf{a}(t)\, dt = \int (\mathbf{j} + 2\mathbf{k})\, dt = t\mathbf{j} + 2t\mathbf{k} + \mathbf{C}$$

where $\mathbf{C} = C_1\mathbf{i} + C_2\mathbf{j} + C_3\mathbf{k}$. Letting $t = 0$ and applying the initial condition $\mathbf{v}(0) = \mathbf{0}$, we have

$$\mathbf{v}(0) = C_1\mathbf{i} + C_2\mathbf{j} + C_3\mathbf{k} = \mathbf{0} \implies C_1 = C_2 = C_3 = 0$$

Thus, the *velocity* at any time t is

$$\mathbf{v}(t) = t\mathbf{j} + 2t\mathbf{k} \qquad \text{\small Velocity vector}$$

Curve:

$$\mathbf{r}(t) = \mathbf{i} + \left(\frac{t^2}{2} + 2\right)\mathbf{j} + t^2\mathbf{k}$$

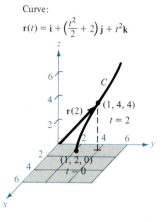

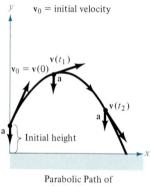

FIGURE 14.10

Now, integrating once more, we have

$$\mathbf{r}(t) = \int \mathbf{v}(t)\, dt = \int (t\mathbf{j} + 2t\mathbf{k})\, dt = \frac{t^2}{2}\mathbf{j} + t^2\mathbf{k} + \mathbf{C}$$

where $\mathbf{C} = C_4\mathbf{i} + C_5\mathbf{j} + C_6\mathbf{k}$. Letting $t = 0$ and applying the initial condition $\mathbf{r}(0) = \mathbf{i} + 2\mathbf{j}$, we have

$$\mathbf{r}(0) = C_4\mathbf{i} + C_5\mathbf{j} + C_6\mathbf{k} = \mathbf{i} + 2\mathbf{j} \implies C_4 = 1,\ C_5 = 2,\ C_6 = 0$$

Thus, the *position function* is

$$\mathbf{r}(t) = \mathbf{i} + \left(\frac{t^2}{2} + 2\right)\mathbf{j} + t^2\mathbf{k} \qquad \text{Position vector}$$

After 2 seconds the location of the object is given by $\mathbf{r}(2) = \mathbf{i} + 4\mathbf{j} + 4\mathbf{k}$, as shown in Figure 14.10.

Motion of a projectile

When we introduced parametric equations in Chapter 12, we began with a description of a projectile moving on a parabolic path. We now return to this problem to see how the path can be derived using a vector model.

We assume that gravity is the only force acting on the projectile after it is launched. Hence, the motion occurs in a vertical plane, which we represent by the xy-coordinate system with the origin as a point on the earth's surface, as shown in Figure 14.11. For a projectile of mass m, the force due to gravity is

$$\mathbf{F} = -mg\mathbf{j} \qquad \text{Force due to gravity}$$

where the gravitational constant is $g = 32$ feet per second per second, or 9.81 meters per second per second. By **Newton's Second Law of Motion,** this same force produces an acceleration $\mathbf{a} = \mathbf{a}(t)$ satisfying the equation $\mathbf{F} = m\mathbf{a}.$ Consequently, the acceleration of the projectile is given by

$$m\mathbf{a} = -mg\mathbf{j}$$

$$\mathbf{a} = -g\mathbf{j} \qquad \text{Acceleration of projectile}$$

With this acceleration, we can derive the position function for the path of a projectile, as shown in the next example.

Parabolic Path of a Projectile

FIGURE 14.11

EXAMPLE 6   *Derivation of the position function for a projectile*

A projectile of mass m is launched from an initial position $\mathbf{r}_0$ with an initial velocity $\mathbf{v}_0$. Find its position vector as a function of time.

Solution: We begin with the acceleration $\mathbf{a}(t) = -g\mathbf{j}$ and integrate twice to obtain

$$\mathbf{v}(t) = \int \mathbf{a}(t)\, dt = \int -g\mathbf{j}\, dt = -gt\mathbf{j} + \mathbf{C}_1$$

$$\mathbf{r}(t) = \int \mathbf{v}(t)\, dt = \int (-gt\mathbf{j} + \mathbf{C}_1)\, dt = -\frac{1}{2}gt^2\mathbf{j} + \mathbf{C}_1 t + \mathbf{C}_2$$

where $\mathbf{C}_1$ and $\mathbf{C}_2$ are constant vectors. Since $\mathbf{v}(0) = \mathbf{v}_0$ and $\mathbf{r}(0) = \mathbf{r}_0$, it follows that

$$\mathbf{v}(0) = 0\mathbf{j} + \mathbf{C}_1 = \mathbf{v}_0 \quad \Longrightarrow \quad \mathbf{C}_1 = \mathbf{v}_0$$
$$\mathbf{r}(0) = 0\mathbf{j} + 0\mathbf{C}_1 + \mathbf{C}_2 = \mathbf{r}_0 \quad \Longrightarrow \quad \mathbf{C}_2 = \mathbf{r}_0$$

Therefore, the position function is given by

$$\mathbf{r}(t) = -\frac{1}{2}gt^2\mathbf{j} + t\mathbf{v}_0 + \mathbf{r}_0 \qquad \square$$

$\|\mathbf{v}_0\| = v_0 = $ initial speed
$\|\mathbf{r}_0\| = h = $ initial height

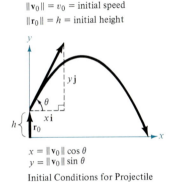

$x = \|\mathbf{v}_0\| \cos\theta$
$y = \|\mathbf{v}_0\| \sin\theta$

Initial Conditions for Projectile

FIGURE 14.12

In many projectile problems the constant vectors $\mathbf{r}_0$ and $\mathbf{v}_0$ are not explicitly given. Often we are given the initial height h, the initial speed v_0, and the angle θ at which the projectile is launched, as shown in Figure 14.12. From the given height, we deduce that

$$h = \|\mathbf{r}_0\| \quad \Longrightarrow \quad \mathbf{r}_0 = h\mathbf{j}$$

Similarly, since the speed gives the magnitude of the initial velocity, we have $v_0 = \|\mathbf{v}_0\|$ and we can write

$$\mathbf{v}_0 = x\mathbf{i} + y\mathbf{j} = (\|\mathbf{v}_0\| \cos\theta)\,\mathbf{i} + (\|\mathbf{v}_0\| \sin\theta)\,\mathbf{j}$$
$$= v_0 \cos\theta\, \mathbf{i} + v_0 \sin\theta\, \mathbf{j}$$

Thus, it follows that the position vector can be written in the form

$$\mathbf{r}(t) = -\frac{1}{2}gt^2\mathbf{j} + t\mathbf{v}_0 + \mathbf{r}_0$$

$$= -\frac{1}{2}gt^2\mathbf{j} + tv_0 \cos\theta\, \mathbf{i} + tv_0 \sin\theta\, \mathbf{j} + h\mathbf{j}$$

$$= (v_0 \cos\theta)t\mathbf{i} + \left[h + (v_0 \sin\theta)t - \frac{1}{2}gt^2\right]\mathbf{j}$$

which is the result listed in the following theorem.

THEOREM 14.3

POSITION FUNCTION FOR A PROJECTILE
The path of a projectile launched from an initial height h with initial speed v_0 and angle of elevation θ is described by the vector function

$$r(t) = (v_0 \cos\theta)t\mathbf{i} + \left[h + (v_0 \sin\theta)t - \frac{1}{2}gt^2\right]\mathbf{j}$$

where g is the gravitational constant.

EXAMPLE 7   *Describing the path of a projectile*

A baseball is hit 3 feet above ground at 100 feet per second and at an angle of $\pi/4$ with respect to the ground, as shown in Figure 14.13. Find the maximum height reached by the baseball. Will it clear a 10-foot-high fence located 300 feet from home plate?

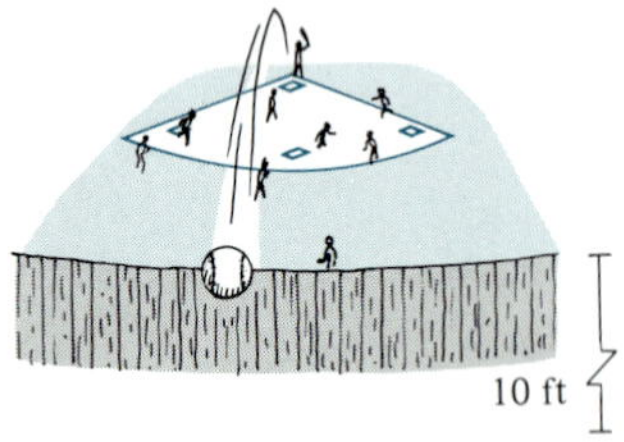

Solution: We are given that $h = 3$, $v_0 = 100$, and $\theta = \pi/4$. Thus, using $g = 32$ feet per second per second, we obtain

$$\mathbf{r}(t) = \left(100 \cos \frac{\pi}{4}\right)t\mathbf{i} + \left[3 + \left(100 \sin \frac{\pi}{4}\right)t - 16t^2\right]\mathbf{j}$$

$$= (50\sqrt{2}t)\,\mathbf{i} + [3 + 50\sqrt{2}t - 16t^2]\,\mathbf{j}$$

$$\mathbf{v}(t) = \mathbf{r}'(t) = 50\sqrt{2}\mathbf{i} + (50\sqrt{2} - 32t)\,\mathbf{j}$$

The maximum height occurs when $y'(t) = 50\sqrt{2} - 32t = 0$, which implies that

$$t = \frac{25\sqrt{2}}{16} \approx 2.21 \text{ sec}$$

Hence, the maximum height reached by the ball is

$$y\left(\frac{25\sqrt{2}}{16}\right) = 3 + 50\sqrt{2}\left(\frac{25\sqrt{2}}{16}\right) - 16\left(\frac{25\sqrt{2}}{16}\right)^2 = \frac{649}{8} \approx 81 \text{ ft}$$

The ball is 300 feet from where it was hit when

$$300 = x(t) = 50\sqrt{2}t \implies t = 3\sqrt{2}$$

At this time the height of the ball is

$$y(3\sqrt{2}) = 3 + 50\sqrt{2}(3\sqrt{2}) - 16(3\sqrt{2})^2 = 303 - 288 = 15 \text{ ft}$$

Therefore, the ball clears the 10-foot fence for a home run.

FIGURE 14.13

10 ft

Section Exercises 14.3

In Exercises 1–8, the position function $\mathbf{r}$ describes the path of an object moving in the xy-plane. Sketch a graph of the path and sketch the velocity and acceleration vectors at the given point.

Position function	*Point*
1. $\mathbf{r}(t) = 3t\mathbf{i} + (t - 1)\,\mathbf{j}$	$(3, 0)$
2. $\mathbf{r}(t) = (6 - t)\,\mathbf{i} + t\mathbf{j}$	$(3, 3)$
3. $\mathbf{r}(t) = t^2\mathbf{i} + t\mathbf{j}$	$(4, 2)$
4. $\mathbf{r}(t) = t^3\mathbf{i} + t^2\mathbf{j}$	$(1, 1)$
5. $\mathbf{r}(t) = 2 \cos t\,\mathbf{i} + 2 \sin t\,\mathbf{j}$	$(\sqrt{2}, \sqrt{2})$
6. $\mathbf{r}(t) = 2 \cos t\,\mathbf{i} + 3 \sin t\,\mathbf{j}$	$(2, 0)$
7. $\mathbf{r}(t) = \langle t - \sin t,\ 1 - \cos t \rangle$	$(\pi, 2)$
8. $\mathbf{r}(t) = \langle e^{-t},\ e^t \rangle$	$(1, 1)$

In Exercises 9–16, the position function $\mathbf{r}$ describes the path of an object moving in space. Find the velocity, speed, and acceleration of the object.

9. $\mathbf{r}(t) = t\mathbf{i} + (2t - 5)\,\mathbf{j} + 3t\mathbf{k}$

10. $\mathbf{r}(t) = 4t\mathbf{i} + 4t\mathbf{j} + 2t\mathbf{k}$

11. $\mathbf{r}(t) = t\mathbf{i} + t^2\mathbf{j} + \dfrac{t^2}{2}\mathbf{k}$

12. $\mathbf{r}(t) = t\mathbf{i} + 3t\mathbf{j} + \dfrac{t^2}{2}\mathbf{k}$

13. $\mathbf{r}(t) = t\mathbf{i} + t\mathbf{j} + \sqrt{9 - t^2}\mathbf{k}$

14. $\mathbf{r}(t) = t^2\mathbf{i} + t\mathbf{j} + 2t^{3/2}\mathbf{k}$

15. $\mathbf{r}(t) = \langle 4t,\ 3 \cos t,\ 3 \sin t \rangle$

16. $\mathbf{r}(t) = \langle e^t \cos t,\ e^t \sin t,\ e^t \rangle$

In Exercises 17–20, use the given acceleration function to find the velocity and position functions. Then find the position at time $t = 2$.

17. $\mathbf{a}(t) = \mathbf{i} + \mathbf{j} + \mathbf{k}$, $\mathbf{v}(0) = \mathbf{0}$, $\mathbf{r}(0) = \mathbf{0}$
18. $\mathbf{a}(t) = \mathbf{i} + \mathbf{k}$, $\mathbf{v}(0) = 5\mathbf{j}$, $\mathbf{r}(0) = \mathbf{0}$
19. $\mathbf{a}(t) = t\mathbf{j} + t\mathbf{k}$, $\mathbf{v}(1) = 5\mathbf{j}$, $\mathbf{r}(1) = \mathbf{0}$
20. $\mathbf{a}(t) = -\cos t\,\mathbf{i} - \sin t\,\mathbf{j}$, $\mathbf{v}(0) = \mathbf{j} + \mathbf{k}$, $\mathbf{r}(0) = \mathbf{i}$

21. A baseball, hit 3 feet above the ground, leaves the bat at an angle of 45° and is caught by an outfielder 300 feet from home plate. What was the initial velocity of the ball, and how high did it rise if it was caught 3 feet above the ground?

22. The nozzle of a hose discharges water with a velocity of 40 feet per second. Determine how high the water rises if the hose makes an angle of 60° with the ground.

23. A child standing 20 feet from the base of a silo attempts to throw a ball into an opening 40 feet from the point of release, as shown in Figure 14.14. Find the minimum initial velocity and the corresponding angle at which the ball must be thrown to go into the opening.

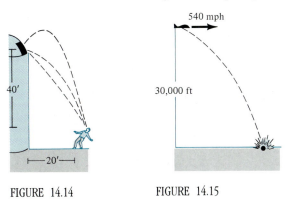

FIGURE 14.14   FIGURE 14.15

24. A bomber is flying at an altitude of 30,000 feet with a speed of 540 miles per hour (792 feet per second), as shown in Figure 14.15. When should the bomb be released to hit the target? (Neglect air resistance, and give your answer in terms of the angle of depression from the plane to the target.) What is the velocity of the bomb at the time of impact?

25. Find the angle at which an object must be thrown to obtain (a) the maximum range and (b) the maximum height.

26. A shot fired from a gun with a muzzle velocity of 1200 feet per second is to hit a target 3000 feet away. Neglecting air resistance, determine the minimum angle of elevation of the gun.

27. A projectile is fired from ground level at an angle of 10° with the horizontal. Find the minimum initial velocity necessary if the projectile is to have a range of 100 feet.

28. Eliminate the parameter t from the position function for the motion of a projectile to show that the rectangular equation is

$$y = -\frac{16 \sec^2 \theta}{v_0^2}x^2 + (\tan \theta)x + h$$

29. The path of a ball is given by the rectangular equation

$$y = x - 0.005x^2$$

Use the result of Exercise 28 to find the position function. Then find the speed and direction of the ball at the point when it has traveled 60 feet horizontally.

In Exercises 30–32, consider the motion of a point on the circumference of a rolling circle. As the circle rolls, it generates the cycloid

$$\mathbf{r}(t) = b(\omega t - \sin \omega t)\,\mathbf{i} + b(1 - \cos \omega t)\,\mathbf{j}$$

where ω is the constant angular velocity of the circle.

30. Find the velocity and acceleration of the particle.
31. Use the results of Exercise 30 to determine the times when the speed of the particle will be (a) zero and (b) maximum.
32. Find the maximum velocity of a point on the circumference of an automobile wheel of radius 1 foot when the automobile is traveling 55 miles per hour. Compare this velocity with the velocity of the automobile.

In Exercises 33–36, consider a particle moving on a circular path of radius b described by

$$\mathbf{r}(t) = b \cos \omega t\,\mathbf{i} + b \sin \omega t\,\mathbf{j}$$

where $\omega = d\theta/dt$ is the constant angular velocity.

33. Find the velocity vector and show that it is orthogonal to $\mathbf{r}(t)$.
34. Show that the speed of the particle is $b\omega$.
35. Find the acceleration vector and show that its direction is always toward the center of the circle.
36. Show that the magnitude of the acceleration vector is $\omega^2 b$.

In Exercises 37 and 38, use the results of Exercises 33–36.

37. A stone weighing 1 pound is attached to a 2-foot string and is whirled horizontally, as shown in Figure 14.16. The string will break under a force of 10 pounds. Find the maximum velocity the stone can attain without breaking the string. (Use $\mathbf{F} = m\mathbf{a}$, where $m = 1/32$.)

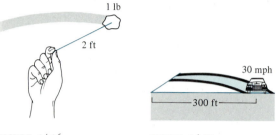

FIGURE 14.16

FIGURE 14.17

38. A 3000-pound automobile is negotiating a circular interchange of radius 300 feet at 30 miles per hour, as shown in Figure 14.17. Assuming the roadway to be level, find the force between the tires and the road so that the car stays on the circular path and does not skid. (Use $\mathbf{F} = m\mathbf{a}$, where $m = 3000/32$.)

39. In Exercise 38, find the angle at which the roadway should be banked so that no lateral friction force is exerted on the tires of the automobile.

40. Determine the relationship between the velocity and acceleration vectors for an object traveling at constant speed.

41. Consider an object moving in a straight line at a constant speed. Prove that the acceleration is zero.

14.4 Tangent vectors and normal vectors

SECTION TOPICS
- Unit tangent vector
- Principal unit normal vector
- Tangential component of acceleration
- Normal component of acceleration

In the previous section we saw that the velocity vector points in the direction of motion. This observation motivates the following definition, which applies to any smooth curve—not just those for which the parameter represents time.

DEFINITION OF UNIT TANGENT VECTOR

Let C be a smooth curve represented by $\mathbf{r}$ on an open interval I. If $\mathbf{r}'(t) \neq \mathbf{0}$, then the **unit tangent vector** $\mathbf{T}(t)$ at t is defined to be

$$\mathbf{T}(t) = \frac{\mathbf{r}'(t)}{\|\mathbf{r}'(t)\|}$$

Remark We define the **tangent line to a curve** at a point $(x(t), y(t), z(t))$ to be the line passing through the point and parallel to the unit tangent vector.

EXAMPLE 1 Finding the tangent line at a point on a curve

Find a set of parametric equations for the tangent line to the curve given by $\mathbf{r}(t) = 2 \sin t\,\mathbf{i} + 2 \cos t\,\mathbf{j} + t\mathbf{k}$ at the point $(\sqrt{2}, \sqrt{2}, \pi/4)$.

Solution: To find the unit tangent vector, we differentiate $\mathbf{r}$:

$$\mathbf{r}(t) = 2 \sin t\,\mathbf{i} + 2 \cos t\,\mathbf{j} + t\mathbf{k} \implies \mathbf{r}'(t) = 2 \cos t\,\mathbf{i} - 2 \sin t\,\mathbf{j} + \mathbf{k}$$

This implies that $\|\mathbf{r}'(t)\| = \sqrt{4 \cos^2 t + 4 \sin^2 t + 1} = \sqrt{5}$, and the unit tangent vector at t is

$$\mathbf{T}(t) = \frac{\mathbf{r}'(t)}{\|\mathbf{r}'(t)\|} = \frac{1}{\sqrt{5}}(2 \cos t\,\mathbf{i} - 2 \sin t\,\mathbf{j} + \mathbf{k})$$

Curve:
$r(t) = 2 \sin t \, \mathbf{i} + 2 \cos t \, \mathbf{j} + t\mathbf{k}$

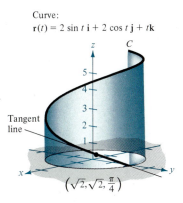

FIGURE 14.18

At $t = \pi/4$, we have

$$\mathbf{T}\left(\frac{\pi}{4}\right) = \frac{1}{\sqrt{5}}(\sqrt{2}\mathbf{i} - \sqrt{2}\mathbf{j} + \mathbf{k})$$

Finally, using the direction numbers $a = \sqrt{2}$, $b = -\sqrt{2}$, and $c = 1$ and the point $(x_1, y_1, z_1) = (\sqrt{2}, \sqrt{2}, \pi/4)$, we obtain the following parametric equations (given with parameter s).

$$x = x_1 + as = \sqrt{2} + \sqrt{2}s$$
$$y = y_1 + bs = \sqrt{2} - \sqrt{2}s$$
$$z = z_1 + cs = \frac{\pi}{4} + s$$

This tangent line is shown in Figure 14.18.

In Example 1, there are infinitely many vectors that are perpendicular to the tangent vector $\mathbf{T}(t)$. Of these, we identify a special one called the **principal unit normal vector.** To find this vector, we consider the following:

$$\mathbf{T}(t) \cdot \mathbf{T}(t) = \|\mathbf{T}(t)\|^2 = 1 \implies D_t[\mathbf{T}(t) \cdot \mathbf{T}(t)] = 0$$

But, by Property 4 of Theorem 14.2, we have

$$D_t[\mathbf{T}(t) \cdot \mathbf{T}(t)] = \mathbf{T}(t) \cdot \mathbf{T}'(t) + \mathbf{T}'(t) \cdot \mathbf{T}(t) = 0 \implies 2[\mathbf{T}(t) \cdot \mathbf{T}'(t)] = 0$$

which implies that $\mathbf{T}(t)$ and $\mathbf{T}'(t)$ are orthogonal for every value of t. By normalizing $\mathbf{T}'(t)$, we arrive at the following definition.

DEFINITION OF PRINCIPAL UNIT NORMAL VECTOR

Let C be a smooth curve represented by $\mathbf{r}$ on an open interval I. If $\mathbf{T}'(t) \neq \mathbf{0}$, then the **principal unit normal vector** at t is defined to be

$$\mathbf{N}(t) = \frac{\mathbf{T}'(t)}{\|\mathbf{T}'(t)\|}$$

EXAMPLE 2   Finding the principal unit normal vector

Find $\mathbf{N}(t)$ for the curve represented by

$$\mathbf{r}(t) = 3t\mathbf{i} + 2t^2\mathbf{j}$$

Solution:

$$\mathbf{r}'(t) = 3\mathbf{i} + 4t\mathbf{j} \implies \|\mathbf{r}'(t)\| = \sqrt{9 + 16t^2}$$

Therefore, the unit tangent vector is

$$\mathbf{T}(t) = \frac{\mathbf{r}'(t)}{\|\mathbf{r}'(t)\|} = \frac{1}{\sqrt{9 + 16t^2}}(3\mathbf{i} + 4t\mathbf{j})$$

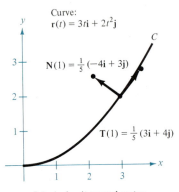

Curve:
$r(t) = 3t\mathbf{i} + 2t^2\mathbf{j}$

$N(1) = \frac{1}{5}(-4\mathbf{i} + 3\mathbf{j})$

$T(1) = \frac{1}{5}(3\mathbf{i} + 4\mathbf{j})$

Principal unit normal vector
points toward the
concave side of the curve.

FIGURE 14.19

Using the Product Rule, we differentiate with respect to t to obtain

$$\mathbf{T}'(t) = \frac{1}{\sqrt{9 + 16t^2}}(4\mathbf{j}) - \frac{16t}{(9 + 16t^2)^{3/2}}(3\mathbf{i} + 4t\mathbf{j})$$

$$= \frac{12}{(9 + 16t^2)^{3/2}}(-4t\mathbf{i} + 3\mathbf{j})$$

$$\|\mathbf{T}'(t)\| = 12\sqrt{\frac{9 + 16t^2}{(9 + 16t^2)^3}} = \frac{12}{9 + 16t^2}$$

$$\mathbf{N}(t) = \frac{\mathbf{T}'(t)}{\|\mathbf{T}'(t)\|} = \frac{1}{\sqrt{9 + 16t^2}}(-4t\mathbf{i} + 3\mathbf{j})$$

The principal unit normal vector can be difficult to evaluate algebraically. For plane curves, we can simplify the algebra by finding

$$\mathbf{T}(t) = x(t)\,\mathbf{i} + y(t)\,\mathbf{j}$$

and observing that $\mathbf{N}(t)$ must be either

$$\mathbf{N}_1(t) = y(t)\,\mathbf{i} - x(t)\,\mathbf{j} \qquad \text{or} \qquad \mathbf{N}_2(t) = -y(t)\,\mathbf{i} + x(t)\,\mathbf{j}$$

Since $\sqrt{[x(t)]^2 + [y(t)]^2} = 1$, it follows that both $\mathbf{N}_1(t)$ and $\mathbf{N}_2(t)$ are unit normal vectors. The *principal* unit normal vector $\mathbf{N}$ is the one that points toward the concave side of the curve. For instance, in Example 2,

$$\mathbf{N}(1) = \frac{1}{5}(-4\mathbf{i} + 3\mathbf{j})$$

which points toward the inside of the parabolic path, as shown in Figure 14.19. This principle also holds for curves in space. That is, for an object moving along a curve C in space, the vector $\mathbf{T}(t)$ points in the direction the object is moving, whereas the vector $\mathbf{N}(t)$ is orthogonal to $\mathbf{T}(t)$ and points in the direction the object is turning as shown in Figure 14.20.

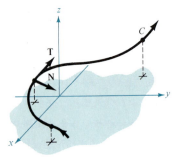

FIGURE 14.20

EXAMPLE 3   *Finding the principal unit normal vector*

Find $\mathbf{N}(t)$ for the curve represented by

$$\mathbf{r}(t) = (t + e^t)\,\mathbf{i} + (t - e^t)\,\mathbf{j} + 2e^t\mathbf{k}$$

Solution: Since

$$\mathbf{r}'(t) = (1 + e^t)\,\mathbf{i} + (1 - e^t)\,\mathbf{j} + 2e^t\mathbf{k}$$

we have

$$\|\mathbf{r}'(t)\| = \sqrt{(1 + e^t)^2 + (1 - e^t)^2 + (2e^t)^2}$$
$$= \sqrt{2 + 6e^{2t}}$$

Therefore, the unit tangent vector is

$$\mathbf{T}(t) = \frac{\mathbf{r}'(t)}{\|\mathbf{r}'(t)\|} = \frac{(1 + e^t)\,\mathbf{i} + (1 - e^t)\,\mathbf{j} + 2e^t\mathbf{k}}{\sqrt{2 + 6e^{2t}}}$$

Now, since

$$\mathbf{T}'(t) = \frac{1}{\sqrt{2 + 6e^{2t}}}(e^t\mathbf{i} - e^t\mathbf{j} + 2e^t\mathbf{k})$$

$$- \frac{6e^{2t}}{(2 + 6e^{2t})^{3/2}}[(1 + e^t)\,\mathbf{i} + (1 - e^t)\,\mathbf{j} + 2e^t\mathbf{k}]$$

$$= \frac{2e^t}{(2 + 6e^{2t})^{3/2}}[(1 - 3e^t)\,\mathbf{i} - (1 + 3e^t)\,\mathbf{j} + 2\mathbf{k}]$$

we have

$$\|\mathbf{T}'(t)\| = \frac{2e^t\sqrt{6 + 18e^{2t}}}{(2 + 6e^{2t})^{3/2}} = \frac{2\sqrt{3}e^t}{2 + 6e^{2t}}$$

and it follows that the principal unit normal vector is

$$\mathbf{N}(t) = \frac{\mathbf{T}'(t)}{\|\mathbf{T}'(t)\|} = \frac{1}{\sqrt{6 + 18e^{2t}}}[(1 - 3e^t)\,\mathbf{i} - (1 + 3e^t)\,\mathbf{j} + 2\mathbf{k}]$$

Tangential and normal components of acceleration

We now return to the problem of describing the motion of an object along a curve. In the previous section we observed that for an object traveling at a *constant speed,* the velocity and acceleration vectors are perpendicular. This seems reasonable, since the speed would not be constant if any acceleration were acting in the direction of motion. (You can verify this observation by showing that $\mathbf{r}''(t) \cdot \mathbf{r}'(t) = 0$ if $\|\mathbf{r}'(t)\|$ is constant.)

However, for an object traveling at a *variable speed,* the velocity and acceleration vectors are not necessarily perpendicular. As a case in point, we saw that the acceleration vector for a projectile always points down, regardless of the direction of motion.

The point is that part of the acceleration (the tangential component) acts in the direction of motion, and part (the normal component) acts perpendicular to the direction of motion. In order to determine these two components, we use the unit vectors $\mathbf{T}(t)$ and $\mathbf{N}(t)$, which serve in much the same way as do $\mathbf{i}$ and $\mathbf{j}$ in representing vectors in the plane. We begin with a theorem that tells us that the acceleration vector lies in the plane determined by $\mathbf{T}(t)$ and $\mathbf{N}(t)$.

THEOREM 14.4   ACCELERATION VECTOR
If $\mathbf{r}(t)$ is the position vector for a smooth curve C, then the acceleration vector $\mathbf{a}(t)$ lies in the plane determined by $\mathbf{T}(t)$ and $\mathbf{N}(t)$.

Proof: To simplify the notation we write $\mathbf{T}$ for $\mathbf{T}(t)$, $\mathbf{T}'$ for $\mathbf{T}'(t)$, and so on. Since $\mathbf{T} = \mathbf{r}'/\|\mathbf{r}'\| = \mathbf{v}/\|\mathbf{v}\|$, it follows that

$$\mathbf{v} = \|\mathbf{v}\|\mathbf{T}$$

Differentiating, we have

$$\mathbf{a} = \mathbf{v}' = D_t[\|\mathbf{v}\|]\mathbf{T} + \|\mathbf{v}\|\mathbf{T}'$$

$$= D_t[\|\mathbf{v}\|]\mathbf{T} + \|\mathbf{v}\|\mathbf{T}'\left(\frac{\|\mathbf{T}'\|}{\|\mathbf{T}'\|}\right)$$

$$= D_t[\|\mathbf{v}\|]\mathbf{T} + \|\mathbf{v}\|\,\|\mathbf{T}'\|\mathbf{N}$$

Since **a** is written as a linear combination of **T** and **N,** it must lie in the plane determined by **T** and **N.**

The coefficients of **T** and **N** in the proof of this theorem are called the **tangential** and **normal components of acceleration** and are denoted by $a_\mathbf{T} = D_t[\|\mathbf{v}\|]$ and $a_\mathbf{N} = \|\mathbf{v}\|\,\|\mathbf{T}'\|$. Thus, we can write

$$\mathbf{a}(t) = a_\mathbf{T}\mathbf{T}(t) + a_\mathbf{N}\mathbf{N}(t)$$

Though $a_\mathbf{T}$ is relatively easy to find, this formula for $a_\mathbf{N}$ is often quite difficult to evaluate. The next theorem gives a convenient alternative.

THEOREM 14.5

TANGENTIAL AND NORMAL COMPONENTS OF ACCELERATION
If $\mathbf{r}(t)$ is the position vector for a smooth curve C, then the tangential and normal components of acceleration are given by

$$a_\mathbf{T} = \mathbf{a} \cdot \mathbf{T} = \frac{\mathbf{v} \cdot \mathbf{a}}{\|\mathbf{v}\|} \quad \text{and} \quad a_\mathbf{N} = \mathbf{a} \cdot \mathbf{N} = \frac{\|\mathbf{v} \times \mathbf{a}\|}{\|\mathbf{v}\|}$$

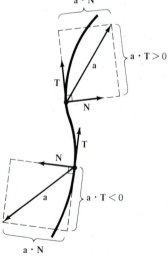

The tangential and normal components of acceleration are obtained by projecting a onto **T** and **N.**

FIGURE 14.21

Proof: Since **a** lies in the plane of **T** and **N,** we can see from Figure 14.21 that for any time t, the projection of the acceleration vector onto **T** is given by $a_\mathbf{T} = \mathbf{a} \cdot \mathbf{T}$, and onto **N** is given by $a_\mathbf{N} = \mathbf{a} \cdot \mathbf{N}$. Moreover, since $\mathbf{a} = \mathbf{v}'$ and $\mathbf{T} = \mathbf{v}/\|\mathbf{v}\|$, we have

$$a_\mathbf{T} = \mathbf{a} \cdot \mathbf{T} = \mathbf{T} \cdot \mathbf{a} = \frac{\mathbf{v}}{\|\mathbf{v}\|} \cdot \mathbf{a} = \frac{\mathbf{v} \cdot \mathbf{a}}{\|\mathbf{v}\|}$$

Furthermore, using $\mathbf{a} = a_\mathbf{T}\mathbf{T} + a_\mathbf{N}\mathbf{N}$, $\mathbf{T} \times \mathbf{T} = \mathbf{0}$, and $\|\mathbf{T} \times \mathbf{N}\| = 1$, we obtain

$$\mathbf{v} \times \mathbf{a} = \|\mathbf{v}\|\mathbf{T} \times (a_\mathbf{T}\mathbf{T} + a_\mathbf{N}\mathbf{N}) = \|\mathbf{v}\|a_\mathbf{T}(\mathbf{T} \times \mathbf{T}) + \|\mathbf{v}\|a_\mathbf{N}(\mathbf{T} \times \mathbf{N})$$

$$= \|\mathbf{v}\|a_\mathbf{N}(\mathbf{T} \times \mathbf{N})$$

$$\|\mathbf{v} \times \mathbf{a}\| = \|\mathbf{v}\|a_\mathbf{N}\|\mathbf{T} \times \mathbf{N}\| = \|\mathbf{v}\|a_\mathbf{N}$$

$$a_\mathbf{N} = \mathbf{a} \cdot \mathbf{N} = \frac{\|\mathbf{v} \times \mathbf{a}\|}{\|\mathbf{v}\|}$$

Remark The normal component of acceleration is also called the **centripetal component of acceleration.**

EXAMPLE 4   *Finding the tangential and normal components of acceleration*

Find the tangential and normal components of acceleration for the position function given by $\mathbf{r}(t) = 3t\mathbf{i} - t\mathbf{j} + t^2\mathbf{k}$.

Solution: We begin by finding the velocity, speed, and acceleration.

$$\mathbf{v}(t) = \mathbf{r}'(t) = 3\mathbf{i} - \mathbf{j} + 2t\mathbf{k}$$

$$\|\mathbf{v}(t)\| = \sqrt{9 + 1 + 4t^2} = \sqrt{10 + 4t^2}$$

$$\mathbf{a}(t) = \mathbf{r}''(t) = 2\mathbf{k}$$

By Theorem 14.5, we find the tangential component to be

$$a_{\mathbf{T}} = \frac{\mathbf{v} \cdot \mathbf{a}}{\|\mathbf{v}\|} = \frac{4t}{\sqrt{10 + 4t^2}}$$

and since

$$\mathbf{v} \times \mathbf{a} = \begin{vmatrix} \mathbf{i} & \mathbf{j} & \mathbf{k} \\ 3 & -1 & 2t \\ 0 & 0 & 2 \end{vmatrix} = -2\mathbf{i} - 6\mathbf{j}$$

we find the normal component to be

$$a_{\mathbf{N}} = \frac{\|\mathbf{v} \times \mathbf{a}\|}{\|\mathbf{v}\|} = \frac{\sqrt{4 + 36}}{\sqrt{10 + 4t^2}} = \frac{2\sqrt{10}}{\sqrt{10 + 4t^2}}$$

Now that we can represent the acceleration vector as a linear combination of the unit vectors $\mathbf{T}$ and $\mathbf{N}$, we can write

$$\|\mathbf{a}\|^2 = \mathbf{a} \cdot \mathbf{a} = (a_{\mathbf{T}}\mathbf{T} + a_{\mathbf{N}}\mathbf{N}) \cdot (a_{\mathbf{T}}\mathbf{T} + a_{\mathbf{N}}\mathbf{N})$$

$$= a_{\mathbf{T}}^2\|\mathbf{T}\|^2 + 2a_{\mathbf{T}}a_{\mathbf{N}}\mathbf{T} \cdot \mathbf{N} + a_{\mathbf{N}}^2\|\mathbf{N}\|^2$$

$$= a_{\mathbf{T}}^2\|\mathbf{T}\|^2 + a_{\mathbf{N}}^2\|\mathbf{N}\|^2 = a_{\mathbf{T}}^2 + a_{\mathbf{N}}^2$$

Therefore, the magnitude of the acceleration is $\|\mathbf{a}\| = \sqrt{a_{\mathbf{T}}^2 + a_{\mathbf{N}}^2}$. Moreover, we can also use this relationship to obtain the following alternative formula for the normal component of acceleration:

$$a_{\mathbf{N}} = \sqrt{\|\mathbf{a}\|^2 - a_{\mathbf{T}}^2} \qquad \text{Alternative formula for } a_{\mathbf{N}}$$

Once you have found $a_{\mathbf{T}}$, you have the option of using this formula to find $a_{\mathbf{N}}$. For instance, in Example 4, we could have used $a_{\mathbf{T}} = 4t/\sqrt{10 + 4t^2}$ to write

$$a_{\mathbf{N}} = \sqrt{\|\mathbf{a}\|^2 - a_{\mathbf{T}}^2} = \sqrt{(2)^2 - \frac{16t^2}{10 + 4t^2}} = \frac{2\sqrt{10}}{\sqrt{10 + 4t^2}}$$

This is further demonstrated in the next example.

EXAMPLE 5   *Finding $a_{\mathbf{T}}$ and $a_{\mathbf{N}}$ for a circular helix*

Find the tangential and normal components of acceleration for the circular helix

$$\mathbf{r}(t) = b \cos t\, \mathbf{i} + b \sin t\, \mathbf{j} + ct\mathbf{k}, \ b > 0$$

Solution: We have

$$\mathbf{v}(t) = \mathbf{r}'(t) = -b \sin t \, \mathbf{i} + b \cos t \, \mathbf{j} + c\mathbf{k}$$

$$\|\mathbf{v}(t)\| = \sqrt{b^2 \sin^2 t + b^2 \cos^2 t + c^2} = \sqrt{b^2 + c^2}$$

$$\mathbf{a}(t) = \mathbf{r}''(t) = -b \cos t \, \mathbf{i} - b \sin t \, \mathbf{j}$$

By Theorem 14.5, we find the tangential component to be

$$a_{\mathbf{T}} = \frac{\mathbf{v} \cdot \mathbf{a}}{\|\mathbf{v}\|} = b^2 \sin t \cos t - b^2 \sin t \cos t + 0 = 0$$

Moreover, since

$$\|\mathbf{a}(t)\| = \sqrt{b^2 \cos^2 t + b^2 \sin^2 t} = b$$

we can use the alternative formula for the normal component to obtain

$$a_{\mathbf{N}} = \sqrt{\|\mathbf{a}(t)\|^2 - a_{\mathbf{T}}^2} = \sqrt{b^2 - 0^2} = b$$

Note that the normal component is equal to the magnitude of the acceleration. In other words, since the speed is constant, the acceleration is perpendicular to the velocity.   □

EXAMPLE 6   *An application*

The position function for the projectile shown in Figure 14.22 is given by

$$\mathbf{r}(t) = (50\sqrt{2}t) \, \mathbf{i} + (50\sqrt{2}t - 16t^2) \, \mathbf{j}$$

Find the tangential component of acceleration when $t = 0$, 1, and $25\sqrt{2}/16$.

Solution: We have

$$\mathbf{v}(t) = 50\sqrt{2}\mathbf{i} + (50\sqrt{2} - 32t) \, \mathbf{j}$$

$$\|\mathbf{v}(t)\| = 2\sqrt{50^2 - 16(50)\sqrt{2}t + 16^2t^2}$$

$$\mathbf{a}(t) = -32\mathbf{j}$$

Therefore, the tangential component of acceleration is

$$a_{\mathbf{T}}(t) = \frac{\mathbf{v}(t) \cdot \mathbf{a}(t)}{\|\mathbf{v}(t)\|} = \frac{-32(50\sqrt{2} - 32t)}{2\sqrt{50^2 - 16(50)\sqrt{2}t + 16^2t^2}}$$

At the specified times, we have

$$a_{\mathbf{T}}(0) = \frac{-32(50\sqrt{2})}{100} = -16\sqrt{2} \approx -22.6$$

$$a_{\mathbf{T}}(1) = \frac{-32(50\sqrt{2} - 32)}{2\sqrt{50^2 - 16(50)\sqrt{2} + 16^2}} \approx -15.4$$

$$a_{\mathbf{T}}\left(\frac{25\sqrt{2}}{16}\right) = \frac{-32(50\sqrt{2} - 50\sqrt{2})}{50\sqrt{2}} = 0$$

We can see from Figure 14.22 that, at the maximum height, the tangential component is zero. This is reasonable since the direction of motion is horizontal at that point and the tangential component is equal to the horizontal component.   □

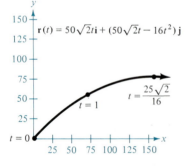

$\mathbf{r}(t) = 50\sqrt{2}t\mathbf{i} + (50\sqrt{2}t - 16t^2)\mathbf{j}$

$t = \dfrac{25\sqrt{2}}{16}$

$t = 1$

$t = 0$

FIGURE 14.22

Note that since the formula for a_N in Theorem 14.5 involves the cross product of two vectors, it can only be applied to three-dimensional vector-valued functions. For vector-valued functions in the plane, the alternative formula provides a useful way to calculate a_N. For instance, in Example 6 the alternative formula yields the following normal component of acceleration

$$a_N = \sqrt{\|\mathbf{a}\|^2 - a_T^2} = \sqrt{32^2 - \frac{32^2(50\sqrt{2} - 32t)^2}{4(50^2 - 16(50)\sqrt{2}t + 16^2t^2)}}$$

$$= \frac{32(50)\sqrt{2}}{2\sqrt{50^2 - 16(50)\sqrt{2}t + 16^2t^2}}$$

When $t = 25\sqrt{2}/16$, we can see that $a_N = 32$ which makes sense since the tangential component is zero at this point and therefore the entire acceleration is directed downward.

Section Exercises 14.4

In Exercises 1–8, find a set of parametric equations for the line tangent to the space curve at the given point.

Function — *Point*

1. $\mathbf{r}(t) = t\mathbf{i} + t^2\mathbf{j} + t\mathbf{k}$ — $(0, 0, 0)$
2. $\mathbf{r}(t) = t\mathbf{i} + t^2\mathbf{j} + \frac{2}{3}\mathbf{k}$ — $\left(1, 1, \frac{2}{3}\right)$
3. $\mathbf{r}(t) = 2\cos t\,\mathbf{i} + 2\sin t\,\mathbf{j} + t\mathbf{k}$ — $(2, 0, 0)$
4. $\mathbf{r}(t) = 3\cos t\,\mathbf{i} + 4\sin t\,\mathbf{j} + \frac{t}{2}\mathbf{k}$ — $\left(0, 4, \frac{\pi}{4}\right)$
5. $\mathbf{r}(t) = \left\langle t, t^2, \frac{2}{3}t^3 \right\rangle$ — $(3, 9, 18)$
6. $\mathbf{r}(t) = \langle t, t, \sqrt{4 - t^2} \rangle$ — $(1, 1, \sqrt{3})$
7. $\mathbf{r}(t) = \langle 2\cos t, 2\sin t, 4 \rangle$ — $(\sqrt{2}, \sqrt{2}, 4)$
8. $\mathbf{r}(t) = \langle 2\sin t, 2\cos t, 4\sin^2 t \rangle$ — $(1, \sqrt{3}, 1)$

In Exercises 9–26, find $\mathbf{T}(t)$, $\mathbf{N}(t)$, a_T, and a_N at the given time t.

Function — *Time*

9. $\mathbf{r}(t) = 4t\mathbf{i}$ — $t = 2$
10. $\mathbf{r}(t) = 4t\mathbf{i} - 2t\mathbf{j}$ — $t = 1$
11. $\mathbf{r}(t) = 4t^2\mathbf{i}$ — $t = 4$
12. $\mathbf{r}(t) = t^2\mathbf{j} + \mathbf{k}$ — $t = 0$
13. $\mathbf{r}(t) = t\mathbf{i} + \frac{1}{t}\mathbf{j}$ — $t = 1$
14. $\mathbf{r}(t) = t\mathbf{i} + t^2\mathbf{j}$ — $t = 1$
15. $\mathbf{r}(t) = 4\cos(2\pi t)\mathbf{i} + 4\sin(2\pi t)\mathbf{j}$ — $t = \frac{1}{8}$
16. $\mathbf{r}(t) = a\cos(\omega t)\mathbf{i} + a\sin(\omega t)\mathbf{j}$ — $t = t_0$
17. $\mathbf{r}(t) = 2\cos(\pi t)\mathbf{i} + 2\sin(\pi t)\mathbf{j}$ — $t = \frac{1}{3}$

18. $\mathbf{r}(t) = a\cos(\omega t)\mathbf{i} + b\sin(\omega t)\mathbf{j}$ — $t = 0$
19. $\mathbf{r}(t) = (e^t\cos t)\mathbf{i} + (e^t\sin t)\mathbf{j}$ — $t = \frac{\pi}{2}$
20. $\mathbf{r}(t) = \langle \omega t - \sin \omega t, 1 - \cos \omega t \rangle$ — $t = t_0$
21. $\mathbf{r}(t) = \langle \cos \omega t + \omega t \sin \omega t, \sin \omega t - \omega t \cos \omega t \rangle$ — $t = t_0$
22. $\mathbf{r}(t) = 4t\mathbf{i} - 4t\mathbf{j} + 2t\mathbf{k}$ — $t = 2$
23. $\mathbf{r}(t) = t\mathbf{i} + t^2\mathbf{j} + \frac{t^2}{2}\mathbf{k}$ — $t = 1$
24. $\mathbf{r}(t) = t\mathbf{i} + 3t^2\mathbf{j} + \frac{t^2}{2}\mathbf{k}$ — $t = 2$
25. $\mathbf{r}(t) = 4t\mathbf{i} + 3\cos t\,\mathbf{j} + 3\sin t\,\mathbf{k}$ — $t = \frac{\pi}{2}$
26. $\mathbf{r}(t) = e^t\cos t\,\mathbf{i} + e^t\sin t\,\mathbf{j} + e^t\mathbf{k}$ — $t = 0$

27. Find the tangential and normal components of acceleration for a projectile fired at an angle θ with the horizontal at an initial speed of v_0. What are the components when the projectile is at its maximum height?

28. A plane flying at an altitude of 30,000 feet and with a speed of 540 miles per hour (792 feet per second) releases a bomb. Find the tangential and normal components of acceleration acting on the bomb.

29. An object is spinning at a constant speed on the end of a string, according to the position function given in Exercise 16.
    (a) If the angular velocity ω is doubled, how is the centripetal component of acceleration changed?
    (b) If the angular velocity is unchanged but the length of the string is halved, how is the centripetal component of acceleration changed?

30. An object of mass m moves at a constant speed v in a circular path of radius r. The force required to produce the centripetal component of acceleration is called the centripetal force and is given by $F = mv^2/r$. Newton's Law of Universal Gravitation is given by $F = GMm/d^2$, where d is the distance between the centers of the two bodies of mass M and m. Use this to show that the speed required for circular motion is $v = \sqrt{GM/r}$.

In Exercises 31–34, use the result of Exercise 30 to find the speed necessary for *the given circular orbit* around the earth. Let $GM = 9.56 \times 10^4$ cubic miles per second per second, and assume the radius of the earth is 4000 miles.

31. A space shuttle 100 miles above the surface of the earth
32. A space shuttle 200 miles above the surface of the earth
33. A heat capacity mapping satellite 385 miles above the surface of the earth, as shown in Figure 14.23.
34. A SYNCOM satellite r miles above the surface of the

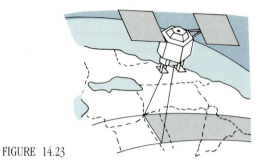

FIGURE 14.23

earth in a geosynchronous orbit. [The satellite completes one orbit per sidereal day (23 hours, 56 minutes), and thus appears to remain stationary above a point on the earth.]

35. If C has a point of inflection at time t, show that $\mathbf{T}' = \mathbf{0}$ at this point.

SECTION TOPICS ▪
Arc length ▪
Arc length as a parameter ▪
Curvature in the plane ▪
Curvature in space ▪

14.5
Arc length and curvature

In Section 12.2, we saw that the arc length of a smooth *plane* curve C given by $x = x(t)$ and $y = y(t)$, $a \le t \le b$, is

$$s = \int_a^b \sqrt{[x'(t)]^2 + [y'(t)]^2} \, dt$$

This formula has a natural extension to a smooth curve in *space*.

THEOREM 14.6   **ARC LENGTH OF A SPACE CURVE**
If C is a smooth curve represented by the equations $x = x(t)$, $y = y(t)$, and $z = z(t)$ on an interval $a \le t \le b$, then the arc length of C on this interval is

$$s = \int_a^b \sqrt{[x'(t)]^2 + [y'(t)]^2 + [z'(t)]^2} \, dt$$

If the curve C is given by a vector-valued function

$$\mathbf{r}(t) = x(t) \, \mathbf{i} + y(t) \, \mathbf{j} + z(t) \, \mathbf{k}$$

then the integrand in Theorem 14.6 can be written as $\|\mathbf{r}'(t)\|$. Thus, **arc length in vector form** is given by

$$s = \int_a^b \|\mathbf{r}'(t)\| \, dt$$

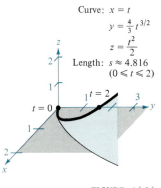

Curve: $x = t$
$\quad\quad y = \frac{4}{3}t^{3/2}$
$\quad\quad z = \frac{t^2}{2}$
Length: $s \approx 4.816$
$\quad\quad (0 \le t \le 2)$

FIGURE 14.24

EXAMPLE 1   Arc length of a space curve

Find the length of the space curve represented by the parametric equations

$$x = t, \qquad y = \frac{4}{3}t^{3/2}, \qquad \text{and} \qquad z = \frac{t^2}{2}$$

from $t = 0$ to $t = 2$, as shown in Figure 14.24.

Solution: Since $x'(t) = 1$, $y'(t) = 2\sqrt{t}$, and $z'(t) = t$, the arc length from $t = 0$ to $t = 2$ is given by

$$s = \int_0^2 \sqrt{[x'(t)]^2 + [y'(t)]^2 + [z'(t)]^2}\, dt$$

$$= \int_0^2 \sqrt{1 + 4t + t^2}\, dt$$

$$= \int_0^2 \sqrt{(t + 2)^2 - 3}\, dt \qquad \text{Integration Formula 26}$$

$$= \left[\frac{t + 2}{2}\sqrt{(t + 2)^2 - 3} - \frac{3}{2}\ln\left|(t + 2) + \sqrt{(t + 2)^2 - 3}\right| \right]_0^2$$

$$= 2\sqrt{13} - \frac{3}{2}\ln(4 + \sqrt{13}) - 1 + \frac{3}{2}\ln 3 \approx 4.816$$

Remark   The formula for arc length given in Theorem 14.6 is independent of the parameter used to represent C. To illustrate this independence, try using the parametric equations

$$x = t^2, \qquad y = \frac{4}{3}t^3, \qquad \text{and} \qquad z = \frac{t^4}{2}$$

to represent the curve given in Example 1. Then find the arc length from $t = 0$ to $t = \sqrt{2}$ and compare the result to that found in Example 1.

EXAMPLE 2   Arc length of a helix

Find the length of one turn of the helix given by

$$\mathbf{r}(t) = b \sin t\, \mathbf{i} + b \cos t\, \mathbf{j} + \sqrt{1 - b^2}\, t\, \mathbf{k}$$

as shown in Figure 14.25.

Solution: Since

$$\mathbf{r}'(t) = b \cos t\, \mathbf{i} - b \sin t\, \mathbf{j} + \sqrt{1 - b^2}\, \mathbf{k}$$

the arc length of one turn is

$$s = \int_0^{2\pi} \|\mathbf{r}'(t)\|\, dt = \int_0^{2\pi} \sqrt{b^2(\cos^2 t + \sin^2 t) + (1 - b^2)}\, dt$$

$$= \int_0^{2\pi} dt = 2\pi$$

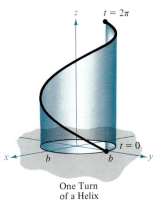

Curve:
$\mathbf{r}(t) = b \sin t\, \mathbf{i} + b \cos t\, \mathbf{j} + \sqrt{1 - b^2}\, t\, \mathbf{k}$

One Turn
of a Helix

FIGURE 14.25

We have seen that curves can be represented by vector-valued functions in different ways, depending on the choice of parameter. For motion along a curve, the convenient parameter is time, t. However, for studying the geometrical properties of a curve, the convenient parameter is arc length, s. The convenience of this **arc length parameter** lies in the fact that as s varies from a to b, the length of the arc traced out by $\mathbf{r}(s)$ matches the length of the interval $[a, b]$. That is,

$$\text{length of arc} = \int_a^b \|\mathbf{r}'(s)\| \, ds = b - a = \text{length of interval}$$

The following theorem can be used to characterize the arc length parameter. You are asked to supply the proof in Exercise 45.

THEOREM 14.7   **ARC LENGTH PARAMETER**

The parameter t for a smooth curve C given by $\mathbf{r}(t) = x(t) \, \mathbf{i} + y(t) \, \mathbf{j} + z(t) \, \mathbf{k}$ is the arc length parameter if and only if

$$\|\mathbf{r}'(t)\| = 1$$

for all t in the domain of $\mathbf{r}$.

Remark   The parameter used for the helix in Example 2 is the arc length parameter since

$$\|\mathbf{r}'(t)\| = \sqrt{b^2 + (1 - b^2)} = 1$$

for every t.

Curvature

An important use of the arc length parameter is to find **curvature**—the measure of how sharply a curve bends. For instance, in Figure 14.26 the curve bends more sharply at P than at Q, and we say its curvature is greater at P than at Q. To measure the bending at a point on a curve, we calculate the rate at which ϕ (the angle of inclination of the unit tangent vector $\mathbf{T}$) changes with respect to arc length s, as shown in Figure 14.27.

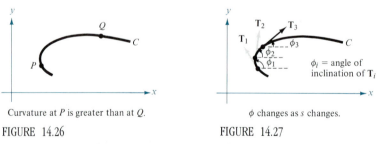

Curvature at P is greater than at Q.

FIGURE 14.26

ϕ changes as s changes.

FIGURE 14.27

DEFINITION OF
CURVATURE
IN THE PLANE

Let C be a smooth curve given by $\mathbf{r}(s) = x(s) \, \mathbf{i} + y(s) \, \mathbf{j}$, where s is the arc length parameter. If ϕ is the angle of inclination of the unit tangent vector $\mathbf{T}$ at s, then the **curvature** at the point $(x(s), y(s))$ is

$$K = \left| \frac{d\phi}{ds} \right|$$

where K is the uppercase Greek letter kappa.

> **Remark** Since a straight line doesn't bend at all, it stands to reason that its curvature should be zero. This conforms to the definition of curvature, since the angle of inclination of a straight line is constant and the derivative of a constant is zero.

A circle has the same curvature at any point on its circumference. Moreover, the curvature and the radius of a circle are inversely related. That is, a circle with a large radius has a relatively small curvature, and a circle with a small radius has a relatively large curvature. This inverse relationship is verified in the following example.

EXAMPLE 3   *The curvature of a circle*

Show that the curvature of a circle of radius r is $K = 1/r$.

Solution: Without loss of generality we consider a circle centered at the origin. Let (x, y) be any point on the circle and s be the length of the arc from $(r, 0)$ to (x, y), as shown in Figure 14.28. We let θ be the central angle of this circular section and let ϕ be the angle of inclination of the tangent line at (x, y). Using the formula for the length of a circular arc, $s = r\theta$, we have

$$\phi = \theta + \frac{\pi}{2} = \frac{s}{r} + \frac{\pi}{2}$$

Consequently, the curvature of the circle at (x, y) is

$$K = \frac{d\phi}{ds} = \frac{1}{r}$$

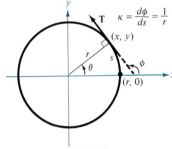

Curvature of a Circle

FIGURE 14.28

From the result of Example 3 we make the following definitions. Let C be a curve with curvature K at the point P. The circle passing through the point P with radius $r = 1/K$ is called the **circle of curvature** if the circle lies on the concave side of the curve and shares a common tangent line with the curve at the point P. We call r the **radius of curvature** at P, and the center of the circle is called the **center of curvature**.

The circle of curvature gives us a nice way to graphically estimate the curvature K at a point P on a curve. Using a compass, we sketch a circle that snuggles up against the concave side of the curve at the point P, as shown in Figure 14.29. If the circle has a radius of r, then we estimate the curvature to be $K = 1/r$.

To calculate curvature for a plane curve given by $y = f(x)$, we use the following theorem.

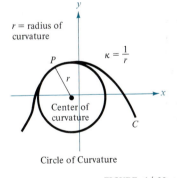

Circle of Curvature

FIGURE 14.29

THEOREM 14.8   **CURVATURE IN RECTANGULAR COORDINATES**
If C is the graph of a twice differentiable function given by $y = f(x)$, then the curvature at the point (x, y) is given by

$$K = \frac{|y''|}{(1 + [y']^2)^{3/2}}$$

Proof: At a point on the graph of $y = f(x)$, we know that $dy/dx = \tan \phi$, which implies that $\phi = \arctan (dy/dx)$. Moreover,

$$\frac{d\phi}{ds} = \left(\frac{d\phi}{dx}\right)\left(\frac{dx}{ds}\right) \implies \text{K} = \left|\frac{d\phi}{ds}\right| = \left|\frac{d\phi/dx}{ds/dx}\right|$$

Now, since the integral definition for arc length

$$s = \int_0^x \sqrt{1 + \left(\frac{dy}{dt}\right)^2}\, dt$$

implies that $ds/dx = \sqrt{1 + (dy/dx)^2}$, and since

$$\frac{d\phi}{dx} = \frac{d}{dx}\left[\arctan \left(\frac{dy}{dx}\right)\right] = \frac{y''}{1 + [y']^2}$$

we have

$$\text{K} = \frac{|y''|}{(1 + [y']^2)^{3/2}}$$

| **Remark** For a smooth curve given by $x = g(y)$, replace y' by x' and y'' by x'' to obtain the formula for K. For a smooth curve given by the parametric equations $x = f(t)$ and $y = g(t)$, the formula is

$$\text{K} = \frac{|f'(t)g''(t) - g'(t)f''(t)|}{([f'(t)]^2 + [g'(t)]^2)^{3/2}}$$

EXAMPLE 4   *Finding curvature in rectangular coordinates*

Find the curvature of the parabola given by

$$y = x - \frac{1}{4}x^2$$

at $x = 2$ and $x = 4$.

Solution:

		At x = 2	*At x = 4*
$y' = 1 - \dfrac{x}{2}$		$y' = 0$	$y' = -1$
$y'' = -\dfrac{1}{2}$		$y'' = -\dfrac{1}{2}$	$y'' = -\dfrac{1}{2}$
$\text{K} = \dfrac{\|y''\|}{[1 + (y')^2]^{3/2}}$		$\text{K} = \dfrac{1}{2}$	$\text{K} = \dfrac{1}{2^{5/2}} \approx 0.177$

| **Remark** Note in Figure 14.30 that at the point (2, 1) the circle of curvature has a radius $r = 1/\text{K} = 2$.

One benefit of the arc length parameter is that it simplifies the formula for the unit tangent vector. Specifically, if $\mathbf{r}(s) = x(s)\, \mathbf{i} + y(s)\, \mathbf{j}$ and s is the arc length parameter, then $\mathbf{r}'(s)$ is a unit vector and

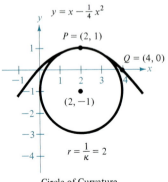

Circle of Curvature

FIGURE 14.30

$$\mathbf{T}(s) = \mathbf{r}'(s) = \cos\phi\,\mathbf{i} + \sin\phi\,\mathbf{j}$$

and

$$\mathbf{T}'(s) = \left(\frac{d\mathbf{T}}{d\phi}\right)\left(\frac{d\phi}{ds}\right) = [-\sin\phi\,\mathbf{i} + \cos\phi\,\mathbf{j}]\frac{d\phi}{ds}$$

This implies that $\|\mathbf{T}'(s)\| = K$, and we use this formula to define curvature in space.

DEFINITION OF CURVATURE IN SPACE

Let C be a smooth space curve given by $\mathbf{r}(s) = x(s)\,\mathbf{i} + y(s)\,\mathbf{j} + z(s)\,\mathbf{k}$, where s is the arc length parameter. Then the curvature at s is given by

$$K = \|\mathbf{T}'(s)\| = \|\mathbf{r}''(s)\|$$

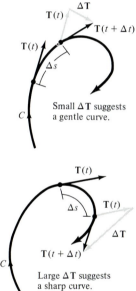

$\mathbf{T}(t)$ $\Delta\mathbf{T}$
$\mathbf{T}(t + \Delta t)$
$\mathbf{T}(t)$
Δs

Small $\Delta\mathbf{T}$ suggests a gentle curve.

C

$\mathbf{T}(t)$
Δs $\mathbf{T}(t)$
$\Delta\mathbf{T}$
$\mathbf{T}(t + \Delta t)$
C

Large $\Delta\mathbf{T}$ suggests a sharp curve.

FIGURE 14.31

For this formula for curvature to be used, the curve must be expressed with arc length as the parameter. For other parameters, we use the following:

$$1.\ \ K = \frac{\|\mathbf{T}'(t)\|}{\|\mathbf{r}'(t)\|} \qquad 2.\ \ K = \frac{\|\mathbf{r}'(t) \times \mathbf{r}''(t)\|}{\|\mathbf{r}'(t)\|^3} \qquad 3.\ \ K = \frac{\mathbf{a}(t)\cdot\mathbf{N}(t)}{\|\mathbf{v}(t)\|^2}$$

Since $\|\mathbf{r}'(t)\| = |ds/dt|$, the first formula implies that curvature is the ratio of the rate of change in the tangent vector $\mathbf{T}$ to the rate of change in arc length. This seems reasonable if we consider that as Δt approaches zero, we can write

$$\frac{\mathbf{T}'(t)}{ds/dt} \approx \frac{[\mathbf{T}(t + \Delta t) - \mathbf{T}(t)]/\Delta t}{[s(t + \Delta t) - s(t)]/\Delta t} = \frac{\mathbf{T}(t + \Delta t) - \mathbf{T}(t)}{s(t + \Delta t) - s(t)} = \frac{\Delta\mathbf{T}}{\Delta s}$$

In other words, for a given Δs, the greater the length of $\Delta\mathbf{T}$, the more bending there is to the curve at t, as shown in Figure 14.31.

EXAMPLE 5   *Finding the curvature of a space curve*

Find the curvature of the curve given by $\mathbf{r}(t) = 2t\mathbf{i} + t^2\mathbf{j} - \frac{1}{3}t^3\mathbf{k}.$

Solution: Since it is not apparent whether or not this parameter is arc length, we use the formula $K = \|\mathbf{T}'(t)\|/\|\mathbf{r}'(t)\|$. Since $\mathbf{r}'(t) = 2\mathbf{i} + 2t\mathbf{j} - t^2\mathbf{k}$, it follows that $\|\mathbf{r}'(t)\| = \sqrt{4 + 4t^2 + t^4} = t^2 + 2$, and we have

$$\mathbf{T}(t) = \frac{\mathbf{r}'(t)}{\|\mathbf{r}'(t)\|} = \frac{2\mathbf{i} + 2t\mathbf{j} - t^2\mathbf{k}}{t^2 + 2}$$

$$\mathbf{T}'(t) = \frac{(t^2 + 2)(2\mathbf{j} - 2t\mathbf{k}) - (2t)(2\mathbf{i} + 2t\mathbf{j} - t^2\mathbf{k})}{(t^2 + 2)^2}$$

$$= \frac{-4t\mathbf{i} + (4 - 2t^2)\mathbf{j} - 4t\mathbf{k}}{(t^2 + 2)^2}$$

$$\|\mathbf{T}'(t)\| = \frac{\sqrt{16t^2 + 16 - 16t^2 + 4t^4 + 16t^2}}{(t^2 + 2)^2} = \frac{2(t^2 + 2)}{(t^2 + 2)^2} = \frac{2}{t^2 + 2}$$

Therefore,

$$K = \frac{\|\mathbf{T}'(t)\|}{\|\mathbf{r}'(t)\|} = \frac{2}{(t^2 + 2)^2}$$

Arc length and curvature are closely related to the tangential and normal components of acceleration. The tangential component of acceleration is the rate of change of the speed, which in turn is the rate of change of the arc length. This component is negative as a moving object slows down and positive as it speeds up—regardless of whether the object is turning or traveling on a straight line. Thus, the tangential component is solely a function of the arc length and is independent of the curvature. On the other hand, the normal component of acceleration is a function of *both* speed and curvature. This component measures the acceleration acting perpendicular to the direction of motion. To see why the normal component is affected by both speed and curvature, imagine that you are driving a car around a turn, as shown in Figure 14.32. If your speed is high and the turn is sharp, you feel yourself thrown against the car door. By lowering your speed *or* taking a more gentle turn, you are able to lessen this sideways thrust.

FIGURE 14.32

The next theorem explicitly states the relationships among speed, curvature, and the components of acceleration.

THEOREM 14.9   ACCELERATION, SPEED, AND CURVATURE
If $\mathbf{r}(t)$ is the position vector for a smooth curve C, then the acceleration vector is given by

$$\mathbf{a}(t) = \frac{d^2s}{dt^2}\mathbf{T} + K\left(\frac{ds}{dt}\right)^2\mathbf{N}$$

where K is the curvature of C and ds/dt is the speed.

Proof: For the position vector $\mathbf{r}(t)$ we have

$$\mathbf{a}(t) = a_\mathbf{T}\mathbf{T} + a_\mathbf{N}\mathbf{N} = D_t[\|\mathbf{v}\|]\mathbf{T} + \|\mathbf{v}\|\,\|\mathbf{T}'\|\mathbf{N}$$

$$= \frac{d^2s}{dt^2}\mathbf{T} + \frac{ds}{dt}(\|\mathbf{v}\|K)\mathbf{N} = \frac{d^2s}{dt^2}\mathbf{T} + K\left(\frac{ds}{dt}\right)^2\mathbf{N}$$

EXAMPLE 6   *Finding the tangential and normal components of acceleration*

Find a_T and a_N for the curve given by $\mathbf{r}(t) = 2t\mathbf{i} + t^2\mathbf{j} - \frac{1}{3}t^3\mathbf{k}$.

Solution: From Example 5 we know that

$$\frac{ds}{dt} = \|\mathbf{r}'(t)\| = t^2 + 2 \qquad \text{and} \qquad K = \frac{2}{(t^2 + 2)^2}$$

Therefore,

$$a_T = \frac{d^2s}{dt^2} = 2t$$

and

$$a_N = K\left(\frac{ds}{dt}\right)^2 = \frac{2}{(t^2+2)^2}(t^2+2)^2 = 2$$

Applications

There are many applications in physics and engineering dynamics that involve the relationships among speed, arc length, curvature, and acceleration. One such application concerns frictional force.

Suppose a moving object with mass m is in contact with a stationary object. The total force required to produce an acceleration $\mathbf{a}$ along a given path is

$$\mathbf{F} = m\mathbf{a} = m\left(\frac{d^2s}{dt^2}\right)\mathbf{T} + mK\left(\frac{ds}{dt}\right)^2\mathbf{N}$$

Force of friction

Force of friction is perpendicular to the direction of motion.

FIGURE 14.33

The portion of this total force that is supplied by the stationary object is called the **force of friction.** For example, if a car is rounding a turn, the roadway exerts a frictional force that keeps the car from sliding off the road. If the car is not sliding, then the frictional force is perpendicular to the direction of motion and has magnitude equal to the normal component of acceleration, as shown in Figure 14.33. The potential frictional force of a road around a turn can be increased by banking the roadway at an angle.

EXAMPLE 7   *An application of the normal component of acceleration*

A 360-kilogram Go-cart is driven at a speed of 60 kilometers per hour around a circular race track of radius 12 meters. To keep the car from skidding off course, what frictional force must the track surface exert on the tires?

Solution: The frictional force must equal the normal component of acceleration. For this circular path we know that the curvature is $K = \frac{1}{12}$. Therefore, the normal component is

$$a_N = mK\left(\frac{ds}{dt}\right)^2 = (360 \text{ kg})\left(\frac{1}{12 \text{ m}}\right)\left(\frac{60{,}000 \text{ m}}{3600 \text{ sec}}\right)^2$$

$$\approx 8333 \text{ (kg)(m)/sec}^2 = 8333 \text{ N}$$

We summarize several of the formulas dealing with vector-valued functions as follows.

SUMMARY OF VELOCITY, ACCELERATION, AND CURVATURE

Let C be a curve given by the position function

$$\mathbf{r}(t) = x(t)\,\mathbf{i} + y(t)\,\mathbf{j} \qquad \text{Curve in the plane}$$

or

$$\mathbf{r}(t) = x(t)\,\mathbf{i} + y(t)\,\mathbf{j} + z(t)\,\mathbf{k} \qquad \text{Curve in space}$$

Velocity vector, speed, and acceleration vector:

$$\mathbf{v}(t) = \mathbf{r}'(t)$$

$$\|\mathbf{v}(t)\| = \frac{ds}{dt}$$

$$\mathbf{a}(t) = \mathbf{r}''(t) = a_{\mathbf{T}}\mathbf{T}(t) + a_{\mathbf{N}}\mathbf{N}(t)$$

Unit tangent vector and principal unit normal vector:

$$\mathbf{T}(t) = \frac{\mathbf{r}'(t)}{\|\mathbf{r}'(t)\|}$$

$$\mathbf{N}(t) = \frac{\mathbf{T}'(t)}{\|\mathbf{T}'(t)\|}$$

Components of acceleration:

$$a_{\mathbf{T}} = \mathbf{a} \cdot \mathbf{T} = \frac{\mathbf{v} \cdot \mathbf{a}}{\|\mathbf{v}\|} = \frac{d^2s}{dt^2}$$

$$a_{\mathbf{N}} = \mathbf{a} \cdot \mathbf{N} = \sqrt{\|\mathbf{a}\|^2 - a_{\mathbf{T}}^2} = K\left(\frac{ds}{dt}\right)^2 = \frac{\|\mathbf{v} \times \mathbf{a}\|}{\|\mathbf{v}\|}$$

Formulas for curvature in the plane:

$$K = \frac{|y''|}{[1 + (y')^2]^{3/2}} \qquad C \text{ given by } y = f(x)$$

$$K = \frac{|x'y'' - y'x''|}{[(x')^2 + (y')^2]^{3/2}} \qquad C \text{ given by } x = x(t),\ y = y(t)$$

Formulas for curvature in the plane or in space:

$$K = \|\mathbf{T}'(s)\| = \|\mathbf{r}''(s)\| \qquad s \text{ is arc length parameter}$$

$$K = \frac{\|\mathbf{T}'(t)\|}{\|\mathbf{r}'(t)\|} = \frac{\|\mathbf{r}'(t) \times \mathbf{r}''(t)\|}{\|\mathbf{r}'(t)\|^3} \qquad t \text{ is general parameter}$$

$$K = \frac{\mathbf{a}(t) \cdot \mathbf{N}(t)}{\|\mathbf{v}(t)\|^2}$$

Cross product formulas apply only to curves in space.

Section Exercises 14.5

In Exercises 1–6, sketch the given curve and find its length over the indicated interval.

Function	Interval
1. $\mathbf{r}(t) = t\mathbf{i} + 3t\mathbf{j}$	[0, 4]
2. $\mathbf{r}(t) = t\mathbf{i} + t^2\mathbf{k}$	[0, 4]
3. $\mathbf{r}(t) = a \cos t\,\mathbf{i} + a \sin t\,\mathbf{j} + bt\mathbf{k}$	$[0, 2\pi]$
4. $\mathbf{r}(t) = a \cos^3 t\,\mathbf{i} + a \sin^3 t\,\mathbf{j}$	$[0, 2\pi]$
5. $\mathbf{r}(t) = \langle \sin t - t \cos t, \cos t + t \sin t, t^2 \rangle$	$\left[0, \dfrac{\pi}{2}\right]$
6. $\mathbf{r}(t) = \langle 4t, 3 \cos t, 3 \sin t \rangle$	$\left[0, \dfrac{\pi}{2}\right]$

In Exercises 7–12, find the curvature and radius of curvature of the plane curve at the indicated point.

Function	Point
7. $y = 3x - 2$	$x = a$
8. $y = mx + b$	$x = a$
9. $y = 2x^2 + 3$	$x = -1$
10. $y = x + \dfrac{1}{x}$	$x = 1$
11. $y = \sqrt{a^2 - x^2}$	$x = 0$
12. $y = \dfrac{3}{4}\sqrt{16 - x^2}$	$x = 0$

In Exercises 13–18, find the curvature K of the given curve at the indicated point.

Function	Point
13. $\mathbf{r}(t) = 4t\mathbf{i}$	$t = 2$
14. $\mathbf{r}(t) = 4t\mathbf{i} - 2t\mathbf{j}$	$t = 1$
15. $\mathbf{r}(t) = 4t^2\mathbf{i}$	$t = 4$
16. $\mathbf{r}(t) = t^2\mathbf{j} + \mathbf{k}$	$t = 0$
17. $\mathbf{r}(t) = t\mathbf{i} + \dfrac{1}{t}\mathbf{j}$	$t = 1$
18. $\mathbf{r}(t) = t\mathbf{i} + t^2\mathbf{j}$	$t = 1$

In Exercises 19–30, find the curvature K of the given curve.

19. $\mathbf{r}(t) = 4 \cos (2\pi t)\,\mathbf{i} + 4 \sin (2\pi t)\,\mathbf{j}$
20. $\mathbf{r}(t) = a \cos (\omega t)\,\mathbf{i} + a \sin (\omega t)\,\mathbf{j}$
21. $\mathbf{r}(t) = 2 \cos (\pi t)\,\mathbf{i} + 2 \sin (\pi t)\,\mathbf{j}$
22. $\mathbf{r}(t) = a \cos (\omega t)\,\mathbf{i} + b \sin (\omega t)\,\mathbf{j}$
23. $\mathbf{r}(t) = e^t \cos t\,\mathbf{i} + e^t \sin t\,\mathbf{j}$
24. $\mathbf{r}(t) = \langle a(\omega t - \sin \omega t), a(1 - \cos \omega t) \rangle$
25. $\mathbf{r}(t) = \langle \cos \omega t + \omega t \sin \omega t, \sin \omega t - \omega t \cos \omega t \rangle$
26. $\mathbf{r}(t) = 4t\mathbf{i} - 4t\mathbf{j} + 2t\mathbf{k}$
27. $\mathbf{r}(t) = t\mathbf{i} + t^2\mathbf{j} + \dfrac{t^2}{2}\mathbf{k}$
28. $\mathbf{r}(t) = t\mathbf{i} + 3t^2\mathbf{j} + \dfrac{t^2}{2}\mathbf{k}$
29. $\mathbf{r}(t) = 4t\mathbf{i} + 3 \cos t\,\mathbf{j} + 3 \sin t\,\mathbf{k}$
30. $\mathbf{r}(t) = e^t \cos t\,\mathbf{i} + e^t \sin t\,\mathbf{j} + e^t\mathbf{k}$

In Exercises 31–34, (a) find the point on the curve at which the curvature K is a maximum and (b) find the limit of K as $x \to \infty$.

31. $y = (x - 1)^2 + 3$ **32.** $y = x^3$
33. $y = x^{2/3}$ **34.** $y = \ln x$

35. Find the circle of curvature of the graph of $y = x + 1/x$ at the point $(1, 2)$.
36. Find all points on the graph of $y = (x - 1)^3 + 3$ at which the curvature is zero.
37. Show that the curvature is greatest at the endpoints of the major axis and least at the endpoints of the minor axis for the ellipse given by

$$x^2 + 4y^2 = 4$$

38. Find all a and b such that the two curves given by

$$y_1 = ax(b - x)$$

and

$$y_2 = \frac{x}{x + 2}$$

intersect at only one point and have a common tangent line and equal curvature at the point. Sketch a graph for each set of values for a and b.

39. The smaller the curvature in a bend of a road, the faster a car can travel. Assume that the maximum speed around a turn is inversely proportional to the square root of the curvature. A car moving on the path $y = \frac{1}{3}x^3$ (x and y measured in miles) can safely go 30 miles per hour at $(1, \frac{1}{3})$. How fast can it go at $(\frac{3}{2}, \frac{9}{8})$?
40. The curve C is given by the polar equation $r = f(\theta)$. Show that the curvature K at the point (r, θ) is

$$K = \frac{|2(r')^2 - rr'' + r^2|}{[(r')^2 + r^2]^{2/3}}$$

(Hint: Represent the curve by $\mathbf{r}(\theta) = r \cos \theta\,\mathbf{i} + r \sin \theta\,\mathbf{j}$.)

In Exercises 41–44, use the result of Exercise 40 to find the curvature of the polar curve.

41. $r = 1 + \sin \theta$ **42.** $r = a \sin \theta$
43. $r = \theta$ **44.** $r = e^{\theta}$

45. Prove Theorem 14.7.

46. Use the definition of curvature in space, $K = \|T'(s)\| = \|r''(s)\|$, to verify the following two formulas:

(a) $K = \dfrac{\|T'(t)\|}{\|r'(t)\|}$ (b) $K = \dfrac{\|r'(t) \times r''(t)\|}{\|r'(t)\|^3}$

Review Exercises for Chapter 14

In Exercises 1–4, sketch the space curve represented by the vector function.

1. $r(t) = i + tj + t^2 k$
2. $r(t) = 2ti + tj + t^2 k$
3. $r(t) = 2 \cos t \, i + tj + 2 \sin t k$
4. $r(t) = i + \sin t \, j + k$

In Exercises 5 and 6, sketch the space curve represented by the intersection of the given surfaces. Find a vector-valued function for the space curve using the indicated parameter.

5. $z = x^2 + y^2,\ x + y = 0,\ t = x$
6. $x^2 + z^2 = 4,\ x - y = 0,\ t = x$

In Exercises 7 and 8, find the indicated limit.

7. $\lim\limits_{t \to 2} (t^2 i + \sqrt{4 - t^2} \, j + k)$

8. $\lim\limits_{t \to 0} \left(\dfrac{\sin 2t}{t} i + e^{-t} j + e^t k \right)$

In Exercises 9–12, (a) find the domain of r and (b) determine the values of t for which the function is discontinuous.

9. $r(t) = ti + \csc t \, k$

10. $r(t) = \sqrt{t} \, i + \dfrac{1}{t - 4} j + k$

11. $r(t) = \ln t \, i + tj + tk$
12. $r(t) = (2t + 1) \, i + t^2 j + tk$

In Exercises 13 and 14, find

(a) $r'(t)$ (b) $r''(t)$
(c) $D_t[r(t) \cdot u(t)]$ (d) $D_t[u(t) - 2r(t)]$
(e) $D_t[\|r(t)\|]$ (f) $D_t[r(t) \times u(t)]$

13. $r(t) = 3ti + (t - 1) \, j$
   $u(t) = ti + t^2 j + \dfrac{2}{3} t^3 k$

14. $r(t) = \sin t \, i + \cos t \, j + tk$
   $u(t) = \sin t \, i + \cos t \, j + \dfrac{1}{t} k$

In Exercises 15–18, find the indefinite integral.

15. $\displaystyle\int (\cos t \, i + t \cos t \, j) \, dt$

16. $\displaystyle\int (\ln t \, i + t \ln t \, j + k) \, dt$

17. $\displaystyle\int \|\cos t \, i + \sin t \, j + tk\| \, dt$

18. $\displaystyle\int (tj + t^2 k) \times (i + tj + tk) \, dt$

In Exercises 19 and 20, find a set of parametric equations for the line tangent to the space curve at the indicated point.

19. $r(t) = 2 \cos t \, i + 2 \sin t \, j + tk,\ t = \dfrac{3\pi}{4}$

20. $r(t) = ti + t^2 j + \dfrac{2}{3} t^3 k,\ t = 3$

In Exercises 21–28, find the velocity, speed, and acceleration at time t. Then, find $a \cdot T$, $a \cdot N$, and the curvature at time t.

21. $r(t) = (1 + 4t) \, i + (2 - 3t) \, j$
22. $r(t) = 5ti$

23. $r(t) = 2(t + 1) \, i + \dfrac{2}{t + 1} j$

24. $r(t) = ti + \sqrt{t} \, j$
25. $r(t) = e^t i + e^{-t} j$
26. $r(t) = t \cos t \, i + t \sin t \, j$

27. $r(t) = ti + t^2 j + \dfrac{1}{2} t^2 k$

28. $r(t) = (t - 1) \, i + tj + \dfrac{1}{t} k$

In Exercises 29 and 30, find the length of the space curve over the indicated interval.

29. $r(t) = \dfrac{1}{2} ti + \sin t \, j + \cos t \, k,\ 0 \le t \le \pi$

30. $r(t) = e^t \sin t \, i + e^t \cos t \, k,\ 0 \le t \le \pi$

31. A projectile is fired from ground level at an angle of elevation of 30°. Find the range of the projectile if the initial velocity is 75 feet per second.

32. The center of a truckbed is 6 feet below and 4 feet horizontally from the end of a horizontal conveyor that is discharging gravel, as shown in Figure 14.34. Determine the speed ds/dt at which the conveyor belt should be moving so that the gravel falls onto the center of the truck bed.

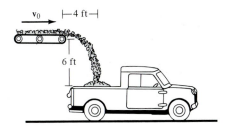

FIGURE 14.34

33. Find the speed necessary for a satellite to maintain a circular orbit 600 miles above the surface of the earth.

34. An automobile is in a circular traffic exchange and its speed is twice that posted. By what factor is the centripetal force increased over that which would occur at the posted speed?

35. A civil engineer designs a highway as indicated in Figure 14.35. BC is an arc of a circle. AB and CD are straight lines tangent to the circular arc. Criticize the design.

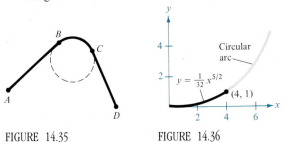

FIGURE 14.35   FIGURE 14.36

36. A highway has an exit ramp that begins at the origin of a coordinate system and follows the curve given by

$$y = \frac{1}{32}x^{5/2}$$

to the point (4, 1), as shown in Figure 14.36. Then it follows a circular path whose curvature is that given by the curve at (4, 1). What is the radius of the circular arc?

15 Functions of several variables

15.1
Introduction to functions of several variables

So far we have dealt only with functions of a single independent variable. Many familiar quantities are functions of two or more independent variables. For instance, the work done by a force ($W = FD$) and the volume V of a right circular cylinder ($V = \pi r^2 h$) are both functions of two variables. The volume of a rectangular solid ($V = lwh$) is a function of three variables. We denote a function of two or more variables by notation similar to that for functions of a single variable. For example,

$$z = \underbrace{f(x, y)}_{\text{2 variables}} = x^2 + xy \qquad \text{and} \qquad w = \underbrace{f(x, y, z)}_{\text{3 variables}} = x + 2y - 3z$$

DEFINITION OF A FUNCTION OF TWO VARIABLES

Let D be a set of ordered pairs of real numbers. If to each ordered pair (x, y) in D there corresponds a real number $f(x, y)$, then f is called a **function of x and y.** The set D is the **domain** of f and the corresponding set of values for $f(x, y)$ is the **range** of f.

Remark Similar definitions can be given for functions of three, four, or n variables, where the domains consist of ordered triples (x_1, x_2, x_3), quadruples (x_1, x_2, x_3, x_4), and n-tuples $(x_1, x_2, \ldots, x_n)$, respectively. In all cases the range is a set of real numbers. We will limit our discussions in this chapter to functions of two and three variables.

For the function given by $z = f(x, y)$, we call x and y the **independent variables** and z the **dependent variable.** Analogous statements are true for a function of three variables.

786

As with functions of one variable, we generally use *equations* to describe functions of several variables, and unless otherwise restricted, we assume the domain to be the set of all points for which the equation is defined. For instance, the domain of the function given by

$$f(x, y) = x^2 + y^2$$

is assumed to be the entire *xy*-plane. In Example 1, we look at two functions whose domains are restricted to regions in the *xy*-plane.

EXAMPLE 1 *Finding the domain of functions of two variables*

Find the domains for the following functions:

(a) $f(x, y) = \dfrac{\sqrt{x^2 + y^2 - 9}}{x}$ (b) $g(x, y) = \dfrac{x}{\sqrt{x^2 + y^2 - 9}}$

Solution:

(a) The function f is defined for all points (x, y) such that $x \neq 0$ and $x^2 + y^2 \geq 9$. Thus, the domain is the set of all points lying on or outside the circle $x^2 + y^2 = 9$, *except* those on the *y*-axis as shown in Figure 15.1.

(b) The function g is defined for all points (x, y) such that $x^2 + y^2 > 9$. Consequently, the domain is the set of all points (x, y) lying outside the circle $x^2 + y^2 = 9$.

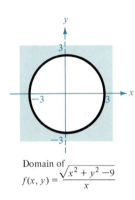

Domain of
$f(x, y) = \dfrac{\sqrt{x^2 + y^2 - 9}}{x}$

FIGURE 15.1

Functions of several variables can be combined in the same ways as functions of a single variable. For two variables, we have

$$f(x, y) \pm g(x, y) \qquad f(x, y)g(x, y) \qquad \frac{f(x, y)}{g(x, y)}$$

Sum or difference   Product   Quotient

The **composite** function given by $(g \circ h)(x, y)$ is defined only if h is a function of x and y and g is a function of a single variable. Then

$$(g \circ h)(x, y) = g(h(x, y))$$

for all (x, y) in the domain of h such that $h(x, y)$ is in the domain of g. For example, the function given by

$$f(x, y) = \sqrt{16 - 4x^2 - y^2}$$

can be viewed as the composite of the function of two variables given by $h(x, y) = 16 - 4x^2 - y^2$ and the function of a single variable given by $g(u) = \sqrt{u}$.

A function that can be expressed as a sum of functions of the form $cx^m y^n$ (where c is a real number and m and n are nonnegative integers) is called a **polynomial function** of two variables. For instance, the functions given by

$$f(x, y) = x^2 + y^2 - 2xy + x + 2 \qquad \text{and} \qquad g(x, y) = 3xy^2 + x - 2$$

are polynomial functions of two variables. A **rational function** is the quotient of two polynomial functions. Similar terminology is used for functions of more than two variables.

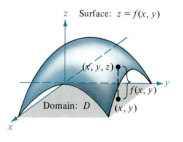

z   Surface: $z = f(x, y)$

(x, y, z)

$f(x, y)$

Domain: *D* (x, y)

x

FIGURE 15.2

The graph of a function of two variables

As with functions of a single variable, we can gain a great deal of insight into the behavior of a function of two variables from sketching its graph. The **graph** of a function f of two variables is the set of all points (x, y, z) for which $z = f(x, y)$ and (x, y) is in the domain of f. This graph can be interpreted geometrically as a *surface in space,* and we already have done much of the preliminary work necessary for sketching such surfaces in Sections 13.5 and 13.6. In Figure 15.2, note that the graph of $z = f(x, y)$ is a surface whose projection onto the *xy*-plane is D, the domain of f. Consequently, to each point (x, y) in D there corresponds a point (x, y, z) on the surface, and, conversely, to each point (x, y, z) on the surface there corresponds a point (x, y) in D.

EXAMPLE 2   Sketching the graph of a function of two variables

Sketch the graph of $f(x, y) = \sqrt{16 - 4x^2 - y^2}$. What is the range of f?

Solution: The domain D implied by the equation for f is the set of all points (x, y) such that $16 - 4x^2 - y^2 \geq 0$. Thus, D is the set of all points lying on or inside the ellipse given by

$$\frac{x^2}{4} + \frac{y^2}{16} = 1$$

as shown in Figure 15.3. The range of f is all values $z = f(x, y)$ such that $0 \leq z \leq \sqrt{16} = 4$. A point (x, y, z) is on the graph of f if and only if

$$z = \sqrt{16 - 4x^2 - y^2}$$
$$z^2 = 16 - 4x^2 - y^2$$
$$4x^2 + y^2 + z^2 = 16$$
$$\frac{x^2}{4} + \frac{y^2}{16} + \frac{z^2}{16} = 1, \quad 0 \leq z \leq 4$$

From our work in Section 13.6, we see that the graph of f is the upper half of an ellipsoid, as shown in Figure 15.4.

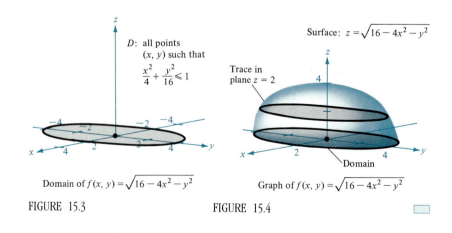

D: all points
(x, y) such that
$\dfrac{x^2}{4} + \dfrac{y^2}{16} \leq 1$

Surface: $z = \sqrt{16 - 4x^2 - y^2}$

Trace in
plane $z = 2$

Domain

Domain of $f(x, y) = \sqrt{16 - 4x^2 - y^2}$

Graph of $f(x, y) = \sqrt{16 - 4x^2 - y^2}$

FIGURE 15.3

FIGURE 15.4

| Remark To sketch a surface in space, it is helpful to use traces in planes parallel to the coordinate planes as shown in Figure 15.4. For example, to find the trace of the surface in the plane $z = 2$, we substitute $z = 2$ in the equation $z = \sqrt{16 - 4x^2 - y^2}$ and obtain

$$2 = \sqrt{16 - 4x^2 - y^2} \implies \frac{x^2}{3} + \frac{y^2}{12} = 1$$

Thus, the trace is an ellipse centered at the point $(0, 0, 2)$ with major and minor axes of lengths $4\sqrt{3}$ and $2\sqrt{3}$, respectively.

Scalar fields and level curves

A second way to visualize a function of two variables is as a **scalar field** in which the scalar $z = f(x, y)$ is assigned to the point (x, y). A scalar field can be characterized by **level curves** (or **contour lines**) along which the value of $f(x, y)$ is constant. For instance, the weather map in Figure 15.5 shows level curves of equal pressure called **isobars.** In weather maps for which the level curves represent points of equal temperature, the level curves are called **isotherms.** Another common use of level curves is in representing electric potential fields. In this type of map the level curves are called **equipotential lines.**

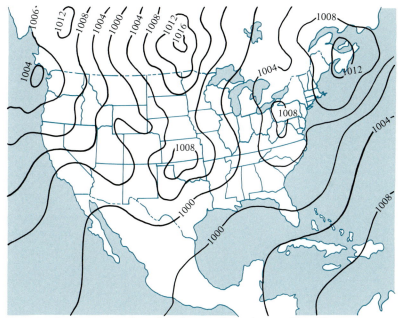

Level curves show lines of equal pressure (isobars) measured in millibars.

FIGURE 15.5

Level curves representing equal heights can be used to describe surfaces in space. For example, in Figure 15.6 the trace of the plane $z = 7$ is projected onto the xy-plane, forming a level curve for the function f. By plotting several level curves corresponding to the constant heights $c_1, c_2, \ldots, c_n$, we can obtain a **contour map** of the surface $z = f(x, y)$, as shown in Figure 15.7.

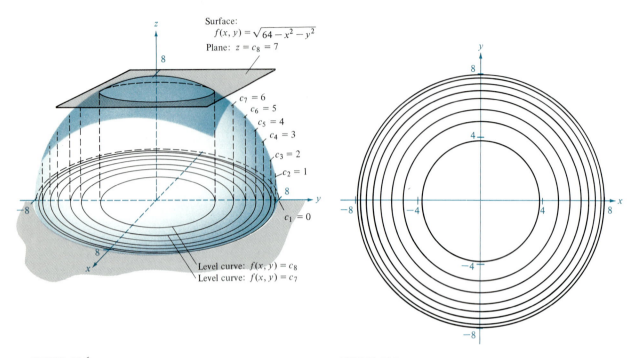

Surface:
$$f(x, y) = \sqrt{64 - x^2 - y^2}$$
Plane: $z = c_8 = 7$

$c_7 = 6$
$c_6 = 5$
$c_5 = 4$
$c_4 = 3$
$c_3 = 2$
$c_2 = 1$

$c_1 = 0$

Level curve: $f(x, y) = c_8$
Level curve: $f(x, y) = c_7$

FIGURE 15.6

FIGURE 15.7

> **Remark** A contour map depicts the variation of z with respect to x and y by the spacing between level curves. Much space between level curves indicates that z is changing slowly, whereas little space indicates a rapid change in z. Furthermore, to give a good three-dimensional illusion in a contour map it is important to choose c-values that are *evenly spaced*.

Contour maps are commonly used to show regions on the earth's surface, with the level curves representing the height above sea level. This type of map, called a **topographic map,** is shown in Figure 15.8.

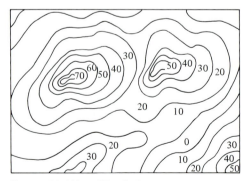

Topographical Map of Two Peaks.
One is over 70 units high and
the other is over 50 units high.

FIGURE 15.8

EXAMPLE 3   Sketching a contour map

The hyperbolic paraboloid given by

$$z = y^2 - x^2$$

is shown in Figure 15.9. Sketch a contour map for this surface.

Solution: For each value of c, we let $f(x, y) = c$ and sketch the resulting level curve in the xy-plane. For this function, each of the level curves ($c \neq 0$) is a hyperbola whose asymptotes are the lines $y = \pm x$. If $c < 0$, the transverse axis is horizontal. For instance, the level curve for $c = -4$ is given by

$$\frac{x^2}{2^2} - \frac{y^2}{2^2} = 1 \qquad \text{Hyperbola with horizontal transverse axis}$$

If $c > 0$, the transverse axis is vertical. For instance, the level curve for $c = 4$ is given by

$$\frac{y^2}{2^2} - \frac{x^2}{2^2} = 1 \qquad \text{Hyperbola with vertical transverse axis}$$

And if $c = 0$, the level curve is the degenerate conic representing the intersecting asymptotes, as shown in Figure 15.10.

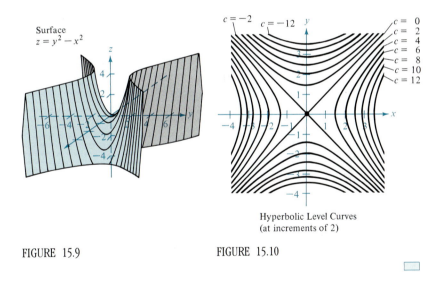

FIGURE 15.9

Hyperbolic Level Curves
(at increments of 2)

FIGURE 15.10

One example of a function of two variables used in economics is the **Cobb-Douglas production function.** This function is used as a model to represent the number of units produced by varying amounts of labor and capital. If x measures the units of labor and y measures the units of capital, then the number of units produced is given by

$$f(x, y) = Cx^a y^{1-a}$$

where C is constant and $0 < a < 1$. We illustrate a property of this function in the next example.

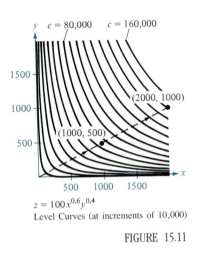

$z = 100x^{0.6}y^{0.4}$

Level Curves (at increments of 10,000)

FIGURE 15.11

EXAMPLE 4   An application from economics

Suppose that a manufacturer estimates a particular production function to be $f(x, y) = 100x^{0.6}y^{0.4}$, where x is the number of units of labor and y is the number of units of capital. Compare the production level when $x = 1000$ and $y = 500$ to the production level when $x = 2000$ and $y = 1000$.

Solution: When $x = 1000$ and $y = 500$, the production level is

$$f(1000, 500) = 100(1000^{0.6})(500^{0.4}) \approx 100(63.10)(12.01) \approx 75,786$$

When $x = 2000$ and $y = 1000$, the production level is

$$f(2000, 1000) = 100(2000^{0.6})(1000^{0.4}) \approx 100(95.64)(15.85) \approx 151,572$$

The graph of the surface $z = f(x, y)$ is shown in Figure 15.11. Note that by doubling *both* x and y we doubled the production level. In Exercise 52 you are asked to show that this is characteristic of the Cobb-Douglas production function.

Level surfaces

The concept of a level curve can be extended by one dimension to define a **level surface.** If f is a function of three variables and c is a constant, then the graph of the equation $f(x, y, z) = c$ is called a **level surface** of the function f, as shown in Figure 15.12. At every point on a given level surface, the function f has a constant value, $f(x, y, z) = c$. For example, if the function $f(x, y, z)$ represents the temperature at (x, y, z) in a region of space, then the level surfaces of equal temperature are called **isothermal surfaces,** as shown in Figure 15.13.

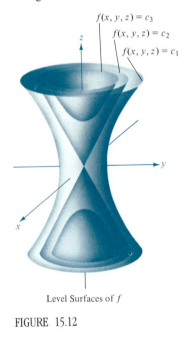

Level Surfaces of f

FIGURE 15.12

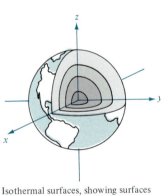

Isothermal surfaces, showing surfaces of equal temperature.

FIGURE 15.13

EXAMPLE 5 Level surfaces

Describe the level surfaces of the function

$$f(x, y, z) = 4x^2 + y^2 + z^2$$

Solution: Each level surface has an equation of the form $4x^2 + y^2 + z^2 = c$. Therefore, the level surfaces are ellipsoids whose cross sections parallel to the yz-plane are circles. As c increases, the radii of the circular cross sections increase according to the square root of c. For example, the level surfaces corresponding to the values $c = 4$ and $c = 16$ are as follows.

$$\text{For } c = 4: \quad \frac{x^2}{1} + \frac{y^2}{4} + \frac{z^2}{4} = 1$$

$$\text{For } c = 16: \quad \frac{x^2}{4} + \frac{y^2}{16} + \frac{z^2}{16} = 1$$

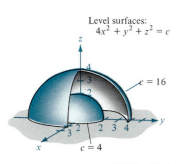

Level surfaces:
$4x^2 + y^2 + z^2 = c$

$c = 16$

$c = 4$

FIGURE 15.14   Two of these level surfaces are shown in Figure 15.14.

Section Exercises 15.1

In Exercises 1–6, determine if z is a function of the variables x and y.

1. $x^2 + y^2 + z^2 = 4$ **2.** $\cos xy - z = 4$
3. $x^2y^2z = 10$ **4.** $x^2yz^2 = 10$
5. $e^{xyz} = 4$ **6.** $x^2 - y^2 = z^2$

In Exercises 7–16, find the functional values.

7. $f(x, y) = \dfrac{x}{y}$

(a) $f(3, 2)$ (b) $f(-1, 4)$
(c) $f(30, 5)$ (d) $f(5, y)$
(e) $f(x, 2)$ (f) $f(5, t)$

8. $f(x, y) = 4 - x^2 - 4y^2$

(a) $f(0, 0)$ (b) $f(0, 1)$
(c) $f(2, 3)$ (d) $f(1, y)$
(e) $f(x, 0)$ (f) $f(t, 1)$

9. $f(x, y) = xe^y$

(a) $f(5, 0)$ (b) $f(3, 2)$
(c) $f(2, -1)$ (d) $f(5, y)$
(e) $f(x, 2)$ (f) $f(t, t)$

10. $g(x, y) = \ln |x + y|$

(a) $g(2, 3)$ (b) $g(5, 6)$
(c) $g(e, 0)$ (d) $g(0, 1)$
(e) $g(2, -3)$ (f) $g(e, e)$

11. $h(x, y, z) = \dfrac{xy}{z}$

(a) $h(2, 3, 9)$ (b) $h(1, 0, 1)$

12. $f(x, y, z) = \sqrt{x + y + z}$

(a) $f(0, 5, 4)$ (b) $f(6, 8, -3)$

13. $f(x, y) = x \sin y$

(a) $f\left(2, \dfrac{\pi}{4}\right)$ (b) $f(3, 1)$

14. $V(r, h) = \pi r^2 h$

(a) $V(3, 10)$ (b) $V(5, 2)$

15. $f(x, y) = \displaystyle\int_x^y (2t - 3)\, dt$

(a) $f(0, 4)$ (b) $f(1, 4)$

16. $g(x, y) = \displaystyle\int_x^y \dfrac{1}{t}\, dt$

(a) $g(4, 1)$ (b) $g(6, 3)$

In Exercises 17–28, describe the region R in the xy-coordinate plane that corresponds to the domain of the given function, and find the range of the function.

17. $f(x, y) = \sqrt{4 - x^2 - y^2}$
18. $f(x, y) = \sqrt{4 - x^2 - 4y^2}$
19. $f(x, y) = \sqrt{x^2 + y^2 - 1}$
20. $f(x, y) = \arcsin(x + y)$
21. $z = \dfrac{x + y}{xy}$ **22.** $z = \dfrac{xy}{x - y}$
23. $f(x, y) = \ln(4 - x - y)$
24. $f(x, y) = \ln(4 - xy)$
25. $f(x, y) = e^{x/y}$ **26.** $f(x, y) = x^2 + y^2$
27. $g(x, y) = \dfrac{1}{xy}$ **28.** $g(x, y) = x\sqrt{y}$

In Exercises 29–36, sketch the graph of the surface specified by the function.

29. $z = 4 - x^2 - y^2$

30. $z = \sqrt{x^2 + y^2}$

31. $f(x, y) = y^2$

32. $z = y^2 - x^2 + 1$

33. $f(x, y) = 6 - 2x - 3y$

34. $f(x, y) = \dfrac{1}{12}\sqrt{144 - 16x^2 - 9y^2}$

35. $f(x, y) = e^{-x}$

36. $f(x, y) = \begin{cases} xy, & x \geq 0, \ y \geq 0 \\ 0, & x < 0 \text{ or } y < 0 \end{cases}$

37. (a) Sketch the graph of the surface given by $f(x, y) = x^2 + y^2$.
   (b) On the surface of part (a), sketch the graphs of $z = f(1, y)$ and $z = f(x, 1)$.

38. (a) Sketch the graph of the surface given by $f(x, y) = xy$ in the first octant.
   (b) On the surface of part (a), sketch the graph of $z = f(x, x)$.

In Exercises 39–48, describe the level curves for each function. Sketch the level curves for the given c-values.

39. $f(x, y) = \sqrt{25 - x^2 - y^2}$, $c = 0, 1, 2, 3, 4, 5$

40. $f(x, y) = x^2 + y^2$, $c = 0, 2, 4, 6, 8$

41. $f(x, y) = xy$, $c = 1, -1, 3, -3$

42. $f(x, y) = 6 - 2x - 3y$, $c = 0, 2, 4, 6, 8, 10$

43. $f(x, y) = \dfrac{x}{x^2 + y^2}$, $c = \pm\dfrac{1}{2}, \pm 1, \pm\dfrac{3}{2}, \pm 2$

44. $f(x, y) = \arctan\dfrac{y}{x}$, $c = 0, \pm\dfrac{\pi}{6}, \pm\dfrac{\pi}{3}, \pm\dfrac{5\pi}{6}$

45. $f(x, y) = \ln(x - y)$, $c = 0, \pm\dfrac{1}{2}, \pm 1, \pm\dfrac{3}{2}, \pm 2$

46. $f(x, y) = \dfrac{x + y}{x - y}$, $c = 0, \pm 1, \pm 2, \pm 3$

47. $f(x, y) = e^{xy}$, $c = 1, 2, 3, 4, 5, \dfrac{1}{2}, \dfrac{1}{3}, \dfrac{1}{4}, \dfrac{1}{5}$

48. $f(x, y) = \cos(x + y)$, $c = 0, \pm\dfrac{1}{2}, \pm 1$

49. The **Doyle Log Rule** is one of several methods used to determine the lumber yield of a log (in board-feet) in terms of its diameter d (in inches) and its length L (in feet). The number of board-feet is given by

$$N(d, L) = \left(\dfrac{d - 4}{4}\right)^2 L$$

   (a) Find the number of board-feet of lumber in a log with a diameter of 22 inches and a length of 12 feet.
   (b) Find $N(30, 12)$.

50. A principal of \$1000 is deposited in a savings account earning r percent interest, compounded continuously. The amount $A(r, t)$ after t years is given by

$$A(r, t) = 1000e^{rt}$$

Use this function of two variables to complete the following table.

		Number of years		
Rate	5	10	15	20
0.08				
0.10				
0.12				
0.14				

51. Queuing model: The average length of time that a customer waits in line for service is given by

$$W(x, y) = \dfrac{1}{x - y}, \quad y < x$$

where y is the average arrival rate, expressed as the number of customers per unit time, and x is the average service rate, expressed in the same units. Evaluate W at the following points.

   (a) $(15, 10)$ (b) $(12, 9)$
   (c) $(12, 6)$ (d) $(4, 2)$

52. Use the Cobb-Douglas production function (see Example 4) to show that if the number of units of labor and the number of units of capital are both doubled, then the production level is also doubled.

53. The temperature T (in degrees Celsius) at any point (x, y) in a circular steel plate of radius 10 feet is given by

$$T = 600 - 0.75x^2 - 0.75y^2$$

where x and y are measured in feet. Sketch some of the isothermal curves.

54. According to the Ideal Gas Law, $PV = kT$, where P is pressure, V is volume, T is temperature, and k is a constant of proportionality. A tank contains 2600 cubic inches of nitrogen at a pressure of 20 pounds per square inch and a temperature of 40°F.

   (a) Determine k.
   (b) Express P as a function of V and T and describe the level curves.

SECTION TOPICS ■
Neighborhoods ■
Limits of functions of two variables ■
Continuity of functions of two variables ■

15.2
Limits and continuity

Sonya Kovalevsky

In this section we discuss limits and continuity as applied to functions of two or three variables. We begin with functions of two variables and then at the end of the section, we indicate how the concepts can be extended to cover functions of three variables.

Before discussing limits and continuity for functions of two variables we need to define some preliminary terms. Much of this terminology was introduced by the German mathematician Karl Weierstrass (1815–1897). His rigorous approach to limits and other topics in calculus gained him the reputation as the "father of modern analysis." Weierstrass was a gifted teacher. One of his best known students was the Russian mathematician Sonya Kovalevsky (1850–1891). Kovalevsky applied many of Weierstrass's techniques to problems in mathematical physics, and became one of the first women to gain acceptance as a research mathematician.

We begin our discussion of the limit of a function of two variables by defining a two-dimensional analog to an interval on the real line. Using the formula for the distance $\delta > 0$ between two points (x, y) and (x_0, y_0) in the plane, we define the **δ-neighborhood** about (x_0, y_0) to be the **disc** centered at (x_0, y_0) with radius δ:

$$\{(x, y): \sqrt{(x - x_0)^2 + (y - y_0)^2} < \delta\}$$

as shown in Figure 15.15.

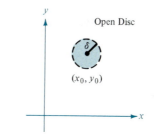

FIGURE 15.15

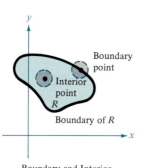

Boundary and Interior
Points of a Region R

FIGURE 15.16

Remark When the formula contains the *less than* inequality, $<$, the disc is said to be **open,** and when it contains the *less than or equal to* inequality, $\leq$, the disc is said to be **closed.** This corresponds to the use of $<$ and $\leq$ to define open and closed intervals.

A point (x_0, y_0) in a plane region R is an **interior point** of R if there exists a δ-neighborhood about (x_0, y_0) that lies entirely in R, as shown in Figure 15.16. If every point in R is an interior point, then we call R an **open region.** A point (x_0, y_0) is a **boundary point** of R if every open disc centered at (x_0, y_0) contains points inside R *and* points outside R. By definition, a region must contain its interior points, but it need not contain its boundary points. If a region contains all of its boundary points, then we say that the region is **closed.** A region that contains some but not all of its boundary points is neither open nor closed.

We are now ready to define the limit of a function of two variables.

DEFINITION OF THE LIMIT OF A FUNCTION OF TWO VARIABLES	Let f be a function of two variables defined, except possibly at (x_0, y_0), on an open disc centered at (x_0, y_0). Then

$$\lim_{(x,y)\to(x_0,y_0)} f(x, y) = L$$

if and only if for each $\varepsilon > 0$, there corresponds a $\delta > 0$ such that

$$|f(x, y) - L| < \varepsilon \quad \text{whenever} \quad 0 < \sqrt{(x - x_0)^2 + (y - y_0)^2} < \delta$$

"For any (x, y) in the circle of radius δ, the value $f(x, y)$ lies between $L + \epsilon$ and $L - \epsilon$."

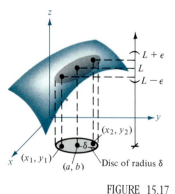

FIGURE 15.17

Remark Graphically, this definition of a limit means that for any point (x, y) in the disc of radius δ, the value $f(x, y)$ lies between $L + \varepsilon$ and $L - \varepsilon$, as shown in Figure 15.17.

The definition of the limit of a function of two variables closely parallels the definition of the limit of a function of a single variable, yet there is a critical difference. To determine whether a function of a single variable possesses a limit, we need only test the approach from two directions—from the left and from the right. If the function approaches the same limit from the right and from the left, we can conclude that the limit exists. However, for a function of two variables, when we write

$$(x, y) \to (x_0, y_0)$$

we mean that the point (x, y) is allowed to approach (x_0, y_0) from any direction. If the value of

$$\lim_{(x,y)\to(x_0,y_0)} f(x, y)$$

is not the same for all possible approaches, or **paths**, to (x_0, y_0), then the limit does not exist. This is demonstrated in the following example.

EXAMPLE 1 Testing for the existence of a limit

Show that the following limit does not exist:

$$\lim_{(x,y)\to(0,0)} \left(\frac{x^2 - y^2}{x^2 + y^2} \right)^2$$

Solution: The domain of the function given by

$$f(x, y) = \left(\frac{x^2 - y^2}{x^2 + y^2} \right)^2$$

consists of all points in the xy-plane except for the point $(0, 0)$. To show that the limit as (x, y) approaches $(0, 0)$ does not exist, we consider approaching along two different paths, as shown in Figure 15.18. Along the x-axis, every point is of the form $(x, 0)$, and the limit along this approach is

$$\lim_{(x,0)\to(0,0)} \left(\frac{x^2 - 0^2}{x^2 + 0^2} \right)^2 = \lim_{(x,0)\to(0,0)} (1)^2 = 1$$

However, if (x, y) approaches $(0, 0)$ along the line $y = x$, we obtain

$$\lim_{(x,x)\to(0,0)} \left(\frac{x^2 - x^2}{x^2 + x^2}\right)^2 = \lim_{(x,x)\to(0,0)} \left(\frac{0}{2x^2}\right)^2 = 0$$

This means that in any open disc centered at $(0, 0)$ there are points (x, y) at which f takes on the values 1 and 0. Hence, f cannot have a limit as $(x, y) \to (0, 0)$.

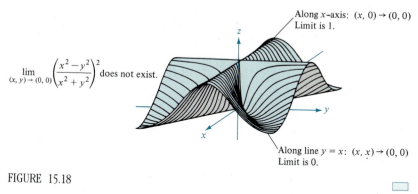

Along x-axis: $(x, 0) \to (0, 0)$
Limit is 1.

$$\lim_{(x, y)\to(0, 0)} \left(\frac{x^2 - y^2}{x^2 + y^2}\right)^2 \text{ does not exist.}$$

Along line $y = x$: $(x, x) \to (0, 0)$
Limit is 0.

FIGURE 15.18

In Example 1 we were able to conclude that the limit did not exist because we found two approaches that yielded different limits. Be sure you see that if the two approaches had yielded the same limit, we still could not conclude that the limit exists. To form such a conclusion, we must show that the limit is the same along all possible approaches. This is demonstrated in the next example.

EXAMPLE 2   *Verifying a limit by the definition*

Show that $\displaystyle\lim_{(x,y)\to(a,b)} x = a$.

Solution: Let $f(x, y) = x$ and $L = a$. We need to show that, for each $\varepsilon > 0$, there exists a δ-neighborhood about (a, b) such that

$$|f(x, y) - L| = |x - a| < \varepsilon$$

whenever $(x, y) \neq (a, b)$ lies in the neighborhood. We first observe that from

$$0 < \sqrt{(x - a)^2 + (y - b)^2} < \delta$$

it follows that

$$|f(x, y) - a| = |x - a| = \sqrt{(x - a)^2} \leq \sqrt{(x - a)^2 + (y - b)^2} < \delta$$

Thus, we can choose $\delta = \varepsilon$, and the limit is verified.

Limits of functions of several variables have the same properties with regard to sums, differences, products, and quotients as do limits of functions of a single variable. (See Theorem 2.3 in Section 2.1.) We make use of these properties in our next example.

EXAMPLE 3   *Verifying a limit*

Evaluate the following limits:

(a) $\lim\limits_{(x,y)\to(1,2)} \dfrac{5x^2y}{x^2+y^2}$ (b) $\lim\limits_{(x,y)\to(0,0)} \dfrac{5x^2y}{x^2+y^2}$

Solution:

(a) Using the properties of limits of products and sums, we have

$$\lim\limits_{(x,y)\to(1,2)} 5x^2y = 5(1^2)(2) = 10$$

and

$$\lim\limits_{(x,y)\to(1,2)} (x^2+y^2) = (1^2+2^2) = 5$$

Since the limit of a quotient is equal to the quotient of the limits, we have

$$\lim\limits_{(x,y)\to(1,2)} \frac{5x^2y}{x^2+y^2} = \frac{10}{5} = 2$$

(b) In this case, the limit of the denominator is 0, and so we cannot determine the existence (or nonexistence) of a limit by taking the limits of the numerator and denominator separately and then dividing. However, from the graph of f in Figure 15.19, it seems reasonable that the limit might be zero. Thus, we try applying the definition to $L = 0$. First, we note that

$$|y| \le \sqrt{x^2+y^2}$$

Then, in a δ-neighborhood about $(0, 0)$, we have $0 < \sqrt{x^2+y^2} < \delta$, and it follows that, for $(x, y) \ne (0, 0)$,

$$|f(x, y) - 0| = \left| \frac{5x^2y}{x^2+y^2} \right| = 5|y|\left(\frac{x^2}{x^2+y^2} \right) \le 5|y| \le 5\sqrt{x^2+y^2} < 5\delta$$

Thus, we choose $\delta = \varepsilon/5$ and conclude that

$$\lim\limits_{(x,y)\to(0,0)} \frac{5x^2y}{x^2+y^2} = 0$$

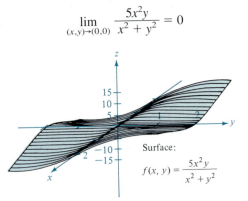

Surface:
$$f(x, y) = \frac{5x^2y}{x^2+y^2}$$

FIGURE 15.19

Notice that in Example 3(a) the limit of $f(x, y)$ as $(x, y) \to (1, 2)$ can be evaluated by direct substitution. That is, the limit is $f(1, 2) = 2$. In such cases, we say that f is **continuous.**

DEFINITION OF CONTINUITY
OF A FUNCTION OF TWO
VARIABLES

A function f of two variables is **continuous** at an interior point (x_0, y_0) of a region R if $f(x_0, y_0)$ is defined and equal to the limit of $f(x, y)$ as (x, y) approaches (x_0, y_0). That is,

$$\lim_{(x,y)\to(x_0,y_0)} f(x, y) = f(x_0, y_0)$$

f is **continuous on R** if f is continuous at every point in R. Functions that are not continuous at (x_0, y_0) are said to be **discontinuous**.

Remark   In Example 3(b), the function f is discontinuous at $(0, 0)$. However, since the limit at this point exists, we can remove the discontinuity by redefining f at $(0, 0)$ to be equal to its limit there. Such a discontinuity is said to be **removable.**

The sum, difference, product, and quotient properties of limits can be used to produce the following two-variable analogue of Theorem 2.8.

THEOREM 15.1

PROPERTIES OF CONTINUOUS FUNCTIONS OF TWO VARIABLES
If k is a real number and f and g are continuous at (x_0, y_0), then the following functions are continuous at (x_0, y_0):

1. Scalar multiple: kf 2. Sum and difference: $f \pm g$
3. Product: fg 4. Quotient: f/g, if $g(x_0, y_0) \neq 0$

Theorem 15.1 establishes the continuity of *polynomial* and *rational* functions at every point in their domains. Furthermore, the continuity of other types of functions can be naturally extended from one to two variables. For instance, the functions given by

$$f(x, y) = \sin xy \qquad \text{and} \qquad g(x, y) = (\cos y^2)e^{-\sqrt{x^2+y^2}}$$

are both continuous at every point in the plane as shown in Figures 15.20 and 15.21.

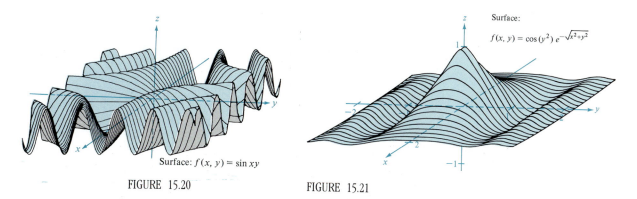

Surface: $f(x, y) = \sin xy$

FIGURE 15.20

Surface:
$f(x, y) = \cos(y^2)\, e^{-\sqrt{x^2+y^2}}$

FIGURE 15.21

THEOREM 15.2

CONTINUITY OF A COMPOSITE FUNCTION

If h is continuous at (x_0, y_0) and g is continuous at $h(x_0, y_0)$, then the composite function given by $(g \circ h)(x, y) = g(h(x, y))$ is continuous at (x_0, y_0). That is,

$$\lim_{(x,y)\to(x_0,y_0)} g(h(x, y)) = g(h(x_0, y_0))$$

| **Remark** Note in Theorem 15.2 that h is a function of two variables and g is a function of one variable.

EXAMPLE 4   Testing for continuity

Discuss the continuity of the following functions:

(a) $f(x, y) = \dfrac{x - 2y}{x^2 + y^2}$ (b) $g(x, y) = \dfrac{2}{y - x^2}$

Solution:

(a) Since a rational function is continuous at every point in its domain, we conclude that f is continuous at each point in the xy-plane except at $(0, 0)$, as shown in Figure 15.22.

(b) The function given by $g(x, y) = 2/(y - x^2)$ is continuous except at the points at which the denominator is zero, $y - x^2 = 0$. Thus, we conclude that the function is continuous at all points except those lying on the parabola $y = x^2$. Inside this parabola, we have $y > x^2$, and the surface represented by the function lies above the xy-plane as shown in Figure 15.23. Outside the parabola, $y < x^2$, and the surface lies below the xy-plane.

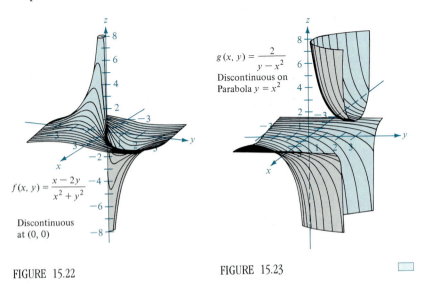

$f(x, y) = \dfrac{x - 2y}{x^2 + y^2}$

Discontinuous at $(0, 0)$

FIGURE 15.22

$g(x, y) = \dfrac{2}{y - x^2}$

Discontinuous on Parabola $y = x^2$

FIGURE 15.23

The preceding definitions of limits and continuity can be extended to functions of three variables by considering points (x, y, z) within the *open sphere*

$$(x - x_0)^2 + (y - y_0)^2 + (z - z_0)^2 = \delta^2$$

of radius δ and center (x_0, y_0, z_0). For example, the function given by

$$f(x, y, z) = \frac{1}{x^2 + y^2 - z}$$

is continuous at each point in space, except at the points on the paraboloid given by $z = x^2 + y^2$.

Section Exercises 15.2

In Exercises 1–4, find the indicated limit by using the limits

$$\lim_{(x,y)\to(a,b)} f(x, y) = 5 \quad \text{and} \quad \lim_{(x,y)\to(a,b)} g(x, y) = 3$$

1. $\displaystyle\lim_{(x,y)\to(a,b)} [f(x, y) - g(x, y)]$

2. $\displaystyle\lim_{(x,y)\to(a,b)} \left[\frac{4f(x, y)}{g(x, y)}\right]$

3. $\displaystyle\lim_{(x,y)\to(a,b)} [f(x, y)g(x, y)]$

4. $\displaystyle\lim_{(x,y)\to(a,b)} \left[\frac{f(x, y) - g(x, y)}{f(x, y)}\right]$

In Exercises 5–14, find the indicated limit and discuss the continuity of the function.

5. $\displaystyle\lim_{(x,y)\to(2,1)} (x + 3y^2)$

6. $\displaystyle\lim_{(x,y)\to(0,0)} (5x + 3xy + y + 1)$

7. $\displaystyle\lim_{(x,y)\to(2,4)} \frac{x + y}{x - y}$

8. $\displaystyle\lim_{(x,y)\to(1,1)} \frac{x}{\sqrt{x + y}}$

9. $\displaystyle\lim_{(x,y)\to(0,1)} \frac{\arcsin (x/y)}{1 + xy}$

10. $\displaystyle\lim_{(x,y)\to(\pi/4,2)} y \sin xy$

11. $\displaystyle\lim_{(x,y)\to(0,0)} e^{xy}$

12. $\displaystyle\lim_{(x,y)\to(1,1)} \frac{xy}{x^2 + y^2}$

13. $\displaystyle\lim_{(x,y,z)\to(1,2,5)} \sqrt{x + y + z}$

14. $\displaystyle\lim_{(x,y,z)\to(2,0,1)} xe^{yz}$

In Exercises 15–18, find the limit of $f(x, y)$ as $(x, y) \to (a, b)$ by finding the limits

$$\lim_{x\to a} g(x) \quad \text{and} \quad \lim_{y\to b} h(y)$$

where $f(x, y) = g(x)h(y)$.

15. $\displaystyle\lim_{(x,y)\to(0,0)} \frac{(1 + \sin x)(1 - \cos y)}{y}$

16. $\displaystyle\lim_{(x,y)\to(0,0)} \frac{\cos x \sin y}{y}$

17. $\displaystyle\lim_{(x,y)\to(1,2)} \frac{2x(y - 1)}{(x + 1)y}$

18. $\displaystyle\lim_{(x,y)\to(1,0)} \frac{xy}{(x - 1)e^y}$

In Exercises 19–28, discuss the continuity of the function and evaluate the indicated limit, if it exists.

19. $\displaystyle\lim_{(x,y)\to(0,0)} \frac{\sin xy}{xy}$

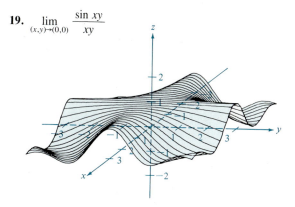

20. $\lim\limits_{(x,y)\to(0,0)} e^{xy}$

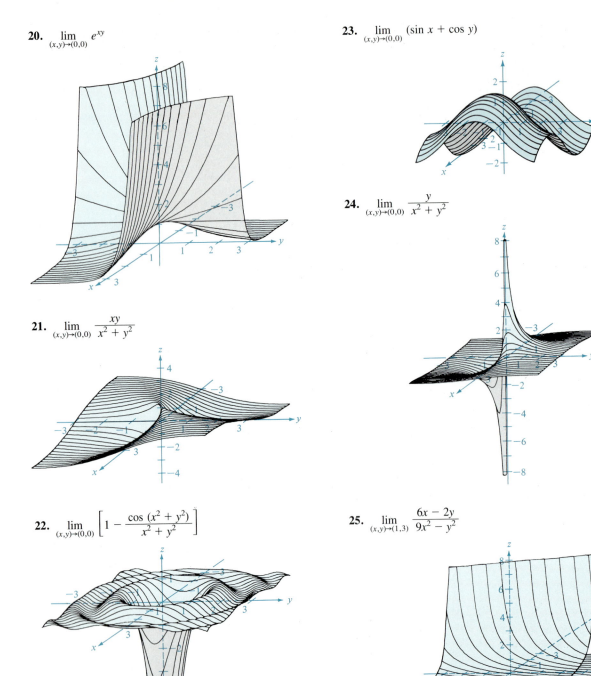

21. $\lim\limits_{(x,y)\to(0,0)} \dfrac{xy}{x^2 + y^2}$

22. $\lim\limits_{(x,y)\to(0,0)} \left[1 - \dfrac{\cos(x^2 + y^2)}{x^2 + y^2} \right]$

23. $\lim\limits_{(x,y)\to(0,0)} (\sin x + \cos y)$

24. $\lim\limits_{(x,y)\to(0,0)} \dfrac{y}{x^2 + y^2}$

25. $\lim\limits_{(x,y)\to(1,3)} \dfrac{6x - 2y}{9x^2 - y^2}$

26. $\lim\limits_{(x,y)\to(0,0)} \dfrac{2x - y^2}{2x^2 + y}$

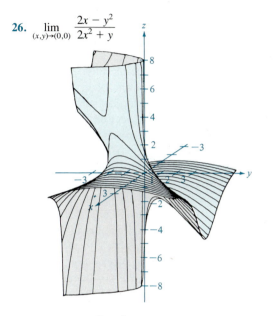

27. $\lim\limits_{(x,y)\to(0,0)} \ln (x^2 + y^2)$

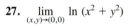

28. $\lim\limits_{(x,y)\to(0,1)} \dfrac{1 - x^2 - y^2}{\sqrt{|1 - x^2 - y^2|}}$

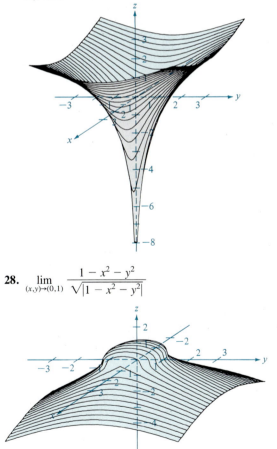

In Exercises 29–32, find the indicated limit by using polar coordinates. (Let $x = r \cos \theta$, $y = r \sin \theta$, and $r \to 0$.)

29. $\lim\limits_{(x,y)\to(0,0)} \dfrac{\sin (x^2 + y^2)}{x^2 + y^2}$

30. $\lim\limits_{(x,y)\to(0,0)} \dfrac{xy^2}{x^2 + y^2}$

31. $\lim\limits_{(x,y)\to(0,0)} \dfrac{x^3 + y^3}{x^2 + y^2}$

32. $\lim\limits_{(x,y)\to(0,0)} \dfrac{x^2 y^2}{x^2 + y^2}$

In Exercises 33–36, discuss the continuity of the function.

33. $f(x, y, z) = \dfrac{1}{\sqrt{x^2 + y^2 + z^2}}$

34. $f(x, y, z) = \dfrac{z}{x^2 + y^2 - 4}$

35. $f(x, y, z) = \dfrac{\sin z}{e^x + e^y}$

36. $f(x, y, z) = xy \sin z$

In Exercises 37–40, discuss the continuity of the composite function $f \circ g$.

37. $f(t) = t^2$, $g(x, y) = 3x - 2y$

38. $f(t) = \dfrac{1}{t}$, $g(x, y) = x^2 + y^2$

39. $f(t) = \dfrac{1}{t}$, $g(x, y) = 3x - 2y$

40. $f(t) = \dfrac{1}{4 - t}$, $g(x, y) = x^2 + y^2$

In Exercises 41–44, find the following limits:

(a) $\lim\limits_{\Delta x \to 0} \dfrac{f(x + \Delta x, y) - f(x, y)}{\Delta x}$

(b) $\lim\limits_{\Delta y \to 0} \dfrac{f(x, y + \Delta y) - f(x, y)}{\Delta y}$

41. $f(x, y) = x^2 - 4y$
42. $f(x, y) = x^2 + y^2$
43. $f(x, y) = 2x + xy - 3y$ **44.** $f(x, y) = \sqrt{y(y + 1)}$

45. Prove that

$$\lim\limits_{(x,y)\to(a,b)} [f(x, y) + g(x, y)] = L_1 + L_2$$

where $f(x, y)$ approaches L_1 and $g(x, y)$ approaches L_2 as $(x, y) \to (a, b)$.

46. Prove that if f is continuous and $f(a, b) < 0$, then there exists a δ-neighborhood about (a, b) such that $f(x, y) < 0$ for every point (x, y) in the neighborhood.

Jean Le Rond d'Alembert

15.3
Partial derivatives

In applications of functions of several variables, the question often arises, "How will the function be affected by a change in one of its independent variables?" We can answer this question by considering the independent variables one at a time. For example, to determine the effect of a catalyst in an experiment, a chemist could conduct the experiment several times using varying amounts of the catalyst, while keeping constant other variables such as temperature and pressure. We follow a similar procedure to determine the rate of change of a function f with respect to one of its several independent variables. That is, we take the derivative of f with respect to one independent variable at a time, while holding the others constant. This process is called **partial differentiation,** and the result is referred to as the **partial derivative** of f with respect to the chosen independent variable.

   The introduction of partial derivatives followed Newton and Leibniz's work in calculus by several years. Between 1730 and 1760, Leonhard Euler and Jean Le Rond d'Alembert (1717–1783) separately published several papers on dynamics, in which they established much of the theory of partial derivatives. These papers used functions of two or more variables to study problems involving equilibrium, fluid motion, and vibrating strings.

DEFINITION OF PARTIAL DERIVATIVES OF A FUNCTION OF TWO VARIABLES

If $z = f(x, y)$, then the **first partial derivatives** of f with respect to x and to y are the functions f_x and f_y defined by

$$f_x(x, y) = \lim_{\Delta x \to 0} \frac{f(x + \Delta x, y) - f(x, y)}{\Delta x}$$

$$f_y(x, y) = \lim_{\Delta y \to 0} \frac{f(x, y + \Delta y) - f(x, y)}{\Delta y}$$

provided the limits exist.

This definition indicates that if $z = f(x, y)$, then to find f_x we *consider y constant* and differentiate with respect to x. Similarly, to find f_y, we *consider x constant* and differentiate with respect to y.

EXAMPLE 1   Finding partial derivatives

Find f_x and f_y for $f(x, y) = 3x - x^2y^2 + 2x^3y$.

Solution: Considering y to be constant and differentiating with respect to x, we have

$$f_x(x, y) = 3 - 2xy^2 + 6x^2y$$

Considering x to be constant and differentiating with respect to y, we have

$$f_y(x, y) = -2x^2y + 2x^3$$

There are several notations for first partial derivatives. We list the common ones here, along with the notation for a partial derivative evaluated at the point (a, b).

NOTATION FOR FIRST PARTIAL DERIVATIVES

For $z = f(x, y)$, the partial derivatives f_x and f_y are denoted by

$$\frac{\partial}{\partial x} f(x, y) = f_x(x, y) = z_x = \frac{\partial z}{\partial x}$$

$$\frac{\partial}{\partial y} f(x, y) = f_y(x, y) = z_y = \frac{\partial z}{\partial y}$$

The first partials evaluated at the point (a, b) are denoted by

$$\frac{\partial z}{\partial x}\bigg|_{(a,b)} = f_x(a, b) \qquad \text{and} \qquad \frac{\partial z}{\partial y}\bigg|_{(a,b)} = f_y(a, b)$$

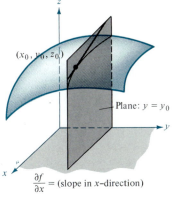

$\dfrac{\partial f}{\partial x}$ = (slope in x–direction)

FIGURE 15.24

EXAMPLE 2   Finding and evaluating partial derivatives

For $f(x, y) = xe^{x^2 y}$, find f_x and f_y, and evaluate each at the point $(1, \ln 2)$.

Solution: Since

$$f_x(x, y) = xe^{x^2 y}(2xy) + e^{x^2 y}$$

we have

$$f_x(1, \ln 2) = e^{\ln 2}(2 \ln 2) + e^{\ln 2} = 4 \ln 2 + 2$$

Since

$$f_y(x, y) = xe^{x^2 y}(x^2) = x^3 e^{x^2 y}$$

we have

$$f_y(1, \ln 2) = e^{\ln 2} = 2$$

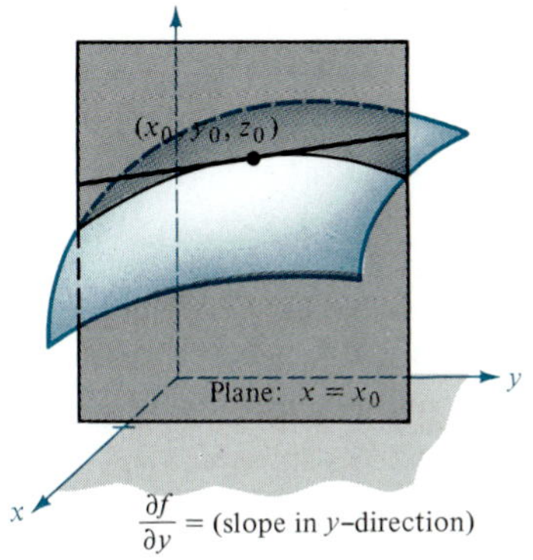

The partial derivatives of a function of two variables, $z = f(x, y)$, have a useful geometric interpretation. If $y = y_0$, then $z = f(x, y_0)$ represents the curve formed by intersecting the surface $z = f(x, y)$ with the plane $y = y_0$, as shown in Figure 15.24. Therefore,

$$f_x(x_0, y_0) = \lim_{\Delta x \to 0} \frac{f(x_0 + \Delta x, y_0) - f(x_0, y_0)}{\Delta x}$$

represents the slope of this curve at the point $(x_0, y_0, f(x_0, y_0))$. (Note that both the curve and the tangent line lie in the plane $y = y_0$.) Similarly,

$$f_y(x_0, y_0) = \lim_{\Delta y \to 0} \frac{f(x_0, y_0 + \Delta y) - f(x_0, y_0)}{\Delta y}$$

represents the slope of the curve given by the intersection of $z = f(x, y)$ and the plane $x = x_0$ at $(x_0, y_0, f(x_0, y_0))$, as shown in Figure 15.25.

Plane: $x = x_0$

$\dfrac{\partial f}{\partial y}$ = (slope in y–direction)

FIGURE 15.25

Informally, we say that the values of $\partial f/\partial x$ and $\partial f/\partial y$ at the point (x_0, y_0, z_0) denote the **slope of the surface in the x and y directions,** respectively.

EXAMPLE 3   *Finding the slope of a surface in the x and y directions*

Find the slope of the surface given by

$$f(x, y) = -\frac{x^2}{2} - y^2 + \frac{25}{8}$$

at the point $(\frac{1}{2}, 1, 2)$ in the x direction and in the y direction.

Solution: In the x direction, the slope is given by

$$f_x(x, y) = -x \implies f_x\left(\frac{1}{2}, 1\right) = -\frac{1}{2} \qquad \text{Figure 15.26}$$

In the y direction, the slope is given by

$$f_y(x, y) = -2y \implies f_y\left(\frac{1}{2}, 1\right) = -2 \qquad \text{Figure 15.27}$$

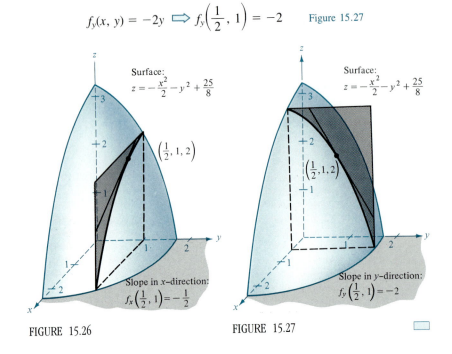

FIGURE 15.26   FIGURE 15.27

The concept of a partial derivative can be extended in a natural way to functions of three or more variables. For instance, if $w = f(x, y, z)$, then there are three partial derivatives, each of which is formed by considering two of the variables to be constant. That is, to define the partial derivative of w with respect to x, we consider y and z to be constant and write

$$\frac{\partial w}{\partial x} = f_x(x, y, z) = \lim_{\Delta x \to 0} \frac{f(x + \Delta x, y, z) - f(x, y, z)}{\Delta x}$$

To define the partial derivative of w with respect to y, we consider x and z to be constant and write

$$\frac{\partial w}{\partial y} = f_y(x, y, z) = \lim_{\Delta y \to 0} \frac{f(x, y + \Delta y, z) - f(x, y, z)}{\Delta y}$$

To define the partial derivative of w with respect to z, we consider x and y to be constant and write

$$\frac{\partial w}{\partial z} = f_z(x, y, z) = \lim_{\Delta z \to 0} \frac{f(x, y, z + \Delta z) - f(x, y, z)}{\Delta z}$$

In general, if $w = f(x_1, x_2, \dots, x_n)$, then there are n partial derivatives denoted by

$$\frac{\partial w}{\partial x_k} = f_{x_k}(x_1, x_2, \dots, x_n), \quad k = 1, 2, \dots, n$$

To find the partial derivative with respect to one of the variables, we consider the others to be constant and differentiate with respect to the given variable.

EXAMPLE 4   *Finding partial derivatives*

Find the indicated partial derivatives.
(a) For $f(x, y, z) = xy + yz^2 + xz$, find $f_z(x, y, z)$.
(b) For $f(x, y, z) = z \sin (xy^2 + 2z)$, find $f_z(x, y, z)$.
(c) For $f(x, y, z, w) = (x + y + z)/w$, find $f_w(x, y, z, w)$.

Solution:

(a) To find the partial derivative of f with respect to z, we consider x and y to be constant and obtain

$$\frac{\partial}{\partial z}[xy + yz^2 + xz] = 2yz + x$$

(b) To find the partial derivative of f with respect to z, we consider x and y to be constant and using the Product Rule, we obtain

$$\frac{\partial}{\partial z}[z \sin (xy^2 + 2z)] = (z)\frac{\partial}{\partial z}[\sin (xy^2 + 2z)] + \sin (xy^2 + 2z)\frac{\partial}{\partial z}[z]$$

$$= (z)[\cos (xy^2 + 2z)](2) + \sin (xy^2 + 2z)$$

$$= 2z \cos (xy^2 + 2z) + \sin (xy^2 + 2z)$$

(c) To find the partial derivative of f with respect to w, we consider x, y, and z to be constant and obtain

$$\frac{\partial}{\partial w}\left[\frac{x + y + z}{w}\right] = -\frac{x + y + z}{w^2}$$

No matter how many variables are involved, partial derivatives of several variables can be interpreted as *rates of change*, just as in the single variable case. This is illustrated in Example 5.

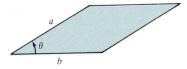

FIGURE 15.28

EXAMPLE 5   *Using partial derivatives to find rates of change*

The area of a parallelogram with adjacent sides a and b and included angle θ is given by $A = ab \sin \theta$, as shown in Figure 15.28.

(a) Find the rate of change of A with respect to a when $a = 10$, $b = 20$, and $\theta = \pi/6$.

(b) Find the rate of change of A with respect to θ when $a = 10$, $b = 20$, and $\theta = \pi/6$.

Solution:

(a) To find the rate of change of the area with respect to a, we hold b and θ constant and differentiate with respect to a to obtain

$$\frac{\partial A}{\partial a} = b \sin \theta$$

When $a = 10$, $b = 20$, and $\theta = \pi/6$, we have $\partial A/\partial a = 20 \sin (\pi/6) = 10$.

(b) To find the rate of change of the area with respect to θ, we hold a and b constant and differentiate with respect to θ to obtain

$$\frac{\partial A}{\partial \theta} = ab \cos \theta$$

When $a = 10$, $b = 20$, and $\theta = \pi/6$, we have $\partial A/\partial \theta = 200 \cos (\pi/6) = 100\sqrt{3}$.

Higher-order partial derivatives

As is true for ordinary derivatives, it is possible to take second, third, and higher partial derivatives of a function of several variables, provided such derivatives exist. We denote higher-order derivatives by the order in which the differentiation occurs. For instance, there are four different ways to find a second partial derivative for $z = f(x, y)$.

1. Differentiate twice with respect to x:

$$\frac{\partial}{\partial x}\left(\frac{\partial f}{\partial x}\right) = \frac{\partial^2 f}{\partial x^2} = f_{xx}$$

2. Differentiate twice with respect to y:

$$\frac{\partial}{\partial y}\left(\frac{\partial f}{\partial y}\right) = \frac{\partial^2 f}{\partial y^2} = f_{yy}$$

3. Differentiate first with respect to x and then with respect to y:

$$\frac{\partial}{\partial y}\left(\frac{\partial f}{\partial x}\right) = \frac{\partial^2 f}{\partial y \partial x} = f_{xy}$$

4. Differentiate first with respect to y and then with respect to x:

$$\frac{\partial}{\partial x}\left(\frac{\partial f}{\partial y}\right) = \frac{\partial^2 f}{\partial x \partial y} = f_{yx}$$

The third and fourth cases are called **mixed partial derivatives.** You should notice that the two types of notation for mixed partials have different conventions for indicating the order of differentiation. For instance, the partial

$$\frac{\partial}{\partial y}\left(\frac{\partial f}{\partial x}\right) = \frac{\partial^2 f}{\partial y \partial x} \qquad \text{Right-to-left order}$$

indicates differentiation with respect to x first, but the partial

$$(f_y)_x = f_{yx} \qquad \text{Left-to-right order}$$

indicates differentiation with respect to y first. You can remember this by observing that in both notations, you differentiate first with respect to the variable "nearest" f.

EXAMPLE 6   *Finding second partial derivatives*

Find the second partial derivatives of $f(x, y) = 3xy^2 - 2y + 5x^2y^2$, and determine the value of $f_{xy}(-1, 2)$.

Solution: We begin by finding the first partial derivatives with respect to x and y:

$$f_x(x, y) = 3y^2 + 10xy^2 \qquad \text{and} \qquad f_y(x, y) = 6xy - 2 + 10x^2y$$

Then, by differentiating each of these with respect to x and y, we have

$$f_{xx}(x, y) = 10y^2 \qquad \text{and} \qquad f_{yy}(x, y) = 6x + 10x^2$$
$$f_{xy}(x, y) = 6y + 20xy \qquad \text{and} \qquad f_{yx}(x, y) = 6y + 20xy$$

Finally, $f_{xy}(-1, 2) = 12 - 40 = -28$. ☐

Notice in Example 6 that the two mixed partials are equal. This is often the case, as indicated in the next theorem, which we list without proof.

THEOREM 15.3   EQUALITY OF MIXED PARTIAL DERIVATIVES
If f is a function of x and y such that f, f_x, f_y, f_{xy}, and f_{yx} are continuous on an open region R, then for every (x, y) in R,

$$f_{xy}(x, y) = f_{yx}(x, y)$$

Theorem 15.3 also applies to a function f of *three or more variables* so long as f and all of its first and second partial derivatives are continuous. For example, if $w = f(x, y, z)$ and f and all of its first and second partial derivatives are continuous in an open region R, then at each point in R the order of differentiation in the mixed second partial derivatives is immaterial. That is,

$$f_{xy}(x, y, z) = f_{yx}(x, y, z)$$
$$f_{xz}(x, y, z) = f_{zx}(x, y, z)$$
and
$$f_{yz}(x, y, z) = f_{zy}(x, y, z)$$

If in addition, the third partial derivatives of f are continuous, then the order of differentiation of the mixed third partial derivatives is immaterial. For example, $f_{xzz} = f_{zxz} = f_{zzx}$, as demonstrated in the next example.

EXAMPLE 7   *Finding higher-order partial derivatives*

Show that $f_{xz} = f_{zx}$ and $f_{xzz} = f_{zxz} = f_{zzx}$ for the function given by

$$f(x, y, z) = ye^x + x \ln z$$

Solution: First partials:

$$f_x(x, y, z) = ye^x + \ln z, \qquad f_z(x, y, z) = \frac{x}{z}$$

Second partials:

$$f_{xz}(x, y, z) = \frac{1}{z}, \qquad f_{zx}(x, y, z) = \frac{1}{z}, \qquad f_{zz}(x, y, z) = -\frac{x}{z^2}$$

(Note that the first two are equal.)
Third partials:

$$f_{xzz}(x, y, z) = -\frac{1}{z^2}, \qquad f_{zxz}(x, y, z) = -\frac{1}{z^2}, \qquad f_{zzx}(x, y, z) = -\frac{1}{z^2}$$

(Note that all three are equal.)

Applications

There are many applications involving the rates of change of related variables. An important application in physical chemistry involves **Boyle's Ideal Gas Law,** named after the English chemist Robert Boyle (1627–1691). This law states that for a so-called ideal gas at a given temperature T, the product of the pressure P and volume V is constant. In other words, the product PV is a function of T, and we can write

$$PV = f(T) \implies V = \frac{f(T)}{P}$$

In order to determine the function f for a particular gas and liquid, chemists use two ratios, called **thermal expansivity** and **isothermal compressibility,** which are defined as

$$\alpha = \frac{\partial V / \partial T}{V} \qquad \text{Thermal expansivity}$$

and

$$\beta = -\frac{\partial V / \partial P}{V} \qquad \text{Isothermal compressibility}$$

These two ratios are used to measure percentage changes in volume per change in temperature and pressure. For example, the thermal expansivity and isothermal compressibility of water are given in Table 15.1. [Listings for α

and β are in units of 10^{-5}. Pressure is measured in atmospheres (Newtons per square meter). Temperature is measured in degrees Celsius. Volume is measured in cubic meters.]

TABLE 15.1

T	0°	10°	20°	30°	40°	50°	60°	70°	80°	90°	100°
α	5.01	4.78	4.58	4.46	4.41	4.40	4.43	4.49	4.57	4.68	4.80
β	−6.81	9.00	20.68	30.32	38.53	45.76	52.31	58.37	64.11	69.62	75.01

Section Exercises 15.3

In Exercises 1–22, find the first partial derivatives with respect to x and with respect to y.

1. $f(x, y) = 2x - 3y + 5$ **2.** $f(x, y) = x^2 - 3y^2 + 7$

3. $f(x, y) = xy$ **4.** $f(x, y) = \dfrac{x}{y}$

5. $z = x\sqrt{y}$ **6.** $z = x^2 - 3xy + y^2$
7. $z = x^2 e^{2y}$ **8.** $z = xe^{x/y}$
9. $z = \ln (x^2 + y^2)$ **10.** $z = \ln \sqrt{xy}$
11. $z = \ln \dfrac{x + y}{x - y}$ **12.** $z = \dfrac{x^2}{2y} + \dfrac{4y^2}{x}$

13. $h(x, y) = e^{-(x^2+y^2)}$
14. $g(x, y) = \ln \sqrt{x^2 + y^2}$
15. $f(x, y) = \sqrt{x^2 + y^2}$

16. $f(x, y) = \dfrac{xy}{x^2 + y^2}$

17. $z = \sin (2x - y)$ **18.** $z = \sin 3x \cos 3y$
19. $z = e^y \sin xy$ **20.** $z = \cos (x^2 + y^2)$

21. $f(x, y) = \displaystyle\int_x^y (t^2 - 1)\, dt$

22. $f(x, y) = \displaystyle\int_x^y (2t + 1)\, dt + \int_y^x (2t - 1)\, dt$

In Exercises 23–26, evaluate f_x and f_y at the indicated point.

23. $f(x, y) = \arctan \dfrac{y}{x}$, $(2, -2)$

24. $f(x, y) = \arcsin xy$, $(1, 0)$

25. $f(x, y) = \dfrac{xy}{x - y}$, $(2, -2)$

26. $f(x, y) = \dfrac{4xy}{\sqrt{x^2 + y^2}}$, $(1, 0)$

In Exercises 27–32, find the first partial derivatives with respect to x, y, and z.

27. $w = \sqrt{x^2 + y^2 + z^2}$

28. $w = \dfrac{xy}{x + y + z}$

29. $F(x, y, z) = \ln \sqrt{x^2 + y^2 + z^2}$

30. $G(x, y, z) = \dfrac{1}{\sqrt{1 - x^2 - y^2 - z^2}}$

31. $H(x, y, z) = \sin (x + 2y + 3z)$
32. $f(x, y, z) = 3x^2y - 5xyz + 10yz^2$

In Exercises 33–40, find the second partial derivatives

$$\frac{\partial^2 z}{\partial x^2}, \quad \frac{\partial^2 z}{\partial y^2}, \quad \frac{\partial^2 z}{\partial y \partial x}, \quad \text{and} \quad \frac{\partial^2 z}{\partial x \partial y}$$

33. $z = x^2 - 2xy + 3y^2$
34. $z = x^4 - 3x^2y^2 + y^4$
35. $z = e^x \tan y$ **36.** $z = 2e^{xy^2}$

37. $z = \arctan \dfrac{y}{x}$ **38.** $z = \sin (x - 2y)$

39. $z = \sqrt{x^2 + y^2}$ **40.** $z = \dfrac{xy}{x - y}$

In Exercises 41–46, show that

$$\frac{\partial^2 z}{\partial y \partial x} = \frac{\partial^2 z}{\partial x \partial y}$$

41. $z = x^3 + 3x^2y$ **42.** $z = \ln (x - y)$
43. $z = x \sec y$ **44.** $z = \sqrt{9 - x^2 - y^2}$
45. $z = xe^{-y^2}$ **46.** $z = xe^y + ye^x$

In Exercises 47–50, show that the mixed partials f_{xyy}, f_{yxy}, and f_{yyx} are equal.

47. $f(x, y, z) = xyz$

48. $f(x, y, z) = x^2 - 3xy + 4yz + z^3$

49. $f(x, y, z) = e^{-x} \sin yz$ **50.** $f(x, y, z) = \dfrac{x}{y + z}$

In Exercises 51–54, show that each fraction satisfies **Laplace's equation**

$$\frac{\partial^2 z}{\partial x^2} + \frac{\partial^2 z}{\partial y^2} = 0$$

51. $z = 5xy$

52. $z = \dfrac{1}{2}(e^y - e^{-y}) \sin x$

53. $z = e^x \sin y$

54. $z = \arctan \dfrac{y}{x}$

In Exercises 55 and 56, show that the function satisfies the **wave equation**

$$\frac{\partial^2 z}{\partial t^2} = c^2 \frac{\partial^2 z}{\partial x^2}$$

55. $z = \sin (x - ct)$

56. $z = \sin \omega ct \sin \omega x$

In Exercises 57 and 58, show that the function satisfies the **heat equation**

$$\frac{\partial z}{\partial t} = c^2 \frac{\partial^2 z}{\partial x^2}$$

57. $z = e^{-t} \cos \dfrac{x}{c}$ **58.** $z = e^{-t} \sin \dfrac{x}{c}$

In Exercises 59–62, use the limit definition of partial derivatives to find $f_x(x, y)$ and $f_y(x, y)$.

59. $f(x, y) = 2x + 3y$

60. $f(x, y) = \dfrac{1}{x + y}$

61. $f(x, y) = \sqrt{x + y}$

62. $f(x, y) = x^2 - 2xy + y^2$

In Exercises 63–66, sketch the curve formed by the intersection of the given surface and plane. Find the slope on the curve at the given point.

63. $z = \sqrt{49 - x^2 - y^2}$, $x = 2$, $(2, 3, 6)$

64. $z = x^2 + 4y^2$, $y = 1$, $(2, 1, 8)$

65. $z = 9x^2 - y^2$, $y = 3$, $(1, 3, 0)$

66. $z = 9x^2 - y^2$, $x = 1$, $(1, 3, 0)$

67. A company manufactures two types of wood-burning stoves: a free-standing model and a fireplace-insert model. The cost function for producing x free-standing and y fireplace-insert stoves is

$$C = 32\sqrt{xy} + 175x + 205y + 1050$$

Find the marginal costs ($\partial C/\partial x$ and $\partial C/\partial y$) when $x = 80$ and $y = 20$.

68. Let $x = 1000$ and $y = 500$ in the Cobb-Douglas production function

$$f(x, y) = 100x^{0.6}y^{0.4}$$

(a) Find the marginal productivity of labor, $\partial f/\partial x$.
(b) Find the marginal productivity of capital, $\partial f/\partial y$.

69. Let N be the number of applicants to a university, p the charge for food and housing at the university, and t the tuition. Suppose that N is a function of p and t such that $\partial N/\partial p < 0$ and $\partial N/\partial t < 0$. How would you interpret the fact that both partials are negative?

70. The range of a projectile fired at an angle θ above the horizontal with velocity v_0 is

$$R = \frac{v_0^2 \sin 2\theta}{32}$$

Evaluate $\partial R/\partial v_0$ and $\partial R/\partial \theta$ when $v_0 = 2000$ feet per second and $\theta = 5°$.

71. The temperature at any point (x, y) in a steel plate is given by

$$T = 500 - 0.6x^2 - 1.5y^2$$

where x and y are measured in feet. At the point $(2, 3)$, find the rate of change of the temperature with respect to the distance moved along the plate in the directions of the x- and y-axes, respectively.

72. According to the Ideal Gas Law, $PV = kT$, where P is pressure, V is volume, T is temperature, and k is a constant of proportionality. Find (a) $\partial P/\partial T$ and (b) $\partial V/\partial P$.

SECTION TOPICS ▪
Increments ▪
Differentials ▪
Approximation by differentials ▪
Differentiability ▪

15.4
Differentials

In this section we generalize the concepts of increments and differentials to functions of two or more variables. Recall from Section 4.10 that for $y = f(x)$ we defined the differential of y to be $dy = f'(x) \, dx$. For a function of two variables, given by $z = f(x, y)$, we use similar terminology. We call Δx and Δy the **increments of x and y,** and the **increment of z** is given by

$$\Delta z = f(x + \Delta x, \, y + \Delta y) - f(x, \, y)$$

The differentials dx, dy, and dz are defined as follows.

DEFINITION OF TOTAL DIFFERENTIAL

If $z = f(x, \, y)$ and Δx and Δy are increments of x and y, then the **differentials** of the independent variables x and y are

$$dx = \Delta x \quad \text{and} \quad dy = \Delta y$$

and the **total differential** of the dependent variable z is

$$dz = \frac{\partial z}{\partial x} \, dx + \frac{\partial z}{\partial y} \, dy = f_x(x, \, y) \, dx + f_y(x, \, y) \, dy$$

Remark This definition can be extended to a function of three or more variables. For instance, if $w = f(x, \, y, \, z, \, u)$, then $dx = \Delta x$, $dy = \Delta y$, $dz = \Delta z$, $du = \Delta u$, and the total differential of w is

$$dw = \frac{\partial w}{\partial x} \, dx + \frac{\partial w}{\partial y} \, dy + \frac{\partial w}{\partial z} \, dz + \frac{\partial w}{\partial u} \, du$$

EXAMPLE 1   *Finding the total differential*

Find dz for the function given by $z = 2x \sin y - 3x^2 y^2$.

Solution:

$$dz = \frac{\partial z}{\partial x} \, dx + \frac{\partial z}{\partial y} \, dy = (2 \sin y - 6xy^2) \, dx + (2x \cos y - 6x^2 y) \, dy$$

In Section 4.10 we saw that for a *differentiable* function given by $y = f(x)$, we can use the differential $dy = f'(x) \, dx$ as an approximation (for small Δx) to the value $\Delta y = f(x + \Delta x) - f(x)$. When a similar approximation is possible for a function of two variables, we say that it is **differentiable.** This is stated explicitly in the following definition.

DEFINITION OF DIFFERENTIABILITY

A function f given by $z = f(x, \, y)$ is **differentiable** at $(x, \, y)$ provided Δz can be expressed in the form

$$\Delta z = f_x(x, \, y) \Delta x + f_y(x, \, y) \Delta y + \varepsilon_1 \Delta x + \varepsilon_2 \Delta y$$

where both ε_1 and $\varepsilon_2 \to 0$ as $(\Delta x, \, \Delta y) \to (0, \, 0)$. The function f is said to be **differentiable in a region R** if it is differentiable at each point of R.

Be sure you see that the term differentiable is used differently for functions of two variables than for one variable. A function of one variable is differentiable at a point if its derivative exists at the point. However, for a function of two variables the existence of the partial derivatives f_x and f_y does not guarantee that the function is differentiable (see Example 4). In the following theorem, we present a *sufficient* condition for differentiability of a function of two variables.

THEOREM 15.4   **SUFFICIENT CONDITION FOR DIFFERENTIABILITY**
If f is a function of x and y, where f, f_x, and f_y are continuous in an open region R, then f is differentiable on R.

Proof: Let S be the surface defined by $z = f(x, y)$, where f, f_x, and f_y are continuous at (x, y) in the domain of f. Let A, B, and C be points on surface S, as shown in Figure 15.29. From this figure we can see that the change in f from point A to point C is given by

$$\Delta z = f(x + \Delta x, y + \Delta y) - f(x, y)$$
$$= [f(x + \Delta x, y) - f(x, y)] + [f(x + \Delta x, y + \Delta y) - f(x + \Delta x, y)]$$
$$= \Delta z_1 + \Delta z_2$$

Between A and B, y is fixed while x changes. Hence, by the Mean Value Theorem, there is a value x_1 between x and $x + \Delta x$ such that

$$\Delta z_1 = f(x + \Delta x, y) - f(x, y) = f_x(x_1, y)\Delta x$$

Similarly, between B and C, x is fixed while y changes, and there is a value y_1 between y and $y + \Delta y$ such that

$$\Delta z_2 = f(x + \Delta x, y + \Delta y) - f(x + \Delta x, y) = f_y(x + \Delta x, y_1)\Delta y$$

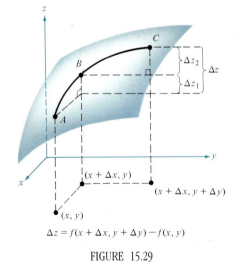

$\Delta z = f(x + \Delta x, y + \Delta y) - f(x, y)$

FIGURE 15.29

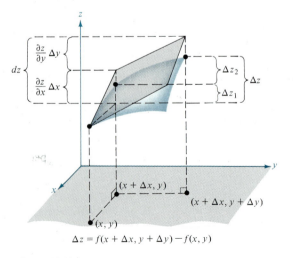

$\Delta z = f(x + \Delta x, y + \Delta y) - f(x, y)$

FIGURE 15.30

By combining these two results, we can write

$$\Delta z = \Delta z_1 + \Delta z_2 = f_x(x_1, y)\Delta x + f_y(x + \Delta x, y_1)\Delta y$$

Now, if we define ε_1 and ε_2 by

$$\varepsilon_1 = f_x(x_1, y) - f_x(x, y) \qquad \text{and} \qquad \varepsilon_2 = f_y(x + \Delta x, y_1) - f_y(x, y)$$

then by the continuity of f_x and f_y and the fact that $x \le x_1 \le x + \Delta x$ and $y \le y_1 \le y + \Delta y$, it follows that $\varepsilon_1 \to 0$ and $\varepsilon_2 \to 0$ as Δx and $\Delta y \to 0$. Thus, we can write

$$\Delta z = \Delta z_1 + \Delta z_2 = [\varepsilon_1 + f_x(x, y)]\Delta x + [\varepsilon_2 + f_y(x, y)]\Delta y$$
$$= [f_x(x, y)\Delta x + f_y(x, y)\Delta y] + \varepsilon_1\Delta x + \varepsilon_2\Delta y$$

Theorem 15.4 tells us that we can choose $(x + \Delta x, y + \Delta y)$ close enough to (x, y) to make $\varepsilon_1\Delta x$ and $\varepsilon_2\Delta y$ insignificant. In other words, for small Δx and Δy, we can use the approximation

$$\Delta z \approx dz$$

This approximation is illustrated graphically in Figure 15.30. Recall that the partial derivatives $\partial z/\partial x$ and $\partial z/\partial y$ can be interpreted as the slope of the surface in the x and y directions, respectively. This means that

$$dz = \frac{\partial z}{\partial x}\Delta x + \frac{\partial z}{\partial y}\Delta y$$

represents the change in height of a plane that is tangent to the surface at the point $(x, y, f(x, y))$. Since a plane in space is represented by a linear equation in the variables x, y, and z, we call the approximation of Δz by dz a **linear approximation.** We will say more about this geometrical interpretation in Section 15.7.

EXAMPLE 2 · *Using a differential as an approximation*

Use the differential dz to approximate the change in $z = \sqrt{4 - x^2 - y^2}$ as (x, y) moves from the point $(1, 1)$ to $(1.01, 0.97)$. Compare this approximation to the exact change in z.

Solution: Letting $(x, y) = (1, 1)$ and $(x + \Delta x, y + \Delta y) = (1.01, 0.97)$, we have

$$dx = \Delta x = 0.01 \qquad \text{and} \qquad dy = \Delta y = -0.03$$

Thus, the change in z can be approximated by

$$\Delta z \approx dz = \frac{\partial z}{\partial x}\,dx + \frac{\partial z}{\partial y}\,dy = \frac{-x}{\sqrt{4 - x^2 - y^2}}\,\Delta x + \frac{-y}{\sqrt{4 - x^2 - y^2}}\,\Delta y$$

When $x = 1$ and and $y = 1$, we have

$$\Delta z \approx -\frac{1}{\sqrt{2}}(0.01) - \frac{1}{\sqrt{2}}(-0.03) = \frac{0.02}{\sqrt{2}} = \sqrt{2}(0.01) \approx 0.0141$$

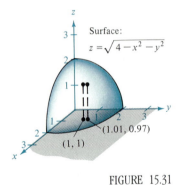

Surface:
$z = \sqrt{4 - x^2 - y^2}$

$(1.01, 0.97)$

$(1, 1)$

FIGURE 15.31

In Figure 15.31 we see that the exact change corresponds to the difference in the heights of two points on the surface of a hemisphere. This difference is given by

$$\Delta z = f(1.01, 0.97) - f(1, 1)$$
$$= \sqrt{4 - (1.01)^2 - (0.97)^2} - \sqrt{4 - 1^2 - 1^2} \approx 0.0137 \qquad \square$$

A function of three variables, $w = f(x, y, z)$, is **differentiable** at (x, y, z) provided $\Delta w = f(x + \Delta x, y + \Delta y, z + \Delta z) - f(x, y, z)$ can be expressed in the form

$$\Delta w = f_x \Delta x + f_y \Delta y + f_z \Delta z + \varepsilon_1 \Delta x + \varepsilon_2 \Delta y + \varepsilon_3 \Delta z$$

where ε_1, ε_2, and $\varepsilon_3 \to 0$ as $(\Delta x, \Delta y, \Delta z) \to (0, 0, 0)$. With this definition of differentiability, Theorem 14.4 has the following extension for functions of three variables. If f is a function of x, y, and z, where f, f_x, f_y, and f_z are continuous in an open region R, then f is differentiable on R.

In Section 4.10, we used differentials to approximate the propagated error introduced by an error in measurement. We demonstrate this application in the following example involving a differential approximation with a function of three variables.

EXAMPLE 3   An application of differentials

The possible error involved in measuring each dimension of a rectangular box is ± 0.1 millimeter. The dimensions of the box are $x = 50$ centimeters, $y = 20$ centimeters, and $z = 15$ centimeters. Use dV to estimate the propagated error and the relative error in the calculated volume of the box.

Solution: The volume of the box is given by

$$V = xyz \qquad \text{Volume of box}$$

and thus

$$dV = \frac{\partial V}{\partial x}\, dx + \frac{\partial V}{\partial y}\, dy + \frac{\partial V}{\partial z}\, dz = yz\, dx + xz\, dy + xy\, dz$$

Now, since 0.1 millimeter $= 0.01$ centimeter, we have $dx = dy = dz = \pm 0.01$, and the propagated error is approximately

$$dV = (20)(15)(\pm 0.01) + (50)(15)(\pm 0.01) + (50)(20)(\pm 0.01)$$
$$= 300(\pm 0.01) + 750(\pm 0.01) + 1000(\pm 0.01)$$
$$= 2050(\pm 0.01) = \pm 20.5 \text{ cm}^3$$

Since the measured volume is

$$V = (50)(20)(15) = 15{,}000 \text{ cm}^3$$

the relative error, $\Delta V/V$, is approximately

$$\frac{\Delta V}{V} \approx \frac{dV}{V} = \frac{20.5}{15{,}000} \approx 0.14\% \qquad \square$$

As is true for a function of a single variable, if a function in two or more variables is differentiable at some point in its domain, then it is also continuous there. This is established in the proof of Theorem 15.5.

THEOREM 15.5   **DIFFERENTIABILITY IMPLIES CONTINUITY**
If a function of x and y is differentiable at (x_0, y_0), then it is continuous at (x_0, y_0).

Proof: Let f be differentiable at (x_0, y_0), where $z = f(x, y)$. Then

$$\Delta z = [f_x(x_0, y_0) + \varepsilon_1]\Delta x + [f_y(x_0, y_0) + \varepsilon_2]\Delta y$$

where both ε_1 and $\varepsilon_2 \to 0$ as $(\Delta x, \Delta y) \to (0, 0)$. However, by definition, we know that Δz is given by

$$\Delta z = f(x_0 + \Delta x, y_0 + \Delta y) - f(x_0, y_0)$$

Thus, letting $x = x_0 + \Delta x$ and $y = y_0 + \Delta y$, we obtain

$$f(x, y) - f(x_0, y_0) = [f_x(x_0, y_0) + \varepsilon_1]\Delta x + [f_y(x_0, y_0) + \varepsilon_2]\Delta y$$
$$= [f_x(x_0, y_0) + \varepsilon_1](x - x_0) + [f_y(x_0, y_0) + \varepsilon_2](y - y_0)$$

Now, taking the limit as $(x, y) \to (x_0, y_0)$, we have

$$\lim_{(x,y)\to(x_0,y_0)} f(x, y) = f(x_0, y_0)$$

which means that f is continuous at (x_0, y_0).

We already pointed out that the existence of f_x and f_y is not sufficient to guarantee differentiability. In the next example we use Theorem 15.5 to validate this claim.

EXAMPLE 4   A function that is not differentiable

Show that $f_x(0, 0)$ and $f_y(0, 0)$ both exist but that f is not differentiable at $(0, 0)$ where f is defined as

$$f(x, y) = \begin{cases} \dfrac{-3xy}{x^2 + y^2}, & \text{if } (x, y) \neq (0, 0) \\ 0, & \text{if } (x, y) = (0, 0) \end{cases}$$

Solution: Using Theorem 15.5 we can show that f is not differentiable at $(0, 0)$ by showing that it is not continuous at this point. To see that f is not continuous at $(0, 0)$, we look at the values of $f(x, y)$ along two different approaches to $(0, 0)$, as shown in Figure 15.32. Along the line $y = x$, the limit is

$$\lim_{(x,x)\to(0,0)} f(x, y) = \lim_{(x,x)\to(0,0)} \frac{-3x^2}{2x^2} = -\frac{3}{2}$$

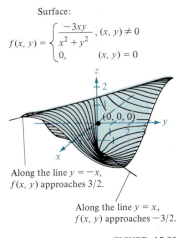

Surface:
$$f(x, y) = \begin{cases} \dfrac{-3xy}{x^2 + y^2}, & (x, y) \neq 0 \\ 0, & (x, y) = 0 \end{cases}$$

Along the line $y = -x$,
$f(x, y)$ approaches $3/2$.

Along the line $y = x$,
$f(x, y)$ approaches $-3/2$.

FIGURE 15.32

whereas along $y = -x$ we have

$$\lim_{(x,-x)\to(0,0)} f(x, y) = \lim_{(x,-x)\to(0,0)} \frac{3x^2}{2x^2} = \frac{3}{2}$$

Thus, the limit of $f(x, y)$ as $(x, y) \to (0, 0)$ does not exist, and we conclude that f is not continuous at $(0, 0)$. Therefore, by Theorem 15.5 we know that f is not differentiable at $(0, 0)$. On the other hand, by the definition of the partial derivatives f_x and f_y, we have

$$f_x(0, 0) = \lim_{\Delta x \to 0} \frac{f(\Delta x, 0) - f(0, 0)}{\Delta x} = \lim_{\Delta x \to 0} \frac{0 - 0}{\Delta x} = 0$$

and

$$f_y(0, 0) = \lim_{\Delta y \to 0} \frac{f(0, \Delta y) - f(0, 0)}{\Delta y} = \lim_{\Delta y \to 0} \frac{0 - 0}{\Delta y} = 0$$

Thus, the partial derivatives at $(0, 0)$ exist. ☐

Section Exercises 15.4

In Exercises 1–10, find the total differential.

1. $z = 3x^2y^3$

2. $z = \dfrac{x^2}{y}$

3. $z = \dfrac{-1}{x^2 + y^2}$

4. $z = e^x \sin y$

5. $z = x \cos y - y \cos x$

6. $z = \dfrac{1}{2}(e^{x^2+y^2} - e^{-x^2-y^2})$

7. $w = 2z^3y \sin x$

8. $w = e^x \cos y + z$

9. $w = \dfrac{x + y}{z - 2y}$

10. $w = x^2yz^2 + \sin yz$

In Exercises 11–16, (a) evaluate $f(1, 2)$ and $f(1.05, 2.1)$ and calculate Δz and (b) use the total differential dz to approximate Δz.

11. $f(x, y) = 9 - x^2 - y^2$

12. $f(x, y) = \sqrt{x^2 + y^2}$

13. $f(x, y) = x \sin y$

14. $f(x, y) = xy$

15. $f(x, y) = 3x - 4y$

16. $f(x, y) = \dfrac{x}{y}$

17. The radius r and height h of a right-circular cylinder are measured with a possible error of 4 percent and 2 percent, respectively. Approximate the maximum possible percentage error in measuring the volume.

18. The centripetal acceleration of a particle moving in a circle is $a = v^2/r$, where v is the velocity and r is the radius of the circle. Approximate the maximum percentage error in measuring the acceleration due to errors of 2 percent in v and 1 percent in r.

19. Electrical power P is given by $P = E^2/R$, where E is voltage and R is resistance. Approximate the maximum percentage error in calculating power if 200 volts is applied to a 4000-ohm resistor and the possible percentage errors in measuring E and R are 2 percent and 3 percent, respectively.

20. A triangle is measured and two adjacent sides are found to be 3 and 4 inches in length, with an included angle of $\pi/4$. The possible errors in measurement are $\frac{1}{16}$ inch in the sides and 0.02 radian in the angle. Approximate the maximum possible error in the computation of the area.

21. The inductance L (in micro-henrys) of a straight nonmagnetic wire in free space is given by

$$L = 0.00021\left(\ln \frac{2h}{r} - 0.75\right)$$

where h is the length of the wire in millimeters and r is the radius of a circular cross section. Approximate L when $r = 2 \pm \frac{1}{16}$ millimeters and $h = 100 \pm \frac{1}{100}$ millimeters.

22. A right-circular cone of height $h = 6$ and radius $r = 3$ is constructed, and in the process errors Δr and Δh are made in the radius and height, respectively. Complete the accompanying table to show the relationship between ΔV and dV for the given errors.

Δr	Δh	dV	ΔV	$\Delta V - dV$
0.1	0.1			
0.1	-0.1			
0.001	0.002			
-0.0001	0.0002			

SECTION TOPICS ▪

Chain Rules for functions
of several variables ▪

Implicit partial differentiation ▪

15.5
The Chain Rule

The work with differentials in the previous section provides the basis for the extension of the Chain Rule to functions of two variables. In this extension, we consider two cases. The first case involves w as a function of x and y, where x and y are functions of a single independent variable t.

THEOREM 15.6

CHAIN RULE: ONE INDEPENDENT VARIABLE

Let $w = f(x, y)$, where f is a differentiable function of x and y. If $x = g(t)$ and $y = h(t)$, where g and h are differentiable functions of t, then w is a differentiable function of t, and

$$\frac{dw}{dt} = \frac{\partial w}{\partial x}\frac{dx}{dt} + \frac{\partial w}{\partial y}\frac{dy}{dt}$$

Proof: Since g and h are differentiable functions of t, we know that both Δx and Δy approach zero as Δt approaches zero. Moreover, since f is a differentiable function of x and y, we know that

$$\Delta w = \frac{\partial w}{\partial x}\Delta x + \frac{\partial w}{\partial y}\Delta y + \varepsilon_1 \Delta x + \varepsilon_2 \Delta y$$

where both ε_1 and $\varepsilon_2 \to 0$ as $(\Delta x, \Delta y) \to (0, 0)$. Thus for $\Delta t \neq 0$, we have

$$\frac{\Delta w}{\Delta t} = \frac{\partial w}{\partial x}\frac{\Delta x}{\Delta t} + \frac{\partial w}{\partial y}\frac{\Delta y}{\Delta t} + \varepsilon_1 \frac{\Delta x}{\Delta t} + \varepsilon_2 \frac{\Delta y}{\Delta t}$$

from which it follows that

$$\frac{dw}{dt} = \lim_{\Delta t \to 0} \frac{\Delta w}{\Delta t} = \frac{\partial w}{\partial x}\frac{dx}{dt} + \frac{\partial w}{\partial y}\frac{dy}{dt} + 0\left(\frac{dx}{dt}\right) + 0\left(\frac{dy}{dt}\right)$$

$$= \frac{\partial w}{\partial x}\frac{dx}{dt} + \frac{\partial w}{\partial y}\frac{dy}{dt}$$

EXAMPLE 1   Using the Chain Rule with one independent variable

Let $w = x^2 y - y^2$, where $x = \sin t$ and $y = e^t$. Find dw/dt when $t = 0$.

Solution: By the Chain Rule for one independent variable, we have

$$\frac{dw}{dt} = \frac{\partial w}{\partial x}\frac{dx}{dt} + \frac{\partial w}{\partial y}\frac{dy}{dt} = 2xy(\cos t) + (x^2 - 2y)e^t$$

When $t = 0$, $x = 0$, and $y = 1$, it follows that

$$\frac{dw}{dt} = 0 - 2 = -2$$

The Chain Rules presented in this section provide alternative solution techniques for many problems in single-variable calculus. For instance, in Example 1, we could have used single-variable techniques to find dw/dt by first writing w as a function of t,

$$w = x^2y - y^2 = (\sin t)^2(e^t) - (e^t)^2 = e^t \sin^2 t - e^{2t}$$

and then differentiating as usual.

The Chain Rule in Theorem 14.6 can be extended to any number of variables. For example, if each x_i is a differentiable function of a single variable t, then for $w = f(x_1, x_2, \ldots, x_n)$ we have

$$\frac{dw}{dt} = \frac{\partial w}{\partial x_1}\frac{dx_1}{dt} + \frac{\partial w}{\partial x_2}\frac{dx_2}{dt} + \cdots + \frac{\partial w}{\partial x_n}\frac{dx_n}{dt}$$

In the next example, we apply this Chain Rule to a function of four variables.

EXAMPLE 2   An application of a Chain Rule to related rates

Two objects are traveling in elliptical paths given by the following parametric equations:

$$x_1 = 4 \cos t \qquad \text{and} \qquad y_1 = 2 \sin t \qquad \text{First object}$$

$$x_2 = 2 \sin 2t \qquad \text{and} \qquad y_2 = 3 \cos 2t \qquad \text{Second object}$$

At what rate is the distance between the two objects changing when $t = \pi$?

Solution: From Figure 15.33 we see that the distance s between the two objects is given by

$$s = \sqrt{(x_2 - x_1)^2 + (y_2 - y_1)^2}$$

and when $t = \pi$, we have $x_1 = -4$, $y_1 = 0$, $x_2 = 0$, $y_2 = 3$, and

$$s = \sqrt{(0 + 4)^2 + (3 - 0)^2} = 5$$

When $t = \pi$, the partial derivatives of s are

$$\frac{\partial s}{\partial x_1} = \frac{-(x_2 - x_1)}{\sqrt{(x_2 - x_1)^2 + (y_2 - y_1)^2}} = -\frac{1}{5}(0 + 4) = -\frac{4}{5}$$

$$\frac{\partial s}{\partial y_1} = \frac{-(y_2 - y_1)}{\sqrt{(x_2 - x_1)^2 + (y_2 - y_1)^2}} = -\frac{1}{5}(3 - 0) = -\frac{3}{5}$$

$$\frac{\partial s}{\partial x_2} = \frac{(x_2 - x_1)}{\sqrt{(x_2 - x_1)^2 + (y_2 - y_1)^2}} = \frac{1}{5}(0 + 4) = \frac{4}{5}$$

$$\frac{\partial s}{\partial y_2} = \frac{(y_2 - y_1)}{\sqrt{(x_2 - x_1)^2 + (y_2 - y_1)^2}} = \frac{1}{5}(3 - 0) = \frac{3}{5}$$

Moreover, when $t = \pi$, the derivatives of x_1, y_1, x_2, and y_2 are

$$\frac{dx_1}{dt} = -4 \sin t = 0 \qquad \frac{dy_1}{dt} = 2 \cos t = -2$$

$$\frac{dx_2}{dt} = 4 \cos 2t = 4 \qquad \frac{dy_2}{dt} = -6 \sin 2t = 0$$

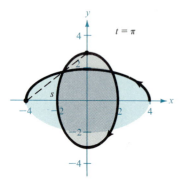

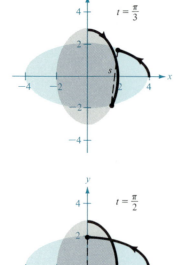

FIGURE 15.33

Therefore, using the appropriate Chain Rule, we know that the distance is changing at the rate of

$$\frac{ds}{dt} = \frac{\partial s}{\partial x_1}\frac{dx_1}{dt} + \frac{\partial s}{\partial y_1}\frac{dy_1}{dt} + \frac{\partial s}{\partial x_2}\frac{dx_2}{dt} + \frac{\partial s}{\partial y_2}\frac{dy_2}{dt}$$

$$= \left(-\frac{4}{5}\right)(0) + \left(-\frac{3}{5}\right)(-2) + \left(\frac{4}{5}\right)(4) + \left(\frac{3}{5}\right)(0)$$

$$= \frac{22}{5}$$

In Example 2, note that s is the function of four *intermediate* variables, x_1, y_1, x_2, and y_2, which in turn are each functions of a single variable t. Another type of composite function is one in which the intermediate variables are themselves functions of more than one variable. For instance, if

$$w = f(x, y)$$

where

$$x = g(s, t) \qquad \text{and} \qquad y = h(s, t)$$

then it follows that w is a function of s and t, and we can consider the partial derivatives of w with respect to s and t. One way to find these partial derivatives is to write w as a function of s and t explicitly by substituting the equations $x = g(s, t)$ and $y = h(s, t)$ into the equation $w = f(x, y)$. Then, we can find the partial derivatives in the usual way, as demonstrated in the next example.

EXAMPLE 3   *Finding partial derivatives by substitution*

Find $\partial w/\partial s$ and $\partial w/\partial t$ for $w = 2xy$, where $x = s^2 + t^2$ and $y = s/t$.

Solution: We begin by substituting $x = s^2 + t^2$ and $y = s/t$ into the equation $w = 2xy$ to obtain

$$w = 2xy = 2(s^2 + t^2)\left(\frac{s}{t}\right) = 2\left(\frac{s^3}{t} + st\right)$$

Then, to find $\partial w/\partial s$ we hold t constant and differentiate with respect to s:

$$\frac{\partial w}{\partial s} = 2\left(\frac{3s^2}{t} + t\right) = \frac{6s^2 + 2t^2}{t}$$

Similarly, to find $\partial w/\partial t$ we hold s constant and differentiate with respect to t to obtain

$$\frac{\partial w}{\partial t} = 2\left(-\frac{s^3}{t^2} + s\right) = 2\left(\frac{-s^3 + st^2}{t^2}\right) = \frac{2st^2 - 2s^3}{t^2}$$

Theorem 15.7 gives an alternative method for finding the partial derivatives in Example 3—without explicitly writing w as a function of s and t.

THEOREM 15.7   CHAIN RULE: TWO INDEPENDENT VARIABLES

Let $w = f(x, y)$, where f is a differentiable function of x and y. If $x = g(s, t)$ and $y = h(s, t)$ such that the first partials $\partial x/\partial s$, $\partial x/\partial t$, $\partial y/\partial s$, and $\partial y/\partial t$ all exist, then $\partial w/\partial s$ and $\partial w/\partial t$ exist and are given by

$$\frac{\partial w}{\partial s} = \frac{\partial w}{\partial x} \frac{\partial x}{\partial s} + \frac{\partial w}{\partial y} \frac{\partial y}{\partial s}$$

and

$$\frac{\partial w}{\partial t} = \frac{\partial w}{\partial x} \frac{\partial x}{\partial t} + \frac{\partial w}{\partial y} \frac{\partial y}{\partial t}$$

Proof: For $\partial w/\partial s$ we hold t constant, which means that g and h are differentiable functions of s. Thus, we can apply Theorem 15.6 to obtain the desired result. Similarly, for $\partial w/\partial t$ we hold s constant, which means that g and h are differentiable functions of t alone, and apply Theorem 15.6.

The "chains" referred to in Theorems 15.6 and 15.7 can be represented schematically as follows.

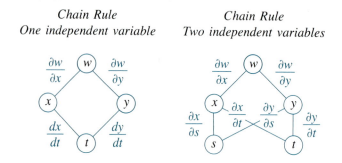

Now, let's use the Chain Rule given in Theorem 15.7 to find the same partial derivatives we found in Example 3.

EXAMPLE 4   Using the Chain Rule with two independent variables

Use the Chain Rule to find $\partial w/\partial s$ and $\partial w/\partial t$ for $w = 2xy$, where $x = s^2 + t^2$ and $y = s/t$.

Solution: If we hold t fixed, then, by Theorem 15.7, we have

$$\frac{\partial w}{\partial s} = \frac{\partial w}{\partial x} \frac{\partial x}{\partial s} + \frac{\partial w}{\partial y} \frac{\partial y}{\partial s} = 2y(2s) + 2x\left(\frac{1}{t}\right)$$

$$= 4\left(\frac{s^2}{t}\right) + \frac{2s^2 + 2t^2}{t} = \frac{6s^2 + 2t^2}{t}$$

Similarly, holding s fixed gives us

$$\frac{\partial w}{\partial t} = \frac{\partial w}{\partial x}\frac{\partial x}{\partial t} + \frac{\partial w}{\partial y}\frac{\partial y}{\partial t} = 2y(2t) + 2x\left(\frac{-s}{t^2}\right)$$

$$= 4s - \frac{2s^3 + 2st^2}{t^2} = \frac{4st^2 - 2s^3 - 2st^2}{t^2}$$

$$= \frac{2st^2 - 2s^3}{t^2}$$

The Chain Rule in Theorem 15.7 can also be extended to any number of variables. For example, if w is a differentiable function of the n variables $x_1, x_2, \ldots, x_n$ where each x_i is a differentiable function of the m variables $t_1, t_2, \ldots, t_m$, then for $w = f(x_1, x_2, \ldots, x_n)$ we have

$$\frac{\partial w}{\partial t_1} = \frac{\partial w}{\partial x_1}\frac{\partial x_1}{\partial t_1} + \frac{\partial w}{\partial x_2}\frac{\partial x_2}{\partial t_1} + \cdots + \frac{\partial w}{\partial x_n}\frac{\partial x_n}{\partial t_1}$$

$$\frac{\partial w}{\partial t_2} = \frac{\partial w}{\partial x_1}\frac{\partial x_1}{\partial t_2} + \frac{\partial w}{\partial x_2}\frac{\partial x_2}{\partial t_2} + \cdots + \frac{\partial w}{\partial x_n}\frac{\partial x_n}{\partial t_2}$$

$$\vdots$$

$$\frac{\partial w}{\partial t_m} = \frac{\partial w}{\partial x_1}\frac{\partial x_1}{\partial t_m} + \frac{\partial w}{\partial x_2}\frac{\partial x_2}{\partial t_m} + \cdots + \frac{\partial w}{\partial x_n}\frac{\partial x_n}{\partial t_m}$$

EXAMPLE 5   *Using the Chain Rule for a function of three variables*

Find $\partial w/\partial s$ and $\partial w/\partial t$ when $s = 1$ and $t = 2\pi$ for the function given by

$$w = xy + yz + xz$$

where $x = s \cos t$, $y = s \sin t$, and $z = t$.

Solution: By extending the result of Theorem 15.7, we have

$$\frac{\partial w}{\partial s} = \frac{\partial w}{\partial x}\frac{\partial x}{\partial s} + \frac{\partial w}{\partial y}\frac{\partial y}{\partial s} + \frac{\partial w}{\partial z}\frac{\partial z}{\partial s}$$

$$= (y + z)(\cos t) + (x + z)(\sin t) + (y + x)(0)$$

When $s = 1$ and $t = 2\pi$, we have $x = 1$, $y = 0$, and $z = 2\pi$. Therefore,

$$\frac{\partial w}{\partial s} = 2\pi(1) + (1 + 2\pi)(0) + 0 = 2\pi$$

Furthermore,

$$\frac{\partial w}{\partial t} = \frac{\partial w}{\partial x}\frac{\partial x}{\partial t} + \frac{\partial w}{\partial y}\frac{\partial y}{\partial t} + \frac{\partial w}{\partial z}\frac{\partial z}{\partial t}$$

$$= (y + z)(-s \sin t) + (x + z)(s \cos t) + (y + x)(1)$$

and for $s = 1$ and $t = 2\pi$ it follows that

$$\frac{\partial w}{\partial t} = (0 + 2\pi)(0) + (1 + 2\pi)(1) + (0 + 1)(1) = 2 + 2\pi$$

Implicit partial differentiation

We conclude this section with an application of the Chain Rule in determining the derivative of a function defined *implicitly*. Suppose that x and y are related by the equation $F(x, y) = 0$, where it is assumed that $y = f(x)$ is a differentiable function of x. To find dy/dx, we could use the techniques discussed in Section 3.6. However, we will see that the Chain Rule provides a convenient alternative. If we consider the function given by

$$w = F(x, y) = F(x, f(x))$$

then we can apply Theorem 14.6 to obtain

$$\frac{dw}{dx} = F_x(x, y)\,\frac{dx}{dx} + F_y(x, y)\,\frac{dy}{dx}$$

Since $w = F(x, y) = 0$ for all x in the domain of f, we know that $dw/dx = 0$ and we have

$$F_x(x, y)\,\frac{dx}{dx} + F_y(x, y)\,\frac{dy}{dx} = 0$$

Now, if $F_y(x, y) \neq 0$, we can conclude that

$$\frac{dy}{dx} = -\,\frac{F_x(x, y)}{F_y(x, y)}$$

A similar procedure can be used to find the partial derivatives of functions of several variables that are defined implicitly. For example, if the equation $F(x, y, z) = 0$ defines z as a function of x and y, then the partial derivatives of z with respect to x and y are given by

$$\frac{\partial z}{\partial x} = -\,\frac{F_x(x, y, z)}{F_z(x, y, z)} \quad \text{and} \quad \frac{\partial z}{\partial y} = -\,\frac{F_y(x, y, z)}{F_z(x, y, z)}$$

provided that $F_z(x, y, z) \neq 0$. We summarize these two results in Theorem 15.8.

THEOREM 15.8   CHAIN RULE: IMPLICIT DIFFERENTIATION
If the equation $F(x, y) = 0$ defines y implicitly as a differentiable function of x, then

$$\frac{dy}{dx} = -\,\frac{F_x(x, y)}{F_y(x, y)}, \quad F_y(x, y) \neq 0$$

If the equation $F(x, y, z) = 0$ defines z implicitly as a differentiable function of x and y, then

$$\frac{\partial z}{\partial x} = -\,\frac{F_x(x, y, z)}{F_z(x, y, z)} \quad \text{and} \quad \frac{\partial z}{\partial y} = -\,\frac{F_y(x, y, z)}{F_z(x, y, z)}, \quad F_z(x, y, z) \neq 0$$

This theorem can be extended to differentiable functions defined implicitly with any number of variables.

EXAMPLE 6   Finding a derivative implicitly

Find dy/dx, given $y^3 + y^2 - 5y - x^2 + 4 = 0$.

Solution: We define a function F by

$$F(x, y) = y^3 + y^2 - 5y - x^2 + 4$$

Then, using Theorem 15.8, we have

$$F_x(x, y) = -2x$$
$$F_y(x, y) = 3y^2 + 2y - 5$$

and it follows that

$$\frac{dy}{dx} = -\frac{F_x(x, y)}{F_y(x, y)} = \frac{-(-2x)}{3y^2 + 2y - 5} = \frac{2x}{3y^2 + 2y - 5} \qquad \square$$

Remark   Compare the solution in Example 6 to that given in Example 2 of Section 3.6.

EXAMPLE 7   Finding partial derivatives implicitly

Find $\partial z/\partial x$ and $\partial z/\partial y$, given $3x^2z - x^2y^2 + 2z^3 + 3yz - 5 = 0$.

Solution: To apply Theorem 15.8 we let $F(x, y, z) = 3x^2z - x^2y^2 + 2z^3 + 3yz - 5$. Then

$$F_x(x, y, z) = 6xz - 2xy^2$$
$$F_y(x, y, z) = -2x^2y + 3z$$
$$F_z(x, y, z) = 3x^2 + 6z^2 + 3y$$

Therefore,

$$\frac{\partial z}{\partial x} = -\frac{F_x}{F_z} = \frac{2xy^2 - 6xz}{3x^2 + 6z^2 + 3y}$$

$$\frac{\partial z}{\partial y} = -\frac{F_y}{F_z} = \frac{2x^2y - 3z}{3x^2 + 6z^2 + 3y} \qquad \square$$

Section Exercises 15.5

In Exercises 1–6, find dw/dt using the appropriate Chain Rule.

1. $w = x^2 + y^2$
   $x = e^t,\ y = e^{-t}$

2. $w = \sqrt{x^2 + y^2}$
   $x = \sin t,\ y = e^t$

3. $w = x \sec y$
   $x = e^t,\ y = \pi - t$

4. $w = \ln \dfrac{y}{x}$
   $x = \cos t,\ y = \sin t$

5. $w = x^2 + y^2 + z^2$
   $x = e^t \cos t,\ y = e^t \sin t,\ z = e^t$

6. $w = xy \cos z$
   $x = t,\ y = t^2,\ z = \arccos t$

In Exercises 7–10, find $\partial w/\partial s$ and $\partial w/\partial t$ using the appropriate Chain Rule, and evaluate each partial derivative at the indicated values of s and t.

Function	Point
7. $w = x^2 + y^2$ $x = s + t,\ y = s - t$	$s = 2,\ t = -1$
8. $w = y^3 - 3x^2y$ $x = e^s,\ y = e^t$	$s = 0,\ t = 1$
9. $w = x^2 - y^2$ $x = s \cos t,\ y = s \sin t$	$s = 3,\ t = \dfrac{\pi}{4}$

10. $w = \sin (2x + 3y)$
$x = s + t, \; y = s - t$ $s = 0, \; t = \dfrac{\pi}{2}$

In Exercises 11–13, find dw/dt (a) by the appropriate Chain Rule and (b) by converting w to a function of t before differentiating.

11. $w = xy$
$x = 2 \sin t, \; y = \cos t$

12. $w = \cos (x - y)$
$x = t^2, \; y = 1$

13. $w = xy + xz + yz$
$x = t - 1, \; y = t^2 - 1, \; z = t$

In Exercises 14–16, find $\partial w/\partial r$ and $\partial w/\partial \theta$ (a) by the appropriate Chain Rule and (b) by converting w to a function of r and θ before differentiating.

14. $w = \sqrt{4 - 2x^2 - 2y^2}$
$x = r \cos \theta, \; y = r \sin \theta$

15. $w = \arctan \dfrac{y}{x}$
$x = r \cos \theta, \; y = r \sin \theta$

16. $w = \dfrac{xy}{z}$
$x = r + \theta, \; y = r - \theta, \; z = \theta^2$

In Exercises 17–20, differentiate implicitly to find the first partial derivatives of z.

17. $x^2 + y^2 + z^2 = 25$ **18.** $xz + yz + xy = 0$
19. $\tan (x + y) + \tan (y + z) = 1$
20. $z = e^x \sin (y + z)$

In Exercises 21 and 22, differentiate implicitly to find the first partial derivatives of w.

21. $xyz + xzw - yzw + w^2 = 5$
22. $x^2 + y^2 + z^2 + 6xw - 8w^2 = 5$

In Exercises 23 and 24, differentiate implicitly to find all first and second partial derivatives of z.

23. $x^2 + 2yz + z^2 = 1$
24. $x + \sin (y + z) = 0$

25. The dimensions of a rectangular chamber are increasing at the following rates: length 3 feet per minute, width 2 feet per minute, and depth $\frac{1}{2}$ foot per minute. What is the rate of change of the volume and the surface area when the length, width, and depth are 10 feet, 6 feet, and 4 feet, respectively?

26. The radius of a right-circular cylinder is increasing at the rate of 6 inches per minute, and the height is decreasing at the rate of 4 inches per minute. What is the rate of change of the volume and surface area when the radius is 12 inches and the height is 36 inches?

27. Repeat Exercise 26 for a right-circular cone.

28. The two radii of the frustrum of a right-circular cone are increasing at the rate of 4 centimeters per minute, and the height is increasing at the rate of 12 centimeters per minute, as shown in Figure 15.34. Find the rate at which the volume and surface area are changing when the two radii are 15 centimeters and 25 centimeters and the height is 10 centimeters.

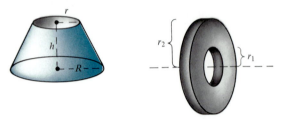

FIGURE 15.34   FIGURE 15.35

29. An annular cylinder has an inside radius of r_1 and an outside radius of r_2, as shown in Figure 15.35. The moment of inertia is given by

$$I = \frac{1}{2} m(r_1^2 + r_2^2)$$

where m is the mass. Find the rate at which I is changing at the instant the radii are 6 centimeters and 8 centimeters, respectively, if the two radii are increasing at the rate of 2 centimeters per second.

30. The Ideal Gas Law is $pV = RT$, where R is a constant. If p and V are functions of time, find dT/dt, the rate at which the temperature changes with respect to time.

In Exercises 31–35, the given function is **homogeneous of degree n.** That is,

$$f(tx, ty) = t^n f(x, y)$$

Find the degree of the given function, and show that

$$x f_x(x, y) + y f_y(x, y) = n f(x, y)$$

31. $f(x, y) = x^3 - 3xy^2 + y^3$

32. $f(x, y) = \dfrac{xy}{\sqrt{x^2 + y^2}}$

33. $f(x, y) = e^{x/y}$

34. $f(x, y) = 2x^3 - 3xy^2$

35. $f(x, y) = \dfrac{x^2}{\sqrt{x^2 + y^2}}$

36. Show that if $f(x, y)$ is homogeneous of degree n, then

$$x f_x(x, y) + y f_y(x, y) = n f(x, y)$$

[Hint: Let $g(t) = f(tx, ty) = t^n f(x, y)$. Find $g'(t)$ and then let $t = 1$.]

37. Show that

$$\frac{\partial w}{\partial u} + \frac{\partial w}{\partial v} = 0$$

for $w = f(x, y)$, $x = u - v$, and $y = v - u$.

38. Demonstrate the result of Exercise 37 for the function given by

$$w = (x - y) \sin (y - x)$$

39. Show that

$$\left(\frac{\partial w}{\partial x}\right)^2 + \left(\frac{\partial w}{\partial y}\right)^2 = \left(\frac{\partial w}{\partial r}\right)^2 + \left(\frac{1}{r^2}\right)\left(\frac{\partial w}{\partial \theta}\right)^2$$

for $w = f(x, y)$, $x = r \cos \theta$, and $y = r \sin \theta$.

40. Demonstrate the result of Exercise 39 for the function given by

$$w = \arctan \frac{y}{x}$$

15.6
Directional derivatives and gradients

Suppose you were standing on the hillside pictured in Figure 15.36 and wanted to determine the hill's incline toward the z-axis. If the hill were represented by $z = f(x, y)$, then you would already know how to determine the slope in two different directions—the slope in the y direction would be given by the partial derivative $f_y(x, y)$, and the slope in the x direction would be given by the partial derivative $f_x(x, y)$. In this section, you will see that these two partial derivatives can be used to find the slope in *any* direction.

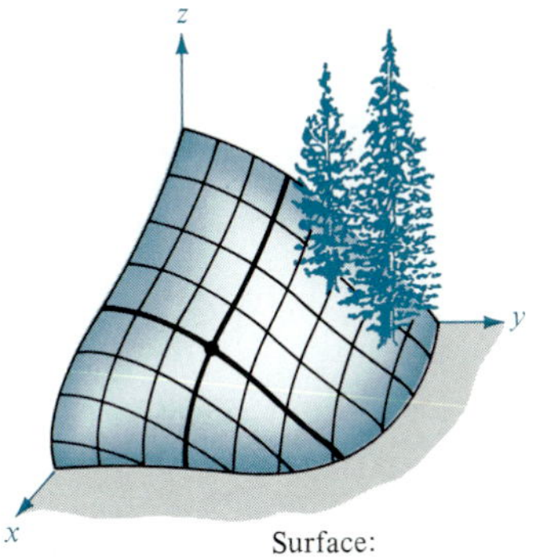

To determine the slope at a point on a surface, we define a new type of derivative called a **directional derivative.** We begin by letting $z = f(x, y)$ be a *surface* and $P = (x_0, y_0)$ a *point* in the domain of f, as shown in Figure 15.37. We specify *direction* by a unit vector $\mathbf{u} = \cos \theta\, \mathbf{i} + \sin \theta\, \mathbf{j}$, where θ is the angle the vector makes with the positive x-axis. Now, to find the desired slope, we reduce the problem to two dimensions by intersecting the surface with a vertical plane passing through the point P and parallel to $\mathbf{u}$, as shown in Figure 15.38. This vertical plane intersects the surface to form a curve C, and we define the slope of the surface at $(x_0, y_0, f(x_0, y_0))$ to be the slope of the curve C at that point.

Surface:
$z = f(x, y)$

FIGURE 15.36

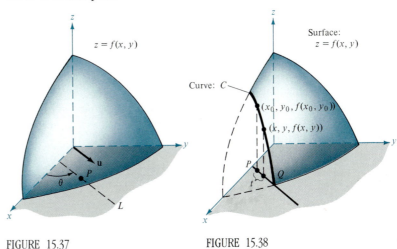

FIGURE 15.37   FIGURE 15.38

We can write the slope of the curve C as a limit that looks much like those used in single-variable calculus. The vertical plane used to form C intersects the xy-plane in a line L, represented by the parametric equations

$$x = x_0 + t \cos \theta \qquad \text{and} \qquad y = y_0 + t \sin \theta$$

so for any value of t, the point $Q = (x, y)$ lies on the line L. For each of the points P and Q, there is a corresponding point on the surface

$$(x_0, y_0, f(x_0, y_0)) \qquad \text{Point above } P$$

and

$$(x, y, f(x, y)) \qquad \text{Point above } Q$$

Moreover, since the distance between P and Q is

$$\sqrt{(x - x_0)^2 + (y - y_0)^2} = \sqrt{(t \cos \theta)^2 + (t \sin \theta)^2} = t$$

we can write the slope of the secant line through $(x_0, y_0, f(x_0, y_0))$ and $(x, y, f(x, y))$ as

$$\frac{f(x, y) - f(x_0, y_0)}{t} = \frac{f(x_0 + t \cos \theta, y_0 + t \sin \theta) - f(x_0, y_0)}{t}$$

Finally, by letting t approach zero, we arrive at the following definition.

DEFINITION OF DIRECTIONAL DERIVATIVE

Let f be a function of two variables x and y and let $\mathbf{u} = \cos \theta \, \mathbf{i} + \sin \theta \, \mathbf{j}$ be a unit vector. Then the **directional derivative of f in the direction of u,** denoted by $D_{\mathbf{u}} f$, is

$$D_{\mathbf{u}} f(x, y) = \lim_{t \to 0} \frac{f(x + t \cos \theta, y + t \sin \theta) - f(x, y)}{t}$$

Calculating directional derivatives by this definition is comparable to finding the derivative of a function of one variable by the four-step process. A simpler "working" formula for finding directional derivatives involves the partial derivatives f_x and f_y.

THEOREM 15.9

DIRECTIONAL DERIVATIVE

If f is a differentiable function of x and y, then the directional derivative of f in the direction of the unit vector $\mathbf{u} = \cos \theta \, \mathbf{i} + \sin \theta \, \mathbf{j}$ is

$$D_{\mathbf{u}} f(x, y) = f_x(x, y) \cos \theta + f_y(x, y) \sin \theta$$

Proof: For a fixed point (x_0, y_0), we let $x = x_0 + t \cos \theta$ and $y = y_0 + t \sin \theta$. Then we let $g(t) = f(x, y)$. Since f is differentiable, we can apply the Chain Rule given in Theorem 15.7 to obtain

$$g'(t) = f_x(x, y) \frac{\partial x}{\partial t} + f_y(x, y) \frac{\partial y}{\partial t}$$

$$= f_x(x, y) \cos \theta + f_y(x, y) \sin \theta$$

If $t = 0$, then $x = x_0$ and $y = y_0$ so

$$g'(0) = f_x(x_0, y_0) \cos \theta + f_y(x_0, y_0) \sin \theta$$

By definition, it is also true that

$$g'(0) = \lim_{t \to 0} \frac{g(t) - g(0)}{t}$$

$$= \lim_{t \to 0} \frac{f(x_0 + t \cos \theta, y_0 + t \sin \theta) - f(x_0, y_0)}{t}$$

Consequently, $D_\mathbf{u} f(x_0, y_0) = f_x(x_0, y_0) \cos \theta + f_y(x_0, y_0) \sin \theta$.

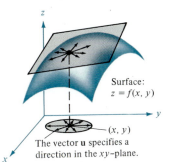

Surface:
$z = f(x, y)$

(x, y)

The vector **u** specifies a
direction in the *xy*-plane.

FIGURE 15.39

Be sure that you see that there are infinitely many directional derivatives to a surface at a given point—one for each direction specified by **u,** as indicated in Figure 15.39. Two of these turn out to be the partial derivatives f_x and f_y.

1. Direction of positive *x*-axis ($\theta = 0$): $\mathbf{u} = \cos 0 \ \mathbf{i} + \sin 0 \ \mathbf{j} = \mathbf{i}$

    $$D_\mathbf{i} f(x, y) = f_x(x, y) \cos 0 + f_y(x, y) \sin 0 = f_x(x, y)$$

2. Direction of positive *y*-axis $\left(\theta = \dfrac{\pi}{2}\right)$: $\mathbf{u} = \cos \dfrac{\pi}{2} \mathbf{i} + \sin \dfrac{\pi}{2} \mathbf{j} = \mathbf{j}$

    $$D_\mathbf{j} f(x, y) = f_x(x, y) \cos \frac{\pi}{2} + f_y(x, y) \sin \frac{\pi}{2} = f_y(x, y)$$

EXAMPLE 1   *Finding a directional derivative*

Find the directional derivative of $f(x, y) = 4 - x^2 - \frac{1}{4}y^2$ at $(1, 2)$ in the direction of $\mathbf{u} = \cos (\pi/3) \ \mathbf{i} + \sin (\pi/3) \ \mathbf{j}$.

Solution:

$$D_\mathbf{u} f(x, y) = f_x(x, y) \cos \theta + f_y(x, y) \sin \theta$$

$$= (-2x) \cos \theta + \left(-\frac{y}{2}\right) \sin \theta$$

Evaluating at $\theta = \pi/3$, $x = 1$, and $y = 2$, we have

$$D_\mathbf{u} f(1, 2) = (-2)\left(\frac{1}{2}\right) + (-1)\left(\frac{\sqrt{3}}{2}\right) = -1 - \frac{\sqrt{3}}{2} \approx -1.866 \quad \square$$

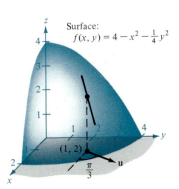

Surface:
$f(x, y) = 4 - x^2 - \frac{1}{4}y^2$

$(1, 2)$

$\dfrac{\pi}{3}$ **u**

FIGURE 15.40

| **Remark** Note in Figure 15.40 that we can interpret the directional derivative as giving the slope of the surface at the point $(1, 2, 2)$ in the direction of the unit vector **u**.

We have been specifying direction by a unit vector **u**. If the direction is given by a vector whose length is not 1, then we must normalize the vector before applying the formula in Theorem 15.9.

EXAMPLE 2   *Finding a directional derivative*

Find the directional derivative of $f(x, y) = x^2 \sin 2y$ at $(1, \pi/4)$ in the direction of $\mathbf{v} = 3\mathbf{i} - 4\mathbf{j}$.

Solution: We begin by finding a unit vector in the direction of $\mathbf{v}$:

$$\mathbf{u} = \frac{\mathbf{v}}{\|\mathbf{v}\|} = \frac{3}{5}\mathbf{i} - \frac{4}{5}\mathbf{j} = \cos\theta\,\mathbf{i} + \sin\theta\,\mathbf{j}$$

Now, using this unit vector, we have

$$D_{\mathbf{u}}f(x, y) = (2x \sin 2y)(\cos\theta) + (2x^2 \cos 2y)(\sin\theta)$$

$$D_{\mathbf{u}}f\left(1, \frac{\pi}{4}\right) = \left(2 \sin \frac{\pi}{2}\right)\left(\frac{3}{5}\right) + \left(2 \cos \frac{\pi}{2}\right)\left(-\frac{4}{5}\right)$$

$$= (2)\left(\frac{3}{5}\right) + (0)\left(-\frac{4}{5}\right) = \frac{6}{5}$$

The directional derivative $D_{\mathbf{u}}f(x, y)$ can be expressed as the dot product of the unit vector

$$\mathbf{u} = \cos\theta\,\mathbf{i} + \sin\theta\,\mathbf{j}$$

and the vector

$$f_x(x, y)\,\mathbf{i} + f_y(x, y)\,\mathbf{j}$$

This vector is important and has a variety of uses. We call it the **gradient of** f.

DEFINITION OF GRADIENT

If $z = f(x, y)$, then the **gradient** of f, denoted by $\nabla f(x, y)$, is the vector

$$\nabla f(x, y) = f_x(x, y)\,\mathbf{i} + f_y(x, y)\,\mathbf{j}$$

We read ∇f as "del f."

Remark We do not assign a value to the symbol ∇ by itself. It is an operator in the same sense that d/dx is an operator. When ∇ operates on $f(x, y)$ it produces the vector $\nabla f(x, y)$. [Another notation for the gradient is **grad** $f(x, y)$.]

Since the gradient of f is a vector, we can write the directional derivative of f in the direction of $\mathbf{u}$ as

$$D_{\mathbf{u}}f(x, y) = [f_x(x, y)\,\mathbf{i} + f_y(x, y)\,\mathbf{j}] \cdot [\cos\theta\,\mathbf{i} + \sin\theta\,\mathbf{j}]$$

In other words, the directional derivative is the dot product of the gradient and the direction vector. This convenient formula is worth memorizing:

$$D_{\mathbf{u}}f(x, y) = \nabla f(x, y) \cdot \mathbf{u}$$

EXAMPLE 3   *Using $\nabla f(x, y)$ to find a directional derivative*

Find the directional derivative of $f(x, y) = 3x^2 - 2y^2$, at $(-1, 3)$ in the direction from $P = (-1, 3)$ to $Q = (1, -2)$.

Solution: A vector in the specified direction is

$$\overrightarrow{PQ} = \mathbf{v} = (1 + 1)\,\mathbf{i} + (-2 - 3)\,\mathbf{j} = 2\mathbf{i} - 5\mathbf{j}$$

and a unit vector in this direction is

$$\mathbf{u} = \frac{\mathbf{v}}{\|\mathbf{v}\|} = \frac{2}{\sqrt{29}}\,\mathbf{i} - \frac{5}{\sqrt{29}}\,\mathbf{j}$$

Since $\nabla f(x, y) = f_x(x, y)\,\mathbf{i} + f_y(x, y)\,\mathbf{j} = 6x\mathbf{i} - 4y\mathbf{j}$, the gradient at $(-1, 3)$ is

$$\nabla f(-1, 3) = -6\mathbf{i} - 12\mathbf{j}$$

Consequently, at $(-1, 3)$ the directional derivative is

$$D_{\mathbf{u}} f(-1, 3) = \nabla f(x, y) \cdot \mathbf{u} = \frac{-12}{\sqrt{29}} + \frac{60}{\sqrt{29}} = \frac{48}{\sqrt{29}}$$

We have already seen that there are many directional derivatives at the point (x, y) on a surface. In many applications we would like to know in which direction to move so that $f(x, y)$ increases most rapidly. We call this the direction of steepest ascent, and it is given by the gradient, as stated in the following theorem.

THEOREM 15.10   **PROPERTIES OF THE GRADIENT**
Let f be differentiable at the point (x, y).

1. If $\nabla f(x, y) = \mathbf{0}$, then $D_{\mathbf{u}} f(x, y) = 0$ for all $\mathbf{u}$.
2. The direction of *maximum* increase of f is given by $\nabla f(x, y)$. The maximum value of $D_{\mathbf{u}} f(x, y)$ is $\|\nabla f(x, y)\|$.
3. The direction of *minimum* increase of f is given by $-\nabla f(x, y)$. The minimum value of $D_{\mathbf{u}} f(x, y)$ is $-\|\nabla f(x, y)\|$.

Proof: If $\nabla f(x, y) = \mathbf{0}$, then for any direction (any $\mathbf{u}$), we have

$$D_{\mathbf{u}} f(x, y) = \nabla f(x, y) \cdot \mathbf{u} = (0\mathbf{i} + 0\mathbf{j}) \cdot (\cos \theta\,\mathbf{i} + \sin \theta\,\mathbf{j}) = 0$$

If $\nabla f(x, y) \neq \mathbf{0}$, then let ϕ be the angle between $\nabla f(x, y)$ and a unit vector $\mathbf{u}$. Using the dot product, we can apply Theorem 13.6 to conclude that

$$D_{\mathbf{u}} f(x, y) = \nabla f(x, y) \cdot \mathbf{u} = \|\nabla f(x, y)\|\,\|\mathbf{u}\|\,\cos \phi = \|\nabla f(x, y)\|\,\cos \phi$$

and it follows that the maximum value $D_{\mathbf{u}} f(x, y)$ will occur when $\cos \phi = 1$. Thus, $\phi = 0$, and the maximum value for the directional derivative occurs when $\mathbf{u}$ has the same direction as $\nabla f(x, y)$. Moreover, this largest value for $D_{\mathbf{u}} f(x, y)$ is precisely

$$\|\nabla f(x, y)\|\,\cos \phi = \|\nabla f(x, y)\|$$

Similarly, the minimum value of $D_{\mathbf{u}} f(x, y)$ can be obtained by letting $\phi = \pi$ so that $\mathbf{u}$ points in the direction opposite that of $\nabla f(x, y)$, as indicated in Figure 15.41.

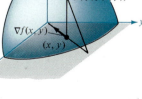

FIGURE 15.41

To visualize one of the properties of the gradient, consider a skier coming down a mountainside. If $f(x, y)$ denotes the altitude of the skier, then $-\nabla f(x, y)$ indicates the *compass direction* the skier should take to ski the path of steepest descent. (Remember that the gradient indicates direction in the xy-plane and does not itself point up or down the mountainside.)

As another illustration of the gradient, consider the temperature $T(x, y)$ at any point (x, y) on a flat metal plate. In this case, $\nabla T(x, y)$ gives the direction of greatest temperature increase at point (x, y), as illustrated in the next example.

EXAMPLE 4   *Finding the direction of maximum increase*

The temperature in degrees Celsius on the surface of a metal plate is given by

$$T(x, y) = 20 - 4x^2 - y^2$$

where x and y are measured in inches. In what direction from $(2, -3)$ does the temperature increase most rapidly? What is this rate of increase?

Solution: The gradient is

$$\nabla T(x, y) = T_x(x, y)\, \mathbf{i} + T_y(x, y)\, \mathbf{j} = -8x\mathbf{i} - 2y\mathbf{j}$$

It follows that the direction of maximum increase is given by

$$\nabla T(2, -3) = -16\mathbf{i} + 6\mathbf{j}$$

as shown in Figure 15.42, and the rate of increase is

$$\|\nabla T(2, -3)\| = \sqrt{256 + 36} = \sqrt{292} \approx 17.09° \text{ per inch}$$

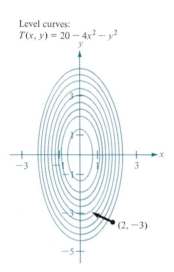

Level curves:
$T(x, y) = 20 - 4x^2 - y^2$

(2, −3)

Direction of Most Rapid
Increase in Temperature
at $(2, -3)$

FIGURE 15.42

The solution presented in Example 4 can be misleading. Although the gradient points in the direction of maximum temperature increase, it does not point toward the hottest spot on the plate. In other words, the gradient provides a local solution to finding an increase relative to the temperature at the point $(2, -3)$. *Once we leave that position, the direction of maximum increase may change.*

EXAMPLE 5   *Finding the path of a heat-seeking particle*

A heat-seeking particle is located at the point $(2, -3)$ on a metal plate whose temperature at (x, y) is $T(x, y) = 20 - 4x^2 - y^2$. Find the path of the particle as it continuously moves in the direction of maximum temperature increase.

Solution: We let the path be represented by the position function

$$\mathbf{r}(t) = x(t)\, \mathbf{i} + y(t)\, \mathbf{j}$$

A tangent vector at each point $(x(t), y(t))$ is given by

$$\mathbf{r}'(t) = \frac{dx}{dt}\mathbf{i} + \frac{dy}{dt}\mathbf{j}$$

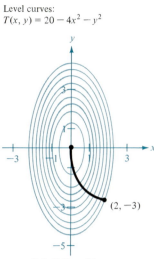

Level curves:
$T(x, y) = 20 - 4x^2 - y^2$

$(2, -3)$

Path Followed by
Heat–Seeking Particle

FIGURE 15.43

Since the particle seeks maximum temperature increase, the directions of $\mathbf{r}'(t)$ and $\nabla T(x, y) = -8x\mathbf{i} - 2y\mathbf{j}$ are the same at each point of the path. Thus,

$$-8x = \frac{dx}{dt} \quad \text{and} \quad -2y = \frac{dy}{dt}$$

These differential equations represent exponential growth (see Theorem 7.18), and the solutions are

$$x(t) = C_1 e^{-8t} \quad \text{and} \quad y(t) = C_2 e^{-2t}$$

Now, since the particle starts at $(2, -3)$, it follows that $2 = x(0) = C_1$ and $-3 = y(0) = C_2$. Thus, the path is represented by

$$\mathbf{r}(t) = x(t)\,\mathbf{i} + y(t)\,\mathbf{j} = 2e^{-8t}\mathbf{i} - 3e^{-2t}\mathbf{j}$$

Eliminating the parameter t, we obtain

$$x = 2e^{-8t} = \frac{2}{81}(-3e^{-2t})^4 = \frac{2}{81}y^4 \implies x = \frac{2}{81}y^4$$

The path is shown in Figure 15.43.

In Figure 15.43, the path of the particle (determined by the gradient at each point) appears to be orthogonal to each of the level curves. This becomes clear when we consider the fact that the temperature $T(x, y)$ is constant along a given level curve. Hence, at any point (x, y) on the curve the rate of change of T in the direction of a unit tangent vector $\mathbf{u}$ is 0, and we can write

$$\nabla f(x, y) \cdot \mathbf{u} = D_{\mathbf{u}}T(x, y) = 0 \qquad \text{\small \textbf{u} is a unit tangent vector}$$

Since the dot product of $\nabla f(x, y)$ and $\mathbf{u}$ is zero, we conclude that they must be orthogonal. This result is stated in the following theorem.

THEOREM 15.11   **GRADIENT IS NORMAL TO LEVEL CURVES**
If f is differentiable at (x_0, y_0), and $\nabla f(x_0, y_0) \neq \mathbf{0}$, then $\nabla f(x_0, y_0)$ is normal to the level curve through (x_0, y_0).

EXAMPLE 6   Finding a normal vector to a level curve

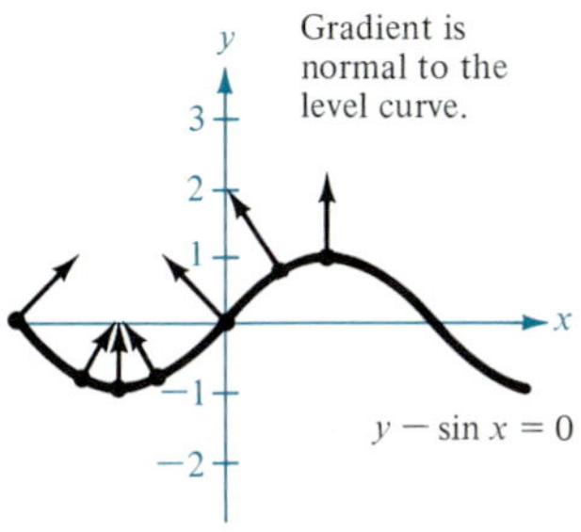

Gradient is
normal to the
level curve.

$y - \sin x = 0$

FIGURE 15.44

Sketch the level curve corresponding to $c = 0$ for the function given by

$$f(x, y) = y - \sin x$$

and find a normal vector at several points on the curve.

Solution: The level curve for $c = 0$ is given by

$$0 = y - \sin x \implies y = \sin x$$

as shown in Figure 15.44. Since the gradient vector of f at (x, y) is

$$\nabla f(x, y) = f_x(x, y)\,\mathbf{i} + f_y(x, y)\,\mathbf{j} = -\cos x\,\mathbf{i} + \mathbf{j}$$

we can use Theorem 15.11 to conclude that $\nabla f(x, y)$ is normal to the level curve at the point (x, y). Some gradient vectors are

$$\nabla f(-\pi, 0) = \mathbf{i} + \mathbf{j}, \qquad \nabla f\left(-\frac{2\pi}{3}, -\frac{\sqrt{3}}{2}\right) = \frac{1}{2}\mathbf{i} + \mathbf{j},$$

$$\nabla f\left(-\frac{\pi}{2}, -1\right) = \mathbf{j}$$

Several others are shown in Figure 15.44.

Functions of three variables

The definitions of the directional derivative and the gradient can be naturally extended to functions of three or more variables. As often happens, some of the geometrical interpretation is lost in the generalization from functions of two variables to three variables. For example, we do not interpret the directional derivative of a function of three variables to represent slope.

We list the definitions and properties of the directional derivative and the gradient of a function of three variables in the following summary.

DIRECTIONAL DERIVATIVE AND GRADIENT FOR A FUNCTION OF THREE VARIABLES

Let f be a function of x, y, and z, with continuous first partial derivatives. The **directional derivative of f** in the direction of a unit vector $\mathbf{u} = \cos \alpha\, \mathbf{i} + \cos \beta\, \mathbf{j} + \cos \gamma\, \mathbf{k}$ is given by

$$D_{\mathbf{u}} f(x, y, z) = f_x(x, y, z) \cos \alpha + f_y(x, y, z) \cos \beta + f_z(x, y, z) \cos \gamma$$

The **gradient of f** is defined to be

$$\nabla f(x, y, z) = f_x(x, y, z)\, \mathbf{i} + f_y(x, y, z)\, \mathbf{j} + f_z(x, y, z)\, \mathbf{k}$$

Properties of the gradient are as follows:

1. $D_{\mathbf{u}} f(x, y, z) = \nabla f(x, y, z) \cdot \mathbf{u}$
2. If $\nabla f(x, y, z) = \mathbf{0}$, then $D_{\mathbf{u}} f(x, y, z) = 0$ for all $\mathbf{u}$.
3. The direction of *maximum* increase of f is given by $\nabla f(x, y, z)$. The maximum value of $D_{\mathbf{u}} f(x, y, z)$ is $\|\nabla f(x, y, z)\|$.
4. The direction of *minimum* increase of f is given by $-\nabla f(x, y, z)$. The minimum value of $D_{\mathbf{u}} f(x, y, z)$ is $-\|\nabla f(x, y, z)\|$.

EXAMPLE 7   *Finding the gradient for a function of three variables*

Find $\nabla f(x, y, z)$ for the function given by

$$f(x, y, z) = x^2 + y^2 - 4z$$

and find the direction of maximum increase of f at the point $(2, -1, 1)$.

Solution: The gradient vector is given by

$$\nabla f(x, y, z) = f_x(x, y, z)\, \mathbf{i} + f_y(x, y, z)\, \mathbf{j} + f_z(x, y, z)\, \mathbf{k}$$
$$= 2x\mathbf{i} + 2y\mathbf{j} - 4\mathbf{k}$$

Hence, it follows that the direction of maximum increase at $(2, -1, 1)$ is

$$\nabla f(2, -1, 1) = 4\mathbf{i} - 2\mathbf{j} - 4\mathbf{k}$$

Section Exercises 15.6

In Exercises 1–12, find the directional derivative of the function at P in the direction of $\mathbf{v}$.

1. $f(x, y) = 3x - 4xy + 5y$,

$P = (1, 2)$, $\mathbf{v} = \dfrac{1}{2}(\mathbf{i} + \sqrt{3}\mathbf{j})$

2. $f(x, y) = x^2 - y^2$,

$P = (4, 3)$, $\mathbf{v} = \dfrac{\sqrt{2}}{2}(\mathbf{i} + \mathbf{j})$

3. $f(x, y) = xy$,
$P = (2, 3)$, $\mathbf{v} = \mathbf{i} + \mathbf{j}$

4. $f(x, y) = \dfrac{x}{y}$,
$P = (1, 1)$, $\mathbf{v} = -\mathbf{j}$

5. $g(x, y) = \sqrt{x^2 + y^2}$,
$P = (3, 4)$, $\mathbf{v} = 3\mathbf{i} - 4\mathbf{j}$

6. $g(x, y) = \arcsin xy$,
$P = (1, 0)$, $\mathbf{v} = \mathbf{i} + 5\mathbf{j}$

7. $h(x, y) = e^x \sin y$,
$P = \left(1, \dfrac{\pi}{2}\right)$, $\mathbf{v} = -\mathbf{i}$

8. $h(x, y) = e^{-(x^2+y^2)}$,
$P = (0, 0)$, $\mathbf{v} = \mathbf{i} + \mathbf{j}$

9. $f(x, y, z) = xy + yz + xz$,
$P = (1, 1, 1)$, $\mathbf{v} = 2\mathbf{i} + \mathbf{j} - \mathbf{k}$

10. $f(x, y, z) = x^2 + y^2 + z^2$,
$P = (1, 2, -1)$, $\mathbf{v} = \mathbf{i} - 2\mathbf{j} + 3\mathbf{k}$

11. $h(x, y, z) = x \arctan yz$,
$P = (4, 1, 1)$, $\mathbf{v} = \langle 1, 2, -1 \rangle$

12. $h(x, y, z) = xyz$,
$P = (2, 1, 1)$, $\mathbf{v} = \langle 2, 1, 2 \rangle$

In Exercises 13–16, find the directional derivative of the function in the direction $\mathbf{u} = \cos \theta \, \mathbf{i} + \sin \theta \, \mathbf{j}$.

13. $f(x, y) = x^2 + y^2$, $\theta = \dfrac{\pi}{4}$

14. $f(x, y) = \dfrac{y}{x + y}$, $\theta = -\dfrac{\pi}{6}$

15. $f(x, y) = \sin (2x - y)$, $\theta = -\dfrac{\pi}{3}$

16. $g(x, y) = xe^y$, $\theta = \dfrac{2\pi}{3}$

In Exercises 17–20, find the directional derivative of the given function at the point P in the direction of Q.

17. $f(x, y) = x^2 + 4y^2$,
$P = (3, 1)$, $Q = (1, -1)$

18. $f(x, y) = \cos (x + y)$,
$P = (0, \pi)$, $Q = \left(\dfrac{\pi}{2}, 0\right)$

19. $h(x, y, z) = \ln (x + y + z)$,
$P = (1, 0, 0)$, $Q = (4, 3, 1)$

20. $g(x, y, z) = xye^z$,
$P = (2, 4, 0)$, $Q = (0, 0, 0)$

In Exercises 21–30, find the gradient of the function and the maximum value of the directional derivative at the indicated point.

Function	Point
21. $f(x, y) = x^2 - 3xy + y^2$	$(4, 2)$
22. $f(x, y) = y\sqrt{x}$	$(4, 2)$
23. $h(x, y) = x \tan y$	$\left(2, \dfrac{\pi}{4}\right)$
24. $h(x, y) = y \cos (x - y)$	$\left(0, \dfrac{\pi}{3}\right)$
25. $g(x, y) = \ln \sqrt[3]{x^2 + y^2}$	$(1, 2)$
26. $g(x, y) = ye^{-x^2}$	$(0, 5)$
27. $f(x, y, z) = \sqrt{x^2 + y^2 + z^2}$	$(1, 4, 2)$
28. $f(x, y, z) = xe^{yz}$	$(2, 0, -4)$
29. $w = \dfrac{1}{\sqrt{1 - x^2 - y^2 - z^2}}$	$(0, 0, 0)$
30. $w = xy^2z^2$	$(2, 1, 1)$

In Exercises 31–38, use the function given by

$$f(x, y) = 3 - \dfrac{x}{3} - \dfrac{y}{2}$$

31. Sketch the graph of f in the first octant and plot the point $(3, 2, 1)$.

32. Find $D_{\mathbf{u}}f(3, 2)$ where $\mathbf{u} = \cos \theta \, \mathbf{i} + \sin \theta \, \mathbf{j}$ and
(a) $\theta = \dfrac{\pi}{4}$ (b) $\theta = \dfrac{2\pi}{3}$

33. Find $D_{\mathbf{u}}f(3, 2)$ where $\mathbf{u} = \cos \theta \, \mathbf{i} + \sin \theta \, \mathbf{j}$ and
(a) $\theta = \dfrac{4\pi}{3}$ (b) $\theta = -\dfrac{\pi}{6}$

34. Find $D_{\mathbf{u}}f(3, 2)$ where $\mathbf{u} = \mathbf{v}/\|\mathbf{v}\|$ and
(a) $\mathbf{v} = \mathbf{i} + \mathbf{j}$ (b) $\mathbf{v} = -3\mathbf{i} - 4\mathbf{j}$

35. Find $D_{\mathbf{u}}f(3, 2)$ where $\mathbf{u} = \mathbf{v}/\|\mathbf{v}\|$ and
(a) $\mathbf{v}$ is the vector from $(1, 2)$ to $(-2, 6)$
(b) $\mathbf{v}$ is the vector from $(3, 2)$ to $(4, 5)$

36. Find $\nabla f(x, y)$.

37. Find the maximum value of the directional derivative at $(3, 2)$.

38. Find a unit vector **u** orthogonal to $\nabla f(3, 2)$ and calculate $D_{\mathbf{u}} f(3, 2)$. Discuss the geometric meaning of the result.

In Exercises 39–42, use function given by

$$f(x, y) = 9 - x^2 - y^2$$

39. Sketch the graph of f in the first octant and plot the point $(1, 2, 4)$ on the surface.

40. Find $D_{\mathbf{u}} f(1, 2)$ where $\mathbf{u} = \cos \theta \, \mathbf{i} + \sin \theta \, \mathbf{j}$ and

(a) $\theta = -\dfrac{\pi}{4}$ (b) $\theta = \dfrac{\pi}{3}$

41. Find $\nabla f(1, 2)$ and $\|\nabla f(1, 2)\|$.

42. Find a unit vector **u** orthogonal to $\nabla f(1, 2)$, and calculate $D_{\mathbf{u}} f(1, 2)$. Discuss the geometric meaning of the result.

In Exercises 43–46, find a normal vector to the level curve $f(x, y) = c$ at P.

43. $f(x, y) = x^2 + y^2$,   **44.** $f(x, y) = 6 - 2x - 3y$,
  $c = 25$, $P = (3, 4)$ $c = 6$, $P = (0, 0)$

45. $f(x, y) = \dfrac{x}{x^2 + y^2}$,   **46.** $f(x, y) = xy$,
        $c = -3$, $P = (-1, 3)$
  $c = \dfrac{1}{2}$, $P = (1, 1)$

In Exercises 47–50, use the gradient to find a unit normal vector to the graph of the equation at the indicated point. Sketch your results.

Equation	*Point*
47. $4x^2 - y = 6$	$(2, 10)$
48. $3x^2 - 2y^2 = 1$	$(1, 1)$
49. $9x^2 + 4y^2 = 40$	$(2, -1)$
50. $xe^y - y = 5$	$(5, 0)$

51. The temperature at the point (x, y) on a plate is given by

$$T = \frac{x}{x^2 + y^2}$$

Find the direction of greatest increase in heat from the point $(3, 4)$.

52. The surface of a mountain is described by the equation

$$h(x, y) = 4000 - 0.001x^2 - 0.004y^2$$

Suppose that a mountain climber is at the point $(500, 300, 3390)$. In what direction should the climber move in order to ascend at the greatest rate?

In Exercises 53 and 55, find the path followed by a heat-seeking particle placed at point P on a metal plate with a temperature field given by $T(x, y)$.

53. $T(x, y) = 400 - 2x^2 - y^2$, $P = (10, 10)$
54. $T(x, y) = 50 - x^2 - 2y^2$, $P = (0, 0)$

55. Let P be a point on the ellipse given by

$$\frac{x^2}{a^2} + \frac{y^2}{b^2} = 1$$

where **T** is a tangent vector to the ellipse at P and $f(x, y) = d_1 + d_2$ is the sum of the distances from the foci to P, as shown in Figure 15.45. Show that $T \cdot \nabla f(x, y)$ is zero, and give a geometric interpretation of this result.

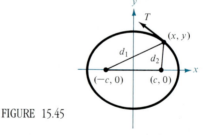

FIGURE 15.45

15.7
Tangent planes and normal lines

Up to this point, we have been representing surfaces in space primarily by equations of the form

$$z = f(x, y) \qquad \text{Equation of a surface } S$$

In the development to follow, it is convenient to use the more general representation $F(x, y, z) = 0$. For a surface S given by $z = f(x, y)$ we can easily convert to the general form by defining F as

$$F(x, y, z) = f(x, y) - z$$

Then, since $z = f(x, y)$, we have $f(x, y) - z = 0$, which means that we can consider S to be the level surface of F given by

$$F(x, y, z) = 0 \qquad \text{Alternative equation for surface } S$$

EXAMPLE 1   Equations of surfaces in space

Identify a function F so that the given surface is represented by an equation of the form $F(x, y, z) = 0$.

(a) $z = x^2 + 2y^2 + 4$ Paraboloid
(b) $y = x^2 + x + z^2 - 2$ Paraboloid
(c) $2x^2 + 4y^2 + z^2 - 2z = 6$ Ellipsoid

Solution:

(a) For the equation $z = x^2 + 2y^2 + 4$, we let

$$F(x, y, z) = x^2 + 2y^2 + 4 - z$$

Then the surface is given by $F(x, y, z) = 0$.

(b) For the equation $y = x^2 + x + z^2 - 2$, we let

$$F(x, y, z) = x^2 + x + z^2 - 2 - y$$

Then the surface is given by $F(x, y, z) = 0$.

(c) For the equation $2x^2 + 4y^2 + z^2 - 2z = 6$, we let

$$F(x, y, z) = 2x^2 + 4y^2 + z^2 - 2z - 6$$

Then the surface is given by $F(x, y, z) = 0$.

Tangent plane and normal line to a surface

We have seen many examples of the usefulness of normal lines in applications involving curves. Normal lines are equally important in the analysis of surfaces and solids. For example, consider the collision of two billiard balls. When a stationary ball is struck at a point P on its surface, it moves along the **line of impact** determined by P and the center of the ball. The impact can occur in *two* ways. If the cue ball is moving along the line of impact, it stops dead and imparts all of its momentum to the stationary ball, as shown in Figure 15.46. This kind of shot requires precision, since the line of impact must coincide exactly with the direction of the cue ball. More often, the cue ball is deflected to one side or the other and retains part of its momentum. That part of the momentum that is transferred to the stationary ball occurs along the line of impact, *regardless* of the direction of the cue ball, as shown in Figure 15.47. We call this line of impact the **normal line** to the surface of the ball at the point P.

In the process of finding a normal line to a surface, we are also able to solve the problem of finding a **tangent plane** to the surface. Let S be a surface given by $F(x, y, z) = 0$, and let $P = (x_0, y_0, z_0)$ be a point on S. Let C be a curve on S through P that is defined by the vector-valued function $\mathbf{r}(t) = x(t)\,\mathbf{i} + y(t)\,\mathbf{j} + z(t)\,\mathbf{k}$. Then, for all t,

$$F(x(t), y(t), z(t)) = 0$$

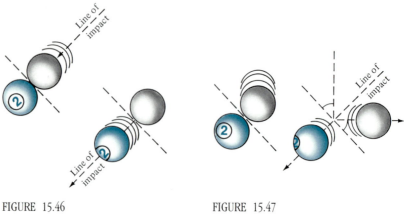

FIGURE 15.46   FIGURE 15.47

Surface S:
  $F(x, y, z) = 0$

$P = (x_0, y_0, z_0)$

∇F

Tangent Plane to Surface S at P

FIGURE 15.48

If F is differentiable and $x'(t)$, $y'(t)$, and $z'(t)$ all exist, it follows from the Chain Rule that

$$0 = F'(t) = F_x(x, y, z)x'(t) + F_y(x, y, z)y'(t) + F_z(x, y, z)z'(t)$$

At (x_0, y_0, z_0), the equivalent vector form is

$$0 = \nabla F(x_0, y_0, z_0) \cdot \mathbf{r}'(t_0) = (\text{gradient}) \cdot (\text{tangent vector})$$

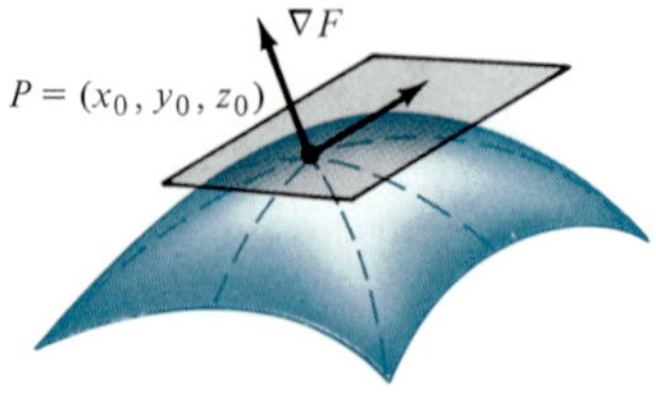

This result means that the gradient at P is orthogonal to the tangent vector of every curve on S through P. Thus, all tangent lines at P lie in a plane that is normal to $\nabla F(x_0, y_0, z_0)$ and contains P, as shown in Figure 15.48. We call this plane the **tangent plane to S at P** and we call the line passing through P in the direction of $\nabla F(x_0, y_0, z_0)$ the **normal line to S at P.**

DEFINITION OF TANGENT
PLANE AND NORMAL LINE

Let F be differentiable at the point $P = (x_0, y_0, z_0)$ on the surface S given by $F(x, y, z) = 0$ such that $\nabla F(x_0, y_0, z_0) \neq \mathbf{0}.$

1. The plane through P that is normal to $\nabla F(x_0, y_0, z_0)$ is called the **tangent plane** to S at P.
2. The line through P having the direction of $\nabla F(x_0, y_0, z_0)$ is called the **normal line** to S at P.

Remark In the remainder of this section we will assume $\nabla F(x_0, y_0, z_0)$ to be nonzero unless we state otherwise.

To find an equation for the tangent plane to S at (x_0, y_0, z_0), we let (x, y, z) be an arbitrary point in the tangent plane. Then the vector

$$\mathbf{u} = (x - x_0)\,\mathbf{i} + (y - y_0)\,\mathbf{j} + (z - z_0)\,\mathbf{k}$$

lies in the tangent plane. Since $\nabla F(x_0, y_0, z_0)$ is normal to the tangent plane at

(x_0, y_0, z_0), it must be orthogonal to every vector in the tangent plane and we have

$$\nabla F(x_0, y_0, z_0) \cdot \mathbf{u} = 0$$

which leads to the result in the following theorem.

THEOREM 15.12   **EQUATION OF TANGENT PLANE**
If F is differentiable at (x_0, y_0, z_0), then an equation of the tangent plane to the surface given by $F(x, y, z) = 0$ at (x_0, y_0, z_0) is

$$F_x(x_0, y_0, z_0)(x - x_0) + F_y(x_0, y_0, z_0)(y - y_0) + F_z(x_0, y_0, z_0)(z - z_0) = 0$$

EXAMPLE 2   Finding an equation of a tangent plane

Find an equation of the tangent plane to the hyperboloid given by

$$z^2 - 2x^2 - 2y^2 - 12 = 0$$

at the point $(1, -1, 4)$.

Solution: Considering

$$F(x, y, z) = z^2 - 2x^2 - 2y^2 - 12 = 0$$

we have

$$F_x(x, y, z) = -4x \qquad F_y(x, y, z) = -4y \qquad F_z(x, y, z) = 2z$$

and at the point $(1, -1, 4)$ the partial derivatives are

$$F_x(1, -1, 4) = -4 \qquad F_y(1, -1, 4) = 4 \qquad F_z(1, -1, 4) = 8$$

Therefore, an equation of the tangent plane at $(1, -1, 4)$ is

$$-4(x - 1) + 4(y + 1) + 8(z - 4) = 0$$
$$-4x + 4y + 8z - 24 = 0$$
$$x - y - 2z + 6 = 0$$

Figure 15.49 shows a portion of the hyperboloid and tangent plane.

Surface:
$z^2 - 2x^2 - 2y^2 - 12 = 0$

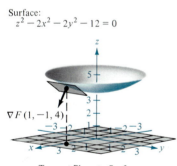

Tangent Plane to Surface

FIGURE 15.49

To find the equation of the tangent plane at a point on a surface given by $z = f(x, y)$, we define the function F by

$$F(x, y, z) = f(x, y) - z$$

Then S is given by the level surface $F(x, y, z) = 0$, and by Theorem 15.12 an equation of the tangent plane to S at the point (x_0, y_0, z_0) is

$$f_x(x_0, y_0)(x - x_0) + f_y(x_0, y_0)(y - y_0) - (z - z_0) = 0$$

This form of the tangent plane equation is demonstrated in the next example.

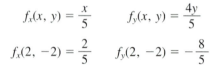

EXAMPLE 3   *Finding an equation for the tangent plane to $z = f(x, y)$*

Find the equation of the tangent plane to the paraboloid

$$z = \frac{x^2 + 4y^2}{10}$$

at the point $(2, -2, 2)$.

Solution: From $z = f(x, y) = (x^2 + 4y^2)/10$, we obtain

$$f_x(x, y) = \frac{x}{5} \qquad f_y(x, y) = \frac{4y}{5}$$

$$f_x(2, -2) = \frac{2}{5} \qquad f_y(2, -2) = -\frac{8}{5}$$

Therefore, an equation of the tangent plane at $(2, -2, 2)$ is

$$f_x(2, -2)(x - 2) + f_y(2, -2)(y + 2) - (z - 2) = 0$$

$$\frac{2}{5}(x - 2) - \frac{8}{5}(y + 2) - (z - 2) = 0$$

$$2x - 8y - 5z - 10 = 0$$

This tangent plane is shown in Figure 15.50.

Surface:
$$f(x, y) = \frac{1}{10}(x^2 + 4y^2)$$

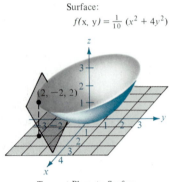

Tangent Plane to Surface

FIGURE 15.50

| Remark   You don't need to memorize the alternative formula used for the tangent plane equation in Example 3, since it can easily be derived from the level surface equation $F(x, y, z) = f(x, y) - z = 0$.

The gradient $\nabla F(x, y, z)$ gives a convenient way to find equations of normal lines. For instance, to find the normal line to the surface given by $z = (x^2 + 4y^2)/10$ at the point $(2, -2, 2)$, we let

$$F(x, y, z) = \frac{x^2 + 4y^2}{10} - z$$

Then the gradient is

$$\nabla F(x, y, z) = \frac{x}{5}\mathbf{i} + \frac{4y}{5}\mathbf{j} - \mathbf{k}$$

and at the point $(2, -2, 2)$, we have

$$\nabla F(2, -2, 2) = \frac{2}{5}\mathbf{i} - \frac{8}{5}\mathbf{j} - \mathbf{k}$$

This vector is normal to the surface and has direction numbers $2, -8$, and -5. Therefore, a set of symmetric equations for the normal line is

$$\frac{x - 2}{2} = \frac{y + 2}{-8} = \frac{z - 2}{-5}$$

EXAMPLE 4   *Finding an equation for a normal line to a surface*

Find a set of symmetric equations for the normal line to the surface given by $xyz = 12$ at the point $(2, -2, -3)$.

Solution: We let

$$F(x, y, z) = xyz - 12$$

Then, the gradient is given by

$$\mathbf{\nabla}F(x, y, z) = F_x(x, y, z)\,\mathbf{i} + F_y(x, y, z)\,\mathbf{j} + F_z(x, y, z)\,\mathbf{k}$$
$$= yz\mathbf{i} + xz\mathbf{j} + xy\mathbf{k}$$

and at the point $(2, -2, -3)$, we have

$$\mathbf{\nabla}F(2, -2, -3) = (-2)(-3)\,\mathbf{i} + (2)(-3)\,\mathbf{j} + (2)(-2)\,\mathbf{k}$$
$$= 6\mathbf{i} - 6\mathbf{j} - 4\mathbf{k}$$

Therefore, the normal line at $(2, -2, -3)$ has direction numbers 6, -6, and -4. The corresponding set of symmetric equations is

$$\frac{x - 2}{6} = \frac{y + 2}{-6} = \frac{z + 3}{-4}$$

Knowing that the gradient $\mathbf{\nabla}F(x, y, z)$ is normal to the surface given by $F(x, y, z) = 0$ allows us to solve a variety of problems dealing with surfaces and curves in space. A typical problem is shown in the next example.

EXAMPLE 5   *Finding the equation of a tangent line to a curve*

Find symmetric equations for the tangent line to the curve of intersection of the ellipsoid given by

$$x^2 + 4y^2 + 2z^2 = 27 \qquad \text{Ellipsoid}$$

and the hyperboloid given by

$$x^2 + y^2 - 2z^2 = 11 \qquad \text{Hyperboloid}$$

at the point $(3, -2, 1)$, as shown in Figure 15.51.

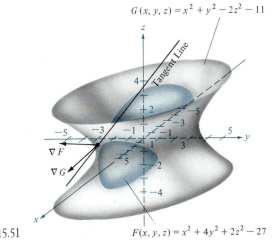

$$G(x, y, z) = x^2 + y^2 - 2z^2 - 11$$

$$F(x, y, z) = x^2 + 4y^2 + 2z^2 - 27$$

FIGURE 15.51

Solution: For the ellipsoid, we let $F(x, y, z) = x^2 + 4y^2 + 2z^2 - 27$. Then, we have

$$\nabla F(x, y, z) = 2x\mathbf{i} + 8y\mathbf{j} + 4z\mathbf{k}$$
$$\nabla F(3, -2, 1) = 6\mathbf{i} - 16\mathbf{j} + 4\mathbf{k}$$

For the hyperboloid, we let $G(x, y, z) = x^2 + y^2 - 2z^2 - 11$ and obtain

$$\nabla G(x, y, z) = 2x\mathbf{i} + 2y\mathbf{j} - 4z\mathbf{k}$$
$$\nabla G(3, -2, 1) = 6\mathbf{i} - 4\mathbf{j} - 4\mathbf{k}$$

The cross product of these two gradient vectors is

$$\nabla F(3, -2, 1) \times \nabla G(3, -2, 1) = \begin{vmatrix} \mathbf{i} & \mathbf{j} & \mathbf{k} \\ 6 & -16 & 4 \\ 6 & -4 & -4 \end{vmatrix} = 8(10\mathbf{i} + 6\mathbf{j} + 9\mathbf{k})$$

Now, since the cross product is orthogonal to both gradient vectors, we can conclude that it is tangent to both surfaces at the point $(3, -2, 1)$. Hence, the direction numbers 10, 6, and 9 give the direction of the required tangent line. Symmetric equations for this tangent line at $(3, -2, 1)$ are

$$\frac{x - 3}{10} = \frac{y + 2}{6} = \frac{z - 1}{9}$$

Remark The two surfaces in Example 5 intersect to form a curve in space. Moreover, at the point $(3, -2, 1)$ the two surfaces have different tangent planes. If two surfaces have a common tangent plane at a point (x_0, y_0, z_0), then we say they are **tangent** at the point.

Another use of the gradient $\nabla F(x, y, z)$ is in determining the angle of inclination of the tangent plane to a surface. The **angle of inclination** of a plane is defined to be the angle θ between the given plane and the xy-plane. (The angle of inclination of a horizontal plane is defined to be zero.) Since the vector $\mathbf{k}$ is normal to the xy-plane, we can use the formula for the cosine of the angle between two planes (given in Section 13.5) to conclude that the angle of inclination of a plane with normal vector $\mathbf{n}$ is given by

$$\cos \theta = \frac{|\mathbf{n} \cdot \mathbf{k}|}{\|\mathbf{n}\| \, \|\mathbf{k}\|} = \frac{|\mathbf{n} \cdot \mathbf{k}|}{\|\mathbf{n}\|} \qquad \text{Angle of inclination of a plane}$$

EXAMPLE 6 *Finding the angle of inclination of a tangent plane*

Find the angle of inclination of the tangent plane to the ellipsoid given by

$$\frac{x^2}{12} + \frac{y^2}{12} + \frac{z^2}{3} = 1$$

at the point $(2, 2, 1)$.

Solution: If we let

$$F(x, y, z) = \frac{x^2}{12} + \frac{y^2}{12} + \frac{z^2}{3} - 1$$

then the gradient of F at the point $(2, 2, 1)$ is given by

$$\nabla F(x, y, z) = \frac{x}{6}\mathbf{i} + \frac{y}{6}\mathbf{j} + \frac{2z}{3}\mathbf{k}$$

$$\nabla F(2, 2, 1) = \frac{1}{3}\mathbf{i} + \frac{1}{3}\mathbf{j} + \frac{2}{3}\mathbf{k}$$

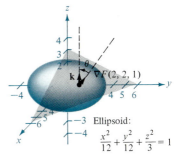

Now since $\nabla F(2, 2, 1)$ is normal to the tangent plane and $\mathbf{k}$ is normal to the xy-plane, it follows that the angle of inclination of the tangent plane is given by

$$\cos\theta = \frac{|\nabla F(2, 2, 1) \cdot \mathbf{k}|}{\|\nabla F(2, 2, 1)\|\,\|\mathbf{k}\|} = \frac{2/3}{\sqrt{(1/3)^2 + (1/3)^2 + (2/3)^2}} = \sqrt{2/3}$$

FIGURE 15.52

which implies that $\theta = \arccos\sqrt{2/3} \approx 35.3°$, as shown in Figure 15.52. ◻

| **Remark**   A special case of the procedure shown in Example 6 is worth noting. The angle of inclination θ of the tangent plane to the surface $z = f(x, y)$ at (x_0, y_0, z_0) is given by

$$\cos\theta = \frac{1}{[f_x(x_0, y_0)]^2 + [f_y(x_0, y_0)]^2 + 1}$$

In Section 15.4, we saw that the total differential

$$dz = f_x(x, y)\,dx + f_y(x, y)\,dy$$

can be used to approximate the increment Δz. (See Figure 15.30 in Section 15.4.) We now give a geometric interpretation of this approximation in terms of the tangent plane to a surface. The tangent plane to the surface $z = f(x, y)$ at (x_0, y_0, z_0) is given by

$$f_x(x_0, y_0)(x - x_0) + f_y(x_0, y_0)(y - y_0) - (z - z_0) = 0$$

or equivalently,

$$z = z_0 + f_x(x_0, y_0)(x - x_0) + f_y(x_0, y_0)(y - y_0)$$

Thus, at (x_0, y_0) the *tangent plane* has height z_0 and at $(x_0 + dx, y_0 + dy)$ it has height

$$z = z_0 + f_x(x_0, y_0)(x_0 + dx - x_0) + f_y(x_0, y_0)(y_0 + dy - y_0)$$
$$= z_0 + f_x(x_0, y_0)\,dx + f_y(x_0, y_0)\,dy$$

That is, $z = z_0 + dz$. We can see from Figure 15.53 that for points near (x_0, y_0, z_0) the increments $dx = \Delta x$ and $dy = \Delta y$ are small and $dz \approx \Delta z$.

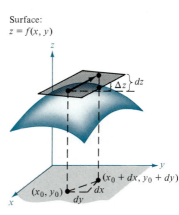

Surface:
$z = f(x, y)$

When dx and dy are small, $dz \approx \Delta z$.

FIGURE 15.53

A comparison of the gradients $\nabla f(x, y)$ and $\nabla F(x, y, z)$

We conclude this section with a comparison of the gradients $\nabla f(x, y)$ and $\nabla F(x, y, z)$. In the previous section we discovered that the gradient of a function f of two variables is normal to the *level curves* of f. Specifically, in Theorem 15.11, we stated that if f is differentiable at (x_0, y_0) and $\nabla f(x_0, y_0) \neq \mathbf{0}$, then $\nabla f(x_0, y_0)$ is normal to the level curve through (x_0, y_0). From our development of normal lines to surfaces we can now extend this result to a function of three variables.

THEOREM 15.13	**GRADIENT IS NORMAL TO LEVEL SURFACES** If F is differentiable at (x_0, y_0, z_0) and $\nabla F(x_0, y_0, z_0) \neq \mathbf{0}$, then $\nabla F(x_0, y_0, z_0)$ is normal to the level surface through (x_0, y_0, z_0).

When working with the gradients $\nabla f(x, y)$ and $\nabla F(x, y, z)$, be sure that you remember that $\nabla f(x, y)$ is a vector in the xy-plane and $\nabla F(x, y, z)$ is a vector in space.

Section Exercises 15.7

In Exercises 1–10, find a unit normal vector to the given surface at the indicated point.

Surface	*Point*
1. $x + y + z = 4$	$(2, 0, 2)$
2. $x^2 + y^2 + z^2 = 11$	$(3, 1, 1)$
3. $z = \sqrt{x^2 + y^2}$	$(3, 4, 5)$
4. $z = x^3$	$(2, 1, 8)$
5. $x^2 y^4 - z = 0$	$(1, 2, 16)$
6. $x^2 + 3y + z^3 = 9$	$(2, -1, 2)$
7. $z - x \sin y = 4$	$\left(6, \dfrac{\pi}{6}, 7\right)$
8. $ze^{x^2 - y^2} - 3 = 0$	$(2, 2, 3)$
9. $\ln\left(\dfrac{x}{y - z}\right) = 0$	$(1, 4, 3)$
10. $\sin(x - y) - z = 2$	$\left(\dfrac{\pi}{3}, \dfrac{\pi}{6}, -\dfrac{3}{2}\right)$

In Exercises 11–24, find an equation of the tangent plane to the given surface at the indicated point.

Surface	*Point*
11. $f(x, y) = 25 - x^2 - y^2$	$(3, 1, 15)$
12. $f(x, y) = \sqrt{x^2 + y^2}$	$(3, 4, 5)$
13. $f(x, y) = \dfrac{y}{x}$	$(1, 2, 2)$
14. $f(x, y) = 2 - \dfrac{2}{3}x - y$	$(3, -1, 1)$
15. $g(x, y) = x^2 - y^2$	$(5, 4, 9)$
16. $g(x, y) = \arctan\dfrac{y}{x}$	$(1, 0, 0)$
17. $z = e^x(\sin y + 1)$	$\left(0, \dfrac{\pi}{2}, 2\right)$
18. $z = x^3 - 3xy + y^3$	$(1, 2, 3)$
19. $h(x, y) = \ln\sqrt{x^2 + y^2}$	$(3, 4, \ln 5)$
20. $h(x, y) = \cos y$	$\left(5, \dfrac{\pi}{4}, \dfrac{\sqrt{2}}{2}\right)$
21. $x^2 + 4y^2 + z^2 = 36$	$(2, -2, 4)$

22. $x^2 + 2z^2 = y^2$	$(1, 3, -2)$
23. $xy^2 + 3x - z^2 = 4$	$(2, 1, -2)$
24. $y = x(2z - 1)$	$(4, 4, 1)$

In Exercises 25–30, find an equation for the tangent plane and find symmetric equations for the normal line to the given surface at the indicated point.

Surface	*Point*
25. $x^2 + y^2 + z = 9$	$(1, 2, 4)$
26. $x^2 + y^2 + z^2 = 9$	$(1, 2, 2)$
27. $xy - z = 0$	$(-2, -3, 6)$
28. $x^2 + y^2 - z^2 = 0$	$(5, 12, 13)$
29. $z = \arctan\dfrac{y}{x}$	$\left(1, 1, \dfrac{\pi}{4}\right)$
30. $xyz = 10$	$(1, 2, 5)$

In Exercises 31–34, find the angle of inclination θ of the tangent plane to the given surface at the indicated point.

Surface	*Point*
31. $3x^2 + 2y^2 - z = 15$	$(2, 2, 5)$
32. $xy - z^2 = 0$	$(2, 2, 2)$
33. $x^2 - y^2 + z = 0$	$(1, 2, 3)$
34. $x^2 + y^2 = 5$	$(2, 1, 3)$

In Exercises 35–40, (a) find symmetric equations of the tangent line to the curve of intersection of the given surfaces at the indicated point and (b) find the cosine of the angle between the gradient vectors at this point. State whether the surfaces are orthogonal at the point of intersection.

Surface	*Point*
35. $x^2 + y^2 = 5$ $z = x$	$(2, 1, 2)$
36. $z = x^2 + y^2$ $z = 4 - y$	$(2, -1, 5)$

37. $x^2 + z^2 = 25$ $(3, 3, 4)$
$\quad\;\; y^2 + z^2 = 25$

38. $z = \sqrt{x^2 + y^2}$ $(3, 4, 5)$
$\quad\;\; 2x + y + 2z = 20$

39. $x^2 + y^2 + z^2 = 6$ $(2, 1, 1)$
$\quad\;\; x - y - z = 0$

40. $z = x^2 + y^2$ $(1, 2, 5)$
$\quad\;\; x + y + 6z = 33$

In Exercises 41 and 42, find the point on the surface where the tangent plane is horizontal.

41. $z = 3 - x^2 - y^2 + 6y$
42. $z = 3x^2 + 2y^2 - 3x + 4y - 5$

In Exercises 43 and 44, find the path of a heat-seeking particle in the temperature field T, starting at the specified point.

43. $T(x, y, z) = 400 - 2x^2 - y^2 - 4z^2$, $(4, 3, 10)$
44. $T(x, y, z) = 100 - 3x - y - z^2$, $(2, 2, 5)$

In Exercises 45 and 46, show that the tangent plane to the quadric surface at the point (x_0, y_0, z_0) can be written in the given form.

45. Ellipsoid: $\dfrac{x^2}{a^2} + \dfrac{y^2}{b^2} + \dfrac{z^2}{c^2} = 1$

      Plane: $\dfrac{x_0 x}{a^2} + \dfrac{y_0 y}{b^2} + \dfrac{z_0 z}{c^2} = 1$

46. Hyperboloid: $\dfrac{x^2}{a^2} + \dfrac{y^2}{b^2} - \dfrac{z^2}{c^2} = 1$

      Plane: $\dfrac{x_0 x}{a^2} + \dfrac{y_0 y}{b^2} - \dfrac{z_0 z}{c^2} = 1$

47. Show that the tangent plane to the cone $z^2 = a^2 x^2 + b^2 y^2$ passes through the origin.
48. Show that any line normal to a sphere passes through the center of the sphere.
49. Prove Theorem 15.13.

15.8
Extrema of functions of two variables

Karl Weierstrass

In Chapter 4 we studied techniques for finding the extreme values of a function of a single variable. In this section we extend these techniques to functions of two variables. For example, in Theorem 15.14 we extend the Extreme Value Theorem for a function of a single variable to a function of two variables. This theorem is difficult to prove in both the single and two variable cases. Although the theorem had been used by earlier mathematicians, the first person to provide a rigorous proof was the German mathematician Karl Weierstrass (1815–1897). This was not the only case in which Weierstrass provided a rigorous justification for mathematical results already in common use, and we are indebted to him for much of the solid logical foundation upon which modern calculus is built.

The values $f(a, b)$ and $f(c, d)$ such that $f(a, b) \le f(x, y) \le f(c, d)$ for all (x, y) in R are called the **absolute minimum** and **absolute maximum** of f in the region R, as shown in Figure 15.54. As in single-variable calculus, we distinguish between absolute extrema and **relative extrema**. Several relative extrema are shown in Figure 15.55.

THEOREM 15.14   **EXTREME VALUE THEOREM**
Let f be a continuous function of two variables x and y defined on a *closed bounded* region R in the xy-plane.

1. There is at least one point in R where f takes on a minimum value.
2. There is at least one point in R where f takes on a maximum value.

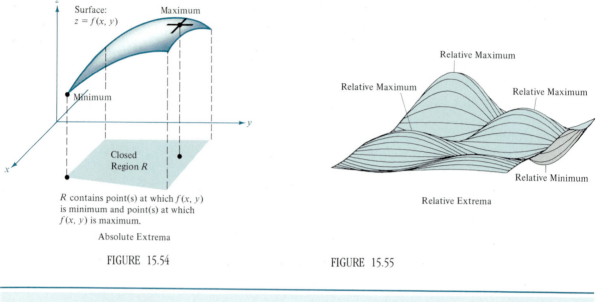

R contains point(s) at which $f(x, y)$
is minimum and point(s) at which
$f(x, y)$ is maximum.

Absolute Extrema

FIGURE 15.54

Relative Extrema

FIGURE 15.55

**DEFINITION OF
RELATIVE EXTREMA**

Let f be a function defined on a region R containing (x_0, y_0).

1. $f(x_0, y_0)$ is a **relative minimum** of f if $f(x, y) \geq f(x_0, y_0)$ for all (x, y)
   in an *open* disc containing (x_0, y_0).
2. $f(x_0, y_0)$ is a **relative maximum** of f if $f(x, y) \leq f(x_0, y_0)$ for all (x, y)
   in an *open* disc containing (x_0, y_0).

To say that $z_0 = f(x_0, y_0)$ is a relative maximum of f means that the
point (x_0, y_0, z_0) is at least as high as all nearby points on the graph of $z =
f(x, y)$. Similarly, $z_0 = f(x_0, y_0)$ is a relative minimum if (x_0, y_0, z_0) is at least
as low as all nearby points on the graph.

To locate relative extrema of f, we investigate the points at which its
gradient is zero or undefined. We call such points **critical points** of f.

**DEFINITION OF
CRITICAL POINT**

Let f be defined on an open region R containing (x_0, y_0). We call (x_0, y_0) a
critical point of f if one of the following is true.

1. $f_x(x_0, y_0) = 0$ and $f_y(x_0, y_0) = 0$
2. $f_x(x_0, y_0)$ or $f_y(x_0, y_0)$ does not exist.

Recall from Theorem 15.10 that if f is differentiable and

$$\nabla f(x_0, y_0) = f_x(x_0, y_0) \, \mathbf{i} + f_y(x_0, y_0) \, \mathbf{j} = 0\mathbf{i} + 0\mathbf{j}$$

then every directional derivative at (x_0, y_0) must be zero. In Exercise 35 you
are asked to show that this implies that the function has a horizontal tangent

plane at the point (x_0, y_0), as shown in Figure 15.56. It appears that such a point is a likely location of a relative extremum.

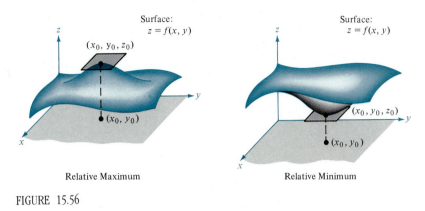

FIGURE 15.56

THEOREM 15.15   **RELATIVE EXTREMA OCCUR ONLY AT CRITICAL POINTS**
If $f(x_0, y_0)$ is a relative extremum of f on an open region R, then (x_0, y_0) is a critical point of f.

Proof: If either f_x or f_y does not exist at (x_0, y_0), then it is a critical point. Thus, we assume both first partial derivatives exist at (x_0, y_0), and we consider the function of one variable, $g(x) = f(x, y_0)$. By hypothesis it has a relative extremum at x_0 and is differentiable there. Hence,

$$g'(x_0) = \lim_{\Delta x \to 0} \frac{f(x_0 + \Delta x, y_0) - f(x_0, y_0)}{\Delta x} = f_x(x_0, y_0) = 0$$

Similarly, $h(y) = f(x_0, y)$ has a local extremum at y_0, and being differentiable there, it satisfies the equation

$$h'(y_0) = \lim_{\Delta y \to 0} \frac{f(x_0, y_0 + \Delta y) - f(x_0, y_0)}{\Delta y} = f_y(x_0, y_0) = 0$$

Thus, $f_x(x_0, y_0) = f_y(x_0, y_0) = 0$, and (x_0, y_0) is a critical point of f.

EXAMPLE 1   *Finding a relative extremum*

Determine the relative extrema of

$$f(x, y) = 2x^2 + y^2 + 8x - 6y + 2x$$

Solution: We begin by finding the critical points of f. Since

$$f_x(x, y) = 4x + 8 \qquad \text{and} \qquad f_y(x, y) = 2y - 6$$

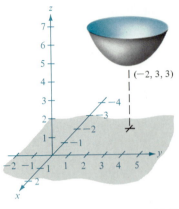

FIGURE 15.57

are defined for all x and y, the only critical points are those for which both first partial derivatives are zero. To locate these points, let $f_x(x, y)$ and $f_y(x, y)$ be zero, and solve the system of equations

$$4x + 8 = 0 \qquad \text{and} \qquad 2y - 6 = 0$$

to obtain the critical point $(-2, 3)$. By completing the square, we can conclude that for all $(x, y) \neq (-2, 3)$,

$$f(x, y) = 2(x + 2)^2 + (y - 3)^2 + 3 > 3$$

Therefore, a relative *minimum* of f occurs at $(-2, 3)$. The value of the relative minimum is $f(-2, 3) = 3$, as shown in Figure 15.57.

Example 1 shows a relative minimum occurring at one type of critical point—the type for which both $f_x(x, y)$ and $f_y(x, y)$ are zero. In the next example we look at a relative maximum that occurs at the other type of critical point—the type for which either $f_x(x, y)$ or $f_y(x, y)$ is undefined.

EXAMPLE 2   *Finding a relative extremum*

Determine the relative extrema of

$$f(x, y) = 1 - (x^2 + y^2)^{1/3}$$

Solution: Since

$$f_x(x, y) = -\frac{2x}{3(x^2 + y^2)^{2/3}} \qquad \text{and} \qquad f_y(x, y) = -\frac{2y}{3(x^2 + y^2)^{2/3}}$$

we see that both partial derivatives are defined for all points in the xy-plane except for $(0, 0)$. Moreover, this is the only critical point, since the partial derivatives cannot both be zero unless both x and y are zero. In Figure 15.58 we see that $f(0, 0)$ is 1. For all other (x, y) it is clear that

$$f(x, y) = 1 - (x^2 + y^2)^{1/3} < 1$$

Therefore, $f(0, 0)$ is a relative *maximum* of f.

Surface:
$$f(x, y) = 1 - (x^2 + y^2)^{1/3}$$

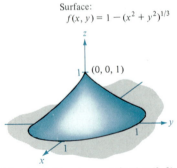

$f_x(x, y)$ and $f_y(x, y)$ are undefined at $(0, 0)$.

FIGURE 15.58

| **Remark**   In Example 2, $f_x(x, y) = 0$ for every point on the x-axis other than $(0, 0)$. However, since $f_y(x, y)$ is nonzero, these points are not critical points. Remember that *one* of the partials must be undefined or *both* must be zero in order to yield a critical point.

Theorem 15.15 tells us that to find relative extrema we need only examine values of $f(x, y)$ at critical points. However, as is true for a function of one variable, the critical points of a function of two variables do not always yield relative maxima or minima. Some critical points yield **saddle points,** which are neither relative maxima nor relative minima. For example, the saddle point shown in Figure 15.59 is not a relative extremum, since in any open disc centered at $(0, 0)$ the function takes on both negative values (along the x-axis) *and* positive values (along the y-axis).

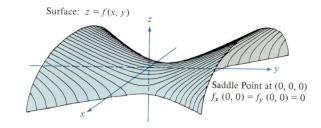

FIGURE 15.59

For the functions in Examples 1 and 2, it is relatively easy to determine the relative extrema, because each function was either given or able to be written in completed square form. For more complicated functions, algebraic arguments are not so fruitful, and we rely on the more analytic means presented in the following Second-Partials Test. This is the two-variable counterpart of the Second-Derivative Test for functions of one variable. The proof of this theorem is best left to a course in advanced calculus.

THEOREM 15.16   **SECOND-PARTIALS TEST**

Let f have continuous first and second partial derivatives on an open region containing a point (a, b) for which $f_x(a, b) = 0$ and $f_y(a, b) = 0$. To test for relative extrema of f, we define the quantity

$$d = f_{xx}(a, b)f_{yy}(a, b) - [f_{xy}(a, b)]^2$$

1. If $d > 0$ and $f_{xx}(a, b) > 0$, then $f(a, b)$ is a **relative minimum.**
2. If $d > 0$ and $f_{xx}(a, b) < 0$, then $f(a, b)$ is a **relative maximum.**
3. If $d < 0$, then $(a, b, f(a, b))$ is a **saddle point.**
4. The test gives no information if $d = 0$.

| **Remark** If $d > 0$, then $f_{xx}(a, b)$ and $f_{yy}(a, b)$ must have the same signs. This means that you can replace $f_{xx}(a, b)$ by $f_{yy}(a, b)$ in the first two parts of the test.

A convenient device for remembering the formula for d in the Second-Partials Test is given by the 2×2 determinant

$$d = \begin{vmatrix} f_{xx}(a, b) & f_{xy}(a, b) \\ f_{yx}(a, b) & f_{yy}(a, b) \end{vmatrix}$$

where $f_{xy}(a, b) = f_{yx}(a, b)$ by Theorem 15.3.

EXAMPLE 3   Using the Second-Partials Test

Find the relative extrema of $f(x, y) = -x^3 + 4xy - 2y^2 + 1$.

Solution: We begin by finding the critical points of f. Since

$$f_x(x, y) = -3x^2 + 4y \quad \text{and} \quad f_y(x, y) = 4x - 4y$$

are defined for all x and y, the only critical points are those for which both first partial derivatives are zero. To locate these points, we let $f_x(x, y)$ and $f_y(x, y)$ be zero and obtain the following system of equations:

$$-3x^2 + 4y = 0$$
$$4x - 4y = 0$$

From the second equation we see that $x = y$, and, by substitution into the first equation, we obtain two solutions: $y = x = 0$ and $y = x = \frac{4}{3}$. Since

$$f_{xx}(x, y) = -6x, \quad f_{yy}(x, y) = -4, \quad \text{and} \quad f_{xy}(x, y) = 4$$

it follows that for the critical point $(0, 0)$,

$$d = f_{xx}(0, 0)f_{yy}(0, 0) - [f_{xy}(0, 0)]^2 = 0 - 16 < 0$$

and, by the Second-Partials Test, we conclude that $(0, 0, 1)$ is a saddle point of f. Furthermore, for the critical point $(\frac{4}{3}, \frac{4}{3})$,

$$d = f_{xx}\left(\frac{4}{3}, \frac{4}{3}\right)f_{yy}\left(\frac{4}{3}, \frac{4}{3}\right) - \left[f_{xy}\left(\frac{4}{3}, \frac{4}{3}\right)\right]^2 = -8(-4) - 16 = 16 > 0$$

and since

$$f_{xx}\left(\frac{4}{3}, \frac{4}{3}\right) = -8 < 0$$

we conclude that $f(\frac{4}{3}, \frac{4}{3})$ is a relative maximum, as shown in Figure 15.60.

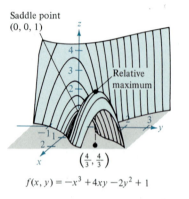

Saddle point
$(0, 0, 1)$

Relative maximum

$(\frac{4}{3}, \frac{4}{3})$

$f(x, y) = -x^3 + 4xy - 2y^2 + 1$

FIGURE 15.60

The Second-Partials Test can fail to find relative extrema in two ways. If either of the first partial derivatives is undefined, then we cannot use the test. Also, if

$$d = f_{xx}(a, b)f_{yy}(a, b) - [f_{xy}(a, b)]^2 = 0$$

the test fails. In such cases, we must rely on a sketch or some other approach, as demonstrated in the next example.

EXAMPLE 4   *Failure of the Second-Partials Test*

Find the relative extrema of $f(x, y) = x^2y^2$.

Solution: Since

$$f_x(x, y) = 2xy^2 \quad \text{and} \quad f_y(x, y) = 2x^2y$$

we see that both partial derivatives are zero if $x = 0$ or $y = 0$. That is, every point along the x- or y-axis is a critical point. Now, since

$$f_{xx}(x, y) = 2y^2, \quad f_{yy}(x, y) = 2x^2, \quad \text{and} \quad f_{xy}(x, y) = 4xy$$

we see that if either $x = 0$ or $y = 0$, then

$$d = f_{xx}(x, y)f_{yy}(x, y) - [f_{xy}(x, y)]^2 = 4x^2y^2 - 16x^2y^2 = -12x^2y^2 = 0$$

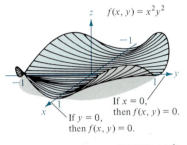

$f(x, y) = x^2y^2$

If $x = 0,$
then $f(x, y) = 0.$
If $y = 0,$
then $f(x, y) = 0.$

FIGURE 15.61

Thus, the Second-Partials Test fails. However, since $f(x, y) = 0$ for every point along the x- or y-axis, and since $f(x, y) = x^2y^2 > 0$ for all other points, we can conclude that each of these critical points yield an absolute minimum, as shown in Figure 15.61.

Absolute extrema of a function can occur in two ways. First, some relative extrema also happen to be absolute extrema. For instance, in Example 1, $f(-2, 3)$ is an absolute minimum of the function. (On the other hand, the relative maximum found in Example 3 is not an absolute maximum of the function.) Second, absolute extrema can occur at a boundary point of the domain. This is illustrated in the next example.

EXAMPLE 5   *Finding absolute extrema*

Find the absolute extrema of the function $f(x, y) = \sin xy$ on the closed region given by $0 \le x \le \pi$ and $0 \le y \le 1$.

Solution: From the partial derivatives

$$f_x(x, y) = y \cos xy \qquad \text{and} \qquad f_y(x, y) = x \cos xy$$

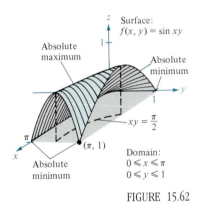

Surface:
$f(x, y) = \sin xy$

Absolute
maximum

Absolute
minimum

$xy = \dfrac{\pi}{2}$

$(\pi, 1)$

Domain:
$0 \le x \le \pi$
$0 \le y \le 1$

Absolute
minimum

FIGURE 15.62

we see that each point lying on the hyperbola given by $xy = \pi/2$ is a critical point. Moreover, these points each yield the value $f(x, y) = \sin(\pi/2) = 1$, which we know is the absolute maximum, as shown in Figure 15.62. The only other critical point of f *lying in the given region* is $(0, 0)$. It yields an absolute minimum of 0, since

$$0 \le xy \le \pi \implies 0 \le \sin xy \le 1$$

If we did not sketch a graph, we might conclude that we had located all the absolute extrema. With the graph shown in Figure 15.62, we can see that we missed an infinite number of points that yield the minimum value. Specifically, $\sin xy = 0$ at all points on the x-axis and the y-axis and at the point $(\pi, 1)$.

The concepts of relative extrema and critical points can be extended to functions of three or more variables. If all first partial derivatives of $w = f(x_1, x_2, x_3, \ldots, x_n)$ exist, then it can be shown that a relative maximum or minimum can occur at $(x_1, x_2, x_3, \ldots, x_n)$ only if every first partial derivative is zero at that point. This means the critical points are obtained by solving the system of equations given by

$$f_{x_1}(x_1, x_2, x_3, \ldots, x_n) = 0$$
$$f_{x_2}(x_1, x_2, x_3, \ldots, x_n) = 0$$
$$\vdots$$
$$f_{x_n}(x_1, x_2, x_3, \ldots, x_n) = 0$$

The extension of Theorem 15.16 to three or more variables is also possible, though we will not consider such an extension in this text.

Section Exercises 15.8

In Exercises 1–20, examine each function for relative extrema and saddle points.

1. $f(x, y) = 2x^2 + 2xy + y^2 + 2x - 3$
2. $f(x, y) = -x^2 - 5y^2 + 8x - 10y - 13$
3. $f(x, y) = -5x^2 + 4xy - y^2 + 16x + 10$
4. $f(x, y) = x^2 + 6xy + 10y^2 - 4y + 4$
5. $z = 2x^2 + 3y^2 - 4x - 12y + 13$
6. $z = -3x^2 - 2y^2 + 3x - 4y + 5$
7. $h(x, y) = x^2 - y^2 - 2x - 4y - 4$
8. $h(x, y) = x^2 - 3xy - y^2$
9. $g(x, y) = xy$
10. $g(x, y) = 120x + 120y - xy - x^2 - y^2$
11. $f(x, y) = x^3 - 3xy + y^3$

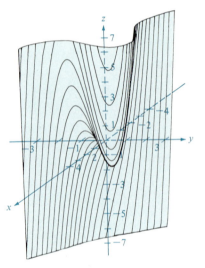

12. $f(x, y) = 4xy - x^4 - y^4$

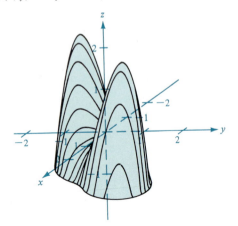

13. $f(x, y) = \dfrac{3x^2 + 1}{2} - x(x^2 + y^2)$

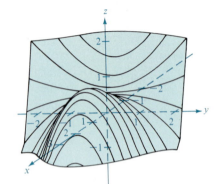

14. $f(x, y) = y^3 - 3yx^2 - 3y^2 - 3x^2 + 1$

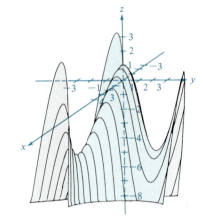

15. $z = (x^2 + 4y^2)e^{1-x^2-y^2}$

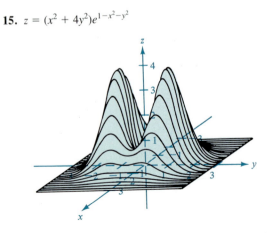

16. $z = e^{-x} \sin y$

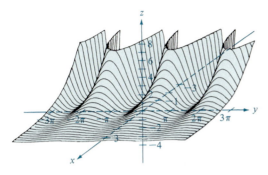

17. $z = \arctan \dfrac{1}{x^2 + y^2}$

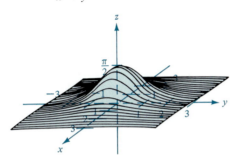

18. $z = e^{-(x^2+y^2)}$

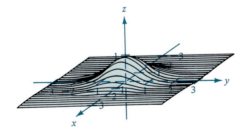

19. $z = \left(\dfrac{1}{2} - x^2 + y^2 \right) e^{1-x^2-y^2}$

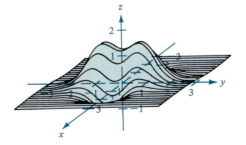

20. $z = \dfrac{-4x}{x^2 + y^2 + 1}$

In Exercises 21–24, find the absolute extrema of the function over the region R.

21. $f(x, y) = x^2 + xy$,
   $R = \{(x, y): |x| \le 2, |y| \le 1\}$
22. $f(x, y) = x^2 + 2xy + y^2$,
   $R = \{(x, y): |x| \le 2, |y| \le 1\}$
23. $f(x, y) = x^2 + 2xy + y^2$,
   $R = \{(x, y): x^2 + y^2 \le 8\}$
24. $f(x, y) = x^2 - 4xy$,
   $R = \{(x, y): 0 \le x \le 4, 0 \le y \le \sqrt{x}\}$

In Exercises 25–30, find the critical points and test for relative extrema. List the critical points for which the Second-Partials Test fails.

25. $f(x, y) = x^3 + y^3$
26. $f(x, y) = x^3 + y^3 - 3x^2 + 6y^2 + 3x + 12y + 7$
27. $f(x, y) = (x - 1)^2(y + 4)^2$
28. $f(x, y) = \sqrt{(x - 1)^2 + (y + 2)^2}$
29. $f(x, y) = x^{2/3} + y^{2/3}$
30. $f(x, y) = (x^2 + y^2)^{2/3}$

In Exercises 31–34, find the relative extrema of the given function.

31. $f(x, y, z) = x^2 + (y - 3)^2 + (z + 1)^2$
32. $f(x, y, z) = 4 - [x(y - 1)(z + 2)]^2$
33. $f(x, y, z) = x^2 + y^2 + 2xz - 4yz + 10z$
34. $f(x, y, z) = x^2 - y^2 + yz - x^2z$

35. Prove that if f is a differentiable function such that $\nabla f(x_0, y_0) = \mathbf{0}$, then the tangent plane at (x_0, y_0) is horizontal.

15.9
Applications of extrema of functions of two variables

There are many applications of extrema of functions of two (or more) variables, and we survey a few in this section.

EXAMPLE 1   Finding maximum volume

A rectangular box is resting on the xy-plane with one vertex at the origin. Find the maximum volume of the box if its vertex opposite the origin lies in the plane $6x + 4y + 3z = 24$, as shown in Figure 15.63.

Solution: Since one vertex of the box lies in the plane $6x + 4y + 3z = 24$, we have

$$z = \frac{1}{3}(24 - 6x - 4y)$$

and we can write the volume, xyz, of the box as a function of two variables:

$$V(x, y) = (x)(y)\left[\frac{1}{3}(24 - 6x - 4y)\right] = \frac{1}{3}(24xy - 6x^2y - 4xy^2)$$

By setting the first partial derivatives equal to zero,

$$V_x(x, y) = \frac{1}{3}(24y - 12xy - 4y^2) = \frac{y}{3}(24 - 12x - 4y) = 0$$

$$V_y(x, y) = \frac{1}{3}(24x - 6x^2 - 8xy) = \frac{x}{3}(24 - 6x - 8y) = 0$$

we obtain the critical points $(0, 0)$ and $(\frac{4}{3}, 2)$. At $(0, 0)$ the volume is zero, so we apply the Second-Partials Test to the point $(\frac{4}{3}, 2)$.

$$V_{xx}(x, y) = -4y, \qquad V_{yy}(x, y) = \frac{-8x}{3}, \qquad V_{xy}(x, y) = \frac{1}{3}(24 - 12x - 8y)$$

Since

$$V_{xx}\left(\frac{4}{3}, 2\right)V_{yy}\left(\frac{4}{3}, 2\right) - \left[V_{xy}\left(\frac{4}{3}, 2\right)\right]^2 = (-8)\left(-\frac{32}{9}\right) - \left(-\frac{8}{3}\right)^2$$

$$= \frac{64}{3} > 0$$

and

$$V_{xx}\left(\frac{4}{3}, 2\right) = -8 < 0$$

we conclude from the Second-Partials Test that the maximum volume is

$$V(x, y) = \frac{1}{3}\left[24\left(\frac{4}{3}\right)(2) - 6\left(\frac{4}{3}\right)^2(2) - 4\left(\frac{4}{3}\right)(2^2)\right] = \frac{64}{9} \text{ cubic units}$$

(Note that the volume is zero at the boundary points of the triangular domain of V.)

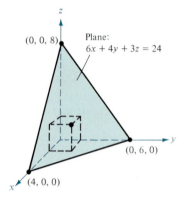

Plane:
$6x + 4y + 3z = 24$

$(0, 0, 8)$

$(0, 6, 0)$

$(4, 0, 0)$

FIGURE 15.63

| Remark   In many applied problems the domain of the function to be optimized is a closed bounded region. To find minimum or maximum points, you must, in addition to testing critical points, consider the value of the function at points on the boundary.

In Section 4.8 we looked at several applications of extrema in economics and business. In practice, such applications often involve more than one independent variable. For instance, a company may produce several models of a single type of product. The price per unit and profit per unit are usually different for each model. Moreover, the demand for each model is often a function of the prices of the other models (as well as its own price). The next example illustrates an application involving two products.

EXAMPLE 2   *Finding the maximum profit*

The profit obtained by producing x units of product A and y units of product B is approximated by the model

$$P(x, y) = 8x + 10y - (0.001)(x^2 + xy + y^2) - 10,000$$

Find the production level that produces a maximum profit.

Solution: We have

$$P_x(x, y) = 8 - (0.001)(2x + y)$$

and

$$P_y(x, y) = 10 - (0.001)(x + 2y)$$

By setting these partial derivatives equal to zero, we obtain the following system of equations.

$$8 - (0.001)(2x + y) = 0 \implies 2x + y = 8000$$
$$10 - (0.001)(x + 2y) = 0 \implies x + 2y = 10,000$$

Solving this system, we have $x = 2000$ and $y = 4000$. The second partial derivatives of P are

$$P_{xx}(x, y) = -0.002$$
$$P_{yy}(x, y) = -0.002$$
$$P_{xy}(x, y) = -0.001$$

Moreover, since $P_{xx} < 0$ and

$$P_{xx}(x, y)P_{yy}(x, y) - [P_{xy}(x, y)]^2 = (-0.002)(-0.002) - (-0.001)^2 > 0$$

we conclude that the production level of $x = 2000$ units and $y = 4000$ units yields a *maximum* profit.   ☐

| Remark   In Example 2 we made the assumption that the plant is able to produce the required number of units to yield a maximum profit. In actual practice, the production would be bounded by physical constraints. We will consider such constrained optimization problems in the next section.

The method of least squares

We have spent some time discussing **mathematical models.** For instance, in Example 2 we gave a quadratic model for profit. We now look at one method for obtaining such a model. In constructing a model to represent a particular phenomenon, the goal is to be as *simple* and as *accurate* as possible. Of course, these two goals often conflict. For instance, a simple linear model for the points in Figure 15.64 is

$$y = 1.8566x - 5.0246$$

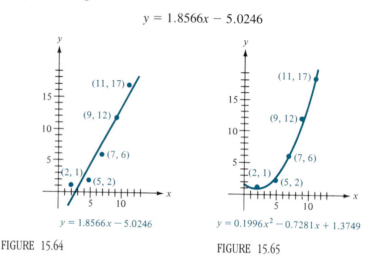

FIGURE 15.64   FIGURE 15.65

However, Figure 15.65 shows that by choosing the slightly more complicated quadratic model

$$y = 0.1996x^2 - 0.7281x + 1.3749$$

we can achieve greater accuracy.

As a measure of how well the model $y = f(x)$ fits the collection of points

$$\{(x_1, y_1), (x_2, y_2), (x_3, y_3), \ldots, (x_n, y_n)\}$$

we add the squares of the differences between the actual y-values and the values given by the model to obtain the **sum of the squared errors**

$$S = \sum_{i=1}^{n} [f(x_i) - y_i]^2$$

Graphically, S can be interpreted as the vertical distances between the graph of f and the given points in the plane, as shown in Figure 15.66. If the model is perfect, then $S = 0$. However, when perfection is infeasible, we settle for a model that minimizes S. Statisticians call the *linear model* that minimizes S the **least squares regression line.** The proof that this line actually minimizes S involves the minimum of a function of two variables.

The method of least squares was introduced by the French mathematician Adrien-Marie Legendre (1752–1833). Legendre is best known for his work in geometry. In fact, his text *Elements of Geometry* was so popular in the United States that it continued to be used for 33 editions, spanning a period of over one hundred years.

Adrien-Marie Legendre

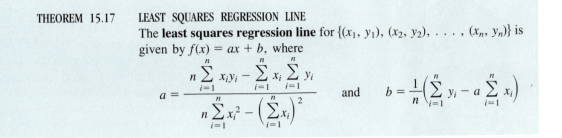

THEOREM 15.17

LEAST SQUARES REGRESSION LINE
The **least squares regression line** for $\{(x_1, y_1), (x_2, y_2), \ldots, (x_n, y_n)\}$ is given by $f(x) = ax + b$, where

$$a = \frac{n\sum_{i=1}^{n} x_i y_i - \sum_{i=1}^{n} x_i \sum_{i=1}^{n} y_i}{n\sum_{i=1}^{n} x_i^2 - \left(\sum_{i=1}^{n} x_i\right)^2} \quad \text{and} \quad b = \frac{1}{n}\left(\sum_{i=1}^{n} y_i - a\sum_{i=1}^{n} x_i\right).$$

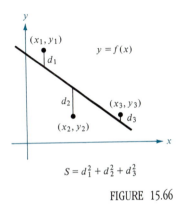

$$S = d_1^2 + d_2^2 + d_3^2$$

FIGURE 15.66

Proof: We begin by letting $S(a, b)$ represent the sum of the squared errors for the model $f(x) = ax + b$ and the given set of points. That is,

$$S(a, b) = \sum_{i=1}^{n} [f(x_i) - y_i]^2 = \sum_{i=1}^{n} [ax_i + b - y_i]^2$$

where the points (x_i, y_i) represent constants. Since S is a function of a and b, we can use the methods discussed in the previous section to find the minimum value of S. Specifically, the first partial derivatives of S are

$$S_a(a, b) = \sum_{i=1}^{n} 2x_i[ax_i + b - y_i] = 2a\sum_{i=1}^{n} x_i^2 + 2b\sum_{i=1}^{n} x_i - 2\sum_{i=1}^{n} x_i y_i$$

$$S_b(a, b) = \sum_{i=1}^{n} 2[ax_i + b - y_i] = 2a\sum_{i=1}^{n} x_i + 2nb - 2\sum_{i=1}^{n} y_i$$

By setting these two partial derivatives equal to zero, we obtain the values for a and b that are listed in the theorem. We leave it up to you to apply the Second-Partials Test to verify that these values of a and b yield a minimum.

Remark If the x-values are symmetrically spaced about the origin, then $\Sigma x_i = 0$ and the formulas for a and b simplify to

$$a = \frac{\sum_{i=1}^{n} x_i y_i}{\sum_{i=1}^{n} x_i^2} \quad \text{and} \quad b = \frac{1}{n}\sum_{i=1}^{n} y_i.$$

This simplification is often possible with a translation of the x-values. For instance, if the x-values in a data collection consisted of the years 1980, 1981, 1982, 1983, and 1984, we could let 1982 be represented by 0.

Complex zeros of polynomial functions

One of the most important theorems in mathematics, the **Fundamental Theorem of Algebra,** states that every nth-degree polynomial function has at least one zero in the set of complex numbers. (By factorization we find that this is equivalent to saying that every nth-degree polynomial has precisely n roots, some of which may be repeated roots.) The theorem was first proved by a man

Carl Friedrich Gauss

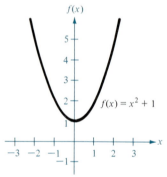

Quadratic Function with No
x-Intercepts

FIGURE 15.67

who is recognized, along with Newton and Archimedes, as one of the three greatest mathematicians in history, Carl Friedrich Gauss (1777–1855). Gauss published the proof at the age of twenty-two, as part of his doctoral dissertation. In the proof he used an analytical argument in which the complex number $x + yi$ corresponds to the point (x, y) in the plane. Today, we can use this same correspondence, coupled with computer graphics, to graphically represent the n roots of an nth-degree polynomial as maximum points on a surface.

Before discussing this graphing procedure, we review some properties of complex numbers. Every **complex number** can be written in the form

$$w = x + yi \qquad \text{Complex number}$$

where x and y are real numbers and

$$i = \sqrt{-1} \qquad \text{Imaginary number}$$

The **real part** of w is x, and the **imaginary part** of w is yi. (This use of the letter i is unrelated to the use of **i** as a unit vector.) Addition and multiplication of complex numbers possess the standard commutative, associative, and distributive properties. The absolute value of a complex number is

$$|x + yi| = \sqrt{x^2 + y^2} \qquad \text{Absolute value}$$

With standard two-dimensional graphs, we visualize the real zeros of a polynomial function as the intercepts with the horizontal axis—however, the complex zeros are not evident. For example, in Figure 15.67, the graph of $f(x) = x^2 + 1$ does not cross the x-axis. From this we know that it must have two complex zeros, but we cannot estimate their value graphically.

To construct a three-dimensional model that pictures both the real and complex zeros of a polynomial function f, we define a new function g such that

$$g(x, y) = -|f(x + yi)|^2$$

Since the absolute value of the complex number $a + bi$ is zero if and only if $a = b = 0$, we know that the graph of g is a surface lying below the xy-plane with maximum points corresponding precisely to the zeros of f. The real zeros correspond to the maximum points lying on the x-axis. (In order to picture these zeros more clearly, the computer-generated graphs in this section were programmed with a logarithmic scale on the z-axis. The same effect can be obtained by graphing $z = -|f(x + yi)|^{2/n}$ where n is the degree of f.)

EXAMPLE 3   *The complex zeros of a polynomial function*

For $f(w) = w^2 + 1$, sketch the surface given by $g(x, y) = -|f(x + yi)|^2$ and show that the surface has two maximum points corresponding to the two complex zeros of f.

Solution: Since

$$f(x + yi) = (x + yi)^2 + 1 = (x^2 + 1 - y^2) + 2xyi$$

the function g is given by

$$g(x, y) = -|f(x + yi)|^2 = -[(x^2 + 1 - y^2)^2 + 4x^2y^2]$$

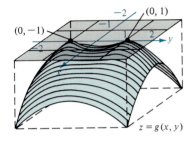

FIGURE 15.68

Using a computer with three-dimensional graphics capability, we obtain the surface shown in Figure 15.68. The maximum points appear to be located at $(0, \pm1)$, and we verify this analytically by finding the critical points of g. The partial derivatives of g are

$$g_x(x, y) = -[-4x(x^2 + 1 - y^2) + 8xy^2] = -4x(x^2 + 1 + y^2)$$

and

$$g_y(x, y) = -[-4y(x^2 + 1 - y^2) + 8x^2y] = -4y(x^2 - 1 + y^2)$$

Since $g_x(x, y) = 0$ only if $x = 0$ and $g_y(0, y) = 0$ only if $y = 0, \pm1$, we see that there are three critical points, $(0, 0)$, $(0, -1)$, and $(0, 1)$. Moreover, since

$$g(0, 0) = -1, \qquad g(0, -1) = 0, \qquad \text{and} \qquad g(0, 1) = 0$$

it follows that $(0, -1)$ and $(0, 1)$ are maximum points on the surface. These two points correspond to the two complex zeros of f.

$$(x, y) = (0, -1) \implies w = x + yi = -i$$
$$(x, y) = (0, 1) \implies w = x + yi = i$$

We used partial derivatives to find the critical points in Example 3 only for the purpose of illustration; we are not recommending that you use this procedure to find the zeros of a polynomial as simple as $f(w) = w^2 + 1$. The point is that a computer can easily be programmed to plot the function $g(x, y) = -|f(x + yi)|^2$, producing a graphical picture of both the real and complex zeros. A practical use of the procedure is demonstrated in Example 4. In this example we use the fact that the complex zeros of a polynomial function with real coefficients must occur in *conjugate pairs* $a + bi$ and $a - bi$.

EXAMPLE 4   An application involving partial fractions

Evaluate the integral

$$\int \frac{2w^2 - 2w - 3}{w^4 + 2w^2 + w + 2} \, dw$$

Solution: Since the integrand is a rational function, we can integrate using partial fractions, *provided* we are able to factor the denominator. After some investigation, we determine that the polynomial

$$f(w) = w^4 + 2w^2 + w + 2$$

has no *real* zeros and we try to find a factorization of the form

$$f(w) = w^4 + 2w^2 + w + 2 = (w^2 + aw + b)(w^2 + cw + d)$$

where $bd = 2$. To find the coefficients a, b, c, and d, we use a computer to plot the function given by $g(x, y) = -|f(x + yi)|^2$. From Figure 15.69, it appears that the zeros occur at approximately $(-0.5, \pm0.85)$ and

$(x, y) \approx (-0.5, 0.85)$ $(x, y) \approx (0.5, 1.3)$

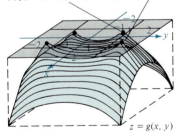

FIGURE 15.69

(0.5, ± 1.3). Moreover, by multiplying the factors corresponding to complex conjugates, we have

$$[w - (-0.5 + 0.85i)][w - (-0.5 - 0.85i)] = w^2 + w + 0.9725$$
$$[w - (0.5 + 1.3i)][w - (0.5 - 1.3i)] = w^2 - w + 1.94$$

From this, we guess that

$$w^2 + aw + b = w^2 + w + 1$$

and

$$w^2 + cw + d = w^2 - w + 2$$

Multiplication of these two polynomials confirms our guess and we have

$$f(w) = w^4 + 2w^2 + w + 2 = (w^2 + w + 1)(w^2 - w + 2)$$

Therefore, we can obtain the partial fraction decomposition

$$\frac{2w^2 - 2w - 3}{w^4 + 2w^2 + w + 2} = \frac{2w - 1}{w^2 - w + 2} - \frac{2w + 1}{w^2 + w + 1}$$

and the integral evaluates as

$$\int \frac{2w^2 - 2w - 3}{w^4 + 2w^2 + w + 2}\, dw = \ln (w^2 - w + 2) - \ln (w^2 + w + 1) + C$$

| Remark   The method in Example 4 can be used as the basis of a numerical technique for approximating the complex zeros of polynomials, much in the same way we use Newton's Method to approximate real zeros.

Section Exercises 15.9

In Exercises 1 and 2, find the minimum distance from the point to the plane given by $2x + 3y + z = 12$.

1. (0, 0, 0)

2. (1, 2, 3)

In Exercises 3 and 4, find the minimum distance from the given point to the paraboloid given by $z = x^2 + y^2$.

3. (5, 5, 0)

4. (5, 0, 0)

In Exercises 5–8, find three positive numbers x, y, and z satisfying the given conditions.

5. The sum is 30 and product is maximum.
6. The sum is 32 and $P = xy^2z$ is maximum.
7. The sum is 30 and the sum of the squares is minimum.
8. The sum is 1 and the sum of the squares is minimum.

9. The sum of the length and the girth (perimeter of a cross section) of packages carried by parcel post cannot exceed 108 inches. Find the dimensions of the rectangular package of largest volume that may be sent by parcel post.

10. The material for constructing the base of an open box costs 1.5 times as much as the material for constructing the sides. For a fixed amount of money C, find the dimensions of the box of largest volume that can be made.

11. The volume of an ellipsoid

$$\frac{x^2}{a^2} + \frac{y^2}{b^2} + \frac{z^2}{c^2} = 1$$

is $4\pi abc/3$. For fixed $a + b + c$, show that the ellipsoid of maximum volume is a sphere.

12. Show that the rectangular box of maximum volume inscribed in a sphere of radius r is a cube.

13. Show that the rectangular box of given volume and minimum surface area is a cube.

14. An eaves trough with trapezoidal cross sections is formed by turning up the edges of a 10-inch-wide sheet of aluminum. Find the cross section of maximum area.

15. Repeat Exercise 14 for a sheet of aluminum that is w inches wide.

16. A company manufactures two products. The total revenue from x_1 units of product 1 and x_2 units of product 2 is

$$R = -5x_1{}^2 - 8x_2{}^2 - 2x_1x_2 + 42x_1 + 102x_2$$

Find x_1 and x_2 so as to maximize the revenue.

17. A retail outlet sells two competitive products, the prices of which are p_1 and p_2. Find p_1 and p_2 so as to maximize total revenue, where

$$R = 500p_1 + 800p_2 + 1.5p_1p_2 - 1.5p_1{}^2 - p_2{}^2$$

18. A corporation manufactures a product at two locations. The cost of producing x_1 units at location 1 is

$$C_1 = 0.02x_1{}^2 + 4x_1 + 500$$

and the cost of producing x_2 units at location 2 is

$$C_2 = 0.05x_2{}^2 + 4x_2 + 275$$

If the product sells for \$15 per unit, find the quantity that should be produced at each location to maximize the profit, $P = 15(x_1 + x_2) - C_1 - C_2$.

19. Find a system of equations whose solution yields the coefficients a, b, and c for the least squares regression quadratic $y = ax^2 + bx + c$ for the points

$$(x_1, y_1), (x_2, y_2), \ldots, (x_n, y_n)$$

by minimizing the sum

$$S(a, b, c) = \sum_{i=1}^{n} (y_i - ax_i{}^2 - bx_i - c)^2$$

20. Use the Second-Partials Test to verify that the formulas for a and b given in Theorem 15.17 yield a minimum.

In Exercises 21–24, (a) find the least squares regression line and (b) calculate S, the sum of the squared errors.

21. **22.**

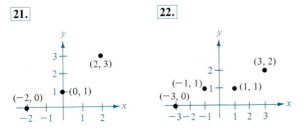

23. **24.**

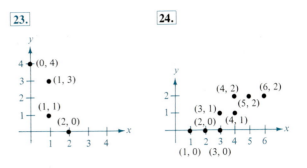

In Exercises 25 and 26, use the result of Exercise 19 to find the least squares regression quadratic for the given points.

25. $(-2, 0)$, $(-1, 0)$, $(0, 1)$, $(1, 2)$, $(2, 5)$

26. $(0, 10)$, $(1, 9)$, $(2, 6)$, $(3, 0)$

27. An agronomist used four test plots to determine the relationship between the wheat yield (in bushels per acre) and the amount of fertilizer (in hundreds of pounds per acre). The results are in the following table.

Fertilizer (x)	1.0	1.5	2.0	2.5
Yield (y)	32	41	48	53

Find the least squares regression line for this data, and estimate the yield for a fertilizer application of 160 pounds per acre.

28. The following table gives the estimated world population (in billions) for four different years. (Let $x = 0$ represent the year 1960.) Find the least squares quadratic for the data.

Year (x)	1960	1970	1974	1976
Population (y)	3.0	3.6	4.0	4.1

29. If $f(w) = aw^2 + bx + c$, show that $f(x + yi)$ is given by

$$\left[f(x) - \frac{f''(x)y^2}{2!} \right] + f'(x)yi$$

30. If $f(w) = aw^3 + bw^2 + cw + d$, show that $f(x + yi)$ is given by

$$\left[f(x) - \frac{f''(x)y^2}{2!} \right] + \left[f'(x)y - \frac{f'''(x)y^3}{3!} \right]i$$

In Exercises 31 and 32, use partial fractions to evaluate the integral

$$\int \frac{h(w)}{f(w)} \, dw$$

In each case, the accompanying figure represents the surface given by

$$g(x, y) = -|f(x + yi)|^2$$

31. $\displaystyle\int \frac{3w^3 + 8w^2 + 7w + 4}{w^4 + 4w^3 + 6w^2 + 4w + 5} \, dw$

32. $\displaystyle\int \frac{w^3 + 2w - 2}{w^4 - 2w^3 + 6w^2 - 8w + 8} \, dw$

15.10
Lagrange multipliers

Many optimization problems have restrictions or **constraints** on the values that can be used to produce the optimal solution. Such constraints tend to complicate optimization problems because the optimal solution can easily occur at a boundary point of the domain. In this section we look at an ingenious technique for solving such problems. It is called the **Method of Lagrange Multipliers,** after the French mathematician Joseph Louis Lagrange (1736–1813). Lagrange first introduced the method in his famous paper on mechanics, written when he was nineteen years old.

We begin with a straightforward constrained optimization problem. Suppose we want to find the rectangle of maximum area that can be inscribed in the ellipse given by

$$\frac{x^2}{3^2} + \frac{y^2}{4^2} = 1$$

Let (x, y) be the vertex of the rectangle in the first quadrant, as shown in Figure 15.70. Then, since the rectangle has sides of length $2x$ and $2y$, its area is given by

$$f(x, y) = 4xy \qquad \text{Objective function}$$

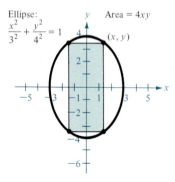

FIGURE 15.70

We want to find x and y such that $f(x, y)$ is a maximum. Our choice of (x, y) is restricted to points that lie on the ellipse.

$$\frac{x^2}{3^2} + \frac{y^2}{4^2} = 1 \qquad \text{Constraint}$$

To see how this problem can be solved by Lagrange multipliers, we consider the constraint equation to be a fixed level curve of

$$g(x, y) = \frac{x^2}{3^2} + \frac{y^2}{4^2}$$

Now, the level curves f represent a family of hyperbolas

$$f(x, y) = k$$

and in this family, the level curves that meet the given constraint correspond to the hyperbolas that intersect the ellipse. Moreover, to maximize $f(x, y)$, we want to find the hyperbola that just barely satisfies the constraint. The level curve that does this is the one that is *tangent* to the ellipse, as shown in Figure 15.71.

To find the appropriate hyperbola, we use the fact that two curves are tangent at a point if and only if their gradient vectors are parallel. This means that $\nabla f(x, y)$ must be a scalar multiple of $\nabla g(x, y)$ at the point of tangency. In the context of constrained optimization problems, we denote this scalar by λ and write

$$\nabla f(x, y) = \lambda \nabla g(x, y)$$

The scalar λ is called a **Lagrange multiplier.** Theorem 15.18 gives the necessary conditions for the existence of such multipliers.

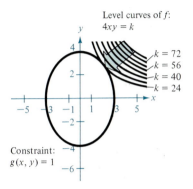

Level curves of f:
$4xy = k$

$k = 72$
$k = 56$
$k = 40$
$k = 24$

Constraint:
$g(x, y) = 1$

FIGURE 15.71

THEOREM 15.18

LAGRANGE'S THEOREM

Let f and g have continuous first partial derivatives such that f has an extremum at a point (x_0, y_0) on the constraint curve $g(x, y) = c$. If $\nabla g(x_0, y_0) \neq 0$, then there is a real number λ such that

$$\nabla f(x_0, y_0) = \lambda \nabla g(x_0, y_0)$$

Proof: The Implicit Function Theorem, discussed in advanced calculus, guarantees that the constraint curve $g(x, y) = c$ is smooth. That is, it can be represented by a vector-valued function

$$\mathbf{r}(t) = x(t)\, \mathbf{i} + y(t)\, \mathbf{j}$$

where x' and y' are continuous on an open interval I. If we define the function h by $h(t) = f(x(t), y(t))$, then, since $f(x_0, y_0)$ is an extreme value of f, we know that

$$h(t_0) = f(x(t_0), y(t_0)) = f(x_0, y_0)$$

is an extreme value of h. This implies that $h'(t_0) = 0$, and, by the Chain Rule,

$$h'(t_0) = f_x(x_0, y_0)x'(t_0) + f_y(x_0, y_0)y'(t_0) = \nabla f(x_0, y_0) \cdot \mathbf{r}'(t_0) = 0$$

Therefore, $\nabla f(x_0, y_0)$ is orthogonal to $\mathbf{r}'(t_0)$. Moreover, by Theorem 15.11, $\nabla g(x_0, y_0)$ is also orthogonal to $\mathbf{r}'(t_0)$. Consequently, the gradients $\nabla f(x_0, y_0)$ and $\nabla g(x_0, y_0)$ are parallel, and there must exist a scalar λ such that

$$\nabla f(x_0, y_0) = \lambda \nabla g(x_0, y_0)$$

| **Remark** Lagrange's Theorem can be shown to be true for functions of three variables, using a similar argument with level surfaces and Theorem 15.13.

The Method of Lagrange Multipliers uses Theorem 15.18 to find the extreme values of a function f subject to a constraint.

METHOD OF LAGRANGE MULTIPLIERS

If f and g satisfy the hypothesis of Lagrange's Theorem and f has a minimum or maximum subject to the constraint $g(x, y) = 0$, then it will occur at one of the critical points of the function F given by

$$F(x, y, \lambda) = f(x, y) - \lambda g(x, y)$$

| **Remark** For functions of three variables, F has the form

$$F(x, y, z, \lambda) = f(x, y, z) - \lambda g(x, y, z)$$

For our first example of the use of Lagrange multipliers, we return to the optimization problem discussed in the beginning of this section.

EXAMPLE 1   Using Lagrange multipliers for optimization with one constraint

Find the maximum value of

$$f(x, y) = 4xy, \quad x > 0, \ y > 0$$

subject to the constraint $(x^2/3^2) + (y^2/4^2) = 1$.

Solution: First, we let

$$g(x, y) = \frac{x^2}{3^2} + \frac{y^2}{4^2} - 1$$

Then, we define F by

$$F(x, y, \lambda) = f(x, y) - \lambda g(x, y) = 4xy - \lambda \left(\frac{x^2}{3^2} + \frac{y^2}{4^2} - 1 \right)$$

By setting the first partial derivatives of F equal to zero, we obtain the system of equations

$$F_x = 4y - \frac{2x\lambda}{9} = 0 \qquad \Rightarrow \qquad 18y - x\lambda = 0$$

$$F_y = 4x - \frac{2y\lambda}{16} = 0 \qquad \Rightarrow \qquad 32x - y\lambda = 0$$

$$F_\lambda = -\frac{x^2}{3^2} - \frac{y^2}{4^2} + 1 = 0 \quad \Rightarrow \quad 16x^2 + 9y^2 = 144$$

By solving for λ in the first equation on the right, we have $\lambda = 18y/x$. Then, by substitution in the second equation, we have

$$32x - y\left(\frac{18y}{x}\right) = 0 \quad \Rightarrow \quad 16x^2 - 9y^2 = 0$$

Now, by subtracting this from the third equation on the right, we have

$$18y^2 = 144 \quad \Rightarrow \quad y^2 = 8$$

Thus, $y = \pm 2\sqrt{2}$. Since it is required that $y > 0$, we choose the positive value and find that

$$16x^2 - 9(8) = 0 \quad \Rightarrow \quad x^2 = \frac{9}{2}$$

Therefore, the only critical point of F is $(3/\sqrt{2}, 2\sqrt{2})$. We know that f has a maximum, since the introductory statement defined the problem as that of finding the area of a rectangle inscribed in an ellipse. Thus, the maximum of f is

$$f\left(\frac{3}{\sqrt{2}}, 2\sqrt{2}\right) = 4\left(\frac{3}{\sqrt{2}}\right)(2\sqrt{2}) = 24 \qquad \qquad \square$$

Note that in Example 1 we were even able to solve for x and y without actually solving for λ. In Example 2 we look at a problem in which it is useful to solve for λ.

EXAMPLE 2   *A business application*

The Cobb-Douglas production function (see Example 4, Section 15.1) for a particular manufacturer is given by

$$f(x, y) = 100x^{3/4}y^{1/4} \qquad \text{Objective function}$$

where x represents the units of labor (at \$150 per unit) and y represents the units of capital (at \$250 per unit). The total cost of labor and capital is limited to \$50,000. Find the maximum production level for this manufacturer.

Solution: From the limit on the cost of labor and capital we have the constraint

$$150x + 250y = 50,000 \qquad \text{Constraint}$$

Using Lagrange's method, we let $g(x, y) = 150x + 250y - 50,000$ and write

$$F(x, y, \lambda) = 100x^{3/4}y^{1/4} - \lambda(150x + 250y - 50,000)$$

Then, the partial derivatives of F are

$$F_x(x, y, \lambda) = 75x^{-1/4}y^{1/4} - 150\lambda = 0$$
$$F_y(x, y, \lambda) = 25x^{3/4}y^{-3/4} - 250\lambda = 0$$
$$F_\lambda(x, y, \lambda) = -150x - 250y + 50{,}000 = 0$$

Solving for λ in the first equation,

$$\lambda = \frac{75x^{-1/4}y^{1/4}}{150} = \frac{x^{-1/4}y^{1/4}}{2}$$

and substituting into the second equation, we have

$$25x^{3/4}y^{-3/4} - 250\left(\frac{x^{-1/4}y^{1/4}}{2}\right) = 0 \qquad \text{Multiply by } x^{1/4}y^{3/4}$$
$$25x - 125y = 0$$

Therefore, $x = 5y$, and, by substituting into the third equation, we have

$$50{,}000 - 150(5y) - 250y = 0$$
$$1000y = 50{,}000$$
$$y = 50 \text{ units of capital}$$
$$x = 250 \text{ units of labor}$$

Finally, the maximum production is

$$f(250, 50) = 100(250)^{3/4}(50)^{1/4} \approx 16{,}719 \text{ product units}$$

Economists call the Lagrange multiplier obtained in a production function the **marginal productivity of money.** For instance, in Example 2 the marginal productivity of money at $x = 250$ and $y = 50$ is

$$\lambda = \frac{x^{-1/4}y^{1/4}}{2} = \frac{(250)^{-1/4}(50)^{1/4}}{2} \approx 0.334$$

which means that for each additional dollar spent on production, 0.334 additional units of the product can be produced.

EXAMPLE 3   *Using Lagrange multipliers with functions of three variables*

Find the minimum value of

$$f(x, y, z) = 2x^2 + y^2 + 3z^2 \qquad \text{Objective function}$$

subject to the constraint $2x - 3y - 4z = 49$.

Solution: Let

$$g(x, y, z) = 2x - 3y - 4z - 49$$

and let F be defined by

$$F(x, y, z, \lambda) = f(x, y, z) - \lambda g(x, y, z)$$
$$= 2x^2 + y^2 + 3z^2 - \lambda(2x - 3y - 4z - 49)$$

To find the critical points of F, we solve the system

$$F_x(x, y, z, \lambda) = 4x - 2\lambda = 0$$
$$F_y(x, y, z, \lambda) = 2y + 3\lambda = 0$$
$$F_z(x, y, z, \lambda) = 6z + 4\lambda = 0$$
$$F_\lambda(x, y, z, \lambda) = -2x + 3y + 4z + 49 = 0$$

$\Rightarrow$ $3x + y = 0$
$\Rightarrow$ $4x + 3z = 0$
$\Rightarrow$ $2x - 3y - 4z = 49$

to obtain $x = 3$, $y = -9$, and $z = -4$. Therefore, the minimum value of f is

$$f(3, -9, -4) = 2(3)^2 + (-9)^2 + 3(-4)^2 = 147$$

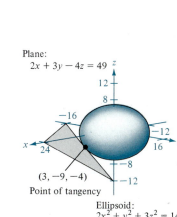

Plane:
$2x + 3y - 4z = 49$

$(3, -9, -4)$
Point of tangency

Ellipsoid:
$2x^2 + y^2 + 3z^2 = 147$

FIGURE 15.72

At the beginning of this section we gave a graphical interpretation of constrained optimization problems in two variables. In three variables, the interpretation is similar, except that we use level surfaces instead of level curves. For instance, in Example 3, the level surfaces of f are ellipsoids centered at the origin, and the constraint $2x - 3y - 4z = 49$ is a plane. The minimum value of f is represented by the ellipsoid that is tangent to the constraint plane, as shown in Figure 15.72.

For optimization problems with *two* constraints, we introduce a second Lagrange Multiplier, μ (the lowercase Greek letter mu), as illustrated in the next example.

EXAMPLE 4   Optimization with two constraints

Let

$$T(x, y, z) = 20 + 2x + 2y + z^2 \qquad \text{Objective function}$$

represent the temperature at each point on the sphere $x^2 + y^2 + z^2 = 11$. Find the extreme temperatures on the curve formed by the intersection of the plane $x + y + z = 3$ and the sphere.

Solution: In this case we have two constraints,

$$g(x, y, z) = x^2 + y^2 + z^2 - 11 = 0 \qquad \text{Constraint 1}$$
$$h(x, y, z) = x + y + z - 3 = 0 \qquad \text{Constraint 2}$$

We let the function F be defined by

$$F(x, y, z, \lambda, \mu)$$
$$= T(x, y, z) - \lambda g(x, y, z) - \mu h(x, y, z)$$
$$= 20 + 2x + 2y + z^2 - \lambda(x^2 + y^2 + z^2 - 11) - \mu(x + y + z - 3)$$

Then, by differentiating, we have the system

$$F_x(x, y, z, \lambda, \mu) = 2 - 2x\lambda - \mu = 0$$
$$F_y(x, y, z, \lambda, \mu) = 2 - 2y\lambda - \mu = 0$$
$$F_z(x, y, z, \lambda, \mu) = 2z - 2z\lambda - \mu = 0$$
$$F_\lambda(x, y, z, \lambda, \mu) = -x^2 - y^2 - z^2 + 11 = 0$$
$$F_\mu(x, y, z, \lambda, \mu) = -x - y - z + 3 = 0$$

$\Rightarrow$ $\lambda(x - y) = 0$
$\Rightarrow$ $2z(1 - \lambda) - \mu = 0$
$\Rightarrow$ $x^2 + y^2 + z^2 = 11$
$\Rightarrow$ $x + y + z = 3$

From the first equation on the right we conclude that $\lambda = 0$ or $x = y$. We consider these two cases separately. If $\lambda = 0$, then the first equation on the left yields $\mu = 2$, and, by substituting into the second equation on the right, we have $z = 1$. The last two equations on the right now become

$$x^2 + y^2 = 10 \qquad \text{and} \qquad x + y = 2$$

and, by substitution, we obtain

$$x^2 + (2 - x)^2 - 10 = 0$$
$$2x^2 - 4x - 6 = 0$$
$$x^2 - 2x - 3 = 0$$

with solutions $x = 3$ and $x = -1$. The corresponding y-values are $y = -1$ and $y = 3$, respectively. Thus, if $\lambda = 0$, the critical points are $(3, -1, 1)$ and $(-1, 3, 1)$. If $\lambda \neq 0$, then $x = y$, and the last two equations become

$$2x^2 + z^2 = 11 \qquad \text{and} \qquad 2x + z = 3$$

and, by substitution, we obtain

$$2x^2 + (3 - 2x)^2 = 11 \implies 3x^2 - 6x - 1 = 0$$

The solutions are $x = (3 \pm 2\sqrt{3})/3$, and the corresponding values of z are $z = (3 \mp 4\sqrt{3})/3$, which yield two additional critical points. Finally, to find the optimal solutions, we compare the temperatures at the four critical points

$$T(3, -1, 1) = T(-1, 3, 1) = 25$$

$$T\left(\frac{3 - 2\sqrt{3}}{3}, \frac{3 - 2\sqrt{3}}{3}, \frac{3 + 4\sqrt{3}}{3}\right) = \frac{91}{3} \approx 30.33$$

$$T\left(\frac{3 + 2\sqrt{3}}{3}, \frac{3 + 2\sqrt{3}}{3}, \frac{3 - 4\sqrt{3}}{3}\right) = \frac{91}{3} \approx 30.33$$

Thus, $T = 25$ is the minimum temperature and $T = \frac{91}{3}$ is the maximum temperature on the curve.

The system of equations that arises in the method of Lagrange multipliers is not, in general, a linear system, and the solution of the system often requires some ingenuity.

Section Exercises 15.10

In Exercises 1–20, use Lagrange multipliers to find the indicated extrema. In each case, assume that x, y, and z are positive.

1. Maximize $f(x, y) = xy$
   Constraint: $x + y = 10$
2. Maximize $f(x, y) = xy$
   Constraint: $2x + y = 4$

3. Minimize $f(x, y) = x^2 + y^2$
   Constraint: $x + y - 4 = 0$
4. Minimize $f(x, y) = x^2 + y^2$
   Constraint: $2x - 4y + 5 = 0$
5. Maximize $f(x, y) = x^2 - y^2$
   Constraint: $y - x^2 = 0$
6. Maximize $f(x, y) = x^2 - y^2$
   Constraint: $x - 2y + 6 = 0$

7. Maximize $f(x, y) = 2x + 2xy + y$
   Constraint: $2x + y = 100$
8. Maximize $f(x, y) = 3x + y + 10$
   Constraint: $x^2 y = 6$
9. Maximize $f(x, y) = \sqrt{6 - x^2 - y^2}$
   Constraint: $x + y - 2 = 0$
10. Minimize $f(x, y) = \sqrt{x^2 + y^2}$
    Constraint: $2x + 4y - 15 = 0$
11. Maximize $f(x, y) = e^{xy}$
    Constraint: $x^2 + y^2 - 8 = 0$
12. Minimize $f(x, y) = 2x + y$
    Constraint: $xy = 32$
13. Minimize $f(x, y, z) = x^2 + y^2 + z^2$
    Constraint: $x + y + z - 6 = 0$
14. Maximize $f(x, y, z) = xyz$
    Constraint: $x + y + z - 6 = 0$
15. Maximize $f(x, y, z) = xyz$
    Constraints: $x + y + z = 32$
    $\qquad\qquad x - y + z = 0$
16. Minimize $f(x, y, z) = x^2 + y^2 + z^2$
    Constraints: $x + 2z = 4$
    $\qquad\qquad x + y = 8$
17. Minimize $f(x, y, z) = x^2 + y^2 + z^2$
    Constraint: $x + y + z = 1$
18. Maximize $f(x, y, z) = xy + yz$
    Constraints: $x + 2y = 6$
    $\qquad\qquad x - 3z = 0$
19. Maximize $f(x, y, z) = xyz$
    Constraints: $x^2 + z^2 = 5$
    $\qquad\qquad x - 2y = 0$
20. Minimize $f(x, y) = x^2 - 8x + y^2 - 12y + 48$
    Constraint: $x + y = 8$

21. Find the dimensions of the rectangular package of largest volume subject to the constraint that the sum of the length and the girth cannot exceed 108 inches. (Maximize $V = xyz$ subject to the constraint $x + 2y + 2z = 108$.)
22. The material for the base of an open box costs 1.5 times as much as the material for the sides. Find the dimensions of the box of largest volume that can be made for a fixed cost C. (Maximize $V = xyz$ subject to $1.5xy + 2xz + 2yz = C$.)
23. Let

$$T(x, y, z) = 100 + x^2 + y^2$$

represent the temperature at each point on the sphere $x^2 + y^2 + z^2 = 50$. Find the maximum temperature on the curve formed by the intersection of the sphere and the plane $x - z = 0$.
24. Repeat Exercise 23 using the plane given by $x + y + z = 5$.

In Exercises 25–28, find the minimum distance from the curve or surface to the specified point. (Hint: In Exercise 25, minimize $z = x^2 + y^2$ subject to the constraint $2x + 3y = -1$.)

Curve	Point
25. Line: $2x + 3y = -1$	$(0, 0)$
26. Circle: $(x - 4)^2 + y^2 = 4$	$(0, 10)$

Surface	Point
27. Plane: $x + y + z = 1$	$(2, 1, 1)$
28. Cone: $z = \sqrt{x^2 + y^2}$	$(4, 0, 0)$

In Exercises 29 and 30, find the maximum production level if the total cost of labor (at \$48 per unit) and capital (at \$36 per unit) is limited to \$100,000.

29. $P(x, y) = 100x^{0.25}y^{0.75}$
30. $P(x, y) = 100x^{0.6}y^{0.4}$

In Exercises 31 and 32, find the minimum cost of producing 20,000 product units, where x is the number of units of labor (at \$48 per unit) and y is the number of units of capital (at \$36 per unit).

31. $P(x, y) = 100x^{0.25}y^{0.75}$ 32. $P(x, y) = 100x^{0.6}y^{0.4}$

In Exercises 33 and 34, find the highest point on the curve of intersection of the given surfaces.

33. Sphere: $\quad x^2 + y^2 + z^2 = 36$
    Plane: $\qquad 2x + y - z = 2$
34. Cone: $\quad x^2 + y^2 - z^2 = 0$
    Plane: $\quad x + 2z = 4$

35. Find the dimensions of the rectangular box of maximum volume that can be inscribed (with edges parallel to the coordinate axes) in the ellipsoid

$$\frac{x^2}{a^2} + \frac{y^2}{b^2} + \frac{z^2}{c^2} = 1$$

36. Prove that the product of three positive numbers whose sum is S is maximum if the three numbers are equal.

Review Exercises for Chapter 15

In Exercises 1–4, discuss the continuity of the function and evaluate the limit, if it exists.

1. $\displaystyle\lim_{(x,y)\to(1,1)} \frac{xy}{x^2 + y^2}$

2. $\displaystyle\lim_{(x,y)\to(1,1)} \frac{xy}{x^2 - y^2}$

3. $\displaystyle\lim_{(x,y)\to(0,0)} \frac{x^2 y}{x^4 + y^2}$

4. $\displaystyle\lim_{(x,y)\to(0,0)} \frac{y + xe^{-y^2}}{1 + x^2}$

In Exercises 5–8, sketch several level curves for the given function.

5. $f(x, y) = e^{x^2 + y^2}$

6. $f(x, y) = \ln xy$

7. $f(x, y) = x^2 - y^2$

8. $f(x, y) = \dfrac{x}{x + y}$

In Exercises 9–18, find all first partial derivatives.

9. $f(x, y) = e^x \cos y$

10. $f(x, y) = \dfrac{xy}{x + y}$

11. $z = xe^y + ye^x$

12. $z = \ln (x^2 + y^2 + 1)$

13. $g(x, y) = \dfrac{xy}{x^2 + y^2}$

14. $w = \sqrt{x^2 + y^2 + z^2}$

15. $f(x, y, z) = z \arctan \dfrac{y}{x}$

16. $f(x, y, z) = \dfrac{1}{\sqrt{1 - x^2 - y^2 - z^2}}$

17. $u(x, t) = ce^{-n^2 t} \sin nx$

18. $u(x, t) = c \sin (akx) \cos kt$

In Exercises 19 and 20, find $\partial z/\partial x$ and $\partial z/\partial y$.

19. $x^2 y - 2xyz - xz - z^2 = 0$

20. $xz^2 - y \sin z = 0$

In Exercises 21–24, find all second partial derivatives and verify that the second mixed partials are equal.

21. $f(x, y) = 3x^2 - xy + 2y^3$

22. $h(x, y) = \dfrac{x}{x + y}$

23. $h(x, y) = x \sin y + y \cos x$

24. $g(x, y) = \cos (x - 2y)$

In Exercises 25–28, show that the function satisfies the **Laplace equation**

$$\frac{\partial^2 z}{\partial x^2} + \frac{\partial^2 z}{\partial y^2} = 0$$

25. $z = x^2 - y^2$

26. $z = x^3 - 3xy^2$

27. $z = \dfrac{y}{x^2 + y^2}$

28. $z = e^x \sin y$

In Exercises 29 and 30, find dz.

29. $z = x \sin \dfrac{y}{x}$

30. $z = \dfrac{xy}{\sqrt{x^2 + y^2}}$

In Exercises 31 and 32, find the indicated derivatives (a) by the Chain Rule and (b) by substitution before differentiating.

31. $u = x^2 + y^2 + z^2,\ \left(\dfrac{\partial u}{\partial r}, \dfrac{\partial u}{\partial t}\right)$

$x = r \cos t,\ y = r \sin t,\ z = t$

32. $u = y^2 - x,\ \left(\dfrac{du}{dt}\right)$

$x = \cos t,\ y = \sin t$

In Exercises 33–36, find the directional derivative at the given point in the direction of **v**.

33. $f(x, y) = x^2 y$
   $\mathbf{v} = \mathbf{i} - \mathbf{j},\ (2, 1)$

34. $f(x, y) = \dfrac{1}{4} y^2 - x^2$
   $\mathbf{v} = 2\mathbf{i} + \mathbf{j},\ (1, 4)$

35. $f(x, y, z) = y^2 + xz$
   $\mathbf{v} = 2\mathbf{i} - \mathbf{j} + 2\mathbf{k},\ (1, 2, 2)$

36. $f(x, y, z) = 6x^2 + 3xy - 4y^2 z$
   $\mathbf{v} = \mathbf{i} + \mathbf{j} - \mathbf{k},\ (1, 0, 1)$

In Exercises 37–40, find the gradient and the maximum value of the directional derivative of the function at the specified point.

Function	Point
37. $z = \dfrac{y}{x^2 + y^2}$	$(1, 1)$
38. $z = \dfrac{x^2}{x - y}$	$(2, 1)$
39. $z = e^{-x} \cos y$	$\left(0, \dfrac{\pi}{4}\right)$
40. $z = x^2 y$	$(2, 1)$

In Exercises 41–44, find an equation of the tangent plane and equations for the normal line to the given surface at the specified point.

Surface	Point
41. $f(x, y) = x^2 y$	$(2, 1, 4)$
42. $f(x, y) = \sqrt{25 - y^2}$	$(2, 3, 4)$
43. $z = -9 + 4x - 6y - x^2 - y^2$	$(2, -3, 4)$
44. $z = \sqrt{9 - x^2 - y^2}$	$(1, 2, 2)$

In Exercises 45 and 46, find symmetric equations of the tangent line to the curve of intersection of the given surfaces at the indicated point.

Surfaces	Point

45. $z = x^2 - y^2, z = 3$ (2, 1, 3)

46. $z = 25 - y^2, y = x$ (4, 4, 9)

In Exercises 47–50, locate and classify any extrema of the function.

47. $f(x, y) = x^3 - 3xy + y^2$

48. $f(x, y) = 2x^2 + 6xy + 9y^2 + 8x + 14$

49. $f(x, y) = xy + \dfrac{1}{x} + \dfrac{1}{y}$

50. $z = 50(x + y) - (0.1x^3 + 20x + 150)$
$$- (0.05y^3 + 20.6y + 125)$$

In Exercises 51 and 52, locate and classify any extrema of the function by using Lagrange multipliers.

51. $z = x^2y$
Constraint: $x + 2y = 2$

52. $w = xy + yz + xz$
Constraint: $x + y + z = 1$

53. The legs of a right triangle are measured and found to be 5 inches and 12 inches, with a possible error of $\frac{1}{16}$ inch. Approximate the maximum possible error in computing the length of the hypotenuse. Approximate the maximum percentage error.

54. To determine the height of a tower, the angle of elevation to the top of the tower was measured from a point $100 \pm \frac{1}{2}$ foot from the base. The angle is measured at $33°$, with a possible error of $1°$. Assuming the ground to be horizontal, approximate the maximum error in determining the height of the tower.

55. The volume of a right circular cone is $V = \frac{1}{3}\pi r^2 h$. Find the approximate error in the volume due to possible errors of $\frac{1}{8}$ inch in the measured values of r and h, if these are found to be 2 and 5 inches, respectively.

56. Approximate the error in the lateral surface area of the cone of Exercise 55. (The lateral surface area is given by $A = \pi r \sqrt{r^2 + h^2}$.)

57. A corporation manufactures a product at two locations. The cost functions for producing x_1 units at location 1 and x_2 units at location 2 are given by

$$C_1 = 0.05x_1{}^2 + 15x_1 + 5400$$
$$C_2 = 0.03x_2{}^2 + 15x_2 + 6100$$

and the total revenue function is

$$R = [225 - 0.4(x_1 + x_2)](x_1 + x_2)$$

Find the production levels at the two locations that will maximize the profit $P(x_1, x_2) = R - C_1 - C_2$.

58. A manufacturer has an order for 1000 units that can be produced at two locations. Let x_1 and x_2 be the number of units produced at the two locations. Find the number that should be produced at each to meet the order and minimize cost, if the cost function is

$$C = 0.25x_1{}^2 + 10x_1 + 0.15x_2{}^2 + 12x_2$$

59. The production function for a manufacturer is

$$f(x, y) = 4x + xy + 2y$$

Assume that the total amount available for labor and capital is $2000, and that units of labor and capital cost $20 and $4, respectively. Find the maximum production level for this manufacturer.

60. Show that a triangle is equilateral if the product of the sines of its angles is maximum.

16 Multiple integration

16.1
Iterated integrals and area in the plane

In the last three chapters of this text, we will survey a variety of applications of integration involving functions of several variables. This first chapter is much like Chapter 6 in that we will use integration to find plane areas, volumes, surface areas, moments, and centers of mass.

In Chapter 15 we saw that it is meaningful to differentiate functions of several variables with respect to one variable while holding the other variables constant. We can *integrate* functions of several variables by a similar procedure. For example, if we are given the partial derivative $f_x(x, y) = 2xy$, then, by considering y constant, we can integrate with respect to x to obtain

$$f(x, y) = \int 2xy \, dx = y \int 2x \, dx = y(x^2) + C(y) = x^2y + C(y)$$

Note that the "constant" of integration, $C(y)$, is actually a function of y. In other words, by integrating with respect to x, we are able to recover $f(x, y)$ only partially. The total recovery of a function of x and y from its partial derivatives is a topic we will study in Chapter 16. For now, we are more concerned with the extension of definite integrals to functions of several variables. For instance, by considering y constant, we can apply the Fundamental Theorem of Calculus to evaluate

$$\int_1^{2y} 2xy \, dx = x^2y \Big]_1^{2y} = (2y)^2y - (1)^2y = 4y^3 - y$$

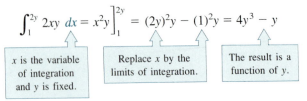

x is the variable of integration and y is fixed.

Replace x by the limits of integration.

The result is a function of y.

872

Notice that we omit the constant of integration, just as we do for the definite integral of a function of one variable.

We can integrate with respect to y in a similar way, holding x fixed, and we summarize both procedures as follows:

$$\int_{h_1(y)}^{h_2(y)} f_x(x,\ y)\ dx = f(x,\ y)\Big]_{h_1(y)}^{h_2(y)} = f(h_2(y),\ y) - f(h_1(y),\ y)$$

$$\int_{g_1(x)}^{g_2(x)} f_y(x,\ y)\ dy = f(x,\ y)\Big]_{g_1(x)}^{g_2(x)} = f(x,\ g_2(x)) - f(x,\ g_1(x))$$

EXAMPLE 1   *Integrating with respect to y*

Evaluate the integral

$$\int_1^x (2x^2y^{-2} + 2y)\ dy$$

Solution: Considering x to be constant and integrating with respect to y, we obtain

$$\int_1^x (2x^2y^{-2} + 2y)\ dy = \left[\frac{-2x^2}{y} + y^2\right]_1^x$$

$$= \left(\frac{-2x^2}{x} + x^2\right) - \left(\frac{-2x^2}{1} + 1\right) = 3x^2 - 2x - 1 \quad \square$$

Notice in Example 1 that the integral defines a function of x and can *itself* be integrated, as shown in the next example.

EXAMPLE 2   *The integral of an integral*

Evaluate the integral

$$\int_1^2 \left[\int_1^x (2x^2y^{-2} + 2y)\ dy\right] dx$$

Solution: Using the results from Example 1, we have

$$\int_1^2 \left[\int_1^x (2x^2y^{-2} + 2y)\ dy\right] dx = \int_1^2 (3x^2 - 2x - 1)\ dx$$

$$= \left[x^3 - x^2 - x\right]_1^2 = 2 - (-1) = 3 \quad \square$$

Iterated integrals

Integrals of the form

$$\int_a^b \left[\int_{g_1(x)}^{g_2(x)} f(x,\ y)\ dy\right] dx \qquad \text{and} \qquad \int_c^d \left[\int_{h_1(y)}^{h_2(y)} f(x,\ y)\ dx\right] dy$$

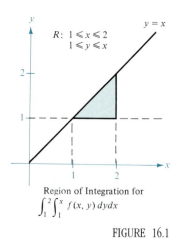

Region of Integration for
$$\int_1^2 \int_1^x f(x, y)\, dy\, dx$$

FIGURE 16.1

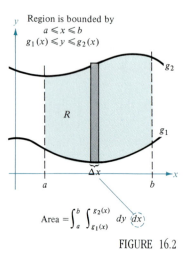

$$\text{Area} = \int_a^b \int_{g_1(x)}^{g_2(x)} dy\ (dx)$$

FIGURE 16.2

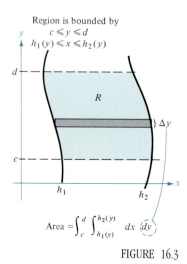

$$\text{Area} = \int_c^d \int_{h_1(y)}^{h_2(y)} dx\ (dy)$$

FIGURE 16.3

are called **iterated integrals**, because the integration is repeated twice. We normally omit the brackets and write simply

$$\int_a^b \int_{g_1(x)}^{g_2(x)} f(x, y)\, dy\, dx \qquad \text{and} \qquad \int_c^d \int_{h_1(y)}^{h_2(y)} f(x, y)\, dx\, dy$$

The **inside limits of integration** can be variable with respect to the outer variable of integration. However, the **outside limits of integration** must be constant with respect to both variables of integration. After performing the inside integration, we are left with a standard definite integral, and the second integration produces a real number. The limits of integration for an iterated integral identify two sets of boundary intervals for the variables. For instance, in Example 2, the outside limits indicate that x lies in the interval $1 \le x \le 2$ and the inside limits indicate that y lies in the interval $1 \le y \le x$. Taken together, these two intervals determine the **region of integration** R of the iterated integral, as shown in Figure 16.1.

| **Remark** An iterated integral is really just a special type of definite integral—one in which the integrand is also an integral. Hence, we can use the properties of definite integrals to evaluate iterated integrals.

The area of a plane region

In the remainder of this section we take a new look at an old problem—the problem of finding the area of a plane region. Consider the plane region R bounded by $a \le x \le b$ and $g_1(x) \le y \le g_2(x)$, as shown in Figure 16.2. The area of R is given by the definite integral

$$\int_a^b [g_2(x) - g_1(x)]\, dx$$

Using the Fundamental Theorem of Calculus, we can rewrite the integrand $g_2(x) - g_1(x)$ itself as a definite integral. Specifically, if we consider x to be fixed and let y vary from $g_1(x)$ to $g_2(x)$, we can write

$$\int_{g_1(x)}^{g_2(x)} dy = y \Big]_{g_1(x)}^{g_2(x)} = g_2(x) - g_1(x)$$

Combining these two integrals, we can write the area of the region R as an iterated integral

$$\int_a^b \int_{g_1(x)}^{g_2(x)} dy\, dx = \int_a^b y \Big]_{g_1(x)}^{g_2(x)} dx = \int_a^b [g_2(x) - g_1(x)]\, dx$$

Placing a representative rectangle in the region R helps determine both the order and the limits of integration. A vertical rectangle implies the order $dy\, dx$, with the inside limits corresponding to the upper and lower bounds of the rectangle, as shown in Figure 16.2. We call this type of region **vertically simple**, since the outside limits of integration represent the vertical lines $x = a$ and $x = b$. Similarly, a horizontal rectangle implies the order $dx\, dy$, with the inside limits determined by the left and right bounds of the rectangle, as shown in Figure 16.3. This type of region is called **horizontally simple**, since the outside limits represent the horizontal lines $y = c$ and $y = d$. We summarize the iterated integrals for these two types of simple regions as follows.

AREA OF A REGION IN THE PLANE

1. If R is defined by $a \le x \le b$ and $g_1(x) \le y \le g_2(x)$, where g_1 and g_2 are continuous on $[a, b]$, then the area of R is given by

$$A = \int_a^b \int_{g_1(x)}^{g_2(x)} dy \, dx$$

2. If R is defined by $c \le y \le d$ and $h_1(y) \le x \le h_2(y)$, where h_1 and h_2 are continuous on $[c, d]$, then the area of R is given by

$$A = \int_c^d \int_{h_1(y)}^{h_2(y)} dx \, dy$$

EXAMPLE 3   Finding area by an iterated integral

Use an iterated integral to find the area of the region bounded by the graphs of $f(x) = \sin x$ and $g(x) = \cos x$ between $x = \pi/4$ and $x = 5\pi/4$.

Solution: Since f and g are given as functions of x, a vertical representative rectangle is convenient, and we choose $dy \, dx$ as the order of integration, as shown in Figure 16.4. The outside limits of integration are $\pi/4 \le x \le 5\pi/4$. Moreover, since the rectangle is bounded above by $f(x) = \sin x$ and below by $g(x) = \cos x$, we have

$$\text{area of } R = \int_{\pi/4}^{5\pi/4} \int_{\cos x}^{\sin x} dy \, dx = \int_{\pi/4}^{5\pi/4} y \Big]_{\cos x}^{\sin x} dx$$

$$= \int_{\pi/4}^{5\pi/4} (\sin x - \cos x) \, dx$$

$$= \Big[-\cos x - \sin x \Big]_{\pi/4}^{5\pi/4}$$

$$= 2\sqrt{2}$$

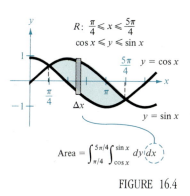

$R: \dfrac{\pi}{4} \le x \le \dfrac{5\pi}{4}$

$\cos x \le y \le \sin x$

$y = \cos x$

$y = \sin x$

$\text{Area} = \int_{\pi/4}^{5\pi/4} \int_{\cos x}^{\sin x} dy \, dx$

FIGURE 16.4

| Remark   In Figure 16.4, we can see that the region of integration of an iterated integral need not have linear boundaries. If all four limits of integration happen to be constant, the region of integration is rectangular.

It often happens that one of the orders of integration will produce a simpler integration problem than the other order. For instance, try reworking Example 3 with the order $dx \, dy$—you may be surprised to see that the task is quite formidable! However, if you succeed, you will see that the answer is the same. In other words, the order of integration affects the ease of integration, but not the value of the integral. This is further demonstrated in the next example.

EXAMPLE 4   *Comparing different orders of integration*

Sketch the region whose area is represented by the integral

$$\int_0^2 \int_{y^2}^4 dx\, dy$$

Then find another iterated integral using the order $dy\, dx$ to represent the same area.

Solution: From the given limits of integration, we know that

$$y^2 \le x \le 4 \qquad \text{Inner limits of integration}$$

which means that the region R is bounded on the left by the parabola $x = y^2$ and on the right by the line $x = 4$. Furthermore, since

$$0 \le y \le 2 \qquad \text{Outer limits of integration}$$

we have the region shown in Figure 16.5. Now, to change the order of integration to $dy\, dx$, we place a vertical rectangle in the region, as shown in Figure 16.6. From this we see that the constant bounds $0 \le x \le 4$ serve as the outer limits of integration. Solving for y in the equation $x = y^2$, we conclude that the inner bounds are $0 \le y \le \sqrt{x}$. Therefore, the area of the region can also be represented by

$$\int_0^4 \int_0^{\sqrt{x}} dy\, dx$$

Evaluating these two integrals, we see that they both yield the same value.

$$\int_0^2 \int_{y^2}^4 dx\, dy = \int_0^2 x\Big]_{y^2}^4 dy = \int_0^2 (4 - y^2)\, dy = \left[4y - \frac{y^3}{3}\right]_0^2 = \frac{16}{3}$$

$$\int_0^4 \int_0^{\sqrt{x}} dy\, dx = \int_0^4 y\Big]_0^{\sqrt{x}} dx = \int_0^4 \sqrt{x}\, dx = \frac{2}{3}x^{3/2}\Big]_0^4 = \frac{16}{3}$$

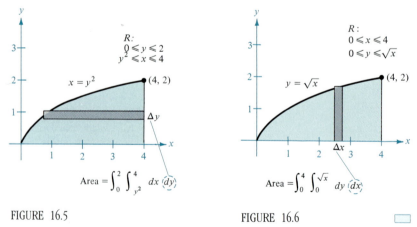

FIGURE 16.5   FIGURE 16.6

It is not always possible to calculate the area of a region with a single iterated integral. In such cases we divide the region into subregions such that the area of each subregion can be calculated by an iterated integral. The total area is then the sum of the iterated integrals. We illustrate this case in Example 5.

EXAMPLE 5 An area represented by two iterated integrals

Find the area of the region R that lies below the parabola $y = 4x - x^2$, above the x-axis, and above the line $y = -3x + 6$, as shown in Figure 16.7.

Solution: We begin by dividing R into the two subregions R_1 and R_2 shown in Figure 16.7. In both regions it is convenient to use vertical rectangles, and we have

$$\text{area} = \int_1^2 \int_{-3x+6}^{4x-x^2} dy \, dx + \int_2^4 \int_0^{4x-x^2} dy \, dx$$

$$= \int_1^2 (4x - x^2 + 3x - 6) \, dx + \int_2^4 (4x - x^2) \, dx$$

$$= \left[\frac{7x^2}{2} - \frac{x^3}{3} - 6x \right]_1^2 + \left[2x^2 - \frac{x^3}{3} \right]_2^4$$

$$= \left(14 - \frac{8}{3} - 12 - \frac{7}{2} + \frac{1}{3} + 6 \right) + \left(32 - \frac{64}{3} - 8 + \frac{8}{3} \right) = \frac{15}{2}$$

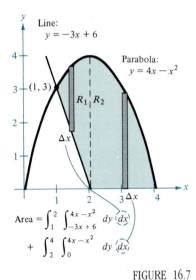

Line:
$y = -3x + 6$

Parabola:
$y = 4x - x^2$

$(1, 3)$

$R_1 \quad R_2$

Δx

$3 \, \Delta x$

$\text{Area} = \int_1^2 \int_{-3x+6}^{4x-x^2} dy \, dx$

$+ \int_2^4 \int_0^{4x-x^2} dy \, dx$

FIGURE 16.7

| Remark In Examples 3, 4, and 5, be sure you see the benefit of sketching the region of integration. We strongly recommend that you develop the habit of making sketches as an aid to determining the limits of integration for all iterated integrals in this chapter.

At this point you may be wondering why we need iterated integrals. After all, we already know how to use conventional integration to find the area of a region in the plane. The need for iterated integrals will become clear in the next section. In this section we have chosen to give primary attention to procedures for finding the limits of integration of the region of an iterated integral, and the following exercise set is designed to develop your skill in this important procedure.

Section Exercises 16.1

In Exercises 1–10, evaluate the given integral.

1. $\int_0^x (2x - y) \, dy$

2. $\int_x^{x^2} \frac{y}{x} \, dy$

3. $\int_1^{2y} \frac{y}{x} \, dx$

4. $\int_0^{\cos y} y \, dx$

5. $\int_0^{\sqrt{4-x^2}} x^2 y \, dy$

6. $\int_{x^2}^{\sqrt{x}} (x^2 + y^2) \, dy$

7. $\int_{e^y}^y \frac{y \ln x}{x} \, dx$

8. $\int_{-\sqrt{1-y^2}}^{\sqrt{1-y^2}} (x^2 + y^2) \, dx$

9. $\int_0^{x^3} y e^{-y/x} \, dy$

10. $\int_y^{\pi/2} \sin^3 x \cos y \, dx$

In Exercises 11–20, evaluate the given iterated integral.

11. $\int_0^1 \int_0^2 (x + y) \, dy \, dx$

12. $\int_0^1 \int_0^x \sqrt{1 - x^2} \, dy \, dx$

13. $\int_1^2 \int_0^4 (x^2 - 2y^2 + 1)\, dx\, dy$

14. $\int_0^1 \int_y^{2y} (1 + 2x^2 + 2y^2)\, dx\, dy$

15. $\int_0^1 \int_0^{\sqrt{1-y^2}} (x + y)\, dx\, dy$

16. $\int_0^2 \int_{3y^2-6y}^{2y-y^2} 3y\, dx\, dy$

17. $\int_0^2 \int_0^{\sqrt{4-y^2}} \dfrac{2}{\sqrt{4-y^2}}\, dx\, dy$

18. $\int_0^{\pi/2} \int_0^{2\cos\theta} r\, dr\, d\theta$

19. $\int_0^{\pi/2} \int_0^{\sin\theta} \theta r\, dr\, d\theta$ **20.** $\int_0^{\infty} \int_0^{\infty} xye^{-(x^2+y^2)}\, dx\, dy$

In Exercises 21–28, sketch the region R whose area is given by the iterated integral. Then switch the order of integration and show that both orders yield the same area.

21. $\int_0^1 \int_0^2 dy\, dx$ **22.** $\int_1^2 \int_2^4 dx\, dy$

23. $\int_0^1 \int_{-\sqrt{1-y^2}}^{\sqrt{1-y^2}} dx\, dy$

24. $\int_0^2 \int_0^x dy\, dx + \int_2^4 \int_0^{4-x} dy\, dx$

25. $\int_0^2 \int_{x/2}^1 dy\, dx$ **26.** $\int_0^4 \int_{\sqrt{x}}^2 dy\, dx$

27. $\int_0^1 \int_{y^2}^{\sqrt[3]{y}} dx\, dy$ **28.** $\int_{-2}^2 \int_0^{4-y^2} dx\, dy$

In Exercises 29–34, use a double integral to find the area of the specified region.

29.   **30.**

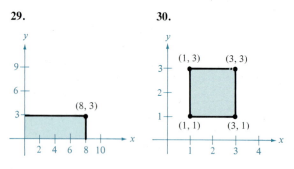

In Exercises 35–40, use an iterated integral to find the area of the region bounded by the graphs of the given equations.

35. $\sqrt{x} + \sqrt{y} = 2$, $x = 0$, $y = 0$
36. $y = x^{3/2}$, $y = x$
37. $2x - 3y = 0$, $x + y = 5$, $y = 0$
38. $xy = 9$, $y = x$, $y = 0$, $x = 9$
39. $\dfrac{x^2}{a^2} + \dfrac{y^2}{b^2} = 1$
40. $y = x$, $y = 2x$, $x = 2$

31.   **32.**   **33.**   **34.**

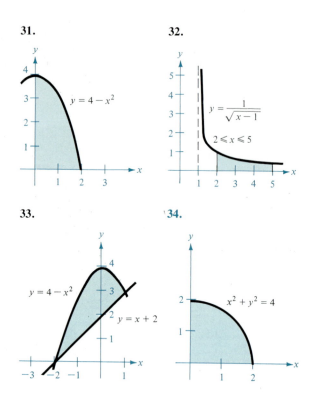

SECTION TOPICS ▪
Double integral ▪
Volume of a solid region ▪
Properties of double integrals ▪
Evaluation of double integrals ▪

16.2
Double integrals and volume

The essence of the definite integral over an *interval* is that it uses a limit process to assign measure to quantities such as area, volume, arc length, and mass. In this section we will see that a similar process can be used to define the **double integral** of a function of two variables over a *region in the plane*.

Later, we will define other types of integrals—triple integrals over solid regions, line integrals over curves, and surface integrals over surfaces. In each case, we follow the pattern established in Section 5.4. That is, we partition the region into subregions, evaluate the function at a point in each subregion, and take the limit of the resulting Riemann sum. To help visualize the process for double integrals, we use a geometric model—the volume of the solid region lying under a surface.

Surface:
$z = f(x, y)$

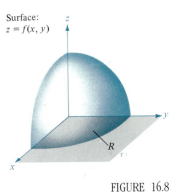

FIGURE 16.8

Consider a continuous function f such that $f(x, y) \geq 0$ for all (x, y) in a region R in the xy-plane. Our goal is to find the volume of the solid region lying between the surface given by $z = f(x, y)$ and the xy-plane, as shown in Figure 16.8. To begin, we superimpose a rectangular grid over the region, as shown in Figure 16.9. The rectangles lying entirely within R form an **inner partition**, whose **norm** $\|\Delta\|$ is defined to be the length of the longest diagonal of the n rectangles. Next, we choose a point (x_i, y_i) in each rectangle and form the rectangular prism whose height is $f(x_i, y_i)$, as shown in Figure 16.10. Since the area of the ith rectangle is $\Delta A_i = (\Delta x_i)(\Delta y_i)$, it follows that the volume of the ith prism is

$$f(x_i, y_i)\Delta A_i = f(x_i, y_i)\Delta x_i \Delta y_i \qquad \text{Volume of } i\text{th prism}$$

and we can approximate the volume of the region by the Riemann sum of the volumes of all n prisms,

$$\sum_{i=1}^{n} f(x_i, y_i)\Delta x_i \Delta y_i \qquad \text{Riemann sum}$$

as shown in Figure 16.11.

Surface:
$z = f(x, y)$

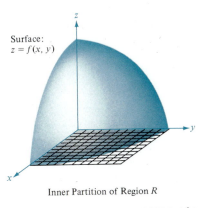

Inner Partition of Region R

FIGURE 16.9

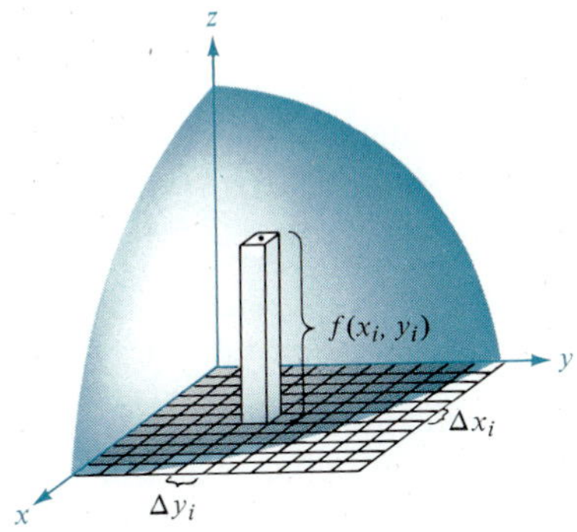

FIGURE 16.10

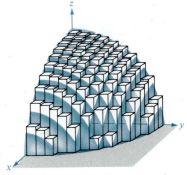

Volume Approximated by Rectangular Prisms

FIGURE 16.11

This approximation can be improved by tightening the mesh of the grid to form smaller and smaller rectangles. That is, we take the limit as $\|\Delta\| \to 0$. If the limit of this sum exists, then we have

$$\text{volume} = \lim_{\|\Delta\| \to 0} \sum_{i=1}^{n} f(x_i, y_i)\Delta x_i \Delta y_i$$

The Riemann sum (with index running from $i = 1$ to $i = n$) is convenient for theoretical purposes. However, for actual computation it is simpler to label the rectangular grid by its rows and columns and use a double sum, as illustrated in the next example.

EXAMPLE 1   *Approximating the volume of a solid*

Approximate the volume of the solid lying between the paraboloid

$$f(x, y) = 4 - x^2 - 2y^2$$

and the square region R given by $0 \le x \le 1$, $0 \le y \le 1$. Use a partition made up of squares whose edges have a length of $\frac{1}{4}$.

Solution: We form the specified partition of R and choose the centers of the subregions as the points at which to evaluate $f(x, y)$. To facilitate the organization of the sum, we use a double subscript notation (like that used to form matrices), so that the point (x_{ij}, y_{ij}) represents the square in the ith row and jth column. Thus, we have

$$(x_{ij}, y_{ij}) = \left(\frac{2i - 1}{8}, \frac{2j - 1}{8} \right), \quad i, j = 1, 2, 3, 4$$

and the value of f at (x_{ij}, y_{ij}) is

$$f(x_{ij}, y_{ij}) = 4 - \left(\frac{2i - 1}{8} \right)^2 - 2\left(\frac{2j - 1}{8} \right)^2$$

Since the area of each square is $\frac{1}{16}$, we approximate the volume by the double sum

$$\sum_{i=1}^{4} \sum_{j=1}^{4} f(x_{ij}, y_{ij})\left(\frac{1}{16} \right) = \sum_{i=1}^{4} \sum_{j=1}^{4} \left[4 - \frac{(2i - 1)^2}{64} - \frac{(2j - 1)^2}{32} \right]\left(\frac{1}{16} \right)$$

$$\approx 3.016$$

This approximation is shown graphically in Figure 16.12. (With a grid of squares with sides of length $\frac{1}{10}$, the approximation would be 3.003.)

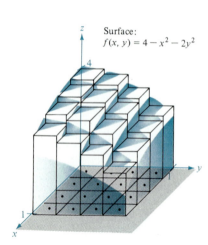

Surface:
$f(x, y) = 4 - x^2 - 2y^2$

FIGURE 16.12

| Remark   The double summation shown in Example 1 is comparable to a nested DO-loop in FORTRAN and a nested FOR-NEXT statement in BASIC.

Using the limit of a Riemann sum to define volume is a special case of using the limit to define a **double integral.** In the general case we do not require that the function be positive or continuous.

DEFINITION OF DOUBLE INTEGRAL

If f is defined on a closed, bounded region R in the xy-plane, then the **double integral of f over R** is given by

$$\iint\limits_{R} f(x, y)\, dA = \lim_{\|\Delta\| \to 0} \sum_{i=1}^{n} f(x_i, y_i)\Delta x_i \Delta y_i$$

provided the limit exists.

Remark Having defined a double integral, we will occasionally refer to a definite integral as a **single integral.**

The definition of the double integral of f on the region R is qualified by the phrase "provided the limit exists." A precise statement of the conditions necessary for this limit to exist is not appropriate in an introductory calculus course. However, we can give some sufficient conditions that cover almost all of the uses of double integrals in engineering and physics. One condition deals with the plane region R. For the remainder of this chapter, we will assume that R can be written as the union of a finite number of subregions that are vertically or horizontally simple, as discussed in Section 16.1. If f is continuous on such a region R, then the double integral of f over R exists.

Double integrals share many of the properties of single integrals, and we summarize some of these properties in the following theorem.

THEOREM 16.1   **PROPERTIES OF DOUBLE INTEGRALS**
Let f be continuous over a closed, bounded plane region R.

1. $\displaystyle \iint\limits_{R} cf(x, y)\, dA = c \iint\limits_{R} f(x, y)\, dA$

2. $\displaystyle \iint\limits_{R} [f(x, y) \pm g(x, y)]\, dA = \iint\limits_{R} f(x, y)\, dA \pm \iint\limits_{R} g(x, y)\, dA$

3. $\displaystyle \iint\limits_{R} f(x, y)\, dA \geq 0 \text{ if } f(x, y) \geq 0$

4. $\displaystyle \iint\limits_{R} f(x, y)\, dA \geq \iint\limits_{R} g(x, y)\, dA \text{ if } f(x, y) \geq g(x, y)$

5. $\displaystyle \iint\limits_{R} f(x, y)\, dA = \iint\limits_{R_1} f(x, y)\, dA + \iint\limits_{R_2} f(x, y)\, dA$

where R is the union of two nonoverlapping subregions R_1 and R_2.

Remark In the fifth property, the statement that R_1 and R_2 are nonoverlapping means that their intersection is a set to which we assign no area. For example, we do not assign area to a line segment.

Evaluation of double integrals

Evaluation of a double integral by the limit definition is difficult, partly because the definition does not specify any particular order for the summation. Example 1 gives a clue to simplifying the process. There we used a double (or nested) sum, holding the outer index of summation constant while we summed the inner index. We follow the same pattern to evaluate a double

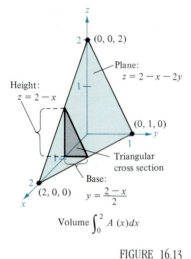

Height:
$z = 2 - x$

Plane:
$z = 2 - x - 2y$

(0, 0, 2)

(0, 1, 0)

Triangular
cross section

Base:
$y = \dfrac{2 - x}{2}$

(2, 0, 0)

Volume $\displaystyle\int_0^2 A(x)\, dx$

FIGURE 16.13

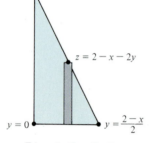

$z = 2 - x - 2y$

$y = 0$ $y = \dfrac{2 - x}{2}$

Triangular Cross Section

FIGURE 16.14

integral. That is, we use an iterated integral and hold the outer variable constant while integrating the inner variable. To see how this is accomplished, we return to our geometric model of a double integral as the volume of a solid.

Consider the solid region bounded by the plane $z = f(x, y) = 2 - x - 2y$ and the three coordinate planes, as shown in Figure 16.13. Each vertical cross section taken parallel to the y-axis is a triangular region whose base has a length of $y = (2 - x)/2$ and whose height is $z = 2 - x$. This implies that for a fixed value of x, the area of the triangular cross section is

$$A(x) = \frac{1}{2}\,(\text{base})(\text{height}) = \frac{1}{2}\left(\frac{2 - x}{2}\right)(2 - x) = \frac{(2 - x)^2}{4}$$

By the formula for the volume of a solid with known cross sections (Section 6.2), the volume of the solid is

$$\text{volume} = \int_a^b A(x)\, dx = \int_0^2 \frac{(2 - x)^2}{4}\, dx$$

$$= -\frac{(2 - x)^3}{12}\bigg]_0^2 = \frac{8}{12} = \frac{2}{3}$$

This procedure works no matter how $A(x)$ is obtained. In particular, we can find $A(x)$ by integration, as indicated in Figure 16.14. That is, we consider x to be constant, and integrate $z = 2 - x - 2y$ from 0 to $(2 - x)/2$ to obtain

$$A(x) = \int_0^{(2-x)/2} (2 - x - 2y)\, dy$$

$$= (2 - x)y - y^2 \bigg]_0^{(2-x)/2} = \frac{(2 - x)^2}{4}$$

Finally, combining these results, we have the *iterated integral*

$$\text{volume} = \iint_R f(x, y)\, dA = \int_0^2 \int_0^{(2-x)/2} (2 - x - 2y)\, dy\, dx$$

To better understand this procedure, it helps to imagine the integration as two sweeping motions. For the inner integration, a vertical line sweeps out the area of a cross section. For the outer integration, the triangular cross section sweeps out the volume, as shown in Figure 16.15.

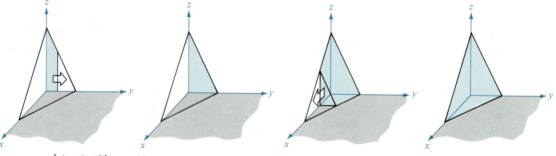

Integrate with respect to y
to obtain the area of the cross section.

Integrate with respect to x to
obtain the volume of the solid.

FIGURE 16.15

The following theorem was proved by the Italian mathematician Guido Fubini (1879–1943). The theorem states that if R is a vertically or horizontally simple region and f is continuous on R, then the double integral of f on R is equal to an iterated integral.

THEOREM 16.2   **FUBINI'S THEOREM**

Let f be continuous on a plane region R.

1. If R is defined by $a \le x \le b$ and $g_1(x) \le y \le g_2(x)$, where g_1 and g_2 are continuous on $[a, b]$, then

$$\iint_R f(x, y) \, dA = \int_a^b \int_{g_1(x)}^{g_2(x)} f(x, y) \, dy \, dx$$

2. If R is defined by $c \le y \le d$ and $h_1(x) \le x \le h_2(x)$, where h_1 and h_2 are continuous on $[c, d]$, then

$$\iint_R f(x, y) \, dA = \int_c^d \int_{h_1(y)}^{h_2(y)} f(x, y) \, dx \, dy$$

EXAMPLE 2   *Evaluating a double integral as an iterated integral*

Evaluate

$$\iint_R (4 - x^2 - 2y^2) \, dA$$

where R is the region given by $0 \le x \le 1$, $0 \le y \le 1$.

Solution: Since the region R is a simple square, it is both vertically and horizontally simple, and we can use either order of integration. Suppose we choose $dy \, dx$ by placing a vertical representative rectangle in the region, as shown in Figure 16.16. This implies that the double integral can be written as

$$\iint_R (4 - x^2 - 2y^2) \, dA = \int_0^1 \int_0^1 (4 - x^2 - 2y^2) \, dy \, dx$$

$$= \int_0^1 \left[(4 - x^2)y - \frac{2y^3}{3} \right]_0^1 dx$$

$$= \int_0^1 \left(\frac{10}{3} - x^2 \right) dx = \left[\frac{10}{3}x - \frac{x^3}{3} \right]_0^1 = 3 \ \square$$

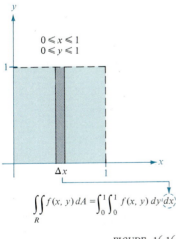

$0 \le x \le 1$
$0 \le y \le 1$

$$\iint_R f(x, y) \, dA = \int_0^1 \int_0^1 f(x, y) \, dy \, dx$$

FIGURE 16.16

Remark   Note that the double integral evaluated in Example 2 represents the volume of the solid region approximated in Example 1. Note also that the approximation obtained in Example 1 is quite good (3.016 vs. 3.000), even though we used a partition consisting of only sixteen squares. The reason we obtained so little error is that we used the centers of the square subregions as the points in the approximation. This is comparable to the Midpoint Rule approximation of a single integral.

The difficulty of evaluating a single integral $\int_a^b f(x)\,dx$ depends basically upon the function f, and not upon the interval $[a, b]$. This is one of the major differences between single and double integrals. In the next example we integrate the same function we did in Example 1. Notice that a change in the region R produces a much more difficult integration problem.

EXAMPLE 3   Finding volume by a double integral

Find the volume of the solid region bounded by the paraboloid $z = 4 - x^2 - 2y^2$ and the xy-plane.

Solution: By letting $z = 0$, we see that the base of the region in the xy-plane is the ellipse $x^2 + 2y^2 = 4$, as shown in Figure 16.17. This plane region is both vertically and horizontally simple, and we choose the order $dy\,dx$.

Variable bounds for y: $-\sqrt{\dfrac{(4 - x^2)}{2}} \le y \le \sqrt{\dfrac{(4 - x^2)}{2}}$

Constant bounds for x: $-2 \le x \le 2$

Therefore, the volume is given by

$$V = \int_{-2}^{2} \int_{-\sqrt{(4-x^2)/2}}^{\sqrt{(4-x^2)/2}} (4 - x^2 - 2y^2)\,dy\,dx$$

$$= \int_{-2}^{2} \left[(4 - x^2)y - \frac{2y^3}{3} \right]_{-\sqrt{(4-x^2)/2}}^{\sqrt{(4-x^2)/2}} dx$$

$$= \frac{4}{3\sqrt{2}} \int_{-2}^{2} (4 - x^2)^{3/2}\,dx$$

$$= \frac{4}{3\sqrt{2}} \int_{-\pi/2}^{\pi/2} 16 \cos^4 \theta\,d\theta \qquad\qquad x = 2 \sin \theta$$

$$= \frac{64}{3\sqrt{2}}(2) \int_{0}^{\pi/2} \cos^4 \theta\,d\theta = \frac{128}{3\sqrt{2}} \left(\frac{3\pi}{16} \right) = 4\sqrt{2}\pi \qquad \text{Wallis's Formula}$$

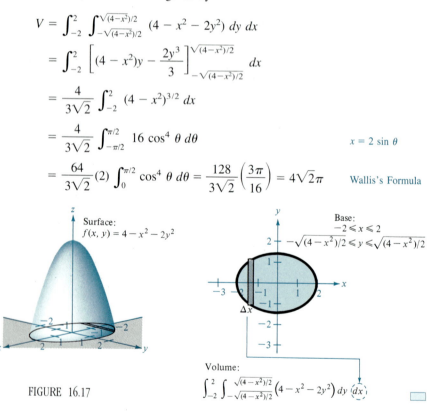

Surface:
$f(x, y) = 4 - x^2 - 2y^2$

Base:
$-2 \le x \le 2$
$-\sqrt{(4 - x^2)/2} \le y \le \sqrt{(4 - x^2)/2}$

Volume:
$$\int_{-2}^{2} \int_{-\sqrt{(4-x^2)/2}}^{\sqrt{(4-x^2)/2}} (4 - x^2 - 2y^2)\,dy\,(dx)$$

FIGURE 16.17

Remark In Example 3, note the usefulness of Wallis's Formula. You may want to review this formula in Section 9.3.

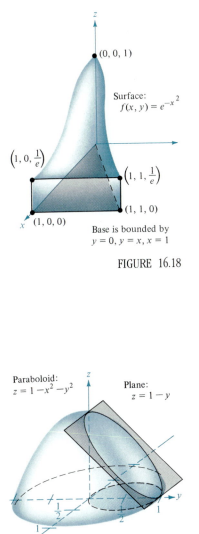

Surface:
$f(x, y) = e^{-x^2}$

$\left(1, 0, \frac{1}{e}\right)$

$\left(1, 1, \frac{1}{e}\right)$

$(1, 1, 0)$

$(1, 0, 0)$

Base is bounded by
$y = 0, y = x, x = 1$

FIGURE 16.18

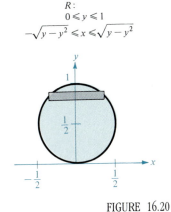

Paraboloid:
$z = 1 - x^2 - y^2$

Plane:
$z = 1 - y$

$R:$
$0 \leqslant y \leqslant 1$
$-\sqrt{y - y^2} \leqslant x \leqslant \sqrt{y - y^2}$

FIGURE 16.20

In Examples 2 and 3, the order of integration was optional, since the regions were both vertically and horizontally simple. Moreover, had we used the order *dx dy*, we would have obtained integrals of comparable difficulty. There are, however, some occasions when one order of integration is much more convenient than the other. Example 4 shows such a case.

EXAMPLE 4   *Comparing different orders of integration*

Find the volume of the solid region R bounded by the surface $f(x, y) = e^{-x^2}$ and the planes $y = 0$, $y = x$, and $x = 1$, as shown in Figure 16.18.

Solution: The base of R in the *xy*-plane is bounded by the lines $y = 0$, $x = 1$, and $y = x$. The two possible orders of integration are given in Figure 16.19.

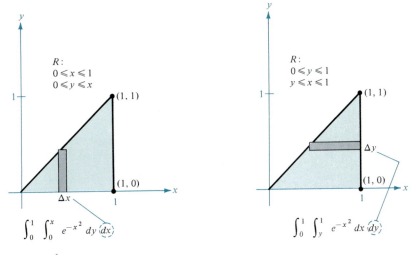

FIGURE 16.19

By setting up the corresponding iterated integrals, we discover that the order *dx dy* requires the antiderivative $\int e^{-x^2} dx$, which we know is not an elementary function. On the other hand, the order *dy dx* produces the integral

$$\int_0^1 \int_0^x e^{-x^2} \, dy \, dx = \int_0^1 e^{-x^2} y \Big]_0^x \, dx = \int_0^1 x e^{-x^2} \, dx$$

$$= -\frac{1}{2} e^{-x^2} \Big]_0^1 = -\frac{1}{2}\left(\frac{1}{e} - 1\right)$$

$$= \frac{e - 1}{2e} \approx 0.316$$

EXAMPLE 5   *Finding volume of a region bounded by two surfaces*

Find the volume of the solid region R bounded above by the paraboloid $z = 1 - x^2 - y^2$ and below by the plane $z = 1 - y$, as shown in Figure 16.20.

Solution: Equating *z*-values, we find that the intersection of the two surfaces occurs on the right-circular cylinder given by

$$1 - y = 1 - x^2 - y^2 \implies x^2 = y - y^2$$

Since the volume of R is the difference between the volume under the parabo-loid and the volume under the plane, we have

$$\text{volume} = \int_0^1 \int_{-\sqrt{y-y^2}}^{\sqrt{y-y^2}} (1 - x^2 - y^2) \, dx \, dy - \int_0^1 \int_{-\sqrt{y-y^2}}^{\sqrt{y-y^2}} (1 - y) \, dx \, dy$$

$$= \int_0^1 \int_{-\sqrt{y-y^2}}^{\sqrt{y-y^2}} (y - y^2 - x^2) \, dx \, dy$$

$$= \int_0^1 \left[(y - y^2)x - \frac{x^3}{3} \right]_{-\sqrt{y-y^2}}^{\sqrt{y-y^2}} dy = \frac{4}{3} \int_0^1 (y - y^2)^{3/2} \, dy$$

$$= \left(\frac{4}{3}\right)\left(\frac{1}{8}\right) \int_0^1 [1 - (2y - 1)^2]^{3/2} \, dy$$

$$= \frac{1}{6} \int_{-\pi/2}^{\pi/2} \frac{\cos^4 \theta}{2} \, d\theta \qquad\qquad 2y - 1 = \sin \theta$$

$$= \frac{1}{6} \int_0^{\pi/2} \cos^4 \theta \, d\theta = \left(\frac{1}{6}\right)\left(\frac{3\pi}{16}\right) = \frac{\pi}{32} \qquad \text{Wallis's Formula} \quad \square$$

EXAMPLE 6   *Reversing the order of integration*

Let $f(x, y) = (\sin x)/x$ for $x \neq 0$ and $f(0, y) = 1$. Evaluate the integral

$$\int_0^1 \int_y^{\sqrt{y}} f(x, y) \, dx \, dy$$

by reversing the order of integration.

Solution: From the boundaries, $y \leq x \leq \sqrt{y}$ and $0 \leq y \leq 1$, we obtain the sketch shown in Figure 16.21. Solving for y in the equations $x = y$ and $x = \sqrt{y}$, we obtain $y = x$ and $y = x^2$, so the boundaries for R are $x^2 \leq y \leq x$ and $0 \leq x \leq 1$. Thus, the integral is

$$\int_0^1 \int_y^{\sqrt{y}} \frac{\sin x}{x} \, dx \, dy = \int_0^1 \int_{x^2}^x \frac{\sin x}{x} \, dy \, dx$$

$$= \int_0^1 \left(\frac{\sin x}{x}\right) y \Big]_{x^2}^x dx$$

$$= \int_0^1 \left(\frac{\sin x}{x}\right)(x - x^2) \, dx$$

$$= \int_0^1 (1 - x)(\sin x) \, dx$$

$$= \int_0^1 (\sin x - x \sin x) \, dx$$

$$= \left[-\cos x + (x \cos x - \sin x) \right]_0^1$$

$$= 1 - \sin 1 \approx 0.159 \qquad \square$$

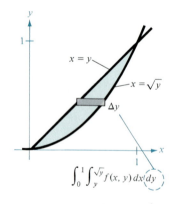

$$\int_0^1 \int_y^{\sqrt{y}} f(x, y) \, dx\, dy$$

FIGURE 16.21

Section Exercises 16.2

In Exercises 1–6, sketch the region R and evaluate the double integral

$$\iint_R f(x, y)\, dA$$

1. $\int_0^2 \int_0^1 (1 + 2x + 2y)\, dy\, dx$

2. $\int_0^\pi \int_0^{\pi/2} \sin^2 x \cos^2 y\, dy\, dx$

3. $\int_0^6 \int_{y/2}^3 (x + y)\, dx\, dy$

4. $\int_0^1 \int_y^{\sqrt{y}} x^2 y^2\, dx\, dy$

5. $\int_{-a}^a \int_{-\sqrt{a^2 - x^2}}^{\sqrt{a^2 - x^2}} (x + y)\, dy\, dx$

6. $\int_0^1 \int_{y-1}^0 e^{x+y}\, dx\, dy + \int_0^1 \int_0^{1-y} e^{x+y}\, dx\, dy$

In Exercises 7–12, set up the integral for both orders of integration, and use the most convenient order to evaluate the integral over the region R.

7. $\iint_R xy\, dA$

   R: rectangle with vertices $(0, 0)$, $(0, 5)$, $(3, 5)$, $(3, 0)$

8. $\iint_R \sin x \sin y\, dA$

   R: rectangle with vertices $(-\pi, 0)$, $(\pi, 0)$, $(\pi, \pi/2)$, $(-\pi, \pi/2)$

9. $\iint_R \dfrac{y}{x^2 + y^2}\, dA$

   R: triangle bounded by $y = x$, $y = 2x$, $x = 2$

10. $\iint_R \dfrac{y}{1 + x^2}\, dA$

   R: region bounded by $y = 0$, $y = \sqrt{x}$, $x = 4$

11. $\iint_R x\, dA$

   R: sector of a circle in the first quadrant bounded by $y = \sqrt{25 - x^2}$, $3x - 4y = 0$, $y = 0$

12. $\iint_R (x^2 + y^2)\, dA$

   R: semicircle bounded by $y = \sqrt{4 - x^2}$, $y = 0$

In Exercises 13–24, use a double integral to find the volume of the specified solid.

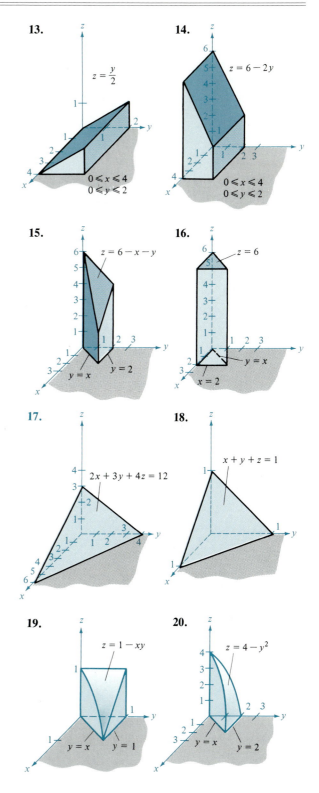

21.

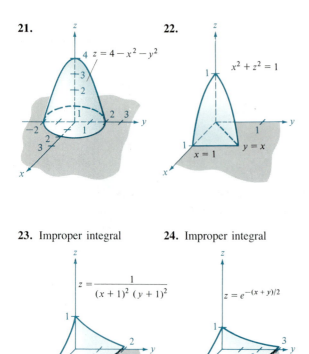

$z = 4 - x^2 - y^2$

22.

$x^2 + z^2 = 1$

$y = x$

$x = 1$

23. Improper integral

$z = \dfrac{1}{(x+1)^2 (y+1)^2}$

$0 \leqslant x < \infty$
$0 \leqslant y < \infty$

24. Improper integral

$z = e^{-(x+y)/2}$

$0 \leqslant x < \infty$
$0 \leqslant y < \infty$

In Exercises 25–32, use a double integral to find the volume of the solid bounded by the graphs of the equations.

25. $z = xy$, $z = 0$, $y = x$, $x = 1$ (first octant)
26. $y = 0$, $z = 0$, $y = x$, $z = x$, $x = 0$, $x = 5$
27. $z = 0$, $z = x^2$, $x = 0$, $x = 2$, $y = 0$, $y = 4$
28. $x^2 + y^2 + z^2 = r^2$
29. $x^2 + z^2 = 1$, $y^2 + z^2 = 1$ (first octant)
30. $y = 1 - x^2$, $z = 1 - x^2$ (first octant)
31. $z = x + y$, $x^2 + y^2 = 4$ (first octant)
32. $z = \dfrac{1}{1 + y^2}$, $x = 0$, $x = 2$, $y \geq 0$

In Exercises 33–36, use Wallis's Formula as an aid to finding the volume of the solid bounded by the graph of the equations.

33. $z = 4 - x^2 - y^2$, $z = 0$
34. $x^2 = 9 - y$, $z^2 = 9 - y$ (first octant)
35. $z = x^2 + y^2$, $x^2 + y^2 = 4$, $z = 0$
36. $z = \sin^2 x$, $z = 0$, $0 \leq x \leq \pi$, $0 \leq y \leq 5$

In Exercises 37–40, find the average value of $f(x, y)$ over the region R where

$$\text{average} = \frac{1}{A} \iint\limits_{R} f(x, y)\, dA$$

where A is the area of R.

37. $f(x, y) = x$
    R: rectangle with vertices $(0, 0)$, $(4, 0)$, $(4, 2)$, $(0, 2)$
38. $f(x, y) = xy$
    R: rectangle with vertices $(0, 0)$, $(4, 0)$, $(4, 2)$, $(0, 2)$
39. $f(x, y) = x^2 + y^2$
    R: square with vertices $(0, 0)$, $(2, 0)$, $(2, 2)$, $(0, 2)$
40. $f(x, y) = e^{x+y}$
    R: triangle with vertices $(0, 0)$, $(0, 1)$, $(1, 1)$

41. For a particular company, the Cobb-Douglas production function is

$$f(x, y) = 100x^{0.6}y^{0.4}$$

Estimate the average production level if the number of units of labor varies between 200 and 250 and the number of units of capital varies between 300 and 325.

42. A firm's profit in marketing two products is given by

$$P = 192x + 576y - x^2 - 5y^2 - 2xy - 5000$$

where x and y represent the number of units of each product. Estimate the average weekly profit if x varies between 40 and 50 units and y varies between 45 and 60 units.

In Exercises 43–46, approximate the integral

$$\iint\limits_{R} f(x, y)\, dA$$

by dividing the rectangle R with vertices $(0, 0)$, $(4, 0)$, $(4, 2)$, and $(0, 2)$ into 8 equal squares and finding the sum

$$\sum_{i=1}^{8} f(x_i, y_i)\Delta x_i \Delta y_i$$

Let (x_i, y_i) be the center of the ith square. Evaluate the double integral and compare it to the approximation.

43. $\displaystyle\int_0^4 \int_0^2 (x + y)\, dy\, dx$

44. $\displaystyle\int_0^4 \int_0^2 xy\, dy\, dx$

45. $\displaystyle\int_0^4 \int_0^2 (x^2 + y^2)\, dy\, dx$

46. $\displaystyle\int_0^4 \int_0^2 \frac{1}{(x+1)(y+1)}\, dy\, dx$

16.3
Change of variables: Polar coordinates

Some double integrals are *much* easier to evaluate in polar form than in rectangular form. This is especially true for regions such as circles, cardioids, and petal curves, and for integrands that involve the sum $x^2 + y^2$. We begin with the definition of a double integral in the polar coordinate system.

Consider a polar region R bounded by the graphs of $r = g_1(\theta)$ and $r = g_2(\theta)$ and by the lines $\theta = \alpha$ and $\theta = \beta$. We begin by superimposing on the region a polar grid made up of rays and circular arcs, as shown in Figure 16.22. The polar sectors R_i lying entirely within R form an **inner polar partition**, whose **norm** $\|\Delta\|$ is defined to be the length of the longest diagonal of the n polar sectors. If we choose a point (r_i, θ_i) in R_i such that r_i is the average radius of R_i, then the area of R_i is

$$\Delta A_i = r_i \Delta r_i \Delta \theta_i \qquad \text{Area of } R_i$$

where θ is measured in radians. Now, if f is a continuous function of r and θ on the region R, then by taking the limit, as $\|\Delta\| \to 0$, of the sum

$$\sum_{i=1}^{n} f(r_i, \theta_i) r_i \Delta r_i \Delta \theta_i$$

it can be shown that we obtain the following polar form of a double integral.

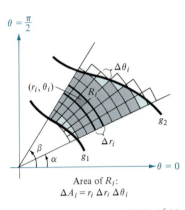

Area of R_i:
$\Delta A_i = r_i \, \Delta r_i \, \Delta \theta_i$

FIGURE 16.22

**DOUBLE INTEGRAL IN
POLAR COORDINATES**

If f is a continuous function of r and θ on a closed bounded plane region R, then the **double integral of f over R** in polar coordinates is given by

$$\iint\limits_{R} f(r, \theta) \, dA = \lim_{\|\Delta\| \to 0} \sum_{i=1}^{n} f(r_i, \theta_i) r_i \Delta r_i \Delta \theta_i = \iint\limits_{R} f(r, \theta) r \, dr \, d\theta$$

As with rectangular coordinates, we restrict the region R to two basic types, which we call **r-simple** and **θ-simple** regions, as in Figure 16.23.

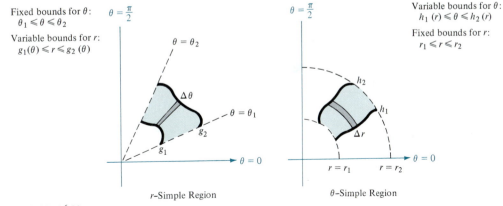

Fixed bounds for θ:
$\theta_1 \leqslant \theta \leqslant \theta_2$

Variable bounds for r:
$g_1(\theta) \leqslant r \leqslant g_2(\theta)$

Variable bounds for θ:
$h_1(r) \leqslant \theta \leqslant h_2(r)$

Fixed bounds for r:
$r_1 \leqslant r \leqslant r_2$

r–Simple Region

θ–Simple Region

FIGURE 16.23

We evaluate a double integral in polar form by an iterated integral, as stated in the following polar form of Fubini's Theorem.

THEOREM 16.3   **POLAR FORM OF FUBINI'S THEOREM**

Let f be continuous on a plane region R.

1. If R is defined by $\theta_1 \leq \theta \leq \theta_2$ and $g_1(\theta) \leq r \leq g_2(\theta)$, where g_1 and g_2 are continuous on $[\theta_1, \theta_2]$, then

$$\iint_R f(r, \theta) \, dA = \int_{\theta_1}^{\theta_2} \int_{g_1(\theta)}^{g_2(\theta)} f(r, \theta) r \, dr \, d\theta$$

2. If R is defined by $r_1 \leq r \leq r_2$ and $h_1(r) \leq \theta \leq h_2(r)$, where h_1 and h_2 are continuous on $[r_1, r_2]$, then

$$\iint_R f(r, \theta) \, dA = \int_{r_1}^{r_2} \int_{h_1(r)}^{h_2(r)} f(r, \theta) r \, d\theta \, dr$$

Remark   Note that the integrand contains a factor r, which arose from the area of the ith polar sector $\Delta A_i = r_i \Delta r_i \Delta \theta_i$.

EXAMPLE 1   *Evaluating a double polar integral*

Evaluate

$$\iint_R \sin \theta \, dA$$

where R is the first-quadrant region lying inside the circle given by $r = 4 \cos \theta$ and outside the circle given by $r = 2$.

Solution: From Figure 16.24 we see that R is an r-simple region, and we sketch a representative sector whose bounds yield the following limits of integration:

$$0 \leq \theta \leq \frac{\pi}{3} \qquad \text{Fixed bounds on } \theta$$

$$2 \leq r \leq 4 \cos \theta \qquad \text{Variable bounds on } r$$

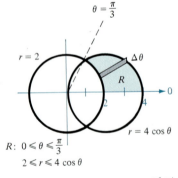

$\theta = \dfrac{\pi}{3}$

$r = 2$ $\Delta \theta$

R

$r = 4 \cos \theta$

$R: 0 \leqslant \theta \leqslant \dfrac{\pi}{3}$

$2 \leqslant r \leqslant 4 \cos \theta$

FIGURE 16.24

Thus, we obtain

$$\iint_R \sin \theta \, dA = \int_0^{\pi/3} \int_2^{4\cos\theta} (\sin \theta) r \, dr \, d\theta = \int_0^{\pi/3} (\sin \theta) \frac{r^2}{2} \bigg]_2^{4\cos\theta} d\theta$$

$$= \frac{1}{2} \int_0^{\pi/3} (\sin \theta)(16 \cos^2 \theta - 4) \, d\theta$$

$$= 2 \int_0^{\pi/3} [4 \cos^2 \theta \, (\sin \theta) - \sin \theta] \, d\theta$$

$$= 2 \left[-\frac{4 \cos^3 \theta}{3} + \cos \theta \right]_0^{\pi/3} = \frac{4}{3}$$

EXAMPLE 2   Finding area of polar regions

Use a double integral to find the area enclosed by the graph of $r = 3 \cos 3\theta$.

Solution: We consider R to be one petal of the curve shown in Figure 16.25. This region is r-simple, and the boundaries are

$$-\frac{\pi}{6} \le \theta \le \frac{\pi}{6} \qquad \text{Fixed bounds on } \theta$$

$$0 \le r \le 3 \cos 3\theta \qquad \text{Variable bounds on } r$$

Thus, the area of one petal is

$$\frac{1}{3} A = \iint_R dA = \int_{-\pi/6}^{\pi/6} \int_0^{3\cos 3\theta} r \, dr \, d\theta$$

$$= \int_{-\pi/6}^{\pi/6} \frac{r^2}{2} \Big]_0^{3\cos 3\theta} d\theta$$

$$= \frac{9}{2} \int_{-\pi/6}^{\pi/6} \cos^2 3\theta \, d\theta$$

$$= \frac{9}{4} \int_{-\pi/6}^{\pi/6} (1 + \cos 6\theta) \, d\theta$$

$$= \frac{9}{4} \left[\theta + \frac{1}{6} \sin 6\theta \right]_{-\pi/6}^{\pi/6} = \frac{3\pi}{4}$$

Therefore, the total area is $A = 9\pi/4$.

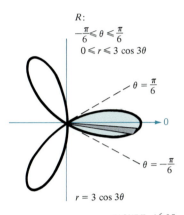

R:

$$-\frac{\pi}{6} \le \theta \le \frac{\pi}{6}$$

$$0 \le r \le 3 \cos 3\theta$$

$\theta = \frac{\pi}{6}$

$\theta = -\frac{\pi}{6}$

$r = 3 \cos 3\theta$

FIGURE 16.25

Change of variables to polar form

We now come to the reason for introducing double integrals in polar form. With sufficient restrictions on a function f and a region R, we make the following change of variables

$$x = r \cos \theta, \qquad y = r \sin \theta$$
$$r^2 = x^2 + y^2 \qquad dA = r \, dr \, d\theta$$

to convert a double integral in rectangular coordinates into one in polar coordinates, as stated in the following theorem. We postpone a discussion of the proof of this theorem until the next section.

THEOREM 16.4   CHANGE OF VARIABLES TO POLAR FORM
Let R be a plane region consisting of all points $(x, y) = (r \cos \theta, r \sin \theta)$ satisfying the condition

$$0 \le g_1(\theta) \le r \le g_2(\theta), \quad \theta_1 \le \theta \le \theta_2$$

where $0 < (\theta_2 - \theta_1) \le 2\pi$. If g_1 and g_2 are continuous on $[\theta_1, \theta_2]$ and f is continuous on R, then

$$\iint_R f(x, y) \, dA = \int_{\theta_1}^{\theta_2} \int_{g_1(\theta)}^{g_2(\theta)} f(r \cos \theta, r \sin \theta) r \, dr \, d\theta$$

If $z = f(x, y)$ is nonnegative on R, then the integral in Theorem 16.4 can be interpreted as the *volume* of the solid region between the graph of f and the region R. We use this interpretation in our next example.

EXAMPLE 3   Change of variables to polar coordinates

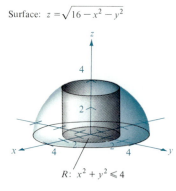

Surface: $z = \sqrt{16 - x^2 - y^2}$

$R: x^2 + y^2 \le 4$

FIGURE 16.26

Use polar coordinates to find the volume of the solid region bounded above by the hemisphere $z = \sqrt{16 - x^2 - y^2}$ and below by the circular region R given by $x^2 + y^2 = 4$, as shown in Figure 16.26.

Solution: From Figure 16.26 we can see that R has the bounds

$$-\sqrt{4 - y^2} \le x \le \sqrt{4 - y^2}, \quad -2 \le y \le 2$$

and that $0 \le z \le \sqrt{16 - x^2 - y^2}$. In polar coordinates the bounds are

$$0 \le r \le 2 \quad \text{and} \quad 0 \le \theta \le 2\pi$$

with height $z = \sqrt{16 - x^2 - y^2} = \sqrt{16 - r^2}$. Consequently, the volume V is given by

$$V = \iint_R f(x, y)\, dA = \int_0^{2\pi} \int_0^2 \sqrt{16 - r^2}\, r\, dr\, d\theta$$

$$= -\frac{1}{3} \int_0^{2\pi} (16 - r^2)^{3/2} \Big]_0^2 d\theta$$

$$= -\frac{1}{3} \int_0^{2\pi} (24\sqrt{3} - 64)\, d\theta$$

$$= -\frac{8}{3}(3\sqrt{3} - 8)\theta \Big]_0^{2\pi}$$

$$= \frac{16\pi}{3}(8 - 3\sqrt{3}) \approx 46.98$$

Remark   To see the remarkable benefit of polar coordinates in Example 3, you should try to evaluate the corresponding rectangular double integral

$$\int_{-2}^{2} \int_{-\sqrt{4 - y^2}}^{\sqrt{4 - y^2}} \sqrt{16 - x^2 - y^2}\, dx\, dy$$

EXAMPLE 4   Change of variables to polar coordinates

Let R be the annular region lying between the two circles $x^2 + y^2 = 1$ and $x^2 + y^2 = 5$, as shown in Figure 16.27. Evaluate the integral

$$\iint_R (x^2 + y)\, dA$$

Solution: The polar boundaries are $1 \le r \le \sqrt{5}$ and $0 \le \theta \le 2\pi$. Thus, we have

$$\iint_R (x^2 + y)\, dA = \int_0^{2\pi} \int_1^{\sqrt{5}} (r^2 \cos^2 \theta + r \sin \theta)\, r\, dr\, d\theta$$

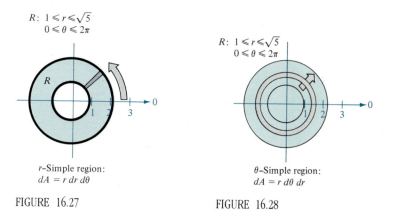

r-Simple region:
$dA = r \, dr \, d\theta$

FIGURE 16.27

θ-Simple region:
$dA = r \, d\theta \, dr$

FIGURE 16.28

For the sake of illustration, we change the order of integration, as indicated in Figure 16.28, and obtain

$$\iint\limits_R (x^2 + y) \, dA = \int_1^{\sqrt{5}} \int_0^{2\pi} r(r^2 \cos^2 \theta + \sin \theta) \, d\theta \, dr$$

$$= \int_1^{\sqrt{5}} r^2 \left[\int_0^{2\pi} \left(\frac{r}{2} + \frac{r \cos 2\theta}{2} + \sin \theta \right) d\theta \right] dr$$

$$= \int_1^{\sqrt{5}} r^2 \left[\frac{r\theta}{2} + \frac{r \sin 2\theta}{4} - \cos \theta \right]_0^{2\pi} dr$$

$$= \int_1^{\sqrt{5}} r^2 (\pi r) \, dr = \pi \frac{r^4}{4} \Bigg]_1^{\sqrt{5}} = 6\pi$$

Try the order $dA = r \, dr \, d\theta$ to see which order is more convenient.

Section Exercises 16.3

In Exercises 1–6, evaluate the double integral $\iint\limits_R f(r, \theta) \, dA$, and sketch the region R.

1. $\int_0^{2\pi} \int_0^6 3r^2 \sin \theta \, dr \, d\theta$

2. $\int_0^{\pi/4} \int_0^4 r^2 \sin \theta \cos \theta \, dr \, d\theta$

3. $\int_0^{\pi/2} \int_2^3 \sqrt{9 - r^2} \, r \, dr \, d\theta$

4. $\int_0^{\pi/2} \int_0^3 re^{-r^2} \, dr \, d\theta$

5. $\int_0^{\pi/2} \int_0^{1+\sin\theta} \theta \, dr \, d\theta$

6. $\int_0^{\pi/2} \int_0^{1-\cos\theta} \sin \theta \, dr \, d\theta$

In Exercises 7–12, use a double integral to find the area of the indicated region.

7.

8.

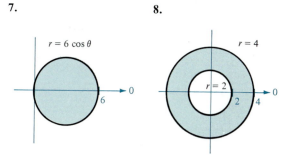

$r = 6 \cos \theta$

$r = 4$

$r = 2$

9.

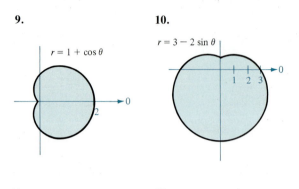

$r = 1 + \cos\theta$

10.

$r = 3 - 2\sin\theta$

11.

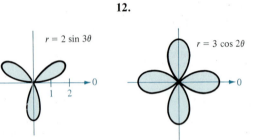

$r = 2\sin 3\theta$

12.

$r = 3\cos 2\theta$

In Exercises 13–18, evaluate the double integral by changing to polar coordinates.

13. $\displaystyle\int_0^a \int_0^{\sqrt{a^2 - y^2}} y\, dx\, dy$ **14.** $\displaystyle\int_0^a \int_0^{\sqrt{a^2 - x^2}} x\, dy\, dx$

15. $\displaystyle\int_0^2 \int_0^x \sqrt{x^2 + y^2}\, dy\, dx$

$\qquad + \displaystyle\int_2^{2\sqrt{2}} \int_0^{\sqrt{8 - x^2}} \sqrt{x^2 + y^2}\, dy\, dx$

16. $\displaystyle\int_0^{2\sqrt{2}} \int_y^{\sqrt{8 - y^2}} \sqrt{x^2 + y^2}\, dx\, dy$

17. $\displaystyle\int_0^2 \int_0^{\sqrt{2x - x^2}} xy\, dy\, dx$

18. $\displaystyle\int_0^4 \int_0^{\sqrt{4y - y^2}} x^2\, dx\, dy$

In Exercises 19–22, use polar coordinates to evaluate the double integral $\displaystyle\iint_R f(x, y)\, dA$.

19. $f(x, y) = x + y$
$\quad R: x^2 + y^2 \le 4,\ 0 \le x,\ 0 \le y$
20. $f(x, y) = e^{-(x^2 + y^2)}$
$\quad R: x^2 + y^2 \le 4,\ 0 \le x,\ 0 \le y$
21. $f(x, y) = \arctan \dfrac{y}{x}$
$\quad R: x^2 + y^2 \le 1,\ 0 \le x,\ 0 \le y$
22. $f(x, y) = 9 - x^2 - y^2$
$\quad R: x^2 + y^2 \le 9,\ 0 \le x,\ 0 \le y$

In Exercises 23–28, use a double integral in polar coordinates to find the volume of the solid bounded by the graphs of the equations.

23. $z = xy,\ x^2 + y^2 = 1$ (first octant)
24. $z = x^2 + y^2 + 1,\ z = 0,\ x^2 + y^2 = 4$
25. $z = \sqrt{x^2 + y^2},\ z = 0,\ x^2 + y^2 = 25$
26. $z = \sqrt{x^2 + y^2},\ z = 0,$
$\qquad x^2 + y^2 \ge 4$ and $x^2 + y^2 \le 16$
27. Inside the hemisphere

$$z = \sqrt{16 - x^2 - y^2}$$

and the cylinder

$$x^2 + y^2 - 4x = 0$$

28. Inside the hemisphere

$$z = \sqrt{16 - x^2 - y^2}$$

and outside the cylinder

$$x^2 + y^2 = 1$$

29. Find a so that the volume inside the hemisphere

$$z = \sqrt{16 - x^2 - y^2}$$

and outside the cylinder

$$x^2 + y^2 = a^2$$

is one-half the volume of the hemisphere.
30. The integral

$$I = \int_{-\infty}^{\infty} e^{-x^2/2}\, dx$$

is important in the study of normal distributions. Use polar coordinates to evaluate I by finding the double integral

$$I^2 = \left(\int_{-\infty}^{\infty} e^{-x^2/2}\, dx \right)\left(\int_{-\infty}^{\infty} e^{-y^2/2}\, dy \right)$$

$$= \int_{-\infty}^{\infty} \int_{-\infty}^{\infty} e^{-(x^2 + y^2)/2}\, dA$$

16.4
Change of variables: Jacobians

For the single integral $\int_a^b f(x)\ dx$, we can change variables by letting $x = g(u)$, so that $dx = g'(u)\ du$, and obtain

$$\int_a^b f(x)\ dx = \int_c^d f(g(u))g'(u)\ du$$

where $a = g(c)$ and $b = g(d)$. Note that the change of variables process introduces an additional factor $g'(u)$ into the integrand. This also occurs in the case of double integrals

$$\iint_R f(x,\ y)\ dA = \iint_S f(g(u,\ v),\ h(u,\ v))\underbrace{\left| \frac{\partial x}{\partial u} \frac{\partial y}{\partial v} - \frac{\partial y}{\partial u} \frac{\partial x}{\partial v} \right|}_{\text{Jacobian}} du\ dv$$

where the change of variables $x = g(u,\ v)$ and $y = h(u,v)$ introduces a factor called the **Jacobian** of x and y with respect to u and v, named after the German mathematician Carl Gustav Jacobi (1804–1851). Jacobi is known for his work in many areas of mathematics, but his interest in integration stemmed from the problem of finding the circumference of an ellipse. In defining the Jacobian, it is convenient to use the following determinant notation.

Carl Gustav Jacobi

DEFINITION OF THE JACOBIAN	If $x = g(u,\ v)$ and $y = h(u,\ v)$, then the **Jacobian** of x and y with respect to u and v, denoted by $\partial(x,\ y)/\partial(u,\ v)$, is

$$\frac{\partial(x,\ y)}{\partial(u,\ v)} = \begin{vmatrix} \dfrac{\partial x}{\partial u} & \dfrac{\partial y}{\partial u} \\ \dfrac{\partial x}{\partial v} & \dfrac{\partial y}{\partial v} \end{vmatrix} = \frac{\partial x}{\partial u} \frac{\partial y}{\partial v} - \frac{\partial y}{\partial u} \frac{\partial x}{\partial v}$$

EXAMPLE 1   Finding the Jacobian for rectangular-to-polar conversion

Find the Jacobian for the change of variables defined by

$$x = r \cos \theta \qquad \text{and} \qquad y = r \sin \theta$$

Solution: From the definition, we evaluate the determinant

$$\frac{\partial(x,\ y)}{\partial(r,\ \theta)} = \begin{vmatrix} \dfrac{\partial x}{\partial r} & \dfrac{\partial y}{\partial r} \\ \dfrac{\partial x}{\partial \theta} & \dfrac{\partial y}{\partial \theta} \end{vmatrix} = \begin{vmatrix} \cos \theta & \sin \theta \\ -r \sin \theta & r \cos \theta \end{vmatrix} = r \cos^2 \theta + r \sin^2 \theta = r$$

Example 1 points out that the change of variables from rectangular to polar coordinates for a double integral can be written as follows.

$$\iint_R f(x,\ y)\ dx\ dy = \iint_S f(r\ \cos\ \theta,\ r\ \sin\ \theta)\left|\frac{\partial(x,\ y)}{\partial(r,\ \theta)}\right|\ dr\ d\theta$$

where S is the region R, described in polar coordinates. The following theorem extends this result to a general change of variables.

THEOREM 16.5

CHANGE OF VARIABLES FOR DOUBLE INTEGRALS
Let R and S be regions in the xy- and uv-planes that are related by the equations $x = g(u,\ v)$ and $y = h(u,\ v)$ such that each point in R is the image of a unique point in S. If f is continuous on R and g and h have continuous partial derivatives on S and $\partial(x,\ y)/\partial(u,\ v)$ is nonzero on S, then

$$\iint_R f(x,\ y)\ dA = \iint_S f(g(u,\ v),\ h(u,\ v))\left|\frac{\partial(x,\ y)}{\partial(u,\ v)}\right|\ du\ dv$$

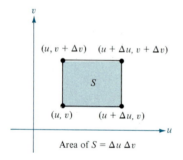

Area of $S = \Delta u\ \Delta v$

FIGURE 16.29

Proof: We will discuss a proof for the case in which S is a rectangular region in the uv-plane with vertices $(u,\ v)$, $(u + \Delta u,\ v)$, $(u + \Delta u,\ v + \Delta v)$, and $(u,\ v + \Delta v)$, as shown in Figure 16.29. The images of these vertices in the xy-plane are

$$M = (g(u,\ v),\ h(u,\ v))$$
$$N = (g(u + \Delta u,\ v),\ h(u + \Delta u,\ v))$$
$$P = (g(u + \Delta u,\ v + \Delta v),\ h(u + \Delta u,\ v + \Delta v))$$
$$Q = (g(u,\ v + \Delta v),\ h(u,\ v + \Delta v))$$

If Δu and Δv are small, then the continuity of g and h implies that R is approximately a parallelogram determined by the vectors $\overrightarrow{MN}$ and $\overrightarrow{MQ}$, as shown in Figure 16.30. Thus, the area ΔA of R is given by

$$\Delta A \approx \|\overrightarrow{MN} \times \overrightarrow{MQ}\|$$

Moreover, for small Δu and Δv, the partial derivatives of g and h with respect to u can be approximated by

$$g_u(u,\ v) \approx \frac{g(u + \Delta u,\ v) - g(u,\ v)}{\Delta u} \quad \text{and} \quad h_u(u,\ v) \approx \frac{h(u + \Delta u,\ v) - h(u,\ v)}{\Delta u}$$

Consequently,

$$\overrightarrow{MN} = [g(u + \Delta u,\ v) - g(u,\ v)]\ \mathbf{i} + [h(u + \Delta u,\ v) - h(u,\ v)]\ \mathbf{j}$$
$$\approx [g_u(u,\ v)\Delta u]\ \mathbf{i} + [h_u(u,\ v)\Delta u]\ \mathbf{j} = \frac{\partial x}{\partial u}\Delta u\mathbf{i} + \frac{\partial y}{\partial u}\Delta u\mathbf{j}$$

Similarly, we can approximate $\overrightarrow{MQ}$ as

$$\overrightarrow{MQ} \approx \frac{\partial x}{\partial v}\Delta v\mathbf{i} + \frac{\partial y}{\partial v}\Delta v\mathbf{j}$$

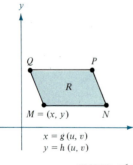

$x = g(u,\ v)$
$y = h(u,\ v)$

FIGURE 16.30

which implies that

$$\overrightarrow{MN} \times \overrightarrow{MQ} \approx \begin{vmatrix} \mathbf{i} & \mathbf{j} & \mathbf{k} \\ \dfrac{\partial x}{\partial u}\Delta u & \dfrac{\partial y}{\partial u}\Delta u & 0 \\ \dfrac{\partial x}{\partial v}\Delta v & \dfrac{\partial y}{\partial v}\Delta v & 0 \end{vmatrix} = \begin{vmatrix} \dfrac{\partial x}{\partial u} & \dfrac{\partial y}{\partial u} \\ \dfrac{\partial x}{\partial v} & \dfrac{\partial y}{\partial v} \end{vmatrix} \Delta u \Delta v \mathbf{k}$$

It follows that, in Jacobian notation,

$$\Delta A \approx \|\overrightarrow{MN} \times \overrightarrow{MQ}\| \approx \left| \frac{\partial(x, y)}{\partial(u, v)} \right| \Delta u \Delta v$$

Finally, since this approximation improves as Δu and Δv approach 0, the limiting case can be written as

$$dA \approx \|\overrightarrow{MN} \times \overrightarrow{MQ}\| \approx \left| \frac{\partial(x, y)}{\partial(u, v)} \right| du\ dv$$

The next two examples in this section show how a change of variables can simplify the integration process. The simplification occurs in *two* ways. We can make a change of variables to simplify either the *region R* or the *integrand f(x, y)*.

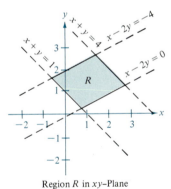

Region R in xy-Plane

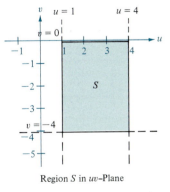

Region S in uv-Plane

FIGURE 16.31

EXAMPLE 2   *A change of variables to simplify the region of integration*

Let R be the region bounded by the lines $x - 2y = 0$, $x - 2y = -4$, $x + y = 4$, and $x + y = 1$, as shown in Figure 16.31. Evaluate the double integral

$$\iint\limits_{R} 3xy\ dA$$

Solution: If we let $u = x + y$ and $v = x - 2y$, the bounds for the region S in the uv-plane are

$$u = x + y = 1, \qquad v = x - 2y = 0$$
$$u = x + y = 4, \qquad v = x - 2y = -4$$

as shown in Figure 16.31. Solving for x and y in terms of u and v, we have

$$y = \frac{1}{3}(u - v) \qquad \text{and} \qquad x = \frac{1}{3}(2u + v)$$

Thus, the Jacobian is

$$\frac{\partial(x, y)}{\partial(u, v)} = \begin{vmatrix} \dfrac{\partial x}{\partial u} & \dfrac{\partial y}{\partial u} \\ \dfrac{\partial x}{\partial v} & \dfrac{\partial y}{\partial v} \end{vmatrix} = \begin{vmatrix} \dfrac{2}{3} & \dfrac{1}{3} \\ \dfrac{1}{3} & -\dfrac{1}{3} \end{vmatrix} = -\frac{2}{9} - \frac{1}{9} = -\frac{1}{3}$$

Therefore, by Theorem 16.5, we obtain

$$\iint_R 3xy \, dA = \iint_S 3\left[\frac{1}{3}(2u + v)\frac{1}{3}(u - v)\right]\left|\frac{\partial(x, y)}{\partial(u, v)}\right| dv \, du$$

$$= \int_1^4 \int_{-4}^0 \frac{1}{9}(2u^2 - uv - v^2) \, dv \, du$$

$$= \frac{1}{9} \int_1^4 \left[2u^2v - \frac{uv^2}{2} - \frac{v^3}{3}\right]_{-4}^0 du$$

$$= \frac{1}{9} \int_1^4 \left(8u^2 + 8u - \frac{64}{3}\right) du = \frac{1}{9}\left[\frac{8u^3}{3} + 4u^2 - \frac{64}{3}u\right]_1^4$$

$$= \frac{164}{9}$$

EXAMPLE 3   A change of variables to simplify the integrand

Let R be the region bounded by the square with vertices $(0, 1)$, $(1, 2)$, $(2, 1)$, and $(1, 0)$. Evaluate the integral

$$\iint_R (x + y)^2 \sin^2 (x - y) \, dA$$

Solution: First, we note that the sides of R lie on the lines $x + y = 1$, $x - y = 1$, $x + y = 3$, and $x - y = -1$, as shown in Figure 16.32. Letting $u = x + y$ and $v = x - y$, we find the bounds for region S in the uv-plane to be

$$u = x + y = 1, \qquad v = x - y = 1$$
$$u = x + y = 3, \qquad v = x - y = -1$$

Solving for x and y in terms of u and v, we have

$$x = \frac{1}{2}(u + v) \qquad y = \frac{1}{2}(u - v)$$

Thus, the Jacobian is

$$\frac{\partial(x, y)}{\partial(u, v)} = \begin{vmatrix} \frac{1}{2} & \frac{1}{2} \\ \frac{1}{2} & -\frac{1}{2} \end{vmatrix} = -\frac{1}{4} - \frac{1}{4} = -\frac{1}{2}$$

By Theorem 16.5, it follows that

$$\iint_R (x + y)^2 \sin^2 (x - y) \, dA = \int_{-1}^1 \int_1^3 u^2 \sin^2 v \left(\frac{1}{2}\right) du \, dv$$

$$= \frac{1}{2} \int_{-1}^1 (\sin^2 v)\frac{u^3}{3}\bigg]_1^3 dv$$

$$= \frac{13}{3} \int_{-1}^1 \sin^2 v \, dv$$

$$= \frac{13}{6} \int_{-1}^1 (1 - \cos 2v) \, dv$$

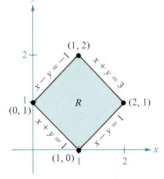

FIGURE 16.32

$$= \frac{13}{6}\left[v - \frac{1}{2}\sin 2v\right]_{-1}^{1}$$

$$= \frac{13}{6}\left[2 - \frac{1}{2}\sin 2 + \frac{1}{2}\sin(-2)\right]$$

$$= \frac{13}{6}[2 - \sin 2] \approx 2.363$$

Section Exercises 16.4

In Exercises 1–8, find the Jacobian for the indicated change of variables.

1. $x = -\frac{1}{2}(u - v)$, $y = \frac{1}{2}(u + v)$

2. $x = au + bv$, $y = cu + dv$

3. $x = u - v^2$, $y = u + v$

4. $x = u - uv$, $y = uv$

5. $x = u \cos\theta - v \sin\theta$, $y = u \sin\theta + v \cos\theta$

6. $x = u + a$, $y = v + a$

7. $x = e^u \sin v$, $y = e^u \cos v$

8. $x = \dfrac{u}{v}$, $y = u + v$

In Exercises 9–14, use the indicated change of variables to evaluate the double integral.

9. $\displaystyle\iint_R 48xy\, dx\, dy$

$x = \frac{1}{2}(u + v)$

$y = \frac{1}{2}(u - v)$

10. $\displaystyle\iint_R 4(x^2 + y^2)\, dx\, dy$

$x = \frac{1}{2}(u + v)$

$y = \frac{1}{2}(u - v)$

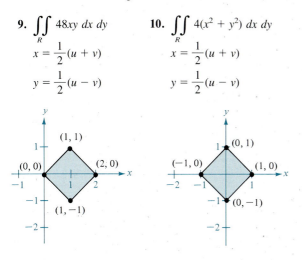

11. $\displaystyle\iint_R 4(x + y)e^{x-y}\, dy\, dx$

$x = \frac{1}{2}(u + v)$

$y = \frac{1}{2}(u - v)$

12. $\displaystyle\iint_R y(x - y)\, dx\, dy$

$x = u + v$, $y = u$

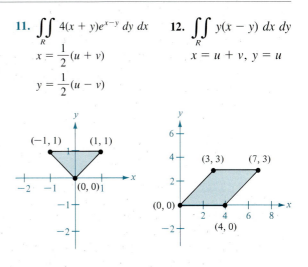

13. $\displaystyle\iint_R \sqrt{x^2 + y^2}\, dy\, dx$

$x = u$, $y = uv$

R: triangle with vertices $(0, 0)$, $(4, 0)$, $(4, 4)$

14. $\displaystyle\iint_R y \sin xy\, dy\, dx$

$x = \dfrac{u}{v}$, $y = v$

R: region lying between the graphs of $xy = 1$, $xy = 4$, $y = 1$, $y = 4$

In Exercises 15–20, use a change of variables to find the volume of the solid region lying below the surface $z = f(x, y)$ and above the plane region R.

15. $f(x, y) = (x + y)e^{x-y}$

R: region bounded by the square with vertices $(4, 0)$, $(6, 2)$, $(4, 4)$, $(2, 2)$

16. $f(x, y) = (x + y)^2 \sin^2(x - y)$

R: region bounded by the square with vertices $(\pi, 0)$, $(3\pi/2, \pi/2)$, (π, π), $(\pi/2, \pi/2)$

17. $f(x, y) = \sqrt{(x - y)(x + 4y)}$
   R: region bounded by the parallelogram with vertices
   $(0, 0)$, $(1, 1)$, $(5, 0)$, $(4, -1)$

18. $f(x, y) = (3x + 2y)^2\sqrt{2y - x}$
   R: region bounded by the parallelogram with vertices
   $(0, 0)$, $(-2, 3)$, $(2, 5)$, $(4, 2)$

19. $f(x, y) = \sqrt{x + y}$
   R: region bounded by the triangle with vertices $(0, 0)$,
   $(a, 0)$, $(0, a)$ where $0 \leq a$

20. $f(x, y) = \dfrac{xy}{1 + x^2y^2}$
   R: region bounded by the graphs of $xy = 1$, $xy = 4$,
   $x = 1$, $x = 4$ (Hint: Let $x = u$, $y = v/u$.)

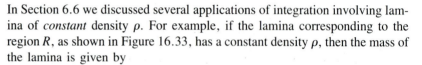

16.5
Center of mass and moments of inertia

SECTION TOPICS ▪
Mass ▪
Moments about x- and y-axes ▪
Center of mass ▪
Moments of inertia ▪

In Section 6.6 we discussed several applications of integration involving lamina of *constant* density ρ. For example, if the lamina corresponding to the region R, as shown in Figure 16.33, has a constant density ρ, then the mass of the lamina is given by

$$\text{mass} = \rho A$$

$$= \rho \int_a^b [g_2(x) - g_1(x)] \, dx$$

$$= \rho \iint_R dA$$

$$= \iint_R \rho \, dA \qquad \text{Constant density}$$

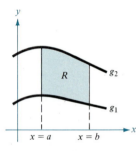

Lamina of Constant Density

FIGURE 16.33

The use of a double integral suggests a natural extension of the formula for finding the mass of a lamina of *variable* density, where the density at (x, y) is given by the **density function** ρ.

DEFINITION OF MASS
OF A PLANAR LAMINA
OF VARIABLE DENSITY

If ρ is a continuous density function on the lamina corresponding to a plane region R, then the mass m of the lamina is given by

$$m = \iint_R \rho(x, y) \, dA \qquad \text{Variable density}$$

| Remark   Density is normally expressed as mass per unit volume. However, for a planar lamina we consider density as mass per unit surface area.

EXAMPLE 1   Finding the mass of a planar lamina

Find the mass of the triangular lamina with vertices $(0, 0)$, $(0, 3)$, and $(2, 3)$, given that the density at (x, y) is

$$\rho(x, y) = 2x + y \qquad \text{Density function}$$

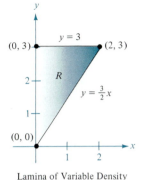

y = 3

(0, 3)   (2, 3)

R

$y = \frac{3}{2}x$

(0, 0)

Lamina of Variable Density

FIGURE 16.34

Solution: As shown in Figure 16.34, region R has the boundaries $x = 0$, $y = 3$, and $y = 3x/2$. Therefore, the mass of the lamina is

$$m = \iint_R (2x + y)\, dA = \int_0^3 \int_0^{2y/3} (2x + y)\, dx\, dy = \int_0^3 \left[x^2 + xy \right]_0^{2y/3} dy$$

$$= \frac{10}{9} \int_0^3 y^2\, dy = \frac{10}{9} \frac{y^3}{3} \Big]_0^3 = 10$$

Remark Note in Figure 16.34 that we shade a plane lamina of variable density so that the darkest shading corresponds to the densest part.

EXAMPLE 2   Finding mass by polar coordinates

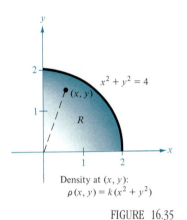

$x^2 + y^2 = 4$

(x, y)

R

Density at (x, y):
$\rho(x, y) = k(x^2 + y^2)$

FIGURE 16.35

Find the mass of the lamina corresponding to the first-quadrant portion of the circle $x^2 + y^2 = 4$, where the density at the point (x, y) is proportional to the distance between the point and the origin, as shown in Figure 16.35.

Solution: At any point (x, y) the density of the lamina is

$$\rho(x, y) = k\sqrt{(x - 0)^2 + (y - 0)^2} = k\sqrt{x^2 + y^2}$$

Since $0 \le x \le 2$ and $0 \le y \le \sqrt{4 - x^2}$, the mass is given by

$$m = \iint_R k\sqrt{x^2 + y^2}\, dA = \int_0^2 \int_0^{\sqrt{4-x^2}} k\sqrt{x^2 + y^2}\, dy\, dx$$

To simplify the integration, we change to polar coordinates, using the bounds $0 \le \theta \le \pi/2$ and $0 \le r \le 2$. Thus, we have

$$m = \iint_R k\sqrt{x^2 + y^2}\, dA = \int_0^{\pi/2} \int_0^2 k\sqrt{r^2}\, r\, dr\, d\theta$$

$$= \int_0^{\pi/2} \int_0^2 kr^2\, dr\, d\theta = \int_0^{\pi/2} \frac{kr^3}{3} \Big]_0^2 d\theta$$

$$= \frac{8k}{3} \int_0^{\pi/2} d\theta = \frac{8k}{3} \theta \Big]_0^{\pi/2} = \frac{4\pi k}{3}$$

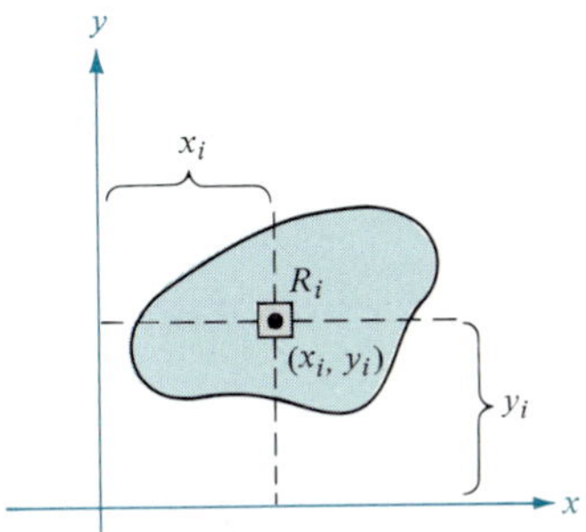

Moments

x_i

R_i

(x_i, y_i)

y_i

$M_x = (\text{mass})(y_i)$
$M_y = (\text{mass})(x_i)$

FIGURE 16.36

For a lamina of variable density, moments of mass are defined in a manner similar to that used for the uniform density case. For a partition Δ of a lamina corresponding to a plane region R, we consider the ith rectangle R_i of area ΔA_i, as shown in Figure 16.36. Assuming that the mass of R_i is concentrated at one of its interior points (x_i, y_i), it follows that the moment of mass of R_i with respect to the x-axis is approximated by

$$(\text{mass})(y_i) \approx [\rho(x_i, y_i)\Delta A_i](y_i)$$

Similarly, the moment of mass with respect to the y-axis is approximated by

$$(\text{mass})(x_i) \approx [\rho(x_i, y_i)\Delta A_i](x_i)$$

By forming the Riemann sum of all such products and taking the limit as the norm of Δ approaches zero, we obtain the following formulas for moments of mass with respect to the x- and y-axes.

DEFINITION OF MOMENTS AND CENTER OF MASS OF A VARIABLE DENSITY PLANAR LAMINA

If ρ is a continuous density function on the lamina corresponding to a plane region R, then the **moments of mass** with respect to the x- and y-axes are

$$M_x = \iint_R y\rho(x,\, y)\, dA \qquad \text{and} \qquad M_y = \iint_R x\rho(x,\, y)\, dA$$

Furthermore, if m is the mass of the lamina, then the **center of mass** is

$$(\bar{x},\, \bar{y}) = \left(\frac{M_y}{m},\, \frac{M_x}{m} \right)$$

| **Remark** If R represents a simple plane region rather than a lamina, the point $(\bar{x},\, \bar{y})$ is called the **centroid** of the region.

Sometimes we can tell that either $\bar{x} = 0$ or $\bar{y} = 0$ by the form of ρ and the symmetry of the region R, as illustrated in the next example.

EXAMPLE 3   Finding the center of mass

Find the center of mass of the lamina corresponding to the parabolic region $0 \le y \le 4 - x^2$, where the density at the point $(x,\, y)$ is proportional to the distance between $(x,\, y)$ and the x-axis, as shown in Figure 16.37.

Solution: Since the lamina is symmetric with respect to the y-axis and $\rho(x,\, y) = ky$, the center of mass will lie on the y-axis. Thus, $\bar{x} = 0$. To find $\bar{y}$, we first find the mass of the lamina

$$\text{mass} = \int_{-2}^{2} \int_{0}^{4-x^2} ky\; dy\; dx = \frac{k}{2} \int_{-2}^{2} y^2 \Big]_{0}^{4-x^2} dx$$

$$= \frac{k}{2} \int_{-2}^{2} (16 - 8x^2 + x^4)\; dx$$

$$= \frac{k}{2} \left[16x - \frac{8x^3}{3} + \frac{x^5}{5} \right]_{-2}^{2}$$

$$= k\left(32 - \frac{64}{3} + \frac{32}{5} \right) = \frac{256k}{15}$$

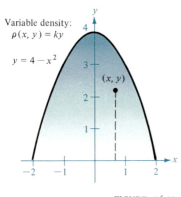

Variable density:
$\rho(x,\, y) = ky$

$y = 4 - x^2$

$(x,\, y)$

FIGURE 16.37

Next, we find the moment about the x-axis.

$$M_x = \int_{-2}^{2} \int_{0}^{4-x^2} (y)(ky)\; dy\; dx = \frac{k}{3} \int_{-2}^{2} y^3 \Big]_{0}^{4-x^2} dx$$

$$= \frac{k}{3} \int_{-2}^{2} (64 - 48x^2 + 12x^4 - x^6)\; dx$$

$$= \frac{k}{3}\left[64x - 16x^3 + \frac{12x^5}{5} - \frac{x^7}{7}\right]_{-2}^{2}$$

$$= \frac{4096k}{105}$$

Thus,

$$\bar{y} = \frac{M_x}{m} = \frac{4096k/105}{256k/15}$$

$$= \frac{16}{7}$$

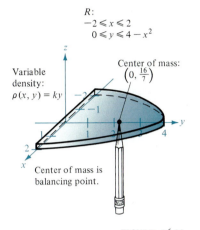

R:
$-2 \leqslant x \leqslant 2$
$0 \leqslant y \leqslant 4 - x^2$

Variable density:
$\rho(x, y) = ky$

Center of mass:
$\left(0, \frac{16}{7}\right)$

Center of mass is balancing point.

FIGURE 16.38

and the center of mass is $(0, \frac{16}{7})$.

Although we think of the moments M_x and M_y as measuring the tendency to rotate about the x- or y-axis, the calculation of moments is usually an intermediate step toward a more tangible goal. The use of the moments M_x and M_y in Example 3 is typical—to find the center of mass. Determination of the center of mass is useful in a variety of applications that allow us to treat a lamina as if its mass were concentrated at just one point. Intuitively, we can think of the center of mass as the balancing point of the lamina. For instance, the lamina in Example 3 should balance on the point of a pencil placed at $(0, \frac{16}{7})$, as shown in Figure 16.38.

Moments of inertia

The moments M_x and M_y used in determining the center of mass of a lamina are sometimes referred to as the **first moments** about the x- and y-axes. In each case the moment is the product of a mass times a distance.

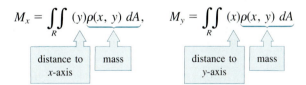

$$M_x = \iint_R (y)\rho(x, y)\, dA, \qquad M_y = \iint_R (x)\rho(x, y)\, dA$$

| distance to x-axis | mass | | distance to y-axis | mass |

We now look at another type of moment referred to as the **second moment,** or the **moment of inertia** of a lamina about a line. In the same way that mass is a measure of the tendency of matter to resist a change in straight line motion, the moment of inertia about a line is a *measure of the tendency of matter to resist a change in rotational motion.* For example, if a particle of mass m has a distance of d from a fixed line, then its moment of inertia about the line is defined to be

$$I = md^2 = (\text{mass})(\text{distance})^2$$

As with moments of mass, we can generalize this concept to obtain the moments of inertia about the x- and y-axes of a lamina of variable density. We denote these second moments by I_x and I_y, and in each case the moment is the product of a mass times the square of a distance.

$$I_x = \iint\limits_{R} (y^2)\rho(x,\, y)\ dA, \qquad I_y = \iint\limits_{R} (x^2)\rho(x,\, y)\ dA$$

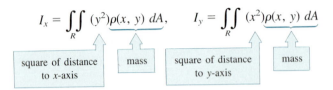

square of distance to x-axis   mass   square of distance to y-axis   mass

The sum of the moments I_x and I_y is called the **polar moment of inertia** and is denoted by I_0. For a lamina in the xy-plane, I_0 represents the moment of inertia of the lamina about the z-axis. The term polar moment of inertia stems from the fact that the square of the polar distance r is used in the calculation.

$$I_0 = \iint\limits_{R} (x^2 + y^2)\rho(x,\, y)\ dA = \iint\limits_{R} r^2\rho(x,\, y)\ dA$$

EXAMPLE 4   *Finding the moment of inertia*

Find the moment of inertia about the x-axis of the lamina described in Example 3.

Solution: From the definition of moment of inertia we have

$$I_x = \int_{-2}^{2} \int_{0}^{4-x^2} y^2(ky)\ dy\ dx = \frac{k}{4} \int_{-2}^{2} y^4 \Big]_{0}^{4-x^2}\ dx$$

$$= \frac{k}{4} \int_{-2}^{2} (256 - 256x^2 + 96x^4 - 16x^6 + x^8)\ dx$$

$$= \frac{k}{4} \left[256x - \frac{256x^3}{3} + \frac{96x^5}{5} - \frac{16x^7}{7} + \frac{x^9}{9} \right]_{-2}^{2} = \frac{32{,}768k}{315}$$

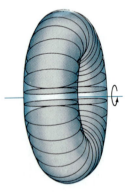

The moment of inertia I of a revolving lamina can be used to measure its kinetic energy. For example, suppose a planar lamina is revolving about a line with an **angular speed** of ω radians per second, as shown in Figure 16.39. The kinetic energy of the revolving lamina is given by

$$E = \frac{1}{2}I\omega^2 \qquad \text{\textcolor{blue}{Kinetic energy for rotational motion}}$$

On the other hand, the kinetic energy of a mass m moving in a straight line at a velocity v is given by

$$E = \frac{1}{2}mv^2 \qquad \text{\textcolor{blue}{Kinetic energy for linear motion}}$$

FIGURE 16.39

Thus, the kinetic energy of a mass moving in a straight line is proportional to its mass, but the kinetic energy of a mass revolving about an axis is proportional to its moment of inertia.

The **radius of gyration** $\bar{\bar{r}}$ of a revolving mass m with moment of inertia I is defined to be

$$\bar{\bar{r}} = \sqrt{\frac{I}{m}}$$

If the entire mass were located at a distance $\bar{\bar{r}}$ from its axis of revolution, it would have the same moment of inertia and, consequently, the same kinetic energy. For instance, the radius of gyration of the lamina in Example 4 about the x-axis is given by

$$\bar{\bar{y}} = \sqrt{\frac{I_x}{m}} = \sqrt{\frac{32{,}768k/315}{256k/15}} = \sqrt{\frac{128}{21}} \approx 2.47$$

EXAMPLE 5   Finding the polar moment of inertia

Find the radius of gyration about the y-axis for the lamina corresponding to the region

$$R: 0 \le y \le \sin x, \quad 0 \le x \le \pi$$

where the density at (x, y) is given by $\rho(x, y) = x$.

Solution: The region R is shown in Figure 16.40. The mass is given by

$$m = \int_0^\pi \int_0^{\sin x} x \, dy \, dx = \int_0^\pi xy \Big]_0^{\sin x} dx$$

$$= \int_0^\pi x \sin x \, dx = \left[-x \cos x + \sin x \right]_0^\pi = \pi$$

The moment of inertia about the y-axis is

$$I_y = \int_0^\pi \int_0^{\sin x} x^3 \, dy \, dx = \int_0^\pi x^3 y \Big]_0^{\sin x} dx = \int_0^\pi x^3 \sin x \, dx$$

$$= \left[(3x^2 - 6)(\sin x) - (x^3 - 6x)(\cos x) \right]_0^\pi = \pi^3 - 6\pi$$

Thus, the radius of gyration about the y-axis is

$$\bar{\bar{x}} = \sqrt{\frac{I_y}{m}} = \sqrt{\frac{\pi^3 - 6\pi}{\pi}} = \sqrt{\pi^2 - 6} \approx 1.97$$

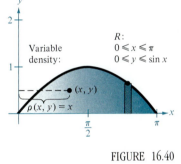

R:
$0 \le x \le \pi$
$0 \le y \le \sin x$

Variable density:

(x, y)

$\rho(x, y) = x$

FIGURE 16.40

Section Exercises 16.5

In Exercises 1–4, find the mass and center of mass for the lamina of specified density.

1. R: rectangle with vertices $(0, 0)$, $(a, 0)$, $(0, b)$, (a, b)
   (a) $\rho = k$ (b) $\rho = ky$
2. R: rectangle with vertices $(0, 0)$, $(a, 0)$, $(0, b)$, (a, b)
   (a) $\rho = kxy$ (b) $\rho = k(x^2 + y^2)$
3. R: triangle with vertices $(0, 0)$, $(b/2, h)$, $(b, 0)$
   (a) $\rho = k$ (b) $\rho = ky$
4. R: triangle with vertices $(0, 0)$, $(0, a)$, $(a, 0)$
   (a) $\rho = k$ (b) $\rho = x^2 + y^2$

In Exercises 5–20, find the mass and center of mass of the lamina bounded by the graphs of the given equations and of the specified density. (Hint: Some of the integrals are simpler in polar coordinates.)

5. $y = \sqrt{a^2 - x^2}$, $y = 0$
   (a) $\rho = k$ (b) $\rho = k(a - y)y$
6. $x^2 + y^2 = a^2$, $0 \le x$, $0 \le y$
   (a) $\rho = k$ (b) $\rho = k(x^2 + y^2)$
7. $y = \sqrt{x}$, $y = 0$, $x = 4$
   $\rho = kxy$

8. $y = x^2$, $y = 0$, $x = 4$
$\rho = kx$

9. $y = e^{-x}$, $y = 0$, $x = 0$, $x = 2$
$\rho = ky$

10. $y = \ln x$, $y = 0$, $x = 1$, $x = e$
$\rho = k/x$

11. $y = \dfrac{1}{1 + x^2}$, $y = 0$, $x = -1$, $x = 1$
$\rho = k$

12. $xy = 4$, $x = 1$, $x = 4$
$\rho = kx^2$

13. $x = 16 - y^2$, $x = 0$
$\rho = kx$

14. $y = 9 - x^2$, $y = 0$
$\rho = ky^2$

15. $y = \sin \dfrac{\pi x}{L}$, $y = 0$, $x = 0$, $x = L$
$\rho = ky$

16. $y = \cos \dfrac{\pi x}{L}$, $y = 0$, $x = 0$, $x = \dfrac{L}{2}$
$\rho = k$

17. $y = \sqrt{a^2 - x^2}$, $y = 0$, $y = x$
$\rho = k$

18. $y = \sqrt{a^2 - x^2}$, $y = 0$, $y = x$
$\rho = k\sqrt{x^2 + y^2}$

19. $r = 2 \cos 3\theta$ (one petal)
$\rho = k$

20. $r = 1 + \cos \theta$
$\rho = k$

In Exercises 21–26, verify the given moments of inertia and find $\bar{\bar{x}}$ and $\bar{\bar{y}}$. Assume each lamina has a density of $\rho = 1$.

21. Rectangle

22. Right Triangle

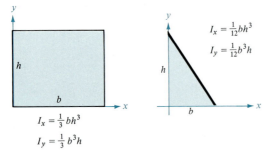

$I_x = \frac{1}{3}bh^3$

$I_y = \frac{1}{3}b^3h$

$I_x = \frac{1}{12}bh^3$

$I_y = \frac{1}{12}b^3h$

23. Circle

24. Semicircle

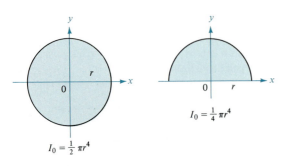

$I_0 = \frac{1}{2}\pi r^4$

$I_0 = \frac{1}{4}\pi r^4$

25. Quarter Circle

26. Ellipse

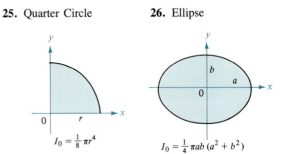

$I_0 = \frac{1}{8}\pi r^4$

$I_0 = \frac{1}{4}\pi ab\,(a^2 + b^2)$

In Exercises 27–34, find I_x, I_y, I_0, $\bar{\bar{x}}$, and $\bar{\bar{y}}$ for the lamina bounded by the graphs of the given equations.

27. $y = 0$, $y = b$, $x = 0$, $x = a$
$\rho = ky$

28. $y = \sqrt{a^2 - x^2}$, $y = 0$
$\rho = ky$

29. $y = 4 - x^2$, $y = 0$
$\rho = kx$

30. $y = x$, $y = x^2$
$\rho = kxy$

31. $y = \sqrt{x}$, $y = 0$, $x = 4$
$\rho = kxy$

32. $y = x^2$, $y^2 = x$
$\rho = x^2 + y^2$

33. $y = x^2$, $y^2 = x$
$\rho = kx$

34. $y = x^3$, $y = 4x$
$\rho = ky$

In Exercises 35–40, find the moment of inertia I of the lamina bounded by the graphs of the given equations about the line.

35. $x^2 + y^2 = b^2$
$\rho = k$, line: $x = a$ $(a > b)$

36. $y = 0$, $y = 2$, $x = 0$, $x = 4$
$\rho = k$, line: $x = 6$

37. $y = \sqrt{x}$, $y = 0$, $x = 4$
$\rho = kx$, line: $x = 6$

38. $y = \sqrt{a^2 - x^2}$, $y = 0$
$\rho = ky$, line: $y = a$

39. $y = \sqrt{a^2 - x^2}$, $y = 0$, $0 \le x$
$\rho = k(a - y)$, line: $y = a$

40. $y = 4 - x^2$, $y = 0$
$\rho = k$, line: $y = 2$

SECTION TOPIC •
Surface area •

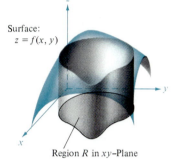

Surface:
$z = f(x, y)$

Region R in xy–Plane

FIGURE 16.41

16.6
Surface area

At this point we know a great deal about the solid region lying between a surface and a closed and bounded region R in the xy-plane, as shown in Figure 16.41. For example, we know how to find

1. The extrema of f on R (Section 15.8)
2. The area of the base R of the solid (Section 16.1)
3. The volume of the solid (Section 16.2)
4. The centroid of the base R (Section 16.5)

In this section we show how to find the upper **surface area** of the solid. In later sections we will find the centroid (center of mass) of the solid (Section 16.7) and the lateral surface area (Section 17.2).

We begin with a surface given by $z = f(x, y)$ defined over a region R. To find the surface area, we construct an inner partition of R consisting of n rectangles, where the area of the ith rectangle R_i is $\Delta A_i = \Delta x_i \Delta y_i$, as shown in Figure 16.42. For a point (x_i, y_i) in R_i, there corresponds a point (x_i, y_i, z_i) on the surface S at which we construct a tangent plane T_i. From Section 15.7 we know that the angle of inclination of this tangent plane is given by

$$\cos \theta_i = \frac{1}{\sqrt{1 + [f_x(x_i, y_i)]^2 + [f_y(x_i, y_i)]^2}}$$

or

$$\sec \theta_i = \sqrt{1 + [f_x(x_i, y_i)]^2 + [f_y(x_i, y_i)]^2}$$

From trigonometry we know that the portion of the tangent plane T_i that lies directly above R_i has an area of $\Delta T_i = \sec \theta_i \, \Delta A_i$. Moreover, we can use the area of this small section of the tangent plane to approximate the area of the portion of the surface that lies directly above R_i, as follows.

$$\Delta S_i \approx \Delta T_i = \sec \theta_i \, \Delta A_i = \sqrt{1 + [f_x(x_i, y_i)]^2 + [f_y(x_i, y_i)]^2} \, \Delta A_i$$

The total area of S can be approximated by the Riemann sum

$$\text{surface area of } S \approx \sum_{i=1}^{n} \Delta S_i \approx \sum_{i=1}^{n} \sqrt{1 + [f_x(x_i, y_i)]^2 + [f_y(x_i, y_i)]^2} \, \Delta A_i$$

Finally, by taking the limit as $\|\Delta\|$ approaches zero, we have the following double integral formula for surface area.

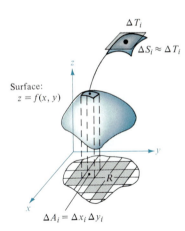

ΔT_i

$\Delta S_i \approx \Delta T_i$

Surface:
$z = f(x, y)$

R

$\Delta A_i = \Delta x_i \Delta y_i$

FIGURE 16.42

DEFINITION OF
SURFACE AREA

If f and its first partial derivatives are continuous on the closed region R in the xy-plane, then the **area of the surface** $z = f(x, y)$ over R is given by

$$\text{surface area} = \iint\limits_{R} dS = \iint\limits_{S} \sqrt{1 + [f_x(x, y)]^2 + [f_y(x, y)]^2} \, dA$$

| Remark   As an aid to remembering the double integral for surface area, it is helpful to note its similarity to the integral for arc length.

Length on x-axis: $\displaystyle\int_a^b dx$

Arc length in xy-plane: $\displaystyle\int_a^b ds = \int_a^b \sqrt{1 + [f'(x)]^2}\, dx$

Area in xy-plane: $\displaystyle\iint_R dA$

Surface area in space: $\displaystyle\iint_R dS = \iint_R \sqrt{1 + [f_x(x, y)]^2 + [f_y(x, y)]^2}\, dA$

EXAMPLE 1   Finding surface area

Find the area S of that portion of the hemisphere $x^2 + y^2 + z^2 = 9$ that lies above the region R bounded by the ellipse $x^2 + 4y^2 = 9$.

Solution: We begin by solving for z as a function of x and y.

$$f(x, y) = z = \sqrt{9 - x^2 - y^2}$$

The first partial derivatives of f are

$$f_x(x, y) = \frac{-x}{\sqrt{9 - x^2 - y^2}} \qquad \text{and} \qquad f_y(x, y) = \frac{-y}{\sqrt{9 - x^2 - y^2}}$$

and from the formula for surface area we have

$$\sqrt{1 + [f_x(x, y)]^2 + [f_y(x, y)]^2} = \frac{3}{\sqrt{9 - x^2 - y^2}}$$

Since f_x and f_y are not continuous at the vertices of the ellipse, $(\pm 3, 0)$, we cannot apply the formula for surface area directly. So, we begin by integrating between the lines $x = \pm a$, where $0 < a < 3$, as shown in Figure 16.43.

$$S = \iint_R \frac{3}{\sqrt{9 - x^2 - y^2}}\, dA = \int_{-a}^a \int_{-\sqrt{9-x^2}/2}^{\sqrt{9-x^2}/2} \frac{3}{\sqrt{(9 - x^2) - y^2}}\, dy\, dx$$

$$= 3 \int_{-a}^a \left. \arcsin \frac{y}{\sqrt{9 - x^2}} \right]_{-\sqrt{9-x^2}/2}^{\sqrt{9-x^2}/2} dx$$

$$= 3 \int_{-a}^a \frac{\pi}{3}\, dx = \pi x \Big]_{-a}^a = 2a\pi$$

Now, by taking the limit as a approaches 3, we obtain a surface area of 6π.

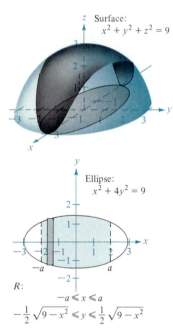

Surface:
$x^2 + y^2 + z^2 = 9$

Ellipse:
$x^2 + 4y^2 = 9$

R:

$-a \leqslant x \leqslant a$

$-\frac{1}{2}\sqrt{9 - x^2} \leqslant y \leqslant \frac{1}{2}\sqrt{9 - x^2}$

FIGURE 16.43

| Remark   The procedure used in Example 1 can be extended to find the surface area of a sphere using the ellipse

$$\frac{x^2}{3^2} + \frac{y^2}{b^2} = 1, \quad 0 < b < 3$$

The surface area over this ellipse is $36 \arcsin (b/3)$. By taking the limit as b approaches 3 and doubling this result, we obtain a total area of 36π. (The surface area of a sphere of radius r is $S = 4\pi r^2$.)

Like integrals for arc length, integrals for surface area are often very difficult to evaluate. However, one type that is easy is demonstrated in the next example.

EXAMPLE 2   *The surface area of a plane region*

Find the surface area of that portion of the plane $z = 2 - x - y$ that lies above the circle $x^2 + y^2 = 1$ in the first quadrant, as shown in Figure 16.44.

Solution: Since $f_x(x, y) = -1$ and $f_y(x, y) = -1$, the surface area is given by

$$S = \iint_R \sqrt{1 + [f_x(x, y)]^2 + [f_y(x, y)]^2} \; dA$$

$$= \iint_R \sqrt{3} \; dA = \sqrt{3} \iint_R dA$$

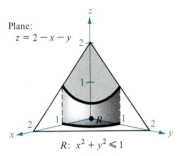

Plane:
$z = 2 - x - y$

$R: \; x^2 + y^2 \leqslant 1$

FIGURE 16.44

Now, since the integral on the right is simply $\sqrt{3}$ times the area of the region R, we have

$$S = \sqrt{3}(\text{area of } R) = \sqrt{3}\left(\frac{\pi}{4}\right) = \frac{\sqrt{3}\,\pi}{4}$$

Another way that integrals for surface area can be simplified is by making a change of variables. In the next example we demonstrate a change of variables to polar coordinates.

EXAMPLE 3   *Change of variables to polar coordinates*

Find the surface area of the paraboloid $z = 1 + x^2 + y^2$ that lies above the unit circle, as shown in Figure 16.45.

Solution: Since $f_x(x, y) = 2x$ and $f_y(x, y) = 2y$, we have

$$S = \iint_R \sqrt{1 + [f_x(x, y)]^2 + [f_y(x, y)]^2} \; dA$$

$$= \iint_R \sqrt{1 + 4x^2 + 4y^2} \; dA$$

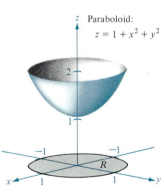

Paraboloid:
$z = 1 + x^2 + y^2$

$R: \; x^2 + y^2 \leqslant 1$

FIGURE 16.45

We convert to polar coordinates by letting $x = r \cos \theta$ and $y = r \sin \theta$. Then, since the region R is bounded by

$$0 \leq r \leq 1 \qquad \text{and} \qquad 0 \leq \theta \leq 2\pi$$

we have

$$S = \int_0^{2\pi} \int_0^1 \sqrt{1 + 4r^2} \; r \, dr \, d\theta = \int_0^{2\pi} \frac{1}{12}(1 + 4r^2)^{3/2} \Big]_0^1 \; d\theta$$

$$= \int_0^{2\pi} \frac{5\sqrt{5} - 1}{12} \; d\theta$$

$$= \frac{\pi(5\sqrt{5} - 1)}{6} \approx 5.33$$

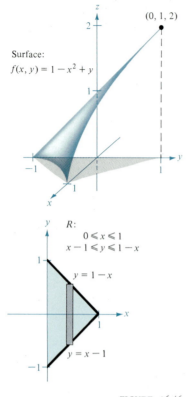

Surface:
$f(x, y) = 1 - x^2 + y$

R:
$0 \leqslant x \leqslant 1$
$x - 1 \leqslant y \leqslant 1 - x$

$y = 1 - x$

$y = x - 1$

FIGURE 16.46

EXAMPLE 4   Finding surface area

Find the area of that portion of the surface $f(x, y) = 1 - x^2 + y$ that lies above the triangular region with vertices $(1, 0, 0)$, $(0, -1, 0)$, and $(0, 1, 0)$, as shown in Figure 16.46.

Solution: Since $f_x(x, y) = -2x$ and $f_y(x, y) = 1$, the surface area is given by

$$S = \iint_R \sqrt{1 + [f_x(x, y)]^2 + [f_y(x, y)]^2} \, dA = \iint_R \sqrt{1 + 4x^2 + 1} \, dA$$

From Figure 16.46 we see that the bounds for R are

$$0 \leq x \leq 1 \qquad \text{and} \qquad x - 1 \leq y \leq 1 - x$$

Thus, the integral becomes

$$S = \int_0^1 \int_{x-1}^{1-x} \sqrt{2 + 4x^2} \, dy \, dx = \int_0^1 y\sqrt{2 + 4x^2} \Big]_{x-1}^{1-x} dx$$

$$= \int_0^1 (2\sqrt{2 + 4x^2} - 2x\sqrt{2 + 4x^2}) \, dx$$

$$= \left[x\sqrt{2 + 4x^2} + \ln(2x + \sqrt{2 + 4x^2}) - \frac{(2 + 4x^2)^{3/2}}{6} \right]_0^1$$

$$= \sqrt{6} + \ln(2 + \sqrt{6}) - \sqrt{6} - \ln\sqrt{2} + \frac{1}{3}\sqrt{2}$$

$$\approx 1.618$$

We can use Simpson's Rule or the Trapezoidal Rule to approximate the value of a double integral, *provided* we can get through the first integration. This is demonstrated in the next example.

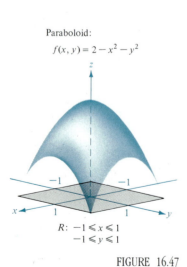

Paraboloid:
$f(x, y) = 2 - x^2 - y^2$

R: $-1 \leqslant x \leqslant 1$
$-1 \leqslant y \leqslant 1$

FIGURE 16.47

EXAMPLE 5   Approximating surface area by Simpson's Rule

Find the area of the surface of the paraboloid $f(x, y) = 2 - x^2 - y^2$ that lies above the square region bounded by $-1 \leq x \leq 1$ and $-1 \leq y \leq 1$, as shown in Figure 16.47.

Solution: Considering the partial derivatives $f_x(x, y) = -2x$ and $f_y(x, y) = -2y$, we have a surface area of

$$S = \iint_R \sqrt{1 + [f_x(x, y)]^2 + [f_y(x, y)]^2} \, dA = \iint_R \sqrt{1 + 4x^2 + 4y^2} \, dA$$

In polar coordinates, the line $x = 1$ is given by $r = \sec \theta$, and we determine from Figure 16.48 that one-fourth of the region R is bounded by

$$0 \leq r \leq \sec \theta \qquad \text{and} \qquad -\frac{\pi}{4} \leq \theta \leq \frac{\pi}{4}$$

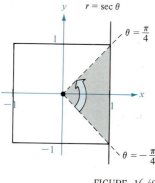

FIGURE 16.48

Now, letting $x = r \cos \theta$ and $y = r \sin \theta$, we have

$$\frac{1}{4}S = \int_{-\pi/4}^{\pi/4} \int_0^{\sec\theta} \sqrt{1 + 4r^2}\, r\, dr\, d\theta$$

$$= \int_{-\pi/4}^{\pi/4} \frac{1}{12}(1 + 4r^2)^{3/2} \Big]_0^{\sec\theta}\, d\theta$$

$$= \frac{1}{12} \int_{-\pi/4}^{\pi/4} [(1 + 4\sec^2\theta)^{3/2} - 1]\, d\theta$$

Now, using Simpson's Rule with $n = 10$, we approximate this *single* integral to be

$$S = \frac{1}{3} \int_{-\pi/4}^{\pi/4} [(1 + 4\sec^2\theta)^{3/2} - 1]\, d\theta \approx 7.45$$

Section Exercises 16.6

In Exercises 1–18, find the area on the surface given by $z = f(x, y)$ over the region R. (Hint: Some of the integrals are simpler in polar coordinates.)

1. $f(x, y) = 2x + 2y$
   R: triangle with vertices $(0, 0, 0)$, $(2, 0, 0)$, $(0, 2, 0)$
2. $f(x, y) = 10 + 2x - 3y$
   R: square with vertices $(0, 0, 0)$, $(2, 0, 0)$, $(0, 2, 0)$, $(2, 2, 0)$
3. $f(x, y) = 8 + 2x + 2y$
   $R = \{(x, y): x^2 + y^2 \leq 4\}$
4. $f(x, y) = 10 + 2x - 3y$
   $R = \{(x, y): x^2 + y^2 \leq 9\}$
5. $f(x, y) = 9 - x^2$
   R: square with vertices $(0, 0, 0)$, $(3, 0, 0)$, $(0, 3, 0)$, $(3, 3, 0)$
6. $f(x, y) = y^2$
   R: square with vertices $(0, 0, 0)$, $(3, 0, 0)$, $(0, 3, 0)$, $(3, 3, 0)$
7. $f(x, y) = 2y + x^2$
   R: triangle with vertices $(0, 0, 0)$, $(1, 0, 0)$, $(1, 1, 0)$
8. $f(x, y) = 2x + y^2$
   R: triangle with vertices $(0, 0, 0)$, $(2, 0, 0)$, $(0, 2, 0)$
9. $f(x, y) = 2 + x^{3/2}$
   R: quadrangle with vertices $(0, 0, 0)$, $(0, 4, 0)$, $(3, 4, 0)$, $(3, 0, 0)$
10. $f(x, y) = 2 + \frac{2}{3}x^{3/2}$
   $R = \{(x, y): 0 \leq x \leq 1, 0 \leq y \leq 1 - x\}$
11. $f(x, y) = \ln |\sec x|$
   $R = \{(x, y): 0 \leq x \leq \frac{\pi}{4}, 0 \leq y \leq \tan x\}$
12. $f(x, y) = 4 + x^2 - y^2$
   $R = \{(x, y): x^2 + y^2 \leq 1\}$

13. $f(x, y) = 4 - x^2 - y^2$
   $R = \{(x, y): 0 \leq f(x, y)\}$
14. $f(x, y) = x^2 + y^2$
   $R = \{(x, y): 0 \leq f(x, y) \leq 16\}$
15. $f(x, y) = \sqrt{x^2 + y^2}$
   $R = \{(x, y): 0 \leq f(x, y) \leq 1\}$
16. $f(x, y) = xy$
   $R = \{(x, y): x^2 + y^2 \leq 16\}$
17. $f(x, y) = \sqrt{a^2 - x^2 - y^2}$
   $R = \{(x, y): x^2 + y^2 \leq b^2, b < a\}$
18. $f(x, y) = \sqrt{a^2 - x^2 - y^2}$
   $R = \{(x, y): x^2 + y^2 \leq a^2\}$

19. Find the surface area of the solid of intersection of the cylinders $x^2 + z^2 = 1$ and $y^2 + z^2 = 1$, as shown in Figure 16.49.

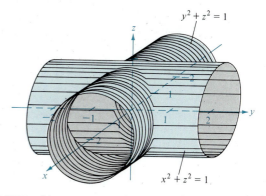

FIGURE 16.49

20. Show that the surface area of the cone

$$z = k\sqrt{x^2 + y^2}, \quad k > 0$$

over a region R in the xy-plane is $A\sqrt{k^2 + 1}$, where A is the area of region R, as shown in Figure 16.50.

In Exercises 21–26, set up the double integral that gives the area of the surface on the graph of f over the region R.

21. $f(x, y) = x^3 - 3xy + y^3$
    R: square with vertices $(1, 1, 0)$, $(-1, 1, 0)$,
    $(-1, -1, 0)$, $(1, -1, 0)$

22. $f(x, y) = e^{-x} \sin y$
    $R = \{(x, y): 0 \le x \le 4, 0 \le y \le x\}$

23. $f(x, y) = e^{-x} \sin y$
    $R = \{(x, y): x^2 + y^2 \le 4\}$

24. $f(x, y) = x^2 - 3xy - y^2$
    $R = \{(x, y): 0 \le x \le 4, 0 \le y \le x\}$

25. $f(x, y) = e^{xy}$
    $R = \{(x, y): 0 \le x \le 4, 0 \le y \le 10\}$

26. $f(x, y) = \cos(x^2 + y^2)$
    $R = \left\{(x, y): x^2 + y^2 \le \dfrac{\pi}{2}\right\}$

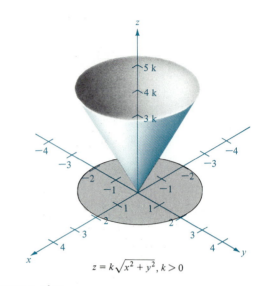

$z = k\sqrt{x^2 + y^2}, k > 0$

FIGURE 16.50

16.7
Triple integrals and applications

The procedure used to define a **triple integral** follows that used for double integrals. First, we assume that f is a continuous function of three variables, defined over a bounded solid region Q. Next we encompass Q with a network of boxes and form the **inner partition** consisting of all boxes lying entirely within Q, as shown in Figure 16.51. The volume of the ith box is

$$\Delta V_i = \Delta x_i \Delta y_i \Delta z_i \qquad \text{Volume of } i\text{th box}$$

and we define the **norm** $\|\Delta\|$ of the partition to be the length of the longest diagonal of the n boxes in the partition. Then we choose a point (x_i, y_i, z_i) in each box and form the Riemann sum

$$\sum_{i=1}^{n} f(x_i, y_i, z_i)\Delta V_i$$

Finally, by taking the limit as $\|\Delta\|$ approaches zero, we obtain the following definition.

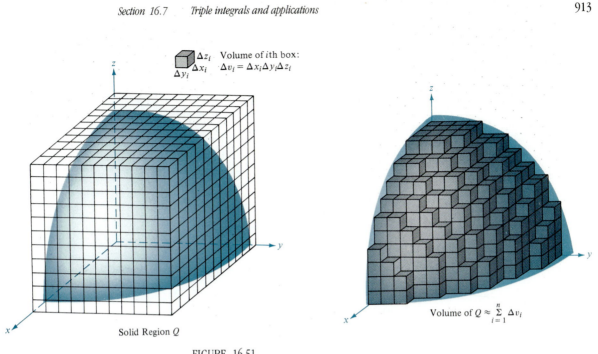

Solid Region Q

FIGURE 16.51

DEFINITION OF TRIPLE INTEGRAL

If f is continuous over a bounded solid region Q, then the **triple integral of f over Q** is defined to be

$$\iiint\limits_{Q} f(x, y, z) \, dV = \lim_{\|\Delta\| \to 0} \sum_{i=1}^{n} f(x_i, y_i, z_i) \Delta V_i$$

provided the limit exists.

| Remark   In the special case where $f(x, y, z) = 1$ in the solid region Q, the triple integral represents the **volume** of Q. That is,

$$\text{volume of } Q = \iiint\limits_{Q} dV$$

Each of the properties of double integrals in Theorem 16.1 can be restated in terms of triple integrals. For example,

1. $\displaystyle\iiint\limits_{Q} c \, f(x, y, z) \, dV = c \iiint\limits_{Q} f(x, y, z) \, dV$

2. $\displaystyle\iiint\limits_{Q} [f(x, y, z) \pm g(x, y, z)] \, dV = \iiint\limits_{Q} f(x, y, z) \, dV \pm \iiint\limits_{Q} g(x, y, z) \, dV$

3. $\displaystyle\iiint\limits_{Q} f(x, y, z) \, dV = \iiint\limits_{Q_1} f(x, y, z) \, dV + \iiint\limits_{Q_2} f(x, y, z) \, dV$

where Q is the union of two nonoverlapping subregions Q_1 and Q_2.

We can evaluate a triple integral with an iterated integral, *provided* the solid region Q is simple with respect to one of the six orders of integration

$$dx\ dy\ dz \qquad dy\ dx\ dz \qquad dz\ dx\ dy$$
$$dx\ dz\ dy \qquad dy\ dz\ dx \qquad dz\ dy\ dx$$

In the following version of Fubini's Theorem, we describe a region that is considered simple with respect to the order $dz\ dy\ dx$. Similar descriptions can be given for the other five orders.

THEOREM 16.6

EVALUATION BY ITERATED INTEGRALS
Let f be continuous on a solid region Q defined by

$$a \le x \le b, \qquad h_1(x) \le y \le h_2(x), \qquad g_1(x,\ y) \le z \le g_2(x,\ y)$$

where h_1, h_2, g_1, and g_2 are continuous functions. Then,

$$\iiint\limits_{Q} f(x,\ y,\ z)\ dV = \int_a^b \int_{h_1(x)}^{h_2(x)} \int_{g_1(x,y)}^{g_2(x,y)} f(x,\ y,\ z)\ dz\ dy\ dx$$

To evaluate a triple iterated integral in the order $dz\ dy\ dx$, we hold *both* x and y constant for the innermost integration, and then hold x constant for the second integration. This is demonstrated in the first example.

EXAMPLE 1   Evaluating a triple iterated integral

Evaluate the iterated integral

$$\int_0^2 \int_0^x \int_0^{x+y} e^x(y + 2z)\ dz\ dy\ dx$$

Solution: Holding x and y constant, we have

$$\int_0^2 \int_0^x \int_0^{x+y} e^x(y + 2z)\ dz\ dy\ dx = \int_0^2 \int_0^x e^x(yz + z^2)\Big]_0^{x+y}\ dy\ dx$$

$$= \int_0^2 \int_0^x e^x(x^2 + 3xy + 2y^2)\ dy\ dx$$

Now, holding x constant, we have

$$\int_0^2 \int_0^x e^x(x^2 + 3xy + 2y^2)\ dy\ dx = \int_0^2 \left[e^x\left(x^2y + \frac{3xy^2}{2} + \frac{2y^3}{3} \right) \right]_0^x\ dx$$

$$= \frac{19}{6} \int_0^2 x^3 e^x\ dx$$

$$= \frac{19}{6} \left[e^x(x^3 - 3x^2 + 6x - 6) \right]_0^2$$

$$= 19\left(\frac{e^2}{3} + 1 \right)$$

Solid region Q lies between two surfaces.

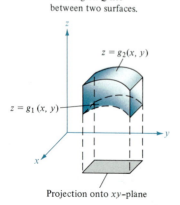

Projection onto xy-plane

FIGURE 16.52

$0 \leq z \leq 2\sqrt{4 - x^2 - y^2}$

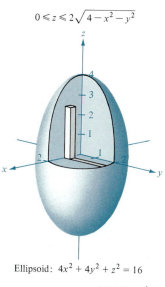

Ellipsoid: $4x^2 + 4y^2 + z^2 = 16$

FIGURE 16.53

In Example 1, we demonstrated the integration order $dz\ dy\ dx$. For other orders, we follow a similar procedure. For instance, to evaluate a triple iterated integral in the order $dx\ dy\ dz$ we hold both y and z constant for the innermost integration and integrate with respect to x. Then, for the second integration we hold z constant and integrate with respect to y. Finally, for the third integration, we integrate with respect to z.

To find the limits for a particular order of integration, it is generally advisable to first determine the innermost limits, which may be functions of the outer two variables. Then, by projecting the solid Q onto the coordinate plane of the outer two variables, we can determine their limits of integration by the methods used for double integrals. For instance, to evaluate

$$\iiint_Q f(x,\ y,\ z)\ dz\ dy\ dx$$

first determine the limits for z, and then the integral has the form

$$\iint \left[\int_{g_1(x,y)}^{g_2(x,y)} f(x,\ y,\ z)\ dz \right] dy\ dx$$

Now, by projecting the solid Q onto the xy-plane, we can determine the limits for x and y as we did for double integrals, as in Figure 16.52.

EXAMPLE 2   *Finding volume by triple integrals*

Find the volume of the ellipsoidal solid given by $4x^2 + 4y^2 + z^2 = 16$.

Solution: Since x, y, and z play similar roles in the equation, the order of integration is probably immaterial, and we arbitrarily choose $dz\ dy\ dx$. Moreover, we can simplify the calculation by considering only that portion of the ellipsoid lying in the first octant, as shown above in Figure 16.53. From the order $dz\ dy\ dx$, we first determine the bounds for z.

$$0 \leq z \leq 2\sqrt{4 - x^2 - y^2}$$

Then, from Figure 16.54 we see that the boundaries for y and x are

$$0 \leq x \leq 2 \qquad \text{and} \qquad 0 \leq y \leq \sqrt{4 - x^2}$$

so the volume of the ellipsoid is

$$V = \iiint_Q dV = 8 \int_0^2 \int_0^{\sqrt{4-x^2}} \int_0^{2\sqrt{4-x^2-y^2}} dz\ dy\ dx$$

$$= 8 \int_0^2 \int_0^{\sqrt{4-x^2}} z \Big]_0^{2\sqrt{4-x^2-y^2}} dy\ dx$$

$$= 16 \int_0^2 \int_0^{\sqrt{4-x^2}} \sqrt{(4 - x^2) - y^2}\ dy\ dx$$

$$= 8 \int_0^2 \left[y\sqrt{4 - x^2 - y^2} + (4 - x^2) \arcsin\left(\frac{y}{\sqrt{4 - x^2}} \right) \right]_0^{\sqrt{4-x^2}} dx$$

$$= 8 \int_0^2 (4 - x^2)\left(\frac{\pi}{2} \right) dx = 4\pi \left[4x - \frac{x^3}{3} \right]_0^2 = \frac{64\pi}{3}$$

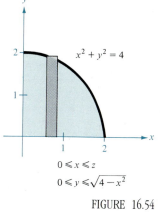

$x^2 + y^2 = 4$

$0 \leq x \leq z$
$0 \leq y \leq \sqrt{4 - x^2}$

FIGURE 16.54

Example 2 is a little unusual in that each of the six possible orders of integration produce integrals of comparable difficulty. Try setting up some other possible orders of integration to find the volume of the ellipsoid. For instance, the order $dx\ dy\ dz$ yields the integral

$$V = 8 \int_0^4 \int_0^{\sqrt{16-z^2}/2} \int_0^{\sqrt{16-4y^2-z^2}/2} dx\ dy\ dz$$

If you solve this integral you will obtain the same volume obtained in Example 2. This is always the case—the order of integration does not affect the value of an integral. However, it often happens that the order of integration does affect the complexity of the integral. In Example 3 the given order of integration is not convenient and we change the order to simplify the problem.

EXAMPLE 3   *Changing the order of integration*

Evaluate

$$\int_0^{\sqrt{\pi/2}} \int_x^{\sqrt{\pi/2}} \int_1^3 \sin y^2\ dz\ dy\ dx$$

Solution: Note that after one integration in the given order, we would encounter the integral $2 \int \sin (y^2)\ dy$, which is not an elementary function. To avoid this problem we change the order of integration to $dz\ dx\ dy$, so that y is the outer variable.

The solid region Q is given by

$$0 \le x \le \sqrt{\frac{\pi}{2}}, \qquad x \le y \le \sqrt{\frac{\pi}{2}}, \qquad 1 \le z \le 3$$

as shown in Figure 16.55, and the projection of Q in the xy-plane yields the bounds

$$0 \le y \le \sqrt{\frac{\pi}{2}} \qquad \text{and} \qquad 0 \le x \le y$$

Therefore, we have

$$V = \iiint_Q dV = \int_0^{\sqrt{\pi/2}} \int_0^y \int_1^3 \sin (y^2)\ dz\ dx\ dy$$

$$= \int_0^{\sqrt{\pi/2}} \int_0^y z \sin (y^2) \Big]_1^3\ dx\ dy$$

$$= 2 \int_0^{\sqrt{\pi/2}} \int_0^y \sin (y^2)\ dx\ dy$$

$$= 2 \int_0^{\sqrt{\pi/2}} x \sin (y^2) \Big]_0^y\ dy$$

$$= 2 \int_0^{\sqrt{\pi/2}} y \sin (y^2)\ dy$$

$$= -\cos (y^2) \Big]_0^{\sqrt{\pi/2}}$$

$$= 1$$

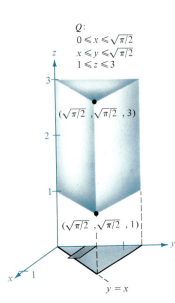

Q:
$0 \le x \le \sqrt{\pi/2}$
$x \le y \le \sqrt{\pi/2}$
$1 \le z \le 3$

$(\sqrt{\pi/2}, \sqrt{\pi/2}, 3)$

$(\sqrt{\pi/2}, \sqrt{\pi/2}, 1)$

$y = x$

FIGURE 16.55

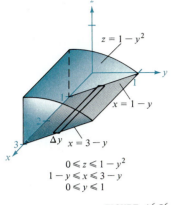

$$z = 1 - y^2$$

$$x = 1 - y$$

$$x = 3 - y$$

$$\Delta y$$

$$0 \leq z \leq 1 - y^2$$
$$1 - y \leq x \leq 3 - y$$
$$0 \leq y \leq 1$$

FIGURE 16.56

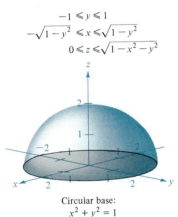

$$-1 \leq y \leq 1$$
$$-\sqrt{1 - y^2} \leq x \leq \sqrt{1 - y^2}$$
$$0 \leq z \leq \sqrt{1 - x^2 - y^2}$$

Circular base:
$$x^2 + y^2 = 1$$

FIGURE 16.57

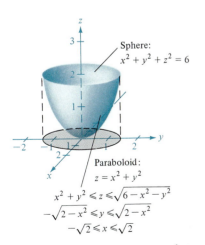

Sphere:
$$x^2 + y^2 + z^2 = 6$$

Paraboloid:
$$z = x^2 + y^2$$

$$x^2 + y^2 \leq z \leq \sqrt{6 - x^2 - y^2}$$
$$-\sqrt{2 - x^2} \leq y \leq \sqrt{2 - x^2}$$
$$-\sqrt{2} \leq x \leq \sqrt{2}$$

FIGURE 16.58

EXAMPLE 4 *Determining the limits of integration*

Set up a triple integral for the volume of each of the following solid regions.
(a) The region in the first octant bounded above by the cylinder $z = 1 - y^2$ and lying between the vertical planes given by $x + y = 1$ and $x + y = 3$.
(b) The upper hemisphere given by $z = \sqrt{1 - x^2 - y^2}$.
(c) The region bounded below by the paraboloid $z = x^2 + y^2$ and above by the sphere $x^2 + y^2 + z^2 = 6$.

Solution:
(a) From Figure 16.56, we see that the solid is bounded below by the xy-plane ($z = 0$) and above by the cylinder $z = 1 - y^2$. Therefore, we have

$$0 \leq z \leq 1 - y^2$$

Now, by projecting the region onto the xy-plane, we obtain a parallelogram. Since two sides of the parallelogram are parallel to the x-axis, we set up the bounds

$$1 - y \leq x \leq 3 - y \qquad \text{and} \qquad 0 \leq y \leq 1$$

Therefore, the volume of the region is given by

$$V = \iiint_Q dV = \int_0^1 \int_{1-y}^{3-y} \int_0^{1-y^2} dz\, dx\, dy$$

(b) For the upper hemisphere given by $z = \sqrt{1 - x^2 - y^2}$, we have

$$0 \leq z \leq \sqrt{1 - x^2 - y^2}$$

In Figure 16.57, we see that the projection of the hemisphere onto the xy-plane is the circle given by $x^2 + y^2 = 1$, and we can use either order $dx\, dy$ or $dy\, dx$. Choosing the first, we have

$$-\sqrt{1 - y^2} \leq x \leq \sqrt{1 - y^2} \qquad \text{and} \qquad -1 \leq y \leq 1$$

which implies that the volume of the region is given by

$$V = \iiint_Q dV = \int_{-1}^1 \int_{-\sqrt{1-y^2}}^{\sqrt{1-y^2}} \int_0^{\sqrt{1-x^2-y^2}} dz\, dx\, dy$$

(c) For the region bounded below by the paraboloid $z = x^2 + y^2$ and above by the sphere $x^2 + y^2 + z^2 = 6$ we have

$$x^2 + y^2 \leq z \leq \sqrt{6 - x^2 - y^2}$$

In Figure 16.58, we can see that the projection of this region onto the xy-plane is the circle given by $x^2 + y^2 = 2$. Using the order $dy\, dx$, we have

$$-\sqrt{2 - x^2} \leq y \leq \sqrt{2 - x^2} \qquad \text{and} \qquad -\sqrt{2} \leq x \leq \sqrt{2}$$

which implies that the volume of the region is given by

$$V = \iiint_Q dV = \int_{-\sqrt{2}}^{\sqrt{2}} \int_{-\sqrt{2-x^2}}^{\sqrt{2-x^2}} \int_{x^2+y^2}^{\sqrt{6-x^2-y^2}} dz\, dy\, dx$$

Center of mass and moments of inertia

In the remainder of this section we look at two important applications of triple integrals. We consider a solid region Q whose density at (x, y, z) is given by the **density function** ρ. The **center of mass** of a solid region Q of mass m is given by $(\bar{x}, \bar{y}, \bar{z})$, where

$$m = \iiint_Q \rho(x, y, z)\, dV \qquad M_{xz} = \iiint_Q y\rho(x, y, z)\, dV$$

$$M_{yz} = \iiint_Q x\rho(x, y, z)\, dV \qquad M_{xy} = \iiint_Q z\rho(x, y, z)\, dV$$

and

$$\bar{x} = \frac{M_{yz}}{m}, \qquad \bar{y} = \frac{M_{xz}}{m}, \qquad \bar{z} = \frac{M_{xy}}{m}$$

The quantities M_{yz}, M_{xz}, and M_{xy} are called the **first moments** of region Q about the yz-, xz-, and xy-planes, respectively.

EXAMPLE 5   Finding the center of mass of a solid region

Find the center of mass of the unit cube shown in Figure 16.59, given that the density at a point (x, y, z) is proportional to the square of its distance from the origin.

Solution: Since the density at (x, y, z) is proportional to the square of the distance between $(0, 0, 0)$ and (x, y, z), we have

$$\rho(x, y, z) = k(x^2 + y^2 + z^2)$$

Now, using this density function, we begin by finding the mass of the cube. Because of the symmetry of the region, the order of integration is immaterial, and we have

$$m = \int_0^1 \int_0^1 \int_0^1 k(x^2 + y^2 + z^2)\, dz\, dy\, dx$$

$$= k \int_0^1 \int_0^1 \left[(x^2 + y^2)z + \frac{z^3}{3} \right]_0^1 dy\, dx$$

$$= k \int_0^1 \int_0^1 \left(x^2 + y^2 + \frac{1}{3} \right) dy\, dx = k \int_0^1 \left[\left(x^2 + \frac{1}{3} \right)y + \frac{y^3}{3} \right]_0^1 dx$$

$$= k \int_0^1 \left(x^2 + \frac{2}{3} \right) dx = k \left[\frac{x^3}{3} + \frac{2x}{3} \right]_0^1 = k$$

The first moment about the yz-plane is

$$M_{yz} = k \int_0^1 \int_0^1 \int_0^1 x(x^2 + y^2 + z^2)\, dz\, dy\, dx$$

$$= k \int_0^1 x \left[\int_0^1 \int_0^1 (x^2 + y^2 + z^2)\, dz\, dy \right] dx$$

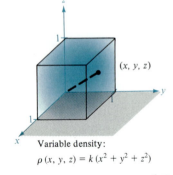

Variable density:

$\rho(x, y, z) = k(x^2 + y^2 + z^2)$

FIGURE 16.59

Note that x can be factored out of the two inner integrals, since it is constant with respect to y and z. After factoring, the two inner integrals are the same as for the mass m; hence we have

$$M_{yz} = k \int_0^1 x\left(x^2 + \frac{2}{3}\right) dx = k\left[\frac{x^4}{4} + \frac{x^2}{3}\right]_0^1 = \frac{7k}{12}$$

Therefore,

$$\bar{x} = \frac{M_{yz}}{m} = \frac{7k/12}{k} = \frac{7}{12}$$

Finally, from the nature of ρ and the symmetry of x, y, and z in this solid region, we have $\bar{x} = \bar{y} = \bar{z}$, and the center of mass is $(\frac{7}{12}, \frac{7}{12}, \frac{7}{12})$. □

The first moments for solid regions are taken about a plane, whereas the second moments for solids are taken about a line. The **second moments** (or **moments of inertia**) about the x-, y-, and z-axes are given by

$$I_x = \iiint\limits_Q (y^2 + z^2)\rho(x, y, z) \, dV \qquad \text{Moment of inertia about } x\text{-axis}$$

$$I_y = \iiint\limits_Q (x^2 + z^2)\rho(x, y, z) \, dV \qquad \text{Moment of inertia about } y\text{-axis}$$

$$I_z = \iiint\limits_Q (x^2 + y^2)\rho(x, y, z) \, dV \qquad \text{Moment of inertia about } z\text{-axis}$$

For problems requiring the calculation of all three of these moments, you can often save considerable effort by applying the additive property of triple integrals and writing

$$I_x = I_{xz} + I_{xy}, \qquad I_y = I_{yz} + I_{xy}, \qquad \text{and} \qquad I_z = I_{yz} + I_{xz}$$

where

$$I_{xy} = \iiint\limits_Q z^2\rho(x, y, z) \, dV$$

$$I_{xz} = \iiint\limits_Q y^2\rho(x, y, z) \, dV$$

and

$$I_{yz} = \iiint\limits_Q x^2\rho(x, y, z) \, dV$$

EXAMPLE 6   Moments of inertia for a solid region

Find the moments of inertia about the x- and y-axes for the solid region lying between the hemisphere $z = \sqrt{4 - x^2 - y^2}$ and the xy-plane, given that the density at (x, y, z) is proportional to the distance between (x, y, z) and the xy-plane.

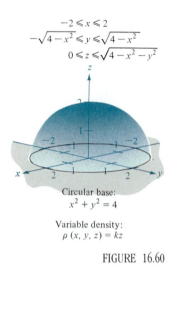

$$-2 \leqslant x \leqslant 2$$
$$-\sqrt{4-x^2} \leqslant y \leqslant \sqrt{4-x^2}$$
$$0 \leqslant z \leqslant \sqrt{4-x^2-y^2}$$

Circular base:
$x^2 + y^2 = 4$

Variable density:
$\rho\,(x,\,y,\,z) = kz$

FIGURE 16.60

Solution: The density of the region is given by $\rho(x, y, z) = kz$. Considering the symmetry of this problem, we know that $I_x = I_y$, and we need to compute only one moment, say I_x. From Figure 16.60 we choose the order $dz\,dy\,dx$ and write

$$I_x = \iiint_Q (y^2 + z^2)\rho(x, y, z)\,dV$$

$$= \int_{-2}^{2} \int_{-\sqrt{4-x^2}}^{\sqrt{4-x^2}} \int_{0}^{\sqrt{4-x^2-y^2}} (y^2 + z^2)(kz)\,dz\,dy\,dx$$

$$= k \int_{-2}^{2} \int_{-4-x^2}^{\sqrt{4-x^2}} \left[\frac{y^2 z^2}{2} + \frac{z^4}{4} \right]_0^{\sqrt{4-x^2-y^2}} dy\,dx$$

$$= k \int_{-2}^{2} \int_{-\sqrt{4-x^2}}^{\sqrt{4-x^2}} \left[\frac{y^2(4 - x^2 - y^2)}{2} + \frac{(4 - x^2 - y^2)^2}{4} \right] dy\,dx$$

$$= \frac{k}{4} \int_{-2}^{2} \int_{-\sqrt{4-x^2}}^{\sqrt{4-x^2}} [(4 - x^2)^2 - y^4]\,dy\,dx$$

$$= \frac{k}{4} \int_{-2}^{2} \left[(4 - x^2)^2 y - \frac{y^5}{5} \right]_{-\sqrt{4-x^2}}^{\sqrt{4-x^2}} dx$$

$$= \frac{k}{4} \int_{-2}^{2} \frac{8}{5}(4 - x^2)^{5/2}\,dx = \frac{4k}{5} \int_{0}^{2} (4 - x^2)^{5/2}\,dx \qquad x = 2\sin\theta$$

$$= \frac{4k}{5} \int_{0}^{\pi/2} 64\cos^6\theta\,d\theta = \left(\frac{256k}{5}\right)\left(\frac{5\pi}{32}\right) = 8k\pi \qquad \text{Wallis's Formula}$$

Thus, $I_x = 8k\pi = I_y$.

Section Exercises 16.7

In Exercises 1–10, evaluate the triple integral.

1. $\int_0^3 \int_0^2 \int_0^1 (x + y + z)\,dx\,dy\,dz$

2. $\int_{-1}^1 \int_{-1}^1 \int_{-1}^1 x^2 y^2 z^2\,dx\,dy\,dz$

3. $\int_0^1 \int_0^x \int_0^{xy} x\,dz\,dy\,dx$

4. $\int_0^4 \int_0^\pi \int_0^{1-x} x\sin y\,dz\,dy\,dx$

5. $\int_1^4 \int_0^1 \int_0^x 2ze^{-x^2}\,dy\,dx\,dz$

6. $\int_1^4 \int_1^{e^2} \int_0^{1/xz} \ln z\,dy\,dz\,dx$

7. $\int_0^9 \int_0^{y/3} \int_0^{\sqrt{y^2 - 9x^2}} z\,dz\,dx\,dy$

8. $\int_0^{\sqrt{2}} \int_0^{\sqrt{2-x^2}} \int_{2x^2+y^2}^{4-y^2} y\,dz\,dy\,dx$

9. $\int_0^2 \int_{-\sqrt{4-x^2}}^{\sqrt{4-x^2}} \int_0^{x^2} x\,dz\,dy\,dx$

10. $\int_0^{\pi/2} \int_0^{y/2} \int_0^{1/y} \sin y\,dz\,dx\,dy$

In Exercises 11–14, sketch the solid whose volume is given by the triple integral and rewrite the integral with the specified order of integration.

11. $\int_0^4 \int_0^{(4-x)/2} \int_0^{(12-3x-6y)/4} dz\,dy\,dx$

Rewrite using the order $dy\,dx\,dz$.

12. $\int_0^4 \int_0^{\sqrt{16-x^2}} \int_0^{10-x-y} dz \, dy \, dx$

Rewrite using the order $dz \, dx \, dy$.

13. $\int_0^1 \int_y^1 \int_0^{\sqrt{1-y^2}} dz \, dx \, dy$

Rewrite using the order $dz \, dy \, dx$.

14. $\int_0^2 \int_{2x}^4 \int_0^{\sqrt{y^2-4x^2}} dz \, dy \, dx$

Rewrite using the order $dx \, dy \, dz$.

In Exercises 15–20, use a triple integral to find the volume of the solid bounded by the graphs of the given equations.

15. $x = 4 - y^2$, $z = 0$, $z = x$
16. $z = xy$, $z = 0$, $x = 0$, $x = 1$, $y = 0$, $y = 1$
17. $x^2 + y^2 + z^2 = r^2$
18. $z = 9 - x^2 - y^2$, $z = 0$
19. $z = 4 - x^2$, $y = 4 - x^2$ (first octant)
20. $z = 9 - x^2$, $y = -x + 2$, $y = 0$, $z = 0$, $x \geq 0$

In Exercises 21–24, find the mass and the indicated coordinates of the center of mass of the solid of specified density bounded by the graphs of the equations.

21. Find $\bar{x}$ using $\rho(x, y, z) = k$.
   $Q: 2x + 3y + 6z = 12$, $x = 0$, $y = 0$, $z = 0$
22. Find $\bar{y}$ using $\rho(x, y, z) = ky$.
   $Q: 2x + 3y + 6z = 12$, $x = 0$, $y = 0$, $z = 0$
23. Find $\bar{z}$ using $\rho(x, y, z) = kx$.
   $Q: z = 4 - x$, $z = 0$, $y = 0$, $y = 4$
24. Find $\bar{y}$ using $\rho(x, y, z) = k$.
   $Q: \dfrac{x}{a} + \dfrac{y}{b} + \dfrac{z}{c} = 1$ $(a, b, c > 0)$, $x = 0$, $y = 0$, $z = 0$

In Exercises 25 and 26, find the mass and the center of mass of the solid bounded by the graphs of the given equations.

25. $x = 0$, $x = b$, $y = 0$, $y = b$, $z = 0$, $z = b$,
   $\rho(x, y, z) = kxy$
26. $x = 0$, $x = a$, $y = 0$, $y = b$, $z = 0$, $z = c$,
   $\rho(x, y, z) = kz$

In Exercises 27–30, find the centroid of the solid region bounded by the graphs of the given equations. (Assume uniform density and find the center of mass.)

27. $z = \dfrac{h}{r}\sqrt{x^2 + y^2}$, $z = h$

28. $y = \sqrt{4 - x^2}$, $y = 0$, $z = y$, $z = 0$
29. $z = \sqrt{4^2 - x^2 - y^2}$, $z = 0$

30. $z = \dfrac{1}{y^2 + 1}$, $z = 0$, $x = -2$, $x = 2$, $y = 0$, $y = 1$

In Exercises 31–34, find I_x, I_y, and I_z for the indicated solid of specified density.

31. (a) $\rho = k$
   (b) $\rho = kxyz$

32. (a) $\rho(x, y, z) = k$
   (b) $\rho(x, y, z) = k(x^2 + y^2)$

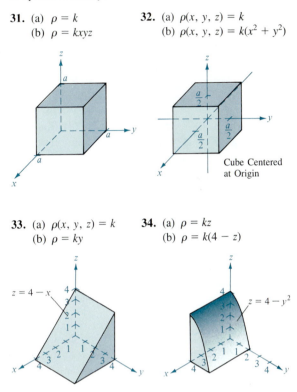

Cube Centered at Origin

33. (a) $\rho(x, y, z) = k$
   (b) $\rho = ky$

34. (a) $\rho = kz$
   (b) $\rho = k(4 - z)$

In Exercises 35 and 36, verify the moments of inertia for the solids of uniform density.

35. $I_x = \dfrac{1}{12}m(a^2 + b^2)$

$I_y = \dfrac{1}{12}m(b^2 + c^2)$

$I_z = \dfrac{1}{12}m(a^2 + c^2)$

36. $I_x = I_z = \dfrac{1}{12}m(3a^2 + L^2)$

$I_y = \dfrac{1}{2}ma^2$

(Hint: Use Wallis's Formula.)

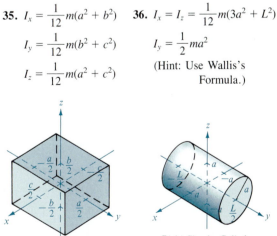

Right Circular Cylinder

16.8
Triple integrals in cylindrical and spherical coordinates

Many common solid regions such as spheres, ellipsoids, cones, and paraboloids can yield difficult triple integrals in rectangular coordinates. In fact, it is precisely this difficulty that led to the introduction of nonrectangular coordinate systems. In this section, we show how *cylindrical* and *spherical* coordinates can be used to evaluate triple integrals. One of the first to use such a system was the French mathematician Pierre Simon de Laplace (1749–1827). Laplace has been called the "Newton of France," and he published many important works in mechanics, differential equations, and probability.

We begin by looking at triple integrals in cylindrical coordinates. Recall that the rectangular conversion equations for cylindrical coordinates are

$$x = r \cos \theta, \qquad y = r \sin \theta, \qquad z = z$$

In this coordinate system, the simplest solid region is a cylindrical block bounded by

$$r_1 \leq r \leq r_2$$
$$\theta_1 \leq \theta \leq \theta_2$$

and

$$z_1 \leq z \leq z_2$$

as shown in Figure 16.61.

To obtain the cylindrical coordinate form of a triple integral, we let f be a continuous function of r, θ, and z defined over a bounded solid region Q. To begin, we encompass the solid by a network of cylindrical blocks and form the **inner partition** Δ, consisting of all blocks lying entirely within Q. The **norm** $\|\Delta\|$ of the partition is the length of the longest diagonal of the n blocks in Δ. If we choose a point (r_i, θ_i, z_i) in the ith block such that r_i is the average radius of the block, then the block's volume is

$$\Delta V_i = (\text{area of base})(\text{height}) = (r_i \Delta \theta_i \Delta r_i) \Delta z_i$$

By forming the sum

$$\sum_{i=1}^{n} f(r_i, \theta_i, z_i) r_i \Delta r_i \Delta \theta_i \Delta z_i$$

and taking the limit as $\|\Delta\| \to 0$, it can be shown that we obtain the following cylindrical coordinate form of a triple integral.

Pierre Simon de Laplace

$\theta = 0$ Volume of cylindrical block:
$\Delta V_i = r_i \Delta r_i \Delta \theta_i \Delta z_i$

FIGURE 16.61

| TRIPLE INTEGRAL IN CYLINDRICAL COORDINATES | If f is a continuous function of r, θ, and z on a bounded solid region Q, then in cylindrical coordinates the **triple integral of f over Q** is |

$$\iiint\limits_{Q} f(r, \theta, z) \, dV = \lim_{\|\Delta\| \to 0} \sum_{i=1}^{n} f(r_i, \theta_i, z_i) r_i \Delta r_i \Delta \theta_i \Delta z_i$$

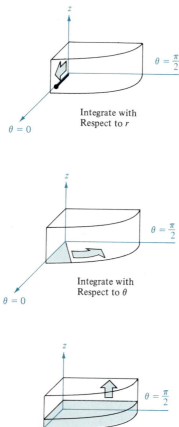

$\theta = \dfrac{\pi}{2}$

Integrate with
Respect to r

$\theta = 0$

$\theta = \dfrac{\pi}{2}$

Integrate with
Respect to θ

$\theta = 0$

$\theta = \dfrac{\pi}{2}$

Integrate with
Respect to z

$\theta = 0$

FIGURE 16.62

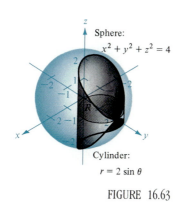

Sphere:
$x^2 + y^2 + z^2 = 4$

Cylinder:
$r = 2 \sin \theta$

FIGURE 16.63

We can evaluate a triple integral in cylindrical coordinates, provided the solid region Q is simple with respect to one of the six possible orders of integration. For example, if the region is bounded above and below by $h_1(r, \theta) \le z \le h_2(r, \theta)$, then we can write

$$\iiint_Q f(r, \theta, z)\, dV = \iint_R \left[\int_{h_1(r,\theta)}^{h_2(r,\theta)} f(r, \theta, z)\, dz \right] r\, dr\, d\theta$$

where the double integral over R is evaluated in polar coordinates. (That is, R is a plane region that is either r-simple or θ-simple, as discussed in Section 16.3.) Now, if R is r-simple, then the iterated form is

$$\int_{\theta_1}^{\theta_2} \int_{g_1(\theta)}^{g_2(\theta)} \int_{h_1(r,\theta)}^{h_2(r,\theta)} f(r, \theta, z) r\, dz\, dr\, d\theta$$

To visualize a particular order of integration, it helps to view the iterated integral in terms of three sweeping motions—each adding another dimension to the solid. For instance, in the order $dr\, d\theta\, dz$, the first integration occurs in the r direction as a point sweeps out a ray. Then, as θ increases, the line sweeps out a sector. Finally, as z increases, the sector sweeps out a solid wedge, as shown in Figure 16.62.

EXAMPLE 1   Finding volume by cylindrical coordinates

Find the volume of the solid region Q cut from the sphere $x^2 + y^2 + z^2 = 4$ by the cylinder $r = 2 \sin \theta$, as shown in Figure 16.63.

Solution: Since $x^2 + y^2 + z^2 = r^2 + z^2 = 4$, the bounds on z are

$$-\sqrt{4 - r^2} \le z \le \sqrt{4 - r^2}$$

We let R be the circular projection of the solid onto the $r\theta$-plane. Then the bounds on R are given by

$$0 \le r \le 2 \sin \theta \qquad \text{and} \qquad 0 \le \theta \le \pi$$

Thus, the volume of Q is

$$V = \int_0^\pi \int_0^{2\sin\theta} \int_{-\sqrt{4-r^2}}^{\sqrt{4-r^2}} r\, dz\, dr\, d\theta$$

$$= 2 \int_0^{\pi/2} \int_0^{2\sin\theta} 2r\sqrt{4 - r^2}\, dr\, d\theta$$

$$= 2 \int_0^{\pi/2} -\frac{2}{3}(4 - r^2)^{3/2} \Big]_0^{2\sin\theta} d\theta$$

$$= \frac{4}{3} \int_0^{\pi/2} [8 - 8\cos^3\theta]\, d\theta$$

$$= \frac{32}{3} \int_0^{\pi/2} [1 - (\cos\theta)(1 - \sin^2\theta)]\, d\theta$$

$$= \frac{32}{3} \left[\theta - \sin\theta + \frac{\sin^3\theta}{3} \right]_0^{\pi/2} = \frac{16}{9}(3\pi - 4)$$

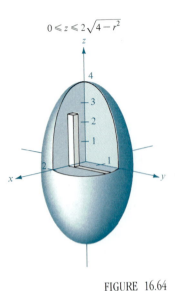

$0 \leq z \leq 2\sqrt{4 - r^2}$

FIGURE 16.64

EXAMPLE 2   Finding mass by cylindrical coordinates

Find the mass of the ellipsoidal solid Q given by $4x^2 + 4y^2 + z^2 = 16$, where the density at a point is proportional to the distance between the point and the xy-plane.

Solution: The density function $\rho(r, \theta, z) = kz$. The bounds on z are

$$0 \leq z \leq \sqrt{16 - 4x^2 - 4y^2} = 2\sqrt{4 - r^2}$$

with $0 \leq r \leq 2$ and $0 \leq \theta \leq 2\pi$, as shown in Figure 16.64. Thus, the mass of the solid is

$$m = \int_0^{2\pi} \int_0^2 \int_0^{\sqrt{16-4r^2}} kzr \, dz \, dr \, d\theta$$

$$= \frac{k}{2} \int_0^{2\pi} \int_0^2 z^2 r \Big]_0^{\sqrt{16-4r^2}} \, dr \, d\theta$$

$$= \frac{k}{2} \int_0^{2\pi} \int_0^2 (16 - 4r^3) \, dr \, d\theta$$

$$= \frac{k}{2} \int_0^{2\pi} \left[8r^2 - r^4 \right]_0^2 \, d\theta$$

$$= 8k \int_0^{2\pi} d\theta = 16\pi k$$

In Examples 1 and 2 we used the integration order $dz \, dr \, d\theta$. In the next example a different order of integration is more convenient.

EXAMPLE 3   Finding a moment of inertia

Find the moment of inertia about the axis of symmetry of the solid bounded by the paraboloid $z = x^2 + y^2$ and the plane $z = 4$, as shown in Figure 16.65. The density at each point is proportional to the distance between the point and the z-axis.

Solution: Since the z-axis is the axis of symmetry, and since $\rho(x, y, z) = k\sqrt{x^2 + y^2}$, it follows that

$$I_z = \iiint_Q k(x^2 + y^2)\sqrt{x^2 + y^2} \, dV$$

In cylindrical coordinates, $0 \leq r \leq \sqrt{x^2 + y^2} = \sqrt{z}$. Therefore, we have

$$I_z = k \int_0^4 \int_0^{2\pi} \int_0^{\sqrt{z}} r^2(r)r \, dr \, d\theta \, dz$$

$$= k \int_0^4 \int_0^{2\pi} \frac{r^5}{5} \Big]_0^{\sqrt{z}} \, d\theta \, dz$$

$$= k \int_0^4 \int_0^{2\pi} \frac{z^{5/2}}{5} \, d\theta \, dz$$

$$= k \int_0^4 \frac{z^{5/2}}{5}(2\pi) \, dz = k\left[\left(\frac{2\pi}{5}\right)\left(\frac{2}{7}\right)z^{7/2} \right]_0^4 = \frac{512k\pi}{35}$$

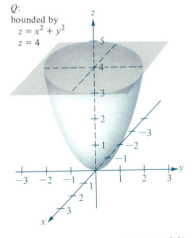

Q:
bounded by
$z = x^2 + y^2$
$z = 4$

FIGURE 16.65

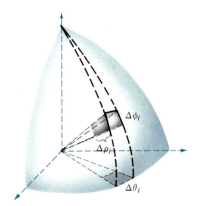

Spherical block:
$$\Delta V_i = (\bar{\rho}_i)^2 \sin \phi_i \, \Delta \rho_i \, \Delta \theta_i \, \Delta \phi_i$$

FIGURE 16.66

Triple integrals in spherical coordinates

Recall that the rectangular conversion equations for spherical coordinates are

$$x = \rho \sin \phi \cos \theta$$
$$y = \rho \sin \phi \sin \theta$$

and

$$z = \rho \cos \phi$$

For solids in spherical coordinates, the fundamental element of volume is a spherical block bounded by $\rho_1 \leq \rho \leq \rho_2$, $\theta_1 \leq \theta \leq \theta_2$, and $\phi_1 \leq \phi \leq \phi_2$, as shown in Figure 16.66. If (ρ, θ, ϕ) is a point in the interior of such a block, then the volume of the block can be approximated as follows:

$$\Delta V \approx \rho^2 \sin \phi \, \Delta \rho \, \Delta \theta \, \Delta \phi$$

Using the usual inner partition-summation-limit process, we can develop the following version of a triple integral in spherical coordinates.

TRIPLE INTEGRAL IN SPHERICAL COORDINATES

If f is a continuous function of ρ, θ, and ϕ on a bounded solid region Q, then in spherical coordinates the **triple integral of f over Q** is

$$\iiint\limits_{Q} f(\rho, \theta, \phi) \, dV = \lim_{\|\Delta\| \to 0} \sum_{i=1}^{n} f(\rho_i, \theta_i, \phi_i)\rho_i^2 \sin \phi_i \, \Delta\rho_i \Delta\theta_i \Delta\phi_i$$

Remark The use of the Greek letter ρ in spherical coordinates is not related to density. Rather, it is the three-dimensional analogue of the r used in polar coordinates. For problems involving spherical coordinates and a density function, we will use a different symbol to denote density.

Like triple integrals in cylindrical coordinates, triple integrals in spherical coordinates are evaluated with iterated integrals, as demonstrated in the next two examples.

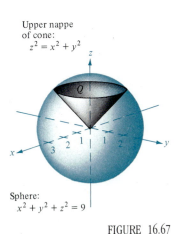

Upper nappe
of cone:
$$z^2 = x^2 + y^2$$

Sphere:
$$x^2 + y^2 + z^2 = 9$$

FIGURE 16.67

EXAMPLE 4   Finding volume in spherical coordinates

Find the volume of the solid region Q bounded below by the upper nappe of the cone $z^2 = x^2 + y^2$ and above by the sphere $x^2 + y^2 + z^2 = 9$, as shown in Figure 16.67.

Solution: In spherical coordinates, the equation of the sphere is

$$\rho^2 = x^2 + y^2 + z^2 = 9 \implies \rho = 3$$

Furthermore, the sphere and cone intersect when

$$(x^2 + y^2) + z^2 = (z^2) + z^2 = 9 \implies z = \frac{3}{\sqrt{2}}$$

and since $z = \rho \cos \phi$, it follows that

$$\left(\frac{3}{\sqrt{2}}\right)\left(\frac{1}{3}\right) = \cos \phi \implies \phi = \frac{\pi}{4}$$

Consequently, we can use the integration order $d\rho \, d\phi \, d\theta$, where

$$0 \le \rho \le 3, \qquad 0 \le \phi \le \pi/4, \qquad \text{and} \qquad 0 \le \theta \le 2\pi$$

as shown in Figure 16.68.

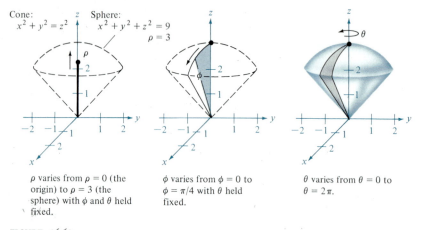

ρ varies from $\rho = 0$ (the origin) to $\rho = 3$ (the sphere) with ϕ and θ held fixed.

ϕ varies from $\phi = 0$ to $\phi = \pi/4$ with θ held fixed.

θ varies from $\theta = 0$ to $\theta = 2\pi$.

FIGURE 16.68

Finally, the volume is

$$V = \iiint_Q dV = \int_0^{2\pi} \int_0^{\pi/4} \int_0^3 \rho^2 \sin \phi \, d\rho \, d\phi \, d\theta$$

$$= \int_0^{2\pi} \int_0^{\pi/4} 9 \sin \phi \, d\phi \, d\theta = 9 \int_0^{2\pi} -\cos \phi \Big]_0^{\pi/4} d\theta$$

$$= 9 \int_0^{2\pi} \left(1 - \frac{\sqrt{2}}{2}\right) d\theta = 9\pi(2 - \sqrt{2}) \approx 16.56 \qquad \square$$

In our next example we demonstrate use of a different order of integration to find the center of mass of the spherical sector described in Example 4.

EXAMPLE 5   *Finding the center of mass of a solid region*

Find the center of mass of the solid region Q of uniform density, bounded below by the upper nappe of the cone $z^2 = x^2 + y^2$ and above by the sphere $x^2 + y^2 + z^2 = 9$.

Solution: Since the density is uniform, we consider the density at the point (x, y, z) to be k. By symmetry, the center of mass lies on the z-axis, and we need only calculate $\bar{z} = M_{xy}/m$, where $m = kV = 9k\pi(2 - \sqrt{2})$ from Example 4. Now, since $z = \rho \cos \phi$, it follows that

$$M_{xy} = \iiint_Q kz \, dV = k \int_0^3 \int_0^{2\pi} \int_0^{\pi/4} (\rho \cos \phi) \rho^2 \sin \phi \, d\phi \, d\theta \, d\rho$$

$$= k \int_0^3 \int_0^{2\pi} \rho^3 \left. \frac{\sin^2 \phi}{2} \right]_0^{\pi/4} d\theta \, d\rho$$

$$= \frac{k}{4} \int_0^3 \int_0^{2\pi} \rho^3 \, d\theta \, d\rho$$

$$= \frac{k\pi}{2} \int_0^3 \rho^3 \, d\rho = \frac{81k\pi}{8}$$

Therefore, we have

$$\bar{z} = \frac{M_{xy}}{m} = \frac{81k\pi/8}{9k\pi(2 - \sqrt{2})} = \frac{9(2 + \sqrt{2})}{16} \approx 1.92$$

and the center of mass is approximately $(0, 0, 1.92)$.

Section Exercises 16.8

In Exercises 1–8, evaluate the triple integral.

1. $\int_0^4 \int_0^{\pi/2} \int_0^2 r \cos \theta \, dr \, d\theta \, dz$

2. $\int_0^{\pi/4} \int_0^2 \int_0^{2-r} rz \, dz \, dr \, d\theta$

3. $\int_0^{\pi/2} \int_0^{2\cos^2 \theta} \int_0^{4-r^2} r \sin \theta \, dz \, dr \, d\theta$

4. $\int_0^4 \int_0^z \int_0^{\pi/2} re^r \, d\theta \, dr \, dz$

5. $\int_0^{\pi} \int_0^{\pi/2} \int_0^2 e^{-\rho^3} \rho^2 \, d\rho \, d\theta \, d\phi$

6. $\int_0^{\pi/2} \int_0^{\pi} \int_0^{\sin \theta} (2 \cos \phi) \rho^2 \, d\rho \, d\theta \, d\phi$

7. $\int_0^{2\pi} \int_0^{\pi/4} \int_0^{\cos \phi} \rho^2 \sin \phi \, d\rho \, d\theta \, d\phi$

8. $\int_0^{\pi/4} \int_0^{\pi/4} \int_0^{\cos \theta} \rho^2 \sin \phi \cos \phi \, d\rho \, d\theta \, d\phi$

In Exercises 9–12, sketch the solid region whose volume is given by the integral and evaluate the integral.

9. $\int_0^{\pi/2} \int_0^3 \int_0^{e^{-r^2}} r \, dz \, dr \, d\theta$

10. $\int_0^{2\pi} \int_0^{\sqrt{3}} \int_0^{3-r^2} r \, dz \, dr \, d\theta$

11. $4 \int_0^{\pi/2} \int_{\pi/6}^{\pi/2} \int_0^4 \rho^2 \sin \phi \, d\rho \, d\phi \, d\theta$

12. $\int_0^{2\pi} \int_0^{\pi} \int_2^5 \rho^2 \sin \phi \, d\rho \, d\phi \, d\theta$

In Exercises 13–16, convert the integral from rectangular coordinates to both cylindrical and spherical coordinates and evaluate the simplest integral.

13. $\int_{-2}^2 \int_{-\sqrt{4-x^2}}^{\sqrt{4-x^2}} \int_{x^2+y^2}^4 x \, dz \, dy \, dx$

14. $\int_0^2 \int_0^{\sqrt{4-x^2}} \int_0^{\sqrt{16-x^2-y^2}} \sqrt{x^2+y^2} \, dz \, dy \, dx$

15. $\int_{-a}^a \int_{-\sqrt{a^2-x^2}}^{\sqrt{a^2-x^2}} \int_a^{a+\sqrt{a^2-x^2-y^2}} x \, dz \, dy \, dx$

16. $\int_0^1 \int_0^{\sqrt{1-x^2}} \int_0^{\sqrt{1-x^2-y^2}} \sqrt{x^2+y^2+z^2} \, dz \, dy \, dx$

In Exercises 17–22, use cylindrical coordinates to find the indicated characteristic of the cone in Figure 16.69.

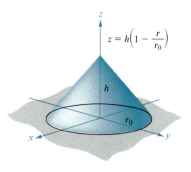

$$z = h\left(1 - \frac{r}{r_0}\right)$$

FIGURE 16.69

17. Find the volume of the cone.

18. Find the centroid of the cone.

19. Find the center of mass of the cone, assuming that its density at any point is proportional to the distance between the point and the axis of the cone.

20. Find the center of mass of the cone, assuming that its density at any point is proportional to the distance between the point and the base.

21. Assume that the cone has uniform density, and show that the moment of inertia about the z-axis is

$$I_z = \frac{3}{10} mr^2$$

22. Assume that the density is given by

$$\rho(x, y, z) = k(x^2 + y^2)$$

and find the moment of inertia about the z-axis.

In Exercises 23 and 24, use cylindrical coordinates to verify the given formula for the moment of inertia of the solid of uniform density.

23. Cylindrical shell: $I_z = \frac{1}{2} m(a^2 + b^2)$

$0 < a \le r \le b, \ 0 \le z \le h$

24. Right circular cylinder: $I_z = \frac{3}{2} ma^2$

$r = 2a \sin \theta, \ 0 \le z \le h$

In Exercises 25 and 26, use cylindrical coordinates to find the volume of the solid.

25. Solid inside both

$$x^2 + y^2 + z^2 = a^2$$

and

$$\left(x - \frac{a}{2}\right)^2 + y^2 = \left(\frac{a}{2}\right)^2$$

26. Solid inside

$$x^2 + y^2 + z^2 = 16$$

and outside

$$z = \sqrt{x^2 + y^2}$$

In Exercises 27 and 28, use spherical coordinates to find the volume of the solid.

27. Solid inside

$$x^2 + y^2 + z^2 = 1$$

and outside

$$z^2 = x^2 + y^2$$

28. Solid between

$$x^2 + y^2 + z^2 = a^2$$

and

$$x^2 + y^2 + z^2 = b^2, \quad b > a$$

and inside

$$z^2 = x^2 + y^2$$

In Exercises 29 and 30, use spherical coordinates to find the mass of the solid.

29. Sphere: $x^2 + y^2 + z^2 = a^2$
The density at any point is proportional to the distance between the point and the origin.

30. Sphere: $x^2 + y^2 + z^2 = a^2$
The density at any point is proportional to the distance of the point from the z-axis.

In Exercises 31 and 32, use spherical coordinates to find the center of mass of the solid of uniform density.

31. Hemispherical solid of radius r

32. Solid lying between two concentric hemispheres of radii r and R, where $r < R$

In Exercises 33 and 34, use spherical coordinates to find the moment of inertia about the z-axis of the solid of uniform density.

33. Solid bounded by the hemisphere $\rho = \cos \phi$, $\pi/4 \le \phi \le \pi/2$ and the cone $0 \le \phi \le \pi/4$

34. Solid lying between two concentric hemispheres of radii r and R, where $r < R$

Review Exercises for Chapter 16

In Exercises 1–14, evaluate the multiple integral. Change coordinate systems when that makes the integration easier.

1. $\displaystyle\int_0^1 \int_0^{1+x} (3x + 2y) \, dy \, dx$

2. $\displaystyle\int_0^2 \int_{x^2}^{2x} (x^2 + 2y) \, dy \, dx$

3. $\displaystyle\int_0^3 \int_0^{\sqrt{9-x^2}} 4x \, dy \, dx$

4. $\int_0^{\sqrt{3}} \int_{2-\sqrt{4-y^2}}^{2+\sqrt{4-y^2}} dx\, dy$

5. $\int_{-2}^4 \int_{y^2/4}^{(4+y)/2} (x-y)\, dx\, dy$

6. $\int_{-2}^2 \int_0^{4-y^2} (8x - 2y^2)\, dx\, dy$

7. $\int_0^h \int_0^x \sqrt{x^2 + y^2}\, dy\, dx$

8. $\int_0^4 \int_0^{\sqrt{16-y^2}} (x^2 + y^2)\, dx\, dy$

9. $\int_{-3}^3 \int_{-\sqrt{9-x^2}}^{\sqrt{9-x^2}} \int_{x^2+y^2}^9 \sqrt{x^2 + y^2}\, dz\, dy\, dx$

10. $\int_{-2}^2 \int_{-\sqrt{4-x^2}}^{\sqrt{4-x^2}} \int_0^{(x^2+y^2)/2} (x^2 + y^2)\, dz\, dy\, dx$

11. $\int_{-1}^1 \int_{-\sqrt{1-x^2}}^{\sqrt{1-x^2}} \int_{-\sqrt{1-x^2-y^2}}^{\sqrt{1-x^2-y^2}} (x^2 + y^2)\, dz\, dy\, dx$

12. $\int_0^5 \int_0^{\sqrt{25-x^2}} \int_0^{\sqrt{25-x^2-y^2}} \frac{1}{\sqrt{x^2 + y^2 + z^2}}\, dz\, dy\, dx$

13. $\int_0^a \int_0^b \int_0^c (x^2 + y^2 + z^2)\, dx\, dy\, dz$

14. $\int_0^2 \int_0^{\sqrt{4-x^2}} \int_0^{\sqrt{4-x^2-y^2}} xyz\, dz\, dy\, dx$

In Exercises 15–22, write the limits to the double integral

$$\iint\limits_R f(x, y)\, dA$$

for both orders of integration. Compute the area of R by letting $f(x, y) = 1$ and integrating.

15. Triangle: vertices $(0, 0)$, $(3, 0)$, $(0, 1)$
16. Triangle: vertices $(0, 0)$, $(3, 0)$, $(2, 2)$
17. The larger area between the graphs of $x^2 + y^2 = 25$, $x = 3$
18. Region bounded by the graphs of $y = 6x - x^2$, $y = x^2 - 2x$
19. Region enclosed by the graph of $y^2 = x^2 - x^4$
20. Region bounded by the graphs of $x = y^2 + 1$, $x = 0$, $y = 0$, $y = 2$
21. Region bounded by the graphs of $x = y + 3$, $x = y^2 + 1$
22. Region bounded by the graphs of $x = -y$, $x = 2y - y^2$

In Exercises 23–28, use an appropriate multiple integral and coordinate system to find the volume of the solid.

23. Solid bounded by the graphs of $z = x^2 - y + 4$, $z = 0$, $x = 0$, $x = 4$
24. Solid bounded by the graphs of $z = x + y$, $z = 0$, $x = 0$, $x = 3$, $y = x$
25. Solid bounded by the graphs of $z = 0$, $z = h$, outside the cylinder $x^2 + y^2 = 1$ and inside the hyperboloid $x^2 + y^2 - z^2 = 1$

26. Solid that remains after drilling a hole of radius b through the center of a sphere of radius $R(b < R)$
27. Solid inside the graphs of $r = 2 \cos \theta$ and $r^2 + z^2 = 4$
28. Solid inside the graphs of $r^2 + z = 16$ and $r = 2 \sin \theta$

In Exercises 29 and 30, find the mass and center of mass of the lamina bounded by the graphs of the equations of specified density.

29. $y = 2x$, $y = 2x^3$, (first quadrant)
    (a) $\rho = kxy$ (b) $\rho = k(x^2 + y^2)$
30. $y = \dfrac{h}{2}\left(2 - \dfrac{x}{L} - \dfrac{x^2}{L^2}\right)$, (first quadrant)
    $\rho = k$

In Exercises 31–34, find the center of mass of the solid of uniform density bounded by the graphs of the given equations.

31. Solid inside the hemisphere $\rho = \cos \phi$, $\pi/4 \le \phi \le \pi/2$ and outside the cone $\phi = \pi/4$
32. Wedge: $x^2 + y^2 = a^2$, $z = cy$ $(0 < c)$, $0 \le y$, $0 \le z$
33. $x^2 + y^2 + z^2 = a^2$ (first octant)
34. $x^2 + y^2 + z^2 = 25$, $z = 16$ (the larger solid)

In Exercises 35 and 36, find the area of the surface on the function $f(x, y)$ over the region R.

35. $f(x, y) = 16 - x^2 - y^2$
    $R = \{(x, y): x^2 + y^2 \le 4\}$
36. $f(x, y) = 16 - x - y^2$
    $R = \{(x, y): 0 \le x \le 2, 0 \le y \le x\}$

In Exercises 37 and 38, find the moment of inertia I_z of the solid of specified density.

37. The solid of uniform density bounded by $z = 16 - x^2 - y^2$, $x^2 + y^2 = 9$, and $z = 0$
38. $x^2 + y^2 + z^2 = a^2$, density is proportional to the distance from the center

39. Give a geometrical interpretation of

$$\int_0^{2\pi} \int_{-\pi/2}^{\pi/2} \int_0^{6\sin\phi} \rho^2 \sin \phi\, d\rho\, d\phi\, d\theta$$

17 Vector analysis

17.1
Vector fields

In Chapter 14 we studied vector-valued functions—functions that assign a vector to a *real number*. There we saw that vector-valued functions of a real number are useful in representing curves and motion along a curve. In this chapter we will study two other types of vector-valued functions—functions that assign a vector to a *point in the plane* or a *point in space*. Such functions are called **vector fields,** and we will see that they are useful in representing various types of **force fields** and **velocity fields.**

**DEFINITION OF
A VECTOR FIELD**

Let M and N be functions of two variables x and y, defined on a plane region R. The function $\mathbf{F}$ defined by

$$\mathbf{F}(x, y) = M\mathbf{i} + N\mathbf{j} \qquad \text{Plane}$$

is called a **vector field over R.**

Let M, N, and P be functions of three variables x, y, and z, defined on a region Q in space. The function $\mathbf{F}$ defined by

$$\mathbf{F}(x, y, z) = M\mathbf{i} + N\mathbf{j} + P\mathbf{k} \qquad \text{Space}$$

is called a **vector field over Q.**

From this definition we see that the *gradient* is one example of a vector field, since it can be written in the form

$$\mathbf{\nabla} f(x, y) = f_x(x, y)\, \mathbf{i} + f_y(x, y)\, \mathbf{j}$$

or

$$\mathbf{\nabla} f(x, y, z) = f_x(x, y, z)\, \mathbf{i} + f_y(x, y, z)\, \mathbf{j} + f_z(x, y, z)\, \mathbf{k}$$

Velocity Field

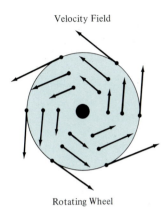

Rotating Wheel

FIGURE 17.1

Air-flow Vector Field

FIGURE 17.2

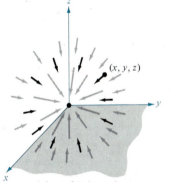

m_1 is located at (x, y, z).
m_2 is located at $(0, 0, 0)$.

Gravitational Force Field

FIGURE 17.3

Some common *physical* examples of vector fields are **velocity fields, gravitational fields,** and **electric force fields.**

Velocity fields are used to describe the motion of a system of particles in the plane or in space. For instance, Figure 17.1 shows the vector field determined by a wheel rotating on an axle. Notice that the velocity vectors are determined by the location of their initial points—the farther a point is from the axle, the greater its velocity. Velocity fields are also determined by the flow of liquids through a container or by the flow of air currents around a moving object, as shown in Figure 17.2.

Gravitational fields are defined by **Newton's Law of Gravitation,** which states that the force of attraction exerted on a particle of mass m_1 located at (x, y, z) by a particle of mass m_2 located at $(0, 0, 0)$ is given by

$$\mathbf{F}(x, y, z) = \frac{-Gm_1m_2}{x^2 + y^2 + z^2}\mathbf{u}$$

where G is the gravitational constant and $\mathbf{u}$ is the unit vector in the direction from the origin to (x, y, z). In Figure 17.3 we can see that the gravitational field $\mathbf{F}$ has the properties that $\mathbf{F}(x, y, z)$ always points toward the origin and that the magnitude of $\mathbf{F}(x, y, z)$ is the same at all points equidistant from the origin. A vector field with these two properties is called a **central force field.** Using the position vector $\mathbf{r} = x\mathbf{i} + y\mathbf{j} + z\mathbf{k}$ for the point (x, y, z), we can express the gravitational field $\mathbf{F}$ as

$$\mathbf{F}(x, y, z) = \frac{-Gm_1m_2}{\|\mathbf{r}\|^2} \cdot \frac{\mathbf{r}}{\|\mathbf{r}\|} = \frac{-Gm_1m_2}{\|\mathbf{r}\|^2}\mathbf{u}$$

Electric force fields are defined by **Coulomb's Law,** which states that the force exerted on a particle with electric charge q_1 located at (x, y, z) by a particle with electric charge q_2 located at $(0, 0, 0)$ is given by

$$\mathbf{F}(x, y, z) = \frac{cq_1q_2}{\|\mathbf{r}\|^2}\mathbf{u}$$

where $\mathbf{r} = x\mathbf{i} + y\mathbf{j} + z\mathbf{k}$, $\mathbf{u} = \mathbf{r}/\|\mathbf{r}\|$, and c is a constant that depends on the choice of units for $\|\mathbf{r}\|$, q_1, and q_2.

Remark Note that an electric force field has the same form as a gravitational field. That is,

$$\mathbf{F}(x, y, z) = \frac{k}{\|\mathbf{r}\|^2}\mathbf{u}$$

where k is a real number and $\mathbf{u} = \mathbf{r}/\|\mathbf{r}\|$. We call this type of vector field an **inverse square field.**

EXAMPLE 1   Sketching a vector field

Sketch some vectors in the vector field given by $\mathbf{F}(x, y) = 2x\mathbf{i} + y\mathbf{j}.$

Solution: We could plot vectors at some random points in the plane. However, it is more enlightening to plot vectors of equal magnitude. This corresponds to finding level curves in scalar fields. In this case, vectors of equal magnitude lie along ellipses given by

$$\|\mathbf{F}\| = \sqrt{(2x)^2 + (y)^2} = c \implies 4x^2 + y^2 = c^2$$

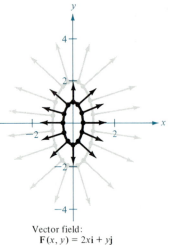

Vector field:
$F(x, y) = 2x\mathbf{i} + y\mathbf{j}$

FIGURE 17.4

For $c = 1$, we sketch several vectors $2x\mathbf{i} + y\mathbf{j}$ of magnitude 1 at points on the ellipse given by

$$4x^2 + y^2 = 1$$

For $c = 2$, we sketch several vectors $2x\mathbf{i} + y\mathbf{j}$ of magnitude 2 at points on the ellipse given by

$$4x^2 + y^2 = 4$$

These vectors are shown in Figure 17.4.

Gradients as vector fields

Notice in Figure 17.4 that the vectors all appear to be normal to the level curve from which they emanate. Since this is a property of gradients, it is natural to ask if the vector field given by $\mathbf{F}(x, y) = 2x\mathbf{i} + y\mathbf{j}$ is the *gradient* for some differentiable function f. Not every vector field is the gradient of a differentiable function—we call those that are **conservative** vector fields.

DEFINITION OF CONSERVATIVE VECTOR FIELD

A vector field $\mathbf{F}$ is called **conservative** if there exists a differentiable function such that

$$\mathbf{F} = \nabla f$$

The function f is called the **potential function** for $\mathbf{F}$.

Many important vector fields, including gravitational fields, magnetic fields, and electric force fields, are conservative. The origin of most of the terminology in this chapter is from physics. For example, the term "conservative" is derived from the classic physical law regarding the conservation of energy. This law states that the sum of the kinetic energy and the potential energy of a particle moving in a conservative force field is constant. (The kinetic energy of a particle is the energy due to its motion, and the potential energy is the energy due to its position in the force field.)

The following important theorem gives a necessary and sufficient condition for a vector field *in the plane* to be conservative.

THEOREM 17.1

TEST FOR CONSERVATIVE VECTOR FIELD IN THE PLANE
Let M and N have continuous first partial derivatives on an *open disc R*. The vector field given by $\mathbf{F}(x, y) = M\mathbf{i} + N\mathbf{j}$ is conservative if and only if

$$\frac{\partial N}{\partial x} = \frac{\partial M}{\partial y}$$

Proof: To prove that the given condition is necessary for **F** to be conservative, we suppose there exists a potential function f such that

$$\mathbf{F}(x,\ y) = \nabla f(x,\ y) = M\mathbf{i} + N\mathbf{j}$$

Then, we have

$$f_x(x,\ y) = M \implies f_{xy}(x,\ y) = \frac{\partial M}{\partial y}$$

$$f_y(x,\ y) = N \implies f_{yx}(x,\ y) = \frac{\partial N}{\partial x}$$

and, by the equivalence of the mixed particles f_{xy} and f_{yx}, we conclude that $\partial N/\partial x = \partial M/\partial y$ for all $(x,\ y)$ in R. We postpone the proof of the sufficiency of the condition until Section 17.4.

| **Remark** Theorem 17.1 requires that the domain of **F** be an open *disc*. If R is simply an open region, then the given condition is necessary but not sufficient to produce a conservative vector field.

EXAMPLE 2   Testing for conservative vector fields in the plane

(a) The vector field given by $\mathbf{F}(x,\ y) = x^2 y\mathbf{i} + xy\mathbf{j}$ *is not* conservative, since

$$\frac{\partial M}{\partial y} = \frac{\partial}{\partial y}[x^2 y] = x^2$$

and

$$\frac{\partial N}{\partial x} = \frac{\partial}{\partial x}[xy] = y$$

(b) The vector field given by $\mathbf{F}(x,\ y) = 2x\mathbf{i} + y\mathbf{j}$ *is* conservative, since

$$\frac{\partial M}{\partial y} = \frac{\partial}{\partial y}[2x] = 0$$

and

$$\frac{\partial N}{\partial x} = \frac{\partial}{\partial x}[y] = 0$$

Theorem 17.1 tells us whether or not a given vector field is conservative. It does not tell us how to find a potential function for **F.** The problem is comparable to antidifferentiation, and sometimes you will be able to find a potential function by simple inspection. For instance,

$$f(x,\ y) = x^2 + \frac{1}{2}y^2$$

has the property that $\nabla f(x,\ y) = 2x\mathbf{i} + y\mathbf{j}$. We will discuss techniques for finding potential functions in detail in Chapter 18 (in the section on exact differential equations). For now, we outline the procedure by way of an example.

EXAMPLE 3   Finding a potential function for **F**(x, y)

Find a potential function for

$$\mathbf{F}(x, y) = 2xy\mathbf{i} + (x^2 - y)\,\mathbf{j}$$

Solution: From Theorem 17.1 it follows that **F** is conservative, since

$$\frac{\partial}{\partial y}[2xy] = 2x \qquad \text{and} \qquad \frac{\partial}{\partial x}[x^2 - y] = 2x$$

Now, if f is a function such that $\nabla f(x, y) = f_x(x, y)\,\mathbf{i} + f_y(x, y)\,\mathbf{j}$, then we have

$$f_x(x, y) = 2xy \qquad \text{and} \qquad f_y(x, y) = x^2 - y$$

To reconstruct the function f from these two partial derivatives, we integrate $f_x(x, y)$ with respect to x and $f_y(x, y)$ with respect to y, as follows:

$$f(x, y) = \int f_x(x, y)\, dx = \int 2xy\, dx = \boxed{x^2y} + \boxed{g(y)} + \boxed{K}$$

$$f(x, y) = \int f_y(x, y)\, dy = \int (x^2 - y)\, dy = x^2y - \frac{y^2}{2} + h(x) + K$$

Now, we reconcile the differences between these two expressions for $f(x, y)$. That is, since $g(y) = -y^2/2$, it follows that $h(x) = 0$, and we have

$$f(x, y) = x^2y + g(y) + K = x^2y - \frac{y^2}{2} + K \qquad \square$$

| Remark   Note that the solution in Example 3 is comparable to that given by an indefinite integral. That is, the solution represents a family of potential functions, each differing by a constant. To find a unique solution we would have to be given an initial condition satisfied by the vector field.

Theorem 17.1 has a counterpart for vector fields in space. Before stating that result, we define the **curl of a vector field** in space.

| DEFINITION OF CURL | The **curl** of $\mathbf{F}(x, y, z) = M\mathbf{i} + N\mathbf{j} + P\mathbf{k}$ is |
| OF A VECTOR FIELD | |

$$\text{curl } \mathbf{F}(x, y, z) = \nabla \times \mathbf{F}(x, y, z)$$

$$= \left(\frac{\partial P}{\partial y} - \frac{\partial N}{\partial z}\right)\mathbf{i} - \left(\frac{\partial P}{\partial x} - \frac{\partial M}{\partial z}\right)\mathbf{j} + \left(\frac{\partial N}{\partial x} - \frac{\partial M}{\partial y}\right)\mathbf{k}$$

| Remark   The cross-product notation used for curl comes from viewing the gradient ∇f as the result of the **differential operator** ∇ acting on the function f. In this context, we can use the following determinant form as an aid to remembering the formula for curl:

$$\nabla \times \mathbf{F}(x, y, z) = \begin{vmatrix} \mathbf{i} & \mathbf{j} & \mathbf{k} \\ \dfrac{\partial}{\partial x} & \dfrac{\partial}{\partial y} & \dfrac{\partial}{\partial z} \\ M & N & P \end{vmatrix}$$

$$= \left(\frac{\partial P}{\partial y} - \frac{\partial N}{\partial z} \right) \mathbf{i} - \left(\frac{\partial P}{\partial x} - \frac{\partial M}{\partial z} \right) \mathbf{j} + \left(\frac{\partial N}{\partial x} - \frac{\partial M}{\partial y} \right) \mathbf{k}$$

Later in this chapter we will assign a physical interpretation to the curl of a vector field. But for now, the primary use we make of curl is in the following test for conservative vector fields in space. The test states that for a vector field whose domain is all of three-dimensional space (or an open sphere), the curl is zero at every point in the domain if and only if **F** is conservative. The proof is similar to that given in Theorem 17.1.

THEOREM 17.2

TEST FOR CONSERVATIVE VECTOR FIELD IN SPACE
Suppose M, N, and P have continuous first partial derivatives on an open sphere Q in space. The vector field given by $\mathbf{F}(x, y, z) = M\mathbf{i} + N\mathbf{j} + P\mathbf{k}$ is conservative if and only if **curl F**$(x, y, z) = \mathbf{0}$. That is, **F** is conservative if and only if

$$\frac{\partial P}{\partial y} = \frac{\partial N}{\partial z}, \qquad \frac{\partial P}{\partial x} = \frac{\partial M}{\partial z}, \qquad \text{and} \qquad \frac{\partial N}{\partial x} = \frac{\partial M}{\partial y}$$

EXAMPLE 4   Testing for conservative vector fields in space

Find the curl of the vector field $\mathbf{F}(x, y, z) = x^3y^2z\mathbf{i} + x^2z\mathbf{j} + x^2y\mathbf{j}$ and show that it is not conservative.

Solution: By definition of curl, we have

$$\mathbf{curl\ F}(x, y, z) = \begin{vmatrix} \mathbf{i} & \mathbf{j} & \mathbf{k} \\ \dfrac{\partial}{\partial x} & \dfrac{\partial}{\partial y} & \dfrac{\partial}{\partial z} \\ x^3y^2z & x^2z & x^2y \end{vmatrix}$$

$$= \begin{vmatrix} \dfrac{\partial}{\partial y} & \dfrac{\partial}{\partial z} \\ x^2z & x^2y \end{vmatrix} \mathbf{i} - \begin{vmatrix} \dfrac{\partial}{\partial x} & \dfrac{\partial}{\partial z} \\ x^3y^2z & x^2y \end{vmatrix} \mathbf{j} + \begin{vmatrix} \dfrac{\partial}{\partial x} & \dfrac{\partial}{\partial y} \\ x^3y^2z & x^2z \end{vmatrix} \mathbf{k}$$

$$= (x^2 - x^2)\,\mathbf{i} - (2xy - x^3y^2)\,\mathbf{j} + (2xz - 2x^3yz)\,\mathbf{k}$$

$$= (x^3y^2 - 2xy)\,\mathbf{j} + (2xz - 2x^3yz)\,\mathbf{k}$$

Since **curl F** $\neq \mathbf{0}$, we conclude by Theorem 17.2 that **F** is not conservative.

For vector fields in space that pass the test for being conservative, we can find a potential function by following the same pattern used in the plane (as demonstrated in Example 3).

EXAMPLE 5   *Finding a potential function for* $\mathbf{F}(x, y, z)$

Find a potential function for $\mathbf{F}(x, y, z) = 2xy\mathbf{i} + (x^2 + z^2)\,\mathbf{j} + 2zy\mathbf{k}$.

Solution: Using Theorem 17.2, we let

$$\mathbf{F}(x, y, z) = 2xy\mathbf{i} + (x^2 + z^2)\,\mathbf{j} + 2yz\mathbf{k} = M\mathbf{i} + N\mathbf{j} + P\mathbf{k}$$

Since

$$\frac{\partial P}{\partial y} = 2z = \frac{\partial N}{\partial z}, \qquad \frac{\partial P}{\partial x} = 0 = \frac{\partial M}{\partial z}, \qquad \text{and} \qquad \frac{\partial N}{\partial x} = 2x = \frac{\partial M}{\partial y}$$

it follows that $\mathbf{F}$ is conservative. Now, if f is a function such that $\mathbf{F}(x, y, z) = \nabla f(x, y, z)$, then

$$f_x(x, y, z) = 2xy, \qquad f_y(x, y, z) = x^2 + z^2, \qquad \text{and} \qquad f_z(x, y, z) = 2zy$$

and by integrating with respect to x, y, and z separately, we obtain

$$f(x, y, z) = \int M\,dx = \int 2xy\,dx = x^2y + g(y, z) + K$$

$$f(x, y, z) = \int N\,dy = \int (x^2 + z^2)\,dy = x^2y + z^2y + h(x, z) + K$$

$$f(x, y, z) = \int P\,dz = \int 2zy\,dz = z^2y + k(x, y) + K$$

Comparing these three versions of $f(x, y, z)$, we conclude that

$$g(y, z) = z^2y, \qquad h(x, z) = 0, \qquad \text{and} \qquad k(x, y) = x^2y$$

Therefore, $f(x, y, z)$ is given by

$$f(x, y, z) = x^2y + z^2y + K$$

Remark Examples 3 and 5 are illustrations of a type of problem called *recovering a function from its gradient,* and we will discuss other methods for solving this problem in Chapter 18. One popular method involves an interplay between successive "partial integrations" and partial differentiations.

Section Exercises 17.1

In Exercises 1–16, sketch several representative vectors in the given vector field.

1. $\mathbf{F}(x, y) = \mathbf{i} + \mathbf{j}$

2. $\mathbf{F}(x, y) = 2\mathbf{i}$

3. $\mathbf{F}(x, y) = x\mathbf{j}$

4. $\mathbf{F}(x, y) = y\mathbf{i}$

5. $\mathbf{F}(x, y) = x\mathbf{i} + y\mathbf{j}$

6. $\mathbf{F}(x, y) = -x\mathbf{i} + y\mathbf{j}$

7. $\mathbf{F}(x, y) = -x\mathbf{i} - y\mathbf{j}$

8. $\mathbf{F}(x, y) = y\mathbf{i} - x\mathbf{j}$

9. $\mathbf{F}(x, y) = x\mathbf{i} + 3y\mathbf{j}$

10. $\mathbf{F}(x, y) = 4x\mathbf{i} + y\mathbf{j}$

11. $\mathbf{F}(x, y) = \dfrac{x}{\sqrt{x^2 + y^2}}\mathbf{i} + \dfrac{y}{\sqrt{x^2 + y^2}}\mathbf{j}$

12. $F(x, y) = i + (x^2 + y^2) j$

13. $F(x, y, z) = 3yj$ **14.** $F(x, y, z) = i + j + k$

15. $F(x, y, z) = xi - yj$

16. $F(x, y, z) = xi + yj + zk$

In Exercises 17–22, find the gradient vector field for the given scalar function. (That is, find the conservative vector field for the given potential.)

17. $f(x, y) = 5x^2 + 3xy + 10y^2$

18. $f(x, y) = \sin 3x \cos 4y$

19. $f(x, y, z) = z - ye^{x^2}$

20. $f(x, y, z) = \dfrac{y}{z} + \dfrac{z}{x} - \dfrac{xz}{y}$

21. $g(x, y, z) = xy \ln (x + y)$

22. $g(x, y, z) = x \arcsin yz$

In Exercises 23–40, apply the test for conservative vector fields and find a potential function for the fields that pass the test.

23. $F(x, y) = 2xyi + x^2j$

24. $F(x, y) = \dfrac{1}{y^2}(yi - 2xj)$

25. $F(x, y) = (2x - 3y) i - 3(x - y^2) j$

26. $F(x, y) = xe^yi + ye^xj$

27. $F(x, y) = xe^{x^2y}(2yi + xj)$

28. $F(x, y) = 2xy^3i + 3y^2x^2j$

29. $F(x, y) = \dfrac{xi + yj}{x^2 + y^2}$

30. $F(x, y) = \dfrac{2y}{x}i - \dfrac{x^2}{y^2}j$

31. $F(x, y) = e^x(\cos y\, i + \sin y\, j)$

32. $F(x, y) = \dfrac{2x}{y}i - \dfrac{x^2}{y^2}j$

33. $F(x, y) = e^x[\sin y\, i + (\cos y + 2)\, j]$

34. $F(x, y) = \dfrac{2xi + 2yj}{(x^2 + y^2)^2}$

35. $F(x, y) = \sin y\, i - x \cos y\, j$

36. $F(x, y, z) = e^z(yi + xj + k)$

37. $F(x, y, z) = e^z(yi + xj + xyk)$

38. $F(x, y, z) = 3x^2y^2zi + 2x^3yzj + x^3y^2k$

39. $F(x, y, z) = \dfrac{1}{y}i - \dfrac{x}{y^2}j + (2z - 1)\, k$

40. $F(x, y, z) = \dfrac{x}{x^2 + y^2}i + \dfrac{y}{x^2 + y^2}j + k$

In Exercises 41–44, find the curl of the vector field F at the indicated point.

Vector field	Point
41. $F(x, y, z) = xyzi + yj + zk$	$(1, 2, 1)$
42. $F(x, y, z) = x^2zi - 2xzj + yzk$	$(2, -1, 3)$

43. $F(x, y, z) = e^x \sin y\, i - e^x \cos y\, j$	$(0, 0, 3)$
44. $F(x, y, z) = e^{-xyz}(i + j + k)$	$(3, 2, 0)$

In Exercises 45–48, find the curl of the vector field F.

45. $F(x, y, z) = \arctan \dfrac{x}{y}i + \ln \sqrt{x^2 + y^2}\, j + k$

46. $F(x, y, z) = \dfrac{yz}{y - z}i + \dfrac{xz}{x - z}j + \dfrac{xy}{x - y}k$

47. $F(x, y, z) = \sin (x - y)\, i + \sin (y - z)\, j + \sin (z - x)\, k$

48. $F(x, y, z) = \sqrt{x^2 + y^2 + z^2}(i + j + k)$

In Exercises 49 and 50, find **curl** $(F \times G)$.

49. $F(x, y, z) = i + 2xj + 3yk$
    $G(x, y, z) = xi - yj + zk$

50. $F(x, y, z) = xi - zk$
    $G(x, y, z) = x^2i + yj + z^2k$

In Exercises 51 and 52, given the vector field F, find **curl (curl F)** $= \nabla \times (\nabla \times F)$.

51. $F(x, y, z) = xyzi + yj + zk$

52. $F(x, y, z) = x^2zi - 2xzj + yzk$

In Exercises 53 and 54, prove the given property for vector fields F and G and scalar function f, whose second partial derivatives are continuous.

53. curl$(F + G) = $ **curl** $F + $ **curl** G

54. curl$(\nabla f) = \nabla \times (\nabla f) = 0$

In Exercises 55–58, let $F(x, y, z) = xi + yj + zk$, and $f(x, y, z) = \|F(x, y, z)\|$.

55. Show that $\nabla(\ln f) = \dfrac{F}{f^2}$.

56. Show that $\nabla\left(\dfrac{1}{f}\right) = -\dfrac{F}{f^3}$.

57. Show that $\nabla f^n = nf^{n-2}F$.

58. The **Laplacian** is the differential operator

$$\nabla^2 = \nabla \cdot \nabla = \frac{\partial^2}{\partial x^2} + \frac{\partial^2}{\partial y^2} + \frac{\partial^2}{\partial z^2}$$

and Laplace's equation is

$$\nabla^2 w = \frac{\partial^2 w}{\partial x^2} + \frac{\partial^2 w}{\partial y^2} + \frac{\partial^2 w}{\partial z^2} = 0$$

Any function that satisfies this equation is called **harmonic.** Show that the function $1/f$ is harmonic.

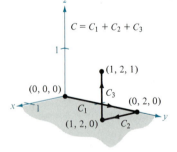

Josiah Willard Gibbs

$C = C_1 + C_2 + C_3$

(1, 2, 1)

(0, 0, 0)

C_3

(0, 2, 0)

C_1

(1, 2, 0) C_2

FIGURE 17.5

17.2
Line integrals

A classic property of gravitational fields is that, subject to certain physical constraints, the work done by gravity on an object moving between two points in the field is independent of the path taken by the object. One of the constraints is that the **path** must be a piecewise smooth curve. Recall that a curve C given by

$$\mathbf{r}(t) = x(t)\,\mathbf{i} + y(t)\,\mathbf{j} + z(t)\,\mathbf{k}, \quad a \le t \le b$$

is **smooth** if dx/dt, dy/dt, and dz/dt are continuous and not simultaneously zero on $[a, b]$. Moreover, a curve C is **piecewise smooth** if the interval $[a, b]$ can be partitioned into a finite number of subintervals, on each of which C is smooth.

| Remark   The parametric representation of a smooth curve is called a **smooth parameterization.**

Many physicists and mathematicians have contributed to the theory and applications described in this chapter—Newton, Gauss, Laplace, Hamilton, Maxwell, and many others. However, the use of vector analysis to describe the results is primarily due to the American mathematical physicist Josiah Willard Gibbs (1839–1903).

EXAMPLE 1   Finding a piecewise smooth parameterization

Find a piecewise smooth parameterization of the curve C shown in Figure 17.5.

Solution: Since C is made up of three smooth line segments C_1, C_2, and C_3, we construct a parameterization for each and piece them together by making the last t-value in C_i correspond to the first t-value in C_{i+1}, as follows.

$$C_1\colon x(t) = 0, \qquad y(t) = 2t, \qquad z(t) = 0, \qquad 0 \le t \le 1$$
$$C_2\colon x(t) = t - 1, \qquad y(t) = 2, \qquad z(t) = 0, \qquad 1 \le t \le 2$$
$$C_3\colon x(t) = 1, \qquad y(t) = 2, \qquad z(t) = t - 2, \quad 2 \le t \le 3$$

Therefore, C is given by

$$\mathbf{r}(t) = \begin{cases} 2t\mathbf{j}, & 0 \le t \le 1 \\ (t - 1)\,\mathbf{i} + 2\mathbf{j}, & 1 \le t \le 2 \\ \mathbf{i} + 2\mathbf{j} + (t - 2)\,\mathbf{k}, & 2 \le t \le 3 \end{cases}$$

Recall from our earlier work that the parameterization of a curve induces an **orientation** to the curve. For instance, in Example 1 the curve is oriented so that the positive direction is from (0, 0, 0), following the curve to (1, 2, 1). Try finding a parameterization that induces the opposite orientation.

Line integrals

We use a definite integral to measure the work done on an object moving through a force field along *a straight line* parallel to one of the coordinate axes. However, to measure the work done on an object moving along a *curve,* we need a new type of integral called a **line integral.** (The terminology is somewhat unfortunate, but of course a curve can be a line segment or a collection of line segments, as shown in Example 1.)

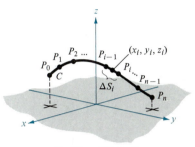

FIGURE 17.6   Partition of Curve *C*

Suppose $f(x, y, z)$ is continuous in some solid region Q containing a smooth space curve C of finite length. We partition the curve C by points $P_0, P_1, \ldots, P_n$ so that it yields n subarcs of length Δs_i, as shown in Figure 17.6. Choosing a point (x_i, y_i, z_i) in each subarc, we form the sum

$$\sum_{i=1}^{n} f(x_i, y_i, z_i)\Delta s_i$$

Letting $\|\Delta\|$ denote the length of the longest subarc and taking the limit of this sum as $\|\Delta\|$ approaches zero, we obtain the following definition.

DEFINITION OF LINE INTEGRAL

If f is defined in a region containing a smooth curve C of finite length, then the **line integral of f along C** is given by

$$\int_C f(x, y)\, ds = \lim_{\|\Delta\| \to 0} \sum_{i=1}^{n} f(x_i, y_i)\Delta s_i \qquad \text{Plane}$$

or

$$\int_C f(x, y, z)\, ds = \lim_{\|\Delta\| \to 0} \sum_{i=1}^{n} f(x_i, y_i, z_i)\Delta s_i \qquad \text{Space}$$

provided this limit exists.

As with the integrals discussed in Chapter 16, evaluation of a line integral is best accomplished by converting to a definite integral. It can be shown that if f is *continuous,* then the above limit exists and is the same for all parameterizations of C that are oriented in the same direction. For a given parameterization, a line integral can be evaluated as shown in Theorem 17.3.

THEOREM 17.3   **EVALUATION OF A LINE INTEGRAL AS A DEFINITE INTEGRAL**
Let f be continuous in a region containing a smooth curve C. If C is given by $\mathbf{r}(t) = x(t) \,\mathbf{i} + y(t) \,\mathbf{j}$, where $a \le t \le b$, then

$$\int_C f(x, y) \, ds = \int_a^b f(x(t), y(t))\sqrt{[x'(t)]^2 + [y'(t)]^2} \; dt$$

If C is given by $\mathbf{r}(t) = x(t) \,\mathbf{i} + y(t) \,\mathbf{j} + z(t) \,\mathbf{k}$ where $a \le t \le b$, then

$$\int_C f(x, y, z) \, ds = \int_a^b f(x(t), y(t), z(t))\sqrt{[x'(t)]^2 + [y'(t)]^2 + [z'(t)]^2} \; dt$$

EXAMPLE 2   Evaluating a line integral

Evaluate

$$\int_C (x^2 - y + 3z) \, ds$$

where C is the line segment shown in Figure 17.7.

Solution: Using a parametric form of the equation of a line, we have

$$x = (1 - 0)t = t, \qquad y = (2 - 0)t = 2t, \qquad \text{and} \qquad z = (1 - 0)t = t$$

where $0 \le t \le 1$, so that the line integral takes the following form.

$$\int_C (x^2 - y + 3z) \, ds = \int_0^1 (t^2 - 2t + 3t)\sqrt{(1)^2 + (2)^2 + (1)^2} \; dt$$

$$= \sqrt{6} \int_0^1 (t^2 + t) \, dt$$

$$= \sqrt{6} \left[\frac{t^3}{3} + \frac{t^2}{2} \right]_0^1 = \frac{5\sqrt{6}}{6}$$

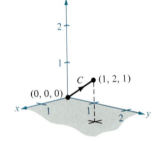

FIGURE 17.7

Suppose C is a path composed of smooth curves $C_1, C_2, \ldots, C_n$. If f is continous on C, then it can be shown that

$$\int_C f(x, y) \, ds = \int_{C_1} f(x, y) \, ds + \int_{C_2} f(x, y) \, ds + \cdots + \int_{C_n} f(x, y) \, ds$$

We use this property in our next example.

EXAMPLE 3   Evaluating a line integral over a path

Evaluate

$$\int_C x \, ds$$

where C is the piecewise smooth curve shown in Figure 17.8.

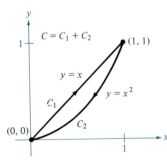

FIGURE 17.8

Solution: We first integrate up the line $y = x$, using the following parameterization.

$$C_1: x = t, \qquad y = t, \quad 0 \le t \le 1$$

Then, $ds_1 = \sqrt{2}\, dt$ and we have

$$\int_{C_1} x \, ds_1 = \int_0^1 t\sqrt{2}\, dt = \frac{\sqrt{2}}{2} t^2 \Big]_0^1 = \frac{\sqrt{2}}{2}$$

Next, we integrate down the parabola $y = x^2$, using the parameterization

$$C_2: x = 1 - t, \qquad y = (1 - t)^2, \quad 0 \le t \le 1$$

Then, $ds_2 = \sqrt{1 + 4(1 - t)^2}\, dt$ and we have

$$\int_{C_2} x \, ds_2 = \int_0^1 (1 - t)\sqrt{1 + 4(1 - t)^2}\, dt$$

$$= -\frac{1}{8}\left[\frac{2}{3}[1 + 4(1 - t)^2]^{3/2}\right]_0^1 = \frac{1}{12}(5^{3/2} - 1)$$

Consequently,

$$\int_C x \, ds = \int_{C_1} x \, ds_1 + \int_{C_2} x \, ds_2 = \frac{\sqrt{2}}{2} + \frac{1}{12}(5^{3/2} - 1) \approx 1.56 \quad \square$$

| **Remark** For parameterizations given by $\mathbf{r}(t) = x(t)\,\mathbf{i} + y(t)\,\mathbf{j} + z(t)\,\mathbf{k}$, it is helpful to remember the form of ds as

$$ds = \|\mathbf{r}'(t)\|\, dt = \sqrt{[x'(t)]^2 + [y'(t)]^2 + [z'(t)]^2}\, dt$$

This is demonstrated in the next example.

EXAMPLE 4   *Evaluating a line integral*

Evaluate

$$\int_C (x + 2) \, ds$$

where C is the curve represented by

$$\mathbf{r}(t) = t\mathbf{i} + \frac{4}{3} t^{3/2}\mathbf{j} + \frac{1}{2} t^2\mathbf{k}, \quad 0 \le t \le 2$$

Solution: Since

$$\|\mathbf{r}'(t)\| = \sqrt{[x'(t)]^2 + [y'(t)]^2 + [z'(t)]^2} = \sqrt{1 + 4t + t^2}$$

it follows that

$$\int_C (x + 2) \, ds = \int_0^2 (t + 2)\sqrt{1 + 4t + t^2}\, dt$$

$$= \frac{1}{2} \int_0^2 2(t + 2)(1 + 4t + t^2)^{1/2}\, dt$$

$$= \frac{1}{3}\left[(1 + 4t + t^2)^{3/2}\right]_0^2 = \frac{1}{3}(13\sqrt{13} - 1) \approx 15.29 \quad \square$$

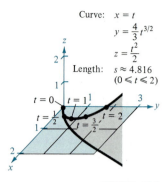

Curve: $x = t$
  $y = \dfrac{4}{3} t^{3/2}$
  $z = \dfrac{t^2}{2}$

Length: $s \approx 4.816$
  $(0 \leqslant t \leqslant 2)$

FIGURE 17.9

| Remark   Compare Example 4 with Example 1 of Section 14.4. If we changed the integrand of the line integral to 1, we would obtain the arc length of curve C, as shown in Figure 17.9. That is,

$$\int_C 1 \, ds = \int_a^b \|\mathbf{r}'(t)\| \, dt = \text{length of curve } C$$

Line integrals possess the usual properties we have come to expect of integrals. We list some of the basic properties in the following summary.

PROPERTIES OF LINE INTEGRALS

1. $\displaystyle\int_C kf(x, y) \, ds = k \int_C f(x, y) \, ds$

2. $\displaystyle\int_C [f(x, y) \pm g(x, y)] \, ds = \int_C f(x, y) \, ds \pm \int_C g(x, y) \, ds$

3. $\displaystyle\int_C f(x, y) \, ds = -\int_{-C} f(x, y) \, ds$ Reversal of orientation

4. *Independence of parameter:* The value of a line integral is independent of the parameter used to represent the curve C, provided the orientation is the same.

Applications of line integrals

Density:
$\rho(x, y, z) = 1 + z$

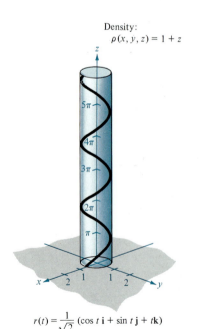

$r(t) = \dfrac{1}{\sqrt{2}} (\cos t\, \mathbf{i} + \sin t\, \mathbf{j} + t\mathbf{k})$

FIGURE 17.10

Line integrals can be used to measure most of the same quantities measured by definite integrals: area, arc length, mass, work, and so on.

For our first application of a line integral, consider a thin wire of variable density. We consider the wire to have the shape of curve C, with $\rho(x, y, z)$ representing the linear density (mass per unit length) at any point (x, y, z) on the wire. For a point (x_i, y_i, z_i) in a small segment Δs_i of wire, the product $\rho(x_i, y_i, z_i)\Delta s_i$ is a reasonable approximation of the mass m_i of this segment. Taking the sum of all such products over a partition of C and taking the limit of this sum as $\|\Delta\| \to 0$, we have

$$\text{mass of wire} = \lim_{\|\Delta\| \to 0} \sum_{i=1}^{n} \rho(x_i, y_i, z_i)\Delta s_i = \int_C \rho(x, y, z) \, ds$$

EXAMPLE 5   Finding the mass of a wire

Find the mass of a spring in the shape of the circular helix

$$\mathbf{r}(t) = \frac{1}{\sqrt{2}}(\cos t\, \mathbf{i} + \sin t\, \mathbf{j} + t\mathbf{k}), \quad 0 \le t \le 6\pi$$

where the density of the wire is $\rho(x, y, z) = 1 + z$, as shown in Figure 17.10.

Solution: Since

$$\|\mathbf{r}'(t)\| = \frac{1}{\sqrt{2}} \sqrt{(-\sin t)^2 + (\cos t)^2 + (1)^2} = 1$$

it follows that the mass of the spring is

$$\text{mass} = \int_C (1 + z) \, ds = \int_0^{6\pi} \left(1 + \frac{t}{\sqrt{2}}\right) dt$$

$$= \left[t + \frac{t^2}{2\sqrt{2}}\right]_0^{6\pi}$$

$$= 6\pi\left(1 + \frac{3\pi}{\sqrt{2}}\right) \approx 144.47$$

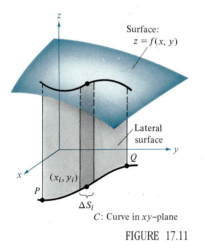

Surface:
$z = f(x, y)$

Lateral
surface

Q

(x_i, y_i)

P

ΔS_i

C: Curve in xy-plane

FIGURE 17.11

Remark   Remember that the orientation of a curve can affect the sign of the line integral. One way to ensure that the sign is determined only by the integrand is to use the arc length parameter to represent C over an interval $a \le s \le b$. For example, the parameter used in Example 5 is actually arc length, since $\|\mathbf{r}'(t)\| = 1$.

Another application of line integrals is to find lateral surface area. Consider a curve C in the xy-plane, as shown in Figure 17.11. If $z = f(x, y)$ represents a surface S lying above C, then the lateral surface (shown in gray) is made up of all points (x, y, z) lying below S and directly above C. To find the area of this lateral surface, we integrate the height $f(x, y)$ along the curve C to obtain

$$\text{lateral surface area} = \int_C f(x, y) \, ds$$

EXAMPLE 6   Finding lateral surface area

Find the lateral surface area of the portion of the circular cylinder $x^2 + y^2 = 1$ that lies between the xy-plane and the surface $f(x, y) = 1 - y^2$, as shown in Figure 17.12.

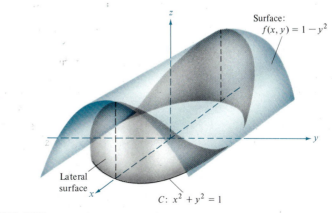

Surface:
$f(x, y) = 1 - y^2$

Lateral
surface

C: $x^2 + y^2 = 1$

FIGURE 17.12

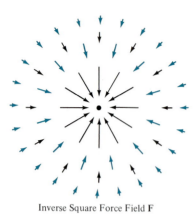

Inverse Square Force Field **F**

Vectors along a Parabolic
Path in the Force Field **F**

FIGURE 17.13

Solution: We represent the circle C parametrically as

$$x = \sin t, \qquad y = \cos t, \quad 0 \le t \le 2\pi$$

so that ds is given by

$$ds = \sqrt{[x'(t)]^2 + [y'(t)]^2} \, dt = \sqrt{(\cos t)^2 + (-\sin t)^2} \, dt = dt$$

Then, substituting $x = \sin t$ and $y = \cos t$ in $f(x, y)$, we have

$$f(x, y) = 1 - y^2 = 1 - \cos^2 t = \sin^2 t$$

Therefore, the lateral surface area is given by integrating $f(x, y)$ around the circle C as follows.

$$\begin{aligned}
\text{lateral surface area} &= \int_C f(x, y) \, ds = \int_0^{2\pi} \sin^2 t \, dt \\
&= \frac{1}{2} \int_0^{2\pi} (1 - \cos 2t) \, dt \\
&= \left[\frac{1}{2} \left(t - \frac{\sin 2t}{2} \right) \right]_0^{2\pi} = \pi \quad \square
\end{aligned}$$

One of the most important physical applications of line integrals is to find the **work** done on an object moving in a force field. For example, Figure 17.13 shows an inverse square force field similar to the gravitational field of the sun. Note that the magnitude of the force along a circular path about the center is constant, whereas the magnitude of the force along a parabolic path varies from point to point.

To see how a line integral can be used to find work done in a force field **F,** consider an object moving along a path C in the field, as shown in Figure 17.14. To determine the work done by the force, we need consider only that part of the force that is acting in the same direction as that in which the object is moving (or the opposite direction). In vector terms this means that at each point on C, we consider the projection $\mathbf{F} \cdot \mathbf{T}$ of the force vector $\mathbf{F}$ onto the unit tangent vector $\mathbf{T}$. On a small subarc of length Δs_i, the increment of work is

$$\Delta W_i = (\text{force})(\text{distance}) \approx [\mathbf{F}(x_i, y_i, z_i) \cdot \mathbf{T}(x_i, y_i, z_i)] \Delta s_i$$

where (x_i, y_i, z_i) is a point in the ith subarc. Consequently, the total work done is given by the integral in the following definition.

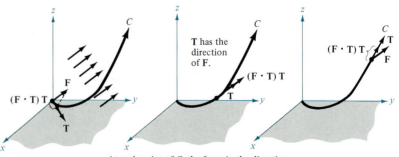

FIGURE 17.14

At each point of C, the force in the direction
of motion is $(\mathbf{F} \cdot \mathbf{T}) \mathbf{T}$.

DEFINITION OF WORK IN A FORCE FIELD	If C is a smooth curve in a force field $\mathbf{F}$ and $\mathbf{T}(x, y, z)$ is the unit tangent vector to C at (x, y, z), then the **work W done by $\mathbf{F}$ along C** is

$$W = \int_C \mathbf{F}(x, y, z) \cdot \mathbf{T}(x, y, z) \, ds$$

Remark In order to obtain the proper sign of the integral for work, C must be oriented in the direction of motion through the force field.

EXAMPLE 7   Work done by a force

Find the work done by the force field

$$\mathbf{F}(x, y, z) = x\mathbf{i} - xy\mathbf{j} + z^2\mathbf{k}$$

in moving a particle along the helix given by $\mathbf{r}(t) = \cos t\, \mathbf{i} + \sin t\, \mathbf{j} + t\mathbf{k}$ from the point $(0, 0, 0)$ to $(-1, 0, 3\pi)$.

Solution: Since $\mathbf{r}(t) = x(t)\, \mathbf{i} + y(t)\, \mathbf{j} + z(t)\, \mathbf{k} = \cos t\, \mathbf{i} + \sin t\, \mathbf{j} + t\mathbf{k}$, we have

$$\mathbf{T}(x, y, z) = \frac{\mathbf{r}'(t)}{\|\mathbf{r}'(t)\|} = \frac{1}{\sqrt{2}}(-\sin t\, \mathbf{i} + \cos t\, \mathbf{j} + \mathbf{k})$$

$$\mathbf{F}(x, y, z) = \cos t\, \mathbf{i} - \cos t \sin t\, \mathbf{j} + t^2\mathbf{k}$$

Therefore,

$$W = \int_C \mathbf{F} \cdot \mathbf{T} \, ds$$

$$= \int_0^{3\pi} (\cos t\, \mathbf{i} - \cos t \sin t\, \mathbf{j} + t^2\mathbf{k}) \cdot (-\sin t\, \mathbf{i} + \cos t\, \mathbf{j} + \mathbf{k}) \, dt$$

$$= \int_0^{3\pi} (-\sin t \cos t - \sin t \cos^2 t + t^2) \, dt$$

$$= \left[\frac{\cos^2 t}{2} + \frac{\cos^3 t}{3} + \frac{t^3}{3} \right]_0^{3\pi} = 9\pi^3 - \frac{2}{3}$$

Remark In each of the applications in the last three examples, you should try using a different parameterization of the curve to see that the value of the integral is unchanged (provided you don't reverse the orientation).

Alternative forms of line integrals

For line integrals involving work, a common alternative notation for $\mathbf{F} \cdot \mathbf{T} \, ds$ is $\mathbf{F} \cdot d\mathbf{r}$. This produces the following **vector form** of a line integral for work.

$$W = \int_C \mathbf{F} \cdot \mathbf{T} \, ds = \int_a^b \mathbf{F} \cdot \underbrace{\left(\frac{\mathbf{r}'(t)}{\|\mathbf{r}'(t)\|} \right) \underbrace{\|\mathbf{r}'(t)\| \, dt}_{ds}}_{\mathbf{T}} = \int_a^b \mathbf{F} \cdot \frac{d\mathbf{r}}{dt} \, dt = \int_C \mathbf{F} \cdot d\mathbf{r}$$

For instance, if we had used this notation in the solution of Example 7, we would have written

$$\mathbf{r}(t) = \cos t\,\mathbf{i} + \sin t\,\mathbf{j} + t\mathbf{k}$$

$$d\mathbf{r} = r'(t)\,dt = (-\sin t\,\mathbf{i} + \cos t\,\mathbf{j} + \mathbf{k})\,dt$$

$$\mathbf{F} = \cos t\,\mathbf{i} - \cos t \sin t\,\mathbf{j} + t^2\mathbf{k}$$

Then, using the vector form of the line integral for work, we would obtain

$$\int_C \mathbf{F} \cdot d\mathbf{r} = \int_0^{3\pi} (-\sin t \cos t - \sin t \cos^2 t + t^2)\,dt$$

A second commonly used form of line integral is derived from the vector field notation used in the previous section. If $\mathbf{F}$ is a vector field of the form $\mathbf{F}(x, y) = M\mathbf{i} + N\mathbf{j}$, and C is given by $\mathbf{r}(t) = x(t)\,\mathbf{i} + y(t)\,\mathbf{j}$, then $\mathbf{F} \cdot d\mathbf{r}$ is often written as $M\,dx + N\,dy$, as follows.

$$\int_C \mathbf{F} \cdot d\mathbf{r} = \int_C \mathbf{F} \cdot \frac{d\mathbf{r}}{dt}\,dt = \int_a^b (M\mathbf{i} + N\mathbf{j}) \cdot (x'(t)\,\mathbf{i} + y'(t)\,\mathbf{j})\,dt$$

$$= \int_a^b \left(M\,\frac{dx}{dt} + N\,\frac{dy}{dt}\right)dt$$

$$= \int_C (M\,dx + N\,dy)$$

This **differential form** can be extended to three variables. We often omit the parentheses and write

$$\int_C M\,dx + N\,dy \qquad \text{and} \qquad \int_C M\,dx + N\,dy + P\,dz$$

For curves represented by $y = g(x)$, $a \le x \le b$, we can let $x = t$ and obtain the parametric form

$$x = t \qquad \text{and} \qquad y = g(t), \quad a \le t \le b$$

Since $dx = dt$ for this form, we have the option of evaluating the line integral in variable x or t. This is demonstrated in the next example.

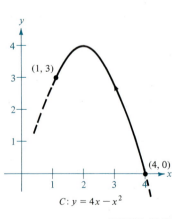

$C: y = 4x - x^2$

FIGURE 17.15

EXAMPLE 8   Evaluating a line integral in differential form

Evaluate

$$\int_C y\,dx + x^2\,dy$$

where C is the parabolic arc given by $y = 4x - x^2$ from $(4, 0)$ to $(1, 3)$, as shown in Figure 17.15.

Solution: Rather than converting to the parameter t, we simply retain the variable x and write

$$y = 4x - x^2 \implies dy = (4 - 2x)\,dx$$

Then, in the direction from $(4, 0)$ to $(1, 3)$, the line integral is

$$\int_C y \, dx + x^2 \, dy = \int_4^1 [(4x - x^2) \, dx + x^2(4 - 2x) \, dx]$$

$$= \int_4^1 [4x + 3x^2 - 2x^3] \, dx$$

$$= \left[2x^2 + x^3 - \frac{x^4}{2} \right]_4^1 = \frac{69}{2}$$

Remark Another parameterization (with the same orientation) for the line integral in Example 8 is

$$x(t) = 4 - t \qquad \text{and} \qquad y(t) = (4 - t)t, \quad 0 \le t \le 3$$

You should verify that this parameterization yields the same result.

Section Exercises 17.2

In Exercises 1–6, find a piecewise smooth parameterization of the path C.

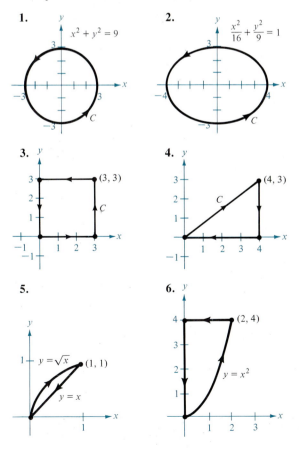

1. $x^2 + y^2 = 9$

2. $\dfrac{x^2}{16} + \dfrac{y^2}{9} = 1$

3. $(3, 3)$

4. $(4, 3)$

5. $y = \sqrt{x}$; $(1, 1)$; $y = x$

6. $(2, 4)$; $y = x^2$

In Exercises 7–14, evaluate

$$\int_C (x^2 + y^2) \, ds$$

along the given path.

7. C: x-axis from $x = 0$ to $x = 3$
8. C: y-axis from $y = 1$ to $y = 10$
9. C: line from $(0, 0)$ to $(1, 1)$
10. C: line from $(0, 0)$ to $(3, 9)$
11. C: counterclockwise around the triangle with vertices $(0, 0)$, $(1, 0)$, $(0, 1)$
12. C: counterclockwise around the square with vertices $(0, 0)$, $(1, 0)$, $(1, 1)$, $(0, 1)$
13. C: counterclockwise around the circle $x^2 + y^2 = 1$ from $(1, 0)$ to $(0, 1)$
14. C: counterclockwise around the circle $x^2 + y^2 = 4$ from $(2, 0)$ to $(0, 2)$

In Exercises 15–20, evaluate the integral

$$\int_C (2x - y) \, dx + (x + 3y) \, dy$$

along the given path.

15. C: x-axis from $x = 0$ to $x = 5$
16. C: y-axis from $y = 0$ to $y = 2$
17. C: line segments from $(0, 0)$ to $(3, 0)$ and from $(3, 0)$ to $(3, 3)$
18. C: line segments from $(0, 0)$ to $(0, -3)$ and from $(0, -3)$ to $(2, -3)$
19. C: parabolic path $x = t$, $y = 2t^2$, from $(0, 0)$ to $(2, 8)$

20. C: elliptic path $x = 4 \sin t$, $y = 3 \cos t$, from $(0, 3)$ to $(4, 0)$

In Exercises 21–26, evaluate

$$\int_C \mathbf{F} \cdot d\mathbf{r}$$

where C is represented by $\mathbf{r}(t)$.

21. $\mathbf{F}(x, y) = xy\mathbf{i} + y\mathbf{j}$
C: $\mathbf{r}(t) = 4t\mathbf{i} + t\mathbf{j}$,
$0 \le t \le 1$

22. $\mathbf{F}(x, y) = xy\mathbf{i} + y\mathbf{j}$
C: $\mathbf{r}(t) = 4 \cos t\,\mathbf{i} + 4 \sin t\,\mathbf{j}$,
$0 \le t \le \dfrac{\pi}{2}$

23. $\mathbf{F}(x, y) = 3x\mathbf{i} + 4y\mathbf{j}$
C: $\mathbf{r}(t) = 2 \cos t\,\mathbf{i} + 2 \sin t\,\mathbf{j}$,
$0 \le t \le \dfrac{\pi}{2}$

24. $\mathbf{F}(x, y) = 3x\mathbf{i} + 4y\mathbf{j}$
C: $\mathbf{r}(t) = t\mathbf{i} + \sqrt{4 - t^2}\,\mathbf{j}$,
$-2 \le t \le 2$

25. $\mathbf{F}(x, y, z) = x^2 y\mathbf{i} + (x - z)\mathbf{j} + xyz\mathbf{k}$
C: $\mathbf{r}(t) = t\mathbf{i} + t^2\mathbf{j} + 2\mathbf{k}$,
$0 \le t \le 1$

26. $\mathbf{F}(x, y, z) = x^2\mathbf{i} + y^2\mathbf{j} + z^2\mathbf{k}$
C: $\mathbf{r}(t) = \sin t\,\mathbf{i} + \cos t\,\mathbf{j} + t^2\mathbf{k}$,
$0 \le t \le \dfrac{\pi}{2}$

In Exercises 27 and 28, find the total mass of two coils of a spring with density ρ in the shape of the circular helix

$$\mathbf{r}(t) = 3 \cos t\,\mathbf{i} + 3 \sin t\,\mathbf{j} + 2t\mathbf{k}$$

27. $\rho(x, y, z) = \dfrac{1}{2}(x^2 + y^2 + z^2)$

28. $\rho(x, y, z) = 2$

In Exercises 29–36, find the area of the lateral surface over the curve C in the xy-plane and under the surface $z = f(x, y)$.

29. $f(x, y) = h$
C: line from $(0, 0)$ to $(3, 4)$
30. $f(x, y) = y$
C: line from $(0, 0)$ to $(4, 4)$
31. $f(x, y) = xy$
C: $x^2 + y^2 = 1$ from $(1, 0)$ to $(0, 1)$
32. $f(x, y) = x + y$
C: $x^2 + y^2 = 1$ from $(1, 0)$ to $(0, 1)$
33. $f(x, y) = h$
C: $y = 1 - x^2$ from $(1, 0)$ to $(0, 1)$

34. $f(x, y) = y + 1$
C: $y = 1 - x^2$ from $(1, 0)$ to $(0, 1)$
35. $f(x, y) = xy$
C: $y = 1 - x^2$ from $(1, 0)$ to $(0, 1)$
36. $f(x, y) = x^2 - y^2 + 4$
C: $x^2 + y^2 = 4$

In Exercises 37–42, find the work done by the force field $\mathbf{F}$ on an object moving along the specified path.

37. $\mathbf{F}(x, y) = -x\mathbf{i} - 2y\mathbf{j}$
C: $y = x^3$ from $(0, 0)$ to $(2, 8)$
38. $\mathbf{F}(x, y) = x^2\mathbf{i} - xy\mathbf{j}$
C: $x = \cos^3 t$, $y = \sin^3 t$ from $(1, 0)$ to $(0, 1)$
39. $\mathbf{F}(x, y) = 2x\mathbf{i} + y\mathbf{j}$
C: counterclockwise around the triangle whose vertices are $(0, 0)$, $(1, 0)$, $(1, 1)$
40. $\mathbf{F}(x, y) = -y\mathbf{i} - x\mathbf{j}$
C: counterclockwise along the semicircle $y = \sqrt{4 - x^2}$ from $(2, 0)$ to $(-2, 0)$
41. $\mathbf{F}(x, y, z) = x\mathbf{i} + y\mathbf{j} - 5z\mathbf{k}$
C: $\mathbf{r}(t) = 2 \cos t\,\mathbf{i} + 2 \sin t\,\mathbf{j} + t\mathbf{k}$,
$0 \le t \le 2\pi$
42. $\mathbf{F}(x, y, z) = yz\mathbf{i} + xz\mathbf{j} + xy\mathbf{k}$
C: line from $(0, 0, 0)$ to $(5, 3, 2)$

In Exercises 43–46, demonstrate the property that

$$\int_C \mathbf{F} \cdot d\mathbf{r} = 0$$

regardless of the initial and terminal points of C, if the tangent vector $\mathbf{r}'(t)$ is orthogonal to the force vector $\mathbf{F}$.

43. $\mathbf{F}(x, y) = y\mathbf{i} - x\mathbf{j}$
C: $\mathbf{r}(t) = t\mathbf{i} - 2t\mathbf{j}$
44. $\mathbf{F}(x, y) = -3y\mathbf{i} + x\mathbf{j}$
C: $\mathbf{r}(t) = t\mathbf{i} + t^3\mathbf{j}$
45. $\mathbf{F}(x, y) = (x^3 - 2x^2)\,\mathbf{i} + \left(x - \dfrac{y}{2}\right)\mathbf{j}$
C: $\mathbf{r}(t) = t\mathbf{i} + t^2\mathbf{j}$
46. $\mathbf{F}(x, y) = x\mathbf{i} + y\mathbf{j}$
C: $\mathbf{r}(t) = 3 \sin t\,\mathbf{i} + 3 \cos t\,\mathbf{j}$

47. The roof on a building has a height above the floor given by $z = 20 + \frac{1}{4}x$, and one of the walls follows a path given by $y = x^{3/2}$. Find the surface area of the wall if $0 \le x \le 40$.

17.3
Conservative vector fields and independence of path

We began the previous section by stating that in a gravitational field the work done by gravity on an object moving between two points in the field is independent of the path taken by the object. In this section we will prove an important generalization of this result. The generalization is called the **Fundamental Theorem of Line Integrals.**

We begin with an example in which we evaluate the line integral of a *conservative vector field* over three different paths.

EXAMPLE 1   Line integral of conservative vector field over different paths

Find the work done by the force field $F(x, y) = 4xy\mathbf{i} + 2x^2\mathbf{j}$ in moving a particle from $(0, 0)$ to $(1, 1)$ along the following paths:
(a) $C_1: y = x$ (b) $C_2: x = y^2$ (c) $C_3: y = x^3$

Solution:

(a) Let $\mathbf{r}(t) = t\mathbf{i} + t\mathbf{j}$ for $0 \le t \le 1$, so that

$$d\mathbf{r} = \mathbf{r}'(t)\, dt = (\mathbf{i} + \mathbf{j})\, dt \qquad \text{and} \qquad F(x, y) = 4t^2\mathbf{i} + 2t^2\mathbf{j}$$

Then,

$$\int_{C_1} F \cdot d\mathbf{r} = \int_0^1 6t^2\, dt = 2t^3 \Big]_0^1 = 2$$

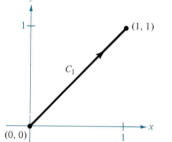

(b) Let $\mathbf{r}(t) = t\mathbf{i} + \sqrt{t}\mathbf{j}$ for $0 \le t \le 1$, so that

$$d\mathbf{r} = \mathbf{i} + \frac{1}{2\sqrt{t}}\mathbf{j} \qquad \text{and} \qquad F(x, y) = 4t^{3/2}\mathbf{i} + 2t^2\mathbf{j}$$

Then,

$$\int_{C_2} F \cdot d\mathbf{r} = \int_0^1 5t^{3/2}\, dt = 2t^{5/2} \Big]_0^1 = 2$$

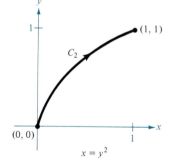

(c) Let $\mathbf{r}(t) = t\mathbf{i} + t^3\mathbf{j}$ for $0 \le t \le 1$, so that

$$d\mathbf{r} = \mathbf{i} + 3t^2\mathbf{j} \qquad \text{and} \qquad F(x, y) = 4t^4\mathbf{i} + 2t^2\mathbf{j}$$

Then,

$$\int_{C_3} F \cdot d\mathbf{r} = \int_0^1 10t^4\, dt = 2t^5 \Big]_0^1 = 2$$

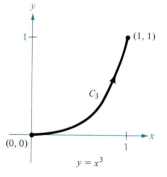

FIGURE 17.16

| Remark   The three paths in Example 1 are shown in Figure 17.16. Note that the vector field $F(x, y) = 4xy\mathbf{i} + 2x^2\mathbf{j}$ is conservative, since

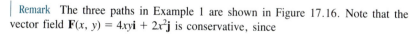

$$\frac{\partial}{\partial y}[4xy] = 4x = \frac{\partial}{\partial x}[2x^2]$$

THEOREM 17.4

FUNDAMENTAL THEOREM OF LINE INTEGRALS

Let C be a piecewise smooth curve lying in an open region R and given by

$$\mathbf{r}(t) = x(t)\,\mathbf{i} + y(t)\,\mathbf{j}, \quad a \leq t \leq b$$

If $\mathbf{F}(x, y) = M\mathbf{i} + N\mathbf{j}$ is conservative in R and M and N are continuous in R, then

$$\int_C \mathbf{F} \cdot d\mathbf{r} = \int_C \nabla f \cdot d\mathbf{r} = f(x(b), y(b)) - f(x(a), y(a))$$

where f is a potential function of $\mathbf{F}$. That is, $\mathbf{F}(x, y) = \nabla f(x, y)$.

Proof: We provide a proof only for a smooth curve. For piecewise smooth curves, the procedure is carried out separately on each smooth portion. Since $\mathbf{F}(x, y) = \nabla f(x, y) = f_x(x, y)\,\mathbf{i} + f_y(x, y)\,\mathbf{j}$, it follows that

$$\int_C \mathbf{F} \cdot d\mathbf{r} = \int_a^b \mathbf{F} \cdot \frac{d\mathbf{r}}{dt}\,dt = \int_a^b \left[f_x(x, y)\,\frac{dx}{dt} + f_y(x, y)\,\frac{dy}{dt} \right] dt$$

and, by the Chain Rule (Theorem 15.6), we have

$$\int_C \mathbf{F} \cdot d\mathbf{r} = \int_a^b \frac{d}{dt}[f(x(t), y(t))]\,dt = f(x(b), y(b)) - f(x(a), y(a))$$

The last step is an application of the Fundamental Theorem of Calculus.

| Remark   In space, the Fundamental Theorem takes the following form. Let C be a piecewise smooth curve lying in an open region Q and given by $\mathbf{r}(t) = x(t)\,\mathbf{i} + y(t)\,\mathbf{j} + z(t)\,\mathbf{k}, \; a \leq t \leq b$. If $\mathbf{F}(x, y, z) = M\mathbf{i} + N\mathbf{j} + P\mathbf{k}$ is conservative and M, N, and P are continuous, then

$$\int_C \mathbf{F} \cdot d\mathbf{r} = \int_C \nabla f \cdot d\mathbf{r} = f(x(b), y(b), z(b)) - f(x(a), y(a), z(a))$$

where $\mathbf{F}(x, y, z) = \nabla f(x, y, z)$.

The Fundamental Theorem of Line Integrals states that if the vector field $\mathbf{F}$ is conservative, then the line integral between any two points is simply the difference in the values of the *potential* function f at these points.

EXAMPLE 2   *Using the Fundamental Theorem of Line Integrals*

Evaluate

$$\int_C \mathbf{F} \cdot d\mathbf{r}$$

where C is a piecewise smooth curve from $(-1, 4)$ to $(1, 2)$ and $\mathbf{F}(x, y) = 2xy\mathbf{i} + (x^2 - y)\,\mathbf{j}.$

Solution: From Example 3 of Section 17.1 we know that **F** is the gradient of *f* where

$$f(x, y) = x^2y - \frac{1}{2}y^2 + K$$

Consequently, **F** is conservative, and by the Fundamental Theorem of Line Integrals, it follows that

$$\int_C \mathbf{F} \cdot d\mathbf{r} = f(1, 2) - f(-1, 4)$$

$$= \left[1^2(2) - \frac{1}{2}(2^2)\right] - \left[(-1)^2(4) - \frac{1}{2}(4^2)\right] = 4$$

Note that it is unnecessary to include *K* as part of *f*, since it is cancelled by subtraction. ▢

Independence of path

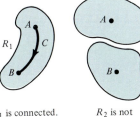

R_1 is connected.

R_2 is not connected.

FIGURE 17.17

From the Fundamental Theorem of Line Integrals it is clear that if **F** is continuous and conservative in an open region *R*, then the value of $\int_C \mathbf{F} \cdot d\mathbf{r}$ is the same for every piecewise smooth curve *C* from one fixed point in *R* to another fixed point in *R*. We describe this result by saying that the line integral $\int_C \mathbf{F} \cdot d\mathbf{r}$ is **independent of path** in the region *R*. In open regions that are *connected*, we will be able to show that the path independence of $\int_C \mathbf{F} \cdot d\mathbf{r}$ is actually equivalent to the condition that **F** is conservative. We say a region in the plane (or in space) is **connected** if any two points in the region can be joined by a piecewise smooth curve lying entirely within the region, as shown in Figure 17.17.

THEOREM 17.5   INDEPENDENCE OF PATH AND CONSERVATIVE VECTOR FIELDS
If **F** is continuous on an open connected region, then the line integral

$$\int_C \mathbf{F} \cdot d\mathbf{r}$$

is independent of path if and only if **F** is conservative.

Proof: If **F** is conservative, then by the Fundamental Theorem of Line Integrals the line integral is independent of path. We establish the converse for a plane region *R*. Let $\mathbf{F}(x, y) = M\mathbf{i} + N\mathbf{j}$, and let (x_0, y_0) be a fixed point in *R*. If (x, y) is any point in *R*, then we choose a piecewise smooth curve *C* running from (x_0, y_0) to (x, y), and define *f* by

$$f(x, y) = \int_C \mathbf{F} \cdot d\mathbf{r} = \int_C M\, dx + N\, dy$$

The existence of *C* in *R* is guaranteed by the fact that *R* is connected. We will show that *f* is a potential function of **F** by considering two different paths

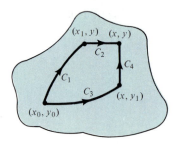

FIGURE 17.18

between (x_0, y_0) and (x, y). For the *first* path, we choose (x_1, y) in R such that $x \neq x_1$. This is possible since R is open. Then we choose C_1 and C_2, as shown in Figure 17.18, and using the independence of path, we write

$$f(x, y) = \int_C M\, dx + N\, dy = \int_{C_1} M\, dx + N\, dy + \int_{C_2} M\, dx + N\, dy$$

$$= g(y) + \int_{C_2} M\, dx$$

Now, since the line integral over C_1 is constant with respect to x and since $dy = 0$ in the line integral over C_2, the partial derivative of f with respect to x is $f_x(x, y) = M$. For the *second* path we choose a point (x, y_1), and following a similar reasoning we conclude that $f_y(x, y) = N$. Therefore,

$$\nabla f(x, y) = f_x(x, y)\, \mathbf{i} + f_y(x, y)\, \mathbf{j}$$

$$= M\mathbf{i} + N\mathbf{j}$$

$$= \mathbf{F}(x, y)$$

and it follows that $\mathbf{F}$ is conservative.

EXAMPLE 3   Finding work in a conservative force field

For the force field given by $\mathbf{F}(x, y, z) = e^x \cos y\, \mathbf{i} - e^x \sin y\, \mathbf{j} + 2\mathbf{k}$, show that $\int_C \mathbf{F} \cdot d\mathbf{r}$ is independent of path, and calculate the work done by $\mathbf{F}$ on an object moving along a curve C from $(0, \pi/2, 1)$ to $(1, \pi, 3)$.

Solution: Writing $\mathbf{F}(x, y, z) = M\mathbf{i} + N\mathbf{j} + P\mathbf{k}$, we have $M = e^x \cos y$, $N = -e^x \sin y$, and $P = 2$, and it follows that

$$\frac{\partial P}{\partial y} = 0 = \frac{\partial N}{\partial z}$$

$$\frac{\partial P}{\partial x} = 0 = \frac{\partial M}{\partial z}$$

and

$$\frac{\partial N}{\partial x} = -e^x \sin y = \frac{\partial M}{\partial y}$$

Hence, $\mathbf{F}$ is conservative. If f is a potential function of $\mathbf{F}$, then

$$f_x(x, y, z) = e^x \cos y$$

$$f_y(x, y, z) = -e^x \sin y$$

and

$$f_z(x, y, z) = 2$$

By integrating with respect to x, y, and z separately, we obtain

$$f(x, y, z) = \int f_x(x, y, z)\, dx = \int e^x \cos y\, dx = e^x \cos y + g(y, z) + K$$

$$f(x, y, z) = \int f_y(x, y, z)\, dy = \int -e^x \sin y\, dy = e^x \cos y + h(x, z) + K$$

$$f(x, y, z) = \int f_z(x, y, z)\, dz = \int 2\, dz = 2z + k(x, y) + K$$

By comparing these three versions of $f(x, y, z)$, we conclude that

$$f(x, y, z) = e^x \cos y + 2z + K$$

Therefore, the work done by **F** along *any* curve C from $(0, \pi/2, 1)$ to $(1, \pi, 3)$ is

$$W = \int_C \mathbf{F} \cdot d\mathbf{r} = \left[e^x \cos y + 2z \right]_{(0,\pi/2,1)}^{(1,\pi,3)}$$

$$= (-e + 6) - (0 + 2) = 4 - e$$

How much work would be done if the object in Example 3 moved from the point $(0, \pi/2, 1)$ to $(1, \pi, 3)$ and then back to the starting point, $(0, \pi/2, 1)$? The Fundamental Theorem tells us that there is zero work done. Remember that by definition work can be negative. Hence, by the time the object gets back to its starting point, the amount of work that registers positively is canceled out by the amount of work that registers negatively. We say that a curve that has the same initial and terminal point is **closed,** as shown in Figure 17.19. By the Fundamental Theorem, we conclude that if **F** is continuous and conservative on an open region R, then the line integral over every closed curve in R is zero. That is,

$$\oint_C \mathbf{F} \cdot d\mathbf{r} = 0 \qquad C \text{ is a closed curve}$$

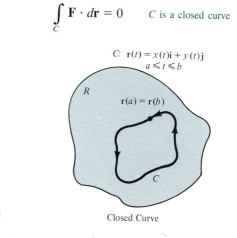

$$C: \ \mathbf{r}(t) = x(t)\mathbf{i} + y(t)\mathbf{j}$$
$$a \leqslant t \leqslant b$$

R

$$\mathbf{r}(a) = \mathbf{r}(b)$$

C

FIGURE 17.19   Closed Curve

One of the benefits of Theorem 17.5 is that for conservative vector fields, we have several options for evaluating line integrals. We can use a potential function, or it might be more convenient to choose a particularly simple path, such as a straight line.

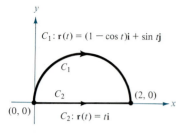

$C_1 : \mathbf{r}(t) = (1 - \cos t)\mathbf{i} + \sin t\mathbf{j}$

C_1

C_2 (2, 0)

(0, 0) $C_2 : \mathbf{r}(t) = t\mathbf{i}$

FIGURE 17.20

EXAMPLE 4   *Evaluating a line integral*

Evaluate

$$\int_{C_1} \mathbf{F} \cdot d\mathbf{r}$$

where $\mathbf{F}(x, y) = (y^3 + 1)\,\mathbf{i} + (3xy^2 + 1)\,\mathbf{j}$ and C_1 is the semicircular path shown in Figure 17.20.

Solution: We have three different options:

(a) We can use the method of the previous section to evaluate the line integral along the *given curve*. To do this, we use the parameterization

$$\mathbf{r}(t) = (1 - \cos t)\,\mathbf{i} + \sin t\,\mathbf{j}, \quad 0 \le t \le \pi$$

so that

$$d\mathbf{r} = \mathbf{r}'(t)\,dt = \sin t\,\mathbf{i} + \cos t\,\mathbf{j}$$

and

$$\mathbf{F}(x, y) = (\sin^3 t + 1)\,\mathbf{i} + [1 + 3\sin^2 t - 3\cos t\sin^2 t]\,\mathbf{j}$$

Thus,

$$\int_{C_1} \mathbf{F} \cdot d\mathbf{r} = \int_0^\pi [\sin t + \sin^4 t + \cos t + 3\sin^2 t\cos t - 3\cos^2 t\sin^2 t]\,dt$$

The sight of this integral dampens our enthusiasm for this option.

(b) We can try to find a *potential function* and evaluate the line integral by the Fundamental Theorem. For example, if f is a potential function of $\mathbf{F}$, then $f_x(x, y) = y^3 + 1$ and $f_y(x, y) = 3xy^2 + 1$, and we have

$$f(x, y) = \int f_x(x, y)\,dx = \int (y^3 + 1)\,dx = xy^3 + x + g(y) + K$$

$$f(x, y) = \int f_y(x, y)\,dy = \int (3xy^2 + 1)\,dy = xy^3 + y + h(x) + K$$

Thus, $f(x, y) = xy^3 + x + y + K$, and, by the Fundamental Theorem, we have

$$W = \int_{C_1} \mathbf{F} \cdot d\mathbf{r} = f(2, 0) - f(0, 0) = 2$$

(c) Knowing that $\mathbf{F}$ is conservative, we have a third option. Since the value of the line integral is independent of path, we can replace the semicircular path with a *simpler path*. Suppose we choose the straight line path C_2 from (0, 0) to (2, 0). Then, we have

$$\mathbf{r}(t) = t\mathbf{i}, \quad 0 \le t \le 2$$

Thus, $d\mathbf{r} = \mathbf{i}$ and $\mathbf{F}(x, y) = (y^3 + 1)\,\mathbf{i} + (3xy^2 + 1)\,\mathbf{j} = \mathbf{i} + \mathbf{j}$, so that

$$\int_{C_1} \mathbf{F} \cdot d\mathbf{r} = \int_{C_2} \mathbf{F} \cdot d\mathbf{r} = \int_0^2 1\,dt = t\Big|_0^2 = 2$$

Of the three options, obviously the third one is the easiest.

Conservation of energy

In 1840, the English physicist Michael Faraday (1791–1867) wrote, ''Nowhere is there a pure creation or production of power without a corresponding exhaustion of something to supply it.'' This statement represents the first formulation of one of the most important laws of physics—the **Law of Conservation of Energy.** Several writers in the philosophy of science have viewed this law as the greatest generalization ever conceived by humankind. Many physicists have contributed to our knowledge of this law. Two early and influential ones were James Prescott Joule (1818–1889) and Hermann Ludwig Helmholtz (1821–1894). In modern terminology, we state the law as follows: *In a conservative force field, the sum of the potential and kinetic energies of an object remains constant from point to point.*

Michael Faraday

We can use the Fundamental Theorem of Line Integrals to derive this law. From physics, the **kinetic energy** of a particle of mass m and speed v is $K = \frac{1}{2}mv^2$. The **potential energy** p of a particle at point (x, y, z) in a conservative vector field $\mathbf{F}$ is defined as $p(x, y, z) = -f(x, y, z)$, where f is the potential function for $\mathbf{F}$. Consequently, the work done by $\mathbf{F}$ along a smooth curve C from A to B is

$$W = \int_C \mathbf{F} \cdot d\mathbf{r} = f(x, y, z)\Big]_A^B = -p(x, y, z)\Big]_A^B = p(A) - p(B)$$

In other words, work W is equal to the difference in the potential energies of A and B. Now, suppose $\mathbf{r}(t)$ is the position vector for a particle moving along C from $A = \mathbf{r}(a)$ to $B = \mathbf{r}(b)$. At any time t, the particle's velocity, acceleration, and speed are $\mathbf{v}(t) = \mathbf{r}'(t)$, $\mathbf{a}(t) = \mathbf{r}''(t)$, and $v(t) = \|\mathbf{v}(t)\|$, respectively. Thus, by Newton's Second Law of Motion, $\mathbf{F} = m\mathbf{a}(t) = m(\mathbf{v}'(t))$, and the work done by $\mathbf{F}$ is

$$W = \int_C \mathbf{F} \cdot d\mathbf{r} = \int_a^b \mathbf{F} \cdot \mathbf{r}'(t)\, dt$$

$$= \int_a^b \mathbf{F} \cdot \mathbf{v}(t)\, dt = \int_a^b [m\mathbf{v}'(t)] \cdot \mathbf{v}(t)\, dt$$

$$= \int_a^b m[\mathbf{v}'(t) \cdot \mathbf{v}(t)]\, dt$$

$$= \frac{m}{2} \int_a^b \frac{d}{dt}[\mathbf{v}(t) \cdot \mathbf{v}(t)]\, dt$$

$$= \frac{m}{2} \int_a^b \frac{d}{dt}[\|\mathbf{v}(t)\|^2]\, dt = \frac{m}{2}\|\mathbf{v}(t)\|^2\Big]_a^b$$

$$= \frac{m}{2}[v(t)]^2\Big]_a^b = \frac{1}{2}m[v(b)]^2 - \frac{1}{2}m[v(a)]^2$$

$$= K(B) - K(A)$$

Equating these two results for W, we obtain

$$p(A) - p(B) = K(B) - K(A) \implies p(A) + K(A) = p(B) + K(B)$$

which implies that the sum of the potential and kinetic energies remains constant from point to point.

Section Exercises 17.3

In Exercises 1–4, show that the value of $\int_C \mathbf{F} \cdot d\mathbf{r}$ is the same for the given parametric representations of C.

1. $\mathbf{F}(x, y) = x^2\mathbf{i} + xy\mathbf{j}$
   (a) $\mathbf{r}_1(t) = t\mathbf{i} + t^2\mathbf{j}, \ 0 \le t \le 1$
   (b) $\mathbf{r}_2(\theta) = \sin\theta\,\mathbf{i} + \sin^2\theta\,\mathbf{j}, \ 0 \le \theta \le \dfrac{\pi}{2}$

2. $\mathbf{F}(x, y) = (x^2 + y^2)\,\mathbf{i} - x\mathbf{j}$
   (a) $\mathbf{r}_1(t) = t\mathbf{i} + \sqrt{t}\,\mathbf{j}, \ 0 \le t \le 4$
   (b) $\mathbf{r}_2(w) = w^2\mathbf{i} + w\mathbf{j}, \ 0 \le w \le 2$

3. $\mathbf{F}(x, y) = y\mathbf{i} - x\mathbf{j}$
   (a) $\mathbf{r}_1(\theta) = \sec\theta\,\mathbf{i} + \tan\theta\,\mathbf{j}, \ 0 \le \theta \le \dfrac{\pi}{3}$
   (b) $\mathbf{r}_2(t) = \sqrt{t+1}\,\mathbf{i} + \sqrt{t}\,\mathbf{j}, \ 0 \le t \le 3$

4. $\mathbf{F}(x, y) = y\mathbf{i} - x^2\mathbf{j}$
   (a) $\mathbf{r}_1(t) = (2 + t)\,\mathbf{i} + (3 - t)\,\mathbf{j}, \ 0 \le t \le 3$
   (b) $\mathbf{r}_2(w) = (2 + \ln w)\,\mathbf{i} + (3 - \ln w)\,\mathbf{j}, \ 1 \le w \le e^3$

In Exercises 5–18, find the value of the line integral $\int_C \mathbf{F} \cdot d\mathbf{r}$ along the given paths.

5. $\mathbf{F}(x, y) = 2xy\mathbf{i} + x^2\mathbf{j}$
   (a) $\mathbf{r}_1(t) = t\mathbf{i} + t^2\mathbf{j}, \ 0 \le t \le 1$
   (b) $\mathbf{r}_2(t) = t\mathbf{i} + t^3\mathbf{j}, \ 0 \le t \le 1$

6. $\mathbf{F}(x, y) = ye^{xy}\mathbf{i} + xe^{xy}\mathbf{j}$
   (a) $\mathbf{r}_1(t) = t\mathbf{i} - \dfrac{3}{2}(t - 2)\,\mathbf{j}, \ 0 \le t \le 2$
   (b) line segments from $(0, 3)$ to $(0, 0)$, and then from $(0, 0)$ to $(2, 0)$

7. $\mathbf{F}(x, y) = y\mathbf{i} - x\mathbf{j}$
   (a) $\mathbf{r}_1(t) = t\mathbf{i} + t\mathbf{j}, \ 0 \le t \le 1$
   (b) $\mathbf{r}_2(t) = t\mathbf{i} + t^2\mathbf{j}, \ 0 \le t \le 1$
   (c) $\mathbf{r}_3(t) = t\mathbf{i} + t^3\mathbf{j}, \ 0 \le t \le 1$

8. $\mathbf{F}(x, y) = xy^2\mathbf{i} + 2x^2y\mathbf{j}$
   (a) $\mathbf{r}_1(t) = t\mathbf{i} + \dfrac{1}{t}\mathbf{j}, \ 1 \le t \le 3$
   (b) $\mathbf{r}_2(t) = (t + 1)\,\mathbf{i} - \dfrac{1}{3}(t - 3)\,\mathbf{j}, \ 0 \le t \le 2$

9. $\displaystyle\int_C y^2\,dx + 2xy\,dy$

(a)   (b)

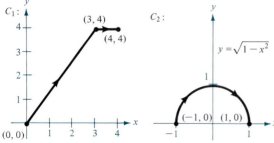

10. $\displaystyle\int_C (2x - 3y + 1)\,dx - (3x + y - 5)\,dy$

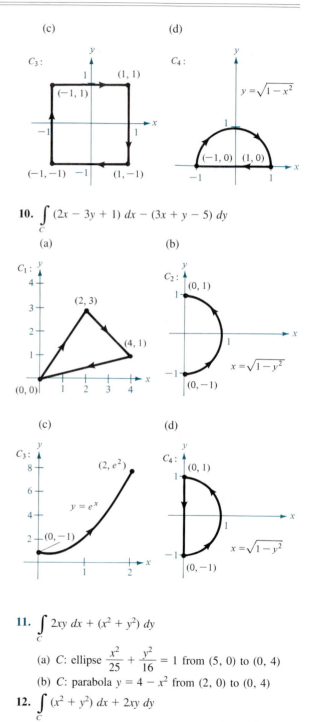

(a)   (b)

(c)   (d)

11. $\displaystyle\int_C 2xy\,dx + (x^2 + y^2)\,dy$
    (a) C: ellipse $\dfrac{x^2}{25} + \dfrac{y^2}{16} = 1$ from $(5, 0)$ to $(0, 4)$
    (b) C: parabola $y = 4 - x^2$ from $(2, 0)$ to $(0, 4)$

12. $\displaystyle\int_C (x^2 + y^2)\,dx + 2xy\,dy$
    (a) $\mathbf{r}_1(t) = t^3\mathbf{i} + t^2\mathbf{j}, \ 0 \le t \le 2$
    (b) $\mathbf{r}_2(t) = 2\cos t\,\mathbf{i} + 2\sin t\,\mathbf{j}, \ 0 \le t \le \dfrac{\pi}{2}$

13. $F(x, y, z) = yz\mathbf{i} + xz\mathbf{j} + xy\mathbf{k}$
    (a) $\mathbf{r}_1(t) = t\mathbf{i} + 2\mathbf{j} + t\mathbf{k}$, $0 \le t \le 4$
    (b) $\mathbf{r}_2(t) = t^2\mathbf{i} + t\mathbf{j} + t^2\mathbf{k}$, $0 \le t \le 2$

14. $F(x, y, z) = \mathbf{i} + z\mathbf{j} + y\mathbf{k}$
    (a) $\mathbf{r}_1(t) = \cos t\ \mathbf{i} + \sin t\ \mathbf{j} + t^2\mathbf{k}$, $0 \le t \le \pi$
    (b) $\mathbf{r}_2(t) = (1 - 2t)\ \mathbf{i} + \pi^2t\mathbf{k}$, $0 \le t \le 1$

15. $F(x, y, z) = (2y + x)\ \mathbf{i} + (x^2 - z)\ \mathbf{j} + (2y - 4z)\ \mathbf{k}$
    (a) $\mathbf{r}_1(t) = t\mathbf{i} + t^2\mathbf{j} + \mathbf{k}$, $0 \le t \le 1$
    (b) $\mathbf{r}_2(t) = t\mathbf{i} + t\mathbf{j} + (2t - 1)^2\mathbf{k}$, $0 \le t \le 1$

16. $F(x, y, z) = -y\mathbf{i} + x\mathbf{j} + 3xz^2\mathbf{k}$
    (a) $\mathbf{r}_1(t) = \cos t\ \mathbf{i} + \sin t\ \mathbf{j} + t\mathbf{k}$, $0 \le t \le \pi$
    (b) $\mathbf{r}_2(t) = (1 - 2t)\ \mathbf{i} + \pi t\mathbf{k}$, $0 \le t \le 1$

17. $F(x, y, z) = e^z(y\mathbf{i} + x\mathbf{j} + xy\mathbf{k})$
    (a) $\mathbf{r}_1(t) = 4\cos t\ \mathbf{i} + 4\sin t\ \mathbf{j} + 3\mathbf{k}$, $0 \le t \le \pi$
    (b) $\mathbf{r}_2(t) = (4 - 8t)\ \mathbf{i} + 3\mathbf{k}$, $0 \le t \le 1$

18. $F(x, y, z) = y\sin z\ \mathbf{i} + x\sin z\ \mathbf{j} + xy\cos z\ \mathbf{k}$
    (a) $\mathbf{r}_1(t) = t^2\mathbf{i} + t^2\mathbf{j}$, $0 \le t \le 2$
    (b) $\mathbf{r}_2(t) = 4t\mathbf{i} + 4t\mathbf{j}$, $0 \le t \le 1$

In Exercises 19–28, evaluate the given line integral using the Fundamental Theorem of Line Integrals.

19. $\displaystyle\int_C (y\mathbf{i} + x\mathbf{j}) \cdot d\mathbf{r}$

    C: smooth curve from $(0, 0)$ to $(3, 8)$

20. $\displaystyle\int_C [2(x + y)\ \mathbf{i} + 2(x + y)\ \mathbf{j}] \cdot d\mathbf{r}$

    C: smooth curve from $(-1, 1)$ to $(3, 2)$

21. $\displaystyle\int_{(0,-\pi)}^{(3\pi/2,\pi/2)} \cos x \sin y\ dx + \sin x \cos y\ dy$

22. $\displaystyle\int_{(1,1)}^{(2\sqrt{3},2)} \frac{y\ dx - x\ dy}{x^2 + y^2}$

23. $\displaystyle\int_C e^x \sin y\ dx + e^x \cos y\ dy$

    C: cycloid $x = \theta - \sin\theta$, $y = 1 - \cos\theta$ from $(0, 0)$ to $(2\pi, 0)$

24. $\displaystyle\int_C \frac{2x}{(x^2 + y^2)^2}\ dx + \frac{2y}{(x^2 + y^2)^2}\ dy$

    C: circle $(x - 4)^2 + (y - 5)^2 = 9$
    clockwise from $(7, 5)$ to $(1, 5)$

25. $\displaystyle\int_C (z + 2y)\ dx + (2x - z)\ dy + (x - y)\ dz$

    (a) C: line segment from $(0, 0, 0)$ to $(1, 1, 1)$
    (b) C: line segments from $(0, 0, 0)$ to $(0, 0, 1)$ to $(1, 1, 1)$
    (c) C: line segments from $(0, 0, 0)$ to $(1, 0, 0)$ to $(1, 1, 0)$ to $(1, 1, 1)$

26. Repeat Exercise 25 using the integral

$$\int_C zy\ dx + xz\ dy + xy\ dz$$

27. $\displaystyle\int_{(0,0,0)}^{(\pi/2,3,4)} -\sin x\ dx + z\ dy + y\ dz$

28. $\displaystyle\int_{(0,0,0)}^{(3,4,0)} 6x\ dx - 4z\ dy - (4y - 20z)\ dz$

In Exercises 29 and 30, find the work done in moving an object from P to Q in the force field F.

29. $F(x, y) = 9x^2y^2\mathbf{i} + (6x^3y - 1)\ \mathbf{j}$
    $P = (0, 0)$, $Q = (5, 9)$

30. $F(x, y) = \dfrac{2x}{y}\mathbf{i} - \dfrac{x^2}{y^2}\mathbf{j}$
    $P = (-1, 1)$, $Q = (3, 2)$

31. A stone weighing 1 pound is attached to the end of a 2-foot string and is whirled horizontally with one end held fixed. It makes one revolution per second. Find the work done by the force F that keeps the stone moving in a circular path.
    [Hint: Use force = (mass) (centripetal acceleration).]

32. If $F(x, y, z) = a_1\mathbf{i} + a_2\mathbf{j} + a_3\mathbf{k}$ is a constant force vector field, show that the work done in moving a particle along any path from P to Q is

$$W = F \cdot \overrightarrow{PQ}$$

33. The kinetic energy of an object moving through a conservative force field is decreasing at a rate of 10 units per minute. At what rate is the potential energy changing?

17.4
Green's Theorem

In this section we discuss a theorem that is named after the English mathematician George Green (1793–1841). Taken by itself, the theorem is fascinating enough—it states that the value of a double integral over a *simply connected* region R is determined by the value of a line integral around the boundary of R. But the result becomes even more impressive when we see some of the loose ends this theorem allows us to tie together.

A plane curve C given by

$$\mathbf{r}(t) = x(t)\,\mathbf{i} + y(t)\,\mathbf{j}, \quad a \le t \le b$$

is **simple** if it does not cross itself. A plane region R is **simply connected** if its boundary consists of *one* simple closed curve, as shown in Figure 17.21.

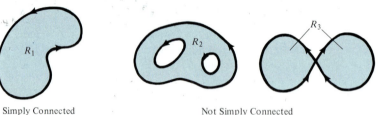

FIGURE 17.21   Simply Connected   Not Simply Connected

THEOREM 17.6

GREEN'S THEOREM
Let R be a simply connected region with boundary C, oriented counterclockwise. If M, N, $\partial M/\partial y$, and $\partial N/\partial x$ are all continuous in an open region containing R, then

$$\int_C M\,dx + N\,dy = \iint_R \left(\frac{\partial N}{\partial x} - \frac{\partial M}{\partial y} \right) dA$$

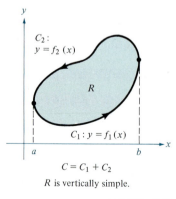

$C = C_1 + C_2$
R is vertically simple.

FIGURE 17.22

Proof: We give a proof only for a region that is both vertically simple and horizontally simple. Thus, R is described by the two forms

$$R: a \le x \le b,\ f_1(x) \le y \le f_2(x) \qquad \text{Figure 17.22}$$

and

$$R: c \le y \le d,\ g_1(y) \le x \le g_2(y) \qquad \text{Figure 17.23}$$

The line integral $\int_C M\,dx$ can be written as

$$\int_C M\,dx = \int_{C_1} M\,dx + \int_{C_2} M\,dx = \int_a^b M(x, f_1(x))\,dx + \int_b^a M(x, f_2(x))\,dx$$

$$= \int_a^b [M(x, f_1(x)) - M(x, f_2(x))]\,dx$$

On the other hand, we have

$$\iint_R \frac{\partial M}{\partial y}\, dA = \int_a^b \int_{f_1(x)}^{f_2(x)} \frac{\partial M}{\partial y}\, dy\, dx = \int_a^b M(x, y)\Big]_{f_1(x)}^{f_2(x)} dx$$

$$= \int_a^b [M(x, f_2(x)) - M(x, f_1(x))]\, dx$$

Consequently,

$$\int_C M\, dx = -\iint_R \frac{\partial M}{\partial y}\, dA$$

In a similar manner we can use $g_1(y)$ and $g_2(y)$ to show that

$$\int_C N\, dy = \iint_R \frac{\partial N}{\partial x}\, dA$$

Thus, we obtain

$$\int_C M\, dx + N\, dy = -\iint_R \frac{\partial M}{\partial y}\, dA + \iint_R \frac{\partial N}{\partial x}\, dA = \iint_R \left(\frac{\partial N}{\partial x} - \frac{\partial M}{\partial y} \right) dA$$

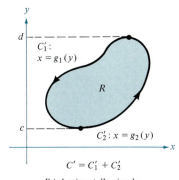

$C' = C_1' + C_2'$

R is horizontally simple.

FIGURE 17.23

Remark By saying that C is oriented counterclockwise, we mean that C is traversed *once* so that region R always lies to the *left*.

EXAMPLE 1   Using Green's Theorem

Use Green's Theorem to evaluate the line integral

$$\int_C y^3\, dx + (x^3 + 3xy^2)\, dy$$

where C is the path from $(0, 0)$ to $(1, 1)$ along the graph of $y = x^3$ and from $(1, 1)$ to $(0, 0)$ along the graph of $y = x$, as shown in Figure 17.24.

Solution: Since $M = y^3$ and $N = x^3 + 3xy^2$, we have

$$\int_C y^3\, dx + (x^3 + 3xy^2)\, dy = \iint_R \left(\frac{\partial N}{\partial x} - \frac{\partial M}{\partial y} \right) dA$$

$$= \int_0^1 \int_{x^3}^x [(3x^2 + 3y^2) - 3y^2]\, dy\, dx$$

$$= \int_0^1 \int_{x^3}^x 3x^2\, dy\, dx = \int_0^1 3x^2 y\Big]_{x^3}^x dx$$

$$= \int_0^1 (3x^3 - 3x^5)\, dx = \left[\frac{3x^4}{4} - \frac{x^6}{2} \right]_0^1 = \frac{1}{4}$$

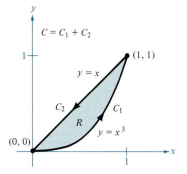

FIGURE 17.24

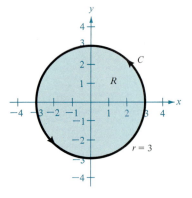

FIGURE 17.25

EXAMPLE 2   Using Green's Theorem to calculate work

While subject to the force $\mathbf{F}(x, y) = y^3\mathbf{i} + (x^3 + 3xy^2)\,\mathbf{j}$, a particle travels once around the circle of radius 3 shown in Figure 17.25. Use Green's Theorem to find the work done by $\mathbf{F}$.

Solution: From Example 1 we know by Green's Theorem that

$$\int_C y^3\,dx + (x^3 + 3xy^2)\,dy = \iint_R 3x^2\,dA$$

In polar coordinates, C is given by $r(t) = 3$ for $0 \le \theta \le 2\pi$, so $dA = r\,dr\,d\theta$, and the work done is

$$W = \iint_R 3x^2\,dA = \int_0^{2\pi} \int_0^3 3(r\cos\theta)^2 r\,dr\,d\theta$$

$$= 3\int_0^{2\pi} \int_0^3 r^3 \cos^2\theta\,dr\,d\theta$$

$$= 3\int_0^{2\pi} \frac{r^4}{4} \cos^2\theta \bigg]_0^3 d\theta$$

$$= 3\int_0^{2\pi} \frac{81}{4} \cos^2\theta\,d\theta = \frac{243}{8} \int_0^{2\pi} (1 + \cos 2\theta)\,d\theta$$

$$= \frac{243}{8} \left[\theta + \frac{\sin 2\theta}{2}\right]_0^{2\pi} = \frac{243\pi}{4}$$

Though a few more steps would be necessary, the line integrals in Examples 1 and 2 could have been evaluated directly, without the aid of Green's Theorem. In the next example the advantage of Green's Theorem is more obvious, since direct evaluation would require four separate line integrals.

EXAMPLE 3   Using Green's Theorem for a piecewise smooth curve

Evaluate

$$\int_C (\arctan x + y^2)\,dx + (\ln y - x^2)\,dy$$

where C is the path enclosing the annular region shown in Figure 17.26.

Solution: In polar coordinates, R is given by $1 \le r \le 3$ for $0 \le \theta \le \pi$. Moreover,

$$\frac{\partial N}{\partial x} - \frac{\partial M}{\partial y} = -2x - 2y = -2(r\cos\theta + r\sin\theta)$$

Thus, by Green's Theorem,

$$\int_C (\arctan x + y^2)\,dx + (\ln y - x^2)\,dy = \iint_R -2(x + y)\,dA$$

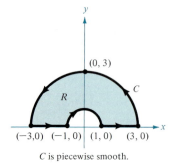

C is piecewise smooth.

FIGURE 17.26

$$= \int_0^\pi \int_1^3 -2r(\cos\theta + \sin\theta)r \, dr \, d\theta$$

$$= \int_0^\pi -2(\cos\theta + \sin\theta)\frac{r^3}{3}\Big]_1^3 \, d\theta$$

$$= \int_0^\pi \left(-\frac{52}{3}\right)(\cos\theta + \sin\theta) \, d\theta$$

$$= -\frac{52}{3}\Big[\sin\theta - \cos\theta\Big]_0^\pi = -\frac{104}{3}$$

In Examples 1–3 we used Green's Theorem to evaluate line integrals as double integrals. We can also use the theorem to evaluate double integrals as line integrals. One useful application occurs when $\partial N/\partial x - \partial M/\partial y = 1$.

$$\int_C M \, dx + N \, dy = \iint_R 1 \, dA = \text{area of region } R$$

Among the many choices for M and N satisfying the stated condition, we choose $M = -y/2$ and $N = x/2$ to obtain the following line integral for the area of region R.

THEOREM 17.7   LINE INTEGRAL FOR AREA

If R is a plane region bounded by a piecewise smooth simple closed curve C, then the area of R is given by

$$A = \frac{1}{2}\int_C x \, dy - y \, dx$$

EXAMPLE 4   Finding area by a line integral

Use a line integral to find the area of the ellipse

$$\frac{x^2}{a^2} + \frac{y^2}{b^2} = 1$$

Solution: Using Figure 17.27, we can induce a counterclockwise orientation to this elliptical path by letting

$$x = a\cos t \quad\text{and}\quad y = b\sin t, \quad 0 \le t \le 2\pi$$

Therefore, we have

$$A = \frac{1}{2}\int_C x \, dy - y \, dx = \frac{1}{2}\int_C [(a\cos t)(b\cos t) \, dt - (b\sin t)(-a\sin t) \, dt]$$

$$= \frac{ab}{2}\int_0^{2\pi} (\cos^2 t + \sin^2 t) \, dt = \frac{ab}{2}t\Big]_0^{2\pi}$$

$$= \pi ab$$

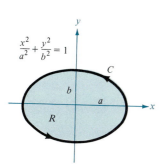

$$\frac{x^2}{a^2} + \frac{y^2}{b^2} = 1$$

FIGURE 17.27

Green's Theorem can be extended to cover regions that are not simply connected. This is demonstrated in the next example.

EXAMPLE 5   Green's Theorem extended to a region with a hole

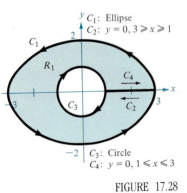

C_1: Ellipse
C_2: $y = 0, 3 \geqslant x \geqslant 1$

C_3: Circle
C_4: $y = 0, 1 \leqslant x \leqslant 3$

FIGURE 17.28

Let R be the region inside the ellipse $(x^2/9) + (y^2/4) = 1$ and outside the circle $x^2 + y^2 = 1$. Evaluate the line integral

$$\int_C 2xy \, dx + (x^2 + 2x) \, dy$$

where $C = C_1 + C_2 + C_3 + C_4$, as shown in Figure 17.28.

Solution: Since the line segments represented by C_2 and C_4 have opposite orientations, the line integrals over these two curves cancel, and we have

$$\int_C 2xy \, dx + (x^2 + 2x) \, dy = \iint_R \left(\frac{\partial N}{\partial x} - \frac{\partial M}{\partial y} \right) dA = \iint_R (2x + 2 - 2x) \, dA$$

$$= 2 \iint_R dA = 2(\text{area of } R)$$

$$= 2[\pi(3)(2) - \pi(1^2)] = 10\pi$$

In Section 17.1 we listed a necessary and sufficient condition for conservative vector fields. At that time, we proved only one direction of the proof. We now outline the other direction, using Green's Theorem. Let $\mathbf{F}(x, y) = M\mathbf{i} + N\mathbf{j}$ be defined on an open disc R. We want to show that if M and N have continuous first partial derivatives and

$$\frac{\partial M}{\partial y} = \frac{\partial N}{\partial x}$$

then $\mathbf{F}$ is conservative. Suppose that C is a closed path forming the boundary of a connected region lying in R. Then, using the fact that $\partial M/\partial y = \partial N/\partial x$, we can apply Green's Theorem to conclude that

$$\int_C \mathbf{F} \cdot d\mathbf{r} = \int_C M \, dx + N \, dy = \iint_R \left(\frac{\partial N}{\partial x} - \frac{\partial M}{\partial y} \right) dA = 0$$

This in turn is equivalent to showing that $\mathbf{F}$ is independent of path in R, and, by Theorem 17.5, we know that $\mathbf{F}$ is conservative.

Alternative forms of Green's Theorem

We conclude this section with the derivation of two vector forms of Green's Theorem for regions in the plane. The extension of these vector forms to three dimensions is the basis for the discussion in the remaining sections of this chapter. If $\mathbf{F}$ is a vector field in the plane, we can write

$$\mathbf{F}(x, y) = M\mathbf{i} + N\mathbf{j} + 0\mathbf{k}$$

so that the curl of $\mathbf{F}$, as described in Section 17.1, is given by

$$\text{curl } \mathbf{F} = \nabla \times \mathbf{F} = \begin{vmatrix} \mathbf{i} & \mathbf{j} & \mathbf{k} \\ \dfrac{\partial}{\partial x} & \dfrac{\partial}{\partial y} & \dfrac{\partial}{\partial z} \\ M & N & 0 \end{vmatrix} = -\dfrac{\partial N}{\partial z}\mathbf{i} - \dfrac{\partial M}{\partial z}\mathbf{j} + \left(\dfrac{\partial N}{\partial x} - \dfrac{\partial M}{\partial y}\right)\mathbf{k}$$

Consequently,

$$(\text{curl } \mathbf{F}) \cdot \mathbf{k} = \left[-\dfrac{\partial N}{\partial z}\mathbf{i} + \dfrac{\partial M}{\partial z}\mathbf{j} + \left(\dfrac{\partial N}{\partial x} - \dfrac{\partial M}{\partial y}\right)\mathbf{k}\right] \cdot \mathbf{k} = \dfrac{\partial N}{\partial x} - \dfrac{\partial M}{\partial y}$$

With appropriate conditions on $\mathbf{F}$, C, and R, we can write Green's Theorem in the vector form

$$\int_C \mathbf{F} \cdot d\mathbf{r} = \iint_R \left(\dfrac{\partial N}{\partial x} - \dfrac{\partial M}{\partial y}\right) dA = \iint_R (\text{curl } \mathbf{F}) \cdot \mathbf{k}\ dA$$

The extension of this vector form of Green's Theorem to surfaces in space produces **Stokes's Theorem,** discussed in Section 17.7.

For the second vector form of Green's Theorem, we assume the same conditions of $\mathbf{F}$, C, and R. Using the arc length parameter for C, we have $\mathbf{r}(s) = x(s)\,\mathbf{i} + y(s)\,\mathbf{j}$, so a unit tangent vector $\mathbf{T}$ to curve C is given by

$$\mathbf{r}'(s) = \mathbf{T} = x'(s)\,\mathbf{i} + y'(s)\,\mathbf{j}$$

From Figure 17.29 we can see that the *outward* unit normal vector $\mathbf{N}$ can then be written as

$$\mathbf{N} = y'(s)\,\mathbf{i} - x'(s)\,\mathbf{j}$$

Consequently, for $\mathbf{F}(x, y) = M\mathbf{i} + N\mathbf{j}$, we can apply Green's Theorem to obtain

$$\int_C \mathbf{F} \cdot \mathbf{N}\ ds = \int_C M\ dy - N\ dx = \iint_R \left(\dfrac{\partial M}{\partial x} + \dfrac{\partial N}{\partial y}\right) dA$$

The extension of this form to three dimensions is called the **Divergence Theorem,** discussed in Section 17.6. The physical interpretation of divergence and curl will be discussed in Sections 17.6 and 17.7.

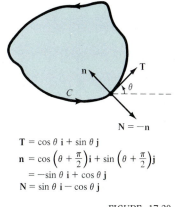

$\mathbf{T} = \cos\theta\,\mathbf{i} + \sin\theta\,\mathbf{j}$

$\mathbf{n} = \cos\left(\theta + \dfrac{\pi}{2}\right)\mathbf{i} + \sin\left(\theta + \dfrac{\pi}{2}\right)\mathbf{j}$

$\quad = -\sin\theta\,\mathbf{i} + \cos\theta\,\mathbf{j}$

$\mathbf{N} = \sin\theta\,\mathbf{i} - \cos\theta\,\mathbf{j}$

FIGURE 17.29

Section Exercises 17.4

In Exercises 1–4, verify Green's Theorem by evaluating both integrals

$$\int_C y^2\ dx + x^2\ dy = \iint_R \left(\dfrac{\partial N}{\partial x} - \dfrac{\partial M}{\partial y}\right) dA$$

for the given path.

1. C: boundary of the square with vertices $(0, 0)$, $(4, 0)$, $(4, 4)$, $(0, 4)$
2. C: boundary of the triangle with vertices $(0, 0)$, $(4, 0)$, $(4, 4)$

3. C: boundary of the region lying between the graphs of $y = x$ and $y = x^2/4$
4. C: $x^2 + y^2 = 1$

In Exercises 5–8, use Green's Theorem to evaluate the integral

$$\int_C (y - x)\ dx + (2x - y)\ dy$$

for the given path.

5. C: boundary of region lying between the graphs of $y = x$ and $y = x^2 - x$

6. $C: x = 2 \cos \theta, \; y = \sin \theta$

7. $C:$ boundary of the region lying inside the rectangle bounded by $x = -5$, $x = 5$, $y = -3$, $y = 3$ and outside the square bounded by $x = -1$, $x = 1$, $y = -1$, $y = 1$ •

8. $C:$ boundary of the region lying inside the circle $x^2 + y^2 = 16$ and outside the circle $x^2 + y^2 = 1$

In Exercises 9–18, use Green's Theorem to evaluate the line integral.

9. $\displaystyle\int_C 2xy \, dx + (x + y) \, dy$

$C:$ boundary of the region lying between the graphs of $y = 0$ and $y = 4 - x^2$

10. $\displaystyle\int_C y^2 \, dx + xy \, dy$

$C:$ boundary of the region lying between the graphs of $y = 0$, $y = \sqrt{x}$, and $x = 4$

11. $\displaystyle\int_C (x^2 - y^2) \, dx + 2xy \, dy$

$C: x^2 + y^2 = a^2$

12. $\displaystyle\int_C (x^2 - y^2) \, dx + 2xy \, dy$

$C: r = 1 + \cos \theta$

13. $\displaystyle\int_C 2 \arctan \frac{y}{x} \, dx + \ln (x^2 + y^2) \, dy$

$C: x = 4 + 2 \cos \theta, \; y = 4 + \sin \theta$

14. $\displaystyle\int_C e^x \sin 2y \, dx + 2e^x \cos 2y \, dy$

$C: x^2 + y^2 = a^2$

15. $\displaystyle\int_C \sin x \cos y \, dx + (xy + \cos x \sin y) \, dy$

$C:$ boundary of the region lying between the graphs of $y = x$ and $y = \sqrt{x}$

16. $\displaystyle\int_C (e^{-x^2/2} - y) \, dx + (e^{-y^2/2} + x) \, dy$

$C:$ boundary of the region lying between the graphs of the circle $x = 5 \cos \theta$, $y = 5 \sin \theta$ and the ellipse $x = 2 \cos \theta$, $y = \sin \theta$

17. $\displaystyle\int_C xy \, dx + (x + y) \, dy$

$C:$ boundary of the region lying between the graphs of $x^2 + y^2 = 1$ and $x^2 + y^2 = 9$

18. $\displaystyle\int_C 3x^2 e^y \, dx + e^y \, dy$

$C:$ boundary of the region lying between the squares with vertices $(1, 1)$, $(-1, 1)$, $(-1, -1)$, $(1, -1)$ and $(2, 2)$, $(-2, 2)$, $(-2, -2)$, $(2, -2)$

In Exercises 19–22, use Green's Theorem to calculate the work done by the force $\mathbf{F}$ in moving a particle around the closed path C.

19. $\mathbf{F}(x, y) = xy\mathbf{i} + (x + y) \mathbf{j}$
$C: x^2 + y^2 = 4$

20. $\mathbf{F}(x, y) = (e^x - 3y) \mathbf{i} + (e^y + 6x) \mathbf{j}$
$C: r = 2 \cos \theta$

21. $\mathbf{F}(x, y) = (x^{3/2} - 3y) \mathbf{i} + (6x + 5\sqrt{y}) \mathbf{j}$
$C:$ boundary of the triangle with vertices $(0, 0)$, $(5, 0)$, $(0, 5)$

22. $\mathbf{F}(x, y) = (3x^2 + y) \mathbf{i} + 4xy^2\mathbf{j}$
$C:$ boundary of the region bounded by the graphs of $y = \sqrt{x}$, $y = 0$, $x = 4$

In Exercises 23–26, use a line integral to find the area of the region R.

23. $R:$ region bounded by the graph of $x^2 + y^2 = a^2$
24. $R:$ triangle bounded by the graphs of $x = 0$, $2x - 3y = 0$, $x + 3y = 9$
25. $R:$ region bounded by the graphs of $y = 2x + 1$, $y = 4 - x^2$
26. $R:$ region inside the loop of the folium of Descartes bounded by the graph of

$$x = \frac{3t}{t^3 + 1}, \quad y = \frac{3t^2}{t^3 + 1}$$

In Exercises 27 and 28, use Green's Theorem to verify the line integral formulas.

27. The centroid of the region having area A bounded by the simple closed path C is

$$\bar{x} = \frac{1}{2A} \int_C x^2 \, dy, \quad \bar{y} = -\frac{1}{2A} \int_C y^2 \, dx$$

28. The area of a plane region bounded by the simple closed path C given in polar coordinates is

$$A = \frac{1}{2} \int_C r^2 \, d\theta$$

In Exercises 29–32, use the result of Exercises 27 to find the centroid of the region.

29. $R:$ region bounded by the graphs of $y = 0$ and $y = 4 - x^2$
30. $R:$ region bounded by the graphs of $y = \sqrt{a^2 - x^2}$ and $y = 0$
31. $R:$ region bounded by the graphs of $y = x^3$, $y = x$, $0 \le x \le 1$
32. $R:$ triangle with vertices $(-a, 0)$, $(a, 0)$, (b, c), where $-a \le b \le a$

In Exercises 33–36, use the result of Exercise 28 to find the area of the region bounded by the graphs of the polar equations.

33. $r = a(1 - \cos \theta)$ **34.** $r = a \cos 3\theta$

35. $r = 1 + 2 \cos \theta$ (inner loop)

36. $r = \dfrac{3}{2 - \cos \theta}$

37. Use Green's Theorem to prove that

$$\int_C f(x) \, dx + g(y) \, dy = 0$$

if f and g are differentiable functions and C is a piecewise smooth, simple closed path.

38. Let $\mathbf{F} = M\mathbf{i} + N\mathbf{j}$, where M and N have continuous first partial derivatives in a simply connected region R. Prove that if C is simple, smooth, and closed and **curl F = 0,** then

$$\int_C \mathbf{F} \cdot d\mathbf{r} = 0$$

17.5
Surface integrals

In the remainder of this chapter we will deal primarily with a type of integral called a **surface integral.** The name comes from the fact that instead of integrating over plane regions bounded by curves, we will integrate over regions in space bounded by surfaces.

Let S be a surface given by $z = g(x, y)$ and R its projection on the xy-plane, as shown in Figure 17.30. Suppose g, g_x, and g_y are continuous at all points in R, and f is defined on S. Employing the procedure used for surface area in Section 16.6, we then evaluate f at (x_i, y_i, z_i) and form the sum

$$\sum_{i=1}^{n} f(x_i, y_i, z_i)\Delta S_i$$

where

$$\Delta S_i \approx \sqrt{1 + [g_x(x_i, y_i)]^2 + [g_y(x_i, y_i)]^2}\,\Delta A_i$$

Providing the limit as $\|\Delta\|$ approaches zero exists, we define the **surface integral of f over S** to be

$$\iint_S f(x, y, z) \, dS = \lim_{\|\Delta\| \to 0} \sum_{i=1}^{n} f(x_i, y_i, z_i)\Delta S_i$$

This integral can be evaluated by the double integral given in the following theorem.

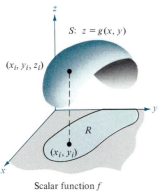

$S: z = g(x, y)$

(x_i, y_i, z_i)

R

(x_i, y_i)

Scalar function f assigns a number to each point of S.

FIGURE 17.30

THEOREM 17.8   **EVALUATING A SURFACE INTEGRAL**

Let S be a surface with equation $z = g(x, y)$ and R its projection on the xy-plane. If g, g_x, and g_y are continuous in R and f is continuous on S, then the surface integral of f over S is

$$\iint_S f(x, y, z) \, dS = \iint_R f(x, y, g(x, y))\sqrt{1 + [g_x(x, y)]^2 + [g_y(x, y)]^2} \, dA$$

For surfaces described by functions of x and z or y and z, we make the following adjustments to Theorem 17.8. If S is the graph of $y = g(x, z)$ and R is its projection onto the xz-plane, then

$$\iint_S f(x, y, z)\, dS = \iint_R f(x, g(x, z), z)\sqrt{1 + [g_x(x, z)]^2 + [g_z(x, z)]^2}\, dA$$

If S is the graph of $x = g(y, z)$ and R is its projection onto the yz-plane, then

$$\iint_S f(x, y, z)\, dS = \iint_R f(g(y, z), y, z)\sqrt{1 + [g_y(y, z)]^2 + [g_z(y, z)]^2}\, dA$$

EXAMPLE 1   Evaluating a surface integral

Evaluate the surface integral

$$\iint_S (y^2 + 2yz)\, dS$$

where S is the first-octant portion of the plane $2x + y + 2z = 6$.

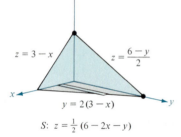

$z = 3 - x$ $z = \dfrac{6 - y}{2}$

$y = 2(3 - x)$

$S:\ z = \tfrac{1}{2}(6 - 2x - y)$

FIGURE 17.31

Solution: By projecting S onto the xy-plane, we can write S as $z = \frac{1}{2}(6 - 2x - y) = g(x, y)$, so that $g_x(x, y) = -1$ and $g_y(x, y) = -\frac{1}{2}$ and obtain

$$\sqrt{1 + [g_x(x, y)]^2 + [g_y(x, y)]^2} = \sqrt{1 + 1 + \left(\frac{1}{4}\right)} = \frac{3}{2}$$

Using Figure 17.31 and Theorem 17.8, we obtain

$$\iint_S (y^2 + 2yz)\, dS = \iint_R f(x, y, g(x, y))\sqrt{1 + [g_x(x, y)]^2 + [g_y(x, y)]^2}\, dA$$

$$= \iint_R \left[y^2 + 2y\left(\frac{1}{2}\right)(6 - 2x - y)\right]\left(\frac{3}{2}\right) dA$$

$$= 3\int_0^3 \int_0^{2(3-x)} y(3 - x)\, dy\, dx = 6\int_0^3 (3 - x)^3\, dx$$

$$= -\frac{3}{2}(3 - x)^4\Big]_0^3 = \frac{243}{2}$$

One alternative solution to Example 1 would be to project S onto the yz-plane, as shown in Figure 17.32. Then, $x = \frac{1}{2}(6 - y - 2z)$, so

$$\sqrt{1 + [g_y(y, z)]^2 + [g_z(y, z)]^2} = \sqrt{1 + (1/4) + 1} = \frac{3}{2}$$

Thus, the surface integral is

$$\iint_S (y^2 + 2yz)\, dS = \iint_R f(g(y, z), y, z)\sqrt{1 + [g_y(y, z)]^2 + [g_z(y, z)]^2}\, dA$$

$$= \int_0^6 \int_0^{(6-y)/2} [y^2 + 2yz]\left(\frac{3}{2}\right) dz\, dy$$

$$= \frac{3}{8}\int_0^6 (36y - y^3)\, dy = \frac{243}{2}$$

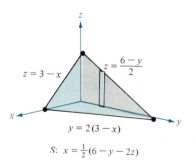

$z = 3 - x$ $z = \dfrac{6 - y}{2}$

$y = 2(3 - x)$

$S:\ x = \tfrac{1}{2}(6 - y - 2z)$

FIGURE 17.32

In Example 1 we could have projected the surface S onto any one of the three coordinate planes. In the next example the surface is a cylinder centered about the z-axis, and we have the option of projecting S onto either the xz-plane or the yz-plane.

EXAMPLE 2   Evaluating a surface integral

Evaluate the surface integral

$$\iint_S (x + z)\, dS$$

where S is the first-octant portion of the cylinder $x^2 + y^2 = 9$ between $z = 0$ and $z = 4$, as shown in Figure 17.33.

Solution: Note that S does not define z as a function of x and y. Hence, we project onto the xz-plane, so that $y = \sqrt{9 - x^2} = g(x, z)$, and obtain

$$\sqrt{1 + [g_x(x, z)]^2 + [g_z(x, z)]^2} = \sqrt{1 + \left(\frac{-x}{\sqrt{9 - x^2}}\right)^2 + (0)^2} = \frac{3}{\sqrt{9 - x^2}}$$

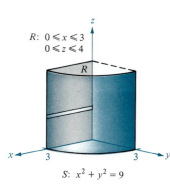

$R:\ 0 \leqslant x \leqslant 3$
$\quad\ 0 \leqslant z \leqslant 4$

R

$x \qquad 3 \qquad\qquad 3 \qquad y$

$S:\ x^2 + y^2 = 9$

FIGURE 17.33

Theorem 17.8 does not apply directly, since g_x is not continuous when $x = 3$. However, we can apply the theorem for $0 < b < 3$, as follows:

$$\iint_S (x + z)\, dS = \int_0^3 \int_0^4 (x + z) \frac{3}{\sqrt{9 - x^2}}\, dz\, dx$$

$$= \int_0^3 \left[\frac{3xz}{\sqrt{9 - x^2}} + \frac{3z^2}{2\sqrt{9 - x^2}} \right]_0^4 dx$$

$$= \int_0^3 \left(\frac{12x}{\sqrt{9 - x^2}} + \frac{24}{\sqrt{9 - x^2}} \right) dx \qquad \text{Improper integral}$$

$$= \lim_{b \to 3^-} \int_0^b \left(\frac{12x}{\sqrt{9 - x^2}} + \frac{24}{\sqrt{9 - x^2}} \right) dx$$

$$= \lim_{b \to 3^-} \left[-12\sqrt{9 - x^2} + 24 \arcsin \frac{x}{3} \right]_0^b$$

$$= \left[-12(0) + 24\left(\frac{\pi}{2}\right) + 12(3) - 0 \right] = 12\pi + 36 \qquad \square$$

If the function f defined on the surface S is simply $f(x, y, z) = 1$, then the surface integral yields the *surface area* of S.

$$\text{area of surface} = \iint_S 1\, dS$$

On the other hand, if S is a lamina of variable density and $\rho(x, y, z)$ is the density at the point (x, y, z), then the *mass* of the lamina is given by

$$\text{mass of lamina} = \iint_S \rho(x, y, z)\, dS$$

We illustrate this application in our next example.

Density:
$\rho(x, y, z) = k\sqrt{x^2 + y^2}$

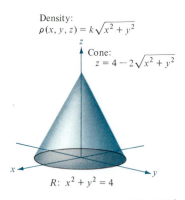

Cone:
$z = 4 - 2\sqrt{x^2 + y^2}$

$R: x^2 + y^2 = 4$

FIGURE 17.34

EXAMPLE 3   Finding the mass of a surface lamina

A surface lamina S has the shape of the cone given by $z = 4 - 2\sqrt{x^2 + y^2}$ between the planes $z = 0$ and $z = 4$, as shown in Figure 17.34. At each point on S, the density is proportional to the distance between the point and the z-axis. Find the mass m of the lamina.

Solution: Projecting S onto the xy-plane, we have

$$S: \quad z = 4 - 2\sqrt{x^2 + y^2} = g(x, y), \quad 0 \le z \le 4$$
$$R: \quad x^2 + y^2 \le 4$$
$$\text{Density:} \quad \rho(x, y, z) = k\sqrt{x^2 + y^2}$$

Using a surface integral, we find the mass to be

$$m = \iint_S \rho(x, y, z) \, dS = \iint_R k\sqrt{x^2 + y^2}\sqrt{1 + [g_x(x, y)]^2 + [g_y(x, y)]^2} \, dA$$

$$= k \iint_R \sqrt{x^2 + y^2}\sqrt{1 + \frac{4x^2}{x^2 + y^2} + \frac{4y^2}{x^2 + y^2}} \, dA$$

$$= k \iint_R \sqrt{5}\sqrt{x^2 + y^2} \, dA$$

In polar coordinates, we have

$$m = k \int_0^{2\pi} \int_0^2 (\sqrt{5}r)r \, dr \, d\theta = \frac{\sqrt{5}k}{3} \int_0^{2\pi} r^3 \Big]_0^2 \, d\theta$$

$$= \frac{8\sqrt{5}k}{3} \Big[\theta\Big]_0^{2\pi} = \frac{16\sqrt{5}k\pi}{3}$$

Orientation of a surface

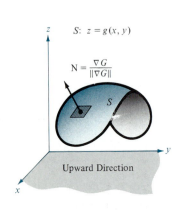

$S: z = g(x, y)$

$N = \dfrac{\nabla G}{\|\nabla G\|}$

S

Upward Direction

$S: z = g(x, y)$

S

$-N = \dfrac{-\nabla G}{\|\nabla G\|}$

Downward Direction

FIGURE 17.35

In using a line integral to find work, we developed the vector form of a line integral

$$\int_C \mathbf{F} \cdot \mathbf{T} \, ds$$

where the unit tangent vector $\mathbf{T}$ points in the direction of positive orientation along C. For every smooth curve, the unit tangent vector is continuous along C. In a similar way, we use unit normal vectors to induce an orientation to a surface S in space. We say that a surface is **orientable** if a unit normal vector $\mathbf{N}$ can be defined at every nonboundary point of S in such a way that the normal vectors vary continuously over the surface S. If this is possible, we call S an **oriented surface,** and we can write the following **vector form** of a surface integral

$$\iint_S \mathbf{F} \cdot \mathbf{N} \, dS \qquad \text{Vector form of surface integral}$$

where $\mathbf{F}$ is a vector field defined on a region containing S.

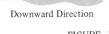

Not all surfaces are orientable (see Exercise 29). However, for many surfaces the gradient vector provides a convenient way to find a unit normal vector.

For a surface S given by $z = g(x, y)$, we let $G(x, y, z) = z - g(x, y)$. Then, S can be oriented by either of the unit normal vectors

$$\mathbf{N} = \frac{\nabla G(x, y, z)}{\|\nabla G(x, y, z)\|} = \frac{-g_x(x, y)\,\mathbf{i} - g_y(x, y)\,\mathbf{j} + \mathbf{k}}{\sqrt{1 + [g_x(x, y)]^2 + [g_y(x, y)]^2}} \qquad \text{Upward unit normal}$$

$$-\mathbf{N} = \frac{-\nabla G(x, y, z)}{\|\nabla G(x, y, z)\|} = \frac{g_x(x, y)\,\mathbf{i} + g_y(x, y)\,\mathbf{j} - \mathbf{k}}{\sqrt{1 + [g_x(x, y)]^2 + [g_y(x, y)]^2}} \qquad \text{Downward unit normal}$$

as shown in Figure 17.35 on the preceding page. In a similar manner, we can orient surfaces given by $y = g(x, z)$ or $x = g(y, z)$, using the gradient vectors

$$\nabla G(x, y, z), \quad G(x, y, z) = y - g(x, z)$$

and

$$\nabla G(x, y, z), \quad G(x, y, z) = x - g(y, z)$$

Flux integrals

One of the principal applications involving the vector form of a surface integral relates to the flow of a fluid through a surface S. Suppose a surface S is submerged in a fluid having a continuous velocity field $\mathbf{F}$. Let ΔS be the area of a small patch of the surface S over which $\mathbf{F}$ is nearly constant. Then the amount of fluid crossing this region per unit of time is approximated by the volume of the column of height $\mathbf{F} \cdot \mathbf{N}$ as shown in Figure 17.36. That is,

$$\Delta V = (\text{height})(\text{area of base}) = (\mathbf{F} \cdot \mathbf{N})\Delta S$$

Consequently, the volume of fluid crossing surface S per unit of time (called the **flux of F across S**) is given by the surface integral in the following definition.

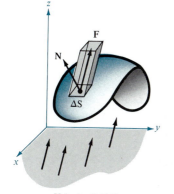

Velocity field $\mathbf{F}$ has direction of fluid flow.

FIGURE 17.36

DEFINITION OF FLUX INTEGRAL	Let $\mathbf{F}(x, y, z) = M\mathbf{i} + N\mathbf{j} + P\mathbf{k}$, where M, N, and P have continuous first partial derivatives on the surface S oriented by a unit normal vector $\mathbf{N}$. The **flux integral of F across S** is given by $$\iint_S \mathbf{F} \cdot \mathbf{N}\, dS$$

Remark   Geometrically, a flux integral is the surface integral over S of the *normal component* of $\mathbf{F}$. If $\rho(x, y, z)$ is the density of the fluid at (x, y, z), then the flux integral

$$\iint_S \rho\mathbf{F} \cdot \mathbf{N}\, dS$$

represents the *mass* of the fluid flowing across S per unit of time.

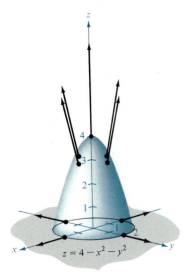

FIGURE 17.37

$z = 4 - x^2 - y^2$

EXAMPLE 4   *Using a flux integral to find the rate of mass flow*

Let S be that portion of the paraboloid $z = 4 - x^2 - y^2$ that lies above the xy-plane, oriented by an upward unit normal vector, as shown in Figure 17.37. A fluid of constant density ρ is flowing through the surface S according to the velocity field $\mathbf{F}(x, y, z) = x\mathbf{i} + y\mathbf{j} + z\mathbf{k}$. Find the rate of mass flow through S.

Solution: Projecting S onto the xy-plane, we have

$$S: \quad z = 4 - x^2 - y^2 \implies G(x, y, z) = z - 4 + x^2 + y^2$$
$$R: \quad x^2 + y^2 \leq 4$$

Unit normal: $\mathbf{N} = \dfrac{\nabla G(x, y, z)}{\|\nabla G(x, y, z)\|} = \dfrac{2x\mathbf{i} + 2y\mathbf{j} + \mathbf{k}}{\sqrt{1 + 4x^2 + 4y^2}}$

$dS: \quad \sqrt{1 + [g_x(x, y)]^2 + [g_y(x, y)]^2}\, dA = \sqrt{1 + 4x^2 + 4y^2}\, dA$

Therefore,

$$\iint_S \rho\mathbf{F} \cdot \mathbf{N}\, dS = \rho \iint_R \left(\frac{2x^2 + 2y^2 + z}{\sqrt{1 + 4x^2 + 4y^2}} \right)\sqrt{1 + 4x^2 + 4y^2}\, dA$$

$$= \rho \iint_R [2x^2 + 2y^2 + (4 - x^2 - y^2)]\, dA$$

$$= \rho \iint_R [4 + x^2 + y^2]\, dA$$

$$= \rho \int_0^{2\pi} \int_0^2 [4 + r^2]r\, dr\, d\theta \qquad \text{Polar coordinates}$$

$$= \rho \int_0^{2\pi} 12\, d\theta = 24\pi\rho$$

Notice the cancellation that occurs in the product

$$\left(\frac{\nabla G(x, y, z)}{\|\nabla G(x, y, z)\|} \right) dS = \nabla G(x, y, z)\, dA$$

By generalizing this result, we can write flux integrals in the following simplified form.

THEOREM 17.9   **EVALUATING A FLUX INTEGRAL**
If S is a surface given by $z = g(x, y)$ and R is its projection on the xy-plane, then

$$\iint_S \mathbf{F} \cdot \mathbf{N}\, dS = \iint_R \mathbf{F} \cdot (-g_x(x, y)\, \mathbf{i} - g_y(x, y)\, \mathbf{j} + \mathbf{k})\, dA$$

Oriented upward

$$\iint_S \mathbf{F} \cdot \mathbf{N}\, dS = \iint_R \mathbf{F} \cdot (g_x(x, y)\, \mathbf{i} + g_y(x, y)\, \mathbf{j} - \mathbf{k})\, dA$$

Oriented downward

EXAMPLE 5   *Finding the flux of an inverse square field*

Find the flux over the sphere $x^2 + y^2 + z^2 = K^2$, where **F** is an inverse square field

$$\mathbf{F}(\mathbf{r}) = \frac{\lambda \mathbf{r}}{\|\mathbf{r}\|^3}$$

Solution: We need only calculate the flux over the upper hemisphere, since the vector field **F** and the surface S have the same symmetry with respect to the origin. Projecting S on the *xy*-plane, we have

$$S: \quad z = \sqrt{K^2 - x^2 - y^2} = g(x, y)$$
$$R: \quad x^2 + y^2 \leq K^2$$

$$\mathbf{F}(\mathbf{r}): \quad \mathbf{F}(x, y, z) = \frac{\lambda(x\mathbf{i} + y\mathbf{j} + z\mathbf{k})}{(x^2 + y^2 + z^2)^{3/2}} = \frac{\lambda(x\mathbf{i} + y\mathbf{j} + z\mathbf{k})}{K^3}$$

Therefore, by Theorem 17.9, we obtain

$$\iint_S \mathbf{F} \cdot \mathbf{N} \, dS = 2 \iint_R \frac{\lambda(x\mathbf{i} + y\mathbf{j} + z\mathbf{k})}{K^3} \cdot \left(\frac{x\mathbf{i} + y\mathbf{j}}{\sqrt{K^2 - x^2 - y^2}} + \mathbf{k} \right) dA$$

$$= 2\lambda \iint_R \left(\frac{x^2 + y^2}{K^3\sqrt{K^2 - x^2 - y^2}} + \frac{z}{K^3} \right) dA$$

$$= 2\lambda \iint_R \left(\frac{x^2 + y^2}{K^3\sqrt{K^2 - x^2 - y^2}} + \frac{\sqrt{K^2 - x^2 - y^2}}{K^3} \right) dA$$

$$= 2\lambda \iint_R \frac{K^2}{K^3\sqrt{K^2 - x^2 - y^2}} \, dA$$

$$= 2\lambda \int_0^{2\pi} \int_0^K \frac{1}{K\sqrt{K^2 - r^2}} r \, dr \, d\theta \qquad \text{Polar coordinates}$$

$$= \frac{2\lambda}{K} \int_0^{2\pi} \left. -\sqrt{K^2 - r^2} \right]_0^K d\theta$$

$$= \frac{2\lambda}{K} 2\pi K = 4\pi\lambda$$

The result in Example 5 shows that the flux across a sphere S in an inverse square field is independent of the radius of S. In particular, if E is an electric field, then the result in Example 5, along with Coulomb's Law, yields one of the basic laws of electrostatics, known as **Gauss's Law:**

$$\iint_S \mathbf{E} \cdot \mathbf{N} \, dS = 4\pi \sum_{i=1}^n q_i$$

where $q_1, q_2, \ldots, q_n$ are the point charges contained in S.

Section Exercises 17.5

In Exercises 1–4, evaluate

$$\iint_S (x - 2y + z)\, dS$$

over the indicated surface S.

1. S: $z = 4 - x$, $0 \le x \le 4$, $0 \le y \le 4$
2. S: $z = 10 - 2x + 2y$, $0 \le x \le 2$, $0 \le y \le 4$
3. S: $z = 10$, $x^2 + y^2 \le 1$
4. S: $z = \dfrac{2}{3}x^{3/2}$, $0 \le x \le 1$, $0 \le y \le x$

In Exercises 5–8, evaluate

$$\iint_S xy\, dS$$

over the indicated surface S.

5. S: $z = 6 - x - 2y$ (first octant)
6. S: $z = xy$, $0 \le x \le 2$, $0 \le y \le 2$
7. S: $z = 9 - x^2$, $0 \le x \le 2$, $0 \le y \le x$
8. S: $z = h$, $0 \le x \le 2$, $0 \le y \le \sqrt{4 - x^2}$

In Exercises 9–14, evaluate

$$\iint_S f(x, y, z)\, dS$$

9. $f(x, y, z) = x^2 + y^2 + z^2$
   S: $z = x + 2$, $x^2 + y^2 \le 1$
10. $f(x, y, z) = \dfrac{xy}{z}$
    S: $z = x^2 + y^2$, $4 \le x^2 + y^2 \le 16$
11. $f(x, y, z) = \sqrt{x^2 + y^2 + z^2}$
    S: $z = \sqrt{x^2 + y^2}$, $x^2 + y^2 \le 4$
12. $f(x, y, z) = \sqrt{x^2 + y^2 + z^2}$
    S: $z = \sqrt{x^2 + y^2}$, $(x - 1)^2 + y^2 \le 1$
13. $f(x, y, z) = x^2 + y^2 + z^2$
    S: $x^2 + y^2 = 9$, $0 \le x \le 3$, $0 \le y \le 3$, $0 \le z \le 9$
14. $f(x, y, z) = x^2 + y^2 + z^2$
    S: $x^2 + y^2 = 9$, $0 \le x \le 3$, $0 \le z \le x$

In Exercises 15–20, find the flux of F through S,

$$\iint_S \mathbf{F} \cdot \mathbf{N}\, dS$$

where $\mathbf{N}$ is the upper unit normal vector to S.

15. $\mathbf{F}(x, y, z) = 3z\mathbf{i} - 4\mathbf{j} + y\mathbf{k}$
    S: $x + y + z = 1$ (first octant)
16. $\mathbf{F}(x, y, z) = x\mathbf{i} + y\mathbf{j}$
    S: $2x + 3y + z = 6$ (first octant)
17. $\mathbf{F}(x, y, z) = x\mathbf{i} + y\mathbf{j} + z\mathbf{k}$
    S: $z = 9 - x^2 - y^2$, $0 \le z$
18. $\mathbf{F}(x, y, z) = x\mathbf{i} + y\mathbf{j} + z\mathbf{k}$
    S: $x^2 + y^2 + z^2 = 16$ (first octant)
19. $\mathbf{F}(x, y, z) = 4\mathbf{i} - 3\mathbf{j} + 5\mathbf{k}$
    S: $z = x^2 + y^2$, $x^2 + y^2 \le 4$
20. $\mathbf{F}(x, y, z) = x\mathbf{i} + y\mathbf{j} - 2z\mathbf{k}$
    S: $z = \sqrt{a^2 - x^2 - y^2}$

In Exercises 21 and 22, find the flux of F over the closed surface. (Let N be the outward unit normal of each surface.)

21. $\mathbf{F}(x, y, z) = 4xy\mathbf{i} + z^2\mathbf{j} + yz\mathbf{k}$
    S: unit cube bounded by $x = 0$, $x = 1$, $y = 0$, $y = 1$, $z = 0$, $z = 1$
22. $\mathbf{F}(x, y, z) = (x + y)\,\mathbf{i} + y\mathbf{j} + z\mathbf{k}$
    S: $z = 1 - x^2 - y^2$, $z = 0$

In Exercises 23 and 24, find the mass of the surface lamina S of density ρ.

23. S: $2x + 3y + 6z = 12$ (first   octant), $\rho(x, y, z) = x^2 + y^2$
24. S: $z = \sqrt{a^2 - x^2 - y^2}$, $\rho(x, y, z) = kz$

In Exercises 25 and 26, use the following formulas for the moments of inertia about the coordinate axes of a surface lamina of density ρ.

$$I_x = \iint_S (y^2 + z^2)\rho(x, y, z)\, dS$$

$$I_y = \iint_S (x^2 + z^2)\rho(x, y, z)\, dS$$

$$I_z = \iint_S (x^2 + y^2)\rho(x, y, z)\, dS$$

25. Show that the moment of inertia of a conical shell about its axis is $\frac{1}{2}ma^2$, where m is the mass and a is the radius.
26. Show that the moment of inertia of a spherical shell of uniform density about its diameter is $\frac{2}{3}ma^2$, where m is the mass and a is the radius.

In Exercises 27 and 28, find I_z for the given lamina with uniform density.

27. $x^2 + y^2 = a^2$, $0 \le z \le h$
28. $z = x^2 + y^2$, $0 \le z \le h$

29. The surface shown in Figure 17.38 is called a Moebius strip. Explain why this surface is not orientable.

30. Determine whether the surface in Figure 17.39 is orientable.

FIGURE 17.38

FIGURE 17.39

SECTION TOPICS ▪
Divergence ▪
Divergence Theorem ▪
Applications ▪

17.6
Divergence Theorem

We begin this section by defining a scalar function called the **divergence** of the vector field **F.** This function can be viewed as a type of derivative of **F** in that, for vector fields representing velocities of moving particles, the divergence measures the rate of particle flow per unit of volume at a point.

DEFINITION OF DIVERGENCE OF A VECTOR FIELD

The **divergence** of $\mathbf{F}(x, y) = M\mathbf{i} + N\mathbf{j}$ is

$$\text{div } \mathbf{F}(x, y) = \nabla \cdot \mathbf{F}(x, y) = \frac{\partial M}{\partial x} + \frac{\partial N}{\partial y} \qquad \text{Plane}$$

The **divergence** of $\mathbf{F}(x, y, z) = M\mathbf{i} + N\mathbf{j} + P\mathbf{k}$ is

$$\text{div } \mathbf{F}(x, y, z) = \nabla \cdot \mathbf{F}(x, y, z) = \frac{\partial M}{\partial x} + \frac{\partial N}{\partial y} + \frac{\partial P}{\partial z} \qquad \text{Space}$$

Remark The dot-product notation used for divergence comes from viewing ∇ as a **differential operator,** as follows.

$$\nabla \cdot \mathbf{F}(x, y, z) = \left[\left(\frac{\partial}{\partial x} \right) \mathbf{i} + \left(\frac{\partial}{\partial y} \right) \mathbf{j} + \left(\frac{\partial}{\partial z} \right) \mathbf{k} \right] \cdot [M\mathbf{i} + N\mathbf{j} + P\mathbf{k}]$$

$$= \frac{\partial M}{\partial x} + \frac{\partial N}{\partial y} + \frac{\partial P}{\partial z}$$

EXAMPLE 1   Finding the divergence of a vector field

Find the divergence at $(2, 1, -1)$ for the vector field

$$\mathbf{F}(x, y, z) = x^3 y^2 z \mathbf{i} + x^2 z \mathbf{j} + x^2 y \mathbf{k}$$

Solution: By definition, we have

$$\text{div } \mathbf{F}(x, y, z) = \frac{\partial}{\partial x}[x^3 y^2 z] + \frac{\partial}{\partial y}[x^2 z] + \frac{\partial}{\partial z}[x^2 y] = 3x^2 y^2 z$$

At the point $(2, 1, -1)$, we have

$$\text{div } \mathbf{F}(2, 1, -1) = 3(2^2)(1^2)(-1) = -12$$

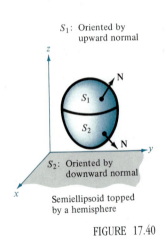

S_1: Oriented by
upward normal

S_2: Oriented by
downward normal

Semiellipsoid topped
by a hemisphere

FIGURE 17.40

At the end of Section 17.4, we pointed out that Green's Theorem can be extended to three dimensions. This important extension is called the **Divergence Theorem** (or **Gauss's Theorem**). Recall from Section 17.4 that an alternative form of Green's Theorem is

$$\int_C \mathbf{F} \cdot \mathbf{N} \, ds = \iint_R \left(\frac{\partial M}{\partial x} + \frac{\partial N}{\partial y} \right) dA = \iint_R \text{div } \mathbf{F} \, dA$$

In an analogous way, the Divergence Theorem shows the relationship between a triple integral over a solid region Q and a surface integral over the surface of Q. In the statement of the theorem, the surface S is **closed** in the sense that it forms the complete boundary of the solid Q. Regions bounded by spheres, hemispheres, ellipsoids, cubes, tetrahedrons, or some combination of these surfaces are typical examples of closed surfaces. We assume that Q is a solid region on which a triple integral can be evaluated, and that the closed surface S is oriented by *outward* unit normal vectors, as shown in Figure 17.40. With these restrictions on S and Q, we state the following theorem.

THEOREM 17.10   **THE DIVERGENCE THEOREM**

Let Q be a solid region bounded by a closed surface S oriented by a unit normal vector directed outward from Q. If $\mathbf{F}$ is a vector field whose component functions have continuous partial derivatives on Q, then

$$\iint_S \mathbf{F} \cdot \mathbf{N} \, dS = \iiint_Q \text{div } \mathbf{F} \, dV$$

Proof: If we let $\mathbf{F}(x, y, z) = M\mathbf{i} + N\mathbf{j} + P\mathbf{k}$, the theorem takes the form

$$\iint_S (M\mathbf{i} \cdot \mathbf{N} + N\mathbf{j} \cdot \mathbf{N} + P\mathbf{k} \cdot \mathbf{N}) \, dS = \iiint_Q \left(\frac{\partial M}{\partial x} + \frac{\partial N}{\partial y} + \frac{\partial P}{\partial z} \right) dV$$

We can prove this result by showing that the following three equations are valid:

$$\iint_S M\mathbf{i} \cdot \mathbf{N} \, dS = \iiint_Q \frac{\partial M}{\partial x} \, dV$$

$$\iint_S N\mathbf{j} \cdot \mathbf{N} \, dS = \iiint_Q \frac{\partial N}{\partial y} \, dV$$

$$\iint_S P\mathbf{k} \cdot \mathbf{N} \, dS = \iiint_Q \frac{\partial P}{\partial z} \, dV$$

Since the verifications of the three equations are similar, we will only discuss the third. We restrict the proof to a **simple solid** region with upper surface $z = g_2(x, y)$ and lower surface $z = g_1(x, y)$, whose projections on the xy-plane

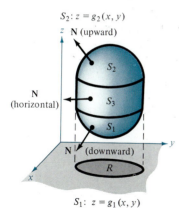

$S_2: z = g_2(x, y)$

N (upward)

S_2

N
(horizontal)

S_3

S_1

N (downward)

R

$S_1: z = g_1(x, y)$

FIGURE 17.41

coincide and form region R. If Q has a lateral surface like S_3 in Figure 17.41, then a normal vector is horizontal, so $P\mathbf{k} \cdot \mathbf{N} = 0$. Consequently, we have

$$\iint_S P\mathbf{k} \cdot \mathbf{N} \, dS = \iint_{S_1} P\mathbf{k} \cdot \mathbf{N} \, dS + \iint_{S_2} P\mathbf{k} \cdot \mathbf{N} \, dS + 0$$

On the upper surface S_2 the outward normal is upward, whereas on the lower surface S_1 the outward normal is downward. Therefore, by Theorem 17.9, we have

$$\iint_{S_1} P\mathbf{k} \cdot \mathbf{N} \, dS = \iint_R P(x, y, g_1(x, y)) \, \mathbf{k} \cdot \left(\frac{\partial g_1}{\partial x} \mathbf{i} + \frac{\partial g_1}{\partial y} \mathbf{j} - \mathbf{k} \right) dA$$

$$= -\iint_R P(x, y, g_1(x, y)) \, dA$$

and

$$\iint_{S_2} P\mathbf{k} \cdot \mathbf{N} \, dS = \iint_R P(x, y, g_2(x, y)) \, \mathbf{k} \cdot \left(-\frac{\partial g_2}{\partial x} \mathbf{i} - \frac{\partial g_2}{\partial y} \mathbf{j} + \mathbf{k} \right) dA$$

$$= \iint_R P(x, y, g_2(x, y)) \, dA$$

Adding these results, we obtain

$$\iint_S P\mathbf{k} \cdot \mathbf{N} \, dS = \iint_R [P(x, y, g_2(x, y)) - P(x, y, g_1(x, y))] \, dA$$

$$= \iint_R \left[\int_{g_1(x,y)}^{g_2(x,y)} \frac{\partial P}{\partial z} \, dz \right] dA = \iiint_Q \frac{\partial P}{\partial z} \, dV$$

EXAMPLE 2   *Using the Divergence Theorem*

Let Q be the solid region bounded by the coordinate planes and the plane $2x + 2y + z = 6$, and let $\mathbf{F} = x\mathbf{i} + y^2\mathbf{j} + z\mathbf{k}$. Find

$$\iint_S \mathbf{F} \cdot \mathbf{N} \, dS$$

Solution: From Figure 17.42 we see that Q is bounded by four subsurfaces, and we would need four surface integrals to evaluate

$$\iint_S \mathbf{F} \cdot \mathbf{N} \, dS$$

However, by the Divergence Theorem, we need only one triple integral. Since

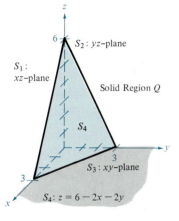

S_2: yz-plane

S_1:
xz-plane

Solid Region Q

S_4

S_3: xy-plane

S_4: $z = 6 - 2x - 2y$

FIGURE 17.42

$$\text{div } \mathbf{F} = \frac{\partial M}{\partial x} + \frac{\partial N}{\partial y} + \frac{\partial P}{\partial z} = 1 + 2y + 1$$

we have

$$\iint_S \mathbf{F} \cdot \mathbf{N}\, dS = \iiint_Q \text{div } \mathbf{F}\, dV$$

$$= \int_0^3 \int_0^{3-y} \int_0^{6-2x-2y} (1 + 2y + 1)\, dz\, dx\, dy$$

$$= \int_0^3 \int_0^{3-y} (2z + 2yz)\Big]_0^{6-2x-2y}\, dx\, dy$$

$$= \int_0^3 \int_0^{3-y} (12 - 4x + 8y - 4xy - 4y^2)\, dx\, dy$$

$$= \int_0^3 \left[12x - 2x^2 + 8xy - 2x^2y - 4xy^2 \right]_0^{3-y}\, dy$$

$$= \int_0^3 (18 + 6y - 10y^2 + 2y^3)\, dy$$

$$= \left[18y + 3y^2 - \frac{10y^3}{3} + \frac{y^4}{2} \right]_0^3 = \frac{63}{2}$$

EXAMPLE 3   Verifying the Divergence Theorem

Let Q be the solid region between the paraboloid $z = 4 - x^2 - y^2$ and the xy-plane. Verify the Divergence Theorem for $\mathbf{F}(x, y, z) = 2z\mathbf{i} + x\mathbf{j} + y^2\mathbf{k}$.

Solution: From Figure 17.43 we see that the outward normal vector for the surface S_1 is $\mathbf{N}_1 = -\mathbf{k}$, whereas the outward normal vector for the surface S_2 is $\mathbf{N}_2 = 2x\mathbf{i} + 2y\mathbf{j} + \mathbf{K}$. Thus, by Theorem 17.9,

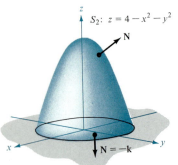

S_2: $z = 4 - x^2 - y^2$

S_1: $x^2 + y^2 \le 4$

FIGURE 17.43

$$\iint_S \mathbf{F} \cdot \mathbf{N}\, dS$$

$$= \iint_{S_1} \mathbf{F} \cdot (-\mathbf{k})\, dS + \iint_{S_2} \mathbf{F} \cdot \mathbf{N}\, dS$$

$$= \iint_R -y^2\, dA + \iint_R \mathbf{F} \cdot (2x\mathbf{i} + 2y\mathbf{j} + \mathbf{k})\, dA$$

$$= -\int_{-2}^2 \int_{-\sqrt{4-y^2}}^{\sqrt{4-y^2}} y^2\, dx\, dy + \int_{-2}^2 \int_{-\sqrt{4-y^2}}^{\sqrt{4-y^2}} (4xz + 2xy + y^2)\, dx\, dy$$

$$= \int_{-2}^2 \int_{-\sqrt{4-y^2}}^{\sqrt{4-y^2}} [4x(4 - x^2 - y^2) + 2xy]\, dx\, dy$$

$$= \int_{-2}^2 \left[8x^2 - x^4 - 2x^2y^2 + x^2y \right]_{-\sqrt{4-y^2}}^{\sqrt{4-y^2}}\, dy = \int_{-2}^2 0\, dy = 0$$

On the other hand, because

$$\text{div } \mathbf{F} = \frac{\partial}{\partial x}[2z] + \frac{\partial}{\partial y}[x] + \frac{\partial}{\partial z}[y^2] = 0$$

we obtain the equivalent result

$$\iiint_Q 0\, dV = 0$$

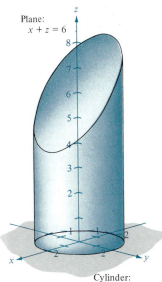

Plane:
$x + z = 6$

Cylinder:
$x^2 + y^2 = 4$

FIGURE 17.44

EXAMPLE 4   Using the Divergence Theorem

Let Q be the solid bounded by the cylinder $x^2 + y^2 = 4$, the plane $x + z = 6$, and the xy-plane, as shown in Figure 17.44. Find

$$\iint_S \mathbf{F} \cdot \mathbf{N} \, dS$$

where $\mathbf{F}(x, y, z) = (x^2 + \sin z) \, \mathbf{i} + (xy + \cos z) \, \mathbf{j} + e^y \mathbf{k}.$

Solution: Direct evaluation of this surface integral would be difficult. However, by the Divergence Theorem, we have

$$\iint_S \mathbf{F} \cdot \mathbf{N} \, dS = \iiint_Q \text{div } \mathbf{F} \, dV = \iiint_Q (2x + x + 0) \, dV$$

$$= \int_0^{2\pi} \int_0^2 \int_0^{6 - r\cos\theta} (3 \, r \cos \theta) r \, dz \, dr \, d\theta \qquad \text{Cylindrical coordinates}$$

$$= \int_0^{2\pi} \int_0^2 (18r^2 \cos \theta - 3r^3 \cos^2 \theta) \, dr \, d\theta$$

$$= \int_0^{2\pi} (48 \cos \theta - 12 \cos^2 \theta) \, d\theta$$

$$= \left[48 \sin \theta - 6\left(\theta + \frac{1}{2} \sin 2\theta \right) \right]_0^{2\pi} = -12\pi \qquad \square$$

Flux and the Divergence Theorem

To increase our understanding of the Divergence Theorem, we consider the two sides of the equation

$$\iint_S \mathbf{F} \cdot \mathbf{N} \, dS = \iiint_Q \text{div } \mathbf{F} \, dV$$

We know from Section 17.5 that the flux integral on the left determines the total fluid flow across surface S per unit of time. This can be approximated by summing the fluid flow across small patches of the surface. The triple integral on the right measures this same fluid flow across S but from a very different perspective—namely, by calculating the flow of fluid into (or out of) small *cubes* of volume ΔV_i. The flux of the ith cube is approximately

$$\text{flux of } i\text{th cube} \approx \text{div } \mathbf{F}(x_i, y_i, z_i) \Delta V_i$$

for some point (x_i, y_i, z_i) in the ith cube. Note that for a cube in the interior of Q, the gain or loss of fluid through any one of its six sides is offset by a corresponding loss or gain through one of the sides of an adjacent cube. After summing over all the cubes in Q, the only fluid flow that is not canceled by adjoining cubes is that on the outside edges of the cubes on the boundary. Thus, the sum

$$\sum_{i=1}^{n} \text{div } \mathbf{F}(x_i, y_i, z_i) \Delta V_i$$

approximates the total flux into (or out of) Q, and hence through the surface S.

To see what is meant by the divergence of $\mathbf{F}$ at a point, we consider ΔV_α to be the volume of a small sphere S_α of radius α and center (x_0, y_0, z_0), contained in region Q. Applying the Divergence Theorem to S_α, we have

$$\text{flux of } \mathbf{F} \text{ across } S_\alpha = \iiint\limits_{Q_\alpha} \text{div } \mathbf{F}\, dV \approx \text{div } \mathbf{F}(x_0, y_0, z_0)\Delta V_\alpha$$

where Q_α is the interior of S_α. Consequently, we have

$$\text{div } \mathbf{F}(x_0, y_0, z_0) \approx \frac{\text{flux of } \mathbf{F} \text{ across } S_\alpha}{\Delta V_\alpha}$$

and, by taking the limit as $\alpha \to 0$, we obtain the divergence of $\mathbf{F}$ at the point (x_0, y_0, z_0).

$$\text{div } \mathbf{F}(x_0, y_0, z_0) = \lim_{\alpha \to 0} \frac{\text{flux of } \mathbf{F} \text{ across } S_\alpha}{\Delta V_\alpha} = \begin{array}{l} \text{flux per unit volume} \\ \text{at } (x_0, y_0, z_0) \end{array}$$

The point (x_0, y_0, z_0) in a vector field is classified as a source, a sink, or incompressible, as follows.

1. **Source,** if div $\mathbf{F} > 0$ (Figure 17.45a)
2. **Sink,** if div $\mathbf{F} < 0$ (Figure 17.45b)
3. **Incompressible** if div $\mathbf{F} = 0$ (Figure 17.45c)

FIGURE 17.45

(a) Source: div $\mathbf{F} > 0$

(b) Sink: div $\mathbf{F} < 0$

(c) Incompressible: div $\mathbf{F} = 0$

EXAMPLE 5   Calculating flux by the Divergence Theorem

Let Q be the region bounded by the sphere $x^2 + y^2 + z^2 = 4$. Find the outward flux of the vector field $\mathbf{F}(x, y, z) = 2x^3\mathbf{i} + 2y^3\mathbf{j} + 2z^3\mathbf{k}$ through the sphere.

Solution: By the Divergence Theorem, we have

$$\text{flux across } S = \iint\limits_S \mathbf{F} \cdot \mathbf{N}\, dS = \iiint\limits_Q \text{div } \mathbf{F}\, dV$$

$$= \iiint\limits_Q 6(x^2 + y^2 + z^2)\, dV$$

$$= 6 \int_0^2 \int_0^\pi \int_0^{2\pi} \rho^4 \sin \phi\, d\theta\, d\phi\, d\rho \qquad \text{Spherical coordinates}$$

$$= 6 \int_0^2 \int_0^\pi 2\pi\rho^4 \sin \phi\, d\phi\, d\rho$$

$$= 12\pi \int_0^2 2\rho^4\, d\rho = 24\pi\left(\frac{32}{5}\right) = \frac{768\pi}{5}$$

Section Exercises 17.6

In Exercises 1–4, find the divergence of the vector field **F**.

1. $F(x, y, z) = 6x^2\mathbf{i} - xy^2\mathbf{j} + (z - 4y)\mathbf{k}$
2. $F(x, y, z) = xe^x\mathbf{i} + ye^y\mathbf{j} + z^2\mathbf{k}$
3. $F(x, y, z) = \sin x\,\mathbf{i} + \cos y\,\mathbf{j} + z^2\mathbf{k}$
4. $F(x, y, z) = \ln(x^2 + y^2)\,\mathbf{i} + xy\mathbf{j} + \ln(y^2 + z^2)\,\mathbf{k}$

In Exercises 5–8, verify the Divergence Theorem by evaluating

$$\iint_S \mathbf{F} \cdot \mathbf{N}\, dS$$

as a surface integral and as a triple integral.

5. $F(x, y, z) = 2x\mathbf{i} - 2y\mathbf{j} + z^2\mathbf{k}$
   S: cube bounded by the planes $x = 0$, $x = a$, $y = 0$, $y = a$, $z = 0$, $z = a$
6. $F(x, y, z) = 2x\mathbf{i} - 2y\mathbf{j} + z^2\mathbf{k}$
   S: cylinder $x^2 + y^2 = 1$, $0 \le z \le h$
7. $F(x, y, z) = (2x - y)\mathbf{i} - (2y - z)\mathbf{j} + z\mathbf{k}$
   S: surface bounded by the plane $2x + 4y + 2z = 12$ and the coordinate planes
8. $F(x, y, z) = xy\mathbf{i} + z\mathbf{j} + (x + y)\mathbf{k}$
   S: surface bounded by the planes $y = 4$, $z = 4 - x$ and the coordinate planes

In Exercises 9–18, use the Divergence Theorem to evaluate

$$\iint_S \mathbf{F} \cdot \mathbf{N}\, dS$$

and find the outward flux of **F** through the surface of the solid bounded by the graphs of the equations.

9. $F(x, y, z) = x^2\mathbf{i} + y^2\mathbf{j} + z^2\mathbf{k}$
   S: $x = 0$, $x = a$, $y = 0$, $y = a$, $z = 0$, $z = a$
10. $F(x, y, z) = x^2z\mathbf{i} - y\mathbf{j} + xyz\mathbf{k}$
    S: $x = 0$, $x = a$, $y = 0$, $y = a$, $z = 0$, $z = a$
11. $F(x, y, z) = x^2\mathbf{i} - 2xy\mathbf{j} + xyz^2\mathbf{k}$
    S: $z = \sqrt{a^2 - x^2 - y^2}$, $z = 0$
12. $F(x, y, z) = xy\mathbf{i} + yz\mathbf{j} - yz\mathbf{k}$
    S: $z = \sqrt{a^2 - x^2 - y^2}$, $z = 0$
13. $F(x, y, z) = x\mathbf{i} + y\mathbf{j} + z\mathbf{k}$
    S: $x^2 + y^2 + z^2 = 4$
14. $F(x, y, z) = xyz\mathbf{j}$
    S: $x^2 + y^2 = 9$, $z = 0$, $z = 4$
15. $F(x, y, z) = x\mathbf{i} + y^2\mathbf{j} - z\mathbf{k}$
    S: $x^2 + y^2 = 9$, $z = 0$, $z = 4$
16. $F(x, y, z) = (xy^2 + \cos z)\,\mathbf{i} + (x^2y + \sin z)\,\mathbf{j} + e^z\mathbf{k}$
    S: $z = \sqrt{x^2 + y^2}$, $z = 4$
17. $F(x, y, z) = x^3\mathbf{i} + x^2y\mathbf{j} + x^2e^y\mathbf{k}$
    S: $z = 4 - y$, $z = 0$, $x = 0$, $x = 6$, $y = 0$
18. $F(x, y, z) = xe^z\mathbf{i} + ye^z\mathbf{j} + e^z\mathbf{k}$
    S: $z = 4 - y$, $z = 0$, $x = 0$, $x = 6$, $y = 0$

19. Use the Divergence Theorem to show that the volume of the solid bounded by a surface S is

$$\iint_S x\, dy\, dz = \iint_S y\, dz\, dx = \iint_S z\, dx\, dy$$

20. Verify the result of Exercise 19 for the cube bounded by $x = 0$, $x = a$, $y = 0$, $y = a$, $z = 0$, and $z = a$.

In Exercises 21 and 22, evaluate

$$\iint_S \text{curl } \mathbf{F} \cdot \mathbf{N}\, dS$$

where S is the closed surface of the solid bounded by the graphs of $x = 4$, $z = 9 - y^2$, and the coordinate planes.

21. $F(x, y, z) = (4xy + z^2)\,\mathbf{i} + (2x^2 + 6yz)\,\mathbf{j} + 2xz\mathbf{k}$
22. $F(x, y, z) = xy\cos z\,\mathbf{i} + yz\sin x\,\mathbf{j} + xyz\mathbf{k}$

23. Show that

$$\iint_S \text{curl } \mathbf{F} \cdot \mathbf{N}\, dS = 0$$

for any closed surface S.

24. For the constant vector field given by

$$F(x, y, z) = a_1\mathbf{i} + a_2\mathbf{j} + a_3\mathbf{k}$$

show that

$$\iint_S \mathbf{F} \cdot \mathbf{N}\, dS = 0$$

where S is any closed surface.

In Exercises 25 and 26, let $F(x, y, z) = x\mathbf{i} + y\mathbf{j} + z\mathbf{k}$.

25. Show that

$$\iint_S \mathbf{F} \cdot \mathbf{N}\, dS = 3V$$

where V is the volume of the solid bounded by the closed surface S.

26. Show that

$$\frac{1}{\|\mathbf{F}\|} \iint_S \mathbf{F} \cdot \mathbf{N}\, dS = \frac{3}{\|\mathbf{F}\|} \iiint_Q dV$$

In Exercises 27–30, find the divergence of the vector field **F** at the indicated point.

27. $F(x, y, z) = xyz\mathbf{i} + y\mathbf{j} + z\mathbf{k}$, $(1, 2, 1)$
28. $F(x, y, z) = x^2z\mathbf{i} - 2xz\mathbf{j} + yz\mathbf{k}$, $(2, -1, 3)$
29. $F(x, y, z) = e^x \sin y\,\mathbf{i} - e^x \cos y\,\mathbf{j}$, $(0, 0, 3)$
30. $F(x, y, z) = e^{-xyz}(\mathbf{i} + \mathbf{j} + \mathbf{k})$, $(3, 2, 0)$

In Exercises 31 and 32, find div ($\mathbf{F} \times \mathbf{G}$).

31. $\mathbf{F}(x, y, z) = \mathbf{i} + 2x\mathbf{j} + 3y\mathbf{k}$
   $\mathbf{G}(x, y, z) = x\mathbf{i} - y\mathbf{j} + z\mathbf{k}$
32. $\mathbf{F}(x, y, z) = x\mathbf{i} - z\mathbf{k}$
   $\mathbf{G}(x, y, z) = x^2\mathbf{i} + y\mathbf{j} + z^2\mathbf{k}$

In Exercises 33 and 34, given the vector field $\mathbf{F}$, find div (**curl F**) = $\nabla \cdot (\nabla \times \mathbf{F})$.

33. $\mathbf{F}(x, y, z) = xyz\mathbf{i} + y\mathbf{j} + z\mathbf{k}$
34. $\mathbf{F}(x, y, z) = x^2z\mathbf{i} - 2xz\mathbf{j} + yz\mathbf{k}$

In Exercises 35 and 36, prove the given property for vector fields $\mathbf{F}$ and $\mathbf{G}$.

35. div ($\mathbf{F} + \mathbf{G}$) = div $\mathbf{F}$ + div $\mathbf{G}$
36. div (**curl F**) = 0

17.7
Stokes's Theorem

George Gabriel Stokes

A second higher-dimension analogue of Green's Theorem is called **Stokes's Theorem,** after the English mathematical physicist George Gabriel Stokes (1819–1903). Stokes was part of a group of English mathematical physicists referred to as the Cambridge School, which included William Thompson (Lord Kelvin) and James Clerk Maxwell. In addition to making contributions to physics, Stokes worked with infinite series and differential equations, as well as with the integration results presented in this section.

Stokes's Theorem gives the relationship between a surface integral over an oriented surface S and a line integral along a closed space curve C forming the boundary of S, as shown in Figure 17.46. The positive direction along C is counterclockwise relative to the normal vector $\mathbf{N}$ to the surface S.

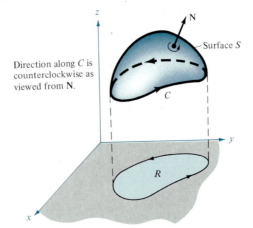

Direction along C is counterclockwise as viewed from $\mathbf{N}$.

FIGURE 17.46

THEOREM 17.11   **STOKES'S THEOREM**
Let S be an oriented surface with unit normal vector $\mathbf{N}$, bounded by a piecewise smooth, simple closed curve C. If $\mathbf{F}$ is a vector field whose component functions have continuous partial derivatives on an open region containing S and C, then

$$\int_C \mathbf{F} \cdot d\mathbf{r} = \iint_S (\textbf{curl F}) \cdot \mathbf{N} \, dS$$

| Remark   The line integral may be expressed in the differential form

$$\int_C M \, dx + N \, dy + P \, dz$$

or in the vector form

$$\int_C \mathbf{F} \cdot \mathbf{T} \, ds$$

EXAMPLE 1   Using Stokes's Theorem

Let C be the oriented triangle lying in the plane $2x + 2y + z = 6$, as shown in Figure 17.47. Evaluate

$$\int_C \mathbf{F} \cdot d\mathbf{r}$$

where $\mathbf{F}(x, y, z) = -y^2\mathbf{i} + z\mathbf{j} + x\mathbf{k}.$

Solution: Using Stokes's Theorem, we begin by finding the curl of $\mathbf{F}$:

$$\mathbf{curl \ F} = \begin{vmatrix} \mathbf{i} & \mathbf{j} & \mathbf{k} \\ \dfrac{\partial}{\partial x} & \dfrac{\partial}{\partial y} & \dfrac{\partial}{\partial z} \\ -y^2 & z & x \end{vmatrix} = -\mathbf{i} - \mathbf{j} + 2y\mathbf{k}$$

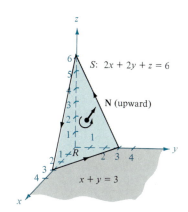

FIGURE 17.47

Considering $z = 6 - 2x - 2y = g(x, y)$ and replacing $\mathbf{F}$ by $\mathbf{curl \ F}$, we can use Theorem 17.9 for an upward normal vector to obtain

$$\iint_S (\mathbf{curl \ F}) \cdot \mathbf{N} \, dS$$

$$= \iint_R (-\mathbf{i} - \mathbf{j} + 2y\mathbf{k}) \cdot [-g_x(x, y) \, \mathbf{i} - g_y(x, y) \, \mathbf{j} + \mathbf{k}] \, dA$$

$$= \iint_R (-\mathbf{i} - \mathbf{j} + 2y\mathbf{k}) \cdot (2\mathbf{i} + 2\mathbf{j} + \mathbf{k}) \, dA$$

$$= \int_0^3 \int_0^{3-y} (2y - 4) \, dx \, dy = \int_0^3 (-2y^2 + 10y - 12) \, dy$$

$$= \left[-\frac{2y^3}{3} + 5y^2 - 12y \right]_0^3 = -9$$

EXAMPLE 2   Verifying Stokes's Theorem

Verify Stokes's Theorem for $\mathbf{F}(x, y, z) = 2z\mathbf{i} + x\mathbf{j} + y^2\mathbf{k}$, where S is the surface of the paraboloid $z = 4 - x^2 - y^2$ and C is the trace of S in the xy-plane, as shown in Figure 17.48.

Solution: As a surface integral, we have $z = g(x, y) = 4 - x^2 - y^2$ and

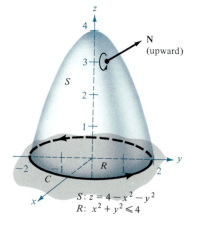

FIGURE 17.48

$$\mathbf{curl \ F} = \begin{vmatrix} \mathbf{i} & \mathbf{j} & \mathbf{k} \\ \dfrac{\partial}{\partial x} & \dfrac{\partial}{\partial y} & \dfrac{\partial}{\partial z} \\ 2z & x & y^2 \end{vmatrix} = 2y\mathbf{i} + 2\mathbf{j} + \mathbf{k}$$

By Theorem 17.9 for an upward normal vector **N**, we obtain

$$\iint_S (\text{curl } \mathbf{F}) \cdot \mathbf{N} \, dS = \iint_R (2y\mathbf{i} + 2\mathbf{j} + \mathbf{k}) \cdot (2x\mathbf{i} + 2y\mathbf{j} + \mathbf{k}) \, dA$$

$$= \int_{-2}^{2} \int_{-\sqrt{4-y^2}}^{\sqrt{4-y^2}} (4xy + 4y + 1) \, dx \, dy$$

$$= \int_{-2}^{2} (8y\sqrt{4 - y^2} + 2\sqrt{4 - y^2}) \, dy$$

$$= \left[-\frac{8}{3}(4 - y^2)^{3/2} + y\sqrt{4 - y^2} + 4 \arcsin \frac{y}{2} \right]_{-2}^{2}$$

$$= 4\pi$$

As a *line integral*, we parameterize C by

$$\mathbf{r}(t) = 2 \cos t \, \mathbf{i} + 2 \sin t \, \mathbf{j} + 0\mathbf{k}, \quad 0 \le t \le 2\pi$$

For $\mathbf{F}(x, y, z) = 2z\mathbf{i} + x\mathbf{j} + y^2\mathbf{k}$, we obtain

$$\int_C \mathbf{F} \cdot d\mathbf{r} = \int_C M \, dx + N \, dy + P \, dz = \int_C 2z \, dx + x \, dy + y^2 \, dz$$

$$= \int_0^{2\pi} [0 + 2 \cos t(2 \cos t) + 0] \, dt$$

$$= 2 \int_0^{2\pi} (1 + \cos 2t) \, dt = 2\left[t + \frac{1}{2} \sin 2t \right]_0^{2\pi} = 4\pi \qquad \square$$

Physical interpretation of curl

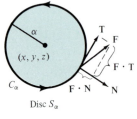

FIGURE 17.49

Stokes's Theorem allows us to gain insight into a physical interpretation of curl. In a vector field **F**, let S_α be a *small* circular disc of radius α, centered at (x, y, z) and with boundary C_α, as shown in Figure 17.49. At each point on the circle C_α, **F** has a normal component $\mathbf{F} \cdot \mathbf{N}$ and a tangential component $\mathbf{F} \cdot \mathbf{T}$. The more closely **F** and **T** are aligned, the greater the value of $\mathbf{F} \cdot \mathbf{T}$, so the fluid tends to move along the circle rather than across it. Consequently, we say that the line integral around C_α measures the **circulation of F around** C_α. That is,

$$\int_{C_\alpha} \mathbf{F} \cdot \mathbf{T} \, ds = \text{circulation of } \mathbf{F} \text{ around } C_\alpha$$

Now consider a small disc S_α to be centered at some point (x, y, z) on the surface S, as shown in Figure 17.50. On such a small disc, **curl F** is nearly constant, since it varies little from its value at (x, y, z). Moreover, **curl F** $\cdot$ **N** is also nearly constant on S_α, since all unit normals to S_α are about the same. Consequently, Stokes's Theorem yields

$$\int_{C_\alpha} \mathbf{F} \cdot \mathbf{T} \, ds = \iint_{S_\alpha} (\text{curl } \mathbf{F}) \cdot \mathbf{N} \, dS$$

$$\approx (\text{curl } \mathbf{F}) \cdot \mathbf{N} \iint_{S_\alpha} dS$$

$$\approx (\text{curl } \mathbf{F}) \cdot \mathbf{N}(\pi\alpha^2)$$

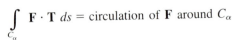

FIGURE 17.50

Therefore,

$$(\textbf{curl } \textbf{F}) \cdot \textbf{N} \approx \frac{\int_{C_\alpha} \textbf{F} \cdot \textbf{T} \, ds}{\pi\alpha^2} = \frac{\text{circulation of } \textbf{F} \text{ around } C_\alpha}{\text{area of disc } S_\alpha}$$

$$= \text{rate of circulation}$$

Assuming conditions are such that the approximation improves over smaller and smaller discs ($\alpha \to 0$), it follows that

$$(\textbf{curl } \textbf{F}) \cdot \textbf{N} = \lim_{\alpha \to 0} \frac{1}{\pi\alpha^2} \int_{C_\alpha} \textbf{F} \cdot \textbf{T} \, ds$$

which is referred to as the **rotation of** F **about** N. That is,

$$\textbf{curl } \textbf{F}(x, y, z) \cdot \textbf{N} = \text{rotation of } \textbf{F} \text{ about } \textbf{N} \text{ at } (x, y, z)$$

In this case, the rotation of **F** is maximum when **F** and **N** have the same direction. Normally, this tendency to rotate will vary from point to point on surface S, and Stokes's Theorem says that the collective measure of this *rotational* tendency taken over the entire surface S (surface integral) is equal to the tendency of the fluid to *circulate* around the boundary C (line integral).

EXAMPLE 3   An application of curl

A liquid is swirling around in a circular container of radius 2, so that its motion is described by the velocity field

$$\textbf{F}(x, y, z) = -y\sqrt{x^2 + y^2}\,\textbf{i} + x\sqrt{x^2 + y^2}\,\textbf{j}$$

as shown in Figure 17.51. Find

$$\iint_S (\textbf{curl } \textbf{F}) \cdot \textbf{N} \, dS$$

where S is the upper surface of the cylindrical tank.

Solution: The curl of **F** is given by

FIGURE 17.51

$$\textbf{curl } \textbf{F} = \begin{vmatrix} \textbf{i} & \textbf{j} & \textbf{k} \\ \dfrac{\partial}{\partial x} & \dfrac{\partial}{\partial y} & \dfrac{\partial}{\partial z} \\ -y\sqrt{x^2 + y^2} & x\sqrt{x^2 + y^2} & 0 \end{vmatrix} = 3\sqrt{x^2 + y^2}\,\textbf{k}$$

Letting $\textbf{N} = \textbf{k}$, we have

$$\iint_S (\textbf{curl } \textbf{F}) \cdot \textbf{N} \, dS = \iint_R 3\sqrt{x^2 + y^2} \, dA$$

$$= \int_0^{2\pi} \int_0^2 (3r)r \, dr \, d\theta$$

$$= \int_0^{2\pi} 8 \, d\theta = 16\pi$$

Remark If **curl F** = **0** throughout region Q, then the rotation of **F** about each unit normal **N** is zero. From earlier work, we know that this is a characteristic of conservative vector fields.

Section Exercises 17.7

In Exercises 1–6, find the curl of the vector field **F**.

1. $\mathbf{F}(x, y, z) = (2y - z)\,\mathbf{i} + xyz\mathbf{j} + e^z\mathbf{k}$
2. $\mathbf{F}(x, y, z) = z^2\mathbf{i} + y^2\mathbf{j} + x^2\mathbf{k}$
3. $\mathbf{F}(x, y, z) = 2z\mathbf{i} - 4x^2\mathbf{j} + \arctan x\ \mathbf{k}$
4. $\mathbf{F}(x, y, z) = x \sin y\ \mathbf{i} - y \cos x\ \mathbf{j} + yz^2\mathbf{k}$
5. $\mathbf{F}(x, y, z) = e^{x^2 + y^2}\mathbf{i} + e^{y^2 + z^2}\mathbf{j} + xyz\mathbf{k}$
6. $\mathbf{F}(x, y, z) = \arcsin y\ \mathbf{i} + \sqrt{1 - x^2}\mathbf{j} + y^2\mathbf{k}$

In Exercises 7–10, verify Stokes's Theorem by evaluating

$$\int_C \mathbf{F} \cdot \mathbf{T}\ dS$$

as a line integral and as a double integral.

7. $\mathbf{F}(x, y, z) = (-y + z)\,\mathbf{i} + (x - z)\,\mathbf{j} + (x - y)\,\mathbf{k}$
   $S: z = \sqrt{1 - x^2 - y^2}$
8. $\mathbf{F}(x, y, z) = (-y + z)\,\mathbf{i} + (x - z)\,\mathbf{j} + (x - y)\,\mathbf{k}$
   $S: z = 4 - x^2 - y^2,\ 0 \le z$
9. $\mathbf{F}(x, y, z) = xyz\mathbf{i} + y\mathbf{j} + z\mathbf{k}$
   $S: 3x + 4y + 2z = 12$ (first octant)
10. $\mathbf{F}(x, y, z) = z^2\mathbf{i} + x^2\mathbf{j} + y^2\mathbf{k}$
    $S: z = x^2,\ 0 \le x \le a,\ 0 \le y \le a$

In Exercises 11–20, use Stokes's Theorem to evaluate

$$\int_C \mathbf{F} \cdot d\mathbf{r}$$

11. $\mathbf{F}(x, y, z) = 2y\mathbf{i} + 3z\mathbf{j} + x\mathbf{k}$
    $C:$ triangle with vertices $(0, 0, 0),\ (0, 2, 0),\ (1, 1, 1)$
12. $\mathbf{F}(x, y, z) = \arctan \dfrac{x}{y}\ \mathbf{i} + \ln \sqrt{x^2 + y^2}\ \mathbf{j} + \mathbf{k}$
    $C:$ triangle with vertices $(0, 0, 0),\ (1, 1, 1),\ (0, 0, 2)$
13. $\mathbf{F}(x, y, z) = z^2\mathbf{i} + x^2\mathbf{j} + y^2\mathbf{k}$
    $S: z = 4 - x^2 - y^2,\ 0 \le z$

14. $\mathbf{F}(x, y, z) = 4xz\mathbf{i} + y\mathbf{j} + 4xy\mathbf{k}$
    $S: z = 4 - x^2 - y^2,\ 0 \le z$
15. $\mathbf{F}(x, y, z) = z^2\mathbf{i} + y\mathbf{j} + xz\mathbf{k}$
    $S: z = \sqrt{4 - x^2 - y^2}$
16. $\mathbf{F}(x, y, z) = x^2\mathbf{i} + z^2\mathbf{j} - xyz\mathbf{k}$
    $S: z = \sqrt{4 - x^2 - y^2}$
17. $\mathbf{F}(x, y, z) = -\ln \sqrt{x^2 + y^2}\mathbf{i} + \arctan \dfrac{x}{y}\mathbf{j} + \mathbf{k}$
    $S: z = 9 - 2x - 3y$ over one petal of $r = 2 \sin 2\theta$ in the first octant
18. $\mathbf{F}(x, y, z) = yz\mathbf{i} + (2 - 3y)\,\mathbf{j} + (x^2 + y^2)\,\mathbf{k}$
    $S:$ the first octant portion of $x^2 + z^2 = 16$ over $x^2 + y^2 = 16$
19. $\mathbf{F}(x, y, z) = xyz\mathbf{i} + y\mathbf{j} + z\mathbf{k}$
    $S: z = x^2,\ 0 \le x \le a,\ 0 \le y \le a$
20. $\mathbf{F}(x, y, z) = xyz\mathbf{i} + y\mathbf{j} + z\mathbf{k}$
    $S:$ the first octant portion of $z = x^2$ over $x^2 + y^2 = a^2$
21. Let f and g be scalar functions with continuous partial derivatives, and let C and S satisfy the conditions of Stokes's Theorem. Show that the following two integrals are equivalent:

$$\int_C f(x, y, z)\nabla g(x, y, z) \cdot d\mathbf{r}$$

and

$$\iint_S [\nabla f(x, y, z) \times \nabla g(x, y, z)] \cdot \mathbf{N}\ dS$$

22. Demonstrate the result of Exercise 21 for the functions $f(x, y, z) = xyz$ and $g(x, y, z) = z$. Let S be the hemisphere $z = \sqrt{4 - x^2 - y^2}$.

Review Exercises for Chapter 17

In Exercises 1 and 2, find a three-dimensional vector field that has the potential function f.

1. $f(x, y, z) = 8x^2 + xy + z^2$
2. $f(x, y, z) = x^2 e^{yz}$

In Exercises 3–10, determine if **F** is conservative. If it is, find the potential function f.

3. $\mathbf{F}(x, y) = \dfrac{1}{y}\mathbf{i} - \dfrac{y}{x^2}\mathbf{j}$ 4. $\mathbf{F}(x, y) = -\dfrac{y}{x^2}\mathbf{i} + \dfrac{1}{x}\mathbf{j}$

5. $\mathbf{F}(x, y) = (6xy^2 - 3x^2)\,\mathbf{i} + (6x^2y + 3y^2 - 7)\,\mathbf{j}$
6. $\mathbf{F}(x, y) = (-2y^3 \sin 2x)\,\mathbf{i} + 3y^2(1 + \cos 2x)\,\mathbf{j}$
7. $\mathbf{F}(x, y, z) = (4xy + z)\,\mathbf{i} + (2x^2 + 6y)\,\mathbf{j} + 2z\mathbf{k}$
8. $\mathbf{F}(x, y, z) = (4xy + z^2)\,\mathbf{i} + (2x^2 + 6yz)\,\mathbf{j} + 2xz\mathbf{k}$
9. $\mathbf{F}(x, y, z) = \dfrac{yz\mathbf{i} - xz\mathbf{j} - xy\mathbf{k}}{y^2 z^2}$
10. $\mathbf{F}(x, y, z) = \sin z\ (y\mathbf{i} + x\mathbf{j} + \mathbf{k})$

In Exercises 11–18, find (a) the divergence and (b) the curl of the vector field **F**.

11. $\mathbf{F}(x, y, z) = x^2\mathbf{i} + y^2\mathbf{j} + z^2\mathbf{k}$
12. $\mathbf{F}(x, y, z) = xy^2\mathbf{j} - zx^2\mathbf{k}$
13. $\mathbf{F}(x, y, z) = (\cos y + y \cos x)\,\mathbf{i}$
$\qquad + (\sin x - x \sin y)\,\mathbf{j} + xyz\mathbf{k}$
14. $\mathbf{F}(x, y, z) = (3x - y)\,\mathbf{i} + (y - 2z)\,\mathbf{j} + (z - 3x)\,\mathbf{k}$
15. $\mathbf{F}(x, y, z) = \arcsin x\,\mathbf{i} + xy^2\mathbf{j} + yz^2\mathbf{k}$
16. $\mathbf{F}(x, y, z) = (x^2 - y)\,\mathbf{i} - (x + \sin^2 y)\,\mathbf{j}$
17. $\mathbf{F}(x, y, z) = \ln(x^2 + y^2)\,\mathbf{i} + \ln(x^2 + y^2)\,\mathbf{j} + z\mathbf{k}$
18. $\mathbf{F}(x, y, z) = \dfrac{z}{x}\mathbf{i} + \dfrac{z}{y}\mathbf{j} + z^2\mathbf{k}$

19. Evaluate

$$\int_C (x^2 + y^2)\,ds$$

for

(a) C: line segment from $(-1, -1)$ to $(2, 2)$
(b) C: $x^2 + y^2 = 16$, one revolution counterclockwise, starting at $(4, 0)$

20. Evaluate

$$\int_C xy\,ds$$

for

(a) C: line segment from $(0, 0)$ to $(5, 4)$
(b) C: counterclockwise around the triangle with vertices $(0, 0)$, $(4, 0)$, $(0, 2)$

21. Evaluate

$$\int_C (2x - y)\,dx + (x + 3y)\,dy$$

for

(a) C: line segment from $(0, 0)$ to $(2, -3)$
(b) C: counterclockwise around the circle $x = 3 \cos t$, $y = 3 \sin t$

22. Evaluate

$$\int_C (2x - y)\,dx + (x + 3y)\,dy$$

for

C: involute given by $x = \cos\theta + \theta \sin\theta$, $y = \sin\theta - \theta \cos\theta$ from $\theta = 0$ to $\theta = \pi/2$

23. Evaluate

$$\int_C (x^2 + y^2)\,ds$$

for

C: $\mathbf{r}(t) = (\cos t + t \sin t)\,\mathbf{i} + (\sin t - t \cos t)\,\mathbf{j}$, $0 \le t \le 2\pi$

24. Evaluate

$$\int_C x\,ds$$

for

C: $\mathbf{r}(t) = (t - \sin t)\,\mathbf{i} + (1 - \cos t)\,\mathbf{j}$, $0 \le t \le 2\pi$

In Exercises 25–30, evaluate

$$\int_C \mathbf{F} \cdot d\mathbf{r}$$

25. $\mathbf{F}(x, y) = xy\mathbf{i} + x^2\mathbf{j}$
$\quad C$: $\mathbf{r}(t) = t^2\mathbf{i} + t^3\mathbf{j}$, $0 \le t \le 1$
26. $\mathbf{F}(x, y) = (x - y)\,\mathbf{i} + (x + y)\,\mathbf{j}$
$\quad C$: $\mathbf{r}(t) = 4 \cos t\,\mathbf{i} + 3 \sin t\,\mathbf{j}$, $0 \le t \le 2\pi$
27. $\mathbf{F}(x, y, z) = x\mathbf{i} + y\mathbf{j} + z\mathbf{k}$
$\quad C$: $\mathbf{r}(t) = 2 \cos t\,\mathbf{i} + 2 \sin t\,\mathbf{j} + t\mathbf{k}$, $0 \le t \le 2\pi$
28. $\mathbf{F}(x, y, z) = (2y - z)\,\mathbf{i} + (z - x)\,\mathbf{j} + (x - y)\,\mathbf{k}$
$\quad C$: curve of intersection of $x^2 + z^2 = 4$ and $y^2 + z^2 = 4$ from $(2, 2, 0)$ to $(0, 0, 2)$
29. $\mathbf{F}(x, y, z) = (y - z)\,\mathbf{i} + (z - x)\,\mathbf{j} + (x - y)\,\mathbf{k}$
$\quad C$: curve of intersection of $z = x^2 + y^2$ and $x + y = 0$ from $(-2, 2, 8)$ to $(2, -2, 8)$
30. $\mathbf{F}(x, y, z) = (x^2 - z)\,\mathbf{i} + (y^2 + z)\,\mathbf{j} + x\mathbf{k}$
$\quad C$: curve of intersection of $z = x^2$ and $x^2 + y^2 = 4$ from $(0, -2, 0)$ to $(0, 2, 0)$

In Exercises 31–36, use Green's Theorem to evaluate each line integral.

31. $\displaystyle\int_C y\,dx + 2x\,dy$

C: boundary of the square with vertices $(0, 0)$, $(0, 2)$, $(2, 0)$, $(2, 2)$

32. $\displaystyle\int_C xy\,dx + (x^2 + y^2)\,dy$

C: boundary of the square with vertices $(0, 0)$, $(0, 2)$, $(2, 0)$, $(2, 2)$

33. $\displaystyle\int_C xy^2\,dx + x^2y\,dy$

C: $x = 4 \cos t$, $y = 2 \sin t$

34. $\displaystyle\int_C (x^2 - y^2)\,dx + 2xy\,dy$

C: $x^2 + y^2 = a^2$

35. $\displaystyle\int_C xy\,dx + x^2\,dy$

C: boundary of the region between the graphs of $y = x^2$ and $y = x$

36. $\displaystyle\int_C y^2\,dx + x^{2/3}\,dy$

C: $x^{2/3} + y^{2/3} = 1$

In Exercises 37 and 38, use the Fundamental Theorem of Line Integrals to evaluate the given integral.

37. $\displaystyle\int_{(0,0,0)}^{(1,4,3)} 2xyz\,dx + x^2z\,dy + x^2y\,dz$

38. $\displaystyle\int_{(0,0,1)}^{(4,4,4)} y\,dx + x\,dy + \dfrac{1}{z}\,dz$

In Exercises 39 and 40, verify the Divergence Theorem by evaluating

$$\int_S \mathbf{F} \cdot \mathbf{N} \, dS$$

as a surface integral and as a triple integral.

39. $\mathbf{F}(x, y, z) = x^2\mathbf{i} + xy\mathbf{j} + z\mathbf{k}$
   Q: solid region bounded by the coordinate planes and the plane $2x + 3y + 4z = 12$

40. $\mathbf{F}(x, y, z) = x\mathbf{i} + y\mathbf{j} + z\mathbf{k}$
   Q: solid region bounded by the coordinate planes and the plane $2x + 3y + 4z = 12$

In Exercises 41 and 42, verify Stokes's Theorem by evaluating

$$\int_C \mathbf{F} \cdot d\mathbf{r}$$

as a line integral and as a double integral.

41. $\mathbf{F}(x, y, z) = (\cos y + y \cos x)\, \mathbf{i}$
                       $+ (\sin x - x \sin y)\, \mathbf{j} + xyz\mathbf{k}$
   S: portion of $z = y^2$ over the square in the xy-plane with vertices $(0, 0)$, $(a, 0)$, (a, a), $(0, a)$

42. $\mathbf{F}(x, y, z) = (x - z)\, \mathbf{i} + (y - z)\, \mathbf{j} + x^2\mathbf{k}$
   S: first octant portion of the plane $3x + y + 2z = 12$

43. Consider the integral

$$\int_C \mathbf{F} \cdot d\mathbf{r}$$

where

$$\mathbf{F}(x, y) = \frac{-y}{x^2 + y^2}\mathbf{i} + \frac{x}{x^2 + y^2}\mathbf{j}$$

and

$$\mathbf{r}(t) = \cos t\, \mathbf{i} + \sin t\, \mathbf{j}$$

Since $\mathbf{F}$ is conservative and C is a circle, it is expected that the line integral will have a value of zero. However, upon direct integration we obtain

$$\int_C \mathbf{F} \cdot d\mathbf{r} = 2\pi$$

Which is correct, and why?

18 *Differential equations*

SECTION TOPICS ▪
Type, order, and degree ▪
General and particular solutions ▪

18.1
Definitions and basic concepts

At several points in the text we have identified physical phenomena that can be described by simple differential equations. For example, we saw that problems involving radioactive decay, population growth, chemical reactions, Newton's Law of Cooling, and gravitational force can all be formulated in terms of differential equations. In this chapter we will see that these are only a few of the many types of problems in physics, chemistry, biology, and engineering that can be solved by the use of equations involving rates of change.

A **differential equation** is an equation involving a function and one or more of its derivatives. If the function has only one independent variable, the equation is called an **ordinary differential equation.** For instance,

$$\frac{d^2y}{dx^2} + 3\,\frac{dy}{dx} - 2y = 0$$

is an ordinary differential equation in which the dependent variable $y = f(x)$ is a twice differentiable function of x. A differential equation involving a function of several variables and its partial derivatives is called a **partial differential equation.** In this chapter we restrict our discussion to ordinary differential equations.

In addition to **type** (ordinary or partial), differential equations are classified by order. The **order** of a differential equation is determined by the highest-order derivative in the equation. Both of these classifications (type and order) are useful in deciding which procedures to use in solving a given differential equation.

EXAMPLE 1   Classifying differential equations

Equation	Type	Order
(a) $y''' + 4y = 2$	Ordinary	3
(b) $\dfrac{d^2s}{dt^2} = -32$	Ordinary	2

(c) $(y')^2 - 3y = e^x$ Ordinary   1

(d) $\dfrac{\partial^2 u}{\partial x^2} + \dfrac{\partial^2 u}{\partial y^2} = 0$ Partial   2

(e) $y - \sin y' = 0$ Ordinary   1

A function $y = f(x)$ is called a **solution** of a differential equation if the equation is satisfied when y and its derivatives are replaced by $f(x)$ and its derivatives, respectively. For example, differentiation and substitution would show that $y = e^{-2x}$, $y = 3e^{-2x}$, and $y = \frac{1}{2}e^{-2x}$ are all solutions to the differential equation

$$y' + 2y = 0$$

The solution $y = Ce^{-2x}$, where C is any real number, is called the **general solution.**

In Section 5.1, Example 8, we saw that the second-order differential equation $s''(t) = -32$ has the general solution

$$s(t) = -16t^2 + C_1 t + C_2$$

that contains two arbitrary constants. It can be shown that a differential equation of order n has a general solution with n arbitrary constants.

EXAMPLE 2   *Verifying solutions*

Determine whether or not the given functions are solutions to the differential equation $y'' - y = 0$.

(a) $y = \sin x$ (b) $y = e^{2x}$ (c) $y = 4e^{-x}$ (d) $y = Ce^x$

Solution:

(a) Since $y = \sin x$, $y' = \cos x$, and $y'' = -\sin x$, we have

$$y'' - y = -\sin x - \sin x = -2 \sin x \neq 0$$

Hence, $y = \sin x$ is *not* a solution.

(b) Since $y = e^{2x}$, $y' = 2e^{2x}$, and $y'' = 4e^{2x}$, we have

$$y'' - y = 4e^{2x} - e^{2x} = 3e^{2x} \neq 0$$

Hence, $y = e^{2x}$ is *not* a solution.

(c) Since $y = 4e^{-x}$, $y' = -4e^{-x}$, and $y'' = 4e^{-x}$, we have

$$y'' - y = 4e^{-x} - 4e^{-x} = 0$$

Hence, $y = 4e^{-x}$ *is* a solution.

(d) Since $y = Ce^x$, $y' = Ce^x$, and $y'' = Ce^x$, we have

$$y'' - y = Ce^x - Ce^x = 0$$

Hence, $y = Ce^x$ *is* a solution for any value of C.

We will see that the general solution to the differential equation in Example 2 is $y = C_1 e^x + C_2 e^{-x}$. A **particular solution** of a differential equation is any solution that is obtained by assigning specific values to the con-

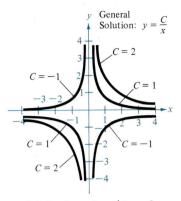

General Solution: $y = \dfrac{C}{x}$

Solution Curves for $xy' + y = 0$

FIGURE 18.1

stants in the general solution. Occasionally a differential equation has other solutions, called **singular solutions.** (Additional information on singular solutions can be found in more advanced texts on differential equations.)

Geometrically, the general solution of a given first-order differential equation represents a family of curves known as **solution curves,** one for each value assigned to the arbitrary constant. For instance, we can easily verify that $y = C/x$ is the general solution to the differential equation $xy' + y = 0$. Figure 18.1 shows some of the solution curves corresponding to different values of C.

In practice, particular solutions to a differential equation are usually obtained from **initial conditions** that give the value of the dependent variable or one of its derivatives for a particular value of the independent variable. The term "initial condition" stems from the fact that often in problems involving time, the value of the dependent variable or one of its derivatives is known at the *initial* time $t = 0$. For instance, the second-order differential equation $s''(t) = -32$ having general solution

$$s(t) = -16t^2 + C_1 t + C_2$$

might have the following initial conditions:

$$s(0) = 80, \qquad s'(0) = 64 \qquad \text{Initial conditions}$$

In this case, the initial conditions yield the particular solution

$$s(t) = -16t^2 + 64t + 80$$

EXAMPLE 3   *Finding a particular solution*

For the differential equation $xy' - 3y = 0$, verify that $y = Cx^3$ is a solution and find the particular solution determined by the initial condition $y = 2$ when $x = -3$.

Solution: We know that $y = Cx^3$ is a solution, because $y' = 3Cx^2$, and

$$xy' - 3y = x(3Cx^2) - 3(Cx^3) = 0$$

Furthermore, the initial condition $y = 2$ when $x = -3$ yields

$$y = Cx^3 \implies 2 = C(-3)^3 \implies C = -\frac{2}{27}$$

so the particular solution is $y = -2x^3/27$.

Remark   In general, for a particular solution to be found, the number of initial conditions must match the number of constants in the general solution.

In the remainder of this chapter we will discuss procedures for finding the general solution to several classes of differential equations. Each section will include practical applications of that particular type of differential equation. We will concentrate mainly on first- and second-order equations and linear equations with constant coefficients. By a linear differential equation with constant coefficients $a_0, a_1, a_2, \ldots, a_n$, we mean an equation of the form

$$a_n y^{(n)} + a_{n-1} y^{(n-1)} + \cdots + a_2 y'' + a_1 y' + a_0 y = F(x)$$

which is linear (of first degree) in y and its derivatives.

Section Exercises 18.1

In Exercises 1–12, classify the differential equation according to type and order.

1. $\dfrac{dy}{dx} + 3xy = x^2$

2. $y'' + 2y' + y = 1$

3. $\dfrac{d^2x}{dt^2} + 2\dfrac{dx}{dt} - 4x = e^t$

4. $\dfrac{d^2u}{dt^2} + \dfrac{du}{dt} = \sec t$

5. $y^{(4)} + 3(y')^2 - 4y = 0$

6. $x^2 y'' + 3xy' = 0$

7. $(y'')^2 + 3y' - 4y = 0$

8. $\dfrac{\partial u}{\partial t} = C^2 \dfrac{\partial^2 u}{\partial x^2}$

9. $\dfrac{\partial u}{\partial t} + \dfrac{\partial u}{\partial y} = 2u$

10. $\dfrac{\partial^2 u}{\partial x \partial y} = \dfrac{\partial u}{\partial y}$

11. $\dfrac{d^2 y}{dx^2} = \sqrt{1 + \left(\dfrac{dy}{dx}\right)^2}$

12. $\sqrt{\dfrac{d^2 y}{dx^2}} = \dfrac{dy}{dx}$

In Exercises 13–18, verify that the given equation represents a solution of the differential equation.

13. $y = Ce^{4x}$, $\dfrac{dy}{dx} = 4y$

14. $x^2 + y^2 = Cy$, $y' = \dfrac{2xy}{x^2 - y^2}$

15. $y = C_1 \cos x + C_2 \sin x$, $y'' + y = 0$

16. $y = C_1 e^{-x} \cos x + C_2 e^{-x} \sin x$, $y'' + 2y' + 2y = 0$

17. $u = e^{-t} \sin bx$, $b^2 \dfrac{\partial u}{\partial t} = \dfrac{\partial^2 u}{\partial x^2}$

18. $u = \dfrac{y}{x^2 + y^2}$, $\dfrac{\partial^2 u}{\partial x^2} + \dfrac{\partial^2 u}{\partial y^2} = 0$

In Exercises 19–24, determine whether the given function is a solution to the differential equation

$$y^{(4)} - 16y = 0$$

19. $y = 3 \cos x$

20. $y = 3 \cos 2x$

21. $y = e^{-2x}$

22. $y = 5 \ln x$

23. $y = C_1 e^{2x} + C_2 e^{-2x} + C_3 \sin 2x + C_4 \cos 2x$

24. $y = 5e^{-2x} + 3 \cos 2x$

In Exercises 25–30, determine if the function is a solution to the differential equation

$$x \dfrac{\partial u}{\partial x} - y \dfrac{\partial u}{\partial y} = 0$$

25. $u = e^{x+y}$

26. $u = 5$

27. $u = x^2 y^2$

28. $u = \sin xy$

29. $u = (xy)^n$

30. $u = x^2 + y^2$

In Exercises 31–36, verify that the general solutions satisfy the differential equation. Then find the particular solution satisfying the given initial condition.

31. $y = Ce^{-2x}$
$y' + 2y = 0$
$y = 3$ when $x = 0$

32. $2x^2 + 3y^2 = C$
$2x + 3yy' = 0$
$y = 2$ when $x = 1$

33. $y = C_1 \sin 3x + C_2 \cos 3x$
$y'' + 9y' = 0$
$y = 2$ and $y' = 1$ when $x = \dfrac{\pi}{6}$

34. $y = C_1 + C_2 \ln x$
$xy'' + y' = 0$
$y = 0$ and $y' = \dfrac{1}{2}$ when $x = 2$

35. $y = C_1 x + C_2 x^3$
$x^2 y'' - 3xy' + 3y = 0$
$y = 0$ and $y' = 4$ when $x = 2$

36. $y = e^{2x/3}(C_1 + C_2 x)$
$9y'' - 12y' + 4y = 0$
$y = 4$ when $x = 0$ and $y = 0$ when $x = 3$

In Exercises 37 and 38, the general solution to the differential equation is given. Sketch the graph of the particular solutions for the given values of C.

37. $4yy' - x = 0$
$2y^2 - x^2 = C$
$C = 0,\ C = \pm1,\ C = \pm4$

38. $yy' + x = 0$
$x^2 + y^2 = C$
$C = 0,\ C = 1,\ C = 4$

In Exercises 39–46, use integration to find a general solution to the differential equation.

39. $\dfrac{dy}{dx} = 3x^2$

40. $\dfrac{dy}{dx} = \dfrac{1}{1 + x^2}$

41. $\dfrac{dy}{dx} = \dfrac{x - 2}{x}$

42. $\dfrac{dy}{dx} = x \cos x$

43. $\dfrac{dy}{dx} = e^x \sin 2x$

44. $\dfrac{dy}{dx} = \tan^2 x$

45. $\dfrac{dy}{dx} = x\sqrt{x - 3}$

46. $\dfrac{dy}{dx} = xe^x$

18.2
Separation of variables in first-order equations

In this section we begin studying techniques for solving specific kinds of ordinary differential equations. We start with a procedure for solving a first-order differential equation that can be written in the form

$$M(x) + N(y) \frac{dy}{dx} = 0$$

where M is a continuous function of x alone and N is a continuous function of y alone. For this type of equation all x terms can be collected with dx and all y terms with dy, and a solution can be obtained by integration. Such equations are said to be **separable,** and the solution procedure is called **separation of variables.** The steps are as follows:

1. Express the given equation in the differential form

$$M(x) \, dx + N(y) \, dy = 0 \qquad \text{or} \qquad M(x) \, dx = -N(y) \, dy$$

2. Integrate to obtain the general solution

$$\int M(x) \, dx + \int N(y) \, dy = C$$

or

$$\int M(x) \, dx = -\int N(y) \, dy + C$$

EXAMPLE 1   *Separation of variables*

Find the general solution to

$$(x^2 + 4) \frac{dy}{dx} = xy$$

Solution: The equation has the differential form

$$(x^2 + 4) \, dy = xy \, dx \qquad \text{Differential form}$$

$$\frac{dy}{y} = \frac{x}{x^2 + 4} \, dx \qquad \text{Separate variables}$$

The general solution is obtained as follows:

$$\int \frac{dy}{y} = \int \frac{x}{x^2 + 4} \, dx \qquad \text{Integrate}$$

$$\ln |y| = \frac{1}{2} \ln (x^2 + 4) + C_1$$

$$= \ln \sqrt{x^2 + 4} + C_1$$

$$|y| = e^{C_1} \sqrt{x^2 + 4}$$

$$y = C\sqrt{x^2 + 4} \qquad \text{General solution}$$

Remark   We encourage you to develop the habit of *checking your solutions* throughout this chapter. For instance, in Example 1 you can check the solution $y = C\sqrt{x^2 + 4}$ by differentiating and substituting into the original equation.

In some cases it is not feasible to write the general solution in the explicit form $y = f(x)$. The next example is a case in point; implicit differentiation can be used to verify this solution.

EXAMPLE 2   *Finding a particular solution by separation of variables*

Given the initial condition $y(0) = 1$, find the particular solution to the equation

$$xy\ dx + e^{-x^2}(y^2 - 1)\ dy = 0$$

Solution: To separate variables we must rid the first term of y and the second of e^{-x^2}. Thus, we multiply by e^{x^2}/y and obtain

$$\left(\frac{e^{x^2}}{y}\right)xy\ dx + \left(\frac{e^{x^2}}{y}\right)e^{-x^2}(y^2 - 1)\ dy = 0$$

$$xe^{x^2}\ dx + \left(y - \frac{1}{y}\right)dy = 0$$

$$\int xe^{x^2}\ dx + \int \left(y - \frac{1}{y}\right)dy = 0$$

$$\frac{1}{2}e^{x^2} + \frac{y^2}{2} - \ln|y| = C_1$$

$$e^{x^2} + y^2 - \ln y^2 = 2C_1 = C$$

From the given initial condition, we use the fact that $y = 1$ when $x = 0$ to obtain $1 + 1 + 0 = 2 = C$. Thus, the particular solution has the implicit form

$$e^{x^2} + y^2 - \ln y^2 = 2$$

EXAMPLE 3   *Finding a particular solution curve*

Find the equation of the curve that passes through the point $(1, 3)$ and has a slope of y/x^2, as shown in Figure 18.2.

Solution: Since the slope of the curve is given by y/x^2, we have

$$\frac{dy}{dx} = \frac{y}{x^2}$$

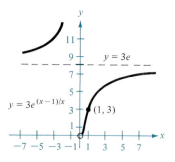

$y = 3e$

$y = 3e^{(x-1)/x}$

$(1, 3)$

FIGURE 18.2

with the initial condition $y(1) = 3$. Separating variables and integrating, we have

$$\int \frac{dy}{y} = \int \frac{dx}{x^2}$$

$$\ln|y| = -\frac{1}{x} + C_1$$

$$y = e^{-(1/x)+C_1} = Ce^{-1/x}$$

Since $y = 3$ when $x = 1$, it follows that $3 = Ce^{-1}$ and $C = 3e$. Therefore, the equation of the specified curve is

$$y = (3e)e^{-1/x} = 3e^{(x-1)/x}$$

Many first-order differential equations are separable. Moreover, some that aren't can be made separable by a change of variables. This is true for differential equations of the form $y' = f(x, y)$, where f is a **homogeneous function.**

DEFINITION OF HOMOGENEOUS FUNCTION

The function given by $z = f(x, y)$ is said to be **homogeneous of degree n** if

$$f(tx, ty) = t^n f(x, y)$$

where n is a real number.

EXAMPLE 4   Verifying homogeneous functions

(a) $f(x, y) = x^2 y - 4x^3 + 3xy^2$ is a homogeneous function of degree 3, since

$$\begin{aligned} f(tx, ty) &= (tx)^2(ty) - 4(tx)^3 + 3(tx)(ty)^2 \\ &= t^3(x^2 y) - t^3(4x^3) + t^3(3xy^2) \\ &= t^3(x^2 y - 4x^3 + 3xy^2) \\ &= t^3 f(x, y) \end{aligned}$$

(b) $f(x, y) = xe^{x/y} + y \sin(y/x)$ is a homogeneous function of degree 1, since

$$f(tx, ty) = txe^{tx/ty} + ty \sin \frac{ty}{tx}$$

$$= t\left(xe^{x/y} + y \sin \frac{y}{x} \right)$$

$$= tf(x, y)$$

(c) $f(x, y) = x + y^2$ is not a homogeneous function, since

$$f(tx, ty) = tx + t^2 y^2 = t(x + ty^2) \neq t^n(x + y^2)$$

(d) $f(x, y) = x/y$ is homogeneous of degree zero, since

$$f(tx, ty) = \frac{tx}{ty}$$

$$= t^0 \frac{x}{y}$$

DEFINITION OF HOMOGENEOUS DIFFERENTIAL EQUATION

A **homogeneous differential equation** is an equation of the form

$$M(x, y)\, dx + N(x, y)\, dy = 0$$

where M and N are homogeneous functions of the same degree.

To solve a homogeneous differential equation by the method of separation of variables, we use the following change of variables theorem.

THEOREM 18.1   CHANGE OF VARIABLES FOR HOMOGENEOUS EQUATIONS
If $M(x, y) \, dx + N(x, y) \, dy = 0$ is homogeneous, then it can be transformed into an equation whose variables can be separated by the substitution

$$y = vx$$

where v is a differentiable function of x.

Proof: Let $y = vx$. Then $dy = v \, dx + x \, dv$, and, by substitution,

$$M(x, y) \, dx + N(x, y) \, dy = M(x, vx) \, dx + N(x, vx)(v \, dx + x \, dv) = 0$$

Since M and N are homogeneous of degree n, it follows that

$$x^n M(1, v) \, dx + x^n N(1, v)(v \, dx + x \, dv) = 0$$
$$M(1, v) \, dx + N(1, v)v \, dx = -N(1, v)x \, dv$$
$$[M(1, v) + vN(1, v)] \, dx = -N(1, v)x \, dv$$
$$\frac{dx}{x} + \frac{N(1, v)}{M(1, v) + vN(1, v)} \, dv = 0$$

provided no denominators are zero. Moreover, if v is a solution to this separable differential equation, we can reverse the steps to show that $y = vx$ is a solution to the given homogeneous differential equation.

EXAMPLE 5   Solving a homogeneous differential equation

Find the general solution to the differential equation

$$(x^2 - y^2) \, dx + 3xy \, dy = 0$$

Solution: Since $(x^2 - y^2)$ and $3xy$ are both homgeneous of degree 2, we let $y = vx$ to obtain $dy = x \, dv + v \, dx$. Then by substitution we have

$$(x^2 - v^2x^2) \, dx + 3x(vx)\overbrace{(x \, dv + v \, dx)}^{dy} = 0$$
$$(x^2 + 2v^2x^2) \, dx + 3x^3v \, dv = 0$$
$$x^2(1 + 2v^2) \, dx + x^2(3vx) \, dv = 0$$

Dividing by x^2 and separating variables, we have

$$(1 + 2v^2) \, dx = -3vx \, dv$$
$$\int \frac{dx}{x} = \int \frac{-3v}{1 + 2v^2} \, dv$$
$$\ln |x| = -\frac{3}{4} \ln (1 + 2v^2) + C_1$$

$$4 \ln |x| = -3 \ln (1 + 2v^2) + \ln |C|$$
$$\ln x^4 = \ln |C(1 + 2v^2)^{-3}|$$
$$x^4 = C(1 + 2v^2)^{-3}$$

Substituting for v, we obtain

$$\left(1 + 2\frac{y^2}{x^2}\right)^3 x^4 = C$$
$$(x^2 + 2y^2)^3 = Cx^2$$

Applications

In the remainder of this section we look at two applications involving differential equations. We begin with one involving exponential growth.

EXAMPLE 6   *Application to food preservation*

In the preservation of food, cane sugar is broken down (inverted) into two simpler sugars: glucose and fructose. In dilute solutions, the inversion rate is proportional to the concentration $y(t)$ of unaltered sugar. If the concentration is $\frac{1}{50}$ when $t = 0$ and $\frac{1}{200}$ after 3 hours, find the concentration of unaltered sugar after 6 hours and after 12 hours.

Solution: Since the rate of inversion is proportional to $y(t)$, we have the differential equation

$$\frac{dy}{dt} = ky$$

Separating the variables and integrating, we get

$$\int \frac{1}{y}\, dy = \int k\, dt$$
$$\ln |y| = kt + C_1$$
$$y = Ce^{kt}$$

From the given boundary conditions, we obtain

$$y(0) = \frac{1}{50} \implies C = \frac{1}{50}$$

$$y(3) = \frac{1}{200} \implies \frac{1}{200} = \left(\frac{1}{50}\right)e^{3k} \implies k = -\frac{\ln 4}{3}$$

Therefore, the concentration of unaltered sugar is given by

$$y(t) = \frac{1}{50} e^{-(\ln 4)t/3} = \frac{1}{50}(4^{-t/3})$$

When $t = 6$ and $t = 12$, we have

$$y(6) = \frac{1}{50}(4^{-2}) = \frac{1}{800} \qquad \text{After 6 hours}$$

$$y(12) = \frac{1}{50}(4^{-4}) = \frac{1}{12{,}800} \qquad \text{After 12 hours}$$

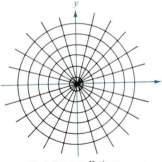

Each line $y = Kx$ is an orthogonal trajectory to the family of circles

FIGURE 18.3

A common problem in electrostatics, thermodynamics, and hydrodynamics involves finding a family of curves, each of which is orthogonal to all members of a given family of curves. For example, Figure 18.3 shows a family of circles

$$x^2 + y^2 = C \qquad \text{Family of circles}$$

each of which intersects the lines in the family

$$y = Kx \qquad \text{Family of lines}$$

at right angles. Two such families of curves are said to be **mutually orthogonal,** and each curve in one of the families is called an **orthogonal trajectory** of the other family. In electrostatics, lines of force are orthogonal to the *equipotential curves.* Similarly, in thermodynamics, the flow of heat across a plane surface is orthogonal to the *isothermal curves.* And in hydrodynamics, the flow (stream) lines are orthogonal trajectories of the *velocity potential curves.*

EXAMPLE 7   *Finding orthogonal trajectories*

Describe the orthogonal trajectories for the family of curves given by $y = C/x$ for $C \neq 0$. Sketch several members of each family.

Solution: First, we solve the given equation for C and write $xy = C$. Then, differentiating implicitly with respect to x, we obtain the differential equation

$$xy' + y = 0 \implies \frac{dy}{dx} = -\frac{y}{x} \qquad \text{Slope of given family}$$

Since y' represents the slope of the given family of curves at (x, y), it follows that the orthogonal family has the negative reciprocal slope, x/y, and we write

$$\frac{dy}{dx} = \frac{x}{y} \qquad \text{Slope of orthogonal family}$$

Now we can find the orthogonal family by separating variables and integrating.

$$\int y \, dy = \int x \, dx$$

$$\frac{y^2}{2} = \frac{x^2}{2} + C_1$$

Therefore, each orthogonal trajectory is a hyperbola given by

$$\frac{y^2}{K} - \frac{x^2}{K} = 1$$

$$2C_1 = K \neq 0$$

The centers are at the origin, and the transverse axes are vertical for $K > 0$ and horizontal for $K < 0$. Several trajectories are shown in Figure 18.4.

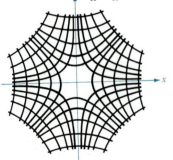

Given family: $xy = C$

Orthogonal family: $\dfrac{y^2}{K} - \dfrac{x^2}{K} = 1$

FIGURE 18.4

Section Exercises 18.2

In Exercises 1–6, find the general solution of the given differential equation.

1. $\dfrac{dy}{dx} = \dfrac{x}{y}$

2. $\dfrac{dy}{dx} = \dfrac{x^2 + 2}{3y^2}$

3. $(2 + x)y' = 3y$

4. $xy' = y$

5. $yy' = \sin x$

6. $\sqrt{1 - 4x^2}\,y' = 1$

In Exercises 7–14, find the particular solution of the differential equation that satisfies the given initial condition.

Differential equation	Initial condition
7. $yy' - e^x = 0$	$y(0) = 4$
8. $\sqrt{x} + \sqrt{y}\,y' = 0$	$y(1) = 4$
9. $y(x + 1) + y' = 0$	$y(-2) = 1$
10. $xyy' - \ln x = 0$	$y(1) = 0$
11. $(1 + x^2)y' - (1 + y^2) = 0$	$y(0) = \sqrt{3}$
12. $\sqrt{1 - x^2}\,y' - \sqrt{1 - y^2} = 0$	$y(0) = 1$
13. $dP - kP\,dt = 0$	$P(0) = P_0$
14. $dT + k(T - 70)\,dt = 0$	$T(0) = 140$

In Exercises 15–20, determine if the function is homogeneous, and if so, determine the degree.

15. $f(x, y) = x^3 - 4xy^2 + y^3$

16. $f(x, y) = \dfrac{xy}{\sqrt{x^2 + y^2}}$

17. $f(x, y) = 2 \ln xy$

18. $f(x, y) = \tan(x + y)$

19. $f(x, y) = 2 \ln \dfrac{x}{y}$

20. $f(x, y) = \tan \dfrac{y}{x}$

In Exercises 21–26, solve the homogeneous differential equation.

21. $y' = \dfrac{x + y}{2x}$

22. $y' = \dfrac{2x + y}{y}$

23. $y' = \dfrac{x - y}{x + y}$

24. $y' = \dfrac{x^2 + y^2}{2xy}$

25. $y' = \dfrac{xy}{x^2 - y^2}$

26. $y' = \dfrac{3x + 2y}{x}$

In Exercises 27–30, find the particular solution that satisfies the given initial condition.

Differential equation	Initial condition
27. $x\,dy - (2xe^{-y/x} + y)\,dx = 0$	$y(1) = 0$
28. $-y^2\,dx + x(x + y)\,dy = 0$	$y(1) = 1$

29. $\left(x \sec \dfrac{y}{x} + y\right) dx - x\,dy = 0 \qquad y(1) = 0$

30. $(y - \sqrt{x^2 - y^2})\,dx - x\,dy = 0 \qquad y(1) = 0$

In Exercises 31–36, find the orthogonal trajectories of the given family and sketch several members of each family.

31. $x^2 + y^2 = C$

32. $2x^2 - y^2 = C$

33. $x^2 = Cy$

34. $y^2 = 2Cx$

35. $y^2 = Cx^3$

36. $y = Ce^x$

In Exercises 37 and 38, find an equation for the curve that passes through the given point and has the specified slope.

Point	Slope
37. $(1, 1)$	$y' = -\dfrac{9x}{16y}$
38. $(8, 2)$	$y' = \dfrac{2y}{3x}$

39. The amount A of an investment P increases at a rate proportional to A at any instant of time t.
   (a) Find an equation for the amount A as a function of t.
   (b) If the initial investment is $1000.00 and the rate is 11 percent, find the amount after 10 years.
   (c) Find the time necessary to double the investment if the rate is 11 percent.

40. The rate of growth of a population of fruit flies is proportional to the size of the population at any instant. If there were 180 flies after the second day of the experiment and 300 after the fourth day, how many flies were there in the original population?

41. The rate of decomposition of radioactive radium is proportional to the amount present at a given instant. Find the percentage that remains of a present amount after 25 years, if the half-life of radioactive radium is 1600 years.

42. In a chemical reaction, a certain compound changes into another compound at a rate proportional to the unchanged amount. If initially there was 20 grams of the original compound, and there is 16 grams after 1 hour, when will 75 percent of the compound be changed?

43. Newton's Law of Cooling states that the rate of change in the temperature of an object is proportional to the difference between its temperature and the temperature of the surrounding air. Suppose a room is kept at a constant temperature of 70°, and an object cooled from 350° to 150° in 45 minutes. At what time will the object cool to a temperature of 80°?

44. If s gives the position of an object moving in a straight line, the velocity and acceleration are given by

$$v = \frac{ds}{dt}, \qquad a = \frac{dv}{dt} = \frac{ds}{dt}\frac{dv}{ds} = v\frac{dv}{ds}$$

Suppose that a boat weighing 640 pounds has a motor that exerts a force of 60 pounds. If the resistance (in pounds) to the motion of the boat is 3 times the velocity (in feet per second) and if the boat starts from rest, find its maximum speed. Use the equation

$$\text{force} = F = ma = mv\frac{dv}{ds} = \frac{w}{32}v\frac{dv}{ds}$$

where w is the weight of the boat.

18.3
Exact first-order equations

In this section we introduce a method for solving the first-order differential equation

$$M(x, y)\, dx + N(x, y)\, dy = 0$$

for the special case in which this equation represents the exact differential of a function $z = f(x, y)$.

DEFINITION OF AN EXACT DIFFERENTIAL EQUATION

The equation

$$M(x, y)\, dx + N(x, y)\, dy = 0$$

is an **exact differential equation** if there exists a function f of two variables x and y having continuous partial derivatives such that

$$f_x(x, y) = M(x, y) \qquad \text{and} \qquad f_y(x, y) = N(x, y)$$

The general solution to the equation is $f(x, y) = C$.

From Section 15.3, we know that if f has continuous second partials, then

$$\frac{\partial M}{\partial y} = \frac{\partial^2 f}{\partial y \partial x} = \frac{\partial^2 f}{\partial x \partial y} = \frac{\partial N}{\partial x}$$

This suggests the following test for exactness.

THEOREM 18.2

TEST FOR EXACTNESS
If M and N have continuous partial derivatives on an open disc R, then the differential equation $M(x, y)\, dx + N(x, y)\, dy = 0$ is exact if and only if

$$\frac{\partial M}{\partial y} = \frac{\partial N}{\partial x}$$

Exactness is a fragile condition in the sense that seemingly minor altera-
tions in an exact equation can destroy its exactness. This is demonstrated in
the following example.

EXAMPLE 1   *Testing for exactness*

(a) The differential equation

$$(xy^2 + x)\, dx + yx^2\, dy = 0$$

is exact, since

$$\frac{\partial}{\partial y}[xy^2 + x] = 2xy = \frac{\partial}{\partial x}[yx^2]$$

But the equation $(y^2 + 1)\, dx + xy\, dy = 0$ is not exact, even though it is
obtained by dividing both sides of the first equation by x.

(b) The differential equation

$$\cos y\, dx + (y^2 - x \sin y)\, dy = 0$$

is exact, since

$$\frac{\partial}{\partial y}[\cos y] = -\sin y = \frac{\partial}{\partial x}[y^2 - x \sin y]$$

But the equation $\cos y\, dx + (y^2 + x \sin y)\, dy = 0$ is not exact, even
though it differs from the first equation only by a single sign.   ☐

| **Remark** Every differential equation of the form $M(x)\, dx + N(y)\, dy = 0$ is exact. In
other words, a separable variables equation is actually a special type of exact equation.

Note that the test for exactness of $M(x, y)\, dx + N(x, y)\, dy = 0$ is the
same as the test for determining if $\mathbf{F}(x, y) = M(x, y)\, \mathbf{i} + N(x, y)\, \mathbf{j}$ is the gradient of a potential function (Theorem 17.1). This means that a general solu-
tion $f(x, y) = C$ to an exact differential equation can be found by the method
used to find a potential function for a conservative vector field. We further
demonstrate this procedure in the next two examples.

EXAMPLE 2   *Solving an exact differential equation*

Show that the differential equation

$$(2xy - 3x^2)\, dx + (x^2 - 2y)\, dy = 0$$

is exact, and find its general solution.

Solution: The given differential equation is exact, since

$$\frac{\partial}{\partial y}[2xy - 3x^2] = 2x = \frac{\partial}{\partial x}[x^2 - 2y]$$

We can obtain the general solution $f(x, y) = C$ as follows:

$$f(x, y) = \int M(x, y)\, dx = \int (2xy - 3x^2)\, dx = x^2y - x^3 + g(y)$$

In Section 17.1 we determined $g(y)$ by integrating $N(x, y)$ with respect to y and reconciling the two expressions for $f(x, y)$. A nice alternative method is to partially differentiate this version of $f(x, y)$ with respect to y and compare the result to $N(x, y)$. In other words,

$$f_y(x, y) = \frac{\partial}{\partial y}[x^2y - x^3 + g(y)] = x^2 + g'(y) = \overbrace{x^2 - 2y}^{N(x,\ y)}$$

$$g'(y) = -2y$$

Thus, $g'(y) = -2y$, and it follows that $g(y) = -y^2 + C_1$. Therefore,

$$f(x, y) = x^2y - x^3 - y^2 + C_1$$

and the general solution is

$$x^2y - x^3 - y^2 = C$$

EXAMPLE 3   *Solving an exact differential equation*

Find the particular solution of

$$(\cos x - x \sin x + y^2)\, dx + 2xy\, dy = 0$$

that satisfies the boundary condition $y = 1$ when $x = \pi$.

Solution: The given equation is exact, since

$$\frac{\partial}{\partial y}[\cos x - x \sin x + y^2] = 2y = \frac{\partial}{\partial x}[2xy]$$

In this case, $N(x, y)$ is simpler than $M(x, y)$, and we proceed as follows:

$$f(x, y) = \int N(x, y)\, dy = \int 2xy\, dy = xy^2 + g(x)$$

$$f_x(x, y) = \frac{\partial}{\partial x}[xy^2 + g(x)] = y^2 + g'(x) = \overbrace{\cos x - x \sin x + y^2}^{M(x,\ y)}$$

$$g'(x) = \cos x - x \sin x$$

Thus, $g'(x) = \cos x - x \sin x$ and

$$g(x) = \int (\cos x - x \sin x)\, dx = x \cos x + C_1$$

which implies $f(x, y) = xy^2 + x \cos x + C_1$ and the general solution is

$$xy^2 + x \cos x = C$$

Applying the given boundary condition, we have $\pi(1)^2 + \pi \cos \pi = C$, which implies that $C = 0$. Hence, the particular solution is

$$xy^2 + x \cos x = 0$$

| **Remark** In Example 3, note that if $z = f(x, y) = xy^2 + x \cos x$, then the total differential of z is given by

$$dz = f_x(x, y)\, dx + f_y(x, y)\, dy = (\cos x - x \sin x + y^2)\, dx + 2xy\, dy$$
$$= M(x, y)\, dx + N(x, y)\, dy$$

In other words, we call $M\, dx + N\, dy = 0$ on *exact* differential equation because $M\, dx + N\, dy$ is exactly the differential of $f(x, y)$.

Integrating factors

If the differential equation $M(x, y)\, dx + N(x, y)\, dy = 0$ is not exact, it may be possible to make it exact by multiplying by an appropriate factor $u(x, y)$, called an **integrating factor** for the differential equation. For instance, if the differential equation

$$2y\, dx + x\, dy = 0 \qquad \text{Not an exact equation}$$

is multiplied by the integrating factor $u(x, y) = x$, the resulting equation

$$2xy\, dx + x^2\, dy = 0 \qquad \text{Exact equation}$$

is exact, and the left side is the total differential of x^2y. Similarly, if the equation

$$y\, dx - x\, dy = 0 \qquad \text{Not an exact equation}$$

is multiplied by the integrating factor $u(x, y) = 1/y^2$, the resulting equation

$$\frac{1}{y}\, dx - \frac{x}{y^2}\, dy = 0 \qquad \text{Exact equation}$$

is exact.

Finding an integrating factor can be a very difficult problem. However, there are two classes of differential equations whose integrating factors can be found in a routine way—namely, those that possess integrating factors that are functions of either x alone or y alone. The following theorem, which we list without proof, outlines a procedure for finding these two special categories of integrating factors.

THEOREM 18.3 **INTEGRATING FACTORS**
For the differential equation $M(x, y)\, dx + N(x, y)\, dy = 0$:

1. If
$$\frac{1}{N(x, y)}[M_y(x, y) - N_x(x, y)] = h(x)$$

   is a function of x alone, then $e^{\int h(x)\, dx}$ is an integrating factor.

2. If
$$\frac{1}{M(x, y)}[N_x(x, y) - M_y(x, y)] = k(y)$$

   is a function of y alone, then $e^{\int k(y)\, dy}$ is an integrating factor.

| Remark If either $h(x)$ or $k(y)$ is constant, Theorem 18.3 still applies. As an aid to remembering the formulas, note that the subtracted partial derivative identifies both the denominator and the variable for the integrating factor.

EXAMPLE 4   Finding an integrating factor

Find the general solution to the differential equation

$$(y^2 - x) \, dx + 2y \, dy = 0$$

Solution: The given equation is not exact, since $M_y(x, y) = 2y$ and $N_x(x, y) = 0$. However, since

$$\frac{M_y(x, y) - N_x(x, y)}{N(x, y)} = \frac{2y - 0}{2y} = 1 = h(x)$$

it follows that $e^{\int h(x) \, dx} = e^{\int dx} = e^x$ is an integrating factor. Multiplying the given differential equation by e^x, we obtain the exact equation

$$(y^2 e^x - x e^x) \, dx + 2y e^x \, dy = 0$$

whose solution is obtained as follows:

$$f(x, y) = \int N(x, y) \, dy = \int 2y e^x \, dy = y^2 e^x + g(x)$$

$$f_x(x, y) = y^2 e^x + g'(x) = \overbrace{y^2 e^x - x e^x}^{M(x,\ y)}$$

$$g'(x) = -x e^x$$

Therefore, $g'(x) = -x e^x$ and $g(x) = -x e^x + e^x + C_1$, which implies that

$$f(x, y) = y^2 e^x - x e^x + e^x + C_1$$

so the general solution is

$$y^2 e^x - x e^x + e^x = C \implies y^2 - x + 1 = C e^{-x}$$

In the next example we show how a differential equation can serve as an aid to sketching a force field $\mathbf{F}(x, y) = M(x, y) \, \mathbf{i} + N(x, y) \, \mathbf{j}$.

EXAMPLE 5   An application to force fields

Sketch the force field given by

$$\mathbf{F}(x, y) = \frac{2y}{\sqrt{x^2 + y^2}} \mathbf{i} - \frac{y^2 - x}{\sqrt{x^2 + y^2}} \mathbf{j}$$

by finding and sketching the family of curves tangent to $\mathbf{F}$.

Solution: At the point (x, y) in the plane, the vector $\mathbf{F}(x, y)$ has a slope of

$$\frac{dy}{dx} = \frac{-(y^2 - x)/\sqrt{x^2 + y^2}}{2y/\sqrt{x^2 + y^2}} = \frac{-(y^2 - x)}{2y}$$

Force field:

$$\mathbf{F}(x, y) = \frac{2y}{\sqrt{x^2 + y^2}} \mathbf{i} - \frac{y^2 - x}{\sqrt{x^2 + y^2}} \mathbf{j}$$

Family of tangent curves to $\mathbf{F}$:
$$y^2 = x - 1 + C e^{-x}$$

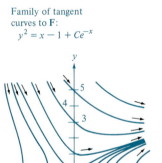

FIGURE 18.5

which in differential form is

$$2y \, dy = -(y^2 - x) \, dx$$
$$(y^2 - x) \, dx + 2y \, dy = 0$$

From Example 4 we know that the general solution to this differential equation is $y^2 - x + 1 = Ce^{-x}$, or

$$y^2 = x - 1 + Ce^{-x}$$

Figure 18.5 on page 1002 shows several representative curves from this family. Note that the force vector at (x, y) is tangent to the curve passing through (x, y).

Section Exercises 18.3

In Exercises 1–10, determine if the differential equation is exact, and if so, find the general solution.

1. $(2x - 3y) \, dx + (2y - 3x) \, dy = 0$
2. $ye^x \, dx + e^x \, dy = 0$
3. $(3y^2 + 10xy^2) \, dx + (6xy - 2 + 10x^2y) \, dy = 0$
4. $2 \cos (2x - y) \, dx - \cos (2x - y) \, dy = 0$
5. $(4x^3 - 6xy^2) \, dx + (4y^3 - 6xy) \, dy = 0$
6. $2y^2e^{xy^2} \, dx + 2xye^{xy^2} \, dy = 0$
7. $\dfrac{1}{x^2 + y^2}(x \, dy - y \, dx) = 0$
8. $e^{-(x^2+y^2)}(x \, dx + y \, dy) = 0$
9. $\dfrac{1}{(x - y)^2}(y^2 \, dx + x^2 \, dy) = 0$
10. $e^y \cos xy \, [y \, dx + (x + \tan xy) \, dy] = 0$

In Exercises 11–16, find the particular solution that satisfies the given initial condition.

11. $\dfrac{y}{x - 1} \, dx + [\ln (x - 1) + 2y] \, dy = 0$
    $y(2) = 4$
12. $\dfrac{1}{\sqrt{x^2 + y^2}}(x \, dx + y \, dy) = 0$
    $y(4) = 3$
13. $\dfrac{1}{x^2 + y^2}(x \, dx + y \, dy) = 0$
    $y(0) = 4$
14. $e^{3x}(\sin 3y \, dx + \cos 3y \, dy) = 0$
    $y(0) = \pi$
15. $(2x \tan y + 5) \, dx + (x^2 \sec^2 y) \, dy = 0$
    $y(0) = 0$
16. $(x^2 + y^2) \, dx + 2xy \, dy = 0$
    $y(3) = 1$

In Exercises 17–26, find the integrating factor that is a function of x or y alone and use it to find the general solution of the differential equation.

17. $y \, dx - (x + 6y^2) \, dy = 0$
18. $(2x^3 + y) \, dx - x \, dy = 0$
19. $(5x^2 - y) \, dx + x \, dy = 0$
20. $(5x^2 - y^2) \, dx + 2y \, dy = 0$
21. $(x + y) \, dx + \tan x \, dy = 0$
22. $(2x^2y - 1) \, dx + x^3 \, dy = 0$
23. $y^2 \, dx + (xy - 1) \, dy = 0$
24. $(x^2 + 2x + y) \, dx + 2 \, dy = 0$
25. $2y \, dx + (x - \sin \sqrt{y}) \, dy = 0$
26. $(-2y^3 + 1) \, dx + (3xy^2 + x^3) \, dy = 0$

In Exercises 27–30, use the given integrating factor to find the general solution of the differential equation.

27. $(4x^2y + 2y^2) \, dx + (3x^3 + 4xy) \, dy = 0$,
    $u(x, y) = xy^2$
28. $(3y^2 + 5x^2y) \, dx + (3xy + 2x^3) \, dy = 0$,
    $u(x, y) = x^2y$
29. $(-y^5 + x^2y) \, dx + (2xy^4 - 2x^3) \, dy = 0$,
    $u(x, y) = x^{-2}y^{-3}$
30. $-y^3 \, dx + (xy^2 - x^2) \, dy = 0$,
    $u(x, y) = x^{-2}y^{-2}$

31. Show that each of the following are integrating factors for the differential equation

    $$y \, dx - x \, dy = 0$$

    (a) $\dfrac{1}{x^2}$ (b) $\dfrac{1}{y^2}$

    (c) $\dfrac{1}{xy}$ (d) $\dfrac{1}{x^2 + y^2}$

32. Show that the differential equation

$$(axy^2 + by)\, dx + (bx^2y + ax)\, dy = 0$$

is exact only if $a = b$. If $a \ne b$, show that $x^m y^n$ is an integrating factor, where

$$m = -\frac{2b + a}{a + b}, \qquad n = -\frac{2a + b}{a + b}$$

In Exercises 33–36, sketch the family of tangent curves to the given force field.

33. $\mathbf{F}(x, y) = \dfrac{y}{\sqrt{x^2 + y^2}}\mathbf{i} - \dfrac{x}{\sqrt{x^2 + y^2}}\mathbf{j}$

34. $\mathbf{F}(x, y) = \dfrac{x}{\sqrt{x^2 + y^2}}\mathbf{i} - \dfrac{y}{\sqrt{x^2 + y^2}}\mathbf{j}$

35. $\mathbf{F}(x, y) = 4x^2y\mathbf{i} - \left(2xy^2 + \dfrac{x}{y^2}\right)\mathbf{j}$

36. $\mathbf{F}(x, y) = (1 + x^2)\,\mathbf{i} - 2xy\mathbf{j}$

In Exercises 37 and 38, find an equation for the curve passing through the given point with the specified slope.

Point	Slope
37. $(2, 1)$	$\dfrac{dy}{dx} = \dfrac{y - x}{3y - x}$
38. $(0, 2)$	$\dfrac{dy}{dx} = \dfrac{-2xy}{x^2 + y^2}$

39. If $y = C(x)$ represents the cost of producing x units in a manufacturing process, then the **elasticity of cost** is defined to be

$$E(x) = \frac{\text{marginal cost}}{\text{average cost}} = \frac{C'(x)}{C(x)/x} = \frac{x}{y}\frac{dy}{dx}$$

Find the cost function if the elasticity function is

$$E(x) = \frac{20x - y}{2y - 10x}$$

where $C(100) = 500$ and $100 \le x$.

18.4
First-order linear differential equations

In this section we will see how integrating factors can be used to solve one of the most important classes of first-order differential equations—first-order *linear* differential equations.

DEFINITION OF FIRST-ORDER LINEAR DIFFERENTIAL EQUATION

A first-order linear differential equation is an equation of the form

$$\frac{dy}{dx} + P(x)y = Q(x)$$

where P and Q are continuous functions of x.

To solve a first-order linear differential equation, we use an integrating factor $u(x)$, which converts the left side into the derivative of the product $u(x)y$. That is, we need a factor $u(x)$ such that

$$u(x)\frac{dy}{dx} + u(x)P(x)y = \frac{d[u(x)y]}{dx}$$

By expanding the right side and using prime notation, we find that

$$u(x)y' + u(x)P(x)y = u(x)y' + yu'(x)$$
$$u(x)P(x)y = yu'(x)$$

$$P(x) = \frac{u'(x)}{u(x)}$$

Integration with respect to x yields

$$\ln |u(x)| = \int P(x)\, dx + C_1 \implies u(x) = Ce^{\int P(x)\, dx}$$

Since we don't need the most general integrating factor, we let $C = 1$. Thus, by multiplying the original equation $y' + P(x)y = Q(x)$ by $u(x)$, we have

$$y'e^{\int P(x)\, dx} + yP(x)e^{\int P(x)\, dx} = Q(x)e^{\int P(x)\, dx}$$

$$\frac{d}{dx}\left[ye^{\int P(x)\, dx}\right] = Q(x)e^{\int P(x)\, dx}$$

whose general solution is given by

$$ye^{\int P(x)\, dx} = \int Q(x)e^{\int P(x)\, dx}\, dx + C$$

THEOREM 18.4

SOLUTION OF A FIRST-ORDER LINEAR DIFFERENTIAL EQUATION
The general solution of $y' + P(x)y = Q(x)$ is

$$ye^{\int P(x)\, dx} = \int Q(x)e^{\int P(x)\, dx}\, dx + C$$

Remark   Rather than trying to memorize this formula, just remember that multiplication by the integrating factor $e^{\int P(x)\, dx}$ converts the left side of the differential equation into the derivative of the product $ye^{\int P(x)\, dx}$.

EXAMPLE 1   Solving a first-order linear differential equation

Find the general solution to $xy' - 2y = x^2$.

Solution: The *standard form* of the given equation is

$$y' - \left(\frac{2}{x}\right)y = x$$

Thus, $P(x) = -2/x$, and we have

$$\int P(x)\, dx = -\int \frac{2}{x}\, dx = -\ln x^2$$

$$e^{\int P(x)\, dx} = e^{-\ln x^2} = \frac{1}{x^2}$$

Therefore, multiplying the standard form by $1/x^2$ yields

$$\frac{d}{dx}\left[y\left(\frac{1}{x^2}\right)\right] = \frac{1}{x}$$

$$\frac{y}{x^2} = \int \frac{1}{x}\, dx = \ln |x| + C$$

$$y = x^2 \ln |x| + Cx^2 \qquad \text{General solution}$$

EXAMPLE 2   Solving a first-order linear differential equation

Find the general solution to

$$y' - y \tan t = 1$$

Solution: Since $P(t) = -\tan t$, we have

$$\int P(t)\, dt = -\int \tan t\, dt = \ln |\cos t|$$

$$e^{\int P(t)\, dt} = e^{\ln |\cos t|} = |\cos t|$$

A quick check would show that $\cos t$ is also an integrating factor. Thus, multiplying $y' - y \tan t = 1$ by $\cos t$, we obtain

$$\frac{d}{dt}[y \cos t] = \cos t$$

$$y \cos t = \int \cos t\, dt = \sin t + C$$

$$y = \tan t + C \sec t \qquad \text{General solution}$$

A well-known *nonlinear* equation that reduces to a linear one with an appropriate substitution is the **Bernoulli equation,** named after James Bernoulli (1654–1705):

$$y' + P(x)y = Q(x)y^n$$

This equation is linear if $n = 0$, and has separable variables if $n = 1$. Thus, in the following development, we assume that $n \neq 0$ and $n \neq 1$. We begin by multiplying by y^{-n} and $(1 - n)$ to obtain

$$y^{-n}y' + P(x)y^{1-n} = Q(x)$$

$$(1 - n)y^{-n}y' + (1 - n)P(x)y^{1-n} = (1 - n)Q(x)$$

$$\frac{d}{dx}[y^{1-n}] + (1 - n)P(x)y^{1-n} = (1 - n)Q(x)$$

which is a linear equation in the variable y^{1-n}. Thus, if we let $z = y^{1-n}$, we obtain the linear equation

$$\frac{dz}{dx} + (1 - n)P(x)z = (1 - n)Q(x)$$

Finally, by Theorem 18.4, the *general solution to the Bernoulli equation* is

$$y^{1-n}e^{\int(1-n)P(x)\, dx} = \int (1 - n)Q(x)e^{\int(1-n)P(x)\, dx}\, dx + C$$

EXAMPLE 3   Solving a Bernoulli Equation

Find the general solution to $y' + xy = xe^{-x^2}y^{-3}$.

Solution: For this Bernoulli equation $n = -3$, and we use the substitution

$$z = y^{1-n} = y^4 \quad \Longrightarrow \quad z' = 4y^3 y'$$

By multiplying the original equation by $4y^3$, we have

$$4y^3y' + 4xy^4 = 4xe^{-x^2}$$

$$z' + 4xz = 4xe^{-x^2} \qquad \text{Linear equation: } z' + P(x)z = Q(x)$$

Now, since this equation is linear in z, we have $P(x) = 4x$ and

$$\int P(x)\,dx = \int 4x\,dx$$

$$= 2x^2$$

which implies that e^{2x^2} is an integrating factor. Multiplying by this factor, we have

$$\frac{d}{dx}[ze^{2x^2}] = 4xe^{x^2}$$

$$ze^{2x^2} = \int 4xe^{x^2}\,dx = 2e^{x^2} + C$$

$$z = 2e^{-x^2} + Ce^{-2x^2}$$

Thus, by substituting $z = y^4$, the general solution is

$$y^4 = 2e^{-x^2} + Ce^{-2x^2} \qquad \text{General solution}$$

At this point we have studied several types of first-order differential equations. Of these, the separable variables case is usually the simplest, and solution by an integrating factor is usually left as a last resort. The following summary lists the various types we have studied.

SUMMARY OF FIRST-ORDER DIFFERENTIAL EQUATIONS	Method	Form of equation
	1. Separable variables	$M(x)\,dx + N(y)\,dy = 0$
	2. Homogeneous:	$M(x, y)\,dx + N(x, y)\,dy = 0$, where M and N are nth-degree homogeneous.
	3. Exact:	$M(x, y)\,dx + N(x, y)\,dy = 0$, where $\partial M/\partial y = \partial N/\partial x$.
	4. Integrating factor:	$u(x, y)M(x, y)\,dx + u(x, y)N(x, y)\,dy = 0$ is exact.
	5. Linear:	$y' + P(x)y = Q(x)$
	6. Bernoulli equation:	$y' + P(x)y = Q(x)y^n$

Applications

One type of problem that can be described in terms of a differential equation involves chemical mixtures, as illustrated in the next example.

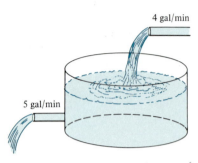

4 gal/min

5 gal/min

FIGURE 18.6

EXAMPLE 4   A mixture application

A tank contains 50 gallons of a solution composed of 90 percent water and 10 percent alcohol. A second solution containing 50 percent water and 50 percent alcohol is added to the tank at the rate of 4 gallons per minute. As the second solution is being added, the tank is being drained at the rate of 5 gallons per minute, as shown in Figure 18.6. Assuming the solution in the tank is stirred constantly, how much alcohol is in the tank after 10 minutes?

Solution: Let y be the number of gallons of alcohol in the tank at any time t. We know that $y = 5$ when $t = 0$. Since the number of gallons of solution in the tank at any time is $50 - t$ and since the tank loses 5 gallons of solution per minute, it must lose

$$\left(\frac{5}{50 - t}\right)y$$

gallons of alcohol per minute. Furthermore, since the tank is gaining 2 gallons of alcohol per minute, the rate of change of alcohol in the tank is given by

$$\frac{dy}{dt} = 2 - \left(\frac{5}{50 - t}\right)y \implies \frac{dy}{dt} + \left(\frac{5}{50 - t}\right)y = 2$$

To solve this linear equation, we let $P(t) = 5/(50 - t)$ and obtain

$$\int P(t)\, dt = \int \frac{5}{50 - t}\, dt = -5 \ln |50 - t|$$

Since $t < 50$, we can drop the absolute value signs and conclude that

$$e^{\int P(t)\, dt} = e^{-5 \ln (50-t)} = \frac{1}{(50 - t)^5}$$

Thus, the general solution is

$$\frac{y}{(50 - t)^5} = \int \frac{2}{(50 - t)^5}\, dt = \frac{1}{2(50 - t)^4} + C$$

$$y = \frac{50 - t}{2} + C(50 - t)^5$$

Since $y = 5$ when $t = 0$, we have

$$5 = \frac{50}{2} + C(50)^5 \implies -\frac{20}{50^5} = C$$

which means that the particular solution is

$$y = \frac{50 - t}{2} - 20\left(\frac{50 - t}{50}\right)^5$$

Finally, when $t = 10$, the amount of alcohol in the tank is

$$y = \frac{50 - 10}{2} - 20\left(\frac{50 - 10}{50}\right)^5$$

$$= 13.45 \text{ gal}$$

which represents a solution containing 33.6 percent alcohol.

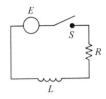

FIGURE 18.7

A simple electric circuit consists of electric current I (in amperes), a resistance R (in ohms), an inductance L (in henrys), and a constant electromotive force E (in volts), as shown in Figure 18.7. According to Kirchhoff's Second Law, if the switch S is closed when $t = 0$, then the applied electromotive force (voltage) is equal to the sum of the voltage drops in the rest of the circuit. This in turn means that the current I satisfies the differential equation

$$L\frac{dI}{dt} + RI = E$$

EXAMPLE 5   *An electric circuit application*

Find the current I as a function of time t (in seconds), given that I satisfies the differential equation $L(dI/dt) + RI = \sin 2t$, where R and L are nonzero constants.

Solution: In standard form the given linear equation is

$$\frac{dI}{dt} + \frac{R}{L}I = \frac{1}{L}\sin 2t$$

Thus, we let $P(t) = R/L$, so that $e^{\int P(t)\,dt} = e^{(R/L)t}$, and, by Theorem 18.4,

$$Ie^{(R/L)t} = \frac{1}{L}\int e^{(R/L)t}\sin 2t\,dt$$

$$= \frac{1}{4L^2 + R^2}e^{(R/L)t}(R\sin 2t - 2L\cos 2t) + C$$

Thus, the general solution is

$$I = e^{-(R/L)t}\left[\frac{1}{4L^2 + R^2}e^{(R/L)t}(R\sin 2t - 2L\cos 2t) + C\right]$$

$$I = \frac{1}{4L^2 + R^2}(R\sin 2t - 2L\cos 2t) + Ce^{-(R/L)t} \qquad \square$$

In most of the falling-body problems discussed so far in the text, we have neglected the air resistance on the falling body. In the next example we include this factor, and we assume that the air resistance of the object falling in the earth's atmosphere is proportional to its velocity v. If g is the gravitational constant, then the downward force F on a falling object of mass m is given by the difference $mg - kv$. But by Newton's Second Law of Motion, we know that $F = ma = m(dv/dt)$, so we obtain the following differential equation.

$$m\frac{dv}{dt} = mg - kv \quad \Longrightarrow \quad \frac{dv}{dt} + \frac{k}{m}v = g$$

EXAMPLE 6   *A falling object with air resistance*

An object of mass m is dropped from a hovering helicopter. Find its velocity as a function of the time t, assuming the resistance due to the air is proportional to the velocity of the object.

Solution: The velocity v satisfies the equation

$$\frac{dv}{dt} + \frac{kv}{m} = g$$

where g is the gravitational constant and k is the constant of proportionality. Letting $b = k/m$, we can *separate variables* to obtain

$$dv = (g - bv)\, dt$$

$$\int \frac{dv}{g - bv} = \int dt$$

$$-\frac{1}{b} \ln |g - bv| = t + C_1$$

$$\ln |g - bv| = -bt - bC_1$$

$$g - bv = Ce^{-bt}$$

Since the object was dropped, $v = 0$ when $t = 0$; thus $g = C$, and it follows that

$$-bv = -g + ge^{-bt} \implies v = \frac{g - ge^{-bt}}{b} = \frac{mg}{k}(1 - e^{-kt/m})$$

Remark Note in Example 6 that the velocity approaches a limit of mg/k. This limit is due to the air resistance. For falling-body problems in which the air resistance is neglected, the velocity increases without bound.

Section Exercises 18.4

In Exercises 1–10, solve the first-order linear differential equation.

1. $\dfrac{dy}{dx} + \left(\dfrac{1}{x}\right) y = 3x + 4$

2. $\dfrac{dy}{dx} + \left(\dfrac{2}{x}\right) y = 3x + 1$

3. $\dfrac{dy}{dx} = e^x - y$

4. $y' + 2y = \sin x$

5. $y' - y = \cos x$

6. $y' + 2xy = 2x$

7. $(3y + \sin 2x)\, dx - dy = 0$

8. $(y - 1) \sin x\, dx - dy = 0$

9. $(x - 1)y' + y = x^2 - 1$

10. $y' + 5y = e^{5x}$

In Exercises 11–16, find the particular solution that satisfies the given boundary condition.

11. $y' \cos^2 x + y - 1 = 0$, $y(0) = 5$

12. $x^3 y' + 2y = e^{1/x^2}$, $y(1) = e$

13. $y' + y \tan x = \sec x + \cos x$, $y(0) = 1$

14. $y' + y \sec x = \sec x$, $y(0) = 4$

15. $y' + \left(\dfrac{1}{x}\right) y = 0$, $y(2) = 2$

16. $y' + (2x - 1)y = 0$, $y(1) = 2$

In Exercises 17–22, solve the Bernoulli differential equation.

17. $y' + 3x^2 y = x^2 y^3$

18. $y' + 2xy = xy^2$

19. $y' + \left(\dfrac{1}{x}\right) y = xy^2$

20. $y' + \left(\dfrac{1}{x}\right) y = x\sqrt{y}$

21. $y' - y = x^3 \sqrt[3]{y}$

22. $yy' - 2y^2 = e^x$

In Exercises 23 and 24, use the differential equation for electrical circuits given by

$$L\frac{dI}{dt} + RI = E$$

23. Solve the differential equation given a constant voltage E_0.

24. Use the result of Exercise 23 to find the equation for the current if $I(0) = 0$, $E_0 = 110$ volts, $R = 550$ ohms, and $L = 4$ henrys. At what time does the current reach 90 percent of its limiting value?

In Exercises 25–28, consider a tank that at time $t = 0$ contains v_0 gallons of a solution, of which, by weight, q_0 pounds is a soluble concentrate. Another solution containing q_1 pounds of the concentrate per gallon is running into the tank at the rate of r_1 gallons per minute. The solution in the tank is kept well stirred and is withdrawn at the rate of r_2 gallons per minute.

25. If Q is the amount of concentrate in the solution at any time t and $Q = q_0$, show that

$$\frac{dQ}{dt} + \frac{r_2 Q}{v_0 + (r_1 - r_2)t} = q_1 r_1$$

26. A 200-gallon tank is full of a solution containing 25 pounds of concentrate. Starting at time $t = 0$, distilled water is admitted to the tank at the rate of 10 gallons per minute, and the well-stirred solution is withdrawn at the same rate.
  (a) Find the amount of the concentrate in the solution as a function of t.
  (b) Find the time when the amount of concentrate in the tank reaches 15 pounds.

27. Repeat Exercise 26, assuming that the solution entering the tank contains 0.05 pound of concentrate per gallon.

28. A 200-gallon tank is half full of distilled water. At time $t = 0$, a solution containing 0.5 pound of concentrate per gallon enters the tank at the rate of 5 gallons per minute, and the well-stirred mixture is withdrawn at the rate of 3 gallons per minute.
  (a) At what time will the tank be full?
  (b) At the time the tank is full, how many pounds of concentrate will it contain?

In Exercises 29–44, solve the first-order differential equation by any appropriate method.

29. $e^{2x+y}\, dx - e^{x-y}\, dy = 0$
30. $(x + 1)\, dx - (y^2 + 2y)\, dy = 0$
31. $(1 + y^2)\, dx + (2xy + y + 2)\, dy = 0$
32. $(1 + 2e^{2x+y})\, dx + e^{2x+y}\, dy = 0$
33. $(y \cos x - \cos x)\, dx + dy = 0$
34. $(x + 1)\, dy + (y - e^x)\, dx = 0$
35. $2xy\, dx + (x^2 + \cos y)\, dy = 0$
36. $y' = 2x\sqrt{1 - y^2}$
37. $(3y^2 + 4xy)\, dx + (2xy + x^2)\, dy = 0$
38. $(x + y)\, dx - x\, dy = 0$
39. $(2y - e^x)\, dx + x\, dy = 0$
40. $(y^2 + xy)\, dx - x^2\, dy = 0$
41. $(x^2 y^4 - 1)\, dx + x^3 y^3\, dy = 0$
42. $y\, dx + (3x + 4y)\, dy = 0$
43. $3y\, dx - (x^2 + 3x + y^2)\, dy = 0$
44. $x\, dx + (y + e^y)(x^2 + 1)\, dy = 0$

18.5
Second-order homogeneous linear equations

SECTION TOPICS ▪
Second-order linear differential equations ▪
Higher-order linear differential equations ▪
Applications ▪

In this and the following section we discuss methods for solving higher-order linear differential equations.

DEFINITION OF A LINEAR DIFFERENTIAL EQUATION OF ORDER n

Let $g_1, g_2, \ldots, g_n$ and f be functions of x with a common domain. An equation of the form

$$y^{(n)} + g_1(x)y^{(n-1)} + g_2(x)y^{(n-2)} + \cdots + g_{n-1}(x)y' + g_n(x)y = f(x)$$

is called a **linear differential equation of order n.** If $f(x) = 0$, the equation is **homogeneous;** otherwise, it is **nonhomogeneous.**

| Remark We will study homogeneous equations in this section and leave the nonhomogeneous case until the next section. (Note that this use of the term homogeneous differs from that in Section 18.2.)

We begin with a definition of linear independence. We say the functions $y_1, y_2, \ldots, y_n$ are **linearly independent** if the *only* solution to the equation

$$C_1 y_1 + C_2 y_2 + \cdots + C_n y_n = 0$$

is the trivial one, $C_1 = C_2 = \cdots = C_n = 0$. Otherwise this set of functions is **linearly dependent.** It can be shown that *two* functions are linearly dependent if and only if one is a constant multiple of the other. For example, $y_1(x) = x$ and $y_2(x) = 3x$ are linearly dependent because $C_1 x + C_2(3x) = 0$ has nonzero solutions $C_1 = -3$ and $C_2 = 1$.

The following theorem points out the importance of linear independence in constructing the general solution of a second-order linear homogeneous differential equation with constant coefficients.

THEOREM 18.5   **GENERAL SOLUTION AS A LINEAR COMBINATION OF LINEARLY INDEPENDENT SOLUTIONS**
If y_1 and y_2 are linearly independent solutions of the differential equation $y'' + ay' + by = 0$, then the general solution is

$$y = C_1 y_1 + C_2 y_2$$

where C_1 and C_2 are constants.

Proof: We prove only one direction of the theorem. If y_1 and y_2 are solutions, we have

$$y_1''(x) + ay_1'(x) + by_1(x) = 0$$
$$y_2''(x) + ay_2'(x) + by_2(x) = 0$$

Multiplying the first equation by C_1, the second by C_2, and adding, we obtain

$$[C_1 y_1''(x) + C_2 y_2''(x)] + a[C_1 y_1'(x) + C_2 y_2'(x)] + b[C_1 y_1(x) + C_2 y_2(x)] = 0$$

which means that $y = C_1 y_1 + C_2 y_2$ is a solution, as desired. The proof that all solutions are of this form is best left to a full course on differential equations.

Remark   This theorem tells us that if we can find two linearly independent solutions, then we can obtain the general solution by forming a **linear combination** of the two solutions.

To find two linearly independent solutions, we note that the nature of the equation $y'' + ay' + by = 0$ suggests that it may have solutions of the form $y = e^{mx}$. If so, then $y' = me^{mx}$ and $y'' = m^2 e^{mx}$. Thus, by substitution, $y = e^{mx}$ is a solution if and only if

$$y'' + ay' + by = 0$$
$$m^2 e^{mx} + ame^{mx} + be^{mx} = 0$$
$$e^{mx}(m^2 + am + b) = 0$$

Since e^{mx} is never zero, $y = e^{mx}$ is a solution if and only if

$$m^2 + am + b = 0$$

This equation is called the **characteristic equation** of the differential equation $y'' + ay' + by = 0$. Note that the characteristic equation can be determined from its differential equation by simply replacing y'' by m^2, y' by m, and y by 1.

EXAMPLE 1   Characteristic equation with distinct real roots

Solve the differential equation $y'' - 4y = 0$.

Solution: In this case the characteristic equation is

$$m^2 - 4 = 0 \qquad \text{Characteristic equation}$$

so $m^2 = 4$ or $m = \pm 2$. Thus, $y_1 = e^{m_1 x} = e^{2x}$ and $y_2 = e^{m_2 x} = e^{-2x}$ are particular solutions of the given differential equation. Furthermore, since these two solutions are linearly independent, we can apply Theorem 18.5 to conclude that the general solution is

$$y = C_1 e^{2x} + C_2 e^{-2x} \qquad \text{General solution}$$

The characteristic equation in Example 1 has two distinct real roots. From algebra we know that this is only one of *three* possibilities for quadratic equations. In general, the quadratic equation $m^2 + am + b = 0$ has roots

$$m_1 = \frac{-a + \sqrt{a^2 - 4b}}{2} \qquad \text{and} \qquad m_2 = \frac{-a - \sqrt{a^2 - 4b}}{2}$$

which fall into one of three cases:

1. Two distinct real roots, $m_1 \neq m_2$
2. Two equal real roots, $m_1 = m_2$
3. Two complex conjugate roots, $m_1 = \alpha + \beta i$ and $m_2 = \alpha - \beta i$

In terms of the differential equation $y'' + ay' + by = 0$, these three cases correspond to three different types of general solutions.

THEOREM 18.6   SOLUTIONS TO $y'' + ay' + by = 0$

Distinct Real Roots: If $m_1 \neq m_2$ are distinct real roots of the characteristic equation, then the general solution is

$$y = C_1 e^{m_1 x} + C_2 e^{m_2 x}$$

Equal Real Roots: If $m_1 = m_2$ are equal real roots of the characteristic equation, then the general solution is

$$y = C_1 e^{m_1 x} + C_2 x e^{m_1 x} = (C_1 + C_2 x) e^{m_1 x}$$

Complex Roots: If $m_1 = \alpha + \beta i$ and $m_2 = \alpha - \beta i$ are complex roots of the characteristic equation, then the general solution is

$$y = C_1 e^{\alpha x} \cos \beta x + C_2 e^{\alpha x} \sin \beta x$$

EXAMPLE 2   *Characteristic equation with complex roots*

Find the general solution of the differential equation $y'' + 6y' + 12y = 0$.

Solution: The characteristic equation

$$m^2 + 6m + 12 = 0$$

has two complex roots:

$$m = \frac{-6 \pm \sqrt{36 - 48}}{2}$$

$$= \frac{-6 \pm \sqrt{-12}}{2}$$

$$= -3 \pm \sqrt{-3} = -3 \pm \sqrt{3}\, i$$

Thus, $\alpha = -3$ and $\beta = \sqrt{3}$, and the general solution is

$$y = C_1 e^{-3x} \cos \sqrt{3}x + C_2 e^{-3x} \sin \sqrt{3}x$$

EXAMPLE 3   *Characteristic equation with repeated roots*

Solve the differential equation $y'' + 4y' + 4y = 0$, subject to the initial conditions $y(0) = 2$ and $y'(0) = 1$.

Solution: The characteristic equation

$$m^2 + 4m + 4 = (m + 2)^2 = 0$$

has two equal real roots. Therefore, the general solution is

$$y = C_1 e^{-2x} + C_2 x e^{-2x} \qquad \text{General solution}$$

Now, since $y = 2$ when $x = 0$, we have

$$2 = C_1(1) + C_2(0)(1) = C_1$$

Furthermore, since $y' = 1$ when $x = 0$, we have

$$y' = -2C_1 e^{-2x} + C_2(-2xe^{-2x} + e^{-2x})$$

$$1 = -2(2)(1) + C_2[-2(0)(1) + 1]$$

$$5 = C_2$$

Therefore, the solution is

$$y = 2e^{-2x} + 5xe^{-2x} \qquad \text{Particular solution}$$

For higher-order homogeneous linear differential equations, we find the general solution in much the same way as we do for second-order equations. That is, we begin by determining the n roots of the characteristic equation, and then, based on these n roots, we form a linearly independent collection of n solutions. The major difference is that with equations of third or higher order, roots of the characteristic equation may occur more than twice. When this happens, the linearly independent solutions are formed by multiplying by increasing powers of x. We demonstrate this in the next two examples.

EXAMPLE 4   *Solving a third-order equation*

Find the general solution to $y''' + 3y'' + 3y' + y = 0$.

Solution: The characteristic equation is

$$m^3 + 3m^2 + 3m + 1 = (m + 1)^3 = 0$$

Since the root $m = -1$ occurs three times, the general solution is

$$y = C_1 e^{-x} + C_2 x e^{-x} + C_3 x^2 e^{-x} \qquad \text{General solution}$$

EXAMPLE 5   *Solving a fourth-order equation*

Find the general solution to $y^{(4)} + 2y'' + y = 0$.

Solution: The characteristic equation

$$m^4 + 2m^2 + 1 = (m^2 + 1)^2 = 0$$

$$m = \pm i$$

Since the roots $m_1 = \alpha + \beta i = 0 + i$ and $m_2 = \alpha - \beta i = 0 - i$ each occur twice, the general solution is

$$y = C_1 \cos x + C_2 \sin x + C_3 x \cos x + C_4 x \sin x \qquad \text{General solution}$$

Applications

One of many applications of linear differential equations is describing the motion of an oscillating spring. According to Hooke's Law (Section 6.4), a spring that is stretched (or compressed) y units from its natural length 1 tends to *restore* itself to its natural length by a force F that is proportional to y. That is, $F(y) = -ky$, where k is called the **spring constant** and indicates the stiffness of a given spring.

Suppose a rigid object of mass m is attached to the end of the spring and causes a displacement, as shown in Figure 18.8. Assume that the mass of the spring is negligible compared to m. Now suppose the object is pulled down and released. The resulting oscillations are a product of two opposing forces— the spring force $F(y) = -ky$ and the weight mg of the object. Under such conditions, we can use a differential equation to find the position y of the object as a function of time t. According to Newton's Section Law of Motion, the force acting on the weight is $F = ma$, where $a = d^2y/dt^2$ is the acceleration. Assuming the motion is **undamped**—that is, there are no other external forces acting on the object—it follows that $m(d^2y/dt^2) = -ky$, and we have

$$\frac{d^2y}{dt^2} + \left(\frac{k}{m}\right)y = 0 \qquad \text{Undamped motion of a spring}$$

We use this equation in our next example.

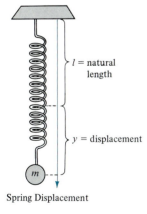

l = natural length

y = displacement

m

Spring Displacement

FIGURE 18.8

EXAMPLE 6   Undamped motion of a spring

Suppose a 4-pound weight stretches a spring 8 inches from its natural length. If the weight is pulled down an additional 6 inches and released with an initial upward velocity of 8 feet per second, find a formula for the position of the weight as a function of time t.

Solution: By Hooke's Law, $4 = k(\frac{2}{3})$, so $k = 6$. Moreover, since the weight w is given by mg, it follows that $m = w/g = \frac{4}{32} = \frac{1}{8}$. Hence the resulting differential equation for this undamped motion is

$$\frac{d^2y}{dt^2} + 48y = 0$$

Since the characteristic equation $m^2 + 48 = 0$ has complex roots $m = 0 \pm 4\sqrt{3}\, i$, the general solution is

$$y = C_1 e^0 \cos 4\sqrt{3}t + C_2 e^0 \sin 4\sqrt{3}t = C_1 \cos 4\sqrt{3}t + C_2 \sin 4\sqrt{3}t$$

Using the initial conditions, we have

$$\frac{1}{2} = C_1(1) + C_2(0) \implies C_1 = \frac{1}{2} \qquad\qquad y(0) = \frac{1}{2}$$

$$y'(t) = -4\sqrt{3}C_1 \sin 4\sqrt{3}t + 4\sqrt{3}C_2 \cos 4\sqrt{3}t$$

$$8 = -4\sqrt{3}\left(\frac{1}{2}\right)(0) + 4\sqrt{3}C_2(1) \implies C_2 = \frac{2\sqrt{3}}{3} \qquad y'(0) = 8$$

Consequently, the position at time t is given by

$$y = \frac{1}{2} \cos 4\sqrt{3}t + \frac{2\sqrt{3}}{3} \sin 4\sqrt{3}t$$

Damped Vibration

FIGURE 18.9

Suppose the object in Figure 18.9 experiences an additional damping or frictional force that is proportional to its velocity. A case in point would be the damping force due to friction and movement through a fluid. With this damping force, $-p(dy/dt)$, taken into account, the differential equation for the oscillations is

$$m\frac{d^2y}{dt^2} = -ky - p\frac{dy}{dt}$$

or, in standard linear form,

$$\frac{d^2y}{dt^2} + \frac{p}{m}\left(\frac{dy}{dt}\right) + \frac{k}{m}y = 0 \qquad \text{Damped motion of a spring}$$

Section Exercises 18.5

In Exercises 1–26, find the general solution of the linear differential equation.

1. $y'' - y' = 0$

2. $y'' + 2y' = 0$

3. $y'' - y' - 6y = 0$

4. $y'' + 6y' + 5y = 0$

5. $2y'' + 3y' - 2y = 0$

6. $16y'' - 16y' + 3y = 0$

7. $y'' + 6y' + 9y = 0$

8. $y'' - 10y' + 25y = 0$

9. $16y'' - 8y' + y = 0$

10. $9y'' - 12y' + 4y = 0$

11. $y'' + y = 0$

12. $y'' + 4y = 0$

13. $y'' - 9y = 0$

14. $y'' - 2y = 0$

15. $y'' - 2y' + 4y = 0$

16. $y'' - 4y' + 21y = 0$

17. $y'' - 3y' + y = 0$ **18.** $3y'' + 4y' - y = 0$
19. $9y'' - 12y' + 11y = 0$ **20.** $2y'' - 6y' + 7y = 0$
21. $y^{(4)} - y = 0$ **22.** $y^{(4)} - y'' = 0$
23. $y''' - 6y'' + 11y' - 6y = 0$
24. $y''' - y'' - y' + y = 0$
25. $y''' - 3y'' + 7y' - 5y = 0$
26. $y''' - 3y'' + 3y' - y = 0$

In Exercises 27 and 28, find the particular solution to the linear differential equation.

27. $y'' - y' - 30y = 0$ **28.** $y'' + 2y' + 3y = 0$
  $y(0) = 1,\ y'(0) = -4$ $y(0) = 2,\ y'(0) = 1$

In Exercises 29–34, describe the motion of a 32-pound weight suspended on a spring. Assume that the weight stretches the spring $\frac{2}{3}$ foot from its natural position.

29. The weight is pulled $\frac{1}{2}$ foot below the equilibrium position and released.
30. The weight is raised $\frac{2}{3}$ foot above the equilibrium position and released.
31. The weight is raised $\frac{2}{3}$ foot above the equilibrium position and started off with a downward velocity of $\frac{1}{2}$ foot per second.
32. The weight is pulled $\frac{1}{2}$ foot below the equilibrium position and started off with an upward velocity of $\frac{1}{2}$ foot per second.

33. The weight is pulled $\frac{1}{2}$ foot below the equilibrium position and released. The motion takes place in a medium that furnishes a damping force of magnitude $\frac{1}{8}$ speed at all times.
34. The weight is pulled $\frac{1}{2}$ foot below the equilibrium position and released. The motion takes place in a medium that furnishes a damping force of magnitude $\frac{1}{4}|v|$ at all times.
35. If the characteristic equation of the differential equation

$$y'' + ay' + by = 0$$

has two equal real roots given by $m = r$, show that

$$y = C_1 e^{rx} + C_2 x e^{rx}$$

is a solution.
36. If the characteristic equation of the differential equation

$$y'' + ay' + by = 0$$

has complex roots given by $m_1 = \alpha + \beta i$ and $m_2 = \alpha - \beta i$, show that

$$y = C_1 e^{\alpha x} \cos \beta x + C_2 e^{\alpha x} \sin \beta x$$

is a solution.

SECTION TOPICS ▪
Nonhomogeneous equations ▪
Method of undetermined coefficients ▪
Variation of parameters ▪

18.6
Second-order nonhomogeneous linear equations

In the previous section we represented damped oscillations of a spring by the *homogeneous* second-order linear equation

$$\frac{d^2y}{dt^2} + \frac{p}{m}\left(\frac{dy}{dt}\right) + \frac{k}{m}y = 0 \qquad \text{Free motion}$$

We call this type of oscillation **free** because it is determined solely by the spring and gravity and is free of the action of other external forces. If such a system is also subject to an external periodic force such as $a \sin bt$, caused by vibrations at the opposite end of the spring, then we call the motion **forced,** and it is characterized by the *nonhomogeneous* equation

$$\frac{d^2y}{dt^2} + \frac{p}{m}\left(\frac{dy}{dt}\right) + \frac{k}{m}y = a \sin bt \qquad \text{Forced motion}$$

In this section we describe two methods for finding general solutions to such nonhomogeneous equations.

The first method is to find the general solution, called y_h, of the corresponding homogeneous equation, and then attempt to find a particular solution, y_p, of the nonhomogeneous equation. If this attempt is successful, the general solution is $y = y_h + y_p$, which is the essence of the following theorem.

THEOREM 18.7   **GENERAL SOLUTION OF NONHOMOGENEOUS LINEAR EQUATION**
Let $y'' + ay' + by = F(x)$ be a second-order nonhomogeneous linear differential equation. If y_p is a particular solution to this equation and y_h is the general solution of the corresponding homogeneous equation, then

$$y = y_h + y_p$$

is the general solution of the nonhomogeneous equation.

Proof: The sum $y = y_h + y_p$ is a solution to the nonhomogeneous equation because

$$
\begin{aligned}
y'' + ay' + by &= (y_h'' + y_p'') + a(y_h' + y_p') + b(y_h + y_p) \\
&= (y_h'' + ay_h' + by_h) + (y_p'' + ay_p' + by_p) \\
&= 0 + F(x) = F(x)
\end{aligned}
$$

Conversely, if y is any solution to the nonhomogeneous equation, then $y - y_p$ is a solution to the homogeneous equation because

$$
\begin{aligned}
(y - y_p)'' + a(y - y_p)' + b(y - y_p) &= (y'' - y_p'') + a(y' - y_p') + b(y - y_p) \\
&= (y'' + ay' + by) - (y_p'' + ay_p' + by_p) \\
&= F(x) - F(x) = 0
\end{aligned}
$$

Consequently, $y - y_p$ is of the form $y - y_p = y_h$, which in turn means that $y = y_p + y_h$ is the general solution of the nonhomogeneous equation.

Method of undetermined coefficients

Since we already have the tools for finding y_h, we focus on ways to find the particular solution y_p. If $F(x)$ in $y'' + ay' + by = F(x)$ consists of sums or products of

$$x^n, \qquad e^{mx}, \qquad \cos \beta x, \qquad \sin \beta x$$

then we can find a particular solution y_p by the method of **undetermined coefficients.** The gist of the method is to guess that the solution y_p is a generalized form of $F(x)$. For instance:

1. If $F(x) = 3x^2$, choose $y_p = Ax^2 + Bx + C$.
2. If $F(x) = 4xe^x$, choose $y_p = Axe^x + Be^x$.
3. If $F(x) = x + \sin 2x$, choose $y_p = (Ax + B) + C \sin 2x + D \cos 2x$.

Then, by substitution, we determine the coefficients for this generalized solution. We illustrate this method in the next three examples.

EXAMPLE 1   *Method of undetermined coefficients*

· Find the general solution of the equation $y'' - 2y' - 3y = 2 \sin x$.

Solution: To find y_h, we solve the characteristic equation

$$m^2 - 2m - 3 = (m + 1)(m - 3) = 0 \implies m = -1 \text{ and } m = 3$$

Thus, $y_h = C_1 e^{-x} + C_2 e^{3x}$. Next, we let y_p be a generalized form of $2 \sin x$. That is, let

$$y_p = A \cos x + B \sin x$$
$$y_p{}' = -A \sin x + B \cos x$$
$$y_p{}'' = -A \cos x - B \sin x$$

Substitution into the given equation yields

$$-A \cos x - B \sin x + 2A \sin x - 2B \cos x - 3A \cos x - 3B \sin x$$
$$= 2 \sin x \quad (-4A - 2B) \cos x + (2A - 4B) \sin x = 2 \sin x$$

Consequently, y_p is a solution, provided the coefficients of like terms are equal. Thus, we obtain the system

$$-4A - 2B = 0 \qquad \text{and} \qquad 2A - 4B = 2$$

with solutions $A = \frac{1}{5}$ and $B = -\frac{2}{5}$. Therefore, the general solution is

$$y = y_h + y_p = C_1 e^{-x} + C_2 e^{3x} + \frac{1}{5} \cos x - \frac{2}{5} \sin x \qquad \square$$

In Example 1, the form of the homogeneous solution $y_h = C_1 e^{-x} + C_2 e^{3x}$ has no overlap with the function $F(x)$ in the equation $y'' + ay' + by = F(x)$. However, suppose the given differential equation in Example 1 were of the form

$$y'' - 2y' - 3y = e^{-x}$$

Now, it would make no sense to guess the particular solution to be $y = Ae^{-x}$, since we know that this solution yields zero. In such cases, we alter the guess by multiplying by the lowest power of x that removes the duplication. For this particular problem, we would guess $y_p = Axe^{-x}$. This is further demonstrated in the next example.

EXAMPLE 2   *Method of undetermined coefficients*

Find the general solution to $y'' - 2y' = x + 2e^x$.

Solution: The characteristic equation $m^2 - 2m = 0$ has solutions $m = 0$ and $m = 2$. Thus,

$$y_h = C_1 + C_2 e^{2x}$$

Since $F(x) = x + 2e^x$, our first choice for y_p would be $(A + Bx) + Ce^x$. However, since y_h *already* contains a constant term C_1, we multiply the *polynomial part* by x and use

$$y_p = Ax + Bx^2 + Ce^x$$
$$y_p{}' = A + 2Bx + Ce^x$$
$$y_p{}'' = 2B + Ce^x$$

Substituting into the differential equation, we have

$$(2B + Ce^x) - 2(A + 2Bx + Ce^x) = x + 2e^x$$
$$(2B - 2A) - 4Bx - Ce^x = x + 2e^x$$

Equating coefficients of like terms yields the system

$$2B - 2A = 0, \qquad -4B = 1, \qquad -C = 2$$

with solutions $A = B = -\frac{1}{4}$ and $C = -2$. Therefore,

$$y_p = -\frac{1}{4}x - \frac{1}{4}x^2 - 2e^x$$

and the general solution is

$$y = C_1 + C_2 e^{2x} - \frac{1}{4}x - \frac{1}{4}x^2 - 2e^x$$

In Example 2 the polynomial part of the initial guess $(A + Bx) + Ce^x$ for y_p overlapped by a constant term with $y_h = C_1 + C_2 e^{2x}$, and it was necessary to multiply the polynomial part by a power of x that removed the overlap. In the next example we further illustrate some choices for y_p that eliminate overlap with y_h. Remember that in all cases the first guess for y_p should match the types of functions occurring in $F(x)$.

EXAMPLE 3   *Choosing the form of the particular solution*

Determine a suitable choice for y_p for the following:

$y'' + ay' + by = F(x)$	y_h
(a) $y'' = x^2$	$C_1 + C_2 x$
(b) $y'' + 2y' + 10y = 4 \sin 3x$	$C_1 e^{-x} \cos 3x + C_2 e^{-x} \sin 3x$
(c) $y'' - 4y' + 4 = e^{2x}$	$C_1 e^{2x} + C_2 x e^{2x}$

Solution:

(a) Since $F(x) = x^2$, the normal choice for y_p would be

$$A + Bx + Cx^2$$

However, since $y_h = C_1 + C_2 x$ already contains a linear term, we multiply by x^2 to obtain

$$y_p = Ax^2 + Bx^3 + Cx^4$$

(b) Since $F(x) = 4 \sin 3x$ and since each term in y_h contains a factor of e^x, we simply let

$$y_p = A \cos 3x + B \sin 3x$$

(c) Since $F(x) = e^{2x}$, the normal choice for y_p would be Ae^{2x}. However, since

$$y_h = C_1 e^{2x} + C_2 x e^{2x} + C_2 x e^{2x}$$

already contains an xe^{2x} term, we multiply by x^2 to get

$$y_p = Ax^2 e^{2x}$$

We can also use the method of undetermined coefficients for nonhomogeneous equations of higher order. The procedure is demonstrated in the next example.

EXAMPLE 4   *Undetermined coefficients with a third-order equation*

Find the general solution to $y''' + 3y'' + 3y' + y = x$.

Solution: From Example 4 in the previous section we know that the homogeneous solution is

$$y_h = C_1e^{-x} + C_2xe^{-x} + C_3x^2e^{-x}$$

Since $F(x) = x$, we let $y_p = A + Bx$ and obtain $y_p' = B$, and $y_p'' = y_p''' = 0$. Thus, by substitution, we have

$$(0) + 3(0) + 3(B) + (A + Bx) = (3B + A) + Bx = x$$

Thus, $B = 1$ and $A = -1/3$, and the general solution is

$$y = C_1e^{-x} + C_2xe^{-x} + C_3x^2e^{-x} - \frac{1}{3} + x$$

Variations of parameters

The method of undetermined coefficients works well if $F(x)$ is made up of polynomials or functions whose successive derivatives have a cyclic pattern. For functions like $1/x$ and $\tan x$, which do not possess such characteristics, we use a more general method called **variation of parameters.** In this method we assume that y_p has the same *form* as y_h, except that the constants in y_h are replaced by variables.

VARIATION OF PARAMETERS   To find the general solution to the equation $y'' + ay' + by = F(x)$:

1. Find $y_h = C_1y_1 + C_2y_2$.
2. Replace the constants by variables to form $y_p = u_1y_1 + u_2y_2$.
3. Solve the following system for u_1' and u_2':

$$u_1'y_1 + u_2'y_2 = 0$$
$$u_1'y_1' + u_2'y_2' = F(x)$$

4. Integrate to find u_1 and u_2. The general solution is $y = y_h + y_p$.

EXAMPLE 5   *Variation of parameters*

Solve the differential equation

$$y'' - 2y' + y = \frac{e^x}{2x}$$

Solution: The characteristic equation $m^2 - 2m + 1 = (m - 1)^2 = 0$ has one solution, $m = 1$. Thus, the homogeneous solution is

$$y_h = C_1 y_1 + C_2 y_2 = C_1 e^x + C_2 x e^x$$

Replacing C_1 and C_2 by u_1 and u_2, we have

$$y_p = u_1 y_1 + u_2 y_2 = u_1 e^x + u_2 x e^x$$

The resulting system of equations is

$$u_1' e^x + u_2' x e^x = 0$$

$$u_1' e^x + u_2' (x e^x + e^x) = \frac{e^x}{2x}$$

Subtracting the second equation from the first, we have $u_2' = 1/2x$. Then, by substitution in the first equation, we have $u_1' = -\frac{1}{2}$. Finally, integration yields

$$u_1 = -\int \frac{1}{2} dx = -\frac{x}{2} \quad \text{and} \quad u_2 = \frac{1}{2} \int \frac{1}{x} dx = \frac{1}{2} \ln x = \ln \sqrt{x}$$

From this result it follows that a particular solution is

$$y_p = -\frac{1}{2} x e^x + (\ln \sqrt{x}) x e^x$$

and the general solution is

$$y = C_1 e^x + C_2 x e^x - \frac{1}{2} x e^x + x e^x \ln \sqrt{x}$$

EXAMPLE 6   *Variation of parameters*

Solve the differential equation $y'' + y = \tan x$.

Solution: Since the characteristic equation $m^2 + 1 = 0$ has solutions $m = \pm i$, the homogeneous solution is

$$y_h = C_1 \cos x + C_2 \sin x$$

Replacing C_1 and C_2 by u_1 and u_2, we have

$$y_p = u_1 \cos x + u_2 \sin x$$

The resulting system of equations is

$$u_1' \cos x + u_2' \sin x = 0$$

$$-u_1' \sin x + u_2' \cos x = \tan x$$

Multiplying the first equation by $\sin x$ and the second by $\cos x$, we have

$$u_1' \sin x \cos x + u_2' \sin^2 x = 0$$

$$-u_1' \sin x \cos x + u_2' \cos^2 x = \sin x$$

Adding these two equations, we have $u_2' = \sin x$, which implies that

$$u_1' = -\frac{\sin^2 x}{\cos x} = \frac{\cos^2 x - 1}{\cos x} = \cos x - \sec x$$

Integration yields

$$u_1 = \int (\cos x - \sec x)\, dx = \sin x - \ln |\sec x + \tan x|$$

$$u_2 = \int \sin x\, dx = -\cos x$$

so that

$$y_p = \sin x \cos x - \cos x \ln |\sec x + \tan x| - \sin x \cos x$$
$$= -\cos x \ln |\sec x + \tan x|$$

and the general solution is

$$y = y_h + y_p = C_1 \cos x + C_2 \sin x - \cos x \ln |\sec x + \tan x|$$

Section Exercises 18.6

In Exercises 1–16, solve the differential equation by the method of undetermined coefficients.

1. $y'' - 3y' + 2y = 2x$
2. $y'' - 2y' - 3y = x^2 - 1$
3. $y'' + y = x^3$
   $y(0) = 1,\ y'(0) = 0$
4. $y'' + 4y = 4$
   $y(0) = 1,\ y'(0) = 6$
5. $y'' + 2y' = 2e^x$
6. $y'' - 9y = 5e^{3x}$
7. $y'' - 10y' + 25y = 5 + 6e^x$
8. $16y'' - 8y' + y = 4(x + e^x)$
9. $y'' + y' = 2 \sin x$
   $y(0) = 0,\ y'(0) = -3$
10. $y'' + y' - 2y = 3 \cos 2x$
   $y(0) = -1,\ y'(0) = 2$
11. $y'' + 9y = \sin 3x$
12. $y'' + 4y' + 5y = \sin x + \cos x$
13. $y''' - 3y' + 2y = 2e^{-2x}$
14. $y''' - y'' = 4x^2$
   $y(0) = 1,\ y'(0) = 1,\ y''(0) = 1$
15. $y' - 4y = xe^x - xe^{4x}$ **16.** $y' + 2y = \sin x$
   $y(0) = \dfrac{1}{3}$ $y\left(\dfrac{\pi}{2}\right) = \dfrac{2}{5}$

In Exercises 17—22, solve the differential equation by the method of variation of parameters.

17. $y'' + y = \sec x$
18. $y'' + y = \sec x \tan x$
19. $y'' + 4y = \csc 2x$
20. $y''' - 4y'' + 4y = x^2 e^{2x}$
21. $y'' - 2y' + y = e^x \ln x$

22. $y'' - 4y' + 4y = \dfrac{e^{2x}}{x}$

In Exercises 23 and 24, use the differential equation

$$5y'' + 80y' + 140y = F(t)$$

to describe the motion of a 5-kilogram mass attached to a spring with a spring constant of 140 newtons per meter and an external driving force $F(t)$. Assume that the mass starts from the equilibrium position with an imparted upward velocity of 2 meters per second, and that the motion takes place in a medium that furnishes a damping force of $-80y'$.

23. $F(t) = 5 \sin 2t$
24. $F(t) = 10 \sin t$

In Exercises 25 and 26, use the electrical circuit differential equation

$$\frac{d^2q}{dt^2} + \left(\frac{R}{L}\right)\frac{dq}{dt} + \left(\frac{1}{LC}\right)q = \left(\frac{1}{L}\right)E(t)$$

where R is the resistance (in ohms), C is the capacitance (in farads), L is the inductance (in henrys), $E(t)$ is the electromotive force (in volts), and q is the charge on the capacitor C (in coulombs). Find the charge q as a function of time for the electrical circuit described. Assume that $q(0) = 0$ and $q'(0) = 0$.

25. $R = 20,\ C = 0.02,\ L = 2$
   $E(t) = 12 \sin 5t$
26. $R = 20,\ C = 0.02,\ L = 1$
   $E(t) = 10 \sin 5t$

18.7
Series solutions of differential equations

We conclude this chapter by showing how power series can be used to solve certain types of differential equations. For the sake of brevity, we will limit our discussion to the statement and demonstration of the method and omit the theoretical development.

We begin with the general **power series solution** method. Recall from Chapter 10 that a power series represents a function f on an interval of convergence, and that we can successively differentiate the power series to obtain a series for f', f'', and so on. For instance,

$$f(x) = a_0 + a_1x + a_2x^2 + a_3x^3 + \cdots = \sum_{n=0}^{\infty} a_nx^n$$

$$f'(x) = a_1 + 2a_2x + 3a_3x^2 + 4a_4x^3 + \cdots = \sum_{n=0}^{\infty} na_nx^{n-1}$$

$$f''(x) = 2a_2 + 6a_{3x} + 12a_4x^2 + 20a_5x^3 + \cdots = \sum_{n=0}^{\infty} n(n-1)a_nx^{n-2}$$

These properties are used in the *power series* solution method demonstrated in the first two examples. For the sake of demonstration, we begin with a simple differential equation that can be solved by other methods.

EXAMPLE 1   Power series solution

Use a power series to find the general solution to the differential equation $y' - 2y = 0$.

Solution: Assume

$$y = \sum_{n=0}^{\infty} a_nx^n$$

is a solution. Then

$$y' = \sum_{n=0}^{\infty} na_nx^{n-1}$$

Substituting for y' and $-2y$, we obtain the following series form for the given differential equation:

$$y' - 2y = \sum_{n=0}^{\infty} na_nx^{n-1} - 2\sum_{n=0}^{\infty} a_nx^n = 0$$

$$\sum_{n=0}^{\infty} na_nx^{n-1} = 2\sum_{n=0}^{\infty} a_nx^n$$

Next we adjust the summation indices so that x^n appears in each series. In this case we need only to replace n by $n + 1$ in the left-hand series to obtain

$$\sum_{n=-1}^{\infty} (n+1)a_{n+1}x^n = 2\sum_{n=0}^{\infty} a_nx^n$$

Now, by equating coefficients of like terms, we obtain the **recursion formula** $(n + 1)a_{n+1} = 2a_n$, which implies that

$$a_{n+1} = \frac{2a_n}{n + 1}, \quad n \geq 0$$

This formula generates the following results in terms of a_0:

$$a_1 = 2a_0$$

$$a_2 = \frac{2a_1}{2} = \frac{2^2 a_0}{2}$$

$$a_3 = \frac{2a_2}{3} = \frac{2^3 a_0}{2 \cdot 3} = \frac{2^3 a_0}{3!}$$

$$a_4 = \frac{2a_3}{4} = \frac{2^4 a_0}{2 \cdot 3 \cdot 4} = \frac{2^4 a_0}{4!}$$

$$\vdots$$

$$a_n = \frac{2^n a_0}{n!}$$

Using these values as the coefficients for the *solution* series, we have

$$y = \sum_{n=0}^{\infty} \frac{2^n a_0}{n!} x^n = a_0 \sum_{n=0}^{\infty} \frac{2^n}{n!} x^n = a_0 e^{2x}$$

The next example *cannot* be solved by any of the methods discussed in previous sections of this chapter.

EXAMPLE 2   *Power series solution*

Use a power series to solve the differential equation $y'' + xy' + y = 0$.

Solution: Assume $\sum_{n=0}^{\infty} a_n x^n$ is a solution. Then

$$y' = \sum_{n=0}^{\infty} n a_n x^{n-1}, \qquad xy' = \sum_{n=0}^{\infty} n a_n x^n, \qquad y'' = \sum_{n=0}^{\infty} n(n - 1)a_n x^{n-2}$$

Substituting for y'', xy', and y in the given differential equation, we obtain the following series:

$$\sum_{n=0}^{\infty} n(n - 1)a_n x^{n-2} + \sum_{n=0}^{\infty} n a_n x^n + \sum_{n=0}^{\infty} a_n x^n = 0$$

$$\sum_{n=0}^{\infty} n(n - 1)a_n x^{n-2} = -\sum_{n=0}^{\infty} (n + 1)a_n x^n$$

To obtain equal powers on x, we adjust the summation indices by replacing n by $n + 2$ in the left-hand sum, to obtain

$$\sum_{n=-2}^{\infty} (n + 2)(n + 1)a_{n+2} x^n = -\sum_{n=0}^{\infty} (n + 1)a_n x^n$$

Now, by equating coefficients, we have $(n + 2)(n + 1)a_{n+2} = -(n + 1)a_n$, from which we get the recursion formula

$$a_{n+2} = -\frac{(n + 1)}{(n + 2)(n + 1)}a_n = -\frac{a_n}{n + 2}, \quad n \geq 0$$

and the coefficients of the solution series are

$$a_2 = -\frac{a_0}{2} \qquad\qquad a_3 = -\frac{a_1}{3}$$

$$a_4 = -\frac{a_2}{4} = \frac{a_0}{2 \cdot 4} \qquad\qquad a_5 = -\frac{a_3}{5} = \frac{a_1}{3 \cdot 5}$$

$$a_6 = -\frac{a_4}{6} = -\frac{a_0}{2 \cdot 4 \cdot 6} \qquad\qquad a_7 = -\frac{a_5}{7} = -\frac{a_1}{3 \cdot 5 \cdot 7}$$

$$\vdots \qquad\qquad\qquad\qquad \vdots$$

$$a_{2n} = \frac{(-1)^n a_0}{2 \cdot 4 \cdot 6 \cdots (2n)} = \frac{(-1)^n a_0}{2^n(n!)} \qquad a_{2n+1} = \frac{(-1)^n a_1}{3 \cdot 5 \cdot 7 \cdots (2n + 1)}$$

Thus, we can represent the general solution as the sum of two series, one for the even-powered terms with coefficients in terms of a_0 and one for the odd-powered terms with coefficients in terms of a_1.

$$y = a_0\left(1 - \frac{x^2}{2} + \frac{x^4}{2 \cdot 4} - \cdots\right) + a_1\left(x - \frac{x^3}{3} + \frac{x^5}{3 \cdot 5} - \cdots\right)$$

$$= a_0 \sum_{n=0}^{\infty} \frac{(-1)^n x^{2n}}{2^n(n!)} + a_1 \sum_{n=0}^{\infty} \frac{(-1)^n x^{2n+1}}{3 \cdot 5 \cdot 7 \cdots (2n + 1)}$$

Note that the solution has two arbitrary constants, a_0 and a_1, as we would expect in the general solution of a second-order differential equation. ☐

A second type of series solution method involves a differential equation with initial conditions and makes use of Taylor's Theorem, as given in Section 10.10.

EXAMPLE 3   Approximation by Taylor's Theorem

Use Taylor's Theorem to find the series solution of

$$y' = y^2 - x$$

given the initial condition $y = 1$ when $x = 0$. Then, use the first six terms of this series solution to approximate values of y for $0 \leq x \leq 1$.

Solution: Recall from Taylor's Theorem that, for $c = 0$,

$$y = y(0) + y'(0)x + \frac{y''(0)}{2!}x^2 + \frac{y'''(0)}{3!}x^3 + \cdots$$

Since $y(0) = 1$ and $y' = y^2 - x$, it follows that

$$y' = y^2 - x \qquad y(0) = 1$$
$$y'' = 2yy' - 1 \qquad y'(0) = 1$$
$$y''' = 2yy'' + 2(y')^2 \qquad y''(0) = 2 - 1 = 1$$
$$y^{(4)} = 2yy''' + 6y'y'' \qquad y'''(0) = 2 + 2 = 4$$
$$y^{(5)} = 2yy^{(4)} + 8y'y''' + 6(y'')^2 \qquad y^{(4)}(0) = 8 + 6 = 14$$
$$\qquad y^{(5)}(0) = 28 + 32 + 6 = 66$$

Therefore, we can approximate the values of the solution from the series

$$y = y(0) + y'(0)x + \frac{y''(0)}{2!}x^2 + \frac{y'''(0)}{3!}x^3 + \frac{y^{(4)}(0)}{4!}x^4 + \frac{y^{(5)}(0)}{5!}x^5 + \cdots$$

$$= 1 + x + \frac{1}{2}x^2 + \frac{4}{3!}x^3 + \frac{14}{4!}x^4 + \frac{66}{5!}x^5 + \cdots$$

Using the first six terms of this series, we compute several values for y in the interval $0 \le x \le 1$, as shown in Table 18.1.

TABLE 18.1

x	0.0	0.1	0.2	0.3	0.4	0.5	0.6	0.7	0.8	0.9	1.0
y	1.00000	1.1057	1.2264	1.3691	1.5432	1.7620	2.0424	2.4062	2.8805	3.4985	4.3000

From our experience with Taylor series, we know that, for a fixed number of terms, the farther we move from the center of convergence (in this case $c = 0$), the less accurate our estimate will be. This observation is supported by comparing the y-values in Table 18.1 with those in Table 18.2, for which we used the first *nine* terms in the series rather than the first *six* terms. Specifically, by continuing the procedure of Example 3, we can compute the first nine terms of the solution to be

$$y = 1 + x + \frac{1}{2}x^2 + \frac{4}{3!}x^3 + \frac{14}{4!}x^4$$
$$+ \frac{66}{5!}x^5 + \frac{352}{6!}x^6 + \frac{2,236}{7!}x^7 + \frac{16,092}{8!}x^8 + \cdots$$

TABLE 18.2

x	0.0	0.1	0.2	0.3	0.4	0.5	0.6	0.7	0.8	0.9	1.0
y	1.00000	1.1057	1.2265	1.3695	1.5462	1.7746	2.0843	2.5232	3.1686	4.1423	5.6316

Convergence of Taylor series

As a final topic in this section, we take another look at a problem introduced in Section 10.10.

EXAMPLE 4   Convergence of a Taylor series

Show that the binomial series

$$y = 1 + kx + \frac{k(k-1)x^2}{2} + \cdots + \frac{k(k-1)\cdots(k-n+1)x^n}{n!} + \cdots$$

converges to $f(x) = (1+x)^k$ in the interval $(-1, 1)$.

Solution: From Example 4 in Section 10.10 we know that this series converges to some function in the interval $(-1, 1)$. Furthermore, in this interval we know that the derivative of y is given by the series

$$y' = k + k(k-1)x + \frac{k(k-1)(k-2)x^2}{2!} + \cdots$$

$$+ \frac{k(k-1)\cdots(k-n+1)x^{n-1}}{(n-1)!} + \cdots$$

and, by multiplying by x, we have

$$xy' = kx + k(k-1)x^2 + \frac{k(k-1)(k-2)x^3}{2!} + \cdots$$

$$+ \frac{k(k-1)\cdots(k-n+1)x^n}{(n-1)!} + \cdots$$

Now, by adding these two series, we have

$$y' + xy' = k + [k(k-1) + k]x + \left[\frac{k(k-1)(k-2)}{2!} + k(k-1)\right]x^2 + \cdots$$

$$= k + k^2x + \frac{k^2(k-1)x^2}{2!} + \cdots$$

$$= k\left[1 + kx + \frac{k(k-1)x^2}{2!} + \cdots\right]$$

$$= ky$$

which yields the differential equation

$$y' + xy' = ky$$

In standard linear form, we have

$$y' - \frac{k}{1+x}y = 0$$

where $P(x) = -k/(1+x)$ and $Q(x) = 0$. Since $\int P(x)\, dx = -\ln(1+x)^k$, the general solution to the equation is

$$ye^{\int P(x)\, dx} = C \implies y(1+x)^{-k}$$

$$= C \implies y = C(1+x)^k$$

Finally, using the initial condition that $y = 1$ when $x = 0$, we determine that $C = 1$ and obtain the particular solution

$$y = (1+x)^k$$

Section Exercises 18.7

In Exercises 1–12, use power series to solve the differential equation.

1. $y' - y = 0$
2. $y' - ky = 0$
3. $y'' - 9y = 0$
4. $y'' - k^2y = 0$
5. $y'' + 4y = 0$
6. $y'' + k^2y = 0$
7. $xy' - 2y = 0$
8. $y' - 2xy = 0$
9. $xy'' - y = 0$
10. $y'' - xy' - y = 0$
11. $(x^2 + 4)y'' + y = 0$
12. $x^2y'' + xy' - y = 0$

In Exercises 13–16, use Taylor's Theorem to find the series solution to the differential equation with the specified initial conditions. Use n terms of the series to approximate y for the given value of x.

13. $y' + (2x - 1)y = 0$, $y(0) = 2$,
$n = 5$, $x = \dfrac{1}{2}$

14. $y' - 2xy = 0$, $y(0) = 1$,
$n = 4$, $x = 1$

15. $y'' - 2xy = 0$, $y(0) = 1$, $y'(0) = -3$,
$n = 6$, $x = \dfrac{1}{4}$

16. $y'' - 2xy' + y = 0$, $y(0) = 1$, $y'(0) = 2$,
$n = 8$, $x = \dfrac{1}{2}$

In Exercises 17–20, verify that the given function converges to its Maclaurin series on the indicated interval.

17. $f(x) = e^x$, $(-\infty, \infty)$
18. $f(x) = \cos x$, $(-\infty, \infty)$
19. $f(x) = \arcsin x$, $(-1, 1)$
20. $f(x) = \arctan x$, $(-1, 1)$

Review Exercises for Chapter 18

In Exercises 1–30, find the general solution of the first-order differential equation.

1. $\dfrac{dy}{dx} - \dfrac{y}{x} = 2 + \sqrt{x}$
2. $\dfrac{dy}{dx} + xy = 2y$
3. $y' - \dfrac{2y}{x} = \dfrac{y'}{x}$
4. $\dfrac{dy}{dx} - 3x^2y = e^{x^3}$
5. $\dfrac{dy}{dx} - \dfrac{y}{x} = \dfrac{x}{y}$
6. $\dfrac{dy}{dx} - \dfrac{3y}{x^2} = \dfrac{1}{x^2}$
7. $y' - 2y = e^x$
8. $y' + \dfrac{2y}{x} = -x^9y^5$
9. $(10x + 8y + 2)\,dx + (8x + 5y + 2)\,dy = 0$
10. $(y + x^3 + xy^2)\,dx - x\,dy = 0$
11. $(2x - 2y^3 + y)\,dx + (x - 6xy^2)\,dy = 0$
12. $3x^2y^2\,dx + (2x^3y + x^3y^4)\,dy = 0$
13. $x\,dy = (x + y + 2)\,dx$
14. $ye^{xy}\,dx + xe^{xy}\,dy = 0$
15. $(1 + y)\ln(1 + y)\,dx + dy = 0$
16. $(2x + y - 3)\,dx + (x - 3y + 1)\,dy = 0$
17. $dy = (y \tan x + 2e^x)\,dx$
18. $y\,dx - (x + \sqrt{xy})\,dy = 0$
19. $(x - y - 5)\,dx - (x + 3y - 2)\,dy = 0$
20. $y' = x^2y^2 - 9x^2$
21. $2xy' - y = x^3 - x$
22. $y' = 2x\sqrt{1 - y^2}$
23. $x + yy' = \sqrt{x^2 + y^2}$
24. $xy' + y = \sin x$
25. $yy' + y^2 = 1 + x^2$

26. $2x\,dx + 2y\,dy = (x^2 + y^2)\,dx$
27. $(1 + x^2)\,dy = (1 + y^2)\,dx$
28. $x^3yy' = x^4 + 3x^2y^2 + y^4$
29. $xy' - ay = bx^4$
30. $y' = y + 2x(y - e^x)$

In Exercises 31–36, find the general solution of the second-order differential equation.

31. $y'' + y = x^3 + x$
32. $y'' + 2y = e^{2x} + x$
33. $y'' + y = 2 \cos x$
34. $y'' + 5y' + 4y = x^2 + \sin 2x$
35. $y'' - 2y' + y = 2xe^x$
36. $y'' + 2y' + y = \dfrac{1}{x^2e^x}$

In Exercises 37 and 38, find the orthogonal trajectories of the given family and sketch several members of each family.

37. $(x - C)^2 + y^2 = C^2$
38. $y - 2x = C$

In Exercises 39 and 40, find the series solution to the differential equation.

39. $(x - 4)y' + y = 0$
40. $x^2y'' + y' - 3y = 0$

Appendices

Proofs of selected theorems

This appendix contains proofs of several theorems that are omitted in the text. We begin with two preliminary theorems dealing with the limit of a composite function and the limit of a reciprocal function.

LIMIT OF A COMPOSITE
FUNCTION

If

$$\lim_{x \to c} g(x) = L \qquad \text{and} \qquad \lim_{x \to L} f(x) = f(L)$$

then

$$\lim_{x \to c} f(g(x)) = f(L)$$

Proof: For a given $\varepsilon > 0$, we must find $\delta > 0$ such that

$$|f(g(x)) - f(L)| < \varepsilon \qquad \text{whenever} \qquad 0 < |x - c| < \delta$$

Since the limit of $f(x)$ as $x \to L$ is $f(L)$, we know there exists $\delta_1 > 0$ such that

$$|f(u) - f(L)| < \varepsilon \qquad \text{whenever} \qquad 0 < |u - L| < \delta_1$$

Moreover, since the limit of $g(x)$ as $x \to c$ is L, we know there exists $\delta > 0$ such that

$$|g(x) - L| < \delta_1 \qquad \text{whenever} \qquad 0 < |x - c| < \delta$$

Finally, letting $u = g(x)$, we have

$$|f(g(x)) - f(L)| < \varepsilon \qquad \text{whenever} \qquad 0 < |x - c| < \delta$$

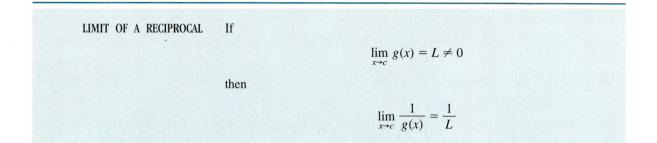

LIMIT OF A RECIPROCAL   If

$$\lim_{x \to c} g(x) = L \neq 0$$

then

$$\lim_{x \to c} \frac{1}{g(x)} = \frac{1}{L}$$

Proof: We let $f(x) = 1/x$ and consider the case for which $c > 0$. Then for $\varepsilon > 0$ we need to find $\delta > 0$ such that

$$\left| \frac{1}{x} - \frac{1}{c} \right| < \varepsilon \qquad \text{whenever} \qquad 0 < |x - c| < \delta$$

Since $c > 0$, we can write

$$\left| \frac{1}{x} - \frac{1}{c} \right| = \left| \frac{c - x}{cx} \right| = \frac{|x - c|}{c|x|}$$

Now, for x such that

$$\frac{c}{2} < x < \frac{3c}{2}$$

it follows that

$$\frac{1}{|x|} < \frac{2}{c}$$

and

$$\left| \frac{1}{x} - \frac{1}{c} \right| < \frac{2}{c^2} |x - c|$$

Therefore, if we choose δ to be the smaller of $c/2$ and $c^2 \varepsilon / 2$, then

$$0 < |x - c| < \delta \qquad \text{implies} \qquad \left| \frac{1}{x} - \frac{1}{c} \right| < \frac{2}{c^2} \left(\frac{c^2 \varepsilon}{2} \right) = \varepsilon$$

Now using the limit of a composite function, we can write

$$\lim_{x \to c} \frac{1}{g(x)} = \lim_{x \to c} f(g(x)) = f(L) = \frac{1}{L}$$

We use the limit of a composite function and the limit of a reciprocal function in the proof of the following theorem.

THEOREM 2.3
(Properties 4 and 5)

If b and c are real numbers, n is an integer, and the functions f and g have limits as $x \to c$, then the following properties are true.

4. $\lim\limits_{x \to c} \dfrac{f(x)}{g(x)} = \dfrac{\lim\limits_{x \to c} f(x)}{\lim\limits_{x \to c} g(x)}$, if $\lim\limits_{x \to c} g(x) \neq 0$

5. $\lim\limits_{x \to c} [f(x)]^n = \left[\lim\limits_{x \to c} f(x) \right]^n$

Proof: For Property 4, let

$$\lim_{x \to c} f(x) = L \quad \text{and} \quad \lim_{x \to c} g(x) = K$$

Using the limit of a product (Property 3 of Theorem 2.3) and the limit of a reciprocal function, we have

$$\lim_{x \to c} \frac{f(x)}{g(x)} = \lim_{x \to c} \left[f(x) \frac{1}{g(x)} \right] = \left[\lim_{x \to c} f(x) \right] \left[\lim_{x \to c} \frac{1}{g(x)} \right] = \frac{L}{K}$$

For Property 5, if n is a positive integer, then the proof can be obtained as a straightforward application of mathematical induction coupled with Property 3. If n is a negative integer, then we can write $n = -k$ where k is a positive integer and using the limit of a reciprocal function, we can write

$$\lim_{x \to c} [f(x)]^n = \lim_{x \to c} \left[\frac{1}{f(x)} \right]^k = \left[\lim_{x \to c} \frac{1}{f(x)} \right]^k$$

$$= \left[\frac{1}{\lim\limits_{x \to c} f(x)} \right]^k = \left[\lim_{x \to c} f(x) \right]^n$$

THEOREM 2.6
(Properties 1 and 2)

If c is a real number, n is an integer, and f is a function whose limit exists at c, then the following properties are true.

1. $\lim\limits_{x \to c} \sqrt[n]{x} = \sqrt[n]{c}$

2. $\lim\limits_{x \to c} \sqrt[n]{f(x)} = \sqrt[n]{f(c)}$

Proof: For Property 1, we consider the case for which $c > 0$ and n is any positive integer. For a given $\varepsilon > 0$, we need to find $\delta > 0$ such that

$$|\sqrt[n]{x} - \sqrt[n]{c}| < \varepsilon \quad \text{whenever} \quad 0 < |x - c| < \delta$$

which is the same as saying

$$-\varepsilon < \sqrt[n]{x} - \sqrt[n]{c} < \varepsilon \qquad \text{whenever} \qquad -\delta < x - c < \delta$$

We assume $\varepsilon < \sqrt[n]{c}$ which implies that $0 < \sqrt[n]{c} - \varepsilon < \sqrt[n]{c}$. Now, we let δ be the smaller of the two numbers.

$$c - (\sqrt[n]{c} - \varepsilon)^n \qquad \text{and} \qquad (\sqrt[n]{c} + \varepsilon)^n - c$$

then we have

$$
\begin{array}{ccc}
-\delta < & x - c & < \delta \\
-[c - (\sqrt[n]{c} - \varepsilon)^n] < & x - c & < (\sqrt[n]{c} + \varepsilon)^n - c \\
(\sqrt[n]{c} - \varepsilon)^n - c < & x - c & < (\sqrt[n]{c} + \varepsilon)^n - c \\
(\sqrt[n]{c} - \varepsilon)^n < & x & < (\sqrt[n]{c} + \varepsilon)^n \\
\sqrt[n]{c} - \varepsilon < & \sqrt[n]{x} & < \sqrt[n]{c} + \varepsilon \\
-\varepsilon < & \sqrt[n]{x} - \sqrt[n]{c} & < \varepsilon
\end{array}
$$

Property 2 follows from Property 1 together with the theorem on the limit of a composite function.

THEOREM 2.9   **CONTINUITY OF A COMPOSITE FUNCTION**
If g is continuous at c and f is continuous at $g(c)$, then the composite function given by $f \circ g(x) = f(g(x))$ is continuous at c.

Proof: Since g is continuous at c and f is continuous at $g(c)$, we can write

$$\lim_{x \to c} g(x) = g(c) \qquad \text{and} \qquad \lim_{u \to g(c)} f(u) = f(g(c))$$

and by applying the theorem on the limit of a composite function, we can conclude that

$$\lim_{x \to c} f(g(x)) = f(g(c))$$

THEOREM 2.11   **VERTICAL ASYMPTOTES**
Let f and g be continuous on an open interval containing c. If $f(c) \neq 0$, $g(c) = 0$, and there exists an open interval containing c such that $g(x) \neq 0$ for all $x \neq c$ in the interval, then the graph of the function given by

$$F(x) = \frac{f(x)}{g(x)}$$

has a vertical asymptote at $x = c$.

Proof: We consider the case for which $f(c) > 0$ and there exists $b > c$ such that $c < x < b$ implies $g(x) > 0$. Then for $M > 0$, we choose δ_1 such that

$$0 < x - c < \delta_1 \qquad \text{implies that} \qquad \frac{f(c)}{2} < f(x) < \frac{3f(c)}{2}$$

and δ_2 such that

$$0 < x - c < \delta_2 \qquad \text{implies that} \qquad 0 < g(x) < \frac{f(c)}{2M}$$

Now we let δ be the smaller of δ_1 and δ_2. Then it follows that

$$0 < x - c < \delta \qquad \text{implies that} \qquad \frac{f(x)}{g(x)} > \frac{f(c)}{2} \left(\frac{2M}{f(c)} \right) = M$$

Therefore, it follows that

$$\lim_{x \to c^+} \frac{f(x)}{g(x)} = \infty$$

and the line $x = c$ is a vertical asymptote of the graph of F.

THEOREM 3.9   **THE CHAIN RULE**
If $y = f(u)$ is a differentiable function of u and $u = g(x)$ is a differentiable function of x, then $y = f(g(x))$ is a differentiable function of x and

$$\frac{dy}{dx} = \frac{dy}{du} \frac{du}{dx}$$

Proof: In Section 3.5, we let $F(x) = f(g(x))$ and used the alternative form of the derivative to show that $F'(c) = f'(g(c))g'(c)$ provided $g(x) \neq g(c)$ for values of x other than c. We now consider a more general proof. We begin by considering the derivative of f.

$$f'(x) = \lim_{\Delta x \to 0} \frac{f(x + \Delta x) - f(x)}{\Delta x} = \lim_{\Delta x \to 0} \frac{\Delta y}{\Delta x}$$

For a fixed value of x, we define a function η such that

$$\eta(\Delta x) = \begin{cases} 0, & \Delta x = 0 \\ \dfrac{\Delta y}{\Delta x} - f'(x), & \Delta x \neq 0 \end{cases}$$

Since the limit of $\eta(\Delta x)$ as $\Delta x \to 0$ doesn't depend upon the value of $\eta(0)$, we have

$$\lim_{\Delta x \to 0} \eta(\Delta x) = \lim_{\Delta x \to 0} \left[\frac{\Delta y}{\Delta x} - f'(x) \right] = 0$$

and we can conclude that η is continuous at 0. Moreover, since $\Delta y = 0$ when $\Delta x = 0$, the equation

$$\Delta y = \Delta x \eta(\Delta x) + \Delta x f'(x)$$

is valid whether Δx is zero or not. Now, by letting $\Delta u = g(x + \Delta x) - g(x)$, we use the continuity of g to conclude that

$$\lim_{\Delta x \to 0} \Delta u = \lim_{\Delta x \to 0} [g(x + \Delta x) - g(x)] = 0$$

which implies that

$$\lim_{\Delta x \to 0} \eta(\Delta u) = 0$$

Finally,

$$\Delta y = \Delta u \eta(\Delta u) + \Delta u f'(u) \implies \frac{\Delta y}{\Delta x} = \frac{\Delta u}{\Delta x} \eta(\Delta u) + \frac{\Delta u}{\Delta x} f'(u), \quad \Delta x \neq 0$$

and taking the limit as $\Delta x \to 0$, we have

$$\frac{dy}{dx} = \frac{du}{dx} \left[\lim_{\Delta x \to 0} \eta(\Delta u) \right] + \frac{du}{dx} f'(u) = \frac{dy}{dx}(0) + \frac{du}{dx} f'(u)$$

$$= \frac{du}{dx} f'(u) = \frac{du}{dx} \frac{dy}{du}$$

THEOREM 7.4   **NATURAL LOGARITHMIC BASE IS e**
The number e is the base for the natural logarithmic function. That is,

$$\ln e = \int_1^e \frac{1}{t} \, dt = 1$$

Proof: We partition the interval $[1, e]$ into n subintervals as follows:

$$x_0 = 1 = \left(\frac{n}{n-1} \right)^0, \quad x_1 = \left(\frac{n}{n-1} \right)^1, \quad \ldots, \quad x_{n-1} = \left(\frac{n}{n-1} \right)^{n-1}, \quad x_n = e$$

The width of the first $n - 1$ subintervals is given by

$$\Delta x_i = x_i - x_{i-1} = \left(\frac{n}{n-1} \right)^i - \left(\frac{n}{n-1} \right)^{i-1} = \left(\frac{n}{n-1} \right)^{i-1} \left(\frac{1}{n-1} \right)$$

The nth subinterval has a width of $\Delta x_n = e - [n/(n-1)]^{n-1}$. Therefore, for $1 \leq i \leq n - 1$

$$\frac{1}{n-1} \leq \left(\frac{n}{n-1} \right)^{i-1} \left(\frac{1}{n-1} \right)$$

Moreover, since

$$\lim_{n \to \infty} \frac{1}{n-1} = 0 \quad \text{and} \quad \lim_{n \to \infty} \left[e - \left(\frac{n}{n-1} \right)^{n-1} \right] = 0$$

it follows that the norm of this partition approaches zero as $n \to \infty$. Now, for $f(t) = 1/t$, we choose c_i to be the left endpoint of the ith subinterval. Then for $i < n$, we have

$$f(c_i) \, \Delta x_i = \left(\frac{1}{c_i}\right)\left(\frac{n}{n-1}\right)^{i-1}\left(\frac{1}{n-1}\right)$$

$$= \left(\frac{n-1}{n}\right)^{i-1}\left(\frac{n}{n-1}\right)^{i-1}\left(\frac{1}{n-1}\right)$$

$$= \frac{1}{n-1}$$

and for the nth subinterval, we have $f(c_{n-1}) \, \Delta x_n = e[(n-1)/n]^{n-1} - 1$. Summing these values, and taking the limit as $n \to \infty$, we have

$$\ln e = \int_1^e \frac{1}{t} \, dt = \lim_{n \to \infty} \sum_{i=1}^{n} f(c_i) \, \Delta x_i$$

$$= \lim_{n \to \infty} \left[\left(\sum_{i=1}^{n-1} \frac{1}{n-1}\right) + e\left(\frac{n-1}{n}\right)^{n-1} - 1\right]$$

$$= \lim_{n \to \infty} \left[1 + e\left(\frac{n-1}{n}\right)^{n-1} - 1\right]$$

$$= e\left[\lim_{n \to \infty} \left(\frac{n-1}{n}\right)^{n-1}\right] = e\left(\frac{1}{e}\right) = 1$$

THE EXTENDED MEAN VALUE THEOREM	If f and g are differentiable on an open interval (a, b) and continuous on $[a, b]$ such that $g(x) \neq 0$ for any x in (a, b), then there exists a point c in (a, b) such that $$\frac{f'(c)}{g'(c)} = \frac{f(b) - f(a)}{g(b) - g(a)}$$

Proof: We can assume that $g(a) \neq g(b)$, since otherwise by Rolle's Theorem it would follow that $g'(x) = 0$ for some x in (a, b). Now, we define $h(x)$ to be

$$h(x) = f(x) - \left(\frac{f(b) - f(a)}{g(b) - g(a)}\right) g(x)$$

Then

$$h(a) = f(a) - \left(\frac{f(b) - f(a)}{g(b) - g(a)}\right) g(a) = \frac{f(a)g(b) - f(b)g(a)}{g(b) - g(a)}$$

and

$$h(b) = f(b) - \left(\frac{f(b) - f(a)}{g(b) - g(a)}\right) g(b) = \frac{f(a)g(b) - f(b)g(a)}{g(b) - g(a)}$$

and by Rolle's Theorem there exists a point c in (a, b) such that

$$h'(c) = f'(c) - \frac{f(b) - f(a)}{g(b) - g(a)} g'(c) = 0$$

which implies that

$$\frac{f'(c)}{g'(c)} = \frac{f(b) - f(a)}{g(b) - g(a)}$$

We can use the Extended Mean Value Theorem to prove L'Hôpital's Rule. Of the several different cases of this rule, we illustrate the proof of only one case and leave the remaining cases for you to prove.

THEOREM 7.19   **L'HÔPITAL'S RULE**

If $\lim f(x)/g(x)$ results in the indeterminate form $0/0$ or ∞/∞, then

$$\lim \frac{f(x)}{g(x)} = \lim \frac{f'(x)}{g'(x)}$$

provided the latter limit exists (or is infinite).

Proof: We consider the case for which

$$\lim_{x \to a^+} f'(x) = 0 \qquad \text{and} \qquad \lim_{x \to a^+} g'(x) = 0$$

Moreover, we assume that

$$\lim_{x \to a^+} \frac{f'(x)}{g'(x)} = L$$

Now, we choose b such that both f and g are differentiable in the interval (a, b), and define two new functions

$$F(x) = \begin{cases} f(x), & x \neq a \\ 0, & x = a \end{cases} \qquad \text{and} \qquad G(x) = \begin{cases} g(x), & x \neq a \\ 0, & x = a \end{cases}$$

It follows that F and G are continuous on $[a, b]$ and we can apply the Extended Mean Value Theorem to conclude that there exists a number c in (a, b) such that

$$\frac{F'(c)}{G'(c)} = \frac{F(b) - F(a)}{G(b) - G(a)} = \frac{F(b) - 0}{G(b) - 0} = \frac{F(b)}{G(b)}$$

Finally, by letting b approach a, we also force c to approach a and it follows that

$$\lim_{b \to a^+} \frac{f(b)}{g(b)} = \lim_{b \to a^+} \frac{F(b)}{G(b)} = \lim_{c \to a^+} \frac{F'(c)}{G'(c)} = \lim_{c \to a^+} \frac{f'(c)}{g'(c)}$$

and we conclude that

$$\lim_{x \to a^+} \frac{f(x)}{g(x)} = \lim_{x \to a^+} \frac{f'(x)}{g'(x)}$$

B Exponential tables

x	e^x	e^{-x}	x	e^x	e^{-x}
0.0	1.0000	1.0000	3.0	20.086	0.0498
0.1	1.1052	0.9048	3.1	22.198	0.0450
0.2	1.2214	0.8187	3.2	24.533	0.0408
0.3	1.3499	0.7408	3.3	27.113	0.0369
0.4	1.4918	0.6703	3.4	29.964	0.0334
0.5	1.6487	0.6065	3.5	33.115	0.0302
0.6	1.8221	0.5488	3.6	36.598	0.0273
0.7	2.0138	0.4966	3.7	40.447	0.0247
0.8	2.2255	0.4493	3.8	44.701	0.0224
0.9	2.4596	0.4066	3.9	49.402	0.0202
1.0	2.7183	0.3679	4.0	54.598	0.0183
1.1	3.0042	0.3329	4.1	60.340	0.0166
1.2	3.3201	0.3012	4.2	66.686	0.0150
1.3	3.6693	0.2725	4.3	73.700	0.0136
1.4	4.0552	0.2466	4.4	81.451	0.0123
1.5	4.4817	0.2231	4.5	90.017	0.0111
1.6	4.9530	0.2019	4.6	99.484	0.0101
1.7	5.4739	0.1827	4.7	109.95	0.0091
1.8	6.0496	0.1653	4.8	121.51	0.0082
1.9	6.6859	0.1496	4.9	134.29	0.0074
2.0	7.3891	0.1353	5.0	148.41	0.0067
2.1	8.1662	0.1225	5.1	164.02	0.0061
2.2	9.0250	0.1108	5.2	181.27	0.0055
2.3	9.9742	0.1003	5.3	200.34	0.0050
2.4	11.023	0.0907	5.4	221.41	0.0045
2.5	12.182	0.0821	5.5	244.69	0.0041
2.6	13.464	0.0743	5.6	270.43	0.0037
2.7	14.880	0.0672	5.7	298.87	0.0033
2.8	16.445	0.0608	5.8	330.30	0.0030
2.9	18.174	0.0550	5.9	365.04	0.0027

EXPONENTIAL TABLES (Continued)

x	e^x	e^{-x}	x	e^x	e^{-x}
6.0	403.43	0.0025	8.0	2980.96	0.0003
6.1	445.86	0.0022	8.1	3294.47	0.0003
6.2	492.75	0.0020	8.2	3640.95	0.0003
6.3	544.57	0.0018	8.3	4023.87	0.0002
6.4	601.85	0.0017	8.4	4447.07	0.0002
6.5	665.14	0.0015	8.5	4914.77	0.0002
6.6	735.10	0.0014	8.6	5431.66	0.0002
6.7	812.41	0.0012	8.7	6002.91	0.0002
6.8	897.85	0.0011	8.8	6634.24	0.0002
6.9	992.27	0.0010	8.9	7331.97	0.0001
7.0	1096.63	0.0009	9.0	8103.08	0.0001
7.1	1211.97	0.0008	9.1	8955.29	0.0001
7.2	1339.43	0.0007	9.2	9897.13	0.0001
7.3	1480.30	0.0007	9.3	10938.02	0.0001
7.4	1635.98	0.0006	9.4	12088.38	0.0001
7.5	1808.04	0.0006	9.5	13359.73	0.0001
7.6	1998.20	0.0005	9.6	14764.78	0.0001
7.7	2208.35	0.0005	9.7	16317.61	0.0001
7.8	2440.60	0.0004	9.8	18033.74	0.0001
7.9	2697.28	0.0004	9.9	19930.37	0.0001
			10.0	22026.47	0.0000

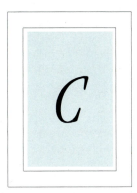

C Natural logarithmic tables

	0.00	0.01	0.02	0.03	0.04	0.05	0.06	0.07	0.08	0.09
1.0	0.0000	0.0100	0.0198	0.0296	0.0392	0.0488	0.0583	0.0677	0.0770	0.0862
1.1	0.0953	0.1044	0.1133	0.1222	0.1310	0.1398	0.1484	0.1570	0.1655	0.1740
1.2	0.1823	0.1906	0.1989	0.2070	0.2151	0.2231	0.2311	0.2390	0.2469	0.2546
1.3	0.2624	0.2700	0.2776	0.2852	0.2927	0.3001	0.3075	0.3148	0.3221	0.3293
1.4	0.3365	0.3436	0.3507	0.3577	0.3646	0.3716	0.3784	0.3853	0.3920	0.3988
1.5	0.4055	0.4121	0.4187	0.4253	0.4318	0.4383	0.4447	0.4511	0.4574	0.4637
1.6	0.4700	0.4762	0.4824	0.4886	0.4947	0.5008	0.5068	0.5128	0.5188	0.5247
1.7	0.5306	0.5365	0.5423	0.5481	0.5539	0.5596	0.5653	0.5710	0.5766	0.5822
1.8	0.5878	0.5933	0.5988	0.6043	0.6098	0.6152	0.6206	0.6259	0.6313	0.6366
1.9	0.6419	0.6471	0.6523	0.6575	0.6627	0.6678	0.6729	0.6780	0.6831	0.6881
2.0	0.6931	0.6981	0.7031	0.7080	0.7129	0.7178	0.7227	0.7275	0.7324	0.7372
2.1	0.7419	0.7467	0.7514	0.7561	0.7608	0.7655	0.7701	0.7747	0.7793	0.7839
2.2	0.7885	0.7930	0.7975	0.8020	0.8065	0.8109	0.8154	0.8198	0.8242	0.8286
2.3	0.8329	0.8372	0.8416	0.8459	0.8502	0.8544	0.8587	0.8629	0.8671	0.8713
2.4	0.8755	0.8796	0.8838	0.8879	0.8920	0.8961	0.9002	0.9042	0.9083	0.9123
2.5	0.9163	0.9203	0.9243	0.9282	0.9322	0.9361	0.9400	0.9439	0.9478	0.9517
2.6	0.9555	0.9594	0.9632	0.9670	0.9708	0.9746	0.9783	0.9821	0.9858	0.9895
2.7	0.9933	0.9969	1.0006	1.0043	1.0080	1.0116	1.0152	1.0188	1.0225	1.0260
2.8	1.0296	1.0332	1.0367	1.0403	1.0438	1.0473	1.0508	1.0543	1.0578	1.0613
2.9	1.0647	1.0682	1.0716	1.0750	1.0784	1.0818	1.0852	1.0886	1.0919	1.0953
3.0	1.0986	1.1019	1.1053	1.1086	1.1119	1.1151	1.1184	1.1217	1.1249	1.1282
3.1	1.1314	1.1346	1.1378	1.1410	1.1442	1.1474	1.1506	1.1537	1.1569	1.1600
3.2	1.1632	1.1663	1.1694	1.1725	1.1756	1.1787	1.1817	1.1848	1.1878	1.1909
3.3	1.1939	1.1969	1.2000	1.2030	1.2060	1.2090	1.2119	1.2149	1.2179	1.2208
3.4	1.2238	1.2267	1.2296	1.2326	1.2355	1.2384	1.2413	1.2442	1.2470	1.2499

NATURAL LOGARITHMIC TABLES (Continued)

	0.00	0.01	0.02	0.03	0.04	0.05	0.06	0.07	0.08	0.09
3.5	1.2528	1.2556	1.2585	1.2613	1.2641	1.2669	1.2698	1.2726	1.2754	1.2782
3.6	1.2809	1.2837	1.2865	1.2892	1.2920	1.2947	1.2975	1.3002	1.3029	1.3056
3.7	1.3083	1.3110	1.3137	1.3164	1.3191	1.3218	1.3244	1.3271	1.3297	1.3324
3.8	1.3350	1.3376	1.3403	1.3429	1.3455	1.3481	1.3507	1.3533	1.3558	1.3584
3.9	1.3610	1.3635	1.3661	1.3686	1.3712	1.3737	1.3762	1.3788	1.3813	1.3838
4.0	1.3863	1.3888	1.3913	1.3938	1.3962	1.3987	1.4012	1.4036	1.4061	1.4085
4.1	1.4110	1.4134	1.4159	1.4183	1.4207	1.4231	1.4255	1.4279	1.4303	1.4327
4.2	1.4351	1.4375	1.4398	1.4422	1.4446	1.4469	1.4493	1.4516	1.4540	1.4563
4.3	1.4586	1.4609	1.4633	1.4656	1.4679	1.4702	1.4725	1.4748	1.4770	1.4793
4.4	1.4816	1.4839	1.4861	1.4884	1.4907	1.4929	1.4951	1.4974	1.4996	1.5019
4.5	1.5041	1.5063	1.5085	1.5107	1.5129	1.5151	1.5173	1.5195	1.5217	1.5239
4.6	1.5261	1.5282	1.5304	1.5326	1.5347	1.5369	1.5390	1.5412	1.5433	1.5454
4.7	1.5476	1.5497	1.5518	1.5539	1.5560	1.5581	1.5602	1.5623	1.5644	1.5665
4.8	1.5686	1.5707	1.5728	1.5748	1.5769	1.5790	1.5810	1.5831	1.5851	1.5872
4.9	1.5892	1.5913	1.5933	1.5953	1.5974	1.5994	1.6014	1.6034	1.6054	1.6074
5.0	1.6094	1.6114	1.6134	1.6154	1.6174	1.6194	1.6214	1.6233	1.6253	1.6273
5.1	1.6292	1.6312	1.6332	1.6351	1.6371	1.6390	1.6409	1.6429	1.6448	1.6467
5.2	1.6487	1.6506	1.6525	1.6544	1.6563	1.6582	1.6601	1.6620	1.6639	1.6658
5.3	1.6677	1.6696	1.6715	1.6734	1.6752	1.6771	1.6790	1.6808	1.6827	1.6845
5.4	1.6864	1.6882	1.6901	1.6919	1.6938	1.6956	1.6974	1.6993	1.7011	1.7029
5.5	1.7047	1.7066	1.7084	1.7102	1.7120	1.7138	1.7156	1.7174	1.7192	1.7210
5.6	1.7228	1.7246	1.7263	1.7281	1.7299	1.7317	1.7334	1.7352	1.7370	1.7387
5.7	1.7405	1.7422	1.7440	1.7457	1.7475	1.7492	1.7509	1.7527	1.7544	1.7561
5.8	1.7579	1.7596	1.7613	1.7630	1.7647	1.7664	1.7681	1.7699	1.7716	1.7733
5.9	1.7750	1.7766	1.7783	1.7800	1.7817	1.7834	1.7851	1.7867	1.7884	1.7901
6.0	1.7918	1.7934	1.7951	1.7967	1.7984	1.8001	1.8017	1.8034	1.8050	1.8066
6.1	1.8083	1.8099	1.8116	1.8132	1.8148	1.8165	1.8181	1.8197	1.8213	1.8229
6.2	1.8245	1.8262	1.8278	1.8294	1.8310	1.8326	1.8342	1.8358	1.8374	1.8390
6.3	1.8405	1.8421	1.8437	1.8453	1.8469	1.8485	1.8500	1.8516	1.8532	1.8547
6.4	1.8563	1.8579	1.8594	1.8610	1.8625	1.8641	1.8656	1.8672	1.8687	1.8703
6.5	1.8718	1.8733	1.8749	1.8764	1.8779	1.8795	1.8810	1.8825	1.8840	1.8856
6.6	1.8871	1.8886	1.8901	1.8916	1.8931	1.8946	1.8961	1.8976	1.8991	1.9006
6.7	1.9021	1.9036	1.9051	1.9066	1.9081	1.9095	1.9110	1.9125	1.9140	1.9155
6.8	1.9169	1.9184	1.9199	1.9213	1.9228	1.9242	1.9257	1.9272	1.9286	1.9301
6.9	1.9315	1.9330	1.9344	1.9359	1.9373	1.9387	1.9402	1.9416	1.9430	1.9445
7.0	1.9459	1.9473	1.9488	1.9502	1.9516	1.9530	1.9544	1.9559	1.9573	1.9587
7.1	1.9601	1.9615	1.9629	1.9643	1.9657	1.9671	1.9685	1.9699	1.9713	1.9727
7.2	1.9741	1.9755	1.9769	1.9782	1.9796	1.9810	1.9824	1.9838	1.9851	1.9865
7.3	1.9879	1.9892	1.9906	1.9920	1.9933	1.9947	1.9961	1.9974	1.9988	2.0001
7.4	2.0015	2.0028	2.0042	2.0055	2.0069	2.0082	2.0096	2.0109	2.0122	2.0136
7.5	2.0149	2.0162	2.0176	2.0189	2.0202	2.0215	2.0229	2.0242	2.0255	2.0268
7.6	2.0281	2.0295	2.0308	2.0321	2.0334	2.0347	2.0360	2.0373	2.0386	2.0399
7.7	2.0412	2.0425	2.0438	2.0451	2.0464	2.0477	2.0490	2.0503	2.0516	2.0528
7.8	2.0541	2.0554	2.0567	2.0580	2.0592	2.0605	2.0618	2.0631	2.0643	2.0656
7.9	2.0669	2.0681	2.0694	2.0707	2.0719	2.0732	2.0744	2.0757	2.0769	2.0782

NATURAL LOGARITHMIC TABLES (Continued)

	0.00	0.01	0.02	0.03	0.04	0.05	0.06	0.07	0.08	0.09
8.0	2.0794	2.0807	2.0819	2.0832	2.0844	2.0857	2.0869	2.0882	2.0894	2.0906
8.1	2.0919	2.0931	2.0943	2.0956	2.0968	2.0980	2.0992	2.1005	2.1017	2.1029
8.2	2.1041	2.1054	2.1066	2.1078	2.1090	2.1102	2.1114	2.1126	2.1138	2.1150
8.3	2.1163	2.1175	2.1187	2.1199	2.1211	2.1223	2.1235	2.1247	2.1258	2.1270
8.4	2.1282	2.1294	2.1306	2.1318	2.1330	2.1342	2.1353	2.1365	2.1377	2.1389
8.5	2.1401	2.1412	2.1424	2.1436	2.1448	2.1459	2.1471	2.1483	2.1494	2.1506
8.6	2.1518	2.1529	2.1541	2.1552	2.1564	2.1576	2.1587	2.1599	2.1610	2.1622
8.7	2.1633	2.1645	2.1656	2.1668	2.1679	2.1691	2.1702	2.1713	2.1725	2.1736
8.8	2.1748	2.1759	2.1770	2.1782	2.1793	2.1804	2.1815	2.1827	2.1838	2.1849
8.9	2.1861	2.1872	2.1883	2.1894	2.1905	2.1917	2.1928	2.1939	2.1950	2.1961
9.0	2.1972	2.1983	2.1994	2.2006	2.2017	2.2028	2.2039	2.2050	2.2061	2.2072
9.1	2.2083	2.2094	2.2105	2.2116	2.2127	2.2138	2.2148	2.2159	2.2170	2.2181
9.2	2.2192	2.2203	2.2214	2.2225	2.2235	2.2246	2.2257	2.2268	2.2279	2.2289
9.3	2.2300	2.2311	2.2322	2.2332	2.2343	2.2354	2.2364	2.2375	2.2386	2.2396
9.4	2.2407	2.2418	2.2428	2.2439	2.2450	2.2460	2.2471	2.2481	2.2492	2.2502
9.5	2.2513	2.2523	2.2534	2.2544	2.2555	2.2565	2.2576	2.2586	2.2597	2.2607
9.6	2.2618	2.2628	2.2638	2.2649	2.2659	2.2670	2.2680	2.2690	2.2701	2.2711
9.7	2.2721	2.2732	2.2742	2.2752	2.2762	2.2773	2.2783	2.2793	2.2803	2.2814
9.8	2.2824	2.2834	2.2844	2.2854	2.2865	2.2875	2.2885	2.2895	2.2905	2.2915
9.9	2.2925	2.2935	2.2946	2.2956	2.2966	2.2976	2.2986	2.2996	2.3006	2.3016

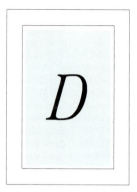

D

Trigonometric tables

1 degree ≈ 0.01745 radians
1 radian ≈ 57.29578 degrees

For $0 \leqslant \theta \leqslant 45$, read from upper left.
For $45 \leqslant \theta \leqslant 90$, read from lower right.
For $90 \leqslant \theta \leqslant 360$, use the identities:

θ	Quadrant II	Quadrant III	Quadrant IV
$\sin \theta$	$\sin(180-\theta)$	$-\sin(\theta-180)$	$-\sin(360-\theta)$
$\cos \theta$	$-\cos(180-\theta)$	$-\cos(\theta-180)$	$\cos(360-\theta)$
$\tan \theta$	$-\tan(180-\theta)$	$\tan(\theta-180)$	$-\tan(360-\theta)$
$\cot \theta$	$-\cot(180-\theta)$	$\cot(\theta-180)$	$-\cot(360-\theta)$

Degrees	Radians	sin	cos	tan	cot		
0° 00′	.0000	.0000	1.0000	.0000	–	1.5708	90° 00′
10	.0029	.0029	1.0000	.0029	343.774	1.5679	50
20	.0058	.0058	1.0000	.0058	171.885	1.5650	40
30	.0087	.0087	1.0000	.0087	114.589	1.5621	30
40	.0116	.0116	.9999	.0116	85.940	1.5592	20
50	.0145	.0145	.9999	.0145	68.750	1.5563	10
1° 00′	.0175	.0175	.9998	.0175	57.290	1.5533	89° 00′
10	.0204	.0204	.9998	.0204	49.104	1.5504	50
20	.0233	.0233	.9997	.0233	42.964	1.5475	40
30	.0262	.0262	.9997	.0262	38.188	1.5446	30
40	.0291	.0291	.9996	.0291	34.368	1.5417	20
50	.0320	.0320	.9995	.0320	31.242	1.5388	10
2° 00′	.0349	.0349	.9994	.0349	28.636	1.5359	88° 00′
10	.0378	.0378	.9993	.0378	26.432	1.5330	50
20	.0407	.0407	.9992	.0407	24.542	1.5301	40
30	.0436	.0436	.9990	.0437	22.904	1.5272	30
40	.0465	.0465	.9989	.0466	21.470	1.5243	20
50	.0495	.0494	.9988	.0495	20.206	1.5213	10
3° 00′	.0524	.0523	.9986	.0524	19.081	1.5184	87° 00′
10	.0553	.0552	.9985	.0553	18.075	1.5155	50
20	.0582	.0581	.9983	.0582	17.169	1.5126	40
30	.0611	.0610	.9981	.0612	16.350	1.5097	30
40	.0640	.0640	.9980	.0641	15.605	1.5068	20
50	.0669	.0669	.9978	.0670	14.924	1.5039	10
4° 00′	.0698	.0698	.9976	.0699	14.301	1.5010	86° 00′
10	.0727	.0727	.9974	.0729	13.727	1.4981	50
20	.0756	.0756	.9971	.0758	13.197	1.4952	40
30	.0785	.0785	.9969	.0787	12.706	1.4923	30
40	.0814	.0814	.9967	.0816	12.251	1.4893	20
50	.0844	.0843	.9964	.0846	11.826	1.4864	10
5° 00′	.0873	.0872	.9962	.0875	11.430	1.4835	85° 00′
10	.0902	.0901	.9959	.0904	11.059	1.4806	50
20	.0931	.0929	.9957	.0934	10.712	1.4777	40
30	.0960	.0958	.9954	.0963	10.385	1.4748	30
40	.0989	.0987	.9951	.0992	10.078	1.4719	20
50	.1018	.1016	.9948	.1022	9.788	1.4690	10
		cos	sin	cot	tan	Radians	Degrees

TRIGONOMETRIC TABLES (Continued)

Degrees	Radians	sin	cos	tan	cot		
6°00′	.1047	.1045	.9945	.1051	9.514	1.4661	84°00′
10	.1076	.1074	.9942	.1080	9.255	1.4632	50
20	.1105	.1103	.9939	.1110	9.010	1.4603	40
30	.1134	.1132	.9936	.1139	8.777	1.4573	30
40	.1164	.1161	.9932	.1169	8.556	1.4544	20
50	.1193	.1190	.9929	.1198	8.345	1.4515	10
7°00′	.1222	.1219	.9925	.1228	8.144	1.4486	83°00′
10	.1251	.1248	.9922	.1257	7.953	1.4457	50
20	.1280	.1276	.9918	.1287	7.770	1.4428	40
30	.1309	.1305	.9914	.1317	7.596	1.4399	30
40	.1338	.1334	.9911	.1346	7.429	1.4370	20
50	.1367	.1363	.9907	.1376	7.269	1.4341	10
8°00′	.1396	.1392	.9903	.1405	7.115	1.4312	82°00′
10	.1425	.1421	.9899	.1435	6.968	1.4283	50
20	.1454	.1449	.9894	.1465	6.827	1.4254	40
30	.1484	.1478	.9890	.1495	6.691	1.4224	30
40	.1513	.1507	.9886	.1524	6.561	1.4195	20
50	.1542	.1536	.9881	.1554	6.435	1.4166	10
9°00′	.1571	.1564	.9877	.1584	6.314	1.4137	81°00′
10	.1600	.1593	.9872	.1614	6.197	1.4108	50
20	.1629	.1622	.9868	.1644	6.084	1.4079	40
30	.1658	.1650	.9863	.1673	5.976	1.4050	30
40	.1687	.1679	.9858	.1703	5.871	1.4021	20
50	.1716	.1708	.9853	.1733	5.769	1.3992	10
10°00′	.1745	.1736	.9848	.1763	5.671	1.3963	80°00′
10	.1774	.1765	.9843	.1793	5.576	1.3934	50
20	.1804	.1794	.9838	.1823	5.485	1.3904	40
30	.1833	.1822	.9833	.1853	5.396	1.3875	30
40	.1862	.1851	.9827	.1883	5.309	1.3846	20
50	.1891	.1880	.9822	.1914	5.226	1.3817	10
11°00′	.1920	.1908	.9816	.1944	5.145	1.3788	79°00′
10	.1949	.1937	.9811	.1974	5.066	1.3759	50
20	.1978	.1965	.9805	.2004	4.989	1.3730	40
30	.2007	.1994	.9799	.2035	4.915	1.3701	30
40	.2036	.2022	.9793	.2065	4.843	1.3672	20
50	.2065	.2051	.9787	.2095	4.773	1.3643	10
12°00′	.2094	.2079	.9781	.2126	4.705	1.3614	78°00′
10	.2123	.2108	.9775	.2156	4.638	1.3584	50
20	.2153	.2136	.9769	.2186	4.574	1.3555	40
30	.2182	.2164	.9763	.2217	4.511	1.3526	30
40	.2211	.2193	.9757	.2247	4.449	1.3497	20
50	.2240	.2221	.9750	.2278	4.390	1.3468	10
13°00′	.2269	.2250	.9744	.2309	4.331	1.3439	77°00′
10	.2298	.2278	.9737	.2339	4.275	1.3410	50
20	.2327	.2306	.9730	.2370	4.219	1.3381	40
30	.2356	.2334	.9724	.2401	4.165	1.3352	30
40	.2385	.2363	.9717	.2432	4.113	1.3323	20
50	.2414	.2391	.9710	.2462	4.061	1.3294	10
14°00′	.2443	.2419	.9703	.2493	4.011	1.3265	76°00′
10	.2473	.2447	.9696	.2524	3.962	1.3235	50
20	.2502	.2476	.9689	.2555	3.914	1.3206	40
30	.2531	.2504	.9681	.2586	3.867	1.3177	30
40	.2560	.2532	.9674	.2617	3.821	1.3148	20
50	.2589	.2560	.9667	.2648	3.776	1.3119	10
15°00′	.2618	.2588	.9659	.2679	3.732	1.3090	75°00′
10	.2647	.2616	.9652	.2711	3.689	1.3061	50
20	.2676	.2644	.9644	.2742	3.647	1.3032	40
30	.2705	.2672	.9636	.2773	3.606	1.3003	30
40	.2734	.2700	.9628	.2805	3.566	1.2974	20
50	.2763	.2728	.9621	.2836	3.526	1.2945	10
		cos	sin	cot	tan	Radians	Degrees

TRIGONOMETRIC TABLES (Continued)

Degrees	Radians	sin	cos	tan	cot		
16°00′	.2793	.2756	.9613	.2867	3.487	1.2915	74°00′
10	.2822	.2784	.9605	.2899	3.450	1.2886	50
20	.2851	.2812	.9596	.2931	3.412	1.2857	40
30	.2880	.2840	.9588	.2962	3.376	1.2828	30
40	.2909	.2868	.9580	.2994	3.340	1.2799	20
50	.2938	.2896	.9572	.3026	3.305	1.2770	10
17°00′	.2967	.2924	.9563	.3057	3.271	1.2741	73°00′
10	.2996	.2952	.9555	.3089	3.237	1.2712	50
20	.3025	.2979	.9546	.3121	3.204	1.2683	40
30	.3054	.3007	.9537	.3153	3.172	1.2654	30
40	.3083	.3035	.9528	.3185	3.140	1.2625	20
50	.3113	.3062	.9520	.3217	3.108	1.2595	10
18°00′	.3142	.3090	.9511	.3249	3.078	1.2566	72°00′
10	.3171	.3118	.9502	.3281	3.047	1.2537	50
20	.3200	.3145	.9492	.3314	3.018	1.2508	40
30	.3229	.3173	.9483	.3346	2.989	1.2479	30
40	.3258	.3201	.9474	.3378	2.960	1.2450	20
50	.3287	.3228	.9465	.3411	2.932	1.2421	10
19°00′	.3316	.3256	.9455	.3443	2.904	1.2392	71°00′
10	.3345	.3283	.9446	.3476	2.877	1.2363	50
20	.3374	.3311	.9436	.3508	2.850	1.2334	40
30	.3403	.3338	.9426	.3541	2.824	1.2305	30
40	.3432	.3365	.9417	.3574	2.798	1.2275	20
50	.3462	.3393	.9407	.3607	2.773	1.2246	10
20°00′	.3491	.3420	.9397	.3640	2.747	1.2217	70°00′
10	.3520	.3448	.9387	.3673	2.723	1.2188	50
20	.3549	.3475	.9377	.3706	2.699	1.2159	40
30	.3578	.3502	.9367	.3739	2.675	1.2130	30
40	.3607	.3529	.9356	.3772	2.651	1.2101	20
50	.3636	.3557	.9346	.3805	2.628	1.2072	10
21°00′	.3665	.3584	.9336	.3839	2.605	1.2043	69°00′
10	.3694	.3611	.9325	.3872	2.583	1.2014	50
20	.3723	.3638	.9315	.3906	2.560	1.1985	40
30	.3752	.3665	.9304	.3939	2.539	1.1956	30
40	.3782	.3692	.9293	.3973	2.517	1.1926	20
50	.3811	.3719	.9283	.4006	2.496	1.1897	10
22°00′	.3840	.3746	.9272	.4040	2.475	1.1868	68°00′
10	.3869	.3773	.9261	.4074	2.455	1.1839	50
20	.3898	.3800	.9250	.4108	2.434	1.1810	40
30	.3927	.3827	.9239	.4142	2.414	1.1781	30
40	.3956	.3854	.9228	.4176	2.394	1.1752	20
50	.3985	.3881	.9216	.4210	2.375	1.1723	10
23°00′	.4014	.3907	.9205	.4245	2.356	1.1694	67°00′
10	.4043	.3934	.9194	.4279	2.337	1.1665	50
20	.4072	.3961	.9182	.4314	2.318	1.1636	40
30	.4102	.3987	.9171	.4348	2.300	1.1606	30
40	.4131	.4014	.9159	.4383	2.282	1.1577	20
50	.4160	.4041	.9147	.4417	2.264	1.1548	10
24°00′	.4189	.4067	.9135	.4452	2.246	1.1519	66°00′
10	.4218	.4094	.9124	.4487	2.229	1.1490	50
20	.4247	.4120	.9112	.4522	2.211	1.1461	40
30	.4276	.4147	.9100	.4557	2.194	1.1432	30
40	.4305	.4173	.9088	.4592	2.177	1.1403	20
50	.4334	.4200	.9075	.4628	2.161	1.1374	10
25°00′	.4363	.4226	.9063	.4663	2.145	1.1345	65°00′
10	.4392	.4253	.9051	.4699	2.128	1.1316	50
20	.4422	.4279	.9038	.4734	2.112	1.1286	40
30	.4451	.4305	.9026	.4770	2.097	1.1257	30
40	.4480	.4331	.9013	.4806	2.081	1.1228	20
50	.4509	.4358	.9001	.4841	2.066	1.1199	10
		cos	sin	cot	tan	Radians	Degrees

TRIGONOMETRIC TABLES (Continued)

Degrees	Radians	sin	cos	tan	cot		
26°00′	.4538	.4384	.8988	.4877	2.050	1.1170	64°00′
10	.4567	.4410	.8975	.4913	2.035	1.1141	50
20	.4596	.4436	.8962	.4950	2.020	1.1112	40
30	.4625	.4462	.8949	.4986	2.006	1.1083	30
40	.4654	.4488	.8936	.5022	1.991	1.1054	20
50	.4683	.4514	.8923	.5059	1.977	1.1025	10
27°00′	.4712	.4540	.8910	.5095	1.963	1.0996	63°00′
10	.4741	.4566	.8897	.5132	1.949	1.0966	50
20	.4771	.4592	.8884	.5169	1.935	1.0937	40
30	.4800	.4617	.8870	.5206	1.921	1.0908	30
40	.4829	.4643	.8857	.5243	1.907	1.0879	20
50	.4858	.4669	.8843	.5280	1.894	1.0850	10
28°00′	.4887	.4695	.8829	.5317	1.881	1.0821	62°00′
10	.4916	.4720	.8816	.5354	1.868	1.0792	50
20	.4945	.4746	.8802	.5392	1.855	1.0763	40
30	.4974	.4772	.8788	.5430	1.842	1.0734	30
40	.5003	.4797	.8774	.5467	1.829	1.0705	20
50	.5032	.4823	.8760	.5505	1.816	1.0676	10
29°00′	.5061	.4848	.8746	.5543	1.804	1.0647	61°00′
10	.5091	.4874	.8732	.5581	1.792	1.0617	50
20	.5120	.4899	.8718	.5619	1.780	1.0588	40
30	.5149	.4924	.8704	.5658	1.767	1.0559	30
40	.5178	.4950	.8689	.5696	1.756	1.0530	20
50	.5207	.4975	.8675	.5735	1.744	1.0501	10
30°00′	.5236	.5000	.8660	.5774	1.732	1.0472	60°00′
10	.5265	.5025	.8646	.5812	1.720	1.0443	50
20	.5294	.5050	.8631	.5851	1.709	1.0414	40
30	.5323	.5075	.8616	.5890	1.698	1.0385	30
40	.5352	.5100	.8601	.5930	1.686	1.0356	20
50	.5381	.5125	.8587	.5969	1.675	1.0327	10
31°00′	.5411	.5150	.8572	.6009	1.664	1.0297	59°00′
10	.5440	.5175	.8557	.6048	1.653	1.0268	50
20	.5469	.5200	.8542	.6088	1.643	1.0239	40
30	.5498	.5225	.8526	.6128	1.632	1.0210	30
40	.5527	.5250	.8511	.6168	1.621	1.0181	20
50	.5556	.5275	.8496	.6208	1.611	1.0152	10
32°00′	.5585	.5299	.8480	.6249	1.600	1.0123	58°00′
10	.5614	.5324	.8465	.6289	1.590	1.0094	50
20	.5643	.5348	.8450	.6330	1.580	1.0065	40
30	.5672	.5373	.8434	.6371	1.570	1.0036	30
40	.5701	.5398	.8418	.6412	1.560	1.0007	20
50	.5730	.5422	.8403	.6453	1.550	.9977	10
33°00′	.5760	.5446	.8387	.6494	1.540	.9948	57°00′
10	.5789	.5471	.8371	.6536	1.530	.9919	50
20	.5818	.5495	.8355	.6577	1.520	.9890	40
30	.5847	.5519	.8339	.6619	1.511	.9861	30
40	.5876	.5544	.8323	.6661	1.501	.9832	20
50	.5905	.5568	.8307	.6703	1.492	.9803	10
34°00′	.5934	.5592	.8290	.6745	1.483	.9774	56°00′
10	.5963	.5616	.8274	.6787	1.473	.9745	50
20	.5992	.5640	.8258	.6830	1.464	.9716	40
30	.6021	.5664	.8241	.6873	1.455	.9687	30
40	.6050	.5688	.8225	.6916	1.446	.9657	20
50	.6080	.5712	.8208	.6959	1.437	.9628	10
35°00′	.6109	.5736	.8192	.7002	1.428	.9599	55°00′
10	.6138	.5760	.8175	.7046	1.419	.9570	50
20	.6167	.5783	.8158	.7089	1.411	.9541	40
30	.6196	.5807	.8141	.7133	1.402	.9512	30
40	.6225	.5831	.8124	.7177	1.393	.9483	20
50	.6254	.5854	.8107	.7221	1.385	.9454	10
		cos	sin	cot	tan	Radians	Degrees

A20   *Appendix D*

TRIGONOMETRIC TABLES (Continued)

Degrees	Radians	sin	cos	tan	cot		
36°00′	.6283	.5878	.8090	.7265	1.376	.9425	54°00′
10	.6312	.5901	.8073	.7310	1.368	.9396	50
20	.6341	.5925	.8056	.7355	1.360	.9367	40
30	.6370	.5948	.8039	.7400	1.351	.9338	30
40	.6400	.5972	.8021	.7445	1.343	.9308	20
50	.6429	.5995	.8004	.7490	1.335	.9279	10
37°00′	.6458	.6018	.7986	.7536	1.327	.9250	53°00′
10	.6487	.6041	.7969	.7581	1.319	.9221	50
20	.6516	.6065	.7951	.7627	1.311	.9192	40
30	.6545	.6088	.7934	.7673	1.303	.9163	30
40	.6574	.6111	.7916	.7720	1.295	.9134	20
50	.6603	.6134	.7898	.7766	1.288	.9105	10
38°00′	.6632	.6157	.7880	.7813	1.280	.9076	52°00′
10	.6661	.6180	.7862	.7860	1.272	.9047	50
20	.6690	.6202	.7844	.7907	1.265	.9018	40
30	.6720	.6225	.7826	.7954	1.257	.8988	30
40	.6749	.6248	.7808	.8002	1.250	.8959	20
50	.6778	.6271	.7790	.8050	1.242	.8930	10
39°00′	.6807	.6293	.7771	.8098	1.235	.8901	51°00′
10	.6836	.6316	.7753	.8146	1.228	.8872	50
20	.6865	.6338	.7735	.8195	1.220	.8843	40
30	.6894	.6361	.7716	.8243	1.213	.8814	30
40	.6923	.6383	.7698	.8292	1.206	.8785	20
50	.6952	.6406	.7679	.8342	1.199	.8756	10
40°00′	.6981	.6428	.7660	.8391	1.192	.8727	50°00′
10	.7010	.6450	.7642	.8441	1.185	.8698	50
20	.7039	.6472	.7623	.8491	1.178	.8668	40
30	.7069	.6494	.7604	.8541	1.171	.8639	30
40	.7098	.6517	.7585	.8591	1.164	.8610	20
50	.7127	.6539	.7566	.8642	1.157	.8581	10
41°00′	.7156	.6561	.7547	.8693	1.150	.8552	49°00′
10	.7185	.6583	.7528	.8744	1.144	.8523	50
20	.7214	.6604	.7509	.8796	1.137	.8494	40
30	.7243	.6626	.7490	.8847	1.130	.8465	30
40	.7272	.6648	.7470	.8899	1.124	.8436	20
50	.7301	.6670	.7451	.8952	1.117	.8407	10
42°00′	.7330	.6691	.7431	.9004	1.111	.8378	48°00′
10	.7359	.6713	.7412	.9057	1.104	.8348	50
20	.7389	.6734	.7392	.9110	1.098	.8319	40
30	.7418	.6756	.7373	.9163	1.091	.8290	30
40	.7447	.6777	.7353	.9217	1.085	.8261	20
50	.7476	.6799	.7333	.9271	1.079	.8232	10
43°00′	.7505	.6820	.7314	.9325	1.072	.8203	47°00′
10	.7534	.6841	.7294	.9380	1.066	.8174	50
20	.7563	.6862	.7274	.9435	1.060	.8145	40
30	.7592	.6884	.7254	.9490	1.054	.8116	30
40	.7621	.6905	.7234	.9545	1.048	.8087	20
50	.7650	.6926	.7214	.9601	1.042	.8058	10
44°00′	.7679	.6947	.7193	.9657	1.036	.8029	46°00′
10	.7709	.6967	.7173	.9713	1.030	.7999	50
10	.7738	.6988	.7153	.9770	1.024	.7970	40
30	.7767	.7009	.7133	.9827	1.018	.7941	30
40	.7796	.7030	.7112	.9884	1.012	.7912	20
50	.7825	.7050	.7092	.9942	1.006	.7883	10
45°00′	.7854	.7071	.7071	1.0000	1.000	.7854	45°00′
		cos	sin	cot	tan	Radians	Degrees

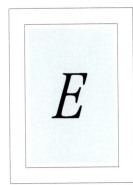

E

Table of square roots and cube roots

n	$\sqrt{n}$	$\sqrt[3]{n}$	n	$\sqrt{n}$	$\sqrt[3]{n}$
1	1.00000	1.00000	43	6.55744	3.50340
2	1.41421	1.25992	44	6.63325	3.53035
3	1.73205	1.44225	45	6.70820	3.55689
4	2.00000	1.58740	46	6.78233	3.58305
5	2.23607	1.70998	47	6.85565	3.60883
6	2.44949	1.81712	48	6.92820	3.63424
7	2.64575	1.91293	49	7.00000	3.65931
8	2.82843	2.00000	50	7.07107	3.68403
9	3.00000	2.08008	51	7.14143	3.70843
10	3.16228	2.15443	52	7.21110	3.73251
11	3.31662	2.22398	53	7.28011	3.75629
12	3.46410	2.28943	54	7.34847	3.77976
13	3.60555	2.35133	55	7.41620	3.80295
14	3.74166	2.41014	56	7.48331	3.82586
15	3.87298	2.46621	57	7.54983	3.84850
16	4.00000	2.51984	58	7.61577	3.87088
17	4.12311	2.57128	59	7.68115	3.89300
18	4.24264	2.62074	60	7.74597	3.91487
19	4.35890	2.66840	61	7.81025	3.93650
20	4.47214	2.71442	62	7.87401	3.95789
21	4.58258	2.75892	63	7.93725	3.97906
22	4.69042	2.80204	64	8.00000	4.00000
23	4.79583	2.84387	65	8.06226	4.02073
24	4.89898	2.88450	66	8.12404	4.04124
25	5.00000	2.92402	67	8.18535	4.06155
26	5.09902	2.96250	68	8.24621	4.08166
27	5.19615	3.00000	69	8.30662	4.10157
28	5.29150	3.03659	70	8.36660	4.12129
29	5.38516	3.07232	71	8.42615	4.14082
30	5.47723	3.10723	72	8.48528	4.16017
31	5.56776	3.14138	73	8.54400	4.17934
32	5.65685	3.17480	74	8.60233	4.19834
33	5.74456	3.20753	75	8.66025	4.21716
34	5.83095	3.23961	76	8.71780	4.23582
35	5.91608	3.27107	77	8.77496	4.25432
36	6.00000	3.30193	78	8.83176	4.27266
37	6.08276	3.33222	79	8.88819	4.29084
38	6.16441	3.36198	80	8.94427	4.30887
39	6.24500	3.39121	81	9.00000	4.32675
40	6.32456	3.41995	82	9.05539	4.34448
41	6.40312	3.44822	83	9.11043	4.36207
42	6.48074	3.47603	84	9.16515	4.37952

n	$\sqrt{n}$	$\sqrt[3]{n}$	n	$\sqrt{n}$	$\sqrt[3]{n}$
85	9.21954	4.39683	143	11.9583	5.22932
86	9.27362	4.41400	144	12.0000	5.24148
87	9.32738	4.43105	145	12.0416	5.25359
88	9.38083	4.44796	146	12.0830	5.26564
89	9.43398	4.46475	147	12.1244	5.27763
90	9.48683	4.48140	148	12.1655	5.28957
91	9.53939	4.49794	149	12.2066	5.30146
92	9.59166	4.51436	150	12.2474	5.31329
93	9.64365	4.53065	151	12.2882	5.32507
94	9.69536	4.54684	152	12.3288	5.33680
95	9.74679	4.56290	153	12.3693	5.34848
96	9.79796	4.57886	154	12.4097	5.36011
97	9.84886	4.59470	155	12.4499	5.37169
98	9.89949	4.61044	156	12.4900	5.38321
99	9.94987	4.62606	157	12.5300	5.39469
100	10.0000	4.64159	158	12.5698	5.40612
101	10.0499	4.65701	159	12.6095	5.41750
102	10.0995	4.67233	160	12.6491	5.42884
103	10.1489	4.68755	161	12.6886	5.44012
104	10.1980	4.70267	162	12.7279	5.45136
105	10.2470	4.71769	163	12.7671	5.46256
106	10.2956	4.73262	164	12.8062	5.47370
107	10.3441	4.74746	165	12.8452	5.48481
108	10.3923	4.76220	166	12.8841	5.49586
109	10.4403	4.77686	167	12.9228	5.50688
110	10.4881	4.79142	168	12.9615	5.51785
111	10.5357	4.80590	169	13.0000	5.52877
112	10.5830	4.82028	170	13.0384	5.53966
113	10.6301	4.83459	171	13.0767	5.55050
114	10.6771	4.84881	172	13.1149	5.56130
115	10.7238	4.86294	173	13.1529	5.57205
116	10.7703	4.87700	174	13.1909	5.58277
117	10.8167	4.89097	175	13.2288	5.59344
118	10.8628	4.90487	176	13.2665	5.60408
119	10.9087	4.91868	177	13.3041	5.61467
120	10.9545	4.93242	178	13.3417	5.62523
121	11.0000	4.94609	179	13.3791	5.63574
122	11.0454	4.95968	180	13.4164	5.64622
123	11.0905	4.97319	181	13.4536	5.65665
124	11.1355	4.98663	182	13.4907	5.66705
125	11.1803	5.00000	183	13.5277	5.67741
126	11.2250	5.01330	184	13.5647	5.68773
127	11.2694	5.02653	185	13.6015	5.69802
128	11.3137	5.03968	186	13.6382	5.70827
129	11.3578	5.05277	187	13.6748	5.71848
130	11.4018	5.06580	188	13.7113	5.72865
131	11.4455	5.07875	189	13.7477	5.73879
132	11.4891	5.09164	190	13.7840	5.74890
133	11.5326	5.10447	191	13.8203	5.75897
134	11.5758	5.11723	192	13.8564	5.76900
135	11.6190	5.12993	193	13.8924	5.77900
136	11.6619	5.14256	194	13.9284	5.78896
137	11.7047	5.15514	195	13.9642	5.79889
138	11.7473	5.16765	196	14.0000	5.80879
139	11.7898	5.18010	197	14.0357	5.81865
140	11.8322	5.19249	198	14.0712	5.82848
141	11.8743	5.20483	199	14.1067	5.83827
142	11.9164	5.21710	200	14.1421	5.84804

F

Basic differentiation rules for elementary functions

1. $\dfrac{d}{dx}[cu] = cu'$

2. $\dfrac{d}{dx}[u \pm v] = u' \pm v'$

3. $\dfrac{d}{dx}[uv] = uv' + vu'$

4. $\dfrac{d}{dx}\left[\dfrac{u}{v}\right] = \dfrac{vu' - uv'}{v^2}$

5. $\dfrac{d}{dx}[c] = 0$

6. $\dfrac{d}{dx}[u^n] = nu^{n-1}u'$

7. $\dfrac{d}{dx}[x] = 1$

8. $\dfrac{d}{dx}[|u|] = \dfrac{u}{|u|}(u'),\ u \neq 0$

9. $\dfrac{d}{dx}[\ln u] = \dfrac{u'}{u}$

10. $\dfrac{d}{dx}[e^u] = e^u u'$

11. $\dfrac{d}{dx}[\sin u] = (\cos u)u'$

12. $\dfrac{d}{dx}[\cos u] = -(\sin u)u'$

13. $\dfrac{d}{dx}[\tan u] = (\sec^2 u)u'$

14. $\dfrac{d}{dx}[\cot u] = -(\csc^2 u)u'$

15. $\dfrac{d}{dx}[\sec u] = (\sec u \tan u)u'$

16. $\dfrac{d}{dx}[\csc u] = -(\csc u \cot u)u'$

17. $\dfrac{d}{dx}[\arcsin u] = \dfrac{u'}{\sqrt{1 - u^2}}$

18. $\dfrac{d}{dx}[\arccos u] = \dfrac{-u'}{\sqrt{1 - u^2}}$

19. $\dfrac{d}{dx}[\arctan u] = \dfrac{u'}{1 + u^2}$

20. $\dfrac{d}{dx}[\text{arccot}\, u] = \dfrac{-u'}{1 + u^2}$

21. $\dfrac{d}{dx}[\text{arcsec}\, u] = \dfrac{u'}{|u|\sqrt{u^2 - 1}}$

22. $\dfrac{d}{dx}[\text{arccsc}\, u] = \dfrac{-u'}{|u|\sqrt{u^2 - 1}}$

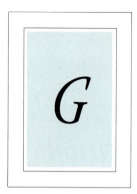

G Integration tables

Forms Involving u^n

1. $\displaystyle\int u^n \, du = \frac{u^{n+1}}{n+1} + C, \; n \neq -1$

2. $\displaystyle\int \frac{1}{u} \, du = \ln |u| + C$

Forms Involving $a + bu$

3. $\displaystyle\int \frac{u}{a+bu} \, du = \frac{1}{b^2}(bu - a \ln |a+bu|) + C$

4. $\displaystyle\int \frac{u}{(a+bu)^2} \, du = \frac{1}{b^2}\left(\frac{a}{a+bu} + \ln |a+bu|\right) + C$

5. $\displaystyle\int \frac{u}{(a+bu)^n} \, du = \frac{1}{b^2}\left[\frac{-1}{(n-2)(a+bu)^{n-2}} + \frac{a}{(n-1)(a+bu)^{n-1}}\right] + C, \; n \neq 1, 2$

6. $\displaystyle\int \frac{u^2}{a+bu} \, du = \frac{1}{b^3}\left(-\frac{bu}{2}(2a - bu) + a^2 \ln |a+bu|\right) + C$

7. $\displaystyle\int \frac{u^2}{(a+bu)^2} \, du = \frac{1}{b^3}\left(bu - \frac{a^2}{a+bu} - 2a \ln |a+bu|\right) + C$

8. $\displaystyle\int \frac{u^2}{(a+bu)^3} \, du = \frac{1}{b^3}\left[\frac{2a}{a+bu} - \frac{a^2}{2(a+bu)^2} + \ln |a+bu|\right] + C$

9. $\displaystyle\int \frac{u^2}{(a+bu)^n} \, du = \frac{1}{b^3}\left[\frac{-1}{(n-3)(a+bu)^{n-3}} + \frac{2a}{(n-2)(a+bu)^{n-2}} - \frac{a^2}{(n-1)(a+bu)^{n-1}}\right] + C,$

$$n \neq 1, 2, 3$$

10. $\displaystyle\int \frac{1}{u(a+bu)} \, du = \frac{1}{a} \ln \left|\frac{u}{a+bu}\right| + C$

11. $\displaystyle\int \frac{1}{u(a + bu)^2}\, du = \frac{1}{a}\left(\frac{1}{a + bu} + \frac{1}{a}\ln\left|\frac{u}{a + bu}\right|\right) + C$

12. $\displaystyle\int \frac{1}{u^2(a + bu)}\, du = -\frac{1}{a}\left(\frac{1}{u} + \frac{b}{a}\ln\left|\frac{u}{a + bu}\right|\right) + C$

13. $\displaystyle\int \frac{1}{u^2(a + bu)^2}\, du = -\frac{1}{a^2}\left[\frac{a + 2bu}{u(a + bu)} + \frac{2b}{a}\ln\left|\frac{u}{a + bu}\right|\right] + C$

Forms Involving $a + bu + cu^2$, $b^2 \neq 4ac$

14. $\displaystyle\int \frac{1}{a + bu + cu^2}\, du = \begin{cases} \dfrac{2}{\sqrt{4ac - b^2}}\arctan\dfrac{2cu + b}{\sqrt{4ac - b^2}} + C, & b^2 < 4ac \\[3ex] \dfrac{1}{\sqrt{b^2 - 4ac}}\ln\left|\dfrac{2cu + b - \sqrt{b^2 - 4ac}}{2cu + b + \sqrt{b^2 - 4ac}}\right| + C, & b^2 > 4ac \end{cases}$

15. $\displaystyle\int \frac{u}{a + bu + cu^2}\, du = \frac{1}{2c}\left(\ln|a + bu + cu^2| - b\int \frac{1}{a + bu + cu^2}\, du\right)$

Forms Involving $\sqrt{a + bu}$

16. $\displaystyle\int u^n\sqrt{a + bu}\, du = \frac{2}{b(2n + 3)}\left[u^n(a + bu)^{3/2} - na\int u^{n-1}\sqrt{a + bu}\, du\right]$

17. $\displaystyle\int \frac{1}{u\sqrt{a + bu}}\, du = \begin{cases} \dfrac{1}{\sqrt{a}}\ln\left|\dfrac{\sqrt{a + bu} - \sqrt{a}}{\sqrt{a + bu} + \sqrt{a}}\right| + C, & 0 < a \\[3ex] \dfrac{2}{\sqrt{-a}}\arctan\sqrt{\dfrac{a + bu}{-a}} + C, & a < 0 \end{cases}$

18. $\displaystyle\int \frac{1}{u^n\sqrt{a + bu}}\, du = \frac{-1}{a(n - 1)}\left[\frac{\sqrt{a + bu}}{u^{n-1}} + \frac{(2n - 3)b}{2}\int \frac{1}{u^{n-1}\sqrt{a + bu}}\, du\right],\ n \neq 1$

19. $\displaystyle\int \frac{\sqrt{a + bu}}{u}\, du = 2\sqrt{a + bu} + a\int \frac{1}{u\sqrt{a + bu}}\, du$

20. $\displaystyle\int \frac{\sqrt{a + bu}}{u^n}\, du = \frac{-1}{a(n - 1)}\left[\frac{(a + bu)^{3/2}}{u^{n-1}} + \frac{(2n - 5)b}{2}\int \frac{\sqrt{a + bu}}{u^{n-1}}\, du\right],\ n \neq 1$

21. $\displaystyle\int \frac{u}{\sqrt{a + bu}}\, du = \frac{-2(2a - bu)}{3b^2}\sqrt{a + bu} + C$

22. $\displaystyle\int \frac{u^n}{\sqrt{a + bu}}\, du = \frac{2}{(2n + 1)b}\left(u^n\sqrt{a + bu} - na\int \frac{u^{n-1}}{\sqrt{a + bu}}\, du\right)$

Forms Involving $a^2 \pm u^2$, $0 < a$

23. $\displaystyle\int \frac{1}{a^2 + u^2}\, du = \frac{1}{a}\arctan\frac{u}{a} + C$

24. $\displaystyle\int \frac{1}{u^2 - a^2}\, du = -\int \frac{1}{a^2 - u^2}\, du = \frac{1}{2a}\ln\left|\frac{u - a}{u + a}\right| + C$

25. $\displaystyle\int \frac{1}{(a^2 \pm u^2)^n}\, du = \frac{1}{2a^2(n - 1)}\left[\frac{u}{(a^2 \pm u^2)^{n-1}} + (2n - 3)\int \frac{1}{(a^2 \pm u^2)^{n-1}}\, du\right],\ n \neq 1$

Forms Involving $\sqrt{u^2 \pm a^2}$, $0 < a$

26. $\displaystyle\int \sqrt{u^2 \pm a^2}\, du = \frac{1}{2}(u\sqrt{u^2 \pm a^2} \pm a^2 \ln |u + \sqrt{u^2 \pm a^2}|) + C$

27. $\displaystyle\int u^2\sqrt{u^2 \pm a^2}\, du = \frac{1}{8}[u(2u^2 \pm a^2)\sqrt{u^2 \pm a^2} - a^4 \ln |u + \sqrt{u^2 \pm a^2}|] + C$

28. $\displaystyle\int \frac{\sqrt{u^2 + a^2}}{u}\, du = \sqrt{u^2 + a^2} - a \ln \left| \frac{a + \sqrt{u^2 + a^2}}{u} \right| + C$

29. $\displaystyle\int \frac{\sqrt{u^2 - a^2}}{u}\, du = \sqrt{u^2 - a^2} - a \, \text{arcsec} \frac{|u|}{a} + C$

30. $\displaystyle\int \frac{\sqrt{u^2 \pm a^2}}{u^2}\, du = \frac{-\sqrt{u^2 \pm a^2}}{u} + \ln |u + \sqrt{u^2 \pm a^2}| + C$

31. $\displaystyle\int \frac{1}{\sqrt{u^2 \pm a^2}}\, du = \ln |u + \sqrt{u^2 \pm a^2}| + C$

32. $\displaystyle\int \frac{1}{u\sqrt{u^2 + a^2}}\, du = \frac{-1}{a} \ln \left| \frac{a + \sqrt{u^2 + a^2}}{u} \right| + C$

33. $\displaystyle\int \frac{1}{u\sqrt{u^2 - a^2}}\, du = \frac{1}{a} \, \text{arcsec} \frac{|u|}{a} + C$

34. $\displaystyle\int \frac{u^2}{\sqrt{u^2 \pm a^2}}\, du = \frac{1}{2}(u\sqrt{u^2 \pm a^2} \mp a^2 \ln |u + \sqrt{u^2 \pm a^2}|) + C$

35. $\displaystyle\int \frac{1}{u^2\sqrt{u^2 \pm a^2}}\, du = \mp \frac{\sqrt{u^2 \pm a^2}}{a^2 u} + C$

36. $\displaystyle\int \frac{1}{(u^2 \pm a^2)^{3/2}}\, du = \frac{\pm u}{a^2\sqrt{u^2 \pm a^2}} + C$

Forms Involving $\sqrt{a^2 - u^2}$, $0 < a$

37. $\displaystyle\int \sqrt{a^2 - u^2}\, du = \frac{1}{2}\left(u\sqrt{a^2 - u^2} + a^2 \arcsin \frac{u}{a}\right) + C$

38. $\displaystyle\int u^2\sqrt{a^2 - u^2}\, du = \frac{1}{8}\left[u(2u^2 - a^2)\sqrt{a^2 - u^2} + a^4 \arcsin \frac{u}{a}\right] + C$

39. $\displaystyle\int \frac{\sqrt{a^2 - u^2}}{u}\, du = \sqrt{a^2 - u^2} - a \ln \left| \frac{a + \sqrt{a^2 - u^2}}{u} \right| + C$

40. $\displaystyle\int \frac{\sqrt{a^2 - u^2}}{u^2}\, du = \frac{-\sqrt{a^2 - u^2}}{u} - \arcsin \frac{u}{a} + C$

41. $\displaystyle\int \frac{1}{\sqrt{a^2 - u^2}}\, du = \arcsin \frac{u}{a} + C$

42. $\displaystyle\int \frac{1}{u\sqrt{a^2 - u^2}}\, du = \frac{-1}{a} \ln \left| \frac{a + \sqrt{a^2 - u^2}}{u} \right| + C$

43. $\displaystyle\int \frac{u^2}{\sqrt{a^2 - u^2}}\, du = \frac{1}{2}\left(-u\sqrt{a^2 - u^2} + a^2 \arcsin \frac{u}{a}\right) + C$

44. $\displaystyle \int \frac{1}{u^2\sqrt{a^2-u^2}}\,du = \frac{-\sqrt{a^2-u^2}}{a^2 u} + C$

45. $\displaystyle \int \frac{1}{(a^2-u^2)^{3/2}}\,du = \frac{u}{a^2\sqrt{a^2-u^2}} + C$

Forms Involving $\sin u$ or $\cos u$

46. $\displaystyle \int \sin u\,du = -\cos u + C$

47. $\displaystyle \int \cos u\,du = \sin u + C$

48. $\displaystyle \int \sin^2 u\,du = \frac{1}{2}(u - \sin u \cos u) + C$

49. $\displaystyle \int \cos^2 u\,du = \frac{1}{2}(u + \sin u \cos u) + C$

50. $\displaystyle \int \sin^n u\,du = -\frac{\sin^{n-1} u \cos u}{n} + \frac{n-1}{n}\int \sin^{n-2} u\,du$

51. $\displaystyle \int \cos^n u\,du = \frac{\cos^{n-1} u \sin u}{n} + \frac{n-1}{n}\int \cos^{n-2} u\,du$

52. $\displaystyle \int u \sin u\,du = \sin u - u \cos u + C$

53. $\displaystyle \int u \cos u\,du = \cos u + u \sin u + C$

54. $\displaystyle \int u^n \sin u\,du = -u^n \cos u + n\int u^{n-1}\cos u\,du$

55. $\displaystyle \int u^n \cos u\,du = u^n \sin u - n\int u^{n-1}\sin u\,du$

56. $\displaystyle \int \frac{1}{1 \pm \sin u}\,du = \tan u \mp \sec u + C$

57. $\displaystyle \int \frac{1}{1 \pm \cos u}\,du = -\cot u \pm \csc u + C$

58. $\displaystyle \int \frac{1}{\sin u \cos u}\,du = \ln|\tan u| + C$

Forms Involving $\tan u$, $\cot u$, $\sec u$, $\csc u$

59. $\displaystyle \int \tan u\,du = -\ln|\cos u| + C$

60. $\displaystyle \int \cot u\,du = \ln|\sin u| + C$

61. $\displaystyle \int \sec u\,du = \ln|\sec u + \tan u| + C$

62. $\displaystyle \int \csc u\,du = \ln|\csc u - \cot u| + C$

63. $\displaystyle\int \tan^2 u \ du = -u + \tan u + C$

64. $\displaystyle\int \cot^2 u \ du = -u - \cot u + C$

65. $\displaystyle\int \sec^2 u \ du = \tan u + C$

66. $\displaystyle\int \csc^2 u \ du = -\cot u + C$

67. $\displaystyle\int \tan^n u \ du = \frac{\tan^{n-1} u}{n-1} - \int \tan^{n-2} u \ du, \ n \neq 1$

68. $\displaystyle\int \cot^n u \ du = -\frac{\cot^{n-1} u}{n-1} - \int (\cot^{n-2} u) \ du, \ n \neq 1$

69. $\displaystyle\int \sec^n u \ du = \frac{\sec^{n-2} u \tan u}{n-1} + \frac{n-2}{n-1} \int \sec^{n-2} u \ du, \ n \neq 1$

70. $\displaystyle\int \csc^n u \ du = -\frac{\csc^{n-2} u \cot u}{n-1} + \frac{n-2}{n-1} \int \csc^{n-2} u \ du, \ n \neq 1$

71. $\displaystyle\int \frac{1}{1 \pm \tan u} \ du = \frac{1}{2}(u \pm \ln|\cos u \pm \sin u|) + C$

72. $\displaystyle\int \frac{1}{1 \pm \cot u} \ du = \frac{1}{2}(u \mp \ln|\sin u \pm \cos u|) + C$

73. $\displaystyle\int \frac{1}{1 \pm \sec u} \ du = u + \cot u \mp \csc u + C$

74. $\displaystyle\int \frac{1}{1 \pm \csc u} \ du = u - \tan u \pm \sec u + C$

Forms Involving Inverse Trigonometric Functions

75. $\displaystyle\int \arcsin u \ du = u \arcsin u + \sqrt{1 - u^2} + C$

76. $\displaystyle\int \arccos u \ du = u \arccos u - \sqrt{1 - u^2} + C$

77. $\displaystyle\int \arctan u \ du = u \arctan u - \ln \sqrt{1 + u^2} + C$

78. $\displaystyle\int \text{arccot } u \ du = u \ \text{arccot } u + \ln \sqrt{1 + u^2} + C$

79. $\displaystyle\int \text{arcsec } u \ du = u \ \text{arcsec } u - \ln |u + \sqrt{u^2 - 1}| + C$

80. $\displaystyle\int \text{arccsc } u \ du = u \ \text{arccsc } u + \ln |u + \sqrt{u^2 - 1}| + C$

Forms Involving e^u

81. $\displaystyle\int e^u \ du = e^u + C$

82. $\displaystyle\int u e^u \ du = (u - 1)e^u + C$

83. $\displaystyle\int u^n e^u \, du = u^n e^u - n \int u^{n-1} e^u \, du$

84. $\displaystyle\int \frac{1}{1 + e^u} \, du = u - \ln(1 + e^u) + C$

85. $\displaystyle\int e^{au} \sin bu \, du = \frac{e^{au}}{a^2 + b^2}(a \sin bu - b \cos bu) + C$

86. $\displaystyle\int e^{au} \cos bu \, du = \frac{e^{au}}{a^2 + b^2}(a \cos bu + b \sin bu) + C$

Forms Involving ln *u*

87. $\displaystyle\int \ln u \, du = u(-1 + \ln u) + C$

88. $\displaystyle\int u \ln u \, du = \frac{u^2}{4}(-1 + 2 \ln u) + C$

89. $\displaystyle\int u^n \ln u \, du = \frac{u^{n+1}}{(n+1)^2}[-1 + (n+1) \ln u] + C, \ n \neq -1$

90. $\displaystyle\int (\ln u)^2 \, du = u[2 - 2 \ln u + (\ln u)^2] + C$

91. $\displaystyle\int (\ln u)^n \, du = u(\ln u)^n - n \int (\ln u)^{n-1} \, du$

Answers to Odd-Numbered Exercises

CHAPTER 1

SECTION 1.1

1. Rational

3. Irrational

5. Rational

7. Rational

9. Rational

11. $\dfrac{4}{11}$

13. $\dfrac{11}{37}$

15. (a) True
    (b) False
    (c) True
    (d) False
    (e) False
    (f) False

17.

Interval Notation	Set Notation	Graph
$[-2, 0)$	$\{x: -2 \le x < 0\}$	
$(-\infty, -4]$	$\{x: x \le -4\}$	
$[3, 11/2]$	$\{x: 3 \le x \le 11/2\}$	
$(-1, 7)$	$\{x: -1 < x < 7\}$	

19. $x \ge 12$

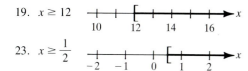

21. $x < -\dfrac{1}{2}$

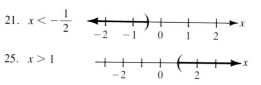

23. $x \ge \dfrac{1}{2}$

25. $x > 1$

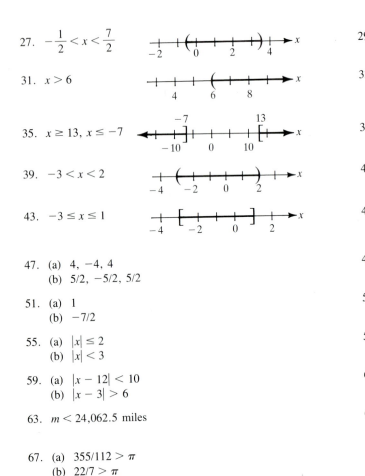

27. $-\dfrac{1}{2} < x < \dfrac{7}{2}$

29. $x < -4$

31. $x > 6$

33. $-1 < x < 1$

35. $x \geq 13,\ x \leq -7$

37. $a - b < x < a + b$

39. $-3 < x < 2$

41. $3 > x > 0$

43. $-3 \leq x \leq 1$

45. $-3 \leq x \leq 2$

47. (a) $4,\ -4,\ 4$
    (b) $5/2,\ -5/2,\ 5/2$

49. (a) $-51,\ 51,\ 51$
    (b) $51,\ -51,\ 51$

51. (a) 1
    (b) $-7/2$

53. (a) 14
    (b) 10

55. (a) $|x| \leq 2$
    (b) $|x| < 3$

57. (a) $|x - 4| \leq 2$
    (b) $|x + 4| < 3$

59. (a) $|x - 12| < 10$
    (b) $|x - 3| > 6$

61. $r > 12.5\%$

63. $m < 24{,}062.5$ miles

65. $x \leq 41$
    $x \geq 59$

67. (a) $355/112 > \pi$
    (b) $22/7 > \pi$

SECTION 1.2

1. $d = 2\sqrt{5}$

3. $d = 2\sqrt{10}$

5. $d = 2\sqrt{37}$

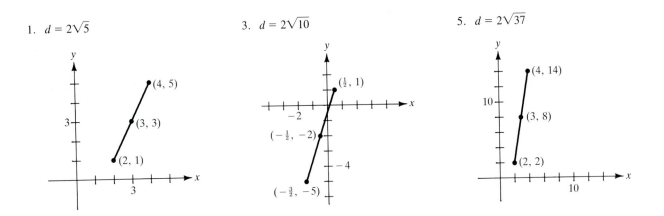

7. $d = \sqrt{8 - 2\sqrt{3}}$

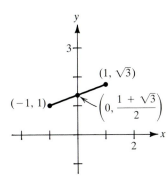

9. Right triangle

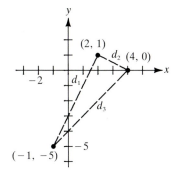

11. Rhombus, the length of each side is $\sqrt{5}$.

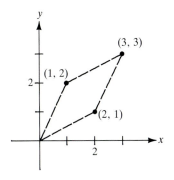

13. Collinear since $d_1 + d_2 = d_3$

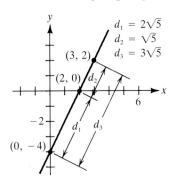

15. Not on a line since $d_1 + d_2 > d_3$

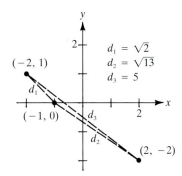

17. $x = \pm 3$

19. $3x - 2y - 1 = 0$

21. $\left(\dfrac{3x_1 + x_2}{4}, \dfrac{3y_1 + y_2}{4}\right), \left(\dfrac{x_1 + x_2}{2}, \dfrac{y_1 + y_2}{2}\right), \left(\dfrac{x_1 + 3x_2}{4}, \dfrac{y_1 + 3y_2}{4}\right)$

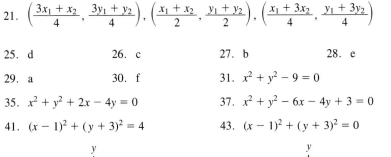

23. (a) $\left(x + \dfrac{5}{2}\right)^2 - \dfrac{25}{4}$

    (b) $(x + 4)^2 - 9$

25. d   26. c

27. b   28. e

29. a   30. f

31. $x^2 + y^2 - 9 = 0$

33. $x^2 + y^2 - 4x + 2y - 11 = 0$

35. $x^2 + y^2 + 2x - 4y = 0$

37. $x^2 + y^2 - 6x - 4y + 3 = 0$

39. $x^2 + y^2 - 6x - 8y = 0$

41. $(x - 1)^2 + (y + 3)^2 = 4$

43. $(x - 1)^2 + (y + 3)^2 = 0$

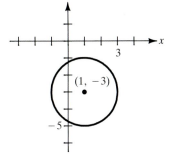

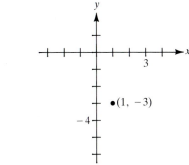

45. $\left(x - \dfrac{1}{2}\right)^2 + \left(y - \dfrac{1}{2}\right)^2 = 2$

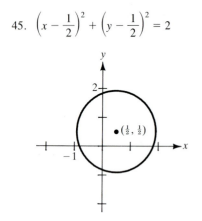

47. $\left(x + \dfrac{1}{2}\right)^2 + \left(y + \dfrac{5}{4}\right)^2 = \dfrac{9}{4}$

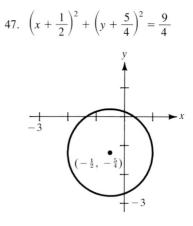

49. $x^2 + y^2 = 26{,}000^2$

51. $(x - 1)^2 + (y + 1)^2 = 25$

53.

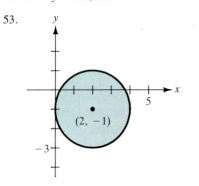

55.

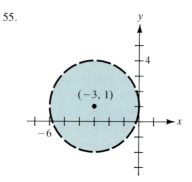

SECTION 1.3

1. c   2. d   3. b

4. f   5. a   6. e

7. Symmetric with respect to the *y*-axis

9. Symmetric with respect to the *y*-axis

11. Symmetric with respect to the *x*-axis

13. Symmetric with respect to the origin

15. Symmetric with respect to the origin

17. $(0, -3)$, $\left(\dfrac{3}{2}, 0\right)$

19. $(0, -2)$, $(-2, 0)$, $(1, 0)$

21. $(0, 0)$, $(-3, 0)$, $(3, 0)$

23. $\left(0, \dfrac{1}{2}\right)$, $(1, 0)$

25. $(0, 0)$

27. $y = x$

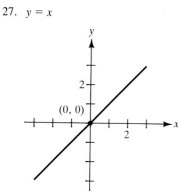

29. $y = x + 3$

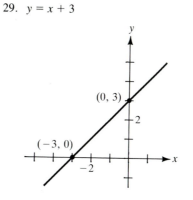

31. $y = -3x + 2$

33. $y = \dfrac{1}{2}x - 4$

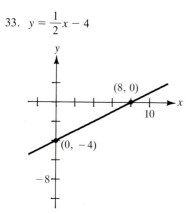

35. $y = 1 - x^2$

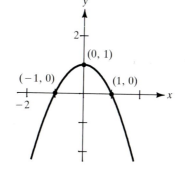

37. $y = -2x^2 + x + 1$

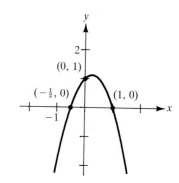

39. $y = x^3 + 2$

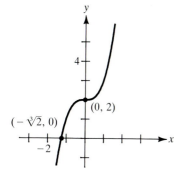

41. $x^2 + 4y^2 = 4$

43. $y = (x + 2)^2$

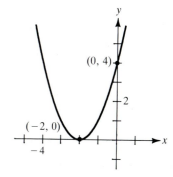

45. $y = \dfrac{1}{x}$

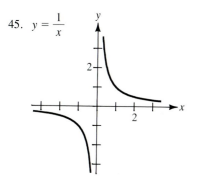

47. $(1, 1)$

49. $(5, 2)$

51. $(-1, -2), (2, 1)$

53. $(-1, -1), (0, 0), (1, 1)$

55. $(-1, -5), (0, -1), (2, 1)$

57. $x \approx 192$ units

59. (1, 2) is not on the graph
    (1, −1) is on the graph
    (4, 5) is on the graph

61. (1, 1/5) is on the graph
    (2, 1/2) is on the graph
    (−1, −2) is not on the graph

63. (a) $k = 4$
    (b) $k = -1/8$
    (c) $k =$ any real number
    (d) $k = 1$

65. (a) Graph
    (b) CPI for 1985: 325.91

67. (a) Graph
    (b) Percentage for
        1990: 2.74%

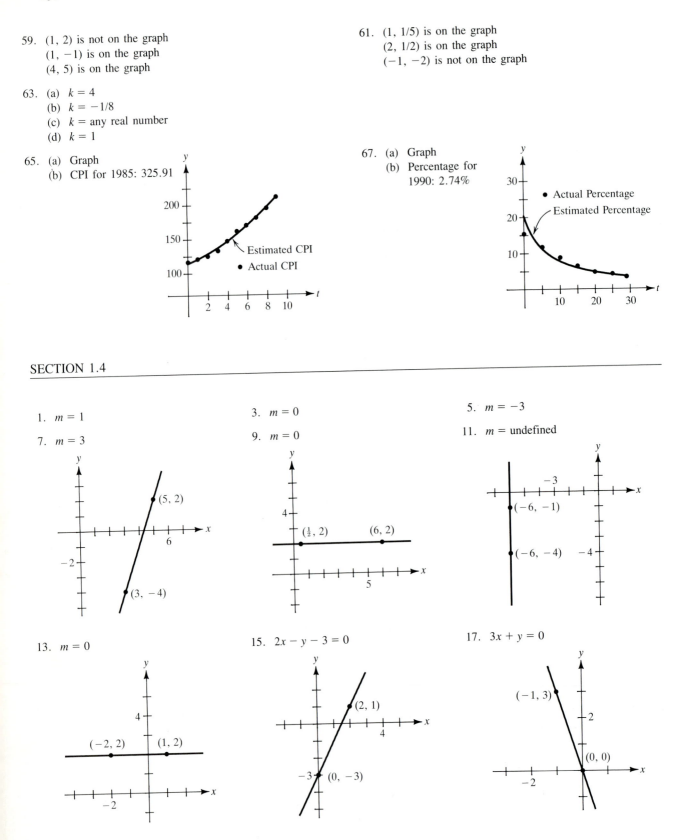

SECTION 1.4

1. $m = 1$

3. $m = 0$

5. $m = -3$

7. $m = 3$

9. $m = 0$

11. $m =$ undefined

13. $m = 0$

15. $2x - y - 3 = 0$

17. $3x + y = 0$

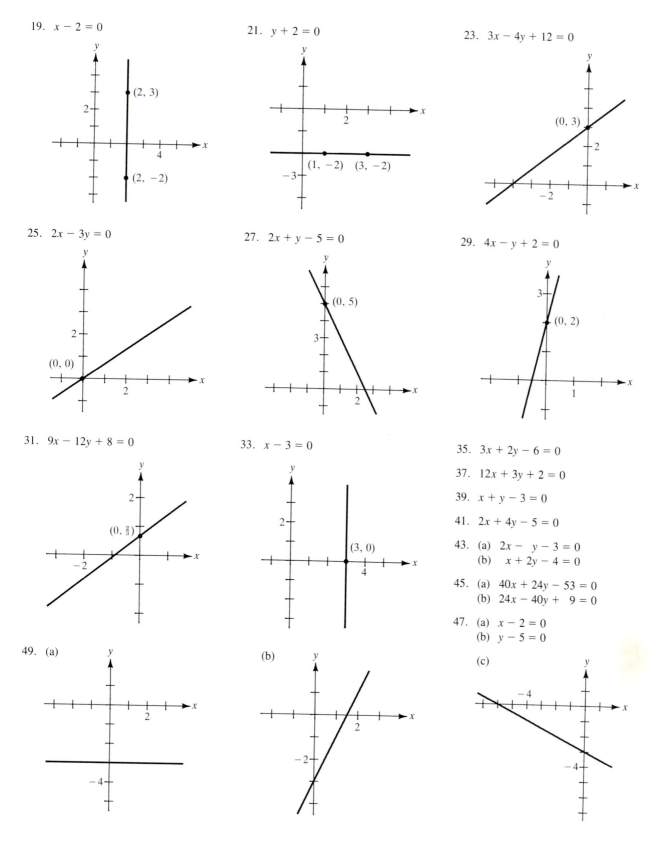

19. $x - 2 = 0$

21. $y + 2 = 0$

23. $3x - 4y + 12 = 0$

25. $2x - 3y = 0$

27. $2x + y - 5 = 0$

29. $4x - y + 2 = 0$

31. $9x - 12y + 8 = 0$

33. $x - 3 = 0$

35. $3x + 2y - 6 = 0$

37. $12x + 3y + 2 = 0$

39. $x + y - 3 = 0$

41. $2x + 4y - 5 = 0$

43. (a) $2x - y - 3 = 0$
    (b) $x + 2y - 4 = 0$

45. (a) $40x + 24y - 53 = 0$
    (b) $24x - 40y + 9 = 0$

47. (a) $x - 2 = 0$
    (b) $y - 5 = 0$

49. (a)   (b)   (c)

51. Collinear, since $m_1 = m_2$

53. Not collinear, since $m_1 \neq m_2$

55. $\left(0, \dfrac{-a^2 + b^2 + c^2}{2c}\right)$

59.

C	−17.8	−10	10	20	32.2	177
F	0		14	50	68	90	350.6

61. $V = 875 - 175t$ where $0 \leq t \leq 5$

63. (a) $x = \dfrac{1}{15}(1030 - p)$
    (b) 45 units
    (c) 49 units

67. 7/5

65. (a) Graph
    (b) Speed of plane = 560 mi/hr

69. 7

71. 9/5

73. 2

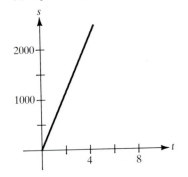

SECTION 1.5

1. (a) −3
    (b) −9
    (c) $2b - 3$
    (d) $2x - 5$

3. (a) 1
    (b) 3
    (c) $\sqrt{c + 3}$
    (d) $\sqrt{x + \Delta x + 3}$

5. (a) 1
    (b) −1
    (c) 1
    (d) $|x - 1|/(x - 1)$

7. $3 + \Delta x$

9. $3x^2 + 3x\,\Delta x + (\Delta x)^2$

11. $\dfrac{-1}{\sqrt{x - 1}(1 + \sqrt{x - 1})}$

13. (a) 0
    (b) 0
    (c) −1
    (d) $\sqrt{15}$
    (e) $\sqrt{x^2 - 1}$
    (f) $x - 1$

15. ± 3

17. 10/7

19. Domain: $[1, \infty)$
    Range: $[0, \infty)$
    $f(x) = \sqrt{x - 1}$

21. Domain: $(-\infty, \infty)$
    Range: $[0, \infty)$
    $f(x) = x^2$

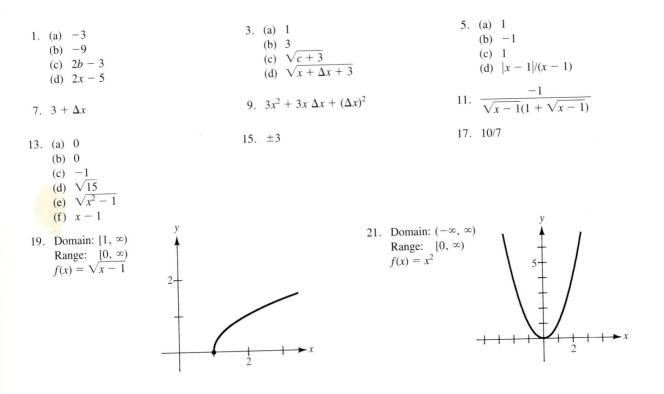

23. Domain: $[-3, 3]$
    Range: $[0, 3]$
    $f(x) = \sqrt{9 - x^2}$

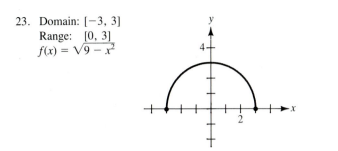

25. Domain: $(-\infty, 0), (0, \infty)$
    Range: $(0, \infty)$
    $f(x) = 1/|x|$

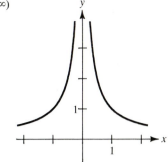

27. Domain: $(-\infty, 0), (0, \infty)$
    Range: -1 and 1
    $f(x) = |x|/x$

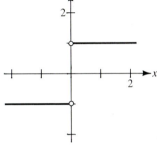

29. y is a function of x

31. y is not a function of x

33. y is a function of x

35. y is a function of x

37. y is not a function of x

39. y is not a function of x

41. y is a function of x

43. y is a function of x

45. y is not a function of x

47. (a) $y = \sqrt{x} + 2$

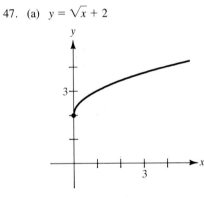

(b) $y = -\sqrt{x}$

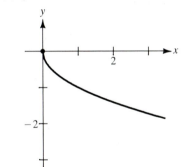

(c) $y = \sqrt{x - 2}$

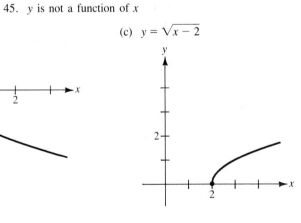

(d) $y = \sqrt{x + 3}$

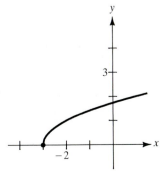

(e) $y = \sqrt{x - 4}$

(f) $y = 2\sqrt{x}$

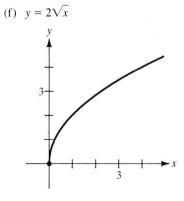

49. (a) $y = (x + 1)\sqrt{x + 4}$
    (b) $y = x\sqrt{x + 3} + 2$
    (c) $y = -x\sqrt{x + 3}$
    (d) $y = -(x - 1)\sqrt{x + 2}$

51. Even

53. Odd

55. Neither

57. y-axis symmetry

63. $A = xy = x\left(\dfrac{100 - 2x}{2}\right)$
    $\quad = x(50 - x)$

65. $V = x(12 - 2x)^2$

67. $V = 108x^2 - 4x^3$

69. $T = \dfrac{\sqrt{x^2 + 4}}{2} + \dfrac{\sqrt{x^2 - 6x + 10}}{4}$

REVIEW EXERCISES FOR CHAPTER 1

1. $-1 \le x \le 5$

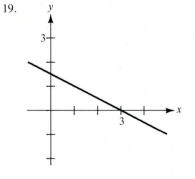

3. $x < -5,\ x > -1$

5. $27/16$

7. $2/3,\ 10/3$

9. $(-1, 3),\ (3, 2),\ (1, 1)$

11. Center: $(-3, 1)$
    Radius $= 3$

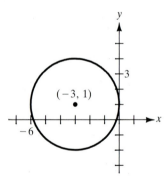

13. Point $(-3, 1)$

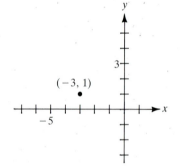

15. $c = -21$

17. $x^2 + y^2 - 2x - 4y = 4$
    (a) on the circle
    (b) inside the circle
    (c) outside the circle
    (d) inside the circle

19.

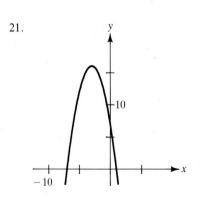

21.

23. The points do not lie on the same line.

25.

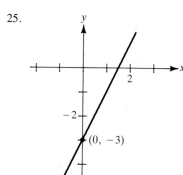

$(0, -3)$

27.

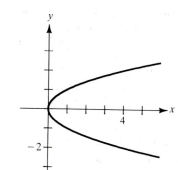

$(0, \frac{6}{5})$

29. (a) $7x - 16y + 78 = 0$
    (b) $5x - 3y + 22 = 0$
    (c) $y + 2x = 0$
    (d) $x + 2 = 0$

31. $(-4, 5)$

33. $(4, 1)$

35. $v = 850a + 300,000$
    Domain: $\{a: a \geq 0\}$

37. $s = 6x^2$
    Domain: $\{x: x \geq 0\}$

39. $d = 45t$
    Domain: $\{t: t \geq 0\}$

41. $R(x) = 4 - \dfrac{x^2}{2}$
    $r(x) = 2$

43. $h(x) = x^2$
    $p(x) = x$

45. $P(x) = 500x - x^2$

47. Function

49. Not a function

51. Function

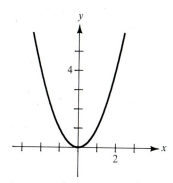

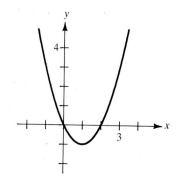

53. (a) $-x^2 + 2x + 2$
    (b) $-x^2 - 2x$
    (c) $-2x^3 - x^2 + 2x + 1$
    (d) $\dfrac{1 - x^2}{2x + 1}$
    (e) $-4x^2 - 4x$
    (f) $3 - 2x^2$

55. $C = 0.25x + 100$

CHAPTER 2

SECTION 2.1

1.

x	1.9	1.99	1.999	2.001	2.01	2.1
$f(x)$	0.3448	0.3344	0.3334	0.3332	0.3322	0.3226

$\lim\limits_{x \to 2} \dfrac{x - 2}{x^2 - x - 2} \approx 0.3333$ (Actual limit is 1/3.)

3.

x	-0.1	-0.01	-0.001	0.001	0.01	0.1
$f(x)$	0.2911	0.2889	0.2887	0.2887	0.2884	0.2863

$$\lim_{x \to 0} \frac{\sqrt{x+3} - \sqrt{3}}{x} \approx 0.2887 \ (\text{Actual limit is } 1/2\sqrt{3}.)$$

5.

x	2.9	2.99	2.999	3.001	3.01	3.1
$f(x)$	-0.0641	-0.0627	-0.0625	-0.0625	-0.0623	-0.0610

$$\lim_{x \to 3} \frac{(1/x + 1) - (1/4)}{x - 3} \approx -0.0625 \ (\text{Actual limit is } -1/16.)$$

7. 1   9. 2   11. Limit does not exist.

13. 4   15. -1 17. -4

19. 2   21. 1   23. 1/2

25. -2 27. (a) 5
                               (b) 6
                               (c) 2/3

SECTION 2.2

1. (a) 1   3. (a) 1   5. (a) 2
    (b) 3   (b) 3   (b) 0

7. -2 9. 1/6   11. 12

13. 2   15. $2x - 2$ 17. 1/10

19. 3/2   21. $\sqrt{3}/6$ 23. $-1/4$

25.

x	-0.1	-0.01	-0.001	0.001	0.01	0.1
$f(x)$	0.358	0.354	0.354	0.354	0.353	0.349

$$\lim_{x \to 0} \frac{\sqrt{x+2} - \sqrt{2}}{x} \approx 0.354 \ \left(\text{Actual limit is } \frac{1}{2\sqrt{2}}.\right)$$

27.

x	-0.1	-0.01	-0.001	0.001	0.01	0.1
$f(x)$	-0.263	-0.251	-0.250	-0.250	-0.249	-0.238

$$\lim_{x \to 0} \frac{1/(2 + x) - (1/2)}{x} \approx -0.250 \ (\text{Actual limit is } -1/4.)$$

29. (a) 1
    (b) 1
    (c) 1

31. (a) 0
    (b) 0
    (c) 0

33. (a) 3
    (b) −3
    (c) Limit does not exist.

35. 1/10

37. Limit does not exist.

39. 2

41. Limit does not exist.

43. Limit does not exist.

45. 2

47. −2

49. 1

SECTION 2.3

1. Continuous for all real x

3. Discontinuous at $x = -1$

5. Discontinuous at $x = 1$

7. Continuous for all real x

9. Nonremovable discontinuity at $x = 1$

11. Continuous for all real x

13. Removable discontinuity at $x = -2$;
    Nonremovable discontinuity at $x = 5$

15. Continuous for all real x

17. Nonremovable discontinuity at $x = 2$

19. Nonremovable discontinuity at $x = -2$

21. Continuous for all real x

23. Nonremovable discontinuities at each integer

25. Continuous for all real x

27. Continuous for all real x

29. Continuous for all real x

31. Removable discontinuity
    at $x = 4$

33. Nonremovable discontinuity
    at each integer

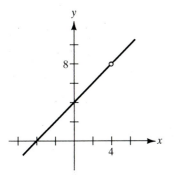

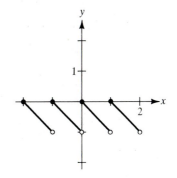

35. $(-\infty, -6), (-6, 6), (6, \infty)$

37. $(-\infty, \infty)$

41. (a) $(0.6, 0.7)$
    (b) $(0.68, 0.69)$

43. $f(3) = 11$

45. $f(2) = 4$

47. $a = 2$

49. Yes, f is continuous on $[-1, 1]$.

51. $C = \begin{cases} 1.04, & 0 < t \le 2 \\ 1.04 + 0.36[t - 1], & t > 2, t \text{ not an integer} \\ 1.04 + 0.36(t - 2), & t > 2, t \text{ is an integer} \end{cases}$

Nonremovable discontinuity at each integer greater than 2

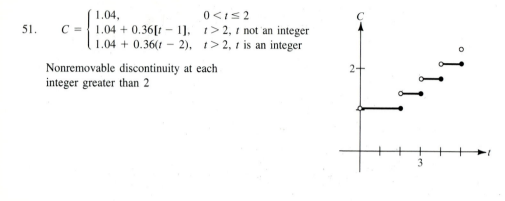

SECTION 2.4

1. $\lim\limits_{x \to -3^+} \dfrac{1}{x^2 - 9} = -\infty$ $\lim\limits_{x \to -3^-} \dfrac{1}{x^2 - 9} = \infty$

3. $\lim\limits_{x \to -3^+} \dfrac{x^2}{x^2 - 9} = -\infty$ $\lim\limits_{x \to -3^-} \dfrac{x^2}{x^2 - 9} = \infty$

5. $\lim\limits_{x \to -2^+} \dfrac{1}{(x + 2)^2} = \infty$ $\lim\limits_{x \to -2^-} \dfrac{1}{(x + 2)^2} = \infty$

7. $x = -1, x = 2$

9. $x = 0$ 11. $x = -2, x = 1$

13. $x = \pm 2$ 15. $x = 0$

17. $x = -2, x = 1$ 19. Removable discontinuity at $x = -1$

21. Vertical asymptote at $x = -1$ 23. $-\infty$

25. ∞ 27. $-\infty$

29. $1/2$ 31. ∞

33. (a) 7/12 ft/sec   35. (a) \$14,117.65
     (b) 3/2 ft/sec   (b) \$80,000
     (c) ∞ (c) \$720,000
                     (d) ∞

SECTION 2.5

1. $\lim\limits_{x \to 2} (3x + 2) = 8$ 3. $\lim\limits_{x \to 2} (x^2 - 3) = 1$

Let $\delta = \dfrac{0.01}{3} \approx 0.0033$ Assume $1 < x < 3$ and let $\delta = \dfrac{0.01}{5} = 0.002$

5. $\lim\limits_{x \to 2} \dfrac{x^2 - 3x + 2}{x - 2} = 1$ 7. $\lim\limits_{x \to 2} (x + 3) = 5$

Let $\delta = 0.01$ Let $\delta = \varepsilon$

9. $\lim\limits_{x \to 6} 3 = 3$
   Any δ will work

11. $\lim\limits_{x \to 0} \sqrt[3]{x} = 0$
   Let $\delta = \varepsilon^3$

13. $\lim\limits_{x \to 0} x^2 = 0$
   Let $\delta = \sqrt{\varepsilon}$

15. $\lim\limits_{x \to 2} \dfrac{x^2 + x - 6}{x - 2} = 5$
   Let $\delta = \varepsilon$

17. $\lim\limits_{x \to 2} \dfrac{1}{x} = \dfrac{1}{2}$
   Let $\delta = 2\varepsilon$

19. $\lim\limits_{x \to 2} (x^2 - 2) = 2$
   Let $\delta = \dfrac{\varepsilon}{5}$

21. $\lim\limits_{x \to 0^+} \sqrt{x} = 0$
   Let $\delta = \varepsilon^2$

23. $\lim\limits_{x \to -1^+} \dfrac{1}{x + 1} = \infty$
   Let $\delta = \dfrac{1}{M}$

25. $\lim\limits_{x \to 2} \dfrac{1}{(x - 2)^2} = \infty$
   Let $\delta = \dfrac{1}{\sqrt{M}}$

27. $\lim\limits_{x \to 3} x^2 = f(3) = 9$

REVIEW EXERCISES FOR CHAPTER 2

1. 7

3. 77

5. 10/3

7. $-1/4$

9. -1

11. 3

13. $-\infty$

15. $-\infty$

17. 1/3

19. 0

21.

x	1.1	1.01	1.001	1.0001
$f(x)$	0.5680	0.5764	0.5772	0.5773

$$\lim\limits_{x \to 1^+} \dfrac{\sqrt{2x + 1} - \sqrt{3}}{x - 1} \approx 0.577$$

23. $1/\sqrt{3}$

25. False

27. False

29. True

31. Nonremovable discontinuity at each integer

33. Removable discontinuity at $x = 1$

35. Nonremovable discontinuity at $x = 2$

37. Nonremovable discontinuity at $x = -1$

39. $c = -1/2$

41. Nonremovable discontinuity every 6 months

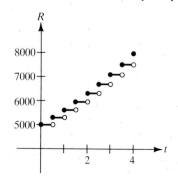

CHAPTER 3

SECTION 3.1

1. (a)   (b)

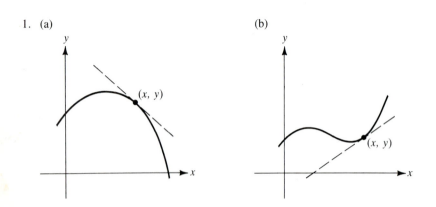

3. (a) $m = 0$
   (b) $m = -2$

5. $f(x) = 3$

   (i) $f(x + \Delta x) = 3$

   (ii) $f(x + \Delta x) - f(x) = 0$

   (iii) $\dfrac{f(x + \Delta x) - f(x)}{\Delta x} = 0$

   (iv) $\lim\limits_{\Delta x \to 0} \dfrac{f(x + \Delta x) - f(x)}{\Delta x} = 0$

7. $f(x) = -5x$

   (i) $f(x + \Delta x) = -5x - 5\Delta x$

   (ii) $f(x + \Delta x) - f(x) = -5\Delta x$

   (iii) $\dfrac{f(x + \Delta x) - f(x)}{\Delta x} = -5$

   (iv) $\lim\limits_{\Delta x \to 0} \dfrac{f(x + \Delta x) - f(x)}{\Delta x} = -5$

9. $f(x) = 2x^2 + x - 1$

   (i) $f(x + \Delta x) = 2x^2 + 4x\Delta x + 2(\Delta x)^2 + x + \Delta x - 1$

   (ii) $f(x + \Delta x) - f(x) = 4x\Delta x + 2(\Delta x)^2 + \Delta x$

   (iii) $\dfrac{f(x + \Delta x) - f(x)}{\Delta x} = 4x + 2\Delta x + 1$

   (iv) $\lim\limits_{\Delta x \to 0} \dfrac{f(x + \Delta x) - f(x)}{\Delta x} = 4x + 1$

11. $f(x) = \dfrac{1}{x - 1}$

   (i) $f(x + \Delta x) = \dfrac{1}{x + \Delta x - 1}$

   (ii) $f(x + \Delta x) - f(x) = \dfrac{-\Delta x}{(x + \Delta x - 1)(x - 1)}$

   (iii) $\dfrac{f(x + \Delta x) - f(x)}{\Delta x} = \dfrac{-1}{(x + \Delta x - 1)(x - 1)}$

   (iv) $\lim\limits_{\Delta x \to 0} \dfrac{f(x + \Delta x) - f(x)}{\Delta x} = \dfrac{-1}{(x - 1)^2}$

13. $f(t) = t^3 - 12t$

   (i) $f(t + \Delta t) = t^3 + 3t^2\Delta t + 3t(\Delta t)^2 + (\Delta t)^3 - 12t - 12\Delta t$

   (ii) $f(t + \Delta t) - f(t) = 3t^2\Delta t + 3t(\Delta t)^2 + (\Delta t)^3 - 12\Delta t$

   (iii) $\dfrac{f(t + \Delta t) - f(t)}{\Delta t} = 3t^2 + 3t\Delta t + (\Delta t)^2 - 12$

   (iv) $\lim\limits_{\Delta t \to 0} \dfrac{f(t + \Delta t) - f(t)}{\Delta t} = 3t^2 - 12$

15. $f'(x) = 2x$

Tangent line:
$y = 4x - 3$

17. $f'(x) = 3x^2$

Tangent line:
$y = 12x - 16$

19. $f'(x) = \dfrac{1}{2\sqrt{x+1}}$

Tangent line:
$4y = x + 5$

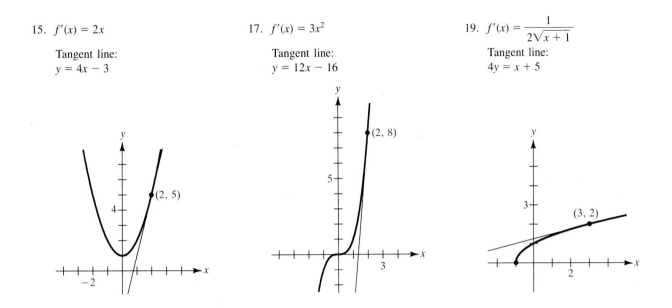

21. $f(x) = x^2 - 1$

$$f'(2) = \lim_{x \to 2} \frac{f(x) - f(2)}{x - 2}$$

$$= \lim_{x \to 2} \frac{(x^2 - 1) - 3}{x - 2}$$

$$= \lim_{x \to 2} (x + 2) = 4$$

23. $f(x) = x^3 + 2x^2 + 1$

$$f'(-2) = \lim_{x \to -2} \frac{f(x) - f(-2)}{x + 2}$$

$$= \lim_{x \to -2} \frac{(x^3 + 2x^2 + 1) - 1}{x + 2}$$

$$= \lim_{x \to -2} x^2 = 4$$

25. $f(x) = (x - 1)^{2/3}$

$$f'(1) = \lim_{x \to 1} \frac{f(x) - f(1)}{x - 1}$$

$$= \lim_{x \to 1} \frac{(x - 1)^{2/3} - 0}{x - 1}$$

$$= \lim_{x \to 1} \frac{1}{(x - 1)^{1/3}}$$

Limit does not exist.
f is not differentiable at $x = 1$.

27. $x = -3$

29. $x = -1$

31. $x = 3$

33. $x = 1$

35. $x = 0$

37. f is not differentiable at $x = 1$.

39. $f'(1) = 0$

41. $y = 3x - 2$
$y = 3x + 2$

43. $y = 2x + 1$
$y = -2x + 9$

SECTION 3.2

1. Average rate: 2
   Instantaneous rates:
   $f'(1) = 2$
   $f'(2) = 2$

3. Average rate: $-1/4$
   Instantaneous rates:
   $f'(0) = -1$
   $f'(3) = -1/16$

5. Average rate: 4.1
   Instantaneous rates:
   $f'(2) = 4$
   $f'(2.1) = 4.2$

7. (a) -48 ft/sec
   (b) $s'(1) = -32$ ft/sec
       $s'(2) = -64$ ft/sec
   (c) $t = 15\sqrt{6}/4 \approx 9.2$ sec
   (d) -293.9 ft/sec

9. $v(5) = 224$ ft/sec
   $v(10) = 64$ ft/sec

11. $v\left(\dfrac{5\sqrt{6}}{2}\right) = -80\sqrt{6}$ ft/sec ≈ -195.96 ft/sec

13. 740 ft

15.

t	0	1	2	3	4
$s(t)$	0	57.75	99	123.75	132
$v(t)$	66	49.5	33	16.5	0
$a(t)$	-16.5	-16.5	-16.5	-16.5	-16.5

17. [0, 1], 57.75 ft/sec
    [1, 2], 41.25 ft/sec
    [2, 3], 24.75 ft/sec
    [3, 4], 8.25 ft/sec

19. $4x$

21. $\dfrac{2}{x^2}$

23. 0

25. $-\$1.91, -\$1.93$

27. (a) -2.145 deg/sec
    (b) -0.953 deg/sec

SECTION 3.3

1. (a) 1/2
   (b) 3/2
   (c) 2
   (d) 3

3. 0

5. 1

7. $2x$

9. $-4t + 3$

11. $3t^2 - 2$

13. -1

15. 5/3

17. 4

19. $2x + \dfrac{4}{x^2}$

21. $3x^2 - 3 + \dfrac{8}{x^5}$

23. $\dfrac{x^3 - 8}{x^3}$

25. $3x^2 + 1$

27. $\dfrac{4}{5x^{1/5}}$

29. $\dfrac{1}{3x^{2/3}} + \dfrac{1}{5x^{4/5}}$

Function	Rewrite	Derivative	Simplify
31. $y = \dfrac{1}{3x^3}$	$y = \dfrac{1}{3}x^{-3}$	$y' = -x^{-4}$	$y' = -\dfrac{1}{x^4}$
33. $y = \dfrac{1}{(3x)^3}$	$y = \dfrac{1}{27}x^{-3}$	$y' = -\dfrac{1}{9}x^{-4}$	$y' = -\dfrac{1}{9x^4}$
35. $y = \dfrac{\sqrt{x}}{x}$	$y = x^{-1/2}$	$y' = -\dfrac{1}{2}x^{-3/2}$	$y' = -\dfrac{1}{2x^{3/2}}$

37. $2x + y - 2 = 0$
39. $(0, 2), (\sqrt{3/2}, -1/4), (-\sqrt{3/2}, -1/4)$

41. No horizontal tangents

43. $y = 2x - 1$ $y = 4x - 4$ 45. 8

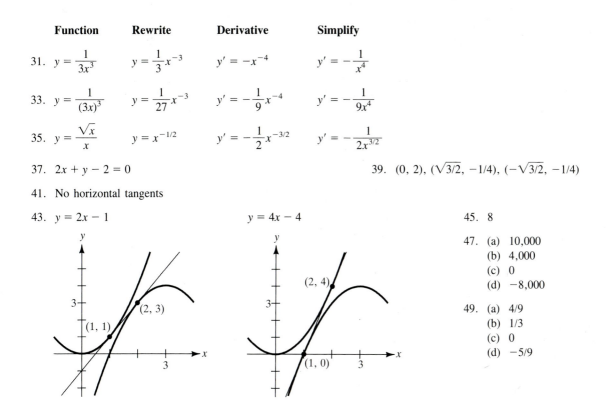

47. (a) 10,000
    (b) 4,000
    (c) 0
    (d) −8,000

49. (a) 4/9
    (b) 1/3
    (c) 0
    (d) −5/9

SECTION 3.4

1. $f'(x) = 2x^2, \; f'(0) = 0$

3. $f'(x) = 5[x^{-2}(1) + (x + 3)(-2x^{-3})] = \dfrac{-5}{x^3}(x + 6), \; f'(1) = -35$

5. $f'(x) = (x^3 - 3x)(4x + 3) + (2x^2 + 3x + 5)(3x^2 - 3) = 10x^4 + 12x^3 - 3x^2 - 18x - 15, \; f'(0) = -15$

7. $f'(x) = (x^5 - 3x)\left(-\dfrac{2}{x^3}\right) + \left(\dfrac{1}{x^2}\right)(5x^4 - 3) = 3x^2 + \dfrac{3}{x^2}, \; f'(-1) = 6$

9. $\dfrac{(2x - 3)(3) - (3x - 2)(2)}{(2x - 3)^2} = -\dfrac{5}{(2x - 3)^2}$

11. $\dfrac{(x^2 - 1)(-2 - 2x) - (3 - 2x - x^2)(2x)}{(x^2 - 1)^2} = \dfrac{2}{(x + 1)^2}$

13. $\dfrac{\sqrt{x}(1) - (x + 1)[1/(2\sqrt{x})]}{x} = \dfrac{x - 1}{2x^{3/2}}$

15. $\dfrac{(t^2 + 2t + 2)(1) - (t + 1)(2t + 2)}{(t^2 + 2t + 2)^2} = \dfrac{-t^2 - 2t}{(t^2 + 2t + 2)^2}$

17. $6s^2(s^3 - 2)$

19. $\left[\dfrac{x+1}{x+2}\right](2) + (2x-5)\left[\dfrac{(x+2)(1)-(x+1)(1)}{(x+2)^2}\right] = \dfrac{2x^2+8x-1}{(x+2)^2}$

21. $15x^4 - 48x^3 - 33x^2 - 32x - 20$

23. $\dfrac{(c^2+x^2)(-2x)-(c^2-x^2)(2x)}{(c^2+x^2)^2} = -\dfrac{4xc^2}{(c^2+x^2)^2}$

25. $\dfrac{3}{\sqrt{x}}$

27. $\dfrac{2}{(x-1)^3}$

Function	Rewrite	Derivative	Simplify
29. $y = \dfrac{x^2+2x}{x}$	$y = x+2$	$y' = 1$	$y' = 1$
31. $y = \dfrac{7}{3x^3}$	$y = \dfrac{7}{3}x^{-3}$	$y' = -7x^{-4}$	$y' = -\dfrac{7}{x^4}$
33. $y = \dfrac{3x^2-5}{7}$	$y = \dfrac{1}{7}(3x^2-5)$	$y' = \dfrac{1}{7}(6x)$	$y' = \dfrac{6x}{7}$

35. $y = -x + 4$

37. $y = -x - 2$

39. $(0, 0), (2, 4)$

41. (a) -0.48
    (b) 0.12
    (c) 0.0149

43. 31.55

SECTION 3.5

$y = f(g(x))$	$u = g(x)$	$y = f(u)$
1. $y = \sqrt{x^2-1}$	$u = x^2 - 1$	$y = \sqrt{u}$
3. $y = (x^2 - 3x + 4)^6$	$u = x^2 - 3x + 4$	$y = u^6$
5. $y = \dfrac{1}{\sqrt{x^2+1}}$	$u = x^2 + 1$	$y = u^{-1/2}$

7. $6(2x-7)^2$

9. $12x(x^2-1)^2$

11. $-\dfrac{1}{(x-2)^2}$

13. $-\dfrac{2}{(t-3)^3}$

15. $-\dfrac{9x^2}{(x^3-4)^2}$

17. $2x(x-2)^3(3x-2)$

19. $-\dfrac{1}{2\sqrt{1-t}}$

21. $\dfrac{t+1}{\sqrt{t^2+2t-1}}$

23. $\dfrac{6x}{(9x^2+4)^{2/3}}$

25. $-\dfrac{2x}{\sqrt{4-x^2}}$

27. $-\dfrac{4x}{3(9-x^2)^{1/3}}$

29. $-\dfrac{1}{2(x+2)^{3/2}}$

31. $-\dfrac{3x^2}{(x^3-1)^{4/3}}$

33. $\dfrac{1}{2\sqrt{x}(\sqrt{x}+1)^2}$

35. $\dfrac{(x^2+1)(1/2)(x^{-1/2})-(\sqrt{x}+1)(2x)}{(x^2+1)^2} = \dfrac{1-3x^2-4x^{3/2}}{2\sqrt{x}(x^2+1)^2}$

37. $\dfrac{(t-1)(3)-(3t+2)}{(t-1)^2} = -\dfrac{5}{(t-1)^2}$

39. $3t^2(-1/2)(t^2 + 2t - 1)^{-3/2}(2t + 2) + (t^2 + 2t - 1)^{-1/2}(6t) = \dfrac{3t(t^2 + 3t - 2)}{(t^2 + 2t - 1)^{3/2}}$

41. $\dfrac{1}{2}\left(\dfrac{x + 1}{x}\right)^{-1/2}\dfrac{x - (x + 1)}{x^2} = -\dfrac{1}{2x^{3/2}\sqrt{x + 1}}$

43. $-\dfrac{2}{3}\left[(2 - t)\dfrac{1}{2\sqrt{1 + t}} + (-1)\sqrt{1 + t}\right] = \dfrac{t}{\sqrt{1 + t}}$

45. $9x - 5y - 2 = 0$ 47. $12(5x^2 - 1)(x^2 - 1)$ 49. $\dfrac{3}{4(x^2 + x + 1)^{3/2}}$

51. (a) 1.461   53. 0.04224
    (b) −1.016

SECTION 3.6

1. $-\dfrac{x}{y}, \quad -\dfrac{3\sqrt{7}}{7}$

3. $-\dfrac{y}{x}, \quad -\dfrac{1}{4}$

5. $-\sqrt{\dfrac{y}{x}}, \quad -\dfrac{5}{4}$

7. $\dfrac{y - 3x^2}{2y - x}, \quad \dfrac{1}{2}$

9. $\dfrac{18x}{(x^2 + 9)^2 y}$, undefined

11. $\dfrac{1 - 3x^2y^3}{3x^3y^2 - 1}, \quad -1$

13. $-\sqrt[3]{\dfrac{y}{x}}, \quad -\dfrac{1}{2}$

15. $\dfrac{4xy - 3x^2 - 3y^2}{2x(3y - x)}, \quad -\dfrac{15}{28}$

17. $y' = -\dfrac{x}{y}$

19. $y' = -\dfrac{9x}{16y}$

21. $\dfrac{10}{x^3}$

23. $-\dfrac{16}{y^3}$

25. $\dfrac{3x}{4y}$

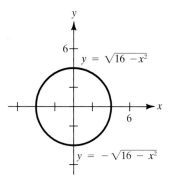

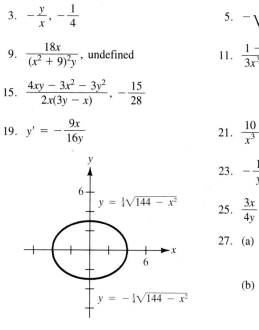

27. (a) At (4, 3)
        Tangent line: $4x + 3y - 25 = 0$
        Normal line: $3x - 4y = 0$
    (b) At (−3, 4)
        Tangent line: $3x - 4y + 25 = 0$
        Normal line: $4x + 3y = 0$

29. Horizontal tangents at (−4, 0), (−4, 10)
    Vertical tangents at (0, 5), (−8, 5)

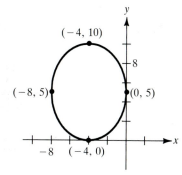

31. At (1, 2):
       Slope of ellipse = −1
       Slope of parabola = 1
     At (1, −2):
       Slope of ellipse = 1
       Slope of parabola = −1

33. At ($\sqrt{2}$, −$\sqrt{2}$):
       Slope of line = −1
       Slope of circle = 1
     At (−$\sqrt{2}$, $\sqrt{2}$):
       Slope of line = −1
       Slope of circle = 1

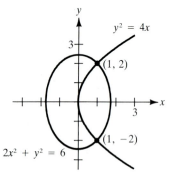

35. $(x − 1 − 2\sqrt{2})^2 + (y − 2 + 2\sqrt{2})^2 = 16$
     $(x − 1 + 2\sqrt{2})^2 + (y − 2 − 2\sqrt{2})^2 = 16$

SECTION 3.7

1. (a) 3/4
    (b) 20

3. (a) −5/8
    (b) 3/2

5. (a) 24π in^2/min
    (b) 96π in^2/min

7. If dr/dt is constant, dA/dt is proportional to r.

9. (a) $5/\pi$ ft/min
    (b) $5/(4\pi)$ ft/min

11. $8/(405\pi)$ ft/min

13. (a) 9 cm^3/sec
     (b) 900 cm^3/sec

15. (a) 0 cm/min
     (b) 12 cm/min

17. (a) 8/25 cm/min
     (b) 0 cm/min
     (c) −8/25 cm/min
     (d) −0.0039 cm/min

19. (a) 24.6%
     (b) 1/64 ft/min

21. (a) −7/12 ft/sec
     (b) −3/2 ft/sec
     (c) −48/7 ft/sec

23. (a) −750 mi/hr
     (b) 20 minutes

25. $28/\sqrt{10} \approx 8.85$ ft/sec

27. (a) 25/3 ft/sec
     (b) 10/3 ft/sec

REVIEW EXERCISES FOR CHAPTER 3

1. $3x(x − 2)$

3. $\dfrac{x + 1}{2x^{3/2}}$

5. $-\dfrac{4}{3t^3}$

7. $\dfrac{3x^2}{2\sqrt{x^3 + 1}}$

9. $2(6x^3 − 9x^2 + 16x − 7)$

11. $5s(s^2 − 1)^{3/2}(s^3 + 5)^{2/3}(2s^3 − s + 5)$

13. $\dfrac{3x^2 + 1}{2\sqrt{x^3 + x}}$

15. $-\dfrac{x^2 + 1}{(x^2 - 1)^2}$

17. $32x - 128x^3$

19. $\dfrac{6x}{(4 - 3x^2)^2}$

21. $\dfrac{x + 2}{(x + 1)^{3/2}}$

23. $\dfrac{5}{6(t + 1)^{1/6}}$

25. $\dfrac{9}{(x^2 + 9)^{3/2}}$

27. $\dfrac{2(t + 2)}{(1 - t)^4}$

29. $\dfrac{2(6x^3 - 15x^2 - 18x + 5)}{(x^2 + 1)^3}$

31. $-\dfrac{2x + 3y}{3(x + y^2)}$

33. $\dfrac{2y\sqrt{x} - y\sqrt{y}}{2x\sqrt{y} - x\sqrt{x}}$

35. $\dfrac{x}{y}$

37. Tangent line: $3x - y + 7 = 0$
    Normal line: $x + 3y - 1 = 0$

39. Tangent line: $x + 2y - 10 = 0$
    Normal line: $2x - y = 0$

41. Tangent line: $2x - 3y - 3 = 0$
    Normal line: $3x + 2y - 11 = 0$

43. (a) $(0, -1), (-2, 7/3)$
    (b) $(-3, 2), (1, -2/3)$
    (c) $(-1 + \sqrt{2}, 2[1 - 2\sqrt{2}]/3)$
        $(-1 - \sqrt{2}, 2[1 + 2\sqrt{2}]/3)$

45. $f(x) = \dfrac{1}{x^2}$

    (i) $f(x + \Delta x) = \dfrac{1}{(x + \Delta x)^2}$

    (ii) $f(x + \Delta x) - f(x) = \dfrac{-2x\Delta x - (\Delta x)^2}{x^2[x^2 + 2x\Delta x + (\Delta x)^2]}$

    (iii) $\dfrac{f(x + \Delta x) - f(x)}{\Delta x} = \dfrac{-2x - \Delta x}{x^2[x^2 + 2x\Delta x + (\Delta x)^2]}$

    (iv) $\lim\limits_{\Delta x \to 0} \dfrac{f(x + \Delta x) - f(x)}{\Delta x} = -\dfrac{2}{x^3}$

47. $f(x) = \sqrt{x + 2}$

    (i) $f(x + \Delta x) = \sqrt{x + \Delta x + 2}$

    (ii) $f(x + \Delta x) - f(x) = \dfrac{\Delta x}{\sqrt{x + \Delta x + 2} + \sqrt{x + 2}}$

    (iii) $\dfrac{f(x + \Delta x) - f(x)}{\Delta x} = \dfrac{1}{\sqrt{x + \Delta x + 2} + \sqrt{x + 2}}$

    (iv) $\lim\limits_{\Delta x \to 0} \dfrac{f(x + \Delta x) - f(x)}{\Delta x} = \dfrac{1}{2\sqrt{x + 2}}$

49. $v(t) = 1 - \dfrac{1}{(t + 1)^2}$, $a(t) = \dfrac{2}{(t + 1)^3}$

51. (a) -18.667
    (b) -7.284
    (c) -3.240
    (d) -0.747

53. 56 ft/sec

55. (a)

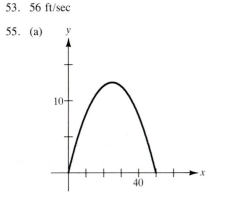

(b) 50
(c) $x = 25$
(d) $y' = 1 - 0.04x$

x	0	10	25	30	50
y'	1	0.6	0	-0.2	-1

(e) $y'(25) = 0$

59. (a) $2\sqrt{2}$ units/sec
    (b) 4 units/sec
    (c) 8 units/sec

61. 2/25 ft/min

63. 31.55 bacteria/hr

67. $f(x) = \begin{cases} x^2 + 4x + 2, & \text{if } x < -2 \\ 1 - 4x - x^2, & \text{if } x \geq -2 \end{cases}$

    (a) nonremovable discontinuity at $x = -2$

    (b) Not differentiable at $x = -2$ because the function is discontinuous there.

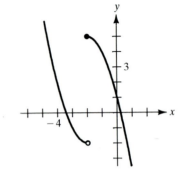

CHAPTER 4

SECTION 4.1

1. (a) Yes
   (b) No

3. (a) No
   (b) Yes

5. (a) Minimum: $(0, -3)$
       Maximum: $(2, 1)$
   (b) Minimum: $(0, -3)$
   (c) Maximum: $(2, 1)$
   (d) No extrema

7. (a) Minimum: $(1, -1)$
       Maximum: $(-1, 3)$
   (b) Maximum: $(3, 3)$
   (c) Minimum: $(1, -1)$
   (d) Minimum: $(1, -1)$
       Maximum: $(3, 3)$

9. (a) No extrema
   (b) Minimum: $(1, 0)$
   (c) Minimum: $(1, 0)$
       Maximum: $(3, 2)$
   (d) Maximum: $(3, 2)$

11. $f'(0) = 0$

13. $f'(4) = 0$

15. $f'(-2)$ is undefined

17. Minimum: $(2, 2)$
    Maximum: $(-1, 8)$

19. Minimum: $(0, 0)$ and $(3, 0)$
    Maximum: $(3/2, 9/4)$

21. Minimum: $(-1, -4)$ and $(2, -4)$
    Maximum: $(0, 0)$ and $(3, 0)$

23. Minimum: $(0, 0)$
    Maximum: $(-1, 5)$

25. Minimum: $(1, -1)$
    Maximum: $(0, -1/2)$

27. Continuous on $[1, 2]$
    Not continuous on $[0, 2]$

29. Maximum: $|f''(0)| = 2$

31. Maximum: $|f''(\sqrt[3]{-10} + \sqrt{108})| \approx 1.47$

33. Maximum: $|f^{(4)}(1/2)| = 360$

35. Maximum: $|f^{(4)}(0)| = 56/81$

37. Maximum: $P(12) = 72$

SECTION 4.2

1. $[0, 2]$, $f'(1) = 0$

3. $[1, 2]$, $f'([6 - \sqrt{3}]/3) = 0$
   $[2, 3]$, $f'([6 + \sqrt{3}]/3) = 0$

5. Not differentiable at $x = 0$

7. Not differentiable at $x = 0$

9. $[-1, 3]$, $f'(-2 + \sqrt{5}) = 0$

11. $f(x) \neq 0$

13. $f'(-1/2) = -1$

15. $f'(8/27) = 1$

17. $f'([-2 + \sqrt{6}]/2) = 2/3$

19. $f'(\sqrt{3}/3) = 1$

21. (a) $f(1) = f(2) = 64$
    (b) velocity $= 0$ for some t in $[1, 2]$

23. (a) -48 ft/sec
    (b) $t = 3/2$ sec

SECTION 4.3

1. Increasing on $(3, \infty)$
   Decreasing on $(-\infty, 3)$

3. Increasing on $(-\infty, -2)$ and $(2, \infty)$
   Decreasing on $(-2, 2)$

5. Increasing on $(-\infty, 0)$
   Decreasing on $(0, \infty)$

7. Critical number: $x = 1$
   Increasing on $(-\infty, 1)$
   Decreasing on $(1, \infty)$
   Relative maximum: $(1, 5)$

9. Critical number: $x = 3$
   Increasing on $(3, \infty)$
   Decreasing on $(-\infty, 3)$
   Relative minimum: $(3, -9)$

11. Critical numbers: $x = -2, 1$
    Increasing on $(-\infty, -2)$ and $(1, \infty)$
    Decreasing on $(-2, 1)$
    Relative maximum: $(-2, 20)$
    Relative minimum: $(1, -7)$

13. Critical numbers: $x = 0, 4$
    Increasing on $(-\infty, 0)$ and $(4, \infty)$
    Decreasing on $(0, 4)$
    Relative maximum: $(0, 15)$
    Relative minimum: $(4, -17)$

15. Critical numbers: $x = 0, 3/2$
    Increasing on $(3/2, \infty)$
    Decreasing on $(-\infty, 3/2)$
    Relative minimum: $(3/2, -27/16)$

17. Critical number: $x = 0$
    Increasing on $(-\infty, \infty)$
    No relative extrema

19. Critical numbers: $x = -1, 1$
    Discontinuity: $x = 0$
    Increasing on $(-\infty, -1)$ and $(1, \infty)$
    Decreasing on $(-1, 0)$ and $(0, 1)$
    Relative maximum: $(-1, -2)$
    Relative minimum: $(1, 2)$

21. Critical number: $x = 0$
    Discontinuities: $x = -3, 3$
    Increasing on $(-\infty, -3)$ and $(-3, 0)$
    Decreasing on $(0, 3)$ and $(3, \infty)$
    Relative maximum: $(0, 0)$

23. Critical numbers: $x = -1, 1$
    Increasing on $(-\infty, -1)$ and $(1, \infty)$
    Decreasing on $(-1, 1)$
    Relative maximum: $(-1, 4/5)$
    Relative minimum: $(1, -4/5)$

25. Critical numbers: $x = -3, 1$
    Discontinuity: $x = -1$
    Increasing on $(-\infty, -3)$ and $(1, \infty)$
    Decreasing on $(-3, -1)$ and $(-1, 1)$
    Relative maximum: $(-3, -8)$
    Relative minimum: $(1, 0)$

27. Moving upward when $0 < t < 3$
    Moving downward when $3 < t < 6$
    Maximum height: $s(3) = 144$ ft

29. $r = 2R/3$

31. Increasing when $0 < t < 84.3388$ minutes
    Decreasing when $84.3388 < t < 120$ minutes

33. Decreasing when $0 < t < 6.02$ days
    Increasing when $6.02 < t < 14$ days

35. $T = 10°$

37. $f(x) = -\dfrac{1}{2}x^3 + \dfrac{3}{2}x^2$

39. $g'(0) < 0$

41. $g'(-6) < 0$

43. $g'(0) > 0$

SECTION 4.4

1. $f'(x)$
   (a) Positive on $(-\infty, 0)$
   (b) Negative on $(0, \infty)$
   (c) Not increasing
   (d) Decreasing on $(-\infty, \infty)$

   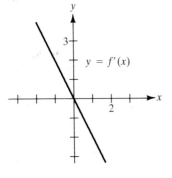

$f(x)$
Increasing on $(-\infty, 0)$
Decreasing on $(0, \infty)$
Not concave upward
Concave downward on $(-\infty, \infty)$

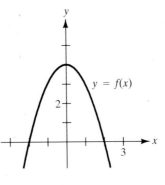

3. $f'(x)$
   (a) Positive on $(0, 2)$
   (b) Negative on $(-\infty, 0)$ and $(2, \infty)$
   (c) Increasing on $(-\infty, 1)$
   (d) Decreasing on $(1, \infty)$

   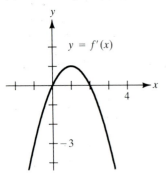

$f(x)$
Increasing on $(0, 2)$
Decreasing on $(-\infty, 0)$ and $(2, \infty)$
Concave upward on $(-\infty, 1)$
Concave downward on $(1, \infty)$

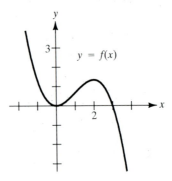

5.

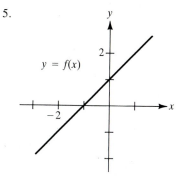

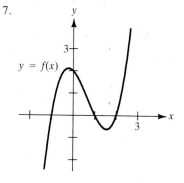

7.

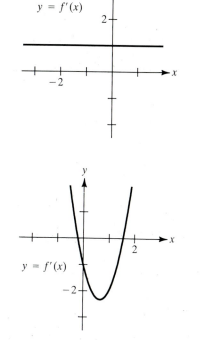

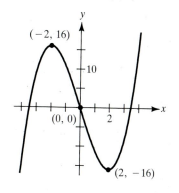

9. Concave upward: $(-\infty, \infty)$

11. Concave upward: $(-\infty, 1)$
    Concave downward: $(1, \infty)$

13. Concave upward: $(-\infty, -1), (1, \infty)$
    Concave downward: $(-1, 1)$

15. Relative maximum: $(3, 9)$

17. Relative minimum: $(5, 0)$

19. Relative maximum: $(0, 3)$
    Relative minimum: $(2, -1)$

21. Relative minimum: $(3, -25)$

23. Relative minimum: $(0, -3)$

25. Relative maximum: $(-2, -4)$
    Relative minimum: $(2, 4)$

27. Relative maximum: $(-2, 16)$
    Relative minimum: $(2, -16)$
    Point of inflection: $(0, 0)$

29. Point of inflection: (2, 0)

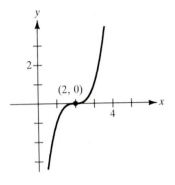

31. Relative minima: $(\pm 2, -4)$
    Relative maximum: $(0, 0)$
    Points of inflection: $(\pm 2/\sqrt{3}, -20/9)$

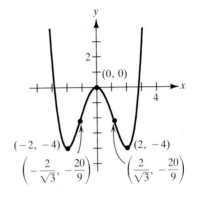

33. Relative minima: $(-1, 2)$, $(1, 2)$

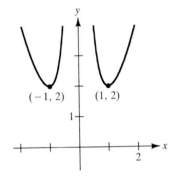

35. Relative minimum: $(-2, -2)$

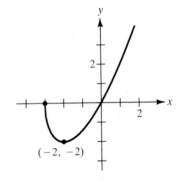

37. Point of inflection: $(0, 0)$

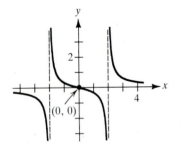

39. Point of inflection: $(2, 0)$

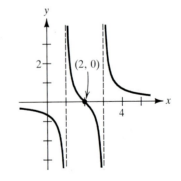

41.

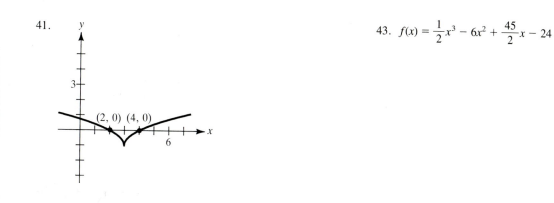

43. $f(x) = \dfrac{1}{2}x^3 - 6x^2 + \dfrac{45}{2}x - 24$

45. Relative extrema: (2, 32), (6, 0)
   Point of inflection: (4, 16)

47. $x = \left(\dfrac{15 - \sqrt{33}}{16}\right)L \approx 0.578\,L$

49. $x = 100$ units

SECTION 4.5

1. h	2. c	3. e
4. a	5. d	6. g
7. b	8. f	9. 2/3
11. 0	13. Limit does not exist.	15. Limit does not exist.
17. 5	19. 0	21. $-1/2$
23. -1	25. 2	27. 1

29.

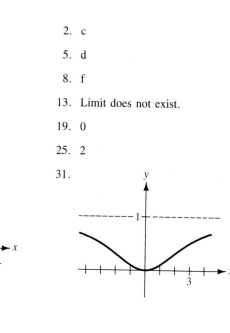

31.

33.

35.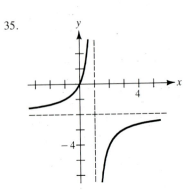

37.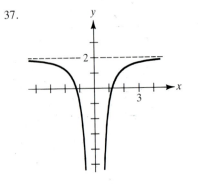

39.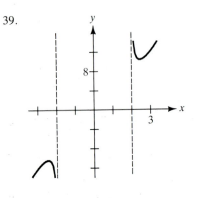

41.

x	1	10	10^2	10^4	10^6
$f(x)$	2.000	0.348	0.101	0.010	0.001

$$\lim_{x \to \infty} \frac{x+1}{x\sqrt{x}} = 0$$

43.

x	1	10	10^2	10^3	10^4	10^5	10^6
$f(x)$	-0.2361	-0.0250	-0.0025	-0.0002	0.0	0.0	0.0

$$\lim_{x \to \infty} (2x - \sqrt{4x^2 + 1}) = 0$$

45. 0.5

47. 100%

SECTION 4.6

1.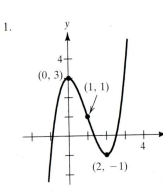

3.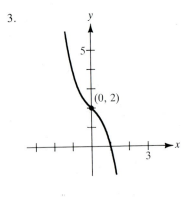

5.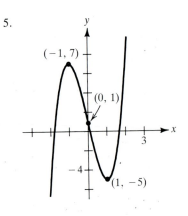

7.

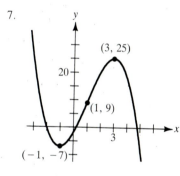

9.

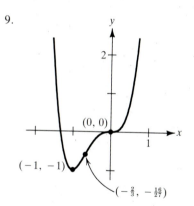

11.

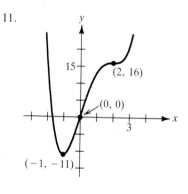

13.

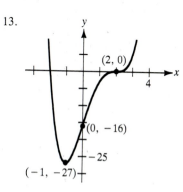

15.

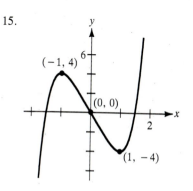

17.

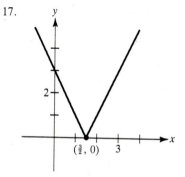

19.

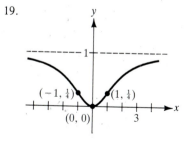

21.

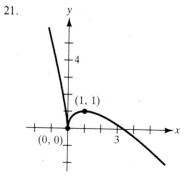

23.

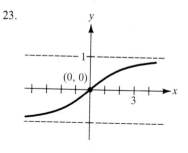

25.

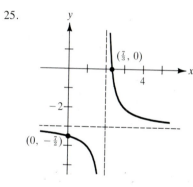

27.

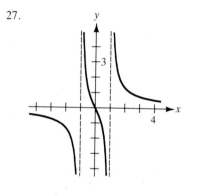

29.

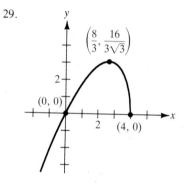

31.   33.   35.

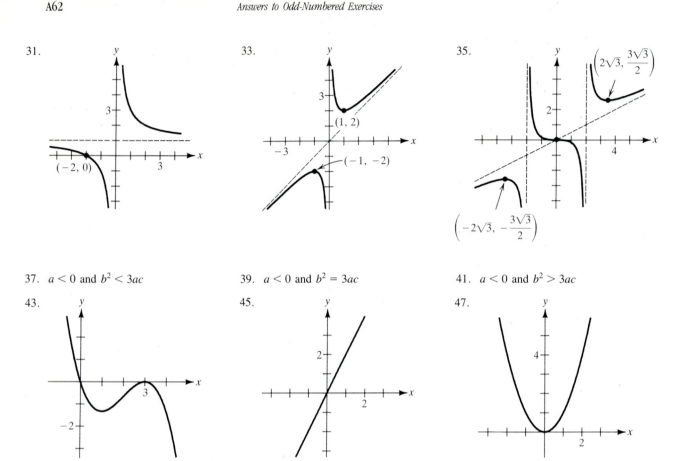

37. $a < 0$ and $b^2 < 3ac$ 39. $a < 0$ and $b^2 = 3ac$ 41. $a < 0$ and $b^2 > 3ac$

43.   45.   47.

49. Relative minimum: $(-3/4, -\sqrt{34}/2)$
    Points of inflection: $(0.19, -1.38)$, $(-1.32, -2.82)$

SECTION 4.7

1. $l = w = 25$ ft

3. 6 and 12

5. 55 and 55

7. $\sqrt{192}$ and $\sqrt{192}$

9. $x = 25$, $y = 100/3$

11. $V = 128$ when $x = 2$

13. $\dfrac{5 - \sqrt{7}}{6}$ ft $\times \dfrac{1 + \sqrt{7}}{3}$ ft $\times \dfrac{4 + \sqrt{7}}{3}$ ft

15. $x = 3$, $y = 3/2$

17. 50 m $\times$ 100/π m

19. $(0, 0)$, $(4, 0)$, $(0, 6)$

21. $x = y = 5\sqrt{2}/2 \approx 3.54$

23. $(7/2, \sqrt{7/2})$

25. $r \approx 1.51$, $h = 2r \approx 3.02$

27. 18 in $\times$ 18 in $\times$ 36 in

29. $32\pi r^3/81$

31. $r = \sqrt[3]{9/\pi} \approx 1.42$ in

33. Side of square: $10\sqrt{3}/(9 + 4\sqrt{3})$
    Side of triangle: $30/(9 + 4\sqrt{3})$

35. (a) $\dfrac{10\pi}{\pi + 3 + 2\sqrt{2}} \approx 3.5$ ft
    (b) 10 ft

37. $W = 8\sqrt{3}$, $H = 8\sqrt{6}$

39. Point 1 mile from the nearest
    point on the coast

SECTION 4.8

1. 4,500

3. 300

5. 200

7. 200

9. $x = 30$, $p = 60$

11. $x = 1,500$, $p = 35$

13. 20

15. 200

17. Line should run from power station to point across the river $3/(2\sqrt{7})$ mi downstream.

19. 8%

21. 3

23. (a) $68.00
    (b) $50.88

25. $92.50

27. (a) $-1/8$
    (b) $x = \sqrt[3]{9/2}$
        $p = \sqrt[3]{9/2}$

SECTION 4.9

1.

n	x_n	$f(x_n)$	$f'(x_n)$	$f(x_n)/f'(x_n)$	$x_n - [f(x_n)/f'(x_n)]$
1	1.700	-0.110	3.400	-0.032	1.732

3.

n	x_n	$f(x_n)$	$f'(x_n)$	$f(x_n)/f'(x_n)$	$x_n - [f(x_n)/f'(x_n)]$
1	1	1	9	0.111	0.889

5. 0.682

7. 1.146

9. 3.317

11. 0.569

13. 1.379

15. $f'(x_1) = 0$

17. $1 = x_1 = x_3 = \ldots$
    $0 = x_2 = x_4 = \ldots$

19. $x_{i+1} = \dfrac{(n-1)x_i^n + a}{nx_i^{n-1}}$

21. 1.565

23. (1.939, 0.240)

25. $x \approx 1.563$ mi

27. $t \approx 4.486$ hr

SECTION 4.10

1. $6x \, dx$

3. $12x^2 \, dx$

5. $-\dfrac{3}{(2x-1)^2} \, dx$

7. $\dfrac{1}{2\sqrt{x}} \, dx$

9. $\dfrac{1 - 2x^2}{\sqrt{1 - x^2}} \, dx$

11.

$dx = \Delta x$	dy	Δy	$\Delta y - dy$	$dy/\Delta y$
1.000	1.000	1.000	0	1
0.500	0.500	0.500	0	1
0.100	0.100	0.100	0	1
0.010	0.010	0.010	0	1
0.001	0.001	0.001	0	1

13.

$dx = \Delta x$	dy	Δy	$\Delta y - dy$	$dy/\Delta y$
1.000	4.000	5.000	1.000	0.800
0.500	2.000	2.250	0.250	0.889
0.100	0.400	0.410	0.010	0.976
0.010	0.040	0.040	0.000	1.000
0.001	0.004	0.004	0.000	1.000

15.

$dx = \Delta x$	dy	Δy	$\Delta y - dy$	$dy/\Delta y$
1.000	80.000	211.000	131.000	0.379
0.500	40.000	65.656	25.656	0.609
0.100	8.000	8.841	0.841	0.905
0.010	0.800	0.808	0.008	0.990
0.001	0.080	0.080	0.000	1.000

17. (a) $dA = 2x\Delta x$, $\Delta A = 2x\Delta x + (\Delta x)^2$

(b)
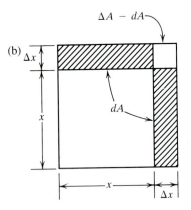

19. $\pm 7\pi$ in^2

21. (a) $\pm 2.88\pi$ in^3
     (b) $\pm 0.96\pi$ in^2
     (c) 1%, 2/3%

23. (a) 1/4%
     (b) 216 sec = 3.6 min

25. $\displaystyle \lim_{\Delta x \to 0} \Delta y = dy$

REVIEW EXERCISES FOR CHAPTER 4

1.
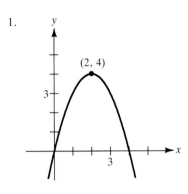

3.
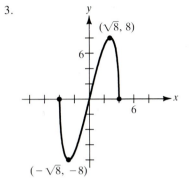

5.
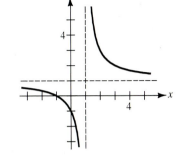

7.
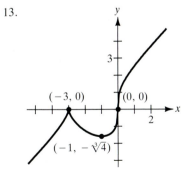

9.
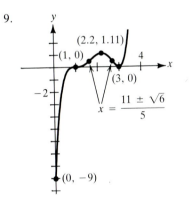

11.
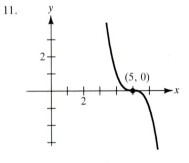

13.
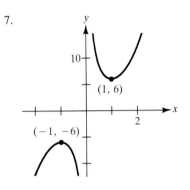

15.
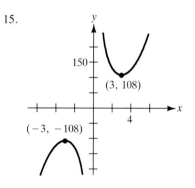

17.
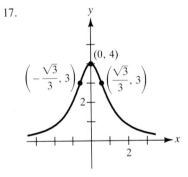

19.
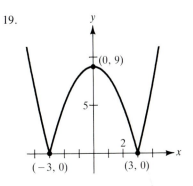

21.
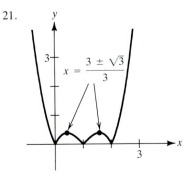

23.
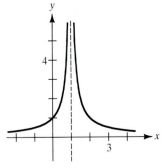

25. Maximum: (1, 3)
    Minimum: (1, 1)

27. $f'([-2 + \sqrt{85}]/3) = -1/17$

29. $f'(2744/729) = 3/7$

31. $f'(2) = \dfrac{5}{4}$

33. No, f is discontinuous at $x = 0$.

35. $c = (x_1 + x_2)/2$

37. $t \approx 4.92 \approx 4:55$ PM
    $d \approx 64$ mi

39. $48

41. $(0, 0), (5, 0), (0, 10)$

43. 120

47. 14.05 ft

49. $x = \sqrt{2Qs/r}$

51. -0.347

53. -0.453

55. (a) 2%   (b) 3%

CHAPTER 5

SECTION 5.1

Given	Rewrite	Integrate	Simplify
1. $\displaystyle\int \sqrt[3]{x}\, dx$	$\displaystyle\int x^{1/3}\, dx$	$\dfrac{x^{4/3}}{4/3} + C$	$\dfrac{3}{4}x^{4/3} + C$
3. $\displaystyle\int \dfrac{1}{x\sqrt{x}}\, dx$	$\displaystyle\int x^{-3/2}\, dx$	$\dfrac{x^{-1/2}}{-1/2} + C$	$-\dfrac{2}{\sqrt{x}} + C$
5. $\displaystyle\int \dfrac{1}{2x^3}\, dx$	$\dfrac{1}{2}\displaystyle\int x^{-3}\, dx$	$\dfrac{1}{2}\left(\dfrac{x^{-2}}{-2}\right) + C$	$-\dfrac{1}{4x^2} + C$

7. $\dfrac{1}{4}x^4 + 2x + C$

9. $\dfrac{2}{5}x^{5/2} + x^2 + x + C$

11. $\dfrac{3}{5}x^{5/3} + C$

13. $-\dfrac{1}{2x^2} + C$

15. $-\dfrac{1}{4x} + C$

17. $\dfrac{2}{15}x^{1/2}(3x^2 + 5x + 15) + C$

19. $x^3 + \dfrac{1}{2}x^2 - 2x + C$

21. $t - \dfrac{2}{t} + C$

23. $\dfrac{2}{7}y^{7/2} + C$

25. $x + C$

27. $y = x^2 - x + 1$

29. $y = x^3 - x + 2$

31. $f(x) = x^2 + x + 4$

33. $f(x) = -4x^{1/2} + 3x = -4\sqrt{x} + 3x$

35. $s(t) = -16t^2 + 1600$
    $t = 10$ sec

37. $v_0 \approx 187.617$ ft/sec

39. (a) $\dfrac{1 + \sqrt{17}}{2} \approx 2.562$ sec
    (b) $-16\sqrt{17} \approx -65.970$ ft/sec

41. (a) $\dfrac{154}{39} \approx 3.95$ ft/sec^2
    (b) $1{,}859/3$ ft ≈ 619.67 ft

43. (a) 300 ft
    (b) 60 ft/sec ≈ 41 mi/hr

47. $C(x) = x^2 - 12x + 50$
    $\overline{C}(x) = x - 12 + \dfrac{50}{x}$

49. $R = x\left(10 - 3x - \dfrac{2}{3}x^2\right)$
    $p = 10 - 3x - \dfrac{2}{3}x^2$

SECTION 5.2

$\int f(g(x))g'(x)\,dx$	$u = g(x)$	$du = g'(x)\,dx$

1. $\int (5x^2 + 1)^2(10x)\,dx$ $5x^2 + 1$ $10x\,dx$

3. $\int \dfrac{x}{\sqrt{x^2 + 1}}\,dx$ $x^2 + 1$ $2x\,dx$

5. $\dfrac{(1 + 2x)^5}{5} + C$

7. $\dfrac{(x^3 - 1)^5}{15} + C$

9. $\dfrac{(x^2 - 1)^8}{16} + C$

11. $4\sqrt{1 + x^2} + C$

13. $\dfrac{15}{8}(1 + x^2)^{4/3} + C$

15. $-3\sqrt{2x + 3} + C$

17. $-\dfrac{1}{2(x^2 + 2x - 3)} + C$

19. $-\dfrac{2}{1 + \sqrt{x}} + C$

21. $-\dfrac{1}{3(1 + x^3)} + C$

23. $\dfrac{1}{2}\sqrt{1 + x^4} + C$

25. $\sqrt{x} + C$

27. $\sqrt{2x} + C$

29. $\dfrac{2}{5}\sqrt{x}(x^2 + 5x + 35) + C$

31. $\dfrac{1}{4}t^4 - t^2 + C$

33. $\dfrac{2}{5}y^{3/2}(15 - y) + C$

35. $\dfrac{1}{6}(2x - 1)^3 + C_1 = \dfrac{4}{3}x^3 - 2x^2 + x - \dfrac{1}{6} + C_1$

37. $\dfrac{2}{5}(x - 3)^{3/2}(x + 2) + C$

$\dfrac{4}{3}x^3 - 2x^2 + x + C_2$, Answers differ by a constant: $C_2 = C_1 - \dfrac{1}{6}$

39. $\dfrac{-2}{105}(1 - x)^{3/2}(15x^2 + 12x + 8) + C$ 41. $\dfrac{\sqrt{2x - 1}}{15}(3x^2 + 2x - 13) + C$

43. $-x - 1 - 2\sqrt{x + 1} + C$ or $-(x + 2\sqrt{x + 1}) + C_1$

45. $\dfrac{1}{3}\sqrt{2x + 1}(x - 1) + C$

47. (b) and (c)

49. $-\dfrac{1}{3}[(1 - x^2)^{3/2} - 5]$

51. (a) $\dfrac{3}{2}[\sqrt{16t + 9} - 3]$

53. (a) $C(x) = \dfrac{3}{2}(12x + 1)^{2/3} + 56.35$

     (b) $\dfrac{3}{2}[\sqrt{1609} - 3] \approx 55.67$ lb.

     (b)

SECTION 5.3

1. 35

3. 158/85

5. 4c

7. 238

9. $\displaystyle\sum_{i=1}^{9} \frac{1}{3i}$

11. $\displaystyle\sum_{j=1}^{8}\left[2\left(\frac{j}{8}\right)+3\right]$

13. $\displaystyle\frac{1}{6}\sum_{k=1}^{6}\left[\left(\frac{k}{6}\right)^2+2\right]$

15. $\displaystyle\frac{2}{n}\sum_{i=1}^{n}\left[\left(\frac{2i}{n}\right)^3-\left(\frac{2i}{n}\right)\right]$

17. $\displaystyle\frac{3}{n}\sum_{i=1}^{n}\left[2\left(1+\frac{3i}{n}\right)^2\right]$

19. 420

21. 2,470

23. $\displaystyle\frac{1015}{n^3}$

25. 8/3

27. 81/4

29. 9

31. $\displaystyle\lim_{n\to\infty}\frac{1}{6}\left[\frac{2n^3-3n^2+n}{n^3}\right]=\frac{1}{3}$

33. $\displaystyle\lim_{n\to\infty}\left[8\left(\frac{n^2+n}{n^2}\right)\right]=8$

35. $\displaystyle\lim_{n\to\infty}2\left(\frac{10n^4+13n^3+4n^2}{n^4}\right)=20$

37. *a*

SECTION 5.4

1. $S\approx0.768$
   $s\approx0.518$

3. $S\approx0.746$
   $s\approx0.646$

5. $S\approx0.859$
   $s\approx0.659$

9. $A=2$

7.

n	5	10	50	100
s(n)	1.6	1.8	1.96	1.98
S(n)	2.4	2.2	2.04	2.02

11. $A=4/3$

13. $A=52/3$

15. $A=3/4$

17. $A = 7/12$

19. $A = 6$

21. 69/8

23. 0.6730

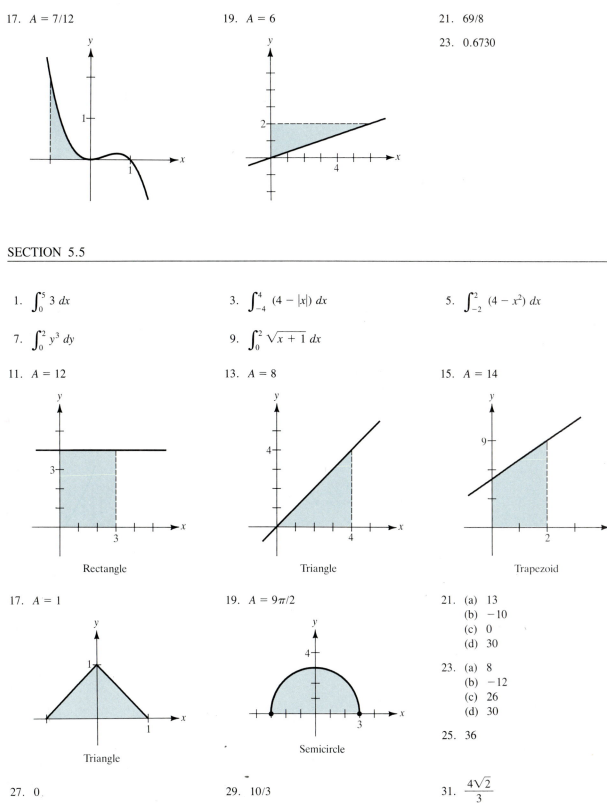

SECTION 5.5

1. $\displaystyle\int_0^5 3 \; dx$

3. $\displaystyle\int_{-4}^4 (4 - |x|) \; dx$

5. $\displaystyle\int_{-2}^2 (4 - x^2) \; dx$

7. $\displaystyle\int_0^2 y^3 \; dy$

9. $\displaystyle\int_0^2 \sqrt{x + 1} \; dx$

11. $A = 12$

13. $A = 8$

15. $A = 14$

Rectangle

Triangle

Trapezoid

17. $A = 1$

19. $A = 9\pi/2$

21. (a) 13
   (b) -10
   (c) 0
   (d) 30

23. (a) 8
   (b) -12
   (c) 26
   (d) 30

25. 36

Triangle

Semicircle

27. 0

29. 10/3

31. $\dfrac{4\sqrt{2}}{3}$

SECTION 5.6

1. 1
3. $-5/2$
5. $-10/3$

7. $1/3$
9. $1/2$
11. 36

13. -4
15. $2/3$
17. $-1/18$

19. $-27/20$
21. 2
23. 0

25. $6 - 3\sqrt[3]{4}/2 \approx 3.619$
27. 1
29. 4

31. $4/15$
33. $144/5$
35. $1,209/28$

37. $7,088/105$
39. $1/6$
41. $8/5$

43. 6
45. 10
47. 6

49. Average $= 8/3$
$x = \pm 2\sqrt{3}/3 \approx \pm 1.155$

51. Average $= 4/3$
$x = \sqrt{2 - (2\sqrt{5}/3)} \approx 0.714$
$x = \sqrt{2 + (2\sqrt{5}/3)} \approx 1.868$

53. Average $= -2/3$
$x = 4 + (2\sqrt{3})/3 \approx 2.488$
$x = 4 - (2\sqrt{3})/3 \approx 0.179$

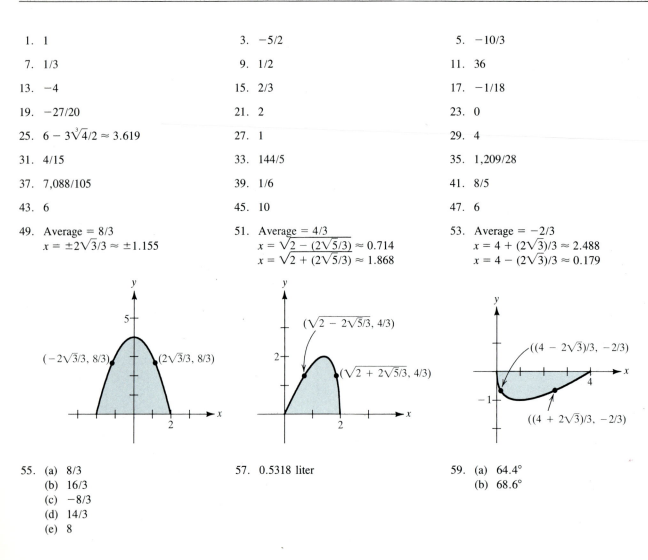

55. (a) $8/3$
    (b) $16/3$
    (c) $-8/3$
    (d) $14/3$
    (e) 8

57. 0.5318 liter

59. (a) $64.4°$
    (b) $68.6°$

SECTION 5.7

	Exact	Trapezoidal	Simpson's		Trapezoidal	Simpson's
1.	2.6667	2.7500	2.6667	11.	1.683	1.622
3.	4.0000	4.2500	4.0000	13.	3.283	3.240
5.	4.0000	4.0625	4.0000	15.	0.342	0.372
7.	0.5000	0.5090	0.5004	17.	1.556	1.571
9.	0.3466	0.3413	0.3468	19.	1.604	1.692

Trapezoidal	Simpson's
21. 0.500	0.000
23. 0.010	0.001
25. 0.021	0.001

27. $n \geq 1033$

29. 8.2

31. 3.14159

33. 89,500 ft^2

REVIEW EXERCISES FOR CHAPTER 5

1. $x^{2/3} + C$

3. $\dfrac{2}{3}x^3 + \dfrac{1}{2}x^2 - x + C$

5. $\dfrac{2\sqrt{x}}{15}(15 + 10x + 3x^2) + C$

7. $\dfrac{2}{3}\sqrt{x^3 + 3} + C$

9. $\dfrac{1}{7}x^7 + \dfrac{3}{5}x^5 + x^3 + x + C$

11. $\dfrac{1}{8}(x^2 + 1)^4 + C$

13. $\dfrac{-1}{12(1 + 4x^2)^3} + C$

15. $-\dfrac{1}{9}\sqrt{25 - 9x^2} + C$

17. $\dfrac{2}{21}(x + 5)^{3/2}(3x^2 - 12x + 40) + C$

19. $\dfrac{-2}{5 + \sqrt{x}} + C$

21. $\displaystyle\sum_{n=1}^{10} (2n - 1)$

23. $\displaystyle\sum_{n=1}^{10} (4n + 2)$

25. $\dfrac{16}{5}$

27. -56

29. 16

31. 0

33. 2

35. 422/5

37. 128/15

39. $\dfrac{2}{3}(6\sqrt{3} - 10)$

41. $y = 2 - x^2$

43. 240 ft/sec

45. (a) 3 sec
    (b) 144 ft
    (c) 3/2 sec
    (d) 108 ft

47. (a) $S = \dfrac{5mb^2}{8}$

    $s = \dfrac{3mb^2}{8}$

    (c) $\dfrac{1}{2}mb^2$

(b) $S(n) = \dfrac{mb^2(n + 1)}{2n}$

    $s(n) = \dfrac{mb^2(n - 1)}{2n}$

    (d) $\dfrac{1}{2}mb^2$

49. 6

51. 10/3

53. 1/4

55. 1/6

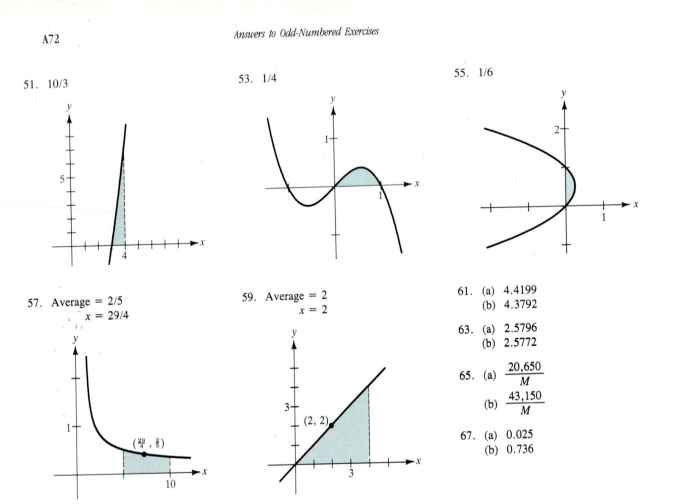

57. Average = 2/5

$x = 29/4$

59. Average = 2

$x = 2$

61. (a) 4.4199
    (b) 4.3792

63. (a) 2.5796
    (b) 2.5772

65. (a) $\dfrac{20,650}{M}$

    (b) $\dfrac{43,150}{M}$

67. (a) 0.025
    (b) 0.736

CHAPTER 6

SECTION 6.1

1. $A = 36$

3. $A = 9$

5. $A = 3/2$

7. $A = 32/3$

9. $A = 9/2$

11. $A = 1$

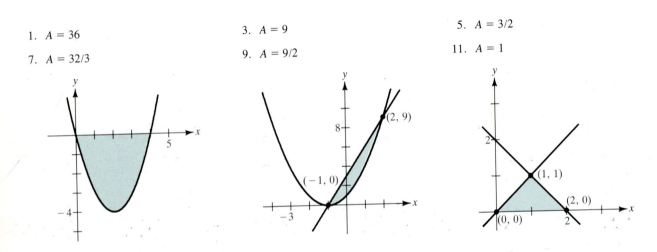

13. $A = 500/27$

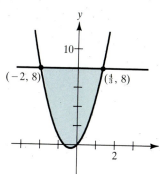

15. $A = 2$

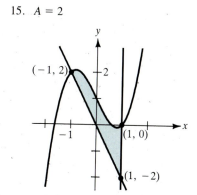

17. $A = 3/2$

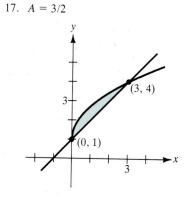

19. $A = 64/3$

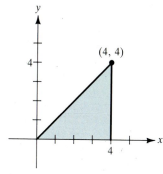

21. $A = 9/2$

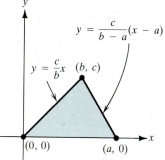

23. $A = 6$

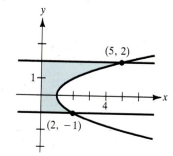

25. $A = 8$

27. $A = \dfrac{1}{2}ac$

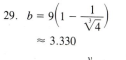

29. $b = 9\left(1 - \dfrac{1}{\sqrt[3]{4}}\right)$

   ≈ 3.330

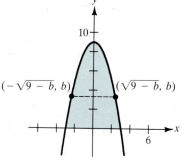

31. $x^4 - 2x^2 + 1 \le 1 - x^2$ on $[-1, 1]$

   $A = \displaystyle\int_{-1}^{1} [(1 - x^2) - (x^4 - 2x^2 + 1)]\, dx = \dfrac{4}{15}$

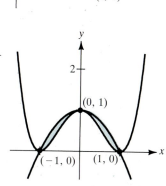

33. \$1.625 billion

35. Consumer surplus $= 1{,}600$
    Producer surplus $= 400$

37. Consumer surplus $= 50{,}000$
    Producer surplus $= 25{,}497$

SECTION 6.2

1. $\pi/3$

3. $16\pi/3$

5. $15\pi/2$

7. $2\pi/35$

9. 8π

11. $\pi/4$

13. (a) 8π
    (b) $128\pi/5$
    (c) $256\pi/15$
    (d) $192\pi/5$

15. (a) $32\pi/3$
    (b) $64\pi/3$

17. $128\pi/15$

19. $3\pi/4$

21. 18π

23. $363\pi/64$

25. $208\pi/3$

27. $384\pi/5$

29. $512\pi/15$

31. 60π

33. 18π

37. $\pi r^2 h\left(1 - \dfrac{h}{H} + \dfrac{h^2}{3H^2}\right)$

39. $\dfrac{4\sqrt{3}}{3}, \dfrac{4\sqrt{6}}{3}$

41. $\pi/30$

43. (a) $128/3$
    (b) $32\sqrt{3}/3$
    (c) $16\pi/3$
    (d) $32/3$

45. (a) $1/10$
    (b) $\pi/80$
    (c) $\sqrt{3}/40$
    (d) $3/80$
    (e) $\pi/20$

47. $2r^3/3$

SECTION 6.3

1. $16\pi/3$

3. $8\pi/3$

5. $128\pi/5$

7. 8π

9. $16\pi/3$

11. 16π

13. 64π

15. $8\pi/3$

17. $192\pi/5$

19. $\dfrac{\pi}{2}$

21. (a) $128\pi/7$
    (b) $64\pi/5$
    (c) $96\pi/5$
    (d) $320\pi/7$

23. (a) $\pi a^3/15$
    (b) $\pi a^3/15$
    (c) $4\pi a^3/15$

25. Diameter $= 2\sqrt{4 - 2\sqrt{3}} \approx 1.464$

29. 2.3

SECTION 6.4

1. $1,000$ ft $\cdot$ lb

3. 300 ft $\cdot$ lb

5. 30.625 in $\cdot$ lb ≈ 2.55 ft $\cdot$ lb

7. 360 in $\cdot$ lb $= 30$ ft $\cdot$ lb

9. 180 in $\cdot$ lb $= 15$ ft $\cdot$ lb

11. 9984 ft $\cdot$ lb

13. $431,308.8\pi$ ft · lb

15. $20,217.6\pi$ ft · lb

17. 2995.2π ft · lb

19. 2457π ft · lb

21. (a) 761.905 mi · ton $\approx 8,046,000,000$ ft · lb
    (b) $1,454.545$ mi · ton $\approx 15,360,000,000$ ft · lb

23. 95.65 mi · ton ≈ 1.01 billion ft · lb

25. 337.5 ft · lb

27. 300 ft · lb

29. 168.75 ft · lb

31. 7987.5 ft · lb

33. $2000 \ln \dfrac{3}{2} \approx 810.93$ ft · lb

SECTION 6.5

1. $1,123.2$ lb

3. 748.8 lb

5. 1064.96 lb

7. 748.8 lb

9. $15,163.2$ lb

11. $2,814$ lb

13. $3,376.8$ lb

15. 94.5 lb

19. 960 lb

21. $46,592$ lb

SECTION 6.6

1. $\bar{x} = -6/7$

3. $\bar{x} = 12$

5. $\bar{x} = 17$

7. $(\bar{x}, \bar{y}) = (10/9, -1/9)$

9. $(\bar{x}, \bar{y}) = (-7/8, -7/16)$

11. $(\bar{x}, \bar{y}) = \left(0, \dfrac{4 + 3\pi}{4 + \pi}\right)$

13. $(\bar{x}, \bar{y}) = (0, 135/34)$

15. $(\bar{x}, \bar{y}) = \left(0, \dfrac{2 + 3\pi}{2 + \pi}\right)$

17. $M_x = 4\rho$, $M_y = 64\rho/5$, $(\bar{x}, \bar{y}) = (12/5, 3/4)$

19. $M_x = 99\rho/5$, $M_y = 27\rho/4$, $(\bar{x}, \bar{y}) = (3/2, 22/5)$

21. $M_x = \rho/12$, $M_y = \rho/15$, $(\bar{x}, \bar{y}) = (2/5, 1/2)$

23. $M_x = 0$, $M_y = 256\rho/15$, $(\bar{x}, \bar{y}) = (8/5, 0)$

25. $M_x = 27\rho/4$, $M_y = -27\rho/10$, $(\bar{x}, \bar{y}) = (-3/5, 3/2)$

27. $(\bar{x}, \bar{y}) = (0, 4/(3\pi))$

29. $(\bar{x}, \bar{y}) = (1/2, 2/5)$

31. $(\bar{x}, \bar{y}) = (b/3, c/3)$

33. $V = 160\pi^2 \approx 1,579.14$

35. $V = 128\pi/3$

SECTION 6.7

1. 13

3. $\dfrac{2}{3}(\sqrt{8} - 1) \approx 1.219$

5. $33/16$

7. $5\sqrt{5} - 2\sqrt{2}$

9. $s = \displaystyle\int_{1}^{3} \dfrac{\sqrt{x^4 + 1}}{x^2}\, dx$

11. $s = \displaystyle\int_{-2}^{1} \sqrt{2 + 4x + 4x^2}\, dx$

13. $s = \displaystyle\int_{0}^{2} \sqrt{1 + 4y^2}\, dy$

15. $2/3$

17. 2.1892

19. $\dfrac{\pi}{9}(82\sqrt{82} - 1) \approx 258.85$ 21. $47\pi/16$ 23. $\dfrac{\pi}{27}(145\sqrt{145} - 10\sqrt{10}) \approx 199.48$

27. $6\pi(3 - \sqrt{5}) \approx 14.40$ 29. Surface area $= \dfrac{\pi}{27}$ in^2

Amount of glass $= \dfrac{\pi}{27}(0.015) \approx 0.0017$ in^3

REVIEW EXERCISES FOR CHAPTER 6

1. $A = 4/5$

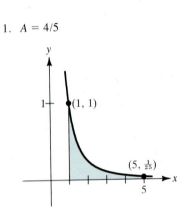

3. $A = \pi/2$

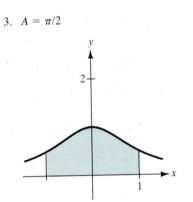

5. $A = 4/3$

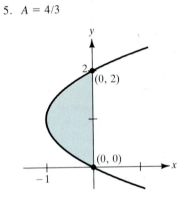

7. $A = 1/2$

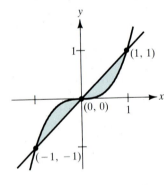

9. $A = 512/3$

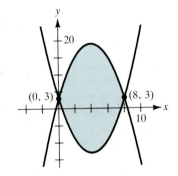

11. $A = 14/3$

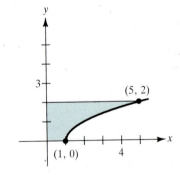

13. $A = 1/6$

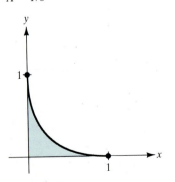

15. (a) $64\pi/3$
    (b) $128\pi/3$
    (c) $64\pi/3$
    (d) $160\pi/3$

17. (a) 64π
    (b) 48π

19. 2.47

21. (a) $512\pi/15$
    (b) 64π

23. 50 in · lb $\approx$ 4.167 ft · lb

25. $104{,}000\pi$ ft · lb $\approx$ 163.4 ft · ton

27. 250 ft · lb

29. 72,800 lb (on side walls)
    62,400 lb (on wall at deep end)
    15,600 lb (on wall at shallow end)

31. $4{,}992\pi$ lb

33. $(\bar{x}, \bar{y}) = (a/5, a/5)$

35. $(\bar{x}, \bar{y}) = (0, 2a^2/5)$

37. $s = \displaystyle\int_0^{\sqrt{3}} \frac{4}{\sqrt{4 - x^2}}\, dx = \frac{4\pi}{3} \approx 4.19$

39. 15π

41. 4/15

43. $32\pi/105$

45. $\dfrac{8}{15}(1 + 6\sqrt{3}) \approx 6.076$

CHAPTER 7

SECTION 7.1

1. $\dfrac{1}{2}x^2 + 2x$

3. $\dfrac{2}{3}(x^{3/2} - 8)$

5. $\dfrac{1}{8}[(x^2 + 1)^4 - 1]$

7. $x^2 - 2x + 5$

9. $\sqrt[4]{x}$

11. $y = \ln 2x$

13. $y = \ln x^2$

15. $f(x) = \ln (x - 1)$

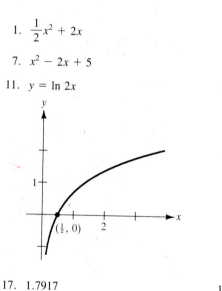

17. 1.7917

19. 4.3944

21. -1.3862

23. 0.8283

25. $\ln 2 - \ln 3$

27. $\ln x + \ln y - \ln z$

29. $\dfrac{3}{2}\ln 2$

31. $3[\ln (x + 1) + \ln (x - 1) - 3 \ln x]$

33. $\ln z + 2 \ln (z - 1)$

35. $\ln \dfrac{x - 2}{x + 2}$

37. $\ln \sqrt[3]{\dfrac{x(x + 3)^2}{x^2 - 1}}$

39. $\ln \dfrac{9}{\sqrt{x^2 + 1}}$

SECTION 7.2

1. 3

3. 2

5. $2/x$

7. $\dfrac{2(x^3 - 1)}{x(x^3 - 4)}$

9. $\dfrac{4(\ln x)^3}{x}$

11. $\dfrac{2x^2 - 1}{x(x^2 - 1)}$

13. $\dfrac{1 - x^2}{x(x^2 + 1)}$

15. $\dfrac{1 - 2 \ln x}{x^3}$

17. $\dfrac{2}{x \ln x^2}$

19. $\dfrac{1}{1 - x^2}$

21. $\dfrac{-4}{x(x^2 + 4)}$

23. $\dfrac{\sqrt{x^2 + 1}}{x^2}$

25. $\dfrac{2x^2 - 1}{\sqrt{x^2 - 1}}$

27. $\dfrac{3x^3 - 15x^2 + 8x}{2(x - 1)^3 \sqrt{3x - 2}}$

29. $\dfrac{(2x^2 + 2x - 1)\sqrt{x - 1}}{(x + 1)^{3/2}}$

31. $xy'' + y' = x\left(\dfrac{-2}{x^2}\right) + \dfrac{2}{x} = 0$

33. Relative minimum:
    $(1, 1/2)$

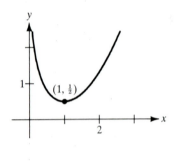

35. Relative minimum:
    $(e^{-1}, -e^{-1})$

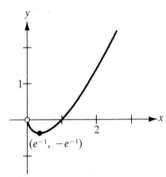

37. Relative minimum:
    (e, e)
    Point of inflection:
    $(e^2, e^2/2)$

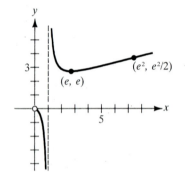

39. 0.567

43. (a) 12.4 primes/100 numbers
    (b) 6.7 primes/100 numbers
    (c) 4.6 primes/100 numbers

SECTION 7.3

1. $\ln |x + 1| + C$

3. $-\dfrac{1}{2} \ln |3 - 2x| + C$

5. $\ln \sqrt{x^2 + 1} + C$

7. $\dfrac{x^2}{2} - 4 \ln |x| + C$

9. $1/4$

11. $7/3$

13. $-\ln 3$

15. $2\sqrt{x + 1} + C$

17. $\dfrac{1}{3} \ln |x^3 + 3x^2 + 9x + 1| + C$

19. $3 \ln |1 + x^{1/3}| + C$

21. $2[\sqrt{x} - \ln (1 + \sqrt{x})] + C$

23. $x + 6\sqrt{x} + 18 \ln |\sqrt{x} - 3| + C$

25. $-\dfrac{2}{3} \ln |1 - x\sqrt{x}| + C$

27. $\ln |x - 1| + \dfrac{1}{2(x - 1)^2} + C$

29. $\dfrac{15}{2} + 8 \ln 2 \approx 13.045$ square units

31. $\pi \ln 4$

33. $\dfrac{\pi}{4}(32 \ln 4 - 3) \approx 32.49$

35. $\dfrac{4\pi}{3}(20 - 9 \ln 3)$

37. $2000 \ln \dfrac{3}{2} \approx 810.93$ ft · lb

39. $P(t) = 1000[12 \ln|1 + 0.25t| + 1]$
    $P(3) \approx 7715$

SECTION 7.4

1.
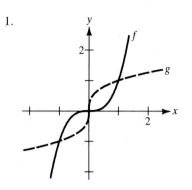

3.
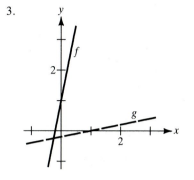

5.
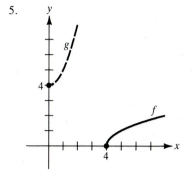

7.
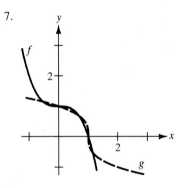

9. $f^{-1}(x) = \dfrac{x + 3}{2}$
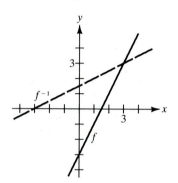

11. $f^{-1}(x) = x^{1/5}$
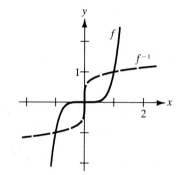

13. $f^{-1}(x) = x^2$
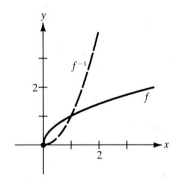

15. $f^{-1}(x) = \sqrt{4 - x^2}$
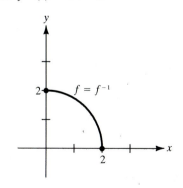

17. $f^{-1}(x) = x^3 + 1$
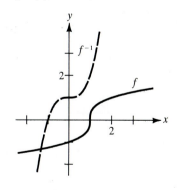

19. $f^{-1}(x) = x^{3/2}$

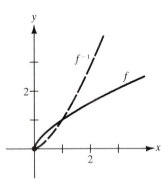

21. $f^{-1}(x) = \dfrac{\sqrt{7}x}{\sqrt{1 - x^2}}$

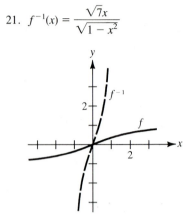

23. Inverse exists.

25. Inverse does not exist.

27. Inverse exists.

29. Inverse exists.

31. Inverse exists.

33. Inverse does not exist.

35. Inverse exists.

37. $f'(x) = 2(x - 4) > 0$ on $(4, \infty)$

39. $f'(x) = -8/x^3 < 0$ on $(0, \infty)$

41. $f'(1/2) = 3/4,\ (f^{-1})'(1/8) = 4/3$

43. $f'(5) = 1/2,\ (f^{-1})'(1) = 2$

45. Not continuous at $x = 0$

SECTION 7.5

1. (a) $\log_2 8 = 3$
   (b) $\log_3 (1/3) = -1$

3. (a) $10^{-2} = 0.01$
   (b) $(1/2)^{-3} = 8$

5. (a) $e^{0.6931\cdots} = 2$
   (b) $e^{2.128\cdots} = 8.4$

7. (a) $x = 3$
   (b) $x = -1$

9. (a) $x = 1/3$
   (b) $x = 1/16$

11. (a) $x = 1/9$
    (b) $x = 3$

13. (a) $x = -1, 2$
    (b) $x = 1/3$

15.

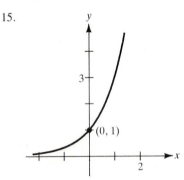

17.

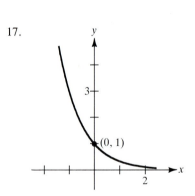

19.

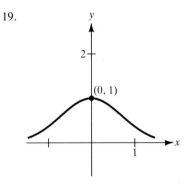

21.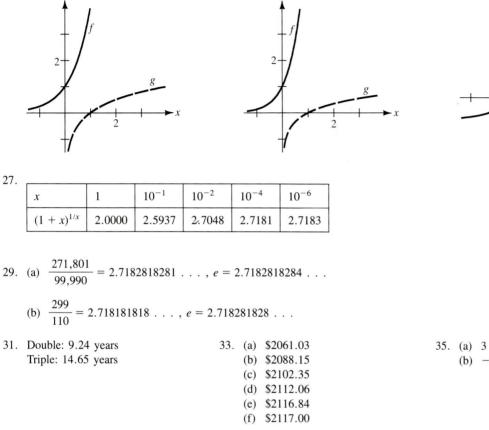

23.

25.

27.

x	1	10^{-1}	10^{-2}	10^{-4}	10^{-6}
$(1 + x)^{1/x}$	2.0000	2.5937	2.7048	2.7181	2.7183

29. (a) $\dfrac{271{,}801}{99{,}990} = 2.7182818281 \ldots$, $e = 2.7182818284 \ldots$

(b) $\dfrac{299}{110} = 2.718181818 \ldots$, $e = 2.718281828 \ldots$

31. Double: 9.24 years
    Triple: 14.65 years

33. (a) \$2061.03
    (b) \$2088.15
    (c) \$2102.35
    (d) \$2112.06
    (e) \$2116.84
    (f) \$2117.00

35. (a) 3
    (b) -3

37. $2e^{2x}$

39. $2(x - 1)e^{-2x+x^2}$

41. $\dfrac{e^{\sqrt{x}}}{2\sqrt{x}}$

43. $3(e^{-x} + e^{x})^2(e^{x} - e^{-x})$

45. $2x$

47. $\dfrac{2e^{2x}}{1 + e^{2x}}$

49. $\dfrac{e^{2x} - 1}{e^{2x} + 1} = \dfrac{e^{x} - e^{-x}}{e^{x} + e^{-x}}$

51. $x^2 e^x$

53. $(\ln 5)5^{x-2}$

55. $e^{-x}\left(\dfrac{1}{x} - \ln x\right)$

57. 1

59. $\dfrac{1}{x(\ln 3)}$

61. $\dfrac{x}{(\ln 5)(x^2 - 1)}$

63. $2x^{(2/x)-2}(1 - \ln x)$

65. $(x - 2)^{x+1}\left[\dfrac{x + 1}{x - 2} + \ln (x - 2)\right]$

67. Relative maximum:
    $(0, 1/\sqrt{2\pi})$
    Points of inflection:
    $(\pm 1, 1/\sqrt{2\pi e})$

69. Relative minimum:
    $(0, 1)$

71. Relative minimum:
    $(0, 0)$
    Relative maximum:
    $(2, 4e^{-2})$
    Points of inflection:
    $(2 \pm \sqrt{2}, (6 \pm 4\sqrt{2})e^{-(2\pm\sqrt{2})})$

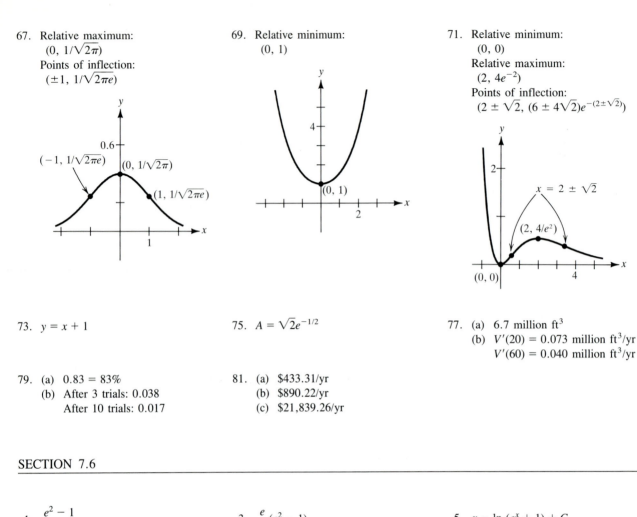

73. $y = x + 1$

75. $A = \sqrt{2}e^{-1/2}$

77. (a) 6.7 million ft^3
    (b) $V'(20) = 0.073$ million ft^3/yr
         $V'(60) = 0.040$ million ft^3/yr

79. (a) $0.83 = 83\%$
    (b) After 3 trials: 0.038
         After 10 trials: 0.017

81. (a) \$433.31/yr
    (b) \$890.22/yr
    (c) \$21,839.26/yr

SECTION 7.6

1. $\dfrac{e^2 - 1}{2e^2}$

3. $\dfrac{e}{3}(e^2 - 1)$

5. $x - \ln(e^x + 1) + C$

7. $\dfrac{1}{2a}e^{ax^2} + C$

9. $\dfrac{e}{3}(e^2 - 1)$

11. $\dfrac{7}{\ln 4}$

13. $-\dfrac{2}{3}(1 - e^x)^{3/2} + C$

15. $\ln|e^x - e^{-x}| + C$

17. $-\dfrac{5}{2}e^{-2x} + e^{-x} + C$

19. 4

21. $e^5 - 1 \approx 147.41$

23. $1 - e^{-1} \approx 0.632$

25. (a) 0.212
    (b) 0.035

27. 0.3414

31. (a) $y = \dfrac{1}{2}e^{0.4605t}$
    (b) $y = 4e^{-0.4159t}$

33. (a) $r = \dfrac{\ln 2}{7.75} \approx 8.94\%$
    (b) $A \approx \$1834.37$

35. \$583,327.77

37. (a) $N = 30(1 - e^{-0.0502t})$
    (b) 36 days

39. (a) $S = 30e^{-1.7918/t}$
    (b) 20,965 units
    (c) S

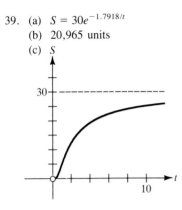

41. 900

45. 95.76%

49. 11.75°

43. 6,015

47. 22.35°

SECTION 7.7

1. 3

3. 0

5. 2

7. $n = 1$: 0
   $n = 2$: 1/2
   $n \geq 3$: ∞

9. 0

11. 3/2

13. ∞

15. 0

17. −3/2

19. 1

21. 0

23. 1

25. 1

27. 0

29. 0

31. 0

33. Limit is not of the form 0/0 or ∞/∞.

35.

x	10	10^2	10^4	10^6	10^8	10^{10}
$\dfrac{(\ln x)^4}{x}$	2.811	4.498	0.720	0.036	0.001	0.000

41. $v = 32t + v_0$

REVIEW EXERCISES FOR CHAPTER 7

1. $f^{-1}(x) = 2x + 6$

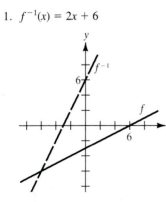

3. $f^{-1}(x) = x^2 - 1$

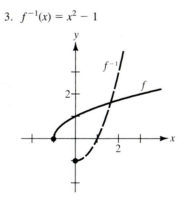

5. $f^{-1}(x) = \sqrt{x + 5}$

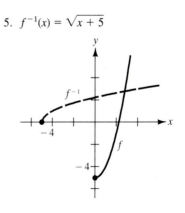

7. $f^{-1}(x) = \sqrt{\dfrac{x}{2} + 4},\ x \geq 4$

9. $x = 3$

11. $x = 5 \pm \sqrt{7}$

13. $1/(2x)$

15. $\dfrac{1 + 2 \ln x}{2\sqrt{\ln x}}$

17. $\dfrac{-y}{x(2y + \ln x)}$

19. $y(1 + \ln x)$

21. $\dfrac{x}{(a + bx)^2}$

23. $\dfrac{1}{x(a + bx)}$

25. $-2x$

27. $xe^x(x + 2)$

29. $\dfrac{e^{2x} - e^{-2x}}{\sqrt{e^{2x} + e^{-2x}}}$

31. $3^{x-1} \ln 3$

33. $\dfrac{y - ye^x - e^y}{e^x + xe^y - x}$

35. ax^{a-1}

37. $x^x(1 + \ln x)$

39. $\dfrac{1}{7} \ln |7x - 2| + C$

41. $\ln |\ln 3x| + C$

43. $\dfrac{x^2}{2} + 3 \ln |x| + C$

45. $\dfrac{1}{3} \ln |x^3 - 1| + C$

47. $2\sqrt{\ln x} + C$

49. $-\dfrac{1}{6}e^{-3x^2} + C$

51. $\dfrac{e^{4x} - 3e^{2x} - 3}{3e^x} + C$

53. $\ln |e^x - 1| + C$

55. $-e^{-x^2/2} + C$

57. $3 + \ln 4$

59. $\ln \dfrac{5}{3}$

61. area $= 2\left[1 - \dfrac{1}{e^8}\right] \approx 1.999$

63. average $= \dfrac{2}{5} \ln \dfrac{3}{2} \approx 0.162,\ x = 1 + \dfrac{5}{2 \ln (3/2)} \approx 7.166$

65. $-\dfrac{1}{2}(e^{-16} - 1) \approx 0.500$

67. (a) $525.64
(b) $824.36
(c) $74,206.58

69. $3499.38

71. (a) 27.73 yrs
(b) 43.94 yrs

73. (a) -24.26%
(b) -14.72%

75. (a) 0.3935
(b) 0.7769
(c) 0.2492
(d) 0.9502

77. (a) 0.60
(b) 0.85

79. $v(t) = \dfrac{1}{K}[32 + (Kv_0 - 32)e^{Kt}]$

81. $v(t) = \dfrac{1}{K}[32 - (20K + 32)e^{Kt}]$

83. $s(t) = \dfrac{32}{K}t - \dfrac{32}{K^2}(1 - e^{Kt}) + s_0$

85. 0

87. ∞

89. 1

91. $1{,}000e^{0.09} \approx 1{,}094.17$

CHAPTER 8

SECTION 8.1

1. (a) $396°, -324°$

   (b) $315°, -405°$

   (c) $240°, -480°$

   (d) $30°, -330°$

3. (a) $\dfrac{19\pi}{9}, -\dfrac{17\pi}{9}$

   (b) $\dfrac{10\pi}{3}, -\dfrac{2\pi}{3}$

   (c) $\dfrac{23\pi}{6}, -\dfrac{\pi}{6}$

   (d) $\dfrac{5\pi}{6}, -\dfrac{19\pi}{6}$

5. (a) $\dfrac{\pi}{6}$

   (b) $\dfrac{5\pi}{6}$

   (c) $\dfrac{7\pi}{4}$

   (d) $\dfrac{2\pi}{3}$

7. (a) $270°$
   (b) $210°$
   (c) $-105°$
   (d) $20°$

9.

r	8 ft	15 in	85 cm	24 in	$\dfrac{12{,}963}{\pi}$ mi
s	12 ft	24 in	63.75π cm	96 in	8642 mi
θ	1.5	1.6	$3\pi/4$	4	$2\pi/3$

11. (a) $\dfrac{5\pi}{12}$

    (b) 7.8125π inches

13. (a) $\sin\theta = 4/5,\ \csc\theta = 5/4$
       $\cos\theta = 3/5,\ \sec\theta = 5/3$
       $\tan\theta = 4/3,\ \cot\theta = 3/4$

    (b) $\sin\theta = -15/17,\ \csc\theta = -17/15$
       $\cos\theta = 8/17,\ \sec\theta = 17/8$
       $\tan\theta = -15/8,\ \cot\theta = -8/15$

15. Quadrant III

17. 2

19. 4/3

21. 17/15

23. (a) $\sin 60° = \sqrt{3}/2$
       $\cos 60° = 1/2$
       $\tan 60° = \sqrt{3}$
    (b) $\sin (2\pi/3) = \sqrt{3}/2$
       $\cos (2\pi/3) = -1/2$
       $\tan (2\pi/3) = -\sqrt{3}$
    (c) $\sin (\pi/4) = \sqrt{2}/2$
       $\cos (\pi/4) = \sqrt{2}/2$
       $\tan (\pi/4) = 1$
    (d) $\sin (5\pi/4) = -\sqrt{2}/2$
       $\cos (5\pi/4) = -\sqrt{2}/2$
       $\tan (5\pi/4) = 1$

25. (a) $\sin 225° = -\sqrt{2}/2$
       $\cos 225° = -\sqrt{2}/2$
       $\tan 225° = 1$
    (b) $\sin (-225°) = \sqrt{2}/2$
       $\cos (-225°) = -\sqrt{2}/2$
       $\tan (-225°) = -1$
    (c) $\sin 300° = -\sqrt{3}/2$
       $\cos 300° = 1/2$
       $\tan 300° = -\sqrt{3}$
    (d) $\sin 330° = -1/2$
       $\cos 330° = \sqrt{3}/2$
       $\tan 330° = -\sqrt{3}/3$

27. (a) 0.1736
    (b) 5.7588

29. (a) 0.3640
    (b) 0.3640

31. (a) $\theta = \pi/4,\ 7\pi/4$
    (b) $\theta = 3\pi/4,\ 5\pi/4$

33. (a) $\theta = \pi/4,\ 5\pi/4$
    (b) $\theta = 5\pi/6,\ 11\pi/6$

35. $\theta = \dfrac{\pi}{4}, \dfrac{3\pi}{4}, \dfrac{5\pi}{4}, \dfrac{7\pi}{4}$

37. $\theta = 0,\ \pi,\ \dfrac{\pi}{4}, \dfrac{5\pi}{4}$

39. $\theta = \dfrac{\pi}{3}, \dfrac{5\pi}{3}$

41. $\theta = 0,\ \pi,\ \dfrac{\pi}{2}$

43. $\dfrac{100\sqrt{3}}{3}$

45. $\dfrac{25\sqrt{3}}{3}$

47. $20\sin 75° \approx 19.32$ feet

49. $\dfrac{150}{\tan 4°} \approx 2145.1$ feet

SECTION 8.2

1. Period: π
   Amplitude: 2

3. Period: $\dfrac{2\pi}{3}$
   Amplitude: 3

5. Period: 4π
   Amplitude: $\dfrac{3}{2}$

7. Period: 6π
   Amplitude: 2

9. Period: $\pi/5$
   Amplitude: 2

11. Period: 1/2
    Amplitude: 3

13. Period: $\pi/2$

15. Period: $2\pi/5$

17.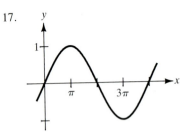

19.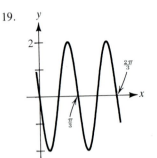

21.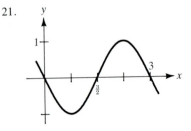

23.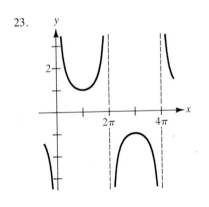

25.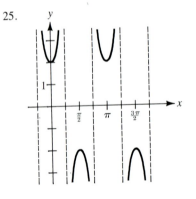

27.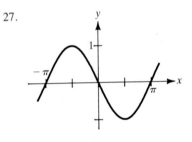

29.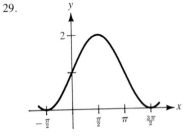

31.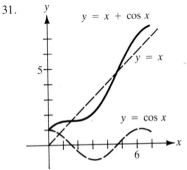

33.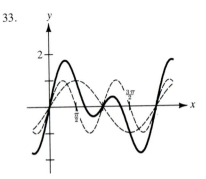

35. 1/5

37. 0

39. 0

41. 1

43. 1

45. 2/3

47. 0

49. Limit does not exist.

51. ∞

53. Continuous for all real x

55. Nonremovable discontinuities at integer multiples of $\pi/2$

57. Continuous for all real x

59. (a) 6 sec
    (b) 10 cycles/min
    (c) v

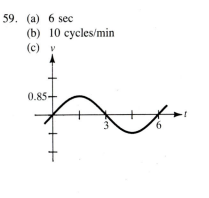

61. (a) 1/440
    (b) 440
    (c) y

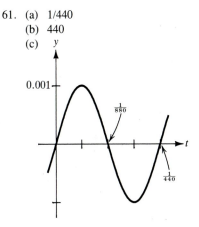

SECTION 8.3

1. $2x + \dfrac{1}{2}\sin x$

3. $-\dfrac{1}{x^2} - 3\cos x$

5. $\dfrac{2}{\sqrt{x}} - 3\sin x$

7. $t(t\cos t + 2\sin t)$

9. $-\dfrac{t\sin t + \cos t}{t^2}$

11. $f'(x) = -1 + \sec^2 x = \tan^2 x$

13. $-5x\csc x\cot x + 5\csc x = 5\csc x(1 - x\cot x)$

15. $\csc x\cot x - \cos x = \cos x\cot^2 x$

17. $t^2\cos t + 2t\sin t - 2t\sin t + 2\cos t - 2\cos t = t^2\cos t$

19. $\pi\cos 2\pi x$

21. $-\dfrac{2\csc x\cot x}{(1 - \csc x)^2}$

23. $-3\sin 3x$

25. $12\sec^2 4x$

27. $\pi\cos \pi x$

29. $\dfrac{1}{4}\sin 2x$

31. $\dfrac{1}{2}\sin 4x$

33. $\dfrac{1}{2}\cot x\sqrt{\sin x}$

35. $6\sec^3 2x\tan 2x$

37. $\csc x$

39. $2e^x\cos x$

41. $\sec^2 xe^{\tan x}$

43. $\sec x \csc x$

45. $\dfrac{\cos x}{2 \sin 2y}$, undefined

47. $-\dfrac{x^2}{x^2 + 1}$, 0

49. $\dfrac{\cot y}{x}$, $\dfrac{1}{2\sqrt{3}}$

53. (a) 1
    (b) 2

55. $4x - 2y = 2 - \pi$

57. $\dfrac{2}{3}$

59. -2

61. $\dfrac{1}{2}$

63. 1

65.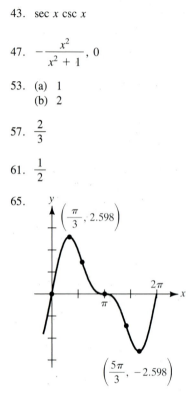

67.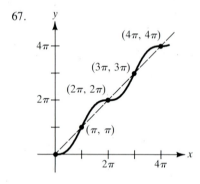

69. -2.03, 0, 2.03

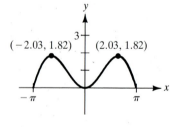

71. (a) $y = 1/4$ in.
       $v = 4$ in./sec
    (c) Period: $\pi/6$
       Frequency: $6/\pi$

73. $\theta = \arctan k$, $F = kW/\sqrt{k^2 + 1}$

75. (a) 1/2 rad/min
    (b) 3/2 rad/min
    (c) 1.8 rad/min

77. (a) 0 ft/sec
    (b) 10π ft/sec
    (c) $10\sqrt{3}\pi$ ft/sec

SECTION 8.4

1. $-2 \cos x + 3 \sin x + C$

3. $t + \csc t + C$

5. $\tan \theta + \cos \theta + C$

7. $-\dfrac{1}{2} \cos 2x + C$

9. $\dfrac{1}{2} \sin x^2 + C$

11. $2 \tan \left(\dfrac{x}{2}\right) + C$

13. $\dfrac{1}{2} \tan^2 x + C$

15. $-\cot x - x + C$

17. $\dfrac{1}{5} \tan^5 x + C$

19. $\dfrac{1}{\pi} \ln |\sin \pi x| + C$

21. $-\dfrac{1}{2} \ln |\csc 2x + \cot 2x| + C$

23. $\ln |\tan x| + C$

25. $\ln |\sec x - 1| + C$

27. $\ln |1 + \sin t| + C$

29. $\ln |\theta - \sin \theta| + C$

31. $\sin e^x + C$

33. $\ln |\cos e^{-x}| + C$

35. $x - \dfrac{1}{4} \cos 4x + C$

37. $3\sqrt{3}/4$

39. $2(\sqrt{3} - 1)$

41. $1/2$

43. $-1 + \sec 1$

45. 2

47. 4

49. $2\left(\dfrac{2\pi}{3} - \ln (2 + \sqrt{3})\right) \approx 6.823$

51. 2π

53. π

55. 3.829

57. (a) 102.352 thousand units
    (b) 102.352 thousand units
    (c) 74.5 thousand units

59. (a) 1.273 amps
    (b) 1.382 amps
    (c) 0 amps

61. $\dfrac{1}{2} \sin^2 x + C_1 = -\dfrac{1}{2} \cos^2 x + C_2$

67. (a) 0.957
    (b) 0.978

69. (a) 0.334
    (b) 0.305

71. (a) 0.194
    (b) 0.186

SECTION 8.5

1. $\pi/6$

3. $\pi/3$

5. $\pi/6$

7. $\pi/4$

9. $\arccos (1/1.269) \approx 0.663$

11. (a) 1/2

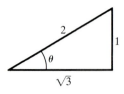

(b) 1/2

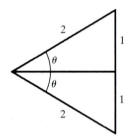

13. (a) 3/5

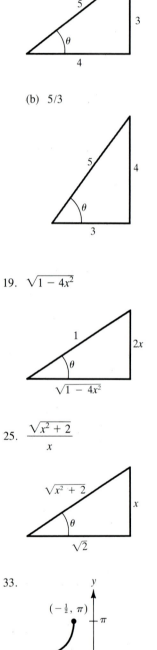

(b) 5/3

15. (a) 12/13

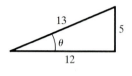

(b) −13/5

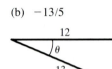

17. x

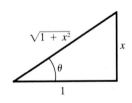

19. $\sqrt{1 - 4x^2}$

20.

21. $\dfrac{\sqrt{x^2 - 1}}{x}$

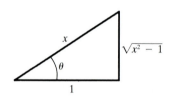

23. $\dfrac{\sqrt{x^2 - 9}}{3}$

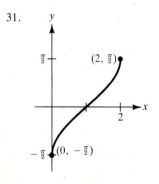

25. $\dfrac{\sqrt{x^2 + 2}}{x}$

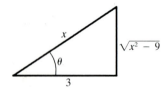

27. $\arcsin\left(\dfrac{9}{\sqrt{x^2 + 81}}\right)$

31.

33.

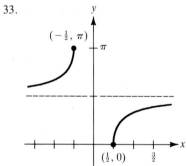

35. $x = \dfrac{1}{3}\left[\sin\left(\dfrac{1}{2}\right) + \pi\right] \approx 1.207$

37. $x = 1/3$

39. $\dfrac{2}{\sqrt{1 - 4x^2}}$

41. $\dfrac{2}{\sqrt{2x - x^2}}$

43. $-\dfrac{3}{\sqrt{4 - x^2}}$

45. $\dfrac{5}{1 + 25x^2}$

47. $\dfrac{1}{|x|\sqrt{x^2 - 1}}$

49. 0

51. $-\dfrac{t}{\sqrt{1 - t^2}}$

53. $\dfrac{1}{2 + 3t^2}$

55. $\dfrac{1}{1 - x^4}$

57. $\arcsin x$

59. (0, 0)

61. Relative maximum: (1.272, −0.606)
    Relative minimum: (−1.272, 3.747)

63. (0.7862, 0.6662)

65. $\sqrt{3}/40$ rad/sec

SECTION 8.6

1. $\pi/18$

3. $\pi/6$

5. $\operatorname{arcsec} 2x + C$

7. $\dfrac{1}{2}x^2 - \dfrac{1}{2}\ln(x^2 + 1) + C$

9. $\arcsin(x + 1) + C$

11. $\dfrac{1}{2}\arcsin t^2 + C$

13. $\dfrac{1}{2}(\arctan x)^2 + C$

15. $\pi^2/32 \approx 0.308$

17. $\dfrac{\sqrt{3} - 2}{2} \approx -0.134$

19. $\arcsin e^x + C$

21. $\dfrac{1}{3}\arctan\left(\dfrac{x - 3}{3}\right) + C$

23. $2\arctan\sqrt{x} + C$

25. $\pi/4$

27. $\pi/2$

29. $\ln|x^2 + 6x + 13| - 3\arctan\left(\dfrac{x + 3}{2}\right) + C$

31. $\arcsin\left(\dfrac{x + 2}{2}\right) + C$

33. $\arcsin(x - 1) + C$

35. $4 - 2\sqrt{3} + \dfrac{\pi}{6} \approx 1.059$

37. $\dfrac{1}{2}\arctan(x^2 + 1) + C$

39. $\dfrac{1}{4}\arcsin(4x - 2) + C$

41. $2(\sqrt{x - 1} - \arctan\sqrt{x - 1}) + C$

43. $2\sqrt{e^t - 3} - 2\sqrt{3}\arctan\left(\dfrac{\sqrt{e^t - 3}}{\sqrt{3}}\right) + C$

45. $\pi/4$

47. $\pi/8$

49. $v = \sqrt{\dfrac{32}{k}}\tan\left[\arctan\left(500\sqrt{\dfrac{k}{32}}\right) - \sqrt{32k}t\right]$

51. (a) $\arcsin x + C$
    (b) $-\sqrt{1 - x^2} + C$
    (c) Does not fit Basic Formulas

53. (a) $\dfrac{2}{3}(x - 1)^{3/2} + C$

    (b) $\dfrac{2}{15}(x - 1)^{3/2}(3x + 2) + C$

    (c) $\dfrac{2}{3}\sqrt{x - 1}(x + 2) + C$

SECTION 8.7

1. (a) 10.018
   (b) −0.964

3. (a) 4/3
   (b) 13/12

5. (a) 1.317
   (b) 0.962

7. $\left(\dfrac{e^x - e^{-x}}{e^x + e^{-x}}\right)^2 + \left(\dfrac{2}{e^x + e^{-x}}\right)^2 = 1$

9. $\left(\dfrac{e^x - e^{-x}}{2}\right)\left(\dfrac{e^y + e^{-y}}{2}\right) + \left(\dfrac{e^x + e^{-x}}{2}\right)\left(\dfrac{e^y + e^{-y}}{2}\right) = \dfrac{e^{(x+y)} - e^{-(x+y)}}{2}$

11. $\left(\dfrac{e^x - e^{-x}}{2}\right)\left[3 + 4\left(\dfrac{e^x - e^{-x}}{2}\right)^2\right] = \dfrac{e^{3x} - e^{-3x}}{2}$

13. $-2x \cosh(1 - x^2)$

15. $\coth x$

17. $\operatorname{csch} x$

19. $\sinh^2 x$

21. $\operatorname{sech} x$

23. $\dfrac{y}{x}[\cosh x + x(\sinh x) \ln x]$

25. $-2e^{-2x}$

27. $\dfrac{3}{\sqrt{9x^2 - 1}}$

29. $|\sec x|$

31. $2 \sec 2x$

33. $2 \sinh^{-1}(2x)$

35. $-\dfrac{1}{2} \cosh(1 - 2x) + C$

37. $\dfrac{1}{3} \cosh^3(x - 1) + C$

39. $\ln |\sinh x| + C$

41. $-\coth \dfrac{x^2}{2} + C$

43. $\operatorname{csch} \dfrac{1}{x} + C$

45. $\dfrac{1}{5} \ln 3$

47. $\pi/4$

49. $\dfrac{1}{2} \arctan(x^2) + C$

51. $-\operatorname{csch}^{-1}(e^x) + C = -\ln\left(\dfrac{1 + \sqrt{1 + e^{2x}}}{e^x}\right) + C$

53. $2 \sinh^{-1} \sqrt{x} + C = 2 \ln(\sqrt{x} + \sqrt{1 + x}) + C$

55. $-\ln\left[\dfrac{1 + \sqrt{(x - 1)^2 + 1}}{|x - 1|}\right] + C$

57. $\dfrac{1}{2\sqrt{6}} \ln \left|\dfrac{\sqrt{2}(x + 1) + \sqrt{3}}{\sqrt{2}(x + 1) - \sqrt{3}}\right| + C$

59. $\dfrac{1}{4} \arcsin\left(\dfrac{4x - 1}{9}\right) + C$

61. $-\dfrac{x^2}{2} - 4x - \dfrac{10}{3} \ln \left|\dfrac{x - 5}{x + 1}\right| + C$

63. Point of inflection: (0, 0)

65. $y''' - y' = a \cosh x - a \cosh x = 0$

67. $-\dfrac{\sqrt{a^2 - x^2}}{x}$

69. 52/31 kg

REVIEW EXERCISES FOR CHAPTER 8

1. $\dfrac{x \cos x - 2 \sin x}{x^3}$

3. $-x \sec^2 x - \tan x$

5. $\cos 4x + 1$

7. $\sin^2 x$

9. $-\dfrac{3}{\sqrt{x}} \csc^3 \sqrt{x} \cot \sqrt{x}$

11. $-\dfrac{\sec^2 \sqrt{1 - x}}{2\sqrt{1 - x}}$

13. $(1 - x^2)^{-3/2}$

15. $\dfrac{x}{|x|\sqrt{x^2 - 1}} + \text{arcsec } x$

17. $(\arcsin x)^2$

19. $\dfrac{1}{2} + x \arctan x$

21. $\dfrac{1}{\cos y} = \dfrac{1}{\sqrt{1 - (x - 2)^2}}$

23. $-\dfrac{e^y + 2x \sin x^2}{xe^y}$

25. $2 \csc^2 x \cot x$

27. $\dfrac{-x^2 \cos x + 2x \sin x + 2 \cos x}{x^3}$

29. $\dfrac{8x}{(1 - 4x^2)^{3/2}}$

31. $\arctan (\sin x) + C$

33. $\dfrac{1}{4} (\arctan 2x)^2 + C$

35. $\dfrac{\tan^{n+1} x}{n + 1} + C$

37. $-\dfrac{1}{2} \ln (1 + e^{-2x}) + C$

39. $\dfrac{1}{2} \arctan (e^{2x}) + C$

41. $-\dfrac{1}{2} \ln |\cos x^2| + C$

43. $\dfrac{1}{2} \arcsin x^2 + C$

45. $\dfrac{1}{2} \ln (16 + x^2) + C$

47. $\dfrac{1}{4} \left[\arctan \dfrac{x}{2} \right]^2 + C$

49. $4 \arcsin \dfrac{x}{2} + \sqrt{4 - x^2} + C$

53. $2\pi x + 4y = \pi^2$

55. 1.122

57. $3(3^{2/3} + 2^{2/3})^{3/2}$ ft

59. $\sqrt{32/L}\sqrt{A^2 + B^2}$

61. (a) $t = 1/2$ (January 15), $Q(1/2) = 7.9$ quads/month
    (b) $t = 13/2$ (July 15), $Q(13/2) = 5.9$ quads/month

63. (a) $C \approx \$9.17$
    (b) $C \approx \$3.14$
        Savings $\approx \$6.03$

65. $\pi^2/4$

CHAPTER 9

SECTION 9.1

1. $\frac{1}{15}(3x - 2)^5 + C$

3. $-\frac{1}{5}(-2x + 5)^{5/2} + C$

5. $\frac{1}{2}v^2 - \frac{1}{6(3v - 1)^2} + C$

7. $-\frac{1}{3}\ln|-t^3 + 9t + 1| + C$

9. $\frac{1}{2}x^2 + x + \ln|x - 1| + C$

11. $\frac{1}{3}\ln\left|\frac{3x - 1}{3x + 1}\right| + C$

13. $-\frac{1}{2}\cos(t^2) + C$

15. $e^{\sin x} + C$

17. $-e^{-t} + 2t + e^t + C$

19. $\frac{1}{3}\sec 3x + C$

21. $2\ln(1 + e^x) + C$

23. $-\cot x - \csc x + C$

25. $\ln(t^2 + 4) - \frac{1}{2}\arctan\frac{t}{2} + C$

27. $3\arctan t + C$

29. $\frac{1}{2}\operatorname{arcsec}\frac{|x|}{2} + C$

31. $-\frac{1}{2}\arcsin(2t - 1) + C$

33. $-\frac{1}{\pi}\csc(\pi x) + C$

35. $\frac{1}{2}\arcsin(t^2) + C$

37. $\frac{1}{2}\arctan\frac{\tan x}{2} + C$

39. $\frac{1}{2}\ln\left|\cos\frac{2}{t}\right| + C$

41. $\frac{x}{15}(12x^4 + 20x^2 + 15) + C$

43. $\frac{1}{2}x^2 + 3x + 3\ln|x| - \frac{1}{x} + C$

45. $\frac{1}{2}e^{2x} + 2e^x + x + C$

47. $3\arcsin\frac{x - 3}{3} + C$

49. $\frac{1}{4}\arctan\frac{2x + 1}{8} + C$

51. $\frac{1}{2}(1 - e^{-1}) \approx 0.316$

53. $\frac{1}{2}$

55. 4

57. $\frac{1}{2}\ln(2)$

59. $\frac{1}{2}\left(\operatorname{arcsec} 4 - \frac{\pi}{3}\right) \approx 0.135$

61. $\frac{4}{3}$

63. $a = \frac{1}{2}$

65. $b = \sqrt{\ln\left(\frac{3\pi}{3\pi - 4}\right)} \approx 0.743$

67. $\dfrac{2}{\arcsin(4/5)} \approx 2.157$

SECTION 9.2

1. $\frac{e^{2x}}{4}(2x - 1) + C$

3. $\frac{1}{2}e^{x^2} + C$

5. $-\frac{1}{4e^{2x}}(2x + 1) + C$

7. $e^x(x^3 - 3x^2 + 6x - 6) + C$

9. $\frac{x^4}{16}(4\ln x - 1) + C$

11. $\frac{1}{4}[2(t^2 - 1)\ln|t + 1| - t^2 + 2t] + C$

13. $x(\ln x)^2 - 2x\ln x + 2x + C$

15. $\frac{(\ln x)^3}{3} + C$

17. $\frac{e^{2x}}{4(2x + 1)} + C$

19. $\frac{2(x - 1)^{3/2}}{15}(3x + 2) + C$

21. $(x - 1)^2 e^x + C$

23. $x\ln x - x + C$

25. $x \sin x + \cos x + C$

27. $x \tan x + \ln |\cos x| + C$

29. $x \arcsin 2x + \frac{1}{2}\sqrt{1 - 4x^2} + C$

31. $x \arctan x - \frac{1}{2} \ln (1 + x^2) + C$

33. $\frac{1}{5}e^{2x}(2 \sin x - \cos x) + C$

35. $-\frac{\pi}{2}$

37. $\frac{e[\sin (1) - \cos (1)] + 1}{2} \approx 0.909$

39. $\frac{\pi}{2} - 1$

41. $\frac{2}{5}(2x - 3)^{3/2}(x + 1) + C$

43. $\frac{1}{3}\sqrt{4 + x^2}(x^2 - 8) + C$

51. $1 - \frac{5}{e^4} \approx 0.908$

53. $\frac{\pi}{1 + \pi^2}\left(\frac{1}{e} + 1\right) \approx 0.395$

55. (a) 1
   (b) $\pi(e - 2) \approx 2.257$
   (c) $\frac{(e^2 + 1)\pi}{2} \approx 13.177$
   (d) $\left(\frac{e^2 + 1}{4}, \frac{e - 2}{2}\right) \approx (2.097, 0.359)$

57. (a) $3.2(\ln 2) - 0.2 \approx 2.018$
   (b) $12.8(\ln 4) - 7.2(\ln 3) - 1.8 \approx 8.035$

59. $\frac{8h}{(n\pi)^2} \sin \left(\frac{n\pi}{2}\right)$

61. For any integrable function,

$$\int f(x)\, dx = C + \int f(x)\, dx$$

but this does not imply that $C = 0$.

SECTION 9.3

1. $-\frac{1}{4} \cos^4 x + C$

3. $\frac{1}{12} \sin^6 2x + C$

5. $-\frac{1}{3} \cos^3 x + \frac{2}{5} \cos^5 x - \frac{1}{7} \cos^7 x + C$

7. $\frac{1}{12}(6x + \sin 6x) + C$

9. $\frac{x}{16} - \frac{\sin 4\pi x}{64\pi} + \frac{\sin^3 2\pi x}{48\pi} + C$

11. $\frac{1}{16}\left(x - \frac{\sin 4x}{4} - \frac{\sin^3 2x}{3}\right) + C$

13. $\frac{1}{8}(2x^2 - 2x \sin 2x - \cos 2x) + C$

15. $\frac{1}{3} \ln |\sec 3x + \tan 3x| + C$

17. $\frac{1}{15} \tan 5x (3 + \tan^2 5x) + C$

19. $\frac{1}{2\pi}(\sec \pi x \tan \pi x + \ln |\sec \pi x + \tan \pi x|) + C$

21. $-\frac{\tan^2 (1 - x)}{2} - \ln |\cos (1 - x)| + C$

23. $\tan^4 \left(\frac{x}{4}\right) - 2 \tan^2 \left(\frac{x}{4}\right) - 4 \ln \left|\cos \frac{x}{4}\right| + C$

25. $\frac{1}{2} \tan^2 x + C$

27. $\frac{\tan^3 x}{3} + C$

29. $\frac{1}{5\pi} \sec^5 \pi x + C$

31. $\frac{\sec^6 4x}{24} + C$

33. $\frac{1}{3} \sec^3 x + C$

35. $\frac{1}{9} \sec^3 3x - \frac{1}{3} \sec 3x + C$

37. $\frac{1}{4}[\ln |\csc^2 (2x)| - \cot^2 (2x)] + C$

39. $-\cot \theta - \frac{1}{3} \cot^3 \theta + C$

41. $\ln |\csc t - \cot t| + \cos t + C$

43. $-\dfrac{1}{10}(\cos 5x + 5\cos x) + C$

45. $\dfrac{1}{8}(2\sin 2\theta - \sin 4\theta) + C$

47. $\ln|\csc x - \cot x| + \cos x + C$

49. $t - 2\tan t + C$

51. π

53. $\dfrac{1}{2}(1 - \ln 2)$

55. $\ln 2$

57. $\dfrac{3\sqrt{2}}{10}$

59. $\dfrac{4}{3}$

61. $\dfrac{\tan^6 3x}{18} + \dfrac{\tan^4 3x}{12} + C_1 = \dfrac{\sec^6 3x}{18} - \dfrac{\sec^4 3x}{12} + C_2$

65. $2\pi\left(1 - \dfrac{\pi}{4}\right) \approx 1.348$

67. (a) $\dfrac{\pi^2}{2}$

   (b) $(\bar{x}, \bar{y}) = \left(\dfrac{\pi}{2}, \dfrac{\pi}{8}\right)$

SECTION 9.4

1. $\dfrac{x}{25\sqrt{25 - x^2}} + C$

3. $5\ln\left|\dfrac{5 - \sqrt{25 - x^2}}{x}\right| + \sqrt{25 - x^2} + C$

5. $\sqrt{x^2 + 9} + C$

7. 2π

9. $\ln|x + \sqrt{x^2 - 9}| + C$

11. $\sqrt{3} - \dfrac{\pi}{3} \approx 0.685$

13. $-\dfrac{(1 - x^2)^{3/2}}{3x^3} + C$

15. $-\dfrac{1}{3}\ln\left|\dfrac{\sqrt{4x^2 + 9} + 3}{2x}\right| + C$

17. $-\dfrac{1}{\sqrt{x^2 + 3}} + C$

19. $\dfrac{1}{15}(x^2 - 4)^{3/2}(3x^2 + 8) + C$

21. $\dfrac{1}{3}(1 + e^{2x})^{3/2} + C$

23. $\dfrac{1}{20}\left[5\arctan\left(\dfrac{1}{2}\right) - 2\right] \approx 0.016$

25. $\dfrac{1}{3}(x^2 + 2x + 2)^{3/2} + C$

27. $\dfrac{1}{2}(\arcsin e^x + e^x\sqrt{1 - e^{2x}}) + C$

29. $\dfrac{1}{4}\left(\dfrac{x}{x^2 + 2} + \dfrac{1}{\sqrt{2}}\arctan\dfrac{x}{\sqrt{2}}\right) + C$

31. $\dfrac{1}{6}(3x\sqrt{9x^2 + 4} + 4\ln|3x + \sqrt{9x^2 + 4}|) + C$

33. $\dfrac{1}{2}(x\sqrt{x^2 - 1} + \ln|x + \sqrt{x^2 - 1}|) + C$

35. $x\,\text{arcsec}\,2x - \dfrac{1}{2}\ln|2x + \sqrt{4x^2 - 1}| + C$

37. $1 - \dfrac{\sqrt{3}\pi}{6} \approx 0.093$

39. $\dfrac{25\pi}{12}$

41. 187.2π lb

43. $6\pi^2$

45. $\ln\left[\dfrac{5(\sqrt{2} + 1)}{\sqrt{26} + 1}\right] + \sqrt{26} - \sqrt{2} \approx 4.367$

47. $100\sqrt{2} + 50\ln\left(\dfrac{\sqrt{2} + 1}{\sqrt{2} - 1}\right) \approx 229.559$

49. $\dfrac{\pi}{32}[102\sqrt{2} - \ln(3 + 2\sqrt{2})] \approx 13.989$

51. 121.3 lb

SECTION 9.5

1. $\dfrac{1}{2} \ln \left| \dfrac{x-1}{x+1} \right| + C$

3. $\ln \left| \dfrac{x-1}{x+2} \right| + C$

5. $\dfrac{3}{2} \ln |2x-1| - 2 \ln |x+1| + C$

7. $5 \ln |x-2| - \ln |x+2| - 3 \ln |x| + C$

9. $x^2 + \dfrac{3}{2} \ln |x-4| - \dfrac{1}{2} \ln |x+2| + C$

11. $\dfrac{1}{x} + \ln |x^4 + x^3| + C$

13. $\dfrac{1}{2}x^2 + 3x + 6 \ln |x-1| - \dfrac{4}{x-1} - \dfrac{1}{2(x-1)^2} + C$

15. $3 \ln |x-3| - \dfrac{9}{x-3} + C$

17. $\ln \left| \dfrac{x^2+1}{x} \right| + C$

19. $\dfrac{1}{6} \left[\ln \left| \dfrac{x-2}{x+2} \right| + \sqrt{2} \arctan \left(\dfrac{x}{\sqrt{2}} \right) \right] + C$

21. $\dfrac{1}{16} \ln \left| \dfrac{4x^2-1}{4x^2+1} \right| + C$

23. $\dfrac{\sqrt{2}}{2} \arctan \dfrac{x}{\sqrt{2}} - \dfrac{1}{2(x^2+2)} + C$

25. $\ln |x+1| + \sqrt{2} \arctan \left(\dfrac{x-1}{\sqrt{2}} \right) + C$

27. $4 \ln |x-2| + \ln (x^2+4) + \dfrac{1}{2} \arctan \dfrac{x}{2} + C$

29. $\ln |x-2| + \dfrac{1}{2} \ln |x^2+x+1| - \sqrt{3} \arctan \left(\dfrac{2x+1}{\sqrt{3}} \right) + C$

31. $\ln 2$

33. $\dfrac{1}{2} \ln \left(\dfrac{8}{5} \right) - \dfrac{\pi}{4} + \arctan (2) \approx 0.557$

35. $10 \ln (2) + \arctan (2) - 4 \arctan (3) + 3 \arctan (4) \approx 7.020$

37. $\ln \left| \dfrac{\cos x}{\cos x - 1} \right| + C$

39. $\dfrac{1}{5} \ln \left| \dfrac{e^x - 1}{e^x + 4} \right| + C$

41. $\ln \left| \dfrac{-1 + \sin x}{2 + \sin x} \right| + C$

45. $6 - \dfrac{7}{4} \ln 7 \approx 2.595$

47. $2\pi \left[\arctan 3 - \dfrac{3}{10} \right] \approx 5.963$

49. $x = \dfrac{n[e^{(n+1)kt} - 1]}{n + e^{(n+1)kt}}$

SECTION 9.6

1. $-\dfrac{1}{2}x(2-x) + \ln |1+x| + C$

3. $-\dfrac{\sqrt{1-x^2}}{x} + C$

5. $\dfrac{1}{9}x^3(-1 + 3 \ln x) + C$

7. $\dfrac{\sqrt{x^2-4}}{4x} + C$

9. $\dfrac{1}{2}e^{x^2} + C$

11. $\dfrac{2}{9} \left(\ln |1-3x| + \dfrac{1}{1-3x} \right) + C$

13. $e^x \arccos (e^x) - \sqrt{1 - e^{2x}} + C$

15. $\dfrac{1}{16}x^4(4 \ln |x| - 1) + C$

17. $\frac{1}{27}\left(3x - \frac{25}{3x - 5} + 10 \ln |3x - 5|\right) + C$

19. $\frac{1}{2}(x^2 + \cot x^2 + \csc x^2) + C$

21. $\arctan (\sin x) + C$

23. $x - \frac{1}{2} \ln (1 + e^{2x}) + C$

25. $\frac{\sqrt{2}}{2} \arctan \left(\frac{1 + \sin \theta}{\sqrt{2}}\right) + C$

27. $-\frac{\sqrt{2 + 9x^2}}{2x} + C$

29. $\frac{1}{2}(e^x\sqrt{e^{2x} + 1} + \ln |e^x + \sqrt{e^{2x} + 1}|) + C$

31. $\frac{1}{16}(6x - 3 \sin 2x \cos 2x - 2 \sin^3 2x \cos 2x) + C$

33. $-2(\cot \sqrt{x} + \csc \sqrt{x}) + C$

35. $(t^4 - 12t^2 + 24) \sin t + (4t^3 - 24t) \cos t + C$

37. $\frac{1}{2}[(x^2 + 1) \text{arcsec} (x^2 + 1) - \ln |(x^2 + 1) + \sqrt{x^4 + 2x^2}|] + C$

39. $\frac{1}{4}(2 \ln x - 3 \ln |3 + 2 \ln x|) + C$

41. $\sqrt{2 - 2x - x^2} - \sqrt{3} \ln \left|\frac{\sqrt{3} + \sqrt{2 - 2x - x^2}}{x + 1}\right| + C$

43. $\frac{1}{2} \arctan (x^2 - 3) + C$

45. $\frac{1}{2} \ln |(x^2 - 3) + \sqrt{x^4 - 6x^2 + 5}| + C$

47. $-\frac{1}{3}\sqrt{4 - x^2}(x^2 + 8) + C$

49. $-\frac{2\sqrt{1 - x}}{\sqrt{x}} + C$

51. $\frac{2}{1 + e^x} - \frac{1}{2(1 + e^x)^2} + \ln |1 + e^x| + C$

59. $\frac{1}{\sqrt{5}} \ln \left|\frac{2 \tan (\theta/2) - 3 - \sqrt{5}}{2 \tan (\theta/2) - 3 + \sqrt{5}}\right| + C$

61. $\ln 2$

63. $\frac{1}{2} \ln (3 - 2 \cos \theta) + C$

65. $2 \sin \sqrt{\theta} + C$

67. $-\csc \theta + C$

69. $1{,}919.145 \text{ ft} \cdot \text{lb}$

SECTION 9.7

1. 4

3. 6

5. 1

7. Diverges

9. 2

11. 1

13. 1/2

15. π

17. $\pi/4$

19. Diverges

21. Diverges

23. 6

25. $-1/4$

27. Diverges

29. $\ln (2 + \sqrt{3})$

31. 0

33. $p > 1$

35. Converges

37. Diverges

39. Converges

41. (a) 1
    (b) $\pi/3$
    (c) Diverges

43. 6

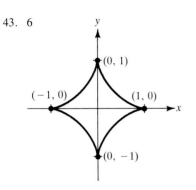

45. (a) $T(1) = 1$
    (b) $T(2) = 1$
    (c) $T(3) = 2$

47. (b) 51.05%
    (c) 7

49. (a) \$743,997.58
    (b) \$795,584.54
    (c) \$858,333.33

REVIEW EXERCISES FOR CHAPTER 9

1. $x + \dfrac{9}{8} \ln |x - 3| - \dfrac{25}{8} \ln |x + 5| + C$

3. $\tan \theta + \sec \theta + C$

5. $\dfrac{e^{2x}}{13}(2 \sin 3x - 3 \cos 3x) + C$

7. $-\dfrac{1}{x}(1 + \ln 2x) + C$

9. $\dfrac{1}{2}\left(4 \arcsin \dfrac{x}{2} + x\sqrt{4 - x^2}\right) + C$

11. $\dfrac{3\sqrt{4 - x^2}}{x} + C$

13. $\dfrac{2}{3}\left[\tan^3 \left(\dfrac{x}{2}\right) + 3 \tan \left(\dfrac{x}{2}\right)\right] + C$

15. $\dfrac{3}{2} \ln \left|\dfrac{x - 3}{x + 3}\right| + C$

17. $\dfrac{1}{4}[6 \ln |x - 1| - \ln (x^2 + 1) + 6 \arctan x] + C$

19. $\dfrac{3}{2} \ln (x^2 + 1) - \dfrac{1}{2(x^2 + 1)} + C$

21. $16 \arcsin \dfrac{x}{4} + C$

23. $\dfrac{1}{2} \arctan \dfrac{e^x}{2} + C$

25. $\dfrac{1}{2} \ln |x^2 + 4x + 8| - \arctan \left(\dfrac{x + 2}{2}\right) + C$

27. $\dfrac{1}{8}(\sin 2\theta - 2\theta \cos 2\theta) + C$

29. $\dfrac{1}{2}(2\theta - \cos 2\theta) + C$

31. $\dfrac{4}{3}[x^{3/4} - 3x^{1/4} + 3 \arctan (x^{1/4})] + C$

33. $2\sqrt{1 - \cos x} + C$

35. $x \ln |x^2 + x| - 2x + \ln |x + 1| + C$

37. $\sin x \ln (\sin x) - \sin x + C$

39. $x - \dfrac{1}{2(x^2 + 1)} + C$

41. $-\dfrac{\sqrt{4 + x^2}}{4x} + C$

43. $\dfrac{1}{3}\sqrt{4 + x^2}(x^2 - 8) + C$

45. $\dfrac{2\pi}{3\sqrt{3}} \approx 1.21$

47. 1.40

49. 0.74

51. (a) $e - 1 \approx 1.72$

    (b) 1

    (c) $\dfrac{1}{2}(e - 1) \approx 0.86$

    (d) 1.46 (Simpson's Rule)

53. $0.015846 < \displaystyle\int_2^\infty \dfrac{1}{x^5 - 1}\, dx < 0.015851$

CHAPTER 10

SECTION 10.1

1. $1 - x + \dfrac{1}{2}x^2 - \dfrac{1}{6}x^3$

3. $1 + 2x + 2x^2 + \dfrac{4}{3}x^3 + \dfrac{2}{3}x^4$

5. $x - \dfrac{1}{6}x^3 + \dfrac{1}{120}x^5$

7. $x + x^2 + \dfrac{1}{2}x^3 + \dfrac{1}{6}x^4$

9. $1 - x + x^2 - x^3 + x^4$

11. $1 + \dfrac{1}{2}x^2$

13. $2 - 3x^3 + x^4$

15. $1 - (x - 1) + (x - 1)^2 - (x - 1)^3 + (x - 1)^4$

17. $-\pi^2 - 2\pi(x - \pi) + \left(\dfrac{\pi^2 - 2}{2}\right)(x - \pi)^2$

19.

x	$\sin x$	$P_1(x)$	$P_3(x)$	$P_5(x)$
0	0	0	0	0
0.25	0.2474	0.25	0.2474	0.2474
0.50	0.4794	0.50	0.4792	0.4794
0.75	0.6816	0.75	0.6797	0.6817
1.00	0.8415	1.00	0.8333	0.8417

21. 0.6042

23. -6.7954

25. For e^x: $P_4(x) = 1 + x + \dfrac{1}{2}x^2 + \dfrac{1}{6}x^3 + \dfrac{1}{24}x^4$

    For xe^x: $Q_4(x) = x + x^2 + \dfrac{1}{2}x^3 + \dfrac{1}{6}x^4$

    $Q_4(x) = xP_4(x) - \dfrac{1}{24}x^5$

27. $-0.3936 < x < 0$

29. 0.00012

31. (a) $P_3(x) = x + \dfrac{x^3}{6}$

(b)

x	-1	-0.75	-0.50	-0.25	0	0.25	0.50	0.75	1
$f(x)$	-1.571	-0.848	-0.524	-0.253	0	0.253	0.524	0.848	1.571
$P_3(x)$	-1.167	-0.820	-0.521	-0.253	0	0.253	0.521	0.820	1.167

(c)

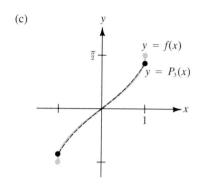

SECTION 10.2

1. 2, 4, 8, 16, 32

3. $-\dfrac{1}{2}, \dfrac{1}{4}, -\dfrac{1}{8}, \dfrac{1}{16}, -\dfrac{1}{32}$

5. $3, \dfrac{9}{2}, \dfrac{27}{6}, \dfrac{81}{24}, \dfrac{243}{120}$

7. $-1, -\dfrac{1}{4}, \dfrac{1}{9}, \dfrac{1}{16}, -\dfrac{1}{25}$

9. $3n - 2$

11. $n^2 - 2$

13. $\dfrac{n + 1}{n + 2}$

15. $\dfrac{(-1)^{n-1}}{2^{n-2}}$

17. $\dfrac{n + 1}{n}$

19. $\dfrac{n}{(n + 1)(n + 2)}$

21. $\dfrac{2^n n!}{(2n)!}$

23. $(-1)^{n(n-1)/2}$

25. Converges to 1

27. Diverges

29. Converges to 3/2

31. Diverges

33. Converges to 0

35. Diverges

37. Converges to 0

39. Diverges

41. Converges to 3

43. Converges to e^k

45. Converges to 0

47. Monotonic

49. Not monotonic

51. Not monotonic

53. Monotonic

55. Not monotonic

57. 5

59. 1/3

61. (a) No

(b)

Month	1	2	3	4	5
Amount	$9,086.25	$9,173.33	$9,261.24	$9,349.99	$9,439.60

Month	6	7	8	9	10
Amount	$9,530.06	$9,621.39	$9,713.59	$9,806.68	$9,900.66

63. (a) $\$2,500,000,000(0.8)^n$
    (b) 1st year: $\$2,000,000,000$
        2nd year: $\$1,600,000,000$
        3rd year: $\$1,280,000,000$
        4th year: $\$1,024,000,000$
    (c) Converges to 0

65. $S_6 = 240$, $S_7 = 440$, $S_8 = 810$, $S_9 = 1,490$, $S_{10} = 2,740$

SECTION 10.3

1. 1, 1.25, 1.361, 1.424, 1.464

3. 3, -1.5, 5.25, -4.875, 10.3125

5. 3, 4.5, 5.25, 5.625, 5.8125

7. $\lim\limits_{n \to \infty} a_n = 1 \neq 0$

9. $\lim\limits_{n \to \infty} a_n = 3 \neq 0$

11. $\lim\limits_{n \to \infty} a_n = 1 \neq 0$

13. Geometric series: $r = \dfrac{3}{2} > 1$

15. Geometric series: $r = 1.055 > 1$

17. $\lim\limits_{n \to \infty} a_n = \dfrac{1}{2} \neq 0$

19. Geometric series: $r = \dfrac{3}{4} < 1$

21. Geometric series: $r = 0.9 < 1$

23. Telescoping series: $a_n = \dfrac{1}{n} - \dfrac{1}{n+1}$

25. 2

27. 2/3

29. $4 + 2\sqrt{2}$

31. 10/9

33. 9/4

35. 3/4

37. 3

39. 1/2

41. $\sum\limits_{n=0}^{\infty} \dfrac{3}{5}(0.1)^n = \dfrac{2}{3}$

43. $\sum\limits_{n=0}^{\infty} \dfrac{3}{40}(0.01)^n = \dfrac{5}{66}$

45. Diverges

47. Converges

49. Diverges

51. Diverges

53. Diverges

55. Diverges

57. $80,000(1 - 0.9^n)$

59. 152.42 ft

61. 1/3

63. $\$11,616.95$

65. $\$235,821.51$

SECTION 10.4

1. Converges

3. Diverges

5. Diverges

7. Converges

9. Converges

11. Diverges

13. Converges

15. Converges

17. Diverges

19. Diverges

21. Diverges

23. Converges

25. Converges

27. Diverges

29. Diverges

31. $R_6 \approx 1.4 \times 10^{-8}$, $\displaystyle\sum_{n=1}^{\infty} (0.075)^n \approx 0.081081066$

33. $R_{10} \approx 0.0997$, $0.9818 < \displaystyle\sum_{n=1}^{\infty} \frac{1}{n^2 + 1} < 1.0815$

35. $R_4 \approx 5.6 \times 10^{-8}$, $\displaystyle\sum_{n=1}^{\infty} ne^{-n^2} \approx 0.404881398$

37. $N = 7$

39. $N = 2$

SECTION 10.5

1. Converges

3. Diverges

5. Converges

7. Diverges

9. Converges

11. Converges

13. Diverges

15. Diverges

17. Converges

19. Diverges

21. Converges

23. Diverges

25. Diverges

27. Diverges, p-Series Test

29. Converges, Comparison Test with $\displaystyle\sum_{n=1}^{\infty} \left(\frac{1}{3}\right)^n$

31. Diverges, nth Term Test

33. Converges, Integral Test

35. $p > 1$

37. Error ≈ 0.000026

39. Diverges

41. Converges

SECTION 10.6

1. Converges

3. Converges

5. Diverges

7. Converges

9. Diverges

11. Converges

13. Diverges

15. Converges

17. Converges

19. Converges

21. Converges

23. Converges absolutely

25. Converges conditionally

27. Converges conditionally

29. Converges absolutely

31. Converges absolutely

33. Converges conditionally

35. Converges absolutely

37. Converges conditionally

39. 0.947

41. 0.368

43. 0.842

45. 0.406

47. At least 499

SECTION 10.7

1. Diverges

3. Converges

5. Converges

7. Converges

9. Diverges

11. Converges

13. Diverges

15. Converges

17. Converges

19. Diverges

21. Converges

23. Diverges

25. Converges

27. Converges

29. Diverges, Ratio Test

31. Converges, Limit Comparison Test with $b_n = \dfrac{1}{2^n}$

33. Converges, Alternating Series Test

35. Converges, Comparison Test with $b_n = \dfrac{1}{2^n}$

37. Converges, Ratio Test

39. Converges, Ratio Test

SECTION 10.8

1. $-2 < x < 2$

3. $-1 < x \le 1$

5. $-\infty < x < \infty$

7. $x = 0$

9. $-4 < x < 4$

11. $0 < x \le 10$

13. $0 < x \le 2$

15. $0 < x < 2c$

17. $-1 \le x \le 1$

19. $-1/2 < x < 1/2$

21. $-\infty < x < \infty$

23. $-1 < x < 1$

25. (a) $-2 < x < 2$
    (b) $-2 < x < 2$
    (c) $-2 < x < 2$
    (d) $-2 \le x < 2$

27. (a) $0 < x \le 2$
    (b) $0 < x < 2$
    (c) $0 < x < 2$
    (d) $0 \le x \le 2$

29. 0.946

31. 0.284

33. (a) For $f(x)$: $-\infty < x < \infty$
    For $g(x)$: $-\infty < x < \infty$
    (d) $f(x) = \sin x$, $g(x) = \cos x$

SECTION 10.9

1. $\displaystyle\sum_{n=0}^{\infty} \frac{x^n}{2^{n+1}}$, $-2 < x < 2$

3. $\displaystyle\sum_{n=0}^{\infty} \frac{(x-5)^n}{(-3)^{n+1}}$, $2 < x < 8$

5. $-3 \displaystyle\sum_{n=0}^{\infty} (2x)^n$, $-\dfrac{1}{2} < x < \dfrac{1}{2}$

7. $-\dfrac{1}{11} \displaystyle\sum_{n=0}^{\infty} \left[\dfrac{2}{11}(x+3)\right]^n$, $-\dfrac{17}{2} < x < \dfrac{5}{2}$

9. $\dfrac{3}{2} \displaystyle\sum_{n=0}^{\infty} \left(\dfrac{x}{-2}\right)^n$, $-2 < x < 2$

11. $\displaystyle\sum_{n=0}^{\infty} \left[\dfrac{1}{(-2)^n} - 1\right] x^n$, $-1 < x < 1$

13. $2 \displaystyle\sum_{n=0}^{\infty} x^{2n}$, $-1 < x < 1$

15. $\displaystyle\sum_{n=1}^{\infty} n(-1)^n x^{n-1}$, $-1 < x < 1$

17. $\displaystyle\sum_{n=0}^{\infty} \dfrac{(-1)^n x^{n+1}}{n+1}$, $-1 < x \le 1$

19. $\displaystyle\sum_{n=0}^{\infty} (-1)^n (2x)^{2n}$, $-\dfrac{1}{2} < x < \dfrac{1}{2}$

21. $\displaystyle\sum_{n=0}^{\infty} \dfrac{x^{2n+1}}{2n+1}$, $-1 < x < 1$

23.

x	$x - \dfrac{x^2}{2}$	$\ln(x+1)$	$x - \dfrac{x^2}{2} + \dfrac{x^3}{3}$
0.0	0.000	0.000	0.000
0.2	0.180	0.182	0.183
0.4	0.320	0.336	0.341
0.6	0.420	0.470	0.492
0.8	0.480	0.588	0.651
1.0	0.500	0.693	0.833

25. 0.245

27. 0.125

SECTION 10.10

1. $\displaystyle\sum_{n=0}^{\infty} \dfrac{(2x)^n}{n!}$

3. $\dfrac{\sqrt{2}}{2} \displaystyle\sum_{n=0}^{\infty} \dfrac{(-1)^{n(n+1)/2}}{n!}\left(x - \dfrac{\pi}{4}\right)^n$

5. $\displaystyle\sum_{n=0}^{\infty} \dfrac{(-1)^n (x-1)^{n+1}}{n+1}$

7. $\displaystyle\sum_{n=0}^{\infty} \dfrac{(-1)^n (2x)^{2n+1}}{(2n+1)!}$

9. $1 + \dfrac{x^2}{2!} + \dfrac{5x^4}{4!} + \left(\dfrac{61x^6}{6!} + \cdots\right)$

11. $\displaystyle\sum_{n=0}^{\infty} (-1)^n (n+1) x^n$

13. $\dfrac{1}{2}\left[1 + \displaystyle\sum_{n=1}^{\infty} \dfrac{(-1)^n 1 \cdot 3 \cdot 5 \cdots (2n-1) x^{2n}}{2^{3n} n!}\right]$

15. $1 + \dfrac{x^2}{2} + \displaystyle\sum_{n=2}^{\infty} \dfrac{(-1)^{n+1} 1 \cdot 3 \cdot 5 \cdots (2n-3)}{2^n n!} x^{2n}$

17. $x - \dfrac{x^3}{3!} + \dfrac{x^5}{5!} - \dfrac{x^7}{7!} + \cdots$

19. $1 + \dfrac{x^2}{2} + \dfrac{x^4}{2^2 2!} + \dfrac{x^6}{2^3 3!} + \cdots$

21. $\displaystyle\sum_{n=0}^{\infty} \dfrac{x^{2n+1}}{(2n+1)!}$

25. $\displaystyle\sum_{n=0}^{\infty} \dfrac{(-1)^n x^{2n}}{(2n+1)!}$

27. 1.3708

29. 0.7040

31. 0.4872

33. 0.2010

35. $\displaystyle\sum_{n=0}^{\infty} \dfrac{(-1)^{n+1} x^{2n+3}}{(2n+3)(n+1)!}$

37. 0.3413

REVIEW EXERCISES FOR CHAPTER 10

1. $a_n = \dfrac{1}{n!}$

3. Converges to 0

5. Diverges

7. Converges to 0

9. Converges to 0

11. 1, 2.5, 4.75, 8.125, 13.3875

13. 0.5, 0.45833, 0.45972, 0.45970, 0.45970

15. 3

17. 1/2

19. 1/11

21. Diverges

23. Converges

25. Diverges

27. Diverges

29. Converges

31. Diverges

33. $1 \le x \le 3$

35. Converges only at $x = 2$

37. $\dfrac{\sqrt{2}}{2} \displaystyle\sum_{n=0}^{\infty} \dfrac{(-1)^{n(n+1)/2}}{n!}\left(x - \dfrac{3\pi}{4}\right)^n$

39. $\displaystyle\sum_{n=0}^{\infty} \dfrac{(x \ln 3)^n}{n!}$

41. $-\displaystyle\sum_{n=0}^{\infty} (x + 1)^n$

43. $\displaystyle\sum_{n=0}^{\infty} \dfrac{(-1)^n x^{2n+1}}{(2n + 1)(2n + 1)!}$

45. $\displaystyle\sum_{n=0}^{\infty} \dfrac{(-1)^n x^{n+1}}{(n + 1)^2}$

47. 0.996

49. 0.560

51.

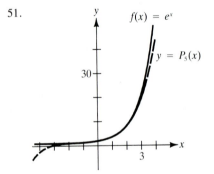

CHAPTER 11

SECTION 11.1

1. e

2. f

3. a

4. c

5. d

6. b

7. Vertex: (0, 0)
    Focus: (0, 1/16)
    Directrix: $y = -1/16$

9. Vertex: (0, 0)
    Focus: (−3/2, 0)
    Directrix: $x = 3/2$

11. Vertex: (0, 0)
    Focus: (0, −2)
    Directrix: $y = 2$

13. Vertex: (1, −2)
    Focus: (1, −4)
    Directrix: $y = 0$

15. Vertex: (5, −1/2)
    Focus: (11/2, −1/2)
    Directrix: $x = 9/2$

17. Vertex: (1, 1)
    Focus: (1, 2)
    Directrix: $y = 0$

19. Vertex: (8, −1)
    Focus: (9, −1)
    Directrix: $x = 7$

21. Vertex: (−2, −3)
    Focus: (−4, −3)
    Directrix: $x = 0$

23. Vertex: (−1, 2)
    Focus: (0, 2)
    Directrix: $x = -2$

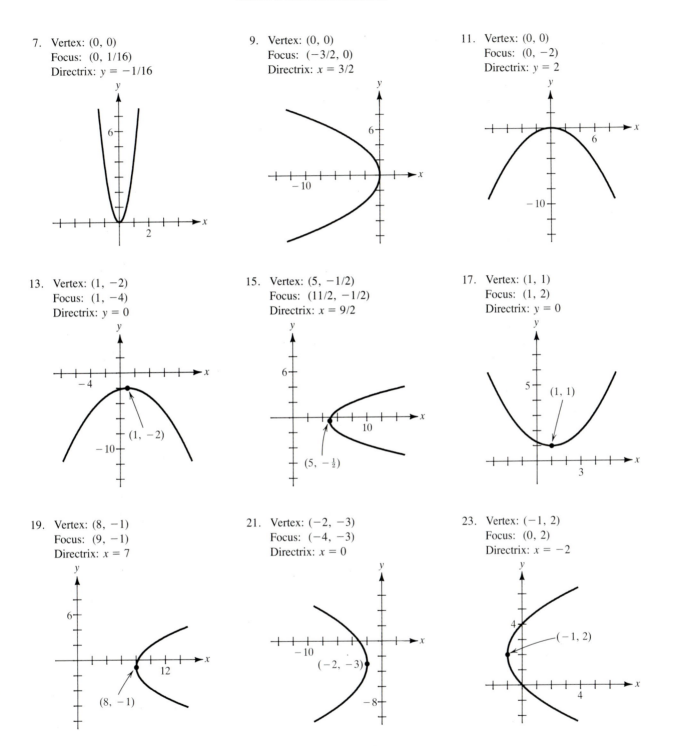

25. Vertex: $(-2, 2)$
    Focus: $(-2, 1)$
    Directrix: $y = 3$

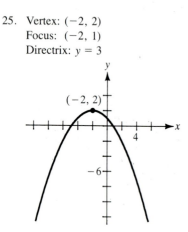

$(-2, 2)$

27. $x^2 + 6y = 0$

29. $y^2 - 4y + 8x - 20 = 0$

31. $x^2 + 24y + 96 = 0$

33. $x^2 + 8y - 16 = 0$

35. $5x^2 - 14x - 3y + 9 = 0$

37. $x^2 + y - 4 = 0$

39. $(x - h)^2 = -8(y + 1)$

41. $3x - 2y^2 = 0$

43. $2[\sqrt{2} + \ln(\sqrt{2} + 1)]$

45. $2\sqrt{5} + \ln(\sqrt{5} + 2)$

47. $100\left[\sqrt{5} + 4\ln\left(\dfrac{1 + \sqrt{5}}{2}\right)\right] \approx 416.1$ ft

49. $10\sqrt{3}$ ft from end of pipe

51. $y = 2ax_0x - ax_0^2$

SECTION 11.2

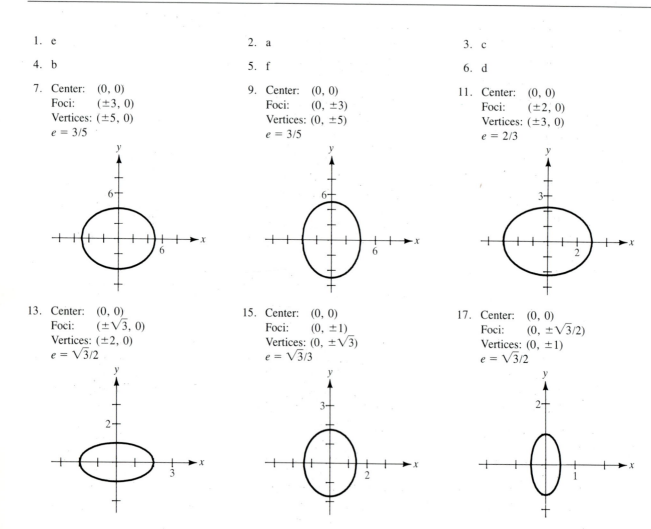

1. e

2. a

3. c

4. b

5. f

6. d

7. Center: $(0, 0)$
   Foci: $(\pm3, 0)$
   Vertices: $(\pm5, 0)$
   $e = 3/5$

9. Center: $(0, 0)$
   Foci: $(0, \pm3)$
   Vertices: $(0, \pm5)$
   $e = 3/5$

11. Center: $(0, 0)$
    Foci: $(\pm2, 0)$
    Vertices: $(\pm3, 0)$
    $e = 2/3$

13. Center: $(0, 0)$
    Foci: $(\pm\sqrt{3}, 0)$
    Vertices: $(\pm2, 0)$
    $e = \sqrt{3}/2$

15. Center: $(0, 0)$
    Foci: $(0, \pm1)$
    Vertices: $(0, \pm\sqrt{3})$
    $e = \sqrt{3}/3$

17. Center: $(0, 0)$
    Foci: $(0, \pm\sqrt{3}/2)$
    Vertices: $(0, \pm1)$
    $e = \sqrt{3}/2$

19. Center: (1, 5)
     Foci: (1, 9), (1, 1)
     Vertices: (1, 10), (1, 0)
     $e = 4/5$

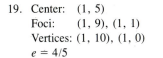

21. Center: $(-2, 3)$
     Foci: $(-2, 3, \pm \sqrt{5})$
     Vertices: $(-2, 6), (-2, 0)$
     $e = \sqrt{5}/3$

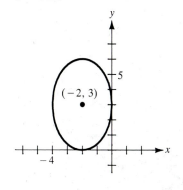

23. Center: $(1, -1)$
     Foci: $(1 \pm 3\sqrt{10}/20, -1)$
     Vertices: $(1 \pm \sqrt{10}/4, -1)$
     $e = 3/5$

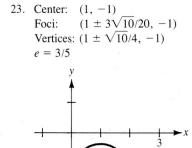

25. Center: $(1/2, -1)$
     Foci: $\left(\dfrac{1}{2} \pm \sqrt{2}, -1\right)$
     Vertices: $\left(\dfrac{1}{2} \pm \sqrt{5}, -1\right)$
     $e = \sqrt{10}/5$

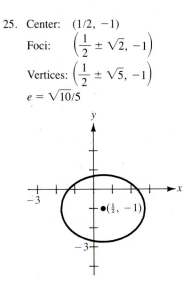

27. $\dfrac{x^2}{9} + \dfrac{y^2}{5} = 1$

31. $\dfrac{(x-2)^2}{4} + \dfrac{(y-2)^2}{1} = 1$

35. $\dfrac{x^2}{24} + \dfrac{y^2}{49} = 1$

41. $(0, 25/3)$

45. 9.91

29. $\dfrac{x^2}{25} + \dfrac{y^2}{16} = 1$

33. $\dfrac{(x-3)^2}{9} + \dfrac{(y-5)^2}{16} = 1$

39. Minor axis: $(-6, -2), (0, -2)$
     Major axis: $(-3, -6), (-3, 2)$

43. 22.10

47. (a) 2π
     (b) Volume $= 8\pi/3$
          Surface area $= 2\pi(9 + 4\sqrt{3}\pi)/9 \approx 21.48$
     (c) Volume $= 16\pi/3$
          Surface area $= 4\pi[6 + \sqrt{3} \ln (2 + \sqrt{3})]/3 \approx 34.69$

49. $\sqrt{2}a \times \sqrt{2}b$

55. 1.5 ft from center

SECTION 11.3

1. e

4. c

2. a

5. d

3. f

6. b

7. Center: (0, 0)
   Vertices: (±1, 0)
   Foci: (±√2, 0)

9. Center: (0, 0)
   Vertices: (0, ±1)
   Foci: (0, ±√5)

11. Center: (0, 0)
    Vertices: (0, ±5)
    Foci: (0, ±13)

13. Center: (0, 0)
    Vertices: (±√3, 0)
    Foci: (±√5, 0)

15. Center: (0, 0)
    Vertices: (0, ±2)
    Foci: (0, ±3)

17. Center: (1, −2)
    Vertices: (−1, −2), (3, −2)
    Foci: (1 ± √5, −2)

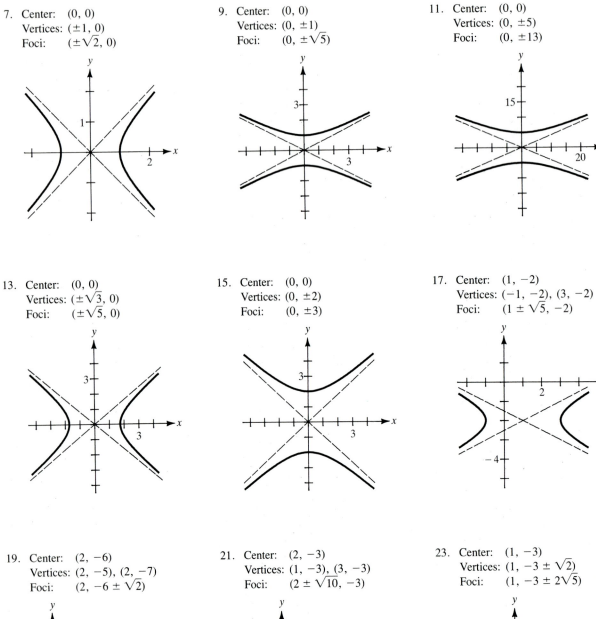

19. Center: (2, −6)
    Vertices: (2, −5), (2, −7)
    Foci: (2, −6 ± √2)

21. Center: (2, −3)
    Vertices: (1, −3), (3, −3)
    Foci: (2 ± √10, −3)

23. Center: (1, −3)
    Vertices: (1, −3 ± √2)
    Foci: (1, −3 ± 2√5)

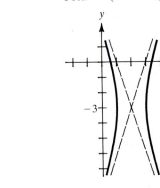

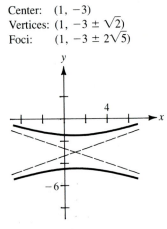

25. Degenerate hyperbola: graph is two intersecting lines with center at $(-1, -3)$

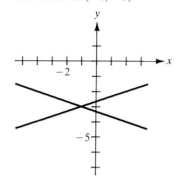

27. $\dfrac{y^2}{4} - \dfrac{x^2}{12} = 1$

29. $\dfrac{x^2}{1} - \dfrac{y^2}{9} = 1$

31. $\dfrac{(x-3)^2}{9} - \dfrac{(y-2)^2}{4} = 1$

33. $\dfrac{y^2}{9} - \dfrac{(x-2)^2}{9/4} = 1$

35. $\dfrac{(x-6)^2}{9} - \dfrac{(y-2)^2}{7} = 1$

37. (a) At $(6, \sqrt{3})$: $2x - 3\sqrt{3}y - 3 = 0$
    At $(6, -\sqrt{3})$: $2x + 3\sqrt{3}y - 3 = 0$

    (b) At $(6, \sqrt{3})$: $2y + 3\sqrt{3}x - 20\sqrt{3} = 0$
    At $(6, -\sqrt{3})$: $2y - 3\sqrt{3}x + 20\sqrt{3} = 0$

39. Volume $= 4\pi/3$

   Surface area $= \pi\left[2\sqrt{7} - 1 + \dfrac{\sqrt{2}}{2}\ln\left(\dfrac{\sqrt{2}+1}{2\sqrt{2}+\sqrt{7}}\right)\right] \approx 11.66$

45. Ellipse

47. Parabola

49. Hyperbola

51. Circle

53. Circle

55. Hyperbola

SECTION 11.4

1. $\dfrac{(y')^2}{2} - \dfrac{(x')^2}{2} = 1$

3. $y' = \dfrac{(x')^2}{6} - \dfrac{x'}{3}$

5. $\dfrac{(x')^2}{1/4} - \dfrac{(y')^2}{1/6} = 1$

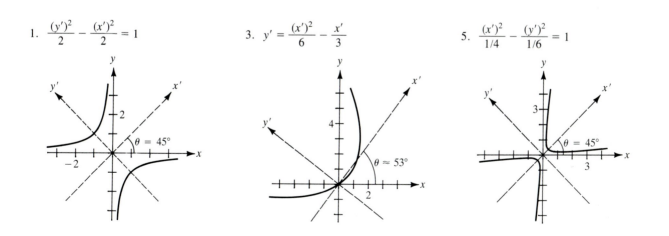

7. $\dfrac{(x' - 3\sqrt{2})^2}{16} - \dfrac{(y' - \sqrt{2})^2}{16} = 1$

9. $\dfrac{(x')^2}{3} + \dfrac{(y')^2}{2} = 1$

11. $x' = -(y')^2$

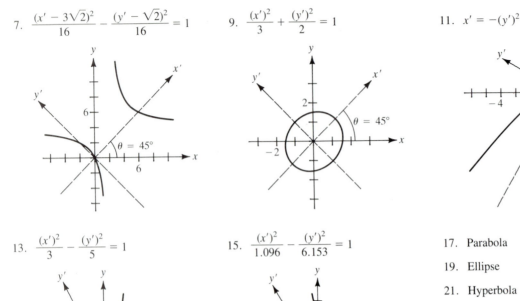

13. $\dfrac{(x')^2}{3} - \dfrac{(y')^2}{5} = 1$

15. $\dfrac{(x')^2}{1.096} - \dfrac{(y')^2}{6.153} = 1$

17. Parabola

19. Ellipse

21. Hyperbola

23. Parabola

REVIEW EXERCISES FOR CHAPTER 11

1. h

2. a

3. e

4. i

5. f

6. b

7. c

8. j

9. g

10. d

11. Circle
Center: (1/2, −3/4)
Radius: 1

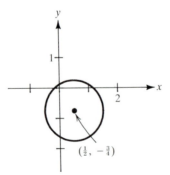

$\left(\tfrac{1}{2}, -\tfrac{3}{4}\right)$

13. Hyperbola
Center: (−4, 3)
Vertices: $(-4 \pm \sqrt{2}, 3)$

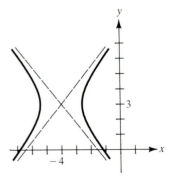

15. Ellipse
Center: $(2, -3)$
Vertices: $\left(2, -3 \pm \dfrac{\sqrt{2}}{2}\right)$

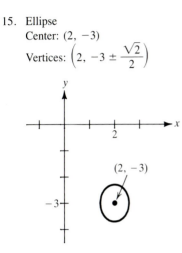

17. Parabola
Vertex: $(3, 0)$

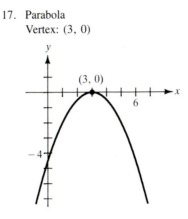

19. Parabola
Vertex: $(-1/\sqrt{8}, 1/\sqrt{8})$

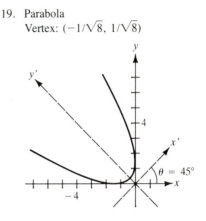

21. Ellipse
Center: $(1, -1/2)$
Vertex: $(1 \pm [3/4], -1/2)$

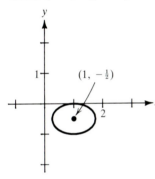

23. $\dfrac{y^2}{1} - \dfrac{x^2}{8} = 1$

25. $\dfrac{(x-2)^2}{25} + \dfrac{y^2}{21} = 1$

27. $x^2 - 2xy + y^2 - 8x - 8y = 0$

29. $\dfrac{x^2}{4} + \dfrac{y^2}{16/3} = 1$

31. $\dfrac{x^2}{4} - \dfrac{y^2}{12} = 1$

33. $4x + 4y - 7 = 0$

35. $a = \sqrt{5}$

41. 15.87

43. 4.212 ft

45. 428.68 ft^2

CHAPTER 12

SECTION 12.1

1. $2x - 3y + 5 = 0$

3. $y = 1 - x^2,\ x \ge 0$

5. $y = (x - 1)^2$

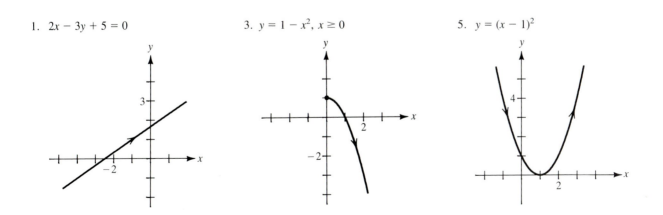

7. $y = \dfrac{1}{2}x^{2/3}$

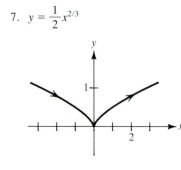

9. $y = \dfrac{x+1}{x}$

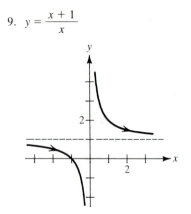

11. $y = \dfrac{|x-4|}{2}$

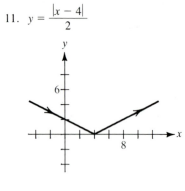

13. $y = \dfrac{1}{x},\ |x| \ge 1$

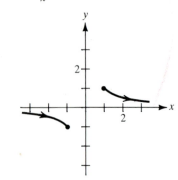

15. $x^2 + y^2 = 9$

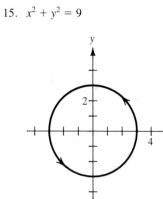

17. $\dfrac{x^2}{16} + \dfrac{y^2}{4} = 1$

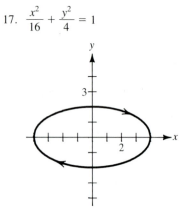

19. $y = 2 - 2x^2$

$-1 \le x \le 1$

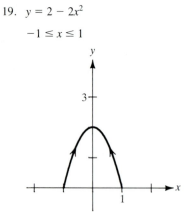

21. $\dfrac{(x-4)^2}{4} + \dfrac{(y+1)^2}{1} = 1$

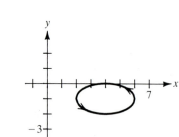

23. $\dfrac{(x-4)^2}{4} + \dfrac{(y+1)^2}{16} = 1$

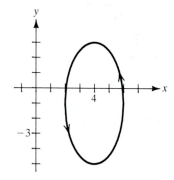

25. $\dfrac{x^2}{16} - \dfrac{y^2}{9} = 1$

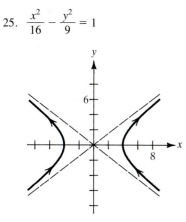

27. $y = \ln x$

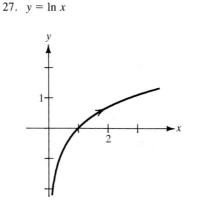

29. $y = \dfrac{1}{x^3},\ x > 0$

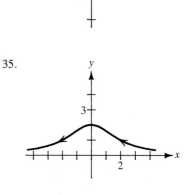

31. Each curve represents a portion of the line given by $y = 2x + 1$.

Domain	**Orientation**
(a) $-\infty < x < \infty$	Up
(b) $-1 \le x \le 1$	Oscillates
(c) $0 < x < \infty$	Down
(d) $0 < x < \infty$	Up

33.

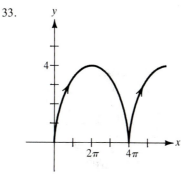

35.

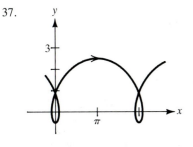

37.

39. $y - y_1 = \dfrac{y_2 - y_1}{x_2 - x_1}(x - x_1)$

41. $(x - h)^2 + (y - k)^2 = r^2$

43. $\dfrac{(x - h)^2}{a^2} - \dfrac{(y - k)^2}{b^2} = 1$

45. $x = 5t,\ y = -2t$
(Solution not unique.)

47. $x = 2 + 4 \cos \theta,\ y = 1 + 4 \sin \theta$
(Solution not unique.)

49. $x = 5 \cos \theta,\ y = 3 \sin \theta$
(Solution not unique.)

51. $x = 4 \sec \theta,\ y = 3 \tan \theta$
(Solution not unique.)

53. $x = t,\ y = t^3$
$x = \tan t,\ y = \tan^3 t$
(Solution not unique.)

55. $x = a\theta - b \sin \theta$
$y = a - b \cos \theta$

SECTION 12.2

1. $\dfrac{dy}{dx} = \dfrac{3}{2},\ \dfrac{d^2y}{dx^2} = 0$

3. $\dfrac{dy}{dx} = 2t + 3,\ \dfrac{d^2y}{dx^2} = 2$

At $t = -1,\ \dfrac{dy}{dx} = 1,\ \dfrac{d^2y}{dx^2} = 2$

5. $\dfrac{dy}{dx} = -\cot \theta,\ \dfrac{d^2y}{dx^2} = -\dfrac{\csc^3 \theta}{2}$

At $\theta = \pi/4,\ \dfrac{dy}{dx} = -1,\ \dfrac{d^2y}{dx^2} = -\sqrt{2}$

7. $\dfrac{dy}{dx} = 2 \csc \theta, \;\; \dfrac{d^2x}{dx^2} = -2 \cot^3 \theta$

    At $t = \pi/6, \; \dfrac{dy}{dx} = 4, \; \dfrac{d^2y}{dx^2} = -6\sqrt{3}$

9. $\dfrac{dy}{dx} = -\tan \theta, \;\; \dfrac{d^2y}{dx^2} = \dfrac{\sec^4 \theta \csc \theta}{3}$

    At $t = \pi/4, \; \dfrac{dy}{dx} = -1, \; \dfrac{d^2y}{dx^2} = \dfrac{4\sqrt{2}}{3}$

11. $2x - y - 5 = 0$

13. $y = 2$

15. $x + 2y - 4 = 0$

17. $(1, 0)$

19. $(0, -2), \; (2, 2)$

21. $(-1, -1), \; (1, 1)$

23. $(4n\pi, 0), \; (2[2n - 1]\pi, 4)$

25. $\sqrt{2}(1 - e^{-\pi/2}) \approx 1.12$

27. $2\sqrt{5} + \ln (2 + \sqrt{5}) \approx 5.916$

29. $\dfrac{1}{12}[\ln (\sqrt{37} + 6) + 6\sqrt{37}] \approx 3.249$

31. $6a$

33. $8a$

35. (a) $32\pi\sqrt{5}$
      (b) $16\pi\sqrt{5}$

37. 32π

39. $12\pi a^2/5$

41. $3\pi/8$

43. $3\pi/2$

45. $(3/4, 8/5)$

47. 36π

49. $2\pi a^2(1 - \cos \theta)$

SECTION 12.3

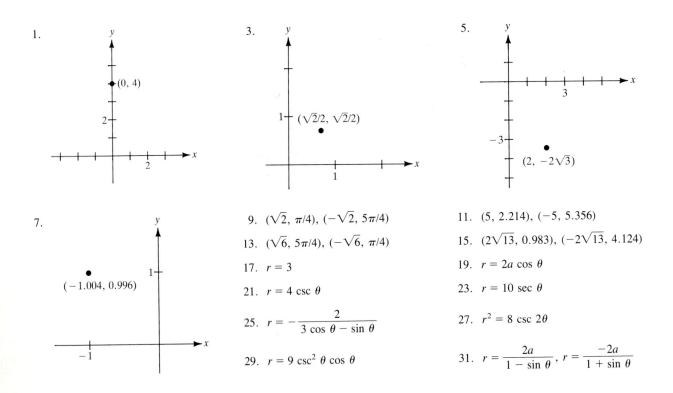

1. $(0, 4)$

3. $(\sqrt{2}/2, \; \sqrt{2}/2)$

5. $(2, -2\sqrt{3})$

7. $(-1.004, 0.996)$

9. $(\sqrt{2}, \pi/4), \; (-\sqrt{2}, 5\pi/4)$

11. $(5, 2.214), \; (-5, 5.356)$

13. $(\sqrt{6}, 5\pi/4), \; (-\sqrt{6}, \pi/4)$

15. $(2\sqrt{13}, 0.983), \; (-2\sqrt{13}, 4.124)$

17. $r = 3$

19. $r = 2a \cos \theta$

21. $r = 4 \csc \theta$

23. $r = 10 \sec \theta$

25. $r = -\dfrac{2}{3 \cos \theta - \sin \theta}$

27. $r^2 = 8 \csc 2\theta$

29. $r = 9 \csc^2 \theta \cos \theta$

31. $r = \dfrac{2a}{1 - \sin \theta}, \; r = \dfrac{-2a}{1 + \sin \theta}$

33. $x^2 + y^2 - 4y = 0$

35. $\sqrt{3}x - 3y = 0$

37. $y = 2$

39. $(x^2 + y^2 + 2y)^2 = x^2 + y^2$

41. $4x^2 - 5y^2 - 36y - 36 = 0$

43. $(x^2 + y^2)^3 = 16y^2$

45. Symmetry: Pole,
Polar axis, $\theta = \pi/2$

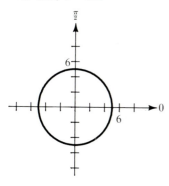

47. Symmetry: $\theta = \pi/2$

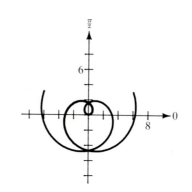

49. Symmetry: $\theta = \pi/2$

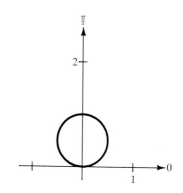

51. Symmetry: Polar axis

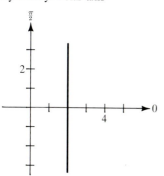

53. Symmetry: $\theta = \pi/2$

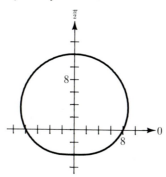

SECTION 12.4

1. $\dfrac{(1 + 2\cos\theta)(1 - \cos\theta)}{\sin\theta\,(2\cos\theta - 1)}$, -1

3. $\dfrac{2\cos\theta\,(3\sin\theta + 1)}{3\cos 2\theta - 2\sin\theta}$, 0

5. $\dfrac{\sin\theta + \theta\cos\theta}{\cos\theta - \theta\sin\theta}$, π

7. $\dfrac{2\sin\theta\cos\theta}{1 - 2\sin^2\theta}$, $-\sqrt{3}$

9. Graph of $r = 2\sec\theta$ is a
vertical line. dy/dx is undefined.

11. Horizontal:
$(2, \pi/2)$, $(1/2, 7\pi/6)$, $(1/2, 11\pi/6)$
Vertical:
$(3/2, \pi/6)$, $(3/2, 5\pi/6)$

13. $(5, \pi/2)$, $(1, 3\pi/2)$

15.

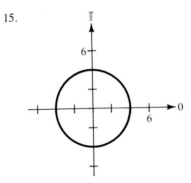

17. $\theta = 0$

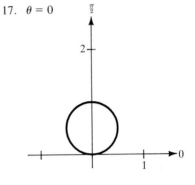

19.

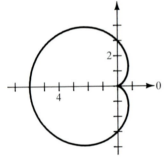

21.

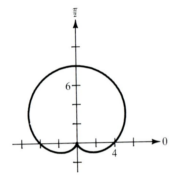

23. $\theta = \arcsin(-2/3)$,
     $\theta = \pi + \arcsin(2/3)$

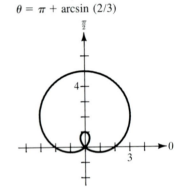

25. $\theta = \arccos\left(\dfrac{3}{4}\right)$

     $\theta = \dfrac{3\pi}{2} + \arccos\left(\dfrac{3}{4}\right)$

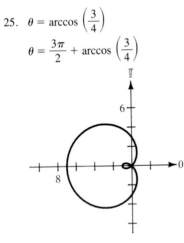

27.

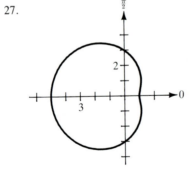

29.

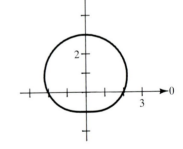

31. $\theta = \pi/6$, $\pi/2$, $5\pi/6$

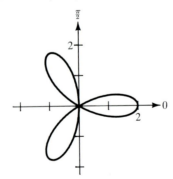

33. $\theta = 0$, $\pi/2$

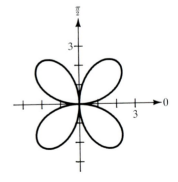

35.

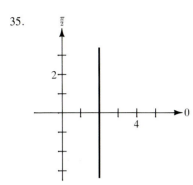

37.

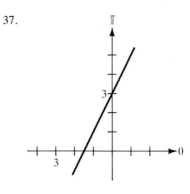

39. $\theta = \pi/4,\ 3\pi/4$

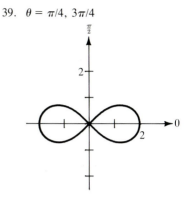

41. $\theta = 0,\ \pi/2$

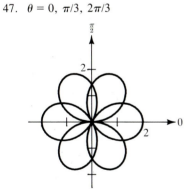

43. $\theta = 0,\ \pi/2$

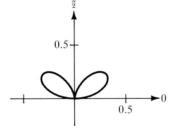

45. $\theta = 0$

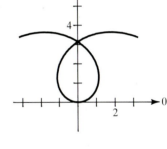

47. $\theta = 0,\ \pi/3,\ 2\pi/3$

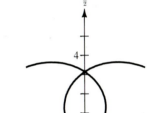

49.

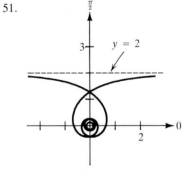

51.

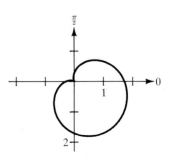

55. (a) $r = 2 - \sin\,[\theta - (\pi/4)]$
     $= 2 - \sqrt{2}(\sin\theta - \cos\theta)/2$
   (b) $r = 2 + \cos\theta$
   (c) $r = 2 + \sin\theta$
   (d) $r = 2 - \cos\theta$

57. (a)

(b)

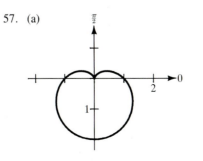

SECTION 12.5

1. Parabola

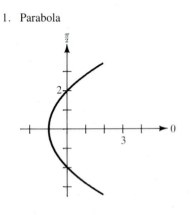

3. Parabola

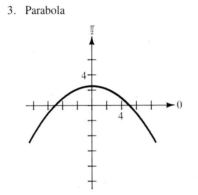

5. Ellipse

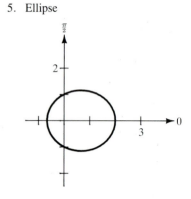

7. Ellipse

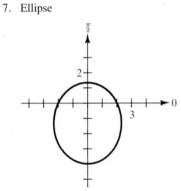

9. Parabola

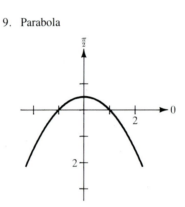

11. Hyperbola

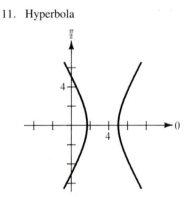

13. Hyperbola

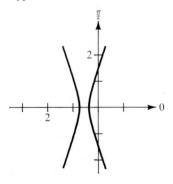

15. Hyperbola

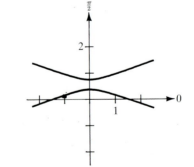

17. $r = \dfrac{1}{2 + \sin \theta}$

19. $r = -\dfrac{1}{1 - \cos \theta}$

21. $r = \dfrac{2}{1 + 2 \cos \theta}$

23. $r = \dfrac{2}{1 - \sin \theta}$

25. $r = -\dfrac{6}{2 - \sin \theta}$

27. $r = -\dfrac{6}{1 - 2 \sin \theta}$

31. $r^2 = \dfrac{9}{1 - (16/25) \cos^2 \theta}$

33. $r^2 = \dfrac{-16}{1 - (25/9) \cos^2 \theta}$

35.

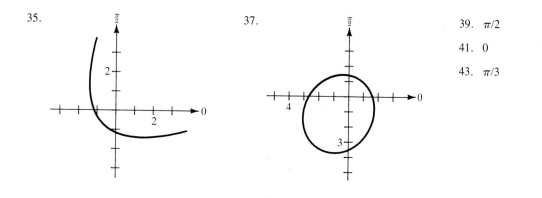

37.

39. $\pi/2$

41. 0

43. $\pi/3$

SECTION 12.6

1. $(1, \pi/2), (1, 3\pi/2), (0, 0)$

3. $\left(\dfrac{2 - \sqrt{2}}{2}, \dfrac{3\pi}{4}\right), \left(\dfrac{2 + \sqrt{2}}{2}, \dfrac{7\pi}{4}\right), (0, 0)$

5. $(3/2, \pi/6), (3/2, 5\pi/6), (0, 0)$

7. $(2, 4), (-2, -4)$

9. $(2, \pi/12), (2, 5\pi/12), (2, 7\pi/12), (2, 11\pi/12),$
   $(2, 13\pi/12), (2, 17\pi/12), (2, 19\pi/12), (2, 23\pi/12)$

11. $(0.581, \pm0.535), (2.581, \pm1.376)$

13. $\pi/3$

15. $\pi/6$

17. $3\pi/2$

19. $2\pi\sqrt{3}$

21. $(2\pi - 3\sqrt{3})/2$

23. $\pi + 3\sqrt{3}$

25. $\dfrac{4}{3}(4\pi - 3\sqrt{3})$

27. $11\pi - 24$

29. $\dfrac{2}{3}(4\pi - 3\sqrt{3})$

31. $5\pi a^2/4$

33. $\dfrac{a^2}{2}(\pi - 2)$

35. $2\pi a$

37. 8

39. $\dfrac{\pi}{4}\sqrt{\pi^2 + 4} + \ln\left(\dfrac{\pi + \sqrt{\pi^2 + 4}}{2}\right)$

41. $\ln\left(\dfrac{\sqrt{4\pi^2 + 1} + 2\pi}{\sqrt{\pi^2 + 1} + \pi}\right) + \dfrac{2\sqrt{\pi^2 + 1} - \sqrt{4\pi^2 + 1}}{2\pi}$

43. 4π

45. $\dfrac{2\pi\sqrt{1 + a^2}}{1 + 4a^2}(e^{\pi a} - 2a)$

47. 21.88

REVIEW EXERCISES FOR CHAPTER 12

1. (a) $\dfrac{dy}{dx} = -\dfrac{3}{4}$

   Horizontal tangents: None

   (b) $y = \dfrac{-3x + 11}{4}$

   (c)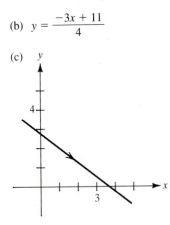

3. (a) $\dfrac{dy}{dx} = -\dfrac{5}{2}\cot\theta$

   Horizontal tangents: $(3, 7)$, $(3, -3)$

   (b) $\dfrac{(x-3)^2}{4} + \dfrac{(y-2)^2}{25} = 1$

   (c)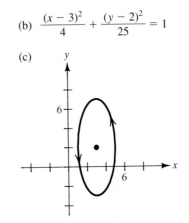

5. (a) $\dfrac{dy}{dx} = -2t^2$

   Horizontal tangents: None

   (b) $y = 3 + \dfrac{2}{x}$

   (c)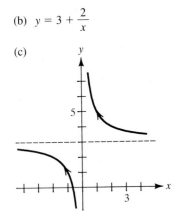

7. (a) $\dfrac{dy}{dx} = -(2t^2 + 2t + 1)$

   Horizontal tangents: None

   (b) $y = \dfrac{1 - x^2}{2x}$

   (c)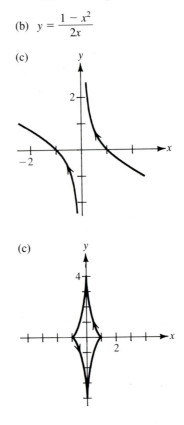

9. (a) $\dfrac{dy}{dx} = -4\tan\theta$

   Horizontal tangents: $(\pm 1, 0)$

   (b) $x^{2/3} + (y/4)^{2/3} = 1$

   (c)

11. $x = -3 + 4 \cos \theta$
     $y = 4 + 3 \sin \theta$

15. $\pi^2 r/2$

17. Cardioid

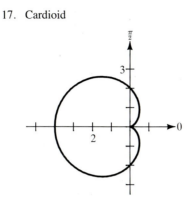

19. Limaçon

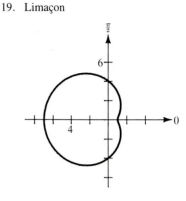

21. Rose curve

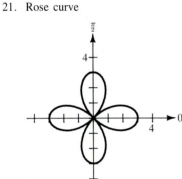

23. Circle

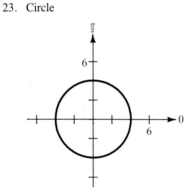

25. Line

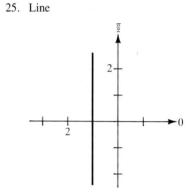

27. Rose curve

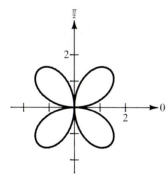

29. Parabola

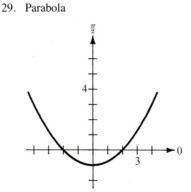

31. Strophoid

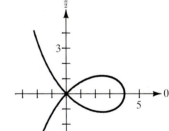

33. $x^2 + y^2 - 3x = 0$

35. $(x^2 + y^2 + 2x)^2 = 4(x^2 + y^2)$

37. $y^2 = x^2 \left(\dfrac{4 - x}{4 + x} \right)$

39. $r = a \cos^2 \theta \sin \theta$

41. $r^2 = a^2 \theta^2$

43. $r = \dfrac{4}{1 - \cos \theta}$

45. $r = \dfrac{7}{3 + 4 \cos \theta}$

47. $r = \dfrac{12}{4 \cos \theta + 3 \sin \theta}$

49. (a) $\pm\pi/3$
    (b) Vertical: $(-1, 0)$, $(3, \pi)$, $(1/2, \pm1.318)$
         Horizontal: $(-0.686, \pm0.568)$, $(2.186, \pm2.206)$
    (c)

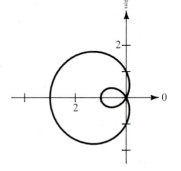

53. $9\pi/2$

55. $\pi/32$

57. a^2

59. $\dfrac{8\pi - 6\sqrt{3}}{3}$

61. $8a$

63. $\arctan\left(\dfrac{2\sqrt{3}}{3}\right)$

CHAPTER 13

SECTION 13.1

1. $\langle 4, 2 \rangle$

3. $\langle -7, 0 \rangle$

5. $\langle 4, 3 \rangle$

7. $\langle -4, -3 \rangle$

9. $\langle 0, 4 \rangle$

11. $\langle -1, 5/3 \rangle$

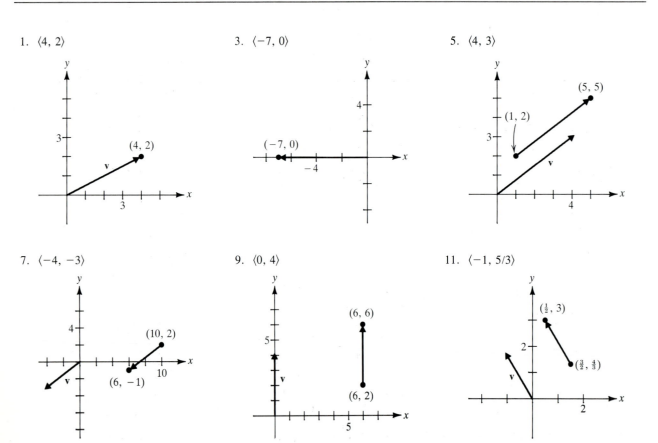

13. (a) $\langle 4, 6 \rangle$ (b) $\langle -6, -9 \rangle$ (c) $\langle 7, 21/2 \rangle$

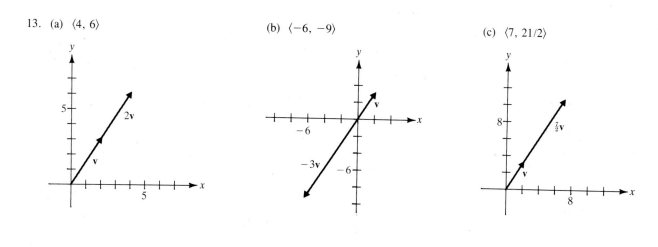

(d) $\langle 4/3, 2 \rangle$ 15. $\langle 3, -3/2 \rangle$ 17. $\langle 4, 3 \rangle$

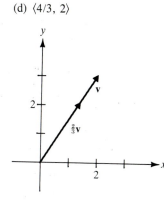   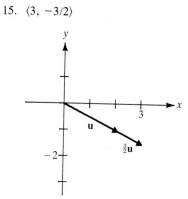   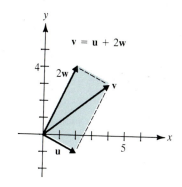

19. $\langle 7/2, -1/2 \rangle$

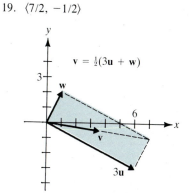

21. $a = 1, b = 1$

23. $a = 1, b = 2$

25. $a = 2/3, b = 1/3$

27. $(3, 5)$

29. 5

31. $\sqrt{61}$

33. 4

35. (a) $\sqrt{5}/2$ (b) $\sqrt{13}$
    (c) $\sqrt{85}/2$ (d) 1
    (e) 1   (f) 1

39. $\langle 2\sqrt{2}, 2\sqrt{2} \rangle$

41. $\langle 1, \sqrt{3} \rangle$

43. (a) $\pm \dfrac{1}{\sqrt{10}} \langle 1, 3 \rangle$

    (b) $\pm \dfrac{1}{\sqrt{10}} \langle 3, -1 \rangle$

45. (a) $\pm \dfrac{1}{5} \langle -4, 3 \rangle$

    (b) $\pm \dfrac{1}{5} \langle 3, 4 \rangle$

47. $\langle 3, 0 \rangle$

49. $\langle -\sqrt{3}, 1 \rangle$

51. $\left(\dfrac{3 + \sqrt{2}}{\sqrt{2}} \right) \mathbf{i} + \left(\dfrac{3}{\sqrt{2}} \right) \mathbf{j}$

53. $(2 \cos 4 + \cos 2) \mathbf{i} + (2 \sin 4 + \sin 2) \mathbf{j}$

55. $\dfrac{-\sqrt{2}}{2}\,\mathbf{i} + \dfrac{\sqrt{2}}{2}\,\mathbf{j}$

57. $F_H = -75\sqrt{3}$ lb

       $F_V = -75$ lb

59. 569 mi/hr

    38° N of W

61. $(-4, -1)$, $(6, 5)$, $(10, 3)$

63. $(3, 3)$, $(5, 4)$

SECTION 13.2

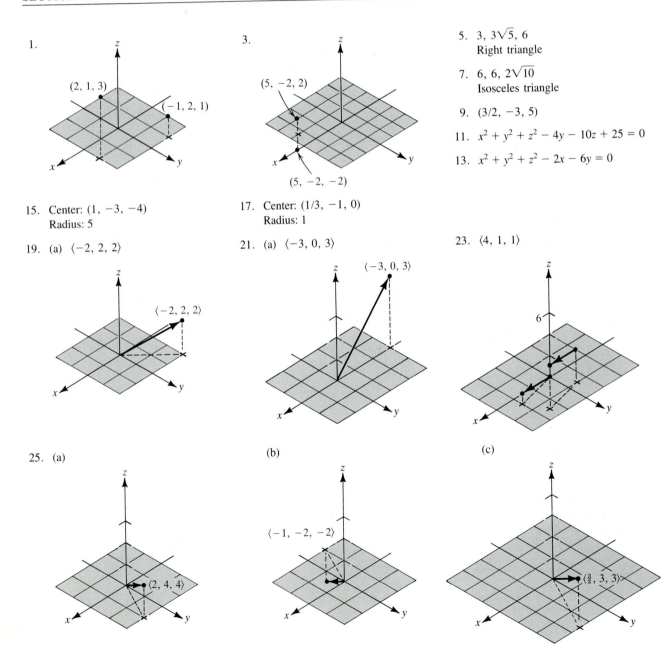

1.

3.

5. $3, 3\sqrt{5}, 6$
   Right triangle

7. $6, 6, 2\sqrt{10}$
   Isosceles triangle

9. $(3/2, -3, 5)$

11. $x^2 + y^2 + z^2 - 4y - 10z + 25 = 0$

13. $x^2 + y^2 + z^2 - 2x - 6y = 0$

15. Center: $(1, -3, -4)$
    Radius: 5

17. Center: $(1/3, -1, 0)$
    Radius: 1

19. (a) $\langle -2, 2, 2 \rangle$

21. (a) $\langle -3, 0, 3 \rangle$

23. $\langle 4, 1, 1 \rangle$

25. (a)

(b)

(c)

25. (d)

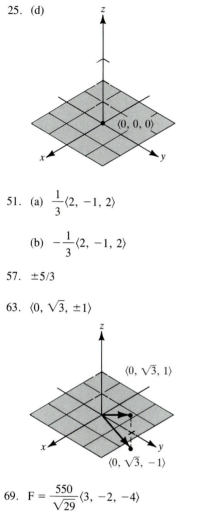

$\langle 0, 0, 0 \rangle$

27. $\langle -1, 0, 4 \rangle$

29. $\langle 6, 12, 6 \rangle$

31. $\langle 7/2, 3, 5/2 \rangle$

33. a and b

35. a

37. Collinear

39. Not collinear

43. $(3, 1, 8)$

45. 0

47. $\sqrt{34}$

49. $\sqrt{14}$

51. (a) $\dfrac{1}{3}\langle 2, -1, 2 \rangle$

   (b) $-\dfrac{1}{3}\langle 2, -1, 2 \rangle$

53. (a) $\dfrac{1}{\sqrt{38}}\langle 3, 2, -5 \rangle$

   (b) $-\dfrac{1}{\sqrt{38}}\langle 3, 2, -5 \rangle$

55. $\pm\sqrt{14}/14$

57. $\pm 5/3$

59. $(2, -1, 2)$

61. $(13/3, 6, 3)$

63. $\langle 0, \sqrt{3}, \pm 1 \rangle$

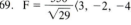

$\langle 0, \sqrt{3}, 1 \rangle$

$\langle 0, \sqrt{3}, -1 \rangle$

65. $\dfrac{\sqrt{3}}{3}\langle 1, 1, 1 \rangle$

67. (c) $a = b = 1$

69. $F = \dfrac{550}{\sqrt{29}}\langle 3, -2, -4 \rangle$

SECTION 13.3

1. (a) -6
   (b) 25
   (c) 25
   (d) $\langle -12, 18 \rangle$
   (e) -12

3. (a) 2
   (b) 29
   (c) 29
   (d) $\langle 0, 12, 10 \rangle$
   (e) 4

5. (a) 1
   (b) 6
   (c) 6
   (d) $\mathbf{i} - \mathbf{k}$
   (e) 2

7. $\pi/2$

9. $\arccos\left(-\dfrac{1}{5\sqrt{2}}\right) \approx 98.1°$

11. $\arccos(\sqrt{2}/3) \approx 61.9°$

13. $\arccos(-8\sqrt{13}/65) \approx 116.3°$

15. Neither

17. Orthogonal

19. Neither

21. Orthogonal

23. $\cos\alpha = 1/3$
    $\cos\beta = 2/3$
    $\cos\gamma = 2/3$

25. $\cos \alpha = 0$
     $\cos \beta = 3/\sqrt{13}$
     $\cos \gamma = -2/\sqrt{13}$

27. (a) $\langle 5/2, 1/2 \rangle$
     (b) $\langle -1/2, 5/2 \rangle$

29. (a) $\langle 0, 33/25, 44/25 \rangle$
     (b) $\langle 2, -8/25, 6/25 \rangle$

31. (a) $\langle 2/3, 1/3, -1/3 \rangle$
     (b) $\langle 1/3, 2/3, 4/3 \rangle$

33. (a) $\langle 5, 0, 0 \rangle$
     (b) $\langle 0, -4, 3 \rangle$

35. (a) 8282.2 lb
     (b) 30,909.6 lb

37. 425 ft · lb

39. 72

41. \$17,139.05

43. (a) $\theta = \pi/2$
     (b) $0 < \theta < \pi/2$
     (c) $\pi/2 < \theta < \pi$

SECTION 13.4

1. $-\mathbf{k}$

3. $\mathbf{i}$

5. $-\mathbf{j}$

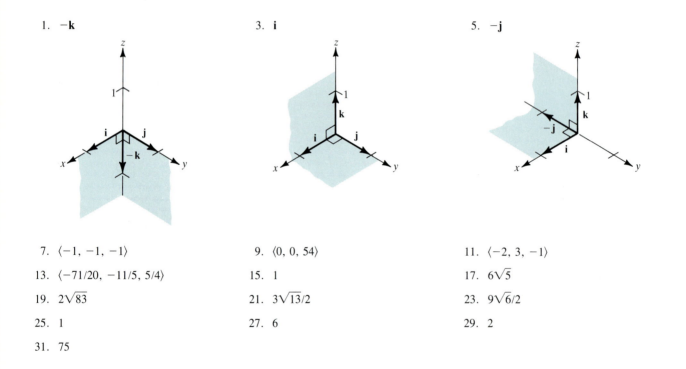

7. $\langle -1, -1, -1 \rangle$

9. $\langle 0, 0, 54 \rangle$

11. $\langle -2, 3, -1 \rangle$

13. $\langle -71/20, -11/5, 5/4 \rangle$

15. 1

17. $6\sqrt{5}$

19. $2\sqrt{83}$

21. $3\sqrt{13}/2$

23. $9\sqrt{6}/2$

25. 1

27. 6

29. 2

31. 75

SECTION 13.5

Parametric Equations	**Symmetric Equations**	**Direction Numbers**
1. $\begin{aligned} x &= t \\ y &= 2t \\ z &= 3t \end{aligned}$	$x = \dfrac{y}{2} = \dfrac{z}{3}$	1, 2, 3
3. $\begin{aligned} x &= -2 + 2t \\ y &= 4t \\ z &= 3 - 2t \end{aligned}$	$\dfrac{x+2}{2} = \dfrac{y}{4} = \dfrac{z-3}{-2}$	2, 4, -2

Parametric Equations	**Symmetric Equations**	**Direction Numbers**

5. $x = 5 + 17t$
$\quad y = -3 - 11t$
$\quad z = -2 - 9t$
$\dfrac{x-5}{17} = \dfrac{y+3}{-11} = \dfrac{z+2}{-9}$ $17, -11, -9$

7. $x = 1 + 3t$
$\quad y = -2t$
$\quad z = 1 + t$
$\dfrac{x-1}{3} = \dfrac{y}{-2} = \dfrac{z-1}{1}$ $3, -2, 1$

9. $x = 2$
$\quad y = 3$
$\quad z = 4 + t$
None $0, 0, 1$

11. a, b, d

13. $(2, 3, 1)$
$\cos \theta = 7\sqrt{17}/51$

15. Nonintersecting

17. $x - 2 = 0$

19. $2x + 3y - z = 10$

21. $3x + 9y - 7z = 0$

23. $4x - 3y + 4z = 10$

25. $z = 3$

27. $x + y + z = 5$

29. $7x + y - 11z = 5$

31. $y - z = -1$

33. Orthogonal

35. $83.5°$

37. Parallel

39. $76.3°$

41.

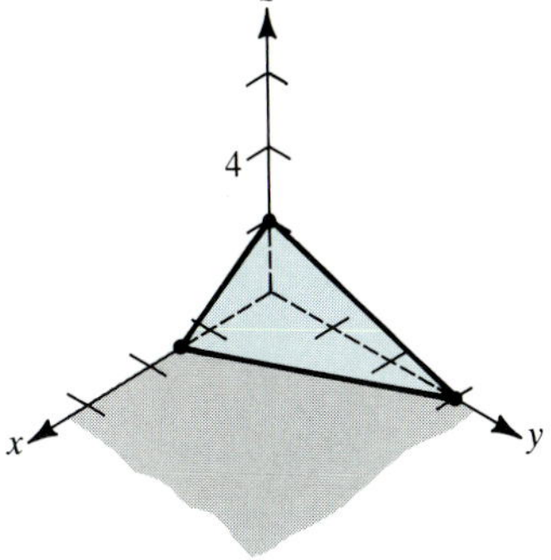

43.

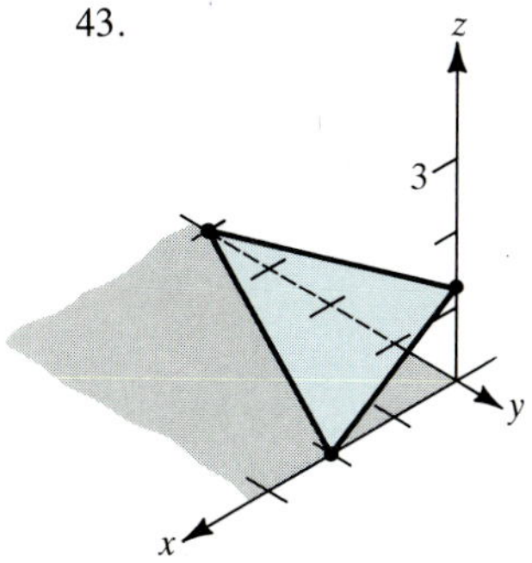

45.

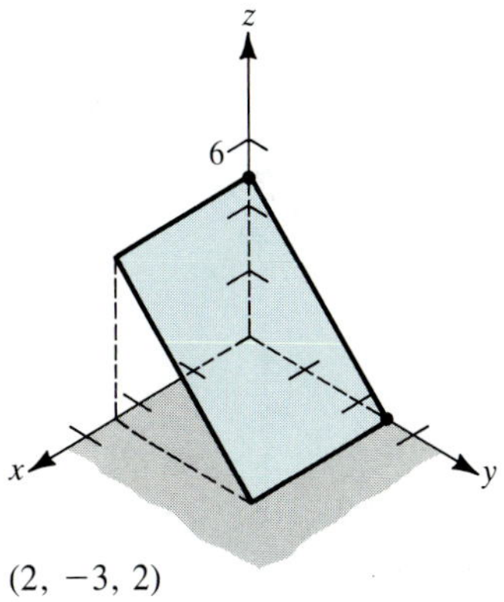

47.

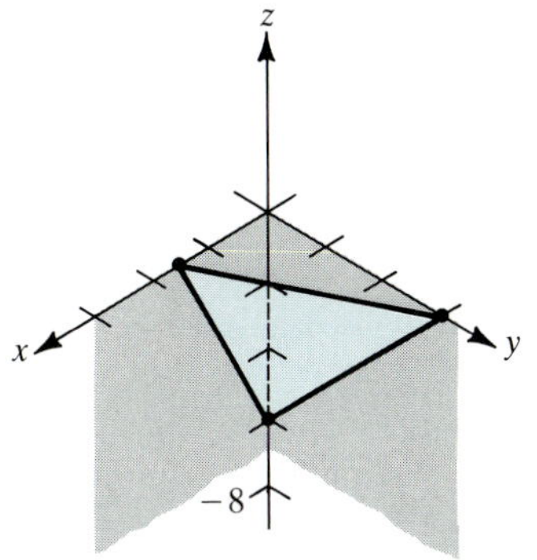

49. $x = 2$
$\quad y = 1 + t$
$\quad z = 1 + 2t$

51. $(2, -3, 2)$

53. Nonintersecting

55. 0

57. $6\sqrt{14}/7$

59. $2\sqrt{26}/13$

61. $10\sqrt{26}/13$

SECTION 13.6

1. c

2. e

3. f

4. g

5. d

6. b

7. a

8. h

9. Plane

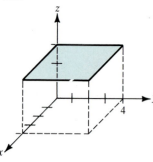

11. Right circular cylinder

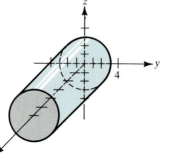

13. Parabolic cylinder

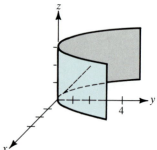

15. Elliptic cylinder

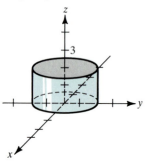

17. Hyperbolic cylinder

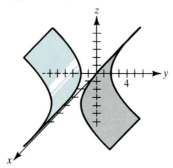

19. Ellipsoid

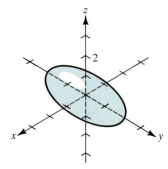

21. Hyperboloid of one sheet

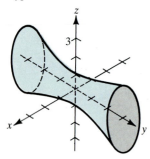

23. Elliptic paraboloid

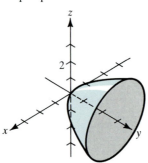

25. Hyperbolic paraboloid

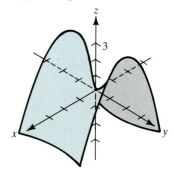

27. Hyperboloid of two sheets

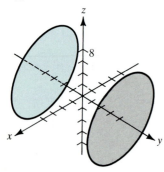

29. Elliptic cone

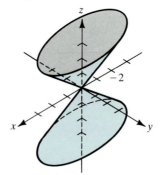

31. Hyperbolic paraboloid

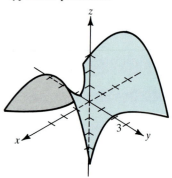

33. Ellipsoid

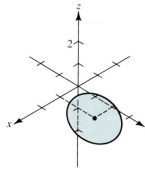

35.

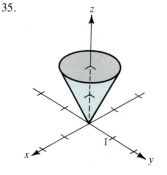

37.

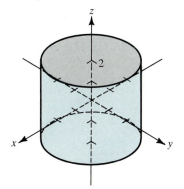

39.

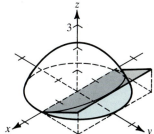

41. $x^2 + z^2 = 4y$

43. $4x^2 + 4y^2 = z^2$

45. $y^2 + z^2 = \dfrac{4}{x^2}$

47. $y = \sqrt{2z}$

49. (a) Major axis: $4\sqrt{2}$
    Minor axis: 4
    Foci: $(0, \pm2, 2)$

(b) Major axis: $8\sqrt{2}$
    Minor axis: 8
    Foci: $(0, \pm4, 8)$

SECTION 13.7

1. $(5, \pi/2, 1)$

3. $(2, \pi/3, 4)$

5. $(2\sqrt{2}, -\pi/4, -4)$

7. $(5, 0, 2)$

9. $(1, \sqrt{3}, 2)$

11. $(-2\sqrt{3}, -2, 3)$

13. $(4, 0, \pi/2)$

15. $(4\sqrt{2}, 2\pi/3, \pi/4)$

17. $(4, \pi/6, \pi/6)$

19. $(\sqrt{6}, \sqrt{2}, 2\sqrt{2})$

21. $(0, 0, 12)$

23. $(5/2, 5/2, -5\sqrt{2}/2)$

25. $(4, \pi/4, \pi/2)$

27. $(2\sqrt{13}, -\pi/6, \arccos[3/\sqrt{13}])$

29. $(13, \pi, \arccos[5/13])$

31. $(10, \pi/6, 0)$

33. $(6\sin 6, -\pi/6, 6\cos 6)$

35. $(4, 7\pi/6, 4\sqrt{3})$

37. d

38. e

39. c

40. a

41. f

42. b

43. $x^2 + y^2 = 4$

45. $x - \sqrt{3}y = 0$

47. $x^2 + y^2 - 2y = 0$

49. $x^2 + y^2 + z^2 = 4$

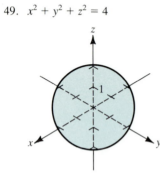

51. $x^2 + y^2 + z^2 = 4$

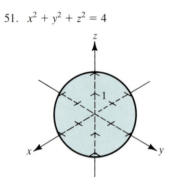

53. $3x^2 + 3y^2 - z^2 = 0$

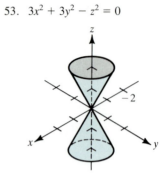

55. $x^2 + y^2 + (z - 2)^2 = 4$

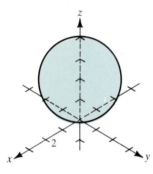

57. $x^2 + y^2 = 1$

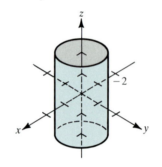

59. (a) $r^2 + z^2 = 16$
    (b) $\rho = 4$

61. (a) $r^2 + (z - 1)^2 = 1$
    (b) $\rho = 2 \cos \phi$

63. (a) $r = 4 \sin \theta$
    (b) $\rho = \dfrac{4 \sin \theta}{\sin \phi} = 4 \sin \theta \csc \phi$

65. (a) $r^2 = \dfrac{9}{\cos^2 \theta - \sin^2 \theta}$
    (b) $\rho^2 = \dfrac{9 \csc^2 \phi}{\cos^2 \theta - \sin^2 \theta}$

67.

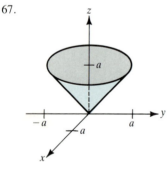

69.

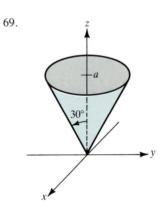

REVIEW EXERCISES FOR CHAPTER 13

1. (a) $\mathbf{u} = 3\mathbf{i} - \mathbf{j}, \mathbf{v} = 4\mathbf{i} + 2\mathbf{j}$
   (b) $2\sqrt{5}$
   (c) 10
   (d) $10\mathbf{i}$
   (e) $2\mathbf{i} + \mathbf{j}$
   (f) $\mathbf{i} - 2\mathbf{j}$

3. (a) $\mathbf{u} = -\mathbf{i} + 4\mathbf{j}, \mathbf{v} = -3\mathbf{i} + 6\mathbf{k}$
   (b) 3
   (c) $24\mathbf{i} + 6\mathbf{j} + 12\mathbf{k}$
   (d) $4x + y + 2z = 20$
   (e) $x = 4 - t, y = 4 + 4t, z = 0$

5. $\theta = \arccos\left(\dfrac{\sqrt{2} + \sqrt{6}}{4}\right)$

7. π

9. $-2\sqrt{2}\mathbf{i} + 2\sqrt{2}\mathbf{j}$

11. $\dfrac{3}{\sqrt{26}}\mathbf{i} - \dfrac{9}{\sqrt{26}}\mathbf{j} + \dfrac{12}{\sqrt{26}}\mathbf{k}$

13. $\sqrt{14}$

19. 4

21. (a) $x = 1$, $y = 2 + t$, $z = 3$
    (b) None

23. (a) $x = t$, $y = -1 + t$, $z = 1$
    (b) $\dfrac{x - 0}{1} = \dfrac{y + 1}{1}$, $z = 1$

25. $x + y + z = 6$

27. $x + 2y = 1$

29. 8/7

31. $\sqrt{35}/7$

33. $\sqrt{6}/6$

35. $\dfrac{\sqrt{3}}{3}$

37.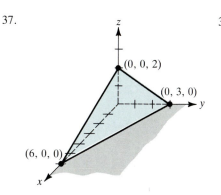

39.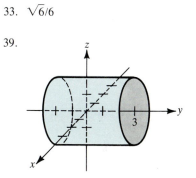

41.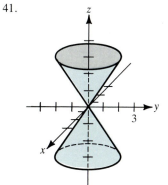

43.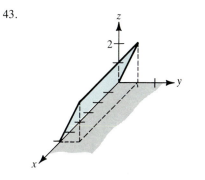

45.

47.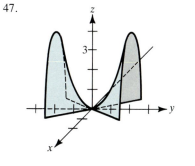

49. (a) $(4, 3\pi/4, 2)$
    (b) $(2\sqrt{5}, 3\pi/4, \arccos(\sqrt{5}/5))$

53. $x^2 - y^2 + z^2 = 1$

51. (a) $r^2 \cos 2\theta = 2z$
    (b) $\rho = 2 \sec 2\theta \cos \phi \csc^2 \phi$

55. $x^2 + y^2 = 1$

CHAPTER 14

SECTION 14.1

1.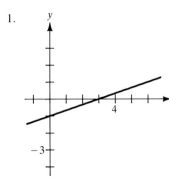

3.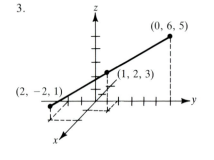

5.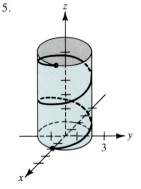

7.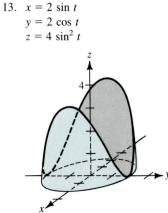

9.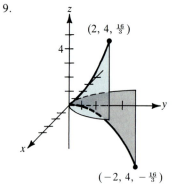

11. $x = t, y = -t, z = 2t^2$

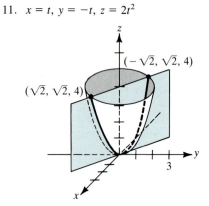

13. $x = 2 \sin t$
    $y = 2 \cos t$
    $z = 4 \sin^2 t$

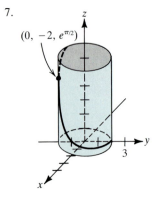

15. $x = 1 + \sin t$
    $y = \pm\sqrt{2} \cos t$
    $z = 1 - \sin t$

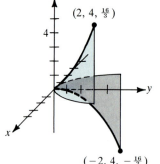

17. $x = t$
    $y = t$
    $z = \sqrt{4 - t^2}$

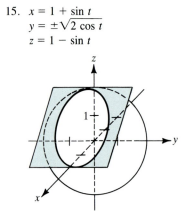

19. $2\mathbf{i} + 2\mathbf{j} + \dfrac{1}{2}\mathbf{k}$

21. $\mathbf{0}$

23. Limit does not exist

25. Discontinuous at $t = 0$

27. Continuous on $[-1, 1]$

29. Discontinuous at $t = \dfrac{\pi}{2} \pm n\pi$

SECTION 14.2

1. $-3a \cos^2 t \sin t \, \mathbf{i} + 3a \sin^2 t \cos t \, \mathbf{j}$

3. $-e^{-t}\mathbf{i}$

5. $(\sin t + t \cos t) \, \mathbf{i} + (\cos t - t \sin t) \, \mathbf{j} + \mathbf{k}$

7. (a) $\mathbf{i} + 3\mathbf{j} + 2t\mathbf{k}$
   (b) $2\mathbf{k}$
   (c) $8t + 9t^2 + 5t^4$
   (d) $-\mathbf{i} + (9 - 2t) \, \mathbf{j} + (6t - 3t^2) \, \mathbf{k}$
   (e) $8t^3\mathbf{i} + (12t^2 - 4t^3) \, \mathbf{j} + (3t^2 - 24t) \, \mathbf{k}$
   (f) $\dfrac{10 + 2t^2}{\sqrt{10 + t^2}}$

9. $t^2\mathbf{i} + t\mathbf{j} + t\mathbf{k} + \mathbf{C}$

11. $e^t\mathbf{i} - \cos t \, \mathbf{j} + \sin t \, \mathbf{k} + \mathbf{C}$

13. $\tan t \, \mathbf{i} + \arctan t \, \mathbf{j} + \mathbf{C}$

15. $2e^{2t}\mathbf{i} + 3(e^t - 1) \, \mathbf{j}$

17. $600\sqrt{3}t\mathbf{i} + (-16t^2 + 600t) \, \mathbf{j}$

19. $\left(\dfrac{2 - e^{-t^2}}{2}\right) \mathbf{i} + (e^{-t} - 2) \, \mathbf{j} + (t + 1) \, \mathbf{k}$

21. $4\mathbf{i} + \dfrac{1}{2}\mathbf{j} - \mathbf{k}$

23. $a\mathbf{i} + a\mathbf{j} + \dfrac{\pi}{2}\mathbf{k}$

SECTION 14.3

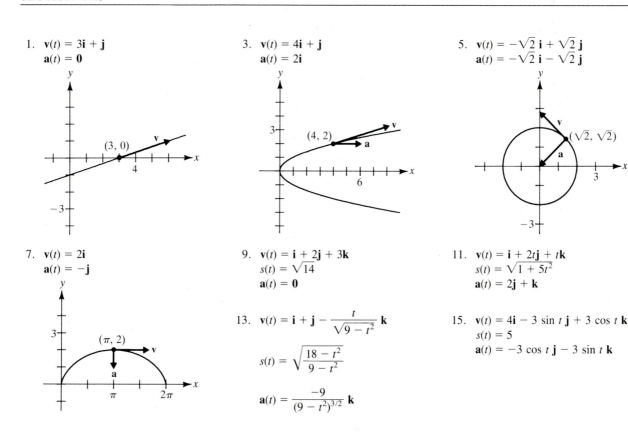

1. $\mathbf{v}(t) = 3\mathbf{i} + \mathbf{j}$
   $\mathbf{a}(t) = \mathbf{0}$

3. $\mathbf{v}(t) = 4\mathbf{i} + \mathbf{j}$
   $\mathbf{a}(t) = 2\mathbf{i}$

5. $\mathbf{v}(t) = -\sqrt{2}\,\mathbf{i} + \sqrt{2}\,\mathbf{j}$
   $\mathbf{a}(t) = -\sqrt{2}\,\mathbf{i} - \sqrt{2}\,\mathbf{j}$

7. $\mathbf{v}(t) = 2\mathbf{i}$
   $\mathbf{a}(t) = -\mathbf{j}$

9. $\mathbf{v}(t) = \mathbf{i} + 2\mathbf{j} + 3\mathbf{k}$
   $s(t) = \sqrt{14}$
   $\mathbf{a}(t) = \mathbf{0}$

11. $\mathbf{v}(t) = \mathbf{i} + 2t\mathbf{j} + t\mathbf{k}$
    $s(t) = \sqrt{1 + 5t^2}$
    $\mathbf{a}(t) = 2\mathbf{j} + \mathbf{k}$

13. $\mathbf{v}(t) = \mathbf{i} + \mathbf{j} - \dfrac{t}{\sqrt{9 - t^2}}\,\mathbf{k}$

    $s(t) = \sqrt{\dfrac{18 - t^2}{9 - t^2}}$

    $\mathbf{a}(t) = \dfrac{-9}{(9 - t^2)^{3/2}}\,\mathbf{k}$

15. $\mathbf{v}(t) = 4\mathbf{i} - 3 \sin t \, \mathbf{j} + 3 \cos t \, \mathbf{k}$
    $s(t) = 5$
    $\mathbf{a}(t) = -3 \cos t \, \mathbf{j} - 3 \sin t \, \mathbf{k}$

17. $\mathbf{v}(t) = t(\mathbf{i} + \mathbf{j} + \mathbf{k})$

   $\mathbf{r}(t) = \dfrac{t^2}{2}(\mathbf{i} + \mathbf{j} + \mathbf{k})$

   $\mathbf{r}(2) = 2\mathbf{i} + 2\mathbf{j} + 2\mathbf{k}$

19. $\mathbf{v}(t) = \left(\dfrac{t^2}{2} + \dfrac{9}{2}\right)\mathbf{j} + \left(\dfrac{t^2}{2} - \dfrac{1}{2}\right)\mathbf{k}$

   $\mathbf{r}(t) = \left(\dfrac{t^3}{6} + \dfrac{9}{2}t - \dfrac{14}{3}\right)\mathbf{j} + \left(\dfrac{t^3}{6} - \dfrac{1}{2}t + \dfrac{1}{3}\right)\mathbf{k}$

   $\mathbf{r}(2) = \dfrac{17}{3}\mathbf{j} + \dfrac{2}{3}\mathbf{k}$

21. $v_0 = 40\sqrt{6}$ ft/sec, 78 ft

23. $v_0 = 4\sqrt{170}$ ft/sec, $\theta \approx 76°$

25. (a) $\theta = \pi/4$
     (b) $\theta = \pi/2$

27. 96.7 ft/sec

29. $\mathbf{r}(t) = 40\sqrt{2}t\,\mathbf{i} + (40\sqrt{2}t - 16t^2)\,\mathbf{j}$
     $s(t) = 8\sqrt{58}$ ft/sec
     Direction: $8\sqrt{2}(5\mathbf{i} + 2\mathbf{j})$

31. (a) When $\omega t = 0, 2\pi, 4\pi, \ldots$
     (b) When $\omega t = \pi, 3\pi, \ldots$

33. $\mathbf{v}(t) = -b\omega \sin \omega t\,\mathbf{i} + b\omega \cos \omega t\,\mathbf{j}$
     $\mathbf{v}(t) \cdot \mathbf{r}(t) = 0$

35. $\mathbf{a}(t) = -b\omega^2[\cos \omega t\,\mathbf{i} + \sin \omega t\,\mathbf{j}] = -\omega^2\mathbf{r}(t)$

37. $8\sqrt{10}$ ft/sec

39. 11.4°

SECTION 14.4

1. $x = t$
    $y = 0$
    $z = t$

3. $x = 2$
    $y = 2t$
    $z = t$

5. $x = 3 + t$
    $y = 9 + 6t$
    $z = 18 + 18t$

7. $x = \sqrt{2} - \sqrt{2}t$
    $y = \sqrt{2} + \sqrt{2}t$
    $z = 4$

9. $\mathbf{T} = \mathbf{i}$
    $\mathbf{N} = \mathbf{0}$
    $\mathbf{a} \cdot \mathbf{T} = 0$
    $\mathbf{a} \cdot \mathbf{N} = 0$

11. $\mathbf{T} = \mathbf{i}$
     $\mathbf{N} = \mathbf{0}$
     $\mathbf{a} \cdot \mathbf{T} = 8$
     $\mathbf{a} \cdot \mathbf{N} = 0$

13. $\mathbf{T} = \dfrac{1}{\sqrt{2}}(\mathbf{i} - \mathbf{j})$

   $\mathbf{N} = \dfrac{1}{\sqrt{2}}(\mathbf{i} + \mathbf{j})$

   $\mathbf{a} \cdot \mathbf{T} = -\sqrt{2}$
   $\mathbf{a} \cdot \mathbf{N} = \sqrt{2}$

15. $\mathbf{T} = \dfrac{1}{\sqrt{2}}(-\mathbf{i} + \mathbf{j})$

   $\mathbf{N} = -\dfrac{1}{\sqrt{2}}(\mathbf{i} + \mathbf{j})$

   $\mathbf{a} \cdot \mathbf{T} = 0$
   $\mathbf{a} \cdot \mathbf{N} = 16\pi^2$

17. $\mathbf{T} = \dfrac{1}{2}(-\sqrt{3}\,\mathbf{i} + \mathbf{j})$

   $\mathbf{N} = -\dfrac{1}{2}(\mathbf{i} + \sqrt{3}\,\mathbf{j})$

   $\mathbf{a} \cdot \mathbf{T} = 0$
   $\mathbf{a} \cdot \mathbf{N} = 2\pi^2$

19. $\mathbf{T} = \dfrac{1}{\sqrt{2}}(-\mathbf{i} + \mathbf{j})$

   $\mathbf{N} = -\dfrac{1}{\sqrt{2}}(\mathbf{i} + \mathbf{j})$

   $\mathbf{a} \cdot \mathbf{T} = \sqrt{2}e^{\pi/2}$
   $\mathbf{a} \cdot \mathbf{N} = \sqrt{2}e^{\pi/2}$

21. $\mathbf{T} = (\cos \omega t_0)\,\mathbf{i} + (\sin \omega t_0)\,\mathbf{j}$

   $\mathbf{N} = (-\sin \omega t_0)\,\mathbf{i} + (\cos \omega t_0)\,\mathbf{j}$

   $\mathbf{a} \cdot \mathbf{T} = \omega^2$
   $\mathbf{a} \cdot \mathbf{N} = \omega Et$

23. $\mathbf{T} = \dfrac{\sqrt{6}}{6}(\mathbf{i} + 2\mathbf{j} + \mathbf{k})$

   $\mathbf{N} = \dfrac{\sqrt{30}}{30}(-5\mathbf{i} + 2\mathbf{j} + \mathbf{k})$

   $\mathbf{a} \cdot \mathbf{T} = 5\sqrt{6}/6$
   $\mathbf{a} \cdot \mathbf{N} = \sqrt{30}/6$

25. $\mathbf{T} = \dfrac{1}{5}(4\mathbf{i} - 3\mathbf{j})$

27. $\mathbf{a} \cdot \mathbf{T} = \dfrac{32(v_0 \sin \theta - 32t)}{\sqrt{v_0{}^2 \cos^2 \theta + (v_0 \sin \theta - 32t)^2}}$, $\mathbf{a} \cdot \mathbf{N} = \dfrac{32v_0 \cos \theta}{\sqrt{v_0{}^2 \cos^2 \theta + (v_0 \sin \theta - 32t)^2}}$

    $\mathbf{N} = -\mathbf{k}$

    At maximum height: Tangential component $= 0$ and Normal component $= 32$

    $\mathbf{a} \cdot \mathbf{T} = 0$

    $\mathbf{a} \cdot \mathbf{N} = 3$

29. (a) Centripetal component is quadrupled

    (b) Centripetal component is halved

31. 4.83 mi/sec

33. 4.67 mi/sec

SECTION 14.5

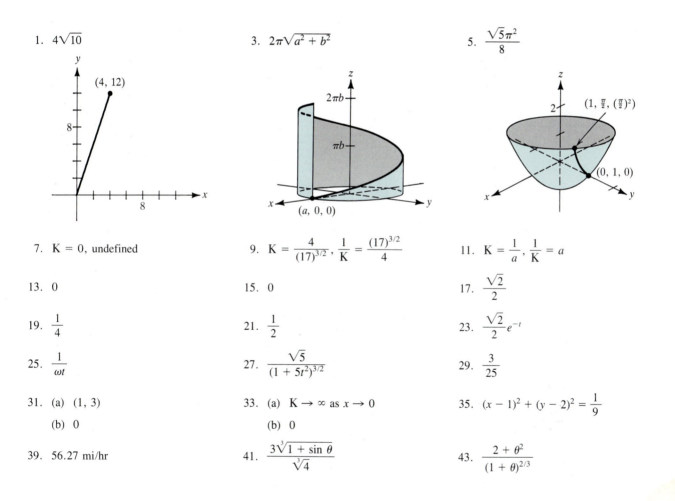

1. $4\sqrt{10}$

3. $2\pi\sqrt{a^2 + b^2}$

5. $\dfrac{\sqrt{5}\pi^2}{8}$

7. $K = 0$, undefined

9. $K = \dfrac{4}{(17)^{3/2}}$, $\dfrac{1}{K} = \dfrac{(17)^{3/2}}{4}$

11. $K = \dfrac{1}{a}$, $\dfrac{1}{K} = a$

13. 0

15. 0

17. $\dfrac{\sqrt{2}}{2}$

19. $\dfrac{1}{4}$

21. $\dfrac{1}{2}$

23. $\dfrac{\sqrt{2}}{2}e^{-t}$

25. $\dfrac{1}{\omega t}$

27. $\dfrac{\sqrt{5}}{(1 + 5t^2)^{3/2}}$

29. $\dfrac{3}{25}$

31. (a) $(1, 3)$

    (b) 0

33. (a) $K \to \infty$ as $x \to 0$

    (b) 0

35. $(x - 1)^2 + (y - 2)^2 = \dfrac{1}{9}$

39. 56.27 mi/hr

41. $\dfrac{3\sqrt[3]{1 + \sin \theta}}{\sqrt[3]{4}}$

43. $\dfrac{2 + \theta^2}{(1 + \theta)^{2/3}}$

REVIEW EXERCISES FOR CHAPTER 14

1.

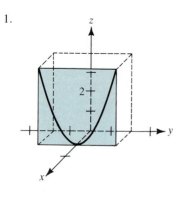

3.

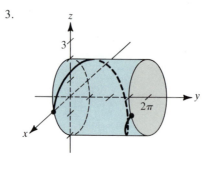

5. $x = t,\ y = -t,\ z = 2t^2$

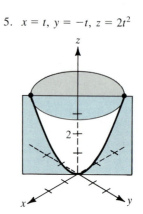

7. $4\mathbf{i} + \mathbf{k}$

9. (a) All reals except $n\pi$, n is an integer
   (b) $t = n\pi$, n is an integer

11. (a) $(0, \infty)$
    (b) Continuous

13. (a) $3\mathbf{i} + \mathbf{j}$
    (b) $\mathbf{0}$
    (c) $4t + 3t^2$
    (d) $-5\mathbf{i} + (2t - 2)\,\mathbf{j} + 2t^2\mathbf{k}$
    (e) $\dfrac{10t - 1}{\sqrt{10t^2 - 2t + 1}}$
    (f) $\left(\dfrac{8}{3}t^3 - 2t^2\right)\mathbf{i} - 8t^3\mathbf{j} + (9t^2 - 2t + 1)\,\mathbf{k}$

15. $(\sin t)\,\mathbf{i} + (t \sin t - \sin t)\,\mathbf{j} + \mathbf{C}$

17. $\dfrac{1}{2}(t\sqrt{1 + t^2} + \ln |t + \sqrt{1 + t^2}|) + C$

19. $x = -\sqrt{2} - \sqrt{2}t$
    $y = \sqrt{2} - \sqrt{2}t$
    $z = \dfrac{3\pi}{4} + t$

21. $\mathbf{v} = 4\mathbf{i} - 3\mathbf{j}$
    $\|\mathbf{v}\| = 5$
    $\mathbf{a} = \mathbf{0}$
    $\mathbf{a} \cdot \mathbf{T} = 0$
    $\mathbf{a} \cdot \mathbf{N} = 0$
    $K = 0$

23. $\mathbf{v} = 2\mathbf{i} - \dfrac{2\mathbf{j}}{(t + 1)^2}$
    $\|\mathbf{v}\| = \dfrac{2\sqrt{(t + 1)^4 + 1}}{(t + 1)^2}$
    $\mathbf{a} = \dfrac{4}{(t + 1)^3}\mathbf{j}$
    $\mathbf{a} \cdot \mathbf{T} = \dfrac{-4}{(t + 1)^3\sqrt{(t + 1)^4 + 1}}$
    $\mathbf{a} \cdot \mathbf{N} = \dfrac{4}{(t + 1)\sqrt{(t + 1)^4 + 1}}$
    $K = \dfrac{(t + 1)^3}{[(t + 1)^4 + 1]^{3/2}}$

25. $\mathbf{v} = e^t\mathbf{i} - e^{-t}\mathbf{j}$
    $\|\mathbf{v}\| = \sqrt{e^{2t} + e^{-2t}}$
    $\mathbf{a} = e^t\mathbf{i} + e^{-t}\mathbf{j}$
    $\mathbf{a} \cdot \mathbf{T} = \dfrac{e^{2t} - e^{-2t}}{\sqrt{e^{2t} + e^{-2t}}}$
    $\mathbf{a} \cdot \mathbf{N} = \dfrac{2}{\sqrt{e^{2t} + e^{-2t}}}$
    $K = \dfrac{2}{(e^{2t} + e^{-2t})^{3/2}}$

27. $\mathbf{v} = \mathbf{i} + 2t\mathbf{j} + t\mathbf{k}$
        $\|\mathbf{v}\| = \sqrt{1 + 5t^2}$
        $\mathbf{a} = 2\mathbf{j} + \mathbf{k}$

   $\mathbf{a} \cdot \mathbf{T} = \dfrac{5t}{\sqrt{1 + 5t^2}}$

   $\mathbf{a} \cdot \mathbf{N} = \dfrac{-5\sqrt{5}}{\sqrt{1 + 5t^2}}$

   $K = \dfrac{\sqrt{5}}{(1 + 5t^2)^{3/2}}$

29. $\dfrac{\sqrt{5}\pi}{2}$

33. 4.56 mi/sec

31. 152 ft

CHAPTER 15

SECTION 15.1

1. Not a function

3. Function

5. Function

7. (a) 3/2
    (b) $-1/4$
    (c) 6
    (d) $5/y$
    (e) $x/2$
    (f) $5/t$

9. (a) 5
    (b) $3e^2$
    (c) $2/e$
    (d) $5e^y$
    (e) xe^2
    (f) te^t

11. (a) 2/3
     (b) 0

13. (a) $\sqrt{2}$
     (b) 3 sin 1

15. (a) 4
     (b) 6

17. Domain: $\{(x, y): x^2 + y^2 \leq 4\}$
     Range: $0 \leq z \leq 2$

19. Domain: $\{(x, y): x^2 + y^2 \geq 1\}$
     Range: $0 \leq z$

21. Domain: $\{(x, y): x \neq 0, y \neq 0\}$
     Range: All real numbers

23. Domain: $\{(x, y): y < -x + 4\}$
     Range: All real numbers

25. Domain: $\{(x, y): y \neq 0\}$
     Range: $0 < z$

27. Domain: $\{(x, y): x \neq 0, y \neq 0\}$
     Range: $0 < |z|$

29.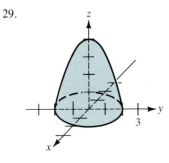

31.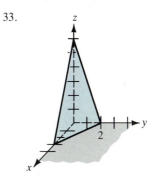

33.

35.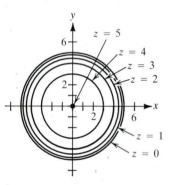

37.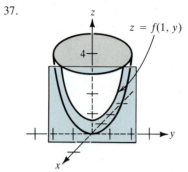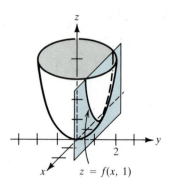

39. Circles centered at
    $(0, 0)$ with radius ≤ 5

41. Hyperbolas: $xy = c$

$z = 0$ $(x$ & y-axes$)$

43. Circles passing through
    $(0, 0)$ centered at $(1/2c, 0)$

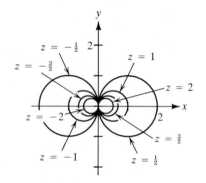

45. Lines with a slope of 1 passing
    through Quadrant IV: $y = x - e^c$

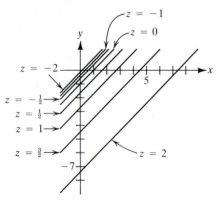

47. Hyperbolas centered at $(0, 0)$ with
    x- and y-axes as asymptotes: $xy = \ln c$

$z = 1$ $(x$ & y-axes$)$

49. (a) 243 board-feet
    (b) 507 board-feet

51. (a) 12 minutes
    (b) 20 minutes
    (c) 10 minutes
    (d) 30 minutes

SECTION 15.2

1. 2

3. 15

5. 5, Continuous

7. -3, Continuous for $x \ne y$

9. 0, Continuous for $xy \ne -1$, $y \ne 0$

11. 1, Continuous

13. $2\sqrt{2}$, Continuous for $x + y + z \ge 0$

15. 0

17. $\dfrac{1}{2}$

19. 1, Continuous for $xy \ne 0$

21. Limit does not exist
    Continuous except at $(0, 0)$

23. 1, Continuous

25. $1/3$, Continuous for $y \ne \pm 3x$

27. $-\infty$, Continuous except at $(0, 0)$

29. 1

31. 0

33. Continuous except at $(0, 0, 0)$

35. Continuous

37. Continuous

39. Continuous for $y \ne 3x/2$

41. (a) $2x$
    (b) -4

43. (a) $2 + y$
    (b) $x - 3$

SECTION 15.3

1. $f_x(x, y) = 2$
   $f_y(x, y) = -3$

3. $f_x(x, y) = y$
   $f_y(x, y) = x$

5. $f_x(x, y) = \sqrt{y}$
   $f_y(x, y) = x/(2\sqrt{y})$

7. $f_x(x, y) = 2xe^{2y}$
   $f_y(x, y) = 2x^2 e^{2y}$

9. $f_x(x, y) = \dfrac{2x}{x^2 + y^2}$

   $f_y(x, y) = \dfrac{2y}{x^2 + y^2}$

11. $f_x(x, y) = \dfrac{-2y}{x^2 - y^2}$

    $f_y(x, y) = \dfrac{2x}{x^2 - y^2}$

13. $f_x(x, y) = -2xe^{-(x^2+y^2)}$
    $f_y(x, y) = -2ye^{-(x^2+y^2)}$

15. $f_x(x, y) = \dfrac{x}{\sqrt{x^2 + y^2}}$

    $f_y(x, y) = \dfrac{y}{\sqrt{x^2 + y^2}}$

17. $f_x(x, y) = 2\cos(2x - y)$
    $f_y(x, y) = -\cos(2x - y)$

19. $f_x(x, y) = ye^y \cos xy$
    $f_y(x, y) = e^y(x \cos xy + \sin xy)$

21. $f_x(x, y) = 1 - x^2$
    $f_y(x, y) = y^2 - 1$

23. $\dfrac{\partial z}{\partial x} = \dfrac{1}{4}$

    $\dfrac{\partial z}{\partial y} = \dfrac{1}{4}$

25. $\dfrac{\partial z}{\partial x} = -\dfrac{1}{4}$

    $\dfrac{\partial z}{\partial y} = \dfrac{1}{4}$

27. $\dfrac{\partial w}{\partial x} = \dfrac{x}{\sqrt{x^2 + y^2 + z^2}}$

    $\dfrac{\partial w}{\partial y} = \dfrac{y}{\sqrt{x^2 + y^2 + z^2}}$

    $\dfrac{\partial w}{\partial z} = \dfrac{z}{\sqrt{x^2 + y^2 + z^2}}$

29. $F_x(x, y, z) = \dfrac{x}{x^2 + y^2 + z^2}$

    $F_y(x, y, z) = \dfrac{y}{x^2 + y^2 + z^2}$

    $F_z(x, y, z) = \dfrac{z}{x^2 + y^2 + z^2}$

31. $H_x(x, y, z) = \cos (x + 2y + 3z)$

$H_y(x, y, z) = 2 \cos (x + 2y + 3z)$

$H_z(x, y, z) = 3 \cos (x + 2y + 3z)$

33. $\dfrac{\partial^2 z}{\partial x^2} = 2, \dfrac{\partial^2 z}{\partial y^2} = 6$

$\dfrac{\partial^2 z}{\partial y \partial x} = \dfrac{\partial^2 z}{\partial x \partial y} = -2$

35. $\dfrac{\partial^2 z}{\partial x^2} = e^x \tan y, \dfrac{\partial^2 z}{\partial y^2} = 2e^x \sec^2 y \tan y$

$\dfrac{\partial^2 z}{\partial y \partial x} = \dfrac{\partial^2 z}{\partial x \partial y} = e^x \sec^2 y$

37. $\dfrac{\partial^2 z}{\partial x^2} = \dfrac{2xy}{(x^2 + y^2)^2}, \dfrac{\partial^2 z}{\partial y^2} = \dfrac{-2xy}{(x^2 + y^2)^2}$

$\dfrac{\partial^2 z}{\partial y \partial x} = \dfrac{\partial^2 z}{\partial x \partial y} = \dfrac{y^2 - x^2}{(x^2 + y^2)^2}$

39. $\dfrac{\partial^2 z}{\partial x^2} = \dfrac{y^2}{(x^2 + y^2)^{3/2}}, \dfrac{\partial^2 z}{\partial y^2} = \dfrac{x^2}{(x^2 + y^2)^{3/2}}$

$\dfrac{\partial^2 z}{\partial y \partial x} = \dfrac{\partial^2 z}{\partial x \partial y} = \dfrac{-xy}{(x^2 + y^2)^{3/2}}$

41. $\dfrac{\partial^2 z}{\partial y \partial x} = \dfrac{\partial^2 z}{\partial x \partial y} = 6x$

43. $\dfrac{\partial^2 x}{\partial y \partial x} = \dfrac{\partial^2 z}{\partial x \partial y} = \sec y \tan y$

45. $\dfrac{\partial^2 z}{\partial y \partial x} = \dfrac{\partial^2 z}{\partial x \partial y} = -2ye^{-y^2}$

47. $f_{xyy}(x, y, z) = f_{yxy}(x, y, z) = f_{yyx}(x, y, z) = 0$

49. $f_{xyy}(x, y, z) = f_{yxy}(x, y, z) = f_{yyx}(x, y, z) = z^2 e^{-x} \sin yz$

51. $\dfrac{\partial^2 z}{\partial x^2} + \dfrac{\partial^2 z}{\partial y^2} = 0 + 0 = 0$

53. $\dfrac{\partial^2 z}{\partial x^2} + \dfrac{\partial^2 z}{\partial y^2} = e^x \sin y - e^x \sin y = 0$

55. $\dfrac{\partial^2 z}{\partial t^2} = -c^2 \sin (x - ct) = \dfrac{\partial^2 z}{\partial y^2}$

57. $\dfrac{\partial z}{\partial t} = -e^{-t} \cos \dfrac{x}{c} = \dfrac{\partial^2 z}{\partial x^2}$

59. $f_x(x, y) = 2$

$f_y(x, y) = 3$

61. $f_x(x, y) = \dfrac{1}{2\sqrt{x + y}}$

$f_y(x, y) = \dfrac{1}{2\sqrt{x + y}}$

63. $-1/2$

65. 18

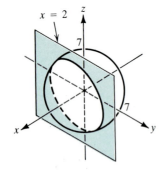

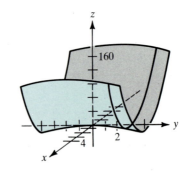

67. $\dfrac{\partial c}{\partial x} = 183, \dfrac{\partial c}{\partial y} = 237$

69. An increase in either price will cause a decrease in demand.

71. $\dfrac{\partial T}{\partial x} = -2.4°/\text{ft}, \dfrac{\partial T}{\partial y} = 9°/\text{ft}$

SECTION 15.4

1. $dz = 6xy^3\,dx + 9x^2y^2\,dy$

3. $dz = \dfrac{2}{(x^2 + y^2)^2}\,(x\,dx + y\,dy)$

5. $dz = (\cos y + y \sin x)\,dy - (x \sin y + \cos x)\,dy$

7. $dw = (2z^3 y \cos x)\,dx + (2z^3 \sin x)\,dy + (6z^2 y \sin x)\,dz$

9. $dw = \dfrac{1}{z - 2y}\,dx + \dfrac{z + 2x}{(z - 2y)^2}\,dy - \dfrac{x + y}{(z - 2y)^2}\,dz$

11. (a) -0.5125
    (b) -0.5

13. (a) -0.00293
    (b) 0.00385

15. (a) -0.25
    (b) -0.25

17. 10%

19. 7%

21. $L \approx 8.096 \times 10^{-4} \pm 6.6 \times 10^{-6}$ micro-henrys

SECTION 15.5

1. $2(e^{2t} - e^{-2t})$

3. $e^t \sec(\pi - t)[1 - \tan(\pi - t)]$

5. $4e^{2t}$

7. $\dfrac{\partial w}{\partial s} = 4s,\ 8,\ \dfrac{\partial w}{\partial t} = 4t,\ -4$

9. $\dfrac{\partial w}{\partial s} = 2s \cos 2t,\ 0$

$\dfrac{\partial w}{\partial t} = -2s^2 \sin 2t,\ -18$

11. $2 \cos 2t$

13. $3(2t^2 - 1)$

15. $\dfrac{\partial w}{\partial r} = 0,\ \dfrac{\partial w}{\partial \theta} = 1$

17. $\dfrac{\partial z}{\partial x} = \dfrac{-x}{z},\ \dfrac{\partial z}{\partial y} = \dfrac{-y}{z}$

19. $\dfrac{\partial z}{\partial x} = \dfrac{-\sec^2(x + y)}{\sec^2(y + z)}$

$\dfrac{\partial z}{\partial y} = -1 - \dfrac{\sec^2(x + y)}{\sec^2(y + z)}$

21. $\dfrac{\partial w}{\partial z} = \dfrac{yw - xy + xw}{xz - yz + 2w}$

$\dfrac{\partial w}{\partial y} = \dfrac{-xz + zw}{xz - yz + 2w}$

$\dfrac{\partial w}{\partial x} = \dfrac{-yz - zw}{xz - yz + 2w}$

23. $\dfrac{\partial z}{\partial x} = \dfrac{-x}{y + z},\ \dfrac{\partial z}{\partial y} = \dfrac{-z}{y + z}$

$\dfrac{\partial^2 z}{\partial x^2} = -\dfrac{(y + z)^2 + x^2}{(y + z)^3}$

$\dfrac{\partial^2 z}{\partial y^2} = \dfrac{z(2y + z)}{(y + z)^3},\ \dfrac{\partial^2 z}{\partial x \partial y} = \dfrac{\partial^2 z}{\partial y \partial x} = \dfrac{xy}{(x + y)^3}$

25. $\dfrac{dV}{dt} = 182$ ft^3/min, $\dfrac{dS}{dt} = 132$ ft^2/min

27. $\dfrac{dV}{dt} = 1536\pi$ in^3/min, $\dfrac{dS}{dt} = \dfrac{36\pi}{5}[20 + 9\sqrt{10}]$ in^2/min
    (surface area includes base)

29. $28\,m$ cm^2/sec

31. 3

33. 0

35. 1

SECTION 15.6

1. $(\sqrt{3} - 5)/2$

3. $5\sqrt{2}/2$

5. $-7/25$

7. $-e$

9. $4/\sqrt{6}$

11. $\dfrac{8 + \pi}{4\sqrt{6}}$

13. $\sqrt{2}(x + y)$

15. $\left(\dfrac{2 + \sqrt{3}}{2}\right) \cos (2x - y)$

17. $-7\sqrt{2}$

19. $7/\sqrt{19}$

21. $(2x - 3y)\,\mathbf{i} + (2y - 3x)\,\mathbf{j},\ 2\sqrt{17}$

23. $\tan y\,\mathbf{i} + x \sec^2 y\,\mathbf{j},\ \sqrt{17}$

25. $\dfrac{1}{3}\left(\dfrac{2x\mathbf{i}}{x^2 + y^2} + \dfrac{2y\mathbf{j}}{x^2 + y^2}\right),\ \dfrac{2\sqrt{5}}{15}$

27. $\dfrac{x\mathbf{i} + y\mathbf{j} + z\mathbf{k}}{\sqrt{x^2 + y^2 + z^2}},\ 1$

29. $\dfrac{x\mathbf{i} + y\mathbf{j} + z\mathbf{k}}{(1 - x^2 - y^2 - z^2)^{3/2}},\ 0$

31.

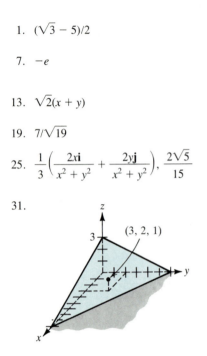

33. (a) $\dfrac{2 + 3\sqrt{3}}{12}$

   (b) $\dfrac{3 - 2\sqrt{3}}{12}$

35. (a) $\dfrac{-1}{5}$

   (b) $\dfrac{-11}{6\sqrt{10}}$

37. $\sqrt{13}/6$

39.

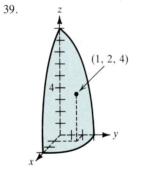

41. $-2\mathbf{i} - 4\mathbf{j},\ 2\sqrt{5}$

43. $6\mathbf{i} + 8\mathbf{j}$

45. $-\dfrac{1}{2}\mathbf{j}$

47. $\dfrac{1}{\sqrt{257}}(16\mathbf{i} - \mathbf{j})$

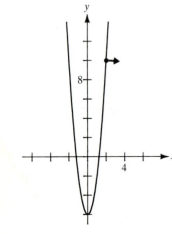

49. $\dfrac{1}{\sqrt{85}}(9\mathbf{i} - 2\mathbf{j})$

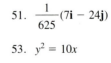

51. $\dfrac{1}{625}(7\mathbf{i} - 24\mathbf{j})$

53. $y^2 = 10x$

SECTION 15.7

1. $\dfrac{1}{\sqrt{3}}(\mathbf{i} + \mathbf{j} + \mathbf{k})$

3. $\dfrac{1}{5\sqrt{2}}(3\mathbf{i} + 4\mathbf{j} - 5\mathbf{k})$

5. $\dfrac{1}{\sqrt{2{,}049}}(32\mathbf{i} + 32\mathbf{j} - \mathbf{k})$

7. $\dfrac{1}{\sqrt{113}}(-\mathbf{i} - 6\sqrt{3}\mathbf{j} + 2\mathbf{k})$

9. $\dfrac{1}{\sqrt{3}}(\mathbf{i} - \mathbf{j} + \mathbf{k})$

11. $6x + 2y + z = 35$

13. $2x - y + z = 2$

15. $10x - 8y - z = 9$

17. $2x - z = -2$

19. $3x + 4y - 25z = 25(1 - \ln 5)$

21. $x - 4y + 2z = 18$

23. $x + y + z = 1$

25. $2x + 4y + z = 14$

$\dfrac{x - 1}{2} = \dfrac{y - 2}{4} = \dfrac{z - 4}{1}$

27. $3x + 2y + z = -6$

$\dfrac{x + 2}{3} = \dfrac{y + 3}{2} = \dfrac{z - 6}{1}$

29. $x - y + 2z = \dfrac{\pi}{2}$

$\dfrac{x - 1}{1} = \dfrac{y - 1}{-1} = \dfrac{z - (\pi/4)}{2}$

31. $86.0°$

33. $77.4°$

35. (a) $\dfrac{x - 2}{1} = \dfrac{y - 1}{-2} = \dfrac{z - 2}{1}$

(b) $\dfrac{\sqrt{10}}{5}$, Not orthogonal

37. (a) $\dfrac{x - 3}{4} = \dfrac{y - 3}{4} = \dfrac{z - 4}{-3}$

(b) $\dfrac{16}{25}$, Not orthogonal

39. (a) $\dfrac{y - 1}{1} = \dfrac{z - 1}{-1}$, $x = 2$

(b) 0, Orthogonal

41. $(0, 3, 12)$

43. $x = 4e^{-4t}$, $y = 3e^{-2t}$, $z = 10e^{-8t}$

SECTION 15.8

1. Relative minimum: $(-1, 1, -4)$

3. Relative maximum: $(8, 16, 74)$

5. Relative minimum: $(1, 2, -1)$

7. Saddle point: $(1, -2, -1)$

9. Saddle point: $(0, 0, 0)$

11. Saddle point: $(0, 0, 0)$
    Relative minimum: $(1, 1, -1)$

13. Relative maximum: $(1, 0, 1)$
    Saddle point: $(0, 0, 1/2)$

15. Relative maxima: $(0, \pm 1, 4)$
    Relative minimum: $(0, 0, 0)$
    Saddle points: $(\pm 1, 0, 1)$

17. No extrema

19. Relative maxima: $(0, \pm\sqrt{2}/2, \sqrt{e})$
    Relative minima: $(\pm\sqrt{6}/2, 0, -e^{-1/2})$
    Saddle point: $(0, 0, e/2)$

21. Absolute minimum: $(0, 0, 0)$
    Absolute maxima: $(2, 1, 6)$, $(-2, -1, 6)$

23. Absolute minima: $(x, -x, 0)$, $|x| \le 2$
    Absolute maxima: $(2, 2, 16)$, $(-2, -2, 16)$

25. Saddle point: (0, 0, 0) [Test fails]

27. Absolute minima: $(1, a, 0)$, $(b, -4, 0)$ [Test fails]

29. Absolute minima: (0, 0, 0) [Test fails]

31. Relative minimum: $(0, 3, -1)$

33. Relative minimum: $(-1, 2, 1)$

SECTION 15.9

1. $6\sqrt{14}/7$

3. 6

5. 10, 10, 10

7. 10, 10, 10

9. $36 \times 18 \times 18$ in

15. Each edge of $w/3$ inches is turned up $60°$ from the horizontal.

17. $p_1 = \$586.67$
$p_2 = \$840.00$

19. $a \sum_{i=1}^{n} x_i^4 + b \sum_{i=1}^{n} x_i^3 + c \sum_{i=1}^{n} x_i^2 = \sum_{i=1}^{n} x_i^2 y_i$

$a \sum_{i=1}^{n} x_i^3 + b \sum_{i=1}^{n} x_i^2 + c \sum_{i=1}^{n} x_i = \sum_{i=1}^{n} x_i y_i$

$a \sum_{i=1}^{n} x_i^2 + b \sum_{i=1}^{n} x_i + cn = \sum_{i=1}^{n} y_i$

21. $y = \dfrac{3}{4}x + \dfrac{4}{3}, \dfrac{1}{6}$

23. $y = -2x + 4, 2$

25. $y = \dfrac{3}{7}x^2 + \dfrac{6}{5}x + \dfrac{26}{35}$

27. $y = 14x + 19$, 41.4 bushels/acre

31. $w^4 + 4w^3 + 6w^2 + 4w + 5 = (w^2 + 1)(w^2 + 4w + 5)$

$\dfrac{1}{2} \ln (w^2 + 1) + \ln (w^2 + 4w + 5) + C$

SECTION 15.10

1. $f(5, 5) = 25$

3. $f(2, 2) = 8$

5. $f(\sqrt{2}/2, 1/2) = 1/4$

7. $f(25, 50) = 2{,}600$

9. $f(1, 1) = 2$

11. $f(2, 2) = e^4$

13. $f(2, 2, 2) = 12$

15. $f(8, 16, 8) = 1{,}024$

17. $f(1/3, 1/3, 1/3) = 1/3$

19. $f(\sqrt{10/3}, \sqrt{5/6}, \sqrt{5/3}) = 5\sqrt{15}/9$

21. $36 \times 18 \times 18$ in

23. 150

25. $\sqrt{13}/13$

27. $\sqrt{3}$

29. $P(3125/6, 6250/3) = 147{,}314$

31. $x = 50\sqrt{2}$, $y = 200\sqrt{2}$,
cost $= \$13{,}576.45$

33. $x = \dfrac{10 + 2\sqrt{265}}{15}$, $y = \dfrac{5 + \sqrt{265}}{15}$, $z = \dfrac{-1 + \sqrt{265}}{3}$

35. $\dfrac{2a}{\sqrt{3}} \times \dfrac{2b}{\sqrt{3}} \times \dfrac{2c}{\sqrt{3}}$

REVIEW EXERCISES FOR CHAPTER 15

1. Limit: 1/2
   Continuous except at $(0, 0)$

3. Limit does not exist
   Continuous except at $(0, 0)$

5.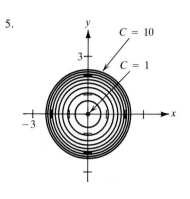

7.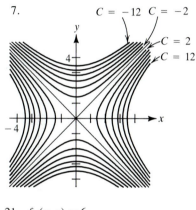

9. $f_x(x, y) = e^x \cos y$
   $f_y(x, y) = -e^x \sin y$

11. $f_x(x, y) = e^y + ye^x$
    $f_y(x, y) = xe^y + e^x$

13. $f_x(x, y) = \dfrac{y(y^2 - x^2)}{(x^2 + y^2)^2}$

    $f_y(x, y) = \dfrac{x(x^2 - y^2)}{(x^2 + y^2)^2}$

15. $f_x(x, y, z) = \dfrac{-yz}{x^2 + y^2}$

    $f_y(x, y, z) = \dfrac{xz}{x^2 + y^2}$

    $f_z(x, y, z) = \arctan \dfrac{y}{x}$

17. $u_x(x, t) = cne^{-n^2 t} \cos nx$
    $u_t(x, t) = -cn^2 e^{-n^2 t} \sin nx$

19. $-\dfrac{2yz + z - 2xy}{2xy + x + 2z}$

21. $f_{xx}(x, y) = 6$
    $f_{yy}(x, y) = 12y$
    $f_{xy}(x, y) = f_{yx}(x, y) = -1$

23. $f_{xx}(x, y) = -y \cos x$
    $f_{yy}(x, y) = -x \sin y$
    $f_{xy}(x, y) = f_{yx}(x, y) = \cos y - \sin x$

29. $\left(\sin \dfrac{y}{x} - \dfrac{y}{x} \cos \dfrac{y}{x} \right) dx + \left(\cos \dfrac{y}{x} \right) dy$

31. $\dfrac{\partial u}{\partial r} = 2r, \ \dfrac{\partial u}{\partial t} = 2t$

33. 0

35. 2/3

37. $\langle -1/2, 0 \rangle$, 1/2

39. $\langle -\sqrt{2}/2, -\sqrt{2}/2 \rangle$, 1

41. $4x + 4y - z = 8, \ \dfrac{x - 2}{4} = \dfrac{y - 1}{4} = \dfrac{z - 4}{-1}$

43. Tangent plane: $z = 4$
    Normal line: $x = 2, \ y = -3$

45. $\dfrac{x - 2}{1} = \dfrac{y - 1}{2}, \ z = 3$

47. Relative minimum: $(3/2, 9/4, -27/16)$
    Saddle point: $(0, 0, 0)$

49. Relative minimum: $(1, 1, 3)$

51. Relative minimum: $(0, 1, 0)$
    Relative maximum: $(4/3, 1/3, 16/27)$

53. 0.082 in, 0.63%

55. π in^3

57. $x_1 = 94, \ x_2 = 157$

59. 49.4, 253

CHAPTER 16

SECTION 16.1

1. $3x^2/2$

3. $y \ln (2y)$

5. $\dfrac{4x^2 - x^4}{2}$

7. $\dfrac{y}{2}[(\ln y)^2 - y^2]$

9. $x^2(1 - e^{-x^2} - x^2 e^{-x^2})$

11. 3

13. 20/3

15. 2/3

17. 4

19. $\dfrac{\pi^2}{32} + \dfrac{1}{8}$

21. $\displaystyle\int_0^1 \int_0^2 dy\, dx = \int_0^2 \int_0^1 dx\, dy = 2$

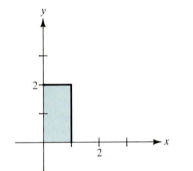

23. $\displaystyle\int_0^1 \int_{-\sqrt{1-y^2}}^{\sqrt{1-y^2}} dx\, dy = \int_{-1}^1 \int_0^{\sqrt{1-x^2}} dy\, dx = \dfrac{\pi}{2}$

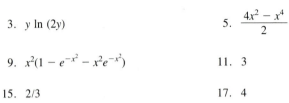

25. $\displaystyle\int_0^2 \int_{x/2}^1 dy\, dx = \int_0^1 \int_0^{2y} dx\, dy = 1$

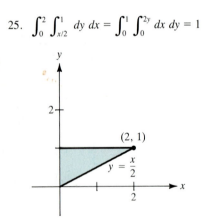

27. $\displaystyle\int_0^1 \int_{y^2}^{\sqrt[3]{y}} dx\, dy = \int_0^1 \int_{x^3}^{\sqrt{x}} dy\, dx = \dfrac{5}{12}$

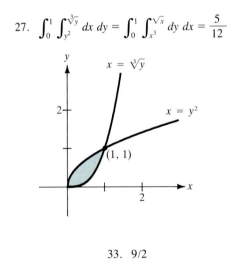

29. 24

31. 16/3

33. 9/2

35. 8/3

37. 5

39. πab

SECTION 16.2

1. 8

3. 36

5. 0

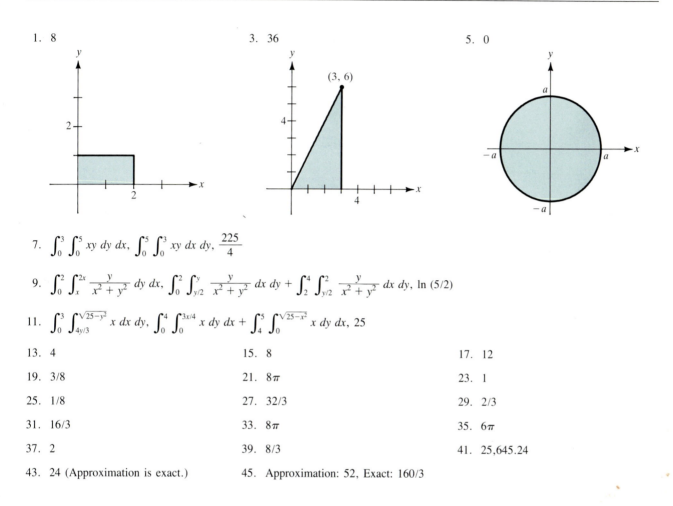

7. $\int_0^3 \int_0^5 xy \, dy \, dx$, $\int_0^5 \int_0^3 xy \, dx \, dy$, $\dfrac{225}{4}$

9. $\int_0^2 \int_x^{2x} \dfrac{y}{x^2 + y^2} \, dy \, dx$, $\int_0^2 \int_{y/2}^y \dfrac{y}{x^2 + y^2} \, dx \, dy + \int_2^4 \int_{y/2}^2 \dfrac{y}{x^2 + y^2} \, dx \, dy$, $\ln (5/2)$

11. $\int_0^3 \int_{4y/3}^{\sqrt{25-y^2}} x \, dx \, dy$, $\int_0^4 \int_0^{3x/4} x \, dy \, dx + \int_4^5 \int_0^{\sqrt{25-x^2}} x \, dy \, dx$, 25

13. 4

15. 8

17. 12

19. 3/8

21. 8π

23. 1

25. 1/8

27. 32/3

29. 2/3

31. 16/3

33. 8π

35. 6π

37. 2

39. 8/3

41. 25,645.24

43. 24 (Approximation is exact.)

45. Approximation: 52, Exact: 160/3

SECTION 16.3

1. 0

3. $\dfrac{5\sqrt{5}\pi}{6}$

5. $\dfrac{\pi^2}{8} + 1$

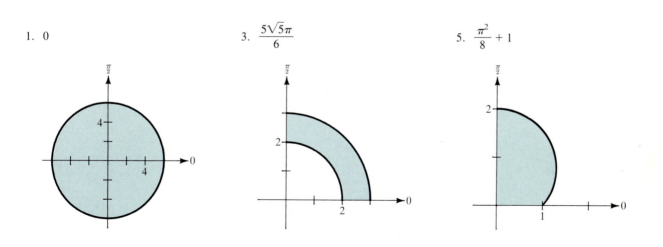

7. 9π

9. $3\pi/2$

11. π

13. $a^3/3$

15. $4\sqrt{2}\pi/3$

17. $2/3$

19. $16/3$

21. $\pi^2/16$

23. $1/8$

25. $250\pi/3$

27. $\dfrac{64}{9}(3\pi - 4)$

29. $2\sqrt{4 - 2\sqrt[3]{2}}$

SECTION 16.4

1. $-1/2$

3. $1 + 2v$

5. 1

7. $-e^{2u}$

9. 0

11. $2(1 - e^{-2})$

13. $\dfrac{32}{3}[\sqrt{2} + \ln(\sqrt{2} + 1)]$

15. $12(e^4 - 1)$

17. $100/9$

19. $\dfrac{2}{5}a^{5/2}$

SECTION 16.5

1. (a) $m = kab, \left(\dfrac{a}{2}, \dfrac{b}{2}\right)$

   (b) $m = \dfrac{kab^2}{2}, \left(\dfrac{a}{2}, \dfrac{2b}{3}\right)$

3. (a) $m = \dfrac{kbh}{2}, \left(\dfrac{b}{2}, \dfrac{h}{3}\right)$

   (b) $m = \dfrac{kh^2b}{6}, \left(\dfrac{b}{2}, \dfrac{h}{2}\right)$

5. (a) $m = \dfrac{k\pi a^2}{2}, \left(0, \dfrac{4a}{3\pi}\right)$

   (b) $m = \dfrac{ka^4}{24}(16 - 3\pi), \left(0, \dfrac{a}{5}\left[\dfrac{15\pi - 32}{16 - 3\pi}\right]\right)$

7. $m = \dfrac{32k}{3}, \left(3, \dfrac{8}{7}\right)$

9. $m = \dfrac{k}{4}(1 - e^{-4}), \left(\dfrac{e^4 - 5}{2(e^4 - 1)}, \dfrac{4}{9}\left[\dfrac{e^6 - 1}{e^6 - e^2}\right]\right)$

11. $m = \dfrac{k\pi}{2}, \left(0, \dfrac{\pi + 2}{4\pi}\right)$

13. $m = \dfrac{8{,}192k}{15}, \left(\dfrac{64}{7}, 0\right)$

15. $m = \dfrac{kL}{4}, \left(\dfrac{L}{2}, \dfrac{16}{9\pi}\right)$

17. $m = \dfrac{k\pi a^2}{8}, \left(\dfrac{4\sqrt{2}a}{3\pi}, \dfrac{4a(2 - \sqrt{2})}{3\pi}\right)$

19. $m = \dfrac{k\pi}{3}, (1.12, 0)$

21. $\bar{\bar{x}} = \dfrac{b}{\sqrt{3}}, \bar{\bar{y}} = \dfrac{h}{\sqrt{3}}$

23. $\bar{\bar{x}} = \dfrac{r}{2}, \bar{\bar{y}} = \dfrac{r}{2}$

25. $\bar{\bar{x}} = \dfrac{r}{2}, \bar{\bar{y}} = \dfrac{r}{2}$

27. $I_x = \dfrac{kab^4}{4}, I_y = \dfrac{kb^2a^3}{6}, I_o = \dfrac{3kab^4 + 2kb^2a^3}{12}, \bar{\bar{x}} = \dfrac{a}{\sqrt{3}}, \bar{\bar{y}} = \dfrac{b}{\sqrt{2}}$

29. $I_x = \dfrac{32k}{3}$, $I_y = \dfrac{16k}{3}$, $I_o = 16k$, $\bar{\bar{x}} = \dfrac{2\sqrt{3}}{3}$, $\bar{\bar{y}} = \dfrac{2\sqrt{6}}{3}$

31. $I_x = 16k$, $I_y = \dfrac{512k}{5}$, $I_o = \dfrac{592k}{5}$, $\bar{\bar{x}} = \dfrac{4\sqrt{15}}{5}$, $\bar{\bar{y}} = \dfrac{\sqrt{6}}{2}$

33. $I_x = \dfrac{3k}{56}$, $I_y = \dfrac{k}{18}$, $I_o = \dfrac{55k}{504}$, $\bar{\bar{x}} = \dfrac{\sqrt{30}}{9}$, $\bar{\bar{y}} = \dfrac{\sqrt{70}}{14}$

35. $\dfrac{k\pi b^2}{4}(b^2 + 4a^2)$

37. $\dfrac{42{,}752k}{315}$

39. $a^5\left(\dfrac{7\pi}{16} - \dfrac{17}{15}\right)k$

SECTION 16.6

1. 6

3. 12π

5. $\dfrac{3}{4}[6\sqrt{37} + \ln(\sqrt{37} + 6)]$

7. $\dfrac{27 - 5\sqrt{5}}{12}$

9. $\dfrac{4}{27}(31\sqrt{31} - 8)$

11. $\sqrt{2} - 1$

13. $\dfrac{\pi}{6}(17\sqrt{17} - 1)$

15. $\sqrt{2}\pi$

17. $2\pi a(a - \sqrt{a^2 - b^2})$

19. 16

21. $\displaystyle\int_{-1}^{1}\int_{-1}^{1} \sqrt{1 + 9(x^2 - y)^2 + 9(y^2 - x)^2}\, dy\, dx$

23. $\displaystyle\int_{-2}^{2}\int_{-\sqrt{4-x^2}}^{\sqrt{4+x^2}} \sqrt{1 + e^{-2x}}\, dy\, dx$

25. $\displaystyle\int_{0}^{4}\int_{0}^{10} \sqrt{1 + e^{2xy}(x^2 + y^2)}\, dy\, dx$

SECTION 16.7

1. 18

3. 1/10

5. $\dfrac{15}{2}\left(1 - \dfrac{1}{e}\right)$

7. 729/4

9. 128/15

11. $\displaystyle\int_{0}^{3}\int_{0}^{(12-4z)/3}\int_{0}^{(12-4z-3x)/6} dy\, dx\, dz$

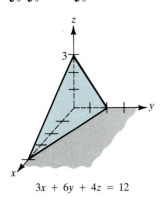

$3x + 6y + 4z = 12$

13. $\displaystyle\int_{0}^{1}\int_{0}^{x}\int_{0}^{\sqrt{1-y^2}} dz\, dy\, dx$

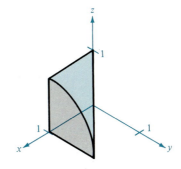

15. 256/15

17. $4\pi r^3/3$

19. 256/15

21. $m = 8k,\ \bar{x} = 3/2$

23. $m = 128k/3,\ \bar{z} = 1$

25. $m = kb^5/4,\ (2b/3,\ 2b/3,\ b/2)$

27. $(0,\ 0,\ 3h/4)$

29. $(0,\ 0,\ 3/2)$

31. (a) $I_x = 2ka^5/3$ (b) $I_x = ka^8/8$
$I_y = 2ka^5/3$ $I_y = ka^8/8$
$I_z = 2ka^5/3$ $I_z = ka^8/8$

33. (a) $I_x = 256k$ (b) $I_x = 2048k/3$
$I_y = 512k/3$ $I_y = 1024k/3$
$I_z = 256k$ $I_z = 2048k/3$

SECTION 16.8

1. 8

3. 52/45

5. $\dfrac{\pi^2}{6}(1 - e^{-8})$

7. 0

9. $\dfrac{\pi}{4}(1 - e^{-9})$

11. $\dfrac{64\sqrt{3}\pi}{3}$

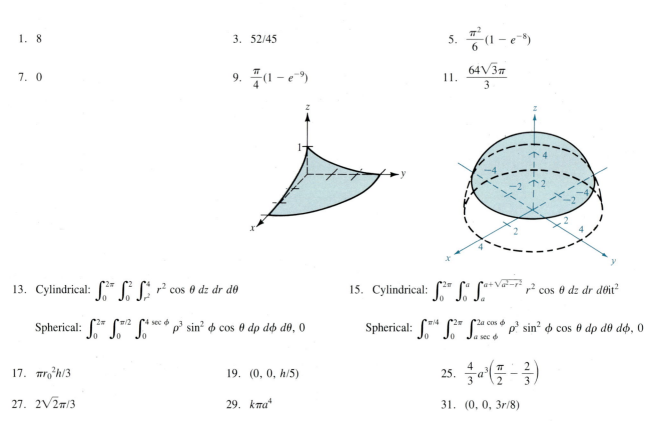

13. Cylindrical: $\displaystyle\int_0^{2\pi}\int_0^2\int_{r^2}^4 r^2 \cos\theta\ dz\ dr\ d\theta$

Spherical: $\displaystyle\int_0^{2\pi}\int_0^{\pi/2}\int_0^{4\sec\phi} \rho^3 \sin^2\phi \cos\theta\ d\rho\ d\phi\ d\theta,\ 0$

15. Cylindrical: $\displaystyle\int_0^{2\pi}\int_0^a\int_a^{a+\sqrt{a^2-r^2}} r^2 \cos\theta\ dz\ dr\ d\theta\text{it}^2$

Spherical: $\displaystyle\int_0^{\pi/4}\int_0^{2\pi}\int_{a\sec\phi}^{2a\cos\phi} \rho^3 \sin^2\phi \cos\theta\ d\rho\ d\theta\ d\phi,\ 0$

17. $\pi r_0^2 h/3$

19. $(0,\ 0,\ h/5)$

25. $\dfrac{4}{3}a^3\left(\dfrac{\pi}{2} - \dfrac{2}{3}\right)$

27. $2\sqrt{2}\pi/3$

29. $k\pi a^4$

31. $(0,\ 0,\ 3r/8)$

33. $k\pi/192$

REVIEW EXERCISES FOR CHAPTER 16

1. 29/6

3. 36

5. 27/5

7. $\dfrac{h^3}{6}[\ln(\sqrt{2} + 1) + \sqrt{2}]$

9. $324\pi/5$

11. $8\pi/15$

13. $\dfrac{abc}{3}(a^2 + b^2 + c^2)$

15. $\displaystyle\int_0^3 \int_0^{(3-x)/3} dy\, dx = \int_0^1 \int_0^{3-3y} dx\, dy = \dfrac{3}{2}$

17. $\displaystyle\int_{-5}^3 \int_{-\sqrt{25-x^2}}^{\sqrt{25-x^2}} dy\, dx = \int_{-5}^{-4} \int_{-\sqrt{25-y^2}}^{\sqrt{25-y^2}} dx\, dy + \int_{-4}^4 \int_{-\sqrt{25-y^2}}^3 dx\, dy + \int_4^5 \int_{-\sqrt{25-y^2}}^{\sqrt{25-y^2}} dx\, dy$

$$= \dfrac{25\pi}{2} + 12 + 25\,\arcsin\dfrac{3}{5} \approx 66.36$$

19. $\displaystyle 4\int_0^1 \int_0^{x\sqrt{1-x^2}} dy\, dx = 4\int_0^{1/2} \int_{\sqrt{(1-\sqrt{1-4y^2})/2}}^{\sqrt{(1+\sqrt{1-4y^2})/2}} dx\, dy = \dfrac{4}{3}$

21. $\displaystyle\int_2^5 \int_{x-3}^{\sqrt{x-1}} dy\, dx + 2\int_1^2 \int_0^{\sqrt{x-1}} dy\, dx = \int_{-1}^2 \int_{y^2+1}^{y+3} dx\, dy = \dfrac{9}{2}$

23. $3{,}296/15$ 25. $\pi h^3/3$

27. $\dfrac{16}{3}\left(\dfrac{\pi}{2} - \dfrac{2}{3}\right)$

29. (a) $m = \dfrac{k}{4}, \left(\dfrac{32}{45}, \dfrac{64}{55}\right)$

   (b) $m = \dfrac{17k}{30}, \left(\dfrac{936}{1{,}309}, \dfrac{784}{663}\right)$

31. $(0,\, 0,\, 1/4)$ 33. $\left(\dfrac{3a}{8}, \dfrac{3a}{8}, \dfrac{3a}{8}\right)$

35. $\dfrac{\pi}{6}(65\sqrt{65} - 1)$ 37. $833k\pi/3$

39. Volume of torus: Formed a circle of radius 3, centered at $(0,\, 3,\, 0)$, revolved about the z-axis

CHAPTER 17

SECTION 17.1

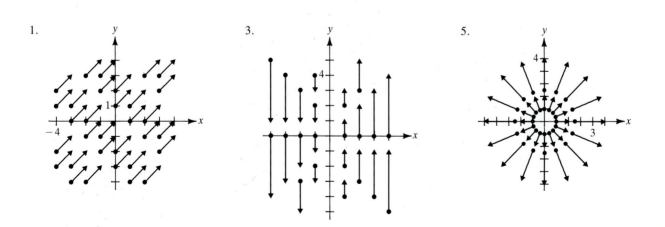

1.   3.   5.

7.
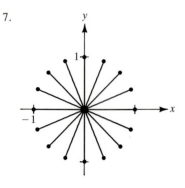

9.
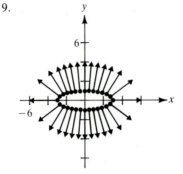

11.
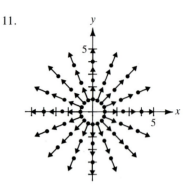

13.
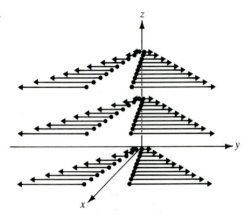

15.
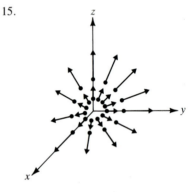

17. $(10x + 3y)\,\mathbf{i} + (3x + 20y)\,\mathbf{j}$

19. $-2xye^{x^2}\mathbf{i} - e^{x^2}\mathbf{j} + \mathbf{k}$

21. $\left[\dfrac{xy}{x + y} + y\,\ln(x + y)\right]\mathbf{i} + \left[\dfrac{xy}{x + y} + x\,\ln(x + y)\right]\mathbf{j}$

23. Conservative: $f(x, y) = x^2 y + K$

25. Conservative: $f(x, y) = x^2 + y^3 - 3xy + K$

27. Conservative: $f(x, y) = e^{x^2 y} + K$

29. Conservative: $f(x, y) = \dfrac{1}{2}\ln(x^2 + y^2) + K$

31. Not conservative

33. Not conservative

35. Not conservative

37. Conservative: $f(x, y, z) = xye^z + K$

39. Conservative: $f(x, y, z) = \dfrac{x}{y} + z^2 - z + K$

41. $2\mathbf{j} - \mathbf{k}$

43. $-2\mathbf{k}$

45. $\dfrac{2x\mathbf{k}}{x^2 + y^2}$

47. $\cos(y - z)\,\mathbf{i} + \cos(z - x)\,\mathbf{j} + \cos(x - y)\,\mathbf{k}$

49. $6x\mathbf{j} - 3y\mathbf{k}$

51. $z\mathbf{j} + y\mathbf{k}$

SECTION 17.2

1. $3\cos t\,\mathbf{i} + 3\sin t\,\mathbf{j},\ 0 \le t \le 2\pi$

3. $\mathbf{r}(t) = \begin{cases} t\mathbf{i}, & 0 \le t \le 3 \\ 3\mathbf{i} + (t-3)\mathbf{j}, & 3 \le t \le 6 \\ (9-t)\mathbf{i} + 3\mathbf{j}, & 6 \le t \le 9 \\ (12-t)\mathbf{j}, & 9 \le t \le 12 \end{cases}$

5. $\mathbf{r}(t) = \begin{cases} t\mathbf{i} + \sqrt{t}\,\mathbf{j}, & 0 \le t \le 1 \\ (2-t)\mathbf{i} + (2-t)\mathbf{j}, & 1 \le t \le 2 \end{cases}$

7. 9

9. $\dfrac{2\sqrt{2}}{3}$

11. $\dfrac{2}{3}(1 + \sqrt{2})$

13. $\pi/2$

15. 25

17. 63/2

19. 316/3

21. 35/6

23. 2

25. $-17/15$

27. $\dfrac{2\sqrt{13}\pi}{3}(27 + 64\pi^2) \approx 4{,}973.8$

29. $5h$

31. 1/2

33. $\dfrac{h}{4}[2\sqrt{5} + \ln(2 + \sqrt{5})]$

35. $\dfrac{1}{120}[25\sqrt{5} - 11]$

37. -66

39. 0

41. $-10\pi^2$

47. $\dfrac{850{,}304\sqrt{91} - 7184}{1215} \approx 6670 \text{ ft}^2$

SECTION 17.3

1. 11/15

3. $-\ln(2 + \sqrt{3})$

5. 1

7. (a) 0
    (b) $-1/3$
    (c) $-1/2$

9. (a) 64
    (b) 0
    (c) 0
    (d) 0

11. (a) 64/3
     (b) 64/3

13. (a) 32
     (b) 32

15. (a) 2/3
     (b) 17/6

17. (a) 0
     (b) 0

19. 24

21. -1

23. 0

25. (a) 2
     (b) 2
     (c) 2

27. 11

29. 30,366

31. 0

33. Increasing: 10 units/min

SECTION 17.4

1. 0

3. 32/15

5. 4/3

7. 56

9. 32/3

11. 0

13. 0

15. 1/12

17. 8π

19. 4π

21. 225/2

23. πa^2

25. 32/3

29. (0, 8/5)

31. (8/15, 8/21)

33. $3\pi a^2/2$

35. $\pi - \dfrac{3\sqrt{3}}{2}$

SECTION 17.5

1. 0

3. 10π

5. $27\sqrt{6}/2$

7. $\dfrac{391\sqrt{17} + 1}{240}$

9. $\dfrac{19\sqrt{2}\pi}{4}$

11. $32\pi/3$

13. 486π

15. $-4/3$

17. $243\pi/2$

19. 20π

21. 5/2

23. 364/3

27. $2\pi a^3 h$

29. If a normal vector at a point P on the surface is moved around the mobius strip once, it will point in the opposite direction.

SECTION 17.6

1. $12x - 2xy + 1$

3. $\cos x - \sin y + 2z$

5. a^4

7. 18

9. $3a^4$

11. 0

13. 32π

15. 0

17. 2,304

21. 0

27. 4

29. 0

31. $2z + 3x$

33. 0

SECTION 17.7

1. $-xy\mathbf{i} - \mathbf{j} + (yz - 2)\,\mathbf{k}$

3. $\left(2 - \dfrac{1}{1 + x^2}\right)\mathbf{j} - 8x\mathbf{k}$

5. $z(x - 2e^{y^2+z^2})\,\mathbf{i} - yz\mathbf{j} - 2ye^{x^2+y^2}\,\mathbf{k}$

7. 2π

9. 0

11. 1

13. 0

15. 0

17. $\dfrac{8}{3}$

19. $a^5/4$

REVIEW EXERCISES FOR CHAPTER 17

1. $(16x + y)\mathbf{i} + x\mathbf{j} + 2z\mathbf{k}$

3. Not conservative

5. Conservative: $f(x, y) = 3x^2y^2 - x^3 + y^3 - 7y + K$

7. Not conservative

9. Conservative: $f(x, y) = \dfrac{x}{yz} + K$

11. (a) div $\mathbf{F} = 2x + 2y + 2z$
    (b) curl $\mathbf{F} = \mathbf{0}$

13. (a) div $\mathbf{F} = -y \sin x - x \cos y + xy$
    (b) curl $\mathbf{F} = xz\mathbf{i} - yz\mathbf{j}$

15. (a) div $\mathbf{F} = \dfrac{1}{\sqrt{1 - x^2}} + 2xy + 2yz$
    (b) curl $\mathbf{F} = z^2\,\mathbf{i} + y^2\mathbf{k}$

17. (a) div $\mathbf{F} = \dfrac{2x + 2y}{x^2 + y^2} + 1$
    (b) curl $\mathbf{F} = \dfrac{2x - 2y}{x^2 + y^2}\,\mathbf{k}$

19. (a) $6\sqrt{2}$
    (b) 128π

21. (a) $35/2$
    (b) 18π

23. $2\pi^2(1 + 2\pi^2)$

25. $5/7$

27. $2\pi^2$

29. $64/3$

31. 4

33. 0

35. $1/12$

37. 12

39. 66

41. $-2a^6/5$

43. $\displaystyle\int_C \mathbf{F} \cdot d\mathbf{r} = 2\pi$, $\mathbf{F}$ is undefined at the point $(0, 0)$.

CHAPTER 18

SECTION 18.1

	Type	Order
1.	Ordinary	1
3.	Ordinary	2
5.	Ordinary	4
7.	Ordinary	2
9.	Partial	1
11.	Ordinary	2

19. Not a solution

21. Solution

23. Solution

25. Not a solution

27. Solution

29. Solution

31. $y = 3e^{-2x}$

33. $y = 2 \sin 3x - \dfrac{1}{3} \cos 3x$

35. $y = -2x + \dfrac{1}{2}x^3$

37.

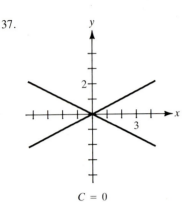

$C = 0$

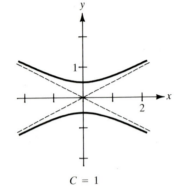

$C = 1$

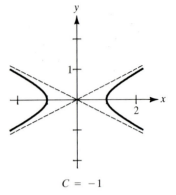

$C = -1$

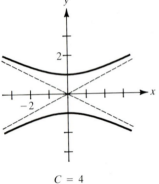

$C = 4$

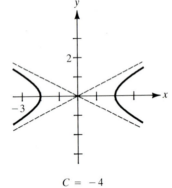

$C = -4$

39. $y = x^3 + C$

41. $y = x - \ln x^2 + C$

43. $y = \dfrac{e^x}{5}(\sin 2x - 2 \cos 2x) + C$

45. $y = \dfrac{2}{5}(x - 3)^{5/2} + 2(x - 3)^{3/2} + C$

SECTION 18.2

1. $y^2 - x^2 = C$

3. $y = C(x + 2)^3$

5. $y^2 = C - 2 \cos x$

7. $y^2 = 2e^x + 14$

9. $y = e^{-(x^2+2x)/2}$

11. $\dfrac{y - x}{1 + xy} = \sqrt{3}$

13. $P = P_0 e^{kt}$

15. Homogeneous of degree 3

17. Not homogeneous

19. Homogeneous of degree 0

21. $x = C(x - y)^2$

23. $y^2 + 2xy - x^2 = C$

25. $y = Ce^{-x^2/2y^2}$

27. $e^{y/x} = 1 + \ln x^2$

29. $x = e^{\sin (y/x)}$

31. Circles: $x^2 + y^2 = C$
     Lines: $y = Kx$

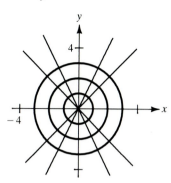

33. Parabolas: $x^2 = Cy$
     Ellipses: $x^2 + 2y^2 = K$

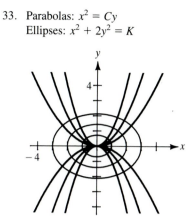

35. Curves: $y^2 = Cx^3$
     Ellipses: $2x^2 + 3y^2 = K$

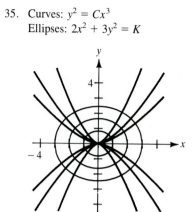

37. $9x^2 + 16y^2 = 25$

39. (a) $A = Pe^{kt}$
     (b) $A = \$3,004.17$
     (c) $t = 6.3$ yrs

41. 98.9% of original amount

43. $t = 119.7$ min

SECTION 18.3

1. $x^2 - 3xy + y^2 = C$

3. $3xy^2 + 5x^2y^2 - 2y = C$

5. Not exact

7. $\arctan \dfrac{x}{y} = C$

9. Not exact

11. $y \ln (x - 1) + y^2 = 16$

13. $x^2 + y^2 = 16$

15. $x^2 \tan y + 5x = 0$

17. Integrating factor: $1/y^2$

     $\dfrac{x}{y} - 6y = C$

19. Integrating factor: $1/x^2$

     $\dfrac{y}{x} + 5x = C$

21. Integrating factor: $\cos x$

     $y \sin x + x \sin x + \cos x = C$

23. Integrating factor: $1/y$

     $xy - \ln y = C$

25. Integrating factor: $1/\sqrt{y}$

     $x\sqrt{y} + \cos \sqrt{y} = C$

27. Integrating factor: xy^2

     $x^4y^3 + x^2y^4 = C$

29. Integrating factor: $1/x^2y^3$

     $\dfrac{y^2}{x} + \dfrac{x}{y^2} = C$

33. $x^2 + y^2 = C$

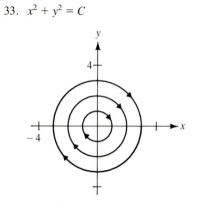

35. $2x^2y^4 + x^2 = C$

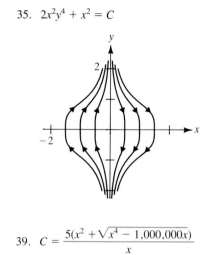

37. $x^2 - 2xy + 3y^2 = 3$

39. $C = \dfrac{5(x^2 + \sqrt{x^4 - 1,000,000x})}{x}$

SECTION 18.4

1. $y = x^2 + 2x + C/x$

3. $y = \dfrac{1}{2}e^x + Ce^{-x}$

5. $y = \dfrac{1}{2}(\sin x - \cos x) + Ce^x$

7. $y = -\dfrac{1}{13}(3 \sin 2x + 2 \cos 2x) + Ce^{3x}$

9. $y = \dfrac{x^3 - 3x + C}{3(x - 1)}$

11. $y = 1 + 4e^{-\tan x}$

13. $y = \sin x + (x + 1) \cos x$

15. $xy = 4$

17. $\dfrac{1}{y^2} = Ce^{2x^3} + \dfrac{1}{3}$

19. $y = \dfrac{1}{Cx - x^2}$

21. $y^{2/3} = Ce^{2x/3} - \dfrac{1}{4}(4x^3 + 18x^2 + 54x + 81)$

23. $y = \dfrac{E_0}{R} + Ce^{-Rt/L}$

27. (a) $Q = 10 + 15e^{-.05t}$

29. $2e^x + e^{-2y} = C$

   (b) $t = -20 \ln \dfrac{1}{3} \approx 21.97$ min

31. $x + xy^2 + \dfrac{1}{2}y^2 + 2y = C$

33. $y = Ce^{-\sin x} + 1$

35. $x^2y + \sin y = C$

37. $x^3y^2 + x^4y = C$

39. $y = \dfrac{e^x(x - 1) + C}{x^2}$

41. $x^4y^4 - 2x^2 = C$

43. $3 \arctan \dfrac{x}{y} - y = C$

SECTION 18.5

1. $y = C_1 + C_2 e^x$

3. $y = C_1 e^{3x} + C_2 e^{-2x}$

5. $y = C_1 e^{x/2} + C_2 e^{-2x}$

7. $y = C_1 e^{-3x} + C_2 x e^{-3x}$

9. $y = C_1 e^{x/4} + C_2 x e^{x/4}$

11. $y = C_1 \sin x + C_2 \cos x$

13. $y = C_1 e^{3x} + C_2 e^{-3x}$

15. $y = e^x[C_1 \cos \sqrt{3}x + C_2 \sin \sqrt{3}x]$

17. $y = C_1 e^{(3+\sqrt{5})x/2} + C_2 e^{(3-\sqrt{5})x/2}$

19. $y = e^{2x/3}\left[C_1 \sin \dfrac{\sqrt{7}x}{3} + C_2 \cos \dfrac{\sqrt{7}x}{3}\right]$

21. $y = C_1 e^x + C_2 e^{-x} + C_3 \sin x + C_4 \cos x$

23. $y = C_1 e^x + C_2 e^{2x} + C_3 e^{3x}$

25. $y = C_1 e^x + e^x[C_2 \sin 2x + C_3 \cos 2x]$

27. $y = \dfrac{1}{11}(e^{6x} + 10e^{-5x})$

29. $y = \dfrac{1}{2} \cos (4\sqrt{3}t)$

31. $y = \dfrac{2}{3} \cos 4\sqrt{3}t - \dfrac{1}{8\sqrt{3}} \sin 4\sqrt{3}t$

33. $y = \dfrac{e^{-t/16}}{2}\left[\cos \dfrac{\sqrt{12{,}287}t}{16} + \dfrac{1}{\sqrt{12{,}287}} \sin \dfrac{\sqrt{12{,}287}t}{16}\right]$

SECTION 18.6

1. $y = C_1 e^x + C_2 e^{2x} + x + \dfrac{3}{2}$

3. $y = \cos x + 6 \sin x + x^3 - 6x$

5. $y = C_1 + C_2 e^{-2x} + \dfrac{2}{3}e^x$

7. $y = (C_1 + C_2 x)e^{5x} + \dfrac{3}{8}e^x + \dfrac{1}{5}$

9. $y = -1 + 2e^{-x} - \cos x - \sin x$

11. $y = \left(C_1 - \dfrac{x}{6}\right) \cos 3x + C_2 \sin 3x$

13. $y = C_1 e^x + C_2 x e^x + \left(C_3 + \dfrac{2x}{9}\right)e^{-2x}$

15. $y = \left(\dfrac{4}{9} - \dfrac{1}{2}x^2\right)e^{4x} - \dfrac{1}{9}(1 + 3x)e^x$

17. $y = (C_1 + \ln |\cos x|) \cos x + (C_2 + x) \sin x$

19. $y = \left(C_1 - \dfrac{x}{2}\right) \cos 2x + \left(C_2 + \dfrac{1}{4} \ln |\sin 2x|\right) \sin 2x$

21. $y = (C_1 + C_2 x)e^x + \dfrac{x^2 e^x}{4}(\ln x^2 - 3)$

23. $y = \dfrac{1}{1{,}200}(225e^{-2t} - 201e^{-14t}) + \dfrac{1}{200}(3 \sin 2t - 4 \cos 2t)$

25. $q = \dfrac{3}{25}(e^{-5t} + 5te^{-5t} - \cos 5t)$

SECTION 18.7

1. $y = a_0 \displaystyle\sum_{n=0}^{\infty} \dfrac{x^n}{n!} = a_0 e^x$

3. $y = a_0 \sum\limits_{n=0}^{\infty} \dfrac{(3x)^{2n}}{(2n)!} + \dfrac{a_1}{3} \sum\limits_{n=0}^{\infty} \dfrac{(3x)^{2n+1}}{(2n+1)!} = C_0 \sum\limits_{n=0}^{\infty} \dfrac{(3x)^n}{n!} + C_1 \sum\limits_{n=0}^{\infty} \dfrac{(-3x)^n}{n!} = C_0 e^{3x} + C_1 e^{-3x}$

Where $C_0 + C_1 = a_0$ and $C_0 - C_1 = \dfrac{1}{3} a_1$

5. $y = a_0 \sum\limits_{n=0}^{\infty} \dfrac{(-1)^n (2x)^{2n}}{(2n)!} + \dfrac{a_1}{4} \sum\limits_{n=0}^{\infty} \dfrac{(-1)^n (2x)^{2n+1}}{(2n+1)!} = C_0 \cos 2x + C_1 \sin 2x$

7. $y = a_2 x^2$

9. $y = a_0 + a_1 \sum\limits_{n=1}^{\infty} \dfrac{x^n}{n!(n-1)!}$

11. $y = a_0 \left(1 - \dfrac{x^2}{8} + \dfrac{x^4}{128} - \cdots \right) + a_1 \left(x - \dfrac{x^3}{24} + \dfrac{7x^5}{1{,}920} - \cdots \right)$

13. $y = 2 + \dfrac{2x}{1!} - \dfrac{2x^2}{2!} - \dfrac{10x^3}{3!} + \dfrac{2x^4}{4!} + \ldots$, when $x = \dfrac{1}{2}$, $y \approx 2.552$

15. $y = 1 - \dfrac{3x}{1!} + \dfrac{2x^3}{3!} - \dfrac{12x^4}{4!} + \dfrac{16x^6}{6!} - \dfrac{120x^7}{7!} + \ldots$, when $x = \dfrac{1}{4}$, $y \approx 0.253$

REVIEW EXERCISES FOR CHAPTER 18

1. $y = x \ln x^2 + 2x^{3/2} + C$

3. $y = C(1 - x)^2$

5. $y^2 = x^2 \ln x^2 + Cx^2$

7. $y = Ce^{2x} - e^x$

9. $5x^2 + 8xy + 2x + \dfrac{5}{2} y^2 + 2y = C$

11. $xy - 2xy^3 + x^2 = C$

13. $y = x \ln |x| - 2 + Cx$

15. $\ln |1 + y| = Ce^{-x}$

17. $y = e^x (1 + \tan x) + C \sec x$

19. $x^2 - 2xy - 10x - 3y^2 + 4y = C$

21. $y = \dfrac{1}{5} x^3 - x + C\sqrt{x}$

23. $y^2 = 2Cx + C^2$

25. $y^2 = x^2 - x + \dfrac{3}{2} + Ce^{-2x}$

27. $\dfrac{y - x}{1 + xy} = C$

29. $y = \dfrac{bx^4}{4 - a} + Cx^a$

31. $y = C_1 \sin x + C_2 \cos x - 5x + x^3$

33. $y = (C_1 + x) \sin x + C_2 \cos x$

35. $y = \left(C_1 + C_2 x + \dfrac{x^3}{3} \right) e^x$

37. Family of circles:
$x^2 + (y - K)^2 = K^2$

39. $y = a_0 \sum\limits_{n=0}^{\infty} \dfrac{x^n}{4^n}$

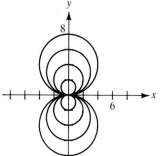

Index

FORMULAS FROM GEOMETRY

Triangle:

$h = a \sin \theta$

$\text{Area} = \dfrac{1}{2}bh$

(Law of Cosines)

$c^2 = a^2 + b^2 - 2ab \cos \theta$

Right Triangle:

(Pythagorean Theorem)

$c^2 = a^2 + b^2$

Equilateral Triangle:

$h = \dfrac{\sqrt{3}s}{2}$

$\text{Area} = \dfrac{\sqrt{3}s^2}{4}$

Parallelogram:

$\text{Area} = bh$

Trapezoid:

$\text{Area} = \dfrac{h}{2}(a + b)$

Circle:

$\text{Area} = \pi r^2$

$\text{Circumference} = 2\pi r$

Sector of Circle:

(θ in radians)

$\text{Area} = \dfrac{\theta r^2}{2}$

$s = r\theta$

Circular Ring:

(p = average radius,

w = width of ring)

$\text{Area} = \pi(R^2 - r^2)$

$\quad = 2\pi pw$

Sector of Circular Ring:

(p = average radius,

w = width of ring,

θ in radians)

$\text{Area} = \theta pw$

Ellipse:

$\text{Area} = \pi ab$

$\text{Circumference} \approx 2\pi \sqrt{\dfrac{a^2 + b^2}{2}}$

Cone:

(A = area of base)

$\text{Volume} = \dfrac{Ah}{3}$

Right Circular Cone:

$\text{Volume} = \dfrac{\pi r^2 h}{3}$

$\text{Lateral Surface Area} = \pi r \sqrt{r^2 + h^2}$

Frustum of Right Circular Cone:

$\text{Volume} = \dfrac{\pi(r^2 + rR + R^2)h}{3}$

$\text{Lateral Surface Area} = \pi s(R + r)$

Right Circular Cylinder:

$\text{Volume} = \pi r^2 h$

$\text{Lateral Surface Area} = 2\pi rh$

Sphere:

$\text{Volume} = \dfrac{4}{3}\pi r^3$

$\text{Surface Area} = 4\pi r^2$

Wedge:

(A = area of upper face,

B = area of base)

$A = B \sec \theta$

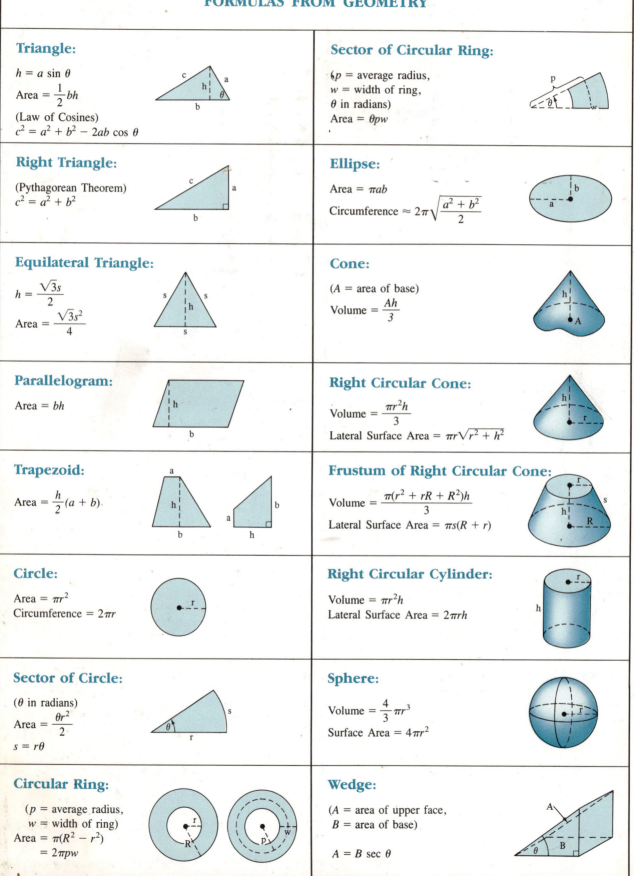